BASIC FORMULAS

Class Width = Range/(# of classes) and roundup

Population Mean: $\mu = \dfrac{\Sigma x}{N}$

Sample Mean: $\bar{x} = \dfrac{\Sigma x}{n}$

Sample variance: $s^2 = \dfrac{\Sigma(x-\bar{x})^2}{n-1}$

Population Variance: $\sigma^2 = \dfrac{\Sigma(x-\mu)^2}{N}$

Population Standard Deviation:

$$\sigma = \sqrt{\dfrac{\Sigma(x-\mu)^2}{N}}; \quad \sigma = \sqrt{\dfrac{\Sigma x^2 - \dfrac{(\Sigma x)^2}{N}}{N}}$$

Sample Standard Deviation:

$$s = \sqrt{\dfrac{\Sigma(x-\bar{x})^2}{n-1}}; \quad s = \sqrt{\dfrac{\Sigma x^2 - \dfrac{(\Sigma x)^2}{n}}{n-1}}$$

z Score Formula: $z = \dfrac{x-\mu}{\sigma}; \quad z = \dfrac{x-\bar{x}}{s}$

Raw Score Formula: $x = \mu + z\sigma; \quad x = \bar{x} + z(s)$

Percentile Rank Formula:

PR of x = $\dfrac{[B + (\frac{1}{2})E]}{N}$ (100)

COEFFICIENT OF VARIATION

$$\text{Coefficient of Variation} = \dfrac{\text{Standard deviation}}{\text{mean}}(100\%)$$

EMPIRICAL RULE (OR 68-95-99.7) RULE

The Empirical Rule states that for a mound or bell-shaped distribution:
- Approximately 68% of the data will lie within one standard deviation of the mean.
- Approximately 95% of the data will lie within two standard deviations of the mean.
- Approximately 99.7% of the data will lie within three standard deviations of the mean.

CHEBYSHEV'S THEOREM

Chebyshev's Theorem states for any distribution, regardless of its shape:

at least $1 - \dfrac{1}{k^2}$ of the data values will lie within k standard deviations of the mean,
where k is greater than one.

5-NUMBER SUMMARY

A **5-number summary** uses the following five numbers to describe a data set:
1. The smallest data value
2. The first quartile (Q_1)
3. The median
4. The third quartile (Q_3)
5. The largest data value

INTER-QUARTILE RANGE (IQR)

$IQR = Q_3 - Q_1$

PROBABILITY

Permutation Rule: $_nP_n = n!$
Permutation Rule for n objects taken s at a time:

$$_nP_s = n(n-1)(n-2) \cdots (n-s+1)$$

Permutation Rule of N objects with k alike objects
Given N objects where n_1 are alike, n_2 are alike,..., n_k are alike, then the number of permutations of these N objects is:

$$\dfrac{N!}{(n_1!)(n_2!)\ldots(n_k!)}$$

Number of Combinations of n objects taken s at a time:

$$_nC_s = \dfrac{n!}{s!(n-s)!}$$

Classical Probability Definition for Equally Likely Outcomes:

$$P(\text{Event E}) = \dfrac{\text{number of outcomes satisfying event E}}{\text{total \# of outcomes in sample space}}$$

Relative Frequency Approach to Calculating Probability:

$$P(\text{Event A}) = \dfrac{\text{number of times event A occurred}}{\text{total \# of times the experiment was repeated}}$$

The Addition Rule:
$P(A \text{ or } B) = P(A) + P(B) - P(A \text{ and } B)$

The Complement Rule:
$P(E) + P(E') = 1$

Multiplication Rule for: Independent Events
$P(A \text{ and } B) = P(A) \bullet P(B)$

Multiplication Rule for: Dependent Events
$P(A \text{ and } B) = P(A) \bullet P(B \mid A)$

Conditional Probability Formula

$$P(A \mid B) = \dfrac{P(A \text{ and } B)}{P(B)}$$

DISCRETE RANDOM VARIABLE X:

Mean: $\mu = \sum_{\text{all possible } X \text{ values}} [X \cdot P(X)]$

Standard Deviation:

$$\sigma = \sqrt{\sum_{\text{all possible } X \text{ values}} [(X-\mu)^2 \cdot P(X)]}$$

Binomial Probability Formula

P(s successes in n trials) = $_nC_s p^s q^{(n-s)}$

Binomial Distribution:
Mean: $\mu_s = np$

Standard Deviation: $\sigma_s = \sqrt{n \cdot p \cdot (1-p)}$

Poisson Probability Formula: $P(x=k) = \dfrac{e^{-\lambda} \lambda^k}{k!}$

Mean of a Poisson Distribution: $\mu = \lambda$

Standard Deviation of a Poisson Distribution: $\sigma = \sqrt{\lambda}$

Geometric Probability Formula

$$P(X=k) = q^k p, \quad k = 0, 1, 2, \ldots$$

Mean of a Geometric Distribution

$$\mu = E(X) = \dfrac{q}{p}$$

Variance and Standard Deviation of a Geometric Distribution

$$\sigma^2 = Var(X) = \dfrac{q}{p^2} \quad \sigma = \sqrt{VarX} = \sqrt{\dfrac{q}{p^2}}$$

Negative Binomial Probability Formula

$$P(X=k) = \binom{r+k-1}{r-1} q^k p^r, \quad k = 0, 1, 2, 3\ldots$$

Mean of a Negative Binomial Distribution

$$\mu = E(X) = \dfrac{rq}{p}$$

Variance and Standard Deviation of a Negative Binomial Distribution

$$VarX = \dfrac{rq}{p^2} \quad \sigma = \sqrt{VarX} = \sqrt{\dfrac{rq}{p^2}}$$

SAMPLING DISTRIBUTION OF THE MEAN:

Mean: $\mu_{\bar{x}} = \mu$

Standard Error: $\sigma_{\bar{x}} = \dfrac{\sigma}{\sqrt{n}}$

estimate of the standard error: $s_{\bar{x}} = \dfrac{s}{\sqrt{n}}$

Test Statistic Formula:
a) for normal distribution: $z = \dfrac{\bar{x} - \mu_{\bar{x}}}{\sigma_{\bar{x}}}$

b) for t distribution: $t = \dfrac{\bar{x} - \mu_{\bar{x}}}{s_{\bar{x}}}$

To find t_c, need to determine degrees of freedom using formula:

$$df = n - 1$$

PROPORTION

population proportion: $p = \dfrac{X}{N}$

sample proportion: $\hat{p} = \dfrac{x}{n}$

SAMPLING DISTRIBUTION OF THE PROPORTION:

Mean: $\mu_{\hat{p}} = p$

Standard Error: $\sigma_{\hat{p}} = \sqrt{\dfrac{p(1-p)}{n}}$

Test Statistic Formula is: $z = \dfrac{\hat{p} - \mu_{\hat{p}}}{\sigma_{\hat{p}}}$

ESTIMATION: POPULATION MEAN:

known population standard deviation
90% Confidence Interval:

$$\bar{x} - (1.65)\dfrac{\sigma}{\sqrt{n}} \text{ to } \bar{x} + (1.65)\dfrac{\sigma}{\sqrt{n}}$$

95% Confidence Interval

$$\bar{x} - (1.96)\dfrac{\sigma}{\sqrt{n}} \text{ to } \bar{x} + (1.96)\dfrac{\sigma}{\sqrt{n}}$$

99% Confidence Interval:

$$\bar{x} - (2.58)\dfrac{\sigma}{\sqrt{n}} \text{ to } \bar{x} + (2.58)\dfrac{\sigma}{\sqrt{n}}$$

margin of error:
(known population standard)

$$E = (z)\left(\frac{\sigma}{\sqrt{n}}\right)$$

sample size:

$$n = \left[\frac{z\sigma}{E}\right]^2$$

POPULATION MEAN:

unknown population standard deviation
90% Confidence Interval:

$$\bar{x} - (t_{95\%})\frac{s}{\sqrt{n}} \text{ to } \bar{x} + (t_{95\%})\frac{s}{\sqrt{n}}$$

95% Confidence Interval

$$\bar{x} - (t_{97.5\%})\frac{s}{\sqrt{n}} \text{ to } \bar{x} + (t_{97.5\%})\frac{s}{\sqrt{n}}$$

99% Confidence Interval:

$$\bar{x} - (t_{99.5\%})\frac{s}{\sqrt{n}} \text{ to } \bar{x} + (t_{99.5\%})\frac{s}{\sqrt{n}}$$

POPULATION PROPORTION:

90% Confidence Interval:

$$\hat{p} - (1.65)\, s_{\hat{p}} \text{ to } \hat{p} + (1.65)\, s_{\hat{p}}$$

95% Confidence Interval

$$\hat{p} - (1.96)\, s_{\hat{p}} \text{ to } \hat{p} + (1.96)\, s_{\hat{p}}$$

99% Confidence Interval:

$$\hat{p} - (2.58)\, s_{\hat{p}} \text{ to } \hat{p} + (2.58)\, s_{\hat{p}}$$

margin of error:

$$E = (z)\left(\sqrt{\frac{\hat{p}(1-\hat{p})}{n}}\right)$$

sample size:

$$n = \frac{z^2(\hat{p})(1-\hat{p})}{E^2}$$

conservative sample size formula:

$$n = \frac{z^2(0.25)}{E^2}$$

SAMPLING DISTRIBUTION OF THE DIFFERENCE BETWEEN TWO PROPORTIONS

Mean: $\mu_{\hat{p}_1-\hat{p}_2} = p_1 - p_2$

Pooled Estimate of the Standard Error:

$$s_{\hat{p}_1-\hat{p}_2} = \sqrt{\frac{\hat{p}(1-\hat{p})}{n_1} + \frac{\hat{p}(1-\hat{p})}{n_2}} \quad \text{or}$$

$$s_{\hat{p}_1-\hat{p}_2} = \sqrt{\hat{p}(1-\hat{p})\left(\frac{1}{n_1} + \frac{1}{n_2}\right)}$$

where $\hat{p} = \dfrac{X_1 + X_2}{n_1 + n_2}$ or $\hat{p} = \dfrac{n_1\hat{p}_1 + n_2\hat{p}_2}{n_1 + n_2}$

Test Statistic Formula is: $z = \dfrac{(\hat{p}_1 - \hat{p}_2) - \mu_{\hat{p}_1-\hat{p}_2}}{s_{\hat{p}_1-\hat{p}_2}}$

THE TWO SAMPLE Z TEST

mean: $\mu_{\bar{x}_1-\bar{x}_2} = \mu_1 - \mu_2 = 0$

standard error: $\sigma_{\bar{x}_1-\bar{x}_2} = \sqrt{\dfrac{\sigma_1^2}{n_1} + \dfrac{\sigma_2^2}{n_2}}$

test statistic formula: $z = \dfrac{(\bar{x}_1 - \bar{x}_2)}{\sigma_{\bar{x}_1-\bar{x}_2}}$

THE POOLED TWO SAMPLE T TEST

mean: $\mu_{\bar{x}_1-\bar{x}_2} = \mu_1 - \mu_2 = 0$

estimate of the standard error:

$$s_{\bar{x}_1-\bar{x}_2} = \sqrt{\frac{(n_1-1)s_1^2 + (n_2-1)s_2^2}{n_1+n_2-2} \cdot \left(\frac{1}{n_1} + \frac{1}{n_2}\right)}$$

test statistic formula: $t = \dfrac{\bar{x}_1 - \bar{x}_2}{s_{x_p}\sqrt{\dfrac{1}{n_1} + \dfrac{1}{n_2}}}$

where, $s_{x_p} = \sqrt{\dfrac{(n_1-1)s_1^2 + (n_2-1)s_2^2}{n_1+n_2-2}}$

degrees of freedom: $df = n_1 + n_2 - 2$

WELCH'S TWO SAMPLE T TEST

mean: $\mu_{\bar{x}_1-\bar{x}_2} = \mu_1 - \mu_2 = 0$

standard error: $s_{\bar{x}_1-\bar{x}_2} = \sqrt{\dfrac{s_1^2}{n_1} + \dfrac{s_2^2}{n_2}}$

test statistic formula: $t = \dfrac{\bar{x}_1 - \bar{x}_2}{s_{\bar{x}_1-\bar{x}_2}}$

degrees of freedom: $df =$ smaller of $(n_1 - 1)$ and $(n_2 - 1)$.

THE PAIRED T TEST

mean: $\mu_d = 0$

standard error: $\dfrac{s_d}{\sqrt{n}}$

test statistic formula: $t = \dfrac{\bar{x}_d - \mu_d}{\dfrac{s_d}{\sqrt{n}}}$

degrees of freedom: $df = n - 1$, where n is the number of data pairs

CHI-SQUARE

Pearson's Chi-Square Statistic

$$X^2 = \sum_{\text{all cells}} \dfrac{(O-E)^2}{E};$$

$$X^2 = \sum_{\text{all cells}} \dfrac{(O)^2}{E} - n$$

degrees of freedom: $df = (r-1)(c-1)$

expected cell frequency $= \dfrac{(RT)(CT)}{n}$

CORRELATION

Pearson's or Sample Correlation Coefficient:

$$r = \dfrac{n\sum(xy) - (\sum x)(\sum y)}{\sqrt{n(\sum x^2) - (\sum x)^2}\sqrt{n(\sum y^2) - (\sum y)^2}}$$

degrees of freedom: $df = n - 2$

Coefficient of Determination: r^2

Regression Line Formula: $y' = a + bx$

$b = \dfrac{n\sum(xy) - \sum x \sum y}{n(\sum x^2) - (\sum x)^2}$; $a = \dfrac{\sum y - b\sum x}{n}$

Alternate Formula for the **sample correlation coefficient, r**:

$r = \dfrac{1}{n-1} \sum z_x z_y$

ANOVA

Sum of Squares $SS = \sum(x - \bar{x})^2$

$SS_{\text{(between treatments)}} + SS_{\text{(within treatments)}} = SS_{\text{(total)}}$

$SS_{\text{(total)}} = \sum x^2 - \dfrac{(\sum x)^2}{N}$

$SS_{\text{(between treatments)}} =$

$(\dfrac{(\sum T_1)^2}{n_1} + \dfrac{(\sum T_2)^2}{n_2} + \ldots + \dfrac{(\sum T_i)^2}{n_i}) - \dfrac{(\sum x)^2}{N}$

$SS_{\text{(within treatments)}} =$

$\sum x^2 - (\dfrac{(\sum T_1)^2}{n_1} + \dfrac{(\sum T_2)^2}{n_2} + \ldots + \dfrac{(\sum T_i)^2}{n_i})$

Degrees of Freedom: $df_{\text{(Between)}} + df_{\text{(within)}} = df_{\text{(Total)}}$

Where: $df_{\text{(Between)}} = k - 1$
$df_{\text{(within)}} = N - k$
$df_{\text{(Total)}} = N - 1$

Mean Square, (denoted MS), formulas:

$MS_{\text{(Between)}} = \dfrac{SS_{bet}}{df_{bet}}$

$MS_{\text{(within)}} = \dfrac{SS_{within}}{df_{within}}$

Test statistic formula: $F = \dfrac{s_1^2}{s_2^2}$

Test statistic: F-statistic

$F = \dfrac{\text{treatment sample variance}}{\text{within sample variance}}$

NONPARAMETRIC TESTS
SIGN TEST for MEDIANS

binomial mean: $\mu_s = np$

binomial std dev: $\sigma_s = \sqrt{np(1-p)}$

If the statistical result 0, is less than n/2,

then test statistic is: $z = \dfrac{(O + \tfrac{1}{2}) - \mu_s}{\sigma_s}$

If the statistical result 0, is greater than n/2,

then the test statistic is: $z = \dfrac{(O - \tfrac{1}{2}) - \mu_s}{\sigma_s}$

MANN-WHITNEY RANK-SUM TEST

When normal distribution can be applied, then test statistic, z, is:

$z = \dfrac{w - \dfrac{n_1(n_1 + n_2 + 1)}{2}}{\sqrt{\dfrac{n_1 n_2 (n_1 + n_2 + 1)}{12}}}$

KRUSKAL-WALLIS H TEST

Kruskal-Wallis H test statistic formula:

$H = \dfrac{12}{N(N+1)}\left[\dfrac{(\sum R_1)^2}{n_1} + \dfrac{(\sum R_2)^2}{n_2} + \ldots + \dfrac{(\sum R_k)^2}{n_k}\right] - 3(N+1)$

BOOTSTRAPPING

Bootstrap Estimate of Bias

$$\text{Bias}(\hat{\Theta}) \approx \frac{1}{B}\sum_{i=1}^{B}\hat{\Theta}_i^* - \hat{\Theta} = \hat{\Theta}^* - \hat{\Theta}$$

Mean of the bootstrap distribution

$$\text{Mean} = \frac{1}{B}\sum \bar{x}^*$$

Bootstrap Standard Error of a Statistic
Standard Error of the Bootstrap Distribution

$$SE_b(\bar{x}) = \sqrt{\frac{1}{B-1} * \sum (\bar{x}^* - \frac{1}{B}\sum \bar{x}^*)^2}$$

Bootstrap Estimate for the Standard Error of the Mean

$$\approx \frac{\text{original sample standard deviation}}{\sqrt{n}}$$

Bootstrap Standard Error t Confidence Interval Formula:

$$\textbf{statistic} \pm (\textbf{t score}) * (\textbf{SE}_b(\textbf{statistic}))$$

When the statistic is a mean,
the Bootstrap Standard Error t Confidence Interval Formula:

$$\bar{x} \pm (\textbf{t score}) * SE_b(\bar{x})$$

Basic Bootstrap Confidence Interval

$$[2\hat{\Theta} - q_{1-\alpha/2}, 2\hat{\Theta} - q_{\alpha/2}]$$

quantile q_α formula
quantile q_α = Integer value of $[\alpha * (B+1)]$

PEARSON — ALWAYS LEARNING

Carmine DeSanto • Michael Totoro • Richard Moscatelli • Rachel Rojas

Introduction to Statistics

Eleventh Custom Edition for Nassau Community College

Cover Art: Courtesy of chbaum/123rf.com, Courtesy of Gecko1968/Shutterstock.

Copyright © 2020 by Pearson Education, Inc.
All rights reserved.
Pearson Custom Edition.

This copyright covers material written expressly for this volume by the editor/s as well as the compilation itself. It does not cover the individual selections herein that first appeared elsewhere. Permission to reprint these has been obtained by Pearson Education, Inc. for this edition only. Further reproduction by any means, electronic or mechanical, including photocopying and recording, or by any information storage or retrieval system, must be arranged with the individual copyright holders noted.

All trademarks, service marks, registered trademarks, and registered service marks are the property of their respective owners and are used herein for identification purposes only.

Pearson Education, Inc., 330 Hudson Street, New York, New York 10013
A Pearson Education Company
www.pearsoned.com

Printed in the United States of America
1 2020

A103000232326
00002733_00000001

EJ

PEARSON   ISBN 10: 0-13-727738-5
ISBN 13: 978-0-13-727738-4

CONTENTS

Preface   xvii

1 Introduction to Statistics   1

- 1.1 Introduction   1
  - What Is Statistics?   1
- 1.2 Why Sample?   8
  - Why Sample?   8
- 1.3 Sampling Techniques   10
  - Random or Probability Sampling Techniques   11
- 1.4 Uses of Statistics   16
- 1.5 Misuses of Statistics   18
  - Misleading Graphs   18
  - Non-Representative Samples   20
  - Inappropriate Comparisons   21
  - The Omission of Variation About an Average   22
- 1.6 Overview and Summary   22

2 Organizing and Presenting Data   33

- 2.1 Introduction   33
- 2.2 Classifications of Data   35
  - Categorical and Numerical Variables   36
  - Continuous and Discrete Data   36
- 2.3 Exploring Data Using the Stem-and-Leaf Display   38
  - Back-to-Back Stem-and-Leaf Display   42
- 2.4 Frequency Distribution Tables   43
  - Frequency Distribution Table for Categorical Data   43
  - Frequency Distribution Table for Numerical Data   44
  - Relative Frequency Distributions   49
  - Cumulative Frequency Distributions   51
- 2.5 Graphs: Bar Graph, Histogram, Frequency Polygon and Ogive   54
  - Bar Graph   54
  - Histogram   58
  - Frequency Polygon and Ogive   64
- 2.6 Specialty Graphs: Pie Chart and Pictograph   68
- 2.7 Identifying Shapes and Interpreting Graphs   69
  - Shapes of Distributions   69
  - Interpreting Graphs   72

3 Numerical Techniques for Describing Data   87

- 3.1 Measures of Central Tendency   88
  - The Mean   88
  - Sample Mean   89

Population Mean 90
Other Calculations Using Σ 91
The Median 94
The Mode 98
The Relationship of the Mean, Median and Mode 102
The Mode and Its Location within the Distribution Shapes 102
The Median and Its Location within the Distribution Shapes 102
The Mean and Its Location within the Distribution Shapes 102
Symmetric Bell-Shaped Distribution 103
Skewed to the Left Distribution 103
Skewed to the Right Distribution 103
Comparing the Mean, the Median and the Mode 104

3.2 Measures of Variability 106
The Range 106
The Variance and Standard Deviation of a Sample 108
Interpretation of the Standard Deviation 111
The Variance and Standard Deviation of a Population 113
Using the Sample Standard Deviation to Estimate the Population Standard Deviation 117

3.3 Applications of the Standard Deviation 118
Chebyshev's Theorem 118
Empirical Rule 120
Usefulness of the Empirical Rule 121
Using the Range to Obtain an Estimate of the Standard Deviation 124

3.4 Measures of Relative Standing 127
z Score 128
Detecting Outliers Using z Scores 132
Converting z Scores to Raw Scores 132
Percentile Rank and Percentiles 133
Deciles and Quartiles 135

3.5 Box-and-Whisker Plot: An Exploratory Data Analysis Technique 138

4 Linear Correlation and Regression Analysis 173

4.1 Introduction 173
4.2 The Scatter Diagram 174
4.3 The Coefficient of Linear Correlation 181
The Computational Formula for Linear Correlation 187
Some Cautions Regarding the Interpretation of Correlation Results 192
4.4 More on the Relationship Between the Correlation Coefficient and the Scatter Diagram 193
4.5 The Coefficient of Determination 193
4.6 Linear Regression Analysis 194
Interpolation versus Extrapolation 199
Explanation and Assumptions for Linear Regression Analysis 201

4.7 Residuals and Analyzing Residual Plots 203
   Analysis of Scatter Diagrams 204
   Homoscedasticity and Heteroscedasticity 204
   Non-Linear Trends 205
   What is a Residual? 205
   Residual Plots 207
   Coefficient of Determination Examined Further 214

5 Probability 231

5.1 Introduction 231
   The Birthday Problem 231
   Chance! Chance! Chance! 232
5.2 Some Terms Used in Probability 232
   Using a Tree Diagram to Construct a Sample Space 234
5.3 Permutations and Combinations 239
   Permutation 239
   Combination 245
   Methods of Selection 249
   Explaining the Difference Between the Idea of a Permutation and a Combination 250
5.4 Probability 251
   Alternate Approaches to Assigning a Probability 252
   Another Approach to Defining Probability 257
   Relative Frequency Concept of Probability or Posteriori Probability 257
   Another Approach to Defining Probability 259
   Subjective or Personal Probability 259
5.5 Fundamental Rules and Relationships of Probability 261
   Probability Problems Using Permutations and Combinations 273
5.6 Conditional Probability 277

6 Random Variables and Discrete Probability Distributions 299

6.1 Introduction 299
6.2 Random Variables 300
6.3 Probability Distribution of a Discrete Random Variable 304
6.4 Mean and Standard Deviation of a Discrete Random Variable 310
   The Mean Value of a Discrete Random Variable 310
   The Variance and Standard Deviation of a Discrete Random Variable 314
6.5 Binomial Probability Distribution 321
   Binomial Probability Formula 324
   An Application of the Binomial Distribution: Acceptance Sampling 336
6.6 The Poisson Distribution 340
   Shape of the Poisson Distribution 345
6.7 The Geometric and Negative Binomial Distributions 346
   The Negative Binomial Distribution 353

7 Continuous Probability Distributions and the Normal Distribution 371

- **7.1** Introduction 371
- **7.2** Continuous Probability Distributions 371
- **7.3** The Normal Distribution 376
- **7.4** Properties of a Normal Distribution 378
  - The Standard Normal Curve 380
- **7.5** Using the Normal Curve Area Table 383
  - Finding a z-score Knowing the Proportion of Area to the Left 396
- **7.6** Applications of the Normal Distribution 401
- **7.7** Percentiles and Applications of Percentiles 406
- **7.8** Probability Applications 417
- **7.9** The Normal Approximation to the Binomial Distribution 422

8 Sampling and Sampling Distributions 445

- **8.1** The Sampling Distribution of the Mean 445
- **8.2** The Mean and Standard Deviation of the Sampling Distribution of the Mean 453
  - Mean of the Sampling Distribution of the Mean 453
  - Standard Deviation of the Sampling Distribution of the Mean 453
  - Interpretation of the Standard Error of the Mean 456
- **8.3** The Finite Correction Factor 457
- **8.4** The Shape of the Sampling Distribution of the Mean 460
  - Sampling from a Normal Population 460
  - Sampling from a Non-Normal Population 463
- **8.5** Calculating Probabilities Using the Sampling Distribution of the Mean 468
- **8.6** The Effect of The Sample Size on the Standard Error of the Mean 474
- **8.7** The Sampling Distribution of the Proportion 476
  - Sampling Error of the Proportion 485
  - Interpretation of the Standard Error of the Proportion 486
  - Shape of the Sampling Distribution of the Proportion 486
  - Calculating Probabilities Using the Sampling Distribution of the Proportion 488

9 Estimation 503

- **9.1** Introduction 503
- **9.2** Point Estimate of the Population Mean and the Population Proportion 504
- **9.3** Interval Estimation 506
- **9.4** Interval Estimation: Confidence Intervals for the Population Mean 507
  - Constructing a Confidence Interval for a Population Mean: When the Population Standard Deviation Is Unknown 515
  - The t Distribution 516
- **9.5** Interval Estimation: Confidence Intervals for the Population Proportion 522

9.6 Determining Sample Size and the Margin of Error   526
   Sample Size for Estimating a Population Mean, μ 526
   Sample Size for Estimating a Population Proportion, p   532
   Summary of Confidence Intervals   536

10 Introduction to Hypothesis Testing   545

10.1 Introduction   545
10.2 Hypothesis Testing   545
   Null and Alternative Hypotheses   546
10.3 The Development of a Decision Rule   554
10.4 p-Values for Hypothesis Testing   569
   Procedure to Calculate the p-Value of a Hypothesis Test   572

11 Hypothesis Testing Involving One Population   583

11.1 Introduction   583
11.2 Hypothesis Testing Involving a Population Proportion   583
   Hypothesis Testing Procedure Involving a Population Proportion   585
11.3 Hypothesis Testing Involving a Population Mean: Population Standard Deviation Known   594
11.4 The t Distribution   599
   Using Table III: Critical Values for the t Distribution   600
11.5 Hypothesis Testing Involving a Population Mean: Population Standard Deviation Unknown   602
11.6 p-Value Approach to Hypothesis Testing Using the TI-84 Calculator   611

12 Hypothesis Testing Involving Two Population Proportions Using Independent Samples   627

12.1 Introduction to Hypothesis Tests Involving a Difference Between Two Population Proportions Using Independent Samples   627
12.2 The Sampling Distribution of the Difference Between Two Proportions   628
12.3 Hypothesis Testing Involving Two Population Proportions Using Large Samples   637
   Hypothesis Testing Procedure Involving the Difference Between the Proportions of Two Populations for Large Samples   637
   The Expected Results of the Sampling Distribution of $\hat{p}_1 - \hat{p}_2$ 640
12.4 Hypothesis Testing Involving Two Population Proportions Comparing Treatment and Control Groups   648
12.5 p-Value Approach to Hypothesis Testing Involving Two Population Proportions Using the TI-84 Calculator   654
12.6 Two Population Hypothesis Testing Summaries Using Independent Samples   658

13 Hypothesis Test Involving Two Population Means Using Independent and Dependent Samples   673

13.1 Introduction   673
13.2 The Sampling Distribution of the Difference Between Two Means   673

- 13.3 Hypothesis Testing Involving two Population Means and known Population Standard Deviations: The Two Sample Z Test 675
    - Two Sample Z Test 675
- 13.4 Hypothesis Testing Involving Two Population Means and Unknown but Equal Population Standard Deviations: The Pooled T-Test 680
    - The Pooled t Test 681
- 13.5 Hypothesis Testing Involving two Population Means and Unknown but Unequal Population Standard Deviations: Welch's T-Test 687
    - Welch's t Test 689
- 13.6 Hypothesis Tests Comparing Treatment and Control Groups 693
- 13.7 The Paired T-Test: A Dependent T-Test 700
- 13.8 p-Value Approach to Hypothesis Testing Involving Two Population Means Using the TI-84 Calculator 711

14 Chi-Square 731

- 14.1 Introduction 731
- 14.2 Properties of the Chi-Square Distribution 733
- 14.3 Chi-Square Hypothesis Test of Independence 734
- 14.4 Assumptions Underlying the Chi-Square Test 745
- 14.5 Test of Goodness-of-Fit 745
- 14.6 p-Value Approach to Chi-Square Hypothesis Test of Independence Using the TI-84 Plus Calculator 752

15 Inferences for Linear Correlation and Regression 767

- 15.1 Introduction 767
- 15.2 Testing the Significance of the Correlation Coefficient 769
    - Procedure to Test the Significance of the Population Correlation Coefficient, ρ 769
- 15.3 Assumptions for Linear Regression Analysis 777
- 15.4 p-Value Approach to Testing the Significance of the Correlation Coefficient Using the TI-84 Calculator 777
- 15.5 Introduction to Multiple Regression 782
    - The Multiple Correlation Coefficient, R 783

16 The F-Distribution and an Introduction to Analysis of Variance (ANOVA) 801 (Only available in the e-book)

- 16.1 Introduction 801
- 16.2 F-Distribution 802
- 16.3 Testing Variances: The F-Test 809
    - F-Test: Hypothesis Testing Procedure to Test if Population Variances are Different 810
- 16.4 Types of Variances 815
- 16.5 One-Way Analysis of Variance 822
- 16.6 A Brief Look at Two-Way ANOVA 849

17 Nonparametric Statistics 865 (Only available in the e-book)

17.1 Introduction 865
17.2 The Sign Test for Medians 867
    The Sign Test for Medians 870
17.3 The Mann-Whitney Rank-Sum Test 875
    The Mann-Whitney Test for Samples of Size of at least 10 878
17.4 The Kruskal-Wallis H Test 884
    The Kruskal-Wallis H Test 886

18 Bootstrapping Concepts & Methods: An Introduction 903 (Only available in the e-book)

18.1 Introduction 903
18.2 Bootstrap Distribution 904
    Plug-In Principle 907
    Justification for the Bootstrap Procedure 907
    Bootstrap Procedure to Generate the Bootstrap Distribution 907
    Bootstrap Notation 908
    Using Bootstrapping to Estimate BIAS 909
    Comments about the Bootstrap Procedure 909
18.3 Estimating Standard Error Using Bootstrapping 910
    Revisiting the Bootstrap Idea 917
    Key Bootstrap Ideas 917
    Useful Characteristics of the Bootstrap Distribution 918
    Comparing the Bootstrap Distribution and the Sampling Distribution 919
18.4 Bootstrap Confidence Intervals 919
    Method 1: Bootstrap Percentile Confidence Interval Method 919
    Cautions about the Bootstrap Percentile Confidence Interval Method 925
    Method 2: Bootstrap Standard Error t Confidence Interval Method 925
    Method 3: Basic Bootstrap Confidence Interval Method
      (or the Reverse Bootstrap Percentile Confidence Interval) 930
    Important Bootstrap Concepts to Remember when
      Applying the Bootstrap Confidence Interval Methods 933
    Bootstrap Confidence Interval Accuracy 934

Answer Section A1
Appendices A41
Index I1

Mohd Suhail/Pearson India Education Services Pvt. Ltd

The credit applies to each copy of the image.

PREFACE

PREFACE TO THE ELEVENTH EDITION

As we all have witnessed, the ubiquitous use of statistical terminology and concepts can be seen across all forms of media. Business and societal representations of statistical analyses are used, not only, in fields such as marketing, manufacturing, economics, science, education and sports, but have become common place in media platforms. A new statistical vocabulary has become part of everyday language. How often does a day go by without a reference to a graph, a study, or a conclusion without the representation of such presentations credited to data analyses? As we developed this eleventh edition, our world is caught in a pandemic that threatens the norms we have grown accustomed to for our entire lives. Today, a day does not go by without a basic presentation of statistical analysis.

Use of the language of statistics is now assumed part of the knowledge set of all world citizens.

We have, in this eleventh edition, continued with the philosophy that student understanding of basic statistical concepts and language is a necessary ingredient of an educated college graduate. We have added some additional topics that reinforce the foundation of elementary statistics. As in all our past editions, the conversational style of communication that has been a major characteristic of our past editions has been maintained. We have updated many applications, case studies and data sets to include variable themes in social media, health, and sports to reflect the changing student population interests. We have also, updated the technology and technological instruction to reflect the new TI-84 Plus calculator. Some changes have been relocated to the MYSTATLAB web component that includes new material, datasets and options for evaluation and student learning. It is our hope that the changes we have made in this edition are welcomed as helpful improvements to elementary statistical education.

FEATURES OF THE ELEVENTH EDITION

In the previous editions, each chapter contained calculator screen shots with basic instructional development to reinforce the understanding of the statistical concepts. In this edition, we continued this effective instructional strategy.

- TI-84plus Calculator screen shots and instructions that reflect new models of the TI-84plus calculator updated throughout.
- New Enriched Case Studies have been added along with updates to previous Case Studies to reflect current data.
- Various updates for text applications to reflect current data.
- Extensive Enrichment on the Usefulness of the Empirical Rule has been added to Chapter 3.
- Alternate Useful Formula to compute Pearson's Correlation Coefficient, r, has been added to Chapter 4
- New Sections on The Geometric and Negative Binomial Distribution were added to Chapter 6 to improve the presentation of probability distributions.
- Margin of Error formula for Estimating a Population Mean when Population Standard Deviation is Unknown was added to Chapter 9.
- The use of Welch's t-Test was added to Chapter 13 to cover a hypothesis testing involving two population means when the population standard deviations are unknown and unequal.
- The paired t-test was added to Chapter 13 to show students how to perform a hypothesis test on data that comes in matched-pairs.
- An extensive new Chapter 18 has been added to introduce Bootstrapping Concepts and Methods. This chapter is available in the e-book website in MYSTATLAB.
- DATABASE end of chapter questions was moved to the MyStatLab.com website.
- The Front Cover and Appendix Formulas have been updated to reflect these changes.

For this 11th Edition, the Website will continue to include the e-book version of the text, along with instructional videos and PDF solutions to selected textbook problems, access to large text data sets, and online assessment tools to help students develop a deeper understanding of the coursework.

As before the Website will maintain features of the student site:
- Videos:
  i. We have provided video instruction for selected problems at the end of each chapter, and instructional videos on basic functions of the TI 84 calculator. These videos can be downloaded or viewed on any personal computer.
  ii. Stat Talk videos featuring statistician, Andrew Vickers, that demonstrate important statistical concepts through interesting stories and real-life events.
- Complete PDF solutions to selected problems
- Printable versions of tables and formulas (Normal, t, Chi-Square distributions etc.)
- Additional Case Studies
- A TI-84plus graphing calculator instruction manual
- Chapters on advanced topics such as ANOVA, Nonparametric Statistics, and the new chapter on Bootstrapping.
- Sample tests with solutions for student practice.
- Detailed solutions to selected problems in each chapter.
- Optional E-Book of the text.

Features of the instructor site:
- An online instructor manual will provide helpful suggestions to the instructor as to the essential topics to be emphasized in each chapter.
- Printable graphs and tables from the text
- Answers to most even numbered problems
- Test Generation software and Test bank
- An E-Book of the text will be available so that instructors will have access to the text from any computer connected to the internet.

ADDITIONAL FEATURES OF THE TEXT

- **Case Studies:** Appearing in all chapters are actual applications of statistical concepts. These case studies are presented as articles, statistical snapshots, or research articles that have been published in newspapers, magazines, or journals. All case studies are based on real data and allow the student to see the relevance of statistics and why a student should study statistics.
- **What Do You Think? Problems:** Each chapter contains a What Do You Think? Section in the chapter exercises. This section provides the student with a set of realistic exercises which enable the student to explore diverse applications. In most cases, the problems will present applications which will require the student to analyze and interpret the information relating to some real practical situations.
- **Projects:** Each chapter exercise section includes a section entitled Projects. Students are presented with suggestions for further exploration of the statistics presented within the chapter.

  This text has been in production and revision for some forty years. We believe that the main features that distinguish it from other introductory statistics texts are:

- **Presentation of the Material:** Our informal "conversational" writing style thoroughly develops and explains each statistical concept. The pedagogical approach of actively learning statistics through discovery and applications makes the subject relevant to the beginning student, and shows each student both the "how to" as well as "the whys" of studying statistics.
- **Design and Format:** All definitions, procedures, theorems, formulas, and other important concepts are highlighted with appropriate headings that include a two-color ink format with appropriate shading where applicable. Most procedures are carefully outlined and highlighted with this two-color enhancement to help students focus on the important aspects of each chapter.
- **Examples:** Every chapter contains an abundant number of examples which cover the concepts for each section. Relevant guided solutions are carefully developed.
- **Solutions:** The solution to every example is presented with a very clear and readable style. Our approach is built on the premise that each step is not trivial. Consequently, we have provided a step-by-step approach in solving each problem.

- **Numerous Diagrams:** The text uses many diagrams to picture many of the concepts visually. Most have been enhanced by using two-color highlighting.
- **Exercises:** The exercises are a special feature of this text. Each chapter contains numerous problems from each chapter section that cover a wide range of applications representing many diverse fields. The content of the problems will interest both the student and the instructor. Most of the problems use real data which have been found in periodicals, reports, newspapers, and journals. These problems provide an insight into the many diverse applications of the use of statistics. Exercise types include: fill-in the blank, multiple choice, true false questions, and word solving problems. The exercises are first keyed to the chapter sections. A chapter review follows along with What Do You Think? database using technology and projects. At the end of each chapter, a chapter test is presented testing the key concepts of the chapter.
- **Key Terms:** At the end of each chapter a glossary of key terms and concepts listing the appropriate page in which the term appears.
- **Database:** Two database files appear in the Appendix. Students are encouraged to use the database with a technological support hardware to investigate the concepts presented in each chapter.
- **How to Use the TI-84plus Calculator: Strategically placed within the text as well as on t**he WEBSITE are detailed step by step approaches to performing the statistical calculations, generating graphs, and hypothesis testing procedures discussed within the text.

ACKNOWLEDGEMENTS

The success of **Introduction to Statistics** is due to the efforts of countless supportive and dedicated individuals.

It is a pleasure to acknowledge the aid, encouragement and helpful suggestions that we have received from our colleagues at Nassau Community College. We would especially like to thank Kira Lariosa for her continued support and participation in the textbook development and our instructional videos. We also thank Heather Huntington for her instructional videos on basic features of the TI 84 calculator. Also, a special thanks to Allison Cramer for her suggestions to include many of the social media variables and problems.

We are also grateful to the following individuals whose excellent suggestions and comments are invaluable in improving the text: Armen Baderian, James Baldwin, George Bruns, Mauro Cassano, Art Cohen, Kathleen Cramer, John Earnest, Eric Girolamo, Les Frimerman, Ron Goodridge, Mark Gwydir, Larry Kaufer, Jack Lubowsky, Abe Mantel, George Miller, Jim Peluso, Rich Silvestri, and Tom Timchek. We continue to extend our sincere thanks and appreciation to Ken Lemp for his valuable suggestions and helpful editing comments. We continue to offer our heartfelt thanks to Joan Tomaszewski and William Porreca for their contributions to our previous editions of the text.

We owe a debt of gratitude to all our students especially their suggestions to improving the text.

Any suggestions for future revisions would be greatly appreciated. These suggestions can be emailed to us at either: carmine.desanto@ncc.edu, richard.moscatelli@ncc.edu, michael.totoro@ncc.edu, or rachel.rojas@ncc.edu.

It is our pleasure to thank all the professionals at Pearson Learning Solutions for their outstanding work on this project. In particular, we would like to thank Matt Vitale, Vicki Pollack, Gina Kayed, Alex Diskin, Ruth Berrie for their assistance, guidance, and dedication to our project. In addition, the technical support and editing assistance provided by Meg Tiedemann.

Finally, and most importantly, we would like to thank our spouses Debbie, Kathleen, Roseann, and Hamilton. Of course, our children Craig, Matthew, Allison, Alexandra, Richard, Katherine, Caleb, Lylah, and Risa for their continuous understanding, patience, and encouragement during the preparation and revision of this project, and for bringing love and happiness to our lives.

Carmine DeSanto
Michael Totoro
Richard Moscatelli
Rachel Rojas

CHAPTER 1

Introduction to Statistics

Contents

1.1 Introduction
    What Is Statistics?
1.2 Why Sample?
    Why Sample?
1.3 Sampling Techniques
    Random or Probability Sampling Techniques
      Simple Random Sampling or Random Sampling
1.4 Uses of Statistics
1.5 Misuses of Statistics
    Misleading Graphs
    Non-Representative Samples
    Inappropriate Comparisons
    The Omission of Variation About an Average
1.6 Overview and Summary

1.1 INTRODUCTION

What Is Statistics?

Your life is a statistic!
    Can you describe a *typical day* in your life?
    For example:

- What do you have for breakfast?
  - eggs or egg beaters?
  - pastry or oat bran muffins?
  - coffee, tea or decaffeinated drink?
  - orange or grapefruit juice?
- Are you a full-time or part-time student?
- How many hours a week do you work?

Consider the following fascinating facts that have been used to describe an "average" day in America. Did you know that on an average day in America . . .

- 128,000,000 people use Facebook.
- 691 people get a nose job (rhinoplasty).
- 20,547 cars are purchased.
- 38 million people eat fast food.
- 87,000 airplanes are in the skies.
- There are 3,693 fires.
- 5.2 billion text messages are sent.
- 144,000 hours of video are uploaded to YouTube.
- The U.S. Postal Service sells 90 million stamps, handles 230 million pieces of mail and delivers 834,000 packages.
- The snack bar at Chicago's O'Hare Airport sells 5,479 hot dogs, covered with 12 gallons of relish and nine gallons of mustard, washed down with 890 gallons of coffee![1]

---

[1] These facts appeared in *American Averages: Amazing Facts of Everyday Life,* by M. Feinsilber and W. B. Mead (Doubleday, New York).

The previous statements represent statistical descriptions of a large collection of information or observations for phenomenon which can be described with a *numerical value* such as a person's height or *non-numerical value* like a person's hair color. Before these numerical and/or non-numerical value's of information or observations are arranged or analyzed in a useful manner they are called raw data values or simply **raw data**. The term **raw** is used because statistical techniques have not as of yet been applied to analyze the data.

We will be primarily concerned with the application of statistical techniques to the collected raw data. The statistical techniques we will examine provides us with ways to summarize and describe the data, to search for patterns or relationships within the data, and to draw inferences or conclusions from the data. Essentially, statistics can be defined as follows.

> **DEFINITION 1.1**
> **Statistics** involves the procedures associated with the data collection process, the summarizing and interpretation of data, and the drawing of inferences or conclusions based upon the analysis of the data.

After raw data is collected, we are usually confronted with a massive list of numbers, names, opinions, dates, measurements, percentages, etc. To make such a collection of raw data more meaningful and useful we need to examine the data set in a formal way.

A formal exploration of raw data may involve the use of:

- graphs to visually display the data, or
- numerical techniques to describe the main characteristics of the data.

These data explorations can also assist us in *discovering patterns or relationships within the data,* which will help *to generate hypotheses or to draw inferences* from the data set.

Essentially the study of statistics involves describing the main characteristics of raw data as well as drawing conclusions from the data based upon the analysis of the data. From this point of view, statistics can thus be subdivided into two branches: *Descriptive Statistics* and *Inferential Statistics*.

> **DEFINITION 1.2**
> **Descriptive Statistics** uses numerical and/or visual techniques to summarize or describe the data in a clear and effective manner.

Let us consider Case Study 1.1 to illustrate the use of descriptive statistics.

> **CASE STUDY 1.1 NHL as a Growth Industry**
>
> The snapshot entitled "NHL a Growth Industry" is shown in Figure 1.1. This snapshot compares the size of the average NHL player of the 1971–72 season to the average NHL player of the 2018–19 season.
>
> According to the snapshot, the *average NHL player* in 1971–72 was 5 feet 11 inches tall and weighed 184 pounds while the *average NHL player* in 2018–19 stands 6 feet 1 inch and weighs in at 201 pounds. The averages used to describe the height and weight of a typical hockey player for the indicated years were determined using a numerical formula. To appreciate the concise manner in which the average NHL player is described, realize that the data from which these averages were computed contained the weights and heights of **ALL** the NHL players for the years 1971–72 and 2018–19. If you were presented with this large data list rather than the average weight and height, you would not find this list helpful in visualizing a typical NHL player. However, presenting the *average weight and height of an NHL player, as shown in Figure 1.1, is a more concise way to numerically describe the "typical"* NHL player which helps us to comprehend the size of a professional hockey player.

CASE STUDY 1.1 NHL as a Growth Industry (continued)

Statistical Snapshot

NHL as a Growth Industry

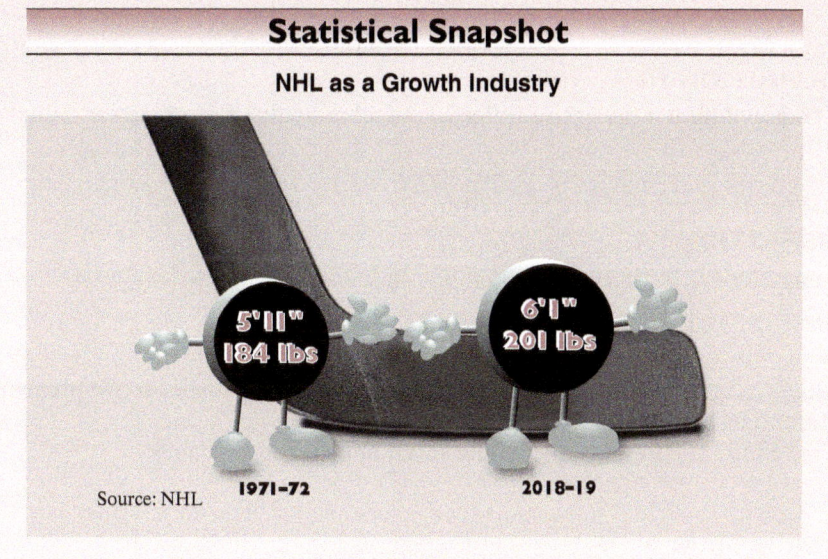

5'11" 184 lbs — 1971-72
6'1" 201 lbs — 2018-19

Source: NHL

FIGURE 1.1

Inferential statistics is the branch of statistics that involves drawing conclusions about a large group, a population, based on the analysis of a smaller group of data, a sample, collected from the population. Let's examine this concept in *Case Study 1.2*.

CASE STUDY 1.2 Time to Relax?

According to a *CNN/Gallup* nationwide telephone poll of 1,000 adults regarding leisure time, *52% of adults say they don't have enough free time.* This poll result is illustrated in Figure 1.2.

Statistical Snapshot

CNN/Gallup Poll States: Most Adults Say They Don't Have Enough Time For Leisure

Enough 48%

Not Enough 52%

FIGURE 1.2

The *CNN/Gallup* pollsters are making an **inference** about the **typical American adult's** leisure time. A conclusion that can be drawn from this survey is that **most U.S. adults** *feel that they don't have enough time for leisure.* Since it is impractical for the pollsters to survey the response of ALL U.S. adults on leisure time, they used the response of only some (1,000) U.S. adults to infer about the leisure time of the majority of ALL U.S. adults. *This conclusion about the leisure time of the majority of ALL U.S. adults based on the response of only 1,000 adults is an example of inferential statistics.*

In Case Study 1.2, the pollsters based their inference about **all U.S. adults** on **the opinions of 1000 adults** during their telephone survey. This is the **essence of Inferential Statistics**. These pollsters are drawing **a conclusion or inference about *ALL* adults using the opinions of only a portion**, (i.e. the 1000 adults telephoned in their nationwide poll), **of U.S. adults**. The pollsters are really interested in the **opinions of ALL U.S. adults**; however it is impractical to poll every U.S. adult. In statistics the entire group of interest is referred to as the **population**, while the smaller group selected from the population for analysis is called the **sample**.

> **DEFINITION 1.3**
> The **population** is the entire collection of all individuals or objects of interest.

> **DEFINITION 1.4**
> The **sample** is the portion of the population that is selected for study.

This concept plays an important role in our study of statistics so we present another case study on the use of inferential statistics.

Case Study 1.3 presents inferential social networking statistics.

CASE STUDY 1.3 Social Networking Statistics

a) Select a graphic within Figures 1.3a and 1.3b name the population(s) being described.
b) Identify the sample used to define one of these populations. How was the sample selected and try to explain what that means to you? What is the sample size—i.e. the number of people selected for the sample?
c) State two conclusions that you can infer from the survey about each population. In each case, identify the data values of interest for the population.
d) Identify the different social medias stated within Figures 1.3a and 1.3.b state some interesting facts associated with each social media.
e) Discuss what conclusions you can draw from the graphic regarding the Daily Active Users and Registered Users of the social medias.
f) Discuss the characteristics of the typical daily active users of the different social medias for males or females.
g) Give some statistics about the users of these different social medias.
h) What conclusion can be drawn about the percentage of smartphone users connected to Facebook?
i) What conclusion can be drawn about the number of photos that are added to Facebook every day?
j) What conclusion can be drawn about the time Internet users spend on Facebook?
k) What conclusion can be drawn about the content of retweets?
l) Compare the percentage of male vs female users for each social media.
m) Compare and draw conclusions of YouTube users to Pinterest users.
n) Compare and draw conclusions of Google+ users to Facebook users.
o) Describe a typical Social Gaming user and give specific statistics to support your response.
p) Describe how businesses use social medias and give specific statistics to support your response.
q) Discuss some mobile use statistics and compare the results for different countries.
r) How likely is a social media user to check their security settings? Give statistics to support your response.
s) What do internet users do to find a good deal on the internet? Give statistics to support your response.
t) Write a composite profile of a typical user for each social media identified within Figures 1.3a and 1.3b based upon the results shown within the figure. For example, for a Facebook user, begin the composite: "the typical Facebook user is more likely to be female, …".
u) Why is this case study an example of inferential statistics? Explain your answer.
v) If we could do the improbable and survey every internet user in the U.S., do you think the results of the information within Figures 1.3a and 1.3b would be exactly the same as the results for ALL the users? Or do you believe there would a difference in the sample results shown in Figures 1.3a and 1.3b the population results? Explain.
w) If additional random samples of internet users were selected, do you believe the results of these new samples would be exactly the same as the results of these samples? Or do you believe there would be a difference? Explain.

CASE STUDY 1.3 Social Networking Statistics

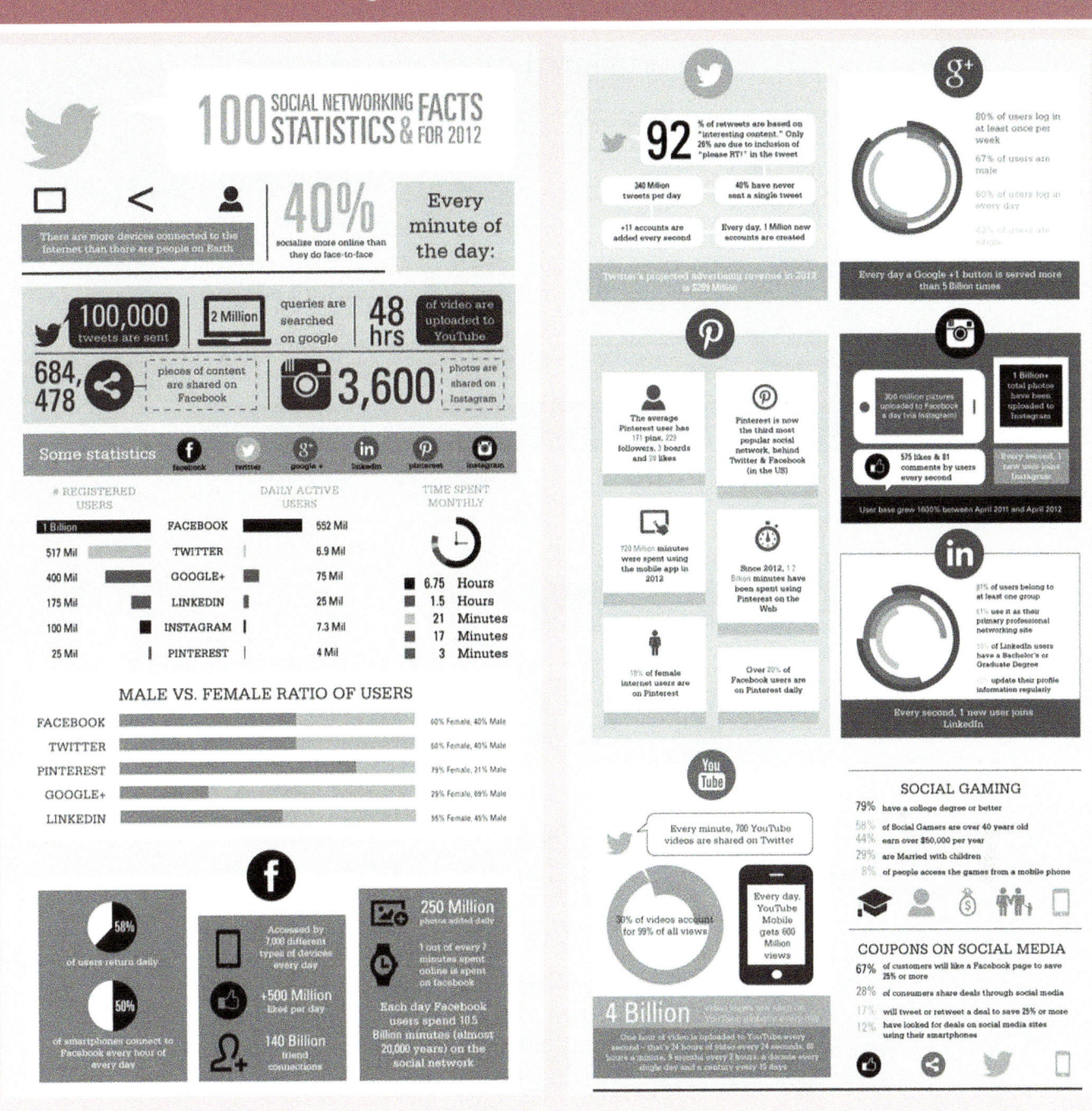

FIGURE 1.3a
Courtesy of Giovanna Bargh/Creotivo.

CASE STUDY 1.3 Social Networking Statistics

FIGURE 1.3b
Courtesy of Giovanna Bargh/Creotivo.

In Case Study 1.2, the pollsters selected **a sample of 1,000 U.S. adults** to poll and used this sample to obtain data regarding the opinions of **the population of ALL U.S. adults**. We may consider *the population to consist of all U.S. adults or the opinions of all U.S. adults.* The **opinions of all the U.S. adults** represent the **data values of the population**. The pollsters used the data values of the sample or opinions of 1,000 adults to draw a conclusion about the entire population. We will now use the notions of population and sample to define inferential statistics.

DEFINITION 1.5
Inferential Statistics is the process of using **sample information** to draw inferences or conclusions about the **population**.

Inferential Statistics is used to help researchers, pollsters and/or decision makers draw conclusions or inferences about a population. The following statements represent situations where sample information can be used to draw conclusions about a population.

- A medical researcher wants to determine if large doses of vitamin C are effective in combating colds.
- Pollsters want to estimate what percentage of the American public approve of the death penalty to curb crime.
- A quality control engineer for a smartphone manufacturer must determine whether their smartphone meets the manufacturers' specifications before shipment.
- The Food and Drug Administration would like to determine if men who use Viagra have undesirable side effects.
- A market researcher wants to know in what quantities the American consumer is willing to purchase a new product.

In most applications of statistics researchers use a sample rather than the entire population because it is usually impractical or impossible to obtain all the population observations or measurements.

For example, suppose an electrical company wanted to determine the average life of their new 40 watt light bulb. All the 40 watt bulbs the company manufactures would constitute the population. To obtain the average life of the population of light bulbs it would be necessary to compute the life of each bulb. Obviously, this would be impractical since they would not have any bulbs left to sell! In such instances, it becomes necessary to select a *representative sample* (i.e., a sample with similar pertinent characteristics as all the new 40 watt light bulbs), that can be used to make inferences about the entire population.

Thus, the electrical company would try to design a procedure to select a representative sample of these new light bulbs and use this sample data to estimate the population average life of their new 40 watt light bulbs.

DEFINITION 1.6
Representative sample is a sample that has the pertinent characteristics of the population in the same proportion, as they are included in that population.

Since the **primary objective of inferential statistics** is to use sample information to estimate a population characteristic, it is imperative that the researcher try to design a procedure to select a **sample which is representative of the population**. For example, if a population has 60% defective light bulbs, then a representative sample of this population should also contain 60% defective light bulbs. If the sample is **not representative** of the population, then inferences about the population characteristics **may not** be reasonable. The idea of ensuring sampling representativeness is similar to the "toothpick" technique used to determine if a cake is completely baked. A toothpick is inserted in **several areas** of the cake, and if the toothpick comes out free of cake batter each time, we conclude that the **entire** cake is done.

In the previous example about the new 40 watt light bulb, a sample was selected and the **sample average** was used **to estimate** the **average life** for *all* **the new 40 watt light bulbs**. That is, a sample average was used to estimate the population average. Whenever a number is used to describe a characteristic of a sample, such as sample average, this number is called a **statistic**.

> **DEFINITION 1.7**
> A **statistic** is a number that describes a characteristic of a sample.

On the other hand, a number like the population average, which describes a characteristic of a population, is called a **parameter**.

> **DEFINITION 1.8**
> A **parameter** is a number that describes a characteristic of a population.

Thus, a population average is an example of a parameter while a sample average is an example of a statistic. The sample statistic (i.e., sample average for the 40 watt light bulb) is being used to estimate the population parameter (i.e., the average life of the population of new light bulbs). We can say that the concept of inferential statistics is to use a sample statistic to make inferences about a population parameter.

Examples 1.1 and 1.2 illustrate the difference between a parameter and a statistic.

EXAMPLE 1.1

A statistician computes the batting average of all American League players to be .256 and the batting average of 24 players randomly selected from all the American League players to be .267. If the population is defined to be all the baseball players in the American League, then determine which batting average represents a parameter and which one is a statistic?

Solution

Since the population is defined to be the baseball players in the American League, then the .256 batting average is an example of a parameter because it describes the batting average or characteristic of all American League players, the population. Since the 24 players randomly selected from the population is a sample of all American League players, then the .267 batting average is an example of a statistic because it describes the batting average of a sample of American League players.

EXAMPLE 1.2

A politician is interested in determining the proportion of 20,000 voters within her district who will vote for her in the upcoming election. An opinion poll of 1500 potential voters was taken to estimate how all the voters in her district will vote in the upcoming election. The results of the opinion poll indicate that 52% will vote for the politician in the upcoming election. Determine:
a) The population.
b) The sample.
c) Whether the result, 52%, is a parameter or statistic.

Solution

a) The 20,000 voters in her district comprise the population.
b) The 1500 potential voters polled represents the sample.
c) The value 52% is a statistic since it describes a characteristic of the sample of 1500 potential voters.

CASE STUDY 1.4 Conducting Polls

The article in Figure 1.4 appeared in an edition of *The New York Times* explaining how a latest *New York Times*/CBS News Poll was conducted. As you carefully read this article notice how the pollsters took great care in selecting their sample.

How the Poll Was Conducted

The latest *New York Times*/CBS News Poll is based on telephone interviews conducted Wednesday and Thursday with 944 adults throughout the United States. The sample of telephone exchanges was randomly selected by a computer from a complete list of more than 42,000 active residential exchanges across the country.

Within each exchange, random digits were added to form a complete telephone number, thus permitting access to both listed and unlisted numbers. Within each household, one adult was designated by a random procedure to be the respondent for the survey.

The results have been weighted to take account of household size and number of telephone lines into residence and to adjust for variations in the sample relating to geographic region, gender, race, age, and education.

In theory, in 19 cases out of 20 the results based on such samples will differ by no more than three percentage points in either direction from what would have been obtained by seeking out all American adults.

For smaller subgroups the potential sampling error is larger. For example, for those adults who are between 18 and 44 years of age and single it is plus or minus about five percentage points.

In addition to sampling error, the practical difficulties of conducting any survey of public opinion may introduce other sources of error into the poll. Variations in question wording or the order of questions, for instance, can lead to somewhat different results.

FIGURE 1.4

a) Identify the population and the number of people selected for the sample, that is, the sample size.
b) Describe the sampling techniques used by the pollsters in their attempt to obtain a representative sample.
c) What were the pollsters trying to accomplish by using a random procedure to decide who should be designated as the respondent to their poll?
d) In the article the following statement appears:

> In theory, in 19 cases out of 20 the results based on such samples will differ by no more than three percentage points in either direction from what would have been obtained by seeking out all American adults.

If a second sample were selected using this same procedure, do you believe the results of the second sample would be identical to the first? Explain your answer in relation to the previous statement.
e) This case study uses terms that identify some important concepts that will be studied in later chapters of this text. What do the terms **random digits** and **sampling error** mean to you?
f) After reading this article, list some areas you would need to address if you planned to take a survey in your town.

1.2 WHY SAMPLE?

Consider the following scenario.

A shopper, who is interested in purchasing a quart of nonfat yogurt, is debating whether to buy a new exotic flavor called "Hawaiian Pleasure" or the old reliable flavor "Simply Vanilla." Before making a decision, the shopper requests a spoonful of the new flavor from the salesperson. Based upon this small spoonful, the shopper decides to purchase Hawaiian Pleasure.

This typical scenario represents the idea of **Inferential Statistics**. The shopper's decision to purchase the Hawaiian Yogurt was based only on a spoonful, or sample, of this yogurt. Obviously, it was not necessary for the shopper to buy and eat the entire quart of Hawaiian Pleasure Yogurt before determining whether or not this new flavor tasted good enough to purchase. This idea of selecting a portion, or sample, to determine the taste or characteristics of all the yogurt, or population, is the essence of inferential statistics. The process of selecting a sample from a population is called sampling.

> **DEFINITION 1.9**
> **Sampling** is the process of selecting a portion, or sample, of the entire population.

The information acquired from a sample is the basis of inferential statistics. A sample is merely a portion of a population, and, when we have only a portion of a population, any conclusion drawn about the characteristics of a population are inferred from the sample data.

Suppose the editors of a student newspaper at a large University are interested in determining the opinion of all U.S. college students on whether they favor some form of censorship of rock music. Is it necessary for the editors to ask "every" college student within the U.S. before drawing a conclusion about how college students feel about censorship of rock music? If the editors had access to the population of all college students, then they could obtain the opinions of all the college students. Such a survey is called a **census**.

> **DEFINITION 1.10**
> A survey that includes every item or individual of the population is called a **census**.

How practical would it be for the editors to take a census to determine the opinions of all the college students? Trying to interview every college student in the United States would become an impossible task in terms of time and money. In practice, a census is rarely taken because it requires an excessive amount of time and it is too costly to gather all the information about the entire population. In addition, it would be virtually impossible for the editors to try to identify each student of the population. The question then becomes: "Is it really necessary for the editors to interview the population of all college students before a valid conclusion can be drawn about the opinions of college students on censorship?" Statisticians are often faced with similar situations. In such situations, statisticians are forced to take a sample to obtain information about the population. However, if a conclusion or inference about a population is based on sample data, then a statistician must try to insure that the sample is representative of the population from which it is drawn. If conclusions are drawn about a population from a nonrepresentative sample, then these conclusions are likely to be incorrect. On the other hand, if the sample is representative, then inferences about the population based on the sample information are likely to be correct. Thus, the editors should try to select and interview a representative sample of college students about censorship of rock music. They could then use this sample information to draw a conclusion about the views of the population of all the college students.

Why Sample?

Samples are taken when it is either impossible or impractical to examine the entire population. Studying an entire population is impractical when the population is extremely large. For such large populations, trying to select every data value (item or individual) in the population would be costly, time consuming or just impossible.

For example:

1. The quality control department at G.E. must test the production of its light bulbs to determine if they meet company specifications regarding the mean life of the bulbs. To check the life of the bulbs produced,

the quality control manager randomly selects 36 bulbs over the daily production of these bulbs and determines the life of each bulb.
2. The Gallup Poll randomly surveys 1500 registered voters within the U.S. to determine if presidential candidate Don Key will win the upcoming presidential election.
3. A Florida distributor of citrus fruits wants to determine the mean weight of Florida oranges to help estimate the cost of shipping the oranges. A sample of oranges are randomly selected and weighed.
4. A medical researcher wishes to determine the mean cholesterol level of men aged 40–59 years old living in a large Maryland community. She randomly selects 100 men aged 40–59 years old and determines their cholesterol levels.

In each of these previous situations, it would be extremely expensive, if not impossible, to test each bulb, to weigh each Florida orange, contact every registered U.S. voter or to measure the cholesterol level of every 40–59 year old man living within the community. In addition, if the G.E. quality control manager were to test every bulb produced, there would be no bulbs left to sell. This type of testing is called **destructive testing** and requires the use of a sample. Consequently, only a small portion of items or individuals is selected to study rather than the entire group. These examples illustrate the idea of using sample information to try to draw a valid conclusion about the population.

In this chapter, we discussed some common statistical terms such as a population, and a sample. Let's review these statistical terms using the previous examples. Recall, the word **population** is used to refer to the entire group that has been identified for study. A population can refer to items or individuals. A **sample** is a portion of items or individuals chosen from the population. The population for the previous examples was all the bulbs G.E. manufactured per day, all the registered U.S. voters, all the oranges the distributor will ship, and the cholesterol measurements of all the 40–59 year old men living within the community. Notice from these examples, the population can be items such as bulbs, oranges, or individuals like registered U.S voters. In each of these situations, a small portion or sample was selected from the entire group or population to draw a conclusion about the entire population. This is the objective of **Inferential Statistics**. For example,

The quality control manager might observe that the sample of 36 bulbs burn out very quickly. Thus, on the basis of this sample of 36 bulbs, the manager would probably conclude that *all* the G.E. bulbs manufactured during the production process were not satisfactory,

or

A Gallup Poll reports that 63% of the 1500 registered U.S. voters preferred presidential candidate Don Key. Based on the results of this sample of 1500 voters, we would expect the population of registered U.S. voters to elect candidate Don Key president, if the election were held today.

In the Gallup Poll example, the 1500 registered U.S. voters represented the **size** of their **sample,** and all registered U.S. voters represented the **population size.**

DEFINITION 1.11
Sample Size — The size of a sample, denoted by n, is the number of data values in the sample.

DEFINITION 1.12
Population Size — The size of a finite population, denoted by N, is the number of data values in the population.

In each of these situations, it is virtually impossible to study the entire population. If we were to test every bulb to determine how many hours it will last before burning out, there would be no bulbs left to sell. Similarly, opinion polls conducted by such pollsters as the Gallup Poll are given to only a sample of people because it would be very time consuming and extremely expensive or impractical to interview everyone in the population. Consequently, we sample the population because it is frequently impossible to study all the individuals or items of a population.

Let's now summarize some of the reasons a sample is selected rather than study the entire population.

Chapter 1 *Introduction to Statistics*

> ### Reasons for Sampling
>
> 1. **A Census is Impossible**
> At times, a census may be impossible because obtaining information about every item in the population will eventually destroy all the items within the population. For example, a cigarette manufacturer would have to burn every cigarette they manufactured if they were interested in obtaining the average nicotine content of all their cigarettes.
>
> A census may also be impossible because the population is so large or infinite that it is practically impossible to obtain information about all the items or individuals of the population. For example, a population may be defined to be all infants who have ever been born or will ever be born. Such a population is endless and would be impossible to study all the individuals.
>
> 2. **Cost**
> The cost of studying all the items of a population or contacting all the individuals within a population is too expensive. For example, it would be too costly for the Gallup Poll to conduct a telephone survey or mail a questionnaire to the population of all U.S. registered voters regarding their opinions on a presidential race.
>
> 3. **Time**
> One major reason of sampling is that it requires less time to take a sample than to study the entire population. Sample information can be processed much faster the population data. For example, the Gallup Poll can survey 1500 people and analyze the results of its survey much faster than the U.S. Census Bureau can analyze the results of its 10 year census.
>
> 4. **Accuracy**
> The results of a sample may be more accurate than trying to study the entire population since gathering data from fewer items or individuals will probably result in fewer human errors when collecting and analyzing the sample data. A statistician surveying a small portion of the population will be able to do a complete and comprehensive job in gathering and analyzing sample data.

Since populations can rarely be studied completely, we must depend upon selecting a sample to draw a conclusion about the characteristics of the population. Thus, our interest is in making inferences from sample data concerning the characteristics of a population. In the example of the Gallup Poll, although the pollsters surveyed 1500 voters to determine how they intend to vote in the upcoming election, their primary interest is not how these 1500 people will vote, but, rather in estimating how the entire population of voters will vote in the upcoming election.

Some numerical characteristics of a population that we are usually interested in describing using sample information are the average, or the variability of a population. Recall from Section 1.1 that a numerical characteristic of a population such as the **population average** or **variability** is called a **parameter**. While a numerical characteristic of a sample such as the sample average or variability is an example of a **statistic**.

Suppose the average IQ of all the students attending a large community college in California is 112. This numerical value 112 is a population parameter because it represents a characteristic of the population of all students attending the community college. On the other hand, if we were to determine the average IQ of a class of statistics students attending this community college to be 114, then the numerical value 114 describes a characteristic of a sample of students. In this case, 114 would be a statistic since it describes the average IQ of a sample of students attending the community college.

Studying the entire population is usually impractical; consequently, we use sample information to estimate the characteristics of a population. That is, we would use a statistic such as the sample average to make inferences about the corresponding population parameter, the **population average**.

One important aspect of Inferential Statistics is to select a representative sample of the population. Since a sample provides only partial information about the population, there is always a chance that any conclusion about the population, based upon sample information, can be subject to error. One way to minimize this error is to insure that you have selected a representative sample from the population. Let's now discuss ways to select a sample from a population.

1.3 SAMPLING TECHNIQUES

In this section, we will examine methods to select a sample from a population. One objective of statistical inference is to try to select a representative sample from the population. If 60% of a population is female, then a **representative sample** of this population should also contain 60% females.

DEFINITION 1.13
A **representative sample** is a sample containing similar characteristics of the population.

Selecting a **nonrepresentative** or **biased** sample from a population can lead to invalid inferences about the population. A **nonrepresentative** or **biased sample** is a sample that either under selects or over selects from certain groups within the population. For example, if a population consisted of 40% males and 60% females, then a representative sample would also contain 60% females. A sample containing 80% females would be a nonrepresentative or biased sample because the percentage of females within the sample is much greater than within the population.

Nonrepresentative or biased samples may yield results that may not be representative of the population and may lead to a distorted picture of the population. A famous example of a nonrepresentative or biased sample leading to an invalid inference is the 1936 *Literary Digest* Poll.

In 1936, a magazine entitled the *Literary Digest* conducted a poll to try to predict the winner of the presidential election between the incumbent President Franklin Delano Roosevelt and the Republican Governor of Kansas, Alfred M. Landon. Most political analysts predicted President Roosevelt would win reelection.

However, the *Literary Digest* gained attention when they predicted Landon would win easily, 57% to 43%. The *Digest* mailed out 10 million questionnaires to prospective voters. However, a major flaw with their poll was that only 2.4 million people returned the questionnaires. Although a sample of over 2 million people may seem to be a reasonably large sample, the *Digest*'s sample was a nonrepresentative or biased sample of voters. Another flaw was the way the *Literary Digest* selected its sample of voters. The *Digest* selected its sample from lists of telephone numbers, automobile registrations and its own list of subscribers. The *Literary Digest*'s selection process for their sample was biased because it favored Republican voters and was nonrepresentative of the voting population. During the depression of 1936, phones, automobiles and magazine subscriptions were luxury items for many people. Most of the wealthy people who could afford these luxuries favored the Republican candidate Landon. Thus, the *Digest*'s sampling procedure led to the selection of a biased sample since it was not representative of the voter population. Although, the *Literary Digest* had correctly predicted the previous five presidential elections, their prediction for the 1936 election was completely inaccurate. In fact, President Roosevelt won reelection by a landslide margin of 62% to 38%. Thus, although, the *Literary Digest* had selected a very large sample their prediction was incorrect because their sampling procedure led to the selection of a nonrepresentative or biased sample. Not too long after their inaccurate prediction, the *Digest* went bankrupt.

The key to inferential statistics is to select a representative sample. We will discuss a sampling technique that is more likely to produce a representative sample. Notice we said "more likely to produce a representative sample." There is always a possibility that a sample selected by this sampling technique may be a nonrepresentative or biased sample. The sampling procedure we will consider is referred to as **random** or **probability sampling technique.** In random or probability sampling, every data value of the population has a chance of being chosen in the sample.

DEFINITION 1.14
A **random** or **probability sample** is a sample which is selected in such a manner that each data value of the population has a non-zero probability of being selected for the sample.

Random or Probability Sampling Techniques

In a random or probability sample, the items or individuals are selected by chance alone. Since each item or individual of the population has a chance of being selected for the sample, the probabilities can be calculated and used to make statistical inferences about the population. Such statements as "we are 90% confident that Candidate Don Key will receive 62% of the votes, plus or minus 3.5 percentage points" can be made. Although there are four random sampling techniques:

1. simple random sampling
2. systematic sampling
3. stratified sampling
4. cluster sampling

We will only consider simple random sampling in the text. Systematic, stratified and cluster sampling techniques are discussed on the web at MyStatLab.com.

Simple Random Sampling or Random Sampling

A **simple random sampling** or **random sampling** is the most fundamental statistical method of selecting a sample. Simple random sampling is a procedure which provides each item or individual of the population an equal chance of being selected. In addition, every possible sample has the same probability of being selected.

> **DEFINITION 1.15**
> A **simple random sample** is a sample of data values selected from a population in such a way that every sample of size n has an equal probability of being selected and every data value of the population has the same chance of being selected for the sample.

EXAMPLE 1.3

Determine if each of the following sampling procedures will produce a random sample.
a) The population is all females between the ages of seventeen and thirty-five living in New York State. The sampling procedure is devised to select a sample of 100 female college students within New York State.
b) The population is defined to be all the people who shop at the Green Acres Mall during the week before Christmas. The sampling procedure is devised to select every 25th shopper who enters the Mall at only one of its ten entrances.
c) The population is all the faculty members teaching at the College of Bill and Mary. The sampling procedure will select a sample of 25 faculty members using the following procedure: The name of each faculty member is placed in a large drum. The drum is rotated several times before each of the 25 faculty names are selected.
d) An advertising agency decides to conduct a marketing poll of people living in a New England community. The sampling procedure will draw a sample of people living within this community by selecting the names of people from the community telephone directory.

Solution

a) The sampling procedure of selecting 100 female college students within New York State is *not* a random sampling procedure because the females aged 17 to 35 years old who are not *college* students do not have a chance of being selected.
b) The sampling procedure of selecting every 25th shopper who enters the Mall at only one of its entrances is not a random sampling procedure because the shoppers who use a different entrance to the Mall do not have a chance of being selected.
c) The sampling procedure of selecting the 25 faculty members *is* a random sampling procedure because every faculty member has an equal chance of being chosen and every possible sample of size 25 had an *equal* chance of being selected.
d) The sampling procedure of selecting the names of the people from the community telephone directory is not a random sampling procedure because the people who have an unlisted telephone number will not have a chance of being selected.

How can we choose a random sample? Let's illustrate one way to select a random sample.
Suppose there are 200 students in a large lecture class and the instructor wants to select 10 students from this class for an experiment.
One way to select a random sample of 10 students from the population of 200 students is to write the name of each student on a slip of paper, put the 200 slips in a box, thoroughly mix the slips within the box, and select the first slip. Repeat this process until this sample of 10 students is selected. The names of the students on the chosen slips represent a random sample. This procedure is acceptable if the slips are the same size, the population size is reasonably small and the slips are mixed thoroughly. If the population size is extremely large, this procedure can be very tedious. A more convenient procedure of selecting a random sample is to use random numbers. These numbers can be generated either by a computer program such as MINITAB or by using a table of random numbers such as Table I in Appendix C: Statistical Tables. Table I contains a table of random numbers that have been generated by some random process in which each of the ten digits from 0 to 9 have the same probability of being selected. For example, the numbers 187 and 024 have the same chance of being selected using this table of random numbers.

To use a table of random numbers to select a random sample of 10 students, you must first assign an identifying number to each of the 200 students in the class. Suppose we assign every student in the class a 3 digit number from 000 to 199. The random number table, Table I, can now be used to choose the sample of 10 students. Table I in Appendix C has 50 rows and 10 columns of 5 digits each. Notice in Table I the random numbers are recorded in blocks of 5 digits. To randomly select the 10 students we can start anywhere within the random number table. One way would be to close your eyes and place your finger anywhere on the page and start at this point. From this starting point, you can move in any direction, that is, to the left, to the right, up, down or even diagonally across the table. Suppose we start at the first block of random numbers in the 36th row from the top of Table I. The random number in the 36th row and the first column is 04230. The 5 rows starting with the 36th row from Table I are shown in Figure 1.5.

Beginning with the random number 04230 as the starting point, we need to select 10 three digit numbers. Since each random number is a five-digit number, we will only use the first 3 digits of the random number to identify the first student from the population. Thus, the random number *04*230 will be used to identify the student number 042 as the first student selected from the random sample. To continue selecting students, we can move in any direction. Suppose we decide to move to the right. The random number to the right of 04230 is *16*831. The first 3 digits of this second random number is 168. Thus, the student identified by number 168 will be the second student selected for our sample. The next random number to the right is *69*085. The first 3 digits of this third random number is 690. We cannot use 690 because the largest student number in the population is 199. This random number is skipped and we continue moving to the right to select our next random number. However, the next two random numbers *30*802 and *65*559 must also be skipped. The random number 09205 identifies the student with the number 092. We continue this process until the 10 students are randomly selected. This sample of 10 students is called a *simple random sample*. Figure 1.6 contains the 3 digit numbers observed in the random number table using our sampling procedure.

Portion of Random Number Table: Table I
Starting with 36th Row

	1st student	2nd student	3rd student	4th student	5th student	6th student	7th student	8th student	9th student	10th student
Starting point	04230	16831	69085	30802	65559	09205	71829	06489	85650	38707
	94879	56606	30401	02602	57658	70091	54986	41394	60437	03195
	71446	15232	66715	26385	91518	70566	02888	79941	39684	54315
	32886	05644	79316	09819	00813	88407	17461	73925	53037	91904
	62048	33711	25290	21526	02223	75947	66466	06232	10913	75366

FIGURE 1.5

Results of the Random Sampling Technique

3-digit number	outcome	3-digit number	outcome
042	1st student selected	413	skipped
168	2nd student selected	604	skipped
690	skipped	031	6th studet selected
308	skipped	714	skipped
655	skipped	152	7th student selected
092	3rd student selected	667	skipped
718	skipped	263	skipped
064	4th student selected	915	skipped
856	skipped	705	skipped
387	skipped	028	8th student selected
948	skipped	799	skipped
566	skipped	396	skipped
304	skipped	543	skipped
026	5th student selected	328	skipped
576	skipped	056	9th student selected
700	skipped	793	skipped
549	skipped	098	10th student selected

FIGURE 1.6

EXAMPLE 1.4

The Zoo, a New York FM radio station, decides to hold a contest to raffle 20 rock concert tickets. To enter the contest, each contestant must submit a postcard containing the answer to the Zoo's special music question. Each postcard the radio station receives with the correct response to their question is numbered. The postcards are numbered using a four digit number ranging from 0001 to 7126. Using the following portion of a 5-digit Random Number Table, starting at the random number 63271 and moving down each column, determine the twenty winning numbers which would be selected using this random sampling procedure.

Portion of Random Number Table

63271	59986	71744	51102	15141	80714	58683	93108	13554	79945
88547	09896	95436	79115	08303	01041	20030	63754	08459	28364
55957	57243	83865	09911	19761	66535	40102	26646	60147	15702
46276	87453	44790	67122	45573	84358	21625	16999	13385	22782
55363	07449	34835	15290	76616	67191	12777	21861	68689	03263

Solution

Beginning with the random number 63271 as the starting point, we need to select 20 *four* digit numbers. Since each random number is a five-digit number, we will only use the first 4 digits of the random number to identify the twenty winning numbers from the random number table.

Using this sampling procedure, the random number *63271* identifies the first winning postcard number 6327. To continue selecting winning numbers, we move down the column. The random number below the starting number is *88547*. The first 4 digits of this second random number is 8854. Since this second random number is outside the range of the postcard numbers 0001 to 7126, we skip this random number. The first four digits of the next random number 55957 is 5595. Thus, 5595 represents a winning number since it falls within the range of postcard numbers. Continuing with this sampling procedure, the twenty winning numbers are:

6327, 5595, 4627, 5536, 5998, 0989, 5724, 0744, 4479, 3483,
5110, 0991, 6712, 1529, 1514, 0830, 1976, 4557, 0104, 6653

CASE STUDY 1.5 Using Simple Random Sampling

The article entitled "Using Simple Random Sample to Study Larger Populations" shown in Figure 1.7 appeared in Investopedia. This article discusses the importance of using simple random sampling to draw conclusions from a population.

Using Simple Random Sample to Study Larger Populations
By GREG DEPERSIO (Investopedia: updated Feb 13, 2018)

Simple random sampling is a method used to cull a smaller sample size from a larger population and use it to research and make generalizations about the larger group. It is one of several methods statisticians and researchers use to extract a sample from a larger population; other methods include stratified random sampling and probability sampling. The advantages of a simple random sample include its ease of use and its accurate representation of the larger population.

How a Simple Random Sample Is Generated

Researchers generate a simple random sample by obtaining an exhaustive list of a larger population and then selecting, at random, a certain number of individuals to comprise the sample. With a simple random sample, every member of the larger population has an equal chance of being selected.

Researchers have two ways to generate a simple random sample. One is a manual lottery method. Each member of the larger population group is assigned a number. Next, numbers are

CASE STUDY 1.5 Using Simple Random Sampling (continued)

drawn at random to comprise the sample group. if a simple random sample were to be taken of 100 students in a high school with a population of 1,000, then every student should have a one in 10 chance of being selected.

The manual lottery method works well for smaller populations, but it isn't feasible for larger ones. In these situations, researchers prefer computer-generated selection. It works via the same principle, but a sophisticated computer system, rather than a human being, assigns numbers and selects them at random.

Room for Error

With a simple random sample, there has to be room for error represented by a plus and minus variance. For example, if in that same high school a survey were to be taken to determine how many students are left-handed, random sampling can determine that eight out of the 100 sampled are left-handed. The conclusion would be that 8% of the student population of the high school are left-handed, when in fact the global average would be closer to 10%.

The same is true regardless of the subject matter. A survey on the percentage of the student population that has green eyes or is physically incapacitated would result in a high mathematical probability based on a simple random survey, but always with a plus or minus variance. The only way to have a 100% accuracy rate would be to survey all 1,000 students which, while possible, would be impractical.

Advantages of Random Sampling

Simple random sample advantages include ease of use and accuracy of representation. No easier method exists to extract a research sample from a larger population than simple random sampling. There is no need to divide the population into sub-populations or take any steps further than plucking the number of research subjects needed at random from the larger group. Again, the only requirements are that randomness governs the selection process and that each member of the larger population has an equal probability of selection.

Selecting subjects completely at random from the larger population also yields a sample that is representative of the group being studied. Even sample sizes as small as 40 can exhibit low sampling error when simple random sampling is performed correctly. For any type of research on a population, using a representative sample to make inferences and generalizations about the larger group is critical; a biased sample can lead to incorrect conclusions being drawn about the larger population.

Simple random sampling is as simple as its name indicates, and it is accurate. These two characteristics give simple random sampling a strong advantage over other sampling methods when conducting research on a larger population.

FIGURE 1.7
Courtesy of Investopedia.

a) Explain the concept of a simple random sample.
b) Discuss different ways that a simple random sample can be selected.
c) Will the results of a simple random sample always accurately represent the population from which it was selected? Give examples to support your answer.
d) Why does the author prefer the simple random sampling process over other sampling techniques?
e) Explain the idea of a sampling error.
f) Why is a representative sample a necessary requirement when drawing a conclusion about a population from a simple random sample? Give examples to support your answer.
g) State two important characteristics about a simple random sample that support using it to draw a conclusion about the population it was selected from.

16 *Chapter 1 Introduction to Statistics*

1.4 USES OF STATISTICS

The role of statistics in our lives is increasing and some may even find it disturbing. Statements such as, "Your family's socio-economic index is too high to qualify for financial aid," or "Your cholesterol level is too high, therefore you are advised to change your diet immediately," are all by-products of the advances in data collection and processing which occurred during the twentieth century.

Today, even at an elementary level, it is impossible to understand economics, finance, psychology, sociology, or the physical sciences without some understanding of the meaning and interpretation of *average, variability, correlation, inference, charts or tables.*

Extensive data collection and distribution activities are performed by the federal and other governmental and private agencies in areas such as education, employment, health, crime, prices, housing, medical care, manufacturing, agriculture, construction, transportation, etc.

Furthermore, the development of the electronic computer has led to a revolution in the area of data collection and analysis. Statistical analysis can now be applied to vast amounts of data accurately and quickly. Such diverse problems as weather forecasting, economic stabilization, and disease control are today being solved using statistical analysis.

The following examples represent some applications of statistics.

EXAMPLE 1.5 "Census Bureau Decides Not To Use Statistical Sampling for the 2010 Census"

It is important for the Census Bureau to provide an accurate count of every person who lives in the United States. Census numbers are used to allocate billions of dollars in federal funds for schools, hospitals, highways, mass transit, etc. They are also used to draw political lines and apportion seats in Congress. The Census Bureau begins a census by mailing out questionnaires to every household in the country. However, most households do not mail the forms back.

Figure 1.8 shows the Census Mail Response Rates for the census years 1970 to 2010. The Census Bureau reported that the rate of nonresponse to their census forms had been increasing until 2010 as shown by Figure 1.8. This increasing rate of nonresponse forces the Census Bureau to send interviewers to visit those households who didn't return the census form. In spite of the many attempts to count every person who lives in the United States, the census usually results in an undercounting of minorities, children and the poor. This is the reason the Census Bureau wanted to use a *statistical sampling method* to estimate people who are too difficult to reach. Proponents of the statistical sampling method believe it will produce more accurate counts of the population, especially for the traditionally undercounted groups such as minorities living in urban areas. The plan was to use a mail questionnaire to count 90% of the population followed by a 10% probability sample of the remaining 10% to complete the enumeration phase. The Supreme Court ruled in 1999 that sampling could not under any circumstances be used to reapportion U.S. House seats. The Supreme Court left open the possibility that sampling could be used for other purposes, such as redrawing state legislature districts or

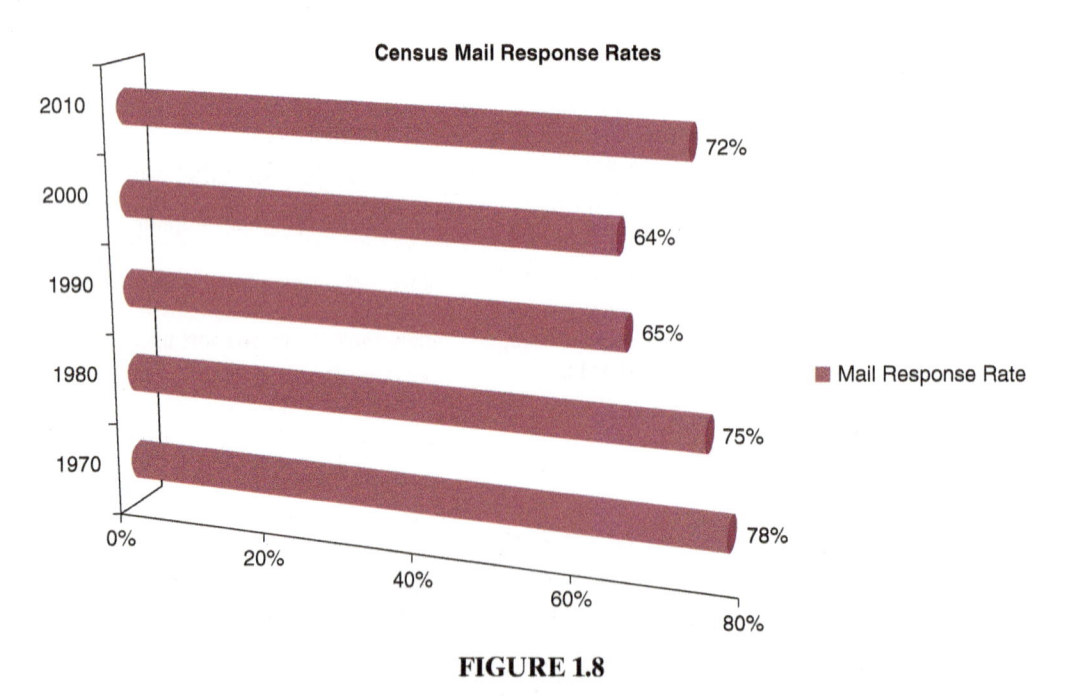

Census Mail Response Rates

Year	Mail Response Rate
2010	72%
2000	64%
1990	65%
1980	75%
1970	78%

FIGURE 1.8

allocating federal funds to cities and states. However, for the 2010 census, Robert Groves, the nominee to head the Census Bureau, ruled out using statistical sampling to adjust the results of the 2010 census. The census is an example of a vital descriptive statistic that may impact the disbursement of federal and state aid as well as the boundaries of the voting districts.

EXAMPLE 1.6 "Dewey Defeats Truman"

In the 1948 presidential election the Gallup Poll incorrectly predicted the defeat of Truman by Dewey. The Gallup Poll uses statistical methods in selecting samples for their public opinion surveys and applies inferential techniques to make predictions. Their method of sampling has produced presidential election predictions which have had an error averaging around 2% since 1948. (They failed to pick the winner in 1948 but have been correct ever since.)

Examples 1.7 and 1.8 discuss descriptive economic measures to reflect changes in the economy.

EXAMPLE 1.7 "Twitter is Behind Google Plus But Growing Faster"

GlobalWebIndex's latest figures show that while Google Plus is the second-most popular social networking service after Facebook, Twitter is actually growing at a slightly faster pace, increasing from 206 million users at the end of June last year to about 297 million today, a rate of 44%.

EXAMPLE 1.8 "Dow Gains More Than 850 Points as the Stock Market Soars"

The Dow Jones Industrial Average(DJIA) is an index of stock prices that is used to reflect the movement of stock prices on the New York Stock Exchange. It is used by the media to indicate the level of stock prices in the United States. For example, a dramatic increase in the Dow represents a major upward movement in the overall stock prices in the stock market. You might ask the question: "how accurate has the Dow Jones Industrial Average been in reflecting the performance of the stock market?" The Dow's tracking of the stock market movements over the long run closely parallels the Standard and Poor's 500 Stock Index, commonly called the S&P 500, which consists of the prices of 500 stocks. Thus, when we are asked: "What did the stock market do today?", it would be appropriate to use the Dow Jones Industrial Average to answer the question even though it is based on only 30 stocks.

EXAMPLE 1.9 "How Many Words Did Shakespeare Know?"

In a 1976 study by Efron & Thisted, they tried to estimate the number of words that Shakespeare knew. Efron and Thisted explored the detailed structure of the word counts from all the plays of Shakespeare. Their study revealed that: *Shakespeare wrote 31,534 words, of which 14,376 appear only once, 4,343 twice, and so on. They concluded that Shakespeare knew at least 35,000 more words.*

EXAMPLE 1.10 "There Is a Hurricane Watch in Effect for the Florida Coast"

Such forecasts are issued by the U.S. Weather Bureau which has the ability to gather weather data from all over the world, quickly analyze this vast amount of information, and produce relatively accurate predictions. Early warnings issued by the bureau are now responsible for the savings of thousands of lives and millions of dollars in property.

In Examples 1.11 and 1.12 the researchers analyzed the data for possible patterns or relationships that might exist. This type of analysis is called **Exploratory Data Analysis**.

EXAMPLE 1.11 "Diet Soda Study: Diet Soda May Be Linked to Health Problems from Obesity to Diabetes to Heart Disease"

A Purdue University study has found that people who drank sodas containing artificial sweeteners were more likely to gain weight than those who drank non-diet sodas. In addition, other studies reported that those who drank diet soda had twice the risk of developing a metabolic syndrome that is often a precursor to heart disease. Some of these studies suggested that drinking diet soda may be just as unhealthy as drinking non-diet sodas. These studies included drinks that contained the artificial sweeteners aspartame, sucralose and saccharin.

EXAMPLE 1.12 "Rise in Rate of Breast Cancer on Long Island"

Medical researchers are trying to determine what factors have led to an abnormally high incidence of breast cancer in women on Long Island, NY. To aid in this study, the researchers will look for any patterns or relationships which may exist among the vast amounts of data. For example, the researchers will examine such factors as race, diet, age, occupation, body fat, family history, age at first child's birth, etc, and the relationship of these factors to the incidence of breast cancer.

EXAMPLE 1.13 "Morning Workout Burns More Fat"

In January 2013, *Cosmopolitan* reported on a study conducted at Northumbria University in the UK which found that males who work out in the morning, and on an empty stomach, burn 20 percent more body fat than those who eat before workouts. The study was based on a comparison of twelve active males who either already ate breakfast, or fasted since the night before. The two groups ran on a treadmill in the mid-morning hours. The researchers hypothesized that burning more fat was based on the male fasting before the workout.

This study illustrates an example of inferential statistics since the researcher uses a sample of 12 men to draw a conclusion about the general population of men.

EXAMPLE 1.14 "Federal Literacy Study: 1 in 7 U.S. Adults Can Not Read"

A new federal National Assessment of Adult Literacy study found that an estimated 32 million adults in the United States (or approximately one in seven) have such low literacy skills that they would have a difficult time reading anything more challenging than a children's picture book or to understand a medication's side effects listed on a pill bottle. The study surveyed more than 19,000 Americans aged 16 years or older. This new study was a follow up to a similar 1992 study on adult literacy.

The previous examples have illustrated the many uses of statistics. Some of the examples emphasized the use of descriptive statistics and exploratory data analysis, while others illustrated the use of inferential statistics in forecasting, predicting and the testing of hypotheses. In the next section, we will discuss some of the misuses of statistics.

1.5 MISUSES OF STATISTICS

Disraeli, a British Statesman, was quoted as saying, "There are three kinds of lies: lies, damned lies, and statistics." Disraeli's statement attests to the fact that there are many instances where statistics are abused.

In this section, our objective is to briefly discuss some of the misuses of statistics. Hopefully, the illustrations presented will make you aware of the need to become statistically literate. After reading this section, we believe you will be more inclined to agree with H. G. Wells' statement that:

Statistical thinking will one day be as necessary for efficient citizenship as the ability to read and write.

We will examine four techniques to illustrate how information can be presented in a misleading fashion. These techniques are: (I) Misleading Graphs, (II) Non-Representative Samples, (III) Inappropriate Comparisons and (IV) The Omission of Variation About an Average. Let's illustrate how graphs can be misleading.

I Misleading Graphs

> **DEFINITION 1.16**
> A **graph** is a descriptive tool used to visually describe the characteristics and relationships of collections of data quickly and attractively.

However, graphs also provide an excellent opportunity to distort the truth.

EXAMPLE 1.15 "Misleading Line Graphs"

An investment analyst could use a graph to show that the rate of return on retirement annuities are falling. This is shown in Figure 1.9.

However, if the analyst wanted to make the decrease in the rate of return appear more dramatic, she could simply change the units of the vertical scale and delete its lower portion. This is illustrated in Figure 1.10.

Although Figures 1.9 and 1.10 present the same information, Figure 1.10 presents the distorted impression that the rate of return is dropping at an incredible rate.

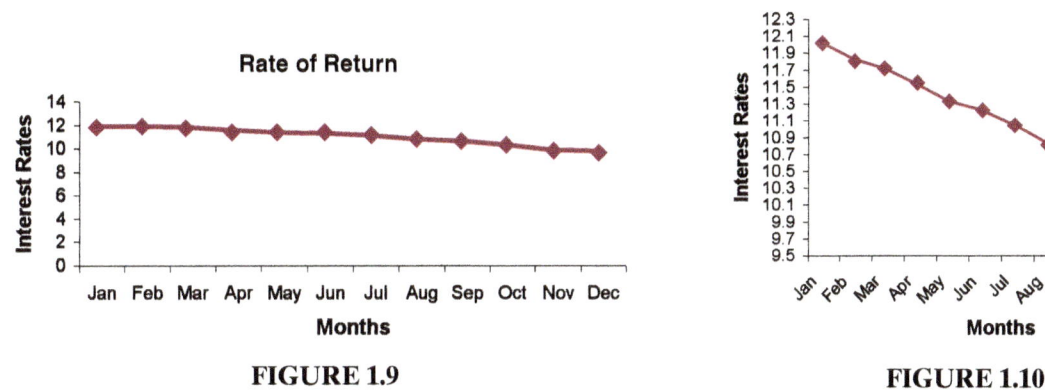

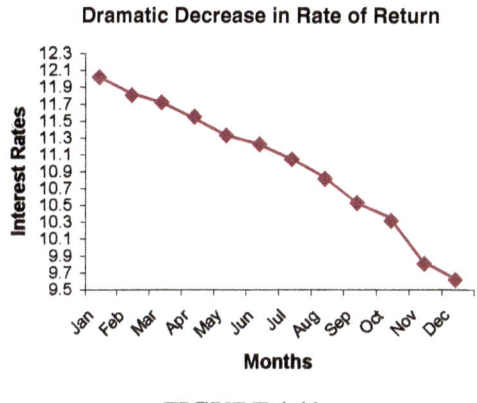

FIGURE 1.9

FIGURE 1.10

Another graph commonly used is the **bar graph** as illustrated in Example 1.16.

EXAMPLE 1.16 "Misleading Bar Graphs"

In order to attract new advertisers, a magazine that wishes to depict dramatic growth in its circulation produced the bar graph shown in Figure 1.11a.

Notice the bars in Figure 1.11a are presented with a portion missing. If not examined carefully this gives the visual impression that the magazine's circulation for 2019 is twice the circulation for 2014. However, when the bars are completely drawn as illustrated in Figure 1.11b, one sees only a slight increase in circulation.

FIGURE 1.11a

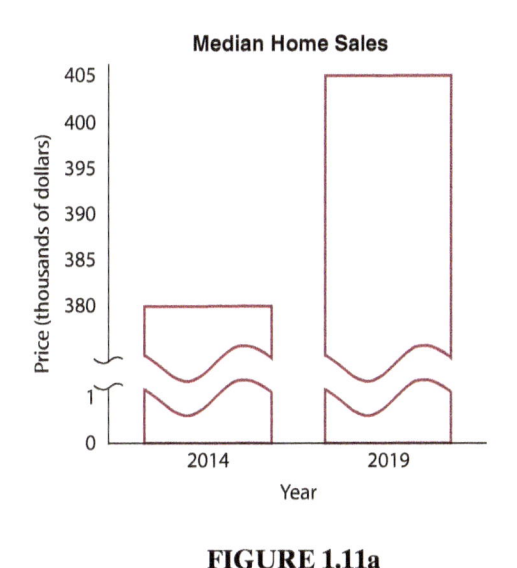

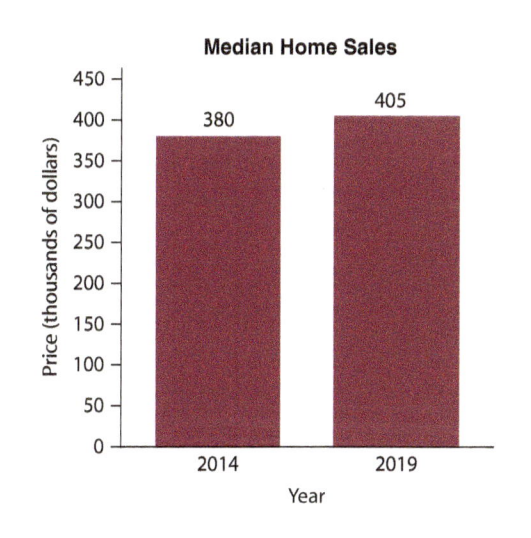

FIGURE 1.11b

A *pictograph* is another popular graphing technique used to distort the truth.

DEFINITION 1.17

A **pictograph** is a graph involving pictures of objects in which the size of the object in the picture represents the relative size of the quantity being represented by the object.

EXAMPLE 1.17 "Misleading Pictographs"

During a recent contract negotiation, the owner of a baseball team decided to dramatize the fact that his players' salaries have doubled from 2015 to 2019 by placing an advertisement in a local newspaper as illustrated by Figure 1.12.

2015   **2019**

FIGURE 1.12

This pictograph is misleading because it creates a visual impression that the baseball players' salaries have more than doubled since 2015. This impression was created by doubling both dimensions of the 2015 picture. This doubling of both dimensions quadrupled the area of the 2019 picture.

II Non-Representative Samples

Since a sample is only a portion of a population, great care must be taken to insure that the sample is representative of the population. Drawing a conclusion from a non-representative sample (or a biased sample) is another device often used to distort the truth.

In Section 1.3 we discussed the 1936 *Literary Digest* Poll. This poll is a perfect example of how a non-representative sample led to an inappropriate inference about the population. The 1936 *Literary Digest* Poll predicted that the Republican Presidential Candidate Alfred M. Landon would defeat the Democratic incumbent Franklin D. Roosevelt by an overwhelming margin. However, to the surprise of the *Literary Digest* pollsters, Roosevelt won the election by a landslide. The *Literary Digest* made a staggering error when it used an unrepresentative sample. The Digest's sampling procedure involved mailing sample ballots to ten million prospective voters of which only 2.3 million ballots were returned.

However, the 2.3 million ballots represented people who had a "relatively intense interest" in expressing their opinion on the election. Thus, this sample was non-representative (biased). In the 1936 election, it seems clear that the minority of anti-Roosevelt voters felt more intense about expressing their opinions about the election than did the pro-Roosevelt majority. In essence, the Digest's error was in basing their prediction of the outcome of the 1936 election on a *biased* voluntary response to its survey. (Incidentally, the *Literary Digest* went bankrupt soon after.)

Examples 1.18, 1.19 and 1.20 contain conclusions based upon samples where the information regarding the sampling selection process is not mentioned. In particular, information such as sample size and representativeness of the population have been conveniently omitted. In such instances an informed citizen must question the procedures used to gather the sample information. For each of the examples we have suggested a possible deception that could have been used in selecting the sample that led to the conclusion.

EXAMPLE 1.18 "Small Non-Representative Samples"

One recent study stated that "90% of the people surveyed were in favor of nuclear power."

What if this survey was based on only 10 workers whose livelihood is dependent upon the construction of nuclear power plants, would you consider the 90% statistic startling?

Probably not, because the study was based on a small non-representative sample!

EXAMPLE 1.19 "Pit-Bull Bias"

You may have noticed an increase in the public's opinion regarding certain dog breeds as dangerous. Certain dog breeds have historical roots in fighting and protection, which would have made breeding for aggression more likely. This brings up a political idea called "Breed-Specific" Legislation.

Say your local city council was looking to enact breed-specific legislation in response to a number of well-publicized incidents involving pit-bulls. A member of the city council attends several PTA meetings at the local schools and polls their opinion on pit-bull owners having to muzzle their dogs when walking them. Would you expect the results of this poll to be biased?

Yes, because parents of school aged children would obviously be interested in the safety of their children. Although some of the PTA members may be pit-bull owners, the negative news coverage regarding incidents involving pit-bulls would bias this poll.

EXAMPLE 1.20 "Questionable Samples"

An independent laboratory test showed that "People who use DPT toothpaste have 30% fewer cavities."

This laboratory made an impressive statement about DPT toothpaste. **But, there are some questions which need to be answered.**

First, how many people were sampled? **Second**, was this the only sample selected? **Third**, what does the statement ". . . . 30% fewer cavities" mean? Fewer than what?! Could this mean fewer than those people who did not use any toothpaste at all?

III Inappropriate Comparisons

The statement "Do Not Compare Apples to Oranges," conveys the essence of this next deceptive technique.

EXAMPLE 1.21 "Inappropriate Unemployment Rates Comparison"

In 2012, Fox News reported a comparison of the unemployment rate in 2009 vs. 2012. Their report is outlined in Figure 1.13 below:

AMERICA'S NOT WORKING

Unemployed, Underemployed, or Have Stopped Looking
23,136,000

Real Unemployment Rate 2009 vs Now
7.8% vs 14.7%

Unemployment Rate for Govt Workers
5.1%

FIGURE 1.13

The reason for the inappropriate comparison is that economists report two different unemployment rates. The first type of unemployment rate is called the Official Unemployment Rate. This is the rate that most people think of when speaking about unemployment. The 7.8% rate in 2009 was the Official Unemployment Rate in 2009. The second type of rate is called a Composite Rate of Unemployed. This rate takes into account those who are actively looking for work, people who are unemployed and discouraged from looking for a new job, part-time workers who prefer full-time employment. The 14.7% rate reported was a composite rate. As you can see this is not an "apples to apples" comparison; and therefore inappropriate. The Official Unemployment Rate in 2012 was in fact 8.1%. Fox News should have used this number instead.

22   Chapter 1   Introduction to Statistics

IV The Omission of Variation About an Average

The omission of pertinent statistical figures concerning variation about an average can result in the public drawing inaccurate conclusions.

EXAMPLE 1.22 "Omission of Information"

New mothers usually read many books about child development. However, these books can cause many unpleasant moments, especially if the reader misinterprets the author's statements concerning a child's development rate.

For instance, a new mother may interpret "A child begins to walk on average around 12 months," to mean that if her child does not walk by 12 months, her baby is abnormal or could be a late developer.

This type of misinterpretation can be avoided if the author provided a **range** of months for the developmental walking stage, such as: "a child begins to walk between 11 and 15 months." To the new mother, the range of ages for walking is much more informative than an average walking age.

1.6 OVERVIEW AND SUMMARY

In this chapter we introduced the two branches of statistics, **descriptive** and **inferential**, and showed how statistics are used and misused. We also discussed the notion of sampling, reasons for sampling and a commonly used sampling technique. The next two chapters will first explore ways to describe and summarize collections of data. We will then consider the concepts and techniques associated with inferential statistics. These include the study of probability, the notion of a sampling distribution, and the statistical models of inference.

Throughout the text, we will present real life examples, case studies, and articles to help you develop an appreciation of the use of statistics. In addition, technology tools such as the graphing calculator and computer statistical software outputs will be illustrated at the end of each chapter. Screen shots from the TI-84 graphing calculator will be presented to illustrate how a statistical graphing calculator can be used to analyze data sets.

For example, consider the following list of 50 randomly selected IQ scores:

				IQ Scores					
95	115	120	105	116	127	101	125	99	104
114	130	130	107	122	110	145	103	103	135
118	105	115	128	131	118	117	113	104	126
104	107	129	134	108	112	104	98	123	105
134	138	121	99	112	127	136	121	102	111

The TI-84 graphing calculator was used to statistically describe the same 50 IQ scores. Figures 1.14 and 1.15 show the graphing calculator's numerical analysis of the IQ scores. The graph in Figure 1.16 represents a histogram of the IQ scores that was generated using the graphing calculator.

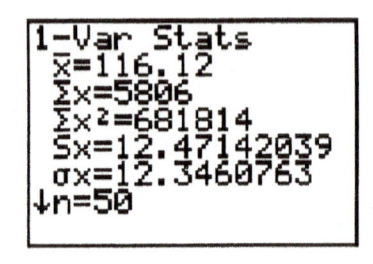

```
1-Var Stats
x̄=116.12
Σx=5806
Σx²=681814
Sx=12.47142039
σx=12.3460763
↓n=50
```

FIGURE 1.14

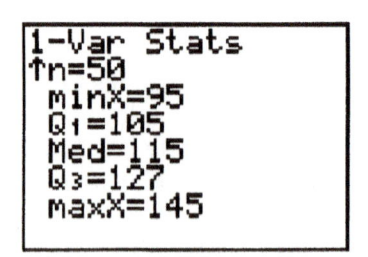

```
1-Var Stats
↑n=50
 minX=95
 Q₁=105
 Med=115
 Q₃=127
 maxX=145
```

FIGURE 1.15

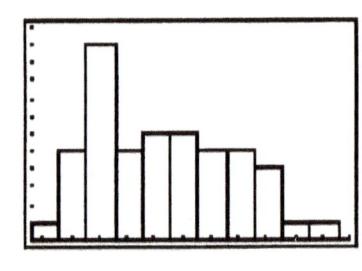

FIGURE 1.16

Statistical software packages can also be used to summarize data. As we move along in the text, computer outputs will be displayed using the software packages Minitab and Excel.

KEY TERMS

Bar Graph 19
Census 8
Descriptive Statistics 2
Destructive Testing 9
Exploratory Data Analysis 17
Graph 18
Inferential Statistics 5
Nonrepresentative Sample 11
Parameter 6

Pictograph 19
Population 4
Population Average 10
Population Size 9
Random or Probability Sampling Techniques 11
Range 22
Raw Data 2
Representative Sample 5, 11

Sample 4
Sample Size 9
Sampling 8
Simple Random Sampling or Random Sampling 12
Statistic 6
Statistics 2

SECTION EXERCISES

Section 1.1

1. Before they are arranged or analyzed information or observations are called _____ data.

2. The collection, presentation, analysis and interpretation of data is called _____.

3. The study of statistics can be separated into two areas: _____ statistics and _____ statistics.

4. The branch of statistics that uses numerical and/or visual techniques to summarize data is called _____.

5. The total collection of individuals or objects under consideration is called the _____.

6. Inferential statistics is the process of using a sample to draw inferences or conclusions about a _____.

7. A number that is used to describe a characteristic of a sample, such as a sample average, is called a _____.

8. A study of all Manhattan apartment renters was conducted to determine their average monthly rent. A sample of 300 renters was selected, and their rents were recorded
   a) The population is _____.
   b) The sample is _____.
   c) The population parameter being studied is _____ _____.

9. A research medical team used data collected from a sample of 4000 dental patients and found that the average cost of a root canal is $2,351. This average is an example of a:
   a) Parameter
   b) Statistic

10. The results of a survey indicate that 23% of the election district will support the proposition. This is an example of:
    a) Descriptive statistics
    b) Inferential statistics

11. Statistics describe the characteristics of a _____.
    a) Population   b) Sample

12. Which of the following is an example of a parameter?
    a) In a sample of 500 adults age 50+, 321 have a smartphone.
    b) Eleven percent of all US senators voted for a cyber-security bill.
    c) Sixty-seven out of 100 chefs surveyed say they prefer vanilla to chocolate.

13. Data exploration techniques are primarily used to help discover errors in the collected information.
    (T/F)

14. A sample represents a portion of the population.
    (T/F)

15. For the following statistical snapshots, I and II.
    a) identify whether the snapshot is an example of inferential or descriptive statistics.

    If the snapshot is an example of inferential statistics, then:
    b) identify the population being sampled
    c) identify the sample size
    d) state one conclusion or inference being drawn and whether or not you are in agreement. Explain.

(I)

Statistical Snapshot

The typical NFL player weighs 246.9 pounds. The heaviest and lightest teams in the NFC and the AFC in average pounds are displayed in the following snapshot graphic.

NFL AVERAGE TEAM WEIGHT (lbs)

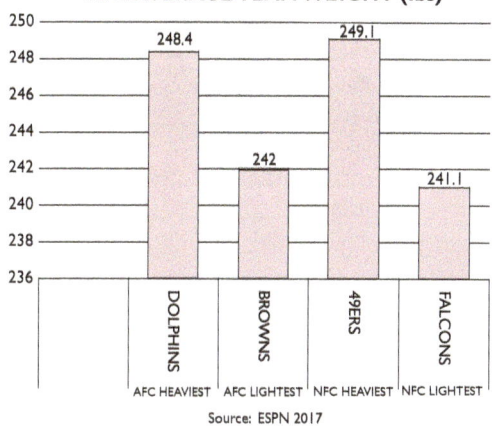

- DOLPHINS (AFC HEAVIEST): 248.4
- BROWNS (AFC LIGHTEST): 242
- 49ERS (NFC HEAVIEST): 249.1
- FALCONS (NFC LIGHTEST): 241.1

Source: ESPN 2017

(II)

Statistical Snapshot

U.S. Mail vs. E-mail

Which do you trust more to send a message: computer E-mail or the U.S. Postal Service?

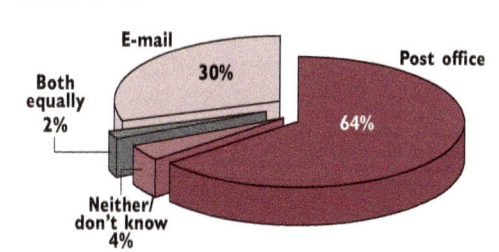

- E-mail: 30%
- Post office: 64%
- Both equally: 2%
- Neither/don't know: 4%

Source: ICR Survey Research Group, survey of 1,014 households.

16. For the following statistical snapshot
    a) identify whether the snapshot is an example of inferential or descriptive statistics.

    If the snapshot is an example of inferential statistics, then:
    b) identify the population being sampled
    c) identify the sample size
    d) state one conclusion or inference being drawn and whether or not you are in agreement. Explain.

Statistical Snapshot

Preferred Source of News

"Which is your preferred media for your source of news?"

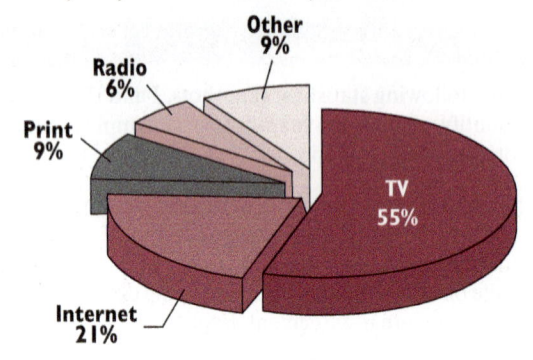

- Other 9%
- Radio 6%
- Print 9%
- TV 55%
- Internet 21%

Source: Gallup - June 2013, poll of 2,048 national adults

Section 1.2

17. The idea of selecting a sample to determine the characteristics of the population is the essence of _____ statistics.
18. Sampling is the process of selecting a _____, or sample, of the entire _____.
19. A survey that includes every item or individual of the population is called a _____.
20. The goal of sampling is to obtain a _____ sample of the population.
21. The objective of Inferential Statistics is to draw a conclusion about the entire _____ using a _____.
22. The size of a sample is denoted by _____.
23. The size of a finite population is denoted by _____.
24. A numerical characteristic of a population such as the population average is called a _____.
25. A numerical characteristic of a sample such as the sample average is a _____.
26. A portion or part of a population is called:
    a) proportion b) subset c) sample d) census

Section 1.3

27. A representative sample is a sample containing similar characteristics of the _____.
28. A random or probability sample is a sample which is selected in such a manner that each data value of the population has a _____ _____ of being selected for the sample.
29. A simple random sample is a sample of data values selected from a population in such a way that every sample of size n has an equal _____ of being selected and every data value of the population has the same chance of being selected for the _____.
30. Random sampling techniques are very important because of their use in formulating valid generalizations about population parameters. (T/F)

31. Consider the problem of drawing a simple random sample of 15 people from a list of 580 people. The 580 people are assigned a number from 001 to 580. Use the following list of random numbers to identify the 15 people selected for the random sample by using the first 3 digits of the 5 digit random numbers:
    a) If you start with the random number 63271 and move to the right of each row.
    b) If you begin with the random number 63271 and move down each column.

63271	59986	71744	51102	15141
80714	58683	93108	13554	79945
88547	09896	95436	79115	08303
01041	20030	63754	08459	28364
55957	57243	83865	09911	19761
66535	40102	26646	60147	15702
46276	87453	44790	67122	45573
84358	21625	16999	13385	22782

32. Using the Random Number Table, Table I in Appendix C: Statistical Tables, start at the top row of the first column of the table and read across the row, select 10 random numbers falling within the values 0000 to 6700.

33. A bank auditor decides to select a random sample, without replacement, of size 15 from the 9,500 five year CD accounts. The auditor decides to use the last four digits of each CD account number as a means of identification. The account numbers are in consecutive order and begin with the number 10–50001 and end at 10–59500. The bank auditor will use the following table of 5-digit random numbers to select the bank accounts. An account number will be selected from these 9,500 accounts when the last four digits of the account number match the first four digits of the random number.

    Use this table to determine the 15 account numbers of the bank customers the auditor will select for the audit if the auditor begins with the number 99041 and reads down each column.

99041	04024	77919	36103	13300	40719
24878	96096	12777	00102	92259	55157
98901	34693	85963	97809	64760	64951
84673	07844	38917	99505	75470	35749
18988	62028	79656	83149	96402	58104

Section 1.4–1.5

34. When a researcher analyzes data for possible patterns or relationships that might exist, this is called _____ Data Analysis.
35. A graph involving pictures of objects in which the size of the object represents the relative size of the quantity being represented by the object is called a _____.
36. The 1936 *Literary Digest* Poll is an instance where a _____ sample led to an inappropriate inference about the population.

37. The following data represents the average daily temperature by year for a city in the Midwest.

Year	2000	2001	2002	2003	2004	2005	2006
Average Daily Temp. (deg F)	50.5	50.8	50.8	50.1	50.0	50.7	50.5

Year	2007	2008	2009	2010	2011	2012	2013
Average Daily Temp. (deg F)	50.0	50.3	50.2	50.4	50.7	50.7	50.8

Using this data, sketch:
a) a graph making the average temperature appear to be stable
b) a graph that exaggerates the rise in average temperature

38. Discuss how the presentation of the statistics within each of the following statements could be misleading.
a) Ivory Soap is 99.44% pure.
b) Nine out of ten dentists recommend chewing sugarless gum for their patients who chew gum.
c) One recent study claimed that 55% of all insured homes in America are underinsured, that is, insured for less than 80% of actual replacement cost.
d) The profits of a foreign owned oil company have doubled in the last two years.

e) A prominent mathematics association journal used the following graph to illustrate its increase in circulation. The ad read:

In Five Years, We'll Have 10 GREAT Years BEHIND us!

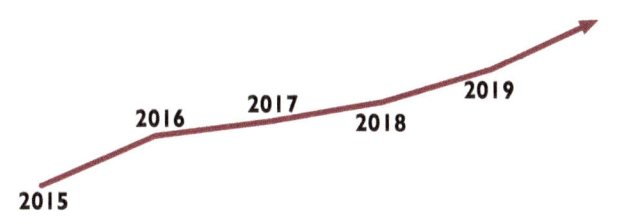

f) Rio Oro, an Arizona land developer, used the following pictographs in their advertisements to emphasize why investors should purchase undeveloped land in Arizona. The ad was stated as:

Here's why so many investors PREFER Arizona land.

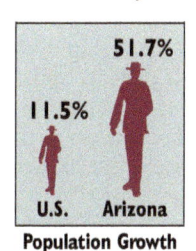

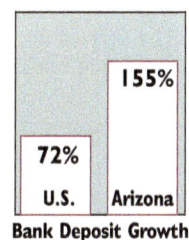

g) The book value of this automobile has increased through the years.

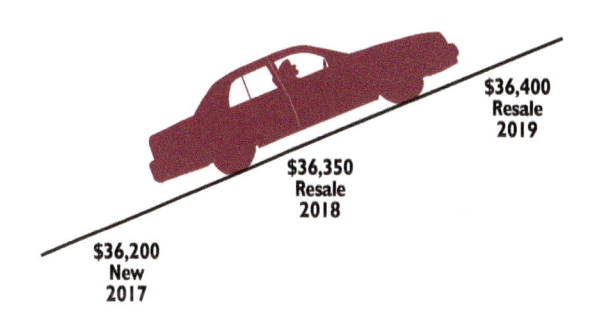

$36,200 New 2017
$36,350 Resale 2018
$36,400 Resale 2019

39. Read the following satiric essay entitled: **I'm Tired**, which comments on how many people work in the United States. Discuss the author's misuse of data and facts within the essay. Within your discussion, comment on:
a) The author's biases on who works and who doesn't work.
b) The author's misuse of facts. For example, are there some groups listed in the nonworking categories who do work? Is it possible to belong to two or more of these categories?
c) What did the author use as the population of the labor force and why it is not as useful as the Census Bureau's definition? (Note: the Census Bureau calculates labor force statistics for persons who are 16 years and over.)
d) The percentage of people that work as stated by the author.
e) What suggestions you would give the anonymous author that would help him/her to use the data more accurately.

I'm Tired

I'm tired. For several years I've been blaming it on middle age, iron poor blood, lack of vitamins, air pollution, obesity, dieting, and a dozen other maladies that make you wonder if life is really worth living. But now I find out, tain't that. I'm tired because I'm overworked.

The population of this country is 200 million. Eighty-four million are retired. That leaves 116 million to do the work. There are 74 million in school, which leaves 42 million to do the work. Of this total, there are 23 million employed by the Federal Government. That leaves 19 million to do the work.

Four million are in the Armed Forces, which leaves 15 million to do the work. Take from that total the 14,800,000 people who work for States and Cities. That leaves 200,000 to do the work. There are 188,000 in hospitals, so that leaves 12,000 to do the work.

Now, there are 11,998 people in prisons. That leaves just two people to do the work. You and me. And you're sitting there reading this.

No wonder I'm tired!!

CHAPTER REVIEW

40. The portion of the population that is selected for study is called a _____.

41. The use of sample information to estimate a population characteristic is the primary objective of _____ statistics. Thus, the researcher must take care to insure that the sample is _____ of the population.

42. A number such as the population average that describes a characteristic of a population is called a _____.

43. Based on an independent poll, 45% of Americans will recycle more this year. This is an example of _____ statistics.

44. A survey of 500 undergraduates at a Long Island university of 16,000 undergraduates was taken to determine the percentage of students who use a laptop or tablet to take notes in class. The survey indicated that 29% of the students use these devices to take notes.
    a) The 500 undergraduates surveyed represent the _____.
    b) The 16,000 undergraduates attending the Long Island university represent the _____.
    c) The percentage value, 29%, represents a _____.

45. Gallup Politics conducted a survey to determine how many registered Democrats trust the government. A sample of 629 registered Democrats were questioned. It was determined that 58% trust the government.
    a) The population is _____.
    b) The sample is _____.
    c) The percentage value, 58%, is a statistic because it describes a characteristic of a _____.

46. Testing research shows that the average math SAT score in New York is 502 (out of 800). This average is an example of a:
    a) Parameter   b) Statistic

47. The average age of those surveyed is 26.73 years. This is an example of
    a) Descriptive statistics   b) Inferential statistics

48. Parameters describe the characteristics of a _____.
    a) Population   b) Sample

49. Inferential statistics helps researchers make generalizations about large populations by using data from a sample of the population. (T/F)

50. Sample characteristics are called statistics. (T/F)

51. In applications of inferential statistics all the observations of the population should be obtained. (T/F)

52. Predictions made from a straw poll are an example of inferential statistics. (T/F)

53. If a sample is not representative of the population, inferences made from the sample information may not be correct. (T/F)

54. a) Define the following terms:
    (i) inferential statistics
    (ii) sampling
    (iii) census
    (iv) destructive testing
    (v) population
    (vi) sample
    (vii) parameter
    (viii) statistic
    (ix) sample size
    (x) population size
    b) Discuss the reasons of why it is necessary to sample.
    c) Identify the symbol for each of the following terms:
    (i) population size
    (ii) sample size

55. A random sample will always be a representative sample. (T/F)

56. Simple random sampling is a technique where each sample does not have an equal chance of being selected. (T/F)

57. A United States senator mailed a questionnaire to all the constituents living in her state. She randomly selected 200 of the questionnaires which were returned to her office. Would this sample represent a random sample of all the constituents living in her state? Explain your answer.

58. Chris, a history instructor at Columbus University, decides to conduct an opinion poll of the students attending the university during the fall semester. Using a random number table, she selects 100 students from all the students attending the morning European History classes.
    a) Does this sample represent a random sample? Explain.
    b) Is this a representative sample of all the students attending the University? Explain.
    c) Identify a population for which this sample could be a representative sample.

59. CCC, a television network, decides to poll its viewers on gun control. The network requests its viewers to call 1-900-GUNSYES if they are against gun control. All viewers who favor gun control are asked to call 1-900-GUNSSNO. Do you believe this sampling procedure represents an appropriate technique in obtaining a random sample of the all the viewers of CCC regarding their views on gun control? Explain.

60. A new home wares department store called Home Arsenal is planning to expand into a new market area. One aspect of its study will be to determine the spending habits of the prospective customers for the type of goods Home Arsenal will sell. Assume the selected marketing area consists of the size of a small town, how would you go about selecting a sample of prospective customers to interview?

61. The service manager of an automobile dealership decides to sample her customers to determine if they are satisfied with the service they receive. The manager prepares a list of 630 customers from the service records of the previous 12 months. Each customer is assigned a number from 000 to 629.

The service managers decides to sample 20 customers. The manager used the following list of random numbers to identify the customers. The sample is selected by using the last 3 digits of each random number. Begin with the random number 08723 and read down each column. Make a list of the 20 customers selected for the sample.

08723	67422	68479	10097	85017	13618	91494
56239	17691	26640	37542	16719	69041	80336
44915	79655	85157	08422	65842	82186	44104
41976	48274	11100	99019	84532	39187	12550
15694	60523	16505	12807	82789	41453	63606

62. Each year Fortune magazine publishes a list of the top 500 U.S. companies based upon their annual earnings. The companies are ranked from 1 to 500. Using the random sampling procedure, identify the numbers of twenty companies from Fortune magazine's list of 500 companies. Start at the 3rd column and 26th row of the Random Number Table, Table I in Appendix C: Statistical Tables, and read across the table to determine the 20 companies.

63. The Daily Satellite, a newspaper company, has received 15,000 entries for its Super Bowl contest. Lois Road and Clark Hent, two newspaper reporters, will randomly select the 10 winners from the entries. If each entry is numbered from 00001 to 15,000, then select the ten winners of the contest by using the Random Number Table, Table I in Appendix C: Statistical Tables, starting at the 31st row of the first column of the table and moving across each row.

64. A bank auditor is planning to take a systematic random sample from a list of bank customer's. The auditor decides to select a sample of 300 bank customers from the bank's list of 60,000 customers. If customer has an account number ranging from 1 to 60,000 and she randomly selects 140 as her starting point, then:
    a) Determine the ratio of the population size to the sample size, and
    b) Identify the customers selected for the audit.

65. The following data represents the trend of U.S. imports of foreign automobiles from 2009 to 2019.

Year	2009	2010	2011	2012	2013	2014
Auto Imports per million Population	35	44	54	65	76	87

Year	2015	2016	2017	2018	2019
Auto Imports per million Population	98	109	120	131	151

Using this data construct:
a) A graph exaggerating the rise in U.S. imports of foreign auto's due to the gasoline shortage.
b) An accurate graph depicting the trend of U.S. imports of foreign autos.

66. Examine the following graph.
    a) Is the Binge Drinkers bar chart in the following figure an example of inferential statistics? Support your answer with an explanation.
    b) Identify the population being represented.
    c) State a conclusion or an inference being drawn.

Binge Drinkers
Percentage of males vs females aged 12 to 20 years who consumed five or more drinks in a row.

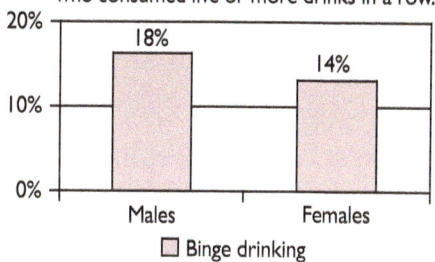

Males: 18%
Females: 14%
☐ Binge drinking

Figure 1: Source: SAMHSA, National Survey on Drug Use and Health

67. For each of the following two media examples of statistics, identify which illustrate inferential statistics.

    For those examples of inferential statistics:
    a) Identify the population being sampled
    b) The sample size
    c) State one conclusion or inference being drawn and whether or not you are in agreement. Explain.

(I)

Statistical Snapshot

Why We Don't Just Do It

Major reasons we don't take part in recreational sports*:

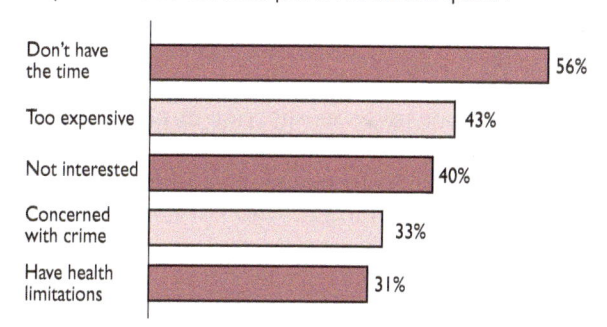

- Don't have the time: 56%
- Too expensive: 43%
- Not interested: 40%
- Concerned with crime: 33%
- Have health limitations: 31%

* Respondents could choose more than one.
Source: The Recreation Roundtable poll of 2,000 persons

(II)

Statistical Snapshot

When Pets Get 'People Food'

How often pet owners say they feed their pet human food:

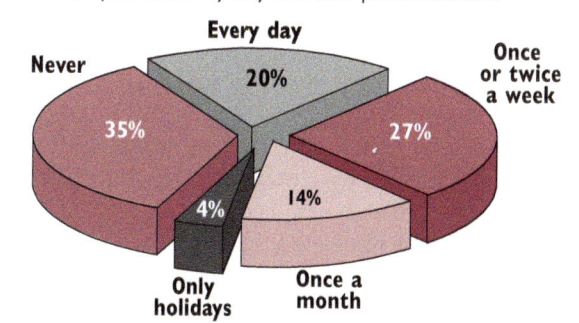

- Every day: 20%
- Once or twice a week: 27%
- Once a month: 14%
- Only holidays: 4%
- Never: 35%

Source: American Animal Hospital Association survey of 1,049 pet owners

68. The article entitled "How Ads Can Bend Statistics" appeared in the *NY Times* on January 17, 1981.

Consumer Saturday

How Ads Can Bend Statistics

"Doctors recommend" "Four out of five people preferred...." "Twice as many repairmen choose...." The slogans and claims are survey advertisements, a technique of selling products that has grown in use over the last five years and one in which the Federal Trade Commission is increasingly interested.

"It is an important technique because consumers give it credibility and they use the information contained in them to make important purchasing decisions," Wallace S. Synder, assistant director for the F.T.C.'s advertising practices division, said in Washington this week. Regulation of advertising is a traditional function of the F.T.C.

Survey advertisements are an effective tool increasingly used in the 60 billion-a-year advertising trade, but such advertisements are not always as scientific or responsible as they may sound.

During the last two weeks, three manufacturers have made consent agreements with F.T.C. to discontinue the use of less-than-substantive claims for their products.

In one, Litton Industries agreed not to advertise its microwave oven as superior to others unless there is a "reasonable basis" for such claims. The company had been claiming that independent technicians preferred their product over competing products, but their survey used only authorized Litton service agencies' technicians who could possibly have a bias toward their product. In addition, there was no effort to make sure that the technicians who preferred Litton microwave ovens were experienced at repairing ovens other than Litton's.

Under F.T.C. regulation, "reasonable basis" means that competent and reliable surveys or tests are conducted by qualified persons. The results, say the regulation, must be objectively evaluated "using procedures that ensure accurate and reliable results." The Litton advertisements, the agency said, did meet such standards.

In a similar case, Standard Brands and Ted Bates & Company, formerly the advertising agency for Standard Brand's Fleischmann's Margarine, agreed not to make claims based on surveys or tests unless they were scientifically conducted and the claims accurately reflected the results.

The agreements between the companies and the F.T.C. cover all the products of Standard Brands, one of the nation's largest food and beverage companies, and every advertisement created by the advertising agency.

At issue were advertisements used between 1974 and 1977. They claimed the doctors preferred and recommended Fleischmann's Margarine. Two of the four advertisements, allegedly based on a national survey of doctors stated:

"When a doctor chooses margarine, chances are it's Fleischmann's."

The F.T.C. alleged that the claims were unproven, charging that only 15.5 percent of the doctors surveyed recommended Fleischmann's and that at least 67.5 percent of the doctors did not recommend any specific name brand.

The companies have now agreed that any surveys or tests featured in advertisements will be designed and analyzed in a reliable scientific manner and must clearly establish any superiority claims.

In another case, Teledyne Inc. and Teledyne Industries Inc. makers of Water Pik, and J. Walter Thompson, its advertising agency, agreed that their advertising claims about the effectiveness of the dental-care product in preventing gum diseases would have a "reasonable basis." The Water Pik is an "oral irrigator" that sprays water into the mouth and that, advertisements say, flushes away debris from beneath the gum line and "reduces the causes of gum disease."

The F.T.C. further alleged that Teledyne falsely claimed that the Water Pik is approved by the American Dental Association. To make such a claim, the endorsement must be made in writing, according to the F.T.C.

C. Collot Guerard, deputy assistant director of the F.T.C.'s advertising division, offered pointers for consumers when they read or see an advertisement:

"Ask yourself," she said, "how many doctors were actually surveyed when you see something like 'five out of six doctors.' Consumers tend to think of big, nationally projected surveys, but it might not be so."

Learn to recognize the "dangling comparatives" in advertisements such as "Product X will get your teeth whiter than ever" and ask yourself just what does that mean. Whiter than the same product did a year ago? Whiter than other products?

Watch for parity claims. "If an advertisement states that 'no hair dryer will dry your hair quicker,'" she said, "it implies that it dries faster than other brands. Usually it means that it dries just as quickly as other products."

Maintain a healthy skepticism.

From *The New York Times*. © 1981 The New York Times Company. All rights reserved. Used under license.

a) This article discusses various problems that arise when claims are made within advertisements. Enumerate several of these problems and discuss some of the possible solutions.

WHAT DO YOU THINK?

69. Examine the snapshot entitled "Learned Believers."
    a) Identify the population being surveyed.
    b) Since there is a greater number of Roman Catholics in the US than Hindus, why don't the Roman Catholics appear as one of the top percentages?
    c) Why are Hindus one of the top percentages?
    d) What information would you like to see that was not included?

Statistical Snapshot

Learned Believers

Religious denominations in the USA with the highest percentage of college graduates in membership:

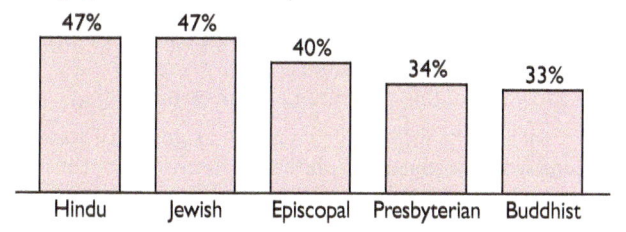

- Hindu: 47%
- Jewish: 47%
- Episcopal: 40%
- Presbyterian: 34%
- Buddhist: 33%

Source: ICR Survey Research Group survey of 113,000 adults for City University of New York Graduate School

70. Examine the graphic entitled "Women Vote More."
    a) Identify the populations presented within the graphic.
    b) What factors might contribute to the comment "Women Vote More"?
    c) What reason can you give for both of these graphs rising and falling every two years? When would you expect the next peak to occur for both graphs? Why?
    d) Do you notice any pattern with respect to certain years for the men and women voters? Explain. In what future year would you expect the number of men voters to be approximately 65 million? Why? In what future year would you expect the number of women voters to be approximately 75 million? Why?
    e) In what year did the number of men and women voters reach their maximum? Can you give an explanation for this?
    f) Would you predict that the number of women and men voters will reach a maximum in the year 2020? Explain.

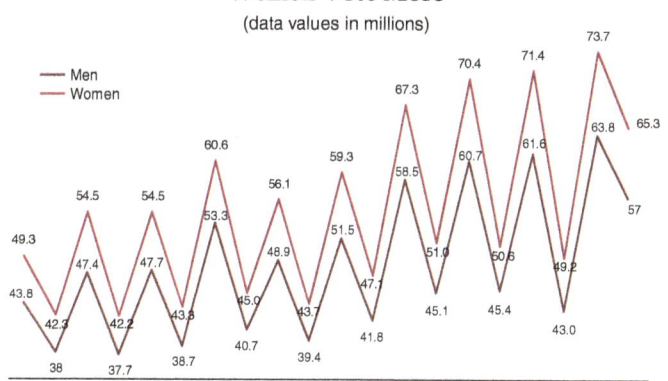

Women Vote More
(data values in millions)

Women (red): 49.3, 47.4, 47.7, 53.3, 54.5, 54.5, 56.1, 60.6, 58.5, 67.3, 60.7, 70.4, 61.8, 71.4, 73.7, 65.3
Men (dark): 43.8, 42.3, 42.2, 43.8, 45.0, 43.7, 48.9, 51.5, 47.1, 51.0, 50.6, 45.1, 45.4, 49.2, 63.8, 57
38, 37.7, 38.7, 40.7, 39.4, 41.8, 43.0

1980 1982 1984 1986 1988 1990 1992 1994 1996 1998 2000 2002 2004 2006 2008 2010 2012 2014 2016 2018

PROJECTS

71. While watching TV pay close attention to the commercials.
    a) Determine the type of statistics (descriptive or inferential) being presented.
    b) Comment as to what information (if any) has been left out that might lead you to a different conclusion.

72. Look in magazines or newspapers for ads using statistics. Comment on the presentation and state if you think any deceptive tricks might have been utilized.

CHAPTER TEST

73. The branch of statistics that uses sample information to draw a conclusion about the population is called _____.

74. A number that is used to describe a characteristic of a population, such as a population proportion, is called a _____.

75. According to a recent cell phone usage poll, 64% of teens have admitted that they have sent and received texts while in the classroom when cell phones are completely banned while 25% of the students have also received or made a call while in the classroom. In addition, the survey indicated that 68% of students use cell phones to make arrangements for after-school activities.
    a) What was the purpose of this survey?
    b) Is this an example of inferential statistics? Explain.
    c) Identify the target population.
    d) Is the numerical value 64% a statistic or a parameter? Explain.
    e) Is the numerical value 25% a statistic or a parameter? Explain.
    f) Based on these survey results, discuss the pros and cons of permitting cell phones in our schools.

76. Which of the following is an example of descriptive statistics?
  a) In a sample of 1,000 college students, their average I.Q. score is 115.
  b) The average height of all NBA basketball players is 6'7".
  c) Nine out of 10 doctors recommend non-aspirin pain reliever.

77. A survey of 2,500 social networking users within the U.S. was randomly conducted to determine the average time a user spends per month on Facebook. According to the survey, the average monthly time on Facebook was 6.75 hours.
  a) Identify the target population.
  b) Identify the sample.
  c) Identify the population parameter being studied.
  d) Is the numerical value 6.75 hours a statistic or a parameter? Explain.

78. Determine whether each of the following sampling methods for selecting a sample from a population will result in a random sample. Explain.
  a) The charge account manager of the department store Lacy's decides to obtain a random sample of all Lacy's customers. The manager proposes to assign a number to all charge account holders. A sample would be selected from this list using a random number table.
  b) To obtain a random sample of customers of the shopping mall Green Pastures, a group of interviewers are placed at all entrances to the mall for 5 consecutive mornings. Each interviewer selects a typical customer carrying packages as they leave the mall.

79. Dear Abby, a newspaper columnist, decides to poll her readers regarding the question: "Would you have children again, if you had to do it all over?". Abby requests her readers to mail in their responses to her survey. Her sample consists of all the responses she receives. Do you believe Abby's sampling procedure represents an appropriate technique in obtaining a random sample of all her readers regarding their views on having children again? Explain.

80. During Wednesday's statistics class, a college instructor requested that her students answer the following questions: Is your birthdate an even or an odd number? If it is even, please answer question (a), but if it is odd, then answer question (b).
  Question (a): Is today Wednesday?
  Question (b): Did you cheat on the last statistics exam?

  Of the 120 responses, 72 had responded yes while 48 responded no. Assuming that even and odd birthdates are equally-likely, what is your estimate of the fraction of the class who cheated on the last exam?

81. Discuss how the presentation of the statistics within each of the following statements could be misleading.
  a) GoodMonth Radial tires stop 45% faster.
  b) Eight out of ten dentists prefer Brand X sugar free gum.

82. Which of the following is the most important criteria for a sample to be considered unbiased?
  a) Representativeness   b) Size   c) Inferential

83. The incorrect prediction made by the *Literary Digest* for the 1936 presidential election was primarily due to the mailing of the ballots to a non-representative sample of 10 million prospective voters.
(T/F)

84. A pictograph is a popular technique used to distort the truth because it can create a visual impression that does not accurately represent the relative size of the quantities being represented.
(T/F)

85. The following advertisement appeared in a wildlife magazine:

Environmental T-Shirts
Support wildlife by wearing environmental t-shirts.
(10% of profits go to environmental groups)

  a) If each t-shirt costs $15 and the manufacturer's profit per t-shirt is $5, then how much of your $15 is being contributed to the environmental groups?
  b) Calculate the percent of the $15 that would be contributed to the environmental groups.
  c) Do you feel this advertisement can be misleading?

86. For each of the following two media examples of statistics, identify which illustrate inferential statistics. For those examples of inferential statistics:
  a) Identify the population being sampled
  b) The sample size
  c) State one conclusion or inference being drawn and whether or not you are in agreement. Explain.

(I) **Long Island Household Statistics**

Number of Households consisting of:
One person	170,273	18%
Two people	284,408	30%
Three people	189,605	20%
Four people	189,605	20%
Five people	94,803	10%
Six people	56,882	6%

Total Households: 948,027
Average household size: 2.96

(II) **Statistical Snapshot**

Busy Days at the Cash Machine
Two-thirds of people say they have a regular day when they're most likely to use an automatic teller machine:

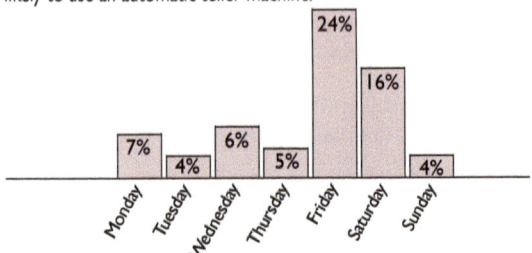

Monday	Tuesday	Wednesday	Thursday	Friday	Saturday	Sunday
7%	4%	6%	5%	24%	16%	4%

Source: AT&T Global Information Systems poll of 1,011

87. For each of the given examples pertaining to Polar Bears,

  i) Identify the variable that is being studied.

ii) Determine if the type of data is quantitative or qualitative.

iii) Is the data discrete or continuous?

  a) A biologist studies bears in a 25 mile radius of a national park and finds there are 7 males and 5 female bears.

  b) A conservationist group tags 25 Polar Bears and finds their weight in the Winter and Summer to make a comparison.

  c) A biologist collects data on 30 Polar Bears in British Columbia in order to find the average number of salmon these bears eat per day.

  d) A camper tracks the number of fatal bear attacks around the Artic in order to determine the safest areas for camping and hiking.

For the answer to this question, please visit www.MyStatLab.com.

CHAPTER 2

Organizing and Presenting Data

Contents

2.1 Introduction
2.2 Classifications of Data
    Categorical and Numerical Values
    Continuous and Discrete Data
2.3 Exploring Data Using the Stem-and-Leaf Display
    Back-to-Back Stem-and-Leaf Display
2.4 Frequency Distribution Tables
    Frequency Distribution Table for Categorical Data
    Frequency Distribution Table for Numerical Data
    Relative Frequency Distributions
    Cumulative Frequency Distributions
2.5 Graphs: Bar Graph, Histogram, Frequency Polygon and Ogive
    Bar Graph
    Histogram
    Frequency Polygon and Ogive
2.6 Specialty Graphs: Pie Chart and Pictograph
2.7 Identifying Shapes and Interpreting Graphs
    Shapes of Distributions
    Interpreting Graphs

2.1 INTRODUCTION

Graphs are commonly used to organize, summarize, and analyze collections of data. In this chapter, we will show how to use tables and graphs to organize, to summarize and to display the important characteristics of a data set. Using a graph to visually present a data set makes it easy to comprehend and to describe the data.

Suppose the student government association of a local college is interested in exploring some characteristics of its student body. They surveyed a sample of 150 students requesting the following information:

- the student's gender
- the number of credits they are currently taking
- the number of hours per week they study
- their current cumulative grade point average (GPA)
- their area of concentration

Table 2.1 shows the raw data collected from this survey. In this form, it is very difficult to observe the characteristics of the data set.

DEFINITION 2.1
Raw Data is data before it has been arranged in a useful manner or analyzed using statistical techniques.

TABLE 2.1 Student Government Survey Data of 150 Students

The column abbreviations are defined as follows:
GENDER: student gender
    M: male
    F: female

HR: study hours per week
GPA: student grade point average
CR: number of credits this semester

MAJ: college major
LIB: liberal arts
BUS: business
CMP: computers or mathematics
SCI: natural and applied sciences
HTH: health sciences
FPA: fine and performing arts

GENDER	MAJ	HR	GPA	CR	GENDER	MAJ	HR	GPA	CR	GENDER	MAJ	HR	GPA	CR	GENDER	MAJ	HR	GPA	CR
M	LIB	15	2.79	14.0	M	BUS	24	3.61	18.0	F	LIB	19	3.05	17.0	F	LIB	8	2.47	15.5
M	BUS	18	2.87	12.5	M	BUS	20	3.45	19.0	M	BUS	31	3.62	15.0	M	LIB	10	2.66	13.0
F	LIB	17	2.71	15.0	F	BUS	15	2.42	15.0	F	CMP	33	3.67	18.0	M	BUS	12	3.28	15.0
M	CMP	20	3.42	17.0	F	LIB	12	2.49	14.0	F	BUS	23	2.25	10.5	F	LIB	15	3.41	17.0
M	BUS	19	2.88	6.0	M	CMP	24	3.41	13.0	M	LIB	20	2.70	15.0	M	LIB	31	3.88	18.0
F	LIB	23	2.91	12.0	F	BUS	29	3.18	12.0	F	LIB	17	2.30	7.5	F	LIB	11	2.32	10.5
M	LIB	8	1.76	18.0	M	CMP	30	3.65	7.5	M	LIB	16	2.51	8.0	F	HTH	20	2.97	16.0
M	SCI	10	2.95	15.0	F	LIB	32	3.55	10.5	M	BUS	26	3.54	18.0	F	BUS	29	3.75	15.5
F	LIB	27	2.82	4.5	F	LIB	30	3.45	12.0	M	BUS	12	1.92	4.5	F	LIB	20	2.78	12.0
M	SCI	28	2.96	12.0	M	LIB	19	2.35	16.0	F	LIB	20	3.15	13.0	F	HTH	15	2.84	16.0
M	BUS	20	2.48	17.0	M	BUS	20	2.37	15.0	F	BUS	11	2.00	3.0	F	HTH	10	3.10	17.0
F	BUS	27	2.24	15.0	F	LIB	21	2.54	17.0	F	LIB	9	1.95	15.0	F	LIB	6	2.49	10.0
F	LIB	36	3.02	14.0	F	BUS	21	2.00	10.0	M	BUS	9	2.88	13.0	F	HTH	10	3.24	15.5
F	LIB	17	2.45	15.5	M	LIB	22	2.59	17.0	F	LIB	9	2.65	15.0	M	BUS	21	2.95	7.5
F	HTH	24	3.39	15.0	F	LIB	17	1.94	14.0	F	LIB	8	1.98	12.0	M	BUS	23	2.81	8.0
F	SCI	26	2.75	17.0	F	BUS	16	2.59	15.5	F	BUS	19	3.21	16.0	F	HTH	10	2.36	17.0
M	BUS	28	2.82	16.5	M	LIB	19	3.16	16.5	M	BUS	7	1.95	12.0	M	CMP	17	3.10	15.0
F	BUS	22	2.29	15.0	F	LIB	18	2.66	7.5	F	LIB	10	2.96	15.0	F	LIB	28	3.78	16.0
F	BUS	16	3.56	12.0	M	BUS	22	3.05	13.0	M	LIB	8	2.95	16.5	M	LIB	15	3.08	17.0
M	LIB	27	3.28	8.0	M	LIB	21	3.16	15.0	M	BUS	18	2.00	10.5	M	BUS	11	3.07	15.0
F	BUS	20	3.06	14.0	M	LIB	24	3.29	12.0	F	LIB	16	2.82	16.0	F	LIB	12	3.15	13.0
M	SCI	24	3.56	7.5	F	LIB	20	2.01	12.0	F	BUS	10	1.95	11.0	F	LIB	15	3.65	12.0
M	BUS	27	3.54	4.5	M	BUS	27	3.51	7.5	F	LIB	12	3.01	13.0	M	BUS	14	3.70	17.0
F	BUS	18	3.07	12.0	F	BUS	18	2.84	15.0	F	HTH	15	2.70	14.0	F	LIB	21	2.73	15.0
M	SCI	28	3.05	12.0	F	LIB	29	3.59	16.0	F	LIB	11	2.67	17.0	M	LIB	17	3.75	17.0
M	BUS	24	2.92	14.0	M	CMP	23	3.25	17.0	M	CMP	9	2.81	16.0	F	LIB	20	3.17	14.0
F	HTH	12	2.48	15.0	M	LIB	25	3.65	7.5	M	LIB	27	3.76	15.5	M	BUS	25	3.56	12.0
M	SCI	24	3.23	13.0	M	CMP	31	3.89	12.0	F	CMP	10	2.67	14.0	F	FPA	30	3.75	17.0
F	LIB	25	3.17	7.5	F	BUS	23	2.96	15.0	F	HTH	7	2.32	14.0	F	BUS	33	3.66	18.0
M	BUS	26	3.57	10.0	F	BUS	18	2.77	15.5	F	BUS	8	2.09	16.0	M	LIB	28	3.85	7.5
F	BUS	28	3.19	11.0	F	LIB	26	3.28	12.0	F	LIB	9	2.42	17.0	M	LIB	9	2.74	17.0
F	LIB	30	3.08	12.0	M	LIB	27	3.13	10.0	F	LIB	10	1.90	6.0	F	BUS	9	2.64	15.0
M	CMP	15	2.47	6.0	F	BUS	33	2.75	7.5	M	LIB	7	1.98	12.0	F	FPA	12	2.93	12.0
F	LIB	12	1.88	7.5	M	CMP	20	2.76	17.0	F	LIB	12	3.15	16.0	M	LIB	24	2.47	16.0
M	BUS	11	2.40	15.0	M	LIB	17	3.17	6.0	F	HTH	14	3.40	17.0	M	FPA	8	2.18	14.0
F	BUS	24	2.89	17.0	M	HTH	15	1.86	16.0	F	BUS	10	2.87	15.0	M	LIB	6	2.43	13.0
F	LIB	25	3.25	14.0	F	BUS	17	2.81	13.0	F	LIB	28	3.51	17.0					
F	LIB	16	2.65	14.0	M	LIB	11	1.87	6.0	F	BUS	8	1.95	12.0					

This table is available on the web in either Microsoft EXCEL or TI-84 format from the textbook's web site at www.MyStatLab.com.

Compare the raw data table to the graph "Major by Gender" in Figure 2.1. This graph depicts a student's major area of concentration to the student's gender. Notice how this graph is a convenient way to display the number of males and females belonging to each of the five listed areas of concentration. The graph makes the data easier to comprehend. From this graph, you can instantly see that there is a greater number of females than males majoring in Liberal Arts, Business, Health Sciences and Fine and Performing Arts.

Major by Gender

LEGEND
LIB: Liberal Arts
BUS: Business
SCI: Natural and Applied Sciences
HTH: Health Sciences
CMP: Computers or Mathematics
FPA: Fine and Performing Arts

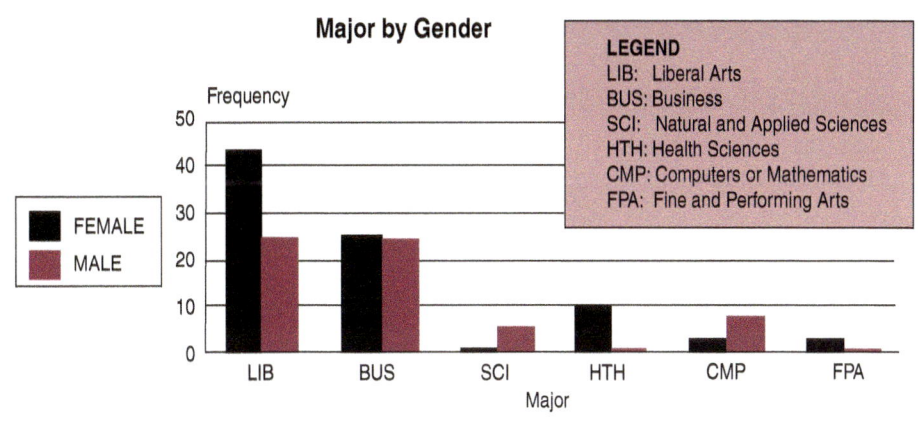

FIGURE 2.1

This chapter will present techniques that will help you organize, summarize, and explore data to determine if patterns or relationships exist. For a given data set, these techniques will help answer questions such as:

- How many credits do most students take for a semester?
- What is the most popular major?
- How many hours do students usually study per week?

or to explore meaningful comparisons such as:

- the number of hours that females study to the number of hours that males study.
- the relationship of the type of major to a student's gender.
- the number of study hours per student to the student's GPA.

This chapter presents a variety of techniques that demonstrate the use of tables and graphs to display the important characteristics of the data. The type of raw data collected will determine the technique. In Section 2.2, we will discuss the different classifications and characteristics of data.

2.2 CLASSIFICATIONS OF DATA

The type of graph you use depends on the type of information about a person or thing you have collected. For example, a person's height or gender, a student's GPA, the circumference of a redwood tree, the closing price of a stock, the pollen count for a given day, or one's preference when buying an automobile illustrate types of information that can be obtained about a person or thing. These types of information are called **variables**.

DEFINITION 2.2

A **variable** is a type of information, usually a property or characteristic of a person or thing, that is measured or observed.

A **specific measurement or observation** for a variable is called the **value** of the variable. Suppose you are interested in the variables height, gender and GPA for the students in your class. If a female student is 63 inches tall, has a GPA of 3.23, then female is the value for the variable gender, 63 inches represents the value for the variable height and 3.23 is the value for the variable GPA. A collection of several data pertaining to one or more variables is called a **data set**.

> **DEFINITION 2.3**
> A **data set** is a collection of several data pertaining to one or more variables.

Table 2.1 represents a data set. In Example 2.1, we will examine the variables of this data set.

EXAMPLE 2.1

Using Table 2.1:
a) Identify the variables.
b) Determine the value of each of these variables for the first student listed in the table.

Solution

a) The variables are Gender, Major, Study Hours, GPA, and Credits.
b) The first student is a Male with a Major in Liberal Arts, who studies 15 hours per week, has a GPA of 2.79 and is taking 14 credits.

Therefore, the variable:
gender has a value of male;
major has a value of Liberal Arts;
study hours has a value of 15 hours per week;
GPA has a value of 2.79; and
Credits has a value of 14.

Categorical and Numerical Variables

There are basically two types of variables: **Categorical** and **Numerical**. A categorical variable yields values that denote categories. For example, the variable gender is a categorical variable since gender is classified by the categories male or female. The variable major is a categorical variable where the values are classified by the categories liberal arts, business, health sciences, computers or mathematics, natural and applied sciences, or fine and performing arts. The values of these categorical variables are referred to as categorical data. Categorical variables are also referred to as **qualitative** variables, and numerical variables are also referred to as **quantitative** variables.

> **DEFINITION 2.4**
> A **categorical (or qualitative) variable** represents categories, and is non-numerical in nature. These values are known as categorical data.

A numerical variable yields numerical values that represent a measurement. For example, the variable GPA is a numerical variable because it represents a numerical measurement of a student's academic performance. The variable study hours is a numerical variable since it numerically measures the number of hours per week a student studies.

> **DEFINITION 2.5**
> A **numerical (or quantitative) variable** is a variable where the value is a number that results from a measurement process. The specific values of numerical variables are called numerical data.

Continuous and Discrete Data

Numerical data can be further classified as either continuous or discrete depending upon the numerical values it can assume.

Numerical data such as height, weight, temperature, and distance are examples of continuous data because they represent measurements that can take on any intermediate value between two numbers. The values of continuous data are usually expressed as a rounded-off number.

DEFINITION 2.6
Continuous data are numerical measurements that can assume any value between two numbers.

A person's weight is an example of continuous data since it is a measurement that can assume any value between two numbers. For example, a person's true weight could be any value between two numbers such as 150.5 and 151.5 pounds. Usually, we represent this person's weight as the *rounded-off weight* of 151 pounds.

Numerical data such as the number of full-time students attending Yale Law School, the number of automobile accidents in Arizona per year, or the number of unemployed people in New York City are examples of discrete data because these data can only take on a limited number of values.

DEFINITION 2.7
Discrete data are numerical values that can assume only a limited number of values.

The closing price of a stock selling between $24 and $25 is considered discrete data since it can only take on a limited number of price values such as: $24, $24⅛, $24¼, $24⅜, $24½, $24⅝, $24¾, $24⅞, $25.

Usually discrete data values represent **count** data and are expressed as whole numbers. However, the main characteristic of discrete data is the **break** between successive discrete values.

EXAMPLE 2.2

For each of the following:
a) Identify each variable and classify as either categorical or numerical, and
b) If the variable is numerical, then further classify it as either continuous or discrete.
  1) a person's height
  2) an individual's political affiliation
  3) the number of times per day an individual updates their status on Facebook
  4) the number of gallons of fuel used to travel from New York to Chicago
  5) a person's hair color
  6) an opinion to the question: "Is the President of the United States doing a good job?"
  7) a shoe size
  8) the number of inches of snowfall per year
  9) the number of followers an individual has on Twitter.
  10) the amount of soda in a 2 liter bottle
  11) a student's credit load for the semester

Solution

1) A person's height is a numerical variable. Since it is a measurement that has been rounded off, it is continuous.
2) Political affiliation is a categorical variable.
3) The number of times per day an individual updates their status on Facebook is a numerical variable. Since this number represents count data, it is discrete.
4) The number of gallons of fuel used to travel from New York to Chicago is a numerical variable. It is a measurement that has been rounded off and is continuous.
5) A person's hair color is a categorical variable.
6) One's opinion is a categorical variable.
7) Shoe size is a numerical variable. Since shoe sizes can only take on a limited number of values, it is discrete.
8) The number of inches of snow is a numerical variable which represents a rounded off measurement. Thus it is continuous.
9) The number of followers an individual has on Twitter represents a numerical variable. Since this number is count data, it is discrete.
10) The amount of soda in a 2 liter bottle is a numerical variable. The amount represents a measurement that has been rounded off. Therefore, it is continuous.
11) The number of credits is a numerical variable. Since the number of credits can only take on a limited number of values, it is discrete.

2.3 EXPLORING DATA USING THE STEM-AND-LEAF DISPLAY

Statistical studies, marketing surveys, scientific and medical experiments all involve collections of data. Before any conclusions can be drawn from these collections, the data must be carefully explored to discover any useful aspects, information, or unanticipated patterns that may exist. This approach to the analysis of data is called **Exploratory Data Analysis**. The idea of exploratory data analysis is to learn as much as possible about the data before conducting any statistical testing of hypotheses or relationships, or drawing any conclusions about the data. The assumption of exploratory data analysis is that the more you know about the data, the more effectively the data can be used to perform any subsequent statistical analysis. The emphasis of exploratory data analysis is to use visual displays, such as tables or graphs, of the data to reveal vital information about the data. To help examine the shape or pattern of the data values, it is necessary to *organize* the data values to form a **distribution**.

> **DEFINITION 2.8**
> A **distribution** of a numerical variable represents the data values of the variable from the lowest to the highest value along with the number of times each data value occurs. The number of times each data value occurs is called its *frequency*.

When examining the data, it is important to know what to look for in the data. The important aspects of the data are called the *characteristics of a distribution*. Let's discuss how some of these characteristics are identified. More characteristics of a distribution will be discussed in Section 2.7.

Identifying Characteristics of a Distribution

Center of the Distribution
Determine the location of the middle value of all the data values.

Overall Shape of the Distribution
Look for the number of peaks within the distribution: Is there one peak or are there several peaks? How are the data values clustered? Are they symmetric about the center or skewed in one direction?
a) A distribution is *symmetric* when the distribution of the data values greater than the center of the display and the data values less than the center of the display are mirror images of each other.
b) A distribution is *skewed* when the distribution of the data values tend to be concentrated toward one end of the display or tail of the distribution, while the data values in the other tail are spread out through extreme values resulting in a longer tail.
c) A *positively skewed* or *skewed to the right distribution* is one in which the data values are concentrated in the left tail (or end of the display containing the smaller data values) with the longer tail of the display on the right.
d) A *negatively skewed* or *skewed to the left distribution* is one in which the data values are concentrated in the right tail (or end of the display containing the larger data values) with the longer tail of the display on the left.

Spread of the Data Within the Distribution
Examine the distribution of data values to determine whether the data values are clustered or spread out about the center. Look for any gap(s) within the distribution or for any individual data value that falls well outside the overall range of the data values, that is, look for any *outliers*.

Outliers can indicate significant deviations from typical data values within a distribution and can provide valuable information about a distribution. Examining the reasons for an outlier are essential to exploring data. There are possible explanations for an outlier. They are:

1. The outlier is due to a recording error.
2. The outlier is due to chance.

3. The outlier is due to an unusual occurrence which may help to provide valuable information about the data set.

In Chapter 3, we will discuss a technique to define what we mean by "a data value that lies far" from the other data values. For now, we will define an outlier as follows.

> **DEFINITION 2.9**
> An **outlier** is an individual data value which lies far (above or below) from most or all of the other data values within a distribution.

A visual display of a distribution of data values helps to highlight these important characteristics of a distribution. One such visual technique used to explore data is the **stem-and-leaf display** developed by Professor John W. Tukey of Princeton University. The stem-and-leaf display helps to provide information about the shape or pattern of the data values of the variable. To help examine the shape or pattern of the data values, it is necessary to organize the data values to form a distribution using a stem-and-leaf display.

> **DEFINITION 2.10**
> A **stem-and-leaf display** is a visual exploratory data analysis technique that shows the shape of a distribution. The display uses the actual values of the variable to present the shape of the distribution of data values.

To see how the stem-and-leaf display is constructed, let's analyze the following data values representing the cholesterol levels of 50 middle-aged men on a regular diet.

Cholesterol Levels (expressed as milligrams of cholesterol per 100 milliliters of blood)

263,	258,	240,	233,	225,	222,	199,	282,	239,	236,	232,	283,	200,
212,	225,	235,	240,	258,	263,	274,	250,	259,	241,	237,	226,	213,
269,	199,	253,	201,	265,	226,	238,	242,	259,	233,	238,	229,	215,
202,	319,	277,	229,	239,	243,	248,	245,	219,	276,	246

To construct a stem-and-leaf display for the 50 cholesterol levels we will use the following procedure. *Each data value is rewritten into two components called a stem and a leaf.*

Step 1 Identify the stem and leaf for your data values. For each cholesterol level, the stem value will pertain to the first two digits of the data value, while the leaf value will indicate the last digit of each data value. Let's discuss how to do this. The first data value of 263 is written as: **26 | 3** within the stem-and-leaf display. The first two leading digits (**26**), appearing to the left of the vertical line, are referred to as the **stem**. The last digit of the data value (**3**), which appears to the right of the vertical line, is called the **leaf**. This is illustrated as follows: **the data value of 263 is displayed as: 26 | 3**. The second data value of 258 would be represented as: **25 | 8** within the stem-and-leaf display, where the 25 represents the stem value and 8 is the leaf value. The stem portion of the third data value of 240 is 24, while the leaf portion is 0. Thus, the cholesterol level of 240 is expressed as: **24 | 0**. To construct a stem-and-leaf display for the 50 cholesterol levels, we will use the following procedure:

Step 2 Start with the smallest stem on top, and end with the largest stem at the bottom, list all the possible stems in order and in a vertical column. Include even the stems with no corresponding leaves. Draw a vertical line to the right of the column of stems as in Figure 2.2. The vertical line separates the stem and leaf of each data value.

Since the smallest data value is 199 and the largest data value is **319**, then the top stem value 19 (which corresponds to the smallest value of 199) and the bottom stem value is **31** (corresponding to the largest value of 319). Thus, the vertical column representing the stem values will include stem values starting with 19 and ending with 31 as in Figure 2.2.

FIGURE 2.2

```
Stem |
-----|
  19 |
  20 |
  21 |
  22 |
  23 |
  24 |
  25 |
  26 |
  27 |
  28 |
  29 |
  30 |
  31 |
```

FIGURE 2.3

```
Stem | Leaf
-----|------
  19 | 9 9
  20 | 0 1 2
  21 | 2 3 5 9
  22 | 5 2 5 6 6 9 9
  23 | 3 9 6 2 5 7 8 3 8 9
  24 | 0 0 1 2 3 8 5 6
  25 | 8 8 0 9 3 9
  26 | 3 3 9 5
  27 | 4 7 6
  28 | 2 3
  29 |
  30 |
  31 | 9
```

Step 3 Place the leaf of each data value with the same stem in the same row corresponding to the stem of the data value. Place this leaf to the right of the vertical line that separates the stem portion (or left side) from the leaf portion (or right side).

For the first data value of 263, the leaf portion 3 will be written within the stem row 26 to the right of the vertical line. This procedure of placing the leaf of each data value within the appropriate stem row is continued until all the data values are included within the stem-and-leaf display. The completed stem-and-leaf display is illustrated in Figure 2.3.

Step 4 Arrange the leaves within each stem row in increasing order from left to right. In practice, it is not necessary to name the stem and leaf portions within the display, so we've omitted them in Figure 2.4.

Stem-and-leaf display of cholesterol levels of 50 middle-aged men on a regular diet

```
19 | 9 9
20 | 0 1 2
21 | 2 3 5 9
22 | 2 5 5 6 6 9 9
23 | 2 3 3 5 6 7 8 8 9 9
24 | 0 0 1 2 3 5 6 8
25 | 0 3 8 8 9 9
26 | 3 3 5 9
27 | 4 6 7
28 | 2 3
29 |
30 |
31 | 9
```

FIGURE 2.4

Stem-and-leaf displays have several important advantages as a first step to exploratory data analysis. Notice that the number of leaves in each stem indicates the number of data values pertaining to that stem. For example, in the second row of Figure 2.4, we have: **20 | 0 1 2**. This row represents the three data values: 200, 201, and 202. In addition, turning the page sideways, shows the shape of the distribution of cholesterol levels. When viewed in this manner the stem-and-leaf display has a likeness to a bar graph. From Figure 2.4, we notice that the greatest number of cholesterol levels are concentrated between the values 222 to 248. **We can see that a major advantage of a stem-and-leaf display is that it shows the shape of the distribution by displaying the number of data values pertaining to each stem using the actual values from the distribution.**

We can also examine the leaves within each stem to determine the distance between the data values, thus providing more information on the shape of the distribution. For example, looking at the two rows for the stems 28 and 31, **28 | 2 3** and **31 | 9**, notice that the largest distance between any two data values within the distribution occurs between data values 283 and 319. In fact, you should notice that the cholesterol level of 319 stands apart from the pattern of all the other cholesterol levels. The cholesterol level of 319 is an **outlier** since it lies well above all the other cholesterol levels within the distribution.

In Figure 2.4, we have seen that the stem-and-leaf display presents the data values of a distribution in a more convenient form that will make it easier to study the shape and important characteristics of a distribution while preserving the actual data values. As we explained before, the stem-and-leaf display is an exploratory data analysis technique that assists one in looking at the data.

Let's now summarize the procedure to construct a stem-and-leaf display.

Procedure to Construct a Stem-and-Leaf Display

Step 1 Identify the stem and leaf portion of your data values. Generally, the stem can have as many digits as needed to represent the beginning digit(s) of each data value. The leaf should only contain the last or terminating digit of the data values.

Step 2 List each possible stem once, in a vertical column starting with the smallest stem on top and ending with the largest stem at the bottom. List even the stems with no corresponding leaves. Draw a vertical line to the right of the column of stems.

Step 3 For each data value, record the leaf within the corresponding stem row and to the right of the vertical line.

Step 4 Arrange the leaves within each stem row in increasing order from left to right. This provides a more informative stem-and-leaf display.

Let's illustrate the use of this procedure in Example 2.3.

EXAMPLE 2.3

For the following data values representing the cholesterol levels of 50 middle-aged men on a low-fat diet and exercise program:
a) Construct a stem-and-leaf display.
b) Describe the characteristics of the distribution. Is it symmetric, skewed to the right, or skewed to the left?
c) Determine the center of the distribution and then decide between which two values on a stem are the greatest concentration of data values located?

Cholesterol Levels (expressed as milligrams cholesterol per 100 milliliters of blood)

161, 172, 193, 190, 205, 214, 217, 205, 168, 176, 188, 195, 209,
219, 189, 209, 220, 211, 200, 195, 184, 175, 198, 209, 211, 228,
199, 188, 172, 165, 177, 197, 202, 211, 198, 184, 171, 200, 218,
220, 210, 189, 231, 176, 180, 193, 230, 200, 181, 190

Solution

a) To construct a stem-and-leaf display, we will use the previously outlined procedure:

Step 1 The first two digits of each cholesterol level are the stem value while the last digit is the leaf value.

Step 2 List all the possible stems once in a vertical column starting with the smallest stem on top and ending with the largest stem at the bottom of the column. Draw a vertical line to the right of the column of stems.

Since the smallest data value is 161 and the largest data value is 231, then the stem values will range from 16 to 23. This is illustrated in Figure 2.5.

Step 3 For each cholesterol value, record the leaf within the corresponding stem row and to the right of the vertical line as shown in Figure 2.6.

42 Chapter 2 Organizing and Presenting Data

Stem	
16	
17	
18	
19	
20	
21	
22	
23	

FIGURE 2.5

Stem	Leaf
16	1 8 5
17	2 6 5 2 7 1 6
18	8 9 4 8 4 9 0 1
19	3 0 5 5 8 9 7 8 3 0
20	5 5 9 9 0 9 2 0 0
21	4 7 9 1 1 1 8 0
22	0 8 0
23	1 0

FIGURE 2.6

Step 4 Arrange the leaves within each stem row in increasing order from left to right as illustrated in Figure 2.7 for a more informative stem-and-leaf display.

**Stem-and-leaf display of cholesterol levels
of 50 middle-aged men on a low-fat diet and exercise program**

16	1 5 8
17	1 2 2 5 6 6 7
18	0 1 4 4 8 8 9 9
19	0 0 3 3 5 5 7 8 8 9
20	0 0 0 2 5 5 9 9 9
21	0 1 1 1 4 7 8 9
22	0 0 8
23	0 1

FIGURE 2.7

b) Rotating the stem-and-leaf display in Figure 2.7 sideways, we see that the shape of the distribution of the men's cholesterol levels is approximately symmetrically bell-shaped.
c) Starting from the lowest data value and counting to the twenty-fifth data value (or to the middle of the data values), we determine that the center of the distribution is between 197 and 198. The greatest concentration of cholesterol levels on a stem falls between 190 and 199, with the majority (more than 50%) of cholesterol levels falling between 171 and 219.

Back-to-Back Stem-and-Leaf Display

At times, it might be of interest to compare two distributions using a stem-and-leaf display with a common stem. This type of display is called a **back-to-back stem-and-leaf display**.

In the next example, we will examine how to construct a back-to-back stem-and-leaf display to compare the distribution of cholesterol levels for middle-aged men who are on a regular diet to those on a low-fat diet and exercise program.

EXAMPLE 2.4

Using the stem-and-leaf display for the cholesterol levels for middle-aged men on a regular diet (shown in Figure 2.4) and the stem-and-leaf display for middle-aged men on a low-fat diet and exercise program (shown in Figure 2.7):
a) Construct a back-to-back stem-and-leaf display.
b) Use the back-to-back stem-and-leaf display to compare the two distributions.

Solution

a) A back-to-back stem-and-leaf display is constructed by writing **one common** stem for both displays. This **common stem** is placed between the leaves of both displays. **This stem must contain all the possible stem values that can exist for both displays.** If you examine Figures 2.4 and 2.7, you will see that the possible stem values range from 16 to 31. Thus, the back-to-back display is constructed using the form shown in Figure 2.8.

Back-to-Back Stem-and-Leaf Display

Leaf	Stem	Leaf
	16	
	17	
	18	
	19	
	20	
	21	
	22	
	23	
	24	
	25	
	26	
	27	
	28	
	29	
	30	

FIGURE 2.8

Leaves for regular diet	Stem	Leaves for low-fat diet
	16	1 5 8
	17	1 2 2 5 6 6 7
	18	0 1 4 4 8 8 9 9
9 9	19	0 0 3 3 5 5 7 8 8 9
2 1 0	20	0 0 0 2 5 5 9 9 9
9 5 3 2	21	0 1 1 1 4 7 8 9
9 9 6 6 5 5 2	22	0 0 8
9 9 8 8 7 6 5 3 3 2	23	0 1
8 6 5 3 2 1 0 0	24	
9 9 8 8 3 0	25	
9 5 3 3	26	
7 6 4	27	
3 2	28	
	29	
	30	
9	31	

FIGURE 2.9

Figure 2.9 shows the back-to-back display comparing the cholesterol levels for middle-aged men on a regular diet to the cholesterol levels of middle-aged men on a low-fat diet and exercise program.

b) Comparing the two distributions shown in the back-to-back stem-and-leaf display of Figure 2.9, we can see that there is a greater number of lower cholesterol levels for the men on the low-fat diet and exercise program. Both distributions are essentially symmetric, but they have different data values representing their centers (239 for the regular diet distribution and 197 or 198 for the low-fat distribution). For the regular diet distribution, the greatest concentration of data values falls between 232 and 239, while the low-fat distribution has the greatest concentration from 190 to 199. The cholesterol level of 319 in the regular diet distribution represents an outlier.

2.4 FREQUENCY DISTRIBUTION TABLES

In many statistical studies, a large quantity of data is usually collected and the statistician is interested in summarizing the large data set by placing it in a more manageable form. The stem-and-leaf display that was developed in Section 2.3 is an example of one way that data can be put into a more manageable form. In this section we will present another form of representing data called a **frequency distribution**. A frequency distribution lists the values of a variable(s) and how often each value occurs. The frequency distribution of a variable can be displayed by a table or graph. In either form we will be able to determine such things as: the highest and lowest values, apparent patterns of a variable, apparent patterns between variables, what values the data may tend to cluster or group around, or which values are the most common. We will first examine the **frequency distribution table** to help determine these characteristics of the data.

> **DEFINITION 2.11**
> A **frequency distribution table** is a table in which a data set has been divided into distinct groups, called classes, along with the number of data values that fall into each class, called the frequency.

Frequency Distribution Table for Categorical Data

The first step in constructing a frequency distribution table is to organize the data into distinctive groups called **classes** (or **class intervals**). If the data collected are categorical, then **each category** will be used to represent a **class** (or **class interval**).

For example, if a biology student is investigating the prevalence of eye color within his college and records the data as either hazel, brown, blue, or black, then each of these four categories of eye color will represent a class.

The second step is to determine the frequency of each class by counting the number of data values that fall within each category or class.

In Example 2.5, we will use this procedure to construct a **frequency distribution table for the categorical data** pertaining to a student's major presented earlier in this chapter in Table 2.1.

EXAMPLE 2.5

Using the raw data pertaining to the Student Government Survey Data of 150 Students listed in Table 2.1,
a) construct a frequency distribution table for the categorical variable **major**, and
b) determine the most popular major for the 150 students.

Solution

a) To construct a frequency distribution table for the categorical variable **major**, we will need to organize the data into classes and then determine the frequency for each class.

Step 1 **Organize the data into distinct classes**. For a categorical variable, the categories of the variable represent the classes. For the variable **major**, these categories or classes are: **liberal arts, business, computers or mathematics, health sciences, natural and applied sciences or fine and performing arts**.

Step 2 **Determine the frequency of each class by counting the number of data values that fall within each class**. From Table 2.1, we count 68 students that are liberal arts majors; therefore, the class labeled liberal arts has a frequency of 68. Continuing this procedure for the remaining five classes, the frequency distribution table for the categorical variable **major** is constructed and shown in Table 2.2.

b) From Table 2.2, one is able to see at a glance that Liberal Arts and Business are the most popular majors for the students within the data set.

TABLE 2.2 Frequency Distribution Table for the Variable Major

Classes (Major)	Frequency
Liberal Arts	68
Business	51
Computers or Mathematics	11
Health Sciences	11
Natural and Applied Sciences	6
Fine and Performing Arts	3

Notice that the frequency distribution table in Table 2.2 displays the data pertaining to majors in a concise and convenient form and makes it easier to interpret the data than the original data set contained in Table 2.1.

Frequency Distribution Table for Numerical Data

Organizing *categorical data* into a frequency distribution table is a relatively simple task since the classes represent the categories of the variable.

On the other hand, constructing a frequency distribution table for *numerical data* is not as obvious. There are two problems that need to be addressed:

(1) How many classes should be used for the data set?
(2) How should these classes be defined?

The choice of the number of classes representing the data set is arbitrary. However, if there are **too many classes**, then it would be difficult to see any apparent pattern within the distribution of the data set. For example, in Table 2.1 the **numerical** variable **study hours** ranges from 6 hours to 36 hours. If we were to construct a frequency distribution for study hours using **each study hour as a separate group or class**, then this procedure would give us too many classes (31 classes, in fact). This is illustrated in Table 2.3. Notice that Table 2.3 would not accomplish our objective of summarizing and examining any patterns that may exist for the study hours.

TABLE 2.3 Frequency Distribution of Study Hours

Study Hours	Frequency	Study Hours	Frequency	Study Hours	Frequency	Study Hours	Frequency
6	2	14	2	22	3	30	4
7	3	15	9	23	5	31	3
8	7	16	5	24	9	32	1
9	7	17	8	25	4	33	3
10	10	18	6	26	4	34	0
11	6	19	5	27	7	35	0
12	9	20	12	28	7	36	1
13	0	21	5	29	3		

On the other hand, using **too few classes**, we would lose much of the information provided by the data. For example, if we were to use only *two* classes as shown in Table 2.4, then the data would be so clustered together that it would be difficult to discover any existing patterns.

TABLE 2.4 Frequency Distribution of Study Hours

Classes (Study Hours)	Frequency
6–21	96
22–37	54

Thus, although there are no hard and fast rules for making a choice on the number of classes to use, we will use the following guidelines to help us in our decision.

Guidelines for Constructing a Frequency Distribution Table for Numerical Data

1. The number of classes should be neither too small nor too large. Generally, the number of classes should be between 5 and 15. The number of classes will depend on the number of data values within the data set as well as how much the data values vary.
2. Define the classes so that they cover the entire range of the data values and that each data value will fall into one and only one class.
3. Whenever possible every class should have an equal range of values, called the class width. A frequency distribution with unequal class widths is more difficult to interpret than one with equal class widths.

We will now outline a procedure to construct a frequency distribution table for the variable study hours using the data set in Table 2.1.

Step 1 Decide on the number of classes into which the data are to be grouped.

Using the guideline of a number between 5 and 15, we will use 7 classes for the variable study hours in Table 2.1. This selection of 7 classes is arbitrary.

Step 2 Define the individual classes.

If we want the 7 classes to accommodate the entire data set and each class to have an equal width, then we must decide upon a class width large enough to include all the data within these 7 classes. You should realize that the number of classes and the class width are dependent upon one another. Since the fewer the number of classes, the larger the class width, and vice versa. To determine an approximate class width, we will use the rule:

$$\text{approximate class width} = \frac{\text{largest data value} - \text{smallest data value}}{\text{number of classes}}$$

Therefore, for the variable study hours in Table 2.1, the approximate width for 7 classes is:

$$\text{approximate class width} = \frac{36-6}{7} = \frac{30}{7} \approx 4.29$$

Thus, the class width must be **at least 4.29 to accommodate all the data into 7 classes**. However, **the actual class width we will use will be dependent upon the number of significant digits used to measure the data values and will always be rounded up to the number of significant digits in the data value.** Since the data values are all expressed as *whole numbers*, we will **round up** this result of 4.29 to **5**. Thus, **our choice for the width of each class will be 5**. (If our data values were expressed with tenths we would have rounded to 4.3.)

Step 3 To construct each class using the appropriate class width we must define the limits of each class so that each data value will fall into one and only one class. Each class must have an upper and lower class limit. The **class limits** represent the **smallest and largest data values** of each class. The **lower class limit is the smallest data value of the class** while the **upper class limit is the largest data value of the class**. The class limits will depend upon the types of data that are being grouped and the way the data were recorded or rounded off.

For example, the study hour data in Table 2.1 represent continuous data that have been rounded off to the nearest hour. Thus, the data values are recorded as whole numbers so the class limits will be expressed as whole numbers. Use the smallest data value of 6 hours as the lower class limit of the first class. Add **one unit** less than the class width to obtain the upper class limit. Since the data values are whole numbers, the unit value is 1. (If tenths were used one unit would be 0.1 or one tenth.) Thus, the upper class limit will be:

$$6 + (5 - 1) = 10.$$

Thus, **the first class represents the data values 6 to 10 and is written as: 6 – 10.**

We can now form the remaining classes by adding the class width to the previous lower and upper class limits. That is:

lower class limit = lower class limit of previous class plus class width

upper class limit = upper class limit of previous class plus class width

Following this procedure for each successive class until the 7 classes have been constructed for our study hour data set, we have:

Classes (Study Hours)
6 – 10
11 – 15
16 – 20
21 – 25
26 – 30
31 – 35
36 – 40

Observe that the class width of 5 is the difference between two consecutive lower class limits and *not* the difference between the upper and lower limits of a class.

DEFINITION 2.12
The **class width** is the difference between two consecutive lower class limits.

Notice all 7 classes have the same class width of five; they do not overlap so every data value will fall into only one class and the classes cover the entire range of data values for the variable study hours. The first class contains the study hours 6 to 10, where the lower class limit is 6 hours and the upper class limit is 10 hours. The second class contains the study hours 11 to 15 where 11 is the lower class limit and 15 is the upper class limit. The last class has a lower class limit of 36 hours and an upper class limit of 40 hours.

Step 4 Determine the frequency of each class.

Tally the number of data values falling within each class. This number is called the **frequency of the class** and is recorded under the frequency column of the frequency distribution table. Table 2.5 represents the **frequency distribution table** for the study hours data of Table 2.1.

TABLE 2.5 Frequency Distribution Table for the Variable Study Hours

Classes (Study Hours)	Frequency
6 – 10	29
11 – 15	26
16 – 20	36
21 – 25	26
26 – 30	25
31 – 35	7
36 – 40	1

Examining Table 2.5, we see that the data values for the distribution of study hours are now organized into a more compact and useful form that tends to highlight the important characteristics of the data. We now can notice that the majority of students study from 6 to 30 hours, with the greatest concentration of students within the class of 16 to 20 study hours. Also, rarely does any student study more than 36 hours. Although, the frequency distribution in Table 2.5 does summarize the data in a more convenient form, information about the original study hours data set is sacrificed. **Unlike a stem-and-leaf display, the actual individual values of the data in a frequency distribution table are no longer apparent.** Notice that one student studied from 36 to 40 hours; however, from Table 2.5 we cannot determine the exact number of hours that this particular student studied. Thus, for each class of a frequency distribution table, we need to **identify a particular value to represent all the data values within that class**. The **midpoint of the class** is used for this purpose, and it is called the **class mark**. The class mark of a class is obtained by calculating the average of the class limits for the class.

DEFINITION 2.13
The **class mark** is the midpoint of a class and is used as the representative value for the class. The formula to calculate the class mark is:

$$\text{Class Mark} = \frac{\text{upper class limit} + \text{lower class limit}}{2}$$

For the frequency distribution in Table 2.5, the class mark of the first class, 6–10, is 8. This class mark is determined by adding the class limits of 6 and 10 and dividing by 2. That is:

$$\text{class mark of first class is: } \frac{10 + 6}{2} = 8$$

The class mark of each class is obtained using this class mark formula. In Table 2.6, the class marks of the frequency distribution of study hours have been added to the frequency distribution table under the column labeled *class mark*.

TABLE 2.6 Frequency Distribution Table for the Variable Study Hours

Classes (Study Hours)	Frequency	Class Mark
6–10	29	8
11–15	26	13
16–20	36	18
21–25	26	23
26–30	25	28
31–35	7	33
36–40	1	38

In examining the class mark column of Table 2.6, you should notice that the difference between each successive class mark is 5. This difference is the class width. **Within a frequency distribution table, the difference between successive class marks will always be equal to the class width.**

Although Table 2.6 represents the frequency distribution for the continuous variable study hours, it is not apparent that the data is continuous. We see that 29 students studied 6 to 10 hours, while 26 students studied 11 to 15 hours. Since study hours represents continuous data, it is possible for a student to have studied 10.7 hours. *In which class would this student be included?* Although the study hours data are continuous, they were recorded and rounded off to the nearest hour. Thus, the 10.7 study hours would be rounded-off to 11 hours, and the student would be included in the class: 11–15. You should understand the class limits shown in Table 2.6 are not the **"real class limits"** or the so-called **"class boundaries"** of the data. For example, consider the first class: 6–10. Since the study hours were originally rounded to the nearest hour, any actual study time greater than or equal to 5.5 hours but less than 10.5 hours is included in the class: 6–10. Thus, the numbers 5.5 and 10.5 are referred to as the **class boundaries** or **real class limits** of the class: 6–10. The class boundaries of the second class: 11–15 are 10.5 and 15.5. The class boundaries of the last class: 36–40 are 35.5 and 40.5.

DEFINITION 2.14
The **class boundary** of a class is the midpoint of the upper class limit of one class and the lower class limit of the next class. The formula to determine the class boundary is:

$$\text{Class Boundary} = \frac{\text{Upper Class Limit of one class} + \text{Lower Class Limit of the next class}}{2}$$

In order to present study hours as continuous data and to identify the real class limits of each class, the frequency distribution table in Table 2.7 contains the class boundaries.

TABLE 2.7 Frequency Distribution Table for the Variable Study Hours

Classes (Study Hours)	Frequency	Class Mark	Class Boundaries
6–10	29	8	5.5–10.5
11–15	26	13	10.5–15.5
16–20	36	18	15.5–20.5
21–25	26	23	20.5–25.5
26–30	25	28	25.5–30.5
31–35	7	33	30.5–35.5
36–40	1	38	35.5–40.5

Notice the classes show a gap of one unit between successive class limits. That is, the first class ends at 10 hours while the second class begins at 11 hours. However, **the class boundaries have eliminated this gap of one unit** by taking one-half of this unit gap and adding it to the upper class limit and subtracting one-half of this unit gap from the lower class limit of the following class. Thus, **class boundaries are numbers midway between consecutive upper and lower class limits that are used to eliminate the gap between the class limits to signify continuous data.** In actual practice, class boundaries are numbers created when a measurement of a continuous variable has been rounded off. Therefore, **when constructing a frequency distribution table for a continuous variable we will include class boundaries as a column within the frequency distribution table**.

On the other hand, **when constructing a frequency distribution table for discrete data, a column for the class boundaries is not required**. Let's now summarize the steps required for constructing a frequency distribution table for numerical data.

Procedure for Constructing a Frequency Distribution Table for Numerical Data

Step 1 Decide on the number of classes for the data set.

Step 2 Determine the class width. The approximate class width formula is:

$$\text{approximate class width} = \frac{\text{largest data value} - \text{smallest data value}}{\text{number of classes}}$$

Always round up the approximate class width result to the next significant digit used to measure the data. This rounded up number is used to represent the class width of each class.

Step 3 Construct the class limits for the first class by:
  a) Selecting the smallest data value within the data set as the lower class limit of the first class, and
  b) Using the formula below to obtain the upper class limit of the first class:

upper class limit = lower class limit plus the class width minus one unit[1]

The remaining class limits of the frequency distribution table can be constructed using the formulas:

lower class limit = lower class limit of previous class plus class width
upper class limit = upper class limit of previous class plus class width

Step 4 To determine the class marks for each class, use the formula:

$$\text{Class Mark} = \frac{\text{Upper Class Limit} + \text{Lower Class Limit}}{2}$$

Step 5 For continuous data, determine the class boundaries by finding the midpoint between two successive class limits using the formula:

$$\text{Class Boundary} = \frac{\text{Upper Class Limit of one class} + \text{Lower Class Limit of the next class}}{2}$$

• Class boundaries are not required for discrete data. •

Step 6 Determine the frequency for each class by finding the number of data values that fall within each class.

Relative Frequency Distributions

Within a frequency distribution table, the frequency of each class is the number of data values within each class. At times, however, it is useful to report the **proportion** or the **percentage** of data values that belong to each class. This is called the **relative frequency** of the class.

---

[1] If the data values are whole numbers, the unit is 1. On the other hand, if the data values are expressed as tenths, then the unit would be 0.1 or one tenth.

DEFINITION 2.15
The **relative frequency** of the class is the proportion of data values within the class. The relative frequency is computed by the formula:

$$\text{Relative Frequency of a Class} = \frac{\text{class frequency}}{\text{total number of data values}}$$

The sum of the relative frequencies of all the classes will always be one except for possible round-off error. If the relative frequencies are multiplied by 100, then they are expressed as percents and are referred to as the **relative percentage frequencies, or relative percentages**.

DEFINITION 2.16
The **relative percentage frequency** of a class is the percent of data values within a class. The formula to calculate the relative percentage frequency is:

$$\text{Relative Percentage Frequency} = (\text{relative frequency})(100\%)$$

The sum of the relative percentage frequencies of all the classes will always equal 100 except for possible round-off error.

A frequency distribution table that displays either the relative frequencies or the relative percentage frequencies of each class is a **relative frequency distribution table**.

Relative frequencies are needed when comparing the frequency distribution tables of two data sets with a different number of data values within each data set.

Let's construct a relative frequency distribution table for the frequency distribution table given in Example 2.6.

EXAMPLE 2.6

Use the frequency distribution table of the weekly earnings (in U.S. dollars) of 50 university students shown in Table 2.8,
a) To construct a relative frequency distribution table.
b) To determine the sum of the relative frequencies and the sum of the relative percentages.
c) To identify the class that has the greatest relative percentage.

**TABLE 2.8 Frequency Distribution Table
Weekly Earnings of the University Students**

Classes (Earnings in U.S. Dollars)	Frequency
52 – 85	5
86 – 119	10
120 – 153	21
154 – 187	10
188 – 221	4

Solution
a) The **total number of students** is determined by summing the frequencies of each class: 5 + 10 + 21 + 10 + 4 = 50. Since there are 50 students that comprise the data set, and there are 5 students that earn from 52 to 85 dollars weekly, then the relative frequency of this class is: 5/50 = 0.10. If we multiply this result by 100%, we get: (0.10)(100%) = 10%. We can state that 10% of the university students earn from 52 to 85 dollars.

The relative frequency of the second class, 86–119, is: $10/50 = 0.20$. Multiplying this result by 100%, we obtain: $(0.20)(100\%) = 20\%$. Therefore, 20% of the university students earn from 86 to 119 dollars per week. The relative frequency of the third class, 120–153, is: $21/50 = 0.42$. As a percent, we have: $(0.42)(100\%) = 42\%$. Thus, 42% of the university students earn from 120 to 153 dollars per week.
The relative frequency of the class: 154–187 is:

$$\frac{10}{50} = 0.20 \text{ or } 20\%,$$

while the relative frequency of the last class, 188–221, is:

$$\frac{4}{50} = 0.08 \text{ or } 8\%.$$

The relative frequencies along with the relative percentages corresponding to each class is shown in Table 2.9. This table is referred to as a **Relative Frequency Distribution Table**.

TABLE 2.9 Relative Frequency Distribution Table Weekly Earnings of the University Students

Classes (Earnings in U.S. dollars)	Frequency	Relative Frequency	Relative Percentages
52 – 85	5	0.10	10
86 – 119	10	0.20	20
120 – 153	21	0.42	42
154 – 187	10	0.20	20
188 – 221	4	0.08	8
	TOTAL = 50	**SUM = 1.00**	**SUM = 100%**

b) **The sum of the relative frequencies is:**
relative frequencies sum $= 0.10 + 0.20 + 0.42 + 0.20 + 0.08$
$= 1.00$
The sum of the relative percentages is:
relative percentages sum $= 10\% + 20\% + 42\% + 20\% + 8\%$
$= 100\%$

c) The greatest percentage of students, which is 42%, earn from 120 to 153 dollars per week. Thus, the class 120–153 has the largest relative percentage. This type of information is very useful in making a comparison between classes or when comparing the 50 university students to another sample of students. ∎

Cumulative Frequency Distributions

Suppose, in considering the information within Table 2.9, we were interested in knowing how many students are earning **less than** a particular weekly salary or how many students are earning **more than** a particular weekly salary. This type of information is displayed in a frequency distribution table called a **cumulative frequency distribution table,** which shows the number of data values that are **less than** or **greater than** a particular data value. There are two types of cumulative frequency distributions, a **"less than" cumulative frequency distribution** and a **"more than" cumulative frequency distribution**.

In a **"less than" cumulative frequency distribution table**, the cumulative frequency of a particular class indicates the number of data values within the class plus the sum of all the data values **less than** this particular class. Such a distribution is constructed by starting with the lowest class and successively summing up all the frequencies within each of the other classes.

In constructing a cumulative frequency distribution table, the manner in which each class is defined is changed. Each class is defined using **the lower class limit**.

Let's examine the technique to construct a "less than" cumulative frequency distribution table for the data in Table 2.10.

EXAMPLE 2.7

Construct a "Less Than" Cumulative Frequency Distribution for the data in Table 2.10.

TABLE 2.10 Frequency Distribution Table Weekly Earnings of the University Students

Classes (Earnings in U.S. Dollars)	Frequency
52 – 85	5
86 – 119	10
120 – 153	21
154 – 187	10
188 – 221	4

Solution

The classes for the "less than" cumulative frequency distribution will be constructed first. Except for the last class of a "less than" cumulative frequency distribution, **each class is constructed by taking the lower class limit within Table 2.10 and expressing the class of the cumulative frequency table in the form**:

less than the lower class limit

For example, the first class is written as: **less than $52**. An extra class must be added so that the data values greater than 188 can be accounted for. To construct the extra or last class of a "less than" cumulative frequency distribution table, **the lower class limit of the previous class**, which is 188, **is added to the class width**, which is 34, to obtain 222 (188 + 34). Using this value of 222, write the last class using the same form as in the previous classes. That is, the last class would be written as **less than $222.** This procedure is illustrated in Table 2.11.

TABLE 2.11 "Less Than" Cumulative Frequency Distribution Table for the Weekly Earnings of the University Students

Classes (Earnings in U.S. Dollars)	Cumulative Frequency
less than $52	
less than $86	
less than $120	
less than $154	
less than $188	
less than $222	

In comparing Table 2.10 to Table 2.11, Table 2.11 has two changes:
(1) there is one additional class.
(2) the classes have been changed to allow for the frequencies to be summed or accumulated. That is, the classes are now written as: "less than 52" dollars, "less than 86" dollars and so on.

The frequencies of any of these classes is the sum of all the frequencies of each class below it. These summed frequencies are listed under the column labeled cumulative frequency. Thus, we simply sum the class frequencies of Table 2.11 as we move down the table. That is, the first class "less than 52" has a frequency of 0 since there are no students who earned less than 52 dollars. The second class "less than 86" has a frequency of 5 since there are 5 students who earned less than 86 dollars. These 5 students represent all the students in the class 52–85 of Table 2.11. The third class "less than 120" has a frequency of 15, which represents the sum of the 5 students of the class 52–85 and the 10 students of the class 86–119 of Table 2.11. Continuing in this manner, the "less than" cumulative frequency distribution is shown in Table 2.12.

In a "Less Than" Cumulative Frequency Distribution, the cumulative frequency of the first class is always zero since there are no data values smaller than its lower class limit. **The cumulative frequency of the** last class always equals the total number of data values, since all data values are smaller than the lower class limit of the last class.

Section 2.4   Frequency Distribution Tables   53

TABLE 2.12 "Less Than" Cumulative Frequency Distribution Table for the Weekly Earnings of the University Students

Classes (Earnings in U.S. Dollars)	Cumulative Frequency
less than $52	**0**
less than $86	**5** (0 + 5)
less than $120	**15** (0 + 5 + 10)
less than $154	**36** (0 + 5 + 10 + 21)
less than $188	**46** (0 + 5 + 10 + 21 + 10)
less than $222	**50** (0 + 5 + 10 + 21 + 10 + 4)

Some observations that can be made from Table 2.12 are:

- all the 50 students made at least $52 per week.
- no student made more than $222 per week.
- the 15 students of the class that earned less than $120 per week includes the 5 students who earned less than $86 per week.
- thirty-six students earned less than $154 per week.

Sometimes cumulative frequency values are expressed using relative percentages of the cumulative frequencies. To represent a cumulative frequency value as a relative percentage, divide each cumulative frequency value by the **total** number of data and **multiply by 100 percent**.

DEFINITION 2.17

$$\text{Relative Percentage Cumulative Frequency} = \frac{(Cumulative\ Frequency)}{(Total\ Number\ of\ Data)} (100\%)$$

Table 2.13 represents the relative percentages for the "less than" cumulative frequency distribution in Table 2.12.

Using the "less than" cumulative percentage distribution in Table 2.13, we can conclude that 72% of the students earned less than $154 per week.

TABLE 2.13 "Less Than" Cumulative Percentage Distribution Table Weekly Earnings of the University Students

Classes (Earnings in U.S. Dollars)	Cumulative Frequency	Relative Percentage Cumulative Frequency $\frac{Cumulative\ Frequency}{Total\ Number\ of\ Data}(100\%)$
Less than 52	0	$\frac{0}{50}(100\%) = 0\%$
Less than 86	5	$\frac{5}{50}(100\%) = 10\%$
Less than 120	15	$\frac{15}{50}(100\%) = 30\%$
Less than 154	36	$\frac{36}{50}(100\%) = 72\%$
Less than 188	46	$\frac{46}{50}(100\%) = 92\%$
Less than 222	50	$\frac{50}{50}(100\%) = 100\%$

In some instances, we are interested in knowing the number of university students that earned $154 or more per week. This type of information is contained within a distribution referred to as a **"more than" cumulative distribution**. In a **"More Than" Cumulative Distribution** the frequencies are summed from the highest class to the lowest class, and the classes are written in the form: **lower class limit or more**. This is the reverse of the "less than" cumulative frequency distribution. Thus, in a **"more than" cumulative frequency distribution table**, the cumulative frequency of a particular class indicates the number of data values within the class plus the sum of all the data values **greater than** this particular class.

Another type of table that is used as a descriptor of categorical variables is called a Contingency Table (We will discuss these tables later in Chapter 14: Chi-Square). This table is also called a Cross *Tabulation Table.*

Basically, these tables are used to summarize count data which helps us to examine the relationship between two variables. For example, suppose you wanted to examine if there is a relationship between gender and attitude for the #*MeToo* movement. One approach would be to survey a defined population, say college students aged 17 to 21 years old from the New York City and ask them, "How frequently do you think about the #*MeToo* movement?"

Let's assume our survey generates non-biased responses. From the survey results you might find that 200 college students between the ages of 17 and 21 think about the #*MeToo* movement in the following way:

- 50 think about it a lot.
- 75 think about it moderately
- 75 do not think about it a lot.

Table 2.14 shows how this information could be placed into a table.

TABLE 2.14

	a lot	moderately	not a lot
College students	50	75	75

Using this table representation, you would not learn anything about how GENDER relates to the ATTITUDE about how the 17 to 21 year old college students think of the #*MeToo* movement.

Suppose, while conducting the survey, we also asked about the college student's gender. Table 2.14 could then be modified as shown below in Table 2.15.

TABLE 2.15

	a lot	moderately	not a lot
Female college student	35	50	15
Male college student	15	25	60

This more informative table representation is an example of a **cross tabulation table** or **contingency table**. By examining the cross tabulation table, do you think you could say that Female college students responding to the survey seem to think about the #*MeToo* movement more frequently than Male college students? As you will learn in Chapter 14, Contingency Tables are often used to help understand how two different variables relate to each other, which will also help us to answer that question.

2.5 GRAPHS: BAR GRAPH, HISTOGRAM, FREQUENCY POLYGON AND OGIVE

A graph is a descriptive tool used to visualize the characteristics and the relationships of the data quickly and attractively. A well-constructed graph will reveal information that may not be apparent from a quick examination of a frequency distribution table. There are many different types of graphs, one is a **Bar Graph**.

Bar Graph

Figures 2.10a and 2.10b display examples of bar graphs commonly seen in a newspaper or magazine. What kind of summary statements can you conclude from the graph in Figure 2.10a? What kind of information can you infer from the graph in Figure 2.10b?

The bar graphs displayed in Figures 2.10a and 2.10b are **using bars** to represent discrete or categorical data. Let's examine how to construct and when to use a bar graph.

> **DEFINITION 2.18**
> A **bar graph** is generally used to depict discrete or categorical data. A bar graph is displayed using two axes. One axis represents the discrete or categorical variable while the other axis usually represents the frequency, or percentage, for each category of the variable.

Section 2.5 *Graphs: Bar Graph, Histogram, Frequency Polygon and Ogive* **55**

Bar graphs are displayed either **horizontally** or **vertically**. A horizontal bar graph entitled "Top Ten Most Visited Websites" presents the top ten most visited websites by search traffic (as of April 2020 – via Ahrefs Site Explorer) is illustrated in Figure 2.10a. Notice Youtube is the most popular website since it has the longest horizontal bar with a monthly traffic of 1,625,928,544 visits. On the other hand, pinterest.com has the shortest horizontal bar since it is the least visited website of the ten websites with 160,008,934 monthly visits.

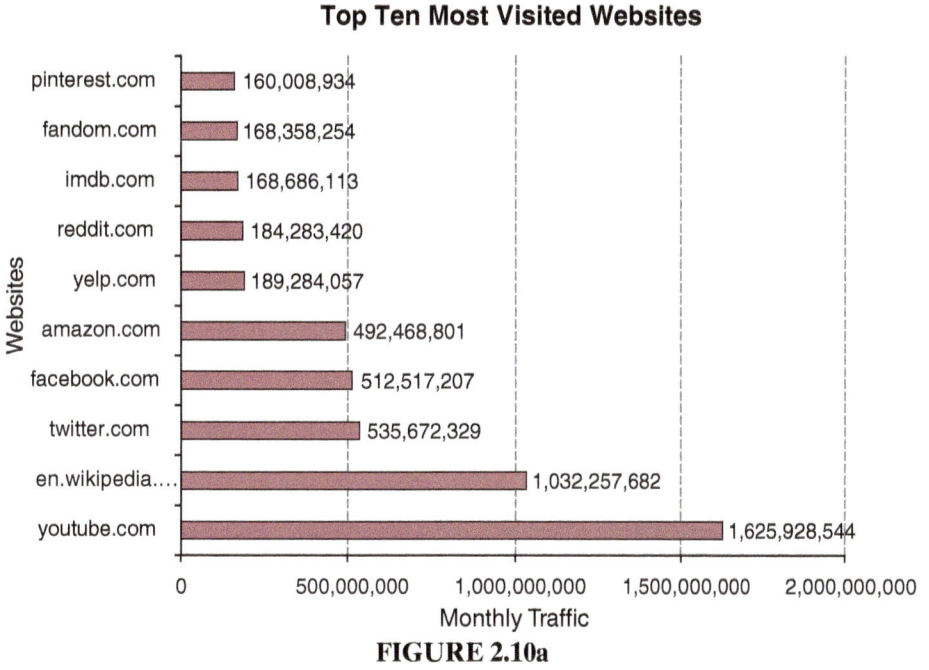

Top Ten Most Visited Websites

Website	Monthly Traffic
pinterest.com	160,008,934
fandom.com	168,358,254
imdb.com	168,686,113
reddit.com	184,283,420
yelp.com	189,284,057
amazon.com	492,468,801
facebook.com	512,517,207
twitter.com	535,672,329
en.wikipedia....	1,032,257,682
youtube.com	1,625,928,544

FIGURE 2.10a

Courtesy of Ahrefs Site Explorer

A **vertical bar graph** entitled: ***Percent of Worldwide Individual Cases That Tested Positive to Covid-19 as of April 30, 2020 for Each of the Top 11 Countries*** is shown in Figure 2.10b. In this bar graph, the horizontal axis listing the top 11 countries represents categorical data while the vertical axis represents the percent of worldwide individual cases that tested positive to Covid-19 as of April 30, 2020. Notice that the bar representing the USA is 32.92%. This means that 32.92% of the total number of individuals worldwide that tested positive to Covid-19 were in the United States. As of April 30, since the worldwide total number of individual cases that tested positive to Covid-19 was 3,352,448, then there were approximately 1,103,961 individuals within the United States that tested positive to Covid-19. The bar heights of the other 10 countries would be interpreted in a similar fashion. Although the countries along the horizontal axis are listed in decreasing percentage order, the order in which the countries are listed along the horizontal axis is not relevant.

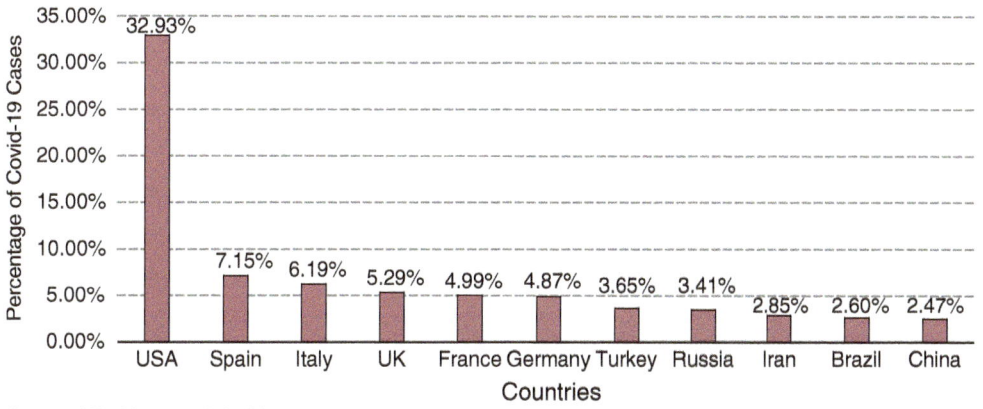

Percent of Worldwide Individual Cases That Tested Positive to Covid-19 as of April 30, 2020 for Each of The Top 11 Countries

Country	Percentage
USA	32.93%
Spain	7.15%
Italy	6.19%
UK	5.29%
France	4.99%
Germany	4.87%
Turkey	3.65%
Russia	3.41%
Iran	2.85%
Brazil	2.60%
China	2.47%

Source: Worldometer.info Site

FIGURE 2.10b

EXAMPLE 2.8

The frequency distribution in Table 2.16 represents the categorical data resulting from a wine merchant's preference survey of 500 customers in the San Francisco Bay Area. The customers were asked to respond to the question: "Which type of wine do you prefer?" The survey responses are shown in Table 2.16. Construct a horizontal bar graph using the data in the frequency distribution table.

TABLE 2.16 Wine Drinker's Preference Survey

Classes (Wine Types)	Frequency
Fume Blanc	35
Pinot Grigio	65
Chardonnay	80
Chianti	45
Cabernet Sauvignon	110
Zinfandel	50
Blended Reds	115

Solution

To construct a horizontal bar graph the vertical axis is labeled "wine type" and the horizontal axis is labeled frequency. The categories of the variable "wine types" are listed in Table 2.16. These wine categories are represented on the vertical axis as illustrated in Figure 2.11. The frequencies in Table 2.16 range from 35 to 115, and the horizontal axis is scaled in units of 20 to represent these frequency values.

Figure 2.11 represents a horizontal bar graph for the categorical data of Table 2.16.

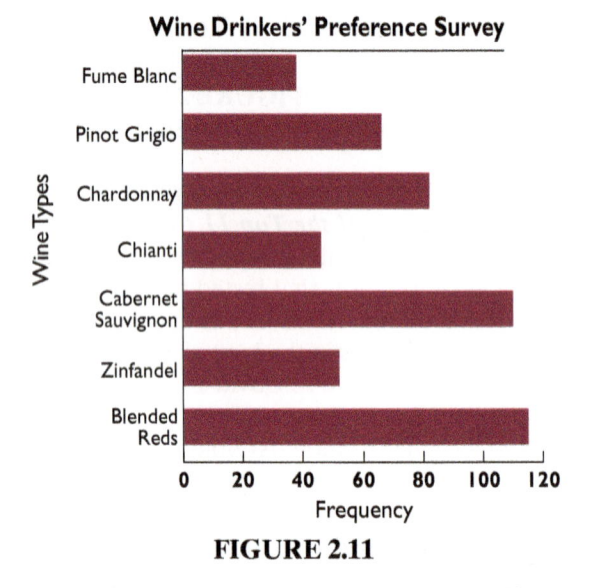

FIGURE 2.11

Notice that the wine types have been placed along the vertical axis while the number of customers preferring each wine type have been placed along the horizontal axis. The length of each horizontal bar represents the number of customers preferring that wine type. The space between the bars emphasizes that the data is discrete.

The second bar graph we will construct is the **vertical** bar graph. This type of bar graph is generally used to represent numerical data that is discrete or data that is classified by time.

EXAMPLE 2.9

The frequency distribution in Table 2.17 represents the rate of marriages per 10,000 United States population for selected years from 1960 to 2015. Construct a vertical bar graph.

TABLE 2.17 U.S. Marriages, 1960–2015

Year	Rate of Marriages (per 10,000 population)
1960	85
1965	93
1970	106
1975	100
1980	106
1985	102
1990	98
1995	89
2000	82
2005	76
2010	68
2015	69

Source: Department of Health & Human Services, National Center for Health Statistics

Solution

In constructing a vertical bar graph, the vertical axis represents the "frequency." For this graph, the vertical axis represents the rate of marriages per 10,000. Since the largest rate of marriages is 106, the vertical axis will be scaled in units of twenty. The horizontal axis depicts the selected years from 1960 to 2015.

Figure 2.12 is a vertical bar graph representing the numerical data in Table 2.17 where the length of each bar corresponds to the rate of marriages per 10,000 for each selected year.

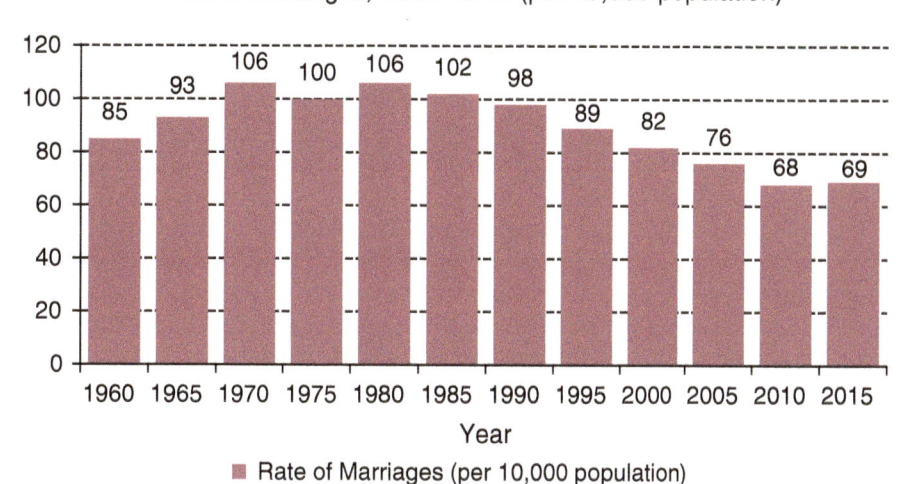

FIGURE 2.12

This vertical bar graph is an example of a **Time-Series Graph** since the value of the variable "Rate of Marriages" on the vertical axis is represented over selected time periods, "years" on the horizontal axis.

DEFINITION 2.19

A **time-series graph** contains information about two variables, where one of the variables is time.

Histogram

The next graph we will consider is the histogram. It is the most common graphical representation of a frequency distribution.

> **DEFINITION 2.20**
> A **histogram** is a vertical bar graph that represents a **continuous** variable. To depict the continuous nature of the data, the rectangles are connected to each other **without any gaps or breaks** between two adjacent rectangles.
>   The width of each rectangle corresponds to the width of each class of the frequency distribution and the height of each rectangle corresponds to the frequency, or relative frequency, of the class.
>   The vertical sides of each rectangle of a histogram correspond to the class boundaries of each class, and so, there are no breaks or gaps between the rectangles.

Let us examine how to construct a histogram using the frequency distribution data in Table 2.18 which represents the average daily temperatures of a South Pacific Island during the month of December.

TABLE 2.18 Average Daily Temperatures (in Degrees Fahrenheit) of a South Pacific Island

Classes (Temperatures in °F)	Frequency	Class Boundaries
74–77	1	73.5–77.5
78–81	7	77.5–81.5
82–85	15	81.5–85.5
86–89	6	85.5–89.5
90–93	1	89.5–93.5
94–97	1	93.5–97.5

In constructing a histogram, the horizontal axis is labeled with the class boundaries. The use of class boundaries for the histogram depicts that the variable temperature is continuous. The class boundaries of each class of daily temperatures are shown in the third column of Table 2.18.

The horizontal axis is scaled from the lowest class boundary of 73.5 to the highest class boundary of 97.5. All the other class boundaries are marked off in equal distances between 73.5 to 97.5, along the horizontal axis. The class frequencies are scaled along the vertical axis. Since the greatest frequency is 15, the vertical axis will be scaled from 0 to 16 in increments of 2.

Rectangles are constructed representing each class of the frequency distribution. The width of each rectangle equals the class width, and the height of each rectangle corresponds to the frequency of each class. In constructing each rectangle, the lower class boundary of each class is used as the beginning point of each rectangle, and the lower class boundary of the next class is the ending point. Thus, the first class, which has the class boundaries 73.5–77.5 and a class width of 4, would be represented by a rectangle with a width of 4

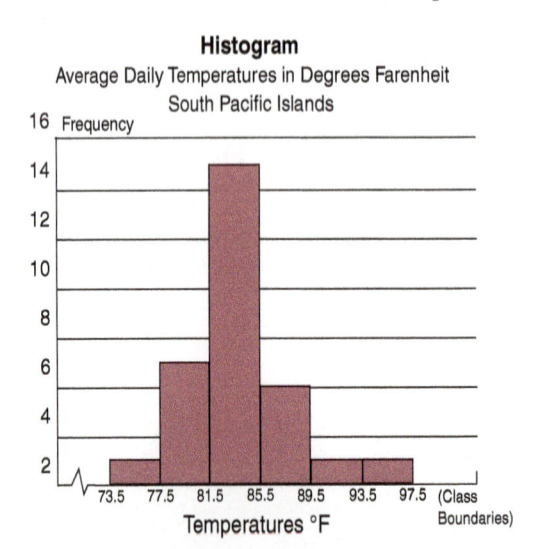

FIGURE 2.13

that begins at the lower class boundary of 73.5 and ends at the class boundary of 77.5. The second class, which has the class boundaries 77.5–81.5 and a class width of 4, would be represented by a rectangle with a width of 4 that begins at the lower class boundary of 77.5 and ends at the class boundary of 81.5. This process continues until the last class 93.5–97.5 is represented by a rectangle with a width of 4 that begins at the lower class boundary of 93.5 and ends at the class boundary of 97.5.

Figure 2.13 illustrates the histogram for the frequency distribution in Table 2.18. Notice that this histogram is basically a vertical bar graph where each rectangle or bar has equal widths and there are no breaks or gaps between the adjacent rectangles or bars. Thus, the histogram portraying the daily temperatures of the South Pacific Island is conveying the fact that the temperature data is **continuous**.

The histogram in Figure 2.13 indicates that the greatest frequency of days during the month of December had an average daily temperature from 81.5 to 85.5 degrees, and rarely does the average daily temperature drop below 77.5 degrees or greater than 89.5 degrees on this South Pacific Island.

Notice that a **break-mark** (√–) has been placed on the horizontal axis in Figure 2.13. This indicates that some values have been omitted. Specifically, the temperatures 0 to 73 have been omitted. This is often done for artistic balance in the presentation of the graph. **In general, a break-mark (√–) is used to indicate the omission of values on either the horizontal or vertical axes.**

Let's summarize the procedure to construct a histogram.

Procedure to Construct a Histogram

Step 1 Organize the data into a frequency distribution table.

Step 2 Label the horizontal axis with the name of the variable. Scale the horizontal axis starting from the lowest class boundary to the highest class boundary, while marking off all the other class boundaries in increments equal to the class width.

Step 3 Label the vertical axis as the frequency, relative frequency, or relative percentage frequency. Determine the largest frequency value and scale the vertical axis using an appropriate increment to represent the frequencies.

Step 4 Construct a rectangle for each class where the class boundaries correspond to the vertical sides of the rectangle, and the height of the rectangle corresponds to the frequency, relative frequency, or relative percentage frequency. Each rectangle should have the same width, and this width is the class width.

EXAMPLE 2.10

In Table 2.1, at the beginning of this chapter, survey data for 150 students was given. Table 2.19 is a frequency distribution representing the study hour data from Table 2.1.

a) Change the frequencies in Table 2.19 to relative percentage frequencies and construct a histogram using the relative percentage frequencies.

b) From the histogram determine the percentage of students who study from 11 to 25 hours.

TABLE 2.19 Study Hours for Students

Classes (Study Hours)	Frequency	Class Boundaries
6–10	29	5.5–10.5
11–15	26	10.5–15.5
16–20	36	15.5–20.5
21–25	26	20.5–25.5
26–30	25	25.5–30.5
31–35	7	30.5–35.5
36–40	1	35.5–40.5

Chapter 2 Organizing and Presenting Data

Solution a

Step 1 Since the class boundaries for each class are included, we need only to determine the relative percentage frequencies for each class. We determine the relative percentage frequencies for each class, using the formula:

$$\text{relative percentage frequency} = \frac{\text{class frequency}}{\text{total number of data}} (100\%)$$

Since the total number of data is 150, divide each class frequency by 150 to determine the relative percentage frequency for each class. For example, the first class representing 6–10 study hours has a frequency of 29. Its relative percentage frequency is:

$$\text{relative percentage frequency for the first class} = \frac{29}{150} (100\%)$$
$$= 19.33\%$$

The relative percentage frequencies for the remaining classes are included in Table 2.20.

Step 2 Label the horizontal axis Study Hours. Scale the axis starting with the lowest class boundary of 5.5 to the largest class boundary of 40.5 while marking off all the other class boundaries in increments equal to the class width of 5. This is illustrated in Figure 2.14.

Step 3 Label the vertical axis Relative Percentage Frequency. To determine an appropriate scale, find the largest percentage. Since the largest percent is 24%, the vertical axis will be scaled in increments of 5%. Examine the scaling of the vertical axis in Figure 2.14.

Step 4 Construct a rectangle for each class. Make the class boundaries correspond to the vertical sides of the rectangle, and the relative percentage frequency of the class corresponds to the height of the rectangle. The class boundaries of the histogram depict the variable study hours as continuous. This is illustrated in Figure 2.14.

TABLE 2.20 Study Hours for Students

Classes (Study Hours)	Relative Percentage	Class Boundaries
6–10	19.33%	5.5–10.5
11–15	17.33%	10.5–15.5
16–20	24.00%	15.5–20.5
21–25	17.33%	20.5–25.5
26–30	16.67%	25.5–30.5
31–35	4.67%	30.5–35.5
36–40	0.67%	35.5–40.5

Histogram
Study Hours

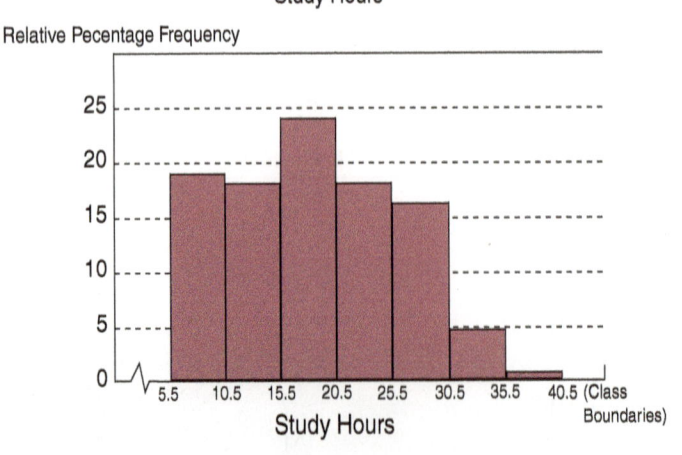

FIGURE 2.14

Solution b

The study hours 11 to 25 are being represented by the three rectangles starting with the class boundary of 10.5 and ending with the boundary of 25.5. To determine the percentage of students corresponding to these study hours we need to sum up the relative percentages of these rectangles. This sum is:

sum = 17.33% + 24.00% + 17.33%
sum = 58.66%

Thus, 58.66% of the students study from 11 to 25 hours.

TI-84 Graphing Calculator Solution for Example 2.10

The TI-84 graphing calculator can also be used to obtain a frequency distribution table and frequency histogram. Before we begin, you should know that this procedure is explained in the graphing calculator manual and in an instructional video on the textbook's website at www.MyStatLab.com.

We will begin by first entering the data into a list in the TI-84 by pressing STAT and **1: EDIT**. This command gets you to the lists in the TI-84 so that data can be entered. Figure 2.15 is a screenshot of the calculator display after pressing those keys:

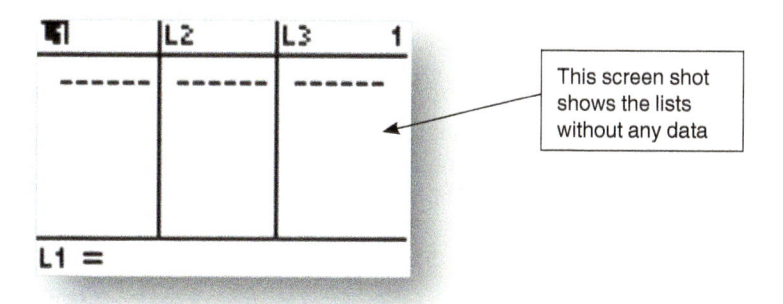

FIGURE 2.15

Now, you must enter the 150 data values for study hours into list 1, L1. This can be quite cumbersome since there are 150 pieces of data so we have put this data set up on the text's website at www.MyStatLab.com so that you can download the data set and import it into your TI-calculator. You will need a graph-link cable and the TI-Connect software to do this. Figure 2.16 shows part of the data entered from list 1 of the TI-84.

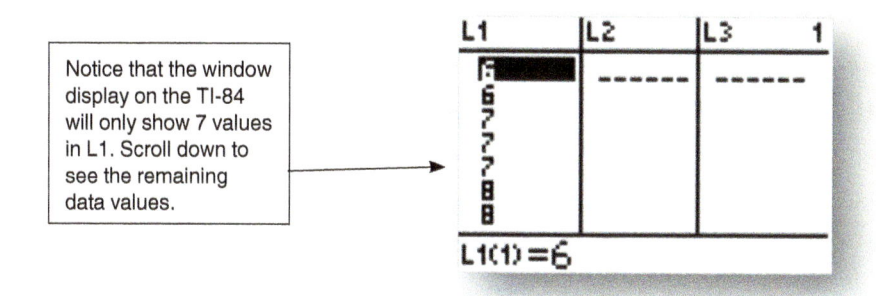

FIGURE 2.16

To create the histogram of study hours, we will have to first set up the graphing calculator window with the appropriate values so that we can see the histogram, and then access the statistical plotting menu in the calculator to enable the calculator to display the histogram for the data in L1. To set the window up, press WINDOW. The class width for study hours was found in Example 2.10 to be 5. This value should be entered into Xscl

62 Chapter 2 Organizing and Presenting Data

which is the *x*-scale or how often you will see a "tick" mark along the horizontal axis. The smallest and largest class boundaries were found to be 5.5 and 40.5. These will be entered into Xmin and Xmax respectively. The frequencies for study hours range from 0 to 36 and these values should be entered into Ymin and Ymax respectively. If the frequencies were very large numbers, you may want to change the scale along the *y*-axis. You would do this by entering a value other than 1 in the Yscl field. We will set Yscl to 2 for this example. The screenshot of this information is shown in Figure 2.17.

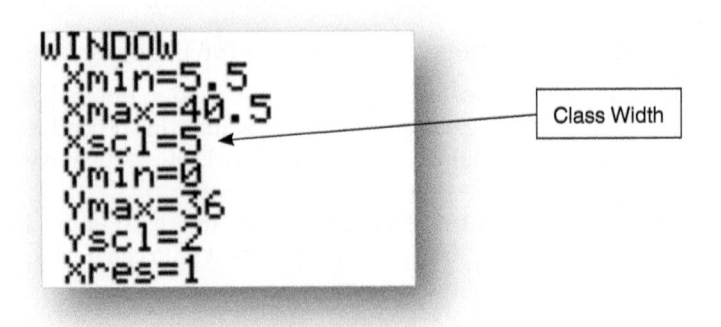

```
WINDOW
 Xmin=5.5
 Xmax=40.5
 Xscl=5      ← Class Width
 Ymin=0
 Ymax=36
 Yscl=2
 Xres=1
```

FIGURE 2.17

To create the histogram, press [2nd] [Y=] and choose option #1 to enter the statistical plotting menu. You will need to turn the data on, select the picture of the histogram in the menu, and then specify what list your data is stored in on the calculator. For our example, we put the data in L1. This is shown in Figure 2.18.

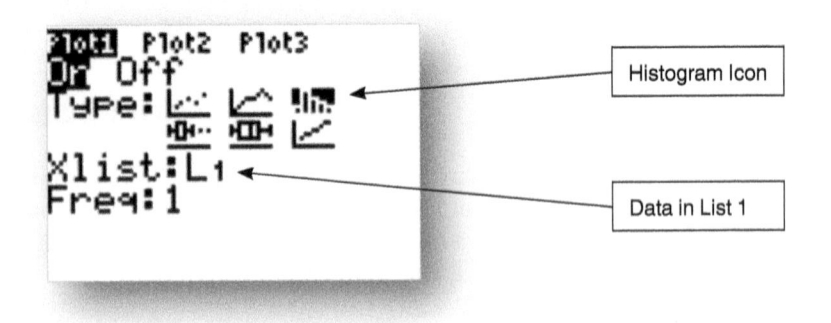

```
Plot1  Plot2  Plot3
On Off
Type: ...  ...  ▥  ← Histogram Icon
      ...  ...  ...
Xlist: L1  ← Data in List 1
Freq: 1
```

FIGURE 2.18

Once this is done, you can press the [GRAPH] key and the histogram will be displayed as shown in Figure 2.19.

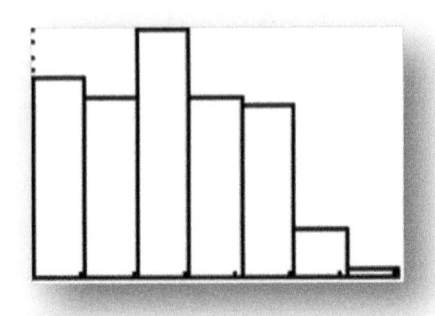

FIGURE 2.19

Once this is done, you can then construct the frequency table by tracing along the bars of the histogram and writing down the information displayed by the calculator. Pressing the TRACE key will give you the image in Figure 2.20.

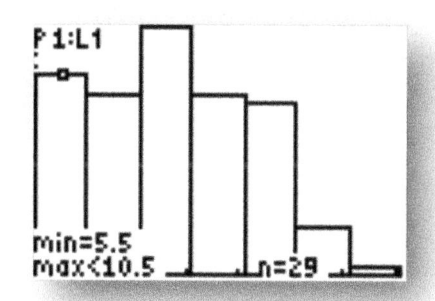

FIGURE 2.20

The dot on top of the first bar of the histogram indicates we are tracing the first bar of the histogram. The meaning of the information given is as follows:

1. Min=5.5 is the lower class boundary of the *first* class.
2. Max<10.5 is the lower class boundary of the *second* class.
3. n=29 is the frequency of the first class.

This will produce the following entry in the frequency table as shown in Figure 2.21.

Class	Class Boundary	Frequency
6–10	5.5–10.5	29

FIGURE 2.21

Pressing ▶ on the calculator once will allow you to trace along the second bar of the histogram. The above procedure can be repeated to create the second entry in your frequency table. Figure 2.22 shows this information. The entry in the frequency table is shown in Figure 2.23.

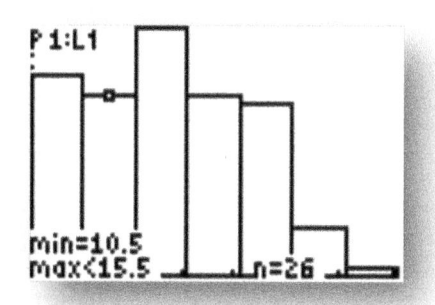

Class	Class Boundary	Frequency
6–10	5.5–10.5	29
11–15	10.5–15.5	26

FIGURE 2.22   **FIGURE 2.23**

This procedure can be repeated four more times to complete the frequency table. This table will have the same results as Table 2.17 in Example 2.10.

CASE STUDY 2.1 Relative Percentage Histogram of Infertility Incidence

Statistical Snapshot

Infertility Incidence

Infertility among women who have never had children:

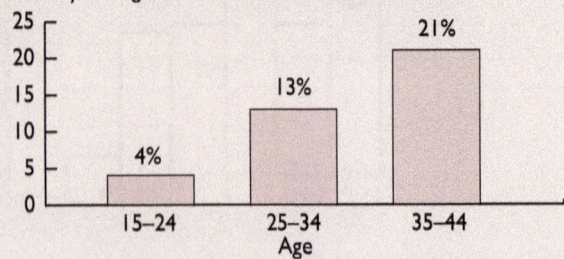

[Bar chart showing: 15–24: 4%, 25–34: 13%, 35–44: 21%, with y-axis from 0 to 25, x-axis labeled "Age"]

Source: National Center for Health Statistics survey of 8,500 women. Incidence represents those who know they are infertile or have had difficulty conceiving.

FIGURE 2.24

The snapshot entitled "Infertility Incidence" shown in Figure 2.24 displays an example of a relative percentage histogram. Using the information contained within the histogram,

a) State the classes used to construct this histogram.
b) What variable do the classes represent? Is this a continuous or discrete variable? Explain.
c) What does the height of each bar within the histogram represent? Interpret the meaning of 4% for the bar representing the ages 15–24.
d) Identify the people who were surveyed. How many people were surveyed? Identify the people who are being represented by the histogram. Does this histogram represent all the people who were surveyed? Explain.
e) Can you determine the total percentage of infertile women within the survey who never had children? If so, what is it? Now determine the total number of infertile women within the survey who never had children.
f) Construct a frequency distribution table—include the classes, frequencies, class boundaries, class marks and relative percentages.
g) Describe the shape of the histogram.
h) Does it seem reasonable that there is no class lower than the class 15–24 or greater than the class 35–44? Explain.
i) Is this an example of descriptive or inferential statistics? Explain.
j) What conclusion can you conclude about the infertility for women who have never had children as their age increases? Explain.

Frequency Polygon and Ogive

A frequency polygon is similar to a histogram. It depicts a frequency distribution by using line segments to connect the midpoints of the bars in a histogram. Frequency polygons are very useful when comparing two or more distributions on the same set of axes.

DEFINITION 2.21

A **frequency polygon** is a line graph that uses the class marks of a frequency distribution to represent continuous data.

Construct a frequency polygon by placing a dot corresponding to the frequency or percentage of each class above each class mark or midpoint. Use straight line segments to connect these dots.

Let's examine how to construct a frequency polygon using the data in Table 2.21. Table 2.21 is the frequency distribution table for the 150 student responses listed in Table 2.1 representing the variable Study Hours. A column for **class marks** has been included in the table since class marks are needed to construct a frequency polygon. Remember, the class mark represents the **midpoint** of each class.

TABLE 2.21 Study Hours for 150 Student Responses

Classes (Study Hours)	Class Mark	Frequency
6–10	8	29
11–15	13	26
16–20	18	36
21–25	23	26
26–30	28	25
31–35	33	7
36–40	38	1

To construct a frequency polygon for these data, a horizontal axis is constructed listing the class marks from Table 2.21 and a vertical axis listing the frequencies represented in the table. A dot is placed directly above each class mark of a corresponding height to the class's frequency (see Figure 2.25a). Use line segments to connect the adjacent dots. Notice that the frequency polygon is not connected to the horizontal axis. To connect the frequency polygon to the horizontal axis, add an **additional class mark to each end of the axis**. These additional class marks have **zero frequency** since no students study less than 6 or more than 40 hours.

Using the additional class marks, the graph is connected to the horizontal axis. Figure 2.25b represents the graph of the frequency polygon.

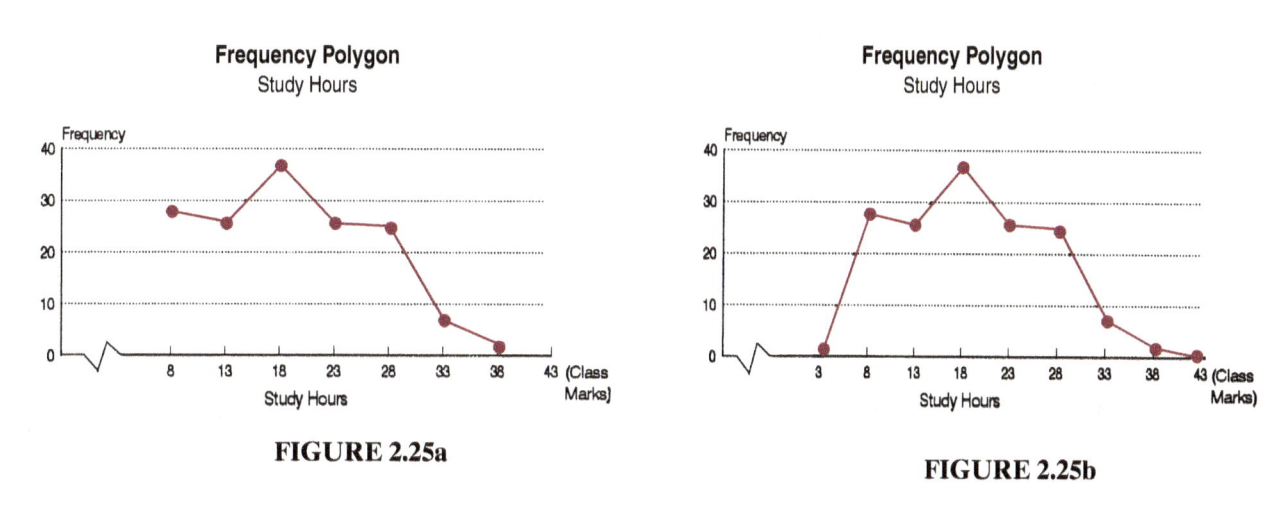

FIGURE 2.25a

FIGURE 2.25b

Let's summarize the steps to construct a frequency polygon.

Procedure to Construct a Frequency Polygon

Step 1 Organize the data into a frequency distribution table and include the class marks.

Step 2 Label the horizontal axis with the name of the variable. The horizontal axis represents the class marks. Scale this axis by marking the class marks from the lowest class mark to the highest class mark. Add two additional class marks calculated by subtracting the class width from the lowest class mark and adding the class width to the highest class mark.

Step 3 Place a dot representing the frequency or percentage of each class above the class mark. Remember, the two additional class marks have zero frequency or percentage, and their dots are placed on the horizontal axis.

Step 4 Connect the adjacent dots using straight line segments. The resulting line graph is called a frequency polygon.

The techniques used to construct frequency polygons can be extended to construct other types of line graphs. These line graphs are especially useful when you wish to compare two or more distributions on the same set of axes. This idea is illustrated in Case Study 2.2.

CASE STUDY 2.2 Social Networking Site Use by Age Group, 2005–2011

Table 2.22 entitled *Social Networking Site Usage by Age Group* presents the percent of social networking usage by specific age groups for the years 2005 to 2011. Inspecting the information within Table 2.22, it is not easy to visualize the social networking usage per age group. Let's use four line graphs to compare the usage per age group.

TABLE 2.22 Social Networking Site Usage by Age Group

YEAR	18–29 years old	30–49 years old	50–64 years old	65+ years old
2005	7%	6%	6%	No Data
2006	35%	8%	4%	0%
2007	49%	9%	6%	1%
2008	67%	25%	7%	7%
2009	76%	48%	25%	13%
2010	86%	61%	47%	26%
2011	83%	70%	51%	33%

The horizontal axis will represent years and be scaled from 2005 to 2011. The vertical axis will represent the percentage of adult internet users in each age group who use social networking sites and will be scaled from 0 to 100%. To construct each line graph, a dot representing the percentage for each age group will be placed above the respective year. Connecting each dot with line segments produced the four line graphs shown in Figure 2.26. These line graphs are time-series graphs since the horizontal line graph represents time represented by years. For time-series graphs, it is inappropriate to extend each line graph to the horizontal axis, since this would indicate the social networking site usage is zero for the years prior to 2005 or after 2011 which may not true.

Notice how quickly these time-series line graphs reveal information about the percentage of adult internet users for the four age groups that was not as apparent from a quick examination of Table 2.22.

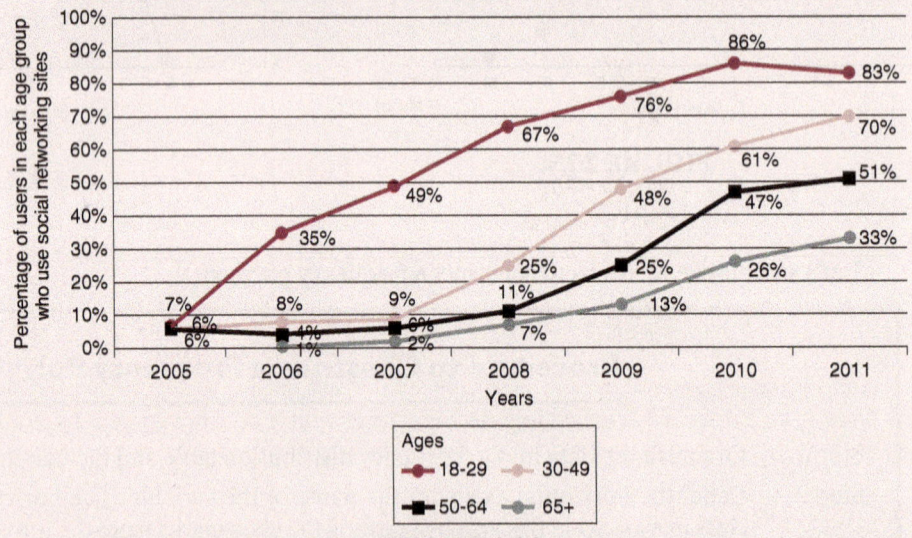

FIGURE 2.26
Courtesy of Pew Internet & American Life Project.

Source: Pew Research Center's Internet & American Life Project Surveys: February 2005, August 2006, May 2008, April 2009, May 2010 and May 2011. (Note: Total n for internet users age 65+ in 2005 was less than 100 and so results for that group are not included.)

In examining the line graph for the age group 18–29, you will notice that the percentage of users who use social networking sites increased dramatically from 2005 to 2007. Then it continued to increase but not at the same dramatic rate until 2010. From 2010 to 2011, the percentage of adult internet users for the 18–29 group dropped slightly.

For the age group 30–49, the percentage of users who used networking sites increased very slightly from 2005 to 2007. However, from 2007 to 2009, the percentage of internet users who use social networking sites increased dramatically. From 2009 to 2011, the percentage increased but at a slower rate.

Examining the line graphs for the age groups 50–64 and 65+, compare the percentage of internet users who use social networking sites.

Based on the trends displayed by these time-series line graphs, which age group would you expect to have the most dramatic increase from 2011 to 2012? Explain.

Section 2.5 Graphs: Bar Graph, Histogram, Frequency Polygon and Ogive

DEFINITION 2.22
A graph of a cumulative frequency distribution is called an **ogive** (pronounced "oh-jive"). It is a line graph of a cumulative frequency distribution that represents the left end of each class on the horizontal axis and the corresponding cumulative frequency on the vertical axis. The ogive of a "less than" cumulative frequency distribution is always increasing while the ogive of a "more than" cumulative distribution is always decreasing.

TABLE 2.23 "Less Than" Cumulative Frequency Distribution Representing Weekly Earnings of 50 University Students

Classes (Earnings in Dollars)	Cumulative Frequency
less than 52	0
less than 86	5
less than 120	15
less than 154	36
less than 188	46
less than 222	50

Table 2.23 is a cumulative frequency distribution representing the weekly earnings of 50 university students.

Let's develop a procedure to construct an ogive using the data in the cumulative frequency distribution of Table 2.23.

Step 1 Create the cumulative frequency distribution table. This is shown in Table 2.23.

Step 2 List the class values on the horizontal axis and label the axis.
From Table 2.23, we will scale the horizontal axis from 52 to 222. The class values are labeled *earnings* (in dollars).

Step 3 Scale the vertical axis using an appropriate increment and list the cumulative frequencies.
Examining Table 2.23, we see that the greatest frequency is 50. An appropriate way to scale the vertical axis would be in units of 10.

Step 4 Place a dot above the value representing the left end of each class at a height corresponding to the cumulative frequency of the class.
Since the first class, less than $52, has a frequency of zero, a dot is placed on the horizontal axis at 52. For the second class, less than $86, a dot is placed above 86 at a frequency height of 5. The remaining dots are placed at a height that corresponds to the frequencies listed in Table 2.23.

Step 5 To construct the ogive, the adjacent dots are connected using line segments.

Figure 2.27 represents the ogive for the "less than" cumulative frequency Table 2.23.

Examine the ogive in Figure 2.27. Between what two successive classes is the change in cumulative frequencies the greatest?

The procedure to construct an ogive is summarized on the next page.

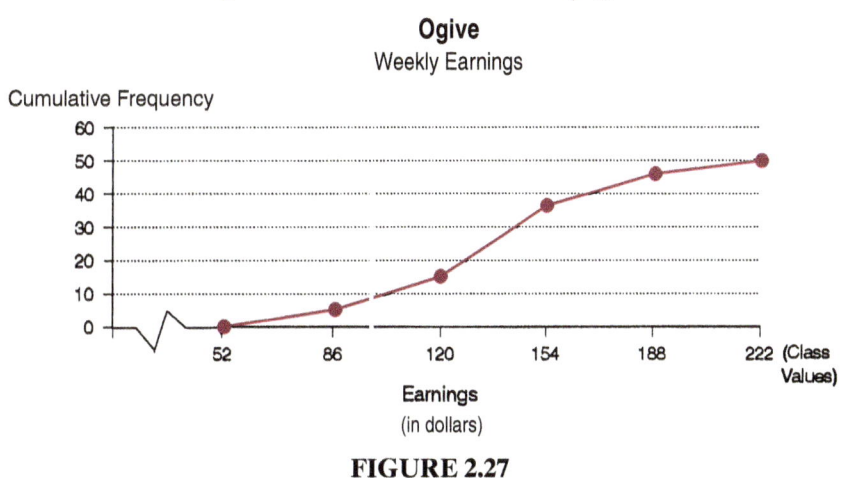

Ogive
Weekly Earnings

FIGURE 2.27

Procedure to Construct an Ogive

Step 1   Organize the data into a Cumulative Frequency Distribution Table.

Step 2   Label the horizontal axis with the name of the variable. Scale the horizontal axis starting from the lowest class value which represents the left end of each class to the highest class value, while marking off all the other class values in equal increments.

Step 3   Label the vertical axis as the cumulative frequency. Determine the largest cumulative frequency value and scale the vertical axis using an appropriate increment to represent these frequencies.

Step 4   Place a dot above the value which represents the left end of each class and that corresponds to the cumulative frequency for each class.

Step 5   Use line segments to connect the adjacent dots. The resulting graph is called an ogive.

2.6 SPECIALTY GRAPHS: PIE CHART AND PICTOGRAPH

Let's examine two specialty graphs called the **pie chart** and the **pictograph**.

DEFINITION 2.23
A **pie chart** is useful in representing categorical data in a circular format. In the pie chart, each category is represented by a sector of the circle or pie that corresponds to the percent or proportion of the data within each category.

Case Study 2.3   How Live-In Partners Fared

The pie chart in Figure 2.28 was constructed from a survey that sampled 8,450 women aged 15–44 regarding the outcomes of unmarried couples who lived together. Each sector represents a different outcome that can be categorized as: successful marriage, marriage dissolved, still living together and broke up. The area of the different sectors of the pie chart reflects the percent of the women who have indicated how their relationship fared. For example, 40% of the women indicated that their relationship resulted in a successful marriage. Thus, 40% of the area of the circle has been sectioned to represent this response.

From Figure 2.28:

a) What percent of the unmarried couples are now not living together?
b) Is there enough information to draw any conclusions regarding whether living together results in a successful marriage? Explain.
c) Do we know the number of women within the survey who were unmarried but lived together as a couple? Can we assume it was all 8,450 women surveyed? Comment.

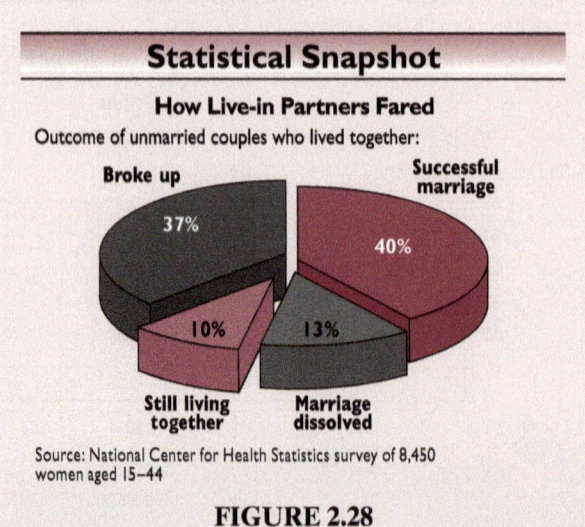

Statistical Snapshot
How Live-in Partners Fared
Outcome of unmarried couples who lived together:
- Broke up: 37%
- Successful marriage: 40%
- Still living together: 10%
- Marriage dissolved: 13%

Source: National Center for Health Statistics survey of 8,450 women aged 15–44

FIGURE 2.28

Section 2.7 *Identifying Shapes and Interpreting Graphs*

> **DEFINITION 2.24**
>
> A **pictograph** uses pictures to represent data in a more attractive and eye catching manner.[2]

Case Study 2.4 displays examples of pictographs.

Case Study 2.4 Data Snapshots

The snapshots entitled "Birthday Party Months," "USA's Stand on Lemonade" and "Kids Who Smoke" use appropriate pictures of objects to present information about birth months, lemonade entrepreneurs, and children smokers. All these snapshots display the data using graphs that were discussed within this chapter but use pictures to replace traditional graph elements.

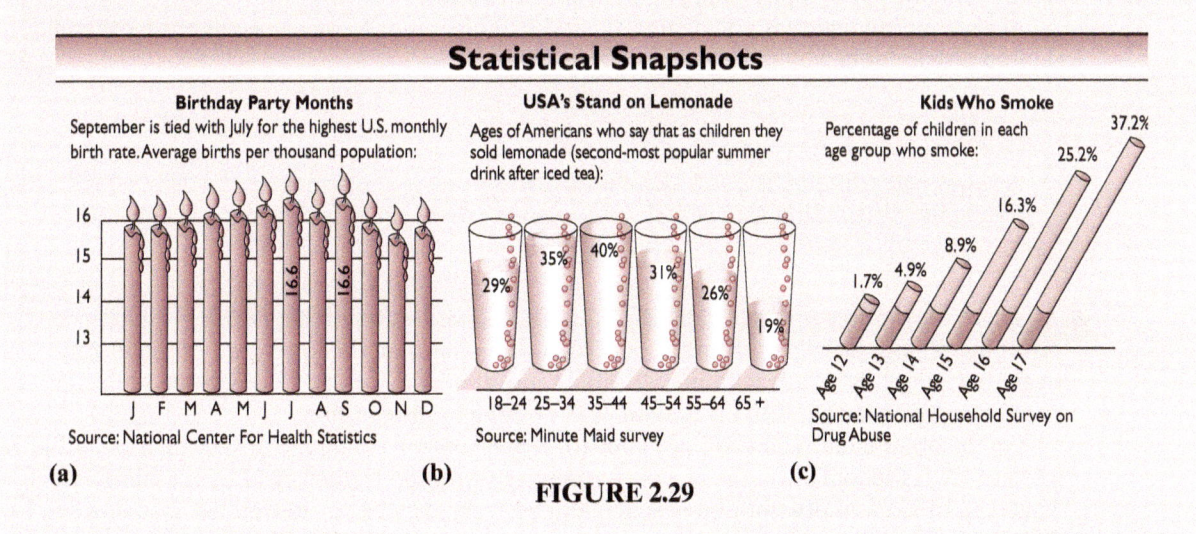

FIGURE 2.29

(a) Birthday Party Months — Source: National Center For Health Statistics
(b) USA's Stand on Lemonade — Source: Minute Maid survey
(c) Kids Who Smoke — Source: National Household Survey on Drug Abuse

For each snapshot within Figures 2.29(a), 2.29(b) and 2.29(c),

a) Identify the variables being displayed. Are they continuous or discrete?
b) Identify the object being displayed and discuss how it is used to present the data, for example, how size plays a role in presenting the data.
c) State the type of graph that is used. For example, is it a histogram, a vertical bar graph or a time-series graph? If it is a histogram, identify the classes. If it is a time-series graph, identify the way time is displayed within the graphic.
d) Discuss the shape of each graphic and how its shape can be used to draw an inference or conclusion from the graphic.

2.7 IDENTIFYING SHAPES AND INTERPRETING GRAPHS

Shapes of Distributions

A frequency distribution can have an unlimited number of shapes. We will discuss and examine the standard shapes and their characteristics for a few different types of distributions. The shapes of these distributions are smooth curves which represent a frequency distribution using very small classes.

The **symmetric bell-shaped distribution**, Figure 2.30, is a familiar distribution that we often encounter in statistics.

> **DEFINITION 2.25**
>
> A **symmetric bell-shaped distribution** has its highest frequency, or peak, at the center with the frequencies steadily decreasing but identically distributed on both sides of the center.

---

[2] A word of caution: Pictographs can be used to misrepresent data. See Section 1.5 for more information on the misuses of statistics.

70 Chapter 2 *Organizing and Presenting Data*

Symmetric bell-shaped curve
FIGURE 2.30

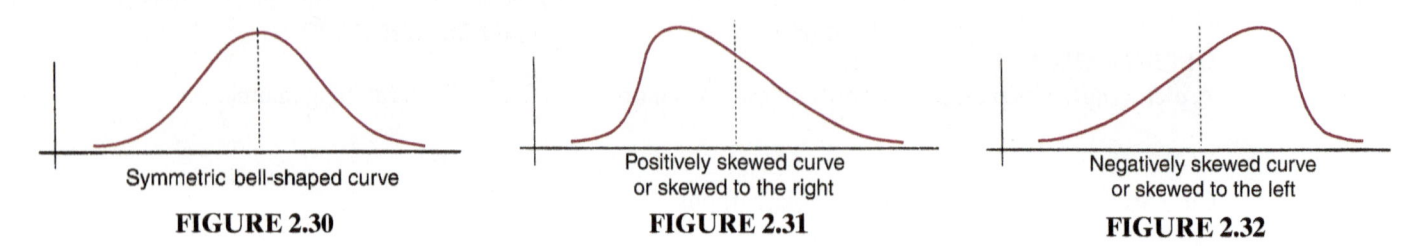

Positively skewed curve or skewed to the right
FIGURE 2.31

Negatively skewed curve or skewed to the left
FIGURE 2.32

If the symmetric bell-shaped distribution were folded along the center, the right and left side would be mirror images of itself as seen in Figure 2.30.

The bell-shaped distribution indicates that there are as many data values in both the low and high classes of the distribution. Data representing adult characteristics such as IQ scores, heights, or weights usually have a symmetric bell-shaped distribution since most adults have IQs, heights or weights that are bunched near the center value with few adults falling at the extreme values.

A distribution which is not symmetric and extends more to one side than the other side is said to be a **skewed distribution**.

> ### DEFINITION 2.26
> A **skewed distribution** is a non-symmetric distribution that has most of the data values falling into one side of the distribution with very few but extreme data values falling in the other side.

There are two types of skewed distributions. A distribution of incomes represents one type of a skewed distribution since usually most of the people have incomes on the lower end of the income scale with a relatively few people who are millionaires falling into the higher end of the income scale. This type of skewed distribution is called **a positively skewed or skewed right distribution**.

On the other hand, a person's age at death represents a second type of skewed distribution since there are a relatively few people who die at a very young age with the frequency of death increasing as a person ages. This type of skewed distribution is called **a negatively skewed or skewed left distribution**. Let's define the characteristics of these two skewed distributions.

> ### DEFINITION 2.27
> A **positively skewed, or skewed right, distribution** has a greater number of relatively low scores and a few extremely high scores. The shape of a skewed right distribution has a longer tail on the right side which represents the few extremely high scores.

A positively skewed or a skewed right distribution is illustrated in Figure 2.31. We notice that the distribution has the longer tail on the right side which represents the few extremely high scores.

> ### DEFINITION 2.28
> A **negatively skewed, or a skewed left, distribution** is a distribution that has a larger number of relatively high scores and a few extremely low scores. Its shape has a longer tail on the left side which represents the few extremely low scores.

The shape of a skewed left distribution is illustrated in Figure 2.32. We notice that the distribution has the longer tail on the left side which represents the few extremely low scores.

Notice that the skewness of a distribution results from the extreme data values falling into one of the tails of the distribution.

If the extreme values occur in the upper or right tail, the distribution has a longer right tail and is referred to as a positively skewed or skewed right distribution.

When the extreme values occur in the lower left tail, then the distribution has a longer left tail and is referred to as negatively skewed or skewed left distribution.

Let's define some other distribution shapes which are illustrated in Figure 2.33.

Uniform

U-Shaped

(a)

(b)

Reverse J-Shaped

Bimodal

(c)

(d)

FIGURE 2.33

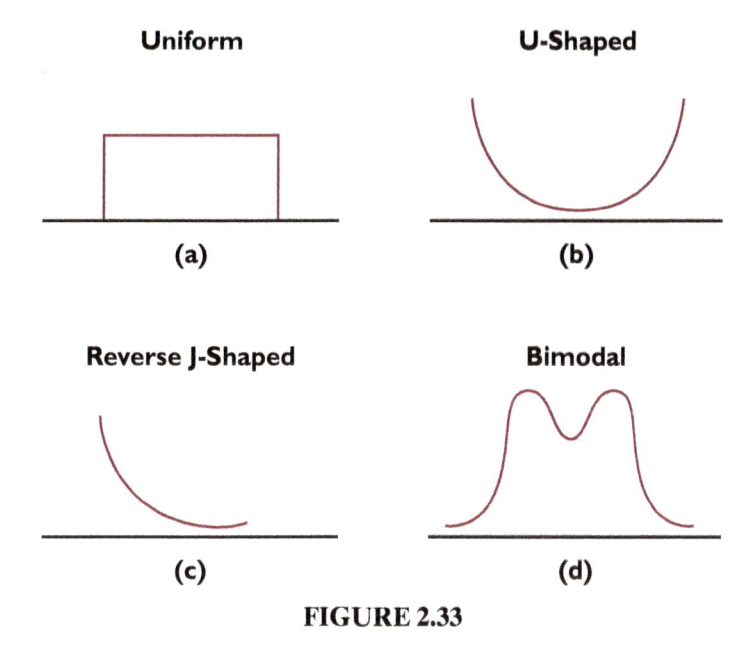

DEFINITION 2.29
A **Uniform** or **Rectangular Distribution** is a distribution where all the classes contain the same number of data values or frequencies.

A uniform distribution is illustrated in Figure 2.33a. A uniform distribution usually occurs when tossing a fair die say 600 times. We would expect the frequency distribution representing the outcome of the die to be the same for each number 1 to 6.

DEFINITION 2.30
A **U-shaped Distribution** is a distribution, as the name suggests, that has a U-shape. That is, it has its two greatest frequencies occurring at each extreme end of the distribution with the lower frequencies of the distribution occurring in the center.

Figure 2.33b illustrates a U-shaped distribution. Death rates by age within the United States form a distribution that is roughly U-shaped.

DEFINITION 2.31
A **Reverse J-Shaped Distribution**, similar to its name, has the greatest frequency of data values occurring at one end of the distribution and then tails off gradually in the opposite direction.

A reverse J-shaped distribution is illustrated in Figure 2.33c. If a pair of fair dice were tossed 720 times, and the outcomes were classified as no sixes, one six, or two sixes, the resulting distribution would resemble a reversed J-shape.

DEFINITION 2.32
A **Bimodal Distribution** is a distribution that has two separate, distinct, and relatively high peaks with the greatest frequencies. It usually indicates the distribution represents two different populations.

Figure 2.33d illustrates a bimodal distribution. A bimodal distribution usually represents the data values of two different groups or populations. For example, a distribution containing both the weights of males and females would be a bimodal distribution where each peak would represent the most frequent weight for each gender.

EXAMPLE 2.11

For the four distributions pictured in Figure 2.34 identify the shape of each distribution and explain what the shape indicates for the variable represented.

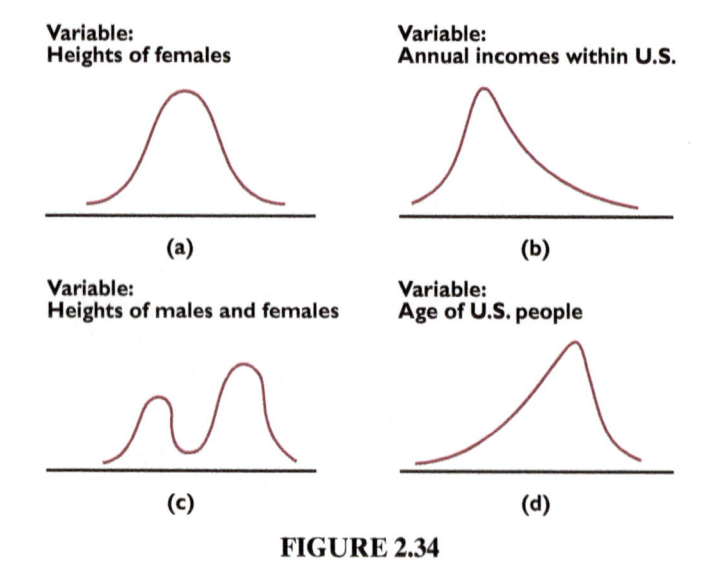

Variable: Heights of females

(a)

Variable: Annual incomes within U.S.

(b)

Variable: Heights of males and females

(c)

Variable: Age of U.S. people

(d)

FIGURE 2.34

Solution

a) The shape of the distribution in Figure 2.34a is bell-shaped. This indicates most of the females have heights that are clustered in the center of the distribution with very few females having extreme heights, i.e. too short or too tall.
b) The shape of the distribution in Figure 2.34b is skewed right. This indicates that few incomes are very high with a greater number of incomes being relatively low.
c) The shape of the distribution in Figure 2.34c is bimodal. Each peak within the distribution represents the most frequent height for each gender.
d) The shape of the distribution in Figure 2.34d is skewed left. This indicates that there are a greater number of older people and relatively fewer younger people.

Interpreting Graphs

Besides identifying the shape of a graph, reading and interpreting the information from a graph are essential. There are no principles taught in this section, it is really just practice. The following example illustrates how to read and interpret information within graphs.

EXAMPLE 2.12

Use the information in the histogram in Figure 2.35 to answer the following questions.

Histogram
Student Ages

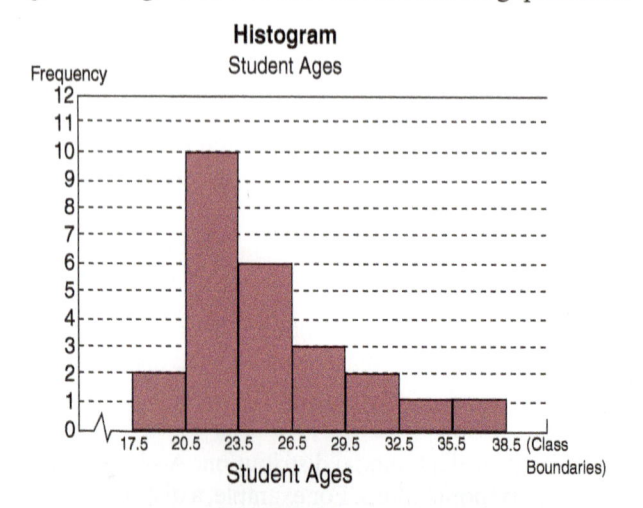

FIGURE 2.35

a) How many students are represented by this histogram?
b) How many students are older than 26?
c) What percent of the students are older than 26?
d) What percent of the students are younger than 21?
e) What percent of the students are older than 20 but younger than 30?
f) Identify the shape that best describes the distribution in Figure 2.35.

Solution

Table 2.24 has been constructed using the histogram in Figure 2.35.

TABLE 2.24 Frequency Distribution of Student Ages

Student Ages	Frequency
18–20	2
21–23	10
24–26	6
27–29	3
30–32	2
33–35	1
36–38	1

a) To calculate the number of students represented by the graph, add the frequencies of all the classes.
   The total number of students is 25.
b) Any student older than 26 must be in one of the classes: 27–29, 30–32, 33–35, and 36–38.
   To calculate the number of students older than 26, add the frequencies of these four classes. Therefore, 7 students are older than 26.
c) To calculate the percent of students older than 26:

Step 1 Divide the number of students older than 26 by the total number of students in the distribution. That is,

$$\frac{\text{The number of students older than 26}}{\text{Total number of students in the dist.}} = \frac{7}{25} = 0.28$$

Step 2 Multiply this decimal by 100% to convert the decimal to a percent.

$$0.28 \times 100\% = 28\%$$

Therefore, 28% of the students in the distribution are older than 26.

d) To calculate the percent of students younger than 21:
   Divide the number of students younger than 21 by the total number of students in the distribution:

$$\frac{\text{Number of students younger than 21}}{\text{Total number of students in the dist.}} = \frac{2}{25} = 0.08$$

Multiply this decimal by 100% to convert the decimal to a percent.

$$0.08 \times 100\% = 8\%$$

Therefore, 8% of the students in the distribution are younger than 21.

e) Calculate the percent of the students older than 20 but younger than 30.
   Using the same procedure illustrated in part d (above), we obtain 19/25 or 76% of the students in the distribution are older than 20 but younger than 30.
f) The shape of the distribution is skewed right.

CASE STUDY 2.5 Living Solo in the USA

Examine Figure 2.36 which represents the percentage of men and women by age who live alone in the USA.

Statistical Snapshot

Living Solo

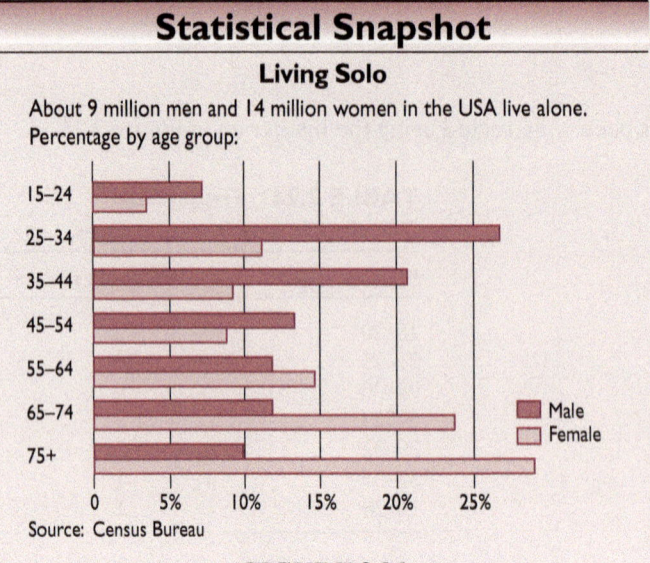

About 9 million men and 14 million women in the USA live alone.
Percentage by age group:

[Bar chart showing percentages by age group 15-24, 25-34, 35-44, 45-54, 55-64, 65-74, 75+ for Male and Female]

Source: Census Bureau

FIGURE 2.36

a) For which age class do you find the greatest number of men living alone?
b) For which age class do you find the greatest number of women living alone?
c) Are there more men or women living alone in the age class 45–54?
d) Identify the shape that best describes the graph representing the women living alone.
e) Identify the shape that best describes the graph representing the men living alone.
f) Why do you think the shapes of the graphs for parts "d" and "e" are different?

Solution

a) In Figure 2.36, the horizontal bar representing the 25–34 years old class for the males has the longest length. Thus, greatest number of men living alone are men 25–34 years old.
b) In Figure 2.36, the class 75+ years old has the longest length for the females. So, the greatest number of women living alone are women over the age of 75.
c) In the 45–54 age class, the horizontal bar for the males appears to represent 13% of the 9 million males. Thirteen percent of the 9 million males is 1.17 million males. While, the horizontal bar for the females in the 45–54 age class appears to represent 9% of the 14 million females. Nine percent of the 14 million females is 1.26 million females. Although, the *percentage* of males is greater, the *number* of females living alone is more than the number of males living alone in the 45–54 age class.
d) The shape of the graph for the male population living alone tends to be skewed to the right.
e) The shape of the graph for the female population living alone tends to be skewed to the left.
f) Women tend to live longer than males.

KEY TERMS

Back-to-Back Stem-and-Leaf Display 42
Bar Graph 54
Bimodal Distribution 71
Categorical Variable 36
Class or Class Interval 43
Class Boundary or Real Class Limit 48
Class Mark 47
Class Width 46
Continuous Data 37
Cumulative Frequency 51
Data Set 35
Discrete Data 37
Distribution 38
Exploratory Data Analysis 38
Frequency Distribution 43
Frequency Distribution Table 43
Frequency Polygon 64
Histogram 58
Horizontal Bar Graph 56
Less Than Frequency Distribution 51
More Than Frequency Distribution 51
Negatively Skewed or Skewed Left 70
Numerical Variable 36
Ogive 67
Outlier 39
Pictograph 68
Pie Chart 68
Positively Skewed or Skewed Right 70
Qualitative 36
Quantitative 36
Raw Data 33
Relative Frequency 49
Relative Percentage 50
Relative Percentage Frequency 50
Reverse J-Shaped Distribution 71
Skewed Distribution 70
Stem-and-Leaf Display 39
Symmetric Bell-Shaped Distribution 69
Time-Series Graph 57
Uniform or Rectangular Distribution 71
U-Shaped Distribution 71
Variable 35
Vertical Bar Graph 55

SECTION EXERCISES

Note: The data sets in these Section Exercises can be downloaded in either Microsoft EXCEL or TI-84 format from the textbook's website at www.MyStatLab.com.

Section 2.2

1. A _____ is a type of information, usually a property or characteristic of a person or thing, that is measured or observed.

2. A specific measurement or observation for a variable is called the _____ of the variable.

3. There are two types of variables: _____ and _____.

4. A categorical variable is a variable where the values represent _____.

5. A variable whose values are the result of a measurement process is called a _____ variable. Its values are referred to as numerical _____.

6. Numerical data can be classified as either _____ or _____.

7. Numerical data such as weight, height, and temperature are examples of _____ data.

8. The values of a variable characterized by descriptive words such as good, poor, or excellent are referred to as categorical data.
(T/F)

9. A discrete variable can be represented using a frequency distribution table.
(T/F)

10. The following table represents a sample of data collected on college students.
Each column abbreviation represents the following information:
gender: student's gender

Student #	Gender	Status	Age	Ht	Wt	Frat	Havg	GPA
5629	f	fr	19	64	110	N	89	2.99
6052	m	jr	22	73	195	N	93	3.45
7679	m	sp	20	71	145	Y	86	2.86

a) Name each variable.
b) Classify each variable as either categorical or numerical.
c) If the variable is numerical then further classify the variable as continuous or discrete.
d) For the student identified by the number 7679, specify the value of each of the variables.

Section 2.3

11. The approach to carefully examining a data set to discover any useful aspects, information, or unanticipated patterns that may exist in the data is called _____.

12. A _____ represents the numerical data values of a variable from the lowest to the highest value along with the number of times each data value occurs.

13. A relatively simple exploratory data analysis technique that helps to visually show the shape of a distribution by using the actual values of the variable is called a _____ display.

14. If a data value within a stem-and-leaf display is written as: 34 | 6, then the stem value is _____ and the leaf value is _____.

15. An individual data value which lies far from most or all of the other data values within a distribution is called an _____.

16. When two distributions are compared using a back-to-back stem-and-leaf display, a common _____ used to display each distribution.

Use the following information for questions 17 and 18: In constructing a Stem-and-Leaf display, a collection of anxiety scores are represented as whole numbers ranging from 15 to 40.

17. What would be the range of values used to construct the Stems?
a) 1 to 4   b) 15 to 40   c) 10 to 100   d) 20 to 35

18. How many digits are used to represent a leaf?
a) 1   b) 2   c) 5   d) 9

19. Express the following data values in a stem-and-leaf format using the given information.

data value	leaf portion
102	units digit
31	units digit
2.4	tenths digit

20. A health club is interested in the distribution of ages for its members that use the club on Friday evenings. The club's personnel director surveys the membership one Friday evening and collects the following data:

   Club Members' Ages in Years
   19 31 52 34 84 63 52 37 24 29
   33 46 19 32 41 49 26 32 46 44
   28 76 49 34 73 63 56 35 74 66
   59 39 61 50 37 29 30 51 54 41

   a) Construct a stem-and-leaf display for the distribution of ages.
   b) Describe the shape of the distribution.
   c) In which age decade are most people?
   d) In which fifteen year age difference are most people?

21. The following distribution represents the 1995 average American College Testing (ACT) scores for state and the District of Columbia. The ACT score is used by some colleges as an admission criteria. The average national ACT score was 20.8 out of a possible 36.

   STATE AVERAGE ACT SCORES
   20.0, 21.0, 21.0, 20.2, 20.9, 21.4, 21.4, 21.6, 18.9, 20.7, 20.2,
   21.8, 21.2, 21.1, 21.2, 21.8, 21.2, 20.1, 19.4, 21.5, 20.6, 20.9,
   21.1, 21.9, 18.8, 21.3, 21.8, 21.4, 21.3, 22.3, 20.4, 20.1, 21.7,
   19.6, 21.2, 21.2, 20.3, 22.6, 20.6, 20.6, 19.1, 21.2, 20.3, 20.1,
   21.4, 21.7, 20.6, 22.2, 20.0, 22.0, 21.3

   a) Construct a stem-and-leaf display for the ACT scores.
   b) Using the stem-and-leaf display, describe the shape of the distribution.

22. A medical study was performed to determine if a special new drug could lower a person's triglycerides. The study was conducted with two groups, an experimental group and a control group. The experimental group received the new drug, while the control group was given a placebo. The triglyceride level (mg/dL) for each participant was measured before the experiment began. The following data represents the triglyceride levels of the participants within each group after the study was conducted.

   Experimental Group
   136, 149, 160, 164, 131, 141, 135, 141, 156, 163, 150, 154, 150
   149, 158, 156, 148, 134, 143, 146, 151, 140, 131, 159, 144

   Control Group
   146, 159, 169, 178, 155, 171, 141, 136, 158, 169, 179, 173, 181
   170, 158, 170, 184, 173, 160, 173, 171, 174, 185, 186, 185

   a) Construct a back to back stem-and-leaf display to compare these two groups.
   b) Does it appear that the new drug had an effect on lowering the triglyceride level of the participants if both groups began with similar triglyceride level distributions?
   c) Identify the distribution shape for the experimental and control groups.

Section 2.4

23. A frequency distribution table helps describe the general characteristics of a distribution. It divides the data into groups, called _____ and the number of data that fall into each class, called the _____.

24. In constructing a frequency distribution table for a categorical variable, the classes represent the _____ of the variable.

25. In constructing a frequency distribution table, the actual class width is dependent upon the number of _____ digits used to measure the data values.

26. The representative value of a class is called a class _____.

27. The class boundary of a class is the _____ of the upper class limit of one class and the lower class limit of the next class.

28. The proportion of data values within a class is called the _____ frequency of the class and is calculated by dividing the class frequency by the _____ _____ of data values.

29. The relative percentage frequency of a class is the _____ of data values within a class. To determine the relative percentage frequency of each class, multiply the relative frequency of each class by _____ %. The sum of all the relative percentage frequencies will always be _____ except for possible round-off error.

30. In a "less than" cumulative frequency distribution table, the cumulative frequency of a particular class indicates the number of data values within the class plus the _____ of all the data values _____ _____ this particular class.

31. In a "more than" cumulative frequency distribution table, the cumulative frequency of a particular class indicates the number of data values within the class plus the _____ of all the data values _____ _____ this particular class.

32. The class boundaries for the class 20–26 are:
   a. 20.5–25.5   b. 19.5–25.5
   c. 20.5–26.5   d. 19.5–26.5

33. The class width for the class 25–30 is:
   a. 7   b. 6   c. 5   d. 4

34. If the lower class limit is 17 and the class width is 7, then the upper class limit is:
   a. 23   b. 24   c. 22   d. 25

35. What is the class mark for the class 10–16?
   a. 10   b. 13   c. 12.5   d. 16

36. *TODAY Moms* and *Parenting.com* surveyed 6,000 parents to determine how much money they were willing to spend on gifts for each one of their children during the holiday season. The following frequency distribution represents the results of the survey:

Classes (Dollar Amount)	Frequency	Relative Percentages
1–100	1134	–
101–200	–	–
201–300	1458	–
301–400	750	–
401–500	468	–
501–and above	588	–

a) Determine the class width.
b) Determine the missing frequency for the second class.
c) Calculate the relative percentages for each class.
d) Is it possible to determine the percentage of parents who are willing to spend:
  i) More than $300 on each child?
  ii) At least $450 on each child?
  iii) No more than $200 on each child?
  iv) Exactly $150 on each child?

37. Class boundaries are used in the construction of a histogram but not in the construction of a bar graph. (T/F)

38. A survey on the number of hours a college student works per week showed that the hours varied from 1 to 40. Determine the class limits, class marks, and class boundaries of a frequency distribution table if the work hours were grouped using 5 classes.

39. The following distribution represents the average high temperature in Central Park for the month of June.

Temperatures In Central Park In June In Degrees Farenheit

67, 71, 76, 80, 83, 81, 75, 74, 72, 72,
71, 79, 83, 85, 86, 84, 80, 82, 86, 88,
83, 80, 79, 74, 77, 76, 83, 80, 81, 82

a) Construct a frequency distribution table using 5 classes.
b) Using only the table, is it possible to determine the number of days the temperature was:
  i) at least 70?
  ii) lower than 80?
  iii) higher than 75?
  iv) exactly 81?

40. According to the National Center for Health Statistics, the most popular birth months are July and September.

Birth Months

October, December, August, May, September, May, May, April, August, January, December, May, July, June, February, November, March, April, November, August, October, September, April, July, November, April, October, August, February, May, June, September, March, November, June, July, March, June, March, August, July, September, June, October, September, January, February, September, January, July, April, November, December, May, June, October, July, May, December, September, June, August, July, September, July

Using the given distribution of birth months for a sample of college students:
a) Identify the variable and specify the type of variable.
b) Construct a frequency distribution table for the birth months.
c) How many classes did you use? Why?

Section 2.5

41. Graphs are used to accurately depict distributions. (T/F)

42. A bar graph is generally used to depict _____ or _____ data.

43. A time-series graph contains information about two variables, where one of the variables is _____.

44. Generally, in a horizontal bar graph, the categories of the categorical data are placed along the _____ axis, while the frequencies are placed along the _____ axis.

45. In a vertical bar graph, the height of each bar represents the _____.

46. A break-mark is used to represent the _____ of values in either a horizontal or vertical axis.

47. A histogram is a graph of a continuous _____. It consists of rectangles connected to each other without any _____ or _____ between two adjacent rectangles thus showing continuity.

48. A frequency polygon is a line graph that uses the class _____ of a frequency distribution to represent a _____ variable.

49. A graph of a cumulative frequency distribution is called an _____.

50. According to a new study on text messaging by The Nielsen Company, teen use of text messages continues to increase. Using the data from the study, the following table displays the number of text messages per month sent or received for the month of January for the seven age groups represented within the following distribution table.

Age Group	The Number of Text Messages Sent or Received for the month of January
12 – 18	3705
19 – 25	1707
26 – 32	758
33 – 39	683
40 – 46	456
47 – 53	349
54 – 60	126

a) What is the class width of these age groups?
b) Add a column representing the class mark of each age group and a second column representing the relative percentages to the given frequency distribution table.
c) Construct a frequency polygon using the class marks and the relative percentages.
d) Identify the shape of the frequency polygon.
e) Does the shape support The Nielsen Company's statement about teen use of text messages? Explain.

51. If two distributions have the same histogram, then the percent of terms greater than the center of the distribution is the same. (T/F)

52. According to a poll, Facebook users enrolled at Nassau Community College make more posts on certain days of the week. Using the following table of information,
a) Name the variable and type of variable.
b) Construct a horizontal bar graph.
c) Calculate the percent for each day.
d) Construct another horizontal bar graph using the percentages for each day.
e) Compare the two bar graphs.

Day of Week	Number of People Who Posted on Facebook
Monday	18,123
Tuesday	19,283
Wednesday	18,391
Thursday	19,484
Friday	19,987
Saturday	17,289
Sunday	18,572

78 Chapter 2 *Organizing and Presenting Data*

53. The National Survey of Families and Households has compiled the following table to represent the percentage of people in certain age groups who have ended a marriage or a cohabiting relationship within the last five years.

Ages	Percentage of People Who Ended a Marriage or Cohabiting Relationship within 5 Years
30–34	28%
35–39	25%
40–44	13%
45–49	11%
50–54	8%
55–59	8%
60–64	7%

a) Use this table to determine the number of people within each age group if the sample size was 1300.
b) What is the width of each class?
c) Construct a frequency table containing classes, frequency, class boundaries, class marks, and relative percentages.
d) Construct a histogram and frequency polygon.
e) Describe the shape of the frequency polygon. What does it seem to indicate about a person's age and ending a relationship?
f) Construct a "Less Than" cumulative frequency distribution table and an "ogive" graph using this table.
g) Construct a "More Than" cumulative frequency distribution table.
h) What percent of the people ended a marriage or a cohabiting relationship when they were less than 45?
i) What percent of the people ended a marriage or a cohabiting relationship within their 30's?
j) What percent of the people ended a marriage or a cohabiting relationship when they were at most 49?

54. The manager of Fast Food Grocery Store is preparing a marketing report to present to his district manager. Construct a horizontal bar graph from the information in the following table.

Cereal	Number of Cases
Bran Chex	20
Raisin Bran	35
Wheat Bran	14
Corn Bran	16

55. According to research from the lifewire.com, the following data represents the number of people who own iPads. Use this table to construct a vertical bar graph.

Growth of iPad Ownership

Year	Number of iPad Owners (in millions)
2010	11.5
2011	32.4
2012	58.1
2013	71.0
2014	68.0
2015	53.9
2016	45.6
2017	43.8

Source: http://www.lifewire.com

Section 2.6

56. A graph used to represent categorical data in a circular format is called a _____ chart.

57. A graph that uses pictures to represent data in an interesting and eye catching way is called a _____.

58. The following pie chart represents last year's expenses of a permanent make-up salon in Queens, New York. The total expenses for the year were $657,234.07.

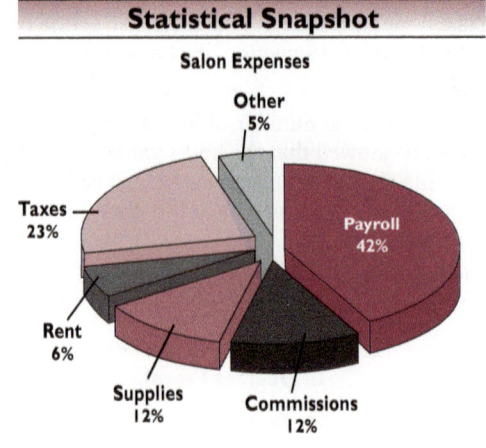

Statistical Snapshot
Salon Expenses
- Other 5%
- Taxes 23%
- Payroll 42%
- Rent 6%
- Supplies 12%
- Commissions 12%

a) In which category was the salon's largest expense?
b) What percent of the salon's expenses were paid towards supplies and rent last year?
c) How much money did the salon pay in taxes last year?
d) How much money did the salon pay in total commissions and payroll expenses last year?

Section 2.7

59. A symmetrically bell-shaped distribution has its highest frequency or peak at the _____ with the frequencies steadily _____ and identically distributed on both sides of the center.

60. A distribution which is not symmetric is a _____ distribution.

61. A positively skewed or skewed _____ distribution has a greater number of relatively _____ scores and a few extremely _____ scores.

62. A negatively skewed or skewed _____ distribution has a larger number of relatively _____ scores and a few extremely _____ scores.

63. A uniform or rectangular distribution is a distribution where all the classes contain the _____ number of data values or frequencies.

64. Use the histogram below to answer the following questions:

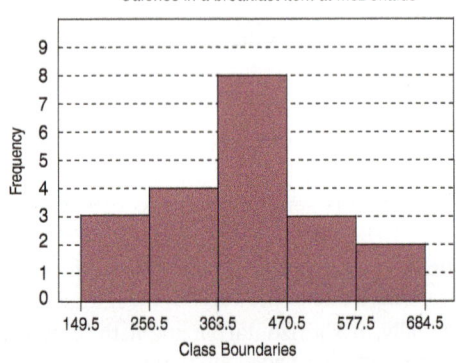

Histogram
Calories in a breakfast item at McDonalds

(Class boundaries: 149.5, 256.5, 363.5, 470.5, 577.5, 684.5; Frequencies: 3, 4, 8, 2, 2)

Source: McDonald's USA Nutrition Facts

a) Find the total number of breakfast items that are represented in the graph.
b) How many breakfast items have more than 470 calories?
c) What percent of the breakfast items are in the first two classes?
d) Which class has the smallest frequency?
e) What percent of the breakfast items have more than 256 calories but less than 578 calories?
f) What percent of the breakfast items have less than 150 calories?

65. A reverse J-shaped distribution has its greatest frequency of data values occurring at _____ end of the distribution and tails off gradually in the _____ direction.
66. A bimodal distribution is a distribution that has _____ _____ greatest frequencies or peaks.
67. If the population of ages of college students is skewed right, then this indicates:
    a) there are more older students within the population
    b) there are just as many younger students as there are older students
    c) as the age of a student increases, the frequency of the ages also increases.
    d) there is a greater number of younger students within the population.

To answer questions 68–74 use the following histogram.

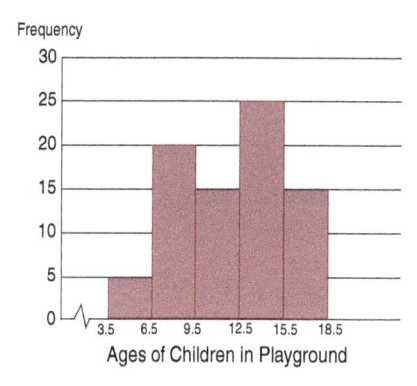

68. Find the total number of children represented in the graph.
    a) 25   b) 80   c) 100   d) 50
69. How many children in the playground are older than 12?
    a) 40   b) 25   c) 15   d) 30
70. What percent of the children are in the last two classes?
    a) 40%   b) 50%   c) 25%   d) 15%
71. Which class has the largest frequency?
    a) 4–6   b) 10–12   c) 13–15   d) 16–18
72. What percent of the children are older than 9 but younger than 16 years old?
    a) 40%   b) 50%   c) 25%   d) 15%
73. What percent of the children are younger than 4 years old?
    a) 40%   b) 50%   c) 25%   d) 0%
74. How many children in the playground are older than 3?
    a) 25   b) 80   c) 100   d) 50

CHAPTER REVIEW

75. Classify the following data as either categorical or numerical. If the data is numerical, then classify it as either continuous or discrete data.
    a) The breaking strength of a fishing line.
    b) The amount of time needed to go to work.
    c) A person's occupation.
    d) The number of bushels of apples produced by a farmer.
76. A statistics instructor has recorded the amount of time students need to complete the final examination. The times, stated to the nearest minute, are:
    50, 70, 62, 55, 38, 42, 49, 75, 80, 79, 48, 45, 53, 58, 64, 77, 48, 49, 50, 61, 72, 74, 10, 95, 120, 79, 75, 48, 72, 37, 35, 79, 72, 75, 77, 32, 30, 77, 75, 79, 45, 70, 39, 75, 72, 45, 47, 73, 72, 71
    a) Construct a stem-and-leaf display.
    b) Describe the shape of the distribution.
    c) Are there any outlier(s) in the data set? If so, give an explanation for the outlier(s)?
    d) What percentage of the class required from 40 to 75 minutes to complete the test?
    e) If the statistics instructor only allowed one hour and 20 minutes for the exam, what percentage of the class would not have had enough time to complete the exam?
77. A medical researcher surveyed one hundred mothers who recently gave birth. The researcher determined whether the mothers smoked during pregnancy and the babies' birth weights. The birth weights were separated into two groups, smoking mothers and non-smoking mothers. The following data represents the researchers findings.

Smoking Mothers
5 lbs 7 ozs, 6 lbs 6 ozs, 7 lbs 2 ozs, 9 lbs 2 ozs, 7 lbs 3 ozs,
8 lbs 2 ozs, 7 lbs 5 ozs, 6 lbs 2 ozs, 5 lbs 3 ozs, 3 lbs 4 ozs,
5 lbs 3 ozs, 7 lbs 6 ozs, 6 lbs 1 oz, 5 lbs 4 ozs, 5 lbs 2 ozs,
6 lbs 8 ozs, 7 lbs 4 ozs, 7 lbs 2 ozs, 9 lbs 3 ozs, 6 lbs 5 ozs,
7 lbs 7 ozs, 8 lbs 2 ozs, 7 lbs 6 ozs, 3 lbs 1 oz, 5 lbs 12 ozs,
6 lbs 2 ozs, 7 lbs 5 ozs, 6 lbs 5 ozs, 8 lbs 8 ozs, 9 lbs 1 oz,
8 lbs 3 ozs, 7 lbs 8 ozs, 6 lbs 1 oz, 5 lbs 9 ozs, 6 lbs 5 ozs,
6 lbs 6 ozs, 8 lbs 3 ozs, 7 lbs 8 ozs, 6 lbs 6 ozs, 5 lbs 12 ozs,
5 lbs 9 ozs, 6 lbs 9 ozs, 6 lbs 8 ozs, 5 lbs 2 ozs, 7 lbs 4 ozs,
6 lbs 9 ozs, 7 lbs 5 ozs, 7 lbs 0 ozs, 8 lbs 0 ozs, 8 lbs 0 ozs

Non-Smoking Mothers
8 lbs 2 ozs, 7 lbs 1 oz, 6 lbs 6 ozs, 5 lbs 9 ozs, 9 lbs 4 ozs,
7 lbs 0 ozs, 8 lbs 5 ozs, 6 lbs 9 ozs, 5 lbs 13 ozs, 8 lbs 4 ozs,
7 lbs 6 ozs, 6 lbs 6 ozs, 5 lbs 11 ozs, 9 lbs 0 ozs, 8 lbs 9 ozs,
7 lbs 1 oz, 6 lbs 8 ozs, 7 lbs 9 ozs, 9 lbs 3 ozs, 8 lbs 8 ozs,
6 lbs 6 ozs, 7 lbs 3 ozs, 7 lbs 14 ozs, 8 lbs 1 oz, 7 lbs 5 ozs,
6 lbs 5 ozs, 7 lbs 6 ozs, 6 lbs 12 ozs, 7 lbs 8 ozs, 7 lbs 9 ozs,
9 lbs 3 ozs, 8 lbs 4 ozs, 7 lbs 11 ozs, 6 lbs 14 ozs, 5 lbs 11 ozs,
8 lbs 6 ozs, 5 lbs 15 ozs, 6 lbs 10 ozs, 7 lbs 3 ozs, 9 lbs 8 ozs,
7 lbs 15 ozs, 7 lbs 2 ozs, 6 lbs 14 ozs, 7 lbs 12 ozs, 7 lbs 10 ozs,
6 lbs 8 ozs, 6 lbs 14 ozs, 11 lbs 2 ozs, 10 lbs 1 oz, 10 lbs 5 ozs,

a) Construct a back-to-back stem-and-leaf display for the birth weights of the babies for the smoking and non-smoking mothers using lbs for the stem portion.
b) Describe the shape of each distribution of birth weights for smoking and non-smoking mothers.
c) Does it appear from the stem-and-leaf display that smoking may have an effect on the birth weight of a newborn? Support your conclusion.

78. The test scores on a statistics exam ranged from 44 to 94. Determine the class limits, class marks, and class boundaries of a frequency distribution if these test scores were grouped using:
a) 8 classes
b) 7 classes
c) How many classes should you use if you want the class marks to be a whole number?

79. The following data set represents the IQ Scores of 120 high school students.

145, 126, 118, 95, 110, 150, 108, 128, 107, 123, 100, 113,
109, 130, 98, 118, 110, 121, 105, 109, 136, 123, 112, 117,
113, 92, 118, 104, 117, 97, 127, 108, 122, 105, 117, 121,
109, 138, 125, 118, 99, 97, 149, 92, 113, 100, 109, 117,
113, 116, 116, 98, 125, 123, 124, 129, 94, 148, 144, 127,
119, 101, 127, 95, 107, 112, 121, 125, 113, 99, 120, 122,
123, 127, 128, 117, 114, 129, 95, 113, 145, 112, 122, 106,
119, 93, 132, 122, 110, 117, 98, 121, 110, 97, 121, 131,
109, 145, 103, 135, 98, 142, 127, 116, 111, 104, 112, 116,
125, 109, 134, 110, 107, 97, 101, 131, 96, 113, 107, 115

a) Construct a frequency distribution table using 10 classes. Within the table include classes, class marks, class boundaries, relative frequencies and percentages.
b) Find the sum of the relative frequencies and percentages.
c) Using this table, determine the percentage of students that have an IQ Score which is:
 1) less than 116
 2) between 128 to 133 inclusive
 3) greater than 121
 4) at most 109
 5) at least 134

80. Fifty people were given a blind taste test to determine their preference for the soft drink colas Sipeps and Coaks. Their responses to the taste test were classified as: S for Sipeps, C for Coaks, and N for Neither. The responses to the taste test were:

S, S, C, S, C, N, N, C, S, C, C, N, C, S, C, C, C,
C, S, S, S, S, C, C, S, N, S, C, C, S, S, C, C, S,
N, C, S, N, C, S, C, S, C, S, C, S, C, C, S, C

a) Construct a frequency distribution table representing the responses to the taste test.
b) Calculate the relative frequencies and relative percents for each of the responses.

81. After the holiday season, 748 people were polled about when they finished their holiday shopping. The results of the poll were as follows:

Finished Shopping By:	Number of Shoppers
November 30	45
December 15	389
December 23	127
December 25	187

a) Determine the percentage of shoppers corresponding to each date.
b) By what date are the majority of the shoppers finished shopping?
c) What percent of the shoppers finish by December 23?
d) What percent of the shoppers are NOT finished by December 23?

82. Pollution indices for a city in the northeast for 100 consecutive days are given in the following table.

59 54 63 60 69 66 42 47 48 41
43 33 48 52 41 48 44 42 40 47
62 70 54 83 69 57 46 58 52 54
58 54 67 49 61 62 54 57 30 44
33 41 52 43 47 54 59 62 68 66
60 57 57 61 58 59 54 48 39 61
60 53 50 44 58 46 57 63 69 72
42 70 54 77 81 84 33 37 44 52
41 43 37 70 72 59 54 61 50 73
61 73 74 72 67 58 49 82 79 68

Construct:
a) A frequency distribution table using seven classes; include class marks and class boundaries within the table.
b) A histogram.
c) A frequency polygon.

83. A survey conducted by Court TV during the O.J. Simpson Trial determined the income of viewers who watch Court TV. The following distribution represents the income of some of the surveyed viewers to the nearest thousands of dollars.

Income
(in thousands of dollars)

19, 33, 49, 65, 43, 73, 62, 50, 37, 25, 44, 55, 70, 65, 54, 35, 40, 57,
61, 70, 56, 68, 38, 42, 55, 69, 72, 58, 36, 24, 70, 41, 64, 28, 60, 33,
63, 54, 20, 66, 32, 27, 49, 62, 42, 22, 37, 52, 67, 15

a) Identify the variable and type of variable.
b) Construct a frequency distribution table using 4 classes. Include within the table: classes, frequencies, class boundaries, class marks, and relative percentages.
c) Construct a "Less Than" cumulative frequency distribution using the frequency distribution table in part (b).
d) What percent of the viewers earned less than $60,000?
e) What percent of the viewers earned $30,000 or more?
f) Is it possible to determine from the distribution table what percent of the viewers earned less than $53,000? Explain.
g) What information is lost when you group data into a frequency distribution table? Is the same information lost when constructing a stem-and-leaf display? Explain.

84. The following data set represents the ages of 110 runners who competed in a city marathon.

44, 23, 19, 51, 22, 69, 31, 30, 41, 21, 52, 27, 20, 32, 23, 71,
37, 60, 29, 54, 27, 31, 23, 27, 44, 29, 58, 43, 49, 38, 60, 32,
34, 50, 28, 18, 29, 54, 41, 22, 37, 36, 37, 35, 33, 37, 20, 26,
22, 27, 31, 29, 41, 43, 67, 55, 35, 29, 48, 19, 25, 69, 60, 43,
26, 33, 24, 22, 35, 34, 18, 28, 65, 58, 21, 60, 59, 45, 27, 26,
42, 34, 52, 61, 56, 72, 19, 29, 19, 28, 44, 56, 29, 43, 50, 49,
19, 22, 23, 27, 30, 58, 64, 59, 24, 55, 48, 18, 40, 21

a) Construct a stem-and-leaf display.
b) Construct a frequency distribution table using 8 classes. Within the table include class marks, class boundaries, and relative percentages.

c) Construct a histogram and frequency polygon for the distribution.
d) Describe the shape of the polygon and classify the shape as either symmetric bell-shaped, skewed right, or skewed left.

85. The following blood pressures are for men and women within the age group 19–25 years.
a) Construct a back-to-back stem-and-leaf display for these blood pressures.
b) Construct a "less than" cumulative frequency distribution table for both the men and women blood pressure readings using 5 classes.
c) Determine the percentage of men who have blood pressures between 104 and 136 inclusive.
d) Determine the percentage of women who have blood pressures between 104 and 136 inclusive.
e) Construct an ogive for the blood pressure readings for both men and women.

Men Blood Pressures
146, 130, 115, 120, 104, 143, 120, 146, 116, 118, 130, 130, 145, 132,
118, 136, 104, 126, 156, 145, 150, 110, 130, 126, 106, 112, 120, 130,
136, 134, 156, 120, 110, 150, 105, 130, 124, 128, 130, 122, 142, 130,
106, 110, 112, 120, 126, 126, 112, 110, 120, 120, 120, 120, 130, 110,
106, 112, 134, 110, 120, 120, 128, 120, 140, 118, 120, 130, 114, 118,
120, 116, 120, 106, 118

Women Blood Pressures
134, 124, 108, 112, 126, 114, 118, 114, 110, 112, 122, 128, 135, 118,
110, 110, 105, 106, 122, 129, 105, 110, 110, 134, 130, 140, 116, 110,
112, 110, 110, 116, 114, 110, 107, 130, 139, 126, 138, 120, 118, 116,
118, 112, 104, 116, 104, 114, 124, 156, 120, 118, 120, 128, 116, 126,
128, 112, 128, 108, 108, 130, 116, 118, 122, 126, 128, 120, 119, 130,
125, 120, 126, 112, 138

86. The following table gives the percentage of deaths in each age group per 100,000 contributed to coronaries and cardiovascular diseases by age group for men and women.

At High Risk
(percentage of adults with serum cholesterol of 240 mg/dl or more)

Age	Men	Women
15-24	6.2%	6.6%
25-34	15.3	11.8
35-44	27.9	20.7
45-54	36.9	40.5
55-64	36.8	52.9
65-74	31.7	51.6

Source: National Center for Health Studies

a) Determine the class width for each class.
b) Find the class mark and class boundaries for each class. Identify the relative percentages for the males and females for each class.
c) Explain why the sum of the relative percents for the males and females exceeds 100.

d) Construct two frequency polygons one representing the males and one representing the females. Compare the two polygons and comment about the differences between the two groups.

87. A business magazine randomly surveyed 210 business owners to determine the age at which they first began their own business. The following distribution represents the survey results:

Age	Frequency
18–22	27
23–27	54
28–32	55
33–37	31
38–42	21
43–47	10
48–52	7
53–57	3
58–62	2

a) Find the class mark, class boundaries, and relative frequencies for each class.
b) Construct a frequency polygon and describe the shape of the distribution.
c) Construct a "less than" cumulative frequency distribution.
d) Construct a "more than" cumulative frequency distribution.
e) Construct an ogive for the "less than" cumulative frequency distribution.
f) What percentage of the business owners started their first business when they were younger than 33 years old?
g) What percentage of the owners started their first business when they were 43 years or older?
h) What percentage of the business owners started their first business when they were 23 to 32 years old?

88. Using the data given in the following table, construct a horizontal bar graph.

Where the Millionaires Live

State	Number of Millionaires
New York	56,096
California	38,691
Illinois	35,545
Ohio	31,202
Florida	29,523
New Jersey	28,613
Indiana	24,880
Idaho	24,738
Minnesota	23,381
Texas	23,002

89. Construct two horizontal bar graphs to represent the results of a poll conducted by the American Communications Group of 1,000 households to determine **how often people watch movies**. The following frequency distribution table represents the results of the poll.

How Often People Watch Movies	Percentage of People Who Watch Movies	
	At Home	At Theater
Less than Once a Month	13%	46%
Once a Month	12%	24%
Twice or More a Month	75%	30%

82　Chapter 2　Organizing and Presenting Data

90. The following bar graph shows the percentage of injuries for different levels of football. Using this information, construct a frequency distribution table to display the number of injuries per level of football for the sample of 400.

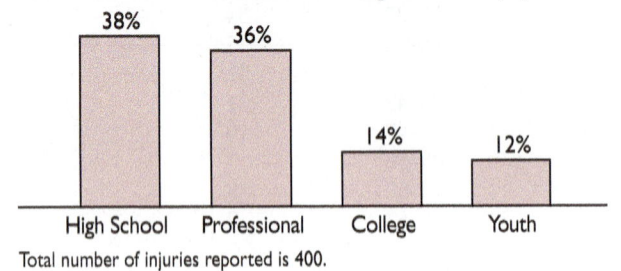

Statistical Snapshot
Sports Injuries
Level of football considered to have the highest risk of injury:

- High School: 38%
- Professional: 36%
- College: 14%
- Youth: 12%

Total number of injuries reported is 400.

91. A writing sample (composition) was submitted by each of 50 students applying for admission to Wrighter's College. Each composition was rated by 3 readers and a composite grade was assigned. The following graph represents the 50 grades.

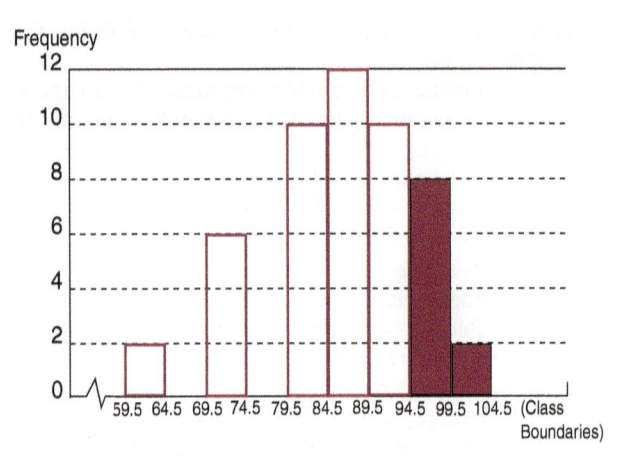

(Histogram with class boundaries 59.5, 64.5, 69.5, 74.5, 79.5, 84.5, 89.5, 94.5, 99.5, 104.5 and frequencies 2, 6, 10, 12, 10, 8, 2)

a) How many compositions were rated 90 – 99 inclusive?
b) What percent of students scored 80 – 84 inclusive?
c) What percent of the total area is in the shaded section?
d) What percent of the students scored 90 – 99 inclusive?
e) What percent of the total area is contained under the graph in the 7th and 8th intervals combined?
f) Did anyone receive a perfect grade of 100?
g) What percent of the total area is contained under the graph in the last interval?

WHAT DO YOU THINK?

92. During the first month of the 1995 baseball season after the strike of the 1994 season, attendance and TV ratings were down compared to the first month of the previous seasons. According to a CNN/Gallup poll, sport fans were asked the question: "What's your favorite sport?" The results were presented within the graphic entitled "Fan Favorites" shown in the following figure.

a) Baseball has been referred to as "America's Pastime." According to this poll, do you agree with this statement? Explain.

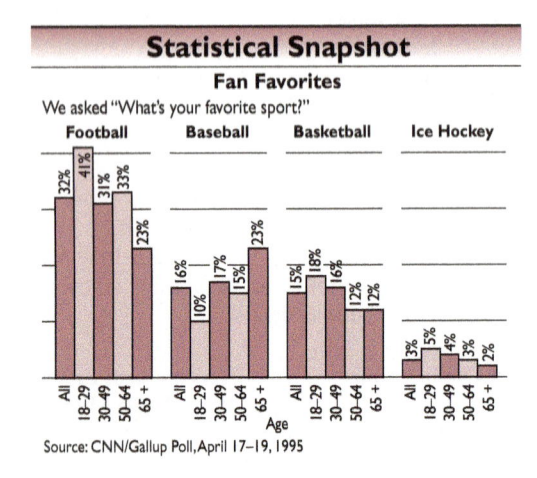

Statistical Snapshot
Fan Favorites
We asked "What's your favorite sport?"

Football: All 32%, 18–29 41%, 30–49 31%, 50–64 33%, 65+ 23%
Baseball: All 16%, 18–29 10%, 30–49 17%, 50–64 15%, 65+ 23%
Basketball: All 15%, 18–29 18%, 30–49 16%, 50–64 12%, 65+ 12%
Ice Hockey: All 3%, 18–29 5%, 30–49 4%, 50–64 3%, 65+ 2%

Source: CNN/Gallup Poll, April 17–19, 1995

b) How could you defend the statement that baseball used to be America's pastime using the results of this poll? Explain.
c) Which sport do you believe has been losing popularity over the years? How could you defend your answer using the results of this poll?
d) Which sport do you believe has been gaining the most popularity over the years? How could you defend your answer using the results of this poll?
e) Examine the football graph and explain how you can have a greater percentage of sport fans aged 18–29 than the sport fans pertaining to the All category.
f) Describe the shape of each graph for the respective sport. If 1500 fans were surveyed, then:
g) How many sport fans overall selected hockey as their favorite sport? baseball? football? basketball? Why doesn't this account for all the fans surveyed? Explain.
h) Can you determine whether there were **more, less or the same number of fans** aged 65 or older who selected football over baseball as their favorite sport? Explain.
i) Can you determine whether there were **more, less or the same number of fans** aged 18–29 who selected basketball over the fans aged 30–49 who selected baseball as their favorite sport? Explain.

PROJECTS

93. Inspect recent issues of your local tabloids and/or business magazines.

Find *one* of each of the following:
a) A frequency distribution table
b) A bar graph
c) A histogram
d) A frequency polygon
e) A pie chart
f) A pictograph

For each example determine:
a) The variables and the variable values
b) The type of data (categorical, discrete or continuous)
c) Comment as to whether or not you believe the example has been depicted appropriately.

94. Survey one hundred students and record for each student the responses to the following questions:

Ques #1: How many hours of television do you watch each week?

Ques #2: Is a Cable TV service installed in your home?
a) Construct a stem-and-leaf display for the student responses to Ques #1.
  1) Describe the shape of the display.
  2) Are there outlier's present? If so, try to explain why the outlier(s) occurred.
b) Construct a back-to-back stem-and-leaf display for the responses to Ques #1 comparing those students that have Cable TV service to those students that do not have Cable TV service.
c) Describe the differences (if any) of each category represented in the back-to-back stem-and-leaf display.
d) Construct a "less than" cumulative frequency table for the responses to Ques #1.
e) Construct an ogive for the information in part d.
f) Formulate three observations using the ogive.

CHAPTER TEST

95. Explain what variable and the value of a variable mean.

96. State the difference between a continuous and discrete data value. Give an example of each.

97. State the difference between a categorical and numerical variable. Give an example of each.

98. Classify the following data as either categorical or numerical. If the data is numerical, then classify it as either continuous or discrete data.
a) A person's weight.
b) A person's eye color.
c) The annual income of a family.
d) The volume of water in a reservoir.
e) An individual's IQ.
f) A person's opinion on the type of job Congress is doing.

99. The following table represents a sample of data collected on college students. Each column abbreviation represents the following information:

gender: student's gender
status: fr = freshman, sp = sophomore, jr = junior, sr = senior
age: student's age
ht: student's height in inches
wt: student's weight in lbs
frat: fraternity or sorority membership: N = No or Y = Yes
havg: high school average
gpa: grade point average

Student #	Gender	Status	Age	Ht	Wt	Frat	Havg	GPA
4233	f	jr	19	61	135	Y	81	2.79
7679	m	sr	24	70	165	Y	95	3.37
6203	m	sp	19	75	195	Y	91	2.94
5924	f	jr	21	60	124	Y	87	3.12

a) Name each variable.
b) Classify each variable as either categorical or numerical.
c) If the variable is numerical then further classify the variable as continuous or discrete.
d) For the student identified by the number 7679, specify the value of each of the variables.

100. Exploratory Data Analysis is used to learn as much as possible about data before conducting statistical tests. (T/F)

101. The difference between two successive class marks is equal to the class _____.

102. The sum of all the relative frequencies of all the classes will always be _____, except for possible round-off error.

103. A survey on the number of hours a college student works per week showed that the hours varied from 1 to 40. Determine the class limits, class marks, and class boundaries of a frequency distribution table if the work hours were grouped using 6 classes.

104. Bar graphs are usually used to represent categorical variables. (T/F)

105. Histograms are usually used to represent continuous variables. (T/F)

106. According to an Opinion Research Corporation survey, the average U.S. shower is 12.2 minutes with men staying in the shower 11.4 minutes on average and women having a longer average time of 13 minutes. Using the following information, construct:
a) A stem-and-leaf display for the all times.
b) Describe the shape of the distribution.
c) A back-to-back stem-and-leaf display for the male and female times.
d) Describe the shape of each distribution.
e) Compare these distributions.

f) Are there any outliers? If so, can you give a possible explanation for the outlier?

Shower times in minutes
(with the gender given in parentheses)

8(F), 14(F), 1(M), 5(M), 4(M), 12(F), 13(F), 12(F),
17(F), 12(F), 6(M), 5(M), 3(M), 11(F), 4(M), 16(F),
5(M), 8(F), 15(F), 12(F), 7(M), 9(M), 9(M), 9(M),
11(F), 17(F), 11(F), 10(M), 15(M), 10(F), 15(M), 13(F),
13(F), 10(F), 12(M), 15(M), 10(F), 18(M), 9(M), 16(M),
17(M), 20(M), 23(M), 18(M), 30(M), 21(F), 23(F)

107. The following distribution represents the IQ scores of 120 high school students.

Classes (IQ Score)	Frequency
81–90	2
91–100	13
101–110	44
111–120	40
121–130	17
131–140	4

Determine, if possible, the number and percent of the students who have IQ scores that are:
a) Less than 101.
b) At most 110.
c) 111 or less.
d) Exactly 121.
e) Between 101 to 120 inclusive.
f) At least 121.

108. According to the latest Census Bureau figures, the following table has been compiled for the given ages of women who have never been married.

Ages	Number of Never Married Women
20–24	5,940,000
25–29	3,356,000
30–34	2,122,000
35–39	1,334,000
40–44	795,000
45–49	744,000
50–54	445,000

Using the table information,
a) Identify the variable and type of variable.
b) Determine the class width.
c) Construct another table containing class boundaries, class marks and relative percentages.
d) Construct a histogram.
e) Construct a frequency polygon.
f) Describe the shape of the frequency polygon.

109. According to the latest Census Bureau figures, the following table represents the estimated lifetime earnings for those people with the stated amount of education. Using this information, construct a horizontal bar graph.

Estimated Lifetime Earnings For Those Who Have:

Education Level	Lifetime Earnings
Professional Degree	2.5 mill
Bachelor's Degree	2.1 mill
Some College	1.5 mill
High School Diploma	1.2 mill
No High School Diploma	1 mill

110. The test scores on a statistics exam ranged from 44 to 94. Determine the class limits, class marks, and class boundaries of a frequency distribution if these test scores were grouped using:
a) 9 classes
b) 10 classes
c) How many classes should you use if you want the class marks to be a whole number?

111. Given the following distribution of statistics test grades: 73, 92, 57, 89, 70, 95, 75, 80, 47, 88, 47, 48, 64, 86, 79, 72, 71, 77, 93, 55, 75, 50, 53, 75, 85, 50, 82, 45, 40, 82, 60, 55, 60, 89, 79, 65, 54, 93, 60, 83, 59

a) Construct a stem-and-leaf display.
b) Describe the shape of the distribution.
c) Construct a frequency distribution table and a histogram using six classes.

Using the histogram, answer the following questions:
d) How many test grades are greater than 89?
e) What percent of the test grades are greater than 79?
f) What percent of the test grades are lower than 70?
g) What percent of the test grades are between 70 and 79 (inclusive)?

112. College students enrolled in a co-ed physical education fitness course must run a mile at the end of the semester as part of their course requirement. The following distribution of mile times (in seconds) represents their run times.

Mile Run Times
(in seconds)

300, 320, 315, 319, 322, 339, 325, 330, 340, 345, 350, 359, 357, 358, 344, 351, 360, 379, 364, 376, 377, 366, 369, 379, 371, 373, 367, 365, 380, 399, 382, 387, 389, 390, 398, 396, 397, 385, 388, 380, 382, 398, 387, 400, 419, 403, 408, 409, 414, 416, 417, 404, 408, 412, 314, 416, 412, 409, 403, 406, 420, 439, 433, 436, 429, 420, 421, 436, 439, 428, 426, 420, 421, 426, 430, 433, 436, 433, 424, 420, 428, 437, 440, 459, 443, 452, 458, 451, 443, 449, 440, 457, 451, 456, 441, 440, 448, 446, 458, 450, 451, 459, 455, 445, 458, 449, 450, 460, 479, 466, 477, 472, 469, 463, 470, 468, 465, 479, 471, 463, 466, 473, 475, 476, 478, 480, 499, 481, 490, 497, 492, 486, 487, 488, 495, 492, 495, 498, 492, 487, 488, 489, 493, 484, 481, 499, 500, 519, 510, 518, 511, 509, 503, 503, 508, 513, 519, 510, 511, 513, 507, 503, 509, 510, 520, 539, 530, 531, 526, 533, 539, 529, 527, 523, 525, 530, 533, 538, 522, 540, 559, 548, 543, 540, 555, 556, 553, 547, 544, 545, 558, 560, 579, 577, 561, 573, 569, 566, 572, 571, 580, 599, 584

Using the mile run data,
a) Construct a stem-and-leaf display.
b) Identify the shape of the stem-and-leaf.
c) If a frequency distribution table of fifteen classes is used to represent this data, and the first class is 300–319 seconds, determine the class width.
d) Using the class width computed in part c, construct a frequency distribution consisting of fifteen classes for the mile run times where the first class is 300-319. Within the frequency distribution table, include: classes, frequencies, class marks, class boundaries and relative percentages.
e) Construct a Histogram using the frequency distribution table.
f) Construct a Frequency Polygon using the frequency distribution table.

113. According to Google Ad Planner, the following histogram based on a survey of 450 Americans represents the age distribution of Americans who use social network sites.

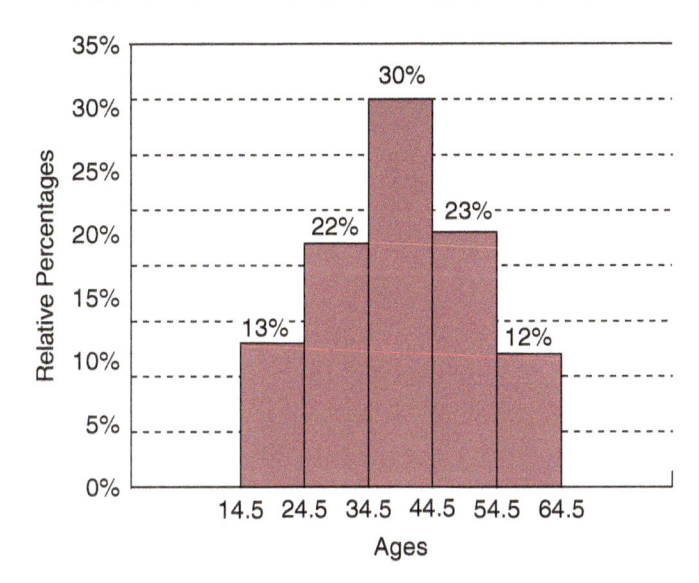

Using the Histogram,
a) Determine the percentage of Americans aged at least 35 years old who use social networking sites.
b) What percentage of Americans aged at most 34 years old use social networking sites?
c) How many Americans aged 45 years old or older use social networking sites?
d) How many Americans less than 55 years old use social networking sites?.
e) Construct a frequency distribution of five classes where the classes represent the ages. Within the table, include the class frequencies, class marks, class boundaries and relative percentages..
f) Construct a Frequency Polygon using the information contained within the frequency distribution table of part (e).
g) Identify the shape of the frequency polygon.

114. A national survey found that the distribution of all drivers that admit to running a red light is skewed right. If the survey results grouped the driver's ages as: 18–25, 26–33, 34–41, 42–49, 50–57 and 58 and older, then:
a) There were a greater percentage of drivers aged 50–57 who ran a red light than those aged 26–33.
b) The age group 18–25 contained the smallest percentage of drivers that ran a red light.
c) As the age of a driver increased, the percentage of the drivers who ran a red light also increased.
d) Younger drivers ran a red light more frequently.

115. *Next Issue Media* surveyed 200 people to determine what topics they are most likely to bond over with a co-worker. The results are summarized in the following pie chart:

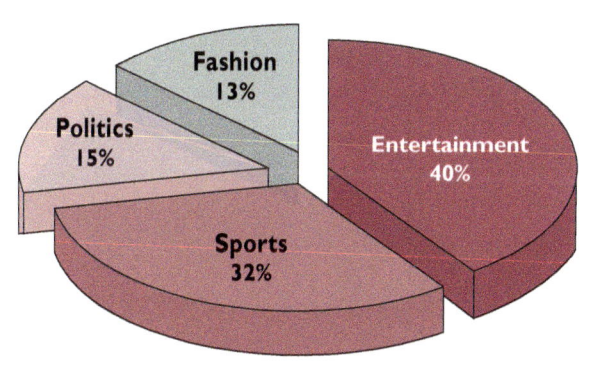

a) What percent are likely to bond over politics?
b) What percent are likely to bond over fashion or entertainment?
c) How many people are likely to bond over sports?
d) How many people are likely to bond over entertainment or sports?

116. The following data set represents the weights (in kg) of 31 Polar Bears that were tracked in North America.

Weights of 31 Polar Bears (kg)

98	224	125
144	116	176
193	160	260
108	226	134
145	117	179
195	169	269
110	245	135
148	124	182
202	173	312
114	257	357
152		

a) Create a frequency table for this data. Use 7 classes.
b) Create a frequency histogram using the table you created in part a.
c) Based on the histogram, what would you say the shape of the distribution is?

For the answer to this question, please visit www.MyStatLab.com.

CHAPTER 3

Numerical Techniques for Describing Data

Contents

3.1 Measures of Central Tendency
    The Mean
    Sample Mean
    Population Mean
    Other Calculations Using Σ
    The Median
    The Mode
    The Relationship of the Mean, Median, and Mode
    The Mode and Its Location within the Distribution Shapes
    The Median and Its Location within the Distribution Shapes
    The Mean and Its Location within the Distribution Shapes
    Symmetric Bell-Shaped Distribution
    Skewed to the Left Distribution
    Skewed to the Right Distribution
    Comparing the Mean, the Median and the Mode
3.2 Measures of Variability
    The Range
    The Variance and Standard Deviation of a Sample
    Interpretation of the Standard Deviation
    The Variance and Standard Deviation of a Population
    Using the Sample Standard Deviation to Estimate the Population Standard Deviation
3.3 Applications of the Standard Deviation
    Chebyshev's Theorem
    Empirical Rule
    Using the Range to Obtain an Estimate of the Standard Deviation
3.4 Measures of Relative Standing
    z Score
    Detecting Outliers Using z Scores
    Converting z Scores to Raw Scores
    Percentile Rank and Percentiles
    Deciles and Quartiles
3.5 Box-and-Whisker Plot: An Exploratory Data Analysis Technique

In Chapter 2 we discussed descriptive methods for organizing and summarizing data into tables and presenting data graphically. This chapter examines *numerical descriptive methods* and techniques to compute a number that describes the main characteristics of a data set. These numerical techniques will allow us to describe the center and the spread of a data set, and the relative position of a specific data value within the data set.

3.1 MEASURES OF CENTRAL TENDENCY

Are you an *average* American?"
If you are, you . . .

. . . would be a 29-year old hermaphrodite (slightly more female than male) who stands about 5'4", weighs about 150 pounds, earns close to $26,000 a year, and eats a hamburger three times a week. He (she? it?) has 1.4 children and watches television about 2hrs 49mins a day.[1]

The previous statement supposedly describes an "average" or a "typical" person for the population of American people. In this section, we will be discussing statistical methods used to depict a **typical or representative value** for all the data values within a distribution. The methods used in determining the average or typical value are trying to give some idea about where the **center** of all of the data values lie. Let's examine this concept using the following quote from a recent baseball negotiation.

After five years as a professional, a player would become a free agent if earning less than the "average" salary. (Owners estimate this at $1,121,000; the players at $1,118,000).

This proposal was one of the principal issues in the negotiations between the baseball club owners and the Major League Players Association. In the previous quote, both groups are trying to describe the "typical" or "representative" salary. However, one might ask why there is a discrepancy of $3,000 between the owners' and players' estimate of the **average** or typical salary if both groups are using the same data. To most people the word "average" has just one meaning. However, in statistics, there are different ways of computing the *average or typical data value* of a distribution. We will examine three such ways. They are the **mean**, the **median**, and the **mode**. These measures are referred to as **Measures of Central Tendency**.

Measures of central tendency are important tools of descriptive statistics because each measure, in its own way, enables a statistician to describe with a single number, the center or typical value of all the data values. However, using a single number to represent an **average** or a **typical data value** may not adequately describe **all the data values** within the distribution. The intent of an average is only to depict where the central or typical data value lies. "How is a typical data value determined?" Should we use the *average value, the middle value* or *the common value* to represent the typical data value? This section presents techniques to compute each of these different measures of central tendencies, examines what each of these measures really describes, and discusses their advantages and disadvantages.

The Mean

The first measure of central tendency that we will discuss is the **mean**. The mean is the most common and useful measure of central tendency.

> **DEFINITION 3.1**
> The **mean** is the number obtained by summing up all the data values and dividing this sum by the total number of data values.
>
> $$\text{Mean} = \frac{\text{sum of all the data values}}{\text{total number of data values}}$$

The notation used for the mean depends upon whether the data values represent a sample or a population. As mentioned in Section 1.1, it is often impractical or impossible to obtain all the data values of a population. In such instances, a sample is selected from the population and the mean of the sample is calculated. The sample mean is used to estimate the mean of the population. We will now discuss the definition and notation for the mean of a sample.

---

[1] This appeared in *The Average American*, by Barry Tarshis.

Sample Mean

In general, the data are represented by $x_1, x_2, x_3, x_4, \ldots, x_n$ where x_1 represents the first data value, x_2 represents the second data value, and x_n represents the last [or nth] data value of the sample. The total number of data values within a sample is called the *sample size* and is represented by the lower case letter n. Using this notation, the sample mean, symbolized by $\bar{x}$ and read *x bar*, can be written as:

Sample Mean Formula

$$\text{sample mean} = \frac{\text{sum of all the sample data values}}{\text{total number of sample data values}}$$

or

$$\bar{x} = \frac{x_1 + x_2 + x_3 + x_4 + \ldots + x_n}{n}$$

In statistics a more concise representation of $x_1 + x_2 + x_3 + x_4 + \ldots + x_n$ is written using the *capital Greek letter Σ (pronounced "sigma") which represents the operation SUM*. Thus, **$x_1 + x_2 + x_3 + x_4 + \ldots + x_n$ can be represented by Σx or "sum of all the data values."** Thus, the formula for the sample mean can now be written as:

$$\bar{x} = \frac{\Sigma x}{n}$$

DEFINITION 3.2
The **mean of a sample** is determined by summing up all the data values of the sample and dividing this sum by the total number of data values.
  The symbol for the sample mean is $\bar{x}$ and the formula to calculate the sample mean is:

$$\bar{x} = \frac{\Sigma x}{n}$$

The **sample mean** can be used as an estimate of the mean of a population.

EXAMPLE 3.1

Let $x_1 = 12$, $x_2 = 25$, $x_3 = 21$ and $x_4 = 8$. Calculate the sample mean, $\bar{x}$.

Solution

1. Calculate Σx, the sum of all the data values.
   $\Sigma x = x_1 + x_2 + x_3 + x_4 = 12 + 25 + 21 + 8$
   $\Sigma x = 66$
2. Determine n, the number of data values.
   $n = 4$

The sample mean is:

$$\bar{x} = \frac{\Sigma x}{n} = \frac{66}{4} = 16.5$$

EXAMPLE 3.2

A quality control inspector of a battery manufacturer would like to estimate the life expectancy of the manufacturer's 9 volt batteries manufactured during the day shift. In order to estimate the life expectancy of all the batteries manufactured during the day shift, a sample of 20 batteries was randomly obtained. The following data represents the battery life in months of the 20 batteries.

$$20, 11, 15, 18, 24, 17, 19, 12, 19, 22$$
$$18, 15, 19, 21, 20, 15, 17, 24, 16, 18$$

a) Calculate the sample mean, $\bar{x}$.
b) What value can the inspector use as an estimate of the average of **all** the 9 volt batteries manufactured?

Solution

a) The sample mean is calculated by using the formula:

$$\bar{x} = \frac{\Sigma x}{n}$$

Since

Σx = 20+11+15+18+24+17+19+12+19+22+18+15+19+21+20+15+17+24+16+18
= 360

and n = 20, then the sample mean is:

$$\bar{x} = \frac{\Sigma x}{n} = \frac{360}{20}$$
$$\bar{x} = 18 \text{ months}$$

b) The quality control inspector can use the sample mean of 18 months as an estimate of the average life of all the batteries.

Population Mean

The formula to compute the mean of a finite population is essentially the same as the sample mean formula. Only the notation is different to indicate we have **all** the data values of the population.

When **all** the data values of a population are obtained, the *Greek letter mu, written* μ, (pronounced "mew") denotes the population mean. The number of data values within the population is symbolized by the uppercase letter N. Using this notation, we can write the formula for the mean of a finite population mean.

Population Mean Formula

The symbol for the population mean is μ. The formula to calculate the mean of a finite population of size N is:

$$\mu = \frac{\Sigma x}{N}$$

As mentioned in Chapter 1, the purpose of inferential statistics is to use sample information to describe the characteristics of a population since it is usually impractical or impossible to obtain all the data values of a population.

Table 3.1 summarizes the notations for the mean of a sample and a population. It is important that you use the proper notation when symbolizing the mean.

TABLE 3.1 Notation

	Sample	Population (finite)
Mean	$\bar{x}$, read x bar	μ, read mu
Size	n	N

In Example 3.3, we consider a small population of data values to illustrate the use of notation and formula for the mean of a population.

EXAMPLE 3.3

According to the U.S. Department of Energy, the following data represent the estimated miles per gallon (mpg) for 2014 Hybrid vehicles under city driving conditions.

$$44, 55, 55, 55, 58, 55, 57, 56, 57, 52, 52, 51, 48, 56, 57, 63, 65, 53, 53$$

Assuming these estimated mpg represent the population of all hybrid vehicles, find the mean of this population.

Solution

To determine the population mean: $\mu = \dfrac{\Sigma x}{N}$, we need to first compute Σx.

$\Sigma x = 44+55+55+55+58+55+57+56+57+52+52+51+48+56+57+63+65+53+53$
$\quad\, = 1042$

Since $N = 19$, then the population mean equals:

$$\mu = \frac{\Sigma x}{N} = \frac{1042}{19}$$
$$\mu = 54.84 \text{ MPG}$$

Thus, the population mean mpg is approximately 54.84 for the hybrid vehicles under city driving conditions.

Other Calculations Using Σ

There are many formulas in statistics involving the use of Σ, and it is important to know how to interpret and evaluate expressions containing Σ. We will consider some of these expressions and discuss how to evaluate them. We will first consider the expression: $(\Sigma x)^2$, the square of the sum of the data values.

Procedure to Calculate $(\Sigma x)^2$
(The Square of the Summed Data Values)

$(\Sigma x)^2$ represents $(x_1 + x_2 + x_3 + x_4 + \ldots + x_n)^2$
1. Sum the data values, (Σx)
2. Square the sum, $(\Sigma x)^2$

The next example will illustrate the procedure used to calculate $(\Sigma x)^2$.

EXAMPLE 3.4

Let $x_1 = 3$, $x_2 = 4$, $x_3 = 1$ and $x_4 = 6$.
Determine $(\Sigma x)^2$.

Solution

Notice the sum of the data values, Σx, is being squared. This means we must first find the sum of all the data values, and then **square** this sum.
That is,
1. sum all the data values: (Σx)
   $\Sigma x = x_1 + x_2 + x_3 + x_4 = 3 + 4 + 1 + 6$
   $\Sigma x = 14$
2. square this sum: $(\Sigma x)^2$
   $(\Sigma x)^2 = (14)^2 = 196$

Now, we will consider the meaning of the expression Σx^2, the sum of the squared data values.

Procedure to Calculate Σx^2
(The Sum of the Squared Data Values)

Σx^2 represents $(x_1)^2 + (x_2)^2 + (x_3)^2 + \ldots + (x_n)^2$
1. Square each data value.
2. Sum these squared data values.

The next example illustrates the calculation of Σx^2.

EXAMPLE 3.5

Let $x_1 = 2$, $x_2 = 4$, $x_3 = 1$ and $x_4 = 3$. Find Σx^2.

Solution:

1. square each data value

   $(x_1)^2 = (2)^2 = 4 \qquad (x_2)^2 = (4)^2 = 16$
   $(x_3)^2 = (1)^2 = 1 \qquad (x_4)^2 = (3)^2 = 9$

2. sum these squared data values

   $\Sigma x^2 = (x_1)^2 + (x_2)^2 + (x_3)^2 + (x_4)^2 = 4 + 16 + 1 + 9$
   $\Sigma x^2 = 30$

Summary for Calculating $(\Sigma x)^2$ and Σx^2

When calculating $(\Sigma x)^2$ you first **add** all the data values and then **square** this sum. When calculating Σx^2 you first **square** each individual data value and then **add** these squared data values.

Statistical calculators have built-in programs to calculate the expressions Σx and Σx^2. You should become familiar with the operation of your calculator and review the examples in this section that refer to these expressions.

An important concept involving the mean is the *deviation from the mean*. The deviation from the sample mean for a particular data value, x, is the amount by which the data value deviates from the mean. This deviation is the difference between the data value and the mean. It is represented as: $x - \bar{x}$. We will now examine the expression $\Sigma(x - \bar{x})$ and how to evaluate it. The expression $\Sigma(x - \bar{x})$ represents the sum of the deviations from the sample mean. To evaluate this expression, we use the following steps.

Procedure to Compute the Sum of the
Deviations from the Sample Mean, $\Sigma(x - \bar{x})$

$\Sigma(x - \bar{x})$ represents $(x_1 - \bar{x}) + (x_2 - \bar{x}) + (x_3 - \bar{x}) + \ldots + (x_n - \bar{x})$
1. Compute the sample mean.
2. Subtract the mean from each data value.
3. Sum these differences.

EXAMPLE 3.6

Let $x_1 = 5, x_2 = 3, x_3 = 7$. Find $\Sigma(x - \bar{x})$, that is, the sum of the deviations from the mean.

Solution

1. Compute the mean: $\bar{x}$.

$$\bar{x} = \frac{\Sigma x}{n} = \frac{15}{3} = 5$$

2. Subtract the mean, $\bar{x}$, from each data value. These differences, $(x - \bar{x})$, are referred to as **deviations from the mean**.

   $(x_1 - \bar{x}) = (5 - 5) = 0$
   $(x_2 - \bar{x}) = (3 - 5) = -2$
   $(x_3 - \bar{x}) = (7 - 5) = +2$

3. Find the sum of the deviations from the mean, $\Sigma(x - \bar{x})$.

   $\Sigma(x - \bar{x}) = 0 + (-2) + 2$
   $= 0$

   The sum of the deviations from the mean is equal to zero, that is:
   $\Sigma(x - \bar{x}) = 0$.

EXAMPLE 3.7

Given the following ages of seven children at a playground: 5, 4, 4, 8, 4, 9, 8
a) Compute the mean age, $\bar{x}$.
b) Evaluate $\Sigma(x - \bar{x})$, the sum of the deviations from the mean.

Solution

a) Calculate the mean, $\bar{x}$.

$$\bar{x} = \frac{\Sigma x}{n} = \frac{42}{7} = 6$$

b) To evaluate $\Sigma(x - \bar{x})$:
   1. Subtract the mean, $\bar{x} = 6$, from each data value.

   $(5 - 6) = -1$ $(4 - 6) = -2$ $(4 - 6) = -2$ $(8 - 6) = 2$
   $(4 - 6) = -2$ $(9 - 6) = 3$ $(8 - 6) = 2$

   2. Find the sum of the deviations from the mean, $\Sigma(x - \bar{x})$.

   $\Sigma(x - \bar{x}) = (-1) + (-2) + (-2) + 2 + (-2) + 3 + 2$
   $= 0$

From the previous two examples we might guess that the **sum of the deviations from the mean is always zero.** That is, $\Sigma(x - \bar{x}) = 0$. In fact, this is true, and it is **a property of the mean. This property indicates that the mean is the balance point within a set of data values.** The mean is like the fulcrum on a seesaw. We will examine this idea using the ages of the seven children from Example 3.7. If the ages of the seven children were marked along a number line on a board, and a one-pound weight is placed on the board corresponding to each child's age, then this board would have its balance point at the mean value of 6. That is, if a fulcrum is placed underneath the board at the position corresponding to the mean value of 6, then the board containing the 7 one-pound weights would be perfectly balanced at the mean! This illustrates the fact that the mean represents the balance point of a distribution. This property is illustrated in Figure 3.1.

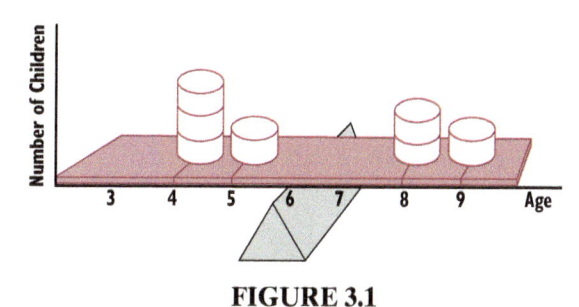

FIGURE 3.1

From Figure 3.1, since the fulcrum of the board is at the mean value of 6, this serves to illustrate that the mean is the point that balances all the values on either side of it. In other words, the sums of the distances from the mean is the same on each side of the mean. This is symbolized as: $\Sigma(x - \bar{x}) = 0$.

Property of the Mean

The sum of the deviations from the mean is always zero.
This is expressed symbolically as:

$$\Sigma(x - \bar{x}) = 0 \text{ for sample data, or}$$
$$\Sigma(x - \mu) = 0 \text{ for population data.}$$

This property states that the mean is the balance point within a distribution. That is, the mean balances the data values on either side of it since the sums of the distances from the mean is the same on each side of the mean.

The Median

Another measure of central tendency is the **median**. The median is the middle value of a data set after the data values have been arranged in numerical order. The procedure to determine the median is dependent upon the number of data values, and not on the actual value of the data.

DEFINITION 3.3
The **median** is the *middle value* of a set of data values after they have been arranged in *numerical order*. The number of data values below and above the position of the median are equal.

The procedure to determine the median is dependent upon whether the number of data values within the data set is odd or even.

Procedure to Determine the Median

1. First arrange the data values in numerical order.
2. a. For an **odd number** of data values, the median is the middle data value.
   b. For an **even number** of data values, the median is the mean of the two middle data values.

EXAMPLE 3.8

For the quiz grades: 8, 7, 9, 9, 9, 6, 10, 7, 5 find the median quiz grade.

Solution

1. Arrange the data values in numerical order.

   5, 6, 7, 7, 8, 9, 9, 9, 10

2. Since there is an odd number of data values (9 data values), the median is the middle grade. Thus, the median quiz grade is 8.

Median's Location within a Distribution

In general, the median's location within a set of n data values that have been arranged in numerical order is determined by the formula:

$$\text{Location of median} = \frac{n+1}{2} \text{th data value}$$

For Example 3.8, the location of the median can be determined using the formula:

$$\text{Location of median} = \frac{n+1}{2} \text{th data value}$$

For n = 9 data values, then the location of the median is:

$$\text{Location of median} = \frac{9+1}{2} = \text{5th data value}$$

In Example 3.8, the **median's location**, as illustrated in Figure 3.2, **is the 5th data value** within the ordered data set.

MEDIAN

5, 6, 7, 7, **8**, 9, 9, 9, 10

5th data value

FIGURE 3.2

EXAMPLE 3.9

Table 3.2 represents the 2017 attendance figures for all the games played by the 10 NCAA football teams listed:

TABLE 3.2

College	2017 Total Attendance
Ohio State	1,254,160
Georgia	1,246,201
Alabama	1,228,376
Auburn	1,162,955
Penn State	1,146,641
Michigan	1,140,358
LSU	1,114,205
Texas A&M	1,093,368
Oklahoma	1,054,046
Tennessee	1,043,298

a) Find the mean and median attendance.
b) Construct a bar graph representing the 10 attendance figures.
c) Determine the percentage of games that are less than the median attendance.
d) Determine the percentage of games that are less than the mean attendance.
e) Construct a bar graph representing the 10 attendance figures, and include a bar for the mean & median attendance figures.
f) Compare the percentage of games below the mean and the median attendances. Explain the differences if any.

Solution

a) 1. The mean attendance is found by summing up the 10 attendances and dividing by 10. Thus the mean attendance is:

$$\frac{1{,}254{,}160 + 1{,}246{,}201 + 1{,}228{,}376 + 1{,}162{,}955 + 1{,}146{,}641 + 1{,}140{,}358 + 1{,}114{,}205 + 1{,}093{,}368 + 1{,}054{,}046 + 1{,}043{,}298}{10}$$

$$\text{mean total attendance} = \frac{11{,}483{,}608}{10} = 1{,}148{,}360.8$$

Hence, the mean attendance is 1,148,360.8.

2. In order to find the median total attendance, begin by arranging the total attendances in numerical order, we have:

1,254,160 1,246,201 1,228,376 1,162,955 1,146,641 1,140,358 1,114,205 1,093,368 1,054,046 1,043,298

3. For a data set of 10 attendances, the location of the median is determined by the formula:

$$\text{Location of median} = \frac{n+1}{2}\text{th data value}$$

For n = 10, the median's position is:

$$\text{Location of median} = \frac{10+1}{2} = \frac{11}{2} = 5.5$$

This result of 5.5 indicates that the median is equal to the mean of the 5th and 6th total attendances which are 1,146,641 and 1,140,358. Thus, we have:

$$\text{Median total attendance} = \frac{1{,}146{,}641 + 1{,}140{,}358}{2} = \frac{2{,}286{,}999}{2} = 1{,}143{,}499.5$$

Thus, the median total attendance is approximately 1,143,500.

b) Figure 3.3 represents the bar graph for the 10 attendance figures.

ATTENDANCE FOR ALL FOOTBALL GAMES PLAYED DURING 2017 SEASON

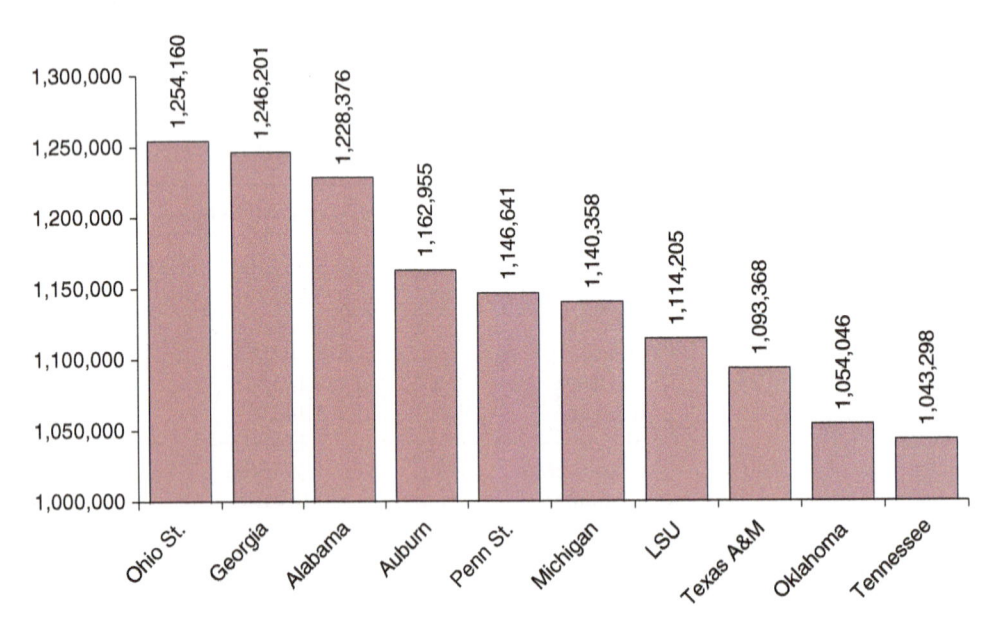

FIGURE 3.3

c) Five of the 10 total attendances were below the median total attendance of 1,143,500. Therefore, 50% of the total attendances are less than the median total attendance.

d) Six of the 10 opening day games had an attendance that was below the mean attendance. Therefore, 60% of the games had attendances that are less than the mean attendance.

e) Figure 3.4 represents the bar graph for the 10 attendance figures including the mean and median.

ATTENDANCE FOR ALL FOOTBALL GAMES PLAYED DURING 2017 SEASON
(Including Mean & Median)

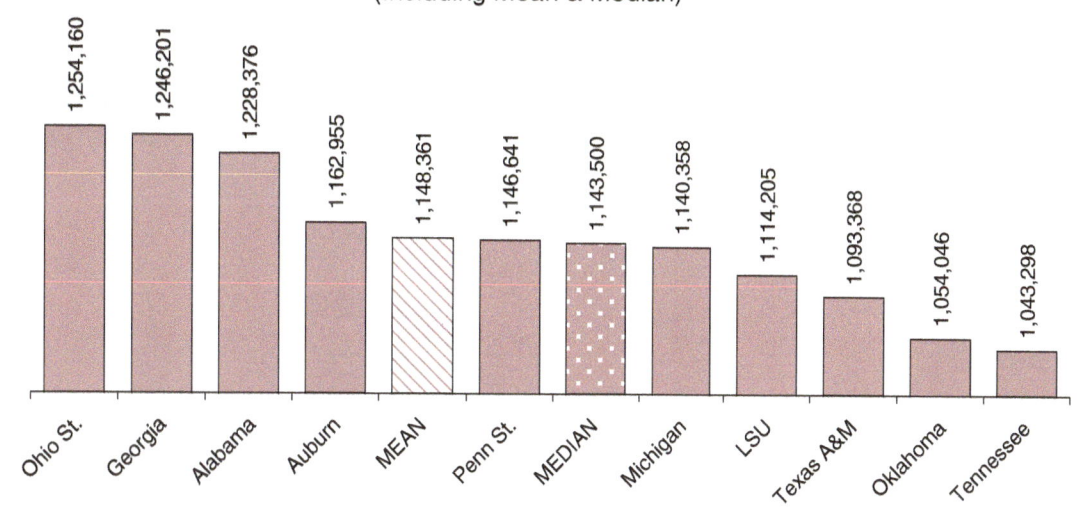

Ohio St.: 1,254,160
Georgia: 1,246,201
Alabama: 1,228,376
Auburn: 1,162,955
MEAN: 1,148,361
Penn St.: 1,146,641
MEDIAN: 1,143,500
Michigan: 1,140,358
LSU: 1,114,205
Texas A&M: 1,093,368
Oklahoma: 1,054,046
Tennessee: 1,043,298

FIGURE 3.4

f) Since the median, by definition, is equal to the middle number, we would expect approximately 50% of the games to be below the median attendance as shown in Figure 3.4. On the other hand, the mean is the balance point so we would **not** expect to have 50% of the games below the mean attendance. In fact, 60% of the games fell below the mean attendance. The reason more than 50% of the games had an attendance figure below the mean attendance is that the mean is influenced by the extreme attendance figure 1,254,160 as shown in Figure 3.4.

TI-84 Graphing Calculator Solution for Example 3.9

The TI-84 could have been used to find the mean and median of the sample data in Example 3.9. To use it, you must first enter the given data into a list such as L1 in the calculator. This is shown in Figure 3.5.

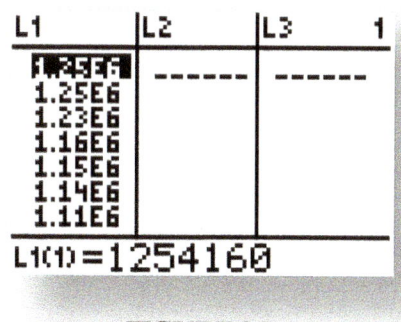

FIGURE 3.5

Next, press STAT ▶ ENTER to select the **1-Var Stats** command. You will then have to specify to the calculator the list your data is stored in. We chose L1 by pressing 2nd then **1**. Press ENTER twice leaving the FreqList field empty. This is shown in Figure 3.6a and Figure 3.6b.

98 Chapter 3 Numerical Techniques for Describing Data

FIGURE 3.6a

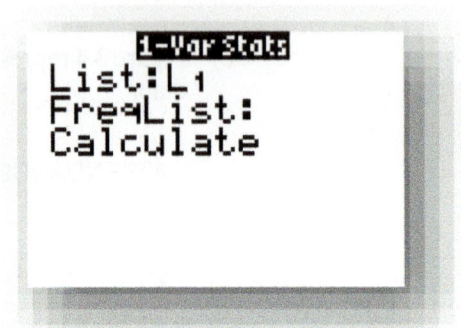

FIGURE 3.6b

Press ENTER, and the results are displayed in Figure 3.7:

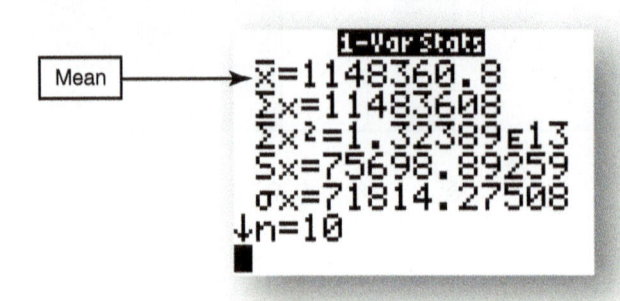

Mean → $\bar{x} = 1148360.8$

FIGURE 3.7

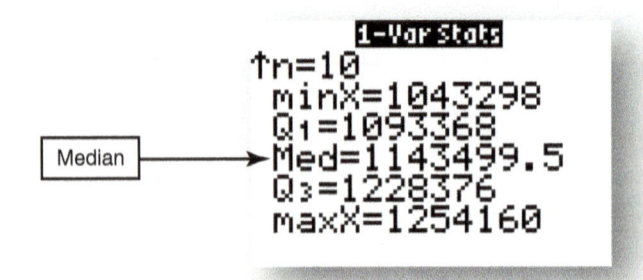

The sample mean is given by $\bar{x}$. Notice this is the same result we calculated in Example 3.9. Pressing the ▼ key will display more statistical information as show in Figure 3.8.

Median → Med = 1143499.5

FIGURE 3.8

Here we can find the mean and median. These are the same answers obtained in Example 3.9.

The Mode

The last measure of central tendency we will examine is the mode. The mode is the easiest measure to compute and the simplest to interpret.

DEFINITION 3.4
The **mode** of a data set is the data value which occurs most frequently. A data set may **not** have a mode or may have **more** than one mode.

EXAMPLE 3.10

Find the modal test score for the following ten test scores: 65, 75, 45, 90, 75, 68, 85, 60, 75, 90

Solution

Since 75 occurs most frequently, it is the modal test score.

EXAMPLE 3.11

Find the mode for the following screw lengths.

screw lengths
(measured in inches)
0.25, 0.33, 0.5, 0.33, 0.5, 0.33, 0.25, 0.16, 0.5, 0.5, 0.33

Solution

Notice there are two data values which occur most frequently. Therefore, this data set has *two* modes. The modes are 0.33 and 0.5.

In the previous example, there were two modes. This type of data set is called *bimodal*.

> **DEFINITION 3.5**
>
> A distribution is a called a **bimodal** distribution if it has two data values which appear with the greatest frequency.

EXAMPLE 3.12

Find the mode for the following distribution:

5, 4, 6, 10, 6, 4, 10, 3, 8, 7

Solution

Since there are more than two data values which occur most frequently, (i.e. 4, 6 and 10 appear twice), we will refer to such a distribution as having *no mode*.

A distribution having *no mode,* as illustrated in the previous example, will be referred to as a *nonmodal distribution*.

> **DEFINITION 3.6**
>
> If a distribution has more than two modes, then the distribution is **nonmodal**.

EXAMPLE 3.13

A college professor randomly selects 36 students from her classes to try to estimate the typical eye color for college students attending the university. The following data set represents her survey results:

brown, blue, hazel, brown, brown, blue, blue, brown, hazel, brown, hazel, brown, brown, blue, brown, brown, blue, blue, brown, brown, brown, blue, hazel, blue, blue, brown, hazel, brown, brown, brown, brown, blue, brown, brown, blue, brown.

Which measure of central tendency should the professor use to estimate the typical college student eye color and what is the typical student eye color for the university?

Solution

Since the mean and median would not be appropriate [WHY?], the professor would need to use the mode. The most frequent eye color is brown. Therefore, the typical college student at the university has brown eyes.

EXAMPLE 3.14

A small community is applying for federal aid. One question on the application requests the *average* family income for the community.

The following data are the family incomes for the five community residents:

$25,000; $20,000; $200,000; $30,000; $15,000.

If the town uses the mean income of $58,000 to represent the average family income, do you believe this number accurately represents the income level of the community?

If not, which measure of central tendency would you use to represent the income level of this community?

Solution

Since the mean income of $58,000 is greater than four of the five family incomes, it does not represent the typical income level of the community. The median income of $25,000 more accurately depicts the typical income level of the residents since it is not sensitive to the extreme values within the community.

CASE STUDY 3.1 Average Costs

Examine the three snapshots shown in Figure 3.9. In each snapshot a measure of central tendency is being used to describe an aspect of a person's life. For each snapshot, state the measure of central tendency being applied and explain why you believe this is or isn't an appropriate measure to use in each instance.

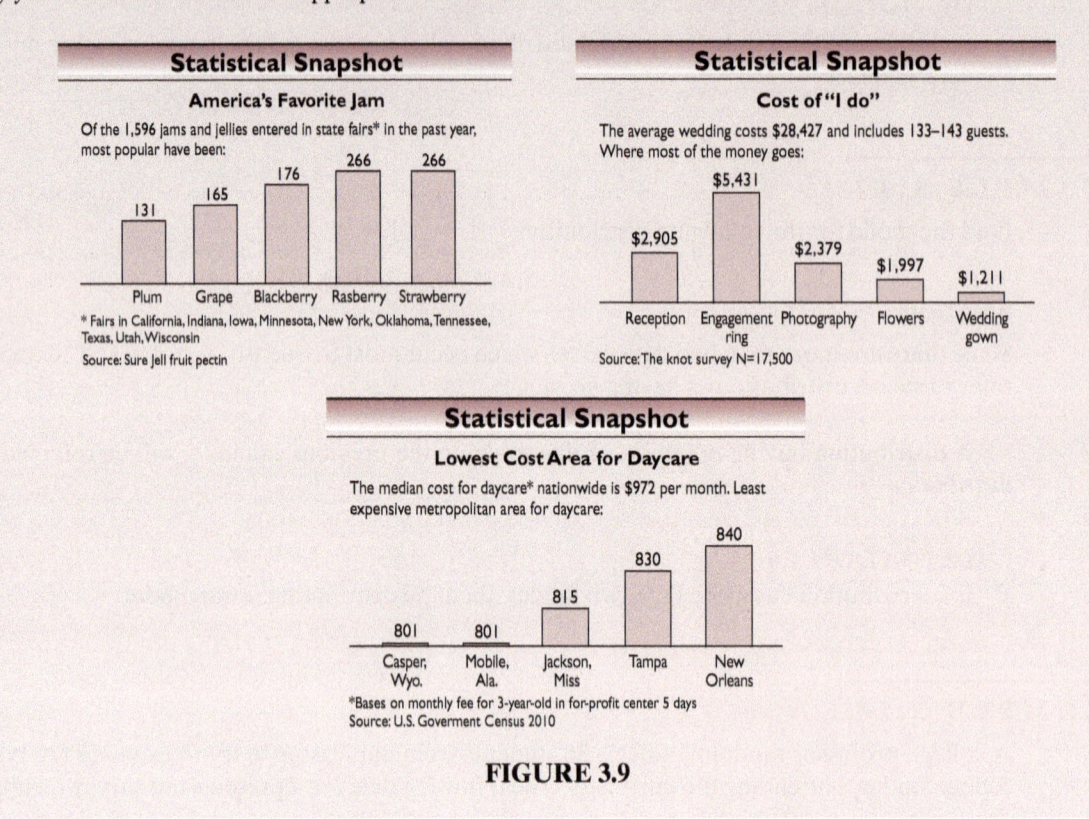

Statistical Snapshot
America's Favorite Jam
Of the 1,596 jams and jellies entered in state fairs* in the past year, most popular have been:
- Plum: 131
- Grape: 165
- Blackberry: 176
- Rasberry: 266
- Strawberry: 266

*Fairs in California, Indiana, Iowa, Minnesota, New York, Oklahoma, Tennessee, Texas, Utah, Wisconsin
Source: Sure Jell fruit pectin

Statistical Snapshot
Cost of "I do"
The average wedding costs $28,427 and includes 133–143 guests. Where most of the money goes:
- Reception: $5,431
- Engagement ring: $2,905
- Photography: $2,379
- Flowers: $1,997
- Wedding gown: $1,211

Source: The knot survey N=17,500

Statistical Snapshot
Lowest Cost Area for Daycare
The median cost for daycare* nationwide is $972 per month. Least expensive metropolitan area for daycare:
- Casper, Wyo.: 801
- Mobile, Ala.: 801
- Jackson, Miss: 815
- Tampa: 830
- New Orleans: 840

*Bases on monthly fee for 3-year-old in for-profit center 5 days
Source: U.S. Goverment Census 2010

FIGURE 3.9

CASE STUDY 3.2 7 Signs You Are Utterly Average

Read the article "7 Signs You Are Utterly Average" and answer the questions displayed in Figure 3.10.

7 Signs You Are Utterly Average
By Devon Delfino (Business Insider: Nov 12, 2018)

Everyone wants to be the best at something. But for the most part, what many people really want is to be "normal," or at least something close to it. That's especially true if you feel that there's some area of your life that's lacking.

Whether that's wealth, health, or general happiness, there's comfort in knowing that you actually aren't struggling more than most, and that you are, in fact, average. However, it can be difficult to tell how you're doing in relation to everyone else when "normal" is edited and distributed through the rose-colored filters of social media.

To help you get a sense for the unedited reality, here are seven areas where you're probably utterly average:

1. **You're not getting as much exercise as you should.**
   Health and nutrition are major issues in the U.S. If you aren't sure you're getting enough exercise, you're likely in the majority. Only about one in three adults gets the recommended amount of physical activity every week, and less than 5% get a half hour of exercise every day, according to the Department of Health and Human Services (HHS). The HHS also found that most Americans have trouble sticking to recommended nutritional rules — 90% of Americans consume more sodium than the CDC recommends for a healthy diet.

CASE STUDY 3.2 7 Signs You Are Utterly Average (continued)

2. **You're dealing with imposter syndrome at work.**
   If you're dealing with imposter syndrome at work, you're not the only one. In fact, the International Journal of Behavioral Science estimated that 70% of people at some point deal with the phenomenon, which it describes as possessing "intense feelings that [your] achievements are undeserved." If you're afraid that you aren't as smart as other people think you are, that's probably a good sign. A Cornell University study confirmed what has long been suspected, since Charles Darwin theorized in 1871 that "ignorance more frequently begets confidence than does knowledge." In other words, people who are actually incompetent are often painfully unaware of that fact, and "grossly overestimate" their abilities compared to their peers.

3. **Your net worth is less than six figures.**
   In 2016, the average net worth (assets minus debt) for families under the age of 35 was about $76,000, according to the Federal Reserve. However, it's important to note that median net worth in that demographic — which was about $11,000 that year — is a more useful figure since it factors out the small percentage of Americans whose wealth falls way outside of "normal" American earnings.

4. **You owe tens of thousands in debt.**
   If you're worried about your debt levels, or that you aren't doing enough to pay it down, it may comfort you to know that the average American owes about $38,000 in personal debt (excluding mortgages) and spends about the same amount of their monthly income on discretionary spending as they do on debt repayment.
   That's according to a study from Northwestern Mutual, which also found that the majority of those in debt are optimistic that they will be able to pay off their debt at some point, stating that their debt has low to no impact on their "ability to achieve financial security."

5. **Your IQ is 100.**
   When it comes to IQ tests, most people score in the 85 to 115 range. In fact, the tests were designed to have an average score of 100. However, it's worth noting that your score can change over time, and there are other signs which can be used to predict higher overall intelligence, like being funny or having the ability to focus and adapt to changes. Not to mention the fact that there has been controversy around the Eurocentric bias of IQ tests, which eugenicists have used to justify arguments around the so-called "superiority" of white people.

6. **You have around six close friends.**
   Everyone has different ideas about what kinds of relationships are best for them, and your unique circumstances will shape your experience. But in general, the average person has 40 friends, including two best friends, four close friends, and five work buddies, according to a recent survey conducted in the UK by One Poll. As for romantic relationships, about half of adult Americans are married, according to the Pew Research Center. That number has been on the decline over the last few decades. And among American singles, only 16% say they are actively looking for a relationship, according to Pew.

7. **You're unhappy.**
   If recent events, or the past few years, have left you feeling a bit down you're not alone. According to the United Nations' World Happiness Report, American happiness has "remained roughly unchanged or has even declined" even as per capita income has "more than doubled" since 1972. Though the U.S. is a major economic power, the report points to life expectancy, the freedom to make choices, perceived corruption, social support, and the generosity of those around you as contributing factors to the decline.
   Though these statistics can provide a useful way to get context around how you're doing, it's important to keep in mind that falling outside the realm of "normal" for any given category isn't necessarily a bad thing. But if you aren't satisfied, there are things you can do to change your circumstances.

A) Which statistical descriptive measure of central tendency discussed in Chapter 3 do you believe best describes what the author means when referring to the term "normal"? Support your answer.
B) Which measure of central tendency do you believe the author is using when she states, " less than 5% get a half hour of exercise every day"? Do you believe, you are in that 5% group? Support your answer.
C) Do you believe you have an "imposter syndrome at college"? In particular, how would you rate your statistical ability to your classmates? And which measure of central tendency would you use to explain your response? Support your answer.
D) When the author discusses net worth, why do you believe the author believes the median is a more useful measure of central tendency than the mean when she states: "it's important to note that median net worth in that demographic — which was about $11,000 that year — is a more useful figure … "? Support your answer.
E) Which measure of central tendency do you believe the author should be using when she states "… the average American owes about $38,000 in personal debt …"? Support your answer by referring to the advantages and disadvantages of the three measures of central measures presented in Chapter 3.
F) Which measure of central tendency do you believe the author is referring to when she states: " … the tests were designed to have an average score of 100"? Support your answer.

G) How would you interpret the author's statement: " … the average person has 40 friends, including two best friends, four close friends, and five work buddies"? Which measure of central tendency do you believe the author is referring to when stating, "the average person"? Support your answer.

H) Which measure of central tendency do you believe should have been used to identify the number of best friends, close friends, and work buddies? Support your response by referring to the advantages and disadvantages of the three measures of central measures presented in Chapter 3.

I) If you were planning to conduct a multiple choice random survey to check on the reliability of the statement: According to the United Nations' World Happiness Report, American happiness has "remained roughly unchanged or has even declined", what type of survey data responses would you give as choices for American happiness? Based on the survey data responses, which measure of central tendency would you use to describe American happiness? Support your answer.

FIGURE 3.10

From *Business Insider.* © 2018 Insider Inc. All rights reserved. Used under license.

The Relationship of the Mean, Median and Mode

In this chapter, we've discussed three ways to describe the center or the "typical" or "representative" value of a distribution. To help decide which measure of central tendency would be appropriate in describing the center of a specific distribution type, we will examine the relationship of the mean, median, and mode within three common distribution shapes: *symmetric bell-shaped*, *skewed left*, and *skewed right*. We will first discuss how to determine where each measure of central tendency will be positioned within each of these distribution shapes.

The Mode and Its Location within the Distribution Shapes

Within each of the distribution shapes illustrated in Figure 3.11, you will notice that the mode, the most frequent value, is located at the point on the curve corresponding to the highest value.

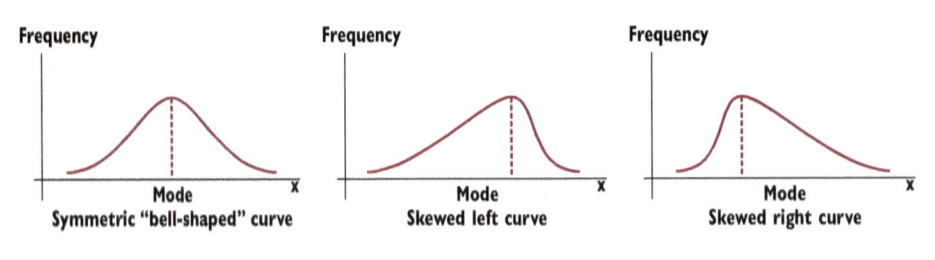

FIGURE 3.11

The Median and Its Location within the Distribution Shapes

The median is located at the point on the curve which divides the area under the curve exactly in half. That is, half or 50% of the area lies to the left of the median value while half or 50% of the area lies to the right of the median value. This is illustrated in Figure 3.12.

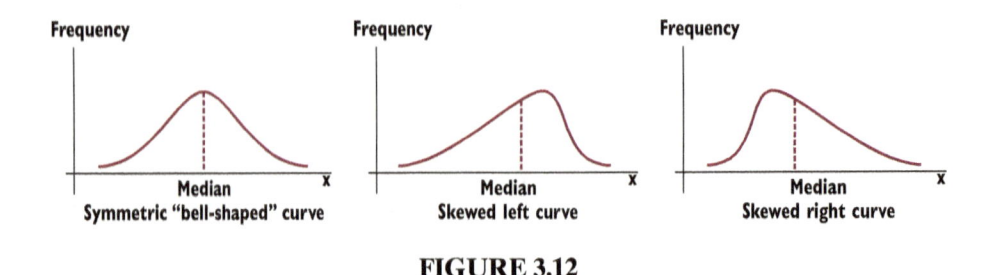

FIGURE 3.12

The Mean and Its Location within the Distribution Shapes

The mean of a distribution is the value which represents the "balance point" of the curve. To understand the idea of a balance point, try to think of the curve as a solid object. The point beneath the curve which will balance the solid curve is the mean. Thus, at the balance point half the weight of the solid curve is to the left of this point and half the weight is to the right of this point. The idea of the balance point for each distribution shape is illustrated in Figure 3.13.

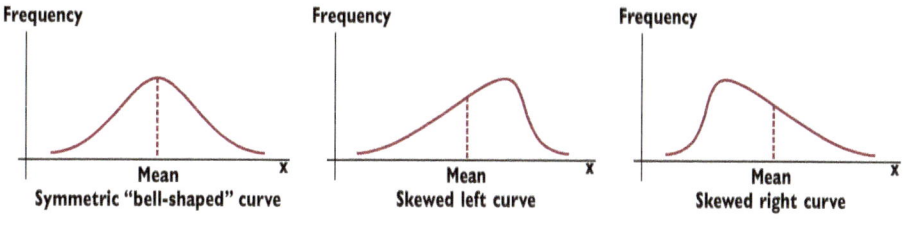

FIGURE 3.13

Now let's examine the position of each measure of central tendency with respect to each other within these three distribution shapes.

Symmetric Bell-Shaped Distribution

The mean, median, and mode are all at the center of the distribution represented by the symmetric bell-shaped curves in Figures 3.11 to 3.13. The mean or balance point is at the center since every data value to the left of the center is balanced by a corresponding data value to the right of the center. The median is located at the center since it divides the area under the curve into two equal halves (50% of the area to the left and 50% of the area to the right). The mode is located at the center since this is the point at which the curve attains its highest value.

Skewed to the Left Distribution

Within a skewed left distribution, the mode is the highest point. Its value is greater than both the median and the mean as illustrated in Figure 3.14.

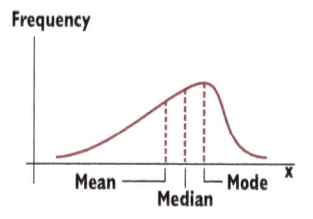

FIGURE 3.14

In a skewed left distribution, the mean has the smallest value of all the measures of central tendency because the mean is influenced the most by the extreme data values that occur in the left tail of the distribution and is *pulled in the direction of the extreme values in the left tail.* The median has a smaller value than the mode since the median is *pulled to the left by the extreme values in the left tail* of the distribution. However, the median will be greater than the mean since the median is not as sensitive or as greatly influenced as the mean to the extreme data values in the left tail.

Skewed to the Right Distribution

Within a skewed right distribution, the mode occurs at the highest point and has a value smaller than the median or the mean as illustrated in Figure 3.15.

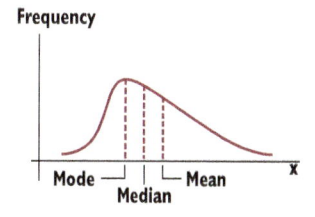

FIGURE 3.15

The mean has the largest value of all the measures of central tendency because the mean is greatly influenced by the extreme data values that occur in the right tail of the distribution. The median has a value greater than the mode since the median is affected by the extreme values within the distribution and *pulled in the direction of the right tail.* However, the median has a value smaller than the mean since the median is not as sensitive or as greatly influenced as is the mean to the extreme data values in the right tail.

EXAMPLE 3.15

Using Figure 3.16, determine the value for:
a) the mode
b) the mean
c) the median

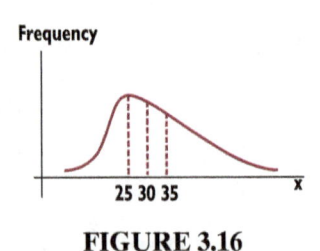

FIGURE 3.16

Solution

a) Within a distribution, the mode occurs at the highest point. Therefore, the mode is 25.
b) Since the distribution in Figure 3.16 is skewed right, then the mean is greater than both the median and the mode. Thus, the mean is 35.
c) In a skewed right distribution, the median is greater than the mode but less than the mean. Therefore, the median is 30.

Comparing the Mean, the Median and the Mode

From the examples and discussions within this section, you should have noticed that there are advantages and disadvantages for each measure of central tendency.

As illustrated in Example 3.14, the mean is a measure of central tendency that is affected by the extreme data values or outliers within a distribution.

> **DEFINITION 3.7**
> A measure is a **resistant measure** if its value is not affected by an outlier or an extreme data value.

Thus we say that **the mean is not a resistant measure of central tendency**, because it is not resistant to the influence of the extreme values or outliers. On the other hand, **the median is a resistant measure of central tendency**. Since the median is resistant to the influence of extreme data values or outliers, its value does not respond strongly to the changes of a few extreme data values regardless of how large the change may be.

Thus, if a distribution has an outlier or is skewed, then the mean and median may differ greatly in value. The mean is sensitive to the extreme data values and will move toward the direction of the long tail in a skewed distribution. On the other hand, the median is resistant to, or unaffected by, outliers or extreme data values. (Figures 3.14–3.15) However, in a symmetric bell-shaped curve (Figures 3.12–3.13), the mean and median will have the same value.

Since only the mean is sensitive to outliers or extreme data values, **a difference between the mean and the median can inform us that an outlier may exist within a distribution**. Once an outlier is detected, the statistician can investigate the reason for such an extreme data value. The statistician can then decide whether to use a measure of central tendency that is resistant to these outliers or to give individual attention to the outliers. Notice in Example 3.14, the existence of the outlier $200,000 was detected by the fact that the mean ($58,000) was much larger than the median ($25,000). In this particular instance, the Community decided to use the median to represent the "average" or center value to ignore the presence of the outlier, since the median value of $25,000 is resistant to the outlier $200,000.

The mode has an advantage over both the mean and median when the data is categorical, since it is not possible to calculate a mean or median for this type of data. This is illustrated in Example 3.13. In addition, the mode usually indicates the location within a large distribution where the data values are concentrated. A shoe retailer would be interested in modal shoe sizes when placing an order for new shoes since the shoe sales will be concentrated around the modal size; while a discount consumers club is interested in the brands of items that most people are purchasing, since these items represent the modal brands.

However, the mode has some disadvantages too. One disadvantage is that the mode cannot always be calculated. For example, if a distribution has all different data values, then the distribution is nonmodal.

Let's now summarize some advantages and disadvantages of the measures of central tendency.

Summarizing the Measures of Central Tendency

Measure of Central Tendency	Definition	Resistant to Outliers	Advantages and Disadvantages
Mean	Sample: $\bar{x} = \dfrac{\Sigma x}{n}$ Finite Population: $\mu = \dfrac{\Sigma x}{N}$	No	Appropriate for use in other statistical methods and for making inferences; affected by outliers; its value takes all data values into account
Median	Middle Data Value	Yes	Excellent choice when outliers exist; its value does not take all data values into account
Mode	Most Frequent Data Value	Yes	Appropriate for categorical data; may not exist or may have more than one mode; its value does not take all data values into account

CASE STUDY 3.3 Major League Baseball Salaries

The graphic in Figure 3.17 entitled the *Major League Baseball Salary* presents a summary of the all the 877 Major League Baseball Players on opening day of the 2019 baseball season. **(2019 MLB Player Salaries Source: USA Today)**

MAJOR LEAGUE BASEBALL SALARY
Who makes what in major league baseball? Based on 2019 opening day rosters of 877 players, the average salary was $4,509,878, the median salary was $1,400,000, and the modal salary was $555,000 while the minimum salary was $555,000. The maximum salary was $42,142,857, earned by Max Scherzer of the Washington Nationals.

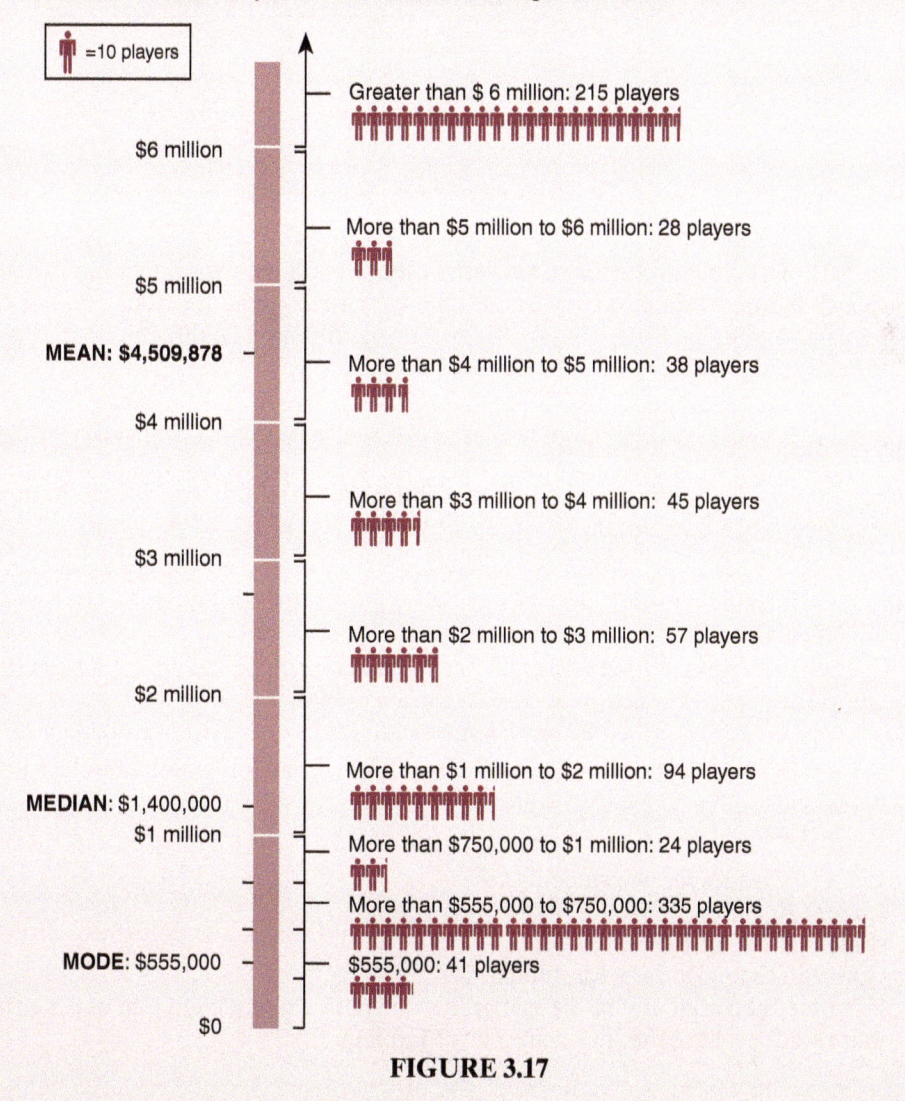

= 10 players

Greater than $6 million: 215 players

$6 million

More than $5 million to $6 million: 28 players

$5 million

MEAN: $4,509,878

More than $4 million to $5 million: 38 players

$4 million

More than $3 million to $4 million: 45 players

$3 million

More than $2 million to $3 million: 57 players

$2 million

More than $1 million to $2 million: 94 players

MEDIAN: $1,400,000
$1 million

More than $750,000 to $1 million: 24 players

More than $555,000 to $750,000: 335 players

MODE: $555,000
$555,000: 41 players

$0

FIGURE 3.17

3.2 MEASURES OF VARIABILITY

A favorite folktale told by the statistics faculty to their students is how a 10 foot fisherman and nonswimmer known as "Big John" drowned while fishing in a lake whose **average depth was only 2 feet!** The purpose of this anecdote is to realize that just knowing a measure of central tendency may not be enough information to adequately describe a data set.

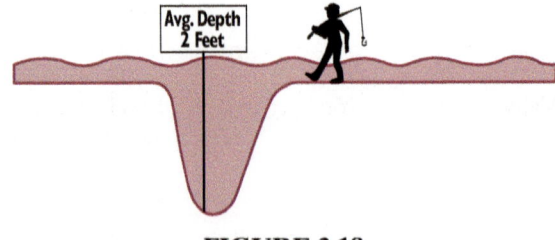

FIGURE 3.18

If the fisherman had been aware of the tremendous variations in depth as illustrated in Figure 3.18, he would have realized that knowing *only* the lake's average depth does not adequately describe its depth everywhere.

As this fable illustrates, the measures of central tendency sometimes fail to adequately describe a distribution. Additional descriptive measures, such as the measures of variability, enhance our ability to describe the main characteristics of a distribution.

Before we discuss the measures of variability, let's discuss the notion of variability by examining the two distributions shown in Table 3.3.

TABLE 3.3

Distribution 1	Distribution 2
4	1
4	22
4	24
4	31
4	100

Notice in Distribution 1, there is **no variability** since all the data values have the same value. While in Distribution 2, there is **variability** because the data values *vary* from 1 to 100.

Let us now examine techniques for measuring variability. The three measures of variability we will discuss are: the range, the variance, and the standard deviation.

The Range

The range provides a quick method of describing the variability of a distribution since it relies only on the extreme values of a distribution.

> **DEFINITION 3.8**
> The **range** of a distribution is the number representing the difference between the largest and smallest data values.
> Range = largest data value − smallest data value

EXAMPLE 3.16

The math test grades for two students are:
student A: 50, 70, 70, 70, 70, 70, 90
student B: 50, 55, 62, 70, 78, 85, 90
a) Compute the range for each student.
b) Examine the variability of the test grades for both students. Do you believe that the grades for both of these students have the same variability? Explain.

Solution

a) For both students, the range is: 90 − 50 = 40.
b) Although the range for both students is the same, the grades for student B vary more than the grades for student A. The variability of student B becomes noticeable when examining all the test grades of each student rather than just their two extreme grades.

In Example 3.16, the range for both students is the same, since the calculation of the range involves only the two extreme data values (the largest and the smallest). Overall, the test grades of student B appear to **vary** more than the test grades of student A. Table 3.4 highlights the two distributions with the box separating the extreme test grade values.

TABLE 3.4

student A:	50	70	70	70	70	70	90
student B:	50	55	62	70	78	85	90

Examining the test grades within the box, you should notice that the variability of the two distributions is different. In this instance, the range may not be a completely adequate way of distinguishing between the variability of these two test grade distributions. Thus, **one disadvantage of the range is that it does not take into consideration the value of all the data values when measuring the variability of a distribution.**

CASE STUDY 3.4 Price Performance of Fonix

The graphic in Figure 3.19 represents the price performance (in U.S. dollars) of the stock Fonix Corporation for the period from March 20th to April 18th.

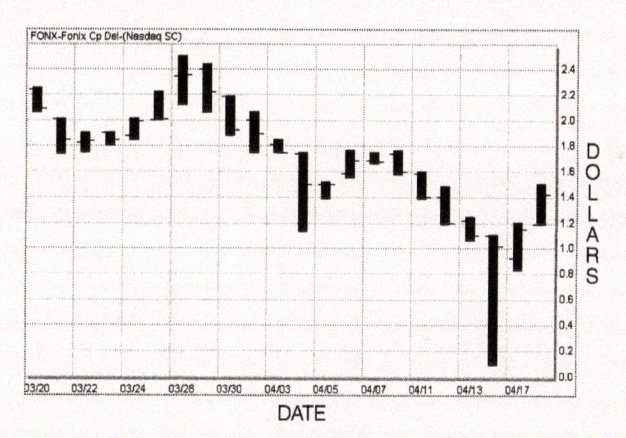

FIGURE 3.19

Each bar represents the price of Fonix for each day displayed within Figure 3.19. The small horizontal lines on each side of the bar indicate the price of Fonix for that day. The line on the left side of each bar indicates the opening price while the line on the right side indicates the closing price.

a) In Figure 3.19, examine the bar representing April 4th. What does the distance between the left and right line for that bar indicate to you about the price of Fonix? Does this distance represent a measure of central tendency or variability? Explain.
b) From Figure 3.19, which day had the shortest bar? the longest bar? What does the length of the bar in Figure 3.19 indicate about the variability of Fonix's price?
c) What does the distance between the daily closing prices measure about the Fonix?
d) Observe the **closing** prices of Fonix for March 25th and March 28th. How much did the closing price of Fonix vary from 3/25 to 3/28? What was the percentage change for this period?
e) On which day could you have bought this stock and sold it the next trading day for the greatest profit? for the greatest loss? for a zero loss?
f) During what 3 day trading period did the closing price of Fonix remain relatively stable?
g) What measure of variability is being presented by the statement *FONIX SURGES 60 CENTS TO CLOSE AT $1.42*? Explain.

108 Chapter 3 Numerical Techniques for Describing Data

Let's examine two additional measures of variability, where *every* data value within the distribution is taken into consideration when computing variability.

The Variance and Standard Deviation of a Sample

The variance and standard deviation are two measures of variability that are determined using all the data values of a distribution. Before we define these measures, let's examine the development of the variance and standard deviation so we can get a better understanding of their definitions and how to compute their formulas.

To develop these new measures of variability, we will first use the mean, as the measure of central tendency, to depict the **typical value** of a distribution, and then determine how far each data value varies from the **mean**.

For the following sample data, find the mean:

$$1, \ 2, \ 3, \ 4, \ 5, \ 6, \ 7$$

$$\bar{x} = \frac{\Sigma x}{n} = \frac{28}{7} = 4$$

Now *determine how far each data value varies from the mean*. We begin by subtracting the mean from each data value. These differences are called **deviations from the mean** and are illustrated in Table 3.5.

TABLE 3.5

Data Value x	Deviation from the Mean $x - \bar{x}$	Explanation
1	1 − 4 = −3	−3 indicates that the value 1 is three units *below* the mean
2	2 − 4 = −2	−2 indicates that the value 2 is two units *below* the mean
3	3 − 4 = −1	−1 indicates that the value 3 is one unit *below* the mean
4	4 − 4 = 0	0 indicates that the value 4 *is* the mean
5	5 − 4 = 1	1 indicates that the value 5 is one unit *above* the mean
6	6 − 4 = 2	2 indicates that the value 6 is two units *above* the mean
7	7 − 4 = 3	3 indicates that the value 7 is three units *above* the mean

Each of these deviations measure the variability of a single data value. Since we would like these new measures to reflect the variability of *all* the data values, we add together all the deviations from the mean that were obtained in Table 3.5.

Notice that the sum of these deviations is:

$$(-3) + (-2) + (-1) + (0) + (1) + (2) + (3) = 0$$

Did we expect this sum to be zero? Indeed we did! WHY? Remember back in Section 3.1, we explained that **the sum of the deviations from the mean**, expressed as $\Sigma(x - \bar{x})$, **is always zero**. Therefore, just finding the sum of the deviations from the mean is not a good way to define a measure of variability since the result will always equal zero!

To remedy this problem we need to consider a technique which will change each deviation to a nonnegative number before calculating the sum. One method of making the deviations nonnegative is to square each deviation. This is illustrated in Table 3.6.

TABLE 3.6

Data Value x	Deviation from the Mean $x - \bar{x}$	Squared Deviation from the Mean $(x - \bar{x})^2$
1	−3	$(-3)^2 = 9$
2	−2	$(-2)^2 = 4$
3	−1	$(-1)^2 = 1$
4	0	$(0)^2 = 0$
5	1	$(1)^2 = 1$
6	2	$(2)^2 = 4$
7	3	$(3)^2 = 9$

We will now use the squared deviations, symbolized as $\Sigma(x-\bar{x})^2$, to define these new measures of variability. Let's now calculate the sum of these squared deviations. This result is:

$$\Sigma(x-\bar{x})^2 = 9 + 4 + 1 + 0 + 1 + 4 + 9 = 28$$

Therefore, the sum of the squared deviations is 28.

Dividing the sum of the squared deviations, $\Sigma(x-\bar{x})^2$, by the number of sample data values minus 1, we obtain the *sample variance*.

$$\text{sample variance} = \frac{\Sigma(x-\bar{x})^2}{n-1}$$

$$= \frac{28}{6}$$

$$\approx 4.67$$

DEFINITION 3.9

The **variance for a sample** of n data values is equal to the sum of the squared deviations from the mean divided by (n – 1). The sample variance is symbolized by the square of the lower-case letter s, written s^2.

The formula for the sample variance is:

$$\text{sample variance} = \frac{\Sigma(\text{data value} - \text{sample mean})^2}{\text{number of data values} - 1}$$

OR

$$s^2 = \frac{\Sigma(x-\bar{x})^2}{n-1}$$

The sample variance uses all the data values to measure the variability of the sample. Although the variance is an important measure of variability, there are instances when the variance is extremely difficult to interpret as a measure of variability. We will explore this disadvantage in Example 3.17.

EXAMPLE 3.17

The following sample data represents the hourly wages of five employees of a small cannoli bakery in Boston.

$$\$6, \quad \$10, \quad \$9, \quad \$12, \quad \$8$$

Compute the sample variance of the hourly wages.

Solution

Step 1 Compute the mean, $\bar{x}$.

$$\bar{x} = \frac{\$45}{5} = \$9$$

Step 2 Subtract the mean of $9 from each wage and then square the deviations from the mean. These results are shown in Table 3.7.

TABLE 3.7

Data Value x	Deviation from the Mean $x - \bar{x}$	Squared Deviation from the Mean $(x-\bar{x})^2$
$ 6	$6 − $9 = −$3	(−$3)² = 9 squared dollars
$10	$10 − $9 = $1	($1)² = 1 squared dollar
$ 9	$9 − $9 = $0	($0)² = 0 squared dollars
$12	$12 − $9 = $3	($3)² = 9 squared dollars
$ 8	$8 − $9 = −$1	(−$1)² = 1 squared dollar

Step 3 To calculate the sample variance, divide the sum of the squared deviations from the mean by the number of data values minus 1.

The sum of the squared deviations is:

$\Sigma(x - \bar{x})^2 = (9 + 1 + 0 + 9 + 1)$ sq. dollars = 20 sq. dollars

and n–1 = 4, thus the sample variance is:

$$\text{sample variance} = \frac{\Sigma(x - \bar{x})^2}{n - 1}$$

$$= \frac{20 \text{ squared dollars}}{4}$$

sample variance = 5 squared dollars

The sample variance is symbolized as:

$$s^2 = \textbf{5 squared dollars}$$

One disadvantage of the variance is that the variance is measured in squared units, while the data values within the distribution are not measured in squared units.

Since the unit measurements for the variance and the data will always be different, it may be difficult to interpret and relate the variance as a measure of variability to the data values within the distribution.

Examining Example 3.17, we see that the variance is expressed in squared dollar units while the data values were given in dollars. We would like to define a new measure of variability, which is expressed in the same unit as the original data values. This can be accomplished by taking the positive square root of the sample variance. This new measure of variability is called the sample standard deviation, and is symbolized by the lower-case letters.

Let's now define this new measure of variability called the sample standard deviation.

DEFINITION 3.10
The **sample standard deviation**, s, is the positive square root of the sample variance.

$$\text{sample standard deviation} = \sqrt{\text{sample variance}}$$

Therefore, the formula for the sample standard deviation is symbolized as:

$$s = \sqrt{\frac{\Sigma(x - \bar{x})^2}{n - 1}}$$

We will refer to this formula for the standard deviation as the *definition formula for the sample standard deviation*.

To calculate the sample standard deviation for Example 3.17, take the positive square root of the sample variance. Since the sample variance is 5 squared dollars, then the sample standard deviation is:

sample standard deviation = $\sqrt{5 \text{ squared dollars}}$

sample standard deviation ≈ 2.24 dollars

This is symbolized as:

s ≈ $2.24

Notice that the standard deviation expresses the variability in the *same* unit of measure as the original data values. That is, the hourly wages and the standard deviation are both expressed in dollars.

Intuitively, we can interpret the standard deviation to be the "average" or "typical" deviation from the mean. Thus, in Example 3.17 we can interpret a standard deviation of $2.24 as indicating that the hourly wages of $6, $10, $9, $12, and $8 vary by an "average" of 2 dollars and 24 cents from the mean hourly wage of $9.

Interpretation of the Standard Deviation

The standard deviation of a set of data values measures the average or typical deviation from the mean. The more the data values are dispersed or spread out, the larger the value of the standard deviation. However, the more the data values are clustered together, the smaller the value of the standard deviation. Thus if there is no variation in the data values, then the standard deviation will equal zero!

Although the variance provides the same information as the standard deviation regarding the variability of a set of data values, we will primarily use the standard deviation because it's expressed in the original units of the data values.

EXAMPLE 3.18

Calculate the standard deviation for the following sample of ages (represented in years) using the definition formula for the sample standard deviation.

$$1, \ 3, \ 5, \ 18, \ 8$$

Solution

Step 1 Calculate the mean, $\bar{x}$.

$$\bar{x} = \frac{\Sigma x}{n} = \frac{35}{5} = 7 \text{ years}$$

Step 2 Subtract the mean of 7 years from each age and then square these deviations. These results are shown in Table 3.8.

TABLE 3.8

Data Value x	Deviation from the Mean $x - \bar{x}$	Squared Deviation from the Mean $(x - \bar{x})^2$
1	−6	$(-6)^2 = 36$
3	−4	$(-4)^2 = 16$
5	−2	$(-2)^2 = 4$
18	11	$(11)^2 = 121$
8	1	$(1)^2 = 1$

Step 3 Divide the squared deviations by the number of data values minus 1. This result is the sample variance.

The sum of the squared deviations is: $\Sigma(x - \bar{x})^2 = 178$

and n − 1 = 4, thus the sample variance is:

$$\text{sample variance} = \frac{\Sigma(x - \bar{x})^2}{n - 1} = \frac{178}{4} = 44.5 \text{ squared years}$$

In symbols, the sample variance is written:

$$s^2 = 44.5 \text{ squared years}$$

Step 4 Take the square root of the sample variance. This is the sample standard deviation.

$$s = \sqrt{\text{sample variance}}$$

$$s = \sqrt{44.5 \text{ squared years}}$$

$$s \approx 6.67 \text{ years}$$

A sample standard deviation of 6.67 years indicates that the ages within this sample vary by an "average" of 6.67 years from the mean age of 7 years.

TI-84 Graphing Calculator Solution for Example 3.18

The TI-84 graphing calculator can be used to find the standard deviation of a sample. To use it, you must first enter the given data into list L1 as shown in Figure 3.20.

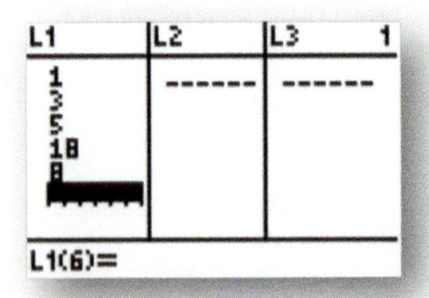

FIGURE 3.20

Next, press STAT ▶ ENTER to select the **1-Var Stats** command. You will then have to specify to the calculator the list your data is stored in. We chose L1 by pressing 2nd then **1**. Press ENTER twice leaving the FreqList field empty. This is shown in Figure 3.21a and Figure 3.21b.

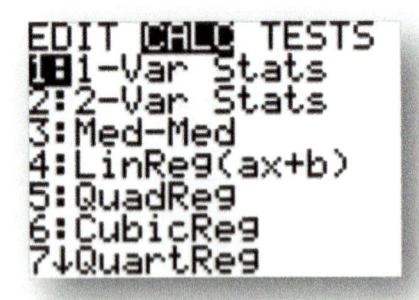

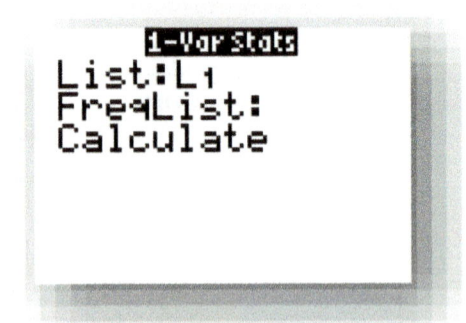

FIGURE 3.21a   **FIGURE 3.21b**

Press ENTER, and the results are displayed in Figure 3.22.

Sample standard deviation → $S_x = 6.670832032$
Population standard deviation → $\sigma x = 5.966573556$

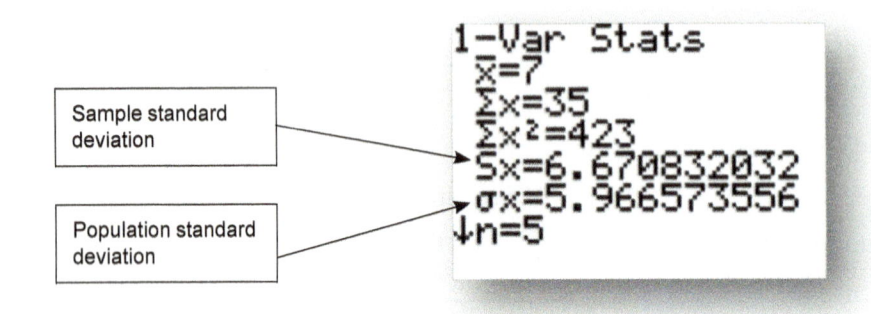

FIGURE 3.22

The sample standard deviation is given by $S_x \approx 6.67$ which is the same result found using the definition formula.

There is a shortcut formula for computing the standard deviation. This formula is called the *computational formula for the sample standard deviation.*

> **The Computational Formula for the Sample Standard Deviation**
>
> $$s = \sqrt{\frac{\Sigma x^2 - \frac{(\Sigma x)^2}{n}}{n-1}}$$

The computational formula is typically a quicker and faster way to compute the standard deviation since it reduces the number of calculations and round-off errors.

The Variance and Standard Deviation of a Population

In the beginning of this section, we developed and discussed the formulas for the variance and standard deviation of a **sample** since, in practice, we usually work with sample data. But remember, as discussed in Chapter 1, the objective of inferential statistics is to use sample data to describe the characteristics of the **population**. Although we rarely ever work with all the data values of a population, we now present the formulas and notations for the population variance and the population standard deviation of a finite population, to help explain how the sample variance and sample standard deviation are used to estimate these population parameters.

Recall the formula to compute the sample variance, s^2, for a sample of n data values is:

$$s^2 = \frac{\Sigma(x - \bar{x})^2}{n-1}$$

The formula to compute the variance of a finite population of N data values is essentially similar to the sample variance formula except the sample mean, $\bar{x}$, is replaced by the Greek symbol for the population mean, μ, and the denominator, n – 1, is replaced by the upper-case letter N, the number of population data values.

> **DEFINITION 3.11**
>
> The **variance for a finite population** of N data values is equal to the sum of the squared deviations from the mean divided by N. The population variance is symbolized by the square of the lower-case Greek letter sigma: σ, written σ^2 and pronounced sigma squared. The formula for the population variance is:
>
> $$\text{population variance} = \frac{\Sigma(\text{data value} - \text{population mean})^2}{\text{number of population data values}}$$
>
> OR
>
> $$\sigma^2 = \frac{\Sigma(x - \mu)^2}{N}$$

In the formula for the population variance, when we divide the sum of the squared deviations from the mean by the total number of data values, N, we are finding the mean of the squared deviations from the mean. Thus, the *population variance is the mean of the squared deviations from the mean.*

The formula to compute the population standard deviation is found by taking the positive square root of the population variance. The population standard deviation is denoted by the lower-case Greek letter sigma, written σ. The following formula for the population standard deviation will be referred to as the definition formula for the population standard deviation.

Chapter 3 Numerical Techniques for Describing Data

DEFINITION 3.12

The **standard deviation for a finite population** of N data values is the positive square root of the population variance. The population standard deviation is symbolized by the lower-case Greek letter sigma, written σ. The formula for the population standard deviation is:

$$\text{population standard deviation} = \sqrt{\frac{\Sigma(\text{data value} - \text{population mean})^2}{\text{number of population data values}}}$$

OR

$$\sigma = \sqrt{\frac{\Sigma(x - \mu)^2}{N}}$$

As mentioned in Chapter 1, the purpose of inferential statistics is to use sample information to describe the characteristics of a population since it is usually impractical or impossible to obtain all the data values of a population. In such instances, it is important that you understand that the sample standard deviation is used to estimate the value of the population standard deviation. It is also important to know and understand the appropriate notation used for the variance and standard deviation of a sample and of a population. Table 3.9 summarizes these notations.

TABLE 3.9 Sample and Population Notation for the Variance and the Standard Deviation

	Sample	Population
Variance	s^2	σ^2, read sigma squared
Standard Deviation	s	σ, read sigma

In Example 3.19, we will treat all the data values of this example as a population to demonstrate the formula and notation for the population standard deviation.

EXAMPLE 3.19

Two high school women basketball players are being considered for an award to be given to the most consistent player with the highest scoring average for the season. The number of points scored per game for **all the games played during the season** is given in Table 3.10.

TABLE 3.10

Points per Game for Player A	Points per Game for Player B
5	18
32	20
45	22
25	15
8	35
10	24
35	29
12	19
37	25
21	23

a. Calculate the population mean and population standard deviation of the number of points scored per game for each player.
b. Using the results of part (a) determine which player should receive the award.

Solution

a. Find the mean number of points scored for each player.
   For player A, the population mean number of points per game is:

   $$\mu = \frac{\Sigma x}{N} = \frac{230}{10} = 23$$

   For player B, the population mean number of points per game is:

   $$\mu = \frac{\Sigma x}{N} = \frac{230}{10} = 23$$

   The population standard deviation of the number of points scored for each player is determined using the formula:

   $$\sigma = \sqrt{\frac{\Sigma(x-\mu)^2}{N}}$$

   For player A, the population standard deviation of the number of points scored is determined by the following steps:

Step 1 Subtract the mean of 23 from each points per game value and then square the deviations from the mean. These results are shown in Table 3.11.

TABLE 3.11

Data Value x	Deviation from the Mean $x - \mu$	Squared Deviation from the Mean $(x-\mu)^2$
5	5 − 23 = −18	$(-18)^2$ = 324
32	32 − 23 = 9	$(9)^2$ = 81
45	45 − 23 = 22	$(22)^2$ = 484
25	25 − 23 = 2	$(2)^2$ = 4
8	8 − 23 = −15	$(-15)^2$ = 225
10	10 − 23 = −13	$(-13)^2$ = 169
35	35 − 23 = 12	$(12)^2$ = 144
12	12 − 23 = −11	$(-11)^2$ = 121
37	37 − 23 = 14	$(14)^2$ = 196
21	21 − 23 = −2	$(-2)^2$ = 4

Step 2 Add the squared deviations from the mean: $\Sigma(x-\mu)^2$

$\Sigma(x-\mu)^2 = 324 + 81 + 484 + 4 + 225 + 169 + 144 + 121 + 196 + 4$

$\Sigma(x-\mu)^2 = 1752$

Step 3 To obtain the population variance, σ^2, divide the sum of the squared deviations from the mean, $(x-\mu)^2$, by the number of data values, N.

population variance $= \frac{\Sigma(x-\mu)^2}{N} = \frac{1752}{10}$

$\sigma^2 = 175.2$

Step 4 To obtain the population standard deviation, σ, take the square root of the variance

population standard deviation $= \sqrt{\text{population variance}}$

$$\sigma = \sqrt{\frac{\Sigma(x-\mu)^2}{N}} = \sqrt{175.2}$$

$\sigma \approx 13.24$ points

The standard deviation for player A is approximately 13.24 points which indicates that player A's points per game varied, on the average, by 13.24 points from her mean of 23 points per game.

TABLE 3.12

Data Value x	Deviation from the Mean $x - \mu$	Squared Deviation from the Mean $(x - \mu)^2$
18	18 − 23 = −5	$(-5)^2$ = 25
20	20 − 23 = 3	$(3)^2$ = 9
22	22 − 23 = −1	$(-1)^2$ = 1
15	15 − 23 = −8	$(-8)^2$ = 64
35	35 − 23 = 12	$(12)^2$ = 144
24	24 − 23 = 1	$(1)^2$ = 1
29	29 − 23 = 6	$(6)^2$ = 36
19	19 − 23 = −4	$(-4)^2$ = 16
25	25 − 23 = 2	$(2)^2$ = 4
23	23 − 23 = 0	$(0)^2$ = 0

For player B, the population standard deviation of the number of points scored is determined by the following steps:

Step 1 Subtract the mean of 23 from each points per game and then square the deviations from the mean. These results are shown in Table 3.12.

Step 2 Add the squared deviations from the mean: $\Sigma(x - \mu)^2$

$$\Sigma(x - \mu)^2 = 25 + 9 + 1 + 64 + 144 + 1 + 36 + 16 + 4 + 0$$

$$\Sigma(x - \mu)^2 = 300$$

Step 3 To obtain the population variance, σ^2, divide the sum of the squared deviations from the mean, $\Sigma(x - \mu)^2$, by the number of data values, N.

$$\text{population variance} = \frac{\Sigma(x - \mu)^2}{N} = \frac{300}{10}$$

$$\sigma^2 = 30$$

Step 4 To obtain the population standard deviation, σ, take the square root of the variance.

$$\text{population standard deviation} = \sqrt{\text{population variance}}$$

$$\sigma = \sqrt{\frac{\Sigma(x - \mu)^2}{N}} = \sqrt{30}$$

$$\sigma \approx 5.48 \text{ points}$$

The standard deviation for player B is approximately 5.48 points which indicates that player B's points per game varied, on the average, by 5.48 points from her mean of 23 points per game.

The mean and standard deviation for each player is shown in Table 3.13.

TABLE 3.13

Player	Mean, μ	Standard Deviation, σ
A	23	13.24
B	23	5.48

b. Although both players had the same average number of points per game as shown in Table 3.13, player B should receive the award for the most consistently outstanding woman's scorer for the season. Player B had the smaller standard deviation. The smaller standard deviation indicates that player B was more consistent because her points per game were closer to the mean than player A.

An equivalent formula to calculate the population standard deviation, σ, is called the computational formula for the population standard deviation. This formula is a quicker and faster way to compute the standard deviation since it reduces the number of calculations and round-off errors.

The Computational Formula for the Population Standard Deviation, σ

$$\sigma = \sqrt{\frac{\Sigma x^2 - \frac{(\Sigma x)^2}{N}}{N}}$$

Using the Sample Standard Deviation to Estimate the Population Standard Deviation

In inferential statistics, a *representative sample* is selected from the population and is used to *estimate* the main characteristics of the population, such as the population variance or the population standard deviation. Usually when a sample variance is computed, we are interested in using this sample variance, s^2, as an estimate of the population variance, σ^2.

Statisticians prefer to use a denominator of n – 1 rather than n in the formula for the sample variance. The resulting sample variance provides an unbiased estimate of the population variance, σ^2, and, on the average, the resulting sample standard deviation, s, is a **better estimate** for the population standard deviation, σ.

One reason we use a denominator of n – 1 in the sample standard deviation formula is that a sample generally has less variability than a population. Thus using n rather than n – 1 in the denominator of the formula for the sample standard deviation, we obtain a smaller result for the sample standard deviation and the estimate of the population standard deviation would tend to be too small. To obtain a better estimate for the population standard deviation, a denominator of n – 1 is used for the sample standard deviation.

In Example 3.20, we illustrate the use of the sample standard deviation to estimate a population standard deviation.

EXAMPLE 3.20

Professor Candida Hart, a college nursing instructor, randomly selects 30 students from the college to estimate the mean and standard deviation of the pulse rate of all the students at the university. She collected the following 30 pulse rates.

pulse rates (measured in heart beats per minute while resting)
52, 74, 76, 70, 75, 45, 70, 50, 76, 65, 73, 78, 74, 73, 66, 59, 70, 62, 65, 72, 55, 70, 60, 72, 76, 66, 73, 76, 68, 79

a. Compute the sample mean, $\bar{x}$
b. Compute the sample standard deviation, s.
c. Determine an estimate of the population mean, μ, and an estimate of the population standard deviation, σ.

Solution

a. Using the sample mean formula,

$$\bar{x} = \frac{\Sigma x}{n}$$

we have:

$$\bar{x} = \frac{2040}{30} = 68 \text{ heart beats per minute}$$

Therefore, 68 heart beats per minute is the sample mean pulse rate for the 30 students.

b. Computing the sample standard deviation, we obtain:

$$s \approx 8.72 \text{ heart beats/minute}$$

The sample standard deviation is approximately 8.72 heart beats per minute.

c. The sample mean of 68 heart beats per minute serves as an estimate of the population mean, μ. That is, 68 heart beats per minute is an estimate of the average heart beat for all the college students. The sample standard deviation (s = 8.72) serves as an estimate of the population standard deviation, σ. Thus, an estimate of the standard deviation of the pulse rates for all the college students is 8.72 heart beats per minute.

118 Chapter 3 *Numerical Techniques for Describing Data*

> ### CASE STUDY 3.5   SAT Score Averages
>
> Examine the sample of critical reading, mathematics, and writing SAT score averages for college-bound high school seniors shown in Figure 3.23. In your observation, which data set of SAT scores seems to have the **more consistent** SAT scores? Would you expect this data to have a smaller or a larger standard deviation?
>
> **SAT SCORE AVERAGES FOR COLLEGE-BOUND HIGH-SCHOOL SENIORS**
>
> | School Year | Critical Reading | Mathematics | Writing |
> |---|---|---|---|
> | 2006 | 503 | 518 | 497 |
> | 2007 | 501 | 514 | 493 |
> | 2008 | 500 | 514 | 493 |
> | 2009 | 499 | 514 | 492 |
> | 2010 | 500 | 515 | 491 |
> | 2011 | 497 | 514 | 489 |
> | 2012 | 496 | 514 | 488 |
> | 2013 | 496 | 514 | 488 |
> | 2014 | 497 | 513 | 487 |
> | 2015 | 495 | 511 | 484 |
> | 2016 | 494 | 508 | 482 |
>
> Source: College Entrance Examination Board.
>
> **FIGURE 3.23**
>
> Calculating the mean and standard deviation for the critical reading, mathematics, and writing SAT averages, we obtain:
>
> | SAT | Critical Reading | Mathematics | Writing |
> |---|---|---|---|
> | Mean | 498 | 513.5 | 489.5 |
> | Standard Deviation | 2.79 | 2.46 | 4.32 |
>
> Using this information, we can see that the more consistent mathematics SAT scores had a smaller standard deviation than the critical reading and writing scores.
>
> Compare the SAT score for the 2006 school year to the other scores within each data set. Do you think the standard deviation of the critical reading, mathematics, and writing SAT scores would decrease, increase or stay the same, if the score for the 2006 school year is omitted from each distribution listed in Figure 3.23?

3.3 APPLICATIONS OF THE STANDARD DEVIATION

In this section, we will look at some rules that illustrate the relationship between the value of the standard deviation and the percentage of data values within a distribution that fall within a certain distance from the mean.

Two measures that help to describe a distribution of data values are the mean and the standard deviation. Remember that the mean is a measure that is used to describe the center of the distribution, while the standard deviation measures the spread of the data values within the distribution. But how can one use the mean and standard deviation to interpret the spread of data values within a distribution? There are two general rules that establish a relationship between the standard deviation and the distribution of data values. These rules are **Chebyshev's Theorem** and the **Empirical Rule**.

Chebyshev's Theorem

For a distribution of data values, regardless of shape, Chebyshev's Theorem states the percentage of data values that **may** lie within various standard deviations from the mean.

> ### Chebyshev's Theorem
>
> Chebyshev's Theorem states for any distribution, regardless of its shape:
>
> at least $1 - \frac{1}{k^2}$ of the data values will lie within k standard deviations of the mean, where k is greater than one.

For example, the percent of data values in a distribution which lie within **two standard deviations of the mean** (k=2) must be at least: $1 - \frac{1}{k^2} = 1 - \frac{1}{2^2} = 1 - \frac{1}{4} = \frac{3}{4}$ or 75%.

This is interpreted to mean that at least 75% of the data values within a distribution will lie within plus and minus 2 standard deviations from the mean.

The percent of data values in a distribution which lie within three standard deviations of the mean (k=3) must be at least: $1 - \frac{1}{k^2} = 1 - \frac{1}{3^2} = 1 - \frac{1}{9} = \frac{8}{9}$ or 89%.

This is interpreted to mean that at least 89% of the data values within a distribution will lie within plus and minus 3 standard deviations from the mean.

Keep in mind that Chebyshev's Theorem only indicates the minimum percentage of data values that will lie within a specific number of standard deviations of the mean. In many distributions, this percentage will be greater than the minimum guaranteed by Chebyshev's Theorem.

In Figure 3.24, the results of Chebyshev's Theorem are shown for the shapes of the following distributions:
a) symmetric bell-shaped shown in Figure 3.24a.
b) skewed left shown in Figure 3.24b.
c) skewed right shown in Figure 3.24c.

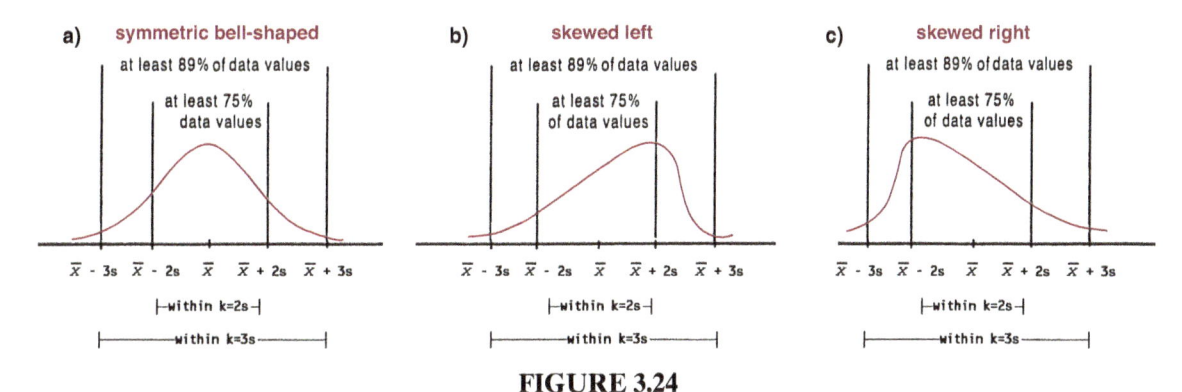

FIGURE 3.24

EXAMPLE 3.21

A statistics instructor has recorded the amount of time students need to complete the final examination. The times, stated to the nearest minute, are:

50, 70, 62, 55, 38, 42, 49, 75, 80, 79, 48, 45, 53, 58, 64, 77, 48, 49, 50, 61, 72, 74, 10, 95, 120, 79, 75, 48, 72, 37, 71, 35, 79, 72, 75, 76, 32, 30, 77, 75, 79, 45, 70, 39, 75, 72, 45, 47, 73, 72

Using this data, determine:
a) $\bar{x}$ and s to the nearest minute.
b) The percent of data values that are within 2 standard deviations of the mean.
c) The percent of data values that are within 3 standard deviations of the mean.
d) Compare the results of parts (b) and (c) to the results of Chebyshev's Theorem.
e) Is there any data value within the distribution that falls outside 3 standard deviations from the mean? If there is such a data value, then can this data value be considered an outlier? Explain.

Solution

a) The mean and standard deviation, to the nearest minute, are $\bar{x}$ = 61 minutes and s = 19 minutes.
b) The time which represents the value that is 2 standard deviations below the mean is:

$$\bar{x} - 2s = 61 - 2(19) = 23 \text{ minutes}$$

While the time which represents the value that is 2 standard deviations above the mean is:

$$\bar{x} + 2s = 61 + 2(19) = 99 \text{ minutes}$$

The number of times that are within 2 standard deviations of the mean, that is from 23 to 99 minutes, is 48. Therefore, the percent of times that are within 2 standard deviations of the mean is:

$$\frac{48}{50}(100\%) = 96\%$$

c) The time which represents the value that is 3 standard deviations below the mean is:

$$\bar{x} - 3s = 61 - 3(19) = 4 \text{ minutes}$$

While the time which represents the value that is 3 standard deviations above the mean is:

$$\bar{x} + 3s = 61 + 3(19) = 118 \text{ minutes}$$

The number of times that are from 4 to 118 minutes, that is within 3 standard deviations of the mean, is 49. Therefore, the percent of times that are within 3 standard deviations of the mean is:

$$\frac{49}{50}(100\%) = 98\%$$

d) Chebyshev's Theorem asserts that the percent of data values that are within 2 standard deviations from the mean is **at least 75% or a minimum of 75%**. From part b, we determined that the percent of actual times that are within 2 standard deviations of the mean is 96%. This is consistent with Chebyshev's Theorem since 96% is **more than the minimum of 75%**.

According to Chebyshev's Theorem, the percent of data values that are within 3 standard deviations from the mean is **at least 89% or a minimum of 89%**. From part c, we determined that the percent of actual times that are within 3 standard deviations of the mean is 98%. This is consistent with Chebyshev's Theorem since 98% is **more than the minimum of 89%**.

e) The data value 120 minutes is outside the range of values representing 3 standard deviations from the mean, 4 to 118 minutes. The data value 120 minutes can be considered to be an outlier since it is considered to lie far from most of the other data values within the distribution.

Another rule that measures with more precision the percentage of data values that fall within a certain distance from the mean for a symmetric bell-shaped distribution is described by the Empirical Rule.

Empirical (or 68-95-99.7) Rule For Bell-Shaped Distributions

Although Chebyshev's Theorem applies to any distribution regardless of shape, the Empirical Rule provides an explanation of how the standard deviation measures the spread of the data values from the mean only for a distribution that is mound or approximately bell-shaped. Figure 3.25 illustrates the Empirical Rule for a symmetric bell-shaped distribution.

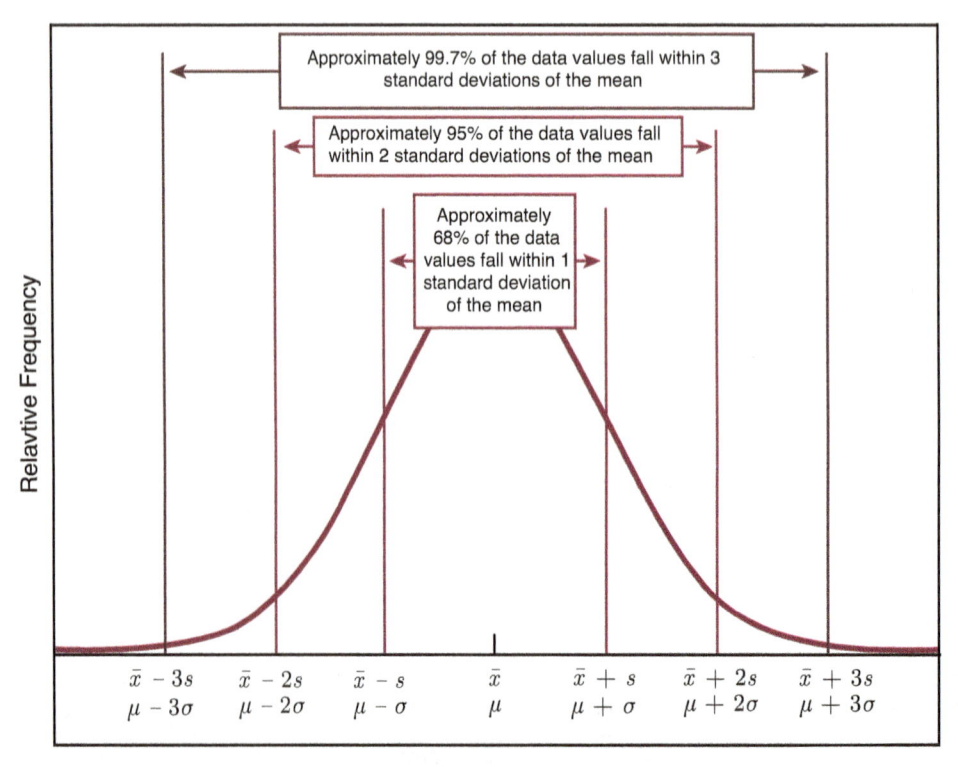

FIGURE 3.25

Empirical (or 68-95-99.7) Rule

The Empirical Rule states that for a mound or bell-shaped distribution nearly all the data will fall within three standard deviations of the mean. The Empirical Rule can be separated into three parts:
- Approximately 68% of the data will lie within one standard deviation of the mean. Using statistical notation, this is represented as $\mu - 1\sigma$ to $\mu + 1\sigma$ for a population. For a sample this is represented as: $\bar{x} - 1s$ to $\bar{x} + 1s$.
- Approximately 95% of the data will lie within two standard deviations of the mean. Using statistical notation, this is represented as $\mu - 2\sigma$ to $\mu + 2\sigma$ for a population. For a sample this is represented as: $\bar{x} - 2s$ to $\bar{x} + 2s$.
- Approximately 99.7% of the data will lie within three standard deviations of the mean. Using statistical notation, this is represented as $\mu - 3\sigma$ to $\mu + 3\sigma$ for a population. For a sample this is represented as: $\bar{x} - 3s$ to $\bar{x} + 3s$.
- This rule is also referred to as the **68-95-99.7 Rule** or the **Three Sigma Rule**.

The Empirical Rule is a widely used rule in real life applications and provides a reasonable estimate for distributions that are approximately bell-shaped. Two essential points of the Empirical Rule are that the distributions must be approximately *bell-shaped* and that the percentage of data values within the specified intervals are only *approximately* true. The Empirical Rule does not apply to distributions that are not bell-shaped and the actual percentage of data values in any of the three intervals specified by the rule could be either greater or less than those given in the rule. The Empirical Rule is especially useful when the mean and standard deviation of a bell-shaped distribution are known but all the data values of the distribution are not known. This type of application of the Empirical Rule is illustrated in Example 3.22.

EXAMPLE 3.22

The results of a standardized achievement test are approximately bell-shaped with $\bar{x} = 350$ and $s = 85$. According to the Empirical Rule, approximately what percentage of test scores would you expect to fall between:
a) 265 to 435?
b) 180 to 520?
c) 95 to 605?
d) If there were a test score equal to 625, could you consider such a score an outlier? Explain.

Solution

a) Since the values 265 and 435 are one standard deviation from the mean, that is, $\bar{x} - 1s = 350 - 85 = 265$ and $\bar{x} + 1s = 350 + 85 = 435$, then according to the Empirical Rule the percentage of test scores between 265 to 435 is approximately 68%.
b) The interval 180 to 520 represents test scores that are two standard deviations from the mean, that is, $\bar{x} - 2s = 180$ and $\bar{x} + 2s = 520$. Therefore, approximately 95% of the scores should lie within this interval.
c) Almost all the test scores, or approximately 99.7%, are expected to lie in the interval 95 to 605 because these test scores are three standard deviations from the mean. That is, $\bar{x} - 3s = 95$ and $\bar{x} + 3s = 605$.
d) The test score 625 is considered an outlier, since an outlier is a test score that lies far from the other test scores. According to the Empirical Rule, the test scores that lie more than three standard deviations from the mean occur less than 1% of the time. This makes the test score 625 a very rare score, therefore, it should be considered an outlier.

Usefulness of the Empirical Rule

The Empirical Rule can be applied to a distribution of data values to confirm whether the graph of the distribution is approximately a bell-shaped curve. Or, at least whether we should reject the notion that the graph of the distribution is a bell-shaped curve. As previously mentioned, the percentages from the data do not have to match the 68-95-99.7 Empirical Rule values exactly in order to conclude that the distribution is approximately bell-shaped. If the percentages are close enough, there is no reason to believe that the distribution of data values is not a bell-shaped distribution. Once we can verify that a distribution's shape is a symmetric bell-shaped curve, we can use the bell-shaped curve to estimate the percentage of data values within the distribution

EXAMPLE 3.23

The following distribution represents the IQ scores of 40 college students.
100, 106, 94, 100, 99, 100, 88, 112, 101, 95, 98, 103, 106, 87, 113, 130, 97, 96, 104, 101, 100, 81, 120, 100, 98, 95, 84, 100, 108, 95, 89, 87, 101, 100, 96, 101, 84, 113, 102, 116

a) Compute the mean, μ, and standard deviation, σ, of the IQ scores to the nearest tenth.
b) Determine the percent of data values within one standard deviation of the mean.
c) Determine the percent of data values within two standard deviations of the mean.
d) Determine the percent of data values within three standard deviations of the mean.
e) Does this distribution of IQ scores adhere to the Empirical Rule?
f) What might you conclude about the shape of the distribution of IQ scores?

Solution:

a) Entering the 40 IQ scores into List L1 of the TI-84 plus calculator and using 1-Var Stats to compute the mean and standard deviation, the mean is $\mu = 100$ and the standard deviation is $\sigma = 9.8$ (to the nearest tenth).

b) First, we need to determine the value representing one standard deviation below the mean. This value is denoted as $\mu - 1\sigma = 100 - 9.8 = 90.2$. Now compute the value representing one standard deviation above the mean. This value is written as $\mu + 1\sigma = 100 + 9.8 = 109.8$. Now we need to count the number of IQ scores that fall between 90.2 and 109.8. There are 27 IQ scores that fall within one standard deviation of the mean. Thus, the percent of IQ scores within one standard deviation of the mean $= \left(\frac{27}{40} \times 100\right) = 67.5\%$.

c) We need to determine the value representing two standard deviations below the mean. This value is denoted as $\mu - 2\sigma = 100 - 2 \times 9.8 = 80.4$. Now compute the value representing two standard deviations above the mean. This value is written as $\mu + 2\sigma = 100 + 2 \times 9.8 = 119.6$. Counting the number of IQ scores that fall between 80.4 and 119.6, there are 38 IQ scores that fall within two standard deviations of the mean. Thus, the percent of IQ scores within two standard deviations of the mean $= \left(\frac{38}{40} \times 100\right) = 95\%$.

d) The value representing three standard deviations below the mean is denoted as $\mu - 3\sigma = 100 - 3 \times 9.8 = 70.6$. Computing the value representing three standard deviations above the mean, this value is written as $\mu + 3\sigma = 100 + 2 \times 9.8 = 129.4$. Counting the number of IQ scores that fall between 70.6 and 129.4, there are 39 IQ scores that fall within three standard deviations of the mean. Thus, the percent of IQ scores within three standard deviations of the mean $= \left(\frac{38}{40} \times 100\right) = 97.5\%$.

e) Since 67.5% of the IQ values fall within one standard deviation of the mean, 95% IQ scores fall within two standard deviations of the mean and 97.5% of the IQ scores fall within three standard deviations of the mean, the distribution of IQ scores adhere to the Empirical Rule.

f) Since the distribution of IQ scores adheres to the Empirical Rule, we could conclude the shape of the distribution is bell-shaped.

Let's consider another application of the Empirical Rule to demonstrate how one can make useful estimations about a bell-shaped distribution just knowing the mean and the standard deviation.

EXAMPLE 3.24

Nine hundred babies were born at Long Island Hospitals during the previous 9 months. The population of the 900 baby weights is bell-shaped with a mean weight $\mu = 7.1$ lbs. and a standard deviation $\sigma = 0.6$ lbs.

a) Determine the value for each of the 7 cutoffs ranging from $\mu - 3\sigma$ to $\mu + 3\sigma$ listed on the bell-shaped diagram shown in Figure 3.26.
b) Use the Empirical Rule to compute the percent of baby weights that one would expect to fall between the 7 cutoffs shown in Figure 3.26.
c) Using Figure 3.27, determine the percent of baby weights that one would expect to be greater than 5.9 lbs.

d) Determine the percent of baby weights that one would expect to fall between 6.5 lbs and 8.3 lbs.
e) Determine the percent of baby weights that one would expect to be less than 5.3 lbs.
f) Determine the number of babies (to a whole number) that one would expect to weigh at least 8.3 lbs.
g) Determine the weight that cuts off the lowest 16% of the baby weights.
h) Determine the percent of baby weights that one would expect to be at most one standard deviation above the mean.

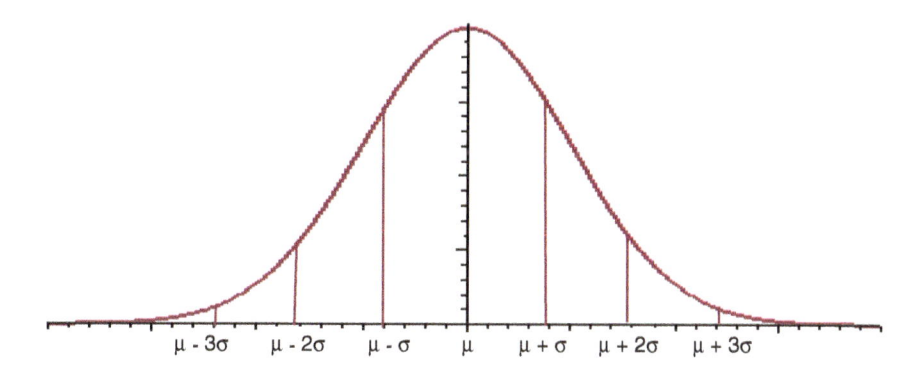

FIGURE 3.26

Solution:

a) The value for each of the 7 cutoffs is: $\mu - 3\sigma = 5.3$ lbs; $\mu - 2\sigma = 5.9$ lbs; $\mu - 1\sigma = 6.5$ lbs; $\mu = 7.1$ lbs; $\mu + 1\sigma = 7.7$ lbs; $\mu + 2\sigma = 8.3$ lbs and $\mu + 3\sigma = 8.9$ lbs. Figure 3.27 displays these cutoff values.
b) The percent of baby weights that one would expect to fall between the 7 cutoff values using the Empirical Rule is shown in Figure 3.27.

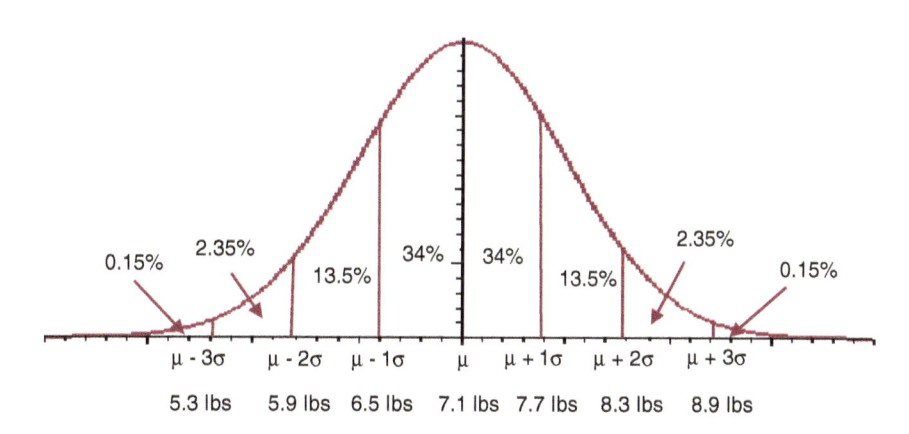

FIGURE 3.27

c) Using the Empirical Rule percentages shown in Figure 3.27, the percent of baby weights that one would expect to be greater than 5.9 lbs (where 5.9 lbs = $\mu - 2\sigma$) is equal to 13.5% + 34% + 34% + 13.5% + 2.35% + 0.15% = 97.5%.
d) Using the Empirical Rule percentages shown in Figure 3.27, the percent of baby weights that one would expect to fall between 6.5 lbs and 8.3 lbs (where 6.5 lbs = $\mu - 1\sigma$ and 8.3 lbs = $\mu + 2\sigma$) is equal to the percent of weights from $\mu - 1\sigma$ to $\mu + 2\sigma$ which equals 34% + 34% + 13.5% = 81.5%.
e) Using the Empirical Rule percentages shown in Figure 3.27, the percent of baby weights that one would expect to be less than 5.3 lbs (where 5.3 lbs = $\mu - 3\sigma$) is equal to the percent of weights below $\mu - 3\sigma$. This result is equal to 0.15%. Therefore, the percent of baby weights that one would expect to be less than 5.3 lbs is 0.15%.

f) Using the Empirical Rule percentages shown in Figure 3.27, the number of babies (to a whole number) that one would expect to weigh at least 8.3 lbs (where 8.3 lbs = μ + 2σ) is determined by finding the expected percent of baby weights greater than μ + 2σ. The percent of baby weights greater than μ + 2σ is equal to 2.35% + 0.15% = 2.5%. Now we need to compute the expected number of baby weights equal to 2.5% of the total 900 baby weights. Thus, the expected number of baby weights greater than 8.3 lbs is 0.025 × 900 = 22.5 or approximately 23 babies are expected to weigh at least 8.3 lbs.

g) Since the sum of the baby weights below μ - 1σ (or 7.1 - 0.6 = 6.5 lbs) is equal to 0.15% + 2.35% + 13.5% = 16% from the Empirical Rule percentages shown in Figure 3.27, the baby weight that cuts off the lowest 16% of the weights is 6.5 lbs.

h) The baby weight that is one standard deviation above the mean is equal to the weight represented by the value μ + 1σ. Using the Empirical Rule percentages shown in Figure 3.27, the percent of baby weights that one would expect to be at most one standard deviation above the mean is equal to the percent of weights less than μ + 1σ. This calculation is equal to: 34% + 34% + 13.5% + 2.35% + 0.15% = 84%.

Using the Range to Obtain an Estimate of the Standard Deviation

According to the Empirical Rule, a bell-shaped distribution would have most of the data values within 2 standard deviations of the mean. This means that the range for the data values will be about 4 standard deviations apart (that is: 2 standard deviations below the mean to 2 standard deviations above the mean). Meanwhile, Chebyshev's Theorem states that for any distribution regardless of shape, approximately all the data values lie within 3 standard deviations of the mean. This results in a range of values that are 6 standard deviations apart (that is: 3 standard deviations below the mean to 3 standard deviations above the mean). This is shown in Figure 3.28.

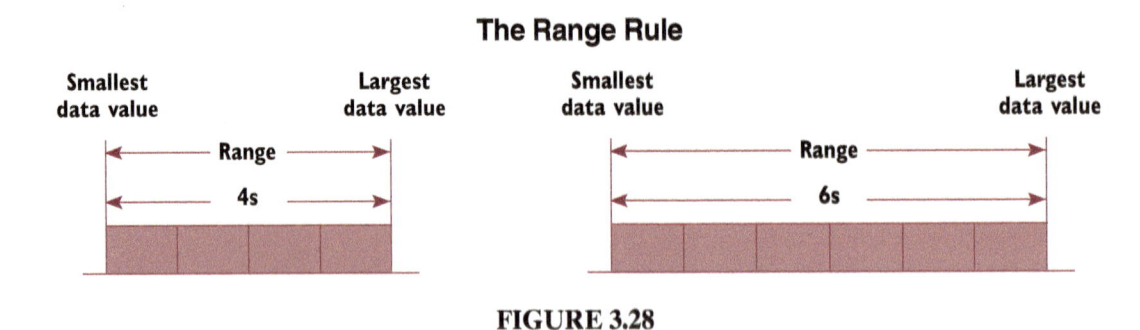

The Range Rule

FIGURE 3.28

From Figure 3.28, we obtain the following relationships:

$$\text{range} \approx 4s \quad \text{or} \quad \text{range} \approx 6s$$

Solving for s in each relationship, we obtain the following estimates for the standard deviation:

$$s \approx \frac{\text{range}}{4} \quad \text{OR} \quad s \approx \frac{\text{range}}{6}$$

From these rules, *the standard deviation should lie within a distance of* $\frac{\text{range}}{6}$ *(the smaller estimate of s) to the* $\frac{\text{range}}{4}$ *(the larger estimate of s).*

In Example 3.21, the standard deviation of the 50 data values was s ≈ 19 minutes (to the nearest minute). For this distribution of data values, the range = 120 - 10 = 110. Using the rules $\frac{\text{range}}{6}$ and $\frac{\text{range}}{4}$ to obtain an estimate of the standard deviation we get an:

over-estimate: $$s \approx \frac{\text{range}}{4} = \frac{110}{4} = 27.5$$

OR

under-estimate: $$s \approx \frac{\text{range}}{6} = \frac{110}{6} \approx 18.33$$

According to these estimates, we would expect the standard deviation to lie between the lower estimate of 18.33 and the upper estimate of 27.5. Notice that the standard deviation s ≈ 19 does lie between these estimates. However, please note that these rules will only provide rough estimates of the standard deviation. **If a distribution has outliers, then the standard deviation may not lie within these two estimates.** In fact, large data sets will usually be more varied since they have a greater chance of containing an outlier. When there are outliers, the range may exceed more than 6 standard deviations from the mean. Therefore, to obtain a better estimate of s using the range, it is necessary to consider the size of the sample. Table 3.14 provides an approximate rule for estimating s based on the sample size.

TABLE 3.14 Rule For Estimating s Using the Range

Sample Size	Rule for Estimating the Standard Deviation, s
5	$s \approx \dfrac{\text{range}}{2.5}$
10	$s \approx \dfrac{\text{range}}{3}$
25	$s \approx \dfrac{\text{range}}{4}$
100	$s \approx \dfrac{\text{range}}{6}$

Please keep in mind that whenever possible you should always calculate the exact value of the standard deviation of data values. These estimates of the standard deviation are only intended to provide you with a rough estimate of the standard deviation if needed.

EXAMPLE 3.25

For a sample containing 30 data values, the range is 47. Using Table 3.14, determine an estimate of the standard deviation, s.

Solution

Notice from Table 3.14, a sample of size n = 30 falls between sample sizes of 25 and 100. We will select the rule: $s \approx \dfrac{\text{range}}{4}$ to estimate the standard deviation, s, since n = 30 is closer to 25. Therefore, an estimate for the sample standard deviation is:

$$s \approx \dfrac{\text{range}}{4} = \dfrac{47}{4} = 11.75$$

Thus an estimate of the sample standard deviation is approximately 11.75.

CASE STUDY 3.6 Security and Exchange Commission

The time-series vertical bar graphs in Figure 3.29 displays the Securities and Exchange Commission's (SEC) staff (Figure 3.29a) and budget (Figure 3.29b) for the five selected years from 2013 to 2017. Based on these bar graphs, which one has more variability over the selected years?

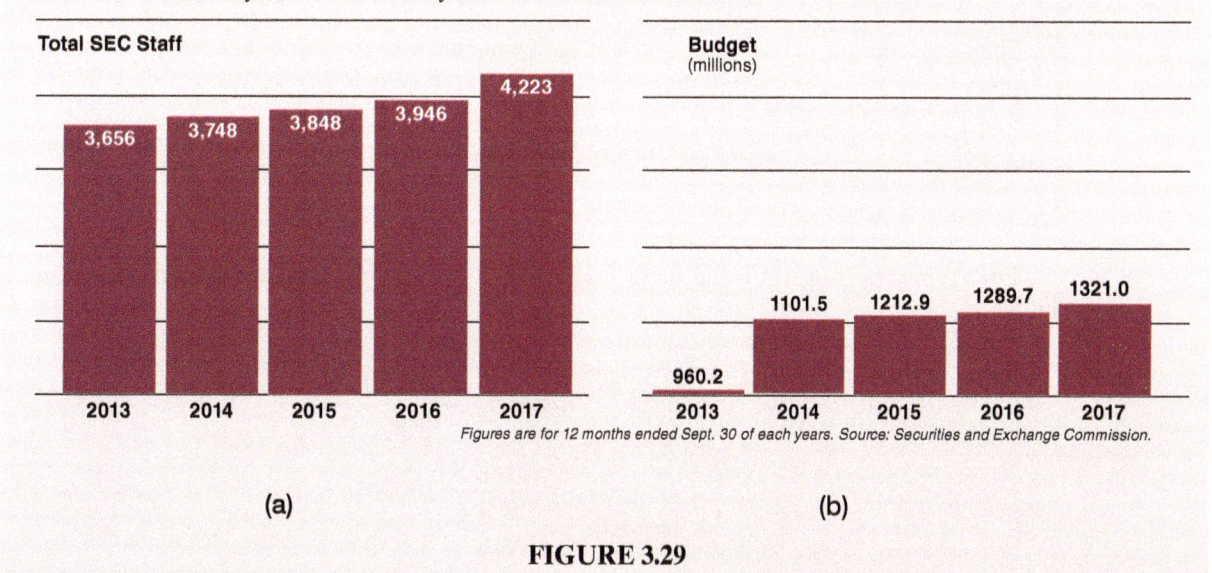

Total SEC Staff

2013	2014	2015	2016	2017
3,656	3,748	3,848	3,946	4,223

Budget (millions)

2013	2014	2015	2016	2017
960.2	1101.5	1212.9	1289.7	1321.0

Figures are for 12 months ended Sept. 30 of each years. Source: Securities and Exchange Commission.

(a)   (b)

FIGURE 3.29

From Figure 3.29, you should observe that the budget bar graph indicates more variability than the SEC staff graph. Using the information in the bar graphs, the mean and standard deviation for the SEC staff and budget are given in Table 3.15.

TABLE 3.15

Statistic	Staff	Budget (millions)
Sample Mean	3884.4	1177.06
Sample Standard Deviation	218.25	147.84
Coefficient of Variation	0.05619	0.1256

Notice that in Table 3.15 the standard deviation for the SEC staff data is greater than the standard deviation for the SEC budget. Why is there a discrepancy between the variability indicated by the graphs in Figure 3.29 verses the variability shown in Table 3.15?

Let's compare this information about the mean and standard deviation to the heights of the bars within the graphs.

For each graph, draw a line on the bar graph that represents the mean. Within which bar graph does there appear to be a greater fluctuation among the heights of the bars with respect to the mean? Explain why the standard deviation reflects the variability among the heights of the bars.

In Case Study 3.6 we observed that when comparing the variability of distributions where the size or magnitude of the data values are very different, the standard deviation **doesn't** take into consideration the magnitude of the data values. That is, the standard deviation is an absolute measure of variability. In such situations we need a **relative measure** of dispersion, that is, a measure that takes into consideration the magnitude and the variability of the data values. This objective can be accomplished by comparing the standard deviation relative to the mean of the distribution. Such a relative measure of variability is called the **coefficient of variation.**

DEFINITION 3.13
The **Coefficient of Variation** is the ratio of the standard deviation to the mean, expressed as a percent.
$$Coefficient\ of\ Variation = \frac{standard\ deviation}{mean}(100\%)$$

When to Use the Coefficient of Variation

The coefficient of variation is a very effective measure to use when one needs to make a meaningful comparison of the dispersion of two distributions under the following two situations.
1. The data values of the two distributions are in different units.
2. The data values of the two distributions are in the same unit, but the distance between their means is large.

Let's reinterpret the information in Table 3.15 by calculating the coefficient of variation for the staff and budget data.

For SEC staff data, the coefficient of variation is:

$$Coefficient\ of\ Variation = \frac{standard\ deviation}{mean}(100\%)$$

$$= \frac{218.25}{3884.4}(100\%)$$

$$\approx 6\%$$

For SEC budget data, the coefficient of variation is

$$= \frac{147.83}{1177.05}(100\%)$$
$$\approx 13\%$$

We find that the SEC staff, which has the greater **absolute** variation (i.e., standard deviation = 218.25), has the less **relative** variation as measured by the coefficient of variation. In this particular situation, the coefficient of variation was an appropriate measure to use because the data values were expressed in different units. The SEC staff data values represent the number of staff members, while the data values of the SEC budget is dollars.

EXAMPLE 3.26

An investor has studied the performance of two stocks for one year. The following data summarizes their performance:

Stock A: mean = $98.38   standard deviation = $9.78
Stock B: mean = $23.50   standard deviation = $6.95

a. Which stock has the greater absolute variation?
b. Which stock has the greater relative variation?
c. If the investor decides to use the standard deviation as a measure of the **risk** of the investment, explain why the coefficient of variation is a more useful measure of risk in comparing these two stocks.

Solution

a. Since the standard deviation is an absolute measure of dispersion, then the larger standard deviation of $9.78 has the greater absolute standard deviation.
b. Using the coefficient of variation as a relative measure of dispersion, we need to calculate the coefficient of variation for each stock. For stock A we have:

$$\text{Coefficient of Variation} = \frac{\text{standard deviation}}{\text{mean}}(100\%)$$

$$= \frac{9.78}{98.30}(100\%) \approx 10\%$$

For stock B we have:

$$\text{Coefficient of Variation} = \frac{\text{standard deviation}}{\text{mean}}(100\%)$$

$$= \frac{6.95}{23.50}(100\%) \approx 30\%$$

Therefore, stock B has the greater relative measure of variability.

c. Notice the standard deviation of the two stocks are relatively close in value. Yet the mean for stock A is about four times larger than the mean of stock B. Since the distance between the means of each stock is large, then the coefficient of variation is a very useful measure to make a meaningful comparison of the dispersion between the two stocks. In particular, a change of one standard deviation from the mean price of either stock will have a more profound effect on the total investment in stock B than in stock A because the standard deviation of stock B represents 30% of the mean as compared to 10% of the mean for stock A. Therefore stock B represents a riskier investment.

3.4 MEASURES OF RELATIVE STANDING

John Gulliver, the famous British traveler, in his fantastic journeys to faraway lands encountered many strange situations. Two of these situations are pictured in Figures 3.30 and 3.31.

As can be seen from the pictures, Gulliver's size is extreme relative to the inhabitants of the kingdoms he visits. In the land of Lilliput, where the inhabitants are only as large as Gulliver's thumb, he is considered a giant (see Figure 3.30). On the other hand, in the land of Brobdingnag, where Gulliver is only as large as an acorn, he is a dwarf among giants (see Figure 3.31).

128 Chapter 3 *Numerical Techniques for Describing Data*

Although Gulliver's height remains constant, throughout his travels, his height is interpreted or ranked differently relative to the heights of the people within each kingdom. For example, in the kingdom of Lilliput, Gulliver is considered a giant, while in the kingdom of Brobdingnag, he is regarded as a dwarf.

In statistics, one encounters numerous situations where it will be necessary to describe the position of a particular data value relative to the other data values of a distribution. In the previous two sections we discussed how the statistical measures of central tendency and variability can be used to describe the center and variability of an entire collection of data values. Now we will examine descriptive techniques which will describe *the position of an individual data value* within a data set. A statistical measure that describes the position of a particular data value relative to the other values of the data set is called a *Measure of Relative Standing*.

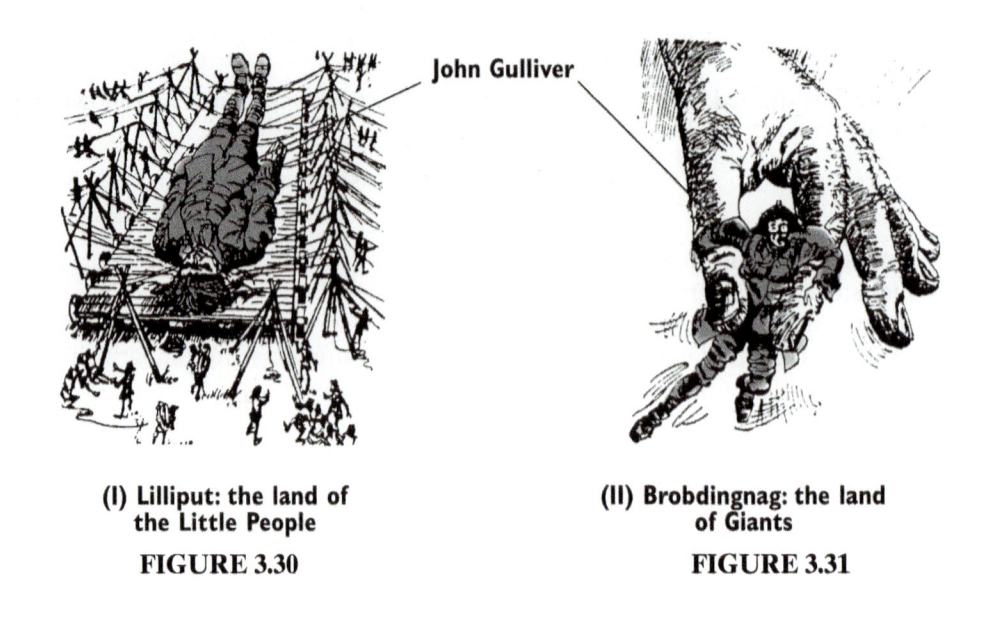

(I) Lilliput: the land of the Little People
FIGURE 3.30

(II) Brobdingnag: the land of Giants
FIGURE 3.31

z Score

In Example 3.27 we will examine a Measure of Relative Standing which uses the mean and the standard deviation to describe the relative location of a particular data value. This Measure of Relative Standing is called the **z score**.

EXAMPLE 3.27

Debbie earned a grade of 87 on her History exam, and a grade of 39 on her English exam. She would like to determine on which exam she did better, relative to the students in each class. How can this be accomplished using the information in Table 3.16?

TABLE 3.16

Exam	Mean	Standard Deviation	Debbie's Grade
History	75	6	87
English	27	4	39

Solution

Using the mean as a typical grade, we can calculate the deviation from the mean for each of Debbie's exams. A deviation from the mean is called a deviation score, and is calculated by the formula:

deviation score = data value – mean.

On the History exam, Debbie's deviation score is:

Test Grade – Mean = Deviation Score
87   –   75   =   12

On the English exam, Debbie's deviation score is:

Test Grade − Mean = Deviation Score
39 − 27 = 12

In both of Debbie's classes, her test grades are twelve units greater than the mean, thereby producing no meaningful comparison. To obtain a meaningful comparison consider the different variations in the test grades for each of her classes. Since the standard deviation reflects the typical deviation from the mean, let's reevaluate her test grades using the standard deviation. Table 3.17 shows the necessary information to reevaluate Debbie's test grades.

TABLE 3.17

	Deviation Score = Test Grade − Mean	Standard Deviation	Number of Standard Deviations from the Mean
History Exam	87 − 75 = 12	6	2
English Exam	39 − 27 = 12	4	3

We can interpret Debbie's History grade of 87 to be **two standard deviations** greater than the mean of 75. That is,

mean + 2(standard deviation) = Debbie's History grade
75 + 2(6) = 87

While Debbie's English grade of 39 is **three standard deviations** greater than the mean of 27. That is,

mean + 3(standard deviation) = Debbie's English grade
27 + 3(4) = 39

We can conclude that Debbie's English test grade of 39 showed a better achievement relative to her English class since it is three standard deviations greater than the mean English grade, whereas her History exam grade of 87 is only two standard deviations greater than the mean History grade.

The statistical Measure of Relative Standing used in Example 3.27 to describe the position of each of Debbie's test grades relative to the test grades of the other students is called a z score. The z score of each of Debbie's grades expresses the relative position of Debbie's grade from the mean grade in terms of *standard deviation units*.

DEFINITION 3.14

The **z score** of a data value indicates the number of standard deviations that the data value deviates from the mean. The formula for the z score of a data value is:

$$\text{z score of a data value} = \frac{\text{data value} - \text{mean}}{\text{standard deviation}}$$

$$\text{z score of a data value} = \frac{\text{deviation score}}{\text{standard deviation}}$$

In symbols, the z score formula for sample data is:

$$z = \frac{x - \bar{x}}{s}$$

The z score formula for population data is:

$$z = \frac{x - \mu}{\sigma}$$

130   Chapter 3   Numerical Techniques for Describing Data

Let's use the z score formula to determine the z score for each of Debbie's exam grades from Example 3.25, and interpret these results. The pertinent data of Example 3.25 is summarized in Table 3.18.

TABLE 3.18

Exam	Mean (μ)	Standard Deviation (σ)	Debbie's Grade (x)
History	75	6	87
English	27	4	39

The z score for the history test grade of 87 is computed using the formula: $z = \frac{x - \mu}{\sigma}$ and the information contained in Table 3.18. Thus, the z score for the data value 87 is:

$$z = \frac{87 - 75}{6} = 2$$

Thus, the test grade of 87 has a z score of 2. This indicates that **87 is two standard deviations greater than the mean of 75**.

The z score for the English test grade of 39 is:

$$z = \frac{39 - 27}{4} = 3$$

Thus, the test grade of 39 has a z score of 3. This indicates that **39 is three standard deviations greater than the mean of 27**. In comparing the two z scores as illustrated in Figure 3.32, we see that Debbie's English grade, because it has a larger z score, has a better relative standing in the English class than her History grade's relative standing within her History class.

Since Debbie's relative standing in the English class is higher than the relative standing in her History class, then we can conclude that her English test grade is better relative to the students in the English class than the History grade is with respect to the students in her History class.

History Exam
Distance from mean to test grade:

← 12 points away from the mean →
or
← 2 standard deviation units →

| 1 std. dev. | 1 std. dev. |

75   87
mean   test grade

English Exam
Distance from mean to test grade:

← 12 points away from the mean →
or
← 3 standard deviation units →

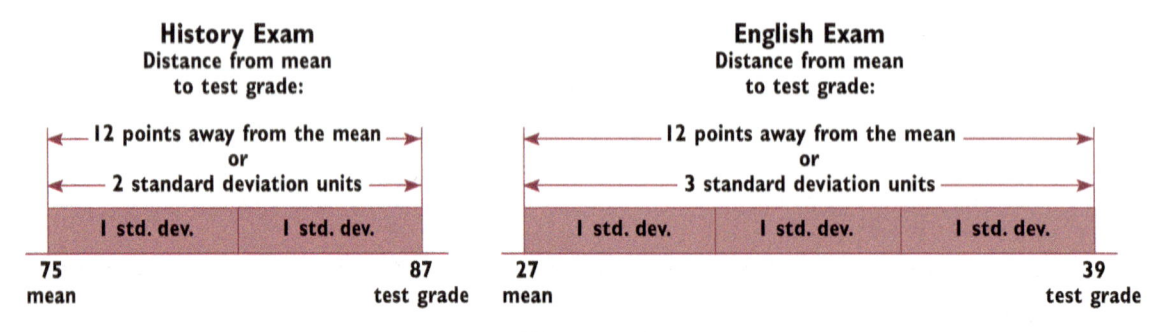

| 1 std. dev. | 1 std. dev. | 1 std. dev. |

27   39
mean   test grade

FIGURE 3.32

EXAMPLE 3.28

Calculate the z scores for each of the following three test grades that were selected from all the college students who took the final exam in Freshman Psychology 101 class, if the population mean is 75 and the population standard deviation is 10.

Psychology test grades:
65, 80, 75

Solution

To calculate the z score of 65, use $z = \frac{x - \mu}{\sigma}$.

Since x = 65, μ = 75 and σ = 10, then z of 65 is:

$$z = \frac{65 - 75}{10} = \frac{-10}{10} = -1$$

To compute the z score of 80, we have:

$$z = \frac{80 - 75}{10} = 0.5$$

Using the z score formula, the test score x = 75 has a z score of:

$$z = \frac{75 - 75}{10} = 0$$

Notice, in Example 3.28 the test grade of 65 is less than the mean and has a negative z score. The test grade of 75 is equal to the mean and has a z score of 0, while the test grade of 80 is greater than the mean and has a positive z score. Notice from the previous results, the sign of the z score indicates the location of the test score relative to the mean. These results are summarized in Table 3.19.

Interpreting the Sign of a z Score

The sign of the z score of a data value indicates the location of the data value relative to the mean as is displayed in Table 3.19.

TABLE 3.19

z Score	Location of Data Value
z score is positive	Data Value is greater than the mean
z score is zero	Data Value is equal to the mean
z score is negative	Data Value is less than the mean

EXAMPLE 3.29

At the end of each Major League baseball season, an award is given to the hitter with the highest batting average (BA). During the 1980 season, George Brett of the Kansas City Royals had the best batting average with a batting average of .390.

At the end of each Masters Tournament, the winner is the golfer with the lowest total golf score. In 1997, Tiger Woods won the Masters Tournament with the lowest total score of 270. Use the following information to answer the questions.

During the 1980 season, the American League hitters had a mean BA of .270 with a standard deviation of 0.031. During the 1997 Masters Tournament, the golfers had a mean total score of 291 with a standard deviation of 6.40.
a) Compute the z score of each player's outstanding performance.
b) Interpret the z scores and determine which player had the more outstanding performance relative to the other players in the category.

Solution
a) Compute the z score for each player.
For George Brett, the z score of his batting average of .390 is:

$$z = \frac{.390 - .270}{.031} = 3.87$$

For Tiger Woods, the z score of his total golf score of 270 is:

$$z = \frac{270 - 291}{6.40} = -3.28$$

b) George Brett's BA of .390 with a z score of 3.87 means that his batting average was 3.87 standard deviations above the mean batting average of .270. Tiger Wood's total golf score of 270 with a z score of -3.28 means that his total score was 3.28 standard deviations below the mean total golf score of 291. Although both players' performances were better than the mean performance, George Brett's performance was more outstanding relative to the other American League hitter since he had a better z score.

Detecting Outliers Using z Scores

As discussed in Chapter 2, an outlier is an extreme data value that falls far from most of the other data values. A z score can be used to identify an outlier since the farther the data value is from a z score of z = 0 the more likely the data value is an outlier. According to the Empirical Rule, any data value within an approximately bell-shaped distribution having a z score less than –3 or greater than +3 can be considered as an outlier since such a data value is considered an extreme data value.

In Example 3.29, Brett's batting average of .390 is almost 4 standard deviations above the mean. According to the Empirical Rule, we would expect almost all of the batting averages to fall within 3 standard deviations of the mean. Since Brett's batting average is almost 4 standard deviations from the mean we can interpret his batting average to be an outlier. That is, it is an extreme batting average that lies well outside the range of all the other batting averages.

Converting z Scores to Raw Scores

We will now develop a formula to convert a z score to a raw score or data value. In Example 3.30, we begin the development of this formula.

EXAMPLE 3.30

A recent medical report claimed that if a male weighs more than one standard deviation from his mean weight, then he is either overweight or underweight. Compute the weights that determine overweight and underweight for a male who stands 5 feet 8 inches, if the mean weight for his height is 155 lbs. and the standard deviation is 10 lbs.

Solution

Step 1 Determine the Overweight.
To determine the overweight we must determine the weight which is one standard deviation greater than the mean. That is, we need to determine the weight corresponding to:

mean + 1 standard deviation

Since the mean is 155 lbs and the standard deviation is 10 lbs then we have:

mean + 1 standard deviation = 155 + 1(10) = 165 lbs.

Therefore, a male who stands 5'8" is considered overweight if he weighs more than 165 lbs.

Step 2 Determine the Underweight.
To determine the underweight we must determine the weight which is one standard deviation less than the mean. That is, we need to find the weight corresponding to:

mean – 1 standard deviation

Using a mean of 155 lbs and a standard deviation of 10 lbs then we have:

mean – 1 standard deviation = 155 – 1(10) = 145 lbs.

Thus, a 5'8" male is considered underweight if he weighs less than 145 lbs.

Consequently, if a 5'8" male has a weight outside the range of 145 to 165 lbs., then the male is either underweight or overweight.

In the previous example, we examined the approach to convert a z score to a raw score, x. Now we present a formula called the **raw score formula** to change a z score to a raw score or data value.

Raw Score Formula

To convert a z Score to a Raw Score

Raw Score = Mean + (z Score)(Standard Deviation)

The raw score formula for a sample is: $x = \bar{x} + zs$

The raw score formula for a population is: $x = \mu + z\sigma$

EXAMPLE 3.31

On each of her student's test papers, a statistics instructor writes the z score rather than the actual test score. The statistics for the test are: $\bar{x} = 72$ and $s = 9.6$. Determine the student's test score (to the nearest whole number) if the test grade:
a) Has a z score of 1.65.
b) Is 2 standard deviations below the mean test grade.
c) Has a z score of 0.

Solution

a) To determine a student's test grade for a z score of 1.65, we will use the raw score formula for a sample: $x = \bar{x} + zs$. Substituting the statistics: $\bar{x} = 72$, $s = 9.6$ and $z = 1.65$ into the raw score formula, we have:
$$x = \bar{x} + zs.$$
$$x = 72 + (1.65)(9.6)$$
$$x = 87.84$$
$$x \approx 88$$
Thus, a student with a z score of 1.65 has a test score approximately equal to 88.

b) A student's test grade which is 2 standard deviations below the mean has a z score of: $z = -2$. To obtain the test grade, substitute the statistics: $\bar{x} = 72$, $s = 9.6$ and $z = -2$.
$$x = \bar{x} + zs.$$
$$x = 72 + (-2)(9.6)$$
$$x = 52.8$$
$$x \approx 53$$
Therefore, a student with a test score of approximately 53 is 2 standard deviations below the mean.

c) A student with a z score of 0 has a test score equal to the mean score of 72.

Percentile Rank and Percentiles

A common Measure of Relative Standing used to describe the position of a particular data value relative to the other data values is its **percentile rank**. The percentile rank of the data value X indicates the percent of data values within the distribution less than the data value X.

For example, Steven scored 687 on a national aptitude exam. How should we interpret Steven's score? Is 687 an exceptional score, an average score or below average score? In order to interpret this score more information is needed. Further investigation of the national results shows that Steven's aptitude score has a percentile rank of 90. We interpret this to mean that 90% of all the students who took the aptitude exam scored below Steven's score of 687 while 10% of the students scored higher than Steven's score. We can thus interpret Steven's score as an exceptional score!

Let's define this percentile rank concept and examine the formula to calculate the percentile rank.

DEFINITION 3.15

The **percentile rank** of a data value, X, is equal to the percentage of data values less than the data value, X, plus one-half the percentage of data values equal to the data value, X, within the distribution.

> **Formula to Compute the Percentile Rank of a Data Value**
>
> After arranging the data values in numerical order, the formula to compute the percentile rank is:
>
> $$\text{PR of the data value } X = \frac{[B+(\tfrac{1}{2})E]}{N}(100)$$
>
> where: PR = percentile rank
> B = the number of data values less than X
> E = the number of data values equal to X
> N = the total number of data values
>
> Percentile ranks are always expressed as whole numbers.

Percentile ranks are meaningful for large data sets. Scores from college boards, I.Q. tests and family incomes in the U.S. represent examples of large data sets where it is practical to report the data values both as a score and as a percentile.

EXAMPLE 3.32

Twenty three people with an Instagram account were asked how many pictures they have posted total. The results are below:

334, 5, 299, 15, 284, 29, 30, 80, 83, 42, 460, 30, 350, 10, 30, 72, 60, 25, 59, 208, 181, 150, 102

Compute the percentile ranks of the following number of pictures:
a) 83
b) 30

Solution

Step 1 First arrange the data in numerical order. We get:

5, 10, 15, 25, 29, 30, 30, 30, 42, 59, 60, 72, 80, 83, 102, 150, 181, 208, 284, 299, 334, 350, 460

Step 2 a) To find the percentile rank of the score 83, use the formula:

$$\text{PR of the data value } X = \frac{[B+(\tfrac{1}{2})E]}{N}(100)$$

Arrange the scores in numerical order, and determine the values of B, E, and N. Examining the ordered scores, we observe that there are thirteen scores less than 83, so B = 13. Since there is only one 83, then E = 1. Finally, N = 23 since there are 23 scores.

Substituting the values of B, E and N into the percentile rank formula, we have:

$$\text{PR of the score } 83 = \frac{[13+(\tfrac{1}{2})1]}{23}100$$

$$\text{PR of the score } 83 = \frac{13.5}{23}(100) \approx 58.7 \approx 59$$

Thus, the PR of 83 is 59. This indicates that approximately 59% of the values are below the value of 83.

b) To compute the percentile rank of 30, determine the values of B, and E for 30. For the score 30, B = 5 and E = 3. Therefore,

$$\text{PR of the score } 30 = \frac{[5+(\frac{1}{2})3]}{23}(100)$$
$$= \frac{6.5}{23}(100) \approx 28.3 \approx 28.$$

The PR of the score 30 is 28. This means that approximately 28% of the values are below the value of 30.

Another concept closely related to percentile rank is that of **percentile**. Let's examine this concept.

> **DEFINITION 3.16**
> The **pth percentile** is the data value within the distribution which has p percent of the data values less than its value and the rest of the data values greater than its data value. 99 percentiles divide the data set into 100 groups.

In Example 3.32 part a, the value 83 has a percentile rank of 59. This indicates that 83 is greater than 59% of the values. Using percentiles, we state that 83 is the 59th percentile. This is written as:

$$P_{59} = 83$$

In Example 3.32 part b, the value 30 has a percentile rank of 28. Therefore, we can state that 30 is the 28th percentile, written as:

$$P_{28} = 30.$$

> **Meaning of Percentile and Percentile Rank**
>
> There is a difference in the use of the words **percentile** and **percentile rank**. In general, the **percentile** is a **data value** of the distribution while a **percentile rank** is a **percentage** value ranging from 0 to 100.

EXAMPLE 3.33

Identify the percentile and the percentile rank in each statement.
a) Exactly 62 percent of the students in a given school system received IQ scores less than 112.
b) In Christopher's third grade class, 70% of the students have heights less than 54 inches.

Solution

a) 112 is the 62nd percentile and 62 is the percentile rank of 112.
b) 54 is the 70th percentile of the student's heights in the third grade class and 70 is the percentile rank of the height of 54 inches.

Deciles and Quartiles

Some important percentiles have individual names. Two such names are *deciles* and *quartiles*. There are nine deciles, and three quartiles. We will first examine the relationship between percentiles and deciles.

The relationship between the nine deciles and the corresponding percentiles are shown in Figure 3.33.

Deciles

The nine deciles which divide the data set into tenths or ten equal parts are:
- The **10th percentile** is called the **1st decile**
- The **20th percentile** is called the **2nd decile**
- The **30th percentile** is called the **3rd decile**
- The **40th percentile** is called the **4th decile**
- The **50th percentile** is called the **5th decile**
- The **60th percentile** is called the **6th decile**
- The **70th percentile** is called the **7th decile**
- The **80th percentile** is called the **8th decile**
- The **90th percentile** is called the **9th decile**

Notice in Figure 3.33, the 1st decile is the value at the tenth percentile.

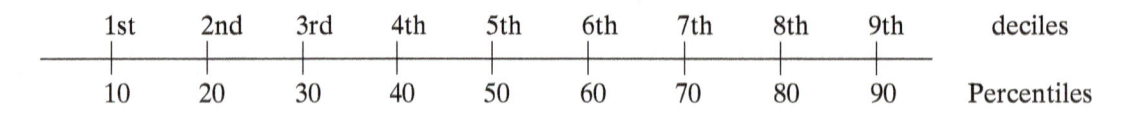

```
1st   2nd   3rd   4th   5th   6th   7th   8th   9th      deciles
 |     |     |     |     |     |     |     |     |
 10    20    30    40    50    60    70    80    90      Percentiles
```

FIGURE 3.33

Let's now examine the relationship between the three quartiles and the corresponding percentiles. The relationship between the three quartiles and the corresponding percentiles are shown in Figure 3.34.

Quartiles

The three quartiles which divide the data set into quarters or four equal parts are:
- The **25th percentile** is called the **1st quartile, Q_1**.
- The **50th percentile** is called the **2nd quartile, Q_2**.
- The **75th percentile** is called the **3rd quartile, Q_3**.

```
         1st           2nd           3rd              quartiles
          |             |             |
          25            50            75              percentiles
```

FIGURE 3.34

EXAMPLE 3.34

Some results from the mathematics reasoning part of the Scholastic Aptitude Test (SAT) are given in Table 3.20. Using Table 3.20 determine:
a) The score which is the 1st decile.
b) The median.
c) The score which has a percentile rank of 90.
d) The percentile rank of the score 380, and state what quartile it represents.
e) The 3rd quartile.
f) The score which is the 6th decile.

TABLE 3.20

Score	Percentile
320	10
365	20
380	25
400	30
410	35
474	50
510	60
570	75
595	80
650	90
750	99

Solution

a) The first decile is another name for the 10th percentile. Since 320 is the 10th percentile, it is also the 1st decile.
b) The median is the 50th percentile. Thus, the median score is 474.
c) A score that has a percentile rank of 90 is the 90th percentile. Thus, the 90th percentile is the score 650.
d) The score 380 is the 25th percentile and has a percentile rank of 25. Thus, 380 represents the 1st quartile, since it has a percentile rank of 25.
e) The 3rd quartile is another name for the 75th percentile. The 75th percentile is 570 and thus, 570 is the 3rd quartile.
f) The sixth decile is the 60th percentile. Since 510 is the 60th percentile then the sixth decile is 510.

CASE STUDY 3.7   Percentile Ranks of SAT Subject Test Scores

SAT Subject Tests Percentile Ranks (2016–2018 Graduating Classes)

Score	Literature	U.S. History	Math Level I	Ecological Biology
800	99	97	99	97
750	91	83	92	88
700	75	64	74	74
650	56	46	57	55
600	40	30	42	37
550	28	19	29	23
500	19	12	19	13
450	12	7	10	8
400	5	3	5	5
350	1	1	1	2
300	1-	1-	1-	1-
250	1-	-	-	1-
200	-	-	-	-
MEAN	607	640	605	618
STANDARD DEV.	115	108	112	110

(SOURCE: 2018 The College Board)

FIGURE 3.35

The table shown in Figure 3.35 lists the percentile ranks of SAT Subject Tests Scores for Literature, U.S. History, Math Level I and Ecological Biology.

a) Interpret the percent of students that scored below a score of 600 on the Literature test.
b) Interpret the percent of students that scored below a score of 650 on the U.S. History test.

c) What percent of students scored above a test score of 450 on the Math Level I test? Above the score of 500 on the Ecological Biology test?

d) What Literature test score represents the 3rd quartile?

e) What Math Level I score represents the 1st decile?

f) Using the mean and standard deviation for the Literature test scores, determine the z score of the Literature test scores: 750, 550, and 400. Interpret these results.

g) Using the mean and standard deviation for the U.S. History test scores, determine the z score of the U.S. History test scores: 700, 550, and 400. Interpret these results.

h) Using the mean and standard deviation for the writing scores, determine the z score of the Math Level I test scores: : 700, 550, and 400. Interpret these results.

i) Using the mean and standard deviation for the Ecological Biology test scores, determine the z score of the Ecological Biology test scores: : 700, 550, and 400. Interpret these results.

j) On which SAT subject test would a score of 750 be considered to be a better test score? Explain your answer using the results of the previous parts: f, g, h and i.

k) On which SAT subject test would a score of 550 be considered to be a better test score? Explain your answer using the results of the previous parts: f, g, h and i.

l) On which SAT subject test would a score of 400 be considered to be a better test score? Explain your answer using the results of the previous parts: f, g, h and i.

3.5 BOX-AND-WHISKER PLOT: AN EXPLORATORY DATA ANALYSIS TECHNIQUE

The Box-and-Whisker Plot (also referred to as a Box Plot) is an Exploratory Data Analysis technique developed by Professor John Tukey of Princeton University to visually display several summary statistics in order to describe the location of the center of the data, the spread and shape of a distribution, and to identify any outliers that may exist within the data. It is also helpful in comparing **two** or **more** distributions.

Before we discuss the construction of a Box-and-Whisker Plot, we will discuss a 5-number summary to describe the characteristics of a data set.

DEFINITION 3.17
A **5-number summary** uses the following five numbers to describe a data set:
1. The smallest data value
2. The first quartile (Q_1)
3. The median
4. The third quartile (Q_3)
5. The largest data value

To provide a **5-number summary** of data set, we must determine the smallest data value, Q_1, the median, Q_3, and the largest data value of the data set. In Section 3.1, we developed a procedure to determine the median of a data set. Remember, the median is the middle value of a data set **after** the data values have been arranged in numerical order.

In a similar manner, the first and third quartiles can be determined. The first quartile, Q_1, is essentially the *median* of the data values below the **position** of the median, while the third quartile, Q_3, is the *median* of the data values above the **position** of the median.

In Example 3.35, we examine how to determine the numbers of a 5-number summary.

EXAMPLE 3.35

Determine the 5-number summary for the following distribution of annual salaries.

Annual Salaries (in Thousands of Dollars)
55 57 60 62 63 64 65 66 67 67 67 70 72 75 78 83 86 87 90

Solution

The 5-number summary of a distribution consists of the smallest data value, Q_1: the first quartile, the median, Q_3: the third quartile, and the largest data value.

Step 1 **Find the smallest and largest data values.**
The smallest data value is 55 while the largest data value is 90.

Step 2 **Determine the median by determining its position in the data set.**

Since the data are arranged in numerical order the median is the $\frac{n+1}{2}$ th data value. For this distribution of 19 data values, the median is the $\frac{19+1}{2}$ or 10th data value which is 67.

Step 3 **Determine the 1st quartile, Q_1.**
The 1st quartile, Q_1, is the median of the data values which are below the position of the median. In Figure 3.36, the median's position is enclosed within a box, and the values below are shown in bold print.

67 at the 10th data value
represents the median's position

55 57 60 62 63 64 65 66 67 | 67 | 67 70 72 75 78 83 86 87 90

FIGURE 3.36

To determine Q_1, we need to find the median of the data values which are below the position of the median. The values below the median are:

55 57 60 62 63 64 65 66 67

The median of these data values is 63, since it is the middle value of these nine numbers. Thus, 63 represents the first quartile, that is $Q_1 = 63$.

Step 4 **Determine the 3rd quartile, Q_3.**
The third quartile, Q_3, is the median of the data values which are above the position of the median. In Figure 3.37, the median's position is enclosed within a box. The values above the median's position are shown in bold print.

67 at the 10th data value
represents the median's position

55 57 60 62 63 64 65 66 67 | 67 | **67 70 72 75 78 83 86 87 90**

FIGURE 3.37

To determine Q_3, we need to find the median of the data values which are above the position of the median. The values above the median are:

67 70 72 75 78 83 86 87 90

The median of these data values is 78, since it is the middle value of these numbers. Thus 78 represents the third quartile, that is $Q_3 = 78$.

Therefore, the 5-number summary for the distribution of annual salaries is:

smallest data value = 55
1st quartile: $Q_1 = 63$
median = 67
3rd quartile: $Q_3 = 78$
largest data value = 90

TI-84 Graphing Calculator Solution for Example 3.33

The TI-84 graphing calculator can be used to find the 5-number summary of the distribution in Example 3.35. First enter the given data into list L1 as shown in Figure 3.38.

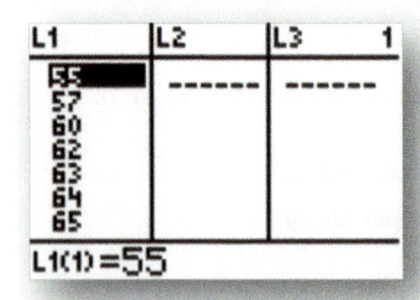

FIGURE 3.38

Next, press STAT ▶ ENTER to select the **1-Var Stats** command. You will then have to specify to the calculator the list your data is stored in. We chose L1 by pressing 2nd then **1.** Press ENTER twice leaving the FreqList field empty. This is shown in Figure 3.39a and Figure 3.39b.

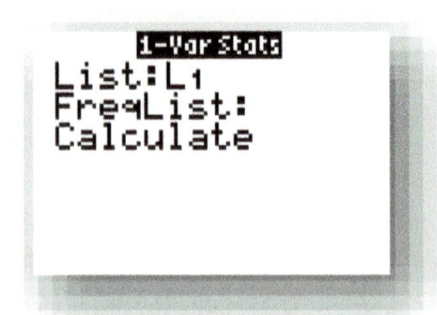

FIGURE 3.39a   **FIGURE 3.39b**

Press ENTER. The calculator results are displayed in Figure 3.40a and Figure 3.40b.

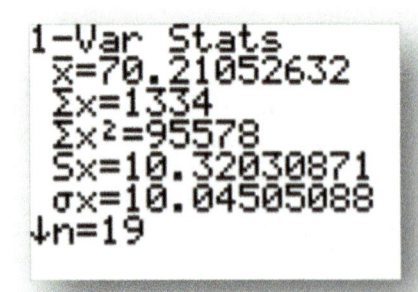

FIGURE 3.40a   **FIGURE 3.40b**

The screenshot in Figure 3.40b is obtained by pressing the ▼ key several times to display the 5-number summary. As you can see from Figure 3.40b, the 5-number summary is given.

Section 3.5 *Box-and-Whisker Plot: An Exploratory Data Analysis Technique* 141

Let's summarize the procedure used to determine the 5-number summary.

Procedure to Determine the 5-Number Summary

Step 1   Find the smallest and largest data values.
Step 2   Find the median.
Step 3   Find the 1st quartile, Q_1.
Step 4   Find the 3rd quartile, Q_3.

We will now examine a visual exploratory data analysis technique called a Box-and-Whisker Plot that displays the 5-number summary of a data set.

DEFINITION 3.18

A **Box-and-Whisker Plot** or **Box Plot** is a visual device that uses a 5-number summary to reveal the characteristics of a data set.

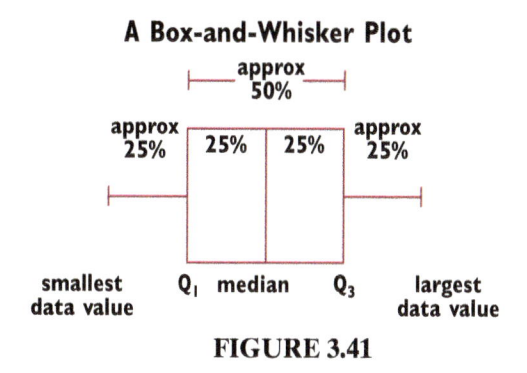

A Box-and-Whisker Plot

approx 50%
approx 25%   25%   25%   approx 25%

smallest data value Q_1 median Q_3 largest data value

FIGURE 3.41

A Box-and-Whisker Plot is a graphical display that can be used when there is a small number of data values in a distribution and therefore the construction of either a stem-and-leaf display, a bar chart, or a histogram may not be appropriate.

The graph in Figure 3.41 consisting of horizontal lines, called the whiskers, and extending outward from a rectangular box is an example of a Box-and-Whisker Plot.

Notice, in Figure 3.41, the values of a 5-number summary, smallest data value, the 1st quartile, the median, the 3rd quartile and the largest data value, are used to construct the rectangular Box and the Whiskers of the box plot. Specifically, the sides of the rectangular box are constructed by using Q_1, the 1st quartile, and Q_3, the 3rd quartile, with a vertical line drawn within the rectangle to represent the median's position. The length of the whiskers (or the horizontal lines) of the box plot are determined by the distance between the quartiles to the smallest and largest data values.

EXAMPLE 3.36

For the following 5-number summary:

smallest data value = 55
the 1st quartile, $Q_1 = 63$
median = 67
the 3rd quartile, $Q_3 = 78$
largest data value = 90

construct a Box-and-Whisker Plot.

Solution

To construct a Box-and-Whisker Plot, the five numerical values from the 5-number summary are graphed and labeled on a horizontal axis.

Step 1 Draw a horizontal axis and scale it appropriately from the smallest to the largest values of the 5-number summary. This is illustrated in Figure 3.42(a).

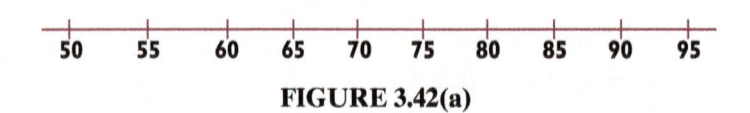

FIGURE 3.42(a)

Step 2 Draw a rectangle above the horizontal scale by drawing vertical lines at the 1st quartile, Q_1, and the 3rd quartile, Q_3, to represent the sides of the rectangle. Thus, the left side of the rectangle is placed at the value of Q_1 which is 63, while the right side of the rectangle is placed at the value of Q_3 which is 78. The height of the rectangle is unimportant. This is illustrated in Figure 3.42(b).

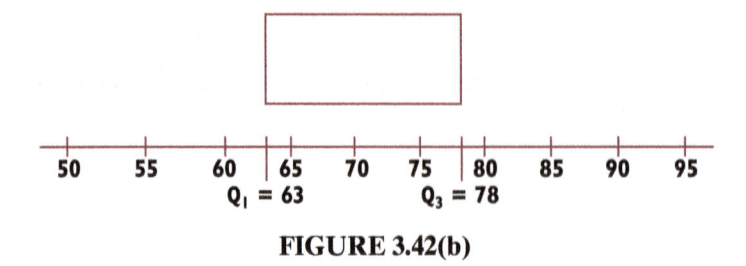

FIGURE 3.42(b)

Step 3 Draw a vertical line within the rectangle to represent the position of the median. Thus, the vertical line representing the median is drawn at 67. This is illustrated in Figure 3.42(c).

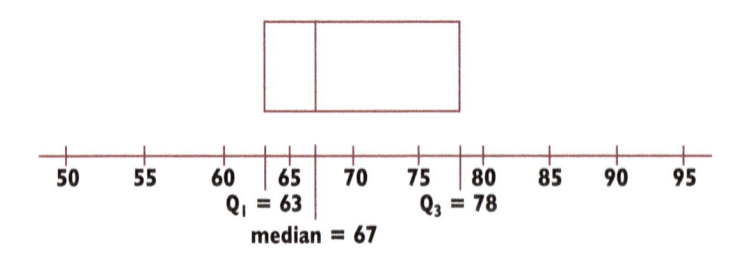

FIGURE 3.42(c)

Step 4 Draw two horizontal lines. One line connects the smallest data value to the left side of the box. The second horizontal line is drawn connecting the largest data value to the right side of the box. These horizontal lines connecting the 1st quartile, Q_1, to the smallest data value and the 3rd quartile, Q_3, to the largest data value are called the Whiskers. The completed Box-and Whisker Plot is illustrated in Figure 3.42(d).

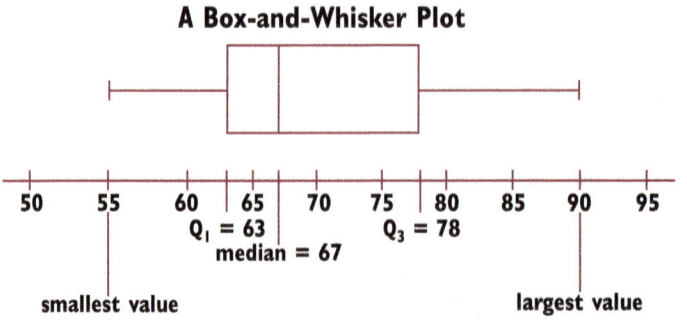

A Box-and-Whisker Plot

FIGURE 3.42(d)

Section 3.5   *Box-and-Whisker Plot: An Exploratory Data Analysis Technique*   143

TI-84 Graphing Calculator Solution for Example 3.36

The TI-84 graphing calculator can be used to display the Box-and-Whisker Plot of the distribution in Example 3.36. Enter the given data into list L1. This is shown in Figure 3.43.

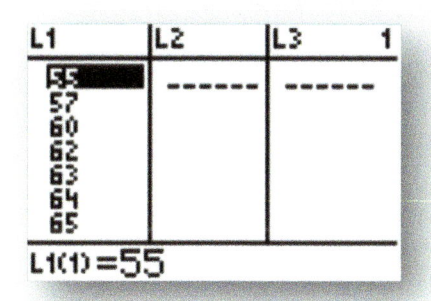

FIGURE 3.43

To create the Box-and-Whisker Plot, press [2nd] [Y=] [ENTER] to open **Stat Plot 1** menu. You will need to turn the plot on, select the icon of the Box and Whisker Plot (5th icon) under **Type**, and then specify the list name your data is stored in. For our example, we have the data in **L1**. This is shown in Figure 3.44.

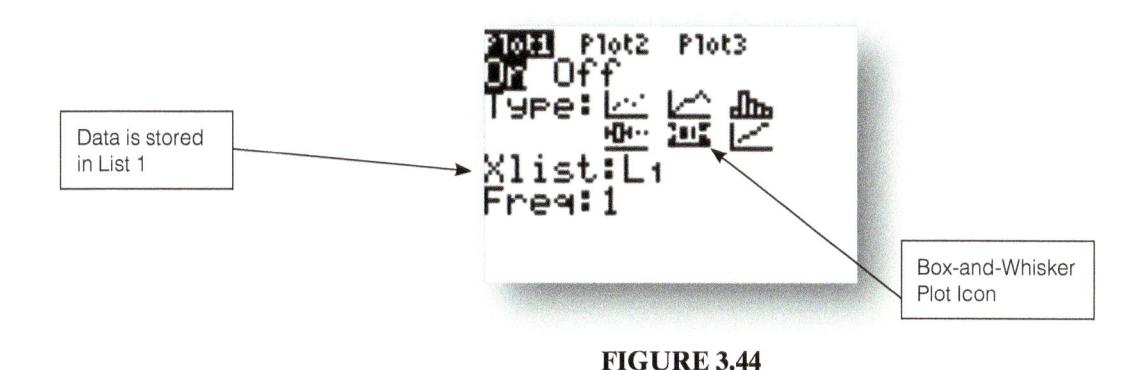

Data is stored in List 1

Box-and-Whisker Plot Icon

FIGURE 3.44

The calculator has a special **ZOOM** feature for statistics in the Zoom menu. Using the **ZoomStat** feature (shown in Figure 3.45a) will fit the Box-and-Whisker Plot to the calculator's display. To activate it press [ZOOM] then scroll to **9:ZoomStat** and press [ENTER] (you could have also just pressed the **9** key). Figure 3.45b shows the resulting Box-and-Whisker Plot.

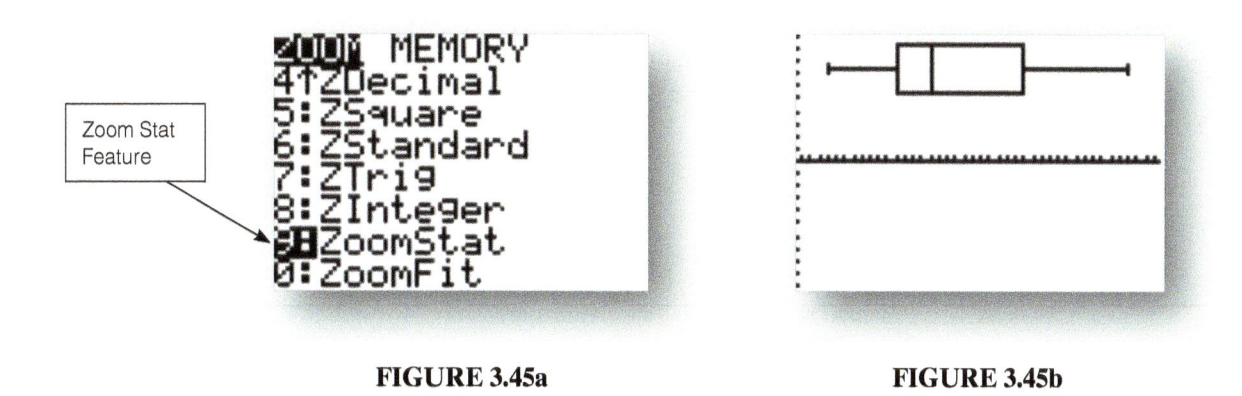

Zoom Stat Feature

FIGURE 3.45a   **FIGURE 3.45b**

To display each value of the 5-number summary, press [TRACE]. Use the [◄] and [►] keys to display the values for the 5-number summary on the Box-and-Whisker Plot. Two values (Q1 and Median) are shown in Figure 3.46a and Figure 3.46b.

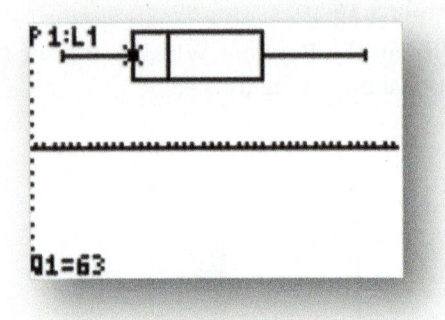

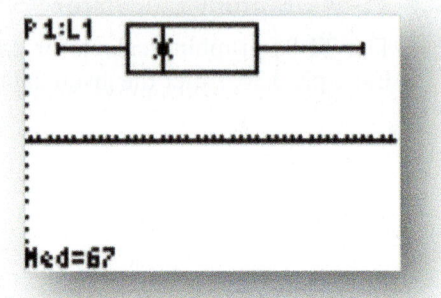

FIGURE 3.46a

FIGURE 3.46b

Let's summarize the procedure to construct a Box-and-Whisker Plot.

Procedure to Construct a Box-and-Whisker Plot

Step 1   Draw a horizontal line and scale it from the smallest to the largest data values.

Step 2   Draw a rectangular box above the horizontal line with the sides at the 1st quartile, Q_1, and the 3rd quartile, Q_3.

Step 3   Draw a vertical line within the box to represent the position of the median.

Step 4   Draw horizontal lines from the box to the smallest and largest data values. These are called the Whiskers of the Box Plot.

A Box-and-Whisker Plot is useful in identifying whether a distribution's shape is symmetric or skewed. Notice that in Figure 3.42(c), the median is not centered within the box. It is positioned closer to Q_1 than to Q_3. Furthermore, notice that the whisker to the right of the third quartile is longer than the whisker to the left of the first quartile. This is an indication that the distribution is not symmetric. In fact it is **skewed** to the right. This suggests that there may be outliers present that are greater than Q_3.

In Figure 3.47 three box plots have been constructed. In Figure 3.47(a) notice that the whisker to the left of Q_1 is longer than the whisker to the right of Q_3. Perhaps there are outliers less than Q_1. Also notice that the median line is positioned closer to Q_3 than to Q_1. This suggests that the distribution is skewed to the left.

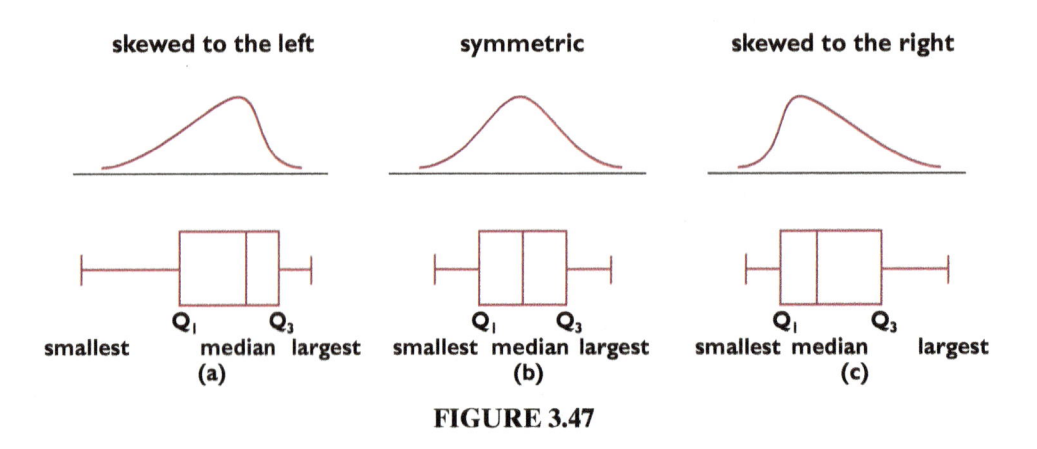

FIGURE 3.47

In Figure 3.47(b) notice that the whisker to the left of Q_1 is the same length as the whisker to the right of Q_3 and that the median line is positioned in the middle of the box. Although this distribution may contain outliers we can certainly conclude that the distribution is symmetric.

In Figure 3.47(c), as was the case in Figure 3.42 (d), the whisker to the left of Q_1 is shorter than the whisker to the right of Q_3 and the median line is closer to Q_1 than to Q_3. This distribution is skewed right.

In summary, a Box Plot is particularly useful in conveying the location of the center, the variability and the shape of a distribution of data. Let's discuss some observations that can be made by examining a Box-and-Whisker Plot.

Box-and-Whisker Plot Interpretations

1. The median represents a measure of the center and is denoted by a vertical line within a rectangular box.
2. Approximately 50% of the data values are contained within the box.
3. Within each component of the box, separated by the median, are approximately 25% of the data values.
4. Approximately the lower 25% of the data values is represented by the whisker connecting the first quartile to the smallest data value.
5. Approximately the upper 25% of the data values is represented by the whisker connecting the third quartile to the largest data value.
6. The length of the box is a measure of the variability of the data set. A longer box indicates more variability, while a shorter box will indicate less variability.
7. The length of the whiskers will provide an idea about the shape of the data set. If one whisker is noticeably longer, then the shape of the data set is probably skewed in the direction of the longer whisker.

EXAMPLE 3.37

The 5-number summaries of the distributions of the calories per serving of several brands of ice cream versus frozen yogurt are given in Table 3.21.
a) Construct a Box Plot for each of the 5-number summaries.
b) Examine the shape of each distribution, and determine what can you conclude about the calories per serving of ice cream compared to yogurt?

TABLE 3.21 Calories Per Serving

Dessert	Smallest Value	Q_1	Median	Q_3	Largest Value
Ice Cream	155	165	175	195	210
Yogurt	125	145	165	185	205

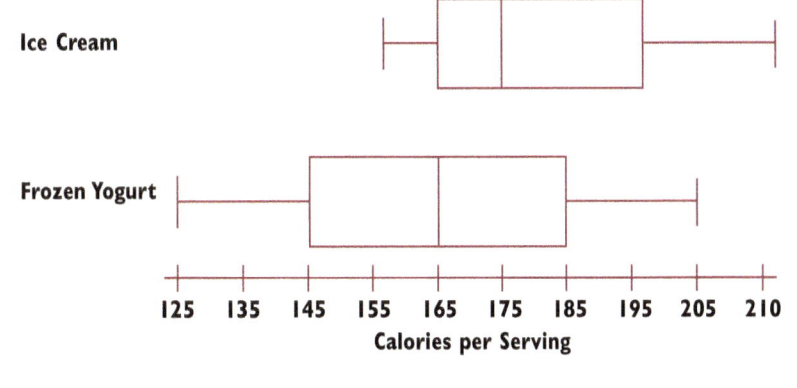

FIGURE 3.48

Solution

a) Using the 5-number summary per each frozen dessert the Box Plots are shown in Figure 3.48:
b) In examining the Box Plots of Figure 3.48, it appears that ice cream tends to have more calories per serving than frozen yogurt. Notice that the median calories per serving for ice cream is 175 while the median

calories per serving for frozen yogurt is 165. The Box Plot for ice cream indicates the shape of the distribution is skewed to the right, since the upper whisker is longer. Whereas the Box Plot for yogurt has a symmetric shaped distribution since the median is in the middle of the box and both whiskers have approximately the same length.   ∎

Box Plots are also useful in detecting the presence of an outlier. Data values that are outside the first and third quartiles, Q_1 and Q_3, may be considered somewhat extreme. In fact, the difference between Q_1 and Q_3, is often used as a guide for determining **outliers**. The difference between Q_1 and Q_3 is called *The Interquartile Range (IQR)*.

DEFINITION 3.19
The **Interquartile Range** (IQR) is the difference between the first quartile, Q_1, and the third quartile, Q_3. It is symbolized as IQR.

$$IQR = Q_3 - Q_1$$

The **Interquartile Range**, IQR, is a measure of the spread within a distribution. Usually, data values that are **at least a distance of 1.5 times the IQR above the third quartile or at least a distance of 1.5 times the IQR below the first quartile** are considered to be potential outliers for a distribution. In fact, Tukey's terminology of **inner and outer fences** to identify potential and definite outliers uses the IQR. Inner fences are located a distance of 1.5*IQR below Q_1 and above Q_3, while outer fences are located a distance of 3*IQR below Q_1 and above Q_3. Using this idea of inner and outer fences, we can now state a rule for detecting an outlier.

Rule for Detecting Outliers

An **outlier** is identified as any data value that falls outside of the outer fences.

A **potential outlier** is identified as any data value that falls between the inner fences and the outer fences.

Using the box portion of the Box Plot to represent the IQR, the relationship between the IQR and the inner and outer fences is illustrated in Figure 3.49.

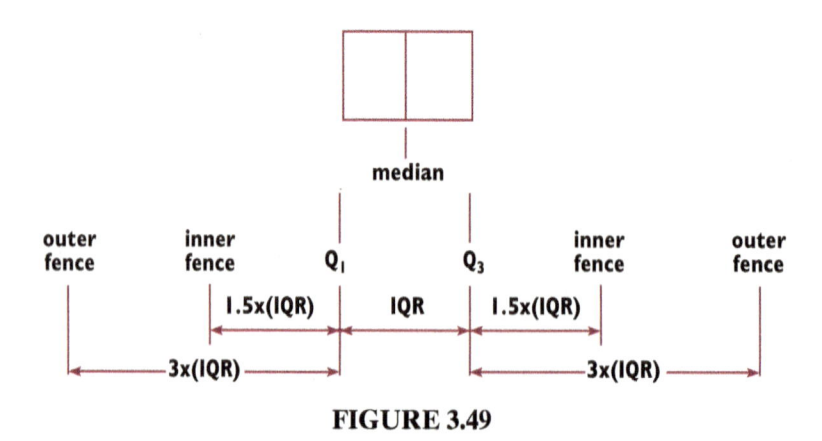

FIGURE 3.49

An outlier is an extreme data value that may be a valid data value or it can be a data value due to an error such as a measurement error or a recording error. Therefore, when an outlier or a potential outlier has been detected it is necessary to determine whether the outlier is due to error or is an unusual data value. If the outlier is due to error, then it should be removed. If the outlier represents an unusual but valid data value, then it should be investigated further since it may reveal useful information about the distribution of data values and may play an important role in subsequent statistical data analysis.

Section 3.5 Box-and-Whisker Plot: An Exploratory Data Analysis Technique

CASE STUDY 3.8 Salary Survey

Examine the Table in Figure 3.50 which represents the fall 2006 to the fall 2012 Salary Survey of Biostatistics and Other Biomedical Statistics Departments and Units, conducted by the American Statistical Association. All the salary figures shown within the Table in Figure 3.50 are for a 12-month period along with the time in rank information based on the responses from 35 departments of U.S. colleges and universities that list programs in statistics plus a few individuals who responded to the survey.

Table 1—Results from the Fall 2012 Salary Survey of Biostatistics and Other Biomedical Statistics Departments and Units

Rank/Years in Rank	Percentile	Fall 2006 (Sample Size)	Fall 2007 (Sample Size)	Fall 2008 (Sample Size)	Fall 2009 (Sample Size)	Fall 2010 (Sample Size)	Fall 2011 (Sample Size)	Fall 2012 (Sample Size)
Assistant 1–3	25th	$82,400	$86,000	$89,200	$89,100	$89.500	$91,900	$92.700
	50th	85,000	88,452	93,600	93,500	93,600	97,400	96,900
	75th	90,000	92,869	98,300	99,100	102,600	104,300	$106,300
	90th			104,900	103,300	111,300	111,500	$117,200
		(112)	(106)	(69)	(82)	(78)	(83)	{73}
4 or more	25th	$84,476	$87,400	$90,500	$89,400	$91,000	$93,000	$96,800
	50th	88,471	92,000	95,500	95,500	97,300	97,200	99,500
	75th	94,819	98,220	106,200	100,700	103,000	104,600	106,300
	90th			118,000	110,500	116,400	113,800	116,400
		(48)	(65)	(62)	(87}	(101)	(97)	(80)
Associate 0–2	25th	$89,937	$102,525	$102,500	$102,400	$103,000	$109,600	$105,800
	50th	100,441	110,493	110,800	114,000	110,100	118,000	$115,500
	75th	113,000	118,900	127,000	129,200	122,400	130,600	$125,000
	90th			140,000	139,000	140,000	143,800	$135,300
		(46)	(50)	(36)	(62)	(46)	(56)	(54)
3 or more	25th	$101,384	$105,000	$109,300	$108,100	$110,100	$112,800	$112,700
	50th	107,981	109,350	118,000	116,800	123,800	125,000	$126,200
	75th	120,000	124,924	130,000	130,000	141,000	140,500	$145,300
	90th			155,000	155,800	153,900	155,900	$159,000
		(65)	(66)	(69)	(69)	(94)	(101)	(114)
Full 0–6	25th	$127,893	$137,991	$130,200	$127,100	$138,500	$141,900	$153,200
	50th	147.488	163,870	145,200	152,300	159,100	157,300	$172.000
	75th	177,840	180,365	174,600	181,100	185,000	185,400	$194,000
	90th			205,500	208,100	208,500	203,000	$220,000
		(54)	(60)	(65)	(84)	{66}	(79)	(78)
7 or more	25th	$139,959	$147,575	$156,200	$145,600	$163,300	$166,900	$160,500
	50th	172.523	180,760	187,000	178,600	194,200	191,500	$191,600
	75th	197,277	209,147	215,500	203,200	222,600	234.500	$232,800
	90th			246,700	224,400	251,800	259,300	$277,000
		(84)	(71)	(77)	(92)	(97)	(109)	(115)
Starting Assistant Professors	25th	$77,500	$84,000	$87,500	$103,200	$85,000	$92,500	$92.000
	50th	82,400	85,000	91,500	107,000	90,000	95,700	$102.200
	75th	85,279	93,150	98,500	110,600	104,100	113,300	$116,000
	90th			NA	NA	NA	120,000	NA
		(27)	(13)	(09)	(12)	(11)	(19)	(15)

FIGURE 3.50
Courtesy of American Statistical Association.

CASE STUDY 3.8 Salary Survey (continued)

a) What salary represents the first quartile for an Associate with 3 or more years in rank for fall 2012?
b) What salary represents the third quartile for an Assistant with 1-3 years in rank for fall 2012?
c) What salary represents the median salary for a Full Professor with 7 or more years in rank for fall 2012?
d) What percent of Starting Assistant Professors during the fall 2012 academic semester earn:
   1) At most $92,000?
   2) At least $116,000?
   3) Between $102,200 and $116,000?
e) For Associate Professors with at least 3 years in rank for the fall 2012 semester:
   1) What salary represents the median salary?
   2) What salary represents the third quartile?
   3) What salary represents the first quartile?
   4) What salary represents the 9th decile?
   5) What is the interquartile rank of their salaries? And, what percent of salaries are represented by the interquartile rank? Approximately how many Associate Professors are represented within the interquartile rank?

KEY TERMS

Bimodal 99
Box-and-Whisker Plot (Box Plot) 141
Chebyshev's Theorem 118
Coefficient of Variation 126
Decile 135
Empirical Rule 121
Five-number Summary 138
Inner Fence 146
Interquartile Range (IQR) 146
Measures of Central Tendency 88

Measures of variability 106
Median 94
Mode 98
Nonmodal 99
Outer Fence 146
Outlier 146
Percentile 135
Percentile Rank, PR 133
Population mean, μ 90
Population standard deviation, σ 114

Potential Outlier 146
Quartile 135
Range 106
Range Rule to Estimates 124
Resistant Measure 104
Sample Mean, $\bar{x}$ 89
Sample standard deviation, s 110
Sample variance, s^2 109
Variance of the population, σ^2 113
z Score, z 128

SECTION EXERCISES

Note: The data sets in these Section Exercises can be downloaded in either Microsoft EXCEL or TI-84 format from the textbook's website at MyStatLab.com.

Section 3.1

1. Three measures of central tendency are _____, _____ and _____.

2. The measures of central tendency usually describe the _____ of a distribution.

3. Since the mean is sensitive to outliers, then it isn't a _____ measure of central tendency.

4. If $\bar{x} = 5$, and n=2, then $\Sigma x = $ _____.

5. A distribution $x_1, x_2, \ldots, x_n$ has a mean of 15. Then for the distribution:
   a) $3x_1, 3x_2, \ldots, 3x_n$, the mean is _____.
   b) each data value is divided by 5, the mean is _____.
   c) $x_1 - 10, x_2 - 10, \ldots, x_n - 10$, the mean is _____.
   d) formed by adding 5 to each data value, the mean is _____.

6. If the data values $x_1, x_2, \ldots, x_n$ have a mean of 13, then $\Sigma(x - 13) = $ _____.

7. If a distribution of 5 data values has $\bar{x} = 11$, and if four of the data values are 5, 10, 20, 25 then the fifth data value is _____.

8. When a distribution with an even number of data values is arranged in numerical order, the median is the _____ of the two middle data values of the distribution.

9. In a distribution, the data value that appears most frequently is called the _____.

10. The sample mean is used as an estimate of the _____ mean.

11. If the data values are categorical, the _____ is the best measure of central tendency to use.

12. In a symmetric bell-shaped distribution, the measures of central tendency are located at the _____ of the distribution.

13. Which of the following is commonly referred to as the "average"?
    a) Mean   b) Median   c) Mode

14. In a skewed to the left distribution, the mean is located to the _____ of the median.
    a) Left   b) Right

15. Since the median is unaffected by an outlier, it is said to be _____ to outliers.

16. The mean, median, and mode are never the same value. (T/F)

17. If the mean of a distribution is equal to 5 and the mode equals 3, then the median must equal 4. (T/F)

18. For the distribution: 6, 2, 4, 4, 5, 3, 5, 3, 1, 7, find:
    a) the mean
    b) the median
    c) the mode
    d) $\Sigma(x-4)$
    e) Σx^2
    f) $(\Sigma x)^2$
    g) $\Sigma(x-4)^2$
    h) $[\Sigma(x-\bar{x})]^2$
    i) $\Sigma x^2 - (\bar{x})^2$

19. The following data values represent a sample of calories burned by a runner during each of her thirty runs last month:

    613 509 405 408 514 508 380 610 475 411
    426 412 608 481 407 425 469 471 614 410
    456 560 495 460 621 409 560 346 708 1334

    a) Determine the mean, median and mode for the number of calories burned.
    b) Which measure seems to be a better indicator of central tendency for this distribution? Explain.

20. A consumer compiled the following prices for a 32 inch LED TV, in an attempt to determine a fair market price:
    $399, $480, $499, $488, $419, $385
    a) Calculate the mean, median and mode.
    b) Which measure(s) of central tendency, if any, do you think best illustrates a fair price based on this data?

21. A statistics instructor gave the following final semester grades to his students:
    86, 82, 100, 96, 91, 94, 85, 95, 78, 80, 87, 92, 85, 81, 97, 98, 89, 90, 84, 92, 91, 87, 98, 90
    a) Compute the mean grade.
    b) Comment on the remark of one of the instructor's statistics students: " On the average, this instructor does not give grades above 90."

22. Given the distribution: 10, 20, 30, 40, 50, 60, 70, 80, 900.
    a) Calculate the mean and the median.
    b) Which measure seems to be a better indicator of the central tendency of this distribution?
    c) For this distribution, change the last data value (900) to 90 and recalculate the mean and the median.
    d) Now which measure seems to be a better indicator of central tendency? Explain.
    From your observation, which measure of central tendency was:
    e) more resistant to the extreme data values?
    f) less resistant to the extreme data values?

23. The student government association is in the process of planning a special awards dinner. To try to determine the menu for the awards dinner the association surveyed the number and types of meals purchased at the campus cafeteria. The following meal table was recorded during the past week.

Meals	Monday	Tuesday	Wednesday	Thursday	Friday
Fried Chicken	125	45	80	100	80
Meat Loaf	40	60	95	70	75
Veal Cutlets	50	35	55	75	45
Spag. & Meatball	100	135	90	125	135
Beef Stew	120	90	80	135	90
Pizza	30	80	70	45	160

   a) What are the most popular daily meals?
   b) What is the most popular weekly meal?
   c) Based on the survey, which meal do you believe the association should choose for the awards dinner?

24. The latest smartphone is claimed by its manufacturer to have a 48-hour battery life, longer than any phone on the market. To test this claim, a technology review group purchases and tests 30 of these phones.

   The number of hours each phone lasted before reaching 0% battery is recorded below:

   Number of Hours
   50, 33, 39, 37, 49, 48, 54, 44, 41, 34, 38, 39, 40, 49, 32, 36, 44, 49, 47, 44, 49, 46, 39, 38, 44, 45, 50, 43, 40, 42

   a) Using the sample information, determine the mean battery life of this new phone. Comment on the manufacturer's claim.
   b) Assume you are a statistician working for the smartphone manufacturer. Using the same information, determine which measure of central tendency you would use to support the manufacturer's claim.

25. If a distribution of study hours is skewed left, then the:
    a) Most frequent study hour equals the median study hour
    b) Most frequent study hour is greater than the median study hour
    c) Most frequent study hour is less than the median study hour
    d) Not enough information

26. Using the following figure which represents a skewed distribution, determine the value for:
    a) the mode
    b) the mean
    c) the median
    d) Identify the shape of the distribution.

e) In which tail would you expect to find an outlier? Explain.
f) Which Measure of Central Tendency would be closest to a possible outlier? Explain.

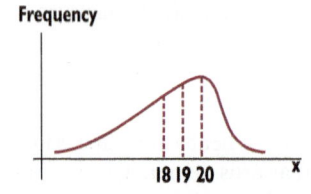

27. For each of the following distribution of data values,
   a) identify the general shape that you believe best describes each distribution, and
   b) identify whether the mean or median would have the larger value within this type of distribution.
   1) The age of people living in the United States.
   2) The amount of soda dispensed into 2 liter bottles.
   3) The amount of soda dispensed by an under-filling malfunctioning machine into 2 liter bottles.
   4) The amount of time to complete an easy exam, if the class is 75 minutes long.
   5) The weights of all the female students at a University.

Section 3.2

28. Three measures of variability are: _____, _____, and _____.

29. Measures of variability describe the _____ of the data values.

30. The range is calculated by subtracting the _____ value of the distribution from the _____ value of the distribution.

31. If the range is 5 and a constant of 6 is added to each data value, then the range of the new data values is _____.

32. If a distribution of five data values has a variance of 100, then the sum of the squared deviations from the mean, $\Sigma(x - \bar{\mu})^2$, is _____.

33. The symbol used to represent the sample standard deviation is _____. The Greek symbol used to represent the population standard deviation is _____.

34. In a sample, if the variance is 144, then s = _____.

35. For a distribution, if $x - \bar{x}$ has a negative value, then x is positioned _____ the mean. If $x - \bar{x}$ has a positive value, then x is positioned _____ the mean. If $x - \bar{x} = 0$ then x = _____.

36. For a distribution where $\bar{x} = 50$ and s = 5, calculate:
    $\bar{x} + s$ _____
    $\bar{x} + 2s$ _____
    $\bar{x} + 3s$ _____
    $\bar{x} - s$ _____
    $\bar{x} - 2s$ _____
    $\bar{x} - 3s$ _____

37. Which measure of variability is always expressed in squared units?
    a) Range   b) Variance   c) Standard deviation

38. Which measure of variability is the square root of another measure of variability?
    a) Range   b) Variance   c) Standard deviation

39. The standard deviation of a distribution is _____.
    a) Never negative
    b) Always a whole number
    c) Always positive

40. The range of a distribution is _____.
    a) Never negative
    b) Always a whole number
    c) Always positive

41. For the same set of data, which formula will yield a larger value:
    a) σ b) s

42. The standard deviation is sensitive to the extreme data values.
    (T/F)

43. If a distribution consists of all negative data values then the range is negative.
    (T/F)

44. In a distribution whose data values all have the same numerical value, the standard deviation is always zero.
    (T/F)

45. A distribution that has a range of one must also have a standard deviation of one.
    (T/F)

46. For the data values: 2, 3, 7, 8, 15 compute:
    a) The range
    b) The sample variance, s^2
    c) The sample standard deviation, s
    d) An estimate of the population standard deviation
    e) The population standard deviation, assuming the data values represent a population.

47. The golf scores of 11 members of a men's and women's college golf teams for the first round of a regional tournament are shown in the following table.

Player	1	2	3	4	5	6	7	8	9	10	11
Men Scores	91	90	88	90	89	84	102	87	88	91	79
Female Scores	99	87	90	88	81	91	90	91	91	80	91

a) Compute the mean (to the nearest whole number) and standard deviation (to 2 decimal places) of the golf scores for both the men and women's team.
b) Which team played a more consistent first round of tournament golf? Explain.
c) Using only the best 10 golf scores for each team, compute the mean (to the nearest whole number) and standard

deviation (to 2 decimal places) of the golf scores for both the men and women's team.

d) Which 10 player team is more consistent? What contributed to this result?

48. Consider the data values: 2, 9, 16, 23, 30, 37, 44.

Calculate the range and sample standard deviation for:
a) The original data values.
b) The data values formed by subtracting 10 from each original data value. Compare this result to the answer for part a.
c) The data values formed by multiplying each original data value by (–4). Compare this result to answer for part a.

49. For the distribution: 42, 30, 24, 48, 60, 36, 54
a) Find the value of $\bar{x}$ and s.
b) Compute $\bar{x}$ + s.
c) Find the data values that are within one standard deviation above the mean.
d) Compute $\bar{x}$ – 1s.
e) Find the data values that are within 1 standard deviation below the mean.
f) Compute $\bar{x}$ – 2s and $\bar{x}$ + 2s.
g) What percent of the data values are within 1 standard deviation from the mean?
h) What percent of the data values are within 2 standard deviations from the mean?

50. A study was conducted by a consumer protection group to determine the effectiveness of two leading pain relievers, Afferin and Banacin. The pain relievers were administered to a sample of 30 subjects suffering from headache pain. Fifteen subjects were administered Afferin while 15 were administered Banacin. The amount of time it took the drug to become effective in relieving the pain was recorded and the results are listed in the following table:

Number of minutes for relief of headache pain

Afferin Group	Banacin Group
45	48
63	54
54	42
37	38
42	42
58	53
47	64
36	44
41	44
39	50
48	52
54	42
55	51
49	52
52	44

a) Compute the sample mean and sample standard deviation for each group.
b) Determine which drug has the more consistent time for relieving headache pain. Explain.

51. Steven, a college professor, computed the mean and the standard deviation of the final exam which he administered to his day and evening statistics classes.

Class	Mean Grade	Standard Deviation
day	66	16
evening	66	2

a) In which class would you expect to find the lowest grade on this test? the highest grade? Explain.
b) If Steven randomly selects a student from each one of these classes, then from which class would Steven have a better chance of selecting a student with a grade closer to the mean grade? Explain.

Section 3.3

52. Two general rules that establish a relationship between the standard deviation and the distribution of data values are _____ Theorem and the _____ Rule.

53. Chebyshev's Theorem states for any distribution, regardless of its shape, at least $1 - \frac{1}{k^2}$ of the data values will lie within k _____ of the mean, where k is greater than one.

54. Empirical Rule states for a bell-shaped distribution that:
a) Approximately ____ % of the data values will lie within 1 standard deviation of the mean.
b) Approximately ____ % of the data values will lie within 2 standard deviations of the mean.
c) Approximately ____ or ____ % to ____ % of the data values will lie within 3 standard deviations of the mean.

55. The Math SAT scores of 2500 students at a local High School is bell-shaped with a mean = 500 and a standard deviation of 75. Use the Empirical Rule to answer the following questions.
a) Determine the percent of Math SAT scores that have a z score of at least 1.
b) Determine the percent of Math SAT scores that have a z score less than 3.
c) Determine the percent of Math SAT scores that have a z score between -1 and 2.
d) Determine the number of students that their Math SAT score has a z score greater than 2.
e) Determine the Math SAT score that represents the 50th percentile.
f) Determine the Math SAT score that represents the 84th percentile.
g) Determine the z score of the Math SAT score that represents the 16th percentile.

56. Which of the following can be applied to a skewed shaped distribution?
a) Chebyshev's Theorem
b) Empirical Rule
c) Schwartz' Rule

57. According to the Empirical Rule, the approximate percentage of data values that are within 2 standard deviations is:
a) 68%   b) 95%   c) 99%   d) 100%

58. For a bell-shaped distribution, which of the following gives a conservative estimate for the percentage of data values that are within two standard deviations of the mean?
a) Chebyshev's Theorem
b) Empirical Rule
c) Schwartz' Rule

59. The following distribution represents the blood pressures of 45 year old females:

108 164 118 136 131 138 141 124 142 131 142 130 145 131 132 148 151 154 135 136 121 139 138 138 124 140 124 141 127 141 129 144 131 132 149 150 133 153 135 114

a) Compute the mean, μ, and standard deviation, σ, of the blood pressures to the nearest whole number.
b) Determine the percent of data values within one standard deviation of the mean.
c) Determine the percent of data values within two standard deviations of the mean.
d) Determine the percent of data values within three standard deviations of the mean.
e) Does this distribution of blood pressures adhere to the Empirical Rule?
f) What might you conclude about the shape of the distribution of blood pressures?

60. The following distribution represents the times (in seconds) it took thirty five seventh grade boys to run a mile:

500 538 347 649 576 515 472 480 423 600 615 520 491 520 460 530 521 520 553 483 455 522 547 498 511 507 520 518 468 438 535 574 550 562 585

a) Compute the mean, μ, and standard deviation, σ, of the times to the nearest second.
b) Determine the percent of values within one standard deviation of the mean.
c) Determine the percent of values within two standard deviations of the mean.
d) Determine the percent of values within three standard deviations of the mean.
e) Does this distribution of times adhere to the Empirical Rule?
f) What might you conclude about the shape of the distribution of times?

61. Three hundred small cups (10 oz) of coffee were sold at a certain coffee shop during the last week. The amount of caffeine in a small cup of coffee follows a bell-shaped distribution with a mean caffeine content of $\mu = 120$ mg and a standard deviation of $\sigma = 4$ mg. Use the Empirical Rule to determine:
a) the percent of small cups of coffee that one would expect to have a caffeine content greater than 124 mg.
b) the percent of small cups of coffee that one would expect to have a caffeine content less than 112 mg.
c) the percent of small cups of coffee that one would expect to have a caffeine content between 116 mg and 128 mg.
d) the number of small cups of coffee (to the nearest cup) that one would expect to have a caffeine content of at most 116 mg.
e) the caffeine content that cuts off the top 16% of the distribution.
f) the percent of small cups of coffee one would expect to be at most two standard deviations above the mean.

62. The resting heart rates of 1000 female college students at a local community college is bell-shaped with a mean of 76 beats per minute (bpm) and a standard deviation of 7 bpm. Use the Empirical Rule to determine:
a) the percent of resting heart rates that have a z score less than 1.
b) the percent of resting heart rates that have a z score greater than –2.
c) the percent of resting heart rates that have a z score between –1 and 3.

63. Use Chebyshev's Theorem to answer the following questions: A machine is set to produce screws with a mean size of 0.42 mm and a standard deviation of 0.003 mm.
a) If screws smaller than 0.411mm or greater than 0.429mm are unusable, at most what percentage of the screws would one expect to be unusable?
b) Would a screw of size 0.435 mm be considered an unusual size? Explain.

64. Assuming the IQ scores of a distribution of 8,000 college-aged students is bell-shaped with a mean of 120 and a standard deviation of 8. Using the Empirical Rule, determine the approximate percentage and number of students who are expected to have IQ scores from:
a) 104 to 136.
b) 112 to 128.
c) 112 to 136.
d) 104 to 144.
e) Would you interpret an IQ score of 150 to be an unusual IQ or a typical IQ? Explain.
f) Would you interpret an IQ score of 90 to be a potential outlier? Explain.

65. The following figures represent the salaries of the chief executives of the given Environmental Groups.

Environmental Group	Salary
National Wildlife Federation	$242,060
Environmental Defense Fund	$213,504
Ducks Unlimited	$194,884
National Parks & Conservation Assoc.	$185,531
The Nature Conservancy	$185,000
World Wildlife Fund	$185,000
National Audubon Society	$171,154
The Wilderness Society	$ 90,896
Sierra Club	$ 77,142
Greenpeace Inc.	$ 65,000

a) Calculate the sample standard deviation of the salaries.
b) Use the appropriate range rule to estimate the standard deviation, s. Compare this estimate to the value computed in part (a).

66. Ellen, a stock broker, is studying the performance of two stocks over a three month period. Stock D had a mean

selling price of $13.38 with a variance of 5.62, while Stock E had a mean selling price of $98.13 with a variance of 61.62.
   a) Which stock has a greater relative variation in the selling price per share?
   b) Why should you consider using a measure of relative dispersion to compare these two stocks?

Section 3.4

67. The statistic that measures the relative position of an individual data value from the mean using standard deviation units is called a _____ .

68. The z score of a sample data value is calculated by the formula: z = ___.
    The z score of a population data value is calculated by the formula: z = ___.

69. A data value that is one standard deviation greater than the mean has a z score of _____.

70. The z score of a data value indicates how many standard deviations the data value is above or below the _____.
    a) Mean   b) Mode   c) Median

71. A data value that has a z score of zero is the _____.

72. For a distribution, $\bar{x}$ = 100 and s = 4.
    a) If x = $\bar{x}$, then z = _____.
    b) If x = 90, then z = _____.
    c) If z = 2, then x = _____.
    d) If z = 1.5, then x = _____.

73. If a raw score has a z score of 3, then the raw score is _____ standard deviations _____ than the mean.

74. The percentile rank formula for a data value, X, is calculated by:

    PR of the data value X = $\frac{[B + (½)E]}{N}$ (100)

    where: B = the number of data values _____ than X.
    E = the number of data values _____ to X.
    N = the _____ number of data values

75. The percentile rank of X indicates the percent of values within the distribution that are _____ than X.

76. A percentile rank is rounded off to a _____ number.

77. The pth percentile is the data value in the distribution which has _____ percent of the data values less than this data value and the rest of the data values greater than it.

78. Lorenzo's IQ score of 120 has a percentile rank of 90. This means that _____ percent of the individuals have IQ scores less than 120 and _____ percent have IQ scores greater than 120.

79. If P_{20} = 75, then we state that 75 is the 20th _____.

80. If the median IQ score is 100, then we can write:
    P__ = 100.

81. The percentile is a _____ of the distribution, while the percentile rank of a data value is a percentage value ranging from _____ to _____.

82. A z score is used to summarize a distribution's variability. (T/F)

83. a) Using the z score formula, calculate and interpret the z scores for each raw score given in the table below.
    b) Using the results of part a, which raw score would you consider to be a potential outlier? Explain.

Raw Score	Mean	Standard Deviation
i) 45	60	9
ii) 165	120	10
iii) 590	400	150
iv) 830	750	100
v) 1470	1470	210

84. a) Using the raw score formula, calculate the raw score for each z score given in the table below.
    b) Using the information in the table and the results of part a, state which raw score might be considered a potential outlier? Explain.

Mean	Standard Deviation	z Score
i) 120	50	+2.33
ii) 500	160	−1.96
iii) 75	25	+0.56
iv) 450	90	−3.58
v) 290	75	+1.65

85. According to FocusDriven, an advocate group for cell-free driving, texting on your cell phone while driving causes drivers to look away from the road for an average of 4.6 seconds. Assume the standard deviation is 0.52 seconds for writing a text while driving at 55 MPH.
    a) Find the z score of a driver who takes 6.03 seconds to write a text while driving at 55 MPH.
    b) Would you consider this an unusually high amount of time to text while driving? Explain.
    c) How many feet per second is a person traveling when driving 55 MPH?
    d) Would a person who is driving 55 MPH travel longer than a football field of 100 yards while looking at their cell phone if the individual took the average time of 4.6 seconds to write a text message? Explain.
    e) What would you tell your best friend about texting while driving?

86. Human body temperatures have a mean of 98.20 degrees F with a standard deviation of 0.62 degrees F. If an individual transported to the emergency room of a hospital has a temperature reading of 101.6 degrees F, would you consider this temperature to be an unusually high temperature? Explain.

87. The length of pregnancies has a mean of 268 days and a standard deviation of 15 days.
    Years ago, a female reader sent a letter to "Dear Abby", a widely syndicated column, claiming that she had given birth 315 days after a brief visit from her husband who was in the armed forces.
    a) Using this information, find the z score of a pregnancy lasting 315 days.
    b) Would you consider a pregnancy lasting 315 days an unusual pregnancy length?
    c) What might this result suggest?

88.
I. The z-score of a student's exam grade is −1.45. If the mean exam grade is 75, which of the following could be the value of the student's grade?
a) 64   b) 85   c) 75   d) cannot be estimated

II. The z-score of a student's exam grade is 0. If the mean exam grade is 65, which of the following is the value of the student's grade?
a) 100   b) 65   c) 0   d) cannot be determined

III. The z-score of a student's exam grade is +2.35. If the mean exam grade is 70, which of the following could be the value of the student's grade?
a) 70   b) 63   c) 87   d) cannot be estimated

89. Some of the ACT test scores of 300 college students are given in the following table:

ACT Score	13	15	21	23	25	31	33
z Score	−2.5	−2.0	−0.5	0	.5	2	2.5
Percentile Rank	10	15	25	40	50	75	99

By inspection of the table, determine the following:
a) mean ACT score of all 300 students
b) median ACT score of all 300 students
c) percent of students who have an ACT score below 23
d) standard deviation of all 300 students
e) ACT score representing the 1st decile
f) ACT score representing the 1st quartile
g) ACT score representing the 3rd quartile

90. Given the sample of temperatures:
90, 100, 79, 70, 95, 85
a) Find the mean temperature.
b) Find the standard deviation.
c) Convert each temperature to a z score.
d) Interpret each z score.

91. Assume you have recently taken a Biology exam and you earned a grade of 85. Your friend, who is in a different biology class, receives a grade of 77. Although you have a higher test score, based upon the information contained in the following table, you feel your friend's test score is more outstanding relative to her class. Using the table information, show that your claim is justified.

Student	Mean Bio Grade	Standard Deviation	Test Score
You	76	4	85
Friend	69	3	77

92. Some of the I.Q. scores of 250 college students are given in the following table.

I.Q. Score	102	104	110	112	114	120	122
z Score	−2.5	−2.0	−0.5	0	0.5	2	2.5
Percentile Rank	10	15	25	40	50	75	99

By inspection of the table, determine:
a) The mean I.Q. score of all 250 students.
b) The median I.Q. score of all 250 students.
c) The percent of students who have an I.Q. score below 112.
d) The standard deviation of all 250 students.
e) The I.Q. score representing the 1st decile.
f) The I.Q. score representing the 1st quartile.
g) The I.Q. score representing the 3rd quartile.

93. What percentage of a population of statistics test scores are:
a) Above the first quartile?
b) Below the third quartile?
c) Greater than the 43rd percentile?
d) Above the 7th decile?
e) Between the first and third quartiles?
f) Between the 3rd and 8th decile?

Section 3.5

94. A 5-number summary consists of the following values: ____, ____, ____, ____, and ____.

95. The Box-and-Whisker Plot is an ____ Data Analysis technique used to visually display several summary statistics to describe the location of the ____ of the data, the ____ and ____ of a distribution.

96. On a Box-and-Whisker Plot, the middle 50% of the data values fall between the ____ and ____ quartiles.

97. The line connecting the 3rd quartile to the largest value of a Box-and-Whisker Plot is called the ____ of the Box Plot.

98. The Interquartile Range (IQR) is the difference between the ____ and the ____ quartile. On a Box Plot, the IQR is represented by the length of the ____. The percent of data values that are represented by the IQR is ____.

99. If the length of the lower and upper whiskers are approximately equal and the median is positioned in the middle of the 1st and 3rd quartiles, then we would expect the shape of the distribution to be ____.

100. The left vertical line of the box of the Box Plot represents the ____.
a) Q_1
b) Q_3
c) Median
d) Lowest data value

101. An outlier is identified as any data value that falls outside of the ____ fences.
a) Extreme   b) Inner   c) Outer

102. An ____ is an extreme data value that may be a valid data value or it can be a data value due to an error.
a) Largest data value
b) Median
c) Mean
d) Quartile
e) Outlier

103. If $Q_1 = 45$ and $Q_3 = 68$, then the IQR is ____.

104. Tukey defines an inner fence to be a distance of ___*IQR from the 1st and 3rd quartiles. An outer fence is defined to be a distance of ___*IQR from the 1st and 3rd quartiles.

105. The Whiskers in a Box-and-Whisker Plot can be used to indicate the presence of data values that are outliers.
(T/F)

106. The 5-number summary uses the mean as a measure of the center of the data set.
(T/F)

107. When comparing the Box Plots of two distributions, then the distribution represented by the larger box has the greater variability within the middle 50% of the data values.
(T/F)

108. The following Box-and-Whisker Plots represent the average monthly temperatures of six different cities in the United States.

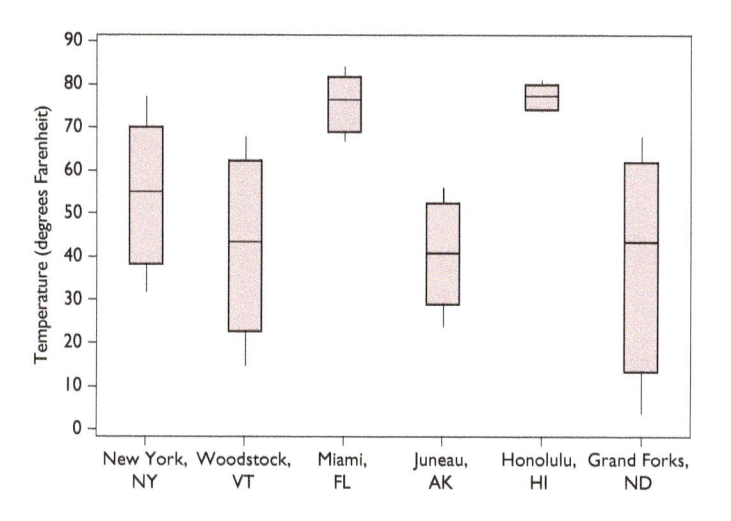

a) Which two cities tend to be the warmest?
b) Which city tends to have the lowest temperatures?
c) Which city has the least variation in temperature?
d) Which city has the most variation in temperature?
e) Identify the shape of the distribution of temperatures for Grand Forks, ND.

109. For the following distribution of statistics test grades:

19, 29, 40, 42, 53, 55, 58, 60, 61, 61, 62, 64, 65, 75, 76, 76, 77, 77, 78, 78, 79, 80, 80, 81, 82, 83, 83, 84, 85, 86, 86, 87, 88, 88, 89, 89, 90, 90, 91, 92, 93, 93, 94, 94, 95, 98, 98

a) Determine the 5-number summary for the statistics grades.
b) Construct a Box-and-Whisker Plot.
c) Describe the shape of the Box-and-Whisker Plot.
d) Calculate the interquartile range (IQR).
e) Determine if there are any outliers within the data set. If there is an outlier, state its value(s).
f) Find the z score of each if any exists and interpret their relative position in terms of standard deviations.
g) Explain what effect an outlier, if any exists, has on the mean and median grade.

110. According to a recent National Survey of Student Engagement, a higher education research center based at Indiana University-Bloomington, the average college freshmen study time is approximately 15 hours a week. A college mathematics instructor at an Arts & Humanities University surveyed her freshmen students to determine the number of weekly hours they spend studying for their introduction to statistics course. The following data set represents the sample weekly student study hour results.

14, 18, 17, 15, 16, 20, 41, 15, 14, 12, 15, 14, 11, 17, 18, 14, 13, 18, 16, 15, 14, 12, 13, 12, 19, 21, 11, 15, 11, 16, 17, 14, 16, 18, 14, 13, 15, 11, 15, 14, 16, 15, 17, 11, 16, 17, 15, 19, 15, 16

a) Compute the three Measures of Central Tendency for the distribution of weekly study hours.
b) Are the median and mean weekly study hours equal or different? What might this indicate about the shape of the distribution?
c) Construct a Box Plot for the weekly student study hours.
d) What is the shape of the distribution of weekly student study hours?
e) Is there an unusual data value? If so, what might you call this data value? Why?
f) How would the Box Plot help to answer part (e)?
g) Compare the National average study hour to the sample average study hour. Now, remove the outlier from the distribution of weekly study hours and compare the National average study hour to the sample mean study hour. How did the outlier impact the sample mean study hour?
h) Compute the standard deviation for the original distribution of weekly study hours.
i) What does the standard deviation tell you about the distribution of weekly study hours?
j) Compute the z score of the largest data value. What does this z score tell you about this data value?
k) How might you use a z score to identify an outlier?
l) Remove the outlier from the distribution of study hours and recalculate the standard deviation. What impact did the outlier have on the standard deviation?

111. The following **samples** of raw scores represent the number of work hours for male and female students per week.

Male Students

45, 23, 27, 23, 30, 27, 0, 30, 18, 36, 15, 60, 30, 38, 26, 30, 36, 8, 55, 50

Female Students

29, 15, 18, 20, 15, 28, 19, 17, 29, 14, 20, 28, 34, 69, 14, 20, 27, 29, 12, 16

For each sample:
a) Construct a Box-and-Whisker Plot.
b) State the shape.
c) Determine the first quartile, the median and the third quartile.
d) Determine if there are any potential outliers. If there are any potential outliers, state their values.
e) Find the z score of each outlier if any exists and interpret its relative position in terms of standard deviations.
f) Explain what effect, if any, these outliers have on the mean and median number of work hours.

CHAPTER REVIEW

112. The mean, median, and mode are known as measures of relative position.
(T/F)
113. The mode of a distribution is the data value which appears:
a) Least   b) Most   c) In the middle
114. The symbol used to represent the mean of a sample is:
a) n   b) x   c) μ d) $\bar{x}$
115. A distribution containing more than two modes is called a _____ distribution.
116. When a distribution with an odd number of data values is arranged in numerical order, the median is the _____ data value of the distribution.
117. The formula for the sample mean is: _____ while the formula for the population mean is _____.
118. $\bar{x}$ is often used to estimate the _____ of a population.
a) Mean   b) Median   c) Mode
119. If a mean is computed using only a portion of the population data, then this statistic is called the _____ mean.
120. The mean is always one of the data values of a distribution.
(T/F)
121. In a distribution with an odd number of data values, the mean is 100, and the median is 75. Fifty percent of the data are less than the data value _____.
122. If $\Sigma(x-3)=0$, then 3 is _____.
123. The mean is a more resistant measure of central tendency than the median.
(T/F)
124. In a skewed to the right distribution, the median is located to the left of the mean.
(T/F)
125. For the following distribution:
25, 20, 8, 7, 9, 6, 8, 4, 5, 8

Which of the three measures of central tendency will change if we remove one of the data values from the above distribution?
a) Mean
b) Median
c) Mode
126. For the sample: 12, 9, 7, 13, 9, 8, 10, 9, 11, 12, find:
1. the mean and state the appropriate symbol to represent this mean.
2. the median
3. the mode
4. $\Sigma(x-10)$
5. Σx^2
6. $(\Sigma x)^2$
7. $\Sigma(x-\bar{x})^2$
8. $[\Sigma(x-\bar{x})]^2$
9. $\Sigma x^2 - (\bar{x})^2$
127. In a population, if all the data values have the same numerical value of 14, then:
Mean = _____
Mode = _____
Median = _____
Range = _____
Variance = _____
Standard deviation = _____

128. The following data values represent a sample of household incomes of Manhattan residents:
$50,000, $19,000, $20,000, $20,000, $800,000, $750,000 $100,000, $90,000, $85,000, $11,000, $20,000
a) Determine the mean, median, and mode for these incomes.
b) Why are the mode and mean not good measures of central tendency for this distribution?
c) What symbol would you use to represent the mean?
129. The following table represents a summary of the final exam scores for all the students at a small community college.

Final Exam	Number of Students
90	2
87	1
86	2
85	5
84	2
83	6
82	8
79	1
76	3
66	7
60	5
20	4
10	5
0	1

a) Determine the mean, median, and modal test grade.
b) What symbol would you use to represent the mean?
c) Which measure of central tendency is less resistant to the extreme data values? Explain.
130. Twenty-four homeowners on Long Island reported the following selling prices for their homes: two homes at $11,000,000, four homes at $5,500,000, four homes at $1,000,000 five homes at $600,000, two homes at $450,000, and nine homes at $350,000.
a) Determine the mean, median, and modal price.
b) Which measure is a poor measure of central tendency? Explain.
131. The following data represents a psychiatrist's rating of fifty severely depressed mental patients' responses to drug therapy.

Response to Drug Therapy	Number of Patients
Excellent	12
Very Good	17
Good	16
No Response	5

The psychiatrist claimed that "very good" was the typical response to this drug. Explain which measure of central tendency she used, and why it is appropriate in this situation.

132. A sociology professor wants to get an estimate of the "average" age of the students in his large lecture. The professor randomly selects 10 students from his class and records their ages. The ages are as follows:

19, 18, 18, 22, 19, 58, 20, 23, 21, 22

a) Compute the mean age for this sample.
b) At the next class meeting the professor states the average age to the class. One of the students is startled at the response to the professor's question: "Who is 24 or older?" Almost no one responded. Can you explain this?

133. A student states that his cumulative grade point average for the past three semesters is 3.0. The student's reasoning is as follows: "I earned a 3.5 average for 4 credits during the summer semester, a 3.0 average for 12 credits in the fall and a 2.5 average for 16 credits in the spring." Is the student correct in his reasoning? Justify your answer.

134. Fourteen psychology exam scores and their respective deviations from the mean of a class of 15 students are listed in the following table. The exam score for student number 7 is missing. Using the information in the table, find the missing psychology score. (Hint: Use the property $\Sigma(x - \bar{x}) = 0$).

Student #	Psychology Score	Deviation from Mean
1	125	2
2	130	7
3	115	−8
4	126	3
5	132	9
6	127	4
7		
8	118	−5
9	128	5
10	134	11
11	126	3
12	119	−4
13	118	−5
14	115	−8
15	125	2

135. The gender and hair and eye color of twenty babies born at a county hospital during the previous week are listed in the following table.

Gender	Hair Color	Eye Color
M	BROWN	BROWN
M	BROWN	BROWN
F	BLONDE	BLUE
F	RED	BROWN
M	BROWN	BROWN
M	BLONDE	BLUE
F	BROWN	BROWN
F	BROWN	BLUE
F	BROWN	BROWN
M	BLACK	BROWN
F	BLONDE	BLUE
M	BLACK	BROWN
M	BROWN	BLUE
M	BROWN	BROWN
F	BROWN	BLUE
F	BROWN	BLUE
F	BLACK	BROWN
F	RED	BLUE
M	BLACK	BROWN
M	BROWN	BLUE

a) Estimate the typical:
  1) hair color.
  2) eye color.
  3) characteristics for a male baby.
  4) characteristics for a female baby.
b) Which measure of central tendency did you use to calculate the answers to the part (a), and explain why you chose this measure?

136. An exercise equipment manufacturer states that their new electronic treadmill will burn an *average of 900 calories for a thirty minute workout*. The following sample data represents the number of calories burned per individual for twenty men aged 18–24 years.

930, 890, 920, 910, 850, 870, 895, 880, 940, 830
825, 890, 925, 855, 960, 860, 915, 830, 920, 875

a) Calculate the sample mean number of calories burned per individual.
b) Does the sample mean calculated in part (a) equal the average claimed by the equipment manufacturer? If there is a difference, what might be an explanation for this difference?
c) If an individual loses one pound for every 3500 calories burned, based on the sample mean calculated in part (a), determine the mean number of pounds lost per individual for a thirty minute workout.
d) If an individual works out 30 minutes a day for 30 days on this treadmill, determine the expected number of pounds lost.

137. A manufacturer of Omega 3 supplements claims that on average, each capsule contains 434 mg. of Omega 3. A health research group decides to test ten capsules to determine if the manufacturer's claim is accurate. The following list represents the group's sample data:

Omega 3 content per capsule (mg)
401, 259, 451, 430, 435, 442, 436, 422, 430, 434

a) Calculate the sample mean amount of Omega 3 content per capsule.
b) Calculate the difference between the sample mean Omega 3 content and the manufacturer's population mean. Does the health research group have reason to dispute the manufacturer's claim?
c) The health research group will decide to dispute the manufacturer's claim if the difference between the sample and population mean is greater than 20 mg. Does the group have reason to dispute the manufacturer's claim?

138. Using the following figure which represents a skewed distribution, identify the letter which represents the position of:
a) The mode
b) The mean
c) The median
d) Identify the shape of the distribution.
e) In which tail would you expect to find an outlier? Explain.

f) Which Measure of Central Tendency would be closest to a possible outlier? Explain.

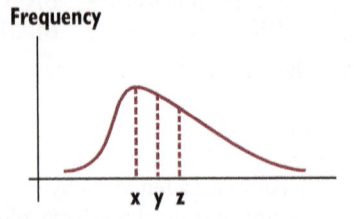

Frequency

x y z

139. For each of the following distribution of data values,
   a) Identify the general shape that you believe best describes each distribution, and
   b) Identify whether the mean or median would have the largest value within this type of distribution.
   (i) The heights of US female adults.
   (ii) The heights of all the male college students playing basketball within the state of North Carolina.
   (iii) The annual income of people living in the United States.
   (iv) The number of text messages sent per day by high school students.

140. Which measure of variability is determined only by the extreme values of a distribution?
   a) Range   b) Variance   c) Standard deviation

141. An estimate of a population standard deviation is denoted by the symbol _____.

142. If the variance of a distribution is 25, and n = 8 then $\Sigma(x - \bar{x})^2 =$ _____.

143. For the data 3, 5, 8, 13, 21 compute:
   a) The range
   b) The sample variance, s^2
   c) The sample standard deviation, s
   d) An estimate of the population standard deviation
   e) The population standard deviation, assuming the data represents a population.

144. If a distribution consists of all negative data values, the standard deviation is also negative. (T/F)

145. Two distributions that have the same range are not necessarily the same. (T/F)

146. If the range of distribution x is denoted by R_x and every data value in this distribution is increased by adding seven to form a new distribution y with a range denoted as R_y, then _____.
   a) R_x is larger than R_y.
   b) R_x equals R_y.
   c) R_y is larger than R_x.

147. A distribution that has a range equal to zero must also have a standard deviation equal to zero. (T/F)

148. A recent survey by The Nielsen Company stated the typical U.S. mobile subscriber sends and receives an average of 12 text messages a day while people within the 18 to 24 age group send and receive an average of 26 text messages. Matthew, a statistics student, decides to randomly text 15 college students during his 75 minute statistics class to determine the average number of text messages college students send and receive on a daily basis. Here are Matthew's survey results.

Text Messages Sent and Received Per Day by College Students:
24, 45, 28, 30, 21, 43, 17, 8, 26, 35, 27, 23, 29, 15, 37
   a) Compute the sample mean (to 1 decimal place) as an estimate of the average text messages sent and received by college students on a daily basis.
   b) Compute the sample standard deviation (to 1 decimal place) to estimate the variability in the number of text messages sent and received by college students per day.
   c) Based on these results, what would you estimate the average number of monthly text messages to be for a college student?

149. A travel agency, trying to promote travel to two ski resorts, base their ad campaign on the following sample temperature data: (data measured in degrees Fahrenheit)

Temperature Data
(in degrees Fahrenheit)

Resort 1	34	37	15	38	42	−16
Resort 2	25	32	16	20	27	30

The ad emphasizes the mean temperature. Observe that the two temperature distributions are not identical.
   a) Calculate the standard deviation for each resort.
   b) Using the results from part (a), explain which resort has the more consistent weather.

150. Craig has recorded prices for a statistical calculator at three discount electronic stores during the last three months. Based on this data, Craig calculated the mean and standard deviation of the calculator prices for each store.

Discount Store	Mean Price	Standard Deviation
store 1	$105.00	$26.25
store 2	$105.00	$12.50
store 3	$105.00	$0

During the past 3 months, in which store did Craig have a better chance to purchase the calculator at a cheaper price?

151. An award is going to be given to the player on a women's college basketball team who has shown the most consistent performance over the entire season. The coach feels that consistent play is a very important factor in team performance. To measure consistency, she decides to use the standard deviation for those players that have scored a mean of 10 points or more per game. What follows are the scoring distributions for all the players on the team:

Games

Player	1	2	3	4	5	6	7	8	9	10
Lucy	7	5	0	16	X	7	4	0	7	8
Allison	12	17	X	10	5	12	0	20	X	7
Lilia	16	18	19	X	15	20	20	14	17	14
Kathleen	9	12	7	14	8	5	4	16	18	7
Debbie	28	18	24	19	22	17	24	12	20	16
Ellen	15	3	9	8	9	13	7	18	9	19
Grace	11	8	20	25	17	16	8	5	10	10

X denotes the player did not play in that game.
   a) Calculate the population mean and standard deviation of each player.
   b) Which player gets the award for the most consistent performance? Explain.

152. A fast food chain samples 10 containers of small french fries from each of two different locations to check the output quality of its production process. The containers are set to have 25 french fries each. The actual number of french fries which were put in each container is recorded in the following table.

Number of Fries at Location One	Number of Fries at Location Two
24	24
24	25
25	25
23	26
26	24
26	23
25	25
25	25
24	24
23	24

a) Using these results, compare the output quality of the two locations by computing the sample mean number of fries given in each container at each location as your standard.
b) Compute the sample standard deviation, s, for each location's output.
c) Using this additional information, compare the output quality of the two locations.

153. If the data values of two distributions have different units, then you should use the standard deviation rather than the coefficient of variation as a measure of variability.
(T/F)

154. In a bell-shaped distribution, if a data value falls beyond 3 standard deviations from the mean, then we might consider this data value to be a:
a) Typical data value
b) Data value smaller than the mean
c) Greater than the mean
d) An outlier

155. Chebyshev's Theorem can only be applied to a bell-shaped distribution.
(T/F)

156. When comparing the variability of two distributions the coefficient of variation is a meaningful measure of variability when the data values are measured in _____ units or when the _____ of the distributions are far apart.

157. During last December in Boston, 40 degrees was the high temperature for the month while 28 degrees was the low temperature for the month. Using this information, an approximation of the variability for the month of December is:
a) 4.8 degrees
b) 4 degrees
c) 3 degrees
d) 2 degrees

158. Use Chebyshev's Theorem to answer the question: A bus company's records indicate that its buses complete their routes 9.3 minutes late on the average with a standard deviation of 2.1 minutes. At least what percentage of its buses complete their routes anywhere between:
1. 5.1 minutes and 13.5 minutes late?
2. 3.0 minutes and 15.6 minutes late?
3. 0.9 minutes and 17.7 minutes late?

159. A credit score agency randomly surveys 15 college students to determine the mean balance for each student's monthly credit card. The result of the survey produced the following sample.
Data values for 15 college students monthly credit card balances (measured in dollars per month)
550, 850, 380, 1035, 870, 985, 1080, 885, 890, 1090, 900, 1050, 1200, 1100, 800
Using these data, calculate:
a) $\bar{x}$ and s.
b) The percent of monthly credit card balances within 2 standard deviations of the mean.
c) Using Chebychev's Theorem, what percent of the data values are within 2 standard deviations of the mean?
d) Compare parts (b) and (c). Are these results consistent with Chebychev's Theorem?

160. During last year's Marathon on Long Island, the average time to run the marathon was 225 minutes with a standard deviation of 25 minutes.
a) What can be said about the percentage of runners that took more than 275 minutes to finish the Marathon, if you have no information about the shape of this distribution?
b) What is the approximate percentage of runners that took more than 275 minutes to finish the Marathon, if the distribution has an approximate bell-shaped distribution?
c) If the distribution is approximately bell-shaped, then which of the following times might be considered a potential outlier: 230, 275 or 375 minutes? Explain.

161. The president of Debbie's Daytime Secretarial school is trying to determine which computer processing training program to use. Two groups were trained for the same task each using a unique training program. Group 1 used training program A; group 2 used training program B. It took the first group an average of 25 hours with a standard deviation of 6 hours to train each student. Whereas, the second group took an average of 16 hours with a standard deviation of 5 hours to train each student. Which training program has less relative variability in its performance?

162. An investor is examining the past performance of three growth mutual funds which invest primarily in small-sized companies. During the past 5 years, the average return on investment and standard deviation of the investment return for each mutual fund are given in the following table.

Mutual Fund	Average Five Year Return	Standard Deviation
Fund A	27.6%	6.47%
Fund B	25.7%	5.91%
Fund C	31.8%	6.92%

a) Which fund had a greater absolute variability in return over this 5 year period?
b) If this investor considers risk to be associated with a greater relative dispersion, then which of these three mutual funds has pursued a riskier strategy?
c) If this investor decides to narrow the selection down to mutual fund B and C because they both belong to excellent mutual fund families, then which fund should the investor choose if he is interested in a less risky investment strategy?

163. The heights of women ages 18 to 24 are approximately bell-shaped with $\bar{x} = 64.5$ inches and s = 2.5 inches. According to the Empirical Rule, approximately what percentage of the women heights would you expect to fall between:
   a) 62 to 67 inches?
   b) 59.5 to 69.5 inches?
   c) 57 to 72 inches?
   d) If there was a height equal to 77 inches, could you consider such a height as an outlier? Explain.

164. For a sample containing 90 data values, the range is 55. Using Table 3.14, determine an estimate of the standard deviation, s.

165. According to the Census Bureau, the following 39 data values represent the average commuting times in metropolitan areas with more than 1 million people.
   30.6, 29.5, 28.1, 26.4, 26.1, 26.0, 26.0, 25.6, 24.4, 24.3, 24.2, 24.1, 24.1, 24.1, 23.4, 23.1, 23.0, 22.9, 22.6, 22.4, 22.2, 22.1, 22.0, 21.9, 21.9, 21.8, 21.8, 21.7, 21.6, 21.6, 21.4, 21.2, 21.1, 20.6, 20.0, 19.8, 19.7, 19.7, 19.4
   a) Calculate the sample standard deviation, s, of the commuting times.
   b) Use the range rule in Table 3.14 to determine an over-estimate and an under-estimate of the standard deviation, s. Compare these estimates to the value computed in part (a).
   c) Determine the percent of data values that are within 2 standard deviations of the mean.
   d) The percent of data values that are within 3 standard deviations of the mean.
   e) Compare the results of parts (c) and (d) to the results of Chebyshev's Theorem.

For questions 166 through 169, use the following information:

At Jefferson High, the final average with a percentile rank of 50 was 82. Mike's average has a percentile rank of 74. Pete's average is equal to the 9th decile. Maria's average has a percentile rank of 94. Joe's average is between the first and second quartile. Susan's average was less than 82. Jeff's average has a percentile rank which is lower than Maria's but greater than Mike's.

166. Of those students mentioned above, who had the highest average?
   a) Joe   b) Maria   c) Mike   d) Susan

167. What is the median average?
   a) 74   b) 94   c) 82   d) 50

168. Of those students mentioned above, how many had an average less than 82?
   a) 0   b) 3   c) 1   d) 2

169. Whose average was equal to the 90th percentile?
   a) Maria   b) Mike   c) Pete   d) Susan

170. If a data value is two standard deviations less than the mean, it has a z score of 2.
   (T/F)

171. The 7th decile is the data value, X, within a distribution that has the property that 70 percent of the data values are less than X and 30 percent of the data values are greater than X.
   (T/F)

172. Which measure of variability is used in calculating the z score of a data value?
   a) Range
   b) Variance
   c) Standard deviation

173. Which of the following indicates the relative position of a data value within its distribution?
   a) Mean
   b) Raw score
   c) z score
   d) Standard deviation

174. The _____ indicates the number of standard deviations a data value is greater than or less than the mean.

175. In a population, if x = 545, μ = 525 and z = 5, then σ = _____.

176. In a distribution, $\bar{x}$ = 75, and s = 5. If z = 3 is the maximum z score value of a data value in the distribution, and z = –2 is the minimum z score value of a data value in the distribution, then, the range of the distribution is equal to _____.

177. If P_{50} = 100, then we state that 100 is the 2nd _____.

178. If a data value has a z score of zero, this indicates that the data value is the _____.
   a) Standard deviation
   b) Median
   c) Mean
   d) None of these

179. If two distributions have equal means, then the data values that have z scores equal to 1 will always be the same.
   (T/F)

180. The 40th percentile is the data value, X, in a distribution that equals 40.
   (T/F)

181. a) Using the z score formula, calculate and interpret the z scores for each raw score given in the table below.
   b) Using the results of part a, identify which raw score might be considered a potential outlier. Explain your reasoning.

Raw Score	Mean	Standard Deviation
i) 24	16	8
ii) 65	100	10
iii) 440	460	15
iv) 875	450	100
v) 1150	1150	120

182. a) Using the raw score formula, calculate the raw score for each z score given in the following table.
   b) Using the table information and the results of part a, state the raw score that might be considered a potential outlier. Explain.

	Mean	Standard Deviation	z Score
i)	20	5	+4.51
ii)	50	6	–1.65
iii)	110	15	+0.95
iv)	500	100	–2.33
v)	90	20	+1.96

183. DEB Homes, Inc., a Long Beach real estate agency, sold three condominiums during the past week. Condo M, a waterfront condo, sold for $305,000 while Condo A, on the canal side of town, sold for $275,000 and Condo T, in the heart of town, sold for $215,000.

In Long Beach, the purchase price of a condo is dependent upon its location. City Hall calculates the mean price of a waterfront condo to be $245,000 (with s = $50,000), while the mean price of a canal side condo is $230,000 (with s = $23,000) and a town condo to have a mean price of $190,000 (with s = $15,000). Based upon its geographic location, which of the three condos was sold for the relatively:
a) Higher purchase price?
b) Lower purchase price?

184. Michael and Lorenzo both took the same calculus test. Michael scored an 87 on the test. Lorenzo's test paper only had the z score of his grade which was +1.2. The instructor gave Lorenzo the following information about the distribution of test grades: the mean test grade is 76 and the standard deviation is 15. Who had the higher grade, Michael or Lorenzo?

185. Two track athletes, a sprinter and a long jumper, break their school's record in their respective events. To determine which feat is more outstanding relative to his respective event, the school's statistician collects the best performances for each event within the United States and lists the necessary data below:

Athlete	Event	Record	Mean	Standard Deviation
sprinter	100 m run	10.3 sec	10.6 sec	0.15 sec
long jumper	long jump	7.82 m	7.62 m	0.13 m

Which of the two athletic records is more outstanding?

186. In certain manufacturing processes, parts must be machined to within specified tolerances. The parts must have a mean length of six inches plus or minus 0.26 inches. A sampling procedure has been established to determine whether the parts are within the specified tolerances. If for a sample of ten parts selected at random, the mean plus 1.5 standard deviations and the mean minus 1.5 standard deviations are within the specified limits of 5.74 inches to 6.26 inches, then the parts are meeting the specified tolerances. This means that a z score of 1.5 and a z score of −1.5 must be within the limits of 5.74 inches and 6.26 inches. The lengths (in inches) representing the ten parts are as follows:

5.94, 6.20, 5.86, 6.08, 6.14, 6.12, 6.01, 5.92, 5.90, 6.23

Procedure:
a) Calculate the mean, $\bar{x}$, and standard deviation, s.
b) Convert z = −1.5 and z = 1.5 into inches (data values).
c) Compare these values with the specified limits of 5.74 inches to 6.26 inches and determine if this sample meets the stated conditions.

187. Using the following collection of 50 IQ scores, compute the percentile rank of the IQ scores and interpret each result:
a) 87 b) 91 c) 105 d) 115 e) 119 f) 121

IQ Scores
81, 85, 85, 86, 87, 87, 88, 89, 89, 90, 90, 90, 91, 93, 93, 94, 95, 97, 98, 99, 100, 101, 102, 104, 105, 105, 107, 107, 110, 112, 112, 113, 114, 115, 115, 115, 115, 119, 120, 121, 121, 121, 122, 123, 125, 125, 126, 126, 127, 135

188. Using the results of problem #187, determine the IQ score representing the:
a) 7th decile
b) 3rd quartile
c) 2nd quartile
d) 1st decile
e) median
f) 1st quartile

189. The reaction time of 250 people is measured after a new drug is administered. Some results of this test are given in the following table.

Reaction Time (sec)	8	12	16	20	28	36	40
z score	−1.5	−1	−0.5	0	1	2	2.5
Percentile Rank	17	25	30	42	50	75	85

By inspection of the table, determine:
a) The mean reaction time for all 250 people.
b) The median reaction time for all 250 people.
c) The standard deviation time for all 250 people.
d) The percent of people who had reaction times 40 secs or below.
e) The reaction time representing the 3rd decile.
f) The reaction time representing the 1st quartile.
g) The reaction time representing the 3rd quartile.

190. The following table indicates the performance of six students on the law boards. Using this information, compute the percentile rank of each student if 10,000 students took the law boards.

Student I.D.	Number of students who scored less than this student	Number of students who scored the same as this student (including student)
L	2,250	500
A	2,900	200
W	4,800	400
Y	7,475	50
E	8,000	100
R	9,000	25

191. To qualify for a scholarship at Richman's College, a student must score at least 1.5 standard deviations above the mean on the entrance exam. Which of the following students qualify if the mean and standard deviation for the entrance examination is 60 and 12 respectively?

Student	Score on Exam
George	75
Alice	78
Barbara	81
Mauro	77
Mike	79

192. A male is considered to be hypertensive for his age group if he has a systolic blood pressure of more than two standard deviations greater than the mean.

The following table contains the mean and standard deviation for systolic blood pressures for some male age groups.

Male systolic blood pressure parameters

Age	Mean	Standard Deviation
18	121	8
25	125	8
35	127	9
45	130	13
60	142	14

The following readings of systolic blood pressure were taken for a sample of males.

Male	Age	Systolic Blood Pressure
Craig	18	134
Jim	25	142
Mike	35	142
Lou	60	150

Determine which (if any) of the males are classified as hypertensive.

193. A 5-number summary is used to describe the shape, center and _____ of a data set.
a) Mean
b) Spread
c) Variance
d) Percentile

194. The percent of data values represented by the IQR is:
a) 25%   b) 50%   c) 75%   d) 100%

195. A sample of the number of minutes spent on Facebook per week of college students is collected. Some of the sample results of this survey are given in the following table.

Minutes Spent on Facebook	35	60	135	160	185	260	285
z Score	−2.5	−2.0	−0.5	0	.5	2	2.5
Percentile Rank	10	25	35	40	50	75	95

By inspection of the table, determine:
a) the mean time spent for all students
b) the median time spent for all students
c) the percent of college students who spend between 135-260 minutes per week
d) the minutes spent representing the 4th decile
e) the number of minutes representing the 1st quartile
f) the number of minutes representing the 3rd quartile
g) the interquartile range
h) the standard deviation number of minutes for all the students
i) the number of minutes corresponding to a z-score of 1.75
j) the number of minutes that is 2.25 standard deviations above the mean

196. The length of the upper whisker represents approximately the top _____ percent of the data values.

197. According to Tukey, a potential outlier is a data value that is less than the _____ fence but greater than the _____ fence.

198. Outer fences are located at (_____)*IQR below Q_1 and above Q_3.
a) 1.5   b) 2   c) 3   d) 4

199. If the IQR = 40 then the outer fences are located a distance of _____ below Q_1 and above Q_3.
a) 60   b) 80   c) 100   d) 120

200. If the length of the lower whisker is longer than the upper whisker and the position of the median is closer to the 3rd quartile, then we would expect the shape of the distribution to be _____.

201. If the largest data value within a distribution is an outlier, then the percent of data values between Q_3 and the largest data value is greater than the percent of data values between Q_1 and the smallest data value.
(T/F)

202. An instructor from the school of business samples the students in his introduction to business class to learn about the amount of dollars students spend per semester on textbooks. The distribution that follows indicates the results of his sample.

Textbook Costs per Student per Semester (in dollars)

65 147 171 142 153 187 195 106 127 178 178 205 175 178 133 186

a) Determine the 5-number summary for the textbook spending data.
b) Construct a Box-and-Whisker Plot.
c) Describe the shape of the Box-and-Whisker Plot.
d) Calculate the interquartile range(IQR).
e) Determine if there are any outliers including potential within the data set. If there is an outlier, state its value(s).
f) Find the z score of each if any exists and interpret their relative position in terms of standard deviations.
g) Explain what effect an outlier, if any exists, has on the mean and median textbook cost.

203. The following distributions represent the number of homeruns that Roger Maris and Babe Ruth hit during the designated years.

The number of homeruns hit by Roger Maris during each baseball season for the years 1957–1968

14, 28, 16, 39, 61, 33, 23, 26, 8, 13, 9, 5

The number of homeruns hit by Babe Ruth during each baseball season for the years 1919–1934

29, 54, 59, 35, 41, 46, 25, 47, 60, 54, 46, 49, 46, 41, 34, 22

For each homerun distribution, determine:
a) The 1st quartile, median, and 3rd quartile.
b) Construct a Box-and-Whisker Plot for each distribution on the same scale.
c) Compare the Box Plots, and discuss the differences between the two distributions.
d) Use the Box Plots to compare the variability of each distribution. Which distribution has the smaller variability? the larger variability? Now, calculate the Interquartile

Range and the standard deviation of each distribution. Compare the IQR and the standard deviation of each distribution. Are these results similar? With respect to variability, what is the relationship between the value of the IQR and the standard deviation of each distribution?
e) Identify any potential outliers for each distribution, and give an explanation for the outlier. Find the z score of any outlier and interpret the meaning of the z score.
f) Describe the shape of each distribution.

204. The following two Box-and-Whisker Plots represent the test results of a statistics final for two different instructors.

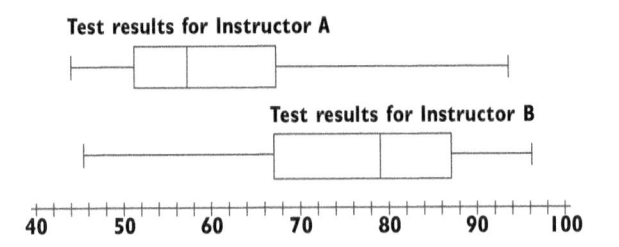

Test results for Instructor A

Test results for Instructor B

From each of the previous two Box Plots representing the test results of instructors A and B, approximate the:
a) 1st quartile test score.
b) 3rd quartile test score.
c) Median test score.
d) Smallest and largest test score.
e) Interquartile range(IQR).
f) Shape of the distribution.
g) Percent of test results to the left of the 3rd quartile.
h) Percent of test results to the right of the median.
i) Percent of test results between the 1st and 3rd quartile.
j) Compare the two test distributions and explain the differences about the test results.
k) Each instructor had a different objective in mind when constructing the final exam. One instructor was trying to determine the top students in the class, while the other instructor was trying to determine the minimum knowledge of the poorest students. Based on the Box Plots, explain which instructor you believe was trying to determine the top students in her class? Which test do you believe was the easiest test? Explain.

205. The following MINITAB output represents a stem-and-leaf display for 42 freshmen enrolled at a small community college. The display represents the number of hours worked per week for both male and female students.

```
MTB  >  Stem-and-Leaf 'hrswrk';
SUBC >  Increment 10.
Stem-and-leaf of hrswrk    N = 42
Leaf Unit = 1.0
    4   0  0000
   12   1  00005555
   21   2  000555555
   21   3  0005555
   14   4  0000005555
    4   5  005
    1   6  5
```

a) Draw the shape of the display using a smooth curve.
b) Indicate, by drawing a line, where the mean, median and mode fall.
c) Discuss the shape of the distribution in terms of the relationship between the location of the mean, median and mode within the distribution.

206. A 2010 study by the Nielsen Company revealed that American women tend to use their mobile phones more frequently than men for talking and texting. To check this claim, a sociologist researcher randomly sampled 30 women and 30 men who use their mobile phones. The following data sets represent the number of voice minutes per month for each sample.

Women Voice Minutes Per Month	Men Voice Minutes Per Month
820	630
836	650
770	550
886	660
844	670
812	670
778	640
868	650
893	645
763	635
788	631
848	649
958	633
798	647
773	628
753	652
748	642
738	638
860	655
826	625
823	610
814	670
797	644
839	636
830	652
805	628
835	631
801	649
827	647
809	633

a) Compute the mean and median voice minutes usage per month for both the women and men.
b) Compute the standard deviation of the voice minute usage per month for both women and men.
c) Compute the z score of the woman who spent 958 minutes on her cell phone that month and interpret the z score. What might you conclude about this data value?
d) Construct a Box Plot of the distribution of voice minutes per month for the women.
e) Identify the shape of the women voice minute usage. What does this shape indicate about a possible outlier?
f) Compute the z score of the man who spent 550 minutes on his cell phone that month and interpret the z score. What might you conclude about this data value?

g) Construct a Box Plot of the distribution of voice minutes per month for the men.
h) Identify the shape of the men voice minute usage. What does this shape indicate about a possible outlier?
i) Compare the Box Plots for both samples and draw a conclusion.
j) On average, what percent of minutes do women spend on their cell phones more than men for the month?

207. The data values in the table compare the amount of dietary fiber for three different cereal types.

Amount of Dietary Fiber for Cereal Type (measured in grams)

Cereal	Whole-Grain Cereal with Bran	All-Bran Cereal
3.0	5.0	13
2.1	4.0	12
2.0	4.0	10
2.0	4.0	9
2.0	3.1	8
1.6	2.9	5

For each cereal type:
a) Construct a Box-and-Whisker Plot and identify its shape.
b) Numerically describe the data.
c) Find the 5-number summary.
d) Identify potential outliers.
e) Write a description comparing the similarities and/or differences for the cereal data.

WHAT DO YOU THINK?

208. The following letter appeared in the *New York Times* on October 3, 1996.

Misusing Statistics

To the Editor:
  In his Sept. 29 letter defending Bob Dole's tax plan, Senator William V. Roth Jr. says that "the average American family pays more in taxes than it spends on food, shelter and clothing." This exemplifies a common misuse of statistics. While it is probably an accurate statement if you add up the taxes paid and divide it by the number of taxpayers, it is not true for typical taxpayers. The more meaningful average is the median, that is, the taxes paid by the taxpayer at the midpoint. That middle taxpayer's taxes do not come close to equaling the cost of food, shelter, and clothing.
  The fallacy of Senator Roth's approach reminds me of the excellent opportunities that were available since "the average employee in our firm makes $100,000 a year." What he didn't tell them was that he made $910,000 and his nine employees made $10,000 each.
      Jerome Trupin
      Scarborough, N.Y., Sept. 29, 1996

Reprinted from the *New York Times*, October 3, 1996, by permission of Jerome Trupin.

a) What does the author mean by the "average American family"?
b) What average does the author believe Senator William V. Roth Jr. used in his article? Why do you think this average is not true for the typical taxpayer in this particular example?
c) What does the author believe is the relationship between the median taxpayer's taxes and the average taxpayer's taxes? Based on the author's opinion, what do you think would be the shape of the distribution of taxes? Explain.
d) If the distribution of taxes has an outlier, then which measure of central tendency do you think is affected the most by the outlier? Would it increase or decrease this measure of central tendency? Explain.
e) Explain the fallacy in the example the author refers to when he states that "the average employee in our firm makes $100,000 a year." Explain which measure of central tendency would have been more appropriate in the example the author gives.
f) Do you think that the standard deviation of this distribution of taxes is a large or a small value? Explain.

209. For each of the following graphics,
a) Define the population and identify the variable being presented.
b) Discuss which measure of central tendency might have been used or was used to describe the variable within the graphic.
c) Explain why you believe this measure of central tendency is or is not the appropriate measure to use to describe the data.
d) State whether the number stated within the graphic represents a statistic or a parameter. Explain.
e) If possible, draw a conclusion using the information within the graphic.

(a) **Statistical Snapshot**

Are You STILL in there?
How long we spend in the bathroom during our lifetime:

- Women: 2.75 years
- Men: 2.1 years

Source: KRC Research and Consulting for Northern Bathroom Tissue (projected based on average life expectancy)

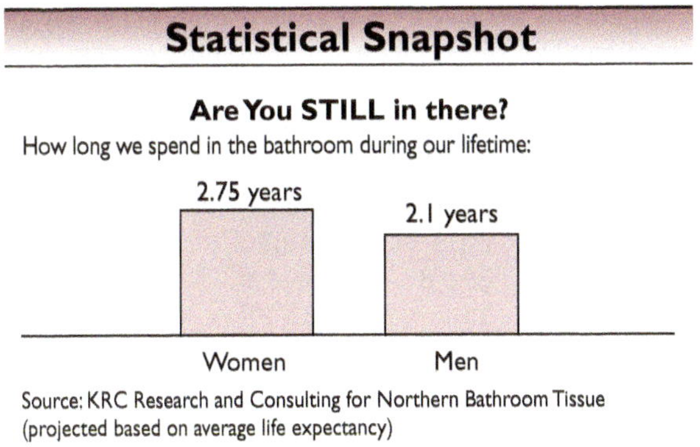

(b) **Statistical Snapshot**

Eye Appeal
About 80% of women say they'd change the color of their eyes to make them sexier. Color they would prefer:

- Green: 35.1%
- Blue: 23.7%
- Violet: 21.1%
- Hazel: 12.7%
- Brown: 7.4%

Source: Advantage Business Research Inc. for Wesley-Jessen

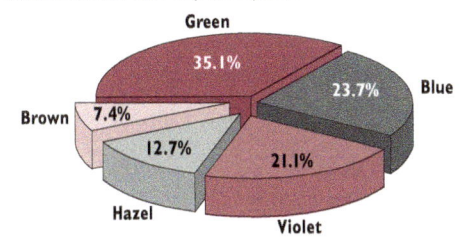

(c) **Statistical Snapshot**

Where Daycare is No Bargain
The median cost for daycare* nationwide is $388 per month. Most expensive metropolitan areas for daycare:

Metro Area	Median Monthly Cost
New York	$589
Boston	$579
Minneapolis	$537
Philadelphia	$503
Washington D.C.	$486

*Based on monthly fee

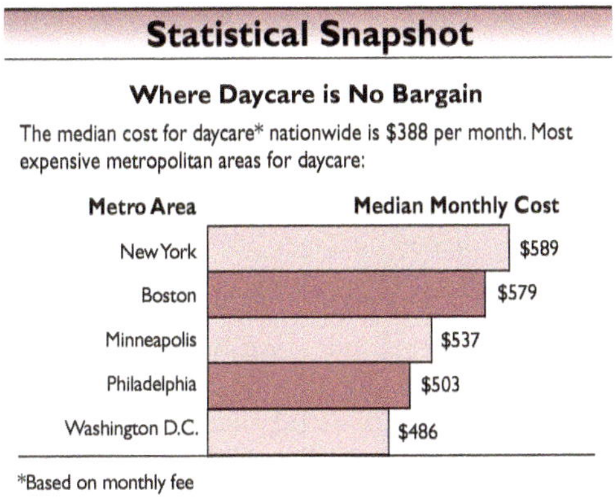

210. Examine the vertical bar graph in the following figure. The graph represents the number of passengers expected to travel by air each day of the Christmas holidays, in millions.
   a) For which of the following three day period: Dec. 19–21st, Dec. 23–25th, or Dec. 28–30th, do you believe the variability of the number of passengers is the largest? the smallest variability? Explain.
   b) From the bar graph, estimate the variability of all the data values using the range. What is a disadvantage in using the range as a measure of variability?
   c) For which any three consecutive day period, is the variability the greatest? the smallest? Explain.

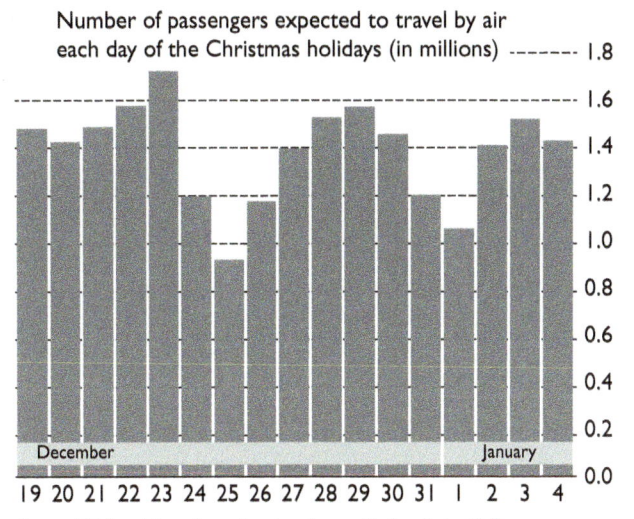

Number of passengers expected to travel by air each day of the Christmas holidays (in millions)

Sources: U.S. Travel Data Center, American Automobile Association, Air Transport Association of America

From *The New York Times.* © 1992 The New York Times Company. All rights reserved. Used under license.

211. The following bar graph represents the time the winning goal was scored in overtime for the 251 National Hockey League playoff overtime games from 1970 to 1993.

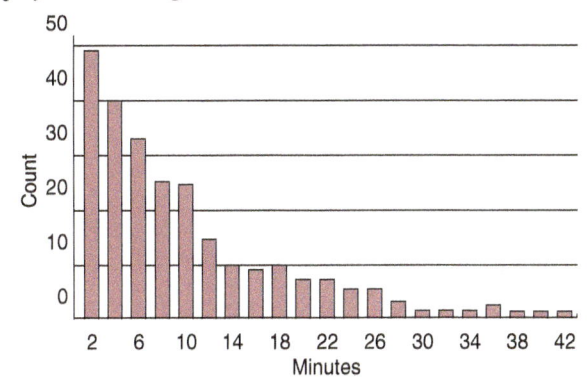

   a) State the shape of the graph.
   b) Using this graph, when can you expect the winning goal to be scored in a playoff overtime game? What measure of central tendency did you use to answer this question?
   c) Place a mark on the graph where you would think the median time would fall? the mean time? Explain your answer.
   d) Calculate the median time and mean time from the information given in the bar graph. Compare these results to your answer to part c. Explain any discrepancies.
   e) Calculate the standard deviation of the times shown in the bar graph. Explain what this statistical measure tells you about the time to score the winning goal in overtime?
   f) Calculate the percentage of goals scored within 1 standard deviation of the mean, 2 standard deviations of the mean and 3 standard deviations of the mean. Compare these results to the expected results using Chebyshev's Theorem.

212. The following information represents the listing of stock quotations for six stocks. Examine the information contained within this listing. Notice that the listing includes the high and low price for each stock during the past 52 weeks, as well as the high, low and closing price for each stock for the previous trade day. The last column represents the net change in the stock price from the previous day's quotation. The stock's name and symbol are also included within the listing.

52 weeks Hi	Lo	Stock	Sym	Hi	Lo	Close	net chg
13 1/2	4	Aileen	AEE	5 1/2	5 1/8	5 1/8	–1/2
43 3/4	25 1/8	AutoZone	AZO	40 7/8	39 5/8	40 5/8	–1/2
15 1/2	12 5/8	WestcstEngy	WE	15 1/4	15	15 1/8	...
28 1/4	11 1/8	Marvel Entn	MRV	28 1/8	24 7/8	27 5/8	+2 5/8
15	9 5/8	Homestake	HM	16 1/4	14 7/8	16	+1 5/8
3 3/4	9 5/8	Ft Nt Film	FNAT	8 9/16	8 7/16	8 7/16	–1/16

Data for exercise 212

a) Which measure of variability is being used to reflect the variability in the stock prices for the year? for the day?
b) Over the year, which stock has been less volatile? more volatile?
c) From the previous trading day, which stock has been less volatile? more volatile?
d) Which stock had the greatest percentage gain in its price for the previous day's trading price? the smallest percentage gain?
e) Which stock had the greatest percentage loss in its price for the previous day's trading price? the smallest percentage loss?
f) The following graph represents the daily prices for the stock First National Film Corp (Symbol: FNAT) for the period from January to April.

First National Film Corp.
Daily stock prices
OTC symbol: FNAT

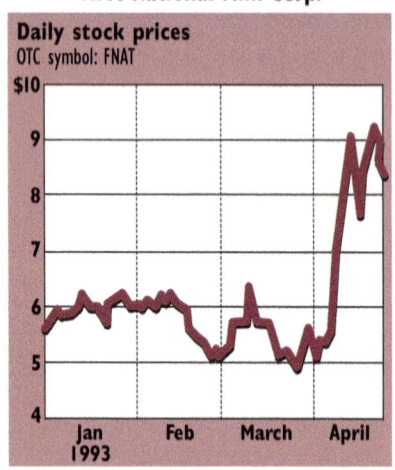

(i) Within which month, was the variability in First National Film Corp. stock's price the smallest? the greatest?
(ii) What information is contained within the graph for First National Film Corp. that is not represented by the stock quotation listed in Table Data for exercise 212? Explain how this shows a disadvantage of the range as a measure of variability as compared to the graphic information for First National Film Corp.

213. Examine the graphic entitled "It was a very wet year in the USA."

Statistical Snapshot

It Was a Very Wet Year in the USA

1996 was USA's seventh wettest year since comprehensive record-keeping began in 1895. The precipitation was driven up by record rainfall in the Northwest and Northeast. The West had its second warmest year, but that was offset by cold in the East and North. The nation's average temperature matched the 102-year average:

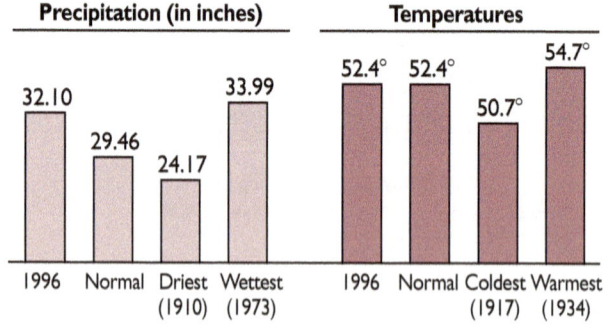

Precipitation (in inches): 32.10 (1996), 29.46 (Normal), 24.17 (Driest 1910), 33.99 (Wettest 1973)

Temperatures: 52.4° (1996), 52.4° (Normal), 50.7° (Coldest 1917), 54.7° (Warmest 1934)

Source: William Brown, National Climate Data Center

a) Define the population and identify the variables being presented.
b) What measure do you think is being used to describe the Normal Precipitation and Temperature? Explain how this is useful to interpret the precipitation and temperature for 1996.
c) What is the purpose of including the information about the driest and wettest? How is this useful when interpreting the precipitation for 1996?
d) What is the purpose of including the information about the coldest and warmest? How is this useful when interpreting the temperature for 1996?
e) Explain why it would be difficult to interpret the precipitation and temperature for 1996 without the information on the driest and wettest precipitation and the coldest and warmest temperature for 1996.

214. A researcher randomly places ten students into two statistics groups. Group I was taught statistics using a calculator

while group II was taught without the use of a calculator. At the end of the semester a final exam was given to each group. The results are as follows.

Group I	Group II
76	73
75	75
77	71
78	74
74	72

a) Determine the mean and variance of each group.
b) Combine the survey data for the two groups. Compute the mean and variance of the combined data. This variance is called the "total" group variance.
c) Using the mean for each group, find the variance of the two means. This variance is called the "between" group variance.
d) Find the mean of the variances of each group. This result is referred to as the "within" group variance.
e) Does the relationship: Total group variance = between group + within group variance hold true?
f) Construct a box plot for each group. Examine the box plots with respect to variability. That is, look at the length of each box plot as an estimate of the "within" variability. Compare this visual representation of the within variability to the numerical computation of variability computed in part a. Comment.
g) Using the box plots of part f, compare the differences between the "average" grade of each group. What variability is indicated by differences in the position of the "average" grade of each group? Comment.
h) The researcher would like to determine if using the calculator results in a higher level of achievement for group I. Examine the value of the means computed in part (a). Do you think it is reasonable for the researcher to conclude that using the calculator to teach statistics results in higher achievement? Explain.
i) Should the researcher take into consideration the variability in the achievement scores for each group? Explain.

215. A health insurance company has conducted a survey that determines the length of hospital stay (in days) for three hospitals in Cincinnati for the same medical procedure. The following data represent the survey results.

Length of Hospital Stay for Patients (in days)

Hospital I	Hospital II	Hospital III
12	20	23
6	14	8
3	5	11
21	17	5
18	2	17
9	11	20
15	8	14

a) Determine the mean and variance length of hospital stay for each hospital.
b) Combine the survey data for all three hospitals. Compute the mean and variance of the combined data. This variance is called the "total" group variance.
c) Using the mean length of stay for each hospital, find the variance of the three means. This variance is called the "between" group variance.
d) Find the mean of the variances of each hospital. This result is referred to as the "within" group variance.
e) Does the relationship: Total group variance = between group + within group variance hold true?
f) Construct a box plot for each hospital. Examine the box plots with respect to variability. That is, look at the length of each box plot as an estimate of the "within" variability. Compare this visual representation of the within variability to the numerical computation of variability computed in part a. Comment.
g) Using the box plots of part f, compare the differences between the "average" length of hospital stay. What variability is indicated by differences in the position of the "average" length of hospital stay? Comment.
h) The insurance company would like to estimate a reasonable length of stay for this medical procedure. Examine the value of the means computed in parts (a) and (b). What do you think is the insurance company's best estimate for the average length of hospital stay for a patient having this medical procedure? Explain.
i) Should the insurance company take into consideration the variability of the length of stay for each hospital? Explain.

PROJECTS

216. Choose 10 stocks arbitrarily from the New York Stock Exchange and compute the mean stock price for these stocks per day for one week. Compare your results per day with the Dow Jones Industrial Stock Average per day by using the chart below:

Day	10 Stock Average	Dollar Change per Day	Dow Jones	Dollar Change per Day
Monday				
Tuesday				
Wednesday				
Thursday				
Friday				

217. Consider the table below:

	Ages			
	16–25	26–35	36–45	46–55
1.				
2.				
3.				
4.				
5.				
6.				
7.				
8.				
9.				
10.				

Chapter 3 Numerical Techniques for Describing Data

For each age group, survey ten people and record their answer to the question, "How many years should a married couple wait before having children?"

a) According to your survey, what is the mean number of years a couple should wait before having children as answered by each group?

b) What is the modal number of years a couple should wait before having children as answered by each group?

c) Consider all the responses as one group. What are the mean and mode of this distribution?

d) Do you think people in different age groups responded with a significantly different waiting period?

www.MyStatLab.com.

218. The following data represents personal information for fifty students.

Student ID	Hair Color	Eye Color	Ht Inchs	Wt Lbs	# Hours Worked Weekly	Gender	IQ Score	# of Sib	Age
8476	BR	HAZEL	66	118	22	M	95	2	19
3834	BK	BROWN	67	107	32	F	115	2	18
3165	BL	BROWN	69	154	35	M	120	5	18
1207	BL	BROWN	72	176	23	M	105	3	23
6784	BR	BROWN	60	103	20	F	116	1	19
6780	BR	HAZEL	67	132	18	M	127	5	20
5380	BK	BLUE	66	123	13	M	101	1	18
4227	BR	BROWN	66	121	40	M	125	1	18
8625	BR	HAZEL	64	105	0	F	99	0	19
0817	BR	BLUE	71	170	35	M	104	3	32
0882	BR	HAZEL	71	175	24	M	114	4	18
0638	BR	HAZEL	67	135	30	M	121	2	19
4390	BK	BROWN	74	206	35	M	130	2	20
5345	BK	BROWN	63	98	0	F	107	2	19
6819	BK	BLUE	74	198	0	M	122	0	20
4002	RD	HAZEL	59	99	10	F	110	2	18
1252	BR	BROWN	68	143	28	F	145	5	18
1031	BL	BROWN	61	115	35	F	109	0	43
7873	BR	HAZEL	60	93	19	F	103	3	19
0194	BR	BROWN	70	160	15	M	135	5	20
9189	BR	BLUE	56	144	23	F	118	5	18
3695	BR	BROWN	64	115	22	F	105	3	18
2444	BK	BROWN	71	172	16	M	115	2	17
2467	BK	BROWN	61	113	19	F	128	2	19
7286	BR	HAZEL	67	98	25	F	131	1	20
7896	BR	BROWN	61	105	24	F	118	2	18
8905	BR	HAZEL	65	135	25	F	117	0	19
7299	BR	BROWN	65	122	0	F	113	3	18
9638	BK	BROWN	67	129	26	M	104	1	29
6798	BR	BROWN	66	118	20	F	126	4	21
4442	BL	BROWN	69	143	44	F	104	1	26
0132	BL	BROWN	73	194	16	M	107	3	18
1884	RD	BROWN	70	157	31	M	129	2	19
7345	BK	BROWN	63	126	9	F	134	1	20
0655	BK	HAZEL	69	149	7	M	108	2	18
4982	BR	HAZEL	73	193	36	M	112	1	24
8825	BL	HAZEL	73	188	17	M	104	3	37
6696	BR	HAZEL	61	122	44	F	98	1	18
9498	BR	BROWN	70	164	10	M	123	3	19
6217	RD	HAZEL	66	133	37	F	105	0	20
6014	BK	BLUE	66	123	24	M	134	1	18
1394	RD	BROWN	69	184	9	M	138	2	17
3199	BR	BROWN	69	190	21	M	121	0	19
2426	BR	HAZEL	66	122	15	M	99	2	19
2033	BR	BROWN	74	207	0	M	112	2	22
2435	BK	HAZEL	68	157	33	M	127	5	20
3173	RD	HAZEL	69	176	28	M	136	2	19
3361	BR	BROWN	70	192	14	M	121	2	36
8886	BR	BROWN	70	125	31	F	102	3	21
3593	BR	BROWN	74	197	41	M	111	2	24

Compute the suitable measure of central tendency for each characteristic and construct a composite description of the typical student at this school. Use the composite of the average American given in the introduction to this chapter as a guide.

219. Select 25 male students and 25 female students and administer the following arithmetic test. Allow 10 minutes for the examination.

The correct answers are (1) b (2) c (3) d (4) d (5) a.

Grade each of the tests and determine which gender had more consistent grades (Calculate the range and standard deviation for the grades for each gender).

The directions: Answer all questions by choosing the letter of the answer of your choice.

1) $\dfrac{4}{3} \div \dfrac{1}{2} = ?$

   a) $\dfrac{2}{3}$ **b)** $\dfrac{8}{3}$ **c)** $\dfrac{4}{5}$ **d)** $\dfrac{3}{8}$

2) $18 + 24 + ? = 60$

   a) 28   **b)** 8   **c)** 18   **d)** 38

3) $(0.03\%) = \dfrac{3}{?}$

   a) 10   **b)** 100   **c)** 1000   **d)** 10,000

4) $\dfrac{2}{5} + \dfrac{3}{4} = ?$

   a) $\dfrac{9}{5}$ **b)** $\dfrac{5}{9}$ **c)** $\dfrac{6}{20}$ **d)** $\dfrac{23}{20}$

5) $7 - 2/5 + 0.12 = ?$

   a) 6.72   **b)** 5.22   **c)** 6.52   **d)** 6.12

220. Select a well-known brand name food product for which there is a comparable store brand. For example, a brand name peanut butter and a store brand peanut butter (same size jar). Get the prices for the brand name product and the store brand product from as many different food stores as you can (at least five). Form two distributions, one from the name brand prices and the other from the store brand prices. Use the standard deviation to determine which price distribution is more variable.

221. Look up the home run records of Babe Ruth, Mickey Mantle, Roger Maris, Mel Ott, Hank Greenberg, Hank Aaron, and Ken Griffey, Jr.
    a) Calculate μ and σ for the number of home runs each player hit during his career.
    b) For each player, take the maximum number of home runs he hit during any one season of his career and determine which player's maximum one year production is most outstanding relative to his career.

222. a) Calculate the mean and standard deviation of the number of electoral votes each Democratic and Republican presidential candidate received in each presidential election since 1900 inclusive.
    b) From your data, choose the Democratic and Republican candidates who received the maximum number of electoral votes and calculate the z score of their electoral votes.
    c) Compare the two z scores and decide which presidential candidate was the higher electoral vote getter relative to his party.

CHAPTER TEST

223. The symbol used for the sample mean is ____ while the Greek symbol used for the population mean is ____.

224. Which measure of central tendency is more resistant to extreme data values or outliers?
    a) Mean
    b) Median

225. The symbol used for the sample standard deviation is ____ and the symbol used to denote the population variance is ____.

226. If a distribution of salaries is skewed right, then the:
    a) most frequent salary equals the median salary
    b) most frequent salary is greater than the median salary
    c) most frequent salary is less than the median salary
    d) not enough information

227. If the range is 10, and the largest data value is 12, then the smallest data value is _____.

228. The Interquartile Range (IQR) is a measurement that indicates the middle 50% of the distribution.
    (T/F)

229. If $Q_1 = 420$ and $Q_3 = 510$, then which one of the following data values is considered an outlier.
    a) 140   b) 190   c) 500   d) 700

230. If the population of ages of college students is skewed right, then this indicates:
    a) There are more older students within the population.
    b) The median age is greater than the mean age.
    c) As the age of a student increases, the frequency of the ages also increases.
    d) There is a greater number of younger students within the population.

231. For each of the following distribution of data values,
    a) identify the general shape that you believe best describes each distribution, and
    b) identify whether the mean or median would have the largest value within this type of distribution.
        (i) The test results of an easy exam.
        (ii) The test results of a difficult exam.
        (iii) The heights of all the female students at a University.

232. For the following sample of IQ scores:
    106, 102, 104, 104, 105, 103, 105, 103, 101, 107, 114, 119, 116, 118, 128, 109, 103, 112, 111, 110, 112, 106, 108, 110, 111, 107
    find:
    a) The mean
    b) The median
    c) The mode
    d) The standard deviation
    e) $\Sigma(x - 109)$
    f) $(\Sigma x)^2$
    g) The sum of the squared IQ values
    h) $\Sigma(x - 109)^2$
    i) $[\Sigma(x - \bar{x})]^2$
    j) The percentile rank of 112 and interpret it.
    k) The z score of 128 and interpret it.
    l) The 5-number summary.
    m) The IQR.
    n) What data value represents the 1st quartile? 3rd quartile?

170 Chapter 3 Numerical Techniques for Describing Data

o) Is there a data value that you consider to be outlier? If an outlier exists, then explain why you believe it is an outlier? Do you believe the outlier is an unusual value or is due to an error? Give an explanation for your answer.
p) Construct a box plot for the IQ scores.
q) State the shape of the box plot.

233. Linda, a sports guru, is interested in comparing the careers of the New York Yankee players Mickey Mantle and Whitey Ford to determine which player had a more consistent career. Mickey Mantle's career batting average was .298 with a standard deviation of 0.033, while Whitey Ford had a career ERA of 3.14 with a standard deviation of 0.46.
  a) If you use the coefficient of variation to compare the consistency of both players' career, which player had the more consistent career?
  b) Why is the coefficient of variation an appropriate measure to use in making this comparison?

234. If the tire life of a certain brand of an all-season radial tire approximates a symmetric bell-shaped distribution with mean = 65,000 miles and standard deviation = 5,000 miles, then use the Empirical Rule to:
  a) Determine the percentage of tires you would expect to last between 60,000 to 70,000 miles?
  b) Determine the percentage of tires you would expect to last between 55,000 to 75,000 miles?
  c) Determine the percentage of tires you would expect to last between 50,000 to 80,000 miles?
  d) Determine the percentage of tires you would expect to last between 60,000 to 75,000 miles?
  e) Can Chebyshev's Theorem be applied to answer part (b) of this question? If it is possible, use Chebyshev's Theorem to answer part (b) and compare the two results.
  f) If a tire lasts 85,000 miles, would you consider this tire wear to be typical or unusual? Explain.

235. A New York Realty Association surveys local banks and mortgage lenders to determine the trends in 30 year fixed and adjustable mortgage rates. Based on their survey, the mean rate for a 30 year fixed mortgage is 10.97% (with s = 0.14%) and the mean rate for a one year adjustable mortgage is 7.79% (with s = 0.17%).

If Matcraig, a mortgage lender, is offering a 30 year fixed mortgage rate of 10.75%(& no points) and a one year adjustable mortgage rate of 7.5%(& no points), which rate would be considered the best deal relative to its type of mortgage?

236. A sample of the number of study hours per week of college students is collected. Some of the sample results of this survey are given in the following table.

Study Hours	4	10	20	24	34	36	42
z Score	−2.5	−1.75	−0.5	0	1.25	1.5	2.25
Percentile Rank	10	25	35	40	50	75	95

By inspection of the table, determine:
  a) The mean study time for all the students.
  b) The median study time for all the students.
  c) The percent of college students who study between 20 and 36 hours per week.
  d) The study time representing the 4th decile.
  e) The study time representing the 1st quartile.
  f) The study time representing the 3rd quartile.
  g) The interquartile range
  h) The standard deviation study time for all the students.
  i) The study time corresponding to a z score of 1.75.
  j) The study time that is 2.5 standard deviations above the mean.

237. Lydia, an Investment analyst, has been plotting the cyclic pattern of the stock of a fast food company. Based upon her analysis, she decides to purchase the stock when it is selling at least one and a half standard deviations below its mean cyclic price. Her plan is to sell the stock when its price is at least 1.65 standard deviations above the mean price. Determine her maximum buying price and her minimum selling price for this stock if its mean cyclic price is $27.65 with a standard deviation of $9.95.

238. What percentage of a population of psychology test scores are:
  a) Above the third quartile?
  b) Below the first quartile?
  c) Greater than the 23rd percentile?
  d) Above the 6th decile?
  e) Between the first and third quartiles?
  f) Between the 2nd and 6th decile?

239. The following data values represent a sample of the heights of female college basketball players.

Heights in Inches

63 71 69 66 73 85 70 69 67 74 75 68 65 63
67 69 68 72 73 75 72 75 73 68 69 75 65 65

  a) Determine (to 2 decimal places) the mean height and standard deviation of the heights.
  b) Determine the z score & percentile rank of the following heights. Interpret these results.
    i) 67   ii) 69   iii) 66   iv) 73   v) 85   vi) 74
  c) Determine the height that represents the:
    i) 1st quartile
    ii) 2nd decile
    iii) 5th decile
    iv) 3rd quartile
    v) 8th decile
  d) Determine the height which represents the 2nd quartile.
  e) Find P_{25} and P_{75}.
  f) Determine the 5-number summary.
  g) Construct a Box Plot for this 5-number summary.
  h) Using the Box Plot, identify the shape of the distribution of the female basketball player's heights.
  i) Calculate the Interquartile Range (IQR). Determine the percentage of heights represented by the IQR? Did you expect this result? Explain.
  j) Identify any outliers. If any data value is identified as an outlier, then indicate whether you believe this data value is an exceptional data value or a data value due to error. Explain. Determine the z score of any possible outlier? Interpret this z score. Explain how the value of a z score may indicate the presence of an outlier.
  k) Determine the percentile rank of the maximum data value. Explain why it isn't 100%.

240. Use this graph to answer the questions a to h.

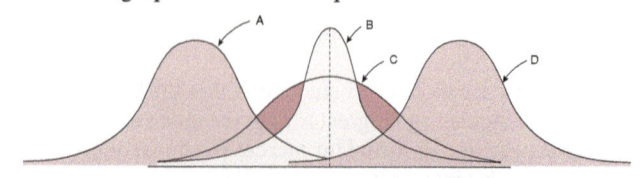

a) Which distributions have the same measures of central tendency but a different standard deviation? Explain.
b) Which one of the distributions identified in part (a) has the smaller standard deviation? Explain.
c) Which distribution has the smallest value for the mode?
d) Which distribution has the largest value for the median?
e) If distribution B had the data value X as an outlier, then in what other distribution would X also be considered an outlier? Explain your answer.
f) For which of these distributions could you use the Empirical Rule? Chebyshev's Theorem? Explain.
g) Would distribution B or C have a larger value for the 1st quartile? 3rd quartile? Explain.
h) Sketch a box plot for each of these distributions using the same horizontal axis.

241. Refer to the accompanying graph that uses Box Plots to compare the number of followers that a person has on Twitter for 4 local colleges. This survey was for college students who had fewer than 300 followers on Twitter. Pick the choice that best represents the answer.

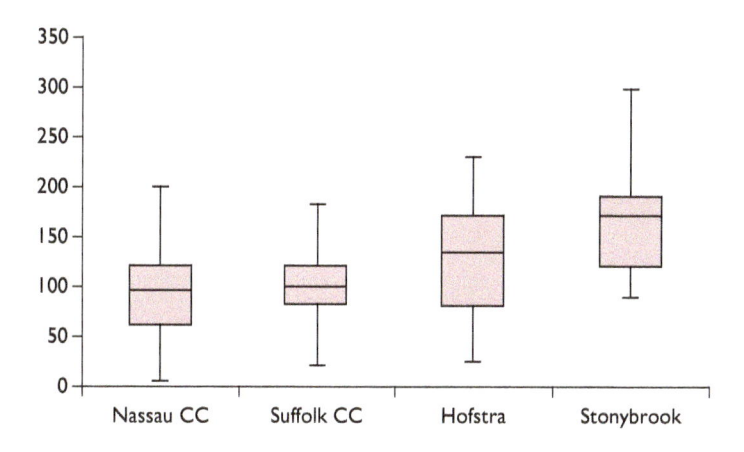

i. Which college would you say contains the largest data value?
   a. Nassau CC
   b. Suffolk CC
   c. Hofstra
   d. Stonybrook

ii. Which college contains the largest value for the standard deviation?
   a. Nassau CC
   b. Suffolk CC
   c. Hofstra
   d. Stonybrook

iii. Which distribution would you say has the most consistent data?
   a. Nassau CC
   b. Suffolk CC
   c. Hofstra
   d. Stonybrook

iv. Which distribution contains the lowest value for its 25th percentile?
   a. Nassau CC
   b. Suffolk CC
   c. Hofstra
   d. Stonybrook

v. What is the shape of the Stonybrook data?
   a. skewed left
   b. uniform
   c. skewed right
   d. approximate Bell

242. The following data set is the weight (in kg) for 15 male and 16 female adult Polar Bears that were tracked in the Arctic.

Males	Females
376	125
379	114
424	108
469	135
395	144
348	124
545	110
653	152
420	134
612	117
605	98
300	116
405	202
605	225
525	169
	160

a) Find the mean & median for each set of data.
b) Find the range, variance and standard deviation of each of the lists.
c) Which of the two genders would you say the data is more varied? Why?
d) Determine the male weight that is in the 75th percentile.
e) What is the percentile rank of the female weight 144 kg?

For the answer to this question, please visit MyStatLab.com.

CHAPTER 4

Linear Correlation and Regression Analysis

Contents

4.1 Introduction
4.2 The Scatter Diagram
4.3 The Coefficient of Linear Correlation
  The Definition Formula for the Correlation Coefficient
  The Computational Formula for the Correlation Coefficient
4.4 More on the Relationship Between the Correlation Coefficient and the Scatter Diagram
4.5 The Coefficient of Determination
4.6 Linear Regression Analysis
  Interpolation versus Extrapolation
  Explanation and Assumptions for Linear Regression Analysis
4.7 Residuals and Analyzing Residual Plots
  Analysis of Scatter Diagrams
  Homoscedasticity and Heteroscedasticity
  Non-Linear Trends
  What is a Residual?
  Residual Plots
  Coefficient of Determination Examined Further

4.1 INTRODUCTION

Most of us have examined a chart in a doctor's office or health magazine that relates height to weight. Imagine the following scenario in a doctor's examination room as illustrated in Figure 4.1

FIGURE 4.1
Courtesy of www.CartoonStock.com.

The patient, a rather plump individual about 5 feet 10 inches, is standing under a measuring height rod scale, facing the doctor. The doctor, dressed in a white examination jacket, turns to the patient after checking a height-weight chart on the wall and informs the patient: "According to the chart, you should be 47 feet tall."

173

The above scenario illustrates the two major concepts that are presented in this chapter. The first concept is known as correlation analysis. In this chapter, we will discuss correlation as a technique that measures the strength (or the degree) of the relationship between two continuous variables. For example, in this scenario, the height and weight of an individual represent the two continuous variables.

Since we are familiar with the notion that there is a strong relationship between the height and weight of an individual, then using the statistical concept of correlation, we could measure the strength of this relationship. We will examine how to determine this measure in a later section.

Remember the chart a physician uses to estimate your "ideal" weight given your respective height? Such charts have been developed using a statistical technique known as regression. Essentially, regression is a statistical technique that produces a model of the relationship (correlation) between the two variables.

The height/weight chart discussed in the previous scenario was generated by a regression formula. This chart represents the model for the relationship between height and weight of an individual. When there is a strong correlation between two variables, such as height and weight, then the regression analysis can be applied to produce a formula (or rule) to model the relationship between the two variables. In this chapter, we will examine two important concepts in elementary statistics: **linear correlation** and **linear regression**.

4.2 THE SCATTER DIAGRAM

There are times when we are interested in determining whether a relationship exists between two variables. In Chapter 14, we will examine whether or not a relationship exists between two classifications where each classification represented a categorical variable. In this chapter, we will examine the relationship between two continuous variables.

To help us begin our examination of the relationship between two variables, let's consider studying the relationship between a female's height and weight for female students attending Bermuda Community College. We will randomly select a sample of 12 female students from this community college and measure each female's height and weight. Thus, *two* measurements are recorded for *each* female student. These two measurements represent the value of each of the two variables, height and weight, for each female student.

If we label the height as the x variable and the weight as the y variable, then for each female student we will have a pair of numbers. The sample data representing the pair of numbers for the twelve female students has been recorded and summarized in Table 4.1.

These pair of numbers can also be written in the following form: **(x,y)**, which is called an **ordered pair** since the value of the x variable is always written first within the parentheses.

For example, the ordered pair for female student number 1 is written: (62, 123), where the first number, 62, within the parentheses represents the height of student number 1 and the second number, 123, within the parentheses represents the weight of student number 1.

TABLE 4.1

Female Student Number	Height (inches) x	Weight (lbs) y
1	62	123
2	58	102
3	64	110
4	69	137
5	61	145
6	70	132
7	59	108
8	60	112
9	63	124
10	72	155
11	71	170
12	68	140

Using the sample information contained in Table 4.1, we can begin to analyze the relationship between a female's height and weight by constructing a graph of these ordered pairs. This type of graph is called a **scatter diagram**.

> **DEFINITION 4.1**
> A **scatter diagram** is a graph representing the ordered pairs of data on a set of axes.

To construct a scatter diagram, we begin by drawing two lines, a horizontal and a vertical line, to represent the two axes. The horizontal line is called the x axis and represents the x values, in this case the heights of the female students. Thus, the x axis is labeled height.

While, the vertical line is called the y axis and represents the y values, in this case the weights of the female students. Thus, the y axis is labeled weight. This is illustrated in Figure 4.2.

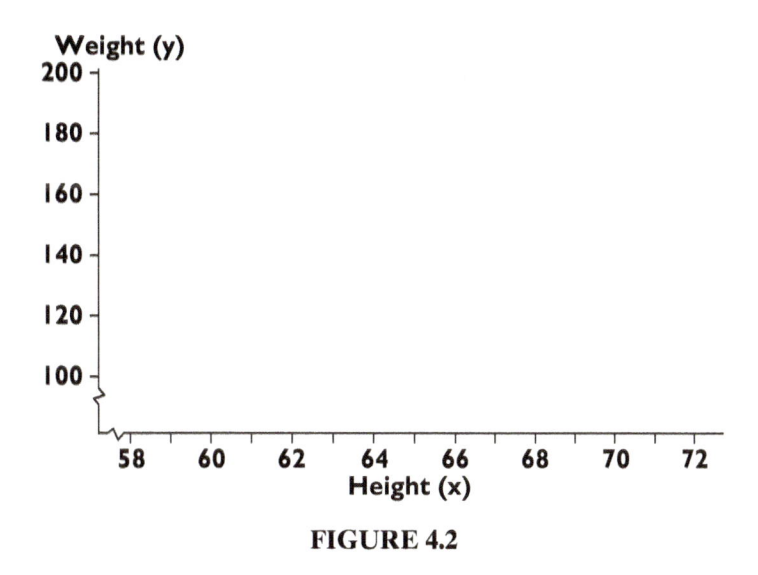

FIGURE 4.2

We use a dot to represent each ordered pair of measurements. The dot is placed directly above the female's height and directly to the right of the female's weight. Thus, for the first female student with a height of 62 inches and a weight of 123 pounds, a dot is placed above 62 on the axis labeled height and to the right of 123 on the axis labeled weight. This dot represents the height and weight of female student number 1. This same procedure is used to plot all the ordered pairs of the sample data of Table 4.1 and Figure 4.3 illustrates the scatter diagram for this sample data.

Visual examination of a scatter diagram can help to determine whether there is an apparent relationship or correlation between the two variables and, if one does exist, to determine the type of correlation that exists. In Figure 4.3, it appears that a relationship does exist between a female's height and her weight as we examine the scatter diagram.

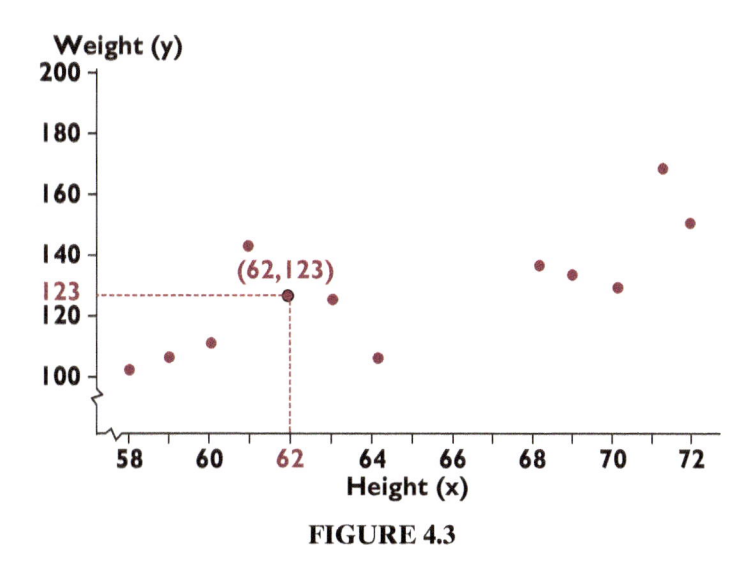

FIGURE 4.3

Let's examine the type of relationship that we notice in Figure 4.3. The scatter diagram of Figure 4.3 seems to show that as a female's height, represented by the x values, increases then her weight, represented by the y values, also increases. This type of correlation is called a **positive correlation**.

DEFINITION 4.2
A **positive correlation** between two variables, x and y, occurs when high measurements for the x variable tend to be associated with high measurements for the y variable and low measurements for the x variable tend to be associated with low measurements for the y variable.

You will also notice that the dots in the scatter diagram tend to follow a straight-line path. This type of correlation is called a **linear correlation**. This can be better visualized by drawing a straight line through the scatter diagram of Figure 4.3, and noticing that the points *tend* to lie very close to the line. This is illustrated in Figure 4.4.

Now, we can state that there is a positive linear correlation between the two variables, height and weight. Another type of correlation that can exist between two variables is a **negative correlation**.

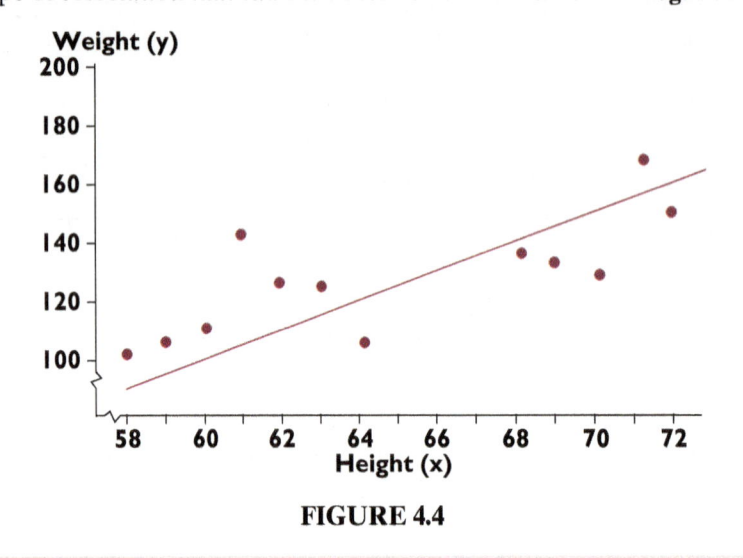

FIGURE 4.4

DEFINITION 4.3
A **negative correlation** between two variables, x and y, occurs when high measurements for the x variable tend to be associated with low measurements for the y variable and low measurements for the x variable tend to be associated with high measurements for the y variable.

Let's examine this idea of negative correlation using the scatter diagram in Figure 4.5. Notice in Figure 4.5, as the values of the x variable increases, the value of the y variable tend to decrease. The two variables tend to vary in opposite directions. Thus, the relationship between the two variables is a negative correlation.

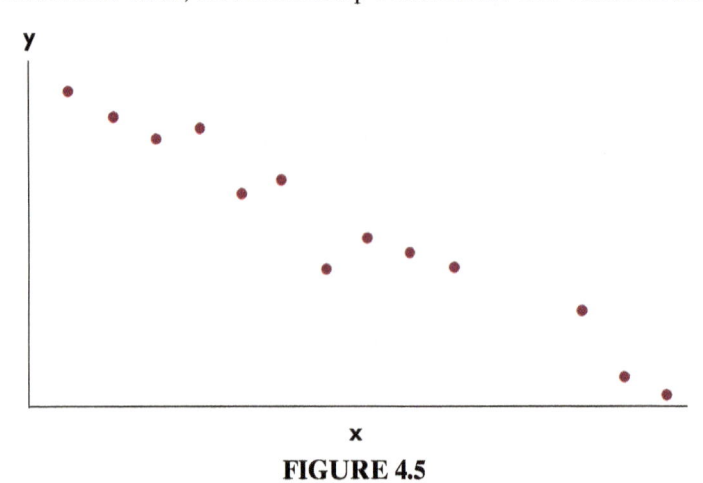

FIGURE 4.5

It is possible that when examining the relationship between two variables, you may notice that there is no linear relationship between the two variables. This is referred to as **no linear correlation**.

DEFINITION 4.4

No linear correlation means there is no linear relationship between the two variables. That is, high and low measurements for the two variables are not associated in any predictable straight line pattern.

Figure 4.6 illustrates a scatter diagram that shows no linear correlation.

Remark: We will only examine linear (or straight line) correlations in this chapter. Although, it is possible that two variables can be related to each other in a curvilinear relationship, this type of correlation is beyond the scope of this textbook. An example of a curvilinear relationship is illustrated by the scatter diagram in Figure 4.7

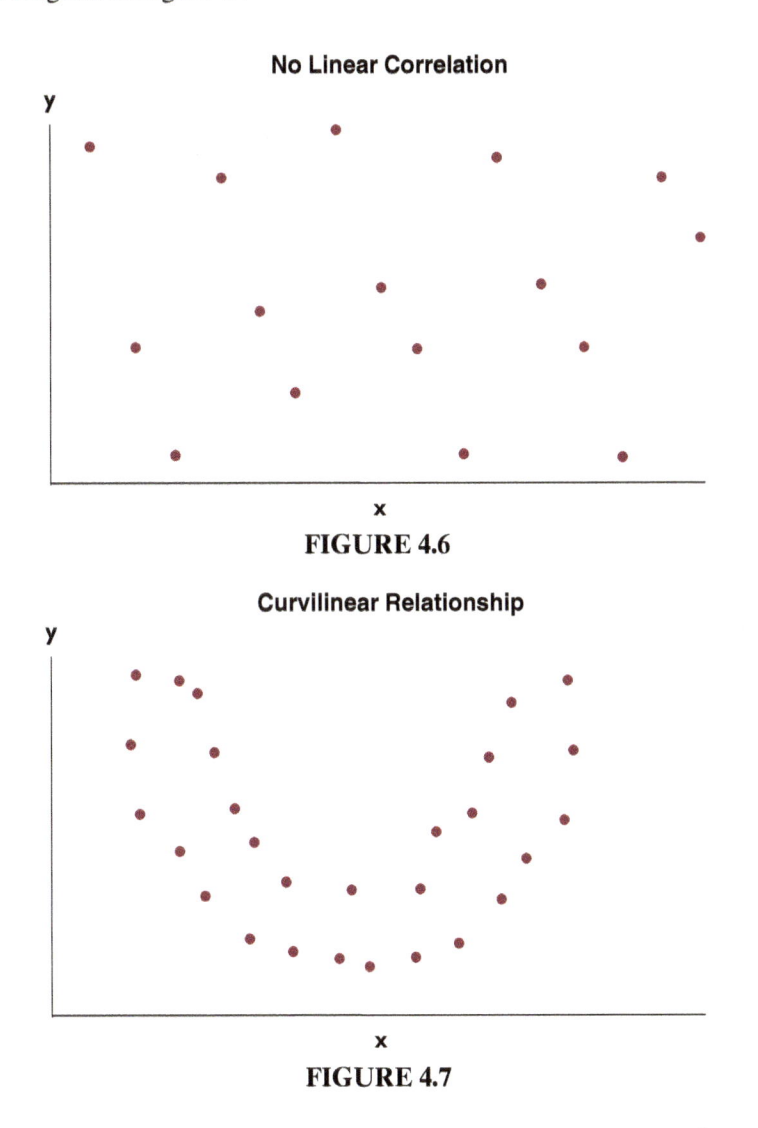

No Linear Correlation

FIGURE 4.6

Curvilinear Relationship

FIGURE 4.7

Let's construct a few scatter diagrams and determine the type of correlation that may exist.

EXAMPLE 4.1

Using the sample data in Table 4.2, construct a scatter diagram and indicate the type of linear correlation, if any exists.

TABLE 4.2

x	1	2	3	5	8
y	4	5	5	6	9

Solution

To construct a scatter diagram, perform the following steps:

Step 1 Draw a horizontal axis and label it as the x axis. Now draw a vertical axis and label it as the y axis.

Step 2 Plot each ordered pair (x,y) of the sample data from Table 4.2 in the appropriate position on the set of axes. Figure 4.8 represents the scatter diagram for the sample data of Table 4.2.

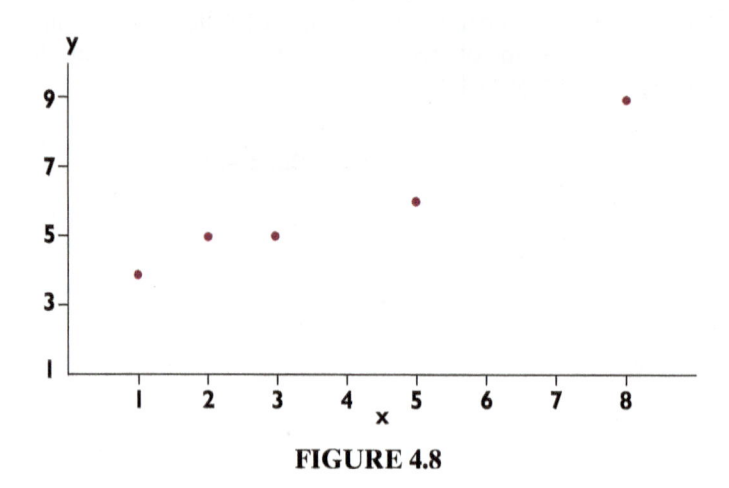

FIGURE 4.8

To determine the type of correlation that may exist, we must examine the pattern that is shown by the sample data pairs in Figure 4.8. As you examine the scatter diagram from left to right, you will notice the pattern of the points are going in an upward direction. That is, as the x values *increase*, the y values also *increase*. Thus, the scatter diagram indicates a pattern that corresponds to a *positive* correlation.

EXAMPLE 4.2

Using the sample data in Table 4.3, construct a scatter diagram and indicate the type of correlation, if any exist.

TABLE 4.3

x	2	3	5	6	8	10
y	11	9	8	6	6	3

Solution

Step 1 Draw a horizontal axis and label it as the x axis. Now draw a vertical axis and label it as the y axis.

Step 2 Plot each ordered pair (x,y) of the sample data from Table 4.3 in the appropriate position on the set of axes. Figure 4.9 represents the scatter diagram for the sample data of Table 4.3.

To determine the type of correlation that may exist, we must examine the pattern that is shown by the sample data pairs in Figure 4.9. As you examine the scatter diagram from left to right, you will notice the pattern of the points are going in a downward direction. That is, as the x variable *increases*, the y variable *decreases*. This scatter diagram indicates a **negative correlation** for the sample data listed in Table 4.3.

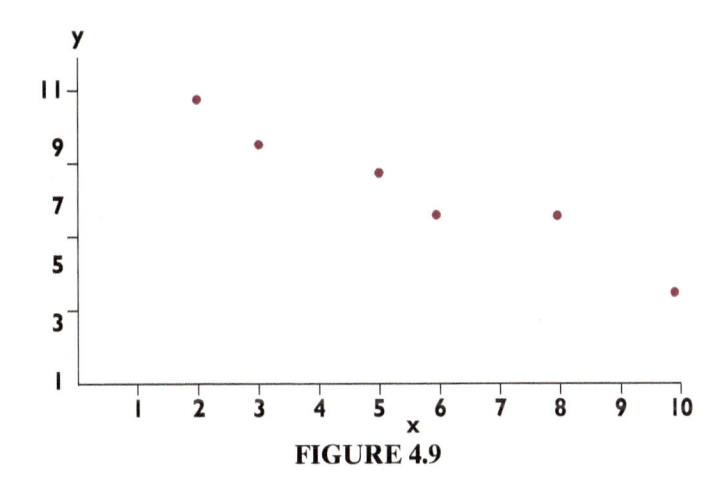

FIGURE 4.9

EXAMPLE 4.3

Using the sample data in Table 4.4, construct a scatter diagram and indicate the type of correlation, if any exist.

TABLE 4.4

x	3	4	5	7	5	3	6	7	8	6
y	1	3	2	5	5	4	4	2	1	3

Solution

Step 1 Draw a horizontal axis and label it as the x axis. Now draw a vertical axis and label it as the y axis.

Step 2 Plot each ordered pair (x,y) of the sample data from Table 4.4 in the appropriate position on the set of axes. Figure 4.10 represents the scatter diagram for the sample data of Table 4.4.

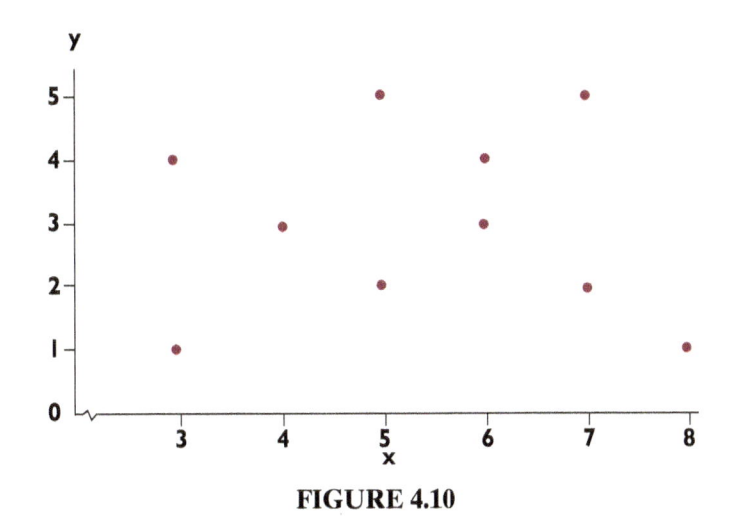

FIGURE 4.10

To determine the type of correlation that may exist, we must examine the pattern that is shown by the sample data pairs in Figure 4.10. As you examine the scatter diagram from left to right, you will notice that there is no definite pattern for the points. This scatter diagram indicates that we have *no linear correlation* for the sample data listed in Table 4.4.

Chapter 4 Linear Correlation and Regression Analysis

TI-84 Graphing Calculator Solution for Example 4.3

The TI-84 graphing calculator can also be used to create a scatterplot. Using the data values from Example 4.3 we will begin by entering the x and y data into lists L1 and L2 by pressing STAT ENTER **[1: EDIT]**. A word of caution... the values of L1 and L2 represent ordered pairs (x, y). This means that each x is paired with a specific y and you must be sure to preserve this relationship when entering the data in L1 and L2. Figure 4.11 displays the ordered pairs in L1 and L2.

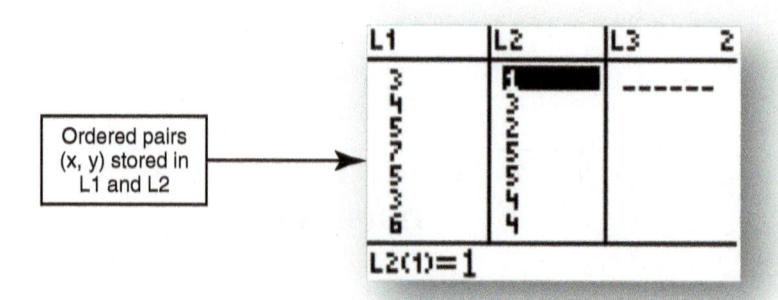

FIGURE 4.11

To create the scatterplot, press 2nd Y= ENTER to open **Stat Plot 1** menu. You will need to turn the Plot on, select the icon of the scatterplot (1st icon) in the **Type** menu, and then specify which lists your data is stored. For our example, we put the x and y data in lists L1 and L2 respectively. You can also specify the type of mark you would like the calculator to use in the scatterplot. This option is useful if you are graphing more than one set of data on the same set of axes, and need to distinguish between the data sets. This is shown in the Figure 4.12.

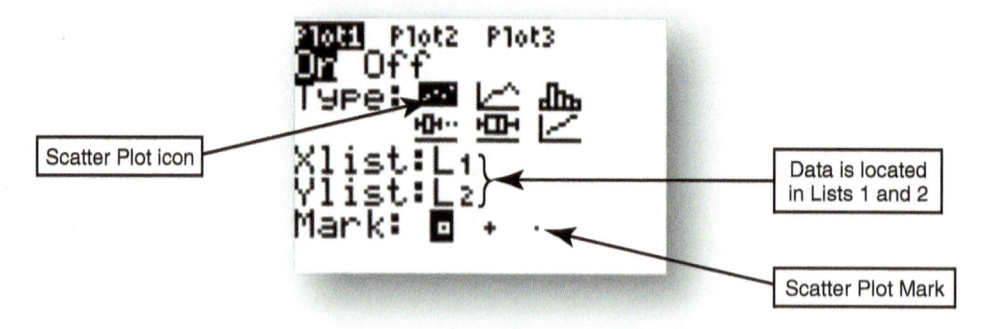

FIGURE 4.12

Remember the calculator has a special ZOOM feature for statistics in the Zoom menu. Using the **Zoom-Stat** feature (shown in Figure 4.13a) will fit the scatter diagram to the calculator's display. To activate this feature press ZOOM then scroll down to **9:ZoomStat** and press ENTER (you could have also just pressed the **9** key). Figure 4.13b shows the resulting scatter diagram.

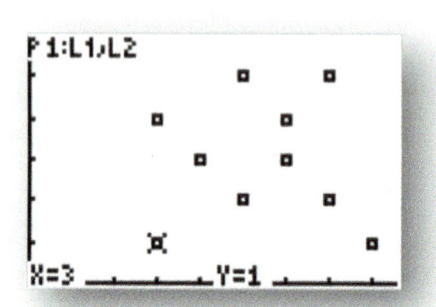

FIGURE 4.13a   **FIGURE 4.13b**

As you can see the scatter diagram is similar to the one given in Figure 4.10.

4.3 THE COEFFICIENT OF LINEAR CORRELATION

When a scatter diagram seems to indicate that there is a linear correlation between two variables, our next step is to measure the strength of the relationship between the two variables. By a **linear correlation**, we mean how closely the points of a scatter diagram approximate a straight-line pattern.

The closer the points of a scatter diagram approximate a straight-line pattern, the stronger the linear correlation between the two variables. The **strength** of a linear correlation between the two variables can be numerically measured by **Pearson's correlation coefficient, r**.

> **DEFINITION 4.5**
>
> The Definition Formula for **Pearson's Correlation Coefficient, r,** measures the strength of a linear relationship between two variables for a sample. This **sample correlation coefficient, r,** is calculated by the following formula:
>
> $$r = \frac{\Sigma(x-\bar{x})(y-\bar{y})}{(n-1)s_x s_y}$$
>
> where: r is the **sample correlation coefficient**
> x represents the data values for the **first variable**.
> y represents the data values for the **second variable**.
> $\bar{x}$ represents the **mean** of the data values for the **first variable, x**
> $\bar{y}$ represents the **mean** of the data values for the **second variable, y**
> s_x represents the **sample standard deviation** for the data values for the **first variable, x**
> s_y represents the **sample standard deviation** for the data values for the **second variable, y**
> n represents the **number of pairs of data values**

The sample correlation coefficient, r, will always be a numerical value between –1 and 1. That is,

$$-1 \leq r \leq 1$$

Let's interpret what the different values of r mean before we examine how to calculate the sample correlation coefficient.

> ### Interpreting the Values of r
>
> A value of **r = 1 represents the strongest positive linear correlation** possible and it indicates a perfect positive linear correlation. This means that all the points of the scatter diagram will lie on a straight line which is sloping upward from left to right.
>
> A value of **r = –1 represents the strongest negative linear correlation** possible and it indicates a perfect negative linear correlation. This means that all the points of the scatter diagram will lie on a straight line which is sloping downward from left to right.
>
> A value of **r = 0 represents no linear correlation** between the two variables.

The closer the value of the correlation coefficient, r, is to –1 or to 1, the stronger the relationship between the two variables. For example, if we say that height, represented by variable one, is strongly correlated to weight, represented by variable two, then we would expect the numerical value of r to be either close to –1 or close to 1. However, a value of r close to zero indicates a weak linear relationship between the two variables.

182 Chapter 4 Linear Correlation and Regression Analysis

Figure 4.14 illustrates the range of values for the correlation coefficient, r.

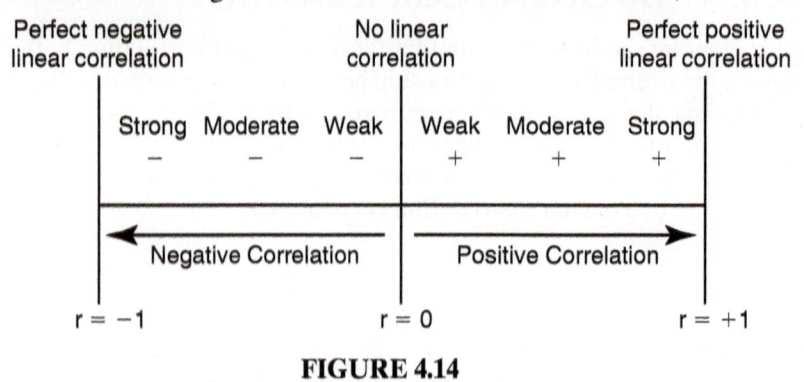

```
Perfect negative          No linear            Perfect positive
linear correlation        correlation          linear correlation

     Strong  Moderate  Weak | Weak  Moderate  Strong
       −        −       −   |   +      +        +

       ←────── Negative Correlation ──│── Positive Correlation ──────→

     r = −1                       r = 0                        r = +1
```

FIGURE 4.14

Several different relationships with their correlation coefficients (r) are illustrated in the following scatter diagrams in Figure 4.15

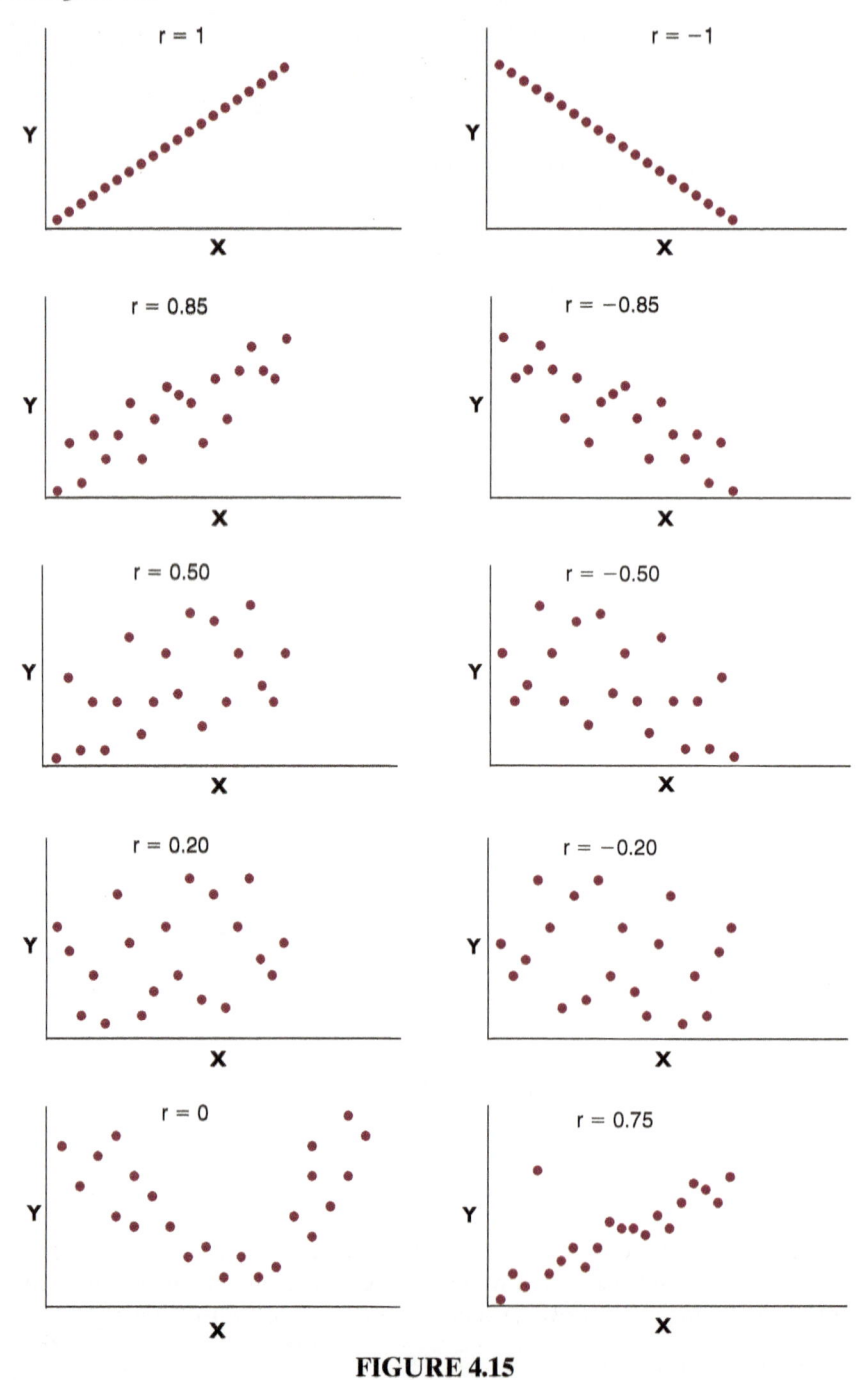

[Scatter diagrams showing: $r = 1$, $r = -1$, $r = 0.85$, $r = -0.85$, $r = 0.50$, $r = -0.50$, $r = 0.20$, $r = -0.20$, $r = 0$, $r = 0.75$]

FIGURE 4.15

Before we examine how to use the sample correlation coefficient formula to calculate the strength of the linear correlation between two variables, let's develop the idea first.

Consider the following scatter diagrams in Figure 4.16. Each scatter diagram is divided into four regions by drawing a vertical line through the mean of the x values, $\bar{x}$, and a horizontal line through the mean of the y values, $\bar{y}$.

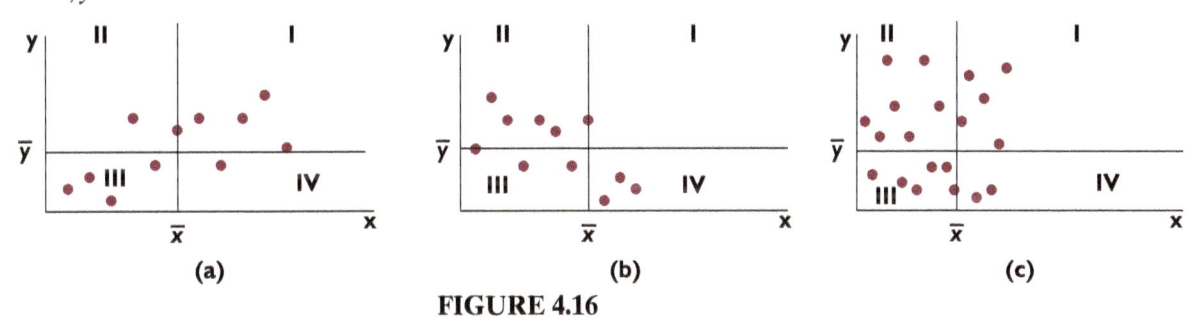

FIGURE 4.16

If we examine the scatter diagram in Figure 4.16 (a), we see that a positive linear relationship exists between x and y. Larger values of x tend to be paired with larger values of y, and smaller values of x tend to be paired with smaller values of y, therefore most of the points lie in regions I and III. The deviations from the mean, $(x-\bar{x})$ and $(y-\bar{y})$, for each data value in region I will be positive and the deviations from the mean in region III will be negative. Now consider taking the product $(x-\bar{x})(y-\bar{y})$ for each ordered pair. The product $(x-\bar{x})(y-\bar{y})$ for each ordered pair in region I is positive, and $(x-\bar{x})(y-\bar{y})$ for each ordered pair in region III is also positive. Notice for the points that lie in regions II and IV, the value of $(x-\bar{x})(y-\bar{y})$ will be negative. Also, the value of $(x-\bar{x})(y-\bar{y})$ for the points that lie on the vertical and horizontal lines for the means will be zero. Therefore if we take the sum of the products, $\Sigma(x-\bar{x})(y-\bar{y})$, the result will be positive since most of the products tend to be positive.

If we examine the scatter diagram in Figure 4.16 (b), we see that a negative linear relationship exists between x and y. Larger values of x tend to be paired with smaller values of y, and smaller values of x tend to be paired with larger values of y, therefore most of the points lie in regions II and IV. The deviations from the mean, $(x-\bar{x})$, for each data value in region II will be negative and the deviations from the mean, $(y-\bar{y})$, in region II will be positive. The deviations from the mean, $(x-\bar{x})$, for each data value in region IV will be positive and the deviations from the mean, $(y-\bar{y})$, in region IV will be negative. Now consider taking the product $(x-\bar{x})(y-\bar{y})$ for each ordered pair. The product $(x-\bar{x})(y-\bar{y})$ for each ordered pair in region II is negative, and the product for each ordered pair in region IV is also negative. Notice for the points that lie in regions I and III, the value of the product $(x-\bar{x})(y-\bar{y})$ will be positive. Also, the value of the product $(x-\bar{x})(y-\bar{y})$ for the points that lie on the vertical and horizontal lines for the means will be zero. Therefore if we take the sum of the products, $\Sigma(x-\bar{x})(y-\bar{y})$, the result will be negative since most of the products tend to be negative.

Since we want the correlation coefficient to be a measure that is unaffected by the particular units of the variables, we divide $\Sigma(x-\bar{x})(y-\bar{y})$ by the product of the sample standard deviations of x and y, s_x and s_y.

Essentially, we are calculating the sum of the products of the z-scores for each pair of data, that is, $\dfrac{(x-\bar{x})}{s_x}$ and $\dfrac{(y-\bar{y})}{s_y}$. Lastly, we divide by $n-1$ to find the average of the products of the z-scores for each pair of data, resulting in the correlation coefficient, r.

$$r = \frac{\Sigma(x-\bar{x})(y-\bar{y})}{(n-1)s_x s_y}$$

Since $\Sigma(x-\bar{x})(y-\bar{y})$ is positive in Figure 4.16(a), the resulting correlation coefficient, r, will be positive. Since $\Sigma(x-\bar{x})(y-\bar{y})$ is negative in Figure 4.16(b), the resulting correlation coefficient, r, will be negative.

In general, if a positive linear correlation exists between x and y, the products $(x-\bar{x})(y-\bar{y})$ will tend to be positive, and $\Sigma(x-\bar{x})(y-\bar{y})$ will also be positive, resulting in a positive correlation coefficient, r. If a negative linear correlation exists between x and y, the products $(x-\bar{x})(y-\bar{y})$ tend to be negative, and $\Sigma(x-\bar{x})(y-\bar{y})$ will also be negative, resulting in a negative correlation coefficient, r.

If you notice that the ordered pairs of the two variables are scattered throughout all four regions as in Figure 4.16 (c), then the resulting correlation coefficient would be *approximately* zero. That is, many of the positive products of $(x-\bar{x})(y-\bar{y})$ would be offset by the negative products of $(x-\bar{x})(y-\bar{y})$ and the sum of the products would be *essentially* zero. This will lead to a correlation coefficient of approximately zero.

Note, the formula for the correlation coefficient, r, can also be written as follows:

$$r = \frac{\Sigma(x-\bar{x})(y-\bar{y})}{(n-1)s_x s_y}$$

$$r = \frac{1}{n-1}\Sigma \frac{(x-\bar{x})}{s_x} \cdot \frac{(y-\bar{y})}{s_y}$$

Substituting $\frac{x-\bar{x}}{s_x} = z_x$ and $\frac{y-\bar{y}}{s_y} = z_y$ into the formula for r we get

$$r = \frac{1}{n-1}\Sigma z_x z_y.$$

Therefore,

$$r = \frac{\Sigma(x-\bar{x})(y-\bar{y})}{(n-1)s_x s_y} = \frac{1}{n-1}\Sigma z_x z_y.$$

This leads us to Definition 4.6, an alternate formula to compute Pearson's Correlation Coefficient, r.

DEFINITION 4.6

An Alternate Formula to compute Pearson's Correlation Coefficient, r, to measure the strength of linear relationship between variables for a sample. This **sample correlation coefficient, r,** is calculated by the following formula:

$$r = \frac{1}{n-1}\Sigma z_x z_y$$

where:
- r is the **sample correlation coefficient**
- n represents the **number of pairs of data values**
- z_x represents the **z score** for the data values for the **first variable, x**
- z_y represents the **z score** for the data values for the **second variable, y**

Now let's examine how to apply the sample correlation coefficient formula in Definition 4.5 to calculate the strength of the linear correlation for the sample data in Example 4.4.

EXAMPLE 4.4

Using the sample data for the number of study hours and exam grade for eight randomly selected students shown in Table 4.5, calculate the sample correlation coefficient, r, using the formula $r = \frac{\Sigma(x-\bar{x})(y-\bar{y})}{(n-1)s_x s_y}$

TABLE 4.5

Student	Number of Study Hours x	Exam Grade y
1	20	85
2	10	65
3	17	75
4	15	70
5	12	75
6	9	72
7	24	86
8	13	72

Section 4.3 The Coefficient of Linear Correlation

Solution

To use the sample correlation coefficient formula: $r = \dfrac{\Sigma(x-\bar{x})(y-\bar{y})}{(n-1)s_x s_y}$ to calculate r, we need to determine the values for n, $\bar{x}$, $\bar{y}$, $\Sigma(x-\bar{x})(y-\bar{y})$, s_x, and s_y. First we will calculate the mean number of study hours, $\bar{x}$, and the mean exam grade, $\bar{y}$.

Step 1 Calculate $\bar{x}$ and $\bar{y}$.

$$\bar{x} = \frac{\Sigma x}{n} = \frac{120}{8} = 15 \text{ and}$$

$$\bar{y} = \frac{\Sigma y}{n} = \frac{600}{8} = 75$$

Remember, you can also find the sample means by using the 2-var stat command in the TI 83/84 calculator. Consider the following scatter diagram in Figure 4.17:

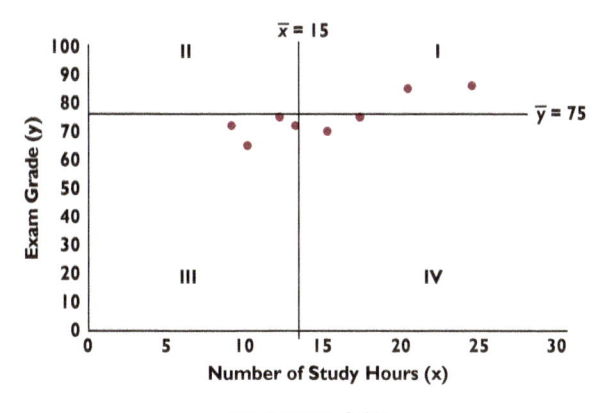

FIGURE 4.17

The vertical line $\bar{x} = 15$ represents the mean number of study hours (x values), and the horizontal line $\bar{y} = 75$ represents the mean exam grade (y values). These lines divide the scatter diagram into four regions. Notice that the ordered pairs tend to lie in regions I and III. Thus the value of r will be positive.

We will now compute the deviations from the mean for each x and y value, that is $(x-\bar{x})$ and $(y-\bar{y})$, as well as the product $(x-\bar{x})(y-\bar{y})$ for each pair of data.

Step 2 Construct a table that has a column for the products $(x-\bar{x})$, $(y-\bar{y})$, and $(x-\bar{x})(y-\bar{y})$.
Now perform each calculation and list the results in the appropriate column, as shown in Table 4.6.

TABLE 4.6

Student	Number of Study Hours x	Exam Grade y	$(x-\bar{x})$	$(y-\bar{y})$	$(x-\bar{x})(y-\bar{y})$
1	20	85	20 − 15 = 5	85 − 75 = 10	(5)(10) = 50
2	10	65	−5	−10	50
3	17	75	2	0	0
4	15	70	0	−5	0
5	12	75	−3	0	0
6	9	72	−6	−3	18
7	24	86	9	11	99
8	13	72	−2	−3	6

Step 3 Using the information in Table 4.6, calculate the quantity $\Sigma(x-\bar{x})(y-\bar{y})$.
Adding the values in the last column we get $\Sigma(x-\bar{x})(y-\bar{y}) = 223$.
Remember, when the ordered pairs tend to be in regions I and III, as illustrated in Figure 4.17, the value of $\Sigma(x-\bar{x})(y-\bar{y})$ will be positive.

Step 4 Calculate S_x and S_y.
Using the 2-var stat command in the TI 83/84 calculator we get:
$s_x \approx 5.13$ and $s_y \approx 7.21$
Recall that **n** is the number of pairs of x and y values. Since there are eight pairs of x and y values, then **n equals 8**.

Step 5 Substitute the calculated values of n, $\Sigma(x-\bar{x})(y-\bar{y})$, s_x, and s_y into the formula and calculate r.

$$r = \frac{\Sigma(x-\bar{x})(y-\bar{y})}{(n-1)s_x s_y}$$

$$r = \frac{223}{(8-1)(5.13)(7.21)}$$

$$r = \frac{223}{258.91}$$

$$r \approx 0.86$$

Thus, the sample correlation coefficient for the data is $r = 0.86$.

You should notice that the sign associated with $r = 0.86$ is *positive*. Whenever the sign associated with a correlation coefficient is positive, this indicates that there is a positive linear correlation between the two variables. Similarly, whenever the sign associated with a correlation coefficient is *negative*, the linear correlation between the two variables is *negative*.

EXAMPLE 4.5

Using the pairs of sample data in Table 4.7, calculate the sample correlation coefficient, r.

TABLE 4.7

x	y
1	7
3	5
5	2
6	3
7	5
8	2

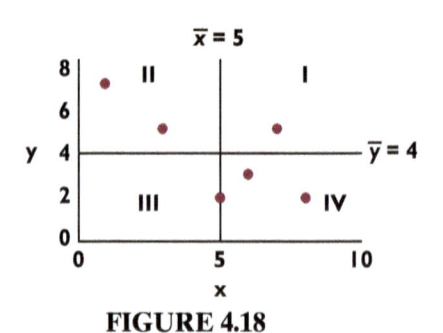

FIGURE 4.18

Solution

Step 1 Calculate $\bar{x}$ and $\bar{y}$

$$\bar{x} = \frac{\Sigma x}{n} = \frac{30}{6} = 5 \text{ and } \bar{y} = \frac{\Sigma y}{n} = \frac{24}{6} = 4$$

In Figure 4.18, the vertical line $\bar{x} = 5$ through the scatter diagram represents the mean of the x values, and the horizontal line $\bar{y} = 4$ represents the mean of the y values of the sample data in Table 4.7. These lines divide the scatter diagram into four regions. Notice that the ordered pairs tend to lie in regions II and IV.

Step 2 Construct a table that has a column for the products $(x-\bar{x})$, $(y-\bar{y})$, and $(x-\bar{x})(y-\bar{y})$.
Now perform each calculation and list the results in the appropriate column. These results are listed in Table 4.8

TABLE 4.8

x	y	$(x-\bar{x})$	$(y-\bar{y})$	$(x-\bar{x})(y-\bar{y})$
1	7	−4	3	−12
3	5	−2	1	−2
5	2	0	−2	0
6	3	1	−1	−1
7	5	2	1	2
8	2	3	−2	−6

Step 3 Using the information in Table 4.8, calculate $\Sigma(x-\bar{x})(y-\bar{y})$.
Adding the values in the last column we get $\Sigma(x-\bar{x})(y-\bar{y}) = -19$.

Step 4 Calculate s_x and s_y.
Using the 2-var stat command in the TI 83/84 calculator we get:
$s_x \approx 2.61$ and $s_y = 2$

Step 5 Substitute the calculated values of $n, \Sigma(x-\bar{x})(y-\bar{y}), s_x,$ and s_y into the formula and calculate r.

$$r = \frac{\Sigma(x-\bar{x})(y-\bar{y})}{(n-1)s_x s_y}$$

$$r = \frac{-19}{(6-1)(2.61)(2)}$$

$$r = \frac{-19}{26.1}$$

$$r \approx -0.73$$

Thus, the sample correlation coefficient for the sample data is $r = -0.73$.

The Computational Formula for Linear Correlation

The definition formula to compute Pearson's Correlation Coefficient, r, defined by Definition 4.5 can be a very cumbersome and tedious formula to use because of all the subtractions. Typically, we calculate the correlation coefficient, r, using a computational formula that involves the use of raw scores rather than z-scores because the calculations tend to be simpler. Definition 4.6 describes the computational formula for calculating Pearson's correlation coefficient, r.

DEFINITION 4.7
The Computational Correlation Coefficient Formula to compute Pearson's Correlation Coefficient, r, is used to measure the strength of a linear relationship between two variables for a sample. This **sample correlation coefficient**, r, is calculated by the following computational formula for r:

$$r = \frac{n\Sigma(xy)-(\Sigma x)(\Sigma y)}{\sqrt{n(\Sigma x^2)-(\Sigma x)^2}\sqrt{n(\Sigma y^2)-(\Sigma y)^2}}$$

where: r is the **sample correlation coefficient**
        x represents the data values for the **first variable**
        y represents the data values for the **second variable**
        n represents the **number of pairs of data values**

Let's apply the computational correlation coefficient formula to calculate the strength of the linear correlation for the sample data in Example 4.4.

EXAMPLE 4.6

a) Using the computational correlation coefficient formula, calculate the sample correlation coefficient, r, for the sample data representing the number of study hours and exam grade for eight randomly selected students in Table 4.9.

TABLE 4.9

Student	Number of Study Hours x	Exam Grade y
1	20	85
2	10	65
3	17	75
4	15	70
5	12	75
6	9	72
7	24	86
8	13	72

b) Compare the sample correlation coefficient value calculated using the computational formula to the value computed using the definition formula in Example 4.4.

Solution

a) To use the computational sample correlation coefficient formula:

$$r = \frac{n\Sigma(xy) - (\Sigma x)(\Sigma y)}{\sqrt{n(\Sigma x^2) - (\Sigma x)^2} \sqrt{n(\Sigma y^2) - (\Sigma y)^2}}$$

to calculate r, we need to determine the values for n, $\Sigma(xy)$, Σx, Σx^2, Σy and Σy^2. To obtain the sums needed to calculate r, we will construct a table that has a column for each of the variables: **x, y, x², y² and xy**. This is the first step in calculating r.

Step 1 Construct a table that lists the variables: x, y, x², y² and xy.
Now, perform the calculation for each of these variables and list the result in the appropriate column. These results are listed in Table 4.10.

TABLE 4.10

x	y	x²	y²	xy
20	85	400	7225	1700
10	65	100	4225	650
17	75	289	5625	1275
15	70	225	4900	1050
12	75	144	5625	900
9	72	81	5184	648
24	86	576	7396	2064
13	72	169	5184	936

Step 2 Using the information in Table 4.10 calculate n, $\Sigma(xy)$, Σx, Σx^2, Σy and Σy^2.
Adding the values within each column, we obtain the following results:

$\Sigma x = 120$ $\Sigma y = 600$ $\Sigma x^2 = 1984$ $\Sigma y^2 = 45364$ $\Sigma(xy) = 9223$

Recall that n is the **number of pairs of x and y values**. Since there are eight pairs of x and y values, then **n equals 8**.

Step 3 Substitute the calculated values of n, Σ(xy), Σx, Σx², Σy and Σy² into the computational formula and calculate r.

$$r = \frac{8(9223)-(120)(600)}{\sqrt{8(1984)-(120)^2}\sqrt{8(45364)-(600)^2}}$$

$$r = \frac{73784-72000}{\sqrt{15872-14400}\sqrt{362912-360000}}$$

$$r = \frac{1784}{\sqrt{1472}\sqrt{2912}}$$

$$r = \frac{1784}{(38.36665)(53.96295)}$$

$$r = \frac{1784}{2070.3776}$$

$$r \approx 0.86$$

b) The sample correlation coefficient calculated using the definition formula in Example 4.4 (r ≈ 0.86) is equal to the sample correlation coefficient value, r ≈ 0.86, calculated using the computational formula.

EXAMPLE 4.7

a) Using the computational correlation coefficient formula, calculate the sample correlation coefficient, r, for the pairs of sample data listed in Table 4.11. Notice this is the same data set as in Example 4.5.

TABLE 4.11

x	y
1	7
3	5
5	2
6	3
7	5
8	2

b) Compare the sample correlation coefficient value calculated using the computational formula to the value computed using the definition formula in Example 4.5.

Solution

a) To use the computational sample correlation coefficient formula:

$$r = \frac{n\Sigma(xy)-(\Sigma x)(\Sigma y)}{\sqrt{n(\Sigma x^2)-(\Sigma x)^2}\sqrt{n(\Sigma y^2)-(\Sigma y)^2}}$$

to calculate r, we need to determine the values for n, Σ(xy), Σx, Σx², Σy and Σy². To obtain the sums required to calculate r, we will construct a table that has a column for each of the variables: **x, y, x², y²** and **xy**. This is the first step in calculating r.

Step 1 Construct a table that lists the quantities: x, y, x², y² and xy.

Now, perform the calculation for each of these variables and list the result in the appropriate column. These results are listed in Table 4.12.

TABLE 4.12

x	y	x²	y²	xy
1	7	1	49	7
3	5	9	25	15
5	2	25	4	10
6	3	36	9	18
7	5	49	25	35
8	2	64	4	16

Step 2 Using the information in Table 4.10 calculate n, $\Sigma(xy)$, Σx, Σx^2, Σy and Σy^2.
Adding the values within each column, we obtain the following results:

$\Sigma x = 30$ $\quad\quad$ $\Sigma y = 24$ $\quad\quad$ $\Sigma x^2 = 184$ $\quad\quad$ $\Sigma y^2 = 116$ $\quad\quad$ $\Sigma(xy) = 101$

Recall that n is the **number of pairs of x and y values**. Since there are six pairs of x and y values, then **n** equals 6.

Step 3 Substitute the calculated values of n, $\Sigma(xy)$, Σx, Σx^2, Σy and Σy^2 into the computational formula and calculate r.

$$r = \frac{6(101)-(30)(24)}{\sqrt{6(184)-(30)^2}\sqrt{6(116)-(24)^2}}$$

$$r = \frac{606-720}{\sqrt{(1104)-(900)}\sqrt{696-576}}$$

$$r = \frac{-114}{\sqrt{204}\sqrt{120}}$$

$$r = \frac{-114}{(14.2829)(10.9545)}$$

$$r = \frac{-114}{156.4620}$$

$$r \approx -0.73$$

b) The sample correlation coefficient calculated using the definition formula in Example 4.5, $r \approx -0.73$, is equal to the sample correlation coefficient value, $r \approx -0.73$, calculated using the computational formula.

TI-84 Graphing Calculator Solution for Example 4.5

The TI-84 graphing calculator can also be used to obtain the sample correlation coefficient, r. We will begin by first entering the x and y data into lists L1 and L2 by pressing STAT ENTER to select **1: EDIT**. Figure 4.19 shows the ordered pairs (x, y) entered in L1 and L2.

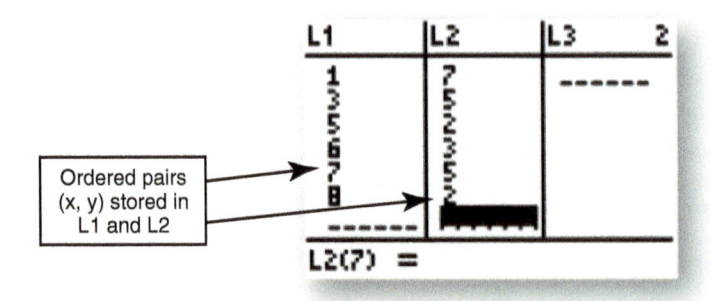

FIGURE 4.19

In order to display the correlation coefficient, r, you must turn on the **DiagnosticOn** function. This is done by pressing 2nd 0 **[Catalog]**, and then scrolling down to the **DiagnosticOn** command. Place the cursor next to this command and press ENTER twice. Figure 4.20a and Figure 4.20b display the calculator's output. (Note: Once the **DiagnosticOn** command is executed, it will remain on unless *you* turn it off.)

Section 4.3 *The Coefficient of Linear Correlation*   191

FIGURE 4.20a   **FIGURE 4.20b**

Press [STAT] [▶] over to CALC, scroll to **8:LinReg(a+bx)** and press [ENTER] as shown in Figure 4.21a. You need to tell the calculator in which lists your x & y data are stored. Enter the list name for the Xlist data by pressing [2nd] 1 [L1] and [ENTER] followed by the list name for the Ylist data by pressing [2nd] 2 [L2] and [ENTER]. Leave the Freqlist and Store RegEQ fields empty by pressing [ENTER] twice as displayed in Figure 4.21b.

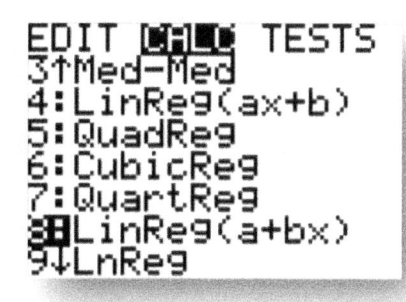

FIGURE 4.21a   **FIGURE 4.21b**

Press [ENTER] and the calculator will display the value of r, as well as other information as shown in Figure 4.22.

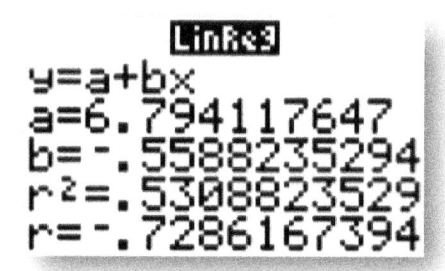

FIGURE 4.22

As you can see the sample correlation coefficient, *r*, to two decimal places is –0.73.

The procedure to calculate the sample coefficient r using the computational formula is outlined in the following box.

Procedure to Calculate the Computational Sample Correlation Coefficient, r

Step 1   Construct a table that contains the variables: x, y, x^2, y^2 and xy.
Perform the calculation for each of these quantities and list the result in the appropriate column of the table.

Step 2   Using the information in the table, calculate:
n, $\Sigma(xy)$, Σx, Σx^2, Σy and Σy^2.

Step 3   Substitute the calculated values of n, $\Sigma(xy)$, Σx, Σx^2, Σy and Σy^2 into the formula and calculate r.
To calculate r, use the computational sample correlation coefficient formula:

$$r = \frac{n\Sigma(xy) - (\Sigma x)(\Sigma y)}{\sqrt{n(\Sigma x^2) - (\Sigma x)^2}\sqrt{n(\Sigma y^2) - (\Sigma y)^2}}$$

When the sample correlation coefficient is greater than zero, that is: r > 0, then there is a positive linear correlation between the two variables.

When the sample correlation coefficient is less than zero, that is: r < 0, then there is a negative linear correlation between the two variables.

Some Cautions Regarding the Interpretation of Correlation Results

There are some common mistakes that people make when they interpret correlation results. To help you avoid these pitfalls, we will discuss two cautions regarding the use of correlation.

Caution #1:
Don't Overlook the Possibility of a Non-Linear Relationship

The correlation coefficient, r, only measures the **linear** relationship between two variables. It is possible for two variables to have a linear correlation near zero, and, yet they could have a significant non-linear relationship. Be careful not to interpret a non-significant linear correlation as meaning there doesn't exist any relationship between the variables.

Caution #2:
Correlation Doesn't Indicate a Cause-and-Effect Relationship

You must be careful not to interpret a significant linear relationship between two variables to imply there exists a *cause-and-effect* relationship. Simply because two variables are correlated does not guarantee a cause-and-effect relationship. Two seemingly non-correlated variables may often be highly correlated.

For example, the number of storks nesting in various European towns in the early 1900's and the number of human babies born in the same towns during this period had a very high correlation. However, we can't conclude that an increase in the number of storks will cause an increase in the number of babies.

Therefore, a significant linear correlation between two variables should not be interpreted to mean that a change in one variable *causes* a change in the other variable, but rather only that changes in one variable are associated with or accompanied by changes in the other variable.

4.4 MORE ON THE RELATIONSHIP BETWEEN THE CORRELATION COEFFICIENT AND THE SCATTER DIAGRAM

We have previously stated that the correlation coefficient, r, measures the strength of a linear relationship between two variables. Furthermore, we have indicated that **−1 ≤ r ≤ 1**. Carefully examine the scatter diagrams in Figures 4.23(a), 4.23(b) and 4.23(c).

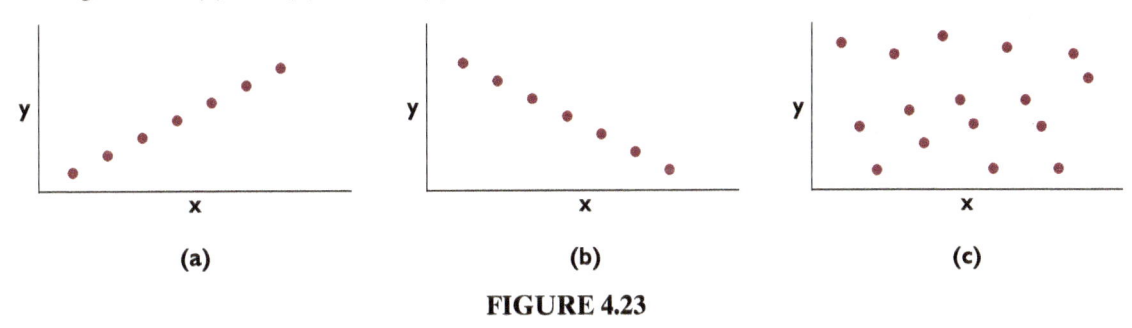

FIGURE 4.23

The scatter diagrams illustrated in Figures 4.23(a) or 4.23(b), have correlation coefficients equal to 1 and −1 respectively. In each case such a value of r would mean that a **perfect linear relationship exists between the x and y variables.** If the value of r is 1, then all the points of the scatter diagram would lie on a line that is sloping upward from left to right as illustrated in Figure 4.23(a) and the relationship between the variables would be referred to as a **perfect positive linear relationship.**

If the value of r is −1, then all the points of the scatter diagram would lie on a line that is sloping downward from left to right as illustrated in Figure 4.23(b) and the relationship between the variables would be referred to as a **perfect negative linear relationship.**

On the other hand if r = 0, then the scatter diagram would indicate no linear correlation as illustrated in Figure 4.23(c). Such a scatter diagram would suggest **no linear** relationship exists between the variables x and y.

In practice, sample correlation coefficient values of r that are 1, or −1, or 0 rarely occur. In the next section we will consider one way to interpret the correlation coefficient specifically for values other than 1, −1, and 0.

4.5 THE COEFFICIENT OF DETERMINATION

An important descriptive statistical measure that can be calculated from the correlation coefficient, r, is called the **Coefficient of Determination**. This statistical measure is used to explain the degree of influence that one variable called the independent variable, usually denoted by x, has on the other variable called the dependent variable, usually denoted by y.

- The coefficient of determination measures how close the data points are to the regression line.
- Thus, the higher the value of the coefficient of determination, the higher percentage of data points the line passes through when the data points and line are plotted together.
- The coefficient of determination, r^2, can be expressed as a percent and the value will be between 0 and 100%.
- A value of 100% means every point on the regression line fits the data; a value of 50% means half of the points lie on the regression line and thus only half (or 50%) of the variation is explained by the regression.
- Thus, the coefficient of determination represents the percentage of variation that can be explained by the regression equation.
- Therefore, the coefficient of determination measures the proportion of the variance of the dependent variable y that can be accounted for by the variance of the independent variable x.

DEFINITION 4.8
The **coefficient of determination** measures the proportion of the variance of the dependent variable **y** that can be accounted for by the variance of the independent variable **x**. It represents the percentage of variation that can be explained by the regression equation and is calculated by squaring the correlation coefficient, r.

Thus, we can write:

$$\text{Coefficient of Determination} = r^2$$

EXAMPLE 4.8

Suppose you are told that a significant linear correlation, r = −0.52, was found between the number of hours per week a child watches television (the independent variable) and the number of hours per week the child spends doing homework (the dependent variable). Using this information, then:
a) Calculate the coefficient of determination, r^2.
b) Interpret the meaning of r^2.

Solution

a) To calculate the coefficient of determination, the correlation coefficient, r, must be determined. Since r was found to be −0.52, the coefficient of determination is found by squaring r. Thus, the coefficient of determination is:

$$r^2 = (-0.52)^2$$
$$r^2 = 0.27$$

b) To convert the coefficient of determination to a percent multiply r^2 by 100%. Thus, in percent form, the coefficient of determination is 27%. This means that 27% of the variance in the number of hours per week a child spends doing homework (the dependent variable) can be accounted for by the variance in the number of hours per week the child watches television (the independent variable).

The coefficient of determination can also be used to suggest the existence of other reasons for the variation of one variable due to the variation in the other. Looking back at Example 4.8, one may conclude that the variation in the hours a child spends doing homework that cannot be accounted for, (i.e., 100% − 27% = 73%) may be attributed to a number of reasons. For example, a child may not be assigned homework on a regular basis. Or, the child is inconsistent in doing homework regularly. Since the coefficient of determination is 27%, we may conclude that there is significant amount of variance in the number of hours per week the child spends doing homework that is **not explained. Thus, 73% reflects this unexplained amount.**

TI-84 Graphing Calculator Solution for Example 4.7

Previously, we outlined how to find the correlation coefficient using the TI-84 graphing calculator. This was done right after Example 4.7. The coefficient of determination was also given as part of the output from the calculator when we found the value of r. Figure 4.24 displays the calculator's output for Example 4.5.

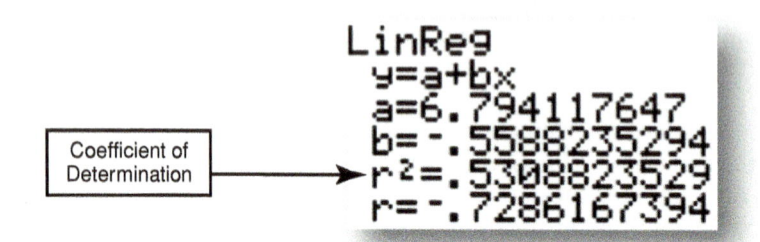

```
LinReg
y=a+bx
a=6.794117647
b=-.5588235294
r²=.5308823529   ← Coefficient of Determination
r=-.7286167394
```

FIGURE 4.24

As you can see the value for the coefficient of determination, r^2, to two decimal places is 0.53. You should also notice that the value for the coefficient of determination is positive although the value for r, the correlation coefficient is negative. The value for the coefficient of determination will always be between 0 and +1 inclusive because of the fact it comes from a number being squared.

4.6 LINEAR REGRESSION ANALYSIS

Once a significant linear correlation has been established between two variables, a **linear model** can be developed to predict a value for the dependent variable, usually denoted as the y variable, given a value for the independent variable, usually denoted as the x variable.

For example, if a significant linear correlation has been established between a father's height (the independent variable) and his son's height (the dependent variable), then it would be useful to develop a linear model that could reasonably *predict* the son's height *given* his father's height. This linear model is used to predict a son's height simply by using the father's height.

You are probably aware that many colleges and universities successfully use a student's standardized test result as a main criteria for acceptance into the school. The schools are able to predict success with a high degree of confidence because research studies have established that there is a strong correlation between the standardized test results and success at the college, as measured by the student's first year grade point average (GPA). To accomplish this, a model is statistically developed that can be used to predict a student's first year GPA from his/her standardized test scores.

These examples show how statistical modeling can be applied to the relationship between two variables. This type of modeling involves **regression analysis**. Regression analysis enables us to develop models so we can make these types of predictions. Linear regression analysis provides us with a linear model or equation that can be used to predict the value of the y or dependent variable given the value of the x or independent variable. The value of y that is predicted by this model is usually not the exact value, however it is a "close" estimate of the actual y value. To determine the linear model that will generate a "close" estimate of the actual y value, we obtain the line that "best fits" all the sample points of the scatter diagram. In regression analysis, the technique used to obtain this best fitting line is called the **method of least squares**. Essentially, this technique selects as the "best fitting" line, the line which minimizes the sum of the squared distances between the predicted y values and the actual or observed y values. This idea is illustrated in Figure 4.25.

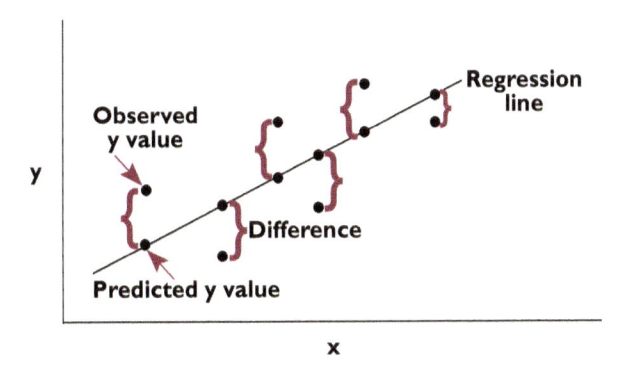

FIGURE 4.25

The best fitting line is called the regression line. It is used to predict the dependent variable (y) based upon the value of the independent variable (x). Let's examine how to obtain the regression line by examining the regression line formula.

Regression Line Formula

$$y' = a + bx$$

where: y' is the predicted value of y, the dependent variable, given the value of x, the independent variable.

and

a and b are the regression coefficients obtained by the formulas:

$$b = \frac{n\Sigma(xy) - \Sigma x \Sigma y}{n(\Sigma x^2) - (\Sigma x)^2}$$

$$a = \frac{\Sigma y - b\Sigma x}{n}$$

Example 4.9 illustrates how to calculate the regression line.

EXAMPLE 4.9

The data in Table 4.13 represents a sample of data pairs collected by a social researcher to examine the relationship between "Annual Family Income" and "Total SAT Score" (out of 2400) for ten students.

Chapter 4 Linear Correlation and Regression Analysis

TABLE 4.13

Family Income (thousands of dollars per year) x	17	31	58	39	135	158	80	75	105	165
SAT (out of 2400) y	1370	1380	1440	1500	1560	1585	1555	1480	1590	1710

Source: Fictitious data, for illustration purposes only

Use the sample data listed in Table 4.13 to determine a regression line to predict the Total SAT score for a student given the student's family income.

Solution

Step 1 Choose the independent (i.e., explanatory) and dependent (i.e., predictive) variables.

To find the regression line, $y' = a + bx$, we must first choose the dependent variable (y). The dependent variable is the variable you want to predict. In this example the Total SAT score out of 2400 is what the regression line equation is trying to predict. Thus, the *Total SAT score out of 2400* is the dependent variable. Therefore, the independent variable (x), is the *Family Income*. This choice of the variable names will produce a regression line model that will *predict* the Total SAT score, y', given the Family Income, x.

Step 2 Calculate the regression constants, a and b.

Of the two regression constants, b **must** be computed **first**, since the calculated value of b is *used* in the formula to calculate a.

To calculate b, use the formula:

$$b = \frac{n\Sigma(xy) - \Sigma x \Sigma y}{n(\Sigma x^2) - (\Sigma x)^2}$$

Since the calculation of b requires Σx, Σy, Σy^2, Σx^2, Σxy and n, use Table 4.14 to help organize the calculations

TABLE 4.14

Family Income (1000 dollars) x	Total SAT score (out of 2400) y	x y	x²
17	1370	23290	289
31	1380	42780	961
58	1440	83520	3364
39	1500	58500	1521
135	1560	210600	18225
158	1585	250430	24964
80	1555	124400	6400
75	1480	111000	5625
105	1590	166950	11025
165	1710	282150	27225
$\Sigma x = 863$	$\Sigma y = 15170$	$\Sigma xy = 1353620$	$\Sigma x^2 = 99599$

Since there are 10 pairs of sample data, n=10.

To calculate the regression coefficient b, substitute the above values into the formula for b:

$$b = \frac{n\Sigma(xy) - \Sigma x \Sigma y}{n(\Sigma x^2) - (\Sigma x)^2}$$

$$b = \frac{10(1353620) - (863)(15170)}{10(99599) - (863)^2}$$

$$b = \frac{(13536200) - (13091710)}{(995990) - (744769)}$$

$$b = \frac{444490}{251221}$$

$$b \approx 1.7693$$

$$b \approx 1.77 \text{ (rounded to two decimal places)}$$

To calculate the regression coefficient a, use the formula:

$$a = \frac{\sum y - b \sum x}{n}$$

Since $\sum x = 863$, $b \approx 1.7693$, $\sum y = 15170$, and $n = 10$

$$a \approx \frac{(15170) - (1.7693)(863)}{10}$$

$$a \approx 1,364.31$$

Step 3 Substitute the values of a and b into the equation for the regression line:

$$y' = a + bx$$

$$y' = 1364.31 + 1.77x$$

TI-84 Graphing Calculator Solution for Example 4.9

The procedure in the TI-84 graphing calculator used to find the values for r, the correlation coefficient and r^2, the coefficient of determination, can also be used to find the regression values a and b for the regression line. We will begin by first entering the x and y data into lists L1 and L2. You do this by pressing STAT ENTER as shown in Figure 4.26.

FIGURE 4.26

Press STAT ▶ over to CALC, scroll to **8:LinReg(a+bx)** and press ENTER as shown in Figure 4.27a. You need to tell the calculator in which lists your x and y data are stored. Enter the list name for the Xlist data by pressing 2nd 1 [L1] and ENTER, followed by the list name for the Ylist data by pressing 2nd 2 [L2] and ENTER. Leave the Freqlist field empty by pressing ENTER. You can store the regression equation by pressing VARS ▶ over to Y-VARS to select 1:Function and select 1:Y1 by pressing ENTER as displayed in Figure 4.27b.

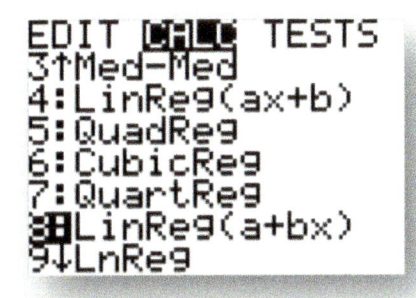

FIGURE 4.27a   **FIGURE 4.27b**

Press ENTER twice. The calculator displays the value of r, the correlation coefficient, r^2, the coefficient of determination and the values for the slope, b, for the regression line and the y-intercept, a, for the regression line as shown in Figure 4.28.

```
LinReg
y=a+bx
a=1364.307801
b=1.769318648
r²=.7987451206
r=.8937254168
```

Regression Coefficients

FIGURE 4.28

Using the regression values a and b (rounded appropriately) displayed in Figure 4.28, the regression line is $y' = 1364.31 + 1.77x$.

Let's summarize the procedure to calculate the regression coefficients a and b, and the regression line $y' = a + bx$

Procedure to Determine the Regression Coefficients and Regression Line

Step 1 Choose the independent and dependent variables.
The dependent variable is the variable you are going to predict.

Step 2 Calculate the regression coefficients, a and b.
Calculate b first. Use the formula:

$$b = \frac{n\Sigma(xy) - \Sigma x \Sigma y}{n(\Sigma x^2) - (\Sigma x)^2}$$

Calculate a. Use the formula:

$$a = \frac{\Sigma y - b\Sigma x}{n}$$

Step 3 Substitute the values of a and b into the equation for the regression line:

$$y' = a + bx$$

EXAMPLE 4.10

In Example 4.9 we found the regression line to be

$$y' = 1364.31 + 1.77x$$

This formula represents a *linear* model of the *best fitting line* for the sample data in Table 4.14. Use this regression line model to predict the SAT score for a student whose family income is:
 a) $60,000
 b) $200,000
 c) $0

Solution

Since we want to predict the student's SAT score, this represents the dependent variable y. Thus, the Annual Family Income in thousands of dollars represents the variable x. The regression line equation,

$$y' = 1364.31 + 1.77x$$

is a linear model to predict the SAT score y' given the student's Annual Family Income (in thousands), x.
a) If x= 60, (the 60 represents $60,000) then

$$y' = 1364.31 + 1.77(60)$$

$y' = 1470.51$ is the predicted SAT score for a student whose Annual Family Income is $60,000.

b) If x= 200, (the 200 represents $200,000) then

$$y' = 1364.31 + 1.77(200)$$

$$y' = 1718.31$$

is the predicted SAT score for a student whose Annual Family Income is $200,000

c) If x =0 dollars of Annual Family Income, then

$$y' = 1364.31 + 1.77(0)$$

$$y' = 1364.31$$

is the predicted SAT score for a student whose Annual Family Income is $0.

Interpolation versus Extrapolation

The regression models we have been studying predict a value of y (dependent variable) for a given value of x (independent variable). **Interpolation** is the name given to the process of making predictions for y within the range of the values for the independent variable x. In Example 4.9, the values for the independent variable, x, "Annual Family Income" ranged between 17 and 165 thousand dollars. Consequently, it is "safe" to use the regression equation to predict a value of y for values of x between 17 and 165 thousand dollars.

Extrapolation is making a prediction for y based on values of x that are outside of the range of values used to determine the regression model. You must be cautious when using the regression model to predict outside of this range since there is no guarantee that the relationship between the two variables will still be linear once you are outside of the range of x values used to determine the regression model.

In Example 4.10, we used the regression equation to predict the SAT score (y') for x values of 60, 200 and 0 thousands of dollars. The values for x that were used to determine the regression model ranged from 17 to 165 thousand dollars. The value 60 is well within the range. The predicted value 1470.54 is an *interpolated* value. The 200 is just outside the range of values of x. Thus, the predicated value 1718.3 is considered an *extrapolation*. Using the regression model for the value 60 would be considered safe. However, a value of 0 is also outside of the range of values for *x* and reflects a prediction of zero family income. So, both 200 and 0 would be considered extrapolated values. Trusting the result of the regression model for extrapolated values would be risky. We would not feel confident that the relationship between Annual Family Income and SAT score has a linear relationship for these extrapolated values.

EXAMPLE 4.11

Graph the regression line equation and the scatter diagram for the sample data pairs from Example 4.9, on the *same* axes. The regression line model equation from Example 4.9 is:

$$y' = 1364.31 + 1.77x.$$

The sample data is listed in Table 4.15

TABLE 4.15

Family Income (thousands of dollars per year) x	17	31	58	39	135	158	80	75	105	165
SAT (out of 2400) y	1370	1380	1440	1500	1560	1585	1555	1480	1590	1710

Solution

The scatter diagram for these data pairs, where the independent variable x represents the Family Income in thousands of dollars per year and the dependent variable y represents the SAT Score, is shown in Figure 4.29. The regression line has also been drawn on the axes. Notice that the regression line has the property that it **"best fits"** the data.

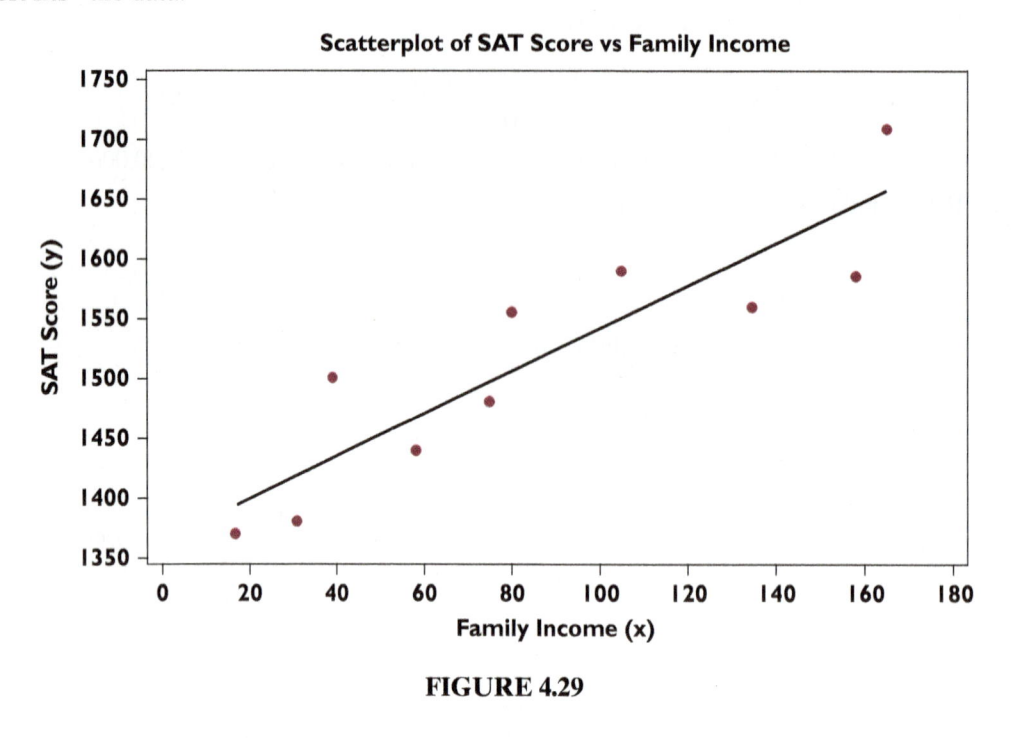

Scatterplot of SAT Score vs Family Income

FIGURE 4.29

TI-84 Graphing Calculator Solution for Example 4.11

We can also graph the regression line along with the scatter diagram for Example 4.11 on the TI-84 graphing calculator. First, we must create the scatter diagram on the calculator. This procedure is outlined at the end of Example 4.3 on page 180. Figure 4.30 displays the scatter diagram for the data in Table 4.15.

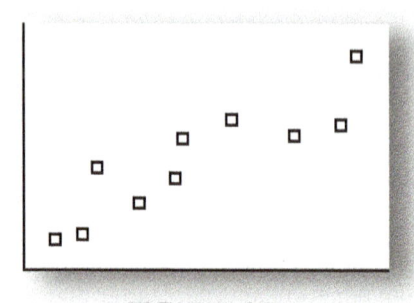

FIGURE 4.30

In Example 4.9, we used the TI-84 graphing calculator to find the regression values a and b for the regression line. We found the equation of the regression line to be $y' = 1364.31 + 1.77x$ and we stored the equation in the Y= editor. Press Y= and you will see the equation for the regression line as shown in Figure 4.31a. Press GRAPH to graph the regression line with the scatter diagram on the same set of axes. This is shown in Figure 4.31b.

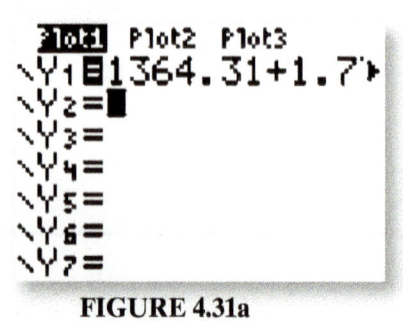

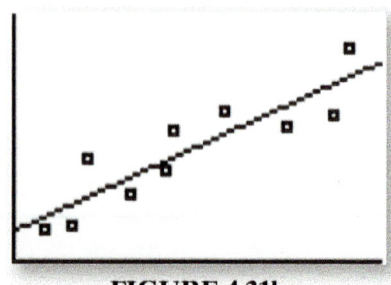

FIGURE 4.31a   **FIGURE 4.31b**

We can also use the graph of the regression line to generate the answers for Example 4.10. Example 4.10a asked us to use the regression equation to predict the SAT score for a student whose family income is $60,000. The calculator has a built-in feature to do this. To use it, press 2nd TRACE ENTER to select the function **1:Value**. Figure 4.32a and Figure 4.32b show this option.

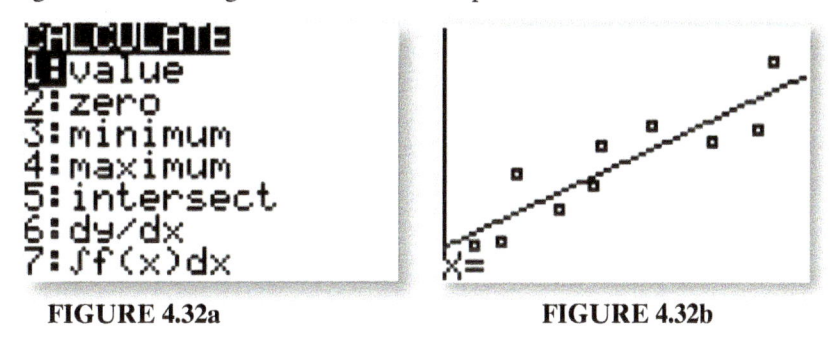

FIGURE 4.32a   **FIGURE 4.32b**

Notice that after you select this option, the calculator's cursor is blinking. It is waiting for you to input an x-value. Input the value 60 for x. Figure 4.33 displays the result given by the calculator for x = 60.

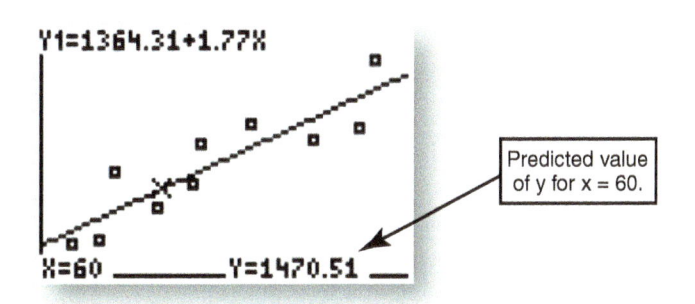

FIGURE 4.33

The predicted value given for y, the SAT score, is 1470.51. This is the same answer obtained in Example 4.10a. Using the same procedure, you can also generate the answers to parts b and c.

Explanation and Assumptions for Linear Regression Analysis

To help explain the idea of Regression Analysis, we will examine two examples of fictitious data sets x and y values. Before we consider these data sets, let's discuss the regression concept. The statistical term "regression," is derived from a Latin root meaning "going back." The symbol r used to represent the sample correlation coefficient means "regression." Francis Galton in his paper "Regression towards Mediocrity in Hereditary Stature" related the heights of children to the average height of their parents.

Galton observed that the heights of the children tended to be more moderate than the heights of their parents. He observed the following phenomenon: if parents were very tall the children tended to be tall but shorter than their parents, i.e. regressed toward the mean. If parents were very short the children tended to be short but taller than their parents, i.e. regressed toward the mean. Galton called his discovery "*regression toward mediocrity*," which in modern terms is referred to as a "*regression to the mean*."

Regression to the mean is a typical occurrence. For example, college students can be expected to score either better or worse on their midterm exams than their performance on their final exams. Their score on a final exam in a course can be *expected* to be less good (or bad) than their score on the midterm exam. Suppose we select a sample of college students whose grades were much higher than the mean on the midterm exam. We will assume their grades were unusually high in part because they were "good students" and in part because they were also "lucky." The fact that they performed so well on the midterm makes it probable that *both* their ability and their luck were better than average. On the final, they may be just as good, but they probably will not be as "lucky". So we would expect that on the final their performance will be closer to the mean grade. Basically, for an imperfect correlation between two variables, it is expected that the more extreme values of one variable to be associated with less extreme values of the other, which is said to 'regress' towards its mean value'.

Regression towards the mean is commonly found when two variables are imperfectly correlated (i.e., have a correlation other than $r = 1$ or $r = -1$). The less correlated the two variables, the larger the effect of regression to the mean. If the correlation between the two variables x and y is zero ($r = 0$), then the best prediction for any x value is the mean of the y values. However, as the correlation (r) increases, a better prediction can be made. If $r = 1.00$, then the prediction is perfect. But as the correlation departs from $r = 1.00$ and approaches $r = 0$, then the prediction from x to y becomes less perfect.

For $r = 1.00$, the x and y values will lie on a straight line which represents a linear relationship of the form: $y' = a + bx$. The closer the value of r to 1.00, the closer the points will lie to the straight line.

To illustrate the idea of statistical regression, we will examine two fictitious examples. Notice both of the following examples contain the same x and y values but the y values are listed in a different order. This will result in the sample correlation coefficient, r, being different for each example.

Example I		Example II	
x	y	x	y
2	3	2	5
4	5	4	9
6	6	6	6
8	8	8	3
10	9	10	8

In Example I, the sample correlation coefficient, r, is 0.99, while the sample correlation coefficient, r, is 0 for Example II. Notice that just changing the order of the y values will change the value of r from $r = 0.99$ to $r = 0$.

The linear regression equation: $y' = a + bx$,

where: x = independent variable
a = intercept constant
b = regression coefficient
y' = predicted y values for each independent variable value x

The regression equation is basically a prediction formula and is used to predict a y value (y') for each x value. The correlation coefficient (r) value will help to determine how "good" the predicted y value (y') is for each x value.

For each of the two examples, the regression equations are:

$$\text{Example I: } y' = 1.7 + 0.75x$$

$$\text{Example II: } y' = 6.2 + 0x$$

Each regression line represented by a straight line has been drawn through the plotted points as shown in Figure 4.34a and 4.34b.

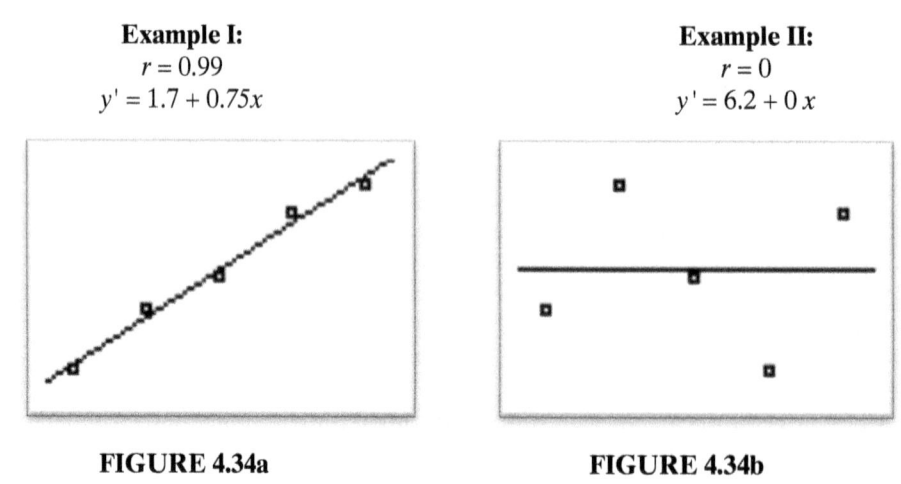

Example I:
$r = 0.99$
$y' = 1.7 + 0.75x$

Example II:
$r = 0$
$y' = 6.2 + 0x$

FIGURE 4.34a **FIGURE 4.34b**

In Figure 4.34a, where r = 0.99, notice the regression line which expresses the regression of y on x lies extremely close to the plotted points. However, in Figure 4.34b, the regression line when $r = 0$, the line does *not* lie close to all the points since the points are in effect randomly dispersed from the line. In Example I, where $r = 0.99$, the points essentially lie on the regression line. On the other hand, when $r = 0$, the points are widely scattered from the regression line. So, the closer the correlation coefficient, r, is to zero, the more the points are scattered. The regression equation for $r = 0$ is:

$$y' = 6.2 + 0x$$

This regression equation indicates that all the predicted y values (y') are 6.2, the mean of all the y values. Consequently, when $r = 0$, the best prediction for the y values is the mean.

However, when $r = 0.99$ (essentially $r = 1$) the predicted y value is determined by adding the value of "$a = 1.7$" to 0.75 of the x value.

That is,

$$y' = 1.7 + 0.75x$$

For example, for $x = 2$, the predicted y value is:

$$y' = 1.7 + 0.75(2) = 3.2$$

If all the predicted y values were computed and plotted with the respective x value, these points would all lie on the regression line: $y' = 1.7 + 0.75x$. That is, the regression line represents the set of all predicted y values for the given x values.

From these two examples, you should notice that that higher the correlation coefficient, r, the more accurate the prediction. The accuracy of the prediction of the two examples can be shown by calculating the differences (d) between the original y values and the predicted y values (y').

That is:

$$d = y - y'$$

and computing the sum of the squares of these differences. The differences (d) are referred to as residuals.

Table 4.16 shows the two sets of residuals, d, and the sums of the squared residuals, d², have also been computed.

TABLE 4.16

		Example I (r = 0.99)					Example II (r = 0)		
x	y	y'	d	d^2	x	y	y'	d	d^2
2	3	3.2	−0.02	0.04	2	5	6.2	−1.2	1.44
4	5	4.7	0.30	0.09	4	9	6.2	2.8	7.84
6	6	6.2	−0.02	0.04	6	6	6.2	−0.2	0.04
8	8	7.7	0.30	0.09	8	3	6.2	−3.2	10.24
10	9	9.2	−0.02	0.04	10	8	6.2	1.8	3.24

The two values of Σd^2 for each example are:

Example I: $\Sigma d^2 = 0.30$ Example II: $\Sigma d^2 = 22.8$

Notice for $r = 0$, $\Sigma d^2 = 22.8$ is much larger than $\Sigma d^2 = .30$ for $r = 0.99$. Consequently, we can conclude the higher the correlation coefficient, r, the smaller the deviations from the predicted y values and the smaller the residuals and thus, the more accurate the prediction.

4.7 RESIDUALS AND ANALYZING RESIDUAL PLOTS

In Section 4.2, we analyzed scatter plots to determine if a data set appeared to have a linear trend. Once we had some "good faith" the data in the scatter plot was in fact linear, we then developed a technique to find the equation of the regression line. The formula for the regression line is given by

$$y' = a + bx$$

where **y'** is the predicted value of y, the dependent variable, given the value of x, the independent variable.

and

a and **b** are the regression coefficients.

This regression equation could then be used to calculate predictions for y within the range of the values for the independent variable x. (Remember Interpolation vs. Extrapolation!)

How "good" were these predictions? One methodology we learned that described the linear association in the data set was the correlation coefficient. Recall that the correlation coefficient *r* measures only linear association; how closely the data fall on a straight line. An *r*-value close to 1 showed a strong positive linear association between the two variables, where an r value of -1 showed a strong negative association, while an *r*-value of 0 shows no linear association between the two variables.

A couple of natural questions arise here. First, is the correlation coefficient the only thing we can use to determine the association (or lack of) between the two variables? Also, how do we know whether the regression we performed is actually an appropriate summary of the given set of data?

Analysis of Scatter Diagrams

One way to answer this question is to have a little more experience in analyzing a scatter diagram. An issue that often arises in a data set is the presence of an outlier. Remember from Chapter 3 that an outlier is a data value that lies several standard deviations from the mean of the data set. The addition of an outlier can significantly change the value of the correlation coefficient, r. For example, Figures 4.35a and 4.35b contain the same data points with the exception of the added outlier in Figure 4.35b.

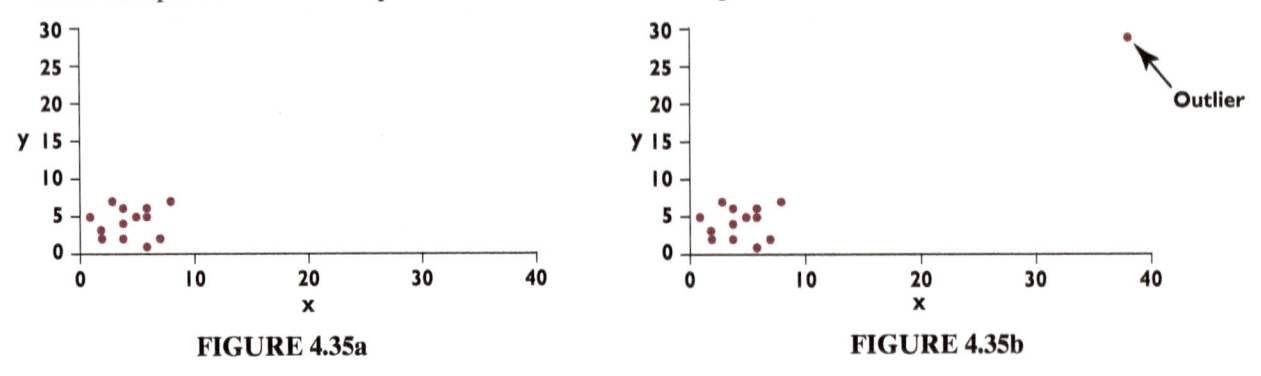

FIGURE 4.35a

FIGURE 4.35b

The computed r value for the scatter diagram in Figure 4.35a is almost 0 (it is actually 0.05). The addition of the outlier in Figure 4.35b significantly changes the value of r to 0.95! Therefore by itself, the correlation coefficient *r* is not a good summary of linear association if the data set has outliers.

Homoscedasticity and Heteroscedasticity

In a scatter diagram, the measure of the horizontal spread is the standard deviation of the variable graphed along the horizontal or x axis, and is denoted by s_x. In a similar fashion, the measure of the vertical spread is the standard deviation of the variable graphed along the vertical or y axis and is denoted s_y. If you can picture vertical slices in a scatter diagram in various locations and the scatter in the y variable (s_y) is approximately the same at all of these slices, then both the data and scatter diagram are said to be **homoscedastic**. Homoscedastic meaning <u>same scatter</u>.

If a scatter diagram is not homoscedastic, then it is referred to as **heteroscedastic**. This idea can be seen in Figures 4.36a and 4.36b.

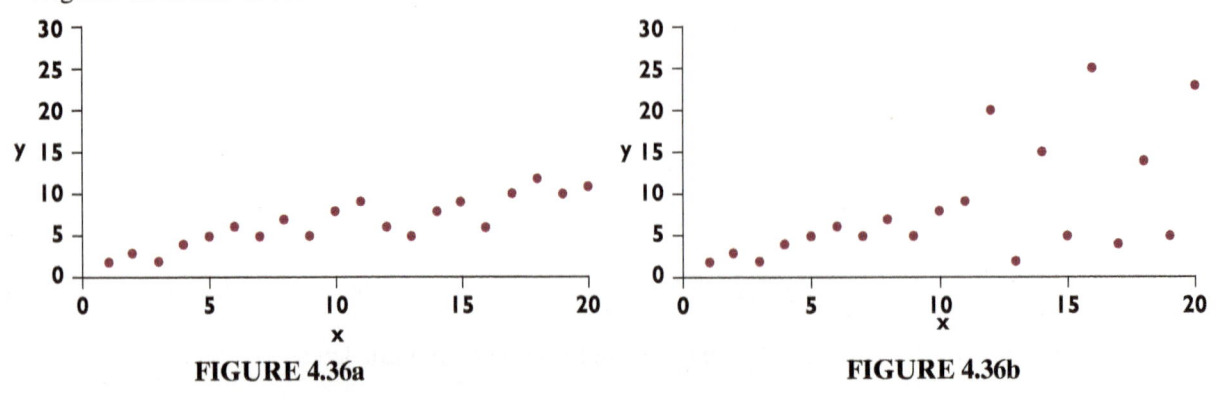

FIGURE 4.36a

FIGURE 4.36b

Figure 4.36a contains a homoscedastic scatter diagram, where the scatter diagram in Figure 4.36b is heteroscedastic. The correlation coefficient of the data in Figure 4.36a is approximately $r = 0.87$, where in Figure 4.36b, r is approximately 0.60. Clearly there is little to no association in the variables of Figure 4.36b after the x value of 10. However, the correlation coefficient of approximately 0.60 tells us otherwise. Because of this ambiguity, the only logical conclusion is that the correlation coefficient r is not a good summary of association *if* the data (and scatter diagram) are heteroscedastic.

Non-Linear Trends

Recall that a data set can be described as curvilinear. Figure 4.37 shows a scatter diagram with a curvilinear relationship.

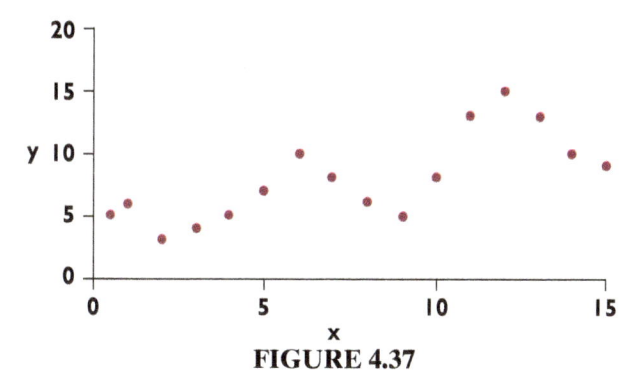

FIGURE 4.37

As you can see from the scatter diagram in Figure 4.37, the relationship is clearly not linear. However, calculating the correlation coefficient gave a value of $r = 0.73$! We must be careful not to misinterpret this value of r. If you look at the overall trend in the data of the scatter diagram, it is increasing which is why we have a positive value for r. However, the correlation coefficient does not indicate how strong the relationship is between the two variables because this trend is not linear. Therefore, if your scatter diagram has a curvilinear relationship, then it makes no sense to perform a linear regression analysis on the data.

> **To summarize, if the data has outliers, are heteroscedastic or have a non-linear trend then the regression line is not a good model of the data, and it is not appropriate to use a linear regression model for predicted values of y.**

What is a Residual?

The vertical distance from a particular data value to the regression line is called a residual, also referred to as a vertical residual. You can think of a residual as the vertical distance by which the regression line "misses" the particular data value. This concept is shown in Figure 4.38.

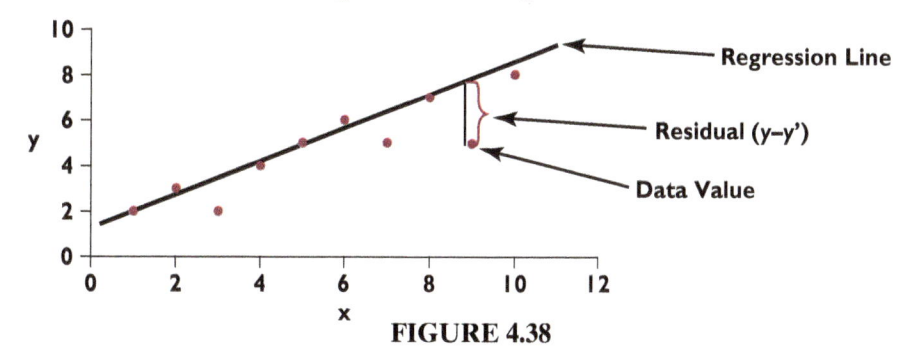

FIGURE 4.38

Remember that the regression equation $y' = a + bx$ gives a predicted value for y given an x value. So a residual can be thought of as the difference between the actual and predicted values for y. Symbolically this is written as $y - y'$. Thus,

Residual = (actual value of y) − (predicted value, y') = $y - y'$

To demonstrate how to calculate residuals, let us revisit **Example 4.9** in Section 4.6.

EXAMPLE 4.12

The data in Table 4.17 represents a sample of data pairs collected by a social researcher to examine the relationship between "Family Income" and "Total SAT Score" (out of 2400) for ten students. Recall that "Family Income" is in thousands of dollars.

TABLE 4.17

Family Income (x)	17	31	58	39	135	158	80	75	105	165
SAT Score (y)	1370	1380	1440	1500	1560	1585	1555	1480	1590	1710

a) Use the linear regression equation to calculate the predicted value, y', for each of the listed Family Income values.

b) Using each pair of actual and predicted SAT scores, calculate the residual.

Solution

a) The regression equation found in **Example 4.9** was given by

$$y' = 1364.31 + 1.77x$$

To find the predicted value, y', substitute each of the given Family Income (x) values into the regression equation we found in **Example 4.9**. This procedure will generate a predicted y value (y') for each x-value. For example, the Family Income of $x = 17$ would give a predicted SAT Score of

$$y' = 1364.31 + 1.77(17) = \mathbf{1394.40}$$

That is,

Family Income (x)	Actual SAT Score (y)	Predicted SAT Score (y')
17	1370	**1394.40**

Continuing to do this for each of the given Family Income values gives the following predicted SAT scores:

Family Income (x)	Actual SAT Score (y)	Predicted SAT Score (y')
17	1370	1394.40
31	1380	1419.18
58	1440	1466.97
39	1500	1433.34
135	1560	1603.26
158	1585	1643.97
80	1555	1505.91
75	1480	1497.06
105	1590	1550.16
165	1710	1656.36

b) Next, we calculate the residual using the formula

Residual = (actual value of y) − (predicted value, y') = $y - y'$

For example, to calculate the residual at the Family Income of $x = 17$, subtract the predicted (y') value *from* the actual(y) value.

$$\text{Residual} = y - y' = 1370 - 1394.40 = \mathbf{-24.40}$$

That is,

Family Income (x)	Actual SAT Score (y)	Predicted SAT Score (y')	Residual (y − y')
17	1370	1394.40	**−24.40**

Continuing to do this for each of the actual and predicted values gives the following residual results:

Family Income (x)	Actual SAT Score (y)	Predicted SAT Score (y')	Residual (y – y')
17	1370	1394.40	–24.40
31	1380	1419.18	–39.18
58	1440	1466.97	–26.97
39	1500	1433.34	66.66
135	1560	1603.26	–43.26
158	1585	1643.97	–58.97
80	1555	1505.91	49.09
75	1480	1497.06	–17.06
105	1590	1550.16	39.84
165	1710	1656.36	53.64

Residual Plots

A natural question you may ask at this point is, *"Why do we need to calculate these residuals?"* We began this section discussing problems that arise with regression. Remember, we discussed earlier that if the data have outliers, are heteroscedastic or have a non-linear trend then the regression line is not a good model of the data, and it is not appropriate to use a linear regression model for predicted values of y.

As it turns out, detecting these "issues" that occur in linear regression is often more difficult than one may think. When using the scatter diagram alone, a statistician may not immediately see an issue such as heteroscedacity or a non-linear trend. In fact, issues with regression are generally easier to detect by creating a **residual plot**.

> **DEFINITION 4.9**
>
> A **residual plot** is a scatter diagram using the original x-values along the horizontal axis, and the residuals along the vertical axis.

As seen in **Example 4.12**, some residuals are positive values where others are negative. It is important to note that if a particular measured y-value falls on the regression line, then the residual would be 0. You may ask, *"What is so special about a residual plot?"* Actually there really shouldn't be *anything* special about your residual plot. A residual plot should not have any special patterns or predictive trends. Residuals should not be extraordinarily high or low. If a data set is in fact linear, then the residual plot for the data set has values that are centered about zero and scattered throughout the range of given data values. Figure 4.39a and Figure 4.39b show a scatter diagram containing a linear trend along with its residual plot.

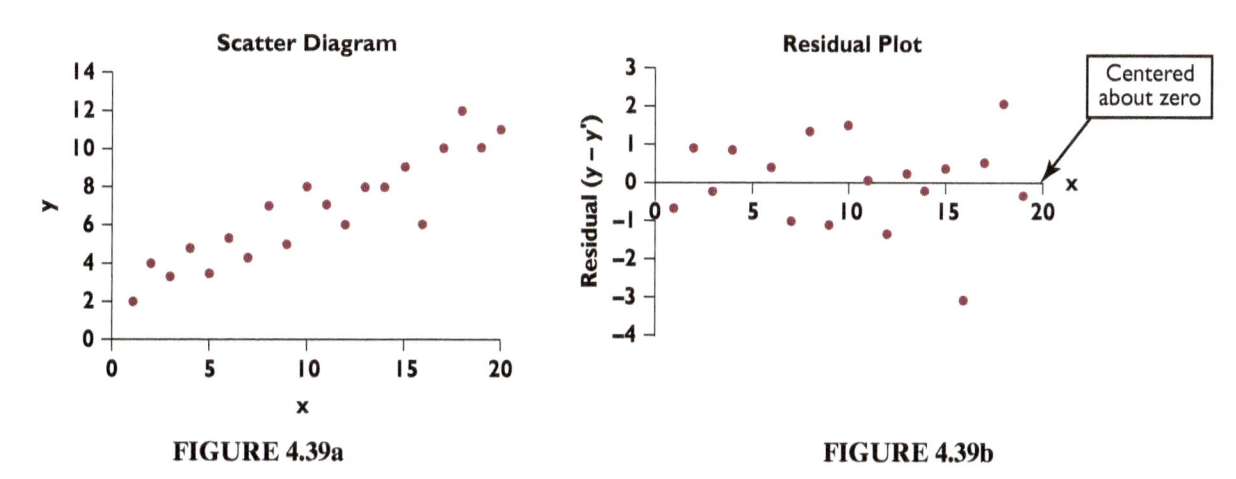

FIGURE 4.39a

FIGURE 4.39b

Figure 4.40 shows a residual plot of a data set that is heteroscedastic. Note how the vertical spread (s_y) of the residuals increase as the x-values increase.

FIGURE 4.40

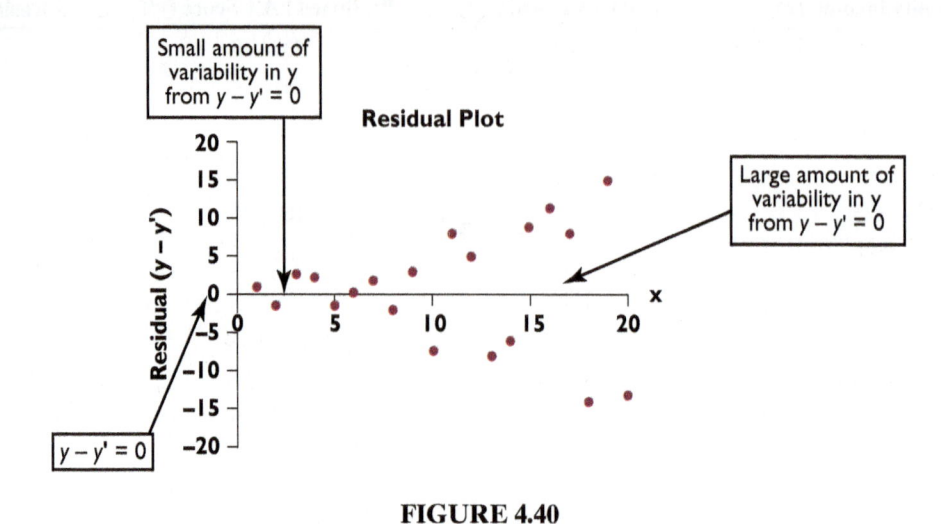

Figure 4.41 shows a residual plot resulting from curvilinear data while Figure 4.42 shows a residual plot of a data set containing outliers.

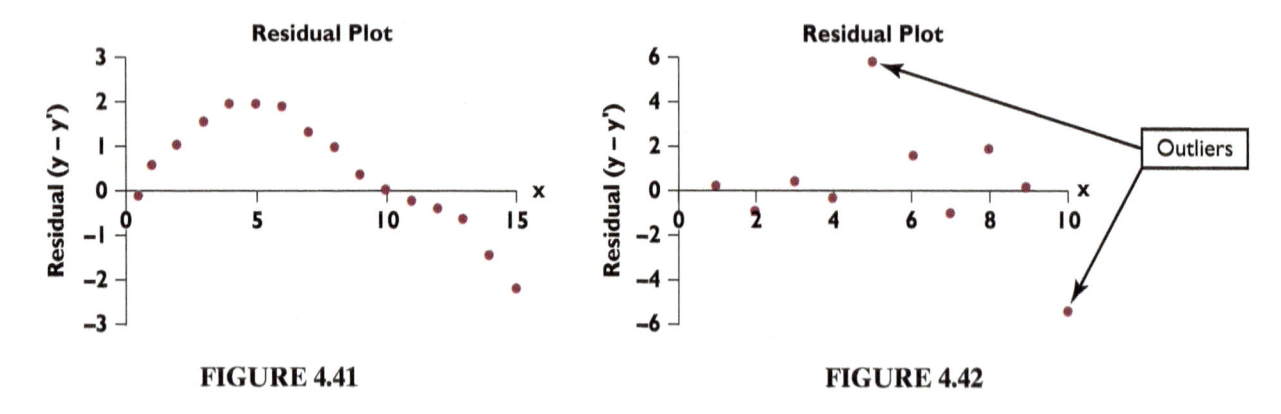

FIGURE 4.41

FIGURE 4.42

It is *important* to remember why we are looking for these situations. The big question is, *"Does the data set we are analyzing in fact have a linear trend?"* If the answer to this question is *no*, then the model we came up with would not be a good summary of the data, and it is not appropriate to use a linear regression model for predicted values of y. Finding the residual plot makes the detection of these *issues* easier.

Let's examine a procedure to create and interpret a residual plot.

Procedure to Create and Interpret a Residual Plot

Step 1 Calculate all of the predicted values for the dependent variable (y') using the regression equation y' = a + bx.

Step 2 Calculate the residuals by subtracting the predicted values calculated in Step 1 from the actual data values that were observed (y − y').

Step 3 Create a scatter diagram using the original x-values observed along the horizontal axis, and the residuals along the vertical axis.

Step 4 If the residual plot has no patterns, then the linear regression model is a good predictor for y. By patterns, we mean the residual plot does not show any outliers, or evidence of heteroscedacity or curvilinear data.

EXAMPLE 4.13 (EXAMPLE 4.9 Revisited)

The data in Table 4.18 represents a sample of data pairs collected by a social researcher to examine the relationship between "Family Income" and "Total SAT Score" (out of 2400) for ten students. Recall that "Family Income" is in thousands of dollars. Create a *residual plot* for the data in Table 4.18.

TABLE 4.18

Family Income (x)	17	31	58	39	135	158	80	75	105	165
SAT Score (y)	1370	1380	1440	1500	1560	1585	1555	1480	1590	1710

Solution

In Example 4.9, we calculated the regression equation and used this regression equation to predict the SAT Score (y) at each of the given Family Income (x) values. We then subtracted each of the predicted SAT Scores (y') from the observed SAT Scores (y) in order to find the residual at each of the given Family Income (x) values. Remember,

$$\text{Residual} = (\text{actual value of } y) - (\text{predicted value, } y') = y - y'$$

The results we previously obtained are given below in Table 4.19.

TABLE 4.19

Family Income (x)	Actual SAT Score (y)	Predicted SAT Score (y')	Residual (y − y')
17	1370	1394.40	−24.40
31	1380	1419.18	−39.18
58	1440	1466.97	−26.97
39	1500	1433.34	66.66
135	1560	1603.26	−43.26
158	1585	1643.97	−58.97
80	1555	1505.91	49.09
75	1480	1497.06	−17.06
105	1590	1550.16	39.84
165	1710	1656.36	53.64

Remember that a residual plot is a scatter diagram formed by using the original x-values observed along the horizontal axis, and the residuals along the vertical axis. The residual plot is given in Figure 4.43.

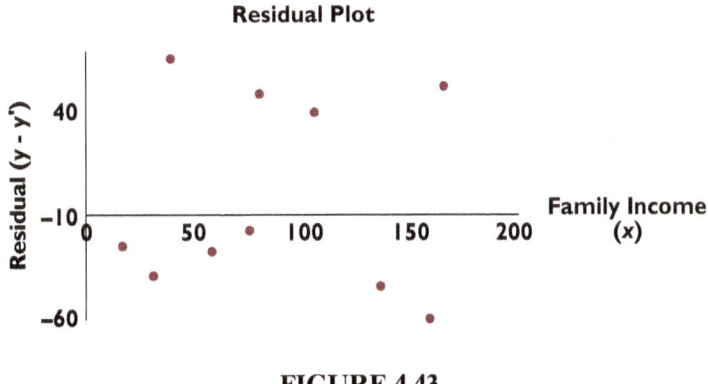

FIGURE 4.43

There are a few observations that should be made regarding this residual plot. First, notice how the residuals are centered about the line $y - y' = 0$. The x-axis isn't at the bottom of the graph as it usually is. Instead, the location of the x-axis is in the center of the graph. Second, notice that there is no pattern such as curvilinear trend in the residual plot as illustrated in Figure 4.43. Also, calculating the 1-variable statistics on the residuals gives a mean of approximately 0, and a standard deviation of approximately 47. Since there are no residuals that are more than 3 standard deviations from the mean, there are no outliers. Consequently, this residual plot isn't giving us any reason to question the linear tendency of the original data set.

TI-84 Graphing Calculator Solution for Example 4.13

The TI-84 graphing calculator can also be used to calculate residuals, and then create a residual plot of a set of data. Previously, we outlined how to use the linear regression model to predict y-values and to calculate the residual. Let's investigate this procedure using the TI-84 graphing calculator.

Recall that in order to find the regression equation, we must first enter the x data from **Example 4.13** into L1 in the calculator and the y data in L2. To access the lists in the calculator, press STAT then **1:EDIT**. Enter the x and y data into **L1** and **L2** respectively. Once the data is in the calculator, then you can perform the linear regression by pressing STAT ▶ over to **CALC** and choose option **8:LinReg(a+bx)**. Remember that this procedure was shown in **Example 4.5** on page 190. Recall that the diagnostics feature of the calculator must be turned on. Figures 4.44a and 4.44b show the output of entering the data and performing the linear regression.

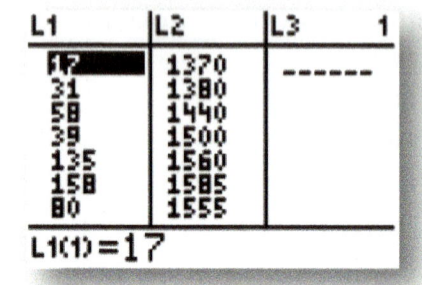

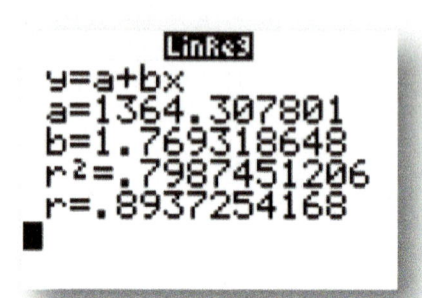

FIGURE 4.44a   **FIGURE 4.44b**

From this output we can see the regression equation, $y' = a + bx$ is given by $y' = 1364.31 + 1.77x$. This was the same answer we came up with in Example 4.9.

We can now have the calculator display the residuals. In the TI-84, when you perform a linear regression, the calculator finds the residuals automatically and stores them in a list called **RESID**. The calculator doesn't display the list automatically, so we must tell it to do so.

Since we already have the x-values in **L1** and the y-values in **L2**, let us put the residuals in **L3**. To do this begin by pressing STAT then **1:EDIT** and move the cursor on top of **L3**. This is shown in Figure 4.45.

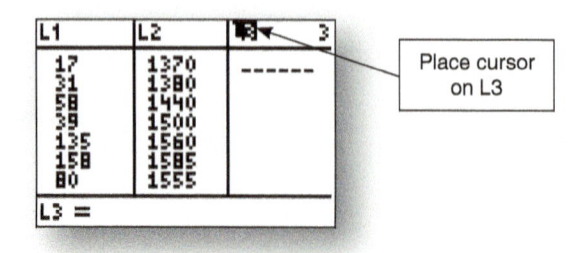

FIGURE 4.45

The second function associated with the STAT key on the calculator is labeled **LIST**. In order to access the list feature containing the residuals we press 2nd STAT. The **LIST** menu is shown below in Figure 4.46.

The residual list is named RESID

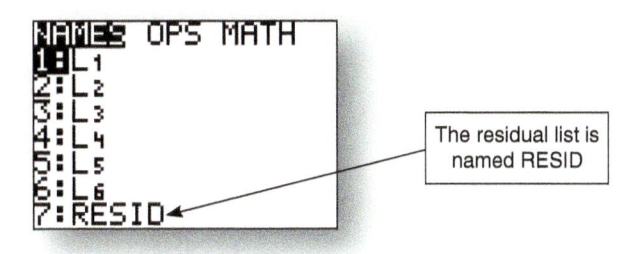

FIGURE 4.46

You can see that option 7 is the list containing the residuals. You may have to scroll down using the down arrow key to find the list containing the residuals on your calculator. To paste the list containing the residuals into **L3**, you must press 7 > ENTER. Figures 4.47a and 4.47b show this feature.

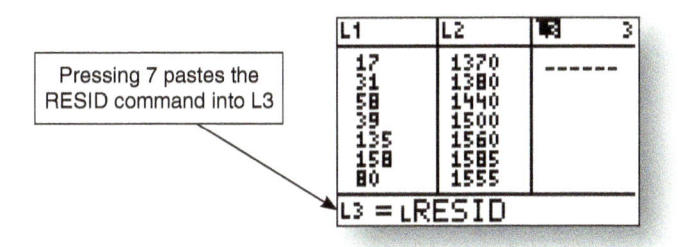

Pressing 7 pastes the RESID command into L3

FIGURE 4.47a

Pressing ENTER pastes the residuals into L3

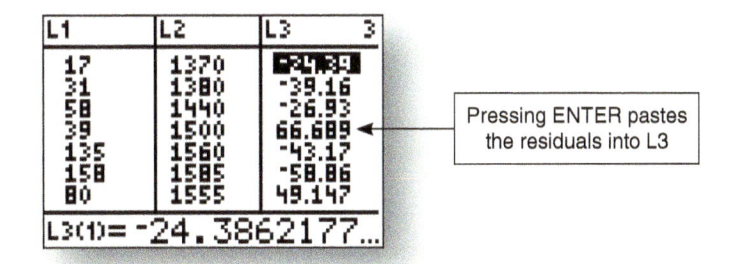

FIGURE 4.47b

You may notice that the values for the residuals are a little different than the ones that we calculated earlier in **Example 4.12** using the regression equation. The reason for this difference is because we rounded the regression coefficients when we wrote down the equation for the regression line. The calculator uses the unrounded regression coefficients when calculating the residuals, so in fact the calculator's answers are a little more accurate. This difference is small, and for predicting SAT scores will not be problematic.

On page 180 we illustrated a procedure to create a scatter diagram on the TI-84 calculator. We will again use this procedure to create the residual plot. Recall that a residual plot is a scatter diagram that uses the original x-values observed along the horizontal axis, and the residuals along the vertical axis. To create the residual plot, we must first set up the STAT PLOT menu on the TI-84. To access the STAT PLOT menu, press 2nd > Y=. Choose option 1 and turn the plot **ON**. For the **X list** enter **L1**, and for the **Y list** enter **RESID**. Remember, you can access **RESID** by pressing 2nd > STAT. This set-up is shown in Figures 4.48a and 4.48b.

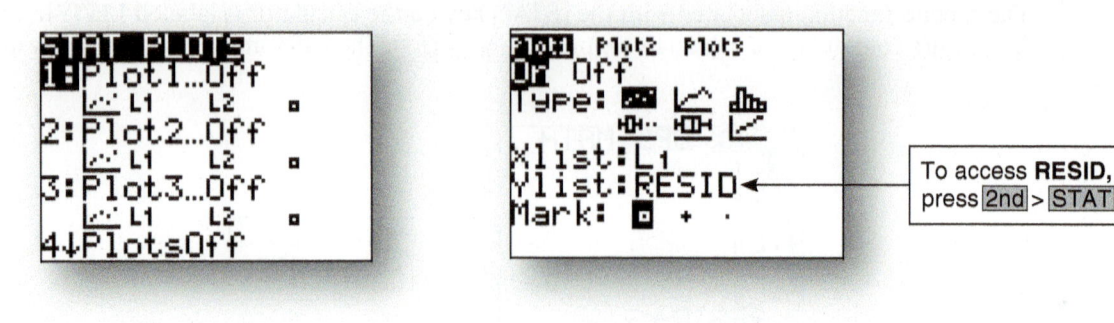

FIGURE 4.48a

FIGURE 4.48b

To access **RESID**, press 2nd > STAT

You can then create the residual plot by pressing ZOOM > 9 : **ZoomStat**. The **ZoomStat** feature is a built-in feature of the calculator that fits data into the calculator's screen. Figures 4.49a and 4.49b show this Zoom menu along with the residual plot.

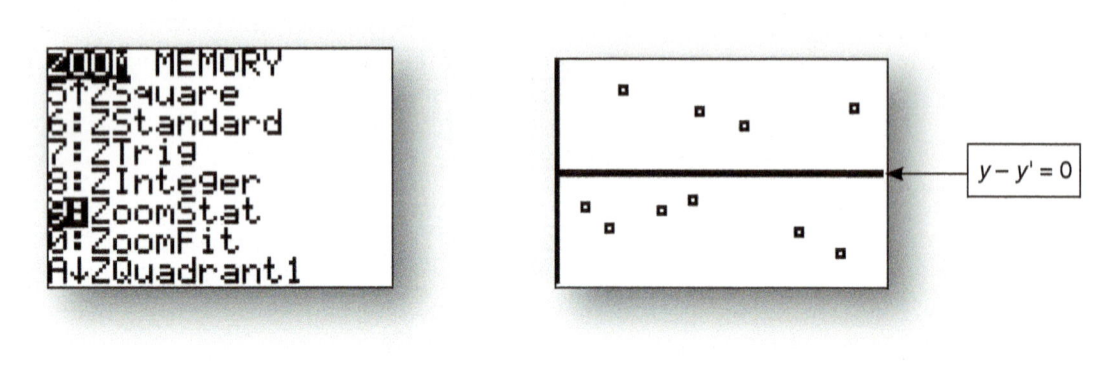

FIGURE 4.49a

FIGURE 4.49b

$y - y' = 0$

The residual plot shown in Figure 4.49b is the same one created earlier in Example 4.13.

It is important to point out that performing a linear regression and analyzing its residual plot normally takes place with a much larger set of data than used in Example 4.13. When looking at a scatter diagram of a large data set, it may be difficult to see issues that would make you question the linear tendency of the data set. Remember that the reason we are creating the residual plot is to rule out these issues that we may not see while analyzing the scatter diagram of the original data set.

Figure 4.50 contains a scatter diagram of 125 data values. The regression line was calculated and is also given as $y' = 0.95 + 0.93x$. Notice how the data values wrap around the regression line in no apparent pattern. The correlation coefficient is calculated as $r = 0.97$, informing us that these variables have high positive correlation.

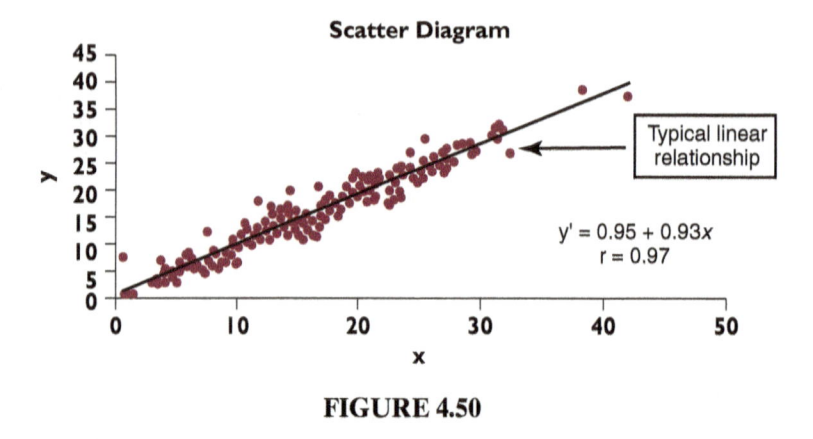

Typical linear relationship

$y' = 0.95 + 0.93x$
$r = 0.97$

FIGURE 4.50

Also, the corresponding residual plot is given in Figure 4.51. Notice how the residual plot has an equal scatter throughout the entire graph, and how this scatter is centered about 0 on the *x*-axis.

Residual Plot

Random scatter supports the linear relationship in FIGURE 4.50

FIGURE 4.51

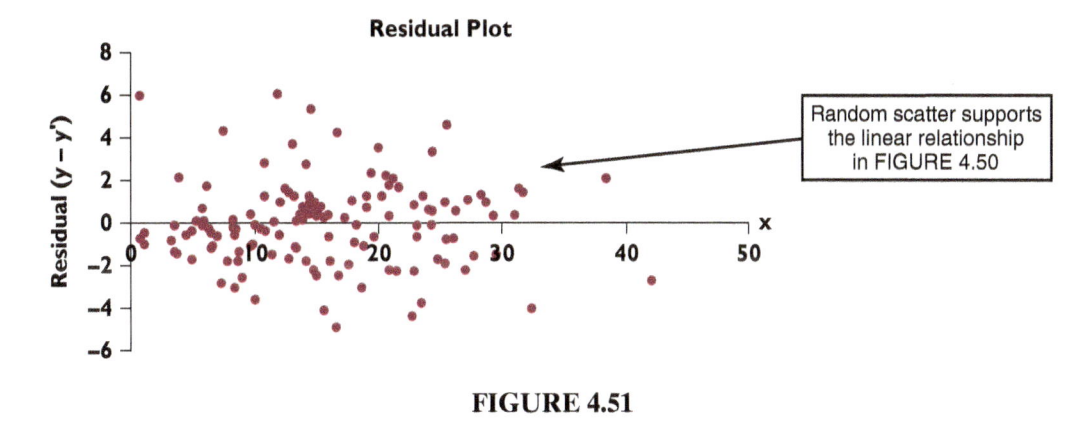

Figure 4.52 contains a scatter diagram of 140 data values along with the regression line and equation. Notice how the data values are tightly packed about the regression line for small values of *x*. However, as the value of *x* increases, so does the scatter in the *y*-direction. This may make an observer feel that the data set is heteroscedastic. However, the correlation coefficient for this data set is $r = 0.964$, giving the appearance that there is a high positive linear correlation between *x* and *y*.

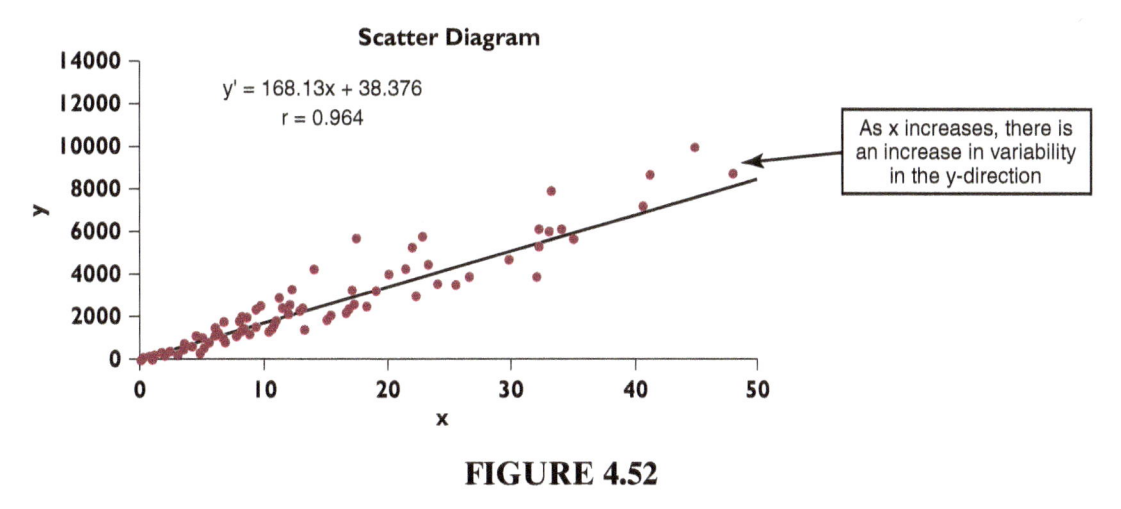

Scatter Diagram

$y' = 168.13x + 38.376$
$r = 0.964$

As x increases, there is an increase in variability in the y-direction

FIGURE 4.52

The question that arises is, "How do we know the increased scatter in the *y*-direction is enough to call the data set heteroscedastic, ruling out the possibility of a linear relationship?" This is precisely why we create the residual plot. Remember, that the residual plot helps to magnify these types of issues, in order to help us make informed decisions. Figure 4.53 contains the residual plot of the data in the scatter diagram in Figure 4.52.

Residual Plot

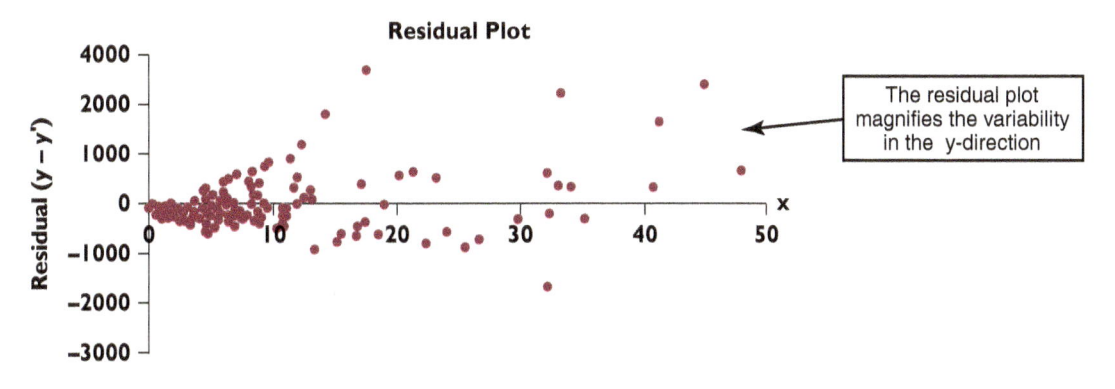

The residual plot magnifies the variability in the y-direction

FIGURE 4.53

The residual plot in Figure 4.53 clearly shows us that the increased scatter in the y-direction is more obvious than in the scatter diagram in Figure 4.52. This would lead us to the conclusion that this data set is heteroscedastic, and *does not* have a linear relationship. We should not use a linear model for this data set.

Let's observe one other scenario that violates an assumption of linearity. Curvilinear data sets often elude inexperienced individuals, leading them to incorrectly conclude that their data set has a linear relationship. Figure 4.54 contains 50 data values along with the regression equation. The correlation coefficient for this data set is $r = 0.995$. This would indicate a near perfect positive correlation between x and y.

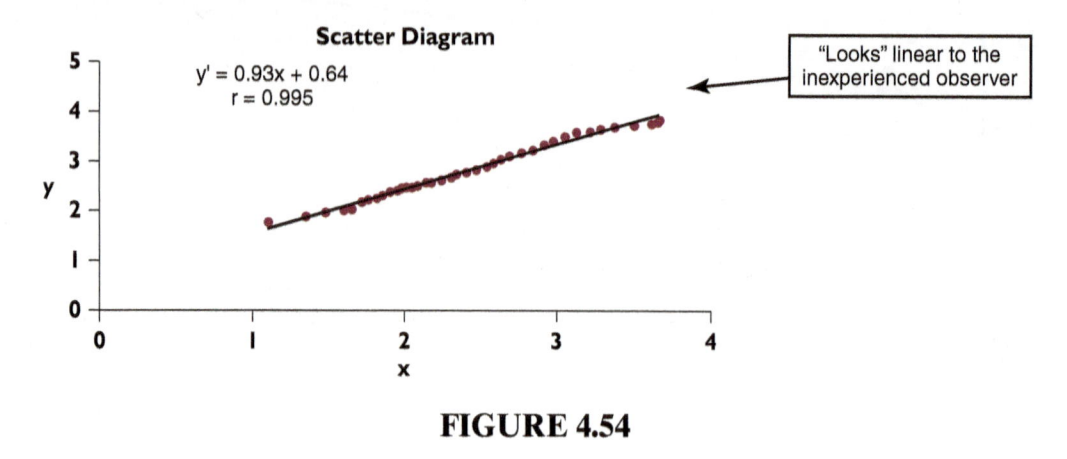

Scatter Diagram

$y' = 0.93x + 0.64$
$r = 0.995$

"Looks" linear to the inexperienced observer

FIGURE 4.54

The residual plot for the data in the scatter diagram in Figure 4.54 is shown in Figure 4.55. It only takes a moment to make the informed decision that this data presented in the scatter diagram is curvilinear.

Residual Plot

The residual plot magnifies the curvilinear relationship of the scatter diagram in FIGURE 4.54

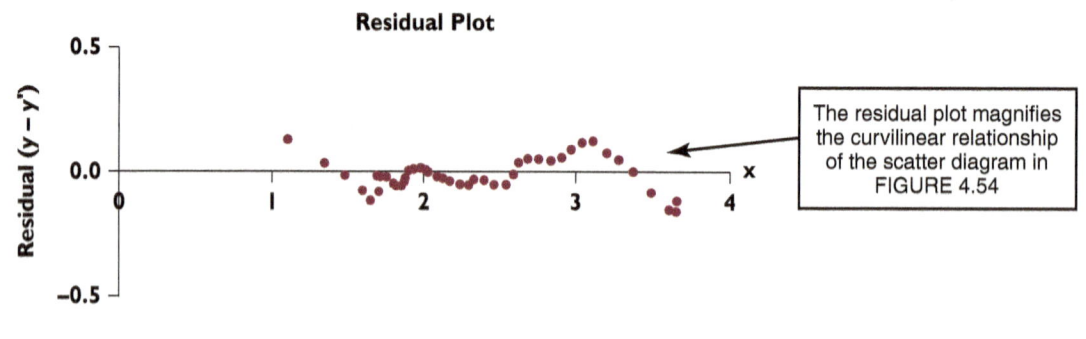

FIGURE 4.55

If we look a bit closer at how the data points scatter around the regression line in Figure 4.54, you can see how the trend in the residual plot follows the trend in the data values about the regression line. The linear assumption violation is not as apparent in the scatter diagram as it is in the residual plot. We should not use a linear model for this data set.

Coefficient of Determination Examined Further

As previously discussed, the coefficient of determination is computed by squaring Pearson's correlation coefficient, r. That is: r^2 = coefficient of determination

To help understand how to interpret the coefficient of determination, we will return to Example 4.4 that examined the relationship between the number of study hours, x, and the exam grade, y, for a random sample of 8 students as shown in Table 4.20.

TABLE 4.20

Student #	Study hours, x	Exam grade, y
1	20	85
2	10	65
3	17	75
4	15	70
5	12	75
6	9	72
7	24	86
8	13	72

The scatter diagram of the sample data is displayed in Figure 4.56.

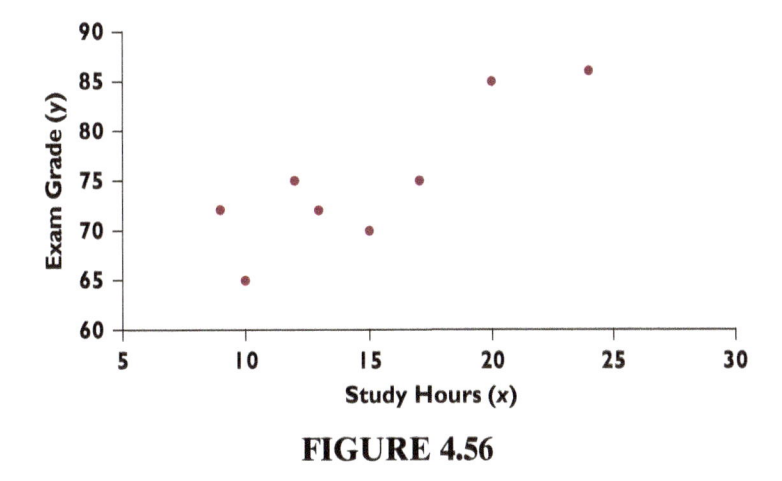

FIGURE 4.56

Since the scatter diagram appears to show a linear relationship between study hours and exam grades, the regression line can be determined using the regression equation:
$y' = a + bx$ where $a = 56.82$ and $b = 1.21$.
We can draw the regression equation $y' = 56.82 + 1.21x$ on the scatter diagram shown in Figure 4.57.

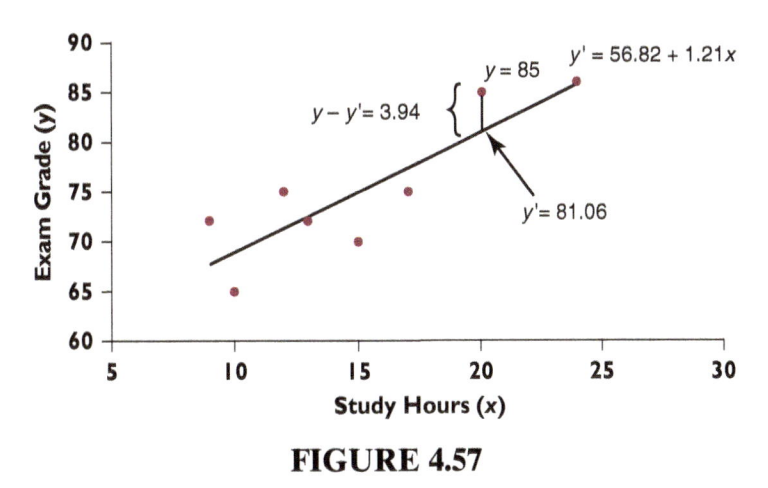

FIGURE 4.57

If we use the regression equation $y' = 56.82 + 1.21x$ to predict a student's exam grade, you will notice there is no prediction (y') that is equal to a student's actual exam grade, y. Consequently, every predicted exam grade will contain an error. For example, for student #1 who studied 20 hours for the exam, the predicted exam grade is 81.06 or approximately 81. Yet, student #1 earned a grade of 85. Thus, for student #1, the error in our prediction is $(y - y') = (85 - 81.06) = 3.94$.

To obtain the total error for all our predictions, we need to compute the squared difference between the predicted y value, y', and the actual y value. Then, these 8 squared differences are summed. That is, $\Sigma(y - y')^2$.

In particular, for student #1, squaring the error in our prediction we get:
$$(y - y')^2 = (85 - 81.06)^2 = (3.94)^2 = 15.52.$$
This variation between the actual y value and the predicted y value is referred to as the unexplained sample variability by the independent variable. This is also referred to as random error. In particular, we cannot explain why student #1's exam grade of y = 85 is 3.94 points above the predicted exam grade of y' = 81.06 solely based on the number of hours student #1 studied. The sum of these squared deviations for all the 8 students, $\Sigma(y - y')^2$, is 93.73 as calculated in Table 4.21.

TABLE 4.21

Student #	Study hours, x	Exam grade, y	y'	y − y'	(y − y')²
1	20	85	81.06	3.94	15.52
2	10	65	68.94	−3.94	15.52
3	17	75	77.42	2.42	5.86
4	15	70	75	−5	25
5	12	75	71.36	3.64	13.25
6	9	72	67.73	4.27	18.23
7	24	86	85.91	0.09	0.0081
8	13	72	72.58	−0.58	0.3368

$$\Sigma (y - y')^2 = 93.73$$

The quantity $\Sigma (y - y')^2 = 93.73$ represents the variation in y (the exam grades) that can **not** be predicted from study hours (the independent variable). This variation in y is referred to as the **unexplained variation in y**.

Now, let's *assume* that we didn't know the study hours (x) for these 8 students. That is, we *only know* the exam grades of the 8 students and we want to predict the exam grade for every student. From Table 4.21, the exam grades are: 85, 65, 75, 70, 75, 72, 86 and 72. If we assume that study hours contributes no information for the prediction of the exam grades, then the best prediction for a value of y is the sample mean of the exam grades, $\bar{y}$. Thus, the best predicted exam grade for every student will be the mean exam grade. That is, mean exam grade: $\bar{y} = \frac{\Sigma y}{n} = \frac{600}{8} = 75$.

Using the mean grade will keep the sum of the squared deviations from the mean grade prediction at a minimum value. Thus, the sum of the squared deviations from the mean = $\Sigma(y - \bar{y})^2$ will be at a minimum value. Table 4.22 contains the necessary calculations to evaluate $\Sigma(y - \bar{y})^2$.

TABLE 4.22

Student #	Exam grade, y	Mean Exam grade, $\bar{y}$	y − $\bar{y}$	Squared Deviations from the mean grade, $(y - \bar{y})^2$
1	85	75	85 − 75 = 10	100
2	65	75	−10	100
3	75	75	0	0
4	70	75	−5	25
5	75	75	0	0
6	72	75	−3	9
7	86	75	11	121
8	72	75	−3	9

$$\Sigma(y - \bar{y})^2 = 364$$

From Table 4.22, the sum of the squared deviations from the mean, $\Sigma(y - \bar{y})^2 = 364$. This quantity $\Sigma(y - \bar{y})^2$ is referred to as the **total variation in y**. Figure 4.58 visually displays this total variation in y.

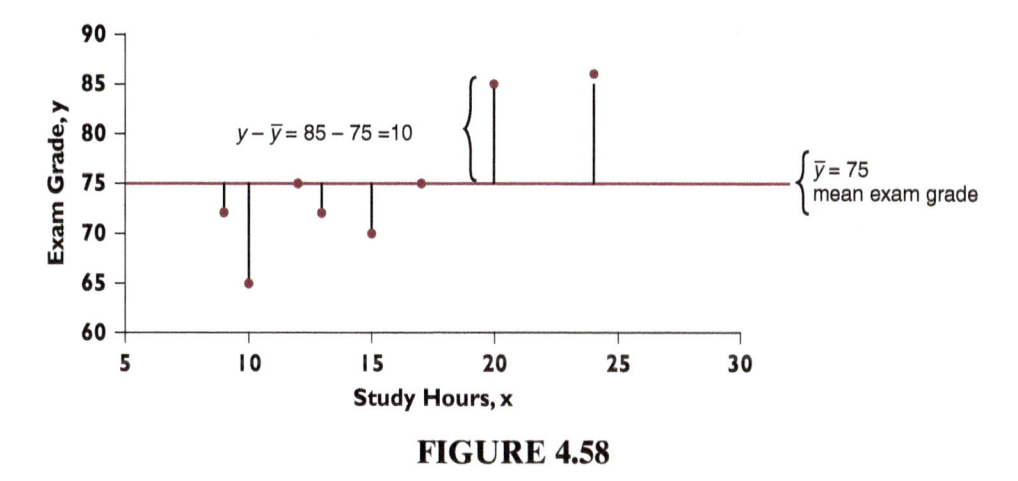

[Figure 4.58: Scatter plot of Exam Grade, y vs Study Hours, x, showing deviations from the mean $\bar{y}=75$. The deviation for exam grade 85 is labeled $y - \bar{y} = 85 - 75 = 10$.]

FIGURE 4.58

In Figure 4.58 the deviation from the mean for the exam grade 85 is illustrated. That is,

$$y - \bar{y} = 85 - 75 = 10.$$

By definition, the total variation in y (exam grades) can be subdivided into the unexplained variation in y and the explained variation in y. That is,

total variation in y (exam grades) = unexplained variation in y + explained variation in y

Solving the equation for the explained variation in y by subtracting the unexplained variation in y from both sides, we get:

explained variation in y = total variation in y − unexplained variation in y

DEFINITION 4.10

The Explained Variation in the dependent variable, y, that can be explained by the Regression Equation is equal to the Total Variation in y minus the Unexplained Variation in y.
This can be expressed by the formula:

Explained variation in y = Total variation in y − Unexplained variation in y

We are stating that by subtracting the unexplained variation in y from the total variation in y, we get the portion of the total variation in y that can be explained by the regression equation.
Now, if we divide both sides of this equation by the total variation in y, we obtain:

$$\frac{\text{explained variation in y}}{\text{total variation in y}} = \frac{\text{total variation in y}}{\text{total variation in y}} - \frac{\text{unexplained variation in y}}{\text{total variation in y}}$$

Or, this can be rewritten as:

$$\frac{\text{explained variation in y}}{\text{total variation in y}} = 1 - \frac{\text{unexplained variation in y}}{\text{total variation in y}}$$

The quantity $\dfrac{\text{explained variation in y}}{\text{total variation in y}}$ represents the proportion of the total variation about the mean value $\bar{y}$ that can be explained by the linear relationship between y and x. This quantity is also referred to as the coefficient of determination. That is,

> **DEFINITION 4.11**
> The Coefficient of Determination, r^2, can now be defined as: $\frac{\text{explained variation in } y}{\text{total variation in } y}$. This can be symbolized as:
>
> $$\text{Coefficient of Determination} = r^2 = \frac{\text{explained variation in } y}{\text{total variation in } y}$$
>
> or
>
> $$\text{Coefficient of Determination} = r^2 = 1 - \frac{\text{unexplained variation in } y}{\text{total variation in } y}$$
>
> This quantity can be expressed as a percentage. As a percentage, r^2 can be interpreted to be the percentage of the total variation that can be explained by the regression equation.

For this example, we have:

$$r^2 = 1 - \frac{\text{unexplained variation in } y}{\text{total variation in } y} = 1 - \frac{93.73}{364} = 1 - 0.2575 = 0.7425.$$

So,

since r^2 is $\frac{\text{explained variation in } y}{\text{total variation in } y}$, then we can write: $r^2 = \frac{\text{explained variation in } y}{\text{total variation in } y} = 0.7425$

Expressing this as a percentage, we get:

$$r^2 = \frac{\text{explained variation in } y}{\text{total variation in } y} = 74.25\%$$

Thus, r^2 can be interpreted to be the percentage of the total variation that can be explained by the regression equation.

Therefore, we can say that 74.25% of the variation in exam grades, y (dependent variable) is determined, or accounted for, by its linear relationship with student study hours, x (independent variable).

KEY TERMS

Best Fitting Line 195
Coefficient of Determination, r^2 193, 218
Dependent Variable, y 217
Extrapolation 199
homoscedastic 204
heteroscedastic 204
Independent Variable, x 199
Interpolation 199

Linear Correlation 174
Linear Regression 174
Method of Least Squares 195
Negative Correlation 176
No Linear Correlation 176
Ordered Pair 174
Pearson's (or Sample) Correlation Coefficient, r 181

Positive Correlation 176
Predicted Value of y, y' 195
Regression Coefficients, a and b 198
Regression Line 195
residual 207
residual plot 207
Scatter Diagram 174

SECTION EXERCISES

Note: The data sets in these Section Exercises can be downloaded in either Microsoft EXCEL or TI-84 format from the textbook's website at MyStatLab.com.

Many exercises require calculations that are long and tedious. Therefore, we recommend that you use a calculator or a computer with statistical application software to help you perform these calculations.

Section 4.2

1. In correlational analysis, the two variables are _____.
2. A pair of numbers that is written in the following form: (x,y) is called an _____ pair since the value of the x variable is always written _____ within the parentheses.
3. A graph representing the ordered pairs of sample data on a set of axes is called a _____ _____.
4. A _____ correlation is a correlation between two variables, x and y, that occurs when high measurements on the x variable tend to be associated with high measurements on the y variable and low measurements on the x variable tend to be associated with low measurements on the y variable.
5. If the points in the scatter diagram tend to follow a straight-line path, then this type of correlation is called a _____.
6. A _____ correlation means there is no linear relationship between the two variables. That is, high and low measurements on the two variables are _____ associated in any predictable way.
7. Which type of correlation is indicated by a scatter diagram whose points appear to be randomly placed in no particular pattern?
   a) Positive
   b) None
   c) Negative
8. A linear correlation between the two variables is a line drawn between the variables. (T/F)
9. Scatter diagrams are graphs that indicate the degree of variation between x and y. (T/F)
10. For each of the following two variables, state whether you would expect a positive, negative or no linear correlation.
    a) An individual's IQ score and annual salary
    b) An individual's age and blood pressure
    c) The number of days a student misses class and final grade
    d) A mother's birthday and her daughter's birthday
    e) An individual's driving speed above 55 mpg and gasoline mileage
    f) An individual's annual income and education level
    g) Average daily temperature during the summer months and the number of air conditioners sold during these months
    h) An adult's shoe size and IQ score
    i) Amount of fertilizer applied to tomato plants and the number of tomatoes per plant
11. In each of the following scatter diagrams, if a linear correlation exists between the two variables, then state whether the type of linear correlation is positive or negative. If no linear correlation exists, then state none.

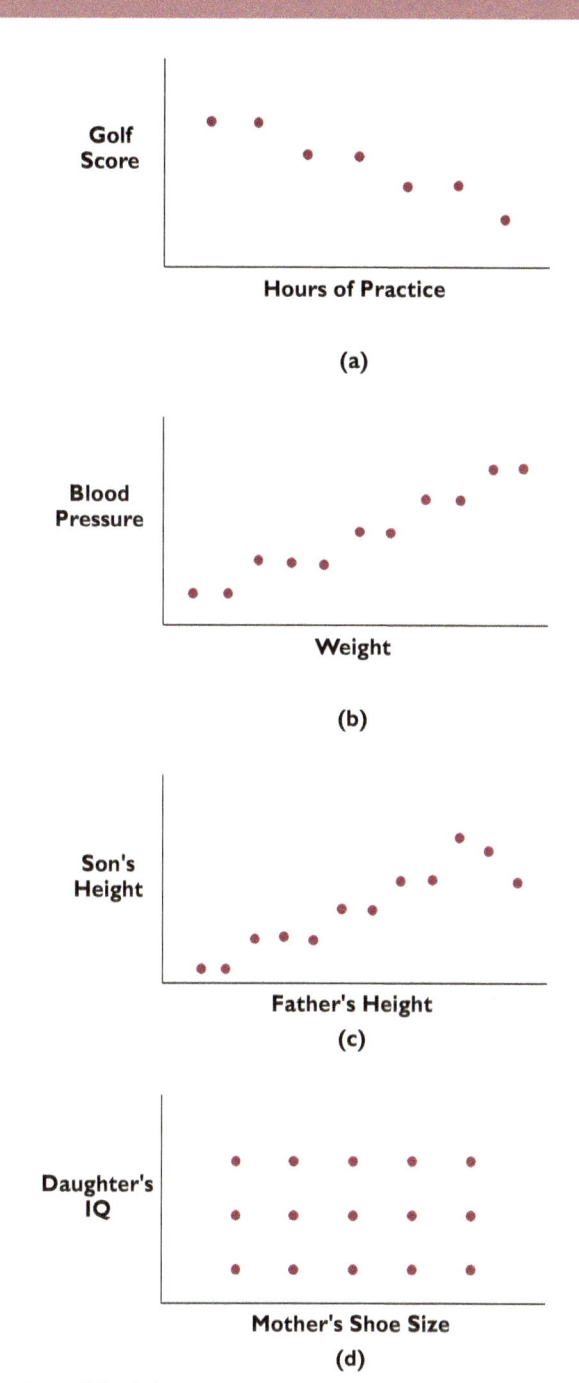

(a) Golf Score vs Hours of Practice

(b) Blood Pressure vs Weight

(c) Son's Height vs Father's Height

(d) Daughter's IQ vs Mother's Shoe Size

Sections 4.3–4.4

12. The closer the points of a scatter diagram approximate a straight-line pattern, the _____ the linear correlation between the two variables.
13. The strength of the linear correlation between the two variables of a sample can be numerically measured by _____.
14. Pearson's correlation coefficient, r, will always be a numerical value between _____ and _____.
15. A value of r = 1 represents the _____ positive correlation possible and it indicates a _____ positive linear correlation. This means that all the points of the scatter diagram will lie on a straight line which is sloping _____ from left to right.

Chapter 4 Linear Correlation and Regression Analysis

16. A value of r = 0 represents no _____ correlation between the two variables.
17. The closer the value of the correlation coefficient, r, is to _____ or to _____ the stronger the linear relationship between the two variables. However, a value of r close to zero indicates a _____ linear relationship between the two variables.
18. To calculate the sample correlation coefficient, we use the formula r = _____.
19. The linear correlation between two variables, x and y, is a number between −1 and 1. (T/F)
20. A linear correlation that is close to zero can be interpreted to mean that no relationship exists between the variables, x and y. (T/F)
21. Rank the following values of Pearson's correlation coefficient, r, from the strongest to weakest linear association:
    $$0.25, -1.00, 0, -0.83, -0.10, 0.50, 0.95$$
22. For the following scatter diagrams:
    a) Identify those that you believe represent a linear relationship between x and y.
    b) For those that you have identified as representing a linear relationship between x and y, determine which are positive relationships and which are negative.
23. Match the following values of Pearson's correlation coefficient, r, with the appropriate scatter diagram:
    Possible r values: I. $r = 0.48$ II. $r = 0.92$ III. $r = 0.78$ IV. $r = 0.0$

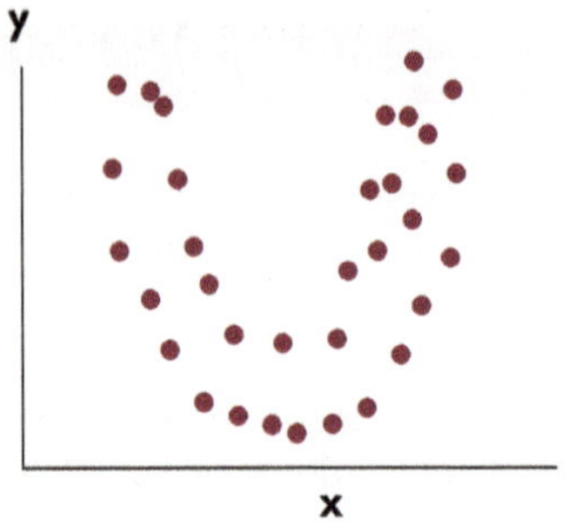

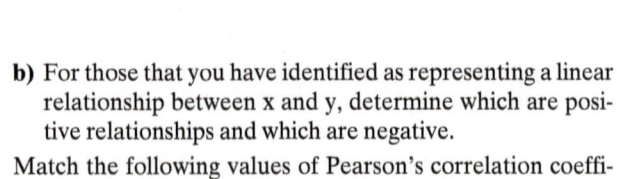

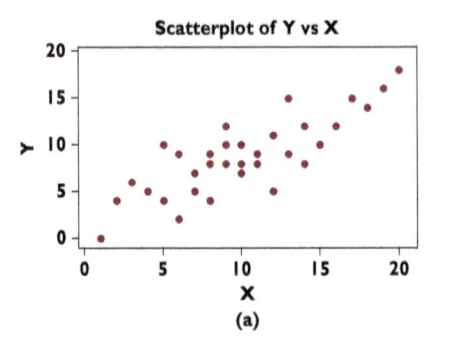

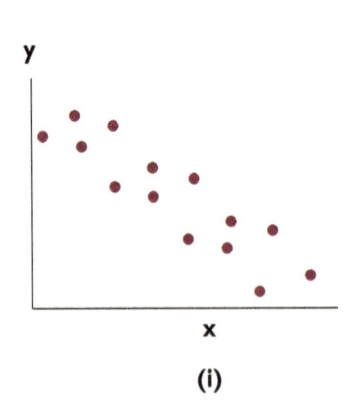

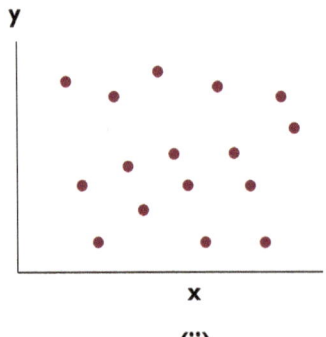

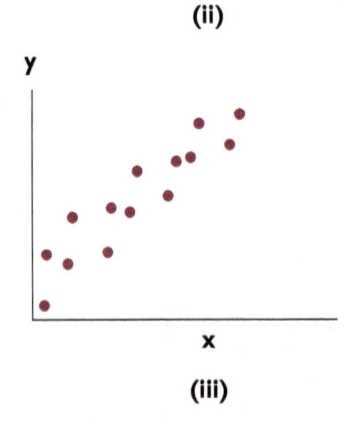

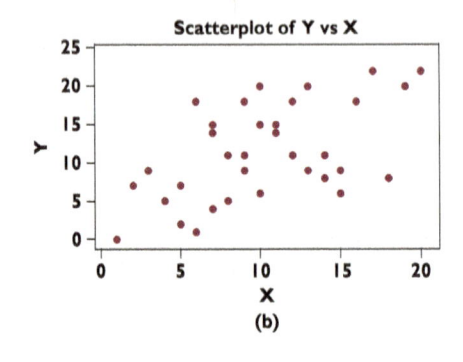

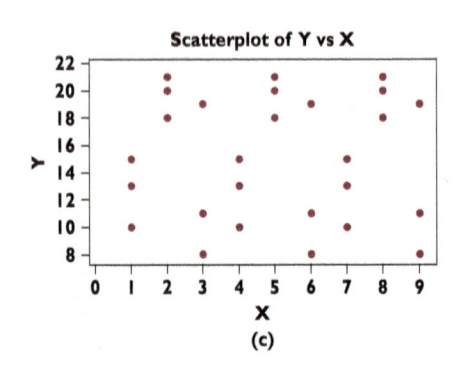

Scatterplot of Y vs X

(d)

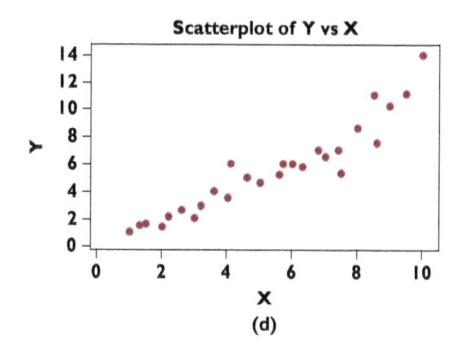

24. Explain the difference between the following pair of r values.
   a) $r = 0.55$ and $r = 0.87$
   b) $r = -0.60$ and $r = 0.60$
   c) $r = 0.05$ and $r = 0.95$

25. Using the sample correlation coefficient formula, calculate the value of r for each of the following data pairs.

a)
x	y
2	4
3	7
4	8
5	11
6	11

b)
x	y
5	11
7	8
9	1
10	6
16	2

c)
x	y
4	9
5	11
6	13
7	15
8	17

d)
x	y
2	3
2	5
3	5
6	6
7	9
4	3
1	4
5	8
3	6
9	10

e)
x	y
11	12
16	9
15	9
21	5
23	4
16	8
26	4
29	3
34	1
28	5

f)
x	y
72	222
62	145
58	125
68	165
68	172
63	135
61	120
73	185
75	210
66	158

Section 4.5

26. If $r = 0.8$, then the coefficient of determination is
   a) 0.88   b) 0.64   c) 1.6   d) 0.40

27. The coefficient of determination, r^2, measures the proportion of variance of:
   a) The independent variable which can be accounted for by dependent variable.
   b) The dependent variable which can be accounted for by the independent variable.
   c) Neither of these.

28. Audrey, an anthropologist, has made a study of the linear relationship between the leg length and total height of prehistoric animals. Using a random sample of 17 prehistoric animal fossils, she obtained a correlation coefficient of 0.48.
   a) Can Audrey conclude that there is a positive correlation between the length and total height of prehistoric animals?
   b) What percent of variability in total height can be accounted for by the variability in the leg length of the prehistoric animals?

Section 4.6

29. In linear regression analysis, the technique used to obtain the best fitting line is called the method of _____ _____. The best fitting line is called the _____ line.

30. The regression line formula is given by: _____. To calculate the regression coefficients a and b, we use the following formulas: b = _____ and a = _____.
The regression line is used to predict the _____ variable given the value of the _____ variable.

31. Craig, a marketing executive for a microbrewery company, wants to determine if there is a positive linear relationship between advertising expenditures and sales for their new lite beer product, Less Ale. He randomly samples data for the past 7 sales years and records the sample data in the following table, where advertising expenditure is measured in thousands and the beer sales in millions of dollars.

Year	Expenditure (thousand $)	Sales (million $)
2013	35	38
2014	47	35
2015	65	49
2016	92	50
2017	55	40
2018	25	35
2019	82	44

   a) Construct a scatter diagram.
   b) Calculate the sample correlation coefficient, r.
   c) Find r^2 and interpret its meaning.
   d) Determine the regression equation, y'.
   e) Using the regression equation, predict the beer sales for the advertising budget expenditure of $50,000.
   f) Is y' an interpolated or extrapolated value? Why?

32. A researcher working for The Sunglass Shack corporation wants to investigate whether there is a correlation between daily temperature and sunglasses sold at its Brooklyn location. The following data pairs represent some sample data collected:

Temperature (degrees F)	Number of Sunglasses Sold
80	131
61	20
56	101
99	62
101	44
95	189

   a) Construct the scatter diagram where T is the independent variable.
   b) Calculate the sample correlation coefficient, r.
   c) Find r^2 and interpret its meaning in the context of this problem.
   d) Determine the regression equation where T is the independent variable.
   e) Using the regression equation, predict the number of sunglasses sold if the temperature was 76 degrees.
   f) Is the predicted value obtained in part e, an interpolated or extrapolated value for y? Why?
   g) Predict the temperature if 105 sunglasses were sold that day.

33. A prominent psychologist wonders if a patient's score on the extrovert scale (those who seek out social environments) is positively correlated with time spent on social network sites online. She gathers data from 20 patients for both variables; the results are as follows:

Extrovert Scale Value	Time on Social Network Sites (min/day)
40	46
45	79
52	33
62	63
31	20
28	18
5	11
83	78
55	63
32	46
47	21
45	55
60	59
13	23
7	30
85	80
38	25
61	26
26	33
3	7

a) Construct the scatter diagram for the data.
b) Calculate the sample correlation coefficient, r.
c) Find r^2 and interpret its meaning.
d) Determine the regression equation, y'.
e) Using the regression equation, predict the time spent online for a person whose extroversion scale is 35.

34. The following data were taken from the information found on the internet on Blood Alcohol Level (BAL) relating to the number of drinks a 100 lb. female consumes in one hour. A drink is defined as: 1.25oz of 80 proof liquor, or 12oz of regular beer, or 5oz of table wine.

Blood Alcohol Level (BAL) for a 100 lb. Female.

# drinks in one hour	1	2	3	4	5	6	7	8
BAL	0.05	0.10	0.15	0.20	0.25	0.30	0.36	0.41

a) Construct a scatter diagram.
b) Calculate the sample correlation coefficient, r.
c) Find r^2 and interpret its meaning in terms of the variables r.
d) Determine the regression equation, y', and draw it on the scatter diagram.
e) Using the regression equation predict the Blood Alcohol Level of a female that weighs 100lbs after consuming 3.5 drinks in one hour.

35. Lylah, a long distance runner wants to know if there is a positive linear relationship between the temperature at the start time of the Chicago Marathon and the time it takes to finish the marathon. She raced in the Chicago Marathon every year for the past ten years and recorded her results in the following table:

Year	Temperature (Fahrenheit)	Finish Time (Minutes)
2010	54	266
2011	51	266
2012	40	265
2013	72	292
2014	64	286
2015	32	267
2016	60	283
2017	59	280
2018	40	272
2019	48	272

a) Construct a scatter diagram.
b) Calculate the sample correlation coefficient, r.
c) Find r^2 and interpret its meaning.
d) Determine the regression equation, y'.
e) Using the regression equation, predict Lylah's time to finish the Chicago Marathon given the temperature at the start of the race is 68° F.
f) Predict Lylah's time to finish the Chicago Marathon given the temperature at the start of the race is 80° F.
g) Is the predicted value obtained in part f) an interpolated or extrapolated value? Why?

Section 4.7

36. If the scatter in the y-direction is approximately the same throughout a scatter diagram, then the data is said to be _____, however if the scatter in the y-direction changes throughout the graph then the data is said to be _____.

37. Three issues that often arise making a statistician question the linear tendency of a data set are _____, _____ and _____.

38. Non-linear trends can always be identified by the scatter diagram.(T/F)

39. To graph a residual plot, create a scatter diagram plotting the original y-values against the predicted y-values. (T/F)

40. To calculate a residual use the formula:
Residual = (_____) − (_____) = y − y'

41. The following table contains actual and predicted values of y. Use this information to complete the table and find the residual.

y	y'	Residual
45	47.5	
34	32.6	
52	50.1	
46	47.8	
35	36.4	

42. The following table contains 10 pairs of data. Use the regression equation y' = 3.47 + 0.26x to complete the table by:

a) computing the predicted value for y at each of the given x-values.

b) computing the residual using the actual and predicted values for y.

x	y	y'	Residual
1	4		
2	3		
3	5		
4	4		
5	5		
6	6		
7	4		
8	7		
9	5		
10	6		

43. You may recall from science class that when water is heated it expands. Using some very accurate instrumentation, the volume of water in a container was measured at various temperatures. The results of this experiment are shown in the following table:

Temperature (C)	Volume (ml)
20	50.124
21	50.146
22	50.169
23	50.193
24	50.215
25	50.237
26	50.265
27	50.292
28	50.321
29	50.349
30	50.376

a) Using your graphing calculator, create a scatter diagram of this data set. Does there appear to be a linear relationship between temperature and volume of water?
b) Calculate the regression equation, y', and the sample correlation coefficient, r.
c) Use the regression equation to calculate the predicted value, y', and use these predicted values to calculate the residuals.
d) Use your graphing calculator to create a residual plot. What does the residual plot reveal to you?

44. The following data were taken from the information found on the internet on Blood Alcohol Level (BAL) relating to the number of drinks a 100 lb. female consumes in one hour. A drink is defined as: 1.25oz of 80 proof liquor, or 12oz of regular beer, or 5oz of table wine.

Blood Alcohol Level (BAL) for a 100 lb. Female.

# drinks in one hour	1	2	3	4	5	6	7	8
BAL	0.05	0.10	0.15	0.20	0.25	0.30	0.36	0.41

a) Using your graphing calculator, create a scatter diagram of this data set. Does there appear to be a linear relationship between drinks per hour and BAL?
b) Calculate the regression equation, y', and the sample correlation coefficient, r.
c) Use the regression equation to calculate the predicted value, y', and use these predicted values to calculate the residuals.
d) Use your graphing calculator to create a residual plot. What does the residual plot reveal to you?

45. Examine each of the following scatter diagrams. Decide if a linear relationship exists between x and y. If you feel that a linear relationship *does not* exist state your reason why (i.e., presence of outliers, curvilinear data or heterscedastic data).

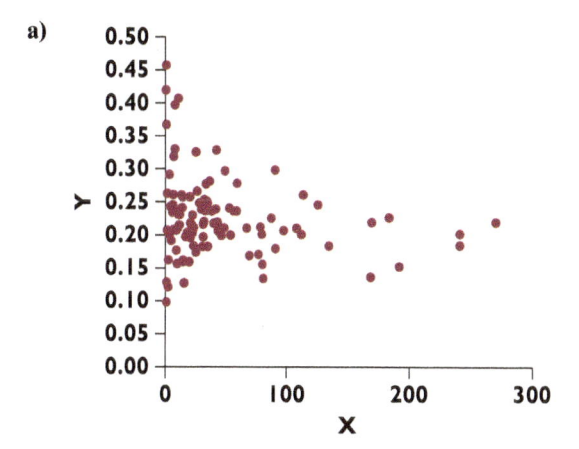

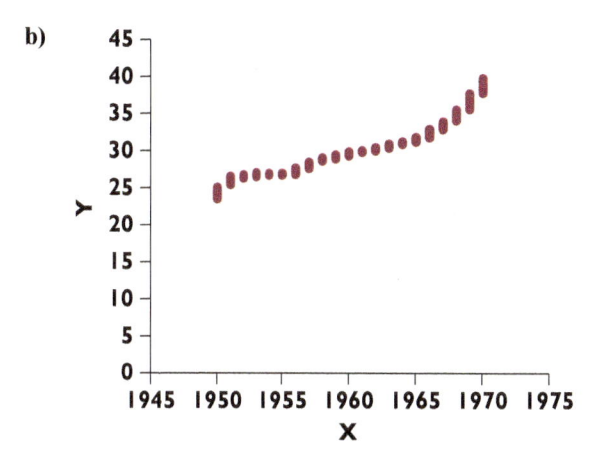

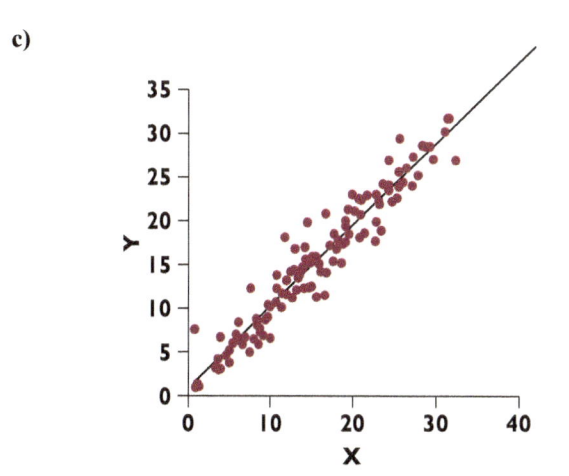

a)

b)

c)

224 Chapter 4 *Linear Correlation and Regression Analysis*

d)

(scatter plot: X-axis 0 to 1, Y-axis 0 to 12, showing strong positive linear relationship)

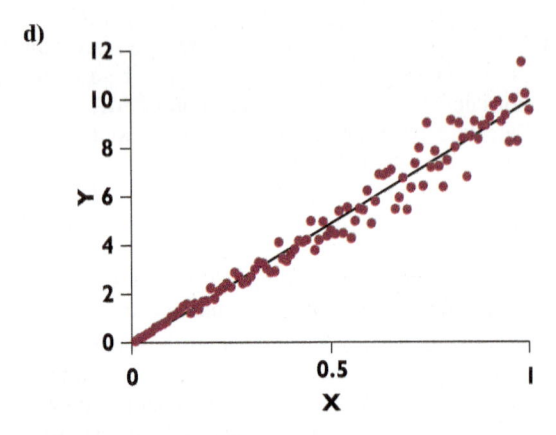

46. A marketing executive is trying to decide the how close the bond is between sales of their product and advertising costs. The table below lists the varying sales according to the amount spent on the advertising of his company's product. X represents Advertising dollars(in thousands), Y represents Sales for the company (in thousands).

	Advertising Dollars (1000 s)	Sales (1000 s)
January	5.5	100
February	5.8	110
March	6	112
April	5.9	115
May	6.2	117
June	6.3	116
July	6.5	118
August	6.6	120
September	6.4	121
October	6.5	120
November	6.7	117
December	6.8	123

a. Use your calculator to make a scatter diagram that shows how advertising dollars explains sales. Does there appear to be a linear relationship? Calculate Pearson's sample correlation coefficient, r. What does r tell you about the strength of the relationship?

b. Find the least squares regression equation, y'. Interpret the slope and the y-intercept in the context of the problem. Be sure to graph this regression equation within your scatter plot to make sure your equations appears to "best fit" the data.

c. Predict the sales for an advertising amount of 6.1.

d. Is the predicted value in part C an interpolated or extrapolated value? Explain.

e. If you were told the sales are 125, what would the model predict for advertising?

f. Would you trust the model's prediction for the value you calculated in part e? Why or why not?

g. In order to help you have "good faith" in your model, create a residual plot using your graphing calculator. Remember the residuals are placed in a List in the TI-84 called "RESID". You must plot the original x-values(Advertising) along with the residuals as y-values(Residual = $y - y'$). Are there any patterns in the residual plot that would make you "second guess" your original assumption of linearity?

47. A doctor knows that muscle mass decreases with age. To help him understand this relationship in women, the doctor selected women aged 40 to 80. The data is given in the following table.

X is age, Y is a measure of muscle mass (the higher the measure, the more muscle mass).

X (Age)	Y (Muscle Mass)	X (Age)	Y (Muscle Mass)
72	80	77	63
62	92	63	86
42	101	46	105
66	69	57	73
55	86	44	95
71	73	51	100
67	79	40	112
57	83	80	75

a. Use your calculator to make a scatter diagram that shows how age helps explain muscle mass. Does there appear to be a linear relationship? Calculate Pearson sample correlation coefficient, r. What does r tell you about the strength of the relationship?

b. Find the least squares regression equation, y'. Interpret the slope and the y-intercept in the context of the problem. Be sure to graph this regression equation within your scatter plot to make sure your equations appears to "best fit" the data.

c. Predict the muscle mass for women aged 60 years.

d. Is the predicted value in part C an interpolated or extrapolated value? Explain.

e. If the muscle mass of a woman was determined to be 58, what age does the model predict she would be?

f. Would you trust the model's prediction for the value you calculated in part e? Why or why not?

g. In order to help you have more "good faith" in your model, create a residual plot using your graphing calculator. Remember the residuals are placed in a List in the TI-84 called "RESID". You must plot the original x-values(AGE) along with the residuals as y-values(Residual = $y - y'$). Are there any patterns in the residual plot that would make you "second guess" your original assumption of linearity?

48. Sydney, An educational researcher, randomly samples 15 college students to determine if there is a correlation between reading and mathematics scores on a standardized educational test. The following data represents the reading and mathematics test results for the 15 students.

Reading Scores X Values	Mathematics Scores Y Values
191	180
103	101
187	173
108	103
180	170
118	113
178	171
127	122
176	168
134	130
165	150
147	145
160	150
157	154
145	130

a. Using your graphing calculator, create a scatter diagram of this data set. Does there appear to be a linear relationship between reading scores and mathematics scores?
b. Calculate the regression equation, y', and the sample correlation coefficient, r.
c. Use the regression equation to calculate the predicted y value, y', and use these predicted values to calculate the residuals.
d. Use your graphing calculator to create a residual plot. What does the residual plot reveal to you?

49. Lylah, a long distance runner wants to know if there is a positive linear relationship between the temperature at the start time of the Chicago Marathon and the time it takes to finish the marathon. She raced in the Chicago Marathon every year for the past ten years and recorded her results in the following table:

Year	Temperature (Fahrenheit)	Finish Time (Minutes)
2010	54	266
2011	51	266
2012	40	265
2013	72	292
2014	64	286
2015	32	267
2016	60	283
2017	59	280
2018	40	272
2019	48	272

a) Using your graphing calculator, create a scatter diagram of this data set. Does there appear to be a linear relationship between Temperature and Finish Time?
b) Calculate the regression equation, y', and the sample correlation coefficient, r.
c) Use the regression equation to calculate the predicted y value, y', and use these predicted values to calculate the residuals.
d) Use your graphing calculator to create a residual plot. What does the residual plot reveal to you?

CHAPTER REVIEW

50. A _____ linear correlation is a correlation between two variables, x and y, that occurs when high measurements on the x variable tend to be associated with low measurements on the y variable and low measurements on the x variable tend to be associated with high measurements on the y variable.

51. If y increases as x decreases this indicates that x and y are correlated:
    a) Positively
    b) Not at all
    c) Negatively

52. A value of r = –1 represents the _____ negative linear correlation possible and it indicates a _____ negative linear correlation. This means that all the points of the scatter diagram will lie on a straight line which is sloping _____ from left to right.

53. For each of the following two variables, state whether you would expect a positive, negative or no linear correlation.
    a) Number of hours practicing golf and golf score
    b) A husband's age and his wife's age
    c) Number of hours practicing bowling and bowling score
    d) Amount of alcohol consumed and reaction time
    e) Mother's birth weight and her child's birth weight
    f) A person's age and number of days of sick-leave
    g) A person's age and life insurance premium
    h) High school average and college grade-point average
    i) Amount of grams of fat consumed daily and cholesterol level

54. A negative value of r indicates:
a) No linear relationship
b) A weak linear relationship
c) A positive linear relationship
d) A negative linear relationship

55. For each of the scatter diagrams below, choose the statement that best describes the type of relationship that exists between the two variables.
1) strong linear correlation.
2) moderate or weak linear correlation.
3) no linear correlation.

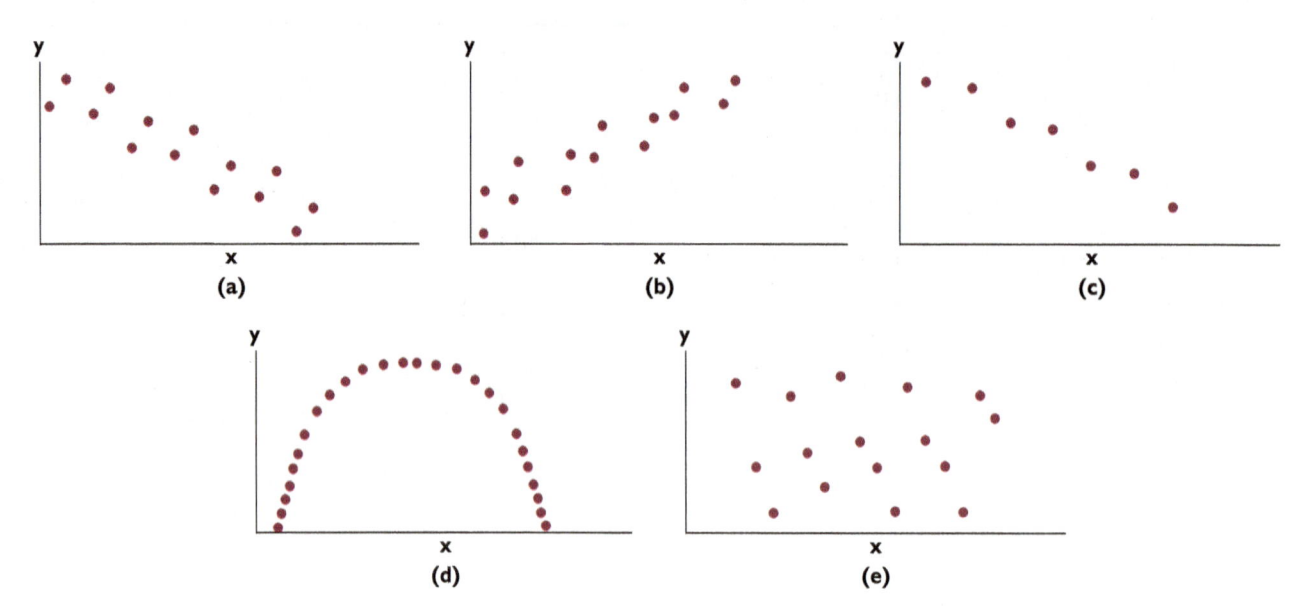

56. Rank the following values of Pearson's correlation coefficient, r, from the strongest to weakest linear association.
−0.95, 0.05, +1.00, 0.69, −0.35, 0, 0.50

57. Match the following values of Pearson's correlation coefficient, r, with the appropriate scatter diagram:
Possible r values:
I. r = 0.48 II. r = −0.85 III. r = 0.80 IV. r = 0.95 V. r = −0.57 VI. r = 0

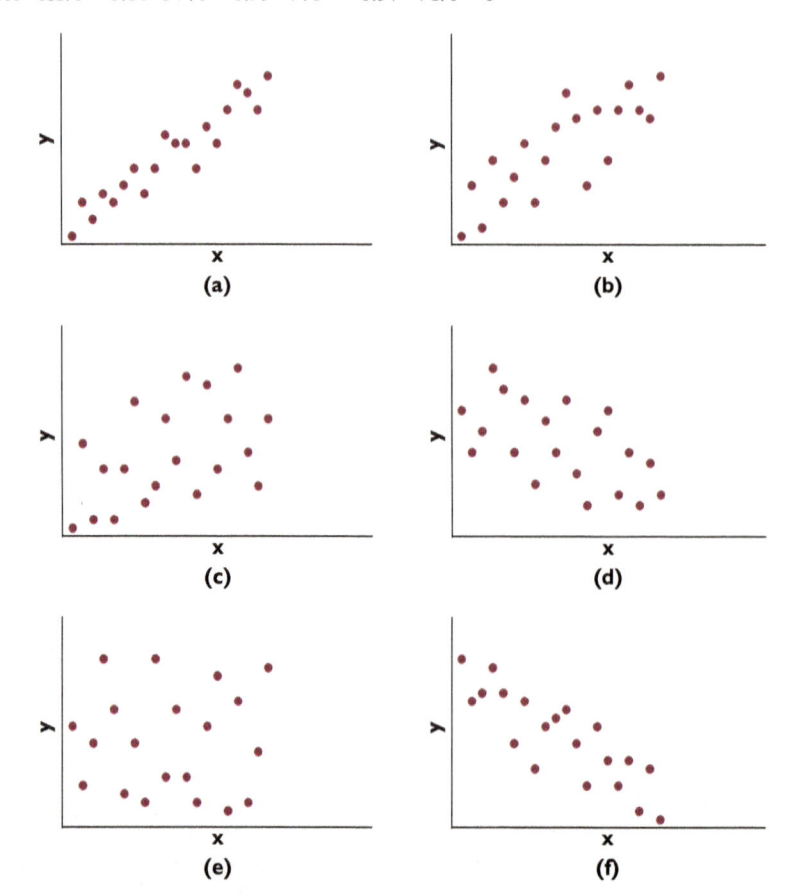

58. A coefficient of determination of 0.5, means that there is a 50% chance that a linear relationship exists between variables x and y.
(T/F)

59. Linear regression can be used to predict a value for the dependent variable from a value of the independent variable.
(T/F)

60. A linear regression model produces a line that best fits the data pairs from a sample.
(T/F)

61. Matthew, a medical researcher, wants to determine if a positive linear relationship exists between the number of pounds a male is overweight and his blood pressure. He randomly selects a sample of 10 males and records the number of pounds that the individual is over his "ideal" weight according to a medical weight chart and his systolic blood pressure. Matthew's sample data is listed in the following table.

Male Subject	Number of Lbs Above Ideal Weight	Systolic Blood Pressure
1	12	150
2	8	142
3	20	165
4	17	152
5	14	147
6	23	158
7	6	135
8	4	128
9	15	135
10	19	153

a) Construct a scatter diagram.
b) Calculate the sample correlation coefficient, r.
c) Find r^2 and interpret its meaning.
d) Determine the regression equation, y'.
e) Using the regression equation, predict a male's systolic blood pressure if the individual is 10 pounds over his ideal weight.
f) Is the predicted value obtained in part e) an interpolated or extrapolated value? Why?

62. Lydia, a sports statistician, wants to test whether there is a negative linear correlation between practice hours and golf score for the golf players who belong to the Arnie Palmer Country Club. She randomly selected ten golf players and determined the number of hours they practiced per week and their golf score on the Country Club Golf Course during the annual tournament. The sample results were:

Golf Score : 92 83 85 87 91 95 80 84 90 88
Practice Hrs.: 10 15 17 20 9 11 26 18 13 22

a) Construct a scatter diagram.
b) Calculate the sample correlation coefficient, r.
c) Find r^2 and interpret its meaning.
d) Determine the regression equation, y'.
e) Using the regression equation, predict the golf score for a golfer who practices 17 hours per week.
f) Predict the golf score if the number of practice hours is 30.
g) Is the predicted value obtained in part f) an interpolated or extrapolated value? Why?

63. Sydney, An educational researcher, randomly samples 15 college students to determine if there is a correlation between reading and mathematics scores on a standardized educational test. The following data represents the reading and mathematics test results for the 15 students.

Reading Scores X Values	Mathematics Scores Y Values
191	180
103	101
187	173
108	103
180	170
118	113
178	171
127	122
176	168
134	130
165	150
147	145
160	150
157	154
145	130

a) Construct a scatter diagram.
b) Using the scatter diagram, do you believe that a linear relationship exits? If so, identify the type of linear relationship?
c) Calculate the sample correlation coefficient, r.
d) Find r^2 and interpret its meaning.
e) Determine the regression equation, y'.
f) Using the regression equation, predict a college student's mathematics test score if the student's reading test score is 182.
g) Is the predicted value obtained in part f an interpolated or extrapolated value? Explain.

64. Nola, a marketing researcher, randomly samples 12 employees to determine if there is a correlation between years of schooling and salary earnings. The following data represents the sample results.

Current Salary	Years of Education
$57,000	15
$40,200	16
$27,500	12
$45,000	15
$23,450	8
$32,100	13
$67,800	16
$37,500	12
$77,600	16
$33,000	13
$55,600	15
$62,500	14

a) Construct a scatter diagram.
b) Using the scatter diagram, do you believe that a linear relationship exits? If so, identify the type of linear relationship?
c) Calculate the sample correlation coefficient, r.
d) Find r^2 and interpret its meaning.
e) Determine the regression equation, y'.
f) Using the regression equation, predict a person's current salary if the employee has 14 years of education.
g) Is the predicted value obtained in part f an interpolated or extrapolated value? Explain.

WHAT DO YOU THINK?

65. Examine the following table which represents the correlations between the SAT (Scholastic Aptitude Test) scores and the GRE (Graduate Record Examination) scores for 22,923 subjects of a study entitled "The Differential Impact Of Curriculum On Aptitude Test Scores" which appeared in the *Journal of Educational Measurement* in 1990.

Intercorrelations Between SAT and GRE Scores for the Total Study Sample
n = 22,923

	SAT Verbal	SAT Mathematical	GRE Verbal	GRE Quantitative	GRE Analytical	Mean	Standard Deviation
SAT Verbal	1.000	.628	.858	.547	.637	518.8	104.7
SAT Mathematical	.628	1.000	.598	.862	.734	556.0	110.2
GRE Verbal	.858	.598	1.000	.560	.649	510.1	107.7
GRE Quantitative	.547	.862	.560	1.000	.730	573.4	125.6
GRE Analytical	.637	.734	.649	.730	1.000	579.7	117.6

Republished with permission of the Journal of Educational Measurement.

The SAT is a test administered by the College Board to college-bound high school students. The results are used by college undergraduate admission departments in making decisions regarding high school applicants. While the Graduate Record Examination (GRE), a College Board administered exam, is taken by college students preparing to apply for admission to graduate schools. The purpose of the study was to determine the correlation between the SAT and GRE scores of students who took these exams the normal times during their academic careers with the typical number of years between exams.

a) What type of correlation exists between the SAT Verbal and the GRE Verbal? Interpret this relationship? That is, in general, what would you expect to happen to the value of the GRE Verbal scores as the scores on the SAT Verbal increases? If a scatter diagram were constructed for the variables SAT Verbal and the GRE Verbal, what would you expect the general pattern to look like?

b) What proportion of the variance in the SAT Verbal scores can be accounted for by the variance in the GRE Verbal scores? What formula is used to determine this proportion, and what is this statistical measure called?

c) What does a value of 1.000 indicate about the relationship between SAT Verbal and SAT Verbal? What type of a relationship is this called?

d) For the SAT Verbal Column, what row variable, other than the SAT Verbal, has the strongest relationship with the SAT Verbal? Explain. What do you believe might cause this to happen?

e) For the SAT Mathematical Column, what row variable, other than the SAT Mathematical, has the strongest relationship with the SAT Mathematical? Explain. What do you think might lead to this strong relationship?

f) What proportion of the variance in the SAT Mathematical scores can be accounted for by the variance in the GRE Analytical? What proportion of the variance in the SAT Mathematical scores can be accounted for by the variance in the GRE Quantitative? Explain what the difference in these two results mean?

g) The authors of the study state: "the correlations between SAT Verbal and GRE Verbal and between the SAT Mathematical and GRE Quantitative, both of which are 0.86, indicating that the linear relationship between SAT and GRE scores explains almost three fourths of the variance in GRE Verbal and GRE Quantitative scores taken four years later." Explain the meaning of this statement and how do you think the authors arrived at this three fourths value? What is an equivalent percentage value for this three fourths figure? What do the authors mean by "linear relationship"? Explain.

h) For the GRE Analytical Column, what row variable has the weakest relationship with the GRE Analytical? Explain. What do you think might lead to this weak relationship?

i) In words, describe the type of relationship and the strength of the relationship between the variables GRE Quantitative and SAT Mathematical.

j) Interpret the information that the mean and standard deviation indicate about the SAT and GRE test results. Which distribution of test scores has the greatest variability?

k) If the distribution of each of these test scores is approximately normal, then what is the median and modal test score for each test?

l) If the distribution of each of these test scores is approximately normal, then what percent of the SAT Mathematical scores are within 110.2 points of 556? within 220.4 points of 556? within 330.6 points of 556?

m) As the scores of the SAT Mathematical increase, what would you expect would happen to the scores on the variable GRE Quantitative? Explain.

66. The following are excerpts from five different studies.

1. In a study of 748 pregnant women, maternal birth weight was significantly related to baby's birth weight. The lower the maternal birth weight, the lower the baby's birth weight.

2. The College Board which administers the SAT's stated that there is a correlation between family income and test performance with low-income students not doing as well on the SAT's as students from high-income families.

3. A research finding indicated that tall children under age 8 tend to do better on intelligence tests than short children.

4. The results of a study support the notion that a child's drive is related to the age of the father. The higher the scores of a student, the younger the father. Older fathers were defined as those over age 30 at the time of their offspring's birth.

5. A researcher cites that time spent on homework is positively related to achievement.
   a) For each excerpt state the variables within the study. Identify which variable is the independent and dependent variable.
   b) Indicate the type of relationship that exists between the two variables.
   c) If a scatter diagram were constructed for each of the variables, what would you expect the general pattern to look like? Explain.

PROJECTS

67. Randomly collect the heights to the nearest inch (x) and the weights to the nearest pound (y) of 20 students at your school.
    a) Construct a scatter diagram.
    b) Find r for your random sample.
    c) What type of linear correlation, if any, do you observe?
    d) Estimate the weight of a student if his height is 74 inches.
    e) Determine the percent of variance in weight that is accounted for by the variance in height.

68. Select two variables that you believe are related and randomly sample the selected population. Using your sample data, follow the **questions** set forth in **problem 49**.

CHAPTER TEST

69. Which type of linear correlation is indicated by a scatter diagram whose points are generally lower as you move from left to right?
    a) Positive
    b) None
    c) Negative

70. If y increases as x increases this indicates that x and y are correlated:
    a) Positively
    b) Not at all
    c) Negatively

71. If x and y are perfectly linearly correlated then r must equal:
    a) 1
    b) –1
    c) 0
    d) 1 or –1

72. A positive value of r indicates:
    a) No linear relationship
    b) A weak linear relationship
    c) A positive linear relationship
    d) A negative linear relationship

73. Match the following values of Pearson's correlation coefficient, r, with the appropriate scatter diagram:
    Possible r values:
    I. r = 0.58   II. r = 0.96   III. r = 0   IV. r = -0.53   V. r = 0.39

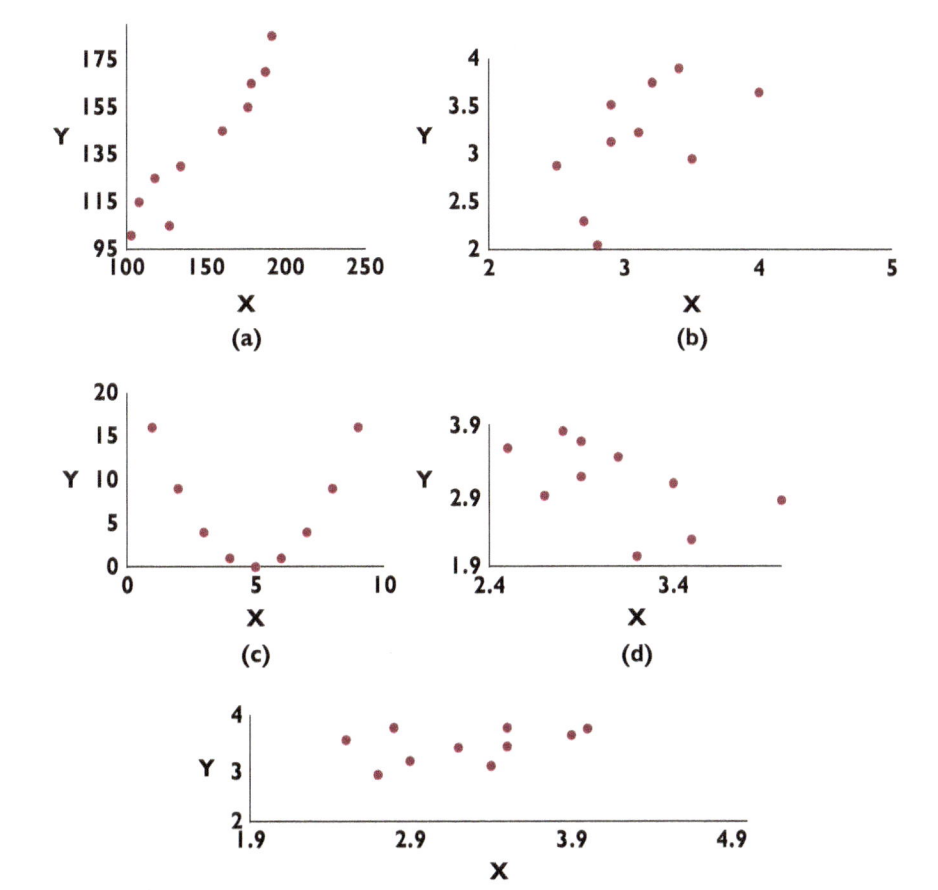

74. When the coefficient of determination is one, this means that there is a perfect positive linear correlation between the x and y variables.
(T/F)

75. I.M.Smart, a statistics nerd, wants to investigate if there is a positive linear relationship between the number of hours a student studies for a statistics exam and the student's test grade. He randomly samples 12 students that are taking statistics at a community college and records the sample data in the following table.

Student	Hours Studying for Exam	Test Grade
1	10	80
2	8	60
3	12	78
4	20	90
5	7	65
6	4	60
7	9	70
8	15	85
9	11	75
10	6	70
11	2	45
12	1	50

a) Construct a scatter diagram.
b) Calculate the sample correlation coefficient, r.
c) Find r^2 and interpret its meaning.
d) Determine the regression equation, y'.
e) Using the regression equation, predict a student's statistics grade if the student spent 14 hours studying for the exam.
f) Predict the test grade if the number of study hours is 25.
g) Is the predicted value obtained in part f) an interpolated or extrapolated value? Why?

76. The following data represents the age (in months) of 19 Polar bear cubs along with their weight at the given age (in pounds).

Age(months)	Weight(pounds)
29	121
104	166
100	220
57	204
53	144
44	140
20	105
9	26
57	125
84	180
57	116
45	182
82	356
70	316
58	202
11	62
83	236
17	76
17	48

a) Draw a scatterplot of the data. Put age on the horizontal axis and weight on the vertical axis.
b) Does there appear to be a linear relationship between these two variables?
c) Find the slope and y-intercept of the "best-fit-line" and write the regression equation.
d) Use the regression equation to predict the weight of a Polar bear cub that is 72 months old.

For the answer to this question, please visit MyStatLab.com.

CHAPTER 5

Probability

Contents

5.1 Introduction
    The Birthday Problem
    Chance! Chance! Chance!
5.2 Some Terms Used in Probability
    Using a Tree Diagram to Construct
      a Sample Space
5.3 Permutations and Combinations
    Permutation
    Combination
    Methods of Selection
    Explaining the Difference Between the Idea
      of a Permutation and a Combination
5.4 Probability
    Alternate Approaches to Assigning
      a Probability
    Another Approach to Defining Probability
    Relative Frequency Concept of Probability
      or Posteriori Probability
    Another Approach to Defining Probability
    Subjective or Personal Probability
5.5 Fundamental Rules and Relationships
      of Probability
    Probability Problems Using Permutations
      and Combinations
5.6 Conditional Probability

5.1 INTRODUCTION

The Birthday Problem

A long long time ago B.C.

FIGURE 5.1 (continues)

200,000 years later A.D.

FIGURE 5.1 (continued)
By permission John Hart Studios, Inc. and Creators Syndicate, Inc

Is Charlie Brown a loser? Lucy thinks so. In every wager Lucy makes with Charlie Brown, Charlie Brown loses. At tomorrow's Christmas party Lucy plans to wager Charlie Brown a brand-new bicycle that "at least two kids have the same birthday" (i.e., born on the same month and day).

Charlie Brown reasons that since there are going to be 50 kids at the Christmas Party and there are 365 possible birthdays, he believes that he is going to win his first bet ever with Lucy.

Do you think Charlie Brown's **chances** are good for winning this bet with Lucy?

Chance! Chance! Chance!

People have always been intrigued with **chance**. In fact an individual's own life involves chance. For example a person's sex is determined by the chromosomes of the parents. Also the purchase of a "lucky" lottery ticket could make one an instant millionaire.

As early as medieval times, man has been concerned with games of chance. In the 16th century, an Italian mathematician, Jerome Cordan (1501–1576), wrote the first book on this subject entitled: *The Book on Games of Chance.*

During the mid-17th century the French mathematician Pierre de Fermat (1601–1660) and Blaise Pascal (1623–1662) were inspired by a gambling friend to study the problem of how one should split the stakes in an interrupted dice game. Their study of this game of chance led to the development of the modern theory of probability.

In today's world probability plays an important role. The French mathematician Pierre Simon de Laplace (1749–1827) in his work, *Theorie Analytique des Probabilities* did a good job defining this role when he stated:

> ... it is remarkable that a science which began with the consideration of games of chance should have become the most important object of human knowledge ... the most important questions of life are, for the most part, really only problems of probability.

Before we investigate problems of probability, let's develop some elementary probability concepts. By the way, Charlie Brown lost the bet! We will examine this problem in detail in Example 5.53.

5.2 SOME TERMS USED IN PROBABILITY

To begin the study of probability, several important terms used in probability need to be defined. The first term we will discuss is what is meant by an *experiment*.

> **DEFINITION 5.1**
> An **experiment** is the process by which an observation is made or obtained. An **outcome** of an experiment is any result that is obtained when performing the experiment.

For example rolling a fair die[1], tossing a fair coin, selecting a card from a deck of cards or selecting the winning numbers of a lottery ticket are all examples of an experiment. For the experiment: selecting a card from a regular deck of 52 playing cards, one possible outcome for this experiment is selecting the three of diamonds.

When performing an experiment, we may be interested in a *particular collection of outcomes.* Such a collection of interested outcomes is called an *event.*

---
[1] A fair die is a die that has the same chance of landing with a 1, 2, 3, 4, 5 or a 6 face up. Similarly, a fair coin is a coin that has the same chance of landing with heads face up as it does with tails face up.

The following illustrations should be helpful in clarifying the terms: an *experiment*, an *outcome* and an *event*.

> **DEFINITION 5.2**
> An **event** is a particular collection of outcomes of an experiment.

Experiment: Rolling a fair die

Possible Outcomes Rolling a one
Rolling a two
Rolling a three
Rolling a four
Rolling a five
Rolling a six

Possible events Rolling an odd number
Rolling a number smaller than 3

Experiment: Selecting a card from a deck of playing cards

Possible Outcomes Selecting a two of clubs
Selecting a three of clubs
.
.
.
Selecting a jack of clubs
Selecting a queen of clubs
Selecting a king of clubs
Selecting an ace of clubs
Selecting a two of diamonds
Selecting a three of diamonds
.
.
.
Selecting a jack of diamonds
Selecting a queen of diamonds
Selecting a king of diamonds
Selecting an ace of diamonds
Selecting a two of hearts
Selecting a three of hearts
.
.
.
Selecting a jack of hearts
Selecting a queen of hearts
Selecting a king of hearts
Selecting an ace of hearts
Selecting a two of spades
Selecting a three of spades
.
.
.
Selecting a jack of spades
Selecting a queen of spades
Selecting a king of spades
Selecting an ace of spades

234 *Chapter 5 Probability*

Possible events   Selecting a red card
Selecting a picture card
Selecting a diamond
.
.
.

Once an experiment is defined, it is necessary to determine *how many* **outcomes** and *what* outcomes are possible when performing the experiment. The collection of all possible outcomes of an experiment is referred to as a **sample space.**

> **DEFINITION 5.3**
> A **sample space** is a representation of all the possible outcomes of an experiment.

In the next few examples, we will illustrate the idea of a sample space and examine different ways to construct a sample space.

EXAMPLE 5.1

A single toss of a coin is an experiment having **two** possible outcomes. Construct a sample space for this experiment.

Solution

The sample space consists of the two possible outcomes: heads and tails. These outcomes are represented in Figure 5.2a.

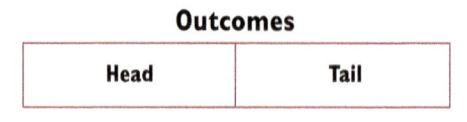

Outcomes

Head	Tail

FIGURE 5.2a

The sample space in Example 5.1 can also be represented using a *tree diagram.*

Using a Tree Diagram to Construct a Sample Space

A **tree diagram** can be very helpful in listing the different possible outcomes of an experiment. A tree diagram for the experiment of a single toss of a coin is illustrated in Figure 5.2b.

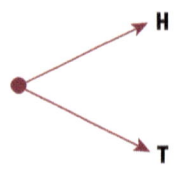

FIGURE 5.2b

Notice that in Figure 5.2b the tree diagram is constructed by first drawing a point or a dot and then drawing lines leading out from the dot, which correspond to the different possible outcomes for the experiment. *These lines represent the branches of the tree.* For this experiment, the two possible outcomes, head and tail, are represented by the two branches.

Since a sample space is a representation of all possible outcomes either Figure 5.2a or Figure 5.2b are ways to display the sample space for the experiment of a single toss of a coin.

EXAMPLE 5.2

Use a tree diagram to construct a sample space for the experiment of tossing a fair coin twice.

Solution

To construct a tree diagram for this experiment, you start by drawing two lines which correspond to the possible outcomes of a head and a tail for the first toss. This is illustrated in the tree diagram of Figure 5.3a. Since the second toss can result in either a head or a tail, start from the tip of each first branch and draw two lines corresponding to the possible outcomes of the second toss. Thus we have two lines branching out from the first possible outcome of a head and two lines from the first possible outcome of a tail. The four possible outcomes of this experiment are illustrated by the tree diagram in Figure 5.3b. Starting from the initial point, one possible outcome of this experiment is determined by moving along one of the initial two branches following by selecting one of the branches of the second toss. This results in the four possible outcomes of tossing a fair coin two times which are a:

head on the first toss and a head on the second toss: HH
head on the first toss and a tail on the second toss: HT
tail on the first toss and a head on the second toss: TH
tail on the first toss and a tail on the second toss: TT

These outcomes are shown to the right of the tree diagram.

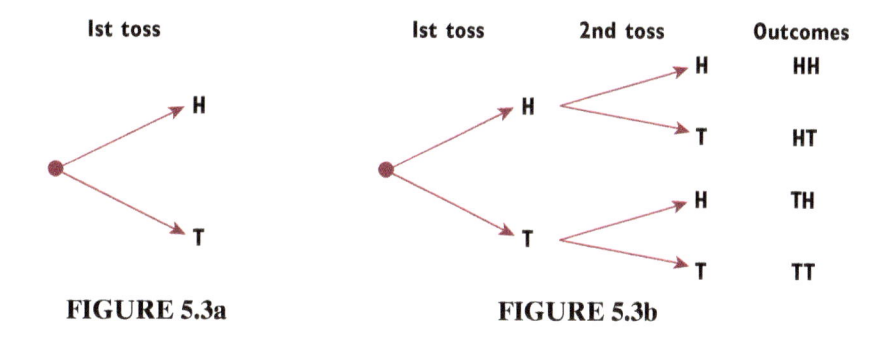

FIGURE 5.3a

FIGURE 5.3b

The sample space for two tosses of a coin can also be represented as shown in Figure 5.3c.

FIGURE 5.3c

In Example 5.2, the sample space consisted of four possible outcomes. For this experiment, the *number of possible outcomes can be determined by* **multiplying** *the number of outcomes for the first toss of the coin by the number of outcomes for the second toss.* That is,

$$\begin{aligned}\text{total number of possible outcomes} &= \left(\begin{array}{c}\text{number of outcomes}\\\text{for first toss}\end{array}\right) * \left(\begin{array}{c}\text{number of outcomes}\\\text{for second toss}\end{array}\right)\\ &= (2)(2)\\ &= 4\end{aligned}$$

Thus, there are four possible outcomes for tossing a coin two times. Furthermore, Example 5.2 is considered a **compound or multi-stage experiment** since the *coin* **is tossed** *more than once*. In fact, this is considered a *two-stage experiment* since the coin is tossed *twice*.

The rule to determine the number of possible outcomes for *a compound or multi-stage experiment* will be referred to as the *Fundamental Counting Principle: Multiplication Rule.*

Fundamental Counting Principle: Multiplication Rule

The total number of possible outcomes for a compound or multi-stage experiment is the product of the number of possible outcomes for each stage of the experiment.

Fundamental Counting Principle
For a Two-Stage and Three-Stage Experiment

Two-Stage Experiment
The total number of possible outcomes for a compound experiment having **two stages**, where there are m ways of doing the first stage and there are n ways of doing the second stage is determined by multiplying the n ways of the first stage by the m ways of the second stage.
   The formula for a two-stage experiment is:

total number of possible outcomes = (m)(n).

Three-Stage Experiment
For a compound experiment with **three stages** where there are m ways for the first stage, n ways for the second stage and p ways for the third stage, the total number of possible outcomes is determined by multiplying these possibilities.
   The formula for a three-stage experiment is:

total number of possible outcomes = (m)(n)(p).

EXAMPLE 5.3

An experiment consists of a single roll of a die and a single toss of a coin.
a) Use the *Fundamental Counting Principle: Multiplication Rule* to determine the total number of possible outcomes for this experiment.
b) Construct a sample space.

Solution

a) This compound experiment is a two-stage experiment where the single roll of the die is the first stage and the single toss of the coin is the second stage. A single roll of a die has **6 possible outcomes** and a single toss of a coin has **2 possible outcomes**. To determine the total number of possible outcomes for this compound experiment using the fundamental counting principle, multiply the number of outcomes for rolling a single die by the number of outcomes for tossing a coin. This can be expressed using the multiplication rule formula:

$$\begin{pmatrix}\text{number of outcomes}\\ \text{for rolling a die}\end{pmatrix} * \begin{pmatrix}\text{number of outcomes}\\ \text{for tossing a coin}\end{pmatrix} = \begin{matrix}\text{total number}\\ \text{of possible}\\ \text{outcomes}\end{matrix}$$

Therefore, there are **6 x 2 = 12 possible outcomes** for this compound experiment.
b) The 12 outcomes can be determined by listing the two outcomes of the coin (head and tail) and the possible six outcomes of the die (1 to 6) along the sides of a table. The 12 outcomes of the sample space are displayed within the body of Table 5.1.

TABLE 5.1 Sample Space for Tossing a Coin and Rolling a Die

Outcomes of the Coin	\multicolumn{6}{c}{Outcomes of the Die}
	1	2	3	4	5	6
Head	H1	H2	H3	H4	H5	H6
Tail	T1	T2	T3	T4	T5	T6

In Example 5.3, a table rather than a tree diagram was used to illustrate another method to construct a sample space. For this experiment, it is appropriate to use a table to construct the sample space since the two objects are each being thrown only once.

EXAMPLE 5.4

Suppose your cat has a litter of three kittens, but you are unaware of the gender of these kittens.
a) Use the Fundamental Counting Principle to determine the number of possible outcomes for the gender of the litter.
b) Construct a sample space for this experiment.
Using the sample space, determine which outcomes satisfy the following events:
c) the first kitten of the litter is male.
d) only one kitten is female.
e) *at least one* kitten is female.
f) *at most one* kitten is male.

Solution

a) Each kitten represents a stage of this three-stage compound experiment, where the kitten's gender represents the outcome for each stage. Using the **Fundamental Counting Principle** to determine the number of possible outcomes in the sample space, we multiply the number of possible outcomes for each stage.
That is,

$$\text{Total number of possible outcomes} = \begin{pmatrix}\text{gender} \\ \text{possibilities} \\ \text{of first kitten}\end{pmatrix} * \begin{pmatrix}\text{gender} \\ \text{possibilities of} \\ \text{second kitten}\end{pmatrix} * \begin{pmatrix}\text{gender} \\ \text{possibilities of} \\ \text{third kitten}\end{pmatrix}$$

$$= [2] * [2] * [2]$$
$$= 8$$

Thus, there are eight possible outcomes for the litter.
b) The eight outcomes in the sample space are displayed by the tree diagram in Figure 5.4.

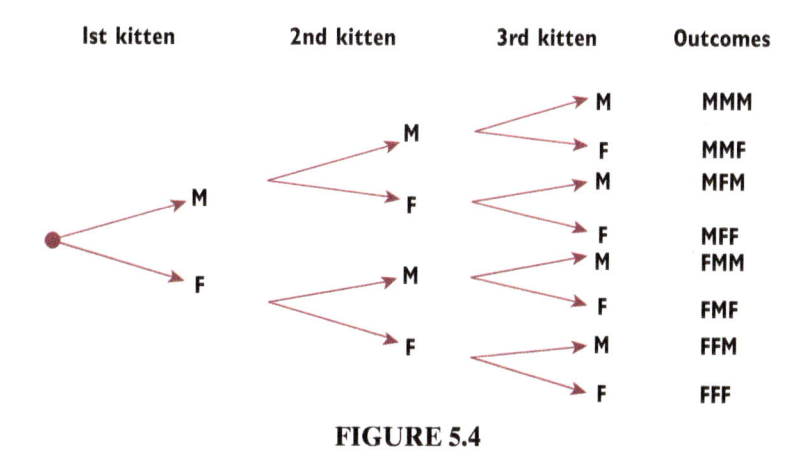

1st kitten	2nd kitten	3rd kitten	Outcomes
M	M	M	MMM
		F	MMF
	F	M	MFM
		F	MFF
F	M	M	FMM
		F	FMF
	F	M	FFM
		F	FFF

FIGURE 5.4

c) In examining the sample space of Figure 5.4, notice that there are **four outcomes** that satisfy **the event that the first kitten of the litter is male**. These outcomes are: **MMM, MMF, MFM and MFF**.
d) From Figure 5.4, there are **three outcomes** that satisfy **the event that only one kitten is female**. The outcomes are: **FMM, MFM and MMF**.
e) **At least one kitten is female** *means* **one or more kittens are female**. Therefore, we are looking for all the outcomes in the litter that contain one or more female kittens. There are seven outcomes satisfying **the event of at least one kitten is female**. As illustrated in Figure 5.4, these outcomes are: **MMF, MFM, MFF, FMM, FMF, FFM, FFF**.
f) **At most one kitten is male** *means* **one male kitten or less than one male kitten**. Therefore, we are looking for all the outcomes that contain *one or less male kittens*. There are four outcomes satisfying **the event at most one male kitten**. These four outcomes are: **FFF, FFM, FMF, MFF**. *Notice the outcome of all female kittens (FFF) satisfies the statement "at most one male kitten" since this outcome (FFF) has no male kittens which is less than one male kitten.*

Example 5.5 contains the phrase *without replacement*. This means when an object is selected it is **not** replaced for the next selection. A detailed explanation is presented in the next section under the heading: Methods of Selection.

EXAMPLE 5.5

Suppose an urn contains 4 balls numbered 1, 2, 3 and 4. Consider the experiment of selecting two balls from the urn *without replacement* and constructing a two-digit number.
a) Use the Fundamental Counting Principle to determine the number of possible outcomes in the sample space for this experiment.
b) Construct a sample space for this experiment.

Using the sample space, determine which outcomes satisfy the following events:
c) both digits of the number are even.
d) only one digit of the number is even.
e) *at least one* digit of the number is even.
f) *at most one* digit of the number is even.

Solution

a) Using the **Fundamental Counting Principle** to calculate the number of possible outcomes in the sample space, we need to multiply the number of possible choices for each selection from the urn. For the first selection, we can choose from any one of the 4 balls in the urn. In a *"without replacement"* experiment, the first selected ball is **not** placed back into the urn. Therefore, for our second selection, we are choosing one ball from the remaining 3 balls left in the urn. Using the **Fundamental Counting Principle**, we have:

total number of possible outcomes = (number of possible outcomes for the 1st selection) * (number of possible outcomes for the 2nd selection)
= (4) * (3)
= 12

Thus, there are 12 possible outcomes for the experiment of selecting two balls from the urn *without replacement*.

b) The 12 outcomes representing the sample space of this experiment are illustrated in Table 5.2. Notice there are cells in Table 5.2 containing an X. This represents the possibilities that are impossible. That is, in a *without replacement experiment*, it is impossible to select the same numbered ball twice such as 1 and 1, 2 and 2, etc.

TABLE 5.2

First Ball Selected	Second Ball Selected			
	1	2	3	4
1	X	12	13	14
2	21	X	23	24
3	31	32	X	34
4	41	42	43	X

c) In examining the sample space of Table 5.2, notice there are **two outcomes** that satisfy the event **both digits of the number are even**. The outcomes are: **24 and 42**.
d) In the sample space of Table 5.2, there are **eight outcomes** that satisfy **the event that only one digit of the number is even**. The outcomes are: **12, 14, 21, 23, 32, 34, 41, and 43**.
e) **At least one digit of the number is even** *means* **one or more digits must be even**. Therefore, we are looking for all the numbers that contain **one or two even digits**. There are 10 outcomes satisfying **the event of at least one digit of the number is even**. As illustrated in Table 5.2, these numbers are: **12, 14, 21, 23, 24, 32, 34, 41, 42, and 43**.
f) **At most one digit of the number is even** *means* **one or less digits are even**. Therefore, we are looking for all the numbers that contain **one or no even digits**. There are 10 numbers satisfying **the event of at most one digit of the number is even**. These numbers are: **12, 13, 14, 21, 23, 31, 32, 34, 41, and 43** as shown in Table 5.2. ∎

The phrases **"at least"** and **"at most"** used to define events in Examples 5.4 and 5.5 are frequently used in probability. It is important that you become familiar with the meaning of these expressions. Let's consider the general interpretation of these phrases.

Interpretation of the Expressions "At Least" and "At Most"

"At least N outcomes" is interpreted as N or more outcomes
"At most N outcomes" means N or less outcomes
(remember to include N = 0 as a possible outcome).

5.3 PERMUTATIONS AND COMBINATIONS

Permutation

When the total number of possible outcomes in a sample space is extremely large, there are techniques that can be used to count the outcomes without having to list every possible outcome. The counting techniques are called **permutations** and **combinations**[2]. **We will first examine the idea of a permutation.**

DEFINITION 5.4
A **permutation** is an arrangement of objects in a *definite* order.

For example, let's examine the number of ways the three symbols: **@, #,** * can be arranged in order. These three objects can be arranged in order six ways as illustrated in Figure 5.5a.

Carefully examining Figure 5.5a, we should notice that within *each* arrangement of the three symbols the **order is different** and there are six different possible arrangements. Each arrangement is referred to as a **permutation**. If we want to determine *how many* arrangements there are without listing them, we could use the following counting technique. Each position in the arrangement is represented by a box as shown in Figure 5.5b.

Since *any one of the three objects* (@,#,*) could have been chosen to be placed in the *first position*, a *three (3)* is written in the box representing position #1 as shown in Figure 5.5b. Once an object has been selected

```
@ * #
@ # *
# @ *
# * @
* @ #
* # @
```

FIGURE 5.5a

1st Position	2nd Position	3rd Position
3	2	1

FIGURE 5.5b

---

[2] Most calculators have the formulas for permutations and combinations built-in. We encourage you to use a calculator when evaluating these formulas.

for the first position, there are only two objects left to choose from for the remaining positions. The two (2) written in position #2 within Figure 5.5b represents the two objects that can be chosen for the second position. Finally, once the first two objects are chosen there is only *one* object left to place in the third position. The one (1) written in the 3rd position in Figure 5.5b represents the last object that is chosen.

To determine the number of different arrangements we *multiply* these three numbers obtaining the result:

$$(3)(2)(1) = 6.$$

Thus, the six represents the number of different arrangements that are possible for the three objects: @,#,*. These were illustrated in Figure 5.5a. In mathematics, we have a shorthand notation for the product (3)(2)(1). The product: *(3)(2)(1)* is written as *3!* and it is read as *3 factorial.* Let's define this new notation.

---

n factorial: n!

In general, *n!* is read as *n "factorial"* and is defined to be:
$$n! = n(n-1)(n-2)(n-3) \cdots 2 \cdot 1$$
where $0! = 1$.

Using this factorial notation, we can define the counting technique that is used to determine the number of ways of arranging n objects in order. We will call this rule as: ***counting rule 1 or the permutation rule***.

DEFINITION 5.5
Counting Rule 1: Permutation Rule. The number of permutations (arrangements) of *n different* objects taken *altogether,* denoted $_nP_n$, is:

$$_nP_n = n!$$

Let's apply this rule to Example 5.6.

EXAMPLE 5.6

How many different arrangements can be found for the following five symbols (objects)?

@, #, %, &, *

Solution
Using the permutation rule: $_nP_n = n!$ for n = 5 objects, we get:

$$\begin{aligned} _5P_5 &= 5! \\ &= (5)(4)(3)(2)(1) \\ &= 120 \end{aligned}$$

Therefore, there are 120 different arrangements for these five different objects.

TI-84 Graphing Calculator Solution for Example 5.6

You can also get the answer for the factorial in Example 5.6 using the TI-84's probability menu. On the TI-84 graphing calculator, you can access the probability menu by pressing MATH then scroll to **PRB**. *Before* you do this, you must be sure to type the value for n in the home screen. To type 5! in the calculator, press the following keystrokes: 5 MATH ◄ over to PRB **4: !** ENTER. Figure 5.6a and Figure 5.6b show these results.

Section 5.3 Permutations and Combinations 241

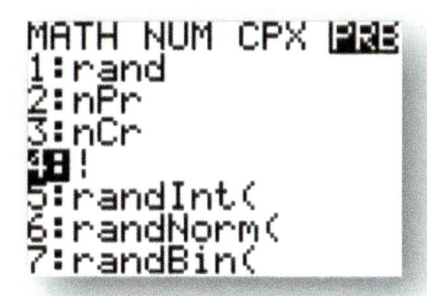

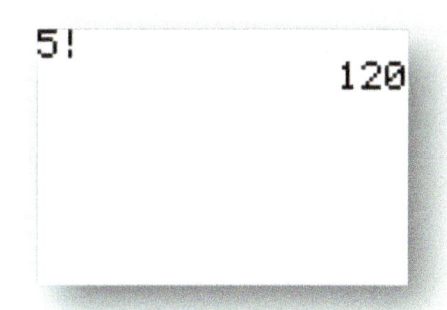

FIGURE 5.6a

FIGURE 5.6b

EXAMPLE 5.7

In how many different ways can the manager of a baseball team arrange the batting order of the nine players in his starting line-up?

Solution

Using the permutation rule: $_nP_n = n!$ for n = 9 players, we have:

$$_9P_9 = 9!$$
$$= (9)(8)(7)(6)(5)(4)(3)(2)(1)$$
$$= 362,880 \text{ different batting orders}$$

In the next example we will consider a slight variation of counting rule 1. We will consider arranging *only some* of the n objects, **not** all of the n objects.

EXAMPLE 5.8

In a *six* horse race how *many different ways of win, place and show* positions (first, second and third position) are possible, assuming no dead heats?

Solution

Although six horses are racing we are only interested in the order of finish for the first three horses as illustrated in Figure 5.7a.

Since any one of the six horses could win the race, there are six possibilities for the first position as displayed in Figure 5.7b.

Once the winner has been determined, any one of the remaining five horses could finish second as shown in Figure 5.7c.

Now that first and second positions have been determined, any one of the remaining four horses could finish third. This is illustrated in Figure 5.7d.

Therefore, the total number of different ways for win, place and show is (6)(5)(4) = 120.

We will now generalize this counting technique and call it counting rule 2 or permutation rule for n objects taken s at a time.

1st Place	2nd Place	3rd Place

FIGURE 5.7a

1st Place	2nd Place	3rd Place
6	5	

FIGURE 5.7c

1st Place	2nd Place	3rd Place
6		

FIGURE 5.7b

1st Place	2nd Place	3rd Place
6	5	4

FIGURE 5.7d

DEFINITION 5.6

Counting Rule 2: Permutation Rule for n objects taken s at a time. The number of permutations of n *different* objects taken s at a time, denoted $_nP_s$, is:

$$_nP_s = n(n-1)(n-2)\cdots(n-s+1)$$

The value of s equals the number of factors for $n(n-1)(n-2)\ldots(n-s+1)$.

The solution to Example 5.8 is determined using the formula for counting rule 2:

$$_nP_s = n(n-1)(n-2)\cdots(n-s+1)$$

The number of arrangements of six horses taken three at a time is expressed as $_6P_3$. The value of n is 6 because there are six horses in the race and the value of s is 3 since we are only interested in the first three finishers. To apply the formula, we need to determine the number of factors required in the expression: $n(n-1)(n-2)\cdots(n-s+1)$. Since the value of s is 3, the expression $_6P_3$ contains 3 factors.

Thus, $_6P_3 = n(n-1)(n-s+1)$
$= (6)(6-1)(6-3+1)$
$= 120$

EXAMPLE 5.9

Using counting rule 2, find the number of permutations (arrangements) there are of seven different objects taken:
a) 3 at a time
b) 4 at a time
c) 7 at a time

Solution

a) The number of permutations of seven different objects taken three at a time is written as:

$$_7P_3, \text{ where: } n = 7 \text{ and } s = 3.$$

Thus, we have: $_7P_3 = (7)(6)(5)$
$= 210$

b) The number of permutations of seven different objects taken four at a time is written as: $_7P_4$.
Thus, we have: $_7P_4 = (7)(6)(5)(4)$
$= 840$

c) The number of permutations of seven different objects taken seven at a time is written as: $_7P_7$.
Thus, we have: $_7P_7 = 7!$
$= (7)(6)(5)(4)(3)(2)(1)$
$= 5040$

TI-84 Graphing Calculator Solution for Example 5.9

Permutations can also be calculated using the TI-84's probability menu. To evaluate $_7P_3$ you must first input 7 on the home screen. To access the probability menu press [MATH] then scroll to **PRB**. To evaluate $_7P_3$, press the following keystrokes: **7** [MATH] ◄ over to PRB **2 :** $_nP_r$. Then type **3** and press [ENTER]. Figure 5.8a and Figure 5.8b show this information.

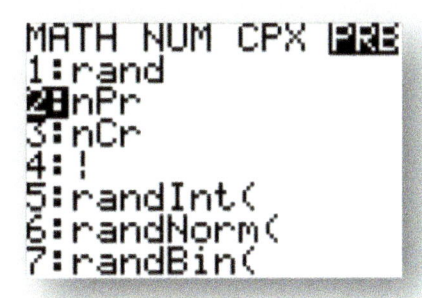

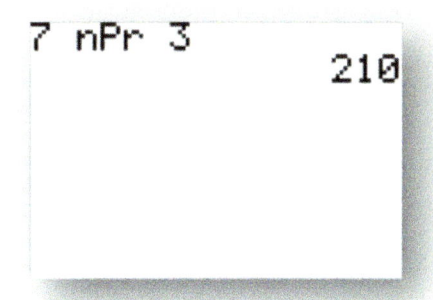

FIGURE 5.8a   **FIGURE 5.8b**

This is the same answer we obtained using the formula in Example 5.9a. The answers for 5.9b and 5.9c can be obtained in the same way as shown in Figure 5.9.

FIGURE 5.9

EXAMPLE 5.10

In how many different ways can a club that has twelve members select a running slate of four officers: *President, Vice President, Treasurer and Secretary?*

Solution

Since a club member can only run for only one office, we are interested in a definite arrangement of the four officers. Therefore, we use counting rule 2, where n = 12 and s = 4. Thus, the number of possible permutations of twelve club members taken four at a time is $_{12}P_4$.

Thus, we have:
$$_{12}P_4 = (12)(11)(10)(9)$$
$$= 11,880$$

EXAMPLE 5.11

List all different arrangements for the letters in the word:
a) *TEA*
b) *TEE*

Solution

a) Since there are three distinct letters to be taken 3 at a time, the number of different arrangements is determined by $_3P_3$.
$$_3P_3 = (3)(2)(1)$$
$$= 6$$

Listing these six arrangements we get:

$$TEA$$
$$TAE$$
$$EAT$$
$$ETA$$
$$ATE$$
$$AET$$

b) The three letters in the word TEE are *not* all distinct. If we take the six arrangements from part a) and change the letter A to E we get:

arrangements from part a)		changing the A to E		
	TEA		TEE	
	TAE		TEE	[repeated]
	EAT		EET	
	ETA		ETE	
	ATE		ETE	[repeated]
	AET		EET	[repeated]

Notice there are only **three *different* arrangements** for the letters T, E, and E. These three arrangements are: *TEE, EET and ETE. There are only three different arrangements since the arrangements containing EE are not distinguishable and thus **reduces the number of distinct arrangements by a factor of 2!**. (This is because the letters E and E can be arranged in 2! ways.)* Consequently, the number of different arrangements of three letters *when two are the same* can be determined by the formula:

$$\frac{3!}{2!} = \frac{(3)(2)(1)}{(2)(1)} = 3$$

Now, we will use Examples 5.12 and 5.13 to help generalize a counting rule for all possible different arrangements that can be made from a group of objects in which **some** objects are **alike**.

EXAMPLE 5.12

a) List all the different arrangements of the letters in the word **FOOT**.
b) Determine the number of different arrangements.

Solution

a) Listing the different arrangements we get:

FTOO	TFOO
OFTO	OTFO
OOFT	OOTF
FOTO	TOFO
OFOT	OTOF
FOOT	TOOF

b) The number of different arrangements of the word FOOT is twelve. The twelve arrangements of the letters in the word FOOT can be determined by the formula:

$$\frac{4!}{2!} = \frac{(4)(3)(2)(1)}{(2)(1)} = 12$$

EXAMPLE 5.13

a) List the different arrangements of the letters in the word *NOON*.
b) Determine the number of different arrangements.

Solution

a) Listing the arrangements we get:

$$\begin{array}{cc} NNOO & NONO \\ ONNO & ONON \\ OONN & NOON \end{array}$$

b) The number of different arrangements of the letters in the word NOON is six. The six different arrangements of the letters in the word NOON can be determined by the formula:

$$\frac{4!}{(2!)(2!)} = \frac{(4)(3)(2)(1)}{(2)(1)(2)(1)} = 6$$

Examine carefully Examples 5.12 and 5.13. In Example 5.12 the word FOOT contained four letters *two of which were the same*. We determined that twelve different arrangements of the letters were possible. In the word NOON, there were also four letters but *both* the letters N and O were the same. That is, in the word NOON there were *two sets of letters that were the same*. It was determined that six arrangements of the letters were possible.

We can now generalize the formula used to calculate the number of permutations of n objects where some objects were the same as illustrated in Examples 5.12 and 5.13. We will call this technique counting rule 3 or the number of permutations of N objects with *k* alike objects.

DEFINITION 5.7

Counting Rule 3: Permutation Rule of N objects with k alike objects. Given N objects where n_1 are alike, n_2 are alike, ..., n_k are alike, then the number permutations of these N objects is:

$$\frac{N!}{(n_1!)(n_2!) \cdots (n_k!)}$$

EXAMPLE 5.14

Use counting rule 3 to determine the number of different arrangements for the letters in each of the following words:
a) BANANA
b) MISSISSIPPI

Solution

a) The word BANANA has six letters with the letter A repeated three times and the letter N repeated twice, therefore the number of different arrangements is:

$$\frac{6!}{3!2!} = \frac{(6)(5)(4)(3)(2)(1)}{(3)(2)(1)(2)(1)} = 60$$

b) The word MISSISSIPPI has eleven letters with the letter I repeated four times, the letter S repeated four times and the letter P repeated twice, therefore the number of different arrangements is:

$$\frac{11!}{4!4!2!} = \frac{(11)(10)(9)(8)(7)(6)(5)(4)(3)(2)(1)}{(4)(3)(2)(1)(4)(3)(2)(1)(2)(1)} = 34,650$$

Combination

In our discussion of permutations we were concerned with the number of ways of arranging s objects selected from n objects in a *particular order*. Now we will consider a selection process where the *order is not important*. This selection process is called a *combination*.

> **DEFINITION 5.8**
> A **combination** is a selection of objects *without* regard to order.

Let's examine this idea of a combination in Example 5.15.

EXAMPLE 5.15

The chief Systems Analyst at H.A.L. Computer Systems, Inc. has to choose two programmers to work overtime on Sunday. If the company employs four programmers, then how many different ways can the Systems Analyst make her selection?

Solution

We will use the letters A, B, C and D to represent the four programmers. The analyst would like all possible arrangements of the four programmers where the ordering of the programmers within each arrangement is *not important*. For example, the selection of programmers A and C where programmer A is chosen first and C second *is the same* as the selection of C and A where programmer C is chosen first and A second. Thus, the *order* of selecting the programmers is **not** *important*.

If **order were important**, then we could use counting rule 2 which determines the number of ways to make a selection of n different objects taken s at a time. For n = 4 and s = 2, counting rule 2 would yield:

$$_4P_2 = (4)(3) = 12.$$

However, the twelve different ways determined by counting rule 2, counts the number of permutations or arrangements of the four programmers if order were important. Within these 12 different arrangements, an arrangement such as AC would be counted as different from arrangement CA. However, the analyst would still end up with the *same* combination of programmers. Therefore to determine the number of different combinations of two programmers, we need to adjust these 12 arrangements since each pair of programmers is *counted twice*. Thus, we must divide the formula $_4P_2$ by 2 to determine the actual number of **different combinations** of 2 programmers.

Thus, there are:

$$\frac{_4P_2}{2} = \frac{(4)(3)}{2} = 6 \text{ different ways}$$

that the systems analyst can make her selection.

We can now generalize the counting technique developed in Example 5.15 to determine the number of ways that s objects can be selected from n objects where **order is not important**. We will call this technique counting rule 4.

> **DEFINITION 5.9**
> **Counting Rule 4: Number of Combinations of n objects taken s at a time.** The number of combinations of n objects taken s at a time, symbolized as $_nC_s$, is:
>
> $$_nC_s = \frac{_nP_s}{s!}$$
>
> The formula for $_nC_s$ in factorial notation is written as:
>
> $$_nC_s = \frac{n!}{s!(n-s)!}$$

EXAMPLE 5.16

Use counting rule 4 to determine the number of combinations of n objects taken s at a time if:
a) n = 5 and s = 3.
b) n = 6 and s = 2.

For part a, use formula: $_nC_s = \dfrac{_nP_s}{s!}$

For part b, use formula: $_nC_s = \dfrac{n!}{s!(n-s)!}$

Solution

To determine the number of combinations of n objects taken s at a time, we will use counting rule 4.

a) If n = 5 and s = 3 then $_5C_3 = \dfrac{_5P_3}{3!} = \dfrac{(5)(4)(3)}{(3)(2)(1)} = 10$

b) If n = 6 and s = 2 then $_6C_2 = \dfrac{6!}{2!4!} = \dfrac{(6)(5)(4)(3)(2)(1)}{(2)(1)(4)(3)(2)(1)} = 15$

EXAMPLE 5.17

Use counting rule 4 to determine the number of combinations for each of the following parts. For parts a and c, use the formula:

$$_nC_s = \dfrac{_nP_s}{s!}$$

For parts b and d, use the formula: $_nC_s = \dfrac{n!}{s!(n-s)!}$

a) the number of combinations of 6 objects taken 4 at a time.
b) the number of ways two books can be chosen from a group of five.
c) the number of ways a college student can choose 3 college courses out of 10 possible choices.
d) the number of ways a student can choose to answer 8 of the 10 questions on an examination.

Solution

a) To determine the number of combinations of 6 objects taken 4 at a time, use the formula:

$$_nC_s = \dfrac{_nP_s}{s!}$$

For n = 6 and s = 4, we have: $_6C_4 = \dfrac{_6P_4}{4!} = 15$

Thus, the number of combinations of 6 objects taken 4 at a time is 15.

b) The number of ways to choose 2 books from a group of five is determined by using formula:

$$_5C_2 = \dfrac{5!}{2!(5-2)!} = 10$$

Therefore, there are 10 ways to choose 2 books from five books.

c) The number of ways a student can select 3 college courses from 10 choices is:

$$_{10}C_3 = \dfrac{_{10}P_3}{3!} = 120$$

Thus, there are 120 ways to select 3 courses from 10 choices.

d) The number of ways a student can choose to answer 8 out of 10 questions is:

$$_{10}C_8 = \frac{10!}{8!(10-8)!} = 45$$

Therefore, there are 45 ways to answer 8 out of 10 questions.

EXAMPLE 5.18

An offer from the Blu-ray Discs of the Month Club to any new member is a choice of 5 Blu-ray Discs for 99 cents from a list of 25 Blu-ray Discs. How many combinations of 5 Blu-ray Discs can be chosen from this list?

Solution

Since we are choosing from 25 Blu-ray Discs, then n = 25. The fact that we are selecting 5 Blu-ray Discs makes s = 5. Therefore, the number of combinations of 25 Blu-ray Discs taken 5 at a time is:

$$_{25}C_5 = \frac{25!}{5!(25-5)!} = 53,130$$

EXAMPLE 5.19

The Holiday Basketball Festival selection committee is considering 20 teams for their 12-team tournament. In how many ways can the 12 teams be selected?

Solution

The fact that the committee is selecting from 20 teams, n = 20. Since they are selecting 12-teams, s = 12. Therefore, the number of combinations of 20 teams taken 12 at a time is:

$$_{20}C_{12} = \frac{20!}{12!(20-12)!} = 125,970$$

TI-84 Graphing Calculator Solution for Example 5.19

Combinations can also be calculated using the TI-84's probability menu. To evaluate $_{20}C_{12}$ you must first input 20 on the home screen. To access the probability menu press [MATH] then scroll to **PRB**. To evaluate $_{20}C_{12}$, press the following keystrokes: **20** [MATH] ◄ over to PRB 3: $_nC_r$. Then type **12** and press [ENTER]. Figure 5.10a and Figure 5.10b show this information.

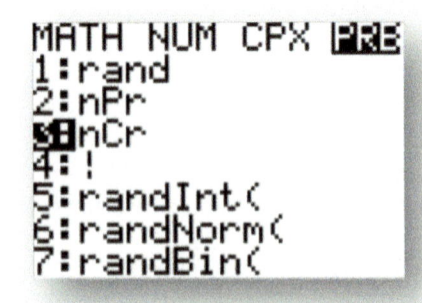

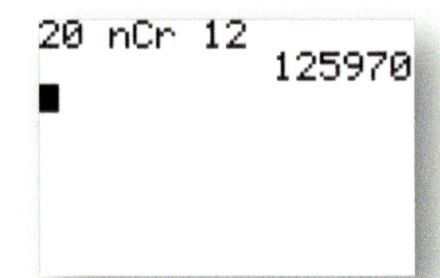

FIGURE 5.10a **FIGURE 5.10b**

This is the same answer we obtained using the formula in Example 5.19.

Methods of Selection

When performing experiments which involve the selection of objects and/or people from a group, it is important that you understand the manner in which the object or person is to be selected from the group. Consequently, it is necessary to briefly discuss some different methods of selection and the wording used in the design of the experiment.

Suppose that you have an urn that contains 5 numbered objects, and the experiment requires you to select 2 objects from this urn. The manner in which the objects are selected depends upon the wording stated within the problem. We will consider two ways the objects can be drawn from the urn.

Methods of Selecting Two Objects from a Group of Objects

Selection without Replacement
If the two objects are selected one at a time from the urn, where the first object *is not* placed back into the urn before the second object is selected, then this type of selection process is called *selection without replacement*.

Selection with Replacement
If the two objects are selected one at a time from the urn, where the first object *is* placed back into the urn before the second object is selected, then this type of selection process is called *selection with replacement*.

Suppose, for example, two balls are selected *without replacement* from an urn containing 5 numbered balls.

a) If order is important, then the selected arrangement of the two balls is considered a permutation with a total of:

$_5P_2$ or $(5)(4) = 20$ **permutations or ordered arrangements of the two numbered balls.**

b) If order is not important, then the selected balls are considered a combination with a total of:

$_5C_2 = \dfrac{5!}{2!(5-2)!} = 10$ **combinations of the two numbered balls.**

On the other hand, if the two balls are selected *with replacement* from the urn containing the 5 numbered balls, then the order in which the balls are selected is usually considered to be of importance. Under these conditions, the number of ways the two balls can be selected is determined by the fundamental counting principle. Since each time a ball is drawn from an urn it contains 5 balls, then there are:

$(5)(5) = 25$ **ways** to select the two numbered balls.

From the previous discussion of the different selection methods used in an experiment, you should realize that *the counting technique employed to determine the number of ways k objects can be selected from a group of n objects is dependent upon the selection method used to choose the k objects, and whether order is an important factor of the experiment.*

EXAMPLE 5.20

An urn has 7 wooden blocks with each block containing only one of the letters: A, C, E, I, M, N, R inscribed on it. If three blocks are selected, then determine:
a) the number of different *arrangements* of the three letters, if the blocks were selected *without* replacement.
b) the number of *combinations* of the three letters, if the blocks were selected *without* replacement.
c) the number of ways to select the three letters, if the blocks are selected *with* replacement.

Solution

a) Since the blocks are being selected **without** replacement, and order **is** important, then this is a permutation problem employing Counting Rule 2 for n **different** objects taken s at a time, denoted $_nP_s$. Because we are selecting from 7 objects, we have n = 7, and since we are choosing 3 objects then s = 3. Substituting this information into the formula for $_nP_s$, we have:

$$_7P_3 = (7)(6)(5) = 210 \text{ different arrangements}$$ of the three letters.

b) Since the blocks are being selected **without** replacement, and order **is not** important, then this is a combination problem using Counting Rule 4 for the number of combinations of n objects taken s at a time, symbolized as $_nC_s$. Substituting n = 7 and s = 3 into the formula, we have:

$$_7C_3 = \frac{7!}{3!4!} = 35 \text{ combinations}$$ of the three letters.

c) Since the blocks are being selected with replacement, and order is usually important, then the number of ways to select the letters is determined by the Fundamental Counting Principle. Since each time a block is drawn from an urn it contains 7 blocks, there are:

$$(7)(7)(7) = 343 \text{ ways}$$ to select three lettered blocks.

Explaining the Difference Between the Idea of a Permutation and a Combination

If the *order of the objects must be different* for each possible outcome then the total number of possible outcomes is determined using a *permutation* formula. For example, the number of different ways that three people Abe, Bill and Carl can be elected for the three different positions: President, Vice President and Secretary of an organization are:

	Abe as President, Bill as VP and Carl as Secretary,
or	Bill as President, Carl as VP and Abe as Secretary,
or	Carl as President, Abe as VP and Bill as Secretary,
or	etc.

Thus, the total number of possible arrangements of these three people into these three positions are determined by the permutation formula:

$$_nP_n = n!$$
$$= 3! = (3)(2)(1)$$
$$= 6 \text{ different arrangements}$$

On the other hand, if *the order of the objects is NOT important,* then the total number of possible outcomes is determined using the *combination* formula. For example, the number of different ways that the three people Abe, Bill and Carl can be selected to form a committee is only one way, since a committee of:

Abe, Bill and Carl is the same as Bill, Carl and Abe.

Thus the total number of ways to form a committee for these three people is determined by the combination formula:

$$_nC_s = \frac{n!}{s!(n-s)!}$$

$$_3C_3 = \frac{3!}{3!(3-3)!}$$

$$= \frac{(3)(2)(1)}{(3)(2)(1)(1)}$$

$$= 1 \text{ way to form a committee for these three people}$$

5.4 PROBABILITY

Each day people make probability statements for such things as the weather, the price of a stock, a baseball team winning the World Series, or a student passing a particular course. For example, a weather forecaster may announce that there is a 50% chance of snow for tomorrow. A stock broker speculates that a particular stock has a 75% chance that its price will double within 6 months. A sportscaster may predict that the likelihood of the National League Baseball Champion beating the American League Baseball Champs in the World Series is 10%. An instructor may indicate that the probability of a student passing his/her course is 90% if the student attends classes every day.

Although you are familiar with such probability statements, how does one go about interpreting these statements? Each of these statements is using a percentage to indicate the likelihood that an event will occur. For example, the statement pertaining to the chance of it snowing tomorrow has the percentage 50% attached to this statement. The weather forecaster is using a percentage (50%) TO REPRESENT THE CHANCE IT WILL SNOW TOMORROW. The percentage 50% or the number 0.50 provide a measure of how likely it is that it will snow tomorrow, and is referred to as the probability of the event that it will snow tomorrow.

DEFINITION 5.10

Probability is a numerical measure of the likelihood that a particular event will occur in the future. The probability of an event can only be a number from zero to one inclusive, or in terms of percent, from zero to 100% inclusive.

The probability or chance of the outcome occurring is expressed as a number ranging from 0 to 1, or as a percentage ranging from 0 to 100%. Probability values can be expressed in the form of a fraction, decimal or percent. For example, one could express the chance of getting a head on a single toss of a fair[3] coin either as: $\frac{1}{2}$ (fraction form) or 0.5 (decimal form) or 50% (percentage form).

A probability value of zero for an outcome is interpreted to mean that the outcome will never happen; while a probability value of one (or 100%) indicates that the outcome will definitely happen. Figure 5.11 shows a scale representing the probability of an outcome and some interpretations of these probabilities.

Probability of an Outcome

form of probability value: interpretation of probability:

fraction or decimal   percent

0 ——— 0%: outcome will never occur

any probability value between 0–50% is unlikely to occur

$\frac{1}{2}$ or 0.50 ——— 50%: outcome is just as likely to occur as not occur

any probability value between 50–100% is likely to occur

1 ——— 100%: outcome will definitely occur

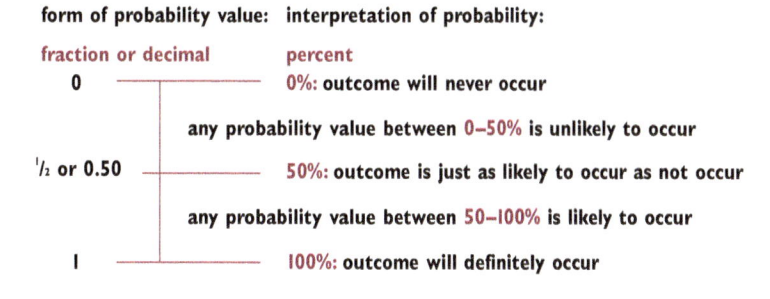

FIGURE 5.11

---

[3] A fair coin is a coin that has the same chance of landing with heads face up as it does with tails face up.

Alternate Approaches to Assigning a Probability

There are three main approaches to assigning a probability value to the likelihood that a particular outcome or event will occur. These commonly used approaches to assigning a probability are:

Approaches to Assigning a Probability

(I) *Classical or a priori probability,*
(II) *Relative frequency concept of probability or posteriori concept*
and (III) *Subjective or personal probability.*

We will first define the **classical** or **a priori approach** to assigning a probability.

DEFINITION 5.11
Classical or *A Priori Probability*: The classical approach to probability is based upon the assumption that each outcome of the experiment is EQUALLY LIKELY.

Consider the experiment of rolling a fair die once. The six outcomes: 1, 2, 3, 4, 5, 6 are equally likely, since the die is fair and each outcome is assumed to have an equal chance of happening.

DEFINITION 5.12
All the possible outcomes of an experiment are **equally likely** if each outcome has the **same chance** of happening or occurring.

If each possible outcome is equally likely, then the probability of an event E can be defined using the Classical Approach.

Classical Probability Definition for Equally Likely Outcomes

Definition: The **probability of an Event E**, written P(Event E), is equal to the number of outcomes satisfying the event E divided by the total number of outcomes in the sample space, provided each outcome within the sample space is equally likely. This formula can be written as:

$$P(\text{Event E}) = \frac{\text{number of outcomes satisfying event E}}{\text{total number of outcomes in the sample space}}$$

In this text, we will primarily use the **Classical Approach of Defining Probability** to assign to an outcome a probability value. Remember, this classical approach can only be used if each outcome of the sample space is equally likely. We will illustrate the classical approach to defining probability in the next few examples.

EXAMPLE 5.21

Consider once again the experiment of tossing a fair coin twice. The sample space for this experiment consists of the four outcomes illustrated in Figure 5.12.

| HH | HT | TH | TT |

FIGURE 5.12

Since the coin is *fair*, each outcome has the *same chance of occurring*. If a fair coin is tossed two times, find the probability of the event that it will land:
a) heads on both tosses, denoted P(HH).
b) heads at most once, denoted P(at most one head).
c) tails at least once, denoted P(at least one tail).

Solution

a) The number of outcomes satisfying the event that the coin will land heads on both tosses is 1. It is HH. Since the total number of outcomes in the sample space is 4 as illustrated in Figure 5.12, the probability of getting two heads is:

$$P(HH) = \frac{\text{number of outcomes satisfying two heads}}{\text{total number of outcomes}} = \frac{1}{4}$$

The probability of two heads is 1/4. This means that if the experiment of tossing a fair coin twice is repeated a large number of times, we would *expect* one-fourth of the outcomes to be two heads.

b) At most one head *means one or less heads*. From Figure 5.12, there are three outcomes which satisfy the event of at most one head. These outcomes are: HT, TH and TT. Since the total number of outcomes in the sample space is 4, the probability of getting at most one head is:

$$P(\text{at most 1 head}) = \frac{\text{number of outcomes satisfying at most one head}}{\text{total number of outcomes}} = \frac{3}{4}$$

The probability of at most one head is 3/4. Therefore, if this experiment were repeated a large number of times, we would *expect* 3/4 of the outcomes to be at most one head.

c) At least one tail *means one or more tails*. The number of outcomes satisfying the event of at least one tail is 3. They are HT, TH and TT. The total number of outcomes in the sample space is 4. Thus, the probability of at least one tail is:

$$P(\text{at least 1 tail}) = \frac{\text{number of outcomes satisfying at least 1 tail}}{\text{total number of outcomes}} = \frac{3}{4}$$

EXAMPLE 5.22

Debbie, the chairperson of the Office Technology Department at a community college, decides to observe the time that two newly hired instructors get to class. Before she performs her experiment, she decides to make a list of all the different possibilities. If she decides to use the notation: E for early to class, O for on time, and L for late to class, then:
a) construct a sample space for this experiment. If each outcome in the sample space is equally likely, then what is the probability that:
b) both instructors will arrive to class late?
c) neither instructor is late for class?
d) at least one instructor is late?
e) both instructors arrive to class at the same time?

Solution

a) Using the Fundamental Counting Principle for this two-stage experiment, the total number of possible outcomes for this sample space can be determined by:

$$\text{number of possible outcomes in sample space} = \left(\begin{array}{c}\text{Number of outcomes}\\ \text{for instructor \#1}\end{array}\right) * \left(\begin{array}{c}\text{Number of outcomes}\\ \text{for instructor \#2}\end{array}\right)$$

Since there are 3 possible outcomes for both instructors, that is: early, on time and late, then:

$$\text{number of possible outcomes in sample space} = (3)(3) \text{ or } 9 \text{ possibilities}$$

The sample space containing these 9 possibilities are illustrated in Table 5.3, using E for early to class, O for on time, and L for late to class.

TABLE 5.3 Sample Space for Two Instructors

	Instructor #2		
Instructor #1	E	O	L
E	EE	EO	EL
O	OE	OO	OL
L	LE	LO	LL

b) There is only one outcome satisfying the event that *both instructors arrive to class late,* which is LL. Since there are 9 outcomes in the sample space and assuming each outcome in the sample space is equally likely, then:

$$P(\text{both instructors arrive to class late}) = \frac{1}{9}.$$

c) The number of outcomes satisfying the event *neither instructor is late for class* is 4. These four outcomes are: EE, EO, OE and OO. Since there are four outcomes out of the 9 possible outcomes satisfying the event, and assuming each outcome in the sample space is equally likely, then:

$$P(\text{neither instructor is late for class}) = \frac{4}{9}.$$

d) The event *at least one instructor is late for class* is satisfied when one or both of the instructors are late. Thus, the outcomes that satisfy this event are: EL, OL, LE, LO, and LL. Since 5 outcomes satisfy this event out of 9 equally likely outcomes, then:

$$P(\text{at least one instructor is late for class}) = \frac{5}{9}.$$

e) There are 3 outcomes satisfying the event *both instructors arrive to class at the same time,* since the event is satisfied by the outcomes: EE, OO and LL. Since 3 outcomes satisfy this event out of 9 equally likely outcomes, then:

$$P(\text{both instructors arrive to class at the same time}) = \frac{3}{9} \text{ or } \frac{1}{3}.$$

∎

EXAMPLE 5.23

An experiment consists of tossing one fair die twice. Find the probability of the following events:
a) the sum of the two tosses is seven, denoted as P(sum is 7).
b) the sum of the two tosses is an odd number, denoted as P(sum is an odd number).
c) the sum of the two tosses is greater than six, denoted as P(sum > 6).

Solution

To enable us to calculate these probabilities, we will use a sample space. But before we construct the sample space we should determine the *total number* of possible outcomes in the sample space by using the Fundamental Counting Principle. Thus, we have:

$$\begin{pmatrix}\text{total number} \\ \text{of} \\ \text{possible outcomes}\end{pmatrix} = \begin{pmatrix}\text{number of} \\ \text{possible outcomes} \\ \text{for first toss}\end{pmatrix} * \begin{pmatrix}\text{number of} \\ \text{possible outcomes} \\ \text{for second toss}\end{pmatrix}$$

$$= (6) * (6)$$
$$= 36$$

The 36 outcomes for this experiment are shown in Table 5.4.

TABLE 5.4

Outcome of First Toss	Outcome of Second Toss					
	1	2	3	4	5	6
1	1,1	1,2	1,3	1,4	1,5	1,6
2	2,1	2,2	2,3	2,4	2,5	2,6
3	3,1	3,2	3,3	3,4	3,5	3,6
4	4,1	4,2	4,3	4,4	4,5	4,6
5	5,1	5,2	5,3	5,4	5,5	5,6
6	6,1	6,2	6,3	6,4	6,5	6,6

a) There are six outcomes that satisfy the event that the sum of the tosses is seven. They are (6,1), (5,2), (4,3), (3,4), (2,5) and (1,6). The number of outcomes in the sample space is 36. Therefore,

$$P(\text{sum is 7}) = \frac{6}{36} \text{ or } \frac{1}{6}.$$

b) The number of outcomes satisfying the event that the sum of the two tosses is odd is 18. Therefore,

$$P(\text{sum is an odd number}) = \frac{18}{36} \text{ or } \frac{1}{2}.$$

c) The number of outcomes satisfying the event that the sum of the two tosses is greater than six is 21. Therefore,

$$P(\text{sum} > 6) = \frac{21}{36} \text{ or } \frac{7}{12}.$$

EXAMPLE 5.24

A four cylinder automobile engine has 2 defective spark plugs. Two plugs are randomly removed from the engine **without replacement**. After each plug is removed, it is tested.
a) Construct a sample space for this experiment.
Assuming each outcome is equally likely, what is the probability that:
b) only the first plug removed will be defective?
c) at most one plug will be defective?
d) both plugs will be good?

Solution

a) To help compute these probabilities, we will construct the sample space for this experiment. To determine the total number of possible outcomes within the sample space, use the Fundamental Counting Principle.

$$\begin{pmatrix}\text{total number}\\\text{of}\\\text{possible}\\\text{outcomes}\end{pmatrix} = \begin{pmatrix}\text{number of possible}\\\text{plugs to choose}\\\text{from for 1st}\\\text{selection}\end{pmatrix} * \begin{pmatrix}\text{number of possible}\\\text{plugs to choose}\\\text{from for 2nd}\\\text{selection}\end{pmatrix}$$

Since the two plugs are being selected **without replacement**, the first selection will be made from an engine having 4 plugs. However, since *the first plug selected is not replaced,* then the second plug selected will be made from an engine containing only 3 plugs. Thus, the total number of possible outcomes is equal to:

$$\text{total number of possible outcomes} = (4) \times (3) = 12$$

To help to distinguish between the 2 good plugs, the good plugs will be labelled as: G_1, and G_2, while the two defective plugs will be labelled as: D_1, and D_2. The 12 equally likely outcomes of the sample space for this experiment are listed in Table 5.5.

TABLE 5.5 The Sample Space for Selecting 2 Plugs Without Replacement from a 4-Cylinder Engine

1st PLUG SELECTED	2nd PLUG SELECTED			
	G_1	G_2	D_1	D_2
G_1	X	G_1G_2	G_1D_1	G_1D_2
G_2	G_2G_1	X	G_2D_1	G_2D_2
D_1	D_1G_1	D_1G_2	X	D_1D_2
D2	D_2G_1	D_2G_2	D_2D_1	X

b) There are four outcomes where *only the first plug removed will be defective.* These outcomes are: D_1G_1, D_1G_2, D_2G_1 and D_2G_2. Since there are 4 outcomes out of 12 equally likely outcomes of the sample space that satisfy this event, then:

$$P(\text{only the first plug removed will be defective}) = \frac{4}{12} \text{ or } \frac{1}{3}.$$

c) The event *at most one plug will be defective* is the same as **one or less plugs** will be defective. For this particular event, it is easier to count the number of outcomes that **do not** satisfy this event. That is, the number of outcomes having two defective plugs. There are only two outcomes where both plugs are defective, thus there are 10 outcomes satisfying the event at most one plug will be defective out of the 12 equally likely outcomes. Therefore,

$$P(\text{at most one plug will be defective}) = \frac{10}{12} \text{ or } \frac{5}{6}.$$

d) The event *both plugs will be good* has only two outcomes satisfying two good plugs. These outcomes are: G_1G_2 and G_2G_1. Since there are 2 outcomes out of 12 equally likely outcomes of the sample space that satisfy this event, then:

$$P(\text{both plugs will be good}) = \frac{2}{12} \text{ or } \frac{1}{6}.$$

EXAMPLE 5.25

Five billiard balls numbered from 1 to 5 are placed in an urn. Two balls are selected in succession from the urn *with replacement,* that is, the first ball selected is returned to the urn before the second ball is selected. Find the probability of selecting:
a) two even-numbered balls
b) two odd-numbered balls
c) one even- and one odd-numbered ball

Solution

To help compute these probabilities, construct the sample space. To determine the total number of possible outcomes within the sample space, use the Fundamental Counting Principle.

$$\begin{array}{c}\text{total number}\\\text{of}\\\text{possible outcomes}\end{array} = \begin{pmatrix}\text{number of possible}\\\text{outcomes for the}\\\text{first selection}\end{pmatrix} * \begin{pmatrix}\text{number of possible}\\\text{outcomes for the}\\\text{second selection}\end{pmatrix}$$

$$= (5) * (5)$$
$$= 25$$

The sample space for this experiment is listed in Table 5.6.

a) The number of outcomes satisfying the event of selecting two even-numbered balls is 4. They are: (4,2),(2,4),(2,2) and (4,4). The number of outcomes in the sample space is 25. Therefore,

$$P(\text{two even-numbered balls}) = \frac{4}{25}$$

TABLE 5.6

Outcome of First Ball	Outcome of Second Ball				
	1	2	3	4	5
1	1,1	1,2	1,3	1,4	1,5
2	2,1	2,2	2,3	2,4	2,5
3	3,1	3,2	3,3	3,4	3,5
4	4,1	4,2	4,3	4,4	4,5
5	5,1	5,2	5,3	5,4	5,5

b) The number of outcomes satisfying the event of selecting two odd-numbered balls is 9. Therefore,

$$P(\text{two odd-numbered balls}) = \frac{9}{25}$$

c) The number of outcomes satisfying the event of selecting one even and one odd numbered ball is 12. Therefore,

$$P(\text{one odd and one even numbered ball}) = \frac{12}{25}$$

Another Approach to Defining Probability

Relative Frequency Concept of Probability or Posteriori Probability

The **relative frequency concept or posteriori probability** is based on observing the number of times an event has occurred over a *long run* in the past and using this information to assign a probability value to the event for the future.

For example, suppose you were interested in assigning a probability value to the following events:

1. the chance that a particular basketball player makes a foul shot.
2. the chance that your favorite coin falls with heads facing up on the next toss.
3. the chance that the next child born at a particular hospital is a boy.
4. the chance that it will rain tomorrow.

The probability for each one of these events **cannot** be calculated using the **classical approach to probability** since the outcomes of these experiments are **not equally likely**. For example, the chance that it will rain tomorrow is not equally likely for each day of the week.

In such cases, **the probability of an event is determined by examining past information on this event or by generating information on this event by performing the experiment a large number of times**. Using this information about the event, the relative frequency of the event can be determined. The formula to compute the relative frequency is:

Relative Frequency Approach to Calculating Probability

For an experiment that is repeated many times, the probability of an event A occurring is:

$$P(\text{Event A}) = \frac{\text{number of times event A occurred}}{\text{total number of times the experiment was repeated}}$$

This relative frequency formula *may not be the exact* probability but it represents an *approximate* probability.

EXAMPLE 5.26

A graduate student at the University of California at Berkeley conducted an experiment to observe the frequency of occurrences for the numerical outcomes of a particular roulette wheel in a Las Vegas Casino. In 1,900 spins of this roulette wheel, the student observed that the number 7 occurred 200 times. Using this information and the relative frequency approach to probability, determine the probability that a 7 will occur on the next spin of this wheel.

Solution

Since the outcome 7 occurred 200 times within a total of 1,900 spins of the roulette wheel, then using the relative frequency approach, the probability of obtaining a 7 on the next spin is equal to:

$$P(\text{obtaining a 7}) = \frac{\text{number of times a 7 occurred}}{\text{total number of times the roulette was spun}}$$

$$P(\text{obtaining a 7}) = \frac{200}{1,900} \approx 0.1053$$

You should realize that the relative frequency approach will most likely not yield an exact probability value. Consider the following case study which illustrates this idea.

CASE STUDY 5.1   Chasing the Perfect Game in Bowling

The snapshot entitled "Elusive Perfect Game" in Figure 5.13 displays the chances an amateur bowler will roll a 300 game for men, women and all bowlers. These probabilities are based upon information available from the American Bowling Congress.

Statistical Snapshot

Elusive Perfect Game

Chances of an amateur bowler rolling a 300 game:

Men	1 in 12,500 games
Women	1 in 644,000 games
All bowlers	1 in 24,000 games

Source: American Bowling Congress

FIGURE 5.13

The probabilities of rolling a 300 game displayed in Figure 5.13 were calculated using the relative frequency approach of defining probability. For example, using the available information, the probability of an amateur male bowler rolling a 300 game was calculated by dividing the *number of times a 300 game was rolled by an amateur male bowler* by the *total number of games an amateur male bowled*.

a) If no male amateur bowler rolls a 300 game during the bowling season following those which produced the information contained in the snapshot, would the relative frequency probability value change for a male bowler rolling a 300 game? Would this relative frequency probability value increase or decrease? Explain.

b) Using the answers to part (a), explain why you would consider the relative frequency probability values in the snapshot to be estimates and not the actual probability values for rolling a 300 game by males, females or all amateur bowlers.

From Case Study 5.1, you should realize that since a relative frequency probability is calculated either from past information or information generated by performing an experiment a large number of times, it provides only an estimate of the actual probability. This is because the relative frequency probability will most likely change as the experiment is repeated more and more times.

Let's examine how a relative frequency probability may change almost each time the experiment is performed. Consider the experiment of tossing your favorite coin 100 times. When you perform this experiment you observe heads 53 out of the 100 tosses; while your friend tosses the coin 100 times and observes 48 heads. You would estimate the probability of getting a head to be 53/100, while your friend would estimate the probability of getting a head as 48/100. Although both you and your friend used the relative frequency approach to calculate the probability of getting a head for this coin, both of you arrived at two different probability values. Consequently, if you were interested in finding the probability of getting a head on the next toss of this coin, which of these two different probability values is the correct one? The idea of the relative frequency approach is not to select one of these probability values, but to realize that both of these serve as estimates of the actual probability value while a better estimate of the actual probability can be determined by tossing this coin a greater number of times. That is, if the coin tossing experiment were repeated a large number of times, say like 1,000 or 10,000 or 100,000 times, the relative frequency probability will begin to stabilize and the relative frequency will begin to approach the actual probability of the event. The law that guarantees this result is called the *Law of Large Numbers*.

Law of Large Numbers

The Law of Large Numbers states that if an experiment is repeated a large number of times, then the probability of an event obtained using the relative frequency formula will approach the actual or theoretical probability of the event.

To learn more about the law of large numbers and the law of averages go to the textbook's website at www.MyStatLab.com.

Another Approach to Defining Probability

Subjective or Personal Probability

There are times when the probability of the outcomes of an experiment cannot be assigned using either the classical definition of probability since the outcomes are not equally likely or approximated by the relative frequency approach because there is no past information generated from repeating the experiment a large number of times. In such instances, the probability may be arrived at *subjectively*. Essentially, this means that the probability is based on a person's judgement, belief, experience or available information. Such a probability is called *Subjective or Personal Probability*.

DEFINITION 5.13

Subjective or Personal probability is the probability assigned to an event based upon an individual's judgement, or available information.

Since, most of the time, there is either little or no information concerning the experiment, the only choice is to use an educated guess at assessing the probability of an event of the experiment. However, even though subjective probability is assigned arbitrarily, this does not imply that the individual assigning the probability randomly selects a number out of the air for this probability assignment. The subjective probability assignment made by the individual reflects the measure of the degree of the individual's personal belief that is based on the knowledge of the situation and the assessment of any other relevant information.

We will now consider some examples which illustrate events of an experiment where the probability assignments rely on the subjective or personal approach to probability. In the following instances, subjective probability is used to estimate the chance that:

1. Steven, a college student, will pass his statistics course this semester.
2. The New York Yankees will win the next World Series.
3. The Centers for Disease Control accurately predicts the percentage of children that will contract chicken pox during the present school year.
4. Tony, a bond-market watcher, will successfully predict when the long-term interest rates will begin to rise this year.
5. Vince, a football coach, will select a first round draft choice that will successfully live up to expectations in the upcoming season.
6. Glenn, an astronaut, estimates the chance that the space shuttle will have a successful launch.

In each of the previous examples, the classical or relative frequency approach to assigning probability are not appropriate since the outcomes of the experiment are not equally likely nor can the experiment be repeated. For example, Vince will select **only one first round draft choice this year**, and the ability of this football player will determine whether he lives up to the coaches expectations in the upcoming season. Therefore, *the relative frequency approach cannot be used* since we cannot repeat this experiment. The *classical approach to probability cannot be used* because **the chance that this player lives up to expectations is not equally likely to the chance that the player does not live up to expectations**. Consequently, in this situation, subjective probability was the approach used to determine the probability of the event that the first round draft choice will successfully live up to expectations since the coaches' personal belief, judgement, information and experience was used in this assignment. Although Vince may assign this event a high probability, it is very likely that Don, another football coach, might assign to this same event a completely different probability value.

CASE STUDY 5.2 Probability Statements

Examine the probability statements contained in each of the three news items of Figure 5.14 and answer the following questions.
  a) For each of the probability statements contained within the news items, explain which probability approach was used to define these probability statements.
  b) Select a probability statement about an event contained within each news item and interpret the meaning of each probability. Explain whether you believe the probability of the event is likely to happen or not likely to happen.
  c) In the news item entitled "Expecting a White Christmas," what do you think historical probability means? Explain.
  d) In "Expecting a White Christmas," interpret the statement: "... there is a 100% probability that those cities will be white on Christmas morning." To which cities does this statement pertain? How do you think they arrived at this statement for these cities? Explain.
  e) In the news item regarding the space shuttle, how do you think they arrived at the overall probability statement: "1 in 131: The Chance of Catastrophe"? That is, do you think they used the other probability information contained within the news item? Explain.
  f) In the lotto graphic, what do you think the overall odds: 1:60 mean? How do you think they arrived at this overall odds? Do you think they used any of the probability information contained within the Lotto graphic? Explain.
  g) Could the classical approach to assigning probability have been used to define the probabilities for "1 in 131: The Chance of Catastrophe" or "Expecting a White Christmas"? Explain.
  h) Could the relative frequency approach to assigning probability have been used to determine the probabilities for "1 in 131: The Chance of Catastrophe"? Explain.
  i) Using the lotto graphic, do you think that if you played one game that you would have an excellent chance of matching all the 6 selected numbers? Explain.

Section 5.5 *Fundamental Rules and Relationships of Probability* **261**

CASE STUDY 5.2 Probability Statements (continued)

(a) Statistical Snapshot

1 in 131: The Chance of Catastrophe

The odds of another shuttle being lost are 1 in 131. Figures in parentheses denote the probability that each component will fail and cause the shuttle to be destroyed. For each component, percentages denote portion of total risk of flying the shuttle.

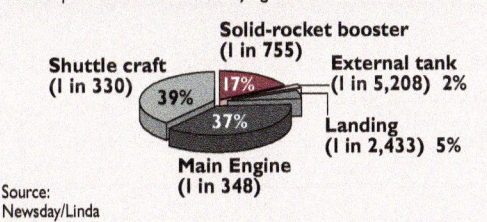

- Solid-rocket booster (1 in 755) — 17%
- Shuttle craft (1 in 330) — 39%
- External tank (1 in 5,208) — 2%
- Main Engine (1 in 348) — 37%
- Landing (1 in 2,433) — 5%

Source: Newsday/Linda McKenney

From *Newsday*. © 1996 Newsday. All rights reserved. Used under license.

(b) Where to go for a holiday snow show

If you're dreaming of a white Christmas, you might want to be in International Falls, Minn., or Sault Ste. Marie, Mich. Although there's no way to accurately predict Christmas weather this far in advance, climatologists say that based on 30 years of records there is a 100% probability that those cities will be white on Christmas morning. "It wouldn't be Christmas without snow," says Jason Thrope, 28, a firefighter in Sault Ste. Marie. "Family is what it's all about, but snow makes it better."
— *Diane Harris*

Statistical Snapshot

Expecting a White Christmas

The historical probability there will be at least an inch of snow on the ground on Christmas in these cities:

City	Chance of Snow	City	Chance of Snow
Albuquerque	9%	Eugene, Ore.	5%
Baltimore	13%	Flagstaff, Ariz.	51%
Billings Mont.	63%	Hartford, Conn.	57%
Boise, Idaho	26%	Indianapolis	30%
Boston	23%	Louisville	13%
Buffalo	67%	Milwaukee	60%
Burlington, VT	77%	Minneapolis	73%
Cheyenne Wyo.	35%	New York	13%
Chicago	40%	Omaha	41%
Cincinnati	17%	Philadelphia	10%
Cleveland	50%	Salt Lake City	52%
Columbus Ohio	23%	Seattle	5%
Denver	42%	St Louis	23%
Des Moines	50%	Tahoe City, Calif.	85%
Detroit	50%	Washington	13%

Source: Northeast Regional Climate Center at Cornell University

(c) Statistical Snapshot

Prizes/Chance of Winning Lotto

Match	Prize	Chance of Winning on a $1 Bet
Match 6 ○○○○○○	JACKPOT**	1: 18,009,460
Match 5 + Bonus Ball ○○○○○ + ○	PARI-MUTUAL**	1: 3,001,577
Match 5 ○○○○○	PARI-MUTUAL**	1: 68,218
Match 4 + Bonus Ball ○○○○ + ○	PARI-MUTUAL**	1: 27,287
Match 4 ○○○○	PARI-MUTUAL**	1: 1,269
Match 3 + Bonus Ball ○○○ + ○	PARI-MUTUAL	1: 952
Match 3 ○○○	FREE PLAY	1: 68

Overall odds of winning a prize are approximately 1 : 60
** Pari-mutual prize is divided equally among multiple winners.

FIGURE 5.14

j) If 18.1 million people played one lotto game each and one of these people were randomly selected, do you think that this individual would have an excellent chance of matching all the 6 selected numbers? Explain.

k) If 18.1 million people played one lotto game each, do you think that there would be an excellent chance that one of these individuals would match all the 6 selected numbers? Explain.

Reprinted by permission of Tribune Media Services.

5.5 FUNDAMENTAL RULES AND RELATIONSHIPS OF PROBABILITY

Now that we have discussed the concept of probability, we are going to study some of the rules which probabilities must obey. The first basic rule states that the probability of an event must range from zero to one.

Probability Rule 1

The **probability of an event E ranges from 0 to 1 inclusive.** Symbolically, this is expressed as:

$$0 \leq P(\text{Event E}) \leq 1$$

This rule is a consequence of the classic definition of the probability of an event. The classic definition states:

$$P(Event\ E) = \frac{number\ of\ outcomes\ satisfying\ event\ E}{total\ number\ of\ outcomes\ in\ the\ sample\ space}$$

Since the number of outcomes satisfying event E and the total number of outcomes cannot be negative numbers, the number P(E) must be greater than or equal to zero. Furthermore since the number of outcomes satisfying event E would always be less than or equal to the total number of outcomes, the number P(E) would have to be less than or equal to one.

> Probability Rule 1 indicates that the smallest probability of an event E is zero. Thus, if an event E cannot occur, then the probability of Event E is zero. That is,
>
> $$P(Event\ E) = 0.$$

EXAMPLE 5.27

An experiment consists of rolling a fair die once. Determine the probability of rolling a nine.

Solution

Of the six possible outcomes there are **no outcomes** corresponding to a nine. Therefore:

$$P(\text{rolling a 9}) = \frac{0}{6} = 0$$

> Probability Rule 1 also indicates that the largest probability of an event E is one. An event with a probability of one is an event which is certain to occur.

EXAMPLE 5.28

An experiment consists of rolling a fair die once. Determine the probability of rolling a number less than seven.

Solution

All six possible outcomes for this experiment are less than seven. Therefore:

$$P(\text{rolling a number less than seven}) = \frac{6}{6} = 1$$

The second probability rule explains the relationship about the sum of the probabilities of all the outcomes within a sample space.

> **Probability Rule 2**
>
> The **sum of the probabilities of all the outcomes in the sample space of an experiment always equals 1.**

EXAMPLE 5.29

For the experiment of rolling a fair die, the sample space consists of the six outcomes illustrated in Figure 5.15. Determine the sum of the probabilities of all the outcomes of this experiment.

Solution

The sample space consists of six equally likely outcomes. The probability for *each* of these six equally likely outcomes is $\frac{1}{6}$. That is,

$$P(\text{rolling a 1}) = P(1) = \frac{1}{6}; \quad P(\text{rolling a 2}) = P(2) = \frac{1}{6}$$

$$P(\text{rolling a 3}) = P(3) = \frac{1}{6}; \quad P(\text{rolling a 4}) = P(4) = \frac{1}{6}$$

$$P(\text{rolling a 5}) = P(5) = \frac{1}{6}; \quad P(\text{rolling a 6}) = P(6) = \frac{1}{6}$$

Observe that the sum of all the probabilities of the outcomes of this experiment is one. That is,

$$P(1) + P(2) + P(3) + P(4) + P(5) + P(6) = 1.$$

This illustrates probability rule 2.

CASE STUDY 5.3

Statistical Snapshot

Odds of Developing Breast Cancer

By age:		By age:	
25	1 in 19,608	60	1 in 24
30	1 in 2,525	65	1 in 17
35	1 in 622	70	1 in 14
40	1 in 217	75	1 in 11
45	1 in 93	80	1 in 10
50	1 in 50	85	1 in 9
55	1 in 33	Ever	1 in 8

Source: National Cancer Institute

FIGURE 5.16

The American Cancer Society announced in January, 1991 that a women's odds of getting breast cancer had risen to 1 in 9. The Cancer Society has been heavily publicizing this 1 in 9 statistic to persuade women to have more regular mammograms and to examine their own breasts. However, this probability figure has terrified many women and made them feel doomed. In fact, many young women have interpreted this 1 in 9 statistic to mean that one of their nine girl friends will get cancer this year. Yet, this is not a correct conclusion based upon the calculated risk of getting breast cancer. Examine the chart shown in Figure 5.16, which indicates a women's risk of getting breast cancer according to the American Cancer Society.

Using the information in Figure 5.16, what is the probability of a woman getting breast cancer:
a) by age 30?
b) by age 40?
c) by age 50?
d) by age 60?
e) by age 70?
f) by age 80?
g) by age 85?
h) By what age does a woman's risk rise to the 1 in 9 chance of getting breast cancer?
i) Would you interpret this 1 in 9 chance as a woman's risk of getting breast cancer for any particular age or the cumulative chance of a woman's risk of getting breast cancer during her lifetime? Explain.
j) Explain what you believe the American Cancer Society means when they say the 1 in 9 chance is meant to be "more of a metaphor than a hard figure."

Chapter 5 Probability

The third probability rule explains how to calculate the probability of the **compound event (A or B)**. This probability rule is called the **addition rule**.

Probability Rule 3: The Addition Rule

The probability of satisfying the event A or the event B is equal to the probability of event A plus the probability of event B minus the probability that both A and B occur at the same time. The addition rule is written as:

$$P(A \text{ or } B) = P(A) + P(B) - P(A \text{ and } B)$$

The key word in knowing when to use the addition rule is **or**. The *word or* within the probability statement P(A **or** B) indicates *addition*. Let's explain and apply the addition rule using the probability problem in Example 5.30.

EXAMPLE 5.30

In the experiment of tossing a fair die twice, what is the probability that the sum of the two tosses is greater than 4 **or** an odd sum?

Solution

The sample space for tossing a fair die twice is shown in Table 5.7. To determine the probability that the sum of the two tosses is greater than 4 **or** an odd sum, the addition rule is used since the probability statement is an **or** statement. If *event A represents the sum is greater than 4* while *event B represents the sum is odd*, then the probability that the sum is greater than 4 (sum > 4) or an odd sum can be written as:

$$P(A \text{ or } B) = P(\text{sum} > 4 \text{ or odd sum}).$$

To determine this probability, we need to calculate the following three probabilities: **P(sum > 4)**, **P(odd sum)** and **P(sum > 4 and odd)**.

TABLE 5.7 Sample Space for Tossing a Fair Die Twice

Outcome of First Toss	\multicolumn{6}{c}{Outcome of Second Toss}
	1	2	3	4	5	6
1	1,1	1,2	1,3	1,4	1,5	1,6
2	2,1	2,2	2,3	2,4	2,5	2,6
3	3,1	3,2	3,3	3,4	3,5	3,6
4	4,1	4,2	4,3	4,4	4,5	4,6
5	5,1	5,2	5,3	5,4	5,5	5,6
6	6,1	6,2	6,3	6,4	6,5	6,6

The probability that the sum is greater than 4 is $30/36$ since there are 30 outcomes out of the 36 equally likely outcomes of the sample space which satisfy the sum is greater than 4. Thus,

$$P(\text{sum} > 4) = \frac{30}{36}$$

The probability that the sum is odd is $18/36$ since there are 18 outcomes out of the 36 equally likely outcomes of the sample space which satisfy the sum is odd. Thus,

$$P(\text{odd sum}) = \frac{18}{36}$$

Notice if you simply tried to add these two probabilities together to determine the probability that the sum > 4 or an odd sum, you would obtain a probability of $30/36 + 18/36 = 48/36$ which is INCORRECT because this would give us a probability that is greater than one which violates probability rule 1. The reason for this error is that the outcomes satisfying **both a sum > 4 and an odd sum have been counted twice** when the probabilities were added together. Consequently, the probability of a sum > 4 **and** an odd sum must be **subtracted** from this probability of $48/36$. Thus, it is necessary to determine the probability of a sum > 4 and an odd sum. That is, P(sum > 4 and odd).

The probability that the sum > 4 and an odd sum is 16/36. That is,

$$P(\text{sum} > 4 \text{ and odd}) = \frac{16}{36}$$

Now, the **P(sum > 4 and odd) can be subtracted from the sum of the probabilities: P(sum > 4) and P(odd sum)** as verified by the addition rule. Thus, using the addition rule to determine the probability that the sum is greater than 4 (sum > 4) or an odd sum, we have:

$$P(\text{sum} > 4 \text{ or odd sum}) = P(\text{sum} > 4) + P(\text{odd sum}) - P(\text{sum} > 4 \text{ and odd}).$$

$$= \frac{30}{36} + \frac{18}{36} - \frac{16}{36}$$

$$P(\text{sum} > 4 \text{ or odd}) = \frac{32}{36} \text{ or } \frac{8}{9}$$

EXAMPLE 5.31

Consider the experiment of selecting one card from an ordinary deck of 52 playing cards. The sample space for this experiment is shown in Figure 5.17.

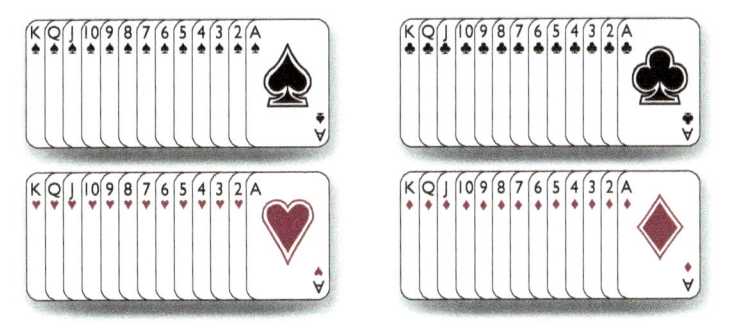

FIGURE 5.17

The first row shows spades and clubs, while the second row shows hearts and diamonds. The club and spade suits are black and the diamond and heart suits are red. Each suit has 13 cards and contains the ranks: Ace, 2, 3, 4, 5, 6, 7, 8, 9, 10, jack, queen and king. Find the probability of selecting:
a) A spade or a diamond
b) A king or a club
c) A picture card or a heart

Solution

a) The probability of selecting a spade **or** a diamond requires the use of the addition rule because we are dealing with an **or** statement. Using the addition rule, we can express the probability of a spade or a diamond as:

$$P(\text{spade or diamond}) = P(\text{spade}) + P(\text{diamond}) - P(\text{spade and diamond})$$

Since the selection of any card from a deck of 52 playing cards is equally likely and there are 13 cards which are spades and 13 cards that are diamonds, then:

$$P(\text{spade}) = \frac{13}{52}, \text{ and } P(\text{diamond}) = \frac{13}{52}$$

Since it is impossible for one card to be both a spade *and* a diamond, then:

$$P(\text{spade and diamond}) = 0.$$

Substituting these probabilities into the addition rule, we have:

$$P(\text{spade or diamond}) = P(\text{spade}) + P(\text{diamond}) - P(\text{spade and diamond})$$

$$P(\text{spade or diamond}) = \frac{13}{52} + \frac{13}{52} - 0 = \frac{26}{52} \text{ or } \frac{1}{2}.$$

b) Using the addition rule for the probability of selecting a king or a club, we can write:

$$P(\text{king or club}) = P(\text{king}) + P(\text{club}) - P(\text{king and club}).$$

Since there are 4 kings within the deck and 13 cards are clubs where one card is the king of clubs, then we have the following probabilities assuming each card has an equal chance of being selected:

$$P(\text{king}) = \frac{4}{52}, P(\text{club}) = \frac{13}{52}, \text{ and } P(\text{king and club}) = \frac{1}{52}$$

Substituting these probabilities into the addition rule, we get:

$$P(\text{king or club}) = \frac{4}{52} + \frac{13}{52} - \frac{1}{52} = \frac{16}{52} \text{ or } \frac{4}{13}$$

c) The probability of selecting a picture card **or** a heart requires the use of the addition rule because we are working with an **or** statement. Using the addition rule, we can express the probability of a picture card or a heart as:

$$P(\text{picture card or heart}) = P(\text{picture card}) + P(\text{heart}) - P(\text{picture card and heart})$$

Since a picture card is either a king, a queen or a jack, then there are 12 chances out of 52 to select a picture card. Thus,

$$P(\text{picture card}) = \frac{12}{52}.$$

There are 13 hearts in the deck, so the chance of selecting a heart is:

$$P(\text{a heart}) = \frac{13}{52}.$$

Since there are only 3 picture cards that are hearts (jack, queen and king of hearts), then the chance of selecting a picture card and a heart is:

$$P(\text{picture card and a heart}) = \frac{3}{52}$$

Substituting these probabilities into the addition rule, we have:

$$P(\text{picture card or heart}) = P(\text{picture card}) + P(\text{heart}) - P(\text{picture card and heart})$$

$$P(\text{picture card or heart}) = \frac{12}{52} + \frac{13}{52} - \frac{3}{52} = \frac{22}{52} \text{ or } \frac{11}{26}$$

∎

If **two events A and B cannot occur at the same time**, then the addition rule can be simplified. These events are said to be **mutually exclusive**. This leads us to the following definition.

DEFINITION 5.14

Two events A and B are **mutually exclusive events** if both events A and B cannot occur at the same time.

For mutually exclusive events, the probability of event A and event B occurring at the same time is zero, that is: P(A and B) = 0. In such instances, the addition rule for two mutually exclusive events A and B is simply the probability of event A plus the probability of event B. This leads us to the following addition rule for mutually exclusive events.

Probability Rule 3A: The Addition Rule for Mutually Exclusive Events

If two events A and B are mutually exclusive, then the probability that event A or event B will occur is equal to the sum of their probabilities. This rule is written as:

$$P(A \text{ or } B) = P(A) + P(B)$$

Section 5.5 *Fundamental Rules and Relationships of Probability*

The next example illustrates the concept of mutually exclusive events.

EXAMPLE 5.32

Which of the following events are mutually exclusive?
a) Rolling a single die once and getting a 2 or an even number.
b) Rolling a single die once and getting a 2 or an odd number.
c) Rolling a single die once and getting a 3 or a 4.
d) Tossing a coin twice and getting two heads or at least one head.
e) Tossing a coin twice and getting at least two heads or at most one head.

Solution

a) The sample space for rolling a single die once is given in Figure 5.18.
   In Figure 5.19, we circled the outcomes that satisfy the event of getting a 2 and placed an X through the outcomes that satisfied the event of getting an even number.
   Notice in Figure 5.19 the outcome 2 has a circle around it and an "X" through it. The events of getting a 2 and an even number can occur at the same time. Therefore, these two events are *not* mutually exclusive.
b) The sample space for rolling a single die once is given in Figure 5.20.
   In Figure 5.21, we circled the outcome(s) that satisfy the event of getting a 2 and placing an "X" through the outcome(s) that satisfy the event of getting an odd number.
   In Figure 5.21 we notice that there are no outcomes that satisfy both events at the same time. Therefore, the two events *are* mutually exclusive.
c) It is impossible to roll a single die once and get the two outcomes 3 and 4 at the same time. Therefore, the two events *are* mutually exclusive.
d) The sample for tossing a coin twice is given in Figure 5.22.
   Circling the outcome(s) that satisfy the event of getting two heads and placing an "X" through the outcome(s) that satisfy the event of getting at least one head, we obtain the results shown in Figure 5.23.
   In Figure 5.23, we notice that there is one outcome that satisfies both events at the same time. Therefore, the two events are *not* mutually exclusive.
e) The sample space for tossing a coin twice is given in Figure 5.24.
   Circling the outcome(s) that satisfy the event of getting at least two heads and placing an "X" through the outcome(s) that satisfy the event of getting at most one head are shown in Figure 5.25.
   In Figure 5.25 we notice that there is no outcome that satisfies both events at the same time. Therefore, the two events *are* mutually exclusive.

1	2	3	4	5	6

FIGURE 5.18

1	②	3	✗4	5	✗6

FIGURE 5.19

1	2	3	4	5	6

FIGURE 5.20

✗1	②	✗3	4	✗5	6

FIGURE 5.21

HH	HT	TH	TT

FIGURE 5.22

(HH)	✗HT	✗TH	TT

FIGURE 5.23

HH	HT	TH	TT

FIGURE 5.24

(HH)	✗HT	✗TH	✗TT

FIGURE 5.25

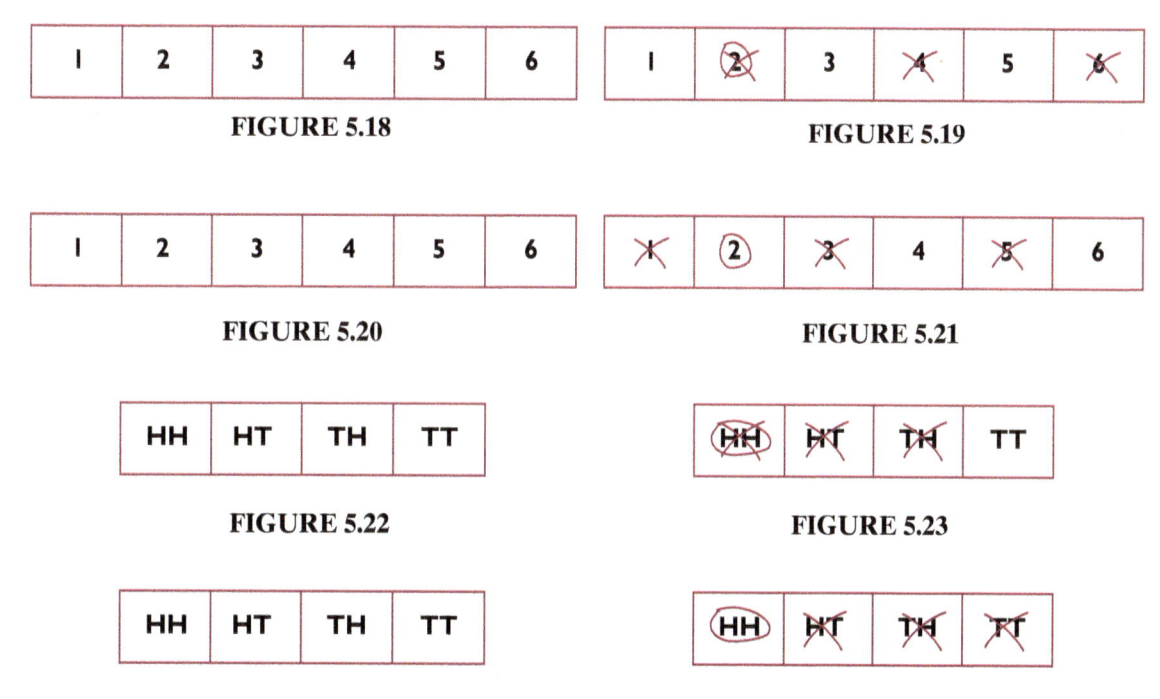

The fourth probability rule explains the relationship between the probability of an event E and the probability of the complement of event E. Let's first define the complement of event E.

> **DEFINITION 5.15**
> Given an event E, the **complement of event E**, symbolized as E' and pronounced as E-complement, represents all the outcomes of an experiment that are not in event E.

If event E represents *it will rain tomorrow*, then event E' (the complement of event E) represents *it will not rain tomorrow*. Therefore, if event E **occurs** (that is, it rains tomorrow), then the complement of event E (it doesn't rain tomorrow) **cannot occur** and vice versa. Thus, if the chance that it will rain tomorrow is 30%, then the chance that it will not rain tomorrow is 70% since the chance it will rain tomorrow (event E) plus the chance it will not rain tomorrow (the complement of event E) must add up to 100% or be equal to a probability of one. This leads us to the Complement Rule.

Probability Rule 4: The Complement Rule

The sum of the probabilities of event E and the complement of event E always equals one. That is,

$$P(\text{Event E}) + P(\text{complement of Event E}) = 1$$

The complement rule can be symbolized as:

$$P(E) + P(E') = 1$$

The next few examples illustrate the application of the complement rule.

EXAMPLE 5.33

Consider the experiment of rolling a fair die once. If event E is rolling a 3, then:
a) identify event E',
   and identify and compute the following probabilities:
b) $P(E)$
c) $P(E')$
d) $P(E) + P(E')$

Solution

a) Since event E represents rolling a 3, the **event E' is *not* rolling a 3**.
b) $P(E)$ is the probability of rolling a 3. Of the six possible outcomes for a single roll of a fair die, only one is a 3. Therefore,

$$P(E) = P(\text{rolling a 3}) = \frac{1}{6}$$

c) $P(E')$ is the probability of not rolling a 3. Since there are five outcomes that are *not* a 3 on the die, then

$$P(E') = P(\text{not rolling a 3}) = \frac{5}{6}$$

d) $P(E) + P(E')$ is the sum of the probabilities of rolling a 3 or not rolling a 3 for a single roll of a fair die. Using parts (b) and (c), we have:

$$P(E) + P(E') = P(\text{rolling a 3}) + P(\text{not rolling a 3})$$

$$= \frac{1}{6} + \frac{5}{6} = 1$$

Thus, the *probability of rolling a 3* **plus** the *probability of not rolling a 3* **must equal one**, because no matter what outcome appears on the die it must satisfy one of the events E or E'. That is,

$$P(E) + P(E') = 1.$$

EXAMPLE 5.34

A young couple planning to have 3 children has eight family possibilities, which are illustrated in Table 5.8.

TABLE 5.8 Sample Space for a 3 Child Family

First Child	Second Child	Third Child
boy	boy	boy
boy	boy	girl
boy	girl	boy
boy	girl	girl
girl	boy	boy
girl	boy	girl
girl	girl	boy
girl	girl	girl

Assuming each outcome is equally likely, compute the following probabilities.
a) P(couple will have 3 boys)
b) P(couple will not have 3 boys)
c) 1 − P(couple will have 3 boys)

Solution

a) Of the eight possible outcomes listed in Table 5.8, only one outcome corresponds to the event the couple will have 3 boys. Therefore,

$$P(\text{couple will have 3 boys}) = \frac{1}{8}$$

b) Of the eight possible outcomes listed in Table 5.8, seven outcomes correspond to the event the couple will not have 3 boys. Therefore,

$$P(\text{couple will not have 3 boys}) = \frac{7}{8}$$

c) From part a, the $P(\text{couple will have 3 boys}) = \frac{1}{8}$

Therefore,

$$1 - P(\text{couple will have 3 boys}) = 1 - \frac{1}{8}$$
$$= \frac{7}{8}$$

From Example 5.34, you should notice that the results of part b and part c are equal. That is,

P(couple will not have 3 boys) = 1 − P(couple will have 3 boys).

This is a direct result from the complement rule since it uses the probability of event E to compute the probability of the complement of event E. In general, this can be written as follows.

> **Computing the Probability of the Complement of Event E Using Event E**
>
> $$P(\text{complement of event E}) = 1 - P(\text{event E})$$
>
> This can be further symbolized as:
>
> $$P(E') = 1 - P(E)$$
>
> Furthermore, we can also compute the probability of event E using the complement of event E. This is expressed as follows:
>
> $$P(\text{event E}) = 1 - P(\text{complement of event E})$$
>
> That is,
>
> $$P(E) = 1 - P(E')$$

The next example illustrates the use of these rules.

EXAMPLE 5.35

If a young couple is planning on having three children, then the P(couple will have 3 girls) $= \frac{1}{8}$.

Using this information, compute the:

$$P(\text{couple will not have 3 girls}).$$

Solution

Using: $P(E') = 1 - P(E)$, we have:

$$P(\text{couple will not have 3 girls}) = 1 - P(\text{couple will have 3 girls})$$

$$= 1 - \frac{1}{8}$$

$$= \frac{7}{8}$$

EXAMPLE 5.36

If a young couple is planning on having three children, then the P(couple will not have any girls) $= \frac{1}{8}$.

Using this information, compute the:

$$P(\text{couple will have at least one girl}).$$

Solution

Using: $P(E) = 1 - P(E')$, we have:

$$P(\text{couple will have at least one girl}) = 1 - P(\text{couple will not have any girls})$$

$$= 1 - \frac{1}{8}$$

$$= \frac{7}{8}$$

The next probability concept we will examine is the idea of independent events.

> **DEFINITION 5.16**
>
> **Independent Events.** Two events A and B are independent if the occurrence or nonoccurrence of event A has no influence on the occurrence or nonoccurrence of event B.

The following example illustrates the concept of independent events.

EXAMPLE 5.37

For each of the following experiments determine which events are independent events.
a) Consider the experiment of rolling a single die twice. Let event A be the outcome that the 1st roll is a 5, while event B represents the outcome that the 2nd roll is a 2.
b) Consider the experiment of tossing a coin twice. Let event A be the outcome that the 1st toss is a head. Let event B be the outcome that the 2nd toss is a head.
c) Consider the experiment of selecting two marbles **without** replacement from an urn containing 5 blue marbles and 1 red marble. Let event A be the outcome that the first marble chosen is red while event B is the outcome that the 2nd marble chosen is red.

Solution

a) When rolling a die twice, does the outcome of the first roll influence the outcome of the second roll? No. Thus, getting a 5 on the first roll has no influence on the second event B, getting a 2 on the second roll. Therefore, events A and B are independent.
b) When tossing a coin twice, does the outcome of the first toss influence the outcome of the second toss? No. Thus, getting a head on the first toss has no influence on getting a head on the second toss. Therefore, events A and B are independent.
c) If the red marble is selected on the first pick, then it cannot be selected on the second pick. Therefore, the two events are **not** independent.

Whenever two events A and B are independent, then the probability of A and B is very simple to calculate. This is illustrated in the next probability rule.

> **Probability Rule 5: Multiplication Rule for Independent Events**
>
> If two events A and B are **independent**, then the probability of A and B equals the probability of A multiplied by the probability of B.
> This rule can be symbolized as:
>
> $$P(A \text{ and } B) = P(A) \cdot P(B)$$

The next few examples illustrate the use of the multiplication rule for independent events.

EXAMPLE 5.38

Consider the experiment of tossing a fair coin twice. Find the probability of the following events:
a) both outcomes are heads
b) both outcomes are tails

Solution

a) Since the successive tosses of a coin are independent, use the multiplication rule for independent events.

$$P(\text{head and head}) = P(\text{head}) \cdot P(\text{head}) = \left(\frac{1}{2}\right)\left(\frac{1}{2}\right) = \frac{1}{4}$$

Therefore, the probability of getting two heads in two tosses is 1/4.

b) Using the multiplication rule for independent events, we have:

$$P(\text{tail and tail}) = P(\text{tail}) \cdot P(\text{tail})$$
$$= \frac{1}{2} \cdot \frac{1}{2} = \frac{1}{4}$$

Therefore, the probability of getting two tails in two tosses is 1/4.

EXAMPLE 5.39

Two cards are selected **with replacement** from an ordinary deck of 52 playing cards. Find the probability of the following events:
a) both cards are clubs.
b) both cards are aces.
c) the first card is a picture card and the second card is an ace.

Solution

a) First notice that in a deck of cards, two successive selections, with replacement of the card after each selection, are independent events. Therefore, we can apply the multiplication rule for independent events. Therefore,

$$P(\text{club and club}) = P(\text{club}) \cdot P(\text{club}) = \frac{13}{52} \cdot \frac{13}{52}$$
$$= \frac{1}{4} \cdot \frac{1}{4}$$
$$= \frac{1}{16}$$

b) Similarly, for the event ace followed by ace with replacement, we can apply the multiplication rule for independent events. Thus,

$$P(\text{ace and ace}) = \frac{4}{52} \cdot \frac{4}{52}$$
$$= \frac{1}{13} \cdot \frac{1}{13}$$
$$= \frac{1}{169}$$

c) Using the multiplication rule for independent events, we have:

$$P(\text{picture card and ace}) = \frac{12}{52} \cdot \frac{4}{52}$$
$$= \frac{3}{13} \cdot \frac{1}{13}$$
$$= \frac{3}{169}$$

EXAMPLE 5.40

Two balls are selected **with replacement** from an urn containing 4 red, 6 green and 2 yellow balls. Find the probability of:
a) selecting 2 yellow balls.
b) selecting first a red ball and then a yellow ball.
c) selecting first a green ball and then a blue ball.

Solution

a) Using the multiplication rule for independent events, we can write:

$$P(\text{yellow and yellow}) = P(\text{yellow}) \cdot P(\text{yellow})$$

$$= \frac{2}{12} \cdot \frac{2}{12}$$

$$= \frac{1}{6} \cdot \frac{1}{6} = \frac{1}{36}$$

b) Using the multiplication rule for independent events, we have:

$$P(\text{red and then yellow}) = P(\text{red}) \cdot P(\text{yellow})$$

$$= \frac{4}{12} \cdot \frac{2}{12}$$

$$= \frac{1}{3} \cdot \frac{1}{6} = \frac{1}{18}$$

c) Using the multiplication for independent events, we can write:

$$P(\text{green and then blue}) = P(\text{green}) \cdot P(\text{blue})$$

$$= \frac{6}{12} \cdot \frac{0}{12} = 0$$

Probability Problems Using Permutations and Combinations

In many instances when the outcomes of a sample space are too large to list, it becomes necessary to use the counting rules to calculate the probability. The next few examples illustrate the use of the counting rules of permutations and combinations in computing probability.

EXAMPLE 5.41

If the four letters A, B, C, K are randomly arranged, what is the probability that the resulting arrangement is the word BACK?

Solution

Since the order of the letters is important this is a permutation problem. First determine the number of ways the letters can be arranged using $_nP_n$, where n=4. Thus,

$$_4P_4 = 4!$$
$$= (4)(3)(2)(1)$$
$$= 24$$

Since only one of these arrangements is the word BACK, the probability that the resulting arrangement is the word BACK is:

$$P(\text{the arrangement spells BACK}) = \frac{1}{24}$$

EXAMPLE 5.42

a) If the five letters A, I, C, R, G are randomly arranged what is the probability that the resulting arrangement is CRAIG?
b) If three of the previous five letters are selected and randomly arranged what is the probability that the result is CAR?

Solution

a) Since the order of the letters is important, this is a permutation problem. There are $_5P_5$ ways to arrange all 5 letters. CRAIG is 1 of these arrangements, therefore:

$$P(\text{the arrangement spells CRAIG}) = \frac{1}{{}_5P_5} = \frac{1}{120}$$

b) Since we are concerned with the order of selecting and arranging only 3 of the 5 letters, there are ${}_5P_3$ or 60 arrangements. Therefore:

$$P(\text{the arrangement is CAR}) = \frac{1}{{}_5P_3} = \frac{1}{60}$$

EXAMPLE 5.43

At a conference 9 executives consisting of 5 men and 4 women are required to sit along one side of a long table as illustrated in Figure 5.26. If they randomly take a seat what is the probability that the women occupy the even places?

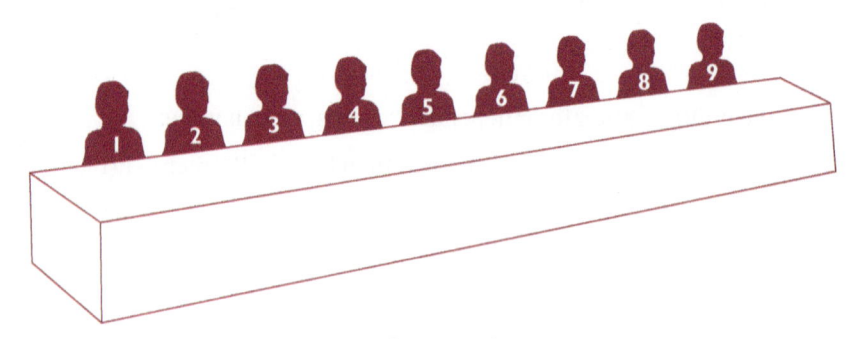

FIGURE 5.26

Solution

If we consider the seats to be numbered from 1 to 9, this arrangement requires that the four women occupy seats 2, 4, 6 and 8. There are ${}_4P_4$ ways that the women can be arranged in these even numbered seats. The 5 men must then occupy seats 1, 3, 5, 7 and 9. There are ${}_5P_5$ ways to arrange them in the odd numbered seats.

Hence using the **fundamental counting principle** we get: $({}_4P_4)({}_5P_5)$ ways that the women can end up in the even numbered seats.

The total number of possible seating arrangements for all nine people is ${}_9P_9$. Thus, the probability that the women occupy the even places is:

$$\frac{({}_4P_4)({}_5P_5)}{{}_9P_9} = \frac{(24)(120)}{362,880}$$

$$= \frac{2880}{362,880}$$

EXAMPLE 5.44

Joe and Laura are part of a group of 5 people waiting for a bus. If as it pulls up they get in line, what is the probability that Joe and Laura are *not* next to each other on line?

Solution

There are ${}_5P_5$ or 120 different ways for the five people to get in line. How many of these *don't* have Joe and Laura next to each other? It's easier to determine the number of ways the people can get in line with Joe and Laura *next* to each and then subtract from 120 (the total number of ways) to get the number we need. For Joe and Laura to be next to each other they can be arranged in two ways either as Joe and Laura or Laura and Joe. If we treat Joe and Laura as **one item**, then there are now ${}_4P_4$ ways to arrange the other three people with the duo of Joe and Laura. Remembering that Joe and Laura can be arranged in 2 ways we get: $2({}_4P_4) = 2(4!) = 48$ ways to form a line with Joe and Laura next to each other. Hence, the number of ways of Joe and Laura *not* being next to each other is:

$${}_5P_5 - 2({}_4P_4) = 120 - 48 = 72$$

Therefore, the probability of Joe and Laura *not* being next to each other on the bus line is:

$$P(\text{Joe and Laura are not next to each other}) = \frac{(_5P_5) - 2(_4P_4)}{_5P_5} = \frac{72}{120}$$

EXAMPLE 5.45

Using a regular deck of 52 playing cards:
a) how many different five card poker hands are possible?
b) what is the probability of being dealt five hearts in a five card poker hand?
c) what is the probability of being dealt a flush (all five cards of the same suit) in a five card poker hand?
d) what is the probability of being dealt 4 kings in a 5 card poker hand?
e) what is the probability of being dealt 4 of a kind in a 5 card poker hand?

Solution

a) Since there are 52 cards in a deck, n=52. A poker hand contains 5 cards, therefore: s=5. Using the formula,

$$_nC_s = \frac{n!}{(n-s)!s!}$$

we have:

$$_{52}C_5 = \frac{52!}{47!5!} = 2{,}598{,}960 \text{ different five card poker hands}$$

b) Since a deck contains thirteen hearts and any five would make a five heart hand, there are $_{13}C_5$ five card heart hands. Therefore,

$$P(\text{five heart poker hand}) = \frac{_{13}C_5}{_{52}C_5} = \frac{1287}{2{,}598{,}960} = 0.0005$$

c) Since there are four suits (diamonds, hearts, clubs and spades) and a flush is possible in any suit, then the probability of being dealt a flush is *four* times the probability of being dealt a flush in any one particular suit. From part (b) we know the probability of a flush in hearts is 0.0005. Thus the probability of a flush in any one suit is also 0.0005. Therefore,

$$\begin{aligned} P(\text{being dealt a flush}) &= 4(\text{probability of a flush in any one particular suit}) \\ &= 4(0.0005) \\ &= 0.002 \end{aligned}$$

d) Since there are only 4 kings, you must be dealt all of them. This can only happen in one way. The fifth card can be any one of the remaining 48 cards. Therefore, there are 48 ways to get a fifth card or 48 five card hands which contain 4 kings. Thus,

$$P(\text{4 kings in a 5 card poker hand}) = \frac{48}{2{,}598{,}960} = 0.0000185$$

Remark: The probability of 4 of any one kind (4 aces, 4 sevens, etc.) is the same as P(4 kings in a 5 card poker hand).

e) Since there are thirteen ranks within any suit, ace through king, and 4 of a kind is possible for any rank (i.e. 4 aces, 4 twos, etc.) thus the

$$\begin{aligned} P(\text{being dealt 4 of a kind for any rank}) &= 13 \,(\text{probability of four of a kind for any particular rank}) \\ &= 13 \,(0.0000185) \\ &= 0.0002405 \end{aligned}$$

EXAMPLE 5.46

If you're a member of an office staff of 25 clerks, what is the probability that you'll be selected as a member of a three person committee which is randomly chosen from the 25 clerks?

Solution

Since there are 25 clerks in the office and 3 clerks are to be chosen where order is not important, then there are $_{25}C_3$ ways to select the committee of 3 clerks. If you're to be included on this committee then only two other clerks from the remaining 24 clerks can be chosen. This is done in $_{24}C_2$ ways. Therefore,

$$P(\text{you are selected to a committee of 3 clerks}) = \frac{_{24}C_2}{_{25}C_2} = \frac{276}{2300} = 0.12$$

CASE STUDY 5.4 Playing to Win

The snapshot in Figure 5.27 indicates the number of $1 tickets necessary to cover every 6-number combination in selected lotto games.

Playing to Win (Tickets in Millions)

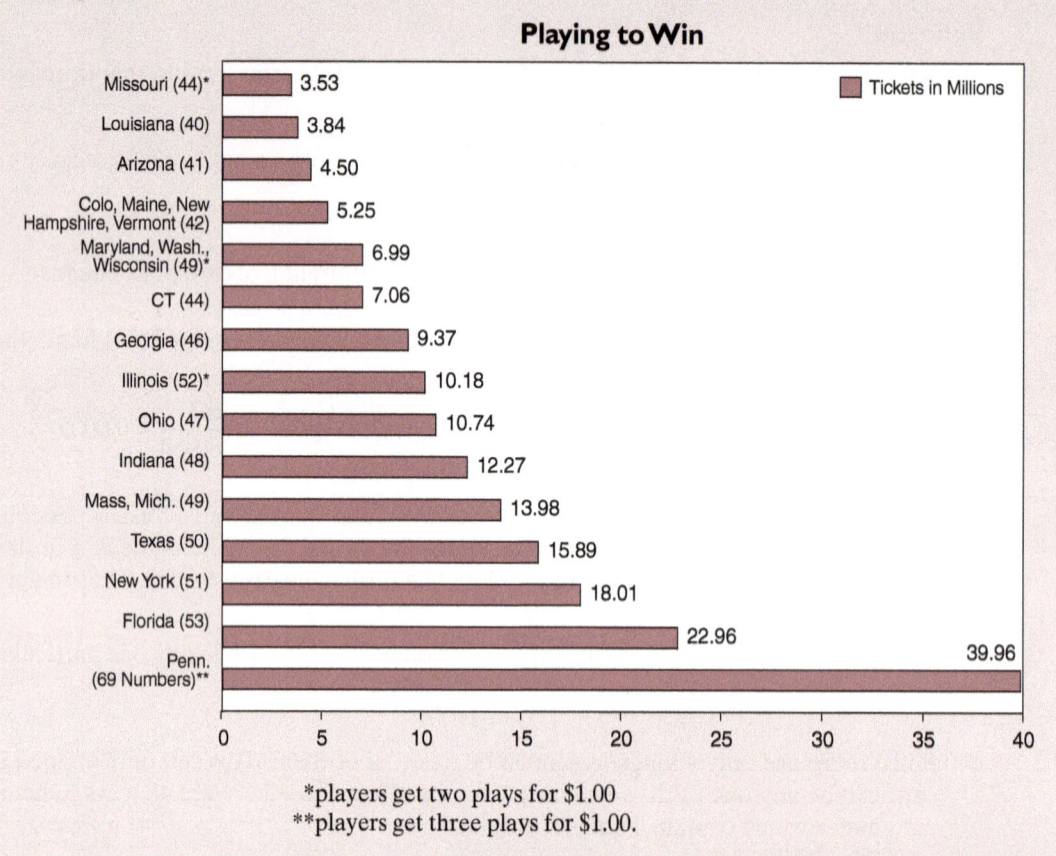

State	Tickets (Millions)
Missouri (44)*	3.53
Louisiana (40)	3.84
Arizona (41)	4.50
Colo, Maine, New Hampshire, Vermont (42)	5.25
Maryland, Wash., Wisconsin (49)*	6.99
CT (44)	7.06
Georgia (46)	9.37
Illinois (52)*	10.18
Ohio (47)	10.74
Indiana (48)	12.27
Mass, Mich. (49)	13.98
Texas (50)	15.89
New York (51)	18.01
Florida (53)	22.96
Penn. (69 Numbers)**	39.96

*players get two plays for $1.00
**players get three plays for $1.00.

FIGURE 5.27

a) What counting technique is used to determine the total number of possible 6-number choices?
b) What is the total number of ways to choose every possible 6-number combination out of 51 numbers? How does this result compare to the value listed in Figure 5.27 for the state of New York?
c) What is the total number of ways to choose every possible 6-number combination out of 49 numbers? How does this result compare to the value listed in Figure 5.27 for the states of Massachusetts and Michigan?
d) For the state of Florida, what is the probability that a person who buys a $1 lotto ticket will win the lottery? For the state of Texas, what is the probability that a person who buys a $1 lotto ticket will win the lottery?
e) For the states Maryland, Washington and Wisconsin, what is the number of ways to choose every possible 6-number combination out of 54 numbers? What is the probability that a person who buys a $1 lotto ticket in Maryland will win the lottery?
f) In which state Texas or Illinois does a person have the maximum number of possible ways to select every possible 6-number combination? Explain.
g) For a $1 ticket, is the probability of winning the lotto greater for Texas or Illinois? Does this result agree with the result of part (f)? Explain.

5.6 CONDITIONAL PROBABILITY

Consider an experiment of selecting two balls in succession from an urn. The selection of the 2 balls can be done in one of two different ways: **with** replacement or **without** replacement.

Selection *with replacement* requires that the first selected ball is returned back to the urn *before* the second ball is selected.

Selection *without replacement* requires that the first ball selected is *not* returned to the urn before the second ball is selected. Let's examine these concepts in the next example.

EXAMPLE 5.47

Let's consider the experiment of selecting two balls in succession from an urn containing 3 balls numbered 1, 2 and 3. First we will examine the experiment where the selection is done *with replacement* and then we will reconsider the experiment *without replacement*.

Find the probability of selecting two odd-numbered balls from this urn and we are:
a) selecting with replacement.
b) selecting without replacement.

Solution

a) The sample space for the experiment in which the selection is done *with replacement* is given in Table 5.9. Using the sample space in Table 5.9, then the:

$$P(\text{selecting two odd numbered balls with replacement}) = \frac{4}{9}$$

Since the first selected ball was returned back to the urn before the second ball was selected, these two selections have no affect on each other. Therefore, they are independent events. Another method of solution is to use the *multiplication rule for independent events* to calculate this probability. Thus,

P(selecting two odd balls with replacement) = P(1st ball selected is odd) · P(2nd ball selected is odd)

$$= \frac{2}{3} \cdot \frac{2}{3}$$

$$= \frac{4}{9}$$

b) The sample space for the experiment in which the selection is done *without replacement* is illustrated in Table 5.10.

TABLE 5.9 Selection of Two Balls with Replacement

Outcome for First Ball	Outcome for Second Ball		
	1	2	3
1	1,1	1,2	1,3
2	2,1	2,2	2,3
3	3,1	3,2	3,3

TABLE 5.10 Selection of Two Balls without Replacement

Outcome for First Ball	Outcome for Second Ball		
	1	2	3
1	X	1,2	1,3
2	2,1	X	2,3
3	3,1	3,2	X

Notice that the sample space does not have the outcomes 1,1; 2,2; and 3,3 since it is impossible to select the same numbered ball twice. Using the sample space in Table 5.10, the

P(selecting two odd numbered balls without replacement) = 2/6 or 1/3

Comparing the answers to parts a and b of Example 5.47, notice that the results are different. This is because in part a, the two selections were *independent*. However in part b, the selection of the second ball **was affected by the outcome of the first selection**. Therefore, we refer to this type of selection process as being *dependent*. Let's now formulate a multiplication rule for **dependent events**.

278 Chapter 5 Probability

> ### Probability Rule 6: Multiplication Rule for Dependent Events: Conditional Probability
>
> **If two events A and B are dependent, then the probability of A *and* B equals the probability of A multiplied by the probability of B given that A has occurred.**
> This can be symbolized as:
>
> $$P(A \text{ and } B) = P(A) \cdot P(B, \text{given A has occurred})$$
>
> This can be further symbolized as:
>
> $$P(A \text{ and } B) = P(A) \cdot P(B|A)$$
>
> where: $P(B|A)$ means $P(B, \text{given A has occurred})$.

We will use the multiplication rule for dependent events to recalculate the P(selecting two odd numbered balls without replacement) as described in Example 5.47 part b. Using Probability Rule 6, the P(selecting two odd numbered balls without replacement) can be written:

P(selecting 2 odd balls without replacement) = P(1st ball is odd) · P(2nd ball is odd | 1st ball was odd)

To compute the P(1st ball is odd), examine the urn pictured in Figure 5.28, where the three numbered balls have now been described as either odd or even. Using Figure 5.28,

$$P(\text{1st ball selected is odd}) = \frac{2}{3}$$

To calculate:

the P(2nd ball selected is odd | 1st ball selected was odd),

the urn in Figure 5.28 must be **modified** to reflect the condition that the 1st ball selected was odd. This modified urn is illustrated in Figure 5.28. Using Figure 5.29, the:

$$P(\text{2nd ball selected is odd | 1st ball selected was odd}) = \frac{1}{2}$$

Therefore,

P(selecting 2 odd balls without replacement) = P(1st ball is odd) · P(2nd ball is odd | 1st ball was odd)

$$= \frac{2}{3} \cdot \frac{1}{2}$$

$$= \frac{2}{6} \text{ or } \frac{1}{3}$$

Notice the previous result obtained using probability rule 6 is the same as the result computed in Example 5.47 part b.

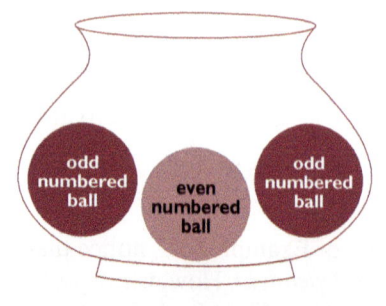

FIGURE 5.28

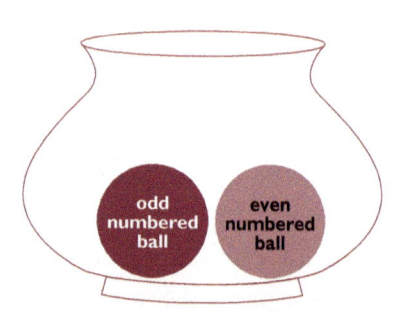

FIGURE 5.29

EXAMPLE 5.48

If two balls are selected *without* replacement from an urn containing two red balls and three white balls, then what is the probability of choosing:
a) 2 white balls?
b) 2 red balls?
c) a red ball and a white ball?

Solution

a) To compute the P(of selecting 2 white balls) without replacement we must use the multiplication rule for **dependent events** since the selection of the 2nd white ball is dependent upon the selection of the 1st white ball.
   Using Probability Rule 6, we have:

P(of selecting 2 white balls) = P(1st ball is white) · P(2nd ball is white | 1st ball was white)

To compute these probabilities, examine the urns shown in Figures 5.30a and .30b.

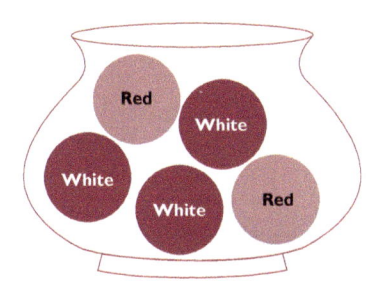

Urn prior to first selection   **Urn prior to second selection**

FIGURE 5.30a   **FIGURE 5.30b**

To determine the P(1st ball selected is white), examine the urn pictured in Figure 5.30a. Therefore,

$$P(\text{1st ball selected is white}) = \frac{3}{5}$$

Examine Figure 5.30b to help calculate the:

P(2nd ball selected is white | 1st ball was white).

Using Figure 5.30b, we have:

$$P(\text{2nd ball selected is white} \mid \text{1st ball was white}) = \frac{2}{4}$$

Now substitute these two results into Probability Rule 6, we get the following result:

P(of selecting 2 white balls) = P(1st ball is white) · P(2nd ball is white | 1st ball was white)

$$= \frac{3}{5} \cdot \frac{2}{4}$$

$$= \frac{6}{20} \text{ or } \frac{3}{10}$$

b) To compute the P(of selecting 2 red balls) without replacement, use the multiplication rule for dependent events. Thus,

P(of selecting 2 red balls) = P(1st ball is red) · P(2nd ball is red | 1st ball was red)

$$= \frac{2}{5} \cdot \frac{1}{4}$$

$$= \frac{2}{20} \text{ or } \frac{1}{10}$$

c) In order to compute the P(of selecting a red ball and a white ball), we must consider the fact that there are two distinct ways of selecting a red and a white ball from the urn. They are:

1st way   or   *2nd way*
selecting a red ball first   selecting a white ball
then a white ball   first then a red ball

That is,

a red ball then white   or   a white ball then red

Therefore, we can express this probability as:

P(of selecting a red ball and a white ball) as

P(a red ball first then a white ball   or   a white ball first then a red ball).

Using the addition rule for mutually exclusive events, Probability Rule 3A, we have:

P(a red and a white ball) = P(first a red ball then a white ball or first a white ball then a red ball)
                        = P(first a red ball then a white ball) + P(first a white ball then a red ball)

First, we will compute the P(of selecting a red ball first and then a white ball), using Probability Rule 6. Thus, we have:

P(1st a red ball then a white ball) = P(1st ball is red) · P(a white ball is selected | 1st ball was red)

$$= \frac{2}{5} \cdot \frac{3}{4}$$

$$= \frac{6}{20} \text{ or } \frac{3}{10}$$

Second, we will compute the:

P(of selecting a white ball first and then a red ball)

using Probability Rule 6. Thus,

P(1st a white ball then a red ball) = P(1st ball is white) · P(a red ball | the 1st ball was white)

$$= \frac{3}{5} \cdot \frac{2}{4}$$

$$= \frac{6}{20} \text{ or } \frac{3}{10}$$

Adding the last two results together, we have:

P(red and a white ball) = P(red ball first then a white ball) + P(white ball first then a red ball)

$$= \frac{3}{10} + \frac{3}{10}$$

$$= \frac{6}{10} \text{ or } \frac{3}{5}$$ ∎

In the previous example, you might have noticed that the P(of selecting a red ball first and then a white ball) and P(of selecting a white first and then a red ball) were equal. This is always true, and is referred to as the commutative probability rule. In general, we can state the commutative probability rule as:

Commutative Probability Rule

For two events A and B, the following probabilities are equal:

$$P(A \text{ and } B) = P(B \text{ and } A).$$

Using Probability Rule 6, we can write:

$$P(A \text{ and } B) = P(A) \cdot P(B \mid A)$$

and also

$$P(B \text{ and } A) = P(B) \cdot P(A \mid B).$$

These two results can be combined to form a new probability rule. This probability rule is called the **general multiplication rule**.

Probability Rule 7: General Multiplication Rule

For any two events A and B, the probability of A and B can be determined by either one of the following formulas:

$$P(A \text{ and } B) = P(A) \cdot P(B \mid A)$$

In words, the probability of A and B is equal to the probability of A multiplied by the probability of B, given that event A has occurred.

OR

$$P(A \text{ and } B) = P(B) \cdot P(A \mid B)$$

In words, the probability of A and B is equal to the probability of B multiplied by the probability of A, given that event B has occurred.

EXAMPLE 5.49

Consider the hopper in Figure 5.31 containing 6 balls where two balls are numbered 1, two balls are numbered 2 and two balls are numbered 3.

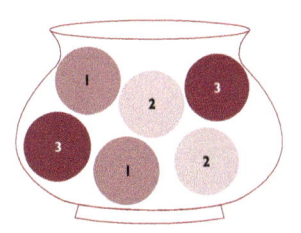

FIGURE 5.31

If this hopper is used to create a two-digit numeral for a lottery game by selecting two balls in succession without replacement, find the probability that:
a) The lottery number 32 is selected
b) The lottery number 33 is selected

Solution

a) To compute the P(the number 32 is selected) use Probability Rule 6 for dependent events since the second numbered ball selected is dependent upon the first selected ball.

$$\begin{aligned}
P(\text{the number 32 is selected}) &= P(\text{first ball selected is a 3 and the second ball is a 2}) \\
&= P(\text{first ball is a 3}) \cdot P(\text{second ball is a 2} \mid \text{first ball is a 3}) \\
&= \frac{2}{6} \cdot \frac{2}{5} \\
&= \frac{4}{30} \text{ or } \frac{2}{15}
\end{aligned}$$

b) To compute P(the number 33 is selected), we will use the Probability Rule 6 for dependent events.

P(the number 33 is selected) = P(first ball selected is a 3 and the second ball is a 3)
= P(first ball is a 3) · P(second ball is a 3 | first ball is a

$$= \frac{2}{6} \cdot \frac{1}{5}$$

$$= \frac{2}{30} \text{ or } \frac{1}{15}$$

EXAMPLE 5.50

Five keys, of which only one works, are tried one after another to unlock a door. What is the probability that:
a) the door does not open on either of the first two tries?
b) the door does open within the first two tries? (HINT: use Probability Rule 4: the complement rule and the result from part a).

Solution

a) The statement that the door does not open on either of the first two tries means that the door did not open on the first try and the door did not open on the second try. Therefore,
P(door doesn't open on either of the first two tries) = P(door doesn't open on 1st try) · P(door doesn't open on 2nd try | door didn't open on 1st try)

Since these two events are dependent, we will use Probability Rule 6 to calculate this probability. Thus,

P(door doesn't open on either of the first two tries) = P(door doesn't open on 1st try) · P(door doesn't open on 2nd try | door didn't open on 1st try)

$$= \frac{4}{5} \cdot \frac{3}{4}$$

$$= \frac{12}{20} \text{ or } \frac{3}{5}$$

b) Use the complement rule: P(E) = 1 – P(E') to compute the P(door opens within the first two tries). Thus,
P(door opens within the first two tries) = 1 – P(door doesn't open on either of the first two tries)
From part a, we have:

$$P(\text{door doesn't open on either of the first two tries}) = \frac{3}{5}$$

Therefore,

$$P(\text{door opens within the first two tries}) = 1 - \frac{3}{5}$$

$$= \frac{2}{5}$$

Using Probability Rule 6: the multiplication rule for dependent events, we can algebraically obtain a formula for the conditional probability of event A, given that event B has occurred.

This result is called **Probability Rule 8: Conditional Probability Formula, denoted P(A | B)**, and is stated as follows:

Probability Rule 8: Conditional Probability Formula

For two events A and B, the probability of A given B has occurred is equal to the probability of A and B divided by the probability of B. This rule is symbolized as:

$$P(A|B) = \frac{P(A \text{ and } B)}{P(B)}$$

where P(B) > 0

EXAMPLE 5.51

In a single toss of a fair die, what is the probability of getting a 1, given the occurrence of an odd number?

Solution

Let's define event A to be the outcome is a 1 and event B to be the outcome is an odd number. Symbolically we want to determine P(A | B), i.e. the probability of getting a one, given that the outcome is odd. To compute this probability, we must first determine P(B) and P(A and B).

Since there are only six possible outcomes of a fair die and three are odd, then:

$$P(B) = P(\text{outcome is odd})$$
$$= \frac{3}{6}$$

Since there is only one outcome satisfying both events A and B, then:

$$P(A \text{ and } B) = P(\text{outcome is a 1 and the outcome is odd})$$
$$= \frac{1}{6}$$

Therefore,

$$P(A \mid B) = \frac{P(A \text{ and } B)}{P(B)}$$
$$= \frac{1/6}{3/6} = \frac{1}{3}$$

EXAMPLE 5.52

Two boxes identical in appearance are on a table. Box 1 contains a nickel and a penny while box 2 contains two nickels. A person randomly picks a box and then randomly selects a coin from the box. If the coin selected is a nickel, what is the probability that the other coin in the box is also a nickel?

Solution

Let's define event A to be the coin remaining in the selected box is a nickel and event B is the selected coin is a nickel. Symbolically we want to determine P(A | B), i.e. the probability the coin remaining in the selected box is a nickel given the selected coin is a nickel. To compute this probability we must first determine P(A and B) and P(B). Thus,

$$P(A \text{ and } B) = P(\text{coin remaining in box is a nickel and selected coin is a nickel})$$

This is equivalent to writing:

$$= P(\text{both coins in the selected box are nickels}).$$

Since there are two boxes and only one box has two nickels, then:

$$P(A \text{ and } B) = \frac{1}{2}$$

Since a nickel can be chosen from either of the two boxes, we have to consider both these possibilities in computing P(B). That is,

$$P(B) = P(\text{coin selected is a nickel})$$
$$= P(\text{box 1 is chosen}) \cdot P(\text{select a nickel} \mid \text{box 1 is chosen})$$
$$+ P(\text{box 2 is chosen}) \cdot P(\text{select a nickel} \mid \text{box 2 is chosen})$$
$$= \frac{1}{2} \cdot \frac{1}{2} + \frac{1}{2} \cdot 1$$
$$= \frac{3}{4}$$

Therefore, the probability the coin remaining in the box is a nickel given that the coin selected was a nickel can be computed using Probability Rule 8:

$$P(A|B) = \frac{P(A \text{ and } B)}{P(B)}$$

$$= \frac{1/2}{3/4}$$

$$= \frac{2}{3}$$

■

EXAMPLE 5.53 The Birthday Problem

In Section 5.1, Lucy wagered Charlie Brown that at least 2 children at their Christmas Party would have the same birthday (month and day).

Let's compute this probability assuming there are 365 days in a year and there were:
a) 3 children at the Christmas Party.
b) 4 children at the Christmas Party.
c) 10 children at the Christmas Party.

Solution

The easiest way to calculate the: P(at least 2 children have the same birthday) is to compute the P(no two children have the same birthday) and use the complement rule: $P(E) = 1 - P(E')$.

Using the complement rule, we can write:

P(at least 2 children have the same birthday) = 1 – P(no two children have the same birthday)

We will use this probability rule to solve the birthday problem.

a) For 3 children at the Christmas Party, we have:

P(no 2 children have the same birthday) = P(all 3 children have different birthdays)
= P(1st child has a birthday)

multiplied by

P(2nd child has a birthday | 2nd child's birthday is different from first child's birthday)

multiplied by

P(3rd child has a birthday | 3rd child's birthday is different from first 2 children's birthdays)

$$= \frac{365}{365} \cdot \frac{364}{365} \cdot \frac{363}{365}$$

= 0.992 (expressed to the nearest thousandths)

Thus,

P(at least 2 children have the same birthday) = 1 – 0.992
= 0.008

This probability is rather small for a group of three children.

b) For 4 children at the Christmas Party, we have:

P(no two children have the same birthday) = P(all 4 children have different birthdays)

$$= \frac{365}{365} \cdot \frac{364}{365} \cdot \frac{363}{365} \cdot \frac{362}{365}$$

= 0.984

Thus,

P(at least 2 children have the same birthday) = 1 – 0.984
= 0.016

c) For 10 children at the Christmas Party, we have:

P(no two children have same birthday) = P(all 10 children will have different birthdays)

$$= \frac{365}{365} \cdot \frac{364}{365} \cdot \frac{363}{365} \cdot \frac{362}{365} \cdot \frac{361}{365} \cdot \frac{360}{365} \cdot \frac{359}{365} \cdot \frac{358}{365} \cdot \frac{357}{365} \cdot \frac{356}{365}$$

$$= 0.883$$

Thus,

P(at least 2 children have the same birthday) = 1 − 0.883
= 0.117

Table 5.11 contains the probability that at least 2 children will have the same birthday for various sized groups. Notice the probability that at least 2 children will have the same birthday for a group of 50 children is 0.970 or 97%. Consequently, Lucy's chances of winning her wager with Charlie Brown (as mentioned in the introduction to this chapter) are extremely good.

TABLE 5.11

Number of People in the Group	Probability of at Least Two People Having the Same Birthday
3	0.008 OR 0.8%
10	0.117 OR 11.7%
20	0.411 OR 41.1%
23	0.507 OR 50.7%
30	0.706 OR 70.6%
40	0.898 OR 89.8%
50	0.970 OR 97%
60	0.994 OR 99.4%
70	0.999 OR 99.9%

EXAMPLE 5.54

Five people are each asked to select a number from 1 to 10 inclusive. What's the probability that at least two of them will select the same number?

Solution

Using the complement rule we get:

P(at least 2 people select the same number) = 1 − P(no two people select the same number)

$$= 1 - \frac{10}{10} \cdot \frac{9}{10} \cdot \frac{8}{10} \cdot \frac{7}{10} \cdot \frac{6}{10}$$

$$= 1 - 0.302$$

$$= 0.698$$

CASE STUDY 5.5

The following article entitled "The Laws of Probability" appeared in *Time* magazine on January 8, 1965.

The Law Trials

The Laws of Probability

Around noon one day last June, an elderly woman was mugged in an alley in San Pedro, Calif. Shortly afterward, a witness saw a blonde girl, her ponytail flying, run out of the alley, get into a yellow car driven by a bearded Negro, and speed away. Police eventually arrested Janet and Malcolm

(continues)

CASE STUDY 5.5 (continued)

Collins, a married couple who not only fitted the description of the fugitive man and woman but also owned a yellow Lincoln. The evidence, though strong, was circumstantial. Was it enough to prove the Collinses guilty beyond a reasonable doubt?

Confidently answering yes, a jury has convicted the couple of second-degree robbery because Prosecutor Ray Sinetar, 30, cannily invoked a totally new test of circumstantial evidence—the laws of statistical probability.

In presenting his case, Prosecutor Sinetar stressed what he felt sure was already in the juror's minds: the improbability that at any one time there could be two couples as distinctive as the Collinses in San Pedro in a yellow car. To "refine the jurors' thinking," Sinetar then explained how mathematicians calculate the probability that a whole set of possibilities will occur at once. Take three abstract possibilities (A, B, C) and assign to each a hypothetical probability factor. A, for example, may have a probability of 1 out of 3: B, 1 out of 10: C, 1 out of 100. The odds against A, B, and C occurring together are the product of their total probabilities (1 out of 3 X 10 X 100), or 3,000 to 1.

After an expert witness approved Sinetar's technique, the young prosecutor asked the jury to consider the six known factors in the Collins case: a blonde white woman, a ponytail hairdo, a bearded man, a Negro man, a yellow car, an interracial couple.

Then he suggested probability factors ranging from 1-to-4 odds that a girl in San Pedro would be blonde to 1-to-1,000 odds that the couple would be Negro-white. Multiplied together, the factors produced odds of 1 to 12 million that the Collinses could have been duplicated in San Pedro on the morning of the crime.

Public Defender Donald Ellertson strenuously objected on the grounds that the mathematics of probability were irrelevant, and that Sinetar's probability factors were inadmissible as assumptions rather than facts. Sinetar, however, merely estimated the factors before inviting the jurors to substitute their own. And the public defender will not appeal because he found no trial errors strong enough to outweigh the strong circumstantial evidence. Convicted by math, Malcolm Collins received a sentence of one year to life. Janet Collins got "not less than one year."

a) Identify each of the six factors that the prosecutor used as evidence to convince the jury that the defendants committed the robbery.
b) What was the probability that the prosecutor assigned to the chance of a girl in San Pedro would have blonde hair? How do you think the prosecutor came up with this probability value? Are you absolutely convinced that this is the correct probability value?
c) What was the probability that the prosecutor assigned to the chance that an interracial couple would be black-white? How do you think the prosecutor came up with this probability value? Are you absolutely convinced that this is the correct probability value?
d) To what event did the prosecutor assign the very convincing probability value of 1 in 12 million?
e) How did the prosecutor arrive at this 1 in 12 million probability value? What assumption did the prosecutor make about the relationship between the six factors before he could multiply each of the respective probability values?
f) Do you agree with the prosecutor's assumption about the relationship among the six factors? Explain.
g) If you were a lawyer who had to defend this couple, what would you say to dispute the prosecutor's mathematical argument?

From *TIME.com*. © 1965 TIME USA LLC. All rights reserved. Used under license.

KEY TERMS

At Least/At Most 239
Chance 232
Classical or A Priori Probability 252
Combination 246
Commutative Probability Rule 280
Complement of an Event 268
Compound or Multi-Stage Experiment 235
Conditional Probability 277

Counting Rule 1: Permutation Rule, $_nP_n$ 240
Counting Rule 2: Permutation Rule for n objects taken s at a time, $_nP_s$ 242
Counting Rule 3: Permutation Rule of N objects with k alike objects 245
Counting Rule 4: number of combinations of n objects taken s at a time, $_nC_s$ 246
Dependent Events 277

Equally Likely 252
Event 233
Experiment 232
Fundamental Counting Principle: Multiplication Rule 236
Independent Events 271
Law of Large Numbers 259
Mutually Exclusive Events 266
n factorial: n! 240

Outcome 232
Permutation 239
Probability 251
Probability Rule 1 261
Probability Rule 2 262
Probability Rule 3: The Addition Rule 264
Probability Rule 3A: The Addition Rule for Mutually Exclusive Events 266
Probability Rule 4: The Complement Rule 268
Probability Rule 5: Multiplication Rule for Independent Events 271
Probability Rule 6: Multiplication Rule for Dependent Events: Conditional Probability 278
Probability Rule 7: General Multiplication Rule 281
Probability Rule 8: Conditional Probability Formula 282
Relative Frequency Concept or Posteriori Probability 257
Sample Space 234
Selection With Replacement 249
Selection Without Replacement 249
Subjective or Personal Probability 259
Three-Stage Experiment 236
Tree Diagram 234
Two-Stage Experiment 236

SECTION EXERCISES

Section 5.2

1. An _____ is the process by which an observation is made or obtained.
2. An _____ is a representation of all possible outcomes of an experiment.
3. The Fundamental Counting Principle is used to determine the number of outcomes in a sample space. T/F
4. Using the Fundamental Counting Principle, the total number of possible outcomes for the experiment of tossing a fair die three times is _____.
5. An _____ is a particular collection of outcomes in the sample space of an experiment.
6. For the experiment of tossing a fair coin four times the event of obtaining at least 3 heads is interpreted as meaning the event of obtaining 3 or _____ heads. The outcomes satisfying this event are _____, _____, _____, _____, and _____.
7. If three dice are tossed, how many different outcomes are in the sample space?
   a) 72   b) 42   c) 216   d) 18
8. If a spinner with four different outcomes is spun and a die is tossed, how many different outcomes are in the sample space?
   a) 40   b) 10   c) 24   d) 1296
9. If two cards are selected from a deck of 52 cards without replacement, how many different outcomes are in the sample space?
   a) 2704   b) 103   c) 1326   d) 2652
10. A *byte* is a computer "word" which consists of a sequence of 0's and 1's. If a particular manufacturer uses a byte of length 8, how many different bytes can be formed?
11. Construct the sample space for the following probability experiments:
    a) Selecting one ball from an urn containing 4 red, 3 yellow and 3 blue balls.
    b) Spinning the following spinner 2 times.
    c) Tossing a fair coin three times.
    d) Tossing a biased coin three times.
    e) Arranging the letters in the word BYE.
    f) Arranging the letters in the word GOOD.
12. Read the following famous nursery rhyme and answer the counting questions pertaining to this rhyme.

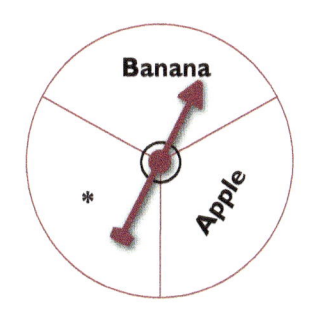

    As I was going to St. Ives,
    I met a man with seven wives,
    Every wife had seven sacks,
    Every sack has seven cats,
    Every cat has seven kits,
    Kits, cats, sacks and wives,
    How many are going to St. Ives?

    Determine the number of:
    a) sacks
    b) cats
    c) kits

Section 5.3

13. Permutations are used as a counting technique when the objects are represented in a definite order. (T/F)
14. 4! is read as four 4 _____ and is a shorthand notation for ____ x ____ x ____ x ____ which has a value of ____.
15. The permutation rule or counting rule 1 is used to determine the number of *permutations* (or arrangements) of n different objects taken altogether, denoted _____. The formula used to calculate this number of permutations is: $_nP_n = $ _____.
16. How many different ways can the letters in the word MOM be arranged?
    a) 6   b) 3   c) 27   d) 9
17. How many different teams containing five members each can be formed from a group of 9 players?
    a) 126   b) 15,120   c) 45   d) 180
18. You've just arrived in Las Vegas for a vacation. You plan to see four dinner shows during your visit. If there are eight shows to select from, then how many ways can you select the four shows you'll see?
    a) 32   b) 8   c) 1680   d) 70
19. The number of different ways four photos can be matched with a list of four names is _____.
20. a) Write the letters of the word DOG in all possible three letter arrangements.
    b) Use the formula $_nP_n=n!$ to compute the number of arrangements.

21. Five textbooks are to be arranged on a shelf. In how many ways can they be arranged?
22. Counting rule 2 is used to determine the number of permutations of n objects taken s at a time. This is denoted by: _____. The formula used to calculate this number of permutations is $_nP_s$ = _____.
23. a) Find the number of permutations that are possible if 5 distinct objects are taken:
    1. one at a time
    2. two at a time
    3. three at a time
    4. four at a time
    5. five at a time
    b) Find the number of combinations that are possible if 5 distinct objects are taken:
    1. one at a time
    2. two at a time
    3. three at a time
    4. four at a time
    5. five at a time
    c) Why are the answers to part (a) larger than the answers to part (b)?
24. The number of different ways to select two numbered balls from an urn containing 5 numbered balls is ____ _____.
25. The chairperson of the scholarship committee wants to select a subcommittee of 2 men and 2 women to evaluate applications. If there are 7 women and 4 men on the committee, then how many different subcommittees can be selected?
    a) 28   b) 333   c) 126   d) 27
26. Counting rule 3 is used to determine the number of permutations of N objects, where: n_1 are alike, n_2 are alike, ..., n_k are alike. The formula to calculate this number of permutations is : _____.
27. In how many different ways can the letters in the following words be arranged?
    a) Calculator
    b) Tennessee
    c) Scissors
    d) Infinity
28. A combination is a selection of objects in which _____ is not important.
29. Counting rule 4 is used to determine the number of combinations of n objects taken s at a time, which is symbolized as _____. The formulas used to calculate this number of combinations are _____ or _____.
30. To determine the number of ways six CDs can be chosen from a selection of 140 one must use the combination formula $_{140}C_6$. (T/F)
31. The number of ways to choose 3 DVD movies from a group of 8 DVDs is _____.
32. In a given semester, a computer science major must select one of five science courses, one of three English courses and one of four psychology courses. How many different programs are available if there are no time conflicts?
33. Maria recently was sent an advertisement to join a record club. As an introductory offer she could choose *any* five selections from the one hundred listed. In how many different ways can Maria make her selections?
34. Twelve college graduates, of which four are female, have applied for seven vacancies in a computer company. If the company has decided to hire three females, how many ways could the company make the seven job offers?
35. On a history test part I has five questions of which three must be answered, and part II has six questions of which four must be answered. In how many ways can a student choose the questions they must answer on this test?
36. When ordering a new car the buyer has the choice of four body styles, five different engines and twelve colors.
    a) In how many different ways can a buyer order one of these cars?
    b) If the buyer also has the option of ordering the car with or without air conditioning, with one of three transmissions, with one of four radio options and with or without a sun roof, in how many ways can the buyer order one of these cars?
37. How many different license plates can be made:
    a) If all six characters are letters?
    b) If all six characters are numerals?
    c) If the first 3 characters are letters and the last 3 characters are numerals?
    d) If all six characters are letters with none repeated?
    e) If all six characters are numerals with none repeated?
38. Four married couples have purchased eight seats in a row for a Broadway show.
    a) In how many different ways can these eight people be seated?
    b) In how many different ways can they be seated if the husband must be seated immediately to the right of his wife?
    c) In how many different ways can they be seated if each couple must sit together?
    d) In how many different ways can they be seated if all the men sit together and all the women sit together?
39. Two attorneys must select nine members for a jury from a group of 15 individuals.
    a) How many different ways can the nine members be selected?
    b) Of the 15 individuals, if 9 are women and 6 are men and the attorneys would like a jury of 5 women and 4 men, how many different ways can the jury be selected?
40. Using the seven points pictured in the following diagram, how many straight lines can be constructed using only two points for each line?

Section 5.4

41. The Classical or a priori probability is based upon the assumption that each outcome of the experiment is _____ _____.
42. The Relative Frequency Concept or Posteriori Probability is based on observing the fraction of the time that an event has occurred over a ____ ____ in the past and using this information to assign a probability value to the event.
43. The probability assigned to an event based upon an individual's personal judgement, or available information is called _____ probability.
44. If an experiment is repeated a large number of times, then the probability of an event obtained using the relative frequency formula will approach the actual or theoretical probability of the event. The law that guarantees this result is called the ____ ____ _____ _____.

45. Assuming a die is fair, and assigning a probability value of ⅙ to a possible outcome is an example of:
    a) Classical probability
    b) Relative frequency probability
    c) Subjective probability

46. If an experiment is repeated a large number of times then the relative frequency of an event will approach the actual theoretical probability.
    (T/F)

47. The law of large numbers guarantees that the probability of an event will equal the theoretical probability for all simulations of an experiment.
    (T/F)

48. Determine whether the scenario is an example of classical, relative frequency, or subjective probability.
    a) Allison does not know the probability of an unbalanced coin landing heads up. She determines that for 100 tosses of the coin the frequency of heads is 65. Thus, she claims the probability that the unbalanced coin lands heads is $^{65}/_{100}$.
    b) Pollsters from *The Arizona Cactus* claim that the probability of getting bitten by a rattlesnake is 1/100,000 for those hiking in the foothills of Tucson.
    c) The probability that a homeless person survives for three years in a major city is ⅔.
    d) Based on a sample of 150,000 high school seniors the probability that a high school senior basketball player makes a professional team is $^{1}/_{2344}$.

49. A fair coin is tossed twenty times and lands with a tail face up on each of these twenty tosses. People who believe that the chance of getting a head on the next toss is not greater than the chance of getting a tail are using the fallacy known as the ____ ____ ____.

50. For equally likely events, the probability of an event E, denoted as P(E), is: P(E) = _____.

51. If we toss a fair coin and roll a fair die at the same time, then:
    a) The probability of getting a 6 on the die and a head on the coin is _____.
    b) The probability of getting an even number on the die and a head on the coin is _____.
    c) The probability of getting a 7 on the die and a tail on the coin is _____.

52. Find the probability of each of the following events:
    a) Selecting the ace of spades by randomly choosing one card from a well shuffled deck containing 52 cards.
    b) Randomly selecting the lottery ticket numbered 0007 from a bin containing lottery tickets numbered 0001 to 9999.
    c) Guessing the month in which your teacher was born.
    d) Randomly selecting an incorrect response to a five item multiple choice question.
    e) Randomly selecting a loser in an eight horse race.

53. Two actors are to be chosen to read the lines of a new play. If there are 17 actors applying, what is the probability that Nick and Jack are randomly chosen?
    a) $^{1}/_{68}$ b) $^{1}/_{136}$ c) $^{1}/_{17}$ d) $^{1}/_{272}$

54. A wine tasting contest is being conducted at the Bowery Hotel. There are 4 varieties of wine labeled: B,M,S, and U to be rated. If one of the contestants has a hangover and has no ability to distinguish a difference in taste between the wines, then what is the probability that he will:
    a) Rank wine B as the least desirable tasting wine?
    b) Rank wine S as the best tasting wine and M as the least desirable tasting wine?
    c) Rank the four wines in the following order: B,U,M,S?

55. a) Construct a sample space for a four child family. Assuming each outcome is equally likely, what is the probability of:
    b) Exactly one boy?
    c) At least one boy?
    d) At least two girls?
    e) At most one boy?
    f) No boy?

56. Derek has the following four coins: A penny, a nickel, a dime and a quarter. Construct a sample space listing all the possible sums of money Derek can form with one or more of these coins. Determine the probability that the sum of money:
    a) Is greater than 10 cents?
    b) Is less than 35 cents?
    c) Is an even amount of money?
    d) Contains a penny?
    e) Contains at least one silver colored coin (nickel, dime, or quarter)?

Section 5.5

57. Probability Rule 1 states that the probability of an event E as a number between ____ and ____ inclusive.

58. Probability Rule 2 states that the sum of the probabilities of all the outcomes in the sample space of an experiment equals _____.

59. It is impossible for P(E) to equal zero (i.e. P(E) = 0).
    (T/F)

60. It is impossible for P(E) to equal one (i.e. P(E) = 1).
    (T/F)

61. It is impossible for P(E) to be greater than one (i.e. P(E) > 1).
    (T/F)

62. Probability Rule 3 is the Addition Rule which states that the probability of satisfying the event A or the event B is equal to the probability of event ____ plus the probability of event B _____ the probability that both events A and B occur at the _____ time. This is symbolized as:
    P(A or B) = _____ + _____ − _____

63. One card is selected at random from an ordinary deck of 52 playing cards. The probability of selecting a queen or a red card is equal to the probability of selecting a _____ plus the probability of selecting a _____ minus the probability of selecting a _____. In symbols, we have:
    P(queen or red card) = P(___) + P(___) − P(___).
    Thus, P(queen or red card) = ___ + ___ − ___ = ___.

64. If the probability a person owns a Chevy is 0.60, the probability a person owns a BMW is 0.40, and the probability a person owns both a Chevy and a BMW is 0.30, then the probability a person owns a Chevy or a BMW is equal to:
    ___ + ___ − ___ = ___.

65. Two events A and B are _____ if both events A and B cannot occur at the same time.

66. The Addition Rule for mutually exclusive events A and B is:
    P(A or B) = _____ + _____.

67. If two events A and B are mutually exclusive then P(A and B) does not equal zero.
    (T/F)

68. Two events that are mutually exclusive must also be independent.
    (T/F)

69. Two events that are independent must also be mutually exclusive.
    (T/F)

70. If two events C and D are mutually exclusive, then:
    a) P(C and D) = _____.
    b) P(C or D) = _____.

71. Consider the experiment of selecting one card at random from an ordinary deck of 52 playing cards. Let event A represent the event of selecting an ace and event B represent the event of selecting a king. Then, events A and B are _____ since a card cannot be both an ace and a king at the _____ time. Therefore, P(ace or king) = P(_____) + P(_____)
    = ___ + _____ = _____.

72. In a single roll of a fair die, the event of getting a 4 and the event of getting a 5 are _____ events, because a die cannot show a _____ and a _____ at the same time.

73. In a single roll of a fair die, the event of getting a 4 and the event of getting an even number are _____ events, because a die can show a ___ and an _____ at the same time. An outcome that satisfies both events is the outcome _____.

74. Probability Rule 4: The Complement Rule states that an event E can either occur or not occur. Thus, the sum of these events always equals _____. This rule is symbolized as: P(_____) + P(_____) = _____.

75. For the experiment of tossing a fair coin four times, the probability of not getting 4 heads can be rewritten using the complement rule as:
    P(not getting 4 heads) = 1 − P(_____).

76. Two events A and B are _____ if the occurrence or nonoccurrence of event A has no influence on the occurrence or nonoccurrence of event B.

77. Probability Rule 5 is the Multiplication Rule for _____ events. This rule states that if two events A and B are independent, then the probability of event A ___ B equals the probability of A _____ the probability of B. This is symbolized as:
    P(_____) = P(_____) _____ P(_____).

78. An urn contains 4 white marbles and 3 red marbles. If two marbles are selected with replacement from this urn, then the probability of selecting two white marbles equals the probability of selecting _____ on the first selection _____ the probability of selecting _____ on the second selection since both events are _____. Therefore, P(two white marbles) = P(white marble) · P(white marble) = _____ · _____ = _____.

79. Construct a sample space for the sum of two dice.
    a) What is the probability of each outcome? (Hint: look at the sample space for Example 5.23.)
    b) What is the probability that one rolls a sum of 7 or 11?
    c) What is the probability that one doesn't roll a sum of 7 or 11?
    d) What is the probability that the sum is less than 9?
    e) What is the probability that the sum is odd?

80. Given five *Scrabble* blocks, each having one of the letters *a, c, e, p, l*:
    a) How many 5 letter arrangements are possible?
    b) How many 3 letter arrangements are possible?
    c) If the 5 blocks are randomly arranged on a table, what is the probability that the word *place* is spelled?

81. A set of ten cards contains two jokers. There are two players, Bob and Audrey. Bob chooses six cards at random and Audrey gets the remaining four cards. What is the probability that Bob has at least one joker?

82. Find the probability that a poker hand will contain:
    a) Full house (1 pair and 1 triple)
    b) 4 of one kind
    c) One pair (hint: more complicated than you may think)
    d) Two pair
    e) A flush (all cards of the same suit)

Section 5.6

83. If Toni, a craps player, rolls a five on her first roll, what is the probability that she makes her point? (In this case, making the point means rolling a five, before rolling a seven).

84. Consider the experiment of selecting two balls without replacement from an urn containing 6 numbered balls. If event A is the first ball selected and event B is the second ball selected, then these two events are said to be _____. This type of probability problem is classified as _____ probability.

85. Probability Rule 6 is the multiplication rule for _____ events. This rule states that if two events A and B are dependent, then the probability of A and B is equal to the probability of A times the probability of _____ given that event A has _____. This is symbolized as: P(_____) = P(A) · P(_____).

86. An urn contains 4 coins: a penny, two nickels and a dime. If two coins are selected without replacement, and event A is first coin selected is a nickel, while event B is second coin selected is a nickel, then events A and B are _____ events. To calculate P(A and B), one can use Probability Rule 6. This is written as: P(A and B) = P(A) · P(_____). Calculating these probabilities, we have:
    P(A) = P(first coin is a nickel) = _____.
    P(B | A) = P(second coin is nickel | first coin is nickel) = _____.
    Thus, P(first coin is a nickel and second coin is a nickel)
    = P(A and B)
    = P(A) · P(B | A)
    = _____.

87. Probability Rule 8: The Conditional Probability Formula states the probability of event A, given that event B has occurred equals the probability of event A ___ B divided by the probability of event _____. This is symbolized as:
    $$P(\_\_\_\_\_) = \frac{P(A \text{ and } B)}{P(\_\_\_\_\_)}$$

88. The six chambers of a handgun randomly were filled with 4 blanks and 2 bullets. If the chambers are spun once, and the gun is fired twice in succession, then to calculate the probability a bullet is fired on the second shot given that the first shot was a blank is determined by using Probability Rule 8: The _____ Probability Formula. Thus,
    P(second shot is a bullet | first shot is a blank)
    $$= \frac{P(\text{first shot is a blank and second shot is a bullet})}{P(_____)} = \frac{8/30}{\_\_\_\_}$$
    = _____.

89. For the Birthday Problem of Example 5.53, the complement rule is used to write: P(at least 2 people have the same birthday) = 1 − P(_____). If there are 35 people in a class, then the probability that at least 2 people have the same birthday is _____.

90. If two balls are selected one at a time from an urn, where the first ball is not placed back into the urn before the next ball is selected, then this selection process is called selection

_____ _____. If the first ball is placed back into the urn before the selection of the next ball, then this selection process is called selection _____ _____.

91. The following table gives a breakdown of the students that attended the graduation picnic.

	BOYS	GIRLS	
5th Grade	24	29	53
6th Grade	18	29	47
	42	58	100

If a student is selected at random to be in charge of coordinating the games, what's the probability of:
a) The student selected being a girl?
b) The student selected is a girl if the student is a 5th grader?
c) The student selected is a 6th grader if the student selected is a boy?

92. From a regular deck of 52 cards, what's the probability of *blackjack* when selecting two cards without replacement? A blackjack has one ace and either a jack, queen, king or ten.

93. If a random drawing is performed and five letters are selected without replacement from the alphabet, determine the probability that:
a) The first three letters in order spell YES.
b) The first five letters in order spell GREAT.

94. A jar contains 5 red, 3 green, 4 blue and 2 yellow marbles. Four marbles are selected without replacement. Determine the probability of the following:
a) One marble of each color is selected.
b) Exactly three red marbles are selected.
c) Exactly two blue marbles are selected.

95. A five card poker hand is dealt to you. Determine the probability of the following:
a) The hand contains exactly 1 club?
b) The hand contains at least 1 diamond?

96. a) If two people are randomly selected and it is known that they are both born in October, what is the probability that they have the same birthday (day only)?
b) If three people are randomly selected and it is known that they are all born in April, what is the probability that their birthdays are all different?
c) What is the probability that at least two of these three people have the same April birthday?

97. Suppose a two child family is moving onto the block where you live.
a) If you learn that not both children are girls, what is the probability that both children are boys?
b) If you know that one child is a boy, what is the probability that the oldest child is a boy?
c) If you learn that both children are of the same sex, what is the probability that both children are girls?

98. Craig, an investment advisor, has a portfolio of 90 stocks: 60 blue-chip and 30 growth stocks. Of the 60 blue-chip stocks, 40 have increased in price during the past month, and 25 of the 30 growth stocks have increased in price. If a stock is selected at random from the portfolio, what is the probability:
a) It will be a growth stock that has not increased in price?
b) If the stock has not increased in price, then it is a blue-chip stock?

99. Vinny, the manager of a Baltimore baseball team, uses probability to help make crucial decisions during the game. In one such instance he estimated that the present pitcher had a 70% chance of getting the next batter out. However, his relief pitcher has a 95% chance of getting the batter out if he is at his best that day, but only a 45% chance if he is not. His pitching coach states that the relief pitcher has a 60% chance of being at his best.
a) What is the probability of the relief pitcher getting the batter out?
b) What should the manager's decision be if he uses probability to make his decisions?

CHAPTER REVIEW

100. The law that guarantees that the relative frequency of an event will begin to approach the actual probability of the event as the experiment is repeated many times is the:
a) Law of averages
b) Law of probability
c) Law of large numbers
d) Law of frequency

101. Find the probability of each of the following events:
a) Selecting an eight by randomly choosing one card from a well shuffled deck containing 52 cards.
b) Randomly selecting a lottery ticket beginning with the digit 5 from a bin containing lottery tickets numbered 00001 to 99999.
c) Selecting the ace of hearts when randomly selecting one card from a deck of 52 cards.
d) Randomly selecting your best friend's birthday from a bin containing all 365 days.
e) Randomly selecting the correct answer to a five choice multiple choice question.
f) Randomly selecting the winner of an eight horse race.

102. In the game of chuck-a-luck a participant wagers on the sum of two die to be above or below seven.
a) What's the probability of the sum being below seven?
b) What's the probability of the sum not being below seven?

103. From a regular deck of 52 cards, what's the probability of
a) Selecting two picture cards with replacement.
b) Selecting two picture cards without replacement.
c) Selecting two cards with replacement whose face values are even numbers.
d) Selecting two cards without replacement whose face values are even numbers?

104. Lorenzo and Luigi play a game where they simultaneously exhibit their right hands with one, two, three or four fingers extended.
a) List all the possible outcomes in the sample space. Determine the probability that:
b) Both players extend the same number of fingers.
c) Both players together extend an even number of fingers.
d) Lorenzo shows an odd number of fingers.

105. A letter of the English alphabet is selected at random. Determine the probability that the letter selected:
a) Is a consonant.
b) Is a vowel (ie. a,e,i,o,u).

c) Follows the letter k.
d) Follows the letter k *and* is a vowel.
e) Follows the letter k *or* is a vowel.
f) Is one of the letters in the word "bushes".
g) Is not in the word "bushes".
h) Is one of the letters in the word "bushes" or the word "tushes".
i) Is one of the letters in both the words "tushes" and "bushes."
j) Is one of the letters in the word "bushes" but not in the word "tushes."

106. Four defective flashlight batteries are mistakenly mixed with two non-defective flashlight batteries which all look identical. If you randomly choose 2 batteries without replacement, what is the probability that:
a) Exactly one is defective?
b) Both are defective?
c) Both are non-defective?
d) At most one is non-defective?

107. Determine whether the scenario is an example of classical, relative frequency, or subjective probability.
a) If three cards: an ace, a deuce and a three are randomly placed face down then the probability that you correctly guess the ace is ⅓.
b) Madeline, a department manager, believes that if you speak quickly to a customer then the probability that you will make a sale is 90%.
c) Debbie, a Bide-A-Wee volunteer, believes that the probability someone will adopt a lovable cat like her cat MEOW is ¹⁄₁₀₀.
d) Based on hospital records, women aged 35–44 have the highest percentage of Caesarean section births in the nation at 29.6%.
e) According to Airline Quality Rating Survey, the chance that Southwest Airlines will depart on-time is 89.5%.

108. a) Write the letters of the word PIG in all possible three letter arrangements.
b) Use the formula: $_nP_n = n!$ to compute the number of arrangements.

109. In how many ways can eight teachers be assigned to eight classrooms?

110. On the way to work Allison must pass through five traffic lights. How many different sequences of green, red could Allison observe as she drives to work?

111. If seven cars are entered in an auto race and only four cars can be placed in the front row, in how many ways can four cars be arranged in the front row?

112. Use counting rule 3 to determine the number of 5 letter arrangements for the letters in the word FUNNY.

113. In how many different ways can the letters in the word TOOTH be arranged?

114. A company wishes to issue ID plates to its employees.
a) If they use a two letter code for each plate, how many different plates can be made?
b) If no letter can be repeated in the same code, how many different plates can be made?
c) If the code consists of two letters and one digit in the last position, now many different plates can be made?
d) If the code consists of two letters and one digit in any order, how many different plates can be made?

115. How many different signals (hanging on a vertical line) can be made with 4 identical green flags, 3 identical blue flags and 3 identical red flags?

116. In a new version of *Master Blaster Brain* one of the players must construct a sequence of 15 colored pegs. There are 2 red, 4 green, 1 blue, 3 white, 2 yellow and 3 black pegs. How many different sequences can be made?

117. Amy has four cassettes to be played at a party.
Cassette 1 contains songs from the early 60's
Cassette 2 contains songs from the late 60's
Cassette 3 contains songs from the 70's
Cassette 4 contains songs from present day rock.
a) How many ways can Amy play the cassettes? (Assuming once started a cassette is played to its completion).
b) What is the probability that if the cassettes were randomly selected from a bag, the order would be cassette 1, cassette 2, cassette 3 and cassette 4?
c) How many ways can Amy play the cassettes if cassette 1 must be played first?
d) What is the probability that after playing cassette 1, cassette 2, 3 and 4 are played in that order?
e) In how many ways can Amy pick two of the four cassettes?
f) What is the probability that she chooses cassette 1 and 2 in that order?

118. If you remember the first five digits of a telephone number but you cannot recall the last two digits, what is the probability that you select the correct missing digits on your first guess?

119. Using six different *Scrabble* blocks with the letters *a, e, o, n, r, s* on them:
a) How many six letter arrangements are possible?
b) How many three letter arrangements are possible?
c) If the 6 blocks are randomly arranged on a table, what is the probability that the word *reason* is spelled?

120. Of seven trees in a line, two are diseased. If any of the seven are equally likely to be affected, what is the probability that the diseased trees are side by side?

121. Calculate:
a) $_4C_3$ b) $_7C_5$ c) $_{12}C_4$ d) $_{52}C_2$ e) $_{52}C_{13}$

122. In how many different ways can the I.R.S. select 7 tax returns to audit out of 50 tax returns?

123. Ten people meet at a pub. Find the number of handshakes that take place if each person shakes hands with everyone else in the group.

124. An urn contains 10 marbles. If there are 6 red marbles and 4 blue marbles, in how many ways can one choose 3 of these balls (one at a time) without replacement so that:
a) None of the red marbles are selected?
b) Exactly one of the blue marbles is selected?
c) 1 red and 2 blue marbles are selected?
d) 3 marbles, regardless of color, are selected?

125. Mary, an intoxicated individual, has five keys on her key ring, one of which opens the front door to her house. If she randomly selected one key at a time and is unable to remember which keys she has already tried, what is the probability that:
a) She selects the correct key for the front door on the first try?
b) She selects the correct key for the front door on the second try?
c) She selects the correct key for the front door on the third try?
d) She selects the correct key for the front door within the first four tries?

WHAT DO YOU THINK?

126. An article entitled "Behind Monty Hall's Doors: Puzzle, Debate and Answer?" appeared in the Sunday *New York Times* on July 21, 1991. The article discussed the debate that was raging among mathematicians, readers of the "Ask Marilyn" column of *Parade Magazine* and the fans of the TV game show "Let's Make a Deal." The argument began in September, 1990 when Ms. Vos Savant, who is listed in the Guinness Book of World Records Hall of Fame for the highest IQ, posed the following question in her "Ask Marilyn" column of *Parade Magazine*:

Suppose you're on a game show, and you're given the choice of three doors: Behind one door is a car; behind the others, goats. You pick a door, say No. 1, and the host, who knows what's behind the other doors, opens another door, say No. 3, which has a goat. He then says to you, 'Do you want to pick door No. 2?' Is it to your advantage to take the switch?

Since Ms. Vos Savant gave her answer, she has received approximately 10,000 letters with the great majority disagreeing with her answer. The most vehement from mathematicians and scientists who have called her the "goat." Her answer has been debated from the halls of the CIA to the mathematicians of MIT. Even Monty Hall conducted a simulation of the problem in his dining room. After which Monty came to the conclusion that Ms. Vos Savant's critics were dead wrong when the host is required to open a door all the time and offer you a switch. However, Monty suggested since he has a choice on the show as to whether he allows the contestant the choice to switch or not, then the question cannot be answered without knowing the motivation of the host. As Monty Hall said: "If he (the host) has the choice whether to allow the switch or not, beware. Caveat emptor. It all depends on his mood. My only advice is, if you can get me to offer you $5,000 not to open the door, take the money and go home."

a) Monty Hall states in the article that the odds on the car being behind door No. 1 are still only 1 in 3 even after he opened another door to reveal a goat. Explain why this is true.
b) Some people argued with Ms. Vos Savant that if one door is shown to be a loser, that information changes the probability of either remaining choice—neither of which has any reason to be more likely—to 1/2. Explain why this is not correct reasoning.
c) Perform a simulation of this game with one of your friends by turning over three playing cards instead of opening doors and using the ace of diamonds as the prize. After doing this simulation 90 times, determine the chance of winning the car when the strategy is to switch doors. What did you get as the chance of winning the car when the strategy was not to switch doors? Are these results close to the actual probabilities?

127. "Tests for AIDS and Drugs: How Accurate Are They?" appeared in the "Ask Marilyn" column of *Parade Magazine* in the March 28, 1993 issue. The question posed was:

Suppose we assume 5% of the people are drug users. A test is 95% accurate, which we'll say means if a person is a user, the result is positive 95% of the time; and if she or he isn't, it's negative 95% of the time. A randomly chosen person tests positive. What is the probability the individual is a drug-user?

To help answer this question, let's assume the population consists of 10,000 people.
a) How many people are drug-users?
b) How many people are non-users?
c) How many of the non-users will test negative?
d) How many of the non-users will test positive?
e) How many of the drug-users will test negative?
f) How many of the drug-users will test positive?
g) How many are positive results?
h) How many are negative results?
i) How many of the positive results are drug-users?
j) How many of the positive results are non-users?
k) What is the probability that given a randomly chosen person tested positive, then the person is a drug-user?
l) What is the probability that given a randomly chosen person tested positive, then the person is a non-user?
m) What would happen to the results of parts (k) and (l) if the percent of drug users in the population was 2%, 1%, 0.5% or 0.25%? What happens as the percentage of drug-users gets closer to 0%?
n) What would happen to the results of parts (k) and (l) if the percent of accuracy for the test is increased from 95% to 98% or 99%? What effect does this have on the results?

128. The following question appeared in the "Ask Marilyn" column of *Parade Magazine* in the July 27, 1997 issue. The question posed was:

A woman and a man (unrelated) each have two children. At least one of the woman's children is a boy, and the man's older child is a boy. Do the chances that the woman has two boys equal the chances that the man has two boys?"

a) What do you think is the probability that the woman has two boys? Explain your answer.
b) What do you think is the probability that the man has two boys? Explain your answer.
c) What approach to defining probability did you use to arrive at the probabilities of part a and b? Explain.
d) This probability question caused much controversy when Marilyn Vos Savant printed her answer to the previous question. Some comments from her readers who felt that the woman and man had the same probability of having two boys were:

"I have lost nearly all my faith in you."

"This is not going to go away until you admit that you are wrong, wrong, wrong!!!"

"You are not the only genius to base your logic on a faulty major premise. Einstein did more than once."

Finally one reader even went as far as to challenge Marilyn to prove her case when the reader wrote:

"I will send $1000 to your favorite charity if you can prove me wrong. The chances of both the woman and the man having two boys are equal."

Marilyn asked her readers to help show why the woman and man have different probabilities of having two boys. She asked her woman readers to write or e-mail her if they have exactly two children (no more), and at least one of them is a boy (either child or both of them) to tell her the sex of both of the children. Marilyn said she would publish the results. What approach to defining probability is Marilyn using to argue her case? Explain.

e) How is this approach different to the one that you used to compute the probabilities in parts a and b?

f) Do you expect the probability results Marilyn will get from her survey to be exactly equal to the probabilities of parts a and b? Explain.

g) Explain how the *law of large numbers* could play an important role in Marilyn's approach to proving her case?

129. In a letter to the editor of the *New York Times* dated September 19, 1997 by Professors Charles Hailey and David Helfand of Columbia University concerning "TWA Flight 800 crash, don't discount meteor." Within their letter, the professors refer to an earlier article on TWA Flight 800 crash which reported that "more than once, senior crash investigators have tried to end the speculation by ranking the possibility of friendly fire at about the same level as that a meteorite destroyed the jet." However, the professors believe this comparison is based on a miscalculation of the probability that a meteorite could have caused the crash of Flight 800. On the other hand, the authors believe:

The odds of a meteor striking TWA Flight 800 or any other single airline flight are indeed small. However, the relevant calculation is not the likelihood of any particular aircraft being hit, but the probability that one commercial airliner over the last 30 years of high-volume air travel would be struck by an incoming meteor with sufficient energy to cripple the plane or cause an explosion.

They state:

Approximately 3,000 meteors a day with the requisite mass strike Earth. There are 50,000 commercial airline takeoffs a day worldwide. Adopting an average flight time of two hours, this translates to more than 3,500 planes in the air; these cover approximately two-billionths of Earth's surface. Multiplying this by the number of meteors per day and the length of the era of modern air travel leads to a 1-in-10 chance that a commercial flight would have been knocked from the sky by meteoric impact.

a) Do you think that the professors' numerical assumptions are reasonable? Explain.

b) What assumptions were made in the professors' conclusion that there is a "1 in 10 chance that a commercial flight would have been knocked from the sky by meteoric impact"?

c) Using the same assumptions the professors did to obtain their "1 in 10 chance" do you come up with the same probability? Explain.

d) Based on the previous information, could you venture a guess at the chance that the Flight 800 was destroyed by friendly fire? That any commercial plane in the last 30 years was destroyed by friendly fire?

In a following letter to the *New York Times* dated September 24, 1996, by Guy Maxtone-Graham, he writes:

As any statistician can tell you, the outcome of past, random events has no bearing on future, unrelated random events. Toss a coin 10 times and the odds of getting heads or tails on the 11th toss are still 50–50. Likewise, calculations based on the number of flights worldwide, the number of takeoffs per day and the number of years that commercial flights have thrived have no bearing on the question of whether a rock from outer space happened to enter the atmosphere to hit one particular airliner on July 17. The odds of such a freak accident downing a specific flight remain small, and the professors' conclusion that "the meteor impact theory deserves more considered attention" is difficult to support.

e) Do you agree with the argument expressed in this letter?

f) In what has been called the "streak of streaks", Joe Dimaggio in 1941 got a hit in 56 successive games. Should you compute the probability that a typical player would achieve such a streak or that sometime in the history of baseball such a streak should occur? How would you estimate these probabilities?

In another letter to the editor of the *New York Times* dated September 28, 1996 from Bill Grassman, he states:

Attempts to prove or disprove the probability that TWA Flight 800 was the victim of a meteor recall the tale of the business executive who, concerned that he might be on a plane with a bomb, commissioned a study to determine the odds of that happening. When the calculations of flights per day, when and where the bombings had occurred and the normal flying patterns of the executive disclosed that the odds of his being on a plane with a bomb were 1 in 13 million, he asked for the probability of his being on a plane with two bombs. On learning that this increased the odds to 1 in 42 billion, he always carried a bomb with him. Statistics!

g) Do you think the writer is criticizing statistics for saying this is true or criticizing those who misunderstand statistics?

h) How would you explain to your Uncle Carmine what is wrong with the executive's reasoning that he should carry a bomb?

130. A classic probability and logic problem: A missionary is captured by a tribe of cannibals who decide to have her for dinner. However, this tribe, being a more sporting group, decide to give the missionary a chance to save her life. They place before her two empty coconuts and twenty berries, ten red berries and ten blue berries. She must distribute *all* the berries as she pleases between the two coconuts, putting at least one berry in each coconut. The head cannibal will randomly choose one berry from the selected coconut. If the berry is blue, the missionary will be set free. However, if it is red, the missionary will be eaten for dinner.

a) How should the missionary distribute the berries to maximize her chances for freedom?

b) Using this distribution, what is the probability that she will be set free?

131. The following legendary problem from India prophesies the end of the world. In a great temple located at the center of the world, rests a brass plate in which are fixed three ivory rods. At the creation of the world, God placed sixty-four discs of pure gold on one of these rods with the largest disc resting on the brass plate and the remaining discs getting smaller and smaller up to the smallest disc at the top. This is referred to as the *Tower of Brahma*. Day and night, the priest on duty transfers the disc from one ivory rod to another adhering to the ancient laws of Brahma. These laws stipulate that the priest must move only one disc at a time, and he must place these discs on the rods so that there never is a smaller disc beneath a larger disc. When all the sixty-four discs have been transferred from the initial ivory rod, on which God placed them, to one of the other rods, the tower and temple will crumble into dust and with a thunderclap the world will vanish.
   a) If there were only two discs on the initial rod, what is the *least* number of moves needed to transfer the discs from the initial rod to one of the other rods?
   b) Determine the *least* number of moves using:
      1) three discs
      2) four discs
      3) five discs
   c) Complete the following table and deduce a general rule to fit the pattern established in the table.

number of discs	least number of moves
1	1
2	
3	
4	
5	
.	
.	
.	
N	

   d) Determine the least number of moves using 64 discs.
   e) If the priests worked day and night, without stopping, making one move every second, how much time would it take for the priests to accomplish their task?

132. The following dialogue took place between the TV personality Monte Hall of *Let's Make a Deal* and a contestant.[4]

[4]This problem was reprinted from The American Statistician, vol. 29. no. 1. February, 1975.

Monte Hall: One of the three boxes labeled A, B, and C contains the keys to that new 1975 Lincoln Continental. The other two are empty. If you choose the box containing the keys, you win the car.
Contestant: Gasp!
Monte Hall: Select one of the boxes.
Contestant: I'll take box B.
Monte Hall: Now box A and box C are on the table and here is box B (contestant grips box B tightly). Is it possible the car keys are in that box! I'll give you $100 for the box.
Contestant: No, thank you.
Monte Hall: How about $200?
Contestant: NO!
Audience: NO!
Monte Hall: Remember that the probability of your box containing the keys to the car is one-third and the probability of your box being empty is two-thirds. I'll give you $500.
Audience: NO!
Contestant: NO, I think I'll keep this box.
Monte Hall: I'll do you a favor and open one of the remaining boxes on the table (he opens box A).
Monte Hall: It's empty! (Audience: applause).
Now either box C or your box B contains the car keys. Since there are two boxes left, the probability of your box containing the keys is now one-half. I'll give you $1,000 cash for your box.
WAIT!!!!

Is Monte right? The contestant knows that at least one of the boxes on the table is empty. He now knows it was box A. Does this knowledge change his probability of having the box containing the keys from 1/3 to 1/2? One of the boxes on the table has to be empty. Has Monte done the contestant a favor by showing him which of the two boxes was empty?
a) Is the probability of winning the car 1/2 or 1/3?
Contestant: I'll trade you my box B for the box C on the table.
Monte Hall: That's weird!!
Do you agree with Monte that the contestant's proposal is "weird," or is the contestant making a logical choice?
b) What is the probability that box C contains the keys?

PROJECTS

133. Reread Example 5.53 *The Birthday Problem*. Ask 20 people to write a whole number between 1 and 100 inclusive onto a piece of paper. Fold the paper in half and put the folded paper into a large envelope. Before examining the results calculate the probability that at least two of the people have the same number.

134. Randomly survey (a) 10, (b) 40, (c) 70, and (d) 100 people within the election district that you live and ask them:
    (1) Whether they are eligible voters within your election district?

(2a) If they answer yes to question number one, ask them to select the appropriate following statement.

I am registered in the following political party:
Democratic
Republican
Conservative
Liberal
Other

or

I am not registered for a political party.

(2b) If they answer no to question number one, exclude them from your survey. Use the probability formula:

$$P(\text{Event E}) = \frac{\text{number of outcomes satisfying Event E}}{\text{total number of outcomes}}$$

and your survey results to calculate the probability that a person chosen at random from your election district is a:
(a) Democrat, (b) Republican, (c) Conservative, (d) Liberal, (e) Not registered for a political party.

Contact the Board of Elections to determine the number of eligible voters in your election district and the actual number of democrats, republicans, conservatives, liberals and voters which are not registered for a political party within your election district. Using the Board of Election information, recalculate the above probabilities. Compare the survey probability results to the Board of Election probability results. Is there any relationship between the two as the survey size increases? Explain this relationship, if any exists.

CHAPTER TEST

135. In New York State's numbers game, an individual chooses a 3 digit number to play the game. This 3 digit number can range from 000 to 999. What is the probability that Tony will win the game, if he selects:
    a) The number 123?
    b) The digits 4,2,7 in all possible arrangements?
    c) All numbers beginning with a 7?
    d) All numbers beginning with 46?
    e) All possible numbers using the digits 1,2,6?

136. An eight cylinder automobile engine has two defective spark plugs. If two plugs are removed from the engine, determine the probability of selecting:
    a) No defective plug.
    b) One defective plug.
    c) Two defective plugs.

137. Millie, a hat check girl, mistakenly mixes up the tickets on three hats. If she randomly places the mixed up tickets on the three hats, what is the probability that:
    a) Exactly one guest will receive the correct hat?
    b) Exactly all 3 guests will receive the correct hats?
    c) Exactly two guests will receive the correct hats?
    d) No one will receive the correct hat?

138. Classify whether each of the following events can be assigned a probability value using either classical, relative frequency or subjective approach to probability.
    a) Getting 2 heads in 3 tosses of a fair coin.
    b) A pregnant woman will give birth to triplets.
    c) The Dow Jones Industrial Stock Index will increase at the close of the next trading day.
    d) The San Francisco 49ers winning the next Super Bowl.
    e) A student will get a passing grade on her first exam in her freshman college introductory psychology course.
    f) Getting 2 tails in 3 tosses of a biased coin.
    g) A heart patient will successfully survive his bypass surgery.
    h) Selecting a red card from an ordinary deck of 52 playing cards.

139. In how many ways can the officers for President, Treasurer and Secretary be filled from the 12 members of the ski club?

140. How many four letter arrangements can be made from ten different letters if:
    a) Repetitions are allowed?
    b) Repetitions are *not* allowed?
    c) A letter can be used at most twice?

141. Matthew, a playboy, is returning to his penthouse on the thirteenth floor late one night. He is concerned that his expected date is already waiting for him in his penthouse. If seven other people get on the elevator with him and no two are together, and he knows that no one in the elevator besides himself is going to the penthouse, what is the probability that:
    a) They all get off at different floors, thus delaying Matthew as long as possible?
    b) All the other passengers get off at the same floor, thus delaying Matthew the least?

142. On the planet Krypton the calendar was almost identical to ours with the exception that every one of the twelve months had 30 days. What's the probability that the day the planet exploded was:
    a) The thirteenth of the month?
    b) A Friday?
    c) A Friday given that it was the thirteenth?
    d) The thirteenth given that it was a Friday?

143. Matt "The Magician" in his "three card Monte" act uses three colored cards. One is red on both sides, one is blue on both sides and the third is red on one side and blue on the other side. If one card is selected at random and the top side is red, what is the probability that the other side is also red?

144. The following data set is the weight (in kg) for 15 male and 16 female adult Polar Bears that were tracked in the Arctic.

Males	Females
376	125
379	114
424	108
469	135
395	144
348	124
445	110
393	152
382	134
345	117
373	98
512	116
557	202
460	225
457	169
	160

Based on the given data find the following:
- **a)** What is the probability that a randomly selected bear has a weight that is greater than 150 kg?
- **b)** What is the probability that a randomly selected bear will have a weight that is between 120 & 170 kg?
- **c)** Given that a female bear is selected, what is the probability that the bear will weigh less than 135 kg?
- **d)** What is the probability that a randomly selected bear will have a weight that is less than 120 kg or greater than 300 kg?

For the answer to this question, please visit www.MyStatLab.com.

CHAPTER 6

Random Variables and Discrete Probability Distributions

Contents

6.1 Introduction
6.2 Random Variables
6.3 Probability Distribution of a Discrete Random Variable
6.4 Mean and Standard Deviation of a Discrete Random Variable
6.5 Binomial Probability Distribution
6.6 The Poisson Distribution
6.7 The Geometric and Negative Binomial Distributions

6.1 INTRODUCTION

In Chapter 2, we discussed how a frequency distribution table was used to organize raw data into a more meaningful form to help describe the important characteristics of a distribution. A frequency distribution represents information pertaining to a situation that has already occurred. In Chapter 5, the concepts of probability allowed us to work with situations that have not yet occurred and where the outcomes are uncertain. In this chapter, we continue our study of probability by introducing the concept of a random variable and a probability distribution which examines all the possible values of an experiment and their associated probabilities. The emphasis of this chapter will be on a discrete random variable and its probability distribution. In particular, the binomial distribution will be presented as an example of an extremely useful discrete probability distribution.

According to the Bureau of the Census, the latest family data pertaining to family size for a small midwestern town, Nomore, is shown in Table 6.1.

Table 6.1 shows that 43.19% of all the families within this town have a family containing 2 people, 22.76% of all family sizes have 3 people, while only 1.51% of the families have a total of 7 people within their family. The listing of all the possible family sizes with the respective proportion of families represents the probability distribution of family size. This probability distribution is useful in answering such questions as:

If a family from this town is selected **at random**, then what is the probability of selecting a 6-person family?

Using the probability distribution in Table 6.1, you can see that there is a 2.83% chance of randomly selecting a 6-person family. Because the family is being selected *at random,* and the size of the family can vary from 2 to 7, then family size is said to be a *random variable.* The proportion or percentage which is associated with each family size is interpreted as the *probability* of each value of family size. The concept of a random variable and a probability distribution will be examined in detail within the next two sections.

TABLE 6.1 Distribution for Family Size

Size of Family	Proportion (or percentage) of Families to the Total Number of Families
2 persons	0.4319 (or 43.19%)
3 persons	0.2276 (or 22.76%)
4 persons	0.2104 (or 21.04%)
5 persons	0.0868 (or 8.68%)
6 persons	0.0283 (or 2.83%)
7 persons	0.0151 (or 1.51%)

6.2 RANDOM VARIABLES

In Chapter 5, we developed techniques that allowed us to calculate the probability of rolling 4 consecutive sevens in four tosses of a pair of dice, or the probability of winning at least once in ten plays of the roulette wheel. In this chapter, we will discuss the numerical values associated with the outcomes of an experiment such as the **number** of heads when tossing a fair coin twice, or the **number** of sixes in five rolls of a die. Other experiments of this type are:

- the number of defective computer chips in a batch of 100 chips,
- the number of correct answers on a 35 question multiple choice test, or
- the life expectancy of a 21 year old female.

In each of these situations, we can associate different numbers with the *uncertain outcomes* of the experiment. For instance, the number of defective computer chips within a batch of 100 chips can be represented by a whole number from 0 to 100, while the number of correct answers on a 35 question multiple choice test can vary from 0 to 35.

The experiment of determining the number of defective chips in a batch of 100 is considered a **random experiment** since *the outcome of this experiment cannot be predicted with any certainty*. Thus, the number of defective chips is determined by chance, and can be any of the following 101 possible values:

$$0, 1, 2, 3, \ldots, 99, 100.$$

If we let *X represent the number of defective computer chips within the batch,* then X can assume any one of these possible 101 values. Until each batch is tested, we are **uncertain** of the actual number of defective chips that will be found so that the number of defective chips may vary from batch to batch. For example, testing one batch of 100 computer chips could result in detecting three defective chips within the batch. This is denoted as X = 3. If the next batch were to produce only one defective chip, then the value of X would change to X = 1. As each batch of 100 chips is tested, the value of X will assume one of the possible 101 values, and probably represent a different value. Consequently, we say that the number of defective chips within the batch, denoted by X, is a *random variable* since **the value of X can vary from batch to batch and is not known until the batch is tested**. The numbers associated with this experiment:

$$0, 1, 2, 3, \ldots, 99, 100$$

represent the values of the random variable.

> **DEFINITION 6.1**
> A **random variable** is a variable that takes on different numerical values which are determined by chance.

The random variables are usually denoted by capital letters such as X, Y, and Z.
Example 6.1 illustrates this idea of a random variable.

EXAMPLE 6.1

For each random experiment, define a random variable and identify the possible values of the variable.
a) If it is assumed that the largest number of children is 6, count the number of children within a family for a small town.
b) Select a radial tire from the production line and determine the life of the tire, assuming no tire has ever lasted less than 20,000 miles or more than 85,000 miles.
c) Count the number of heads in two tosses of a fair coin.

Solution

a) For the random experiment of counting the number of children within a family in this small town, the random variable is the number of children. If we let X denote the number of children, then the seven possible values of X are:

$$X = 0, 1, 2, 3, 4, 5, \text{ or } 6 \text{ children.}$$

b) For the random experiment of selecting a radial tire from the production line, the random variable is the life of the tire. The possible values are infinite, since the tire selected can last any mileage from 20,000 to 85,000 miles including tenths, hundredths etc. of a mile. Letting Y equal the life of the tire, the value of Y can be any number greater than or equal to 20,000 but less than or equal to 85,000. The values of Y can be expressed as:

$$20{,}000 \leq Y \leq 85{,}000.$$

c) For the random experiment of tossing a fair coin twice, the number of heads is the random variable. There are three possible outcomes when counting the number of heads in two tosses, they are:

both tosses land heads (or 2 heads),
only one toss lands heads (1 head), or
neither toss lands heads (0 heads).

Thus, if we let X denote the number of heads, then the possible values of X are: $X = 0, 1,$ or 2 heads.

A random variable can be classified as either a *discrete* or *continuous* random variable depending upon the numerical values that it can assume. In Example 6.1, the number of children in a family is an example of a *discrete random variable* because the values of this variable: 0, 1, 2, 3, 4, 5, and 6 are *finite or can be counted*.

DEFINITION 6.2
A **discrete random variable** is a random variable that can take on a finite or countable number of values.

Some examples of a *discrete* random variable are:

- the number of correct answers on a 35 question multiple choice test
- the number of people attending the performance of a Broadway play
- the number of earrings a person wears
- the number of students who come to class late
- the number of cars involved in an accident at an intersection
- the daily closing price of a stock on the New York Stock Exchange

To determine a value for most of these random variables, you would have to count the number of items or people or things that would satisfy the random variable. For example, to determine the number of correct answers on a 35 question test, you would **count the number correct**, or to determine the number of people attending the performance of a Broadway play, you would **count the number of people** attending the performance. Consequently, we say that the possible numerical values of a discrete random variable usually represent a **count**. Therefore, these possible values of a discrete random variable can only represent **a limited number of values** which have a **break between successive values**. Recall that the number of correct answers a student gets on a 35 question multiple choice test represents a **count** which could result, say, in 20 correct, or possibly 21 correct. However, it is ***impossible*** *for the student to get a number **between** 20 and 21 correct.* The fact that there is *a **break between successive values** of this random variable* (between 20 and 21) is the **characteristic** that distinguishes it as a **discrete random variable**. Not all discrete random variables have values that represent a count. The values of the daily closing price of a stock *are limited and have a break between successive values.* For example, if the closing price of the stock ranged from $42 to $43 for the day, then it could only assume these limited values: $42, $42⅛, $42¼, $42⅜, $42½, $42⅝, $42¾, $42⅞, or $43. Notice the values of this discrete random variable have breaks between successive values such as $42¼ and $42⅜.

On the other hand, a random variable that *measures* the life of a radial tire is an example of a *continuous random variable*. It represents *an uncountable number of values* since it can take on **all values** between any *two possible values* of the variable.

> **DEFINITION 6.3**
>
> A **continuous random variable** is continuous if the value of the random variable can assume any value or an uncountable number of values between any two possible values of the variable.

The values of continuous random variables represent measurements. In measuring the life of a radial tire or the amount of time a person watches MTV, the number of different values that the variable can assume between any two possible values of the variable are uncountable. For example, the number of hours a person watches MTV could possibly be **measured to be any value** between 2 hours and 2.5 hours. Some such possible values are:

$$2.41 \text{ hrs, or } 2.401 \text{ hrs, or } 2.4001 \text{ hrs, or } 2.40001 \text{ hrs, etc.}$$

Since it is impossible to list all these values, we use the inequality symbol < to express the values of a continuous variable. Thus, if we let X = the number of hours a person watches MTV, then to represent all the values between 2.0 and 2.5, we would write: $2.0 < X < 2.5$.

Although we may record the number of hours a person watches MTV as a whole number, the random variable is treated as a continuous variable since **it can assume an uncountable number of values**. Thus, any random variable that can be measured such as height, weight, time, temperature or speed are considered continuous even though they might be recorded as a whole number or to a particular decimal place.

Some examples of a *continuous* random variable are:

- the amount of time needed to travel to college
- the height of a person
- the amount of iced tea in a 16 oz bottle
- the temperature in Central Park at noon
- the amount of rainfall during the year in San Francisco

EXAMPLE 6.2

For each random variable, state whether it is discrete or continuous.
a) The life of a LED 32 inch 4K Ultra High Definition TV.
b) The number of telephone calls arriving at a college switchboard during the day.
c) The number of employees that are absent from work at a steel manufacturing plant during a summer day.
d) The thickness of a multivitamin tablet.

Solution

a) Since the life of a LED 32 inch 4K Ultra High Definition TV can last say 10.1 years or 10.01 or even 10.001 years, it is a continuous variable. The values represent an amount of time which can assume an uncountable number of values.
b) The number of telephone calls is discrete because there can only be a finite or countable number of calls arriving at the switchboard during the day.
c) If there are 300 people that work at this plant, then the possible number of employees that can be absent is: 0, 1, 2, ... , 300. Since these represent a finite or countable number of values, the number of employees that are absent from work is a discrete random variable.
d) The thickness of a multivitamin tablet is a continuous random variable since it represents a measurement of its width which can assume an uncountable number of values.

CASE STUDY 6.1 Random Variables

Examine each graphic or table shown in Figure 6.1. Each graphic or table presents a random variable.

For each graphic or table,
a) Describe each random variable being illustrated.
b) State whether the random variable is discrete or continuous.
c) Indicate two non-consecutive possible values of the random variable. Identify these values on a number line.

To identify all the possible values of the random variable between these two values on the number line, could you highlight these values using separate dots or would you need a darken line to represent all the values? What does this indicate about the type of random variable? Explain your answer.

Statistical Snapshots

(a) The Facts of Long Island Life

How many scoops of ice cream do you typically eat in a week?

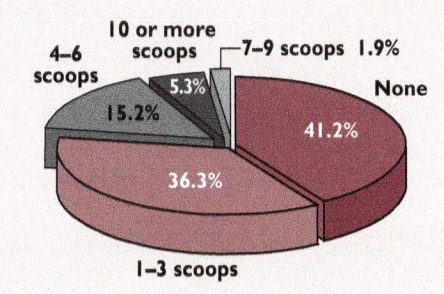

- None: 41.2%
- 1–3 scoops: 36.3%
- 4–6 scoops: 15.2%
- 7–9 scoops: 1.9%
- 10 or more scoops: 5.3%

Source: Responses are from a Newsday-Hofstra poll that was conducted March 6–28 by telephone with 802 Long Island residents. The margin of error is plus or minus 3.5 percentage points.
From *Newsday*. © 1998 Newsday. All rights reserved. Used under license."

(b) How Many Telephones We Have

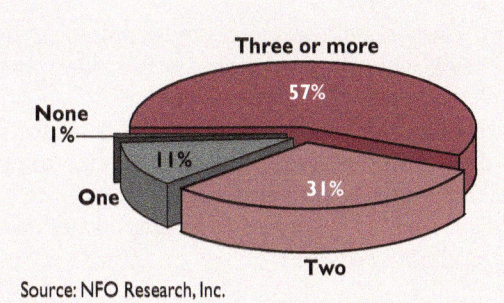

- Three or more: 57%
- Two: 31%
- One: 11%
- None: 1%

Source: NFO Research, Inc.

(c) Age of Foreign-car Shopper

Half of U.S. adults say they would consider buying a foreign-made car. Percentage by age group:

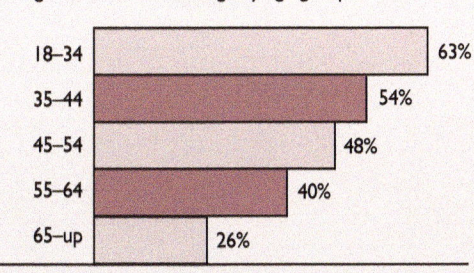

- 18–34: 63%
- 35–44: 54%
- 45–54: 48%
- 55–64: 40%
- 65–up: 26%

*ICR for Japan Automobile Manufacturers Association

(d) Beer Consumption

More than half the adults over the age of 21 had at least one serving of beer during the past month. Beer drinkers average 18.7 servings per month. Servings per month by age:

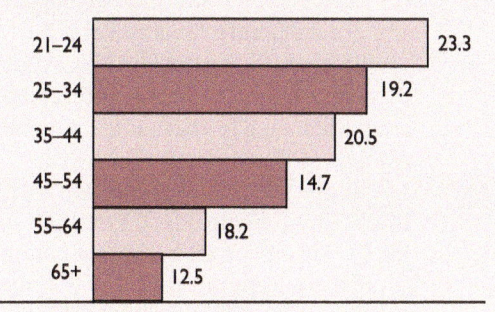

- 21–24: 23.3
- 25–34: 19.2
- 35–44: 20.5
- 45–54: 14.7
- 55–64: 18.2
- 65+: 12.5

Note: A serving equals 12 ounces
*Source: Maritz AmeriPoll

(e) Who's Got Mail?

Average number of e-mail received daily:

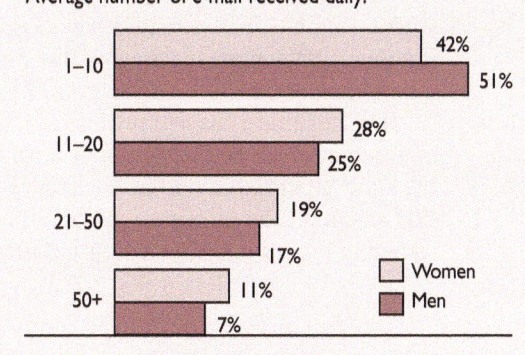

	Women	Men
1–10	42%	51%
11–20	28%	25%
21–50	19%	17%
50+	11%	7%

(f) Who's Losing Sleep

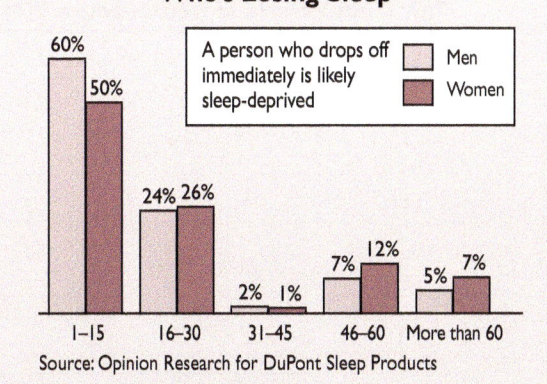

A person who drops off immediately is likely sleep-deprived

	Men	Women
1–15	60%	50%
16–30	24%	26%
31–45	2%	1%
46–60	7%	12%
More than 60	5%	7%

Source: Opinion Research for DuPont Sleep Products

FIGURE 6.1

In most experiments, we are interested in determining all the possible values of a random variable and the probability associated with each value. Such a listing is called a *probability distribution*. In this chapter, we will examine the concepts of a discrete probability distribution. In the next chapter, we will examine the idea of a continuous probability distribution.

6.3 PROBABILITY DISTRIBUTION OF A DISCRETE RANDOM VARIABLE

We will now examine how to assign probabilities to the values of a discrete random variable. A distribution that displays all the possible values of a random variable and the probability associated with each value of the random variable is called a probability distribution.

> **DEFINITION 6.4**
> A **probability distribution** is a distribution which displays the probabilities associated with all the possible values of a random variable.

Let's examine this idea of a probability distribution and determine how to generate a probability distribution for a discrete random variable.

To begin let's consider the experiment of tossing a fair coin twice. There are four possible outcomes for this experiment. If we are interested in counting the number of heads obtained on the two tosses of the fair coin, then each of these four outcomes can be classified within one of the 3 possible values: 0 heads, 1 head or 2 heads as illustrated in Table 6.2.

Notice there is only one way to obtain 2 heads (HH), two ways to get 1 head (HT and TH), and one way to obtain no heads (TT). For a fair coin, the probability of each outcome is ¼, since each outcome is equally-likely. This is shown in the last column of Table 6.2.

Since we are interested in the number of heads obtained when a fair coin is tossed twice, we can construct another table which just lists *the possible number of heads along with their associated probabilities*. For this experiment, the number of heads can be categorized into the following three possible values: **2 heads, 1 head and no heads**. Table 6.3, which displays these possible number of heads along with their associated probabilities, is the **probability distribution of the number of heads in two tosses of a fair coin**. Notice the two column headings of the probability distribution of Table 6.3, the possible values of the number of heads is symbolized by X while the associated probabilities are denoted by P(X).

You should realize that Table 6.3 *doesn't represent the actual outcomes* of tossing a fair coin twice, rather it is a **theoretical model of the possible number of heads** which indicates what we would **expect to happen in the long-run** in two tosses of a fair coin. This is the idea of a probability distribution.

It is helpful to graph a probability distribution of a random variable since this helps to visualize the probability distribution. The graph of a probability distribution is accomplished by scaling the possible values of the random variable on the horizontal axis and the corresponding probability values along the vertical axis. For example, the information contained in Table 6.3 can also be displayed graphically using a histogram as shown in Figure 6.2. This histogram is called a *probability histogram*.

TABLE 6.2 Possible Outcomes of Two Tosses of a Fair Coin Where H Represents a Head and T Is a Tail

First Toss	Second Toss	Number of Heads	Probability
H	H	2	$\frac{1}{4}$
H	T	1	$\frac{1}{4}$
T	H	1	$\frac{1}{4}$
T	T	0	$\frac{1}{4}$

TABLE 6.3 Probability Distribution of Number of Heads in Two Tosses of a Fair Coin

Number of Heads X	Probability P(X)
0	$\frac{1}{4}$ or 0.25
1	$\frac{2}{4}$ or 0.50
2	$\frac{1}{4}$ or 0.25

DEFINITION 6.5
A **probability histogram** is a histogram representing a probability distribution. The values of the discrete random variable are scaled along the horizontal axis, while the corresponding probability values are scaled along the vertical axis.

Notice, in Figure 6.2, that the values on the horizontal axis represent the number of heads, denoted as X, and are placed at the center of the base of each bar. While the height of each bar equals the probability for the value of X. For example, the height of the bar pertaining to the value X = 0 is 0.25 because the probability of 0 heads is 0.25 (that is P(0 heads) = 0.25).

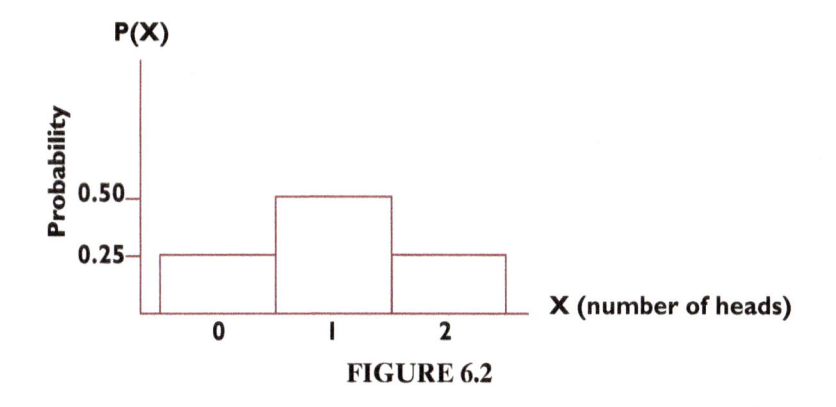

FIGURE 6.2

The probability histogram in Figure 6.2 is a graph of a probability distribution of the number of heads for two tosses of a fair coin. Notice that the probability distribution of the number of heads in two tosses of a fair coin can be displayed by a table as illustrated by Table 6.3 as well as a graph as shown in Figure 6.2. *In general, the probability distribution of a discrete random variable can be defined using a table, a graph or a formula.* We will consider the formula approach when we examine a **discrete probability distribution** called the binomial distribution in a subsequent section of this chapter.

EXAMPLE 6.3

Consider the experiment of tossing a fair coin until a head appears, or until the coin has been tossed three times, whichever occurs first.
a) Construct a probability distribution for the number of tails.
b) Construct a probability histogram for this probability distribution.
c) Using the probability histogram, what is the probability of getting no tails?

Solution

a) There are four possible outcomes for this experiment. They are:

Head on 1st toss, represented as H,
Tail on 1st toss and a Head on 2nd toss, written TH
Tail on 1st two tosses and a Head on 3rd toss, denoted TTH,
Tail on all three tosses, symbolized as TTT.

If X represents the number of tails, then the possible values of X are: 0, 1, 2, or 3 tails.
Since the probability of getting a head is ½, and the probability of getting a tail is also ½, then the probabilities for each of these values of X are shown in Table 6.4.
The probability assigned to the probability of 1 tail, written P(X = 1), is ¼ which was determined by multiplying the probability of a tail by the probability of a head. This is shown in the second row of Table 6.4. The probability assigned to the probability of 2 tails, written P(X = 2), is ⅛ which was determined by multiplying the probability of a tail by the probability of a tail by the probability of a head. This is shown in the third row of Table 6.4. Finally, the probability assigned to the probability of 3 tails, written P(X = 3), is ⅛ which was determined by multiplying the probability of a tail by the probability of a tail by the probability of a tail. This is shown in the last row of Table 6.4. Thus, the probability distribution for the number of tails is given in Table 6.5.

TABLE 6.4 Probability Distribution with Outcomes

Number of Tails X	Possible Outcomes	Probability P(X)
0	H	$\frac{1}{2}$
1	TH	$\frac{1}{2} \cdot \frac{1}{2} = \frac{1}{4}$
2	TTH	$\frac{1}{2} \cdot \frac{1}{2} \cdot \frac{1}{2} = \frac{1}{8}$
3	TTT	$\frac{1}{2} \cdot \frac{1}{2} \cdot \frac{1}{2} = \frac{1}{8}$

TABLE 6.5 Probability Distribution of Number of Tails

Number of Tails X	Probability P(X)
0	$\frac{1}{2}$
1	$\frac{1}{4}$
2	$\frac{1}{8}$
3	$\frac{1}{8}$

b) For this probability distribution, the discrete random variable is the number of tails which is scaled along the horizontal axis from 0 to 3. The probability values corresponding to the number of tails are scaled along the vertical axis from 0 to ½. The value of the discrete random variable, the number of tails, is placed at the center of each bar of the histogram while the bar's height is equal to the probability of the variable. The probability histogram for this probability distribution is shown in Figure 6.3.

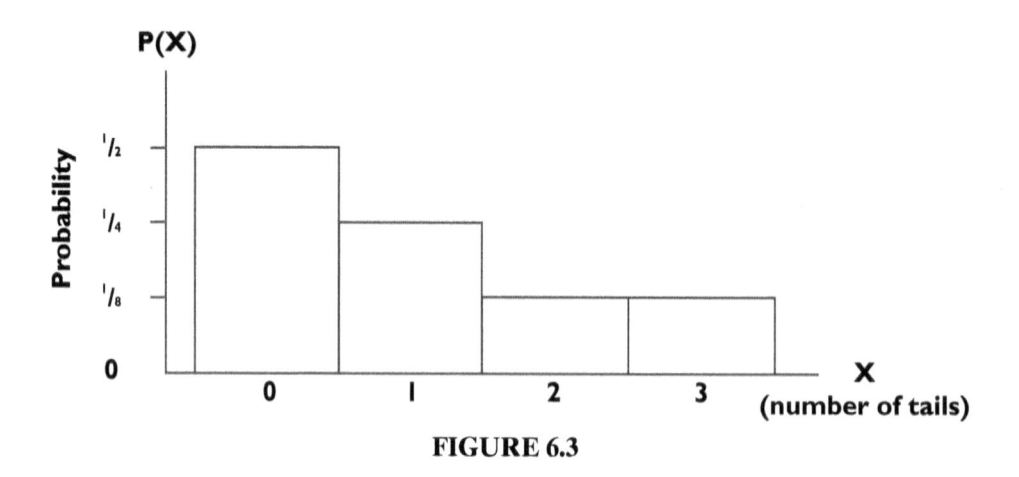

FIGURE 6.3

c) From the probability histogram in Figure 6.3, the probability of getting no tails is equal to the height of the bar pertaining to 0 tails which is ½. Thus, the probability of getting no tails is ½.

In Figure 6.3, notice *the probability assigned to each value of the discrete random variable, number of tails, is a number between 0 and 1.* The probability values of this probability distribution are: ⅛, ¼, and ½. In addition, *notice that the sum of all the probabilities of this discrete probability distribution is equal to one.* From Figure 6.3, the sum of the probabilities of the probability distribution is:

$$\frac{1}{2} + \frac{1}{4} + \frac{1}{8} + \frac{1}{8} = 1.$$

These observations represent two important characteristics of a probability distribution of a discrete random variable. Let's summarize these characteristics of a probability distribution.

Characteristics of a Probability Distribution of a Discrete Random Variable

1. The probability associated with a particular value of a discrete random variable of a probability distribution is always a number between 0 and 1 inclusive.
2. The sum of all the probabilities of a probability distribution must always be equal to one.

EXAMPLE 6.4

The graph below illustrates the lifestyle of a popular newspaper's business travelers while on the road. The study was entitled "What it's like to be a Road Warrior." One pie chart which appeared in the article is shown in Figure 6.4. The pie chart describes the number of times that the business traveler calls home per day. Using this information,

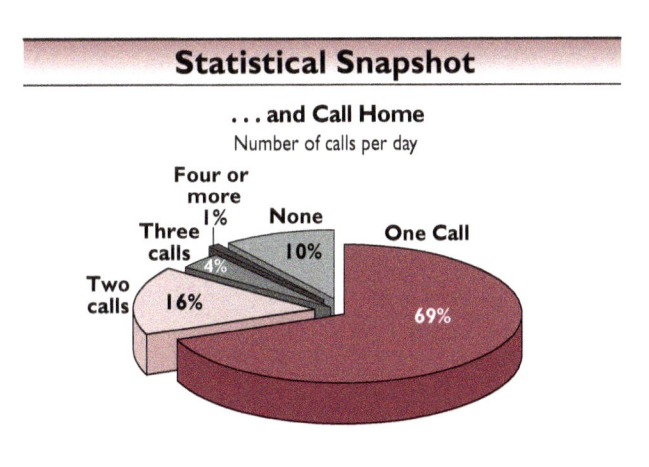

Statistical Snapshot
...and Call Home
Number of calls per day

Four or more 1%
Three calls 4%
None 10%
Two calls 16%
One Call 69%

FIGURE 6.4

a) Identify the discrete random variable being presented in the pie chart,
b) Construct a probability distribution for this variable,
c) Construct a probability histogram for this probability distribution,
d) Find the sum of the probabilities for this probability distribution. Did you expect this result? Explain.

Solution

a) The number of times a business traveler calls home per day is the discrete random variable.
b) To construct a probability distribution for the number of times a business traveler calls home per day, we need to list all the possible values of the variable along with the respective probabilities. According to the pie chart in Figure 6.4 and treating the possibility of 4 or more as 4, the possible values of the variable are: 0, 1, 2, 3, or 4. To determine the probabilities associated to each of these values of the random variable, we can use the percentages which accompany the values since they represent a reasonable estimate of the likelihood that each value will occur in the future. For example, since 69% of the business travelers called home once, then we will use 69% or 0.69 to represent the probability that a business traveler will call home once per day while on a business trip. Thus, the probability distribution for the number of times a business traveler calls home per day is shown in Table 6.6.
c) For this probability distribution, the discrete random variable, which is the number of calls per day, is scaled along the horizontal axis from 0 to 4. The probability corresponding to the number of calls are scaled along the vertical axis from 0 to 0.69. The value of the variable, the number of calls, is placed at the center of each bar of the histogram while the bar's height is equal to the probability associated to each value of the variable. The probability histogram for this probability distribution is shown in Figure 6.5.

TABLE 6.6 Probability Distribution for the Number of Times a Business Traveler Calls Home per Day

Number of Calls to Home per Day X	Probability P(X)
0	0.10
1	0.69
2	0.16
3	0.04
4	0.01

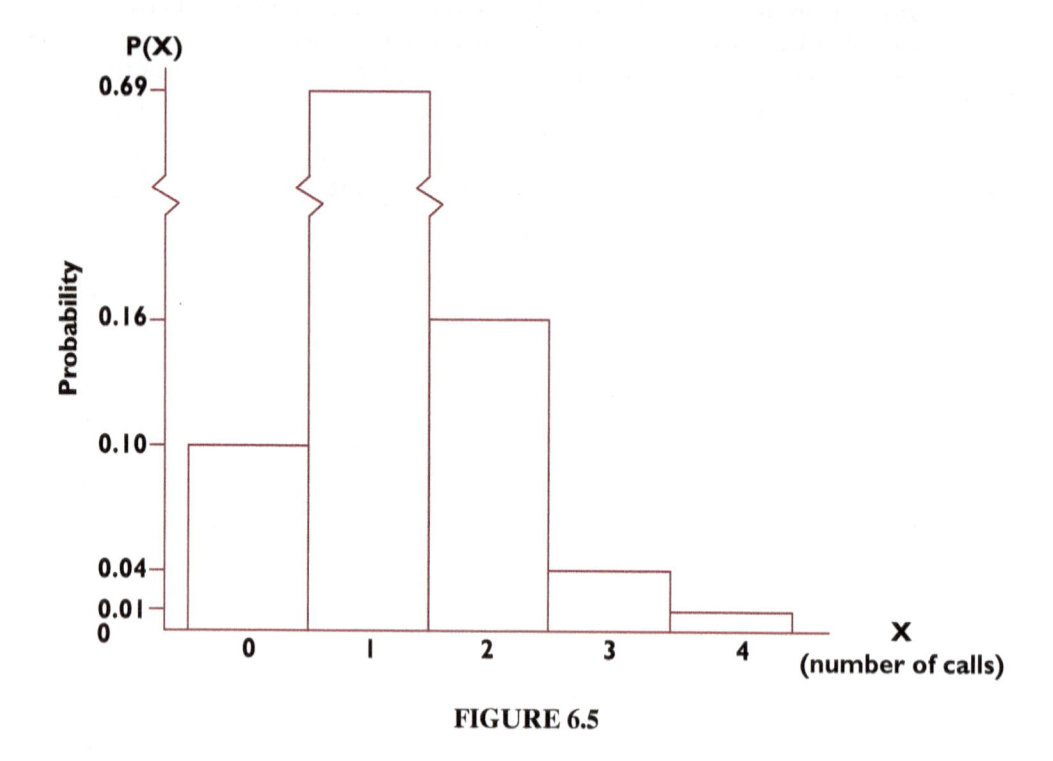

FIGURE 6.5

d) The sum of all the probabilities for the probability distribution are:

sum of the probabilities = 0.10 + 0.69 + 0.16 + 0.04 + 0.01
= 1.00.

This result is expected since one of the properties of a discrete probability distribution is the sum of all the probabilities is equal to one.

Before we discuss other aspects of a probability distribution, you should observe that a probability distribution can be constructed in different ways. One is on the basis of theoretical considerations like the tossing of a fair coin as illustrated in both Table 6.3 and Example 6.3. Another probability distribution is established on the basis of experience and observation as discussed in Example 6.4. In addition, a probability distribution can also be based on subjective judgements of the likelihood of the possible values of the discrete random variable. At this point, we should explain the difference between a frequency distribution and a probability distribution. A frequency distribution is a listing of all the observed values and frequencies of an experiment that was already conducted; whereas, a probability distribution represents a list of all the possible outcomes of an experiment that could result when the experiment is eventually conducted. Probability distributions will serve as models of experiments. If we can find a probability distribution model that accurately fits the experiment then we can use the probability values of the model to draw conclusions or make inferences from the probability distribution.

EXAMPLE 6.5

According to the latest Census Bureau, 40% of all families have only one child. Consider the experiment of randomly selecting two families from the population of all U.S. families. Let X represent the number of families having only one child in this sample.
a) Construct a sample space for this experiment.
b) Determine the probability for each outcome in the sample space.
c) Construct a probability distribution for the discrete random variable X.
d) Construct a probability histogram for the discrete random variable X.
e) What is the probability that both families selected at random will have only one child?

Solution

a) Since X represents the number of families having only one child, then there are four possible outcomes for this sample of two families. These four outcomes are illustrated in Figure 6.6 where H represents a family having only one child while D represents that the family doesn't have only one child.

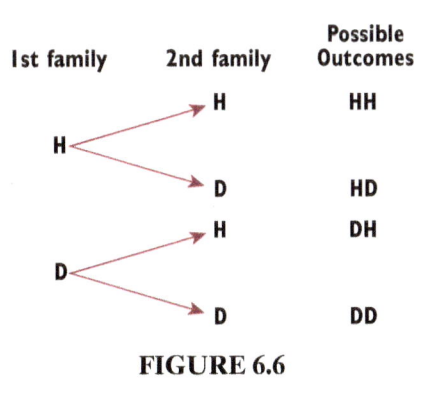

```
                Possible
1st family   2nd family   Outcomes
                 H          HH
         H
                 D          HD
                 H          DH
         D
                 D          DD
```

FIGURE 6.6

From Figure 6.6, the four possible outcomes are:

HH: both families have only one child
HD: first family selected has only one child, but second family selected doesn't have only one child.
DH: second family selected has only one child, but first family selected doesn't have only one child.
DD: both families don't have only one child.

b) To compute the probability for each of these outcomes, we need to use the Census Bureau information that 40% of all families have only one child. Thus, the probability that a family has only one child is 0.40, while the probability that a family doesn't have only one child is 0.60 (or 1 − 0.40). For a sample of two randomly selected families, the number of possible families that have only one child, denoted by X, can be classified into the three values:

$$X = 0, X = 1, \text{ or } X = 2.$$

The probability for each of these possible values of X are calculated as follows.

$P(X = 0) = P(DD) = (0.60)(0.60)$
P(X = 0) = 0.36
$P(X = 1) = P(HD \text{ or } DH)$
$\quad\quad\quad\ = P(HD) + P(DH)$
$\quad\quad\quad\ = (0.40)(0.60) + (0.60)(0.40)$
$\quad\quad\quad\ = 0.24 + 0.24$
P(X = 1) = 0.48
$P(X = 2) = P(HH) = (0.40)(0.40)$
P(X = 2) = 0.16

c) Assigning these probabilities to the respective possible values of X, the probability distribution of the number of families having only one child for a sample of two families is shown in Table 6.7.

TABLE 6.7 Probability Distribution for the Number of Families Having Only One Child

Number of Families Having Only One Child X	Probability P(X)
0	0.36
1	0.48
2	0.16

d) A probability histogram representing the number of families having only one child is constructed where the values of the number of families are scaled along the horizontal axis from 0 to 2. While the corresponding probability values are scaled along the vertical axis from 0 to 0.48. The value of the variable, the number of families having only one child, is placed at the center of each bar of the histogram while the bar's height is equal to the probability of the variable. The probability histogram for this probability distribution is shown in Figure 6.7.

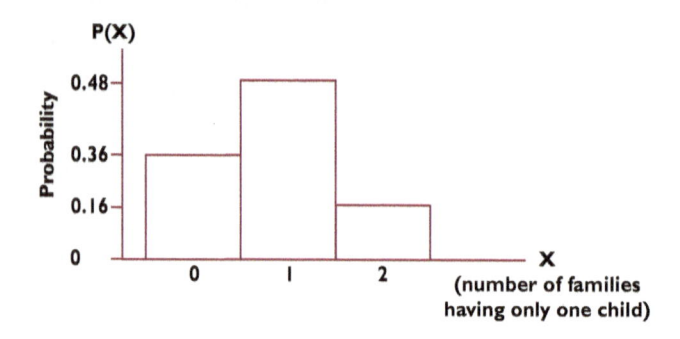

FIGURE 6.7

e) The probability that both families selected at random will have only one child is 0.16 is equal to the height of the bar representing X = 2 as shown in Figure 6.7. Thus, the P(X = 2) = 0.16.

6.4 MEAN AND STANDARD DEVIATION OF A DISCRETE RANDOM VARIABLE

In the previous section, we examined the probability distribution for a discrete random variable. For example, we examined the number of times a business traveler calls home per day or the number of families having only one child in a sample of two families. In each of these probability distributions, a model of all the possible values of the discrete random variable along with their associated probabilities was used. Furthermore, a probability histogram was used to graphically describe the probability distribution. Now, we would like to examine the numerical measures such as the mean and the standard deviation that will help us describe the main characteristics of the distribution. Remember that the mean and the standard deviation are used to describe the center value and the variability or spread of a distribution. In this section, we will learn how to find and interpret the mean and the standard deviation of a discrete probability distribution.

The Mean Value of a Discrete Random Variable

Table 6.8 presents the probability distribution for the number of televisions per family in Poorrich, Kansas.

From the information in Table 6.8, can you estimate the number of televisions a typical family in Poorrich, Kansas owns? Such an estimate can be considered to be a measure of a central tendency of the probability distribution. As discussed in Chapter 3, the *mean* represents a measure of the center of a distribution and serves as the balance point of the distribution. Thus, we can estimate the mean or typical number of televisions by trying to identify the center or balance point of the probability distribution. Examine the histogram of the probability distribution for the number of TV's per family in Poorrich, Kansas in Figure 6.8a and try to estimate the position of the balance point of the probability distribution.

Section 6.4 *Mean and Standard Deviation of a Discrete Random Variable* 311

TABLE 6.8 Probability Distribution of the Number of TV's per Family

Number of TV's per Family X	Probability of the Number of TV's per Family P(X)
0	0.05
1	0.35
2	0.40
3	0.13
4	0.07

Probability Distribution of the Number of TVs Per Family in Poorrich, Kansas

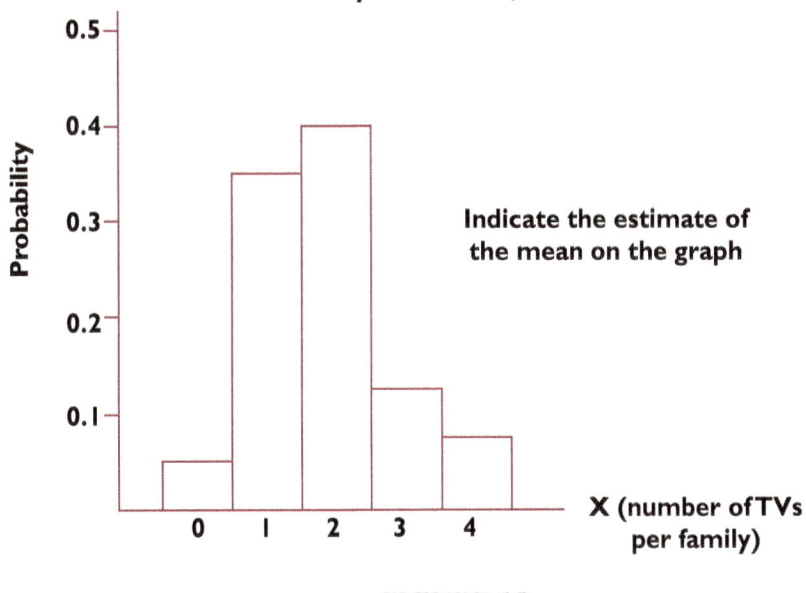

Indicate the estimate of the mean on the graph

FIGURE 6.8a

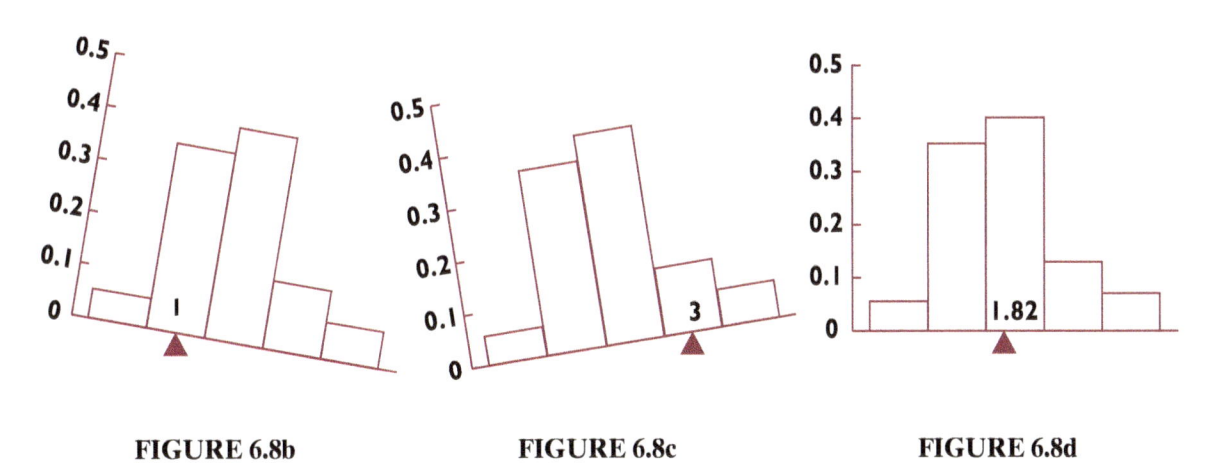

FIGURE 6.8b

FIGURE 6.8c

FIGURE 6.8d

At what value of the random variable, X, do you think the balance point lies? That is, at what value of the random variable, X, could we place a fulcrum which would balance the probability histogram of Figure 6.8a? Figure 6.8b shows the effect of trying to balance the distribution by placing the fulcrum under the value of X = 1 along the x axis. However, this is not the balance point since the distribution tilts down to the right. Figure 6.8c indicates the effect of placing the fulcrum under the value of X = 3. In this case the probability distribution tilts down to the left. Using the histogram, can you estimate a placement for the fulcrum so that the distribution would be balanced? The value of the mean for the probability distribution is the value of X along the horizontal axis that corresponds to the fulcrum's position that balances the histogram.

In trying to determine this balance point for the histogram of Figure 6.8a, the height of each bar which represents the **probability weight** for each value of the random variable has to be taken into consideration. The mean value of a probability distribution is the balance point which takes into account the **weights** or **the probability for each value of the random variable**. Thus, we will refer to the mean value for a discrete random variable as a *weighted mean*.

From the probability histogram of Figure 6.8a, an estimate of the mean value or the balance point of the distribution happens to fall between 1.5 and 2. To compute the actual mean or balance point of a probability distribution, use the following formula.

The Mean of a Discrete Random Variable X

The mean of a discrete random variable X denoted by μ, is computed by *multiplying* each possible X value of the discrete random variable by its probability, P(X), and then *adding* all the resulting products: $X \cdot P(X)$. In symbols, the formula for the mean is:

$$\mu = \sum_{\substack{\text{all possible} \\ X \text{ values}}} \left[(X \text{ value}) \cdot (\text{probability of } X) \right]$$

$$\mu = \sum_{\substack{\text{all possible} \\ X \text{ values}}} \left[X \cdot P(X) \right]$$

Using this formula to determine the mean number of TV's per family for the probability distribution in Table 6.8, we can use the following procedure.

First, substitute each possible value of the random variable X into the formula for the mean:

$$\mu = \sum_{\substack{\text{all possible} \\ X \text{ values}}} \left[X \cdot P(X) \right]$$

Since the possible X values are: 0, 1, 2, 3, and 4, then we have:

$$\mu = 0P(0) + 1P(1) + 2P(2) + 3P(3) + 4P(4)$$

Second, substitute the probability value, P(X), corresponding to each X value into the formula.

$$\mu = 0(0.05) + 1(0.35) + 2(0.40) + 3(0.13) + 4(0.07)$$

Third, calculate the value of each product.

$$\mu = 0 + 0.35 + 0.80 + 0.39 + 0.28$$

Fourth, add up these products to obtain the mean.

$$\mu = 1.82.$$

Thus, **the mean number of TV's per family is 1.82.**

Examining the probability histogram in Figure 6.8d, we now know that the balance point is 1.82, since this is the mean value of the number of televisions. This mean value signifies that the typical family in Poorrich, Kansas has 1.82 televisions. This doesn't suggest that each family has *exactly* 1.82 televisions. Instead, the mean value of 1.82 televisions is the average number of televisions that can be **expected** per family in the long-run. For this reason this value is also referred to as the expected value. That is, if we repeatedly selected families at random living in Poorrich, Kansas and counted the number of televisions they have within their household, then, as the number of families selected gets larger and larger, the average number of televisions will approach 1.82. Therefore, in the long-run, we would expect an average value of 1.82 televisions to occur. The mean value is not necessarily a possible value of the random variable, it is an expected value that measures the typical number of televisions per family. Thus, in this case, some families will have less than 1.82 televisions, while some families will have more than 1.82 televisions, but no families will have exactly 1.82 televisions.

EXAMPLE 6.6

Consider the experiment of tossing a fair coin until a head appears, or until the coin has been tossed three times, whichever occurs first. Table 6.9 lists the outcomes and the probability distribution for this experiment. What is the expected or mean number of tosses of the coin for this experiment?

TABLE 6.9 Probability Distribution with Outcomes

Number of Tosses X	Possible Outcomes	Probability P(X)
1	H	$\frac{1}{2}$
2	TH	$\frac{1}{4}$
3	TTH or TTT	$\frac{1}{8} + \frac{1}{8} = \frac{1}{4}$

Solution

To find the expected or mean number of tosses of the probability distribution in Table 6.9, use the formula for the mean of a discrete variable.

To determine the mean value using the formula:

$$\mu = \sum_{\substack{\text{all possible} \\ X \text{ values}}} [X \cdot P(X)]$$, we will use the following procedure.

First, substitute each value of the random variable X into the formula for the mean:

$$\mu = \sum_{\substack{\text{all possible} \\ X \text{ values}}} [X \cdot P(X)]$$

Since the possible X values are: X = 1, X = 2, and X = 3, then we obtain:

$$\mu = 1P(1) + 2P(2) + 3P(3)$$

Second, substitute the probability value, P(X), corresponding to each X value into the formula.

$$\mu = 1\left(\frac{1}{2}\right) + 2\left(\frac{1}{4}\right) + 3\left(\frac{1}{4}\right)$$

Third, calculate the value of each product, and sum up these products.

$$\mu = \frac{1}{2} + \frac{2}{4} + \frac{3}{4}$$

Fourth, add up these products to obtain the mean.

$$\mu = \frac{7}{4} \text{ or } 1.75 \text{ tosses}$$

Thus, the expected or mean number of tosses for this coin tossing experiment is 1.75. That is, if we were to repeat this coin tossing experiment many, many times then we would expect the long-run average number of tosses to be 1.75.

Let's now summarize the procedure used to calculate the mean of a discrete random variable.

Procedure to Calculate the Mean (or expected value) of a Discrete Random Variable Using the Formula:

$$\mu = \sum_{\substack{\text{all possible} \\ X \text{ values}}} [X \cdot P(X)]$$

Step 1   Substitute each value of the random variable X into the formula for the mean.
Step 2   Substitute the probability value, P(X), corresponding to each X value into the formula for the mean.
Step 3   Calculate the value of each product.
Step 4   Sum up these products to obtain the mean (or expected value).

The Variance and Standard Deviation of a Discrete Random Variable

Although the mean is the balance point which helps to describe the center or the typical value of a probability distribution, it doesn't provide a measure of the spread or variability of the distribution.

Recall in Chapter 3, we used the variance and the standard deviation to measure the variability of a distribution. Remember that both of these measures of variability took into consideration how far each of the data values within a distribution deviated from the mean. Examine the probability distribution of the number of TV's per family shown in Figure 6.8a. We found that the mean number of televisions per family was 1.82. By examining the histogram in Figure 6.8a, you see that there is a greater weight or probability for families having values near the mean of 1.82 televisions such as X = 1, X = 2 and X = 3. In contrast, there is less weight or probability for families having values far from the mean such as X = 0 televisions or X = 4 televisions. Since the values closer to the mean, such as X = 1, X = 2 or X = 3, have a greater chance of occurring than the values far from the mean, X = 0 or X = 4, we should expect the variance and the standard deviation to be relatively small because the standard deviation and variance measure the concentration of the values about the mean. On the other hand, if the values of the random variable which are far from the mean had greater probabilities then we would expect the variance and the standard deviation to be relatively larger since the values further from the mean carry a greater weight or probability.

Thus, the formulas for the *variance and the standard deviation* of a probability distribution must take into consideration the probability associated with each value of the random variable. Let's examine the formulas to compute the variance and standard deviation of a discrete random variable.

Procedure to Calculate the Variance of a Discrete Random Variable Using the Formula:

$$\sigma^2 = \sum_{\substack{\text{all possible} \\ X \text{ values}}} [(X-\mu)^2 \cdot P(X)]$$

Step 1   Subtract the mean from each possible value of the random variable X to obtain the deviation from the mean, written: $(X-\mu)$;
Step 2   Square each deviation and multiply the result by the probability associated with the corresponding X value, symbolized as

$$(X-\mu)^2 \cdot P(X)$$

Step 3   Add these results to obtain the variance, written:

$$\sigma^2 = \sum_{\substack{\text{all possible} \\ X \text{ values}}} [(X-\mu)^2 \cdot P(X)]$$

DEFINITION 6.6

Standard Deviation of a Discrete Random Variable, denoted by σ, is the positive square root of the variance. Thus, the formula for the standard deviation is symbolized as:

$$\sigma = \sqrt{\sum_{\text{all possible } X \text{ values}} [(X - \mu)^2 \cdot P(X)]}$$

Let's calculate the variance, σ^2, and the standard deviation, σ, for the probability distribution in Table 6.10.

TABLE 6.10 Probability Distribution

Number of TV's per Family X	Probability of the Number of TV's per Family P(X)
0	0.05
1	0.35
2	0.40
3	0.13
4	0.07

To calculate the variance and the standard deviation for the discrete random variable X, we must first find the mean, μ. In the previous discussion on the mean, the mean for the probability distribution given in Table 6.10 was: $\mu = 1.82$ televisions.

Table 6.11 contains the necessary computations to calculate the variance and the standard deviation.

TABLE 6.11

X	P(X)	X − μ	(X − μ)²	(X − μ)²P(X)
0	0.05	0 − 1.82 = −1.82	(−1.82)² = 3.3124	(3.3124) (0.05) = 0.16562
1	0.35	1 − 1.82 = −0.82	(−0.82)² = 0.6724	(0.6724) (0.35) = 0.23534
2	0.40	2 − 1.82 = 0.18	(0.18)² = 0.0324	(0.0324) (0.40) = 0.01296
3	0.13	3 − 1.82 = 1.18	(1.18)² = 1.3924	(1.3924) (0.13) = 0.181012
4	0.07	4 − 1.82 = 2.18	(2.18)² = 4.7524	(4.7524) (0.07) = 0.332668

To construct Table 6.11, you need the following five columns:

X: *the values of the random variable,*
P(X): *the probability values associated with each X value*
(X − μ): *the deviations from the mean*
(X − μ)²: *the squared deviations from the mean*
(X − μ)²P(X): *the products of the squared deviation by the probability associated with the corresponding X value*

Using the information in Table 6.11, the variance and the standard deviation are calculated using the following steps.

Step 1 Find the sum of the quantities: $(X-\mu)^2 P(X)$, symbolized as $\sum[(X-\mu)^2 \cdot P(X)]$. This result is the variance, σ^2.

$\sum[(X-\mu)^2 \cdot P(X)] = 0.16562 + 0.23534 + 0.01296 + 0.181012 + 0.332668$

$\sum[(X-\mu)^2 \cdot P(X)] = 0.9276$

Thus, the variance is:

$$\sigma^2 = 0.9276$$

Step 2 Take the square root of the variance to obtain the standard deviation, σ.

That is, $\sigma = \sqrt{\sum_{\substack{\text{all possible} \\ x \text{ values}}} [(X-\mu)^2 \cdot P(X)]}$

Thus, $\sigma = \sqrt{0.9276}$

or

$\sigma \approx 0.96$.

Therefore, the standard deviation for the probability distribution is approximately 0.96 televisions per family. Thus, the standard deviation of 0.96 televisions per family represents a measure of the spread of the possible values of the number of televisions from the mean of 1.82 televisions. Or intuitively, we can say that a standard deviation of 0.96 is approximately the average deviation the values of the random variable are from the mean of 1.82 televisions. Since the standard deviation measures the variability or dispersion in the random variable values from the mean, it describes how spread out a probability distribution is. Thus, a probability distribution with a small standard deviation will have a probability histogram where the bars are clustered together, while a probability distribution having a large standard deviation will have a probability histogram where the bars are more dispersed.

At this point, let's summarize the procedure used to calculate the standard deviation of a discrete random variable.

Procedure to Calculate the Standard Deviation of a Discrete Random Variable Using the Formula:

$$\sigma = \sqrt{\sum_{\substack{\text{all possible} \\ X \text{ values}}} [(X-\mu)^2 \cdot P(X)]}$$

Step 1 Construct a table containing the following five columns:

X: the values of the random variable,

P(X): the probability values associated with each X value

$(X - \mu)$: the deviation from the mean obtained by subtracting the mean from each possible value of the random variable

$(X-\mu)^2$: square each deviation from the mean

$(X-\mu)^2 P(X)$: multiply each square deviation by the probability associated with the corresponding X value

Step 2 Find the sum of the quantities: $(X-\mu)^2 P(X)$, symbolized as $\sum[(X-\mu)^2 \cdot P(X)]$. This result is the variance, σ^2.

Step 3 Take the square root of the variance to obtain the standard deviation, σ.

EXAMPLE 6.7

The probability distribution given in Table 6.12 represents the number of computer systems a salesman named Hal expects to sell during a particular month.
Find:
a) The most likely number of computer systems that Hal will sell during the month.
b) The average monthly number of computer systems that Hal expects to sell. How would you interpret this result?
c) The standard deviation of this probability distribution.
d) The probability that the number of computer systems that Hal sells will be within one standard deviation from the mean.

Section 6.4 Mean and Standard Deviation of a Discrete Random Variable 317

TABLE 6.12 Probability Distribution

Number of Computer Systems Sold X	Probability P(X)
0	0.10
1	0.25
2	0.30
3	0.20
4	0.10
5	0.05

Solution

a) The most likely number of computer systems that Hal will expect to sell during the month is 2, since it has the largest probability.

b) To find the average monthly number of computers that Hal expects to sell, we need to compute the mean of this probability distribution. The mean is calculated using the formula:

$$\mu = \sum_{\substack{\text{all possible} \\ X \text{ values}}} [X \cdot P(X)]$$

That is, each value of the discrete random variable, denoted by X, is multiplied by its probability, denoted by P(X).

To determine the average number of computer systems that Hal expects to sell, use the procedure to calculate the mean from the information in Table 6.12.

Step 1 Substitute each possible value of the random variable X into the formula for the mean:

$$\mu = \sum_{\substack{\text{all possible} \\ X \text{ values}}} [X \cdot P(X)]$$

We have:

$$\mu = 0P(0) + 1P(1) + 2P(2) + 3P(3) + 4P(4) + 5P(5)$$

Step 2 Substitute the probability value, P(X), corresponding to each X value into the formula.

$$\mu = 0(0.10) + 1(0.25) + 2(0.30) + 3(0.20) + 4(0.10) + 5(0.05)$$

Step 3 Calculate the value of each product.

$$\mu = 0 + 0.25 + 0.60 + 0.40 + 0.40 + 0.25$$

Step 4 Sum up these products to obtain the mean.

$$\mu = 2.1$$

Thus, **the average number of computer systems Hal expects to sell is 2.1.** We would interpret this result to mean that over a large number of months, Hal expects to sell 2.1 computer systems.

c) To calculate the standard deviation for the probability distribution we must calculate the mean, μ, and utilize the procedure to compute the standard deviation.

From part (b), the mean is: $\mu = 2.1$.
Using the procedure, we have:

Step 1 Table 6.13 contains the five columns needed to compute the standard deviation.

Step 2 Find the sum of the quantities: $(X - \mu)^2 P(X)$, symbolized as $\Sigma[(X-\mu)^2 \cdot P(X)]$. This result is the variance, σ^2.

$$\Sigma[(X-\mu)^2 \cdot P(X)] = 0.441 + 0.3025 + 0.003 + 0.162 + 0.361 + 0.4205$$

$$\Sigma(X-\mu)^2 \cdot P(X) = 1.69$$

TABLE 6.13

X	P(X)	X−μ	(X−μ)²	(X−μ)²P(X)
0	0.10	0−2.1 = −2.1	(−2.1)² = 4.41	(4.41)(0.10) = 0.441
1	0.25	1−2.1 = −1.1	(−1.1)² = 1.21	(1.21)(0.25) = 0.3025
2	0.30	2−2.1 = 0.1	(0.1)² = 0.01	(0.01)(0.30) = 0.003
3	0.20	3−2.1 = 0.9	(0.9)² = 0.81	(0.81)(0.20) = 0.162
4	0.10	4−2.1 = 1.9	(1.9)² = 3.61	(3.61)(0.10) = 0.361
5	0.05	5−2.1 = 2.9	(2.9)² = 8.41	(8.41)(0.05) = 0.4205

Thus, the variance is:

$$\sigma^2 = 1.69$$

Step 3 Take the square root of the variance to obtain the standard deviation, σ.

That is, $\sigma = \sqrt{\sum_{\substack{\text{all possible} \\ x \text{ values}}} \left[(X - \mu)^2 \cdot P(X)\right]}$

Thus, $\sigma = \sqrt{1.69}$

or

$$\sigma \approx 1.30$$

Therefore the standard deviation for the probability distribution is approximately 1.30 computer systems.

d) The number of computer systems that are within one standard deviation of the mean are from $\mu - \sigma$ to $\mu + \sigma$. Since the mean is: $\mu = 2.1$, and the standard deviation is: $\sigma \approx 1.30$, then:

$$\mu - \sigma = 2.1 - 1.30 = 0.8$$

and

$$\mu + \sigma = 2.1 + 1.30 = 3.4.$$

Thus, we are interested in the probability that the number of computer systems sold will range from 0.8 to 3.4. The possible values of the random variable X that fall within this range are: X = 1, X = 2, and X = 3. To determine the probability that the number of computer systems will fall within this range of 0.84 to 3.4, we must add the corresponding probability for each of these X values. That is:

probability of the number falling from X = 1 to X = 3 is: P(X = 1) + P(X = 2) + P(X = 3)
$$= 0.25 + 0.30 + 0.20$$
$$= 0.75.$$

Thus, the probability that the number of computer systems that Hal sells will be within one standard deviation from the mean is 0.75.

TI-84 Graphing Calculator Solution for Example 6.7

The TI-84 graphing calculator can also be used to obtain mean and standard deviation of a discrete probability distribution. We will begin by entering the x and p(x) data from Example 6.7 into lists L1 and L2 respectively. You do this by pressing STAT ENTER for **1: EDIT**. Figure 6.9 presents the entered data.

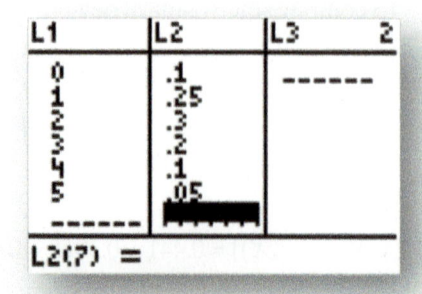

FIGURE 6.9

Section 6.4 Mean and Standard Deviation of a Discrete Random Variable

Obtaining the mean and standard deviation for a discrete probability distribution is similar to how we obtained the mean and standard deviation back in Chapter 3. We will use the **1-VarStats** command in the **STAT** menu. To obtain the statistics, press STAT ▶ over to CALC ENTER for **1-VarStats.** Then, press 2nd 1[L1] and 2nd 2[L2] ENTER. Now press ENTER to calculate the results. Figure 6.10a and Figure 6.10b show the command and results of this command.

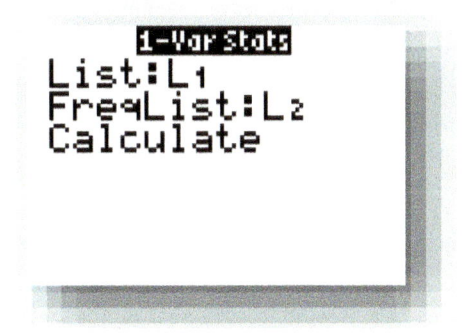

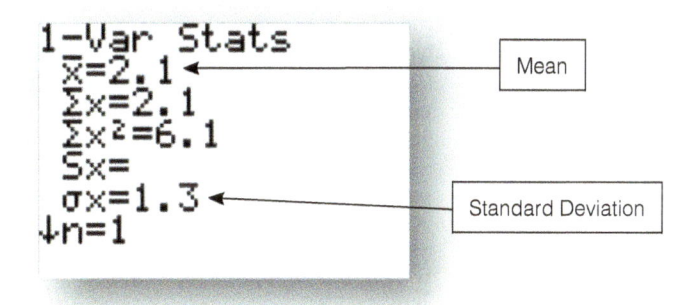

FIGURE 6.10a   **FIGURE 6.10b**

Notice the output indicates the mean is 2.1 and the standard deviation is 1.30.

CASE STUDY 6.2

Examine the graphic entitled "How Long World Series Run" which is shown in Figure 6.11. This graphic represents 111 World Series played from 1905 to 2019 excluding the series played during the years from 1919 to 1921 when they played more than 7 games and 1994 when there was no World Series because of the baseball players strike.

a) Identify the random variable being illustrated in Figure 6.11.
b) Is this a discrete or continuous random variable? Explain your answer.
c) State the possible values of the random variable.
d) State the probability for each possible value. Interpret what each probability indicates about each possible random variable value.
e) Determine the average number of games a World Series will last. Interpret this result.

Statistical Snapshot
HOW LONG WORLD SERIES RUN
Length in Games

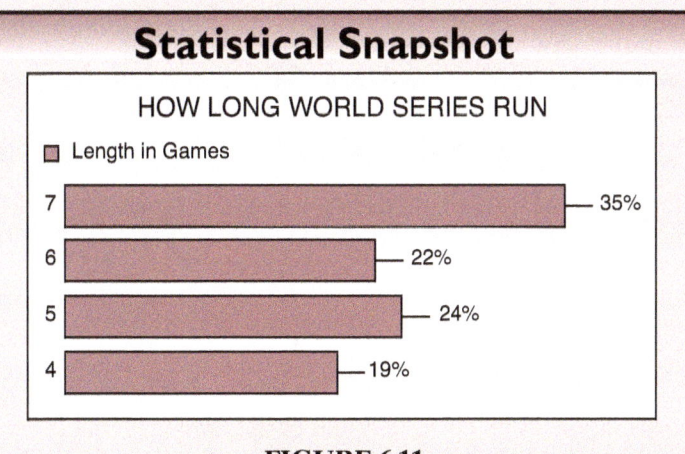

Games	Percent
7	35%
6	22%
5	24%
4	19%

FIGURE 6.11

CASE STUDY 6.2 (continued)

f) Which value of the random variable had the greatest impact on this average? Which value of the random variable has the greatest probability? Compare these answers. Did you expect these results? Using these results, explain what is meant by the statement that "*the mean is a weighted average.*" What quantities represent the "weights"? Explain.
g) Determine the standard deviation of this random variable. Interpret this result.
h) Construct a probability histogram for the probability distribution of this random variable.
i) Examine the probability histogram's spread and compare this to the value of the standard deviation. What can you indicate about the relationship between the histogram's spread and the standard deviation? Explain your answer.

EXAMPLE 6.8

Examine the two probability histograms pictured within Figures 6.12a and 6.12b.

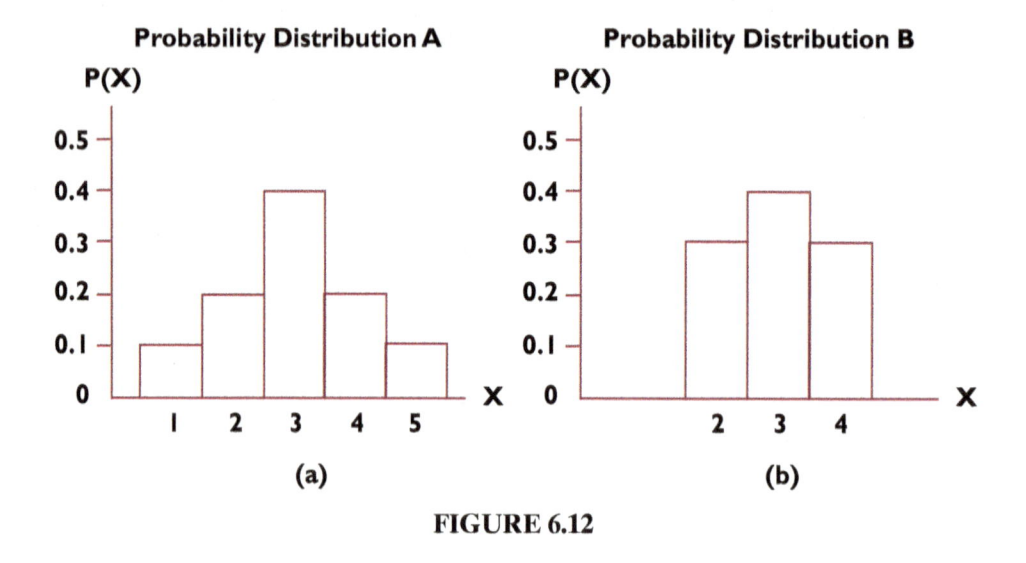

FIGURE 6.12

a) Indicate the possible values of the random variable X for each probability distribution and the corresponding probability.
b) By examining the histograms in Figure 6.12, what is your estimate of the mean of each probability distribution?
c) Within which distribution are the values of the random variable more dispersed or scattered from the mean?
d) One of the probability distributions has a standard deviation of 0.77, while the other probability distribution has a standard deviation of 1.34. Which of the probability distributions shown in Figure 6.12 has a standard deviation of 1.34? Explain your answer.

Solution

a) For Probability Distribution A, the possible values of the random variable are: X = 1, 2, 3, 4, or 5. Their corresponding probabilities are: P(1) = 0.1, P(2) = 0.2, P(3) = 0.4, P(4) = 0.2, and P(5) = 0.1. For Probability Distribution B, the possible values of the random variable are: X = 2, 3, or 4. Their corresponding probabilities are: P(2) = 0.3, P(3) = 0.4, and P(4) = 0.3.
b) Since the mean is the balance point for each distribution, and the distributions are symmetric, then the mean for each probability distribution is 3.
c) Probability Distribution A has values which are more dispersed from the mean of 3.
d) Since Probability Distribution A has values which are more dispersed from the mean of 3, this distribution will have the greater variability. Therefore, Probability Distribution A has the larger standard deviation, which is 1.34. Thus, Probability Distribution B has the smaller standard deviation, which is 0.77.

CASE STUDY 6.3

According to the registrar records of Nassau Community College on Long Island, Table 6.14 represents the number of semesters in which 21,541 students have attended the college during the Fall semester of 1990.

TABLE 6.14   Student Duration of Attendance

Number of Semesters X	Day	Evening	Total
1	5,704	2,059	7,763
2	1,201	843	2,044
3	3,661	969	4,630
4	900	582	1,482
5	1,741	720	2,461
6	438	437	875
7	416	439	855
8	157	269	426
9	114	232	346
10 or more	177	482	659
Total	14,509	7,032	21,541

Replace the last row of the table labelled 10 or more semesters with just 10 semesters, and let X be the number of semesters attending the college. If we imagine selecting one day student at random, then:
  a) construct a probability distribution for X for day students, for evening students and for all students.
  b) What is the probability of selecting a day student who has been in attendance for more than 4 semesters?
  c) What is the probability of selecting an evening student who has been in attendance for less than 5 semesters?
  d) What is the probability of selecting a student at random, day or evening, who has been in attendance for more than 4 semesters?
  e) Construct a probability histogram for X for day, evening and all students.
  f) Find the mean number of semesters attended for day, evening and all students.
  g) Find the standard deviation of the number of semesters attended for day, evening and all students.

6.5   BINOMIAL PROBABILITY DISTRIBUTION

In this section, we will examine a discrete probability distribution called the **Binomial Probability Distribution** or **Binomial Distribution.** One of the characteristics of a binomial distribution is that it involves probability experiments where each outcome can be classified in *one of two possibilities*. For example:

- the toss of a coin results in either *a head or a tail.*
- the selection of a possible answer for a question on a multiple choice test is either *correct or incorrect.*
- the toss of a die results in an outcome which is either *a 5 or not a 5.*
- the selection of a single card from a deck of cards results in either *an ace or not an ace.*
- a new drug will either be *effective or not effective* when administered to a single patient.
- in an ESP experiment using five objects, the subject will either *guess the object or not guess the object correctly.*
- in a quality control procedure, an item tested will either be *defective or non-defective.*

An outcome, in each of these experiments, can be classified in only one of the two defined possibilities. Thus, each outcome is mutually exclusive, that is, both outcomes cannot happen at the same time. For example, when a coin is tossed it will land either heads or tails, but it cannot land both heads and tails at the same time. Also, a tested item can be either defective or not defective, but not both.

Since the outcomes of a binomial distribution can only take on one of two possible outcomes, we usually refer to these two outcomes as either a success or a failure. For example, in the coin tossing experiment, we might classify the *outcome heads as a success*. Thus, the *outcome tails will then be considered a failure*. In the quality control experiment, *a defective item might be classified as a success, while a non-defective item would be considered a failure*. Keep in mind that the use of the words **success** and **failure** are probability terms and should not be confused with your everyday meaning of these words. Consequently, our decision on what outcomes to define a success will be dependent on the probability question. That is, the selection of which one of the two possibilities to identify as a success or a failure is usually dependent upon the question being asked. The term *success is usually defined to be the outcome which is being referred to within the question*, while the word *failure refers to the outcome that is not referred to within the question*. For example, if a binomial probability question were stated as: "What is the probability of guessing 4 *wrong answers* on a 10 question multiple choice test?", then *success would be defined as a wrong answer*, while *a failure is defined to be a correct answer*. Notice, we are identifying the outcome **wrong answer** as a **success** since this is the outcome to which the question refers, while the outcome of a *correct answer* which is not referred within the question is identified as a failure. Thus, the word success is simply a term used to refer to one of the two possibilities of a binomial experiment, and should not be confused with the outcome(s) that you may find to be desirable.

A second characteristic of a binomial distribution is that it is the result of a probability experiment where *the experiment has been repeated a predetermined number of times and each repetition of the experiment is identical*. That is, it is a probability experiment having n **identical trials**, where each repetition of the experiment is referred to as a **trial**. For example, if an experiment is defined as rolling a fair die 20 times, then this experiment consists of repeatedly rolling the same die 20 times. Since the same identical die is being rolled each time, then each repetition of the experiment is said to be identical. The number of times the die will be rolled was predetermined or known prior to performing the actual experiment. Therefore, an experiment consisting of rolling a fair die 20 times is said to consist of 20 trials, since each roll is considered a trial. In addition, since the outcomes of a binomial distribution are values which were arrived at by **counting**, the *binomial probability distribution is a discrete probability distribution*. For example, the number of times a coin lands heads in 10 tosses, or the number of defective items within a batch of 50 items represent *counts,* since the results are determined by *counting* the number of heads in 10 tosses or *counting* the number of defective items within a batch of 50 items.

A third characteristic of a binomial distribution is that *the n identical trials are* **independent.** For example, in the experiment of tossing a coin 10 times, each toss is independent of all the other tosses, since regardless of what happens on any one toss, this result does not affect the outcome of any other toss.

Finally, a fourth characteristic of a binomial distribution is that *the probability of a success must remain the same for each trial.* For example, in the coin tossing experiment, the chance of a fair coin landing heads will be the same for each toss or trial. Thus, the probability of the coin landing heads on the first toss (or first trial) is ½, while the probability of the coin landing heads on the second toss (or second trial) is also ½, and this is true for the third toss (or third trial) and so on.

An experiment that has these four characteristics is classified as a binomial experiment.

DEFINITION 6.7

A **binomial experiment** satisfies the following four conditions.
1) There are *n identical trials*.
2) The n identical trials are *independent*.
3) The outcome for each trial can be classified as either a *success* or a *failure*.
4) The *probability of a success* is the *same* for *each trial*.

Let's now consider an example to explain how to identify whether a probability experiment is binomial.

EXAMPLE 6.9

Consider the experiment of tossing a fair coin five times, where we are interested in getting a head. Can this experiment be classified as a binomial experiment?

Solution

Let's determine if the four conditions for a binomial experiment are satisfied.

1) Are there n identical trials? Yes, since the same coin is being tossed 5 times and each toss is considered a trial, thus there are 5 identical trials.

2) Are the 5 identical trials independent? Yes, since the outcome for the first toss (or trial) does not influence the outcome for the second toss (or trial), or any other toss. In fact, the outcome of any toss has no influence on the outcome of any other toss. Thus, the trials are independent.
3) Can the outcome of each trial be classified as either a success or a failure? Yes, since there are only two possible outcomes heads and tails, we can define the outcome of a head as a success, and the outcome of a tail as a failure.
4) Is the probability of a success the same for each trial? Yes, for each toss (or trial) of the fair coin the probability of a success equals the probability of a head which is always ½.

Therefore, the experiment of tossing a fair coin five times is a binomial experiment.

EXAMPLE 6.10

Consider spinning the spinner shown in Figure 6.13 four times and recording the four outcomes.

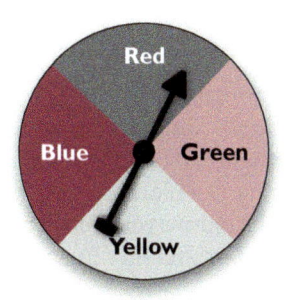

FIGURE 6.13

Can this experiment be considered a binomial, if we are interested in the outcome red?

Solution

Let's check the four conditions of a binomial experiment.

1) Are there n identical trials? Yes, since the same spinner is being spun 4 times and each spin is considered a trial, then the number of identical trials is 4.
2) Are the 4 identical trials independent? Since the outcome of any spin has no influence on the outcome of the any other spin, then the 4 spins are independent of each other.
3) Can the outcome of each trial be classified as either a success or a failure? Since we are interested in the outcome red, then we can define a success to be obtaining a red, while all other outcomes: green, blue and yellow are considered to be a failure.
4) Is the probability of a success the same for each trial? The probability of a success (or a red) for each spin is ¼ since there are four equally-likely outcomes for each spin.

EXAMPLE 6.11

Medical records indicate that 40% of the people living in New York have type O+ blood. Seven people donating blood at a New York hospital are randomly selected to determine their blood type. Is this experiment a binomial experiment?

Solution

We need to determine if this experiment satisfies the four conditions of a binomial experiment. These conditions are:

1) Are there n identical trials? This experiment consists of randomly selecting 7 people and identifying the blood type of each person. Since each person selected is considered a trial, then there are 7 identical trials.
2) Are the 7 identical trials independent? Each trial or selected person is independent since whether or not one person has a O+ type blood has no affect on any other person having O+ blood. We can make this assumption since our population is very large compared to a sample of 7 people.

3) Can the outcome of each trial be classified as either a success or a failure? Since each person selected either will have O+ blood or not have O+ blood, then we can separate each of these possibilities into a success or a failure. That is, we can define a success to be a person with O+ blood and a failure to be a person not having O+ blood.
4) Is the probability of a success the same for each trial? If we define a success to be a person with O+ blood and knowing that 40% of the people have O+ blood type, then the probability of selecting a person with O+ blood type is 0.40. Thus, the probability of a success (or a person with O+ blood) is the same for each person selected.

Binomial Probability Formula

Now that we've illustrated the four conditions of a binomial experiment, let's examine how to calculate the probabilities associated with each outcome of a binomial experiment. For example, the binomial experiment of selecting a person with O+ blood type from a sample of 7 people has 8 possible values of the random variable, X, representing the number of people with O+ blood type. They are:

$$X = 0 \text{ people}, X = 1 \text{ person}, X = 2 \text{ people}, X = 3 \text{ people},$$
$$X = 4 \text{ people}, X = 5 \text{ people}, X = 6 \text{ people and } X = 7 \text{ people}.$$

This random variable which represents the number of people with O+ blood type of a binomial experiment is called a *binomial random variable*, and the probability distribution of the binomial random variable is called the **binomial probability distribution** or simply the **binomial distribution**. To construct a binomial distribution for the number of people having O+ blood type, it becomes necessary to determine the probability for each possible value of this binomial random variable. To determine each of these probabilities, we can use a formula referred to as the **Binomial Probability Formula**.

Binomial Probability Formula

For a binomial experiment, the probability of getting **s** successes in **n** trials is computed using the binomial probability formula. This formula is written:

P(s successes in n trials) = $_nC_s \, p^s \cdot q^{(n-s)}$
where:
 n = number of independent trials
 s = number of successes
 (n–s) = number of failures
 $_nC_s$ = the number of ways "s" successes can occur in "n" trials
 p = probability of a success for one trial
 q = probability of a failure for one trial = 1–p

A discussion of $_nC_s$ can be found in Chapter 5 Section 3. Most statistical calculators have the ability to evaluate $_nC_s$. On page 248, the keystrokes to evaluate $_nC_s$ were illustrated for the TI-84 graphing calculator. Determine the procedure to evaluate $_nC_s$ on your calculator. Using your calculator, answer the following two examples.

To evaluate $_nC_s$, we must determine the number of ways **s** successes can occur in **n** trials. The following two examples illustrate this calculation.

EXAMPLE 6.12

For the binomial experiment of tossing a fair coin three times, where a success is getting a head on a single toss of the coin, determine the number of ways to get:
a) Three successes (or heads) in three tosses of the coin, written as $_3C_3$
b) Two successes (or heads) in three tosses of the coin, written as $_3C_2$
c) One success (or head) in three tosses of the coin written as $_3C_1$
d) No successes (or heads) in three tosses of the coin, written as $_3C_0$

Solution

a) There is only one way to get three successes (or 3 heads) in three tosses of a coin. To obtain 3 successes, you need to get a success (or a head) on the first toss, a success (or a head) on the second toss and a success (or a head) on the third toss. This occurrence of three successes (or 3 heads) is symbolized as: sss.
   Therefore, $_3C_3$, which represents the number of ways of getting 3 successes (or 3 heads) in 3 tosses of a fair coin, is 1.

b) There are three ways to get two successes (or 2 heads) in three tosses of a coin. They are: ssf, sfs, and fss.
   Therefore, $_3C_2$, which represents the number of ways of getting 2 successes (or 2 heads) in 3 tosses of a fair coin, is equal to 3.

c) There are three ways to get one success (or 1 head) in three tosses of a coin. They are: sff, fsf, and ffs.
   Therefore, $_3C_1$, which represents the number of ways of getting 1 success (or 1 head) in 3 tosses of a fair coin, is 3.

d) There is only one way to get no successes (or 0 heads) in three tosses of a coin. It is: fff.
   Therefore, $_3C_0$, which represents the number of ways of getting no successes (or 0 heads) in 3 tosses of a fair coin, is 1.

EXAMPLE 6.13

Consider the binomial experiment of rolling a fair die four times where a success is rolling a 1. Determine:
a) The number of ways to get 4 successes in four rolls of the die.
b) $_4C_2$
c) $_4C_0$

Solution

a) There is only one way to get 4 successes in four rolls or trials of the die. It is: ssss. Therefore $_4C_4 = 1$.
b) $_4C_2$ means the number of ways to get 2 successes in 4 rolls or trials which is 6, since we can get 2 successes in 4 trials six different ways. They are: ssff, sfsf, sffs, fsfs, ffss, and fssf.
c) $_4C_0$ means the number of ways to get no successes in 4 rolls or trials which is 1. Since there is only one way to get no successes in 4 trials. It is: ffff.

In general the number of ways **s** successes can occur in **n** independent trials, written as $_nC_s$, can be determined by the formula: $_nC_s = \dfrac{n!}{s!(n-s)!}$

EXAMPLE 6.14

Use the binomial probability formula:

$$P(s \text{ successes in } n \text{ trials}) = {}_nC_s\, p^s\, q^{(n-s)}$$

to determine the probability of getting 3 successes in 4 trials if p = ⅔.

Solution

1) n = 4 (number of trials)
2) s = 3 (number of successes)
3) n − s = 4 − 3 = 1
4) $_nC_s = {}_4C_3 = 4$
5) p = ⅔
6) q = 1 − p = 1 − ⅔ = ⅓

Substituting this information into the binomial probability formula, we get:

$$P(3 \text{ successes in 4 trials}) = {}_4C_3 \left(\frac{2}{3}\right)^3 \left(\frac{1}{3}\right)^1$$

$$= 4\left(\frac{8}{27}\right)\left(\frac{1}{3}\right)$$

$$= \frac{32}{81} \approx 0.3951$$

326   Chapter 6   Random Variables and Discrete Probability Distributions

EXAMPLE 6.15

Each year the FBI reports the probability of a car being stolen. In a recent report, the FBI states that the probability a new car will be stolen during the year is 1 out of 75. If you and your three friends own new cars, what is the probability that **none** of these cars will be stolen this year?

Solution

We will use the binomial probability formula since the four conditions of a binomial experiment are true. Thus, we assume that:
1. the number of cars represent the number of identical trials.
2. whether your car is stolen or any one of your friends' car is stolen are independent of each other.
3. the possible outcomes can be classified in one of two possibilities: either the car is stolen or the car is not stolen.
4. the probability of a new car being stolen is the same for each automobile.

Since the probability question pertains to the chance that a car will be stolen, then we will define a **success to be a car is stolen**. To use the binomial probability formula to compute the probability that none of these cars are stolen, we need to determine the quantities n, s, p and q. Thus,

n = 4, since the total number of cars is four;
s = 0, because we are interested in zero cars being stolen;
$_nC_s = {_4C_0} = 1$
p = $1/75$, since this is the probability of a car being stolen;
q = $1 - p = 1 - 1/75 = 74/75$

Substituting this information into the binomial probability formula, we have:

$$P(\text{no cars being stolen}) = {_4C_0}\left(\frac{1}{75}\right)^0 \left(\frac{74}{75}\right)^4$$

$$\approx 0.9477$$

Therefore, the probability that none of the cars will be stolen is approximately 0.95.

EXAMPLE 6.16

A student is going to guess at the answers to all the questions on a five question multiple choice test where there are four choices for each question. Calculate the probability of:
a) Guessing three correct answers.
b) Guessing five correct answers.
c) Guessing at most two correct answers.
d) Guessing at least four correct answers.

Solution

a) Since the question pertains to a correct answer, we will define **success to be guessing the correct answer** to a question. Using this definition of success, and considering a **trial to be a question**, then we use the following values to determine the probability of 3 correct:
1. $n = 5$ (since there are 5 questions)
2. $s = 3$ (we are interested in 3 correct answers)
3. $n - s = 5 - 3 = 2$
4. $_nC_s = {_5C_3} = 10$
5. $p = 1/4$ (since the probability of a correct answer is one chance out of four possible choices)
6. $q = 1 - p = 1 - 1/4 = 3/4$

Substituting these values into the binomial probability formula, we have:

$$P(3 \text{ correct answers}) = {_5C_3}\left(\frac{1}{4}\right)^3 \left(\frac{3}{4}\right)^2$$

$$= 10 \cdot \left(\frac{1}{64}\right) \cdot \left(\frac{9}{16}\right)$$

$$= \frac{90}{1024} \approx 0.0879$$

b) To find the probability of guessing five correct answers, we will define success to be guessing the correct answer to a question, and use the following values in the binomial probability formula:
   1. n = 5 (because there are five questions)
   2. s = 5 (since we are interested in 5 correct answers)
   3. n − s = 5 − 5 = 0
   4. $_nC_s = {}_5C_5 = 1$
   5. p = ¼ (since the probability of getting a correct answer for one question is one chance out of four choices)
   6. q = 1 − ¼ = ¾ (using 1 − p)
   7. Substituting these values into the binomial probability formula, and remembering that **any non-zero number raised to a power of zero is one**, we get:

$$P(5 \text{ correct answers}) = {}_5C_5 \left(\frac{1}{4}\right)^5 \left(\frac{3}{4}\right)^0$$

$$= \frac{1}{1024} \approx 0.0010$$

c) Again success is defined to be guessing the correct answer to a question. To determine the probability of guessing **at most two correct answers**, you must interpret this compound statement **to mean guessing two or less correct answers**. Thus, to calculate this probability, we need to determine and add the probabilities: *the probability of getting two correct answers, the probability of getting one correct answer* and *the probability of zero correct answers*. That is:

P(at most 2 correct answers) = P(2 correct) + P(1 correct) + P(0 correct)

To determine the probability of two correct, we use the following values:
1. n = 5 (number of questions)
2. s = 2 (success is two correct answers)
3. n − s = 5 − 2 = 3
4. $_nC_s = {}_5C_2 = 10$
5. p = ¼ (probability of a correct answer)
6. q = 1 − ¼ = ¾ (using 1 − p)

Substituting these values into the binomial probability formula, we have:

$$P(2 \text{ correct answers}) = {}_5C_2 \left(\frac{1}{4}\right)^2 \left(\frac{3}{4}\right)^3$$

$$= \frac{270}{1024}$$

To determine the probability of one correct, use n, p and q from the previous part. However,

$$s = 1 \text{ (success is one correct answer)}$$

$$n - s = 5 - 1 = 4$$

and

$$_nC_s = {}_5C_1 = 5$$

Substituting these values into the binomial probability formula, we have:

$$P(1 \text{ correct}) = {}_5C_1 \left(\frac{1}{4}\right)^1 \left(\frac{3}{4}\right)^4$$

$$= \frac{405}{1024}$$

To calculate the probability of zero correct,

$$s = 0 \text{ (zero correct answers)}$$

$$n - s = 5 - 0 = 5$$

and
$$_nC_s = {}_5C_0 = 1$$

Substituting these values into the binomial probability formula, and using the fact that **any non-zero number raised to a power of zero is one**, we have:

$$P(\text{zero correct}) = {}_5C_0 \left(\frac{1}{4}\right)^0 \left(\frac{3}{4}\right)^5$$

$$= \frac{243}{1024}$$

Therefore, to calculate the probability of at most two correct, we add these three probabilities together. Thus, we have:

$$P(\text{at most 2 correct answers}) = P(2 \text{ correct}) + P(1 \text{ correct}) + P(0 \text{ correct})$$

$$= \frac{270}{1024} + \frac{405}{1024} + \frac{243}{1024}$$

$$= \frac{918}{1024} \approx 0.8965$$

d) Again success is defined to be guessing the correct answer to a question, since this is the outcome to which the question refers.

To determine the probability of guessing **at least four correct answers**, you must interpret this compound statement: **at least four correct answers. At least four correct means guessing four or more correct answers.**

Since there are only five questions, the most correct answers that one can guess is five. Thus, to calculate this probability, we need to add the two probabilities: the *probability of getting four correct answers* and the *probability of five correct answers*. That is,

$$P(\text{at least 4 correct answers}) = P(4 \text{ correct}) + P(5 \text{ correct})$$

To determine the probability of four correct,

$$s = 4 \text{ (four correct answers)}$$

$$n - s = 5 - 4 = 1$$

and

$$_nC_s = {}_5C_4 = 5$$

Substituting these values into the binomial probability formula, we have:

$$P(4 \text{ correct}) = {}_5C_4 \left(\frac{1}{4}\right)^4 \left(\frac{3}{4}\right)^1$$

$$= \frac{15}{1024}$$

To calculate the probability of five correct,

$$s = 5 \text{ (five correct answers)}$$

$$n - s = 5 - 5 = 0$$

and

$$_nC_s = {}_5C_5 = 1$$

Substituting these values into the binomial probability formula, and using the fact that **any non-zero number raised to a power of zero is one**, we have:

$$P(5 \text{ correct}) = {}_5C_5 \left(\frac{1}{4}\right)^5 \left(\frac{3}{4}\right)^0$$

$$= \frac{1}{1024}$$

Therefore, to calculate the probability of at least four correct, we add these two probabilities together. Thus, we have:

$$P(\text{at least 4 correct}) = P(\text{4 correct}) + P(\text{5 correct})$$

$$= \frac{15}{1024} + \frac{1}{1024}$$

$$= \frac{16}{1024} \approx 0.0156$$

TI-84 Graphing Calculator Solution for Example 6.16

The TI-84 graphing calculator has two built-in binomial probability functions found in the distribution menu. To access the distribution menu press [2nd] [VARS] [DISTR]. On the TI-84, the binomial probability functions are in options A and B as shown in Figure 6.14.

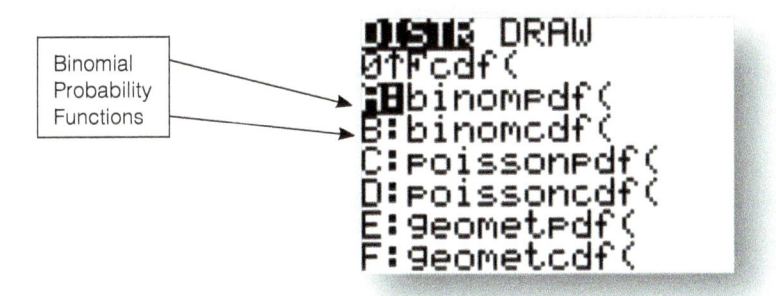

FIGURE 6.14

These functions have a syntax that must be followed in order for them to work correctly. The function **binompdf** has the syntax **binompdf(n, p, r)** and will return the probability of exactly r successes in n trials of an event. The values for the variables n, p and r must be entered in that order for the function to give you the correct answer.

Example 6.16a could have been worked as follows:

$$P(r \text{ successes in n trials}) = \text{binompdf}(n, p, r)$$

$$P(3 \text{ successes in 5 trials}) = \text{binompdf}(5, 1/4, 3)$$

Press [2nd] [VARS] [DISTR] and scroll down to option **A:binompdf(** and press [ENTER]. Now enter the number of trials, the probability of a success, and number of successes in n trials as shown in Figure 6.15a. Press [ENTER] twice to paste and calculate the answer. This probability is shown in Figure 6.15b. The probability of guessing 3 correct answers in 5 questions, rounded to 4 decimal places, is 0.0879.

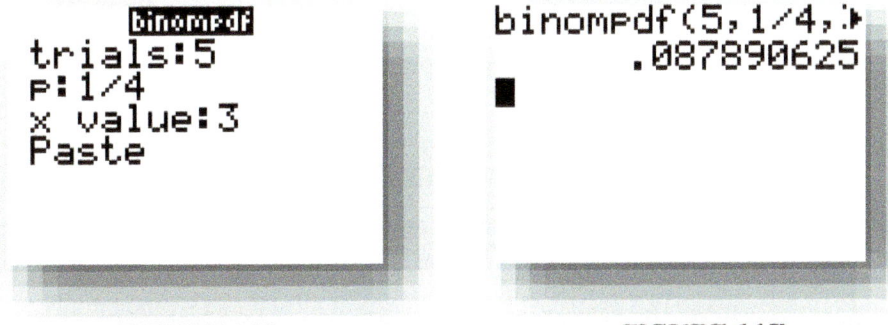

FIGURE 6.15a   **FIGURE 6.15b**

Example 6.16b could have been worked in a similar fashion as shown below:

$$P(5 \text{ successes in 5 trials}) = \text{binompdf}(5, 1/4, 5)$$

Figure 6.16a shows the calculator input and Figure 6.16b shows the probability.

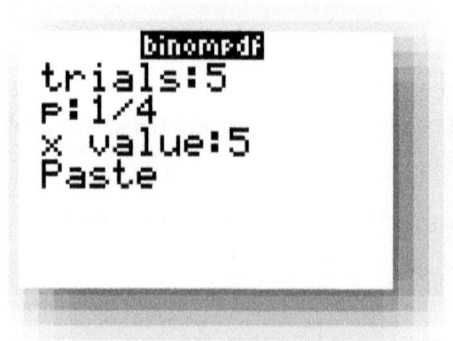

FIGURE 6.16a   **FIGURE 6.16b**

The probability of guessing 5 correct answers in 5 questions is 9.765625E-4, which is scientific notation for 0.0009765625 or 0.0010 to 4 decimal places.

The second binomial function in the TI-84 is used when you need to find the probability of several events. This is called the cumulative probability distribution. This option used for cumulative probabilities is **binomcdf** and its syntax is **binomcdf(n,p,r)**. It returns the probability of *at most* r successes in n trials.

Example 6.16c asked for the probability of at most 2 correct answers in 5 questions. When we found the answer in 6.16c using the formula for the binomial distribution, we used the fact that:

$$P(X \text{ is at most } 2) = P(X \leq 2) = P(X=0) + P(X=1) + P(X=2).$$

We then found each of the three probabilities and then found their sum. Since **binomcdf(n,p,r)** returns the probability of *at most* r successes in n trials of some event, we can just use this function to give us the answer to 6.16c directly. That is,

$$P(X \text{ is at most } 2) = P(X \leq 2) = \text{binomcdf}(5, 1/4, 2)$$

Press [2nd] [VARS] [DISTR] and scroll down to option **B:binomcdf(** and press [ENTER]. Now enter the number of trials, the probability of a success, and maximum number of successes in n trials as shown in Figure 6.17a. Press [ENTER] twice to paste and calculate the answer. This probability is shown in Figure 6.17b.

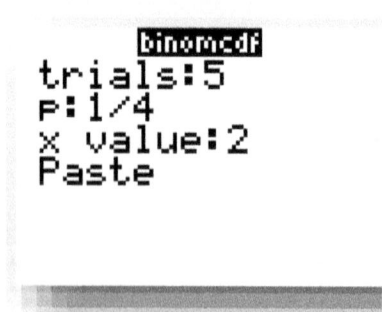

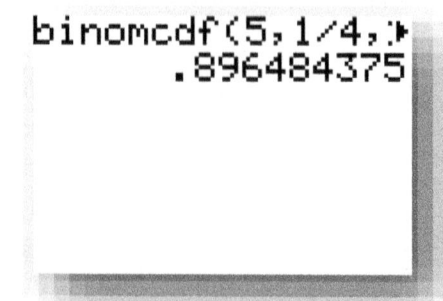

FIGURE 6.17a   **FIGURE 6.17b**

Notice the result when rounded to 4 decimal places is 0.8965.

For Example 6.16d, we can use the fact that:

$$P(X \text{ is at least } 4) = P(X=4) + P(X=5) = 1 - P(X \text{ is at most } 3).$$

Since we know the function **binomcdf** will give us the result to the **at most** probability in the previous statement, this probability can be worked as follows:

$$P(X \text{ is at least } 4) = 1 - P(X \text{ is at most } 3) = 1 - \text{binomcdf}(5, \tfrac{1}{4}, 3).$$

This probability is shown in Figure 6.18.

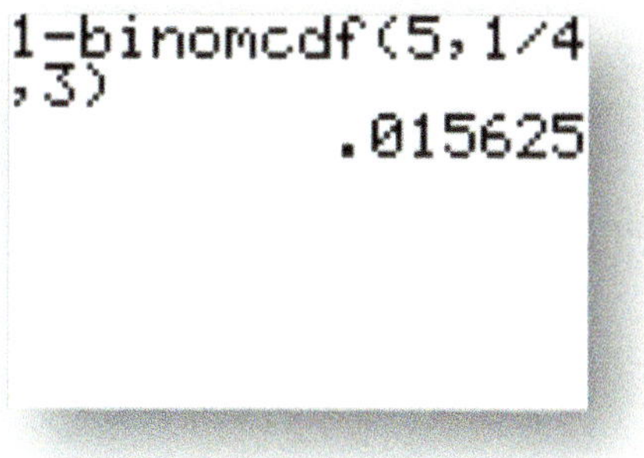

```
1-binomcdf(5,1/4
,3)
            .015625
```

FIGURE 6.18

The result to 4 decimal places is 0.0156, the same as we obtained using the formula for the binomial distribution in Example 6.16d. Notice we could have worked the probabilities P(X=4) and P(X=5) directly using the **binompdf** operation and obtained the same result. This probability is shown in Figure 6.19.

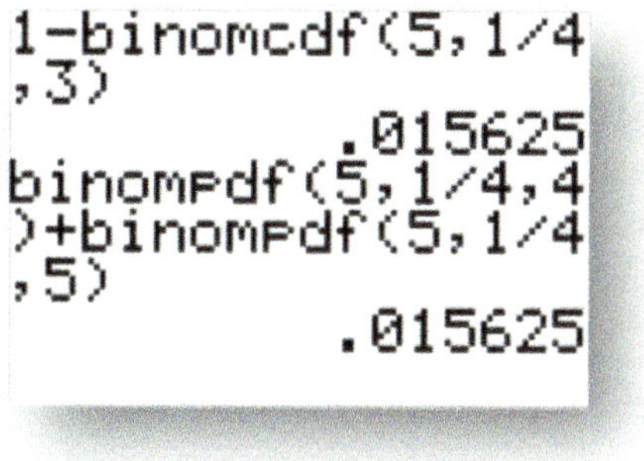

```
1-binomcdf(5,1/4
,3)
            .015625
binompdf(5,1/4,4
)+binompdf(5,1/4
,5)
            .015625
```

FIGURE 6.19

As you can see, the two answers are the same.

EXAMPLE 6.17

After examining her college attendance records for the past few years, Professor Bea Earlee has determined that there is a 30% chance of any one student coming late to her class. Six students are randomly selected from her class rosters. Assuming her students arrive to class independently of one another, what is the probability that:
a) None of the six students arrive to class late?
b) Exactly four students arrive to class late?
c) At most one student doesn't arrive to class late?

Solution

a) To determine the probability that none of the six students arrive to class late, we need to define success and determine the appropriate values for the binomial probability formula. Success is defined as a student is late to class, since the question refers to coming to class late. The appropriate values for the binomial probability formula are:
1. $n = 6$ (since there are 6 students)
2. $s = 0$ (the number of students the question refers to as arriving late to class)
3. $n - s = 6 - 0 = 6$
4. $_nC_s = {}_6C_0 = 1$
5. $p = 0.30$ (since there is a 30% or 0.30 chance that a student arrives to class late)
6. $q = 1 - 0.30 = 0.70$ (using $1 - p$)

Using the binomial probability formula, and the previous information we get:
$$P(\text{none of the 6 students arrive to class late}) = {}_6C_0 \, (0.30)^0 \, (0.70)^6$$
$$\approx 0.12$$

b) To determine the probability that exactly four of the students arrive to class late, we need to define success and determine the appropriate values for the binomial probability formula. Success is still defined as a student arriving late to class, since the question still refers to coming to class late. The appropriate values for the binomial probability formula are:

$$n = 6$$
$$s = 4$$
$$n - s = 6 - 4 = 2$$
$${}_nC_s = {}_6C_4 = 15$$
$$p = 0.30$$
$$q = 1 - 0.30 = 0.70$$

Using the binomial probability formula, and the previous information we get:
$$P(\text{four of the 6 students arrive to class late}) = {}_6C_4 \, (0.30)^4 \, (0.70)^2$$
$$\approx 0.06$$

c) To determine the probability that at most one of the six students *doesn't arrive to class late*, we need to define success and determine the appropriate values for the binomial probability formula. Success is defined as no student arrives late to class, since the question refers to **not** arriving to class late. To determine the number of successes you must interpret the compound statement: **at most one student isn't late to class. At most one student isn't late means one or no students aren't late to class.** Thus, to calculate this probability, we need to add the two probabilities: the *probability of no student isn't late to class* and the *probability of one student isn't late to class.* That is,

$$P(\text{at most 1 student isn't late}) = P(\text{no student isn't late}) + P(\text{1 student isn't late})$$

To determine the probability of no student is late to class, we use the following values in the binomial probability:
1. n = 6 (since there are 6 students)
2. s = 0 (the number of students the question refers to **not** arriving late to class)
3. n − s = 6 − 0 = 6
4. ${}_nC_s = {}_6C_0 = 1$
5. p = 0.70 (since there is a 30% or 0.30 chance that a student *arrives to class late*, then there is a 70% or 0.70 chance that a student *doesn't arrive late to class*)
6. q = 1 − 0.70 = 0.30 (using 1 − p)

Using the binomial probability formula, and the previous information we get:
$$P(\text{no student isn't late}) = {}_6C_0 \, (0.70)^0 \, (0.30)^6$$
$$\approx 0$$

To determine the probability of one student isn't late to class, we use the following values in the binomial probability:

$$n = 6$$
$$s = 1$$
$$n - s = 6 - 1 = 5$$
$${}_nC_s = {}_6C_1 = 6$$
$$p = 0.70$$
$$q = 1 - 0.70 = 0.30$$

Using the binomial probability formula, and the previous information we get:
$$P(\text{one student isn't late}) = {}_6C_1 \, (0.70)^1 \, (0.30)^5$$
$$\approx 0.01$$

Therefore, to calculate the probability of at most one student isn't late to class, we add these two probabilities together. Thus, we have:

P(**at most 1 student isn't late**) = P(*no student isn't late*) + P(**1 student isn't late**)
$$\approx 0 + 0.01$$
$$\approx 0.01$$

In Example 6.16, we examined the binomial experiment of guessing the answers to five multiple choice questions where each question has four choices. For this experiment, if we define the number of correct answers as the discrete random variable X, then there are six possible values. They are:

X = 0 correct, X = 1 correct, X = 2 correct,
X = 3 correct, X = 4 correct, X = 5 correct.

The probability associated for each of these possible values can be computed using the binomial probability formula. These probabilities were computed in Example 6.16 and are listed in Table 6.15 along with the associated values of the random variable X.

Table 6.15 represents a **binomial probability distribution** for the number of correct answers on a five question multiple choice test. Recall that *a binomial probability distribution is a discrete probability distribution.* Remember, one of the characteristics of a discrete probability distribution is that the *sum of all the probabilities is equal to one.* Let's verify that this is true for the binomial distribution in Table 6.15. The sum of all the probabilities for the binomial probability distribution for the number of correct answers is:

sum = 243/1024 + 405/1024 + 270/1024 + 90/1024 + 15/1024 + 1/1024
sum = 1.

The binomial probability distribution can also be displayed using a probability histogram as illustrated in Figure 6.20.

TABLE 6.15 Binomial Probability Distribution for the Number of Correct Answers for a 5 Question Test

Number of Correct Answers X	Probability of the Number of Correct Answers P(X)
0	243/1024
1	405/1024
2	270/1024
3	90/1024
4	15/1024
5	1/1024

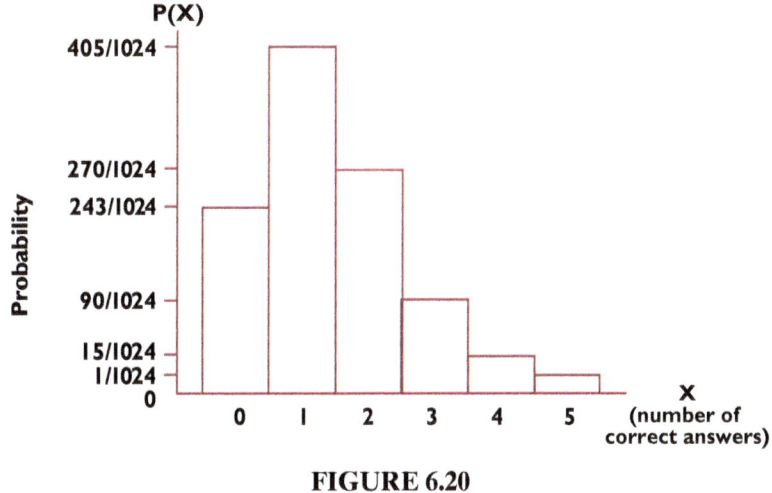

**Probability Histogram For
The Number of Correct Answers For a 5 Question Test**

FIGURE 6.20

334 *Chapter 6* *Random Variables and Discrete Probability Distributions*

Notice that you can answer any questions pertaining to this binomial probability distribution using the probability histogram in Figure 6.20. In Example 6.16 we asked the following questions:
calculate the probability of:
a) Guessing three correct answers.
b) Guessing five correct answers.
c) Guessing at most two correct answers.
d) Guessing at least four correct answers.

The answers to these questions can be easily determined using this probability histogram. For example, *the probability of guessing three correct answers is equal to the height of the bar pertaining to the value of $X = 3$ which is 90/1024. While the probability of guessing at most two correct answers is equal to the sum of the height of the bars pertaining to the values of $X = 0$, $X = 1$ and $X = 2$. That is, the probability of at most two correct answers equals* $243/1024 + 405/1024 + 270/1024 = 918/1024$. By examining the probability histogram of Figure 6.20 and using the idea of the mean as the balance point, what would be your estimate of the mean number of correct answers? To determine the actual value of the mean, we can use the formula for the mean of a discrete random variable. This formula is:

$$\mu = \sum_{\substack{all\ possible \\ X\ values}} [X \cdot P(X)]$$

Using the procedure to calculate the mean as outlined in Section 6.3 and the information in Table 6.15, we have:

First, substitute each possible value of the random variable X into the formula for the mean:

$$\mu = \sum_{\substack{all\ possible \\ X\ values}} [X \cdot P(X)]$$

Since the possible X values are: 0, 1, 2, 3, 4, and 5 then we have:

$$\mu = 0P(0) + 1P(1) + 2P(2) + 3P(3) + 4P(4) + 5P(5)$$

Second, substitute the probability value, P(X), corresponding to each X value into the formula.

$$\mu = 0\left(\frac{243}{1024}\right) + 1\left(\frac{405}{1024}\right) + 2\left(\frac{270}{1024}\right) + 3\left(\frac{90}{1024}\right) + 4\left(\frac{15}{1024}\right) + 5\left(\frac{1}{1024}\right)$$

Third, calculate the value of each product.

$$\mu = 0 + \frac{405}{1024} + \frac{540}{1024} + \frac{270}{1024} + \frac{60}{1024} + \frac{5}{1024}$$

Fourth, add up these products to obtain the mean.

$$\mu = 1280/1024$$

or

$$\mu = 1.25 \text{ correct answers.}$$

Similarly, the standard deviation can be computed using the formula for the standard deviation of a discrete random variable as outlined in Section 6.3. Using the standard deviation formula

$$\sigma = \sqrt{\sum_{\substack{all\ possble \\ X\ values}} [(X - \mu)^2 \cdot P(X)]} \ ,$$

we get: $\sigma \approx 0.97$ correct answers. (Verify this result using the standard deviation formula of a discrete random variable.)

For a binomial distribution, the **mean** and **standard deviation** can be determined using the following simplified formulas.

The Formulas for the Mean and Standard Deviation of a Binomial Distribution

Mean of a Binomial Distribution, denoted by μ_s, is given by the formula:

$$\mu_s = np$$

Standard Deviation of a Binomial Distribution, denoted by σ_s, is given by the formula:

$$\sigma_s = \sqrt{np(1-p)}$$

where:
n = number of identical trials
p = probability of getting a success in one trial

Notice the use of the subscript **s** in the symbols for the mean and standard deviation. The subscript **s** pertains to the idea of a success for a binomial distribution. Thus, the symbol μ_s, read *mu sub s*, denotes the *mean number of successes* for a binomial distribution, while the symbol σ_s, read *sigma sub s*, denotes the *standard deviation of the number of successes* for a binomial distribution.

Using the formulas for the mean and the standard deviation of a binomial distribution, let's verify that the mean is 1.25 and the standard deviation is approximately 0.97 for the binomial experiment of guessing the correct answer on a five question multiple choice test.

To calculate the mean and standard deviation, we need to determine the values of n and p. For this binomial experiment, the value of n is equal to the number of questions which is 5, and the value of p is equal to the probability of guessing a correct answer which is ¼ or 0.25. Thus, substituting the values n = 5 and p = 0.25 into the formula for the mean and standard deviation, we get:

$$\mu_s = np$$
$$= (5)(0.25)$$
$$\mu_s = 1.25 \text{ correct answers}$$

and

$$\sigma_s = \sqrt{np(1-p)}$$
$$= \sqrt{(5) \cdot (0.25) \cdot (0.75)}$$
$$\sigma_s \approx 0.97.$$

Thus, assuming students guess on all 5 questions, the expected number of correct answers is 1.25 with a standard deviation of 0.97.

EXAMPLE 6.18

The following table represents a binomial probability distribution where n = 4 and p = 0.5.
a) Construct a probability histogram and describe its shape.
b) Find the binomial mean, μ_s, and the binomial standard deviation, σ_s.
c) What is the probability that the number of successes is within two standard deviations of the mean?

Number of Successes s	Probability of Success P(s) = p
0	0.0625
1	0.2500
2	0.3750
3	0.2500
4	0.0625

Solution

a) Using the number of successes, s, as discrete values along the horizontal axis and the corresponding P(s) values for the vertical axis, the probability histogram for the binomial experiment is shown in Figure 6.21.

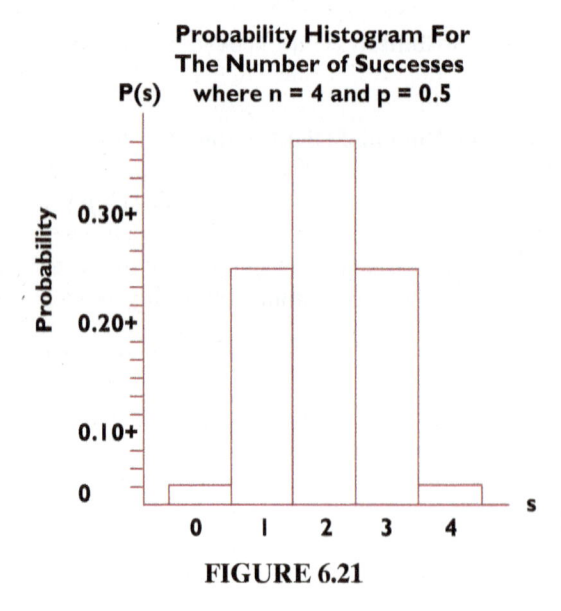

Probability Histogram For The Number of Successes where n = 4 and p = 0.5

FIGURE 6.21

The shape of the probability histogram is symmetric bell-shaped. Can you estimate the balance point along the s axis?

b) To calculate the binomial mean, μ_s, and the binomial standard deviation, σ_s where n = 4 and p = 0.5 we use the formulas:

$$\mu_s = np$$

and

$$\sigma_s = \sqrt{np(1-p)}$$

Thus, $\mu_s = (4)(0.5)$

$\mu_s = 2$

and $\sigma_s = \sqrt{(4) \cdot (0.5) \cdot (0.5)}$

$$\sigma_s = 1$$

c) Since to be within two standard deviations of the mean translates as $\mu_s - 2\sigma_s$ and $\mu_s + 2\sigma_s$ then

$$\mu_s - 2\sigma_s = 2 - 2(1) = 0$$

and

$$\mu_s + 2\sigma_s = 2 + 2(1) = 4$$

By examining the probability histogram in Figure 6.21 you should notice that 100% of the distribution is within two standard deviations of the mean. Therefore, the probability that the number of successes is within two standard deviations of the mean is 100%.

An Application of the Binomial Distribution: Acceptance Sampling

Everyday, both manufacturers and purchasers of certain products need to determine if these products meet certain specifications before the products are sold or purchased. For example, a light bulb manufacturer is interested in knowing whether the thousand bulbs they produce meet their specifications or whether a large percentage are defective before selling these bulbs to the consumer. An automotive manufacturer who orders ten thousand computer chips for their new automobiles is interested in knowing whether all these chips meet their specifications and are not defective before accepting this entire order from the computer chip manufacturer.

In each one of these situations it is either extremely difficult or impossible to inspect each item before shipping the product or accepting the shipment. If the light bulb manufacturer were to test all the bulbs before shipment, there would be no bulbs to sell. For the automotive manufacturer, testing each computer chip prior to purchasing the chip would be extremely time consuming and expensive.

Consequently, in each of these situations, a small sample of these products, relative to the shipment size, are randomly selected and tested. From this sample result, they infer that they would probably obtain similar results if they were to test all the items within the shipment. This procedure is called **acceptance sampling**. The main objective of acceptance sampling is to use a small inexpensive sample of items which will serve as a measuring stick of the quality of the entire shipment. In many instances, the binomial distribution is used to help draw these inferences about the entire population of items from the small sample selected. To use the binomial distribution when doing acceptance sampling, it is assumed that the experiment of selecting the sample items satisfies the four conditions of a binomial experiment. That is, the random selection of the n items for the sample represents n identical and independent trials where the probability of selecting a defective item is the same for each trial.

Example 6.19 examines how the binomial probability distribution can be used to do acceptance sampling.

EXAMPLE 6.19

Suppose an automobile company decides to reject a shipment of 10,000 computer chips based upon the results of a sample of 20 chips. The company's criteria is to reject the entire shipment *if 10% or more of the chips tested are defective*.
a) What is the probability that the company will not reject (i.e. accept) this shipment of 10,000 computer chips if 25% of the chips in this shipment are actually defective?
b) What is the probability that the company will reject this shipment of 10,000 computer chips if 25% of the chips in this shipment are actually defective?

Solution

a) Since we can assume that this is a binomial experiment, we can use the binomial distribution to determine the probability of not rejecting the shipment. The shipment will not be rejected if less than 10% of the chips in the sample of 20 chips are defective. Since 10% of 20 chips is 2, then the shipment will not be rejected if less than 2 chips is determined to be defective. Thus, we need to calculate the probability that less than 2 chips, 1 chip or no chips, will be defective in a sample of 20 chips.

To calculate the probability of less than 2 defective chips, we need to add the probability of 1 defective chip to the probability of no defective chips. That is,

$$P(\text{less than 2 defective chips}) = P(\text{1 defective chip}) + P(\text{no defective chips})$$

To determine each of these probabilities, use the binomial probability formula:

$$P(s \text{ successes in n trials}) = {}_nC_s \, p^s \, q^{(n-s)}$$

and determine:

        the number of trials, n,
        the definition of a success,
        the number of successes, s,
        the probability of a success, p,
and   the probability of a failure, q = 1 – p.

The size of the sample represents the number of trials, thus n = 20.

Since the question pertains to selecting a defective chip, then we will *define success to be selecting a defective chip*, and the number of successes will be:

    s = 1 (for the possibility of 1 defective chip) or
    s = 0 (for the possibility of no defective chips).

The probability of selecting a defective chip or a success has been given to be 25% or 0.25, thus p = 0.25. Therefore, the probability of not selecting a defective chip or a failure, q, is found by: q = 1 – p = 1 – 0.25 = 0.75.

Substituting these values into the binomial probability formula, we obtain:

$$\begin{aligned}P(\text{less than 2 defective chips}) &= P(\text{1 defective chip}) + P(\text{no defective chips}) \\ &= {}_{20}C_1 \, (0.25)^1 \, (0.75)^{19} + {}_{20}C_0 \, (0.25)^0 \, (0.75)^{20} \\ &\approx 0.0211 + 0.0032 \\ &\approx 0.0243 \text{ or } 2.43\%\end{aligned}$$

Thus, the probability that less than 2 chips will be defective in a sample of 20 chips, if 25% of the chips in this shipment are actually defective, is approximately 0.0243 or 2.43%. Therefore, the probability that the company will not reject (or accept) this shipment of 10,000 computer chips is approximately 0.0243 or 2.43%.

b) Since we know from part (a) that the probability that the company **will not reject this shipment** is approximately 0.0243, then the probability that the company **will reject this shipment** can be determined using the complement statement:

$$P(\text{company will reject this shipment}) = 1 - P(\text{company will not reject this shipment})$$
$$= 1 - 0.0243$$
$$= 0.9757$$

Thus, the probability that the company will reject this shipment of 10,000 chips, if 25% of the chips in this shipment are actually defective, is 0.9757 or 97.57%.

CASE STUDY 6.4 Trial by Mathematics

The following article entitled "Trial by Mathematics" appeared in *Time* magazine on April 26, 1968.

Decisions: Trial by Mathematics

After an elderly woman was mugged in an alley in San Pedro, Calif., a witness saw a blonde girl with a ponytail run from the alley and jump into a yellow car driven by a bearded Negro. Eventually tried for the crime, Janet and Malcolm Collins were faced with the circumstantial evidence that she was white, blonde and wore a ponytail while her Negro husband owned a yellow car and wore a beard. The prosecution, impressed by the unusual nature and number of matching details, sought to persuade the jury by invoking a law rarely used in a courtroom—the mathematical law of statistical probability.

The jury was indeed persuaded, and ultimately convicted the Collinses (TIME, Jan. 8, 1965). Small wonder. With the help of an expert witness from the mathematics department of a nearby college, the prosecutor explained the probability of a set of events actually occurring is determined by multiplying together the probabilities of each of the events. Using what he considered "conservative" estimates (for example, that the chances of a car's being yellow were 1 in 10, the chances of a couple in a car being interracial 1 in 1,000), the prosecutor multiplied all the factors together and concluded that the odds were 1 in 12 million that any other couple shared the characteristics of the defendants.

Only One Couple. The logic of it all seemed overwhelming, and few disciplines pay as much homage to logic as do the law and math. But neither works right with the wrong premises. Hearing an appeal of Malcolm Collins' conviction, the California Supreme Court recently turned up some serious defects, including the fact that not even the odds were all they seemed.

To begin with, the prosecution failed to supply evidence that "any of the individual probability factors listed were even roughly accurate." Moreover, the factors were not shown to be fully independent of one another as they must be to satisfy the mathematical law: the factor of a Negro with a beard, for instance, overlaps the possibility that the bearded Negro may be part of an interracial couple. The 12 million to 1 figure, therefore, was just "wild conjecture." In addition, there was not complete agreement among the witnesses about the characteristics in question. "No mathematical equation," added the court, "can prove beyond a reasonable doubt (1) that the guilty couple *in fact* possessed the characteristics described by the witnesses or even (2) that only *one* couple possessing those distinctive characteristics could be found in the entire Los Angeles area."

Improbable Probability. To explain why, Judge Raymond Sullivan attached a four-page appendix to his opinion that carried the necessary math far beyond the relatively simple formula of probability. Judge Sullivan was willing to assume it was unlikely that such a couple as the one described existed. But since such a couple did exist—and the Collinses demonstrably did exist—there was a perfectly acceptable mathematical formula for determining the probability that another such couple existed. Using the formula and the prosecution's figure of 12 million, the judge demonstrated to his own satisfaction and that of five concurring justices that there was a 41% chance that at least one other couple in the area might satisfy the requirements.

CASE STUDY 6.4 Trial by Mathematics (continued)

"Undoubtedly," said Sullivan, "the jurors were unduly impressed by the mystique of the mathematical demonstration but were unable to assess its relevancy or value." Neither could the defense attorney have been expected to know of the sophisticated rebuttal available to them. Janet Collins is already out of jail, has broken parole and lit out for parts unknown. But Judge Sullivan concluded that Malcolm Collins, who is still in prison at the California Conservation Center, had been subjected to "trial by mathematics" and was entitled to a reversal of his conviction. He could be tried again, but the odds are against it.

This robbery was discussed in a Case Study in Chapter 5. Recall, that the interracial couple was convicted of the crime, because they fit the characteristics of the robbers which had been established as a 1-in-12 million chance. The couple however appealed and were exonerated because the judge used a mathematical argument of his own. The judge's argument assumed that the probability p that a couple has all the characteristics of the robbers was $1/12{,}000{,}000$ and that there does exist one such couple in a population of 12,000,000 (the prosecutor's figure) couples. The judge proceeded to calculate the chance that another such couple could exist. The judge's mathematical reasoning used binomial probability and conditional probability. He was trying to arrive at a value for the following probability:

P(2 or more such couples exist | at least one such couple exists).

This is equal to:

$$\frac{P(2 \text{ or more such couples})}{P(\text{at least one such couple})}$$

a) What probability rule was used to write the above statement?
   This in turn equals:

$$\frac{1 - P(\text{one or less such couples})}{1 - P(\text{no such couple})}$$

b) What probability rule allowed us to write the above probability statement?
   If we use p to represent the probability that a couple has all the characteristics of the robbers, and let N be the number of couples in the population, then this probability results in the following formula:

$$\frac{1 - (1-p)^n - Np(1-p)^{n-2}}{1 - (1-p)^n}$$

c) What probability formula did we use to arrive at the above formula?
   If $1/12{,}000{,}000$ is substituted in for the value of p, and N equals 12,000,000 (the prosecutor's figure) couples, then the:

   P(2 or more such couples exist | at least one such couple exists) is equal to approximately 41%.

d) Is the judge's argument based upon an unlikely or a likely chance that such a couple has the stated characteristics? What is this probability value that the judge is using in his argument?
   Table 6.16 contains the probabilities for several values of N couples where p is equal to $1/12{,}000{,}000$.

TABLE 6.16

Number of Couples N (in millions)	Probability	Number of Couples N (in millions)	Probability
1	4.02%	10	34.79%
2	7.86%	15	48.35%
3	11.60%	20	59.59%
4	15.22%	25	68.75%
5	18.75%	30	76.10%
6	22.16%	40	86.44%
7	25.47%	50	92.56%
8	28.68%	75	98.52%
9	31.79%	100	99.73%

(continues)

> **CASE STUDY 6.4   Trial by Mathematics (continued)**
>
> e) Examining this probability table, what conclusion can you draw about the existence of another such couple as the population of N couples increases?
> f) Was it to the judge's advantage or disadvantage to use the prosecutor's 12,000,000 population figure for N couples, if the judge was trying to demonstrate that there was a reasonable chance that another such couple could exist with the same characteristics?
>
> From *TIME.com*. © 1968 TIME USA LLC. All rights reserved. Used under license.

6.6 THE POISSON DISTRIBUTION

Probability distributions are defined by their random variables. In this section we will study a discrete probability distribution called the **Poisson distribution**. The Poisson distribution applies to experiments having a discrete random variable, where the occurrences are random and independent. The probability of an event occurring in a fixed time period, given that we know the average number of times the event occurs during this time period can be computed using the Poisson distribution. The Poisson distribution has a single parameter λ (the Greek letter lambda), which represents the average number of occurrences for a specified time interval. For example:

- The average number of accidents at a busy intersection in a 72 hour period is 1.5. In this case $\lambda = 1.5$ since the average number of occurrences is 1.5.
- If 1000 people are exposed to the flu, on average 10 will show symptoms within 48 hours. Here $\lambda = 10$ because the average number of occurrences is 10.
- Policyholders of an insurance company report claims at a mean rate of 0.4 per year. In this case $\lambda = 0.4$ since the average number of occurrences is 0.4.
- Shoppers enter a mall during the holiday season at a mean rate of 50 per 10 minutes. Here $\lambda = 50$.

It is important to see that the parameter, λ, is for a specified time period. For example, "a mean of 1.5 accidents <u>per 72 hours</u>" **or** "on average 0.4 claims <u>per year</u>." If we change the time interval for any of the above examples, then the mean amount of occurrences must change as well. This will be covered in more detail later on in this section.

The Poisson Distribution

Let X be a discrete random variable representing the number of times an event occurs in a specified time period. It is known that the mean number of occurrences is λ.

The probability of exactly k random and independent occurrences is given by:

$$P(X = k) = \frac{e^{-\lambda} \lambda^k}{k!}, \; k = 0, 1, 2, 3, \ldots$$

where λ represents the mean number of occurrences and $e \approx 2.718$.

For this distribution,

Mean of X, $\mu = \lambda$
Variance of X = $V(X) = \lambda$
Standard Deviation of X = $\sqrt{\lambda}$

In the above formula, the numerator contains the number *e*. This should not be confused with a variable. In mathematics, the number *e* is an irrational number (like π) and is approximately 2.718. Also, notice the denominator contains a factorial which we studied in Chapter 5.

EXAMPLE 6.20

The number of shoppers entering a supermarket follows a Poisson distribution with a mean of 110 per hour.
a) What is the value of λ in this situation?
b) How would the value of λ change if we were only interested in a 30 minute time period?
c) Say you wanted to calculate a probability involving the number of shoppers entering the supermarket between 10:00 A.M. and 11:30 A.M. What would the value of λ be in this case?
d) What is the mean number of shoppers entering the supermarket in 12 minutes?

Solution

a) The mean number of shoppers entering the mall is 110, thus $\lambda = 110$.
b) Since the time interval for λ and k must be equal, we must redefine the time interval for λ. Thus, a mean of 110 per hour is the equivalent of 55 per 30 minutes. So, in this situation $\lambda = 55$.
c) If we realize that this is just three 30 minute intervals, then we can use the answer from part b). If the mean is 55 per 30 minutes, then there are 3*55 or 165 in 150 minutes. So in this case $\lambda = 165$.
d) 12 minutes is just 1/5 of an hour, so we can take 1/5 of the mean.
⅕*110 = 22. Therefore, $\lambda = 22$ shoppers.

Also, we could have just set up a proportion to solve any of these problems.
For example, in part d) above:

$$\frac{110 \; shoppers}{60 \; min} = \frac{\lambda \; shoppers}{12 \; min}$$

$$\left(\frac{110 \; shoppers}{60 \; min}\right) \cdot (12 \; min) = \lambda$$

thus, $\lambda = 22$ shoppers

EXAMPLE 6.21

The number of car accidents that occur on a highway per day follows the Poisson distribution with mean number of accidents equal to 3.
a) What is the probability of exactly 2 car accidents in one day?
b) What is the probability of at most 1 accident in one day?
c) What is the probability of more than 1 accident in one day?

Solution

a) In this case $\lambda = 3$, since this represents the mean number of accidents per day.

Using the formula $P(X = k) = \frac{e^{-\lambda}\lambda^k}{k!}$, where $\lambda = 3$ and $k = 2$ we have:

$$P(X = 2) = \frac{e^{-3} \cdot 3^2}{2!} \approx 0.2240$$

b) This example is a little more complex. The mean is still 3, but we are asked to work a series of probabilities, and not just one probability as in part a). Thus,

$$P(X \leq 1) = P(X = 0) + P(X = 1)$$

$$= \left[\frac{e^{-3} \cdot 3^0}{0!} + \frac{e^{-3} \cdot 3^1}{1!}\right]$$

$$= \left[\frac{e^{-3}}{1} + \frac{3e^{-3}}{1}\right]$$

$$= 0.04978 + 0.14936$$

$$= 0.1991$$

In the calculation of the probability in part b) we used the fact that $x^0 = 1$. This exponent rule is true for any x value not equal to 0. In this example it was $3^0 = 1$ specifically. We also used the fact that by definition $0! = 1$.

c) We can use the answer from part b) to get the answer for this probability. The probability of there being more than one accident in a day is actually the sum of an infinite number of probabilities.

$$P(X > 1) = P(X = 2) + P(X = 3) + P(X = 4) + \ldots$$

We can use the fact that the Poisson distribution is a discrete probability distribution, and therefore has the property that the sum of all probabilities is 1 ($\sum p(x) = 1$). The probability of more than 1 accident in a day is the same as 1 minus the probability of 1 or less accidents in a day.

$$P(X > 1) = 1 - P(X \leq 1) = 1 - [P(X = 0) + P(X = 1)]$$

$$= 1 - \left[\frac{e^{-3} \cdot 3^0}{0!} + \frac{e^{-3} \cdot 3^1}{1!} \right]$$

$$\approx 1 - [0.04978 + 0.14936]$$

$$= 0.8009$$

TI-84 Graphing Calculator Solution for Example 6.21

The TI-84 graphing calculator has two built-in Poisson probability functions in the distribution menu. To access the distribution menu press 2nd VARS [DISTR]. On the TI-84, the Poisson probability functions are in options C and D as shown in Figure 6.22.

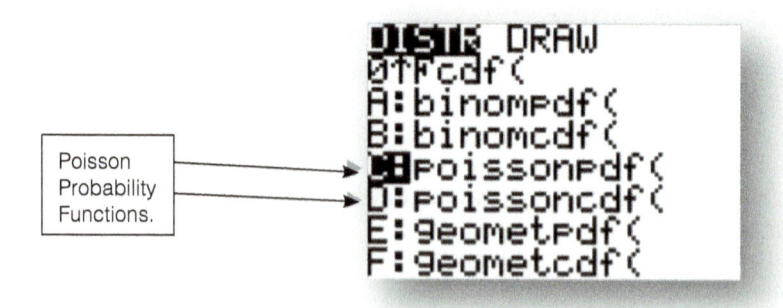

FIGURE 6.22

The function **poissonpdf(λ, k)** gives the probability of exactly k occurrences given a mean of λ. The function **poissoncdf(λ, k)** gives the probability of *at most* k occurrences given a mean of λ. We can use these functions to rework the questions in Example 6.21.

Example 6.21a could have been worked as follows:

$$P(k \text{ given } \lambda) = \text{poissonpdf}(\lambda, k)$$

$$P(k=2 \text{ given } \lambda=3) = \text{poissonpdf}(3, 2)$$

Press 2nd VARS [DISTR] and scroll down to option **C:poissonpdf(** and press ENTER. Now enter the mean and number of occurrences as shown in Figure 6.23a. Press ENTER twice to paste and calculate the answer. This probability is shown in Figure 6.23b.

FIGURE 6.23a **FIGURE 6.23b**

As you can see the answer to 4 decimal places is 0.2240.

Example 6.21b is a probability involving at most, so we will use the cumulative distribution function to work this probability.

$$P(k \text{ given } \lambda) = \text{poissoncdf}(\lambda, k)$$
$$P(k \text{ is at most 1 given } \lambda=3) = \text{poissoncdf}(3, 1)$$

Press 2nd VARS [DISTR] and scroll down to option **D:poissoncdf(** and press ENTER. Now enter the mean and maximum number of occurrences as shown in Figure 6.24a. Press ENTER twice to paste and calculate the answer. This probability is shown in Figure 6.24b.

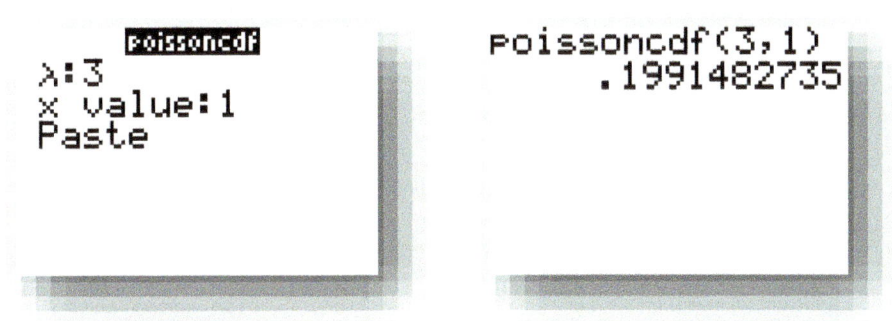

FIGURE 6.24a   **FIGURE 6.24b**

To 4 decimal places the answer is 0.1991.

Example 6.21c is a probability involving *at least*, so we will use the cumulative distribution function to work this probability and subtract the result from 1.

$$P(k \text{ is at least 2 given } \lambda=3) = 1 - P(k \text{ is at most 1 given } \lambda=3) = 1 - \text{poissoncdf}(3, 1).$$

This probability is shown in Figure 6.25.

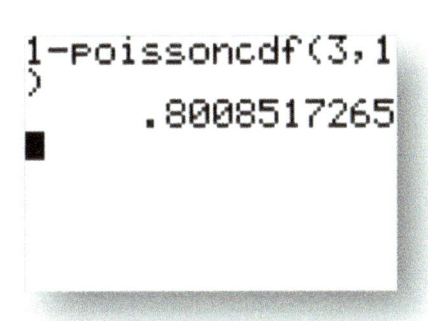

FIGURE 6.25

The result to 4 decimal places is 0.8009.

EXAMPLE 6.22

The number of calls made by a telemarketer follows a Poisson distribution with a mean of 2 calls per minute.
a) What is the probability the telemarketer makes 6 calls in 4 minutes?
b) What is the probability the telemarketer makes 22 calls between 10:00 A.M. & 10:10 A.M?

Solution

a) Since the mean number of calls is defined to be 2 per minute, we must compute the mean number of calls for 4 minutes.

$$\frac{2 \text{ calls}}{1 \text{ min}} = \frac{\lambda \text{ calls}}{4 \text{ min}},$$

$$\left(\frac{2 \text{ calls}}{1 \text{ min}}\right) \cdot (4 \text{ min}) = \lambda$$

so, $\lambda = 8$ calls

To determine the probability the telemarketer makes 6 calls per 4 minutes, we have: $P(X = 6) = \frac{e^{-8} \cdot 8^6}{6!}$ 0.1221

b) Since the mean number of calls is defined to be 2 per minute, we must compute the mean number of calls for 10 minutes.

$$\frac{2 \text{ calls}}{1 \text{ min}} = \frac{\lambda \text{ calls}}{10 \text{ min}},$$

$$\left(\frac{2 \text{ calls}}{1 \text{ min}}\right) \cdot (10 \text{ min}) = \lambda$$

thus, $\lambda = 20$ calls

To determine the probability the telemarketer makes 22 calls per 20 minutes we have: $P(X = 22) = \dfrac{e^{-20} \cdot 20^{22}}{22!} \approx 0.0769$

EXAMPLE 6.23

The number of calls received by the Silver City Police Department follows a Poisson distribution with a mean of 3.5 calls per hour.
a) Construct a probability distribution table for k = 0 calls to k = 6 calls per hour.
b) Construct a probability histogram to determine the shape of the distribution.

Solution

a) A probability distribution table for k = 0 to k = 6 calls requires the calculation of 7 probabilities. Since $\lambda = 3.5$, the general formula in each case will be:

$$P(X = k) = \frac{e^{-3.5}(3.5)^k}{k!} \text{ for } k = 0, 1, 2, 3, 4, 5, 6.$$

It may be useful to use a technology tool for this example. The Minitab v.15 output for this problem is:

k	P(X = k)
0	0.030197
1	0.105691
2	0.184959
3	0.215785
4	0.188812
5	0.132169
6	0.077098

b) Figure 6.26 represents the probability histogram created in Minitab v.15:

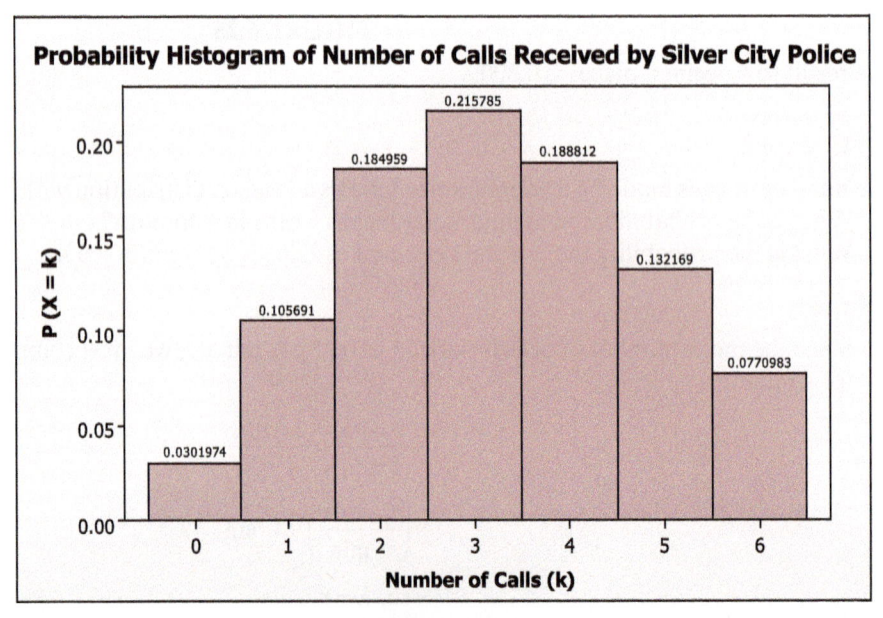

FIGURE 6.26

Shape of the Poisson Distribution

One important concept to note is how the shape of the distribution for the Poisson changes as the value of λ changes. Here is what the distribution looks like for mean (λ) values of 1, 2, 5 and 10.

a. $\lambda = 1$

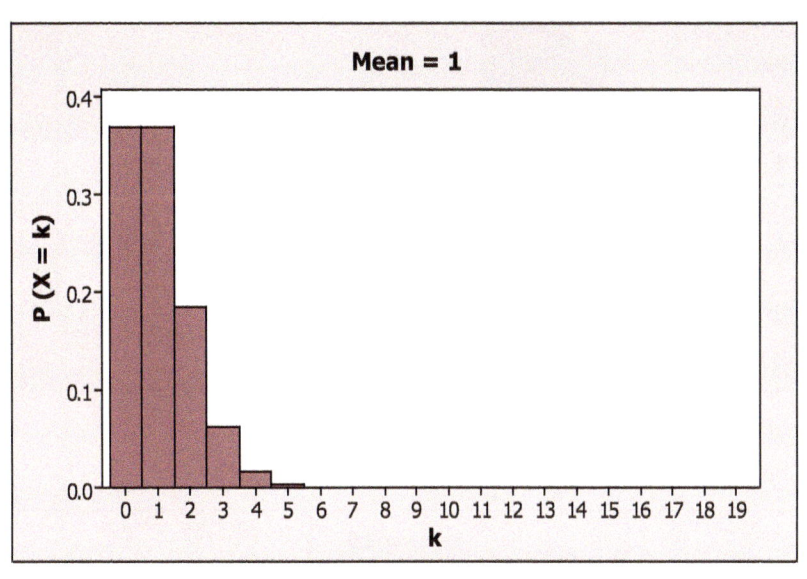

FIGURE 6.27a

As seen in Figures 6.27a and 6.27b, the shape of the distribution appears skewed right for small values of λ. Notice as the mean increases, the shape of the distribution tends towards a symmetrical (bell-shaped) distribution as shown in Figures 6.27c and 6.27d. Also note that the highest value (the mode) is close to λ.

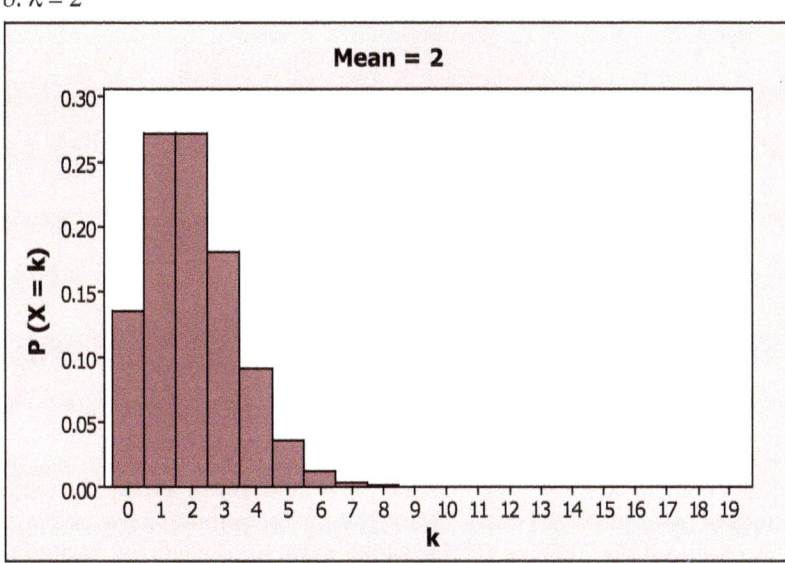

b. $\lambda = 2$

FIGURE 6.27b

c. $\lambda = 5$

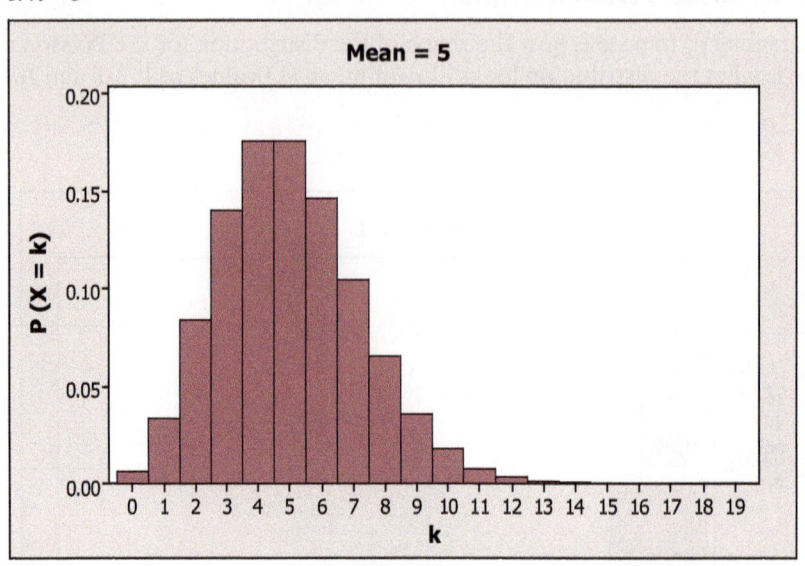

FIGURE 6.27c

d. $\lambda = 10$

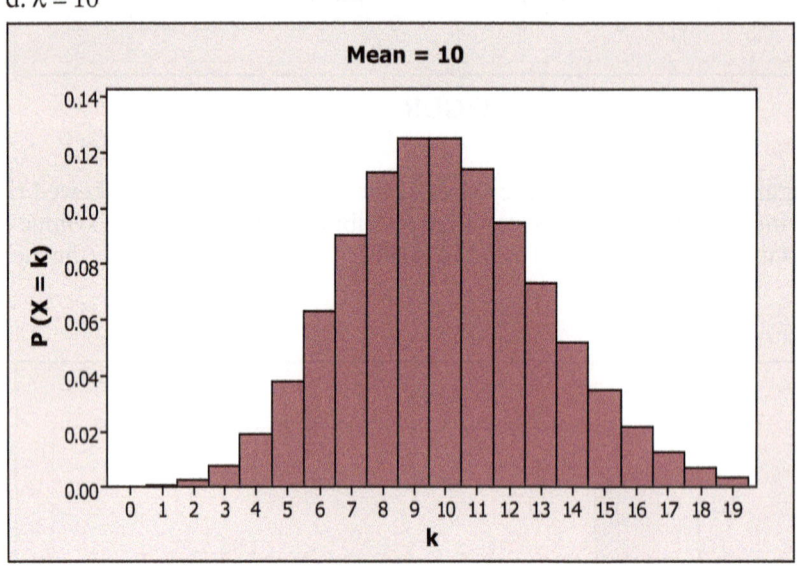

FIGURE 6.27d

6.7 THE GEOMETRIC AND NEGATIVE BINOMIAL DISTRIBUTIONS

Recall, that we have previously discussed that probability distributions are defined by their random variables. For example, there is a big difference between the following two situations:

Situation #1: A gambler sits at a slot machine and plays 20 times in a row. He is interested in the number of times he wins in the 20 trials. The value for the number of wins is a random variable since he hasn't played the 20 times yet. He wonders what his chances of winning *exactly twice* are.

Situation #2: The same gambler wants to win at this slot machine *once* and then he'll quit. He is wondering how many times he will lose *before* he wins once. The number of losses is a random variable because it hasn't happened yet.

You may have to think about these two situations for a while, but although they appear to be similar they in-fact are *very* different. In situation #1, the gambler is interested in X, the number of wins when he plays 20 times in a row. In situation #2, the gambler is interested in Y, the number of losses he must incur *before* he wins for the first time.

Section 6.7 The Geometric and Negative Binomial Distributions

The *formulas* used to calculate probabilities for the two above situations are very different. In this section we will be looking at the second situation. In fact, we've already looked at the random variable in the first situation; it's the **Binomial random variable**! The random variable in the second situation is defined to be the **Geometric random variable**, that is, the random variable describing the number of failures someone incurs before having a success for the first time is defined to be the geometric random variable. The probability distribution that this random variable describes is called the **Geometric Distribution**.

In order to determine a formula to find the probability described in situation #2 above, we must first describe the properties of a geometric experiment.

Properties of a Geometric Experiment

- There are two outcomes, success and failure.
- Successive trials are independent.
- You are interested in X, the number of times you *fail* at this event *before* you have a success for the *first* time.

EXAMPLE 6.24

Suppose the probability of a gambler winning on any try of a slot machine is 2%. Since the gambler has a 0.02 = 2% chance of winning(success) on any given try, then consequently, this implies that the probability of losing(failure) is 1 − 0.02 = 0.98 = 98% on any given try.

Develop a formula that will calculate the probability that he *loses k times* before winning for his *first time*.

Solution

He begins playing the slot machine many times in a row. Most likely he'll fail several times in a row because the probability of a failure is so great, however, *eventually* he will have his first success. The big question is "When will this success happen, and how many times did he fail before he finally won for the first time?" Typically, we use the variable k to represent the number of failures.

This situation would look as follows

Fail	Fail	Fail	Fail	Fail	Fail	...	Fail	**Success**
Try 1	Try 2	Try 3	Try 4	Try 5	Try 6		Try k	**Try k+1**

Again, note that the value for k is a random variable. Now remember that the probability this gambler fails on each try at this slot machine is 98% or 0.98. You can recall from Chapter 5 that when you wish to perform many events in a row and the successive trials are independent of one another then you can find the probability of all those successive events occurring by *multiplying* the probabilities together. This would look as follows:

k successive failures | success on the $k + 1$ trial

(0.98)	(0.98)	(0.98)	(0.98)	(0.98)	(0.98)	....	(0.98)	(0.02)
Try 1	Try 2	Try 3	Try 4	Try 5	Try 6		Try k	**Try k+1**

The probability of a failure, 0.98, appears k-times, where the probability of success, 0.02, appears once.

Mathematically, we write this as $(0.98)^k (0.02)$. You should think of this statement as follows:

> "The probability of failing k times in a row *before* winning for the first time, is equal to the probability of failing (0.98) raised to the number of failures, k, multiplied by the probability of a success (0.02)."

Therefore, let's say that the gambler was interested in the probability of his first success occurring on his 10th trial. The probability the gambler wins for the first time on his 10th trial is equivalent to finding the probability the gambler fails 9 times in a row before winning for the first time. This implies the value of k would be 9. The probability of this event would be written as follows:

$k = 9$ losses

$$P(X = 9) = (0.98)^9 (0.02) \approx 0.0167 \text{ or about } 1.7\%$$

The probability found in the previous example can be calculated using the geometric probability distribution formula.

The Geometric Probability Distribution

Let X be a *discrete* random variable representing the number of failures, k, that occur before the first success. Then X is a geometric random variable whose probability distribution is defined as:

$$P(X = k) = q^k p, \quad k = 0, 1, 2, \ldots$$

Where p is the probability of a success, and q is the probability of a failure. Also, $p + q = 1$.

Furthermore, for this distribution:

$$\mu = E(X) = \frac{q}{p} \text{ failures} \qquad \sigma^2 = Var(X) = \frac{q}{p^2} \qquad \sigma = \sqrt{Var(X)} = \sqrt{\frac{q}{p^2}} \text{ failures}$$

EXAMPLE 6.25

A telemarketer makes successive calls in order to sell a service that their company provides. The probability that the telemarketer will get a callback on any particular call is 5%.

a) What is the probability the telemarketer won't get a callback on a particular call?
b) What is the probability that the first callback the telemarketer gets occurs on their 10th call?
c) What is the probability the telemarketer has exactly 5 failures before the first callback?
d) What is the probability the telemarketer has at most 2 failures before their first callback?
e) What is the mean number of failures the telemarketer must incur before getting their first callback?
f) What are the variance and standard deviation of X?

Solution

a) Since the probability of a success (getting a callback) is 5%, then the chances of failing (not getting a callback) must be 95%. Remember that the probability of a success plus the probability of a failure must be 100% or 1, that is, $p + q = 1$. So, for this problem $p = 0.05$ and $q = 0.95$.

b) If they get their first success (callback) on the tenth trial, then that means they failed 9 times *before* their first success (callback). This gives a k value of 9. So, the probability of 9 failures before their first success is given by:

$$P(X = k) = q^k p$$
$$P(X = 9) = (0.95)^9 (0.05) \approx 0.0315$$

c) In this case $k = 5$, so the probability of 5 failures before the first success is given by:

$$P(X = 5) = (0.95)^5 (0.05) \approx 0.0387$$

d) The phrase "*at most* 2" means that there could be 0, 1 or 2 failures before the first success. So this question is really three questions in one, first working the problem with $k = 0$, then $k = 1$ followed by $k = 2$. Then the final result will be found by adding the three probabilities together.

$$P(X \text{ is at most } 2) = P(X \leq 2) = P(X = 0) + P(X = 1) + P(X = 2)$$
$$= (0.95)^0(0.05) + (0.95)^1(0.05) + (0.95)^2(0.05)$$
$$= 0.05 + 0.0475 + 0.04513$$
$$= 0.14263 \text{ or } \approx 14.3\%$$

e) The formula for the mean of the geometric random variable is:

$$\mu = E(X) = \frac{q}{p} = \frac{0.95}{0.05} = 19 \text{ failures}$$

This tells us that, *on average*, the telemarketer should expect 19 failed attempts (i. e. hang-ups) *before* their first callback.

f) The variance and standard deviation are given by:

$$\text{Var}(X) = \sigma^2 = \frac{q}{p^2} = \frac{0.95}{(0.05)^2} = 380 \qquad \sigma = \sqrt{\text{Var}X} = \sqrt{\sigma^2} = \sqrt{380} \approx 19.49$$

It is important to point out that the values we calculated in *Example 6.25* are part of a bigger picture. In parts b, c & d of that example we found probabilities for the X values of 0, 1, 2, 5 & 9 failures. A portion of the corresponding discrete probability distribution table is shown in Table 6.17:

TABLE 6.17

X	P(X)
0	0.0500
1	0.0475
2	0.0451
3	
4	
5	0.0387
6	
7	
8	
9	0.0315
10	
11	
12	

The table keeps going on for higher values of X, however we are just showing a small piece of the entire table. We could very easily find the probabilities for the other X values in the table by using the defining formulas or the *geometPDF* operation of the TI-84 graphing calculator. This operation on the TI-84 graphing calculator is described in the next couple of pages in Example. 6.25…revisited. Once these values are found, then answering probability questions becomes very easy. We have completed the table and this is shown in Table 6.18. Again, remember that the values of X can keep going but we are only showing a portion of the table.

TABLE 6.18

X	P(X)
0	0.0500
1	0.0475
2	0.0451
3	0.0429
4	0.0407
5	0.0387
6	0.0368
7	0.0349
8	0.0332
9	0.0315
10	0.0299
11	0.0284
12	0.0270

A *probability histogram* of the values shown in Table 6.18 is given below in FIGURE 6.28. Notice how the values for P(X) are getting closer and closer to 0.

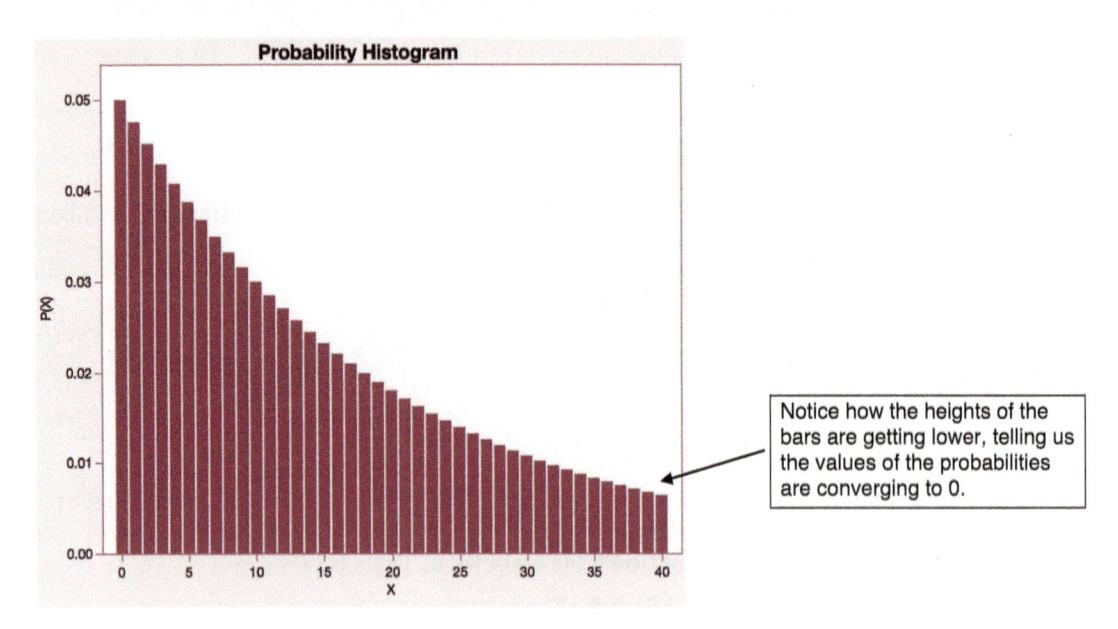

FIGURE 6.28

Some points to consider regarding the output in this table and corresponding probability histogram include:

- The values of X, in theory, go on forever.
- By examining the graph in Figure 6.28, we can see as the value of X increase, the probabilities get closer and closer to 0. Another way to say this is that as the probabilities *converge* to 0. For example, the probability that the first callback will happen on the 10,000th call is *essentially* 0. We *of course* would expect the first callback to happen sooner. We already know from Example 6.25 part e that *on average* there are 19 failures before the first success (the mean).
- If it were actually possible to find all of the values in the table, their sum would be exactly 1 because the geometric distribution is a **discrete probability distribution** and therefore possesses the properties that all probabilities are greater than or equal to 0 and the sum of all of the probabilities in the distribution is 1.
- There are other technology tools out there that will calculate geometric probabilities. These include other hand-held calculators, software packages such as JMP and Minitab, as well as geometric probability applets that you can search for in Google.

TI-84 Graphing Calculator Solution for Example 6.25

The probabilities asked in Example 6.25 parts b, c & d could be found using the geometric distribution feature of the TI-84 graphing calculator.

EXAMPLE 6.25 REVISITED

A telemarketer makes successive calls in order to sell a service that their company provides. Suppose the probability that the telemarketer will get a callback on any particular call is 5%.

a) What is the probability that the first callback the telemarketer gets occurs on their 10th call?
b) What is the probability the telemarketer has exactly 5 failures before the first callback?
c) What is the probability the telemarketer has at most 2 failures before their first callback?

Solution

b) Recall, that the answer to this part of the problem was approximately 0.0315.

To use the TI-84's geometric probability calculator, press **2nd > Vars**. This will bring you to the distribution menu(**DISTR**). This is shown below in Figure 6.29.

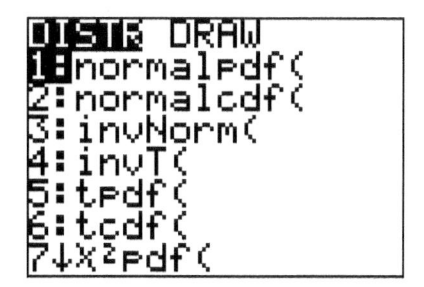

Section 6.7 *The Geometric and Negative Binomial Distributions* 351

FIGURE 6.29

Scroll to the very bottom of the screen and you'll see two geometric functions, **geometPDF** and **geometCDF**. These two functions operate in a very similar manner to the binomPDF, binomCDF, poissonPDF and poissonCDF functions that we worked with in Sections 6.5 and 6.6. The geometPDF operation will work out the probability of a single event X, where the geometCDF operation will work out the cumulative probability of *at most* X.

These two functions are shown below in Figure 6.30.

Geometric probability distribution functions.

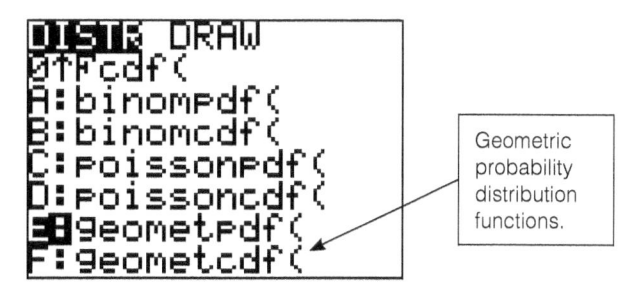

FIGURE 6.30

Previously, we've identified the parameters of success and failure in the question to be $p = 0.05$, $q = 0.95$. In addition, the number of failures was $k = 9$. Next, we must input these values into the TI-84. We'll do this by selecting option **E: geometpdf.** Selecting this option brings you to the parameter input screen.

This is shown below in Figure 6.31.

FIGURE 6.31

Many software packages have a slightly different view of the geometric distribution. (The TI-84 being one of them!) Rather than viewing the geometric random variable X as the number of failures before the first success, the TI-84 views the random variable X as the trial that the first success happens on. In other words, the screen shown in Figure 6.31 looks at everything from a "success" point of view. This screen asks you for p, the probability of a success and for the "x-value" which is the trial the success occurs on. For this part, the probability of a success for the telemarketer is $p = 0.05$ and his first success happens on his 10th trial so, $x = 10$. Asking for the probability of 9 failures before the telemarketer's first success is the same thing as asking, "What is the probability that the telemarketer's first success occurs in his 10th time making a phone call?"

Figures 6.31a and 6.31b show the values of the parameters entered into the TI-84 as well as what occurs after you select the **Paste** option to paste the command into the home screen. The syntax for the geometpdf operation in the TI-84 graphing calculator is geometpdf(p, x) where p is the probability of a success, and x is the trial the success occurs on.

So, for this problem the syntax would be geometpdf(.05, 10). In Figure 6.31b, the value 10 is cut off from view on the TI-84 because it lies off of the screen.

FIGURE 6.31a   **FIGURE 6.31b**

Once the command is pasted into the home screen, you can press **ENTER** in order to have the calculator calculate the probability. This result is shown below in Figure 6.32.

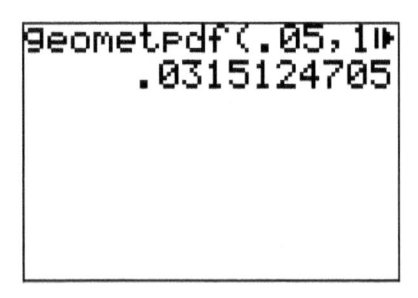

FIGURE 6.32

As you can see the probability is approximately 0.0315, the same value we calculated using the defining formula for the geometric distribution.

c) This question asks us to find the probability of 5 failures before the first success. This is equivalent to asking for the probability the first success occurs on the 6th trial. So, the parameters for this part are $p = 0.05$ and x-value = 6. Figures 6.33a and 6.33b show these values entered into the TI-84 along with the calculated probability.

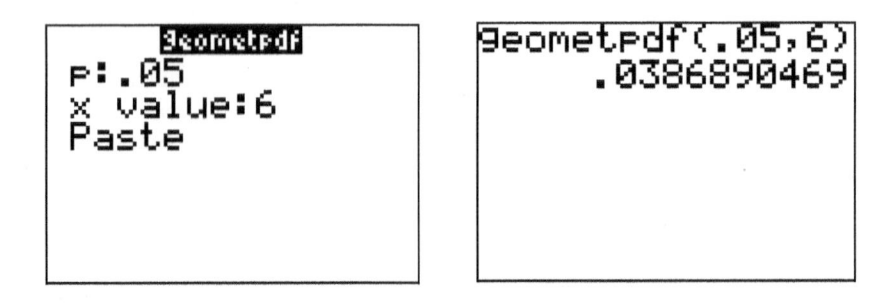

FIGURE 6.33a   **FIGURE 6.33b**

As you can see, the probability is approximately 0.0387 which is the same result obtained using the defining formulas.

d) Recall that this part of the problem asked us for the probability the telemarketer has at most 2 failures before their first callback. Asking for the probability of at most 2 failures before the first success is equivalent to asking, "What is the probability the first success occurs on *at most* the 3rd trial?" Remember that in order to work out a probability involving the phrase "*at most x*", we use the cumulative probability function, which in this case is **geometcdf**. Figures 6.34a and 6.34b show the geometcdf input menu along with the parameters $p = 0.05$ and x-value = 3 entered into the menu.

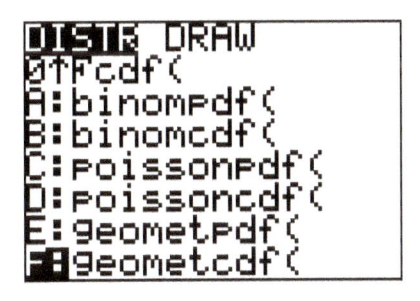

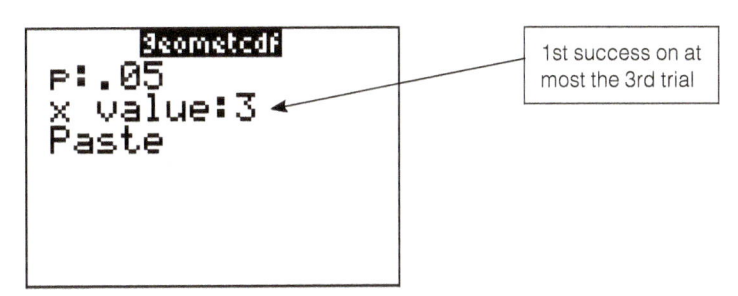

FIGURE 6.34a

FIGURE 6.34b

Pasting the **geometcdf** command into the home screen and calculating the probability are shown in Figures 6.35a and 6.35b.

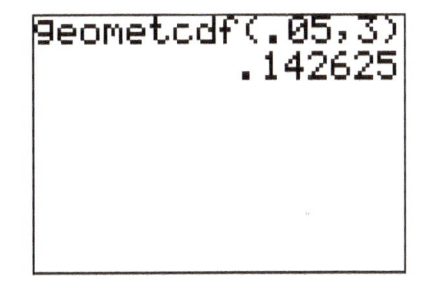

FIGURE 6.35a

FIGURE 6.35b

As you can see, the probability is approximately 0.1426, which is the same answer we obtained by calculating this probability using the defining formulas.

The Negative Binomial Distribution

Next, we will discuss a probability distribution very similar to the geometric distribution. This distribution is called the *negative binomial* distribution. In fact, you'll learn that the geometric distribution is a special case of the negative binomial distribution. The geometric is called a special case of the negative binomial because the negative binomial distribution creates the geometric distribution under some special conditions of the parameters. We will investigate this fact later in the section.

We'll begin by defining the *negative binomial random variable*. Recall that the geometric random variable involved the study of the number of failures, k, that occurred *before* your **first** success. However, what if you wanted to win *more* than one time? Well, that is precisely what the negative binomial distribution studies! In other words, the *negative binomial random variable* involves the study of the number of failures, k, that occur *before* your rth success.

DEFINITION 6.8
Let X be a discrete random variable representing the number of failures k that occur at some event before the rth success. Then X is called the *negative binomial* random variable.

- The **geometric random variable** studies the number of failures that occur before your **first** success.
- The **negative binomial random variable** studies the number of failures that occur before your r^{th} success.

In order to calculate probabilities for the negative binomial random variable, we first must understand the properties of a negative binomial experiment. These include the assumptions we make while using the defining formula to calculate probabilities.

Properties of a Negative Binomial Experiment

- There are two outcomes, success and failure.
- Successive trials are independent.
- You are interested in X, the number of times you fail at this event before you have a success for the rth time.

Let's revisit the telemarketer example covered in Example 6.25 making a slight change...

EXAMPLE 6.25 REVISITED AGAIN

A telemarketer makes successive calls in order to sell a service that their company provides. The probability that the telemarketer will get a callback on any particular call is 5%. What is the probability that the **third** callback(success) occurs on the tenth trial?

To find the probability of this event, we must first reword the question, so we know exactly how many successes(callbacks) and failures(hang-ups) that there are. If the third success occurs on the tenth trial, then that means that 3 of the ten trials were successes and the 7 other trials were failures. So, another way to word this question is as follows:

"What is the probability of the telemarketer having 7 hang-ups(failures) before their 3rd call back(success)?"

The prior statement is equivalent to asking,

"What is the probability that the 3rd callback (success) occurs on the tenth trial?"

It is **very important** that you see that the two prior statements are asking the same thing. When dealing with the negative binomial probability distribution, you need to know a numerical value for how many failures and successes that will occur, so it is often necessary to reword a question.

To help us find the probability of this event, let's look at a few situations that could possibly occur. If the telemarketer will have their third success on the tenth trial, then the combinations of successes and failures will have the following format:

7 failures and 2 successes occur in an unknown order | Trial 10 contains the third success (callback).

?	?	?	?	?	?	?	?	?	success
Try 1	Try 2	Try 3	Try 4	Try 5	Try 6	Try 7	Try 8	Try 9	Try 10

The **only trial** whose event is really known is the 10th (last) trial...it **must** be a success. The other 9 trials are yet to be determined, however we know that there are 7 failures, and 2 more successes to in those places. The issue is that **we don't know** the order of the 7 failures and 2 successes.

Section 6.7 The Geometric and Negative Binomial Distributions

So, a few possibilities that could occur would be:

Possibility	Try 1	Try 2	Try 3	Try 4	Try 5	Try 6	Try 7	Try 8	Try 9	Try 10
#1	F	F	F	S	F	F	F	S	F	S
#2	S	S	F	F	F	F	F	F	F	S
#3	F	F	S	F	F	F	F	F	S	S

$$\vdots$$

I think you can see that we would have to spend quite a while to figure out all the different ways we can mix up the 7 failures with the remaining two successes in the first 9 trials (actually there are 36 ways to do this!) The **one** thing each of these possibilities has *in common* is that **all of them** have 7 failures and 3 successes.

Let's say we want to find the probability that possibility #1 will occur. To find this, we would just multiply the probability of each of the events together; remembering that the probability of a success is 0.05 and the probability of a failure is 0.95. Recall, that the reason we can multiply these probabilities is because we are making the assumption that successive trials are *independent*.

For example, for possibility #1,

Possibility	Try 1	Try 2	Try 3	Try 4	Try 5	Try 6	Try 7	Try 8	Try 9	Try 10
#1	F	F	F	S	F	F	F	S	F	S

$= (0.95) \cdot (0.95) \cdot (0.95) \cdot \mathbf{(0.05)} \cdot (0.95) \cdot (0.95) \cdot (0.95) \cdot \mathbf{(0.05)} \cdot (0.95) \cdot \mathbf{(0.05)}$

mathematically, we write this product as:

$= (0.95)^7 \cdot (0.05)^3$

Now, what about all the *other* possibilities? They too would have the same probability of occurring since they all have 7 failures and 3 successes. Earlier, we mentioned that there are 36 of these possibilities. The question is, "How do we know this to be true?"

To answer this question, consider the following:

Each possibility has 7 failures and 3 successes; however, the last spot in every possibility is a success. So, the question now becomes. "In how many ways can we place the **other 2 successes** into the 9 remaining places?" You can recall from Chapter 5 that given n distinct objects, the number of ways you can choose r of them is

called a combination and found using the formula $_nC_r = \frac{n!}{r!(n-r)!}$. So, for this situation, the number of ways we can choose 2 locations from the 9 locations to place the remaining two successes is given by

$$_9C_2 = \frac{9!}{2!(9-2)!} = \frac{9!}{2!(7)!} = \frac{9 \cdot 8 \cdot 7!}{2! \cdot 7!} = \frac{72}{2} = 36.$$

Now, you do not have to work this out by hand. Almost all calculators have an $_nC_r$ button; you'll just have to locate it in the calculator that you are using. For instance, in the TI-84 the combination key is located in: **MATH > PRB > #3: nCr.**

Therefore, to finally answer the question, there are 36 possibilities **each** with a probability of occurring equal to $(0.95)^7 \cdot (0.05)^3$.

Multiplying by 36 will give us the desired probability:

The probability that the third success comes on the tenth trial

$= 36 \cdot (0.95)^7 \cdot (0.05)^3 \approx 0.0031$ ... not such a good chance of occurrence.

Chapter 6 Random Variables and Discrete Probability Distributions

We summarize the above discussion with the following:

The Negative Binomial Distribution

Let X be a discrete random variable representing the number of failures, k, that occur before the rth success. Then X is a negative binomial random variable whose probability distribution is defined as:

$$P(X = k) = \binom{r+k-1}{r-1} q^k p^r, \quad k = 0, 1, 2, 3...$$

Where k is the number of failures, r is the number of successes, p is the probability of a success, and q is the probability of a failure. Also, $p + q = 1$.

Mean and variance of the Negative Binomial Distribution:

$$\mu = E(X) = \frac{rq}{p} \text{ failures} \qquad VarX = \frac{rq}{p^2} \qquad \sigma = \sqrt{\frac{rq}{p^2}} \text{ failures}$$

NOTE:

We often replace the traditional notation for the combination $_nC_r$ with the notation $\binom{n}{r}$. Also, some prefer to see the above probability formula written as $P(X = k) = {}_{r+k-1}C_{r-1} \cdot q^k \cdot p^r$.

Either way, just realize that $\binom{r+k-1}{r-1} = {}_{r+k-1}C_{r-1}$.

Also, the number of successes, r, plus the number of failures, k, equals the total number of trials, n, of the experiment; $r + k = n$. Therefore, since $n = r + k$ the combination $\binom{r+k-1}{r-1}$ can be rewritten as $\binom{n-1}{r-1}$. The interpretation of this is, "since we know the location of one of the wins in the experiment, in how many ways can we distribute the remaining $r - 1$ successes to the $n - 1$ remaining locations?"

Let's apply the above formulas with an example.

EXAMPLE 6.26

1. An illness has spread throughout a small village and many of its inhabitants have pneumonia. A medical clinic gives its patients an X-ray to test for pnuemonia. It is known that 15% of the patients tested have the disease. Assuming a negative binomial distribution,

   a) What are the values of p & q?
   b) What is the probability the first positive test result occurs on the 9th person to be tested?
   c) What is the probability the fifteenth patient tested will be the third with pneumonia?
   d) What is the probability the tenth patient tested will be the second with pneumonia?
   e) What is the mean number of patients without pneumonia tested before the sixth patient with pneumonia is tested?

Solution

a) Successes don't always have to be *good* things. In this situation, the medical clinic is trying to detect who has pneumonia. The probability of successfully finding someone with the disease is 15% or 0.15. Therefore, the value of p is 0.15 and the probability of detecting someone without the disease is $q = 1 - p = 1 - 0.15 = 0.85$.

b) The probability the first positive test occurs on the 9th person means that you really want the probability of 8 failures *before* the first success. So, the value of $k = 8$ and the value of $r = 1$. Substituting into the formula for the negative binomial gives:

$$P(X=k) = \binom{r+k-1}{r-1} q^k p^r$$

$$P(X=8) = \binom{1+8-1}{1-1} (0.85)^8 (0.15)^1$$

$$P(X=8) = \binom{8}{0} (0.85)^8 (0.15)$$

$$P(X=8) = (0.85)^8 (0.15) \;\leftarrow\; \text{Note: This is the geometric formula } q^k \cdot p^r$$

$$P(X=8) \approx 0.0409$$

*$\binom{8}{0}$ is actually 1 since there is only 1 way to choose no successes from the first 8 trials.

It is very interesting to note that this question asked for the probability of 8 failures before the *first* success... this is *actually a geometric distribution example*. Substituting the value $r = 1$ in to the formula for the negative binomial formula gives,

$$P(X=k) = \binom{r+k-1}{r-1} q^k p^r = \binom{1+k-1}{1-1} q^k p^1 = \binom{k}{0} q^k p = 1 \cdot q^k p = q^k p$$

This shows us the geometric distribution is actually a *special case* of the negative binomial distribution when the value of r, the number of successes is 1!

c) If the 15th patient is the 3rd with pneumonia, then that means you wish to find the probability of 12 failures *before* the 3rd success. This means the value of k, *the number of failures*, is 12 and the value of r, *the number of successes*, is 3.

Substituting these values into the formula for the negative binomial gives:

$$P(X=k) = \binom{r+k-1}{r-1} q^k p^r$$

$$P(X=12) = \binom{3+12-1}{3-1} (0.85)^{12} (0.15)^3$$

$$P(X=12) = \binom{14}{2} (0.85)^{12} (0.15)^3$$

Using your calculator to evaluate gives,

$$P(X=12) \approx 0.0437$$

d) To find the probability the 10th patient is the 2nd with pneumonia, we must find the probability of 8 failures *before* the 2nd success. So, the value of k is 8 and the value of r is 2. Note, it is often popular to write a vertical line in the probability statement that shows both the value of k & r within it. For example, the expression $P(X = 8 | r = 2)$ means what is the probability k is 8 *given* that r is 2 (the vertical line in the statement means the word "given").

$$P(X=k|r) = \binom{r+k-1}{r-1} q^k p^r$$

$$P(X=8|r=2) = \binom{2+8-1}{2-1} (0.85)^8 (0.15)^2$$

$$P(X=8|r=2) = \binom{9}{1} (0.85)^8 (0.15)^2$$

Using your calculator to evaluate gives,

$$P(X=8|r=2) \approx 0.0552$$

e) This question is asking you to find the mean number of failures before the 6th patient with pneumonia is found. The formula for the mean of the negative binomial is given by $\mu = \frac{rq}{p}$. Using $r = 6$, with $p = 0.15$ and $q = 0.85$ gives $\mu = \frac{rq}{p} = \frac{6 \cdot (0.85)}{0.15} = 34$ failures.

This means that, *on average*, we would expect about 34 failures before the 6th person with pneumonia is found.

KEY TERMS

Acceptance Sampling 337
Binomial Experiment 322
Binomial Probability Distribution 321
Binomial Probability Formula 324
Continuous Random Variable 302
Discrete Probability Distribution 305
Discrete Random Variable 302
Failure 322
Geometric Experiment 347
Geometric Probability Distribution 348
Identical Trials 322
Negative Binomial Distribution 356
Negative Binomial Experiment 354
Mean of the Negative Binomial Random Variable 354
Mean of the Poisson Distribution 340

Mean/Expected Value of a Binomial Distribution, μ_s 335
Mean/Expected Value of a Discrete Random Variable 314
Means of the Geometric Random Variable 347
Mutually Exclusive 317
Poisson Probability Distribution 340
Probability Distribution 304
Probability Histogram 305
Random Experiment 300
Random Variable 300
Standard Deviation of a Binomial Distribution, σ_s 335
Standard Deviation of a Discrete Random Variable 315

Standard Deviation of the Geometric Variable 348
Standard Deviation of the Poisson Distribution 345
Success 322
Trial 322
Variance of a Discrete Random Variable 314
Variance of the Geometric Random Variable 347
Variance of the Negative Binomial Random Variable 354

SECTION EXERCISES

Note: The data sets in these Section Exercises can be downloaded in either Microsoft EXCEL or TI-84 format from the textbook's website at www.MyStatLab.com.

Section 6.2

1. A random variable is a variable that takes on different numerical values which are determined by _____.

2. A discrete random variable is a random variable that can take on a _____ or _____ number of values.

3. A random variable is continuous if the value of the random variable can assume _____ or _____ of values between any two possible values of the variable.

4. Which is **not** an example of a discrete random variable?
   a) The number of containers of milk sold at a grocery store during a given month.
   b) The number of customers using a credit card at a restaurant during lunch time.
   c) The height of police rookies for a given entering class.
   d) The amount of vitamins taken each morning for a family of four.

5. Which is **not** an example of a continuous random variable?
   a) The time it takes for an aspirin to take effect.
   b) The amount of cavities found for a visit to the dentist.
   c) The net weight of a box of oat-bran cereal.
   d) The systolic blood pressure level for a 25 year old female.

6. State whether the following random variables are discrete or continuous.
   a) The strength of a 2 × 4 in pounds per square inch.
   b) The height of a person.
   c) The number of conservatives within a family in the state of Virginia.
   d) The arrival time of an Amtrak train.
   e) The number of accidents at a particular stretch of a parkway.
   f) The amount of time it takes a computer to run a program.
   g) The batting average of a baseball player.
   h) The number of people attending a rock concert on a certain day.
   i) The weight of a UPS package.
   j) A person's age.
   k) The number of flight delays per hour at an international airport.
   l) The number of students absent per class each day.
   m) The time to finish the Boston Marathon.
   n) The lifetime of an alkaline size C battery.
   o) The pulse rate of a person after an aerobic class.

7. For each of the following situations, discuss the possible values of the indicated random variable and classify the random variable as discrete or continuous.

Situation	Random Variable
a) Taking a test	The time taken to complete the test
b) CD's sold	The number of CD's sold at Skyway records.
c) Tennis player	The number of volleys per point
d) Movie theater	The number of bags of popcorn sold

Section 6.3

8. A probability distribution is a distribution which displays the _____ associated with all the possible _____ of a random variable.

9. A histogram representing a probability distribution is called a _____. The values of the discrete random variable are scaled along the _____ axis, while the corresponding probability values are scaled along the _____ axis.

10. The probability assigned to a particular value of a discrete random variable of a probability distribution is always a number between _____ and _____.
11. Which is **not** a possible probability associated with a value of a discrete random variable?
    a) 0   b) 0.34   c) 1/19   d) 1.25
12. In a discrete probability experiment with four distinct outcomes, the sum of the first three probabilities is 0.80. What is the probability value of the remaining outcome?
    a) 0.2   b) 0.8   c) 1.8   d) cannot be determined
13. The probability distribution of a discrete random variable can be displayed using a graph. (T/F)
14. The sum of all the probabilities of a probability distribution can be less than one but never less than 0. (T/F)
15. Examine the probability assignments for the tossing of a biased die shown in the following table.

Possible Values of Die	Probability Assignment A	Probability Assignment B	Probability Assignment C	Probability Assignment D
1	2/6	1/9	1/6	1/2
2	1/6	1/9	2/6	1/12
3	0	1/9	1/6	1/12
4	1/6	1/9	2/6	1/6
5	1/6	1/9	−1/6	1/12
6	1/6	1/9	1/6	1/12

   a) State which assignments satisfy the probability characteristics for a probability distribution.
   b) State which assignments do not satisfy the probability characteristics for a probability distribution. Explain which probability rules are being violated.
16. Construct a probability distribution table which lists the possible values of the random variable shown within the following probability histogram along with the assigned probabilities.

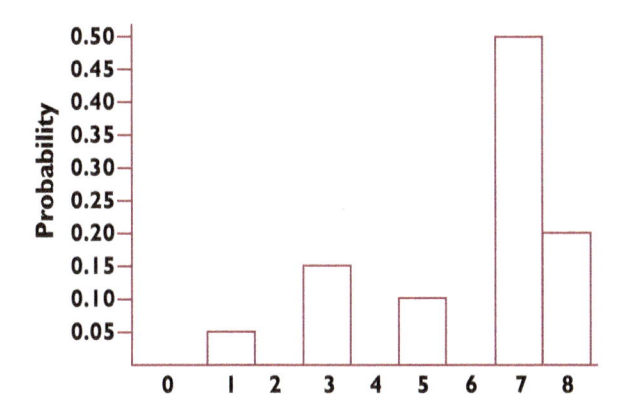

17. For each of the discrete probability distributions labelled I, II, III and IV given below.
    a) Identify the values of the random variable.
    b) Construct a probability histogram and determine its shape.
    c) Determine the sum of all the probabilities. What is the value of this sum? Did you expect this value? Explain.
    d) Find the mean.
    e) Calculate the standard deviation.
    f) Estimate the probability that a random variable is within 1 standard deviation of the mean.
    g) Estimate the probability that a random variable is within 2 standard deviations of the mean.

Probability Distribution I		Probability Distribution II	
x	P(x)	x	P(x)
0	1/4	1	1/2
1	1/4	2	1/4
2	1/4	3	1/8
3	1/4	4	1/8

Probability Distribution III		Probability Distribution IV	
x	P(x)	x	P(x)
−1	1/6	2	1/8
0	2/3	4	1/3
1	1/6	6	13/24

18. Professor Kwizz used her previous years attendance records from her statistics classes of 32 students to construct the following probability distribution. The table lists the number of students who are absent on an exam day and its assigned probability.

Number of Students Absent per Exam	1	2	3	4	5	6	7	8
Percent of Students	15%	25%	20%	15%	5%	10%	5%	5%

   a) Construct a probability distribution for this distribution.
   b) Determine the probability that the number of students absent for Professor's Kwizz class on a given exam day is:
      1) exactly 7.
      2) more than 5.
      3) between 1 and 6.
      4) less than 4.
      5) at most 3.
      6) at least 4.
19. According to the latest figures, the Census Bureau states that 10% of the families have exactly 3 earners within the family. Consider the experiment of randomly selecting 3 families from the population of all U.S. families. If we let X represent the number of families having exactly 3 earners in this sample, then:
    a) Construct a sample space of all the possible outcomes for this experiment.
    b) Determine the probability of each one of the possible outcomes of the sample space, assuming each family has a 10% chance of having 3 earners.
    c) Construct a probability distribution for the number of possible families who have 3 earners in a sample of 3 families.
    d) Determine the most likely value of X.
    e) Determine the probability that at least one family of the 3 families randomly selected has 3 earners.
    f) Determine the probability that more than one family of the 3 families randomly selected has 3 earners.
    g) Determine the probability that none of the 3 families randomly selected has 3 earners.
    h) Determine the probability that at most one family of the 3 families randomly selected has 3 earners.

Section 6.4

20. The mean of a discrete random variable X, denoted by _____, is calculated by the formula: _____, or _____.

21. The standard deviation of a discrete random variable X, denoted by _____, is determined using the formula: _____.

22. If a fair coin is tossed 1000 times, what is the expected number of heads?
 a) 50 b) ½ c) 1 d) 500

23. The mean of a discrete random variable represents:
 a) The value of the next outcome of the experiment.
 b) An actual value of the experiment.
 c) The sum of all the probabilities of the experiment.
 d) The long-run average outcome of the experiment.
 e) A non-weighted average of the experiment.

24. The expected value for a discrete random variable is called the mean.
 (T/F)

25. The standard deviation of a random variable is a measure of chance variation, that is, a measure of what is expected and what actually happens for a probability experiment.
 (T/F)

26. During tax season, the probability distribution of the number of calls per hour to the IRS office in Nassau County regarding tax filing questions is shown in the following table.

Number of Calls X	Probability P(X)	Number of Calls X	Probability P(X)
5	0.01	30	0.10
10	0.01	35	0.10
15	0.02	40	0.15
20	0.05	45	0.23
25	0.08	50	0.25

 a) Determine the mean number of tax filing calls per hour to this IRS office.
 b) Determine the probability that X will be within 5 of its mean value.
 c) Find the standard deviation of X.
 d) What is the probability that the number of calls will be within one standard deviation of the mean value?
 e) What is the probability that the number of calls will be more than two standard deviations from the mean value?

27. The Department of Transportation (D.O.T.) of a small town is studying the request of a civic association to install a traffic light at the busy intersection of Maple and Munster Streets within the town. Based upon the D.O.T.'s investigation, they devised the following probability distribution which models the number of accidents per month at this intersection.

Number of Accidents	2	3	4	5	6	7	8	9	10	11	12	13	14
Probability	.02	.04	.05	.10	.10	.05	.20	.05	.25	.06	.06	0	.02

The D.O.T.'s policy regarding the installation of a traffic light is that the monthly number of accidents must be greater than 8. Based upon this criteria, should the D.O.T. install a traffic light at this intersection?

28. Professor Par Kay, an economics professor at a local college, assigns grades in her *bread and butter* freshman introductory course using the following numerical scale: A = 4.0, B+ = 3.5, B = 3.0, C+ = 2.5, C = 2.0, D+ = 1.5, D = 1.0, and F = 0. Based upon previous class records, the probability distribution of the grades students receive in her introductory economics class is:

Grade X	4.0	3.5	3.0	2.5	2.0	1.5	1.0	0
Probability P(X)	0.13	0.15	0.25	0.20	0.10	0.04	0.06	0.07

What is the probability that a student selected at random from her class will receive:
 a) A grade of B or greater?
 b) A failing grade, if D is the minimum passing grade?
 c) A passing grade?
 d) What is the average grade that a student receives in her class?

Section 6.5

29. For an experiment to be binomial, it must satisfy the following conditions:
 a) There are n _____ trials.
 b) The n identical trials are _____.
 c) The outcome for each trial can be classified as either a _____ or a _____.
 d) The probability of a _____ is the same for each trial.

30. Which of the following probability experiments cannot be modeled by a binomial distribution?
 a) Tossing a biased coin 4 times.
 b) Randomly selecting five 12 oz soda cans from a production line and measuring the amount of soda in each bottle.
 c) Randomly selecting 6 cards with replacement from a deck of 52 playing cards.
 d) Rolling a fair die 7 times.
 e) A basketball player takes 8 foul shots.

31. The number $_nC_s$ gives us the number of ways "s" successes can occur in "n" independent trials.
 (T/F)

32. To calculate the probabilities for a binomial experiment, we use the binomial probability formula. This is symbolized as:
 P(s successes in **n** trials) = $_nC_s\, p^s\, q^{(n-s)}$
 where: n = number of _____ trials
 s = number of _____
 n−s = number of _____
 $_nC_s$ = number of _____ s successes can occur in **n** trials
 p = probability of a _____
 q = 1 − p = probability of a _____

33. For the binomial experiment of tossing a fair coin 4 times the number of ways to get 3 successes in 4 independent trials is: $_4C_3 =$ _____. If you consider a head as a success, the four ways of getting 3 successes would be _____, _____, _____, and _____.

34. For a binomial experiment, where n = 5, s = 3, and p = ⅓, calculate:
 a) P(3 successes in 5 trials) = P(3) = $_5C_3 \left(\tfrac{1}{3}\right)^3 \left(\tfrac{2}{3}\right)^2$
  = ___ ___ ___
  = _____.
 b) P(5 successes in 5 trials) = $_5C_5 \left(\tfrac{1}{3}\right)^5 \left(\tfrac{2}{3}\right)^0$
  P(5) = ___ ___ ___
  = _____.

35. For a binomial experiment, if: P(s = 5) = $_7C_5 \left(\tfrac{1}{4}\right)^5 \left(\tfrac{3}{4}\right)^2$
 then: n = _____, s = _____, p = _____, and q = _____.
 The number of ways to get 5 successes in 7 trials is _____.

The number of successes is _____ .

$_7C_5$ = _____ .

36. A binomial probability distribution is an example of a _____ probability distribution.

37. In a binomial probability distribution, the probability associated for each of the possible values of the discrete random variable can be computed using the _____ _____ _____.

38. In a probability histogram, the probability of a value of the random variable X is equal to the _____ of the bar pertaining to the value of X.

39. The mean of a binomial distribution, denoted by _____, is given by the formula: _____.

40. The standard deviation of a binomial distribution, denoted by _____, is given by the formula: _____.

41. The procedure of using a small inexpensive sample of items which is randomly selected from a large shipment to decide whether to accept or to reject the entire shipment is called _____.

42. If in a binomial experiment n = 20 and p = 0.4 then μ_s =
    a) 8   b) 10   c) 12   d) 20   e) 80

43. If in a binomial probability experiment n = 16 and p = 0.1 then σ_s =
    a) 4
    b) $\sqrt{16(0.1)}$
    c) $\sqrt{16(0.1)(0.1)}$
    d) $\sqrt{(0.1)(0.9)}$
    e) $\sqrt{16(0.1)(0.9)}$

44. If the probability that a company rejects a shipment is 0.10, then the probability that the company will not reject the shipment is:
    a) 0.05   b) 0.10   c) 0.50   d) 0.90
    e) cannot be determined

45. In a binomial probability experiment, the probability of success, P(s), should be less than 0.5.
    (T/F)

46. Acceptance sampling tests all the items in a population to infer acceptance of a sample from the population.
    (T/F)

47. The probability that a company will not reject a shipment is the same as the probability that the company accepts the shipment.
    (T/F)

48. If you guess at the answers to five multiple choice questions where each question has four choices, then what is the probability that:
    a) You guess them all wrong?
    b) You guess at least 4 correct?
    c) You guess less than 3 correct?

49. The probability that a softball player named Casey gets a base hit during a single time at bat is 4/10. If he goes to bat 4 times in his next game (which is played in Mudville), determine:
    a) The probability that he gets 4 hits.
    b) The probability that he gets no hits, causing no joy in Mudville.
    c) The probability that he gets 3 or more hits.

50. Matthew "The Magician" claims his hands are quicker than your eyes. To illustrate this he places a red ball under one of three cups, does some quick manipulations and asks you to tell which cup is covering the red ball. Since his hands are indeed faster than your eyes, calculate the probability that you do not guess the correct cup ten times in succession. (Calculate probability to three decimal places.)

51. A survey of undergraduate college students by NellieMae, a student loan company, stated only 20 percent of undergraduate college students said they pay off their credit cards in full each month. Assume this is true for all credit card carrying undergraduate college students. What is the probability that out of 10 randomly selected undergraduate college students with a credit card statement balance:
    a) none of them pay off in full their entire credit card balance by the due date?
    b) at most 3 pay off in full their entire credit card balance by the due date?
    c) exactly 3 pay off in full their entire credit card balance by the due date?
    d) at least 3 pay off in full their entire credit card balance by the due date?
    e) at least 3 do not pay off in full their entire credit card balance by the due date?

52. From past experience, Caryl, a golfer, knows that she will hit her drive into a sand trap from the tee one-fourth of the time. Using this figure, what is the probability that she will hit the ball into a sand trap on exactly four of the first nine holes?

53. Medical records show that 40% of all persons affected by a certain viral illness recover. Ten people with this illness were selected at random and injected with the vaccine.
    a) What is the probability that eight of the ten people recover?
    b) What is the probability of at most 4 recoveries?

54. In Denmark the probability that a carton of a dozen eggs contains one rotten egg is 1/20. If a boy named Hamlet buys 5 cartons of eggs, what is the probability that at least 1 carton contains a rotten egg (causing Hamlet to mumble: "Something is rotten in Denmark!")?

55. A slot machine has four windows. The wheel behind each window has the same five identical objects. They are: cherry, grape, peach, star and jackpot. After the handle is pulled, the four wheels revolve independently several times before coming to a stop. If the slot machine is played once, what is the probability that:
    a) Four jackpots appear in the windows?
    b) At most 1 grape appears in the windows?
    c) Less than 3 stars appear in the windows?
    d) Only fruits appear in the windows?

56. Four out of five patients who have an artery bypass heart operation are known to survive at least three years. Of five patients who recently had the operation, what is the probability that:
    a) All five will survive at least 3 years?
    b) At most two will *not* survive at least 3 years?
    c) At least four will survive at least 3 years?

57. A psychologist claims that he has trained a rat named Algernon to select the correct path of a maze that leads to food 90% of the time. Based upon this assumption, what is the probability that Algernon will select the correct path six times in eight previously unknown mazes?

58. Anne Marie, a botanist, is studying a new hybrid Kentucky blue grass seed. It is known that these grass seeds have a 95% probability of germinating. If Anne Marie plants six seeds, what is the probability that:
a) Exactly four seeds will germinate?
b) At least five will germinate?
c) At most one will not germinate?

59–69. For each of the binomial probability experiments listed in problems 48 to 58:
a) Determine the binomial probability distribution.
b) Construct a probability histogram. Indicate the shape of the histogram and the value of p. What relationship do you notice between the shape and the value of p?
c) Calculate the binomial mean, μ_s, and the binomial standard deviation, σ_s.

70. Will Short, a fuse inspector for an automobile manufacturer, randomly selects and tests a sample of 20 fuses from a shipment of 10,000 fuses to decide whether to reject or not reject (accept) the entire shipment. If at least 10% of the fuses are defective within the sample, he will reject the entire shipment. Assuming that the defective fuses occur independently and historically, the percent of defective fuses is 15%, then:
a) Calculate the probability that the entire shipment will not be rejected (or accepted)?
b) Calculate the probability that the entire shipment will be rejected?

71. Ms. Kathleen Shape, the manager of a racquet sport complex, has ordered 1000 cans of low quality racquet balls. Based on prior orders, 20% are usually defective. Before accepting the shipment she decides to reject the shipment if 25% or more of a sample of 16 balls are defective.
a) What is the probability that Ms. Shape will accept the order of 1000 cans?
b) What is the probability that Ms. Shape will reject the order of 1000 cans?

Section 6.6

72. The Poisson distribution has 2 parameters, n and p. (T/F)

73. The mean and variance of the Poisson distribution are the same. (T/F)

74. Let X be a Poisson random variable with mean equal to 3 accidents per week. The value for λ in this situation is _____. If the time interval is changed to a year then the value of λ will change to _____.

75. For small values of λ the shape of the Poisson distribution appears to be _____ _____, but as λ gets large the shape of the distribution tends to be _____ _____.

76. The number of calls received by a fire department is Poisson with $\lambda = 2$ per hour.
a) How would the value of λ change if we were only interested in a 30 minute time period?
b) Say you wanted to calculate a probability involving the number of calls received between 10:00 a.m. and 1:30 p.m. What would the value of λ be in this case?
c) What is the mean number of calls received by the fire department in 15 minutes?

77. The distribution of a random variable, X is Poisson. Construct a probability distribution table for k = 0 to k = 3,
a) if $\lambda = 1.25$.
b) if the mean is 9.
c) if the standard deviation is 16

78. For a Poisson random variable with a mean equal to 15.
a) What is the probability that k = 5?
b) What is the probability that k at most 5?
c) What is the probability that k is greater than 5?

79. Suppose the average number of car accidents on the highway in one week is Poisson with mean equal to 2.
a) What is the probability of no car accident in one week?
b) What is the probability of 2 car accidents in three weeks?
c) What is the probability of more than one car accident in one week?

80. An auto insurance company has determined that the average number of claims against the coverage of a policy is Poisson with mean .5 claims per year. What is the probability that a policyholder will file:
a) 1 claim in a year
b) more than 1 claim in a year
c) less than 2 claims in a year
d) 2 claims in two years

81. Suppose the average number of calls made by a telemarketer in 5 minutes is 8.
a) What is the probability of 10 calls in 5 minutes?
b) What is the probability of 20 calls in 15 minutes?

82. Shoppers enter a supermarket at an average of 90 per hour.
a) What is the probability that exactly 10 shoppers will enter the department store between 1:00 and 1:10 p.m.?
b) What is the probability that at least 40 shoppers will enter the department store between 9:00 and 9:30 p.m.?

83. The number of students that call on Professor Deporto's office hours follows a Poisson distribution with a mean of 300 students per semester.
a) What is the value of λ in this situation? What is the value of the standard deviation?
b) How would the value of λ change if we were only interested in a weekly time period (Assume that a semester has 15 weeks)? How about the standard deviation?
c) What is the probability that for a given week the number of students visiting the professor is 12?
d) What is the probability that for a given week the number of students visiting the professor is less than 12?
e) What is the probability that for a given week the number of students visiting the professor is 20?

Section 6.7

84. When working with the geometric probability distribution, if the chances of a success are 22%, then the chances of a failure are _____.

85. The geometric probability formula, $P(X = k) = q^k p$, calculates the probability of _____ failures before your first _____.

86. The negative binomial probability formula, $P(X = k) = \binom{r + k - 1}{r - 1} q^k p^r$, calculates the probability of _____ failures before your _____ success.

87. The formula for the mean of the geometric distribution is given by _____, where the formula for the negative binomial distribution is given by _____.

88. The formulas for the geometric distribution are a special case of the negative binomial distribution formulas when the value of _____ is _____.

89. The probability that a gambler wins on any try for betting on the color BLACK in the game of roulette is 18/38.
a) What is the probability that the gambler losses on a particular bet on BLACK?
b) What is the probability that the gambler losses 5 times in a row before he wins for the first time?

c) What is the probability that the gambler has at most 3 losses before winning on BLACK?
d) What is the mean number of losses that the gambler has before the first win?
e) What are the variance and standard deviation number of losses before the first win?

90. Kathleen likes to play the casino game CRAPS. Let's assume that the probability of winning on the PASS bet is 244/495.
   a) What is the probability that she doesn't win betting on PASS?
   b) What is the probability that she losses 4 times in a row before she wins for the first time?
   c) What is the probability that she loses at most 4 times before she wins for the first time?
   d) What is the mean number of losses that a Kathleen should have before the first win?
   e) What are the variance and standard deviation number of losses before the first win?

91. Oftentimes, when given a choice on test format, students like a multiple-choice type exam. Suppose your instructor offers you a multiple-choice exam where each question has four choices. For this situation each question has a 25% probability of success. Say you decide that you are unprepared and that you must guess on all questions on this exam.
   a. What is the probability that you guess on a question and it is *not* correct?
   b. What is the probability that your first correct guess response occurs on the 10th question?
   c. What is the probability of at *most* 5 failed guesses before your first successful guess?
   d. What is the mean number of incorrect guess before your first successful guess?
   e. What are the variance and standard deviation number of incorrect guesses before the first correct guess?

92. Allison, a college basketball player, has a 30% success rate when practicing a shot from 3-point range. Her warmup regiment is that she moves around the 3-point range when she makes a shot successfully.
   a) What is the probability that she fails to make a 3-point shot?
   b) What is the probability her first successful 3-point shot occurs on her 4th attempt?
   c) What is the probability her first successful 3-point shot occurs no later than her 4th attempt?
   d) What is the mean number of missed shots before her first successful 3-point shot?
   e) What are the variance and standard deviation number of missed shots before the first successful 3-point shot?

For problems 93 to 96, assume a *negative binomial* distribution.

93. Let's revisit the roulette problem in Problem 89: The probability that a gambler wins on any try for betting on the color BLACK in the game of roulette is 18/38.
   a) What are the values of p and q?
   b) What is the probability the gambler wins for the 3rd time on the 10th bet on BLACK?
   c) What is the probability the gamblers 4th win occurs on his 15th bet on BLACK?
   d) What is the mean number of losses the gambler will incur prior to his 20th success?

94. Let's revisit the CRAPS problem in problem 90: Kathleen likes to play the casino game CRAPS. Let's assume that the probability of winning on the PASS bet is 244/495.
   a) What are the values of p and q?
   b) What is the probability that Kathleen wins for the 12th time on the 20th bet on PASS?
   c) What is the probability that Kathleen loses 18 times prior to her 12th win while betting on PASS?
   d) What is the mean number of loses Kathleen will incur before winning for her 10th time when betting on PASS?

95. Let's revisit the multiple-choice question from problem 91. Oftentimes, when given a choice on test format, students like a multiple-choice type exam. Suppose your instructor offers you a multiple-choice exam where each question has four choices. For this situation each question has a 25% probability of success. Say you decide that you are unprepared and that you must guess on all questions on this exam.
   a) What are the values of p and q?
   b) What is the probability that your 20th correct guess occurs on the 50th test question?
   c) What is the probability that you have 7 incorrect guesses prior to your 3rd correct guess?
   d) What is the mean number of incorrect guesses prior to your 30th correct guess?

96. Let's revisit the three point college basketball question from problem 92. Allison, a college basketball player, has a 30% success rate when practicing a shot from 3-point range. Her warmup regiment is that she moves around the 3-point range when she makes a shot successfully.
   a) What are the values of p and q?
   b) What is the probability that Allison's 5th successful 3-point shot occurs on her 10th try?
   c) What is the probability that Allison's 4th shot will be her second successfully shot?
   d) What is the mean number of missed shots Allison will incur before her third successful shot occurs?

CHAPTER REVIEW

97. A random variable can be classified as either a _____ or _____ random variable depending upon the numerical values that it can assume.

98. A discrete random variable is a random variable that can assume any value between any two possible values of the variable.
(T/F)

99. Determine which of the following random variables are discrete or continuous?
   a) A person's chest size.
   b) The number of customer's arriving at a credit union office during any hour.
   c) The number of TV's sold at an appliance store during a given month.
   d) The time required to travel to work.

364 Chapter 6 *Random Variables and Discrete Probability Distributions*

e) The number of hats a person owns.
f) The actual amount of ice tea in a 12 oz. bottle.
g) The number of early arrivals at an airport on a given day.
h) The number of times *00* comes up on a particular roulette wheel during a certain day.
i) The number of points scored in a basketball game.
j) The number of inches of rainfall per month in San Francisco.
k) The number of air bags per car.
l) The number of people who visit Disney World each day.
m) The blood pressure of a middle-aged male.
n) The number of questions asked during a class test.
o) The amount of automobile gas consumed each week.

100. In general, the probability distribution of a discrete random variable can be shown by use of a ____, a ____ or a ____.

101. The sum of all the probabilities of a discrete probability distribution must always be equal to ____.

102. A probability distribution is a distribution which displays the frequencies associated with all the possible values of a random variable.
(T/F)

103. Let X be the number of defective radial tires obtained when randomly selected batch of eight tires is chosen from the production process. Using past quality control records of this production process, the probability distribution of X serves as a model of this production process and is shown in the following table.

Number of Defective Tires in the Batch X	0	1	2	3	4	5	6	7	8
Probability P(X)	0.85	0.10	0.03	0.01	0.005	0.002	0.002	0.001	0.000

What is the probability that a batch of eight tires randomly selected from this production process will have:
a) At most 2 defective tires?
b) At least 5 defective tires?
c) More than 5 defective tires?
d) If a batch of eight tires which were randomly selected from this production process had 7 defective tires, would you consider this an expected result or an unusual result?
e) If you are a quality control engineer in charge of monitoring this production process and this result occurred, what would this result mean to you?

104. The mean value of a discrete random variable is what can be expected the next time the probability experiment is attempted.
(T/F)

105. An investment analyst ranks a mutual fund by analyzing the type of investments made by the fund. According to the analyst's investigation of this fund, she constructs a probability distribution where X is the risk of the investment.

Risk of Investment X	1	2	3	4	5
Probability P(X)	0.15	0.30	0.15	0.25	0.15

a) Determine the average risk of this fund.
b) How would you classify this fund's risk, using the following classifications: lowest risk is 1, medium risk is 3, and highest risk is 5. Explain your answer.

106. Consider the experiment of tossing a fair coin until a tail appears, or until the coin has been tossed four times, whichever occurs first.
a) Construct a probability distribution for the number of heads.
b) What is the probability that less than three heads will occur?
c) What is the probability that at most three heads will occur?
d) What is the probability that at least three heads will occur?
e) What is the probability that more than three heads will occur?
f) What is the probability that a tail will occur on the first toss?
g) What is the probability that a tail will occur on the fourth toss?
h) What is the probability that a tail will not occur?
i) What is the average of number of times the coin will be tossed for this experiment?
j) What is the standard deviation of the number of times the coin will be tossed for this experiment?

107. The number and type of headache pain tablets taken by an individual can vary. Two types of headache pain tablets, aspirin and ibuprofen, are listed in the following table along with the probability for the number of tablets taken by an individual.

	Aspirin	Ibuprofen
X	P(X)	P(X)
1	0.4	0.6
2	0.5	0.3
3	0.1	0.1

For each of the discrete probability distributions:
a) Identify the values of the random variable.
b) Construct a probability histogram and determine its shape.
c) Calculate the sum of all the probabilities. What is the value of this sum? Did you expect this value? Explain.
d) What is the expected number of tablets for each type of pain tablet?
e) Calculate the standard deviation.
f) Is an individual more likely to take more than 2 aspirin or more than 2 ibuprofen? How did you arrive at your conclusion?

108. Craig, a young basketball player, has a 80% chance of making a free throw based on his performance over the season. If he attempts 10 free throws during his next game, then:
a) What is his expected number of misses?
b) What is his expected number of hits?
c) What is the probability that he hits all 10 free throws?

d) What is the probability that he misses all 10 free throws?
e) What is the probability that he hits at least 2 free throws?
f) What is the probability that he misses at most 1 free throw?
g) In the probability histogram of the probability distribution of making free throws, what is the height of the bar representing the probability of making 9 free throws?
h) In the probability histogram of the probability distribution of not making free throws, what is the height of the bar representing the probability of missing 4 free throws?

109. During the football season, the sportswriters Craig, Matthew and Debbie have a 0.2, 0.5, and 0.8 probability of predicting the correct winning team respectively. For the next five football games each sportswriter predicts:
a) Construct a binomial probability distribution of the number of correct predictions for each sportswriter.
b) Construct a probability histogram for each sportswriter.
c) Identify the shape of each histogram. Based on the relationship between the shape of the histogram and the probability of making a correct prediction, what would be the shape of a histogram for a probability of 0.3? 0.7?
d) Estimate the balance point for each histogram.
e) Determine the mean number of correct predictions for each sportswriter.
f) Find the standard deviation of the number of correct predictions for each sportswriter.
g) What is the sum of all the probabilities of each binomial probability distribution? Did you expect these results? Explain.

110. One of the characteristics of a binomial distribution is that it involves probability experiments where each outcome can be classified in one of _____ possibilities.

111. In a binomial experiment, each trial must be:
a) Mutually exclusive
b) Dependent
c) Independent
d) Fixed

112. For a binomial experiment, n = 6 and s = 3, then: $_nC_s = {_6C_3}$ = _____ . Therefore, the number of ways to get _____ successes in _____ independent trials is _____ .

113. For a binomial experiment, if p = 4/5, then q = _____ .

114. For a binomial distribution, the mean is denoted by the symbol:
a) μ **b)** m   **c)** μ_x **d)** σ_x **e)** μ_s

115. For a binomial distribution, the standard deviation is denoted by the symbol:
a) σ **b)** s   **c)** σ_x **d)** s_x **e)** σ_s

116. The mean of a binomial distribution can be determined by the formula:
a) pq   **b)** npq   **c)** n(1–p)   **d)** np
e) $\sqrt{np(1-p)}$

117. The standard deviation of a binomial distribution can be determined by the formula:
a) pq   **b)** npq   **c)** n(1–p)   **d)** np
e) $\sqrt{np(1-p)}$

118. A binomial probability experiment is an example of a discrete probability experiment. (T\F)

119. A *biased* coin is flipped 5 times. If the probability is 1/3 that it will land heads on any toss, calculate the probability that the coin:

a) Lands heads all five times.
b) Lands heads at least three times.
c) Lands tails at most one time.

120. Brooke and Matthew, a young couple, plan to have five children. Assuming that the probability of having a boy is 1/2 determine:
a) The probability of having 2 boys and 3 girls.
b) The probability of having only 1 boy.
c) The probability of having at least 4 boys.

121. If you roll a pair of fair dice, the probability that the sum is seven is 1/6. If the dice are rolled four times, what is the probability that:
a) None of the outcomes are seven?
b) All of the outcomes are seven?
c) At most 1 of the outcomes is a seven?

122. Suppose an urn contains 10,000 balls, where 4,000 are red and 6,000 are blue. If five balls are randomly selected one at a time, with replacement, from this urn find the probability that there were:
a) Exactly two blue balls selected.
b) More than four red balls selected.
c) At most two blue balls selected.
d) No red balls selected.

123. Joe, the vegetable man, has ordered a truckload of grapefruits. He usually finds that 15% of the grapefruits are not acceptable for sale. Joe decides to reject the shipment if a sample of 20 grapefruits has at least 5 bad grapefruits.
a) What is the probability that Joe will accept the truckload of grapefruits?
b) What is the probability that Joe will reject the truckload of grapefruits?

124. The distribution of a random variable, X is Poisson. Construct a probability distribution table for k = 0 to k = 3,
a) if λ = 2.25.
b) if the mean is 5.
c) if the standard deviation is 1.

125. The distribution of a random variable, X is Poisson. Construct a probability distribution table for k = 0 to k = 5,
a) if λ = 8.35.
b) if the mean is 12.
c) if the standard deviation is 9.

126. The distribution of a random variable, X is Poisson with a mean equal to 7.
a) Construct a probability distribution for k = 0 to k = 6.
b) What is the probability that k = 3?
c) What is the probability that k < 3?
d) What is the probability that k is at least 4?

127. During the first week of classes, the mean number of students entering the bookstore per 12 hour shift is 250.
a) What is the value of λ in this situation? What is the value of the standard deviation?
b) How would the value of λ change if we were only interested in an hourly time period? How about the standard deviation?
c) What is the probability that for a given 12 hour shift the number of students visiting the bookstore is 280?
d) What is the probability that for a given week the number of students visiting the bookstore is less than 1800?
e) What is the probability that for a given week the number of students visiting the bookstore is 1800?

128. A couple decides to start having children and is really interested in having boys. Let's assume the probability of a birth of a boy is equal to a birth of a girl, and the couple only has one child per birth (no twins or triplets etc.).
   a. What is the probability the couple's first child is a boy? What is the probability it is a girl?
   b. What is the probability the first-born son occurs on the couple's fourth child?
   c. What is the probability the couple has at most 6 girls before their first-born son?
   d. What is the mean number of girls the couple will have prior to their first-born son?

129. Refer to the couple's situation in question 128.
   a. What is the probability the couple's 4th child is their second son?
   b. What is the probability the couples 4th son occurs on their 6th child born?
   c. What is the probability the couples 5th son occurs on their 5th child? (i.e. They have 5 boys in a row).
   d. What is the mean number of girls the couple will have prior to their 3rd son being born?

WHAT DO YOU THINK?

130. *To Foul or Not to Foul — That Is the Question* — A basketball team is losing by 3 points with only seconds left in the game. The losing team has the ball and is about to attempt a 3-point shot for a possible tie. *Should the winning team intentionally foul the shooter, thereby preventing the 3-point shot?* To answer this question, you must realize that when a basketball player is fouled while attempting a 3-point shot, the shooter is allowed to take 3 free-throws which count 1 point for each free-throw that is made. In addition, if the basketball player who is fouled while attempting the 3-point shot makes the shot, then the shot counts and the player is given one free-throw for a possible 4-point play.

Taking all these factors into consideration: *should the defensive team foul the shooter who is attempting the 3-point shot?* To answer this question, we need some information regarding the probability of a player making a 3-point shot and the chance of a player making a free-throw. According to the 2000 LYCOS Sports Almanac, the top 10 NBA players made approximately 44% of their 3-point shots, and 89% of their free-throws. Assuming the best shooter takes the 3-point attempt, and using the Sports Almanac information, determine:
   a) An estimate for a top NBA shooter making a free-throw and a 3-point shot?
   b) The probability of the player making three consecutive free-throws, assuming independence?
   c) The two possible outcomes of a 3-point shot, and their associated probabilities?
   d) The expected or average value of a 3-point attempt?
   e) The two possible outcomes of a free-throw, and their associated probabilities?
   f) The expected or average value of a free-throw attempt?
   g) The expected or average value of three free-throw attempts?
   h) Compare the expected (or average value) of a 3-point attempt to the three free-throw attempts, and determine which has a greater value? Interpret this comparison.
   i) What would your advice be to player defending the 3-point shooter — *to foul or not to foul*?

131. Examine the following U.S. Bureau of the Census table on Families by Size of Family: 2009.
   a) Identify the random variables within the table.
   b) Discuss the type of random variable and the possible values for each variable.
   c) Construct a probability distribution for the size of family for the categories: Total, Married-Couple families and Other families for United States, Metropolitan and Female Householder Nonfamily. Replace the seven or more persons family size with just seven persons.
   d) Construct a probability histogram for each of the distributions described in part (c). Explain the differences or similarities for these histograms.
   e) Determine the mean and standard deviation for each of the distributions discussed in part (c).
   f) Compare the average family size for each distribution in part (e) and explain any differences between your results to the average size computed by the Census Bureau. Are there any discrepancies between your results and results of the Census Bureau? Do you think that this could be attributed to the fact that you changed the category *7 or more persons to just 7 persons*?
   g) What do you think would happen if we replaced the category 7 or more persons by 8 persons? Re-calculate the average family size using 8 persons instead of 7 persons. How does this average result compare to the Census Bureau's average family size? Which result gave you a better estimate of the average family size, using 7 persons or 8 persons for the category 7 or more persons?
   h) Do you think that a different value, like 9 or 10 or 11 or 12 persons, would provide a closer estimate to the Census Bureau's average family size? Try these values and explain what value you think the Census bureau might have used to compute their average family size?

TABLE 6.17 Households by Type and Tenure of Householder for Selected Characteristics: 2009

(Numbers in thousands.)

		Family Households				Nonfamily Households		
	Total	Total	Married Couple	Male Householder	Female Householder	Total	Male Householder	Female Householder
ALL HOUSEHOLDS	117,181	78,850	59,118	5,252	14,480	38,331	17,694	20,637
SIZE OF HOUSEHOLD								
One member	31,657	0	0	0	0	31,657	13,758	17,899
Two members	39,242	33,879	25,851	2,168	5,860	5,363	3,027	2,336
Three members	18,606	17,785	11,686	1,599	4,500	821	551	270
Four members	16,099	15,761	12,478	875	2,409	338	244	94
Five members	7,406	7,308	5,870	366	1,072	99	79	19
Six members	2,640	2,610	2,081	161	368	30	21	9
Seven or more members	1,529	1,506	1,153	83	270	23	14	9
Total Persons	293,928	246,899	186,579	16,010	44,314	47,034	23,060	23,969
Average per Family	2.508324728	3.131249207	3.156043845	3.048362529	3.060359116	1.227048603	1.303266644	1.161457576

132. The baseball World Series is a best-of-7 game series between the American and National League Champions, that is, the series ending as soon as one team wins 4 games. If we assume that both teams are evenly matched, determine the probability that the World Series will end:
   a) In 7 games.
   b) In 6 games.
   c) In 5 games.
   d) In 4 games.
   e) Construct a probability histogram for the number of games the World Series will last.
   f) Estimate the mean number of games that the World Series will last from the probability histogram.
   g) Determine the expected number of games the World Series will last. Compare this result to the mean result found in Case Study 6.2 and explain why they are not the same? What is the difference between these two probability distributions?
   h) Determine the standard deviation of the number of games the World Series will last. Compare this standard deviation result to the result found in Case Study 6.2 and explain why they are not the same?
   i) Determine the probability that the World Series will last at least 6 games?
   j) Explain why the probability of the World Series ending in 6 games and 7 games are the same?

PROJECTS

133. Randomly select 150 students to survey. Determine for each student the size of their respective family. Using these data,
   a) Construct a probability distribution table.
   b) Construct a probability histogram.
   c) Determine the mean and standard deviation.
   d) State the shape of the histogram.
   Research the Bureau of the Census on the latest data pertaining to the U.S. family size. Using these data,
   e) Construct a probability distribution table.
   f) Construct a probability histogram.
   g) Determine the mean and standard deviation.
   h) State the shape of the histogram.
   Compare your results to the Census Bureau's data.

134. Administer the following test to 32 students who have no previous knowledge about the material within each question.
   a) Complete the following table with your test results.
   b) Construct a histogram using the data of the test results.
   c) Compare the test results and the histogram of the test results to the binomial experiments table and histogram.
   Test: Answer each question True or False.
   1) If $f(x) = e^x$, x is a real number, then:
   $$\int f(x)\,dx = f(x) + C, \text{ where C is a constant.}$$
   2) Set A is an open set of a metric space, M, if for every x belonging to A, there exists a $\delta > 0$ such that the interval $(x - \delta, x + \delta) \in A$.
   3) In 1733, DeMoivre developed the equation for the T distribution.
   4) For all values of x, $e^x = 1 + x + \dfrac{x^2}{2!} + \ldots + \dfrac{x^n}{n!} + \ldots$
   5) If a function f(x) is differentiable at x_0, then it is continuous at x_0.

   Test Results:

Number of Correct Answers	Number of Students per Correct Answer
0	
1	
2	
3	
4	
5	

CHAPTER TEST

135. A random variable having only a limited number of values with breaks between successive possible values of the random variable is:
a) discrete
b) continuous
c) numerical
d) categorical

136. For each of the following situations, discuss the possible values of the indicated random variable and classify the random variable as discrete or continuous.

Situation	Random Variable
a) Airplane tickets	The number of seats on plane
b) Bicycle trip	The number of flat tires
c) Diet soda	The amount of ounces dispensed
d) Restaurant entree	The number of a particular type of entree ordered

137. For a probability distribution of a discrete random variable, the probability of any value of the random variable is always:
a) Less than 0.5
b) Greater than 0
c) Less than 1.0
d) Between 0 and 1.0 inclusive.

138. The mean of a discrete random variable is calculated by multiplying each possible value of the variable by its corresponding:
a) Frequency
b) Percentage
c) Ratio
d) Probability

139. According to the highway department in Praire City, Idaho, the number of passengers per car (including driver) commuting on highway 57 per evening is defined by the probability distribution

Number of Passengers X	Probability P(X)
1	0.35
2	0.35
3	0.15
4	0.10
5	0.05

For this probability distribution:
a) Identify the values of the random variable.
b) Construct a probability histogram and determine its shape.
c) Verify that the sum of all the probabilities is one.
d) What is the expected number of passengers each evening?
e) Calculate the standard deviation.
f) How unusual is it for a car to contain more than 2 passengers? Explain.

140. An experiment that has six possible outcomes cannot be classified as a binomial experiment. This statement is:
a) True   b) False   c) Cannot be determined

141. For a binomial experiment, if $_7C_6 = 7$, then the number of ways to get _____ successes in _____ independent trials is _____. The seven successes are: sssssf, _____, _____, _____, _____, _____, and _____.

142. For a binomial experiment, where p = ½, calculate: P(4 successes in 5 trials) = _____ .

143. If five cards are selected with replacement from a regular deck of 52 playing cards, what's the probability that:
a) None of the cards are clubs?
b) One of the five cards is a diamonds?
c) At least three of the cards are hearts?

144. Larry, an old basketball player, has a 70% chance of making a free throw based on his performance over the season. If he attempts 9 free throws during his next game, then what is the probability that:
a) He makes all nine free throws?
b) He misses all nine free throws?
c) He makes at least one free throw?
d) He will make the 9th free throw given that he has missed the first 8 free throws?

145. The U.S. Department of Education's National Center for Education Statistics (NCES) found that only 36 percent of students graduate from college within four years. Assuming this percentage is true for all first time students seeking a bachelor's degree, then what is the probability that out of seven randomly selected first time students seeking a bachelor's degree:
a) All seven will graduate from college within four years?
b) None will graduate from college within four years?
c) At most two will graduate from college within four years?
d) At least four will not graduate from college within four years?

146. A slot machine has four windows. The wheel behind each window has the same five identical objects. They are: cherry, grape, peach, star and jackpot. After the handle is pulled, the four wheels revolve independently several times before coming to a stop. If the slot machine is played once, what is the probability that:
a) Exactly three cherries appear in the windows?
b) At least three peaches appear in the windows?
c) Only 1 fruit appears in the windows?

147. Medical records indicate that 40% of the people have type O+ blood. What is the probability that of the next seven people donating blood that:
a) At least five have type O+ blood?
b) At most one has type O+ blood?
c) More than five do *not* have type O+ blood?

148. Determine which of the following random variables are discrete and which are continuous.
a) The number of people attending a jazz concert.
b) The actual number of liters of spring water in a gallon container.
c) The number of shares of stock owned by an investor in a mutual fund.
d) The age of an automobile.
e) The amount of time waiting on line to change a college student's schedule.
f) The number of pages in a statistics textbook.
g) The number of new clients opening up a charge card account.
h) The amount of time required to run a 10K race.
i) The number of broken bottles in a case of wine.
j) The price of a college textbook.

149. Michele, the quality control supervisor, at a radial tire manufacturer has her assistant, Lynn, select a batch of 4 radial tires at random times during the daily production process to check this manufacturing operation. Each batch of four tires are inspected to determine the number of defective tires. The assistant records one of five possible numbers: 0, 1, 2, 3, or 4 to indicate the number of defective tires within each batch.

The following table represents their summary of the sampling results of this manufacturing process over a period of 12 months.

Number of Defective Tires Within a Batch	Percent of Defective Tires
0	71%
1	26%
2	2.5%
3	0.5%
4	0%

a) Identify the discrete random variable for this experiment, and the possible values of the random variable.
b) What probability would you associate with each possible value of this discrete random variable? Explain.
c) Construct a probability distribution for the discrete random variable X.
d) Construct a probability histogram for the discrete random variable X.
e) What is the probability that the next batch of radial tires selected from this production process will have 3 or more defective tires?
f) If you were the quality control supervisor and a randomly selected batch of radial tires had 4 defective tires, what might you conclude about the manufacturing process? Explain your answer.

150. Through careful record keeping over thousands of customers, MACBLIMP, a large fast food hamburger restaurant, has determined the probability distribution for the random variable, X, the number of premium sized hamburgers ordered per customers, as shown in Table 6.18.

TABLE 6.18

Number of Premium Sized Hamburgers per Customer X	Probability of the Number of Premium Sized Hamburgers per Customer P(X)
0	0.10
1	0.25
2	0.35
3	0.15
4	0.10
5	0.05

a) Construct a probability histogram for this probability distribution.
b) Determine the mean and standard deviation for this distribution.
c) What is the probability that a customer will order at least 2 premium sized hamburgers?

151. From an urn containing 8 marbles, four of which are colored red, two are colored white, and two blue. Four marbles are selected with replacement. Calculate the probability of:
a) Selecting four red marbles.
b) Selecting at least two blue marbles.

152. The mean of a binomial probability distribution, μ_s, can be calculated using the formula for the mean of a discrete probability distribution.
(T/F)

153. According to a survey, 80% of young women aged 17 to 23 are still living at home. If we assume this statistic is accurate for all young women in the age group, then for a sample of 4 women:
a) Determine the probability distribution for this binomial experiment.
b) Construct a probability histogram and describe its shape.
c) Find the binomial mean, μ_s, and the binomial standard deviation, σ_s.

154. Ms. Pix Els, a quality control engineer for a television manufacturer, must decide whether to accept or reject a shipment of five thousand 32" color picture tubes. She randomly samples and tests 100 of these tubes from the shipment. Ms. Els' criteria is to reject the entire shipment if more than 1% of the tubes tested are defective.
a) What is the probability that the entire shipment will **not be rejected** if 5% of the color tubes in this shipment are actually defective?
b) What is the probability that the entire shipment will **be rejected** if 5% of the color tubes in this shipment are actually defective?

155. The number of accidents that occur at a very dangerous intersection follows a Poisson Distribution with mean number of accidents equal to 9 per year.
a) What is the probability that there will be exactly 8 accidents this year?
b) What is the probability that there will be less than 6 accidents this year?
c) What is the probability that there will be 15 accidents in 2 years?

156. Vehicles arrive at the Queen Burger drive through at a Poisson rate of 20 per hour.
a) What is the probability that exactly 45 cars arrive at the drive-thru window in 2 hours?
b) What is the probability that exactly 18 cars arrive at the drive-thru window in 1 hour?
c) What is the probability that exactly 8 cars arrive at the drive-thru window in 1/2 hour?

157. A quality assurance engineer works at a plant that manufactures bolts for cars. The machines he is responsible for have a 3% defective rate. The morning shift has just produced a batch of bolts and he decides to check on the shift's quality. He randomly selects bolts and checks for defectives. Assuming a geometric/negative binomial distribution,
a. What is the probability on any given try he selects a bolt that is not defective?
b. What is the probability his first defective choice occurs on his 25th selection?
c. What is the probability he finds 35 good bolts prior to finding his first defective bolt?
d. What is the probability his third defective bolt is found on his 60th selection?
e. What is the probability he finds 28 good bolts *before* he finds his second defective bolt?
f. What is the mean number of good bolts selected *prior* to finding his 3rd defective bolt?

158. In the Arctic, the percent of Polar Bear cubs that die during the first year of life is approximately 29%. Twenty five cubs are selected at random and tagged for study by a biologist.

a) What is the probability that the number of cubs who die in the first year is exactly 6?
b) What is the probability of at most 10 cubs dying in the first year of life?
c) What is the probability of at least 18 deaths in the first year of life?
d) What is the probability that there are at most 5 cubs that survive past the first year of life?

For the answer to this question, please visit www.MyStatLab.com.

CHAPTER 7

Continuous Probability Distributions and the Normal Distribution

Contents

7.1 Introduction
7.2 Continuous Probability Distributions
7.3 The Normal Distribution
7.4 Properties of a Normal Distribution
    The Standard Normal Curve
7.5 Using the Normal Curve Area Table
    Finding a z-score Knowing the Proportion of Area to the Left
7.6 Applications of the Normal Distribution
7.7 Percentiles and Applications of Percentiles
7.8 Probability Applications
7.9 The Normal Approximation to the Binomial Distribution

7.1 INTRODUCTION

In the previous chapter, the concept of a discrete probability distribution was introduced. In particular, the binomial and the Poisson probability distributions were examined as models to many real world applications. Recall that *a discrete random variable can only assume a finite or countable number of values with a break between successive values of the variable.* For example, the number of free throws that a basketball player can make when attempting 12 free throws can be only one of thirteen values: 0, 1, 2, 3, . . . , 11, 12. Thus, it is impossible for a player to make 8½ free throws.

In this chapter, the concept of a continuous probability distribution will be examined. We will study a particular continuous probability distribution having a bell-shaped curve called the *normal probability distribution* or simply the *normal distribution.*

7.2 CONTINUOUS PROBABILITY DISTRIBUTIONS

As discussed in Chapter 6, a **continuous random variable** can assume an uncountable number of possible values between any two values of the variable. The values of a continuous random variable represent a measurement. For example, the amount of time a college student needs to get from one class to the next is an example of a continuous random variable. The variable time represents a measurement and its value can assume an uncountable number of possible values. For example, the amount of time a college student needs to travel between classes could possibly take 9.1 minutes or 9.01 minutes or even 9.001 minutes, etc. The accuracy to which the value of the variable can be measured depends upon the precision of the measuring device.

Suppose a continuous random variable X represents the amount of time a college student needs to travel between classes at a small eastern community college. Based on the results of a particular week, let's consider the probability distribution histogram representing the continuous variable X shown in Figure 7.1. The amount of time has been recorded to the nearest minute up to a maximum time of twenty minutes. There is a rectangle associated for each minute to get to class.

The height of each rectangle of the histogram in Figure 7.1 represents the probability that the student will take X minutes to get to the next class. For example, the probability that a student will take 5 minutes to get to the next class is 0.12, since the height of the rectangle corresponding to X = 5 is 0.12. Since each rectangle has

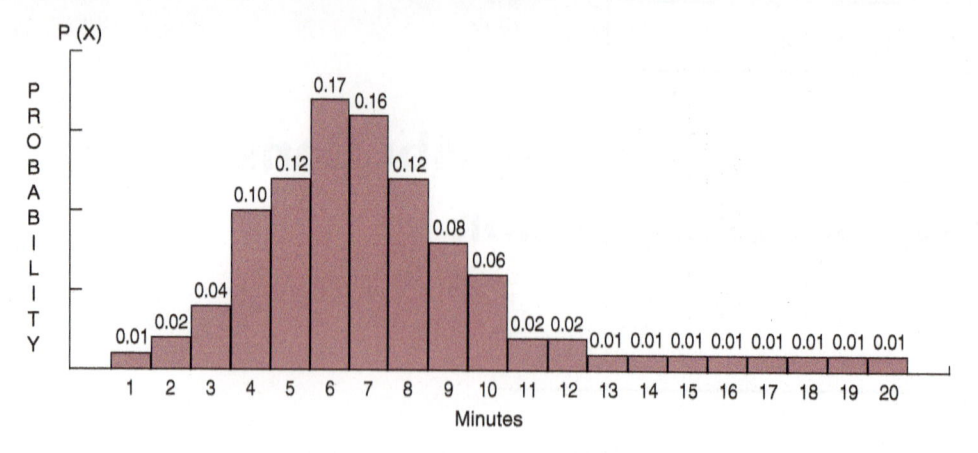

FIGURE 7.1

a width of 1 minute, the area of each rectangle is equal to the probability of the X value occurring. For example, the area of the rectangle for X = 5 minutes is equal to the probability of X = 5 minutes, written as P(X = 5).

$$\text{Area of rectangle for } X = 5 \text{ minutes} = (\text{base})(\text{height})$$
$$= (1)(0.12)$$
$$= 0.12$$
$$= P(X = 5).$$

Thus, **the area of the rectangle for X = 5 minutes is 0.12 and the probability associated with X = 5 minutes is also 0.12**.

Notice that the probability a student will require from 5 to 7 minutes to get to the next class is equal to the sum of the areas corresponding to the shaded rectangles associated with X = 5 minutes, X = 6 minutes, and X = 7 minutes. This is illustrated in Figure 7.2a.

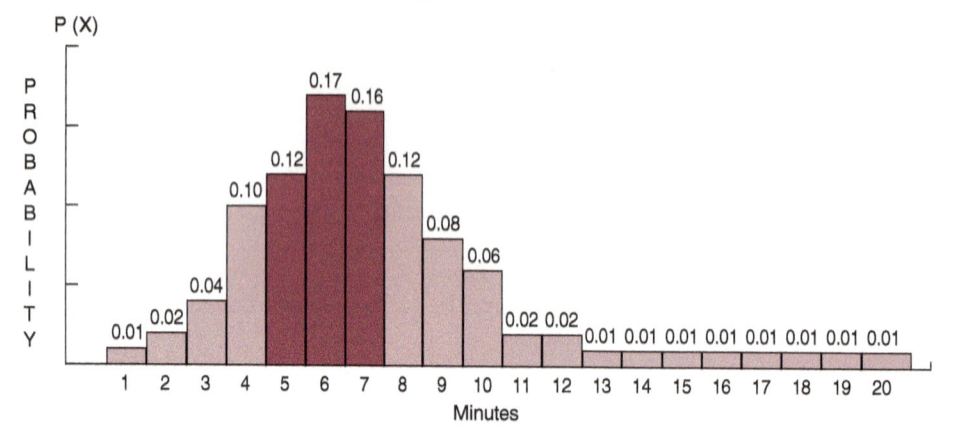

FIGURE 7.2 (a)

Thus,

$$P(\text{student requires from 5 to 7 min.}) = \text{sum of the areas of the rectangles for } X = 5, X = 6, \& X = 7$$
$$= P(X = 5) + P(X = 6) + P(X = 7)$$
$$= 0.12 + 0.17 + 0.16$$
$$= 0.45$$

You should also observe that *the sum of all the probabilities or the sum of all the areas of the rectangles equals one*. Thus,

P(student requires from 1 to 20 minutes to get to the next class) = the sum of the area of ALL the rectangles from 1 to 20
= 1.

Since time is a continuous random variable, it can assume any value between two possible values. If the time a student needs to get to the next class is measured to the nearest tenth of a minute, the amount of time can assume values such as 5 minutes or 5.1 minutes or 5.2 minutes etc. Thus, the probability histogram of Figure 7.2b will have more rectangles and each rectangle will have a narrower width as illustrated in Figure 7.2b. Although the histogram in Figure 7.2b has narrower rectangles, the total area of all these rectangles, which equals the sum of all the probabilities, is still one.

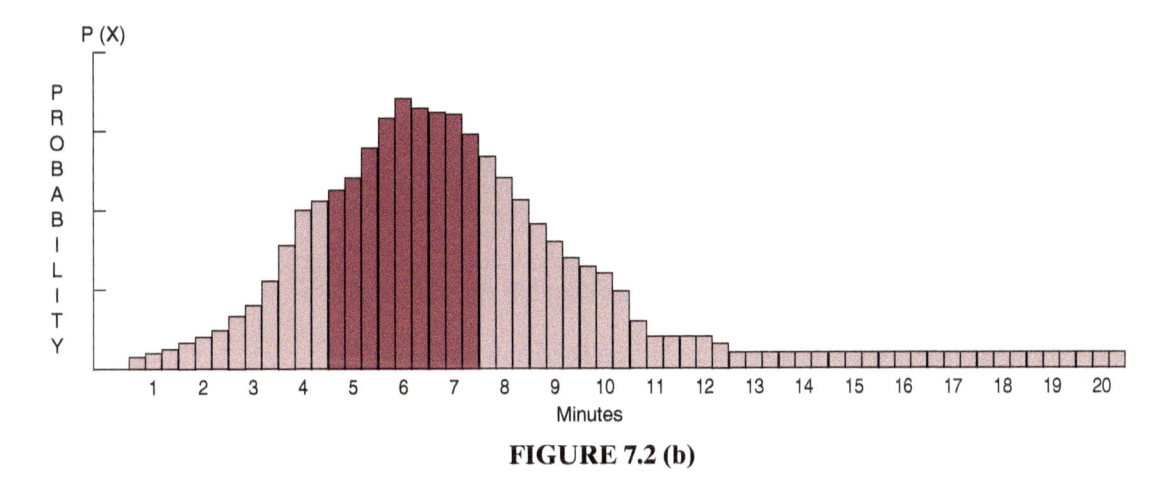

Section 7.2 *Continuous Probability Distributions* 373

FIGURE 7.2 (b)

If we continue this process of measuring the continuous variable time (the amount of time a student needs to get to the next class) to a finer and finer degree of precision, such as a hundredth of a minute or a thousandth of a minute, etc. the number of rectangles will increase while the width of each rectangle gets narrower. As the widths get narrower, the top of the histogram begins to approach a smooth curve as shown in Figure 7.2c.

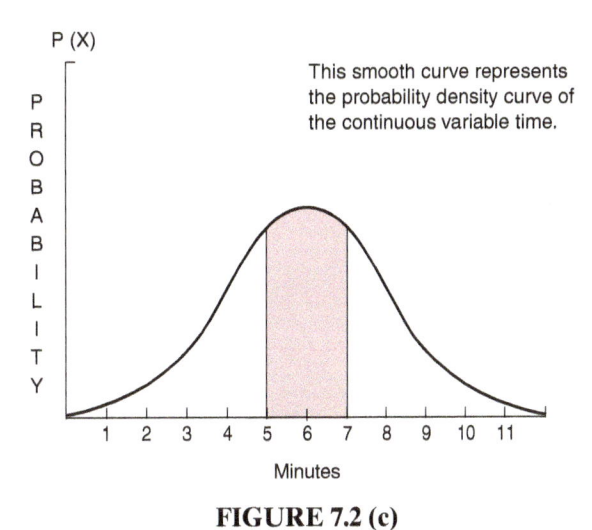

FIGURE 7.2 (c)

The smooth curve of the continuous variable time is called the *probability density curve* of the continuous variable time. The area under the probability density curve between two possible values of the continuous variable represents the probability that the value of the continuous variable will fall between these two values. Thus, the **probability** that the amount of time a student will take from 5 to 7 minutes to get to the next class is equal to the **area** under the probability density curve as illustrated in Figure 7.3.

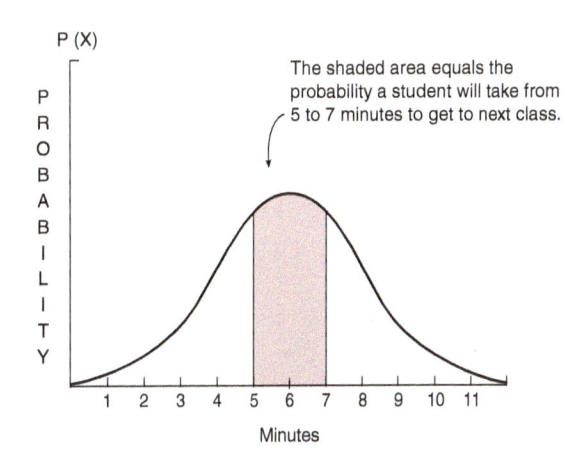

FIGURE 7.3

Figure 7.3 illustrates that for a continuous probability distribution the probability assigned to an event, such as the event a student will take from 5 to 7 minutes to get to the next class, is equal to the area under the density curve satisfying the event. This probability assignment satisfies the characteristics of a probability distribution as discussed in Chapter 6. Let's now state the characteristics of a probability distribution for a continuous variable.

> ### Characteristics of a Continuous Probability Distribution
>
> 1. The **probability** that the continuous random variable X will assume a value between two possible values X = a and X = b of the variable **is equal to the area** under the density curve between the values X = a and X = b.
> 2. The **total area** (or **probability**) under the density curve is equal to **1**.
> 3. The **probability** that the continuous random variable X will assume a value between **any two possible values** X = a and X = b of the variable is **between 0 and 1**.

Let's discuss these characteristics of a continuous probability distribution.

The first characteristic states that the probability a continuous random variable X will have a value between the possible values X = a and X = b is equal to the area under the density curve between these two values. This can be written as:

$$P(X = a \text{ to } X = b) = \text{the area under the curve between } X = a \text{ and } X = b.$$

This is illustrated in Figure 7.4.

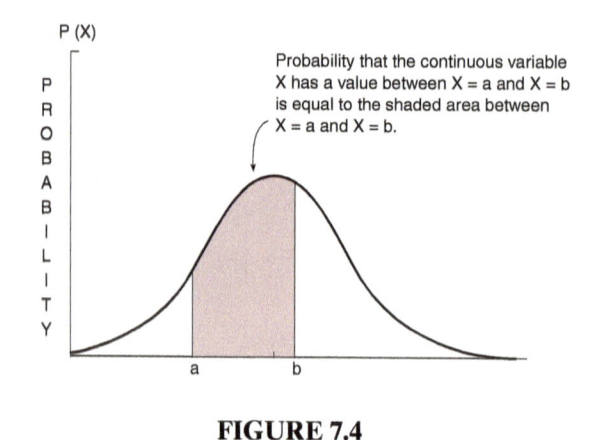

FIGURE 7.4

The second characteristic indicates that the total area or probability under the density curve is 1 or 100% as shown in Figure 7.5.

The third characteristic states that the probability a continuous variable X will have a value between any two values X = a and X = b will be greater than 0 but less than 1 as illustrated in Figure 7.6.

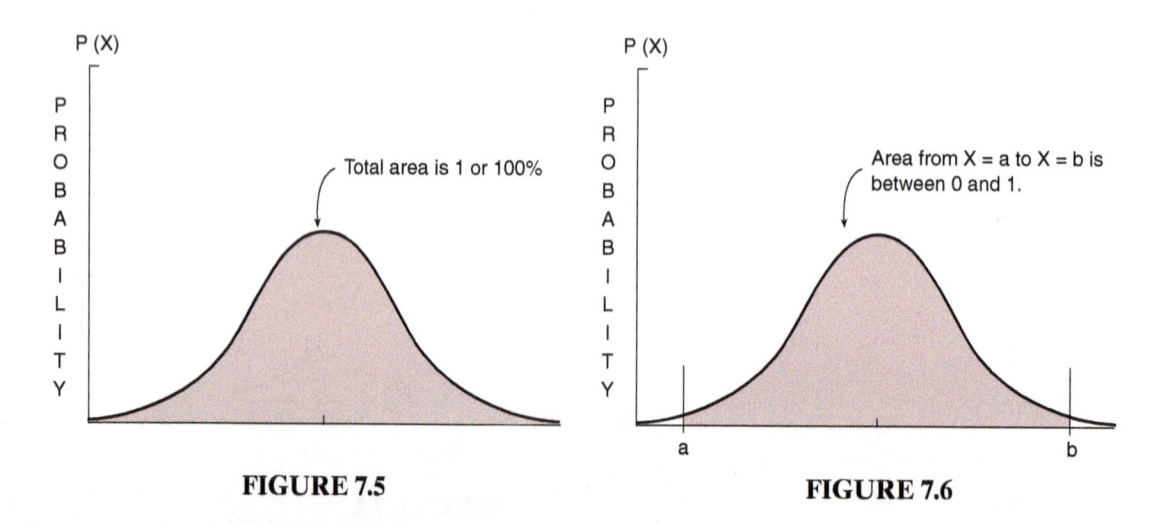

FIGURE 7.5

FIGURE 7.6

This can be written as:

$$0 < P(X = a \text{ to } X = b) < 1.$$

This third characteristic is true for a continuous variable since the largest possible probability is equal to the total area under the density curve which is 1 as indicated by the second characteristic. While the smallest possible probability is the probability assigned to an individual outcome of a continuous variable which is 0. That is, **the probability that a continuous variable will assume a specific value, such as X = a, is zero**, since the probability of a specific value corresponds to the area directly above the point X = a. This can be written as: **P(X = a) = 0**. Because the area of a single point is equal to the area above the point which is a line having no width there is no (or zero) area.

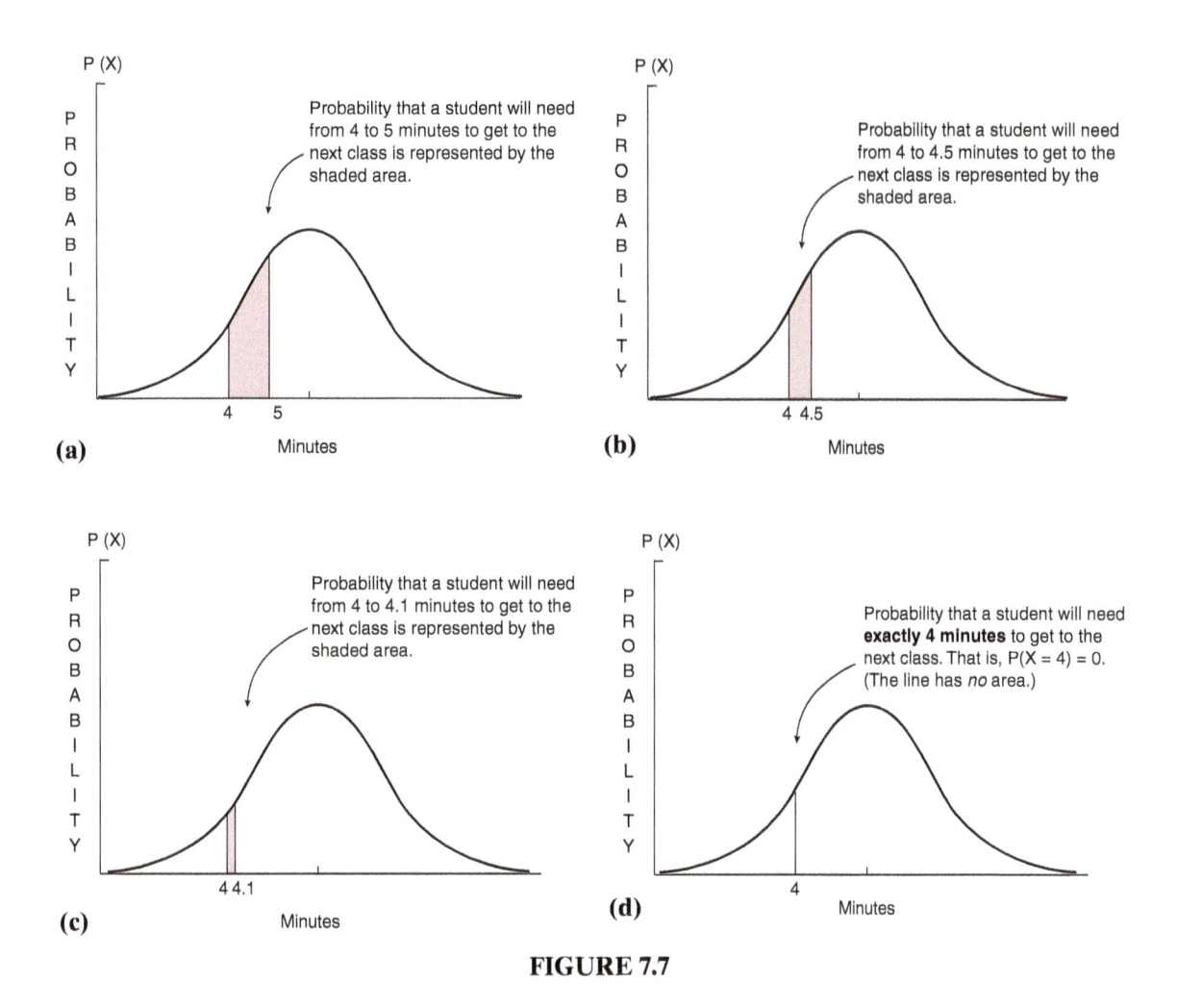

FIGURE 7.7

Let's explain this idea of assigning a probability of zero to a specific value such as X = a by using the previous example pertaining to the amount of time a student needs to get to the next class. Specifically, the probability that a student will need from 4 to 5 minutes to get to the next class is equal to the area under the density curve between 4 and 5 minutes as illustrated in Figure 7.7a.

The probability that a student will need from 4 to 4.5 minutes to get to the next class is the area under the density curve between 4 to 4.5 minutes as shown in Figure 7.7b.

The probability that a student will need from 4 to 4.1 minutes to get to the next class is equal to the area under the density curve from 4 to 4.1 minutes as illustrated in Figure 7.7c.

As you can see from Figures 7.7a to 7.7c, as the number of minutes gets closer to the single value of 4 minutes (starting from 5 to 4.1 minutes), the areas are getting smaller and smaller. Continuing in this manner, you should begin to understand that the probability, which represents the area, that a student will need **exactly 4 minutes to get to the next class is 0**, since there is no (or zero) area associated with 4 minutes. This is illustrated in Figure 7.7d.

Because there is no probability associated with a single value of a continuous variable, then the probability that a continuous variable X assumes a value between a and b, written P(a < X < b), and the probability that a continuous variable X assumes a value from a to b, written P(a ≤ X ≤ b), are equal. Consequently, for a **continuous variable**,

$$P(a < X < b) = P(a \leq X \leq b),$$

since P(X = a) = 0 and P(X = b) = 0. For example, the probability that a student will need between 4 and 5 minutes to get to the next class, written P(4 < X < 5), is equal to the probability that a student will need from 4 to 5 minutes inclusive to get to the next class, written P(4 ≤ X ≤ 5), as shown in Figure 7.8.

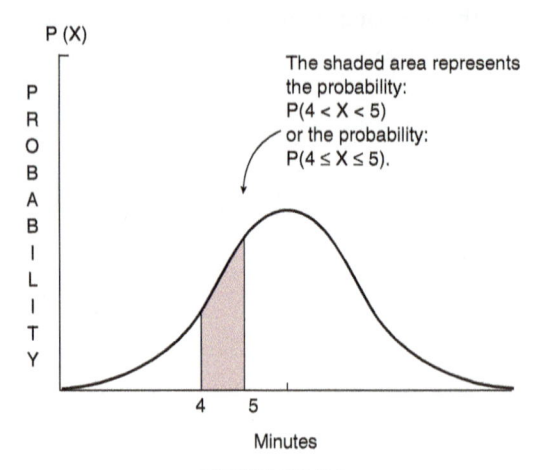

FIGURE 7.8

In summary, the following probabilities are true for a continuous probability distribution.

Probability Statements Associated with a Continuous Probability Distribution

1. The probability of obtaining a specific value X = a of a continuous variable is zero. That is,

$$P(X = a) = 0.$$

2. The probability that a continuous variable X assumes a value between a and b, written P(a < X < b), and the probability that a continuous variable X assumes a value from a to b, written P(a ≤ X ≤ b), are equal. That is,

$$P(a < X < b) = P(a \leq X \leq b).$$

In the next section, we will use these concepts and the characteristics of a continuous variable when we examine one of the most important continuous probability distributions called the normal distribution.

7.3 THE NORMAL DISTRIBUTION

If one uses the pinball demonstrator pictured in Figure 7.9, a collection of balls is released. As the balls fall through the different paths as shown in Figure 7.10, they are constantly hitting other balls and pins. Thus, the paths the balls follow are random. Because the paths are random one might expect to get a *radically different* figure each time the balls are released. However, this is not the case. In fact, a bell-shaped figure similar to the one pictured in Figure 7.11 occurs almost all of the time.

This occurrence was first observed by Sir Frances Galton, in an 1889 publication entitled *Natural Inheritance*, when he used a similar pinball machine to demonstrate that many unknown factors acting together yield a phenomena whose distribution is bell-shaped. The demonstration led to the discovery that the **normal distribution**, pictured in Figure 7.12, could be used as a model for this type of phenomena. The significance of this finding is that it provides an explanation of why many observable phenomena (common occurrences) in nature have *approximately* **normal distributions**. It should be noted that the use of the word *normal* is based upon historical significance which just simply refers to the shape of the distribution. It does not indicate that this particular distribution is more typical or more appropriate than any other distribution.

FIGURE 7.9

FIGURE 7.10

FIGURE 7.11

FIGURE 7.12

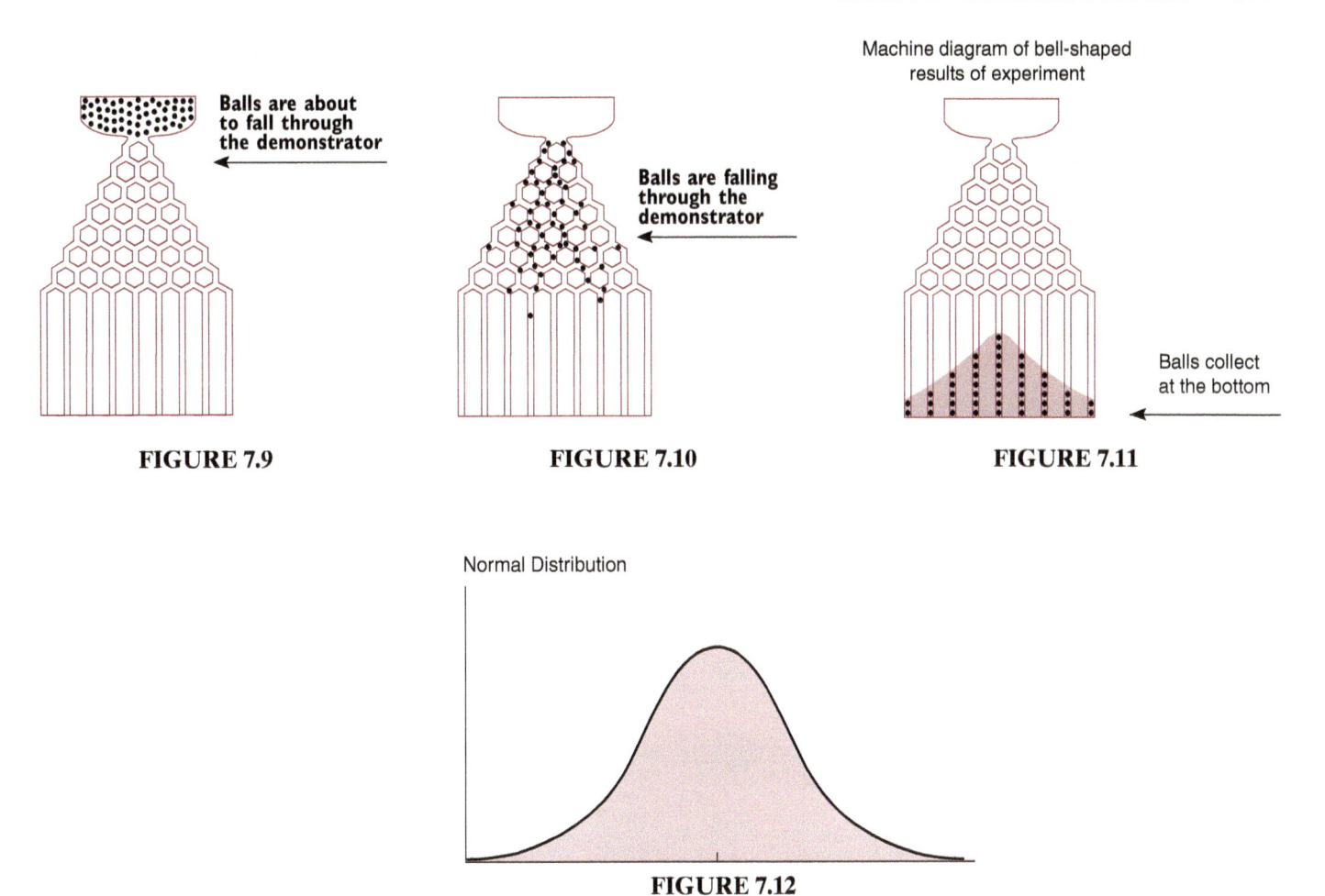

To understand why a phenomena can be modeled or approximated by a normal distribution let's consider the distribution of adult male heights. The distribution of adult male heights can be modeled or approximated by a normal distribution since each height is influenced by many random factors acting together. These factors could be hereditary, environmental, physiological or based on the male's diet. Thus, we can think of the height of an adult male as the result of *a combination of many such factors*. Each factor is perceived as applying a small force or influence toward a male's height which can either increase or decrease the height of a male. Since so many factors influence a male's height, it would be extremely rare to expect **all these factors** exerting either all positive or all negative influences in one direction. Consequently, we rarely expect to find many male adults to be extremely short or exceptionally tall. Instead, we would expect to find more male adults whose heights are positively influenced by some of these factors and negatively influenced by the other factors. Thus, the heights of adult males would tend to be approximately a normal distribution with most of the males clustered around the average height, located at the center of the distribution, with fewer males that are either much shorter or much taller than the average height at the ends of the distribution. Thus, the distribution of adult male heights can be modeled or approximated by the normal bell-shaped curve shown in Figure 7.12.

Other examples of distributions that can be modeled or approximated by the normal distribution include IQ scores of individuals, adult weights, the diameters of tree trunks, blood pressures of men and women, scores on standardized exams measuring an individual's knowledge, tire wear, the size of red blood cells, the time required to get to work, and the actual amount of soda dispensed into a 2 liter soda bottle. The variables that have been observed to have an approximately normal distribution are continuous variables. Thus, *a normal distribution, also referred to as the normal probability distribution, is a distribution that represents the values of a continuous variable.* The importance of a normal distribution is that it serves as a *good model* in approximating many distributions of real-world phenomena. *If a continuous variable is said to be approximately normal, then the normal distribution can be used to model the continuous variable.* The graph of a normal distribution is called a *normal curve* which is a smooth symmetric bell-shaped curve with a single peak at the center. This is illustrated in Figure 7.12. It should be noted that not all bell-shaped curves represent normal distributions. In fact, we will discuss another bell-shaped distribution called a *t distribution* which is not a normal distribution in Chapter 9. In this chapter, however, we will study in detail the normal distribution.

A normal probability distribution, or simply the normal distribution, is a specific kind of bell-shaped curve with certain properties. Distributions that exhibit these properties all belong to the normal distribution family. Let's now discuss the properties of a normal distribution.

7.4 PROPERTIES OF A NORMAL DISTRIBUTION

In order to analyze distributions which are approximately normal, or that can be approximated by the normal distribution, we must discuss the properties that a normal distribution possesses.

Let's now summarize some of the more important properties of the normal distribution.

Properties of a Normal Distribution

The normal distribution and its graph, the normal curve, have the following properties which are illustrated in Figure 7.13.

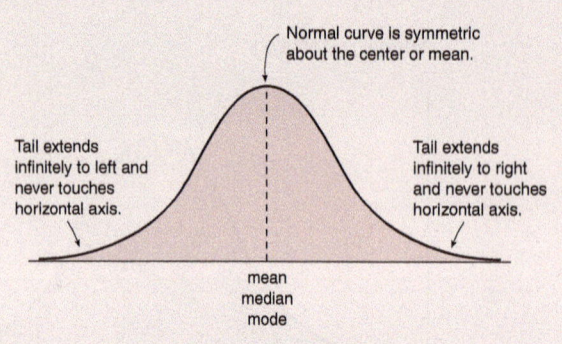

FIGURE 7.13

1. The normal curve is bell-shaped and has a single peak which is located at the center.
2. The mean, median and mode all have the same numerical value and are located at the center of the distribution.
3. The normal curve is symmetric about the mean.
4. The data values in the normal distribution are clustered about the mean.
5. In theory, the normal curve extends infinitely in both directions, always getting closer to the horizontal axis but never touching it. As the normal curve extends out away from the mean and gets closer to the horizontal axis, the area in the tails of the normal curve is getting closer to zero.
6. The total area under the normal curve is 1 which can also be interpreted as probability. Thus, the area under the normal curve represents 100% of all the data values.

Let's examine these properties assuming that a large distribution of IQ scores with a mean of 110 can be approximated by a normal distribution.

Property 1 indicates that the graph representing the distribution of IQ scores is bell-shaped and has a single peak at the center since the IQ scores are approximately a normal distribution. Thus, a graph of all the IQ scores can be modeled by a normal curve as shown in Figure 7.14.

As stated in **property 2**, the mean, median, and mode all have the same value and are located at the center of the normal distribution. Since the IQ scores are approximately a normal distribution, then the mean, median and mode are equal to 110 and are located at the center of the normal curve as illustrated in Figure 7.15.

By definition, the median is equal to the IQ score representing the middle value which is 110. Thus, the median is located at the center of the distribution and one half or 50% of the IQ scores are less than 110 because the median is 110. The mode represents the IQ score having the greatest frequency, thus the mode is also 110 and it is located at the peak which is located at the center of the distribution.

The **symmetric** property of the normal distribution, **property 3**, can be illustrated by examining Figures 7.16a and 7.16b.

If the normal curve shown in Figure 7.16a is folded along the center which is located at the mean value of 110, then the two halves of the normal curve would be identical. This is illustrated by Figure 7.16b.

Figure 7.17 illustrates **property 4** which states that the data values in a normal distribution are clustered about the mean. In examining Figure 7.17, notice that greatest concentration of the area under the normal curve is clustered around the center value of 110 which represents the mean.

Property 5 states in theory that the normal curve extends infinitely in both directions. This means that the graph of the mathematical formula representing a normal curve will result in a bell-shaped curve extending infinitely in both directions. This is illustrated in Figure 7.18.

However, in practical applications, a normal distribution is used as a model to represent a continuous variable such as the IQ of an individual. It is important to realize that it is impossible for an individual's IQ to approach infinity. For real-life applications, we will draw a normal curve extending infinitely in both directions; however, it is important to understand that the normal curve is serving as a model and the values of the continuous variable will not approach infinity.

Property 6 states that the total area under the normal curve is 1. This is a geometric characteristic of the curve. For a normal distribution, the areas under a normal curve can be interpreted as probabilities. This is a characteristic of a continuous probability distribution.

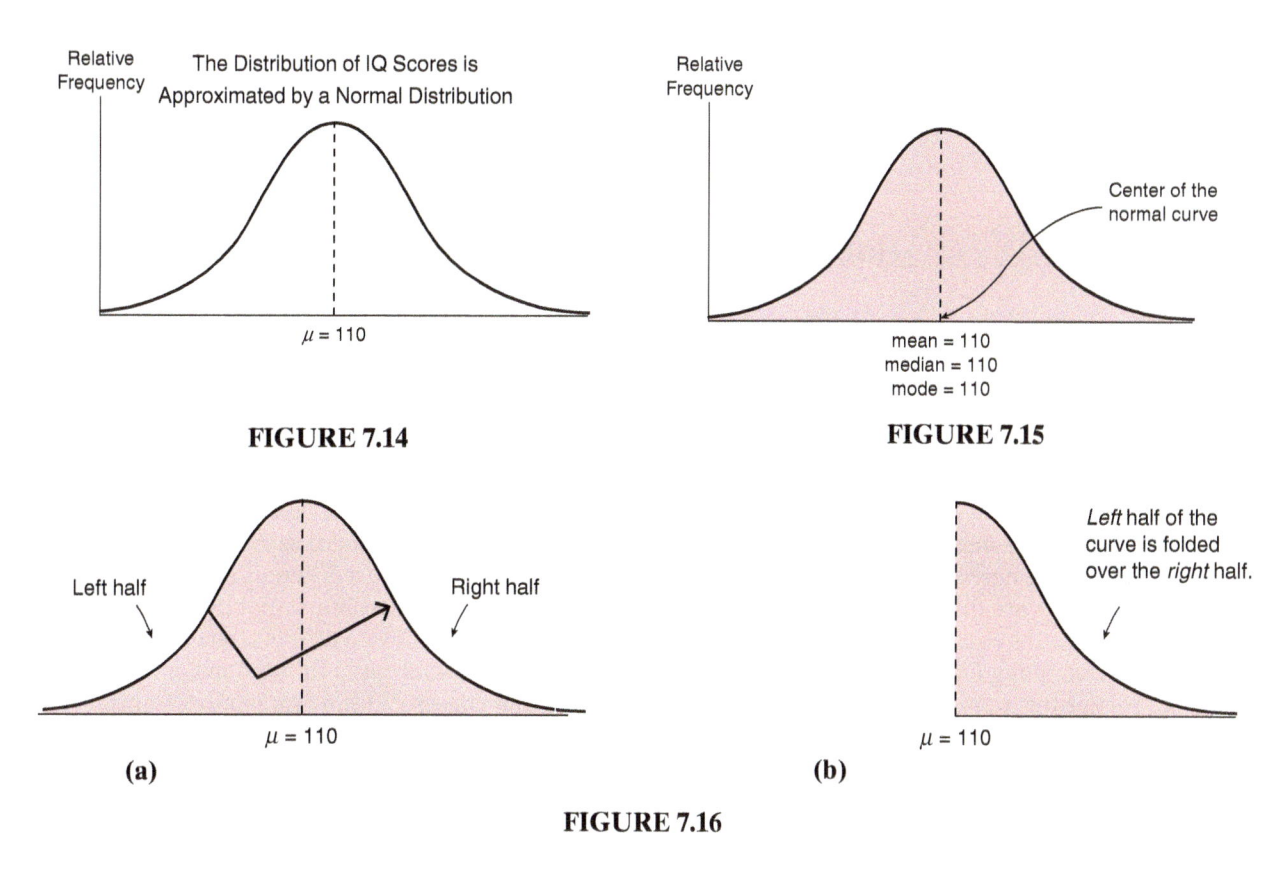

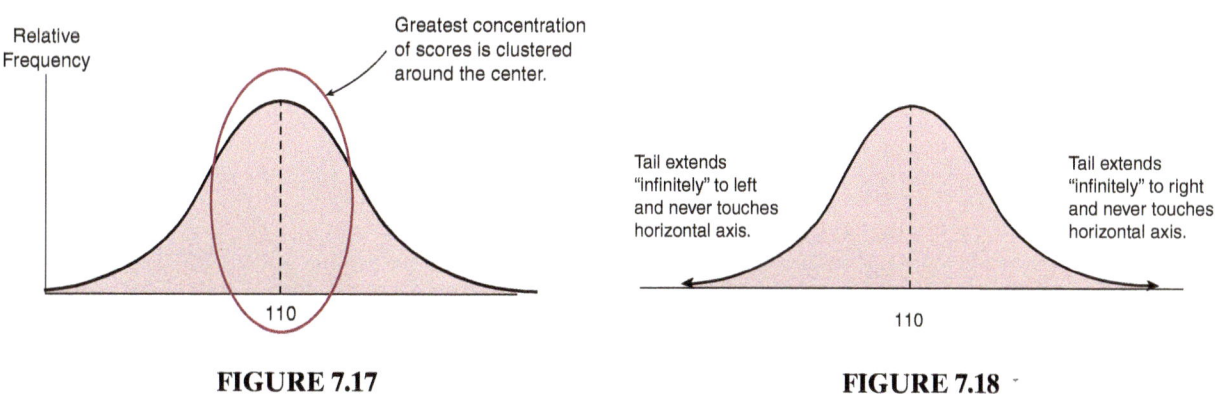

FIGURE 7.14

FIGURE 7.15

FIGURE 7.16

FIGURE 7.17

FIGURE 7.18

The Standard Normal Curve

When we state that the distribution of a continuous variable is *approximately* normal, we are indicating that this continuous variable can be modeled by a normal distribution and its graph would be modeled by a normal curve. Thus, for each continuous variable that can be approximated by a normal distribution with a particular mean and standard deviation, there is a normal curve having the same mean and standard deviation which serves as a model of this variable. Consequently, there is *no single normal curve*, but rather *many normal curves* which are referred to as a **family of normal curves**. Let's examine this idea of a *family of normal curves*.

Consider three normal distributions representing the IQ scores of students at different colleges each of which has the *same mean of 110 but different standard deviations*. The standard deviations are $\sigma = 10$, $\sigma = 20$ and $\sigma = 30$. In Figure 7.19, the curve for each normal distribution is drawn. Although these normal curves vary in shape and size, notice that all three graphs in Figure 7.19 are *normal curves having the same center value* since their means are equal. Observe from Figure 7.19 that the normal distribution with the largest standard deviation, $\sigma = 30$, has a normal curve with greatest spread or dispersion, while the normal distribution with the smallest standard deviation, $\sigma = 10$, has a normal curve with the smallest spread or dispersion.

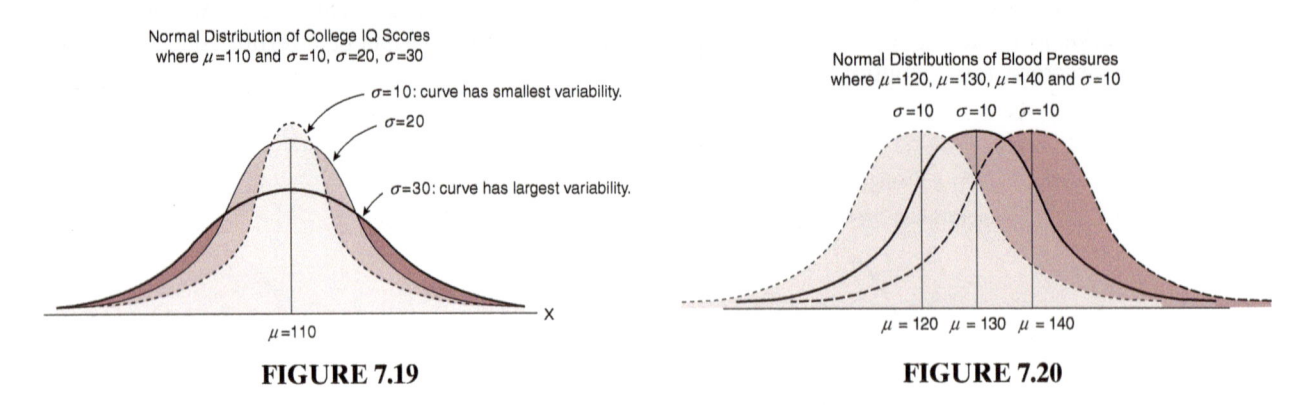

FIGURE 7.19

FIGURE 7.20

In Figure 7.19, notice for each pair of mean and standard deviation, there is a specific normal curve. There is a normal curve representing the IQ scores of the college where the mean is 110 and the standard deviation is 10. There is a different normal curve representing the IQ scores of the college whose mean is 110 and standard deviation is 20. Finally, there is a third normal curve representing the college IQ scores with a mean of 110 and a standard deviation of 30. Thus, for every particular combination of mean and standard deviation, a different normal curve can be generated. Figure 7.19 illustrates the idea of a **family of normal curves**.

In Figure 7.20, we illustrate three normal distributions representing the blood pressures of different aged males where the means vary from 120 to 130 to 140, but the standard deviation for each distribution is the same, $\sigma = 10$.

The normal distributions of blood pressures in Figure 7.20 have a different value at the center because the means are different. However, the normal distributions have the same variability since the standard deviations are equal. Much like a family of siblings where each can have a different shape and size, you can see by examining Figures 7.19 and 7.20 that there is a different normal curve for each mean and standard deviation. For this reason we say that there is a **family of normal curves**. Within this family of normal curves, there is only one normal distribution or curve for each combination of mean and standard deviation.

We have seen that there is a family of normal curves where each curve can have a different mean or a different standard deviation. Consequently, the number of possible normal curves is infinite. For example, there is a normal curve having a mean $\mu = 100$ and standard deviation $\sigma = 10$ which models the IQ scores of college students. Another normal curve with a mean $\mu = 120$ and standard deviation $\sigma = 15$ models the blood pressures of 20 year old males. These normal curves are illustrated in Figures 7.21a and 7.21b.

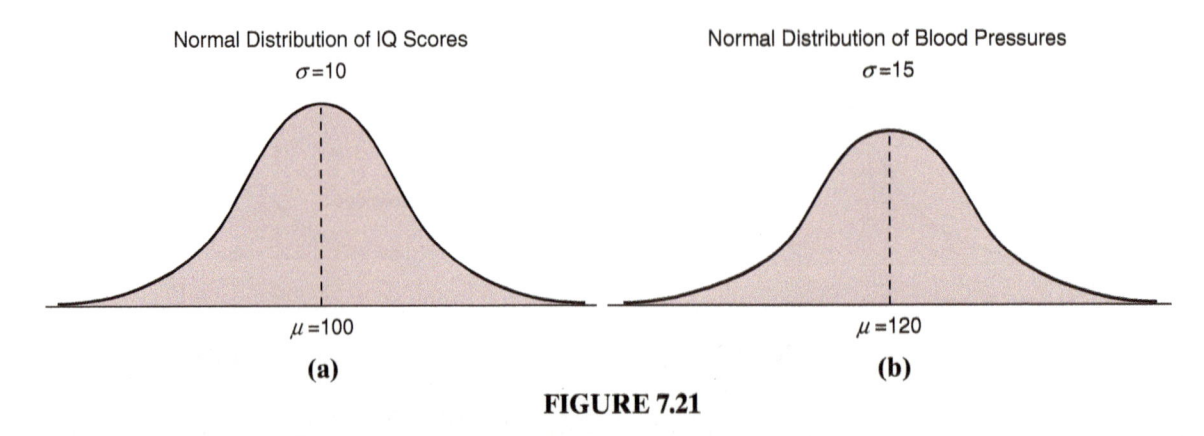

FIGURE 7.21

Section 7.4 *Properties of a Normal Distribution* 381

Let's examine the *similarity between the two normal curves* in Figures 7.22a and 7.22b. In Figure 7.22a, the proportion of area under the normal curve for the distribution of IQ scores between the $\mu = 100$ and 110, which is one standard deviation above the mean, has been shaded. Similarly, in Figure 7.22b, an equivalent proportion of area under the normal curve for blood pressures between the $\mu = 120$ and 135, which is also one standard deviation above the mean, has also been shaded. Although the shaded areas appear to be different, they represent the *same proportion of area under each normal curve*.

This is also true for the proportion of area under every normal curve between the mean and the data value which is **one standard deviation above the mean** regardless of its mean and standard deviation. In general, **for every normal curve the proportion of area under the curve between the mean and the data value which is one**

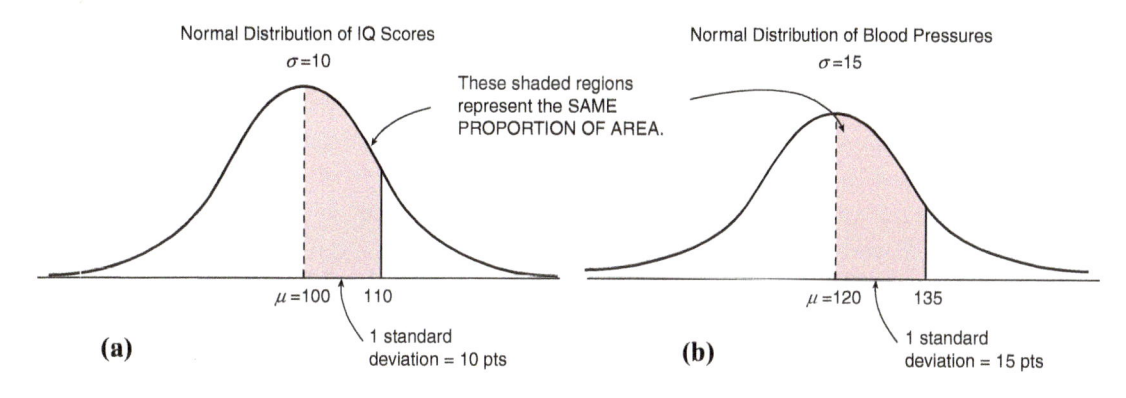

FIGURE 7.22

standard deviation above the mean is the same proportion of area. It is also true that *the proportion of area under each normal curve between the mean and the data value which is* **two standard deviations above the mean** *represents the same proportion of area*. This is illustrated in Figures 7.23a and 7.23b for the distribution of IQ scores and blood pressures.

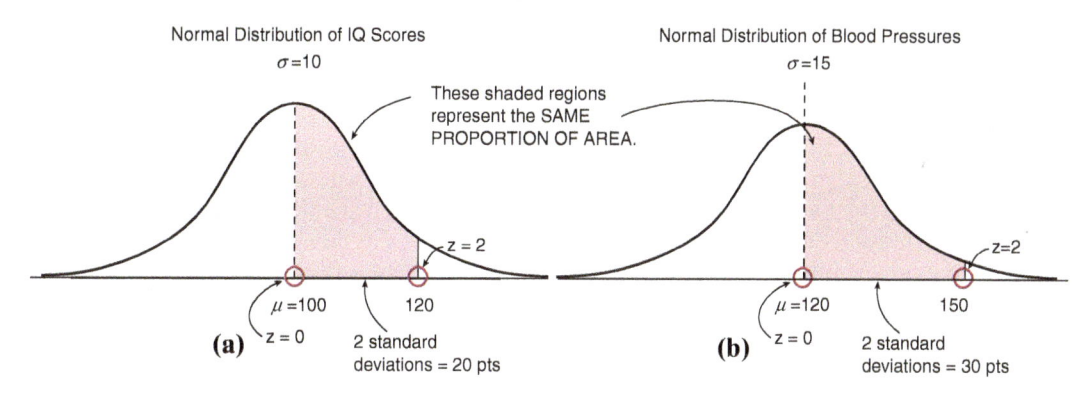

FIGURE 7.23

The proportion of area under every normal curve between the mean and the data value which is **one standard deviation below the mean** *is the same proportion of area.* This is illustrated in Figures 7.24a and 7.24b for the distribution of IQ scores and blood pressures.

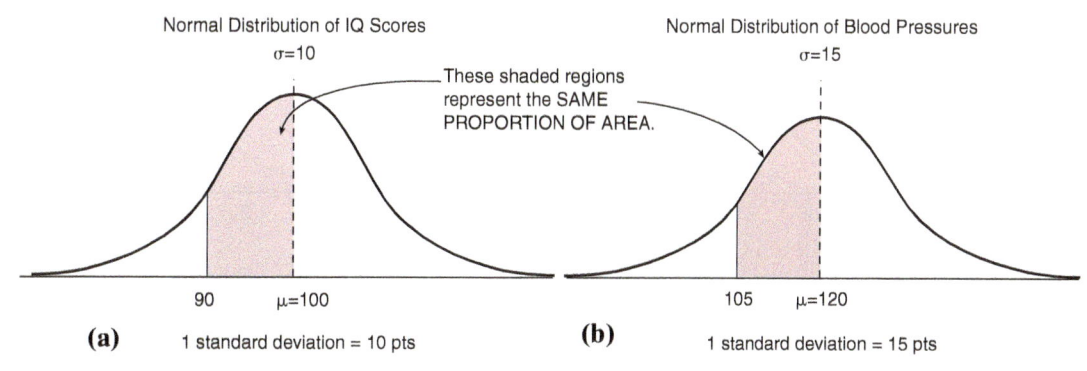

FIGURE 7.24

382 *Chapter 7 Continuous Probability Distributions and the Normal Distribution*

In fact, **the proportion of area under any normal curve between the mean and the data value which is N standard deviations away from the mean is the same for all normal curves.** Consequently, when working with normal distribution applications it is possible only to have to work with one normal curve since each member of the family of normal distribution curves exhibits the same characteristics. Thus, although there are an infinite number of normal curves (or normal distributions), *there is one special member of the normal distribution family* that can be used to serve as the normal distribution model for any application. This special normal distribution is referred to as the **standard normal distribution**. The **standard normal distribution** has a mean equal to zero, $\mu = 0$, and a standard deviation equal to one, $\sigma = 1$. The standard normal distribution model is shown in Figure 7.25.

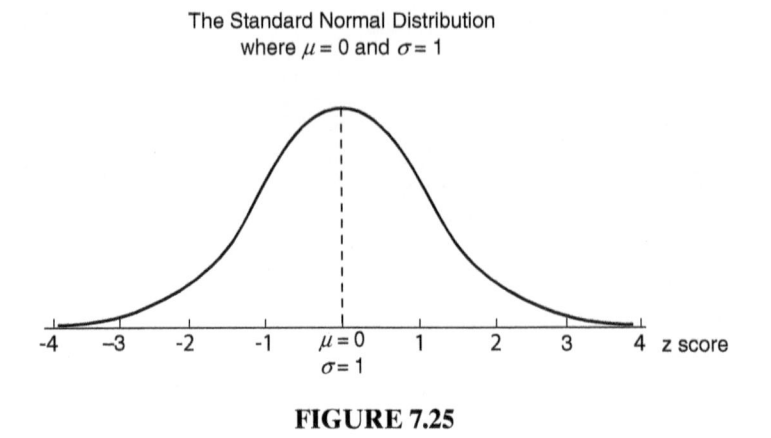

FIGURE 7.25

To utilize this standard normal distribution as a model for any normal distribution, it becomes necessary to convert any normal distribution to the standard normal distribution. This is accomplished by converting the values of the continuous variable to z scores. Recall from Chapter 3, the z score of a data value indicates the number of standard deviations that the data value deviates from the mean. The formula for the z score of a population data value is:

$$\text{z score of a data value x} = \frac{x - \mu}{\sigma}.$$

In essence, the z score formula helps us to convert any normal distribution to the standard normal distribution. This idea is illustrated in Figures 7.26a and 7.26b, where the normal distributions representing the IQ scores and the blood pressures of Figures 7.21a and 7.21b are shown with their respective z scores written underneath some of the values of the continuous variables, IQ scores and blood pressures.

As shown in Figures 7.26a and 7.26b, regardless of the values and units of any normal continuous variable, the z score formula can be used to transform any normal distribution to the standard normal distribution with a mean of 0 and a standard deviation of one. Thus, although both normal curves in Figures 7.26a and 7.26b represent different normal distributions, we can convert each normal curve to the standard normal curve by changing each data value to its z score. Consequently, the normal curves in Figures 7.26a and 7.26b can be modeled by the standard normal curve in Figure 7.27.

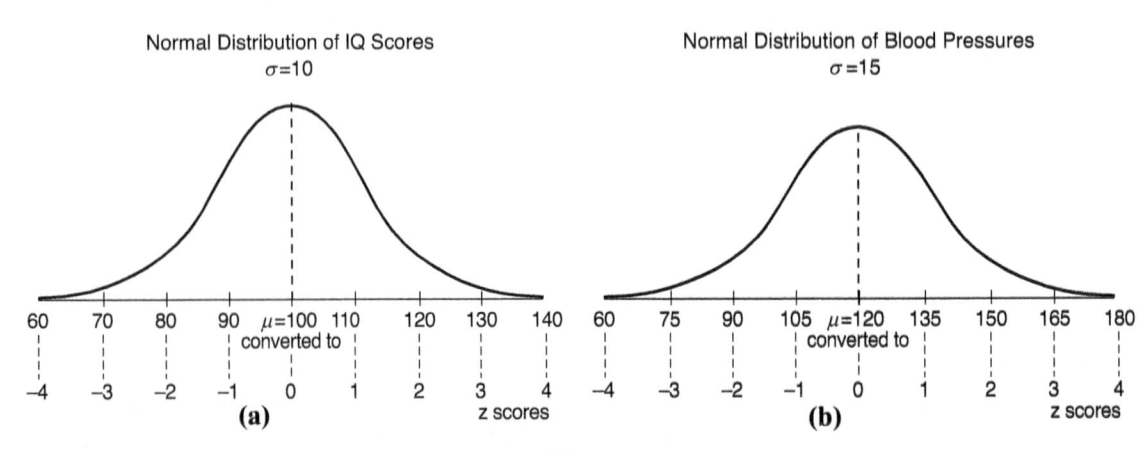

FIGURE 7.26

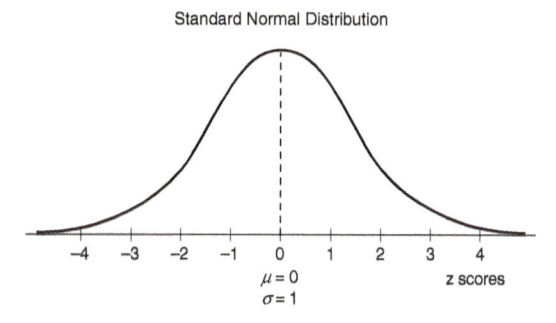

FIGURE 7.27

Since any normal distribution can be converted to the standard normal distribution, we only need to refer to one statistical table to find the area under any normal curve rather than an infinite number of tables for all the possible normal curves. This statistical table[1] is called the **Standard Normal Curve Area Table** and will be referred to as **Table II in Appendix C: Statistical Tables**. Therefore, for any normal continuous variable, *Table II* can be used to find the area (or probability) which falls to the left of a particular z score associated with a normal curve. In Section 7.5, we will explain how to use *Table II: The Normal Curve Area Table* to determine the area or probability associated with any normal curve.

7.5 USING THE NORMAL CURVE AREA TABLE

The **Normal Curve Area Table, Table II,** located in **Appendix C: Statistical Tables** at the back of the text, gives the relationship between a z score and the proportion of area under the normal curve. In particular, when a z score is looked up in Table II, the proportion of area under the normal curve to the **left of this z score** is given. Let's now examine how to use Table II to find the different areas under a normal curve. Notice Table II, located in **Appendix C: Statistical Tables**, consists of two pages. One page pertains to the negative z scores while the other page pertains to the positive z scores. On each page there is a column labelled **z** within the table. When a z score is looked up in the table, it gives the proportion of area under the normal curve which is to the left of the z score.

For example, suppose we wanted to find the proportion of area under the standard normal curve to the left of the z score z = –1.28. The shaded area in Figure 7.28 illustrates the proportion of area to the left of z = –1.28. To determine this area, we need to locate z = –1.28 in Table II. To find this z score, look up –1.2 in the column labeled z on the page pertaining to the negative z scores. The second decimal place of z = –1.**28** (i.e. the value 8) is found by moving across the row labeled –1.2 of the table until you reach the column labeled 8. The entry in Table II corresponding to z = –1.28 is 0.1003. This is illustrated in Figure 7.29. Thus, 0.1003 represents the proportion of area to the **left** of z = –1.28 as shown by the shaded area in Figure 7.28.

In general, the following procedure can be used to find the proportion of area under a normal curve to the left of a z score in Table II.

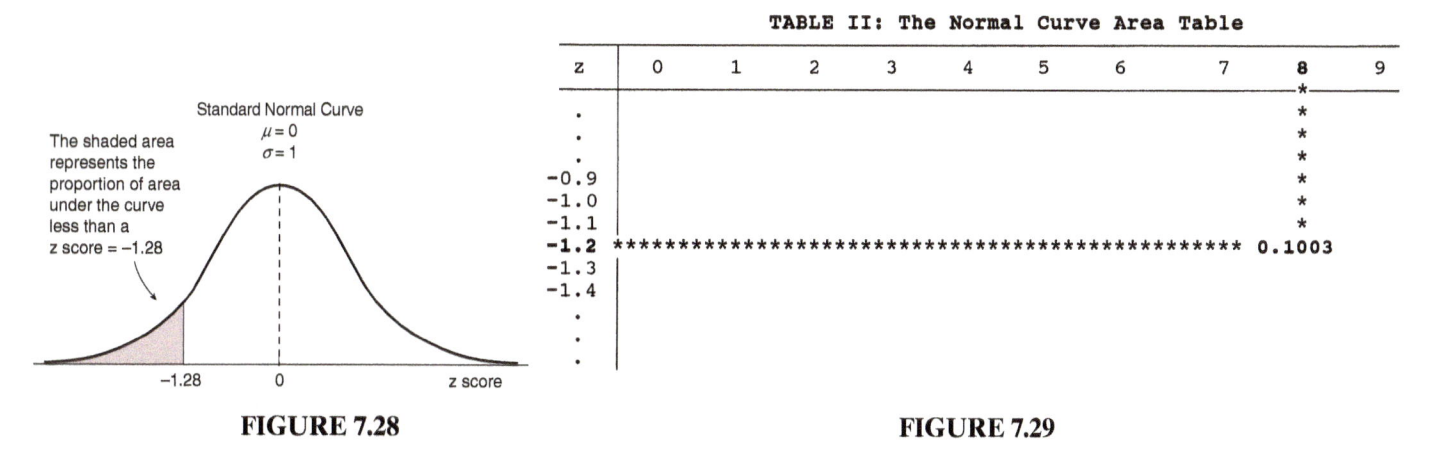

FIGURE 7.28

FIGURE 7.29

---

[1] The TI–84 graphing calculator can find the area under a normal curve associated to a given z-score or raw score.

Procedure to Determine the Proportion of Area Under a Normal Curve to the Left of a z Score in Table II

1. Determine which page of Table II to use.
2. Using the appropriate page of the table, find the row corresponding to the z score under the column labeled z. This row is determined by looking for the integer value and the value of the first decimal place of the z score.
3. Locate the column corresponding to the z score. To determine this column, look for the value of the second decimal place of the z score by moving across the top row of the table until you find the column pertaining to the second decimal place.
4. The entry found in the table where the row and column determined in steps 2 and 3 intersect represents the proportion of area to the left of the z score.

EXAMPLE 7.1

Find the proportion of area under the normal curve:
a) To the left of $z = -0.53$.
b) To the left of $z = 2.56$.

Solution

a) It's helpful to make a sketch of the normal curve and to shade in the proportion of area one needs to determine. In Figure 7.30, the proportion of area under the normal curve to the left of $z = -0.53$ is shaded. To determine this shaded area, look up the z score $z = -0.53$ in Table II using the procedure outlined above.
   1. Since the z score is negative, use the page pertaining to the negative z scores.
   2. Locate the row corresponding to the z score $z = -0.53$ by looking up **–0.5** under the column labeled z.
   3. Locate the column corresponding to the second decimal place, 3, of the z score $z = -0.5\mathbf{3}$.
   4. The entry in Table II corresponding to $z = -0.53$ is 0.2981. This is illustrated in Figure 7.31.

   Thus, 0.2981 represents the proportion of area to the **left** of $z = -0.53$ as shown by the shaded area in Figure 7.30.

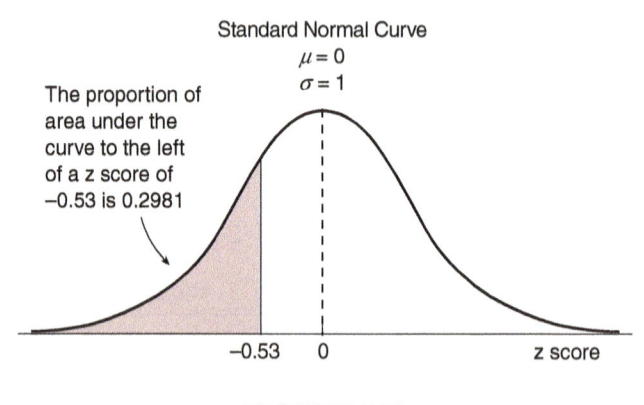

FIGURE 7.30

b) The proportion of area under the normal curve to the left of $z = 2.56$ has been shaded and is shown in Figure 7.32. To determine this shaded area, we need to look up the z score $z = 2.56$ in Table II.
   Using the following procedure, we can locate $z = 2.56$ in Table II.
   1. Since the z score is positive, use the page pertaining to positive z scores.
   2. Locate the row corresponding to the z score $z = \mathbf{2.5}6$ by looking up **2.5** under the column labeled z.
   3. Locate the column corresponding to the second decimal place, 6, of the z score $z = 2.5\mathbf{6}$.
   4. The entry in Table II corresponding to $z = 2.56$ is 0.9948. This is illustrated in Figure 7.33.

   Thus, 0.9948 represents the proportion of area to the **left** of $z = 2.56$ as shown by the shaded area in Figure 7.32.

TABLE II: The Normal Curve Area Table

```
z    | 0   1   2   3   4   5   6   7   8   9
-----|---------------------------------------
                     *
  .                  *
  .                  *
  .                  *
-1.0                 *
-0.9                 *
-0.8                 *
-0.7                 *
-0.6                 *
-0.5*************    0.2981
  .
  .
  .
```

FIGURE 7.31

Standard Normal Curve
$\mu = 0$
$\sigma = 1$

The proportion of area under the curve to the left of z = 2.56 is 0.9948

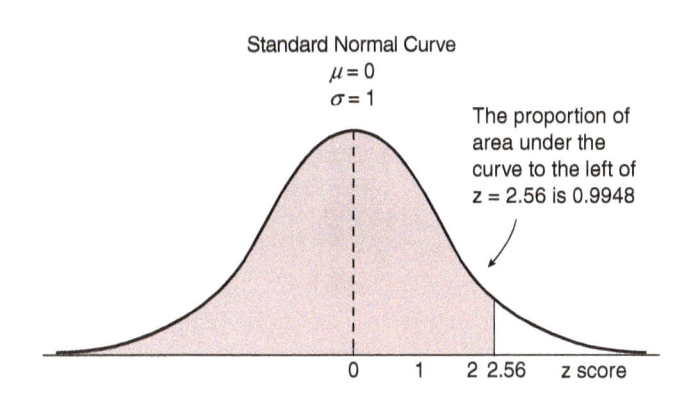

FIGURE 7.32

TABLE II: The Normal Curve Area Table

```
z    | 0   1   2   3   4   5   6   7   8   9
-----|---------------------------------------
                                 *
  .                              *
  .                              *
  .                              *
 2.0                             *
 2.1                             *
 2.2                             *
 2.3                             *
 2.4                             *
 2.5******************************  0.9948
  .
```

FIGURE 7.33

We can generalize the procedure to find the proportion of area to the left of a z score.

Procedure to Find the Proportion of Area to the Left of a z Score

Proportion of Area to the left of z = Entry in Table II corresponding to z

386 Chapter 7 Continuous Probability Distributions and the Normal Distribution

The normalcdf function of the TI-84 calculator can be used to find the area under a normal curve to the left of a z-score. This function can be found in the distribution menu of the TI-84. To access the distribution menu, press 2nd VARS [DISTR]. Figure 7.34 shows the calculator's display when you access the distribution menu.

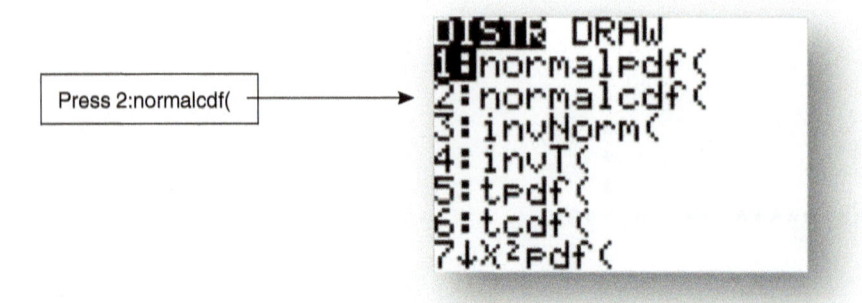

FIGURE 7.34

Press the **2: normalcdf(** key on the calculator and you will be brought to the input screen shown in Figure 7.35.

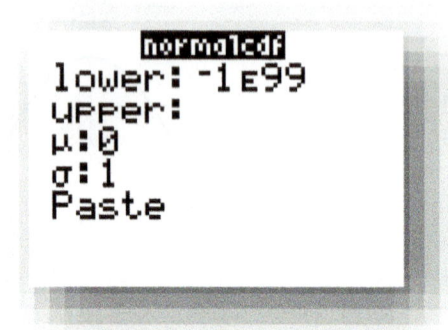

FIGURE 7.35

The normal function requires that you input four parameters in a specific order. The syntax for the parameters is as follows:

normalcdf(lower bound of shaded area, upper bound of shaded area, mean, standard deviation)

In Example 7.1a, we were asked to find the area under the normal curve to the left of a z-score of −0.53. From Figure 7.30, the lower boundary of the shaded region is −∞, and the upper bound of the shaded region is −0.53. For the standard normal curve, the mean and standard deviation are 0 and 1 respectively. Please note that the calculator does not have a symbol for −∞. To input −∞, we must input "−E99" which is scientific notation for -10^{99}, the smallest number the calculator can display. The default value on the calculator for the lower bound is −1E99. To enter this value manually, press (-) 2nd , 99. Now enter the upper bound as shown in Figure 7.36a. Notice the default mean and standard deviation are 0 and 1 respectively. Press the down arrow to highlight Paste and press ENTER twice to paste and calculate the answer. This probability is shown in Figure 7.36b.

FIGURE 7.36a   **FIGURE 7.36b**

Notice the result rounded to 4 decimal places is 0.2981. This is the same answer we found using Table II.

In Example 7.1b, we were asked to find the area to the left of a z-score of 2.56. This can be done in a similar fashion by using –E99 for the lower boundary of the shaded area, 2.56 as the upper bound of the shaded area, 0 for the mean and 1 for the standard deviation. Figure 7.37a shows the calculator input and Figure 7.37b shows the result.

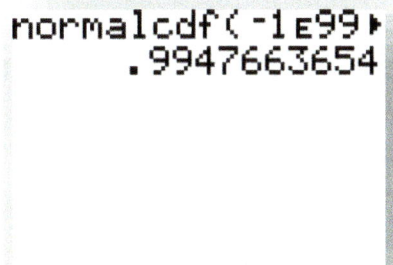

FIGURE 7.37a

FIGURE 7.37b

The result is 0.9948 when rounded to 4 decimal places. This is the same answer as the value obtained from Table II.

Example 7.2 illustrates how to use Table II to find areas under the normal curve to the ***right of a z score.***

EXAMPLE 7.2

Find the proportion of area under the normal curve:
a) To the right of z = –1.37.
b) To the right of z = 0.67.

Solution

a) In Figure 7.38, the proportion of area under the normal curve to the right of z = –1.37 has been shaded.

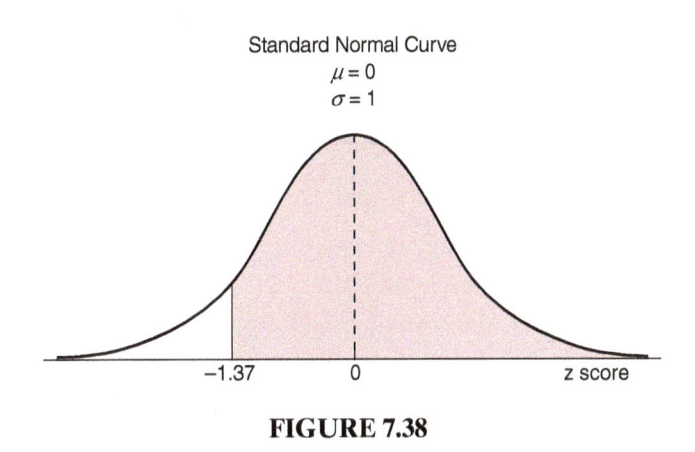

FIGURE 7.38

Find z = –1.37 in Table II. Using the techniques described, we find that the entry in Table II corresponding to z = –1.37 is 0.0853. This is illustrated in Figure 7.39.

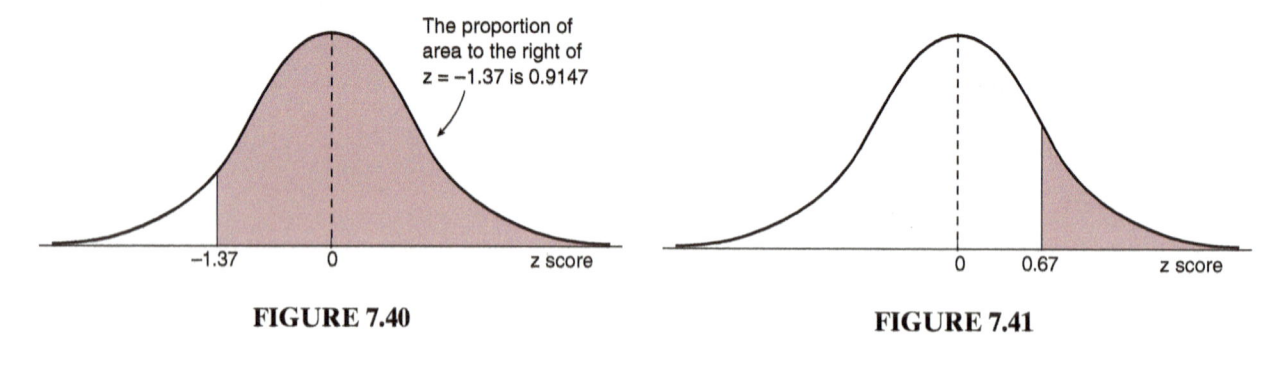

388 Chapter 7 Continuous Probability Distributions and the Normal Distribution

TABLE II: The Normal Curve Area Table

z	0	1	2	3	4	5	6	7	8	9
.								*		
.								*		
.								*		
-1.1								*		
-1.2								*		
-1.3	** 0.0853
-1.4										
.										

FIGURE 7.39

Consequently, 0.0853 of the **total** area is to the left of z = –1.37. *Since the total proportion of area under the normal curve is 1, to find the proportion of area to the right of z = –1.37 subtract 0.0853 from 1.* Thus, the proportion of area to the *right* of z = –1.37 is:

$$\text{proportion of area to the } right \text{ of } z = -1.37 = 1 - 0.0853$$
$$= 0.9147.$$

This is illustrated in Figure 7.40.

b) The proportion of area under the normal curve to the right of z = 0.67 has been shaded and is shown in Figure 7.41. To determine this shaded area, we need to look up the z score z = 0.67 in Table II.

Using Table II, the entry corresponding to z = 0.67 is 0.7486. This is illustrated in Figure 7.42. Consequently, 0.7486 of the **total** area is to the left of z = 0.67.

FIGURE 7.40

FIGURE 7.41

TABLE II: The Normal Curve Area Table

z	0	1	2	3	4	5	6	7	8	9
.								*		
.								*		
.								*		
0.1								*		
0.2								*		
0.3								*		
0.4								*		
0.5								*		
0.6	** 0.7486
.										

FIGURE 7.42

To find the proportion of area to the right of z = 0.67 subtract 0.7486 from 1 since the total proportion of area under the normal curve is 1. Thus, the proportion of area to the *right* of z = 0.67 is:

$$\text{proportion of area to the } right \text{ of } z = 0.67 = 1 - 0.7486$$
$$= 0.2514.$$

This is illustrated in Figure 7.43.

We can now generalize the procedure to find the proportion of area to the right of a z score.

The proportion of area to the right of z = 0.67 is 0.2514

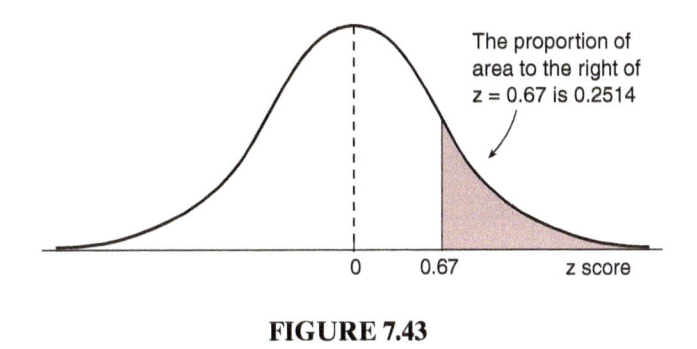

FIGURE 7.43

We can use the normalcdf function of the TI-84 calculator to compute the areas in Example 7.2. To access this function, press 2nd VARS [DISTR] 2:normalcdf(.

Remember that the normalcdf function requires that you input four parameters in a specific order. The syntax for the parameters is as follows:

normalcdf(lower bound of shaded area, upper bound of shaded area, mean, standard deviation)

In Example 7.2a, we were asked to find the area under the normal curve to the right of a z-score of -1.37. Examining Figure 7.38, the lower boundary of the shaded region is -1.37, and the upper bound of the shaded region is ∞(remember to use E99 for ∞). For the standard normal curve, the mean and standard deviation are 0 and 1 respectively. Figure 7.44a shows the calculator input and Figure 7.44b shows the output for these four values.

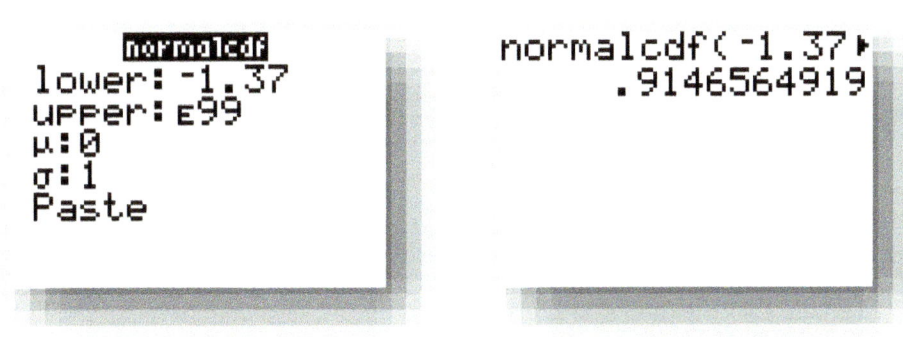

```
normalcdf
lower: -1.37
upper: E99
μ: 0
σ: 1
Paste
```

```
normalcdf(-1.37▶
          .9146564919
```

FIGURE 7.44a **FIGURE 7.44b**

Notice the result rounded to 4 decimal places is 0.9147. This is the same answer we found using Table II.

In Example 7.2b, we were asked to find the area to the right of a z-score of 0.67. This can be computed using 0.67 for the lower boundary of the shaded area, E99 as the upper bound of the shaded area, 0 for the mean and 1 for the standard deviation. Figure 7.45a shows the calculator input and Figure 7.45b shows the result.

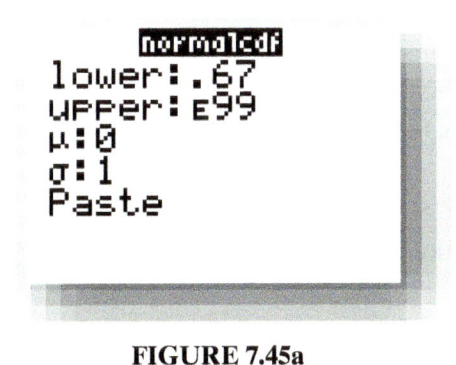

```
normalcdf
lower: .67
upper: E99
μ: 0
σ: 1
Paste
```

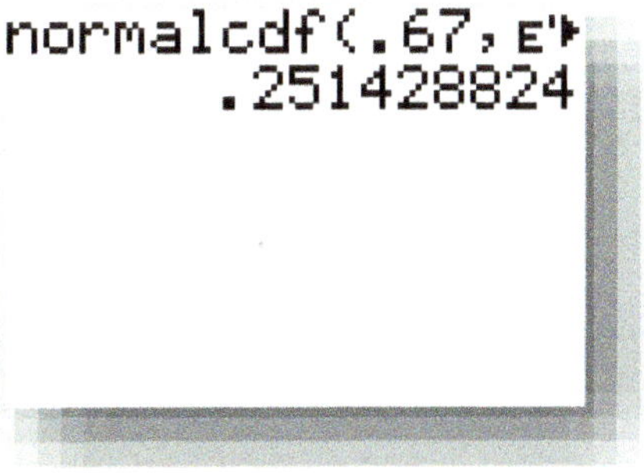

```
normalcdf(.67,E▶
          .251428824
```

FIGURE 7.45a **FIGURE 7.45b**

The result is 0.2514 rounded to 4 decimal places. This is the same as the value using Table II.

390   Chapter 7   Continuous Probability Distributions and the Normal Distribution

Another procedure which can be used to determine the proportion of area to the right of a z score is to utilize the symmetric property of the normal curve (property 3).

> **Procedure to Find the Proportion of Area to the Right of a z Score**
>
> 1 – Proportion of Area to the left of z = Proportion of Area to the right of z

For example, using the alternate procedure to Part (a) of Example 7.2, the proportion of area to the right of z = –1.37 can be found by looking up the proportion of area to the left of z = 1.37, which is 0.9147. In Part (b) of Example 7.2, the proportion of area to the right of z = 0.67 can be found by looking up the proportion of area to the left of z = –0.67, which is 0.2514.

> **Alternate Procedure to Find the Proportion of Area to the Right of a z Score**
>
> Proportion of Area to the right of z = Proportion of Area to the left of –z

Example 7.3 illustrates how to use Table II to find areas under the normal curve **between two z scores**.

EXAMPLE 7.3

Find the proportion of area under the normal curve:
a) Between z = 0 and z = 1.5.
b) Between z = –1.96 and z = 1.96.
c) Between z = –1.25 and z = 1.0.

Solution

a) In Figure 7.46, the proportion of area under the normal curve between z = 0 and z = 1.5 has been shaded. The procedure to determine this shaded region is outlined in the following two steps.

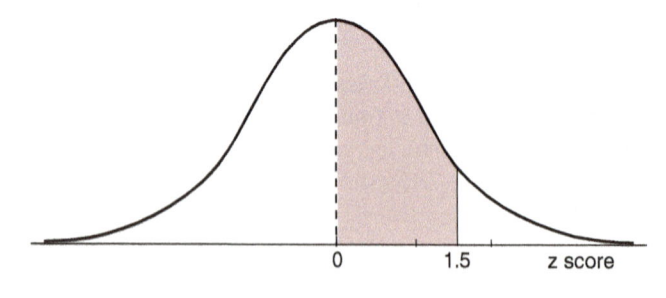

FIGURE 7.46

1. Determine the proportion of area to the left of z = 0 and the proportion of area to the left of z = 1.5. From Table II, the proportion of area to the left each z score is:

z Score	Proportion of Area to the Left
0	0.5000
1.5	0.9332

2. The proportion of area between the two z scores can be found by subtracting the **smaller area** from the **larger area**. Thus, *the proportion of area between z = 0 and z = 1.5 is*:

$$0.9332 - 0.5000 = 0.4332.$$

This is illustrated in Figure 7.47.

b) In Figure 7.48, the proportion of area between z = –1.96 and z = 1.96 has been shaded. To determine this shaded area, use the following two steps.

1. Determine the proportion of area to the left of z = –1.96 and the proportion of area to the left of z = 1.96. From TABLE II, the proportion of area to the left of each z score is:

z Score	Proportion of Area to the Left
–1.96	0.0250
1.96	0.9750

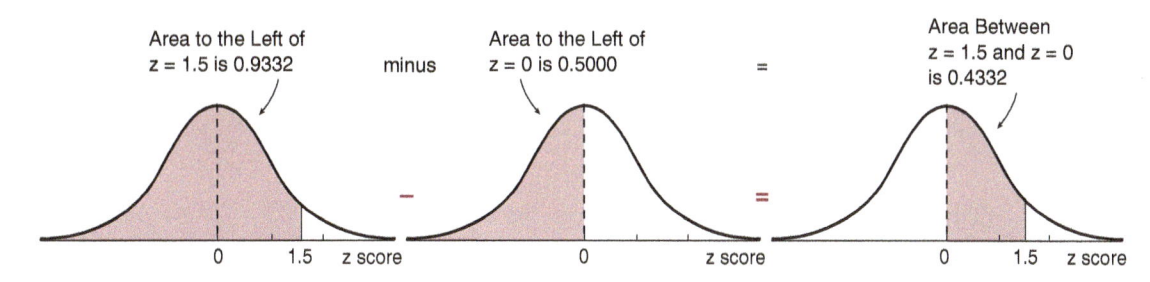

FIGURE 7.47

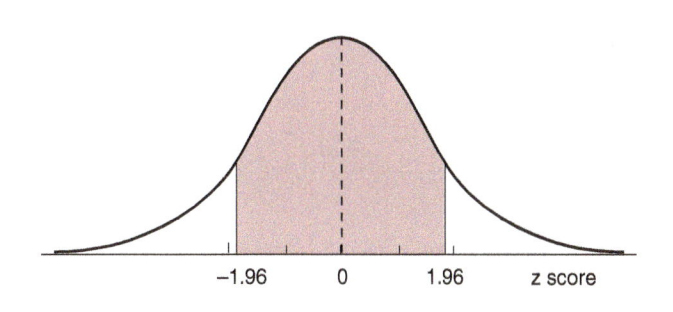

FIGURE 7.48

2. Subtract the **smaller area** from the **larger area**. Thus, *the proportion of area between z = 1.96 and z = –1.96 is*:

$$0.9750 - 0.0250 = 0.9500.$$

This is illustrated in Figure 7.49.

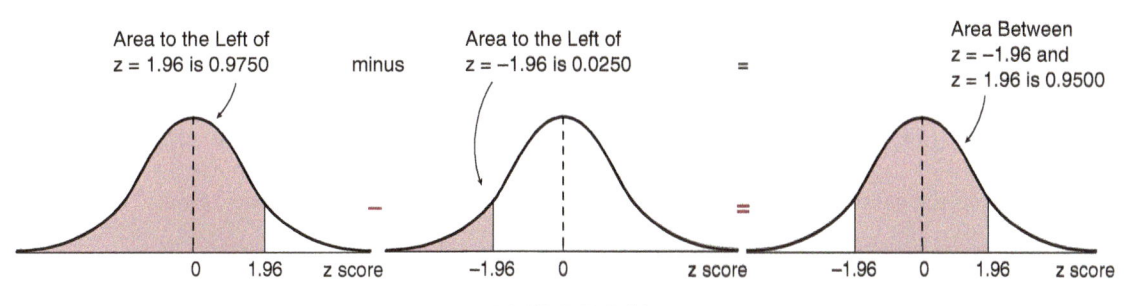

FIGURE 7.49

Chapter 7 *Continuous Probability Distributions and the Normal Distribution*

c) In Figure 7.50, the proportion of area between z = –1.25 and z = 1.0 is shaded. To determine this shaded region, perform the following two steps.

1. Determine the proportion of area to the left of each z score. From Table II, these proportions are:

z Score	Proportion of Area to the Left
–1.25	0.1056
1.00	0.8413

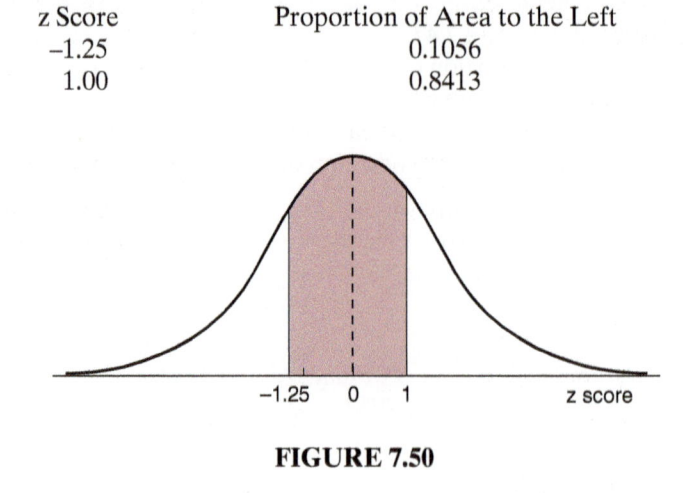

FIGURE 7.50

2. By subtracting the **smaller area** from the **larger area**, *the proportion of area between z = 1.00 and z = –1.25 is*:

$$0.8413 - 0.1056 = 0.7357.$$

This is illustrated in Figure 7.51.

We can now generalize the procedure to find the proportion of area between two z scores.

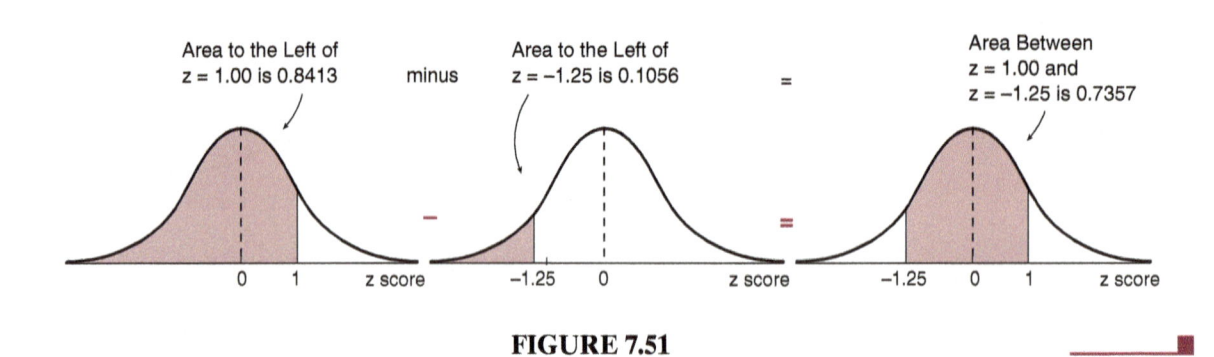

Area to the Left of z = 1.00 is 0.8413 minus Area to the Left of z = –1.25 is 0.1056 = Area Between z = 1.00 and z = –1.25 is 0.7357

FIGURE 7.51

We can use the normalcdf function on the TI-84 calculator to find the areas in Example 7.3. To access this function, press 2nd VARS [DISTR] 2:normalcdf(.

Remember that the normalcdf function requires that you input four parameters in a specific order. The syntax for the parameters is as follows:

normalcdf(lower bound of shaded area, upper bound of shaded area, mean, standard deviation)

In Example 7.3a, we were asked to find the area under the normal curve between z-scores of 0 and 1.5. Examining Figure 7.46, the lower boundary of the shaded region is 0 and the upper bound of the shaded region is 1.5. For the standard normal curve, the mean and standard deviation are 0 and 1 respectively. Figure 7.52a shows the calculator input and Figure 7.52b shows the output for these values.

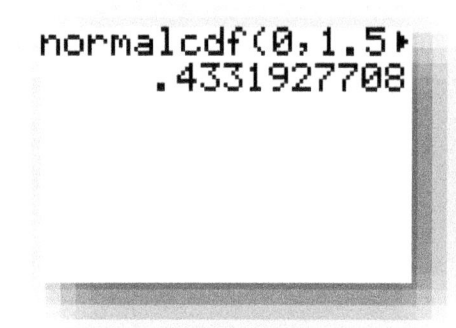

Section 7.5 *Using the Normal Curve Area Table*   **393**

FIGURE 7.52a   **FIGURE 7.52b**

Notice the result rounded to 4 decimal places is 0.4332. This is the same answer we found using Table II.

In Example 7.3b, we were asked to find the area between z-scores of -1.96 and 1.96. This can be done in a similar fashion by using -1.96 for the lower boundary of the shaded area, 1.96 as the upper bound of the shaded area, 0 for the mean and 1 for the standard deviation. Figure 7.53a shows the calculator input and Figure 7.53b shows the result.

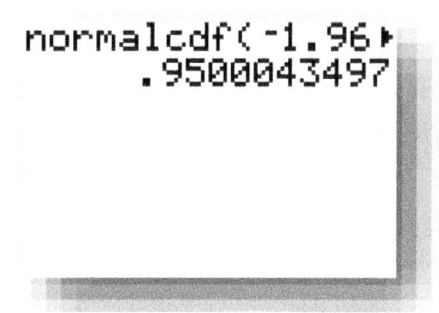

FIGURE 7.53a   **FIGURE 7.53b**

The result is 0.9500 when rounded to 4 decimal places. This is the same result as the value from Table II.

In Example 7.3c, we were asked to find the area between z-scores of -1.25 and 1.0. This can be done by using -1.25 for the lower boundary of the shaded area, 1.0 as the upper bound of the shaded area, 0 for the mean and 1 for the standard deviation. Figure 7.54a shows the calculator input and Figure 7.54b shows the result.

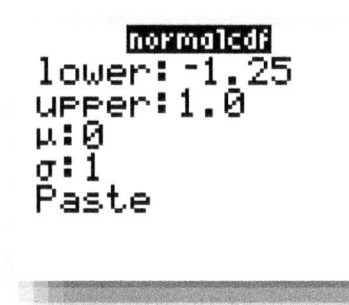

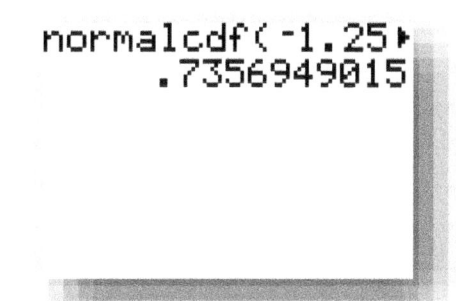

FIGURE 7.54a   **FIGURE 7.54b**

The result is 0.7357 rounded to 4 decimal places. This is the same result as the value from Table II.

394 Chapter 7 *Continuous Probability Distributions and the Normal Distribution*

Since the normal curve represents a continuous variable, then the area associated with a particular z score is zero. This is true because there is no area under a continuous curve and exactly over a z score. Therefore, we have the following rule.

Procedure to Find the Proportion of Area Between Two z Scores

Proportion of Area between two z scores = Proportion of Area to the left of larger z score *minus* Proportion of Area to the left of smaller z score

---

Proportion of Area Associated to a Single z Score

Proportion of area for a particular z score equals zero.

Using the procedure to find the area between two z scores we can determine and verify the percentages associated with the Empirical Rule discussed in Chapter 3. The Empirical Rule states for a bell-shaped distribution that:

1. approximately 68% of the data values will lie within one standard deviation of the mean.
2. approximately 95% of the data values will lie within two standard deviations of the mean.
3. approximately all (99% to 100%) of the data values will lie within three standard deviations of the mean.

In Example 7.4, we will verify the results of the Empirical Rule for a normal distribution.

EXAMPLE 7.4

For a normal distribution, determine the percent of area under a normal curve within:
a) One standard deviation of the mean.
b) Two standard deviations of the mean.
c) Three standard deviations of the mean.
d) Why do the results of the Empirical Rule hold true for a normal distribution?

Solution

a) The percent of area under a normal curve within one standard deviation of the mean is interpreted as the percent of area between the z scores $z = -1$ and $z = 1$. This is illustrated in Figure 7.55. To determine this shaded area, use the following two steps.
   1. Find the proportion of area to the left of each z score. From Table II, these proportions are:

z Score	Proportion of Area to the Left
−1.00	0.1587
1.00	0.8413

   2. Subtract the **smaller area** from the **larger area**. Thus, *the proportion of area between $z = 1.00$ and $z = -1.00$ is*:

      $$0.8413 - 0.1587 = 0.6826.$$

   To determine the percent of area between $z = 1.00$ and $z = -1.00$, multiply the proportion of area by 100%. Thus, the percent of area between $z = 1.00$ and $z = -1.00$ is:

   $$\begin{aligned}\text{percent of area between } z=1.00 \text{ and } z=-1.00 &= \left(\text{proportion of area between } z=1.00 \text{ and } z=-1.00\right)(100\%) \\ &= (0.6826)(100\%) \\ &= 68.26\%\end{aligned}$$

   Thus, the percent of area under a normal curve within one standard deviation of the mean is 68.26%.

Section 7.5 Using the Normal Curve Area Table 395

b) The percent of area under a normal curve within two standard deviations of the mean is interpreted as the percent of area between the z scores z = –2 and z = 2. This is illustrated in Figure 7.56.

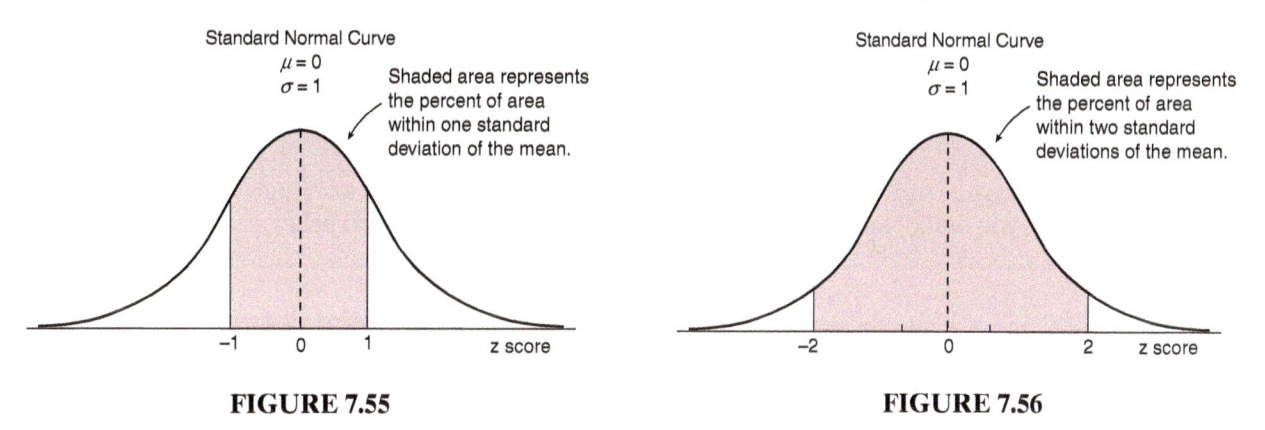

FIGURE 7.55

FIGURE 7.56

To determine this shaded area, use the following two steps.
1. Determine the proportion of area to the left of each z score. From Table II, these proportions are:

z Score	Proportion of Area to the Left
–2.00	0.0228
2.00	0.9772

2. Subtract the **smaller area** from the **larger area**. Thus, *the proportion of area between z = 2.00 and z = –2.00 is*:

$$0.9772 - 0.0228 = 0.9544.$$

To determine the percent of area between z = 2.00 and z = –2.00, multiply the proportion of area by 100%.

$$\text{percent of area between } z = 2.00 \text{ and } z = -2.00 = (0.9544)(100\%)$$
$$= 95.44\%$$

Thus, the percent of area under a normal curve within two standard deviations of the mean is 95.44%.

c) The percent of area under a normal curve within three standard deviations of the mean is interpreted as the percent of area between the z scores z = –3 and z = 3. This is illustrated in Figure 7.57. To determine this shaded area, use the following two steps.

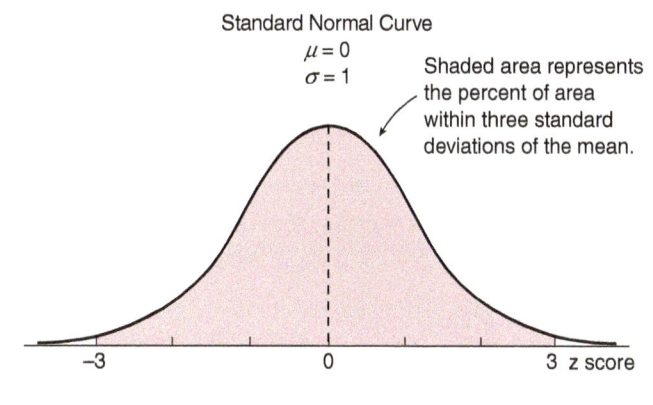

FIGURE 7.57

1. Determine the proportion of area to the left of each z score. From Table II:

z Score	Proportion of Area to the Left
–3.00	0.0013
3.00	0.9987

2. Subtract the **smaller area** from the **larger area**. Thus, *the proportion of area between z = 3.00 and z = –3.00 is*:

$$0.9987 - 0.0013 = 0.9974.$$

To change this proportion to percent, multiply the proportion of area by 100%. Thus:

percent of area between z = 3.00 and z = –3.00 = (0.9974)(100%)
= 99.74%

Thus, the percent of area under a normal curve within three standard deviations of the mean is 99.74%.

d) The results of the Empirical Rule hold true for bell-shaped curves. Since a normal curve is a bell-shaped curve, the results of the Empirical Rule are true for any normal distribution.

The results of Example 7.4 are very useful to remember when working with a normal distribution especially if you don't have Table II handy. These three results are stated below and illustrated in Figure 7.58.

In a Normal Distribution:

a) Approximately 68% of the area or data values under the normal curve is within one standard deviation of the mean. That is, approximately 68% of the area or data values falls between $\mu - 1\sigma$ and $\mu + 1\sigma$.

b) Approximately 95% of the area or data values under the normal curve is within two standard deviations of the mean. That is, approximately 95% of the area or data values falls between $\mu - 2\sigma$ and $\mu + 2\sigma$.

c) Approximately 100% of the area or data values under the normal curve is within three standard deviations of the mean. That is, approximately 100% of the area or data values falls between $\mu - 3\sigma$ and $\mu + 3\sigma$.

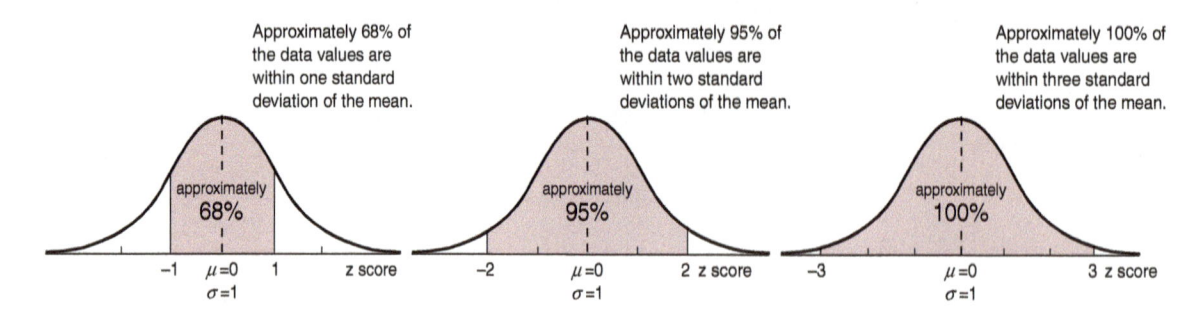

FIGURE 7.58

Finding a z-score Knowing the Proportion of Area to the Left

An important technique associated with the proportion of area and a z-score is to be able to **find** a z score **knowing** information about the proportion of area to the **left** of the z score.

EXAMPLE 7.5

For a normal distribution, find the z score(s) that cut(s) off:
a) The lowest 20% of area.
b) The highest 10% of area.
c) The middle 90% of the area.

Solution

a) The shaded proportion of area in Figure 7.59 represents the lowest 20% of the area in a normal distribution.

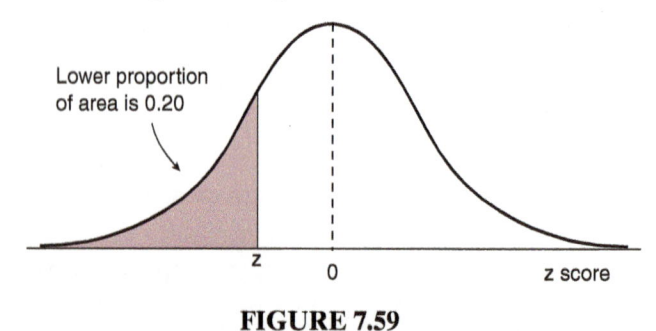

FIGURE 7.59

Notice that this area is positioned to the left of the mean: $\mu = 0$. Thus, this z score will be negative since it is below the mean: $\mu = 0$. To determine the z score that cuts off the lowest 20%, we must use Table II. Since Table II gives the relationship between z scores and the proportion of area to the left of a z score, knowing the proportion of area to the left of this unknown z score will enable us to determine this z score. We can calculate this z score by the using the following procedure.

Step 1 **Determine the proportion of area to the left of the z score, z.**

Since the z score cuts off the lowest 20%, then the proportion of area to the LEFT of the z score is 0.20.

Remark: To convert a percent to a decimal move the decimal point two places to the left. For example:

percent ────────────────→ **decimal**
20.00% (move decimal point two places to left) 0.2000

Step 2 **Locate the proportion of area in Table II which is closest to the area to the left of the z score and determine the z score corresponding to this proportion of area.**

In Table II find the proportion of area closest to 0.20. This is illustrated in Figure 7.60. Thus, the proportion of area closest to 0.2000 which can be found in Table II is 0.2005. The z score that is associated with 0.2005 is z = –0.84. Thus, the z score that cuts off the lowest 20% of area in a normal distribution is z = –0.84.

TABLE II: The Normal Curve Area Table

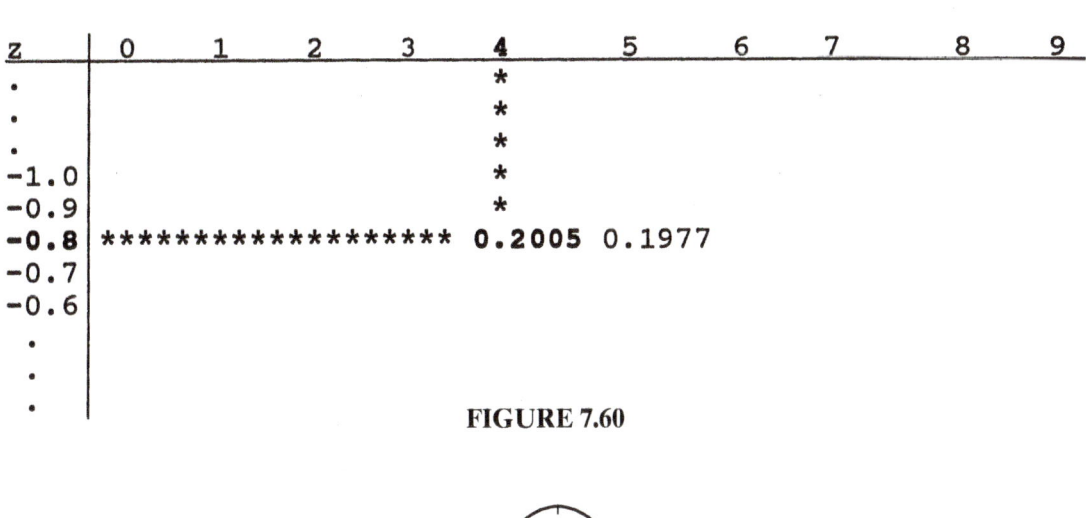

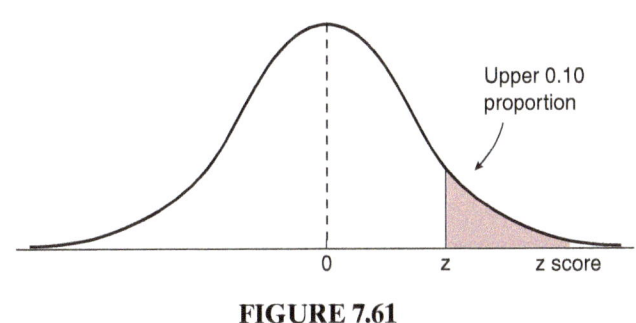

```
z    |  0    1    2    3    4    5    6    7    8    9
 .                              *
 .                              *
 .                              *
-1.0                            *
-0.9                            *
-0.8 |******************  0.2005  0.1977
-0.7
-0.6
 .
 .
 .
```

FIGURE 7.60

FIGURE 7.61

b) The shaded proportion of area in Figure 7.61 represents the highest 10% of area in a normal distribution.

Step 1 **Determine the proportion of area to the left of the z score, z.**

Notice that the proportion of area to the right of the z score that cuts off the highest 10% is 0.10 and it is positioned to the right of the mean: $\mu = 0$. Thus, the z score will be positive and has 0.10 of the proportion of area under the normal curve to the right of this z score. Remember in Table II, each z score is associated with the proportion of area to the **LEFT** of the z score. Thus, we need to compute the proportion of area to the **LEFT** of the z score to use Table II. To find the z score associated with the highest 10% or 0.10 proportion of area, we need to determine the proportion of area to the **LEFT** of the z score.

398 Chapter 7 Continuous Probability Distributions and the Normal Distribution

To compute the proportion of area to the LEFT, we subtract the proportion to the RIGHT of the z score from the **TOTAL AREA** under the normal curve which is 1. That is:

$$\text{proportion of area to the LEFT of z score} = 1.0000 \text{ minus proportion of area to the RIGHT of z score}$$

$$= 1.0000 - 0.10$$
$$= 0.9000$$

Step 2 **Locate the proportion of area in Table II which is closest to the proportion of area that is to the left of the z score and determine the z score corresponding to this proportion of area.**

We must find the z score that corresponds to the proportion closest to 0.9000 in Table II. This is illustrated in Figure 7.62. The proportion of area closest to 0.9000 is 0.8997 and is highlighted in Figure 7.62. The z score that is associated with 0.8997 is z = 1.28. Thus the z score that cuts off the highest 10% of area in a normal distribution is z = 1.28.

TABLE II: The Normal Curve Area Table

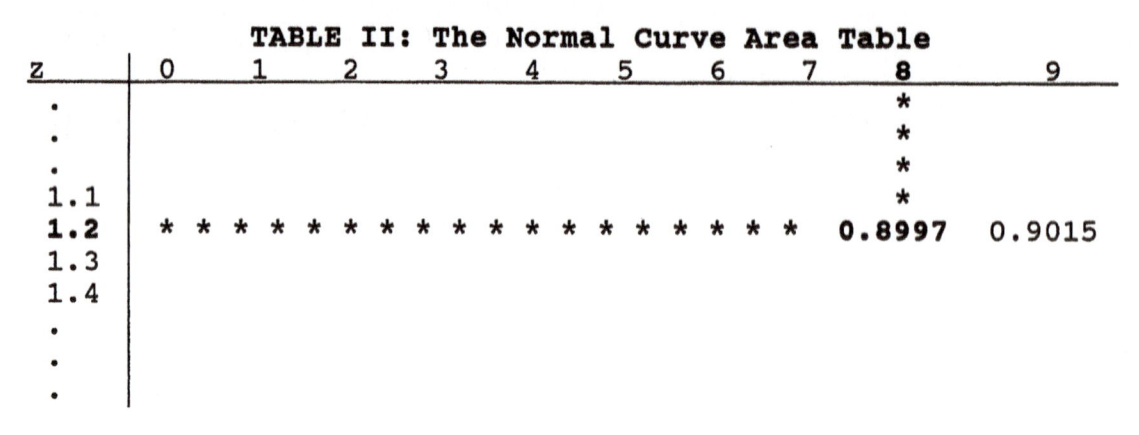

z	0	1	2	3	4	5	6	7	8	9
.									*	
.									*	
.									*	
1.1									*	
1.2	*	*	*	*	*	*	*	*	*	*	*	*	*	*	*	*	0.8997	0.9015
1.3										
1.4										
.										
.										
.										

FIGURE 7.62

c) In a normal distribution, z = 0 is located at the center. Therefore the middle 90% of the z scores will be centered about the z = 0, that is 45% of the z scores will be less than the z = 0 and 45% of the z scores will be greater than the z = 0. This is illustrated in Figure 7.63. By examining Figure 7.63, you should notice that the middle of the normal distribution means the proportion of area is **centered** about z = 0. Thus we are looking for two z scores, which we will symbolize as z_1 and z_2. To determine these z scores, we will follow the procedure outlined in part a of this problem.

Step 1 **Determine the proportion of area to the left of the each z score, z.**

To determine the proportion of area to the left of z_1, observe that 45% or 0.45 of the area is between the z score z_1 and the z = 0. Since the proportion of area to the left of z = 0 is 50% OR 0.50, then the remaining proportion of area to the left of z_1 is:

$$0.50 - 0.45 = 0.05.$$

This is illustrated in Figure 7.64.

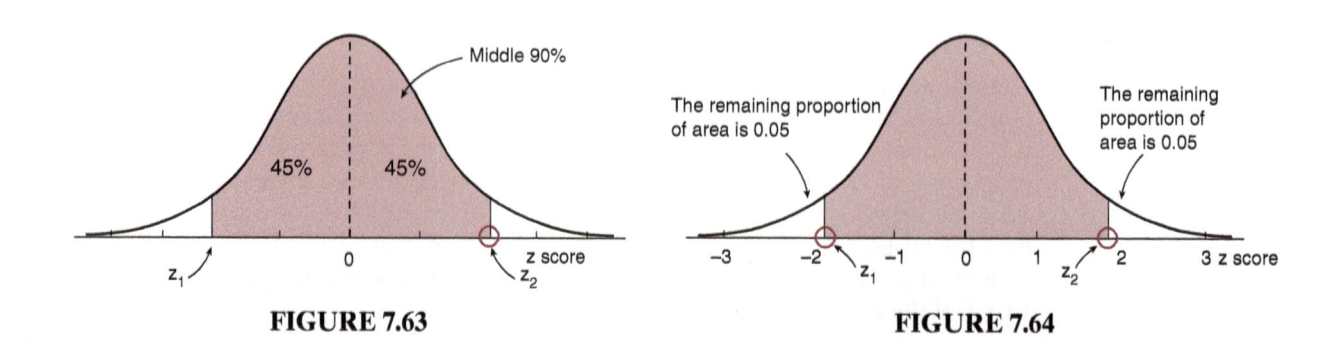

FIGURE 7.63

FIGURE 7.64

Step 2 **Locate the proportion of area in Table II which is closest to the proportion of area that is to the left of the z score and determine the z score corresponding to this proportion of area.**

Since Table II in Appendix C gives the proportion of area to the left of a z score, we are looking for the area closest to 0.0500. The z score closest to 0.0500 can be found in Table II. This is illustrated in Figure 7.65.

Notice that the z scores −1.64 and −1.65 are each 0.0005 away from the proportion 0.0500. When this occurs *either* z score could be chosen. In this text, we will always use the z score **farthest from z = 0**, i.e. the z score with the greater absolute value. Thus, $z_1 = -1.65$ is chosen to represent the z score that cuts off the lower proportion of 0.05 or the lowest 5%. Since the normal distribution is symmetric about z = 0, then the value of z_2 is also 1.65 standard deviations away from the mean of z = 0. However, since z_2 is greater than the mean of z = 0, then z_2 is positive and is equal to $z_2 = 1.65$. Therefore, the z scores that cut off the middle 90% of the normal distribution are $z_1 = -1.65$ and $z_2 = 1.65$. This is illustrated in Figure 7.66.

```
         TABLE II:  The Normal Curve Area Table
  z    |  0    1    2    3    4    5    6    7    8    9
  .    |                              *
  .    |                              *
  .    |                              *
 -1.8  |                              *
 -1.7  |                              *
 -1.6  |  *  *  *  *  *  *  *  * 0.0505 0.0495
 -1.5  |
 -1.4  |
  .
  .
  .
```

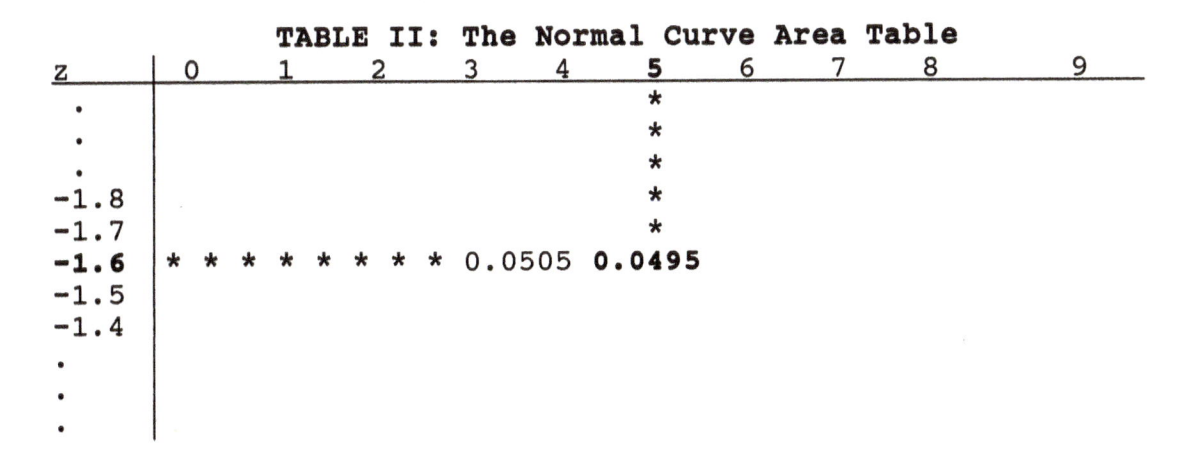

FIGURE 7.65

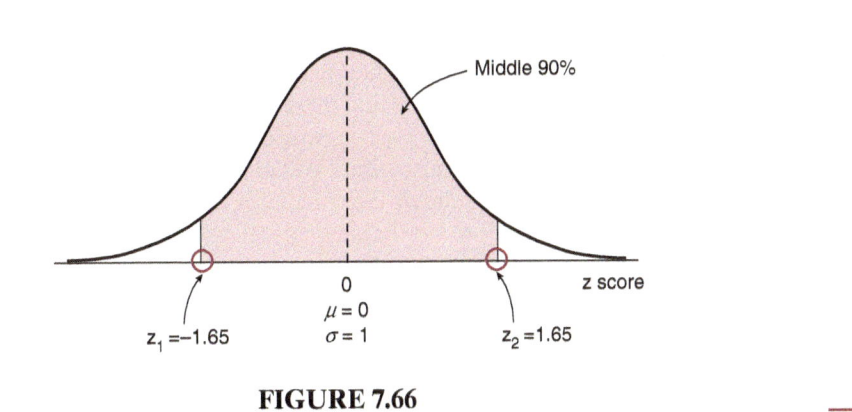

FIGURE 7.66

We can use the InvNorm function of the TI-84 calculator to find the value of a z-score given the area under the normal curve to the left of the z-score. To access this function, press 2nd VARS [DISTR] 3:InvNorm(.

The syntax of the parameters for InvNorm is:

InvNorm(Area to the left of the z-score, mean, standard deviation)

In Example 7.5a, we were asked to find the z-score that cuts off the lower 20% of area under the normal curve. Examining Figure 7.59, the area to the left of the z-score is 20% or 0.20. For the standard normal curve, the mean and standard deviation are 0 and 1 respectively. Figure 7.67a shows the calculator input and Figure 7.67b shows the output for these values.

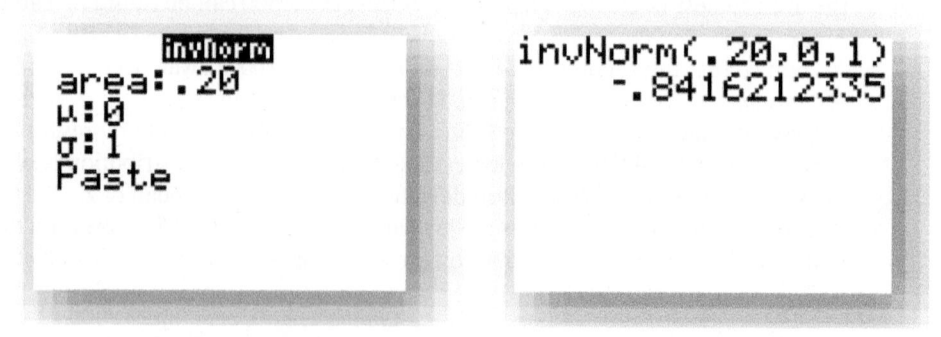

FIGURE 7.67a

FIGURE 7.67b

Notice the result rounded to 2 decimal places is -0.84. This is the same answer we found using Table II.

In Example 7.5b, we were asked to find the z-score that cuts off the upper 10% of area under the normal curve. Examining Figure 7.61, the area to the left of the z-score is 90% or 0.90. Although the question asks for the z-score cutting the upper 10% of area, the calculator function *requires* us to use the area to the left of the z-score for its input so we use 90% or 0.90. For the standard normal curve, the mean and standard deviation are 0 and 1 respectively. Figure 7.68a shows the calculator input and Figure 7.68b shows the output for these three values.

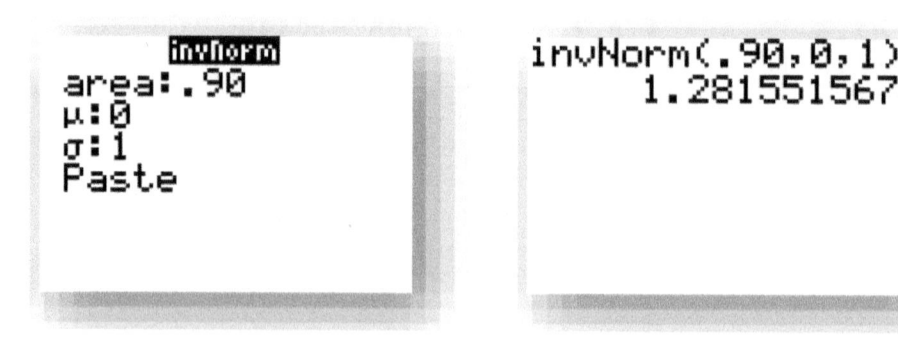

FIGURE 7.68a

FIGURE 7.68b

The result is 1.28 rounded to 2 decimal places. This is the same z-score as the value from Table II.

In Example 7.5c, we were asked to find the z-score that cuts off the middle 90% of area under the normal curve. Remember that there are two z-scores that cut a middle area. Examining Figure 7.64 the area to the left of z_1 is given as 0.05. We can find this z-score by using 0.05 as the area to the left, 0 for the mean and 1 for the standard deviation. Figure 7.69a shows the calculator input and Figure 7.69b shows the result.

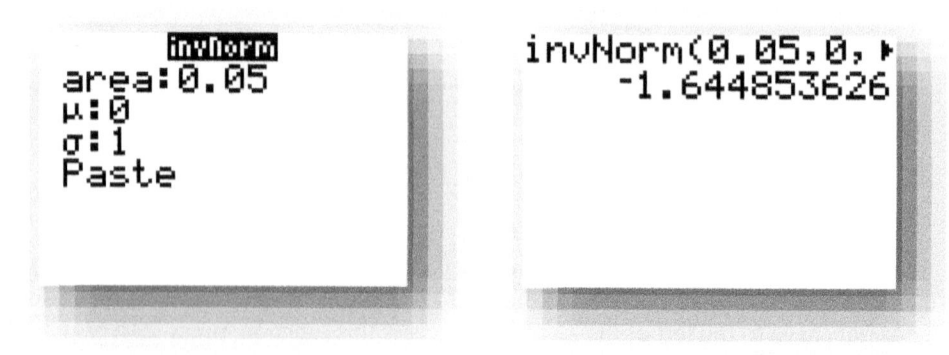

FIGURE 7.69a

FIGURE 7.69b

The result is -1.645 rounded to 3 decimal places. From Table II we found the result to be -1.65. The difference in these two answers is due to round-off error. Since the normal distribution is symmetrical about $z = 0$, the upper value for z_2 must be 1.645.

7.6 APPLICATIONS OF THE NORMAL DISTRIBUTION

Let's now consider some applied problems in which it will be assumed that the distributions under consideration can be approximated by a normal distribution.

EXAMPLE 7.6

Assume the distribution of the annual cost of textbooks for college students is approximately normal with $\mu = \$900$ and $\sigma = \$125$. Find the percent of college students who will be expected to spend:
a) Less than $775 on textbooks for the year.
b) More than $962.50 on textbooks for the year.
c) Between $837.50 and $1087.50 on textbooks for the year.

Solution

Since this distribution is approximately normal, we will assume all of the properties of the normal distribution and use the normal curve as a sketch of the distribution representing the annual cost of textbooks for college students. This is shown in Figure 7.70a.

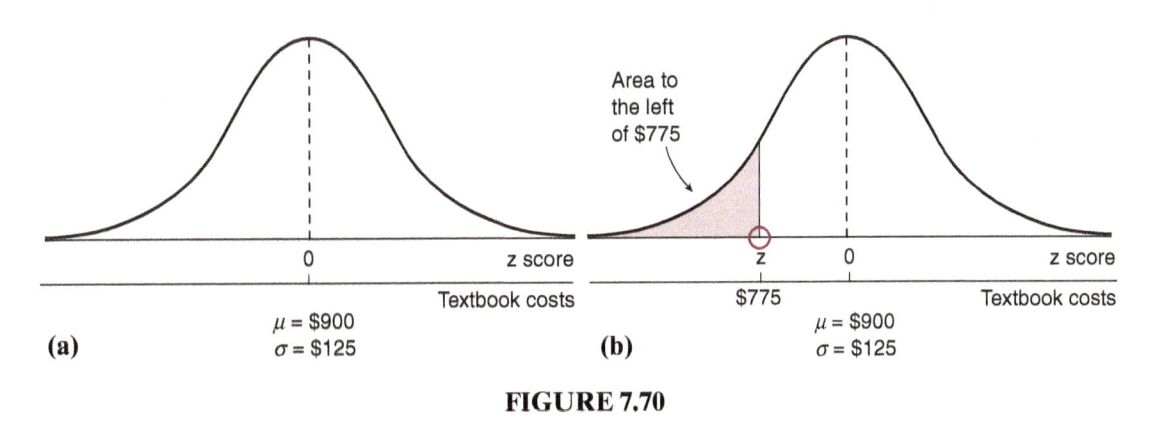

FIGURE 7.70

In examining Figure 7.70a, you will notice that the mean is placed at the center of the normal curve below the z score line and corresponding to the z score of zero.

a) In Figure 7.70b, the shaded area represents the proportion of college students who are expected to spend less than $775 on their textbooks for the year.

To determine the percent of students who are expected to spend less than $775 on their textbooks for the year, we must find the portion of area to the left of $775. To determine this area, we first calculate the z score for $775, using the z score formula:

$$z = \frac{x - \mu}{\sigma}$$

Thus,

$$z = \frac{\$775 - \$900}{\$125} = \frac{-\$125}{\$125} = -1$$

Using Table II: The Normal Curve Area Table, the proportion of area to the left of $z = -1$ is 0.1587. Therefore the percent of students who are expected to spend less than $775 on their textbooks for the year is 15.87%.

b) In Figure 7.71, the shaded area represents the proportion of college students who are expected to spend more than $962.50 on their textbooks for the year. To find the proportion of college students who are expected to spend more than $962.50 on their textbooks for the year, find the proportion of area to the right of $962.50. To determine this area, first calculate the z score of $962.50.

Thus,

$$z = \frac{\$962.50 - \$900}{\$125} = \frac{\$62.50}{\$125} = 0.50$$

Recall that Table II gives the proportion of area to the left of $962.50 or z = 0.50, which is 0.6915. To find the proportion of area to the *right* of $962.50, subtract 0.6915 (the area to the left of z = 0.50) from 1.0000 (the total area under the normal curve).

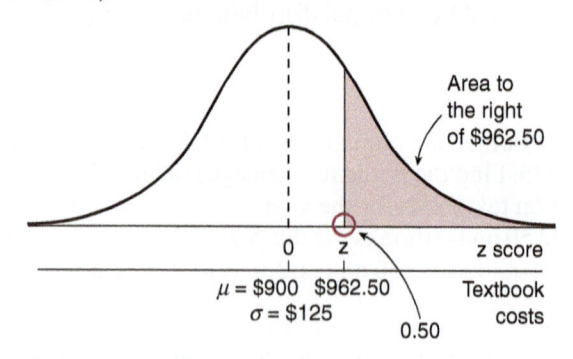

FIGURE 7.71

Thus, the proportion of college students who are expected to spend more than $962.50 on their textbooks for the year is 1.0000 − 0.6915 = 0.3085. Therefore, the percent of college students who are expected to spend more than $962.50 on their textbooks for the year is 30.85%.

c) In Figure 7.72, the shaded area represents the proportion of college students who are expected to spend between $837.50 and $1087.50 on textbooks for the year. To find the area between $837.50 and $1087.50, first calculate their z scores.

$$\text{For z of } \$837.50: z = \frac{\$837.50 - \$900}{\$125}$$

$$z = \frac{-\$62.50}{\$125} = -0.50$$

$$\text{For z of } \$1087.50: z = \frac{\$1087.50 - \$900}{\$125}$$

$$z = \frac{\$187.50}{\$125} = 1.50$$

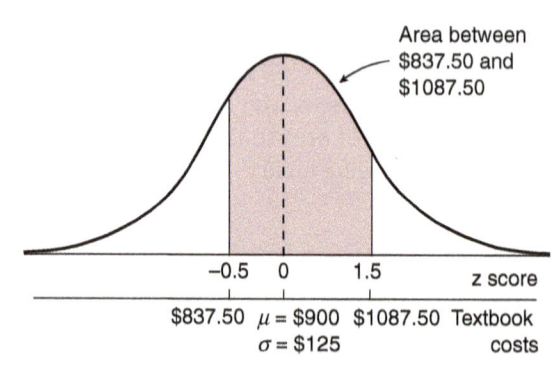

FIGURE 7.72

To find the proportion of area *between* the two z scores of −0.50 and 1.5, determine each of the areas in Table II corresponding to each z score and then subtract the smaller area from the larger area.

Annual Textbook Cost	z Score	Proportion of Area to the Left
$837.50	−0.5	0.3085
$1087.50	1.5	0.9332

shaded area = larger area − smaller area
= 0.9332 − 0.3085
= 0.6247

Thus, the proportion of college students who are expected to spend between $837.50 and $1087.50 on textbooks for the year is 0.6247. Therefore, the percent of college students who are expected to spend between $837.50 and $1087.50 on textbooks for the year is 62.47%.

Section 7.6 *Applications of the Normal Distribution* 403

We can use the normalcdf function of the TI-84 to find the answers to the questions in Example 7.6. Using the calculator does not require us to convert to z-scores and use Table II. Remember, the syntax for the parameters for normalcdf is:

normalcdf(lower bound of shaded area, upper bound of shaded area, mean, standard deviation)

In Example 7.6a, we were asked to find the percentage of college students who expect to spend less than $775 on textbooks for the year. Examining Figure 7.70, the lower boundary of the shaded area is $-\infty$ or –E99 (we can also use 0 as a common sense value because money spent on text books doesn't have a negative value in dollars). The upper boundary of the shaded area is 775. The mean and standard deviation are given as 900 and 125 respectively. Figures 7.73a and 7.73b show the calculator output.

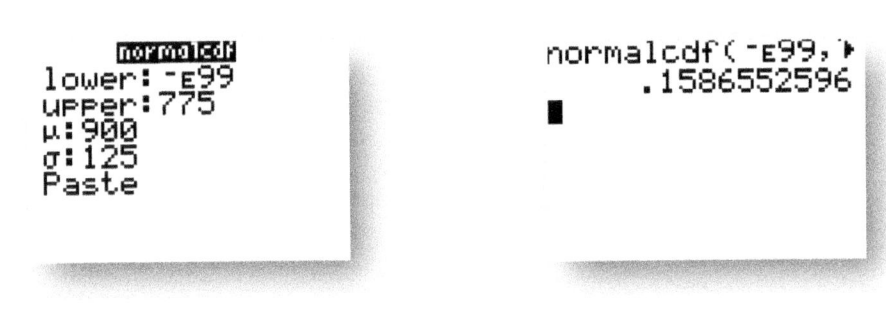

FIGURE 7.73a   **FIGURE 7.73b**

The result is 0.1587 rounded to 4 decimal places.

For Example 7.6b, we are asked to find the percentage of college students who expect to spend more than $962.50 on textbooks for the year. Examining Figure 7.71 the lower boundary of the shaded region is 962.50 and the upper boundary is ∞ or E99. The mean and standard deviation are given as 900 and 125 respectively. Figures 7.74a and 7.74b show the calculator output.

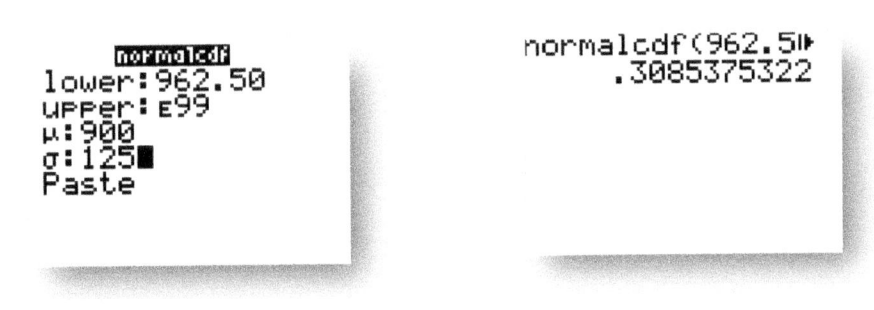

FIGURE 7.74a   **FIGURE 7.74b**

The result is 0.3085 rounded to 4 decimal places.

For Example 7.6c, we are asked to find the percentage of college students who expect to spend between $837.50 and $1,087.50 on textbooks for the year. Examining Figure 7.72 the lower boundary of the shaded region is $837.50 and the upper boundary is $1,087.50. The mean and standard deviation are 900 and 125 respectively. The result is shown in Figures 7.75a and 7.75b.

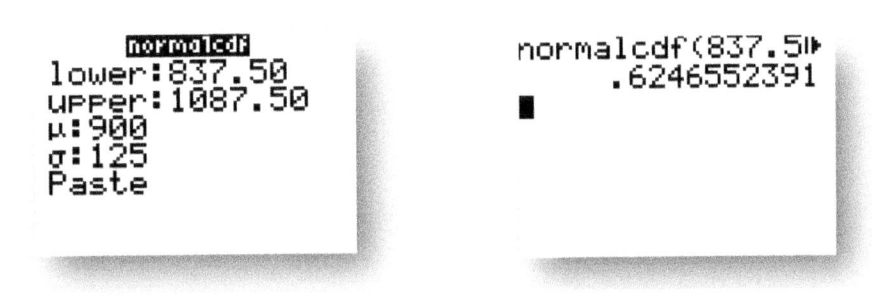

FIGURE 7.75a   **FIGURE 7.75b**

The result is 0.6247 rounded to 4 decimal places.

EXAMPLE 7.7

A distribution of IQ scores is normally distributed with a mean of 100 and a standard deviation of 15. Find the proportion of IQ scores that are:
a) Less than 88.
b) Greater than 127.
c) Less than 70 or greater than 145.
d) Less than one standard deviation from the mean.
e) Within two standard deviations of the mean.

Solution

a) In Figure 7.76, the shaded area represents the proportion of IQ scores less than 88. To determine this area calculate z of 88.

$$z = \frac{88 - 100}{15} = -0.80$$

The proportion of IQ scores less than 88 is 0.2119.

b) In Figure 7.77, the shaded area represents the proportion of IQ scores greater than 127.

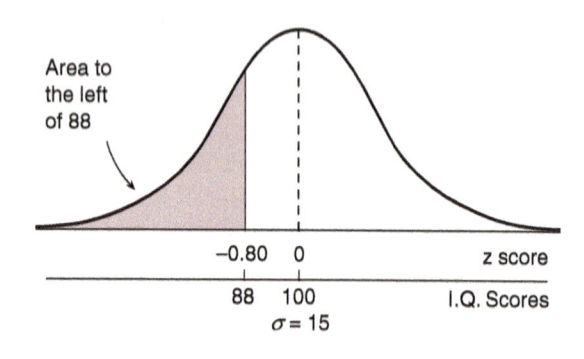

FIGURE 7.76

FIGURE 7.77

To determine this area calculate the z of 127.

$$z = \frac{127 - 100}{15} = 1.80$$

The proportion of area to the left 127 is 0.9641. Therefore the proportion of IQ scores greater than 127 is equal to:

$$1.0000 - 0.9641 = 0.0359.$$

c) In Figure 7.78, the shaded area represents the proportion of IQ scores less than 70 or greater than 145.
As can be seen in Figure 7.78, there are two separate areas to be determined. To calculate the *total* shaded area, calculate each area separately and add the two areas together. For the proportion of area less than 70, calculate the z of 70.

$$z = \frac{70 - 100}{15} = -2.00$$

Using Table II, the proportion of area less than 70 is 0.0228.
For the proportion of area greater than 145, calculate z of 145.

$$z = \frac{145 - 100}{15} = 3.00$$

From Table II, the proportion of area to the right of 145 is 0.0013.
Therefore the total shaded area is the sum of these two areas: $0.0013 + 0.0228 = 0.0241$. Thus the proportion of IQ scores less than 70 or greater than 145 is 0.0241.

d) In Figure 7.79, the shaded area represents the proportion of IQ scores within one standard deviation of the mean.

Less than one standard deviation from the mean would include all IQ scores between z = −1 and z = +1 as shown in Figure 7.79. We need to determine the proportion of area between the z scores of −1 and +1.

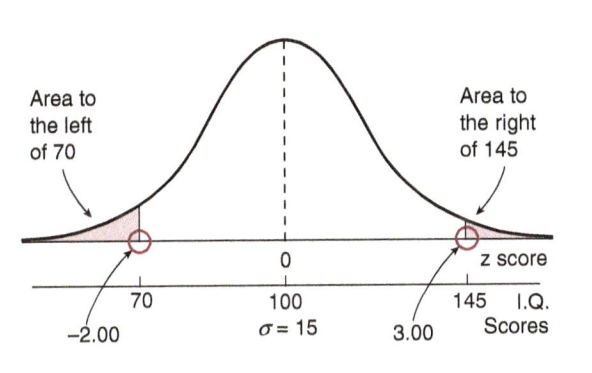

FIGURE 7.78

FIGURE 7.79

Using Table II, we have:

z Score	Proportion of Area to the Left
−1	0.1587
1	0.8413

To find the area between the z scores of −1 and +1 subtract the smaller area from the larger area.

$$\text{shaded area} = \text{larger area} - \text{smaller area}$$
$$= 0.8413 - 0.1587$$
$$= 0.6826.$$

Therefore the proportion of IQ scores less than one standard deviation from the mean is 0.6826.

e) In Figure 7.80, the shaded area represents the proportion of IQ scores within two standard deviations of the mean.

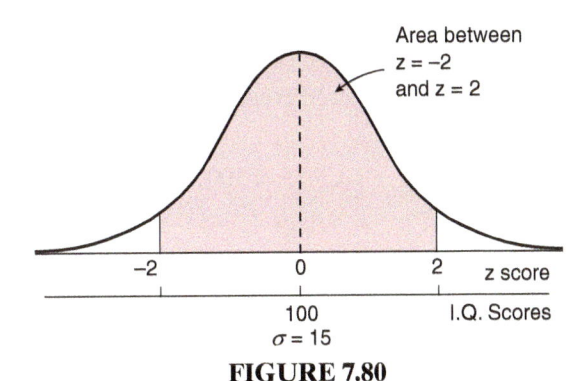

FIGURE 7.80

The IQ scores within two standard deviations from the mean would include all the IQ scores between z = −2 and z = +2 as shown in Figure 7.80. Let's determine the proportion of the area between the z scores of −2 and +2. Using Table II, we find the corresponding areas to the left of each z score:

z Score	Proportion of Area to the Left
−2	0.0228
2	0.9772

Subtracting the smaller area from the larger area, the proportion of IQ scores within two standard deviations of the mean is 0.9544.

EXAMPLE 7.8

One thousand students took a standardized psychology exam. The results were approximately normal with $\mu = 83$ and $\sigma = 8$. Find the *number* of students who scored:
a) Less than 67.
b) Greater than 87.

Solution:

a) In Figure 7.81, the shaded area represents the proportion of students who scored less than 67.

The z score of 67 is: $z = -2$. Using Table II, the proportion of area to the left of 67 is 0.0228. Thus, the number of students who scored less than 67 is:

$$0.0228 \text{ of } 1000 = (0.0228)(1000)$$
$$= 22.8$$

This is rounded off to 23 students.

b) In Figure 7.82, the shaded area represents the proportion of students who scored greater than 87.

The z score of 87 is 0.50. From Table II, the proportion of the area to the left of $z = 0.50$ is 0.6915. Therefore the proportion of area to the right of 87 equals 0.3085.

To find the *number* of students who scored greater than 87 multiply the proportion of area by the total number of students who took the exam. (The *area* to the right of 87 represents the proportion of

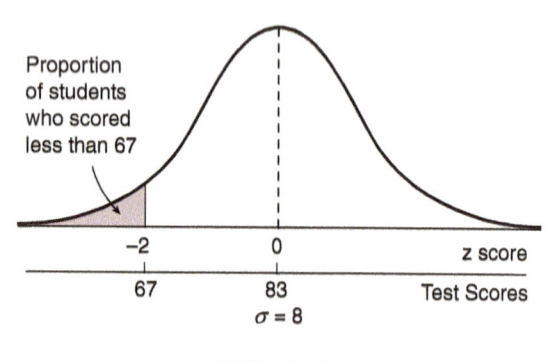

FIGURE 7.81

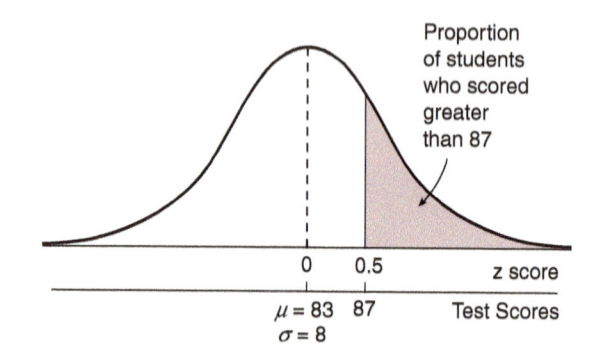

FIGURE 7.82

students in the distribution who scored greater than 87.) Thus, the number of students who scored greater than 87 is equal to:

$$0.3085 \text{ of } 1000 = (0.3085)(1000)$$
$$= 308.5.$$

We round this off to 309 students.

7.7 PERCENTILES AND APPLICATIONS OF PERCENTILES

Percentiles are widely used to describe the position of a data value within a distribution. Let's examine the definition of a **percentile rank** for a data value within a normal distribution.

DEFINITION 7.1

The **percentile rank** of a data value X within a normal distribution is equal to the percent of area under the normal curve to the left of the data value X. Percentile ranks are always expressed as a whole number.

EXAMPLE 7.9

Matthew earned a grade of 87 on his history exam. If the grades in his class were normally distributed with $\mu = 80$ and $\sigma = 7$, find Matthew's percentile rank on this exam.

Solution

The shaded area in Figure 7.83 represents the percentile rank of the grade 87.

To find the proportion of area to the left of 87 we must first calculate the z score of 87.

$$z = \frac{87 - 80}{7} = 1$$

Shaded area represents the percentile rank of 87

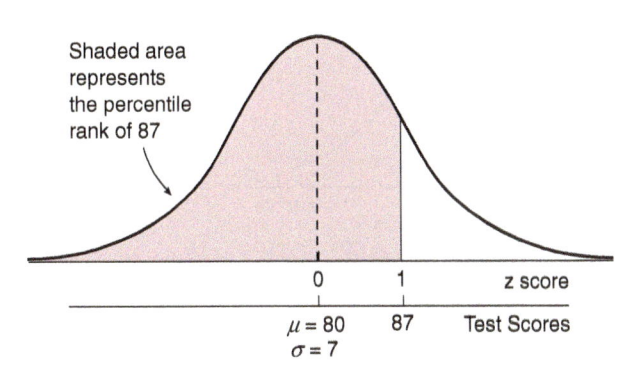

$\mu = 80$
$\sigma = 7$

FIGURE 7.83

From Table II, the proportion of area to the left of 87 is 0.8413. To convert a proportion to a percent move the decimal two places to the right. Thus, the proportion becomes 84.13%. Since percentile ranks are always expressed as whole numbers, 84.13% is rounded off to the nearest whole percent which is 84%. Therefore, the **percentile rank of 87 is 84**. Using Percentile notation, we can express this result as: **$87 = P_{84}$**.

Matthew's grade of 87 has a percentile rank of 84 which means that his grade of 87 was better than 84% of the test grades.

Percentile rank is an important statistical measure that describes the position of a data value within a distribution. It can also be used to compare data values from different distributions. This is illustrated in Example 7.10.

EXAMPLE 7.10

Miss Brooke took two exams last week. Her grades and class results were as follows:

Exam	Her Grade	Class Mean (μ)	Class Standard Deviation (σ)
English	83	80	6
Math	77	74	5

Assuming the results of both exams to be normally distributed, use percentile ranks to determine on which exam Miss Brooke did better relative to her class.

Solution

a) The shaded area in Figure 7.84 represents the percentile rank of Miss Brooke's English grade of 83. To determine this area, calculate the z score of 83.

$$z = \frac{83 - 80}{6} = 0.5$$

From Table II, the proportion of area to the left of 83 is 0.6915. This is expressed in percent form as 69.15%. Since percentile ranks are expressed in whole numbers, 69.15% rounded off to the nearest whole percent is 69. Therefore, the percentile rank of Miss Brooke's English grade is 69.

b) The shaded area in Figure 7.85 represents the percentile rank of Miss Brooke's math grade of 77. To determine this area, compute the z score of 77.

$$z = \frac{77 - 74}{5} = 0.60$$

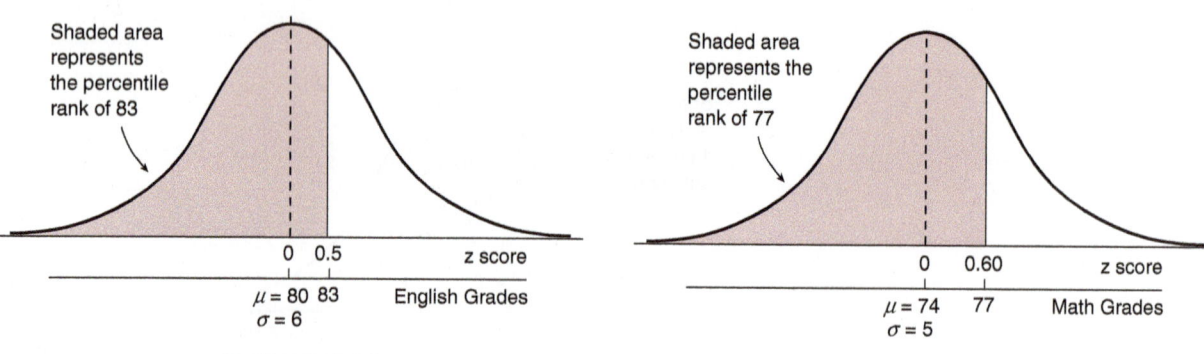

FIGURE 7.84

FIGURE 7.85

The proportion of area corresponding to the z = 0.6 is 0.7257. This is 72.57%. Therefore, the percentile rank of her math exam grade is 73. The percentile rank of her math grade is larger than the percentile rank of her English grade. Thus, *relative* to her class, Miss Brooke did better on her math exam.

EXAMPLE 7.11

One of the entrance requirements at a local college is a percentile rank of at least 45 on the verbal SAT exam. Richard scored 470 on his verbal SAT exam. Does he meet the minimum requirement of this college, if the verbal SAT distribution is normally distributed with $\mu = 500$ and $\sigma = 100$?

Solution

The shaded area in Figure 7.86 represents the percentile rank of the SAT score 470.

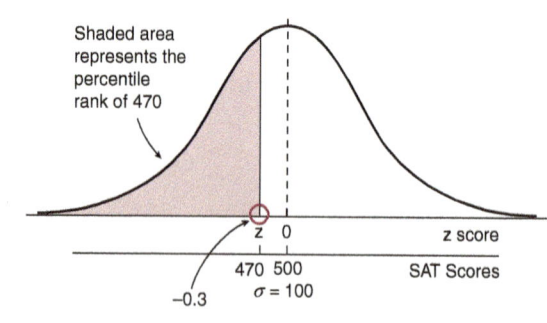

FIGURE 7.86

Find the z score of 470.

$$z = \frac{470 - 500}{100} = -0.30$$

In Table II, the entry corresponding to the z = −0.30 is 0.3821 or 38.21%. Thus, the percentile rank of Richard's SAT score is 38. Therefore, Richard does not meet the entrance requirement.

EXAMPLE 7.12

An entrance requirement at an eastern university is a high school average in the top 25% of one's high school class. Craig would like to know if he qualifies for entrance into this college. His high school average is 85. If the high school averages of his class are normally distributed with a mean of 75 and a standard deviation of 10, does he meet the requirement?

Solution

Being in the top 25% of the graduating class means that Craig's average must be better than **at least 75%** of the graduating class. Therefore the percentile rank of Craig's average must be at least 75. To determine if Craig qualifies for entrance into this college, find the percentile rank of Craig's high school average and compare it to the minimum requirement of a percentile rank of 75. The shaded area in Figure 7.87 represents the percentile rank of Craig's high school average of 85.

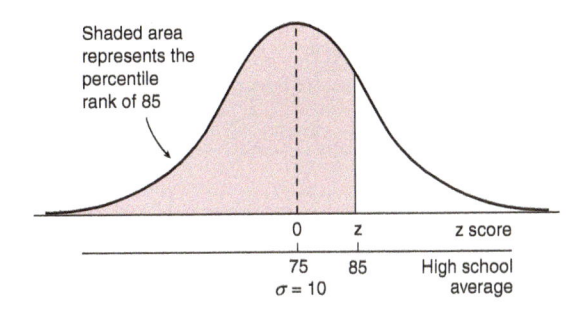

FIGURE 7.87

To determine this area, find the z score of 85.

$$z = \frac{85 - 75}{10} = 1$$

From Table II, the proportion of area corresponding to the z = 1 is 0.8413 which equals 84.13%. Therefore, the percentile rank of Craig's average is 84. Since Craig's percentile rank is greater than 75, he meets this entrance requirement.

CASE STUDY 7.1 SAT Subject Language Test Percentile Rank Data

The information contained in the table of Figure 7.88 represents the percentile ranks of SAT subject language tests administered by the College Board for college bound students who studied a language in high school.

Percentile ranks of SAT subject language tests (for college-bound students)

Score	Chinese with Listening	French	French with Listening	German	German with Listening	Modern Hebrew	Italian	Japanese with Listening	Korean with Listening	Latin	Spanish	Spanish with Listening
800	89	96	93	99	99	97	97	97	88	94	98	99
750	64	87	84	90	96	91	86	76	59	81	88	89
700	46	75	71	79	88	84	70	59	43	66	73	74
650	34	61	53	66	78	76	52	46	29	53	56	55
600	23	47	38	53	67	67	36	33	21	39	39	36
550	15	31	24	39	50	55	24	22	12	25	25	23
500	7	18	14	26	31	40	14	13	7	13	14	14
450	2	8	6	13	18	22	8	6	3	4	7	6
400	1-	2	1	4	8	10	5	3	—	1	2	2
350	—	1-	1-	1	2	1	3	2	—	—	1	1
300	—	—	—	1-	1	—	1	1-	—	—	1-	1-
250	—	—	—	—	—	—	—	1-	—	—	1-	—
200	—	—	—	—	—	—	—	—	—	—	—	—
Nmber	1,413	12,070	2,094	886	487	296	637	787	150	5,846	27,084	2,987
Mean	679	607	627	585	550	549	622	641	689	631	619	621
SD	105	111	108	114	109	124	116	117	103	107	105	103

The SAT Sbject Test percentile rank data on this page are based on collage-bound students in the classes of 2016-2018 who took SAT Subject Tests at any time in their high school career.

FIGURE 7.88

(continues)

CASE STUDY 7.1   SAT Subject Language Test Percentile Rank Data (continued)

a) Do you think that the test results of the SAT subject language tests would be approximately normally distributed when administered to a large group of students across the country?
b) If the Chinese with listening test results were approximately normal, explain what this information would mean to you?
c) Examine the table in Figure 7.88 and determine the mean test score for the French, Germain, Italian and Spanish tests.
d) Approximate the median test score for each of these tests. If these test results are approximately normal, then what value would you expect the median test score to equal?
e) Using the mean and standard deviations of the French, Germain, Italian and Spanish tests, approximate the percent of test scores that are within one, and two standard deviations of the mean.
f) For a normal distribution, what percentage of the data values are within one, two, and three standard deviations of the mean.
g) Compare the results of parts (e) and (f).
h) Which exam(s) seem to approximate the percentages of a normal distribution? Do these results have any effect on the way you answered part (a)?

Another application involves finding the data value that is associated with a certain proportion of the area. This will be illustrated in the following problems.

EXAMPLE 7.13

Consider a normal distribution with a mean of 75 and $\sigma = 5$. Find the data value(s) in the distribution that:
a) Cut(s) off the top 25% of the data values.
b) Represents the 30th percentile.
c) Has a percentile rank of 55.
d) Cut(s) off the middle 40% of the data values.

Solution

a) The shaded area in Figure 7.89 represents the top 25% of the data values in a normal distribution.

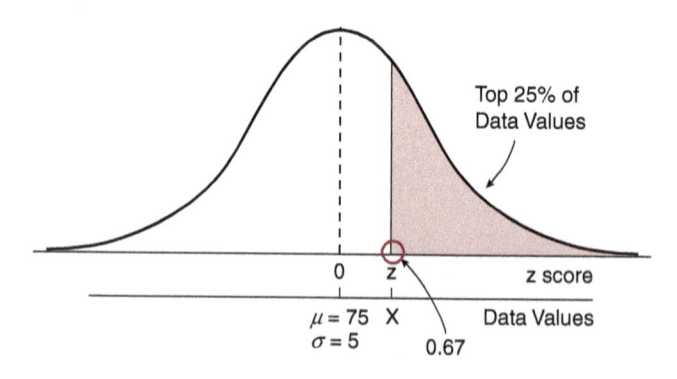

FIGURE 7.89

To determine the data value that cuts off the top 25% of the data values, we must find the data value that has 25% of the data values greater than this value and 75% of the data values less than this value. Since Table II gives the relationship between z scores and the proportion of area to the left of a z score, then knowing the proportion of area to the left of this unknown z score will enable us to determine this z score. We can now begin to calculate this data value by the using the following procedure.

Section 7.7 *Percentiles and Applications of Percentiles* 411

Step 1 **Determine the proportion of area to the left of the data value, X.**
Since 25% is to the right of the data value, X, then 75% is to the left of the data value, X. Recall that 75% corresponds to a proportion of 0.7500.

Step 2 **Find the area in Table II which is closest to the area that is to the left of the data value and determine the z score corresponding to this area.**
In Table II find the area closest to 0.75. This is illustrated in Figure 7.90.

Thus, the closest area to 0.75 which can be found in the table is 0.7486. Thus, the corresponding z score for the data value X is 0.67. Therefore the data value X which cuts off the top 25% has a z score of 0.67.

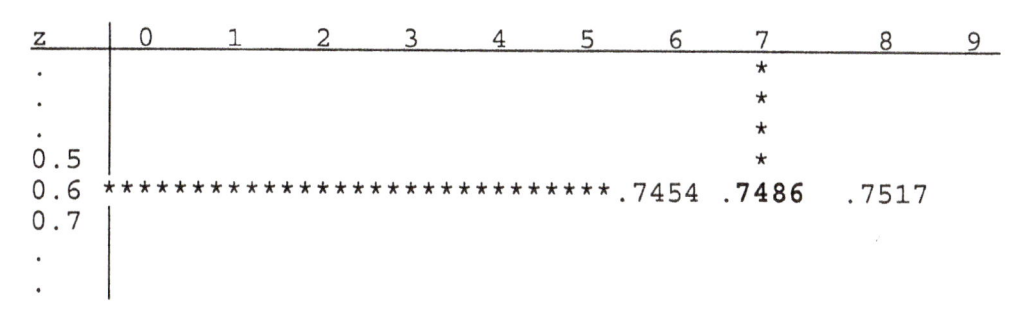

```
z    | 0   1   2   3   4   5   6   7      8      9
.    |                                *
.    |                                *
.    |                                *
0.5  |                                *
0.6  |****************************.7454  .7486  .7517
0.7  |
.    |
.    |
```

FIGURE 7.90

Step 3 **Using the raw score formula, $X = \mu + (z)\sigma$, convert the z score to a data value.**
In order to convert the z score of 0.67 to a data value, we need to use the raw score formula:

$$X = \mu + (z)\sigma$$

The data value for z = 0.67 is:

$$\begin{aligned} X &= \mu + (z)\sigma \\ &= 75 + (0.67)(5) \\ &= 78.35 \end{aligned}$$

Therefore the data value which cuts off the top 25% in this normal distribution is 78.35.

b) To determine the data value that represents the 30th percentile, recall that the 30th percentile is the data value that has 30% of the distribution below it. We will follow the procedure outlined in part a of this question.

Step 1 **Determine the proportion of area to the left of the data value, X.**
We are looking for the data value X that has 30% or 0.30 of the area under the normal curve to the left of this data value.

Step 2 **Find the area in Table II which is closest to the area that is to the left of the data value and determine the z score corresponding to this area.**
We must find the z score that corresponds to the bottom 0.30 of area under the normal curve as illustrated in Figure 7.91.

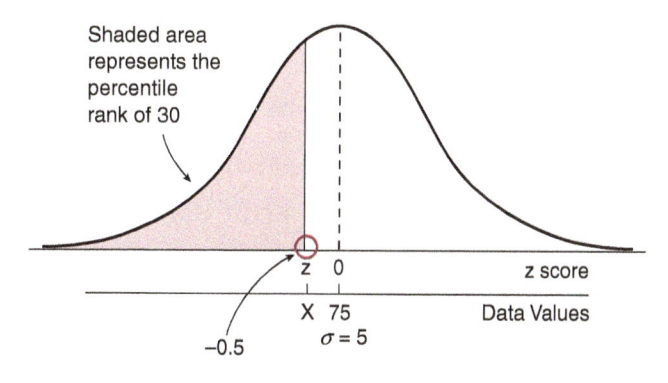

FIGURE 7.91

In Table II find the area closest to 0.30. This is illustrated in Figure 7.92.

TABLE II: Normal Curve Area Table

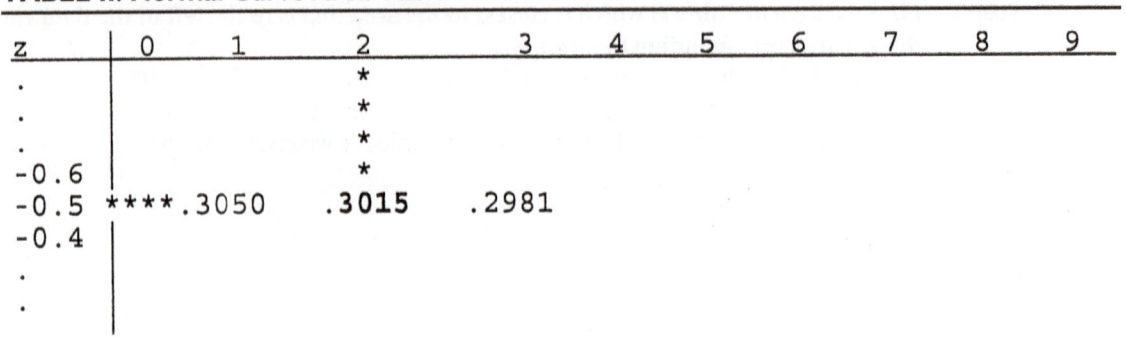

z	0	1	2	3	4	5	6	7	8	9
.			*							
.			*							
.			*							
-0.6			*							
-0.5	****	.3050	.3015	.2981						
-0.4										

FIGURE 7.92

Thus, the closest area to 0.30 is 0.3015. The corresponding z score is –0.52.

Step 3 Using the raw score formula, $X = \mu + (z)\sigma$, convert the z score to a data value.

Using the formula: $X = \mu + (z)\sigma$, we can transform this z score to the data value X.

$$X = 75 + (-0.52)(5)$$
$$= 72.4.$$

Therefore the data value X which represents the 30th percentile in this distribution is 72.4.

c) To determine the data value that has percentile rank of 55, we will follow the procedure outlined in part a of this problem.

Step 1 **Determine the proportion of area to the left of the data value, X.**
The data value that has percentile rank of 55, cuts off the bottom 55% of the data values in the distribution. This is illustrated in Figure 7.93.

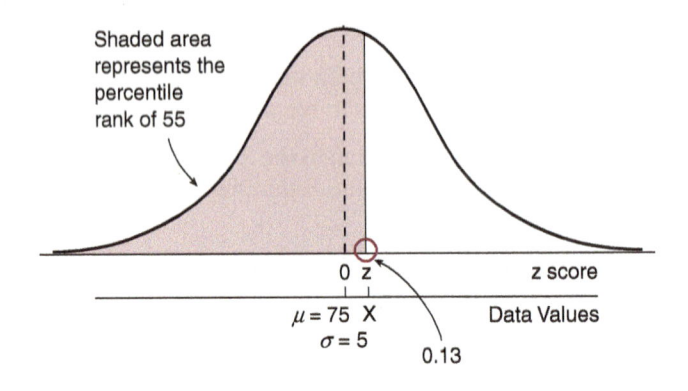

FIGURE 7.93

Step 2 Find the area in Table II which is closest to the area that is to the left of the data value and determine the z score corresponding to this area.
From Table II, the area closest to 0.55 is 0.5517, which corresponds to a z score of 0.13.

Step 3 Using the raw score formula, $X = \mu + (z)\sigma$, convert the z score to a data value.

$$X = 75 + (0.13)(5)$$
$$X = 75.65$$

Therefore, the data value that has a percentile rank of 55 is 75.65.

d) In a normal distribution, the mean is located at the center. Therefore the middle 40% of the data values will be centered about the mean, that is 20% of the data values will be less than the mean and 20% of the data values will be greater than the mean. This is illustrated in Figure 7.94.

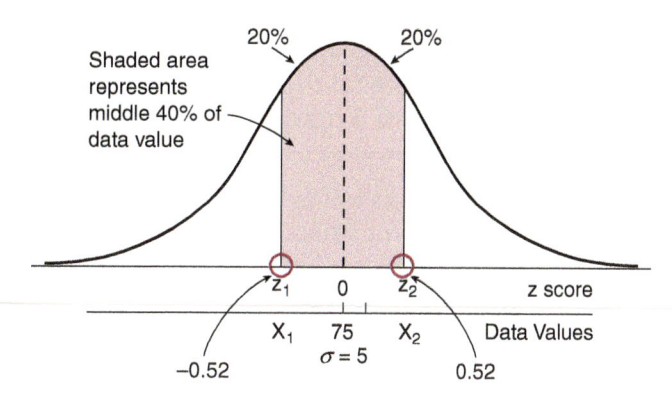

Section 7.7 Percentiles and Applications of Percentiles   413

FIGURE 7.94

Notice in Figure 7.94 that we are looking for two data values, which we will symbolize as X_1 and X_2. To determine these data values, we will follow the procedure outlined in part a of this problem.

Step 1 **Determine the proportion of area to the left of the data value, X.**

To determine the proportion of area to the left of X_1, observe that 0.20 of the area is between the data value X_1 and the mean. Since the total proportion of area to the left of the mean is 0.50, then the remaining area to the left of X_1 is 0.50 − 0.20 = 0.30.

Step 2 **Find the area in Table II which is closest to the area that is to the left of the data value and determine the z score corresponding to this area.**

In Table II the area closest to 0.30 is 0.3015. Its corresponding z score is −0.52.

Step 3 Using the raw score formula, $X = \mu + (z)\sigma$, convert the z score to a data value.

$$X_1 = 75 + (-0.52)(5)$$
$$X_1 = 72.4$$

To determine the value of X_2, we will use the symmetry property of the normal curve.

Step 1 **Use the symmetry property of the normal curve to determine the z score of X_2.**

In examining Figure 7.94, you should notice that the percent of area between the mean and z_1 is the same as the area between the mean and z_2. Using the symmetry property of the normal curve and the value of z_1 is −0.52, the value of z_2 is 0.52. Thus, the z score of X_2 is 0.52.

Step 2. **Using the raw score formula, $X = \mu + (z)\sigma$, convert the z score to a data value.**

$$X_2 = 75 + (0.52)(5)$$
$$X_2 = 77.6$$

Therefore the middle 40% of the distribution is located between the two data values 72.4 and 77.6.

We can now generalize the procedures to determine the data value that cuts off the bottom p% in a normal distribution and the data value that cuts off the top q% in a normal distribution.

Procedure to Find Data Value (or Raw Score) That Cuts Off Bottom p % Within a Normal Distribution

1. Change the percentage p to a proportion.
2. Using Table II, find the z score that has the area closest to this proportion. If there is a tie, choose the z score with the larger absolute value.
3. Substitute this z score into the raw score formula

$$X = \mu + (z)\sigma$$

to determine the data value that corresponds to this z score.

Procedure to Find Data Value (or Raw Score) That Cuts Off Top q % Within a Normal Distribution

1. Calculate the percentage of data values below the raw score by subtracting q from 100: (100 – q).
2. Change this percentage to a proportion.
3. Using Table II, find the z score that has the area closest to this proportion. If there is a tie, choose the z score with the larger absolute value.
4. Substitute this z score into the raw score formula

$$X = \mu + (z)\sigma$$

to determine the data value that corresponds to this z score.

EXAMPLE 7.14

In a large physical education class the grades on the final exam were normally distributed with $\mu = 80$ and $\sigma = 8$. Professor Stevenson assigns letter grades in the following way: the top 10% receive A's, the next 15% receive B's, the lowest 4% receive F's, the middle 50% receive C's, and the rest of the grades are D's. Find the numerical grades that correspond to each of the letter grades.

Solution

The information is interpreted in Figure 7.95.

Using Figure 7.95, let us determine the test grades X_1, X_2, X_3, and X_4 that separate the letter grades.

To determine X_1:

1. Find the area to the left of X_1. This area is 4% or as a proportion 0.04.
2. Find the closest entry to 0.04 in Table II. The closest entry to 0.04 is 0.0401. The z score that corresponds to 0.0401 is –1.75.

To find the value of X_1, we can use the raw score formula:

$$X_1 = \mu + (z)\sigma$$
$$X_1 = 80 + (-1.75)(8)$$
$$X_1 = 66$$

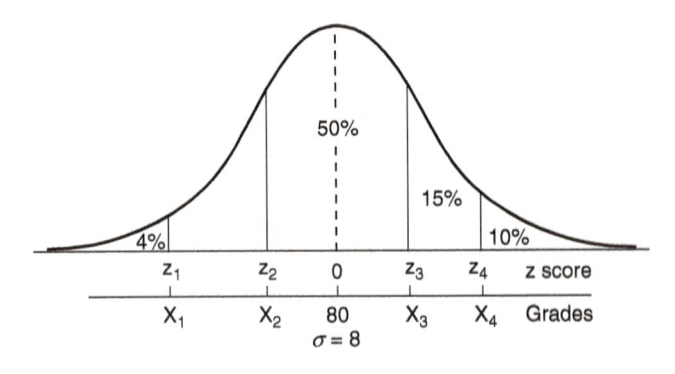

FIGURE 7.95

Therefore, the grade which separates the F's from the D's is 66.

To determine X_2:

1. Find the proportion of area to the left of X_2. From the figure we see that 75% of the area is to the right of X_2. This means that the area to the left is 25% or 0.25.
2. The closest entry in Table II to 0.25 is 0.2514. The z score that corresponds to 0.2514 is –0.67. The formula to determine X_2 is:

$$X_2 = \mu + (z)\sigma$$
$$X_2 = 80 + (-0.67)(8)$$
$$X_2 = 74.64$$

Therefore, the test grade which separate grades C and D is 74.64.
To determine X_3:
1. The proportion of area to the left of X_3 is 0.75.
2. The closest entry in Table II is 0.7486. It has a z score of 0.67. Using the raw score formula, we have:

$$X_3 = \mu + (z)\sigma$$
$$X_3 = 80 + (0.67)(8)$$
$$X_3 = 85.36$$

The test grade which separate grades B and C is 85.36.
Finally, to determine X_4:
1. The proportion of area to the left of X_4 is 0.90.
2. The closest entry in Table II is 0.8997, which has a z score of 1.28. Using the raw score formula, we have:

$$X_4 = \mu + (z)\sigma$$
$$X_4 = 80 + (1.28)(8)$$
$$X_4 = 90.24$$

Thus the test grade which separate grades A and B is 90.24.

EXAMPLE 7.15

To determine the size of the annual Christmas bonus, Grace, the personnel and budget director of a telemarketing firm, decides to use annual sales per salesperson as the criterion. The salespeople in the top 20% of annual sales will receive $3000 in bonus, those in the next 30% will receive $2000 in bonus, those in the next 30% will receive $1000 and the rest will receive $500 in bonus. If the annual sales per salesperson is normally distributed with $\mu = \$120,000$ and $\sigma = \$15,000$ determine the annual sales required to obtain:
a) A bonus of $3000.
b) A bonus of $2000.

Solution

The relevant information is summarized in Figure 7.96.
a) To determine X_2:
1. The percent of area to the left of X_2 is 80% which as a proportion is 0.80.
2. The closest entry in Table II to 0.80 is 0.7995 which has the z score of 0.84. Converting this z score to a raw score, we get $X_2 = \$132,600$.
Thus, annual sales of at least $132,600 will receive a $3000 bonus.

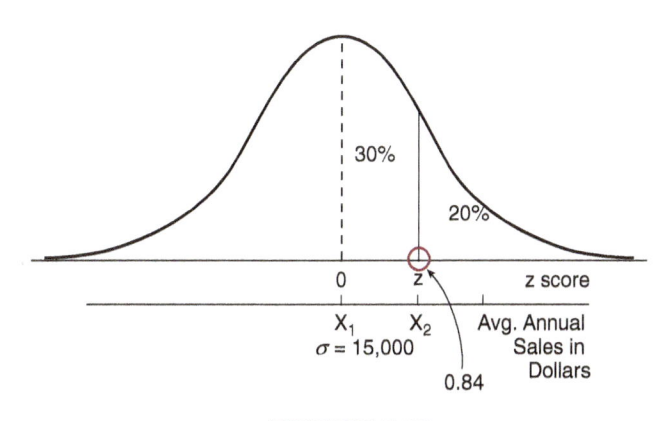

FIGURE 7.96

b) To determine X_1:
1. Notice that X_1 has 50% of the area to the left of it. Thus, X_1 is the mean.
2. Therefore, $X_1 = \$120{,}000$.
   Thus, annual sales greater than or equal to $120,000 but less than $132,600 will receive a $2000 bonus.

EXAMPLE 7.16

A local community college decides to award scholarships based upon the results of a standardized achievement test. If this year's 5,000 test results are approximately normal with $\mu = 137$ and $\sigma = 15$, then determine the minimum test grade needed to win a scholarship if the top:
a) 20 students receive scholarships.
b) 30 students receive scholarships.

Solution

a) To find the minimum test grade needed to win a scholarship if the top 20 students receive scholarships, we must first determine the *proportion* of the students that will win scholarships. Since there are 20 students receiving scholarships out of 5,000 students, the proportion of scholarship winners equals:

$$\frac{20}{5000} = 0.0040$$

This is illustrated in Figure 7.97.

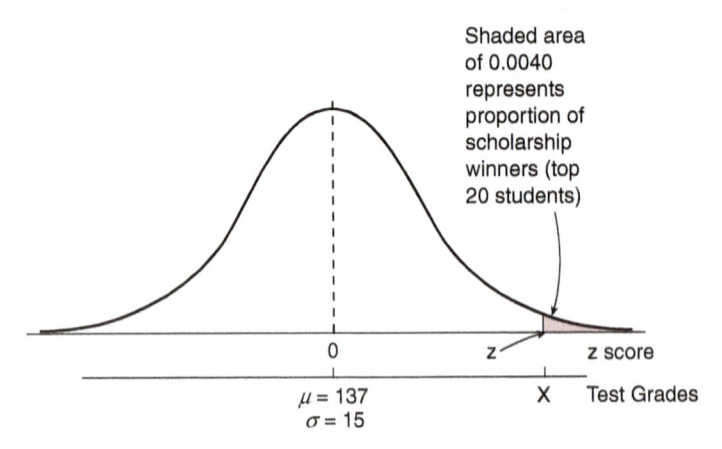

FIGURE 7.97

To determine the test grade X which cuts off the top 0.0040 we need to determine the proportion of area to the left of the test grade X. Since 0.0040 is to the right of the test grade X, then 0.9960 is to the left of X. Using Table II, 0.9960 corresponds to a z score of 2.65.

In order to convert this z score into a test grade, we need to use the raw score formula:

$$X = \mu + (z)\sigma$$
$$X = 137 + (2.65)(15)$$
$$X = 176.75$$

Therefore, the minimum test grade needed to win a scholarship if the top 20 students will receive scholarships is 176.75.

b) To find the minimum test grade needed to win a scholarship if the top 30 students will receive scholarships, we must determine the *proportion* of the students that will win scholarships. Since 30 students out of 5,000 students will win scholarships, the proportion of scholarship winners equals:

$$\frac{30}{5000} = 0.0060$$

This is illustrated in Figure 7.98.

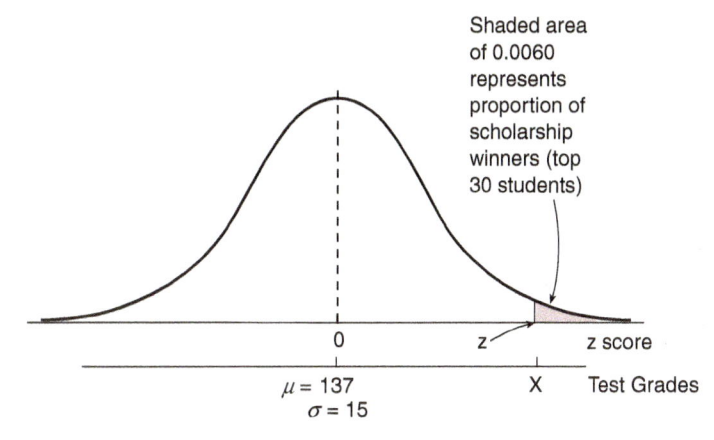

Figure 7.98: Normal curve with shaded area of 0.0060 representing proportion of scholarship winners (top 30 students); μ = 137, σ = 15.

To find the value of the test grade X we need to determine the proportion of area to the left of the test grade X. Since 0.0060 is to the right of X, then the proportion to the left is 0.9940. Using Table II, 0.9940 corresponds to a z score of 2.51.

Using the raw score formula, we have:

$$X = \mu + (z)\sigma$$
$$X = 137 + (2.51)(15)$$
$$X = 174.65$$

Therefore, the minimum test grade needed to win a scholarship if the top 30 students will receive scholarships is 174.65. ∎

7.8 PROBABILITY APPLICATIONS

In Section 7.2, we discussed the characteristics of a probability distribution for a continuous random variable. Since a normal distribution is an example of a continuous probability distribution, the characteristics for a continuous probability distribution hold for the normal distribution. Thus, for a normal distribution:

1. The probability that a normal variable X will assume a value between two possible values X_1 and X_2 equals the area under the normal curve between the values X_1 and X_2. This is illustrated in Figure 7.99.
2. The total area (or probability) under a normal curve is equal to 1. This is illustrated in Figure 7.100.
3. The probability that a normal variable X assumes a value between $X = X_1$ and $X = X_2$ is greater than or equal to 0 but less than or equal to 1. This is written as $0 \leq P(X_1 < X < X_2) \leq 1$. Since probability is expressed as a number from 0 to 1 this property emphasizes three important ideas:
    I. The largest possible probability is 1, which is equal to the total area under a normal curve.
    II. The smallest possible probability is 0.

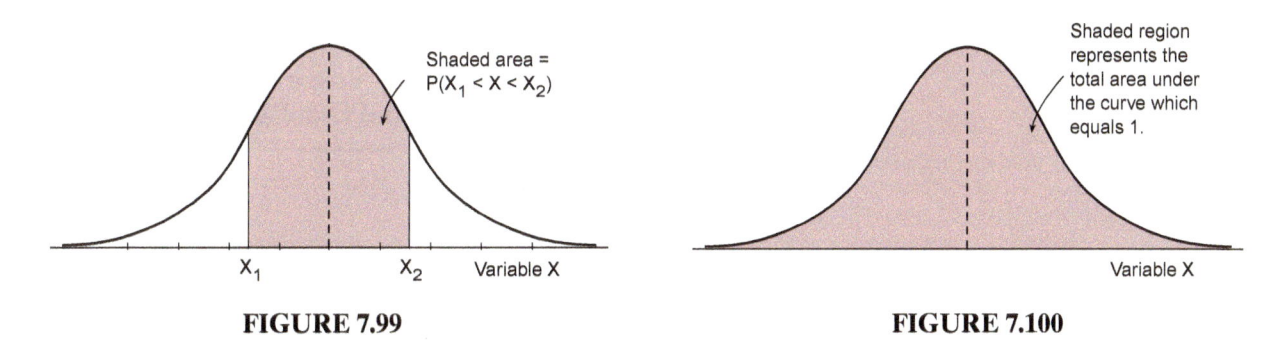

FIGURE 7.99 — Shaded area = $P(X_1 < X < X_2)$.

FIGURE 7.100 — Shaded region represents the total area under the curve which equals 1.

III. The probability that a variable will assume a specific value, say $X = X_1$, is 0 since the area associated with a specific value of the variable is 0. That is,

$$P(X = X_1) = 0.$$

This leads us to the following definition.

DEFINITION 7.2
The **probability of an Event E in a normal distribution** is defined to be:

P(Event E) = Proportion of area under the normal curve satisfying Event E.

This is illustrated in Figure 7.101.

P(EVENT E) equals the shaded region which represents the proportion of area satisfying EVENT E.

EVENT E

Variable X

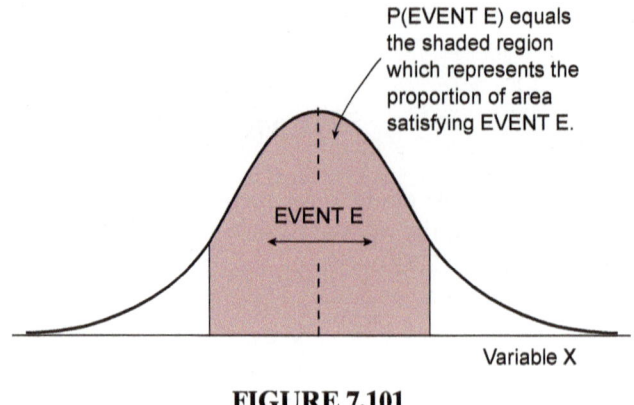

FIGURE 7.101

This leads to the following definition for the probability that a data value falls between two data values X_1 and X_2 within a normal distribution.

DEFINITION 7.3
Probability of a data value X falling between two data values: X_1 and X_2.
The probability that a data value X selected at random from a normal distribution will fall between the two data values X_1 and X_2 is equal to the **proportion of area** between X_1 and X_2. This can be expressed as:

$P(X_1 \leq X \leq X_2)$ = Area under the normal curve between X_1 and X_2

Remember that a normal curve represents the distribution of a continuous variable, and the proportion of area under the normal curve that is associated with a particular data value is zero! Therefore, the probability of random selecting a data value from a normal distribution that is exactly equal to the value X is zero. This can be expressed as:

Probability of Selecting a Data Value Equal to the Value X_1 Within a Normal Distribution

P(Data Value is X_1) = 0

Let's apply these probability concepts to a normal distribution application.

EXAMPLE 7.17
Assume the distribution of the playing careers of major league baseball players can be approximated by a normal distribution with a mean of 8 years and a standard deviation of 4 years. Find the probability that a player selected at random will have a career that will last:

a) Less than 6 years
b) More than 14 years
c) Between 4 and 10 years.

Solution

a) The shaded area in Figure 7.102 represents the probability that a player's career will last less than 6 years.
   To determine the probability that a player selected at random will have a playing career that will last less than 6 years, determine the proportion of area to the left of 6 years. This is represented by the shaded area in Figure 7.102. To find the proportion of area to the left of 6 years, we need to use Table II. In order to use Table II, we must convert 6 years to a z score. The z score of 6 years is:

$$z = \frac{6-8}{4} = -0.5$$

From Table II, the proportion of area to the left of $z = -0.5$ is 0.3085. Therefore, the probability that a player's career will last less than 6 years is 0.3085.

b) The shaded area in Figure 7.103 represents the probability that a player's career will last more than 14 years.

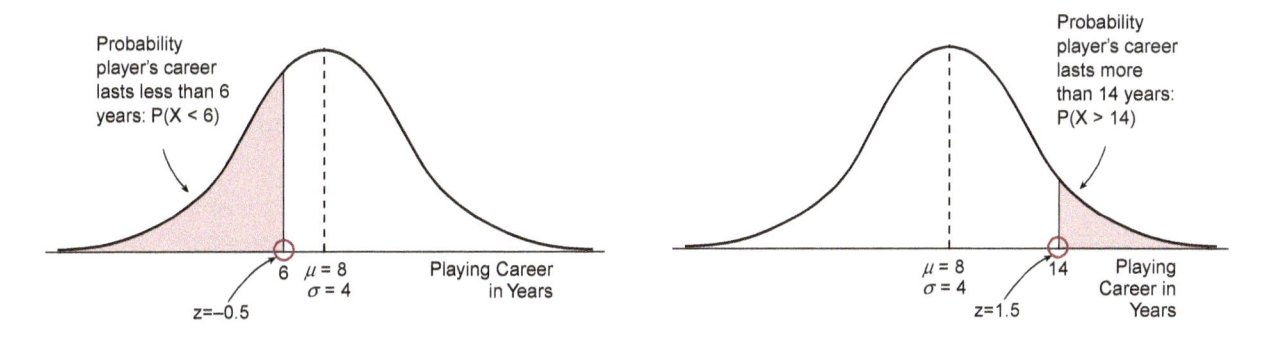

FIGURE 7.102

FIGURE 7.103

To determine the probability that a player selected at random has a career that lasts more than 14 years, determine the proportion of area to the right of 14 years or the shaded area in Figure 7.103. To find the proportion of area to the right of 14 years, we need to use Table II. In order to use Table II, we must convert 14 years to a z score. The z score of 14 years is:

$$z = \frac{14-8}{4} = 1.5$$

From Table II, the proportion of area to the left of $z = 1.5$ is 0.9332. The proportion of area to the right of $z = 1.5$ is:

$$= 1 - 0.9332$$
$$= 0.0668$$

Therefore, the probability that a player's career lasts more than 14 years is 0.0668.

c) The shaded area in Figure 7.104 represents the probability that a player's career will last between 4 and 10 years.

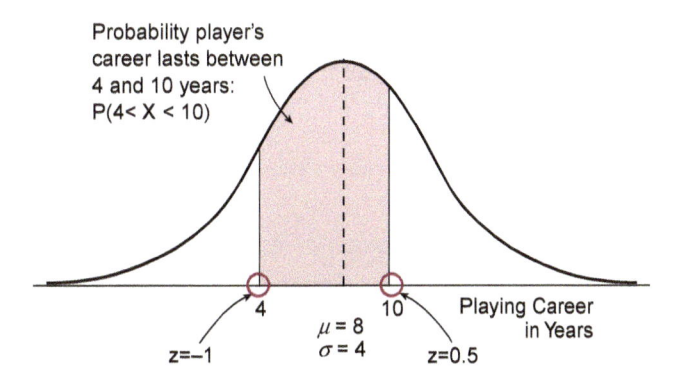

FIGURE 7.104

To determine the probability that a player selected at random will have a career that will last between 4 and 10 years, determine the proportion of area between 4 and 10 years. This is represented by the shaded area in Figure 7.104. To find the proportion of area between 4 and 10 years, we need to use Table II. To use Table II, we must convert both 4 years and 10 years to their respective z scores. The z scores of 4 and 10 years are:

$$z = \frac{4-8}{4} = -1$$

and

$$z = \frac{10-8}{4} = 0.5$$

From Table II, the proportion of area to the left of z = –1 is 0.1587, while the proportion of area to the left of z = 0.5 is 0.6915. Thus, the proportion of area between z = 0.5 and z = –1 is equal to the larger area (0.6915) minus the smaller area (0.1587):

$$\text{proportion of area between } z = 0.5 \text{ and } z = -1 = 0.6915 - 0.1587$$
$$= 0.5328.$$

Therefore, the probability that a player's career will last between 4 and 10 years is 0.5328.

We can use the normalcdf function on the TI-84 to find the answers to the questions in Example 7.17. Remember, the syntax for the parameters for normalcdf is:

normalcdf(lower bound of shaded area, upper bound of shaded area, mean, standard deviation)

For 7.17a, P(X < 6) = normalcdf(-E99, 6, 8, 4) ≈ 0.3085

For 7.17b, P(X > 14) = normalcdf(14, E99, 8, 4) ≈ 0.0668

Figure 7.105 shows the calculator output for 7.17a and 7.17b.

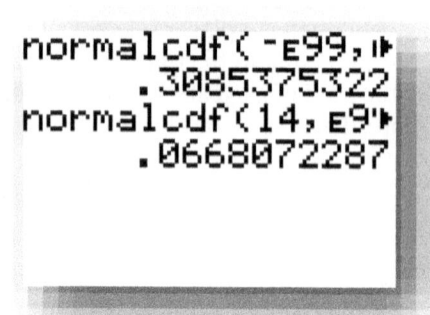

FIGURE 7.105

For 7.17c, P(4 < X < 10) = normalcdf(4, 10, 8, 4) ≈ 0.5328

The calculator's result for 7.17c is shown in Figure 7.106.

FIGURE 7.106

CASE STUDY 7.2 The Curve Moves Higher

Examine the two normal distributions in Figure 7.107 which appeared in a 1997 *New York Times* graphic entitled *The Curve Moves Higher*. Both normal distributions represent the IQ scores of American children as measured by the Stanford-Binet IQ test. The normal distribution labeled 1932 represent the IQ results of American children tested in 1932 while the 1997 normal distribution displays the IQ results of American children tested in 1997.

a) What was the mean IQ score for both distributions?
b) Determine the median IQ score for both distributions.
c) State the modal IQ score for both distributions.
e) Assuming the standard deviation for both distributions is 15, calculate the probability that a student would have scored within the intellectually deficient area for the 1932 group and then calculate it for the 1997 group.

The Curve Moves Higher

As measured by the Stanford-Binet intelligence test, American children seem to be getting smarter. Scores of a group tested in 1932 fell along a bell-shaped curve with half below 100 and half above. Studies show that if children took that same test today, half would score above 120 on the 1932 scale. Very few of them would score in the "intellectually deficient" end, on the left side, and about one-fourth would rank in the "very superior" range.

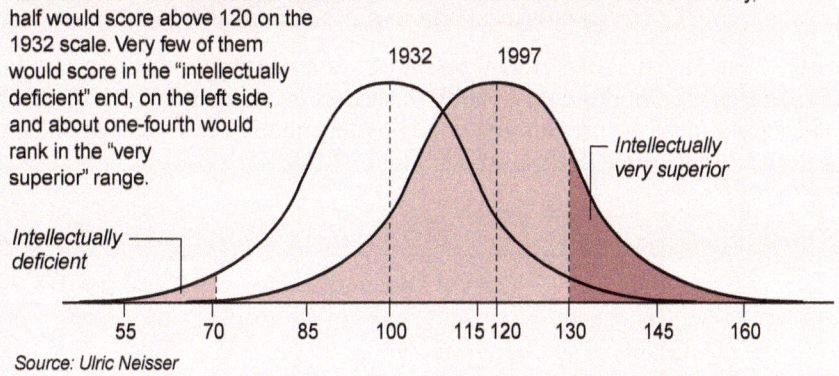

Source: Ulric Neisser

FIGURE 7.107
From *The New York Times*. © 1998 The New York Times Company. All rights reserved. Used under license.

f) Assuming the standard deviation for both distributions is 15, calculate the probability that a student would have scored within the intellectually very superior area for the 1932 group and then calculate it for the 1997 group.
g) Using the results of parts e and f, compare both distributions and explain the differences in intellectual ability as measured by the Stanford-Binet IQ test for both groups.
h) For the 1932 test, determine the percentile rank of the score 115. Now determine the percentile rank of the score 115 for the 1997 test. (Assume the standard deviation for both distributions is 15.) Explain what the difference between these percentile ranks would signify.
i) What IQ score on the 1932 test would represent the 3rd quartile? What IQ score on the 1997 test would represent the 3rd quartile? Explain what this signifies.
j) What IQ score on the 1997 test would be equivalent to the IQ score 70 on the 1932 test? How would you define intellectually deficient for the 1997 if you wanted to use the same standard applied for the 1932 group? (Assume the standard deviation for both distributions is 15.)
k) What IQ score on the 1997 test would be equivalent to the IQ score 130 on the 1932 test? How would you define intellectually very superior for the 1997 if you wanted to use the same standard applied for the 1932 group? (Assume the standard deviation for both distributions is 15.)
l) If we assume that 100,000 American children took the 1932 IQ test, then how many children would you expect to fall within the area defined to be "intellectually very superior"?
m) Assuming that 100,000 American children took the 1997 IQ test, then how many children would you expect to fall within the area defined to be "intellectually very superior" according to the 1932 standards?

7.9 THE NORMAL APPROXIMATION TO THE BINOMIAL DISTRIBUTION

The normal distribution plays an important role in statistics because there are many probability distributions that can be approximated by a normal curve. In this section, we will examine how the normal distribution which represents a continuous probability distribution can serve as an approximation to the **binomial distribution** which represents a discrete probability distribution. However, before we consider this binomial approximation let us re-examine the binomial distribution which was introduced in Chapter 6.

In Chapter 6, we discussed how the binomial distribution serves as a model for a discrete random variable of a probability experiment which satisfies the following four conditions:

1. there are n identical trials.
2. the n identical trials are independent.
3. the outcome for each trial can be classified as either a success or a failure.
4. the probability of success, p, is the same for each trial.

If these conditions are met, then the probabilities of the binomial experiment can be computed using the binomial probability formula:

$$P(s \text{ successes in n trials}) = {}_nC_s \, p^s \, q^{(n-s)}, \text{ where: } q = 1-p.$$

Consider the binomial experiment tossing a fair coin twice. For this binomial experiment, the number of trials or tosses is n = 2 and the probability of success (or a head) is p = ½. Using the binomial probability formula, we can compute the probability for all possible number of successes or heads. The binomial distribution for tossing a fair coin twice is illustrated in Figure 7.108.

Binomial Distribution for n = 2 and p = 1/2	
number of successes or heads "s heads"	Probability of "s" successes or heads $P(s \text{ heads}) = {}_nC_s \, p^s \, q^{(n-s)},$ where: $q = 1 - p$
s = 2	${}_2C_2 \left(\frac{1}{2}\right)^2 \left(\frac{1}{2}\right)^0 = \frac{1}{4}$
s = 1	${}_2C_1 \left(\frac{1}{2}\right)^1 \left(\frac{1}{2}\right)^1 = \frac{2}{4}$
s = 0	${}_2C_0 \left(\frac{1}{2}\right)^0 \left(\frac{1}{2}\right)^2 = \frac{1}{4}$

FIGURE 7.108

Figure 7.108 represents the binomial distribution for n = 2 and p = ½. A histogram and frequency polygon of this binomial distribution is illustrated in Figure 7.109, where the vertical axis represents the probability of **s** successes and the horizontal axis represents the number of **s** successes. As you can observe in Figure 7.109, the histogram and the polygon for this binomial distribution is symmetric and the polygon formed by joining the midpoints of the rectangles is approximately bell-shaped.

In Figure 7.109 the *probability* for each success or head is represented by the *area* of the rectangle associated with this success. Observe that the probability of getting zero successes is represented by the rectangle starting at –0.5 and ending at 0.5. Since the width of this rectangle is 1 and the height or probability is ¼ then using the formula for the area of a rectangle we have:

$$\text{Area of rectangle} = (\text{height})(\text{width})$$
$$= (¼)(1)$$
$$= ¼.$$

The Binomial Distribution for n=2, p=$\frac{1}{2}$

FIGURE 7.109

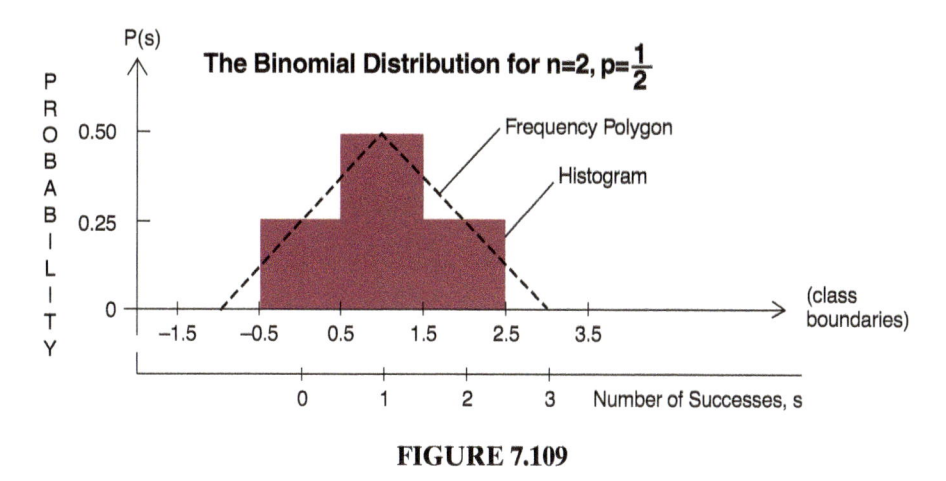

Therefore, the area of the rectangle represents the probability of getting zero successes or heads is ¼. Continuing in this manner, the area of the rectangle representing 1 success or head which represents the probability of getting one head is ½. Likewise, the area of the rectangle representing 2 successes or heads is ¼, which represents the probability of getting two heads.

Like all distributions we have studied the binomial distribution has a mean and standard deviation. Symbolically, the **mean of a binomial distribution** is expressed μ_s and is pronounced **mu sub s**. The **standard deviation of a binomial distribution** is expressed σ_s and is pronounced **sigma sub s.** To calculate the mean and standard deviation of a binomial, we use the following definitions:

DEFINITION 7.4
Mean of the Binomial Distribution.
For a binomial distribution, the mean number of successes, denoted μ_s, is determined by the formula:

$$\mu_s = np$$

where: n = the number of trials
p = the probability of a success in one trial

DEFINITION 7.5
Standard Deviation of the Binomial Distribution.
In a binomial distribution, the standard deviation, denoted σ_s, is determined by the formula:

$$\sigma_s = \sqrt{np(1-p)}$$

where: n = the number of trials
p = the probability of a success in one trial
1−p = the probability of a failure in one trial

For the binomial distribution illustrated in Figure 7.109 where n = 2 and p = ½, the mean is:

$$\begin{aligned}\mu_s &= np \\ &= (2)(½) \\ &= 1 \text{ head}\end{aligned}$$

And the standard deviation of this binomial distribution is:

$$\sigma_s = \sqrt{np(1-p)}$$
$$= \sqrt{(2)(\tfrac{1}{2})(\tfrac{1}{2})}$$
$$= \sqrt{0.5}$$
$$\sigma_s \approx 0.71 \text{ heads}$$

EXAMPLE 7.18

Using the binomial distribution where n = 4 and p = ½ shown in Figure 7.110, compute the mean, μ_s, and the standard deviation, σ_s.

Solution

To compute mean, μ_s, use the formula:

$$\mu_s = np$$

Binomial Distribution for n = 4 and p = 1/2	
number of successes "s"	Probability of "s" successes $P(s \text{ heads}) = {}_nC_s\, p^s\, q^{(n-s)}$, where: q = 1 − p
4	${}_4C_4\left(\tfrac{1}{2}\right)^4\left(\tfrac{1}{2}\right)^0 = \tfrac{1}{16}$
3	${}_4C_3\left(\tfrac{1}{2}\right)^3\left(\tfrac{1}{2}\right)^1 = \tfrac{4}{16}$
2	${}_4C_2\left(\tfrac{1}{2}\right)^2\left(\tfrac{1}{2}\right)^2 = \tfrac{6}{16}$
1	${}_4C_1\left(\tfrac{1}{2}\right)^1\left(\tfrac{1}{2}\right)^3 = \tfrac{4}{16}$
0	${}_4C_0\left(\tfrac{1}{2}\right)^0\left(\tfrac{1}{2}\right)^4 = \tfrac{1}{16}$

FIGURE 7.110

If n = 4 and p = ½, then: $\mu_s = 2.$
To compute standard deviation, σ_s, use the formula:

$$\sigma_s = \sqrt{np(1-p)}$$

For n = 4, p = ½ and 1−p = ½,

$$\sigma_s = \sqrt{4(\tfrac{1}{2})(\tfrac{1}{2})}$$
$$\sigma_s = 1$$

The graph of this binomial distribution is illustrated in Figure 7.111.

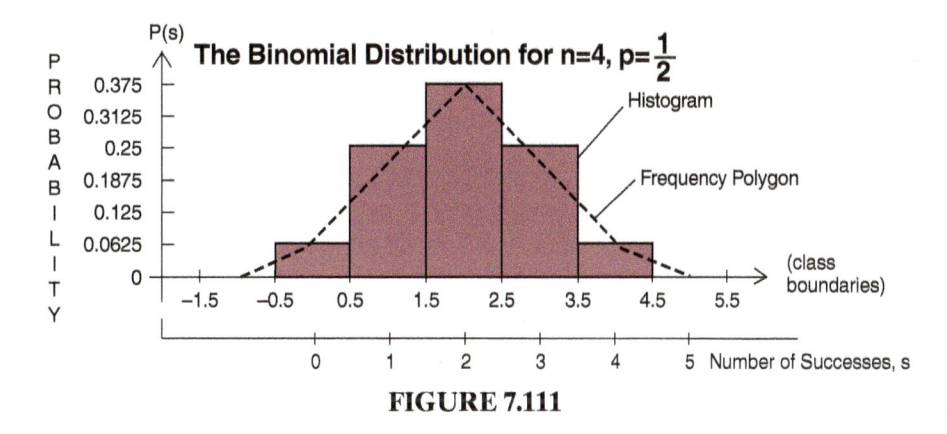

FIGURE 7.111

In many applications of the binomial distribution the number of trials, n, is large. Figure 7.112 illustrates the effect of *increasing the number of trials, n*, on the shape of the binomial distribution with p = ½.

In Figure 7.112, notice that **as n** *increases* **the shape of the binomial distribution gets** *closer* **to the bell-shape of a normal distribution**. In fact, the binomial distribution *can* be approximated by a normal distribution with a mean $\mu_s = np$ and a standard deviation $\sigma_s = \sqrt{np(1-p)}$.

However, this approximation is reasonable only when the following conditions are *both* **true:**

$$np \geq 10 \text{ and } n(1-p) \geq 10.$$

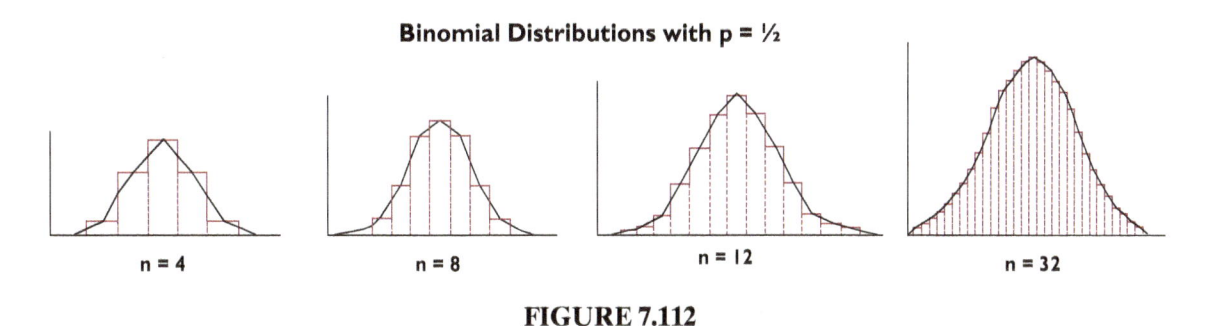

FIGURE 7.112

Let's summarize these conditions.

Conditions of Normal Approximation of a Binomial Distribution

A binomial distribution with a probability of a success p and n trials can be approximated by a normal distribution with a mean $\mu_s = np$ and a standard deviation

$$\sigma_s = \sqrt{np(1-p)}$$

only when both:

np and n(1 − p) are greater than or equal to 10.

Thus, if the number of trials, n, is sufficiently large to make *both np and n(1-p) greater than or equal to 10*, then the binomial distribution can be approximated by a normal curve where the mean is $\mu_s = np$ and the standard deviation is $\sigma_s = \sqrt{np(1-p)}$.

Under these conditions, the areas under the normal curve may be used to approximate the binomial probabilities. Let's illustrate the use of the normal approximation to a binomial distribution for tossing a fair coin 20 times.

Suppose we want to determine the probability of getting at least 14 heads in 20 tosses of a fair coin. The probability of at least 14 heads can be determined using the binomial probability formula:

$$P(s \text{ successes in n trials}) = {}_nC_s \, p^s \, q^{(n-s)}$$

The probability of getting at least 14 heads is equal to the probability of getting 14 or more heads. Thus, we have:

P(at least 14 heads) = P(14 or more heads)
= P(14 heads)+ P(15 heads)+ P(16 heads)+ P(17 heads)+ P(18 heads)+ P(19 heads)+ P(20 heads)

Using the binomial probability formula to compute each of these probabilities, we have:
P(at least 14 heads) = P(14 heads)+ P(15 heads)+ P(16 heads)+ P(17 heads)+ P(18 heads)+ P(19 heads)+ P(20 heads)

$$= {}_{20}C_{14}\left(\frac{1}{2}\right)^{14}\left(\frac{1}{2}\right)^{6} + {}_{20}C_{15}\left(\frac{1}{2}\right)^{15}\left(\frac{1}{2}\right)^{5} + {}_{20}C_{16}\left(\frac{1}{2}\right)^{16}\left(\frac{1}{2}\right)^{4} + \ldots + {}_{20}C_{20}\left(\frac{1}{2}\right)^{20}\left(\frac{1}{2}\right)^{0}$$

P(at least 14 heads) ≈ 0.0370 + 0.0148 + 0.0046 + 0.0011 + 0.0002 + 0.00002 + 0.000001
≈ 0.0577.

Using the binomial probability formula to determine the result of 0.0577, you will experience these calculations to be both long and tedious.

Let's consider an easier approach which uses **the normal curve to approximate this binomial probability**. This approximation is reasonable since both np and n(1–p) are greater than or equal to 10. That is,

$$np = (20)\left(\frac{1}{2}\right) = 10 \quad \text{and} \quad n(1-p) = (20)\left(\frac{1}{2}\right) = 10$$

To use the normal approximation, we need to calculate the mean, μ_s, and standard deviation, σ_s, of this binomial distribution using the formulas:

$$\mu_s = np \text{ and } \sigma_s = \sqrt{np(1-p)},$$

where n = 20, and p = $\frac{1}{2}$

The mean is:
$$\mu_s = np = (20)\left(\frac{1}{2}\right) = 10$$

and the standard deviation is:
$$\sigma_s = \sqrt{np(1-p)}$$
$$\sigma_s = \sqrt{20\left(\frac{1}{2}\right)\left(\frac{1}{2}\right)}$$
$$\sigma_s = \sqrt{5}$$
$$\sigma_s \approx 2.24$$

Thus, the binomial distribution where n = 20 and p = can be approximated by a normal distribution with mean: μ_s = 10 and standard deviation: σ_s = 2.24. Examine Figure 7.113 which shows the binomial distribution for n = 20 and p = $\frac{1}{2}$ and the normal curve superimposed on this binomial distribution.

Section 7.9 The Normal Approximation to the Binomial Distribution

Figure 7.113 shows a normal curve superimposed over the binomial distribution with $\mu_s = 10$ and $\sigma_s = 2.24$. The shaded region represents the probability: $P(14 \leq X \leq 20) \approx 0.0577$. Bar values: .037, .0148, .0046, .0011, .0002, .00002, .000001. Lower class boundary = 13.5.

FIGURE 7.113

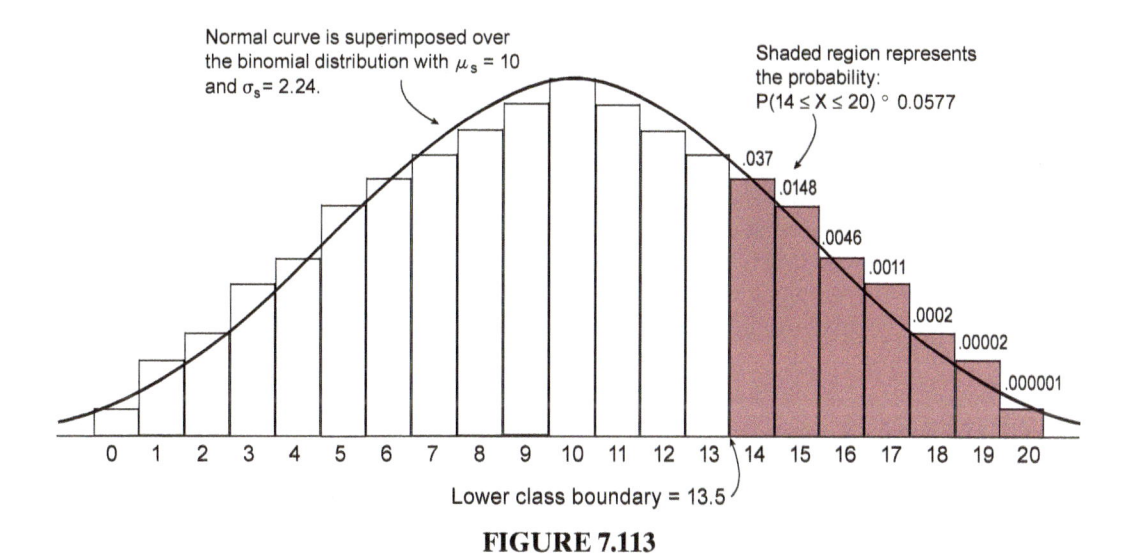

In the probability histogram, as pictured in Figure 7.113, which represents the binomial distribution, the area of each rectangle equals the probability of the X value shown under the respective rectangle. The probability (or the area) of the X values 14 to 20 are shown above each of these rectangles. For example, the probability (or area of the rectangle) of X = 14 heads is 0.0370, while the probability (or area of the rectangle) of X = 15 heads is 0.0148 ... and the probability (or area of the rectangle) X = 20 heads is 0.000001. Thus, the probability of at least 14 heads is the sum of these probabilities (or areas) which equals:

$$P(14 \leq X \leq 20) \approx 0.037 + 0.0148 + 0.0011 + 0.0002 + 0.00002 + 0.000001$$
$$\approx 0.0577.$$

Notice, in Figure 7.113, a normal curve with mean: $\mu_s = 10$ and standard deviation: $\sigma_s = 2.24$ has been superimposed over the binomial distribution. The normal curve can be used to obtain an approximation of these binomial probabilities without performing all the tedious calculations of the binomial formula.

The probability histogram of Figure 7.113 illustrates the binomial distribution where **the shaded region represents the probability of getting at least 14 heads**, since this region represents **the probability of 14 or more heads**. This area can be approximated by **the proportion of area under the normal curve to the *right of* 13.5** which is the lower class boundary of 14. Notice that we are finding the proportion of area to the right of 13.5, the boundary value of 14, and not to the right of 14. We are using the boundary value (13.5) rather than the actual value (14) because the Binomial Distribution is a discrete probability distribution and the Normal Distribution is a continuous probability distribution. We refer to the use of 13.5 rather than 14 as a **correction for continuity**. This **continuity correction** is due to the fact that we cannot find the probability of a specific value, like 14 heads, within a normal distribution. Therefore, the probability of exactly 14 heads corresponds to the proportion of area under the normal curve between the boundary values of 13.5 and 14.5. Therefore, any time we use the normal distribution to approximate a binomial probability, we must make a slight modification which is to either add or subtract 0.5 from the data value.

Therefore, the probability of at least 14 heads corresponds to the proportion of area under the normal curve which is to the right of 13.5. To determine the proportion of area to the right of 13.5, we must first compute the z score for x = 13.5. Using the z score formula,

$$z = \frac{x - \mu_s}{\sigma_s}$$

and substituting 13.5 for x, we have:

$$z = \frac{13.5 - 10}{2.24} = 1.56$$

From Table II, the proportion of area to the left of z = 1.56 is 0.9406. The proportion of area to the right of z = 1.56 is 0.0594.

This area of 0.0594 represents the normal approximation of the binomial probability of getting at least 14 heads in 20 tosses of a fair coin.

Let's compare the actual binomial probability of at least 14 heads to the normal approximation:

Binomial Calculation: **0.0577**

Normal Approximation: **0.0594**

Notice that the normal approximation (**0.0594**) is *very close* to the actual value (**0.0577**), since the difference between these two values is only **0.0017**.

When using the normal distribution to approximate binomial probabilities, the **class boundaries** of a data value are used to get a better approximation of the binomial probability. This is referred to a *continuity correction*, since the binomial distribution, which is a discrete distribution, is being corrected so that it can be approximated by the normal distribution, a continuous distribution.

Continuity Correction When Using the Normal Distribution to Approximate a Binomial Probability

When using the normal distribution to approximate a binomial probability, you must either add or subtract 0.5 from the data value depending upon the binomial probability question.

Let's summarize the procedure to approximate the binomial distribution using the normal distribution.

Normal Approximation Procedure to the Binomial Distribution

1. Determine the values of n, number of trials, and p, probability of a success, needed to solve the binomial probability problem.
2. Calculate the mean and standard deviation of the binomial distribution using the formulas:

$$\mu_s = np$$

$$\sigma_s = \sqrt{np(1-p)}$$

3. Draw a normal curve and a rectangle or rectangles that represent the number of successes. Shade the area under the normal curve representing the solution to the binomial probability problem.
4. Identify the class boundary or boundaries required for the solution to the binomial probability. The boundary value of the data value is found by either adding or subtracting 0.5 from the data value depending upon the binomial probability question.
5. Calculate the z score for each class boundary and use Table II to determine the proportion of area corresponding to the shaded proportion of area identified in step 4. This result represents the normal approximation to the binomial probability.

EXAMPLE 7.19

For each of the following binomial experiments, calculate the mean and standard deviation.
a) n = 100 and p = 0.2
b) n = 400 and p = 0.7

Solution

a) For n = 100 and p = 0.2, the mean of this binomial experiment is:

$$\mu_s = np$$
$$= (100)(0.2)$$
$$\mu_s = 20$$

The standard deviation of this binomial experiment is:

$$\sigma_s = \sqrt{np(1-p)}$$
$$= \sqrt{(100)(0.2)(1-0.2)}$$
$$= \sqrt{(100)(0.2)(0.8)}$$
$$\sigma_s = 4$$

Thus the mean and standard deviation for this binomial experiment are $\mu_s = 20$ and $\sigma_s = 4$.

b) For n = 400 and p = 0.7,

$$\mu_s = (400)(0.7)$$
$$\mu_s = 280$$

and:

$$\sigma_s = \sqrt{(400)(0.7)(0.3)}$$
$$\sigma_s \approx 9.17$$

Thus the mean and standard deviation for this binomial experiment are $\mu_s = 280$ and $\sigma_s \approx 9.17$.

EXAMPLE 7.20

Consider the binomial experiment of tossing a fair coin 100 times. Use the normal approximation to calculate the probability of getting:
a) At least 55 heads.
b) Between 40 and 60 heads.
c) Exactly 54 heads.
d) At most 45 heads.

Solution

To use the normal approximation to the binomial distribution, calculate the mean and standard deviation for the binomial distribution where n = 100, p = ½, and 1 − p = ½.

The mean is:
$$\mu_s = np = 50$$

The standard deviation is:

$$\sigma_s = \sqrt{np(1-p)}$$
$$= \sqrt{(100)(\tfrac{1}{2})(\tfrac{1}{2})}$$
$$= \sqrt{25}$$
$$\sigma_s = 5$$

Thus, the normal distribution with $\mu_s = 50$ and $\sigma_s = 5$ will be used to approximate these binomial probabilities.

a) To calculate the probability of getting at least 55 heads, we must determine the probability of getting 55 or more heads. This is illustrated in Figure 7.114.

To approximate the probability of getting 55 or more heads use the lower boundary of 55 which is 54.5 and compute the z score of 54.5.

$$z = \frac{54.5 - 50}{5} = \frac{4.5}{5} = 0.90$$

Using Table II the area to the left of z = 0.90 is 0.8159. Thus, the area to the right of z = 0.90 is 0.1841. Therefore, the probability of at least 55 heads is approximately 0.1841.

430 *Chapter 7* Continuous Probability Distributions and the Normal Distribution

b) To calculate the probability of getting **between** 40 and 60 heads, compute the probability of getting **more than** 40 heads and **less than** 60 heads. This is illustrated in Figure 7.115.

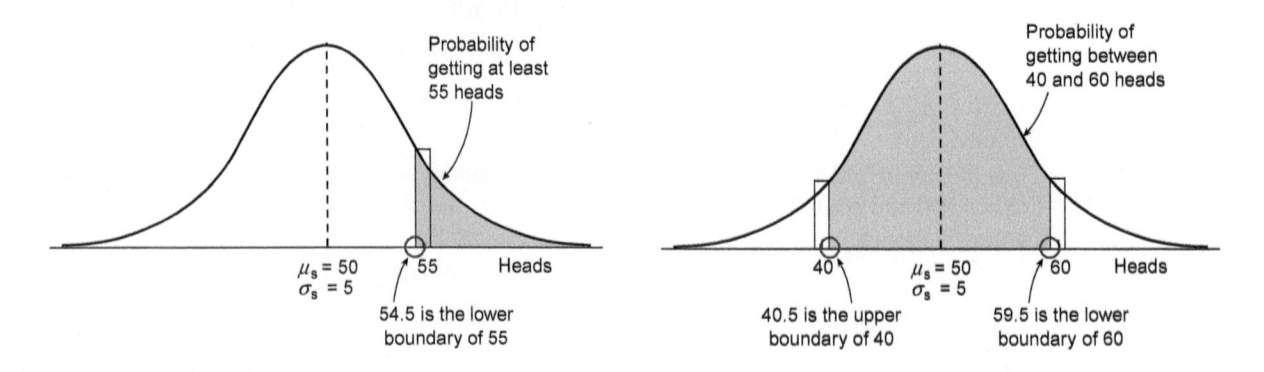

FIGURE 7.114

FIGURE 7.115

To approximate the probability of getting **between 40 and 60 heads**, use the **upper boundary of 40**, which is **40.5**, and the **lower boundary of 60**, which is **59.5**. This is shown in Figure 7.115. To determine the shaded proportion of area in Figure 7.115 compute the z score of 40.5 and the z score of 59.5. We obtain:

$$z \text{ of } 40.5 \text{ is } -1.90$$
$$z \text{ of } 59.5 \text{ is } 1.90$$

From Table II the proportion of area to the left of $z = -1.9$ is 0.0287, and the proportion of area to the left of $z = 1.9$ is 0.9713. Therefore, the area between $z = -1.9$ and $z = 1.9$ is $0.9713 - 0.0287 = 0.9426$. Hence, the probability of getting between 40 and 60 heads is 0.9426.

c) The probability of getting exactly 54 heads is approximated by the shaded proportion of area in Figure 7.116.

To approximate the **probability of exactly 54 heads, use the lower boundary of 54, which is 53.5, and the upper boundary of 54, which is 54.5**.

To determine the shaded area in Figure 7.116 compute the z score of 53.5 and the z score of 54.5. We obtain:

$$z \text{ of } 53.5 = 0.70$$
$$z \text{ of } 54.5 = 0.90$$

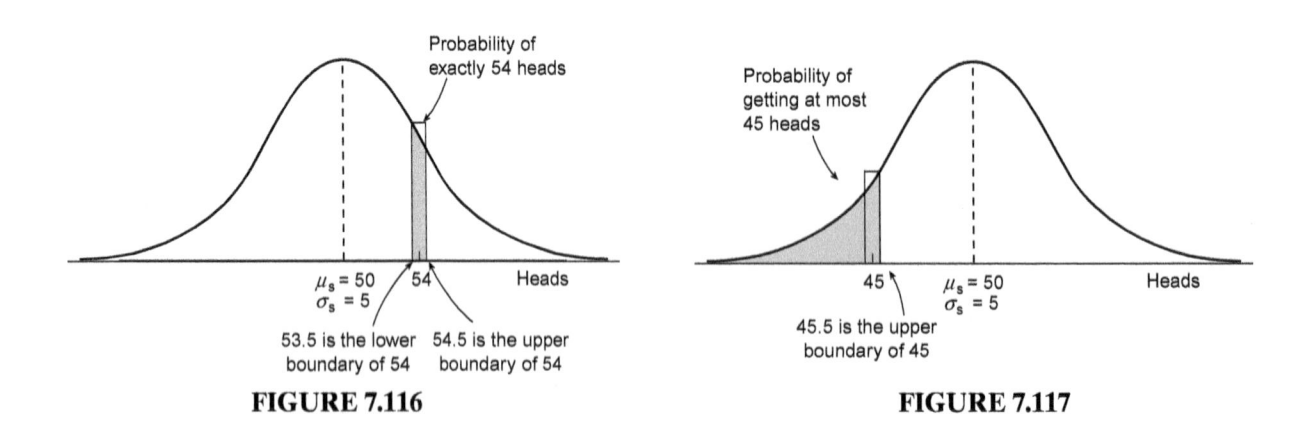

FIGURE 7.116

FIGURE 7.117

Using Table II the proportion of area to the left of $z = 0.70$ is 0.7580, and the proportion of area to the left of $z = 0.90$ is 0.8159. Thus, the area between $z = 0.70$ and $z = 0.90$ is $0.8159 - 0.7580 = 0.0579$. Therefore, the probability of getting exactly 54 heads is approximately 0.0579.

d) To calculate the probability of getting at most 45 heads, we must determine the probability of getting 45 or less heads. This is illustrated in Figure 7.117.

To approximate the probability of getting 45 or less heads use the upper boundary of 45 which is 45.5 and compute the z score of 45.5.

Section 7.9 *The Normal Approximation to the Binomial Distribution* 431

$$z = \frac{45.5 - 50}{5} = \frac{-4.5}{5} = -0.90$$

Using Table II the proportion of area to the left of $z = -0.90$ is 0.1841. Therefore, the probability of at most 45 heads is approximately 0.1841.

In Example 7.20, we examined four different types of probability questions. For each question, we determined the boundary value required to find the appropriate proportion of area under the normal curve to approximate the binomial probability. Let's examine the different probability questions and the boundary values that were used to approximate the probability. Figure 7.118 shows the boundary value or values needed to determine the normal curve area to approximate the different binomial probability problems.

Binomial Probability Problem	Boundary Value Used to Solve Problem	Area Under Normal Curve Corresponding to Binomial Probability
At least 55 heads	Boundary value is 54.5	Proportion of area to the right of 54.5 (see Figure 7.114)
Between 40 and 60 heads	Lower boundary is 40.5 and upper boundary is 59.5	Proportion of area between 40.5 and 59.5 (see Figure 7.115)
Exactly 55 heads	Lower boundary is 54.5 and upper boundary is 55.5	Proportion of area between 54.5 and 55.5 (see Figure 7.116)
At most 45 heads	Boundary value is 45.5	Proportion of area to the left of 45.5 (see Figure 7.117)

FIGURE 7.118

EXAMPLE 7.21

An automobile salesperson is successful in selling new cars to 40% of the customers that visit the showroom. What is the probability that the salesperson will sell a new car to at least 30 of the next 50 customers?

Solution

This problem can be considered to be a binomial problem where the probability that the salesperson successfully sells a new car to the customer is 0.40, and the probability the salesperson does not sell a new car to the customer is: $1 - 0.4 = 0.60$.

To determine if the normal approximation to the binomial distribution can be used to solve this problem, calculate np and $n(1-p)$.

$$np = (50)(0.4) \qquad \text{and} \qquad n(1-p) = (50)(0.6)$$
$$= 20 \qquad \qquad \qquad \qquad = 30$$

Since np and $n(1-p)$ are both greater than or equal to 10 the normal approximation to the binomial distribution can be used to solve this problem.

The mean of the binomial distribution is:

$$\mu_s = np = 20$$

The standard deviation of the binomial distribution is:

$$\sigma_s = \sqrt{np(1-p)}$$
$$\sigma_s = \sqrt{12}$$
$$\sigma_s \approx 3.5$$

Thus, a normal distribution with $\mu_s = 20$ and $\sigma_s \approx 3.5$ will be used to obtain an approximation for the binomial probability of the salesperson selling a new car to at least 30 of the next 50 customers. This is illustrated in Figure 7.119.

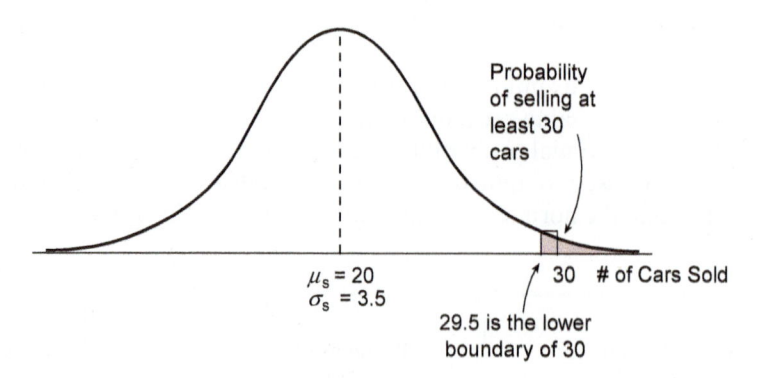

FIGURE 7.119

To approximate the probability of selling at least 30 new cars, we need to calculate the probability of selling 30 or more cars. We will use the lower boundary of 30 which is 29.5 and find the proportion of area to the right of 29.5. This is illustrated by the shaded area in Figure 7.119.

To calculate the proportion of area to the right of 29.5, find the z score of 29.5 which is 2.71. In Table II, we find the proportion of area to the left of $z = 2.71$ is 0.9966.

Thus, the area to the right of $z = 2.71$ is $1 - 0.9966 = 0.0034$. Therefore, the probability that the salesperson successfully sells at least 30 new cars to his next 50 customers is approximately 0.0034.

EXAMPLE 7.22

Tom, a daily commuter to Wall Street, is late to work 20% of the time. Calculate the probability that in his next 300 trips to work Tom will be late more than 25 times.

Solution

Let's consider using the normal approximation to the binomial distribution for this problem, where the probability that Tom gets to work late is 0.20 and the probability that Tom does not get to work late is 0.80.

Calculate np and n(1 – p) to determine if both expressions are greater than or equal to 10.

$$np = (300)(0.20) \quad \text{and} \quad n(1-p) = (300)(0.80)$$
$$= 60 \qquad\qquad\qquad\qquad\qquad = 240$$

Since np and n(1 – p) are both greater than or equal to 10, the *normal* approximation to the binomial distribution can be used to solve this problem.

The *mean* of the *binomial* distribution is:

$$\mu_s = 60$$

The *standard deviation* of the *binomial* distribution is:

$$\sigma_s \approx 6.93$$

Thus, the normal distribution with $\mu_s = 60$ and $\sigma_s \approx 6.93$ will be used to obtain an approximation for the binomial probability that Tom gets to work late more than 25 times in his next 300 trips. This is illustrated in Figure 7.120.

Section 7.9 The Normal Approximation to the Binomial Distribution

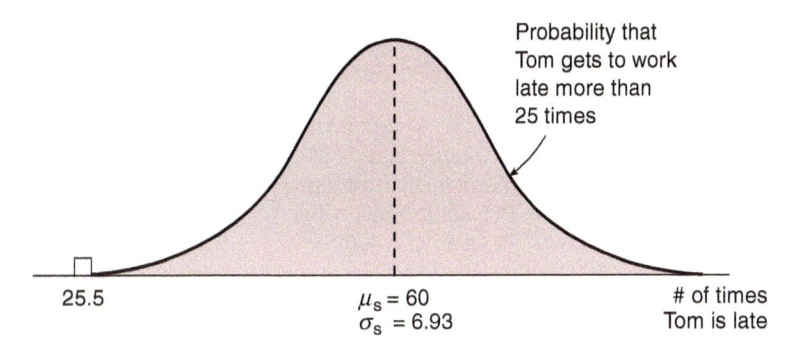

FIGURE 7.120

A normal curve is shown with $\mu_s = 60$ and $\sigma_s = 6.93$. The x-axis is labeled "# of times Tom is late" with the value 25.5 marked to the left. An arrow points to the right tail indicating "Probability that Tom gets to work late more than 25 times."

To approximate this binomial probability, use the *upper* boundary of 25 which is 25.5 since we are trying to find the probability of more than 25. Thus, this probability is determined by finding the proportion of area to the right of 25.5. Finding the z score of 25.5 = –4.98 in Table II, the area to the right of z = –4.98 is essentially 1. Therefore, the probability that Tom gets to work late more than 25 times is approximately 100%.

CASE STUDY 7.3

In Case Study 6.4, the article entitled "Trial By Mathematics" which appeared in *Time* magazine on April 26, 1968 discussed a robbery committed by an interracial couple. The couple accused of the crime had been convicted because the likelihood of a couple having the known characteristics of the robbers had been established at 1 in 12 million. In the couples' appeal, the judge assumed the probability p of a couple having all the characteristics of the robbers to be: p = 1/12,000,000. Assume that there does exist another such couple where each of the N couples in the population serve as a trial of a binomial experiment. The judge then calculated the probability that another couple has the same characteristics of the convicted couple for the value of N = 12,000,000.

a) Would the normal distribution be a good approximation to the binomial distribution? What values of N, p, and (1 – p) would you use, if you could use the normal approximation?
b) What is the mean and standard deviation of this binomial distribution?
c) If you used the normal approximation to the binomial distribution, what boundary value would you use to calculate the probability that two or more such couples within the population of N couples satisfying the characteristics of the robbers?
d) Using the normal distribution, what did you get for the probability that two or more such couples satisfy the characteristics of the robbers? How does this compare to the judges' calculation of approximately 41%?
e) Calculate this probability again for the following different values of N couples (in millions): 15, 20, 25, 30, 40, 50, 75, and 100. Compare your results using the normal approximation to the results of Table 6.16 in Chapter 6 which represent the probabilities calculated using the binomial distribution. Are the results of the normal approximation close to the binomial distribution? What happens to the difference between the normal approximation and the binomial distribution as the value of N gets larger?

Chapter 7 Continuous Probability Distributions and the Normal Distribution

KEY TERMS

Binomial Distribution 422
Class Boundaries 428
Continuous Probability Distribution 371
Continuous Random Variable 371
Mean of a Binomial Distribution 423

Normal Approximation to a Binomial Distribution 428
Normal Distribution 376
Percentile Rank 406
Percentiles 406

Probability 373
Standard Deviation of a Binomial Distribution 423
Standard Normal Curve 383

SECTION EXERCISES

Section 7.2

1. The variable *gender* is not a continuous random variable because it can assume two distinct values.
(T/F)

2. The variable *airplane flight times between New York and Los Angeles* is a continuous random variable because time can assume many possible values.
(T/F)

3. The sum of all the areas of the rectangles in a probability distribution histogram representing a continuous random variable is always equal to one.
(T/F)

4. The smooth curve of the continuous variable X is called the normal distribution.
(T/F)

5. The normal distribution is an example of a probability density curve.
(T/F)

6. The probability that a continuous variable will assume a specific value, such as X = a, is always zero.
(T/F)

7. If the process of measuring a continuous variable is to a finer and finer degree of precision then:
   a) The number of rectangles used to estimate the curve will decrease
   b) The width of each of the rectangles used to estimate the curve will get narrower
   c) The height of the rectangles used to estimate the curve will approach a line
   d) All of the above are correct

8. Which is not a characteristic of a continuous probability distribution?
   a) The probability that the continuous random variable X will assume a value between two possible values, X = a and X = b, is equal to the area under the density curve between X = a and X = b.
   b) The total area under the curve is one.
   c) The probability that the continuous variable X will assume a value between any two values X = a and X = b is between 0 and 1.
   d) The probability that the continuous variable X will assume a value of 0 is 1.

Section 7.3

9. Explain why many observable phenomena have shapes that are approximated by a normal distribution.

10. For each of the following continuous variables estimate whether or not their respective shapes are approximately normally distributed.
    a) The time it takes to travel from JFK International Airport in New York to Logan International Airport in Chicago.
    b) The amount of dollars earned for selling advertising space in the New York Times from 1970 to 1996.
    c) The number of pages in an elementary statistics book.
    d) The number of bacteria present in a culture after one hour.
    e) The weight in pounds of one day old Doberman puppies.
    f) The amount of electricity used to run your refrigerator.

Section 7.4

11. List the properties of a normal distribution.

12. Draw a standard normal distribution from z = –4 to z = 4. On the normal curve, show the proportion of data values that fall between each integer z score.

13. Use the properties of the normal distribution to answer the questions relating to the normal distribution in the following figure.

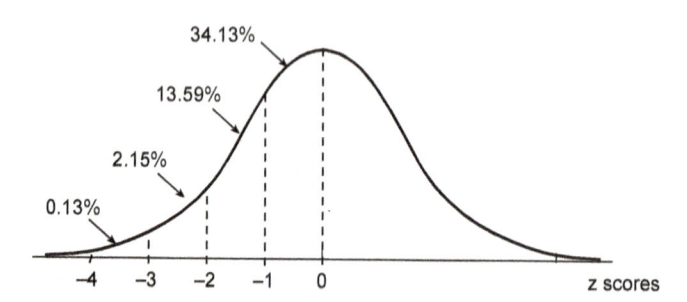

a) What proportion of the distribution is less than a z-score of –3?
b) What proportion of the distribution is less than a z-score of –1?
c) What proportion of the distribution is less than a z-score of 0?
d) What proportion of the distribution is less than a z-score of 2?
e) What proportion of the distribution is less than a z-score of 4?
f) What proportion of the distribution is greater than a z-score of 0?
g) What proportion of the distribution is greater than a z-score of –1?

h) What proportion of the distribution is greater than a z-score of 2?
i) What proportion of the distribution is greater than a z-score of –3?
j) What proportion of the distribution is greater than a z-score of 4?
k) What proportion of the distribution is between a z-score of –1 and 1?
l) What proportion of the distribution is between a z-score of –2 and 2?
m) What proportion of the distribution is between a z-score of –3 and 3?
n) What proportion of the distribution is between a z-score of –4 and 4?

14. The data values in a normal distribution tend to cluster about the _____.
15. In a normal distribution, the mean is located at the _____ of the distribution.
16. The proportion of area under the normal curve which is to the left of the mean is equal to _____.
17. In a normal distribution if the median is 100, then the mean is _____.
18. In a normal distribution, the percent of data values to the left of z = –0.53 is the same as the percent of data values to the right of z = _____.
19. In a normal distribution, the percent of area to the left of z = –1 is ____. Therefore, the percent of area to the right of z = –1 is ____.
20. The shaded area to the right of z = 0 is equal to the shaded area to the left of z = 0.
(T/F)

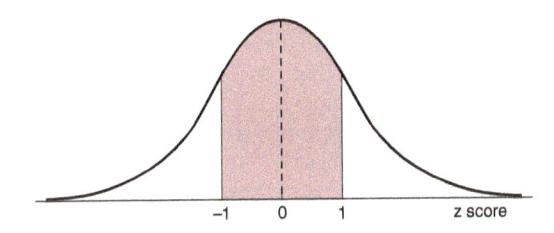

21. The shaded area A1 is equal to the shaded area A2.
(T/F)

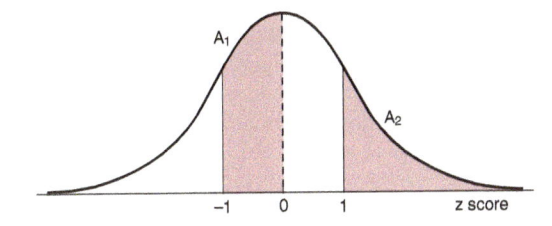

22. The z score of 0 divides the normal distribution into two equal parts.
(T/F)

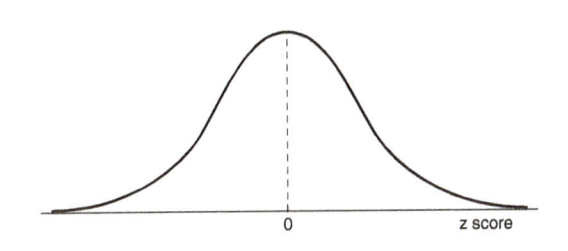

23. The area to the left of a z score of 1 equals the area to the right of a z score of 1.
(T/F)

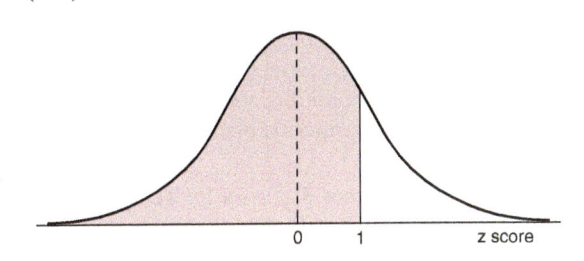

24. The area between the z scores –1 and –2 is equal to the area between the z scores +1 and +2.
(T/F)

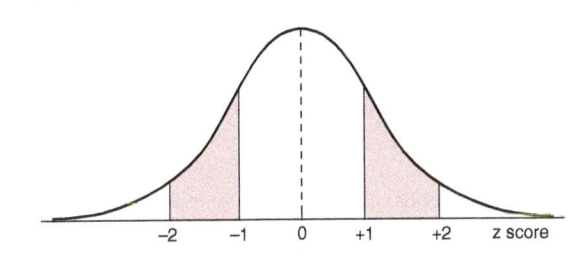

25. The total area under the normal curve is infinite.
(T/F)

Section 7.5

26. In a normal distribution find the proportion of area:
a) Less than a z score of 1.78.
b) Between the z scores of –2.09 and +0.53.
c) Greater than the z score of 2.19.

27. In a normal distribution find the percent of area:
a) Less than a z score of –1.65.
b) Between the z scores of –3 and 3.
c) Greater than the z score of 2.33.

28. In a normal distribution find the proportion of area:
a) Less than a z score of 4.
b) Between the z scores of –4 and +4.
c) Greater than the z score of –4.

29. In a normal distribution find the z score(s) that:
a) Cuts off the bottom 0.40 of the z scores.
b) Cuts off the top 0.01 of the z scores.
c) Cuts off the middle 90% of the z scores.

30. In a normal distribution find the z score(s) that:
a) Cuts off the bottom 10% of the z scores.
b) Cuts off the top 1% of the z scores.
c) Cuts off the middle 95% of the z scores.

Section 7.6

31. In a normal distribution with $\mu = 90$ and $\sigma = 15$ find the proportion of data values which are:
a) Less than 106.
b) Between 78 and 107.
c) Greater than 100.

32. The amount of caffeine in an 8 oz. energy drink is approximately normally distributed with a mean of 101 mg. and a standard deviation of 37.4 mg. Determine:

a) the proportion of 8 oz. energy drinks that are expected to contain less than 71 mg of caffeine.
b) the proportion of 8 oz. energy drinks that are expected to contain between 65 and 125 mg of caffeine.
c) the percent of 8 oz. energy drinks that are expected to contain greater than 150 mg of caffeine.
d) the minimum amount of caffeine expected to be in the top 1% of the distribution of caffeine in an 8 oz. energy drink.

33. According to Internet Statistics by Tech Crunchies, the average American Facebook user spends approximately an average of 7 hours a month on the Facebook site. Assume the amount of user time on the Facebook site can be approximated by a normal distribution with $\mu = 7$ hours per month and $\sigma = 1.65$ hours per month. Determine the percentage of Facebook users who spend:
a) less than 8.5 hours per month on the site?
b) at least 4.5 hours per month on the site?
c) either less than 4.5 hours or more than 10 hours on the site?
d) at most 390 minutes on the site?

34. According to the latest computer internet usage survey of 1500 high school students, the amount of social time per day a student spends on the internet is approximately normally distributed with $\mu = 210$ minutes and $\sigma = 45$ minutes.
Determine:
a) The proportion of students that spend less than 176 minutes on the internet.
b) The number of students that spend between 150 and 300 minutes on the internet.
c) The percent of students that spend more than 4 hours on the internet.
d) The minimum computer usage time spent on the internet by the top 5% of the users.
e) The maximum computer usage time spent on the internet by the lowest 400 users.

35. Debbie, an architect, is designing the interior doors to a new squash club. She wants the doors to be high enough so that 90% of the men using the doors will have at least a one foot clearance. Assuming the heights to be normally distributed with a mean of 69 inches and a standard deviation of 4 inches, how high must Debbie design the doors to satisfy her requirements?

36. Matthew, an experimental psychologist, indicates that the age at which a child learns to walk is normally distributed with a mean age of 10.9 months with a standard deviation of 0.5 months. If Matthew classifies an early walker as a child who is among the youngest 10% to walk, then by what age must a child walk to be identified as an early walker?

37. Body mass index (BMI) is a measure of body fat based on a person's height and weight. The BMI of an American female is approximately normally distributed with a mean of 28.7 and a standard deviation of 1.4. A female is considered obese if her BMI is within the top 18% of the distribution. What is the minimum BMI of a female to be considered obese?

Section 7.7

38. In a normal distribution, if the percent of data values greater than the data value 60 is 84.13 then, the percentile rank of 60 is_____.

39. In a normal distribution if the percentile rank of a data value is 50, then the data value is equal to the _____, and its z score is equal to _____.

40. In a normal distribution what is the percentile rank of the data value which is:
a) Two standard deviations greater than the mean?
b) One standard deviation less than the mean?
c) One-half a standard deviation greater than the mean?

41. In a normal distribution with $\mu = 250$ and $\sigma = 20$ find the percentile rank of the following data values:
a) 286.
b) 265.
c) 225.

42. In a normal distribution with $\mu = 80$ and $\sigma = 10$, if a data value is the 70th percentile, then the data value has a value of _____.

43. In a normal distribution with $\mu = 75$ and $\sigma = 15$ the data value with a percentile rank of 77 is:
a) 77   b) 75   c) 80   d) 86.1

44. In a normal distribution with $\mu = 300$ and $\sigma = 40$ find the data value equal to:
a) P_{85}.
b) P_{35}.
c) P_{70}.
d) The first decile.
e) The third quartile.
f) The sixth decile.

45. Three hundred first year law students at a local law school had a mean of 150 on the LSAT, the law boards, with a standard deviation of 10. Assuming these results are approximately normally distributed, determine:
a) The percentile rank of the LSAT score 163.
b) The percentile rank of the LSAT score 135.
c) The LSAT score that cuts off the top 2% of the LSAT scores (to the nearest whole number).
d) The LSAT score that cuts off the top 50 LSAT scores (to the nearest whole number).
e) The LSAT score that represents the 6th decile (to the nearest whole number).
f) The LSAT score that represents the first quartile (to the nearest whole number).
g) The LSAT score that represents the third quartile (to the nearest whole number).
h) The LSAT score that represents the 15th percentile (to the nearest whole number).
Suppose one of these first year law student's is selected at random, and you have to guess the student's LSAT score. If you guess it correctly to within 12 points, then you will be given ten dollars.
i) What LSAT score should you guess and what is your chance of winning the ten dollars?

Section 7.8

46. In a normal distribution find the probability that a data value picked at random has a z score:
a) Less than z = –1.16.
b) Between z = 1.0 and z = +0.5.
c) Greater than z = 1.75.

47. In a normal distribution which has $\mu = 900$ and $\sigma = 350$ find the probability that a data value picked at random is:
a) Less than 500.
b) Between 800 and 1200.
c) Greater than 1100.

48. According to the McDonald's nutrition guide, a McDonald's sandwich has an average of 500 calories with a standard deviation of 100 calories. Assuming that the calorie content

of a McDonald's sandwich is normally distributed, find the probability that a customer will order a sandwich with:
a) Less than 380 calories.
b) More than 675 calories.
c) Less than 400 or greater than 700 calories.

49. It was recently reported that "people in their 20s change jobs an average of 7 times". If we assume for the population of employed individuals in their 20s the number of times they change jobs is approximately normally distributed with a mean of 7 and standard deviation of 3.2, answer the following:
a) What percent of the population of employed individuals in their 20s change jobs less than 8 times?
b) What is the probability that an employed person in their 20s changes jobs more than 10 times?
c) What percent of the population of employed individuals in their 20s change jobs between 4 and 12 times?
d) Determine the expected number of job changes for an employed person in their 20s that represent the middle 90% of the population.

50. Dr. Nägele determined that the average human pregnancy is 266 days from conception to birth. Assume that the length of human pregnancies can be approximated by a normal distribution with μ = 266 days and σ = 16 days. Find the probability that a pregnancy will last:
a) less than 240 days (approximately 8 months)?
b) between 240 and 270 days (approximately 8 and 9 months)?
c) more than 270 days (approximately 9 months)?
d) Determine how long do the longest 15% of pregnancies last?

51. According to the manufacturer of Crackle natural peach flavored iced tea drinks, the net amount of iced tea in a glass bottle is normally distributed with a mean of 12 ounces and a standard deviation of 0.018 ounces. What is the probability that a bottle selected at random from the manufacturing process contains:
a) Less than 11.98 ounces?
b) Between 11.96 and 12.01 ounces?
c) Greater than 12.03 ounces?
d) If the maximum amount of iced tea that a standard bottle can hold is 12.05 ounces, then what is the probability that a bottle selected at random was overfilled?

52. The average cell phone user interacts with their cell phone 60 times per day with a standard deviation of 20. Assuming the number of interactions follows a normal distribution, find:
a) the probability that a randomly selected cell phone user will look at their phone more than 80 times today.
b) the probability that a randomly selected cell phone user will look at their phone between 40 and 80 times today.
c) the percent of cell phone users who look at their phone less than 25 times per day.
d) the minimum number of interactions by cell phone users in the upper 1%.

53. The amount of time required for Jain to travel to and from home and the office can be approximated by a normal distribution with μ = 55 minutes and σ = 10 minutes.
a) If Jain leaves home at 7:45 a.m., what is the probability that she will arrive at the office no later than 9:00 a.m.?
b) If coffee is served at the office between 8:45–9:00 a.m., what is the probability that Jain will miss the coffee break (arrives after 9:00 a.m.) if she leaves home at 8:05 a.m.?
c) If Jain has an appointment with a client at the office promptly at 9:05 a.m., what is the probability she will make the appointment if she leaves home promptly at 8:15 a.m.?

Section 7.9

54. The normal distribution is a good approximation of the binomial distribution when both np and n(1–p) are greater than or equal to _____.

55. a) For a binomial distribution with n = 10 and p = ⅕, would the normal distribution be a good approximation to the binomial distribution? _____.
b) For a binomial distribution with n = 48 and p = ¾, would the normal distribution be a good approximation to the binomial distribution? _____.

56. The frequency polygon of a binomial distribution looks more like a _____ distribution as the number of trials, n, gets larger.

57. For a binomial distribution, the formula for:
a) The mean of the binomial distribution, μ_s, is _____.
b) The standard deviation of the binomial distribution, σ_s, is _____.

58. For a binomial distribution, if n = 100 and p = ½, then:
a) $\mu_s = np =$ _____.
b) $\sigma_s = \sqrt{np(1-p)}$ = _____.

59. For a binomial distribution with n = 200 and p = 0.37 what is the value of μ_s?
a) 74   b) 200   c) 37   d) 100

60. For a binomial distribution with n = 120 and p = 0.4 what is the value of σ_s?
a) 0.24   b) 28.8   c) 5.37   d) 14.4

61. To get a good approximation of a binomial probability problem by using the normal approximation, real boundaries should be applied.
(T/F)

62. Use the normal distribution to approximate the following binomial probabilities:
a) If n = 100 and p = 0.5, then find the probability that the number of successes "s" is:
1. Greater than 60, i.e. P(s > 60).
2. Equal to 60, i.e. P(s = 60).
3. Greater than or equal to 60, i.e. P(s ≥ 60).
4. Less than 60 or greater than 60, i.e. P(s < 60 or s > 60).
b) If n = 100 and p = 0.1, then find the probability that the number of successes "s" is:
1. Equal to 10, i.e. P(s = 10).
2. Less than 7, i.e. P(s<7).
3. Greater than or equal to 7, i.e. P(s ≥ 7).
4. Less than 7 or greater than or equal to 7, i.e. P(s < 7 or s ≥ 7).

63. For the binomial experiment of tossing a fair coin 400 times, use the normal distribution to approximate the probability that:
a) The coin lands heads at least 217 times.
b) The coin lands tails between 180 and 195 inclusive.
c) The coin lands heads exactly 200 times.
d) The coin lands heads less than 185 or more than 210.

64. If a pitcher has 0.25 probability of striking out a batter what's the probability that this pitcher strikes out 18 or more of the next 40 batters he faces?
a) 0.0031   b) 0.05   c) 0.01   d) 0.9969

65. A multiple choice test consists of 100 questions, each with five possible answers. If a student must answer at least 33 questions correctly in order to pass, what is the probability that the student passes the test? (Assume the student guesses at all the answers.)

438 Chapter 7 Continuous Probability Distributions and the Normal Distribution

66. Over the past year, 60% of the patients waiting to see Dr. Hawkeye, an ophthalmologist, for their scheduled appointment have had to wait at least 45 minutes. What is the probability that out of the next 600 patients:
   a) More than 390
   b) Less than 340
   will have to wait at least 45 minutes for their scheduled appointments?

67. According to a study by Dow Jones & Co., the best time to buy stocks is on a Monday afternoon since the chance of the market rising is 0.43 while the best time to sell is on a Friday since the chance of the market rising is 0.58. Use the results of this study to determine the probability that over the next 100 trading weeks, the market will:
   a) Rise on a Monday afternoon at least 55 times.
   b) Rise on a Friday at most 45 times.
   c) Not rise on Friday more than 40 times.
   d) Not rise on a Monday afternoon less than 60 times.

68. A baseball slugger hits a home run once out of every fifteen times he bats, that is, the probability that he will hit a home run at any given time at bat is $1/15$. If he bats 450 times in a season, what is the probability that he hits:
   a) At least 40 home runs?
   b) Between 15 and 30 home runs (inclusive)?
   c) More than 49 home runs?

69. Last year, the records of a Brooklyn college show that 80% of their graduates who were recommended to U.S. medical schools were admitted. If the records of 100 of these Brooklyn college graduates who were recommended to U.S. medical schools were randomly selected from the school's files, determine the probability that:
   a) At least 90 of these graduates were admitted to a U.S. medical school.
   b) At most 15 of these graduates were not admitted to a U.S. medical school.

70. Medical records show that 25% of all people suffering from "hardening of the arteries" have side effects from a certain type of medication. If 200 people suffering from "hardening of the arteries" are given this medication, what is the probability that:
   a) No more than 35 will have side effects?
   b) At least 160 will not have side effects?

71. Clinical records indicate that 20% of the outpatients in a facility for emotional disorders have been treated in this facility for at least 3 years. What is the probability that a sample of 400 outpatients from this facility will include:
   a) Not more than 60 patients who have been treated at least 3 years?
   b) No fewer than 90 patients who have been treated at least 3 years?

72. The Best Fit Screw Company institutes the following quality control procedure to improve the quality of the screws it uses. Before a shipment of screws is accepted a sample of 100 screws is selected and tested for defects. If more than six of the screws are found to be defective, the shipment is rejected.
   a) If a shipment is 20 percent defective, what is the probability that it will be accepted?
   b) If a shipment is 10 percent defective, what is the probability that it will be accepted?

73. A money market fund company reports that next to individual retirement accounts (IRA's), the tax-shelter most frequently used by their shareholders is the uniform gifts to minors account. If 10% of their shareholders are custodians of an account for a minor, then what is the probability that out of the next 40,000 shareholders opening new accounts no more than 4100 will be custodian accounts for a minor?

CHAPTER REVIEW

74. Use the properties of the normal distribution to answer the questions relating to the normal distribution in the following figure.

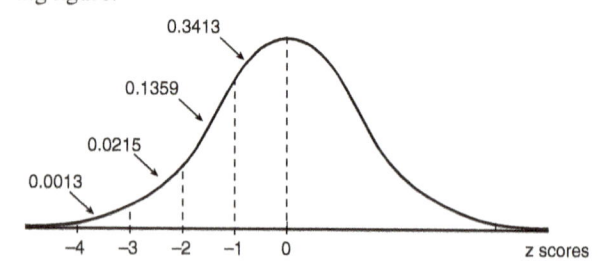

   a) What proportion of the distribution is less than a z-score of –4?
   b) What proportion of the distribution is less than a z-score of –2?
   c) What proportion of the distribution is less than a z-score of 0?
   d) What proportion of the distribution is less than a z-score of 1?
   e) What proportion of the distribution is less than a z-score of 3?
   f) What proportion of the distribution is less than a z-score of 5?
   g) What proportion of the distribution is within 1 standard deviation of the mean?
   h) What proportion of the distribution is within 2 standard deviations of the mean?
   i) What proportion of the distribution is within 3 standard deviations of the mean?

75. The shaded areas (A1 and A2) are equal. (T/F)

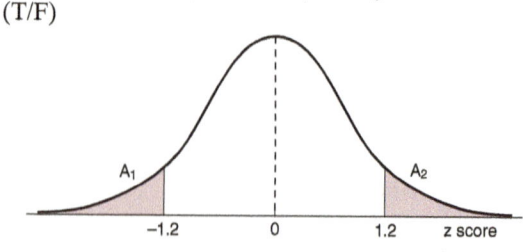

76. The shaded area to the right of a z score of –1 is greater than the area to the left of z score of –1. (T/F)

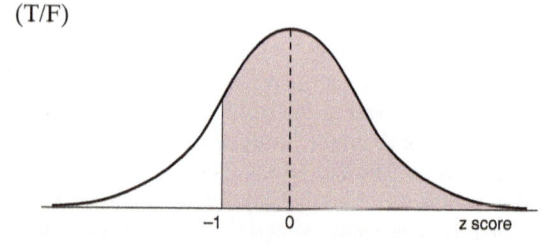

77. For a normal distribution the percent of area between z = 1 and z = −1 is approximately 68.26. Therefore the percent of data values between z = −1 and z = 1 is approximately _____.

78. In a normal distribution, the proportion of data values between z = 0 and z = −2 is the same as the proportion of data values between z = _____ and z = _____.

79. The percent of data values in a normal distribution that are located within:
   a) One standard deviation of the mean (that is, between z = +1 and z = −1) is approximately _____.
   b) Two standard deviations of the mean (that is, between z = 2 and z = −2) is approximately _____.
   c) Three standard deviations of the mean (that is, between z = +3 and z = −3) is approximately _____ _____.

80. In a normal distribution find the z score(s) that:
   a) Cuts off the bottom 84% of the z scores.
   b) Cuts off the top 10% of the z scores.
   c) Cuts off the middle 50% of the z scores.

81. In a normal distribution find the proportion of area:
   a) Less than a z score of −1.65.
   b) Between the z scores of 1.03 and +2.81.
   c) Greater than the z score of 0.30.

82. Suppose that the time to complete the 2013 Long Island Marathon is approximately normally distributed with mean μ minutes and a standard deviation of 44 minutes. Jack raced in the 2013 Long Island Marathon and he completed the race in 293.03 minutes. Also, 20% of the marathon runners took longer than Jack to complete the race. What is the mean time to complete the 2013 Long Island Marathon?

83. The English Mastiff is one of the largest and heaviest dog breeds recognized by the American Kennel Club. Suppose the weight of the male Mastiff is approximately normally distributed with a mean of 190 pounds and a standard deviation of 20 pounds. What is the probability that a male Mastiff selected at random will:
   a) Weigh less than 165 pounds?
   b) Weigh greater than 235 pounds?
   c) Weigh between 178 and 224 pounds?
   d) Determine the weight of a male Mastiff that lies at the 75th percentile.

84. In a normal distribution with $\mu = 50$ and $\sigma = 10$ the proportion of data values greater than 65 is:
   a) 0.0668   b) 0.6915   c) 0.9332   d) 0.1587

85. During the past season the number of points scored per game by the local High School football team was approximately normally distributed with $\mu = 18$ and $\sigma = 10$.

   Use the normal distribution to answer the following questions:
   a) If one assumes this season's play is indicative of next season's play then estimate the percent of next season's games that the team will score more than 31 points?
   b) If the team plays 30 games in approximately how many would they score more than 6 points?
   c) What's the probability that the team will score at least 22 points in any given game?

86. At a local high school 5000 juniors and seniors recently took an aptitude test. The results of the exam were normally distributed with $\mu = 450$ and $\sigma = 50$. Calculate the following:
   a) The percent of students that scored over 575.
   b) The number of students that scored less than 425.
   c) The probability of a student selected at random having scored between 400 and 510.

87. In order to qualify for the Presidential Physical Fitness Award, elementary school students must be *at least* the 85th percentile for sit-ups. If the number of sit-ups completed is normally distributed with $\mu = 27$ and $\sigma = 3$, determine which of the ten students failed to qualify for the award. A random sample of students produced the following results.

Student	# of Sit-Ups
1. Lylah	30
2. Justin	38
3. Hector	20
4. Aaron	40
5. Alexandra	52
6. Hassan	29
7. Katie	12
8. Edward	32
9. Michael	30
10. Caleb	42

88. A new drug developed to immobilize wild life is administered to a herd of four hundred buffaloes which must be transported to another grazing area for environmental reasons. The buffaloes reaction time to this drug is approximately normal with a mean of 17.4 seconds and a standard deviation of 3.4 seconds. What is the probability that a buffalo selected at random will have a reaction time to the drug:
   a) Less than 15 seconds?
   b) Greater than 20 seconds?
   c) Between 13 and 23 seconds?
   Determine:
   d) What is the maximum time required for the ten most *susceptible* buffaloes to react to this drug?

89. The normal distribution can be used as a good approximation to the binomial distribution if np and n(1 − p) are both greater than or equal to:
   a) 10   b) 5   c) 0.5   d) 25

90. For the binomial distribution with n = 30 and p = ⅙, would the normal distribution be a good approximation to the binomial distribution? _____.

91. For a binomial distribution with n = 800 and p = 0.45, calculate the value of:
   a) μ_s?
   b) σ_s?

92. Use the normal distribution to approximate the following binomial probabilities: If n = 10,000 and p = 0.8, then find the probability that the number of successes "s" is:
   a) Greater than 8060, i.e P(s > 8060).
   b) Greater than or equal to 8060, i.e. P(s ≥ 8060).
   c) Less than 7960, i.e. P(s < 7960).
   d) Equal to 7960, i.e. P(s = 7960).
   e) Between 7960 and 8060, i.e. P(7960 < s < 8060).

93. If a fair coin is tossed 100 times what is the probability that the coin lands heads more than 60 times?
   a) 0.9821   b) 0.05   c) 0.0179   d) 0.95

94. A recent survey showed that only 17% of the married women in the labor force have husbands who were earning over $55,000 annually. If a random sample of 100 married women is selected from the labor force, what is the probability that:
   a) More than 25 have husbands earning over $55,000 annually?
   b) Less than 76 have husbands earning less than $55,000 annually?

95. Jacqueline, a sales representative of Day News, a Nassau County newspaper, makes 200 calls a day trying to persuade

potential customers to subscribe to Day News. If she has a 30% chance of selling a new subscription, what is the probability that Jacqueline will sell:
a) At least 78 new subscriptions for the day?
b) Fewer than 40 new subscriptions for the day?

96. The accountant firm of a major credit card company reports that 30% of all the company's delinquent accounts will eventually require legal action to collect the payment due. If this statement is accurate, what is the probability that of the 10,000 delinquent accounts due:
a) At least 3100 accounts will require legal action?
b) At most 6940 accounts will not require legal action?

97. A New York municipal court cites "incompatibility" as the legal cause for 80 percent of all divorce cases. If one hundred divorce cases are randomly selected from the court files, what is the probability that:
a) Not more than 72 cite incompatibility as the grounds for divorce?
b) At least 85 cite incompatibility as the grounds for divorce?
c) Less than 15 do *not* cite incompatibility as the grounds for divorce?

98. Medical researchers at a Boston university report that 20% of babies born to American mothers this year will be delivered by cesarean sections. What is the probability that out of the next 300 deliveries:
a) Not more than 85 will be cesarean sections?
b) At least 260 will not be cesarean sections?

99. A study released by the Bureau of Justice Statistics found that eight out of 10 young adults paroled from prison are rearrested for serious crimes within six years. If 1600 cases of young parolees are reviewed after six years, then what is the probability that:
a) More than 1300 were rearrested for a serious crime?
b) Less than 1320 were rearrested for a serious crime?

WHAT DO YOU THINK?

100. The graphic on the next page entitled "Mind Over Muscle" appeared in *The New York Times* on November 19, 1995.

The vertical bars within the graphic represent the number of people who finished the 1995 New York City Marathon for each ten-minute period starting with 2 hours. Notice that the vertical bar graph can be approximated by a bell-shaped curve.
a) Re-read the introduction to Section 7.3. Using this information, explain why you think the finishing times for the marathon would approximate a bell-shape curve.
b) Notice in the bar graph that there are five bars identified by black tips that do not conform to the bell-shaped curve. What are the times associated with these bars? The author of the article believes that runners strive to run the marathon on the "fast side of the closest round time." That is, a runner pushes to run a 3:57 rather than 4:02 or a 4:58 rather than a 5:03. Do you agree with this explanation? What do you think is the reason for these five spikes?

101. An article entitled "Student Finds Penny IS Tail-Heavy" appeared in *The Washington Post* on November 27, 1965. The article gives an account of a high school senior named Edward who flipped a penny 17,950 times. He determined that the penny was tail-heavy. According to the story, a student flipped a penny 17,950 times and he recorded 464 more heads than tails. From his experiment, he proclaimed that the United States Mint produces tail-heavy pennies.
a) Can this be considered a binomial experiment? If it is, then what are the values of n, p, and (1 − p)?
b) Can you use the normal approximation to the binomial distribution to determine probabilities associated with this experiment? Explain.
c) How many heads and tails did Edward get during his experiment?
d) If we define p to be the probability of getting a head, then what is the expected mean number of heads for this experiment? How does the mean number of heads compare to the result in part (c)?
e) If a coin is fair, would you expect the number of times a coin lands heads to be equal to the mean number of heads found in part (d)? If not, then how many heads might you expect the coin to deviate from this expected mean number?
f) The number of heads can be 9,207 or 9,208, but it can't be any value in between. Is the number of heads a discrete or continuous variable? Since the normal distribution represents values between whole numbers, does it represent a discrete or continuous variable? Why are you able to use the normal distribution to approximate the binomial distribution? Do any adjustments have to be made to perform this approximation? If so, what are these adjustments?
g) What is the probability that a coin will land at least 9,207 heads in 17,950 tosses, using the normal approximation?
h) Is this probability small or large? If the event is rare, then would you expect the probability of the event to be small or large? What can you say about this event?
i) Do you think that the coin is fair or biased?
j) What did you use as the boundary value to calculate the probability in part (g)? Explain why it is necessary to use a boundary value rather than the actual value of 9,207, when using the normal approximation to the binomial distribution?

PROJECTS

102. a) Randomly select 36 females and record their heights (or weights).
b) Construct a frequency table, a histogram and a frequency polygon using your data.
c) Calculate the mean and standard deviation of your data.
d) Find the percent of heights (or weights) that are within 1 standard deviation of the mean.
e) For a normal distribution which has the same mean and standard deviation obtained in part (c) calculate the percent of data values within one standard deviation of the mean.

Mind Over Muscle (for exercise 100)

Running up 13,000-foot volcanoes? Yes, that's a pretty good way to squeeze out your best performance at the New York City Marathon. It worked for German Silva of Mexico, the men's winner last Sunday. But what really lights a runner's fire is the magic of round numbers.

You hear it all the time: "If I can just get under four hours," "Break three hours and I can die in peace." Runners carry these mantras around for months, even years.

Now, statistics would dictate that if you grouped the finishing times in 10-minute chunks, you'd see a tidy bell curve—the rarefied ranks of the speedy bulging smoothly toward the bulk of 4:15 runners in the middle, then falling off just as smoothly.

But no, runners gleefully defy statistics. Hundreds of them regularly leap to the fast side of the closest round time. A 4:02 runner pushes for a 3:57; a 5:03 is pulled by 4:58. This year at the four-hour barrier, the one most accessible to the average runner, the bell curve cracked.

New York runners often say, "Well, I lost 7 minutes at the start, so I *really* did ..." But it's what the clock says that imbeds itself in their souls. Just try telling them a 3:00:10 is the same as a 2:59:50.

Hubert B. Herring

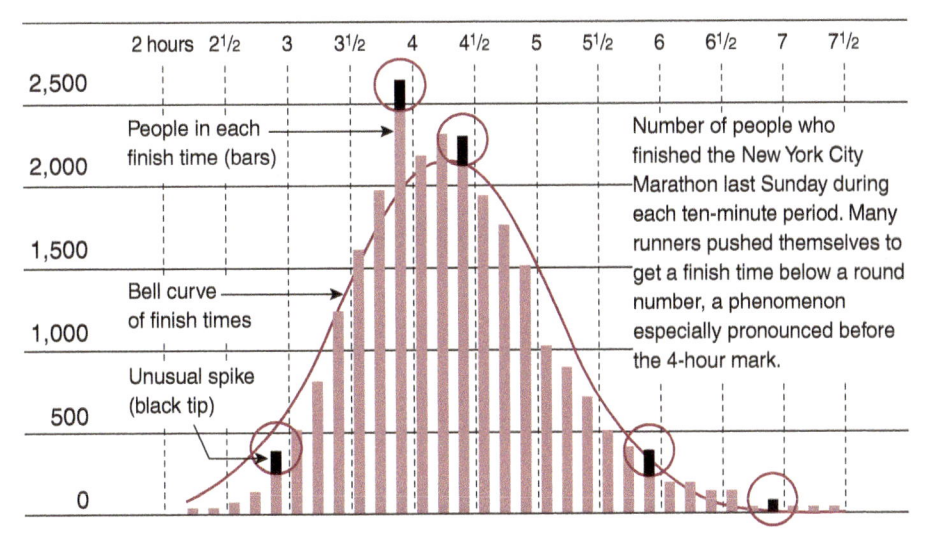

From *The New York Times*. © 1995 The New York Times Company. All rights reserved. Used under license.

The New York Times

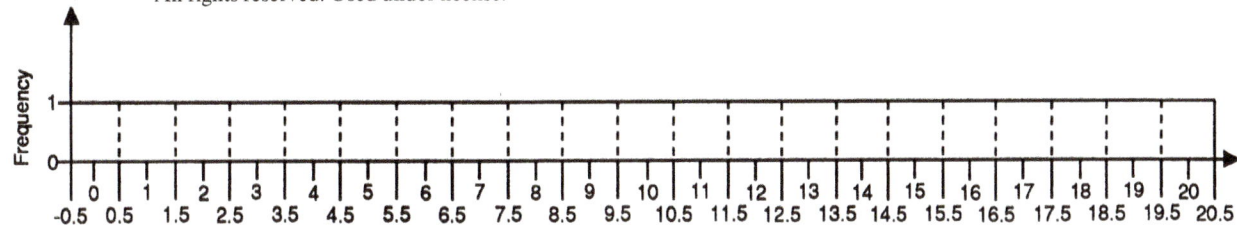

f) Compare the results of parts (d) and (e).
g) Repeat parts (a) through (f), using samples of size 64, 100, and 200.

103. Consider the uniform distribution pictured in the histogram above.
  a) On twenty-one small pieces of paper write the numbers 0 through 20, one number to a piece of paper.
  b) Place these pieces of paper into a container.
  c) Select ten numbers from the container with replacement and record the result of each selection.
  d) Calculate the sum of the ten numbers selected.
  e) Repeat steps (c) and (d) until 100 sums have been calculated.
  f) Construct a histogram and frequency polygon using an interval width of 5 for the 100 sums.
  g) Compare their general shape to that of the normal distribution.

104. Use the random number generator within a technological tool such as MINITAB, EXCEL, or TI–84 Plus Calculator to produce approximately normally distributed data values for N = 10, 100, 200 and 500 data values. Use $\mu = 0$ and $\sigma = 1$. Construct a histogram for each distribution.
  a) Describe the shape of each histogram.
  b) Does the shape of the histogram change as the sample size increases? Explain.
  c) Repeat parts (a) and (b). Are the results the same as before? Explain.
  d) What do you believe would be the shape of the resulting histogram if N = 100,000?

105. Use the random number generator within a technological tool such as MINITAB, EXCEL, or TI–84 Plus Calculator to produce approximately normally distributed data values for N = 10, 100, 200 and 500 data values. Use $\mu = 500$ and $\sigma = 100$. Construct a histogram for each distribution.

a) Describe the shape of each histogram.
b) Does the shape of the histogram change as the sample size increases? Explain.
c) Repeat parts (a) and (b). Are the results the same as before? Explain.
d) For each histogram calculate the proportion of data values that are within two standard deviations of the mean.
e) Compare the results obtained in part (d) to the actual proportion of data values that are within two standard deviations of the mean in a normal distribution.
f) As the sample size is increased, did the proportion of data values within two standard deviations of the mean for the samples, get closer to or farther from the proportion of data values that are within two standard deviations of the mean for a normal distribution?

106. Use a technological tool to produce a binomial distribution along with a table of binomial probability values for the binomial experiment where n = 100 and p = 0.5.
a) Using this information, calculate $P(s \leq 45)$.
b) Compare this probability value to the one obtained by using the normal approximation to the binomial distribution. Are these results very close? Did you expect this to happen? Explain.

107. a) Repeat problem 115 for n = 200, 300 and 400.
b) 1) for n = 200, calculate $P(s \leq 90)$.
Compare this probability value to one obtained by using the normal approximation to the binomial distribution.
2) for n = 300, calculate $P(s \leq 135)$.
Compare this probability value to one obtained by using the normal approximation to the binomial distribution.
3) for n = 400, calculate $P(s \leq 180)$.
Compare this probability value to one obtained by using the normal approximation to the binomial distribution.
c) By analyzing the previous results, write a conclusion regarding the normal approximation to the binomial distribution as n increases?

108. In the casino game roulette the probability of success for the outcome "black" is approximately 0.474. Use a technological tool to generate a table of successes along with the probability of success for choosing "black" and to construct a graph of each distribution for the game being played:
a) n = 10 times
b) n = 100 times
c) n = 1000 times
d) What happens to the probability of success in the game as the number of times the game is played increases?

CHAPTER TEST

109. In a normal distribution approximately two-thirds of the data values are within _____ standard deviation(s) of the mean.
a) 3   b) 1   c) 2   d) 1 and ½

110. In a normal distribution find the proportion of area:
a) Less than a z score of –0.50.
b) Between the z scores of –0.33 and +1.5.
c) Greater than the z score of 2.88.

111. If the distribution of annual family incomes in New York City is normally distributed with a mean of $77,100 and standard deviation of $15,000, then determine the following:
a) The proportion of families with incomes less than $55,000.
b) The percent of families with incomes between $65,000 and $95,000.
c) The probability that a family selected at random living in New York City will have an annual family income greater than $125,000.
d) If there are an estimated 3 million families living in New York City, then how many families would be expected to have family incomes greater than $100,000?
e) The family income that represents the 8th decile.
f) The family income that represents the 1st quartile.
g) The family income that represents the 65th percentile.
h) The family income that cuts off the lowest 20,000 incomes.
i) The family income that cuts off the highest 100,000 incomes.
j) If the upper middle incomes are defined as those families earning in the upper 20% of all family incomes. Find the minimum family income in New York City that is considered upper middle.

112. A woman wrote to Dear Abby the following letter.

Dear Abby,
When my son first married his wife, she would spend a couple of weeks with him and then go visit her parents across the state for a week or so. This pattern kept up for more than a year. My son knew his wife dated other men while on her visits home.

Last June she went for a visit with her folks and stayed almost six weeks. 253 days after the night of her return, she gave birth to a full-term healthy son. According to my information, 270 days must elapse after insemination, or 275 to 280 days after the first day of monthly period. Taking all this into consideration, she would have been impregnated while on her visit at her home.

I don't think my son is aware of these facts. Should I mention to him that even though he is the legal father of the baby he may not be the biological father? Or would it be better not to say anything?
— His Mom.

Although Dear Abby approached the problem from a non-statistical view (she told the mom to mind her own business), we can use the normal distribution to provide an answer using a statistical approach. If we assume that completed pregnancies are normally distributed with a mean of 270 days and a standard deviation of 16 days, then calculate the probability that a pregnancy lasts:
a) At most 253 days (i.e. the wife could've gotten pregnant after she returned home).
b) Longer than 295 days (i.e. the wife could've been pregnant before she left = 253 days + 6 weeks = 295 days).
c) Between 253 and 295 days.
d) What do you think are the chances that the wife was impregnated by her husband? Explain.

113. A person having a Twitter account spends, on average, 21 minutes per month on Twitter. This includes reading and sending Tweets. The standard deviation is 11.5 minutes. Assuming the time spent on Twitter per month is normally distributed, find:
  a) the probability a randomly selected person with a Twitter account will spend more than 30 minutes on Twitter this month.
  b) the probability a randomly selected person with a Twitter account will spend more than 5 minutes on Twitter this month.
  c) the *minimum* amount of time spent on Twitter per month by the upper 2% of users.
  d) the time spent on Twitter per month representing the 90th percentile.

114. If the amount of soda which a dispensing machine puts into a two liter bottle is normally distributed with $\sigma = 0.5$ oz. find the mean amount of soda which the machine will put into a two liter bottle if only 65.54% of the bottles the machine filled will have less than 67.6 oz. of soda.

115. The lifetime of a car battery is found to be normally distributed with $\mu = 5$ years and $\sigma = ¾$ year. If the company which manufactures this battery is only willing to replace 5% of these batteries under its guarantee how many months should they guarantee their batteries?

116. Use the properties of the normal distribution to answer the questions relating to the normal distribution shown in the following figure.

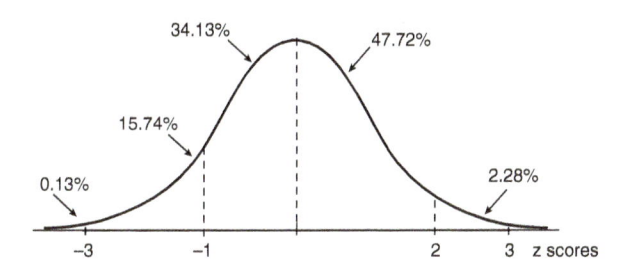

  a) What proportion of the distribution is less than a z-score of 2?
  b) What proportion of the distribution is between a z-score of –1 and 0?
  c) What proportion of the distribution is greater than a z-score of 1?
  d) What proportion of the distribution is less than a z-score of –1?
  e) What proportion of the distribution is between a z-score of 1 and 2?
  f) What proportion of the distribution is between a z-score of –1 and –2?
  g) What proportion of the distribution is between a z-score of –2 and –3?

117. The binomial distribution can be approximated by a normal distribution if either np or n(1 – p) is greater than 5. (T/F)

118. Assuming that the probability that someone will respond positively to a survey question is 0.40 what is the chance that more than half of the 150 people surveyed will respond positively?
  a) 0.40   b) 0.20   c) 0.50   d) 0.0049

119. The probability that a student that has a credit card pays their balance off every month is 55%. What is the probability that for the next 600 students that you encounter that have credit cards, more than 350 will pay their balance off next month?

120. An airline company knows that 90% of the people making flight reservations for a certain flight will show up for the flight. What is the probability that out of the next 10,000 people making flight reservations:
  a) 9075 or less will show up for the flight?
  b) At least 9030 of them will show up for the flight?
  c) Exactly 1000 will not show up for the flight?

121. From past experience a restaurant owner knows that 80% of the drinks ordered will be alcoholic. For the next 400 drinks ordered:
  a) How many would be expected to be alcoholic?
  b) What is the probability that at least 300 drinks will be alcoholic?
On a *good* weekend the owner sells at least $990 worth of alcoholic drinks at $3.00 per drink.
  c) If 400 drinks are ordered next weekend what is the probability that next weekend is good?
  d) The owner estimates that 25,600 drinks will be sold during the next year. He knows that for every 30 alcoholic drinks sold he uses one bottle of liquor. If each bottle of liquor costs him $9.00, how much money should he budget to be 95% confident his liquor costs are covered?

122. A biologist is studying the weights of female Polar Bears in the Arctic and finds that the mean weight is 179 kg with a standard deviation of 63 kg. Assuming the distribution of weights is approximately normally distributed find:
  a) The probability that a randomly selected female bear from this population will weigh greater than 200 kg.
  b) The probability that a randomly selected female bear from this population will weigh between 130 & 180 kg.
  c) The number of female bears that will weigh between 130 & 180 kg given that the size of the bear population in a certain area is 10,000.
  d) The percentile rank of a female bear that weighs 195 kg.

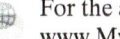

For the answer to this question, please visit www.MyStatLab.com.

CHAPTER 8

Sampling and Sampling Distributions

Contents

8.1 The Sampling Distribution of the Mean
8.2 The Mean and Standard Deviation of the
   Sampling Distribution of the Mean
   Mean of the Sampling Distribution of the Mean
   Standard Deviation of the Sampling Distribution
      of the Mean
   Interpretation of the Standard Error of the Mean
8.3 The Finite Correction Factor
8.4 The Shape of the Sampling Distribution of the Mean
   Sampling from a Normal Population
   Sampling from a Non-Normal Population
   The Central Limit Theorem
8.5 Calculating Probabilities Using the Sampling
      Distribution of the Mean
8.6 The Effect of the Sample Size on the Standard
      Error of the Mean
8.7 The Sampling Distribution of the Proportion
   Sampling Error of the Proportion
   Interpretation of the Standard Error of the
      Proportion
   Shape of the Sampling Distribution of the
      Proportion
   Calculating Probabilities Using the Sampling
      Distribution of the Proportion

8.1 THE SAMPLING DISTRIBUTION OF THE MEAN

As previously mentioned, populations are generally either quite large, have an unlimited number of data values, (that is, infinite) or partly unobtainable. Since a population can rarely be studied completely, a sample randomly selected from the population serves as a convenient and economical procedure to estimate the characteristics of a population. One population characteristic we are interested in estimating is the mean of a population. For example,

- an internist uses samples to estimate the average cholesterol level in a patient's body,
- an economist uses samples to estimate the average interest rate for short-term bonds,
- a tire manufacturer uses samples to estimate the average wear of their new radial tire,
- a light-bulb manufacturer uses samples to estimate the average life of each type of bulb they produce,
- a pool owner uses samples to determine the average pH reading of her pool,
- a pharmaceutical company uses samples to estimate the average time its new pain relieving drug will be effective,
- a college student uses samples to estimate the average textbook cost for all the college students.

In each of these situations, sample information is being used to estimate the mean of a population. When the population mean is being estimated using sample data, the mean of the sample will probably not be equal to the mean of the population. In fact, if a few samples were randomly selected from the same population, *it is very likely that none of these sample means would be exactly equal to the mean of the population.*

Consider, for example, a soft-drink company who manufactures a 2 liter bottle of their new diet drink. Due to variations in the manufacturing process, some bottles will probably contain slightly less than 2 liters while others will contain slightly more. Because of the imperfections in the dispensing process, we really have no idea of what the "true" mean amount of soda, μ, is for the population of soda bottles. The only possible way we could determine the true population mean, μ, is to remove and open every bottle from the production line and determine the amount of soda within each bottle. Practically speaking, however, it is virtually impossible to do this. In such instances, **inferential statistics** is used to estimate the mean of the population. A sample of the bottles are randomly selected from the production line and the mean amount of soda for this sample is computed. Assume a random sample of 15 bottles yields the following amounts of soda.

Amount of soda, in liters, for the random sample of 15 bottles

1.93	1.95	1.94	1.96	2.03
1.96	2.02	1.93	2.04	2.01
2.01	2.01	1.96	1.93	2.02

The sample mean, $\bar{x}$, for these 15 bottles is:

$$\bar{x} = \frac{1.93 + 1.95 + 1.94 + \ldots + 2.02}{15} = 1.98 \text{ liters}$$

It is reasonable to use this sample information to estimate the mean amount of soda for all the bottles manufactured by this company. Thus, the **sample mean of $\bar{x}$ = 1.98 liters** would serve as an **estimate** of the **true mean amount of soda, μ, of the population**. When we use the sample mean, $\bar{x}$, of 1.98 liters as an estimate of the true population mean, μ, we are using inferential statistics, that is, using sample information, $\bar{x}$, to draw an inference about a population parameter, μ.

However, once we use a sample mean to estimate the mean of a population, it is natural to ask the question: *how good of an estimate do we have?* That is, *how close is $\bar{x}$ to μ?* Is it possible that this sample mean, $\bar{x}$, is identically equal to the true population mean, μ? In essence, we are asking: how likely is it that a sample of 15 bottles would yield a mean equal to the mean amount of soda for all the bottles manufactured?

We would have to say it is probably very unlikely that this sample mean would be equal to the population mean. In fact, if we were to select 5 more samples of 15 bottles from the production line, more likely than not, these 5 sample means would not only differ from the population mean, μ, but would probably also differ from one another. That is, all 5 sample means would most likely have a different value for $\bar{x}$, since the samples would contain different bottles randomly selected from the same production line. Therefore, in answer to our question, it is rather unlikely to expect the sample mean, $\bar{x}$, to be identically equal to the population mean, μ.

At this point, you should begin to understand that although there is only one value for the true mean amount of soda for all the bottles manufactured by this company, there are many different values for the mean amount of soda from the samples selected from this population. Therefore, *the estimates of a true population mean will vary from sample to sample due to the data values randomly selected within each sample, that is, by chance alone.* Different samples selected from the same population will possibly yield different sample results because they contain different data values. For example, by chance alone, a sample may contain mostly bottles that are below the mean amount of soda or mostly bottles containing amounts of soda above the mean amount of soda. Consequently, since the samples of bottles vary from one another, by chance alone, the estimates of the true population mean amount of soda will vary.

Suppose we selected a sample and calculated the sample mean to be 1.98 liters. *Our estimate of the true population mean would be 1.98 liters.* However, suppose you selected another sample from the same production line and computed the sample mean to be 2.03 liters. *Your estimate of the same population mean would be 2.03 liters.* A third person selecting a sample from the same production line would most likely obtain a different sample mean result and consequently, a different estimate of the same population mean. *These variations in the estimates of the population mean from sample to sample are due to chance and are called **sampling errors**.* This sampling error is due to the luck of the sampling selection, that is, the random selection of the sample data values chosen from the population. For example, assume, hypothetically, that the true population mean amount of soda is 2 liters. If we used our first sample mean of 1.98 liters to estimate the population mean, our estimate would result in an error. The difference between our estimate of the true population mean (1.98 liters) and the true population mean (2 liters) represents the sampling error. That is,

$$\text{sampling error} = \text{sample mean} - \text{population mean}$$
$$= 1.98 \text{ liters} - 2 \text{ liters}$$
$$= -0.02 \text{ liters.}$$

On the other hand, if your estimate of this same population mean is 2.03 liters, based on the random selection of another sample from the same population, then this estimate would also result in an error. In this instance, the sampling error is:

the sampling error = sample mean − population mean
= 2.03 liters − 2 liters
= 0.03 liters.

These sampling errors are attributed to the fact that the samples contained a different collection of randomly selected bottles from the same production line. So, by chance alone, these different sampling errors occurred. The sampling errors are not the result of any errors which can occur during the:

	selection process (i.e. not selecting a representative sample of bottles),
or	measuring process (i.e. incorrectly measuring the amount of soda per bottle),
or	recording process (i.e. incorrectly recording the amount of soda per bottle),
or	computation process (i.e. incorrectly calculating the sample mean amount of soda).

These type of errors are called **nonsampling errors**.
   Sampling errors are simply errors due to chance or the luck of the selection process.

DEFINITION 8.1
Sampling error is the difference between the value of a sample statistic, such as the sample mean, $\bar{x}$ and the value of the corresponding population parameter, such as the population mean, μ. Thus, the sampling error for the mean is:

sampling error = sample mean − population mean

assuming the sample is random and there are no nonsample errors.
   In symbols,

sampling error of the mean = $\bar{x} - \mu$.

EXAMPLE 8.1

Suppose the population of seven college students on the Student Government Association (SGA) have the following ages:

$$23, 19, 20, 21, 18, 19, 25$$

a) Compute the population mean age, μ, of all the students on the SGA.
   If a random sample of 3 students was selected from this population having the ages: 19, 21, and 18, then
b) Compute the sample mean age for this sample.
c) Determine the sampling error if this sample were used to estimate the population mean age.
If a random sample of 3 students was selected from this population having the ages: 23, 21, and 25, then
d) Compute the sample mean age for this sample.
e) Determine the sampling error if this sample were used to estimate the population mean age.
f) Interpret the meaning of this sampling error.

Solution

a) The population mean age, μ, is:

$$\mu = \frac{23+19+20+21+18+19+25}{7} \approx 20.71 \text{ years}$$

b) The sample mean, $\bar{x}$, of the ages 19, 21, and 18 is:

$$\bar{x} = \frac{19+21+18}{3} \approx 19.33 \text{ years}$$

c) If a sample mean of 19.33 is used to estimate the population mean of 20.71, then the sampling error would be:

$$\text{sampling error} = \bar{x} - \mu.$$
$$= 19.33 - 20.71$$
$$= -1.38 \text{ years}$$

d) The sample mean, $\bar{x}$, of the ages 23, 21, and 25 is:

$$\bar{x} = \frac{23+21+25}{3} = 23 \text{ years}$$

e) If a sample mean of 23 is used to estimate the population mean of 20.71, then the sampling error would be:

$$\text{sampling error} = \bar{x} - \mu.$$
$$= 23 - 20.71$$
$$= 2.29 \text{ years}$$

f) Sampling error represents the error that would be made in estimating the population mean using the mean of a sample selected from this population. This error is due to chance alone. That is, by chance alone, a sample mean of 23 was selected from this population of ages. Thus, if we were to use a sample mean of 23 to estimate the population mean of 20.71, then, by chance alone, our estimate would have an error of 2.29 years.

Sampling error is inevitable because we are using a chosen number of data values which are randomly selected, by chance, from the population. In practice, we will select only one sample from the population to estimate the mean of the population. However, keep in mind, this chosen sample is only one of many possible samples which could have been selected from the population. The value of the sample mean, $\bar{x}$ is dependent upon the sample of data values selected, so the sample mean does not have a constant value like the population mean. That is, a population can have only one mean. Yet, *depending upon which sample is selected from a population, the mean of a sample can vary from sample to sample as different samples of the same size are randomly chosen from the same population.* Thus, the sample mean is a **random variable** because it is dependent upon the particular data values which are randomly selected from the population. In order for us to determine how closely a sample mean estimates the mean of a population, it is necessary for us to have some prior knowledge about all the possible sample means which could have been selected and how these sample means are distributed. That is, we are interested in examining the distribution of all the possible values of the sample mean, $\bar{x}$, for a population. Such a distribution is called the **distribution of sample means**, or more commonly referred to as the **sampling distribution of the mean**.

> **DEFINITION 8.2**
> The **sampling distribution of the mean** is a probability distribution which lists the sample means from all possible samples of the same sample size selected from the same population along with the probability associated with each sample mean.

Let's illustrate the idea of a sampling distribution of the mean by constructing such a distribution. This illustration is, of course, hypothetical because in practice the population would be much larger and we would be unable to obtain all the values of the population.

Suppose the prices of 7 statistical calculators sold in a college bookstore represent the population. The data values of this population are:

$$\$30, \$32, \$34, \$36, \$38, \$40, \$42.$$

For this population, the mean price and standard deviation of these prices are: $\mu = \$36$ and $\sigma = \$4$. Figure 8.1 illustrates how these population values were computed.

We will construct a sampling distribution of the mean by selecting all possible samples of size 2 that can be selected **without replacement** from the population of prices. The total number of possible samples of size 2 can

Population of calculator prices and its characteristics:
the population mean, μ, and standard deviation, σ

X	$X - \mu$	$(X - \mu)^2$
30	$30 - 36 = -6$	$(-6)^2 = 36$
32	$32 - 36 = -4$	$(-4)^2 = 16$
34	$34 - 36 = -2$	$(-2)^2 = 4$
36	$36 - 36 = 0$	$(0)^2 = 0$
38	$38 - 36 = 2$	$(2)^2 = 4$
40	$40 - 36 = 4$	$(4)^2 = 16$
42	$42 - 36 = 6$	$(6)^2 = 36$

$$\mu = \frac{\Sigma X}{N} = \frac{252}{7} \qquad \sigma = \sqrt{\frac{\Sigma(X - \mu)^2}{N}} = \sqrt{\frac{112}{7}} = \sqrt{16}$$

$$\mu = 36 \qquad\qquad\qquad\qquad\qquad \sigma = 4$$

FIGURE 8.1

be determined using counting rule 4: $_nC_s$. Since there are 7 prices which comprise the population and we are going to select samples of size 2, the total number of possible samples that can be selected without replacement is equal to $_7C_2$. Thus, we have:

$$\text{total number of possible samples of size 2} = {_7C_2} = \frac{7!}{2!5!} = 21.$$

Thus, there are 21 distinct samples of size 2. These 21 samples and their respective means are shown in Figure 8.2.

All Possible Samples of Size 2 and Their Sample Means

Sample	Data Values of Selected Sample	Sample Mean, $\bar{x}$
1	$30, $32	$\bar{x} = \frac{30+32}{2} = \$31$
2	$30, $34	$\bar{x} = \frac{30+34}{2} = \$32$
3	$30, $36	$\bar{x} = \frac{30+36}{2} = \$33$
4	$30, $38	$\bar{x} = \frac{30+38}{2} = \$34$
5	$30, $40	$\bar{x} = \frac{30+40}{2} = \$35$
6	$30, $42	$\bar{x} = \frac{30+42}{2} = \$36$
7	$32, $34	$\bar{x} = \frac{32+34}{2} = \$33$
8	$32, $36	$\bar{x} = \frac{32+36}{2} = \$34$
9	$32, $38	$\bar{x} = \frac{32+38}{2} = \$35$
10	$32, $40	$\bar{x} = \frac{32+40}{2} = \$36$
11	$32, $42	$\bar{x} = \frac{32+42}{2} = \$37$
12	$34, $36	$\bar{x} = \frac{34+36}{2} = \$35$
13	$34, $38	$\bar{x} = \frac{34+38}{2} = \$36$
14	$34, $40	$\bar{x} = \frac{34+40}{2} = \$37$
15	$34, $42	$\bar{x} = \frac{34+42}{2} = \$38$
16	$36, $38	$\bar{x} = \frac{36+38}{2} = \$37$
17	$36, $40	$\bar{x} = \frac{36+40}{2} = \$38$
18	$36, $42	$\bar{x} = \frac{36+42}{2} = \$39$
19	$38, $40	$\bar{x} = \frac{38+40}{2} = \$39$
20	$38, $42	$\bar{x} = \frac{38+42}{2} = \$40$
21	$40, $42	$\bar{x} = \frac{40+42}{2} = \$41$

FIGURE 8.2

450 Chapter 8 *Sampling and Sampling Distributions*

Figure 8.2 illustrates that different samples will yield different values of the sample mean. For example, the sample mean can be as small as $31 or as large as $40, depending upon the sample chosen. If a sample is chosen by simple random sampling, then each sample has an equal chance of being selected. Each of the 21 possible samples shown within Figure 8.2 has a probability of 1/21 of being randomly selected. Since there are two samples out of 21 possible samples that yield a sample mean of $33, then the probability is 2/21 that a sample randomly selected from these 21 samples will yield a sample mean of $33. Similarly, the probability is 2/21 that the sample randomly selected from these 21 samples will yield a sample mean of $34. From the information in Figure 8.2, we can construct a probability distribution of these sample means which lists all the possible values which the sample mean $\bar{x}$ can assume along with the probabilities of each of these sample means occurring.

Figure 8.3 represents a probability distribution of the sample means since it lists all the possible sample means and their corresponding probabilities. Since the data values of this probability distribution are the means of all the possible samples, this probability distribution represents the sampling distribution of the mean for random samples of size 2.

A histogram of the sampling distribution of the mean shown in Figure 8.3 is illustrated in Figure 8.4.

An examination of the sampling distribution in Figure 8.4 reveals information about the variations, due to chance, of the mean of random samples of size 2 selected from the population of prices. For instance, the probability that a sample mean will be within $1 of the population mean of μ = $36 is 9/21. This is because a sample mean of $35, $36 or $37 will fall within $1 of the population mean and their corresponding probabilities add up to 9/21.

Sampling Distribution of the Mean
for random samples of size 2

Value of Sample Mean, $\bar{x}$	Probability of selecting Sample Mean, $\bar{x}$
31	1/21
32	1/21
33	2/21
34	2/21
35	3/21
36	3/21
37	3/21
38	2/21
39	2/21
40	1/21
41	1/21

FIGURE 8.3

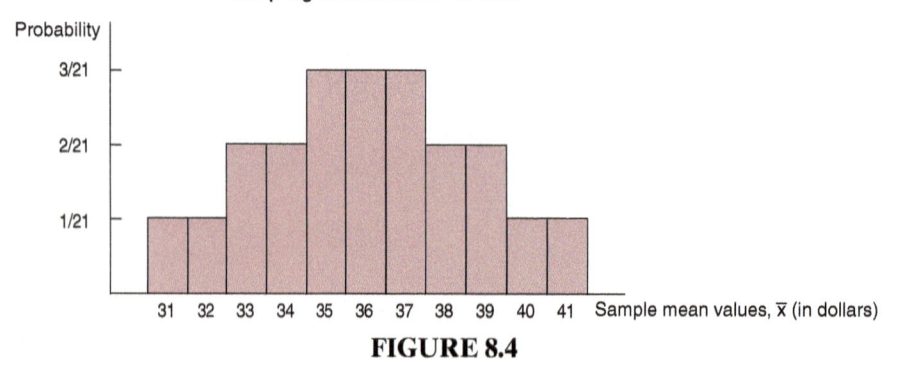

FIGURE 8.4

That is,

$$P\binom{\text{sample mean will be}}{\text{within \$1 of } \mu = \$36} = P\binom{\text{sample mean}}{\text{is \$35}} + P\binom{\text{sample mean}}{\text{is \$36}} + P\binom{\text{sample mean}}{\text{is \$37}}$$

$$= 3/21 + 3/21 + 3/21$$
$$= 9/21.$$

EXAMPLE 8.2

Using the sampling distribution of the mean given in Figure 8.2, what is the probability that a sample randomly selected from the population of calculator prices will result in a sampling error in estimating the population mean of $2 or less?

Solution

Since the population mean is $36, then selecting a sample mean from $34 to $38 would result in a sampling error of $2 or less. That is, the sample mean will deviate from the population mean of $36 by $2 or less. According to the sampling distribution of the mean given in Figure 8.3, the probability of selecting a sample mean from $34 to $38 is equal to the sum of the probabilities associated to these sample means. That is:

$$P(\text{sample mean from \$34 to \$38}) = P(\$34) + P(\$35) + P(\$36) + P(\$37) + P(\$38)$$
$$= 2/21 + 3/21 + 3/21 + 3/21 + 2/21$$
$$= 13/21.$$

Thus, the probability that a sample mean will result in a sampling error of $2 or less is equal to $13/21$.

In describing the main characteristics of a distribution, we typically determine its mean, standard deviation, and the shape of the distribution. Before we discuss the mean and standard deviation of the sampling distribution of the mean we will introduce the notation for these measures.

Notation for the Mean of the Sampling Distribution of the Mean

The **mean of the sampling distribution of the mean** is denoted by $\mu_{\bar{x}}$, read mu sub x bar. Thus,

$\mu_{\bar{x}}$ = mean of all the sample means of the sampling distribution

---

Notation for the Standard Deviation of the Sampling Distribution of the Mean

The **standard deviation of the sampling distribution of the mean** is denoted by $\sigma_{\bar{x}}$, read sigma sub x bar. Thus,

$\sigma_{\bar{x}}$ = standard deviation of all the sample means of the sampling distribution

---

To compute these measures for the sampling distribution of the mean shown in Figure 8.3, we will use the definitions of the population mean and population standard deviation discussed in the probability distribution chapter where the variable X is replaced by $\bar{x}$.

For the sampling distribution of the mean shown in Figure 8.3, the mean of the sampling distribution of the mean is: $\mu_{\bar{x}} = \$36$, and the standard deviation of the sampling distribution of the mean is: $\sigma_{\bar{x}} \approx \$2.58$. Figure 8.5 contains the calculations of $\mu_{\bar{x}}$ and $\sigma_{\bar{x}}$.

Figure 8.6 summarizes the characteristics of the sampling distribution of the mean and the analogous characteristics of the population of prices.

Calculation of Mean and Standard Deviation of the Sampling Distribution of the Mean

$\bar{x}$	$P(\bar{x})$	$\bar{x}P(\bar{x})$	$\bar{x} - \mu_{\bar{x}}$	$(\bar{x} - \mu_{\bar{x}})^2$	$(\bar{x} - \mu_{\bar{x}})^2 P(\bar{x})$
31	1/21	31/21	31 − 36 = − 5	$(-5)^2 = 25$	(25) (1/21) = 25/21
32	1/21	32/21	32 − 36 = − 4	$(-4)^2 = 16$	(16) (1/21) = 16/21
33	2/21	66/21	33 − 36 = − 3	$(-3)^2 = 9$	(9) (2/21) = 18/21
34	2/21	68/21	34 − 36 = − 2	$(-2)^2 = 4$	(4) (2/21) = 8/21
35	3/21	105/21	35 − 36 = − 1	$(-1)^2 = 1$	(1) (3/21) = 3/21
36	3/21	108/21	36 − 36 = 0	$(0)^2 = 0$	(0) (3/21) = 0
37	3/21	111/21	37 − 36 = 1	$(1)^2 = 1$	(1) (3/21) = 3/21
38	2/21	76/21	38 − 36 = 2	$(2)^2 = 4$	(4) (2/21) = 8/21
39	2/21	78/21	39 − 36 = 3	$(3)^2 = 9$	(9) (2/21) = 18/21
40	1/21	40/21	40 − 36 = 4	$(4)^2 = 16$	(16) (1/21) = 16/21
41	1/21	41/21	41 − 36 = 5	$(5)^2 = 25$	(25) (1/21) = 25/21

$$\mu_{\bar{x}} = \Sigma[\bar{x} \cdot P(\bar{x})]$$
$$= 756/21$$
$$\mu_{\bar{x}} = \$36$$

$$\Sigma[(\bar{x} - \mu_{\bar{x}})^2 \cdot P(\bar{x})] = 140/21 \approx 6.67$$
$$\sigma_{\bar{x}} = \sqrt{\Sigma[(\bar{x} - \mu_{\bar{x}})^2 \cdot P(\bar{x})]} \approx \sqrt{6.67}$$
$$\sigma_{\bar{x}} \approx \$2.58$$

FIGURE 8.5

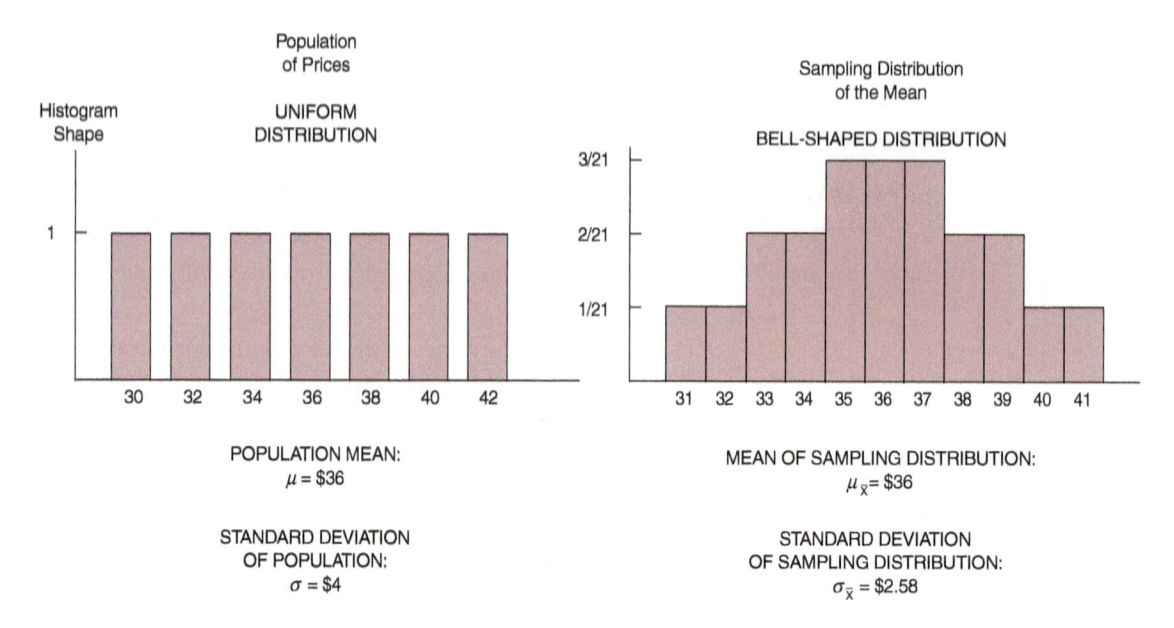

FIGURE 8.6

From Figure 8.6, observe some of the important relationships between the population and the sampling distribution of the mean for this example.

1. The mean of the sampling distribution of the mean, $\mu_{\bar{x}}$, equals the mean of the population, μ.
2. The standard deviation of the sampling distribution of the mean, $\sigma_{\bar{x}}$, is less than the population standard deviation, σ.
3. The shape of the sampling distribution of the mean and the shape of the population are different. The sampling distribution of the mean has a more normal curve shape.

These observations illustrate some very important and fundamental concepts of the sampling distribution of the mean. We will discuss these further in the following sections.

Before we continue, it is important to understand and be able to distinguish between the mean and standard deviation of a population, a sample, and the sampling distribution of the mean. In earlier chapters, we described a population by its mean, μ, and standard deviation, σ. A sample was described by its mean, $\bar{x}$, and standard deviation, s. Now in this chapter, we have discussed the mean and standard deviation of the sampling distribution of the mean. Figure 8.7 contains the symbols for the mean and standard deviation of a population, a sample and the sampling distribution of the mean.

**Notation for Mean and Standard Deviation for a Population,
a Sample and the Sampling Distribution of the Mean**

	Population	Sample	Sampling Distribution of the Mean
Description:	collection of all data values	portion of data values randomly selected from the population	means of all possible samples of size n randomly selected from the population
Mean:	μ	$\bar{x}$	$\mu_{\bar{x}}$
Standard Deviation:	σ	s	$\sigma_{\bar{x}}$

FIGURE 8.7

8.2 THE MEAN AND STANDARD DEVIATION OF THE SAMPLING DISTRIBUTION OF THE MEAN

In the previous section, we examined the concept of the sampling distribution of the mean, using a hypothetical situation of selecting all possible samples of size 2 from a known population of 7 data values. This unrealistic situation was used to demonstrate that when randomly selecting a sample from a population and computing its mean, the value of this sample mean is the mean of only one of many possible samples that might have been selected. In practical applications, we will never know all the data values of a population and it will be impractical or impossible to randomly select all the possible samples of a particular sample size from the population. Consequently, in inferential statistics, we never construct a sampling distribution rather we conceive of the situation of selecting all possible samples of a particular size from a population containing an infinite number of data values. Such a sampling distribution is referred to as a **Theoretical Sampling Distribution**. Since the theoretical sampling distribution is a sampling distribution obtained by theoretical results we must rely on statisticians to provide the theory and formulas needed to help us determine the characteristics of the Theoretical Sampling Distribution of the mean. In this section, we will examine how to determine the mean and standard deviation of the Theoretical Sampling Distribution of the mean by sampling from a list of means.

Mean of the Sampling Distribution of the Mean

As discussed earlier, the mean of all the sample means of the sampling distribution is called the mean of the sampling distribution of the mean and is denoted by $\mu_{\bar{x}}$, read mu sub x bar.

Statistical theory tells us that if we were to take all the possible samples of the same size n from a population, the mean of all the sample means, $\mu_{\bar{x}}$, will be equal to the mean of the population, μ.

> **DEFINITION 8.3**
> **Mean of the Sampling Distribution of the Mean, $\mu_{\bar{x}}$** The mean of the sample means of all possible samples of size n is called the mean of the sampling distribution of the mean, denoted by $\mu_{\bar{x}}$. It is equal to the mean of the population from which the samples were selected. In symbols, this is expressed as:
> $$\mu_{\bar{x}} = \mu$$

Standard Deviation of the Sampling Distribution of the Mean

The standard deviation of the sampling distribution of the mean, however, is not equal to the standard deviation of the population. As we illustrated in Figure 8.6, the standard deviation of the sampling distribution of the mean is smaller than the standard deviation of the population. That is, **the sampling distribution of the mean is less dispersed than the population from which the samples were selected**. According to statistical theory, the standard deviation of the sampling distribution of the mean, denoted by $\sigma_{\bar{x}}$, is equal to the standard deviation of the population, σ, divided by the square root of the sample size, n. The standard deviation of the sampling distribution of the mean is commonly referred to as the **Standard Error of the Mean**.

DEFINITION 8.4

Standard Deviation of the Sampling Distribution of the Mean or Standard Error of the Mean, denoted by $\sigma_{\bar{x}}$ The standard error of the mean is the standard deviation of the sample means of all possible samples of size n of the sampling distribution, denoted by $\sigma_{\bar{x}}$. The standard error of the mean is equal to the standard deviation of the population, σ, divided by the square root of the sample size n. That is,

$$\frac{\text{standard error}}{\text{of the mean}} = \frac{\text{population standard deviation}}{\sqrt{\text{sample size}}}$$

In symbols, this is expressed as:

$$\sigma_{\bar{x}} = \frac{\sigma}{\sqrt{n}}$$

Example 8.3 illustrates the use of these formulas to calculate the mean and standard error of the sampling distribution of the mean.

EXAMPLE 8.3

According to a study of TV viewing habits, the average number of hours an American over the age of 2 watches TV per week is $\mu = 34$ hours with a standard deviation of $\sigma = 6$ hours. If a sample of 64 Americans over the age of 2 is randomly selected from the population, then determine the mean and standard error of the mean of the sampling distribution of the mean.

Solution

Since a random sample of 64 Americans over the age of 2 is being selected from the population, then the sampling distribution of the mean consists of all possible samples of size n = 64. According to statistical theory, the mean of the sampling distribution of the mean, $\mu_{\bar{x}}$, is equal to the population mean, μ. That is, $\mu_{\bar{x}} = \mu$. Since the population mean is $\mu = 34$ hours, then the mean of the sampling distribution of the mean is equal to 34. Thus, $\mu_{\bar{x}} = 34$ hours.

The standard deviation of the sampling distribution of the mean or the standard error of the mean, $\sigma_{\bar{x}}$, is equal to the standard deviation of the population, σ, divided by the square root of the sample size n. That is,

$$\sigma_{\bar{x}} = \frac{\sigma}{\sqrt{n}}$$

Since the population standard deviation is $\sigma = 6$ and the sample size is n = 64, then the standard error of the mean is:

$$\sigma_{\bar{x}} = \frac{\sigma}{\sqrt{n}}$$

$$\sigma_{\bar{x}} = \frac{6}{\sqrt{64}}$$

$$\sigma_{\bar{x}} = \frac{6}{8}$$

$$\sigma_{\bar{x}} = 0.75$$

In Example 8.4, we will discuss what happens to the standard error of the mean when the size of the sample is increased.

EXAMPLE 8.4

The registrar at a large University states that the mean grade point average of all the students is $\mu = 2.95$ with a population standard deviation of $\sigma = 0.20$.
a) Determine the mean and standard error of the sampling distribution if the sampling distribution of the mean consists of all possible sample means from samples of size 25.
b) Determine the mean and standard error of the sampling distribution if the sampling distribution of the mean consists of all possible sample means from samples of size 100.
c) What effect did increasing the sample size have on the mean and standard error of the sampling distribution?

d) In which sampling distribution of the mean, the one with sample size 25 or the one with sample size 100, would you have a better chance of selecting a sample mean which is closer to the population mean grade point average? Explain.

Solution

a) According to statistical theory, $\mu_{\bar{x}} = \mu$. Since the population mean is $\mu = 2.95$, then the mean of the sampling distribution of the mean is equal to 2.95. Thus, $\mu_{\bar{x}} = 2.95$. The formula for the standard error of the mean, is,

$$\sigma_{\bar{x}} = \frac{\sigma}{\sqrt{n}}.$$

Since the population standard deviation is $\sigma = 0.20$ and the sample size is n = 25, then the standard error is:

$$\sigma_{\bar{x}} = \frac{\sigma}{\sqrt{n}}$$

$$\sigma_{\bar{x}} = \frac{0.20}{\sqrt{25}}$$

$$\sigma_{\bar{x}} = \frac{0.20}{5}$$

$$\sigma_{\bar{x}} = 0.04.$$

b) The mean of the sampling distribution of the mean, $\mu_{\bar{x}}$, is equal to the population mean, μ, regardless of the sample size; therefore, the mean of the sampling distribution of the mean is still equal to 2.95. Therefore, $\mu_{\bar{x}} = 2.95$. The standard error of the mean, $\sigma_{\bar{x}}$, however, is affected by the sample size since $\sigma_{\bar{x}}$ is equal to the standard deviation of the population, σ, divided by the square root of the sample size n. That is,

$$\sigma_{\bar{x}} = \frac{\sigma}{\sqrt{n}}.$$

The population standard deviation is still $\sigma = 0.20$ but the sample size has increased to n = 100. Thus, the standard error of the mean is:

$$\sigma_{\bar{x}} = \frac{\sigma}{\sqrt{n}}$$

$$\sigma_{\bar{x}} = \frac{0.20}{\sqrt{100}}$$

$$\sigma_{\bar{x}} = \frac{0.20}{10}$$

$$\sigma_{\bar{x}} = 0.02.$$

c) The results of parts (a) and (b) are shown in the following table.

Sample Size n	Mean of the Sampling Distribution $\mu_{\bar{x}}$	Standard Error of the Sampling Distribution $\sigma_{\bar{x}}$
n = 25	$\mu_{\bar{x}} = 2.95$	$\sigma_{\bar{x}} = 0.04$
n = 100	$\mu_{\bar{x}} = 2.95$	$\sigma_{\bar{x}} = 0.02$

Observe from this table, the mean of the sampling distribution was not affected as the sample size increased from n = 25 to n = 100. This will always be true. That is, *the mean of the sampling distribution is not affected by the sample size.* However, the standard error of the sampling distribution will be affected as the sample size increases. Examining the table, we observe as the sample size increased from 25 to 100, the standard error of the sampling distribution decreased from 0.04 to 0.02. *In general, as the sample size increases, the standard error of the sampling distribution will decrease.*

d) Since the sampling distribution of the mean for all possible samples of size n = 100 has a smaller standard deviation, $\sigma_{\bar{x}} = 0.02$, then *you have a better chance of selecting a sample mean closer to the population mean grade point average from this sampling distribution.* This is true because a smaller standard deviation indicates that samples of size n = 100 will have less dispersion about the mean than samples of size n = 25. In the sampling distribution containing samples of size n = 100, the sample means will be relatively closer to the mean of the population. This will provide us with a better chance of selecting a sample mean that is closer to the population mean grade point average.

Interpretation of the Standard Error of the Mean

The standard deviation of the sampling distribution of the mean is referred to as the Standard Error of the Mean because it is a measure of how much a sample mean is likely to deviate from the population mean, that is, it is a measure of the average sampling error.

Recall when a sample is randomly selected from the population and its mean, $\bar{x}$, is used as an estimate of the population mean, μ, the difference between the sample mean, $\bar{x}$, and the population mean, μ, represents the sampling error. For any distribution, the standard deviation of the distribution provides an approximate measure of the average deviation of the data values from the mean. Thus, in a sampling distribution of the mean, the standard deviation of the sampling distribution of the mean, $\sigma_{\bar{x}}$, will roughly indicate the average deviation of the sample means, $\bar{x}$, from the population mean, μ, or average sampling error. Consequently, the standard deviation of the sampling distribution of the mean, $\sigma_{\bar{x}}$, is a measure of the average sampling error and is referred to as the Standard Error of the Mean.

If the *Standard Error of the Mean*, $\sigma_{\bar{x}}$, is a small number, then the sampling distribution *of the mean has relatively little dispersion* and the *sample means will be relatively close to the population mean.* In such instances, a sample mean which is randomly selected from the sampling distribution will have a relatively large probability of not deviating considerably from the population mean. Consequently, **there is a good chance of a small sampling error in estimating the population mean when $\sigma_{\bar{x}}$ is a small number**.

On the other hand, if the *Standard Error of the Mean*, $\sigma_{\bar{x}}$, is a large number, then *the sampling distribution of the mean has a relatively large dispersion* and the *sample means will be relatively far from the population mean.* In such instances, a sample mean that is randomly selected from the sampling distribution will have a relatively large probability of deviating considerably from the population mean. Consequently, **there is a good chance of a large sampling error in estimating the population mean when $\sigma_{\bar{x}}$ is a large number**.

EXAMPLE 8.5

At a large community college the mean IQ score of the student body is 107 with a standard deviation of 20. Suppose a random sample of size n is selected from the student body, compute the standard error of the mean, $\sigma_{\bar{x}}$, for samples of size:
a) n = 4
b) n = 25
c) n = 400
d) What happens to the standard error of the mean as the sample size increases?
e) Answer the following question based on the results of parts (a) to (c).
   If you were going to select a random sample from this population of IQ scores, which sample size (n = 4, n = 25 or n = 400) would give you a better chance of generating a sample mean closer to the population mean IQ score? Explain.

Solution

The standard error of the mean is determined by the formula:

$$\sigma_{\bar{x}} = \frac{\sigma}{\sqrt{n}}$$

a) Since the standard deviation of the population, σ, is 20 and the sample size, n, is 4, then the standard error of the mean is:

$$\sigma_{\bar{x}} = \frac{\sigma}{\sqrt{n}}$$

$$\sigma_{\bar{x}} = \frac{20}{\sqrt{4}}$$

$$\sigma_{\bar{x}} = 10$$

b) For $\sigma = 20$ and n = 25, the standard error of the mean is:

$$\sigma_{\bar{x}} = \frac{\sigma}{\sqrt{n}}$$

$$\sigma_{\bar{x}} = \frac{20}{\sqrt{25}}$$

$$\sigma_{\bar{x}} = 4$$

c) For a sample size of n = 400 and a population standard deviation, σ, of 20 then the standard error of the mean is:

$$\sigma_{\bar{x}} = \frac{\sigma}{\sqrt{n}}$$

$$\sigma_{\bar{x}} = \frac{20}{\sqrt{400}}$$

$$\sigma_{\bar{x}} = 1$$

d) As the sample size *increases* from n = 4 to n = 400, the standard error of the mean, $\sigma_{\bar{x}}$, *decreases* from $\sigma_{\bar{x}} = 10$ to $\sigma_{\bar{x}} = 1$.

e) A sample size of n = 400 would give you a better chance of generating a sample mean closer to the population mean IQ score because the standard error of the mean has the smaller value of $\sigma_{\bar{x}} = 1$. As the standard error of the mean, $\sigma_{\bar{x}}$, *decreases*, then the sample means will become *closer* to the population mean, and thus *increase* the chances of selecting a sample with a mean IQ score being closer to the population mean IQ score.

8.3 THE FINITE CORRECTION FACTOR

In our previous discussion of the standard error of the mean or the standard deviation of the sampling distribution *of the mean*, $\sigma_{\bar{x}}$, we used the formula:

$$\sigma_{\bar{x}} = \frac{\sigma}{\sqrt{n}}$$

to compute its value. *This formula holds true for when the population contains an infinite number of data values,* (that is, it contains an unlimited number of data values) *or when sampling from a finite population with replacement* (that is, after a data value is selected from the population, it is placed back into the population before the next data value is chosen).

In many practical applications, the populations contain a finite number of data values and the sampling is done without replacement (that is, each data value sampled from the population is not replaced). *If sampling without replacement is being done from a finite population, then the formula to compute the standard deviation of the sampling distribution of the mean must be modified.* This modification is referred to as the *finite correction factor*. The following formula represents the exact formula for the standard deviation of the sampling distribution of the mean when sampling without replacement from a finite population, N.

Formula for the Standard Deviation of the Sampling Distribution of the Mean for a Finite Population

For a finite population, the formula for the standard deviation of the sampling distribution of the mean, $\sigma_{\bar{x}}$, is:

$$\sigma_{\bar{x}} = \frac{\sigma}{\sqrt{n}} \cdot \sqrt{\frac{N-n}{N-1}}$$

Where:

σ is the standard deviation of the population
n is the sample size
N is the population size

the term

$$\sqrt{\frac{N-n}{N-1}}$$

is called the finite correction factor

EXAMPLE 8.6

In a large lecture college psychology class containing 250 students, the mean test grade on the first exam was $\mu = 75$ with a population standard deviation $\sigma = 12$. If the college instructor samples n students from this population without replacement and finds their mean test grade, then determine the standard deviation of the sampling distribution of the mean for the following samples of size n.

a) n = 9
b) n = 49
c) n = 100

Solution

For each sample size the sampling is being done without replacement from a finite population; therefore, we need to apply the formula for the standard deviation of the sampling distribution of the mean for a finite population:

$$\sigma_{\bar{x}} = \frac{\sigma}{\sqrt{n}} \cdot \sqrt{\frac{N-n}{N-1}}$$

a) For a sample size of n = 9 students, a population standard deviation of $\sigma = 12$, and a population size of N = 250, the standard deviation of the sampling distribution of the mean is:

$$\sigma_{\bar{x}} = \frac{\sigma}{\sqrt{n}} \cdot \sqrt{\frac{N-n}{N-1}}$$

$$\sigma_{\bar{x}} = \frac{12}{\sqrt{9}} \cdot \sqrt{\frac{250-9}{250-1}}$$

$$\sigma_{\bar{x}} = \frac{12}{3} \cdot \sqrt{\frac{241}{249}}$$

$$\sigma_{\bar{x}} \approx 4 \cdot \sqrt{0.968}$$

$$\sigma_{\bar{x}} \approx 4 \cdot (0.984)$$

$$\sigma_{\bar{x}} \approx 3.94$$

Thus, for a sample size of n = 9, the standard deviation of the sampling distribution of the mean is approximately 3.94.

b) For a sample size of n = 49 students, a population standard deviation of $\sigma = 12$, and a population size of N = 250, the standard deviation of the sampling distribution of the mean is:

$$\sigma_{\bar{x}} = \frac{\sigma}{\sqrt{n}} \cdot \sqrt{\frac{N-n}{N-1}}$$

$$\sigma_{\bar{x}} = \frac{12}{\sqrt{49}} \cdot \sqrt{\frac{250-49}{250-1}}$$

$$\sigma_{\bar{x}} = \frac{12}{7} \cdot \sqrt{\frac{201}{249}}$$

$$\sigma_{\bar{x}} \approx (1.71) \cdot \sqrt{0.807}$$

$$\sigma_{\bar{x}} \approx (1.71) \cdot (0.898)$$

$$\sigma_{\bar{x}} \approx 1.54$$

Therefore, for a sample size of n = 25, the standard deviation of the sampling distribution of the mean is approximately 1.54.

c) For a sample size of n = 100 students, a population standard deviation of $\sigma = 12$, and a population size of N = 250, the standard deviation of the sampling distribution of the mean for a finite population without replacement is

$$\sigma_{\bar{x}} = \frac{\sigma}{\sqrt{n}} \cdot \sqrt{\frac{N-n}{N-1}}$$

$$\sigma_{\bar{x}} = \frac{12}{\sqrt{100}} \cdot \sqrt{\frac{250-100}{250-1}}$$

$$\sigma_{\bar{x}} = \frac{12}{10} \cdot \sqrt{\frac{150}{249}}$$

$$\sigma_{\bar{x}} \approx (1.2) \cdot \sqrt{0.602}$$

$$\sigma_{\bar{x}} \approx (1.2) \cdot (0.776)$$

$$\sigma_{\bar{x}} \approx 0.93$$

For a sample size of n = 100 students, the standard deviation of the sampling distribution of the mean is approximately 0.93.

In Example 8.6, when the sample size was n = 9, the finite correction factor

$$\sqrt{\frac{N-n}{N-1}}$$

was approximately 0.984, which is very close to 1. As the sample size increased to n = 100, the finite correction factor decreased to a value of approximately 0.776. Thus, as shown from this example, when the sample size, is small, n = 9, relative to the population size, N = 250, the finite correction factor is very close to 1. In such instances, the standard deviation of the sampling distribution of the mean, $\sigma_{\bar{x}}$, is approximately equal to the formula:

$$\frac{\sigma}{\sqrt{n}}$$

since

$$\sqrt{\frac{N-n}{N-1}} \approx 1 \ .$$

In practice, if the sample size, n, is small *relative* to the population size, N, then the finite correction factor,

$$\sqrt{\frac{N-n}{N-1}} \ ,$$

is approximately equal to 1, and thus can be ignored. The general rule used to determine whether the sample is small in relation to the population is:

the sample size, n, is less than 5% of the population size, N.

That is,

$$\frac{\text{sample size}}{\text{population size}} \text{ is less than 5\%.}$$

In symbols, this condition is written:

$$\frac{n}{N} < 5\%.$$

Consequently, we can establish the following general rule for the standard deviation of the sampling distribution when sampling from a finite population.

General Rule for Computing the Standard Deviation of the Sampling Distribution of the Mean when Sampling from a Finite Population

When sampling without replacement from a finite population, and the sample size, n, is less than 5% of the population size, N, then we can use the formula:

$$\sigma_{\bar{x}} = \frac{\sigma}{\sqrt{n}}$$

to determine the standard deviation of the sampling distribution of the mean.

In most applications, *the sample size, n, is relatively small with respect to the population size, N. Consequently, we will generally assume this is true and use the formula*

$$\sigma_{\bar{x}} = \frac{\sigma}{\sqrt{n}}$$

8.4 THE SHAPE OF THE SAMPLING DISTRIBUTION OF THE MEAN

As previously discussed, the sampling distribution of the mean is a distribution of the values of all the possible sample means. Suppose a consumer group selects a random sample of 100 bulbs to test a leading manufacturer's claim that their new bulb More Lite will last an average of 5400 hours. Although, the consumer group will only select one sample of bulbs from the population of bulbs and calculate the mean of the sample, you should realize that this selected sample occurred by chance. This means that the chosen light bulbs were randomly selected, and other samples of the same size, n = 100 bulbs, could just as easily been chosen, each one of which would have yielded its own sample mean. Now, let's conceive of continuing to select random samples of the same size, n = 100 bulbs, from the population of More Lite bulbs manufactured by this company, and computing the mean life for each sample. This is continued until we have completely exhausted the selection of all the possible samples of size n = 100 bulbs from the population. The distribution of all these possible sample means from samples of size n = 100 would represent the sampling distribution of the mean. Of course, in practice, it would be impossible or impractical to construct such a sampling distribution of the mean. In this sense, the sampling distribution of the mean is a theoretical distribution. In general, when we refer to the sampling distribution of the mean we are actually referring to a theoretical probability distribution which relates the value of each possible sample to the probability that such a sample mean will occur.

In practice, since only one sample will be selected from the population, it is impossible to construct the actual sampling distribution of the mean for a population containing an infinite number of data values. Thus, we must rely on statisticians to provide us with the necessary theorems to enable us to construct a theoretical sampling distribution of the mean. We have already discussed the formulas to help us determine the mean, $\mu_{\bar{x}}$, and standard deviation, $\sigma_{\bar{x}}$, of the sampling distribution of the mean if we know the population mean and standard deviation. To completely describe the essential characteristics of the sampling distribution of the mean, we need to determine its shape. We will now present and discuss two important theorems in inferential statistics that will enable us to determine the shape of the sampling distribution of the mean. The first theorem will discuss the shape of the sampling distribution of the mean when sampling from a normal population. The second theorem, called the **Central Limit Theorem**, will provide us with information about the shape of the sampling distribution of the mean when sampling from a population with an unknown shape.

Sampling from a Normal Population

If all possible samples of size n are selected from a population that is normally distributed, then the shape of the sampling distribution of the mean will also be normally distributed regardless of the sample size. This is stated in Theorem 8.1.

THEOREM 8.1   The Shape of the Sampling Distribution when Sampling from a Normal Population

If the population being sampled is a normal distribution, then the sampling distribution of the mean is a normal distribution regardless of the sample size, n.

The sampling distribution of the mean will be a normal distribution even for samples of size n = 2 provided the population from which the samples are drawn is a normal distribution. This is illustrated in Figure 8.8.

Figure 8.8 shows the population from which the samples are selected is a normal distribution. The resulting shapes of the sampling distribution of the mean for different sample sizes ranging from n = 2 to n = 100 are also normal as illustrated in Figures 8.8 (a) to (e). Observe, in Figure 8.8, as the sample size increases from n = 2 to n = 100, the dispersion of the sample means within the sampling distributions decreases.

Shape of the Sampling Distribution of the Mean for Different Sample Sizes, n, Selected from a Normal Population

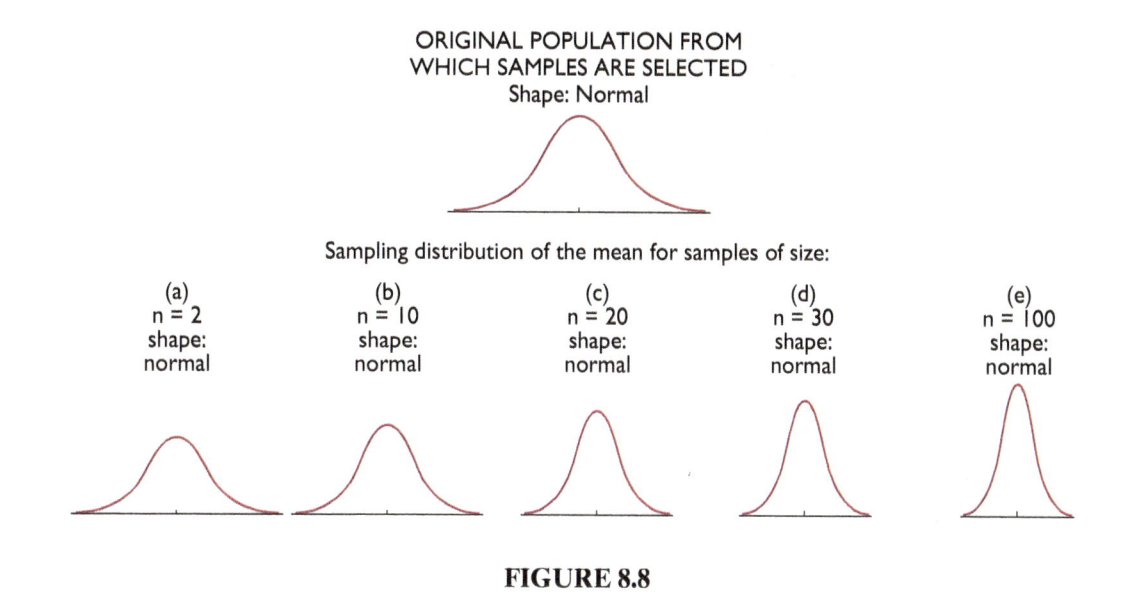

ORIGINAL POPULATION FROM
WHICH SAMPLES ARE SELECTED
Shape: Normal

Sampling distribution of the mean for samples of size:

(a) n = 2 shape: normal
(b) n = 10 shape: normal
(c) n = 20 shape: normal
(d) n = 30 shape: normal
(e) n = 100 shape: normal

FIGURE 8.8

Using the formulas to compute the mean and standard deviation of the sampling distribution discussed in the previous sections, we can now completely describe the sampling distribution of the mean, (that is, determine the mean, standard error and the shape of the sampling distribution), when sampling from a normal population whose mean is μ and a standard deviation is σ. Let's now summarize these characteristics of the sampling distribution of the mean for a normally distributed population.

Characteristics of the Sampling Distribution of the Mean When Sampling from a Normal Population Whose Mean Is μ and Standard Deviation Is σ

If all possible samples of size n are selected from a normal population, then the sampling distribution of the mean has the following three characteristics.
1) The sampling distribution of the mean is a normal distribution, regardless of the sample size, n.
2) The mean of the sampling distribution of the mean, $\mu_{\bar{x}}$, is equal to the mean of the population, μ. Symbolically, this is written:

$$\mu_{\bar{x}} = \mu.$$

3) The standard error of the sampling distribution of the mean, $\sigma_{\bar{x}}$, is equal to the standard deviation of the population, σ, divided by the square root of the sample size, n. Symbolically, this is written:

$$\sigma_{\bar{x}} = \frac{\sigma}{\sqrt{n}}.$$

Figure 8.9 illustrates the conceptual idea of the sampling distribution of the mean, its characteristics, and the relationship to the characteristics of a normal population.

Example 8.9 shows how to determine the characteristics of the sampling distribution of the mean when sampling from a normal population.

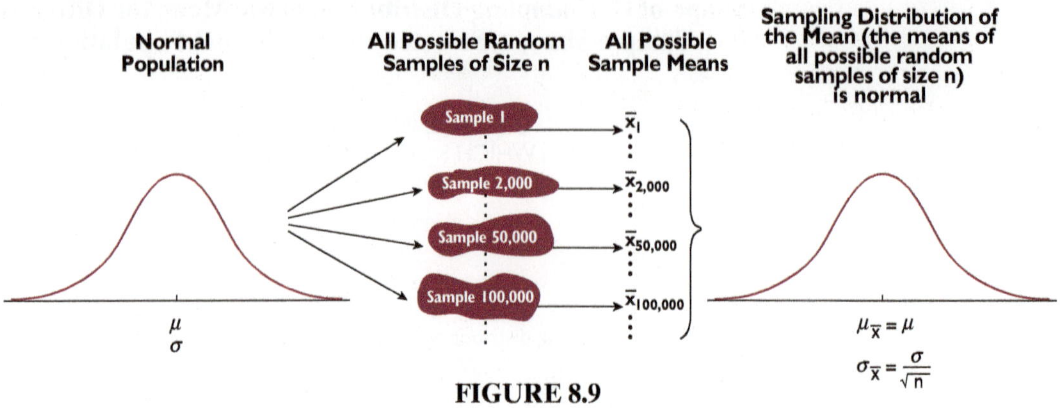

FIGURE 8.9

EXAMPLE 8.7

At a large New England college, the grade point average (GPA) of all attending students is normally distributed with a mean of $\mu = 2.95$ and a population standard deviation of $\sigma = 0.32$. A sample is randomly selected from this population and its sample mean, $\bar{x}$, is calculated. Determine the mean, $\mu_{\bar{x}}$, the standard error, $\sigma_{\bar{x}}$ and identify the shape of the sampling distribution of the mean for samples of size:
a) n = 9
b) n = 49
c) n = 100

Solution

In each case the sampling distribution of the mean will be a normal distribution since the samples were randomly selected from a normal population. The mean of the sampling distribution of the mean is determined by the formula: $\mu_{\bar{x}} = \mu$. Since the population mean is $\mu = 2.95$, the mean of the sampling distribution of the mean is also equal to 2.95. That is, $\mu_{\bar{x}} = 2.95$. These results hold for each sample size. However the standard error of the sampling distribution of the mean, $\sigma_{\bar{x}}$, depends on the sample size and will be calculated for each.

a) Since the sample size is n = 9, and the population standard deviation is $\sigma = 0.32$, then the standard error of the sampling distribution of the mean is determined by the formula:

$$\sigma_{\bar{x}} = \frac{\sigma}{\sqrt{n}}.$$

Substituting, n = 9 and $\sigma = 0.32$ into this formula, we have:

$$\sigma_{\bar{x}} = \frac{0.32}{\sqrt{9}}$$

$$\sigma_{\bar{x}} = \frac{0.32}{3}$$

$$\sigma_{\bar{x}} \approx 0.11.$$

Therefore, the sampling distribution of the mean will be a normal distribution with a mean of $\mu_{\bar{x}} = 2.95$ and a standard error of $\sigma_{\bar{x}} \approx 0.11$.

b) For a sample size of n = 49, and a population standard deviation of $\sigma = 0.32$, then the standard error of the sampling distribution of the mean is calculated by the formula:

$$\sigma_{\bar{x}} = \frac{\sigma}{\sqrt{n}}.$$

Substituting, n = 49 and $\sigma = 0.32$ into this formula, we obtain:

$$\sigma_{\bar{x}} = \frac{0.32}{\sqrt{49}}$$

$$\sigma_{\bar{x}} = \frac{0.32}{7}$$

$$\sigma_{\bar{x}} \approx 0.05$$

Thus, the sampling distribution of the mean will be a normal distribution with a mean of $\mu_{\bar{x}} = 2.95$ and a standard error of $\sigma_{\bar{x}} \approx 0.05$.

c) For a sample size of n = 100, and a population standard deviation of $\sigma = 0.32$, then the standard error of the sampling distribution of the mean is obtained using the formula:

$$\sigma_{\bar{x}} = \frac{\sigma}{\sqrt{n}}.$$

Substituting, n = 100 and $\sigma = 0.32$ into this formula, we obtain:

$$\sigma_{\bar{x}} = \frac{0.32}{\sqrt{100}}$$

$$\sigma_{\bar{x}} = \frac{0.32}{10}$$

$$\sigma_{\bar{x}} \approx 0.03$$

Consequently, the sampling distribution of the mean will be normally distributed with a mean of $\mu_{\bar{x}} = 2.95$ and a standard error of $\sigma_{\bar{x}} \approx 0.03$.

Sampling from a Non-Normal Population

Theorem 8.1 assumes that when sampling from a normal population, the sampling distribution of the mean will also be a normal distribution. However, there are many statistical applications where the population will not be normally distributed. In fact, there are applications where the shape of the population will be unknown. In such instances, how can we determine the shape of the sampling distribution of the mean, if the population is not a normal distribution or has an unknown shape? In these situations, we rely on the most important theorem of inferential statistics called the **Central Limit Theorem**.

THEOREM 8.2   The Central Limit Theorem

For any population, the sampling distribution of the mean approaches a normal distribution as the sample size n becomes large. This is true regardless of the shape of the population being randomly sampled.

The Central Limit Theorem is a very important theorem in inferential statistics because it tells us that the sampling distribution of the mean will always approach the same distribution shape, a normal bell-shaped distribution, as the sample size n becomes larger, regardless of the population shape. Thus, according to the Central Limit Theorem, the sampling distribution of the mean can be approximated by a normal distribution as long as the sample size is large enough regardless of the shape of the population distribution from which the sample is randomly selected. In essence, since the **population can have any shape, it is not even necessary to know the shape of the population provided the sample size is large enough**. Since in most statistical applications the shape of the population is usually not known, this emphasizes the importance of the Central Limit Theorem in the field of statistical inference.

How large the sample size n should be to justify the use of the normal curve as an approximation of the sampling distribution of the mean depends on the shape of the population being sampled. The closer the shape of the sampled population is to a normal or bell-shaped distribution, the smaller the sample size n required for the sampling distribution of the mean to be closely approximated by a normal distribution. **Extremely skewed populations will require a relatively larger sample size n** for the sampling distribution of the mean to achieve a close approximation of the normal distribution. That is, more data values need to be randomly selected for the sample when the population shape is far from a normal population.

Figure 8.10 illustrates the use of the Central Limit Theorem for populations of different shapes and different sample sizes.

In Figure 8.10, the sampling distribution of the mean is shown for samples of size n = 2, n = 5, and n = 30 from four different population shapes: uniform, bimodal, extremely skewed to the right and normal. Notice,

Application of Central Limit Theorem: Sampling Distribution of the Mean for Various Population Shapes and Different Sample Sizes

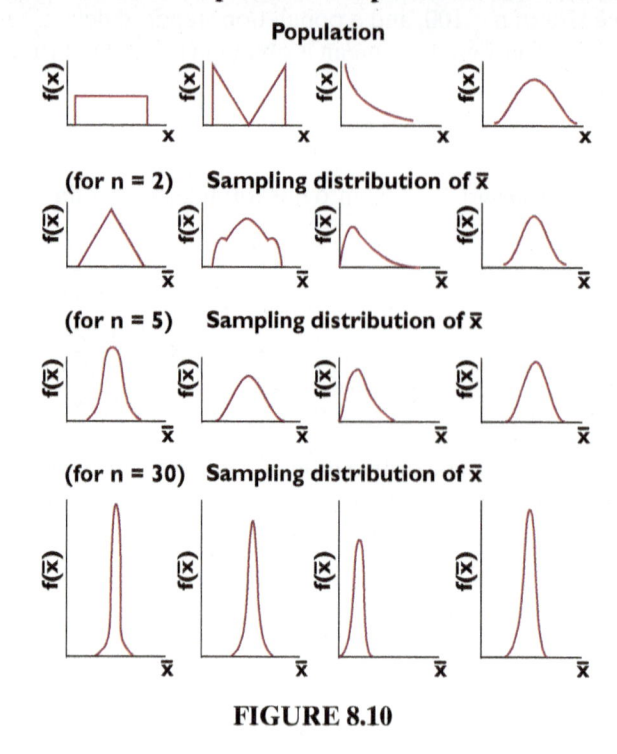

FIGURE 8.10

for samples of size n = 5, the sampling distribution of the mean for all four populations is beginning to approach a normal distribution. For a sample size of n = 30, regardless of the shape of the population, the sampling distributions of the mean are close to a normal distribution. We cannot say precisely how large a sample size, n, is needed to apply the Central Limit Theorem since, in practice, we usually do not know the shape of the population being sampled. Consequently, there is a need for a useful general rule of thumb to follow when a sample is large enough to apply the Central Limit Theorem. The general rule widely accepted and used by statisticians is that the sampling distribution of the mean can be approximated by a normal distribution provided the sample size n is greater than 30 regardless of the shape of the population.

General Rule for Applying the Central Limit Theorem: The n Greater than 30 Rule

For most applications, a sample size n greater than 30 is considered large enough to apply the Central Limit Theorem. Thus, the sampling distribution of the mean can be reasonably approximated by a normal distribution whenever the sample size n is greater than 30.

Although, we will use this general rule as our definition of a large sample throughout this text, remember, the Central Limit Theorem states that the larger the sample size, the better the normal approximation of the sampling distribution of the mean. Consequently, in practical applications, this **n greater than 30 rule** for a large sample should be applied with caution. That is, a larger sample size may be required for a good approximation when the population is far from a normal bell-shaped distribution.

Using the result of the Central Limit Theorem, and the formulas for the mean and standard error of the sampling distribution, we can state the characteristics of the sampling distribution of the mean when sampling from a non-normal population in Theorem 8.3.

Figure 8.11 illustrates the concept of the sampling distribution of the mean, its characteristics, and the relationship to the characteristics of a normal population.

To illustrate Theorem 8.3, consider selecting all possible random samples of *size n* = 100 from each of the populations shown in Figure 8.12. Thus, each of the resulting sampling distributions of the mean will be approximately normal with their respective mean and standard error shown in Figure 8.12.

THEOREM 8.3 Characteristics of the Sampling Distribution of the Mean when Sampling from a Non-Normal Population

If the following three conditions are satisfied:
- given any infinite population with mean, μ, and standard deviation, σ, and
- all possible random samples of size n are selected from the population to form a sampling distribution of the mean, and
- the sample size, n, is large (greater than 30),

then:
1) the sampling distribution of the mean is **approximately normal**.
2) the mean of the sampling distribution of the mean is equal to the mean of the population. This is expressed as:

$$\mu_{\bar{x}} = \mu$$

3) the standard error of the sampling distribution of the mean is equal to standard deviation of the population divided by the square root of the sample size. Symbolically, this is written as:

$$\sigma_{\bar{x}} = \frac{\sigma}{\sqrt{n}}$$

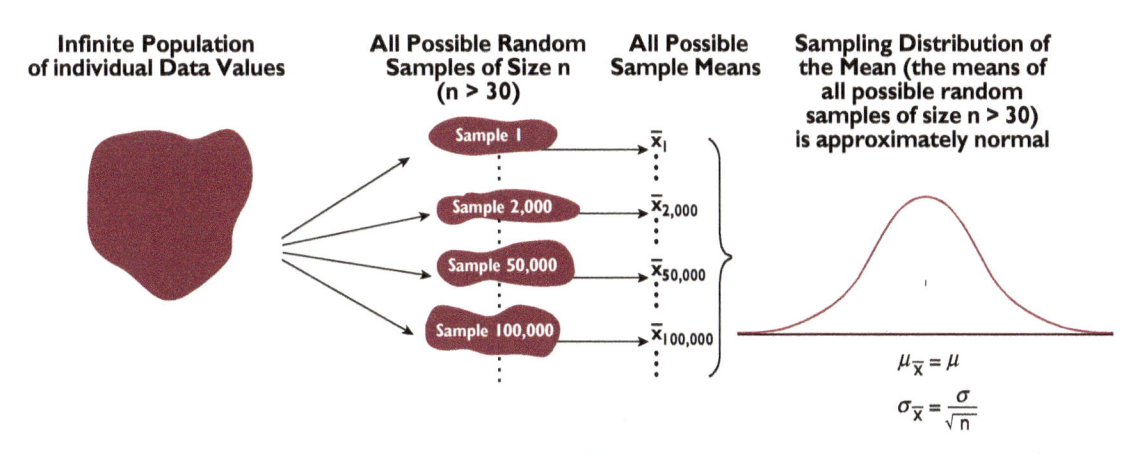

FIGURE 8.11

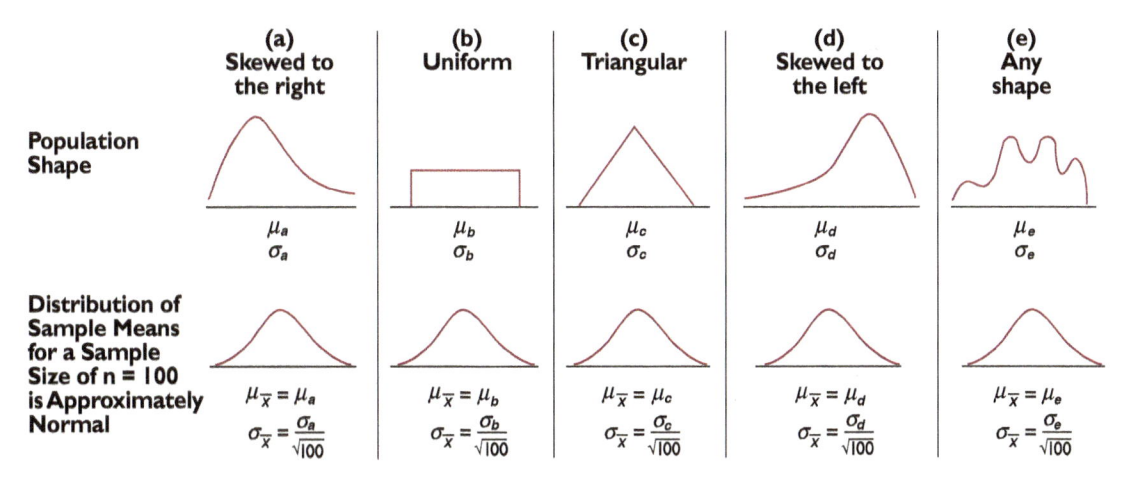

FIGURE 8.12

EXAMPLE 8.8

A leading cell phone carrier reports that the average length of time of mobile telephone calls is $\mu = 1.81$ minutes with a standard deviation of $\sigma = 0.98$ minutes. Determine the mean and the standard error of the sampling distribution of the mean and describe the shape of the sampling distribution of the mean when the sample size is:
a) $n = 36$
b) $n = 100$
c) Compare the sampling distributions of the mean for the sample size of $n = 36$ and $n = 100$.
d) If you had to estimate the mean of the population by either randomly selecting a sample of size $n = 36$ or a sample size of $n = 100$ from the population, then which sample size would give you a better chance of obtaining a smaller sampling error? Explain.

Solution

a) The mean of the sampling distribution of the mean is equal to the population mean and is determined by the formula: $\mu_{\bar{x}} = \mu$. Since the population mean is $\mu = 1.81$ minutes, then the mean of the sampling distribution is:

$$\mu_{\bar{x}} = \mu = 1.81 \text{ minutes}$$

Thus, the mean of the sampling distribution of the mean is: $\mu_{\bar{x}} = 1.81$ minutes.

The standard error of the sampling distribution of the mean is equal to the population standard deviation divided by the square root of the sample size. That is,

$$\sigma_{\bar{x}} = \frac{\sigma}{\sqrt{n}}$$

For a population standard deviation of $\sigma = 0.98$ minutes, and a sample size of $n = 36$, the standard error of the sampling distribution of the mean is:

$$\sigma_{\bar{x}} = \frac{\sigma}{\sqrt{n}}$$

$$\sigma_{\bar{x}} = \frac{0.98}{\sqrt{36}}$$

$$\sigma_{\bar{x}} = \frac{0.98}{6}$$

$$\sigma_{\bar{x}} \approx 0.16 \text{ minutes}$$

Although the shape of the population of length of times of long-distance telephone calls is unknown, the sampling distribution of the mean is approximately a normal distribution because a sample size of $n = 36$ (n greater than 30) can be considered as a large enough sample to apply the Central Limit Theorem.

The sampling distribution of the mean is illustrated in Figure 8.13.

b) The mean of the sampling distribution of the mean is equal to the population mean and is determined by the formula: $\mu_{\bar{x}} = \mu$. Since the population mean is $\mu = 1.81$ minutes, then the mean of the sampling distribution is:

$$\mu_{\bar{x}} = \mu = 1.81 \text{ minutes}$$

Thus, the mean of the sampling distribution of the mean is: $\mu_{\bar{x}} = 1.81$ minutes.

The standard error of the sampling distribution of the mean is equal to the population standard deviation divided by the square root of the sample size. That is,

$$\sigma_{\bar{x}} = \frac{\sigma}{\sqrt{n}}.$$

For a population standard deviation of $\sigma = 0.98$ minutes, and a sample size of $n = 100$, the standard error of the sampling distribution of the mean is:

$$\sigma_{\bar{x}} = \frac{\sigma}{\sqrt{n}}$$

$$\sigma_{\bar{x}} = \frac{0.98}{\sqrt{100}}$$

$$\sigma_{\bar{x}} = \frac{0.98}{10}$$

$$\sigma_{\bar{x}} = 0.098$$

$$\sigma_{\bar{x}} \approx 0.10 \text{ minutes}$$

Sampling Distribution of the Mean for a Sample Size of n = 36

$\mu_{\bar{x}} = 1.81$
$\sigma_{\bar{x}} = 0.16$

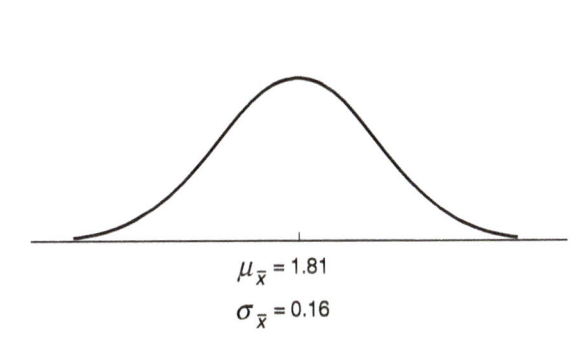

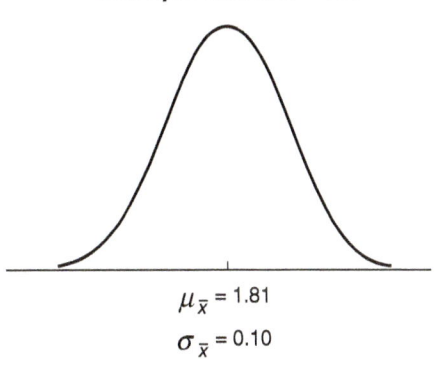

FIGURE 8.13

Sampling Distribution of the Mean for a Sample Size of n = 100

$\mu_{\bar{x}} = 1.81$
$\sigma_{\bar{x}} = 0.10$

FIGURE 8.14

Although the shape of the population of length of times of long-distance telephone calls is unknown, the sampling distribution of the mean is approximately a normal distribution because a sample size of n = 100 (n greater than 30) can be considered as a large enough sample to apply the Central Limit Theorem.

The sampling distribution of the mean is illustrated in Figure 8.14.

c) Comparing the characteristics of the sampling distributions shown in Figures 8.13 and 8.14, we should notice that both sampling distributions can be approximated by a normal distribution since the sample sizes are greater than 30. Also, notice that the mean of both sampling distributions is equal to 1.81 minutes regardless of the sample size. However, the standard error of the sampling distribution for a sample size of n = 100 is smaller than the standard error of the sampling distribution for a sample size of n = 36.

d) Since the standard error of the sampling distribution of the mean for a sample size of n = 100 ($\sigma_{\bar{x}} = 0.10$) is smaller than the standard error of the sampling distribution of the mean for a sample size of n = 36 ($\sigma_{\bar{x}} = 0.16$), then you should select a random sample of size n = 100. A smaller standard deviation of the sampling distribution or standard error of the mean indicates a relatively smaller dispersion among the sample means and thus, the sample means will be relatively closer to the population mean.

Consequently, there is a good chance of a small sampling error in estimating the population mean when the standard error of the sampling distribution is smaller.

Let's now summarize the different possible cases discussed when working with the sampling distribution of the mean. Figure 8.15 contains the assumptions along with the appropriate characteristics of the sampling distribution of the mean.

Characteristics of the Sampling Distribution of the Mean

Case I: Sampling from a Normal Population, Regardless of Sample Size, n

	Population	Sampling Distribution of the Mean
Mean:	μ	$\mu_{\bar{x}} = \mu$
Std Dev:	σ	$\sigma_{\bar{x}} = \dfrac{\sigma}{\sqrt{n}}$
Shape:	Normal	Normal

Case II: Sampling from a Non-Normal Population and the Sample Size, n, is Large (ie., n > 30)

	Population	Sampling Distribution of the Mean
Mean:	μ	$\mu_{\bar{x}} = \mu$
Std Dev:	σ	$\sigma_{\bar{x}} = \dfrac{\sigma}{\sqrt{n}}$
Shape:	Non-Normal	Normal when n > 30

FIGURE 8.15 (continues)

> **Characteristics of the Sampling Distribution of the Mean (continued)**
>
> **Case III: Sampling without Replacement from a Non-Normal Finite Population and Sample Size, n, is Large Relative to the Population Size, N (i.e., n>0.05N)**
>
> | | Population | Sampling Distribution of the Mean |
> |---|---|---|
> | Mean: | μ | $\mu_{\bar{x}} = \mu$ |
> | Std Dev: | σ | $\sigma_{\bar{x}} = \dfrac{\sigma}{\sqrt{n}} \cdot \sqrt{\dfrac{N-n}{N-1}}$ |
> | Shape: | Non-Normal | |

FIGURE 8.15 (continued)

8.5 CALCULATING PROBABILITIES USING THE SAMPLING DISTRIBUTION OF THE MEAN

Using the Central Limit Theorem and the characteristics of the sampling distribution of the mean discussed in Section 8.4, we can now illustrate how to calculate probabilities using the sampling distribution of the mean. To calculate these probabilities, we will need to determine a z score of the sample mean and use Table II: The Normal Curve Area Table. As discussed in an earlier chapter, the z score formula for a data value x is:

$$\text{z of data value} = \frac{\text{data value} - \text{population mean}}{\text{population standard deviation}}.$$

Symbolically, the z score formula was:

$$\text{z of x} = \frac{x - \mu}{\sigma}$$

However, in this chapter, since we are working with the sampling distribution of the mean, the data values are sample means, $\bar{x}$, thus, the z score formula becomes:

$$\text{z score of a sample mean} = \frac{\text{sample mean} - \text{mean of the sampling distribution}}{\text{standard error of the sampling distribution}}$$

Consequently, when working with the sampling distribution of the mean, the z score formula is now symbolized as:

$$\text{z of } \bar{x} = \frac{\bar{x} - \mu_{\bar{x}}}{\sigma_{\bar{x}}}$$

Example 8.9 demonstrates how to calculate probabilities using the sampling distribution of the mean, this z score formula and Table II: The Normal Curve Area Table.

EXAMPLE 8.9

At a large public state college in Virginia, the mean Verbal SAT score of all attending students was $\mu = 600$ with a population standard deviation of $\sigma = 65$. If a random sample of 100 students is selected from the population of students, determine:

a) The probability that mean Verbal SAT score of the selected sample will be less than 615.
b) The probability that mean Verbal SAT score of the selected sample will be within 10 points of the population mean.

Solution

Since a random sample of n = 100 students is being selected from the population, then, according to the Central Limit Theorem, the sampling distribution of the mean can be approximated by a normal distribution because the sample size (n = 100) is greater than 30. To answer parts (a) and (b) we need to determine the mean and standard error of the sampling distribution of the mean using Theorem 8.3. From Theorem 8.3, the mean of the sampling distribution of the mean, $\mu_{\bar{x}}$, is equal to the population mean. Thus, the mean of the sampling distribution is:

$$\mu_{\bar{x}} = \mu$$

$$\mu_{\bar{x}} = 600.$$

The standard error of the sampling distribution of the mean is equal to the population standard deviation divided by the square root of the sample size. Therefore, the standard error of the sampling distribution is:

$$\sigma_{\bar{x}} = \frac{\sigma}{\sqrt{n}}$$

$$\sigma_{\bar{x}} = \frac{65}{\sqrt{100}}$$

$$\sigma_{\bar{x}} = \frac{65}{10}$$

$$\sigma_{\bar{x}} = 6.5$$

Figure 8.16 shows the characteristics of the sampling distribution of the mean: an approximately normal distribution with mean $\mu_{\bar{x}} = 600$, and a standard error $\sigma_{\bar{x}} = 6.5$.

a) To determine the probability that the mean Verbal SAT score of the selected sample will be less than 615, we will use the sampling distribution of the mean illustrated in Figure 8.16. To calculate the probability that the sample of 100 students will be less than 615, we need to find the area to the left of 615. The shaded area shown in Figure 8.17 to the left of 615 represents the probability that the selected sample mean will be less than 615.

To find this shaded area to left of 615, we must first calculate the z score of 615. Since the data values of the sampling distribution are sample means, $\bar{x}$, the z score formula becomes:

$$z \text{ score of a sample mean} = \frac{\text{sample mean} - \text{mean of the sampling distribution}}{\text{standard error of the sampling distribution}}$$

The z score formula is now symbolized as:

$$z \text{ of } \bar{x} = \frac{\bar{x} - \mu_{\bar{x}}}{\sigma_{\bar{x}}}$$

$\mu_{\bar{x}} = 600$
$\sigma_{\bar{x}} = 6.5$

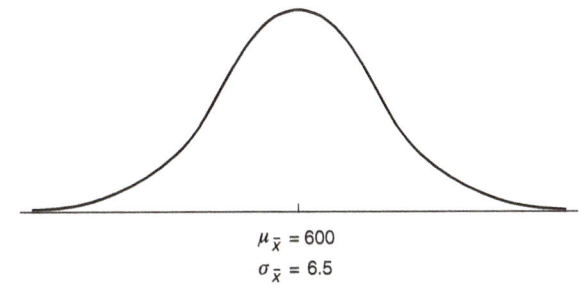

FIGURE 8.16

Probability that the sample mean is less than 615

$\mu_{\bar{x}} = 600$ 615
$\sigma_{\bar{x}} = 6.5$

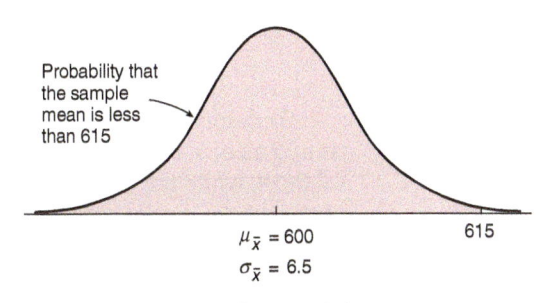

FIGURE 8.17

Substituting $\bar{x} = 615$, $\mu_{\bar{x}} = 600$ and $\sigma_{\bar{x}} = 6.5$, into this z score formula, we have:

$$\text{z of } 615 = \frac{615 - 600}{6.5}$$

$$\text{z of } 615 \approx 2.31$$

From Table II: Normal Curve Area Table in Appendix C, the area to the left of z = 2.31 is 0.9896. Therefore, the probability of selecting a sample of 100 students whose mean Verbal SAT score will be less than 615 is 0.9896.

b) To determine the probability that the mean Verbal SAT of the selected sample will be within 10 points of the population mean, we need to find the sample means that are within 10 points of the mean. Since the mean is 600, the sample means which are 10 points away from the mean are:

$$600 - 10 = 590 \text{ and } 600 + 10 = 610.$$

Thus, to find the probability that the mean Verbal SAT of the selected sample will be within 10 points of the population mean, we need to determine the area between 590 and 610. This probability is represented by the shaded area in Figure 8.18.

To find the shaded area in Figure 8.18, we must find the z scores of 590 and 610, using the z score formula:

$$\text{z of } \bar{x} = \frac{\bar{x} - \mu_{\bar{x}}}{\sigma_{\bar{x}}}, \mu_{\bar{x}} = 600 \text{ and } \sigma_{\bar{x}} = 6.5.$$

Probability of selecting a sample within 10 points of the population mean

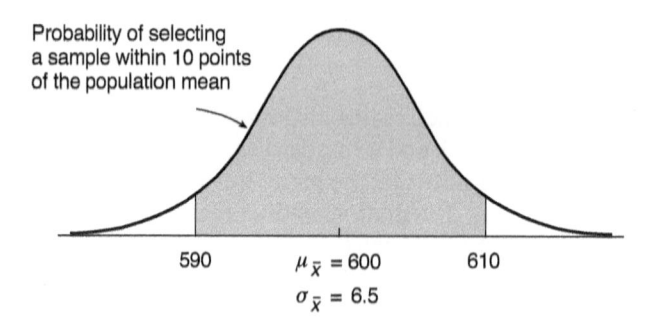

590     $\mu_{\bar{x}} = 600$ 610
         $\sigma_{\bar{x}} = 6.5$

FIGURE 8.18

The z score of $\bar{x} = 590$ is:

$$\text{z of } 590 = \frac{590 - 600}{6.5}$$

$$\text{z of } 590 \approx -1.54$$

While the z score of $\bar{x} = 610$ is:

$$\text{z of } 610 = \frac{610 - 600}{6.5}$$

$$\text{z of } 610 \approx 1.54$$

To determine the area between 590 and 610, we must look up the z scores corresponding to these sample means. From Table II: Normal Curve Area Table in Appendix C, the area to the left of z = –1.54 is 0.0618, while the area to the left of z = 1.54 is 0.9382. To determine the area between 590 and 610, we must subtract the area to the left of z = –1.54 from the area to the left of z = 1.54. Thus, we have:

area between 590 and 610 = area to the left of z = 1.54 − area to the left of z = –1.54
= 0.9382 − 0.0618
= 0.8764

Therefore, the probability that mean Verbal SAT score of the selected sample will be within 10 points of the population mean is 0.8764.

EXAMPLE 8.10

The population mean weight of newborn babies for a western suburb is 7.4 lbs. with a standard deviation of 0.8 lbs. What is the probability that a sample of 64 newborns selected at random will have a mean weight greater than 7.5 lbs.?

Solution

To determine the probability that a random sample of 64 babies will have a mean weight greater than 7.5 lbs, we must compute the characteristics of the sampling distribution. Since a random sample of n = 64 babies is being selected from the population, then, according to the Central Limit Theorem, the sampling distribution of the mean can be approximated by a normal distribution because the sample size (n = 64) is greater than 30. The mean of the sampling distribution of the mean is equal to the population mean. Therefore, mean of the sampling distribution is:

$$\mu_{\bar{x}} = \mu$$
$$\mu_{\bar{x}} = 7.4 \text{ lbs}$$

The standard error of the sampling distribution of the mean is:

$$\sigma_{\bar{x}} = \frac{\sigma}{\sqrt{n}}$$
$$\sigma_{\bar{x}} = \frac{0.8}{\sqrt{64}}$$
$$\sigma_{\bar{x}} = 0.1$$

Figure 8.19 shows the characteristics of the sampling distribution of the mean: an approximately normal distribution with mean $\mu_{\bar{x}} = 7.4$, and a standard error $\sigma_{\bar{x}} = 0.1$.

To find the probability that the sample mean will be greater than 7.5 lbs. we must find the area to the *right* of 7.5 lbs. The shaded area in Figure 8.20 illustrates the probability that the selected sample mean will be greater than 7.5 lbs.

To determine the shaded area in Figure 8.20, compute the z score of 7.5 using the z score formula:

$$z \text{ of } \bar{x} = \frac{\bar{x} - \mu_{\bar{x}}}{\sigma_{\bar{x}}}, \text{ where } \mu_{\bar{x}} = 7.4 \text{ and } \sigma_{\bar{x}} = 0.1.$$

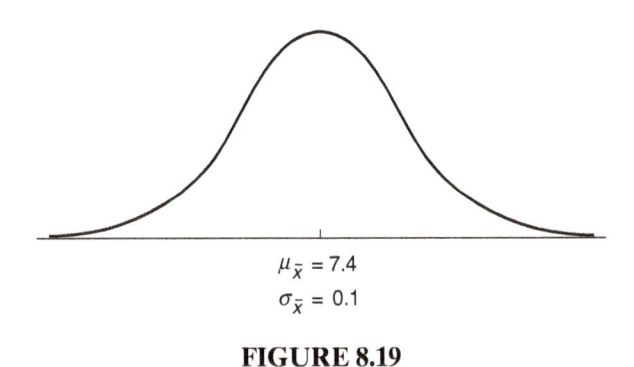

$\mu_{\bar{x}} = 7.4$
$\sigma_{\bar{x}} = 0.1$

FIGURE 8.19

Probability of selecting a sample mean greater than 7.5 lbs.

$\mu_{\bar{x}} = 7.4 \quad 7.5$
$\sigma_{\bar{x}} = 0.1$

FIGURE 8.20

The z score of $\bar{x} = 7.5$ is:

$$z = \frac{7.5 - 7.4}{0.1} = 1$$

Using Table II: Normal Curve Area Table in Appendix C, the area to the left of z = 1 is 0.8413. To obtain the area to the right, subtract the area to the left from 1. Thus, the area to the right of 7.5 lbs is equal to:

$$1 - 0.8413 = 0.1587.$$

Therefore, the probability of randomly selecting a sample of 64 newborns whose mean weight is greater than 7.5 lbs. is 0.1587.

472 Chapter 8 *Sampling and Sampling Distributions*

The TI-84 calculator can be used to find a probability involving the sampling distribution of the mean. To do this we will use the normalcdf function of the TI-84 distribution menu. To access the distribution menu, press 2nd VARS [DISTR] 2: **normalcdf(**. Figure 8.21 shows the calculator's display when you access the distribution menu.

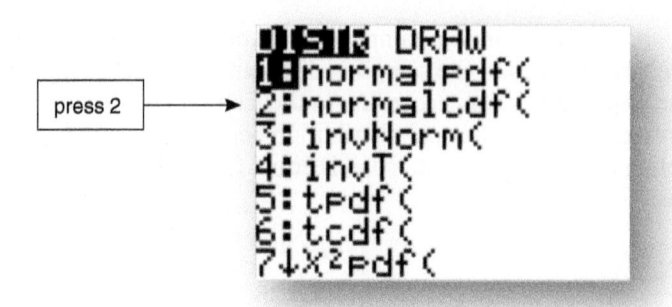

press 2 →

FIGURE 8.21

Press the **2** key on the calculator and this will bring you to the input screen. Remember, the normalcdf function requires that you input four parameters in a specific order.

The syntax for the parameters for this chapter is:

normalcdf(lower bound of shaded area, upper bound of shaded area, $\mu_{\bar{x}}, \sigma_{\bar{x}}$)

In Example 8.10, we were asked to find the probability that 64 randomly selected newborns will have a mean weight greater than 7.5 lbs. In symbols this is written, $P(\bar{X} > 7.5)$. The mean of the sampling distribution was given as $\mu_{\bar{x}} = 7.4$ and the standard error was given as $\sigma_{\bar{x}} = 0.1$. Examining Figure 8.20, the lower boundary of the shaded region is 7.5 and the upper boundary of the shaded region is ∞ or E99. Entering these four parameters into normalcdf as shown in Figure 8.22a gives the output shown in Figure 8.22b.

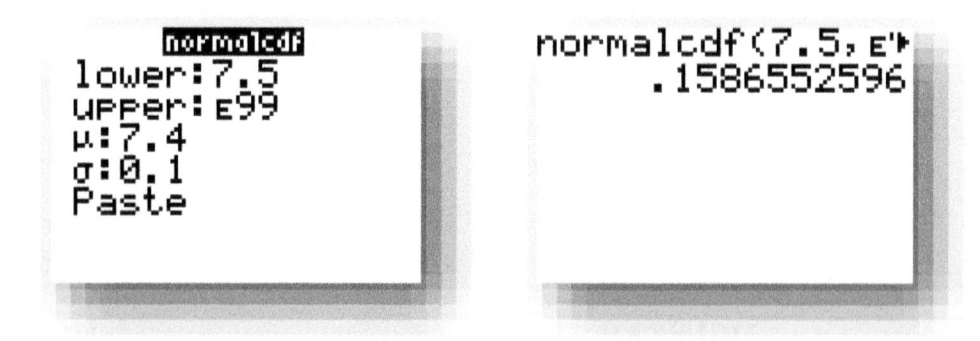

FIGURE 8.22a   **FIGURE 8.22b**

Notice the result of the probability is 0.1587 rounded to 4 decimal places. This is the same result found in Example 8.10.

EXAMPLE 8.11

The population of the ages of all U. S. college students is skewed to the right with a mean age of 27.4 years and a standard deviation of 5.8 years. Determine the probability that a random sample of 49 students selected from the population will have a sample mean age within one year of the population mean age?

Solution

Since the sample size is 49 (greater than 30), then according to the Central Limit Theorem, the sampling distribution of the mean can be approximated by a normal distribution, regardless of the shape of the population. The mean of this sampling distribution is equal to:

$$\mu_{\bar{x}} = \mu$$
$$\mu_{\bar{x}} = 27.4 \text{ years}$$

and, its standard error is equal to:

$$\sigma_{\bar{x}} = \frac{\sigma}{\sqrt{n}}$$
$$\sigma_{\bar{x}} = \frac{5.8}{\sqrt{49}}$$
$$\sigma_{\bar{x}} \approx 0.83$$

To determine the probability that the sample mean will be within one year of the population mean age, we must find the sample ages which are 1 year less than the population mean age of 27.4 years (that is: 27.4 − 1 = 26.4 years) and 1 year greater than the population mean age of 27.4 (that is: 27.4 + 1 = 28.4 years). Thus, we need to find the probability that the selected sample will have a mean between 26.4 years and 28.4 years. This probability is represented by the shaded area shown in Figure 8.23.

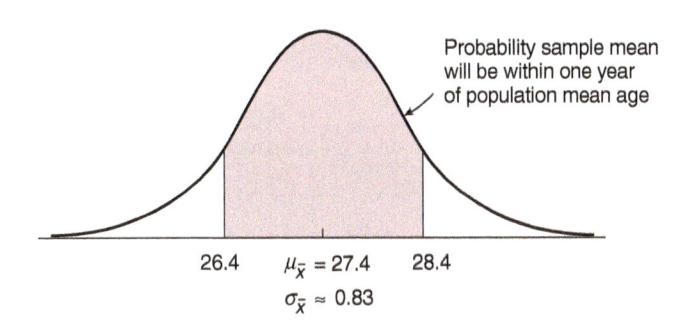

FIGURE 8.23

To determine this shaded area we must compute the z scores for 26.4 years and 28.4 years. Substituting $\mu_{\bar{x}} = 27.4$ and $\sigma_{\bar{x}} = 0.83$ into the z score formula, we can compute the z scores of 26.4 and 28.4. Thus:

$$z \text{ of } 26.4 = \frac{26.4 - 27.4}{0.83} \text{ and } z \text{ of } 28.4 = \frac{28.4 - 27.4}{0.83}$$
$$z \text{ of } 26.4 = -1.20 \text{ and } z \text{ of } 28.4 = 1.20$$

Using Table II: Normal Curve Area Table in Appendix C, the area to the left of z = −1.20 is 0.1151, while the area to the left of z = 1.20 is 0.8849. To obtain the area between 26.4 and 28.4 we must subtract the smaller area, 0.1151, from the larger area, 0.8849, which is:

$$0.8849 - 0.1151 = 0.7698.$$

Therefore, the probability that a random sample of 49 students selected from the population will have a sample mean age within one year of the population mean age is 0.7698.

CASE STUDY 8.1 Average Annual Income for Families with Children

Examine the information on the "Average Annual Income for Families with Children" in Figure 8.24, which appeared in the *New York Times*.

a) Define the population for Figure 8.24(a). What do you believe is the shape of the population? Explain.

For the population stated in Figure 8.24(a):

b) If all possible samples of size 100 are selected from the population of Figure 8.24(a) and the mean is computed for these samples, what distribution shape would you expect for the resulting distribution?
c) What is the name of the resulting distribution?
d) What value would you expect the mean of the resulting distribution to be?
e) If the population standard deviation were σ, then what would be the standard deviation of the distribution containing the mean from all of the possible samples?

For the population stated in Figure 8.24(b):

f) Define the population and what do you think is the shape of this population? Explain.
g) If all possible samples of size 64 are selected from the population of Figure 8.24(b) and the means are computed for these samples, what distribution shape would you expect for the resulting distribution?
h) What is the name of the resulting distribution?
i) What value would you expect the mean of the resulting distribution to be?
j) If the population standard deviation were σ, then what would be the standard deviation of the distribution containing the mean from all of the possible samples?

For the population stated in Figure 8.24(c):

k) Define the population and what do you think is the shape of this population? Explain.
l) If all possible samples of size 36 are selected from the population of Figure 8.24(c) and the mean is computed for these samples, what distribution shape would you expect for the resulting distribution?
m) What is the name of the resulting distribution?
n) What value would you expect the mean of the resulting distribution to be?
o) If the population standard deviation were σ, then what would be the standard deviation of the distribution containing the mean from all of the possible samples?

Average Annual Income for Families with Children

FIGURE 8.24

8.6 THE EFFECT OF THE SAMPLE SIZE ON THE STANDARD ERROR OF THE MEAN

Let's examine the effect of an increase in the sample size, n, on the standard error of the mean, $\sigma_{\bar{x}}$.

EXAMPLE 8.12

Consider the distribution of 5 feet 9 inch American males whose mean weight is 160 lbs. with a standard deviation of 10 lbs. Determine the mean, $\mu_{\bar{x}}$, and standard error, $\sigma_{\bar{x}}$, for the sampling distribution of the mean for samples of size 36, 100, 400 and 1600.

Solution

For each we shall use the formulas for

$$\mu_{\bar{x}} = \mu \text{ and } \sigma_{\bar{x}} = \frac{\sigma}{\sqrt{n}},$$

the population mean $\mu = 160$ and the population standard deviation of $\sigma = 10$.

a) For a sample size of n = 36, then the mean and standard error of the sampling distribution are:

$$\mu_{\bar{x}} = 160 \qquad \sigma_{\bar{x}} = \frac{10}{\sqrt{36}}$$
$$\sigma_{\bar{x}} = 1.67$$

b) For a sample size of n = 100, then the mean and standard error of the sampling distribution are:

$$\mu_{\bar{x}} = 160 \qquad \sigma_{\bar{x}} = \frac{10}{\sqrt{100}} = 1$$

c) For a sample size of n = 400, then the mean and standard error of the sampling distribution are:

$$\mu_{\bar{x}} = 160 \text{ and } \sigma_{\bar{x}} = \frac{10}{\sqrt{400}} = 0.5$$

d) For a sample size of n = 1600, then the mean and standard error of the sampling distribution are:

$$\mu_{\bar{x}} = 160 \text{ and } \sigma_{\bar{x}} = \frac{10}{\sqrt{1600}} = 0.25$$

Using the results of Example 8.12, notice as the sample size increases from n = 36 to n = 1600, the standard error of the mean decreases from $\sigma_{\bar{x}} = 1.67$ to $\sigma_{\bar{x}} = 0.25$. This relationship is illustrated in Figure 8.25.

Also, notice that as the sample size increases the sample means become more clustered about the mean, $\mu_{\bar{x}}$. Figure 8.26 illustrates that the size of the interval of sample means containing the middle 99% of the sample means decreases as the sample size increases.

FIGURE 8.25

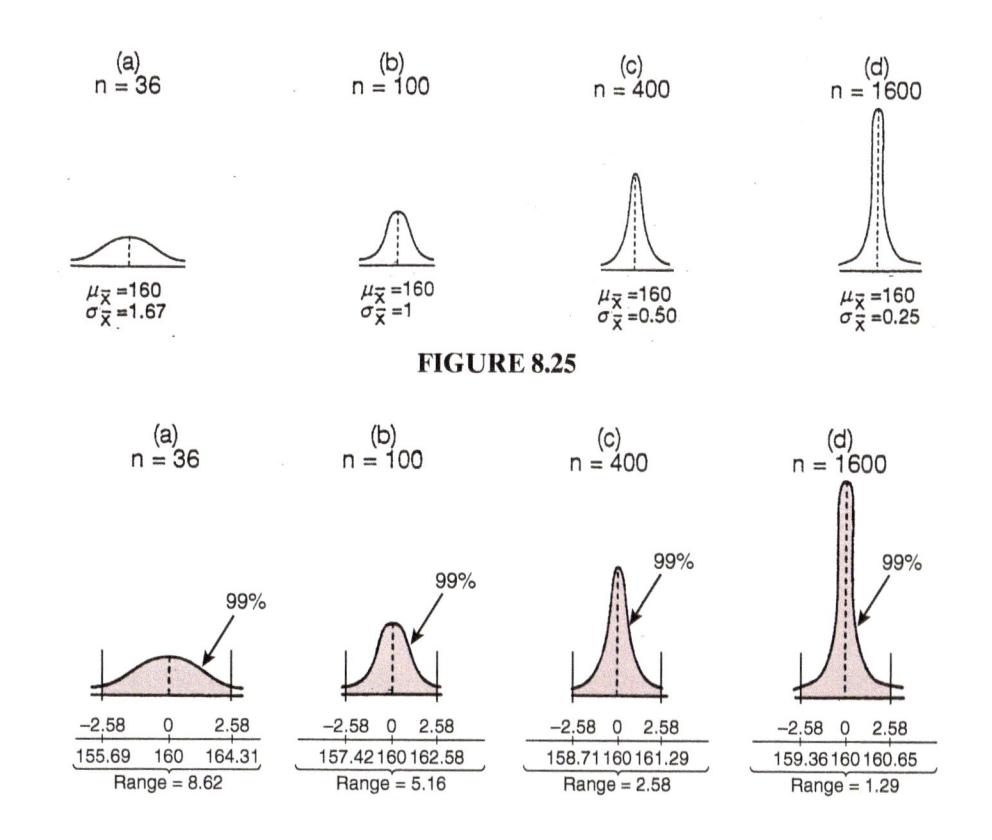

FIGURE 8.26

For instance when n = 36, 99% of the sample means lie within the interval 155.69 to 164.31. Whereas when n = 1600, 99% of the sample means lie within the smaller interval 159.36 to 160.65.

Suppose you didn't know the mean weight for 5 foot 9 inch American males discussed previously and you were going to estimate the population mean by selecting a random sample from the population of 5 foot 9 inch American males. Using Figure 8.26, which sample size would you use if you wanted to be 99% confident that the sample mean estimate would be within one pound of the actual population mean?

You should choose a sample size of n = 1600 since 99% of the sample means are within one pound of the actual population mean. In fact for this sample size, the sample mean is within 0.65 lbs. of the actual population mean. Thus, we can now summarize the effect of the sample size on the standard error of the sampling distribution of the mean and the accuracy of using a sample mean to estimate the population mean.

The Effect of the Sample Size on the Standard Error of the Sampling Distribution of the Mean

As the sample size increases, the standard error of the sampling distribution of the mean decreases. In general, when using a sample mean to estimate a population mean, **as the sample size increases the more confidence you have in the accuracy of the sample mean as an estimate of the population mean**.

8.7 THE SAMPLING DISTRIBUTION OF THE PROPORTION

Our discussion so far in this chapter has centered around the mean of a population. However, sometimes we are interested in the **proportion** or fraction of a population that possesses a particular characteristic.

DEFINITION 8.5
A **proportion** is a fraction, ratio or percentage that indicates the part of a population or sample that possesses a particular characteristic.

For example, statisticians, pollsters or researchers might be interested in inferences about the proportion of:

1. U.S. adults who wear seat belts.
2. individuals not catching a cold who take large doses of vitamin C.
3. children who live with only one parent within the U.S.
4. people who will vote for the Democratic candidate in the upcoming senate election.
5. defective computer memory chips produced using a new innovative manufacturing process.
6. patients that recover from a certain illness who are treated with a new drug.

In each of these cases, we are interested in the proportion of a population. To determine this proportion, it becomes necessary to count the number of data values within the population that possess a particular characteristic. In the earlier sections of this chapter, we worked with data that was measurable. Now we will be working with non-measurable data, that is, data that can only be counted. For example, the population data will be categorized and counted according to a particular characteristic such as wearing seat belts, taking vitamin supplements, living with only one parent, voting for a particular candidate, gender, race, religion, marital status, educational level, state of health, and so on. When dealing with count data, the **proportion** or **fraction** of a finite **population** that possesses each of these characteristics can be determined using the following definition. This proportion is a parameter because it represents a population value.

The concept of a proportion is similar to the probability of a success of a binomial distribution discussed in Chapter 6. This is because the probability of a success for a binomial experiment represents the proportion of the population that possesses a particular characteristic. Example 8.13 illustrates the calculation of a population proportion.

Section 8.7 *The Sampling Distribution of the Proportion* **477**

> **DEFINITION 8.6**
> The **population (or true) proportion** of a finite population is the ratio of the number of occurrences (or observations) in the population that possess the particular characteristic to the population size. The population proportion, denoted by p, can be expressed by the formula:
>
> $$p = \frac{X}{N}$$
>
> where: p = population proportion
> X = number of occurrences (or observations) in the population possessing the particular characteristic
> N = population size
>
> The population proportion is often expressed in an equivalent percentage form.

EXAMPLE 8.13

In the 2019 New York City Marathon, there were 53,638 runners who finished the race. The following table shows the number of finishers by their gender.

Gender	Number who finished the Marathon
Men	30,893
Women	22,745

Determine the population proportion (to 4 decimal places) of:
a) men who completed the marathon.
b) women who completed the marathon.

Solution

a) The proportion of all runners who completed the marathon who were men is:

$$p = \frac{X}{N}$$

where X = number of men runners who completed the marathon
N = total number of all runners who completed marathon

$$p = \frac{X}{N} = \frac{\text{number of men runners who completed the marathon}}{\text{total number of runners who completed marathon}}$$

Using the table information, $p = \frac{X}{N} = \frac{30{,}893}{53{,}638} = 0.5760$ (rounded to 4 decimal places).

The population proportion of men who completed the marathon is 0.5760 or 57.60%.

b) The proportion of all runners who completed the marathon who were women is:

$$p = \frac{X}{N}$$

where X = number of women runners who completed the marathon
N = total number of all runners who completed marathon

$$p = \frac{X}{N} = \frac{\text{number of women runners who completed the marathon}}{\text{total number of runners who completed marathon}}$$

Using the table information, $p = \frac{X}{N} = \frac{22{,}745}{53{,}638} = 0.4240$ (rounded to 4 decimal places).

The population proportion of women who completed the marathon is 0.4240 or 42.40%.

There are many instances where it is either impractical or impossible to calculate the population proportion. For example, suppose you wanted to determine the proportion of blue colored plain M&M's manufactured by the Mars Corporation. Since it is virtually impossible to determine the actual population proportion of blue M&M's manufactured by Mars Corporation (although it might be fun trying to accomplish this task), it becomes necessary to **estimate the population proportion of blue M&M's**. In such instances, a random sample is selected and the proportion of the sample, symbolized by $\hat{p}$ and read as p-hat, is used to estimate the true population proportion, p. The sample proportion represents a statistic and is computed using the following definition.

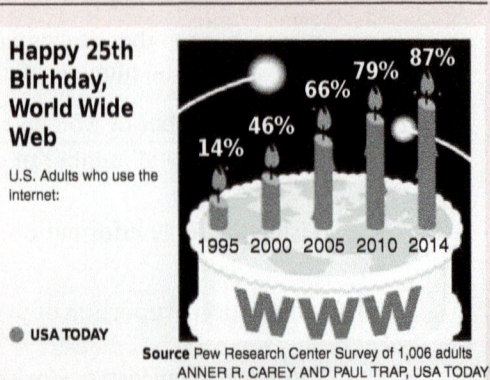

> **DEFINITION 8.7**
> The **sample proportion** is the ratio of the number of occurrences (or observations) in the sample that possess the particular characteristic to the sample size.
>   Thus, the sample proportion, denoted by $\hat{p}$ (read p-hat), can be expressed by the formula:
>
> $$\hat{p} = \frac{x}{n}$$
>
> where: $\hat{p}$ = sample proportion
>   x = number of occurrences (or observations) in the sample possessing the particular characteristic
>   n = sample size

EXAMPLE 8.14

A 2019 semiannual U.S. teen survey claimed that 1,833 teens stated they own an iPhone. If a random sample of 2,350 teens were surveyed, then determine the sample proportion of U.S. teens who own an iPhone.

Solution

The sample proportion of U.S. teens who own an iPhone is:

$$\hat{p} = \frac{x}{n} = \frac{\text{Number of U.S. teens within sample who own an iPhone}}{\text{sample size}}$$

$$\hat{p} = \frac{x}{n} = \frac{1{,}833}{2{,}350} = 0.78$$

Thus, the sample proportion of U.S. teens who own an iPhone is 0.78 or 78%.

The concept of a population and a sample proportion are discussed within Case Study 8.2.

> **CASE STUDY 8.2   Population and Sample Proportion**
>
> Examine the two snapshots entitled "Cars with best resale value" and "Happy 25th Birthday World Wide Web" shown in Figures 8.27a and 8.27b respectively.
> a) What variable is being represented within each snapshot?
> b) Is the variable numeric or categorical? Explain.
> c) Identify the populations being represented within each snapshot.
> d) For the proportions stated within each snapshot, explain why you believe the proportion is a population proportion or a sample proportion. Write the symbol you would use for each proportion.

Statistical Snapshot

Cars with best resale value
By percentage of sticker price after 5 years:

Toyota FJ Cruiser	69%
Chevrolet Corvette Stingray	53%
Subaru Impreza WRX Hatchback	49%
Mercedes-Benz GL350	49%
Lexus ES 350	47%

USA TODAY
Source Kiplinger's Personal Finance/Kelly Blue Book
JAEYANG AND PAUL TRAP, USA TODAY

FIGURE 8.27a
From *USA Today.* © 2014 Gannett-USA Today.
All rights reserved. Used under license.

Statistical Snapshot

Happy 25th Birthday, World Wide Web
U.S. Adults who use the internet:

1995: 14%
2000: 46%
2005: 66%
2010: 79%
2014: 87%

USA TODAY
Source Pew Research Center Survey of 1,006 adults
ANNER R. CAREY AND PAUL TRAP, USA TODAY

FIGURE 8.27b
From *USA Today.* © 2014 Gannett-USA Today.
All rights reserved. Used under license.

Let's return to the task of estimating the population proportion of blue M&M's. Suppose a sample of 500 M&M plain candies is randomly selected and is found to contain 43 blue M&M's. Using this sample data, the sample proportion of blue M&M's is:

$$\hat{p} = \frac{\text{number of blue M\&M's within the sample}}{\text{total number of M\&M's within the sample}}$$

$$\hat{p} = \frac{x}{n} = \frac{43}{500}$$

$$\hat{p} = 0.086$$

Consequently, the proportion of blue M&M's within the sample is 0.086 or 8.6%. Keep in mind that this sample proportion of 8.6% blue M&M's is a **statistic** and serves as an **estimate** of the unknown population parameter p, the population proportion of blue M&M's. At this time, it is appropriate to ponder the following question: Do you think that 8.6% is the exact proportion of blue M&M's within the population? Probably not! And what would happen if we decided to take a second sample to estimate the population proportion? Do you think that we would obtain the same result? That is, if a second random sample of 500 M&M's was selected from this same population, do you think this second sample would contain exactly 8.6% blue M&M's? In all likelihood, we would expect a second sample to yield a different proportion of blue M&M's. Let's consider taking a third sample. Would you expect a third sample to contain the same proportion of blue M&M's as either the first or second sample? Hopefully, you should begin to realize that the proportion of blue M&M's can vary from sample to sample even though there is only one true value for the proportion of blue M&M's for all the plain M&M's manufactured. Thus, just by chance alone, the estimate of a population proportion will vary from sample to sample due to the data values randomly selected within each sample. Therefore, the sample proportion is called a random variable because it is dependent upon the particular data values which are randomly selected from the population. For example, by chance alone, a sample may vary from containing no blue M&M's to one containing only blue M&M's. In order for us to determine how close a sample proportion estimates the population proportion, it is necessary for us to have information about **all** the possible sample proportions which can be selected from the population and how these sample proportions are distributed. Such a distribution of all sample proportions is referred to as the **sampling distribution of the proportion**.

DEFINITION 8.8
The sampling distribution of the proportion is a probability distribution which lists all the possible values of the sample proportions of the same sample size selected from a population along with the probability associated with each value of the sample proportion.

In Example 8.15, we will construct a sampling distribution of the proportion to facilitate the understanding of this concept.

EXAMPLE 8.15

Suppose a small bag of plain M&M's contains 2 brown, 1 red, 1 yellow, 1 blue, 1 green and 1 orange. We will assume this bag represents our population. Suppose we take all possible samples of 5 M&M's each.
a) Determine the population proportion p of brown M&M's.
b) Determine the total number of samples of size 5 that can be selected from this population.
c) Construct a frequency distribution table containing all the possible samples of size 5 and the proportion of brown M&M's for each of those samples.

Solution
a) The population proportion of brown M&M's is:

$$p = \frac{\text{number of brown M\&M's within the population}}{\text{population size}}$$

$$p = \frac{X}{N} = \frac{2}{7} = 0.2857$$

480 Chapter 8 *Sampling and Sampling Distributions*

b) The number of samples of size 5 that can be selected from a population containing 7 M&M's can be computed using the combination formula: $_nC_s$. Since n = 7 and s = 5 we have $_7C_5 = 21$. (This counting technique is discussed in Section 5.3.) Thus, there are 21 possible samples of size 5 that can be selected from the population of 7 M&M's.

c) The 21 possible samples of size 5 are illustrated in the following table.

All Possible Samples of Size 5 and the Sample Proportion of Brown M&M's for Each Sample.

All Samples of Size 5	Sample Proportion of Brown M&M's Using the Formula: $\hat{p} = \dfrac{x}{n}$
brown, brown, red, yellow, blue	$\hat{p} = \dfrac{x}{n} = \dfrac{2}{5} = 0.40$
brown, brown, red, yellow, orange	$\hat{p} = \dfrac{x}{n} = \dfrac{2}{5} = 0.40$
brown, brown, red, yellow, green	$\hat{p} = \dfrac{x}{n} = \dfrac{2}{5} = 0.40$
brown, brown, yellow, blue, green	$\hat{p} = \dfrac{x}{n} = \dfrac{2}{5} = 0.40$
brown, brown, yellow, blue, orange	$\hat{p} = \dfrac{x}{n} = \dfrac{2}{5} = 0.40$
brown, brown, red, yellow, green	$\hat{p} = \dfrac{x}{n} = \dfrac{2}{5} = 0.40$
brown, brown, blue, green, orange	$\hat{p} = \dfrac{x}{n} = \dfrac{2}{5} = 0.40$
brown, brown, green, orange, red	$\hat{p} = \dfrac{x}{n} = \dfrac{2}{5} = 0.40$
brown, brown, green, orange, yellow	$\hat{p} = \dfrac{x}{n} = \dfrac{2}{5} = 0.40$
brown, brown, green, orange, red	$\hat{p} = \dfrac{x}{n} = \dfrac{2}{5} = 0.40$
brown, red, yellow, blue, green	$\hat{p} = \dfrac{x}{n} = \dfrac{1}{5} = 0.20$
brown, red, yellow, blue, green	$\hat{p} = \dfrac{x}{n} = \dfrac{1}{5} = 0.20$
brown, red, yellow, blue, orange	$\hat{p} = \dfrac{x}{n} = \dfrac{1}{5} = 0.20$
brown, red, yellow, blue, orange	$\hat{p} = \dfrac{x}{n} = \dfrac{1}{5} = 0.20$
brown, red, blue, green, orange	$\hat{p} = \dfrac{x}{n} = \dfrac{1}{5} = 0.20$
brown, red, blue, green, orange	$\hat{p} = \dfrac{x}{n} = \dfrac{1}{5} = 0.20$
brown, yellow, blue, green, orange	$\hat{p} = \dfrac{x}{n} = \dfrac{1}{5} = 0.20$

(continues)

Sample Proportion of Brown M&M's Using the Formula:

$$\hat{p} = \frac{x}{n}$$

All Samples of Size 5	
brown, yellow, blue, green, orange	$\hat{p} = \frac{x}{n} = \frac{1}{5} = 0.20$
brown, red, yellow, green, orange	$\hat{p} = \frac{x}{n} = \frac{1}{5} = 0.20$
brown, red, yellow, green, orange	$\hat{p} = \frac{x}{n} = \frac{1}{5} = 0.20$
red, yellow, blue, green, orange	$\hat{p} = \frac{x}{n} = \frac{0}{5} = 0$

Using this table we can construct the following frequency distribution table of the sample proportions.

Sample Proportion of Brown M&M's: $\hat{p}$	Frequency
0.40	10
0.20	10
0	1

Before we discuss the characteristics of the sampling distribution of the proportion, we will define the notation for the mean and standard deviation of the sampling distribution of the proportion.

Notation for the Mean of the Sampling Distribution of the Proportion

The mean of the sampling distribution of the proportion is denoted by $\mu_{\hat{p}}$, read mu sub p hat. Thus,

$\mu_{\hat{p}}$ = mean of all the sample proportions of the sampling distribution

---

Notation for the Standard Deviation of the Sampling Distribution of the Proportion
Also Known as the Standard Error of the Proportion

The standard deviation of the sampling distribution of the proportion or the standard error of the proportion is denoted by $\sigma_{\hat{p}}$, read sigma sub p hat. Thus,

$\sigma_{\hat{p}}$ = standard deviation of all the sample proportions of the sampling distribution

---

Using this notation, we will now present the characteristics of the sampling distribution of the proportion. The mean of the sampling distribution of the proportion is equal to the population proportion, p. You should recall that this result is similar to the relationship between the mean of the sampling distribution of the mean which was equal to the population mean.

The Mean of the Sampling Distribution of the Proportion

The mean of the sampling distribution of the proportion, $\mu_{\hat{p}}$, is equal to the population proportion, p. Thus,

$$\mu_{\hat{p}} = p$$

The standard deviation of the sampling distribution of the proportion is referred to as the standard error of the proportion because it is a measure of how much a sample proportion is likely to deviate from the population proportion. It is also a measure of the average sampling error which was discussed in Section 8.2 with respect to the sampling distribution of the mean.

The Standard Deviation of the Sampling Distribution of the Proportion also known as the Standard Error of the Proportion

The standard deviation of the sampling distribution of the proportion or the standard error of the proportion, $\sigma_{\hat{p}}$, is defined as:

$$\sigma_{\hat{p}} = \sqrt{\frac{p(1-p)}{n}}$$

This formula for the standard error of the proportion is appropriate provided that the sample size is small as compared to the population size. Recall from Section 8.3, the general rule used to determine whether the sample is small in relation to the population is:

the sample size, n, is less than 5% of the population size, N.

That is,

$$\frac{sample\ size}{population\ size} \text{ is less than 5\%.}$$

In symbols, this condition is written:

$$\frac{n}{N} < 5\%.$$

If this condition doesn't hold true, then the finite correction factor as discussed in Section 8.3 should appear in the formula for the standard error of the proportion. That is, the formula for the standard error of the proportion becomes:

$$\sigma_{\hat{p}} = \sqrt{\frac{p(1-p)}{n}} \sqrt{\frac{N-n}{N-1}}.$$

In this chapter, we will assume when working with the sampling distribution of the proportion that the population is large relative to sample size so **we will not apply the finite correction factor** except for Example 8.17.

EXAMPLE 8.16

According to a survey in *Men's Health Magazine,* (about) 37% of adults aged 18 to 54 have taken a herbal diet supplement. If $\hat{p}$ represents the proportion of adults aged 18 to 54 within a random sample of 200 such adults, then:
a) State the mean of the sampling distribution of the proportion using the appropriate notation.
b) State the standard error of the proportion using the appropriate notation.

Solution

a) Let's define p as the proportion of all adults aged 18 to 54 who have taken a herbal diet supplement. Thus,

$$p = 0.37$$

and

$$1 - p = 1 - 0.37 = 0.63.$$

The mean of the sampling distribution of the proportion is:

$$\mu_{\hat{p}} = p$$
$$\mu_{\hat{p}} = 0.37$$

b) Using the information, p = 0.37, 1 − p = 0.63 and the sample size, n, is 200 then the standard error of the proportion is:

$$\sigma_{\hat{p}} = \sqrt{\frac{p(1-p)}{n}}$$

$$= \sqrt{\frac{(0.37)(0.63)}{200}}$$

$$\sigma_{\hat{p}} = 0.0341 \text{ (to 4 decimal places)}$$

In Example 8.17, we will illustrate the use of the finite correction factor when computing the standard deviation of the sampling distribution of the proportion.

EXAMPLE 8.17

Using the information contained in the frequency distribution table found in part (c) of Example 8.15, then:
a) Compute the mean and standard deviation of all the sample proportions of brown M&M's determined in part (c) of Example 8.15.
b) Using the formulas for the mean and standard deviation of the sampling distribution of the proportion, compute the mean and standard deviation of the sampling distribution constructed in Example 8.15.
c) Compare parts (a) and (b).

Solution

a) The following table represents the frequency distribution table of all the possible sample proportions of brown M&M's determined in Example 8.15.

Sample Proportion of Brown M&M's: $\hat{p}$	Frequency
0.40	10
0.20	10
0	1

The mean of all the sample proportions of brown M&M's is determined by the formula:

$$\text{mean} = \frac{\Sigma(\hat{p} * \text{frequency})}{\text{sum of the frequencies}}$$

Thus, the mean of all the sample proportions of brown M&M's is:

$$\text{mean} = \frac{(0.40)(10) + (0.20)(10) + 0(1)}{21} = 0.2857 \text{ (to 4 decimal places)} = \mu_{\hat{p}}$$

The standard deviation of all the sample proportions of brown M&M's is determined by the formula:

$$\text{standard deviation} = \sqrt{\frac{\sum (\hat{p} - \text{mean})^2 * \text{frequency}}{\text{sum of the frequencies}}}$$

The standard deviation of all the sample proportions of brown M&M's is:

$$\text{standard deviation} = \sqrt{\frac{(0.40 - 0.2857)^2 * 10 + (.20 - 0.2857)^2 * 10 + (0 - 0.2857)^2 * 1}{21}}$$

$$\text{standard deviation} \approx 0.1166 \text{ (to 4 decimal places)} = \sigma_{\hat{p}}$$

b) Using the formula for the mean of the sampling distribution of the proportion and the result from part (a) of Example 8.15, we have:

From Example 8.15 part (a), the population proportion of brown M&M's is:

$$p = \frac{X}{N} = \frac{2}{7} = 0.2857$$

Using the formula for the mean of the sampling distribution of the proportion, we have:

$$\mu_{\hat{p}} = p$$
$$\mu_{\hat{p}} = 0.2857$$

Since $\frac{n}{N}$ is greater than 5%, the **finite correction factor formula** for the standard deviation of the sampling distribution of the proportion must be used.

For $p = 0.2857$, $N = 7$ and $n = 5$, we have:

$$\sigma_{\hat{p}} = \sqrt{\frac{p(1-p)}{n}} \sqrt{\frac{N-n}{N-1}}$$

$$= \sqrt{\frac{(0.2857)(1-0.2857)}{5}} \sqrt{\frac{7-5}{7-1}}$$

$$\sigma_{\hat{p}} \approx 0.1166 \text{ (to 4 decimal places)}$$

c) The results for parts (a) and (b) are exactly the same.

The purpose of Example 8.17 was to verify the formulas for the mean and the standard deviation of the sampling distribution of the proportion. We will always use the formulas for the mean and standard deviation of the sampling distribution of the proportion because, in practice, we will be sampling from a very large or infinite population which makes it impractical or impossible to construct the sampling distribution of the proportion.

CASE STUDY 8.3   Top States for Caesarean Delivery

Examine the snapshot entitled "Top Ten U.S. States that had the highest percentage of Caesarean Deliveries at 39 Weeks" shown in Figure 8.28.
  a) What variable is being represented within the snapshot?
  b) Is this variable numeric or categorical? Explain.
  c) If we define the population to be all births in the USA for the year 2011, then state whether each of the percentages shown in Figure 8.28 represent a population or a sample proportion.
  d) Using your answer to part (c), rewrite each percentage as a decimal and use the appropriate notation to represent this proportion.
  e) Explain why this would not be considered a sampling distribution of the proportion for the population stated in part (c).
  f) Explain how you might go about constructing an appropriate sampling distribution of the proportion for the stated population. Within your explanation, identify how you would select each sample, your sample size, and so on.
  g) For the sampling distribution of the proportion you identified in part(f), state the value of the mean and standard error of the proportion using the information shown in the snapshot and the information you presented in part(f). Use appropriate notation when stating these values.

(continues)

CASE STUDY 8.3 Top States for Caesarean Delivery (continued)

Statistical Snapshot

Top Ten U.S. States that had the highest percentage of Caesarean Deliveries at 39 Weeks

In 2011, 33.7% of all births in the USA were Caesarean Delivery at 39 weeks.

State	Percentage
New York	36.10%
Mississippi	36.20%
Maryland	36.20%
West Virginia	36.90%
Kentucky	37.10%
Rhode Island	37.50%
Connecticut	37.70%
Florida	38.60%
Louisiana	39.20%
New Jersey	41.40%

Source: CDC/NCHS, National Vital Statistics System

FIGURE 8.28

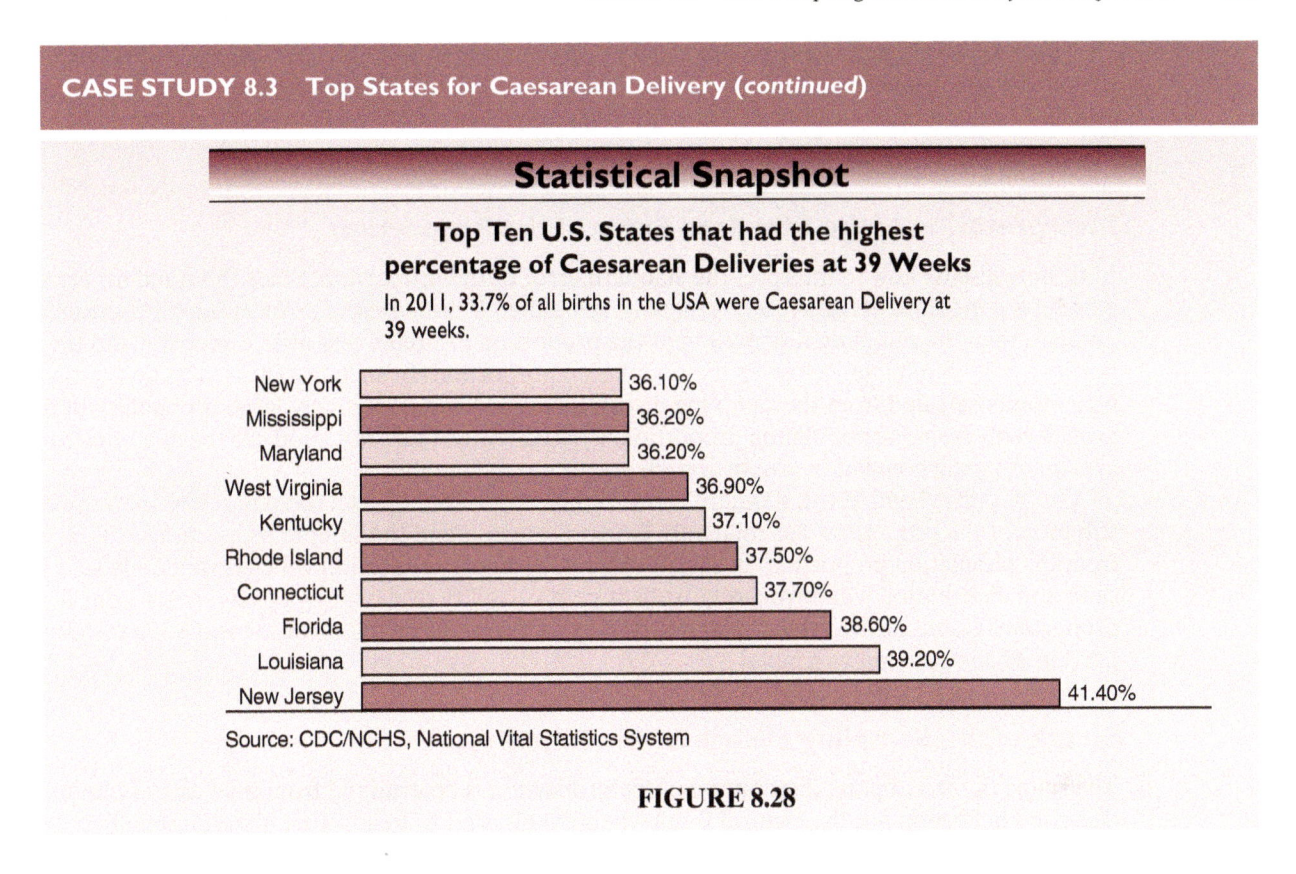

Sampling Error of the Proportion

You should realize that Example 8.15 is unusual because very rarely will we ever have any knowledge of the actual value of the population proportion. In fact that is why we need to select a random sample to estimate the value of the population proportion. Consequently, whenever we estimate the population proportion we will be subjected to committing a sampling error.

Recall from Section 8.1 that sampling errors are simply errors due to chance or the luck of the selection process. That is, *we can expect the estimates of a population proportion to vary from sample to sample just due to chance only.* This difference between the sample proportion, $\hat{p}$, and the value of the corresponding population proportion, p, is called the sampling error of the proportion.

DEFINITION 8.9
Sampling error of the proportion is the difference between the value of a sample statistic, such as the sample proportion, $\hat{p}$, and the value of the corresponding population proportion, p.

Thus, the *sampling error for the proportion* is:

sampling error = sample proportion − population proportion

assuming the sample is random and there are no nonsample errors.

In symbols,

sampling error of the proportion = $\hat{p}$ − p.

Recall in Example 8.15, the population proportion of brown M&M's was p = 0.2857. However, if we were to randomly select any of the 21 possible samples of size 5 to estimate this population proportion then we would make a sampling error. For example, suppose we had selected one of the samples with proportion $\hat{p}$ = 0.40 then the *sampling error for the proportion* would be:

sampling error = sample proportion − population proportion
= 0.40 − 0.2857
sampling error = 0.1143

Thus, we would make a sampling error of 0.1143 if the sample proportion of 0.40 was used to estimate the population proportion of brown M&M's which is p = 0.2857. What would be the sampling error if we had selected a sample with proportion $\hat{p}$ = 0.20? Hopefully, you computed the sampling error to be –0.0857! Why is the sampling error negative?

Interpretation of the Standard Error of the Proportion

Let's now discuss how to interpret the standard error of the proportion. Since the standard error of the proportion is a measure of the average sampling error, then if the *standard error* of the proportion, $\sigma_{\hat{p}}$, is a small number, then the sampling distribution of the proportion has relatively little dispersion and the sample proportions will be relatively close to the population proportion. In such instances, a sample proportion which is randomly selected from the sampling distribution will have a relatively large probability of not deviating considerably from the population proportion. Consequently, there is a good chance of a *small* sampling error in estimating the population proportion when $\sigma_{\hat{p}}$ is a *small* number.

On the other hand, if the standard error of the proportion, $\sigma_{\hat{p}}$, is a **large** number, then the sampling distribution of the proportion has relatively larger dispersion and the sample proportions will be relatively far from the population proportion. In such instances, a sample proportion which is randomly selected from the sampling distribution will have a relatively large probability of deviating considerably from the population proportion. Consequently, there is a good chance of a *large* sampling error in estimating the population proportion when $\sigma_{\hat{p}}$ is a large number.

Shape of the Sampling Distribution of the Proportion

The shape of the sampling distribution of the proportion is determined from the Central Limit Theorem. As discussed in Section 8.4, the Central Limit Theorem allowed us to conclude that the sampling distribution of the mean can be approximated by a normal distribution whenever the sample size is large (i.e. greater than 30). We can now apply the Central Limit Theorem to the sampling distribution of the proportion to obtain a statement about its shape. This is stated in Theorem 8.4.

THEOREM 8.4 Central Limit Theorem for the Sampling Distribution of the Proportion

If a random sample of size n is selected from a population with a population proportion p, then the sampling distribution of the proportion can be approximated by a normal distribution whenever the sample size is sufficiently large. A sample size n can be considered sufficiently large whenever np and n(1 – p) are both greater than 10. That is,

$$np \geq 10$$

and

$$n(1-p) \geq 10.$$

You should recall that these were the same conditions that were required for the normal approximation of the binomial distribution as discussed in Section 7.9.

EXAMPLE 8.18

According to the U.S. Census Bureau, 20% of men aged from 25 to 29 are still living at home. Assume this proportion is true for the population of all men aged 25 to 29. If a random sample of size 100 is selected from this population, then determine whether the sampling distribution of the proportion can be approximated by a normal distribution.

Solution

Since p is defined to be the proportion of all men aged 25 to 29 who still live at home, then:

$$p = 0.20$$

and

$$1 - p = 1 - 0.20 = 0.80.$$

According to the Central Limit Theorem, we need to calculate the value of np and n(1 − p) to determine if they are both greater than or equal to 10. Since n = 100, p = 0.20 and 1 − p = 0.80, we have:

$$np = 100(0.20) = 20$$

and

$$n(1 - p) = 100(0.80) = 80.$$

Since both np and n(1 − p) are greater than or equal to 10, then according to the Central Limit Theorem the sampling distribution of the proportion can be approximated by a normal distribution. ∎

Using the result of the Central Limit Theorem for the sampling distribution of the proportion and the formula for its mean and standard error, we state the characteristics of the sampling distribution of the proportion in Theorem 8.5.

THEOREM 8.5 Characteristics of the Sampling Distribution of the Proportion

If the following three conditions are satisfied:
- given a large population with population proportion p
- all possible random samples of size n are selected from the population to form the sampling distribution of the proportion, and,
- the sample size, n, is sufficiently large, i.e. np ≥ 10 and n(1 − p) ≥ 10,

then:
1) the sampling distribution of the proportion can be approximated by a normal distribution
2) the mean of the sampling distribution of the proportion is equal to the population proportion, p. This is expressed as:

$$\mu_{\hat{p}} = p$$

3) the standard deviation of the sampling distribution of the proportion or the standard error of the proportion, $\sigma_{\hat{p}}$, is:

$$\sigma_{\hat{p}} = \sqrt{\frac{p(1-p)}{n}}$$

Figure 8.29 illustrates the conceptual idea of the sampling distribution of the proportion, its characteristics, and the relationship to the characteristics of the population.

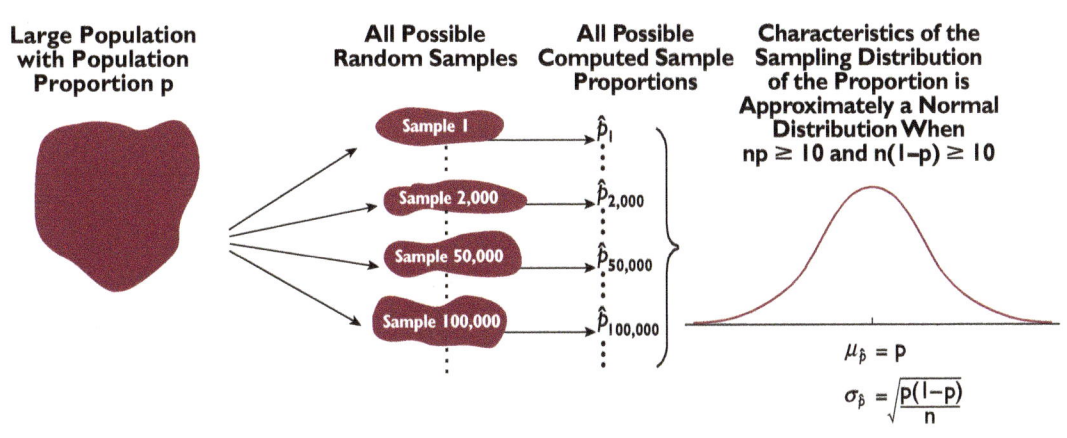

FIGURE 8.29

EXAMPLE 8.19

According to the National Highway Traffic Safety Administration, 70% of all motor vehicle drivers within the state of California use their seat belts. Assume this proportion is true for the population of all motor vehicle drivers in California who use their seat belts. If a random sample of size 200 is selected from this population, then determine the:

a) Mean of the sampling distribution of the proportion
b) Standard error of the proportion (to 4 decimal places)
c) Shape of the sampling distribution of the proportion

Solution

a) Let p be defined as the proportion of all motor vehicle drivers within the state of California who use their seat belts. Since p = 0.70 then the mean of the sampling distribution of the proportion is:

$$\mu_{\hat{p}} = p$$
$$\mu_{\hat{p}} = 0.70$$

b) The standard error of the proportion, $\sigma_{\hat{p}}$, is:

$$\sigma_{\hat{p}} = \sqrt{\frac{p(1-p)}{n}}$$

$$\sigma_{\hat{p}} = \sqrt{\frac{(0.70)(1-0.70)}{200}}$$

$$\sigma_{\hat{p}} = \sqrt{\frac{(0.70)(0.30)}{200}}$$

$$\sigma_{\hat{p}} = 0.0324$$

c) Since the values of np and n(1 – p) are

$$np = 200(0.70) = 140$$

and

$$np = 200(0.30) = 60$$

are both greater than or equal to 10, then the sampling distribution of the proportion can be approximated by a normal distribution with mean $\mu_{\hat{p}} = 0.70$, and standard error of the proportion $\sigma_{\hat{p}} = 0.0324$. This is illustrated in Figure 8.30.

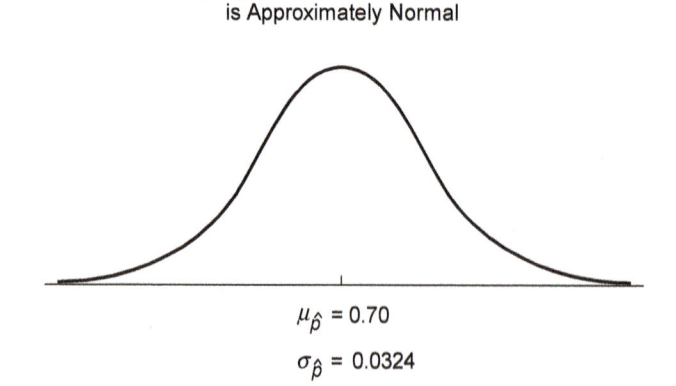

Sampling Distribution of the Proportion is Approximately Normal

$\mu_{\hat{p}} = 0.70$

$\sigma_{\hat{p}} = 0.0324$

FIGURE 8.30

Calculating Probabilities Using the Sampling Distribution of the Proportion

Later in this text we will need to make an inference or draw a conclusion about a population proportion. We will usually select only one random sample from the population and draw our inference or conclusion on the basis of this one sample. In the following examples, we will show how the sampling distribution of the proportion can be used to calculate probabilities about the value of a sample proportion $\hat{p}$ based upon the results of only one random sample.

EXAMPLE 8.20

According to Princeton Survey Research, 47% of drivers in the Midwest say they drive at or below the speed limit. Assume this percentage represents the true population proportion of all Midwest drivers who drive at or below the speed limit. Determine the probability that more than 53% in a random sample of 200 Midwest drivers will say they drive at or below the speed limit.

Solution

Let p be defined as the proportion of all drivers in the Midwest who say they drive at or below the speed limit. Thus

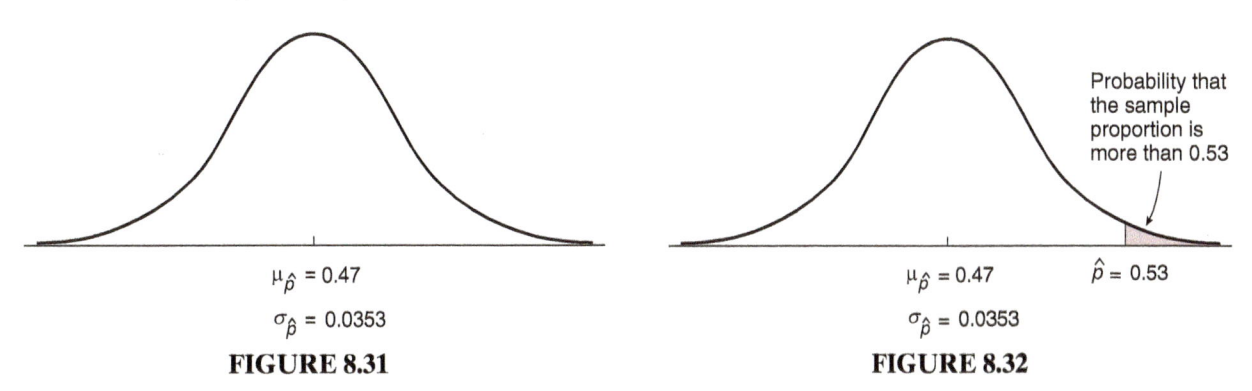

$$p = 0.47.$$

Using Theorem 8.5, the mean of the sampling distribution of the proportion is:

$$\mu_{\hat{p}} = p = 0.47.$$

The standard error of the proportion (or the standard deviation of the sampling distribution) $\sigma_{\hat{p}}$, is computed by the formula:

$$\sigma_{\hat{p}} = \sqrt{\frac{p(1-p)}{n}}$$

Since n = 200, p = 0.47 and 1 − p = 0.53, we have:

$$\sigma_{\hat{p}} = \sqrt{\frac{0.47(0.53)}{200}}$$

$$\sigma_{\hat{p}} \approx 0.0353 \text{ (to 4 decimal places)}$$

According to the Central Limit Theorem, since

$$np = 200(0.47) = 94$$

and

$$n(1-p) = 200(0.53) = 106$$

are both greater than or equal to 10, we can conclude that the sampling distribution of the proportion is approximately a normal distribution. Thus the sampling distribution of the proportion for this example is shown in Figure 8.31.

To calculate the probability that more than 53% in a random sample of 200 Midwest drivers will say they drive at or below the speed limit we must determine the shaded area under the normal curve to the right of $\hat{p} = 0.53$. This is illustrated in Figure 8.32.

Since the sampling distribution of the proportion is approximately normal, we can use Table II: The Normal Curve Area Table to obtain the probability associated with the shaded area in Figure 8.32. To compute this probability, we will need to determine a z score of the sample proportion $\hat{p} = 0.53$. As discussed in Chapter 3, the z score formula for a data value x is:

$$z \text{ of data value} = \frac{\text{data value} - \text{mean}}{\text{standard deviation}}.$$

Sampling Distribution of the Proportion is Approximately Normal

$\mu_{\hat{p}} = 0.47$

$\sigma_{\hat{p}} = 0.0353$

FIGURE 8.31

$\mu_{\hat{p}} = 0.47$ $\hat{p} = 0.53$

$\sigma_{\hat{p}} = 0.0353$

Probability that the sample proportion is more than 0.53

FIGURE 8.32

In this section, since we are working with the sampling distribution of the proportion, the data values are sample proportions, $\hat{p}$, thus, the z score formula becomes:

$$z \text{ score of } \hat{p} = \frac{\text{sample proportion} - \text{mean of the sampling distribution}}{\text{standard deviation of the sampling distribution}}.$$

Therefore, when working with the sampling distribution of the proportion, the z score formula is now symbolized as:

$$z \text{ of } \hat{p} = \frac{\hat{p} - \mu_{\hat{p}}}{\sigma_{\hat{p}}}.$$

Thus, the z score of $\hat{p} = 0.53$ is:

$$z = \frac{0.53 - 0.47}{0.0353}$$

z of $\hat{p} = 0.53$ is 1.70

From Table II, the Normal Curve Area Table found in Appendix C: Statistical Tables, the area to the left of z = 1.70 is 0.9554. The area to the right of z = 1.70 is 1 − 0.9554 = 0.0446.

Therefore, the probability that more than 53% in a random sample of 200 Midwest drivers will say they drive at or below the speed limit is 0.0446. ∎

The TI-84 calculator can be used to find a probability involving the sampling distribution of the proportion. To do this we will use the **normalcdf** function. To access this function, press 2nd VARS [DISTR] **2:Normalcdf**. Remember, the **normalcdf** function requires that you input four parameters in a specific order. The syntax for the parameters for this chapter is:

normalcdf(lower bound of shaded area, upper bound of shaded area, $\mu_{\hat{p}}$, $\sigma_{\hat{p}}$)

In Example 8.20, we were asked to find the probability that more than 53% of Midwest drivers report that they drive at or below the speed limit in a sample of 200. The mean of the sampling distribution of the proportion was given as $\mu_{\hat{p}} = 0.47$ and the standard error of the proportion was given as $\sigma_{\hat{p}} = 0.0353$. Referring to Figure 8.32 shows us that the lower boundary of the shaded region is 0.53 and the upper boundary of the shaded region is ∞ or E99. Entering these four parameters into normalcdf as shown in Figure 8.33a gives the output shown in Figure 8.33b.

FIGURE 8.33a

FIGURE 8.33b

Notice the result of the probability is 0.0446 rounded to 4 decimal places. This is the same result found in Example 8.20. ∎

EXAMPLE 8.21

According to a survey by the Better Sleep Council, (nearly) one-third of people admit to dozing off at their workplace. Assume this proportion represents the true population proportion of all workers who doze off at their workplace. If a random sample of 100 workers is selected from this population, then determine the probability that the proportion of workers who admit to dozing off at work falls between 0.24 and 0.42.

Solution

Let p be defined as the proportion of all workers who doze off at their workplace. Thus,

$$p \approx 0.33.$$

Using Theorem 8.5, the mean of the sampling distribution of the proportion is:

$$\mu_{\hat{p}} = p = 0.33.$$

The standard error of the proportion (or the standard deviation of the sampling distribution), $\sigma_{\hat{p}}$, is computed by the formula:

$$\sigma_{\hat{p}} = \sqrt{\frac{p(1-p)}{n}}$$

For n = 100, p = 0.33 and 1 − p = 0.67, we have:

$$\sigma_{\hat{p}} = \sqrt{\frac{0.33(0.67)}{100}}$$

$$\sigma_{\hat{p}} \approx 0.0470 \text{ (to 4 decimal places)}$$

According to the Central Limit Theorem, since

$$np = 100\,(0.33) = 33$$

and

$$n(1-p) = 100\,(0.67) = 67$$

are both greater than or equal to 10, we can conclude that the sampling distribution of the proportion is approximately a normal distribution. Thus the sampling distribution of the proportion for this example is shown in Figure 8.34.

To calculate the probability that the proportion of workers within the sample who admit to dozing off at work falls between 0.24 and 0.42, we must determine the shaded area under the normal curve between $\hat{p} = 0.24$ and $\hat{p} = 0.42$. This is illustrated in Figure 8.35.

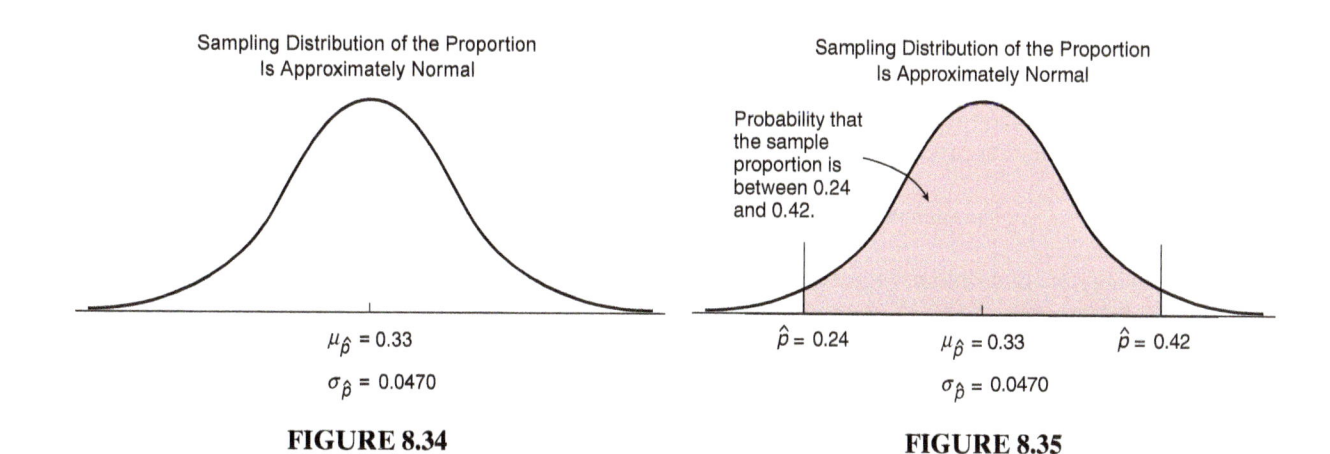

FIGURE 8.34

FIGURE 8.35

Since the sampling distribution of the proportion is approximately normal, we will use Table II: The Normal Curve Area Table to obtain the probability associated with the shaded area in Figure 8.35. To compute this probability, we will need to determine a z score for the sample proportions: $\hat{p} = 0.24$ and $\hat{p} = 0.42$. Thus, the z scores are:

$$z = \frac{0.24 - 0.33}{0.0470} \quad \text{and} \quad z = \frac{0.42 - 0.33}{0.0470}$$

z of $\hat{p} = 0.24$ is -1.91 and z of $\hat{p} = 0.42$ is 1.91

From Table II: the Normal Curve Area Table, the area associated with each of these z scores is:

z score	area to the left
−1.91	0.0281
1.91	0.9719

Subtract the smaller area from the larger area to compute the shaded area shown in Figure 8.35. This result is:

$$0.9719 - 0.0281 = 0.9438.$$

Therefore, the probability is 0.9438 that between 0.24 and 0.42 of the workers in a random sample of 100 workers will admit to dozing off at work. ∎

KEY TERMS

Bell-shaped Distribution 452
Central Limit Theorem, Theorem 8.2 463
Central Limit Theorem for the Sampling Distribution of the Proportion, Theorem 8.4 486
Characteristics of the Sampling Distribution of the Mean, Theorem 8.3 465
Characteristics of the Sampling Distribution of the Proportion, Theorem 8.5 487
Distribution of Sample Means or the Sampling Distribution of the Mean 448
Nonsampling Errors 447
Population Proportion 477

Proportion 476
Random Variable 448
Standard Deviation of the Sampling Distribution of the Mean for a Finite Population 457
Sampling Distribution of the Proportion 479
Sampling Error 447
Sampling Error of the Proportion 485
Sample Proportion 478
The Finite Correction Factor for the Standard Deviation of the Sampling Distribution of the Proportion 484

The Mean of the Sampling Distribution of the Mean, $\mu_{\bar{x}}$ 453
The Mean of the Sampling Distribution of the Proportion, $\mu_{\hat{p}}$ 482
The Standard Deviation of the Sampling Distribution of the Mean or Standard Error of the Mean, $\sigma_{\bar{x}}$ 454
The Standard Deviation of the Sampling Distribution of the Proportion or Standard Error of the Proportion, $\sigma_{\hat{p}}$ 482
Theoretical Sampling Distribution of the Mean 453
Uniform Distribution 452
Without Replacement 448

SECTION EXERCISES

Section 8.1

1. An error that occurs during the selection, measuring, recording or computation process of sampling is called an _____ error.
2. The variation in a sample estimate of the population mean is called _____ _____.
3. The sampling error of the mean is equal to $\bar{x} -$ _____.
4. The sampling distribution of the mean is a probability distribution which lists all the possible values of the _____ _____ of the same sample size selected from a population along with the _____ associated with each value of the sample mean.
5. The mean of all the sample means of the sampling distribution of the mean is symbolized as _____.
6. The standard deviation of all the sample means of the sampling distribution of the mean is symbolized as _____.

7. The difference between the population mean and a sample mean is called the:
   a) Standard deviation   b) Variance
   c) Sampling error   d) Finite correction factor
8. The sampling distribution of the mean is made up of the means of *many* random samples of size n selected from an infinite population.
   (T/F)
9. Suppose the following ten I.Q. scores represent a population of I.Q. scores:

   123, 130, 106, 117, 127, 103, 118, 133, 124, 131

   a) Compute the population mean I.Q. score, μ.
   For each of the following random samples of I.Q. scores selected from this population
   i) compute the sample mean.
   ii) determine the sampling error if this sample were used to estimate the population mean.
   b) 130, 103, 133   c) 124, 118, 117   d) 127, 124, 106

10. The following list of hourly wages represent the population of all employees working at a small local computer store
$10, $8, $9.50, $7.50, $12, $15
   a) Calculate the mean hourly wage for all the employees.
   b) Determine all the possible samples of size n = 2 that can be selected from this population without replacement.
   c) Compute all the possible sample means of size n = 2 that can be made and used to estimate the population mean.

11. Consider the finite population of six individuals who are identified by the first letters of their surnames:
    M, A, T, H, E, W. If we are going to randomly select two people from this group to form a committee, then:
   a) Identify all 15 possible pairs of people that can be randomly selected such as: MA, MT, MH, etc.
   b) Determine the probability of selecting one of these pairs of people.
   c) If each letter is assigned the following number:
      M is assigned #1, A is assigned #2, T is assigned #3, H is assigned #4, E is assigned #5, AND W is assigned #6, then list the two pairs of people who would be randomly selected using the following random numbers: 94879 16831.

Section 8.2

12. a) The mean of a sample is symbolized by _____.
    b) The mean of the population is symbolized by _____.
    c) The standard deviation of the population is symbolized by _____.
    d) The mean of the sampling distribution of the mean is symbolized by _____.
    e) The standard error of the sampling distribution of the mean is symbolized by _____.

13. The sampling distribution of the mean is made up of the means of _____ possible random samples of size n selected from a population.

14. The mean of the sampling distribution of the mean, denoted by _____ and is equal to the mean of the _____ from which the samples were selected. In symbols, this is expressed as: _____.

15. The standard deviation of the sampling distribution of the mean is called the _____ _____ of the mean.

16. The standard error of the mean is the _____ _____ of all the sample means of size n of the sampling distribution, denoted by _____. The standard error of the mean is equal to the standard deviation of the population, σ, divided by the square root of the _____ _____. In symbols, this is expressed as: _____.

17. If random samples of size 25 are selected from a population with a mean of 45 and a standard deviation of 15, then the mean of all the sample means will be:
    a) 9   b) 45   c) 60   d) 15

18. Suppose a random sample of size n = 64 is selected from a population with mean, μ, and standard deviation, σ.
    For each of the following values of μ and σ, find the value of $\mu_{\bar{x}}$ and $\sigma_{\bar{x}}$.
   a) $\mu = 20$, and $\sigma = 2$
   b) $\mu = 90$, and $\sigma = 5$
   c) $\mu = 150$, and $\sigma = 10$
   d) $\mu = 200$, and $\sigma = 50$

19. Suppose you are sampling from a population with $\mu = 850$ and $\sigma = 95$. If a random sample of size n = 81 is selected from this population, determine the mean and variance of the sampling distribution of the mean.

20. Suppose a random sample of size n is selected from a population with $\mu = 470$ and $\sigma^2 = 625$. For each of the following values of n, state the mean and standard error of the sampling distribution of the mean.
    a) 25   b) 36   c) 81   d) 200   e) 1,000

Section 8.3

21. When sampling without replacement from a finite population, the formula to compute the standard deviation of the sampling distribution of the mean is: _____.

22. When sampling without replacement from a finite population, and the sample size, n, is less than 5% of the population size, N, then a _____ _____ factor must be used. In such instances, the standard error of the sampling distribution of the mean can be computed using the formula: _____.

23. The finite correction factor is approximately one when _____.

24. Many cell phone providers are reporting the population mean length of a cell phone call to be $\mu = 1.8$ minutes with a population standard deviation of $\sigma = 0.98$ minutes. If a sample of 100 cell phone users from our community of 10,000 is selected without replacement and the mean length of time for cell phone calls for this sample is computed, then:
   a) Determine the standard error of the mean.
   b) What is the value of the finite correction factor?
   c) Compute the ratio of n/N. Is this value less than 5%? If it is, what should this indicate about the usage of the finite correction factor?

25. As part of its annual Economy and Personal Finance Survey, the Gallup Organization reported that as of 2013 the population mean retirement age has increased to 61 with a population standard deviation of 8. A random sample of 1,000 recent retirees from Nassau County's population of 200,000 retirees is selected without replacement and the mean retirement age was computed.
   a) Determine the standard error of the mean.
   b) Find the value of the finite correction factor.
   c) Calculate the value of the ratio n/N. Is this value less than 5%? If it is, what should this tell you about the usage of the finite correction factor?

Section 8.4

26. The Central Limit Theorem states that the sampling distribution of the mean can be closely approximated by a _____ distribution if the sample size is _____. That is, usually greater than _____.

27. The importance of the Central Limit Theorem is that it guarantees that the sampling distribution of the mean is approximately normal as long as the sample size is large enough regardless of the _____ of the population from which the samples were selected.

28. If all possible random samples of size 100 are selected from an infinite population whose mean is 130 and standard deviation is 10, then:
   a) The sampling distribution of the mean is approximately _____. This is a result of the _____ _____ Theorem.
   b) The mean of the sampling distribution of the mean is _____. In symbols, we write: $\mu_{\bar{x}} = $ _____.
   c) The standard error of the mean is _____. In symbols, we have: $\sigma_{\bar{x}} = $ _____.

494 Chapter 8 *Sampling and Sampling Distributions*

29. The sampling distribution of the mean is normal for all sample sizes, even as small as n = 2, if the original population, from which the samples were selected, is a _____ distribution.

30. Describe the characteristics (its mean, standard error and shape) of the sampling distribution of the mean if sampling is done from a normal population where $\mu = 110$, and $\sigma = 10$ for samples of size:
    a) n = 4
    b) n = 25
    c) n = 100

31. If the standard error of the mean, $\sigma_{\bar{x}}$, is a small number, then the sampling distribution of the mean has relatively _____ dispersion and the sample means will be relatively _____ to the population mean.

32. If random samples of size 10 are selected from a normal distribution, then the sampling distribution of the mean will be:
    a) Unknown
    b) Any shape
    c) Uniform
    d) Normal

33. When sampling from a non-normal distribution, the sampling distribution of the mean will be approximately normal when n < 30.
    (T/F)

34. For each of the following parts, you are given the approximate *population* shape, the population mean, the population standard deviation and are asked to form a sampling distribution of the mean by selecting all possible random samples of size n. Determine the:
    1) Approximate shape (normal or unknown) for the sampling distribution of the mean
    2) Mean of the sampling distribution of the mean, $\mu_{\bar{x}}$
    3) Standard error of the mean, $\sigma_{\bar{x}}$.
    a) shape: uniform
       mean, $\mu = 160$
       standard deviation,
       $\sigma = 8$
       sample size, n = 64
    b) shape: triangular
       mean, $\mu = 16$
       standard deviation,
       $\sigma = 3$
       sample size, n = 81

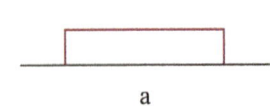

    c) shape: any shape
       mean, $\mu = 4000$
       standard deviation,
       $\sigma = 200$
       sample size, n = 25
    d) shape: normal
       mean, $\mu = 1200$
       standard deviation,
       $\sigma = 400$
       sample size, n = 16

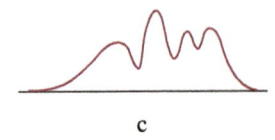

    e) shape: unknown
       $\mu = 32$
       $\sigma = 5$
       n = 100
    f) shape: bimodal
       $\mu = 500$
       $\sigma = 100$
       n = 400

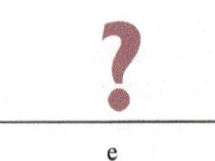

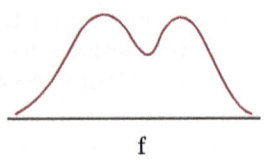

    g) shape: any shape
       $\mu = 20,000$
       $\sigma = 4500$
       n = 225
    h) shape: unknown
       $\mu = 8$
       $\sigma = 1.6$
       n = 16

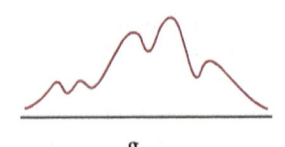

    i) shape: normal
       $\mu = 1000$
       $\sigma = 100$
       n = 4
    j) shaped: skewed
       $\mu = 24,000$
       $\sigma = 6000$
       n = 9

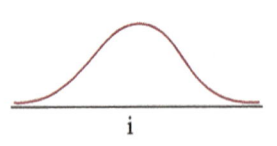

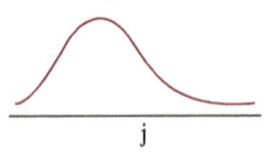

35. Suppose you are sampling from a normal population with $\mu = 525$ and $\sigma = 100$, what is the smallest sample size (greater than one) required for the sampling distribution of the mean to be normal?

Section 8.5

36. When working with the sampling distribution of the mean, the formula used for the z score is _____.

37. The mean systolic blood pressure of young female adults is 125 with a standard deviation of 24. Find the probability that a random sample of 100 young female adults will have a mean systolic blood pressure:
    a) Less than 122.
    b) Between 121 and 127.
    c) Greater than 126.

38. The population of IQ scores from a large community college has a mean of 110 with a standard deviation of 10. Find the probability that a random sample of size
    a) 36
    b) 81
    c) 100
    selected from this population will have a sample mean between 108 and 112.
    d) What happens to the probability as the sample size increases? How would this help someone who is interested in estimating the population mean?

39. The mean weight of newborn babies in a Long Island Community is 7.5 lbs. with a standard deviation of 1.4 lbs. What is the probability that a random sample of 49 babies has a mean weight of at least 7.2 lbs?

40. A National Fatherhood Initiative survey concluded that the average age at first marriage for American women is 26 years. Suppose the standard deviation for this population

is 2.5 years. If a random sample of 36 married American women is selected, what is the probability that their mean age at first marriage was less than 25 years?

41. At a sports arena, twelve professional wrestlers squeeze into an elevator on the third floor going down to the wrestling ring for the Royal Rumble. The weights of professional wrestlers are normally distributed with $\mu = 265$ lbs and $\sigma = 30$ lbs. If the elevator's weight capacity is 3400 lbs, calculate the probability that the elevator will not be able to carry the 12 wrestlers to the Royal Rumble?

42. Miss Peld, a statistics instructor, announces to her class that the number of words per page in their 800 page statistics textbook is normally distributed with $\mu = 437$ wpp and $\sigma = 46$ wpp. Angelo, a statistics student, decides to test Miss Peld's claim by randomly selecting 16 pages from the textbook and determining the mean number of words per page. What is the probability that the mean number of words per page for Angelo's sample will be greater than 460?

43. In 2014, the average number of emails received by an individual's business email account is 81 with a standard deviation of 13. If 50 business email accounts are selected at random, find the probability that the mean number of emails received is greater than 85. Is this a likely event?

44. The birthrate for a county in Kentucky is stated to be 1.45 per women. Assume the population standard deviation is 0.30. If a random sample of 400 women is selected from this county, what is the probability that the mean birthrate of the sample will fall between 1.20 and 1.70?

45. Last year, the mean batting average of all major league baseball players was .275 with a standard deviation of .018. What is the probability that the mean of a random sample of 36 baseball players is at most .280?

46. The Nielsen Company in a recent report on Internet usage in the U.S. stated the average U.S. Internet user spent an estimated 68 mean monthly number of hours online (both at home and at work). Assume the standard deviation is 10 hours per month for this population. If a random sample of 49 Internet U.S. users is selected, what is the probability that the mean monthly hours online (for both at home and at work) will be at most 65 hours (assuming The Nielsen Company's average is the true population mean)?

47. The Environmental Protection Agency's (EPA) rating of an automobile manufacturer's Z car is 27 mpg with a standard deviation of 6 mpg Suppose the Miami Police Department has randomly purchased a sample of 36 Z cars from this manufacturer. If the police officers have driven these cars exclusively in the city and have recorded a mean of 24 mpg, then:
   a) What are the chances of getting a random sample mean of at most 24 mpg?
   b) Would you question the EPA's rating for this model car based upon this sample result? Explain.

48. A light bulb manufacturer states that its energy saver 60 watt light bulb has a mean life of 1850 hours with a standard deviation of 240 hours. If random samples of size 64 of these light bulbs are selected and tested, then within what interval (centered about the mean) would you expect 95% of these sample means to fall?

49. The production characteristics for a small engine part are as follows: mean is 2.36", standard deviation is 0.04" and normally distributed.
   a) What percent of these parts fail to meet the engineering specifications of 2.35" plus or minus 0.1"?
   b) If sixteen of these parts are selected at random what is the probability that the average length is less than 2.35"?
   c) A quality control procedure requires that a random sample of size 4 be selected and that the sample mean be in the interval 2.32" to 2.38". If not, then the case of parts is declared defective. What percent of these samples will fail to meet this test?

50. According to National Express Credit the average college graduate owes about $19,000 in student loan debt, with a standard deviation of $4,000.
   a) What is the probability that a random sample of 100 graduates will have a credit card debt greater than 20,000 dollars?
   b) For a sample of size 200, what percent of graduates would you expect to have a credit card debt less than $18,000?

Section 8.6

51. As the sample size increases, the standard error of the mean _____.

52. The standard deviation of the sampling distribution of the mean gets smaller as the size of the sample
   a) Increases
   b) Remains the same
   c) Decreases
   d) Not enough information

53. The sampling distribution of the mean will have _____ dispersion as the population.
   a) Less
   b) The same
   c) More

54. As the sample size increases, the standard error of the mean
   a) Increases
   b) Decreases
   c) Remains the same
   d) Cannot be determined

55. For a particular country club of 500 members, the standard deviation of the ages is 16. If a survey is conducted to estimate the mean age by randomly sampling 81 members from this population without replacement, then determine the standard error of the mean.

56. The standard error of the mean will be reduced by 50%, if the sample size is:
   a) Doubled
   b) Tripled
   c) Quadrupled
   d) Halved

57. At a midwestern college, the mean height of all the women students is 5' 3" with a standard deviation of 2.4". Determine the mean, $\mu_{\bar{x}}$, and standard error, $\sigma_{\bar{x}}$, for the sampling distribution of the mean for samples of size:
   a) 49
   b) 81
   c) 144
   d) 900
   e) What affect does increasing the sample size have on the standard error of the mean? Explain.

Section 8.7

58. The Central Limit Theorem of the sampling distribution of the proportion states that the sampling distribution of the proportion can be approximated by a normal distribution whenever the sample size is sufficiently _____. That is, when _____ ≥ 10 and _____ ≥ 10.

59. When working with the sampling distribution of the proportion, the formula used for the z score is _____.

60. A _____ is a fraction, ratio or percentage that indicates the part of a population or sample that possesses a particular characteristic.

61. The formula to compute the population (or true) proportion of a finite population is _____.

62. The formula to compute the sample proportion is _____.

63. The notation for the mean of the sampling distribution of the proportion is _____.

64. The notation for the standard deviation of the sampling distribution of the proportion is _____.

65. The formula for the mean of the sampling distribution of the proportion is _____.

66. The formula for the standard error of the proportion is _____.

67. The formula to compute the sampling error for the proportion is: sampling error = _____ − _____.

68. If 14 out of every 35 people in the population interviewed answered yes to the survey question, then the population proportion, p, would be:
    a) 0.14
    b) 0.20
    c) 0.50
    d) 0.40

69. If a random sample of size 40 is selected from an infinite population with a proportion of women equal to 0.25, what is the standard error of the proportion of women?
    a) 0.0610   b) 0.0801   c) 0.0685   d) 0.0488

70. If the sample size is increased, then the standard deviation of the sampling distribution of the proportion will:
    a) Decrease
    b) Increase
    c) Remain the same
    d) Not enough information

71. The formula for $\mu_{\hat{p}}$ is given by:
    a) $\mu_{\hat{p}} = np$
    b) $\mu_{\hat{p}} = p$
    c) $\mu_{\hat{p}} = p/n$
    d) $\mu_{\hat{p}} = n(1-p)$

72. Determine the sample proportion, $\hat{p}$, (to 4 decimal places) for each of the following statements.
    a) Thirty out of 100 incoming freshman do not meet the minimum college requirements in reading.
    b) A family counselor finds that 180 out of 200 patients are lonely because they have forgotten how to have fun.
    c) Four out of 37 stock investors purchase stocks based upon presumed inside information (i.e. a hot tip).

73. Determine the mean and standard error of the sampling distribution of the proportion and whether or not the sampling distribution of the proportion can be approximated by a normal distribution, if:
    a) p = 0.40, and n = 50
    b) The population proportion is 0.65 and the sample size is 200.
    c) A random sample of 400 was selected from a population composed of 70% democratic.

74. Suppose a small bag of plain M&M's contains 2 brown, 2 red, 1 yellow, 1 blue, 1 green and 1 orange. We will assume this bag represents our population. Suppose we take all possible samples of 3 M&M's each.
    a) Determine the population proportion p of red M&M's.
    b) Determine the total number of samples of size 3 that can be selected from this population.
    c) Construct a frequency distribution table containing all the possible samples of size 3 and the proportion of red M&M's for each of those samples.
    d) Compute the mean and standard deviation of all the sample proportions of red M&M's.
    e) Using the formulas for the mean and standard deviation of the sampling distribution of the proportion, compute the mean and standard error of the sampling distribution. (Hint: use the finite correction factor when computing the standard deviation.)
    f) Compare parts (d) and (e).

75. According to a National Sleep Foundation Survey, 70% of adults say it takes more than 30 minutes to fall asleep at night. Let's assume that this percentage is true for the population of all U.S. adults. If a random sample of 250 adults is selected from this population, then determine the probability that the sample proportion, $\hat{p}$, of adults who take more than 30 minutes to fall asleep will be:
    a) Less than 0.66.
    b) Greater than 0.74.
    c) Between 0.64 and 0.76.

76. According to the Princeton Survey Research Group, 42% of U.S. drivers who live in the South say they drive at or below the speed limit. Assume this proportion is true for the population of all U.S. drivers living in the South. If a random sample of 300 drivers is selected from this population, then determine the probability that the sample proportion, $\hat{p}$, of drivers living in the South who say they drive at or below the speed limit is:
    a) At most 39%.
    b) At least 47%.
    c) Between 36% to 49%.

77. According to a recent study, 40% of children under the age of 2 have used a mobile device as a form of entertainment (i.e., play video games, watch movies, listen to music, etc.). If a random sample of 200 children under the age of 2 is selected, determine the probability that:
    a) At least 45% have used a mobile device as a form of entertainment.
    b) Between 37% and 48% have used a mobile device as a form of entertainment.
    c) At least 55% have **not** used a mobile device as a form of entertainment.

CHAPTER REVIEW

78. An infinite population has a $\mu = 120$ and $\sigma = 15$. A random sample of size n = 16 is selected from the population. Determine the mean, $\mu_{\bar{x}}$, and the standard error of the mean, $\sigma_{\bar{x}}$.

79. A very large population has a mean = 1540 and a variance of 40,000. A random sample of n = 100 is selected from the population. Find the mean, $\mu_{\bar{x}}$, of the sampling distribution of the mean and the standard error, $\sigma_{\bar{x}}$, of the mean.

80. The general rule widely accepted and used by statisticians is that the sampling distribution of the mean can be approximated by a normal distribution provided the sample size n is greater than 30 regardless of the shape of the population. (T/F)

81. A Mobile Marketing Company surveyed all 2,000 college students at a small private college. They reported that college students spend a mean time of 2.7 hours per day socializing on their mobile device with a standard deviation of 0.79 hours. If a sample of n students is randomly selected from this college without replacement and the mean number of hours a student spends socializing on their mobile device for the sample is computed, then determine the standard error of the sampling distribution of the mean for the following samples of size n.
  a) n = 16
  b) n = 81
  c) n = 400
  d) Explain what happens to the value of the finite correction factor as the sample size increased from 16 to 400.

82. A small calculus class of 8 students received the following scores on their midterm: 91, 84, 73, 82, 69, 66, 78, 87.

Assuming these scores represent a population, then:
  a) Compute the population mean test score, μ.

For each of the following random samples of test scores selected from this population
  i) compute the sample mean.
  ii) determine the sampling error if this sample were used to estimate the population mean.
  b) 84, 69, 87
  c) 91, 82, 73
  d) 66, 91, 66

83. The mean balance that college students owe on their credit cards is $896 with a standard deviation of $104. If all possible random samples of size 169 are taken from this population, determine:
  a) The mean and standard error of the sampling distribution of the mean.
  b) The percent of sample means that are less than $880.
  c) The probability that sample means fall between $884 and $900.
  d) Below what sample mean can we expect to find the lowest 20% of all the sample means?

84. According to a National Health Survey, the mean cholesterol level for men aged 18 to 24 is 197. Assuming that this is the mean of the population of cholesterol levels for men aged 18 to 24 with a standard deviation of $\sigma = 20$, then determine the mean and standard error of the sampling distribution of the mean for samples of size:
  a) n = 49
  b) n = 400
  c) What effect did increasing the sample size have on the mean and standard error of the sampling distribution?

85. A college instructor has determined that the mean time it takes all the students to complete the final exam is $\mu = 92.5$ minutes with a standard deviation of $\sigma = 17.6$ minutes. Determine the mean and the standard error of the sampling distribution of the mean and describe the shape of the sampling distribution of the mean when the sample size is:
  a) n = 49
  b) n = 144

86. Determine the sample proportion, $\hat{p}$, (to 4 decimal places) for each of the following statements.
  a) A nationwide survey of 1200 people found that 800 favored the death penalty for murder.
  b) A government agency has determined that 1850 out of a sample of 5000 student borrowers have defaulted on their federal student loans.

87. Determine the mean and standard error of the sampling distribution of the proportion and whether or not the sampling distribution of the proportion can be approximated by a normal distribution, if:
  a) p = 0.10, and n = 30
  b) The population proportion is 0.80 and the sample size is 100.
  c) A random sample of 150 was selected from a population composed of 20% unemployed people.

88. A National Survey conducted by The Social Science Research Center at Old Dominion University found that 56% of all drivers admitted they run red lights. Assume this proportion is true for the population of all U.S. drivers. If a random sample of 500 drivers is selected from this population, then determine the probability that the sample proportion, $\hat{p}$, of drivers who admit to running red lights is:
  a) Less than 59%.
  b) Between 53% and 58%.
  c) Greater than 61%.
What would be the sampling error if we had selected a sample:
  d) With proportion $\hat{p} = 59\%$ of drivers who admit to running a red light.
  e) With proportion $\hat{p} = 51\%$ of drivers who admit to running a red light.
  f) Is it possible for the sampling error to be negative? Explain.

89. According to a Gallup Poll, 60% of adults say they experience heartburn. Assume this proportion is true for the population of all adults. If a random sample of 300 adults is selected from this population, then determine the probability that:
  a) Less than 172 adults within the sample experience heartburn.
  b) More than 191 adults within the sample experience heartburn.
  c) Between 164 to 198 adults within the sample experience heartburn.
What would be the sampling error if we had selected a sample:
  d) With proportion $\hat{p} = 59\%$ of adults who experienced heartburn.
  e) With proportion $\hat{p} = 51\%$ of adults who experienced heartburn.
  f) Is it possible for the sampling error to be negative? Explain.

WHAT DO YOU THINK?

90.I Each of the following statements represent an excerpt from different studies which indicate how the sample for the study was selected. For each sampling procedure, indicate whether you think the procedure to select the sample results is biased or unbiased. Explain the reason for your answer, in some cases you may wish to answer "not enough information given." In such cases, explain why.

a) "A market research survey was mailed to a random sample of 250 owners of new automobiles. A total of 85 completed surveys was received yielding a return rate of 34%."

b) "Forty-six rats, all coming from one of several litters, were separated from their mother when they were three weeks old. One half (Group A) were put into individual cages; the other half (Group B) were left together to see what effect a rat's early environment has on its behavior later in life."

c) "A total of 356 persons were interviewed, 160 men and 196 women, to test whether men have nightmares as often as women."

d) "The subjects in this sample were 90 children, ranging in age from 6 to 12, who were selected from a list of students attending an elementary school. Three children were selected at random from each of the classes within the school."

e) "A telephone survey was conducted to determine current attitudes toward birth control education in the public high schools. The respondents were 1,345 men and women of voting age; all were white."

f) "A total of 199 persons, all complaining of frequent headaches, agreed to participtttate in a study to compare the effectiveness of four different headache remedies. The participants were divided into four groups (I, II, III, and IV)."

90.II The article entitled "Poll Involved Queries to 1,540" describes the procedure used to select adults for a New York Times/CBS News Public Opinion Poll.

Read the procedure that was used to conduct this public opinion poll.

a) Define the population for this poll.

b) What do you think are some problems with selecting a random sample of adults for a national poll using a telephone interview?

c) What does the statement "the exchanges were chosen in such a way as to insure that each region of the country was represented in proportion to its population" mean? What is the purpose of doing this?

d) What does the statement "the results have been weighted to take account of household size and to adjust for variations in the sample relating to region, race, gender, age and education"? What do you believe is the purpose for doing this?

e) How would you translate the statement "that in 95 cases out of 100 the results based on the entire sample differ by no more than 3 percentage points in either direction from what would be obtained by interviewing all adult Americans"?

Poll Involved Queries to 1,540

The latest New York Times/CBS News Poll is based on telephone interviews conducted from Jan. 11 through Jan. 15 with 1,540 adults around the United States.

The sample of telephone exchanges called was selected by a computer from a complete list of exchanges in the country. The exchanges were chosen in such a way as to insure that each region of the country was represented in proportion to its population. For each exchange, the telephone numbers were formed by random digits, thus permitting access to both listed and unlisted residential numbers.

The results have been weighted to take account of household size and to adjust for variations in the sample relating to region, race, gender, age and education.

In theory, it can be said that in 95 cases out of 100 the results based on the entire sample differ by no more than 3 percentage points in either direction from what would have been obtained by interviewing all adult Americans. The error for smaller subgroups is larger, depending on the number of sample cases in the subgroup.

The theoretical errors do not take into account a margin of additional error resulting from the various practical difficulties in taking any survey of public opinion.

From *The New York Times*. © 1982 The New York Times Company. All rights reserved. Used under license.

91. Was The 1969 Draft Lottery a Random Selection?

Background:

On November 29, 1969, President Nixon issued an executive order prescribing the Selective Service draft process to be a random selection. The purpose for a random draft selection was to address perceived inequities in the draft system.

On December 1, 1969, the first draft lottery since World War II was held, at the Selective Service National headquarters in Washington, D.C. The draft process determined the "order of call" for 1970 for all men of draft age (approximately 850,000 men), which included all men born in the years 1944 through 1950.

Draft Lottery Procedure:

Navy Capt. William S. Pascoe, chief of public information for the Selective Service System, was assigned the task of conducting the lottery. This is a summary description of the actual procedure.

Before the December 1st draft lottery, 366 cylindrical capsules for each of the 366 days of the calendar year including February 29th were setup. Thirty-once capsules were selected and slips of paper with the 31 January dates were inserted within the capsules. The January capsules were then placed in a large, square wooden box and pushed to one side with a cardboard divider, leaving part of the box empty. Then, the 29 February capsules were then poured into the empty portion of the box, counted again, and then pushed with the divider into the January capsules. This same process was followed with each subsequent month, counting the capsules into the empty side of the box and then pushing them with the divider into the capsules of the previous months. Under this process,

The 1969 Draft Numbers Listed By Day and Month

DAY OF THE MONTH	Jan	Feb	March	April	May	June	July	Aug	Sept	Oct	Nov	Dec
1	305	86	108	32	330	249	93	111	225	359	19	129
2	159	144	29	271	298	228	350	45	161	125	34	328
3	251	297	267	83	40	301	115	261	49	244	348	157
4	215	210	275	81	276	20	279	145	232	202	266	165
5	101	214	293	269	364	28	188	54	82	24	310	56
6	224	347	139	253	155	110	327	114	6	87	76	10
7	306	91	122	147	35	85	50	168	8	234	51	12
8	199	181	213	312	321	366	13	48	184	283	97	105
9	194	338	317	219	197	335	277	106	263	342	80	43
10	325	216	323	218	65	206	284	21	71	220	282	41
11	329	150	136	14	37	134	248	324	158	237	46	39
12	221	68	300	346	133	272	15	142	242	72	66	314
13	318	152	259	124	295	69	42	307	175	138	126	163
14	238	4	354	231	178	356	331	198	1	294	127	26
15	17	89	169	273	130	180	322	102	113	171	131	320
16	121	212	166	148	55	274	120	44	207	254	107	96
17	235	189	33	260	112	73	98	154	255	288	143	304
18	140	292	332	90	278	341	190	141	246	5	146	128
19	58	25	200	336	75	104	227	311	177	241	203	240
20	280	302	239	345	183	360	187	344	63	192	185	135
21	186	363	334	62	250	60	27	291	204	243	156	70
22	337	290	265	316	326	247	153	339	160	117	9	53
23	118	57	256	252	319	109	172	116	119	201	182	162
24	59	236	258	2	31	358	23	36	195	196	230	95
25	52	179	343	351	361	137	67	286	149	176	132	84
26	92	365	170	340	357	22	303	245	18	7	309	173
27	355	205	268	74	296	64	289	352	233	264	47	78
28	77	299	223	262	308	222	88	167	257	94	281	123
29	349	285	362	191	226	353	270	61	151	229	99	16
30	164		217	208	103	209	287	333	315	38	174	3
31	211		30		313		193	11		79		100

FIGURE 8.36

the January capsules were mixed with the other capsules 11 times, the February capsules 10 times and so on, with the November capsules intermingled with others only twice and the December ones only once. Finally, the box was closed, and the entire box was shook several times. The box was then carried down to the room where the drawing take place. In public view, the 366 capsules were poured from the black box into a two-foot-deep clear bowl. Once in the bowl, the capsules were *not* stirred since it was a stationary bowl.

No one knows which end of the box containing the 366 capsules were poured into the bowl. If the end where the capsules with the early months had been repeatedly shoved, these capsules might have fallen to the bottom of the bowl. Conversely, if the other end was poured from, the later months could have fallen to the bottom. This conjecture is assuming that the shoving and shaking procedure did not adequately mix the capsules.

Generally, it was noted that the individuals who selected the capsules picked the capsules from the top, however, occasionally an individual would pick from the middle or the bottom of the bowl.

Interpretation of the Draft number:

A person's draft number corresponds with the order in which his birthday was selected. For example, September 14th was the first date selected and men with that birthday were assigned the draft number 1. As opposed to June 8th which was the last birthday selected and men with that birthday were assigned the draft number 366. Thus, the lower the draft number the greater the probability the individual would be drafted. Figure 8.36 displays the 1969 Draft Numbers by Day and Month.

There has been much discussion that the 1969 draft lottery was **not** a random draft selection process. Over the years, statistical studies of the supposedly random selection process indicated that the later dates of the year received disproportionately low draft numbers, due possibly to insufficient mixing of the capsules. Regardless, the results of the lottery were not changed.

One argument that the draft selection process was not random was based on the monthly mean average draft numbers. For example, the mean average draft number for men born in January was 201 while the mean average draft number for men born in December was 122.

a) Compute the monthly mean average draft numbers for the months February to November.
b) Which months have the highest monthly mean average draft number? Which months have the lowest monthly mean average draft number? Do you notice any difference in the monthly mean average draft number for the later months of the year compared to the beginning months of the year?
c) How would you interpret the highest monthly mean average draft number vs the lowest monthly mean average draft number regarding the months with the greater chance to be drafted?
d) Use the mean of a discrete random variable formula (found in Chapter 6) to compute the expected the mean monthly draft value.
e) Compare the monthly mean average draft number for each of the 12 months to the expected mean monthly draft value to determine which months have an expected mean monthly draft number value below the expected value and which months have an expected mean monthly draft number value above the expected value.
f) Using the results of part (e), what did you notice about the later months July to December? What did you notice about the beginning months January to June? Using this information, what do you believe this indicates about months with the greater probability to be drafted?
g) Construct a Box Plot of the draft numbers for each month and place the Box Plots vertically next to each other starting with January to the last month December. What do you notice about the median value of each Box Plot? Which months have the largest median value? Which months have the smallest median value? After examining the Box Plots, do you believe regardless of the month an individual was born an individual would have the same chance of being drafted?
h) If you recall, the bowl containing the capsules was stationary so the capsules could not be rotated. Furthermore, if the capsules containing the earlier months were placed into the bowl first and the capsules containing the later months were placed on top of the earlier months, how does this information coincide with the results of part (e)? How does this information coincide with the information displayed in the 12 box plots of part (g)?
i) If the lottery selection process was random, would you expect the results of part(e) or the Box Plot results of part (g)?
j) So, based on the statistical information you computed: "Do You Think the 1969 Draft Lottery selection process was random?"

PROJECTS

92. a) Take a set of 50 blank plastic chips and write the following numbers on the chips:

Number on Chips	Frequency
10	6
9	5
8	6
7	4
6	7
5	7
4	4
3	5
2	5
1	1

b) Construct a histogram for the above distribution using a class width of 1.
c) Calculate:
   1) the mean (μ) of the population.
   2) the standard deviation (σ) of the population.
d) Place the 50 chips in a box. Shake the box rigorously and randomly choose 20 samples of size 9 with replacement. (Be sure to shake the box each time before a new sample is chosen.) For *each* sample chosen, calculate the sample mean, $\bar{x}$, and indicate the results in the table below.

Sample	$\bar{x}$
1	
2	
3	
.	
.	
.	
20	

e) Using the 20 sample means, construct a histogram and frequency polygon.
f) Calculate the mean of all the sample means, $\mu_{\bar{x}}$, and the standard deviation of the sample means, $\sigma_{\bar{x}}$. Place these values in the following table.
g) Calculate the expected mean of the sample means, and the expected standard deviation of the sample means using the formulas:

$$\mu_{\bar{x}} = \mu, \text{ and } \sigma_{\bar{x}} = \frac{\sigma}{\sqrt{n}}.$$

Place these values in the following table.

Computed Values	Expected Values
$\mu_{\bar{x}} =$	$\mu_{\bar{x}} = \mu$
$\sigma_{\bar{x}} =$	$\sigma_{\bar{x}} = \dfrac{\sigma}{\sqrt{n}}$

h) Compare the computed values to the expected values of $\mu_{\bar{x}}$ and $\sigma_{\bar{x}}$.
i) Repeat the above procedure for twenty random samples of size 25 with replacement. Compare and interpret the results of both experiments.

93. The formulas of $\mu_{\bar{x}}$ and $\sigma_{\bar{x}}$ are dependent upon the values of the population mean, μ, and population standard deviation, σ, and the size, n, of the sample selected. Given the population: 1, 2, 3, 4, 5.
   a) Calculate μ and σ for this population.
   b) Form a distribution by selecting *with replacement* all samples of size n = 2.
   c) Calculate the mean of each sample found in part (b).
   d) Form a sampling distribution of the mean from the sample means found in part (c).
   e) Find the mean and standard deviation of the sample means found in part (c). (i.e. find $\mu_{\bar{x}}$ and $\sigma_{\bar{x}}$).
   f) Form a distribution by selecting *without replacement* all samples of size n = 2.
   g) Calculate the mean of each sample found in part (f).
   h) Form a sampling distribution of the mean from the sample means found in part (g).
   i) Find the mean and standard deviation of the sample means found in part (g).
   j) Compute the theoretical values for $\mu_{\bar{x}}$ and $\sigma_{\bar{x}}$ using the formulas for the mean and standard deviation of the sampling distribution of the mean.
   k) Compare the results of parts (e) and (j).
   l) Compare the results of parts (i) and (j). Notice that the results of parts (i) and (j) will not be equal unless the finite correction factor is applied. Verify. Explain why the finite correction factor is necessary for parts (i) and (j).

CHAPTER TEST

94. To apply the Central Limit Theorem for means the shape of the original population must be known. (T/F)

95. The sampling distribution of the mean will be approximately normally distributed when the sample size is greater than:
   a) 25   b) 30   c) 50   d) 100

96. A random sample of size n = 8 is selected from a population. Under what conditions will the sampling distribution of the mean be a normal distribution?

97. According to the registrar at a local community college, the average mathematics SAT score for all the students is 514. A random sample of 100 students selected from this student body yields an average math SAT score of 529. Determine the sampling error if this sample were used to estimate the population mean mathematics SAT score.

98. For a normally distributed population, what sample size is required to insure that the sampling distribution of the mean is also normally distributed?
   a) n = 2   b) n = 10   c) n = 30   d) n = 31

99. Recently, it was reported that a baby born in 2014 is expected to live to a mean age of 79 years. If the standard deviation for this population is 10 years, use the Central Limit Theorem to answer questions about a sampling distribution of the mean formed by taking from the population of babies born in 2014, all possible random samples of size 100.
   a) Find $\mu_{\bar{x}}$
   b) Determine $\sigma_{\bar{x}}$
   c) Can this sampling distribution of the mean be approximated by a normal distribution?
   d) If the population of life expectancies for those born in 2014 is approximately normal, what is the probability that a baby born in 2014 will live to: at least age 78? at least age 82?
   e) If a random sample of 100 babies born in 2014 were selected, what is the probability that these babies will live to at least a mean age 78? to at least a mean age 82?
   f) Why are the answers to parts (d) and (e) different?

100. Consider the following population of data values:

   29, 31, 33, 35, 37, 39, 41

   a) Find the mean, μ, and the standard deviation, σ.
   b) Construct the sampling distribution of the mean for random samples of size n = 2 from this population with replacement by listing all the possible sample means along with the probability associated with each sample mean.
   c) Calculate $\mu_{\bar{x}}$ and $\sigma_{\bar{x}}$ using the information in part (b). Compare these values to the population values computed in part (a). Discuss your observations.
   d) Compute all the possible sampling errors that could be made if any of these samples were used to estimate the population mean. In addition, determine the probability of making such an error.

101. Suppose a population is infinite with $\sigma = 90$. Calculate the standard error of the mean for random samples of size:
   a) 64
   b) 100
   c) 144
   d) 1000
   e) 10,000
   f) Explain what happens to the standard error of the mean as the sample size, n, increases?
   g) If you wanted to estimate the mean of the population, which of the previous sample sizes would you use to minimize the sampling error? Explain.

102. At a university sports complex, eleven football players squeeze into the elevator on the seventh floor. The weights of the university's football players is normally distributed with $\mu = 235$ lbs and $\sigma = 30$ lbs. If the elevator's weight capacity is 2500 lbs, calculate the probability that the elevator will not be able to carry this heavy load?

103. The mean tire life of a certain brand of radial tires is 40,000 miles with a standard deviation of 2000 miles. If the life of this brand of tires is normally distributed then:
   a) What is the probability a tire will last longer than 38,000 miles?
   b) What is the probability that the mean life of a random *sample* of 100 tires is greater than 40,400 miles?
   c) What is the probability that the mean life of a random *sample* of 400 tires is less than 40,200 miles?

104. The battery life for a smartphone is approximately normally distributed with a mean number of discharge/recharge cycles of 400, and a standard deviation of 50 discharge/recharge cycles. Find the probability that a random sample of size
   a) 36
   b) 81
   c) 100
   selected from the manufacturing process has a sample mean between 390 and 410.
   d) What happens to the probability as the sample size increases?
   e) How would this help someone who is interested in estimating the population mean?

105. If you are sampling from a normal population, would it be advantageous to take a large sample? Explain.

106. Air travel from JFK New York to Sarasota, Florida takes an average 150 minutes with σ = 8 minutes. Use the Central Limit Theorem to determine the characteristics and to answer questions about a sampling distribution of the mean formed by selecting from this population random samples of size 64.
   a) Find $\mu_{\bar{x}}$.
   b) Determine $\sigma_{\bar{x}}$.
   c) Can this sampling distribution of the mean be approximated by a normal distribution?
   d) What is the probability that a random sample of 64 flights from JFK to Sarasota will average (mean average) more than 10 minutes late?
   e) If the population of air travel time from JFK to Sarasota is normally distributed, what is the probability that a flight from JFK to Sarasota is more than 10 minutes late?
   f) Why are the answers to parts (d) and (e) different?

107. Suppose you are sampling from a population with a population variance σ^2 = 1600. If you would want the standard error of the mean to be at most 8, what is the minimum sample size you should use?

108. For the following population shapes where the population standard deviation is known,
   (I) a normal population
   (II) a bell-shaped population: close to a normal
   (III) a slightly skewed population
   (IV) a highly skewed population
   determine which type of population shape would require:
   a) The smallest sample size
   b) The largest sample size
   to use the normal distribution as an approximation of the sampling distribution of the mean?

109. A CNN survey reported that 803 of women drivers and 874 of men drivers pump their own gas. If this survey was based upon a random sample of 1005 male drivers and a random sample of 1100 female drivers, then determine the sample proportion of (and state result using the appropriate notation):
   a) Male drivers who pump their own gas.
   b) Male drivers who do not pump their own gas.
   c) Women drivers who pump their own gas.
   d) Women drivers who do not pump their own gas.

110. Suppose a population is infinite with p = 0.54. Calculate the standard error of the proportion for random samples of size:
   a) 100
   b) 200
   c) 400
   d) 800
   e) 1200
   f) Explain what happens to the standard error of the proportion as the sample size, n, increases?
   g) Assuming you didn't know the population proportion and you wanted to estimate p, which of the previous sample sizes would you use to minimize the sampling error? Explain.

111. According to a Fox News/Opinion Dynamics Poll, 44% of adults believe there is intelligent life on other planets. Let's assume that this percentage is true for the population of all U.S. adults. If a random sample of 200 adults is selected from this population, then determine the probability that the sample proportion, $\hat{p}$, of adults who believe there is intelligent life on other planets is:
   a) Between 46% and 51%.
   b) At least 50%.
   Determine the probability that:
   c) At most 95 adults sampled will indicate they believe in aliens.
   d) More than 105 adults sampled will indicate they believe in aliens.
   Determine the sampling error if we had selected a sample:
   e) With proportion $\hat{p}$ = 49% of adults who believe there is intelligent life on other planets.

112. A biologist is studying the weights of female Polar Bears in the Arctic and finds that the mean weight is 190 kg with a standard deviation of 35 kg.
   a) One female Polar bear is chosen at random from the population. What is the probability that the randomly selected bear will have a weight that is greater than 200 kg?
   b) What is the probability that a sample of 50 randomly selected female Polar bears will have a mean weight that is greater than 200 kg?
   c) What is the probability that a sample of 50 randomly selected female Polar bears will have a mean weight that is between 185 and 195 kg?
   d) Compare the answers you computed in parts a. and b. Is it easier for 1 individual bear's weight to deviate far from the mean; or is it easier for a group of bears average weight to deviate far from the mean?

For the answer to this question, please visit www.MyStatLab.com.

CHAPTER 9

Estimation

Contents

9.1 Introduction
9.2 Point Estimate of the Population Mean and the Population Proportion
9.3 Interval Estimation
9.4 Interval Estimation: Confidence Intervals for the Population Mean
    Constructing a Confidence Interval for a Population Mean: When the Population Standard Deviation Is Unknown
    The t Distribution
9.5 Interval Estimation: Confidence Intervals for the Population Proportion
9.6 Determining Sample Size and the Margin of Error
    Sample Size for Estimating a Population Mean, μ
    Sample Size for Estimating a Population Proportion, p
    Summary of Confidence Intervals

9.1 INTRODUCTION

"What time is it?"

The response to this question is an **estimate** of the actual time. Estimates like this are made everyday. In fact, occasions arise in business, in social science, in medicine and in science when it becomes necessary to estimate a population parameter such as the population mean, the population proportion or the population standard deviation. For instance:

- A sociologist is interested in the mean age of women when delivering their first child.
- The census bureau is trying to determine the proportion of American workers who use computers at work.
- A quality control engineer wants to monitor the variability of the sugar content of soft drinks.
- An environmentalist needs to know the mean amount of toxins in the city's water supply.
- A student is interested in the proportion of students who pass a statistics course with a particular instructor.
- A commuter wishes to know the variability of the amount of time necessary to travel to work using mass transit.
- A company president wants to determine the proportion of people that would purchase a new line of shavers.

In each of these examples, the researcher is trying to determine the *true* or *actual population parameter*. In practical applications, the true population parameter is unattainable since populations are large and it would be either impractical or impossible to obtain the entire population data to calculate the true population parameter. In such instances, it becomes necessary to select a random sample from the population and use the sample data to estimate the population parameter. The process of using sample data to make an inference about a population parameter is called statistical inference. In this chapter, we will discuss one major area of statistical inference called **estimation**.

9.2 POINT ESTIMATE OF THE POPULATION MEAN AND THE POPULATION PROPORTION

In this section, we will introduce procedures that enable us to estimate the population mean or the population proportion using information from a sample. A procedure that assigns a numerical value to a population parameter based upon sample information is called estimation.

> **DEFINITION 9.1**
> **Estimation** is the statistical procedure which uses sample information to estimate the value of a population parameter such as the population mean, population standard deviation or population proportion.

Suppose a sociologist is interested in determining the population mean age of southern California women when delivering their first child. How would the sociologist obtain an *estimate* of the mean age of this population?

The sociologist could *randomly* select a sample of the birth records of first time mothers from the southern California area. Using this sample of records, the sample mean age of the first time mothers can be calculated. This *sample mean* is used to *estimate* the population mean age of first time mothers. Such an estimate is called a **point estimate of the population mean**.

> **DEFINITION 9.2**
> **Point Estimate** is a sample estimate of a population parameter, such as, a population mean or population proportion. The point estimate is expressed by a single number.

CASE STUDY 9.1 Average Shower Time

Examine the snapshot entitled "Long, hot shower?" shown in Figure 9.1. Notice that there are estimates for the average U.S. shower time for all adults, men, women, 18–24 age group and 45–54 age group. These averages represent *point estimates* of the *true* population shower times for these different population groups.

From this snapshot, it is estimated that men spend an average of 11.4 minutes taking a shower while women spend an average of 13 minutes in the shower.

On the other hand, adults aged 18–24 spend an average of 16.4 minutes taking a shower while adults aged 45–54 spend an average of 10.4 minutes in the shower.

Overall, an adult spends an average of 12.2 minutes taking a shower. This average represents **a point estimate** of the mean shower time of **all U. S. adults**.

Statistical Snapshot

Long, Hot Showers?

The average U.S. shower is 12.2 minutes. Variations by gender and age:

Minutes

Group	Minutes
Men	11.4
Women	13
18–24	16.4
45–54	10.4

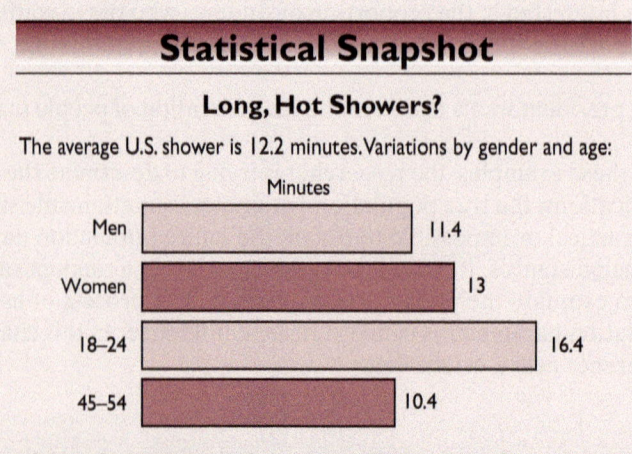

Source: Opinion Research Corporation of Teldyne Water Pik

FIGURE 9.1

CASE STUDY 9.2 Do You Believe in Aliens?

Examine the pie chart graphics shown in the snapshot of Figure 9.2. The graphic provides an estimate of the proportion of U.S. men and women who believe in aliens.

Statistical Snapshot
Do You Believe in Aliens?

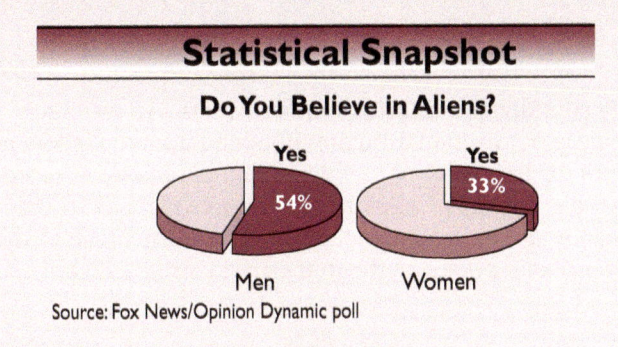

Men: Yes 54%
Women: Yes 33%

Source: Fox News/Opinion Dynamic poll

FIGURE 9.2

According to the information contained within the snapshot in Figure 9.2, the *point estimate* of the **proportion** of **all U.S. men** who believe in intelligent life on other planets is 54% while the *point estimate* of the **proportion** of **all U.S. women** who believe in intelligent life on other planets is 33%.

In Case Studies 9.1 and 9.2, a point estimate was used to estimate a population parameter. In Case Study 9.1, the population mean was estimated by a sample mean, while in Case Study 9.2, the population proportion was estimated by a sample proportion. **A sample mean, $\bar{x}$, is one method used to obtain a point estimate of the population mean, μ; while the sample proportion, $\hat{p}$, is used to obtain a point estimate of the population proportion, p. The sample mean and the sample proportion formulas are called estimators.** Any sample statistic formula that is used to estimate a population parameter is called an estimator.

DEFINITION 9.3
An **Estimator** is a method or formula used in estimating a population parameter.

It is important to distinguish between an *estimate* and an *estimator*. If a sample mean is used to **estimate** the population mean, then the **numerical value obtained** from the sample mean formula is the **estimate** while the **estimator** is the **sample mean** *formula* which was used to compute this point estimate.

EXAMPLE 9.1

The Student Government Association (SGA) at Nassau Community College is interested in estimating the mean number of hours that all full-time students work during a school week. If the SGA randomly samples 200 students and calculates the sample mean number of hours students work to be 22.4 hours, then determine the:
a) Population
b) Population parameter being estimated
c) Estimator used to estimate the population parameter
d) Point estimate for the population parameter.

Solution

a) The population is all full-time students attending Nassau Community College.
b) The population parameter to be estimated is the **population mean** number of hours that a full-time student works during a school week.
c) The estimator is the *sample mean formula* used to calculate the **mean number of hours** a full-time student works during a school week for the **sample of 200 students**.
d) The point estimate for the population mean number of hours is 22.4 hours.

A question that needs to be addressed at this point is:

How do we know that the sample mean, $\bar{x}$, is a good estimator for the population mean, μ?

The criteria for a **good estimator** are:

- Unbiasedness
- Efficiency
- Consistency
- Sufficiency

We will **only** consider the criteria of *unbiasedness*. An estimator of a population parameter is said to be **unbiased** if the mean of the point estimates obtained from the independent samples selected from the population will *approach* the true value of the population parameter as more and more samples are selected.

Based on our discussion in Chapter 8, the sample mean is an unbiased estimator of the population mean because the mean of all the sample means is equal to the mean of the population. It can also be shown that the sample proportion, $\hat{p}$, possesses the four criteria discussed above to be a good estimator of the population proportion, p.

9.3 INTERVAL ESTIMATION

In the previous section, we considered the concept of a point estimate. In practice, we will only select **one** sample to compute a point estimate. However, we must realize that the sample we select is **only one** of the many possible samples that could have been selected from the population.

For example, if you were interested in estimating the mean GPA of the population of 26,000 community college students attending Nassau Community College, you would *randomly* select a sample of students and calculate the mean GPA of the sample. Suppose you determine that the mean GPA for the selected sample is 2.63. Although this **point estimate** of 2.63 would represent our best guess for the true population mean GPA, you should realize that this estimate of 2.63 is *probably* **not going to be exactly equal to the true population mean**! In fact, if you selected a second random sample of students and calculated the mean GPA of this second sample, you probably would have arrived at a result that is *different* from the first sample mean GPA of 2.63.

Since samples do vary, one major disadvantage of using one sample mean as a point estimate of the population mean is that we don't know how close or far the point estimate is from the *true* population mean. Thus, when using a point estimate to estimate a population parameter, we are not sure of the extent of the *error* involved in the estimate.

An estimate of a population parameter would be more useful if we could provide a measure of the error associated with the estimate of a population parameter. In fact many opinion polls contain a measure of the extent of error associated with estimating a population parameter. Case Study 9.3 illustrates the idea of a "sampling error" associated with a point estimate of a population proportion. This error is referred to as the **margin of error**.

CASE STUDY 9.3 Do Politicians Take Advantage of Us?

The Opinion Poll conducted by an Independent Poll in Figure 9.3 indicates that 67% is a point estimate for the proportion of the people who say that they feel that politicians in power take advantage of the general population.

Statistical Snapshot

Public Opinion of Politicians

Percentage of people who say they feel those in power try to take advantage of the general population:

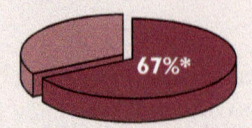

67%*

*Opinion poll result from a sample size of 983. The margin of error is plus or minus three percentage points

FIGURE 9.3

CASE STUDY 9.3   Do Politicians Take Advantage of Us? (continued)

Notice the statement attached to the point estimate of 67% includes a warning that the sample result of 67% in estimating the true population proportion has a **margin of error** *of plus or minus of 3 percentage points*. This "margin of error" is due to "sampling error" because not everyone in the population has been polled! The sampling error means that if a second random sample were selected from this population and a point estimate of the population proportion is determined, we would **expect** this sample proportion to be different from 67%.

Since the proportion of a sample can vary from sample to sample, then the point estimates determined from different samples can yield **different estimates for the population proportion**. Thus, the margin of error of ±3 percentage points reflects the sampling error in estimating the true population proportion. So, we interpret the opinion poll as stating that the true population proportion lies within the interval of: 67% ± 3%. The interval is estimating that the true population proportion will lie between:

$$(67\% - 3\%) \text{ to } (67\% + 3\%)$$

or,

$$64\% \text{ to } 70\%$$

This **range of percentage values from 64% to 70% is referred to as an interval estimate of the true population proportion**. Thus, an interval estimate uses a point estimate in constructing a range of values within which one can be reasonably sure that the true population parameter will lie. In particular, the interval estimate consists of two components:
1) **67%:** the value of the sample proportion which represents a point estimate of the population proportion.
2) **3 percentage points:** the margin of error which represents the range allowed for the anticipated sampling error.

From Case Study 9.3, we can see that an **interval estimate** is a technique that uses a point estimate along with an associated sampling error to construct an interval to estimate a population parameter.

DEFINITION 9.4
Interval Estimate is an estimate that specifies a *range of values* within which the population parameter is likely to fall.

Let's go back to the previous example of trying to estimate the population mean GPA of Nassau Community College students using the point estimate of 2.63. If we instead stated our estimate as: the mean GPA may lie between the values: 2.43 and 2.83, then the range of values from 2.43 to 2.83 would represent an interval estimate of Nassau's mean GPA.

In comparison to a point estimate which uses a single value to estimate the population parameter and is very unlikely to be exactly equal to the population parameter, an interval estimate uses a range of values to estimate the population parameter. A probability level can be assigned to the interval estimate which will indicate how confident we are that the interval will include the value of the population parameter. The probability level that is associated with an interval estimate is referred to as the confidence level. This type of interval estimate is called a confidence interval. In Section 9.4, we will examine how to construct a confidence interval for estimating the mean of a population.

9.4 INTERVAL ESTIMATION: CONFIDENCE INTERVALS FOR THE POPULATION MEAN

Let's illustrate the procedure to construct an interval estimate of a population mean by considering an example. Suppose we want to determine an interval estimate for the mean GPA of Nassau Community College students.

The first step of the procedure in constructing an interval estimate of a population mean is to select a random sample and calculate the sample mean, $\bar{x}$. Suppose a random sample of 36 students was selected from the population of Nassau students and the sample mean is determined to be: $\bar{x} = 2.63$. Remember, though, this sample mean is only one of many possible sample means that could have been determined. That is, this is one of the possible sample means within the sampling distribution of the mean. Since the sample size, n, is greater than 30 (for this example, n = 36), then according to Theorem 8.2: The Central Limit Theorem, the sampling distribution of the mean is approximately normal with:

$$\mu_{\bar{x}} = \mu$$
$$\sigma_{\bar{x}} = \frac{\sigma}{\sqrt{n}},$$

The Sampling Distribution of the Mean is illustrated in Figure 9.4.

Sampling Distribution of the Mean
(is approximately a Normal Distribution)

$\mu_{\bar{x}} = \mu$
$\sigma_{\bar{x}} = \frac{\sigma}{\sqrt{n}}$

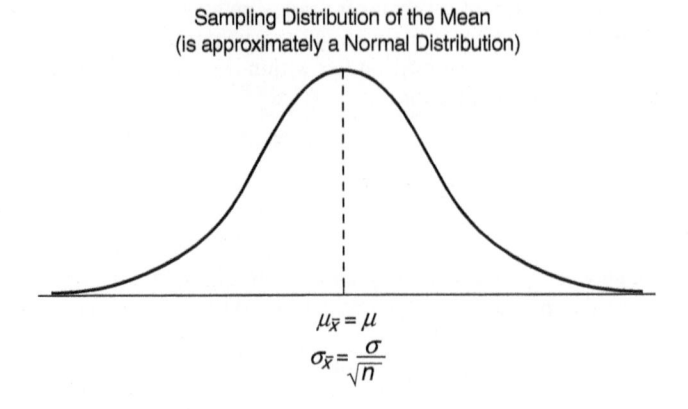

FIGURE 9.4

Within this normal distribution of sample means, we would expect 95% of the sample means, $\bar{x}$, to lie within 1.96 standard deviations[1] of the Population Mean. Therefore, we expect 95% of the sample means to lie between the values:

population − (1.96)(standard deviation) and population mean + (1.96)(standard deviation)

This can be expressed as:

$$\mu - (1.96)\sigma_{\bar{x}} \quad \text{and} \quad \mu + (1.96)\sigma_{\bar{x}}$$

Using

$$\sigma_{\bar{x}} = \frac{\sigma}{\sqrt{n}},$$

we can then rewrite the values between which we would expect 95% of the sample means to fall. These values are expressed as:

$$\mu - (1.96)\frac{\sigma}{\sqrt{n}} \quad \text{and} \quad \mu + (1.96)\frac{\sigma}{\sqrt{n}}$$

These two values are illustrated in Figure 9.5.

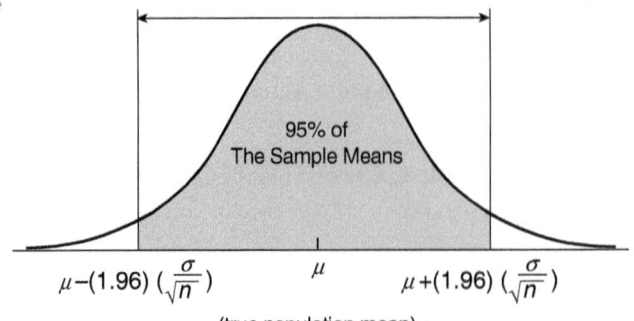

95% of
The Sample Means

$\mu - (1.96)\left(\frac{\sigma}{\sqrt{n}}\right)$ μ $\mu + (1.96)\left(\frac{\sigma}{\sqrt{n}}\right)$

(true population mean)

FIGURE 9.5

[1] Please note: The standard deviation used in this statement refers to the standard deviation of the sampling distribution of the mean, $\sigma_{\bar{x}}$.

Section 9.4 *Interval Estimation: Confidence Intervals for the Population Mean* 509

If the sample mean, $\bar{x}$, that we select from the population, happens to be one of the sample means that falls within the 95% region illustrated in Figure 9.5, and we construct an interval around the sample mean by:

subtracting 1.96 standard deviations from the sample mean, $\bar{x}$,

expressed as:

$$\bar{x} - (1.96)\frac{\sigma}{\sqrt{n}}$$

and

adding 1.96 standard deviations to the sample mean, $\bar{x}$,

expressed as:

$$\bar{x} + (1.96)\frac{\sigma}{\sqrt{n}}$$

then the interval, written as:

$$\bar{x} - (1.96)\frac{\sigma}{\sqrt{n}} \text{ to } \bar{x} + (1.96)\frac{\sigma}{\sqrt{n}}$$

will contain the *true* population mean, μ. This is illustrated in Figure 9.6.

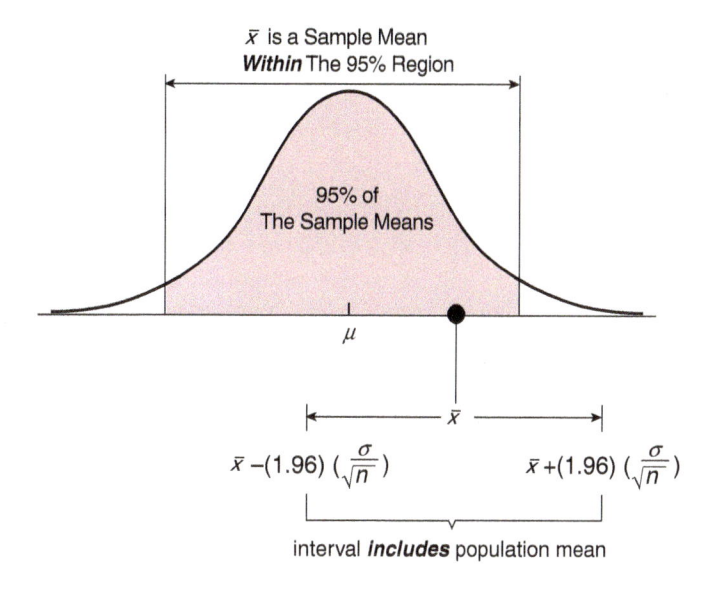

FIGURE 9.6

However, if the sample mean, $\bar{x}$, that we select is a sample mean that comes from one of the regions outside the 95% region illustrated in Figure 9.6, then the interval constructed using the formula:

$$\bar{x} - (1.96)\frac{\sigma}{\sqrt{n}} \text{ to } \bar{x} + (1.96)\frac{\sigma}{\sqrt{n}}$$

will not contain the true population mean. This is illustrated in Figure 9.7.

Therefore, if all possible samples of size n (for n greater than 30) are selected, and an interval is constructed using the formula:

$$\bar{x} - (1.96)\frac{\sigma}{\sqrt{n}} \text{ to } \bar{x} + (1.96)\frac{\sigma}{\sqrt{n}}$$

for each possible sample mean, then 95% of such intervals are expected to contain the true population mean, μ. This interval is referred to as a **95% Confidence Interval for the Population Mean**, and is defined as follows.

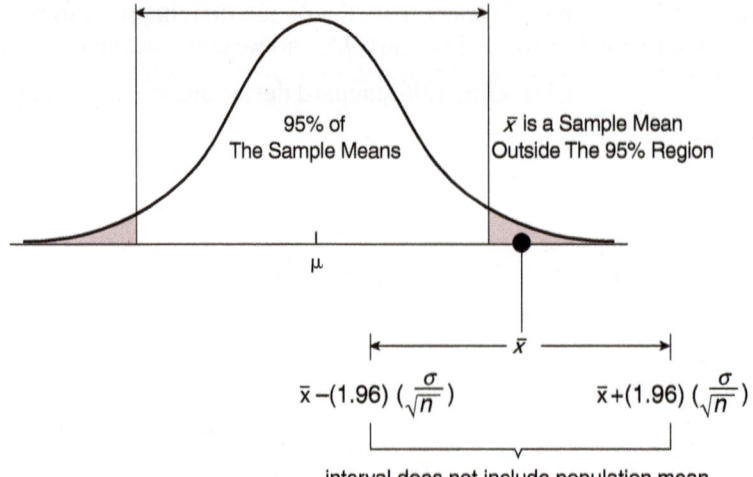

Chapter 9 *Estimation*

FIGURE 9.7

DEFINITION 9.5
95% Confidence Interval for the Population Mean, μ, where the Population Standard Deviation is Known.

A 95% Confidence Interval for the population mean, μ, (with known population standard deviation) is an interval expected to capture the true population mean, μ, 95% of the time and is constructed by the formula:

$$\bar{x} - (1.96)\frac{\sigma}{\sqrt{n}} \text{ to } \bar{x} + (1.96)\frac{\sigma}{\sqrt{n}}$$

where:

- $\bar{x}$ is the sample mean,
- σ is the population standard deviation,
- n is the sample size greater than 30

Using definition 9.5, a 95% Confidence Interval to estimate the population mean GPA of Nassau students can be constructed using the formula:

$$\bar{x} - (1.96)\frac{\sigma}{\sqrt{n}} \text{ to } \bar{x} + (1.96)\frac{\sigma}{\sqrt{n}}$$

If a sample of 36 students has a mean GPA of 2.63 and we assume that the population standard deviation, σ, is 0.6, then a 95% confidence interval is:

$$\bar{x} - (1.96)\frac{\sigma}{\sqrt{n}} \text{ to } \bar{x} + (1.96)\frac{\sigma}{\sqrt{n}}$$

$$(2.63) - (1.96)\frac{0.6}{\sqrt{36}} \text{ to } (2.63) + (1.96)\frac{0.6}{\sqrt{36}}$$

$$(2.63) - (1.96)\frac{0.6}{6} \text{ to } (2.63) + (1.96)\frac{0.6}{6}$$

$$(2.63) - (1.96)(0.1) \text{ to } (2.63) + (1.96)(0.1)$$

$$2.43 \text{ to } 2.83$$

Therefore, a 95% Confidence Interval is: 2.43 to 2.83. That is, we estimate the mean GPA of all Nassau Community College students to fall within the interval: 2.43 to 2.83.

The interval 2.43 to 2.83 is called the confidence interval, and the endpoints of the interval are called the confidence limits. The smaller endpoint of the confidence interval is called the lower confidence limit, while the larger endpoint of the confidence interval is referred to as the upper confidence limit.

DEFINITION 9.6
The endpoints of a confidence interval are called the **confidence limits** of the interval. The smaller endpoint of the interval is referred to as the **lower confidence limit**. The larger endpoint of the interval is referred to as the **upper confidence limit.**

For this interval, the lower confidence limit is 2.43, while 2.83 represents the upper confidence limit. The probability value expressed as a percentage which is associated with the confidence interval is called the confidence level or degree of confidence.

DEFINITION 9.7
The **Confidence Level** of a confidence interval is the probability value, expressed as a percentage, that is associated with an interval estimate. The probability value represents the chance that the procedure used to construct the confidence interval will give an interval that will include the population parameter. Thus, the higher the probability value, the greater the confidence that the interval estimate will include the population parameter.

The confidence level for the interval: 2.43 to 2.83 is 95%. This confidence level of 95% indicates that: in the long run, the percentage of all intervals constructed using this procedure will probably contain the population mean 95% of the time. In fact, if we were to repeat this process of constructing such an interval for all possible samples, we would be confident that 95% of all possible intervals constructed would include the population mean.

Because of this, the interval 2.43 to 2.83 is referred to as a 95% Confidence Interval. This statement doesn't indicate whether one particular interval, like the interval, 2.43 to 2.83, will actually contain the true population parameter. It indicates that we are 95% confident that the true population mean is contained in the interval 2.43 to 2.83.

Furthermore, it is interpreted to mean: that if a 95% confidence interval for the population mean was constructed for each possible sample of size n selected from the population, we would expect that 95% of these confidence intervals to include the population mean and 5% of these confidence intervals not to include the population mean.

In Figure 9.8, a 95% confidence interval is shown for four different possible samples. Notice from Figure 9.8 that if a sample mean falls within the 95% shaded area (like $\bar{x}_1$ and $\bar{x}_3$) then the confidence interval will contain the population mean. While if the sample mean falls outside this 95% region (like $\bar{x}_2$ and $\bar{x}_4$), then the confidence

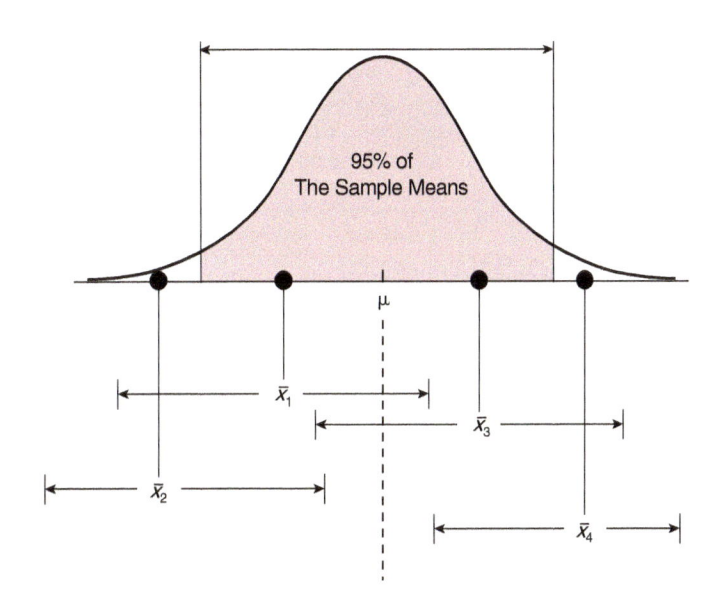

FIGURE 9.8

interval will not contain the population mean. Again, keep in mind that a 95% confidence interval doesn't tell you whether a particular interval will contain the population mean. But rather it means that: in the long run, the percentage of all intervals constructed using this procedure will probably contain the population mean 95% of the time.

Although, any confidence level can be chosen to construct a confidence interval, the most common confidence levels are: 90%, 95% and 99%. The formulas to construct 90% and 99% Confidence Intervals are similar to a 95% Confidence Interval formula except for the value of the z score which is used in the formula. In a 90% Confidence Interval, the z score 1.65 is used to construct the interval estimate, while in a 99% Confidence Interval, the appropriate z score is 2.58 in constructing this interval estimate. Thus, formulas for 90% and 99% Confidence Intervals are defined as follows.

DEFINITION 9.8
90% Confidence Interval for the Population Mean, μ, where the Population Standard Deviation is Known. A 90% Confidence Interval for the population mean, μ, (with known population standard deviation) is an interval that is expected to capture the population mean, μ, 90% of the time and is constructed by the formula:

$$\bar{x} - (1.65)\frac{\sigma}{\sqrt{n}} \text{ to } \bar{x} + (1.65)\frac{\sigma}{\sqrt{n}}$$

where:

$\bar{x}$ is the sample mean
σ is the population standard deviation
n is the sample size greater than 30.

Similarly, we can define a **99% Confidence Interval for the Population Mean** to be:

DEFINITION 9.9
99% Confidence Interval for the Population Mean, μ, where the Population Standard Deviation is Known. A 99% Confidence Interval for the population mean, μ, (with known population standard deviation) is an interval that is expected to capture the population mean, μ, 99% of the time and is constructed by the formula:

$$\bar{x} - (2.58)\frac{\sigma}{\sqrt{n}} \text{ to } \bar{x} + (2.58)\frac{\sigma}{\sqrt{n}}$$

where:

$\bar{x}$ is the sample mean
σ is the population standard deviation
n is the sample size greater than 30

Notice that in the formula for a 90% confidence interval, the values, –1.65 and 1.65, represent the appropriate z-scores associated with a 90% Confidence Interval, while the z-scores, –2.58 and +2.58, are used to construct a 99% Confidence Interval. The z-scores, –1.65 and +1.65, are used for a 90% confidence interval, since the sampling distribution of the mean, which is approximately a normal distribution for a sample size greater than 30 and a known population standard deviation, has approximately 90% of the sample means within 1.65 standard deviations of the true population mean. Similarly, 99% of the sample means are expected to lie within 2.58 standard deviations of the population mean. This is illustrated in Figure 9.9.

Let's calculate a 99% confidence interval for the population mean GPA for Nassau students, using $\bar{x} = 2.63$, n= 36 and $\sigma = 0.6$. From definition 9.9, a 99% confidence interval can be constructed using the formula:

$$\bar{x} - (2.58)\frac{\sigma}{\sqrt{n}} \text{ to } \bar{x} + (2.58)\frac{\sigma}{\sqrt{n}}$$

Section 9.4 Interval Estimation: Confidence Intervals for the Population Mean

99% of The Sample Means

$\bar{x} - (2.58)\left(\frac{\sigma}{\sqrt{n}}\right)$ μ $\bar{x} + (2.58)\left(\frac{\sigma}{\sqrt{n}}\right)$

(true population mean)

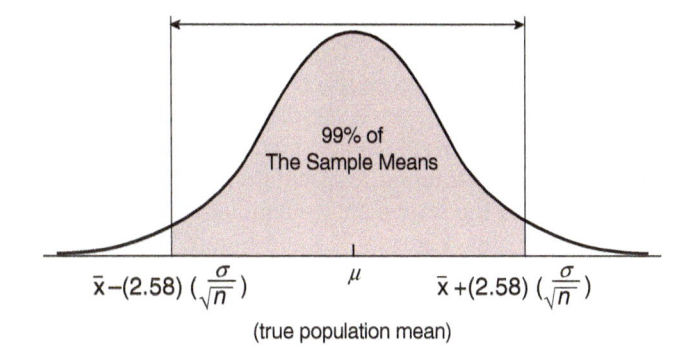

FIGURE 9.9

Substituting the sample mean, $\bar{x} = 2.63$, sample size, n = 36, and the standard deviation, $\sigma = 0.6$, we have:

$$(2.63) - (2.58)\frac{0.6}{\sqrt{36}} \text{ to } (2.63) + (2.58)\frac{0.6}{\sqrt{36}}$$

$$(2.63) - (2.58)\frac{0.6}{6} \text{ to } (2.63) + (2.58)\frac{0.6}{6}$$

$$(2.63) - (2.58)(0.1) \text{ to } (2.63) + (2.58)(0.1)$$

$$2.37 \text{ to } 2.89$$

Thus, a 99% confidence interval is 2.37 to 2.89.

That is, we are 99% confident that the mean GPA of all Nassau Community College students falls within the interval:

$$2.37 \text{ to } 2.89$$

In summary: we have computed both a 95% confidence interval and a 99% confidence interval for the population mean GPA of the Nassau Community College students to be:

95% Confidence Interval:	Width of Interval is:
2.43 to 2.83	2.83 − 2.43 = 0.40
99% Confidence Interval:	Width of Interval is:
2.37 to 2.89	2.89 − 2.37 = 0.52

You should notice that as the confidence level increases from 95% to 99%, the width of the confidence interval also increases.

In general, as the confidence level increases, the width of the confidence interval also increases.

The ZInterval function of the TI-84 calculator can be used to find the confidence interval for a population mean when the population standard deviation is known. The ZInterval function is found in the TESTS menu. To access the ZInterval function of the TESTS menu, press STAT ▶ ▶ [TESTS] 7: ZInterval. This is shown in Figure 9.10.

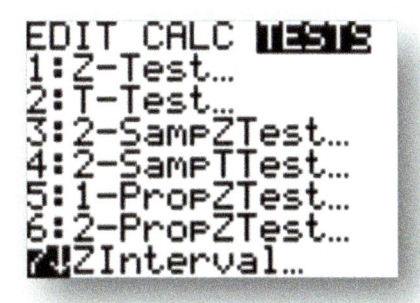

```
EDIT CALC TESTS
1: Z-Test…
2: T-Test…
3: 2-SampZTest…
4: 2-SampTTest…
5: 1-PropZTest…
6: 2-PropZTest…
7↓ZInterval…
```

FIGURE 9.10

After you select option **7: ZInterval**, you will then have to set up the menu by inputting the parameters given in the question. The menu requires that you have the sample size, n, the sample mean $\bar{x}$, the population standard deviation σ and the confidence level.

In the GPA example we were given a sample size of n = 36, the sample mean $\bar{x}$ = 2.63 and the population standard deviation σ = 0.6. We were asked to calculate 95% and 99% confidence intervals. We must enter these values in the menu for **ZInterval** to calculate the 95% confidence interval. Be sure to select STAT in the input menu at the top of the screen and to highlight Calculate and press the ENTER key after you input the values to obtain the results. Figure 9.11a and Figure 9.11b show the **ZInterval** menu and the 95% confidence interval result.

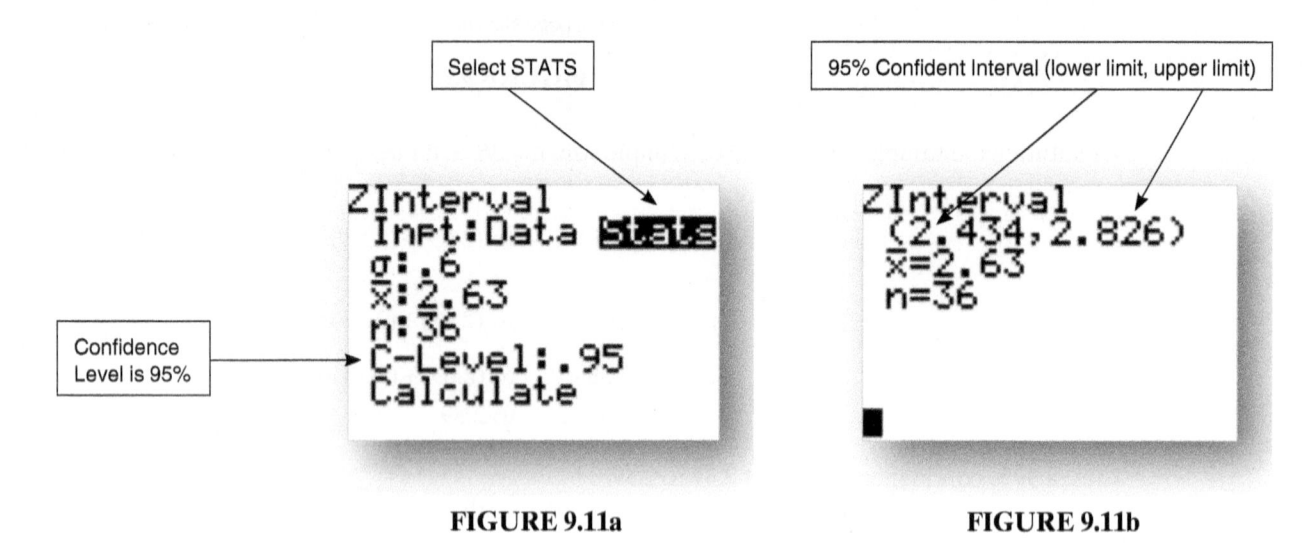

FIGURE 9.11a   **FIGURE 9.11b**

Notice, the calculator displays the result of a confidence interval as an ordered pair (lower limit, upper limit). The 95% confidence interval for the population mean is given as 2.43 to 2.83 after you round to the nearest hundredth.

We can also compute the 99% confidence interval by changing the confidence level from 95% to 99% in the confidence interval menu of the calculator. The other inputs for n, $\bar{x}$ and σ will remain the same. This is shown in Figure 9.12a and Figure 9.12b.

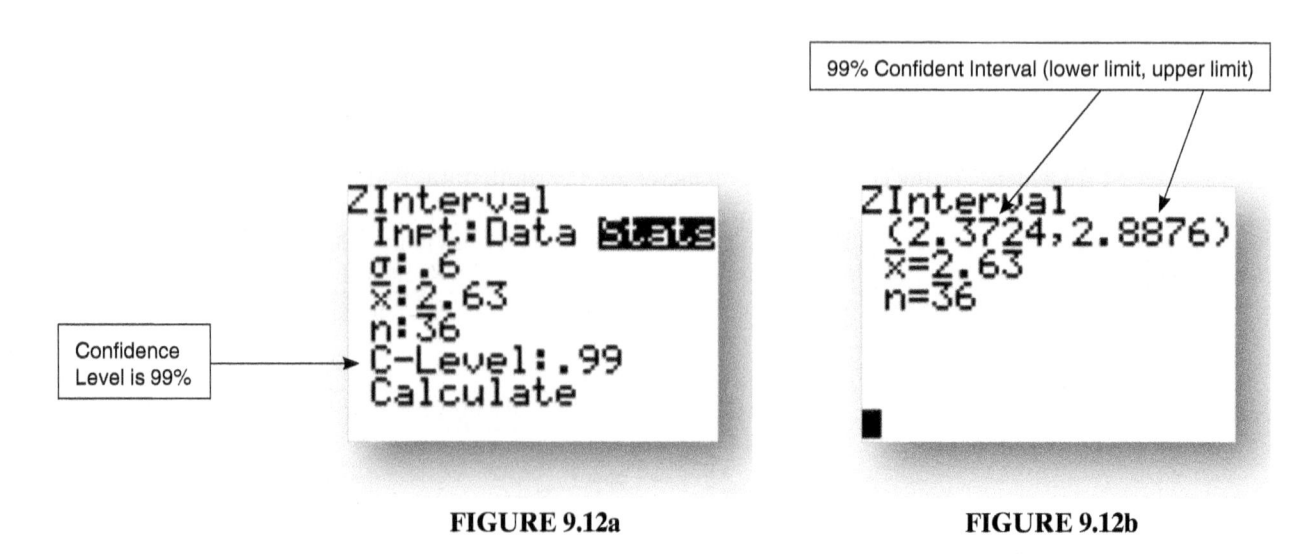

FIGURE 9.12a   **FIGURE 9.12b**

The 99% confidence interval for the population mean is 2.37 to 2.89. This is the same result as obtained using the confidence interval formulas.

Constructing a Confidence Interval for a Population Mean: When the Population Standard Deviation Is Unknown

In constructing a Confidence Interval for a population mean, it was assumed that the standard deviation of the population was known. In most practical applications, the population standard deviation is rarely known. Therefore, a more realistic approach in constructing an interval estimate of the true population mean is to consider the situation where the **population standard deviation is unknown**.

When the population standard deviation, σ, is unknown, the sample standard deviation, s, is used to estimate the population standard deviation. When we estimate σ, we cannot be confident that our estimated confidence interval will contain the population mean, μ. Sometimes, the confidence interval will not contain μ—not because the sample mean is off, but because the estimate of the **width** of the confidence interval is off. This uncertainty with the width of the confidence interval is due to the fact that the z score can no longer be used to construct a confidence interval for a population mean when the population standard deviation is unknown.

In the early 1900's, W. S. Gosset faced a similar problem when he worked for Guinness Breweries as a chemist. He had to make conclusions about the mean quality of various brews where the population standard deviation was unknown. Gosset's work led him to the discovery that when s, the sample standard deviation, was used to approximate the population standard deviation, σ, and was substituted for σ into the formula for the z statistic

$$z = \frac{\bar{x} - \mu_{\bar{x}}}{\sigma/\sqrt{n}}$$

the new statistic

$$\frac{\bar{x} - \mu_{\bar{x}}}{s/\sqrt{n}}$$

generated by Gosset was **not** normally distributed. However, Gosset showed that this new statistic which he called a t *statistic* possessed a distribution that was similar to a normal distribution but was more varied. In fact, the smaller the sample size, n, the greater the spread in the distribution for t. Gosset published his work on the t *distribution* in 1908. Since the policy of the Guinness Brewing Company required that he not publish his discoveries under his own name, he used the pen name "Student." His unique choice of this pen name has resulted in the t *score* statistic

$$t = \frac{\bar{x} - \mu_{\bar{x}}}{s/\sqrt{n}}$$

also being called a *Student's* t *score*, and the t *distribution* also called *"Student's* t*" distribution* in his honor.

Thus, the **t Distribution** rather than the normal distribution is used to construct confidence intervals for the population mean under the following conditions.

Conditions to Use the t Distribution When Constructing a Confidence Interval for the Mean of a Population

The **t Distribution** is used to construct a confidence interval for the population mean when the following conditions hold:
- The population standard deviation, σ, is unknown, and is estimated by the sample standard deviation, s.
- The population from which the sample is selected is approximately a normal distribution (or bell-shaped).

The t Distribution

Before we discuss the procedure to construct a confidence interval for the population mean using the t distribution, we need to discuss the properties of a t distribution.

You should notice that a t statistic is very similar to the z statistic. A z score

$$z = \frac{\bar{x} - \mu_{\bar{x}}}{\sigma/\sqrt{n}}$$

measures the difference between the sample mean, $\bar{x}$, and its mean $\mu_{\bar{x}}$ in units of the standard error

$$\frac{\sigma}{\sqrt{n}},$$

while the *t* score

$$t = \frac{\bar{x} - \mu_{\bar{x}}}{s/\sqrt{n}}$$

measures the difference between the sample mean, $\bar{x}$, and its mean $\mu_{\bar{x}}$ in units of the **estimated standard error**

$$\frac{s}{\sqrt{n}}.$$

The properties of a t distribution are similar to the normal distribution in some respects. Both distributions are symmetric bell-shaped curves about the mean and never touch the horizontal axis. The total area under the t curve is 1.0 or 100% as is the case with the normal curve. For small samples, the t distribution is more dispersed than the standard normal distribution. That is, it has more area in the tails and less in the center than the normal distribution. Thus, in appearance, a t distribution is lower and flatter than the normal distribution. This is due to the extra variability caused by substituting the varying sample standard deviations s for the fixed value parameter σ. However as the sample size increases, the t distribution approaches the standard normal distribution. This reflects the fact that as the sample size increases, the value of s approaches the value of σ. The t distribution is composed of a family of t curves, where there is a different t curve for each sample size. Thus, the area under a particular t curve is dependent upon the concept of **degrees of freedom (df)**. The degrees of freedom for this t statistic is equal to the sample size minus one, that is,

$$df = n - 1.$$

The number of degrees of freedom can be thought of as the number of observations within the sample that can be chosen freely. For example, suppose that a sample of n = 5 observations has a mean of 120. This sample would have (5 – 1) or 4 degrees of freedom since the 5 deviations must always have a sum of 0. That is,

$$(x_1 - 120) + (x_2 - 120) + (x_3 - 120) + (x_4 - 120) + (x_5 - 120) = 0.$$

Therefore, if we know any 4 of these 5 deviations then the remaining deviation **cannot** be free to vary. So we subtract 1 from 5 because we lose one degree of freedom since we think of 4 of these deviations as being **free to change**. Hence, the number of degrees of freedom for this sample is:

degrees of freedom: $df = n - 1 = 5 - 1 = 4$.

DEFINITION 9.10

Degrees of Freedom of a statistic are the number of free choices used in computing the statistic. The degrees of freedom, denoted by *df,* for each sample of size n, is one less than the sample size. Thus, for a sample of size n, the degrees of freedom are given by formula:

$$df = n - 1$$

Figure 9.13 shows the relationship between the normal distribution and two particular t distributions of sample sizes n = 5 and n = 12, or degrees of freedom **df** = 4 and **df** = 11, respectively.

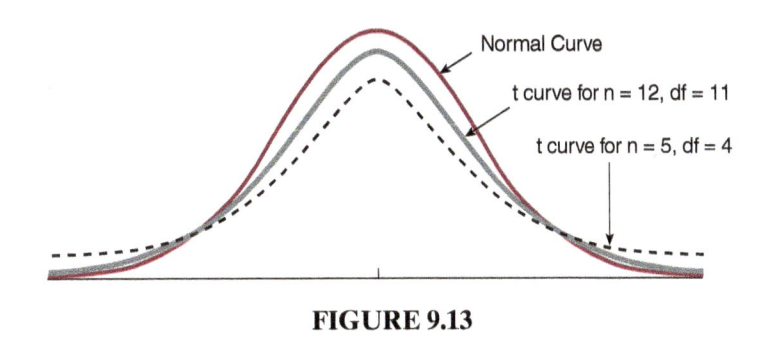

FIGURE 9.13

Notice, from Figure 9.13, the t distributions are **flatter and more dispersed** than the normal distribution, and, as the degrees of freedom *increase,* the t distribution is getting closer to the normal distribution. For most practical purposes, whenever the degrees of freedom are greater than 30, the t and normal distributions are considered to be sufficiently close.

Let's summarize the properties of the t distribution.

Properties of the t Distribution

- The t distribution is bell-shaped and is symmetric about t = 0.
- The t distribution is more varied and flatter than the standard normal distribution.
- There is a different t distribution for each sample size. A particular t distribution is specified by giving the degrees of freedom, $df = n - 1$.
- As the number of degrees of freedom increases, the t distribution approaches the standard normal distribution. They are sufficiently close when the degrees of freedom are greater than 30.

When the **t Distribution** is used to construct a confidence interval for the population mean, then the formulas for a 90%, 95% and a 99% confidence interval will include:

- the sample standard deviation, s, rather than the population standard deviation, σ, and
- t scores rather than z scores.

Since the t distribution is composed of a family of t curves that are dependent upon degrees of freedom, the table representing the area associated with the t distribution is given with degrees of freedom. Hence, to determine the specific t score to use for a confidence interval formula, the degrees of freedom, *df*, associated with the t score must be determined using the formula: $df = n - 1$. Knowing the degrees of freedom, *df*, and the confidence level, then the appropriate t score for a confidence interval is found using Table III: Critical values for the t Distributions found in Appendix C.

Substituting s for σ, and replacing the z score by the appropriate t score, we can now define a **90%, 95% or a 99% Confidence Interval for the Population Mean,** *where the population standard deviation is unknown*.

DEFINITION 9.11
90% Confidence Interval for the Population Mean, μ, where Population Standard Deviation is Unknown. A 90% Confidence Interval for the population mean, μ, when the population standard deviation is *unknown*, is constructed by the formula:

$$\bar{x} - (t_{95\%})\frac{s}{\sqrt{n}} \text{ to } \bar{x} + (t_{95\%})\frac{s}{\sqrt{n}}$$

where:
- $\bar{x}$ is the sample mean
- n is the sample size
- $t_{95\%}$ is the positive critical *t* score with $df = n - 1$ that corresponds to a **one-tailed test at $\alpha = 5\%$**
- s is the sample standard deviation.

DEFINITION 9.12
95% Confidence Interval for the Population Mean, μ, where Population Standard Deviation is Unknown. A 95% Confidence Interval for the population mean, μ, when the population standard deviation is *unknown*, is constructed by the formula:

$$\bar{x} - (t_{97.5\%})\frac{s}{\sqrt{n}} \text{ to } \bar{x} + (t_{97.5\%})\frac{s}{\sqrt{n}}$$

where:
- $\bar{x}$ is the sample mean
- n is the sample size
- $t_{97.5\%}$ is the **positive** critical *t* score with $df = n - 1$ that corresponds to a **two-tailed test at $\alpha = 5\%$**
- s is the sample standard deviation.

DEFINITION 9.13
99% Confidence Interval for the Population Mean, μ, where Population Standard Deviation is Unknown. A 99% Confidence Interval for the population mean, μ, when the population standard deviation is *unknown*, is constructed by the formula:

$$\bar{x} - (t_{99.5\%})\frac{s}{\sqrt{n}} \text{ to } \bar{x} + (t_{99.5\%})\frac{s}{\sqrt{n}}$$

where:
- $\bar{x}$ is the sample mean
- n is the sample size
- $t_{99.5\%}$ is the **positive** critical t score with $df = n - 1$ that corresponds to a **two-tailed test at $\alpha = 1\%$**
- s is the sample standard deviation.

Notice that the t score for a 90% confidence interval is found in Table III under the column listed $t_{95\%}$, while the appropriate t score for a 95% confidence interval is found under the column heading $t_{97.5\%}$, and the t score for a 99% confidence interval is found under the column for $t_{99.5\%}$. The subscript for each of these t scores refers to the area under the t distribution that is to the LEFT of each positive t score. This is illustrated in Figure 9.14.

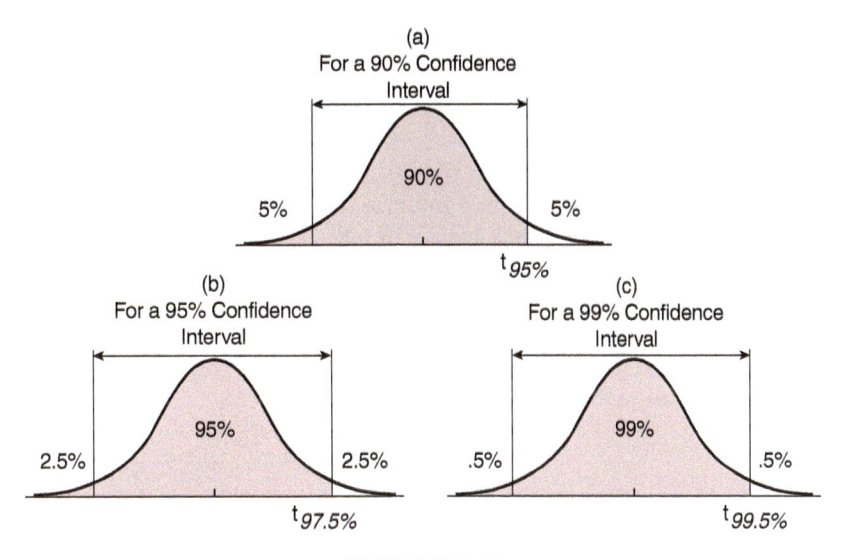

FIGURE 9.14

(a) For a 90% Confidence Interval: 5% | 90% | 5%, with $t_{95\%}$

(b) For a 95% Confidence Interval: 2.5% | 95% | 2.5%, with $t_{97.5\%}$

(c) For a 99% Confidence Interval: .5% | 99% | .5%, with $t_{99.5\%}$

In Figure 9.14a, for a 90% confidence interval, the area to the left of the positive t score is 95%, therefore, this t score is referred to as: $t_{95\%}$. From Figure 9.14b, for a 95% confidence interval, the area to the left of the positive t score is 97.5%, therefore, this t score is referred to as: $t_{97.5\%}$. Examining Figure 9.14c, for a 99% confidence interval, the area to the left of the positive t score is 99.5%, therefore, this t score is referred to as: $t_{99.5\%}$.

EXAMPLE 9.2

A random sampling of 16 students was selected from the student body at a local college to determine the average amount of money that students carry with them. If the sample mean, $\bar{x}$, was equal to $28, with a sample standard deviation of s = 6, then construct a:
a) 95% Confidence Interval for estimating the mean amount of money all students carry on them.
b) 99% Confidence Interval for estimating the mean amount of money all students carry on them.

Solution

a) Since the population standard deviation is unknown and is estimated by s, then according to definition 9.12 the formula to compute a 95% Confidence Interval for the population mean is:

$$\bar{x} - (t_{97.5\%}) \frac{s}{\sqrt{n}} \text{ to } \bar{x} + (t_{97.5\%}) \frac{s}{\sqrt{n}}$$

To determine a 95% Confidence Interval, we need to determine the appropriate t score by finding the degrees of freedom using the formula: df = n − 1. For a sample size of 16, we have: df = 16 − 1 = 15. Using Table III, the appropriate t score is found under the column $t_{97.5\%}$ and along the row for 15 degrees of freedom. Thus, the t score, $t_{97.5\%}$, for 15 degrees of freedom is: 2.13. Using the following values:

$$\bar{x} = 28, t_{97.5\%} \text{ (with } df = 15) = 2.13, s = 6 \text{ and } n = 16$$

in the formula for a 95% Confidence Interval, we have:

$$(28) - (2.13)\frac{6}{\sqrt{16}} \text{ to } 28 + (2.13)\frac{6}{\sqrt{16}}$$

$$(28) - (2.13)\frac{6}{4} \text{ to } 28 + (2.13)\frac{6}{4}$$

$$(28) - (2.13)(1.50) \text{ to } 28 + (2.13)(1.50)$$

$$(28) - (3.20) \text{ to } 28 + (3.20)$$

$$24.81 \text{ to } 31.20$$

Thus, a 95% Confidence Interval for the mean amount of money all students carry on them is: $24.81 to $31.20.

b) To construct a 99% Confidence Interval for the population mean, where the population standard deviation is **unknown**, use the formula:

$$\bar{x} - (t_{99.5\%})\frac{s}{\sqrt{n}} \text{ to } \bar{x} + (t_{99.5\%})\frac{s}{\sqrt{n}}$$

To determine a 99% Confidence Interval, we need to determine the appropriate t score by finding the degrees of freedom using the formula: df = n − 1. For a sample size of 16, we have: df = 16 − 1 = 15. Using Table III, the appropriate t score is found under the column $t_{99.5\%}$ and along the row for 15 degrees of freedom. Thus, the t score, $t_{99.5\%}$, for 15 degrees of freedom is: 2.95. Using the following values:

$$\bar{x} = 28, t_{99.5\%} \text{ (with } df = 15) = 2.95, s = 6 \text{ and } n = 16$$

in the formula for a 99% Confidence Interval, we have:

$$(28) - (2.95)\frac{6}{\sqrt{16}} \text{ to } 28 + (2.95)\frac{6}{\sqrt{16}}$$

$$(28) - (2.95)\frac{6}{4} \text{ to } 28 + (2.95)\frac{6}{4}$$

$$(28) - (2.95)(1.50) \text{ to } 28 + (2.95)(1.50)$$

$$(28) - (4.43) \text{ to } 28 + (4.43)$$

$$\$23.58 \text{ to } \$32.43$$

Thus, a 99% Confidence Interval for the mean amount of money all students carry on them is: $23.58 to $32.43.

EXAMPLE 9.3

A professor in the English department would like to estimate the average number of typing errors per page in term papers for all students enrolled in liberal arts English courses. For a random sample of 36 term papers, the professor found the mean number of typing errors per page was 4.6 with s = 1.4. Find the 95% Confidence Interval for the mean number of typing errors per page for all students enrolled in liberal arts English courses.

Solution

Since the population standard deviation is **unknown** and is estimated by s, then the formula to compute a 95% Confidence Interval for the population mean is:

$$\bar{x} - (t_{97.5\%})\frac{s}{\sqrt{n}} \text{ to } \bar{x} + (t_{97.5\%})\frac{s}{\sqrt{n}}$$

To determine a 95% Confidence Interval, we need to determine the appropriate t score by finding the degrees of freedom using the formula: df = n −1. For a sample size of 36, we have: df = 36 − 1 = 35. Using Table III, the appropriate t score is found under the column $t_{97.5\%}$. Since df = 35 does not appear in the table and is exactly halfway between df = 30 and df = 40, we will choose the t score value that is farthest from zero. Thus in this case we will choose the t score for 30 degrees of freedom. Thus, the t score, $t_{97.5\%}$, for 30 degrees of freedom is: 2.04. Using the following values:

$$\bar{x} = 4.6, t_{97.5\%} \text{ (with } df = 30) = 2.04, s = 1.4 \text{ and } n = 36$$

in the formula for a 95% Confidence Interval, we have:

$$(4.6) - (2.04)\frac{1.4}{\sqrt{36}} \text{ to } 4.6 + (2.04)\frac{1.4}{\sqrt{36}}$$

$$(4.6) - (2.04)\frac{1.4}{6} \text{ to } 4.6 + (2.04)\frac{1.4}{6}$$

$$(4.6) - (2.04)(0.233) \text{ to } 4.6 + (2.04)(0.233)$$

$$4.6 - (0.476) \text{ to } 4.6 + (0.476)$$

$$4.12 \text{ to } 5.08$$

Thus, a 95% Confidence Interval for the mean number of typing errors per page for all students enrolled in liberal arts English courses is: 4.12 to 5.08.

Section 9.4 *Interval Estimation: Confidence Intervals for the Population Mean*

The TInterval function of the TI-84 calculator can be used to find the confidence interval for the population mean when the population standard deviation is unknown. The TInterval function can be found in the TESTS menu. To access the TInterval function of the TESTS menu, press [STAT] ▶ ▶ [TESTS] 8: TInterval. This is shown in Figure 9.15.

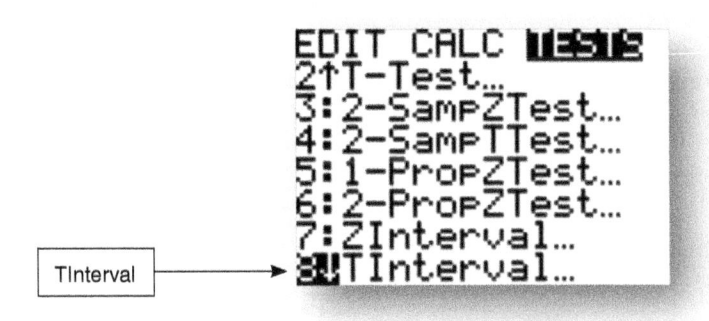

FIGURE 9.15

After you select option **8: TInterval**, you will then have to set up the menu by inputting the statistics given in the question. The menu requires that you have the sample size n, the sample mean $\bar{x}$, the sample standard deviation s and the confidence level.

In Example 9.3 we were given that the sample size n = 36, the sample mean $\bar{x}$ = 4.6 and the sample standard deviation s = 1.4. We were asked to calculate a 95% confidence interval. We must enter these values into the menu for TInterval to calculate the 95% confidence interval. Be sure to select [STAT] in the input menu at the top of the screen and to highlight Calculate and press [ENTER] to obtain the result. Figure 9.16a and Figure 9.16b show the TInterval menu and the 95% confidence interval result.

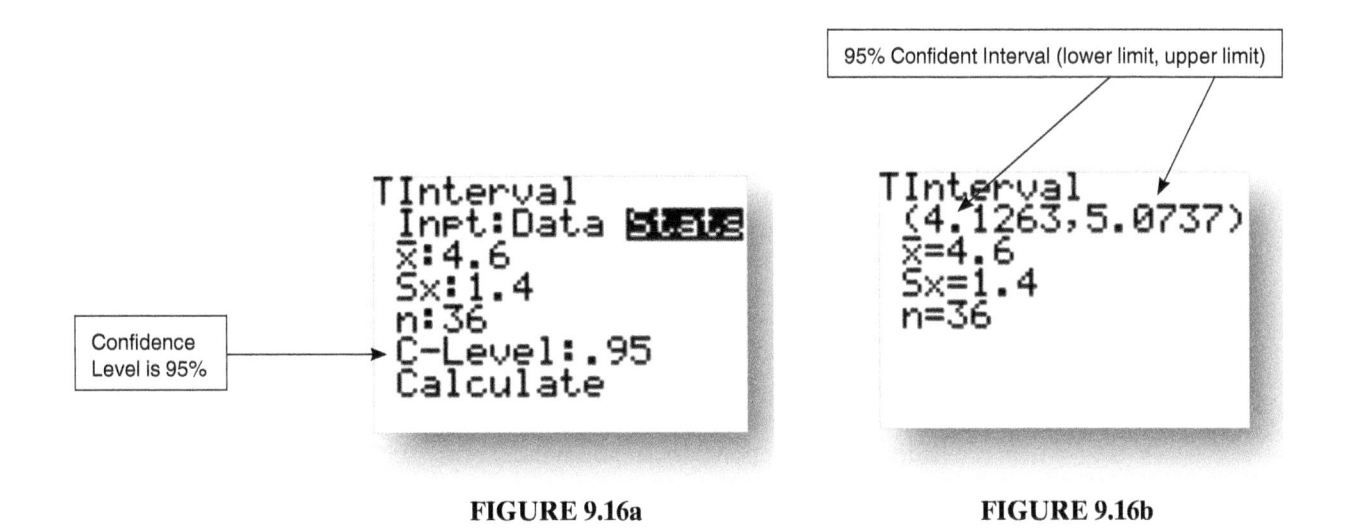

FIGURE 9.16a   **FIGURE 9.16b**

The calculator displays the confidence interval result as an ordered pair (lower limit, upper limit). Notice, the 95% confidence interval for the population mean is 4.13 to 5.07 rounded to the nearest hundredth. This is approximately the same result obtained by using the formula for the 95% confidence interval. The difference in the two answers is due to calculator precision.

9.5 INTERVAL ESTIMATION: CONFIDENCE INTERVALS FOR THE POPULATION PROPORTION

Very often, you see the results of an opinion poll stated in a newspaper or a magazine. Polls are conducted to determine the proportion or percentage of people who favor a particular issue like gun control, the proportion of people who favor a certain candidate, the percentage of people who prefer a particular soft drink, the proportion of people who don't drink and drive. In each instance, the purpose of the opinion poll is to estimate the proportion or percentage of a population with a particular characteristic or opinion.

For example, suppose an opinion poll is conducted to determine the proportion of all American adults who disapprove of a new Congressional law. If an opinion poll of 1500 American adults indicates that 1170 adults disapprove of the new law, we can use this sample information to estimate the proportion of **all** American adults who disapprove of the new law by calculating the sample proportion, $\hat{p}$. This sample proportion of people is determined by the formula:

$$\hat{p} = \frac{\text{number of adults disapproving of the new Congressional law}}{\text{total number of adults polled}}$$

Thus,

$$\hat{p} = \frac{1170}{1500}$$

or

$$\hat{p} = 0.78 \text{ or } 78\%$$

Since this opinion poll has determined that 78% of the American adults polled disapprove of the new Congressional law, then we can use this sample proportion as a point estimate of the proportion of all American adults who disapprove of the new law. Thus, the question becomes: is it reasonable to conclude from this sample proportion, $\hat{p}$, that 78% of all American adults disapprove of the new law? That is, how confident can one be that this poll of 1500 American adults reflects the true feelings of the entire population of American adults or the population proportion, p?

Suppose a second opinion poll was conducted for another 1500 adults, would one expect the same percentage of American adults to disapprove of the new law? Probably not! In fact, if more and more samples were selected, the sample proportion, $\hat{p}$, would vary from sample to sample. Some of these sample proportions will be smaller than the population proportion, p, and some will be larger. Therefore, this sample proportion, $\hat{p}$, which serves as a point estimate of the population proportion, p, does not indicate how accurate this estimate is.

Thus, in order to provide a level of confidence in our estimate of the population proportion, we need to construct an interval estimate of the population proportion, p. To construct an interval estimate of the population proportion, we need to examine the distribution of all possible sample proportions. This distribution is referred to as the sampling distribution of the proportion.

A detailed discussion regarding the sampling distribution of the proportion was previously discussed in Section 8.9. The **Characteristics of the Sampling Distribution of the Proportion, p,** are:

- the sampling distribution of the proportion is approximately normal when the size of the sample n is large, that is, when both np and n(1 − p) are greater than or equal to 10.
- the mean of the sampling distribution of the proportion, denoted $\mu_{\hat{p}}$, is: $\mu_{\hat{p}} = p$
- the standard deviation of the sampling distribution of the proportion (or the standard error of the proportion), denoted $\sigma_{\hat{p}}$, is:

$$\sigma_{\hat{p}} = \sqrt{\frac{p(1-p)}{n}}$$

In the same way that the sample mean was used to construct a confidence interval for the population mean, the sample proportion, $\hat{p}$, will be used to construct 90%, 95% and a 99% confidence intervals for the population proportion, p.

When constructing an interval estimate of the population proportion, the actual value of the population proportion, p, is **not known**. Since the best estimate that we have for the population proportion, p, is the sample proportion, $\hat{p}$, we will use the sample proportion, $\hat{p}$, as an estimate of the population proportion, p, in the formula for σ_p.

When we replace the population proportion, p, by the sample proportion, $\hat{p}$, in the formula for the standard deviation of the sampling distribution of the proportion, σ_p, we have an **estimate of the standard deviation of the sampling distribution of the proportion** (also referred to as an **estimate of the standard error of the proportion**) and it is denoted by the symbol: $s_{\hat{p}}$.

DEFINITION 9.14
The Estimate of Standard Deviation of the Sampling Distribution of the Proportion, $s_{\hat{p}}$ (also referred to as the **estimate of the standard error of the proportion**) The value of $s_{\hat{p}}$, is an estimate of the standard deviation of the sampling distribution of the proportion, σ_p, and is calculated by the formula:

$$s_{\hat{p}} = \sqrt{\frac{\hat{p}(1-\hat{p})}{n}}$$

where:

$\hat{p}$ is the sample proportion.
n is the sample size.

EXAMPLE 9.4

Determine an estimate of the standard error of the sampling distribution of the proportion, σ_p, using the formula for $s_{\hat{p}}$, if a random sample of 49 data values has a sample proportion of: $\hat{p} = 0.4$.

Solution

An estimate of σ_p is found using the formula for $s_{\hat{p}}$, which is:

$$s_{\hat{p}} = \sqrt{\frac{\hat{p}(1-\hat{p})}{n}}$$

Substitute the value of the sample proportion: $\hat{p} = 0.4$ and the sample size: n = 49 into the formula for $s_{\hat{p}}$, we have:

$$s_{\hat{p}} = \sqrt{\frac{(0.4)(0.6)}{49}}$$

$$s_{\hat{p}} = \sqrt{0.0049}$$

$$s_{\hat{p}} = 0.07$$

Under the assumption that the sampling distribution of the population proportion, p, is approximately a normal distribution, $\hat{p}$ serves as a *point estimate of the population proportion,* and $s_{\hat{p}}$ serves as an estimate of the standard error of the sampling distribution of the proportion, we can use the reasoning that was developed in constructing confidence intervals for population means to construct confidence intervals for the population proportion. Thus, we can define the *90%, 95% and 99% confidence intervals for the population proportion, p* as follows.

> **DEFINITION 9.15**
> **Confidence Intervals for the Population Proportion, p, Where n Is Large**
> 1. A 90% Confidence Interval for the Population Proportion, p, is constructed by the formula:
> $$\hat{p} - (1.65)\, s_{\hat{p}} \quad \text{to} \quad \hat{p} + (1.65)\, s_{\hat{p}}$$
> 2. A 95% Confidence Interval for the Population Proportion, p, is constructed by the formula:
> $$\hat{p} - (1.96)\, s_{\hat{p}} \quad \text{to} \quad \hat{p} + (1.96)\, s_{\hat{p}}$$
> 3. A 99% Confidence Interval for the Population Proportion, p, is constructed by the formula:
> $$\hat{p} - (2.58)\, s_{\hat{p}} \quad \text{to} \quad \hat{p} + (2.58)\, s_{\hat{p}}$$
>
> where:
> $\hat{p}$ is the point estimate of the population proportion,
> $s_{\hat{p}}$ is the estimate of the standard error of the sampling distribution of the proportion, and is computed by the formula:
> $$s_{\hat{p}} = \sqrt{\frac{\hat{p}(1-\hat{p})}{n}}$$
> n   is the sample size

EXAMPLE 9.5

According to a Mobile Marketing Company random survey of 1,000 respondents, 61% of the respondents stated that they use their mobile phones to play games. Using this information, find a:

a) 95% confidence interval for the population proportion, p, of mobile phone users that play games on their phones.
b) 99% confidence interval for the population proportion, p, of mobile phone users that play games on their phones.

Solution

a) To find a confidence interval for a population proportion, p, we must first calculate $s_{\hat{p}}$ using the formula:

$$s_{\hat{p}} = \sqrt{\frac{\hat{p}(1-\hat{p})}{n}}$$

Since the point estimate of the sample proportion is $\hat{p} = 0.61$, and the sample size is n = 1000, then substituting these values into the formula for $s_{\hat{p}}$ we obtain:

$$s_{\hat{p}} = \sqrt{\frac{\hat{p}(1-\hat{p})}{n}} = \sqrt{\frac{0.61(1-0.61)}{1000}} = \sqrt{\frac{0.61(0.39)}{1000}} = 0.0154$$

A 95% Confidence Interval for the population proportion p is found using the formula:

$$\hat{p} - (1.96)\, s_{\hat{p}} \quad \text{to} \quad \hat{p} + (1.96)\, s_{\hat{p}}$$

Substituting the values: $\hat{p} = 0.61$, and $s_{\hat{p}} = 0.0154$ into the formula, we have:

$$0.61 - (1.96)(0.0154) \quad \text{to} \quad 0.61 + (1.96)(0.0154)$$

$$0.580 \quad \text{to} \quad 0.640$$

Therefore, a 95% confidence interval for the population proportion, p, of mobile phone users that play games on their phones is:

58% to 64%.

b) A 99% Confidence Interval for the population proportion p is found using the formula:

$$\hat{p} - (2.58)\, s_{\hat{p}} \text{ to } \hat{p} + (2.58)\, s_{\hat{p}}$$

Substituting the values: $\hat{p} = 0.61$, and $s_{\hat{p}} = 0.0154$ into the formula, we have:

$$0.61 - (2.58)(0.0154) \text{ to } 0.61 + (2.58)(0.0154)$$

$$0.570 \text{ to } 0.650$$

Therefore, a 95% confidence interval for the population proportion, p, of mobile phone users that play games on their phones is:

57% to 65%.

In Example 9.5 we were given that the sample size, n, was 1000 and the sample percentage $\hat{p} = 61\%$. Since $x = n\hat{p}$, then $x = (1000)(0.61) = 610$. We can enter these values into the menu for **1-PropZInt** to calculate the 95% confidence interval in part a. Figure 9.17a and Figure 9.17b display the **1-PropZInt** menu and the 95% confidence interval.

The calculator displays the result of the confidence interval as an ordered pair (lower limit, upper limit). The 95% confidence interval for the population proportion is 0.58 to 0.64, or as percentages 58% to 64%.

We can also work part b of Example 9.5 by changing the confidence level from 95% to 99% in the confidence interval menu for the calculator. The inputs for x and n will remain the same. This is shown in Figure 9.18a and Figure 9.18b.

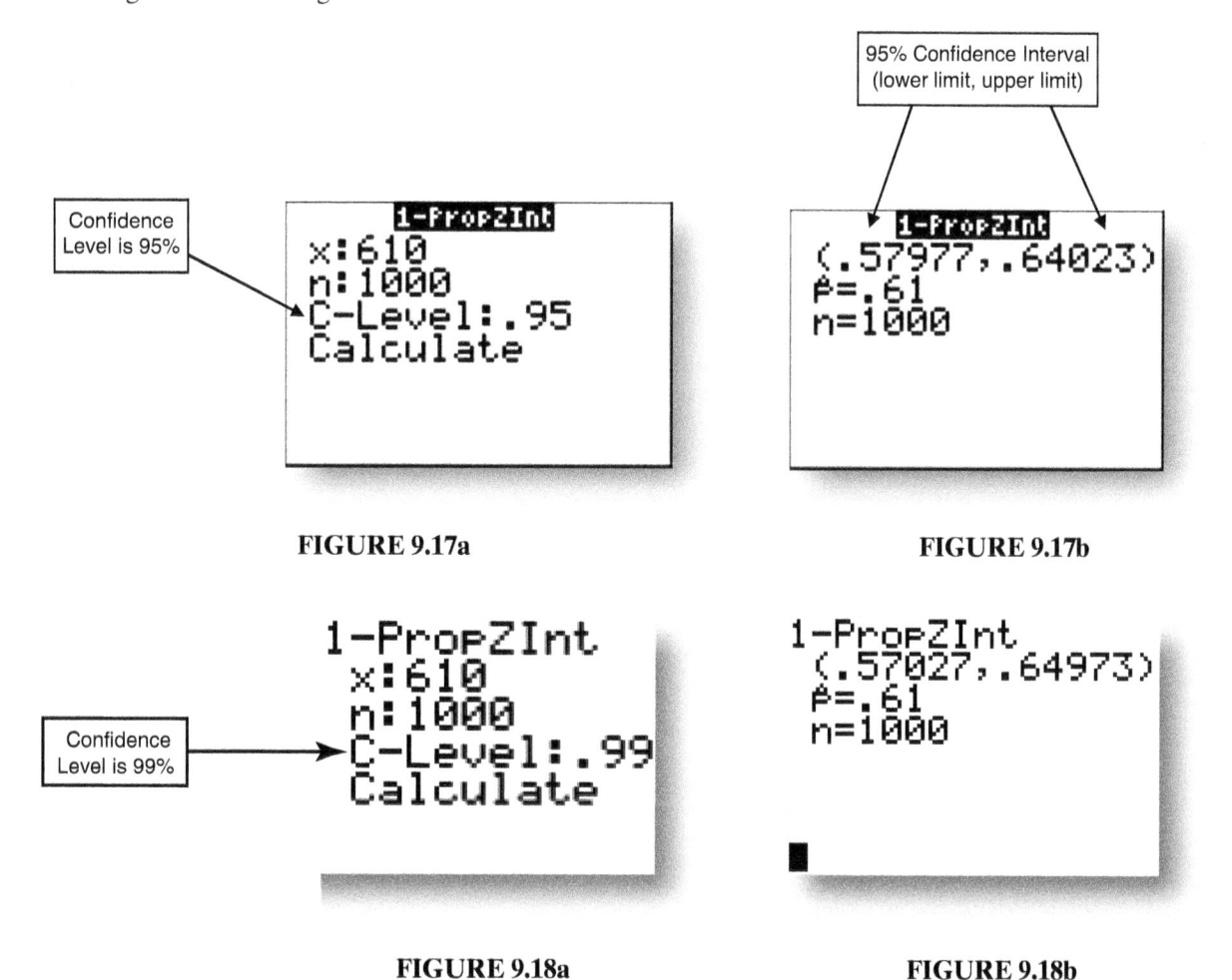

FIGURE 9.17a

FIGURE 9.17b

FIGURE 9.18a

FIGURE 9.18b

The 99% confidence interval for the population proportion is 0.57 to 0.65, or as percentages 57% to 65%.

9.6 DETERMINING SAMPLE SIZE AND THE MARGIN OF ERROR

In the previous two sections, we developed the formulas to construct a confidence interval for a population mean and a population proportion. In those sections, we arbitrarily chose a sample size. In practice, it is important to determine an appropriate sample size for a desired confidence level. Selecting a sample size larger than necessary will involve a greater cost. Thus, we must answer the question, "how does one decide on how large a sample size, n, is required?"

One factor that affects the sample size, n, is the *level of precision* that is required. If a high level of precision (that is, a better estimate or more accuracy) is required for the confidence interval, then the sample size must be increased since a larger sample will decrease the sampling error. Therefore, the more precision you need for a confidence interval, the larger the sample size you will be required to take.

Let's examine some procedures that are used to determine the sample size for a specified precision level and confidence level. First, we will consider the procedure for determining the sample size for estimating a population mean.

Sample Size for Estimating a Population Mean, μ

A confidence interval for the population mean, μ, where the population standard deviation, σ, is known has the form:

$$\bar{x} \pm (z \text{ score})\left(\frac{\sigma}{\sqrt{n}}\right)$$

Thus, the confidence interval for a specific level of confidence has two components:

1. $\bar{x}$: which represents a point estimate of the population mean

2. $(z \text{ score})\left(\frac{\sigma}{\sqrt{n}}\right)$: which represents the **margin of error**

This **margin of error** is the maximum possible error of estimate for the population mean using the sample mean, $\bar{x}$. We will denote the margin of error by E.

> **DEFINITION 9.16**
> **Margin of Error for Estimating the Population Mean, where the Population Standard Deviation is Known.** The margin of error, denoted by E, for the population mean is the maximum possible error of estimate for the population mean, μ, using the sample mean, $\bar{x}$. The formula to determine this margin of error is:
>
> $$E = (z)\left(\frac{\sigma}{\sqrt{n}}\right)$$
>
> where:
>
> E = margin of error
> z = z score corresponding to the level of confidence
> σ is the population standard deviation
> n is the sample size which is greater than 30

The margin of error, E, represents one-half the width of the confidence interval as illustrated in Figure 9.19.

For a fixed sample size and population standard deviation, as the level of confidence is increased, the margin of error also increases. This is illustrated in Table 9.1.

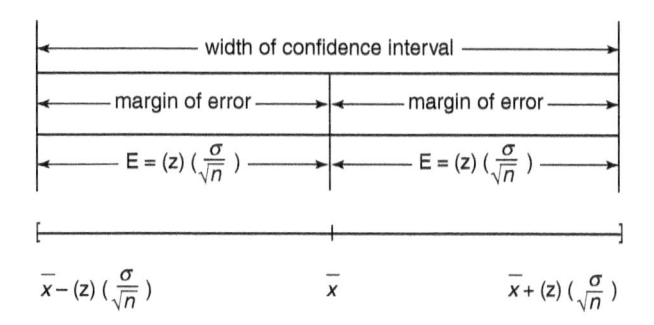

FIGURE 9.19

TABLE 9.1

Level of Confidence	z Score	Margin of Error
90%	1.65	$(1.65)(\frac{\sigma}{\sqrt{n}})$
95%	1.96	$(1.96)(\frac{\sigma}{\sqrt{n}})$
99%	2.58	$(2.58)(\frac{\sigma}{\sqrt{n}})$

The size of the margin of error determines the precision or accuracy of the estimate of the population mean. A precise estimate has a small margin of error. We would like our estimate of the population mean to be precise and also to have a high level of confidence. However, as shown in Table 9.1, as the confidence level increases from 90% to 99%, the margin of error increases, which causes a decrease in the precision level.

In examining the margin of error formula, you should notice that there are three quantities within the formula. These quantities are the z score, the population standard deviation, and the sample size. All three have an effect on the margin of error, but, since the population standard deviation represents the variability of **all** the data values, it is a fixed value over which we have no control. On the other hand, the z score and the sample size are quantities over which we have control.

Our objective will be to reduce the margin of error without reducing the confidence level, so that the precision of the estimate of the population mean can be improved. If we hold the confidence level fixed, a simple way to reduce the margin of error is to increase the sample size. This is true because the sample size is in the denominator of the margin of error formula; therefore, as the sample size increases, the margin of error will decrease. This concept is illustrated in Example 9.6.

EXAMPLE 9.6

During the past year, the flight times between two cities was reported to have a population standard deviation of 20 minutes. Determine the margin of error for estimating the mean flight time between the two cities for a confidence level of 95% for each of the following sample sizes:
a) 36
b) 64
c) 100
d) Compare the results of the previous three parts and state what happens to the margin of error as the sample size is increased.

Solution

To calculate the margin of error, we use the formula:

$$E = (z)\left(\frac{\sigma}{\sqrt{n}}\right)$$

Since the population standard deviation is given to be 20 minutes, then $\sigma = 20$. For a 95% level of confidence, the z score is: 1.96. Substituting these values into the margin of error formula, we have:

$$E = (1.96)\left(\frac{20}{\sqrt{n}}\right)$$

a) For a sample size of 36, the margin of error is:

$$E = (1.96)\left(\frac{20}{\sqrt{36}}\right)$$
$$E \approx (1.96)(3.33)$$
$$E \approx 6.53 \text{ minutes}$$

b) For a sample size of 64, the margin of error is:

$$E = (1.96)\left(\frac{20}{\sqrt{64}}\right)$$
$$E = (1.96)(2.5)$$
$$E = 4.9 \text{ minutes}$$

c) For a sample size of 100, the margin of error is:

$$E = (1.96)\left(\frac{20}{\sqrt{100}}\right)$$
$$E = (1.96)(2)$$
$$E = 3.92 \text{ minutes}$$

d) Notice from previous results, as the sample size increased from 36 to 100, the margin of error, E, decreased from 6.53 minutes to 3.92 minutes. Thus, a smaller margin of error can be achieved by increasing the sample size.

As illustrated in Example 9.6, selecting a larger sample size for a fixed confidence level will reduce the margin of error. As the margin of error is decreased, the width of the confidence interval will also be reduced. This will result in an increase in the precision or accuracy of estimating the population mean.

Since the margin of error and the size of a sample are related, we can develop a formula that determines the required sample size, n, for a predetermined margin of error (or precision requirement) and confidence level. Using the margin of error formula and solving for n, we can obtain a formula that will determine the sample size for a given margin of error.

The Formula to Determine the Sample Size for a Given Margin of Error in Estimating the Population Mean, μ

For a given confidence level, a known population standard deviation, and a given margin of error, the formula to determine the sample size, n, to satisfy these requirements in estimating the population mean is:

$$n = \left[\frac{z\sigma}{E}\right]^2$$

If the resulting value of n is not a whole number, the next largest whole number should be taken for the required sample size.

where:
- n = the required sample size for the specified conditions
- E = predetermined margin of error
- z is the z score corresponding to the level of confidence
- σ is the population standard deviation

EXAMPLE 9.7

A light-bulb manufacturer wants to determine how large a random sample the quality control department should take to be 99% confident that the sample mean will be within 25 hours of estimating the population mean.

Determine the sample size required to satisfy these requirements if we assume that the population standard deviation is 250 hours.

Solution

To determine the sample size, n, for the given conditions, we need to use the formula:

$$n = \left[\frac{z\sigma}{E}\right]^2$$

For a confidence level of 99%, the z score is 2.58. The population standard deviation is given to be 250 hours, therefore, $\sigma = 250$. Since the quality control department wants to be within 25 hours of the population mean, then the margin of error is 25 hours. Substituting this information into the sample size formula, we have:

$$n = \left[\frac{(2.58)(250)}{25}\right]^2$$
$$n = 665.64$$

Thus, the sample size required to meet the requirements of the quality control department is 666.

If the population standard deviation, σ, is unknown, then the conventional procedure is to take a preliminary sample (of any size greater than 30) and to find the sample standard deviation, s. The value of s is used to replace the value of σ in the formula for the sample size. However, it should be noted that when s is used to estimate the value of σ, then the sample size determined using s may yield a margin of error that can be either smaller or larger than the predetermined margin of error. Why?

> **DEFINITION 9.17**
> **Margin of Error for Estimating the Population Mean, where the Population Standard Deviation is Unknown.** The margin of error, denoted by E, for the population mean is the maximum possible error of estimate for the population mean, μ, using the sample mean, $\bar{x}$. The formula to determine this margin of error is:
>
> $$E = (t)\frac{s}{\sqrt{n}}$$
>
> where:
>
> E = margin of error
> t = t score corresponding to the level of confidence
> s is the sample standard deviation
> n is the sample size

The margin of error, E, represents one-half the width of the confidence interval as illustrated in Figure 9.20.

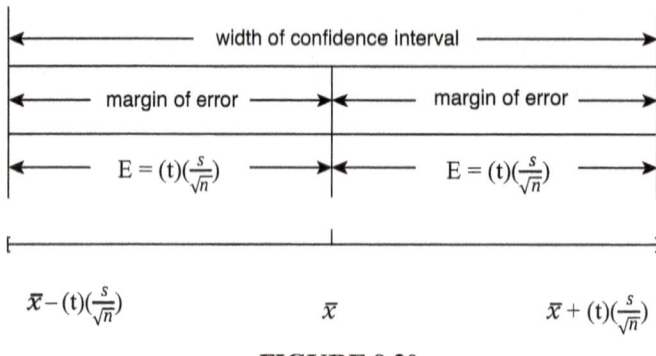

FIGURE 9.20

For a fixed sample size and sample standard deviation, as the level of confidence is increased, the margin of error also increases. This is illustrated in Table 9.2.

TABLE 9.2

Level of Confidence	t Score	Margin of Error
90%	$t_{95\%}$	$(t_{95\%})\frac{s}{\sqrt{n}}$
95%	$t_{97.5\%}$	$(t_{97.5\%})\frac{s}{\sqrt{n}}$
99%	$t_{99.5\%}$	$(t_{99.5\%})\frac{s}{\sqrt{n}}$

Where:
  n is the sample size
  s is the sample standard deviation.
  $t_{95\%}$ is the positive critical t score with $df = n - 1$ that corresponds to a one-tailed test at $\alpha = 5\%$
  $t_{97.5\%}$ is the positive critical t score with $df = n - 1$ that corresponds to a two-tailed test at $\alpha = 5\%$
  $t_{99.5\%}$ is the positive critical t score with $df = n - 1$ that corresponds to a two-tailed test at $\alpha = 1\%$

Section 9.6 *Determining Sample Size and the Margin of Error* 531

EXAMPLE 9.8

The US Department of Agriculture (USDA) randomly sampled 25 U.S. consumers to determine the average amount of red meat consumed daily per person. The following sample data (in ounces) represents the daily red meat consumption of 25 individuals.

| 11.8 | 10.5 | 12.5 | 10.2 | 11.2 | 13.5 | 11.7 | 9.6 | 6.5 | 12.0 | 11.3 | 10.8 |
| 9.9 | 12.2 | 10.3 | 12.6 | 11.6 | 8.5 | 11.0 | 9.2 | 8.3 | 10.1 | 11.2 | 7.5 | 8.5 |

a) Find the sample mean of the daily red meat consumption.
b) Find the sample standard deviation of the daily red meat consumption.
c) If the USDA decides to construct a confidence interval for the population mean daily red meat consumption, what distribution would you use and why?
d) Determine a 99% confidence interval estimate for the population mean daily red meat consumption.
e) Determine the margin of error for the confidence interval you found in part d?
f) If an individual told you the mean daily red meat consumption per person is 14 oz. Would you believe the individual? Support your answer using the 99% confidence interval information.

Solution

a) Entering the 25 data values into a list of the TI-84 plus calculator and using 1-Var Stats, the sample mean daily red meat consumption is $\bar{x} = 10.5$ oz..
b) Using the 1-Var Stats calculator function again, the sample standard daily red meat consumption is s = 1.7 oz. (to the nearest tenth).
c) The USDA would use a t distribution to construct a confidence interval for the population mean daily red meat consumption since the population standard deviation is unknown.
d) To construct a 99% Confidence Interval for the population mean, where the population standard deviation is unknown, we use the formula:

$$\bar{x} - (t_{99.5\%}) \frac{s}{\sqrt{n}} \quad \text{to} \quad \bar{x} + (t_{99.5\%}) \frac{s}{\sqrt{n}}$$

To determine a 99% Confidence Interval, we need to first determine the appropriate t score by computing the degrees of freedom using the formula: df = n − 1. For a sample size of 25, we have: df = 25 − 1 = 24 degrees of freedom. Using Table III, the appropriate t score is found under the column $t_{99.5\%}$ and across the row for 24 degrees of freedom. The t score, $t_{99.5\%}$, for 24 degrees of freedom is: 2.80.

Substituting the values: $\bar{x} = 10.5$, $t_{99.5\%} = 2.80$, s = 1.7 and n = 25 into the formula for the 99% Confidence Interval, we get:

$$\bar{x} - (t_{99.5\%}) \frac{s}{\sqrt{n}} \quad \text{to} \quad \bar{x} + (t_{99.5\%}) \frac{s}{\sqrt{n}}$$

$10.5 - (2.80)(\frac{1.7}{\sqrt{25}})$	to	$10.5 + (2.80)(\frac{1.7}{\sqrt{25}})$
10.5 − (2.80(0.34)	to	10.5 + (2.80)(0.34)
10.5 − 0.952	to	10.5 + 0.952
9.548	to	11.452
9.55 oz.	to	11.45 oz.

Therefore, a 99% Confidence Interval for the population mean daily red meat consumption per person is 9.55 to 11.45 oz.

e) The margin of error formula is: $E = (t)\frac{s}{\sqrt{n}}$.

For t = 2.80, s = 1.7 and n = 25, the margin of error is: $E = (2.80)(\frac{1.7}{\sqrt{25}}) = (2.80)(0.34) = 0.952$ oz..

Thus, the margin of error is E = 0.95 oz. (to two decimal places).

f) You should not believe the individual who states the mean daily red meat consumption per person is 14 oz. since this value exceeds the upper value of the 99% Confidence Interval.

Sample Size for Estimating a Population Proportion, p

A confidence interval for the population proportion, p, where the sample size is large, and the population proportion, p, is unknown, and is estimated by the sample proportion, $\hat{p}$, has the form:

$$\hat{p} \pm (z \text{ score})\left(\sqrt{\frac{\hat{p}(1-\hat{p})}{n}}\right)$$

Thus, the confidence interval for a specific level of confidence has two components:

1. $\hat{p}$: which represents a point estimate of the population proportion
2. $(z \text{ score})\left(\sqrt{\frac{\hat{p}(1-\hat{p})}{n}}\right)$: which represents the margin of error

To determine this margin of error, we need to know the value of the sample proportion, $\hat{p}$, and the z score corresponding to the confidence level. This margin of error is the maximum possible error of estimate for the population proportion using the sample proportion, $\hat{p}$.

We will denote the margin of error by E. Opinion Polls usually express the point estimate, the sample proportion, of the poll as a percentage and the margin of error of the poll as percentage points. Care must be taken to distinguish between percentages and percentage points. Percentage refers to a proportion value multiplied by 100%, while percentage points represent units on the percent scale. The margin of error for opinion polls is always stated as a certain number of percentage points, and not as a percentage.

EXAMPLE 9.9

The Nielsen television ratings used a random sample of 2,000 homes to estimate with 95% confidence that 45% of the proportion of all homes watched a football game on Monday Night Football. Determine the margin of error for the Nielsen's estimate of the proportion of all homes watching this football game on Monday Night Football.

Solution

To calculate the margin of error, we use the formula:

$$E = (z)\sqrt{\frac{\hat{p}(1-\hat{p})}{n}}$$

For a 95% level of confidence, the z score is: 1.96. Since a random sample of 2,000 homes was selected, then n = 2,000. Given that the sample proportion of homes watching the football game is 45%, then $\hat{p}$ = 0.45, and $1 - \hat{p} = 1 - 0.45 = 0.55$. Substituting these values into the margin of error formula, we have:

$$E = (1.96)\left(\sqrt{\frac{(0.45)(0.55)}{2,000}}\right)$$

$$E = (1.96)\left(\sqrt{\frac{(0.2475)}{2,000}}\right)$$

$$E = (1.96)(\sqrt{0.0001238})$$

$$E = (1.96)(0.0111)$$

$$E \approx 0.0218$$

Thus, the margin of error is approximately 0.0218. This can also be expressed as approximately **2.18 percentage points.**

CASE STUDY 9.4 Phone Survey Methods

The following *New York Times* article entitled "Method Used in Taking Survey in Phone Interview" describes the technique used in selecting the sample for a *New York Times*/CBS News Public Opinion Poll.
  a) How many people were selected for this poll?
  b) Name some procedures that the pollsters did when selecting the sample to insure that it was representative of the population?
  c) Explain what is meant by the statement: "in theory, one can say with 95% certainty that the results based on the entire sample differ by no more than 3 percentage points in either direction from what would have been obtained by interviewing all adult Americans"? What confidence interval are they referring to within this statement? What is the margin of error for this poll? What is the population of people for this poll?
  d) What additional errors are the pollsters referring to in the statement: "the theoretical errors do not take into account a margin of additional errors resulting from the various practical difficulties in taking any survey of public opinion"?

Method Used in Taking Survey in Phone Interview

The latest New York Times/CBS News Poll is based on telephone interviews conducted from last Wednesday through last Sunday with 1,536 adult men and women around the United States. Of this total, 542 said that they were Democrats, 495 said they were Republicans and 499 said they were independents.

The sample of telephone exchanges called was selected by a computer from a complete list of exchanges in the country. The exchanges were chosen in a way that would insure that each region of the country was represented in proportion to its population. For each exchange, the telephone numbers were formed by random digits, thus permitting access to both listed and unlisted residential numbers.

The results have been weighted to take account of household size and to adjust for variations in the sample relating to region, race, sex, age and education.

Higher Rate for Republicans

Republicans were sampled at a higher rate than others to insure a sufficiently large number from that party. The groups of voters, by party identification, were then weighted to reflect their proper proportion in the voting population.

In theory, one can say with 95 percent certainty that the results based on the entire sample differ by no more than 3 percentage points in either direction from what would have been obtained by interviewing all adult Americans. The error for smaller subgroups is larger, depending on the number of sample cases in the subgroup.

The theoretical errors do not take into account a margin of additional error resulting from the various practical difficulties in taking any survey of public opinion.

FIGURE 9.21
From *The New York Times*. © 1980 The New York Times Company. All rights reserved. Used under license.

As shown earlier in this section, when estimating the mean of a population, the margin of error decreases as the sample size increases. This also holds true when estimating a population proportion. As the margin of error is decreased, then the precision or accuracy of estimating the population proportion will increase.

Since the margin of error and the size of a sample are related, we can develop a formula that determines the required sample size, n, for a predetermined margin of error (or precision requirement) and confidence level. Using the margin of error formula and solving for n, we can obtain a formula that will determine the sample size for a given margin of error when estimating a population proportion, p. When the margin of error formula is solved for n, we get:

$$n = \frac{z^2(\hat{p})(1-\hat{p})}{E^2}$$

The Formula to Determine the Sample Size for a Given Margin of Error in Estimating the Population Proportion, p

For a given confidence level, a point estimate of the population proportion, $\hat{p}$, and a given margin of error, the formula to determine the sample size, n, to satisfy these requirements in estimating the population proportion is:

$$n = \frac{z^2(\hat{p})(1-\hat{p})}{E^2}$$

If the resulting value of n is not a whole number, the next largest whole number should be taken for the required sample size.

where:
- n = the required sample size for the specified conditions
- E = predetermined margin of error
- z = the z score corresponding to the level of confidence
- $\hat{p}$ = a point estimate of the population proportion

Thus, to calculate for a sample size, n, we need to determine the values for z, E and $\hat{p}$. The value for z is the z score that corresponds to the confidence level used in estimating the population proportion, while the value for E is the predetermined margin of error for estimating the population proportion. However, the value of $\hat{p}$, the sample proportion, is *unknown* since we would need to select a sample to determine $\hat{p}$.

Thus, our problem is: the formula for the sample size requires that we have a value for $\hat{p}$, but in order for us to determine this value for $\hat{p}$, we need to decide upon how large a sample to select, which is the purpose of using this sample size formula.

Consequently, we don't have a value for $\hat{p}$ and thus, we cannot use the formula for the sample size, n. We can alleviate this problem in one of two ways:

1. Take a pilot sample (of any sample size that is sufficiently large) and calculate $\hat{p}$ for this sample, or use the value of $\hat{p}$ from a previous sample. This value of $\hat{p}$ is used to find the sample size, n.
2. Determine the maximum possible product of $\hat{p}$ and $1 - \hat{p}$, and simply replace $\hat{p}$ and $1 - \hat{p}$ with values that produce the maximum possible product of $\hat{p}$ and $1 - \hat{p}$. This procedure generates the **conservative estimate of the sample size.**

To help find this conservative estimate of the sample size, we need to determine the maximum product of $\hat{p}$ and $1 - \hat{p}$.

Table 9.3 displays the product of $\hat{p}$ and $1 - \hat{p}$ for values of $\hat{p}$ from 0.1 to 0.9.

TABLE 9.3 Table of Values of $\hat{p}$, $(1 - \hat{p})$, and Their Product

$\hat{p}$	$(1 - \hat{p})$	$(\hat{p})(1 - \hat{p})$
0.1	0.9	0.09
0.2	0.8	0.16
0.3	0.7	0.21
0.4	0.6	0.24
0.5	0.5	0.25
0.6	0.4	0.24
0.7	0.3	0.21
0.8	0.2	0.16
0.9	0.1	0.09

From Table 9.3, you should notice that the maximum value of the product of $\hat{p}$ and $1 - \hat{p}$ occurs when $\hat{p}$ is equal to 0.5. Thus, the maximum possible product of $\hat{p}$ and $1 - \hat{p}$ will be 0.25, no matter what value of $\hat{p}$ is chosen. Therefore, $\hat{p}$ will always be less than $\hat{p} = 0.25$. Substituting 0.25 for the product: $(\hat{p})(1 - \hat{p})$ into the formula for the sample size, n, we obtain:

$$n = \frac{z^2(0.25)}{E^2}$$

This formula for the sample size, n, based upon using 0.5 for the value of $\hat{p}$ will give a conservative estimate of the sample size. This is referred to as a conservative estimate because it gives the maximum sample size regardless of the possible value of $\hat{p}$.

> **Conservative Estimate of Determining the Sample Size for a Given Margin of Error in Estimating the Population Proportion, p**
>
> For a given confidence level, and a given margin of error, the formula to determine the conservative sample size, n, to satisfy these requirements in estimating the population proportion is:
>
> $$n = \frac{z^2(0.25)}{E^2}$$
>
> If the resulting value of n is not a whole number, the next largest whole number should be taken for the required sample size.
>
> where:
>
> n = the required sample size for the specified conditions
> E = predetermined margin of error
> z is the z score corresponding to the level of confidence

EXAMPLE 9.10

An automobile club would like to estimate the proportion of all drivers who wear seat belts while driving. If the automobile club wants the margin of error to be within 2 percentage points of the true population proportion for a 99% confidence interval, then determine:
a) A conservative estimate for the sample size.
b) An estimate of the sample size if a previous study showed that 65% wear seat belts.

Solution

a) To determine a conservative estimate for the sample size of drivers who wear seat belts, we use the formula:

$$n = \frac{z^2(0.25)}{E^2}$$

To calculate the sample size, n, using this formula, we need to determine the z score and a margin of error for these requirements. A confidence level of 99% uses the z score, z = 2.58. Since the automobile club wants to be within 2 percentage points, then the margin of error is: E = 0.02. Substituting these values into the formula for n, we have:

$$n = \frac{(2.58)^2(0.25)}{(0.02)^2}$$

$$n = \frac{1.6641}{0.0004}$$

$$n = 4160.25$$

Since the resulting value of n is not a whole number, the next largest whole number is taken to be: 4161. Thus, the automobile club's conservative estimate of the sample size is 4161 drivers. Therefore, if the club wants to be within two percentage points of the population proportion of drivers who wear seat belts, and 99% confident of their estimate, they should take a sample of 4161 drivers.

b) To determine an estimate of the sample size, n, using a sample proportion, $\hat{p}$, determined from a previous study, we use the formula:

$$n = \frac{z^2(\hat{p})(1-\hat{p})}{E^2}$$

To calculate the sample size, n, using this formula, we need to determine the z score, a margin of error for these requirements, and a sample proportion, $\hat{p}$. A confidence level of 99% uses a z score, z = 2.58. Since the automobile club wants to be within 2 percentage points, then the margin of error is: E = 0.02. From the previous study, we can use 0.65 for the value for $\hat{p}$, so $\hat{p}=0.65$ and $1-\hat{p}=0.35$. Substituting these values into the formula for the sample size, n, we have:

$$n = \frac{(2.58)^2(0.65)(0.35)}{(0.02)^2}$$

$$n = \frac{1.5143}{0.0004}$$

$$n = 3785.75$$

Since the resulting value of n is not a whole number, the next largest whole number is taken: 3786. Thus, if the automobile club takes a sample of 3786 drivers, then the estimate of the population of all drivers that wear seat belts will be within 2 percentage points of the population proportion.

In Example 9.9, it is important to realize that the sample size of 3786 drivers was determined using the sample proportion, $\hat{p}=0.65$, and that the automobile club wanted to be within 2 percentage points of the population proportion. However, you should realize that if the sample proportion for the new sample of 3786 drivers happens to be less than 0.65, the margin of error will not be within the requirement of 2 percentage points.

Therefore, if you want to be cautious and more conservative in your estimate, you may want to select a sample size that is greater than 3786 drivers. For example, perhaps you would use the sample size of 4161 that was determined in *part a* of Example 9.9.

Summary of Confidence Intervals

Table 9.4 contains a summary of the confidence intervals and the appropriate conditions.

TABLE 9.4 Summary of Confidence Intervals

Parameter Being Estimated	Conditions of Estimate	Point Estimate	Confidence Interval	Sampling Distribution Is:
μ	Known σ	$\bar{x}$	$\bar{x} - (z)(\frac{\sigma}{\sqrt{n}})$ to $\bar{x} + (z)(\frac{\sigma}{\sqrt{n}})$ where: z=1.65 for 90% z=1.96 for 95% z=2.58 for 99%	Approximately a Normal Distribution
μ	Unknown σ	$\bar{x}$	$\bar{x} - (t)(\frac{s}{\sqrt{n}})$ to $\bar{x} + (t)(\frac{s}{\sqrt{n}})$ where: $t = t_{95\%}$ for 90% $t = t_{97.5\%}$ for 95% $t = t_{99.5\%}$ for 99%	A t Distribution with: df = n − 1
p	$n\hat{p} \geq 10$ and $n(1-\hat{p}) \geq 10$	$\hat{p}$	$\hat{p} - (z)(s_{\hat{p}})$ to $\hat{p} + (z)(s_{\hat{p}})$ where: z = 1.65 for 90% z = 1.96 for 95% z = 2.58 for 99% $s_{\hat{p}} = \sqrt{\frac{\hat{p}(1-\hat{p})}{n}}$	Approximately a Normal Distribution

KEY TERMS

Confidence Level 511
Confidence Limits 511
Conservative Estimate of the Sample Size 532
Degrees of Freedom, df 516
Estimate 503
Estimate of the Standard Deviation of the Sampling Distribution of the Proportion or the Estimate of the Standard Error of the Proportion, $s_{\hat{p}}$ 523
Estimation 503
Estimator 504
Good Estimator 506

Interval Estimate 507
Margin of Error 506, 526
90% Confidence Interval for the Population Mean, μ 512
90% Confidence Interval for Population Proportion, p 524
95% Confidence Interval for the Population Mean, μ 510
95% Confidence Interval for Population Proportion, p 524
99% Confidence Interval for Population Mean, μ 512

99% Confidence Interval for Population Proportion, p 525
Point Estimate 504
Point Estimate of the Population Proportion, p 505
Precision 526
Sampling Distribution of the Proportion 523
t Distribution 515
Unbiased 506

SECTION EXERCISES

Note: The data sets in these Section Exercises can be downloaded in either Microsoft EXCEL or TI-84 format from the textbook's website at MyStatLab.com.

Section 9.2

1. The mean, proportion or standard deviation of a population is called a population _____.
2. To estimate a population parameter, a random _____ is selected from the population.
3. The statistical procedure where sample information is used to estimate the value of a population parameter such as the population mean, population standard deviation or population proportion is called _____.
4. A sample estimate using a single number to estimate a population parameter, such as, a population mean or population proportion is called a _____ estimate.
5. A method that is used to obtain a point estimate of the population proportion, p, is the _____.
6. An estimator of a population parameter is said to be _____, if the mean of the point estimates obtained from the independent samples selected from the population will approach the true value of the population parameter as more and more samples are selected.
7. Which symbol represents a point estimate of the population mean?
   a) p   b) $\mu_{\hat{p}}$ c) $\hat{p}$ d) $\bar{x}$ e) $E\mu$
8. Which symbol represents a point estimate of the population proportion?
   a) p   b) $\mu_{\hat{p}}$ c) $\hat{p}$ d) $\bar{x}$ e) $E\mu$
9. An example of an estimator is the formula for the standard error of the proportion.
   (T/F)
10. One major disadvantage of using a sample mean point estimate to estimate the true population mean is that we are never sure how far the point estimate is from the true population mean.
    (T/F)

Section 9.3

11. The margin of error usually given in an opinion poll is due to _____ error because not everyone in the population was selected for the poll.
12. An _____ estimate is an estimate that specifies a range of values that the population parameter is likely to fall within.
13. The first step of the procedure in constructing an interval estimate of a population parameter is to select a random sample and calculate a _____ estimate.
14. When constructing a 99% confidence interval for the population mean where the population standard deviation is not known, the sample standard deviation is used to construct such an interval.
    (T/F)
15. State the four criteria required for a good estimator.
16. A CNN/Gallup Poll of 496 adults nationwide stated that 39% of respondents nationally say the federal government has become so large and powerful it poses an immediate threat to the rights and freedoms of ordinary citizens. Percentage of various demographic groups who hold that opinion were:

CNN/Gallup Poll Results

Gender:	
Men:	35%
Women:	41%
Education:	
No College:	45%
Some College:	40%
College Grad:	26%

If the margin of error was ± 5 percentage points, then: give an interval estimate for the percentage of:
a) Men who fear big government.
b) Women who fear big government.
c) Adults with no college education who fear big government.

d) Adults with some college education who fear big government.
e) College graduates who fear big government.

17. Mikayla Inc. is a large company that manufactures dirt bike tires. Many professional dirt bike riders feel that when ⅛ of the tread on a tire has worn, the tire has exceeded its lifetime. Mikayla Inc. needs to determine the average lifetime for their most popular line of dirt bike tires for advertisement purposes. A random sample of 100 tires is tested and the mean lifetime is found to be 170 miles. Assume the lifetime of the tires is normally distributed with an unknown population mean and a population standard deviation of 30 miles. Find the 95% confidence interval for the population mean lifetime of these tires.

Section 9.4

18. The 95% confidence interval for the population mean, μ, where the population standard deviation is **known** and the sample size is greater than 30, is constructed by the formula: _____ to _____ .

19. The 99% confidence interval for the population mean, μ, where the population standard deviation is **known** and the sample size is greater than 30, is constructed by the formula: _____ to _____ .

20. The 90% confidence interval for the population mean, μ, when the population standard deviation is **unknown**, is constructed by the formula: _____ to _____ .

21. The 99% confidence interval for the population mean, μ, when the population standard deviation is **unknown**, is constructed by the formula: _____ to _____ .

22. The confidence level of a confidence interval is the _____ value, expressed as percentage, that is associated with an interval estimate, which represents the chance that the _____ used to construct the confidence interval will give an interval that will include the population _____. Thus, the higher the probability value, the greater the _____ that the interval estimate will include the population parameter.

23. The t Distribution is used to construct a confidence interval for the population mean when the following conditions hold.
    a) The population standard deviation, σ is _____, and is estimated by the _____ .
    b) The population from which the sample is selected is approximately a _____ distribution.

24. According to a CNN random sample of 64 couples, the mean length of dating before marriage is 540 days with a sample standard deviation of 90 days.
    a) Construct a 90% confidence interval to estimate the mean number of dating days for the population.
    b) State the lower and upper confidence limits for this interval.

25. If a sample of size 36 is selected, then the degrees of freedom for the t distribution is:
    a) 30   b) 37   c) 6   d) 35

26. To insure that the sampling distribution of the mean can be approximated by the normal distribution, the sample size must be less than 30.
    (T/F)

27. Degrees of freedom is a concept used to determine critical t scores.
    (T/F)

28. Which formula is used in the calculation of a confidence interval for the mean when the population standard deviation is **unknown**?
    a) $\dfrac{\sigma}{\sqrt{n}}$
    b) $\dfrac{s}{\sqrt{n}}$
    c) $\sqrt{np(1 \pm p)}$
    d) $E\mu$
    e) np

29. Which formula is used to calculate the estimate of the standard error of the proportion?
    a) $\dfrac{\sigma}{\sqrt{n}}$
    b) $\dfrac{s}{\sqrt{n}}$
    c) $\sqrt{np(1-p)}$
    d) $\sqrt{\dfrac{\hat{p}(1-\hat{p})}{n}}$
    e) np

30. Queuing theory is the study of waiting. Many companies are interested in the mean amount of time their customers spend waiting on line. A local bank reports that the average wait time for their customers is 3 minutes with a population standard deviation of 1.2 minutes. To support this claim the bank takes a random sample of 1000 people, and finds that the average wait time on line for the sample is 3.2 minutes.
    a) Construct a 99% confidence interval estimate for the population mean waiting time at this bank.
    b) What distribution did you use to calculate the confidence interval in part a and why did you use it?
    c) Write a statement to interpret this confidence interval.
    d) Does the bank's claim that the average wait time is 3 minutes seem reasonable? Why or why not?

31. The concentration of caffeine in a popular domestic caffeinated Diet Cola soft drink is known to be approximately normally distributed. A random sample of 25 Diet Cola 12-ounce cans was selected from the production process. The concentration of caffeine in these 25 Diet Cola soft drinks was measured by the method of high-pressure liquid chromatography. The mean caffeine content was determined to be 45.6 milligrams with s = 6.5 milligrams. Construct a 95% confidence interval for the mean caffeine content of this domestic Diet Cola soft drink.

32. The registration time for students at a community college is approximately normally distributed. If a random sample of 49 students has a mean registration time of 92 minutes with a standard deviation of 11.2 minutes, construct a 90% confidence interval for the mean registration time at this community college.

33. A sociologist is interested in estimating the educational level of a community. She randomly interviews 25 adult residents of the community and determines the number of years they attended school. Her sample had a mean of 12.5 years with s = 3.8 years. Use a 95% confidence interval

to estimate the educational level of all the adults living in this community.

34. The mean cholesterol level of a random sample of 36 adults aged 45–54 was 215 milligrams of cholesterol with s = 19.84. Construct a 95% confidence interval for the mean cholesterol level of all adults aged 45–54.

35. A bride-to-be is interested in estimating the cost of a wedding on Long Island. A random sample of 25 recent Long Island weddings has a mean cost of $40,000 and a sample standard deviation of $12,000. Construct a 95% confidence interval to estimate the population mean cost of a wedding on Long Island.

36. A survey of 40 NFL players found the average weight to be 245.3 pounds with s = 15.7 lbs, with an average height of 6 feet 2 inches with s = 3.8 inches. Find:
    a) A 95% confidence interval to estimate the mean weight of all NFL players.
    b) A 90% confidence interval to estimate the mean height of all NFL players.

Section 9.5

37. The value of $s_{\hat{p}}$, which is an estimate of the standard error of the population proportion, $\sigma_{\hat{p}}$, is determined by the formula: $s_{\hat{p}}$ = _____

38. The 95% confidence interval for the population proportion is constructed by the formula: _____ to _____ .

39. Which formula is used in the calculation of a confidence interval for the population proportion?
    a) p   b) $\mu_{\hat{p}}$ c) $\hat{p}$ d) $\bar{x}$ e) $E\mu$

40. When the sampling distribution of the population proportion is approximated by a normal distribution, which z-score is used to construct the 95% confidence interval?
    a) 0.95   b) 1.65   c) 1.96   d) 2.33   e) 2.58

For problems 41 to 46, when computing $s_{\hat{p}}$, round off to four decimal places.

41. Determine a 90%, 95% and 99% confidence interval for the following sample data:
    a) $\hat{p}$ = 0.30, and n=100
    b) $\hat{p}$ = 0.30, and n = 1600
    c) $\hat{p}$ = 0.30, and n = 100,000
    d) Explain the relationship between the width of each confidence interval and the margin of error as the sample size increases.

42. According to a Facebook Opinion Poll of 100 randomly selected users, 61 users stated that they do not install Facebook applications that require the user to invite others. Construct a 99% confidence interval to estimate the population proportion of Facebook users who do not install Facebook applications that require the user to invite others.

43. According to a new survey by the Harvard Institute of Politics, Democrats love Google+ and Twitter, while Republicans are more interested in sharing on Pinterest. The poll asked 3,058 Americans ages 18 to 29 which social sites they preferred to use. The results showed that Democrats use Google+ 52% of the time, while Republicans only use it 36% of the time. Determine the margin of error for the Harvard Institute of Politics estimate of the proportion of all Democrats who use Google+ if this estimate is based on a 95% confidence level.

44. A Department of Transportation survey of 1480 American drivers stated that 873 drivers wear seat belts. Construct:
    a) A 95% confidence interval to estimate the true proportion of all American drivers that wear seat belts.
    b) A 99% confidence interval to estimate the true proportion of all American drivers that wear seat belts.

45. Given the statement: *a 95% confidence interval for the proportion of people owning an answering machine is: 45% ± 3 percentage points,* explain why the following interpretation of this 95% confidence interval statement is incorrect.

    This confidence interval is interpreted to mean that you are 95% certain that the true proportion of all people that own an answering machine is within 3 percentage points of 45%.

Section 9.6

46. To increase the precision of a confidence interval with a fixed confidence level, you must _____ the sample size.

47. A confidence interval for a specific level of confidence consists of two components: a point estimate of the _____, and the _____ of error.

48. The margin of error for a confidence interval to estimate the population mean represents the _____ possible error of estimate for the population mean using the _____ .

49. The margin of error is denoted by the symbol _____. The margin of error represents one-half the _____ of the confidence interval.

50. As the level of confidence is increased, where the sample size and population standard deviation remain fixed, the margin of error _____ .

51. To reduce the margin of error while holding the confidence level fixed, then the sample size must be _____ .

52. The width of a confidence interval is twice the _____ .

53. For a given confidence level, a known population standard deviation, and a given margin of error, the formula to determine the sample size, n, to satisfy these requirements in estimating the population mean is: n = _____ .

54. For a given confidence level, a point estimate of the population proportion, $\hat{p}$, and a given margin of error, the formula to determine the sample size, n, to satisfy these requirements in estimating the population proportion is: n = _____ .

55. When determining the margin of error, the sample standard deviation is the one measure you can control.
    (T/F)

56. Determine a **conservative estimate** of the sample size for estimating the population proportion using a 95% confidence interval where the margin of error is to be:
    a) 0.01
    b) 0.10
    c) Explain one advantage in computing **conservative estimate**.

57. A manufacturing company wants to estimate the proportion of defective squash balls that are produced by a machine to be within 0.03 of the population proportion for a 99% confidence level. How large of a sample size, n, must the company select to accomplish this using the **conservative estimate** of the sample size formula?

58. a) How large a sample is required if one would like to be 95% confident that the estimate of the population mean

height of Americans will be within 0.5 inches of the population mean? (Assume that the population standard deviation is 6 inches.)
b) How large a sample is required for a 99% confidence interval of the population mean height of Americans? (Assume that the population standard deviation is 6 inches.)

59. How many times would a fair coin have to be randomly tossed to be 99% confident that between 40% and 60% of the tosses will land heads?

60. An opinion poll estimates the percentage of voters who favor candidate P to be 48% with a margin of error of 3 percentage points. Candidate P is not pleased with this poll and decides to take another poll that will have a higher precision level of 2 percentage points for the margin of error. If the candidate is willing to accept a 95% confidence level, then determine:
a) The number of voters that should be interviewed for this new poll.
b) The cost of the new poll, if the cost per interview is $1.50.

CHAPTER REVIEW

61. A Gallup survey of 1000 adults stated 49% of American adults suffer sleep-related problems such as insomnia. In addition, 52% of women say they have sleeping difficulties while only 45% of men stated they had sleeping difficulties. Whereas 40% of adults say they sometimes doze off when bored. The margin of error was 3 percentage points for this survey. Using this information, state an interval estimate for each of the stated percentages within this Gallup survey.

62. A random sample of 100 American workers showed that the average number of hours an employee works per week is 36.3 with the sample standard deviation equal to 4.2 hours. Find a 95% confidence interval for the mean number of hours that all American employees work per week.

63. A sociology professor is interested in estimating the mean work hours that a student spends at a job for which the student gets a monetary reward. The professor randomly selects 400 day students and calculates that the sample mean number of hours spent at a job is 18.5 hrs with s = 4. Determine the:
a) Population
b) Population parameter being estimated
c) Estimator used to estimate the population parameter
d) Point estimate
e) Margin of error for the 90%, 95%, and 99% confidence level
f) 90% confidence interval for the population mean, μ
g) 95% confidence interval for the population mean, μ
h) 99% confidence interval for the population mean, μ

64. There has been an increasing presence of adult students age 25 and older on U.S. college campuses. A recent study determined that 189 out of 1500 undergraduates surveyed were age 25 and older. Construct a 90% confidence interval to estimate the population proportion of U.S. college undergraduates that are age 25 and older.

65. A random sample of n college students is selected from the student body to estimate the mean IQ score of the student population. Assuming that the sample mean is 108, and the population standard deviation is equal to 15, construct a 95% confidence interval for samples of size:
a) 36
b) 100
c) 400
d) Determine the width of the confidence intervals found in parts a, b & c.
e) What is the effect on the width of the confidence interval as the sample size is increased while the confidence level remains the same?

For problems 66 to 70, when computing s_p, round off to four decimal places.

66. DPT Enterprises, Inc. is about to market a new kind of shaver, a shaver with six blades. In order to determine if they should proceed with their manufacturing plans, the marketing department of the company decides to survey a cross section of the potential buyers as to their feelings regarding their purchasing of the new shaver. The results of the survey were:

152 potential buyers said they would buy the shaver

sample size = 7569

Determine the:
a) Population
b) Population parameter being estimated
c) Estimator used to estimate the population parameter
d) Point estimate
e) Margin of error for the 90%, 95%, and 99% confidence level
f) 90% confidence interval for the population proportion.
g) 95% confidence interval for the population proportion.
h) 99% confidence interval for the population proportion.

67. Suppose you are the polling consultant for a state senator. You find that a straw poll of 625 registered voters indicates that 263 registered voters name drugs within the schools as the most important issue facing the inner city schools. Determine a 95% confidence interval for the proportion of all voters who hold this opinion. What is the margin of error associated with this estimate?

68. A technology research firm published that Android now has 79% of the market share for smartphone users. Determine the:
a) population
b) population parameter being estimated
c) estimator used to estimate the population parameter
d) point estimate
e) 90% confidence interval and margin of error for the population proportion if the sample size used is
1) 100
2) 500
3) 110
f) 95% confidence interval and margin of error for the population proportion if the sample size used is
1) 100
2) 500
3) 1100

g) 99% confidence interval and margin of error for the population proportion if the sample size used is
1) 100
2) 500
3) 1100
h) Discuss what happens to the margin of error as the sample size increases for each confidence interval.
i) Discuss what happens to the margin of error as the confidence level increases.

69. Mike's housemate Kathleen is trying to convince Mike to get a tattoo. Mike is very sensitive to popular opinion, and will only get a tattoo if "everyone else is doing it". Mike will get the tattoo if Kathleen can show that more than 50% of students at Mike's college have one. Kathleen wastes no time in taking a random sample of 120 people from Mike's campus. Her sample of 120 revealed 72 people who admitted to having a tattoo.
   a) Find a 95% confidence interval for the estimate of the population proportion of students that have a tattoo.
   b) What is the margin of error for this confidence interval?
   c) Write a statement to interpret this confidence interval.
   d) Will Mike get the tattoo? Why or why not?

70. A Roper Starch survey of 1350 men aged 18 years and older found that 486 men were single, while a random sample of 1560 women aged 18 years and older showed that 624 were single. Construct:
   a) A 95% confidence interval to estimate the true proportion of all single women that are 18 years or older.
   b) A 99% confidence interval to estimate the true proportion of all single men that are 18 years or older.

71. A random sample of 49 college male students indicates that the mean pulse rate, while at rest, is 68.3 beats per minute. Assuming the population standard deviation is 9.6 beats per minute, determine:
   a) A 95% confidence interval for the mean pulse rate, while at rest, for the population of college male students.
   b) The width of this 95% confidence interval.
   c) The margin of error for this 95% confidence interval.
   d) How large a sample of college students would one need to select to be 95% confident that the margin of error is within 1 heart beat of the mean pulse rate for all college male students?

72. You are hired by a cell phone provider to determine the number of times per day their customers interact with their smartphone. You take a random sample of 28 customers and find the average number of interactions to be 60 times per day with a sample standard deviation of 6.
   a) If you are to find a confidence interval for the population mean number of interactions, what distribution would you use and why?
   b) Find a 99% confidence interval estimate for the population mean number of interactions.
   c) What is the margin of error for the confidence interval you found in part b?
   d) A representative for the cell phone provider told you they believe the mean number of interactions per day was 65. What would you say to the company representative? Explain.

73. The formula used to calculate the margin of error when estimating the population mean, where the population standard deviation is known, is: E = _____ .

74. The margin of error formula used to determine the maximum possible error of estimate for the population proportion using the sample proportion, $\hat{p}$, is: E = _____ .

75. For a given confidence level, and a given margin of error, the formula to determine the **conservative** sample size, n, to satisfy these requirements in estimating the population proportion is: n = _____ .

76. A medical student is interested in estimating with a 90% confidence level the proportion of babies that are born between the hours of: 6AM and 6PM, called a "*morning baby*." If the student wants the margin of error to be within 4 percentage points, determine:
   a) A **conservative estimate** for the sample size required to meet these requirements.
   b) An estimate of the sample size required to satisfy these conditions, if a previous sample showed the proportion of morning babies to be 0.62.

WHAT DO YOU THINK?

77. The following article entitled "Survey Finds High Fear of Crime" discusses the results of a Gallup Poll on crime.
   a) Define the population for this survey.
   b) What type of estimate does the percentage value of 76% represent? What population parameter is it being used to estimate? Explain.
   c) What was the margin of error for this survey? How would you interpret this margin of error?
   d) State an interval that you believe might contain the true population proportion of Americans who are afraid to go out alone at night within one mile of their homes. How did you construct such an interval?
   e) What is meant by the statement "the difference in these figures is statistically insignificant because the survey's margin of error is plus or minus three percentage points"?
   f) State a sample point estimate for the proportion of women who are afraid to go out alone at night within a mile of their homes. State an interval estimate for the proportion of women who are afraid to go out alone at night within a mile of their homes. What is the difference between these two estimates?

Survey Finds High Fear of Crime

Fear of crime continues to pervade American society, especially in urban areas, where 76 percent of women fear walking alone at night in their neighborhood, according to the latest Gallup Poll.

Over all, 45 percent of Americans are afraid to go out alone at night within a mile of their homes, the researchers reported. Thirteen percent had that fear in the day. Sixty-two percent of the women surveyed had that fear, as did 26 percent of the men.

The survey also found that crime had affected an average of 25 percent of American households in the last 12 months.

Comparing surveys over the last decade, the poll reported that the incidence of crime and fear of crime had remained about the same.

Another statistic that has remained constant is the number of people who do not feel safe at home at night. Sixteen percent said they felt unsafe, as did 16 percent in 1991, 15 percent in 1987 and 20 percent in 1985. The difference in these figures is statistically insignificant because the survey's margin of error is plus or minus three percentage points.

The one area in which a marked change was found was the number of people who thought there was "more crime in this area than there was a year ago". In 1992, 47 percent felt there was more crime than there had been a year before, in contrast to 37 percent in the latest poll.

The poll said its findings showed that "the actual crime situation in this country is more serious than official Government figures" indicate because "many incidents are not reported to the police."

The findings were based on interviews conducted in person with 1,555 adults from Jan. 28-31.

PROJECTS

78. Poll your fellow students on an issue of concern for your school. For your poll, identify the:
a) Population
b) Population parameter being estimated
c) Estimator used to estimate the population parameter
d) Point estimate
e) Margin of error
f) 90% confidence interval for the population parameter
g) 95% confidence interval for the population parameter
h) 99% confidence interval for the population parameter

79. Randomly select a page from this text and count the number of words on the page.
a) Estimate the number of words in this book using this information.
b) Construct a 95% confidence interval using this information.

80. Use a technological tool to find the 90%, 95%, and 99% confidence intervals for the population mean for the sample data randomly selected from the population whose standard deviation is 20, if the data values are:

104 75 59 73 71 74 39
93 84 80 62 120 50 80
56 62

a) What is the point estimate?
b) For each confidence interval determine the margin of error.

81. Use a technological tool to find the 90%, 95%, and 99% confidence intervals for the population proportion for a random sample of 400 data values selected from a population that yields a point estimate of 0.6.
For each confidence interval determine the margin of error.

CHAPTER TEST

82. The numerical value obtained from the sample mean (or sample proportion) formula is referred to as an _____ of the population mean (or sample proportion). The sample mean (or sample proportion) formula which is used to find a point estimate of the population mean (or population proportion) is called an _____ .

83. The 95% confidence interval for the population mean suggests that 5% of these confidence intervals constructed for all possible sample means will not contain the true population mean.
(T/F)

84. A poll conducted for the National Heart, Lung and Blood Institute determined that 49% of adults know their own cholesterol level. If the survey's margin of error was 2.5 percentage points, then:
a) State an interval estimate for the percentage of adults who know their cholesterol level.
b) If 1,583 adults were surveyed, then state whether the interval estimate of part (a) is either a 90%, 95% or 99% interval estimate. Explain your reasoning.

85. Determine a 90%, 95% and 99% confidence interval and the margin of error for the following sample data:
a) $\bar{x} = 78$, s = 4 and n = 16
b) $\bar{x} = 78$, s = 4 and n = 36
c) $\bar{x} = 78$, s = 4 and n = 49
d) $\bar{x} = 78$, s = 4 and n = 100
e) What happens to the width of each confidence interval and the margin of error as the sample size increases?

86. In general, as the confidence level increases, the width of the confidence interval _____ .

87. For a sample of size n, the degrees of freedom is equal to the sample size minus _____ . In symbols, $df = $ _____ .

88. A *New England Journal of Medicine* diet study stated that an interval estimate of total caloric intake per day for the control group was 1581 to 3127. While the interval estimate of total caloric intake per day for the treatment group was 1555 to 3409.
a) Determine the margin of error for each group.
b) Which group has the wider interval estimate? How does this compare to the size of the margin of error?
c) Is there a relationship between the width of an interval estimate and the margin of error? Explain.

89. Using the information: A random sample of 24 Twitter users averaged 6 minutes on the site per day with a sample standard deviation of 1 minute, construct:
   a) a 90% confidence interval for the population mean.
   b) a 95% confidence interval for the population mean.
   c) a 99% confidence interval for the population mean.

In addition,

   d) Determine the width of each confidence interval by subtracting the lower confidence limit from the upper confidence limit.
   e) Compare and discuss the relationship of the width of each confidence interval with the confidence level.
   f) What happens to the width as the confidence level increases?

90. Which of the following is *not* a property of the t distribution?
   a) Less dispersed than the normal distribution
   b) Symmetric with respect to t = 0
   c) Bell-shaped
   d) Different t distribution for each sample size

91. A random sample of 14 Harbor High School seniors sent a mean of 60 text messages daily, with sample standard deviation of 7. Construct a 90% confidence interval for the mean number of text messages sent for all Harbor High School seniors.

92. The 99% confidence interval for the population proportion is constructed by the formula: _____ to _____ .

93. When the sampling distribution of the population proportion is approximated by a normal distribution, which z-score is used to construct the 99% confidence interval?
   a) 0.95   b) 1.65   c) 1.96   d) 2.33   e) 2.58

94. A CNN telephone survey of 1050 married adults indicated that 724 adults preferred to celebrate their anniversary by eating at a romantic restaurant. Construct a 90% confidence interval to estimate the true proportion of all married adults who prefer to celebrate their anniversary by eating at a romantic restaurant.

95. As the sample size increases, the margin of error increases. (T/F)

96. A political activist wants to estimate the proportion of all college students who are registered to vote. Determine:
   a) A **conservative estimate** of the sample size that would produce a margin of error of 3 percentage points for a 95% confidence interval.
   b) An estimate for the size of a sample for a 95% confidence interval, if a previous sample indicated that the proportion of college students who are registered to vote was 28%.

97. A biologist is studying the weights of 16 female Polar Bears in the Arctic and finds that the mean weight is 200 kg with a standard deviation of 35 kg. She wishes to construct a 95% confidence interval for the true population mean weight for female Polar bears in the Arctic.

   a) In order to construct the confidence interval, which distribution will you be using? What assumption about the distribution do you have to make in order to use it?
   b) For a 95% confidence interval, what is the value of the t-score? How many degrees of freedom do you need to use?
   c) Construct the 95% confidence interval estimate for the true population mean weight for female Polar bears in the Arctic. Interpret this confidence interval verbally.
   d) Would it be reasonable to say that the mean weight for female Polar bears in the Arctic is 150 kg? What about 220 kg? Explain your answers.

For the answer to this question, please visit www.MyStatLab.com.

CHAPTER 10

Introduction to Hypothesis Testing

Contents

10.1 Introduction
10.2 Hypothesis Testing
   Null and Alternative Hypotheses
10.3 The Development of a Decision Rule
10.4 p-Values for Hypothesis Testing

10.1 INTRODUCTION

Hypothesis testing is an important statistical decision-making tool that has applications in a variety of fields such as education, economics, medicine, psychology, marketing, sociology, management, etc. The following situations represent typical instances where a decision is required and the hypothesis testing procedure can be applied to help with this decision.

- A new drug is being tested to determine its effectiveness and safety. These results will be compared to the effectiveness of the drug currently in use. Based on the comparison of the test results, a decision will be made either to market the new drug or to continue marketing the present drug.
- The college registrar is testing a new registration procedure. The effectiveness of the new procedure is to be determined by comparing data collected under the existing system with the data collected using the new registration system. The registrar's decision as to which procedure to use will depend upon the analysis of the test results.
- A new manufacturing procedure is being tested to determine its effectiveness in reducing the number of defective finished products. These test results will be compared to the number of defective products under the present manufacturing procedure. Based on the comparison of the test results, a decision will be made to either implement the new procedure or to retain the present procedure.
- The developer of a special reflective coating for windows claims that this coating applied to the windows of a house will reduce air conditioning costs by at least 40%. Before a manufacturer will agree to purchase the manufacturing and marketing rights of this product a test will be conducted to determine if this product is as effective as claimed by its developer. Based on the test results the manufacturer will decide if it should purchase the special reflective coating.
- Many companies use "**acceptance sampling**" to determine if the quality of parts received from suppliers meet their specifications. Before the company accepts the supplier's shipment of parts, a sample of the parts is selected from the shipment and tested to determine if the parts meet the specifications. Based on the analysis of the test results, the company will either accept or reject the supplier's shipment.

10.2 HYPOTHESIS TESTING

Hypothesis testing is an example of inferential statistics since sample information is used to draw a conclusion about the population from which the sample was selected. For example, suppose an automotive club researcher decides to test the claim that "70% of Americans use seat belts" made by the National Highway Traffic Safety Administration by using a sample selected from the population. The National Highway's claim states that the population proportion of Americans who use a seat belt is 70%. If the researcher samples 1500 Americans and determines 67.5% indicate that they use a seat belt, then based on this statistical result can the researcher conclude that National Highway's claim about the population proportion is too high? A hypothesis test is an appropriate statistical technique to answer this question because the researcher is using a sample proportion to draw a conclusion about the population proportion of Americans who use a seat belt. The hypothesis testing procedure will enable the researcher to determine if the sample proportion of 67.5% is significantly less than the stated population proportion of 70% or whether this difference happened just by chance alone.

It is also important to realize that we cannot be 100% certain about the conclusions we draw about the population based upon one sample since a different sample selected from the same population could conceivably lead one to draw a different conclusion about the same population. Consequently, it is necessary to use probability to help explain how reliable our conclusions are. The inferences we make will be most reliable when the sample data is randomly selected from the population. Therefore, when we apply the hypothesis testing procedure we will always assume that the sample data was randomly selected from the population.

In this chapter, we will examine the reasoning used in the hypothesis testing procedure. The essence of hypothesis testing is to evaluate the evidence provided by a sample regarding a claim made about a population parameter. The following simplified example regarding gender prediction illustrates the reasoning associated with a hypothesis test.

EXAMPLE 10.1 Hypothesis Test of Gender Prediction

A doctor asserts she has a method that is 75% accurate in predicting, several months before birth, the gender of an unborn child. She is challenged to convince us that her claim is true. To test her claim, she predicts the gender of 36 unborn children and is correct in 17 of her predictions. Do we have any basis to support the doctor's claim?

Solution

We cannot support the doctor's claim, since we can argue that if her claim of 75% accuracy is true, she would almost never guess ONLY 17 out of 36 newborns. The reasoning behind rejecting her claim is supported by the extremely small probability that she would correctly predict as few as 17 out of 36 newborns if she really can predict the gender 75% of the time. The probability of such an occurrence is only 0.0002. In other words, if her claim of 75% is true, then we would expect her to correctly predict as few as 17 out of 36 newborns only twice in 10,000 times over the long run. This extremely small probability convinces us that her claim cannot be supported and should be rejected.

The basic idea behind hypothesis testing as illustrated by this example is simply a sample result that would rarely happen if a claim were true is good evidence to reject the claim. Clearly, we started to suspect the doctor's claim when the outcome of 17 correct predictions out of 36 newborns was shown to be a very unusual or rare event. (That is, a probability of 0.0002!) Thus, the more unusual or rare the sample outcome, the more doubtful we are that the sample outcome couldn't just happen by chance alone. In such instances, we believe that the sample outcome is enough evidence "beyond a reasonable doubt" to reject the claim!

Null and Alternative Hypotheses

A hypothesis test begins by stating the hypotheses or claims of the statistical test. There are two hypotheses that are formulated. One is referred to as the **null hypothesis** while the other is called the **alternative hypothesis**. The null hypothesis is a statement which hypothesizes a value for a population parameter. The alternative hypothesis is a statement that claims the population parameter is a value **different** than the value stated within the null hypothesis. In the hypothesis testing procedure we try to gather evidence to **reject** the null hypothesis and thus assert the alternative hypothesis.

In many ways, the hypothesis testing procedure is similar to the U.S. judicial process. In our development of hypothesis testing we will draw an analogy to the judicial process to provide a rationale and better understanding of the procedure used in the statistical testing of a hypothesis.

Within the judicial system, once an individual has been accused of committing a crime, a trial is conducted to determine if there is enough evidence to convict the individual. Throughout the trial, the fundamental assumption is that the individual is innocent. However, it is the objective of the prosecutor to present enough evidence to convince the jury to reject the assumption of innocence and find the individual guilty as charged.

In hypothesis testing the researcher is confronted with a similar problem. The researcher begins by formulating a fundamental assumption referred to as the null hypothesis. Similar to the judicial system where it is the prosecutor's task to present evidence to cause the rejection of the assumption of innocence, in the hypothesis testing procedure it is the objective of the researcher to present sample data (evidence) to cause the rejection of the null hypothesis.

Thus, the only way the prosecutor can obtain the objective of a guilty verdict is by convincing the jury to reject the assumption of innocence. This last statement is the essence of hypothesis testing.

In the hypothesis testing procedure, the researcher formulates *two hypotheses:* the **null hypothesis** and an **alternative hypothesis**. The alternative hypothesis is stated in such a way that by *rejecting the null hypothesis* the researcher is able to support his or her belief, the alternative hypothesis. The null hypothesis is symbolized by H_o and the alternative hypothesis is symbolized by H_a. Let's now define these two hypotheses.

Section 10.2 Hypothesis Testing

DEFINITION 10.1
Null Hypothesis, symbolized by **H₀**, is the statement being tested which states a value for a population parameter and is initially **assumed to be true**. The null hypothesis is stated as to indicate **"no change"** in the status quo or **"no difference"** or **"no relationship."** It is formulated for the sole purpose of trying to reject it by seeking evidence against it.

Within the hypothesis testing procedure, the researcher will be testing whether the claim stated about the population parameter within the null hypothesis is reasonable based upon the sample information. If the sample information leads to the rejection of the null hypothesis, then this will convince the researcher to accept the alternative hypothesis. The alternative hypothesis is a statement about the population parameter that the researcher is trying to find evidence to support it.

DEFINITION 10.2
Alternative Hypothesis, symbolized by **Hₐ**, is a statement about the value of a population parameter which is different from the value stated in the null hypothesis. The alternative hypothesis is stated to indicate that the status quo is **"not true"** or **"there is a difference"** or **"there is a relationship."** The alternative hypothesis is suspected to be true and is supported by the rejection of the null hypothesis.

The alternative hypothesis is always stated in a manner whereby the rejection of the null hypothesis will lead to the acceptance of the alternative hypothesis.

EXAMPLE 10.2

State the null hypothesis and the alternative hypothesis for the following survey.

According to a recent Edison Research media poll, 56% of Americans have a profile on a social networking site. A social media researcher who decides to test this statement would like to show that more than 56% of Americans have a social networking profile.

Solution

The social researcher will need to find evidence to support her claim that more than 56% of Americans have a profile on a social networking site. This statement about the population parameter of Americans who have a profile on a social networking site is the alternative hypothesis, Hₐ. In order to support the *alternative* hypothesis, Hₐ, the social researcher will need to perform a hypothesis test and to *gather sufficient evidence to reject* the statement that 56% of Americans have a profile on a social networking site. Since this statement about the population of Americans who have a profile on a social networking site represents "no change" in the status quo, then this is the *null* hypothesis, H₀. Thus, the null and alternative hypotheses can be stated as:

H₀: The population proportion of the Americans who have a profile on a social networking site is 56%.
Hₐ: The population proportion of the Americans who have a profile on a social networking site is **more than 56%**.

EXAMPLE 10.3

State the null hypothesis and the alternative hypothesis for the following research.

A college newspaper claims that full-time college students work an average of 20 hours a week. A marketing professor who believes this claim is too high decides to conduct a study to test the newspaper's claim.

Solution

The professor would like to find evidence to show that full-time college students work less than an average of 20 hours per week. This statement about the population parameter is the *alternative* hypothesis, Hₐ. In order to support the alternative hypothesis, Hₐ, the professor must *test* and *try to gather evidence to reject* the statement that full-time college students work an average of 20 hours per week which represents the status quo. This is the *null* hypothesis, H_o. The hypotheses are:

H_o: The population average number of hours that full-time college students work **is 20 hours per week**.
H_a: The population average number of hours that full-time college students work **is less than 20 hours per week**.

In the **Examples 10.2 and 10.3**, each **alternative** hypothesis contained words such as: **more than** or **less than**. These words indicate the *direction* of the difference from the value of the population parameter as stated in the null hypothesis. Such hypotheses are called *directional* alternative hypotheses.

DEFINITION 10.3
A **directional alternative hypothesis** is an alternative hypothesis that considers **only one specified direction of difference** away from the value stated in the null hypothesis. A directional alternative hypothesis is stated using words equivalent to: **less than** or **greater than**.

EXAMPLE 10.4

State the null hypothesis and the alternative hypothesis for the following research.

The New York Times reported that the mean private law school graduate's debt last year was $125,000. A researcher decides to test this claim against the suspicion that the mean debt has changed this year.

Solution

The hypothesis to be tested is: the population mean private law school graduate's debt this year is $125,000. That is, there is no change in the graduate's debt this year. Since the researcher would like to determine whether the debt has changed (that is, the debt has decreased or increased this year) both possibilities must be considered when stating the alternative hypothesis. This is accomplished by stating the alternative hypothesis as: the mean debt this year is not $125,000. Thus, the null and alternative hypotheses can be stated:

H_o: The population mean private law school graduate's debt this year is $125,000.
H_a: The population mean private law school graduate's debt this year is not $125,000.

When we need to consider **both** directions for the alternative hypothesis, that is, when phrases such as: too high, too low, more than, less than, etc. are **not** explicitly used or implied then the alternative hypothesis is a *nondirectional* alternative hypothesis.

DEFINITION 10.4
A **nondirectional alternative hypothesis** is a hypothesis that considers **both directions** away from the value specified in the null hypothesis. A **nondirectional** alternative hypothesis is stated using phrases equivalent to: **is not**, or **is not equal to**.

After the hypotheses are formulated the researcher's task is to design a procedure to test the null hypothesis. This statistical procedure consists of five steps and is referred to as the **hypothesis testing procedure**.

Hypothesis Testing Procedure

The hypothesis testing procedure consists of the five steps.
- **Step 1** Formulate the hypotheses.
- **Step 2** Determine the model to test the null hypothesis.
- **Step 3** Formulate a decision rule.
- **Step 4** Analyze the sample data.
- **Step 5** State the conclusion.

To help you understand the logic of the hypothesis testing procedure, we will outline the hypothesis testing procedure and illustrate how it is similar to the procedure followed within the judicial system.

A Comparison of the Hypothesis Testing Procedure to the Judicial System

Hypothesis Testing Procedure	Application of Hypothesis Testing Procedure to the Judicial System
Step 1 Formulate the hypotheses. State the null hypothesis, H_o, and alternative hypothesis, H_a.	**Step 1** H_o: *Individual is innocent.* This statement represents the status quo. H_a: *Individual is guilty.* The statement the prosecutor will try to convince the jury to accept by presenting evidence to **reject the null hypothesis.**
Step 2 Determine the model to test the null hypothesis. In this step the researcher identifies an appropriate distribution to serve as a model and calculates its mean and standard error. The mean and standard error values are referred to as the **expected results** of the test and are computed based upon the **assumption that the null hypothesis is true**.	**Step 2** The model is how the trial will be conducted. This includes the selection of a jury and the strategies formulated by both the defense attorney and the prosecutor. **This trial is based upon the assumption that the accused is presumed to be innocent** (i.e. the null hypothesis is assumed to be true).
Step 3 Formulate a decision rule. A decision rule is a statement formulated by the researcher which defines the criteria necessary for the rejection of the null hypothesis.	**Step 3** Each juror, in his/her own mind, will determine how much evidence is required to convict the individual (i.e. reject the assumption of innocence.)
Step 4 Analyze the sample data. After the data has been collected, the researcher analyzes the sample data by calculating the **test statistic**. A **test statistic** is a quantity computed from sample information that is used to make a decision about the null hypothesis.	**Step 4** The jury must analyze the evidence presented by both the prosecutor and the defense attorney.
Step 5 State the conclusion. The researcher compares the **test statistic** to the **decision rule** and determines the conclusion. The conclusion is either: a) **Reject the Null Hypothesis, H_o** OR b) **Fail to Reject the Null Hypothesis, H_o.**	**Step 5** The jury determines whether the evidence presented at the trial is enough beyond a reasonable doubt to convict the individual. Their verdict is either: a) **Individual is guilty** (this is equivalent to: **reject the null hypothesis**) OR b) **Individual is not guilty** (this is equivalent to: **fail to reject the null hypothesis**)

When a researcher draws a conclusion using the hypothesis testing procedure an important question to be considered is: will this procedure always lead to the correct conclusion? Unfortunately, the answer is no. Let's explore why by examining the possible outcomes of a trial. Everyone should be aware that the jury's verdict at the end of a trial is not always correct. The possible outcomes of a trial are shown in Figure 10.1.

	Possible Outcomes of a Trial	
	Truth: Person Is	
Jury's Verdict:	Not Guilty	Guilty
Guilty	Wrong Decision	Correct Decision
Not Guilty	Correct Decision	Wrong Decision

FIGURE 10.1

By examining Figure 10.1 we see that there are four possible outcomes. In two of these outcomes the correct decision is rendered but in the other two outcomes the wrong decision is reached. These wrong decisions are to *convict* an innocent person or to *free* a guilty person. When the jury convicts an innocent person they've committed an error. In statistics, this type of error is referred to as a **Type I error**. The error the jury commits when setting a guilty person free in statistics is referred to as a **Type II error**. A good procedure tries to *minimize* the probabilities of making both types of errors or incorrect decisions. In hypothesis testing the errors a researcher can make which are similar to the incorrect decisions of a jury are summarized in Figure 10.2.

	Possible Outcomes to a Statistical Hypothesis Test	
	Reality:	
Conclusion:	H_o is True	H_o is False
Reject H_o	Type I Error	Correct Decision
Fail to Reject H_o	Correct Decision	Type II Error

FIGURE 10.2

In examining Figure 10.2 notice that there are two types of errors that a researcher can make when performing a hypothesis test. These errors are referred to as a *Type I Error* and a *Type II Error*.

> **DEFINITION 10.5**
> A **Type I Error** is the error that is made when a **true null hypothesis is incorrectly rejected**.

For example, suppose a researcher is testing the null hypothesis that "70% of Americans use seat belts." A Type I error would occur if 70% of Americans use a seat belt is true for the entire population but the researcher *wrongfully* rejects this null hypothesis because a sample with a proportion much less than 70% was selected from the population.

> **DEFINITION 10.6**
> A **Type II Error** is the error that is made when a **false null hypothesis is not rejected**.

Let's return to the example where the researcher is testing the null hypothesis that 70% of Americans use seat belts. A Type II Error occurs when this null hypothesis is actually false (that is, 70% of Americans do not use seat belts) yet the researcher selects a sample, just by chance, with a proportion that is extremely close to 70% and *incorrectly* concludes that the null hypothesis (70% of Americans use seat belts) should not be rejected.

Consequently, when performing a hypothesis test, it is necessary to attempt to minimize the effects of a Type I and Type II Error. Such techniques will be discussed in Section 10.3.

Now, we will illustrate the hypothesis testing procedure through the following two examples.

EXAMPLE 10.5

Suppose you have a favorite coin which you use to make decisions every day. For example, you might ask, "should I attend my statistics class tomorrow? Heads, I attend, tails I don't." Over a period of time, you notice that you have a perfect attendance record and you begin to question the fairness of your favorite coin. How would you go about testing whether or not your favorite coin is biased? You would need to obtain a sample of the behavior of your favorite coin and draw a conclusion concerning its possible bias. In essence, you would need to conduct a hypothesis test to determine whether your favorite coin is biased (i.e. the probability of heads, p, is not equal to ½: $p \neq ½$). We will illustrate the hypothesis testing procedure to test your favorite coin for bias.

To help us concentrate just on the concepts of hypothesis testing, the calculations for the mean, standard error (SE) and distribution model usually calculated in step 2 and the calculations for the decision rule(s) usually calculated in step 3 are deliberately omitted within this discussion. You will **NOT** be responsible for these calculations within this chapter. However, these calculations will be discussed and included in subsequent chapters.

Solution

Step 1 Formulate the hypotheses.
We will **assume** that the coin is fair. Thus, the probability of a head, p, is ½: $p = ½$. This statement represents the null hypothesis.

Null Hypothesis, H_o: $p = ½$ (coin is fair).

Since we suspect that our favorite coin is not fair then we state the alternative hypothesis as:

Alternative Hypothesis, H_a: $p \neq ½$ (coin is biased).

Step 2 Determine the model to test the null hypothesis.
To test the null hypothesis we need to sample the behavior of our favorite coin. We will do this by tossing the coin 100 times. *Assuming the null hypothesis is true,* we expect the sampling distribution to have the following characteristics:
1) approximately a normal distribution
2) mean = 0.50
3) standard error: SE = 0.05
These results are shown in Figure 10.3.

Step 3 Formulate the decision rule.
Before one can formulate a decision rule, it is necessary to define a criteria for the rejection of the null hypothesis. This criteria will **always** be based upon the assumption that the null hypothesis is **true**.

Based upon the assumption that the coin is fair, one would expect a sample of 100 tosses to land heads about *50* times. We might decide to consider an *unusual* sample as a statistical result containing more than 60 heads or fewer than 40 heads. For each case, we would conclude that the coin is biased. Thus, we would write our *decision rule* as: *reject the null hypothesis,* H_o, if, the statistical result is either *more than 60% heads or less than 40% heads*. This decision rule is illustrated in Figure 10.4.

Step 4 Analyze the sample data.
The coin is tossed 100 times and the proportion of heads is recorded. This proportion is referred to as the **statistical result**.

Suppose the coin lands heads 67 times in 100 tosses. The statistical result is:

sample proportion = 67/100 = 0.67

552   *Chapter 10*   *Introduction to Hypothesis Testing*

FIGURE 10.3 — Sampling distribution model is approximately normal; mean = 0.50, standard error = 0.05

FIGURE 10.4 — REJECT H₀ if the statistical result is less than 0.40; REJECT H₀ if the statistical result is greater than 0.60; mean = 0.50, standard error = 0.05

FIGURE 10.5 — REJECT H₀ (below 0.40); REJECT H₀ (above 0.60); mean = 0.50; Statistical result = 0.67

FIGURE 10.6 — FAIL TO REJECT H₀ if the statistical result falls within 0.40 and 0.60; mean = 0.50

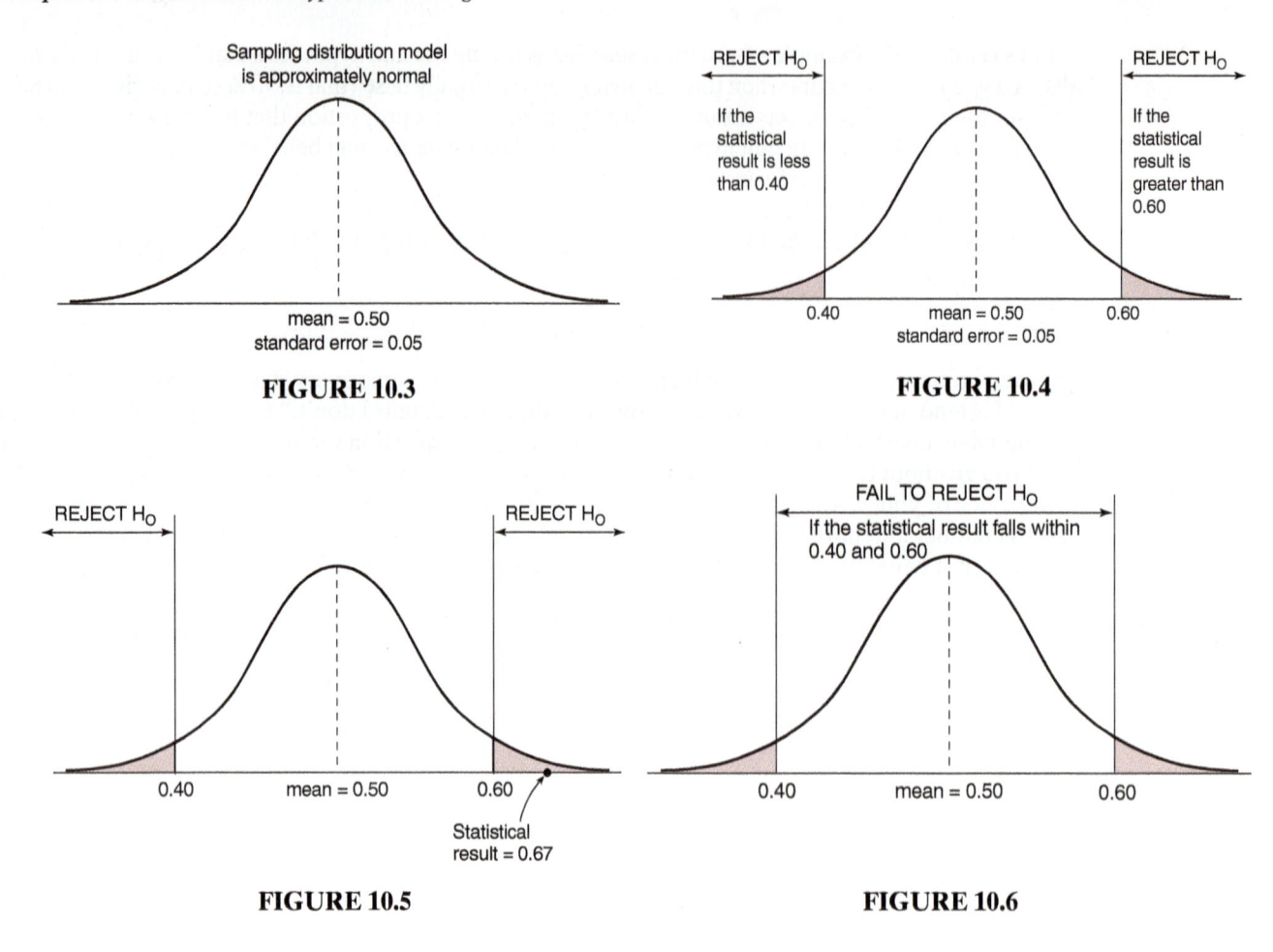

Step 5   State the conclusion.
Compare the statistical result to the *decision rule* and determine the conclusion.

For a sample proportion of 0.67, the conclusion is: reject the null hypothesis. This is illustrated in Figure 10.5.

The rejection of the null hypothesis means that one does *not* agree with the assumption that the coin is fair. Therefore, this leads to the conclusion that the coin is biased. Thus, in statistical terms, a rejection of the null hypothesis leads to the acceptance of the alternative hypothesis. Furthermore, a rejection of the null hypothesis means that the *difference* between the statistical result and the expected result (the mean) is significant enough to support the alternative hypothesis. We can then say that this difference is **statistically significant**.

If the statistical result fell between 0.40 and 0.60, the conclusion is *not* to reject the null hypothesis or, in other words, to *fail to reject* the null hypothesis since the statistical result is not greater than 0.60 or less than 0.40. This is illustrated in Figure 10.6.

When the statistical result does *not* fall within the rejection region of the decision rule, the conclusion is stated as: *fail to reject* the null hypothesis.

The interpretation of the conclusion (i.e. fail to reject the null hypothesis) means that, according to the decision rule, the statistical result is *not* significantly different from the mean. Therefore, although there is a difference, this difference is **not statistically significant** (not large enough) to conclude that the coin is biased. We are attributing this difference to chance alone. That is, if the coin is truly fair, we would not expect the coin to always land heads 50% of the time for every sample of 100 tosses. We would expect some of the statistical results to vary from 50% heads. This sample variability is often referred to as "**variability due to chance alone.**"

In general, when the researcher *fails* to reject the null hypothesis, this means that the difference between the **statistical result** and the **expected result** (the mean) is **not statistically significant**. ∎

Let's us use the hypothesis testing procedure to test the claim about the new iPad stated in Example 10.6.

EXAMPLE 10.6

A manufacturer claims that 90% of its new iPads operate three years before repairs are necessary. A consumer group feels that the 90% figure is too high and decides to test the claim.

Solution

Step 1 Formulate the hypotheses.

The claim being tested is that the proportion of all new iPads that operate three years before needing repair is 90%. Since this statement represents the status quo, the null hypothesis is:

H_o: the population proportion of new iPads that operate three years before needing repair is 90%: p = 0.90.

On the other hand, the consumer group believes that the 90% figure is too high, thus the alternative hypothesis is stated as:

H_a: the population proportion of new iPads that operate three years before needing repair is *less than 90%*: p < 0.90.

Step 2 Determine the model to test the null hypothesis.

To test the null hypothesis the consumer group randomly selects 100 people who have owned one of the manufacturer's new iPads for at least three years. Assuming the *null* hypothesis is true we would expect 90% of the new iPads to last three years before needing repair with a standard error, SE, of 3% or 0.03.

(Note: As indicated in Example 10.5 all calculations necessary to obtain the mean, standard error (SE), model distribution and decision rule(s) are deliberately omitted to simplify the presentation.)

The sampling distribution which is approximately normal with the expected results shown in Figure 10.7 serves as the model.

Step 3 Formulate the decision rule.

Based upon the assumption that the manufacturer's claim is correct one would expect 90% of the iPod Touches to have lasted three years before needing repair. Suppose the consumer group formulates its decision rule to be: **reject the null hypothesis if less than 85% of the new iPads have operated at least three years before needing repair.** Thus, if the sample proportion is less than 0.85, then the consumer group will reject the null hypothesis. This decision rule is illustrated in Figure 10.8.

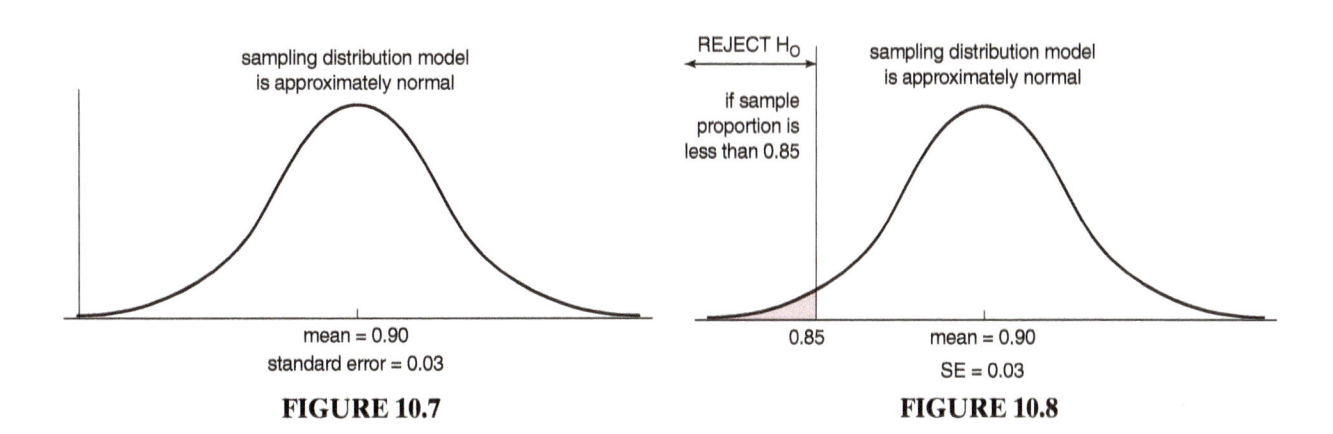

FIGURE 10.7 — sampling distribution model is approximately normal; mean = 0.90, standard error = 0.03

FIGURE 10.8 — sampling distribution model is approximately normal; REJECT H_o if sample proportion is less than 0.85; mean = 0.90, SE = 0.03

Step 4 Analyze the sample data.

The 100 randomly selected people are interviewed to determine how long their new iPads operated before needing repair.

If the statistical result is:

87% of the new iPads operated at least three years before needing repair. Thus, the sample proportion of new iPads needing repair is 87% or 0.87.

Step 5 State the conclusion.

For a statistical result of 0.87, the conclusion is *fail to reject* H_o since the sample proportion of 0.87 is greater than the decision rule of 0.85. This is illustrated in Figure 10.9.

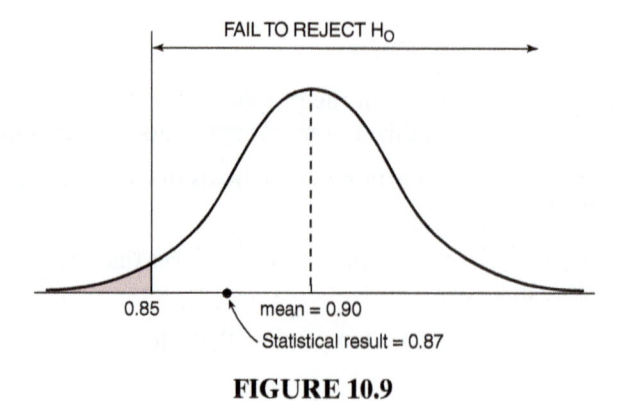

FIGURE 10.9

Our conclusion is *fail to reject* H_o since the sample proportion is *not* significantly less than the mean. Consequently, our conclusion (*fail to reject* H_o) means that the consumer group's claim that less than 90% of the New iPads operate less than 3 years before needing repair is not supported.

10.3 THE DEVELOPMENT OF A DECISION RULE

The objective of a hypothesis test is to use sample data to decide if one should reject or fail to reject the null hypothesis. Researchers often use a predetermined decision rule to decide whether the sample data supports or refutes the null hypothesis. To decide whether to reject H_o or fail to reject H_o, the researcher compares the statistical result to the value(s) of the decision rule, referred to as the critical value(s). We examine this idea of a decision rule through the use of Figure 10.10.

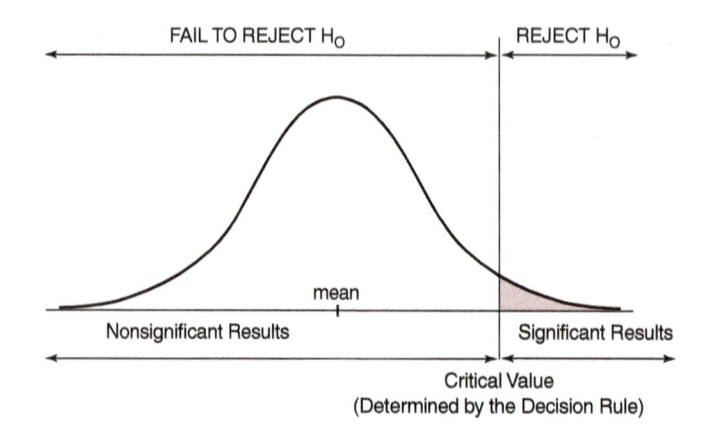

FIGURE 10.10

According to the decision rule pictured in Figure 10.10 one *rejects* the null hypothesis if the statistical result is *greater than* the critical value. One *fails to reject* the null hypothesis if the statistical result is *less than* the critical value. Any statistical result falling *beyond* the critical value (shown by the shaded region) is referred to as a **significant** result because it leads to the *rejection* of the null hypothesis. For the decision rule illustrated in Figure 10.10, any statistical result less than the critical value is referred to as a *nonsignificant result* since it leads one to *fail to reject the null hypothesis*.

The shaded area to the right of the critical value is referred to as the **level of significance**, since any statistical result falling within this shaded area is considered *significantly* greater than the expected result (the mean) and will lead to a rejection of the null hypothesis. Let's now define the level of significance.

DEFINITION 10.7
The **level of significance** (definition one) is the probability of a statistical result falling beyond the critical value, assuming that the null hypothesis is true. The level of significance is represented by the lower-case Greek letter α, pronounced alpha.

This is illustrated in Figure 10.11.

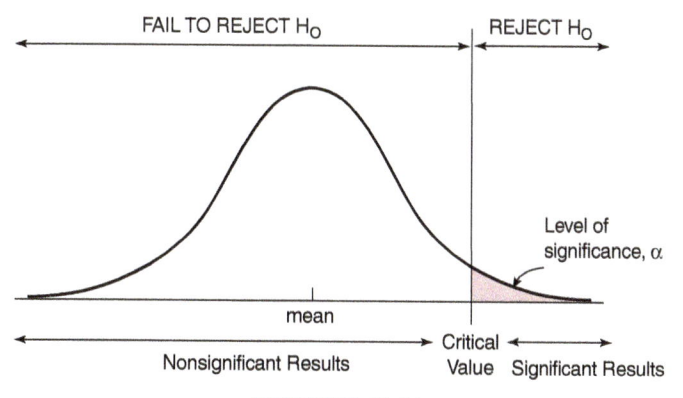

FIGURE 10.11

EXAMPLE 10.7

Using the decision rule from Example 10.6, illustrated in Figure 10.12, determine the level of significance, α.

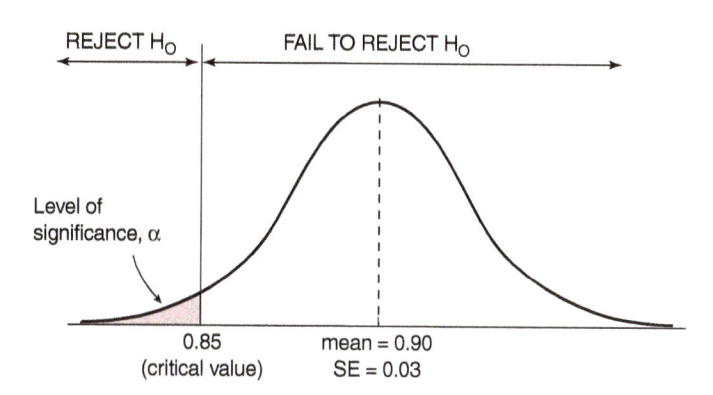

FIGURE 10.12

Solution

For this decision rule, the level of significance, α, is the area to the left of the critical value of 0.85 because any sample proportion falling below 0.85 is considered a significant result. This sampling distribution is approximated by the normal distribution. We can determine the level of significance by computing the z score of 0.85 and by using Table II: The Normal Curve Area Table.

$$z = \frac{0.85 - 0.90}{0.03} = -1.67$$

Using Table II the area to the left of $z = -1.67$ is 0.0475 or 4.75%. Therefore the level of significance is 4.75%, written $\alpha \approx 4.75\%$.

EXAMPLE 10.8

The decision rule in Example 10.5 is illustrated in Figure 10.13. Determine the significance level for this decision rule.

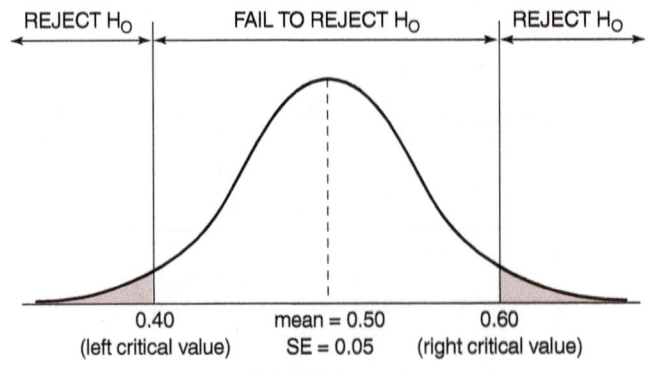

FIGURE 10.13

Solution

From the diagram in Figure 10.13, the decision rule has two areas which must be considered in the evaluation of the significance level. These areas are: the area that exceeds the right critical value of 0.60, and the area that is less than the left critical value of 0.40. If the statistical result falls into **either** one of the shaded areas then this result is **considered significant enough to reject the null hypothesis**.

The level of significance, α, is the **sum** of the two shaded areas in Figure 10.13. We can determine the level of significance by computing the z scores of 0.40 and 0.60. The z score of 0.40 is:

$$z = \frac{0.40 - 0.50}{0.05} = -2.00$$

From Table II, the area to the left of z = –2.00 is 0.0228 or 2.28%. Calculating the z score of 0.60, we get:

$$z = \frac{0.60 - 0.50}{0.05} = 2.00$$

From Table II, the area to the left of z = 2.00 is 0.9772 or 97.72%. Therefore the area to the right of z = 2.00 is 2.28% (see Figure 10.14).

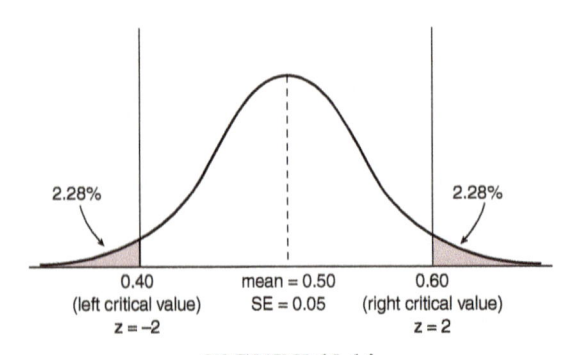

FIGURE 10.14

Since the significance level, α, represents the two shaded areas illustrated in Figure 10.14, then α is equal to the sum of these two shaded areas. Thus, we have:

α = the area to the left of 0.40 plus the area to the right of 0.60
α = area to the left of z = –2 plus area to the right of z = 2
α = 2.28% + 2.28%
α = 4.56%

Let's take another look at Type I and Type II errors and their relationship to the significance level, α. Recall that a Type I error is made when a true null hypothesis is incorrectly rejected. Therefore **whenever** the null hypothesis is rejected the possibility of a Type I error occurs. Consequently, a Type I error will occur whenever the null hypothesis is true *and* a statistical result falls beyond the critical value, leading to the rejection of the true null hypothesis. This is illustrated in Figure 10.15.

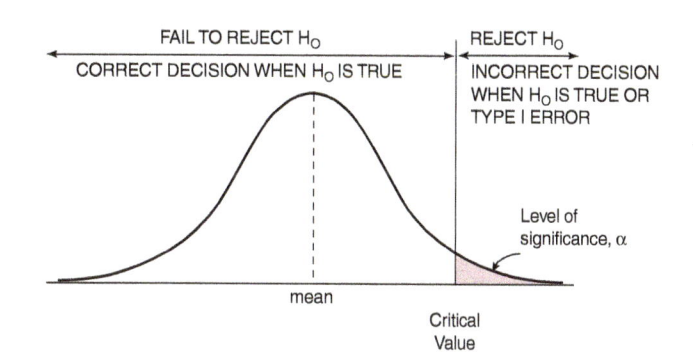

FIGURE 10.15

As can be seen from the diagram in Figure 10.15, when the null hypothesis is true the probability of committing a Type I error equals the level of significance, α.

Thus an alternative definition of the level of significance can be defined as follows:

DEFINITION 10.8
The **level of significance** (definition two) is the probability of committing a Type I error or the probability of rejecting a true null hypothesis.

The decision rule determines the level of significance. Thus, by choosing different decision rules the level of significance can be increased or decreased. Since the level of significance is the probability of committing a Type I error one might ask: "why not choose a decision rule that will have the *smallest* possible significance level"? Example 10.9 addresses this question.

EXAMPLE 10.9

Suppose a hypothesis test is conducted to determine if a coin is biased. If the coin is tossed 100 times consider the following three decision rules that are illustrated in Figures 10.16, 10.17 and 10.18.

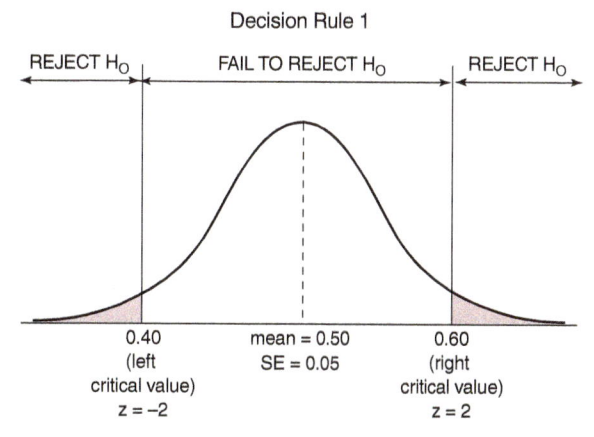

FIGURE 10.16

Decision Rule 2

REJECT H₀ | FAIL TO REJECT H₀ | REJECT H₀

0.35
(left
critical value)
z = −3

mean = 0.50
SE = 0.05

0.65
(right
critical value)
z = 3

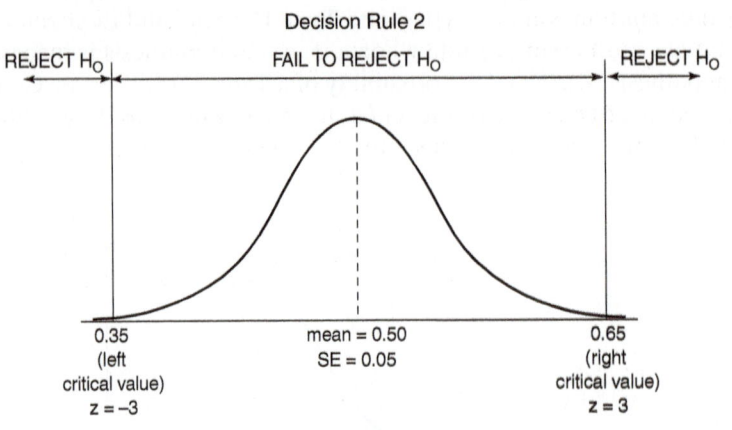

FIGURE 10.17

Decision Rule 3

REJECT H₀ | FAIL TO REJECT H₀ | REJECT H₀

0.1
(left
critical value)
z = −8

mean = 0.50
SE = 0.05

0.9
(right
critical value)
z = 8

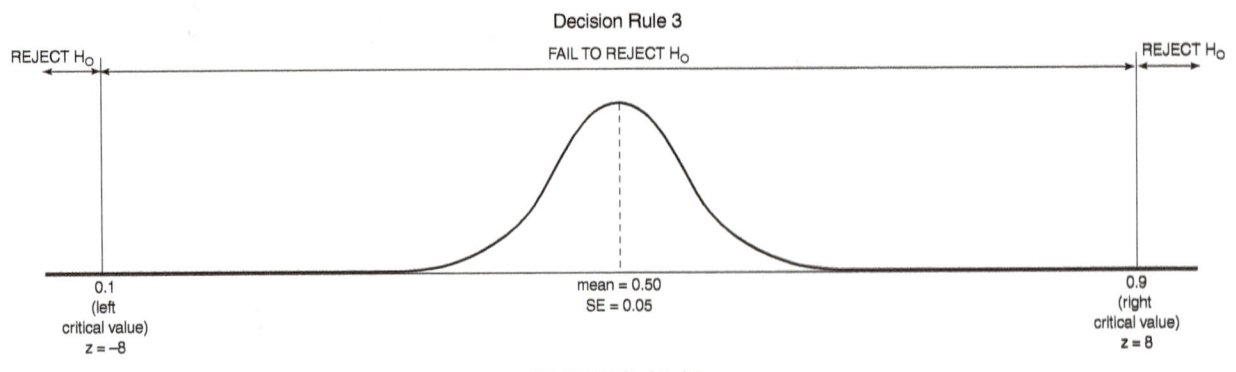

FIGURE 10.18

Solution

Notice in Decision Rule 1 illustrated in Figure 10.16, a coin will be considered biased if the sample proportion of heads is either *less than 0.40 or greater than 0.60*. In Decision Rule 2 illustrated in Figure 10.17, a coin will be considered biased if the sample proportion is either *less than 0.35 or greater than 0.65*. And finally, using Decision Rule 3 illustrated in Figure 10.18, a coin is considered biased if the sample proportion is *less than 0.1 or greater than 0.9*.

Notice as the critical values for the decision rules are chosen ***farther*** from the mean of 0.50, the significance level, represented by the shaded areas, ***decreases***. Since no one ever knows whether the coin being tossed is fair or biased, let's consider how choosing the critical values of our decision rule farther from the mean value affects the probabilities of a Type I and Type II error for a coin that is in *reality* fair (possibility #1) verses biased (possibility #2).

Possibility #1:   Coin is Fair

Assuming the coin being tossed is *fair*, then the effect of choosing the critical values **farther** from the mean ***decreases*** the chance of concluding that a fair coin is **biased**. Thus, the chance of committing a *Type I has decreased*.

Possibility #2:   Coin is Biased

Assuming the coin being tossed is *biased*, then the effect of choosing the critical values farther from the mean *decreases* the chance of concluding that the coin is biased. Thus, the chance of concluding that this biased coin is actually fair *has increased*. Therefore, the chance of committing a *Type II error has increased*.

To illustrate the relationship between choosing the critical values farther from the mean and the chance of committing a Type II error let's consider tossing a ***biased*** coin (possibility #2) 100 times. If the coin lands heads 85 times out of 100 tosses (i.e. the sample proportion is 0.85) let's examine the possible conclusions for each of the three previously stated decision rules shown in Figures 10.16, 10.17 and 10.18.

Using the Decision Rule 1 (Figure 10.16) and Decision Rule 2 (Figure 10.17) we would conclude that the coin is biased since the sample proportion 0.85 falls beyond the right critical value. This happens to be the correct decision.

However, using Decision Rule 3 (Figure 10.18), which have the critical values farther from the mean, would lead to the *incorrect conclusion that the biased coin is fair,* since the sample proportion 0.85 *does not* fall beyond the right critical value of 0.90.

Now let's reconsider the question posed prior to Example 10.9: "why not choose a decision rule that will have the smallest possible significance level"?

We can conclude from the previous discussion that *as the significance level decreases the chance of committing a Type II error increases.* Figure 10.19 illustrates this relationship. The probability of committing a Type II error will be denoted by the lower-case Greek letter ß, pronounced beta.

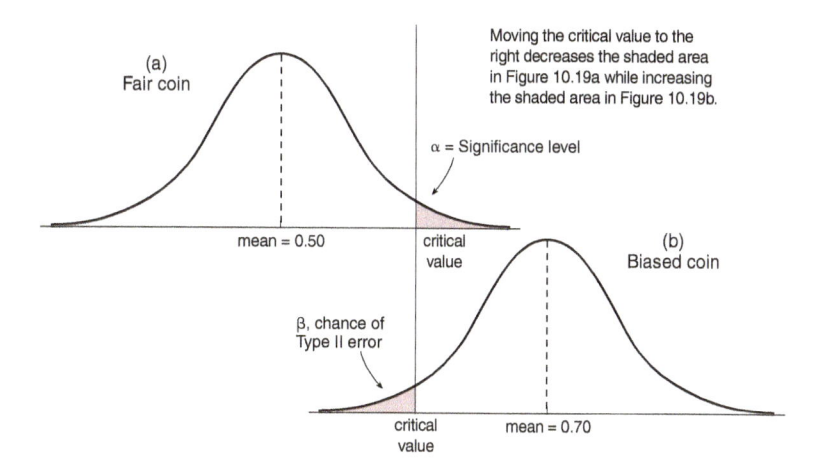

FIGURE 10.19

Before we begin, you should note that in Figure 10.19b, 0.70 will be used as the mean of the biased coin. However, *any* value other than 0.50 could have been used to represent the mean of the biased coin.

By carefully examining Figure 10.19, we observe that as the critical value line is moved to the right, the significance level, α, decreases *but* the chance of committing a Type II error, ß, *increases*. On the other hand, if the critical value line is moved to the left, the significance level, α, increases while the chance of committing a Type II error, ß, *decreases*.

Now that we have examined the relationship between the significance level, α, and the probability of committing a Type II error, ß, one might ask: "what *is* an appropriate value for the significance level"?

As can be seen from the previous discussion, if the significance level, α, is chosen too small then this increases the risk of committing a Type II error. Usually a significance level, α, of less than 1% is generally not used. On the other hand, values greater than 10% are generally not chosen for the significance level, α, since this would allow for a greater chance of committing a Type I error. Consequently in practice the two most commonly used values for the significance level, α, are 1% and 5%.

After the researcher has selected a suitable significance level, the critical z-score value(s) of the decision rule can be found in Table II.[1]

In the following examples we will illustrate the technique used to identify the critical z-score value(s) for the decision rule.

EXAMPLE 10.10

A consumer testing agency is conducting a hypothesis test to determine the validity of an advertised claim made by the Eye Saver Light Bulb Co. that the mean life of its new 60 watt bulb is 1800 hours. To test this claim the agency randomly selects 400 bulbs. Determine the critical z-score value(s) for the decision rule if:
a) The consumer agency believes the claim is too high and selects a significance level of 5%.
b) The consumer agency wants to test the claim using a nondirectional test and selects a significance level of 5%.

---
[1] Assuming the Distribution is Normal. Other Distributions will be discussed in subsequent chapters.

Chapter 10 Introduction to Hypothesis Testing

Solution to Example 10.10(a):

a) To determine the decision rule, we will use the first three steps in the hypothesis testing procedure.

Step 1 Formulate the hypotheses.
The null hypothesis is:

H_o: The population mean life of the new 60 watt bulb is 1800 hrs: $\mu = 1800$.

Since the consumer agency believes that the claim is *too high,* the alternative hypothesis is:

H_a: The population mean life of the new 60 watt bulb is *less than* 1800 hrs: $\mu < 1800$.

Step 2 Determine the model to test the null hypothesis.
(Note: all calculations necessary to obtain the mean and standard error (SE) are deliberately omitted to simplify the presentation.) To test this claim, the consumer agency randomly selects 400 bulbs from the population.

Assuming the null hypothesis is true implies that the mean life for samples of 400 bulbs is 1800 hrs with a standard error of 5 hrs.

The sampling distribution which is approximately normal with the expected results shown in Figure 10.20 serves as the model.

Step 3 Formulate the decision rule.
First, we must determine whether we have a directional or nondirectional alternative hypothesis. Since the alternative hypothesis is the mean life is *less than* 1800 hours, we have a *directional alternative hypothesis.* Therefore the critical z-score value of the decision rule will be located to the *left* of the mean. This is illustrated in Figure 10.21.

In this example the consumer agency selected a significance level of 5%. The significance level of 5% is represented by the shaded region in Figure 10.22. Using this information we can determine the critical z score, written as z_c. Since we have a normal distribution and the area to the left of the critical z score, z_c, is 5%, we can find the value of z_c associated with the area 0.05 in Table II. The value for z_c is –1.65. This is illustrated in Figure 10.22.

For the critical z-score, $z_c = -1.65$, we formulate the following decision rule: **reject the null hypothesis if the z-score of the sample mean life is less than the critical z-score of –1.65.** This is illustrated in Figure 10.23.

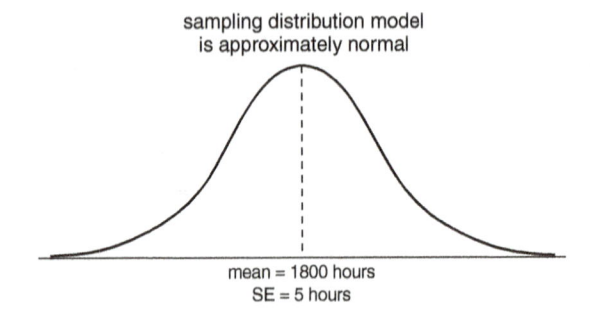

FIGURE 10.20

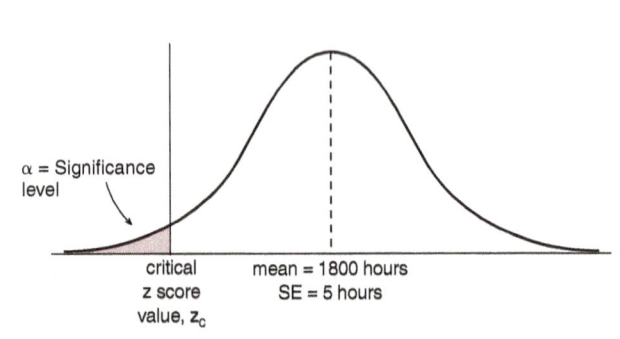

FIGURE 10.21

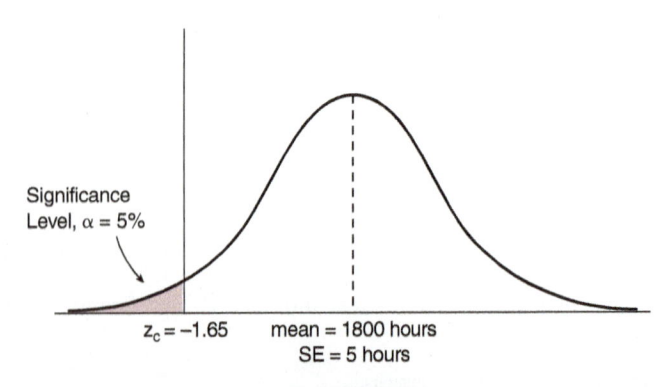

FIGURE 10.22

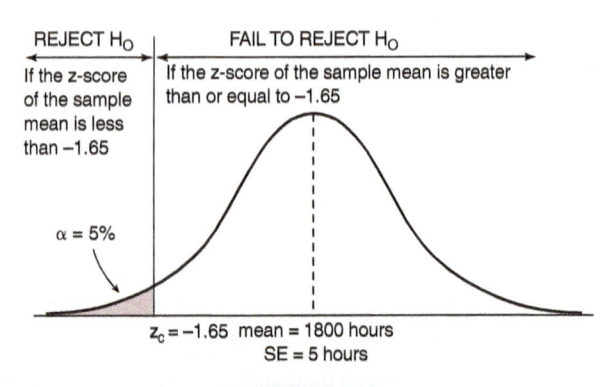

FIGURE 10.23

Notice from Figure 10.23, the z score of the sample mean enables us to make a decision about the null hypothesis. The z score of the sample mean is commonly referred to as the **test statistic**.

> **DEFINITION 10.9**
> The **test statistic** is a quantity computed from sample information that is used to decide whether to reject or fail to reject the null hypothesis.

In this chapter, the test statistic is determined by the formula:

$$z = \frac{\text{sample statistic} - \text{mean}}{\text{SE}}$$

Since the region for rejecting the null hypothesis is only on the left side (or in the left tail) of the sampling distribution this type of hypothesis test is referred to as a *one-tailed test*, abbreviated **1TT**.

> **One-Tailed Test, 1TT**
>
> A **one-tailed test** is conducted when the alternative hypothesis is directional.

A one-tailed test is illustrated in Figure 10.24.

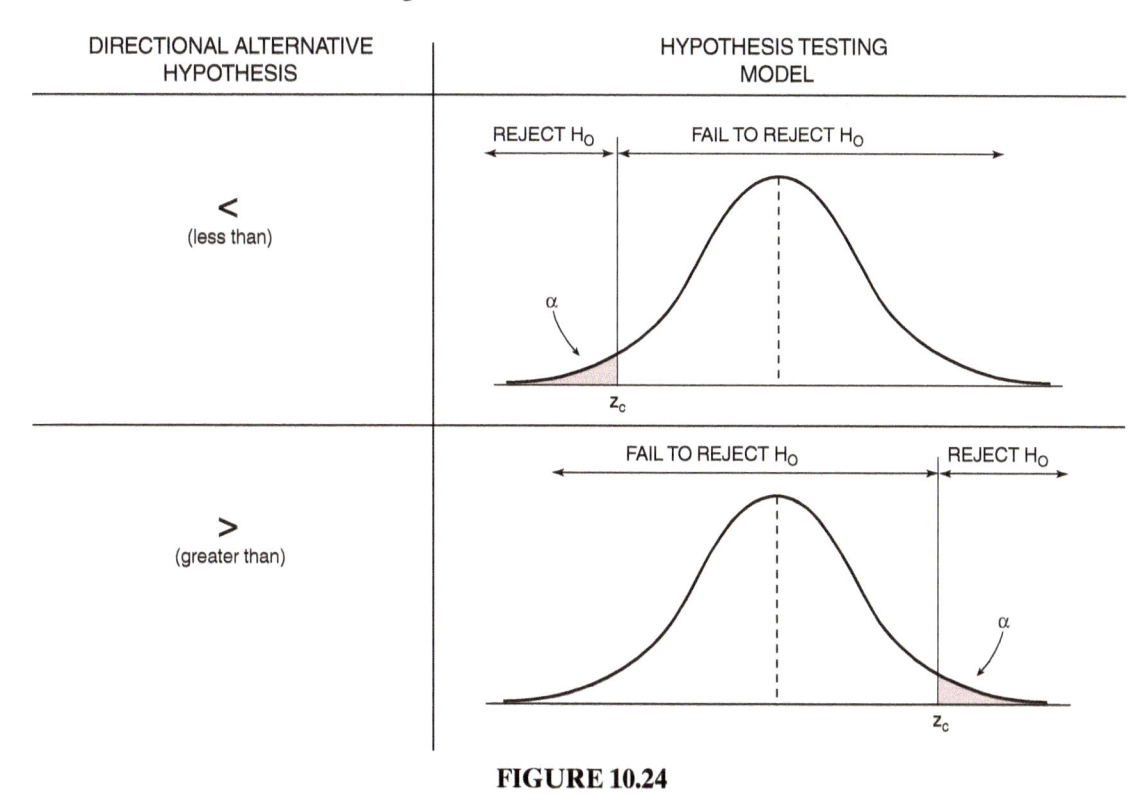

FIGURE 10.24

b) As in part (a), we will use the first three steps in the hypothesis testing procedure to determine the decision rule.

Step 1 Formulate the hypotheses.
The null hypothesis is:

H_o: The population mean life of the new 60 watt bulb is 1800 hrs: $\mu = 1800$.

562   Chapter 10   Introduction to Hypothesis Testing

Since the consumer agency wants to test the null hypothesis using a nondirectional test, the alternative hypothesis is:

H$_a$: The population mean life of the new 60 watt bulb is *not* 1800 hrs: $\mu \neq 1800$.

Step 2 Determine the model to test the null hypothesis.
(Note all calculations necessary to obtain the mean and standard error (SE) are deliberately omitted to simplify the presentation.) To test the null hypothesis the consumer agency randomly selects 400 bulbs. Assuming the null hypothesis is true implies that the mean life for all samples of 400 bulbs is 1800 hrs with a standard error (SE) of 5 hrs. The sampling distribution which is approximately normal with the expected results shown in Figure 10.25 serves as the model.

Step 3 Formulate the decision rule.
First we must determine whether we have a directional or nondirectional alternative hypothesis. Since the alternative hypothesis is nondirectional we must determine *two* critical z-score values, one located to the left of the mean and one located to the right of the mean. The critical z-score value located to the left of the mean is referred to as the **left critical z-score value**, denoted z_{LC}, and the critical value located to the right of the mean is referred to as the **right critical z-score value**, denoted z_{RC}. This is illustrated in Figure 10.26.

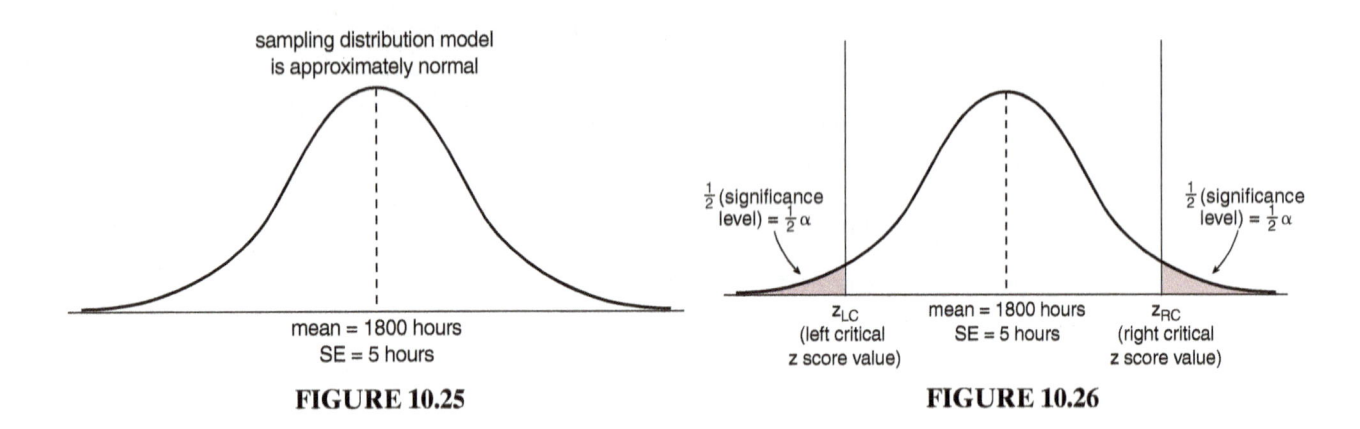

FIGURE 10.25

FIGURE 10.26

Since the significance level for this example is 5%, then the sum of the two shaded regions in Figure 10.26 must total 5%. The conventional procedure is to **divide the significance level *equally*** between these two shaded regions. Therefore the area of each shaded region is half the significance level (i.e. ½ α). Since the significance level is 5% the area in each shaded region (or tail) is:

$$\tfrac{1}{2}\alpha = \tfrac{1}{2}(5\%)$$
$$= 2.5\%$$

This is illustrated in Figure 10.27.

The decision rule requires two critical z-score values: the left critical value (referred to as the **left critical z score**, denoted z_{LC}) and the right critical value (referred to as the **right critical z score**, denoted z_{RC}). Using the information in Figure 10.27 and Table II, we can find the value of the two critical z scores, z_{LC} and z_{RC}.

To find z_{LC} we must determine the z score which has 2.5% or 0.025 of the area to the left. From Table II, $z_{LC} = -1.96$. Similarly to find z_{RC} we must determine the z score which has 97.5% or 0.9750 of the area to the left. From Table II, $z_{RC} = 1.96$. This is illustrated in Figure 10.28.

Using these critical z score values we formulate the decision rule: **reject the null hypothesis if the z score of the sample mean life (the test statistic) is either less than the left critical z score value of −1.96 or more than the right critical z score value of 1.96.** This is illustrated in Figure 10.29.

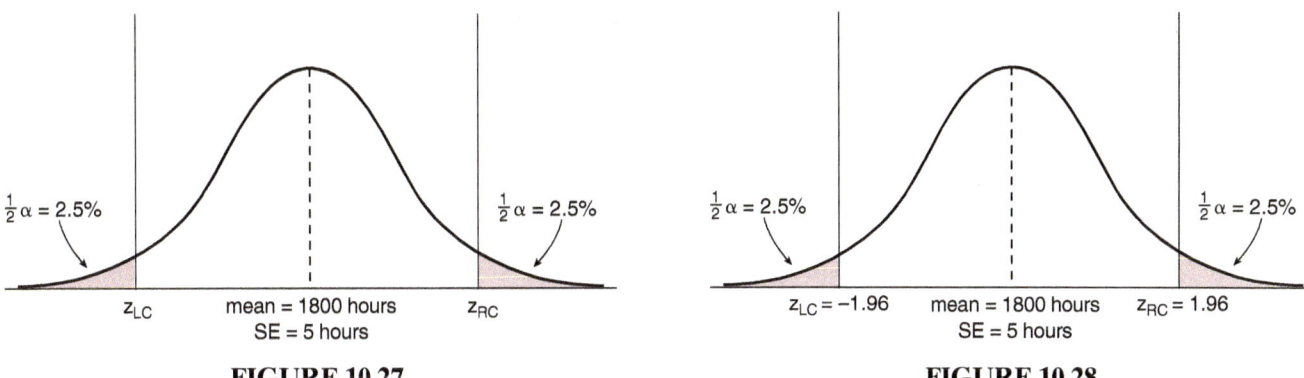

Section 10.3 *The Development of a Decision Rule* **563**

FIGURE 10.27

$\frac{1}{2}\alpha = 2.5\%$ $\frac{1}{2}\alpha = 2.5\%$

z_{LC} mean = 1800 hours z_{RC}
SE = 5 hours

FIGURE 10.28

$\frac{1}{2}\alpha = 2.5\%$ $\frac{1}{2}\alpha = 2.5\%$

$z_{LC} = -1.96$ mean = 1800 hours $z_{RC} = 1.96$
SE = 5 hours

REJECT H_O | FAIL TO REJECT H_O | REJECT H_O

If test statistic is less than −1.96   If test statistic is greater than 1.96

$\frac{1}{2}\alpha = 2.5\%$ $\frac{1}{2}\alpha = 2.5\%$

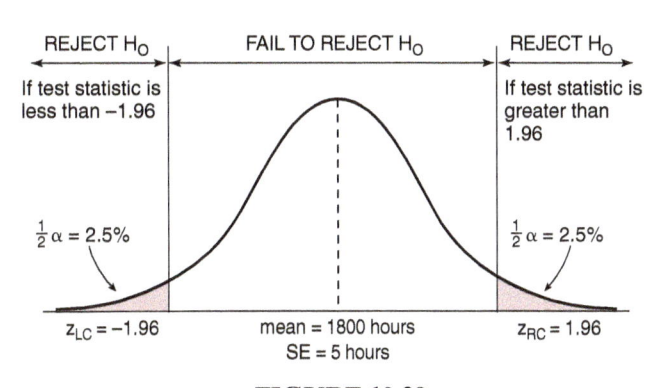

$z_{LC} = -1.96$ mean = 1800 hours $z_{RC} = 1.96$
SE = 5 hours

FIGURE 10.29

In the previous example there were two regions on the normal curve for rejecting the null hypothesis. One rejection region was in the left tail and the other rejection region was in the right tail of the curve. This type of hypothesis test is referred to as a **two-tailed test**, abbreviated **2TT**.

Two-Tailed Test, 2TT

A **two-tailed test** is conducted when the alternative hypothesis is nondirectional.

In a two-tailed test, the level of significance, α, is divided equally between the two tails. Therefore, the area in each tail is half the significance level, $\frac{1}{2}\alpha$.

A two-tailed test is illustrated in Figure 10.30.

NONDIRECTIONAL ALTERNATIVE HYPOTHESIS	HYPOTHESIS TEST MODEL
$\neq$ (not equal)	REJECT H_O | FAIL TO REJECT H_O | REJECT H_O $\frac{1}{2}\alpha$ $\frac{1}{2}\alpha$ z_{LC} z_{RC}

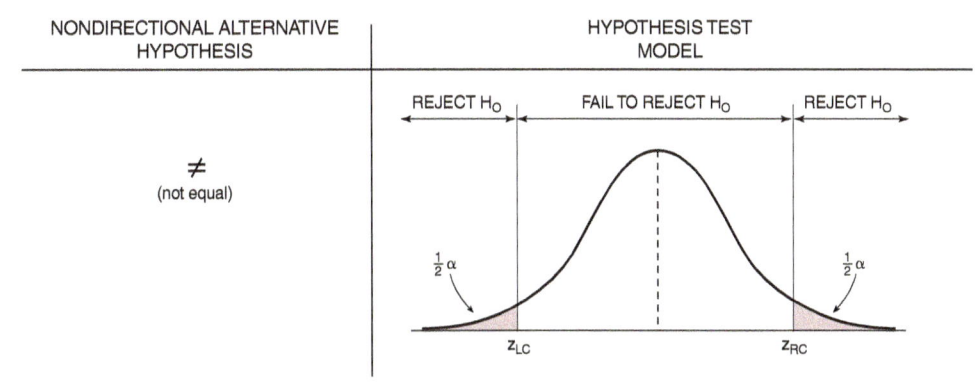

FIGURE 10.30

EXAMPLE 10.11

A national department store chain has determined that in previous years 60% of its customers used the department store's special holiday credit card to make Christmas purchases. This year because of changing economic conditions the department store management would like to determine if the 60% figure is too low. To test the 60% figure, 150 customers are randomly selected during the week of December 10th and asked whether they were using the store's special holiday credit card. Determine the decision rule for the hypothesis test. Use a significance level of 1%.

Assume the sampling distribution is approximately normal with a mean of 0.60 and a standard error of: SE = 0.04.

Test the null hypothesis:

H_o: The population proportion of the customers who use the special holiday credit card is 60%: p = 0.60.

Against the alternative hypothesis:

H_a: The population proportion of customers who use the special holiday credit card is *more than* 60%: p > 0.60

Solution

Since the first two steps of the hypothesis testing procedure are given we begin with Step 3.

Step 3 Formulate the decision rule.

We have a directional alternative hypothesis since it is written as *more than 60%* of the customers use the special holiday credit card. Therefore the critical value will be located to the *right* of the mean. This is illustrated in Figure 10.31.

Using the information in Figure 10.31 and Table II: the Normal Curve Area Table, we can determine the critical z score, z_c. Since the percent of the area to the left of z_c is 99% then critical z score is $z_c = 2.33$.

The decision rule is: reject the null hypothesis if the z score of the sample proportion (test statistic) is *more than* the critical z score of 2.33. This is illustrated in Figure 10.32.

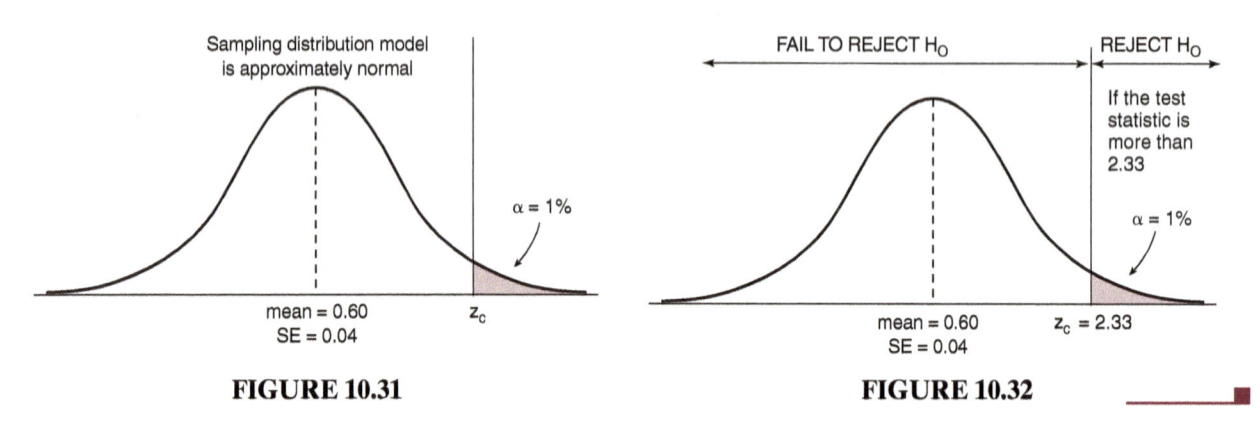

FIGURE 10.31

FIGURE 10.32

From the previous examples, we can now generalize the procedure to formulate the decision rule.

Procedure to Formulate the Decision Rule

a) Determine the type of alternative hypothesis (directional or nondirectional).
b) Determine the type of test (1TT or 2TT).
c) Identify the significance level ($\alpha = 5\%$ or $\alpha = 1\%$).
d) Construct the appropriate hypothesis test model.
e) Find the critical z score(s) using Table II for the appropriate level of significance.
f) State the decision rule.

Tables 10.1 and 10.2 summarize the procedure for formulating the decision rule for the significance levels of 5% and 1%.

TABLE 10.1

Formulating the Decision Rule Using 5% Significance Level

Type of Alternative Hypothesis, H_a	Directional (less than) <	Directional (greater than) >	Nondirectional (not equal to) ≠
Type of Test	One-tailed (1TT)	One-tailed (1TT)	Two-tailed (2TT)
Hypothesis Test Model	REJECT H_o | FAIL TO REJECT H_o; $\alpha = 5\%$; $z_C = -1.65$, mean	FAIL TO REJECT H_o | REJECT H_o; $\alpha = 5\%$; mean, $z_C = 1.65$	REJECT H_o | FAIL TO REJECT H_o | REJECT H_o; $\frac{1}{2}\alpha = 2.5\%$, $\frac{1}{2}\alpha = 2.5\%$; $z_{LC} = -1.96$, mean, $z_{RC} = 1.96$
Critical Z Score(s)	$z_C = -1.65$	$z_C = 1.65$	$z_{LC} = -1.96$ and $z_{RC} = 1.96$
Decision Rules	Reject H_O, if the test statistic is less than −1.65	Reject H_O, if the test statistic is greater than 1.65	Reject H_O, if the test statistic is either less than −1.96 or greater than 1.96

TABLE 10.2

Formulating the Decision Rule Using 1% Significance Level

Type of Alternative Hypothesis, H_a	Directional (less than) <	Directional (greater than) >	Nondirectional (not equal to) ≠
Type of Test	One-tailed (1TT)	One-tailed (1TT)	Two-tailed (2TT)
Hypothesis Test Model	REJECT H_o | FAIL TO REJECT H_o; $\alpha = 1\%$; $z_C = -2.33$, mean	FAIL TO REJECT H_o | REJECT H_o; $\alpha = 1\%$; mean, $z_C = 2.33$	REJECT H_o | FAIL TO REJECT H_o | REJECT H_o; $\frac{1}{2}\alpha = 0.5\%$, $\frac{1}{2}\alpha = 0.5\%$; $z_{LC} = -2.58$, mean, $z_{RC} = 2.58$
Critical Z Score(s)	$z_C = -2.33$	$z_C = 2.33$	$z_{LC} = -2.58$ and $z_{RC} = 2.58$
Decision Rules	Reject H_O, if the test statistic is less than −2.33	Reject H_O, if the test statistic is greater than 2.33	Reject H_O, if the test statistic is either less than −2.58 or greater than 2.58

Hypothesis Testing Procedure

Step 1 Formulate the two hypotheses, H_o and H_a.

Step 2 Determine the model to test the null hypothesis, H_o.
Under the assumption H_o is true, calculate the expected results: (**The calculation of the expected results will be discussed in the subsequent chapters.**)
 a) Identify the appropriate sampling distribution used as the hypothesis testing model.
 b) Mean of the sampling distribution
 c) Standard error (SE) of the sampling distribution

Step 3 **Formulate the Decision Rule**:
 a) Determine type of alternative hypothesis, directional or nondirectional.
 b) Determine type of test, 1TT or 2TT.
 c) Identify the significance level, $\alpha = 1\%$ or $\alpha = 5\%$.
 d) Construct the appropriate hypothesis test model.
 e) Find the critical z score(s) using Table II for the appropriate level of significance.
 f) State the decision rule.

Step 4 **Analyze the sample data.**
Collect the necessary sample data to calculate the test statistic using the formula

$$z = \frac{\text{sample statistic} - \text{mean}}{\text{SE}}$$

Step 5 **State the conclusion.**
Compare the test statistic to the decision rule and either:
a) **Reject H_o and accept H_a at α,**
or
b) **Fail to reject H_o at α.**

Example 10.12 illustrates the use of the hypothesis testing procedure to perform a hypothesis test.

EXAMPLE 10.12

Last year, 75% of adults within the USA stated that they believed a college degree was very important. This year a national poll randomly selected 1,020 American adults and found that 794 adults believed that a college degree is very important. Can you conclude that the percent of adults within the USA who believe a college degree is very important has significantly increased this year? Use mean = 0.75 and standard error: SE = 0.0136 to perform a hypothesis test at $\alpha = 5\%$ to answer this question.

Solution

We will use the hypothesis testing procedure to perform this hypothesis test. The first step is to formulate the hypotheses.

Step 1 Formulate the two hypotheses, H_o and H_a.
The hypothesis being tested is that 75% of the adults within the USA believe a college degree is important. Therefore, the null hypothesis is:

H_o: The population proportion of adults within the USA who believe a college degree is important is 75%: $p = 0.75$.

Since we are trying to show that the percent of adults who believe a college degree is important *has increased*, the alternative hypothesis is:

H_a: The population proportion of adults within the USA who believe a college degree is important is *greater than 75%*: $p > 0.75$.

Section 10.3 The Development of a Decision Rule   567

The second step involves determining the model to test the null hypothesis.

Step 2 Determine the model to test the null hypothesis, H_o.
To test the null hypothesis, a national poll of 1,020 American adults was randomly selected to determine their belief about the importance of a college degree. Under the assumption H_o is true, we need to calculate the expected results. As we have mentioned repeatedly throughout Chapter 10, the expected results are given so we can concentrate on the concepts of the hypothesis testing procedure. Thus, the sampling distribution is approximately normal with a mean = 0.75 and a standard error, SE = 0.0136. These expected results are shown in Figure 10.33.

Now that we have calculated the expected results based upon the assumption that the null hypothesis is true, we need to formulate a decision rule.

Step 3 Formulate the decision rule.
a) Determine type of alternative hypothesis.
Since the alternative hypothesis is in the form of a *greater than statement*, it is a directional hypothesis.
b) Determine the type of test.
A directional alternative hypothesis is a one tail test (1TT). Since it is in the form of a *greater than statement*, the critical value will be located to the *right* of the mean.
c) Identify the significance level.
The significance level is $\alpha = 5\%$.
d) Construct the appropriate hypothesis test model.
This is illustrated in Figure 10.34.
e) Calculate the critical z score using Table II:
Using the information in Figure 10.34 and Table II: the Normal Curve Area Table, we can determine the critical z score, z_c. Since the percent of the area to the left of z_c is 95%, then $z_c = 1.65$.
The decision rule is: reject the null hypothesis if the z score of the sample proportion (test statistic) is *greater* than the critical z score of 1.65. This is illustrated in Figure 10.35.

Now that the decision rule has been formulated, we need to calculate the test statistic.

Step 4 Analyze the sample data.
The experiment is the process of collecting the sample information from the population. The national poll of 1,020 randomly selected American adults represents the sample data. Since 794 adults within the sample said that a college degree is very important, then the sample proportion is: $794/1020 \approx 0.7784$. The test statistic for the sample proportion is:

$$z = \frac{0.7784 - 0.75}{0.0136}$$

$$z = 2.09 \text{ (test statistic)}$$

Now we compare the test statistic to the decision rule, so we can draw a conclusion.

Step 5 State the conclusion.
Since the test statistic 2.09 is greater than the critical z score of 1.65 then the conclusion is *reject H_o and accept H_a* at $\alpha = 5\%$. This is illustrated in Figure 10.36.

Therefore, we can conclude that more than 75% of adults within USA believe a college degree is very important at $\alpha = 5\%$.

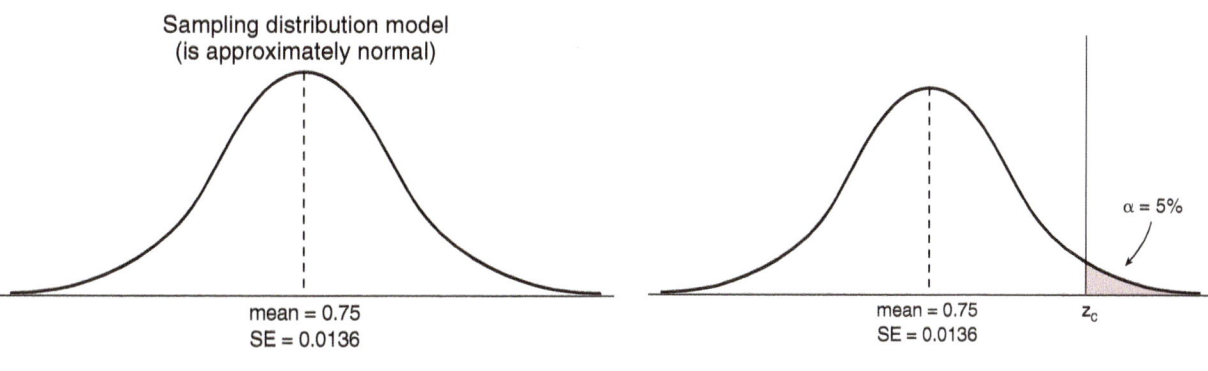

FIGURE 10.33

Sampling distribution model (is approximately normal)
mean = 0.75
SE = 0.0136

FIGURE 10.34

mean = 0.75
SE = 0.0136
$\alpha = 5\%$
z_c

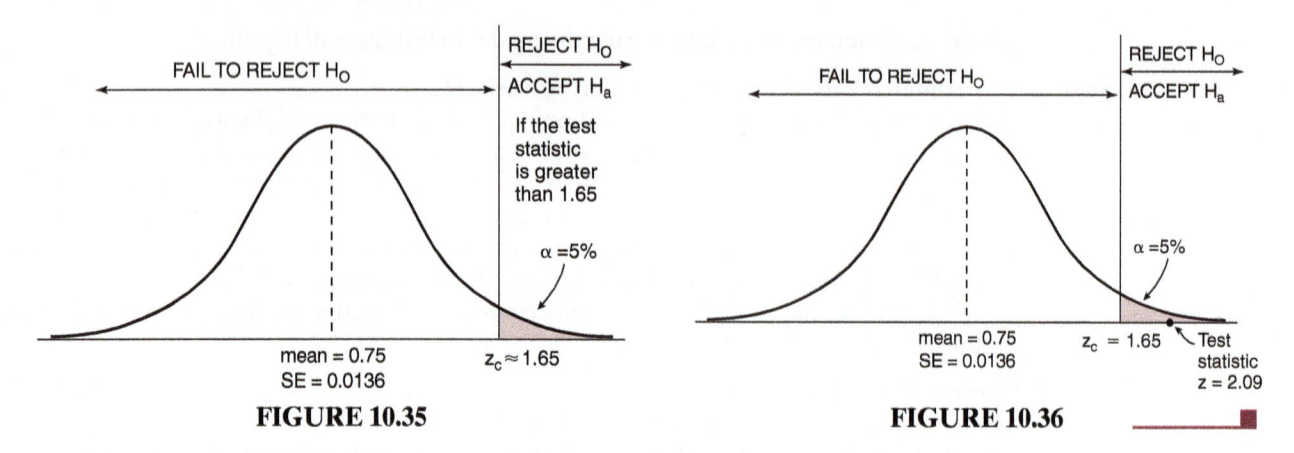

568 Chapter 10 Introduction to Hypothesis Testing

FIGURE 10.35

FAIL TO REJECT H₀ | REJECT H₀ ACCEPT Hₐ
If the test statistic is greater than 1.65
α = 5%
mean = 0.75
SE = 0.0136
$z_c \approx 1.65$

FIGURE 10.36

FAIL TO REJECT H₀ | REJECT H₀ ACCEPT Hₐ
α = 5%
mean = 0.75
SE = 0.0136
$z_c = 1.65$
Test statistic z = 2.09

CASE STUDY 10.1 Trying to Kick the Habit

Using the information contained in the snapshot entitled "Trying to Kick the Habit" shown in Figure 10.37, answer the following questions.

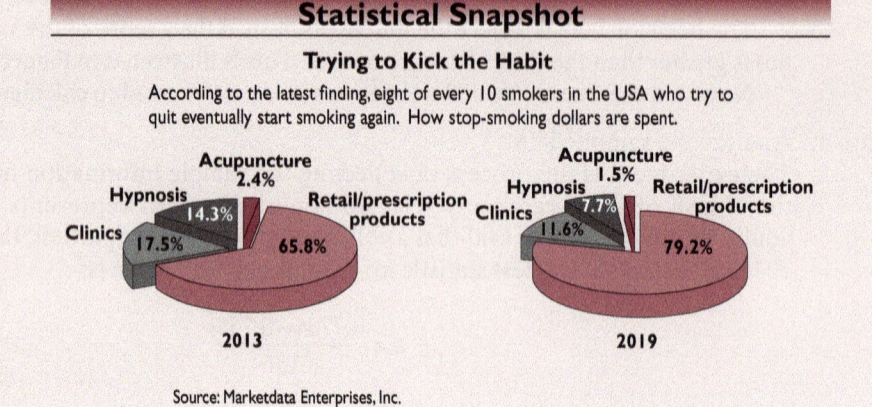

Statistical Snapshot
Trying to Kick the Habit

According to the latest finding, eight of every 10 smokers in the USA who try to quit eventually start smoking again. How stop-smoking dollars are spent.

2013
Acupuncture 2.4%
Hypnosis 14.3%
Clinics 17.5%
Retail/prescription products 65.8%

2019
Acupuncture 1.5%
Hypnosis 7.7%
Clinics 11.6%
Retail/prescription products 79.2%

Source: Marketdata Enterprises, Inc.

FIGURE 10.37

a) Define the population of people being illustrated in Figure 10.37.
b) Define the variable(s) being discussed in the snapshot in Figure 10.37. If you use the information pertaining to the year 2013 as representing the status quo, and the information regarding the year 2019 as the direction you believe the percentages for each category are headed, then:
c) Formulate a null hypothesis and an alternative hypothesis for each of the four categories shown in Figure 10.37.
d) Indicate the type of alternative hypothesis.
e) Determine the type of test (i.e. 1TT or 2TT), and on which side of the curve the critical z values would be positioned.
f) Describe how you would select an appropriate sample to test the null hypothesis.
g) When calculating the expected results in step 2 of the hypothesis testing procedure, what assumption are you making? Explain why it is necessary to make this assumption.
h) For the hypothesis test you outlined regarding Acupuncture, what would be the conclusion to your test if the test statistic were to fall to the left of your critical z score value?
i) For the hypothesis test you outlined regarding Hypnosis, what would be the conclusion to your test if the test statistic were to fall to the right of your critical z score value?

10.4 p-VALUES FOR HYPOTHESIS TESTING

When performing a hypothesis test, the last step or step 5 of the hypothesis testing procedure requires that the test statistic be compared to the critical z score value z_c of the decision rule to determine whether to either reject the null hypothesis or fail to reject the null hypothesis. In Example 10.12, a one-tailed test (1TT) on the right side was performed to determine if we could reject the null hypothesis at a 5% level of significance. In step 5 of the hypothesis test, the test statistic of 2.09 was compared to the critical z score $z_c = 1.65$. Since the test statistic was *greater than* the critical z score, $z_c = 1.65$, the *null hypothesis H_o was rejected at the 5% level of significance* (that is, $\alpha = 5\%$). This is illustrated in Figure 10.38.

A question that might be asked is: *could we have rejected the null hypothesis for a different significance level, such as $\alpha = 4\%$ or $\alpha = 3\%$ or $\alpha = 2\%$ or even $\alpha = 1\%$?*

For example, if the significance level of Example 10.12 was $\alpha = 4\%$, would the null hypothesis still have been rejected? To answer this question, we need to determine the critical z score z_c for a significance level of **$\alpha = 4\%$**, and a 1TT on the right side. Since the percent of area to the right of the critical z score z_c is 4%, then the percent of area to the left of the critical z score, z_c, is 96%. Using Table II: The Normal Curve Area Table, the nearest area is 0.9599, thus the critical z score is: $z_c = 1.75$.

Since the test statistic of z = 2.09 **is still greater than** this critical z score of $z_c \approx 1.75$ for a significance level of 4%, the null hypothesis is rejected at the significance level of **$\alpha = 4\%$**. This is illustrated in Figure 10.39.

Using this same procedure to determine the critical z-score z_c for each of the significance levels **$\alpha = 3\%$** and **$\alpha = 2\%$**, we obtain a critical z score of $z_c \approx 1.88$ for **$\alpha = 3\%$**, and a critical z score of $z_c \approx 2.05$ for **$\alpha = 2\%$**. Since the test statistic of z = 2.09 is still greater than both of the critical z scores, $z_c \approx 1.88$ (for **$\alpha = 3\%$**) and $z_c \approx 2.05$ (**$\alpha = 2\%$**), then the null hypothesis is rejected at each of these significance levels. This is illustrated in Figure 10.40(a) and Figure 10.40(b) respectively.

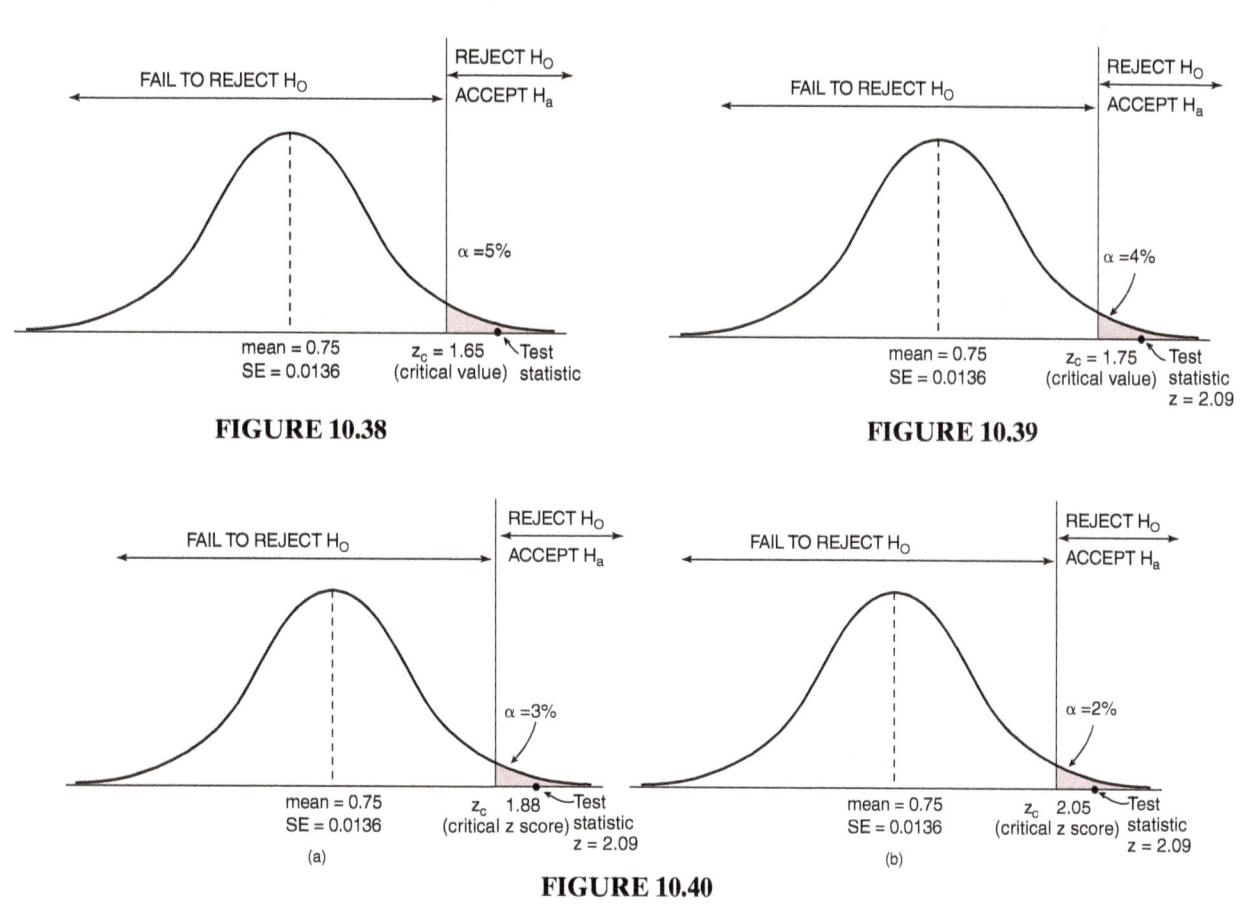

FIGURE 10.38

FIGURE 10.39

FIGURE 10.40

However, at the significance level of **$\alpha = 1\%$,** the null hypothesis **is not rejected**, since the test statistic z = 2.09 is now **less than** the critical z score of $z_c = 2.33$. This is illustrated in Figure 10.41.

Examining the results for the different significance levels, we can begin to obtain an estimate to the answer of the question: *what is the **smallest significance level, α, at which the null hypothesis can be rejected?*** Answer: the null hypothesis of Example 10.12 can be rejected using a significance level of $\alpha = 2\%$ or greater. This is illustrated in Figure 10.42.

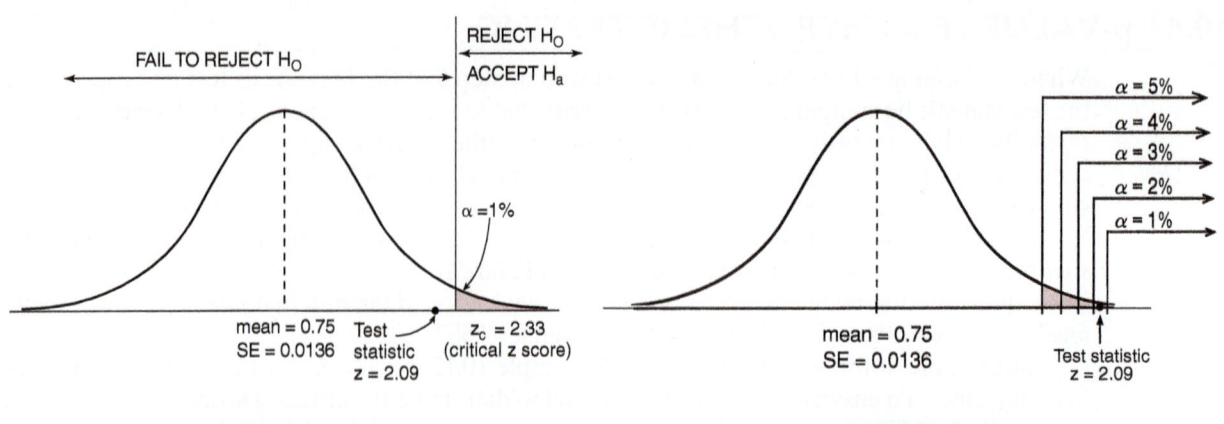

FIGURE 10.41

FIGURE 10.42

From Figure 10.42, notice it is possible to use a significance level smaller than 2% but greater than 1% that would still allow us to reject the null hypothesis. In essence, we are looking for the *true level of significance* for this hypothesis test or the *smallest value* of α for which the null hypothesis can be rejected. The smallest value of α is called the **probability-value** or **p-value of a hypothesis test**. The p-value of a hypothesis test is also referred to as the **observed significance level of a hypothesis test**. The p-value of a hypothesis test is usually expressed in decimal form rather than a percent. We will express the p-value as both a decimal and a percent.

DEFINITION 10.10

The **p-value**, or observed significance level, of a hypothesis test is the **smallest level of significance** (α) at which the null hypothesis H_o can be rejected using the test statistic.

To find the smallest value of α at which the null hypothesis of Example 10.12 can be rejected, that is, the p-value, it is necessary to find the area to the right of the test statistic of 2.09. This is illustrated in Figure 10.43.

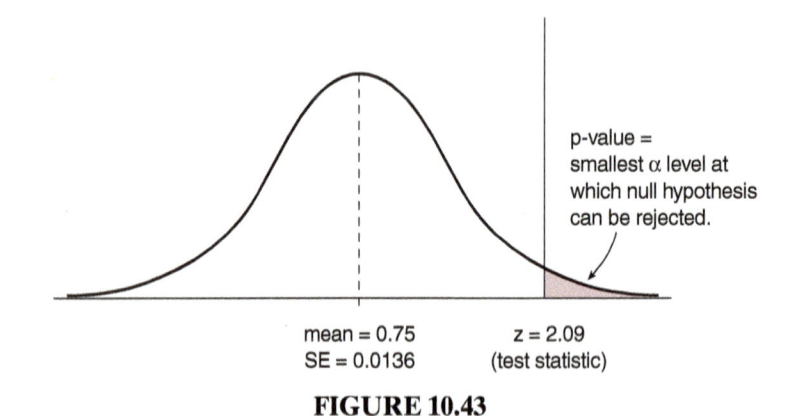

FIGURE 10.43

Using Table II: The Normal Curve Area Table, the area to the left of $z \approx 2.09$ is 0.9817, thus the area to the right which represents the p-value is:

$$\text{p-value} = 1 - 0.9817 = 0.0183.$$

Thus, the p-value of this hypothesis test is 0.0183 or 1.83%. Consequently, the smallest value of α at which the null hypothesis of Example 10.12 can be rejected is 1.83%. This is illustrated in Figure 10.44.

Therefore, we say that this test statistic of 2.09 is "extremely significant," since the p-value or probability of observing a sample proportion of at least 0.7784 is only 0.0183 or 1.83%. **Reporting the p-value of a hypothesis test is more informative than stating the conclusion of a hypothesis test.** Suppose instead of stating the conclusion of the hypothesis test in Example 10.12 as reject the null hypothesis at $\alpha = 5\%$, the p-value of 0.0183 or 1.83% was reported. This knowledge about the p-value of the hypothesis test conveys more information to the

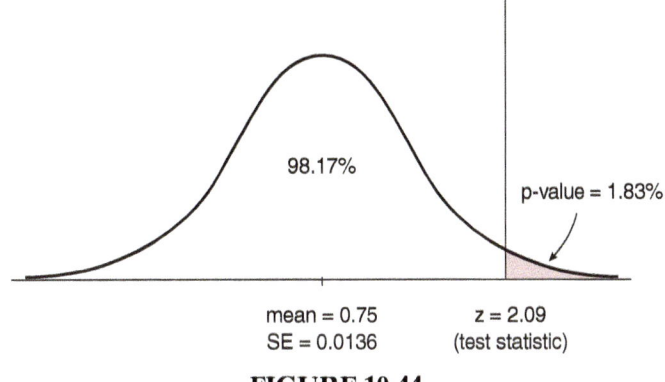

[Figure 10.44: Normal curve showing 98.17% area to left, p-value = 1.83% in right tail, mean = 0.75, SE = 0.0136, z = 2.09 (test statistic)]

FIGURE 10.44

reader. It tells the reader that the null hypothesis can be rejected at $\alpha = 4\%$ and that it can be rejected even at smaller values of α such as $\alpha = 2\%$. In fact, this p-value of 0.0183 or 1.83% indicates that the null hypothesis can be rejected at values of α as small as $\alpha = 1.83\%$. Compare the content information indicated by the statement "the p-value is 0.0183 or 1.83%" with the very limited information contained by the statement "reject the null hypothesis at $\alpha = 5\%$."

Computer software programs, like MINITAB, carry out hypothesis tests by calculating the p-value for you. In fact, many research articles containing statistical tests publish p-values. Thus, instead of selecting a significance level, α, before conducting the experiment as outlined in the hypothesis testing procedure, the researcher or the computer software reports the p-value of the hypothesis test along with the statistical result. It is left to the reader to interpret the significance of the hypothesis test. That is, the reader must determine whether to reject the null hypothesis based upon the reported p-value. The null hypothesis is rejected if the p-value is less than the significance level, α, that the reader decides to choose. Thus, the smaller the p-value, the stronger the evidence against the null hypothesis provided by the statistical result. To decide whether to reject the null hypothesis when the p-value of a hypothesis test is reported, you can use the following guideline.

Deciding Whether to Reject the Null Hypothesis When a p-Value Is Reported

Compare the p-value of the hypothesis test to the chosen significance level, α, that you are willing to tolerate. If the p-value is less than or equal to α, then the hypothesis test is significant at a selected significance level, α. That is,

if p-value $\leq \alpha$,

then reject H_o at significance level, α.

Remember, however, that a p-value conveys more information than just reporting whether or not to reject the null hypothesis at a given significance level, α. Thus, the p-value indicates how unusual the statistical result is assuming that the null hypothesis is true. Therefore, the p-value will provide an indication as to whether or not the statistical result is very significant to not significant. The following rules of thumb established by researchers will help you to interpret p-values regarding the significance of the statistical result.

Interpreting p-Values

1. If the p-value is less than 0.01 (or 1%), the statistical result is **very significant**.
2. If the p-value is between 0.01 (or 1%) and 0.05 (or 5%), the statistical result is **significant**.
3. If the p-value is between 0.05 (or 5%) and 0.10 (or 10%), the statistical result is **marginally or not significant**.
4. If the p-value is greater than 0.10 (or 10%), the statistical result is **not significant**.

EXAMPLE 10.13

A statistical hypothesis test was conducted using a significance level of $\alpha = 5\%$. If the p-value of this test is equal to 0.06 (or 6%), can the null hypothesis be rejected?

Solution

Since the p-value of a hypothesis test is the smallest level of significance (α) at which the null hypothesis H_o can be rejected, then for this test the null hypothesis can not be rejected because the p-value of 0.06 (or 6%) is greater than the significance level of $\alpha = 5\%$.

EXAMPLE 10.14

The null hypothesis of a statistical hypothesis test was rejected at a significance level of $\alpha = 5\%$. If the p-value of this test is equal to 0.025 (or 2.5%), then what is the smallest value of α for which the null hypothesis can be rejected?

Solution

Since the p-value of a hypothesis test is the smallest level of significance (α) at which the null hypothesis H_o can be rejected, then the smallest value of α is equal to 2.5%.

Procedure to Calculate the p-Value of a Hypothesis Test

Let's discuss how to compute the p-value of a hypothesis test for the following three possible cases.

Case One: A one-tailed test on the left side
For a one-tailed test on the left side, the p-value of a hypothesis test is equal to the percent of area in the tail to the left of the test statistic of the test. This is illustrated in Figure 10.45.

Case Two: A one-tailed test on the right side
For a one-tailed test on the right side, the p-value of a hypothesis test is equal to the percent of area in the tail to the right of the test statistic of the test. This is illustrated in Figure 10.46.

Case Three: A two-tailed test
For a two-tailed test, the p-value of a hypothesis test is equal to *twice* the percent of area in the one tail beyond the direction of the test statistic of the test. This is illustrated in Figure 10.47.

FIGURE 10.45

FIGURE 10.46

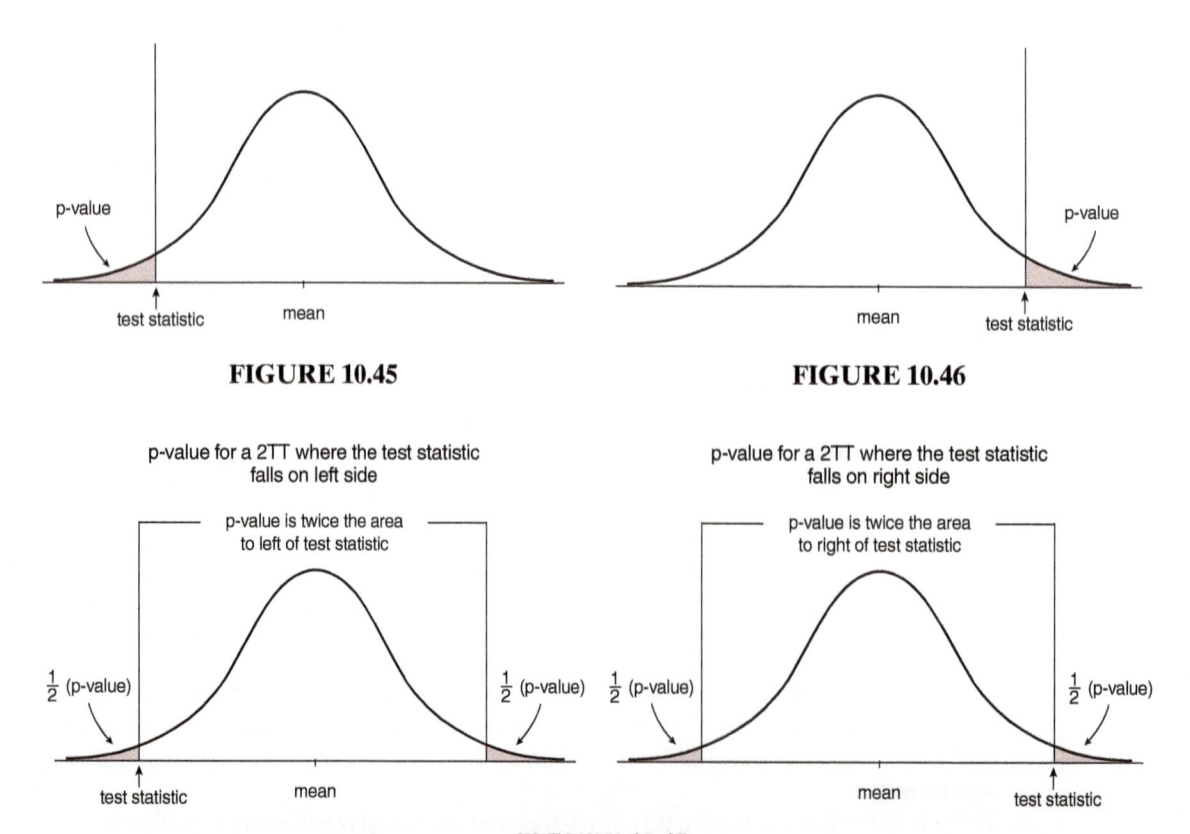

FIGURE 10.47

Section 10.4 p-Values for Hypothesis Testing   573

EXAMPLE 10.15

Use the information contained within the given hypothesis test to answer the following questions. The first two steps of the hypothesis testing procedure are given.

A consumer testing agency is conducting a hypothesis test to determine the validity of an advertised claim made by the Eye Saver Light Bulb Co. that the mean life of its new 60 watt bulb is 1800 hours. The consumer agency believes the claim is too high and randomly selects 400 bulbs to test the company's claim and determines the mean life of these 400 bulbs to be 1788 hours.

Step 1   Formulate the hypotheses.

H_o: The population mean life of the new 60 watt bulb is 1800 hrs: $\mu = 1800$ hrs.
H_a: The population mean life of the new 60 watt bulb is *less than* 1800 hrs: $\mu < 1800$ hrs.

Step 2   Determine the model to test the null hypothesis.
To test the null hypothesis, the consumer agency randomly selects 400 bulbs. Assuming the null hypothesis is true implies that the mean life for all possible samples of 400 bulbs is 1800 hrs with a standard error (SE) of 5 hrs.

The sampling distribution is approximately normal with the expected results shown in Figure 10.48 serving as the model.

a) Calculate the p-value of this hypothesis test.
b) Can the consumer agency reject the null hypothesis at $\alpha = 5\%$? at $\alpha = 1\%$?
c) What is the smallest level of significance at which the null hypothesis H_o can be rejected?

Solution

a) To calculate the p-value, we need to find the area to the left of the sample mean of 1788 hours, since this is a 1TT on the left side. This is illustrated in Figure 10.49.
To determine this area find the z score of 1788. This is the test statistic. Using the z score formula, the test statistic is:

$$z \text{ of } 1788 = \frac{1788 - 1800}{5} = -2.40.$$

Using Table II: The Normal Curve Area Table, the area to the left of $z = -2.40$ is 0.0082. Thus the p–value is 0.0082 or 0.82%.

b) Yes it can be rejected at both significance levels, since the p-value of 0.0082 or 0.82% is less than both 5% and 1%.

c) The smallest level of significance at which the null hypothesis can be rejected is equal to the p-value which is 0.0082 (or 0.82%).

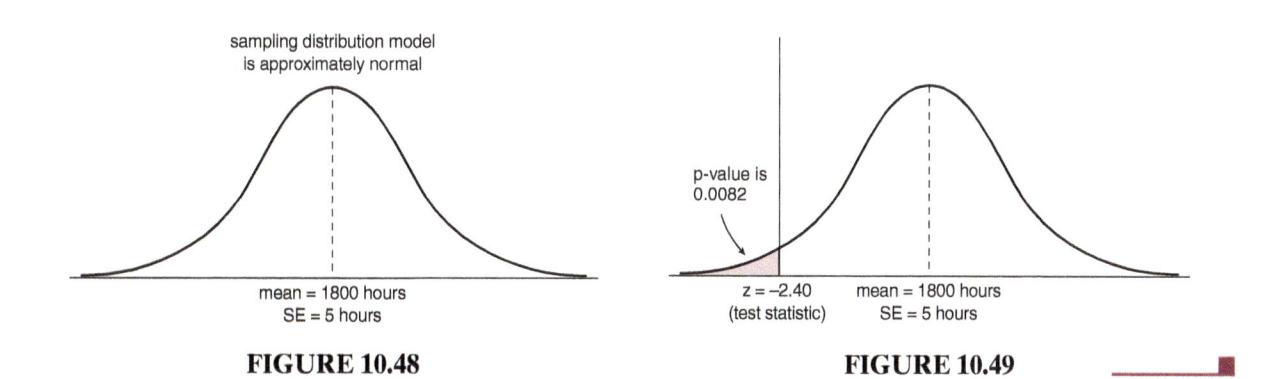

FIGURE 10.48 — sampling distribution model is approximately normal; mean = 1800 hours, SE = 5 hours

FIGURE 10.49 — p-value is 0.0082; z = –2.40 (test statistic); mean = 1800 hours, SE = 5 hours

p-Value Approach to the Hypothesis Testing Procedure

Step 1 **Formulate the two hypotheses, H_o and H_a.**

Step 2 **Determine the model to test the null hypothesis, H_o.**
Under the assumption H_o is true, calculate the expected results: **(The calculation of the expected results will be discussed in the subsequent chapters.)**
  a) Identify the appropriate sampling distribution used as the hypothesis testing model.
  b) Mean of the sampling distribution.
  c) Standard error (SE) of the sampling distribution.

Step 3 **Formulate the Decision Rule:**
  a) Determine type of alternative hypothesis, directional or nondirectional.
  b) Determine type of test, 1TT or 2TT.
  c) Identify the significance level, $\alpha = 1\%$ or $\alpha = 5\%$.
  d) Construct the appropriate hypothesis test model.
  e) State the decision rule as: *Reject H_o if the p-value of the test statistic is less than or equal to α.*

Step 4 **Analyze the sample data.**
Collect the necessary sample data to calculate the test statistic. Determine the p-value of the test statistic.

Step 5 **Determine the conclusion.**
Compare the p-value of the test statistic to the decision rule and either:
a) **Reject H_o and accept H_a at α,**
or
b) **Fail to reject H_o at α.**

KEY TERMS

Alternative Hypothesis, H_a 546
Chance of Type II Error, ß 555
Critical z Score, z_c 560
Directional Alternative Hypothesis 548
Expected Result (mean) 552
Hypothesis Testing Procedure 549
Left Critical z Score, z_{LC} 562
Level of Significance, α 554
Marginal or Not Significant p-Value 571
Nondirectional Alternative Hypothesis 548
Not Significant p-Value 567
Null Hypothesis, H_o 546
One-Tailed Test, 1TT 561
p-Value 569
Right Critical z Score, z_{RC} 562
Significant p-Value 591
Statistically Significant 552
Test Statistic 561
Two-Tailed Test, 2TT 563
Type I Error 550
Type II Error 550
Very Significant p-Value 571

SECTION EXERCISES

Section 10.2

1. The null hypothesis is the hypothesis to be tested that is initially assumed to be _____ and formulated for the sole purpose of trying to _____. The null hypothesis is symbolized as _____.

2. The alternative hypothesis is the hypothesis which is supported by the rejection of the _____ hypothesis. The alternative is symbolized as _____.

3. When a false null hypothesis is accepted, one has committed a Type _____ error.

4. If a false alternative hypothesis is accepted, then:
   a) A Type I error has been made
   b) A Type II error has been made
   c) No error has been made

5. A Type I error is an error that is made when a true null hypothesis is incorrectly rejected.
   (T/F)

6. A rejection of the null hypothesis means that the difference between the test statistic and the _____ result, the mean, is statistically _____.
7. If the null hypothesis is rejected then the alternative hypothesis is:
   a) Accepted
   b) Rejected
   c) Neither accepted nor rejected
8. "Statistically significant" means that the sample statistic is different enough from the expected result to support the alternative hypothesis.
   (T/F)

Section 10.3

9. The level of significance is the probability of committing a Type _____ error.
10. The Greek letter used to symbolize a Type II error is _____.
11. Whenever the alternative hypothesis is directional we have a _____ tailed test.
12. In a two-tailed test, the significance level is divided _____ between the two tails. Therefore, for $\alpha = 1\%$, the percent of the area in each tail is _____.
13. The critical z score for a one-tailed test at $\alpha = 5\%$ is either _____ or _____.
14. The two critical z scores for a two-tailed test at $\alpha = 1\%$ are _____ and _____.
15. In a hypothesis test, if the test statistic does not fall beyond the critical z score, then the conclusion is stated as _____ H_o.
16. If a nondirectional alternative hypothesis is being tested then:
    a) A one-tailed test is performed
    b) A two-tailed test is performed
    c) Not enough information is given
17. When the test statistic does not fall beyond the critical z score(s) then the decision is to "fail to reject H_o." What type of error may have been made?
    a) No error
    b) Type I error
    c) Type II error
18. As the significance level increases, the chance of committing a Type II error decreases.
    (T/F)
19. The decision rule is usually determined after a suitable significance level is chosen.
    (T/F)
20. A directional alternative hypothesis results from a two-tailed test.
    (T/F)

For problems 21–26, state the null and alternative hypotheses assuming you are the individual or agency performing the hypothesis test. Also indicate whether these tests are directional or nondirectional.

21. The Department of Consumer Affairs is going to conduct a study to determine if the following claim made by Weight Examiners Inc. is an exaggeration. "The mean weight loss during the first 10 days is 12 pounds!"
22. The Environmental Protection Agency is going to conduct a series of tests to determine if V & D Inc. is exceeding the allowable pollutant levels. The allowable pollutant level is a mean of 13 ppm.
23. According to a research study linking maternal smoking and childhood asthma, 10% of the children whose mothers smoked during pregnancy are asthmatic. A local official from a public health agency has doubts about this statement and decides to test this claim against her suspicion that it is not 10%.
24. The National Education Association (NEA) states that the national mean salary of public school teachers in the United States is $45,890. A teacher's union wants to show that their mean salary is less than the national mean.
25. A consumer protection group is interested in testing a new energy-saving refrigerator because of their suspicion that the new refrigerator uses more than the advertised mean consumption of 4.9 kilowatt hours of electricity per day.
26. According to the National Center for Education Statistics, the national average daily rate of attendance for kindergarten through 12th grade is 94%. The school superintendent of a city school district believes her district has a daily average attendance rate that is higher than the national average.

For problems 27–29, you are given the null hypothesis (H_o), the alternative hypothesis (H_a), a diagram indicating the sampling distribution model, the mean, standard error (SE), critical z score(s) and a test statistic.

27. $H_o: \mu = 100$
    $H_a: \mu \neq 100$

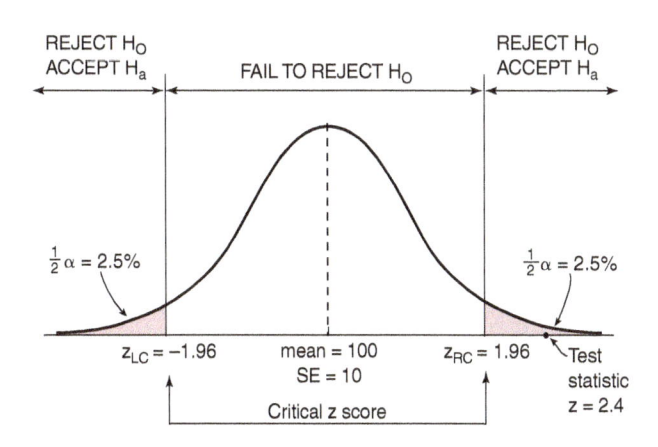

    REJECT H_O ACCEPT H_a | FAIL TO REJECT H_O | REJECT H_O ACCEPT H_a
    $\frac{1}{2}\alpha = 2.5\%$ ... $\frac{1}{2}\alpha = 2.5\%$
    $z_{LC} = -1.96$ | mean = 100, SE = 10 | $z_{RC} = 1.96$
    Critical z score
    Test statistic z = 2.4

    a) State the decision rule.
    b) Do you reject the null hypothesis?
    c) Do you accept the alternative hypothesis?

28. $H_o: p = 0.50$
    $H_a: p > 0.50$

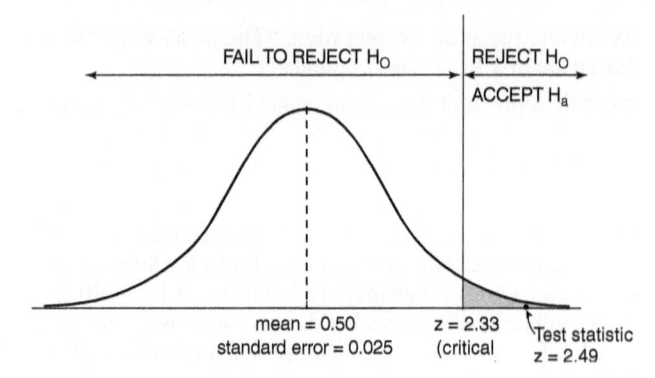

576 Chapter 10 Introduction to Hypothesis Testing

[Diagram: normal curve with mean = 0.50, standard error = 0.025; FAIL TO REJECT H₀ region to left of z = 2.33 (critical); REJECT H₀ ACCEPT Hₐ region to right; Test statistic z = 2.49]

a) State the decision rule.
b) Do you reject the null hypothesis?
c) Do you reject the alternative hypothesis?

29. $H_o: \mu = 114$
$H_a: \mu < 114$
a) State the decision rule. If the test statistic is z=2, then:
b) Do you reject the null hypothesis?
c) Do you accept the alternative hypothesis?

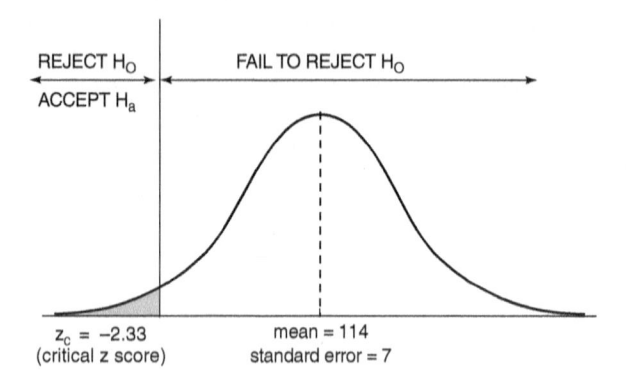

[Diagram: normal curve with mean = 114, standard error = 7; REJECT H₀ ACCEPT Hₐ region to left of $z_c = -2.33$ (critical z score); FAIL TO REJECT H₀ region to right]

If the test statistic is z = −3, then:

d) Do you reject the null hypothesis?
e) Do you accept the alternative hypothesis?

For problems 30–32, use the given information to:
  a) Label each diagram with critical z score(s), reject and fail to reject areas and the test statistic.
  b) Determine the conclusion.
  c) Is the test statistic statistically significant?

30. $H_o: p = 0.90$
$H_a: p > 0.90$
Decision Rule: Reject H₀ if test statistic is greater than z = 2.33.
The test statistic is z = 2.

[Diagram: normal curve, mean = 0.90, SE = 0.03]

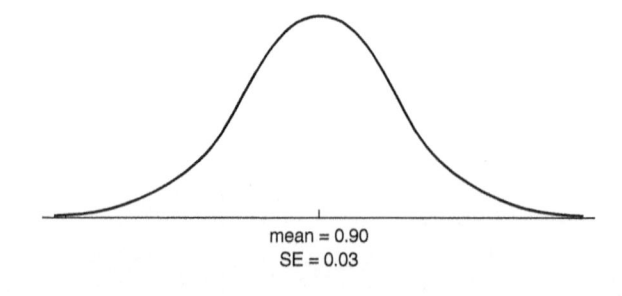

31. $H_o: \mu = 190$
$H_a: \mu \neq 190$

Decision Rule: Reject H₀ if the test statistic is either less than −1.96 or greater than 1.96.
The test statistic is z = 2.2.

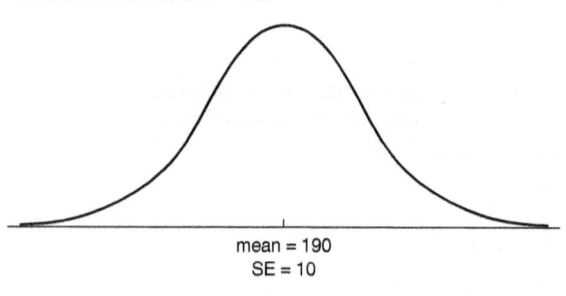

[Diagram: normal curve, mean = 190, SE = 10]

32. $H_o: \mu = 14.5$
$H_a: \mu < 14.5$

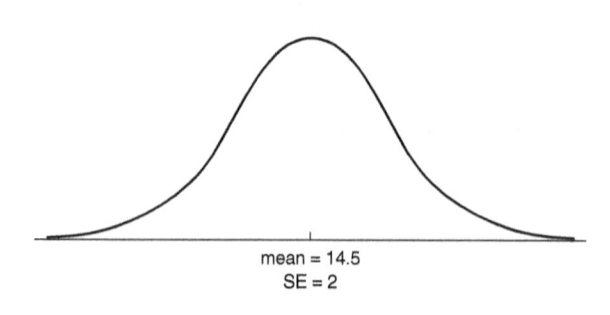

[Diagram: normal curve, mean = 14.5, SE = 2]

Decision Rule: Reject H₀ if the test statistic is less than −1.65.
The test statistic is z = −1.9.

For problems 33–35, use problems 30–32 and:

  a) Determine the chance of a Type I error.

For problems 36–39, the null and alternative hypotheses, the mean and standard error (SE), the significance level, α, and a diagram are given.
Calculate the critical z score(s) and formulate the decision rule. Label the diagram with the mean, standard error (SE), significance level, critical z score(s) and reject and fail to reject areas. Answer the appropriate questions.

36. $H_o: \mu = 2500$
$H_a: \mu > 2500$
Mean = 2500
Standard Error (SE) = 100
Significance Level, $\alpha = 5\%$

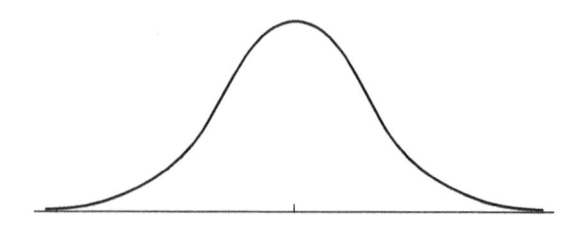

[Diagram: normal curve]

a) If the test statistic is:
  1) 2.15   2) 1.00   3) −1.00   4) −13
then determine the conclusion.
b) Which of these results are statistically significant?

37. $H_o: \mu = 18$
$H_a: \mu < 18$
Mean = 18
Standard Error (SE) = 3
Significance Level, $\alpha = 1\%$

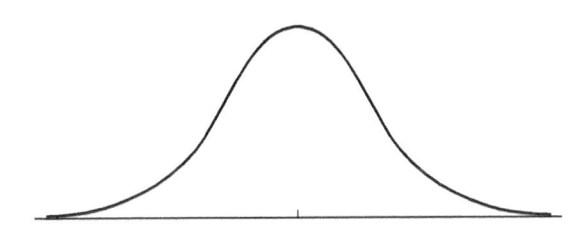

a) If the test statistic is:
1) –2   2) 1.00   3) –2.35   4) 3.33
then determine the conclusion.
b) Which of these results are statistically significant?

38. $H_o: p = 0.80$
$H_a: p < 0.80$
Mean = 0.80
Standard Error (SE) = 0.02
Significance Level, $\alpha = 1\%$

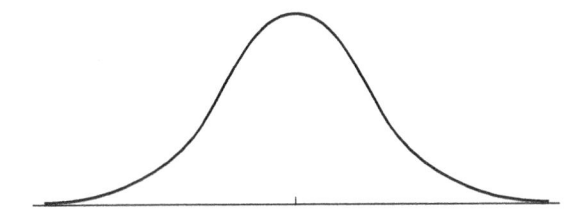

a) If the test statistic is:
1) –2.74   2) –0.58   3) –2.03   4) –2.68
then determine the conclusion.
b) Which of these results are statistically significant?

39. $H_o: p = 0.40$
$H_a: p \neq 0.40$
Mean = 0.40
Standard Error (SE) = 0.0315
Significance Level, $\alpha = 5\%$

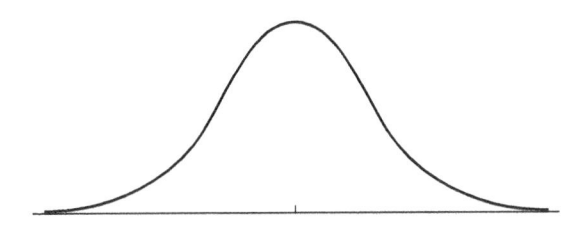

a) If the test statistic is:
1) –1.85   2) 2.36   3) –2.49   4) 1.34
then determine the conclusion.
b) Which of these results are statistically significant?

Section 10.4

40. The p-value, or observed significance level, of a hypothesis test is the _____ level of significance at which the null hypothesis H_o can be rejected using the test statistic.

41. When interpreting p-values, a rule of thumb that can be followed is:
a) If the p-value is less than 1%, the test statistic is _____ significant.
b) If the p-value is between 1% and 5%, the test statistic is _____.
c) If the p-value is between 5% and 10%, test statistic is _____ significant.
d) If the p-value is greater than 10%, the test statistic is _____ significant.

42. The p-value, or observed significance level is the largest level of significance, α, at which the null hypothesis H_o can be rejected.
(T/F)

43. P-values can be represented as either percents or decimals.
(T/F)

44. If the p-value for a test statistic is greater than 10%, the test statistic is not significant.
(T/F)

45. An unemployment agency claims that the mean age of recipients of unemployment benefits is 37 years. A trade union association believes this claim is not correct. The association randomly interviews 400 recipients of unemployment benefits and obtains a mean age of 35 years. The researcher computes the decision rule for this test to be: reject H_o if the test statistic is greater than 1.96 or less than –1.96. Use the following model to perform this hypothesis test, and:
a) State the null and alternative hypotheses.
b) Label the diagram with the critical z-scores, z_c, reject and fail to reject areas, and the test statistic.

mean = 37
standard error = 0.25

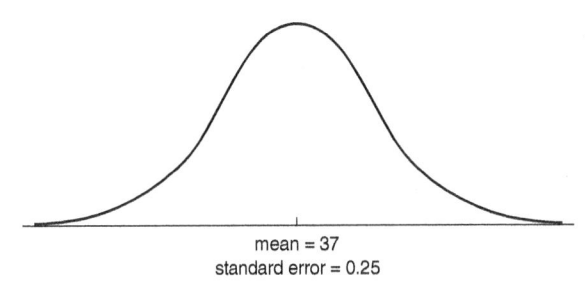

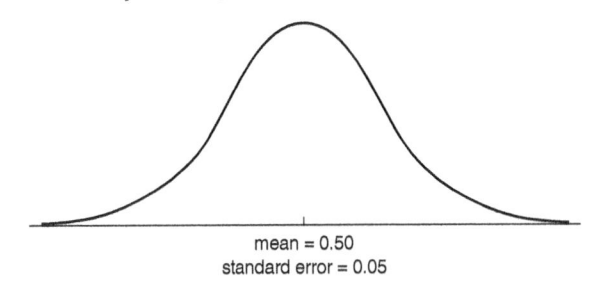

c) Determine the p-value for the test statistic.
d) State your conclusion.

46. According to a recent census report 50% of U.S. families earn more than $20,000 per year. A sociologist from a northeastern university believes this percent is too low. He randomly selects 100 families and determines that 64 have incomes of more than $20,000. The decision rule is to reject H_o if the test statistic is more than 2.33.
a) State the null and alternative hypotheses.
b) Label the diagram with the critical z-scores, z_c, reject and fail to reject areas, and the test statistic.

mean = 0.50
standard error = 0.05

c) Determine the p-value for the test statistic.
d) State your conclusion.

For problems 47–50:

a) Formulate the null and alternative hypotheses.
b) Use the mean, standard error (SE), and significance level, α, given in the problem to calculate the critical z-score value(s) and formulate the decision rule.
c) Label the diagram with the mean, standard error (SE), significance level, reject and fail to reject areas.
d) Determine the p-value for the test statistic.
e) Use the test statistic to determine the conclusion.
f) State whether the test statistic is statistically significant.
g) Answer the questions in each problem.

47. An independent research group is interested in showing that the percent of babies delivered by Cesarean Section is increasing. Last year, 20% of the babies born were delivered by Cesarean Section. The research group randomly inspects the medical records of 100 recent births and finds that 25 of the births were by Cesarean Section. Can the research group conclude that the percent of births by Cesarean section has increased? Use mean = 0.20, standard error (SE) = 0.04 and α = 5%.

48. According to the U.S. Dept. of Labor, as of May 2014 unemployment is at 6.3%. The Congresswoman from the third district believes that the unemployment rate is lower in her district. To test her belief she interviews 200 residents of her district and finds 7 of them to be unemployed. Is the Congresswoman's belief correct? Use mean = 0.063, standard error (SE) = 0.0172 and α = 5%.

49. Social networking websites such as Facebook and Twitter have contributed to an increase in time spent on the Internet. Last year, a study conducted by The Nielsen Company indicates that the average American Facebook user spends a mean of 7 hours per month on the Facebook site. If a recent sample of 64 randomly selected American Facebook users average 8.1 hours for the month on the Facebook site, can one conclude that monthly average has increased this year? Use mean = 7 hours, standard error (SE) = 0.8 hours and α = 1%.

50. A beverage company has a machine that is supposed to fill soda bottles with two liters of soda. To check that the machine is operating correctly, (i.e., not underfilling or overfilling the bottles), a sample of 81 bottles was selected and the contents of each bottle was determined. The sample mean was 1.93 liters of soda. Does this indicate that the machine is improperly filling the bottles? Use mean = 2.0 liters, standard error (SE) = 0.02 liter and α = 1%.

51. The college newspaper at Rich Man's College states the average time that students study per week is 8.5 hours. A skeptical student decides to test the claim to determine if the claim is true. The student selects a random sample of 400 students from the student body and determines the average study time of the sample to be 9.2 hours. Use a mean = 8.5 and a standard error (SE) = 0.12.
a) What is the p-value for this hypothesis test?
b) Can the null hypothesis be rejected at α = 5%?
c) What is the smallest level of significance for which the null hypothesis can be rejected?

CHAPTER REVIEW

52. If a false null hypothesis is rejected then
a) A Type I error has been made
b) A Type II error has been made
c) No error has been made

53. If the test statistic is beyond the critical z-score, z_c, then the result is considered:
a) To be a Type I error
b) Statistically significant
c) To be a Type II error

54. The Greek letter used to symbolize the level of significance is _____.

55. Whenever the alternative hypothesis is nondirectional we have a _____ tailed test.

56. The general formula used to calculate a test statistic, z = _____.

57. The two critical z scores for a two-tailed test at α = 5% are _____ and _____.

58. The critical z score for a one-tailed test at α = 1% is either _____ or _____.

59. When the test statistic falls beyond the critical z-score(s) then the decision is to:
a) Reject H_o and accept H_a
b) Fail to reject H_o
c) Seek more information

60. When a directional hypothesis is being tested and the critical z-score is to the right of the mean then the alternative hypothesis must be of the form:
a) Less than
b) More than
c) Not equal to

61. When a nondirectional hypothesis is being tested and the significance level is 5% then the critical z scores are:
a) ± 1.96
b) ± 2.58
c) ± 1.65

62. In a hypothesis test, the null hypothesis is rejected if the p-value is _____ or _____ the significance level, α. Thus, the smaller the p-value, the stronger the evidence against the _____ hypothesis provided by the test statistic.

63. a) If the p-value of a statistical hypothesis test is equal to 3%, can the null hypothesis be rejected at α = 5%?
b) For this test, the smallest significance level in which the null hypothesis can be rejected is _____.

64. In a two-tailed test, the p-value of a hypothesis test is equal to _____ the percent of area in the tail beyond the direction of the test statistic of the test.

For problems 65–66, state the null and alternative hypotheses assuming you are the individual or agency performing the hypothesis test. Also indicate whether these tests are directional or nondirectional.

65. In a recent issue of Psychology for Tomorrow researchers report that 20% of the wives who hold full-time jobs now earn

more than their husbands. A sociologist from the midwest believes this claim is too high and decides to test the claim.

66. A national magazine recently claimed that 3 out of every 10 women between the ages of 16 and 40 years have a form of herpes virus. Researchers representing the New World Medical Society believe the claim is not true and decide to test the claim.

67. You are given the null hypothesis (H_o), the alternative hypothesis (H_a), a diagram indicating the sampling distribution model, the mean, standard error (SE), critical z-score(s) and a test statistic.

$H_o: \mu = 12$
$H_a: \mu < 12$

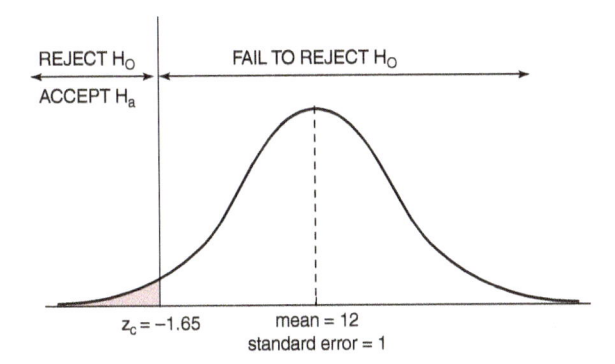

a) State decision rule.
If the test statistic is z = –2, then:
b) Do you reject the null hypothesis?
c) Do you accept the alternative hypothesis?
If the test statistic is z = 3, then:
d) Do you reject the null hypothesis?
e) Do you accept the alternative hypothesis?

68. Use the given information to:
a) Label the diagram with critical z-score(s), reject and fail to reject areas and the test statistic.
b) Determine the conclusion.
c) Is the test statistic statistically significant?
$H_o: p = 0.80$
$H_a: p < 0.80$
Decision Rule: Reject H_o if the test statistic is less than $z_c = -1.65$.
The test statistic is z = –1.88.

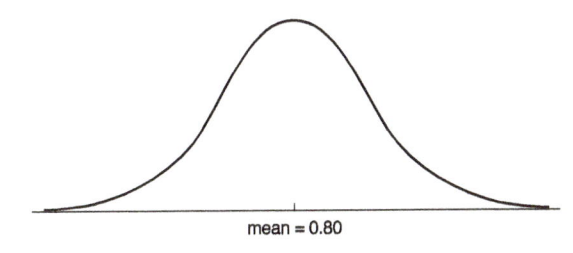

69. Use the information in problem 68 to calculate the chance of a Type I error.

For problems 70–71, the null and alternative hypotheses, the mean and standard error (SE), the significance level, α, and a diagram are given.

Calculate the critical z-score(s) and formulate the decision rule. Label the diagram with the mean, standard error (SE), significance level, critical z-score(s) and reject and fail to reject areas. Answer the appropriate questions.

70. $H_o: \mu = 90$
$H_a: \mu \neq 90$
Mean = 90
Standard Error (SE) = 10
Significance Level, $\alpha = 5\%$

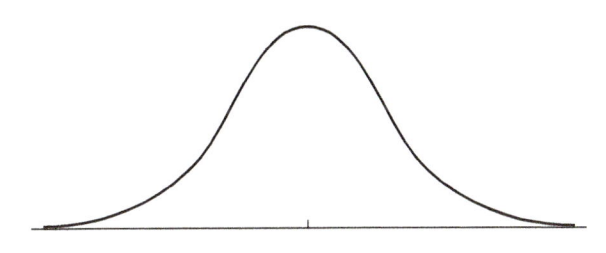

a) If the test statistic is:
1) –0.7   2) –1.10   3) 2.00   4) 1.30
then determine the conclusion.
b) Which of these results are statistically significant?

71. $H_o: p = 0.30$
$H_a: p > 0.30$
Mean = 0.30
Standard Error (SE) = 0.0229
Significance Level, $\alpha = 5\%$

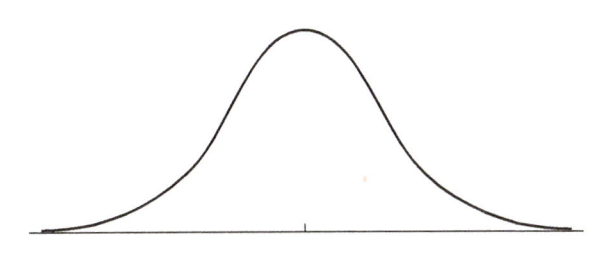

a) If the test statistic is:
1) 0.93   2) 1.61   3) 0.91   4) –0.09
then determine the conclusion.
b) Which of these results are statistically significant?

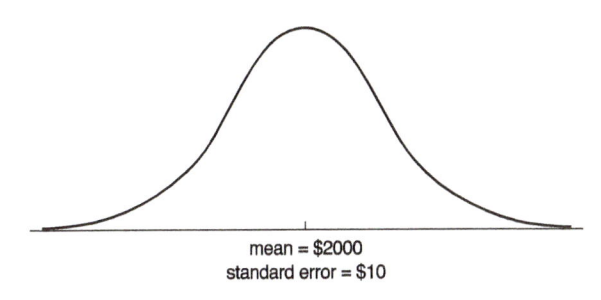

72. According to a recent insurance company claim an American family of four spends a mean of $2000 per year for medical care. The medical association believes this claim is too high. They randomly survey 3600 families and find the mean expenditure to be $1990. Their decision rule is to reject H_o if the test statistic is less than $z_c = -2$.
Use the following model to perform this hypothesis test, and:
a) State the null and alternative hypotheses.
b) Label the diagram with the critical z-score(s), reject and fail to reject areas, and the test statistic.

c) Determine the p-value for the test statistic.
d) State your conclusion.

73. A psychiatrist wants to determine if a small amount of alcohol decreases the reaction time in adults. The mean reaction time for a specified test is 0.20 seconds. A sample of 64 people were given a small amount of alcohol and their reaction time averaged 0.17 seconds. Does this sample result support the psychiatrist's hypothesis? Use mean = 0.20 sec., standard error (SE) = 0.01 sec. and $\alpha = 1\%$.
a) Formulate the null and alternative hypotheses.
b) Use the mean, standard error (SE), and significance level, α, given in the problem to calculate the critical z-score(s) and formulate the decision rule.
c) Label the diagram with the mean, standard error (SE), significance level, reject and fail to reject areas.
d) Determine the p-value for the test statistic.
e) Use the test statistic to determine the conclusion.
f) State whether the test statistic is statistically significant.
g) Answer the question.

WHAT DO YOU THINK?

74. Examine the graph entitled "Spending for the Holidays."

Statistical Snapshot
Spending for the Holidays

Men claim they'll spend an average $653 this holiday season; women, $580. Spending plans by age groups:

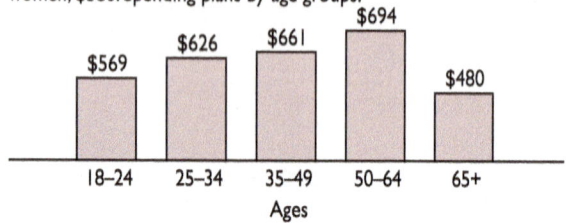

18–24	25–34	35–49	50–64	65+
$569	$626	$661	$694	$480

Ages

Source: Opinion Research Corporation, survey of 508 minority members and a survey of 1,009 members of the general population

a) What different populations are being represented within the snapshot?
b) Does the average amount of money spent by each age group represent both males and females?
c) What claim do the men make regarding their spending during the holidays?
d) What claim do the women make regarding their spending during the holidays?
e) How many people were surveyed to determine their spending plans for the holiday? Do you have any information regarding their break down with respect to their age and gender? Why do think that information might be helpful in interpreting the results stated in the snapshot?

If you are planning to test the statement about what men spend for the holidays and your belief is that the figure within the graph is too high, then:
f) What would the null and alternative hypotheses be for your test?
g) Indicate the type of alternative hypothesis for each population.
h) Determine the type of test (i.e. 1TT or 2TT), and on which side of the curve the critical z-score would be positioned.
i) Describe how you would select an appropriate sample to test the null hypothesis.
j) What assumption are you making when calculating the expected results in step 2 of the hypothesis testing procedure? Explain why it is necessary to make this assumption to test the null hypothesis.
k) What would the conclusion be to your test if the test statistic fell to the left of your critical z-score?

If you are planning to test the statement about what women spend for the holidays and your belief is that the figure within the graph is too low:
l) What would the null and alternative hypotheses be for your test?
m) Indicate the type of alternative hypothesis for each population.
n) Determine the type of test (i.e. 1TT or 2TT), and on which side of the curve the critical z-score would be positioned.
o) Describe how you would select an appropriate sample to test the null hypothesis.
p) What assumption are you making when calculating the expected results in step 2 of the hypothesis testing procedure? Explain why it is necessary to make this assumption to test the null hypothesis.
q) What would the conclusion be to your test if the test statistic fell to the right of your critical z-score?

If you are planning to test the statement about what people within the age group 25–34 spend for the holidays and your belief is that the figure within the graph is too low:
r) What would the null and alternative hypotheses be for your test?
s) Indicate the type of alternative hypothesis for each population.
t) Determine the type of test (i.e. 1TT or 2TT), and on which side of the curve the critical z-score would be positioned.
u) Describe how you would select an appropriate sample to test the null hypothesis.
v) What assumption are you making when calculating the expected results in step 2 of the hypothesis testing procedure? Explain why it is necessary to make this assumption to test the null hypothesis.
w) What would the conclusion be to your test if the test statistic fell to the right of your critical z-score?

PROJECTS

75. Read a research article from a professional journal in a field such as medicine, education, marketing, science, social science, etc. Write a summary of the research article which includes the following:
   a) Name of article, author and journal (include volume, number and date).
   b) The statement of the research problem.
   c) The null and alternative hypotheses.
   d) How the experiment was conducted (include the sample size and the sampling technique).
   e) Determine the test statistic and the level of significance.
   f) State the conclusion in terms of the null and alternative hypotheses.

76. Select a claim found in a magazine, newspaper or stated on television or radio. Formulate an opinion regarding this claim, and devise a hypothesis testing procedure to test the claim. Be sure to include the following:
   a) The source from which you found this claim.
   b) The null and alternative hypotheses.
   c) Define the population and how you would go about selecting a sample from this population.
   d) State the method you would use to obtain your data from the sample.

CHAPTER TEST

77. When a true null hypothesis is rejected, one has committed a Type _____ error.

78. If a test statistic is statistically significant, then the _____ hypothesis is accepted.

79. As the significance level decreases the chance of committing a Type I error _____, while the chance of committing a Type II error _____.

80. In a hypothesis test, if the test statistic falls beyond the critical z-score, then the conclusion is stated as _____ H_o and _____ H_a.

81. The significance level, α, is the probability of:
   a) Rejecting a false null hypothesis
   b) Making a Type II error
   c) Rejecting a true null hypothesis

82. Reporting a p-value of 0.89% indicates that the null hypothesis can be rejected at values of α as small as $\alpha =$ _____ %.

For problems 83–84, state the null and alternative hypotheses assuming you are the individual or agency performing the hypothesis test. Also indicate whether these tests are directional or nondirectional.

83. The Center for Student Credit reports that only 17 percent of students on college campuses do not have credit cards. An economics professor believes this claim is too high and decides to test it.

84. The U.S. Navy is going to conduct a series of tests to determine if an automatic opening device for parachutes meets the manufacturer's claim of a mean opening time of 6 seconds.

85. You are given the null hypothesis (H_o), the alternative hypothesis (H_a), a diagram indicating the sampling distribution model, the mean, standard error (SE), critical value(s) and a statistical result.

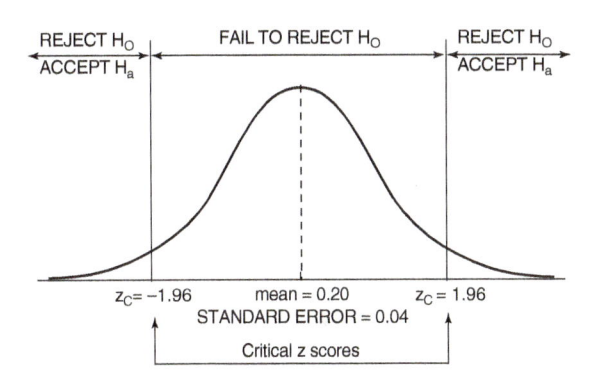

[Diagram: normal curve with REJECT H_O / ACCEPT H_a regions on both tails, FAIL TO REJECT H_O in middle; $z_C = -1.96$, mean = 0.20, $z_C = 1.96$, STANDARD ERROR = 0.04, Critical z scores]

H_o: p = 0.20
H_a: p ≠ 0.20
   a) State the decision rule.
   b) If the test statistic is –0.03, state your conclusion.
   c) If the test statistic is 2.23, state your conclusion.
   d) If the test statistic is 1.95, state your conclusion.

86. The null and alternative hypotheses, the mean and standard error (SE), the decision rule and a diagram are given.

H_o: $\mu = 520$
H_a: $\mu > 520$

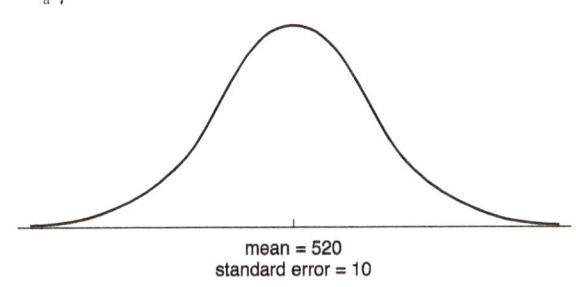

[Diagram: normal curve with mean = 520, standard error = 10]

Decision Rule: Reject H_o if the test statistic is greater than 2.33.

The test statistic is 2.9.

87. A manufacturer wants to advertise a product in *Playperson* magazine. The product is specifically designed for a person aged 21 through 35 years. The magazine has stated that 70% of its subscribers are in this age group. The marketing supervisor for the manufacturer believes the claim is too high

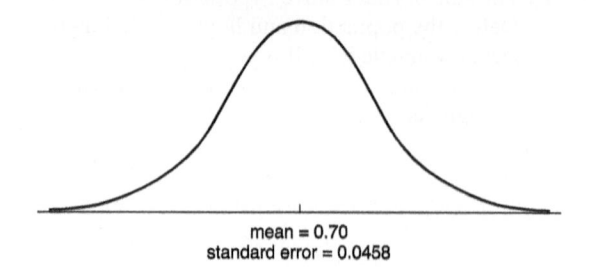

mean = 0.70
standard error = 0.0458

and randomly surveys 100 of the magazine's subscribers and determines that 64 are in this age group. The decision rule is reject H_o if the test statistic is less than -1.75.
a) State the null and alternative hypotheses.
b) Label the diagram with the critical z-score(s), reject and fail to reject areas, and the test statistic.
c) Determine the p-value for the test statistic.
d) State your conclusion.

88. A pharmaceutical company claims to have developed a new antibiotic which is more effective against type A bacteria than its current product. The current product is 90% effective against type A bacteria. To test its claim a random sample of 400 individuals who were infected with this bacteria are treated with the new antibiotic and 380 recover. Do these results support the pharmaceutical company's claim? Use mean = 0.90, standard error (SE) = 0.015 and $\alpha = 1\%$.
a) Formulate the null and alternative hypotheses.
b) Use the mean, standard error (SE), and significance level, α, given in the problem to calculate the critical z score(s) and formulate the decision rule.
c) Label the diagram with the mean, standard error (SE), significance level, reject and fail to reject areas.
d) Determine the p-value for the test statistic.
e) Use the test statistic to determine the conclusion.
f) State whether the test statistic is statistically significant.
g) Answer the question.

89. For the following examples, determine the null and alternative hypotheses both verbally and in symbolic form.
a) The published mean age of death (due to natural causes) for adult male Polar bears is known to be 19.7 years. A local biologist believes that due to the decline in food in many areas, this age is no longer valid and that the mean age is in fact lower.
b) The proportion of Polar bear cubs that survive the first year of life is known to be 30%. A local statistician feels this proportion is outdated and disagrees with this claim.

For the answer to this question, please visit www.MyStatLab.com.

CHAPTER 11

Hypothesis Testing Involving One Population

Contents

11.1 Introduction
11.2 Hypothesis Testing Involving a Population Proportion
11.3 Hypothesis Testing Involving a Population Mean: Population Standard Deviation Known
11.4 The t Distribution
11.5 Hypothesis Testing Involving a Population Mean: Population Standard Deviation Unknown
11.6 p-value Approach to Hypothesis Testing
Using the TI-84 Calculator

11.1 INTRODUCTION

The general concept of hypothesis testing was introduced in Chapter 10. In this chapter we will examine in detail the procedure to test hypotheses (or claims) made about a population mean or population proportion **using information obtained from a sample drawn from the population**.

To illustrate the use of this procedure, consider the following story about the Precision Tool and Die Company, which was the principal employer in a southern Nevada town. This company has been credited with building up the town not only from an economic point of view, but also in a physical sense, since they supplied precision manufactured screws used to assemble many of the town's building projects. In fact, today the company still boasts: "we're the company that screwed up this town."

The success of this company can be attributed to the quality control department whose motto is: "we manufacture only precision screws." The quality control manager is responsible for testing every batch of screws before they leave the plant to insure they meet the manufacturer's specifications.

For example, if the quality control manager needed to determine if the entire batch of 4" screws have been manufactured to the required length specification, she would randomly select a sample of the screws and compute the mean length of this sample. Her decision to either reject or not reject the entire batch is based upon a comparison of the mean of this sample to the manufacturer's length specification. If the difference between the sample mean length and the manufacturer's length specification is **statistically significant**, then the entire batch will be rejected. On the other hand, if the difference is **not statistically significant**, then the manager is **confident** that the entire batch meets the required length specification based upon the sample tested and the entire batch would not be rejected.

In this chapter, we will examine the statistical procedure used by the quality control manager to make her decision. The underlying theorems which support this statistical test are the Central Limit Theorem and the Theorem on the Properties of Sampling Distributions (discussed in Chapter 8). The development of the hypothesis testing procedure will incorporate these theorems into the general hypothesis testing procedure outlined in Chapter 10. We will start by developing the hypothesis testing procedure involving a population proportion.

11.2 HYPOTHESIS TESTING INVOLVING A POPULATION PROPORTION

Are the following claims startling? What are your feelings about these claims?

EXAMPLE 11.1

The Bureau of Statistics claims that 75% of new small businesses will go bankrupt within ten months.

FIGURE 11.1 Success / Failure

EXAMPLE 11.2

A sociologist claims that two out of three couples that wed this year will divorce within five years.

FIGURE 11.2 Newlyweds / Divorce Court

EXAMPLE 11.3

Educational Research Associates (ERA) claims that at least 40% of the male graduates of the State University have a reading level below 9th grade.

FIGURE 11.3 Entering / Graduating

In Chapter 8 we developed the sampling distribution for the population proportion. We will now use this model to test claims such as the ones made in Examples 11.1–11.3. In this section we will develop the **hypothesis testing procedure for hypotheses involving a population proportion** using the five step general hypothesis testing procedure discussed in Chapter 10. Throughout this section, we will be assuming that the sampling distribution of the proportion satisfies the conditions of Theorem 8.5.

Hypothesis Testing Procedure Involving a Population Proportion

Step 1 Formulate the two hypotheses, H_o and H_a.
Null Hypothesis:
The null hypothesis has the following form:

Null Hypothesis Form:

H_o: The population proportion, p, is claimed to be **equal to** the proportion p_o.
   This is symbolized as:

$$H_o: p = p_o$$

Alternative Hypothesis:
Since the alternative hypothesis, H_a, can be stated as either *greater than*, *less than*, or *not equal to* the numerical value, p_o, then the alternative hypothesis can have one of three forms:

Form (a): Greater Than Form for H_a

H_a: The population proportion, p, is claimed to be **greater than** the numerical value, p_o.
   This is symbolized as:

$$H_a: p > p_o$$

Form (b): Less Than Form for H_a

H_a: The population proportion, p, is claimed to be **less than** the numerical value, p_o.
   This is symbolized as:

$$H_a: p < p_o$$

Form (c): Not Equal Form for H_a

H_a: The population proportion, p, is claimed to be **not equal** to the numerical value, p_o.
   This is symbolized as:

$$H_a: p \neq p_o$$

Step 2 Determine the model to test the null hypothesis, H_o.
Under the assumption H_o is true, calculate the expected results. Since we are testing a hypothesis about a population proportion using a sample proportion, $\hat{p}$, we use as our model the sampling distribution of the proportion and should choose a sample size which will make both np and n(1 − p) greater than or equal to 10.
Using Theorem 8.5 (the sampling distribution of the proportion was discussed in Section 8.7) and under the assumption H_o is true(i.e., $p = p_o$), then:
the **expected results** for this hypothesis testing model are:
a) The sampling distribution of the proportion is *approximately normal* and serves as the hypothesis testing model.
b) The mean of the sampling distribution of the proportion is:

$$\mu_{\hat{p}} = p_o,$$

586 Chapter 11 *Hypothesis Testing Involving One Population*

(i.e. the mean $\mu_{\hat{p}}$ equals the claimed population proportion, p_o.)

c) The standard error of the proportion is:

$$\sigma_{\hat{p}} = \sqrt{\frac{p_o(1-p_o)}{n}}$$

This hypothesis testing model is illustrated in Figure 11.4.

Hypothesis Testing Model

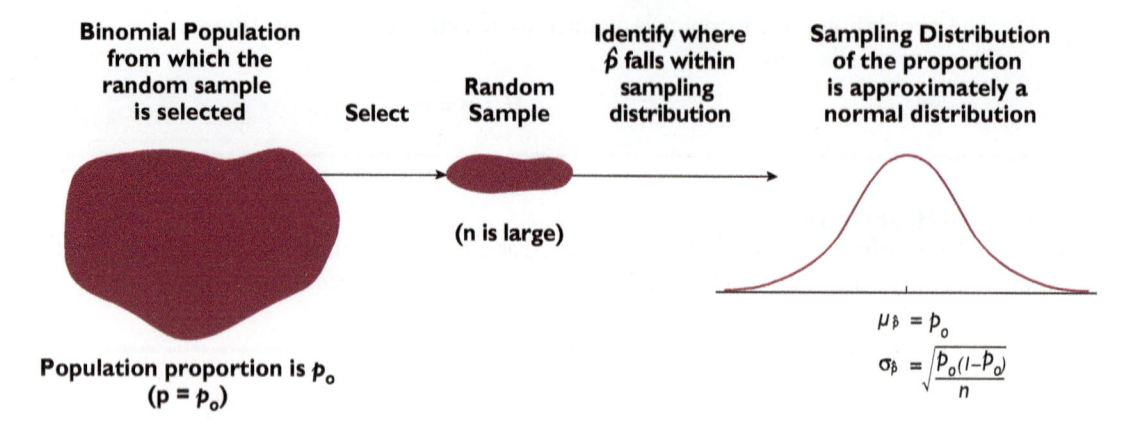

FIGURE 11.4

In theory, the binomial population being sampled must be infinite to use the standard error formula:

$$\sigma_{\hat{p}} = \sqrt{\frac{p(1-p)}{n}}$$

However, it is appropriate to use this formula when the population is *finite* and the population size, N, is large relative to the sample size, n. The rule used to determine whether the population is large relative to the sample size is: the sample size, n, is less than 5% of the population size, N. That is, $n/N < 5\%$.

Step 3 Formulate the decision rule.
  a) Determine the type of alternative hypothesis: directional or non-directional.
  b) Determine the type of test: 1TT or 2TT.
  c) Identify the significance level: $\alpha = 1\%$ or $\alpha = 5\%$.
  d) Using Table II, find the critical z score(s) z_c.
  e) Construct the hypothesis test model.
  f) State the decision rule.

Step 4 Analyze the sample data.
Randomly select the necessary sample data from the binomial population and calculate the sample proportion, $\hat{p}$, using the formula:

$$\hat{p} = \frac{\text{number of occurrences possessing particular characteristic}}{\text{sample size}}$$

Calculate the z score of the sample proportion using the formula:

$$z = \frac{\hat{p} - \mu_{\hat{p}}}{\sigma_{\hat{p}}}$$

This is referred to as the **test statistic**.

Step 5 State the conclusion.
Compare the test statistic, z, to the critical z score(s), z_c, of the decision rule and draw one of the following conclusions:

a) **Reject H₀ and Accept Hₐ at α** or b) **Fail to reject H₀ at α**

In summary the procedure just outlined to perform a hypothesis test involving a population proportion, p, contains three refinements to the general hypothesis testing procedure developed in Chapter 10. They are:

I. The sampling distribution of the proportion is used as the appropriate hypothesis testing model.
II. The properties of the sampling distribution of the proportion are used to calculate the expected values.
III. The formula: $z = \dfrac{\hat{p} - \mu_{\hat{p}}}{\sigma_{\hat{p}}}$ is used to calculate the test statistic, z.

EXAMPLE 11.4

According to a recent social media statistical study, 47% of Americans stated that Facebook has the greatest influence on their buying decisions. A marketing researcher decides to test this claim against the suspicion that the social media claim is too high. To test this population proportion, the researcher randomly surveys 400 Americans who use social media before purchasing a product. The researcher finds that 176 of those surveyed stated Facebook had a major impact on their purchase. Based upon this sample proportion, is the researcher's claim reasonable at α = 5%?

Solution

Step 1 Formulate the hypotheses.

H₀: The population proportion of Americans who are influenced by Facebook with their buying decisions is 47%:

In symbols, we have: H₀: p = 0.47

Since the marketing researcher is interested in determining if the population proportion is lower than 0.47, then the alternative hypothesis is directional and is stated as:

Hₐ: The population proportion of Americans who are influenced by Facebook with their buying decisions is less than 47%.

In symbols, we have: Hₐ: p < 0.47

Step 2 Determine the model to test the null hypothesis, H₀.
Under the assumption H₀ is true, calculate the expected results. Using Theorem 8.5 and the assumption H₀ is true, the expected results are:
a) The sampling distribution of the proportion is approximately a normal distribution since both np and n(1 − p) are greater than 10.
b) For p = 0.47, then the mean of the sampling distribution of the proportion is: $\mu_{\hat{p}} = 0.47$
c) Since n = 400, p = 0.47 and 1 − p = 0.53, then the standard error of the proportion is:

$$\sigma_{\hat{p}} = \sqrt{\dfrac{p(1-p)}{n}} = \sqrt{\dfrac{0.47(1-0.47)}{400}} = 0.0250$$

Figure 11.5 illustrates the sampling distribution of the proportion as the Hypothesis Testing model.

Hypothesis Testing Model: sampling distribution of the proportion (is approximately a normal distribution)

$\mu_{\hat{p}} = 0.47$
$\sigma_{\hat{p}} = 0.0250$

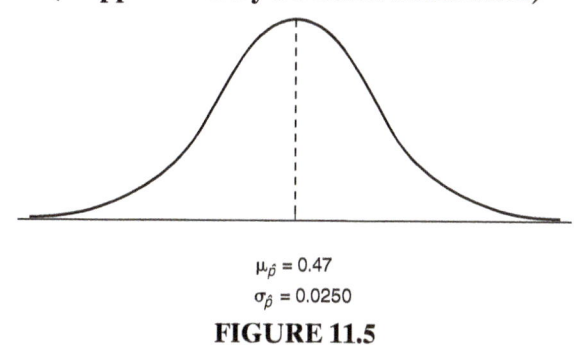

FIGURE 11.5

588 Chapter 11 Hypothesis Testing Involving One Population

Step 3 Formulate the decision rule.
a) The alternative hypothesis is directional since the researcher wants to determine if the population proportion is significantly lower than 0.47.
b) The type of hypothesis test is one-tailed on the left side.
c) The significance level is: $\alpha = 5\%$.
d) The critical z score, z_c, for a one-tailed test on the left side with $\alpha = 5\%$ is $z_c = -1.65$.
e) Figure 11.6(a) illustrates the decision rule for this hypothesis test.
f) The decision rule is: Reject H_o if the test statistic, z, is less than -1.65.

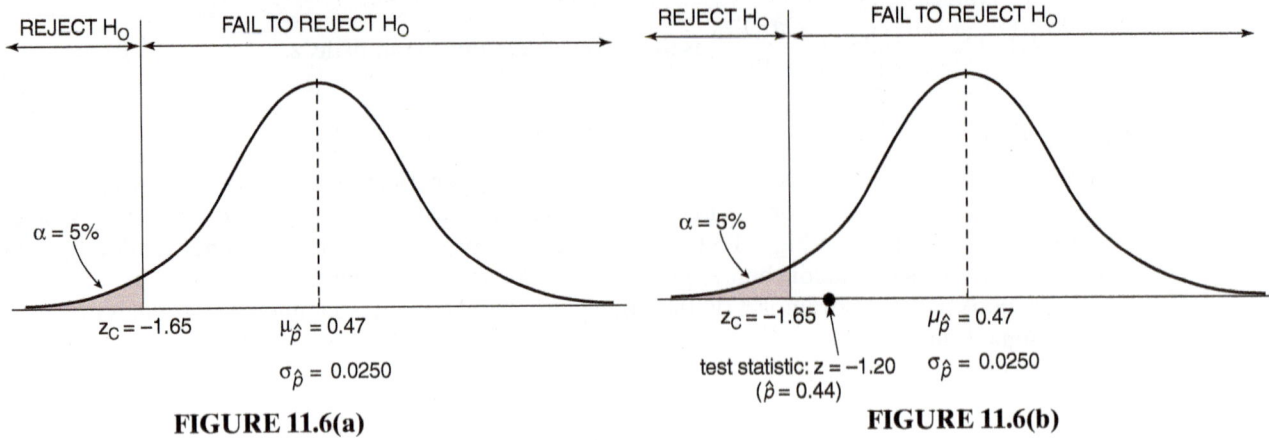

FIGURE 11.6(a)

$z_c = -1.65$, $\mu_{\hat{p}} = 0.47$, $\sigma_{\hat{p}} = 0.0250$, $\alpha = 5\%$

FIGURE 11.6(b)

$z_c = -1.65$, $\mu_{\hat{p}} = 0.47$, $\sigma_{\hat{p}} = 0.0250$, $\alpha = 5\%$, test statistic: $z = -1.20$ ($\hat{p} = 0.44$)

Step 4 Analyze the sample data.
First, using the survey data, compute the sample proportion, $\hat{p}$, using the formula:

$$\hat{p} = \frac{\text{number of Americans who use social media before purchasing a product}}{\text{sample size}}$$

$$\hat{p} = \frac{176}{400} = 0.44$$

Second, compute the test statistic, z, using the formula: $z = \dfrac{\hat{p} - \mu_{\hat{p}}}{\sigma_{\hat{p}}}$

where: $\hat{p} = 0.44$, $\mu_{\hat{p}} = 0.47$ and $\sigma_{\hat{p}} = 0.0250$.

Thus, the test statistic is: $z = \dfrac{0.44 - 0.47}{0.0250} = -1.20$

Step 5 Step 5 State the conclusion.
From Figure 11.6(b), since the test statistic $z = -1.20$ is not less than the critical z score $z_c = -1.65$, then we fail to reject H_o at $\alpha = 5\%$. Therefore, the marketing researcher cannot conclude that the social media claim is too high at $\alpha = 5\%$.

EXAMPLE 11.5

The IRS stated that last year 20% of the Federal Income Tax Returns contained arithmetic errors. A random sample of 500 of this year's Federal Returns found 130 returns with arithmetic errors. Can the IRS conclude that the proportion of Federal Returns containing arithmetic errors has significantly changed at $\alpha = 5\%$?

Solution

Step 1 Formulate the hypotheses.

H_o: For this year, the population proportion of Federal Income Tax Returns with arithmetic errors is 0.20.

Section 11.2 Hypothesis Testing Involving a Population Proportion 589

In symbols, we have:

H_o: p = 0.20

H_a: For this year, the population proportion of Federal Income Tax Returns with arithmetic errors is **not** 0.20.

In symbols, we have:

H_a: p ≠ 0.20

Step 2 Determine the model to test the null hypothesis, H_o.
Under the assumption H_o is true and using Theorem 8.5, the expected results are:
a) The sampling distribution of the proportion is approximately a normal distribution since both np and n(1 − p) are greater than 5.
b) Since p = 0.20, then the mean is: $\mu_{\hat{p}}$ = 0.20.
c) For n = 500, p = 0.20 and 1 − p = 0.80, then the standard error is:

$$\sigma_{\hat{p}} = \sqrt{\frac{p(1-p)}{n}}$$

$$\sigma_{\hat{p}} = 0.0179$$

Figure 11.7 illustrates the sampling distribution of the proportion which serves as the hypothesis testing model.

**sampling distribution of the proportion
(is approximately a normal distribution)**

$\mu_{\hat{p}}$ = 0.20

$\sigma_{\hat{p}}$ = 0.0179

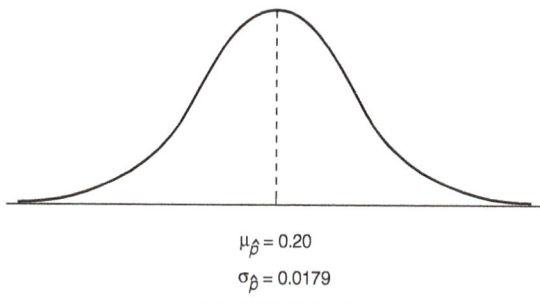

FIGURE 11.7

Step 3 Formulate the decision rule.
a) The alternative hypothesis is **non-directional** since the IRS is trying to determine if the proportion of Federal Returns containing arithmetic errors for this year is **significantly different** from 0.20.
b) The type of hypothesis test is two-tailed.
c) The significance level is: α = 5%.
d) The critical z scores, z_{LC} and z_{RC}, for a two-tailed test with α = 5% are z_{LC} = −1.96 and z_{RC} = +1.96.
e) Figure 11.8(a) illustrates the decision rule for this hypothesis test.
The decision rule is:
f) Reject H_o if the test statistic, z, is either *less than* z_{LC} = −1.96 or *greater than* z_{RC} = 1.96.

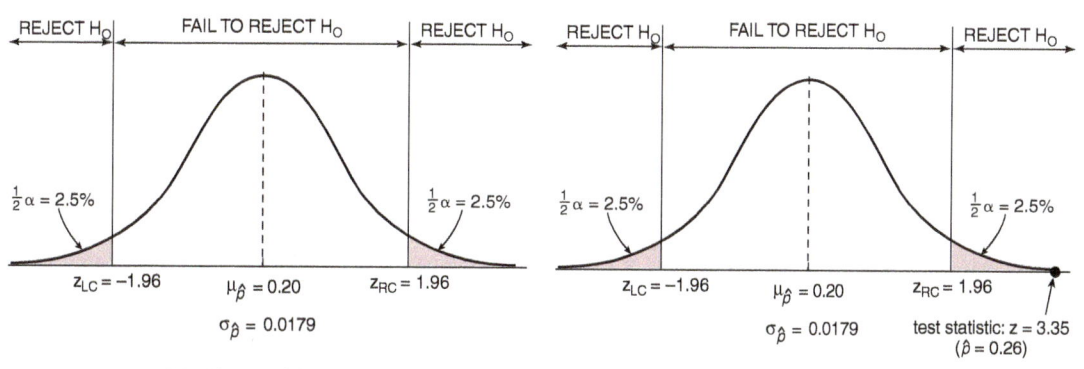

FIGURE 11.8(a)   **FIGURE 11.8(b)**

(test statistic: z = 3.35, $\hat{p}$ = 0.26)

Step 4 Analyze the sample data.
First, compute the sample proportion, $\hat{p}$.

$$\hat{p} = \frac{\text{number of returns in the sample containing arithmetic errors}}{\text{sample size}}$$

$$\hat{p} = \frac{130}{500}$$

$$\hat{p} = 0.26$$

Second, compute the test statistic, z for, $\hat{p} = 0.26$.

$$z = \frac{0.26 - 0.20}{0.0179} \approx 3.35$$

Step 5 State the conclusion.
As shown in Figure 11.8(b), since the test statistic z = 3.35, is *greater than* the right critical z score $z_{RC} = 1.96$, we reject H_o and accept H_a at $\alpha = 5\%$. Therefore, the IRS can conclude that the proportion of this year's Federal Returns containing arithmetic errors has significantly changed at $\alpha = 5\%$.

EXAMPLE 11.6

A biochemist has developed a new drug which he claims is more effective in the treatment of a skin disorder than the existing drug. Using the existing drug only 30 percent of the people contracting this skin disorder recover completely. To test his claim, he administers this new drug to a sample of 300 people who have contracted this skin disorder. The biochemist determines that 110 patients in his sample show a complete recovery. Based upon this sample result, can the biochemist conclude at $\alpha = 1\%$ that his new drug is more effective than the existing drug in the treatment of this skin disorder?

Solution

Step 1 Formulate the hypotheses.

H_o: The population proportion of people who completely recover from this skin disorder using the new drug is 0.30.

In symbols, we have:

H_o: p = 0.30

H_a: The population proportion of people who completely recover from this skin disorder using the new drug is *greater than* 0.30.

In symbols, we have:

H_a: p > 0.30

Step 2 Determine the model to test the null hypothesis, H_o.
Under the assumption H_o is true, and using Theorem 8.5, the expected results are:
a) The sampling distribution of the proportion is approximately a normal distribution since both np and n(1 – p) are greater than 5.
b) Since p = 0.30 the mean is: $\mu_{\hat{p}} = 0.30$.
c) For n = 300, p = 0.30 and 1 – p = 0.70, the standard error is:

$$\sigma_{\hat{p}} = \sqrt{\frac{p(1-p)}{n}}$$

$$\sigma_{\hat{p}} = 0.0265$$

Figure 11.9 illustrates the sampling distribution of the proportion as the hypothesis testing model.

**sampling distribution of the proportion
(is approximately a normal distribution)**

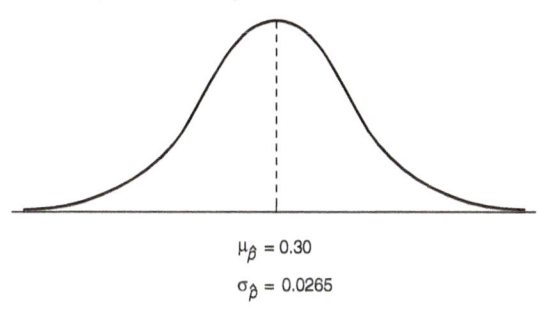

$\mu_{\hat{p}} = 0.30$

$\sigma_{\hat{p}} = 0.0265$

FIGURE 11.9

Step 3 Formulate the decision rule.
  a) The alternative hypothesis is *directional* since the biochemist wants to determine if the proportion of patients completely recovering using the new drug is *greater than* 0.30.
  b) The type of hypothesis test is one-tailed on the right side.
  c) The significance level is: $\alpha = 1\%$.
  d) The critical score, z_c, for a one-tailed test on the right side with $\alpha = 1\%$ is 2.33.
  e) Figure 11.10(a) illustrates the decision rule for this hypothesis test.

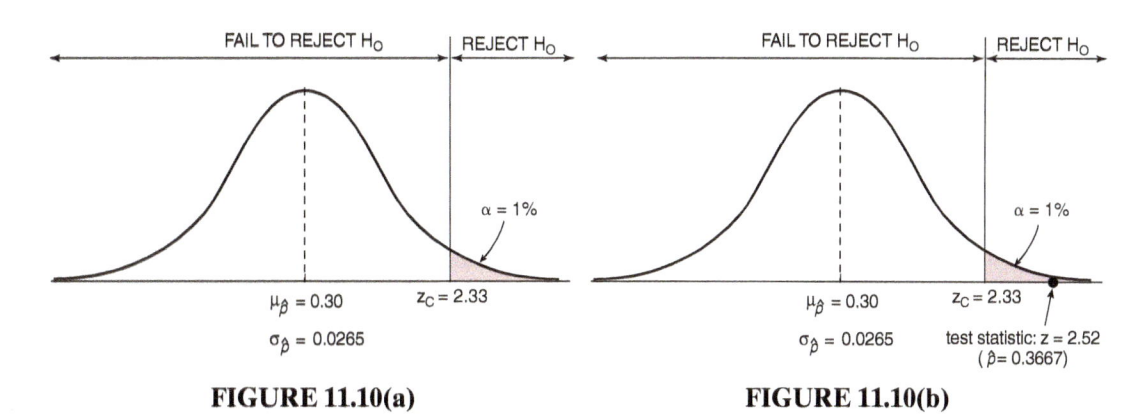

FAIL TO REJECT H₀ | REJECT H₀

$\alpha = 1\%$

$\mu_{\hat{p}} = 0.30$ $z_C = 2.33$

$\sigma_{\hat{p}} = 0.0265$

FIGURE 11.10(a)

FAIL TO REJECT H₀ | REJECT H₀

$\alpha = 1\%$

$\mu_{\hat{p}} = 0.30$ $z_C = 2.33$

$\sigma_{\hat{p}} = 0.0265$ test statistic: z = 2.52
($\hat{p}$= 0.3667)

FIGURE 11.10(b)

The decision rule is:
  f) Reject H₀ if the test statistic, z, is greater than $z_c = 2.33$.

Step 4 Analyze the sample data.
First, compute the sample proportion, $\hat{p}$.

$$\hat{p} = \frac{\text{number of people in the sample who recover using new drug}}{\text{sample size}}$$

$$\hat{p} = \frac{110}{300}$$

$$\hat{p} = 0.3667$$

Second, compute the test statistic, z, for $\hat{p} = 0.3667$.
$z \approx 2.52$

Step 5 State the conclusion.
As shown in Figure 11.10(b), since the test statistic, z = 2.52, is greater than the critical z score, $z_c = 2.33$, we reject H₀ and accept Hₐ at $\alpha = 1\%$. Therefore, the biochemist can conclude that his new drug is more effective than the existing drug in the treatment of this skin disorder at a significance level of 1%.

CASE STUDY 11.1 PROGRESS OR PLATEAU? The State of Women in IT Leadership

Examine the information contained in the Snapshot entitled "Percentage of Female Chief Information Officers (CIOs) in Fortune 500 Companies Over Time" shown in Figure 11.11.

a) What parameter is represented by the graph in Figure 11.11?
b) What type of graph is used in Figure 11.11 to present the percentage values?
c) Define the population described in Figure 11.11.
d) What variable is represented by the graph in Figure 11.11?
e) Is this variable numeric or categorical? Explain.
f) According to Figure 11.11, what population proportion of female workers were CIOs in 1995?
g) According to Figure 11.11, what population proportion of female workers were CIOs in 2018?

Statistical Snapshot

Percentage of Female Chief Information Officers (CIOs) in Fortune 500 Companies Over Time

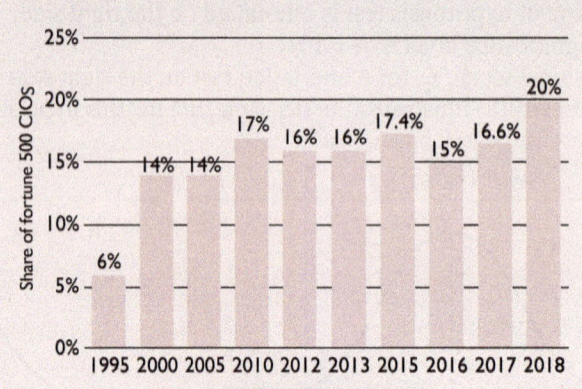

Share of fortune 500 CIOs:
- 1995: 6%
- 2000: 14%
- 2005: 14%
- 2010: 17%
- 2012: 16%
- 2013: 16%
- 2015: 17.4%
- 2016: 15%
- 2017: 16.6%
- 2018: 20%

FIGURE 11.11

h) Using the population proportion of female CIOs for 2015, design a hypothesis test assuming you believe that the trend shown in Figure 11.11 will continue to the present time. Within your hypothesis test, state:
i) The null hypothesis, using the information pertaining to 2015 as the status quo.
j) An alternative hypothesis based on your belief.
k) Is the alternative hypothesis directional or nondirectional? A 1 TT or a 2 TT? Explain.
l) What population parameter value are you assuming you know to perform this hypothesis test?
m) State the Sampling Distribution that will serve as the model for this test. Can you use a z test to perform the hypothesis test? Explain.
n) State the mean and standard error of the Sampling Distribution.
o) In which tail(s) of the curve would you place the critical value for your test?
p) If you decide to use a significance level of 5%, then state the critical value notation required to perform the test and determine the critical value.
q) If the test statistic (or sample result) falls to the left of the critical z score value of your test, what would be the conclusion to this test? Based upon this conclusion, would you agree that the population proportion of female CIOs has significantly changed? If you can, then explain the type of significant change for the population proportion you can conclude from this test. What can you conclude about the population proportion?
r) If the test statistic (or sample result) falls to the right of the critical z score value of your test, what would be the conclusion to this test? Based upon this conclusion, would you agree that the population proportion of female CIOs has significantly changed? If you can, then explain the type of significant change for the population proportion you can conclude from this test. What can you conclude about the population proportion?
s) If the p-value of the hypothesis test is 4% (or 0.04) then can you reject the null hypothesis at $\alpha = 5\%$? Or at $\alpha = 1\%$? For each of these significant levels, does your conclusion indicate that the population proportion of female CIOs has significantly changed? Explain what you can conclude about the population proportion of female CIOs for each significant level?
t) If the p-value of the hypothesis test is 0.9% (or 0.009) and $\alpha = 1\%$, then can you conclude that the population proportion of female CIOs has significantly changed? Explain.

Summary of Test Statistic Approach to Hypothesis Test Involving a Population Proportion

The following outline is a summary of the 5 step hypothesis testing procedure for a hypothesis test involving one population proportion.

Step 1 Formulate the two hypotheses, H_o and H_a.
Null Hypothesis, H_o: $p = p_o$
Alternative Hypothesis, H_a: has one of the following three forms:
H_a: $p < p_o$ or H_a: $p > p_o$ or H_a: $p \neq p_o$

Step 2 Determine the model to test the null hypothesis, H_o.
Under the assumption H_o is true, the expected results are:
a) The sampling distribution of the proportion is approximately normal when both np and $n(1-p)$ are greater than or equal to 10.
b) The mean of the Sampling Distribution of The Proportion, denoted $\mu_{\hat{p}}$, is: $\mu_{\hat{p}} = p_o$
c) the standard error of the proportion, denoted $\sigma_{\hat{p}}$, is:

$$\sigma_{\hat{p}} = \sqrt{\frac{p_o(1-p_o)}{n}}$$

Step 3 Formulate the decision rule.
a) Alternative hypothesis: directional or non-directional.
b) Type of test: 1TT or 2TT.
c) Significance level: $\alpha = 1\%$ or $\alpha = 5\%$.
d) Find the critical z score(s), z_c.
e) Construct the hypothesis test model.
f) State the decision rule.

Step 4 Analyze the sample data.
a) Compute the sample proportion, $\hat{p}$, using the formula:

$$\hat{p} = \frac{\text{number of occurrences possessing particular characteristic}}{\text{sample size}}$$

b) Compute test statistic, z, using: $z = \dfrac{\hat{p} - \mu_{\hat{p}}}{\sigma_{\hat{p}}}$

Step 5 State the conclusion.
Compare the test statistic, z, to the critical z score(s), z_c, of the decision rule and draw one of the following conclusions:
either: (a) Reject H_o and Accept H_a at α
or (b) Fail to Reject H_o at α

Table 11.1 contains a condensed summary of the hypothesis testing procedure involving a population proportion.

TABLE 11.1 Summary of Hypothesis Test for Testing the Value of a Population Proportion

Form of the Null Hypothesis	Conditions of Test	Test Statistic Formula	Sampling Distribution is:
$p = p_o$	$np \geq 10$ and $n(1-p) \geq 10$	$z = \dfrac{\hat{p} - \mu_{\hat{p}}}{\sigma_{\hat{p}}}$ where $\mu_{\hat{p}} = p_o$ $\sigma_{\hat{p}} = \sqrt{\dfrac{p_o(1-p_o)}{n}}$	Approximately a Normal Distribution

11.3 HYPOTHESIS TESTING INVOLVING A POPULATION MEAN: POPULATION STANDARD DEVIATION KNOWN

We are now going to outline the hypothesis testing procedure for testing the mean of a population by modifying the general five step hypothesis testing procedure developed in Chapter 10. The first step of the hypothesis testing procedure is to formulate the null and alternative hypotheses. Let's examine how to state these hypotheses for a test involving a population mean.

Step 1 Formulate the two hypotheses, H_o and H_a.
Null Hypothesis: In general, the null hypothesis has the following form.

Null Hypothesis Form

H_o: The mean of the population, μ, is claimed to be equal to a numerical value which will be symbolized as: μ_o.
   Thus, the null hypothesis can be symbolized as:

$$H_o: \mu = \mu_o$$

Alternative Hypothesis:
   Since the alternative hypothesis, H_a, can be stated as either *greater than, less than,* or *not equal* to the numerical value, μ_o, then the alternative hypothesis can have one of three forms:

Form (a): Greater Than Form for H_a

H_a: The mean of the population, μ, is claimed to be **greater than** the numerical value, μ_o.
This is symbolized as:

$$H_a: \mu > \mu_o$$

Form (b): Less Than Form for H_a

H_a: The mean of the population, μ, is claimed to be **less than** the numerical value, μ_o.
This is symbolized as:

$$H_a: \mu < \mu_o$$

Form (c): Not Equal Form for H_a

H_a: The mean of the population, μ, is claimed to be **not equal** to the numerical value, μ_o.
This is symbolized as:

$$H_a: \mu \neq \mu_o$$

Step 2 Determine the model to test the null hypothesis, H_o.
   Two important design considerations are choosing a sample size, n, and determining the sampling distribution model. Since we are testing a hypothesis about a population mean using a sample mean, $\bar{x}$, we use as the hypothesis test model the sampling distribution of the mean and its properties, which were discussed in Chapter 8. Using this information, we can draw the following conclusions.
   Under the following assumptions that:
   a) The null hypothesis, H_o, is true, i.e. $\mu = \mu_o$, and,
   b) The standard deviation of the population, σ, is **known**, and,
   c) The sample size, n, is greater than 30,

then we can conclude that:
a) The sampling distribution of the mean is approximately **normal** and serves as the hypothesis test model,
b) The mean of the sampling distribution of the mean is:

$$\mu_{\bar{x}} = \mu_o$$

(i.e. the mean, $\mu_{\bar{x}}$, is equal to the claimed population mean, μ_o).
c) The standard error of the mean, $\sigma_{\bar{x}}$, is:

$$\sigma_{\bar{x}} = \frac{\sigma}{\sqrt{n}}$$

$\sigma_{\bar{x}}$ is also the standard deviation of the sampling distribution of the mean. This is illustrated in Figure 11.12.

Hypothesis Testing Model

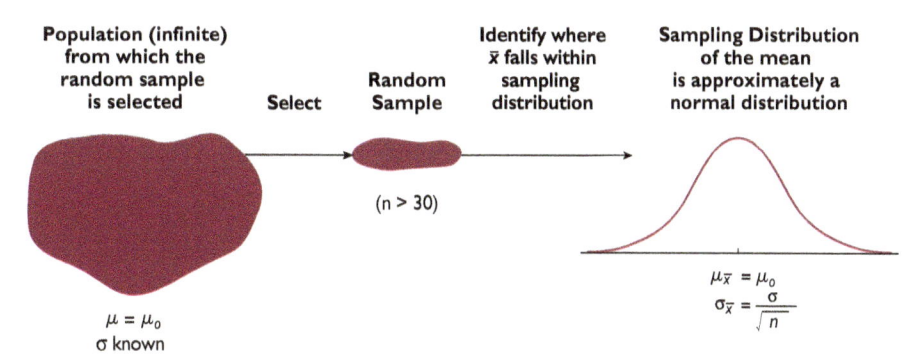

FIGURE 11.12

In theory, the population being sampled must be infinite to use the standard error of the mean formula:

$$\sigma_{\bar{x}} = \frac{\sigma}{\sqrt{n}}$$

However it is appropriate to use this formula when the population is *finite* and the population size, N, is large relative to the sample size, n. The rule used to determine whether the population is large relative to the sample is: the sample size, n, is less than 5% of the population size, N. That is, $\frac{n}{N} < 5\%$.

Step 3 Formulate the decision rule.
  a) Determine the type of alternative hypothesis (i.e., is the alternative hypothesis **directional** or **nondirectional**?)
  b) Determine the type of hypothesis test (i.e., is the test **1TT** or **2TT**?)
  c) Identify the significance level (i.e., is $\alpha = 1\%$ or $\alpha = 5\%$?)
  d) Find the critical z score(s), z_c, using Table II.
  e) Construct the appropriate **hypothesis testing model**.
  f) State the decision rule.

Step 4 Analyze the sample data.
  a) Compute the sample mean, $\bar{x}$.
  b) Compute the test statistic, z, using: $z = \dfrac{\bar{x} - \mu_{\bar{x}}}{\sigma_{\bar{x}}}$

Step 5 State the conclusion.
Compare the test statistic to the **critical z score**, z_c, of the decision rule and draw one of the following conclusions:

(a) **Reject H_o and Accept H_a at α**   or   (b) **Fail to reject H_o at α**

In summary, the procedure just outlined to perform a hypothesis test involving a population mean, μ, where the **population standard deviation, σ, is known** and **n is greater than 30**, contains three refinements to the general hypothesis testing procedure developed in Chapter 10. They are:

    I The sampling distribution of the mean is used as the appropriate hypothesis testing model.
    II The properties of the sampling distribution of the mean are used to calculate the expected values.
    III The formula: $z = \dfrac{\bar{x} - \mu_{\bar{x}}}{\sigma_{\bar{x}}}$ is used to calculate the test statistic, z.

Example 11.7 demonstrates the use of the hypothesis testing procedure to test a population mean.

EXAMPLE 11.7

An automatic opening device for parachutes has a stated mean release time of 10 seconds with a population standard deviation of 3 seconds. A local parachute club decides to test this claim against the alternative hypothesis that the release time is not 10 seconds. The club purchases 36 of these devices and finds that the mean release time is 8.5 seconds. Does this sample result indicate that the opening device does not have a mean release time of 10 seconds? Use $\alpha = 1\%$.

Solution

Step 1 Formulate the hypotheses.

H_o: The population mean release time is 10 seconds.

This is symbolized as: $H_o: \mu = 10$ secs.

H_a: The population mean release time *is not* 10 seconds.

This is symbolized as: $H_a: \mu \neq 10$ secs.

Step 2 Determine the model to test the null hypothesis, H_o.

To test the null hypothesis, the club decides to purchase 36 parachutes.

Since the sample size, n, is greater than 30, we can use the Central Limit Theorem (previously discussed in Chapter 8). Using the results of this theorem, the sampling distribution of the mean is a normal distribution and *serves as the hypothesis testing model.*

Assuming the null hypothesis is true, the *expected results* for the sampling distribution of the mean are:

$$\mu_{\bar{x}} = \mu$$
$$\mu_{\bar{x}} = 10 \text{ seconds}$$

and

$$\sigma_{\bar{x}} = \frac{\sigma}{\sqrt{n}}$$
$$\sigma_{\bar{x}} = \frac{3}{\sqrt{36}}$$
$$\sigma_{\bar{x}} = 0.50 \text{ seconds}$$

Figure 11.13 illustrates the sampling distribution of the mean hypothesis testing model.

The Sampling Distribution of the Mean (is a normal distribution)

$\mu_{\bar{x}} = 10$ sec.
$\sigma_{\bar{x}} = 0.50$ sec.

FIGURE 11.13

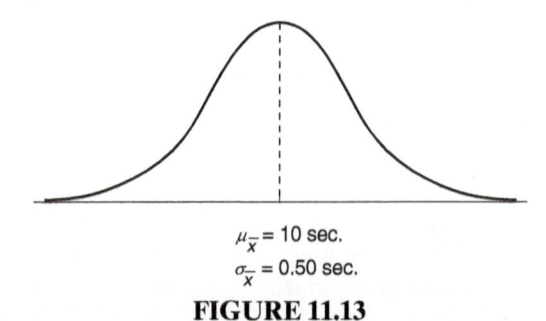

Step 3 Formulate the decision rule.
a) The alternative hypothesis is nondirectional.
b) The type of test is two-tailed, 2TT.
c) The significance level is: $\alpha = 1\%$.
d) Using Table II: The Normal Curve Area Table, for a 2TT at $\alpha = 1\%$. The two critical z scores, z_{LC} and z_{RC}, are:

$$z_{LC} = -2.58 \text{ and } z_{RC} = 2.58$$

e) Figure 11.14(a) illustrates the decision rule for this hypothesis test.

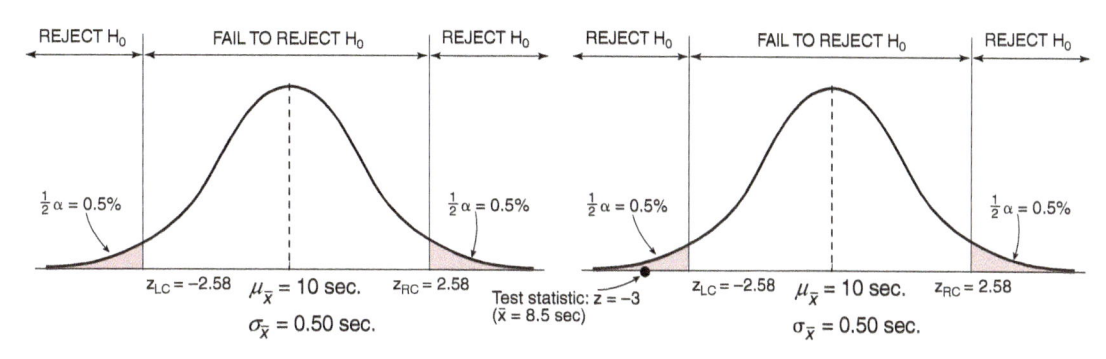

FIGURE 11.14(a)   **FIGURE 11.14(b)**

The decision rule is:
f) Reject the null hypothesis if the test statistic, z, is less than -2.58, or more than 2.58.

Step 4 Analyze the sample data.
a) The **sample mean** is: $\bar{x} = 8.5$ seconds.
b) Compute the test statistic, z, using $z = \frac{\bar{x} - \mu_{\bar{x}}}{\sigma_{\bar{x}}}$. For $\bar{x} = 8.5$, $\mu_{\bar{x}} = 10$ and $\sigma_{\bar{x}} = 0.5$, the test statistic is $z = -3$.

Step 5 State the conclusion.
From Figure 11.14(b), since the test statistic $z = -3$ is less than the critical z score $z_{LC} = -2.58$; we *reject* H_o and *accept* the alternative hypothesis, H_a, at $\alpha = 1\%$. Therefore, the mean release time for the automatic opening device is **significantly different** than 10 seconds.

EXAMPLE 11.8

A tire manufacturer advertises that its brand of radial tires has a mean life of 40,000 miles with a population standard deviation of 1,500 miles.

A consumer's research team decides to investigate this claim after receiving several complaints from people who believe this advertisement is false (i.e. the mean life of 40,000 miles is too high).

If the research team tests 100 of these radial tires and obtains a sample mean tire life of 39,750, is the advertisement legitimate? Use $\alpha = 5\%$.

Solution

Step 1 Formulate the hypotheses.

H_o: The population mean tire life is 40,000 miles.

This is symbolized as: $H_o: \mu = 40{,}000$ miles
H_a: The population mean tire life is less than 40,000 miles.

This is symbolized as: $H_a: \mu < 40{,}000$ miles

Step 2 Determine the model to test the null hypothesis, H_o.
Under the assumption that the null hypothesis is true and the sample size, n, is 100, then the *expected results* are:

$$\mu_{\bar{x}} = \mu$$
$$\mu_{\bar{x}} = 40{,}000 \text{ miles}$$

and

$$\sigma_{\bar{x}} = \frac{\sigma}{\sqrt{n}} = \frac{1500}{\sqrt{100}}$$

$$\sigma_{\bar{x}} = 150 \text{ miles}$$

From the Central Limit Theorem, the sampling distribution of the mean is approximately normal and serves as the hypothesis testing model. This is illustrated in Figure 11.15.

$\mu_{\bar{x}} = 40,000$
$\sigma_{\bar{x}} = 150$

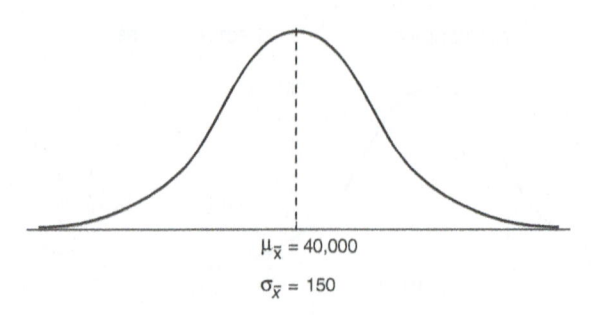

FIGURE 11.15

Step 3 Formulate the Decision Rule.
a) The alternative hypothesis is directional, since the research team is trying to show that the mean tire life is too high.
b) The type of test is one-tailed on the left side.
c) The significance level is: $\alpha = 5\%$.
d) For a one-tailed test on the left side at $\alpha = 5\%$, the critical z score, z_c, using Table II, (the Normal Curve Area Table), is: $z_c = -1.65$.
e) Figure 11.16(a) illustrates the decision rule for this hypothesis test.

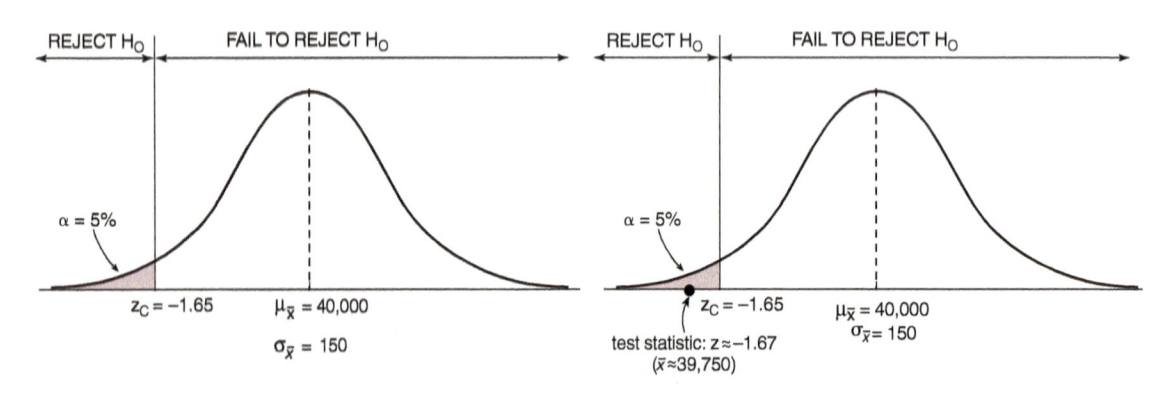

FIGURE 11.16(a)   **FIGURE 11.16(b)**

The decision rule is:
f) Reject the null hypothesis if the test statistic is less than $z_c = -1.65$.

Step 4 Analyze the sample data.
a) The sample mean is: $\bar{x} = 39,750$ miles.
b) Compute the test statistic, z. The test statistic is $z \approx -1.67$.

Step 5 State the conclusion.
From Figure 11.16(b), since the test statistic $z = -1.67$ is less than the critical z_c of -1.65 we *reject* H_o and accept H_a at $\alpha = 5\%$. Thus, the research team can conclude that the advertisement was *not* legitimate.

11.4 THE t DISTRIBUTION

There are many practical instances when one performs a hypothesis test involving a population mean, μ, where **the population standard deviation, σ, is unknown**.

For example:

- An automobile manufacturer claims that its new subcompact averages 50 mpg for highway driving.
- A cereal company claims that their family-sized cereal box has an average net weight of 15 oz.
- An archaeologist claims that the dinosaur species Tyrannosaurus Rex has an average length of 50 feet.

In such instances, when the population standard deviation is unknown, it becomes necessary to obtain a good estimate of the population standard deviation. A good estimate of the population standard deviation is determined by using the standard deviation of a random sample selected from the population.

When confronted with having to estimate the population standard deviation when performing a hypothesis test involving a population mean, we need to use a distribution (previously discussed in Section 9.4) developed by a statistician named William S. Gosset called the **Student's t distribution** or simply, the t distribution rather than the normal distribution as our model for the sampling distribution of the mean. Gosset's research work was done while he was employed by the famous Guinness Brewery in Ireland and was published under his pen-name "Student." Gosset's development of the *t* distribution allows us to test hypotheses involving a population mean when the population standard deviation, σ, is unknown. Let's re-examine the **t** distribution and its properties.

The t Distribution

If samples of size n are randomly selected from a **normal or approximately normal population** with an unknown standard deviation, then when σ is approximated by s the test statistic

$$z = \frac{\bar{x} - \mu_{\bar{x}}}{\sigma/\sqrt{n}} \text{ becomes } \frac{\bar{x} - \mu_{\bar{x}}}{s/\sqrt{n}} \text{ and is called a t score.}$$

The distribution of all the t scores is referred to as the t distribution.

Let's summarize the properties of the **t** distribution that were discussed in Chapter 9.

Properties of the t Distribution

- The **t** distribution is bell-shaped.
- The **t** distribution is symmetric about the mean.
- There is a different **t** distribution for each sample size, n, or degrees of freedom, $df = n - 1$.
- As the number of degrees of freedom increases, the **t** distribution approaches the normal distribution. They are sufficiently close when the degrees of freedom are greater than 30.

It is important to realize that when you are performing a hypothesis test involving a population mean where the population is normal and s is used as an estimator of the unknown population standard deviation, then the distribution of the test statistic $t = \frac{\bar{x} - \mu_{\bar{x}}}{s/\sqrt{n}}$ is a **t** distribution.

Unlike the normal distribution, the t distribution is composed of a family of t curves that are dependent upon sample size or degrees of freedom. Thus, for varying degrees of freedom, a different t curve is required to perform a hypothesis test. As the degrees of freedom change, the shape of the t curve shifts and the critical values called critical t scores, denoted t_c, change as well. The critical t scores for significance levels of 5% and 1% are found in Table III: Critical Values for the t distribution in Appendix C: Statistical Tables. Let's examine Table III and discuss how to use it.

Using Table III: Critical Values for the t Distribution

To find the **critical t score**, t_c, in Table III needed to perform a hypothesis test, you need the following information:

- The significance level, α.
- Type of hypothesis test. (i.e. 1TT or 2TT)
- Degrees of freedom, df.

Example 11.3 illustrates how to use Table III, given the previous information.

EXAMPLE 11.9

Use Table III to determine the critical **t** score, t_c, for a hypothesis test where:
1) $\alpha = 1\%$.
2) one-tailed test on the right side.
3) $df = 16$.

Solution

To determine the critical **t** score on the right side, t_c, examine Table III. A portion of Table III is shown in Table 11.2.

TABLE 11.2 Table III: Critical Values for the t Distribution

$\alpha = 1\%$

	One Tail			Two Tail	
df	Critical t left (t_c)	Critical t right (t_c)		Critical t left (t_{LC})	Critical t right (t_{RC})
		$t_{99\%}$			$t_{99.5\%}$
1		*			
2		*			
3		*			
.		*			
.		*			
.		*			
15		*			
16	* * * * * * * * * * * * * * *	2.58			
17					
.					
.					
.					
Normal Distribution					

From Table 11.2, the critical **t** score on the right side, t_c, is 2.58, because it is found under the column for one tail test at $\alpha = 1\%$, critical **t** right and in the row for $df = 16$.

EXAMPLE 11.10

Find the critical **t** score, t_c, for the **t** distribution in Figure 11.17 if:
1) $\alpha = 5\%$
2) One-tailed test on the left side.
3) $df = 30$.

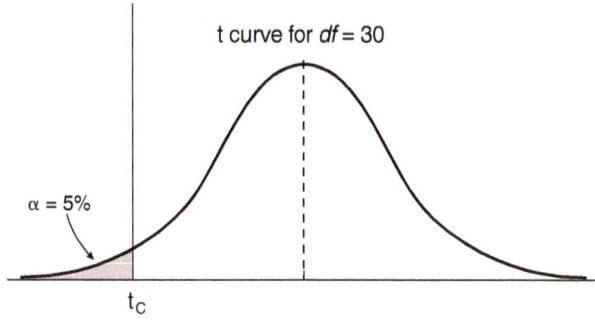

FIGURE 11.17

Solution

From Table III, the $t_c = -1.70$ since it is in the column for one tail test at $\alpha = 5\%$, critical **t** left and in the row for $df = 30$.

EXAMPLE 11.11

Find the critical **t** scores, t_{LC} and t_{RC}, for a two-tailed test, $\alpha = 5\%$ and $df = 9$ as illustrated in Figure 11.18.

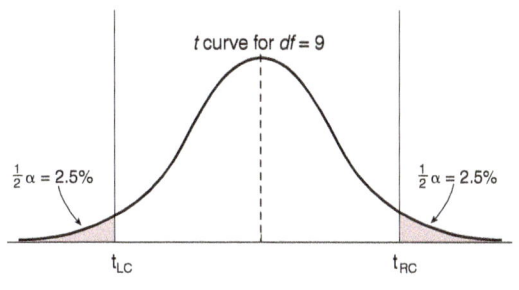

FIGURE 11.18

Solution

From Table III, the critical **t** scores are $t_{LC} = -2.26$ and $t_{RC} = 2.26$.

EXAMPLE 11.12

Using the information in Figure 11.19, find the critical **t** score, t_c.

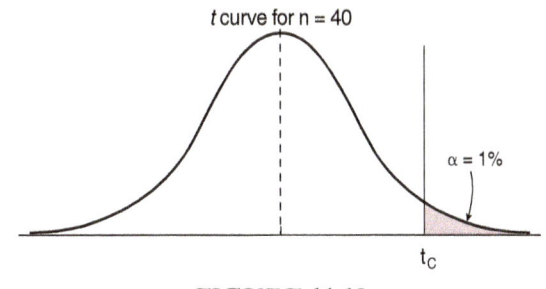

FIGURE 11.19

Solution

Before Table III can be used, the degrees of freedom must be calculated using $df = n - 1$. For $n = 40$, $df = 39$. Since there is no row for 39 degrees of freedom in Table III, we use the row for 40 degrees of freedom because it is *closest* to 39 df. Therefore, for $\alpha = 1\%$, $t_c = 2.42$.

Procedure to Locate a Critical t Score in Table III

To locate a critical t score in Table III, use the following procedure.
1. Select the appropriate α value column.
2. Identify the type of hypothesis test:
   one or two tail test
3. Select the appropriate *df* row using the following guidelines:
   (i) If the *df* value is found in the table, use this row to locate the critical t score.
   (ii) If the exact *df* value does not appear in the table, then choose the row corresponding to the closest *df* value to locate the critical t score.
   (iii) If the *df* value does not appear in the table but falls exactly halfway between two *df* values, then choose the row that corresponds to the critical *t* score that is farthest from the t score: t = 0.

11.5 HYPOTHESIS TESTING INVOLVING A POPULATION MEAN: POPULATION STANDARD DEVIATION UNKNOWN

The **t** distribution is important in hypothesis tests involving a population mean where the population is approximately normal and the population standard deviation is *unknown*. In these instances the population standard deviation, σ, is estimated by s, the sample standard deviation, and the distribution of the test statistic is a **t** distribution. Therefore, the **t** distribution is used in hypothesis testing under the following conditions.

When the Distribution of the Test statistic: $\dfrac{\bar{x} - \mu_{\bar{x}}}{s_{\bar{x}}}$ is a t Distribution

The **t** distribution is used while performing a hypothesis test involving the population mean, when the following two conditions are true:
1. The population is approximately normal
2. The population standard deviation is unknown

The five step hypothesis testing procedure has to be modified when performing a hypothesis test involving a population mean when the **t** distribution serves as the model for the distribution of the test statistic. Let's discuss these minor changes in the hypothesis testing procedure.

Hypothesis Testing Procedure Changes When the Distribution of the Test Statistic is a t Distribution

In performing a hypothesis test where the distribution of the test statistic is a **t** distribution, the hypothesis testing procedure using a **t** distribution is similar to the hypothesis testing procedure using a normal distribution. The same five step procedure is used with the only differences being:
1. $\sigma_{\bar{x}}$ has been estimated by $s_{\bar{x}}$, where

$$s_{\bar{x}} = \dfrac{s}{\sqrt{n}}$$

2. The test statistic z is replaced by the test statistic t. The formula for the test statistic is:

$$t = \dfrac{\bar{x} - \mu_{\bar{x}}}{s_{\bar{x}}}.$$

Section 11.5 Hypothesis Testing Involving a Population Mean

3. The use of the t distribution in place of the normal distribution for the test statistic and serves as the hypothesis testing model.
4. The critical z score, z_c, is replaced by a critical t score, t_c. Critical t scores are found in Table III: Critical Values for the t distribution.

Let's now consider a few examples of hypothesis testing involving a population mean where the **t** distribution serves as the model for the sampling distribution of the mean.

EXAMPLE 11.13

An automobile manufacturer claims that their hybrid car averages 55 mpg for highway driving. An engineer representing a leading automotive magazine believes this claim is too high. To test the manufacturer's claim, the engineer randomly selects sixteen of the hybrid cars and finds that they averaged 53 mpg with s = 2.5 mpg. Does this sample result indicate that the manufacturer's claim is too high at $\alpha = 1\%$?

Solution

Step 1 Formulate the hypotheses.

H_o: The population mean mpg for the hybrid car is 55.

In symbols, $H_o: \mu = 55$ mpg.

H_a: The population mean mpg for the hybrid car is less than 55.
In symbols, $H_a: \mu < 55$ mpg.

Step 2 Determine the model to test the null hypothesis, H_o.
To test the null hypothesis, a random sample of n = 16 cars was selected and tested. **Since σ is unknown, it is necessary to estimate σ using s, thus, the distribution of the test statistic is a t distribution** and serves as the hypothesis testing model.
For the sampling distribution of the mean, the expected results are:

$$\mu_{\bar{x}} = \mu$$

$$\mu_{\bar{x}} = 55 \text{ mpg}$$

and, **since s is an estimate for σ, it is necessary to estimate $\sigma_{\bar{x}}$ using the formula:**

$$s_{\bar{x}} = \frac{s}{\sqrt{n}}$$

$$s_{\bar{x}} = \frac{2.5}{\sqrt{16}}$$

$$s_{\bar{x}} = 0.625 \text{ mpg}$$

For n = 16, the degrees of freedom, *df*, is 15.
Thus, the test statistic distribution with df = 15 is illustrated in Figure 11.20.

The distribution of the test statistic with *df* = 15 (is a t Distribution)

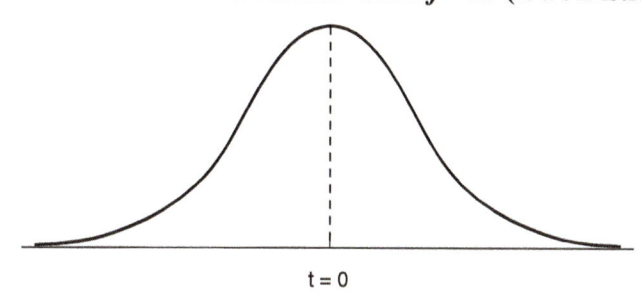

t = 0

FIGURE 11.20

Step 3 Formulate the decision rule.
a) The alternative hypothesis is directional since the engineer believes that the manufacturer's claim is *too high*.
b) The type of hypothesis test is one-tailed (1TT) on the left side.
c) The significance level is: $\alpha = 1\%$.
d) Using Table III, the critical t score, t_c, for a one-tailed test on the left side with $\alpha = 1\%$ and $df = 15$, is: $t_c = -2.60$.
e) Figure 11.21(a) illustrates the decision rule for this hypothesis test.

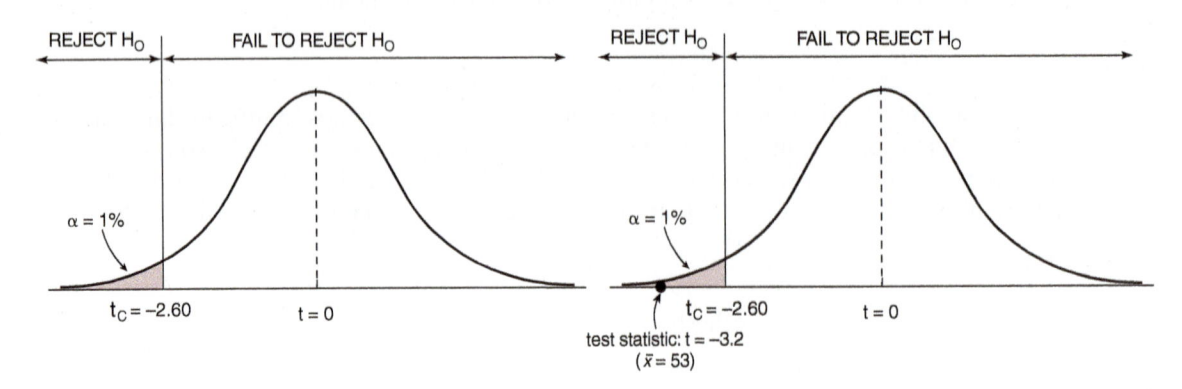

FIGURE 11.21(a)

FIGURE 11.21(b)

The decision rule is:
f) reject the null hypothesis if the test statistic t is less than $t_c = -2.60$.

Step 4 Analyze the sample data.
a) The sample mean is: $\bar{x} = 53$ mpg
b) Compute the test statistic, t, using $t = \dfrac{\bar{x} - \mu_{\bar{x}}}{s_{\bar{x}}}$.

For $\bar{x} = 53$, $\mu_{\bar{x}} = 55$ and $s_{\bar{x}} = 0.625$, $t = -3.2$.

Step 5 State the conclusion.
Using Figure 11.21(b), since the test statistic $t = -3.2$ is less than the critical t score $t_c = -2.60$, we reject H_o and accept H_a at $\alpha = 1\%$. Thus, the sample mean is *significantly different* than the manufacturer's claim of 55 mpg. Therefore, the engineer can conclude that the manufacturer's claim is too high at $\alpha = 1\%$.

EXAMPLE 11.14

A medical research team decides to study whether pressure exerted on a person's upper arm will increase bleeding time. The average bleeding time for a pricked finger where no pressure is applied to the upper arm is 1.6 minutes. To test their claim the research team randomly selects 64 subjects. They find that the average bleeding time is 1.9 minutes with s = 0.80 minutes when 50 mm of pressure is applied to the upper arm.

Does this sample result support the claim that pressure applied to the upper arm will increase bleeding time? (use $\alpha = 5\%$)

Solution

Step 1 Formulate the hypotheses.

H_o: The population mean bleeding time when 50 mm of pressure is applied to the upper arm is 1.6 minutes.

In symbols, H_o: $\mu = 1.6$ minutes.

H_a: The population mean bleeding time when 50 mm of pressure is applied to the upper arm is greater than 1.6 minutes.

In symbols, H_a: $\mu > 1.6$ minutes.

Step 2 Determine the model to test the null hypothesis, H_o.

To test the null hypothesis, a random sample of n = 64 subjects was selected and tested. **Since σ is unknown, s was used to estimate σ, and s was found to be 0.8 minutes.**

Under the assumption H_o is true, the sampling distribution of the mean has the expected results:

$$\mu_{\bar{x}} = \mu \qquad \text{and} \qquad s_{\bar{x}} = \frac{s}{\sqrt{n}}$$

$$\mu_{\bar{x}} = 1.6 \text{ minutes} \qquad s_{\bar{x}} = 0.1 \text{ minutes}$$

The distribution of the test statistic is a t distribution with df = 63 and serves as the hypothesis testing model. This is illustrated in Figure 11.22.

The distribution of the test statistic with df = 63 (is a t distribution)

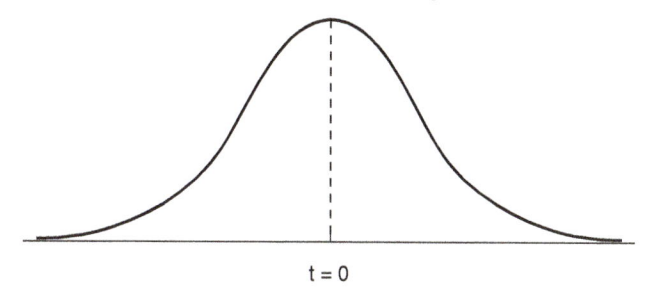

FIGURE 11.22

Step 3 Formulate the decision rule.
  a) The alternative hypothesis is directional since the research team believes that pressure applied to the upper arm will *increase* bleeding time.
  b) The type of hypothesis test is one-tailed on the right side.
  c) The significance level is: α = 5%.
  d) For a one-tailed test on the right side with α = 5% and df = 63, we need to find the critical **t** score on the right, t_c, in Table III using the closest df table entry.
     From Table III, t_c = 1.67
  e) Figure 11.23(a) illustrates the decision rule for this hypothesis test.

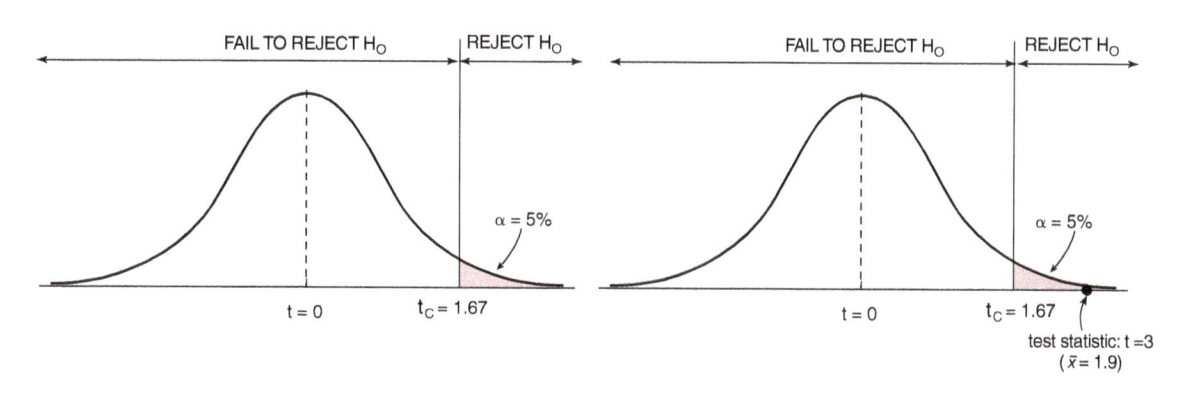

FIGURE 11.23(a)   FIGURE 11.23(b)

The decision rule is:
  f) Reject the null hypothesis if the test statistic, t, is greater than t_c = 1.67.

Step 4 Analyze the sample data.
  a) The sample mean is: $\bar{x}$ = 1.9 minutes.
  b) Compute the test statistic, t. The test statistic is $t = \frac{1.9 - 1.6}{0.1} = 3$.

Step 5 State the conclusion.
Using Figure 11.23(b), since the test statistic t = 3 is *greater than* the t score, t_c = 1.67, we reject H_o and accept H_a at α = 5%.

EXAMPLE 11.15

American Computer Machines, a large computer firm, has branch offices in several major cities of the world. From past experience, ACM knows that its employees relocate on the average once every ten years. Due to recent population trends, the company wants to determine if the average relocation time has changed (that is, is the average relocation time *different* from ten years?). To determine if a change has occurred, a random sample of twenty-five employees were interviewed and it was found that their mean relocation time was 9.5 years with s = 4.5 years.

Does this sample result indicate a change has occurred in the mean relocation time? (use $\alpha = 5\%$)

Solution

Step 1 Formulate the hypotheses.

H_o: The population mean relocation time is 10 years.

In symbols, $H_o: \mu = 10$ years.

H_a: The population mean relocation is not 10 years.

In symbols, $H_a: \mu \neq 10$ years.

Step 2 Determine the model to test the null hypothesis, H_o.

To test the null hypothesis, a random sample of n = 25 employees were interviewed. **Since σ is unknown, s was used to estimate σ and determined to be 4.5 years.** Under the assumption H_o is true, the expected results for the sampling distribution of the mean are:

$$\mu_{\bar{x}} = 10 \text{ years}$$

and

$$s_{\bar{x}} = 0.9 \text{ years}$$

The hypothesis testing model is the distribution of the test statistic which is a *t* distribution with **df = 24**. This is illustrated in Figure 11.24.

The distribution of the test statistic with df = 24 (is a *t* Distribution)

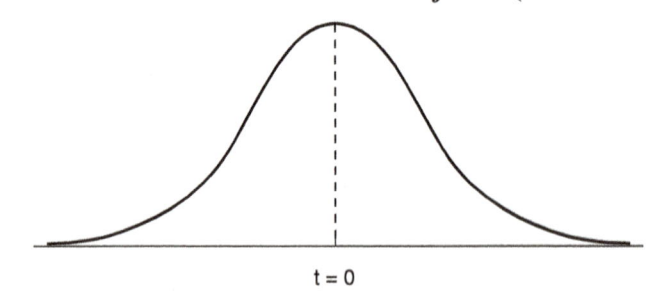

t = 0

FIGURE 11.24

Step 3 Formulate the Decision Rule.
a) The alternative hypothesis is *nondirectional* since the company is testing whether the mean relocation time has decreased or increased.
b) The type of hypothesis test is two-tailed.
c) The significance level is: $\alpha = 5\%$.
d) Using Table III, the critical values for a two-tailed test with $\alpha = 5\%$ and df = 24 are:

$$t_{LC} = -2.06 \text{ and } t_{RC} = 2.06$$

e) Figure 11.25(a) illustrates the decision rule for this hypothesis test.
The decision rule is:
f) Reject the null hypothesis if the test statistic t is either less than –2.06 or greater than 2.06.

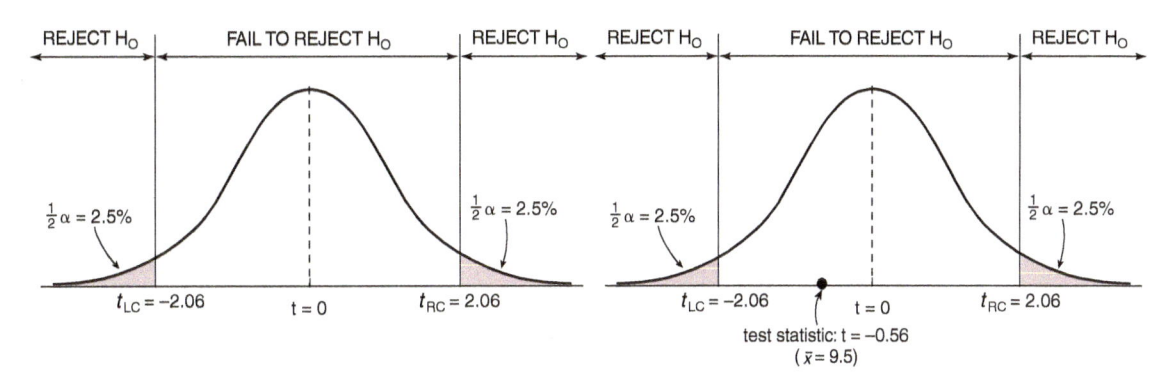

FIGURE 11.25(a)   FIGURE 11.25(b)

Step 4 Analyze the sample data.
  a) The sample mean is: $\bar{x} = 9.5$ years.
  b) The test statistic $t = \dfrac{9.5 - 10}{0.9} \approx -0.56$.

Step 5 State the conclusion.
  Using Figure 11.25(b), since the test statistic t = –0.56 falls between the critical t scores of –2.06 and 2.06, we *fail to reject* H_o at $\alpha = 5\%$.
  Although there is a *difference* between the sample mean of 9.5 years and the hypothesized mean of 10 years, this difference is ***not statistically significant***. Therefore, the company cannot conclude that the mean relocation time of 10 years has changed.

EXAMPLE 11.16

The IRS claims that the mean amount of time it takes to transfer tax information from a tax return to a computer record is 35 minutes. Before they will invest into a new system, they decide to do a test run to determine if the new system is significantly better than the present system.

Four hundred tax returns are randomly selected and the data is transferred using the new system. The mean amount of time it takes to transfer the data from a tax return to a computer record is 34.3 minutes with s = 4.2 minutes. Based upon this sample result should the IRS invest in the new system? (use $\alpha = 5\%$).

Solution

Step 1 Formulate the hypotheses.
  In hypothesis testing problems dealing with a comparison between a present (or existing) system and a new system, the null hypothesis is *always* stated as: the mean of the new system is *equal* to the mean of the present (or existing) system. The null hypothesis is stated in this form because if the null hypothesis is rejected, then this would *indicate* the new system is significantly better than the present system.
  On the other hand, if the null hypothesis *cannot* be rejected, then the new system is *not* considered to be significantly better. Thus, in this instance, the present system would *not* be replaced by the new system. Therefore, the null hypothesis is:

  H_o: Under the new transfer system, the population mean time to transfer a tax return to a computer record is 35 minutes.

  In symbols, H_o: $\mu = 35$ minutes.

  H_a: Under the new transfer system, the population mean time to transfer a tax return to a computer record is less than 35 minutes.

  In symbols, H_a: $\mu < 35$ minutes.

Step 2 Determine the model to test the null hypothesis, H_o.
  To test the null hypothesis, a random sample of n = 400 tax returns were transferred to computer records using the new transfer system. Since the population standard deviation, σ, is unknown, s was

used to estimate σ. The value of s was determined to be 4.2 minutes. Under the assumption H_o is true, the expected results for the sampling distribution of the mean are:

$$\mu_{\bar{x}} = 35 \text{ minutes}$$

and

$$s_{\bar{x}} = 0.21 \text{ minutes}$$

The hypothesis testing model is the distribution of the test statistic with $df = 399$ which is a **t** distribution. This is illustrated in Figure 11.26.

distribution of the test statistic with $df = 399$ (is a t Distribution)

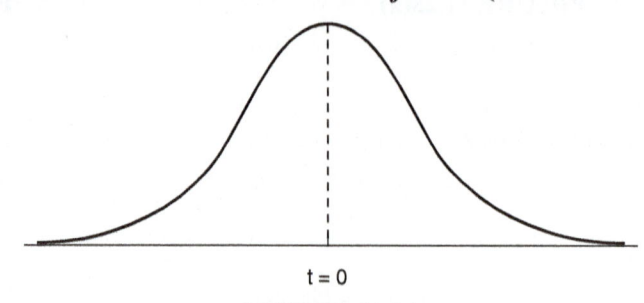

t = 0

FIGURE 11.26

Step 3 Formulate the decision rule.
a) The alternative hypothesis, H_a, is directional.
b) The type of hypothesis test is one-tailed on the left side.
c) The significance level is: $\alpha = 5\%$.
d) The critical **t** score for a one-tailed test on the left side with $\alpha = 5\%$ and $df = 399$ is found in Table III. It is $t_c = -1.65$.
e) Figure 11.27(a) illustrates the decision rule for this hypothesis test.

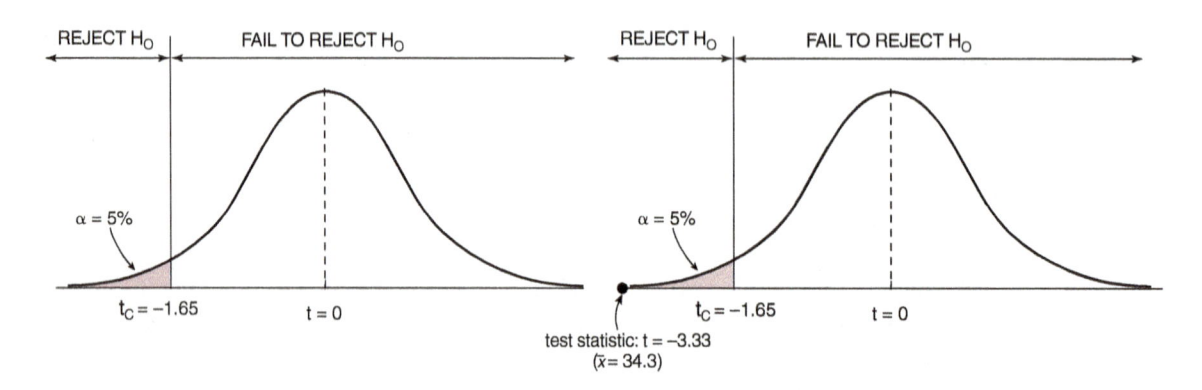

FIGURE 11.27(a)   **FIGURE 11.27(b)**

The decision rule is:
f) Reject the null hypothesis if the test statistic t is less than $t_c = -1.65$.

Step 4 Analyze the sample data.
a) The sample mean is: $\bar{x} = 34.3$ minutes.
b) The test statistic $t = \dfrac{34.3 - 35}{0.21} \approx -3.33$.

Step 5 State the conclusion.

Using Figure 11.27(b), since the test statistic t = −3.33 falls below the critical t score of −1.65, we reject H₀ and accept Hₐ at α = 5%. Thus, there is a significant difference between the sample mean time of 34.3 minutes and the expected mean time of 35 minutes. Therefore, the IRS should consider investing into the new data transferring system.

CASE STUDY 11.2   Average Minutes on Different Social Medias

The statistical snapshot entitled "*Time Flies: Average minutes per visitor to social-media sites in January*" shown in Figure 11.28 shows the average time spent per visitor to most of the popular social media sites.

For each social media site shown in Figure 11.28,
  a) Define the population.
  b) Define the variable being displayed.
  c) Is the variable discrete or continuous?
  d) For the Facebook site, what does the value 405 represent about the population parameter?
  e) For the Pinterest site, what does the value 89 represent about the population parameter?

Statistical Snapshot
Time Flies | Average minutes per visitor to social-media sites in January

- tumblr. 89
- Pinterest 89
- LinkedIn 17
- facebook 405 minutes per visitor
- myspace 8
- twitter 21
- 3 Google+

Notes: World-wide data. Does not include mobile usage. *Twitter.com data only
Source: comScore   The Wall Street Journal

FIGURE 11.28
Reprinted by permission from the *Wall Street Journal*.

Assuming you are an individual who believes the population parameter value for the Facebook site is too high, and decides to perform a hypothesis test to test the value 405 minutes.
  f) What is the null and alternative hypothesis for this test? Is the alternative hypothesis directional or non-directional? Explain.
  g) Which population parameter are you assuming you know? What is the value of the population parameter?
  h) Which population parameter are you missing to perform this test? How might you go about obtaining an estimate of this parameter?
  i) What is the name of the Sampling Distribution that will serve as the model to perform this hypothesis test? Is the Sampling Distribution of the test statistic a normal or a t distribution, if we assume that the distribution of minutes spent on a particular social media site is normally distributed? Explain.
  j) What is the mean of the Sampling Distribution? What formula would you use to estimate the standard error of the Sampling Distribution?
  k) Which tail of the curve would the critical score value of the test be placed?
  l) If the test statistic falls to the left of the critical score, what is the conclusion to this test? Would this sample result indicate the population average minutes per visit for the Facebook site in January is more or less than 405 minutes? Explain.
  m) If the p-value of the hypothesis test is 3.96% (or 0.0396) and α = 1%, then can you reject the null hypothesis? What would you say about the average minutes per visit on the Facebook site? Explain.
  n) If the p-value of the hypothesis test is 0.98% (or 0.0098) and α = 1%, then can you conclude that the average minutes per visit on the Facebook site is less than 405 minutes?

Summary of Test Statistic Approach to Hypothesis Testing Involving a Population Mean

The following outline is a summary of the 5 step hypothesis testing procedure for a hypothesis test involving one population mean.

Step 1 Formulate the two hypotheses, H_o and H_a.
Null Hypothesis, H_o: $\mu = \mu_o$
Alternative Hypothesis, **H_a**: has one of the following three forms:
H_a: $\mu < \mu_o$ or H_a: $\mu > \mu_o$ or H_a: $\mu \neq \mu_o$

Step 2 Determine the model to test the null hypothesis, H_o.
Under the assumption H_o is true, the expected results are:
a) The distribution of the test statistic is the hypothesis testing model and is either:
  (i) approximately a normal distribution, if the standard deviation of the population is known and the sample size is greater than 30,
  or
  (ii) a t Distribution with degrees of freedom: $df = n - 1$ if the standard deviation of the population is unknown.
b) The mean of the sampling distribution of the mean is given by the formula:

$$\mu_{\bar{x}} = \mu_o$$

(since H_o is true :i.e. $\mu = \mu_o$)

c) For a normal distribution: the standard error of the mean, denoted by $\sigma_{\bar{x}}$ is:

$$\sigma_{\bar{x}} = \frac{\sigma}{\sqrt{n}}$$

or

for a t distribution: the estimate of the standard error of the mean, written $s_{\bar{x}}$, is given by the formula:

$$s_{\bar{x}} = \frac{s}{\sqrt{n}}$$

Step 3 Formulate the decision rule.
a) Alternative hypothesis: directional or non-directional.
b) Type of test: 1TT or 2TT.
c) Significance level: $\alpha = 1\%$ or $\alpha = 5\%$.
d) Critical score:
  for a normal distribution: z_c is found in Table II.
  for a t distribution: t_c is found in Table III where $df = n - 1$.
e) Construct the hypothesis testing model.
f) State the decision rule.

Step 4 Analyze the sample data.
a) Compute the sample mean: $\bar{x}$.
b) for a normal distribution: test statistic is: $z = \frac{\bar{x} - \mu_{\bar{x}}}{\sigma_{\bar{x}}}$.
for a t distribution: test statistic is: $t = \frac{\bar{x} - \mu_{\bar{x}}}{s_{\bar{x}}}$.

Step 5 State the conclusion.
Compare the test statistic to the critical score of the decision rule and state the conclusion:
either: (a) Reject H_o and Accept H_a at α
or (b) Fail to Reject H_o at α

Table 11.3 contains a condensed summary of the hypothesis tests involving a population mean.

TABLE 11.3 Summary of Hypothesis Tests for Testing the Value of a Population Mean

Form of the Null Hypothesis	Conditions of Test	Test Statistic Formula	Distribution of the Test Statistic Is a:
$\mu = \mu_o$	KNOWN σ and n > 30	$z = \dfrac{\bar{x} - \mu_{\bar{x}}}{\sigma_{\bar{x}}}$ where: $\sigma_{\bar{x}} = \dfrac{\sigma}{\sqrt{n}}$	Normal Distribution
$\mu = \mu_o$	UNKNOWN σ	$t = \dfrac{\bar{x} - \mu_{\bar{x}}}{s_{\bar{x}}}$ where: $s_{\bar{x}} = \dfrac{s}{\sqrt{n}}$	t Distribution where: df = n − 1

11.6 p-VALUE APPROACH TO HYPOTHESIS TESTING USING THE TI-84 CALCULATOR

The p-value approach will require refinements in steps 3, 4 and 5 of the hypothesis testing procedure. In step 3, the decision rule will be stated as: Reject H_o if the p-value of the test statistic is less than α. In step 4, the p-value will be determined using the TI-84 calculator. While in step 5, the conclusion of the hypothesis test is determined by comparing the p-value to the level of significance, α.

Let's reexamine the hypothesis test shown in Example 11.16 using a p-value approach with the aid of the TI-84 Calculator.

EXAMPLE 11.17

A biochemist has developed a new drug which he claims is **more effective** in the treatment of a skin disorder than the existing drug. Using the existing drug only 30 percent of the people contracting this skin disorder recover completely. To test his claim, he administers this new drug to a sample of 300 people who have contracted this skin disorder. The biochemist determines that 110 patients in his sample show a complete recovery.

Based upon this sample result, can the biochemist conclude at $\alpha = 1\%$ that his new drug is more effective than the existing drug in the treatment of this skin disorder?

Solution

Step 1 *Formulate the hypotheses.*

H_o: The population proportion of people who completely recover from this skin disorder using the new drug is 0.30.

In symbols, we have:

H_o: p = 0.30

H_a: The population of people who completely recover from this skin disorder using the new drug is *greater than* 0.30.

In symbols, we have:
H_a: p > 0.30

Step 2 *Determine the model to test the null hypothesis, H_o.*

Under the assumption H_o is true, and using Theorem 8.5, the expected results are:
a) The Sampling Distribution of The Proportion is approximately a normal distribution since both np and n(1 − p) are greater than or equal to 10.
b) Since p = 0.30 the mean is: $\mu_{\hat{p}} = 0.30$.
c) For n = 300, p = 0.30 and 1 − p = 0.70, the standard error is:

$$\sigma_{\hat{p}} = \sqrt{\dfrac{p(1-p)}{n}}$$

$$\sigma_{\hat{p}} = 0.0265$$

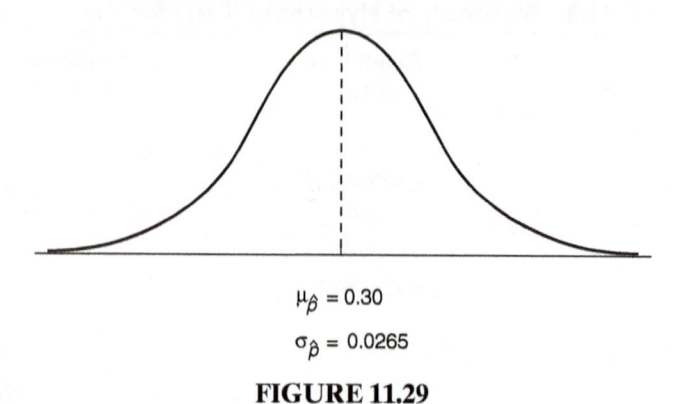

**sampling distribution of the proportion
(is approximately a normal distribution)**

$\mu_{\hat{p}} = 0.30$

$\sigma_{\hat{p}} = 0.0265$

FIGURE 11.29

Figure 11.29 illustrates the Sampling Distribution of The Proportion Hypothesis Testing model.

Step 3 *Formulate the decision rule.*
   a) the alternative hypothesis is *directional* since the biochemist wants to determine if the proportion of patients completely recovering using the new drug is *greater than* 0.30.
   b) the type of hypothesis test is one-tailed on the right side.
   c) the significance level is: $\alpha = 1\%$.
   d) Figure 11.30 illustrates the decision rule for this hypothesis test.

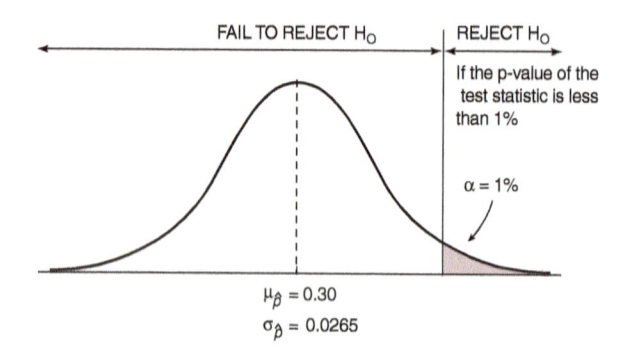

FAIL TO REJECT H₀ | REJECT H₀

If the p-value of the test statistic is less than 1%

$\alpha = 1\%$

$\mu_{\hat{p}} = 0.30$

$\sigma_{\hat{p}} = 0.0265$

FIGURE 11.30

The decision rule is:
   e) Reject H₀ if the p-value of the test statistic is less than $\alpha = 1\%$.

Step 4 *Analyze the sample data.*
The 1-PropZTest function can be used to perform a hypothesis test for the population proportion. To determine the p-value of the test statistic, the sample statistics must be entered into the TI-84's TESTS menu. To access the 1-PropZTest function in the TESTS menu press [STAT] ▶ ▶ [TESTS]. This is shown in Figure 11.31.

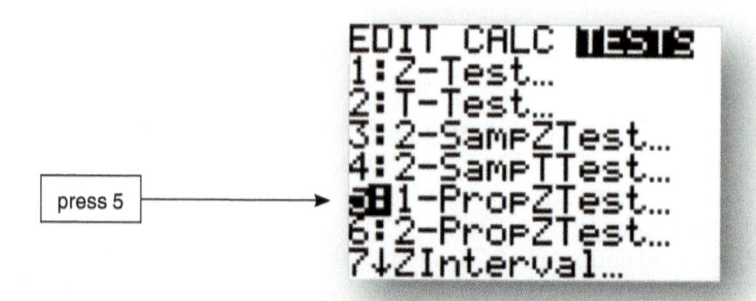

```
EDIT CALC TESTS
1:Z-Test...
2:T-Test...
3:2-SampZTest...
4:2-SampTTest...
5:1-PropZTest...
6:2-PropZTest...
7↓ZInterval...
```

press 5

FIGURE 11.31

Section 11.6 *p-Value Approach to Hypothesis Testing Using the TI-84 Calculator*   613

Press **5** to access the 1-PropZTest menu. Enter the value of the hypothesized population proportion along with the sample statistics into the menu. This is shown in Figure 11.32.

```
1-PropZTest
 p0:.3
 x:110
 n:300
 prop≠p0  <p0  >p0
 Calculate  Draw
```
Null Hypothesis → p0:.3
Alternative Hypothesis → prop≠p0 <p0 **>p0**

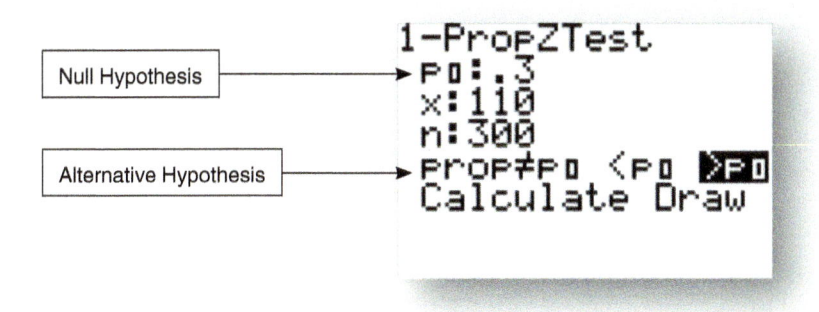

FIGURE 11.32

Highlight **Calculate** and press ENTER to obtain the output screen for this test. This is illustrated in Figure 11.33.

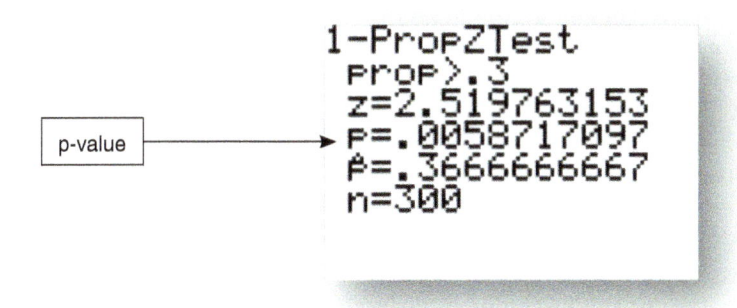

```
1-PropZTest
 prop>.3
 z=2.519763153
 p=.0058717097
 p̂=.3666666667
 n=300
```
p-value → p=.0058717097

FIGURE 11.33

From the output screen, the p-value of the test statistic $z \approx 2.52$ is 0.0059. Using this p-value we can position the test statistic on the hypothesis testing model in Figure 11.34.

FAIL TO REJECT H_o | REJECT H_o

If the p-value of the test statistic is less than 1%

$\alpha = 1\%$

test statistic: $z \approx 2.52$
(p-value = 0.0059)

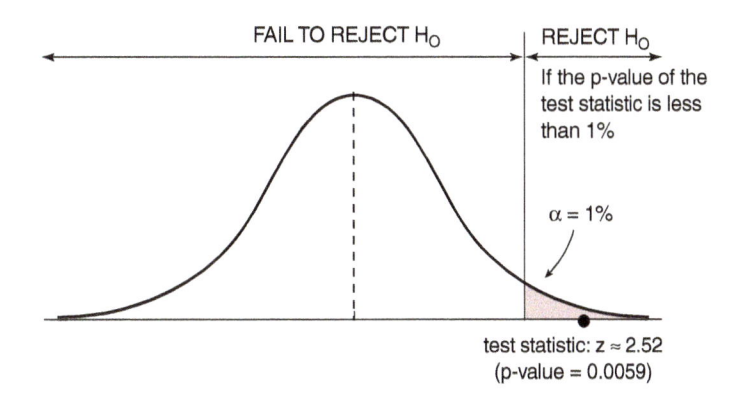

FIGURE 11.34

Step 5   *State the conclusion.*
Since the p-value (0.0059) of the test statistic is less than 1%, our conclusion is reject H_o and accept H_a at a p-value of 0.0059. Therefore, the biochemist can conclude that his new drug is more effective than the existing drug in the treatment of this skin disorder at a significance level of 1%. ∎

Now we will summarize this procedure.

Summary of the p-Value Approach to Hypothesis Testing Involving a Population Proportion Using the TI-84 Plus Calculator

Step 1 *Formulate the two hypothesis, H_o and H_a.*
Null Hypothesis, **H_o: p = p_o**
Alternative Hypothesis, **H_a**: has one of the following three forms:
H_a: p < p_o or **H_a: p > p_o** or **H_a: p ≠ p_o**

Step 2 *Determine the model to test the null hypothesis, H_o.*
Under the assumption H_o is true, the expected results are:
a) the Sampling Distribution of The Proportion is approximately normal when **both** np and n(1 – p) are greater than or equal to 10.
b) the mean of the Sampling Distribution of The Proportion, denoted $\mu_{\hat{p}}$, is: $\mu_{\hat{p}} = p_o$.
c) the standard error of the proportion denoted $\sigma_{\hat{p}}$, is:

$$\sigma_{\hat{p}} = \sqrt{\frac{p_o(1-p_o)}{n}}$$

Step 3 *Formulate the decision rule.*
a) alternative hypothesis: directional or non-directional.
b) type of test: 1TT or 2TT.
c) significance level: $\alpha = 1\%$ or $\alpha = 5\%$.
d) construct the appropriate hypothesis test model.
e) state the decision rule as *Reject H_o if the p-value of the test statistic is less than or equal to α.*

Step 4 **Analyze the sample data.**
Enter the necessary sample data into the TI-84 calculator to determine the p-value of the test statistic. Use 1-PropZTest.

Step 5 **Determine the conclusion.**
Compare the p-value of the test statistic to the decision rule and either:
a) **Reject H_o and accept H_a at α,**
or
b) **Fail to reject H_o at α.**

EXAMPLE 11.18

The IRS claims that the mean amount of time it takes to transfer tax information from a tax return to a computer record is 35 minutes. Before they will invest into a new system, they decide to do a test run to determine if the new system is significantly better than the present system.

Four hundred tax returns are randomly selected and the data is transferred using the new system. The mean amount of time it takes to transfer the data from a tax return to a computer record is 34.3 minutes with s = 4.2 minutes. Based upon this sample result should the IRS invest in the new system? (use $\alpha = 5\%$).

Solution

Step 1 Formulate the hypotheses.
In hypothesis testing problems dealing with a comparison between a present (or existing) system and a new system, the null hypothesis is *always* stated as: the mean of the new system is *equal* to the mean of the present (or existing) system. The null hypothesis is stated in this form because if the null hypothesis is rejected, then this would *indicate* the new system is significantly better than the present system.

On the other hand, if the null hypothesis *cannot* be rejected, then the new system is *not* considered to be significantly better. Thus, in this instance, the present system would *not* be replaced by the new system. Therefore, the null hypothesis is:

H_o: Under the new transfer system, the population mean time to transfer a tax return to a computer record is 35 minutes.

In symbols, H_o: $\mu = 35$ minutes.

H_a: Under the new transfer system, the population mean time to transfer a tax return to a computer record is less than 35 minutes.

In symbols, H_a: $\mu < 35$ minutes.

Step 2 Determine the model to test the null hypothesis, H_0.
To test the null hypothesis, a random sample of n = 400 tax returns were transferred to computer records using the new transfer system. Since the population standard deviation, σ, is unknown, s was used to estimate σ. The value of s was determined to be 4.2 minutes. Under the assumption H_0 is true, the expected results for the sampling distribution of the mean are:

$$\mu_{\bar{x}} = 35 \text{ minutes}$$
and
$$s_{\bar{x}} = 0.21 \text{ minutes}$$

The hypothesis testing model is the distribution of the test statistic which is a t distribution with df = 399. This is illustrated in Figure 11.35.

distribution of the test statistic (is a t distribution with df = 399)

t = 0

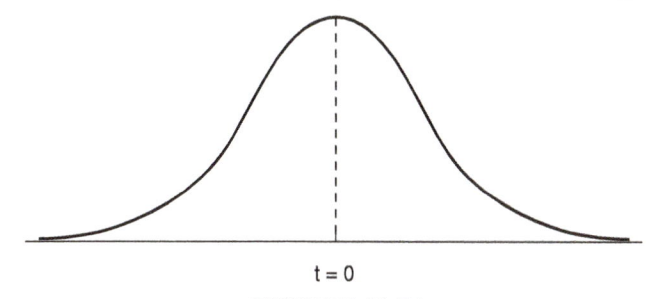

FIGURE 11.35

Step 3 Formulate the decision rule.
a) The alternative hypothesis, H_a, is directional.
b) The type of hypothesis test is one-tailed on the left side.
c) The significance level is: $\alpha = 5\%$.
d) Figure 11.36 illustrates the decision rule for this hypothesis test.
e) The decision rule is: Reject H_0 if the p-value of the test statistic is less than $\alpha = 5\%$.

REJECT H_0 | FAIL TO REJECT H_0

If the p-value of the test statistic is less than 5%

$\alpha = 5\%$

t_c t = 0

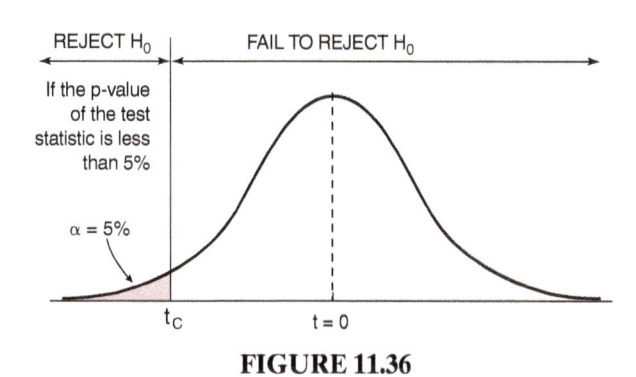

FIGURE 11.36

Step 4 Analyze the sample data.
The T-Test function can be used to perform a hypothesis test for the population mean when the population standard deviation is unknown. To determine the p-value of the test statistic, the sample statistics must be entered into the TI-84's TESTS menu. To access the T-Test function in the TESTS menu press STAT ▶ ▶ [TESTS]. This is shown in Figure 11.37.

616 Chapter 11 Hypothesis Testing Involving One Population

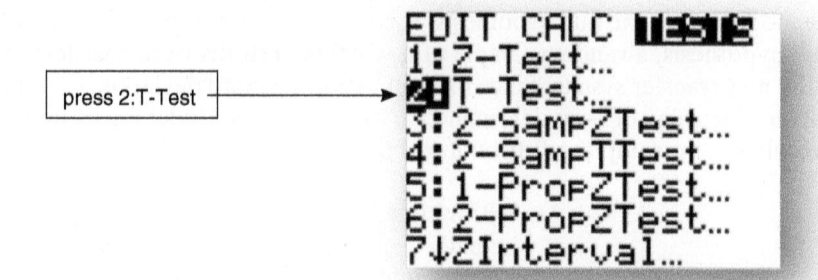

[press 2:T-Test]

FIGURE 11.37

Press **2** to access the T-Test menu. Highlight **Stats** and press ENTER. Now enter the value of the hypothesized population mean along with the sample statistics into the menu. This is shown in Figure 11.38.

[Null Hypothesis] [Highlight Stats]
[Alternative Hypothesis]

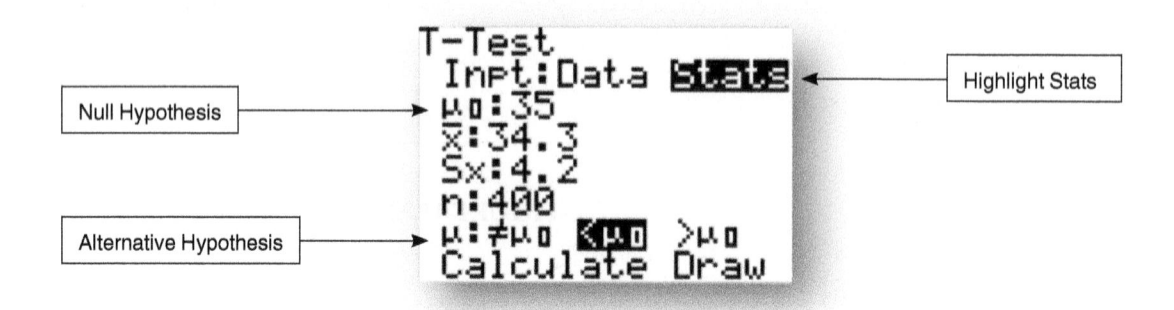

FIGURE 11.38

Highlight **Calculate** and press ENTER to obtain the output screen for this test. This is illustrated in Figure 11.39.

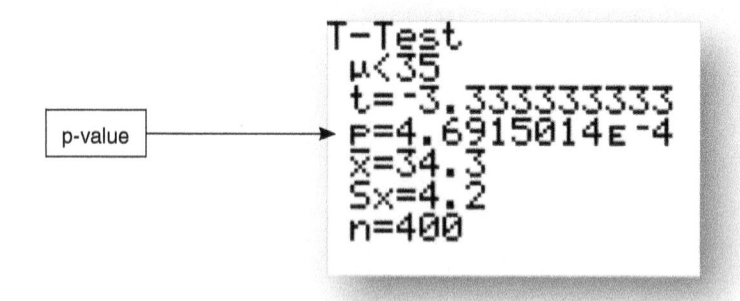

[p-value]

FIGURE 11.39

From the output screen, the p-value of the test statistic $t \approx -3.33$ is approximately 4.692 E-4. The p-value of 4.692 E-4 can also be written as 0.0004692. Using this p-value we can now position the test statistic on the hypothesis testing model shown in Figure 11.40.

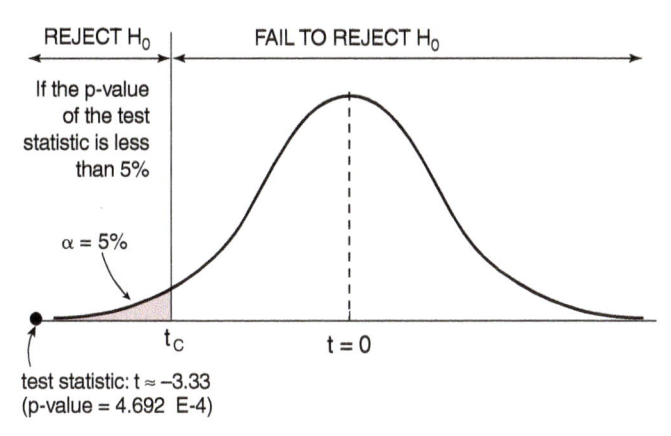

FIGURE 11.40

Step 5 State the conclusion.
Since the p-value (4.692 E-4) of the test statistic is less than $\alpha = 5\%$, we reject H_0 and accept H_a at a p-value of 4.692 E-4. Thus, there is a significant difference between the sample mean time of 34.3 minutes and the expected mean time of 35 minutes. Therefore, the IRS should consider investing into the new data transferring system.

Now we will summarize this procedure.

Summary of the p-Value Approach to Hypothesis Testing Involving a Population Mean Using the TI-84 Plus Calculator

Step 1 Formulate the two hypotheses, H_0 and H_a.
Null Hypothesis, H_0: $\mu = \mu_o$
Alternative Hypothesis, H_a: has one of the following three forms:
H_a: $\mu < \mu_o$ or H_a: $\mu > \mu_o$ or H_a: $\mu \neq \mu_o$

Step 2 Determine the model to test the null hypothesis, H_0.
Under the assumption H_0 is true, the expected results are:
a) the distribution of the test statistic is either:
  (i) approximately a normal distribution, if the standard deviation of the population is known and the sample size is greater than 30,
  or,
  (ii) a t Distribution with degrees of freedom: $df = n - 1$ if the standard deviation of the population is unknown.
b) the mean of the sampling distribution of the mean is given by the formula:

$$\mu_{\bar{x}} = \mu_0$$

(since H_o is true: i.e. $\mu = \mu_o$)

c) for a normal distribution: the standard error of the mean, denoted by $\sigma_{\bar{x}}$, is:

$$\sigma_{\bar{x}} = \frac{\sigma}{\sqrt{n}}$$

or
for a t distribution: the estimate of the standard error of the mean, written $s_{\bar{x}}$, is given by the formula:

$$s_{\bar{x}} = \frac{s}{\sqrt{n}}$$

(continues)

618 Chapter 11 Hypothesis Testing Involving One Population

> Step 3 Formulate the decision rule.
>   a) alternative hypothesis: directional or non-directional.
>   b) type of test: 1TT or 2TT.
>   c) significance level: $\alpha = 1\%$ or $\alpha = 5\%$.
>   d) construct the hypothesis test model.
>   e) state the decision rule as: *Reject H_o if the p-value of the test statistic is less than or equal to α.*
>
> Step 4 Analyze the sample data.
>   Enter the necessary sample data into the TI-84 calculator to determine the p-value of the test statistic. Use Z-Test if σ is given or T-Test if s is given.
>
> Step 5 Determine the conclusion.
>   Compare the p-value of the test statistic to the decision rule and either:
>   a) Reject H_o and accept H_a at α,
>   or
>   b) Fail to reject H_o at α.

KEY TERMS

critical t score, t_c 602
degrees of freedom, df 599
distribution of the test statistic 602
Estimate of the Population Standard Deviation, s 599
Estimate of the Standard Error of the Mean (Estimate of the Standard Deviation of the Sampling Distribution of the Mean), $s_{\bar{x}}$ 602
Expected Results 585
Hypothesis Testing Model 586
Hypothesis Testing Procedure Involving a Population Mean 594, 602, 610, 611

Hypothesis Testing Procedure Involving a Population Proportion 585
Mean of Samples Distribution of the Proportion, $\mu_{\hat{p}}$ 585
Mean of The Sampling Distribution of The Proportion, $\mu_{\hat{p}}$ 585
Mean of the Sampling Distribution of the Mean, $\mu_{\bar{x}}$ 595
Population Proportion, p 585
Sample Mean, $\bar{x}$ 597
Sample Proportion, $\hat{p}$ 586
Sampling Distribution of The Proportion 586

Standard Error of the Mean (Standard Deviation of the Sampling Distribution of the Mean), $\sigma_{\bar{x}}$ 595
Standard Error of the Proportion, $\sigma_{\hat{p}}$ 586
Standard Error of the Proportion or The Standard Deviation of the Sampling Distribution of The Proportion, $\sigma_{\hat{p}}$ 589
t Distribution 599
t score 600
Test Statistic 586
test statistic t 602
test statistic z 586

SECTION EXERCISES

Section 11.2

(Round off ALL answers to four decimal places.)

1. To perform a hypothesis test involving a population proportion, the sampling distribution of the _____ is used as the hypothesis testing model.

2. The null hypothesis of a hypothesis test involving a population proportion is stated as: The population proportion, symbolized _____, is _____ to the proportion p_o. In symbols, H_o: $p = p_o$.

3. The alternative hypothesis of a hypothesis test involving a population proportion has one of the following three forms:
   a) the population proportion, symbolized _____, is greater than the proportion p_o. In symbols, H_a: p ___ p_o.
   b) The _____ proportion, p, is less than the proportion p_o. In symbols, H_a: p ___ p_o.
   c) The population _____, p, is not equal to the proportion p_o. In symbols, H_a: p ___ p_o.

4. The formula for $\mu_{\hat{p}}$ is given by:
   a) $\mu_{\hat{p}} = np$
   b) $\mu_{\hat{p}} = p$
   c) $\mu_{\hat{p}} = p/n$
   d) $\mu_{\hat{p}} = n(1-p)$

5. If the alternative hypothesis is H_a: $p > p_o$, then the researcher is trying to show:
   a) p is less than p_o
   b) p equals p_o
   c) p is not equal to p_o
   d) p is greater than p_o

6. The formula for the test statistic, z, for a hypothesis test involving a population proportion is: z = _____ .

7. In a hypothesis test the value used for p in calculating $\mu_{\hat{p}}$ and $\sigma_{\hat{p}}$ is obtained from:
   a) The statistical result
   b) The alternative hypothesis
   c) The null hypothesis
   d) The standard error

8. To determine the conclusion of a hypothesis test involving a population proportion, the test statistic is computed and compared to the _____ value(s).

For problems 9 to 18:

  a) State the Null and Alternative Hypotheses.
  b) Calculate the expected results for the hypothesis test assuming the null hypothesis is true and determine the hypothesis test model.
  c) Formulate the decision rule.
  d) Determine the test statistic.
  e) Determine the conclusion and answer the question(s) posed in the problem, if any.
  f) Round off all calculations to four decimal places.

9. A car manufacturer claims that less than 10% of their cars have major defective parts. A skeptical consumer believes the claim is too low and randomly surveys 100 owners of the manufacturer's cars and finds 16 cars had major defective parts. Would this indicate that the manufacturer's claim was incorrect at $\alpha = 5\%$?

10. A gasoline lawn mower manufacturer claims that their power lawn mowers start up on the first try 95% of the time. A consumer's protection group feels the manufacturer claim is too high and randomly selects 200 of the manufacturer's mowers and finds that 178 start on the first try. Would this indicate that the manufacturer was indeed exaggerating at $\alpha = 1\%$?

11. A survey conducted by an Institute for Social Research claims that 40% of the American people say life in the United States is getting worse. Would one disagree with this claim at $\alpha = 5\%$ if an independent poll of 400 Americans shows that 175 believe life in the U.S. is getting worse?

12. Some people are concerned that tougher standards and high-stakes examinations may increase the high school dropout rate. The National Center for Education Statistics reported that the high school dropout rate for the year 2013 was 7.9%. One school district on Long Island, whose dropout rate has always been very close to the national average, reports that 157 of their 1782 students dropped out last year. Is their experience evidence that the dropout rate is increasing? Test at 5% significance.

13. According to a study released by a New York newspaper 25% of Long Island shoppers are being charged sales tax on non-taxable items in supermarkets because of the complexity of the state's tax law. Mr. Jessel, a consumer advocate, decides to test this claim against his suspicion that the claim is too low. In a random survey of supermarket shoppers, Mr. Jessel finds that 65 out of 200 shoppers were improperly charged tax on their purchases. Does this sample result support Mr. Jessel's claim at $\alpha = 1\%$?

14. Within the latest statistics on Federal Financial Aid, the U.S. Education Department reported that 57% of college undergraduates are receiving Financial Aid. A researcher in the Office of Institutional Research at a Northeastern Community College decides to test this claim against the suspicion that this percentage is too low for the students who attend her community college. To test this population proportion, the researcher randomly surveys the records of 200 students attending the community college. The researcher determines that 131 students are receiving Financial Aid. Based upon this sample proportion, is the researcher's claim reasonable at $\alpha = 5\%$?

15. A census bureau study of divorce rates reports that 68% of the men who wed before age 22 were divorced within 20 years. The Institute of Family Relations decided to test the validity of this report by randomly selecting and reviewing 250 marriages where the men wed before age 22. If 186 of these marriages ended in divorce within 20 years, can the Institute of Family Relations conclude the census bureau's claim is incorrect at $\alpha = 5\%$?

16. A new 3D printing company will only continue production of 3D printers if they can get a test market acceptance rate of at least 81%. If the test market survey of their latest 3D printer model received 360 favorable responses out of a random sample of 450 interviews, should the company continue producing their new printer? Test at $\alpha = 1\%$.

17. According to the latest National Crime Poll 75% of all women fear walking alone at night in their neighborhoods. A sociologist feels that this proportion is too high for women living in a middle class suburb on Long Island. To test her claim, the sociologist randomly samples 250 women living in this Long Island suburb and determines that 162 fear walking alone at night in their neighborhoods. Does this sample data support the sociologist's claim at $\alpha = 5\%$?

18. A national survey on "The Evolving Role of the Secretary in the Information Age" reported that 60% of the secretaries experienced eye strain when using word processors with a CRT display screen. An executive secretary decided to test this claim. She randomly surveys fifty secretaries who work with a CRT and determines that 35 experience eye strain. Based upon this sample data can the executive secretary reject the national claim at $\alpha = 1\%$?

Section 11.3

For questions 19–20, complete the following statements with the word: directional or nondirectional.

19. If the null hypothesis has the form: $H_o: \mu = \mu_o$, and the alternative hypothesis has the form: $H_a: \mu < \mu_o$, then this is a _____ hypothesis test.

20. If the null hypothesis has the form: $H_o: \mu = \mu_o$, and the alternative hypothesis has the form: $H_a: \mu \neq \mu_o$, then this is a _____ hypothesis test.

21. The expected results of a hypothesis test are calculated under the assumption that the null hypothesis is _____.

22. Given the null hypothesis is $H_o: \mu = \mu_o$, the standard deviation of the population is σ, and the size of the random sample selected from the population is greater than 30, then:
  a) the sampling distribution of the mean is _____.
  b) the mean of the sampling distribution of the mean is given by the formula: $\mu_{\bar{x}} =$ _____, and
  c) the standard error of the mean is given by the formula: $\sigma_{\bar{x}} =$ _____.

23. The alternative hypothesis of a hypothesis test involving a population mean can have one of the following three forms. They are:
  a) the population mean, μ, is *greater than* the numerical value, _____. This is symbolized as: $H_a: \mu$ _____ μ_o.

b) the population ____, μ, is *less than* the numerical value ____. This is symbolized as: $H_a: \mu \_\_ \mu_o$.
c) the ____ mean, μ, is *not equal* to the numerical value, μ_o. This is symbolized as: $H_a: \mu \_\_ \mu_o$.

24. The formula for the test statistic for a hypothesis test involving a population mean is:
   a) when the distribution of the test statistic is approximately normal, and the standard deviation of the population, σ, is known:
   $$z = \_\_\_\_.$$
   b) when the distribution of the test statistic is a *t* distribution:
   $$t = \_\_\_\_.$$

25. If the population is normal with $\sigma = 8$, and samples of size 16 are selected, then the standard error of the mean is:
   a) 8   b) 2   c) 0.5   d) 1

26. If the alternative hypothesis, H_a, is $\mu < 27$, then the hypothesis test uses a:
   a) 1TT on right
   b) 2TT
   c) 1TT on left
   d) 2TT on left

27. Which of the following could be an alternative hypothesis?
   a) $\mu \neq 17$ b) $\mu = 26$ c) $\mu = 4.3$ d) $\mu = 0.10$

28. The sample mean is used as the mean for the sampling distribution of the mean. (T/F)

29. When doing a hypothesis test about a population mean, μ, and the population standard deviation, σ, is known, then the sampling distribution of the mean is always normal, regardless of the size, n, of the random sample. (T/F)

For problems 30 through 37:
For each of the hypothesis testing problems, find the following:

   a) State the null and alternative hypotheses.
   b) Calculate the expected results for the hypothesis test assuming the null hypothesis is true and determine the hypothesis test model.
   c) Formulate the decision rule.
   d) Determine the statistical result and test statistic.
   e) Determine the conclusion and answer the question(s) posed in the problem, if any.

30. The mean systolic blood pressure for twenty-five year old females is given to be 124 with a population standard deviation of 12. A sociologist believes that twenty-five year old females who live in the southwestern section of the United States will have significantly lower blood pressures. To test her claim, the sociologist randomly selects 36 of these females and determines their mean systolic blood pressure to be 120. Do these sample results support the sociologist's claim at $\alpha = 5\%$?

31. A sports equipment company claims to have developed a new manufacturing process which will increase the mean breaking strength of their new tennis racquet. The present manufacturing process produces a tennis racquet with a mean breaking strength of 90 lbs. with a population standard deviation of 6 lbs. Larry K., a wealthy tennis perfectionist who is known for his intense play that usually results in a broken racquet, employs a product testing company to test the manufacturer's claim. The testing company randomly purchases sixty-four of these new racquets and tests them. If the testing company determines the mean breaking strength to be 93 lbs., can Larry K. agree with the manufacturer's claim? (use $\alpha = 1\%$)

32. An unemployment agency claims that the average age of recipients of unemployment benefits are 37 years old with a population standard deviation of 5 years. A trade union association wants to test this claim. They randomly interview 200 recipients of unemployment benefits and obtain an average age of 36 years. Using a 1% significance level, can the trade union reject the unemployment agency's claim?

33. The LIRR carries a mean of 125,000 passengers daily with a population standard deviation of 4000. The railroad has averaged 126,000 passengers daily for the last 36 snowstorms. Does this represent a significant increase over the mean passenger load at $\alpha = 5\%$?

34. According to the U.S. Bureau of the Census, the average age of brides marrying for the first time is 23.9 years with a population standard deviation of 4.2 years. A sociologist believes that young women are delaying marriage and marrying at a later age. The sociologist randomly samples 100 marriage records and determines the average age of the first time brides is 24.9. Is there sufficient evidence to support the sociologist's claim at $\alpha = 5\%$?

35. A Florida citrus grower claims to have devised a new method for harvesting oranges which is faster and just as efficient as the method currently used. The current method takes a mean of 5 hours to harvest 1 acre of orange trees with a population standard deviation of 45 minutes. One hundred acres harvested by the new method averaged 4 hours and 45 minutes per acre. The average yield for the new method was equal to that of the current method. Do the sample results support the citrus grower's claim? (Use a significance level of 1%).

36. A traffic engineer believes that the time needed to drive from the downtown area out to the suburbs has significantly decreased due to increased use of public transportation. A study done five years ago determined that it took a mean of 45 minutes for this trip with a population standard deviation of 7 minutes. Recently, the mean time for a sample of 49 trips was 42 minutes. Does this data support the traffic engineer? Use a significance level of 5%.

37. The Fort Knox Federal Savings Bank claims that their regular passbook savings accounts have an average balance of $1236.45 with a population standard deviation of $317.68. A group of federal bank auditors believes this claim is too high. To test the bank's claim the auditors randomly selected 400 accounts and obtained a mean balance of $1217.69. Do these sample results indicate that the bank's claim was too high? (use $\alpha = 5\%$)

38. Why state the conclusion as "Fail to Reject H_o" rather than Accept H_o? In the hypothesis testing procedure, when the null hypothesis cannot be rejected we state the conclusion as "fail to reject H_o" rather than "accept H_o". This problem examines the reason why we prefer to state the conclusion as "fail to reject H_o".

Two researchers, Maria and Larry, (who often don't agree on what the null hypothesis should be) decide to perform a hypothesis test on the same population of students to determine if the population mean IQ score has increased at the 5% level of significance.

Maria states her null hypothesis as: "The population mean is 120." Larry states his null hypothesis as: "The population mean is 122." Both researchers agree that the population standard deviation is 12. Their hypothesis tests are shown side by side in the following table.

Maria's Test

1. Formulate the hypotheses:
   H_o: population mean is 120
   $\mu = 120$
   H_a: population mean is greater than 120
   $\mu > 120$

Larry's Test

1. Formulate the hypotheses:
   H_o: population mean is 122
   $\mu = 122$
   H_a: population mean is greater than 122
   $\mu > 122$

2. Determine the model to test the null hypothesis, H_o.

To test their null hypotheses, Maria and Larry select a random sample of 36 students. The expected results for their respective tests are:

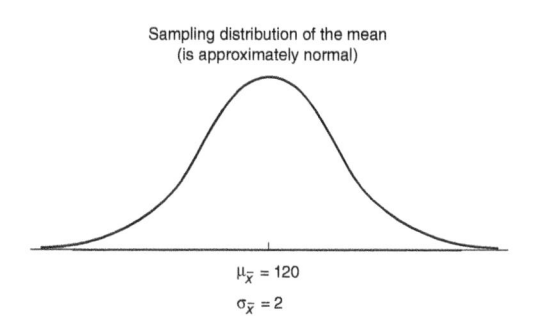

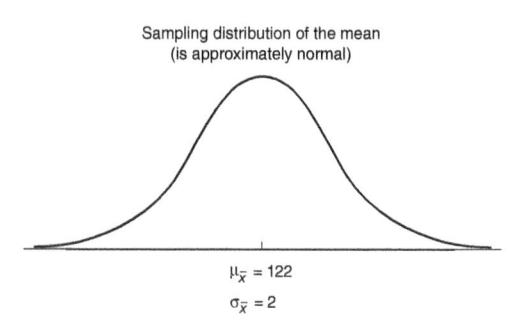

Sampling distribution of the mean (is approximately normal)
$\mu_{\bar{x}} = 120$
$\sigma_{\bar{x}} = 2$

Sampling distribution of the mean (is approximately normal)
$\mu_{\bar{x}} = 122$
$\sigma_{\bar{x}} = 2$

3. Formulate the decision rule.

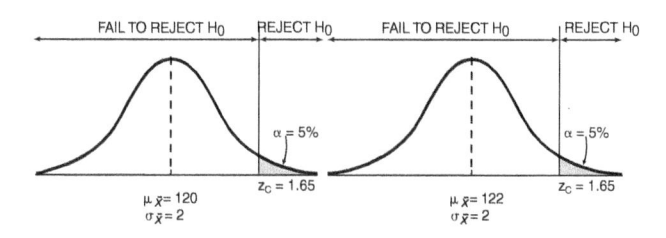

FAIL TO REJECT H_0 | REJECT H_0
$\alpha = 5\%$
$z_c = 1.65$
$\mu_{\bar{x}} = 120$
$\sigma_{\bar{x}} = 2$

FAIL TO REJECT H_0 | REJECT H_0
$\alpha = 5\%$
$z_c = 1.65$
$\mu_{\bar{x}} = 122$
$\sigma_{\bar{x}} = 2$

Maria's decision rule:
Reject H_o if the test statistic is greater than 1.65.

Larry's decision rule:
Reject H_o if the test statistic is greater than 1.65.

4. Analyze the sample data.

Maria and Larry randomly selected 36 students and found the sample mean IQ (the statistical result) to be $\bar{x} = 123$.

The test statistic is $z = 1.5$ for Maria's test and $z = 0.5$ for Larry's test.

Using these test statistics, state:
a) The conclusion to Maria's hypothesis test.
b) The conclusion to Larry's hypothesis test.
c) Was the test statistic result statistically significant for either Maria's or Larry's hypothesis test?
d) If Maria had decided to state her conclusion as **accept H_o** rather than **fail to reject H_o**, what would she be accepting as the population mean IQ score?
e) If Larry had decided to state his conclusion as **accept H_o** rather than **fail to reject H_o**, what would he be accepting as the population mean IQ score?
f) Using the ideas of this problem, explain what is wrong with the stating the conclusion as: **accept H_o**.

Section 11.4

39. The properties of a **t** distribution are:
    a) The graph of a **t** distribution is _____.
    b) The **t** distribution is symmetric about the _____.
    c) The percent of area to the left of a **t** score is dependent upon the degrees of _____.

40. If samples of size 36 are selected, then the degrees of freedom for the **t** distribution is:
    a) 30   b) 37   c) 6   d) 35

41. The formula used to determine the *df* with one sample is:
    a) $df = n$
    b) $df = n + 1$
    c) $df = n - 1$
    d) $df = s - 1$

42. The **t** distribution is used as a model for the normal distribution whenever the sample size is less than 30. (T/F)

43. Degrees of freedom is a concept used to determine critical **t** scores. (T/F)

44. Use Table III to determine the critical **t** score(s) t_c, for a hypothesis test where:
    a) $\alpha = 5\%$, one-tail test on the left side, and $df = 25$.
    b) $\alpha = 1\%$, one-tail test on left side, and $df = 11$.
    c) $\alpha = 5\%$, one-tail test on right side, and $df = 22$.
    d) $\alpha = 1\%$, 1TT on right side, and $df = 17$.
    e) $\alpha = 5\%$, 2TT, and $df = 43$.
    f) $\alpha = 1\%$, 2TT, and $df = 56$.

45. Use the information in each of the Figures a through f to determine the critical **t** score(s), t_c, from Table III.

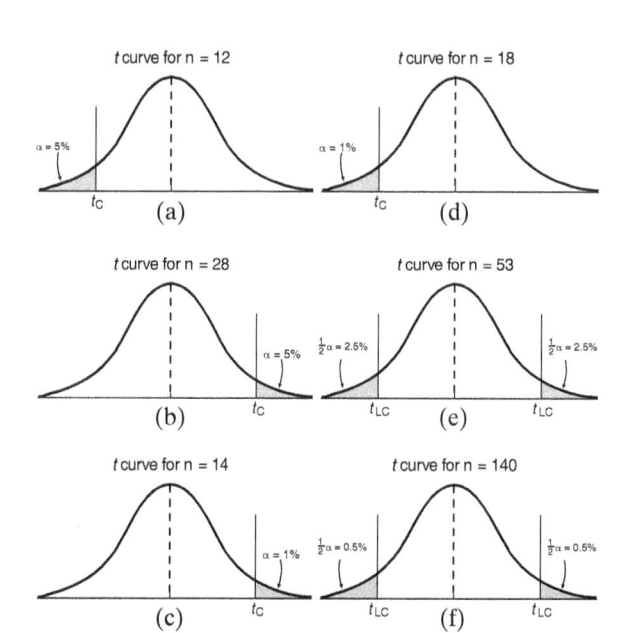

t curve for n = 12, $\alpha = 5\%$, t_c (a)

t curve for n = 18, $\alpha = 1\%$, t_c (d)

t curve for n = 28, $\alpha = 5\%$, t_c (b)

t curve for n = 53, $\frac{1}{2}\alpha = 2.5\%$, t_{LC} t_{LC} (e)

t curve for n = 14, $\alpha = 1\%$, t_c (c)

t curve for n = 140, $\frac{1}{2}\alpha = 0.5\%$, t_{LC} t_{LC} (f)

Section 11.5

For problems 46 through 55:

If the population standard deviation is unknown, assume that the population is approximately normal.

For each of the hypothesis testing problems, find the following:

a) State the null and alternative hypotheses.
b) Calculate the expected results for the hypothesis test assuming the null hypothesis is true and determine the hypothesis test model.
c) Formulate the decision rule.
d) Determine the statistical result and test statistic.
e) Determine the conclusion and answer the question(s) posed in the problem, if any.

46. A recent National Survey of Student Engagement (NSSE) study reported that first year full-time students at four year U.S. colleges spent an average of 14 hours per week preparing for class. An educational researcher at a small private four year college decides to test this claim against the suspicion that the average is significantly higher for the students attending his institution. To test this population mean, the researcher randomly surveys 25 students and determines that the students within the sample spent average of 14.8 hours per week preparing for class with a sample standard deviation of 1.95 hours. Based upon this sample information, is the researcher's claim reasonable at $\alpha = 1\%$?

47. A certain type of oak tree has a mean growth of 14.3 inches in 4 years. A forestry biologist believes that a new variety will have a greater mean growth during this same length of time. A sample of 36 trees of this new variety are studied. If it is found that the mean growth is 15.4 inches with a sample standard deviation of 1.8 inches, can one conclude the biologist is correct? (use $\alpha = 5\%$)

48. Facebook claims that its users have an average of 49 photos of themselves posted on their profile. A social media researcher thinks this number is too low. He surveys 50 Facebook users to test this claim, and finds that the mean number of photos is 59 with a standard deviation of 11 photos. Do these findings support his claim? (use $\alpha = 5\%$)

49. Facebook.com claims that users of the site spend a mean of 55 minutes per day on the site. A communication professor believes this claim is too low. To test the claim she surveys 49 users of the site and finds the mean time spent on the site per day is 58 minutes with a standard deviation of 10 minutes. Do these findings support the professor's belief? (Use $\alpha = 5\%$)

50. An ecologist claims that because of environmental influences, the mean life of Louisiana shellfish is 27 months. Feeling the claim is too low, a congressman decides to test the ecologist's claim at $\alpha = 5\%$. If the mean life of a random sample of 16 shellfish is 28.05 months with a standard deviation of 2.8, do the sample results indicate that the ecologist's claim is too low?

51. According to the McKinley Health Center at the University of Illinois in Champlain, a college student gets an average of 6 hours of sleep a night. A recent random survey of 49 students at a local college produced a mean of 5.5 hours of sleep per night with $s = 1.4$ hours. Can one conclude that these sample results indicate that the mean number of hours of sleep has decreased at a 1% level of significance?

52. A pharmaceutical company claims that their muscle relaxing tablets have a mean effective period of 4.4 hours. Researchers representing the Food and Drug Administration suspect the company's claim is too high. The researchers administer the muscle relaxing tablets to a random sample of 121 patients and find that the mean effective period is 4.2 hours with $s = 1.1$ hours. Do the sample results support the researcher's suspicions? (Use $\alpha = 1\%$)

53. The commerce department in a California County claims that the mean rental for a 3 room apartment is $850. A real estate company believes the commerce department's claim is not correct and randomly surveys 144 apartment dwellers and finds that their average rental is $872, with a standard deviation of $90. Do the sample data indicate that the commerce department's claim is *not* correct at $\alpha = 1\%$?

54. A pharmaceutical company manufactures a blood pressure drug which they claim will lower blood pressure an average of 20 points over a three week period. Dr. Zorba administers this drug to 36 patients for three weeks and records the following drops in their blood pressure: 22, 18, 19, 17, 24, 22, 15, 17, 21, 18, 15, 19, 24, 20, 14, 16, 19, 23, 24, 13, 18, 22, 21, 17, 24, 19, 15, 17, 21, 19, 22, 17, 19, 21, 17, 20. Do these sample data indicate that the pharmaceutical company's claim is incorrect at $\alpha = 5\%$?

55. A prominent psychologist at a New York hospital claims that a new form of psychotherapy for a certain type of mental disorder will help a patient to be ready for outpatient treatment in an average of twelve weeks. A resident psychologist decides to test this claim under the suspicion that the claim is too low. The resident psychologist uses this new form of psychotherapy on a random sample of 40 patients. The time required for these patients to reach outpatient status were (in days): 91, 88, 86, 95, 90, 93, 87, 89, 83, 91, 93, 86, 92, 91, 89, 83, 92, 88, 87, 90, 93, 87, 94, 86, 84, 92, 87, 85, 90, 88, 84, 92, 94, 86, 91, 89, 84, 89, 85, 95. Do these sample data support the claim made by the resident psychologist at $\alpha = 1\%$?

Section 11.6

For problems 56 to 78, use questions 9 to 18 of Section exercises 11.2, questions 30 to 32 of Section 11.3 and 46 to 55 of Section 11.5 exercises and apply the p-value approach to hypothesis testing using the TI-84 calculator. For each of these problems, use the 5 step procedure to perform a hypothesis test at **both** $\alpha = 1\%$ and $\alpha = 5\%$. Include the following within your test:

a) State the null and alternative hypotheses.
b) Calculate the expected results for the experiment assuming the null hypothesis is true and state the hypothesis test model.
c) Formulate the decision rule.
d) Identify the test statistic and its p-value.
e) State the conclusion and answer any question(s) posed in the problem for each alpha level.
f) Determine the smallest level of significance for which the null hypothesis can be rejected.

CHAPTER REVIEW

79. If the null hypothesis has the form: $H_o: \mu = \mu_o$, and the alternative hypothesis has the form: $H_a: \mu > \mu_o$, then this is a _____ hypothesis test.

80. To perform a hypothesis test involving a population mean, the distribution of the _____ is used as the hypothesis testing model.

81. The null hypothesis for a hypothesis test involving a population mean is stated as: H_o: the population mean, μ, is _____ to the numerical value, μ_o. In symbols this would be, $H_o: \mu$ ___ μ_o.

82. When performing a hypothesis test about a population mean where H_o is $\mu = 75$, the expected result $\mu_{\bar{x}} = 75$ is determined under the assumption that _____ is true.

83. When the mean of a sample, $\bar{x}$, is used to test the null hypothesis, $\mu = \mu_o$, one must use the sampling distribution of the mean and its properties.
(T/F)

84. To insure that the sampling distribution of the mean can be approximated by the normal distribution, the sample size must be less than 30.
(T/F)

85. When the standard deviation of the population is known then $\sigma_{\bar{x}}$ is equal to the standard deviation of the population.
(T/F)

86. For a sample of size n, the degrees of freedom is equal to the sample size minus _____. In symbols, $df =$ _____.

87. The statistic, s, is a good estimate for:
a) μ b) $\sigma_{\bar{x}}$ c) $\bar{x}$ d) σ

88. Whenever the population standard deviation, σ, is unknown, a good estimate of the population standard deviation is obtained from a random sample taken from the population.
(T/F)

89. The **t** distribution is generally less dispersed than the normal distribution.
(T/F)

90. a) Find the critical **t** score, t_c, for $\alpha = 5\%$, and a one-tailed test on left side for each value of n given in the following table.

n	t_c
5	
10	
50	
100	
500	

b) From Table II: The Normal Distribution Table, find the critical z score, z_c, for $\alpha = 5\%$ and a one-tailed test on the left side.

c) Compare the values of the critical **t** scores, t_c, to the critical z scores, z_c, as the sample size, n, increases.

For problems 91 through 98:
If the population standard deviation is unknown, assume that the population is approximately normal.

For each of the hypothesis testing problems, find the following:

a) State the null and alternative hypotheses.
b) Calculate the expected results for the hypothesis test assuming the null hypothesis is true and determine the hypothesis test model.
c) Formulate the decision rule.
d) Determine the statistical result and test statistic.
e) Determine the conclusion and answer the question(s) posed in the problem, if any.

91. The population norms (i.e. parameters μ, and σ) for a mathematics anxiety test are known to be $\mu = 6$ and $\sigma = 1.6$. A teacher wants to determine if the students at her school will yield significantly different test results from the population norm. She administers the test to a random sample of 100 students and determines the sample mean test score to be 5.73. Can the teacher conclude that these results are different from the population norm at $\alpha = 5\%$?

92. The personnel director of a Wall Street brokerage house believes that the executives working in the "Big Apple" tend to have higher IQ's than the national executive mean IQ. The national mean IQ score for executives is 112 with a population standard deviation of 18. If a random sample of 81 "Big Apple" executives yields a mean IQ score of 116, do these sample results support the personnel director's claim at $\alpha = 1\%$?

93. According to the Centers for Disease Control and Prevention (CDC), 18% of adults age 18 and over in the U.S. smoke cigarettes. A medical doctor believes that the CDC is incorrect and decides to test this claim. The doctor selected a random sample of 500 adults age 18 and over and found that 111 smoked cigarettes. Does this sample result support the doctor's belief at $\alpha = 1\%$?

94. Presently, the mean life expectancy of a rare strain of bacteria is 12 hours. A scientist claims she has developed a medium that will increase the mean life of the bacteria. The scientist tests 16 cultures of the newly treated bacteria and finds that they have a mean life of 13 hours with s = 1 hour. Do these results show that the medium is effective in increasing the bacteria's life expectancy? (Use $\alpha = 1\%$)

95. A psychologist believes that a child has a mean of one hundred words in their vocabulary by the age of two. A pediatrician suspects that the psychologist's claim is too high and decides to survey the parents of 25 two-year-olds and finds that for a sample 49 children the mean number of words in their vocabulary to be 95 with a sample standard deviation of 20 words. Do the sample results support the pediatrician's claim at $\alpha = 5\%$?

96. The Media Research Institute reports that iPhone users have an average of 19 apps on their phone. An Apple employee believes that the customers of his store have fewer apps than that on average. He randomly selects 30 customers from his store and determines the mean number of apps on an iPhone is 17 with a sample standard deviation of 4 apps. Do these sample results support the Apple employee's claim at $\alpha = 5\%$?

97. A National Research Group reports that children between the ages of 6 and 10 years watch television an average of

21 hours per week. A sociologist believes that the children within her upper-middle class neighborhood watch television less than an average of 21 hours per week. She randomly selects 25 children aged 6 to 10 years from her neighborhood and determines the mean number of hours that the children watch television is 20 hours with a sample standard deviation of 3 hours. Do these sample results support the sociologist's claim at $\alpha = 5\%$?

98. An educational psychologist claims that 80% of the recent high school graduates that enter college do so primarily because of parental pressure. Researchers from the California University system believe this claim is too high when applied to entering freshman in the California University system. The researchers randomly sample 1600 of their most recent entering freshman and find that 1248 entered college primarily because of parental pressure. Does this sample result significantly support the researcher's belief at $\alpha = 1\%$?

WHAT DO YOU THINK?

99. Examine the two graphs shown in the figure entitled "A Shifting Burden" which represents the average student loan debt is leveling off.
    a) What parameter is represented by both graphs? Define the variable being displayed in the graphs. Is the variable discrete or continuous?
    For each loan debt, design a hypothesis test that includes:
    b) A null hypothesis. Use the information contained in the graph for the 2017–2018 year as the status quo.
    c) An alternative hypothesis. Use the direction that the line graph is heading as to your belief regarding the true value of the population parameter for this year.
    d) Is the alternative hypothesis directional or nondirectional? Is this a 1TT or a 2TT? Explain.
    e) Which population parameter are you assuming you know? What is the value of this parameter?
    f) Which population parameter are you missing? How would you go about obtaining an estimate of this parameter?
    g) State the name of the sampling distribution that will serve as the model for this test. Is the distribution of the test statistic a normal or a t distribution, if we assume that the distribution of student-aids is a normal distribution? Explain.
    h) What is the mean of this sampling distribution? What formula would you use to estimate the standard error of the sampling distribution?
    i) Which tail of the curve would the critical z or t score value of the test be placed?
    j) If you decide to use a significance level of 5% for each hypothesis test, then determine the test statistic.
    k) If the test statistic falls to the right of the critical z or t score value for the student loan dept hypothesis test, what is the conclusion to this test? Could you agree that

Statistical Snapshot
A Shifting Burden
The average student's dept load is leveling off, a new analysis of federal loan data shows. That's not true for many parents.

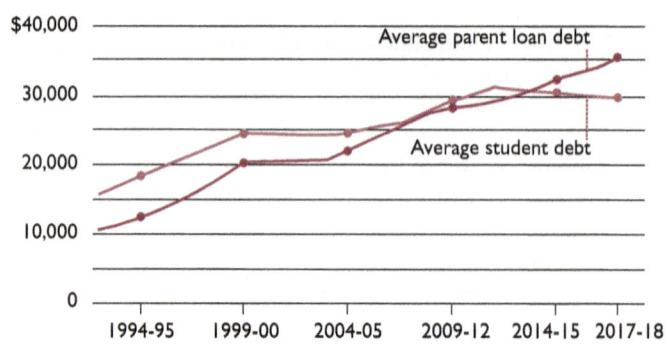

Source: Mark Kantrowitz (Saving for college.com)

the conclusion to this hypothesis test follows the trend shown in the graph? Explain.
l) If the p-value of the hypothesis test is 6% (or 0.06), then can you reject the null hypothesis at $\alpha = 5\%$? Does this indicate that the student loan debt average is following the trend shown in the graph? Explain.
m) If the p-value of the hypothesis is 2.90% (or 0.290) and $\alpha = 2\%$, could you conclude that the student loan debt average is following the trend shown in the graph? Explain.

PROJECTS

100. Hypothesis Testing Project
    A. Selection of a topic
       1. Select a reported fact or claim about a population mean made in a local newspaper, magazine or other source.
       2. Suggested topics:
          a) Test the claim that the average number of hours worked per week by a student is 20 hours.
          b) Test the claim that the cumulative grade point average is higher today then it was ten years ago. The cumulative grade point average ten years ago was 2.64.
          c) Test the claim that the average SAT score on the mathematics section is greater than 480.
          d) Test the claim that the average pulse rate at rest for a woman between the ages of 16 and 23 inclusive is lower than 80.
    B. Use the following procedure to test the claim selected in part A.
       1. State the claim, identify the population referred to in the statement of the claim and indicate where the claim was found.
       2. State your opinion regarding the claim (i.e., do you feel it is too high, too low, or simply don't agree with the claim) and clearly identify the population you plan to sample to support your opinion.

3. State the null and alternative hypotheses.
4. Develop and state a procedure for selecting a random sample from your population. Indicate the technique you are going to use to obtain your sample data.
5. Compute the expected results, indicate the hypothesis testing model, and choose an appropriate level of significance.
6. Determine the critical z or t score value(s) and state the decision rule.
7. Calculate the statistical result, $\bar{x}$, and the test statistic.
8. Construct a hypothesis testing model and place the test statistic on the model.
9. Formulate the appropriate conclusion and interpret the conclusion with respect to your opinion (as stated in step 2).

101. Read a research article from a professional journal in a field such as medicine, education, marketing, science, social science, etc. Write a summary of the research article that includes the following:
  a. Name of article, author and journal (include volume number and date).
  b. Statement of the research problem.
  c. The null and alternative hypotheses.
  d. How the experiment was conducted (give the sample size; and the sampling technique).
  e. Determine the test statistic and the level of significance.
  f. State the conclusion in terms of the null and alternative hypotheses.

CHAPTER TEST

102. Which of the following statements could be used for the null hypothesis, H_o?
  a) $\mu < \mu_o$ b) $\mu > \mu_o$ c) $\mu \neq \mu_o$ d) $\mu = \mu_o$

103. Which of the following could be the alternative hypothesis for a 1TT on the left?
  a) $\mu < 56$ b) $\mu > 173$ c) $\mu \neq 93$ d) $\mu = 63$

104. The hypothesis testing model for a hypothesis test involving a population mean where n = 25 and the population standard deviation, σ, is unknown, is a t distribution with _____ degrees of freedom.

105. Which of the following is *not* a property of the t distribution?
  a) Less dispersed than the normal distribution
  b) Symmetric with respect to the mean
  c) Bell-shaped
  d) Different t distribution for each sample size

106. When σ is unknown and the formula
$$s = \sqrt{\frac{\sum (x - \bar{x})^2}{n - 1}}$$
is used as an estimate for σ, then s is said to be a good estimator.
(T/F)

107. The critical t score for a 1TT on the right with $\alpha = 5\%$ and $df = 17$ is:
  a) 1.74   b) 2.11   c) 2.57   d) 2.90

For the hypothesis testing problem, find the following:
  a) State the null and alternative hypotheses.
  b) Calculate the expected results for the hypothesis test assuming the null hypothesis is true and determine the hypothesis test model.
  c) Formulate the decision rule.
  d) Determine the statistical result and test statistic.
  e) Determine the conclusion and answer the question(s) posed in the problem, if any.

108. An automobile manufacturer claims that their leading compact car averages 36 mpg in the city. The population standard deviation is 4 mpg for city driving. Suppose a city police department purchases 64 cars from this auto manufacturer. If these cars were driven exclusively under city conditions and averaged 33 mpg, can one argue that the manufacturer's claim is too high at $\alpha = 5\%$?

109. Facebook.com claims that college users have a mean of 300 friends on the site. Professor K believes this claim is too low. She randomly samples 16 students in her college and finds that they have a mean of 320 friends on the site with a standard deviation of 144. Do these results support the professor's claim? (Use $\alpha = 5\%$)

110. A new chemical process has been developed for producing gasoline. The company claims that this new process will increase the octane rating of the gasoline. Sixteen samples of the gasoline produced with the new process are selected at random and their octane ratings were: 94, 93, 97, 92, 96, 94, 95, 91, 98, 95, 92, 91, 95, 96, 97, 93. If the mean octane using the existing process is 93, is the company's claim correct? (Use $\alpha = 5\%$)

111. Based on a recent survey, the New York Off-Track Betting Corporation (O.T.B.) stated that 57% of its single-male bettors earned over $38,000 per year. If 300 randomly selected single male O.T.B. bettors were interviewed and 178 earned over $38,000 per year, would one disagree with the O.T.B.'s claim at $\alpha = 5\%$?
  a) state the Null and Alternative Hypotheses.
  b) calculate the expected results for the hypothesis test assuming the null hypothesis is true and determine the hypothesis test model.
  c) formulate the decision rule.
  d) determine the test statistic.
  e) determine the conclusion and answer the question(s) posed in the problem, if any.
  f) round off all calculations to four decimal places.

112. In 2019 a study across various sites near the Arctic circle found that the percent of the population of Polar bears that are female is 54%. A statistician feels that this number has declined in recent years and decides to test this claim. She randomly collects a sample of 50 bears throughout the various sites and finds that 24 of them are female. Does the data back up the statisticians claim? Test at the .01 alpha level.

For the answer to this question, please visit www.MyStatLab.com.

CHAPTER 12

Hypothesis Testing Involving Two Population Proportions Using Independent Samples

Contents

12.1 Introduction to Hypothesis Tests Involving a Difference Between Two Population Proportions Using Independent Samples
12.2 The Sampling Distribution of the Difference Between Two Proportions
12.3 Hypothesis Testing Involving Two Population Proportions Using Large Samples
    Hypothesis Testing Procedure Involving the Difference Between the Proportions of Two Populations for Large Samples
12.4 Hypothesis Testing Involving Two Population Proportions Comparing Treatment and Control Groups
12.5 p-value Approach to Hypothesis Testing Involving Two Population Proportions Using the TI-84 Calculator
12.6 Two Population Hypothesis Testing Summaries Using Independent Samples

12.1 INTRODUCTION TO HYPOTHESIS TESTS INVOLVING A DIFFERENCE BETWEEN TWO POPULATION PROPORTIONS USING INDEPENDENT SAMPLES

In this chapter, we will present a hypothesis test involving the comparison of two population proportions to determine if the populations are statistically different. For example, a medical researcher might be interested in determining whether there is a significant difference in the proportion of individuals catching the common cold for the individuals who take large doses of vitamin C and those individuals who do not take vitamin C. Such a study would involve the comparison of two population proportions. We will now consider the procedure to conduct a hypothesis test that can be used to compare two population proportions where an independent random sample is selected from each population.

In order to conduct a hypothesis test involving two populations, we will randomly select a sample from each population. Before we can discuss the procedure to perform such a hypothesis test it is necessary to consider how each sample is selected from each population. The two samples selected can be either **independent** or **dependent** depending on the selection process.

> **DEFINITION 12.1**
> **Independent Samples.** Two samples are independent if the sample data values selected from one population are unrelated to the sample data values selected from the second population.

In this chapter we will only consider hypothesis tests where the samples selected are **independent samples**. In Chapter 11, we examined hypothesis tests involving a single population proportion. We will now discuss hypothesis tests that involve the comparison of the proportions from two populations to determine if the population proportions are statistically different. For example:

- *Is there a statistically significant difference in the proportion of individuals catching the common cold among individuals who take large doses of vitamin C and those individuals who do not take vitamin C?*
  This test involves the comparison of two population proportions: the proportion of individuals taking vitamin C who caught a common cold and the proportion of individuals not taking vitamin C who caught a common cold.

- *Is the difference between the opinion of men and women regarding their attitude towards purchasing non-designer jeans statistically significant?*

    This test involves the comparison of two population proportions: the proportion of women who will purchase non-designer jeans and the proportion of men who will purchase non-designer jeans.

- *Is there a statistically significant difference in the proportion of defective memory chips produced using a new innovative manufacturing process and the traditional manufacturing process?*

    This test involves the comparison of two population proportions: the proportion of defective memory chips produced using the new innovative manufacturing process and the proportion of defective memory chips produced using the traditional manufacturing process.

- *Is the difference between the recovery rate of patients with a certain illness who are treated with a new drug (the treatment group) and the recovery rate of patients with this same illness who are given a placebo (the control group) statistically significant?*

    This test involves the comparison of two population proportions: the proportion of patients who recover within the treatment group using the new drug treatment and the proportion of patients who recover within the control group who take the *placebo*.

Statistical questions such as these involve the study of two populations, instead of just one population, and the comparison of the proportion from each of these two populations. The appropriate sampling distribution to perform a hypothesis test involving two population proportions is called the *Sampling Distribution of the Difference Between Two Proportions*. In Section 12.2, we will develop the concept underlying this sampling distribution.

12.2 THE SAMPLING DISTRIBUTION OF THE DIFFERENCE BETWEEN TWO PROPORTIONS

Studies involving population proportions will produce **counts** rather than measurements. For example, the question:

Is there a statistically significant difference in the proportion of individuals catching the common cold among individuals who take high doses of vitamin C and those individuals who do not take vitamin C?

involves the comparison of the population of individuals taking vitamin C and the population of individuals not taking vitamin C. To determine the effectiveness of vitamin C, the researcher will **count** the number of vitamin C users who catch a cold and **count** the number of people not taking vitamin C who catch a cold. These **counts** will be used to calculate the **proportion** of individuals who caught the common cold within each group.

As mentioned in Chapter 11, the symbol **p** is used to represent the population proportion of items or individuals possessing a particular characteristic. In this chapter, however, we will need to attach subscripts 1 and 2 to the population proportion symbol p, because we need to distinguish between the proportions of two populations. For example, if we define *population 1 as the population of individuals taking vitamin C*, and *population 2 as the population of individuals not taking vitamin C*, then:

p_1 (read p sub one) = *the population proportion of individuals taking vitamin C who caught a cold.*

p_2 (read p sub two) = *the population proportion of individuals not taking vitamin C who caught a cold.*

In practice, the size of each population is either too large or infinite so that it will be impractical or impossible to determine the true proportion, p, of each population. Consequently, *an independent random sample is selected from each population and the proportion of each sample is computed*. Recall, from Chapter 11, the sample proportion, symbolized $\hat{p}$ and read *p hat*, is the ratio of the number of items or individuals in the sample that possess a particular characteristic to the sample size. That is,

$$\hat{p} = \frac{\text{number of items or individuals possessing a particular characteristic}}{\text{sample size}}$$

Thus, in the example on the effectiveness of vitamin C on the incidence of catching a cold, an independent random sample of vitamin C users is selected and the *number of individuals who caught a cold is counted*. If

Section 12.2 The Sampling Distribution of the Difference Between Two Proportions

we let n_1 represent the size of the random sample selected while x_1 equals the number of individuals within the sample who caught a cold using vitamin C, then the sample proportion of individuals who caught a cold using vitamin C, symbolized $\hat{p}_1$ and read *p hat sub 1*, is determined using:

$$\hat{p}_1 = \frac{\text{number of individuals who caught a cold using vitamin C}}{\text{sample size}}$$

In symbols, we write:

$$\hat{p}_1 = \frac{x_1}{n_1}$$

To determine the effectiveness of vitamin C, we need to compare this sample result, $\hat{p}_1$, to a group of individuals not taking vitamin C. Therefore, another independent random sample of individuals who do not use vitamin C is selected and the number of individuals who caught a cold would be counted. Letting n_2 represent the sample size and x_2 equal the number of individuals within the sample who caught a cold not using vitamin C, then the sample proportion of individuals who caught a cold not using vitamin C, symbolized $\hat{p}_2$ and read *p hat sub 2*, is determined using:

$$\hat{p}_2 = \frac{\text{number of individuals who caught a cold not using vitamin C}}{\text{sample size}}$$

In symbols, we have:

$$\hat{p}_2 = \frac{x_2}{n_2}$$

Suppose a medical researcher decides to conduct such a study and she randomly selects 500 people for the experiment. From these 500 people, she randomly selects 250 who are given high doses of vitamin C while the other 250 people are given a placebo (that is, a fake ineffective tablet similar in appearance to the vitamin C tablet.) At the conclusion of the experiment, the researcher determines the number of individuals within each group who caught a cold. Examine the results of the study shown in Table 12.1.

TABLE 12.1

	Vitamin C Group	Placebo Group
Number of individuals who caught a cold	$x_1 = 104$	$x_2 = 158$
Sample Size	$n_1 = 250$	$n_2 = 250$

Using the results shown in Table 12.1, the researcher can determine the sample proportion of individuals within each group who caught a cold. These sample proportions are:

$\hat{p}_1$ = sample proportion of vitamin C users who caught a cold

$$\hat{p}_1 = \frac{x_1}{n_1} = \frac{104}{250} \approx 0.4160 \text{ (computed to 4 decimal places)},$$

while,

$\hat{p}_2$ = sample proportion of placebo users who caught a cold

$$\hat{p}_2 = \frac{x_2}{n_2} = \frac{158}{250} \approx 0.6320 \text{ (computed to 4 decimal places)}.$$

The researcher will use the *sample proportions* to decide if there is *a significant difference between the proportion of all individuals taking high doses of vitamin C who catch a cold and the proportion of all individuals* not taking vitamin C who catch a cold. To determine if there is **a significant difference between the two population proportions**, the researcher will need to compute *the difference between the two sample proportions: $\hat{p}_1 - \hat{p}_2$.* That is, the researcher will subtract the two sample proportions, $\hat{p}_1 - \hat{p}_2$, to determine if this sample difference, $\hat{p}_1 - \hat{p}_2$, is large enough to infer that the difference between the proportion of all individuals taking vitamin C and the proportion of all individuals not taking vitamin C is statistically significant. For this example, the difference between the sample proportions is:

$$\hat{p}_1 - \hat{p}_2 = 0.4160 - 0.6320 = -0.2160.$$

Consequently, the researcher must now determine if the *difference between the **sample** proportions $\hat{p}_1 - \hat{p}_2 = -0.216$ is **large** enough* to infer that there is *a significant difference between the two **population** proportions* **or** she may conclude that this *difference* between the sample proportions *is due solely to chance.*

If the researcher concludes that the difference is attributed to chance alone, then she is simply stating that it is *likely* to select an independent random sample from each of the two populations having equal proportions (i.e. $p_1 = p_2$), and obtain a difference between their sample proportions, $\hat{p}_1$ and $\hat{p}_2$ respectively. As explained in Chapter 11, when selecting random samples from a population, the values of the sample proportions can vary from sample to sample. Consequently, when selecting an independent random sample from two populations which have equal proportions (i.e. $p_1 = p_2$), their sample proportions, $\hat{p}_1$ and $\hat{p}_2$ respectively, can vary and thus exhibit a difference between the sample proportions $\hat{p}_1 - \hat{p}_2$ by chance alone.

For example, if the medical researcher were to conduct a similar study by randomly selecting another independent sample from each of these same two populations, the researcher could obtain different sample proportions, such as:

$$\hat{p}_1 = 0.5890 \text{ and } \hat{p}_2 = 0.7450,$$

which would result in the following *different* difference between the sample proportions:

$$\hat{p}_1 - \hat{p}_2 = 0.5890 - 0.7450 = -0.1560.$$

Thus the researcher needs to determine if the difference between the sample proportions, $\hat{p}_1 - \hat{p}_2$, is large enough and not just attributed to chance alone before she can conclude that the difference between the proportion of **all** individuals taking high doses of vitamin C who catch a cold and the proportion of **all** individuals not taking vitamin C who catch a cold is statistically significant. To determine if the difference between the sample proportions, $\hat{p}_1 - \hat{p}_2$, is large enough, it is necessary to understand the characteristics of the **distribution of all possible sample differences of $\hat{p}_1 - \hat{p}_2$**. Such a distribution is called the *Sampling Distribution of the Difference Between Two Proportions. This distribution is also referred to as the Sampling Distribution of $\hat{p}_1 - \hat{p}_2$.* The researcher will need **to use the Sampling Distribution of the Difference Between Two Proportions to determine if the difference between the sample proportions $\hat{p}_1 - \hat{p}_2 = -0.2160$** is significant (or large enough) to infer that there is a statistically significant difference between the population proportions, $\hat{p}_1 - \hat{p}_2$. To make such a decision, the researcher must know the shape, mean, and standard deviation *of the Sampling Distribution of the Difference Between Two Proportions.* These characteristics will be discussed in the next section. Before we leave this section, examine Table 12.2 which summarizes the notation which will be used in this chapter.

TABLE 12.2 Summary of Notation for Population and Sample Proportions

Population	Population Proportion	Sample Size	Count or Number of Occurrences or Observations Possessing a Particular Characteristic or Number of Successes	Sample Proportion
1	p_1	n_1	x_1	$\hat{p}_1 = \dfrac{x_1}{n_1}$
2	p_2	n_2	x_2	$\hat{p}_2 = \dfrac{x_2}{n_2}$

Let's consider Example 12.1, which will examine the idea and notation of a population and a sample proportion.

EXAMPLE 12.1

Craig, a psychological researcher, is trying to determine if there is a significant difference between the proportion of women and men who frequently have nightmares when taking statistics at the Big Red University. Craig randomly selected 195 women and 180 men who had taken statistics at the University to perform his study. Within his study, Craig determined that 69 of the women and 58 of the men had frequent nightmares when taking statistics at the University.

Using this information and letting the women represent population 1 and the men as population 2,
a) identify the two population proportions being compared and the symbols used to symbolize these proportions.
b) determine the sample proportion (to 4 decimal places) of students who had frequent nightmares for each population and the symbols used to symbolize these proportions.
c) calculate the difference between the sample proportions and identify the symbol to represent this difference.

Solution

a) The researcher is interested in comparing the *proportion of all University women who have frequent nightmares when taking statistics* to the *proportion of all University men who have frequent nightmares when taking statistics*. Letting the college women who took statistics at the University represent population 1, and the college men who took statistics at the University represent population 2, then:

p_1 = *the population proportion of all University women who have frequent nightmares when taking statistics*

p_2 = *the population proportion of all University men who have frequent nightmares when taking statistics*

b) The sample proportion of women who had frequent nightmares when taking statistics is symbolized by $\hat{p}_1$ and is equal to:

$$\hat{p}_1 = \frac{\text{number of women having frequent nightmares}}{\text{total number of women in the sample}},$$

or

$$\hat{p}_1 = \frac{x_1}{n_1}$$

where:
x_1 = number of women having frequent nightmares
   = 69,
and
n_1 = total number of women in the sample
   = 195.

Thus, the sample proportion of women who had frequent nightmares when taking statistics is:

$$\hat{p}_1 = \frac{69}{195} \approx 0.3538$$

The sample proportion of men who had frequent nightmares when taking statistics is symbolized by $\hat{p}_2$ and is equal to:

$$\hat{p}_2 = \frac{\text{number of men having frequent nightmares}}{\text{total number of men in the sample}}$$

or

$$\hat{p}_2 = \frac{x_2}{n_2},$$

where:
x_2 = number of men having frequent nightmares
   = 58,
and
n_2 = total number of men in the sample
   = 180.

Thus, the sample proportion of men who had frequent nightmares when taking statistics is:

$$\hat{p}_1 = \frac{58}{180} \approx 0.3222$$

c) The difference between the sample proportions is symbolized by $\hat{p}_1 - \hat{p}_2$ and is equal to:

$$\hat{p}_1 - \hat{p}_2 = 0.3538 - 0.3222$$
$$= 0.0316.$$

We are interested in studying the relationship of the proportions of two populations, rather than just one population proportion as discussed in Chapter 11. The appropriate sampling distribution used to compare the proportions of two populations is called the *Sampling Distribution of the Difference Between Two Proportions*.

DEFINITION 12.2
The Sampling Distribution of the Difference Between Two Proportions or the Sampling Distribution of $\hat{p}_1 - \hat{p}_2$. The Sampling Distribution of $\hat{p}_1 - \hat{p}_2$ is a theoretical distribution of all the possible differences between the values of the sample proportions selected from each of the two populations.

Figure 12.1 will help you to visualize the conceptual idea behind this sampling distribution.

Notice at the top of Figure 12.1, the two populations, identified as *Population 1* and *Population 2*, are shown with their respective population proportions p_1 and p_2. Suppose an independent random sample of size n_1 is selected from population 1 and an independent random sample of size n_2 is selected from population 2. Underneath each population, the Sampling Distribution of the Proportion is shown. The theoretical Sampling Distribution of the Proportion corresponding to population 1 represents all the possible samples of size n_1 that can be selected from population 1. While the theoretical Sampling Distribution of the Proportion corresponding to population 2 represents all the possible samples of size n_2 that can be selected from population 2. As discussed in Chapter 11, each Sampling Distribution of the Proportion is approximately a normal distribution when the sample size is large. Recall, *a sample is large if both np and n(1 − p) are greater than or equal to 10*. Using this rule, the theoretical Sampling Distribution of the Proportion corresponding to population 1 is approximately normal when both $n_1 p_1$ and $n_1(1 - p_1)$ are greater than or equal to 10, while theoretical Sampling Distribution of the Proportion corresponding to population 2 is approximately normal when both $n_2 p_2$ and $n_2(1 - p_2)$ are greater than or equal to 10.

Now, consider selecting an independent random sample and determining the sample proportion for each of these two populations. These sample proportions are symbolized as $\hat{p}_1$ and $\hat{p}_2$ for sample 1 and sample 2 respectively. Subtracting these two sample proportions, we have $\hat{p}_1 - \hat{p}_2$ **which represents the difference between the sample proportions**. *This difference will be negative if the sample proportion from population one $\hat{p}_1$ is smaller than the sample proportion from population two $\hat{p}_2$, and positive if $\hat{p}_1$ is larger than $\hat{p}_2$.* Theoretically, if we continue this process of randomly selecting an independent sample from each of the two populations and calculating the difference between the sample proportions $\hat{p}_1 - \hat{p}_2$, then the distribution of **all the possible sample differences of $\hat{p}_1 - \hat{p}_2$** represents the **Distribution of the Difference Between Two Sample Proportions** or simply the *Sampling Distribution of the Difference Between Two Proportions*. This concept is illustrated at the bottom of Figure 12.1.

Since it is impossible or impractical to construct such a Sampling Distribution of the Difference Between Two Proportions, this sampling distribution is a *theoretical probability distribution* which relates the value of each possible difference $\hat{p}_1 - \hat{p}_2$ to the probability that such a difference will occur. Since, in practice, only one sample will be randomly selected from each population, and it is impossible to construct this theoretical sampling distribution, we must rely on statisticians to provide us with the necessary theorems to completely describe the essential characteristics of this sampling distribution. These theorems will help us to determine the shape, mean and standard deviation of the Sampling Distribution of the Difference Between Two Proportions. The first theorem will discuss the shape of the Sampling Distribution of $\hat{p}_1 - \hat{p}_2$, while the second theorem will provide us with formulas to compute the mean and standard deviation of the Sampling Distribution of $\hat{p}_1 - \hat{p}_2$.

As mentioned in Chapter 11, *when sampling from one population, the Sampling Distribution of The Proportion is approximately normal when the sample size n is large. A sample is defined to be large when both:*

np is greater than or equal to 10
and
n(1 − p) is greater than or equal to 10.

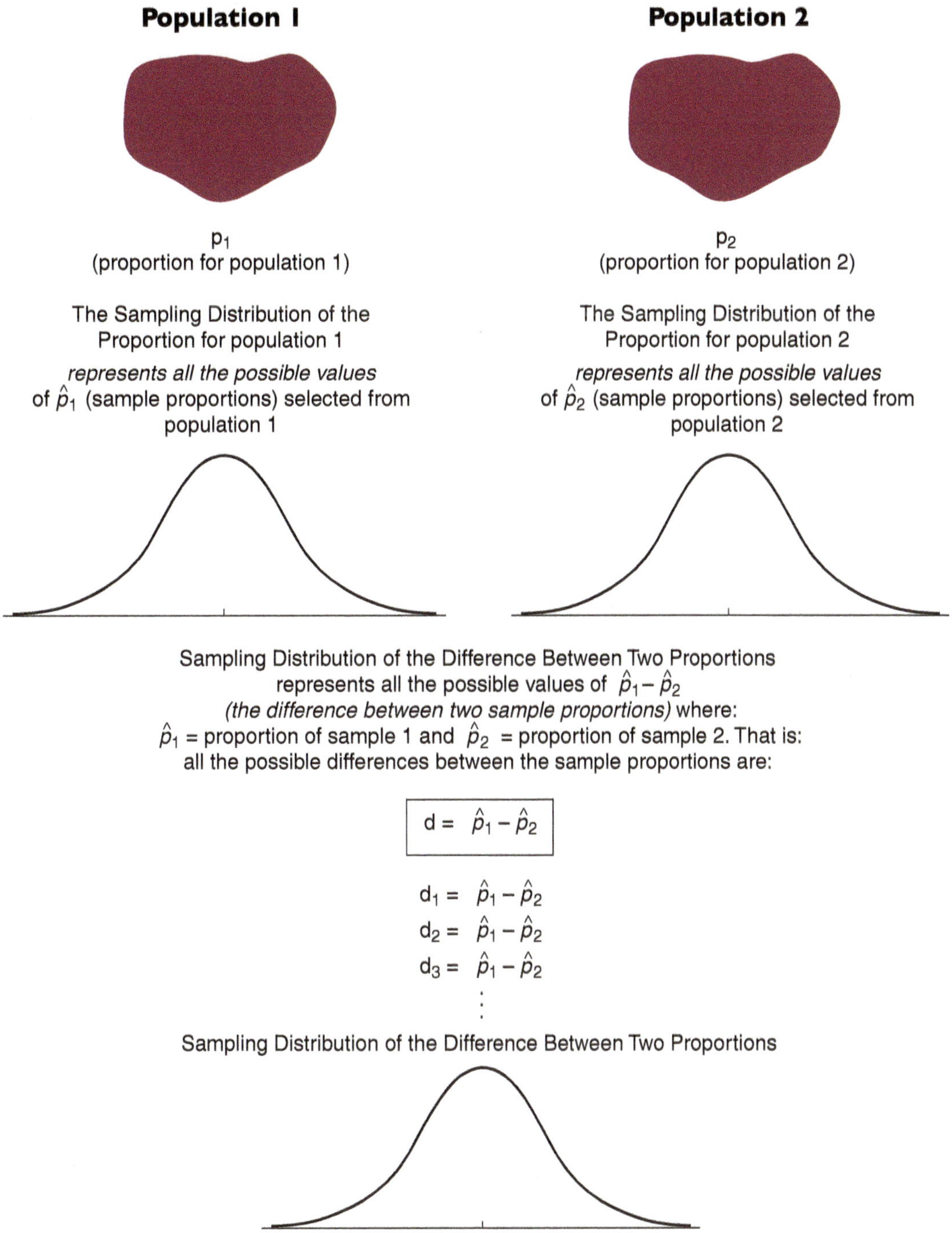

Section 12.2 The Sampling Distribution of the Difference Between Two Proportions

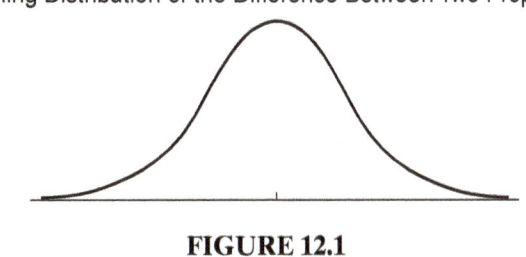

Population 1

p_1
(proportion for population 1)

The Sampling Distribution of the Proportion for population 1

represents all the possible values of $\hat{p}_1$ (sample proportions) selected from population 1

Population 2

p_2
(proportion for population 2)

The Sampling Distribution of the Proportion for population 2

represents all the possible values of $\hat{p}_2$ (sample proportions) selected from population 2

Sampling Distribution of the Difference Between Two Proportions represents all the possible values of $\hat{p}_1 - \hat{p}_2$ *(the difference between two sample proportions)* where: $\hat{p}_1$ = proportion of sample 1 and $\hat{p}_2$ = proportion of sample 2. That is: all the possible differences between the sample proportions are:

$$d = \hat{p}_1 - \hat{p}_2$$

$$d_1 = \hat{p}_1 - \hat{p}_2$$
$$d_2 = \hat{p}_1 - \hat{p}_2$$
$$d_3 = \hat{p}_1 - \hat{p}_2$$
$$\vdots$$

Sampling Distribution of the Difference Between Two Proportions

FIGURE 12.1

In the case of sampling from two populations, *the Sampling Distribution of The Difference Between Two Proportions is approximately normal when the independent random sample selected from each of the two populations are both large*. Therefore, in the case of two samples, **both samples are considered to be large if the quantities $n_1 p_1$, $n_1(1 - p_1)$, $n_2 p_2$ and $n_2(1 - p_2)$ are all greater than or equal to 10.** Theorem 12.1 describes the shape of the Sampling Distribution of The Difference Between Two Proportions for large samples.

Theorem 12.2 discusses the formulas to compute the mean and standard deviation of the Sampling Distribution of $\hat{p}_1 - \hat{p}_2$.

Theorem 12.1. Shape of the Sampling Distribution of the Difference Between Two Proportions for Large Samples.

If an independent random sample of size n_1 is selected from population 1 having a proportion p_1 and an independent random sample of size n_2 is selected from population 2 having a proportion p_2, then the Sampling Distribution of the Difference Between Two Proportions is approximately a normal distribution when both samples are large. (That is, when the quantities $n_1 p_1$, $n_1(1-p_1)$, $n_2 p_2$ and $n_2(1-p_2)$ are all greater than or equal to 10.)

Theorem 12.2. Characteristics of the Sampling Distribution of the Difference Between Two Proportions.

Given two populations, referred to as population 1 and population 2, having proportions p_1 and p_2 respectively. If a large independent random sample of size n_1 is selected from population 1 and a second large independent random sample of size n_2 is selected from population 2, then the Sampling Distribution of The Difference Between Two Proportions has the following characteristics.

1. The **mean** of the Sampling Distribution of the Difference Between Two Proportions, denoted by $\mu_{\hat{p}_1-\hat{p}_2}$, is:

$$\mu_{\hat{p}_1-\hat{p}_2} = p_1 - p_2$$

2. The **standard deviation** of the Sampling Distribution of the Difference Between Two Proportions or commonly called the **standard error** of the difference between two proportions, denoted by $\sigma_{\hat{p}_1-\hat{p}_2}$, is:

$$\sigma_{\hat{p}_1-\hat{p}_2} = \sqrt{\frac{(p_1)(1-p_1)}{n_1} + \frac{(p_2)(1-p_2)}{n_2}}$$

EXAMPLE 12.2

Based upon the latest Census Bureau's information, 70% of single women who are the heads of households own a car while 82% of households headed by single men own cars. Assuming the Census Bureau's percentages represent the proportions of the indicated populations and if an independent random sample of 1800 households is selected from each population, then:
a) define the two populations, and the proportions of these populations.
   Determine the:
b) shape of the Sampling Distribution of $\hat{p}_1 - \hat{p}_2$
c) mean of the Sampling Distribution of $\hat{p}_1 - \hat{p}_2$
d) standard deviation of the Sampling Distribution of $\hat{p}_1 - \hat{p}_2$
e) standard error of the difference between two proportions

Solution

a) We will define the population of households headed by single women as population 1 while the population of households headed by single men will be defined as population 2. According to the Census Bureau, the proportion of single women car owners who serve as the head of a household is 70%. This percentage refers to population 1 and is denoted as $p_1 = 0.70$. While the proportion of single men car owners who head a household is 82%. This percentage refers to population 2 and is symbolized as $p_2 = 0.82$.
b) The shape of the Sampling Distribution of $\hat{p}_1 - \hat{p}_2$ is dependent upon the size of the two samples selected. For population 1, we have a population proportion of $p_1 = 0.70$ and a sample size of $n_1 = 1800$, while for population 2, we have a population proportion of $p_2 = 0.82$ and a sample size of $n_2 = 1800$.

To determine if the two samples are large, we need to determine if:

$$n_1 p_1, n_1(1 - p_1), n_2 p_2 \text{ and } n_2(1 - p_2)$$
are all greater than or equal to 10.

For population 1, we have:
$n_1 p_1$ = (1800)(0.70) = 1260,
and
$n_1(1 - p_1)$ = (1800)(1 − 0.70) = (1800)(0.30) = 540,
which are both greater than or equal to 10.

For population 2, we have:
$n_2 p_2$ = (1800)(0.82) = 1476,
and
$n_2(1 - p_2)$ = (1800)(1 − 0.82) = (1800)(0.18) = 324,
which are also both greater than or equal to 10.

According to Theorem 12.1, since the quantities $n_1 p_1$, $n_1(1 - p_1)$, $n_2 p_2$ and $n_2(1 - p_2)$ are all greater than or equal to 10, then **the two samples are large** and **the Sampling Distribution of the Difference Between Two Proportions is approximately a normal distribution.**

c) According to Theorem 12.2, the mean of the Sampling Distribution of $\hat{p}_1 - \hat{p}_2$, denoted by $\mu_{\hat{p}_1 - \hat{p}_2}$, is determined by the formula:

$$\mu_{\hat{p}_1 - \hat{p}_2} = p_1 - p_2.$$

For population 1, the proportion is $p_1 = 0.70$, while for population 2, the proportion is $p_2 = 0.82$, thus the mean of the Sampling Distribution of $\hat{p}_1 - \hat{p}_2$ is:

$$\mu_{\hat{p}_1 - \hat{p}_2} = 0.70 - 0.82$$

$$\mu_{\hat{p}_1 - \hat{p}_2} = -0.12.$$

d) Using Theorem 12.2, the **standard deviation** of the Sampling Distribution of $\hat{p}_1 - \hat{p}_2$, denoted by $\sigma_{\hat{p}_1 - \hat{p}_2}$, is computed using the formula:

$$\sigma_{\hat{p}_1 - \hat{p}_2} = \sqrt{\frac{(p_1)(1-p_1)}{n_1} + \frac{(p_2)(1-p_2)}{n_2}}.$$

For population 1, the proportion is $p_1 = 0.70$ and the sample size is $n_1 = 1800$, while for population 2, the proportion is $p_2 = 0.82$, and the sample size is $n_2 = 1800$. Substituting these values into the formula for the standard deviation, we have:

$$\sigma_{\hat{p}_1 - \hat{p}_2} = \sqrt{\frac{(p_1)(1-p_1)}{n_1} + \frac{(p_2)(1-p_2)}{n_2}}$$

$$\sigma_{\hat{p}_1 - \hat{p}_2} = \sqrt{\frac{(0.70)(1-0.70)}{1800} + \frac{(0.82)(1-0.82)}{1800}}$$

$$\sigma_{\hat{p}_1 - \hat{p}_2} = \sqrt{\frac{(0.21)}{1800} + \frac{(0.1476)}{1800}}$$

$$\sigma_{\hat{p}_1 - \hat{p}_2} = \sqrt{\frac{0.3576}{1800}}$$

$$\sigma_{\hat{p}_1 - \hat{p}_2} = \sqrt{0.000198621}$$

$$\sigma_{\hat{p}_1 - \hat{p}_2} \approx 0.0141$$

e) The standard error of the difference between two proportions is equal to the standard deviation of the Sampling Distribution of $\hat{p}_1 - \hat{p}_2$. Therefore, the standard error is:

$$\sigma_{\hat{p}_1 - \hat{p}_2} \approx 0.0141.$$

In most practical applications, **the true population proportions p₁ and p₂ are not known**. When the population proportions are unknown, the standard error of the difference between two proportions cannot be calculated. In these instances, a large independent sample is selected from each population and the sample proportions $\hat{p}_1$ and $\hat{p}_2$ are computed. The sample proportion $\hat{p}_1$ will serve as an **estimate** for the population proportion p₁ while the sample proportion $\hat{p}_2$ will serve as an **estimate** for the population proportion p₂. Hence, within the formula for the standard error, the sample proportion $\hat{p}_1$ replaces p₁ while the sample proportion $\hat{p}_2$ replaces p₂. Thus, the formula for the estimated value of the standard error becomes:

$$\sqrt{\frac{(\hat{p}_1)(1-\hat{p}_1)}{n_1} + \frac{(\hat{p}_2)(1-\hat{p}_2)}{n_2}}$$

This formula is called *the* **estimate** *of the standard error of the difference between two proportions* and is symbolized as $s_{\hat{p}_1 - \hat{p}_2}$.

Formula for the Estimate of the Standard Error of the Difference Between Two Proportions

When the population proportions are not known, the formula for **the estimate of the standard error of the difference between two proportions** is used. The *estimate of the standard error*, symbolized by $s_{\hat{p}_1 - \hat{p}_2}$, is:

$$s_{\hat{p}_1 - \hat{p}_2} = \sqrt{\frac{(\hat{p}_1)(1-\hat{p}_1)}{n_1} + \frac{(\hat{p}_2)(1-\hat{p}_2)}{n_2}}$$

EXAMPLE 12.3

Recent studies have indicated that children who were enrolled in pre-school tended to not require Social Services later in life. A school sociologist wishes to test the validity of this statement and randomly samples children from his school's population. For the population of children who did not attend pre-school, he randomly sampled 55 children and determined that 43 required Social Services later on in life. For the population of children who attended pre-school, he randomly sampled 60 children and found 37 required the use of Social Services later in life. Use this information to answer the following:
a) define the two populations, and the proportions of these populations
b) state the sample proportions
c) compute the estimate of the standard error of the difference between two proportions
d) compute the estimate of the standard deviation of the Sampling Distribution of the Difference Between Two Proportions

Solution

a) One population represents the children who attended pre-school, and the second population represents the children who did not attend pre-school. Of the two populations identified by this sociologist, one proportion represents all children who did not attend pre-school and needed Social Services later in life, while the second proportion represents all children who did attend pre-school and needed Social Services later in life.

b) If we let the sample proportion of children who did not attend pre-school and needed Social Services later in life equal $\hat{p}_1$, then $\hat{p}_1 = \dfrac{x_1}{n_1} = \dfrac{43}{55} \approx 0.78$. If we let the sample proportion of children who did attend pre-school and needed Social Services later in life equal $\hat{p}_2$, then $\hat{p}_2 = \dfrac{x_2}{n_2} = \dfrac{37}{60} \approx 0.62$.

c) Since the population proportions are not known, we must estimate the standard error of the difference between two proportions using the formula:

$$s_{\hat{p}_1-\hat{p}_2} = \sqrt{\frac{\hat{p}_1(1-\hat{p}_1)}{n_1} + \frac{\hat{p}_2(1-\hat{p}_2)}{n_2}}$$

For the population of children who did not attend pre-school, the sample of 55 students has a $\hat{p}_1 = 0.78$ and $n_1 = 55$. For the population of children who did attend pre-school, the sample of 60 has a $\hat{p}_2 = 0.62$ and $n_2 = 60$. Substituting this sample information into the formula for $s_{\hat{p}_1-\hat{p}_2}$, we have:

$$s_{\hat{p}_1-\hat{p}_2} = \sqrt{\frac{\hat{p}_1(1-\hat{p}_1)}{n_1} + \frac{\hat{p}_2(1-\hat{p}_2)}{n_2}}$$

$$s_{\hat{p}_1-\hat{p}_2} = \sqrt{\frac{(0.78)(1-0.78)}{55} + \frac{(0.62)(1-0.62)}{60}}$$

$$s_{\hat{p}_1-\hat{p}_2} = \sqrt{\frac{(0.78)(0.22)}{55} + \frac{(0.62)(0.38)}{60}}$$

$$s_{\hat{p}_1-\hat{p}_2} = \sqrt{\frac{0.1716}{55} + \frac{0.2356}{60}}$$

$$s_{\hat{p}_1-\hat{p}_2} = \sqrt{0.003120 + 0.003927}$$

$$s_{\hat{p}_1-\hat{p}_2} = \sqrt{0.007047}$$

$$s_{\hat{p}_1-\hat{p}_2} \approx 0.0839$$

Thus, the estimate of the standard error of the difference between two proportions is approximately 0.0839.

d) The estimate of the standard deviation of the Sampling Distribution of the Difference Between Two Proportions is equal to the estimate of the standard error of the difference between two proportions. Thus, the estimate of the standard deviation of the sampling distribution is:

$$s_{\hat{p}_1-\hat{p}_2} \approx 0.0839$$

12.3 HYPOTHESIS TESTING INVOLVING TWO POPULATION PROPORTIONS USING LARGE SAMPLES

Within this section, we will discuss the procedure to perform a hypothesis test involving two population proportions using a large independent random sample selected from each population. This procedure involves the general 5-step procedure initially discussed in Chapter 10. To perform this hypothesis test involving two population proportions, a large independent random sample from each of the two populations is selected. The difference between the sample proportions, $\hat{p}_1 - \hat{p}_2$, is computed. If the difference between the sample proportions, $\hat{p}_1 - \hat{p}_2$, is statistically significant then we will conclude that the two population proportions are statistically different. To determine if the difference between the two sample proportions, $\hat{p}_1 - \hat{p}_2$ is large enough to indicate that the two population proportions are statistically different, we will use the Sampling Distribution of $\hat{p}_1 - \hat{p}_2$ as the hypothesis testing model.

Let's now examine the hypothesis testing procedure involving the difference between the proportions of two populations.

Hypothesis Testing Procedure Involving the Difference Between the Proportions of Two Populations for Large Samples

Step 1 Formulate the two hypotheses: H_0 and H_a.

In most hypothesis tests involving a comparison of two population proportions, the **null hypothesis assumes that the difference between the population proportions is equal to zero**. That is, *there is no*

> **Null Hypothesis Form**
>
> H_0: There is *no difference* between the proportion of population 1, written p_1, and the proportion of population 2, written p_2.
>
> The null hypothesis is symbolized as:
>
> $$H_0: p_1 - p_2 = 0$$

difference between the proportions of the two populations. In such hypothesis tests, the null hypothesis, H_0, has the following form:

By stating the null hypothesis, H_0, as *there is no difference between the population proportions, p_1 and p_2*, we are making the assumption that p_1 equals p_2. Consequently, only by rejecting the null hypothesis, H_0, can we conclude that there is a *statistically significant difference* between the two population proportions: p_1 and p_2. Therefore, *by rejecting H_0, we may conclude one of the three possible forms of the alternative hypothesis*:

a) The *difference* between the proportion of population 1, p_1, and the proportion of population 2, p_2, *is less than zero*.
b) The *difference* between the proportion of population 1, p_1, and the proportion of population 2, p_2, *is greater than zero*.
c) The *difference* between the proportion of population 1, p_1, and the proportion of population 2, p_2, *is not equal to zero*.

Consequently, the statement and symbolic representation for each of the three types of alternative hypothesis forms when testing the difference between two population proportions are:

> **Alternative Hypothesis Forms**
>
> | Less than Form | Greater than Form | Not Equal to Form |
> |---|---|---|
> | H_a: the proportion of population 1, p_1, is *less than* the proportion of population 2, p_2. | H_a: the proportion of population 1, p_1, is *greater than* the proportion of population 2, p_2. | H_a: the proportion of population 1, p_1, is *not equal to* the proportion of population 2, p_2. |
> | This form is symbolized as: | This form is symbolized as: | This is form symbolized as: |
> | $H_a: p_1 - p_2 < 0.$ | $H_a: p_1 - p_2 > 0.$ | $H_a: p_1 - p_2 \neq 0.$ |
> | That is, the *difference $p_1 - p_2$ is significantly less than zero*. | That is, the *difference $p_1 - p_2$ is significantly greater than zero*. | That is, the *difference $p_1 - p_2$ is significantly different than zero*. |

Step 2 Determine the model to test the null hypothesis, H_0. Under the assumption H_0 is true, determine the expected results of the Sampling Distribution of $\hat{p}_1 - \hat{p}_2$.

Since the difference between the two sample proportions, $\hat{p}_1 - \hat{p}_2$, is used to determine if $p_1 - p_2$ is significantly different from zero we will use *the Sampling Distribution of The Difference Between Two Proportions, or* **the Sampling Distribution of $\hat{p}_1 - \hat{p}_2$** as our hypothesis testing model.

To determine the **expected results** of the sampling distribution, we must identify **the shape** and calculate **the mean and standard error** of the Sampling Distribution of $\hat{p}_1 - \hat{p}_2$. For large samples, Theorem 12.1 states that *the shape of the Sampling Distribution of $\hat{p}_1 - \hat{p}_2$ is approximately a normal distribution* (i.e. the quantities $n_1\hat{p}_1, n_1(1-\hat{p}_1), n_2\hat{p}_2$ and $n_2(1-\hat{p}_2)$ are all greater than or equal to 10).

The mean of the Sampling Distribution of $\hat{p}_1 - \hat{p}_2$ is defined by the formula: $\mu_{\hat{p}_1-\hat{p}_2} = p_1 - p_2$, according to Theorem 12.2. Under the assumption that the null hypothesis, H_0, is true, the difference between the population proportions is zero (i.e., $p_1 - p_2 = 0$). Thus, *the mean of the Sampling Distribution, $\mu_{\hat{p}_1-\hat{p}_2}$, of $\hat{p}_1 - \hat{p}_2$ is equal to zero*. That is, assuming H_0 is true, we have:

Section 12.3 *Hypothesis Testing Involving Two Population Proportions Using Large Samples*

$$\mu_{\hat{p}_1-\hat{p}_2} = p_1 - p_2$$
$$\mu_{\hat{p}_1-\hat{p}_2} = 0$$

From Theorem 12.2, the *estimate of the standard error of the Sampling Distribution*, $s_{\hat{p}_1-\hat{p}_2}$, of $\hat{p}_1 - \hat{p}_2$ is given by:

$$s_{\hat{p}_1-\hat{p}_2} = \sqrt{\frac{(\hat{p}_1)(1-\hat{p}_1)}{n_1} + \frac{(\hat{p}_2)(1-\hat{p}_2)}{n_2}}$$

Again using the assumption that the null hypothesis, H_0, is true we are assuming that there is *no difference between the population proportions p_1 and p_2, or* **$p_1 - p_2 = 0$**. Thus, it follows that $p_1 = p_2$ when H_0 is true. Whenever we make the assumption that H_0 is true, we are assuming we have **a common population proportion value.** *p will be used to symbolize this common value of p_1 and p_2, that is:* $p_1 = p_2 = p$. Thus, *p represents the proportion of successes common to both populations.*

However, this common hypothesized population proportion p is unknown, since the population values of p_1 and p_2 are unknown. Each of these population values p_1 and p_2 must be approximated by the sample statistics $\hat{p}_1$ and $\hat{p}_2$. Consequently, we will use the sample information collected from both populations to obtain the *best estimate for the common population proportion value, p*. If we denote this *best estimate for the value of p*, by $\hat{p}$, then the formula used to compute $\hat{p}$, is determined by **pooling (i.e. combining) the sample information collected from both populations**. This *pooled formula is referred to as the pooled sample proportion and is denoted by $\hat{p}$.* The value for this pooled sample proportion can be determined by using one of the following formulas, depending upon the sample information available.

Formulas for the Pooled Sample Proportion, $\hat{p}$:
The Best Estimate for the Common Population Proportion Value

$$\hat{p} = \frac{x_1 + x_2}{n_1 + n_2} \quad \text{or} \quad \hat{p} = \frac{n_1\hat{p}_1 + n_2\hat{p}_2}{n_1 + n_2}$$

where:
$\hat{p}$ is the pooled sample proportion (the best estimate for common population proportion)
$\hat{p}_1$ = sample proportion selected from population 1.
x_1 = number of successes within the sample selected from population 1.
n_1 = size of sample selected from population 1.
$\hat{p}_2$ = sample proportion selected from population 2.
x_2 = number of successes within the sample selected from population 2.
n_2 = size of sample selected from population 2.

Substituting the *pooled sample proportion result, $\hat{p}$* into the formula for the estimated standard error of the difference between two proportions, we obtain **the pooled formula for the estimated standard error of the Sampling Distribution of $\hat{p}_1 - \hat{p}_2$, $s_{\hat{p}_1-\hat{p}_2}$.** Under the assumption that the null hypothesis, H_0, is true (i.e. $p_1 - p_2 = 0$), then the pooled formula for the estimated standard error of the Sampling Distribution of $\hat{p}_1 - \hat{p}_2$, denoted by $s_{\hat{p}_1-\hat{p}_2}$, is determined by the following formula.

Pooled Formula for the Estimated Standard Error of the Sampling Distribution of $\hat{p}_1 - \hat{p}_2$, $s_{\hat{p}_1-\hat{p}_2}$:

$$s_{\hat{p}_1-\hat{p}_2} = \sqrt{\frac{\hat{p}(1-\hat{p})}{n_1} + \frac{\hat{p}(1-\hat{p})}{n_2}}$$

or

(continues)

Chapter 12 *Hypothesis Testing Involving Two Population Proportions Using Independent Samples*

**Pooled Formula for the Estimated Standard Error
of the Sampling Distribution of $\hat{p}_1 - \hat{p}_2$, $s_{\hat{p}_1 - \hat{p}_2}$: (continued)**

$$s_{\hat{p}_1 - \hat{p}_2} = \sqrt{\hat{p}(1-\hat{p})\left(\frac{1}{n_1} + \frac{1}{n_2}\right)}$$

where: $\hat{p}$ is the proportion obtained by combining (pooling) both samples and is computed using one of the following formulas:

$$\hat{p} = \frac{x_1 + x_2}{n_1 + n_2} \quad \text{or} \quad \hat{p} = \frac{n_1\hat{p}_1 + n_2\hat{p}_2}{n_1 + n_2}$$

Thus, for a large sample hypothesis test involving two population proportions, **the expected results of the Sampling Distribution of $\hat{p}_1 - \hat{p}_2$,** are summarized as follows.

The Expected Results of the Sampling Distribution of $\hat{p}_1 - \hat{p}_2$ Under the Assumption That H_0 Is True

1. *The shape of the sampling distribution is approximately a normal distribution for large samples.*
2. *The mean of the Sampling Distribution of $\hat{p}_1 - \hat{p}_2$ is equal to zero. That is:*

$$\mu_{\hat{p}_1 - \hat{p}_2} = p_1 - p_2 = 0$$

3. *The pooled formula for the estimate of the standard error of the Sampling Distribution of $\hat{p}_1 - \hat{p}_2$,*

   is:

$$s_{\hat{p}_1 - \hat{p}_2} = \sqrt{\frac{\hat{p}(1-\hat{p})}{n_1} + \frac{\hat{p}(1-\hat{p})}{n_2}}$$

   or

$$s_{\hat{p}_1 - \hat{p}_2} = \sqrt{\hat{p}(1-\hat{p})\left(\frac{1}{n_1} + \frac{1}{n_2}\right)}$$

   where

$$\hat{p} = \frac{x_1 + x_2}{n_1 + n_2} \quad \text{or} \quad \hat{p} = \frac{n_1\hat{p}_1 + n_2\hat{p}_2}{n_1 + n_2}$$

The expected results are shown in Figure 12.2.

The Expected Results of the Sampling Distribution of $\hat{p}_1 - \hat{p}_2$

The Sampling Distribution of $\hat{p}_1 - \hat{p}_2$ is approximately normal for large samples.

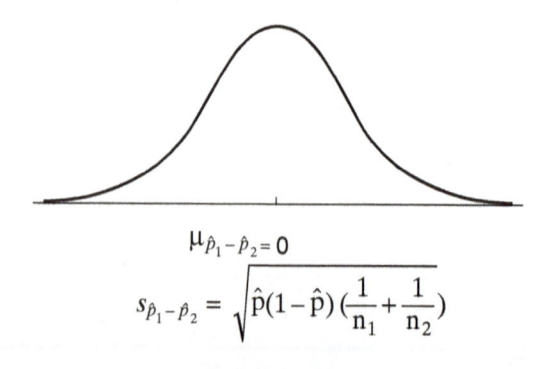

$$\mu_{\hat{p}_1 - \hat{p}_2} = 0$$

$$s_{\hat{p}_1 - \hat{p}_2} = \sqrt{\hat{p}(1-\hat{p})\left(\frac{1}{n_1} + \frac{1}{n_2}\right)}$$

FIGURE 12.2

Step 3 Formulate the decision rule.
  a) Determine the type of alternative hypothesis: *directional* H_a or *non-directional* H_a.
  b) Determine the type of test: *1TT or 2TT*.
  c) Identify the significance level: $\alpha = 1\%$ or $\alpha = 5\%$.
  d) Find the critical z score(s).
  e) Construct the appropriate hypothesis test model.
  f) State the decision rule.

Step 4 Analyze the sample data.
Randomly select a large sample from each population, and determine the proportion for each sample, denoted $\hat{p}_1$ and $\hat{p}_2$. *Calculate the z score of the difference between the two sample proportions using the formula:* $z = \dfrac{(\hat{p}_1 - \hat{p}_2) - \mu_{\hat{p}_1 - \hat{p}_2}}{s_{\hat{p}_1 - \hat{p}_2}}$. This is referred to as the test statistic.

Step 5 State the conclusion.
Compare the test statistic to the critical z score(s), z_c, of the decision rule and draw one of the following conclusions:

Conclusion (a)	or	Conclusion (b)
Reject H_0 and		Fail to Reject H_0
Accept H_a at α		at α

In summary, the procedure just outlined to perform a hypothesis test involving two population proportions for large samples included three refinements to the general hypothesis testing procedure developed in Chapter 10. These refinements were:

1. The appropriate hypothesis testing model is the Sampling Distribution of the Difference Between Two Proportions or the Sampling Distribution of $\hat{p}_1 - \hat{p}_2$.
2. Under the assumption that the null hypothesis, H_0, is true, the expected results of the Sampling Distribution of $\hat{p}_1 - \hat{p}_2$ are:
  a) *The shape of the sampling distribution is approximately a normal distribution for large samples.*
  b) *The mean of the Sampling Distribution of $\hat{p}_1 - \hat{p}_2$ is equal to zero.* That is:
  $$\mu_{\hat{p}_1 - \hat{p}_2} = p_1 - p_2 = 0$$
  c) *The pooled formula for the estimate of the standard error of the Sampling Distribution of $\hat{p}_1 - \hat{p}_2$, $s_{\hat{p}_1 - \hat{p}_2}$, is:*
  $$s_{\hat{p}_1 - \hat{p}_2} = \sqrt{\dfrac{\hat{p}(1-\hat{p})}{n_1} + \dfrac{\hat{p}(1-\hat{p})}{n_2}}$$
  or
  $$s_{\hat{p}_1 - \hat{p}_2} = \sqrt{\hat{p}(1-\hat{p})\left(\dfrac{1}{n_1} + \dfrac{1}{n_2}\right)}$$

3. The formula used to calculate the test statistic is:
$$z = \dfrac{(\hat{p}_1 - \hat{p}_2) - \mu_{\hat{p}_1 - \hat{p}_2}}{s_{\hat{p}_1 - \hat{p}_2}}.$$

Example 12.4 illustrates the use of the hypothesis testing procedure to perform a hypothesis test involving two population proportions.

PLEASE NOTE: FOR CHAPTER 12, ALL CALCULATIONS WILL BE ROUNDED OFF TO FOUR DECIMAL PLACES IN ALL THE EXAMPLES AND ANSWERS.

EXAMPLE 12.4

According to the Center for Disease Control and Prevention, a random survey of 1500 American men indicated that 435 smoke while a random sample of 1450 American women showed that 363 women smoke.

Chapter 12 Hypothesis Testing Involving Two Population Proportions Using Independent Samples

Perform a hypothesis test at $\alpha = 5\%$ to answer the question:

Is the proportion of U.S. men that smoke significantly greater than the proportion of U.S. women that smoke?

Solution

Step 1 Formulate the null hypothesis, H_o, and the alternative hypothesis, H_a.

If we let population 1 represent American men while population 2 represents American women, then the null hypothesis can be stated as:

H_o: There is **no difference** in the population proportion of American men who smoke and the population proportion of American women who smoke.

In symbols, we state the null hypothesis as:

H_o: $p_1 - p_2 = 0$.

Since we are trying to determine if there is a **greater** proportion of American males who smoke than females who smoke then the alternative hypothesis is *directional* and is stated using a *greater than* statement. Consequently, the alternative hypothesis can be stated as:

H_a: The population proportion of American men who smoke is greater than the population proportion of American women who smoke.

The alternative hypothesis is symbolized as:

H_a: $p_1 - p_2 > 0$.

Step 2 Determine the model to test the null hypothesis, H_o. Under the assumption H_o is true, determine the expected results of the Sampling Distribution of $\hat{p}_1 - \hat{p}_2$.

To test the null hypothesis, a large random independent sample was selected from each population and the sample results are summarized in Table 12.3.

TABLE 12.3

Population 1 Represents American Men Who Smoke	Population 2 Represents American Women Who Smoke
Sample information: $x_1 = 435$ $n_1 = 1500$	Sample information: $x_2 = 363$ $n_2 = 1450$

In conducting this hypothesis test, we will need to compute the difference between the two sample proportions, $\hat{p}_1 - \hat{p}_2$, to determine if this difference is *significantly greater than zero*. Consequently, we will use the **Sampling Distribution of** $\hat{p}_1 - \hat{p}_2$ as the hypothesis testing model when conducting this hypothesis test. To determine the expected results of the Sampling Distribution of $\hat{p}_1 - \hat{p}_2$, we need to use Theorems 12.1 and 12.2. Under the assumption that H_0 **is true**, we have the following expected results:

The Expected Results of the Sampling Distribution of $\hat{p}_1 - \hat{p}_2$

1. *The shape of the sampling distribution is approximately a normal distribution for large samples.*
2. *The mean of the Sampling Distribution of $\hat{p}_1 - \hat{p}_2$ is equal to zero since we are assuming that H_0 is true.* That is:

$$\mu_{\hat{p}_1 - \hat{p}_2} = p_1 - p_2 = 0$$

3. *To find the estimate of the standard error of the Sampling Distribution of $\hat{p}_1 - \hat{p}_2$, $s_{\hat{p}_1 - \hat{p}_2}$, we must first compute the pooled sample proportion, $\hat{p}$. Under the assumption that H_0 is true, i.e. p_1 equals p_2, the pooled sample proportion, $\hat{p}$, is determined by* **combining (i.e., pooling) the sample information collected from both populations**.

We will use the formula: $\hat{p} = \dfrac{x_1 + x_2}{n_1 + n_2}$ since we have the values $x_1 = 435$, $n_1 = 1500$, $x_2 = 363$ and $n_2 = 1450$.

The pooled sample proportion value, $\hat{p}$, is:

Section 12.3 *Hypothesis Testing Involving Two Population Proportions Using Large Samples*

$$\hat{p} = \frac{435 + 363}{1500 + 1450} = \frac{798}{2950}$$

$$\hat{p} = 0.2705 \text{ (rounded off to 4 decimal places)}$$

Substituting this pooled sample proportion result:
$\hat{p} \approx 0.2705$, the sample size for sample one: $n_1 = 1500$, and the sample size for sample two: $n_2 = 1450$ into the formula for the **pooled estimate of the standard error** of the Sampling Distribution of The Difference Between Two Proportions:

$$s_{\hat{p}_1 - \hat{p}_2} = \sqrt{\hat{p}(1-\hat{p})\left(\frac{1}{n_1} + \frac{1}{n_2}\right)}, \text{ the \textbf{pooled estimate of the standard error} equals:}$$

$$s_{\hat{p}_1 - \hat{p}_2} = \sqrt{(0.2705)(1 - 0.2705)\left(\frac{1}{1500} + \frac{1}{1450}\right)}$$

$$s_{\hat{p}_1 - \hat{p}_2} = \sqrt{(0.2705)(0.7295)\left(\frac{1}{1500} + \frac{1}{1450}\right)}$$

$$s_{\hat{p}_1 - \hat{p}_2} = \sqrt{(0.2705)(0.7295)(0.0013562)}$$

$$s_{\hat{p}_1 - \hat{p}_2} = \sqrt{0.0002676}$$

$$s_{\hat{p}_1 - \hat{p}_2} \approx 0.0164 \text{ (rounded off to 4 decimal places)}$$

Thus, the **pooled estimate of the standard error** of the Sampling Distribution of The Difference Between Two Proportions is: $s_{\hat{p}_1 - \hat{p}_2} \approx 0.0164$.

In summary, the Sampling Distribution of $\hat{p}_1 - \hat{p}_2$ will serve as our hypothesis testing model with the following expected results:

1. the Sampling Distribution can be approximated by the normal distribution for large samples.
2. the mean of the Sampling Distribution of $\hat{p}_1 - \hat{p}_2$ is:

$$\mu_{\hat{p}_1 - \hat{p}_2} = 0$$

3. the pooled estimate of the standard error, $s_{\hat{p}_1 - \hat{p}_2}$, is:

$$s_{\hat{p}_1 - \hat{p}_2} \approx 0.0164$$

This hypothesis testing model is illustrated in Figure 12.3.

The Sampling Distribution of $\hat{p}_1 - \hat{p}_2$ is the hypothesis testing model and is approximately a normal distribution.

$\mu_{\hat{p}_1 - \hat{p}_2} = 0$

$s_{\hat{p}_1 - \hat{p}_2} \approx 0.0164$

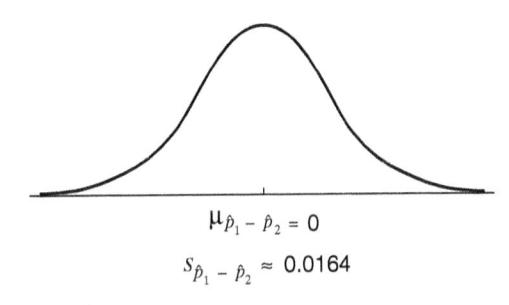

FIGURE 12.3

Step 3 Formulate the decision rule.
a) The alternative hypothesis is directional since we are trying to determine if the proportion of American men who smoke is *greater than* the proportion of American women who smoke.
b) The type of hypothesis test is one-tailed on the right side.
c) The significance level is: $\alpha = 5\%$.
d) To find the critical score, z_c, use Table II. For a one-tailed test on the right side with $\alpha = 5\%$, the critical z score, z_c, is: $z_c = 1.65$.
e) Figure 12.4 illustrates the decision rule for this hypothesis test.
f) Thus, *the decision rule is:* **Reject H_0 if the test statistic is greater than 1.65.**

FAIL TO REJECT H_0 | REJECT H_0

$\mu_{\hat{p}_1-\hat{p}_2} = 0$ $z_c = 1.65$

$s_{\hat{p}_1-\hat{p}_2} \approx 0.0164$

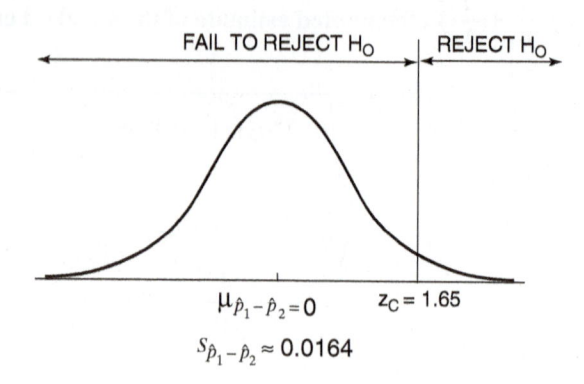

FIGURE 12.4

Step 4 Analyze the sample data.
Collect the sample data and calculate the test statistic using the formula: $z = \dfrac{(\hat{p}_1 - \hat{p}_2) - \mu_{\hat{p}_1-\hat{p}_2}}{s_{\hat{p}_1-\hat{p}_2}}$.

To determine the test statistic, we must first determine the values of $\hat{p}_1$ and $\hat{p}_2$.

$$\hat{p}_1 = \frac{435}{1500} \quad \text{and} \quad \hat{p}_2 = \frac{363}{1450}$$

$$\hat{p}_1 \approx 0.29 \quad \text{and} \quad \hat{p}_2 \approx 0.2503$$

Thus, the difference between the sample proportions is:

$$\hat{p}_1 - \hat{p}_2 = 0.29 - 0.2503$$

$$\hat{p}_1 - \hat{p}_2 = 0.0397$$

Since $\mu_{\hat{p}_1-\hat{p}_2} = 0$ and $s_{\hat{p}_1-\hat{p}_2} \approx 0.0164$, the test statistic is:

$$z = \frac{0.0397 - 0}{0.0164}$$

$$z \approx 2.42.$$

Step 5 State the conclusion.
Since the test statistic, $z \approx 2.42$, is greater than the critical z score, $z_c = 1.65$, our conclusion is reject H_0 and accept H_a at $\alpha = 5\%$. Therefore, *we can conclude that the proportion of American men who smoke is significantly greater than the proportion of American women who smoke.*

In Example 12.4, we considered a directional hypothesis test. We will now consider a non-directional hypothesis test in Example 12.5. Within this example, we will use the formula

$$\hat{p} = \frac{n_1 \hat{p}_1 + n_2 \hat{p}_2}{n_1 + n_2}$$ to estimate the pooled proportion value, $\hat{p}$.

EXAMPLE 12.5

According to a recent survey released by the National Highway Traffic and Safety Administration, 63% of women drivers use their seat belts while 60% of men drivers wear seat belts. Assume these estimates are based upon a random sample of 1150 women drivers and 1350 male drivers.

Perform a hypothesis test at $\alpha = 1\%$ to determine if there is a significant difference between the proportion of men drivers and the proportion of women drivers who use seat belts.

Solution

Step 1 Formulate the null hypothesis, H_0, and the alternative hypothesis, H_a.

If we let population 1 be the women drivers and population 2 be the men drivers then the null hypothesis is:

H_0: *There is no difference in the population proportion of women drivers who use seat belts and the population proportion of men drivers who use seat belts.*

In symbols, we have:

H_0: $p_1 - p_2 = 0$.

Since we are interested in determining if there is a significant difference in the proportions for the two populations then the alternative hypothesis is nondirectional and is stated as:

H_a: *There is a significant difference in the population proportion of women drivers who use seat belts and the population proportion of men drivers who use seat belts.*

Symbolically, the alternative hypothesis is stated:

H_a: $p_1 - p_2 \neq 0$.

Step 2 Determine the model to test the null hypothesis, H_0. Under the assumption H_0 is true, determine the expected results of the Sampling Distribution of $\hat{p}_1 - \hat{p}_2$.

To test the null hypothesis, a random sample was selected from each population and the sample results are summarized in Table 12.4.

TABLE 12.4

	Population 1: Women Drivers	**Population 2: Men Drivers**
	Sample information: Sample size is: $n_1 = 1150$ Proportion of women drivers wearing seat belts is: $\hat{p}_1 = 0.63$	Sample information: Sample size is: $n_2 = 1350$ Proportion of men drivers wearing seat belts is: $\hat{p}_2 = 0.60$

To determine if we can infer that $p_1 - p_2$ is significantly different from zero, we will need to decide if the difference between the two sample proportions, $\hat{p}_1 - \hat{p}_2$, is significantly different from zero. This decision will be based on our hypothesis testing model: the Sampling Distribution of Difference Between Two Proportions. Since we are working with large samples, the **expected results of the Sampling Distribution of $\hat{p}_1 - \hat{p}_2$**, are calculated using Theorems 12.1 and 12.2. Using the assumption that **H_0 is true**, the characteristics of this sampling distribution are determined using the following expected results:

The Expected Results of the Sampling Distribution of $\hat{p}_1 - \hat{p}_2$

1. *The shape of the sampling distribution is approximately a normal distribution for large samples.*
2. *Assuming that H_0 is true, the mean of the Sampling Distribution of $\hat{p}_1 - \hat{p}_2$ is equal to zero.*

That is: $\mu_{\hat{p}_1 - \hat{p}_2} = p_1 - p_2 = 0$.

3. Using the assumption that H_0 is true, the pooled formula for the estimate of the standard error of the sampling distribution of $\hat{p}_1 - \hat{p}_2$, $s_{\hat{p}_1 - \hat{p}_2}$, is found using the formula:

$$s_{\hat{p}_1 - \hat{p}_2} = \sqrt{\hat{p}(1-\hat{p})\left(\frac{1}{n_1} + \frac{1}{n_2}\right)}$$

Under the assumption that H_0 is true, i.e. p_1 equals p_2, the pooled sample proportion, $\hat{p}$, is determined by **combining (i.e. pooling) the sample information collected from both populations** using the formula:

$$\hat{p} = \frac{n_1 \hat{p}_1 + n_2 \hat{p}_2}{n_1 + n_2}$$

Substituting the sample information, $n_1 = 1150$, $\hat{p}_1 = 0.63$, $n_2 = 1350$, and $\hat{p}_2 = 0.60$ into the pooled sample proportion formula we obtain:

$$\hat{p} = \frac{(1150)(0.63) + (1350)(0.60)}{1150 + 1350}$$

$$\hat{p} = \frac{724.5 + 810}{2500}$$

$$\hat{p} = \frac{1534.5}{2500}$$

$$\hat{p} = 0.6138$$

Using this pooled sample proportion result, $\hat{p} = 0.6138$, the **pooled estimate of the standard error** of the Sampling Distribution of The Difference Between Two Proportions is determined using the formula:

$$s_{\hat{p}_1 - \hat{p}_2} = \sqrt{\hat{p}(1-\hat{p})\left(\frac{1}{n_1} + \frac{1}{n_2}\right)} .$$

Using the information: $\hat{p} = 0.6138$, $n_1 = 1150$, and $n_2 = 1350$, the **pooled estimate of the standard error** is:

$$s_{\hat{p}_1 - \hat{p}_2} = \sqrt{(0.6138)(1 - 0.6138)\left(\frac{1}{1150} + \frac{1}{1350}\right)}$$

$$s_{\hat{p}_1 - \hat{p}_2} = \sqrt{(0.6138)(0.3862)\left(\frac{1}{1150} + \frac{1}{1350}\right)}$$

$$s_{\hat{p}_1 - \hat{p}_2} \approx \sqrt{(0.6138)(0.3862)(0.0016102)}$$

$$s_{\hat{p}_1 - \hat{p}_2} \approx \sqrt{(0.0003928)}$$

$$s_{\hat{p}_1 - \hat{p}_2} \approx 0.0195 \text{ (rounded off to 4 decimal places)}$$

Thus, the **pooled estimate of the standard error** of the Sampling Distribution of the difference between two proportions is:

$$s_{\hat{p}_1 - \hat{p}_2} \approx 0.0195.$$

In summary, the Sampling Distribution of $\hat{p}_1 - \hat{p}_2$ will serve as our hypothesis testing model with the following expected results:

1. the Sampling Distribution can be approximated by the normal distribution for large samples.
2. the mean of the Sampling Distribution of $\hat{p}_1 - \hat{p}_2$ is:

$$\mu_{\hat{p}_1 - \hat{p}_2} = 0.$$

Section 12.3 *Hypothesis Testing Involving Two Population Proportions Using Large Samples*

3. the pooled estimate of the standard error, $s_{\hat{p}_1 - \hat{p}_2}$, is:

$$s_{\hat{p}_1 - \hat{p}_2} \approx 0.0195.$$

This hypothesis testing model is illustrated in Figure 12.5.

The Sampling Distribution of $\hat{p}_1 - \hat{p}_2$ is the hypothesis testing model and is approximately a normal distribution.

$\mu_{\hat{p}_1 - \hat{p}_2} = 0$

$s_{\hat{p}_1 - \hat{p}_2} \approx 0.0195$

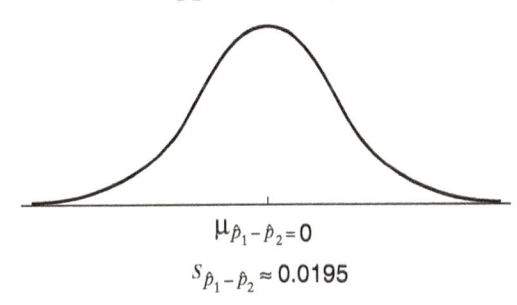

FIGURE 12.5

Step 3 Formulate the decision rule.
   a) The alternative hypothesis is nondirectional because we are trying to determine if there is a **significant difference** between the proportion of women drivers and the proportion of men drivers who use seat belts.
   b) The type of hypothesis test is two-tailed (2TT).
   c) The significance level is: $\alpha = 1\%$.
   d) To find two critical z scores, z_{LC} and z_{RC}, use Table II. From Table II, the critical z scores for $\alpha = 1\%$ are:

$$z_{LC} = -2.58 \text{ and } z_{RC} = 2.58$$

   e) Figure 12.6 illustrates the decision rule for this hypothesis test.

REJECT H₀ | FAIL TO REJECT H₀ | REJECT H₀

$z_{LC} = -2.58$ $\quad z_{RC} = 2.58$
$\mu_{\hat{p}_1 - \hat{p}_2} = 0$
$s_{\hat{p}_1 - \hat{p}_2} \approx 0.0195$

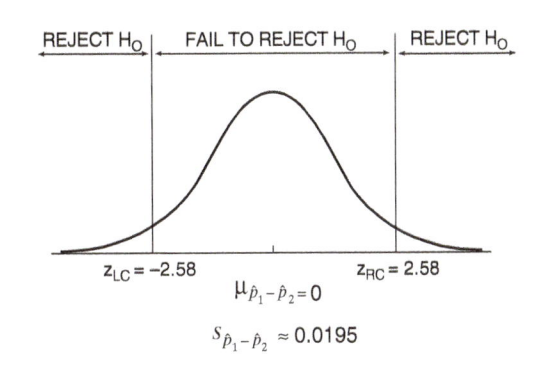

FIGURE 12.6

   f) Thus, *the decision rule* is: **Reject H₀ if the test statistic is either less than −2.58 or greater than 2.58.**

Step 4 Analyze the sample data.
Collect the sample data and calculate the test statistic using the formula: $z = \dfrac{(\hat{p}_1 - \hat{p}_2) - \mu_{\hat{p}_1 - \hat{p}_2}}{s_{\hat{p}_1 - \hat{p}_2}}$.

The difference between the sample proportions is:

$$\hat{p}_1 - \hat{p}_2 = 0.63 - 0.60$$
$$\hat{p}_1 - \hat{p}_2 = 0.03.$$

Since $\mu_{\hat{p}_1 - \hat{p}_2} = 0$ and $s_{\hat{p}_1 - \hat{p}_2} \approx 0.0195$, the test statistic is:

$$z = \frac{0.03 - 0}{0.0195}$$

$$z \approx 1.54.$$

Step 5 State the conclusion.
Since the test statistic, $z \approx 1.54$, falls between the critical z scores, $z_{LC} = -2.58$ and $z_{RC} = 2.58$, our conclusion is fail to reject H_0 at $\alpha = 1\%$. Therefore, *we can conclude that there is no significant difference in the proportion of women drivers and the proportion of men drivers who wear seat belts at $\alpha = 1\%$*.

CASE STUDY 12.1

Examine the graphic entitled *'No' for Women in Combat* in Figure 12.7. The Roper Organization surveyed 7,206 military men and women to determine their opinion of the Pentagon's policy of not assigning women to any direct combat position. Notice that results indicate that overall 57% favor the Pentagon's policy, 57% of the men are in agreement with the Pentagon's policy while only 42% of the women agree.

Statistical Snapshot

"No" for women in combat

Military people were asked their opinion of the Pentagon's current policy not to assign women to any direct combat position. The percentage who favored and opposed the no-combat policy:

■ Favor policy   □ Opposed policy

Overall
57%
42%

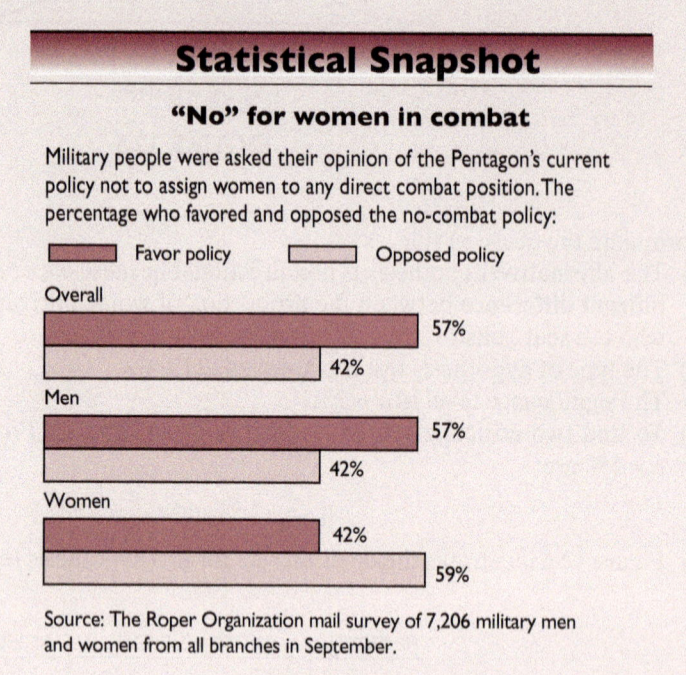

Men
57%
42%

Women
42%
59%

Source: The Roper Organization mail survey of 7,206 military men and women from all branches in September.

FIGURE 12.7

a) In Figure 12.7, is the sample data discrete or continuous? Does it represent a measurement or is it count data? Explain.

If we identify population 1 as the military men and population 2 represents the military women, then:

b) state the sample proportion of men and the sample proportion of women who favor the Pentagon's policy.
c) explain the relationship between the concept of a pooled sample proportion which was discussed in this section and the overall percentage favoring the policy reported in Figure 12.7.
d) For this case study, assume that the number of men participating in the survey was 6,206 while the number of women polled was 1,000. Use these sample results and the sample proportion results of part (b) to compute the pooled sample proportion. How does this pooled sample result compare to the overall percentage value?
e) Using this information, perform a hypothesis test to determine if there is a significant difference in the proportion of men who favor the policy and the proportion of women who favor the policy at $\alpha = 5\%$.

12.4 HYPOTHESIS TESTING INVOLVING TWO POPULATION PROPORTIONS COMPARING TREATMENT AND CONTROL GROUPS

A statistical study concerning the comparison of two groups of subjects is usually called an *experimental study* and it involves a *treatment group* and a *control group*. Usually, the subjects are randomly assigned to each group. The purpose of using a random assignment of subjects to both groups is to start the experiment with the treatment and control groups being as similar as possible with respect to all characteristics that are critical to the experiment as well as other factors which can influence the results of the experiment.

For example, if a medical researcher is interested in determining the effectiveness of aspirin in preventing heart attacks, the researcher would design an experiment by randomly assigning subjects to the treatment group who WILL be administered aspirin during the study. A second group of subjects will be randomly assigned to the control group who will NOT be administered aspirin during the study. The control group will resemble the treatment group as much as possible except they will not receive the aspirin. However, a placebo (i.e. a fake pill or an ineffective pill) is given to the subjects within the control group to insure that the subjects of the experiment aren't aware of whether they belong to the treatment or the control group. A placebo is administered to the control group as a precaution against *treatment effect*. That is, the psychological effect of the treatment group feeling better not because of the actual treatment they receive (the aspirin) but because they feel they were selected for some special treatment. In an experimental study where the subjects are unaware of whether they belong to the treatment or control group is called a *single-blind experiment*.

DEFINITION 12.3
Single-blind experiment. A single-blind experiment is a study in which the subjects of the experiment do not know who is receiving the treatment and who is receiving the placebo.

The treatment and control groups are compared to determine if the treatment (i.e., the aspirin) makes any difference in the subjects having a heart attack. To determine the effectiveness of the aspirin, the researcher will compare the *proportion of subjects having a heart attack* within the treatment and the control groups.

In a single-blind *experimental study,* the subjects are unaware of whether they belong to either the treatment group or the control group. However, if the researchers conducting the study are also unaware of whether the subjects belong to the treatment group or the control group, then the experiment is called a *double-blind experiment*.

DEFINITION 12.4
Double-blind experiment. Double-blind experiment is a study in which neither the subjects nor the researchers involved in the experiment know which group is the treatment or control group.

In an experimental study, the researcher's objective is to conduct a study where the treatment group and the control group are as similar as possible except for the treatment (i.e., the aspirin). Since all the subjects are randomly assigned to either the treatment or the control group and each subject receives a pill (the aspirin or the placebo) during the study, then the subjects of both groups are assumed to begin the experiment as similar as possible. *In essence, the only difference between the treatment and the control groups is the effect of the aspirin.* Thus, if the researcher finds a significant difference between the proportion of subjects who have a heart attack within the treatment group and the control group, then the researcher can conclude that this difference may be attributed to the aspirin (or treatment) rather than some other influence or factor.

Example 12.6 illustrates the comparison of a treatment versus a control group using a hypothesis test involving two population proportions.

EXAMPLE 12.6

According to the New England Journal of Medicine, a double-blind experiment was conducted to determine whether taking an aspirin tablet regularly helps prevent heart attacks. The medical researcher randomly assigned six thousand 40-year-old males to take an aspirin while another six thousand 40-year-old males were given a placebo. Perform a hypothesis test to determine if the following sample results indicate that taking aspirin will significantly lower the incidence of having a heart attack at $\alpha = 1\%$.

After 6 years, 78 of the males within the treatment group and 112 males in the control group suffered heart attacks.

Solution

Step 1 Formulate the null hypothesis, H_0, and the alternative hypothesis, H_a.

If we let population 1 represent the 40 year old males taking an aspirin daily while population 2 represents the 40 year old males taking a placebo daily, then the null hypothesis can be stated as:

H_0: There is **no difference** in the proportion of heart attacks between the population of 40 year old men who take an aspirin daily and the population of 40 year old men who take a placebo daily.

In symbols, we state the null hypothesis as:

$H_0: p_1 - p_2 = 0$.

Since we are trying to determine if the men taking the aspirin will have a *lower proportion of heart attacks* then the alternative hypothesis is *directional* and is stated using a *less than* statement. Consequently, the alternative hypothesis can be stated as:

H_a: The population proportion of heart attacks for 40 year old men taking an aspirin daily is less than the population proportion of heart attacks for 40 year old men taking a placebo daily.

The alternative hypothesis is symbolized as:

$H_a: p_1 - p_2 < 0$.

Step 2 Determine the model to test the null hypothesis, H_0. Under the assumption H_0 is true, determine the expected results of the Sampling Distribution of $\hat{p}_1 - \hat{p}_2$.

To test the null hypothesis, the medical researcher selected a large random independent sample from each population and the sample results are summarized in Table 12.5.

TABLE 12.5

Population 1 Represents 40 Year Old Men Taking Aspirin	Population 2 Represents 40 Year Old Men Taking a Placebo
Sample information: $x_1 = 78$ $n_1 = 6{,}000$	Sample information: $x_2 = 112$ $n_2 = 6{,}000$

To conduct this hypothesis test, we will need to compute the difference between the two sample proportions, $\hat{p}_1 - \hat{p}_2$, to determine if this difference is *significantly less than zero*. Consequently, we will use the **Sampling Distribution of $\hat{p}_1 - \hat{p}_2$ as the hypothesis testing model when conducting this hypothesis test.** To determine the expected results of the Sampling Distribution of $\hat{p}_1 - \hat{p}_2$, we need to use Theorems 12.1 and 12.2. Under the assumption that H_0 **is true**, we expect the following results:

The Expected Results of the Sampling Distribution of $\hat{p}_1 - \hat{p}_2$

1. *The shape of the sampling distribution is approximately a normal distribution for large samples.*
2. *The mean of the Sampling Distribution of $\hat{p}_1 - \hat{p}_2$ is equal to zero since we are assuming that H_0 is true.*

$$\mu_{\hat{p}_1 - \hat{p}_2} = p_1 - p_2 = 0.$$

3. *The estimate of the standard error of the sampling distribution of $\hat{p}_1 - \hat{p}_2$, $s_{\hat{p}_1 - \hat{p}_2}$, is found using the pooled formula:*

$$s_{\hat{p}_1 - \hat{p}_2} = \sqrt{\hat{p}(1-\hat{p})\left(\frac{1}{n_1} + \frac{1}{n_2}\right)}$$

Using the assumption that H_0 is true, i.e., p_1 equals p_2, the pooled sample proportion, $\hat{p}$, is determined by **combining (i.e., pooling) the sample information collected from both populations** and calculated using the formula: $\hat{p} = \dfrac{x_1 + x_2}{n_1 + n_2}$.

Section 12.4 *Hypothesis Testing Involving Two Population Proportions Comparing Treatment and Control Groups* 651

Using the sample information, $x_1 = 78$, $n_1 = 6000$, $x_2 = 112$ and $n_2 = 6000$, the pooled sample proportion value, $\hat{p}$, is:

$$\hat{p} = \frac{78 + 112}{6000 + 6000} = \frac{190}{12000}$$

$\hat{p} \approx 0.0158$ (rounded off to 4 decimal places)

The pooled estimate of the standard error of the Sampling Distribution of The Difference Between Two Proportions is determined using the formula:

$$s_{\hat{p}_1 - \hat{p}_2} = \sqrt{\hat{p}(1-\hat{p})\left(\frac{1}{n_1} + \frac{1}{n_2}\right)}$$

Using the information: $\hat{p} \approx 0.0158$, $n_1 = 6000$, and $n_2 = 6000$, the pooled estimate of the standard error is:

$$s_{\hat{p}_1 - \hat{p}_2} = \sqrt{(0.0158)(1 - 0.0158)\left(\frac{1}{6000} + \frac{1}{6000}\right)}$$

$$s_{\hat{p}_1 - \hat{p}_2} = \sqrt{(0.0158)(0.9842)\left(\frac{1}{6000} + \frac{1}{6000}\right)}$$

$$s_{\hat{p}_1 - \hat{p}_2} = \sqrt{(0.0158)(0.9842)(0.0003333)}$$

$$s_{\hat{p}_1 - \hat{p}_2} = \sqrt{(0.0000051)}$$

$s_{\hat{p}_1 - \hat{p}_2} \approx 0.0023$ (rounded off to 4 decimal places)

Thus, the pooled estimate of the standard error of the Sampling Distribution of The Difference Between Two Proportions is:

$$s_{\hat{p}_1 - \hat{p}_2} \approx 0.0023$$

In summary, the Sampling Distribution of $\hat{p}_1 - \hat{p}_2$ will serve as our hypothesis testing model with the following expected results:

1. the Sampling Distribution can be approximated by the normal distribution for large samples.
2. the mean of the Sampling Distribution of $\hat{p}_1 - \hat{p}_2$ is:

$$\mu_{\hat{p}_1 - \hat{p}_2} = 0.$$

3. the pooled estimate of the standard error, $s_{\hat{p}_1 - \hat{p}_2}$, is:

$$s_{\hat{p}_1 - \hat{p}_2} \approx 0.0023$$

This hypothesis testing model is illustrated in Figure 12.8.

The Sampling Distribution of $\hat{p}_1 - \hat{p}_2$ is the hypothesis testing model and is approximately a normal distribution.

$\mu_{\hat{p}_1 - \hat{p}_2} = 0$
$s_{\hat{p}_1 - \hat{p}_2} \approx 0.0023$

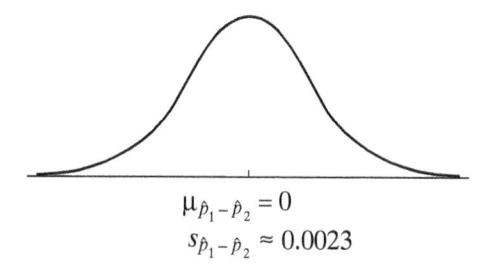

FIGURE 12.8

Step 3 Formulate the decision rule.
a) The alternative hypothesis is directional since we are trying to determine if the proportion of 40 year old men taking aspirin have a *lower* heart attack rate than the proportion of 40 year old men taking a placebo.
b) The type of hypothesis test is one-tailed on the left side.
c) The significance level is: $\alpha = 1\%$.
d) To find the critical score, z_c, use Table II.
For a one-tailed test on the left side with $\alpha = 1\%$, the critical z score, z_c, is: $z_c = -2.33$.
e) Figure 12.9 illustrates the decision rule for this hypothesis test.
f) Thus, *the decision rule* is: **reject H_0 if the test statistic is less than -2.33.**

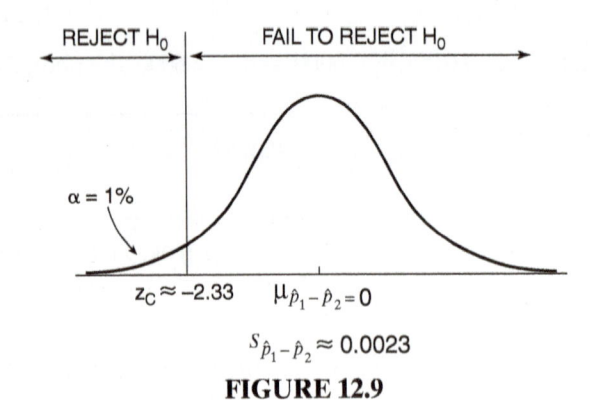

$z_c \approx -2.33 \quad \mu_{\hat{p}_1-\hat{p}_2}=0$

$s_{\hat{p}_1-\hat{p}_2} \approx 0.0023$

FIGURE 12.9

Step 4 Analyze the sample data.
Collect the sample data and calculate the test statistic using the formula: $z = \dfrac{(\hat{p}_1 - \hat{p}_2) - \mu_{\hat{p}_1-\hat{p}_2}}{s_{\hat{p}_1-\hat{p}_2}}$.
We must first determine the values of $\hat{p}_1$ and $\hat{p}_2$.

$$\hat{p}_1 = \frac{78}{6000} \quad \text{and} \quad \hat{p}_2 = \frac{112}{6000}$$

$$\hat{p}_1 \approx 0.0130 \quad \text{and} \quad \hat{p}_2 = 0.0187.$$

Thus, the difference between the sample proportions is:

$$\hat{p}_1 - \hat{p}_2 = 0.0130 - 0.0187$$
$$\hat{p}_1 - \hat{p}_2 = -0.0057$$

Since $\mu_{\hat{p}_1-\hat{p}_2} = 0$ and $s_{\hat{p}_1-\hat{p}_2} \approx 0.0023$, the test statistic is:

$$z = \frac{-0.0057 - 0}{0.0023}$$

$$z \approx -2.48$$

Step 5 State the conclusion.
Since the test statistic, $z \approx -2.48$, is less than the critical z score, $z_c = -2.33$, our conclusion is reject H_0 and accept H_a at $\alpha = 1\%$. Therefore, *we can conclude that the proportion of 40 year old men taking an aspirin daily who have heart attacks is significantly less than the proportion of 40 year old men who don't take aspirin.*

CASE STUDY 12.2

The data in Table 12.6 represents the findings from the aspirin component of a Physicians Health Study which appeared in the January, 1988 issue of *The New England Journal of Medicine*. The Physicians Health Study was a randomized, double-blind, placebo-controlled trial testing two hypotheses. One hypothesis was formulated to determine whether 325 mg of aspirin taken every other day reduces mortality from cardiovascular disease. Another hypothesis was formulated to determine the benefits of aspirin on the fatality of strokes.

CASE STUDY 12.2 (continued)

TABLE 12.6 Data on the Randomized Aspirin Component of the Study Regarding Myocardial Infarction and Stroke

End Point	Aspirin	Placebo
Myocardial Infarction		
Fatal	5	18
Nonfatal	99	171
Total	104	189
Stroke		
Fatal	6	2
Nonfatal	74	68
Total	80	70

The study involved male physicians aged 40 to 84 years of age residing in the United States at the beginning of the study in 1982. The selection of the physicians for the study was based on many factors. Many physicians were not selected for the study because of their previous history of myocardial infarction or stroke, or their current use of aspirin. Those physicians with liver or renal disease, gout, or peptic ulcer were also excluded. After the physicians were finally selected for the study, there was a trial run for 18 weeks. During this trial period, all the participants were given calendar packs containing active aspirin and the placebo. At the end of the trial, physicians were excluded from the study if they had changed their minds or demonstrated inadequate compliance with taking their medication. The physicians selected for the actual study were randomly selected to one of two groups: those assigned to the active aspirin group and those assigned to the aspirin placebo. Every six months, the physicians were sent a supply of monthly calendar packs with white tablets which were either aspirins or placebos. An independent group called the Data Monitoring Board met twice a year to review the accumulated data and to monitor the progress of the study. The Board recommended to terminate the randomized aspirin component of the study based upon the data shown in Table 12.6.

If we define the physicians taking aspirin as population 1 and the physicians taking the placebo as population 2, then:

a) determine the sample proportion of physicians taking aspirin who had a fatal myocardial infarction and the sample proportion of physicians taking the placebo who had a fatal myocardial infarction.
b) determine the sample proportion of physicians taking aspirin who had a fatal stroke and the sample proportion of physicians taking the placebo who had a fatal stroke.
c) Why do you believe many physicians were excluded from the study because of their personal medical history or their use of aspirin?
d) Why do you think a trial run of the study was conducted?
e) Explain why you think that the placebo was a white tablet?
f) What was the purpose of randomly assigning the physicians to the aspirin or the placebo group?
g) How did the trial run help to make this a better study?
h) Explain what you believe to be the meaning of a randomized, double-blind, placebo-controlled trial study?
i) Which group would you call the treatment group and which group would you identify as the control group? Explain.
j) What do you believe is the purpose of the Data Monitoring Board?
k) Conduct a hypothesis test to determine if aspirin taken every other day reduces mortality from cardiovascular disease at $\alpha = 1\%$.
l) Conduct a hypothesis test to determine if aspirin taken every other day reduces mortality from strokes at $\alpha = 1\%$.
m) Based on the results of your hypothesis test, would you conclude that aspirin helps to prevent cardiovascular disease?
n) Based upon your analysis, would it be advisable to have the general population take aspirin to prevent cardiovascular disease? If not, to whom would you advise to take aspirin for the prevention of myocardial infarction? Explain.
o) Why do you think the monitoring board decided to terminate the study earlier than had been scheduled and release the results in a preliminary report through the New England Journal of Medicine?
p) Based on your analysis of the data, do you think that the board was correct in terminating the study?

12.5 p-VALUE APPROACH TO HYPOTHESIS TESTING INVOLVING TWO POPULATION PROPORTIONS USING THE TI-84 CALCULATOR

The p-value approach will require refinements in steps 3, 4 and 5 of the hypothesis testing procedure. In step 3, the decision rule will be stated as: Reject the H_0 if the p-value of the test statistic is less than α. In step 4, the p-value will be determined using the TI-84 calculator. While in step 5, the conclusion of the hypothesis test is determined by comparing the p-value to the level of significance, α.

Let's reexamine the hypothesis test shown in Example 12.6 using a p-value approach with the aid of the TI-84 calculator.

EXAMPLE 12.7

According to the New England Journal of Medicine, a double-blind experiment was conducted to determine whether taking an aspirin tablet regularly helps prevent heart attacks. The medical researcher randomly assigned six thousand 40 year old males to take an aspirin while another six thousand 40 year old males were given a placebo. Perform a hypothesis test to determine if the following sample results indicate that taking aspirin will significantly lower the incidence of having a heart attack at $\alpha = 1\%$.

After 6 years, 78 of the males within the treatment group while 112 males in the control group suffered heart attacks.

Solution

Step 1 Formulate the null hypothesis, H_0, and the alternative hypothesis, H_a.

If we let population 1 represent the 40 year old males taking an aspirin daily while population 2 represents the 40 year old males taking a placebo daily, then the null hypothesis can be stated as:

H_0: There is **no difference** in the proportion of heart attacks between the population of 40 year old men who take an aspirin daily and the population of 40 year old men who take a placebo daily.

In symbols, we state the null hypothesis as:

H_0: $p_1 - p_2 = 0$.

Since we are trying to determine if the men taking the aspirin will have a *lower proportion of heart attacks* then the alternative hypothesis is *directional* and is stated using a *less than* statement. Consequently, the alternative hypothesis can be stated as:

H_a: The population proportion of heart attacks for 40 year old men taking an aspirin daily is less than the population proportion of heart attacks for 40 year old men taking a placebo daily.

The alternative hypothesis is symbolized as:

H_a: $p_1 - p_2 < 0$.

Step 2 Determine the model to test the null hypothesis, H_0. Under the assumption H_0 is true, determine the expected results of the Sampling Distribution of $\hat{p}_1 - \hat{p}_2$.

To test the null hypothesis, the medical researcher selected a large random independent sample from each population and the sample results are summarized in Table 12.7.

TABLE 12.7

Population 1 Represents 40 Year Old Men Taking Aspirin	Population 2 Represents 40 Year Old Men Taking a Placebo
Sample information: $x_1 = 78$ $n_1 = 6{,}000$	Sample information: $x_2 = 112$ $n_2 = 6{,}000$

To conduct this hypothesis test, we will need to compute the difference between the two sample proportions, $\hat{p}_1 - \hat{p}_2$, to determine if this difference is *significantly less than zero*. Consequently, we will use the **Sampling Distribution of $\hat{p}_1 - \hat{p}_2$** as the hypothesis testing model when conducting this hypothesis test. To determine the **expected results of the Sampling Distribution of $\hat{p}_1 - \hat{p}_2$**, we need to use Theorems 12.1 and 12.2. Under the assumption that H_0 **is true**, we expect the following results:

The Expected Results of the Sampling Distribution of $\hat{p}_1 - \hat{p}_2$

1. *The shape of the sampling distribution is approximately a normal distribution for large samples.*
2. *The mean of the Sampling Distribution of $\hat{p}_1 - \hat{p}_2$ is equal to zero since we are assuming that H_0 is true. That is:*

$$\mu_{\hat{p}_1 - \hat{p}_2} = p_1 - p_2 = 0$$

3. *The estimate of the standard error of the sampling distribution of $\hat{p}_1 - \hat{p}_2$, $s_{\hat{p}_1 - \hat{p}_2}$, is found using the pooled formula:*

$$s_{\hat{p}_1 - \hat{p}_2} = \sqrt{\hat{p}(1-\hat{p})\left(\frac{1}{n_1} + \frac{1}{n_2}\right)}$$

Using the assumption that H_0 is true, i.e. p_1 equals p_2, the pooled sample proportion, $\hat{p}$, is determined by **combining (i.e. pooling) the sample information collected from both populations** and calculated using the formula:

$$\hat{p} = \frac{x_1 + x_2}{n_1 + n_2}$$

Using the sample information, $x_1 = 78$, $n_1 = 6000$, $x_2 = 112$ and $n_2 = 6000$, the pooled sample proportion value, $\hat{p}$, is:

$$\hat{p} = \frac{78 + 112}{6000 + 6000} = \frac{190}{12000}$$

$$\hat{p} \approx 0.0158 \text{ (rounded off to 4 decimal places)}$$

The pooled estimate of the standard error of the Sampling Distribution of The Difference Between Two Proportions is determined using the formula:

$$s_{\hat{p}_1 - \hat{p}_2} = \sqrt{\hat{p}(1-\hat{p})\left(\frac{1}{n_1} + \frac{1}{n_2}\right)}.$$

Using the information: $\hat{p} \approx 0.0158$, $n_1 = 6000$, and $n_2 = 6000$, the pooled estimate of the standard error is:

$$s_{\hat{p}_1 - \hat{p}_2} = \sqrt{(0.0158)(1 - 0.0158)\left(\frac{1}{6000} + \frac{1}{6000}\right)}$$

$$s_{\hat{p}_1 - \hat{p}_2} = \sqrt{(0.0158)(0.9842)\left(\frac{1}{6000} + \frac{1}{6000}\right)}$$

$$s_{\hat{p}_1 - \hat{p}_2} = \sqrt{(0.0158)(0.9842)(0.0003333)}$$

$$s_{\hat{p}_1 - \hat{p}_2} = \sqrt{(0.0000051)}$$

$$s_{\hat{p}_1 - \hat{p}_2} \approx 0.0023 \text{ (rounded off to 4 decimal places)}$$

Thus, the pooled estimate of the standard error of the Sampling Distribution of The Difference Between Two Proportions is: $s_{\hat{p}_1 - \hat{p}_2} \approx 0.0023$.

In summary, the Sampling Distribution of $\hat{p}_1 - \hat{p}_2$ will serve as our hypothesis testing model with the following expected results:

1. the Sampling Distribution can be approximated by the normal distribution for large samples.
2. the mean of the Sampling Distribution of $\hat{p}_1 - \hat{p}_2$ is:

$$\mu_{\hat{p}_1 - \hat{p}_2} = 0$$

3. the pooled estimate of the standard error, $s_{\hat{p}_1 - \hat{p}_2}$, is:

$$s_{\hat{p}_1 - \hat{p}_2} \approx 0.0023$$

This hypothesis testing model is illustrated in Figure 12.10.

The Sampling Distribution of $\hat{p}_1 - \hat{p}_2$ is the hypothesis testing model and is approximately a normal distribution.

$\mu_{\hat{p}_1 - \hat{p}_2} = 0$

$s_{\hat{p}_1 - \hat{p}_2} \approx 0.0023$

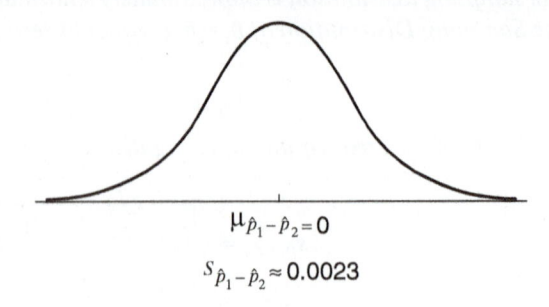

FIGURE 12.10

Step 3 Formulate the decision rule.
a) The alternative hypothesis is directional since we are trying to determine if the proportion of 40 year old men taking aspirin have a *lower* heart attack rate than the proportion of 40 year old men taking a placebo.
b) The type of hypothesis test is one-tailed on the left side.
d) The significance level is: $\alpha = 1\%$.
d) Figure 12.11 illustrates the decision rule for this hypothesis test.
e) Thus, *the decision rule* is: **reject H_0 if the p-value of the test statistic is less than $\alpha = 1\%$.**

REJECT H_0 | FAIL TO REJECT H_0

If the p-value of the test statistic is less than 1%

$\alpha = 1\%$

$\mu_{\hat{p}_1 - \hat{p}_2} = 0$

$s_{\hat{p}_1 - \hat{p}_2} \approx 0.0023$

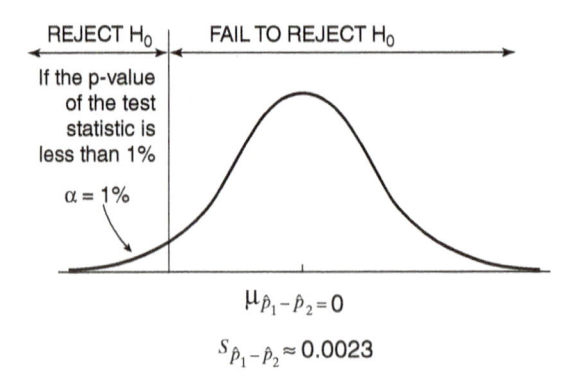

FIGURE 12.11

Step 4 Analyze the sample data.
The **2-PropZTest** function can be used to perform a hypothesis test involving two population proportions using independent samples. To determine the p-value of the test statistic, the sample statistics must be entered into the TI-84's TESTS menu. To access the **2-PropZTest** function in the TESTS menu press [STAT] ▶ ▶ [TESTS]. This is shown in Figure 12.12.

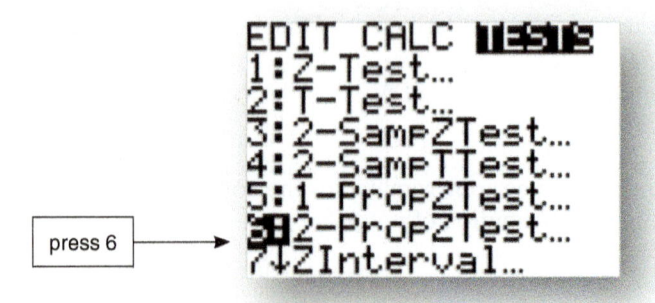

```
EDIT CALC TESTS
1:Z-Test…
2:T-Test…
3:2-SampZTest…
4:2-SampTTest…
5:1-PropZTest…
6:2-PropZTest…
7↓ZInterval…
```
press 6

FIGURE 12.12

Section 12.5 *p-Value Approach to Hypothesis Testing Involving Two Population Proportions Using the TI-84 Calculator*   **657**

Press **6** to access the **2-PropZTest** menu. Enter the sample statistics into the menu. This is shown in Figure 12.13.

Alternative Hypothesis →

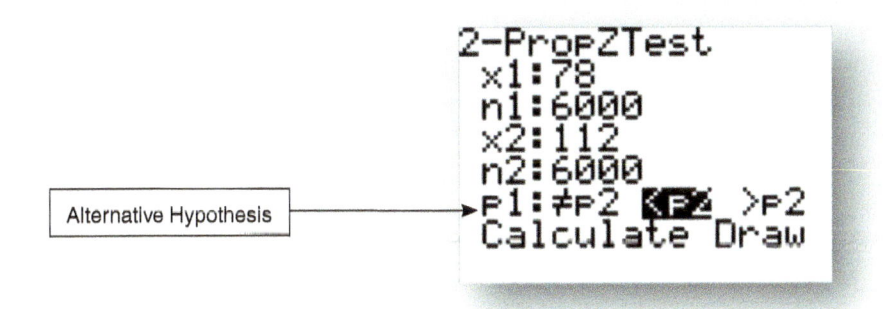

```
2-PropZTest
x1:78
n1:6000
x2:112
n2:6000
p1:≠p2  <p2  >p2
Calculate Draw
```

FIGURE 12.13

Highlight **Calculate** and press ENTER to obtain the output screens for this test. These are illustrated in Figure 12.14a and Figure 12.14b.

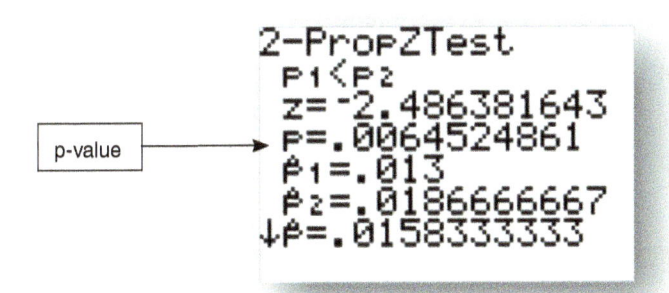

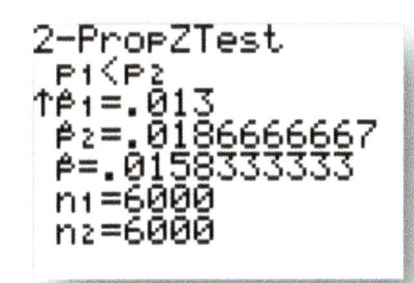

p-value →

```
2-PropZTest
p1<p2
z=-2.486381643
p=.0064524861
p̂1=.013
p̂2=.018666667
↓p̂=.0158333333
```

```
2-PropZTest
p1<p2
↑p̂1=.013
p̂2=.018666667
p̂=.0158333333
n1=6000
n2=6000
```

FIGURE 12.14a   **FIGURE 12.14b**

From the output screen, the p-value of the test statistic $z \approx -2.49$ is 0.0065. Using this p-value we can position the test statistic on the hypothesis testing model shown in Figure 12.15.

REJECT H_0 | FAIL TO REJECT H_0

If the p-value of the test statistic is less than 1%

$\alpha = 1\%$

test statistic $z \approx -2.49$ $\mu_{\hat{p}_1 - \hat{p}_2} = 0$
(p-value ≈ 0.0065) $s_{\hat{p}_1 - \hat{p}_2} \approx 0.0023$

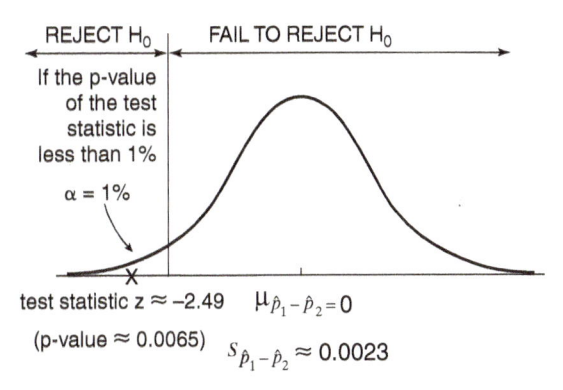

FIGURE 12.15

Step 5 State the conclusion.
Since the p-value (0.0065) of the test statistic is less than $\alpha = 1\%$, our conclusion is reject H_0 and accept H_a at a p-value of 0.0065. Therefore, *we can conclude that the proportion of 40 year old men taking an aspirin daily who have heart attacks is significantly less than the proportion of 40 year old men who don't take aspirin.*

12.6 TWO POPULATION HYPOTHESIS TESTING SUMMARIES USING INDEPENDENT SAMPLES

The following outline is a summary of the 5-step hypothesis testing procedure for a hypothesis test involving two population proportions for large independent samples using the test statistic.

Test Statistic Approach to Hypothesis Tests Involving Two Population Proportions for Large Independent Samples

STEP 1. Formulate the hypotheses, H_0 and H_a.
Null Hypothesis, H_0: $p_1 - p_2 = 0$
Alternative Hypothesis, H_a: *has one of the following three forms:*

H_a: $p_1 - p_2 < 0$ or
H_a: $p_1 - p_2 > 0$ or
H_a: $p_1 - p_2 \neq 0$

STEP 2. Determine the model to test the null hypothesis, H_0. Under the assumption H_0 is true, the expected results of the Sampling Distribution of $\hat{p}_1 - \hat{p}_2$

are:

(1) *The shape of the sampling distribution is approximately a normal distribution for large samples.*
(2) *The mean of the Sampling Distribution of $\hat{p}_1 - \hat{p}_2$ is equal to zero. That is:*

$$\mu_{\hat{p}_1 - \hat{p}_2} = p_1 - p_2 = 0$$

(3) *The pooled formula for the estimate of the standard error of the Sampling Distribution of $\hat{p}_1 - \hat{p}_2$, symbolized $s_{\hat{p}_1 - \hat{p}_2}$, is:*

$$s_{\hat{p}_1 - \hat{p}_2} = \sqrt{\hat{p}(1-\hat{p})\left(\frac{1}{n_1} + \frac{1}{n_2}\right)}$$

where $\hat{p}$ represents the pooled sample proportion and is computed by one of the following formulas:

$$\hat{p} = \frac{x_1 + x_2}{n_1 + n_2} \quad or \quad \hat{p} = \frac{n_1 \hat{p}_1 + n_2 \hat{p}_2}{n_1 + n_2}$$

STEP 3. Formulate the decision rule.
(a) alternative hypothesis, H_a: directional or non-directional.
(b) type of test: 1TT or 2TT.
(c) significance level: $\alpha = 1\%$ or $\alpha = 5\%$.
(d) find critical z scores using Table II.
(e) construct the hypothesis testing model.
(f) state the decision rule.

STEP 4. Analyze the sample data.
Calculate the test statistic using the formula:

$$z = \frac{(\hat{p}_1 - \hat{p}_2) - \mu_{\hat{p}_1 - \hat{p}_2}}{s_{\hat{p}_1 - \hat{p}_2}}$$

STEP 5. State the conclusion.
Compare the test statistic to the critical z score(s), z_c, of the decision rule and either:

Reject H_0 and Accept H_a at α

or

Fail to Reject H_0 at α

Table 12.8 contains a condensed version of the hypothesis test involving two population proportions for large samples.

Section 12.6 Two Population Hypothesis Testing Summaries Using Independent Samples 659

TABLE 12.8 Summary of Hypothesis Tests Involving Two Population Proportions for Large Independent Samples

Form of the Null Hypothesis	Conditions of Test	Test Statistic Formula	Sampling Distribution Is a:
$H_0: p_1 - p_2 = 0$	the two samples are large, if: $n_1 p_1$, $n_1(1-p_1)$, $n_2 p_2$ and $n_2(1-p_2)$ are all greater than or equal to 10.	$z = \dfrac{(\hat{p}_1 - \hat{p}_2) - \mu_{\hat{p}_1 - \hat{p}_2}}{s_{\hat{p}_1 - \hat{p}_2}}$ where: $s_{\hat{p}_1 - \hat{p}_2} = \sqrt{\hat{p}(1-\hat{p})\left(\dfrac{1}{n_1} + \dfrac{1}{n_2}\right)}$ and $\hat{p} = \dfrac{x_1 + x_2}{n_1 + n_2}$ or $\hat{p} = \dfrac{n_1 \hat{p}_1 + n_2 \hat{p}_2}{n_1 + n_2}$	Normal Distribution

The following outline is a summary of the 5-step hypothesis testing procedure for a hypothesis test involving two population proportions for large independent samples using p-values.

---

p-Value Approach to Hypothesis Tests Involving Two Population Proportions for Large Independent Samples Using the TI-84 Calculator

STEP 1. Formulate the hypotheses, H_0 and H_a.
Null Hypothesis, $H_0: p_1 - p_2 = 0$
Alternative Hypothesis, H_a: *has one of the following three forms:*
 $H_a: p_1 - p_2 < 0$ or
 $H_a: p_1 - p_2 > 0$ or
 $H_a: p_1 - p_2 \neq 0$

STEP 2. Determine the model to test the null hypothesis, H_0. Under the assumption H_0 is true, the expected results of the Sampling Distribution of $\hat{p}_1 - \hat{p}_2$ are:

(1) *The shape of the sampling distribution is approximately a normal distribution for large samples.*
(2) *The mean of the Sampling Distribution of $\hat{p}_1 - \hat{p}_2$ is equal to zero. That is:*
$$\mu_{\hat{p}_1 - \hat{p}_2} = p_1 - p_2 = 0$$
(3) *The pooled formula for the estimate of the standard error of the Sampling Distribution of $\hat{p}_1 - \hat{p}_2$, symbolized $s_{\hat{p}_1 - \hat{p}_2}$, is:*
$$s_{\hat{p}_1 - \hat{p}_2} = \sqrt{\hat{p}(1-\hat{p})\left(\dfrac{1}{n_1} + \dfrac{1}{n_2}\right)}$$

where $\hat{p}$ represents the pooled sample proportion and is computed by one of the following formulas:
$$\hat{p} = \dfrac{x_1 + x_2}{n_1 + n_2} \quad \text{or} \quad \hat{p} = \dfrac{n_1 \hat{p}_1 + n_2 \hat{p}_2}{n_1 + n_2}$$

STEP 3. Formulate the decision rule.
(a) alternative hypothesis, H_a: directional or non-directional.
(b) type of test: 1TT or 2TT.
(c) significance level: $\alpha = 1\%$ or $\alpha = 5\%$.
(d) construct the hypothesis testing model.
(e) state the decision rule as: Reject the H_0 if the p-value of the test statistic is less than or equal to α.

STEP 4. Analyze the sample data.
Enter the necessary sample data into the TI-84 calculator to determine the p-value of the test statistic. Use 2-PropZTest.

STEP 5. State the conclusion.
Compare the p-value of the test statistic to the decision rule and either:

Reject H_0 and Accept H_a at α

or

Fail to Reject H_0 at α

KEY TERMS

Common hypothesized population proportion, p 639
Control group 648
Count data 628
Difference between sample proportions, $\hat{p}_1 - \hat{p}_2$ 630
Double-blind experiment 649
Estimate of the standard error of the difference between two proportions, $\sigma_{\hat{p}_1 - \hat{p}_2}$ 634
Independent Samples 627
Mean of the Sampling Distribution of the Difference Between Two Proportions, $\mu_{\hat{p}_1 - \hat{p}_2}$ 634

Placebo 649
Pooled estimate of the standard error of the Sampling Distribution of $\hat{p}_1 - \hat{p}_2$, $s_{\hat{p}_1 - \hat{p}_2}$ 639
Pooled sample proportion, $\bar{p}$ 639
Population proportion, 628
Sample proportion, $\hat{p}$ 628
Single-blind experiment 649
Standard Deviation of the Sampling Distribution of the Difference Between Two Proportions or Standard error of the difference between two proportions, $\sigma_{\hat{p}_1 - \hat{p}_2}$ 634

Test statistic 641
The Sampling Distribution of the Difference Between Two Proportions or the Sampling Distribution of $\hat{p}_1 - \hat{p}_2$ 632
Theoretical probability distribution 632
Treatment effect 649
Treatment group 648

SECTION EXERCISES

Note: The data sets in these Section Exercises can be downloaded in either Microsoft EXCEL or TI-84 format from the textbook's website at www.MyStatLab.com.

Sections 12.1–12.2

1. Studies consisting of population means involve measurements while studies dealing with population proportions will produce _____.

2. The symbol used to represent the population proportion of items or individuals possessing a particular characteristic is _____.

3. When dealing with two populations, the symbol used to represent the population proportion for population 2 is _____.

4. A sample proportion is symbolized by _____ and read _____.

5. The symbol p_1 represents _____.
   The symbol n_1 represents _____.
   The symbol x_1 represents _____.

6. The sampling distribution used to compare the proportions of two populations is called the _____.

7. The symbol used to represent the difference between the sample proportions is _____.

8. In the case of sampling from two populations, the Sampling Distribution of The Difference Between Two Proportions is approximately normal when the independent random sample selected from each of the two populations are both _____. Therefore, in the case of two samples, both samples are considered to be large if the quantities _____, _____, _____, and _____ are all greater than 5.

9. The mean of the Sampling Distribution of the Difference Between Two Proportions, denoted by $\mu_{\hat{p}_1 - \hat{p}_2}$, is equal to _____.

10. The standard deviation of the Sampling Distribution of the Difference Between Two Proportions or commonly called the standard error of the differences between two proportions, denoted by $\sigma_{\hat{p}_1 - \hat{p}_2}$, is equal to _____.

11. For large samples, the shape of the Sampling Distribution of $\hat{p}_1 - \hat{p}_2$ is approximately a _____ _____.

12. According to Theorem 12.2, the mean of the Sampling Distribution of $\hat{p}_1 - \hat{p}_2$ is determined by the formula: _____ = _____. Under the assumption that the null hypothesis, H_0, is true, then $\mu_{\hat{p}_1 - \hat{p}_2}$ = _____.

13. The estimate of the standard error of the Sampling Distribution, $s_{\hat{p}_1 - \hat{p}_2}$, of $\hat{p}_1 - \hat{p}_2$ is determined by the formula: _____ = _____.

14. If the sample data values selected from one population are unrelated to the sample data values selected from the second population, then the two sample are _____.

15. The survey information within the following table represents the number of men and women aged 18 and over who never eat breakfast. Use the table information to determine the sample proportion of men and women aged 18 and over who never eat breakfast.

	Groups	
	Men 18 Years and Over	Women 18 Years and Over
Sample Size	$n_1 = 1450$	$n_2 = 1500$
Number of individuals who don't eat breakfast	$x_1 = 389$	$x_2 = 327$

16. According to a recent survey of 760 men aged 25–34 and 790 women aged 25–34 produced the following results: 26.3% of the men live alone while 12% of the women live alone. Determine:
   a) the sample proportion of men aged 25–34 who do not live alone
   b) the sample proportion of women aged 25–34 who do not live alone
   c) the number of men within the survey who indicated that they lived alone
   d) the number of women within the survey who indicated that they lived alone
   e) the number of men within the survey who indicated that they do not live alone
   f) the number of women within the survey who indicated that they do not live alone

17. Identify the notation and state the formula for:
    a) the mean of the Sampling Distribution of the Difference Between Two Proportions.
    b) the standard deviation of the Sampling Distribution of the Difference Between Two Proportions.
    c) the standard error of the difference between two proportions.
    d) the estimate of the standard error of the difference between two proportions.

18. Based upon the latest Reality Poll, 16% of home owners have an annual maintenance expenditure of between $1000–$2000, while 38% of condo owners have an annual maintenance expenditure of between $1000–$2000. Assuming these percentages represent the proportions of the indicated populations and if an independent random sample of 900 owners is selected from each population, then:
    a) define the two populations, and the proportions of these populations.
    Determine the:
    b) shape of the Sampling Distribution of $\hat{p}_1 - \hat{p}_2$
    c) mean of the Sampling Distribution of $\hat{p}_1 - \hat{p}_2$
    d) standard deviation of the Sampling Distribution of $\hat{p}_1 - \hat{p}_2$
    e) standard error of the difference between two proportions

19. a) Identify the notation and formula used to compute the estimate of the standard error of the difference between two proportions.
    b) State the conditions of when this estimate of the standard error can be applied.
    c) Identify the notation and formula used to compute the estimate of the standard deviation of the Sampling Distribution of the Difference Between Two Proportions.

20. Using the following sample values:
    $$\hat{p}_1 = 0.38 \text{ and } n_1 = 270$$
    $$\hat{p}_2 = 0.46 \text{ and } n_2 = 350.$$
    compute:
    a) the estimate of the standard error of the difference between two proportions.
    b) the estimate of the standard deviation of the Sampling Distribution of the Difference Between Two Proportions.

21. Using the following sample values:
    $$x_1 = 360 \text{ and } n_1 = 1500$$
    $$x_2 = 388 \text{ and } n_2 = 1250$$
    compute:
    a) the proportions for each of the two samples.
    b) the estimate of the standard error of the difference between two proportions.
    c) the estimate of the standard deviation of the Sampling Distribution of the Difference Between Two Proportions.

22. In a PC magazine for dummies, a poll of 1500 readers indicated that 78% of men and 83% of the women said they preferred at most one technical article on computer hardware in each monthly issue. Assuming that there were 970 men and 530 women sampled in this poll, answer the following questions:
    a) identify the two population proportions
    b) identify each sample proportion
    c) state the value of each sample proportion
    d) compute the estimate of the standard error of the difference between two proportions
    e) compute the estimate of the standard deviation of the Sampling Distribution of the Difference Between Two Proportions.

23. According to the Nielsen Media Research, the share of the T.V. viewing audience of the Networks: CBS, ABC and NBC was 61% for 2001. The share of the T.V. viewing audience of the Networks was 70% in 2000. If the Nielsen Media Research surveyed 1500 households in 2000 and 2000 households in 2001 then answer the following questions:
    a) define the two populations, and the proportions of these populations
    b) identify each sample proportion
    c) compute the value of each sample proportion
    d) compute the estimate of the standard error of the difference between two proportions
    e) compute the estimate of the standard deviation of the Sampling Distribution of the Difference Between Two Proportions.

24. According to a study by the Better Sleep Council which appeared in the August 8, 1994 issue of *USA Today*, nearly one-third of people say their job has been affected by a lack of sleep. Of those people who were surveyed, 26% of the men and 13% of the women admit to dozing off at their workplace. If this study involved 975 men and 895 women, then:
    a) define the two populations, and the proportions of these populations
    b) identify each sample proportion
    c) compute the value of each sample proportion
    d) compute the estimate of the standard error of the difference between two proportions
    e) compute the estimate of the standard deviation of the Sampling Distribution of the Difference Between Two Proportions.

25. The problems in this chapter require the selection of two samples from:
    a) the same population
    b) two different populations
    c) either the same or different populations.

26. In practice, it is extremely easy to compute the proportion of a population.
    (T/F)

27. The Sampling Distribution of the Differences Between Sample Proportions is used to determine if the difference between the sample proportions, $\hat{p}_1 - \hat{p}_2$, is significant to infer that there is a statistically significant difference between the population proportions, $\hat{p}_1 - \hat{p}_2$.
    (T/F)

28. The estimate of the standard deviation of the Sampling Distribution of the Difference Between Two Proportions, $s_{\hat{p}_1 - \hat{p}_2}$ can be obtained using:
    $$s_{\hat{p}_1 - \hat{p}_2} = \sqrt{\hat{p}(1-\hat{p})\left(\frac{1}{n_1} + \frac{1}{n_2}\right)}$$
    only when we assume $p_1 = p_2$.
    (T/F)

29. One way to insure that $n_1 p_1$, $n_2 p_2$, $n_1(1 - p_1)$ and $n_2(1 - p_2)$ are all greater than 5 is to choose the sample sizes that are large enough.
    (T/F)

30. A study was conducted to determine if the effectiveness of two vaccines in the prevention of a new flu strain. The results of the study are shown in the following table.

	Flu Vaccines	
	Vaccine A	Vaccine B
Number of individuals who didn't catch the flu	$x_1 = 432$	$x_2 = 270$
Sample size	$n_1 = 650$	$n_2 = 580$

Use the information in the previous table to determine the sample proportion of individuals receiving:
a) vaccine A
b) vaccine B
c) vaccine A who caught the flu.
d) vaccine B who caught the flu.

31. Last year, a random sample of 978 people making dinner reservations, 392 people requested the nonsmoking section. Recently, a random sample 1014 people showed that 729 people prefer the nonsmoking section when making dinner reservations. Using this information, determine the sample proportion of:
a) people who preferred the nonsmoking section last year.
b) people who preferred the nonsmoking section this year.
c) people who did not prefer the nonsmoking section last year.
d) people who did not prefer the nonsmoking section this year.

32. A current newspaper article reported that 73% of women drivers and 87% of men drivers pump their own gas. If this survey was based upon a random sample of 1005 male drivers and a random sample of 1100 female drivers, then determine:
a) the sample proportion of male drivers who do not pump their own gas
b) the sample proportion of women drivers who do not pump their own gas
c) the number of men drivers within the survey who indicated that they pump their own gas
d) the number of women drivers within the survey who indicated that they pump their own gas
e) the number of men within the survey who indicated that they do not pump their own gas
f) the number of women within the survey who indicated that they do not pump their own gas

33. Using the following information:

$$p_1 = 0.73, n_1 = 1230, \text{ and } x_1 = 901$$
$$p_2 = 0.62, n_2 = 1570, \text{ and } x_2 = 989$$

determine:
a) the proportion of the sample selected from population 1
b) the proportion of the sample selected from population 2
c) if the samples are considered large
d) the difference between the sample proportions: $\hat{p}_1 - \hat{p}_2$
e) the shape of the Sampling Distribution of $\hat{p}_1 - \hat{p}_2$
f) mean of the Sampling Distribution of $\hat{p}_1 - \hat{p}_2$
g) standard deviation of the Sampling Distribution of $\hat{p}_1 - \hat{p}_2$
h) standard error of the difference between two proportions

34. According to Economist magazine, 55% of singles and 68% of married people report being happy with life. If 1050 of single and 1325 of married respondents report being happy with life, answer the following questions:

a) define the two populations, and the proportions of these populations.
b) compute the value of each sample proportion.
c) compute the estimate of the standard error of the difference between two proportions.
d) compute the estimate of the standard deviation of the Sampling Distribution of the Difference Between Two Proportions.

Section 12.3

35. The common population proportion value is symbolized by _____. The best estimate for this common population proportion value $\hat{p}$, is determined by _____ the sample information collected from both populations.

36. The formula for the pooled sample proportion, $\hat{p}$, is determined by using one of the following formulas:

$$\hat{p} = \underline{} \text{ or } \hat{p} = \underline{}.$$

37. The pooled formula for the estimated standard error of the sampling distribution of $\hat{p}_1 - \hat{p}_2$, $s_{\hat{p}_1 - \hat{p}_2}$, is:

$$s_{\hat{p}_1 - \hat{p}_2} = \underline{}$$

38. To perform a hypothesis test involving two population proportions, the Sampling Distribution of the Difference Between _____ _____ is used as the hypothesis testing model.

39. The notation $\hat{p}_1 - \hat{p}_2$ represents the difference between the _____ _____, where $\hat{p}_1$ represents the _____ of sample 1 and $\hat{p}_2$ represents the _____ of sample 2.

40. The null hypothesis of a hypothesis test involving two population proportions, p_1 and p_2, is stated as: There is _____ difference between the _____ of population 1 and the _____ of population 2. In symbols, H_0 is written as: _____ = 0.

41. The alternative hypothesis of a hypothesis test involving two population proportions, p_1 and p_2, has one of three possible forms. They are:
a) The proportion of population 1 is greater than the _____. In symbols, H_a is written as $H_a: p_1 - p_2$ _____ 0.
b) The _____ of population 1 is less than the proportion of _____. In symbols, H_a is written as $H_a: p_1 - p_2$ _____ 0.
c) The proportion of _____ 1 is not equal to the _____ of population _____. In symbols, H_a is written as: $H_a: p_1 - p_2$ _____ 0

42. The formula for the critical z score, z_c, for a hypothesis test involving two population proportion is: $z_c =$ _____.

43. To determine the conclusion of a hypothesis test involving two population proportions, the test statistic, denoted _____, is compared to the critical z score(s), z_c.

44. The pooled sample proportion value, $\hat{p}$, is used when:
a) computing $\mu_{\hat{p}_1 - \hat{p}_2}$
b) stating H_0
c) computing $\sigma_{\hat{p}_1 - \hat{p}_2}$
d) stating H_a

45. When computing the critical z score(s), z_c, a value from which distribution is utilized?
a) normal distribution
b) binomial distribution
c) t distribution
d) skewed distribution

46. If 40 out of 100 men and 50 out of 150 women agreed with the proposal, then what is the pooled sample proportion value, $\hat{p}$?
a) 0.40   b) 0.60   c) 0.365   d) 0.36

47. If the pooled sample value is $\bar{p} = 0.40$, $n_1 = 200$ and $n_2 = 150$, then $\sigma_{\hat{p}_1 - \hat{p}_2}$ equals:
   a) 0.003   b) 0.002   c) 0.053   d) 529

48. If the critical z score is 1.65 and the test statistic is: z = 1.96, then the conclusion for a 1TT on the right is:
   a) Reject H_0 and Accept H_a
   b) Fail to reject H_0
   c) Accept H_0
   d) Reject H_a

49. The form of the null hypothesis, H_0, for testing the difference between two population proportions is:
   a) $H_0: p_1 - p_2 = 0$
   b) $H_0: p_1 - p_2 < 0$
   c) $H_0: p_1 - p_2 > 0$
   d) $H_0: p_1 - p_2 \neq 0$

50. The standard error of the difference between the two proportions is the standard deviation of the Sampling Distribution of the Difference Between Two Proportions. (T/F)

51. The null hypothesis for tests that compare the proportions from two populations is usually expressed as $p_1 - p_2 = 0$, where p_1 represents the proportion of population 1 and p_2 represents the proportion of population 2. (T/F)

52. If the difference between the sample proportions $\hat{p}_1 - \hat{p}_2$ is different from zero, then we can conclude that there is a statistically significant difference between p_1 and p_2. (T/F)

53. If p_1 and p_2 are unknown, then one cannot perform a hypothesis test for the difference between two proportions. (T/F)

54. $\bar{p}$, the proportion obtained by combining results from both samples is based upon the assumption that n_1 and n_2 are equal. (T/F)

55. If the test statistic is z = 0, then the null hypothesis is correct. (T/F)

56. When doing a hypothesis test for a comparison of population proportions, it is very important to choose random samples. (T/F)

57. Under the assumption that the null hypothesis, H_0, is true, what is the value of the mean of the Sampling Distribution of $\hat{p}_1 - \hat{p}_2$?

58. State the pooled formulas for the best estimate for the common population proportion value, p. What is the name of each of these formulas?

59. The following information concerning hypothesis tests for two proportions. For each
   (i) define the two population proportions.
   (ii) state the null and alternative hypotheses.
   (iii) identify the type of test: 1TT or 2TT.
   a) A researcher would like to determine if married women are more apt to express anger than unmarried women.
   b) The Naturalizer Company would like to determine if there is a significantly smaller percentage of males compared to females that believe wearing miniskirts is inappropriate office attire for women.
   c) The National Association of Home Builders is interested in determining if the percentage of new homes with garages large enough for two or more cars in 2010 is different compared to new homes in 1990.
   d) A pharmaceutical company is preparing to do a study to investigate whether the proportion of college educated mothers who breast feed their babies in the hospital is higher than non-college educated mothers.
   e) The Motion Picture Association of America would like to study whether the percentage of the U.S. population that attends the movies every week has decreased in 2010 compared to 2000.

60. State the pooled formula for $s_{\hat{p}_1 - \hat{p}_2}$, the estimated standard error of the Sampling Distribution of $\hat{p}_1 - \hat{p}_2$.

61. Compute the pooled sample proportion, $\bar{p}$, for the sample information in parts (a) to (h):
   a) $x_1 = 540$, $n_1 = 1175$, $x_2 = 415$ and $n_2 = 950$
   b) $x_1 = 663$, $n_1 = 850$, $x_2 = 537$ and $n_2 = 1060$
   c) $n_1 = 1050$, $\hat{p}_1 = 0.46$, $n_2 = 1170$, and $\hat{p}_2 = 0.58$
   d) $n_1 = 750$, $\hat{p}_1 = 0.74$, $n_2 = 675$, and $\hat{p}_2 = 0.69$
   e) $x_1 = 930$, $n_1 = 1260$, $x_2 = 685$ and $n_2 = 1075$
   f) $x_1 = 786$, $n_1 = 1150$, $x_2 = 846$ and $n_2 = 1390$
   g) $n_1 = 1510$, $\hat{p}_1 = 0.86$, $n_2 = 1425$, and $\hat{p}_2 = 0.79$
   h) $n_1 = 870$, $\hat{p}_1 = 0.37$, $n_2 = 990$, and $\hat{p}_2 = 0.45$

62. (a) to (h): For each part in question 61, use the sample information to determine the pooled estimate of the standard error of the Sampling Distribution of The Difference Between Two Proportions.

For the problems 63–77, use the 5 step procedure to perform a hypothesis test. Include the following within your test:

   a) state the null and alternative hypotheses.
   b) calculate the expected results for the experiment assuming the null hypothesis is true and state the hypothesis test model.
   c) formulate the decision rule.
   d) compute the test statistic.
   e) state the conclusion and answer any question(s) posed in the problem.

63. According to a *Psychology Today* survey, 29% of females like to take chances while 35% of males prefer to take chances. Assume these estimates are based upon a random sample of 750 females and 730 males. Perform a hypothesis test at $\alpha = 1\%$ to determine if there is a significant difference between the proportion of females and the proportion of males who like to take chances.

64. According to a survey by Opinion Research Corporation, 38% of 1050 smokers and 61% of 1125 non-smokers said that they strongly agree that smoking should be banned in all public places. Perform a hypothesis test at $\alpha = 1\%$ to determine if the proportion of non-smokers who strongly agree that smoking should be banned in all public places is significantly greater than that of smokers.

65. A marriage counselor believes that couples that live together prior to getting married are more likely to get divorced than couples that did not live together prior to marriage. To test his claim, the marriage counselor randomly surveys 1050 couples that lived together prior to marriage and 1000 couples that did not live together prior to marriage and determines that 376 of the couples that lived together prior to marriage got divorced, and 287 of the couples that did not live together prior to marriage got divorced. Can the marriage counselor conclude that the

proportion of couples that got divorced that lived together prior to marriage is significantly higher than the proportion of couples that got divorced that did not live together prior to marriage at $\alpha = 1\%$?

66. A recent Pew Research Center study compared the percent of internet users in different income brackets on each social media network. According to the study, 17% of Twitter internet users earned at least $75,000 while 14% of Twitter internet users earned from $50,000 to $75,000. A social researcher believes that high income earners use Twitter more frequently. Perform a hypothesis test at $\alpha = 5\%$ to determine if the proportion of Twitter users earning at least $75,000 is significantly different than the proportion of Twitter users earning from $50,000 to $75,000. Assume these sample results are based on a random sample of 2000 internet users earning from $50,000 to $75,000 and 1500 internet users earning at least $75,000. Can the social researcher conclude that high income earners use Twitter more frequently at $\alpha = 1\%$?

67. A poll of women aged 18–59 years conducted by *New Woman* determined that 71% of the women in the Northwest say they are attractive without makeup while only 57% of the women in the Midwest say they are attractive without makeup. Perform a hypothesis test at $\alpha = 0.05$ to determine if the proportion of Northwest women who say they are attractive without makeup is significantly greater than the proportion of Midwest women who say they are attractive without makeup. Assume these estimates are based on a random sample of 1070 Northwest women and 1180 Midwest women.

68. The National Highway Traffic Safety Administration has released the results of a recent survey which indicate that people who wear seat belts are more likely to survive and be uninjured in car crashes than people not wearing seat belts. According to their survey, 21.1% of people not wearing seat belts are uninjured in passenger car crashes compared to 58.8% of people wearing a seat belt. Perform a hypothesis test at $\alpha = 0.01$ to determine if the proportion of people wearing seat belts who are uninjured in passenger car crashes is significantly greater than the proportion of people not wearing seat belts. Assume these estimates are based on a random sample of 1025 passenger car crashes where the passengers wore seat belts and a random sample of 1185 passenger car crashes involving passengers who didn't wear seat belts.

69. The Opinion Research Corporation released a study stating that 83% of non-smokers strongly believe that nicotine is an addictive substance while only 79% of smokers believe that nicotine is an addictive substance.

Perform a hypothesis test at $\alpha = 0.05$ to determine if the proportion of smokers is significantly less than the proportion of non-smokers who strongly believe that nicotine is an addictive substance. Assume these estimates are based on a random sample of 990 smokers and a random sample of 1040 non-smokers.

70. According to the *American Journal of Sports Medicine*, the following table represents the frequency of knee injuries for grass and AstroTurf for the National Football League from 1980–1989.

	MCL Sprains[1]	ACL Sprains[2]	Other	Total number of Injuries
AstroTurf	395	59	123	577
Natural Grass	342	55	107	504
Total Number of Injuries	737	114	230	1081

[1]Medical Collateral Knee; [2]Anterior Cruciate Knee Ligament

Using the table information, perform a hypothesis test to determine:
i) if there is a significant greater proportion of MCL Sprains on AstroTurf than Natural Grass at $\alpha = 0.05$.
ii) if there is a significant greater proportion of ACL Sprains on AstroTurf than Natural Grass at $\alpha = 0.05$.
iii) if there is a significant greater proportion of total injuries on AstroTurf than Natural Grass at $\alpha = 0.05$.

71. Researchers at the University of Singapore believe that diet is a critical factor in coronary heart disease in men. To test this claim, a sample of case histories of men was randomly chosen from society A where the main protein component of their diet is red meat. A second sample of case histories of men was randomly chosen from society B where the main protein component of their diet is fish. Researchers found that of 200 case histories from society A, 26 have experienced some form of coronary heart disease and of 200 case histories from society B, 14 have experienced some form of coronary heart disease. Do these results indicate that the researchers belief is correct at $\alpha = 1\%$?

72. A study in the New England Journal of Medicine reported a test to determine if there was a higher percentage of women who still had significant levels of a sleep enhancer Zolpidem in their system after 8 hours of sleep than did men. In the study, of the 150 women who took the drug, 50 still had significant levels of Zolpidem in their systems. Of the 125 men who took the drug, 31 still had significant levels of Zolpidem in their systems after 8 hours of sleep.

Can one conclude that the levels of Zolpidem that remained in the women's system after 8 hours of sleep is significantly higher than it is for men? Test at an alpha level of 5%.

73. According to a recent Nielsen study on "Why Americans Use Social Media," one of the top reasons was to stay in touch with family and friends. The study indicated that 64% of women and 59% of men use social media to stay in touch with family and friends. Sydney, a social researcher, believes that more women use social media to stay in touch with family and friends. Perform a hypothesis test at $\alpha = 1\%$ to determine if the proportion of women who state they use social media to stay in touch with family and friends is significantly greater than the proportion of men who use social media to stay in touch with family and friends. Assume these sample results are based on a random sample of 1500 women and 1200 men. Can the social researcher conclude that more women use social media to stay in touch with family and friends at $\alpha = 1\%$?

74. An economist believes that women are more likely than men to carry a credit card balance from month to month. To test her claim, the economist randomly surveys 800 female credit card holders and 750 male credit card holders and

determines that 480 females and 413 males carry a credit card balance from month to month. Can the economist conclude that the proportion of women that carry a credit card balance from month to month is significantly higher than the proportion of males that carry a credit card balance from month to month at $\alpha = 5\%$?

75. A researcher wishes to determine if there is a significant difference in the proportion of men and women who approve of mothers working when their children are less than five years old. Two random samples of men and women are independently drawn. On the basis of the sample data below, can the researcher show that there is a significant difference in the men's and women's attitude toward working mothers at $\alpha = 5\%$?

Group	Sample Size	Number Who Approve of Working Mothers
Men	300	120
Women	200	90

76. A quality control computer engineer would like to determine if a new manufacturing process will significantly reduce the proportion of defective computer chips. In a random sample of 400 computer chips manufactured using the new process, 31 were defective while in a random sample of 400 computer chips manufactured using the present process, 47 were defective.

Based upon the sample data, can the quality control engineer state the new manufacturing process is significantly better than the present process at $\alpha = 0.05$?

77. A medical researcher is studying the effects of two alternative drugs for the treatment of high blood pressure. Drug A has no apparent side-effect whereas Drug B sometimes has a side-effect of mild headaches. The medical researcher is willing to recommend the use of Drug B if the proportion of people cured is significantly higher than the proportion of people cured using Drug A. The medical researcher used the drugs experimentally on two random samples of people suffering from high blood pressure. Drug A was administered to a group of 100 patients while Drug B was administered to a group of 105 patients. At the end of the experimental period, 62 of those treated with Drug A were classified as "cured" while 69 of those treated with Drug B were also classified as "cured." Can the medical researcher recommend the use of Drug B over Drug A at the 0.01 level of significance?

Section 12.4

78. Statistical studies concerning the comparison of two groups of subjects usually consists of an experiment involving a _____ and _____.
79. An ineffective pill administered to the control group is called a _____.
80. An experiment where the subjects of the experiment do not know who is receiving the treatment and who is receiving the placebo is called a ____-____ ____.
81. An experiment where neither the subjects nor the researchers involved in the experiment know which group of subjects is the treatment or the control group is called a ____-____ ____.

82. Hypotheses tests that involve the comparison of the proportions from two populations use the Sampling Distribution of the Difference Between Two Proportions as the hypothesis testing model. (T/F)
83. Explain the difference between a treatment and control group.
84. What is a placebo?
85. What is meant by a treatment effect?
86. Define:
    a) a single-blind experiment.
    b) a double-blind experiment.

For each of the following studies stated in problems 87 to 88, define:

a) the treatment group
b) the control group
c) the null and alternative hypotheses.

87. The results of a study conducted at Thomas Jefferson University indicated that a cancer vaccine treatment made from a patient's own tumor cells was administered to a group of patients who were operated on for malignant melanoma while another group of patients received the standard treatment of surgery and a placebo vaccine. The results of the study showed that the group receiving the cancer vaccine and the surgery had a greater percentage of patients remaining disease free after three years.

88. The consumer report publication, *Health Letter*, reported the following research to determine whether acupuncture significantly relieves pain. In one acupuncture study, a group of randomly selected people suffering from back pain and headache received genuine acupuncture while a similar group of people also suffering from back pain and headache received a "sham" acupuncture in which needles were placed at nonacupuncture points.

For problems 89–92, use the 5 step procedure to perform a hypothesis test. Include the following within your test:

a) state the null and alternative hypotheses.
b) calculate the expected results for the experiment assuming the null hypothesis is true and state the hypothesis test model.
c) formulate the decision rule.
d) compute the test statistic.
e) state the conclusion and answer any question(s) posed in the problem.

89. A study in the *New England Journal of Medicine* reported that an experimental drug significantly reduces life-threatening blockages after the artery-opening procedure angioplasty one year later. In the study, for a group of 150 patients who received angioplasty and the drug CentoRx 125 patients didn't have life-threatening blockages after the angioplasty procedure. While the second group of 75 patients who received the artery-opening procedure and was administered a fake ineffective pill 50 patients didn't have life-threatening blockages after the angioplasty procedure.

Can one conclude that the experimental drug significantly reduces life-threatening blockages after the artery-opening procedure angioplasty at $\alpha = 1\%$?

90. A study conducted by Brigham and Women's Hospital reported that regular aspirin use substantially decreases the risk of colon cancer. For a group of 250 subjects who were administered two or more aspirins a week during the study 20 contracted colon cancer while a second group of 225 subjects who were administered a fake pill similar in appearance to an aspirin 32 contracted colon cancer. Can one conclude that regular use of aspirin can significantly reduce the risk of colon cancer $\alpha = 1\%$?

91. Findings released by the North American Association for the Study of Obesity showed that a cream treated with an asthma drug may melt away fat. The group of 265 women in the study applied the cream five days a week for six weeks to one thigh, while a placebo cream was applied to the other thigh. At the end of the study, the researchers measured each thigh to determine the proportion of women who lost at least an average of half an inch. Perform a hypothesis test using the data in the following table to determine if the proportion of thighs which lost an average of half an inch of fat is significantly greater for the thighs administered the treated cream than for thighs treated with the placebo cream at $\alpha = 0.01$.

Thigh Applied with the	Number of Thighs Measuring an Average of Half an Inch Less
Treated Cream	169
Placebo Cream	96

92. Harvard researchers conducted a study on 456 elderly women to determine if cranberry juice can help women avoid bladder infections. Half the women were randomly assigned to the group who was administered one cup of cranberry juice while the remaining women were given a berry-free imitation liquid similar in appearance to cranberry juice. After eight weeks, the results of the survey showed that 15% of the women consuming the cranberry juice had bladder infections while 28% of the berry-free group contracted bladder infections. Perform a hypothesis test at $\alpha = 0.01$ to determine if the women on the cranberry juice suffered a significantly less proportion of bladder infections.

93. Within the advertisement for Hismanal the ad states that Hismanal causes no more drowsiness than a placebo. Do you support this statement? Explain.

94. Within the advertisement for Hismanal the ad states that Hismanal causes no more dry mouth or nervousness than with a sugar pill. Do you support this statement? Explain.

(For problems 93 and 94)

Percent of Patients Reporting in Controlled Clinical Studies*

ADVERSE EVENT	HISMANAL (n = 1630)	Placebo (n = 1109)
Central Nervous System		
Drowsiness	7.1	6.4
Headache	6.7	9.2
Fatigue	4.2	1.6
Appetite Increase	3.9	1.4
Weight Increase	3.6	0.7
Nervousness	2.1	1.2
Dizzy	2.0	1.8
Gastrointestinal System		
Nausea	2.5	2.9
Diarrhea	1.8	2.0
Abdominal Pain	1.4	1.2
Eye, Ear, Nose, and Throat		
Mouth Dry	5.2	3.8
Pharyngitis	1.7	2.3
Conjunctivitis	1.2	1.2
Other		
Arthralgia	1.2	1.6

*Duration of treatment in Controlled Studies ranged from 7 to 182 days

Section 12.5

For problems 95–113, refer to problems 63 to 77 in Section exercises 12.3 and problems 89 to 92 in Section exercises 12.4, and use the p-value approach to hypothesis testing using the TI-84 calculator. For each of these problems, use the 5 step procedure to perform a hypothesis test at **both** $\alpha = 1\%$ and $\alpha = 5\%$. Include the following within your test:

a) state the null and alternative hypotheses.
b) calculate the expected results for the experiment assuming the null hypothesis is true and state the hypothesis test model.
c) formulate the decision rule.
d) identify the test statistic and its p-value.
e) state the conclusion and answer any question(s) posed in the problem for each alpha level.
f) determine the smallest level of significance for which the null hypothesis can be rejected.

CHAPTER REVIEW

114. When the sample size is large, the Sampling Distribution of the Proportion is approximately a _____ distribution. A sample is large if both _____ and _____ are greater than 5.

115. Studies involving population means deal with measurements while studies involving population proportions produce _____.

116. Identify what each of the following symbols represent:
   a) n_1
   b) n_2
   c) $\hat{p}_1$
   d) $\hat{p}_2$
   e) p_1
   f) p_2

117. The name of the Sampling Distribution which is used to determine if the difference between the sample proportions, $\hat{p}_1 - \hat{p}_2$, is large enough to conclude that the difference between the population proportions, $p_1 - p_2$, is statistically significant is: _____.

118. The mean of the Sampling Distribution of the Difference Between Two Proportions, denoted by $\mu_{\hat{p}_1 - \hat{p}_2}$,
   a) $\mu_{\hat{p}_1 - \hat{p}_2} = 0$
   b) $\mu_{\hat{p}_1 - \hat{p}_2} = p_1 + p_2$
   c) $\mu_{\hat{p}_1 - \hat{p}_2} = p_1 - p_2$

119. To determine if the distribution of difference between sample proportions can be approximated by a normal distribution, both n_1p_1 and n_2p_2 must be greater than 5. (T/F)

120. A random sample of 625 male drivers showed that 376 wear seatbelts, while a random sample of 758 female drivers revealed that 492 wear seatbelts. Using this information, determine the sample proportion of:
a) men drivers who wear seatbelts
b) women drivers who wear seatbelts
c) men drivers who do not wear seatbelts
d) women drivers who do not wear seatbelts

121. Define:
a) the Sampling Distribution of the Difference Between Two Proportions.
b) the conditions under which the Sampling Distribution of the Proportion can be approximated by a normal distribution.
c) the conditions under which the Sampling Distribution of The Difference Between Two Proportions can be approximated by a normal distribution.

122. Using the following information:

$$p_1 = 0.56, n_1 = 1860, \text{ and } x_1 = 1172$$
$$p_2 = 0.82, n_2 = 1945, \text{ and } x_2 = 1537$$

determine:
a) the proportion of the sample selected from population 1
b) the proportion of the sample selected from population 2
c) if the samples are considered large
d) the difference between the sample proportions: $\hat{p}_1 - \hat{p}_2$
e) the shape of the Sampling Distribution of $\hat{p}_1 - \hat{p}_2$
f) mean of the Sampling Distribution of $\hat{p}_1 - \hat{p}_2$
g) standard deviation of the Sampling Distribution of $\hat{p}_1 - \hat{p}_2$
h) standard error of the difference between two proportions

123. A marketing research agency conducted a survey to determine the spending plans for the forthcoming holiday season of the consumers in the Green Acres Shopping Mall on Long Island. Out of 1952 people surveyed, 650 responded that they planned to spend less than last year's holiday season. According to a similar survey in 2010, 809 of the 1924 consumers surveyed indicated that they planned to spend less during the upcoming holiday season. Using this information:
a) define the two populations, and the proportions of these populations
b) identify each sample proportion
c) compute the value of each sample proportion
d) compute the estimate of the standard error of the difference between two proportions
e) compute the estimate of the standard deviation of the Sampling Distribution of the Difference Between Two Proportions.

124. A survey by Simmons Research for Interep Radio Store reported on the level of education and the percentage of sports watchers or listeners. According to the survey, 33% of college graduates listen to sports on the radio while 27% of college graduates watch sports on TV. If the Simmons Research polled 830 college graduates who listen to the radio and 910 college graduates who watch TV, then:
a) define the two populations, and the proportions of these populations
b) identify each sample proportion
c) compute the value of each sample proportion
d) compute the estimate of the standard error of the difference between two proportions

e) compute the estimate of the standard deviation of the Sampling Distribution of the Difference Between Two Proportions.

125. For a hypothesis test involving two population proportions, p_1 and p_2, state, in words and in symbols, the general form of the:
a) null hypothesis
b) alternative hypothesis

126. Which of the following would not be suitable as an alternative hypothesis:
a) $p_1 - p_2 < 0$
b) $p_1 - p_2 > 0$
c) $p_1 - p_2 \neq 0$
d) $p_1 - p_2 = 0$

127. For large samples, state the expected results of the Sampling Distribution of $\hat{p}_1 - \hat{p}_2$. That is, identify the shape and give the formulas needed to calculate the mean and the estimate of the standard error of the Sampling Distribution of $\hat{p}_1 - \hat{p}_2$.

128. When performing a hypothesis test involving two population proportions, the expected result $\mu_{\hat{p}_1 - \hat{p}_2} = p_1 - p_2 = 0$ is determined under the assumption that _____ is true.

129. Give the notation and formulas for the pooled sample proportion, and state under what conditions you would use these formulas.

130. Using the assumption that H_0 is _____, p_1 _____ p_2 and then we can pool the two samples together to obtain the pooled estimate of the standard error formula, $s_{\hat{p}_1 - \hat{p}_2}$, is:

$$s_{\hat{p}_1 - \hat{p}_2} = \sqrt{(\quad)(\quad)(\quad)(\frac{1}{n_1} + \frac{1}{n_2})}$$

131. The following information concerning hypothesis tests for two proportions. For each
(i) define the two population proportions.
(ii) state the null and alternative hypotheses.
(iii) identify the type of test: 1TT or 2TT.
a) American Express is interested in determining if the percentage of no-shows for dinner reservations is greater on the weekend.
b) *Psychology Today* is interested in determining whether there is a difference in the percentage of males who pray and the percentage of females who pray.

132. Compute the pooled sample proportion, $\hat{p}$, for the sample information in parts (a) to (b):
a) $x_1 = 450, n_1 = 975, x_2 = 615$ and $n_2 = 1050$
b) $n_1 = 1250, \hat{p}_1 = 0.76, n_2 = 1370,$ and $\hat{p}_2 = 0.85$

133. a) to b): For each part in question 139, use the sample information to determine the pooled estimate of the standard error of the Sampling Distribution of The Difference Between Two Proportions.

For problems 134–138, use the 5 step procedure to perform a hypothesis test. Include the following within your test:
a) state the null and alternative hypotheses.
b) calculate the expected results for the experiment assuming the null hypothesis is true and state the hypothesis test model.
c) formulate the decision rule.
d) compute the test statistic.
e) state the conclusion and answer any question(s) posed in the problem.

134. A hospital administrator is interested in comparing the survival rates for patients who have had a kidney transplant at his hospital to the survival rate at a competing hospital. He believes the survival rate of patients receiving a kidney transplant at his hospital is greater than the hospital's closest competitor. At his hospital, of the last 200 patients who received a kidney from a deceased donor, 161 lived for at least five years. At the competing hospital, the last 100 patients who received a kidney from a deceased donor had 76 who lived for at least five years. Is this hospital administrator correct in stating that the population proportion of kidney transplant patients who live at least 5 years is greater at his hospital? Test at the 5% level of significance.

135. According to a League of Women Voters survey, 63% of males said that people have very little influence on government while 52% females said that people have very little influence on government. If there were 1050 women and 1275 males in the survey, perform a hypothesis test to determine if there is a significantly different proportion of males than females who believe that people have very little influence on government at $\alpha = 0.01$.

136. An educational researcher believes that students from small classes (i.e. class size less than 35) receive more A's and B's as compared to students from large classes (i.e. class size at least 35). To test this claim he randomly samples students from large and small classes and finds that of 400 students from small classes 150 received A's and B's, whereas 75 of the 350 students from large classes received A's and B's. Do these results support the researcher's claim at $\alpha = 5\%$?

137. According to a new survey by the Wine Market Council, more American adults than ever are drinking wine. Wine is increasingly being chosen as an accompaniment to meals in 'casual chain' restaurants, and at home when all the family dine together, the survey found.

Mr. Lorenzo, an Italian sociologist, wants to determine if the proportion of male wine drinkers is greater than the proportion of female wine drinkers who prefer red wine. To test this claim, the sociologist randomly surveys each population. Based on random samples of 150 males and 120 females, the sociologist determines that 115 males prefer red wine while 76 females prefer red wine. Do these results confirm the sociologist's belief at $\alpha = 1\%$?

138. According to research conducted by Brigham and Women's Hospital, two groups of women were studied to determine if the rate of heart attack was significantly different among those who took the vitamin E supplement versus the women who didn't take any vitamin E supplements. One group of 375 women who took vitamin E supplements of 100 or more units a day, while a second group of 425 women who didn't take any vitamin E supplements. The number of heart attacks for the women who took the vitamin E supplement was 15 while 32 of the women who didn't take the vitamin E supplement had heart attacks.

Do these sample results indicate there is a significant difference in the number of heart attacks among women who take vitamin E supplements versus women who do not take any vitamin E supplements at $\alpha = 5\%$?

WHAT DO YOU THINK?

139. Read the following article which appeared in the June 15, 1993 USA TODAY issue.

The placebo effect: Gauging the mind's role in healing

From *USA Today*. © 1993 Gannett-USA Today.
All rights reserved. Used under license.

Long ignored and disparaged, the so-called placebo effect may finally be getting the respect it deserves. A new study out today in the *Annals of Internal Medicine* underlines the power the placebo effect — a person's expectation about the outcome of an illness or therapy and suggests that the success of many therapies may really spring from our own minds, experts say.

"The placebo effect is greatly under appreciated," says Dr. David Sobel, co-author of the *Healing Brain*, (Simon & Schuster). "When we treat patients with various drugs or procedures, we often forget that a sizable portion of the improvement nay not be due to therapy but to the positive expectations of the patient." The new study examined the effectiveness of increasingly popular mental techniques used for lowering high blood pressure. It found that the techniques — medication, biofeedback and relaxation — were a little better than doing nothing, but no better than sham or placebo techniques. The finding may be bad news for people who paid a lot of money to lower blood pressure with the methods. But it also demonstrates that everyone has a natural ability to help themselves, and all that's needed is something to believe in. "I would say placebo, which reflects patient expectations, is very powerful medicine indeed," says Dr. Thomas L. Delbanco, Beth Israel Hospital, Boston, who helped conduct the study. "What we've shown, which is very important, is that the cognitive therapies have a small impact when you compare them to sham therapy, or placebo, then you find no difference at all. So, it is possible that the meditation and relaxation response basically reflects the placebo effect."

Pharmaceutical companies, academic researchers and the Food and Drug Administration are well aware of the placebo effect. New drugs and many other therapies approved in the USA, including those for cancer and heart disease, must be compared to a placebo. Carefully conducted studies show placebos alleviate symptoms in 10% to 70% of people, depending on their condition. On average, if you take 100 people with certain illnesses and give them a sugar pill, about a third would be expected to improve. If the new drug is better than the placebo, then it's considered effective. But, even though placebos work, doctors tend to view the effect as a nuisance, says Dr. Margaret Caudill, Mind/Body Medical Institute of New England Deaconess Hospital and Harvard Medical School, Boston. "Placebo is the X factor and we have to account for it because it keeps appearing," Caudill says. "But we often say, 'The effect is not real or meaningful, it's just placebo.' I say if it works, who cares."

Parents know the power of placebo and practice it every time they make a child's scrape feel better by kissing it. We might take offense if doctors took the same approach, but they could learn a few lessons. The first thing a parent does is show concern, compassion and thoughtfulness. And in their authority as a parent, they inspire confidence in the child when they tell them a kiss will stop the pain.

Sobel says studies show that doctors spend less than 18 seconds listening to a patient talk about 'where it hurts' before interrupting. And when treatment is prescribed, the emphasis is usually on technology or drugs.

"Some of the most powerful prescriptions are those filled in the natural pharmacy of the brain," says Sobel, who teaches doctors to use the placebo effect by improving the doctor-patient relationship and communication better with their patient's expectation that they will feel better after having seen the doctor.

Until the explosion of medical technology in the 20th century, placebo power was, by necessity, the foundation of medicine for thousands of years. And to maximize its benefit, many cultures mastered the use of symbols and rituals to trigger the response.

"The placebo effect is the triggering of our intrinsic healing mechanisms. You can do it with a pill, a surgical scalpel or headdress," Sobel says. "No treatment is immune to the effect of the positive expectation and this is why medicine has been so successful despite treatments throughout history which themselves were ineffective or harmful."

And even modern medicine makes use of symbols to exploit the placebo effect, the experts say.

In the U.S., we give magical powers to the doctor's white lab coat, the stethoscope and the laboratory test. Other cultures may use a wooden mask, a headdress and a rattle.

"We don't like to think that we use primitive symbols, but the lab coat and stethoscope definitely are," Caudill says. "Patients demand certain stereotypes."

Delbanco agrees: "Magic has always played an enormously important role in virtually every type of society. While we pride ourselves on our scientific foundations, there's no doubt in my mind that medicine will always be mixture of art and science. Part of that art is magic."

A small number of doctors are beginning to appreciate the placebo phenomenon as a tool for everyday medical practice, Sobel says.

By understanding its potential as well as its limitations, it should be possible to harness the power of placebo and boost the effectiveness of any medical therapy or procedure.

"Making the maximum use of placebo power in the modern doctor's office may mean displaying these symbols, or whatever is relevant, to particular types of patients," Delbanco says. "Some may prefer blue jeans to a lab coat."

Taking time to learn about a patient's particular belief system also may be a critical element and key part of communication.

Says Sobel: "A placebo is not a good substitute for an effective antibiotic or a life-saving surgery. But words can be scalpels. They can cut and damage, and when used effectively, they can also heal."

a) Summarize the article and explain what is meant by the "placebo effect."
b) Pharmaceutical companies purposely use certain colors for different type pills. For example, pills used for depression are usually small yellow pills, sedative pills are blue, and stimulants are red or pink pills. Using the ideas discussed within the article, explain why you think the pharmaceutical companies might purposely use these different color pills for the indicated ailments?
c) For each of the following excerpts from different studies:
   — identify the treatment and placebo groups within the first two studies.
   — discuss what you think contributed to the unusual results in the studies.
   (i) In the late 1950's, surgeons were using a new operation to relieve chest pain by making an incision in the patient's chest and tieing off the mammary artery. In the 1960's, Dr. E. Grey Diamond of the University of Kansas Medical Center who was suspicious of the benefit of this procedure decided to do a study to test the procedure. Dr. Diamond divided a group of patients into two small groups. One group received the actual surgery while the other group had sham surgery, that is, an incision was made on their chest but the artery was left intact. Both groups had identical results.
   (ii) Studies on the use of beta blockers have shown that these drugs can reduce the risk of heart attack in certain patients. In 1990, Dr. R. Horowitz published in Lancet the following results on beta blocker studies. Dr. Horowitz indicated that the patients who failed to take the beta blockers had a 2.6 times higher incidence of death than the patients who regularly took their medication. He also noted there was a higher death rate among the patients who failed to take the fake pill than the patients who took the fake pill regularly.
   (iii) In an article in a 1950 issue of the Journal of Clinical Investigation Dr. Sobel reported that a group of doctors were having difficulty treating a pregnant woman who was suffering from severe nausea and vomiting. As a last resort, the doctors gave her syrup of ipecac and told her it was a powerful new drug that would help her. Although ipecac is actually used to induce vomiting in people who have swallowed poison, the pregnant woman's nausea and vomiting stopped 20 minutes after taking the syrup of ipecac.
d) With respect to the discussed ideas in the previous article, the unusual results mentioned in the previous studies and the concepts discussed in this chapter, what factors would you consider if you were conducting an experiment which involved a treatment and a control group?

140. **What Do You Think?**
A study entitled Personality Characteristics of Nightmare Sufferers appeared in an issue of The Journal of Nervous and Mental Disease contained in Table 12.9.

TABLE 12.9

	Frequent		Sometimes		Seldom		Never	
Groups	N	%	N	%	N	%	N	%
Males	14	9	41	25	38	24	67	42
Females	12	6	48	25	48	25	84	44
Total	26	7	89	25	86	25	151	43

Table 12.9 reports the percentages of nightmares in the 352 subjects of the study. In this study, a nightmare was operationally defined as a frightening dream that awakens the dreamer. The categories of the nightmares were defined as follows:

frequent — at least once a week;
sometimes — at least once a month;
seldom — at least once a year;
never — it never happens.

a) How many males participated in this study? How many females participated in this study?
b) Determine the sample proportion of the males that have experienced a nightmare and the sample proportion of females that have experienced a nightmare.
c) What percentage of the participants have experienced a nightmare? How does this result compare to the results you found in part (b)? Explain the relationship between a pooled sample proportion which was discussed in this chapter and this overall percentage of participants who experienced a nightmare.

d) Conduct a hypothesis test at $\alpha = 0.05$ to determine if the proportion of males who have experienced a nightmare is significantly different from the proportion of males who have never experienced a nightmare. Does your conclusion indicate that there is a significant difference? Explain.

e) Conduct a hypothesis test at $\alpha = 0.05$ to determine if the proportion of females who have experienced a nightmare is significantly greater than the proportion of males who have experienced a nightmare. Does your conclusion indicate that the females experience a significant greater proportion of nightmares than males? Explain.

PROJECTS

141. Hypothesis Testing Project
   A. Selection of a topic.
      1. Select a reported fact or claim about two population proportions made in a local newspaper, magazine or other source.
      2. Suggested topics:
         a) More men smoke than women! Test the claim that the proportion of men who smoke is higher than the proportion of women who smoke.
         b) More men vote than women! Test the claim that the proportion of men who vote is higher than the proportion of women who vote.
         c) Women get better grades! Test the claim that the proportion of women whose G.P.A. is higher than 3.3 is greater than the proportion of men with a G.P.A. higher than 3.3.
         d) According to the Mars Company, each package of M&M's should contain 30% brown, 20% yellow, 20% red, and 10% each of green, orange and blue. Test the claim that the proportion of brown M&M's is significantly greater than the proportion of red M&M's.
   B. Use the following procedure to test the claim selected in Part A.
      1. State the claim, and identify the populations referred to in the statement of claim, also indicate the source.
      2. State your opinion regarding the claim (True or False) and clearly identify the populations you plan to sample to support your opinion.
      3. State the null and alternative hypotheses.
      4. Develop and state a procedure for the selection of a random sample from each population. Indicate the technique you are going to use to obtain your sample data.
      5. Perform the experiment and compute the following:
         a) $\mu_{\hat{p}_1 - \hat{p}_2}$
         b) p (pooled value for p)
         c) $s_{\hat{p}_1 - \hat{p}_2}$
         d) Critical z score(s)
      6. State the Decision Rule.
      7. Calculate the test statistic: z.
      8. Construct a hypothesis testing model and place the test statistic on the model.
      9. Formulate the appropriate conclusion and interpret the conclusion with respect to your opinion (as stated in STEP 2).

142. Read a research article from a professional journal in a field such as medicine, education, marketing, science, social science, etc. Write a summary of the research article that includes the following:
   a) Name of article, author, and journal (include volume number and date).
   b) Statement of research problem.
   c) The null and alternative hypotheses.
   d) How the experiment was conducted (give the sample sizes and how the sampling was done).
   e) Determine the test statistic statistic, p-value and the level of significance.
   f) State the conclusion in terms of the null and alternative hypotheses.

CHAPTER TEST

143. Write the notation used to represent each of the following:
   a) a population proportion
   b) the proportion for population 2
   c) a sample proportion
   d) the proportion for a sample selected from population 1
   e) the size of a sample
   f) the sample size for a sample selected from population 2
   g) the difference between two sample proportions
   h) the number of occurrences or observations possessing a particular characteristic within a sample
   i) the formula for the proportion of a sample selected from population 1

144. Using the following information:

$$p_1 = 0.48, n_1 = 970, \text{ and } x_1 = 470$$
$$p_2 = 0.56, n_2 = 1060, \text{ and } x_2 = 670$$

determine:
   a) the proportion of the sample selected from population 1
   b) the proportion of the sample selected from population 2
   c) if the samples are considered large
   d) the difference between the sample proportions: $\hat{p}_1 - \hat{p}_2$
   e) the shape of the Sampling Distribution of $\hat{p}_1 - \hat{p}_2$
   f) mean of the Sampling Distribution of $\hat{p}_1 - \hat{p}_2$
   g) standard deviation of the Sampling Distribution of $\hat{p}_1 - \hat{p}_2$
   h) standard error of the Difference Between Two Proportions

145. According to a poll conducted by a national health magazine, 295 of 1200 male readers and 694 of the 1275 female readers surveyed were on a diet at the time of its poll. Using this information:
  a) define the two populations, and the proportions of these populations.
  b) identify each sample proportion.
  c) identify each sample size.
  d) determine if the two samples are considered large.
  e) compute the value of each sample proportion.
  f) compute the estimate of the standard error of the difference between two proportions.
  g) compute the estimate of the standard deviation of the Sampling Distribution of the Difference Between Two Proportions.

146. The Sampling Distribution of the Difference Between Two Proportions has the following characteristics:
  a) the mean of the Sampling Distribution of the Difference Between Two Proportions, denoted _____, is determined by the formula:

  $$\mu_{\hat{p}_1 - \hat{p}_2} = \underline{\qquad} - \underline{\qquad},$$

  where p_1 represents the _____ proportion of _____ 1 and p_2 represents the _____ proportion of _____ 2.
  b) The formula for the estimate of the standard deviation of the Sampling Distribution of the Difference Between Two Proportions, denoted by _____, is given by the formula:

  $$s_{\hat{p}_1 - \hat{p}_2} = \underline{\qquad}$$

  c) The Sampling Distribution of the Difference Between Two Proportions is approximately normal when:

  n_1 _____, $n_1(1 - p_1)$, n_2 _____ and n_2 _____ are all greater than _____.

147. One correct formula for the pooled sample proportion, $\hat{p}$, is:

  a) $\dfrac{\hat{p}_1 + \hat{p}_2}{2}$ b) $\dfrac{x_1 + x_2}{n_1 + n_2}$ c) $\dfrac{\hat{p}_1 + \hat{p}_2}{x_1 + x_2}$ d) $\dfrac{\hat{p}_1 + \hat{p}_2}{n_1 + n_2}$

148. In a random sample of size 200, $x_1 = 120$ and in another independent random $ss_{\hat{p}_1 - \hat{p}_2}$ of size 300, $x_2 = 140$, then the "pooled" formula for ____ is approximately equal to ____.

149. The pooled sample proportion formulas are: $\hat{p} = $ ____ or ____.

150. The following information concerning hypothesis tests for two proportions. For each
  (i) define the two population proportions.
  (ii) state the null and alternative hypotheses.
  (iii) identify the type of test: 1TT or 2TT.
  a) The League of Women Voters would like to determine if a greater proportion of females than males believe that people can have at least some influence on government.
  b) The American Association for Marriage and Family Therapy wants to conduct a study to determine whether the percentage of divorce is smaller for couples who don't live together before marriage than for those who do live together before marriage.

151. According to a national credit organization, the percent of college students who do not have a credit card in their name is the same for males as it is for females. Using the 5 step hypothesis testing procedure, perform a hypothesis test of this claim using an alpha level of 5%. Sample data revealed the following:

	Males	Females
Number who do not have a credit card	$x_1 = 24$	$x_2 = 20$
Sample size	$n_1 = 48$	$n_2 = 56$

Based on the sample data, can we conclude that the percent who do not have a credit card is the same for both male and female college students?

For each of the following problems, use the 5 step procedure to perform a hypothesis test. Include the following within your test:

  a) state the null and alternative hypotheses.
  b) calculate the expected results for the experiment assuming the null hypothesis is true and state the hypothesis test model.
  c) formulate the decision rule.
  d) compute the test statistic.
  e) state the conclusion and answer any question(s) posed in the problem.

152. According to the Department of Transportation (DOT), a significantly greater proportion of Southwest airline flights arrived on time as compared to on time flights for Alaska airlines. Perform a hypothesis test at $\alpha = 5\%$, if 84% of a sample of 1250 flights for Southwest airlines arrived on time and 80.9% of a sample size of 1175 flights for Alaska airlines arrived on time. Do you agree with the DOT's claim?

153. In a double-blind clinical study of 725 subjects, a prescription medication for seasonal allergy called Hismanal was administered to a group of 400 subjects while a second group of 325 subjects received a sugar pill for their allergies. The number of subjects suffering from drowsiness while taking Hismanal was 120 while the number of subjects who reported drowsiness taking the sugar pill was 75. Based on the sample data, can one conclude that there was a significant difference in the proportion of drowsiness for the subjects on Hismanal and the sugar pill at $\alpha = 1\%$?

154. A study of the Arctic Polar bears was conducted in order to find the proportion of the bear population that lived in coastal areas. Of the 370 males sampled in the Arctic, 41 inhabited coastal regions. Of the 482 females sampled, 97 inhabited coastal regions. Does this data provide sufficient evidence that the proportion of female Polar bears that inhabit coastal regions is significantly greater than the proportion of male Polar bears that inhabit coastal regions? Test at the .05 alpha level.

For the answer to this question, please visit www.MyStatLab.com.

CHAPTER 13

Hypothesis Test Involving Two Population Means Using Independent and Dependent Samples

Contents

13.1 Introduction
13.2 The Sampling Distribution of the Difference Between Two Means
13.3 Hypothesis Testing Involving Two Population Means and Known Population Standard Deviations: The Two Sample z Test
13.4 Hypothesis Testing Involving Two Population Means and Unknown But Equal Population Standard Deviations: The Pooled Two Sample t Test
13.5 Hypothesis Testing Involving Two Population Means and Unknown But Unequal Population Standard Deviations: Welch's Two Sample t Test
13.6 Hypothesis Tests Comparing Treatment and Control Groups
13.7 The Paired t Test: a Dependent t-test
13.8 p-Value Approach to Hypothesis Testing Involving Two Population Means Using the TI-84 Calculator

13.1 INTRODUCTION

In Chapter 11, we discussed hypothesis tests involving one population mean. In this chapter we will study hypothesis tests that involve the comparison of two population means. We will first discuss hypothesis tests involving two population means where an independent random sample is selected from each population to determine if the population means are statistically different.

For example:

- Is there a difference in the mean physical fitness scores of third grade girls and boys? This involves the comparison of two populations: the population of physical fitness scores for the girls and the population of physical fitness scores for the boys.
- Will Goodday automobile tires give better average mileage wear than Goodpoor tires? (Neither company owns a blimp!) The mileage data for Goodday tires would comprise one of the populations and the mileage data for the Goodpoor tires would comprise the other population.
- Does a toothpaste that contains fluoride significantly lower the mean number of cavities as compared to the toothpaste without fluoride? One population represents those people that used the toothpaste with fluoride and the other population represents those people that used the toothpaste without fluoride.

To answer these questions about whether there is a significant difference between the means of two independent populations, we have to select a random sample from each population, calculate the mean of each sample and determine the difference between the sample means. We would then have to test whether the **difference between the two sample means** is significant enough to conclude that the two population means are different. To help us perform this test, we need to examine the **sampling distribution of the difference between two means**.

13.2 THE SAMPLING DISTRIBUTION OF THE DIFFERENCE BETWEEN TWO MEANS

To perform a hypothesis test involving two population means, an **independent random sample**[1] from each population will be selected and compared to determine whether **the difference** between the two sample means is significantly different. This is illustrated in Figure 13.1.

---

[1] By independent random samples, we mean that the selection of one random sample has no influence upon the selection of the other random sample. See chapter 12 for definition of independent samples.

674 *Chapter 13* Hypothesis Test Involving Two Population Means Using Independent and Dependent Samples

FIGURE 13.1

To determine if the difference between the sample means, denoted by d = $\bar{x}_1 - \bar{x}_2$, is statistically significant, we must compare this difference to a distribution formed by calculating the difference between the means of all possible random sample pairs, where the first random sample with size n_1 for the pair is selected from an infinite population 1 and the second random sample with size n_2 of the pair is selected from an infinite population 2. This sampling distribution is referred to as the **sampling distribution of the difference between two means** or the **sampling distribution of $\bar{x}_1 - \bar{x}_2$**. This is illustrated in Figure 13.2.

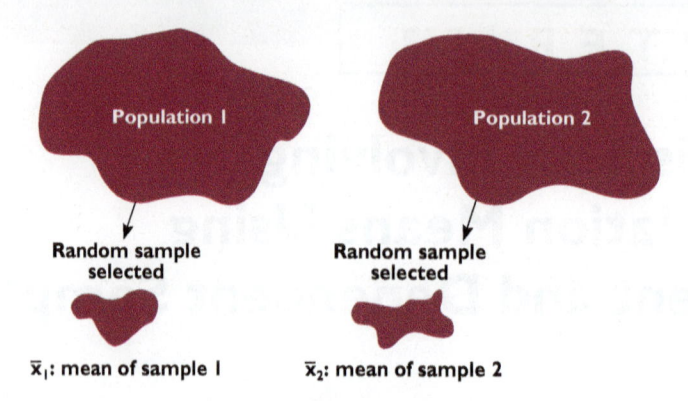

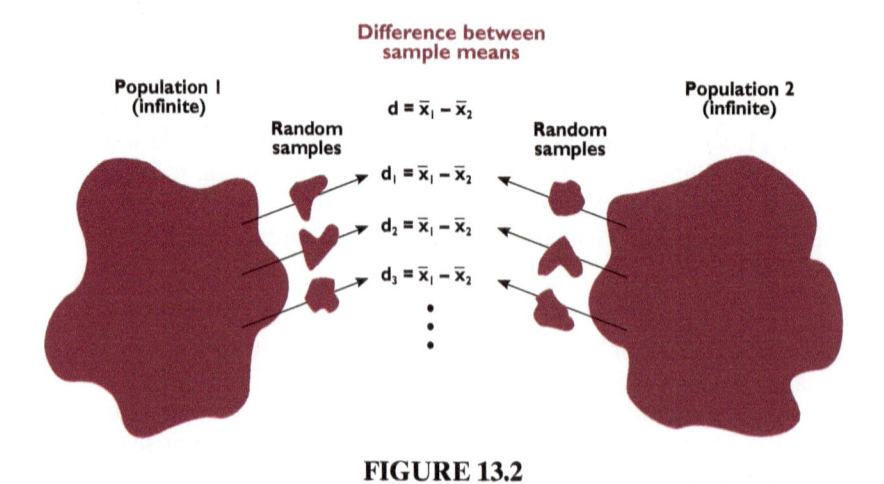

FIGURE 13.2

The distribution of the differences between the means of all possible independent random sample pairs form the sampling distribution of the difference between two means.

Let's now examine Theorem 13.1 which explains the properties of the sampling distribution of the differences between two means where the population standard deviations are known and the sample sizes are greater than 30.

Theorem 13.1   Characteristics of the Sampling Distribution of The Difference Between Two Means with Population Standard Deviations Known

Suppose you are given two populations, referred to as population 1 and population 2, with means μ_1 and μ_2 respectively and standard deviations σ_1, and σ_2, respectively. If all possible independent random samples of size n_1 and n_2 (where **both sample sizes are greater than 30**) are selected from population 1 and population 2 respectively, the mean of all these pairs of samples is determined, and the differences between the pairs of sample means ($\bar{x}_1 - \bar{x}_2$) are computed, then the sampling distribution of the difference between two means is formed and has the following characteristics.

1. The **mean of the sampling distribution of the difference between two means**, denoted by $\mu_{\bar{x}_1 - \bar{x}_2}$, is:

$$\mu_{\bar{x}_1 - \bar{x}_2} = \mu_1 - \mu_2$$

2. The **standard deviation** of the sampling distribution of the difference between two means, denoted by $\sigma_{\bar{x}_1 - \bar{x}_2}$, is:

$$\sigma_{\bar{x}_1 - \bar{x}_2} = \sqrt{\frac{\sigma_1^2}{n_1} + \frac{\sigma_2^2}{n_2}}$$

This standard deviation is also referred to as **the standard error of the difference between two means**.

3. The sampling distribution of the difference between two means is **approximately normal**.

13.3 HYPOTHESIS TESTING INVOLVING TWO POPULATION MEANS AND KNOWN POPULATION STANDARD DEVIATIONS: THE TWO SAMPLE Z TEST

Applying the general five-step hypothesis testing procedure discussed in Chapter 10 to hypotheses involving two population means, where the population standard deviations are known, we will develop a hypothesis testing procedure that will allow us to determine if there is a significant difference between the means of two independent populations. This type of hypothesis test is often referred to as a two-sample z test.

The Two Sample z Test

Step 1 Formulate the two hypotheses, H_0 and H_a.
Null Hypothesis, H_0:
The **Null Hypothesis**, H_0, in testing the difference between two population means has the following form:

Null Hypothesis Form

H_0: There is no difference between the mean of population 1, written μ_1, and the mean of population 2, written μ_2.

The null hypothesis can be symbolized as:
$$H_0: \mu_1 - \mu_2 = 0$$

Alternative Hypothesis, H_a: Since the alternative hypothesis, H_a, can be stated as either *greater than*, *less than*, or *not equal to* 0, then the alternative hypothesis can have one of the following three forms.

Alternative Hypothesis Form

Form (a): Less Than Form for H_a

H_a: the mean of population 1, written μ_1, is **less** than the mean of population 2, written μ_2. This can also be stated as the difference between the population means ($\mu_1 - \mu_2$) is **significantly less than 0**.

This is symbolized as:
$$H_a: \mu_1 - \mu_2 < 0$$

> #### Form (b): Greater Than Form for H_a
>
> H_a: the mean of population 1, written μ_1, is **greater** than the mean of population 2, written μ_2. This can also be stated as the difference between the population means ($\mu_1 - \mu_2$) is **significantly greater than 0**.
>
> This is symbolized as:
> $$H_a : \mu_1 - \mu_2 > 0$$
>
> #### Form (c): Not Equal to Form for H_a
>
> H_a: the mean of population 1, written μ_1, is **not equal to** the mean of population 2, written μ_2. This can also be stated as the difference between the population means ($\mu_1 - \mu_2$) is **significantly different from 0**.
>
> This is symbolized as:
> $$H_a : \mu_1 - \mu_2 \neq 0$$

Step 2 Determine the model to test the null hypothesis, H_0. Under the assumption H_0 is true, calculate the expected results.

Since we are testing the hypothesis that the **difference between two population means** ($\mu_1 - \mu_2$) is significantly different from zero by **using the difference between two sample means**, $\bar{x}_1 - \bar{x}_2$, then we will use the **sampling distribution of the difference between two means as the hypothesis testing model.** Using Theorem 13.1, and under the assumption H_0 is true (that is: $\mu_1 - \mu_2 = 0$ is true), then:

the expected results for the Two Sample z test are:

a) The mean of the sampling distribution of the difference between two means is given by the formula:

$$\mu_{\bar{x}_1 - \bar{x}_2} = \mu_1 - \mu_2$$

Under the assumption H_0 is true (i.e. $\mu_1 - \mu_2 = 0$ is true), then we can state that:

$$\mu_{\bar{x}_1 - \bar{x}_2} = 0$$

b) For the two sample z-test, the standard error of the difference between two means, written $\sigma_{\bar{x}_1 - \bar{x}_2}$, is computed by the formula:

$$\sigma_{\bar{x}_1 - \bar{x}_2} = \sqrt{\frac{\sigma_1^2}{n_1} + \frac{\sigma_2^2}{n_2}}$$

Step 3 Formulate the Decision Rule:
a) Determine type of alternative hypothesis, directional or non-directional.
b) Determine type of test, 1TT or 2TT.
c) Identify the significance level, $\alpha = 1\%$ or $\alpha = 5\%$.
d) Construct the appropriate hypothesis test model.
e) Find the critical z score(s) using Table II for the appropriate level of significance.
f) State the decision rule.

Step 4 Analyze the sample data
a) Randomly select an independent sample from each population and calculate the mean of each sample, denoted $\bar{x}_1$ and $\bar{x}_2$.
b) **Determine the difference between the two sample means, the statistical result, using the formula:** $d = \bar{x}_1 - \bar{x}_2$.

Section 13.3 Hypothesis Testing Involving Two Population Means and Known Population Standard Deviations

c) Calculate the z score of the difference between the two sample means using the formula:

$$z = \frac{(\bar{x}_1 - \bar{x}_2)}{\sigma_{\bar{x}_1 - \bar{x}_2}}$$

This z score is referred to as the **test statistic**.

Step 5 **State the conclusion.**
Compare the test statistic to the decision rule and either:
a) Reject H_0 and accept H_a at α,
or
b) Fail to reject H_0 at α.

To summarize, in the procedure used to outline a two sample z-test, there were three refinements to the general hypothesis testing procedure developed in Chapter 10. They were:

1. The test statistic hypothesis testing model is a normal distribution.
2. The properties of the sampling distribution of the difference between two means are used to calculate the expected values.
3. The formula to calculate the test statistic z is:

$$z = \frac{(\bar{x}_1 - \bar{x}_2)}{\sigma_{\bar{x}_1 - \bar{x}_2}}$$

Example 13.1 illustrates the use of the Two Sample z Test to perform a hypothesis test involving two population means with known population standard deviations and independent random samples.

EXAMPLE 13.1

An educational researcher believes that the mean SAT Verbal scores for female students is significantly greater than the SAT verbal scores for male students. She decides to test her claim. It is known that the verbal portion of the SAT has a population standard deviation of 100. A random sample of 50 female SAT verbal scores had a mean of 525, and a random sample of 50 male SAT verbal scores had a mean of 497. Do these sample results support the high school teacher's claim at $\alpha = 5\%$

Since the two population standard deviations are known and both independent random sample sizes are greater than 30 we'll use the Two sample z-test.

Step 1 Formulate the two hypotheses, H_0 and H_a.
If we let the female verbal SAT scores represent population 1 and population 2 represents the male verbal SAT scores, then the **Null Hypothesis** is:

H_0: There is no difference between the population mean verbal SAT score for females and males.

In symbols, $H_0: \mu_1 - \mu_2 = 0$.

We are trying to show the population mean verbal SAT score for females is greater than the population mean verbal SAT score for males.

Thus the alternative hypothesis is directional and is stated as:

H_a: The difference between the population mean verbal SAT score for females and males is **significantly greater** than *zero*. This is symbolized as:

$$H_a: \mu_1 - \mu_2 > 0$$

Step 2 Determine the model to test the null hypothesis, H_0.

To test the null hypothesis an independent random sample is selected from each population. The sample results are summarized in Table 13.1

Table 13.1 Summary of sample data selected from each population

Random Sample Selected from Population 1 (Female)	Random Sample Selected from Population 2 (Male)
$\bar{x}_1 = 525$	$\bar{x}_2 = 497$
$\sigma_1 = 100$	$\sigma_2 = 100$
$n_1 = 50$	$n_2 = 50$

Since we are testing the hypothesis that the **difference between two population means** $\mu_1 - \mu_2$ is significantly greater than zero **using the difference between two sample means,** $\bar{x}_1 - \bar{x}_2$, then we will use the **sampling distribution of the difference between two means as the hypothesis testing model.** Using Theorem 13.1, and under the assumption H_0 is true (that is: $\mu_1 - \mu_2 = 0$ is true), then:

the expected results for the Two Sample z test are:

a) The mean of the sampling distribution of the difference between two means is given by the formula:

$$\mu_{\bar{x}_1-\bar{x}_2} = \mu_1 - \mu_2$$

Under the assumption H_0 is true (i.e. $\mu_1 - \mu_2 = 0$ is true), then we can state that:

$$\mu_{\bar{x}_1-\bar{x}_2} = 0$$

b) For the two sample z-test, the standard error of the difference between two means, written $\sigma_{\bar{x}_1-\bar{x}_2}$, is computed by the formula:

$$\sigma_{\bar{x}_1-\bar{x}_2} = \sqrt{\frac{\sigma_1^2}{n_1} + \frac{\sigma_2^2}{n_2}}$$

Substituting $n_1 = 50, n_2 = 50, \sigma_1 = 100, \sigma_2 = 100$ into the formula for $\sigma_{\bar{x}_1-\bar{x}_2}$, we have

$$\sigma_{\bar{x}_1-\bar{x}_2} = \sqrt{\frac{\sigma_1^2}{n_1} + \frac{\sigma_2^2}{n_2}}$$

$$= \sqrt{\frac{100^2}{50} + \frac{100^2}{50}}$$

$$= \sqrt{200 + 200}$$

$$= \sqrt{400}$$

$$\sigma_{\bar{x}_1-\bar{x}_2} = 20$$

Figure 13.3 Illustrates the hypothesis testing model.

**Hypothesis Testing Model: Sampling distribution of the differences between two means
(is approximately a normal distribution)**

$\mu_{\bar{x}_1-\bar{x}_2} = 0$
$\sigma_{\bar{x}_1-\bar{x}_2} = 20$

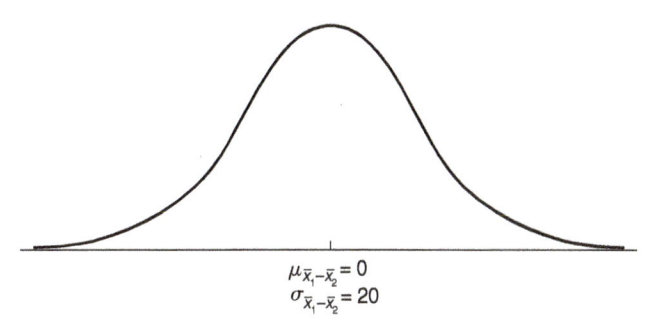

FIGURE 13.3

Step 3 **Formulate the Decision Rule**:

a) The alternative hypothesis is directional since the researcher wants to show the population mean verbal SAT score for females is greater than the population mean verbal SAT score for males.
b) The hypothesis test is one-tailed on the right side.
c) The significance level is $\alpha = 5\%$.
d) The critical z score, z_c, for a one tailed test on the right side with $\alpha = 5\%$ is $z_c = 1.65$.
e) Figure 13.4a illustrates the decision rule for this hypothesis test.

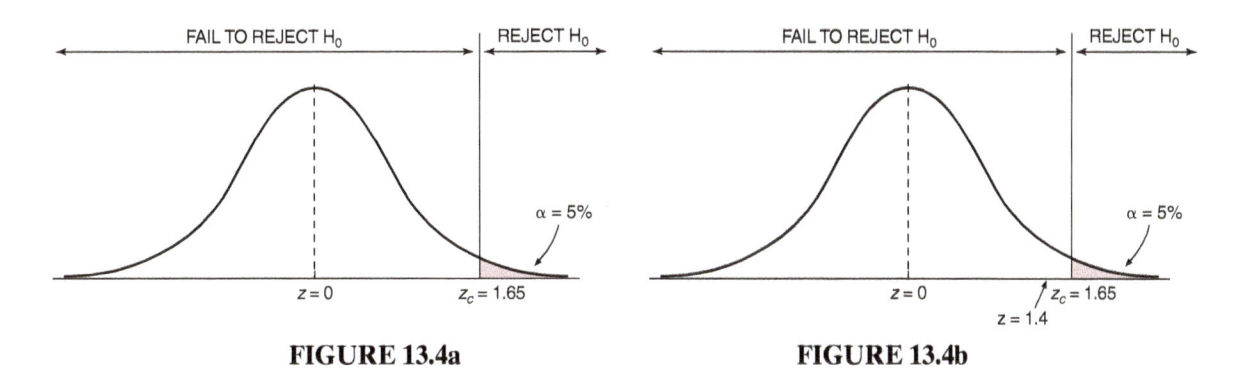

FIGURE 13.4a   **FIGURE 13.4b**

f) The decision rule is Reject H_0 if the test statistic, z, is greater than 1.65.

Step 4 **Analyze the sample data**

a) Randomly select an independent sample from each population and calculate the mean of each sample, denoted $\bar{x}_1$ and $\bar{x}_2$.
b) **Determine the difference between the two sample means, using the formula: $d = \bar{x}_1 - \bar{x}_2$.**
c) Calculate the z score of the difference between the two sample means using the formula:

$$z = \frac{(\bar{x}_1 - \bar{x}_2)}{\sigma_{\bar{x}_1-\bar{x}_2}}$$

$$z = \frac{(525 - 497)}{20}$$

$$z = \frac{28}{20}$$

$$z = 1.4$$

The z score $z = 1.4$ is referred to as the **test statistic**.

Step 5 State the conclusion.
From Figure 13.4b, since the test statistic $z = 1.4$ is less than the critical z score $z_c = 1.65$, then we Fail to Reject H_0 at $\alpha = 5\%$. Therefore, the educational researcher cannot conclude the population mean verbal SAT score for females is greater than the population mean verbal SAT score for males.

13.4 HYPOTHESIS TESTING INVOLVING TWO POPULATION MEANS AND UNKNOWN BUT EQUAL POPULATION STANDARD DEVIATIONS: THE POOLED T-TEST

In most instances the standard deviations of the two populations are **unknown**. Thus, the **estimates** of the two population standard deviations, s_1 and s_2 respectively, are used to approximate the two population standard deviations, σ_1 and σ_2 respectively, and the Central Limit Theorem may not apply. That is, the sampling distribution of the difference between two means may not be approximated by a normal distribution. However, if the two populations are approximately normal with equal standard deviations, and the sample standard deviations are used to estimate the population standard deviations, then the distribution of the test statistic can be approximated by a **t** distribution. This result and the characteristics of the sampling distribution of the difference between two means with population standard deviations unknown and independent random samples is stated in Theorem 13.2.

Theorem 13.2 Characteristics of the Sampling Distribution of the Difference Between Two Means with Population Standard Deviations Unknown and Assumed Equal

Suppose you are given two approximately normal populations, referred to as population 1 and population 2, with means μ_1 and μ_2 and approximately **equal standard deviations** σ_1 and σ_2 respectively. If all possible independent random samples (i.e. the selection of one random sample has no influence upon the selection of the other random sample) of size n_1 and n_2 are selected from population 1 and population 2 respectively, the means of all these pairs of samples are determined, and the differences between the pairs of sample means $(\bar{x}_1 - \bar{x}_2)$ are computed, then the sampling distribution of the difference between two means has the following characteristics:

1. The mean of the sampling distribution of the difference between two means, denoted by $\mu_{\bar{x}_1 - \bar{x}_2}$, is:

$$\mu_{\bar{x}_1 - \bar{x}_2} = \mu_1 - \mu_2$$

2. The standard error of the difference between two means cannot be determined since the population standard deviations, σ_1 and σ_2, are unknown. However, since the two sample standard deviations, s_1 and s_2, provide good estimates of the population standard deviations, σ_1 and σ_2, we can combine the sample standard deviations to obtain **an estimate of the standard error of the difference between two means**. The process of combining the two independent sample results to provide an estimate of this standard error is referred to as **pooling**. The formula to calculate this pooled estimate for the standard error, written $s_{\bar{x}_1 - \bar{x}_2}$, is:

$$s_{\bar{x}_1 - \bar{x}_2} = \sqrt{\frac{(n_1 - 1)s_1^2 + (n_2 - 1)s_2^2}{n_1 + n_2 - 2} \cdot \left(\frac{1}{n_1} + \frac{1}{n_2}\right)}$$

where:
- s_1 is the estimate of the standard deviation of population 1.
- n_1 is the sample size for the sample selected from population 1.
- s_2 is the estimate of the standard deviation of population 2.
- n_2 is the sample size for the sample selected from population 2.

and $\sqrt{\dfrac{(n_1 - 1)s_1^2 + (n_2 - 1)s_2^2}{n_1 + n_2 - 2}}$ is the pooled standard deviation.

3. The distribution of the test statistic t defined as:

$$t = \frac{(\bar{x}_1 - \bar{x}_2) - \mu_{\bar{x}_1 - \bar{x}_2}}{s_{\bar{x}_1 - \bar{x}_2}}$$

is approximated by a t distribution with degrees of freedom given by: $df = n_1 + n_2 - 2$.

Applying the general five step hypothesis testing procedure discussed in Chapter 10 to hypotheses involving **two population means**, where the population standard deviations are **unknown**, we will develop a hypothesis testing procedure that will allow us to determine if there is a significant difference between the means of two independent populations. This type of hypothesis test is often referred to as a **pooled two sample t test** or simply the **pooled t-test**.

The Pooled t Test

Step 1 Formulate the two hypotheses, H_o and H_a.
Null Hypothesis, H_o:
The **Null Hypothesis**, H_o, in testing the difference between two population means has the following form:

Null Hypothesis Form

H_o: There is **no difference** between the mean of population 1, written μ_1, and the mean of population 2, written μ_2.
This is symbolized as:

$$H_o: \mu_1 - \mu_2 = 0$$

By stating H_o as there is no difference between the population means, μ_1 and μ_2, we are making the assumption that μ_1 **equals** μ_2. Consequently, only by **rejecting the null hypothesis**, H_o, can we conclude that there is a *statistically significant difference* between μ_1 and μ_2. Therefore, by rejecting H_o, we can conclude that the alternative hypothesis is true.

Alternative Hypothesis, H_a
The **Alternative Hypothesis, H_a**, in testing the difference between two population means, has one of the following three forms:

Form (a): Less Than Form for H_a

H_a: the mean of population 1, written μ_1, is **less than** the mean of population 2, written μ_2.
This can also be stated as: **the difference $(\mu_1 - \mu_2)$ is significantly less than zero.**
This is symbolized as:

$$H_a: \mu_1 - \mu_2 < 0$$

Form (b): Greater Than Form for H_a

H_a: the mean of population 1, written μ_1, is **greater than** the mean of population 2, written μ_2. This can also be stated as: **the difference $(\mu_1 - \mu_2)$ is significantly greater than zero**.

This is symbolized as:

$$H_a: \mu_1 - \mu_2 > 0$$

Form (c): Not Equal Form for H_a

H_a: the mean of population 1, written μ_1, is **not equal to** the mean of population 2, written μ_2. This can also be stated as: **difference $(\mu_1 - \mu_2)$ is significantly different from zero**. This is symbolized as:

$$H_a: \mu_1 - \mu_2 \neq 0$$

Step 2 Determine the model to test the null hypothesis, H_o. Under the assumption H_o is true, calculate the expected results.

Since we are testing the hypothesis that the **difference between two population means $(\mu_1 - \mu_2)$** is significantly different from zero by **using the difference between two sample means, $\bar{x}_1 - \bar{x}_2$**, then we will use the **sampling distribution of the difference between two means as the hypothesis test model**.

Using Theorem 13.2, and under the assumption H_o is true (that is: $\mu_1 - \mu_2 = 0$ is true), then:

the expected results for a pooled two sample t test are:

a) The test statistic hypothesis testing model is approximately a t Distribution with degrees of freedom:

$$df = n_1 + n_2 - 2$$

b) The mean of the sampling distribution of the difference between two means is given by the formula:

$$\mu_{\bar{x}_1 - \bar{x}_2} = \mu_1 - \mu_2$$

Under the assumption H_o is true (i.e. $\mu_1 - \mu_2 = 0$ is true), then we can state that:

$$\mu_{\bar{x}_1 - \bar{x}_2} = 0$$

c) For a pooled t-test, the pooled estimate of the standard error of the difference between two means, written $s_{\bar{x}_1 - \bar{x}_2}$, is given by the formula:

$$s_{\bar{x}_1 - \bar{x}_2} = \sqrt{\frac{(n_1 - 1)s_1^2 + (n_2 - 1)s_2^2}{n_1 + n_2 - 2} \cdot \left(\frac{1}{n_1} + \frac{1}{n_2}\right)}$$

Step 3 Formulate the decision rule.
  a) Determine the type of alternative hypothesis: directional or non-directional.
  b) Determine the type of test: 1TT or 2TT.
  c) Identify the significance level: $\alpha = 1\%$ or $\alpha = 5\%$.
  d) Find the critical t score(s).
  e) Construct the hypothesis test model.
  f) State the decision rule.

Step 4 Analyze the sample data.
Randomly select an independent sample from each population and calculate the mean of each sample, denoted $\bar{x}_1$ and $\bar{x}_2$. **Determine the difference between the two sample means, using the formula: $d = \bar{x}_1 - \bar{x}_2$**. Calculate the t score of the difference between the two sample means using the formula:

$$t = \frac{(\bar{x}_1 - \bar{x}_2)}{s_{\bar{x}_1 - \bar{x}_2}}$$

Section 13.4 *Hypothesis Testing Involving Two Population Means and Unknown but Equal Population Standard Deviations*

This is referred to as the **test statistic**.

Step 5 State the conclusion.
Compare the test statistic, t, to the critical t score(s), t_c, of the decision rule and draw one of the following conclusions:

(a) Reject H_o and Accept H_a at α

or

(b) Fail to reject H_o at α

To summarize, in the procedure used to outline a **pooled two sample t test**, there were three refinements to the general hypothesis testing procedure developed in Chapter 10. They were:

1. The test statistic hypothesis testing model is a *t* distribution, with df = $n_1 + n_2 - 2$.
2. The properties of the sampling distribution of the difference between two means are used to calculate the expected values.
3. The formula to calculate the test statistic t is:

$$t = \frac{(\bar{x}_1 - \bar{x}_2)}{s_{\bar{x}_1 - \bar{x}_2}}$$

Example 13.2 illustrates the use of a pooled two sample t test to perform a hypothesis test involving two population means using independent random samples.

EXAMPLE 13.2

A computer programming aptitude test was given to 50 men and 60 women. The women's sample mean test score was 84 with s = 7, while the men's mean test score was 82 with s = 5. Can one conclude that there is a significant difference between the mean test scores for the population of men and the population of women on this test? Assume equal population standard deviations and use $\alpha = 1\%$.

Solution

Since we have unknown but equal population standard deviations we will use the pooled t-test.

Step 1 Formulate the hypotheses.

Let population 1 represent the men's computer aptitude scores and population 2 represent the women's computer aptitude scores:

Since we are trying to show that there is a significant difference between the population mean test scores of the men and women we state the null hypothesis in such a way that by *rejecting* H_o we can conclude that there *is* a difference between the population means. Therefore, we will always state the null hypothesis as "there is no difference between the mean of population 1 and the mean of population 2." This is symbolized as: $\mu_1 - \mu_2 = 0$.

H_o: There is **no difference** between the population means of the computer programming aptitude test scores for men and women.

In symbols, H_o: $\mu_1 - \mu_2 = 0$

We are trying to show that there is a difference between the population mean for the men's and women's aptitude test scores. Thus, the alternative hypothesis is non-directional and is stated as:

H_a: The difference between the population mean computer aptitude test scores for the men and women **is significantly different** from *zero*.

This is symbolized as: H_a: $\mu_1 - \mu_2 \neq 0$

Step 2 Determine the model to test the null hypothesis, H_o.

To test the null hypothesis, an independent random sample is selected from each population. The sample results are summarized in Table 13.2.

Since the difference between the two sample means, $\bar{x}_1 - \bar{x}_2$, will be used to determine if $(\mu_1 - \mu_2)$ is **significantly different** from **zero, the expected results** are:

a) The test statistic is approximated by a t distribution with degrees of freedom:

$$df = n_1 + n_2 - 2 = 108.$$

TABLE 13.2 Summary of the sample data selected from each population

Random Sample Selected from Population 1 (Men)	Random Sample Selected from Population 2 (Women)
$\bar{x}_1 = 82$	$\bar{x}_2 = 84$
$s_1 = 5$	$s_2 = 7$
$n_1 = 50$	$n_2 = 60$

b) The mean of the sampling distribution of the difference between two means is:

$$\mu_{\bar{x}_1-\bar{x}_2} = \mu_1 - \mu_2 = 0$$

c) For the pooled t-test, the estimate of the pooled standard error of the difference between two means formula is:

$$s_{\bar{x}_1-\bar{x}_2} = \sqrt{\frac{(n_1-1)s_1^2 + (n_2-1)s_2^2}{n_1+n_2-2} \cdot \left(\frac{1}{n_1}+\frac{1}{n_2}\right)}$$

Using Table 13.2, substitute $s_1=5$, $n_1=50$, $s_2=7$ and $n_2=60$ into the formula for $s_{\bar{x}_1-\bar{x}_2}$, we have:

$$s_{\bar{x}_1-\bar{x}_2} = \sqrt{\frac{(n_1-1)s_1^2 + (n_2-1)s_2^2}{n_1+n_2-2} \cdot \left(\frac{1}{n_1}+\frac{1}{n_2}\right)}$$

$$= \sqrt{\frac{(50-1)5^2 + (60-1)7^2}{50+60-2} \cdot \left(\frac{1}{50}+\frac{1}{60}\right)}$$

$$= \sqrt{\frac{(49)25 + (59)49}{108} \cdot \left(\frac{1}{50}+\frac{1}{60}\right)}$$

$$\approx \sqrt{(38.11111)(0.03666)}$$

$$s_{\bar{x}_1-\bar{x}_2} \approx 1.18$$

Figure 13.5 illustrates the hypothesis testing model.

distribution of the test statistic (is a t Distribution with df = 108)

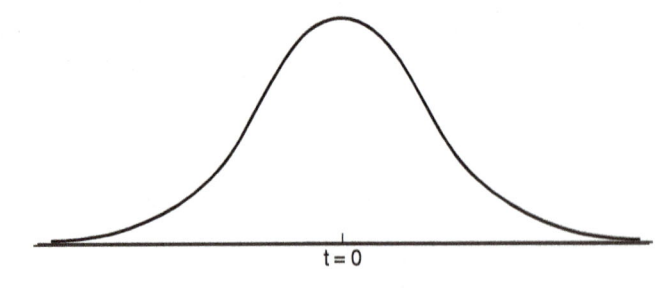

t = 0

FIGURE 13.5

Step 3 Formulate the decision rule.
   a) The alternative hypothesis is non-directional since we are trying to show that there is a significant difference between the population mean scores of the men and women.
   b) The type of hypothesis test is two-tailed (2TT).
   c) The significance level is: $\alpha = 1\%$.
   d) To determine the critical t scores t_{LC} and t_{RC}, for a two-tailed test with $\alpha = 1\%$ and $df = 108$ use Table III.

From Table III, the critical t scores are:

$t_{LC} = -2.63$ and $t_{RC} = 2.63$

e) Figure 13.6(a) illustrates the decision rule for this hypothesis test.

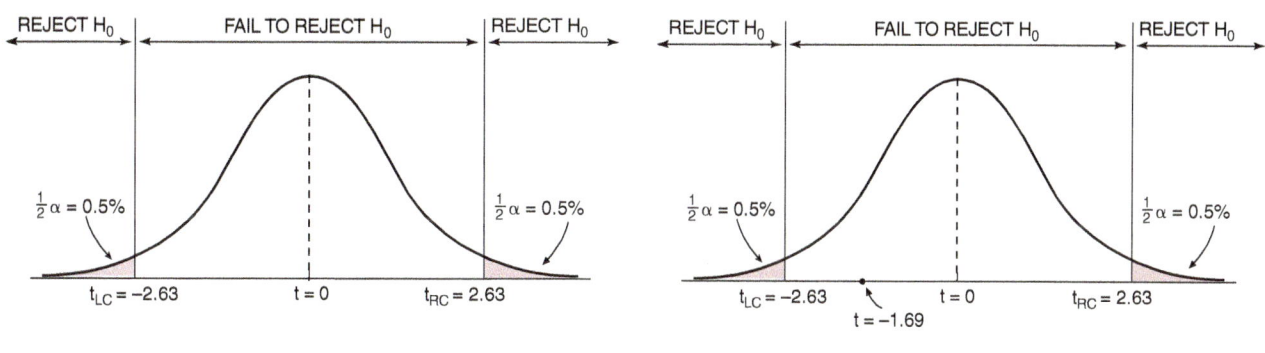

FIGURE 13.6(a)

FIGURE 13.6(b)

The decision rule is:
f) Reject H_o if the test statistic t is either less than $t_{LC} = -2.63$ or more than $t_{RC} = 2.63$.

Step 4 Analyze the sample data.
Calculate the test statistic t using

$$t = \frac{(\bar{x}_1 - \bar{x}_2)}{s_{\bar{x}_1 - \bar{x}_2}}$$

The difference between the two sample means, $d = \bar{x}_1 - \bar{x}_2$, is:

$$d = \bar{x}_1 - \bar{x}_2 = 82 - 84 = -2$$

Since $s_{\bar{x}_1 - \bar{x}_2} = 1.18$, the test statistic t is:

$$t = \frac{(\bar{x}_1 - \bar{x}_2)}{s_{\bar{x}_1 - \bar{x}_2}}$$

$$t = \frac{-2}{1.18}$$

$$t \approx -1.69$$

Step 5 State the conclusion.
Using Figure 13.6(b), since the test statistic $t = -1.69$ falls between the critical t scores -2.63 and 2.63, we fail to reject H_o at $\alpha = 1\%$. Therefore, we conclude that the difference between the population mean aptitude test scores for men and women is **not statistically significant** at $\alpha = 1\%$.

EXAMPLE 13.3

A sociologist claims that women in the South marry at a younger age than women in the North. An independent random sample of 40 recently married southern women had a mean age of 21 years with $s = 2$ years. An independent random sample of 40 recently married northern women had a mean age of 23.5 years with $s = 3$ years. Do these sample results support the sociologist's claim? Assume equal population standard deviations, independent random samples and use $\alpha = 5\%$.

Solution

Since we have unknown but equal population standard deviations we will use the pooled t-test.

Step 1 Formulate the hypotheses.
If we let population 1 represent the ages of the southern brides and population 2 represent the ages of the northern brides then the null hypothesis is:

H_o: there is **no difference** in the population mean age of southern and northern brides.

In symbols we have, $H_o: \mu_1 - \mu_2 = 0$.

H$_a$: the population mean age for southern brides is **less than** the population mean age of northern brides.

In symbols we have, H$_a$: $\mu_1 - \mu_2 < 0$.

Step 2 Determine the model to test the null hypothesis, H$_o$.

To test the null hypothesis, an independent random sample is selected from each population and the sample results are summarized in Table 13.3.

TABLE 13.3 Summary of the Data for the Random Samples Selected from Each Population

Population 1 (South)	Population 2 (North)
$\bar{x}_1 = 21$	$\bar{x}_2 = 23.5$
$s_1 = 2$	$s_2 = 3$
$n_1 = 40$	$n_2 = 40$

Since the differences between the two sample means, $\bar{x}_1 - \bar{x}_2$, will be used to determine if $(\mu_1 - \mu_2)$ is significantly less than zero, the expected results are:

a) The test statistic is approximated by a t distribution with $df = 78$.
b) The mean of the sampling distribution of the difference between two means is:

$$\mu_{\bar{x}_1 - \bar{x}_2} = \mu_1 - \mu_2$$
$$= 0$$

c) For the pooled t-test, the estimate of the pooled standard error of the difference between two means is:

$$s_{\bar{x}_1 - \bar{x}_2} = \sqrt{\frac{(n_1-1)s_1^2 + (n_2-1)s_2^2}{n_1 + n_2 - 2} \cdot \left(\frac{1}{n_1} + \frac{1}{n_2}\right)}$$

Using Table 13.3, substitute s$_1$=2, n$_1$=40, s$_2$=3 and n$_2$=40 into $s_{\bar{x}_1 - \bar{x}_2}$, we have:

$$s_{\bar{x}_1 - \bar{x}_2} = \sqrt{\frac{(n_1-1)s_1^2 + (n_2-1)s_2^2}{n_1 + n_2 - 2} \cdot \left(\frac{1}{n_1} + \frac{1}{n_2}\right)}$$

$$= \sqrt{\frac{(40-1)2^2 + (40-1)3^2}{40 + 40 - 2} \cdot \left(\frac{1}{40} + \frac{1}{40}\right)}$$

$$= \sqrt{\frac{(39)4 + (39)9}{78} \cdot \left(\frac{1}{40} + \frac{1}{40}\right)}$$

$$\approx \sqrt{(6.5)(0.05)}$$

$$s_{\bar{x}_1 - \bar{x}_2} \approx 0.57$$

Figure 13.7 illustrates the hypothesis testing model.

distribution of the test statistic
(is a t Distribution with $df = 78$)

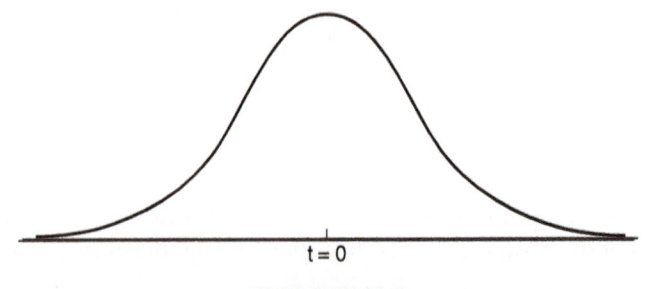

FIGURE 13.7

Step 3 Formulate the decision rule.
a) The alternative hypothesis is directional since we are trying to show that the population mean age of southern brides is significantly less than the population mean age of northern brides.
b) The type of hypothesis test is one-tailed (i.e. 1TT) on the **left side** since we are trying to show that $(\mu_1 - \mu_2)$ is **less than** zero.
c) The significance level is: $\alpha = 5\%$.
d) For a one-tailed test on the left side with $\alpha = 5\%$ and $df = 78$ the critical t score, t_c, using Table III is $t_c = -1.66$.
e) Figure 13.8(a) illustrates the decision rule for this hypotheses test.
The decision rule is:
f) Reject H_o if the test statistic t is less than $t_c = -1.66$.

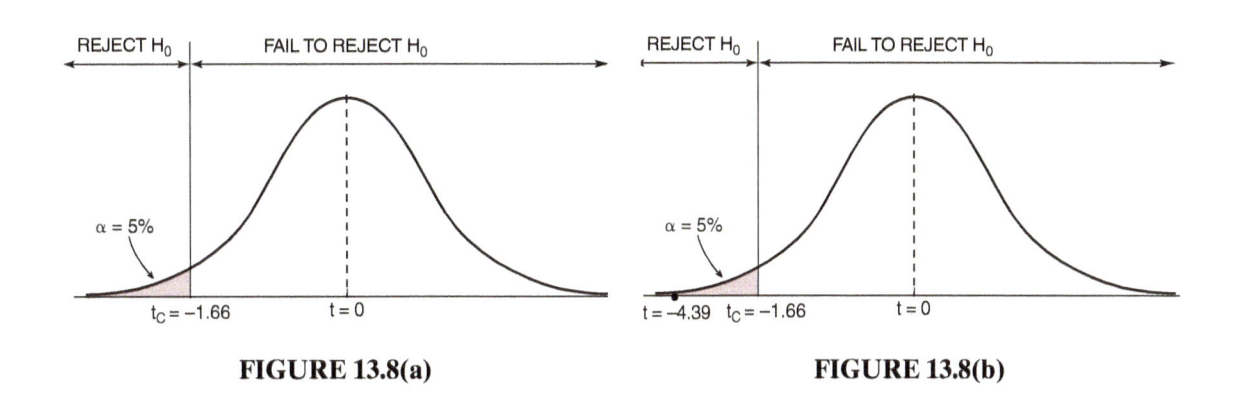

FIGURE 13.8(a)

FIGURE 13.8(b)

Step 4 Analyze the sample data.
Calculate the test statistic t where:

$d = \bar{x}_1 - \bar{x}_2 = 21 - 23.5 = -2.5$ and $s_{\bar{x}_1 - \bar{x}_2} \approx 0.57$

The test statistic is $t = \dfrac{-2.5}{0.57} \approx -4.39$.

Step 5 State the conclusion.
Examine Figure 13.8(b). The test statistic $t = -4.39$ is *less than* the critical t score -1.66. We *reject H_o and accept H_a* at $\alpha = 5\%$. Therefore, we can agree with the sociologist's claim that women in the south marry at a younger mean age than women in the north.

13.5 HYPOTHESIS TESTING INVOLVING TWO POPULATION MEANS AND UNKNOWN BUT UNEQUAL POPULATION STANDARD DEVIATIONS: WELCH'S T-TEST

Theorem 13.2 outlined the characteristics of the sampling distribution of the difference between two population means with unknown population standard deviations. Although unknown, these population standard deviations were **assumed to be equal**. In the previous section, we described a procedure to test the difference between two population means when the population standard deviations were unknown but *assumed* to be equal and independent random samples. This hypothesis test is called the **pooled t-test.**

One obvious question that we must ask ourselves is, "How do we perform this two sample t-test if the population standard deviations are still unknown but **assumed to be unequal**?" The simple answer to this question is we use a technique developed by the statistician Bernard Lewis Welch. This hypothesis test is named **Welch's t-test** after its creator. Before we can discuss how to perform the test, we must first outline the characteristics of the sampling distribution of the difference between two population means with *unknown* and *unequal* population standard deviations and independent random samples. These characteristics are stated in Theorem 13.3.

> **Theorem 13.3 Characteristics of the Sampling Distribution of the Difference Between Two Means ith Population Standard Deviations Unknown and Assumed Unequal**
>
> Suppose you are given two approximately normal populations, referred to as population 1 and population 2, with means μ_1 and μ_2 and *unequal* standard deviations σ_1 and σ_2 respectively. If all possible independent random samples (i.e. the selection of one random sample has no influence upon the selection of the other random sample) of size n_1 and n_2 are selected from population 1 and population 2 respectively, the means of all these pairs of samples are determined, and the differences between the pairs of sample means $(\bar{x}_1 - \bar{x}_2)$ are computed, then the sampling distribution of the difference between two means has the following characteristics:
>
> 1. The mean of the sampling distribution of the difference between two means, denoted by $\mu_{\bar{x}_1-\bar{x}_2}$, is
>
> $$\mu_{\bar{x}_1-\bar{x}_2} = \mu_1 - \mu_2$$
>
> 2. The standard error of the difference between two means cannot be determined since the population standard deviations, σ_1 and σ_2, are unknown. However, since the two sample standard deviations, s_1 and s_2, provide good estimates of the population standard deviations, σ_1 and σ_2, we can combine the sample standard deviations to obtain **an estimate of the standard error of the difference between two means**. This formula for the estimate of the standard error is referred to as **Welch's approximation** denoted by $s_{\bar{x}_1-\bar{x}_2}$, and is:
>
> $$s_{\bar{x}_1-\bar{x}_2} = \sqrt{\frac{s_1^2}{n_1} + \frac{s_2^2}{n_2}}$$
>
> where:
> - s_1 is the estimate of the standard deviation of population 1.
> - n_1 is the sample size selected from population 1.
> - s_2 is the estimate of the standard deviation of population 2.
> - n_2 is the sample size selected from population 2.
>
> 3. The distribution of the test statistic t is defined as
>
> $$t = \frac{(\bar{x}_1 - \bar{x}_2) - \mu_{\bar{x}_1-\bar{x}_2}}{s_{\bar{x}_1-\bar{x}_2}}$$
>
> and is approximated by a *t* distribution with degrees of freedom given by:
>
> $$df = \text{smaller of } (n_1 - 1) \text{ and } (n_2 - 1).*$$

*Note: The actual formula for the degrees of freedom for this test was developed by the statistician Franklin Satterthwaite and is $df = \dfrac{\left(\dfrac{s_1^2}{n_1} + \dfrac{s_2^2}{n_2}\right)^2}{\dfrac{1}{n_1-1}\left(\dfrac{s_1^2}{n_1}\right)^2 + \dfrac{1}{n_2-1}\left(\dfrac{s_2^2}{n_2}\right)^2}$. As you can see, this formula for the actual degrees of freedom would be quite cumbersome to work with. It is because of this that we often use the simpler and more *conservative method* described in Theorem 13.3 to calculate the degrees of freedom for this test. The method described in Theorem 13.3 is called a more *conservative method* because it will almost always yield a value for the degrees of freedom that is smaller than the value from Satterthwaite's equation. Smaller values for degrees of freedom yield larger critical *t* scores(t_c) which in turn makes it more difficult for the test to reject the null hypothesis. This is a common practice in many introductory texts, and one that we'll practice as well. We should also point out that many software programs have the actual formula for degrees of freedom programmed and this conservative two approach is often not needed if technology is being used to help perform the hypothesis test.

Section 13.5 *Hypothesis Testing Involving Two Population Means and Unknown but Unequal Population Standard Deviations*

Let's apply the general five step hypothesis testing procedure discussed in Chapter 10 to the hypothesis test involving two population means, where the population standard deviations are unknown and assumed to be unequal and independent random samples. This hypothesis test is often referred to as ***Welch's t-Test***.

Welch's t Test

Step 1   Formulate the two hypotheses, H_0 and H_a.
Null Hypothesis, H_0:
The **Null Hypothesis**, H_0, in testing the difference between two population means has the following form:

Null Hypothesis Form

H_0: There is no difference between the mean of population 1, written μ_1, and the mean of population 2, written μ_2.

The null hypothesis can be symbolized as:

$$H_0 : \mu_1 - \mu_2$$

Alternative Hypothesis, H_a: Since the alternative hypothesis, H_a, can be stated as either *greater than*, *less than*, or *not equal to* 0, then the alternative hypothesis can have one of the following three forms.

Alternative Hypothesis Form

Form (a): Less Than Form for H_a

H_a: the mean of population 1, written μ_1, is **less** than the mean of population 2, written μ_2. This can also be stated as the difference between the population means $(\mu_1 - \mu_2)$ is **significantly less than 0**.

This is symbolized as:

$$H_a: \mu_1 - \mu_2 < 0$$

Form (b): Greater Than Form for H_a

H_a: the mean of population 1, written μ_1, is **greater** than the mean of population 2, written μ_2. This can also be stated as the difference between the population means $(\mu_1 - \mu_2)$ is **significantly greater than 0**.

This is symbolized as:

$$H_a: \mu_1 - \mu_2 > 0$$

Form (c): Not Equal to Form for H_a

H_a: the mean of population 1, written μ_1, is **not equal to** the mean of population 2, written μ_2. This can also be stated as the difference between the population means $(\mu_1 - \mu_2)$ is **significantly different from 0**.

This is symbolized as:

$$H_a: \mu_1 - \mu_2 \neq 0$$

Step 2 Determine the model to test the null hypothesis, H_0. Under the assumption H_0 is true, calculate the expected results.

Since we are testing the hypothesis that the **difference between two population means** $(\mu_1 - \mu_2)$ is significantly different from zero by **using the difference between two sample means,** $\bar{x}_1 - \bar{x}_2$, then we will use the **sampling distribution of the difference between two means as the hypothesis testing model**. Using Theorem 13.3, and under the assumption H_0 is true (that is: $\mu_1 - \mu_2 = 0$ is true), then:

the expected results for Welch's t test are:

a) The test statistic hypothesis testing model is approximately a t distribution with degrees of freedom given by:

$$df = \text{smaller of } (n_1 - 1) \text{ and } (n_2 - 1).$$

b) The mean of the sampling distribution of the difference between two means is given by the formula:

$$\mu_{\bar{x}_1 - \bar{x}_2} = \mu_1 - \mu_2$$

Under the assumption H_0 is true (i.e. $\mu_1 - \mu_2 = 0$ is true), then we can state that:

$$\mu_{\bar{x}_1 - \bar{x}_2} = 0$$

c) For Welch's t-test, the estimate of the standard error of the difference between two means, written $s_{\bar{x}_1 - \bar{x}_2}$, is computed by the formula:

$$s_{\bar{x}_1 - \bar{x}_2} = \sqrt{\frac{s_1^2}{n_1} + \frac{s_2^2}{n_2}}$$

Step 3 **Formulate the Decision Rule:**
a) Determine type of alternative hypothesis, directional or nondirectional.
b) Determine type of test, 1TT or 2TT.
c) Identify the significance level, $\alpha = 1\%$ or $\alpha = 5\%$.
d) Construct the appropriate hypothesis test model.
e) Find the critical t score(s) using Table III for the appropriate level of significance.
f) State the decision rule.

Step 4 **Analyze the sample data**
a) Randomly select an independent sample from each population and calculate the mean of each sample, denoted $\bar{x}_1$ and $\bar{x}_2$.
b) **Determine the difference between the two sample means, using the formula:** $d = \bar{x}_1 - \bar{x}_2$.
c) Calculate the t score of the difference between the two sample means using the formula:

$$t = \frac{(\bar{x}_1 - \bar{x}_2)}{s_{\bar{x}_1 - \bar{x}_2}}$$

Step 5 **State the conclusion.**
Compare the test statistic to the decision rule and either:
a) Reject H_0 and accept H_a at α,
or
b) Fail to reject H_0 at α.

Example 13.4 illustrates the use of **Welch's t test** to perform a hypothesis test involving two population mean with unknown and unequal population standard deviations and independent random samples.

EXAMPLE 13.4

A market analyst is trying to convince her boss to advertise their new perfume for women on social media platforms. The analyst believes that Facebook would be a great match because she is convinced that women spend significantly more time each day on Facebook than men. To support this decision, the analyst randomly

Section 13.5 *Hypothesis Testing Involving Two Population Means and Unknown but Unequal Population Standard Deviations*

samples 26 women and 22 men in the 18-28 demographic range who have Facebook accounts. The summary statistics for the number of minutes spent per day by the two groups are given in Table 13.4.

TABLE 13.4 Summary Statistics for the Random Samples of Facebook times

Gender	Sample size	Sample mean (mins/day)	Sample standard deviation
Female	26	32.3	26.5
Male	22	28.7	22.8

At the 5% level of significance, perform a hypothesis test to test the claim that females spend significantly more time on Facebook each day than males. Assume that the population standard deviations are unequal, and use Welch's *t*-Test.

Solution

Since we are assuming the population standard deviations are unequal and unknown we will use Welch's t-test.

Step 1 Formulate the two hypotheses, H_0 and H_a.

If we let population 1 represent the female time spent on Facebook per day and population 2 represent the male time spent on Facebook per day, then the **Null Hypothesis** is:

H_0: There is no difference between the population mean time spent on Facebook per day for females and males.

In symbols we have, $H_0: \mu_1 - \mu_2 = 0$.

The **Alternative Hypothesis** is:

H_a: The population mean time spent on Facebook per day for females is greater than the population mean time spent on Facebook per day by males.

In symbols we have, $H_a: \mu_1 > \mu_2$. This is equivalent to stating that the difference between the population mean times for females and males is greater than 0. Therefore, this alternative hypothesis can also be symbolized as $H_a: \mu_1 - \mu_2 > 0$.

Step 2 Determine the model to test the null hypothesis, H_0.

To test the null hypothesis, the analyst selected an independent random sample from each population and the sample results are summarized in Table 13.5

TABLE 13.5 Summary Statistics for the Random Samples of Facebook times

Population 1(female)	Population 2(male)
$n_1 = 26$	$n_2 = 22$
$\bar{x}_1 = 32.3$	$\bar{x}_2 = 28.7$
$s_1 = 26.5$	$s_2 = 22.8$

Since the differences between the two sample means, $d = \bar{x}_1 - \bar{x}_2$, will be used to determine if $\mu_1 - \mu_2$ is significantly less than zero, the expected results are:

a) The test statistic hypothesis testing model is approximately a t distribution with degrees of freedom given by:

df = smaller of $(n_1 - 1)$ and $(n_2 - 1)$, where $n_1 = 26$ and $n_2 = 22$
df = smaller 25 and 21.

Therefore, $df = 21$.

b) The mean of the sampling distribution of the difference between two means is given by the formula:

$$\mu_{\bar{x}_1 - \bar{x}_2} = \mu_1 - \mu_2 = 0$$

c) For Welch's t-test, the estimate of the standard error of the difference between two means, written $s_{\bar{x}_1-\bar{x}_2}$, is given by the formula:

$$s_{\bar{x}_1-\bar{x}_2} = \sqrt{\frac{s_1^2}{n_1} + \frac{s_2^2}{n_2}}$$

$$= \sqrt{\frac{26.5^2}{26} + \frac{22.8^2}{22}}$$

$$\approx \sqrt{50.6387}$$

$$s_{\bar{x}_1-\bar{x}_2} \approx 7.1161$$

The distribution of the test statistic is a t distribution with $df = 21$. This is illustrated in Figure 13.9.

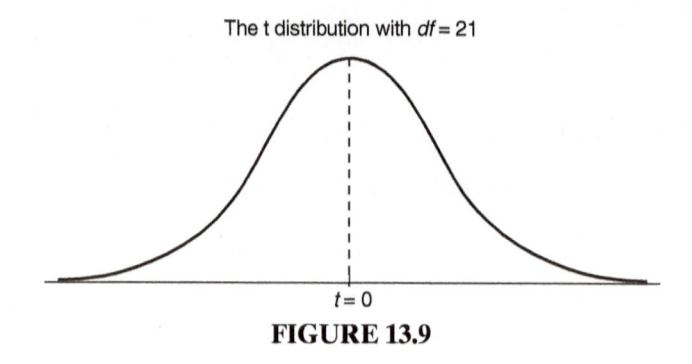

The t distribution with $df = 21$

FIGURE 13.9

Step 3 **Formulate the Decision Rule**:
a) The alternative hypothesis is directional since we are trying to show the population mean time per day for the females is greater than the population mean time per day for the males. That is: $\mu_1 - \mu_2 > 0$.
b) The type of test is a one tailed (1TT) on the right side since we are trying to show that $\mu_1 - \mu_2 > 0$.
c) The significance level is $\alpha = 5\%$.
d) For a one-tailed test on the right side with $\alpha = 5\%$ and $df = 21$, the critical t-score, t_c, using Table III is $t_c = 1.72$.
e) Figure 13.10 Illustrates the decision rule for this hypothesis test.
The decision rule is:
Reject H_0 if the test statistic is greater than $t_c = 1.72$.

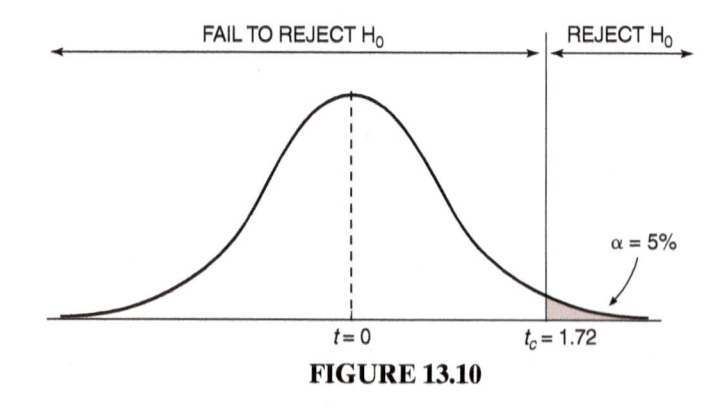

FAIL TO REJECT H_0 | REJECT H_0

$\alpha = 5\%$

$t = 0$ $t_c = 1.72$

FIGURE 13.10

Step 4 **Analyze the sample data**
Calculate the t score of the difference between the two sample means using the formula:

$$t = \frac{(\bar{x}_1 - \bar{x}_2)}{s_{\bar{x}_1-\bar{x}_2}}$$

$$t = \frac{32.3 - 28.7}{7.1161}$$

$$t \approx 0.51$$

Thus, the test statistic is $t \approx 0.51$.

Step 5 **State the conclusion.**
The test statistic $t \approx 0.51$ is less than the critical t score of 1.72 as shown in Figure 13.11. Thus, we fail to reject H_0 at $\alpha = 5\%$. Therefore, the analyst cannot conclude that the female population spends significantly more time than the male population on Facebook per day. A new marketing strategy will need to be discussed.

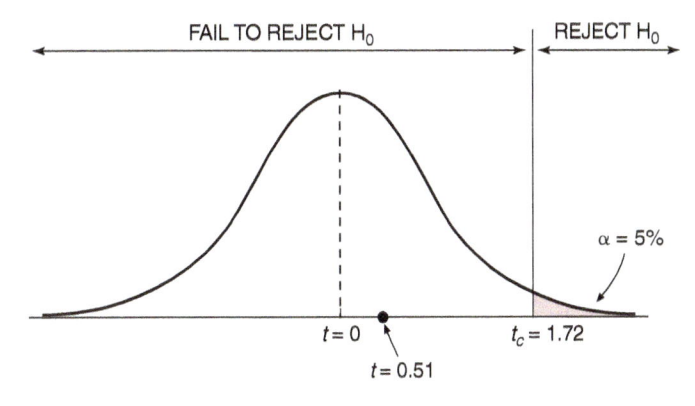

FIGURE 13.11

Pooled t-test vs. Welch's t-Test

One major difference between the pooled t-test and Welch's t-test is that when using the pooled t-test, the population variances (and standard deviations) are assumed to be *equal*, where in Welch's t-test the population variances (and standard deviations) are assumed to be *unequal*. One last question regarding these two different interpretations of the two sample t-tests we have to ask is, "How could we possibly know if the population standard deviations are equal or not?" As it turns out, there is a hypothesis test that will test this equal/unequal variance assumption called the **F-Test for Variances**. This hypothesis test is covered later in the text in Chapter 16: The F-Distribution and Introduction to Analysis of Variance(ANOVA).

13.6 HYPOTHESIS TESTS COMPARING TREATMENT AND CONTROL GROUPS

Frequently, an experimenter wants to compare two populations or groups where one group receives a treatment called the **treatment group**, while a second group called the **control group** does not receive the treatment. Ideally, the members of both groups are similar in every way except that the members of the control group are not given the treatment.

For example, a new medication is tested to determine if it is effective. An independent random sample of the population called the treatment group is selected and given this new medication. Another independent random sample of the population called the control group is selected and does not receive the medication. However, the control group is given a placebo, that is, a pill which looks like the real medication but does not contain the medication. This is done to guard against the psychological effect of the treatment group feeling better not because of the new medication but because they were picked for the special treatment. This type of experimental design is called a **single-blind study**, because the *subjects* do **not** know who is receiving the *real* medication. Sometimes it has been shown that studies have been affected by the *doctors* who are administering the medication, since *they* know who is and who is not receiving the placebo. Consequently, it becomes necessary to conduct a study where *both* the subjects and doctors are *not* aware of who is receiving the medication and who is receiving the placebo. This type of experimental design is called a **double-blind study**.

Example 13.5 illustrates the comparison of treatment versus control groups using a **pooled two sample *t* test**.

EXAMPLE 13.5

A pharmaceutical company has developed a diet pill called, Effective Anti-Hunger Tablet (E.A.T.), which the company believes will significantly reduce an individual's weight at the end of one week. To support their claim, the company selects 30 overweight individuals and randomly divides them into two independent groups: a **treatment** group of 15 subjects who are administered the diet pill E.A.T. and a **control** group of 15 subjects who are given a *placebo* tablet (i.e. **a pill that has *no effect* on weight**).

After one week, the treatment group had a sample mean weight loss of $\bar{x} = 7.1$ lbs. with s = 3.6 lbs. while the control group had a sample mean weight loss of $\bar{x} = 4.2$ lbs. with s = 2.3 lbs.

Can the company conclude that their diet pill E.A.T. will significantly reduce an individual's weight within one week? Assume equal population standard deviations and an $\alpha = 5\%$.

Remark: In this study the administration of the diet pill E.A.T. to the treatment group is the only difference between the two groups. Therefore, if we can show a significant difference between the treatment and control groups then this *difference can be attributed only to the treatment (the diet pill)*.

Solution

Since we are assuming the population standard deviations are unknown but equal, we will use the pooled t-test.

Step 1 Formulate the hypotheses.

If we let population 1 represent the treatment group and population 2 represent the control group then the null hypothesis is:

H_o: there is **no difference** in the population mean weight loss for the treatment group and the control group.

In symbols we have, $H_o: \mu_1 - \mu_2 = 0$

H_a: the population mean weight loss for the treatment group is **greater than** the population mean weight loss for the control group.

In symbols we have, $H_a: \mu_1 - \mu_2 > 0$.

Step 2 Determine the model to test the null hypothesis, H_o.

To test the null hypothesis the company selects 30 overweight individuals and randomly divides them into two independent groups and the sample results are summarized in Table 13.6.

TABLE 13.6 Summary of the Data for the Random Samples Selected from Each Population

Population 1 (Treatment)	Population 2 (Control)
$\bar{x}_1 = 7.1$	$\bar{x}_2 = 4.2$
$s_1 = 3.6$	$s_2 = 2.3$
$n_1 = 15$	$n_2 = 15$

Since the differences between the two sample means, $d = \bar{x}_1 - \bar{x}_2$, will be used to determine if $(\mu_1 - \mu_2)$ is significantly greater than zero, the expected results are:
a) The test statistic is approximated by a t distribution with $df = 28$.
b) The mean of the sampling distribution of the difference between two means is:

$$\mu_{\bar{x}_1 - \bar{x}_2} = \mu_1 - \mu_2$$
$$= 0$$

c) For the pooled t-test, the estimate of the pooled standard error of the difference between two means is:

$$s_{\bar{x}_1 - \bar{x}_2} = \sqrt{\frac{(n_1 - 1)s_1^2 + (n_2 - 1)s_2^2}{n_1 + n_2 - 2} \cdot \left(\frac{1}{n_1} + \frac{1}{n_2}\right)}$$

Using Table 13.6, substitute $s_1 = 3.6$, $n_1 = 15$, $s_2 = 2.3$ and $n_2 = 15$ into $s_{\bar{x}_1 - \bar{x}_2}$, we have:

$$s_{\bar{x}_1 - \bar{x}_2} = \sqrt{\frac{(15-1)(3.6)^2 + (15-1)(2.3)^2}{15 + 15 - 2}\left(\frac{1}{15} + \frac{1}{15}\right)}$$

$$= \sqrt{\frac{181.44 + 74.06}{28}(0.13)}$$

$$\approx \sqrt{(9.1250)(0.1333)}$$

$$s_{\bar{x}_1 - \bar{x}_2} \approx 1.10$$

Figure 13.12 illustrates the hypothesis testing model.

**distribution of the test statistic
(is a t Distribution with *df* = 28)**

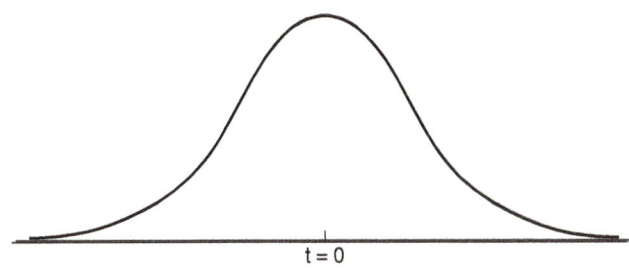

FIGURE 13.12

Step 3 Formulate the decision rule.
  a) The alternative hypothesis is **directional** since we are trying to show that the population mean weight loss for the treatment group is **significantly greater** than the population mean weight loss for the control group.
  b) The type of hypothesis test is one-tailed (1TT) on the **right side** since we are trying to show that $(\mu_1 - \mu_2)$ is **greater than** zero.
  c) The significance level is: $\alpha = 5\%$.
  d) For a one-tailed test on the right with $\alpha = 5\%$ and $df = 28$, the critical t score, t_c, using Table III is $t_c = 1.70$.
  e) Figure 13.13(a) illustrates the decision rule for this hypothesis test.

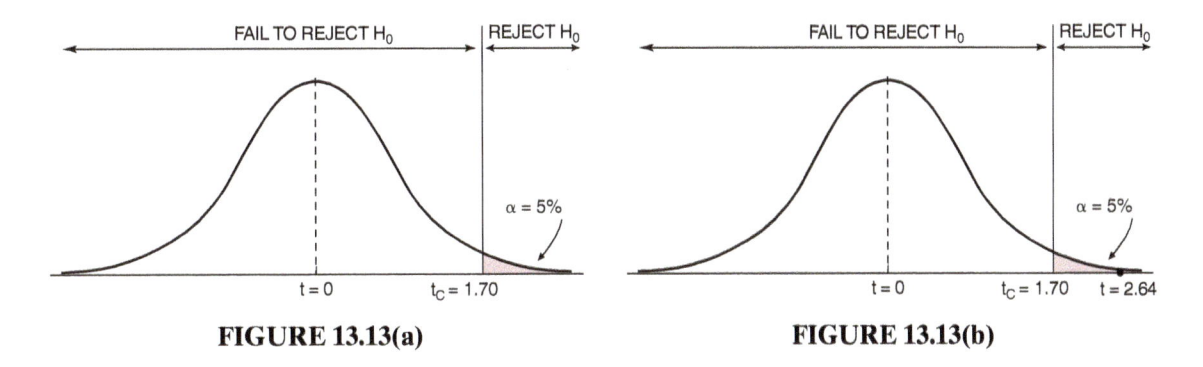

FIGURE 13.13(a) **FIGURE 13.13(b)**

The decision rule is:
  f) Reject H_o if the test statistic t is greater than $t_c = 1.70$.

Step 4 Analyze the sample data.
  Calculate the test statistic t where

$$\bar{x}_1 - \bar{x}_2 = 7.1 - 4.2 = 2.9 \text{ and } s_{\bar{x}_1 - \bar{x}_2} \approx 1.10$$

The test statistic t is: $t = \dfrac{2.9}{1.10} \approx 2.64$.

Step 5 State the conclusion.

Using Figure 13.13(b), since the test statistic t ≈ 2.64, is greater than the critical t score $t_c = 1.70$, *reject* H_o and *accept* H_a at $\alpha = 5\%$.

Therefore, the company can conclude that their diet pill (E.A.T.) will **significantly reduce** an individual's weight at the end of one week at $\alpha = 5\%$.

CASE STUDY 13.1 Academic Differences

A research article entitled "Differences in Extracurricular Activity Participation, Achievement, And Attitudes Toward School Between Ninth-grade Students Attending Junior High School And Those Attending Senior High School" appeared in *Adolescence*. The following table represents a comparison of the Mean Number of Total Extracurricular Activities Participated, Overall GPA, and Mean Attitude Toward Self and School for Ninth Graders in Junior High (n = 771) and Senior High Settings (n = 825).

Groups	Mean	Standard Deviation	t value
Extracurricular Activities			
Ninth Graders in JHS	2.68	2.30	6.20*
Ninth Graders in SHS	1.99	2.14	
Overall GPA			
Ninth Graders in JHS	2.59	0.89	6.91*
Ninth Graders in SHS	2.24	1.11	
Attitude Toward Self and School			
Ninth Graders in JHS	77.87	8.07	0.60
Ninth Graders in SHS	78.12	8.48	

*p < 0.01

Let's define population 1 as: Ninth Graders in Junior High School (JHS) and population 2 is defined as: Ninth Graders in Senior High School (SHS).

a) If you are trying to show that there is a significant difference in the mean number of extracurricular activities for these two populations, then state the null hypothesis and alternative hypothesis for this two-sample t test.

b) If you want to show that there is a significant difference in the mean overall GPA for these two populations, then state the null and alternative hypotheses for this two-sample t test.

c) If you would like to show that there is a significant difference in the mean attitude toward self and school for these two populations, then state the null and alternative hypotheses for this two-sample t test.

d) What is the mean of the sampling distribution which serves as a model for this test? Is the test statistic distribution a normal or a *t* distribution? Explain.

e) For the Extracurricular Activities Study, which group had the greater variability? For the Overall GPA Study, which group had the less variability? For the Attitude Study, which group had the greater variability? Do these measures of variability represent a parameter or a statistic? Explain.

f) Was there a significant difference between the populations with respect to the mean number of extracurricular activities? Explain.

g) If the significance level for the extracurricular activities had been 0.5%, would the null hypothesis be rejected? Explain.

h) Was there a significant difference between the populations with respect to the mean overall GPA? Was the null hypothesis rejected? Explain.

i) Would the null hypothesis have been rejected for the overall GPA study if $\alpha = 1\%$? Explain.

j) Was there a significant difference between the populations with respect to the mean attitude? Was the null hypothesis rejected? Explain.

k) Would the null hypothesis have been rejected for the attitude study if $\alpha = 2\%$? Explain.

CASE STUDY 13.2 Potency of the Placebo Effect

In Figure 13.14 is an article entitled *Study says 'placebo effect' is potent* appeared in the Orange County Register, a California newspaper, on July 1, 1993.

Study says 'placebo effect' is potent

Health: Research shows people with certain diseases get better 70 percent of the time with dummy treatment.

By Lee Siegel
The Associated Press

LOS ANGELES—People with certain diseases get better 70 percent of the time even when they receive dummy treatments, according to a study that suggests the "placebo effect" can be a powerful healer.

"Even if a treatment is not effective, large numbers of people will feel they've been helped if both the therapist and the patient believe in its effectiveness." said Alan H. Roberts, principal author of the study, being published Thursday in Clinical Psychology Review.

The research is "consistent with what we knew, but documents it better than before," said William Jarvis, a preventive medicine professor at Loma Linda University and president of the National Council Against Health Fraud.

The findings suggest doctors as well as patients may be too quick to use experimental treatments that seem promising but ultimately prove ineffective, said Roberts, chief psychologist at Scripps Clinic and Research Foundation in La Jolla.

He said the research also indicates people seek unconventional, unproven therapies because they often seem to be helped by them, but that they would do even better with scientifically proven treatments.

That would allow them to benefit from both real medicine and the "placebo effect"—the improvement patients get from inert drugs or ineffective treatments purely because of belief the treatments will work, he said.

Roberts and colleagues at San Diego State University and the University of Michigan, Ann Arbor, analyzed dozens of old studies on five treatments that were medically accepted in the 1960s or 1970s but later found ineffective.

The treatments were glomectomy, an asthma-relief surgery; briefly freezing the stomach lining in an attempt to treat peptic ulcers; and three drugs applied to herpes simplex sores.

Of the 6,931 patients treated in the old studies, the outcome was excellent for 40 percent, good for 30 percent and poor for only 30 percent.

That shows "a lot of healing from medicine—and probably from psychotherapy—is the result of what are generally called placebo effects," Roberts said.

The most frequently cited research on the subject has been a 1955 study that found the placebo effect accounted for one-third of the improvement in people who got better after being given either a medicine or a placebo for a variety of ailments, Roberts said.

Read the article and comment on the following questions.

FIGURE 13.14
Reprinted by permission of Wright's Media.

a) What does the term 'dummy treatment' mean to you? Is there another name for this?
b) What does the author mean by the statement: "the 'placebo effect' can be a powerful healer"?
c) What does the "placebo effect" mean? Give examples from the article to illustrate this effect.
d) How did the researchers come to the conclusion of a placebo effect? What did they use to support their claim?

Summary of the Test Statistic Approach to Hypothesis Testing Involving Two Population Means Using Independent Random Samples

The following outline is a summary of the 5 step hypothesis testing procedure for a hypothesis test involving two population means.

Hypothesis Testing Involving Two Population Means

Step 1 Formulate the two hypotheses, H_o and H_a.

Null Hypothesis, $H_o: \mu_1 - \mu_2 = 0$

Alternative Hypothesis, H_a: has one of the following three forms:

$$H_a: \mu_1 - \mu_2 < 0 \quad \text{or} \quad H_a: \mu_1 - \mu_2 > 0 \quad \text{or} \quad H_a: \mu_1 - \mu_2 \neq 0$$

Step 2 Determine the model to test the null hypothesis, H_0.

Under the assumption H_0 is true, the expected results are:

a) The distribution of the test statistic is either:
   (i) a normal distribution, if the standard deviations of both populations are known and both sample sizes are greater than 30,

   or (ii) is a t distribution with degrees of freedom: $df = n_1 + n_2 - 2$, if the standard deviations of both populations are unknown and assumed to be **equal**.

   or (iii) is a t distribution with degrees of freedom: $df =$ smaller of $(n_1 - 1)$ and $(n_2 - 1)$, if the standard deviations of both populations are unknown and assumed to be **unequal**.

b) The mean of the sampling distribution of the difference between two means is given by the formula: $\mu_{\bar{x}_1 - \bar{x}_2} = \mu_1 - \mu_2 = 0$ (since H_0 is true: i.e. $\mu_1 - \mu_2 = 0$)

c) I. For a test statistic having a normal distribution: the standard error of the difference between two means, denoted by $\sigma_{\bar{x}_1 - \bar{x}_2}$ is:

$$\sigma_{\bar{x}_1 - \bar{x}_2} = \sqrt{\frac{\sigma_1^2}{n_1} + \frac{\sigma_2^2}{n_2}},$$

and the **two sample z Test** is applied.

II. For a test statistic having a t distribution: the **estimate of the standard error** of the difference between two means is denoted $s_{\bar{x}_1 - \bar{x}_2}$, and has two cases. The formula is dependent on the appropriate condition stated below.

1) If the population standard deviations are unknown and assumed to be **equal**, then the estimate of the standard error is given by

$$s_{\bar{x}_1 - \bar{x}_2} = \sqrt{\frac{(n_1 - 1)s_1^2 + (n_2 - 1)s_2^2}{n_1 + n_2 - 2} \cdot \left(\frac{1}{n_1} + \frac{1}{n_2}\right)},$$

and the **pooled t Test** is applied.

Note: $s_{\bar{x}_1 - \bar{x}_2} = s_{x_p} \sqrt{\left(\frac{1}{n_1} + \frac{1}{n_2}\right)}$, where s_{x_p} is called the pooled standard deviation and is computed by the formula $s_{x_p} = \sqrt{\frac{(n_1 - 1)s_1^2 + (n_2 - 1)s_2^2}{n_1 + n_2 - 2}}$.

2) If the population standard deviations are unknown and assumed to be **unequal**, then the formula for the estimate of the standard error is given by

$$s_{\bar{x}_1 - \bar{x}_2} = \sqrt{\frac{s_1^2}{n_1} + \frac{s_2^2}{n_2}},$$

and **Welch's t Test** is applied.

Step 3   Formulate the decision rule.
  a) Alternative hypothesis: directional or nondirectional.
  b) Type of test: 1TT or 2TT.
  c) Significance level: $\alpha = 1\%$ or $\alpha = 5\%$.
  d) Critical score:
      for a normal distribution: z_c is found in Table II
      for a t distribution: t_c is found in Table III where $df = n_1 + n_2 - 2$ in the case of the pooled t-test or $df =$ smaller of $(n_1 - 1)$ or $(n_2 - 1)$ in the case of Welch's t-test.
  e) Construct the hypothesis testing model.
  f) State the decision rule.

Step 4   Analyze the sample data.
  a) Compute the difference between the two sample means: $\bar{x}_1 - \bar{x}_2$.
  b) For a normal distribution: test statistic: $z = \dfrac{\bar{x}_1 - \bar{x}_2}{\sigma_{\bar{x}_1 - \bar{x}_2}}$

  For a t distribution: test statistic: $t = \dfrac{\bar{x}_1 - \bar{x}_2}{s_{\bar{x}_1 - \bar{x}_2}}$

Step 5   State the conclusion.
  Compare the test statistic to the critical score of the decision rule and state the conclusion: either: (a) Reject H_o and Accept H_a at α or (b) Fail to Reject H_o at α

Table 13.7 contains a condensed version of the hypothesis tests involving two population means.

TABLE 13.7 Summary of Hypothesis Tests Involving Two Population Means Using Independent Random Samples

Type of Test	Form of the Null Hypothesis	Conditions of Test	Test Statistic Formula	Distribution of the Test Statistic Is a:
Two sample z Test	$\mu_1 - \mu_2 = 0$	KNOWN: σ_1 and σ_2 and n_1 and n_2 are both greater than 30	$z = \dfrac{\bar{x}_1 - \bar{x}_2}{\sigma_{\bar{x}_1 - \bar{x}_2}}$ where: $\sigma_{\bar{x}_1 - \bar{x}_2} = \sqrt{\dfrac{\sigma_1^2}{n_1} + \dfrac{\sigma_2^2}{n_2}}$	Normal Distribution
Pooled t Test	$\mu_1 - \mu_2 = 0$	UNKNOWN: σ_1 and σ_2 and $\sigma_1 \approx \sigma_2$	$t = \dfrac{\bar{x}_1 - \bar{x}_2}{s_{\bar{x}_1 - \bar{x}_2}}$ where: $s_{\bar{x}_1 - \bar{x}_2} = \sqrt{\dfrac{(n_1 - 1)s_1^2 + (n_2 - 1)s_2^2}{n_1 + n_2 - 2} \cdot \left(\dfrac{1}{n_1} + \dfrac{1}{n_2}\right)}$	t Distribution with: $df = n_1 + n_2 - 2$
Welch's t Test	$\mu_1 - \mu_2 = 0$	UNKNOWN: σ_1 and σ_2 and $\sigma_1 \neq \sigma_2$	$t = \dfrac{\bar{x}_1 - \bar{x}_2}{s_{\bar{x}_1 - \bar{x}_2}}$ where: $s_{\bar{x}_1 - \bar{x}_2} = \sqrt{\dfrac{s_1^2}{n_1} + \dfrac{s_2^2}{n_2}}$	t Distribution with: $df =$ smaller of $(n_1 - 1)$ and $(n_2 - 1)$.

13.7 THE PAIRED T-TEST: A DEPENDENT T-TEST

In Example 13.3 of Section 13.6, we discussed how a pharmaceutical company could test a new medication to determine its effectiveness. The company's researcher performed that hypothesis test by selecting an independent random sample from the population, dividing the sample in two groups, and administered to one group the medication and measured its effectiveness. This first group is called the **experimental group**. The second group did not receive the medication, but instead, received a pill which looks like the real medication but does not actually contain the medication. This group was called the **control group**. We then learned how to perform a 2-sample t-test to compare the difference in the population mean effectiveness between the experimental and control groups. The 2-sample t-test determines if there is a statistical difference between the population means of these two groups.

We should point out a couple of facts about the situation mentioned above. First, the sample sizes of the two groups need not be the same, and in many cases are not. Second, the members of each population are not necessarily related, in fact, they're considered to be *independent*. For example, a person observed in the control group has no significant relation to a person observed in the experimental group. However, there are situations where there is a relation between the observed values within groups and populations.

Consider the following situations:

- A biologist wishes to know if there has been a decline in the clam population in several areas off Long Island that commercial fisherman harvest clams. The biologist measures and estimates the clam population at ten popular clamming locations during the summer of 2019. The biologist then **repeats** this estimate at *the same* 10 clamming locations during the summer of 2020. He compares the results of the 2019 and 2020 measurements to determine if there is a statistically significant decrease in the clam population.
- A dietician has developed a balanced diet low in sugar and believes if followed, her clients will be very successful at achieving their weight loss goals. She weighs twenty of her clients and has them follow this diet. These *same* twenty clients are weighed again three months later to determine if there is a statistically significant amount of weight loss.

Notice that in both situations mentioned above there is a relation between observations. In the first example, the population of clams was measured at the *same* location in 2019 and 2020 and those two values were compared. In the second example, the weight of an individual was compared at the beginning and end of a three-month time period. In each situation, each entity or subject was measured twice and resulted in pairs of observations. These observations are *matched pairs* and are also called a *repeated measure* on the subject. In this section, we'll discuss a hypothesis test that will help us to compare the mean difference in these matched pairs of data. This test is called the *paired t-test* or sometimes referred to as the dependent sample t-test. In Section 11.3, we learned a procedure to test the population mean using the t distribution. That test is also referred to as a single sample t-test. As it turns out, the paired t-test is actually a single sample t-test *in disguise*. To motivate this fact, let's begin with an example that will introduce some new terminology.

EXAMPLE 13.6

A dietician has developed a new low-carb diet and believes if followed, her clients will be very successful at achieving their weight loss goals. She randomly selects eight of her clients who agree to participate in this new diet. To test her hypothesis, the dietician weighs the eight participants before administering the diet, and again after administering the diet. The **before** and **after** results of the eight clients are recorded below in Table 13.8.

TABLE 13.8 Before and After Weights (in lbs.) For Each Client

Client	Before	After
1	161	166
2	171	135
3	182	144
4	166	158
5	171	142
6	175	160
7	160	144
8	181	152

What are the sample mean and sample standard deviation of the differences in the weights of the eight participants?

Solution

In order to find the sample mean and sample standard deviation of the *differences* in the **before** and **after** weights, we must first compute the differences. We do this by creating a new column in the table and subtracting the before and after weights. The differences are shown in Table 13.9.

TABLE 13.9 Before, After and Differences in Weights (in lbs.) For Each Client

Client	Before	After	Difference
1	161	166	−5
2	171	135	36
3	182	144	38
4	166	158	8
5	171	142	29
6	175	160	15
7	160	144	16
8	181	152	29

Paired Difference = Before − After

We should point out that the "−5" difference is actually a weight gain. All other differences were positive, indicating a weight loss. Let's find the sample mean and sample standard deviation of these paired differences. Finding the sample mean and sample standard deviation by hand is a bit cumbersome, however we can use the TI-84's **1–VarStats** command to do this. Recall, you can access the lists in the TI-84 by pressing [STAT] > **1:Edit**. Figures 13.15a, b, c & d show the values entered into L1 of the TI-84 as well as the results of running the **1–VarStats** command.

As you can see the sample mean difference of the weight loss was 20.75 lbs. and the standard deviation of the deferences in weight loss is approximately 14.87 lbs. As we've done frequently throughout this text, we

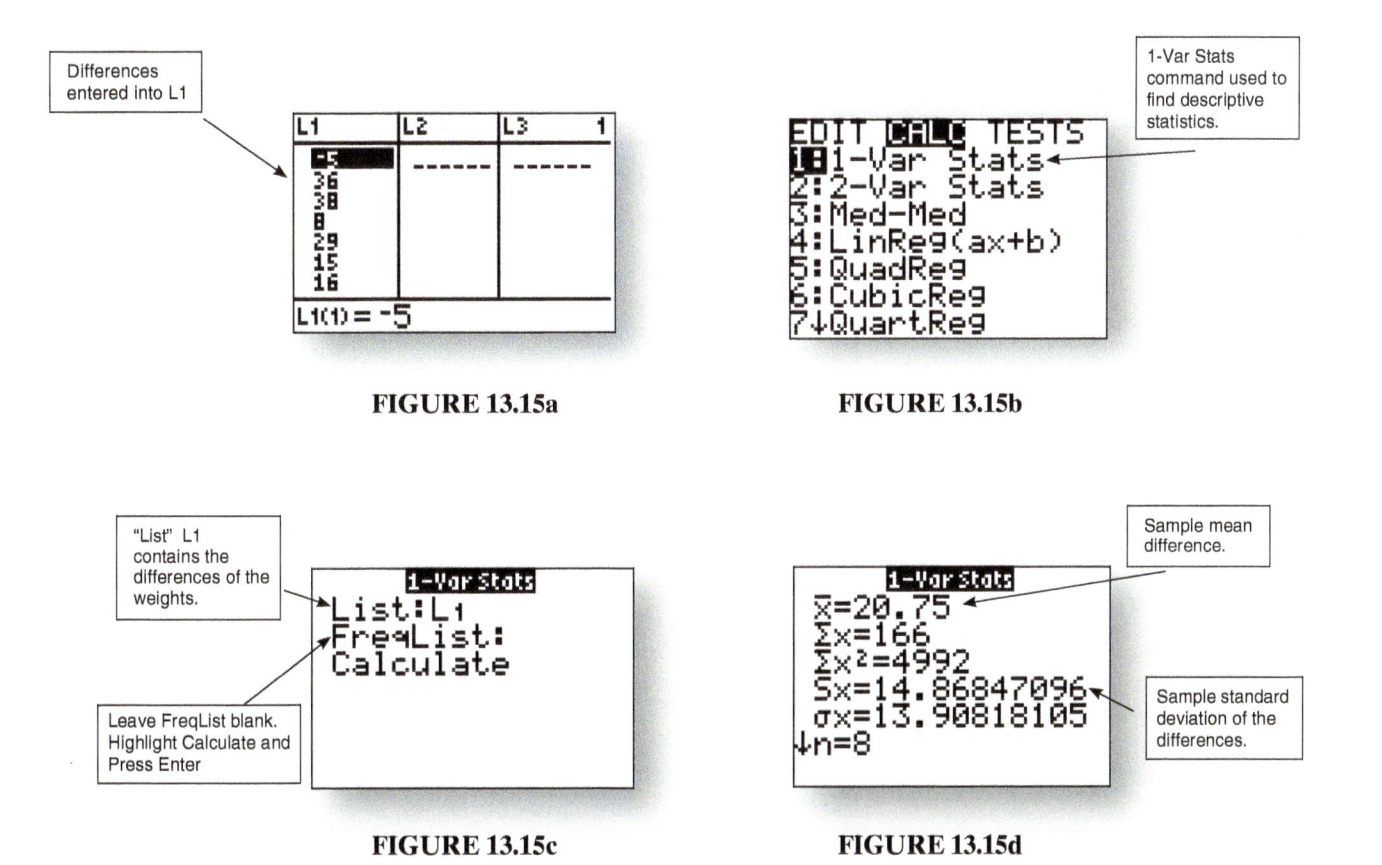

FIGURE 13.15a

Differences entered into L1

FIGURE 13.15b

1-Var Stats command used to find descriptive statistics.

FIGURE 13.15c

"List" L1 contains the differences of the weights.

Leave FreqList blank. Highlight Calculate and Press Enter

FIGURE 13.15d

Sample mean difference.

$\bar{x} = 20.75$
$\Sigma x = 166$
$\Sigma x^2 = 4992$
$Sx = 14.86847096$
$\sigma x = 13.90818105$
$n = 8$

Sample standard deviation of the differences.

will use helpful notation that describe our statistics that are either standard to most statistical texts or are just more appropriate to the section we are working within.

Notation For The Paired T-Test

It is because of this fact that instead of using the symbol $\bar{x}$ for the sample mean of the differences, we'll use $\bar{x}_d$. In addition, we'll use s_d for the sample standard deviation of the differences rather than s. The subscript of "d" on each of those statistical symbols just indicates they are a mean and standard deviation for *differences of the paired observations*.

Using this Paired t-test notation, we have:

- The sample mean difference of the paired observations in weight loss, $\bar{x}_d = 20.75$ lbs.
- The sample standard deviation of the differences of the paired observations in weight loss, $s_d = 14.87$ lbs.

We are now going to outline the hypothesis testing procedure for the **paired t-test** by modifying the general five step hypothesis testing procedure developed in Chapter 10. The first step of the hypothesis testing procedure is to formulate the null and alternative hypotheses. Let's examine how to state these hypotheses for the paired t-test.

Step 1 Formulate the two hypotheses H_0 and H_a.

Null Hypothesis: In general, the null hypothesis states that the population mean difference of the paired observations, which is symbolized by μ_d, is 0 and has the following form.

Null Hypothesis Form

H_0: The population mean difference of the paired observations, μ_d, is equal to 0.

The null hypothesis can be symbolized as:

$$H_0: \mu_d = 0$$

Alternative Hypothesis: Since the alternative hypothesis, H_a, can be stated as either *greater than*, *less than*, or *not equal to* 0, then the alternative hypothesis can have one of the following three forms.

Alternative Hypothesis Form

Form (a): Greater Than Form for H_a

H_a: The population mean difference of the paired observations, μ_d, is greater than 0.

This is symbolized as:

$$H_a: \mu_d > 0$$

Form (b): Less Than Form for H_a

H_a: The population mean difference of the paired observations, μ_d, is less than 0.

This is symbolized as:

$$H_a: \mu_d < 0$$

Form (c): Not Equal to Form for H_a

H_a: The population mean difference of the paired observations, μ_d, is not equal to 0.

This is symbolized as:

$$H_a: \mu_d \neq 0$$

Step 2 Determine the model to test the null hypothesis, H_0.

To test the null hypothesis, a random sample of n data pairs will be selected and tested. **Since the population standard deviation of the differences of the paired observations, σ_d, is unknown, it is necessary to estimate σ_d using s_d, thus, the distribution of the test statistic is a t distribution with $n-1$ degrees of freedom** and serves as the hypothesis testing model.

The mean of the sampling distribution of the differences of the paired observations is equal to μ_d.

Also, since s_d is an estimate of σ_d it is necessary to estimate the standard deviation of the differences of the paired observations. This can also be referred to as the estimate of the standard error of the differences of the paired observations.

For the paired t-test, the Estimate of the Standard Error of the differences of the paired observations equals $\frac{s_d}{\sqrt{n}}$, where n is the number of data pairs.

Thus, the test statistic's distribution is a t distribution with $df = n-1$, where n is the number of data pairs.

Step 3 **Formulate the Decision Rule**:
a) Determine type of alternative hypothesis, directional or nondirectional.
b) Determine type of test, 1TT or 2TT.
c) Identify the significance level, $\alpha = 1\%$ or $\alpha = 5\%$.
d) Construct the appropriate hypothesis test model.
e) Find the critical t score(s) using Table III for the appropriate level of significance.
f) State the decision rule.

Step 4 **Analyze the sample data**
Collect the necessary sample data to calculate the test statistic using the formula

$$t = \frac{\bar{x}_d - \mu_d}{\frac{s_d}{\sqrt{n}}}$$

Step 5 **State the conclusion.**
Compare the test statistic to the decision rule and either:
a) Reject H_0 and accept H_a at α,
or
b) Fail to reject H_0 at α.

In Example 13.7 we'll revisit Example 13.6 and apply the five-step hypothesis testing procedure to perform a paired t-test.

EXAMPLE 13.7

A dietician has developed a new low-carb diet and believes if followed, her clients will be very successful at achieving their weight loss goals. She randomly selects eight of her clients who agree to participate in this new diet. To test her hypothesis, the dietician weighs the eight participants before administering the diet, and again after administering the diet. The **before** and **after** results of the eight clients were recorded in Table 13.10.

TABLE 13.10 Before and After Weights (in lbs.) For Each Client

Client	Before	After
1	161	166
2	171	135
3	182	144
4	166	158
5	171	142
6	175	160
7	160	144
8	181	152

At the 5% level of significance, perform a *paired t-test* to test the claim that there is a significant amount of weight loss in lbs. for the participants in this new diet.

Solution

Since we are conducting a hypothesis test involving matched pairs, we will use a paired t-test.

To begin, let's remind ourselves that in Example 13.6 we have already calculated the sample mean of the paired differences, $\bar{x}_d$, and the sample standard deviation of the paired differences, s_d.

- The sample mean difference of the paired observations in weight loss is $\bar{x}_d = 20.75$ lbs.
- The sample standard deviation of the differences of the paired observations in weight loss is $s_d = 14.87$ lbs.

We are trying to determine if this observed sample mean difference of the paired observations of 20.75 pounds is significantly greater than 0 (i.e., a positive amount of weight loss).

Step 1 Formulate the two hypotheses H_0 and H_a.
H_0: The population mean difference of the paired observations in weight loss is 0.
In symbols, $H_0: \mu_d = 0$.

H_a: The population mean difference of the paired observations in weight loss is greater than 0.
In symbols, $H_a: \mu_d > 0$.

Step 2 Determine the model to test the null hypothesis, H_0.
To test the null hypothesis, a random sample of $n = 8$ data pairs was selected and tested. **Since the population standard deviation of the differences of the paired observations, σ_d, is unknown, it is necessary to estimate σ_d using s_d, thus, the distribution of the test statistic is a t distribution with $n - 1 = 7$ degrees of freedom** and serves as the hypothesis testing model.

The mean of the sampling distribution of the differences of the paired observations is equal to μ_d. Under the assumption that the null hypothesis, H_0, is true, $\mu_d = 0$.

Also, since s_d is an estimate of σ_d it is necessary to estimate the standard deviation of the differences of the paired observations. This is also referred to as the estimate of the standard error of the differences of the paired observations.

For the paired t-test,

the Estimate of the Standard Error of the differences of the paired observations is $\frac{s_d}{\sqrt{n}}$, The Estimate of the Standard Error of the differences of the paired observations = $\frac{14.87}{\sqrt{8}} \approx 5.26$ lbs.

The distribution of the test statistic is a t distribution with $df = 7$. This is illustrated in Figure 13.16.

**Distribution of the test statistic
(is a t distribution with $df = 7$)**

t = 0

FIGURE 13.16

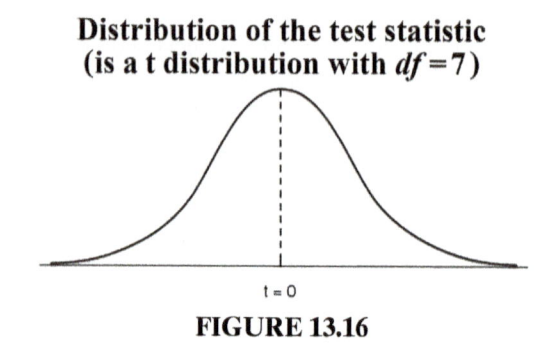

Step 3 Formulate the decision rule.
a) The alternative hypothesis is directional since we are testing for a positive amount of weight loss.
b) The type of hypothesis test is one-tailed (1TT) on the right side.
c) The significance level is: $\alpha = 5\%$.

d) Using Table III, the critical t score, t_c, for a one-tailed test on the right side with $\alpha = 5\%$ and $df = 7$, is: $t_c = 1.89$.
e) Figure 13.17 illustrates the decision rule for this hypothesis test.

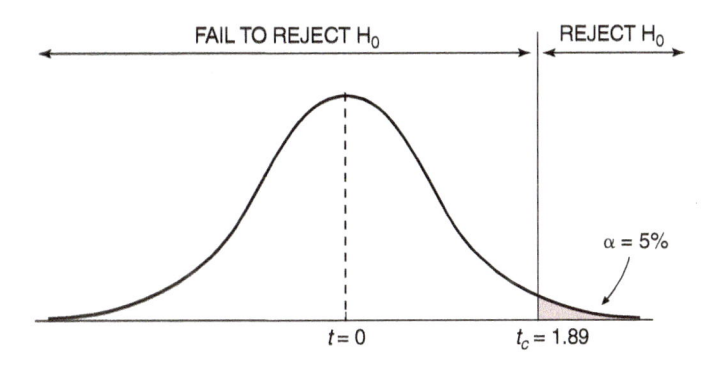

FIGURE 13.17

The decision rule is:
  f) reject the null hypothesis if the test statistic is greater than $t_c = 1.89$.

Step 4 Analyze the sample data
  a) The sample mean difference of the paired observations in weight loss is $\bar{x}_d = 20.75$ lbs. The sample standard deviation of the differences of the paired observations in weight Loss is $s_d = 14.87$ lbs.
  b) Compute the test statistic, t using $t = \dfrac{\bar{x}_d - \mu_d}{\dfrac{s_d}{\sqrt{n}}}$. Remember, μ_d is assumed to be 0.

$$t = \dfrac{\bar{x}_d - \mu_d}{\dfrac{s_d}{\sqrt{n}}} = \dfrac{(20.75 - 0)}{\left(\dfrac{14.87}{\sqrt{8}}\right)} \approx 3.95.$$

Thus, the test statistic t is approximately 3.95.

Step 5 State the conclusion.
Using Figure 13.18, since the test statistic t = 3.95 is greater than the critical t score, $t_c = 1.89$, we reject H_0 and accept H_a at $\alpha = 5\%$.

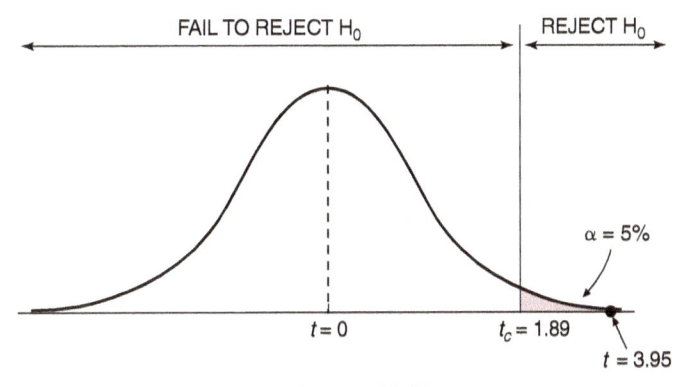

FIGURE 13.18

Therefore, the dietician can conclude that her diet does yield significant weight loss results at $\alpha = 5\%$.

Using p-values for the paired t-test

The p-value approach will require refinements in steps 3, 4 and 5 of the hypothesis testing procedure. In step 3, the decision rule will be stated as: Reject H_0 if the p-value of the test statistic is less than α. In step 4, the p-value will be determined using the TI-84 calculator. While in step 5, the conclusion of the hypothesis test is determined by comparing the p-value to the level of significance, α. Let's reexamine the hypothesis test shown in Example 13.7 using a p-value approach with the aid of the TI-84 Calculator.

TI-84 Graphing Calculator Solution for Example 13.7

A dietician has developed a new low-carb diet and believes if followed, her clients will be very successful at achieving their weight loss goals. She randomly selects eight of her clients who agree to participate in this new diet. To test her hypothesis, the dietician weighs the eight participants before administering the diet, and again after administering the diet. The **before** and **after** results of the eight clients were recorded below in Table 13.11.

TABLE 13.11 Before and After Weights (in lbs.) For Each Client

Client	Before	After
1	161	166
2	171	135
3	182	144
4	166	158
5	171	142
6	175	160
7	160	144
8	181	152

At the 5% level of significance, perform a paired t-test to test the claim that there is a significant amount of weight loss for the participants in this new diet.

Solution

Since we are conducting a hypothesis test involving matched pairs, we will use the Dependent t-test called the paired t-test.

Step 1 Formulate the two hypotheses H_0 and H_a.

H_0: The population mean difference of the paired observations in weight loss is 0.
    In symbols, $H_0: \mu_d = 0$.

H_a: The population mean difference of the paired observations in weight loss is greater than 0.
    In symbols, $H_0: \mu_d > 0$.

Step 2 Determine the model to test the null hypothesis, H_0.

To test the null hypothesis, a random sample of $n = 8$ data pairs was selected and tested. **Since the population standard deviation of the differences of the paired observations, σ_d, is unknown, it is necessary to estimate σ_d using s_d, thus, the distribution of the test statistic is a t distribution with $n - 1 = 7$ degrees of freedom** and serves as the hypothesis testing model.

The mean of the sampling distribution of the differences of the paired observations is equal to μ_d. Under the assumption that the null hypothesis, H_0, is true, $\mu_d = 0$.

Also, since s_d is an estimate of σ_d it is necessary to estimate the standard deviation of the differences of the paired observations. This is also referred to as the estimate of the standard error of the differences of the paired observations.

For the paired t-test,

the Estimate of the Standard Error of the differences of the paired observations is $\frac{s_d}{\sqrt{n}}$, The Estimate of the Standard Error of the differences of the paired observations = $\frac{14.87}{\sqrt{8}} \approx 5.26$

The distribution of the test statistic is a t distribution with $df = 7$. This is illustrated in Figure 13.19.

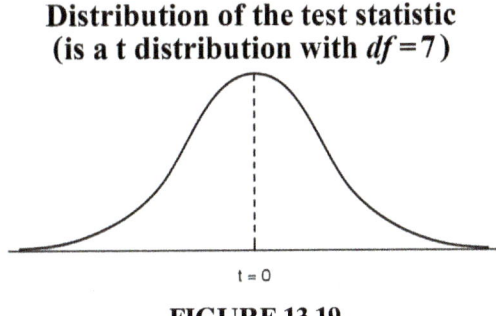

**Distribution of the test statistic
(is a t distribution with $df = 7$)**

FIGURE 13.19

Step 3 Formulate the decision rule.
a) The alternative hypothesis is directional since we are testing for a positive amount of weight loss.
b) The type of hypothesis test is one-tailed (1TT) on the right side.
c) The significance level is: $\alpha = 5\%$.
d) Figure 13.20 illustrates the decision rule for this hypothesis test.

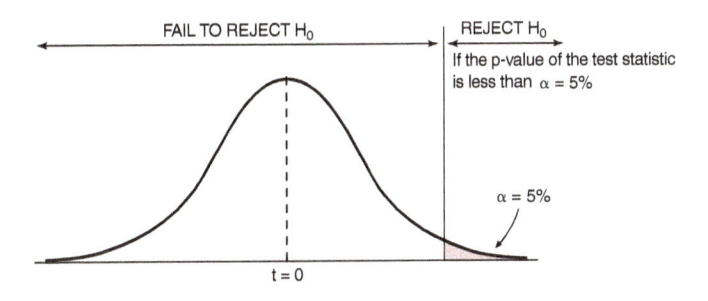

FAIL TO REJECT H_0 | REJECT H_0
If the p-value of the test statistic is less than $\alpha = 5\%$

$\alpha = 5\%$

$t = 0$

FIGURE 13.20

The decision rule is:
f) reject the null hypothesis if the p-value for the test statistic is smaller than $\alpha = 5\%$.

Step 4 **Analyze the sample data**
The **T-Test** function can be used to perform a hypothesis test involving one population mean when the population standard deviation is unknown. To determine the p-value of the test statistic, the sample data must be entered into the TI-84's lists. To access the lists in the TI-84, press STAT > **1:EDIT.** Figures 13.21a and 13.21b show the edit menu along with the before data entered into L1 and the after data entered into L2.

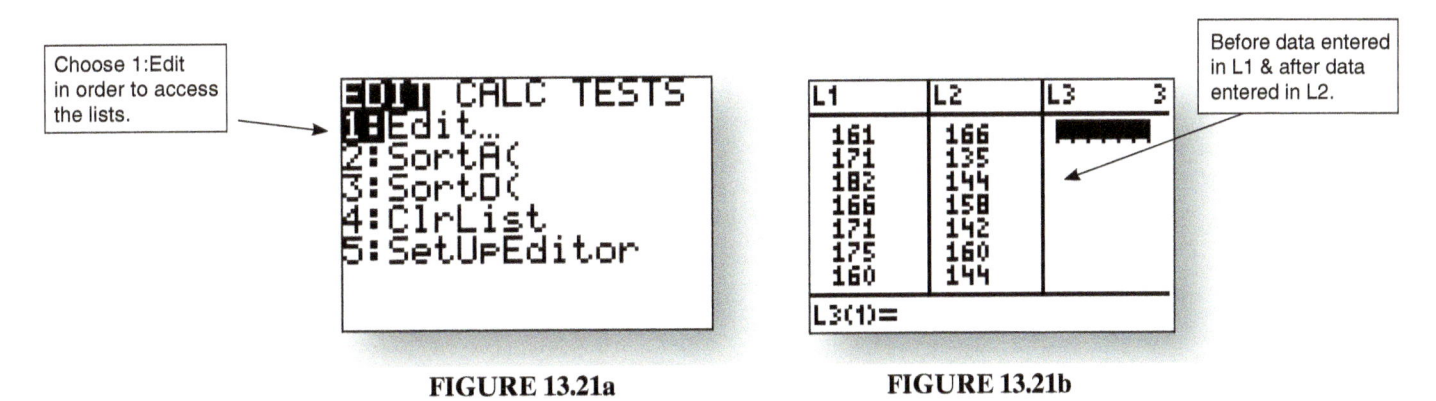

Choose 1:Edit in order to access the lists.

Before data entered in L1 & after data entered in L2.

FIGURE 13.21a   FIGURE 13.21b

Next, we can have the TI-84 calculate the differences for us. Use the arrow keys and move the cursor so that L3 is highlighted. This is shown below in Figure 13.22.

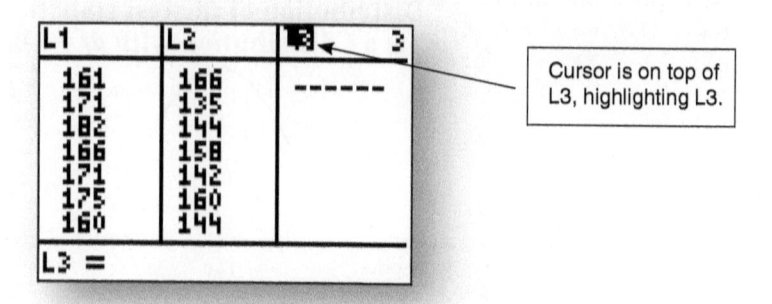

708 Chapter 13 *Hypothesis Test Involving Two Population Means Using Independent and Dependent Samples*

FIGURE 13.22

Next, we are going to tell the TI-84 to subtract the values that are in L1 & L2 and place those results in L3. This will be done by pressing 2nd ▶ 1 ▶ − ▶ 2nd ▶ 2. This is shown in Figure 13.23.

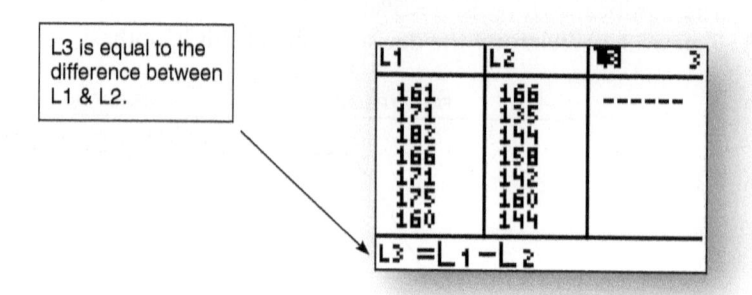

FIGURE 13.23

Pressing ENTER will calculate the differences of the paired observations and paste them into L3. This is shown in Figure 13.24.

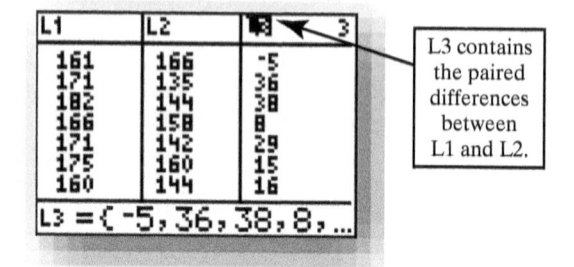

FIGURE 13.24

Now that the paired difference data is entered into the calculator, we can now utilize the **T-Test** function in the TI-84 to help us find the p-value. To access the **T-Test** menu, press **STAT** and then the right arrow key twice to access the **TESTS** menu. This is shown in Figure 13.25.

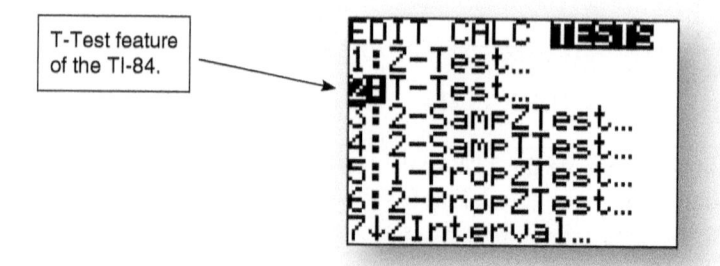

FIGURE 13.25

Next, press option **2** to access the **T-Test** menu. In the **Inpt** field, highlight **Data**. Also, remember that the null hypothesis form states that the population mean difference of the paired observations, μ_d, is 0. The TI-84 shows this hypothesized mean as μ_0. The paired differences data was calculated and pasted in L3, and this goes in the "List" field. Also, recall that this is a 1TT on the right-hand side, so the format for the alternative hypothesis is greater than (>). Figure 13.26 shows this menu with all the fields completed.

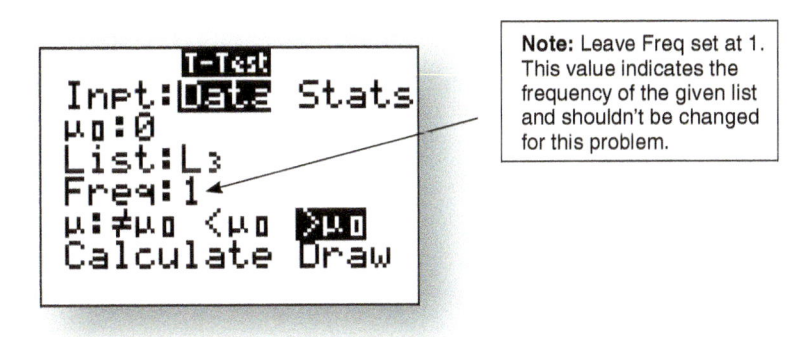

Note: Leave Freq set at 1. This value indicates the frequency of the given list and shouldn't be changed for this problem.

FIGURE 13.26

To execute the **T-Test** command, highlight **Calculate** and press **ENTER**. The results screen is shown in Figure 13.27.

The test statistic, t = 3.95 calculated earlier by hand.

The p-value of approximately 0.003.

The sample mean and standard deviation of the paired differences.

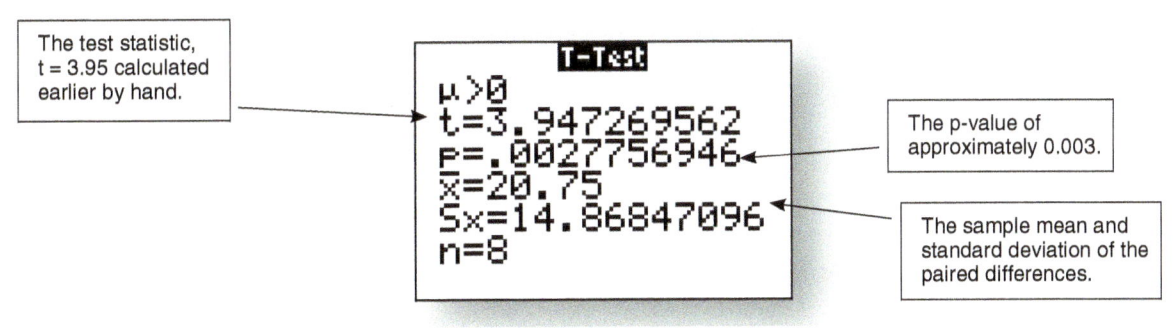

FIGURE 13.27

Some points to note regarding this output. A few statistics that we calculated earlier in this section are listed. These include the test statistic $t = 3.95$, as well as the sample mean difference of the paired observations in weight loss, $\bar{x}_d = 20.75$, and the sample standard deviation of the paired differences in the difference in weight loss, $s_d = 14.87$. These values were used when performing this hypothesis test using the traditional method. Right now, we are using a p-value approach and will focus on that.

Observe that the p-value for the test statistic of t = 3.95 is approximately 0.003.

Step 5 **State the conclusion**
Using the p-value, we can position it along with the test statistic on the hypothesis testing model. This is shown in Figure 13.28.

FAIL TO REJECT H_0 | REJECT H_0
If the p-value of the test statistic is less than $\alpha = 5\%$

$\alpha = 5\%$

$t = 0$

test statistic: $t \approx 3.95$
(p-value 0.003)

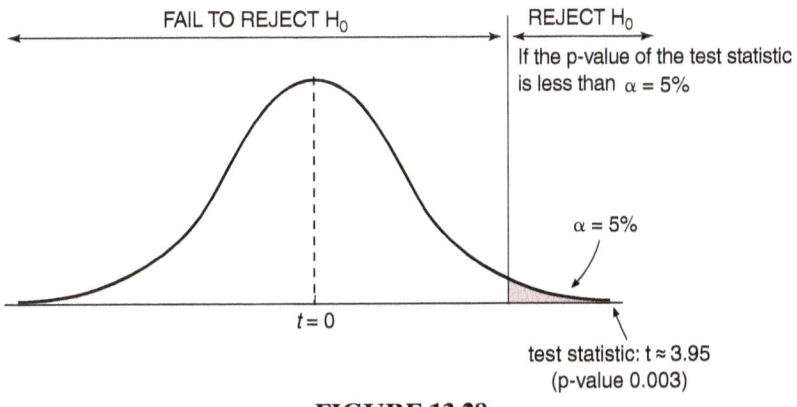

FIGURE 13.28

Therefore, since the p-value of 0.003 is smaller than $\alpha = 0.05$, we reject the null hypothesis, H_0, and accept the alternative hypothesis, H_a. The dietician can conclude that her new diet does yield significant weight loss results at $\alpha = 0.05$

It is important to note that since the paired t-test is *essentially* a one sample t-test on the paired differences column, the paired t-test has the same underlying assumptions as the one sample t-test. These assumptions are outlined below.

Assumptions For The Paired t-Test

The *t*-distribution is used while performing a paired *t*-test. The following conditions must be met for this test to be valid:

- Each data pair sampled from the population are taken randomly and the pairs are independent.
- The base population we are sampling from is approximately normally distributed.
- The population standard deviation of the differences of the paired observations, σ_d, is unknown.
- For each sample, the observations (subjects) are measured twice for the same variable to produce data pairs.
- Just because the data are related in pairs does not mean you can use the paired t-test. The pairing should be part of the experimental design and not something you do as an afterthought once you've collected data.

The following outline is a summary of the 5-step hypothesis testing procedure for a paired *t* Test.

Summary of the Test Statistic Approach to the Paired t Test

Step 1 **Formulate the two hypotheses H_0 and H_a.**
Null Hypothesis: $H_0: \mu_d = 0$

Alternative Hypothesis, H_a, has one of three forms:
$H_a: \mu_d < 0$ or $H_a: \mu_d > 0$ or $H_a: \mu_d \neq 0$

Step 2 **Determine the model to test the null hypothesis, H_0.**
To test the null hypothesis, a random sample of *n* data pairs will be selected and tested. **The distribution of the test statistic is a *t* distribution with $n - 1$ degrees of freedom** and serves as the hypothesis testing model.

The mean of the sampling distribution of the differences of the paired observations is denoted by μ_d, and $\mu_d = 0$.

The estimate of the Standard Error of the differences of the paired observations equals $\frac{s_d}{\sqrt{n}}$.

Step 3 **Formulate the Decision Rule**:
a) Determine the type of alternative hypothesis, directional or nondirectional.
b) Determine the type of test, 1TT or 2TT.
c) Identify the significance level, $\alpha = 1\%$ or $\alpha = 5\%$.
d) Construct the appropriate hypothesis test model.
e) Find the critical *t* score(s) using Table III for the appropriate level of significance.
f) State the decision rule.

Step 4 **Analyze the sample data**
Collect the necessary sample data to calculate the test statistic using the formula

$$t = \frac{\bar{x}_d - \mu_d}{\frac{s_d}{\sqrt{n}}}$$

Step 5 **State the conclusion.**
Compare the test statistic to the decision rule and either:
a) Reject H_0 and accept H_a at α,
or
b) Fail to reject H_0 at α.

13.8 p-VALUE APPROACH TO HYPOTHESIS TESTING INVOLVING TWO POPULATION MEANS USING THE TI-84 CALCULATOR

The p-value approach will require refinements in steps 3, 4 and 5 of the hypothesis testing procedure. In step 3, the decision will be stated as: Reject H_O if the p-value of the test statistic is less than α. In step 4, the p-value will be determined using the TI-84 calculator. While in step 5, the conclusion of the hypothesis test is determined by comparing the p-value to the level of significance, α.

Let's reexamine the hypothesis test shown in Example 13.5 using a p-value approach with the aid of the TI-84 calculator.

EXAMPLE 13.8

A pharmaceutical company has developed a diet pill called, Effective Anti-Hunger Tablet (E.A.T.), which the company believes will significantly reduce an individual's weight at the end of one week. To support their claim, the company selects 30 overweight individuals and randomly divides them into independent groups: a **treatment** group of 15 subjects who are administered the diet pill E.A.T. and a *control* group of 15 subjects who are given a *placebo* tablet (i.e. **a pill that has *no effect* on weight**).

After one week, the treatment group had a sample mean weight loss of $\bar{x}$ = 7.1 lbs. with s = 3.6 lbs. while the control group had a sample mean weight loss of $\bar{x}$ = 4.2 lbs. with s = 2.3 lbs.

a) Use the 5 step hypothesis testing procedure to answer the question: can the company conclude that their diet pill E.A.T. will significantly reduce an individual's weight within one week? Assume equal population standard deviations and use α = 5%.
b) Can we conclude that their diet pill E.A.T. will significantly reduce an individual's weight within one week if we use α = 1%? Assume equal population standard deviations.
c) What is the smallest α level that can be used to support the pharmaceutical company's claim?

Remark: In this study the administration of the diet pill E.A.T. to the treatment group is the only difference between the two groups. Therefore, if we can show a significant difference between the treatment and control groups then this *difference can be attributed only to the treatment (the diet pill)*.

Solution
part (a):
Since we are assuming the population standard deviations are unknown but equal, we will use the pooled t-test.

Step 1 Formulate the hypotheses.
If we let the population 1 represent the treatment group and population 2 represent the control group then the null hypothesis is:

H_0: there is **no difference** in the population mean weight loss for the treatment group and the control group.

In symbols we have, H_0: $\mu_1 - \mu_2 = 0$

H_a: the population mean weight loss for the treatment group is **greater than** the population mean weight loss for the control group.

In symbols we have $H_a: \mu_1 - \mu_2 > 0$.

Step 2 Determine the model to test the null hypothesis, H_0.

To test the null hypothesis the company selects 30 overweight individuals and randomly divides them into two independent groups and the sample results are summarized in Table 13.12.

TABLE 13.12 Summary of the Data for the Random Samples Selected from Each Population

Population 1 (Treatment)	Population 2 (Control)
$\bar{x}_1 = 7.1$	$\bar{x}_2 = 4.2$
$s_1 = 3.6$	$s_2 = 2.3$
$n_1 = 15$	$n_2 = 15$

Since the differences between the two sample means, $d = \bar{x}_1 - \bar{x}_2$, will be used to determine if $\mu_1 - \mu_2$ is significantly greater than zero, **the expected results** are:

a) The distribution of the test statistic is approximated by a t distribution with $df = 28$.
b) The mean of the Sampling Distribution of the Difference Between Two Means is:

$$\mu_{\bar{x}_1 - \bar{x}_2} = \mu_1 - \mu_2 = 0$$

c) For the pooled t-test, the estimate of the pooled standard error of the difference between two means is:

$$s_{\bar{x}_1 - \bar{x}_2} = \sqrt{\frac{(n_1 - 1)s_1^2 + (n_2 - 1)s_2^2}{n_1 + n_2 - 2} \cdot \left(\frac{1}{n_1} + \frac{1}{n_2}\right)}$$

Substituting $s_1 = 3.6$, $n_1 = 15$, $s_2 = 2.3$ and $n_2 = 15$ into $s_{\bar{x}_1 - \bar{x}_2}$, we have:

$$s_{\bar{x}_1 - \bar{x}_2} = \sqrt{\frac{(15-1)(3.6)^2 + (15-1)(2.3)^2}{15 + 15 - 2} \left(\frac{1}{15} + \frac{1}{15}\right)}$$

$$= \sqrt{\frac{181.44 + 74.06}{28}(0.13)}$$

$$\approx \sqrt{(9.1250)(0.1333)}$$

$$s_{\bar{x}_1 - \bar{x}_2} \approx 1.10$$

Figure 13.29 illustrates hypothesis testing model.

distribution of the test statistic
(is a t Distribution with $df = 28$)

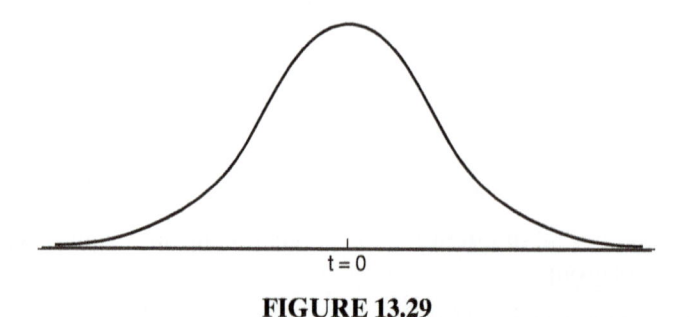

t = 0

FIGURE 13.29

Section 13.8 *p-value Approach to Hypothesis Testing Involving Two Population Means Using the TI-84 Calculator* 713

Step 3 Formulate the decision rule.
  a) the alternative hypothesis is **directional** since we are trying to show that the population mean weight loss for the treatment group is **significantly greater** than the population mean weight loss for the control group.
  b) the type of hypothesis test is one-tailed (1TT) on the **right side** since we are trying to show that $(\mu_1 - \mu_2)$ is **greater than** zero.
  c) the significance level is: $\alpha = 5\%$.
  d) Figure 13.30 illustrates the decision rule for this hypothesis test.
  The decision rule is:
  e) Reject H_0 if the p-value of the test statistic t is less than $\alpha = 5\%$.

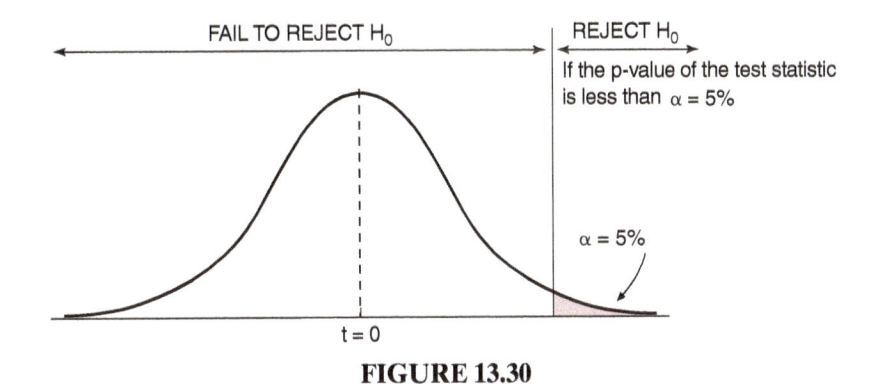

FIGURE 13.30

Step 4 Analyze the sample data.

The **2-SampTTest** function can be used to perform a hypothesis test involving two population means when the population standard deviations are unknown. To determine the p-value of the test statistic, the sample statistics must be entered into the TI-84's TESTS menu. To access the **2-SampTTest** function in the TESTS menu press [STAT] ▶ ▶ [TESTS]. This is shown in Figure 13.31.

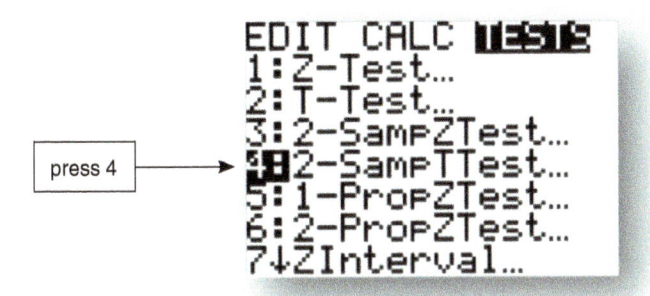

FIGURE 13.31

Press **4** to access the **2-SampTTest** menu. Highlight **Stats**, enter the sample statistics and highlight Yes for Pooled. This is shown in Figure 13.32a and Figure 13.32b.

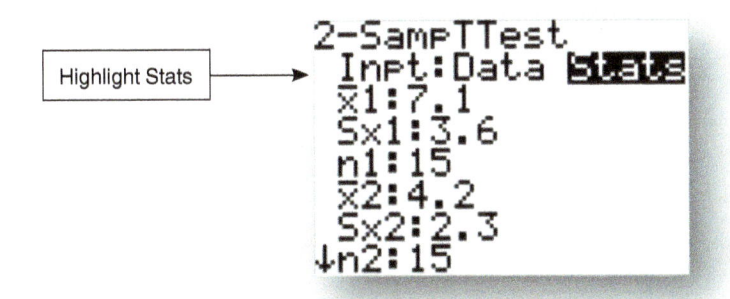

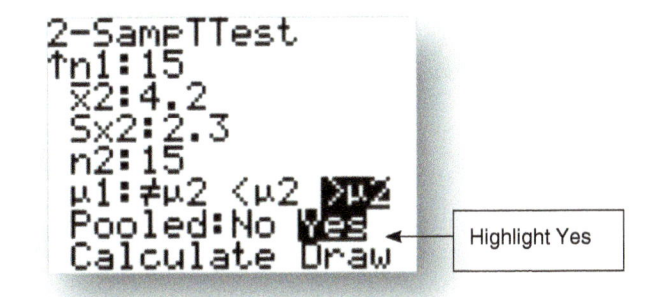

FIGURE 13.32a   **FIGURE 13.32b**

Highlight **Calculate** and press ENTER to obtain the output screens for this test. These are illustrated in Figure 13.33a and Figure 13.33b.

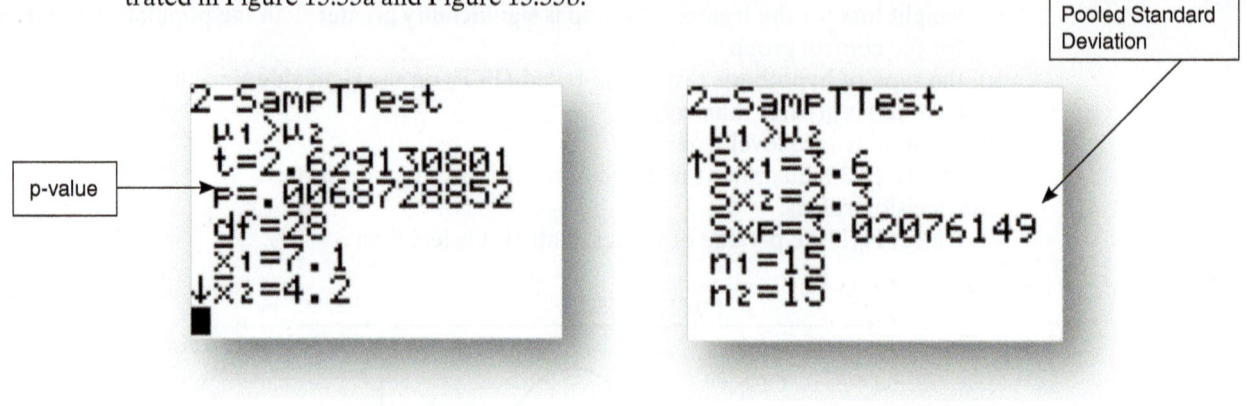

p-value →

FIGURE 13.33a (2-SampTTest: μ1>μ2, t=2.629130801, p=.0068728852, df=28, x̄1=7.1, ↓x̄2=4.2)

Pooled Standard Deviation →

FIGURE 13.33b (2-SampTTest: μ1>μ2, ↑Sx1=3.6, Sx2=2.3, SxP=3.02076149, n1=15, n2=15)

From the output screen shown in Figure 13.33a, the p-value of the test statistic $t \approx 2.63$ is approximately 0.0069. Using the p-value (0.0069) we can position the test statistic on the hypothesis testing model shown in Figure 13.34.

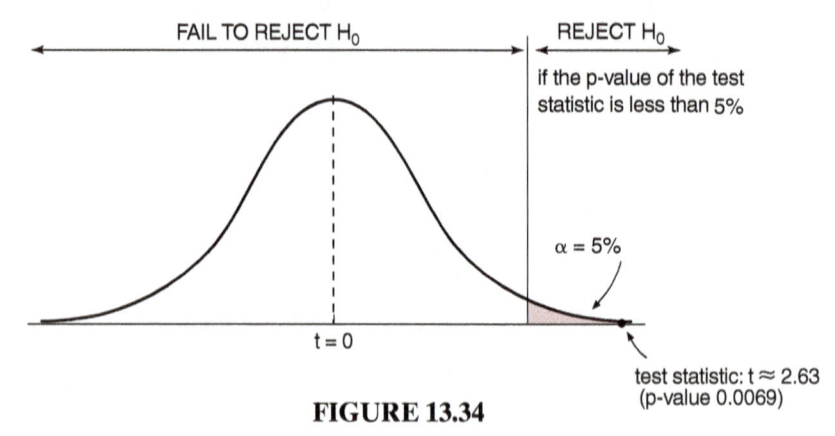

FAIL TO REJECT H_0 | REJECT H_0 if the p-value of the test statistic is less than 5%

$\alpha = 5\%$

t = 0

test statistic: $t \approx 2.63$
(p-value 0.0069)

FIGURE 13.34

Step 5 State the conclusion.

Since the p-value (0.0069) of the test statistic t is less than $\alpha = 5\%$ *reject* H_0 and *accept* H_a at a p-value of 0.0069. Therefore, the company can conclude that their diet pill (E.A.T.) will **significantly reduce** an individual's weight at the end of one week at $\alpha = 5\%$.

part (b): Since the p-value of 0.0069 is still less than $\alpha = 1\%$, we can support the pharmaceutical company's claim as shown in Figure 13.35.

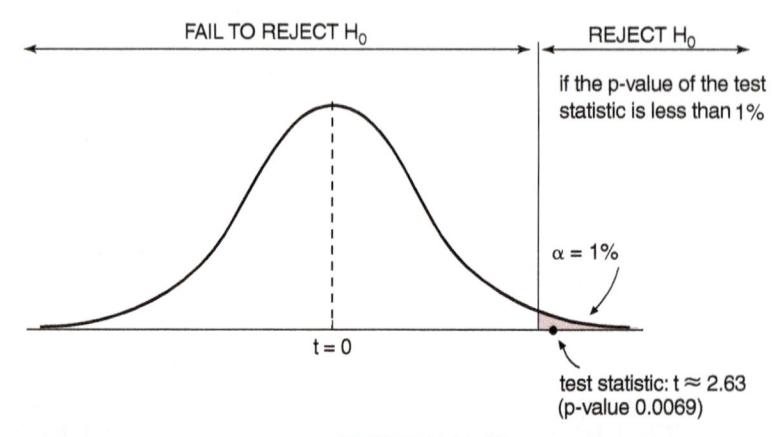

FAIL TO REJECT H_0 | REJECT H_0 if the p-value of the test statistic is less than 1%

$\alpha = 1\%$

t = 0

test statistic: $t \approx 2.63$
(p-value 0.0069)

FIGURE 13.35

part (c): The smallest α level that can be used to support the pharmaceutical company's claim is equal to the p-value, 0.0069. ∎

Let's reexamine Welch's t-test shown in Example 13.4 using a p-value approach with the aid of the TI-84 Calculator.

TI-84 Graphing Calculator Solution for Example 13.9

A market analyst is trying to convince her boss to advertise their new perfume for women on social media platforms. The analyst believes that Facebook would be a great match because she is convinced that women spend significantly more time each day on Facebook than men. To support this decision, the analyst randomly samples 26 women and 22 men in the 18-28 demographic range who have Facebook accounts. The summary statistics for the number of minutes spent per day by the two groups are given below in Table 13.13.

TABLE 13.13 Summary Statistics for Random Samples of Facebook times

Gender	Sample size	Sample mean (mins/day)	Sample standard deviation
Female	26	32.3	26.5
Male	22	28.7	22.8

At the 5% level of significance, perform a hypothesis test to test the claim that females spend significantly more time on Facebook each day than males. Assume that the population standard deviations are unequal, and use Welch's *t*-Test.

Solution

Since we are assuming the population standard deviations are unequal and unknown, we will use Welch's t-test.

Step 1 If we let population 1 represent the female time spent on Facebook per day and population 2 represent the male time spent on Facebook per day, then the **Null Hypothesis** is:

H_0: There is no difference between the population mean time spent on Facebook per day for females and males.
In symbols we have, $H_0 : \mu_1 - \mu_2 = 0$.

The **Alternative Hypothesis** is:

H_a: The population mean time spent on Facebook per day for females is greater than the population mean time spent on Facebook per day by males.
In symbols we have, $H_a : \mu_1 > \mu_2$. This is equivalent to stating that the difference between the population mean times for females and males is greater than 0. Therefore, this alternative hypothesis can also be symbolized as $H_a : \mu_1 - \mu_2 > 0$

Step 2 Determine the model to test the null hypothesis, H_0.
To test the null hypothesis, the analyst selected a random sample from each population and the sample results are summarized in Table 13.14

TABLE 13.14 Summary Statistics for the Random Samples of Facebook times

Population 1(female)	Population 2(male)
$n_1 = 26$	$n_2 = 22$
$\bar{x}_1 = 32.3$	$\bar{x}_2 = 28.7$
$s_1 = 26.5$	$s_2 = 22.8$

Since the differences between the two sample means, $d = \bar{x}_1 - \bar{x}_2$, will be used to determine if $\mu_1 - \mu_2$ is significantly less than zero, the expected results are:

a) The test statistic hypothesis testing model is approximately a t distribution with degrees of freedom given by:
df = smaller of $(n_1 - 1)$ and $(n_2 - 1)$, where $n_1 = 26$ and $n_2 = 22$
df = smaller 25 and 21.
Therefore, $df = 21$.

b) The mean of the sampling distribution of the difference between two means is given by the formula:
$$\mu_{\bar{x}_1 - \bar{x}_2} = \mu_1 - \mu_2 = 0$$

c) For Welch's t-test, the estimate of the standard error of the difference between two means, written $s_{\bar{x}_1 - \bar{x}_2}$, is given by the formula:

$$s_{\bar{x}_1 - \bar{x}_2} = \sqrt{\frac{s_1^2}{n_1} + \frac{s_2^2}{n_2}}$$

$$= \sqrt{\frac{26.5^2}{26} + \frac{22.8^2}{22}}$$

$$\approx \sqrt{50.6387}$$

$s_{\bar{x}_1 - \bar{x}_2} \approx 7.1161$

The distribution of the test statistic is a t distribution with $df = 21$. This is illustrated in Figure 13.36.

Distribution of the test statistic (is a t distribution with $df = 21$)

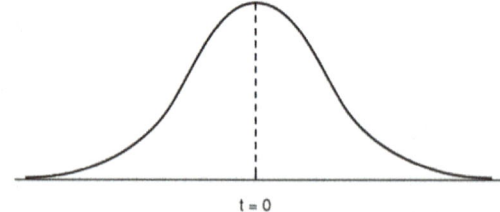

FIGURE 13.36

Step 3 Formulate the Decision Rule:
a) The alternative hypothesis is directional since we are trying to show the population mean time per day for the females is greater than the population mean time per day for the males. That is: $\mu_1 - \mu_2 > 0$.
b) The type of test is a one tailed (1TT) on the right side since we are trying to show that $\mu_1 - \mu_2 > 0$.
c) The significance level is $\alpha = 5\%$.

Section 13.8 *p-value Approach to Hypothesis Testing Involving Two Population Means Using the TI-84 Calculator*   717

d) For a one-tailed test on the right side with $\alpha = 5\%$ and $df = 21$, the critical t-score, t_c, using Table III is $t_c = 1.72$.

e) Figure 13.37 illustrates the decision rule for this hypothesis test. The decision rule is: Reject H_0 if the p-value is less than $\alpha = 5\%$.

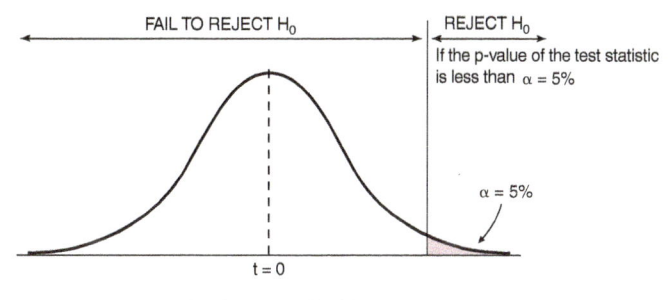

FIGURE 13.37

Step 4 **Analyze the sample data**

The **2-SampTTest** function can be used to perform a hypothesis test involving two population means when the population standard deviations are unknown. To determine the p-value of the test statistic, the sample statistics must be entered into the TI-84's TESTS menu. To access the **2-SampTTest** function in the TESTS menu press STAT ▶ ▶. This is shown in Figure 13.38.

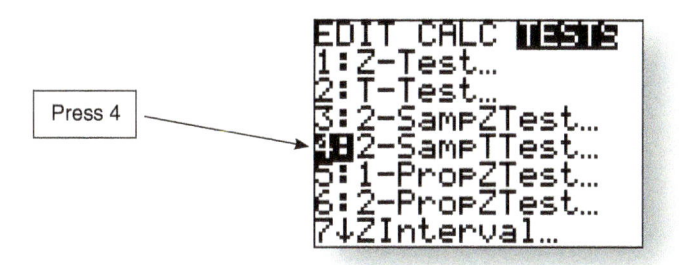

FIGURE 13.38

Press **4** to access the **2-SampTTest** menu. Highlight **Stats**, enter the sample statistics and highlight NO for Pooled. *Selecting NO for pooling ensures that you'll be using Welch's t-Test and not the pooled t-Test.* This is shown in Figures 13.39a and 13.39b.

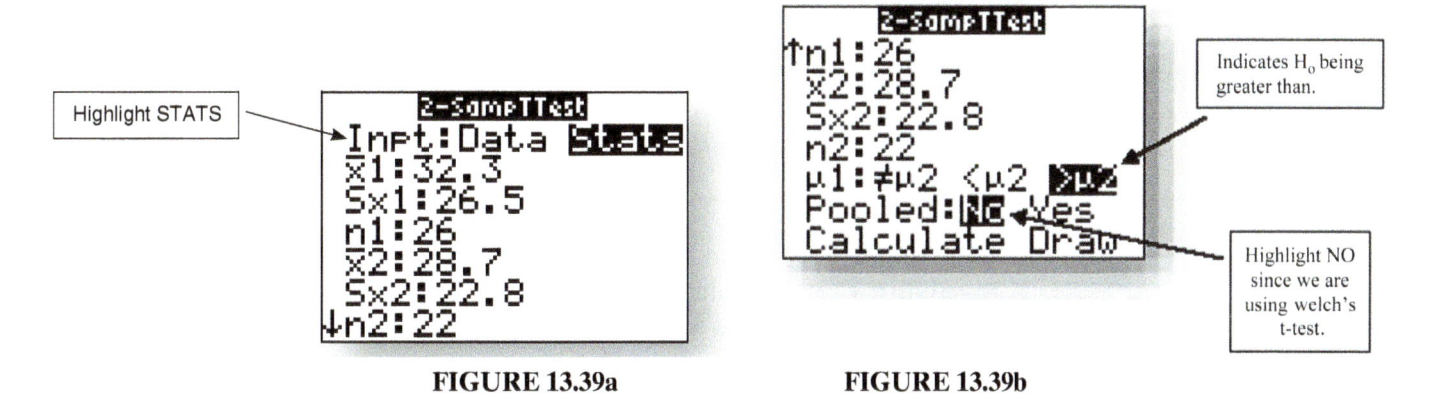

FIGURE 13.39a   **FIGURE 13.39b**

Highlight **Calculate** and press ENTER to obtain the output screens for this test. These are illustrated in Figures 13.40a and 13.40b.

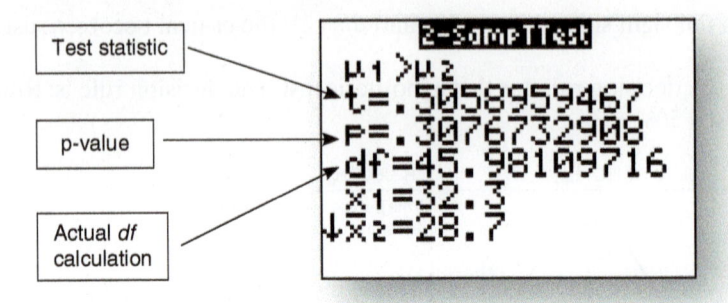

718 Chapter 13 Hypothesis Test Involving Two Population Means Using Independent and Dependent Samples

- Test statistic → t = .5058959467
- p-value → P = .3076732908
- Actual df calculation → df = 45.98109716
- x̄1 = 32.3
- x̄2 = 28.7

2-SampTTest
μ1 > μ2
↑x̄2 = 28.7
Sx1 = 26.5
Sx2 = 22.8
n1 = 26
n2 = 22

FIGURE 13.40a 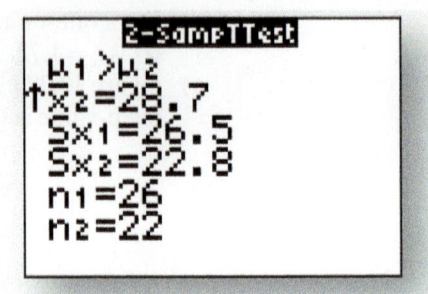   **FIGURE 13.40b**

From the output screen the p-value of the test statistic $t \approx 0.51$ is approximately 0.308. Notice the TI-84 calculated the actual degrees of freedom using Satterthwaite's formula for the degrees of freedom. Earlier in Section 13.3 we used the conservative value of df = 22 − 1 = 21. Note the conservative degrees of freedom calculation of 21 is smaller than the actual degrees of freedom calculation of approximately 46. This will normally be the case when comparing the conservative and actual values for the degrees of freedom.

Using the p-value we can position the test statistic on the hypothesis testing model shown in Figure 13.41.

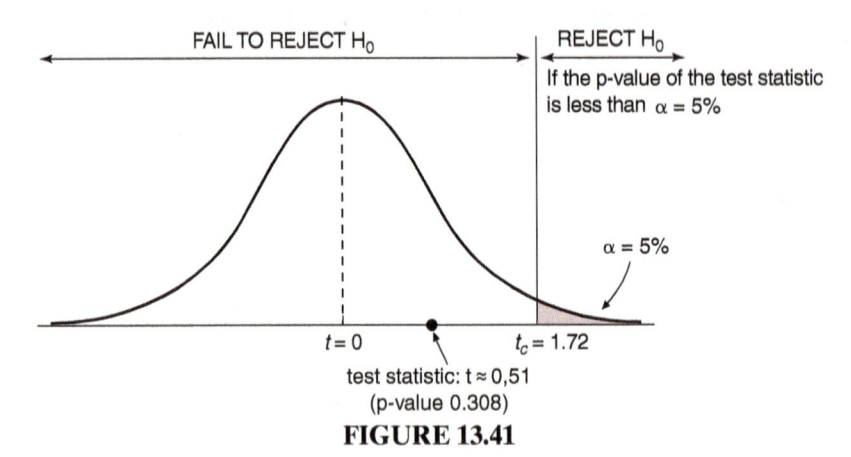

FAIL TO REJECT H₀ | REJECT H₀
If the p-value of the test statistic is less than α = 5%
α = 5%
t = 0, t_c = 1.72
test statistic: t ≈ 0.51
(p-value 0.308)

FIGURE 13.41

Step 5   **State the conclusion**
The p-value (0.308) of the test statistic t is greater than 5% as shown in Figure 13.41. Thus, we fail to reject H_0 at $\alpha = 5\%$. Therefore, the analyst cannot conclude that the female population spends significantly more time than males on Facebook per day. A new marketing strategy will need to be discussed.

The following outline is a summary of the 5 step hypothesis testing procedure for a hypothesis test involving two population means using p-values.

Summary of the p-Value Approach to Hypothesis Testing Involving Two Population Means and Independent Random Samples Using the TI-84 Calculator

Step 1   Formulate the two hypotheses, H_o and H_a.

Null Hypothesis, $H_o: \mu_1 - \mu_2 = 0$

Alternative Hypothesis, H_a: has one of the following three forms:

$H_o: \mu_1 - \mu_2 < 0$ or $H_o: \mu_1 - \mu_2 > 0$ or $H_o: \mu_1 - \mu_2 \neq 0$

Step 2 Determine the model to test the null hypothesis, H_0.
Under the assumption H_0 is true, the expected results are:

a) The distribution of the test statistic is either:
   (i) a normal distribution, if the standard deviations of both populations are known and both sample sizes are greater than 30,
   or (ii) is a t distribution with degrees of freedom: $df = n_1 + n_2 - 2$, if the standard deviations of both populations are unknown and assumed to be **equal**.
   or (iii) is a t distribution with degrees of freedom: $df =$ smaller of $(n_1 - 1)$ and $(n_2 - 1)$, if the standard deviations of both populations are unknown and assumed to be **unequal**.

b) The mean of the sampling distribution of the difference between two means is given by the formula: $\mu_{\bar{x}_1 - \bar{x}_2} = \mu_1 - \mu_2 = 0$ (since H_0 is true: i.e. $\mu_1 - \mu_2 = 0$)

c) I. For a test statistic having a normal distribution: the standard error of the difference between two means, denoted by $\sigma_{\bar{x}_1 - \bar{x}_2}$ is:

$$\sigma_{\bar{x}_1 - \bar{x}_2} = \sqrt{\frac{\sigma_1^2}{n_1} + \frac{\sigma_2^2}{n_2}}.$$

and the **two sample z Test** is applied.

II. For a test statistic having a t distribution: the **estimate of the standard error** of the difference between two means is denoted $s_{\bar{x}_1 - \bar{x}_2}$, and has two cases. The formula is dependent on the appropriate condition stated below.

1) If the population standard deviations are unknown and assumed to be **equal**, then the estimate of the standard error is given by

$$s_{\bar{x}_1 - \bar{x}_2} = \sqrt{\frac{(n_1 - 1)s_1^2 + (n_2 - 1)s_2^2}{n_1 + n_2 - 2} \cdot \left(\frac{1}{n_1} + \frac{1}{n_2}\right)},$$

and the **pooled t Test** is applied.

Note: $s_{\bar{x}_1 - \bar{x}_2} = s_{x_p}\sqrt{\frac{1}{n_1} + \frac{1}{n_2}}$, where s_{x_p} is called the pooled standard deviation and is computed by the formula $s_{x_p} = \sqrt{\frac{(n_1 - 1)s_1^2 + (n_2 - 1)s_2^2}{n_1 + n_2 - 2}}$.

2) If the population standard deviations are unknown and assumed to be **unequal**, then the formula for the estimate of the standard error is given by

$$s_{\bar{x}_1 - \bar{x}_2} = \sqrt{\frac{s_1^2}{n_1} + \frac{s_2^2}{n_2}},$$

and **Welch's t Test** is applied.

Step 3 Formulate the decision rule.
a) Alternative hypothesis: directional or nondirectional.
b) Type of test: 1TT or 2TT.
c) Significance level: $\alpha = 1\%$ or $\alpha = 5\%$.
d) Construct the hypothesis testing model.
e) State the decision rule as: Reject H_o if the p-value of the test statistic is less than or equal to α.

Step 4 Analyze the sample data.
Enter the necessary sample data into the TI-84 calculator to determine the p-value of the test statistic. Use 2-SampleZTest if population standard deviations are known or 2-SampleT-Test if population standard deviations are unknown.

Step 5 Determine the conclusion.
Compare the p-value of the test statistic to the decision rule and either:
(a) Reject H_o and Accept H_a at α or (b) Fail to Reject H_o at α

KEY TERMS

Assumptions of Paired t-test 710
Control Group 693
Difference between the sample means 673
Difference of Paired Observations 703
Double-Blind Study 693
Independent Random Samples 688
Matched Pairs 700
Mean/Standard Error of the Difference between Paired Observations 675
p-value 711
p-values for Paired t-test 705
Paired t-test 700
Placebo 694
Pooled Standard deviation 680
Pooled t-test 693
Repeated Measure 700
Sampling Distribution of the Difference Between Two Means 674
Single-Blind Study 693
Test statistic 683
Treatment Group 693
Two Sample t Test 681, 683
Welch's t-test 687
$\sigma_{\bar{x}_1-\bar{x}_2}$, the Standard Error of the Difference Between Two Means 675
$s_{\bar{x}_1-\bar{x}_2}$, Estimate of the Standard Error of the Difference Between Two Means 680
$\mu_{\bar{x}_1-\bar{x}_2}$, Mean of the Sampling Distribution of the Difference Between Two Means 675

SECTION EXERCISES

Note: The data sets in these Section Exercises can be downloaded in either Microsoft EXCEL or TI-84 format from the textbook's website at www.MyStatLab.com.

Section 13.2

1. To perform a hypothesis test involving two population means, the distribution of the _____ _____ is used as the hypothesis testing model.

2. The notation $\bar{x}_1 - \bar{x}_2$ represents the difference between the _____ _____, where: $\bar{x}_1$ represents the _____ of sample 1 and $\bar{x}_2$ represents the _____ of sample 2.

3. The sampling distribution of the difference between two means has the following characteristics:
   a) The mean of the sampling distribution of the difference between two means, denoted _____, is determined by the formula: $\mu_{\bar{x}_1-\bar{x}_2}$ = _____ – _____, where: μ_1 represents the mean of _____ one and μ_2 represents the mean of _____ two.
   b) The formula for the estimate of the standard error of the difference between two means, denoted $s_{\bar{x}_1-\bar{x}_2}$, is dependent upon the sample standard deviations and the sample sizes. The formula for $s_{\bar{x}_1-\bar{x}_2}$ is _____.
   c) Under Theorem 13.2, the distribution of the test statistic is approximated by a _____ distribution with df, degrees of freedom, given by the formula: df = _____.

4. If $n_1 = 35$, $n_2 = 40$, $\sigma_1 = 4$ and $\sigma_2 = 6$, then $\sigma_{\bar{x}_1-\bar{x}_2}$ = _____.

Section 13.3 and 13.4

5. The null hypothesis of a hypothesis test involving two population means, μ_1 and μ_2, is stated as: There is _____ difference between the _____ of population 1 and the _____ of population 2. In symbols, H_o is written as: _____ = 0.

6. The alternative hypothesis of a hypothesis test involving two population means, μ_1, and μ_2, has one of three possible forms. They are:
   a) The mean of population 1 is greater than the _____. In symbols, this H_a is written: _____.
   b) The _____ of population 1 is less than the mean of _____. In symbols, this H_a is written: _____.
   c) The mean of _____ 1 is not equal to the _____ of population _____. In symbols, this H_a is written: _____.

7. The null hypothesis used in most comparisons involving two population means is based upon the assumption that mean of population 1 equals the mean of population 2.
(T/F)

8. The formula for the test statistic t, for a hypothesis test involving two population means, is:

   t = _____.

   Under the assumption H_o is true, $\mu_{\bar{x}_1-\bar{x}_2}$ = _____, and thus, the formula for the test statistic is reduced to: t = _____.

9. To determine the conclusion of a hypothesis test involving two population means, the _____ _____ is compared to the critical score(s).

10. The formula for the degrees of freedom of a t Distribution for testing the difference between two population means using two independent samples is:
    a) $df = n - 1$ b) $df = n_1 + n_2 - 2$ c) $df = n_1 - n_2$

11. When performing a two-tailed hypothesis test, if the test statistic t falls between the two critical values, t_{LC} and t_{RC}, then the conclusion is:
    a) Reject H_o
    b) Reject H_a
    c) Fail to Reject H_o
    d) Accept H_a

12. To determine if the difference between the sample means is significantly different from zero the test statistic t is compared to a critical t score, t_c, that is determined by:

    $$t = \frac{\bar{x}_1 - \bar{x}_2}{s_{\bar{x}_1 - \bar{x}_2}}$$

    (T/F)

13. If the difference between the sample means, $\bar{x}_1 - \bar{x}_2$, is different from zero then there is a significant difference between the sample means.
    (T/F)

For each of the problems 14 to 20:
   a) Define population 1 and population 2.
   b) Formulate the null and alternative hypotheses.
   c) Calculate the expected results for the hypothesis test assuming the null hypothesis is true, and determine the hypothesis test model.
   d) Formulate the decision rule.
   e) Determine the test statistic.
   f) Determine the conclusion and answer the question(s) posed in the problem, if any.

For problems where the population standard deviations are unknown, assume that they **are equal** and use the Pooled t Test to perform the hypothesis test using the 5-step hypothesis testing procedure.

14. A psychologist who needs two groups of students for a learning experiment decides to select a random sample of 100 female college students and 81 male college students. Before the experiment, the psychologist administers an IQ test to both groups to determine if there is a significant difference in their mean IQ scores. The population standard deviation for this IQ test is known to be 12. If the sample mean IQ score of the 100 female students is 116.9 and the sample mean IQ score of the 81 male students is 120.8, can the psychologist conclude that there is a significant difference in the mean IQ score of both groups at $\alpha = 1\%$? (Use Theorem 13.1 and the hypothesis testing procedure)

15. A high school teacher believes that there is a significant difference in the mean SAT math scores of male and female students and decides to test her claim. A random sample of 60 female SAT math scores had a mean of 487 and a standard deviation of 101, and a random sample of 50 male SAT math scores had a mean of 527 and a standard deviation of 93. Do these sample results support the high school teacher's claim at $\alpha = 5\%$?

16. A Pew Research Center analysis of the American Time Use Survey which was based on adults aged 18 to 64 who were currently married or living with a partner indicated that the mean number of weekly hours that a mother spends on child care was 12 hours while the mean number of weekly hours that a father spends on child care is 9.9 hours. Lydia, a social worker, decides to perform a hypothesis test at $\alpha = 1\%$ to determine if mothers spend significantly more time on child care per week. Can the social worker conclude that mothers do spend significantly more weekly time on child care if a random sample of 81 mothers produced a sample standard deviation of 3.4 weekly child care hours while a random sample of 36 fathers generated a sample standard deviation of 3.6 weekly child care hours?

17. A recent report issued by The Nielsen Company stated that the average U.S. Internet user spent an estimated 68 hours online (both at home and at work). Ms. Maria, a social researcher, wants to compare the 18 to 24 year old users to the 25 to 34 year old users to determine if there is a significant difference in the mean number of hours that the groups spend on the Internet per month. An independent random sample of fifty 18 to 24 year old users had a monthly mean of 22.5 hours online with a standard deviation of 3.5 hours. An independent random sample of forty 25 to 34 year old users had a monthly mean of 24.5 hours online with a standard deviation of 4.2 hours. Can the researcher conclude that there is a significant difference in the mean online usage between the two age groups at $\alpha = 5\%$?

18. A farmer believes that his special fertilizer will significantly increase his tomato crop yield. To test his belief, he randomly divides 200 acres into two groups. In Group I, he administers his special fertilizer. In Group II, he administers the usual tomato fertilizer. The following table summarizes the tomato yields for this experiment.

	Group I	Group II
Number of acres, n	100	100
sample average bushels per acre, $\bar{x}$	830	805
sample standard deviation, s	40	36

Do these sample results support the farmer's claim at $\alpha = 1\%$?

19. A sociologist claims that marriage lowers the cumulative average of graduate students. If the cumulative average of 16 randomly selected married graduate students was 3.35 with a standard deviation of 0.3 and the cumulative average of 9 unmarried graduate students was 3.56 with a standard deviation of 0.5, do you agree with the sociologist's claim at $\alpha = 1\%$?

20. After many years of experience, a marriage counselor feels that the mean education level for married men is higher than the mean education level for divorced men. In order to back-up her belief, she decides to perform a 2-sample t-test. She randomly samples from her clients 25 married men and 25 divorced men and coded their education level as a 0, 1 or 2. The meaning for the education level is as follows:
    0 : less than 12 years of education,
    1 : 12 to 15 years of education,
    2 : 16 or more years of education.

    For the married men, the mean education level was 1.16 with a sample standard deviation of 0.73. For the divorced men, the mean education level was 0.71 with a sample standard deviation of 0.68. Perform a 2-sample t-test to test the sociologist's claim that the population mean education level for married men is higher than the population mean education level for divorced men. Is the sociologist correct? Assume equal variances. Test at the 5% level of significance.

Section 13.5

For problems 21–26, redo problems 15–20 under the assumption that the population standard deviations are *unknown* and **unequal**; use Welch's t Test to perform the hypothesis test using the 5-step hypothesis testing procedure.

Section 13.6

27. To study whether a new pill will significantly reduce the blood pressure of patients, forty patients are randomly selected and divided randomly into two equal sized groups, A and B. Group A is administered the new blood pressure pill, while Group B is administered a placebo. Then, the control group is Group _____, while the treatment group is Group _____.

For each of the problems 22-23, assume that the population standard deviations are equal, and use the Pooled t Test. Additionally, be sure to:
   a) Define population 1 and population 2.
   b) Formulate the null and alternative hypotheses.
   c) Calculate the expected results for the hypothesis test assuming the null hypothesis is true and determine the hypothesis test model.
   d) Formulate the decision rule.
   e) Determine the test statistic.
   f) Determine the conclusion and answer the question(s) posed in the problem, if any.

28. A school nutritionist believes that her special diet given along with a rigorous training routine will significantly increase stamina. She randomly selects 30 students and divides them into two equal independent groups. Group I, called the treatment group, is put on the special diet and the rigorous training routine. Group II, called the control group, is just given the rigorous training routine with no special diet. After six months the increase in stamina level of all subjects is measured. The results are given in the following table.

Treatment Group	Control Group
$n_1 = 15$	$n_2 = 15$
$\bar{x}_1 = 26$	$\bar{x}_2 = 18$
$s_1 = 4$	$s_2 = 10$

Do these sample results support the nutritionist's belief at $\alpha = 5\%$?

29. A medical researcher wants to study whether a diet deficient in Vitamin E affects the mean amount of Vitamin A stored in mice. She randomly selects two samples of fifteen mice each. Group I called the control group is fed a normal diet while the second group called the treatment group is fed a diet deficient in vitamin E. She collected the following data:

Control Group I.U. of Vitamin A	Treatment Group I.U. of Vitamin A
3850	3300
3700	3250
3650	2900
3500	2700
3450	2000
3450	2550
3900	2400
3700	2600
2900	2300
3400	2450
3200	1900
3150	2100
3850	2500
3000	2350
2800	2800

Can the researcher conclude that the mean amount of Vitamin A for mice on the deficient diet is less than the mean amount of Vitamin A for mice on a normal diet at $\alpha = 1\%$?

Section 13.7

30. In order to perform a paired t Test, we must first compute the _____ of the before and after data.
31. The sample mean and sample standard deviation of the differences of the paired observations in a paired t Test are denoted by _____ and _____.
32. When performing a paired t Test, the symbol used for the population mean difference is denoted by _____.
33. In a paired t Test, $\dfrac{s_d}{\sqrt{n}}$ represents the _____ _____ of the differences of the paired observations.
34. For a paired t Test, the Null Hypothesis states the population mean difference of the paired observations is _____. In symbols, this is represented as _____.

For each of the problems 35 to 39:
  a) Identify the Before and After populations.
  b) Formulate the null and alternative hypotheses.
  c) Calculate the expected results for the hypothesis test assuming the null hypothesis is true, and determine the hypothesis test model.
  d) Formulate the decision rule.
  e) Determine the test statistic.
  f) Determine the conclusion and answer the question(s) posed in the problem, if any.

35. It is believed that adding the supplement, Co-X 15 daily to your diet for 20 weeks will effectively lower LDL cholesterol in fifty-year-old males. A medical researcher tests this hypothesis for a random sample of 10 fifty-year-old males. The results of the before and after the twenty-week study is given in the following table:

TABLE: LDL Level in Fifty-Year-Old Males

Subject	Before adding Co-X 15 daily to diet for 20 weeks	After adding Co-X 15 daily to diet for 20 weeks
1	130	122
2	110	112
3	145	143
4	150	155
5	135	120
6	120	120
7	112	115
8	155	145
9	125	120
10	115	120

Perform a paired t Test to determine if these sample results support the medical researcher's claim at $\alpha = 5\%$.

36. A college women's softball coach believes that adding a weekly weight training program to player's exercises will increase the throwing distance for softball athletes. She randomly selects 12 softball athletes and measures their throwing distances before and after the training program. The results are given in the following table:

TABLE: Throwing Distances in Feet

Subject	Throwing Distance BEFORE Training (ft)	Throwing Distance AFTER Training (ft)
1	122	130
2	112	110
3	143	145
4	155	150
5	120	135
6	120	120
7	115	112
8	145	155
9	120	125
10	120	115

Perform a paired t Test to determine if these results indicate a statistically significant improvement in the throwing distances for the softball athletes. (use $\alpha = 1\%$)

37. A sociologist believes that turning off cell phones for at least five hours prior to sleeping for fourteen-year-old females will result in their spending more time interacting with their parents. She randomly selects 200 fourteen-year-old females to participate in a study. The BEFORE turning off cell phones and AFTER turning off cell phones for each

fourteen-year-old is found. The mean difference between the BEFORE and AFTER paired subjects is found to be 0.48 hours and the standard deviation of differences between the BEFORE and AFTER paired subject is 3.05 hours. Perform a paired t Test to determine if these results indicate that the sociologist's claim is statistically significant at $\alpha = 1\%$.

38. A biologist wishes to know if there has been a decline in the clam population in several areas off Long Island that commercial fisherman harvest clams. The biologist measures the number of clams per square meter and estimates the clam population at ten popular clamming locations during the summer of 2018. The biologist then **repeats** this estimate at *the same* 10 clamming locations during the summer of 2019. The results of this study have been summarized in the following table.

TABLE: Number of Clams per Square Meter

Location	Clam Population in 2019	Clam Population in 2020
1	17.1	10.4
2	41.7	29.2
3	.92	.73
4	29.9	21.9
5	11.6	9.3
6	32.6	31.5
7	25.4	22.7
8	19.2	21.9
9	45.7	41.3
10	22.6	21.8

Perform a paired t Test to determine if the results of the 2019 and 2020 measurements indicate there is a statistically significant decrease in the clam population. Use $\alpha = 1\%$.

39. A psychologist wishes to test the effectiveness of the prescription medication she has been prescribing to her patients for their depression. She has developed a wellbeing scale that measures her patient's mood on a 0 to 10 scale. She randomly selects 10 of her new patients and measures their mood *prior* to taking the medication. She then gives the medication to these 10 individuals for 30 days, and then measures their mood again. The results of her measurements have been summarized in the following table.

TABLE: Wellbeing Score

Patient #	Mood pre medicine	Mood post medicine
1	0	1
2	2	5
3	5	4
4	4	5
5	6	7
6	1	3
7	3	5
8	3	3
9	1	1
10	2	4

Perform a paired t Test to determine if the psychologist can conclude there is a statistically significant increase in the mood of her patients after taking the medication for 30 days. Use $\alpha = 1\%$.

Section 13.8

For problems 40–45, use questions 15 & 17 of Section 3.3/13.4, questions 22 & 24 of Section 13.5 and questions 35 & 36 of Section 13.7 and apply the p-value approach to hypothesis testing using the TI-84 graphing calculator. For each of these problems, use the 5-step procedure to perform a hypothesis test at **both** $\alpha = 1\%$ and $\alpha = 5\%$. Include the following:

a) state the null and alternative hypotheses.
b) calculate the expected results for the experiment assuming the null hypothesis is true and state the hypothesis test model.
c) formulate the decision rule.
d) identify the test statistic and its p-value.
e) state the conclusion and answer any question(s) posed in the problem for each alpha level.
f) determine the smallest level of significance for which the null hypothesis can be rejected.

CHAPTER REVIEW

46. Symbolically, the difference between two sample means is expressed as:
    a) $\bar{x}_1 - \bar{x}_2$
    b) $\mu_1 - \mu_2$
    c) $\mu_{\bar{x}_1 - \bar{x}_2}$
    d) $p_1 - p_2$

47. The formula for the mean of the sampling distribution of the difference between two means is:
    a) $\mu_{\bar{x}_1 - \bar{x}_2} = \mu_1 + \mu_2$
    b) $\mu_{\bar{x}_1 - \bar{x}_2} = \mu_1 - \mu_2$
    c) $\mu_{\bar{x}_1 - \bar{x}_2} = \bar{x}_1 - \bar{x}_2$
    d) $\mu_{\bar{x}_1 - \bar{x}_2} > 0$

48. The formula
$$\sigma_{\bar{x}_1 - \bar{x}_2} = \sqrt{\frac{\sigma_1^2}{n_1} + \frac{\sigma_2^2}{n_2}}$$
can be used only if:
    a) Both population standard deviations are known
    b) Both population standard deviations are unknown
    c) Both s_1 and s_2 are known

49. If $n_1 = 25$, $n_2 = 50$, $s_1 = 3$ and $s_2 = 5$, then the estimate of the standard error under Welch's t-test equals _____.

50. The sampling distribution of the difference between two means is formed by finding the differences between all

possible independent sample means of the form $\bar{x}_1 - \bar{x}_2$, where $\bar{x}_1$ represents the mean of an independent random sample selected from population 1 and $\bar{x}_2$ represents the mean of an independent random sample selected from population 2.
(T/F)

51. The mean of the sampling distribution of the difference between two means is computed by the formula: $\bar{x}_1 - \bar{x}_2$.
(T/F)

52. The standard error of the difference between two means is estimated by the formula: $s_{\bar{x}_1 - \bar{x}_2} = s_1 - s_2$.
(T/F)

53. When performing a hypothesis test involving two population means, the expected result $\mu_{\bar{x}_1 - \bar{x}_2} = 0$ is determined under the assumption that _____ is true.

54. When using the **t** distribution with $n_1=16$ and $n_2=20$ to approximate the sampling distribution of the difference between two means, the degrees of freedom for the **t** distribution is _____.

55. The null hypothesis, H$_o$, for testing the difference between two population means can be represented as:
a) H$_o$: $\mu_1 = 0$
b) H$_o$: $\mu_1 = \mu_2$
c) H$_o$: $\mu_1 - \mu_2 > 0$
d) H$_o$: $\mu_1 - \mu_2 = 0$

56. When the experimenter shows no significant difference between the sample means, then the mean of population 1 is said to equal the mean of population 2.
(T/F)

57. Only by rejecting H$_o$ can you conclude that there is a difference between the population means.
(T/F)

58. The treatment group is frequently compared to the:
a) Control group
b) Null group
c) Test group
d) Pseudo group

For each of the problems 59 to 63:
a) Define population 1 and population 2.
b) Formulate the null and alternative hypotheses.
c) Calculate the expected results for the hypothesis test assuming the null hypothesis is true and determine the hypothesis test model.
d) Formulate the decision rule.
e) Determine the test statistic.
f) Determine the conclusion and answer the question(s) posed in the problem, if any.

59. An educational researcher wants to determine if the mean score of urban sixth grade students is significantly greater than the mean score of rural sixth grade students on a standardized reading comprehension test. From past experience, the population standard deviation of the scores on this standardized test is 10. If a random sample of 64 urban students had a mean test score of 78.4, and a random sample of 36 rural students had a mean test score of 73.9, can the educational researcher conclude that the urban sixth grade students performed significantly better on the standardized test at $\alpha = 5\%$? (Use Theorem 13.1 and the hypothesis testing procedure)

60. Recently, there have been a number of articles stating that women spend, on average, more time on social media sites than men. A local sociologist disagrees with these articles and decides to test this claim. He samples 55 males and finds the average time spent on social media per day to be 36.3 minutes with a sample standard deviation of 48.2 minutes. A sample of 60 females gave a mean of 42.5 minutes per day with a sample standard deviation of 51.4 minutes. Perform a 2-sample t-test in order to test the claim that the population mean time spent on social media sites for men is not significantly different than the population mean time spent on social media sites for women. Assume **equal** population standard deviations and use $\alpha = 5\%$.

61. A financial analyst claims that women tend to have less debt than men. A random sample of forty women had a mean debt of $11,119 with a standard deviation of $6,857, and a random sample of fifty five men had a mean debt of $13,456 with a standard deviation of $5,983. Do these sample results support the financial analyst's claim? Assume **unequal** population standard deviations and use $\alpha = 5\%$.

62. An insurance company claims that male drivers under age 25 have more accidents per year than male drivers 25 years of age or older. An independent sample of size 40 is randomly selected from each age group. The mean number of accidents per year for the male drivers under age 25 was found to be 1.2 with a standard deviation of 0.1 while the mean number of accidents per year for the drivers 25 years of age or older was found to be 0.94 with a standard deviation of 0.2. Do you believe the claim of the company is correct? Assume **equal** population standard deviations and use $\alpha = 5\%$.

63. An educational researcher wants to determine whether the Scholastic Aptitude Test (SAT) scores of students attending an Ivy League College is significantly different than the SAT scores of students attending a Military Academy. An independent random sample is chosen from each population and the SAT scores for the verbal, the mathematics and the combined sections are summarized in the following table.

Verbal SAT's		Mathematics SAT's		Combined SAT's	
Ivy	Military	Ivy	Military	Ivy	Military
$\bar{x} = 680$	$\bar{x} = 640$	$\bar{x} = 675$	$\bar{x} = 710$	$\bar{x} = 1355$	$\bar{x} = 1350$
s = 50	s = 60	s = 70	s = 50	s = 86	s = 78
n = 50	n = 50	n = 50	n = 50	n = 50	n = 50

Although the population standard deviations are unknown, the educational researcher believes them to be **equal** and wishes to use a pooled t Test. Using this sample data, can the educational researcher conclude that the:

(I) Population mean verbal score is higher for the Ivy League students at $\alpha = 5\%$?

(II) Population mean math score is higher for the military academy students at $\alpha = 5\%$?

(III) Population mean score of the combined SAT scores is significantly different from the Ivy League and military academy students at $\alpha = 5\%$?

64. A botanist wants to test whether the new product "Miraculous Growth" would significantly increase the number of flowers per plant over an eight week period. Two trays of 12 plants were prepared. Tray I received the "Miraculous Growth" during the test period while Tray II served as the **control group**. The plants were grown in a controlled environment for this eight week period. At the end of this period, the number of

flowers per plant was recorded. The results are summarized in the following table:

Tray I (treatment group)		Tray II (control group)	
14	17	12	9
16	19	10	10
20	20	13	15
19	18	14	12
16	17	15	13
15	16	12	14

Can the botanist conclude that the plants treated with "Miraculous Growth" had a greater number of flowers per plant than the plants in the control group? Assume **equal** population standard deviations and use $\alpha = 5\%$.

65. Residents in a local community have been complaining of increased trace amounts of lead in their drinking water. The local water authority claims that trace amounts of lead in the area's drinking water are at or below the State's Department of Health Regulations. The water authority takes samples at six of its storage tanks in the community in question and measures the amount of lead in each sample. Since the Department of Health keeps records of their measurements from year to year, they compare the results of this survey to their results at these locations from the prior year. The results of this survey have been recorded in the following table. Note, lead levels are measure in parts per billion, *ppb*.

TABLE: Lead Measurements

Tank #	Amount of Lead in the Prior Year(*ppb*)	Amount of Lead in the Current Year(*ppb*)
1	10	11
2	7	8
3	8	7
4	6	8
5	6	5
6	4	6

Perform a paired *t* Test to determine if the water authority can conclude there is a statistically significant increase in the amount of lead in this communities drinking water. Use $\alpha = 1\%$.

66. The following two hypothesis tests should be conducted using Welch's t-Test. Thus, it is assumed that the populations for each variable are approximately normal, but the population standard deviations are unknown and assumed to be *not* equal. When doing these problems you should do them two ways. **First**, follow the 5-step hypothesis testing procedure and use Welch's approximation for the estimate of the standard error

$$s_{\bar{x}_1 - \bar{x}_2} = \sqrt{\frac{s_1^2}{n_1} + \frac{s_2^2}{n_2}}.$$

Then, use that estimate of the standard error when computing the test statistic $t = \frac{(\bar{x}_1 - \bar{x}_2)}{s_{\bar{x}_1 - \bar{x}_2}}$. When looking up the critical test statistic, t_c, use the smaller of $n_1 - 1$ and $n_2 - 1$ for the *degrees of freedom*.

Second, use the TI-84 calculator to calculate the testing statistic, p-value and actual degrees of freedom. Recall that when using the TI-84 calculator, the output screen will use the Satterthwaite formula to calculate the actual *degrees of freedom*. Also, when using the calculator, you must select "NO" under the "Pooled" option as indicated in Example 13.9 on page 717.

1. An educational researcher would like to compare scores on mathematics placement exam given at their college for students under age 25 and students 25 years old or older. She randomly selects 150 students, ninety under the age of 25 and sixty 25 years old or older. For the students under age 25, the mean mathematics placement score is 55 with a standard deviation of 12. Students 25 years old or older, the mean placement score is 48 with a standard deviation of 18. Can the researcher conclude that the group's performance is significantly different on the mathematics placement scores? (Use $\alpha = 5\%$

2. A sociologist believes that for people under the age of 30, men sleep longer per night on the weekend than women. 24 people were surveyed. Test at alpha =5%. The following summary statistics were determined from the sample.

	Men	Women
Mean	11.45	10.18
Standard Deviation	1.95	2.81
Sample size, *n*	11	13

Do the sample result indicate that the sociologist's claim is correct?

WHAT DO YOU THINK?

67. A research study explored the relationship between the personality characteristics and concepts and skills necessary for leadership development in students. The study involved 53 girls and 42 boys attending a Leadership Studies Program for one week on a university campus. Figure 13.42 (see next page) presents the means and standard deviations for the girls and boys on the Leadership Skills Inventory subscale scores and for each of the High School Personality Questionnaire factors along with the *t* values of the comparison of the mean scores. The purpose of administering these instruments was to assess the strengths and weaknesses of the students' leadership concepts and skills.

a) Define the two populations within the study. By examining the t value can you determine which group has been defined as population 1? Explain.
b) What variables are being measured? How are they being measured?
c) On which of the variables are the results of the t test significant?
d) Would these results be significant at $\alpha = 1\%$? Explain.

Means, Standard Deviations, and t Ratios for Subscale Scores on Leadership Skills Inventory and Factors (Sten Scores) of High School Personality Questionnaire for Boys and Girls in Leadership Studies Program.

Variable	All Subjects M	All Subjects SD	Boys M	Boys SD	Girls M	Girls SD	t*
1. Fundamentals of Leadership	50.04	11.09	49.61	12.02	50.37	10.40	−.33
2. Written Communication Skills	55.27	7.98	53.66	6.88	56.45	8.60	−1.77
3. Speech Communication Skills	53.21	9.48	52.19	9.30	54.01	9.64	−.93
4. Values Clarification	52.96	6.78	51.35	6.08	54.24	7.09	−2.10*
5. Decision-making Skills	54.10	7.48	53.09	7.43	54.92	7.50	−1.18
6. Group Dynamic Skills	53.58	7.53	51.97	6.72	54.86	7.97	−1.88
7. Problem-solving Skills	55.08	9.17	54.11	10.38	55.84	8.10	−.91
8. Personal Development Skills	54.55	6.39	53.26	6.51	55.58	6.16	−1.78
9. Planning Skills	55.94	7.20	54.90	7.55	56.77	6.88	−1.26
10. A Warmth	5.98	1.93	5.88	2.34	6.07	1.56	−.48
11. B Intelligence	6.52	1.85	6.40	1.72	6.62	1.95	−.57
12. C Emotional Stability	6.26	1.82	5.81	1.89	6.62	1.71	−2.20*
13. D Excitability	5.69	1.89	5.85	2.07	5.56	1.74	.74
14. E Dominance	5.89	2.16	5.33	1.85	6.34	2.30	−2.30*
15. F Cheerfulness	5.41	2.10	5.54	2.23	5.30	2.00	.56
16. G Conformity	6.35	1.89	6.38	1.88	6.34	1.91	.11
17. H Boldness	6.07	2.16	5.95	2.23	6.17	2.11	−.49
18. I Sensitivity	6.45	2.21	6.45	2.10	6.45	2.31	.00
19. J Withdrawal	5.90	1.95	5.59	1.93	6.15	1.94	−1.38
20. O Apprehension	4.26	2.14	4.40	2.16	4.15	2.13	.57
21. Q_2 Self-sufficiency	5.36	1.95	5.00	1.87	5.66	1.99	−1.65
22. Q_3 Self-discipline	6.11	1.86	5.88	1.86	6.30	1.86	−1.09
23. Q_4 Tension	5.15	1.90	5.28	2.06	5.05	1.78	.58
24. Extraversion	6.90	1.60	7.11	1.70	6.74	1.51	1.09
25. Anxiety	4.71	1.86	5.13	1.85	4.38	1.83	1.97
26. Tough Poise	5.13	2.19	5.09	2.20	5.16	2.20	−.16
27. Independence	6.30	1.93	5.71	1.60	6.77	2.06	−2.71*
28. Delinquency	6.49	2.04	6.22	1.84	6.70	2.17	−1.15
29. Accident Proneness	4.55	1.91	4.73	1.73	4.41	2.05	.81
30. Creativity	6.75	1.98	6.77	1.99	6.82	1.99	−.36
31. Leadership Potential	6.52	1.98	6.74	2.09	6.34	1.89	.95
32. School Achievement	6.51	2.10	6.25	1.95	6.70	2.20	−1.08
33. Vocational Growth	6.62	2.26	6.22	2.21	6.93	2.28	−1.52
34. Vocational Success	5.91	2.20	5.91	2.15	5.91	2.27	−.02

*p = .05.

FIGURE 13.42

e) State the null and alternative hypotheses for each of these significant t tests if the alternative hypothesis was a nondirectional hypothesis.

f) What is the name and mean of the sampling distribution that serves as the model for this test?

g) Did the boys or girls score significantly higher on the Dominance Factor?

h) Who scored significantly higher on the Independence Factor?

i) For the Dominance and Independence Factor, was the null hypothesis rejected at $\alpha = 5\%$? at $\alpha = 1\%$? Explain.

j) For the Fundamentals of Leadership Skill, was the null hypothesis rejected at $\alpha = 5\%$? at $\alpha = 1\%$? Explain.

k) Does the t test analysis indicate that there is a significant or nonsignificant difference between the girls and boys with respect to Speech Communication Skills at $\alpha = 5\%$? Explain.

l) Does the t test analysis indicate that there is a significant or nonsignificant difference between the girls and boys with respect to Values Clarification at $\alpha = 5\%$? at $\alpha = 1\%$? Explain.

68. The following excerpt is from a recent research article from the *New England Journal of Medicine* regarding a study concerning the effects of oat bran on cholesterol. Read this excerpt and answer the following questions.

A Harvard study from the *New England Journal of Medicine* found no significant difference on cholesterol levels between people who ate oat bran and those that ate a placebo supplement, which consisted of just plain flour. In both cases, the levels decreased slightly, mainly because the volunteers ate grains instead of fatty foods. In the study, 20 healthy people from the staff of Brigham and Women's Hospital in Boston were put for six weeks on an oat bran high-fiber diet, two weeks on their normal diets, and then six weeks on a diet high in white flour, which is low in fiber. The study was double-blinded.

a) Define the two groups contained within the study.
b) State the purpose of the study, and a null and an alternative hypothesis for this study.
c) What was the result of the study? Was the null hypothesis rejected? Explain.
d) What does "placebo supplement" mean?
e) What is the purpose of giving a group a placebo? How does this help to conduct the study?
f) Explain what is meant by the statement: "the study was double-blinded"?
g) Comment on the following statement about this study: "twenty volunteers are too few to be meaningful and that the participants had cholesterol levels of 186, which is less than the 235 cutoff point set by the American Heart Association for serious concern about heart disease."
h) Comment on the following statement about this study: "Because the cholesterol levels dropped slightly, but equally, regardless of whether the volunteers ate high- or low fiber foods, the oat bran hype is clearly overblown. This is a classic case of where the oat bran industry has run ahead of science."

PROJECTS

69. Hypothesis Testing Project:

    A. Selection of a topic.
    1. Select a reported fact or claim involving two population means made in a local newspaper, magazine or other source.
    2. Suggested topics:
        a) Compare the cumulative averages of males and females at your school and test the hypothesis that there is no significant difference between their cumulative averages.
        b) Compare the mean number of traffic violations incurred during the past 3 years for drivers under 25 against the average for drivers over 25 to determine if a significant difference exists between these averages.
        c) Compare the average pulse rate of football players and cheerleaders at your school and test the hypothesis that there is no significant difference between their average pulse rates.
        d) Compare the average number of part-time hours worked by freshmen and sophomores at your school to determine if there is a significant difference between these averages.

    B. Use the following procedure to test the claim selected in Part A.
    1. State the claim, identify the two populations referred to in the statement of the claim, and indicate where the claim was found.
    2. State your opinion regarding the claim of the two population means, μ_1 and μ_2 (i.e. $\mu_1 > \mu_2$, $\mu_1 < \mu_2$ or $\mu_1 \neq \mu_2$).
    3. State the null and alternative hypotheses.
    4. Develop and state a procedure for selecting an independent random sample from each population. Indicate the technique you are going to use to obtain your sample data.
    5. Compute the expected results, indicate the hypothesis testing model, and choose an appropriate level of significance.
    6. Calculate the critical score(s) and state the decision rule.
    7. Calculate the test statistic.
    8. Construct a hypothesis testing model and place the test statistic on the model.
    9. Formulate the appropriate conclusion and interpret the conclusion with respect to your opinion (as stated in Step 2).

70. Read a research article from a professional journal in a field such as medicine, education, marketing, science, social science, etc. Write a summary of the research article that includes the following:
    a) Name of article, author, and journal (include volume number and date).
    b) Statement of the research problem.
    c) The null and alternative hypotheses.
    d) How the study was conducted (give the definition of the treatment group, treatment used and the sample size; give the definition of the control group and the sample size. Were placebos used? If so, what were they?)
    e) Determine the test statistic and the level of significance.
    f) State the conclusion in terms of the null and alternative hypotheses.

CHAPTER TEST

71. The distribution used to approximate the sampling distribution of the differences between two means, when the population standard deviations are unknown, is the:
    a) Binomial distribution
    b) t distribution
    c) Normal distribution

72. If the sample sizes, n_1 and n_2, are increased and the sample standard deviations, s_1 and s_2, remain the same then the value of $s_{\bar{x}_1-\bar{x}_2}$ will:
    a) Increase
    b) Remain the same
    c) Decrease

73. If the test statistic t is beyond the critical t score, then the conclusion is:
    a) Reject H_o
    b) Reject H_a
    c) Fail to Reject H_o
    d) Accept H_a

74. If the conclusion of a hypothesis test involving two population means is that there is no significant difference between the population means then the experimenter has failed to reject H_o.
    (T/F)

75. If the test statistic of the difference between the sample means, $\bar{x}_1 - \bar{x}_2$, does not fall beyond t_c then the null hypothesis, H_o, is assumed true.
    (T/F)

For each of the problems 76 to 84:
   a) Define population 1 and population 2.
   b) Formulate the null and alternative hypotheses.
   c) Calculate the expected results for the hypothesis test assuming the null hypothesis is true and determine the hypothesis test model.
   d) Formulate the decision rule.
   e) Determine the test statistic.
   f) State the conclusion and answer the question(s) posed in the problem, if any.

76. A psychologist administered a perception test to two randomly selected groups. The first group of 10 subjects were asked to estimate the length of an object dangling from a rod at a distance of 20 feet. These subjects had both eyes open while observing the object. The second group of 12 subjects estimated the length of the object with one eye covered. The results are summarized in the following table.

Group I (using both eyes)	Group II (one eye covered)
$n_1 = 10$	$n_2 = 12$
$\bar{x}_1 = 12.5$ inches	$\bar{x}_2 = 14.25$ inches
$s_1 = 1.25$ inches	$s_2 = 2.5$ inches

Can the psychologist (with both eyes open) conclude that there is a significant difference in the perception of the two groups? Assume **equal** population standard deviations and use $\alpha = 1\%$.

77. A national credit organization reports the mean amount of credit card debt is the same for male college students as it is for female college students. Student government representatives believe men have a lower mean credit card debt than women. The representatives randomly select independent samples from men and women regarding credit card balances and report the following results:

	Males	Females
Mean	$450	$519
Standard Deviation	$65	$75
Sample size	15	15

Can the research group conclude, based on the sample data, male college students have a lower mean credit card debt than female college students? Assume **equal** population standard deviations and use $\alpha = 5\%$.

78. Many college professors believe that full-time male students work more hours per week than full-time female students. An independent random sample of 400 full-time college students found that 210 full-time male students worked a mean of 25.7 hours per week with a standard deviation of 6.3 hours, and, the remaining female students worked 24.2 hours per week with a standard deviation of 7.9 hours.
    What is the largest p-value for which these results support the claim that fulltime male students work more hours per week than fulltime female students? Assume **equal** population standard deviations and the pooled t Test.

79. Is there a difference in the 40 hour work week base salary for working moms as compared to the 40 hour work week for stay at home moms? Salary.com, a website that measures various salary factors, released a comparison of these two groups. It was reported that on average a working mom's annual salary is $40,583 while a stay at home mom's annual salary is $38,126. Let's assume that the sample used to determine these salaries contained 15,000 participants, where 7,500 were working moms and 7,500 were stay at home moms. Furthermore, let's assume that the sample standard deviation for the sample of working mom salaries was $8,500 while the stay at home mom salaries sample standard deviation was $6,500. Do these results indicate that there is a significant difference in working versus stay at home mom salaries? Assume **equal** population standard deviations and use $\alpha = 1\%$.

80. A recent report issued by The Nielsen Company stated men aged 18 to 34 years watch TV a daily average of 244 minutes while women aged 18 to 34 years view TV a daily average of 254 minutes. Nola, a social researcher, decides to test if women aged 18 to 34 years spend significantly more time watching TV than men aged 18 to 34 years. If a random sample of 49 women yielded a sample standard deviation of 25 minutes and a random sample of 64 men generated a sample standard deviation of 18 minutes, test to determine if the social researcher can conclude that women aged 18 to 34 years spend significantly more time watching TV? Assume **unequal** population standard deviations and use $\alpha = 1\%$.

81. A new toothpaste containing an anti-cavity substance DPT claims to significantly reduce cavities in children. To test this claim, researchers from a dental school randomly select two groups of children. To one group, called the treatment group, the children are given the toothpaste containing DPT. To the other group, called the control group, the children are given the toothpaste *without* DPT. After one year the mean number of cavities for each group is determined. The results for each group are summarized in the following table:

Treatment Group	Control Group
$n_1 = 64$	$n_2 = 100$
$\bar{x}_1 = 3$	$\bar{x}_2 = 3.2$
$s_1 = 0.4$	$s_2 = 0.6$

Do these sample results indicate that DPT significantly reduces cavities? Assume **unequal** population standard deviations and use $\alpha = 1\%$.

82. The following two hypothesis tests should be conducted using Welch's t-Test. Thus, it is assumed that the populations for each variable are approximately normal, but the population standard deviations are unknown and assumed to be *not* equal. When doing these problems you should do them two ways. **First**, follow the 5-step hypothesis testing procedure and use Welch's approximation for the estimate of the standard error

$$s_{\bar{x}_1-\bar{x}_2} = \sqrt{\frac{s_1^2}{n_1} + \frac{s_2^2}{n_2}}.$$

Then, use that estimate of the standard error when computing the test statistic $t = \frac{(\bar{x}_1 - \bar{x}_2)}{s_{\bar{x}_1-\bar{x}_2}}$. When looking up the critical test statistic, t_c, use the smaller of $n_1 - 1$ and $n_2 - 1$ for the *degrees of freedom*.

Second, use the TI-84 calculator to calculate the testing statistic, p-value and actual degrees of freedom. Recall that when using the TI-84 calculator, the output screen will use the Satterthwaite formula to calculate the actual *degrees of freedom*. Also, when using the calculator, you must select "NO" under the "Pooled" option as indicated in Example 13.9 on page 717.

A marketing professor believes that students between the ages of 18 and 24 that attend local 4-year colleges work less hours per week at their part time jobs than students who attend local 2-year colleges. To test this hypothesis the marketing professor random selects 100 students from the local colleges in her state. Forty-five students were surveyed from local 4-year colleges and fifty-five students were surveyed from local 2-year colleges. The summary statistics for her findings are summarized in the following table:

TABLE: Random Sample of Hours Worked per Week

Population 1 (4-year college students)	Population 2 (2-year college students)
$n_1 = 45$	$n_2 = 55$
$\bar{x}_1 = 21.3 hr/wk$	$\bar{x}_2 = 24.73 hr/wk$
$s_1 = 6.53 hr/wk$	$s_2 = 8.83 hr/wk$

Do these results support the marketing professor's belief that students attending 4-year institutions work fewer hours per week than students attending 2-year institutions? (Use $\alpha = 1\%$)

83. A popular manufacturer for pro-level ice hockey sticks has developed a new composite material that they believe will increase the speed of a player's slapshot. Eight professional ice hockey players had their slapshot speeds measured by first using their current stick, and then again while using the new stick from the manufacturer. The speeds of the slapshots using their current and manufacturer's new sticks were recorded in the following table.

TABLE: Slapshot speed (*mph*)

Player	Slapshot speed using current stick	Slapshot speed using the new stick
1	91	90
2	87	89
3	82	86
4	93	92
5	101	103
6	88	89
7	90	93
8	85	87

Perform a paired t Test to determine if the manufacturer can conclude there is a statistically significant increase in the speed of the player's slapshots when using sticks manufactured with this new composite material. Use $\alpha = 5\%$.

84. A recent study of 35 female Polar bears in Canadian territories found the mean litter size to be 1.63 cubs with a standard deviation of .77 cubs. A recent study of 21 brown bears found that the mean litter size was 2.01 with a standard deviation of .8. Can one conclude that there is a significant difference in the mean litter size between the two species of bears? Test at the .01 alpha level. Assume **equal** population standard deviations and use $\alpha = 1\%$.

For the answer to this question, please visit www.MyStatLab.com.

CHAPTER 14

Chi-Square

Contents

14.1 Introduction
14.2 Properties of the Chi-Square Distribution
14.3 Chi-Square Hypothesis Test of Independence
14.4 Assumptions Underlying the Chi-Square Test
14.5 Test of Goodness-of-Fit
14.6 p-Value Approach to Chi-Square Hypothesis Test of Independence Using the TI-84 Calculator

14.1 INTRODUCTION

A population can be divided many ways. For example, it can be divided according to gender, age group, type of personality, religion, marital status, annual income, political affiliation, or education. Divisions such as these are usually referred to as classifications or factors of the population.

In this chapter, we will present a technique that determines whether two classifications of a population are **independent** (i.e., the classifications are not related) or **dependent** (i.e., the classifications are related). For example,

- are categories of annual income (low, medium, high) and education level dependent? If the classifications of annual income and education level are **dependent**, then one might *expect* to find individuals with a low education level to have a low annual income, and, those with a high education level to have a high annual income. On the other hand, if the classifications are **independent**, then knowledge of one's education level is *not* indicative of that individual's annual income level.
- are movie preference and an individual's gender dependent? If the classifications of movie preference and gender are **dependent** then a relationship between the classifications might be that females generally prefer romance type movies and males generally prefer adventure type movies. Whereas, if the classifications are **independent** then knowing one's gender would *not* give you a clue to their movie preference.
- are male personality types and incidence of heart attack dependent? If the classifications of age and incidence of heart attack are **dependent** then we might expect to find type A personality more susceptible to heart attack than type B personality. If the classifications are **independent** then one's personality type would have no relation to the incidence of heart attack.

All of these examples involve the analysis of **qualitative** data to determine if there is a statistically significant relationship between the population classifications. That is, are the classifications dependent?

A statistical technique used to investigate whether the classifications of a population are related is the **chi-square test of independence**. To apply the chi-square test, a random sample is selected from the population and the sample data is separated into the different categories for each classification of the population. The **frequency** (or number) of responses falling into the distinct categories for each classification are recorded in a table. These sample results are called *observed frequencies*.

Table 14.1 illustrates the observed frequencies for a random sample of 200 co-eds selected from a midwestern university. The sample is classified by gender and movie preference. This table is referred to as a **contingency table**.

> **DEFINITION 14.1**
> A **contingency table** is an arrangement of sample count data into a two-way classification.

TABLE 14.1 Contingency Table

Gender	Romance	Comedy	Drama	Adventure	Mystery	Row Total
Male	ᵃ 10	ᵇ 20	ᶜ 5	ᵈ 40	ᵉ 15	90
Female	ᶠ 50	ᵍ 25	ʰ 5	ⁱ 15	ʲ 15	110
Column Total	60	45	10	55	30	200

Movie Preference

Contingency Table 14.1 has the row classification (i.e., the Gender classification) separated into two categories and the column classification (i.e., the Movie Preference classification) separated into five categories. A contingency table with 2 row classifications and 5 column classifications is referred to as a *2 by 5 contingency table*.

In general, an *r by c contingency table* is a table where the row classification has been separated into **r** rows and the column classification has been separated into **c** columns.

Each individual box in the table is referred to as a **cell**. The number within each cell is the **observed cell frequency**. For example, in cell d of Table 14.1 the observed cell frequency is 40. That is, the number of males within this sample that prefer adventure movies is 40. Similarly, in cell f the observed frequency is 50 which represents the number of females within the sample that prefer a romantic movie.

To determine whether the two classifications, gender and movie preference, are related we will compare the observed frequencies of the random sample to the frequencies we would expect if there is no relationship between the classifications gender and movie preference. If the differences between the **observed** and these **expected frequencies** are *small*, then we will conclude that the two classifications are not related or independent. If the differences between the observed and expected frequencies are *large*, then we will conclude that the two classifications are related or dependent. The statistic used to determine whether the differences between the observed and expected frequencies for all the cells of the contingency table are **large** or **small** is known as **Pearson's chi-square statistic, denoted by** X^2 (where X is the capital Greek letter Chi, pronounced kye which rhymes with hi). Pearson's chi-square statistic is also referred to as the **test statistic**.

The Definition Formula for Computing Pearson's Chi-Square Test Statistic, (X^2)

Pearson's Chi-Square Test Statistic = $\sum_{\text{all cells}} \dfrac{(Observed - Expected)^2}{Expected}$

This formula is symbolized as:

$$X^2 = \sum_{\text{all cells}} \dfrac{(O-E)^2}{E}$$

where: O = Observed Cell Frequency
E = Expected Cell Frequency

If the classifications are *truly independent*, then this would be indicated by a situation where each expected cell frequency, symbolized by E, is *close* or *equal* to its corresponding observed cell frequency, symbolized by O. Consequently, in the case of independence, the numerical value of X^2 would be *small* or *zero*.

When the observed and expected frequencies are the same (i.e., there are *no differences* between the observed and expected frequencies), the value of X^2 is zero. Whereas, *any difference* between the observed and expected frequencies will result in a positive value for X^2. As the differences between the observed and expected frequencies get larger, the value of X^2 will increase.

It then follows that a *large* value of X^2 would tend to indicate a **significant** relationship exists between the two classifications. Thus, we need to determine how *large* a value of X^2 would indicate that the two classifications are dependent. To determine when a value of Pearson's chi-square statistic is considered large, we need to examine a distribution that consists of all the possible values of the chi-square statistic. This distribution is called the sampling distribution of Pearson's chi-square statistic. The sampling distribution of Pearson's chi-square statistic is approximated by the **chi-square distribution**, denoted as χ^2 **distribution** (where χ is the lowercase Greek letter chi, pronounced kye and rhymes with hi). Let us now examine the properties of the **chi-square, χ^2, distribution**.

14.2 PROPERTIES OF THE CHI-SQUARE DISTRIBUTION

Figure 14.1 illustrates a typical chi-square distribution.

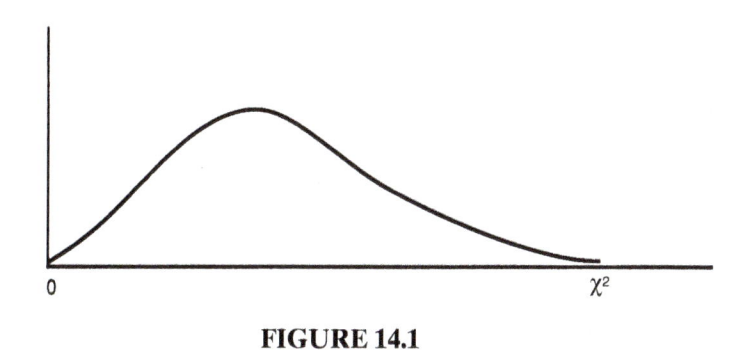

FIGURE 14.1

Properties of the Chi-Square Distribution

1. The value of χ^2 is always non-negative; it is zero or positively valued.
2. The graph of a chi-square distribution is *not symmetric*, it is *skewed* to the right and extends infinitely to the right of zero.
3. There is a family of chi-square distributions similar to the family of t distributions as illustrated in Figure 14.2. Each chi-square distribution is dependent upon the number of degrees of freedom, *df*, determined by the formula:

$$df = (r-1)(c-1),$$

where: r represents the number of row classifications within the contingency table, and
        c represents the number of column classifications within the contingency table.

As the number of degrees of freedom, *df*, increases, the shape of the chi-square distribution approaches the shape of the normal distribution.

Various Chi-Square Distributions

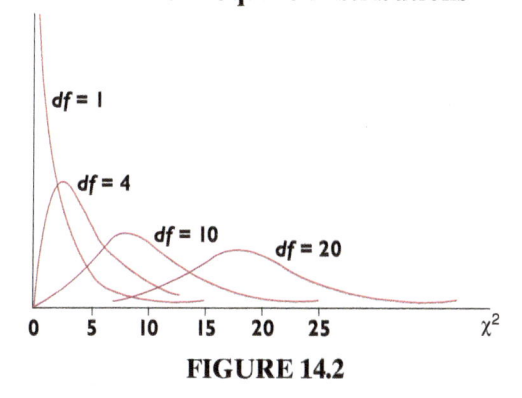

FIGURE 14.2

14.3 CHI-SQUARE HYPOTHESIS TEST OF INDEPENDENCE

Let's apply the general five step hypothesis testing procedure to test the independence of two population classifications. This hypothesis testing procedure is called the *Chi-Square Test of Independence*.

Step 1 Formulate the two Hypotheses, H_o and H_a.
These hypotheses have the following form:
Null Hypothesis Form
H_o: The two population classifications are **independent**.

This means that the population classifications are *not* related.
Alternative Hypothesis Form
H_a: The two population classifications are **dependent**.

This means that the population classifications *are* related.

By stating the null hypothesis as "the two population classifications are independent" we are making the assumption that there is *no relationship* between the classifications. Consequently, by *rejecting* the null hypothesis we can conclude that the two population classifications are *dependent*. This infers that there is a relationship between the two population classifications.

Step 2 Determine the model to test the null hypothesis, H_o.
The chi-square distribution with $df = (r-1)(c-1)$ is the hypothesis testing model (see Figure 14.2). Under the assumption that H_o is true, calculate the expected cell frequencies. To calculate the expected frequency for a particular cell, we multiply its row total (RT) by its column total (CT) and divide this product by the sample size (n).

Expected Frequency Cell Formula

The expected frequency formula for a particular cell is:

$$\text{expected cell frequency} = \frac{(RT)(CT)}{n}$$

where:
- RT is the row total for the cell
- CT is the column total for the cell
- n is the sample size

Step 3 Formulate the decision rule.
All chi-square tests of independence will be one-tailed tests on the right, since a significant Pearson's X^2 statistic will always be a large positive number.

To determine the critical chi-square value necessary to formulate the decision rule, we perform the following steps:

Procedure to Formulate the Decision Rule

a) Identify the significance level, $\alpha = 1\%$ or $\alpha = 5\%$.
b) Calculate the degrees of freedom, *df*, using:
df = (number of rows minus one)(number of columns minus one)
In symbols, $df = (r-1)(c-1)$.
c) Determine the **critical χ^2 value, denoted χ^2_α**, using Table IV: Critical Values for the Chi-Square Distribution, χ^2_α, found in Appendix C: Statistical Tables, and draw a chi-square distribution model with the χ^2_α value appropriately placed on the curve. This hypothesis testing model is illustrated in Figure 14.3.
d) State the decision rule.
The decision rule is: reject H_o if X^2 test statistic is greater than χ^2_α.

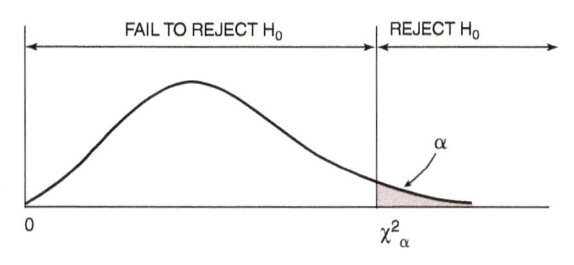

FIGURE 14.3

Step 4 Analyze the sample data.
Calculate Pearson's X^2 test statistic.
There are two equivalent formulas, the definition formula and the computational formula, that are used to calculate Pearson's X^2 statistic. Let's examine each of these two formulas for the X^2 statistic.

Pearson's Chi-Square Test Statistic Formulas

The definition formula for Pearson's chi-square test statistic is:

$$X^2 = \sum_{\text{all cells}} \frac{(O-E)^2}{E}$$

The computational formula for Pearson's chi-square test statistic is:

$$X^2 = \sum_{\text{all cells}} \left(\frac{O^2}{E}\right) - n$$

where: O represents the observed frequency of a particular cell,
E represents the expected frequency for the same cell,
n represents the sample size.

Step 5 State the conclusion.
Compare the value of Pearson's chi-square test statistic, X^2, to the critical value of chi-square, χ^2_α, and choose the appropriate conclusion.

If the value of the X^2 test statistic is greater than the critical value of chi-square, χ^2_α, then the conclusion is:

Reject H_o and Accept H_a at α and we would conclude that there is a significant relationship between the two population classifications. This means that the two population classifications are dependent.

Similarly, if the value of the X^2 test statistic is less than or equal to the critical value of chi-square, χ^2_α, then the conclusion is:

Fail to Reject H_o at α and we would conclude that there is no significant relationship between the two population classifications. This means that the two population classifications are independent.

In summary, we have:

Conclusions for the Chi-Square Test of Independence

(a)
If $X^2 > \chi^2_\alpha$, then:
Reject H_o
and
Accept H_a at α

(b)
If $X^2 \leq \chi^2_\alpha$, then:
Fail to Reject H_o
at α

These conclusions are illustrated in Figure 14.4.

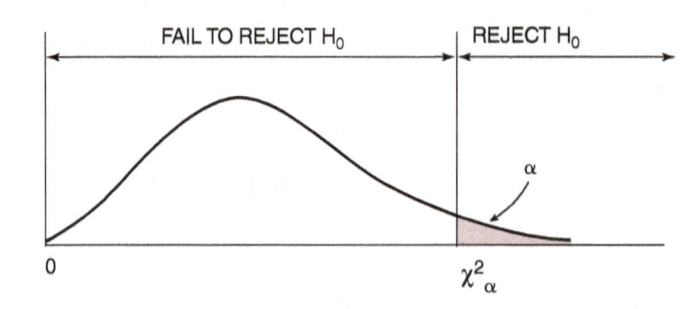

FIGURE 14.4

EXAMPLE 14.1

A study was conducted at the 1% level of significance to determine if there is a relationship between gender and one's view on capital punishment. A random sample of 100 adults was selected and asked their opinion on the question:

"Do you believe that the death penalty should be given to those convicted of first degree murder?"

The sample results which represent the observed frequencies are summarized in Table 14.2.

TABLE 14.2 Table of Observed Frequencies

Gender	View on Capital Punishment			Row Totals
	Favor	Against	Undecided	
Male	[a] 20	[b] 25	[c] 15	60
Female	[d] 10	[e] 25	[f] 5	40
Column Totals	30	50	20	100

Based on these sample results can we conclude that there is a relationship between gender and one's view on capital punishment at $\alpha = 1\%$?

Solution

Step 1 Formulate the two hypotheses, H_o and H_a.

H_o: The population classifications gender and one's view on capital punishment are **independent.**

H_a: The population classifications gender and one's view on capital punishment are **dependent.**

Step 2 Determine the model to test the null hypothesis, H_o.
The hypothesis testing model is a chi-square distribution with degrees of freedom:
$df = (r-1)(c-1)$.
 For this contingency table the row classification of **gender** has a row value of 2; the column classification of **view on capital punishment** has a column value of 3.

$$df = (2-1)(3-1)$$
$$df = 2$$

Thus, this 2 by 3 contingency table has **two degrees of freedom.**
 Thus, the chi-square distribution with $df = 2$ is the hypothesis testing model shown in Figure 14.5. Table 14.2 is a contingency table of the observed frequencies. *To test H_o, the expected frequencies for each cell must be computed.* To compute these expected frequencies, we assume H_o is true. That is, we assume that the classifications of gender and one's view on capital punishment are independent. This

Section 14.3 Chi-Square Hypothesis Test of Independence

FIGURE 14.5 Chi-Square Distribution with df = 2

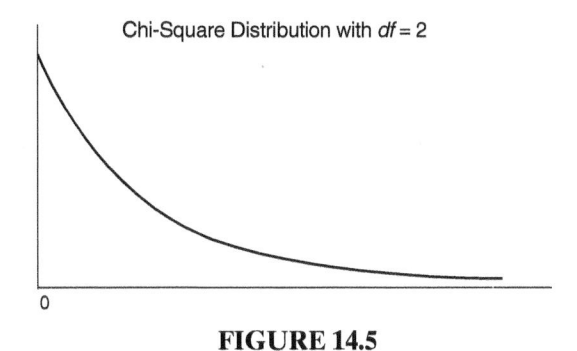

means that since 30 of the 100 people sampled favor capital punishment then we would expect 30% of the 60 males to favor capital punishment. Similarly, we would also expect 30% of the 40 females to favor capital punishment. Therefore, the expected number of males favoring capital punishment is

$$\frac{(30)}{(100)}(60) = 18$$

and the expected number of females favoring capital punishment is

$$\frac{(30)}{(100)}(40) = 12$$

Table 14.3 represents the *expected frequencies* for each of the cells a through f. Notice that the value in cell (a) in Table 14.3 is 18. This cell value represents the number of males that are *expected* to favor capital punishment. Also the value in cell (d) in Table 14.3 is 12. This cell value represents the number of females that are expected to favor capital punishment.

Each of the remaining cell values are obtained based upon the assumption that there is no relationship between gender and one's view on capital punishment. The expected frequency for a particular cell may be calculated by multiplying its row total by its column total and then dividing the product by the sample size. Therefore, the value in cell (a), the expected frequency for cell (a), could have been obtained by the formula:

$$\text{expected frequency for cell a} = \frac{\left(\begin{array}{c}\text{row total}\\\text{for cell a}\end{array}\right)\left(\begin{array}{c}\text{column total}\\\text{for cell a}\end{array}\right)}{\text{sample size}}$$

$$= \frac{(60)(30)}{100}$$

$$= 18$$

Therefore, each of the remaining expected cell values can be obtained using the formula:

$$\text{expected frequency for a cell} = \frac{\left(\begin{array}{c}\text{row total}\\\text{for the cell}\end{array}\right)\left(\begin{array}{c}\text{column total}\\\text{for the cell}\end{array}\right)}{\text{sample size}}$$

TABLE 14.3   Table of Expected Frequencies View on Capital Punishment

Gender	View on Capital Punishment			
	Favor	Against	Undecided	Row Totals
Male	[a] 18	[b] 30	[c] 12	60
Female	[d] 12	[e] 20	[f] 8	40
Column Totals	30	50	20	100

The observed and expected frequencies are conveniently summarized in Table 14.4. In each cell, the expected frequency is shown in the upper right corner and the observed cell frequency is in the center of the cell.

TABLE 14.4 Table of Observed and Expected Frequencies View on Capital Punishment

Gender	View on Capital Punishment			Row Totals
	Favor	Against	Undecided	
Male	[a]20 18	[b]25 30	[c]15 12	60
Female	[d]10 12	[e]25 20	[f]5 8	40
Column Totals	30	50	20	100

Step 3 Formulate the decision rule.
  a) The significance level is $\alpha = 1\%$.
  b) The degrees of freedom is:

$$df = 2$$

  Thus, this 2 by 3 contingency table has *two degrees of freedom*.
  c) The critical value, χ^2_α, is found in Table IV. For $\alpha = 1\%$ and $df = 2$,

$$\chi^2_\alpha = 9.21$$

  The hypothesis testing model is a chi-square distribution with $df = 2$ and $\chi^2_\alpha = 9.21$.
  d) The decision rule is: reject H_o if X^2 test statistic is greater than 9.21. This is illustrated in Figure 14.6.

Step 4 Analyze the sample data.
Calculate the X^2 test statistic. For this example we will use the definition formula to calculate Pearson's chi-square test statistic. The definition formula is:

$$X^2 = \sum_{\text{all cells}} \frac{(O-E)^2}{E}$$

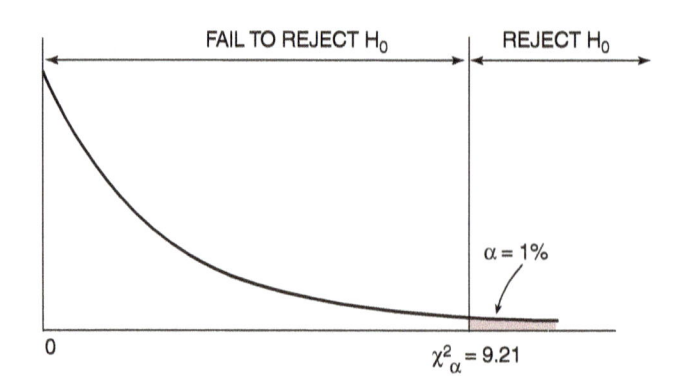

FIGURE 14.6

Table 14.5 has been constructed to aid in the computation of this formula. The observed and expected cell frequency values were obtained from Table 14.4.

TABLE 14.5

Cell	Observed Frequency (O)	Expected Frequency (E)	(O – E)	(O – E)²	$\frac{(O-E)^2}{E}$
a	20	18	2	4	0.22
b	25	30	–5	25	0.83
c	15	12	3	9	0.75
d	10	12	–2	4	0.33
e	25	20	5	25	1.25
f	5	8	–3	9	1.13

$$\sum_{\text{all cells}} \frac{(O-E)^2}{E} = 4.51$$

To determine the X^2 test statistic sum the values in the last column labeled:

$$\frac{(O-E)^2}{E}.$$

Thus, the test statistic is:

$$X^2 = 4.51$$

Pearson's X^2 test statistic could have been calculated using the computational formula:

$$X^2 = \sum_{\text{all cells}} \frac{O^2}{E} - n$$

We will illustrate the use of the computational formula in Example 14.2.

Step 5 State the conclusion.
Since X^2 test statistic value of 4.51 is less than χ^2_α value of 9.21 we *fail to reject the null hypothesis at* $\alpha = 1\%$. Thus, there is no statistically significant relationship between gender and one's view on capital punishment. (i.e., the two classifications are independent). This is illustrated in Figure 14.7.

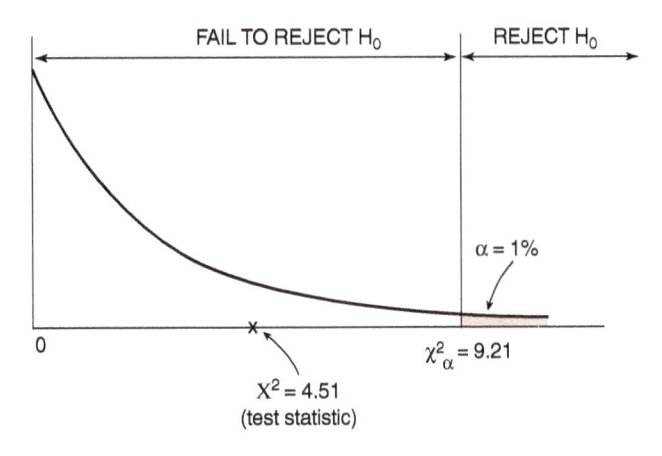

FIGURE 14.7

EXAMPLE 14.2

A study was conducted to determine if there is a significant relationship between the frequency of times meat is served as a main meal per month for individuals living in the eastern, central, and western United States.

A questionnaire was administered to a random sample of 300 families and the results are summarized in Table 14.6.

Do these sample results indicate that there is a significant relationship between the frequency of meat served as a main meal per month and geographic living area at $\alpha = 5\%$?

TABLE 14.6 Table of Observed Frequencies

Living Area	Frequency of Meat as Main Meal per Month			
	Less than 11	11 to 20	More than 20	Row Totals
Eastern US	[a] 46	[b] 36	[c] 20	102
Central US	[d] 40	[e] 40	[f] 26	106
Western US	[g] 16	[h] 38	[i] 38	92
Column Totals	102	114	84	300

Solution

Step 1 Formulate the two hypotheses, H_o and H_a.

H_o: The population classifications frequency of meat served as a main meal per month and geographic living area are independent.

H_a: The population classifications frequency of meat served as a main meal per month and geographic living area are dependent.

Step 2 Determine the model to test the null hypothesis, H_o.
The hypothesis testing model is a chi-square distribution with degrees of freedom: $df = (r-1)(c-1)$. The degrees of freedom for a 3 by 3 contingency table is:

$$df = (3-1)(3-1)$$
$$df = 4$$

Thus, this 3 by 3 contingency table has **four degrees of freedom**.
The chi-square distribution with $df = 4$ is the hypothesis testing model as shown in Figure 14.8.

chi-square distribution with $df = 4$

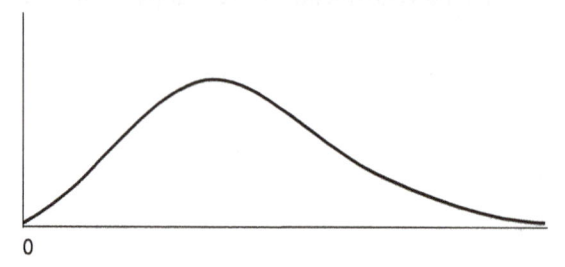

FIGURE 14.8

The results of the sample are summarized in Table 14.6. To calculate the chi-square statistic, an expected frequency table representing the expected frequency for each cell must be developed. The expected frequency table is based upon the assumption that the null hypothesis is true. Each of the cell entries is found by the formula:

$$\text{expected frequency for a cell} = \frac{\begin{pmatrix}\text{row total}\\ \text{for the cell}\end{pmatrix}\begin{pmatrix}\text{column total}\\ \text{for the cell}\end{pmatrix}}{\text{sample size}}$$

The expected and observed cell frequencies are summarized in Table 14.7.

TABLE 14.7 Table of Observed and Expected Frequencies

Living Area	Frequency of Meat as Main Meal per Month			
	Less than 11	11 to 20	More than 20	Row Totals
Eastern US	a 46 34.68	b 36 38.76	c 20 28.56	102
Central US	d 40 36.04	e 40 40.28	f 26 29.68	106
Western US	g 16 31.28	h 38 34.96	i 38 25.76	92
Column Totals	102	114	84	300

Step 3 Formulate the decision rule.
   a) The significance level is $\alpha = 5\%$.
   b) The degrees of freedom, df, is:

$$df = 4$$

   c) Using Table IV, the critical χ_α^2 value for $\alpha = 5\%$ and $df = 4$ is:

$$\chi_\alpha^2 = 9.49$$

The hypothesis testing model is a chi-square distribution with $df = 4$ and $\chi_\alpha^2 = 9.49$.
d) The decision rule is: reject H_o if X^2 test statistic is greater than 9.49.
Figure 14.9 illustrates the decision rule for this hypothesis test.

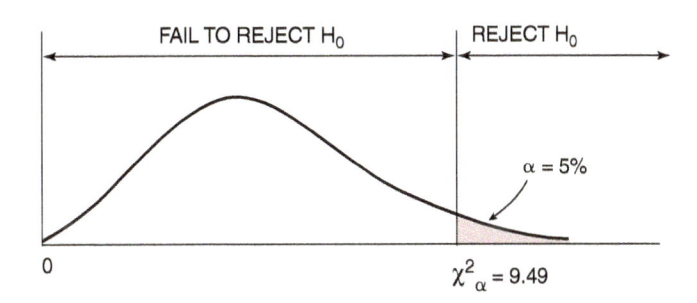

FIGURE 14.9

Step 4 Analyze the sample data.
Calculate X^2 test statistic.
For this example we will use the computational formula for Pearson's chi-square test statistic.

$$X^2 = \sum_{\text{all cells}} \left(\frac{O^2}{E}\right) - n$$

Table 14.8 has been constructed to aid in the computation of this formula.

TABLE 14.8

Cell	O	O^2	E	$\dfrac{O^2}{E}$
a	46	2116	34.68	61.01
b	36	1296	38.76	33.44
c	20	400	28.56	14.01
d	40	1600	36.04	44.40
e	40	1600	40.28	39.72
f	26	676	29.68	22.78
g	16	256	31.28	8.18
h	38	1444	34.96	41.30
i	38	1444	25.76	56.06

$n = 300$ $\qquad \sum\limits_{\text{all cells}} \dfrac{(O^2)}{E} = 320.90$

To compute X^2 test statistic using the computational formula:

$$X^2 = \sum_{\text{all cells}} \frac{(O^2)}{E} - n$$

1) Sum the values in the last column labeled

$$\frac{(O^2)}{E}$$

in Table 14.8. Thus,

$$\sum_{\text{all cells}} \frac{(O^2)}{E} = 320.90$$

2) The sample size, n, is 300. Therefore,

$$X^2 = \sum_{\text{all cells}} \frac{(O^2)}{E} - n$$
$$= 320.90 - 300$$
$$X^2 = 20.90$$

Step 5 State the Conclusion.

Since X^2 test statistic value of 20.90 is greater than the χ_α^2 value of 9.49, we reject H_o and accept H_a at $\alpha = 5\%$. These results are illustrated in Figure 14.10.

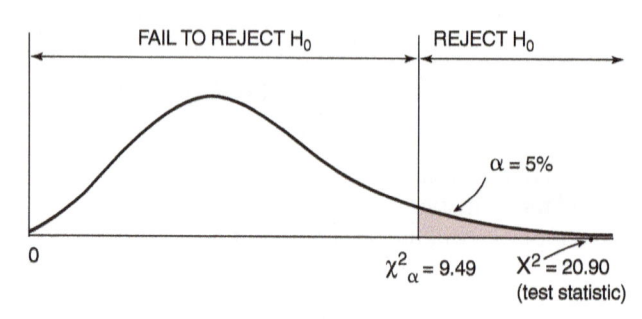

FIGURE 14.10

Thus, there is a **statistically significant** relationship between the frequency of times meat is served as a main meal and one's geographic living location at $\alpha = 5\%$.

CASE STUDY 14.1 Music That Makes You Dumb

Examine the graphic within the statistical snapshot Figure 14.11 entitled "Music That Makes You Dumb". This graphic was created by Mr. Virgil Griffith, a grad student from Cal Tech, who researched the relationship between musical tastes and the SAT scores of people who listen to that type of music. To conduct his research, Mr. Griffith used Facebook statistics to identify the favorite music preferences of the students who attend a particular college and the average SAT/ACT scores for the students attending that college. Based on this information, Mr. Griffith generated the graphic entitled "Music That Makes You Dumb".

a) What data values did Mr. Griffith plot along the horizontal axis of the graphic?
b) Using the graphic's legend, what do the colored boxes within the graphic signify?
c) If we wanted to use a Chi-Square Test of Independence to test whether there is a relationship between the music preferences of college students who use Facebook and the average SAT/ACT scores for the students who attend those colleges, then state the two population classifications you might use to conduct this hypothesis test.
d) If you were to perform such a Chi-Square Hypothesis Test of Independence, then how would you state the null hypothesis?
e) How would you state the alternative hypothesis?
f) When the null hypothesis is assumed to be true, what does this indicate about the relationship between the two population classifications?
g) In conducting a Chi-Square Test of Independence to determine whether there is a relationship between the music preferences of college students who use Facebook and the average SAT/ACT scores for the students who attend those colleges, then what distinct categorical sub-classifications might you use to represent the population classification Music Preferences?
h) In conducting a Chi-Square Test of Independence to test whether there is a relationship between the music preferences of college students who use Facebook and the average SAT/ACT scores for the students who attend those colleges, then what distinct three categorical sub-classifications might you use to represent the population classification SAT/ACT scores of the college students as qualitative data?
i) Based on your answers to the *previous questions g* and *h*, what would be the size of the Contingency Table?

Section 14.3 *Chi-Square Hypothesis Test of Independence*

CASE STUDY 14.1   Music That Makes You Dumb (continued)

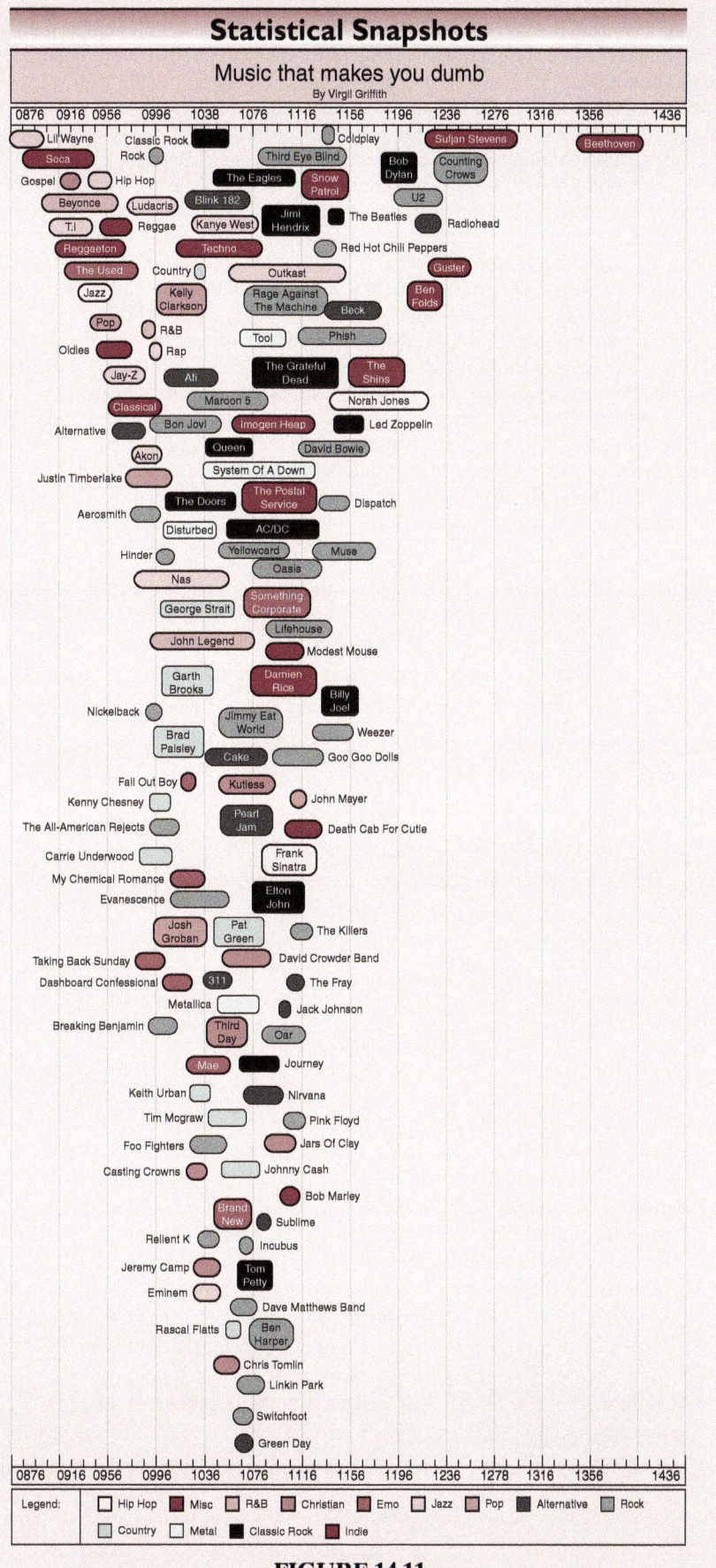

FIGURE 14.11
Reprinted by permission of Virgil Griffith.

(continues)

CASE STUDY 14.1 Music for the Ages (continued)

j) Using your answer to the *previous question i*, what would be the degrees of freedom for this Chi-Square Test?

k) Determine the critical Chi-Square value if the test were conducted at a significance level of 5%? At 1%?

l) When you examine the graphic, you should notice that the students with the highest SAT/ACT scores (average SAT of 1371) preferred to listen to Beethoven while the students with the lowest SAT scores (average SAT of 889) preferred to listen to Lil' Wayne. Do you believe Mr. Griffith was trying to determine if there is a relationship between a college student's musical taste and intelligence? Explain.

m) There were several journalists who were upset with the title "Music That Makes You Dumb" that Mr. Virgil Griffith used for his research graphic. One journalist suggested that the proper title should have been "Music Preferences of Facebook-Using College Students Based on School Average SAT/ACT Scores". What is your opinion and explain your reasoning?

n) Mr. Griffith was trying to determine if intelligence is related to a student's taste in music. Can you think of any other qualitative factors that might influence a student's taste in music?

o) If you were considering a follow-up study to Mr. Griffith's research and wanted to show that a different factor other than intelligence might influence a student's musical taste, would you consider using a student's personality as a population classification rather than a student's intelligence? Explain.

p) Would any of these factors be reasonable to consider: a student's race? A student's ethnicity? A student living in a particular region within the U.S.? A student's political party affiliation?

q) If you wanted to consider the continuous data such as a person's age or a person's IQ score, how might you consider transforming this data into qualitative categories to conduct the Chi-Square test of Independence? Do you believe musical taste might be related to age or an IQ score? Explain.

Summary of the Test Statistic Approach to the Chi-Square Hypothesis Test of Independence

The following outline is a summary of the 5 step hypothesis testing procedure for The Chi-Square Test of Independence.

Step 1 Formulate the two hypotheses, H_o and H_a.

H_o: The two population classifications are independent.

H_a: The two population classifications are dependent.

Step 2 Determine the model to test the null hypothesis, H_o.

The chi-square distribution with the $df = (r-1)(c-1)$ is the hypothesis testing model.

Under the assumption that H_o is true, calculate the expected cell frequencies using the formula:

$$\text{expected cell frequency} = \frac{(RT)(CT)}{n}$$

Step 3 Formulate the decision rule.

All chi-square tests of independence will be one-tailed tests on the right, since a significant Pearson's X^2 test statistic will always be a large positive number. To determine the critical chi-square value, perform the following steps:

a) Identify the significance level, $\alpha = 1\%$ or $\alpha = 5\%$.

b) Calculate the degrees of freedom: $df = (r-1)(c-1)$.

c) Determine the critical χ^2 value, denoted χ^2_α, using Table IV: Critical Values for the Chi-Square Distribution, χ^2_α, found in Appendix C: Statistical Tables, and draw a chi-square distribution model with the χ^2_α value appropriately placed on the curve.

d) State the decision rule.

The decision rule is: reject H_o if X^2 test statistic is greater than χ^2_α.

Step 4 Analyze the sample data.

Calculate Pearson's X^2 test statistic, using either of the following formulas:

$$X^2 = \sum_{\text{all cells}} \frac{(O-E)^2}{E} \quad \text{or} \quad X^2 = \sum_{\text{all cells}} \frac{(O^2)}{E} - n$$

> **Summary of the Test Statistic Approach to the**
> **Chi-Square Hypothesis Test of Independence (continued)**
>
> Step 5   State the conclusion.
> Compare the value of Pearson's chi-square test statistic, X^2, to the critical value of chi-square, χ^2_α, and either:
> a) Reject H_o and Accept H_a at α, if X^2 is greater than the critical value of chi-square, χ^2_α.
> or
> b) Fail to Reject H_o at α, if X^2 is less than or equal to the critical value of chi-square, χ^2_α.

14.4 ASSUMPTIONS UNDERLYING THE CHI-SQUARE TEST

When using the χ^2 distribution to perform a hypothesis test, several conditions must be satisfied. These conditions are referred to as the **assumptions underlying the chi-square test.** Let's now discuss these assumptions.

> **Assumptions Underlying the Chi-Square Test**
>
> - Each of the observations must be **independent of all other observations**.
> - Each observation must be **classified into one and only one cell**. Care must be exercised to avoid designing classifications that allow an observation to fall into more than one cell.
> - Perhaps the most important condition necessary to perform a χ^2 test is that the sample size, n, be sufficiently large. To be sufficiently large means that the **smallest expected cell frequency is at least five**.

14.5 TEST OF GOODNESS-OF-FIT

Another important application of the chi-square distribution is the determination of how well sample data fits a particular theoretical probability model. This type of Chi-Square Test is referred to as a **Goodness-of-Fit Test**.

The Goodness-of-Fit Test is used to determine whether a particular population can be described by a known theoretical distribution model. Throughout the text, we have always assumed that the distribution we were working will fit a particular theoretical distribution. For example, in several chapters, we assumed that certain distributions were normal. Using the Goodness-of-Fit Test, we can use the chi-square distribution to determine whether a particular distribution is approximately normal by comparing a random sample of data selected from the distribution to a theoretical normal distribution.

The chi-square test statistic is used to compare the sample data to the theoretical distribution model, like a normal distribution, to see if the model provides a "good fit" for the data. This type of statistical test is called a Goodness-of-Fit Test. The procedure to perform a Goodness-of-Fit Test involves the same five step hypothesis testing procedure that we've used in the preceding chapters with a few modifications. Example 14.3 outlines the procedure for a Goodness-of-Fit Test.

EXAMPLE 14.3

A market researcher wants to determine if the color of a clients' packaging material is an important factor when a consumer is selecting their product. To answer this question the researcher selects 800 consumers at random and gives them four different colored packages of the same product. However, the consumers are told that the four packages have different chemical properties and that the different color wrappers are to be used only for identification. Six weeks later each person in the sample is asked to complete an order form indicating which one of the four packages they preferred. To insure that the consumer selects the color they truly prefer, each consumer will be sent a free months supply of the product of their choice. The results of this survey are given in Table 14.9.

TABLE 14.9 Survey Results

Color of Wrapper	Pink	Blue	Brown	Grey
Number of Consumers Selecting Wrapper	250	300	100	150

Remark: In this example the market researcher is trying to determine if the consumer prefers a particular colored package. If color is *not* a factor in the consumer's selection then we would expect an *equal* number of consumers choosing each color package. Whenever a statistician expects an equal number of observations in each category, the theoretical distribution model is called the **uniform distribution**. The Goodness-of-Fit test can now be used to determine if the observed data in Table 14.9 fits the theoretical uniform distribution.

Solution

We will modify the five step hypothesis testing procedure used to perform a chi-square test of independence to outline the Goodness-of-Fit test.

Step 1 Formulate the two hypotheses H_o and H_a.
In a Goodness-of-Fit test, the **null hypothesis** is stated as:

Null Hypothesis Form

H_o: the sample data fits a particular distribution

In this example we are trying to determine if there is an equal preference for each color. Therefore, the null hypothesis, H_o, is stated:

H_o: The sample data fits a uniform distribution.

In a *Goodness-of-Fit* test, the **alternative hypothesis** is stated as:

Alternative Hypothesis Form

H_a: the sample data does *not* fit the particular distribution

---

If color is a significant factor in consumer selection of this product then the alternative hypothesis, H_a, is:

H_a: The sample data does not fit a uniform distribution.

Step 2 Determine the model to test the null hypothesis, H_o
The chi-square distribution with $df = k-1$ is the hypothesis testing model.
In this example the sample data is only classified by color. This is referred to as a single classification case. **In a Goodness-of-Fit test, the degrees of freedom are dependent upon the number of categories within a single classification case.**

The Degrees of Freedom, *df*, for a Single Classification Case

$$df = k - 1$$

where: k equals the number of categories within the single classification.

Since there are four categories of color, then k = 4.

$$\text{Thus, } df = 4 - 1$$
$$df = 3$$

Thus, the chi-square distribution with $df = 3$ is the hypothesis testing model as shown in Figure 14.12.

chi-square distribution with $df = 3$

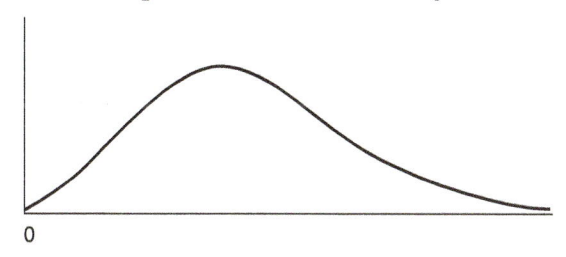

FIGURE 14.12

To calculate the expected frequency for each cell, we assume the null hypothesis to be true. Thus **if we assume that the sample data fits a uniform distribution, then all the expected frequencies would be equal**. Thus, the expected frequency for each cell of the four cells is 200, since 800 consumers responded to the survey. The observed and expected frequencies are summarized in Table 14.10.

TABLE 14.10 Observed and Expected Frequencies

Color of Wrapper	Pink	Blue	Brown	Grey	Total
Number of Consumers Selecting Wrapper	a 250 200	b 300 200	c 100 200	d 150 200	800

Step 3 Formulate the decision rule.
  a) In this example we will use a significance level $\alpha = 1\%$.
  b) In this example the sample data is only classified by color.

Since there are four categories of color, then k = 4.

$$\text{Thus, } df = 3$$

  c) The critical χ^2 value for this test can be found in Table IV where $\alpha = 1\%$ and $df = 3$. The critical χ^2 value is:

$$\chi^2_\alpha = 11.34$$

  d) The decision rule is: reject H_o if Pearson's X^2 test statistic is greater than 11.34. The hypothesis testing model is illustrated in Figure 14.13.

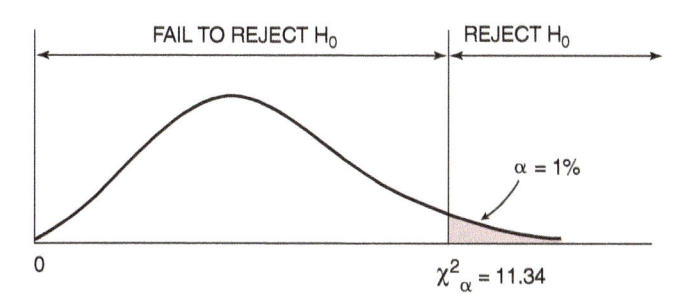

FIGURE 14.13

Step 4 Analyze the sample data.
Calculate the X^2 test statistic using the definition formula:

$$X^2 = \sum_{\text{all cells}} \frac{(O-E)^2}{E}$$

Table 14.11 is used for the computation of the X^2 test statistic.

TABLE 14.11

Cell	Observed Frequency (O)	Expected Frequency (E)	(O – E)	(O – E)²	$\frac{(O-E)^2}{E}$
a	250	200	50	2500	12.50
b	300	200	100	10000	50.00
c	100	200	–100	10000	50.00
d	150	200	–50	2500	12.50
					$X^2 = 125.00$

Step 5 State the conclusion.
Since the X^2 test statistic of 125.00 is greater than χ_α^2 value of 11.34, we reject H_o and accept H_a at $\alpha = 1\%$. **Thus the sample data does not fit a uniform distribution.** This conclusion is illustrated in Figure 14.14.

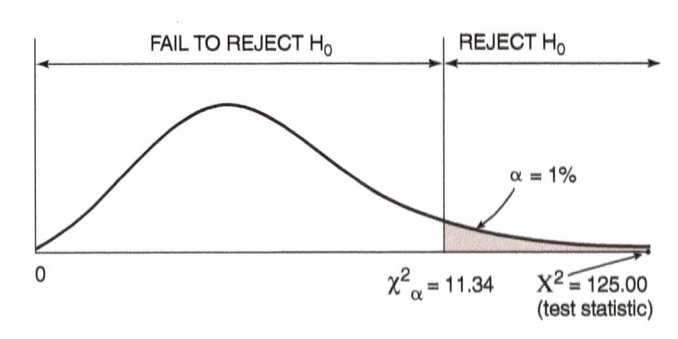

FIGURE 14.14

Thus, color is a significant factor in consumer selection of this product.

EXAMPLE 14.4

Table 14.12 indicates the proportion of viewers per network that watched the six o'clock news last year.

This year a random sample of 400 viewers was conducted to determine if the number of viewers watching the six o'clock news on each of the networks had significantly changed from last years results. The results of the sample data is listed in Table 14.13.

TABLE 14.12 Percent of Viewers

Network	Percent of Viewers
ABC	30
CBS	25
NBC	30
Independent	15

TABLE 14.13 Sample of Viewers

Network	Number of Viewers
ABC	110
CBS	95
NBC	130
Independent	65

Based on the results of this sample data can we conclude that this year's viewing audience has significantly changed from last year's? Use $\alpha = 5\%$.

Solution

The Goodness-of-Fit Test will be used to test whether this year's sample data fits last year's viewing distribution.

Step 1 Formulate the two hypotheses H_o and H_a.

H_o: The sample data fits last year's viewing distribution.

H_a: The sample data does not fit last year's viewing distribution.

Step 2 Determine the model to test the null hypothesis, H_o.

Thus, the chi-square distribution with $df = 3$ is the hypothesis testing model as shown in Figure 14.15.

Chi-Square Distribution with $df = 3$

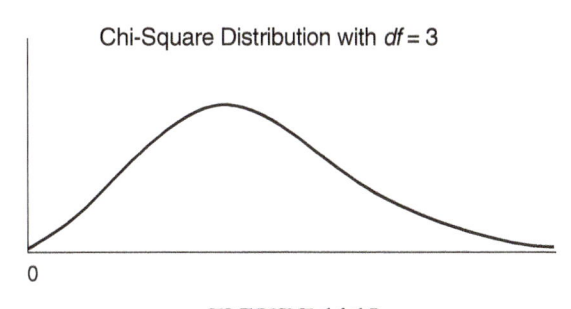

FIGURE 14.15

In this example the sample data is only classified by network. The degrees of freedom, df, for a single classification case is $df = k - 1$, where k is the number of categories within the single classification. Since there are four networks, then k = 4.

$$\text{Thus, } df = 4 - 1$$
$$df = 3$$

To calculate the expected frequency for each cell, we assume the null hypothesis to be true. Thus if we assume that the sample data fits last year's viewing distribution, then the expected frequencies are calculated from last year's viewing audience. For example, 30 percent of last year's viewing audience watched the six o'clock news on the ABC network. Thus, for this year's sample of 400 viewers, we would expect that 30 percent of the viewers to have watched the six o'clock news on the ABC network. Therefore we would expect 120 viewers to watch ABC's six o'clock news. The expected and the observed frequencies are summarized in Table 14.14.

TABLE 14.14 Observed and Expected Frequencies

Network	ABC	CBS	NBC	Independent	Total
Number of Viewers	a 110 120	b 95 100	c 130 120	d 65 60	400

Step 3 Formulate the decision rule.
   a) The significance level is $\alpha = 5\%$.
   b) In this example the sample data is only classified by network. Since there are four networks, then k = 4.

$$\text{Thus, } df = 3$$

   c) The critical χ^2 value for this test can be found in Table IV where $\alpha = 5\%$ and $df = 3$. The critical χ^2 value is:

$$\chi^2_\alpha = 7.82$$

Chapter 14 Chi-Square

d) The decision rule is: reject H_o if the X^2 test statistic is greater than 7.82. The hypothesis testing model is illustrated in Figure 14.16.

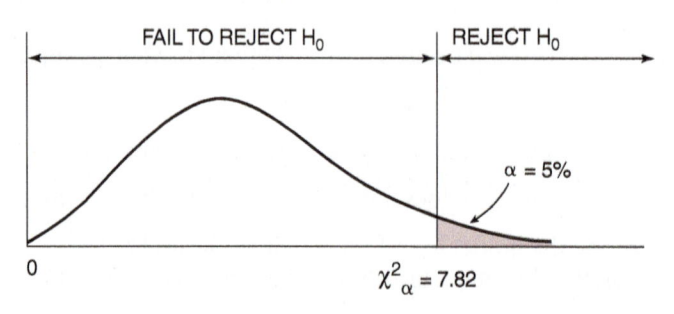

FIGURE 14.16

Step 4 Analyze the sample data.
Calculate the X^2 test statistic using the definition formula:

$$X^2 = \sum_{\text{all cells}} \frac{(O-E)^2}{E}$$

Table 14.15 is used for the computation of the X^2 statistic.

TABLE 14.15

Cell	Observed Frequency (O)	Expected Frequency (E)	(O − E)	(O − E)²	$\frac{(O-E)^2}{E}$
a	110	120	−10	100	0.83
b	95	100	−5	25	0.25
c	130	120	10	100	0.83
d	65	60	5	25	0.42
					$X^2 = 2.33$

Step 5 State the conclusion.
Since the X^2 test statistic (2.33) is less than χ^2_α (7.82) we fail to reject H_o at $\alpha = 5\%$. This conclusion is illustrated in Figure 14.17.

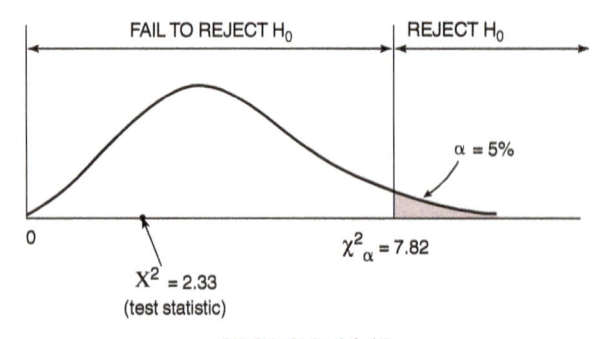

FIGURE 14.17

Thus, we can conclude that this year's viewing distribution has not significantly changed from last year's.

Section 14.5 Test of Goodness-of-Fit   751

CASE STUDY 14.2 Special K

A research article entitled "The Rekurring Kase of the Special K" examined whether the letter K occurs with more frequency as the initial letter in top brand names than it does as a first letter in our language in general. This study appeared in the October/November 1990 issue of the *Journal of Advertising Research*. One aspect of this research was to examine the eight most frequently occurring initial letters were: C, S, A, B, M, P, D and T in top name brands to determine if they occur more frequently as the initial letter in top brand names than they appear as a first letter in our language in general.

These eight most frequently occurring letters were selected by the researchers from a composite list of unduplicated top brand names. From this list, only unique names were selected. Using only unique names eliminated over-representation of certain letters. The expected frequency for each of these letters were based on *Webster's Ninth New Collegiate Dictionary* (1987) model. The result of this chi-square test is shown in Figure 14.18. The letter K does not appear on this list since it was not among the most frequent occurring initial letters.

Chi-Square Analysis of Most Frequently Occurring Initial Letters of Top Brand Names

Initial letter	Observed frequency	Expected* frequency	Cumulative percent Observed	Cumulative percent Expected
C	51	52.4	9.43	9.68
S	46	67.3	17.93	22.12
A	42	32.2	25.69	28.08
B	42	29.2	33.45	33.47
M	40	29.5	40.84	38.92
P	40	47.5	48.23	47.70
D	30	28.2	53.78	52.91
T	29	32.0	59.14	58.83

X^2 = 20.677   df = 7   p < .005

*Based on Webster's Ninth New Collegiate Dictionary (1987), distribution of words beginning with each letter.

FIGURE 14.18
Republished with permission of the *Journal of Advertising Research*.

The second aspect of this research was to determine what happens to the analysis when the letter K is added to this list. This aspect of the hypothesis test will be examined in the: What Do You Think? question set of the exercise section at the end of this chapter.

a) Define the population for this test.
b) What variable is being examined within this test?
c) Is this variable numeric or categorical? Explain.
d) What type of chi-square test is being illustrated in this research?
e) What distribution serves as the theoretical distribution?
f) What serves as the sample data for this test?
g) What would serve as the null hypothesis for this test?
h) What would serve as the alternative hypothesis?
i) What assumption was used to calculate the expected frequencies within the table? Which hypothesis does this assumption represent?
j) How many classifications are there for this test? State the classification(s) for this test. What does each category within a classification represent for this test?
k) What formula was used to calculate the number of degrees of freedom for this test? What were the degrees of freedom for this test?
l) What is the value of $\frac{(O-E)^2}{E}$ for each initial letter of the information contained within the table of Figure 14.18, where O is the Observed frequency and E is the Expected frequency? What does the sum of the values of $\frac{(O-E)^2}{E}$ represent?

(continues)

CASE STUDY 14.2 Special K (continued)

m) What are the critical chi-square values for this test at $\alpha = 5\%$ and $\alpha = 1\%$?
n) What is the conclusion of this hypothesis test for each of these significance levels? Can you reject the null hypothesis for any of these significance levels? Explain.
o) What is the conclusion to this test for a significance level of 0.75%? What information did you use to answer this question? Explain.
p) What is the smallest level of significance that you can use to obtain a significant result?
q) State the conclusion to the chi-square test shown in Figure 14.18.
r) Was the chi-square result significant for the hypothesis test shown in Figure 14.18? Explain.

14.6 p-VALUE APPROACH TO CHI-SQUARE HYPOTHESIS TEST OF INDEPENDENCE USING THE TI-84 PLUS CALCULATOR

The p-value approach will require refinements in steps 3, 4 and 5 of the hypothesis testing procedure. In step 3, the decision rule will be stated as: Reject the H_o if the p-value of the X^2 test statistic is less than α. In step 4, the p-value will be determined using the TI-84 Plus calculator. While in step 5, the conclusion of the hypothesis test is determined by comparing the p-value to the level of significance, α.

Let's reexamine the hypothesis test shown in Example 14.2 using a p-value approach with the aid of the TI-84 Plus calculator.

EXAMPLE 14.5

A study was conducted to determine if there is a significant relationship between the frequency of times meat is served as a main meal per month for individuals living in eastern, central, and western United States.

A questionnaire was administered to a random sample of 300 families and the results are summarized in Table 14.16.

TABLE 14.16 Frequency of Meat as Main Meal per Month

Living Area	Less than 11	11 to 20	More than 20	Row Totals
Eastern US	a 46	b 36	c 20	102
Central US	d 40	e 40	f 26	106
Western US	g 16	h 38	i 38	92
Column Totals	102	114	84	300

a) Do these sample results indicate that there is a significant relationship between the frequency of meat served as a main meal per month and geographic living are at $\alpha = 5\%$?
b) Can we conclude that there is a significant relationship between the frequency of meat served as a main meal per month and geographic living area at $\alpha = 1\%$?
c) What is the smallest α level that can be used to show a significant relationship between the frequency of times meat is served as a main meal per month for individuals living in eastern, central, and western United States?

Solution

part (a):

Step 1 Formulate the two hypothesis, H_o and H_a.

H_o: The population classifications frequency of meat served as a main meal per month and geographic living area are independent.

H_a: The population classifications frequency of meat served as a main meal per month and geographic living area are dependent.

Section 14.6 *p-Value Approach to Chi-Square Hypothesis Test of Independence Using the TI-84 Plus Calculator*

Step 2 Determine the model to test the null hypothesis, H_o.

The hypothesis testing model is a chi-square distribution with $df = (r - 1)(c - 1)$. Since $r = 3$ and $c = 3$, then $df = 4$. The chi-square distribution with $df = 4$ is the hypothesis testing model as shown in Figure 14.19.

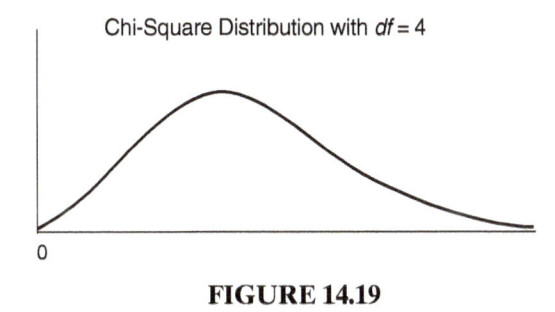

FIGURE 14.19

Step 3 Formulate the decision rule.
a) The significance level is $\alpha = 5\%$.
b) The decision rule is: reject H_o if the p-value of the X^2 test statistic is less than $\alpha = 5\%$. Figure 14.20 illustrates the decision rule for this hypothesis test.

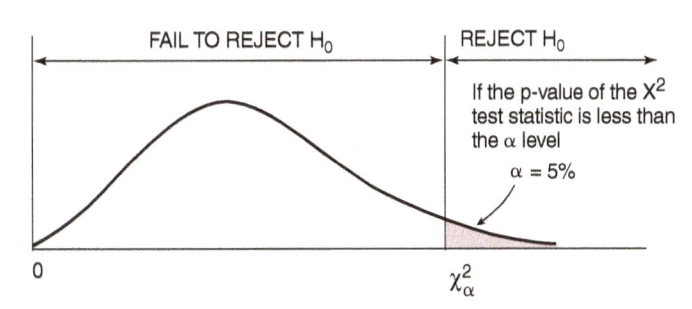

FIGURE 14.20

Step 4 Analyze the sample data.

Before we can calculate the p-value of the Chi-Square statistic the sample data must be entered into the TI-84 calculator using the MATRIX function. To access the MATRIX function on the TI-84 press 2nd x^{-1} [MATRIX] ▶ ▶. This is shown in Figure 14.21.

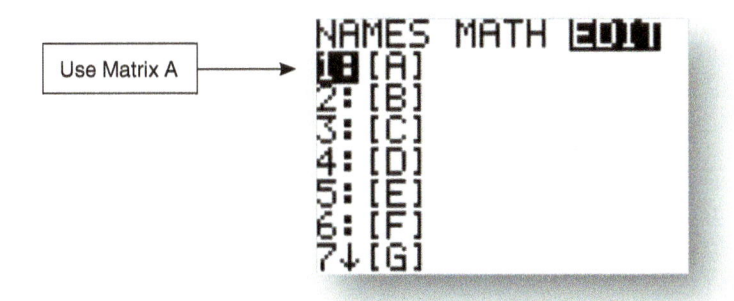

FIGURE 14.21

754 Chapter 14 Chi-Square

To enter the observed values of the 3 × 3 contingency table into Matrix [A], press ENTER. The result is shown in Figure 14.22a. At this point, we must change the 1 × 1 in Figure 14.22a to 3 × 3 and then press ENTER to obtain a 3 × 3 MATRIX. This is shown in Figure 14.22b.

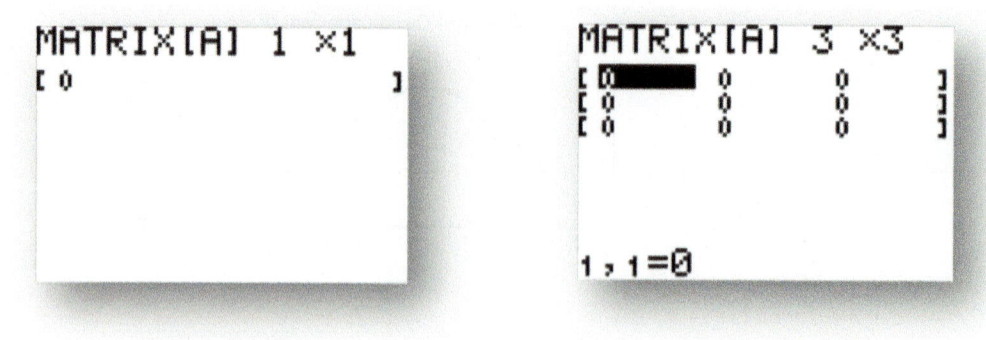

FIGURE 14.22a **FIGURE 14.22b**

Input the values for the 3 × 3 contingency table into the MATRIX. Be sure to press ENTER after you type each value in the matrix. This is shown in Figure 14.23.

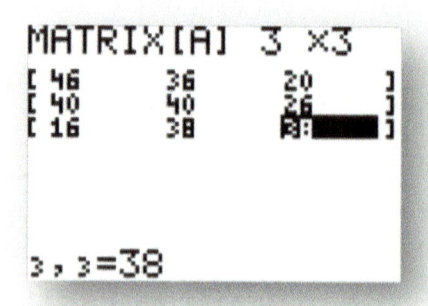

FIGURE 14.23

To perform the Chi-Square test, press STAT ▶ ▶ [TESTS] and highlight C: X^2-Test. This is shown in Figure 14.24.

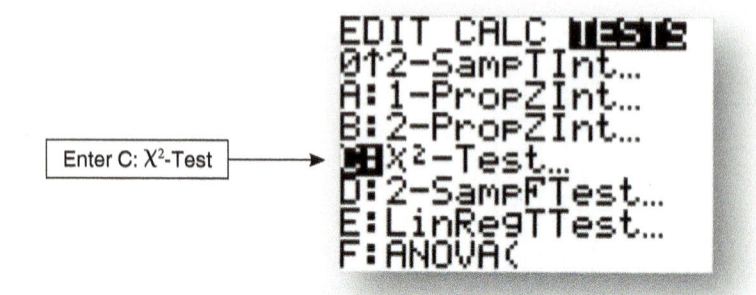

Enter C: X^2-Test

FIGURE 14.24

Section 14.6 *p-Value Approach to Chi-Square Hypothesis Test of Independence Using the TI-84 Plus Calculator*   **755**

Press ENTER to obtain the Chi-Square testing menu. This is shown in Figure 14.25.

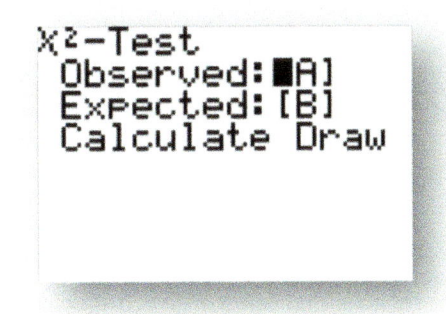

```
X²-Test
 Observed:■A]
 Expected: [B]
 Calculate Draw
```

FIGURE 14.25

Figure 14.25 shows that the observed frequencies are located in Matrix [A], and the expected frequencies will be outputted into Matrix [B]. Highlight Calculate and press ENTER to obtain the results of the Chi-Square test. This is shown in Figure 14.26.

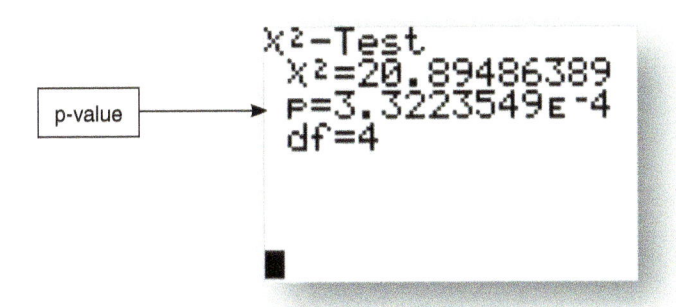

```
X²-Test
 X²=20.89486389
 P=3.3223549E-4    ← p-value
 df=4
```

FIGURE 14.26

Notice from the output screen, the p-value of the test statistic ($X^2 \approx 20.89$) is 3.22 E-4.

Step 5 State the Conclusion.

Since the p-value of this X^2 test statistic is less than $\alpha = 5\%$, we reject H_o and accept H_a at a p-value of 0.000322. These results are illustrated in Figure 14.27.

FAIL TO REJECT H_0 | REJECT H_0

$\alpha = 5\%$

p-value: $p \approx 0.000322$

χ^2_α X^2 test statistic ≈ 20.89

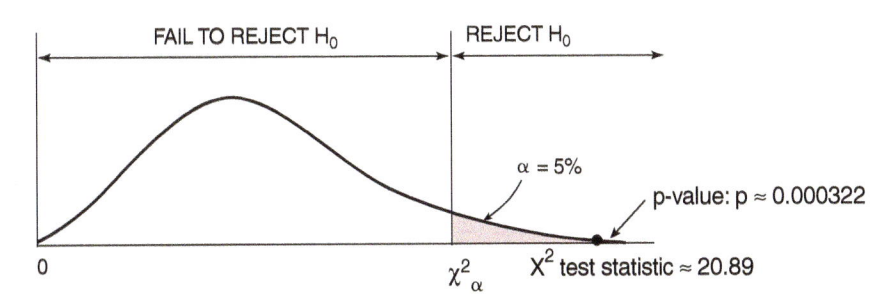

FIGURE 14.27

part (b): Since the p-value of 0.000322 is less than $\alpha = 1\%$, we can conclude that there is a significant relationship between the frequency of meat is served as a main meal per month and geographic living area at $\alpha = 1\%$. This is shown in Figure 14.28.

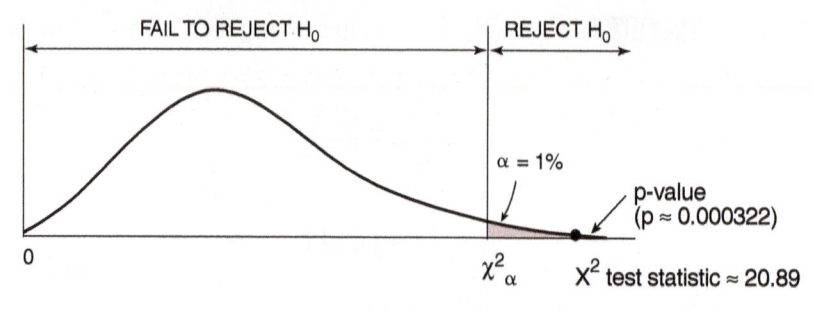

FIGURE 14.28

part (c): The smallest α level that can be used to show a significant relationship between the frequency of times meat is served as a main meal per month for individuals living in eastern, central, and western United States is equal to the p-value, 0.000322.

The expected and observed cell frequencies are summarized in Table 14.17.

TABLE 14.17 Frequency of Meat as Main Meal per Month

Living Area	Less than 11	11 to 20	More than 20	Row Totals
Eastern US	a 46 34.68	b 36 38.76	c 20 28.56	102
Central US	d 40 36.04	e 40 40.28	f 26 29.68	106
Western US	g 16 31.28	h 38 34.96	i 38 25.76	92
Column Totals	102	114	84	300

To display the expected results on the TI-84 calculator, press [2nd] [x^{-1}] [MATRIX] [▶] [▶] [▼]. This is shown in Figure 14.29.

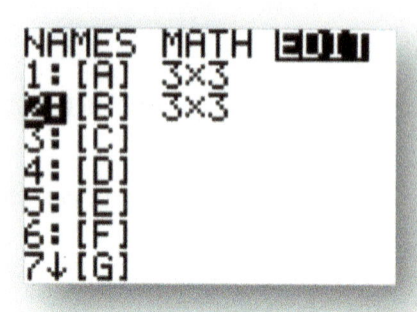

FIGURE 14.29

Press [ENTER] to display Matrix [B] which contains the expected results of this hypothesis test. This is shown in Figure 14.30.

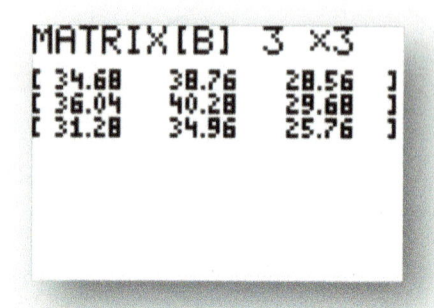

FIGURE 14.30

Summary of the p-value Approach to the Chi-Square Hypothesis Test of Independence Using the TI-84 Calculator

The following outline is a summary of the 5 step hypothesis testing procedure for the p-value approach to the Chi-Square Test of Independence Using the TI-84 Calculator

Step 1. *Formulate the two Hypothesis, H_o and H_a.*

H_o: The two population classifications are independent.

H_a: The two population classifications are dependent.

Step 2. *Determine the model to test the null hypothesis, H_o.*
   a) Construct the chi-square distribution with $df = (r-1)(c-1)$ that serves as the hypothesis testing model.
   b) Using H_o *is true* and the Matrix function of the TI-84 calculator, determine the expected cell frequencies.

Step 3. *Formulate the decision rule.*
   a) Identify the significance level, $\alpha = 1\%$ or $\alpha = 5\%$.
   b) State the decision rule.
   The decision rule is: reject H_o if the p-value of the X^2 test statistic is less than α.

Step 4. *Analyze the sample data.*
Enter the sample data into the TI-84 using the Matrix function and determine the p-value of the X^2 test statistic using the χ^2 test.

Step 5. *State the conclusion.*
Compare the p-value of the X^2 test statistic to the significance level α and either:
   a) Reject H_o and Accept H_a at the computed p-value, if the p-value of the X^2 test statistic is less than or equal to α.
   b) Fail to Reject H_o at the computed p-value, if the p-value of the X^2 test statistic is not less than α.

KEY TERMS

Cell 732
Chi-Square Distribution 733
Chi-Square Test of Independence 731
Computational Formula for Pearson's X^2 732
Contingency Table 731
Critical Chi-Square, χ_α^2 734
Definition Formula for Pearson's X^2 732
Degrees of Freedom df 736
Dependent 731
Expected Cell Frequency 732
Frequency 731
Goodness-of-Fit Test 745
Independent 731
Observed Cell Frequency 732
Pearson's Chi-Square Test Statistic, X^2 732
Population Classifications 731
Qualitative Data 731
Skewed 733
Symmetric 733
Uniform Distribution 746

SECTION EXERCISES

Note: The data sets in these Section Exercises can be downloaded in either Microsoft EXCEL or TI-84 format from the textbook's website at www.MyStatLab.com.

Section 14.1

1. A population can be classified by gender, age, religion, marital status, etc. These divisions are called _____.

2. The sample data falling into each of the cells of the population classifications are called the _____ frequencies.

3. A 4 × 3 contingency table has _____ rows and _____ columns.

4. Pearson's Chi-Square test statistic is computed by comparing the observed frequencies to the _____ frequencies of each cell within the contingency table.

5. How many degrees of freedom does a 3 by 4 contingency table have?
   a) 3   b) 12   c) 6   d) 4

6. Chi-Square tests of independence involve the analysis of quantitative continuous data.
   (T/F)

Section 14.2

7. The value of the Chi-Square test statistic is always _____.

8. There are a family of Chi-Square distributions where each chi-square distribution is dependent upon the number of _____. The formula to calculate df is given by: $df = $ _____.

9. The shape of the chi-square distribution is similar to the shape of the normal distribution. (T/F)

Section 14.3

10. A statistical technique used to investigate whether the classifications of a population are related is the Chi-Square Test of _____.

11. a) The definition formula for computing Pearson's Chi-Square test statistic is: $X^2 = $ _____.
    b) The computational formula for computing Pearson's Chi-Square test statistic is: $X^2 = $ _____.

12. In a Chi-Square test of Independence, the null hypothesis is stated as: the two population classifications are _____. The alternative hypothesis is stated as: the two population classifications are _____.

13. In a Chi-Square test of Independence, the expected frequency for a particular cell is calculated by multiplying its row _____ by its _____ total and dividing this product by the _____.

14. In formulating the conclusion of a Chi-Square test of Independence when:
    a) The value of the X^2 test statistic is less than or equal to the critical value of chi-square, χ^2_α, then the conclusion is _____.
    b) The value of the X^2 test statistic is greater than the critical value of chi-square, χ^2_α, then the conclusion is _____.

15. What is the critical value for a χ^2 test of independence for a 2 by 5 contingency table using $\alpha = 1\%$?
    a) 13.28   b) 11.07   c) 5.99   d) 9.49

16. In a χ^2 test for independence the null hypothesis is always
    a) The distribution is normal
    b) The population classifications are independent
    c) The population classifications are dependent
    d) The distribution is uniform

17. The expected frequency for a particular cell is calculated by multiplying its row total by its column total and dividing by
    a) The degrees of freedom
    b) The critical χ^2
    c) The sample size
    d) Number of classifications

18. If the value of the X^2 test statistic is greater than the critical value of chi-square, χ^2_α, then the conclusion is to reject H_o and accept H_a at α. (T/F)

19. If the conclusion is that the classifications are independent then the Pearson's chi-square test statistic, X^2, is greater than the χ^2_α. (T/F)

Problems 20–28 use the chi-square test of independence.

For all of these problems:
a) Formulate the hypotheses.
b) State the decision rule.
c) Calculate Pearson's chi-square test statistic.
d) Determine the conclusion, and answer the question(s) posed in the problem, if any.

20. A study was undertaken to determine whether or not membership in campus fraternity groups is related to grades. A random sample of 100 students was selected and the results are summarized in the following table:

Fraternity Group	Less than 2.00	From 2.00 to 3.00	Greater than 3.00	Row Totals
Member	[a] 16	[b] 17	[c] 7	40
Non-Member	[d] 9	[e] 28	[f] 23	60
Column Totals	25	45	30	100

Grade Point Average(GPA)

Based on these sample results can we conclude that there is a relationship between membership in a campus fraternity group and GPA at $\alpha = 1\%$?

21. A study was conducted to determine if a relationship exists between gender and movie preference. Two hundred students were questioned regarding their movie preference. The results of the survey are listed in the following table:

Movie Preference

Gender	Romance	Comedy	Drama	Adventure	Mystery	Row Totals
Male	[a] 10	[b] 20	[c] 5	[d] 40	[e] 15	90
Female	[f] 50	[g] 25	[h] 5	[i] 15	[j] 15	110
Column Totals	60	45	10	55	30	200

Based on the sample data can we conclude that gender is related to movie preference at $\alpha = 5\%$?

22. Caffeine is one of the ingredients now being included in many of the weight-loss supplements because of its energy enhancement, appetite suppressant, and "fat-burning" properties. Ms. JoAnne, a medical researcher, decides to conduct a study to monitor the impact of a green tea-caffeine combination on weight loss. Two hundred randomly selected participants were divided into two categories: those who consume low levels of caffeine (≤ 300 mg/day) and high-caffeine consumers (>300 mg/day). After a six month study, the researcher's results are summarized in the following table.

Caffeine Consumption Level

Weight Loss	Low levels (≤ 300 mg/day)	High levels (>300 mg/day)	Row Totals
Less than 6 lbs	36	14	50
6 to 10 lbs	37	43	80
11 to 16 lbs	22	33	55
More than 16 lbs	5	10	15
Column Totals	100	100	200

Based upon the sample results, can the researcher conclude that caffeine consumption level and weight loss are independent at $\alpha = 1\%$?

23. In an experiment designed to compare the effectiveness of different brands of pain relievers, a group of 225 individuals who complained of chronic pain were randomly assigned to three daily treatments: three tablets of Brand A (prescription drug), three tablets of Brand B (non-prescription drug), or three placebos.

After the experiment, each subject was questioned to determine the effectiveness of their treatment. Their responses are classified in the following table.

Result of Treatment	Drug Treatment			Row Totals
	Brand A	Brand B	Placebo	
Relief from Pain	a 17	b 15	c 13	45
Some Relief from Pain	d 33	e 34	f 38	105
No Relief from Pain	g 25	h 26	i 24	75
Column Totals	75	75	75	225

Based upon these sample results can one conclude that drug treatment and relief from pain are independent at $\alpha = 5\%$?

24. A PhD student decided to test whether a person's music preference and their IQ score are related. The student researcher believes that individuals with a high IQ are drawn to classical music. To conduct her study, the researcher randomly surveyed 400 individuals who were asked to rate their enjoyment of Classical Music. Her sample results are recorded in the following contingency table.

Classical Music Rating	Low IQ	Med IQ	High IQ
Does not enjoy Classical Music	56	42	30
Mixed Feelings	42	40	45
Enjoys Classical Music	37	48	60

Perform a chi-square test of independence to determine if a person's rating of Classical Music and their IQ score are dependent at $\alpha = 1\%$. Based upon the sample results, can the PhD student researcher conclude that an individual's IQ level and their Classical Music rating are independent at $\alpha = 1\%$?

25. A sociologist is researching the question: "Is there a relationship between the level of education and job satisfaction?" For a random sample of 270 workers she determines each individual's education level and level of job satisfaction. The results are recorded in the following table:

Education Level	Job Satisfaction Level		
	Low	Medium	High
College	a 20	b 70	c 30
High School	d 15	e 60	f 25
Grade School	g 15	h 25	i 10

Perform a chi-square test of independence to determine if educational level and job satisfaction are dependent at $\alpha = 5\%$.

Section 14.4

26. The assumptions underlying the Chi-Square test of Independence are:
   a) Each of the observations must be _____ of all the other observations.
   b) Each observation must be classified into one and only one _____.
   c) The sample size must be sufficiently large so that the smallest expected cell frequency is at least _____.

27. When doing an χ^2 test what must be true about the sample size?
   a) At least 50
   b) No more than 300
   c) Large enough so that the smallest expected cell frequency is at least 5
   d) Large enough so that each cell in the expected frequency is at least 30

Section 14.5

28. The Goodness-of-Fit Chi-Square Test is used to determine if the sample data fits a particular _____ distribution model.

29. In a Goodness-of-Fit Test, the null hypothesis is stated as the _____ data fits a particular theoretical distribution model.

30. In a Goodness-of-Fit Test, the degrees of freedom of a single classification case with k categories is: $df =$ _____.

31. The test of Goodness-of-Fit refers to a type of chi-square test that can be used to determine how well sample data fits the normal distribution.
   (T/F)

32. A Goodness-of-Fit test is used to determine if sample data fits a uniform distribution. The conclusion is to reject H_o and accept H_a at α. This conclusion means that the sample data does not fit a uniform distribution.
   (T/F)

Problems 33–37 use the test of Goodness-of-Fit. For all of the problems:
   a) Formulate the hypotheses.
   b) State the decision rule.
   c) Calculate Pearson's chi-square test statistic.
   d) Determine the conclusion, and answer the question(s) posed in the problem, if any.

33. A marketing research agency conducted a random survey to determine the opinions of prospective customers for a new product. The results of the survey were:

Opinion	Number
Outstanding	97
Excellent	98
Good	88
Fair	67
Not interested	70

If the opinions are uniformly distributed among the five categories then the marketing agency will not recommend the manufacturer to mass produce the new product. Test the null hypothesis that the sample data (opinions of the customers) fits a uniform distribution at $\alpha = 5\%$. Should the marketing agency recommend mass production of the new product to the manufacturer?

34. A market researcher wants to determine if the shape of a client's packaging material is an important factor in a consumer's selection of the product. The researcher randomly selects 2000 consumers and gives each of them five differently shaped packages of the same product. One month later each consumer is asked to indicate a preference for one of the shaped packages. The results of the survey are given in the table below.

Number of Consumers Selecting Package	#1	#2	#3	#4	#5
	430	450	385	365	370

Use a Goodness-of-Fit Test to test the null hypothesis that sample data fits a uniform distribution at $\alpha = 1\%$. Can you conclude that the sample data indicates a preference for packaging shape?

35. According to recently published census figures the socio-economic stratification for a small northeast town is: 20% upper, 50% middle and 30% lower. One hundred and fifty residents of this town are supposedly chosen at random. To check this the sample data is sorted and summarized to be:

Socio-Economic Stratification	Number of Residents
Lower	37
Middle	86
Upper	27

Should we conclude that the sample was not randomly selected from the town using $\alpha = 1\%$?

36. Victoria, a student at a local commuter college, wants to determine if the day of the week is independent of time it takes to locate a parking space at the college. A random sample of times it took to locate a parking space produced the following data:

Day of the Week	Time (minutes) to Locate a Parking Space
Monday	14
Tuesday	22
Wednesday	16
Thursday	28
Friday	6

Victoria states the null hypothesis as: the sample data (time to locate a parking space) fits a uniform distribution. Perform a Goodness-of-Fit test at $\alpha=5\%$ and determine if Victoria can conclude that the Time to Locate a Parking Space is not evenly distributed throughout the work week.

37. The Pit Boss at the Mirage casino can't believe her eyes when she notices that a particular pair of dice rarely rolls certain numbers. She decides to test if the dice are "loaded." She tosses the dice 1080 times and records the sum for each toss. The data for her experiment is listed below:

Sum of the Pair of Dice	Number of Times
2	24
3	70
4	77
5	128
6	156
7	156
8	148
9	136
10	111
11	48
12	26

Can the Pit Boss conclude that the dice are indeed "loaded" using $\alpha = 5\%$?

The following problem outlines a Goodness-of-Fit Test to determine if the sample data fits a normal distribution.

38. It has been claimed that the number of hours worked per week by a full-time college student is normally distributed. To test this claim a random sample of 600 full-time college students were selected. It was determined that the sample data had a mean number of hours worked per week per student of 21.7 hrs with s = 6.7 hrs. Six intervals were used to summarize the data and the sample results are given in the following table.

Interval Boundaries	Frequency
Below 12.32	31
12.32–17.01	96
17.01–21.70	173
21.70–26.39	147
26.39–31.08	112
Above 31.08	41

Complete the following steps which outline a Goodness-of-Fit Test to test the null hypothesis that the number of hours worked per week by a full-time student is normally distributed at $\alpha = 5\%$.

Step 1. Formulate the two hypotheses H_o and H_a.
H_o
H_a

Step 2. Determine the model to test the null hypothesis, H_o. To calculate the expected frequency for each interval we assume the null hypothesis to be true, that is, we assume a normal distribution with $\mu \approx$ sample mean $= 21.70$ and $\sigma \approx s = 6.70$. The expected frequencies for each of the interval boundaries are calculated by:
 1) converting each boundary into a z score.
 2) obtaining the corresponding probabilities for the z scores from Table II:
 3) calculating each expected frequency using these probabilities and the sample size.

Complete the following table using this procedure.

Interval Boundaries	z Score	Expected Probability	Frequency
Below 12.32	–1.40	0.0808	48.48
12.32 – 17.01			
17.01 – 21.70			
21.70 – 26.39			
26.39 – 31.08			
Above 31.08			

Step 3. Formulate the decision rule.
 1) In this example we will use a significance level $\alpha =$ _____.

2) The degrees of freedom, *df*, for a single classification case is *df* = k – 1, where k is the number of categories within the single classification. However, since the mean and standard deviation were estimated then the degrees of freedom must be reduced by one for each estimated parameter used.

Thus, *df* = k – 3

df = _____

3) The critical chi-square value for this test can be found in Table IV where α = 5% and *df* = 3. The critical chi-square value is:

χ_α^2 = _____

4) The decision rule is: _____.

Step 4. Calculate Pearson's chi-square test statistic.

X^2 = _____

Step 5. State the conclusion.
State whether the number of hours worked per week by a full-time student fits a normal distribution.

Section 14.6

For problems 39–44, use questions 20 to 25 of Section 14.2 exercises and apply the p-value approach to hypotheses testing using the TI-84 Plus calculator. For each of these problems, use the 5 step procedure to perform a hypothesis test at **both** α = 1% and α = 5%. Include the following within your test:

a) state the null and alternative hypotheses.
b) formulate the chi-square decision rule.
c) identify the test statistic and its p-value.
d) state the conclusion and answer any question(s) posed in the problem for each alpha level.
e) determine the smallest level of significance for which the null hypothesis can be rejected.

CHAPTER REVIEW

45. If the test statistic is a very large positive number, then the probability is small that the null hypothesis will be rejected. (T/F)

46. An arrangement of sample count data into a two-way classification is called a _____ table.

47. Each individual box in the contingency table is referred to as a _____.

48. The graph of a Chi-Square distribution is not symmetric, but it is skewed to the _____.

49. In a Chi-Square Test of Independence the sample data represents qualitative or _____ data.

50. For a sample of 200, what is the expected frequency for a cell whose row total is 70 and its column total is 30?
    a) 10.5   b) 21   c) 105   d) 2100

51. All chi-square tests of independence are one-tailed tests on the right. (T/F)

52. If the value of Pearson's chi-square test statistic is 6.79 and χ_α^2 = 9.49, what is the conclusion of this test?
    a) Reject H_o and accept H_a
    b) Fail to reject H_o
    c) Reject H_o and reject H_a
    d) Fail to reject H_a

53. If the test statistic X^2 is less than the critical χ_α^2, then the conclusion is: Fail to reject the null hypothesis at α. (T/F)

54. What type of test is used to determine if the sample data conforms to a theoretical distribution model?
    a) Sample mean hypothesis test
    b) Goodness-of-Fit test
    c) Chi-square test of independence
    d) A normal distribution test

55. A Goodness-of-Fit Test is used to determine if sample data fits a uniform distribution. The conclusion is fail to reject H_o at α. This means that the data fits a normal distribution. (T/F)

Problems 56–59 use the chi-square test of independence. For each problem:
a) Formulate the hypotheses.
b) State the decision rule.
c) Calculate Pearson's chi-square test statistic.
d) Determine the conclusion, and answer the question(s) posed in the problem, if any.

56. A study was conducted to determine if gender is related to the number of songs downloaded on iTunes per year. A random sample of 200 iTunes users were interviewed and their responses are recorded in the following table:

Gender	0 – 50	51 – 100	101 and higher	Row Totals
Male	40	50	30	120
Female	20	50	10	80
Column Totals	60	100	40	200

Number of songs downloaded on iTunes per year

Based on the sample data, can we conclude that gender is related to the number of songs downloaded on iTunes per year at α = 5%?

57. An author of a popular blog site feels her readers are especially interested in topics involving social media. For a future topic, she decides to focus on the possibility that there is a relationship between the "Relationship Status (RS)" of an individual on Facebook and the "Number of Friends (NF)" that individual has on Facebook. She performs a survey, and her results are summarized in the contingency table below.

NF \ RS	0 < NF ≤ 100	100 < NF ≤ 500	NF > 500	Row Totals
Single	4	25	14	43
Married	8	15	3	26
Divorced	4	6	1	11
Column Totals	16	46	18	80

a) Perform a Chi-Square test of independence to see if the variables Relationship Status and Number of Friends are dependent or independent. Test at the 5% level of significance.

b) Now that you have completed this test, do you trust the results? It appears that the one of the chi-square assumptions has been violated. Which one?

58. A study was conducted to determine if gender is related to the number of tweets a Twitter user makes per year. A random sample of 415 Twitter users was taken with the following results:

Gender	Number tweets per year			
	0–100	101–200	201 and higher	Row Totals
Male	52	61	50	163
Female	58	89	105	252
Column Totals	110	150	155	415

Based on the sample data, can we conclude that gender is related to the number of tweets a Twitter user makes per year? Test at an alpha level of 1%. Would the conclusion be different if the test were run with an alpha level of 5%? Why?

59. A study was conducted on a random sample of 500 married couples to determine if there is a significant relationship between the frequency of times per year a married couple goes to the movies and their educational background. The results of the study are shown here:

Frequency of Movie Attendance per Year	Educational Background (Highest Level Achieved)			
	Neither Went to College	Each Has Some College	Only One Grad College	Both Grad College
0–6	a 55	b 20	c 35	d 17
6–12	e 65	f 30	g 30	h 8
Above 12	i 55	j 75	k 45	l 65

Do these sample results indicate at $\alpha = 5\%$ that there is a significant relationship between frequency of times married couples go to the movies and their educational background?

60. A travel group wants to determine whether people from different areas of the country have valid passports. A survey of 233,831 people from across the country was taken to determine whether they have valid passports.

The table below shows the numbers of people in 4 geographical areas who do and do not hold valid passports.

Region / Valid Passport	Northeast	South	Midwest	West	Row Totals
Yes	19105	26572	17816	23342	86835
No	23603	60666	33033	29694	146996
Column Totals	42708	87238	50849	53036	233831

Do the results suggest that people from different areas of the country hold valid passports in different percentages? Use an alpha level of 5%.

61. A social science researcher intends to conduct a hypothesis test to determine if there is a relationship between the number of children and a mother's education level for women aged 22 to 44 years old. The researcher determines the mother's education level and the number of children for a random sample of 1,000 mothers aged 22 to 44 years old. The sample results are recorded in the following contingency table.

Education level / Number of children	No HS Diploma	HS Diploma	Some College Education	Bachelor's Degree or Higher
none	55	70	78	124
1 to 2	67	65	70	114
3 or more	95	85	82	95

Perform a chi-square test of independence to determine if a mother's educational level and number of children are dependent at $\alpha = 5\%$.

62. Prerequisite requirements for college courses have been a topic of conversation amongst students, educators, and administrators for several years. Many students feel that they can still perform at an acceptable level in a college level class, even though they do not have the required prerequisite course. A statistics professor at our college decides to see if there is a relationship between passing/failing a course and having the required elementary algebra prerequisite. The professor allows 60 students to register for their course, and waived the prerequisite requirement. Of the 60 students who registered, 35 had the elementary algebra prerequisite, and 25 did not. The results of their grades are summarized in the contingency table below.

	Pass	Fail	Total
Have Algebra Pre-req	29	6	35
Don't have Algebra Pre-req	10	15	25
Total	39	21	60

Does it appear that students are more likely to pass their statistics class if they have the elementary algebra prerequisite? Test at the 1% level of significance.

Problems 63–64 use the test of Goodness-of-Fit. For each problem:

a) Formulate the hypotheses.

b) State the decision rule.

c) Calculate Pearson's chi-square test statistic.

d) Determine the conclusion, and answer the question(s) posed in the problem, if any directions, then.

63. "Let it Ride" Jack decides to test the fairness of a die. He rolls it 120 times with the results shown in the following table:

Number Rolled	Frequency
1	24
2	18
3	27
4	17
5	16
6	18

Use a Goodness-of-Fit Test to determine if the die is fair at $\alpha = 5\%$.

64. The student government at Fernworth College has been told that the student population consists of 58% female and 42% male. A random sample of 400 students consists of 54% female and 46% male. Are the sample data consistent with the population proportions at $\alpha = 5\%$?

WHAT DO YOU THINK?

65. Examine the snapshot entitled "Would You Date a Co-worker" in the Figure 14.31.
    a) If you wanted to determine if there was a relationship between the gender of a worker and the response to the question: Would you date a co-worker, then identify the population classifications presented within the snapshot.
    b) If you were to perform a hypothesis test to determine if there is a relationship between the population classifications, how would the null hypothesis be stated?
    c) How would the alternative hypothesis be stated?
    d) When the null hypothesis is assumed to be true, what does that indicate about the relationship between the two population classifications?
    e) If we classify the response to the question as either: **yes**, **no** and **undecided**, then what would be the size of the contingency table for this type of chi-square test?
    f) What is the sample size for this information contained within the snapshot? If three hundred of the sample were women, then determine the number of people that would fall within each category of the contingency table.
    g) What are these values within the contingency table called?
    h) Determine the expected frequencies for this hypothesis test.
    i) What is the number of degrees of freedom for this test?
    j) Determine the critical value for the 5% and 1% level of significance.
    k) State a decision rule for each level of significance.
    l) Calculate Pearson's Chi-Square test statistic.

Statistical Snapshot

Would You Date a Co-worker?

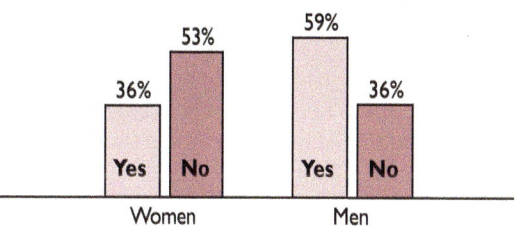

Women: 36% Yes, 53% No
Men: 59% Yes, 36% No

Source: Gallup Organization poll of 679 employed adults

FIGURE 14.31

    m) What is the conclusion for this test, if $\alpha = 5\%$? Can the null hypothesis be rejected at $\alpha = 5\%$? Can you conclude that the two population classifications are independent?
    n) Is there a statistically significant relationship between the gender of a worker and the response to the question: *Would you date a co-worker* at $\alpha = 1\%$? Can the null hypothesis be rejected at $\alpha = 1\%$? Can you conclude that the two population classifications are dependent?
    o) Can you conclude that the p-value of this hypothesis test is less than 5%? Explain.
    p) Can you conclude that the p-value of this hypothesis test is less than 1%? Explain.

PROJECTS

66. Chi-Square Test of Independence Hypothesis Testing Project.
    A. Selection of a topic
       1. Select an article from a local newspaper, magazine or other source that can be used to perform a chi-square test of independence.
       2. Suggested topics:
          a) Test the claim that there is a relationship between student views on course requirements for their major and the major area of concentration.
          b) Test the claim that there is a relationship between gender and student views on capital punishment.
    B. Use the following procedure to test the claim selected in part A.
       1. Identify the classifications of the population referred to in the statement of the claim and indicate where the claim was found.
       2. State the null and alternative hypotheses.
       3. Develop and state a procedure for selecting a random sample from your population. Indicate the technique you are going to use to obtain your sample data.
       4. Indicate the hypothesis testing model, and choose an appropriate level of significance.
       5. Collect the data and summarize the observed frequencies in a contingency table.
       6. Find the degrees of freedom and determine the appropriate critical chi-square value.
       7. Calculate the expected frequencies and determine Pearson's chi-square test statistic.
       8. Construct a chi-square distribution model and place Pearson's chi-square test statistic on the model.
       9. Formulate the appropriate conclusion and interpret the conclusion with respect to your opinion (as stated in step 2).

67. Get hold of a pair of dice from any board game. Do an experiment by tossing the dice 1000 times and record the number of times each sum appears. Compare the experimental results to the results that one would expect if the dice were fair. Use $\alpha = 5\%$.

CHAPTER TEST

68. Chi-square tests of independence involve the analysis of qualitative data.
(T/F)

69. The graph of a chi-square distribution is not symmetric, but it is skewed left.
(T/F)

70. If the conclusion is to reject H_o and accept H_a at α, then the chi-square test of independence has indicated that the population classifications are independent.
(T/F)

71. In designing a contingency table, care must be taken so that each observation falls into
a) At least one cell
b) No more than two cells
c) Only one cell
d) None of these choices

72. What are the degrees of freedom for a Goodness-of-Fit Test with a single classification of six categories?
a) 4   b) 5   c) 6   d) 3

73. The test of Goodness-of-Fit refers to a type of chi-square test that can be used to determine how well sample data fits a chi-square distribution.
(T/F)

Problems 74–75 use the chi-square test of independence.
For each problem:
a) Formulate the hypotheses.
b) State the decision rule.
c) Calculate Pearson's chi-square test statistic.
d) Determine the conclusion, and answer the question(s) posed in the problem, if any.

74. A recent study conducted by The Nielsen Company shows that the USA is among countries with the highest average time spent per person on social networking sites. Ms. Lydia Scarlet, a social science researcher, decided to investigate the relationship between gender and the average time 18–24 year olds spend on social networking websites per month. The researcher randomly sampled 500 18 to 24 year olds and summarized the results in the following table.

Average Monthly Time 18–24 Year Olds Spend on Social Networking Websites

Gender	Less Than 10 Hours	10 to 30 Hours	31 to 50 Hours	More Than 50 Hours	Row Totals
Female	35	90	65	60	250
Male	55	100	55	40	250
Column Totals	90	190	120	100	500

Based upon the sample results, can the researcher conclude that gender and monthly average time 18–24 year olds spend on social networking websites are independent at $\alpha = 5\%$?

75. A statistics instructor decides to investigate the relationship between a student's grade in the course and class attendance. The instructor randomly samples the records of 120 students and the results are listed in the following table.

Number of Absences	Grade in Course F–D+	C–B	B+–A
0–1	a 10	b 18	c 12
2–5	d 25	e 17	f 8
above 5	g 20	h 5	i 5

Perform a chi-square test of independence to determine if a student's grade in the statistics course and class attendance are dependent at $\alpha = 1\%$?

76. Use the test of Goodness-of-Fit.
a) Formulate the hypotheses.
b) State the decision rule.
c) Calculate Pearson's chi-square test statistic.
d) Determine the conclusion, and answer the question posed in the problem, if any.

A mathematics professor has analyzed the results of her statistics final exam for the past few years. She has determined that the grade distribution for her final exam is:

Grade	Percent of Students Receiving
A	10%
B	15%
C	40%
D	25%
F	10%

This year the professor administers the exam to 200 students. The number of students receiving the letter grades A, B, C, D, and F are 24, 20, 89, 40, and 27 respectively. Use the Goodness-of-Fit Test to determine if this year's final exam results fit her previous grade distribution at $\alpha = 5\%$.

77. Cell phones are frequently broken due to a misfortunate happening. A survey of 40000 cell phone users who reported that their cell phone broke due to a misfortunate happening was examined to determine if gender was related to the way in which the phone broke.

A summary of the top six reasons for breakage by gender is listed in the table that follows:

Breakage / Gender	Dropped	Spilled Water	Run Over	Thrown	Toilet/bath	Other misfortune	Row Totals
Men	7500	5800	2400	2880	700	1900	19200
Women	8500	3000	1600	720	2900	2100	20800
Column Totals	16000	8800	4000	3600	3600	4000	40000

Do the results of this table suggest that there is a significant difference in the way men and women break their cell phones? Use $\alpha = 5\%$.

78. Five hundred randomly selected British Columbians were polled regarding their attitude towards the hunting of Polar Bears.

The 500 respondents were asked the same questions, once with no knowledge about Polar Bear reproduction and once with knowledge about Polar Bear reproduction.

Survey Question #1: Do you believe the government should or should *not* continue to permit the hunting of Polar Bears?

Survey Question #2: Given that the population of Polar Bears can increase at a rate of 8% per year after hunting occurs. Do you believe the government should or should *not* continue to permit the hunting of Polar Bears?

The results of the two survey questions are listed in the following contingency table.

Knowledge of Reproduction Rate	Responses to Survey Question		
	Should Permit Hunting	Should *Not* Permit Hunting	Don't Know
Question w/no knowledge	245	230	25
Question with knowledge	285	185	30

a) Based on the sample results, can one conclude that knowledge of the Polar Bear rate of reproduction and responses to the survey questions are independent at $\alpha = 5\%$?

b) Based upon the results of the Chi-Square Test of Independence, do you believe that as the respondents became more aware of the Polar Bear rate of reproduction they were more inclined to allow the hunting of Polar Bears?

For the answer to this question, please visit www.MyStatLab.com.

CHAPTER 15

Inferences for Linear Correlation and Regression

Contents

15.1 Introduction
15.2 Testing the Significance of the Correlation Coefficient
Procedure to Test the Significance of the
Population Correlation Coefficient, ρ
15.3 Assumptions for Linear Regression Analysis
15.4 p-Value Approach to Testing the Significance
of the Correlation Coefficient Using the TI-84
Calculator
15.5 Introduction to Multiple Regression

15.1 INTRODUCTION

In Chapter 4, we introduced linear regression analysis as a descriptive statistical technique used to measure the strength of a linear relationship between a dependent variable (y') and one independent variable x. As discussed in Chapter 4, the dependent variable, y', is the variable that is predicted by the independent variable, x. The linear regression equation is a prediction formula.

The general form of the linear regression equation was found to be: $y' = a + bx$

where:
y' = the predicted value of the dependent variable y given the x value
x = the independent variable
b = the slope of the line (or regression coefficient)
a = the y' intercept of the line (or the intercept constant)

The strength of the relationship between the variables was measured using Pearson's sample correlation coefficient, r. Remember that values of r close to –1 or 1 were said to have a high linear correlation. If the variables had little or no linear correlation, then the value for r would be approximately 0.

For example, suppose a statistician wants to measure the correlation between the dependent variable College GPA and the independent variable number of study hours per week for the entire population of college students. Suppose a linear correlation exists between the variables for this population of college students. This population correlation coefficient is designated by the Greek letter rho, ρ, pronounced "row." Figure 15.1 shows this relationship.

Since it is impossible for the statistician to collect data for the entire population of college students, the statistician must select a random sample. Using the sample data, the sample correlation coefficient r will be used as an estimate for rho, ρ. Twenty five college students are randomly sampled from the population and the sample correlation coefficient is calculated to be r = 0.74. This is represented in Figure 15.2.

How can the statistician use the sample correlation coefficient r = 0.74 calculated from the 25 students to draw a conclusion about the population correlation coefficient ρ?

In this chapter we will develop a hypothesis testing procedure to determine if the population correlation coefficient, ρ, is significantly different than 0. The value for the sample correlation coefficient r, will be involved in this hypothesis testing procedure. After we calculate the sample correlation coefficient r, we will use this hypothesis test to determine if this sample value of r indicates that the unknown population correlation coefficient ρ, is a significant correlation. That is, is the value of the population correlation coefficient ρ significantly different from a correlation of 0 (no linear correlation)?

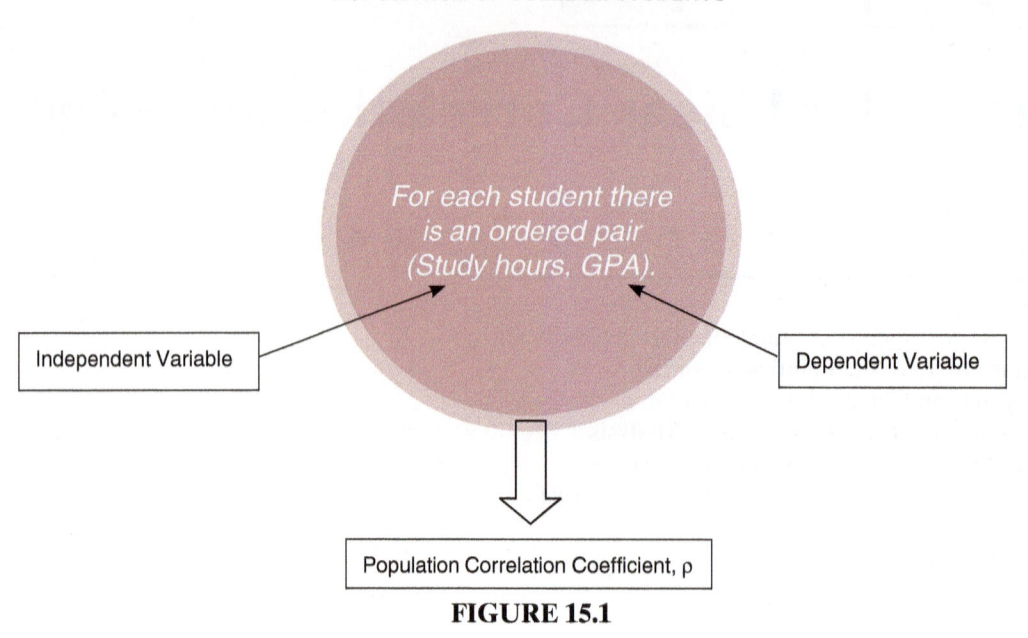

Chapter 15 *Inferences for Linear Correlation and Regression*

POPULATION OF COLLEGE STUDENTS

For each student there is an ordered pair (Study hours, GPA).

Independent Variable

Dependent Variable

Population Correlation Coefficient, ρ

FIGURE 15.1

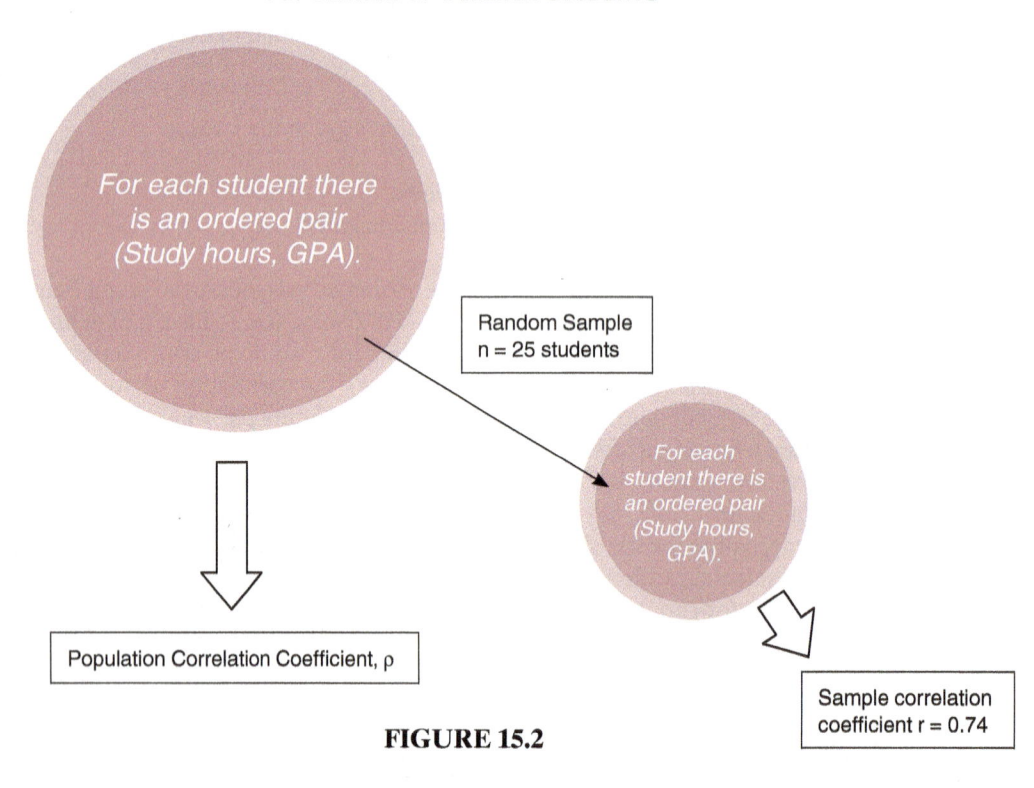

POPULATION OF COLLEGE STUDENTS

For each student there is an ordered pair (Study hours, GPA).

Random Sample n = 25 students

For each student there is an ordered pair (Study hours, GPA).

Population Correlation Coefficient, ρ

Sample correlation coefficient r = 0.74

FIGURE 15.2

We will now examine how to test a population correlation coefficient for statistical significance. The statistical procedure needed to perform this test involves a modification of the general five step hypothesis testing procedure developed in Chapter 10. Let's develop this procedure to test the significance of a population correlation coefficient.

15.2 TESTING THE SIGNIFICANCE OF THE CORRELATION COEFFICIENT

There are many practical situations where it is necessary to determine whether a *significant* correlation exists between two variables. For example:

- Is there a significant correlation between High School average and success in college? or,
- Is there a significant correlation between the number of hours of television a five year old watches per week and their IQ score? or,
- Is there a significant correlation between a person's age and their blood pressure? or,
- Is there a significant correlation between a company's advertising expenditures and their sales for one of their new products?

We will now **examine how to test a population correlation coefficient for statistical significance**. The statistical procedure needed to perform this test involves a modification of the general five step hypothesis testing procedure that was developed in Chapter 10. Let's apply this procedure to test the significance of a population correlation coefficient.

Procedure to Test the Significance of the Population Correlation Coefficient, ρ

This test is based upon the assumption that both variables x and y are normally distributed.

Step 1 Formulate the two hypotheses, H_o and H_a.
Null Hypothesis:
In general, the null hypothesis for testing the significance of the population correlation coefficient has the following form.

Null Hypothesis Form

H_o: The population correlation coefficient is equal to zero. That is, there is no linear correlation between the two variables.
Thus, the null hypothesis is symbolized as:

$$H_o: \rho = 0$$

Alternative Hypothesis:
Since the alternative hypothesis, H_a, can be stated as the population correlation coefficient is either: *greater than*, *less than*, or *not equal to zero*, then the alternative hypothesis can have one of three forms:

Form (a): Positive or Greater than Form for H_a

H_a: The population correlation coefficient, ρ, is claimed to be greater than zero. That is, there is a positive linear correlation between the two variables. This is symbolized as:

$$H_a: \rho > 0$$

Form (b): Negative or Less than Form for H_a

H_a: The population correlation coefficient, ρ, is claimed to be less than zero. That is, there is a negative linear correlation between the two variables. This is symbolized as:

$$H_a: \rho < 0$$

(continues)

Form (c): Not Equal Form for H_a

H_a: The population correlation coefficient, ρ, is claimed to be not equal to zero. That is, there is a significant linear correlation between the two variables. This is symbolized as:

$$H_a: \rho \neq 0$$

Step 2 Determine the model to test the null hypothesis, H_o.

An important consideration to the design is the assumption that the variables x and y are both normally distributed. Under the assumption the null hypothesis is true, that is: $\rho = 0$, the distribution of r values is bell-shaped and symmetric about r = 0 and is referred to as **the sampling distribution of the correlation coefficients.** This distribution will be used as the **model for the hypothesis test**. Figure 15.3 illustrates different sampling distributions of the correlation coefficients for various degrees of freedom.

DEFINITION 15.1 DEGREES OF FREEDOM, df
The **degrees of freedom, df,** for testing the correlation coefficient is calculated by subtracting two from the number of pairs of sample data, n. Thus, the formula for df is: $df = n - 2$

Hypothesis Testing Model
the sampling distribution of the correlation coefficients
when $\rho = 0$

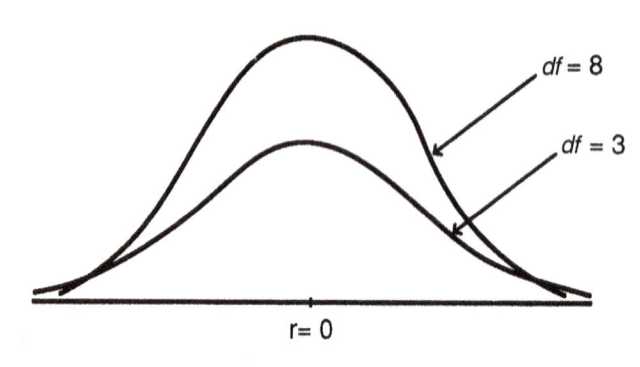

FIGURE 15.3

Step 3 Formulate the decision rule.
The decision rule is based on the following information.
a) Determine the type of alternative hypothesis: directional or nondirectional.
b) Determine the type of hypothesis test: 1TT or 2TT.
c) Identify the significance level: $\alpha = 1\%$ or $\alpha = 5\%$.
d) Find the critical r value(s), r_α, using Table V: Critical Values of the Correlation Coefficient, r_α, found in Appendix C: Statistical Tables.
Table V contains the Critical Values of the Correlation Coefficient, r_α. To determine the appropriate critical r value, denoted as r_α, from Table V for the decision rule, we need to determine the degrees of freedom for testing the correlation coefficient using the formula: $df = n - 2$.
e) Construct the appropriate hypothesis testing model.

Figure 15.4 illustrates the appropriate sampling distribution of the correlation coefficients that are used as the hypothesis testing model. Notice that the actual model chosen is dependent upon the form of the alternative hypothesis.

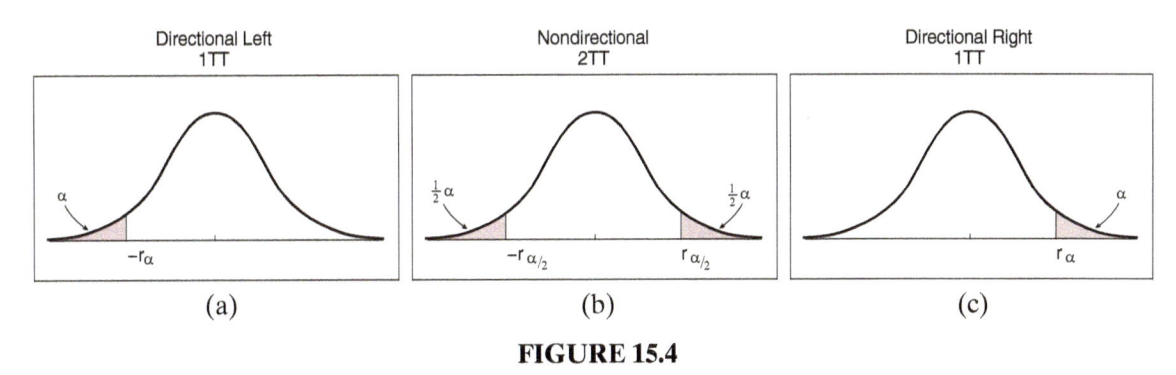

Section 15.2 Testing the Significance of the Correlation Coefficient

Directional Left
1TT
(a)

Nondirectional
2TT
(b)

Directional Right
1TT
(c)

FIGURE 15.4

Step 4 Analyze the sample data.
Randomly select the necessary data from the population to calculate the sample correlation coefficient, r, for the sample data. This r value is called the **test statistic**. To calculate the sample correlation coefficient, r, use the formula:

$$r = \frac{n\Sigma(xy) - (\Sigma x)(\Sigma y)}{\sqrt{n(\Sigma x^2) - (\Sigma x)^2} \sqrt{n(\Sigma y^2) - (\Sigma y)^2}}$$

Step 5 State the conclusion.
Compare the **test statistic, (i.e., the sample correlation coefficient r)** to the **critical r value, r_α**, of the decision rule and draw one of the following conclusions:

a) Reject H_o and Accept H_a at α

b) Fail to reject H_o at α

In summary, the general procedure used to perform a hypothesis test about a population correlation coefficient, ρ, is similar to the general hypothesis testing procedure developed in Chapter 10. However, there were some refinements to this general procedure. They were:

- The variables x and y are each normally distributed and the sampling distribution of the correlation coefficients which is bell-shaped and symmetric about r = 0 when $\rho = 0$ is used as the appropriate hypothesis testing model.
- The test statistic is calculated by using the sample correlation coefficient, r, for the sample data.
- The critical value, r_α, for the decision rule is found in Table V using degrees of freedom, df, calculated by the formula: $df = n - 2$, the α value and the form of the alternative hypothesis.

Example 15.1 illustrates the use of this hypothesis testing procedure to test the significance of a correlation coefficient.

EXAMPLE 15.1

A Parents' Association in a Mission Viejo, California school district interviewed and collected data from a random sample of twenty parents whose children attend the elementary school. The data indicated the number of hours per week that each child watches television and the number of hours spent doing homework. The Association believes there is a negative correlation between the number of hours per week a child watches television and the number of hours per week the child spends doing homework. The sample correlation coefficient, r, was calculated to be –0.52. Test the hypothesis that there is a negative linear relationship between the number of hours a child watches television and spends doing homework. Use $\alpha = 5\%$.

Solution

Step 1 Formulate the hypotheses.
Null Hypothesis:

H_o: The population correlation coefficient of the number of hours a child watches TV and spends doing homework is zero.

This is symbolized as: $H_o: \rho = 0$.
A population correlation coefficient of zero is interpreted to mean that there is no linear relationship between the number of hours per week a child watches television and the number of hours per week the child spends doing homework.
Alternative Hypothesis:
   Since the association believes that there is a negative linear correlation between the number of hours per week a child watches television and the number of hours per week the child spends doing homework, then they are trying to show that the population correlation coefficient is *less than* zero.

H_a: The population correlation coefficient of the number of hours a child watches TV and spends doing homework is *less than* zero.

This is symbolized as: $H_a: \rho < 0$.

Step 2 Determine the model to test the null hypothesis, H_o.
To test the null hypothesis, the association randomly selected 20 parents whose children attend the elementary school and determines the number of hours per week each child watched television and the number of hours per week each child dedicated to homework. To test the null hypothesis the sampling distribution of the correlation coefficients is used as the hypothesis testing model and both variables are assumed to be normally distributed.
   Assuming the null hypothesis is true, i.e. $\rho = 0$, the distribution of r values is bell-shaped and symmetric about $r = 0$ and serves as the hypothesis testing model. This is illustrated in Figure 15.5.

Hypothesis Testing Model
the sampling distribution of the correlation coefficients
when $\rho = 0$

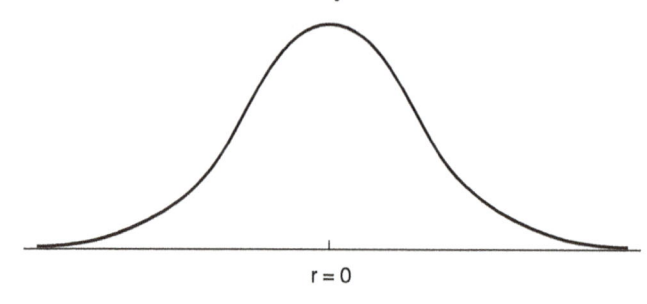

r = 0

FIGURE 15.5

Step 3 Formulate the decision rule.
   a) The alternative hypothesis is directional, since the association is trying to show that ρ is less than zero.
   b) The type of hypothesis test is one-tailed on the left side.
   c) The significance level is: $\alpha = 5\%$.
   d) To determine the critical r value, r_α, we need to compute the degrees of freedom, *df*, using:

$$df = n - 2.$$

For n = 20, we have: df = 20 − 2 = 18. Thus, the critical r value, r_α, for a 1TT on the left with *df* = 18 is found in Table V. From Table V, the critical value is: $r_\alpha = -0.38$.
   e) Thus, the decision rule is: *Reject H_o*, if the test statistic, r, is *less than* −0.38.
Figure 15.6 illustrates the decision rule for this hypothesis test.

Hypothesis Testing Model
the sampling distribution of the correlation coefficients
when $\rho = 0$

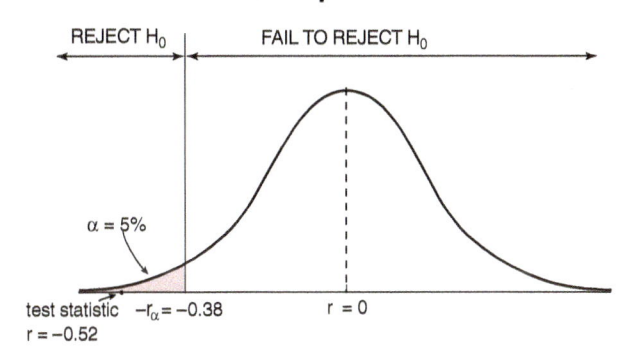

FIGURE 15.6

Step 4 Analyze the sample data.
The Parents' Association calculated the *test statistic* using the sample correlation coefficient formula. The test statistic is:

$$r = -0.52.$$

Step 5 State the conclusion.
Since the test statistic, $r = -0.52$, is *less than* the critical value, $r_\alpha = -0.38$, then the conclusion is *Reject H_o* and *Accept H_a* at $\alpha = 5\%$. Therefore, the Parents' Association found that there is a significant negative linear correlation between the number of hours per week a child watches television and the number of hours per week the child spends doing homework at $\alpha = 5\%$.

EXAMPLE 15.2

A personnel director for ECHO publishing, Inc. believes that there is a significant correlation between the distance (in miles) an employee travels to work and the number of minutes per month the employee is late. Data from a random sampling of 10 employees obtained a sample correlation coefficient of: $r = 0.65$. Test the significance of the population correlation coefficient at $\alpha = 1\%$.

Solution

Step 1 Formulate the hypotheses.
Null Hypothesis:

H_o: The population correlation coefficient of the number of miles traveled to work and number of late minutes per month is zero.

This is symbolized as: H_o: $\rho = 0$.

Alternative Hypothesis:

Since the personnel director did not indicate a direction as to his belief about the relationship between the number of miles an employee travels to work and the number of minutes per month the employee is late, the alternative hypothesis is *nondirectional*.

Therefore, the alternative hypothesis is:

H_a: The population correlation coefficient of the number of miles traveled to work and number of late minutes per month is *not equal* to zero.

This is symbolized as: H_a: $\rho \neq 0$.

Step 2 Determine the model to test the null hypothesis, H_o.

The personnel director will randomly select 10 employees and determine the number of miles each employee travels to work and the number of minutes per month the employee is late. The sampling distribution of the correlation coefficients is used as the hypothesis testing model and both variables are assumed to be normally distributed. Under the assumption that the null hypothesis is true, i.e. $\rho = 0$, the distribution of values is bell-shaped and symmetric about r = 0 and serves as the hypothesis testing model. This is illustrated in Figure 15.7.

Hypothesis Testing Model
the sampling distribution of the correlation coefficients
when $\rho = 0$

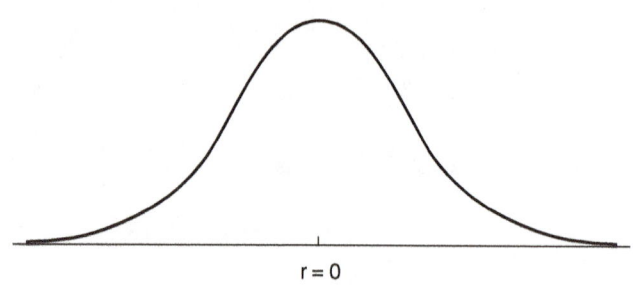

r = 0

FIGURE 15.7

Step 3 Formulate the decision rule.
  a) The alternative hypothesis is nondirectional, since the director is trying to show that ρ is not equal to zero.
  b) The type of hypothesis test is *two-tailed*.
  c) The significance level is: $\alpha = 1\%$.
  d) To determine the critical r values, $r_{\alpha/2}$, we need to compute the degrees of freedom, df, using $df = n - 2$. For n = 10, we have: df = 10 − 2 = 8. Thus, the critical r value, $r_{\alpha/2}$, for a 2TT with $df = 8$ and $\alpha = 1\%$ is found in Table V. From Table V, the critical r values are: $r_{\alpha/2} = \pm 0.76$.
  e) Thus, the decision rule is to Reject H_o, if the test statistic, r, is either: less than the critical r value, $r_{\alpha/2} = -0.76$ or r is greater than the critical value, $r_{\alpha/2} = 0.76$. Figure 15.8 illustrates the appropriate hypothesis testing model.

Hypothesis Testing Model
the sampling distribution of the correlation coefficients
when $\rho = 0$

REJECT H_0 | FAIL TO REJECT H_0 | REJECT H_0

$\frac{1}{2}\alpha = 0.5\%$ $\frac{1}{2}\alpha = 0.5\%$

$-r_{\alpha/2} = -0.76$ r = 0 $r_{\alpha/2} = 0.76$
test statistic:
r = 0.65

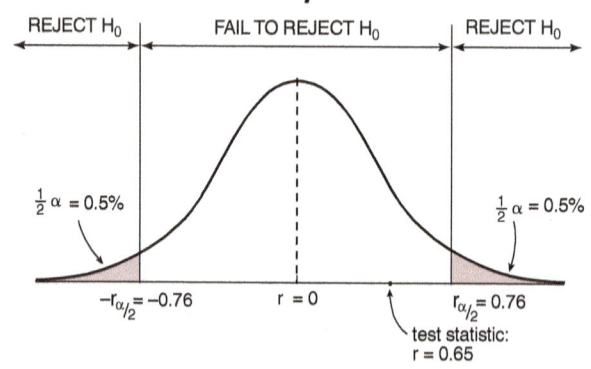

FIGURE 15.8

Step 4 Analyze the sample data.
For this study, ten employees were surveyed. The sample correlation coefficient, r, was calculated to be 0.65. Thus, the test statistic is: r = 0.65.

Step 5 State the conclusion.
Since the test statistic, r = 0.65, is neither less than the critical value of $r_{\alpha/2} = -0.76$ nor greater than the critical value of $r_{\alpha/2} = 0.76$, then the personnel director *fails to reject* H_o at $\alpha = 1\%$. Thus, the linear correlation between the number of miles travelled to work and the number of minutes late per month is *not statistically significant at $\alpha = 1\%$*.

CASE STUDY 15.1 Dropping Out or Missing School

The research article entitled "Dropping Out And Absenteeism in High School" appeared in *Psychological Reports*. The study investigated the relationships among dropping out, absenteeism days of school year, and size of school enrollment. The data were collected from the records of a North Central Kansas High School District. The table in Figure 15.9 represents the Pearson correlations for the variables.

Pearson Correlations For Variables

	1	2	3	4	5	6	7
1. Days of school in the year							
2. Enrollment	.80*						
3. Total days students absent less semester test days	.81*	.92*					
4. Number of students dropping out	.53†	.74*	.48†				
5. Percentage of dropouts							
6. Total number of graduating seniors	.07	.72*	.66*	.29			
7. Total number of seniors not graduating	−.74†	−.15	−.37	−.18		.09	

* $p<.01$. † $p<.05$.

FIGURE 15.9

a) Identify the different variables that were investigated within the study.
b) If a scatter diagram were constructed for the variables enrollment and total number of seniors not graduating, what would you expect the general pattern to look like?
c) If a scatter diagram were constructed for the variables enrollment and total days students absent less semester test days, what would you expect the general pattern to look like?
d) As enrollment increases, what would you expect would happen to the value of the variable number of students dropping out?
e) Interpret the meaning of 0.80 within the second row and first column of the table? What variables are being compared? Is this a positive or negative relationship between the variables? Explain. Is this a significant relationship at $\alpha = 5\%$? at $\alpha = 1\%$? Explain.
f) Interpret the meaning of 0.07 within the sixth row and first column? What variables are being compared? Is this a positive or negative relationship between the variables? Explain. Is this a significant result at 5%? at $\alpha = 1\%$? Explain.
g) What happens to the value of the variables on enrollment, days of school of the school year, and absences as the value of the variable on the number of students dropping out increases? Are these relationships positive or negative?
h) Can the null hypothesis be rejected for the variables number of students dropping out and days of the school year at $\alpha = 5\%$? at $\alpha = 1\%$? Explain. What is the smallest level of significance that would cause the rejection of the null hypothesis? Explain.
i) What is proportion of the variance in enrollment that can be accounted for by the variance in the number of students dropping out? What formula is used to determine this proportion, and what is this statistical measure called?
j) What is proportion of the variance in total number of seniors not graduating that can be accounted for by the variance in the days of school in the year?

CASE STUDY 15.2 Sociodemographic Variables

The research article entitled "Suicide Rates, Handgun Control Laws, and Sociodemographic Variables" appeared in *Psychological Reports*. One aspect of the study examined the relationships between suicide rates, and gender, age, percent black, percent metropolitan population, population density, and rates of population change (increase or decrease), divorce, crime, and unemployment. The Pearson correlation coefficient obtained among suicide rates, handgun control laws (identified as Attribute I: Restrictions on the seller of handguns and Attribute II: Restrictions on the buyer of handguns), and all sociodemographic variables are presented in the table of Figure 15.10.

(continues)

CASE STUDY 15.2 Sociodemographic Variables (continued)

Pearson Correlations (N = 51)

Variable	1	2	3	4	5	6
1. Suicide rate, 1985						
2. Percent male, 1980	.55‡					
3. Percent ages 35-64, 1986	−.29*	−.35†				
4. Percent black, 1985	−.42†	−.52‡	.23			
5. Percent metropolitan, 1985	−.31*	−.35†	.54‡	.32*		
6. Percent per square mile, 1985	−.37†	−.42†	.21	.67‡	.32*	
7. Percent population change, 1980-1985	.52‡	.63‡	−.17	−.14	.03	−.15
8. Divorce rate, 1985	.69‡	.41†	−.05	−.12	−.08	−.11
9. Crime rate, 1985	.19	.02	.19	.26	.60‡	.34*
10. Unemployment rate, 1985	.15	.08	−.20	.22	−.18	.04
11. Restrictions on seller (Attribute I), 1987	−.45‡	−.25	.45‡	−.02	.48‡	−.07
12. Restrictions on buyer (Attribute II), 1987	−.54‡	−.30*	.30*	.37†	.35†	.44‡
M	12.80	48.90	31.80	10.80	64.10	351.30
SD	3.20	1.10	1.90	12.50	22.50	1386.40

	7	8	9	10	11	12
8. Divorce rate, 1985	.58‡					
9. Crime rate, 1985	.51‡	.32*				
10. Unemployment rate, 1985	−.01	.22	.03			
11. Restrictions on seller (Attribute I), 1987	−.40†	−.35†	−.06	−.09		
12. Restrictions on buyer (Attribute II), 1987	−.31*	−.29*	.18	−.03	.21	
M	6.00	5.20	48.20	7.10	1.30	0.50
SD	5.70	1.80	13.60	1.90	1.20	0.90

*p<.05.   †p<.01.   ‡p < .001.

FIGURE 15.10

a) Examine the table in Figure 15.10 and identify all the variables examined within this study.
b) What type of relationship exists between the suicide rate and the restrictions of the seller? Is this a significant relationship at $\alpha = 5\%$? at $\alpha = 1\%$? Can one conclude from this relationship that high suicide rates had significantly less stringent handgun control laws? Explain. What proportion of the variance in the suicide rate that can be accounted for by the variance in the restrictions of the seller? What formula is used to determine this proportion, and what is this statistical measure called?
c) What type of relationship exists between the suicide rate and the percent of males? Is this a significant relationship at $\alpha = 5\%$? at $\alpha = 1\%$? What proportion of the variance in the suicide rate that can be accounted for by the variance in the percent of males? What formula is used to determine this proportion, and what is this statistical measure called?
d) If a scatter diagram were constructed for the variables suicide rate and divorce rate, what would you expect the general pattern to look like?
e) As suicide rates increase, what would you expect would to happen to the variable persons per square mile?
f) As crime rates increase, what would you expect would to happen to the variable persons per square mile?
g) With what variables would suicide rates be significantly correlated? Complete the following statements:
high suicide rates had significantly _____ percent of blacks.
high suicide rates had significantly _____ divorce rates.
high crime rates had significantly _____ persons per square mile.
For each of the other variables, write a similar statement.
h) Can the null hypothesis be rejected for the variables suicide rate and percentage of persons in the 35 to 64 age group at $\alpha = 5\%$? at $\alpha = 1\%$? Explain. What is the smallest level of significance that would cause the rejection of the null hypothesis? Explain.

Summary of the Test Statistic Approach to the Hypothesis Testing Procedure to Test the Significance of a Correlation Coefficient

The following outline is a summary of the test statistic approach to the 5 step hypothesis testing procedure for the test of a correlation coefficient.

Step 1 Formulate the two hypotheses, H_o and H_a.
Null Hypothesis: $H_o: \rho = 0$

Alternative Hypothesis, H_a: can have one of following three forms:

$$H_a: \rho > 0 \text{ or } H_a: \rho < 0 \text{ or } H_a: \rho \neq 0$$

Step 2 Determine the model to test the null hypothesis, H_o.
Under the assumption the null hypothesis is true, that is: $\rho = 0$, the distribution of r values is bell-shaped and symmetric about r = 0 and is referred to as the:

sampling distribution of the correlation coefficients.

This is the hypothesis testing model.

Step 3 Formulate the decision rule.
 a) alternative hypothesis: directional or nondirectional
 b) type of test: 1TT or 2TT
 c) significance level: $\alpha = 1\%$ or $\alpha = 5\%$
 d) determine the critical r value, r_α, using Table V with degrees of freedom: $df = n - 2$
 e) draw diagram.
 f) state the decision rule.

Step 4 Analyze the sample data.
Calculate the sample correlation coefficient, r. This is the test statistic.

Step 5 State the conclusion.
Compare the test statistic, (i.e., the sample correlation coefficient r) to the critical r value, r_α, and state the conclusion:

either: a) Reject H_o and Accept H_a at α
or   b) Fail to reject H_o at α

15.3 ASSUMPTIONS FOR LINEAR REGRESSION ANALYSIS

Beside the assumption of a **linear relationship** between the two variables **x** and **y**, the following three assumptions must be satisfied in order to apply the linear regression model.

Linear Regression Model Assumptions

For each value of x under consideration, there exists a population of y values and they must conform to the following conditions:
- these y populations must have a normal distribution.
- the means of all these populations must lie on a straight line called the population regression line.
- the standard deviation of all the y populations must be equal.

Furthermore, when selecting a value for the independent variable x, this x value must be within the range of the sample x data.

15.4 p-VALUE APPROACH TO TESTING THE SIGNIFICANCE OF THE CORRELATION COEFFICIENT USING THE TI-84 CALCULATOR

The p-value approach will require refinements in steps 3, 4 and 5 of the hypothesis testing procedure. In step 3, the decision rule will be stated as: Reject the H_0 if the p-value of the test statistic, r, is less than α. In step 4, the p-value will be determined using the TI-84 calculator. While in step 5, the conclusion of the hypothesis test is determined by comparing the p-value to the level of significance, α.

Let's reexamine the hypothesis test shown in Example 15.2 using a p-value approach with the aid of the TI-84 calculator.

EXAMPLE 15.3

A personal director of ECHO publishing, Inc. believes that there is a significant correlation between the distance (in miles) an employee travels to work and the number of minutes per month the employee is late. The sample data from a random sampling of 10 employees is shown in Table 15.1.

TABLE 15.1

miles traveled to work:	5	10	15	20	25	30	35	40	45	50
minutes late per month:	3	6	6	5	2	4	7	7	12	8

Perform a hypothesis test at $\alpha = 1\%$ to determine if there is a significant correlation between the distance an employee travels to work and the number of minutes late per month.

Solution

Step 1 Formulate the hypotheses.
Null Hypothesis:

H_o: The population correlation coefficient of the number of miles traveled to work and the number of minutes late per month is zero.

This is symbolized as: $H_o: \rho = 0$.

Alternative Hypothesis:

Since the personnel director did not indicate a direction as to his belief about the relationship between the number of miles an employee travels to work and the number of minutes per month the employee is late, the alternative hypothesis is *nondirectional*.

Therefore, the alternative hypothesis is:

H_a: The population correlation coefficient of the number of miles traveled to work and the number of minutes late per month is *not equal* to zero.

This is symbolized as: $H_a: \rho \neq 0$.

Step 2 Determine the model to test the null hypothesis, H_o.

The personnel director will randomly select 10 employees and determine the number of miles each employee travels to work and the number of minutes per month the employee is late. The sampling distribution of correlation coefficients is used as the hypothesis testing model and both variables are assumed to be normally distributed. Under the assumption that the null hypothesis is true, i.e. $\rho = 0$, the distribution of r values is bell-shaped and symmetric about $r = 0$ and serves as the hypothesis testing model. This is illustrated in Figure 15.11.

Hypothesis Testing Model
the sampling distribution of the correlation coefficients
when $\rho = 0$

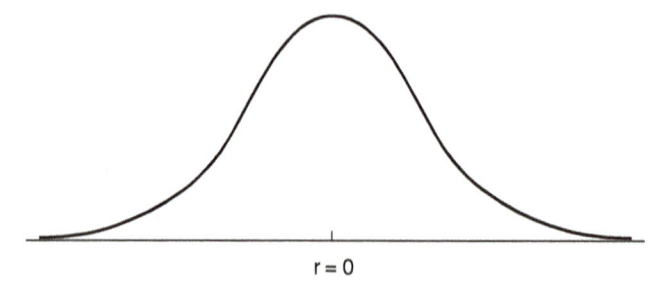

FIGURE 15.11

Section 15.4 *p-Value Approach to Testing the Significance of the Correlation Coefficient Using the TI-84 Calculator* 779

Step 3 Formulate the decision rule.
  a) The alternative hypothesis is nondirectional, since the director is trying to show that ρ is not equal to zero.
  b) The type of hypothesis test is two-tailed.
  c) The significance level is: $\alpha = 1\%$.
  d) The decision rule is: Reject H$_o$ if the p-value of the test statistic, r, is less than $\alpha = 1\%$. Figure 15.12 illustrates the decision rule.

**Hypothesis Testing Model
the sampling distribution of the correlation coefficients
when $\rho = 0$**

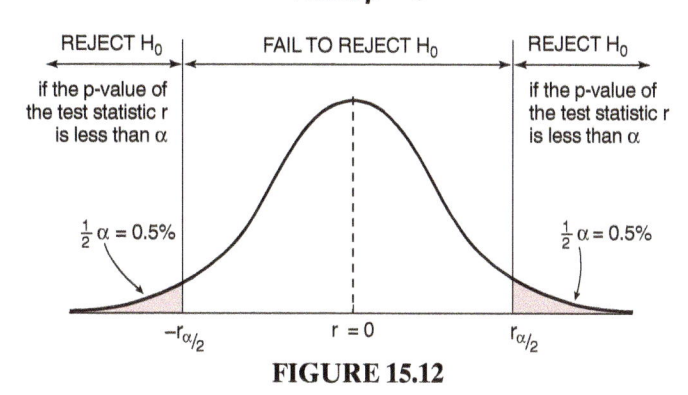

REJECT H$_0$ if the p-value of the test statistic r is less than α

FAIL TO REJECT H$_0$

REJECT H$_0$ if the p-value of the test statistic r is less than α

$\frac{1}{2}\alpha = 0.5\%$

$\frac{1}{2}\alpha = 0.5\%$

$-r_{\alpha/2}$ r = 0 $r_{\alpha/2}$

FIGURE 15.12

Step 4 Analyze the sample data.
  Enter the sample data of Table 15.1 into the TI-84 calculator. Input the values for miles traveled to work in L1 and minutes late per month in L2. This is shown in Figure 15.13.

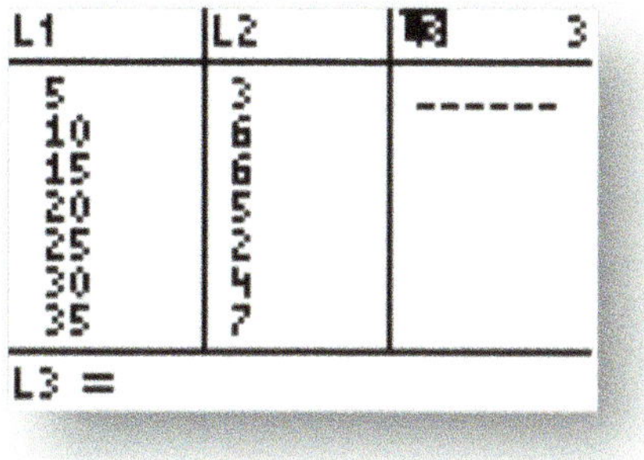

FIGURE 15.13

To determine the p-value of the test statistic, r, we must use the **LinRegTTest** function found in the TESTS menu. To access the **LinRegTTest** function press STAT ▶ ▶ . Scroll down to highlight **E: LinRegTTest**. This is shown in Figure 15.14.

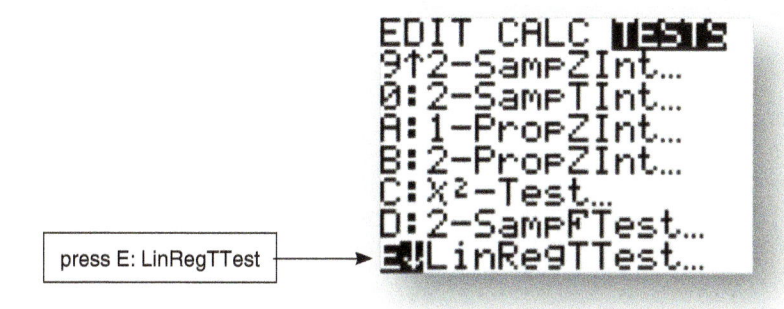

press E: LinRegTTest

FIGURE 15.14

780 Chapter 15 Inferences for Linear Correlation and Regression

Press ENTER to access the **LinRegTTest** menu. Enter the location of the independent and dependent variables and the form of the alternative hypothesis. This is shown in Figure 15.15.

L1 contains data for the independent variable.

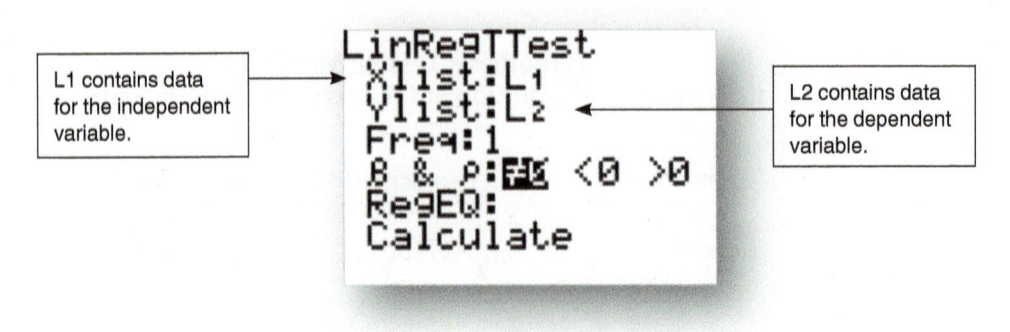

```
LinRegTTest
Xlist:L1
Ylist:L2
Freq:1
β & ρ:≠0  <0  >0
RegEQ:
Calculate
```

L2 contains data for the dependent variable.

FIGURE 15.15

To perform the hypothesis test, highlight Calculate and press ENTER. The output screens are shown in Figure 15.16a and Figure 15.16b

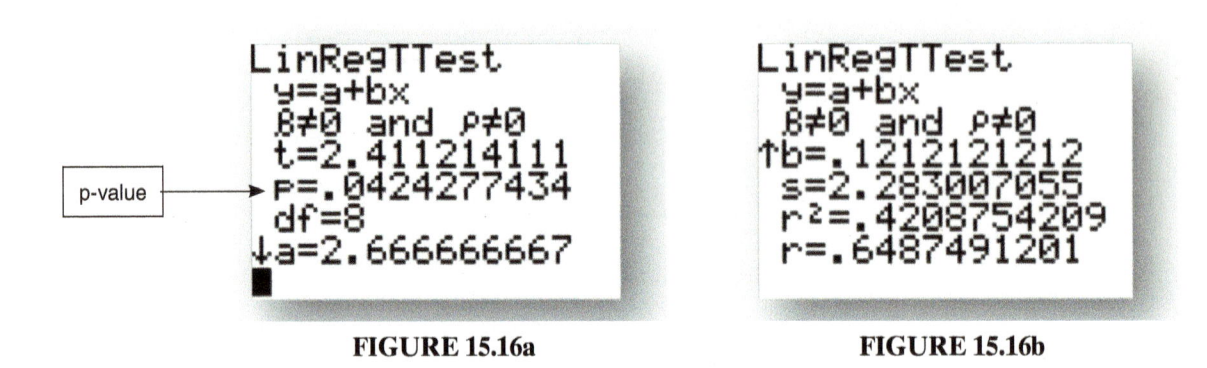

p-value

```
LinRegTTest
y=a+bx
β≠0 and ρ≠0
t=2.411214111
p=.0424277434
df=8
↓a=2.666666667
```

```
LinRegTTest
y=a+bx
β≠0 and ρ≠0
↑b=.1212121212
s=2.283007055
r²=.4208754209
r=.6487491201
```

FIGURE 15.16a   **FIGURE 15.16b**

Notice from the output screen Figure 15.16a, the p-value of the test statistic (r = 0.65) is approximately 0.0424.

Step 5 Determine the conclusion.
   Since the p-value (0.0424) of the test statistic is greater then $\alpha = 1\%$, the personnel director *fails to reject* H_o at $\alpha = 1\%$. This is illustrated in Figure 15.17. Notice for a 2TT, the position of the test statistic (r ≈ 0.65) was determined by taking half of the p-value, i.e. ½(0.0424) = 0.0212, as illustrated is Figure 15.17. Thus, the linear correlation between the number of miles traveled to work and the number of minutes late per month is *not statistically significant at* $\alpha = 1\%$.

Hypothesis Testing Model
the sampling distribution of the correlation coefficients
when $\rho = 0$

```
        REJECT H₀  |   FAIL TO REJECT H₀   |  REJECT H₀
```

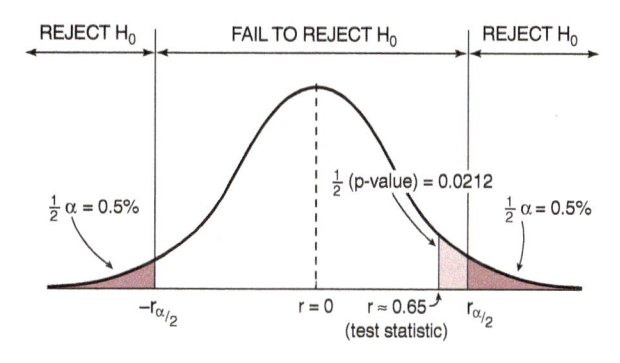

$\frac{1}{2}\alpha = 0.5\%$ $\frac{1}{2}$(p-value) = 0.0212 $\frac{1}{2}\alpha = 0.5\%$

$-r_{\alpha/2}$ $r = 0$ $r \approx 0.65$ (test statistic) $r_{\alpha/2}$

FIGURE 15.17

The following outline is a summary of the 5 step hypothesis testing procedure for the p-value approach to test the significance of a correlation coefficient using the TI-84 calculator.

Summary of p-Value Approach to the Hypothesis Testing Procedure to Test the Significance of a Correlation Coefficient Using the TI-84 Calculator

Step 1 Formulate the two hypotheses, H_o and H_a.
Null Hypothesis: H_o: $\rho = 0$
Alternative Hypothesis, H_a: can have one of following three forms:

$$H_a: \rho > 0 \text{ or } H_a: \rho < 0 \text{ or } H_a: \rho \neq 0$$

Step 2 Determine the model to test the null hypothesis, H_o.
Under the assumption the null hypothesis is true, that is: $\rho = 0$, the distribution of r values is bell-shaped and symmetric about r = 0 and is referred to as the:

sampling distribution of the correlation coefficients.

Step 3 Formulate the decision rule.
a) alternative hypothesis: directional or nondirectional
b) type of test: 1TT or 2TT
c) significance level: $\alpha = 1\%$ or $\alpha = 5\%$
d) state the decision rule as Reject H_o if the p-value of the test statistic, r, is less than α.

Step 4 Analyze the sample data.
Enter the sample data into the TI-84 calculator and use LinRegTTest to determine the p-value of the test statistic, r.

Step 5 State the conclusion.
Compare the p-value of the test statistic, (i.e., the sample correlation coefficient r) to α and state the conclusion as either:
   a) Reject H_o and Accept H_a at the computed p-value, if the p-value of the test statistic r is less than or equal to α

or

   b) Fail to reject H_o at the computed p-value, if the p-value of the test statistic r is not less than α

15.5 INTRODUCTION TO MULTIPLE REGRESSION

Up to this point, we have only covered the concept of a linear regression equation involving one independent variable, x, and one dependent variable, y'. This linear regression equation has the form:

$$y' = a + bx \qquad \text{(equation 1)}$$

where "a" is the intercept of the y' equation and b is the slope of the y' equation. In a **multiple regression equation,** there are several independent variables and one dependent variable. For example, the multiple regression equation involving two independent variables has the form:

$$y' = a + b_1 x_1 + b_2 x_2 \qquad \text{(equation 2)}$$

where x_1 and x_2 represent the independent variables. The variable y' still represents the dependent variable and the constant "a" still represents the y'-intercept. Similarly, the multiple regression equation involving three independent variables would have the form:

$$y' = a + b_1 x_1 + b_2 x_2 + b_3 x_3 \qquad \text{(equation 3)}$$

This idea can be expanded for any number of independent variables and is given in definition 15.2.

DEFINITION 15.2   MULTIPLE REGRESSION EQUATION
The multiple regression equation has the form

$$y' = a + b_1 x_1 + b_2 x_2 + \cdots + b_k x_k$$

where:
- y' is the dependent variable,
- a is the y'-intercept,
- $b_1 \ldots b_k$ are the partial regression coefficients,
- and $x_1 \ldots x_k$ are the independent variables.

To put this concept into perspective, suppose a psychologist is interested in determining if there is a relationship between a student's GPA, the number of hours per week the student spends studying and the age of the student. The psychologist could conduct a study using the dependent variable GPA, and two independent variables x_1, for number of study hours and x_2, for age of the student. This multiple regression equation would have the form of equation 2 above.

As another example, suppose a college admissions officer is trying to determine the relationship that exists between GPA, SAT score, ACT score and quality of recommendation letters. Once again, GPA is the dependent variable and SAT score, ACT score and quality of recommendation letters are the independent variables. This multiple regression equation would have the form of equation 3.

The calculations for the equations of multiple regression analysis are very complicated; as a result, in this section we will use technology to find the multiple regression equations. The technological output will be examined to interpret the results. Before we begin with the multiple regression analysis, you should be familiar with the multiple regression assumptions.

Multiple Regression Assumptions

1. Normality and equal variances between y and each of the independent variables $x_1 \ldots x_k$.
2. There is a linear relationship between the dependent variable y and each of the independent variables $x_1 \ldots x_k$.
3. There is no correlation between independent variables.

The Multiple Correlation Coefficient, R

In simple linear regression, we identified a statistic to measure the strength of the bond that existed (or didn't exist) between the two variables we were studying. This statistic was called the sample correlation coefficient and was designated by r. In multiple regression analysis, a similar statistic exists as well. It is called the sample multiple correlation coefficient and is designated by R. In multiple regression there exists a number of different simple linear regressions between all the variables. For example, take the case where there are two independent variables x_1 and x_2 and the dependent variable y. There are three different simple regressions that we could perform; y with x_1, y with x_2, and x_1 with x_2. Each of these simple regressions would produce a sample correlation coefficient, r. We will denote these simple regression correlation coefficients as r_{yx_1}, r_{yx_2} and $r_{x_1x_2}$ respectively. These values are used in the calculation for the multiple correlation coefficient, R. Let's examine the formula for R for two independent variables.

Formula for the Multiple Correlation Coefficient, R, for Two Independent Variables

$$R = \sqrt{\frac{r_{yx_1}^2 + r_{yx_2}^2 - 2r_{yx_1} \cdot r_{yx_2} \cdot r_{x_1x_2}}{1 - r_{x_1x_2}^2}}$$

where:

r_{yx_1} is the value of the correlation coefficient between variables y and x_1,

r_{yx_2} is the value of the correlation coefficient between variables y and x_2, and

$r_{x_1x_2}$ is the value of the correlation coefficient between variables x_1 and x_2,

---

If the number of independent variables increases, the formula for R will change.

Interpretation of the Multiple Regression Coefficient

In Chapter 4 we discussed the correlation coefficient for simple linear regression. Remember, the values for the sample correlation coefficient r, ranged from –1 to +1. The closer r was to either –1 or +1, the stronger the bond between the two variables. We have a similar situation for the multiple regression correlation coefficient. However, since R is defined by a square root, the only possible values for R would range from 0 to +1. Therefore, the interpretation used for R is that the closer the value of R is to +1, the stronger the bond is between the variables being studied. So, an R value of 0.92 would represent a strong relationship among the variables being studied, where an R value of 0.24 would represent a weak bond among the variables.

Example 15.4 illustrates how to calculate and interpret a multiple correlation coefficient, R, for two independent variables.

EXAMPLE 15.4

A psychologist is interested in determining if a relationship exists between a student's GPA, the number of hours per week the student spends studying and the age of the student. She lets y represent the dependent variable for GPA, x_1 represents the independent variable for number of study hours per week and x_2 represents the independent variable for the age of the student. She performs three simple linear regressions and computes the following correlation coefficients:

$$r_{yx_1} = 0.863, \ r_{yx_2} = 0.787 \text{ and } r_{x_1x_2} = 0.391.$$

a. Using this information, compute the value of R.
b. Would you describe the relationship between these variables as strong, weak or no correlation?

Solution

Step 1 Substituting the values into the formula for the multiple correlation coefficient gives

$$R = \sqrt{\frac{r_{yx_1}^2 + r_{yx_2}^2 - 2r_{yx_1} \cdot r_{yx_2} \cdot r_{x_1x_2}}{1 - r_{x_1x_2}^2}}$$

$$= \sqrt{\frac{(0.863)^2 + (0.787)^2 - 2 \cdot (0.863)(0.787)(0.391)}{1 - (0.391)^2}}$$

$$= \sqrt{\frac{0.833018}{0.847119}}$$

$$R = 0.9916$$

Step 2 An R value close to +1 represents a strong relationship between the variables.

The Coefficient of Determination for a Multiple Regression Equation

As in simple linear regression, R^2 denotes the coefficient of determination. Since there is more than one independent variable in multiple regression, we refer to R^2 as the **multiple coefficient of determination.** The interpretation is the same as in simple linear regression. Remember that R^2 denotes the amount of explained variance given by the regression model. The amount of unexplained variance is given by $1 - R^2$ and is called the **residual error**. For example, the R value of 0.9916 found in Example 15.4 has an explained variance of $R^2 = 0.983$, with a residual error of $1 - R^2 = 0.017$.

The multiple coefficient of determination, R^2 is an increasing function based on the number of independent variables. This means, as the number of independent variables increase, the value of R^2 also increases. This statement is true regardless of the importance of the variables in the multiple regression model and is a huge flaw for multiple regression. Because of this another statistic is needed. This statistic is called the adjusted R^2 and is denoted by R_{adj}^2. The adjusted R^2 takes into account the sample size, n, and the number of independent variables, k, of the multiple regression model.

Formula for the Adjusted R^2

The formula for the adjusted R^2 is given by:

$$R_{adj}^2 = 1 - \frac{(1 - R^2)(n - 1)}{n - k - 1}$$

One way to interpret this formula is that it adjusts R^2 for the number of degrees of freedom in the multiple regression model. The resulting adjusted R^2 value will be smaller than the original R^2. In multiple regression analysis both R^2 and R_{adj}^2 are usually reported and taken into consideration when interpreting the output.

This is illustrated in Example 15.5.

EXAMPLE 15.5

Using the calculated value of $R = 0.9916$ from Example 15.4 and a sample size of n = 50, calculate the adjusted R_{adj}^2.

Solution

There were two independent variables given in Example 15.4 so the value of k = 2. The R value was calculated to be 0.9916. For n = 50, the adjusted R^2 is:

$$R^2_{adj} = 1 - \frac{(1-R^2)(n-1)}{n-k-1}$$

$$= 1 - \frac{(1-0.9916^2)(50-1)}{50-2-1}$$

$$R^2_{adj} = 0.9826$$

As you can see, the value for the adjusted R^2 is smaller than the original R^2. This means that the explained variance due to the multiple regression model is smaller when adjusted for the number of independent variables and sample size. Consequently, this resulted in an increase in the residual error or unexplained variance.

Testing the Significance of the Population Multiple Correlation Coefficient, ρ

Just as with simple linear regression, the Null and Alternative hypotheses are given by:

$$H_o: \rho = 0 \quad \text{and} \quad H_a: \rho \neq 0$$

where ρ represents the value of the population multiple correlation coefficient. The Null Hypothesis $\rho = 0$ states that the population multiple correlation coefficient is 0 and that there is no correlation between the variables being studied. The Alternative Hypothesis $\rho \neq 0$ means that the population multiple correlation coefficient is different than 0 and states that there is a correlation among the variables being studied. The hypothesis testing model is given by the F Distribution and uses what we call an F Test. The F Test is covered in the e-book on the textbook's website at www.MyStatLab.com. For simplicity, we'll only look at the p-value approach to this hypothesis test. Remember, if the p-value is between 0.01 and 0.05, then the test is considered to be significant, and if the p-value is smaller than 0.01, then the test is very significant. We will demonstrate this approach using the technological tool MINITAB as outlined in Example 15.6.

EXAMPLE 15.6

Table 15.2 gives the information for 25 college student's college GPA, High School GPA, SAT score and a rating of the quality of the student's letters of recommendations. Use MINITAB to:

a. Find and interpret the multiple regression correlation coefficient, R.
b. Find the values for R^2 and the adjusted R^2.
c. Use the p-value approach to test the significance of the population multiple correlation coefficient.
d. Identify the multiple regression equation.

TABLE 15.2 Data for 25 College Students

College GPA	HS GPA	SAT	Quality of Letters	College GPA	HS GPA	SAT	Quality of Letters
2.04	2.01	1070	5	3.11	3.12	1246	6
2.56	3.4	1254	6	1.92	2.14	1106	4
3.75	3.68	1466	6	0.81	2.6	790	5
1.1	1.54	706	4	1.01	1.9	954	4
3	3.32	1160	5	3.66	3.06	1500	6
0.05	0.33	756	3	2	1.6	1046	5
1.38	0.36	1058	2	2.05	1.96	1054	4
1.5	1.97	1008	7	2.6	1.96	1198	6
1.38	2.03	1104	4	2.55	1.56	940	3
4.0	2.05	1200	7	0.38	1.6	456	6
1.5	2.13	896	7	2.48	1.92	1150	7
1.29	1.34	848	3	2.74	3.09	636	6
1.9	1.51	958	5				

Data obtained from: http://davidmlane.com/hyperstat/Multiple_regression_exampl.html

Solution

Technological tools such as MINITAB are used to perform multiple regression analysis. To perform the regression analysis, first you must import the data into a spreadsheet in MINITAB. A snapshot of this can be seen in Figure 15.18.

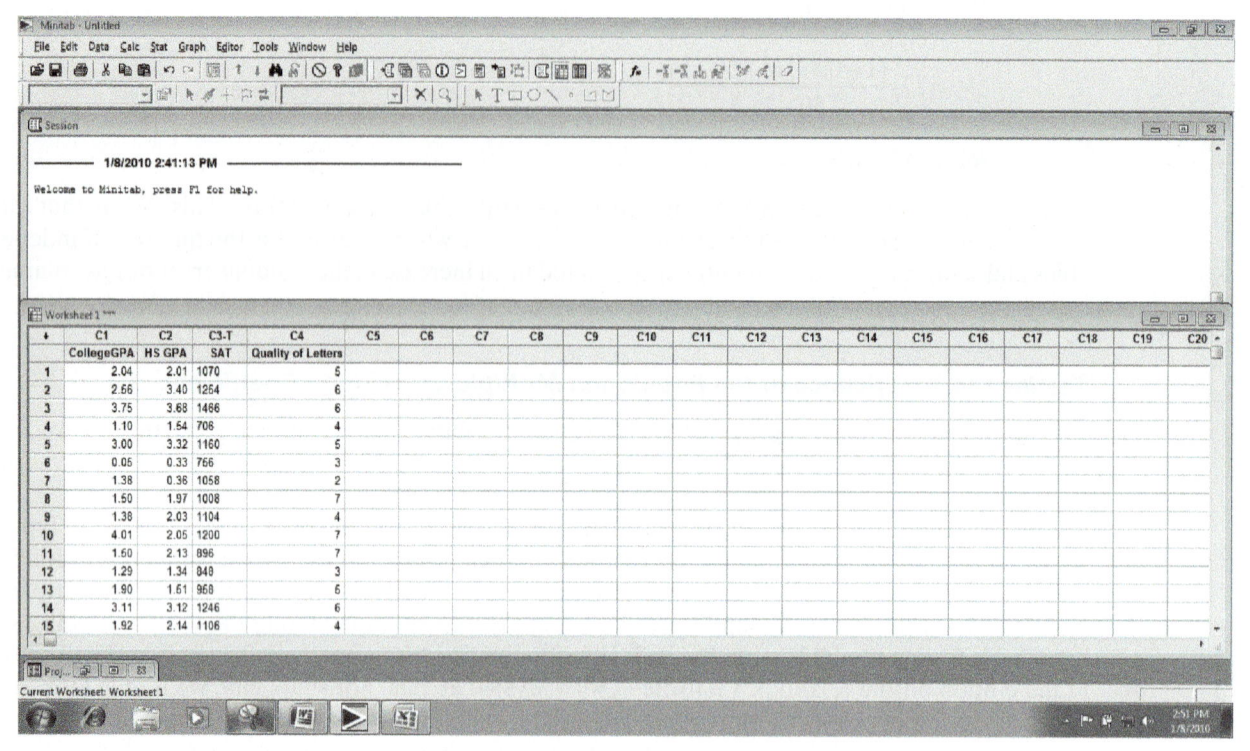

FIGURE 15.18

To perform the regression, follow these steps:

Step 1 Click Stat>Regression>Regression.

Step 2 As shown in the Figure 15.19, you should then place College GPA in the Response variable field (the dependent variable), and HS GPA, SAT and Quality of Letters in the Predictors field (the independent variables).

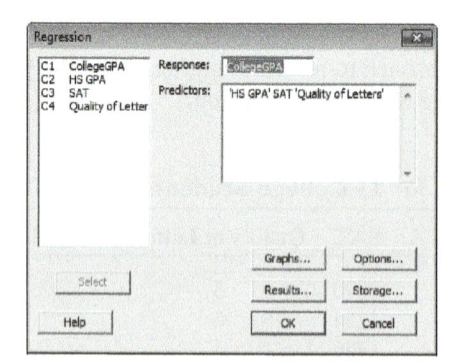

FIGURE 15.19

Step 3 Click OK. The results are posted in the session window of MINITAB. The results of the regression are given in Figure 15.20.

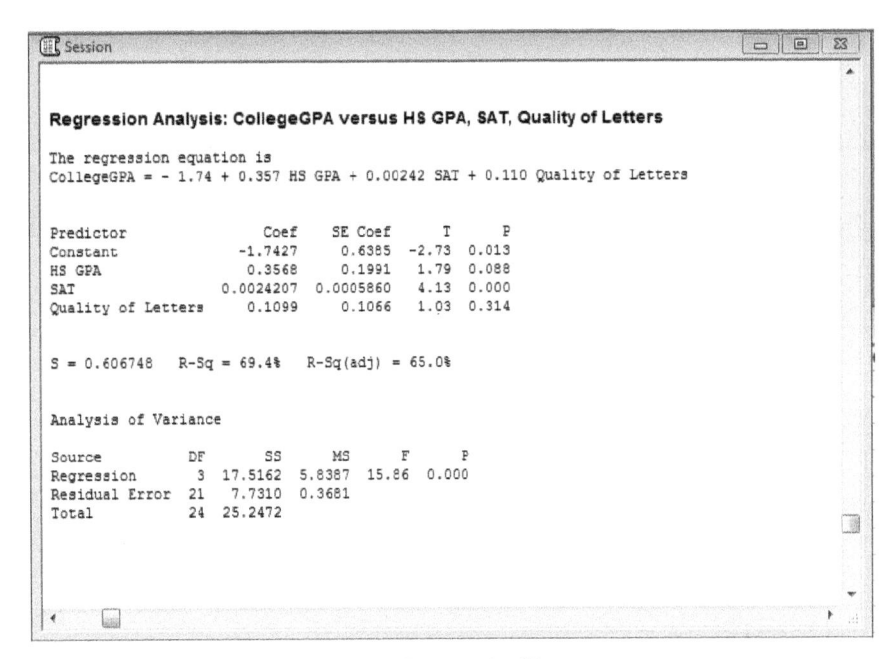

```
Session

Regression Analysis: CollegeGPA versus HS GPA, SAT, Quality of Letters

The regression equation is
CollegeGPA = - 1.74 + 0.357 HS GPA + 0.00242 SAT + 0.110 Quality of Letters


Predictor              Coef     SE Coef       T       P
Constant            -1.7427      0.6385   -2.73   0.013
HS GPA               0.3568      0.1991    1.79   0.088
SAT              0.0024207    0.0005860    4.13   0.000
Quality of Letters   0.1099      0.1066    1.03   0.314


S = 0.606748   R-Sq = 69.4%   R-Sq(adj) = 65.0%


Analysis of Variance

Source          DF       SS       MS       F       P
Regression       3  17.5162   5.8387   15.86   0.000
Residual Error  21   7.7310   0.3681
Total           24  25.2472
```

FIGURE 15.20

a. The value for R is not given in the output for MINITAB. However the value for R^2 is given. To find R, take the square root of R^2.

So, $\sqrt{R^2} = \sqrt{0.694} \approx 0.833$. Therefore, the value of the multiple regression correlation coefficient is 0.833. A value for R = 0.833 shows a good relationship exists among the variables.

b. Directly from the output, the values for R^2 and the adjusted R^2 are given as 69.4% and 65% respectively.

c. We are testing the Null Hypothesis H_o: $\rho = 0$ versus the Alternative Hypothesis H_a: $\rho \neq 0$. There are a number of different p-values given in the MINITAB output. One p-value is given for each of the independent variables with regard to the dependent variable. If we refer to the line in the MINITAB output entitled "Regression," the p-value for the test statistic with regard to all of the independent variables is given in the MINITAB output as being 0.000, which is a p-value of *essentially* 0. Since the p-value for the test statistic is smaller than 0.01, this test is very significant and we can reject the Null Hypothesis. To interpret the result of this test, since we are rejecting the null hypothesis, the decision would be that there exists a significant relationship among the variables student's college GPA, High School GPA, SAT score and the rating of the quality of the student's letters of recommendations.

d. The regression equation is the first result listed in the MINITAB output and is given as:

CollegeGPA = –1.74 + 0.357 HS GPA + 0.00242 SAT + 0.110 Quality of Letters

This same output can be obtained in EXCEL using the Data Analysis tool. Figure 15.21 shows the data set along with the output the *Regression* command in the Data Analysis tool in EXCEL gives.

As you can see from the output, the results obtained in EXCEL are the same as the ones given by MINITAB. The p-value for the hypothesis test is given under the caption "Significance F" in Figure 15.21.

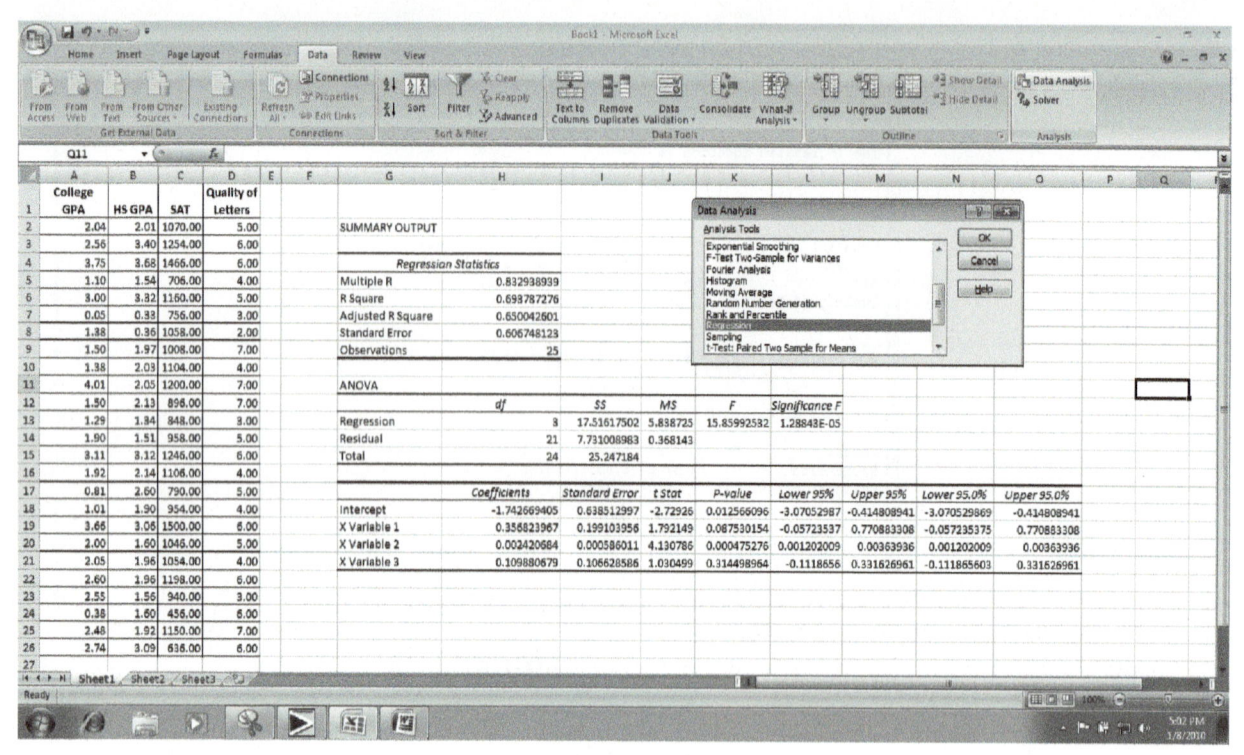

Chapter 15 Inferences for Linear Correlation and Regression

FIGURE 15.21

The given p-value is 1.28843E-05 which is scientific notation for 0.0000128843 or essentially 0 as displayed by MINITAB. Also, the regression equation isn't given in EXCEL as it was in MINITAB. EXCEL only gives the coefficient for the y'-intercept along with the coefficients for each of the independent variables. This information is listed in the third table under the Coefficients column.

You should also note that in addition to the information we discussed, both MINITAB and EXCEL give other information for multiple regression analysis that was not covered in this chapter. We limited our discussion to these topics in order to just introduce this concept to you. Further study would be required in an Intermediate Statistics course.

KEY TERMS

Adjusted R^2, R^2_{adj} 784
Assumptions for Multiple Regression 782
Coefficient of Determination for a Multiple Regression Equation, R^2 784
Critical Value of the Correlation Coefficient, r_α 771
Dependent Variable 767
Independent Variable, x 767
Linear Regression 777
Multiple Correlation Coefficient, R 783
Multiple Regression Equation 782
Predicted Value of y, y' 767
Regression Coefficients, a and b 782
Residual Error 784
Sampling Distribution of Correlation Coefficients 770
y'-intercept 782

SECTION EXERCISES

Note: The data sets in these Section Exercises can be downloaded in either Microsoft EXCEL or TI-84 format from the textbook's website at www.MyStatLab.com.

Many exercises require calculations that are long and tedious. Therefore, we recommend that you use a calculator or a computer with statistical application software to help you perform these calculations.

Section 15.2

1. The population correlation coefficient is symbolized by the lower case Greek letter _____, pronounced as row.

2. We use the sample correlation coefficient, r, to calculate the correlation coefficient for the _____ data. The correlation coefficient, r, for the sample is used as _____ of the population correlation coefficient, ρ.

3. **a)** The null hypothesis for testing the significance of the population correlation coefficient is stated as: the population correlation coefficient is equal to _____. That is, there is _____ linear correlation between the two variables. This is symbolized as: $H_o: \rho =$ _____.
    b) The alternative hypothesis can be stated in one of the following three forms:
    form a: The population correlation coefficient, ρ, is claimed to be greater than zero. That is, there is a _____ linear correlation between the two variables. This is symbolized as: $H_a: \rho$ _____ 0.
    form b: The population correlation coefficient, ρ, is claimed to be less than zero. That is, there is a _____ linear correlation between the two variables.
    This is symbolized as: $H_a: \rho$ _____ 0
    form c: The population correlation coefficient, ρ, is claimed to be not equal to zero. That is, there is a _____ linear correlation between the two variables.
    This is symbolized as: $H_a: \rho$ _____ 0.
4. The test of the significance of the correlation coefficient is based upon the assumption that both variables x and y are _____ distributed. This test uses as its hypothesis testing model: the sampling distribution of the _____ _____.
5. The critical value of the correlation coefficient is denoted as _____ . To determine this critical r value, you need to calculate the degrees of freedom for the test using the formula: $df =$ _____ .
6. The coefficient of determination measures the proportion of the variance of the _____ variable that can be accounted for by the variance of the _____ variable. To calculate the coefficient of determination, you need to square the value of _____.
7. Negative correlations are usually not significant. (T/F)
8. In a to f, there is a list of sample information, Pearson's correlation coefficient, type of test, and alpha level. Use this information and Table V to determine r_α, and to test if r is significant.
    a) n = 15, r = 0.60, 2TT, $\alpha = 1\%$
    b) n = 25, r = −0.60, 1TT, $\alpha = 1\%$
    c) n = 30, r = 0.60, 1TT, $\alpha = 1\%$
    d) $df = 10$, r = 0.42, 2TT, $\alpha = 5\%$
    e) $df = 20$, r = −0.42, 1TT, $\alpha = 5\%$
    f) $df = 28$, r = 0.42, 1TT, $\alpha = 5\%$

Section 15.3

In problems requiring the testing of the significance of the correlation coefficient, assume that both the independent and dependent variables are normally distributed.

9. Don, the grounds keeper at the Sunview Golf course, wants to know if there is a positive linear relationship between the density of seed spread on a newly landscaped fairway and the density of new grass six months later. The following data pairs represent some sample plantings made by Don.

Seed Density (lbs/400ft²)	Grass Density (seedlings/ft²)
1.0	170
2.0	200
3.0	240
4.0	300
5.0	310
6.0	290
7.0	290

a) Calculate the sample correlation coefficient, r.
b) Use your answer from part a and perform a hypothesis test to determine if a significant positive correlation exists between seed density and grass density. Perform this test at $\alpha = 5\%$.

10. A prominent psychologist wonders if a patient's score on the extrovert scale (those who seek out social environments) is positively correlated with time spent on social network sites online. She gathers data from 20 patients for both variables; the results are as follows:

Extrovert Scale Value	Time on Social Network Sites (min/day)
40	46
45	79
52	33
62	63
31	20
28	18
5	11
83	78
55	63
32	46
47	21
45	55
60	59
13	23
7	30
85	80
38	25
61	26
26	33
3	7

a) Calculate the sample correlation coefficient, r.
b) Use your answer from part a and perform a hypothesis test to determine if, a significant positive correlation exists between a patient's extrovert scale score and time spent on social networking sites. Perform this test at $\alpha = 1\%$.

11. Grace, the personnel director at Weight Lookers International, believes there is a negative linear relationship between the number of sick days taken per year by an employee and the percent of annual salary increase the employee receives. A random sample of 20 employees yielded the following data pairs.

Number of Sick Days per Year	% of Annual Salary Increase
4	12
16	8
11	8
14	7
6	14
20	0
9	10
12	9
13	7
1	14
0	12
17	2
18	2
7	13
9	10
15	7
15	8
13	9
10	8
5	12

a) Calculate the sample correlation coefficient, r.
b) Use your answer from part a and perform a hypothesis test to determine if a significant negative correlation exists between number of sick days per year and percent of annual salary increase. Perform this test at $\alpha = 5\%$.

12. Professor DePorto believes there is a positive linear relationship between Midterm examination grades and Final examination grades for students taking his introduction to statistics course. A random sample of 26 students who have taken both exams produced the following data pairs.

Midterm Exam Grade	Final Exam Grade
39	17
28	42
32	30
35	24
30	15
42	39
34	16
36	15
22	19
33	32
22	39

(Continued)

Midterm Exam Grade	Final Exam Grade
38	20
29	25
50	42
16	36
41	37
43	35
44	40
39	21
30	44
24	11
16	17
29	19
33	15
44	29
39	35

a) Calculate the sample correlation coefficient, r.
b) Use your answer from part a and perform a hypothesis test to determine if a significant positive correlation exists between a student's midterm exam grade and final exam grade. Perform this test at $\alpha = 5\%$.

13. Craig, a marketing executive for a microbrewery company, wants to determine if there is a positive linear relationship between advertising expenditures and sales for their new lite beer product, Less Ale. He randomly samples data for the past 7 sales years and records the sample data in the following table, where advertising expenditure is measured in thousands and the beer sales in millions of dollars.

Year	Expenditure (thousand $)	Sales (million $)
2013	35	38
2014	47	35
2015	65	49
2016	92	50
2017	55	40
2018	25	35
2019	82	44

a) Construct a scatter diagram.
b) Calculate the sample correlation coefficient, r.
c) Determine if r is significant at $\alpha = 1\%$.
   If r is significant at $\alpha = 1\%$, then do parts d,e,f.
d) Find r^2 and interpret its meaning.
e) Determine the regression equation, y'.
f) Using the regression equation, predict the beer sales for the advertising budget expenditure of $50,000.

14. A nutritionist believes that there is a positive correlation between the total amount of fat (g) and total amount of cholesterol (mg) in a fast food item. She randomly selects 15 items from a fast food menu and records the following sample data:

Item	Fat (g)	Cholesterol (mg)
1	9	25
2	12	40
3	11	0
4	4	5
5	22	45
6	9	65
7	16	35
8	19	40
9	29	75
10	26	95
11	26	70
12	12	25
13	2	5
14	7	25
15	9	20

a) Calculate the sample correlation coefficient, r.

b) Perform a hypothesis test at $\alpha = 1\%$ to determine if there is a significant positive linear relationship between the total amount of fat (g) and total amount of cholesterol (mg) in a fast food item.

15. Many educators believe that there is a linear relationship between the number of hours a student works per week versus the course grade. Some think that the relationship is positively correlated while others believe the relationship is negatively correlated. The following data pairs represent the number of hours students work per week and their final course average in a liberal arts statistics class.

Number of hours worked per week (x)	Final course average (y)
21	75
10	81
5	57
9	84
40	60
0	85
45	55
12	96
8	88
15	76
55	50
23	69
35	75
60	50
18	87
16	75
10	70
48	60
35	80
25	83

a) Calculate the sample correlation coefficient, r.

b) Use your answer from part a and perform a hypothesis test to determine if a significant negative correlation exists between number of hours a student spends working per week and the student's final course average. Perform this test at $\alpha = 5\%$.

16. A researcher working for The Sunglass Shack Corporation wants to investigate whether there is a significant positive correlation between daily temperature and sunglasses sold at its Brooklyn location. The following data pairs represent some sample data collected:

Temperature (degrees F)	Number of Sunglasses Sold
80	131
61	20
56	101
99	62
101	44
95	189

a) Calculate the sample correlation coefficient, r.

b) Use your answer from part a and perform a hypothesis test to determine if a significant positive correlation exists between the temperature outside and number of sunglasses sold by Sunglass Shack. Perform this test at $\alpha = 5\%$.

17. The following data were taken from the information found on the internet on Blood Alcohol Level (BAL) relating to the number of drinks a 100 lb female consumes in one hour. A drink is defined as: 1.25 oz of 80 proof liquor, or 12 oz of regular beer, or 5 oz of table wine.

Blood Alcohol Level (BAL) for a 100 lb. Female.

# drinks in one hour	BAL
1	0.05
2	0.10
3	0.15
4	0.20
5	0.25
6	0.30
7	0.36
8	0.41

a) Calculate the sample correlation coefficient, r.

b) Use your answer from part a and perform a hypothesis test to determine if a significant positive correlation exists between the number of drinks a 100 lb female student consumes in one hour and her blood alcohol level. Perform this test at $\alpha = 5\%$.

18. Lylah, a long distance runner, wants to know if there is a positive linear relationship between the temperature at the start time of the Chicago Marathon and the time it takes her to finish the marathon. She raced in the Chicago Marathon every year for the past ten years and recorded her results in the following table:

Year	Temperature (Fahrenheit)	Finish Time (minutes)
2004	54	266
2005	51	266
2006	40	265
2007	72	292
2008	64	286
2009	32	267
2010	60	283
2011	59	280
2012	40	272
2013	48	272

a) Calculate the sample correlation coefficient, r.
b) Use your answer from part a and perform a hypothesis test to determine if a significant positive correlation exists between the temperature outside and her finish time for the Chicago Marathon. Perform this test at $\alpha = 1\%$.

19. Nola, a marketing researcher, randomly samples 12 employees to determine if there is a correlation between years of schooling and salary earnings. The following data represents the sample results.

Current Salary	Years of Education
$57,000	15
$40,200	16
$27,500	12
$45,000	15
$23,450	8
$32,100	13
$67,800	16
$37,500	12
$77,600	16
$33,000	13
$55,600	15
$62,500	14

a) Calculate the sample correlation coefficient, r.
b) Use your answer from part a and perform a hypothesis test to determine if a significant positive correlation exists between number of years of education and current salary. Perform this test at $\alpha = 1\%$.

Section 15.4

For problems 20–30, use questions 9 to 19 of Section 15.3 exercises and apply the p-value approach to hypothesis testing using the TI-84 calculator. For each of these problems, use the 5 step procedure to perform a hypothesis test at **both** $\alpha = 1\%$ and $\alpha = 5\%$. Include the following within your test:

a) state the null and alternative hypotheses.
b) calculate the expected results for the experiment assuming the null hypothesis is true and state the hypothesis test model.
c) formulate the decision rule.
d) identify the test statistic and its p-value.
e) state the conclusion and answer any question(s) posed in the problem for each alpha level.
f) determine the smallest level of significance for which the null hypothesis can be rejected.

Section 15.5

Note: The data files for this problem set can be downloaded in EXCEL or TI-84 format from the textbook's website at www.MyStatLab.com.

31. A multiple regression equation has the form $y' = a + b_1x_1 + b_2x_2 + \cdots + b_kx_k$ where _____ is the dependent variable, _____ is the y'-intercept, _____ are the partial regression coefficients, and _____ are the independent variables.

32. A multiple regression equation containing two independent variables has the form _____.

33. A multiple regression equation has the form $y' = a + b_1x_1 + b_2x_2 + b_3x_3$. How many independent variables are there? _____

34. In a multiple regression analysis, there exists a strong relationship among the independent variables (T/F) _____.

35. In multiple regression analysis, it is assumed that there is normality among the dependent variable and each of the independent variables (T/F) _____.

36. The multiple correlation coefficient, R will always take on a value between _____ and _____.

37. A value of +1 for R represents a _____ bond among the variables being studied.

38. To calculate the multiple correlation coefficient, R, use the formula _____.

39. A multiple regression equation contains two independent variables. The simple regression correlation coefficients are given as $r_{yx_1} = 0.854$, $r_{yx_2} = 0.790$ and $r_{x_1x_2} = 0.398$. What is the value of R, the multiple regression correlation coefficient?

40. A doctor is studying the effects of smoking and high cholesterol on his patients. He is trying to decide if a strong relationship exists between the age (y) his patient, number of years the patient was a smoker (x_1) and average cholesterol level of his patient over the last 5 physicals (x_2). Using a simple linear regression model among the variables, the doctor determines the following correlations: $r_{yx_1} = 0.801$, $r_{yx_2} = 0.850$ and $r_{x_1x_2} = 0.458$.
a) What is the value of the multiple regression correlation coefficient, R?
b) Would you classify the relationship among these variables as strong or weak? Explain.

41. Use your result from problem 39 for a sample size of n = 50 to find:
a) the value for R^2 and the adjusted R^2.
b) Interpret the meaning of R^2. Which value is smaller?
c) Calculate the residual error.

42. Use your result from problem 40 for a sample size of n = 100 to find:
a) the value for R^2 and the adjusted R^2.
b) Interpret the meaning of R^2. Which value is smaller?
c) Calculate the residual error.

43. The value for R in a multiple regression equation with 5 independent variables was given as R = 0.914.
a) Find and interpret the value of R^2.
b) A statistician feels a bit "uncomfortable" relying only on R^2 for her analysis. Why might this be? What other statistic could she compute in order to settle her comfort level?
c) Knowing that the sample size n = 75, use your result from part (a) to find the value of the adjusted R^2.

44. The following table gives the values for 13 football players' left leg strength (in lbs.), right leg strength and punting distance (in feet). A football analyst is trying to decide if there is a strong correlation among left and right leg strength and punting distance. He decides to use punting distance for his dependent variable. For the independent variables he used left and right leg strength.

Left (lbs)	Right (lbs)	Punt (ft)
170	170	162.5
130	140	144
170	180	174.5
160	160	163.5
150	170	192
150	150	171.75
180	170	162
110	110	104.83
110	120	105.67
120	130	117.58
140	120	140.25
130	140	150.17
150	160	165.17

Source for data: http://www.sci.usq.edu.au/staff/dunn/Datasets/applications/popular/punting.html

Use a technological tool such as MINITAB or EXCEL to:

a) Find and interpret the values of R, R^2 and the adjusted R^2. Does there appear to be a relationship among these variables?
b) Use the p-value approach to test the significance of the multiple correlation coefficient.
c) Identify the multiple regression equation.
d) Use the multiple regression equation to predict the punting distance for a punter whose left leg strength is 145 lbs and right leg strength is 162 lbs.

45. Back in the days when ice cream could have been considered a luxury item, people spent a relatively "large portion" of their weekly income when purchasing ice cream. The following data gives the ice cream consumption over 30 four-week periods from March 18, 1950 to July 11, 1953. The variables given are consumption (in pints), price of ice cream (in dollars), weekly family income (in dollars) and temperature (in degrees). Use consumption as the dependent variable.

Consumption (pt)	Price ($)	Income ($)	Temp (degrees)
0.386	0.27	78	41
0.374	0.28	79	56
0.393	0.28	81	63
0.425	0.28	80	68
0.406	0.27	76	69
0.344	0.26	78	65
0.327	0.28	82	61
0.288	0.27	79	47
0.269	0.27	76	32
0.256	0.28	79	24
0.286	0.28	82	28
0.298	0.27	85	26
0.329	0.27	86	32
0.318	0.29	83	40

(Continued)

Consumption (pt)	Price ($)	Income ($)	Temp (degrees)
0.381	0.28	84	55
0.381	0.29	82	63
0.47	0.28	80	72
0.443	0.28	78	72
0.386	0.28	84	67
0.342	0.28	86	60
0.319	0.29	85	44
0.307	0.29	87	40
0.284	0.28	94	32
0.326	0.29	92	27
0.309	0.28	95	28
0.359	0.27	96	33
0.376	0.27	94	41
0.416	0.27	96	52
0.437	0.27	91	64
0.548	0.26	90	71

Data source: http://www.sci.usq.edu.au/staff/dunn/Datasets/Books/Hand/Hand-R/icecream-R.html

Use a technological tool such as MINITAB or EXCEL to:

a) Find and interpret the values of R, R^2 and the adjusted R^2. Does there appear to be a relationship among these variables?
b) Use the p-value approach to test the significance of the multiple correlation coefficient.
c) Identify the multiple regression equation.
d) Use the multiple regression equation to predict the consumption given the price of the ice cream was $0.25, the weekly family income was $82 and temperature was 68 degrees.

46. The following data set contains data for 38 college students IQ, brain size, height (inches) and weight (in lbs). A school psychologist would like to know if there exists a significant relationship between these variables and decides to perform a multiple regression analysis of the data. She decides to use IQ as the dependent variable.

IQ	Brain	Height	Weight	IQ	Brain	Height	Weight
124	81.69	64.5	118	84	90.59	76.5	186
150	103.84	73.3	143	134	79.06	62	122
128	96.54	68.8	172	128	95.5	68	132
134	95.15	65	147	102	83.18	63	114
110	92.88	69	146	131	93.55	72	171
131	99.13	64.5	138	84	79.86	68	140
98	85.43	66	175	110	106.25	77	187
84	90.49	66.3	134	72	79.35	63	106
147	95.55	68.8	172	124	86.67	66.5	159
124	83.39	64.5	118	132	85.78	62.5	127
128	107.95	70	151	137	94.96	67	191
124	92.41	69	155	110	99.79	75.5	192
147	85.65	70.5	155	86	88	69	181
90	87.89	66	146	81	83.43	66.5	143
96	86.54	68	135	128	94.81	66.5	153
120	85.22	68.5	127	124	94.94	70.5	144
102	94.51	73.5	178	94	89.4	64.5	139
84	80.8	66.3	136	74	93	74	148
86	88.91	70	180	89	93.59	75.5	179

Data source: http://www.stat.psu.edu/~lsimon/stat501wc/sp05/data/iqsize.txt

Use a technological tool such as MINITAB or EXCEL to:
a) Find and interpret the values of R, R^2 and the adjusted R^2. Does there appear to be a relationship among these variables?
b) Use the p-value approach to test the significance of the multiple correlation coefficient.
c) Identify the multiple regression equation.
d) Use the multiple regression equation to predict the IQ of a student given a brain size of 86.6, a height of 71.5 inches and a weight of 182.5 lbs.

47. The accompanying table gives the gender, height and weight for 92 people. A gender value of 1 indicates male and 2 indicates female.

Gender	Height	Weight	Gender	Height	Weight
1	66	140	1	71	150
1	72	145	1	68	155
1	73.5	160	1	69.5	150
1	73	190	1	73	180
1	69	155	1	75	160
1	73	165	1	66	135
1	72	150	1	69	160
1	74	190	1	66	130
1	72	195	1	73	155
1	71	138	1	68	150
1	74	160	1	74	148
1	72	155	1	73.5	155
1	70	153	1	70	150
1	67	145	1	67	140
1	71	170	1	72	180
1	72	175	1	75	190
1	69	175	1	68	145
1	73	170	1	69	150
1	74	180	1	71.5	164
1	66	135	1	71	140
1	71	170	1	72	142
1	70	157	1	69	136
1	70	130	1	67	123
1	75	185	1	68	155
2	61	140	2	66	130
2	66	120	2	65.5	120
2	68	130	2	66	130
2	68	138	2	62	131
2	63	121	2	62	120
2	70	125	2	63	118
2	68	116	2	67	125
2	69	145	2	65	135
2	69	150	2	66	125
2	62.75	112	2	65	118
2	68	125	2	65	122
1	74	190	2	65	115
1	71	155	2	64	102
1	69	170	2	67	115
1	70	155	2	69	150
1	72	215	2	68	110
1	67	150	2	63	116
1	69	145	2	62	108
1	73	155	2	63	95

Gender	Height	Weight	Gender	Height	Weight
1	73	155	2	64	125
2	67	150	2	68	133
2	61.75	108	2	62	110

a) Is there a significant relationship between the weight of a person (dependent variable) and that person's height and gender (independent variables)? Use the p-value method to perform this test with a technological tool.
b) The p-value for this test is extremely small. Do you feel it is appropriate to perform a multiple regression analysis using these data? What can you conclude by making a comparison of the independent variables?

48. A doctor wishes to determine if a significant relationship exists between the recovery time of a patient after surgery (dependent variable), the size of the dose of anesthetic and average blood pressure during the surgery (independent variables).

Dose	BP	Recovery	Dose	BP	Recovery
3.96	59	8	4.42	69	12
4.01	68	26	4.58	72	25
3.68	63	16	4.58	63	45
4.95	65	23	5.41	56	72
5.2	72	7	4.14	70	25
3.8	58	11	5.43	69	28
3.75	69	8	3.66	60	10
5.53	70	14	4.84	51	25
6.22	73	39	4.14	61	44
4.37	56	28	5.2	66	7
6.4	83	12	4.17	52	10
5.23	67	60	4.1	72	18
4.01	84	10	3.55	67	4
6.03	68	60	4.74	69	10
4.14	64	22	4.01	71	13
4.17	60	21	5.89	88	21
3.64	62	14	5.27	68	12
5.55	76	4	4.14	59	9
3.8	60	27	5.34	73	65
5.16	60	26	4.7	68	20
3.91	59	28	4.33	58	31
5.64	84	15	2.72	61	23
3.96	66	8	4.79	68	22
5.46	68	46	3.91	69	13
5.13	65	24	4.01	55	9

Data source: http://www.oxfordjournals.org

Use a technological tool such as MINITAB or EXCEL to:
a) Find and interpret the values of R, R^2 and the adjusted R^2. Does there appear to be a relationship among these variables?
b) Use the p-value approach to test the significance of the multiple correlation coefficient.
c) Identify the multiple regression equation.
d) Use the multiple regression equation to predict the recovery time for a patient who had a dose size of 4.75 and a blood pressure during surgery of 70.

(*Continued*)

CHAPTER REVIEW

49. For a 1TT on the left side, as the value of the test statistic decreases, then the p-value increases.
(T/F)

50. A _____ linear correlation is a correlation between two variables, x and y, that occurs when high measurements on the x variable tend to be associated with low measurements on the y variable and low measurements on the x variable tend to be associated with high measurements on the y variable.

51. If y increases as x decreases this indicates that x and y are correlated:
a) Positively   b) Not at all
c) Negatively

52. A value of r = –1 represents the _____ negative linear correlation possible and it indicates a _____ negative linear correlation. This means that all the points of the scatter diagram will lie on a straight line which is sloping _____ from left to right.

53. For each of the following two variables, state whether you would expect a positive, negative or no linear correlation.
a) Number of hours practicing golf and golf score
b) A husband's age and his wife's age
c) Number of hours practicing bowling and bowling score
d) Amount of alcohol consumed and reaction time
e) Mother's birth weight and her child's birth weight
f) A person's age and number of days of sick-leave
g) A person's age and life insurance premium
h) High school average and college grade-point average
i) Amount of grams of fat consumed daily and cholesterol level

54. When the sample correlation coefficient is greater than zero, that is: r > 0, then there is a _____ linear correlation between the two variables. When the sample correlation coefficient is less than zero, that is: r < 0, then there is a _____ linear correlation between the two variables.

55. A negative value of r indicates:
a) No linear relationship
b) A weak linear relationship
c) A positive linear relationship
d) A negative linear relationship

56. In a to f, there is a list of sample information, Pearson's correlation coefficient, type of test, and alpha level. For each part, use the information and Table V to determine r_α, and to test if r is significant.
a) n = 18, r = 0.35, H_a: $\rho \neq 0$, $\alpha = 1\%$
b) n = 28, r = 0.47, H_a: $\rho > 0$, $\alpha = 1\%$
c) n = 30, r = –0.46, H_a: $\rho < 0$, $\alpha = 1\%$
d) df = 19, r = 0.58, H_a: $\rho \neq 0$, $\alpha = 5\%$
e) df = 25, r = –0.34, H_a: $\rho < 0$, $\alpha = 5\%$
f) df = 13, r = 0.40, H_a: $\rho > 0$, $\alpha = 5\%$

57. Rank the following values of Pearson's correlation coefficient, r, from the strongest to weakest linear association.
–0.95, 0.05, +1.00, 0.69, –0.69, 0, 0.50

58. When testing the significance of the correlation coefficient, r, the null hypothesis, H_o, is always symbolized as:
a) $\rho > 0$ b) $\rho < 0$
c) $\rho = 0$ d) $\rho \neq 0$

59. When the test statistic, r, is significant, then the coefficient of determination is greater than 50%.
(T/F)

60. For each of the scatter diagrams in Figure 15.22, choose the statement that best describes the type of relationship that exists between the two variables.
1) strong linear correlation.
2) moderate or weak linear correlation.
3) no linear correlation.

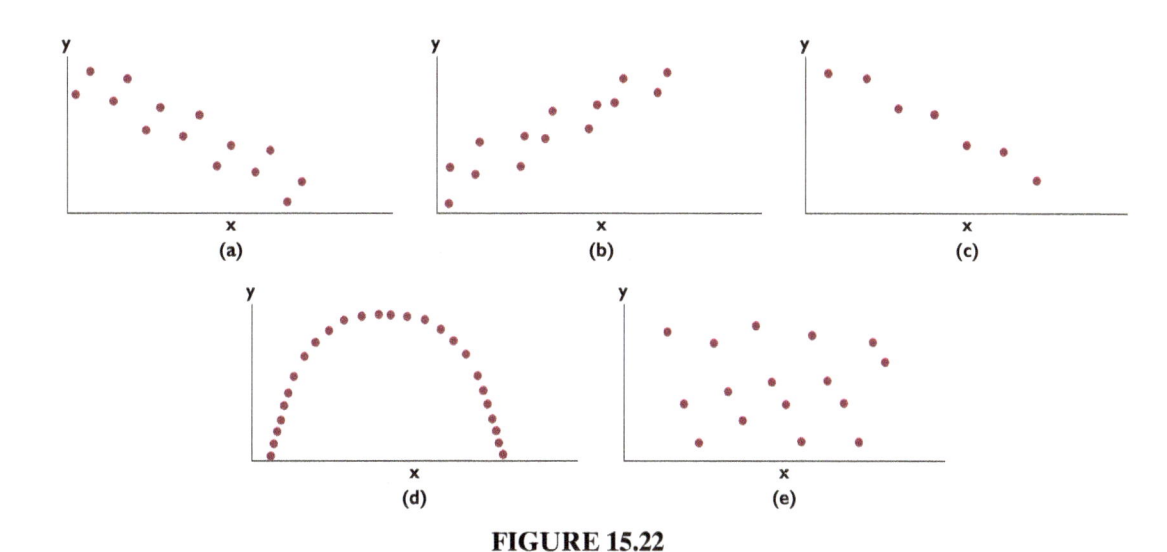

FIGURE 15.22

61. A coefficient of determination of 0.5, means that there is a 50% chance that a linear relationship exists between variables x and y.
(T/F)

62. Linear regression can be used to predict a value for the dependent variable from a value of the independent variable.
(T/F)

63. A linear regression model produces a line that best fits the data pairs from a sample.
(T/F)

64. Matthew, a medical researcher, wants to determine if a positive linear relationship exists between the number of pounds a male is overweight and his blood pressure. He randomly selects a sample of 10 males and records the number of pounds that the individual is over his "ideal" weight according to a medical weight chart and his systolic blood pressure. Matthew's sample data is listed in the following table.

Male Subject	Number of Lbs Above Ideal Weight	Systolic Blood Pressure
1	12	150
2	8	142
3	20	165
4	17	152
5	14	147
6	23	158
7	6	135
8	4	128
9	15	135
10	19	153

a) Calculate the sample correlation coefficient, r.
b) Use your answer from part a and perform a hypothesis test to determine if a significant positive correlation exists between the number of pounds above your idea weight and blood pressure. Perform this test at $\alpha = 1\%$.

65. A sports statistician wants to test whether there is a negative linear correlation between practice hours and golf score for the golf players who belong to the Arnie Palmer Country Club. She randomly selected ten golf players and determined the number of hours they practiced per week and their golf score on the Country Club Golf Course during the annual tournament. The sample results were:

Golf Score: 92 83 85 87 91 95 80 84 90 88
Practice Hrs.: 10 15 17 20 9 11 26 18 13 22

a) Calculate the sample correlation coefficient, r.
b) Use your answer from part a and perform a hypothesis test to determine if a significant negative correlation exists between a players golf score and number of practice hours. Perform this test at $\alpha = 1\%$.

66. Sydney, an educational researcher, randomly samples 15 college students to determine if there is a correlation between reading and mathematics scores on a standardized educational test. The following data represents the reading and mathematics test results for the 15 students.

Reading Scores X Values	Mathematics Scores Y Values
191	180
103	101
187	173
108	103
180	170
118	113
178	171
191	180

(Continued)

Reading Scores X Values	Mathematics Scores Y Values
176	168
134	130
165	150
147	145
160	150
157	154
145	130

a) Calculate the sample correlation coefficient, r.
b) Use your answer from part a and perform a hypothesis test to determine if a significant correlation exists between reading and mathematics scores on this standardized test. Perform this test at $\alpha = 1\%$.

67. Back in the early 1990's, psychologists studied the relationship between an individual's physical attractiveness and penalties given to them for misdemeanor crimes by judges. The psychologists found that the more attractive the individual was, the lower the fine on the misdemeanor crime. You wish to determine if this negative correlation still exists today and study 15 recent convicted individual's shoplifting fines. These individuals were all first time offenders in NYC. Photos of the 15 individuals convicted of shoplifting were shown to 100 random people in a separate survey in order to determine their level of attractiveness. Attractiveness was rated on a 1 – 10 scale, one being least attractive and 10 being highest. The attractiveness scores were averaged. The table below shows the results of the survey along with the actual fines.

Average Attractiveness Rating	Fine Given by Judge
2.5	$500
3.2	$400
4.5	$450
4.7	$400
5.6	$425
5.8	$225
6.1	$350
6.5	$300
6.7	$225
7.2	$250
7.6	$350
8.8	$200
9.1	$225
9.2	$350
9.5	$300

a) Calculate the sample correlation coefficient, r.
b) Use your answer from part a and perform a hypothesis test to determine if a significant negative correlation exists between average attractiveness and fine given by a judge for the first time offense of a misdemeanor shoplifting crime. Perform this test at $\alpha = 1\%$.

WHAT DO YOU THINK?

68. Examine the following table which represents the correlations between the SAT (Scholastic Aptitude Test) scores and the GRE (Graduate Record Examination) scores for 22,923 subjects of a study entitled "The Differential Impact Of Curriculum On Aptitude Test Scores" which appeared in the *Journal of Educational Measurement* in 1990.

The population of interest for the study consisted of examinees who took the SAT and also the GRE General Test at the normal times in their academic careers, with the typical number of years intervening.

Intercorrelations Between SAT and GRE Scores for the Total Study Sample
n = 22,923

	SAT Verbal	SAT Mathematical	GRE Verbal	GRE Quantitative	GRE Analytical	Mean	Standard Deviation
SAT Verbal	1.000	.628	.858	.547	.637	518.8	104.7
SAT Mathematical	.628	1.000	.598	.862	.734	556.0	110.2
GRE Verbal	.858	.598	1.000	.560	.649	510.1	107.7
GRE Quantitative	.547	.862	.560	1.000	.730	573.4	125.6
GRE Analytical	.637	.734	.649	.730	1.000	579.7	117.6

The SAT is a test administered by the College Board to college-bound high school students. The results are used by college undergraduate admission departments in making decisions regarding high school applicants. While the Graduate Record Examination (GRE), a College Board administered exam, is taken by college students preparing to apply for admission to graduate schools. The purpose of the study was to determine the correlation between the SAT and GRE scores of students who took these exams the normal times during their academic careers with the typical number of years between exams.

a) What type of correlation exists between the SAT Verbal and the GRE Verbal? Interpret this relationship. That is, in general, what would you expect to happen to the value of the GRE Verbal scores as the scores on the SAT Verbal increases? If a scatter diagram were constructed for the variables SAT Verbal and the GRE Verbal, what would you expect the general pattern to look like?

b) What proportion of the variance in the SAT Verbal scores can be accounted for by the variance in the GRE Verbal scores? What formula is used to determine this proportion, and what is this statistical measure called?

c) What does a value of 1.000 indicate about the relationship between SAT Verbal and SAT Verbal? What type of a relationship is this called?

d) For the SAT Verbal Column, what row variable, other than the SAT Verbal, has the strongest relationship with the SAT Verbal? Explain. What do you believe might cause this to happen?

e) For the SAT Mathematical Column, what row variable, other than the SAT Mathematical, has the strongest relationship with the SAT Mathematical? Explain. What do you think might lead to this strong relationship?

f) What proportion of the variance in the SAT Mathematical scores can be accounted for by the variance in the GRE Analytical? What proportion of the variance in the SAT Mathematical scores can be accounted for by the variance in the GRE Quantitative? Explain what the difference in these two results mean?

g) The authors of the study state: "the correlations between SAT Verbal and GRE Verbal and between the SAT Mathematical and GRE Quantitative, both of which are 0.86, indicating that the linear relationship between SAT and GRE scores explains almost three fourths of the variance in GRE Verbal and GRE Quantitative scores taken four years later." Explain the meaning of this statement and how do you think the authors arrived at this three fourths value? What is an equivalent percentage value for this three fourths figure? What do the authors mean by "linear relationship"? Explain.

h) For the GRE Analytical Column, what row variable has the weakest relationship with the GRE Analytical? Explain. What do you think might lead to this weak relationship?

i) In words, describe the type of relationship and the strength of the relationship between the variables GRE Quantitative and SAT Mathematical.

j) Interpret the information that the mean and standard deviation indicate about the SAT and GRE test results. Which distribution of test scores has the greatest variability?

k) If the distribution of each of these test scores is approximately normal, then what is the median and modal test score for each test?

l) If the distribution of each of these test scores is approximately normal, then what percent of the SAT Mathematical scores are within 110.2 points of 556? within 220.4 points of 556? within 330.6 points of 556?

m) As the scores of the SAT Mathematical increase, what would you expect would happen to the scores on the variable GRE Quantitative? Explain.

69. The following are excerpts from five different studies.

1. In a study of 748 pregnant women, maternal birth weight was significantly related to baby's birth weight. The lower the maternal birth weight, the lower the baby's birth weight.

2. The College Board which administers the SAT's stated that there is a correlation between family income and test performance with low-income students not doing as well on the SAT's as students from high-income families.

3. A research finding indicated that tall children under age 8 tend to do better on intelligence tests than short children.

4. The results of a study support the notion that a child's drive is related to the age of the father. The higher the scores of a student, the younger the father. Older fathers were defined as those over age 30 at the time of their offspring's birth.

5. A researcher cites that time spent on homework is positively related to achievement.
   a) For each excerpt state the variables within the study. Identify which variable is the independent and dependent variable.
   b) Indicate the type of relationship that exists between the two variables.
   c) If a scatter diagram were constructed for each of the variables, what would you expect the general pattern to look like? Explain.

PROJECTS

70. Randomly collect the heights to the nearest inch (x) and the weights to the nearest pound (y) of 20 students at your school.
    a) Construct a scatter diagram.
    b) Find r for your random sample.
    c) What type of linear correlation, if any, do you observe?
    d) Test $H_o: \rho = 0$, against $H_a: \rho > 0$ at $\alpha = 5\%$.
    e) If the test is significant at $\alpha = 5\%$, determine the regression line model and plot it on your scatter diagram.
    f) Estimate the weight of a student if his height is 74 inches.
    g) Determine the percent of variance in weight that is accounted for by the variance in height.

71. Select two variables that you believe are related and randomly sample the selected population. Using your sample data follow the procedure set forth in project 70.

CHAPTER TEST

72. Which type of linear correlation is indicated by a scatter diagram whose points are generally lower as you move from left to right?
    a) Positive
    b) None
    c) Negative

73. If y increases as x increases this indicates that x and y are correlated:
    a) Positively
    b) Not at all
    c) Negatively

74. If x and y are perfectly linearly correlated then r must equal:
    a) 1
    b) –1
    c) 0
    d) 1 or –1

75. A positive value of r indicates:
    a) No linear relationship
    b) A weak linear relationship
    c) A positive linear relationship
    d) A negative linear relationship

76. When the coefficient of determination is one, this means that there is a perfect positive linear correlation between the x and y variables. (T/F)

77. When testing the significance of r the degrees of freedom is given by:
    a) n   b) n – 2   c) n – 1   d) 2n

78. When testing the significance of the correlation coefficient, if there are ten pairs of data, $\alpha = 5\%$ and it's a 1TT, the $r_\alpha = $ _____.
    a) 0.54   b) 0.83   c) 0.62   d) 0.71

79. MaryRose, a nutritionist, has conducted a study to determine the linear relationship between oat bran consumption and cholesterol level.
    a) If MaryRose obtains a correlation coefficient of –0.56 from a random sample of 24 people, can she conclude that there is a negative linear correlation between the amount of oat bran consumed and cholesterol level at $\alpha = 1\%$?
    b) What percent of the variability in cholesterol level can be accounted for by the variability in the amount of oat bran consumed?

80. I.M.Smart, a statistics nerd, wants to investigate if there is a positive linear relationship between the number of hours a student studies for a statistics exam and the student's test grade. He randomly samples 12 students that are taking statistics at a community college and records the sample data in the following table.

Student	Hours Studying for Exam	Test Grade
1	10	80
2	8	60
3	12	78
4	20	90
5	7	65
6	4	60
7	9	70
8	15	85
9	11	75
10	6	70
11	2	45
12	1	50

a) Calculate the sample correlation coefficient, r.
b) Use your answer from part a and perform a hypothesis test to determine if a significant positive correlation exists between the number of hours a student spends studying for their statistics exam and the student's test grade. Perform this test at $\alpha = 1\%$.

81. It has been suspected for many years that muscle mass decreases with an individual's age. A doctor wishes to test the significance of this relationship in a certain group of his own patients. To help him understand this relationship in his female patients, the doctor selected a subset of his women patients aged 40 to 80. The data is given in the table below.

X is age, Y is a measure of muscle mass (the higher the measure, the more muscle mass).

X (Age)	Y (Muscle Mass)	X (Age)	Y (Muscle Mass)
72	80	77	63
62	92	63	86
42	101	46	105
66	69	57	73
55	86	44	95
71	73	51	100
67	79	40	112
57	83	80	75

a) Calculate the sample correlation coefficient, r.
b) Use your answer from part a and perform a hypothesis test to determine if a significant negative correlation exists between a patient's age and their muscle mass. Perform this test at $\alpha = 1\%$.

82. Rachel Roe Hass, a sports dietitian who counsels athletes about optimal nutrition to enhance their performance, believes there is a positive linear correlation between calcium intake and knowledge about calcium benefits for sports science students. Rachel randomly selects 20 sports science students and administers a quiz to measure their knowledge about the benefits of taking calcium. In addition, she also measures their calcium intake in mg/day. Rachel's sample data results are listed in the following table.

Student	Knowledge Score (out of a possible 50)	Calcium Intake (mg/day)	Student	Knowledge Score (out of a possible 50)	Calcium Intake (mg/day)
1	12	450	11	39	740
2	42	1050	12	25	733
3	38	990	13	48	985
4	15	575	14	28	763
5	22	710	15	22	583
6	37	854	16	45	850
7	40	800	17	18	798
8	14	493	18	24	854
9	26	630	19	30	805
10	32	894	20	43	1085

a) Calculate the sample correlation coefficient, r.
b) Perform a hypothesis test to determine if there is a significant positive linear correlation between calcium intake and knowledge about calcium benefits for sports science students at $\alpha = 1\%$.

83. The following data represents the age (in months) of 19 female Polar bears along with their weight at the given age (in pounds).

Age(months)	Weight(pounds)
29	121
104	166
100	220
57	204
53	144
44	140
20	105
9	26
57	125
84	180
57	116
45	182
82	356
70	316
58	202
11	62
83	236
17	76
17	48

a) Calculate the sample correlation coefficient, r.
b) Use your answer from part a and perform a hypothesis test to determine if a significant positive correlation exists between the age in months of a Polar bear and weight of the bear in pounds. Perform this test at $\alpha = 1\%$.

For the answer to this question, please visit www.MyStatLab.com.

CHAPTER 16

The F-Distribution and an Introduction to Analysis of Variance (ANOVA)

Contents

16.1 Introduction
16.2 F-Distribution
16.3 Testing Variances: The F-Test
16.4 Types of Variances
16.5 One-Way Analysis of Variance
16.6 A Brief Look at Two-Way ANOVA

16.1 INTRODUCTION

We have studied various hypothesis testing procedures. Some of these procedures involved population proportions, while others involved population means. You should be aware that each application of an hypothesis testing procedure requires a specific model distribution whose *critical values* are obtained from a table listed in Appendix D. For example, when a hypothesis test involved two population means where the population standard deviations were unknown, the specific model distribution was called the sampling distribution of the difference between two means. Critical values for the distribution of the test statistic for this model were found using the *t* distribution table listed in Appendix C.

In this chapter we will present a new distribution called the *F-Distribution*. This distribution will be used to answer questions about variability or variance. Let's consider the following two examples. As you read through each example you will notice each one raises a different question. However, both questions will be answered by using the **same** distribution, that is, by using the F-Distribution.

EXAMPLE 16.1

In traveling to work Paulette likes to drive different routes on different days. One day she might drive through the side streets and back roads and the next day, she would ride on the highway. A sample of her travel times, measured in minutes, appears in Table 16.1.

TABLE 16.1 Paulette's Travel Times to Work During a Two Week Time Period

Back Roads	Highway
10	12
12	13
14	14
16	15
18	16
$\bar{x} = 14$	$\bar{x} = 14$
$s^2 = 8$	$s^2 = 2$

Only available in the e-book.

Notice that **both** mean values are the same but the sample variances are **different** (Remember that variance is a measure of variability and that the positive square root of the variance is the standard deviation). For these data, the variance for highway driving is **smaller** than the variance for back road driving. Can Paulette infer from these sample results that the population variances have a similar relationship?

To answer this question an *F-Test* can be used. However to use an F-Test, a distribution called the F-Distribution must be explored.

EXAMPLE 16.2

Jim is interested in buying a highly rated computer. He checks a PC magazine and selects three manufacturers. From these manufacturers he draws a sample of computers rated by the magazine. Table 16.2 displays the samples with their respective means and variances.

TABLE 16.2 Ratings of Computers From a PC Magazine

Computers	Compost	Dovell	ACE Power
	29.2	33.6	24.4
	21.4	44.4	42.0
	47.1	25.2	25.2
	34.0	33.6	34.9
	37.1	46.4	52.9
$\bar{x}$	33.76	36.64	35.88
s^2	90.63	76.21	143.52

Jim would like to determine which manufacturer has a **significantly** higher mean rating. Notice that this question is similar to those posed when using an hypothesis testing procedure like a t-test. However the t-test was only used with hypotheses involving one population mean or at most two population means. In this instance we are comparing three population means. To answer this question we would use a procedure called *Analysis of Variance* or to use the acronym, *ANOVA*. As in Example 16.1 the ANOVA procedure also makes use of the F-Distribution. Thus, in order for us to answer the questions posed in Examples 16.1 and 16.2, we must first explore the F-Distribution.

16.2 F-DISTRIBUTION

When we studied the differences of two population means we used the t-test for independent samples and assumed the means were calculated from data randomly selected from two distinct approximately normal populations. We also assumed that **the populations had equal variances** and consequently equal standard deviations. However, at the time we did not have the tools to test the assumption of equal population variances. The F-Distribution and its corresponding F-Test, can be used to test if two population variances are equal. Let us first consider the F-Distribution.

> **DEFINITION 16.1**
> **F-Distribution.** The F-Distribution is a family of distributions composed of the ratio of the variances for samples drawn from two distinct normal populations with equal variances. This is illustrated in Figure 16.1.

To form an F-Distribution, we theoretically select independent random samples from two normal populations having equal variances, one from population one and one from population two. For each

Section 16.2 F-Distribution

$$\sigma_1^2 = \sigma_2^2$$

Normal Population 1   **Normal Population 2**

Random sample selected → s_1^2 : variance of sample 1

Random sample selected → s_2^2 : variance of sample 2

Form Ratio $F = \dfrac{s_1^2}{s_2^2}$

FIGURE 16.1

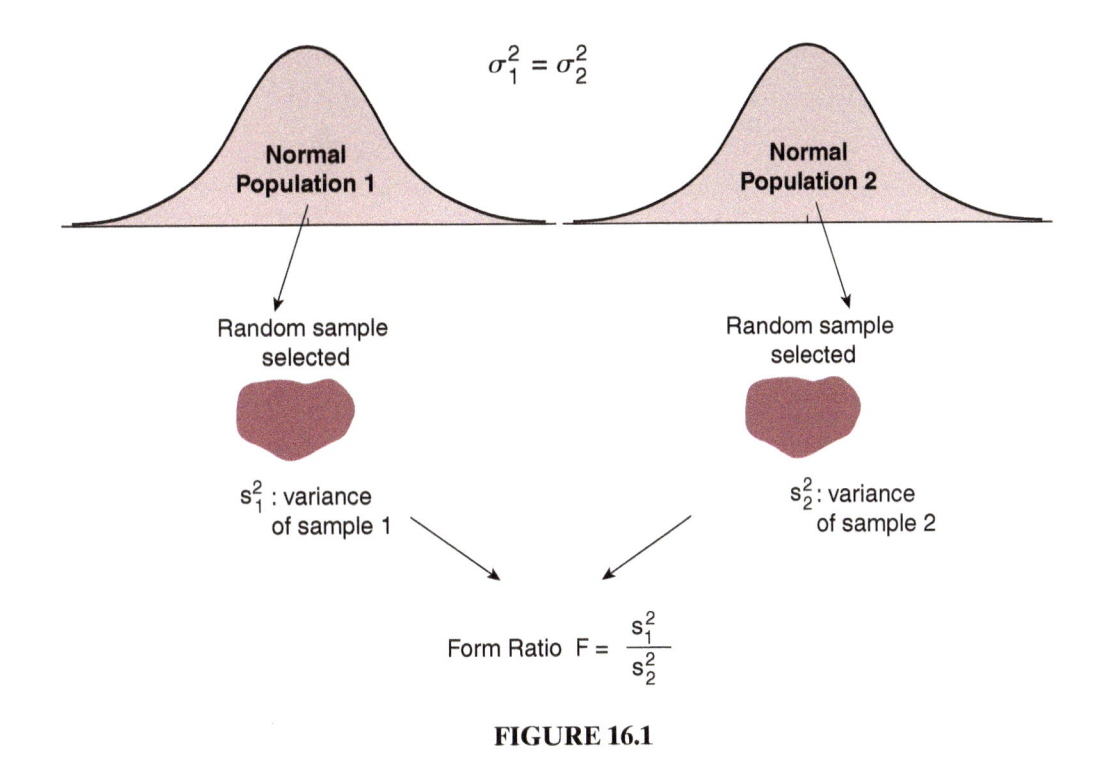

random sample, the sample variance is computed. The ratio (fraction) of each pair of sample variances is formed. Each ratio is called an F-ratio.

Formula to Compute an F-ratio, Denoted by F

$$F = \frac{s_1^2}{s_2^2}$$

where s_1^2 represents a sample variance for a random sample of size n_1 selected from population one,

and

       s_2^2 represents a sample variance for a random sample of size n_2 selected from population two.

If this process is continued for all possible random sample pairs the resulting distribution of ratios is called an F-Distribution. This is shown in Figure 16.2.

To illustrate the creation of the F-Distribution further consider Table 16.3. The three samples displayed in Table 16.3 were randomly selected from two populations, identified as population one and population two, having equal population variances of 2.87. Within Table 16.3, the sample variances of each sample are given along with the corresponding F-ratio

$$F = \frac{s_1^2}{s_2^2} .$$

804 *Chapter 16* *The F-Distribution and an Introduction to Analysis of Variance (ANOVA)*

FIGURE 16.2

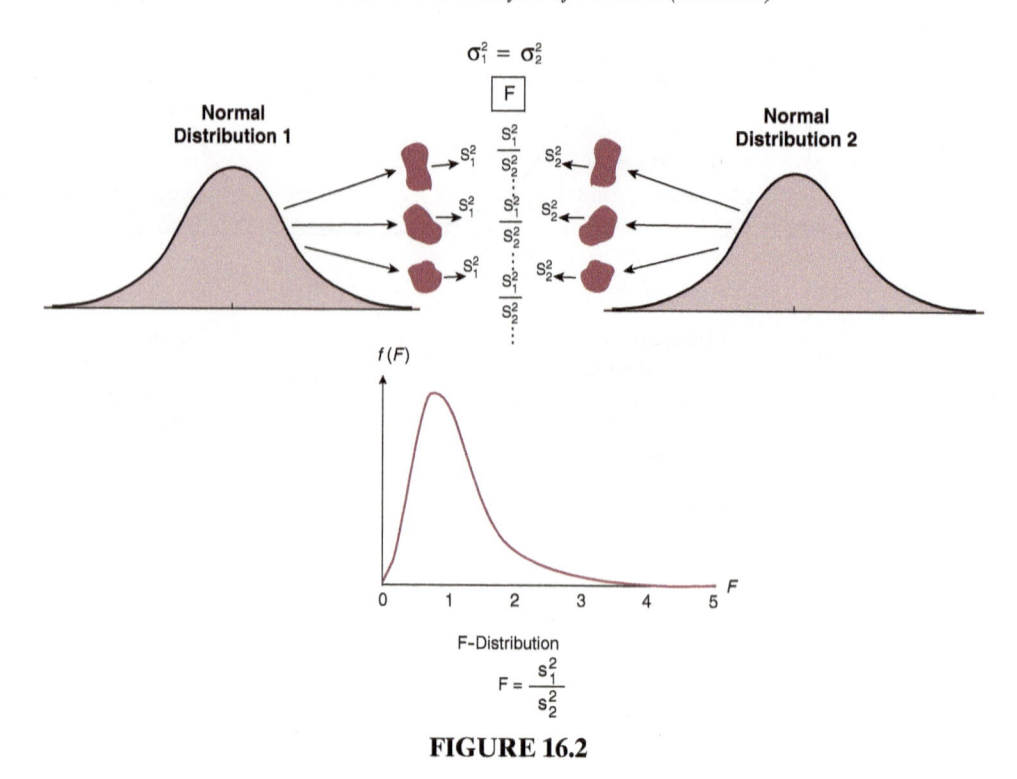

The three F-ratios represent just three possible F-statistics that might appear in a F-Distribution. Although the random samples in this illustration each contain ten data values, in general the random samples need not contain ten data values, nor do they need to be of equal size.

TABLE 16.3   SAMPLES DRAWN FROM TWO POPULATIONS WITH EQUAL VARIANCES: $\sigma_1^2 = \sigma_2^2$

	Population 1 $\sigma_1^2 = 2.87$	Population 2 $\sigma_2^2 = 2.87$
1st Sample n = 10	7 8 5 1 4 7 6 8 8 3	14 16 10 11 17 11 17 12 11 15
Sample Variances	$s_1^2 = 5.76$	$s_2^2 = 7.38$
F-ratio	$F = \dfrac{s_1^2}{s_2^2} = \dfrac{5.76}{7.38} \approx 0.78$	
2nd Sample n = 10	8 3 1 3 5	13 15 15 10 16

TABLE 16.3 (continued)

	Population 1 $\sigma_1^2 = 2.87$	Population 2 $\sigma_2^2 = 2.87$
	7	19
	4	16
	9	16
	2	19
	5	10
Sample Variances	$s_1^2 = 6.90$	$s_2^2 = 9.88$
F-ratio	$F = \dfrac{s_1^2}{s_2^2} = \dfrac{6.90}{9.88} \approx 0.70$	
3rd Sample n=10	4	19
	3	14
	0	13
	9	13
	3	16
	4	16
	2	13
	4	18
	9	10
	8	11
Sample Variances	$s_1^2 = 9.38$	$s_2^2 = 8.46$
F-ratio	$F = \dfrac{s_1^2}{s_2^2} = \dfrac{9.38}{8.46} \approx 1.11$	

By examining the F-ratios in Table 16.3 you should notice that each are positive and *close to the value one.* In fact, all F-ratios are non-negative since they are composed of variances which are always non-negative. **In this illustration the F-ratios of the sample variances were near one because the population variances were equal.** In fact, the ratio of the population variances is equal to one. Since the sample variances are approximations of the population variances the ratio of the sample variances should also be near one. Thus, if a sampling distribution of all possible F-ratios of sample variances where the population variances are equal were constructed, we might expect the mean value of the F-ratios to be 1.

Let's examine the possible values of the F-ratios if the population variances are not equal. In Table 16.4 the samples displayed are drawn from two populations with *unequal* population variances. Notice that population 1 has a variance of 32.94 while the second population has a variance of 8.24.

TABLE 16.4 SAMPLES DRAWN FROM POPULATIONS WITH UNEQUAL POPULATION VARIANCES $\sigma_1^2 \neq \sigma_2^2$

	Population 1 $\sigma_1^2 = 32.94$	Population 2 $\sigma_2^2 = 8.24$
1st Sample n=10	8	7
	12	8
	0	5
	2	1
	14	4
	2	7
	14	6
	4	8

(*continues*)

Chapter 16 The F-Distribution and an Introduction to Analysis of Variance (ANOVA)

TABLE 16.4 *(continued)*

		Population 1 $\sigma_1^2 = 32.94$	Population 2 $\sigma_2^2 = 8.24$
	n=10	2	8
		10	3
Sample Variances		$s_1^2 = 29.16$	$s_2^2 = 5.76$
F-ratio		$F = \dfrac{s_1^2}{s_2^2} = \dfrac{29.16}{5.76} \approx 6.06$	
	2nd Sample	6	8
	n = 10	10	3
		10	1
		0	3
		12	5
		18	7
		12	4
		12	9
		18	2
		0	5
Sample Variances		$s_1^2 = 39.51$	$s_2^2 = 6.90$
F-ratio		$F = \dfrac{s_1^2}{s_2^2} = \dfrac{39.51}{6.90} \approx 5.73$	
	3rd Sample	18	4
	n = 10	8	3
		6	0
		6	9
		12	3
		12	4
		6	2
		16	4
		0	9
		2	8
Sample Variances		$s_1^2 = 33.83$	$s_2^2 = 9.38$
F-ratio		$F = \dfrac{s_1^2}{s_2^2} = \dfrac{33.83}{9.38} \approx 3.61$	

By examining the F-ratios in Table 16.4 you should notice that they are all larger than one and in fact, they are larger than any F-ratio in Table 16.3. Since the variance of population 1 is larger than the variance of population 2, the ratio of the population variances is:

$\sigma_1^2/\sigma_2^2 = 32.94/8.24 = 4$, which is greater than one.

Since the sample variances approximate their respective population variances, we would expect the F-ratios of the sample variances in Table 16.4 to be approximately 4.

Now let us consider the effects of *sample size on the value of the sample variances*. When we use relatively small sample sizes we are not selecting many data values from the population. Thus extreme data values selected within a small sample can cause their sample variances to vary greatly. However, if we selected larger sample sizes extreme data values would not affect the sample variances as much, and, thus the sample variances would vary less. Consequently, the F-ratios attained from smaller sample sizes would vary more than those F-ratios computed from larger sample sizes. **Hence the shape of the resulting F-Distribution would change relative to the *sizes* of the samples drawn.** Thus the F-Distribution *is a family of curves,* similar to the concept of family of curves for the *t*-distribution or the chi-square distribution. As you recall with the *t*-distribution we used the degrees of freedom associated with the sample size to determine the appropriate critical *t* value(s). In a similar way, to determine the appropriate F-Distribution curve, the degrees of freedom, *df*, for *each* sample is used. The degrees of freedom associated with the sample sizes are placed in parenthesis (***df for sample variance one in the numerator, df for sample variance two in the denominator***). This is written as F (dfs_1^2, dfs_2^2). The first number in the parenthesis represents the degrees of freedom for sample variance one of the **numerator**. It is calculated using $df = n_1 - 1$. The **second** number in the parenthesis represents the degrees of freedom for the sample variance used in the **denominator** of the F-ratio. The degrees of freedom for this sample is $df = n_2 - 1$. Figure 16.3 displays the F-Distribution curve for different size samples.

The F-Distribution Family of Curves for Different Degrees of Freedom, *df*

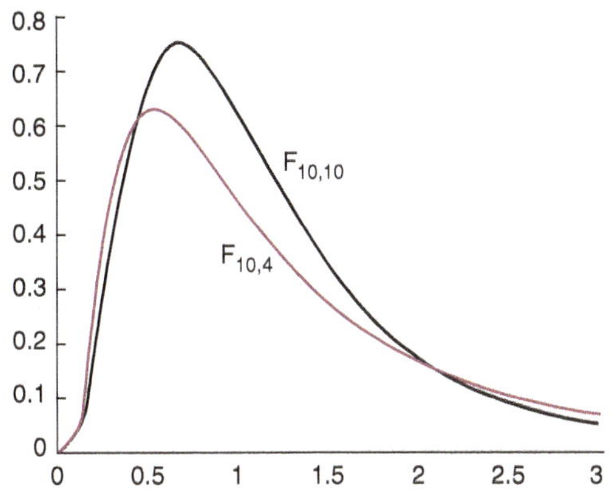

FIGURE 16.3

The family of F-Distribution curves is based on two basic assumptions.

Assumptions of the F-Distribution

1. The two populations from which the samples are drawn must be independent.
2. The two populations from which the samples are selected are normally distributed and have equal variances.

The first assumption implies that the data values in one population must not depend in any way on the data values in the second population. The second assumption indicates that the two populations must be normal distributions. This last assumption is **critical** to the F-Distribution. If the original populations are not normal other statistical models would have to be used instead of the F-Distribution. Let's examine the main characteristics of a F-Distribution.

Characteristics of the F-Distribution

1. The F-Distribution is composed of statistics called F-statistics which are ratios formed from the sample variances calculated from two independent and random samples selected from two normal populations having equal variances. Each F-statistic is determined by the formula:

(continues)

Characteristics of the F-Distribution (continued)

$$F - statistic = \frac{s_1^2}{s_2^2}$$

where s_1^2 represents a sample variance for a random sample of size n_1 selected from population one,

and s_2^2 represents a sample variance for a random sample of size n_2 selected from population two.

2. The F-Distribution is a family of curves dependent upon the degrees of freedom values of the sample variances associated with an F-ratio. Each pair of degrees of freedom for the F-Distribution is determined by the degrees of freedom of the variance in the numerator and the degrees of freedom of the variance in the denominator.
3. The F-Distribution is skewed to the right. As the degrees of freedom increases, the F-Distribution approaches the shape of a normal distribution.
4. The peak or modal value of the F-Distribution is about 1. Thus, values *far* from 1 in either direction provide evidence *against* a hypothesis of equal population standard deviations.

The F-Distribution is a probability distribution. Therefore, the probability that a given F-statistic will be above a certain critical F-value can be computed. Statisticians have constructed tables representing critical F-values. In Appendix C: Table VII you will find such a table. Table VII is called *Critical Values of the F-Distribution*. The critical values of the F-Distribution table is set up to give values of F which must be **exceeded** at a given level of significance, α, **to provide the evidence that the population standard deviations are not equal.** To determine the critical F-value, denoted $F_\alpha(dfs_1^2, dfs_2^2)$, you must calculate the degrees of freedom for the sample variance (dfs_1^2) of the numerator and degrees of freedom of the sample variance of the denominator (dfs_2^2) for the F-statistic.

Let's discuss how to use **Appendix C: Table VII Critical Values of the F-Distribution** to find a critical F-value.

EXAMPLE 16.3

Find the critical F-value if $dfs_1^2 = 8$, $dfs_2^2 = 10$ and $\alpha = 0.05$.

Solution

A portion of Table VII Critical Values of the F-Distribution found in Appendix C is shown in Table 16.5. You should notice that to use the F-Distribution table you must identify the degrees of freedom for the numerator, the degrees of freedom for the denominator and the level of significance.

TABLE 16.5 Critical Values of F-Distribution F_α (dfs_1^2, dfs_2^2)

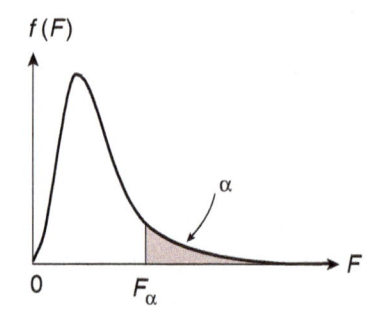

TABLE 16.5 (continued)

s_2^2 \ s_1^2	1	2	3	4	5	6	7	8	9
				Numerator Degrees of Freedom					
1	161.4	199.5	215.7	224.6	230.2	234.0	236.8	**238.9**	240.5
2	18.51	19.00	19.16	19.25	19.30	19.33	19.35	**19.37**	19.38
3	10.13	9.55	9.28	9.12	9.01	8.94	8.89	**8.85**	8.81
4	7.71	6.94	6.59	6.39	6.26	6.16	6.09	**6.04**	6.00
5	6.61	5.79	5.41	5.19	5.05	4.95	4.88	**4.82**	4.77
6	5.99	5.14	4.76	4.53	4.39	4.28	4.21	**4.15**	4.10
7	5.59	4.74	4.35	4.12	3.97	3.87	3.79	**3.73**	3.68
8	5.32	4.46	4.07	3.84	3.69	3.58	3.50	**3.44**	3.39
9	5.12	4.26	3.86	3.63	3.48	3.37	3.29	**3.23**	3.18
10	**4.96**	**4.10**	**3.71**	**3.48**	**3.33**	**3.22**	**3.14**	3.07	**3.02**
11	4.84	3.98	3.59	3.36	3.20	3.09	3.01	2.95	2.90
12	4.75	3.89	3.49	3.26	3.11	3.00	2.91	2.85	2.80
13	4.67	3.81	3.41	3.18	3.03	2.92	2.83	2.77	2.71

(Denominator Degrees of Freedom on left axis)

Since the $dfs_1^2 = 8$ represents the degrees of freedom for the numerator, $dfs_2^2 = 10$ represents the degrees of freedom for the denominator, and the level of significance, $\alpha = 0.05$, the critical F-value, $F_\alpha(dfs_1^2, dfs_2^2)$, expressed as $F_{0.05}(8,10)$ has a table value of 3.07. This is highlighted in Table 16.5 and is denoted as

$$F_{0.05}(8,10) = 3.07$$

16.3 TESTING VARIANCES: THE F-TEST

In the previous section, we discussed the F-Distribution and mentioned one of its uses is to test for the equality of variances from two independent and **normal** populations having equal variances. Now we will present the procedure to perform this test for equality.

In the chapter introduction we proposed the following example:

In traveling to work Paulette likes to drive different routes on different days. One day she might drive through the side streets and back roads and the next day, she would ride on the highway. A sample of her travel times, measured in minutes, are shown in Table 16.6.

TABLE 16.6 Paulette's Travel Times to Work During a Two Week Time Period

Back Roads	Highway
10	12
12	13
14	14
16	15
18	16
$\bar{x} = 14$	$\bar{x} = 14$
$s^2 = 8$	$s^2 = 2$

Can we conclude that there is a significant difference in the population variances for the given sample results?

To answer this question we can *compare* the sample variances by forming a ratio of the **larger** sample variance to the **smaller** sample variance. This ratio is called an F-statistic and is expressed as:

$$\text{F-Statistic} = \frac{\text{Larger Variance}}{\text{Smaller Variance}} = \frac{8}{2} = 4$$

Since the larger sample variance has been substituted in the numerator of the F ratio we would expect the F-statistic value to be greater than one. The question is whether this F-statistic is *significantly* greater than one. If we can conclude that the F-statistic is significantly greater than one then we can infer that the population variances from which the samples were randomly selected are *different*. The question to be answered lends itself to an analysis for which our *five step hypothesis testing procedure* can be applied. The hypothesis testing procedure summarized below explains how the procedure can be applied to this problem. This application is called the *F-Test*.

F-Test: Hypothesis Testing Procedure to Test if Population Variances are Different

Step 1 *Formulate the two hypotheses, H_0 and H_a.*
The Null Hypothesis, H_0, in testing the difference between two population variances has the following form:

Null Hypothesis Form

H_0: The variance of population 1, written σ_1^2, and the variance of population 2, written σ_2^2 are equal.

In symbols:

$$\mathbf{H_0}:\ \sigma_1^2 = \sigma_2^2$$

The Alternative Hypothesis, H_a, for the F-test has the following form:

Alternative Hypothesis Form

H_a: The variance of population 1, written σ_1^2, and the variance of population 2, written σ_2^2, are **not** equal.

This is symbolized:

$$\mathbf{H_a}:\ \sigma_1^2 \neq \sigma_2^2$$

Step 2 *Determine the model to test the null hypothesis, H_0.*
The appropriate sampling distribution is the F-Distribution. Assuming the null hypothesis of equality of population variances is true, the F-ratio equals one.

Step 3 *Formulate the Decision Rule.*
Since the form of the alternative hypothesis is nondirectional, to test the calculated F-statistic we would need an F-Distribution Table that contains two critical F-values, a left and right critical value. This is illustrated in Figure 16.4.

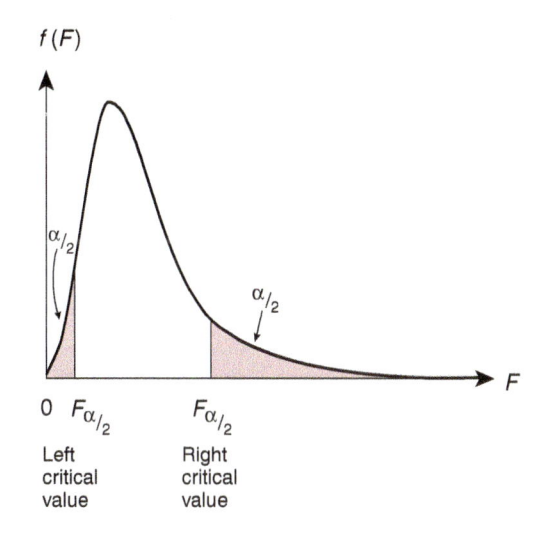

FIGURE 16.4

We can simplify the F-Distribution table by calculating the F-statistic, also known as the *test statistic*, by always placing the **larger sample variance in the numerator.** *This procedure will insure that the F-test statistic is always greater than or equal to one. Consequently we will only need the right hand side of the F-Distribution which contains the critical F-value. This is shown in Figure 16.5.*

Thus the computed F-test statistic will be significant when its value is **greater** than the critical F-value. Thus, if the computed F-test statistic is greater than the critical F-value, $F_{\alpha/2}$ of Table VII, we will **reject** the null hypothesis, H_0, and accept the alternative hypothesis, H_a.

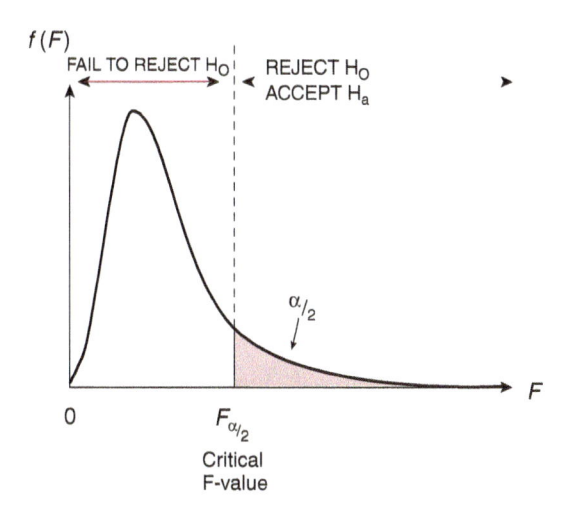

FIGURE 16.5

In order to complete the analysis, an alpha level, α, must be chosen. But since we have eliminated the need for critical F-values less than one, to use the F-Distribution Table VII in Appendix C, look up the critical F-value using the chosen alpha level divided by 2.

For this example, $\alpha = 0.05$, thus $\alpha/2 = 0.05/2$ or 0.025. To determine if the F-statistic is significantly greater than one, we use the F-Table located in Appendix C: Table VII. Remember to use the F-Table we must first find the degrees of freedom for the two sample variances. Table 16.7 contains a summary of the sample variances, the sample sizes and the corresponding degrees of freedom for the sample data values given in Table 16.6.

TABLE 16.7 Summary Table to Compute the F-statistic

	Numerator of F-Statistic Larger s^2	Denominator of F-Statistic Smaller s^2
Sample variance, s^2	8	2
Sample size, n	5	5
Degrees of freedom, $df = n - 1$	4	4

Thus we are using a F-Distribution with (4,4) degrees of freedom where $\alpha/2 = 0.025$. Therefore, the critical F-value from Table VII is:

$$F_{\alpha/2}(dfs_1^2, dfs_2^2) = F_{0.025}(4,4) = 9.60.$$

The decision rule is: Reject H_o, if the F-statistic is greater than 9.60.

Step 4 *Analyze the sample data.*
Calculate the **F-Statistic** using the test statistic formula:

$$F = \frac{\text{larger } s^2}{\text{smaller } s^2}$$

The F-Statistic is also referred to as the **test statistic.**
Using sample variances in Table 16.7, the calculated F-Statistic is

$$F = \frac{8}{2}$$

$$F = 4$$

Step 5 *State the conclusion.*
In Figure 16.6 the calculated F-statistic (F = 4) and the critical F-value (9.60) are indicated on the graph. Since the test statistic is **not** greater than the critical F-value, we fail to reject the null hypothesis, H_0. Thus, although there is a difference in the sample variances, the difference is not statistically significant at an alpha level of 0.05.

Therefore, Paulette can be assured, based on the results of this hypothesis test, the two routes to work are approximately equal in their variability of travel times.

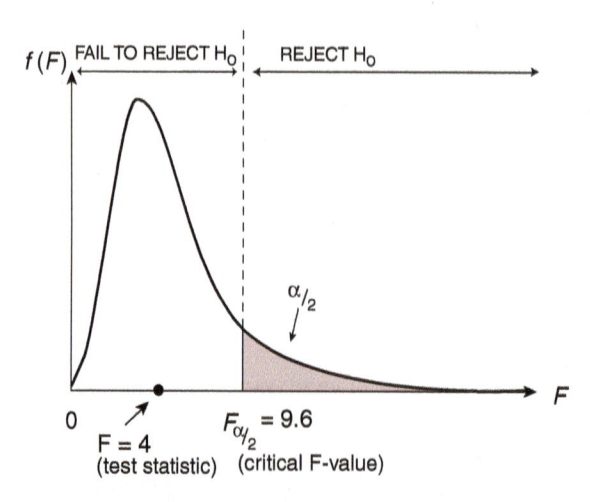

FIGURE 16.6

Let's now summarize the F-Test used to test the equality of two population variances.

A Summary of the F-Test:
A hypothesis test to compare variances from two NORMAL populations

Step 1 Formulate the two hypotheses, H_0 and H_a.

H_0: The variance of population 1, written σ_1^2, and the variance of population 2, written σ_2^2 are equal.

$$H_0: \sigma_1^2 = \sigma_2^2$$

H_a: The variance of population 1, written σ_1^2, and the variance of population 2, written σ_2^2 are **not** equal.

$$H_a: \sigma_1^2 \neq \sigma_2^2$$

Step 2 Determine the model to test the null hypothesis, H_0.
The appropriate sampling distribution is the F-Distribution. Assuming the null hypothesis of equality of population variances is true, the F-ratio equals one.

Step 3 Formulate the Decision Rule.
a) Choose a significance level, α, and divide by 2: $\alpha/2$.
b) Determine the degrees of freedom for the larger sample variance, and the degrees of freedom for the smaller sample variance. Look up in Appendix C: Table VII, the critical F-value using the notation:

$$F_{\alpha/2}(dfs_1^2, dfs_2^2)$$

where s_1^2 represents the **larger** sample variance for a random sample of size n_1 selected from population one,
and s_2^2 represents the **smaller** sample variance for a random sample of size n_2 selected from population two.
c) State the decision rule.

Step 4 Analyze the sample data.
Calculate the F-statistic using the test statistic formula:

$$F = \frac{\text{larger } s^2}{\text{smaller } s^2}$$

Step 5 State the conclusion.
If the test statistic, F, is greater than the critical F-value, symbolized $F_{\alpha/2}(dfs_1^2, dfs_2^2)$, then the population variances are significantly different at the significance level, α. Otherwise, the population variances are not significantly different at the level of significance, α. This is shown in Figure 16.7.

EXAMPLE 16.4

(a) Compute an F-statistic for the following sample variances.

$$s_1^2 = 14 \qquad s_2^2 = 4$$
$$n_1 = 10 \qquad n_2 = 11$$

(b) Conduct an F-Test for the equality of the population variances using $\alpha = 0.10$.

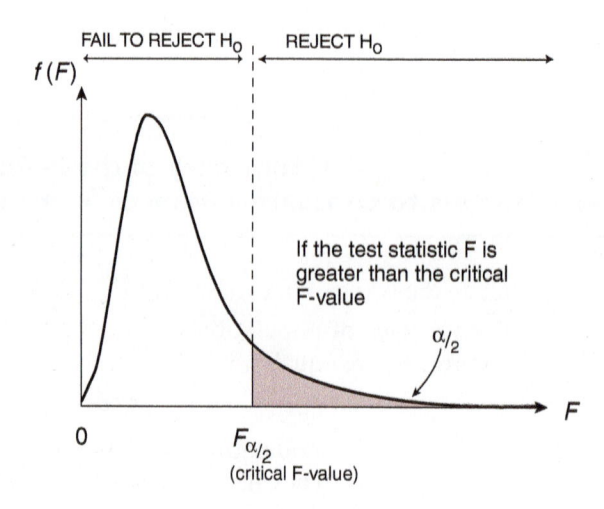

Chapter 16 The F-Distribution and an Introduction to Analysis of Variance (ANOVA)

FIGURE 16.7

Solution

(a) To compute an F-statistic use the test statistic formula

$$F = \frac{\text{larger } s^2}{\text{smaller } s^2}$$

Since the larger of the two sample variances is $s_1^2 = 14$, and the smaller is $s_2^2 = 4$, then the F-statistic is

$$F = \frac{14}{4} = 3.5$$

(b) To conduct a hypothesis test for the equality of the population variances, we will follow the F-Test procedure.

Step 1 *Formulate the two hypotheses, H_0 and H_a.*
H_0: The population variances are equal.
This is symbolized as:

$$H_o: \sigma_1^2 = \sigma_2^2$$

and
H_a: The population variances are not equal.
This is symbolized as:

$$H_a: \sigma_1^2 \neq \sigma_2^2$$

Step 2 *Determine the model to test the null hypothesis, H_0.*
Since the sample size of the larger sample variance is $n_1 = 10$ and the sample size of the smaller sample variance is $n_2 = 11$ the F-Distribution with degrees of freedom $df = (9,10)$ will be used as the hypothesis testing model. The F-Distribution model is shown in Figure 16.8.

Step 3 *Formulate the Decision Rule.*
a) This F-Test is to be conducted using an $\alpha = 0.10$ therefore, $\alpha/2 = 0.1/2 = 0.05$.
b) The degrees of freedom for the larger sample variance is $dfs_1^2 = 10 - 1 = 9$, and for the smaller sample variance is $dfs_2^2 = 11 - 1 = 10$.

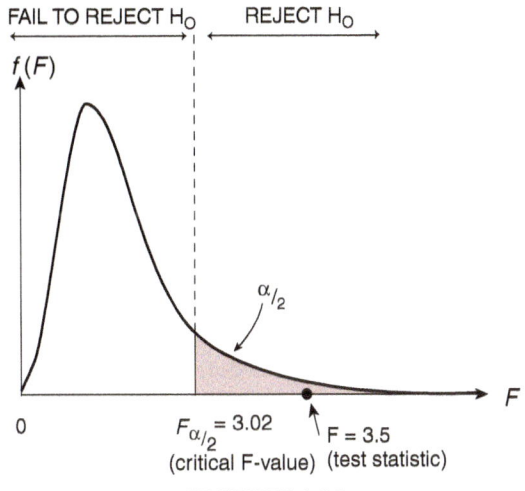

FIGURE 16.8

Using Table VII, the critical F-value is

$$F_{\alpha/2}(dfs_1^2,\ dfs_2^2) = F_{0.05}(9,10)$$

$$F_{0.05}(9,10) = 3.02$$

c) The decision rule is: Reject H_0 if the test statistic F is greater than 3.02.

Step 4 *Analyze the sample data.*
The test statistic is, F = 3.5. (see part a.)

Step 5 *State the conclusion.*
Since the test statistic, F = 3.5, is greater than the critical F-value, $F_{0.05}(9,10) = 3.02$, reject the null hypothesis and accept the alternative hypothesis. Thus, the population variances are significantly different at $\alpha = 0.10$.

16.4 TYPES OF VARIANCES

In this chapter, we have discussed the concept of variance. In the previous two sections, population and sample variances were revisited and an application to inferential statistics was presented. That is, sample variances were used to make inferences about population variances. Let us now consider variance in a different manner.

The hypothesis testing procedure is used to analyze an experiment that is designed and conducted around a test of a null hypothesis. The experiment is set up so that by rejecting the null hypothesis at a given significance level, an alternative hypothesis is accepted. Various applications to population parameters have been explored. For example, we have tested hypotheses about one or two population proportions. And, we have tested hypotheses about one or two population means.

This section is concerned with a hypothesis testing procedure designed to compare **more than two** population means. Its application is used to analyze many problems. For example, let's consider the following problems:

1. Which of the life saving *procedures* will produce the longest life expectancy in someone who is deathly ill with the xyz virus: *drug therapy, a surgical procedure,* or *laser therapy*?
2. Which *method* of instruction will increase the achievement levels of students: *directed method, undirected method* or *a combination of both*?

In each of these problems there is a need to find the best alternative among those given. Many times the "best" (i.e. best method or procedure) is measured by using the mean, such as mean life expectancy, or mean achievement level.

As we have seen in the other hypothesis testing procedures, besides a comparison of means we also must account for the differences in variability. We have done this by examining standard deviations as a measure of variability. In the application to compare many population means we will use variance as a measure of variability.

Let's examine the problem about the effect of various medical procedures on life expectancy. The variable being measured is the length of life. Each individual (subject) in the experiment will live a number of years. Besides being able to calculate a *mean* lifetime, a *variation* of lifetime among the **total** participants in the experiment can be calculated. Also, **within** each treatment, not every person will live the same number of years. Thus there is a mean lifetime as well as variability in length of life when considering those people within each medical treatment. Finally, if we look at the mean number of years people live under one medical treatment in comparison to people within each of the other medical treatments, we will notice differences in those mean times as well as the variability **between** the means of each treatment group.

This first example illustrates the different ways means and variability can be examined when performing an experiment. What are the causes for the differences in variability and how we can explain them is a major objective of the **analysis of variance**. The analysis of the variability has been used extensively in many research scenarios. First there is the variability *within each of the treatment* procedures: drug therapy, surgical procedures and laser therapy. Everyone given the *same* treatment will *not* live the same number of years. This variance is referred to as **Within Group Variance**. Second, there is variability in length of life *among all the participants regardless* of which treatment procedure they were given. This variance is referred to as **Total Variance**. Finally, there is variability *between* the mean length of life among *the three treatment groups*. This variability is referred to as **Between Group Variance**.

DEFINITION 16.2
Within Group Variance. The variance due to the differences among the individual data values within each group is called Within Group Variance.

This concept of within group variance is examined in Table 16.8.

TABLE 16.8 No Within Group Variance

Groups:	I	II	III
	30	45	60
	30	45	60
	30	45	60
	30	45	60
	30	45	60
Means:	30	45	60

Examining Table 16.8, you will notice that there is **no Within Group Variance** since every data value within each group is the same. However, there is variation among the groups since the means are different. This variation among the groups is called **Between Group Variance**.

DEFINITION 16.3
Between Group Variance. The variance due to the differences among the groups or the differences among the sample means is called Between Group Variance.

The concept of between group variance is examined in Table 16.9.

TABLE 16.9   No Between Group Variance

Groups:	I	II	III
	32	25	30
	40	35	43
	48	45	49
	50	55	47
	55	65	56
Means:	45	45	45

Examining Table 16.9, you will notice that there is **no Between Group Variance** since the mean of each group is the *same*. However, there *is* variation *within* each group since the data values within each group are *not* the same.

In the second example regarding *method of instruction* and *achievement* we would notice the **same** breakdown of variances. The students who receive only one method of instruction or treatment will still have some variability in their achievement scores. This variance is the **Within Group Variance.** Since within group variance is oftentimes due to *individual differences* among subjects that is not explainable, the **Within Group Variance** is also called **error variance**.

There will be variability in achievement scores regardless of how the students are assigned into their treatment subgroups. This variance is the **Total Variance.**

Finally there will be differences in the variability of the mean achievement scores of the groups, that is, there will be differences in the variability of the mean achievement scores of those students given the **Within Group Variance** directed method versus those given the non-directed method versus those given the combined method. This variance is the **Between Group Variance.** In most experiments the experimenter is interested in the effects of different methods of treatment or the effects of different methods of instruction. Thus, the **Between Group Variance is also referred to as Experimental or Treatment Variance.**

This breakdown of variance into Within Group Variance, Total Variance and Between Group Variance can be discussed in terms of the population variance or the sample variances. When the population variance is broken down by the three variances then the following relationship holds:

Total Variance = Between Group Variance + Within Group Variance

Pictorially we can represent the relationship with the following circle (see Figure 16.9). The total circle is the total population variance. The two subdivisions are the Between Group Variance and the Within Group Variance.

To illustrate the relationship in Figure 16.9 assume we have a **total population** of 15 students and their scores on an achievement test. The 15 students were subdivided into three groups, called treatment 1, treatment 2, and treatment 3. The achievement scores of the students, the means and standard deviations of the scores within each treatment group are displayed in Table 16.10.

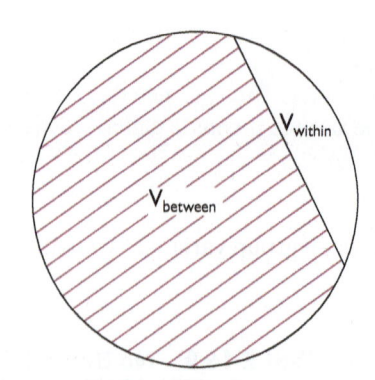

Chapter 16 The F-Distribution and an Introduction to Analysis of Variance (ANOVA)

Variance Total = Variance between + Variance within
Vtotal = Vbetween + Vwithin

FIGURE 16.9

TABLE 16.10 Achievement Scores of a Population of 15 Students Subdivided into Three Groups

	Treatment 1	Treatment 2	Treatment 3
	10	28	10
	20	29	29
	30	30	30
	40	31	31
	50	32	50
Population means μ	30	30	30
Population variances **Within** each treatment σ^2	200	2	160.4

You should notice that the means for each treatment is the same, whereas each of the population variances are different. You could interpret this to indicate that although three different treatments have been administered, the effects due to the treatments were equal. However since the variability in each treatment group is different you could attribute this effect to the differences among individual subjects. Let's examine the different types of variability that are displayed in the example.

Given the population variances within each group the *average within variance* can be found by **pooling** the three population variances. Pooling the three population variances is accomplished by summing the three population variances within each treatment and dividing by three:

$$\text{Sum of the population variances} = 200 + 2 + 160.4 = 362.4$$

Dividing by the sum by three

$$362.4/3 = 120.8$$

yields

$$\text{Within Variance} = 120.8$$

Thus, the Within Group variance, or simply the variance due to differences among the individuals is 120.8

We can compute the *Between Group Variance* by using the population variance formula with the values of the group means: 30, 30, and 30. Since there is no variability between the group means, the Between Group Variance is zero.

$$\text{Between Group Variance} = 0$$

Thus, the Between Group Variance, or simply the variance due to the treatment (i.e. the experiment) is 0.

Finally to compute the *Total Variance* regardless of how the students are grouped we use the population variance formula for all 15 achievement scores. Performing that task we find the

$$\text{Total Variance to be } 120.8$$

(To get a greater appreciation of this result, it might be helpful to actually calculate the Total Variance of the data in Table 16.10.)

Thus, Between Variance + Within Variance = Total Variance
                0   +   120.8   =   120.8

In the above example the **means of the treatment were all equal** and thus the Total Variance was determined **solely** by the Within Variance. You might ask:

Can the same relationship be demonstrated when the means are *not equal*, and, as a result, the Between Group Variance would *not* be equal to zero? Let's examine this question in Example 16.5.

EXAMPLE 16.5

Consider the data in Table 16.11.

TABLE 16.11 Achievement Scores of a Population of 21 Students Subdivided into Three Groups

	Treatment 1	Treatment 2	Treatment 3
	10	17	24
	11	18	25
	12	19	26
	13	20	27
	14	21	28
	15	22	29
	16	23	30
Population means μ	13	20	27
Population variances **Within** each treatment σ^2	4	4	4

a) Using the within group variance for each group calculate the pooled Within Variance.
b) Compute the Between Group Variance using the population variance formula on the group means.
c) Calculate the Total Group Variance for all 21 data values in Table 16.11.
d) Verify the relationship

$$\text{Between Variance} + \text{Within Variance} = \text{Total Variance}$$

Solution

a) Given the Within Variances for each group the pooled Within Variance can be found by summing the three variances and dividing by three:

$$4 + 4 + 4 = 12$$

$$12/3 = 4$$

$$\text{Within Variance} = 4$$

b) To compute the Between Group Variance we use the population variance formula for the values of the group means : 13, 20, 27.
First, the mean of the group means is

$$\mu = \frac{\Sigma X}{N} = \frac{13+20+27}{3} = 20$$

Second, compute the population variance using the formula:

$$\sigma^2 = \frac{\Sigma(X-\mu)^2}{N},$$

to obtain the Between Group Variance.

Table 16.12 summarizes the relevant data necessary for the calculation of the Between Group Variance.

TABLE 16.12 Between Group Variance

	X	$X - \mu$	$(X - \mu)^2$
	13	−7	49
	20	0	0
	27	7	49
Σ	60	0	98

Substituting $\Sigma (X - \mu)^2 = 98$ and $N = 3$ into the formula for the population variance

$$\sigma^2 = \frac{\Sigma(X-\mu)^2}{N}$$

$$\sigma^2 = \frac{98}{3}$$

$$\sigma^2 \approx 32.67$$

Thus, the Between Group Variance ≈ 32.67.

c) Finally to compute the Total Variance regardless of how the students are grouped we use the population variance formula for all 21 achievement scores. Performing this task we find the

Total Group Variance = 36.67

Table 16.13 summarizes the relevant data necessary for the calculation of the Total Group Variance. For these data the Total Group Mean = 20.

TABLE 16.13 Total Group Variance
All Achievement Scores for a Population of 21 Students

X	$X - \mu$	$(X - \mu)^2$
10	−10	100
11	−9	81
12	−8	64
13	−7	49
14	−6	36
15	−5	25
16	−4	16
17	−3	9
18	−2	4
19	−1	1
20	0	0
21	1	1
22	2	4
23	3	9
24	4	16
25	5	25
26	6	36
27	7	49
28	8	64
29	9	81
30	10	100
Σ 420	0	9170

Substituting $\Sigma(X - \mu) = 9170$ and $N = 21$ into the formula for the population variance

$$\sigma^2 = \frac{\Sigma(X - \mu)^2}{N}$$

$$\sigma^2 = \frac{9170}{21}$$

$$\sigma^2 = 36.67$$

d) Since the values of the Between Variance, Within Variance and Total Variance have been computed, the relationship

Between Variance + Within Variance = Total Variance
32.67   +   4   =   36.67

is verified.

Most of the time when performing statistical experiments we do not have the population data values available. That is, we only have sample data to use. And, as a result, we are unable to use population parameters, such as μ and σ. Thus, statistics are computed using other formulas, and the values obtained by these statistics are used for inferential conclusions about the parameters.

As you recall, the sample variances are *estimates* of the population variance. The relationship

$$\text{Between Variance} + \text{Within Variance} = \text{Total Variance}$$

does not hold for **sample variances.**

A formula that can be used in the calculation of the sample variance is

$$s^2 = \frac{\Sigma(X - \overline{X})^2}{n - 1}$$

The expression

$$SS = \Sigma(X - \overline{X})^2$$

is referred to as the **Sums of Squares**. The degrees of freedom, df, associated with the sample is n − 1. Thus the sample variance

$$s^2 = \frac{\Sigma(X - \overline{X})^2}{n - 1}$$

can be rewritten as:

$$s^2 = \frac{\text{Sums of squares}}{\text{degrees of freedom}}$$

or simply,

$$s^2 = \frac{SS}{df}$$

It turns out that when we *only consider* the Sums of Squares, SS, then the relationship

$$\begin{array}{c}\text{Sums of Squares} \\ \text{Between Treatments}\end{array} + \begin{array}{c}\text{Sums of Squares} \\ \text{Within Treatments}\end{array} = \begin{array}{c}\text{Sums of Squares} \\ \text{Total Subjects}\end{array}$$

will hold all the time. This relationship can also be denoted as,

$$SS_{\text{(between treatments)}} + SS_{\text{(within treatments)}} = SS_{\text{(total)}}.$$

16.5 ONE-WAY ANALYSIS OF VARIANCE

A statistical procedure that accounts for differences between treatment effects by studying variability as measured by variance is called *Analysis of Variance*.

In this section an Analysis of Variance procedure for one-way ANOVA, will be presented.

DEFINITION 16.4
ANOVA. ANOVA is a statistical procedure that involves the analysis of one or more independent variables called factors with two or more levels for each factor. When only one independent variable is used in the ANOVA it is called *one-way* ANOVA.

The ANOVA procedure compares variances under the *assumption* that the variance due to experimental manipulation or treatment is the same. This implies that all sample means in an experiment must be equal or, at best, differences between the sample means are not significant. Thus, the differences are simply due to chance (i.e. error) variance. Since the F-Distribution compares two estimates of population variances, we will use the ANOVA procedure to find estimates of the population variance. In fact, the *within group* variance and the *between group* variance that was discussed in Section 16.3 are the two estimates of the population variance that the analysis of variance procedure uses and compares via an F-ratio.

A basic fundamental question that ANOVA explores concerns variability is:

Basic Fundamental Question:
Is the variation among individual sample means due purely to chance or are these differences due to different experimental conditions?

Let's consider the following example.

A market researcher decides to test market a new coffee product using a random sample of 15 supermarkets. Besides the quality of the coffee product, the predominant color used in the packaging design is a major concern. Three colors are chosen to be used as the predominant color. They are Gold, Silver, and Blue as shown in Figure 16.10.

FIGURE 16.10

Of the total of 15 stores, five are randomly chosen to carry the coffee packaged in Gold, five are randomly chosen to carry the coffee packaged in Silver, and five are randomly chosen to carry the coffee packaged in Blue.

After one month of sales the following results were found and summarized in Table 16.14. The results lists the number of units sold (i.e. the independent variable) from each supermarket, along with the mean number of units sold, and the corresponding sample variance s^2, (i.e. the population variance estimates).

TABLE 16.14 Units of Coffee Product Sold After 1 Month
N = 15 stores participating

	Gold package	Silver package	Blue package	
	7	19	8	
	15	12	12	
	20	21	19	
	18	23	11	
	19	15	16	
sample mean, $\bar{x}$	15.8	18	13.2	overall mean = 15.67
sample variance, s^2	27.7	20	18.7	

The basic question that the market researcher would like answered is whether the *variation* among the individual sample means for the number of units sold is due purely to chance or are these differences due mainly to packaging color?

The ANOVA technique *compares variances* under the *assumption* that the variance among all levels, that is packaging colors, are the same. Let's modify our hypothesis testing procedure and perform an analysis of variance to determine if the market researcher can indeed show a statistical difference in sales due to packaging differences.

Based on the hypothesis testing procedure:

Step 1 *Formulate the two hypotheses, H_0 and H_a.*

In ANOVA, all population means are assumed equal. For this example, we will assume the color of packaging used to sell the coffee is *not* related to coffee product sales. The null hypothesis can also be stated as: there is no difference in the population mean number of units sold for the coffee product packaged in Gold, Silver or Blue.

Symbolically, if μ_G = the population mean number of units of the coffee product sold when packaged in Gold, μ_S = the population mean number of units of the coffee product sold when packaged in Silver, and μ_B = the population mean number of units of the coffee product sold when packaged in Blue, then the null hypothesis, H_0, is expressed

$$H_0: \mu_G = \mu_S = \mu_B$$

whereas, the alternative hypothesis, H_a, is expressed

$$H_a: \text{all three means are } not \text{ equal.}$$

The alternative hypothesis implies that there is at least one population mean that is different from the other population means. That is, the population mean number of units sold for at least one of the packaging coloring techniques is different from one of the others.

In general the null and alternative hypothesis forms for this type of ANOVA test are:

Null Hypothesis Form

H_0: There is no difference between the population means.

In the case where there are only three population means we write H_0 as:

$$H_0: \mu_1 = \mu_2 = \mu_3$$

The general alternative hypothesis form is:

H_a: There is a difference between the population means.

or, in this case,

H_a: all three population means are not equal.

Step 2 *Determine the model to test the null hypothesis, H_0.*
The appropriate sampling distribution is the F-Distribution based on the null hypothesis of equality of population means and variances. To use the ANOVA technique, the following assumptions **must be true.**
(a) Each sample is selected from a population that is approximately normally distributed.
(b) Each sample is selected from a population that has the same population variance, σ^2.
(c) When samples are drawn from different populations the samples must be random and independent.

In the example we are examining, to use ANOVA, we are assuming that for the three package coloring techniques the population of data values for each technique is approximately normally distributed. Second, we are assuming that the population of data values for each packaging technique has the same variance, σ^2. Third, for each sample used in the procedure, the samples are independent and were randomly selected from each population.

Step 3 *Formulate the Decision Rule.*
To do the one-ways ANOVA test, the right hand side of the F-Distribution is always used. This is illustrated in Figure 16.11. The critical F-value, $F_\alpha(df$ between, df within) is determined from Appendix C: Table VII, by using α, df between and df within. These three values are indicated in the experimental design for the problem and are generated using the sample data. Specifically,
(a) the *df associated with the numerator and denominator* for the F-value are found using the degrees of freedom associated with the sample variances. These two estimates are called the *between group* variance and the *within group* variance.
For the *df* associated with the numerator we use the *df* for the between group variance.
df(between) = k-1,
where k is the number of levels of the factor.
For the *df* associated with the denominator, use the *df* for the within group variance.
df(within) = N − k, where N is the total number of data for all samples and k is the number of levels of the factor.
(b) the level of significance, α.
(c) State the decision rule.

ANOVA: HYPOTHESIS TESTING MODEL

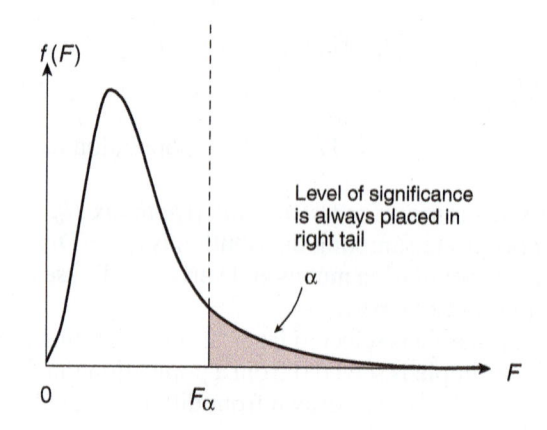

FIGURE 16.11

Thus when the ANOVA test is applied, a *computed* test statistic F will be significant if its value is **greater** than the *critical* F-value, F_α. When this occurs we will **reject** the null hypothesis. This is shown in Figure 16.12.

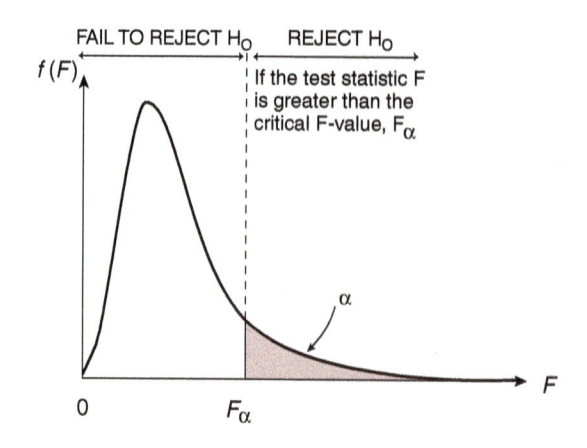

FIGURE 16.12

In this example, let us choose $\alpha = 0.05$. To determine the critical F-value, $F\alpha(df \text{ between}, df \text{ within})$, the degrees of freedom for the between group variance and the degrees of freedom for the within group variance must be obtained.

The degrees of freedom for the between group variance, *df(between groups)*, is determined by the **number of levels, k,** of the factor being compared minus 1. In this example, the 3 different packaging colors represent the 3 levels of the factor: packaging colors. Thus k = 3 and the:

$$df(between\ groups) = k - 1$$
$$= 3 - 1$$
$$df(between\ groups) = 2$$

The *df(between groups)* is often referred to as *df(between treatments)*. Thus, we can say,

$$df(between\ treatments) = k - 1 = 2$$

The degrees of freedom for the within group variance, *df(within groups)* is determined by the **total** number of sample data, N, minus the number of levels of the factor being compared, k. In this example, we have fifteen data values representing our total number of sample data, so N = 15, and we are comparing these data values by separating them into three packaging techniques, therefore, k = 3, thus the

$$df(within\ groups) = N - k$$
$$= 15 - 3$$
$$df(within\ groups) = 12$$

The Critical F-value, $F_\alpha(df\ between, df\ within) = F_{.05}(2,12)$ is determined from Appendix C: Table VII. Using the Table VII,

$$F_{0.05}(2,12) = 3.89$$

Therefore, the decision rule is: if the test statistic F is **greater** than the *critical* F-value, $F_{.05}(2,12)=3.89$, we reject H_0. This decision rule is illustrated in Figure 16.13.

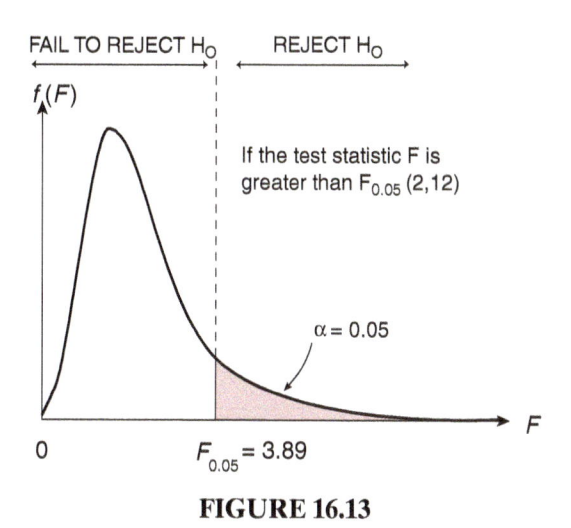

FIGURE 16.13

Step 4 *Analyze the sample data.*

The ANOVA test requires the calculation of a test statistic F by forming a ratio of the variance due to treatment (also called the *between* group variance) **to** the variance within each group (called the *within* sample variance). This test statistic F is computed by the formula:

$$F = \frac{treatment\ sample\ variance}{within\ sample\ variance}$$

Let's calculate the test statistic F for the data in Table 16.14. We have reprinted the data values of Table 16.14 and represent these data in Table 16.15.

Chapter 16 The F-Distribution and an Introduction to Analysis of Variance (ANOVA)

TABLE 16.15 Units of Coffee Product Sold After 1 Month N = 15 stores participating

	Gold package	Silver package	Blue package	
	7	19	8	
	15	12	12	
	20	21	19	
	18	23	11	
	19	15	16	
sample mean, $\bar{x}$	15.8	18	13.2	overall mean = 15.67
sample variance, s^2	27.7	20	18.7	

To calculate the ANOVA test statistic F, a *special ANOVA Table* which is shown in Table 16.16 has been used to summarize the Analysis of Variance results.

TABLE 16.16 ANOVA TABLE

Source of Variation	Sum of Squares SS	Degrees of Freedom df	Mean Square MS	Calculated F-statistic
Between (Treatment)				
Within				
Total				

Notice that the **sources of variation** have been *partitioned* into three distinct cells called Between, Within and Total. Recall that in Section 16.3 we observed that the **sums of squares** of variation due to the Between, Within and Total variation were related as follows:

Sum of Squares Between Treatments + Sum of Squares Within Treatments = Sum of Squares Total Subjects

or

$$SS_{(between\ treatments)} + SS_{(within\ treatments)} = SS_{(total)}$$

The second column of the ANOVA Table is labelled Sum of Squares, denoted SS. To determine the values needed to help complete the column for Sum of Squares in the ANOVA table, formulas are needed for each of the sum of squares. These formulas are:

$$SS_{(total)} = \Sigma X^2 - \frac{(\Sigma X)^2}{N}$$

$$SS_{(between\ treatments)} = \left(\frac{(\Sigma T_1)^2}{n_1} + \frac{(\Sigma T_2)^2}{n_2} + ... + \frac{(\Sigma T_i)^2}{n_i} \right) - \frac{(\Sigma X)^2}{N}$$

$$SS_{(within\ treatment)} = \Sigma X^2 - \left(\frac{(\Sigma T_1)^2}{n_1} + \frac{(\Sigma T_2)^2}{n_2} + \ldots + \frac{(\Sigma T_i)^2}{n_i} \right)$$

where X = a data value
n_1 = the size of sample 1
n_2 = the size of sample 2
n_i = the size of sample i

N = the total number of data values

T_1 = the data values of sample 1
T_2 = the data values of sample 2
T_i = the data values of sample i

Let's compute the various Sums of Squares for the market research problem using the data of Table 16.17.

Step 1 Compute $SS_{(total)} = \Sigma X^2 - \dfrac{(\Sigma X)^2}{N}$

To compute $SS_{(total)}$ you must:
(a) calculate ΣX^2, the sum of the squared data values for **all** the data of Table 16.17. This is,

$$\Sigma X^2 = (7)^2 + (15)^2 + (20)^2 + (18)^2 + \ldots + (11)^2 + (16)^2 = 4005$$

(b) calculate $(\Sigma X)^2$, the sum of all the data values squared. This is,

$$(\Sigma X)^2 = [7 + 15 + 20 + \ldots + 11 + 16]^2 = (235)^2$$

(c) divide $(\Sigma X)^2$ by N, that is, compute $\dfrac{(\Sigma X)^2}{N}$.

$$\frac{(\Sigma X)^2}{N} = \frac{(235)^2}{15}$$

Therefore,

$$SS_{(total)} = 4005 - \frac{(235)^2}{15}$$
$$= 4005 - 3681.67$$
$$SS_{(total)} = 323.33$$

Step 2 *Compute* $SS_{(between\ treatments)} = \left(\dfrac{(\Sigma T_1)^2}{n_1} + \dfrac{(\Sigma T_2)^2}{n_2} + \dfrac{(\Sigma T_3)^2}{n_3} \right) - \dfrac{(\Sigma X)^2}{N}$

(a) To compute $SS_{(between\ treatments)}$ the

$$\frac{(\Sigma T)^2}{n}$$

needs to be calculated for each **level** of treatment.

Thus, for the gold packaging,

$$\frac{(\Sigma T)^2}{n} = \frac{(7+15+20+18+19)^2}{5} = \frac{(79)^2}{5}$$

For the Silver packaging,

$$\frac{(\Sigma T)^2}{n} = \frac{(19+12+21+23+15)^2}{5} = \frac{(90)^2}{5}$$

And, for the Blue packaging,

$$\frac{(\Sigma T)^2}{n} = \frac{(8+12+19+11+16)^2}{5} = \frac{(66)^2}{5}$$

Since

$$\frac{(\Sigma X)^2}{N}$$

was computed in Step one, where the value of

$$\frac{(\Sigma X)^2}{N}$$

was found to be

$$\frac{(235)^2}{15} \approx 3681.67$$

Therefore,

$$SS_{(between\ treatments)} = \frac{(79)^2}{5} + \frac{(90)^2}{5} + \frac{(66)^2}{5} - \frac{235^2}{15}$$

$$= 3739.4 - 3681.67$$

$$SS_{(between\ treatments)} = 57.73$$

Step 3 Compute

$$SS_{(within\ treatment)} = \Sigma X^2 - \left(\frac{(\Sigma T_1)^2}{n_1} + \frac{(\Sigma T_2)^2}{n_2} + \frac{(\Sigma T_3)^2}{n_3} \right)$$

Since both ΣX^2 and

$$\left(\frac{(\Sigma T_1)^2}{n_1} + \frac{(\Sigma T_2)^2}{n_2} + \frac{(\Sigma T_3)^2}{n_3} \right)$$

have been computed in Steps one and two respectively, we have

$$SS_{(within\ treatment)} = 4005 - \frac{(79)^2}{5} + \frac{(90)^2}{5} + \frac{(66)^2}{5}$$

$$= 4005 - 3739.4$$

$$SS_{(within\ treatment)} = 265.6$$

You should notice that the relationship

$$SS_{(between\ treatments)} + SS_{(within\ treatments)} = SS_{(total)}$$

holds for the calculated SS values

$$57.73 + 265.6 = 323.33.$$

When doing ANOVA calculations it is **helpful to remember the relationship for the Sum of Squares:**

$$SS_{(between\ treatments)} + SS_{(within\ treatments)} = SS_{(total)}$$

because knowing any two of the calculations will yield the third.

Thus, the Sum of Squares column in the ANOVA table can be filled in as shown in Table 16.17.

TABLE 16.17   ANOVA TABLE

Source of Variation	Sum of Squares SS	Degrees of Freedom df	Mean Square MS	Calculated F-statistic
Between	57.73			
Within	265.6			
Total	323.33			

The third column of the ANOVA Table represents the *Degrees of Freedom, df,* for each Sum of Squares. The Degrees of Freedom of each Sum of Squares is computed by the formulas:

$df_{(Between)} = k - 1$,

where k = the number of levels of factors (or treatments).

$df_{(within)} = N - k$,

where N = total number of data values,

and   k = the number of treatment levels.

$df_{(Total)} = N - 1$

where N = total number of data values.

For the color packaging data of Table 16.15 we have a total of 15 data values (N = 15) and 3 levels (k = 3) of treatment (i.e. package design), thus,

$df_{(Between)} = k - 1$,
       $= 3 - 1$
       $= 2$

$df_{(within)} = N - k$,
       $= 15 - 3$
       $= 12$

$df_{(Total)} = N - 1$
       $= 15 - 1$
       $= 14$

Chapter 16 The F-Distribution and an Introduction to Analysis of Variance (ANOVA)

Notice that $df_{(Between)} + df_{(within)} = df_{(Total)}$

$$2 + 12 = 14.$$

This is always true!

In Table 16.18, the Degrees of Freedom column in the ANOVA table has been completed.

TABLE 16.18   ANOVA TABLE

Source of Variation	Sum of Squares SS	Degrees of Freedom df	Mean Square MS	Calculated F-statistic
Between	57.73	2		
Within	265.6	12		
Total	323.33	14		

The fourth column of the ANOVA table represents the Mean Square, (denoted MS), which can be determined by the formulas:

$$MS_{(between)} = \frac{SS_{bet}}{df_{bet}}$$

$$MS_{(within)} = \frac{SS_{within}}{df_{within}}$$

To calculate the Mean Square for each sum of square, you need only to use the ANOVA table since the values needed for each calculation is contained within the table. Using Table 16.18, we have:

$$MS_{(between)} = \frac{SS_{bet}}{df_{bet}}$$

$$MS_{(between)} = \frac{57.73}{2}$$

$$MS_{(between)} = 28.87$$

and

$$MS_{(within)} = \frac{SS_{within}}{df_{within}}$$

$$MS_{(within)} = \frac{265.6}{12}$$

$$MS_{(within)} = 22.13$$

Substituting these values into the ANOVA Table, produces Table 16.19.

TABLE 16.19   ANOVA TABLE

Source of Variation	Sum of Squares SS	Degrees of Freedom df	Mean Square MS	Calculated F-statistic
Between	57.73	2	28.87	
Within	265.6	12	22.13	
Total	323.33	14		

Mean Square is an alternative way of referring to the estimates of the population variances. Recall that a sample variance was computed using the formula:

$$\text{sample variance} = \frac{\Sigma(X-\overline{X})^2}{n-1} = \frac{\text{Sum of Squares}}{\text{degrees of freedom}} = \frac{SS}{df}$$

In ANOVA, the terminology, Mean Square (or MS) is used to represent the term variance. Thus, in ANOVA,

Mean Square = Variance

In general, the Mean Square is computed using the formula

$$MS = \frac{SS}{df} = \text{sample variance}$$

The

$$\text{Between sample variance} = \frac{SS_{between}}{df} = MS_{Between}$$

and

$$\text{Within sample variance} = \frac{SS_{within}}{df} = MS_{Within}$$

Thus, once they are calculated, the test statistic F can be determined. Therefore, the **test statistic** is computed by the formula:

$$F = \frac{MS_{bet}}{MS_{within}}$$

$$F = \frac{28.87}{22.13}$$

$$F \approx 1.30$$

Therefore the test statistic for the data representing the three different coffee packaging colors is F= 1.30.

Step 4 *State the conclusion.*

The test statistic (F=1.30) and the critical F-value, $F_{0.05}(2,12) = 3.89$, is indicated in Figure 16.14. Since the test statistic is not greater than the critical F-value, then our conclusion is that we fail to reject H_0. Thus, although there is a difference between the sample means, the difference is not statistically significant at an alpha level of 0.05.

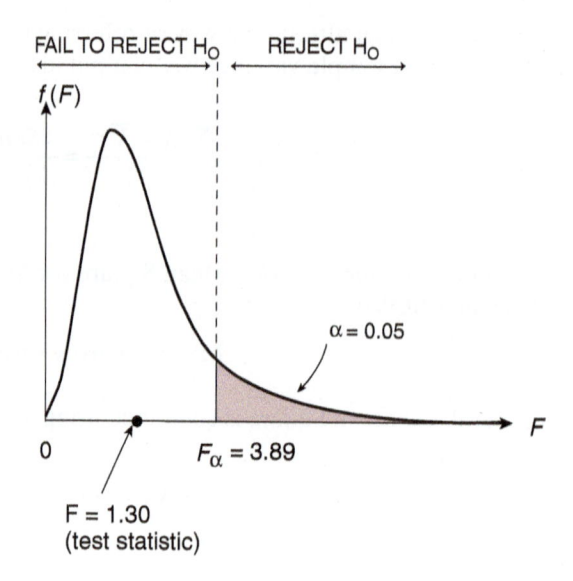

834 *Chapter 16* The F-Distribution and an Introduction to Analysis of Variance (ANOVA)

[Figure showing F-distribution curve with FAIL TO REJECT H₀ and REJECT H₀ regions, α = 0.05, $F_\alpha = 3.89$, and F = 1.30 (test statistic)]

FIGURE 16.14

In practice, when conducting an ANOVA procedure **two** summary tables are needed. The first lists the raw data along with the appropriate statistics ultimately needed for the calculation of the test statistic F. Table 16.20 includes the appropriate sums, sums of squares, k, n, and N needed for the ANOVA procedure. These are highlighted in Table 16.20.

TABLE 16.20 Units of Coffee Product Sold After 1 Month
N = 15 stores participating

	Gold package	Silver package	Blue package	
	7	19	8	
	15	12	12	
	20	21	19	
	18	23	11	
	19	15	16	
sample mean, $\bar{x}$	15.8	18	13.2	overall mean = 15.67
sample variance, s^2	27.7	20	18.7	
ΣT	79	90	66	
n	5	5	5	

k (levels) = 3

$\Sigma X = 235$
$\Sigma X^2 = 4005$
N = 15 data values

The second table that is necessary to compute the test statistic F is called **the ANOVA Table.** The ANOVA Table is pictured in Table 16.21.

TABLE 16.21 The ANOVA TABLE

Source of Variation	Sum of Squares SS	Degrees of Freedom df	Mean Square MS	Calculated test statistic F
Between				
Within				
Total				

Table 16.22 lists the Formulas necessary to complete the ANOVA Table. As you can see, the highlighted statistics of Table 16.20 are needed in the formulas for **sums of squares** and **degrees of freedom** shown in Table 16.22.

TABLE 16.22 ANOVA FORMULAS

Formulas for Sums of Squares

$$SS_{(total)} = \Sigma X^2 - \frac{(\Sigma X)^2}{N}$$

$$SS_{(between\ treatments)} = \left(\frac{(\Sigma T_1)^2}{n_1} + \frac{(\Sigma T_2)^2}{n_2} + ... + \frac{(\Sigma T_i)^2}{n_i} - \frac{(\Sigma X)^2}{N} \right)$$

$$SS_{(within\ treatment)} = \Sigma X^2 - \left(\frac{(\Sigma T_1)^2}{n_1} + \frac{(\Sigma T_2)^2}{n_2} + ... + \frac{(\Sigma T_i)^2}{n_i} \right)$$

where X = a data value
n_1 = the size of sample 1
n_2 = the size of sample 2
n_i = the size of sample i

N = the total number of data values

T_1 = the data values of sample 1
T_2 = the data values of sample 2
T_i = the data values of sample i

Formulas for Degrees of Freedom

$df_{(between)} = k - 1$,
   where k = the number of treatment levels.
$df_{(within)} = N - k$,
   where N = total number of data values,
   and   k = the number of treatment levels.
$df_{(total)} = N - 1$
   where N = total number of data values.

TABLE 16.22   ANOVA FORMULAS (continued)

Formulas for Mean Square

$$MS_{(between)} = \frac{SS_{bet}}{df_{bet}}$$

$$MS_{(within)} = \frac{SS_{within}}{df_{within}}$$

Calculated test statistic F

$$F = \frac{MS_{bet}}{MS_{within}}$$

Therefore, to conduct an Analysis of Variance application, the five step hypothesis testing procedure that we have used throughout the text can be **modified** to include the two summary tables. The hypothesis testing procedure for ANOVA is summarized below.

ANOVA
Hypothesis Testing Procedure

Step 1 Formulate the two hypotheses, H_0 and H_a.
In general the null and alternative hypothesis forms in this type of ANOVA test are:

Null Hypothesis Form

H_0: There is no difference between the population means.
In the case where there are only three population means we have:

$$H_0: \mu_1 = \mu_2 = \mu_3$$

Alternative Hypothesis Form

H_a: There is a difference between the population means.
or
In the case of three population means:
   H_a: all three population means are not equal.

Step 2 *Determine the model to test the null hypothesis, H_0.*
The appropriate sampling distribution is the F-Distribution based on the null hypothesis of equality of population means and variances. To use the ANOVA technique, the following assumptions **must be true.**

Assumptions for ANOVA

(a) Each sample is selected from a normal population.
(b) Each sample is selected from a population that has the same population variance, σ^2.
(c) When samples are drawn from different populations the samples must be random and independent.

Step 3 *Formulate the Decision Rule.*

In performing the ANOVA test, the right hand side of the F-Distribution is always used as shown in Figure 16.15. The critical F-value, F_α(*df* between, *df* within) is determined from Appendix C: Table VII, by using α, *df* between and *df* within. The α value is determined by the researcher. The *df* between is computed by the number of levels of the factor (treatment), k, minus 1. And, the *df* within by the number of data values in the experiment, N, minus k.

Specifically,
(a) the *df associated with the numerator and denominator* for the critical F-value are found using the degrees of freedom associated with the two estimates of population variance from the sample data. These two estimates are called the *between group* variance and the *within group* variance.

For the *df* associated with the numerator we use the *df* for the between group variance.
df(between) = k − 1,
where k is the number of levels of the factor.

For the *df* associated with the denominator, use the *df* for the within group variance.
df(within) = N − k, where N is the total number of data for all samples and k is the number of levels of the factor.
(b) the level of significance, α, is assumed or given by the researcher.
(c) State the Decision Rule. The Decision Rule is shown in Figure 16.15.

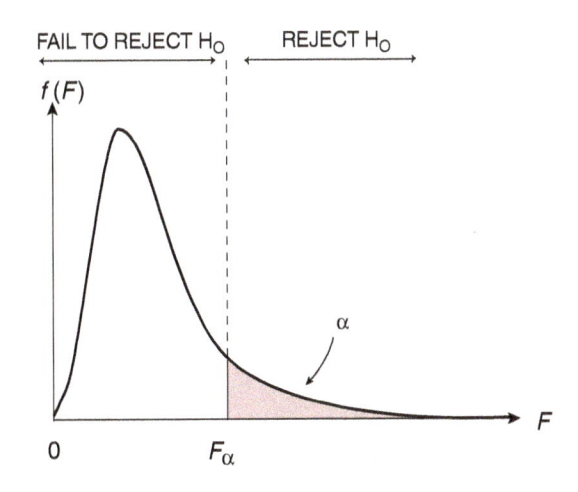

FIGURE 16.15

Thus when the ANOVA test is applied, a test statistic F will be shown to be significant if its value is **greater** than the *critical* F-value. When this occurs we **reject** the null hypothesis. This is shown in Figure 16.16.

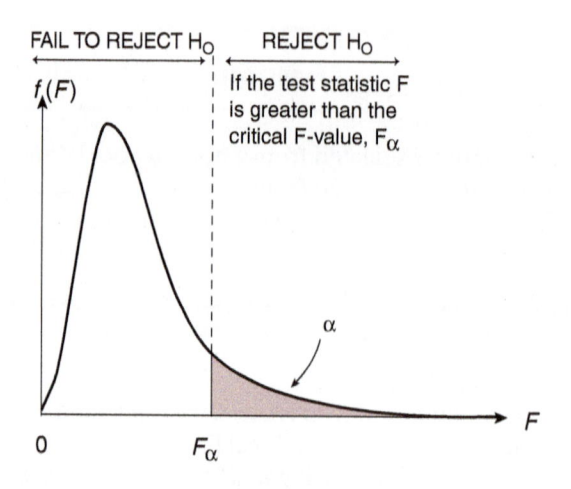

Chapter 16 The F-Distribution and an Introduction to Analysis of Variance (ANOVA)

FIGURE 16.16

A graph of $f(F)$ showing the F-distribution with critical value F_α marking the boundary between "FAIL TO REJECT H_O" and "REJECT H_O" regions. If the test statistic F is greater than the critical F-value, F_α, reject H_O. The shaded area in the right tail represents α.

Step 4 *Analyze the sample data.*

The ANOVA test requires the calculation of a test statistic F by comparing the variance due to treatment, (or the *between* group variance) **to** the variance within each group, (or the *within* sample variance). This test statistic F is expressed as:

$$F = \frac{MS_{between}}{MS_{within}}$$

Using the sample data complete the following list of calculations:

- ΣT for each level of sample data.
- n for each level of sample data.
- ΣX = Sum of ALL the data values using all the samples.
- ΣX^2 = Sum of ALL the Squared data values using all the samples.
- N = Total number of data values using all the samples.
- k = Number of Treatment Levels.

Formulas for Sums of Squares

$$SS_{(total)} = \Sigma X^2 - \frac{(\Sigma X)^2}{N}$$

$$SS_{(between\ treatments)} = \left(\frac{(\Sigma T_1)^2}{n_1} + \frac{(\Sigma T_2)^2}{n_2} + ... + \frac{(\Sigma T_i)^2}{n_i} - \frac{(\Sigma X)^2}{N} \right)$$

$$SS_{(within\ treatment)} = \Sigma X^2 - \left(\frac{(\Sigma T_1)^2}{n_1} + \frac{(\Sigma T_2)^2}{n_2} + ... + \frac{(\Sigma T_i)^2}{n_i} \right)$$

where X = a data value
n_1 = the size of sample 1
n_2 = the size of sample 2
n_i = the size of sample i

N = the total number of data values
T₁ = the data values of sample 1
T₂ = the data values of sample 2
Tᵢ = the data values of sample i

Formulas for Degrees of Freedom

$df_{(Between)} = k - 1$,
  where k = the number of treatment levels.
$df_{(within)} = N - k$,
  where N = total number of data values,
  and k = the number of treatment levels.
$df_{(Total)} = N - 1$
  where N = total number of data values.

Formulas for Mean Square

$$MS_{(between)} = \frac{SS_{bet}}{df_{bet}}$$

$$MS_{(within)} = \frac{SS_{within}}{df_{within}}$$

Calculated test statistic F

$$F = \frac{MS_{bet}}{MS_{within}}$$

Substitute the calculated values into the ANOVA TABLE shown in Table 16.23.

TABLE 16.23 ANOVA TABLE

Source of Variation	Sum of Squares SS	Degrees of Freedom df	Mean Square MS	Calculated F-statistic
Between				
Within				
Total				

Step 5 *State the conclusion.*
Figure 16.17 is used to model the possible conclusions.

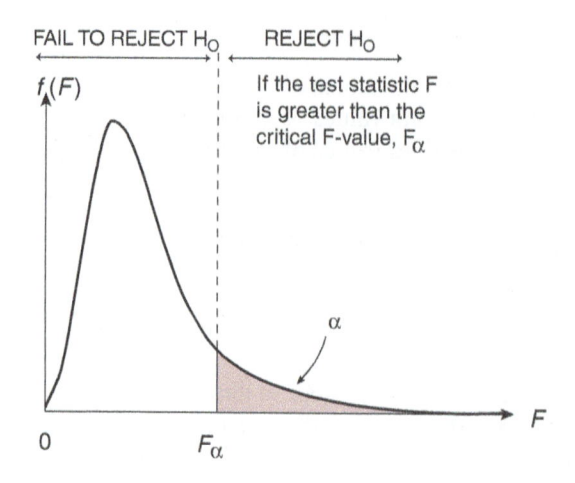

840 Chapter 16 *The F-Distribution and an Introduction to Analysis of Variance (ANOVA)*

FIGURE 16.17

CASE STUDY 16.1

The 1994 baseball season began with Ray Lankford of the St. Louis Cardinals hitting a home run. During the 1994 baseball season, the month of April produced more runs per game since 1930 and more homers per game than ever before. This led to the speculation that the 1994 baseballs were "juiced" or "livelier." Independent tests were conducted on a sample of '94 baseballs to determine if the balls were livelier which would account for the 1994's offensive firepower. One test conducted on the baseballs was to determine the average bounce of the balls when they were heated, cold or regular. Figure 16.18 presents the results of the test.

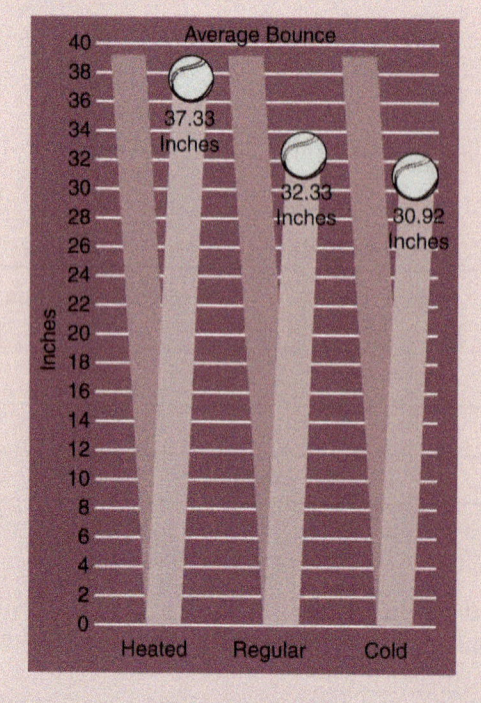

FIGURE 16.18

a) What was the average bounce for the heated baseballs? for the cold baseballs? for the regular baseballs?
b) What is the independent variable? Name the different levels of the variable.
c) Interpret the within group variance and between group variance for this test.
d) Does the temperature of the ball seem to have an effect on the average bounce?
e) What additional information would you need to determine whether the temperature of the ball plays a significant role in the distance the ball will travel or bounce?
f) If you were to conduct an ANOVA test to determine whether there was a significant difference in the distance the balls bounced, then what assumptions would be made regarding the sample of balls selected for the experiment and the population from which the balls were selected?

Let's consider an ANOVA where the samples are **not** of equal size n.

EXAMPLE 16.6

A medical researcher is interested in the comparison of the effectiveness of four different exercise programs for infants. The researcher randomly selects 23 infants, and randomly assigns them to four treatment groups. After administering to each treatment group, one of the four exercise programs, she measures the age at which each infant first walks without support. Table 16.24 lists the relevant data values.

TABLE 16.24 Age at which infant first walks without support (in months)

Treatment Group	A	B	C	D
	9.50	11.25	11.75	13.25
	9.25	10.50	12.50	11.50
	9.75	9.75	9.50	12.25
	10.00	9.50	11.50	13.25
	12.75	11.75	13.50	11.50
	9.50	10.25	13.25	

Do these data indicate that an exercise treatment program is related to the number of months an infant will walk without support at $\alpha = 0.05$?

Solution

Step 1 *Formulate the two hypotheses, H_0 and H_a.*

H_0: There is no difference between the population means for infants to walk without support.

$$H_0: \mu_A = \mu_B = \mu_C = \mu_D$$

H_a: Not all four population means are equal.

Step 2 *Determine the model to test the null hypothesis, H_0.*

The appropriate sampling distribution is the F-Distribution.

Step 3 *Formulate the Decision Rule.*

To perform the ANOVA test, the right hand side of the F-Distribution is always used as shown in Figure 16.19. The critical F-value, $F_\alpha(df$ between, df within) is determined from

Appendix C: Table VII, by using α, *df* between and *df* within.

For this problem, $\alpha = 0.05$. Since there are four levels of treatment, k = 4 and:

df(between) = k − 1,
df(between) = 3

Since there are 23 data values, N = 23. Thus:

df(within) = N − k
df(within) = 23 − 4
df(within) = 19

Using **df(between) = 3, df(within) = 19,** and Appendix C: Table VII, the critical F-value is $F_{0.05}(3,19) = 3.13$.

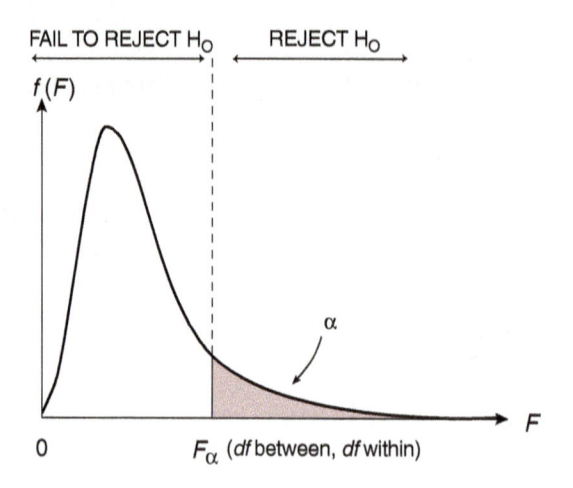

FIGURE 16.19

In this ANOVA test the **decision rule** is: *if the test statistic F is **greater** than the critical F-value, $F_{0.05}(3,19) = 3.13$, **reject** the null hypothesis.* This decision rule is illustrated in Figure 16.20.

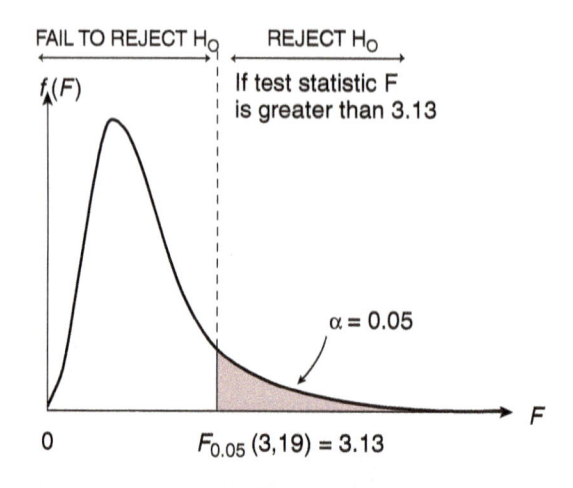

FIGURE 16.20

Step 4 Analyze the sample data.

The ANOVA test requires the calculation of the test statistic F by comparing the variance due to treatment, (or the *between* group variance) **to** the variance within each group, (or the *within* sample variance). This test statistic F is expressed as

$$F = \frac{MS_{between}}{MS_{within}}$$

Using the sample data complete the following list of calculations:
- ΣT for each level of sample data.
- n for each level of sample data.
- ΣX = Sum of ALL the data values using all the samples.
- ΣX^2 = Sum of ALL the Squared data values using all the samples.
- N = Total number of data values using all the samples.
- k = Number of Treatment Levels.

Table 16.25 contains all the data values of Example 16.6 along with the necessary calculations required to determine the sums of squares expressions.

TABLE 16.25 Age, in months, at which infant first walks without support

Treatment Group	A	B	C	D
	9.50	11.25	11.75	13.25
	9.25	10.50	12.50	11.50
	9.75	9.75	9.50	12.25
	10.00	9.50	11.50	13.25
	12.75	11.75	13.50	11.50
	9.50	10.25	13.25	
ΣT	60.75	63	72	61.75
n	6	6	6	5

$\Sigma X = 257.5$
$\Sigma X^2 = 2929.25$
N = 23 data values
k = 4 levels

Using the calculated values listed in Table 16.25 we can apply the formulas for the Sums of Squares, degrees of freedom, Means Square and the calculated test statistic F.

Calculations for Sums of Squares

$$SS_{(total)} = \Sigma X^2 - \frac{(\Sigma X)^2}{N} = 2929.25 - \frac{(257.5)^2}{23}$$

$$SS_{(total)} = 46.37$$

$$SS_{(between\ treatments)} = \left(\frac{(\Sigma T_1)^2}{n_1} + \frac{(\Sigma T_2)^2}{n_2} + \ldots + \frac{(\Sigma T_i)^2}{n_i}\right) - \frac{(\Sigma X)^2}{N}$$

since there are four treatments we have,

$$= \left(\frac{(60.75)^2}{6} + \frac{(63)^2}{6} + \frac{(72)^2}{6} + \frac{(61.75)^2}{5} \right) - \frac{(257.5)^2}{23}$$

$$= 2903.21 - 2882.88$$

$$SS_{(between\ treatments)} = 20.33$$

$$SS_{(within\ treatment)} = \Sigma X^2 - \left(\frac{(\Sigma T_1)^2}{n_1} + \frac{(\Sigma T_2)^2}{n_2} + ... + \frac{(\Sigma T_i)^2}{n_i} \right)$$

$$= 2929.25 - 2903.21$$

$$SS_{(within\ treatment)} = 26.04$$

Calculations for Degrees of Freedom

$df_{(between)} = k - 1,$ where k = the number of treatment levels.
$df_{(between)} = 4 - 1$
$df_{(between)} = 3$

$df_{(within)} = N - k,$ where N = total number of data values, and k = the number of treatment levels.

$df_{(within)} = 23 - 4$
$df_{(within)} = 19$

$df_{(total)} = N - 1,$ where N = total number of data values.
$df_{(total)} = 23 - 1$
$df_{(total)} = 19$

Calculations for Mean Square

$$MS_{(between)} = \frac{SS_{bet}}{df_{bet}} = \frac{20.33}{3}$$

$$MS_{(between)} = 6.78$$

$$MS_{(within)} = \frac{SS_{within}}{df_{within}} = \frac{26.04}{19}$$

$$MS_{(within)} = 4.95$$

Calculated test statistic F

$$F = \frac{MS_{bet}}{MS_{within}}$$

$$F = \frac{6.78}{1.37}$$

$$F = 4.95$$

Substitute the calculated values into the ANOVA TABLE shown in Table 16.26.

TABLE 16.26 ANOVA TABLE

Source of Variation	Sum of Squares SS	Degrees of Freedom df	Mean Square MS	Calculated test statistic F
Between	20.33	3	6.78	4.95
Within	26.04	19	1.37	
Total	46.37	22		

Step 5 *State the conclusion.*

The calculated test statistic F is 4.95, and the critical F-value, $F_{0.05}(3,19)= 3.13$. Since the test statistic F is **greater than** the critical F-value, the **null hypothesis is rejected.** This is illustrated in Figure 16.21. Thus, the alternative hypothesis is **accepted** at $\alpha = 0.05$. Therefore, the population means are **not** all equal. This implies that at least one of the exercise treatment programs affects the initial month that an infant will begin to walk without support.

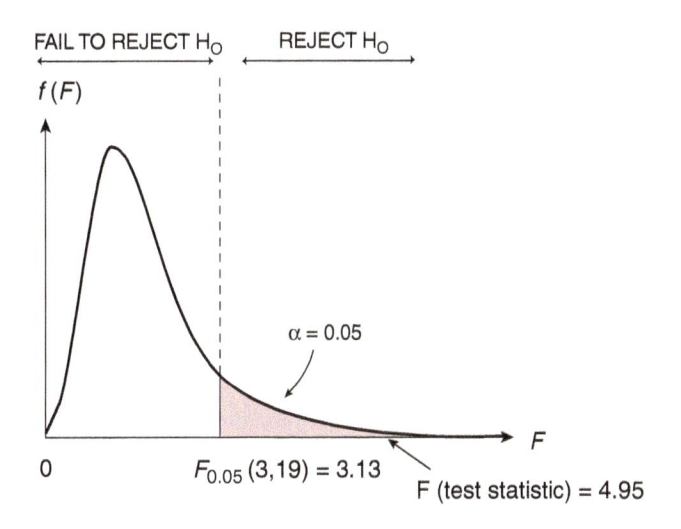

FIGURE 16.21

CASE STUDY 16.2

A study entitled Teacher Job Satisfaction and Teacher Job Stress: School Size, Age and Teaching Experience appeared in *Education* (volume 112 issue number 2). The purpose of this study was to determine the effect that school size, age and teaching experience has on job satisfaction and job stress.

The subjects selected for the study were 229 North Florida and South Georgia urban secondary school physical education teachers. Two instruments were used to measure the job satisfaction and job related stress of the teachers. A high score on the Job Satisfaction Scale indicates a high level of job satisfaction while a high score on the Job-Related Stress Scale indicates a high level of job stress. Some of the results of this study are shown in Table 16.27 to Table 16.29.

(continued)

846 Chapter 16 The F-Distribution and an Introduction to Analysis of Variance (ANOVA)

TABLE 16.27 Differences in Job Stress Scores by School Size

Size of School	Number of Subjects	Mean Scores	Standard Deviation	F-statistic	Probability Level
500–1000	45	34.87	12.21	5.33	0.00*
1001–1500	69	37.80	12.51		
1501–2000	60	44.01	15.02		
2001+	55	44.35	14.24		
Total	229	40.29			

* Exceeds the 0.05 alpha level

TABLE 16.28 Differences in Job Satisfaction Scores by Age Group

Size of School	Number of Subjects	Mean Scores	Standard Deviation	F-statistic	Probability Level
30 or less	52	46.50	11.13	0.82	0.48
31–45	88	45.77	13.35		
46–60	66	43.40	12.24		
61+	23	47.22	16.72		
Total	229	44.39			

TABLE 16.29 Differences in Job Stress Scores by Age Group

Size of School	Number of Subjects	Mean Scores	Standard Deviation	F-statistic	Probability Level
30 or less	52	42.38	13.74	0.010	0.99
31–45	88	42.62	14.17		
46–60	66	42.50	16.09		
61+	23	43.00	11.82		
Total	229	42.57			

For Table 16.27:
a) identify the independent variable. State the different levels of this variable.
b) compare the mean job stress score for a school of size 500–1000 to the mean job stress score for a school size of 2001+. Interpret the meaning of a each mean score.
c) Within which group are the job stress scores more variable? less variable? How would you interpret each of these results?
d) What would you compare to determine the between group variance?
e) What are the degrees of freedom for the within groups and the degrees of freedom for the between groups?
f) As the size of the school increases, what appears to happen to the job stress scores?
g) Explain the meaning of the probability level for the given F-statistic.
h) Based upon the results of Table 16.27, can you conclude that the largest schools are significantly more stressful? Explain.

For Table 16.28:
a) identify the independent variable. State the different levels of this variable.
b) compare the mean job satisfaction score for teachers aged 61+ to the mean job satisfaction score for teachers aged 46–60. Interpret the meaning of each mean score.
c) Within which group are the job satisfaction scores more variable? less variable? How would you interpret each of these results?
d) What would you compare to determine the between group variance?
e) Explain the meaning of the probability level for the given F-statistic.
f) What are the degrees of freedom for the within groups and the degrees of freedom for the between groups?

(continued)

g) Based upon the results of Table 16.28, can you conclude that the age of teachers is a significant factor in their job satisfaction? Explain.
h) Identify the age group that had the lowest job satisfaction and the age group with the highest satisfaction. Can you conclude at an alpha level of 0.01 that this difference is significant?

For Table 16.29:
a) identify the independent variable. State the different levels of this variable.
b) compare the mean job stress score for teachers aged 30 or less to the mean job stress score for teachers aged 61+. Interpret the meaning of each mean score.
c) Within which group are the job stress scores more variable? less variable? How would you interpret each of these results?
d) What would you compare to determine the between group variance?
e) Explain the meaning of the probability level for the given F-statistic.
f) What are the degrees of freedom for the within groups and the degrees of freedom for the between groups?
g) Based upon the results of Table 16.29, can you conclude that the age of teachers is a significant factor in their job stress? Explain.
h) Identify the age group that had the lowest job stress and the age group with the highest job stress. Can you conclude at an alpha level of 0.05 that this difference is significant?

CASE STUDY 16.3

The study entitled Factors In Children's Attitudes Toward Pets which appeared in the journal *Psychological Reports*, 1990, issue 66, investigated the parental attitudes, family size, structure, and presence or absence of household pets as influences on children's attitudes toward pets. The researchers, Kidd & Kidd, believe that attachment to pet animals begin as early as 18 months. Since a child-pet attachment begins at such an early age, Kidd and Kidd decided to investigate the factors which may influence feelings and attitudes toward pets. The researchers administered a Parent Questionnaire to the parents of the 700 children who were the subjects of the study. These subjects were composed of 339 boys and 361 girls between the ages of 6 months and 18 years: 150 boys and 150 girls from two-parent pet-owning homes, 43 boys and 55 girls from one-parent owning homes, 97 boys and 105 girls from two-parent, nonpet-owning homes, and 49 boys and 51 girls from one-parent, nonpet-owning homes.

The Parent Questionnaire were scored for the children's Activities with Pets and Interest in Pet's scales whether or not the family owned a pet. They were scored for a child's Responsibility for Pet Care scale only if the family owned a pet or indicated that the child actually helped with responsible care for a neighbor's or a friend's pet. One aspect of the study was to determine whether the age of the child significantly influenced the scores on each of the three scales. The researchers believed that involvement in activities with and interest in pets were good measures of attachment to pets, who showed more interest in them, and who took more responsibility for them also were reported as having a greater affection for their pets than children who were scored lower on these scales. A higher score on each of these scales is interpreted as displaying more activities with the pets, showing more interest in the pets and more responsibility with the pets. The results of the nonpet-owning children on the three scales are given in Table 16.30. The results of the pet-owning children on the three scales are discussed in the What Do You Think Exercises at the end of the chapter.

TABLE 16.30 Mean and Standard Deviations For Nonpet-Owning Children On Three Scales

	n	Mean	Std Dev
Activities Scale			
Ages			
0–5	110	15.26	7.10
Grade School	122	18.64	9.38
High School	70	18.11	10.16

(continued)

TABLE 16.30 (continued)

	n	Mean	Std Dev
Interest Scale			
Ages			
0–5	110	9.92	7.80
Grade School	122	12.89	9.19
High School	70	9.03	7.96

TABLE 16.31 One-Way Analysis of Variance of Activities With Pets, Interest In Pets: Scores of Children Without Household Pets By Age Groups

Source of Variance	SS	df	MS	F
Activities With Pets				
SS Between Age Groups	622.75	2	311.38	4.54*
SS Within Groups	20505.89	299	68.58	
Interest In Pets				
SS Between Age Groups	725.42	2	362.71	4.66*
SS Within Groups	23250.09	299	77.76	

*$p<0.05$

Using the information on the Activities Scale in Table 16.30 and Table 16.31:
- a) Identify the independent variable. State the different levels of this variable.
- b) Which age group should the greatest activity with pets? the smallest activity with pets?
- c) Which age group had the greatest variability in the activities scale with pets? the smallest variability?
- d) Which age group had the greatest mean score on the Activities scale? the smallest mean score?
- e) Determine the difference between the mean scores of part d? Do you think this difference is significant? If it is significant, state the alpha level and explain how you came to this conclusion.
- f) What are the degrees of freedom for the within groups and the degrees of freedom for the between age groups?
- g) Determine the Sums of Square due to Factor variance. Determine the Sums of Square due to Error variance.
- h) Explain what the MS = 311.38 indicates about the between age groups for the activities scale.
- i) Explain what the MS = 68.58 indicates about the within groups for the activities scale.
- j) Identify the Mean Square population variance estimate for the variance Between the groups.
- k) Identify the Mean Square population variance estimate for the variance within the groups.
- l) Based upon the results of Table 16.30, can you conclude that the age of a child is a significant factor in heir activities with pets at an alpha level of 0.10? at an alpha level of 0.01? Explain.

Using the information on the Interest Scale in Table 16.30 and Table 16.31:
- m) Identify the independent variable. State the different levels of this variable.
- n) Which age group should the greatest interest with pets? the smallest interest with pets?
- o) Which age group had the greatest variability in the interest scale with pets? the smallest variability?
- p) Which age group had the greatest mean score on the Interest scale? the smallest mean score?
- q) Determine the difference between the mean scores of part p? Do you think this difference is significant? If it is significant, state the alpha level and explain how you came to this conclusion.
- r) What are the degrees of freedom for the within groups and the degrees of freedom for the between age groups?

(continued)

s) Determine the Sums of Square due to Factor variance. Determine the Sums of Square due to Error variance.
t) Explain what the MS = 362.71 indicates about the between age groups for the interest scale.
u) Explain what the MS = 77.76 indicates about the within groups for the interest scale.
v) Identify the Mean Square population variance estimate for the variance Between the groups.
w) Identify the Mean Square population variance estimate for the variance within the groups.
x) Based upon the results of Table 16.31, can you conclude that the age of a child is a significant factor in their interest with pets at an alpha level of 0.10? at an alpha level of 0.01? Explain.

16.6 A BRIEF LOOK AT TWO-WAY ANOVA

Single factor or one-way ANOVA examines populations that have been classified with one categorical variable or factor. When populations have been classified using two categorical variables or two factors, each with its own levels, the ANOVA model is called *Two-Way ANOVA*.

Consider an experiment designed to study the effects of diet and aerobic exercise on blood pressure. In such an experiment, each of the categorical variables (factors) could have its own levels. For example, the categorical variable diet could be classified into three fat-content levels, such as low, medium or high. The other categorical variable aerobic exercise could be classified into the levels of low, normal or high aerobic activities. Table 16.32 diagrams the way the categories and levels are examined for the experiment.

TABLE 16.32 Diet (Fat Content)

	Low	Medium	High
EXERCISE			
Low	LL	LM	LH
Normal	NL	NM	NH
High	HL	HM	HH

In this two way design there are **nine** combinations of experimental treatment conditions. Using the first letter of each category level we have:

LL or Low-Low to be a combination where a subject is placed in a low aerobic exercise program and a low fat diet.

NL or Normal-Low to be a combination where a subject is placed in a normal aerobic exercise program and a low fat diet.

HL or High-Low to be a combination where a subject is placed in a high aerobic exercise program and a low fat diet.

These combinations of treatment conditions continue to HH, where HH indicates a high aerobic exercise program and a high fat diet.

All levels of exercise are combined with **all levels of diet.** The efficiency and control obtained in combining and analyzing the treatment conditions in this design are major advantages for using the two-way design.

Further development for a two-way design are beyond the scope of this text. However, many of the key concepts in the two-way design are similar to those of the one-way ANOVA.

KEY TERMS

F-Distribution 802
F-Test 809
Test statistic 812
Test statistic F 812
Critical F-Value 808
Analysis of Variance, ANOVA 816

One-way ANOVA 822
Between Variance 816
Within Variance 816
Total Variance 816
Treatment Variance 817
Error Variance 817

Experimental Variance 817
Pooling Variance 818
Sums of Squares, SS 828
Mean Square, MS 822
ANOVA Table 831
Two-Way ANOVA 849

SECTION EXERCISES

Sections 16.1–16.2

1. ANOVA is an acronym for _____.
2. The F-Distribution is a family of distributions composed of the value of the variances for samples drawn from two distinct _____ populations with equal _____.
3. The symbol used to represent the variance of a sample selected from population 2 is _____.
4. The formula to compute the test statistic F is given by: F = _____.
5. If $s_1^2 = 12.93$ and $s_2^2 = 8.47$, then F = _____.
6. The shape of the F-Distribution is _____. However, as the degrees of freedom increases, the shape approaches a _____.
7. The peak or modal value of the F-Distribution is approximately _____.
8. In the expression $F_{0.01}(10, 20)$, the quantity 0.01 represents the _____, the value 10 represents the _____, while the value 20 represents _____.
9. From Table VII, the value of $F_{0.01}(10, 20)$ is _____.
10. If a test statistic F is around one, then the population variances are approximately equal. (T/F)
11. Unlike standard deviations, the test statistics F are sometimes negative. (T/F)
12. There is a different F-Distribution for each set of degrees of freedom associated with the numerator and denominator of the test statistic F. (T/F)

Use **Table VII: Critical Values of the F-Distribution in Appendix C** to answer questions 13–17.

13. Find the critical F-value if $df_{numerator} = 8$, $df_{denominator} = 9$ and $\alpha = 0.01$.
14. Find the critical F-value if $dfs_1^2 = 24$, $dfs_2^2 = 22$ and $\alpha = 0.01$.
15. Find the critical F-value, $F_\alpha(dfs_1^2, dfs_2^2)$, expressed as $F_{0.01}(13, 20)$.
16. Find the critical F-value, $F_{0.10}(35, 45)$.
17. Place the critical F-values on the following F-Distribution graph.
    (a) $F_{0.10}(30, 40)$
    (b) $F_{0.05}(30, 40)$
    (c) $F_{0.025}(30, 40)$
    (d) $F_{0.01}(30, 40)$

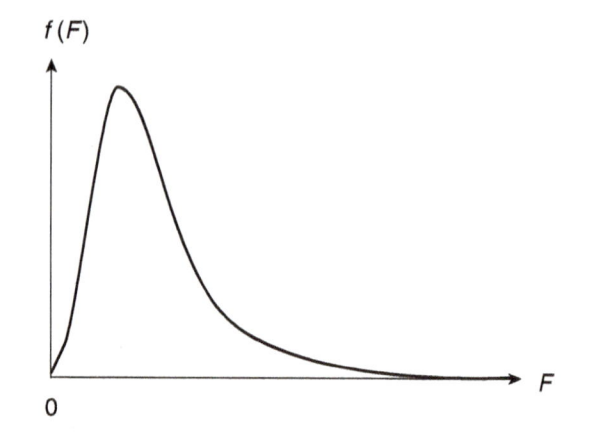

Section 16.3

18. In an F-test, the null hypothesis: the variance of population 1 and the variance of population 2 are equal is symbolized as _____.
    The alternative hypothesis: the variance of population 1, and the variance of population 2 are **not** equal is symbolized as _____.
19. In an F-test, the test statistic is computed using the formula: F = _____.
20. In an F-test, if the test statistic, F, is greater than the critical F-value, symbolized $F\alpha/2(dfs_1^2, dfs_2^2)$, then the population variances are _____ _____ at the significance level, α.

21. To perform an F-test, which represents the null hypothesis?
    (a) $H_0: \sigma_1^2 = \sigma_2^2$ (b) $H_0: \sigma_1 = \sigma_2$ (c) $H_0: \mu_1 = \mu_2$
    (d) $H_0: p_1 = p_2$

For questions 22 to 26:

If $F_{0.025}(12,20)$, is the critical F-value for an F-Test, then which number represents:

22. the degrees of freedom for the larger variance?
    (a) 21   (b) 12   (c) 20   (d) 13
23. the degrees of freedom for the smaller variance?
    (a) 21   (b) 12   (c) 20   (d) 13
24. the sample size for the larger variance?
    (a) 21   (b) 12   (c) 20   (d) 13
25. the sample size for the smaller variance?
    (a) 21   (b) 12   (c) 20   (d) 13
26. the value for the critical F-statistic?
    (a) 2.68   (b) 2.28   (c) 3.07   (d) 2.54
27. The F-test is a hypothesis testing procedure to determine if sample variances are different. (T/F)
28. (a) Compute the test statistic F for the following sample variances.
    $s_1^2 = 10$ $s_2^2 = 6$
    $n_1 = 22$ $n_2 = 25$
    (b) Conduct an F-Test for the equality of the population variances using an $\alpha = 0.05$.
29. (a) Compute the test statistic F for the following sample variances.
    $s_1^2 = 6.25$ $s_2^2 = 9.85$
    $n_1 = 50$ $n_2 = 50$
    (b) Conduct an F-Test for the equality of the population variances using an $\alpha = 0.10$.
30. (a) Compute the test statistic F for the following sample variances.
    $s_1^2 = 18$ $s_2^2 = 16$
    $n_1 = 12$ $n_2 = 15$
    (b) Conduct an F-Test for the equality of the population variances using an $\alpha = 0.05$.
31. (a) Compute the test statistic F for the following sample variances.
    $s_1^2 = 49$ $s_2^2 = 18$
    $n_1 = 21$ $n_2 = 22$
    (b) Conduct an F-Test for the equality of the population variances using an $\alpha = 0.05$.
32. (a) Compute the test statistic F for the following sample variances.
    $s_1^2 = 4.25$ $s_2^2 = 9.45$
    $n_1 = 40$ $n_2 = 40$
    (b) Conduct an F-Test for the equality of the population variances using an $\alpha = 0.10$.
33. (a) Compute the test statistic F for the following sample variances.

    $s_1^2 = 1.6$ $s_2^2 = 3.8$
    $n_1 = 100$ $n_2 = 100$
    (b) Conduct an F-Test for the equality of the population variances using an $\alpha = 0.10$.
34. (a) Compute the test statistic F for the following samples.

sample 1	sample 2
2.4	5.6
7.8	6.7
8.9	5.9
5.8	8.05
6.7	8.93
8.8	7.34

    (b) Conduct an F-Test for the equality of the population variances using an $\alpha = 0.05$.
35. (a) Compute the test statistic F for the following samples.

sample 1	sample 2
112	111
146	137
134	88
156	109
126	98
114	122
	149
	159

    (b) Conduct an F-Test for the equality of the population variances using an $\alpha = 0.02$.
36. (a) Compute the test statistic F for the following samples.

sample 1	sample 2
3	11
7	13
9	8
23	10
5	9
18	12
	14
	15

    (b) Conduct an F-Test for the equality of the population variances using an $\alpha = 0.02$.
37. Kathleen, an educational psychologist, is interested in the academic achievement of third graders enrolled in a special computer assisted problem solving program. Prior to the beginning of the school year, the total group of third graders and the total group of teachers are randomly assigned to two treatment groups within their school. Treatment Group 1 contains third graders not enrolled in the special computer assisted problem solving program, whereas, Treatment Group 2 contains students that are enrolled in the special computer assisted problem solving program. Kathleen would like to pool the sample variances but first she must perform an F-Test. Using the data within the table determine if Kathleen could pool the sample variances. ($\alpha = 0.10$).

Treatment Group	Mean Achievement	Variance	Sample Size
Group 1	74.6	75.9	31
Group 2	115.8	38.4	41

38. Good neighbor Dave runs a limousine service from Hempstead, New York to JFK airport. Dave is considering one of two routes. The first is via Sunrise Highway and the other is the Southern State Parkway. Dave decided to conduct a study of both routes and compares and finds that the mean times for the routes are very similar, but there is more variation in the Sunrise Highway route, since it contains more stoplights than the Southern State Parkway route. But the Parkway route is somewhat longer in miles. The statistics recorded by Dave for his study are listed in the following table. Use the F-test to determine if there really is a difference in the variation of the two routes at $\alpha = 0.05$.

Route	Mean Time(MIN)	Variance	Sample Size
Sunrise Highway	25	169	8
Southern State Pkwy	28	25	8

39. A medical researcher decides to perform an experiment comparing the effects of calcium and a placebo on the blood pressure of males between the ages of 40 and 55. The researcher wants to use a two sample t test where a pooled standard deviation is to be calculated. Since the procedure requires equal population standard deviations, an F-Test is called upon to determine if the population standard deviations are equal. The experimental data produce two standard deviations. The larger of the two is s = 8.45 from 10 observations, while the smaller is s = 6.14 using 11 observations. Do these results support the use of the two sample t test with a pooled standard deviation? (use $\alpha = 0.02$)

Section 16.4

40. The variability in the differences among the individual data values of each group is called _____.
41. The variability in the differences among the sample means is called _____.
42. In an experiment, if all the subjects were grouped into one overall sample, the variability in the differences among all the subjects is called _____.
43. Another name for error variance is _____.
44. The relationship of the three variances: within, between and total is: _____ variance equals _____ variance + _____ variance.
45. The symbol SS is an abbreviation for _____ and is computed using the expression _____.
46. If the $SS_{(within\ treatments)} = 140$, and the value of $SS_{(total)} = 190$ then $SS_{(between\ treatments)}$ is:
    (a) 50 (b) 140 (c) 330 (d) not enough information

47. If the $SS_{(between\ treatments)} = 33$, and the value of $SS_{(total)} = 117$ then $SS_{(within\ treatments)}$ is:
    (a) 33 (b) 84 (c) 150 (d) not enough information
48. Between Group Variance is often referred to as Experimental Variance. (T/F)
49. If the variance due to treatment is zero, then the Within Group Variance must equal the Total Variance. (T/F)
50. Consider the following information about three different populations:

	Population one	Population two	Population three
Population means μ	23	23	23
Population variances Within each treatment σ^2	4	4	4

   (a) Identity the population variance within each group.
   (b) Find the average within group variance.
   (c) Calculate the between group variance.
   (d) Determine the total group variance.
   (e) Is the Total group variance due solely to the Within group variance? Explain.

51. Consider the following information about three different populations:

	Population one	Population two	Population three
Population means μ	100	150	200
Population variances Within each treatment σ^2	4	4	4

   (a) Identity the population variance within each group.
   (b) Find the average within group variance.
   (c) Calculate the between group variance.
   (d) Determine the total group variance.
   (e) Is the Total group variance due solely to the Within group variance? Explain.

52. Consider the following information about three different populations:

	Population one	Population two	Population three
Population means μ	350	400	450
Population variances Within each treatment σ^2	100	100	100

   (a) Identity the population variance within each group.
   (b) Find the average within group variance.
   (c) Calculate the between group variance.
   (d) Determine the total group variance.
   (e) Is the total group variance due solely to the within group variance? Explain.

53. Data Values of a Population of 21 Students Subdivided into Three Groups

Treatment 1	Treatment 2	Treatment 3
10	24	38
12	26	40
14	28	42
16	30	44
18	32	46
20	34	48
22	36	50

Population means
μ 16   30   44
Population variances
Within each treatment
σ^2 16   16   16

(a) Calculate $SS_{(within\ treatments)}$.
(b) Calculate $SS_{(between\ treatments)}$.
(c) Calculate $SS_{(total)}$.
(d) Is the relationship
$SS_{(between\ treatments)} + SS_{(within\ treatments)} = SS_{(total)}$
True?
(e) Identify the population variance within each group.
(f) Find the average within group variance.
(g) Calculate the between group variance.
(h) Determine the total group variance.
(i) Is the Total group variance due solely to the Within group variance? Explain.

54. Data Values of a Population of 15 Students Subdivided into Three Groups

Treatment 1	Treatment 2	Treatment 3
23	34	45
26	37	48
28	39	50
35	46	57
39	50	61

Population means
μ 30.2   41.2   52.2
Population variances
Within each treatment
σ^2 34.96   34.96   34.96

(a) Calculate $SS_{(within\ treatments)}$.
(b) Calculate $SS_{(between\ treatments)}$.
(c) Calculate $SS_{(total)}$.
(d) Is the relationship
$SS_{(between\ treatments)} + SS_{(within\ treatments)} = SS_{(total)}$
True?
(e) Identify the population variance within each group.
(f) Find the average within group variance.
(g) Calculate the between group variance.
(h) Determine the total group variance.
(i) Is the Total group variance due solely to the Within group variance? Explain.

55. Data Values of a Population of 15 Scores Subdivided into Three Groups

Treatment 1	Treatment 2	Treatment 3
115	121	112
118	124	115
121	127	118
124	130	121
127	133	124

Population means
μ 121   127   118
Population variances
Within each treatment
σ^2 18   18   18

(a) Calculate $SS_{(within\ treatments)}$.
(b) Calculate $SS_{(between\ treatments)}$.
(c) Calculate $SS_{(total)}$.
(d) Is the relationship
$SS_{(between\ treatments)} + SS_{(within\ treatments)} = SS_{(total)}$
True? Verify your results.
(e) Identity the population variance within each group.
(f) Find the average within group variance.
(g) Calculate the between group variance.
(h) Determine the total group variance.
(i) Is the total group variance due solely to the within group variance? Explain.

Section 16.5

56. In this section we have presented the one-way analysis of variance procedure. Why is the procedure called "one way"?

57. What are the assumptions necessary to use the one-way ANOVA procedure?

58. Which distribution is used to test the calculated results?

59. One-way ANOVA involves the analysis of one independent variable with two or more treatment levels. (T/F)

60. To use one-way ANOVA, how may independent variables are compared?
(a) 1   (b) 2   (c) at least 2
(d) not enough information

61. The ANOVA procedure uses the normal distribution as the model distribution to determine differences between experimental conditions.
(T/F)

62. The ANOVA procedure assumes that the variation between the sample means is zero.
(T/F)

63. The test statistic F is found by dividing the within group variance by the between group variance.
(T/F)

64. The ANOVA technique compares variances under the assumption that the variance among all levels is the same.
(T/F)

65. In ANOVA, as the value of between group variance increases, the value of the test statistic F will _____. As the value of the within group variance increases, the value of the test statistic F will _____.

66. The formula for the degrees of freedom for the between group variance is: df(between groups) = k − 1, where k represents _____. The formula for df(between treatments) = _____.

67. The formula for the degrees of freedom for the within group variance is: df(within groups) = N − k, where k represents _____ and N is _____.

68. If a sample one contains 15 data values, and sample two contains 18 data values, then the degrees of freedom for the F-value is
    (a) df = 33
    (b) df = 31
    (c) df = (15,18)
    (d) df = (14,17)

69. In the ANOVA Hypothesis Test, the null hypothesis is stated as: the _____ means are _____. The alternative hypothesis states that there is at least _____ mean of the _____ which is _____.

70. In the ANOVA Hypothesis Test, assuming the null hypothesis is true, then we would expect the value of the test statistic F to be around the value _____.

71. Which null hypothesis is used for the ANOVA procedure?
    (a) all sample means are the same.
    (b) not all sample means are equal.
    (c) all population variances are equal.
    (d) all population means are equal.

72. If $\alpha = 0.05$, the $df_{between\ groups}$ is 4 and $df_{within\ groups}$ is 26, then the critical F-value is:
    (a) 5.77 (b) 3.78 (c) 2.74 (d) 2.17

73. If the critical F-value is greater than the test statistic (i.e. the calculated F-statistic) then what is the appropriate conclusion.
    (a) fail to reject H_0
    (b) reject H_0
    (c) reject H_a
    (d) not enough information

74. In the ANOVA hypothesis test, if the critical F-value is $F_{0.01}(7,22)$, then the df(within groups) is _____ and the df(between treatments) = _____.

75. In the ANOVA hypothesis test, if the calculated test statistic F is greater than the critical F-value, then the decision is to _____.

76. The formula to compute the $SS_{(total)}$ is:
    $SS_{(total)} = $ _____ + _____.

77. The formula to compute the $SS_{(between\ treatments)}$ for four levels of a factor is: $SS_{(between\ treatments)} = $ _____ − _____.

78. The formula to compute the $SS_{(within\ treatments)}$ for four levels of a factor is: $SS_{(within\ treatments)} = $ _____ − _____.

79. Identify what each of the following notations represent:
    $n_4 = $ _____
    $N = $ _____
    $T_3 = $ _____

80. If $\Sigma T_1 = 3$, $\Sigma T_2 = 6$, $\Sigma T_3 = 12$, $\Sigma T_4 = 15$, $N = 12$, $\Sigma X^2 = 154$, $n_1 = n_2 = n_3 = n_4 = 3$ then:
    $\Sigma X = $ _____,
    $SS_{(total)} = $ _____,
    $SS_{(between\ treatments)} = $ _____
    $SS_{(within\ treatments)} = $ _____
    $k = $ _____
    $df_{(between)} = $ _____
    $df_{(within)} = $ _____
    $df_{(total)} = $ _____
    $MS_{(between)} = $ _____
    $MS_{(within)} = $ _____
    Calculated test statistic F = _____
    Critical F-value = _____

81. If the $SS_{(total)} = 87$ and the $SS_{(between\ treatments)} = 53$, then the $SS_{(within\ treatments)} = $ _____.
    If the $df_{(between)} = 4$, and $df_{(total)} = 40$, then $df_{(within)} = $ _____.
    The $MS_{(between)} = $ _____, $MS_{(within)} = $ _____ and the calculated F-statistic = _____.

82. Use the information within the following table:

Treatment #1	Treatment #2	Treatment #3	Treatment #4	Treatment #5
n = 7	n = 6	n = 7	n = 8	n = 7
$\Sigma T_1 = 182$	$\Sigma T_2 = 162$	$\Sigma T_3 = 182$	$\Sigma T_4 = 224$	$\Sigma T_5 = 175$
$\Sigma T_1^2 = 4754$	$\Sigma T_2^2 = 4396$	$\Sigma T_3^2 = 4764$	$\Sigma T_4^2 = 6294$	$\Sigma T_5^2 = 4407$

to determine the quantities:
a) $\Sigma X^2 = $ _____
b) $(\Sigma X)^2 = $ _____
c) $N = $ _____
d) $SS_{(total)} = $ _____
e) $SS_{(between\ treatments)} = $ _____
f) $SS_{(within\ treatments)} = $ _____
g) $df_{(between)} = $ _____
h) $df_{(total)} = $ _____
i) $df_{(within)} = $ _____
j) $MS_{(between)} = $ _____
k) $MS_{(within)} = $ _____
l) calculated test statistic F = _____

Use the F-Distribution Table located in Appendix C: Table VII to answer questions 83 – 91.

83. Find the critical F-value
    if $df_{numerator} = 10$, $df_{denominator} = 10$ and $\alpha = 0.05$.

84. Find the critical F-value
 if $df_{numerator} = 18, df_{denominator} = 19$ and $\alpha = 0.01$.
85. Find the critical F-value
 if $dfs_1^2 = 17, dfs_2^2 = 15$ and $\alpha = 0.05$.
86. Find the critical F-value
 if $dfs_1^2 = 22, dfs_2^2 = 24$ and $\alpha = 0.01$.
87. Find the critical F-value,
 $F_\alpha(dfs_1^2, dfs_2^2)$, expressed as $F_{0.05}(19,20)$.
88. Find the critical F-value,
 $F_\alpha(dfs_1^2, dfs_2^2)$, expressed as $F_{0.01}(23,20)$.
89. Find the critical F-value, $F_{0.025}(15,15)$.
90. Find the critical F-value, $F_{0.10}(25,35)$.
91. Place the critical F-values on the graph.
 (a) $F_{0.10}(20,20)$
 (b) $F_{0.05}(20,20)$
 (c) $F_{0.025}(20,20)$
 (d) $F_{0.01}(20,20)$

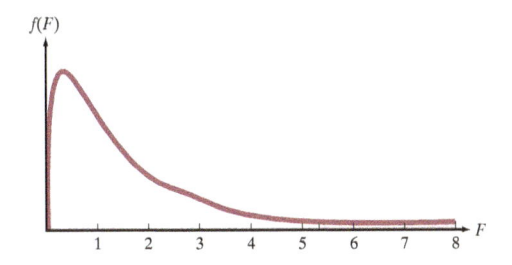

92. Examine the ANOVA table. Notice that you are given information about Sums of Squares for the Between group variance and the Within group variance along with their respective degrees of freedom.
 (a) Complete the table.
 (b) Determine if the calculated test statistic F is significant at $\alpha = 0.05$.

Source of Variation	Sum of Squares SS	Degrees of Freedom df	Mean Square MS	Calculated test statistic F
Between	15.51	3		
Within	27.43	16		
Total				

93. Examine the ANOVA table. Notice that you are given information about Sums of Squares for the Between group variance and the Within group variance along with their respective degrees of freedom.
 (a) Complete the table.
 (b) Determine if the calculated test statistic F is significant at $\alpha = 0.01$.

Source of Variation	Sum of Squares SS	Degrees of Freedom df	Mean Square MS	Calculated test statistic F
Between	39.2	1		
Within	392.6	18		
Total				

94. Examine the ANOVA table. Notice that you are given information about Sums of Squares for the Between group variance and the Within group variance along with their respective degrees of freedom.
 (a) Complete the table.
 (b) Determine if the calculated test statistic F is significant at $\alpha = 0.01$.

Source of Variation	Sum of Squares SS	Degrees of Freedom df	Mean Square MS	Calculated test statistic F
Between	14.79	3		
Within	43.84	21		
Total				

95. Fill in the missing values, a through e in the following ANOVA table. Determine the critical F-value $F_{0.05}(df\text{ between}, df\text{ within})$. Is the result statistically significant?

The ANOVA TABLE

Source of Variation	Sum of Squares SS	Degrees of Freedom df	Mean Square MS	Calculated test statistic F
Between	456	2	c	e
Within	140	b	d	
Total	a	11		

96. Fill in the missing values, a through e in the following ANOVA table. Determine the critical $F_{0.01}(df\text{ between}, df\text{ within})$. Is the result statistically significant?

The ANOVA TABLE

Source of Variation	Sum of Squares SS	Degrees of Freedom df	Mean Square MS	Calculated test statistic F
Between	896	b	c	e
Within	a	18	d	
Total	1931	21		

97. For the following sample data. Test the hypothesis that the means are equal at the 0.05 significance level.

Treatment One	Treatment Two	Treatment Three
6	3	4
8	2	5
9	1	4
9	2	5

a) State the null and alternative hypotheses.
b) Identify the model distribution.
c) Determine the decision rule.
d) Compute all Sum of Squares and Degrees of Freedom.
e) Complete an ANOVA Table.
f) State the conclusion regarding the null hypothesis.

98. For the following sample data. Test the hypothesis that the means are equal at the 0.05 significance level.

Treatment 1	Treatment 2	Treatment 3	Treatment 4
13	14	19	13
11	12	13	15
15	10	17	17
13	11		

a) State the null and alternative hypotheses.
b) Identify the model distribution.
c) Determine the decision rule.
d) Compute all Sum of Squares and Degrees of Freedom.
e) Complete an ANOVA Table.
f) State the conclusion regarding the null hypothesis.

99. A study to compare the playing weights of female athletes in the sports of basketball, track and soccer produced a Mean Square for treatment of 239.5 and a Mean Square Within of 45.3.
(a) State the null and alternative hypotheses of interest.
(b) Calculate the test statistic F for this test.
(c) If the study used 10 females per sport do the results indicate a significant result at $\alpha = 0.01$?.

100. The human resource manager of Bigco Corporation is investigating employee down time on the job. Four types of employees were examined. Of the four types of employees random samples of 20 were chosen. A Mean Square Between the groups was calculated to be 151.4 and a Mean Square Within the groups produced a value of 73.8.
(a) State the null and alternative hypotheses of interest.
(b) Calculate the test statistic F for this test.
(c) Do the results indicate a significant result at $\alpha = 0.05$?

101. A medical technician wants to compare the water holding properties of three different brands of diapers. The technician devises a test that measures the maximum amount of fluid that each diaper can hold. The following results (in fluid ounces) were obtained:

Brand 1	Brand 2	Brand 3
5.6	6.1	4.5
5.8	6.0	5.5
5.1	5.9	4.9
6.1	6.2	4.8
5.7	5.8	5.2

Use the one-way ANOVA procedure to determine whether the differences among the means of each sample is significant at $\alpha = 0.05$.

102. To study the effectiveness of four kinds of packaging, a processor of yogurt foods obtained the following data on the numbers of sales for five different weeks.

Packaging Type (number of Sales per week in hundreds of units)

one	two	three	four
6.0	5.5	5.5	7.1
5.2	6.6	5.6	6.5
5.6	6.8	7.0	6.0
5.2	5.7	5.8	5.9
6.5	5.5	5.6	6.2

Use the one-way ANOVA procedure to determine whether the differences among the means of each sample is significant at $\alpha = 0.01$.

103. The ages and the mercury levels of 12 Loons inspected in a Lake in Southern Maine are listed in the following table.

Mercury Levels (ppm)

Young of the year	Yearlings	Two years or older
0.54	0.65	1.02
0.59	0.82	0.83
0.57	0.83	0.78
0.44	0.77	0.95

Using the 1% significance level, use the five step hypothesis testing procedure to test the null hypothesis that the mean mercury level for the Loons are all equal.

104. The police commissioner of the City of Oz has four districts. She wants to determine if there is a difference in the mean number of traffic tickets served in the four districts. The following data reflects the number of tickets served for a random sample of five business days.

Can the police commissioner conclude there is a significant difference with $\alpha = 0.05$ in the mean number of tickets served?

Number of Tickets Served

District 1	District 2	District 3	District 4
23	31	22	26
25	23	24	27
24	28	25	28
25	29	23	25
24	28	22	30

CHAPTER REVIEW

105. If $s_1^2 = 14.78$ and $s_2^2 = 7.93$, then F = _____.

106. An F-curve is dependent upon the degrees of freedom of the sample _____.

107. The F-Distribution is a family of distributions composed of the value of the standard deviations for samples drawn from two distinct populations. (T/F)

108. If a test statistic F is much greater than one, then the population variances are approximately equal. (T/F)

109. Which is **not** a characteristic of an F-Distribution?
 (a) Each sample used in the test statistic F is independent.
 (b) The populations for each sample is assumed to be normally distributed.
 (c) The F-Distribution is assumed to be approximately normally distributed.
 (d) The modal value of the F-Distribution is assumed to be 1.

Use **Table VII: Critical Values of the F-Distribution in Appendix C** to answer questions 110–113.

110. Find the critical F-value if $df_{numerator} = 15$, $df_{denominator} = 20$ and $\alpha = 0.05$.

111. Find the critical F-value if $dfs_1^2 = 7$, $dfs_2^2 = 10$ and $\alpha = 0.05$.

112. Find the critical F-value, $F\alpha(dfs_1^2, dfs_2^2)$, expressed as $F_{0.05}(20,19)$.

113. Find the critical F-value, $F_{0.025}(15,17)$.

For questions 114 to 118:
If $F_{0.01}(9,25)$, is the critical F-value for an F-Test, then which number represents:

114. the degrees of freedom for the larger variance?
 (a) 9 (b) 12 (c) 25 (d) 1

115. the degrees of freedom for the smaller variance?
 (a) 9 (b) 12 (c) 25 (d) 1

116. the sample size for the larger variance?
 (a) 10 (b) 34 (c) 26 (d) 35

117. the sample size for the smaller variance?
 (a) 10 (b) 34 (c) 26 (d) 35

118. the value for the critical F-value?
 (a) 4.73 (b) 3.22 (c) 2.28 (d) 1.89

119. (a) Compute the test statistic F for the following sample variances.
 $s_1^2 = 2.9$ $s_2^2 = 1.08$
 $n_1 = 19$ $n_2 = 22$
 (b) Conduct an F-Test for the equality of the population variances using an $\alpha = 0.05$.

120. (a) Compute the test statistic F for the following sample variances.
 $s_1^2 = 136$ $s_2^2 = 398$
 $n_1 = 100$ $n_2 = 100$
 (b) Conduct an F-Test for the equality of the population variances using an $\alpha = 0.10$.

121. (a) Compute the test statistic F for the following samples.

sample 1	sample 2
4	5
7	3
6	4
3	8
6	8
8	12

 (b) Conduct an F-Test for the equality of the population variances using an $\alpha = 0.05$.

122. Mark of East Bluff Glass replaces car windshields. During the past year Mark has rejected an average of 9 windshield replacement glass per month with a variance of 2. For the same period, Lori who also replaces windshields, has rejected an average of 8.5 windshield replacement glass per month with a variance of 1.5. At the 0.05 significance level, can we conclude that there is more variation in the number of rejects per day attributed to Mark?

123. Experimental or Treatment Variance is also referred to as _____.

124. Using the expressions SS and df, the formula for the sample variance is: $s^2 =$ _____.

125. Using the following expressions: $SS_{(between\ treatments)}$, $SS_{(total)}$ and $SS_{(within\ treatments)}$, write a relationship that is always true. _____ = _____ + _____.

126. If $SS_{(between\ treatments)} = 14$, the $SS_{(within\ treatments)} = 160$, then the value of $SS_{(total)}$ is:
 (a) 146 (b) 160 (c) 174
 (d) not enough information

127. Consider the following information about three different populations:

	Population one	Population two	Population three
Population means μ	100	100	100
Population variances Within each treatment σ^2	25	25	25

(a) Identity the population variance within each group.
(b) Find the average within group variance.
(c) Calculate the between group variance.
(d) Determine the total group variance.
(e) Is the total group variance due solely to the within group variance? Explain.

128. Consider the following information about three different populations:

	Population one	Population two	Population three
Population means μ	200	250	300
Population variances Within each treatment σ^2	7	7	7

(a) Identify the population variance within each group.
(b) Find the average within group variance.
(c) Calculate the between group variance.
(d) Determine the total group variance.
(e) Is the Total group variance due solely to the Within group variance? Explain.

129. Data Values of a Population of 21 Students Subdivided into Three Groups

	Treatment 1	Treatment 2	Treatment 3
	20	34	48
	22	36	50
	24	38	52
	26	40	54
	28	42	56
	30	44	58
	32	46	60
Population means μ	26	40	54
Population variances Within each treatment σ^2	16	16	16

(a) Calculate $SS_{(within\ treatments)}$.
(b) Calculate $SS_{(between\ treatments)}$.
(c) Calculate $SS_{(total)}$.
(d) Is the relationship
$SS_{(between\ treatments)} + SS_{(within\ treatments)} = SS_{(total)}$
True?
(e) Identify the population variance within each group.
(f) Find the average within group variance.
(g) Calculate the between group variance.
(h) Determine the total group variance.
(i) Is the Total group variance due solely to the Within group variance? Explain.

130. Data Values of a Population of 12 scores Subdivided into Three Groups

	Treatment 1	Treatment 2	Treatment 3
	70	74	78
	74	78	82
	78	82	86
	82	86	90
Population means μ	76	80	84
Population variances Within each treatment σ^2	20	20	20

(a) Calculate $SS_{(within\ treatments)}$.
(b) Calculate $SS_{(between\ treatments)}$.
(c) Calculate $SS_{(total)}$.
(d) Is the relationship
$SS_{(between\ treatments)} + SS_{(within\ treatments)} = SS_{(total)}$
True? Verify your answer.
(e) Identify the population variance within each group.
(f) Find the average within group variance.
(g) Calculate the between group variance.
(h) Determine the total group variance.
(i) Is the total group variance due solely to the within group variance? Explain.

131. Identify what each of the following notations represent:
$n_6 = $ _____
$N = $ _____
$T_5 = $ _____

132. If $\Sigma T_1 = 6$, $\Sigma T_2 = 18$, $\Sigma T_3 = 24$, $\Sigma T_4 = 30$, $N = 12$, $\Sigma X^2 = 612$, $n_1 = n_2 = n_3 = n_4 = 3$ then:
$\Sigma X = $ _____,
$SS_{(total)} = $ _____,
$SS_{(between\ treatments)} = $ _____
$SS_{(within\ treatments)} = $ _____
$k = $ _____
$df_{(between)} = $ _____
$df_{(within)} = $ _____
$df_{(total)} = $ _____
$MS_{(between)} = $ _____
$MS_{(within)} = $ _____
Calculated test statistic $F = $ _____
Critical F-statistic = _____

133. Examine the ANOVA table. Notice that you are given information about Sums of Squares for the Between group variance and the Within group variance along with their respective degrees of freedom.
(a) Complete the table.
(b) Determine if the calculated test statistic F is significant at $\alpha = 0.05$.

Source of Variation	Sum of Squares SS	Degrees of Freedom df	Mean Square MS	Calculated test statistic F
Between	150.51		4e	
Within	87.43	26		
Total				

134. Debbie wants to compare the effectiveness of three methods of teaching word processing. Method 1 is a traditional lecturing technique without instructor accessing a computer. Method 2 is in a laboratory setting with instructor using a computer to illustrate techniques. Method 3 uses no traditional lecturing but rather individual paced learning material with instructor giving individual attention to students having special difficulty. Of the many students taught by one of the three methods the following random samples representing achievement test scores in word processing were obtained:

Method 1	Method 2	Method 3
69	70	88
73	75	78
63	74	84
67	77	82

Use the one-way ANOVA procedure to determine whether the differences among the means of each sample is significant at $\alpha = 0.01$.

135. The following table lists the hourly salaries of certain randomly selected college students working part time representing colleges from the Northeast, Midwest and Southwest portions of the United States.

Northeast	Midwest	Southwest
7.80	6.90	7.60
8.65	7.10	6.90
7.75	6.40	7.85
9.40	6.90	8.10
8.80	6.75	6.85

Using the $\alpha = 0.05$ significance level, use the five step hypothesis testing procedure to test the null hypothesis that the mean hourly wage for colleges students working part time employment from colleges in the Northeast, Midwest and Southwest portions of the United States are all equal.

WHAT DO YOU THINK?

136. In Case Study 16.2, we discussed the effect of school size and age of teachers on job satisfaction and job stress for the study entitled Teacher Job Satisfaction and Teacher Job Stress: School Size, Age and Teaching Experience. The subjects selected for the study were 229 North Florida and South Georgia urban secondary school physical education teachers. Two instruments were used to measure the job satisfaction and job related stress of the teachers. A high score on the Job Satisfaction Scale indicates a high level of job satisfaction while a high score on the Job-Related Stress Scale indicates a high level of job stress. The following tables present the results of this study which examined the relationship of number of years of teaching with respect to job satisfaction and job stress.

Table A
Differences in Job Satisfaction Scores by Number of Years of Teaching Experience

Years of Teaching	Number of Subjects	Mean Scores	Standard Deviation	F-statistic	Probability Level
1–5	36	44.42	10.71	0.65	0.63
6–10	62	46.77	11.95		
11–15	66	44.21	14.35		
16–20	36	44.30	13.78		
21+	29	47.65	13.02		
Total	229	45.39			

Table B
Differences in Job Stress Scores by Number of Years of Teaching Experience

Years of Teaching	Number of Subjects	Mean Scores	Standard Deviation	F-statistic	Probability Level
1–5	36	42.65	13.62	0.18	0.95
6–10	62	42.90	15.07		
11–15	66	42.33	14.46		
16–20	36	41.13	13.78		
21+	29	44.06	14.99		
Total	229	42.57			

For Table A:

a) identify the independent variable. State the different levels of this variable.

b) compare the mean job satisfaction score for teachers with 11–15 years of teaching experience to the mean job satisfaction score for teachers with 21+ years of teaching experience. Interpret the meaning of each mean score.

c) Within which group are the job satisfaction scores more variable? less variable? How would you interpret each of these results?

d) What would you compare to determine the between group variance?

e) Explain the meaning of the probability level for the given F-statistic.

f) What are the degrees of freedom for the within groups and the degrees of freedom for the between groups?

g) Based upon the results of Table A, can you conclude that the number of years of teaching experience is a significant factor in their job satisfaction? Explain.

h) Identify the number of years of teaching experience group that had the lowest job satisfaction and the number of years of teaching experience group with the highest job satisfaction. Can you conclude at an alpha level of 0.10 that this difference is significant?

For Table B:

i) identify the independent variable. State the different levels of this variable.

j) compare the mean job stress score for teachers with 11–15 years of teaching experience to the mean job stress score for teachers with 21+ years of teaching experience. Interpret the meaning of each mean score.

k) Within which group are the job stress scores more variable? less variable? How would you interpret each of these results?

l) What would you compare to determine the between group variance?

m) Explain the meaning of the probability level for the given F-statistic.

n) What are the degrees of freedom for the within groups and the degrees of freedom for the between groups?

o) Based upon the results of Table B, can you conclude that the number of years of teaching experience is a significant factor in their job stress? Explain.

p) Identify the number of years of teaching experience group that had the lowest job stress and the number of years of teaching experience group with the highest job stress. Can you conclude at an alpha level of 0.10 that this difference is significant?

137. The study entitled Factors In Children's Attitudes Toward Pets which appeared in the journal *Psychological Reports,* 1990, issue 66, investigated the parental attitudes, family size, structure, and presence or absence of household pets as influences on children's attitudes toward pets. The researchers, Kidd & Kidd, believe that attachment to pet animals begin as early as 18 months. Since a child-pet attachment begins at such an early age, Kidd and Kidd decided to investigate the factors which may influence feelings and attitudes toward pets. The researchers administered a Parent Questionnaire to the parents of the 700 children who were the subjects of the study. These subjects were composed of 339 boys and 361 girls between the ages of 6 months and 18 years: 150 boys and 150 girls from two-parent pet-owning homes, 43 boys and 55 girls from one-parent owning homes, 97 boys and 105 girls from two-parent, nonpet-owning homes, and 49 boys and 51 girls from one-parent, nonpet-owning homes.

The Parent Questionnaire were scored for the children's Activities with Pets and Interest in Pet's scales whether or not the family owned a pet. They were scored for a child's Responsibility for Pet Care scale only if the family owned a pet or indicated that the child actually helped with responsible care for a neighbor's or a friend's pet. One aspect of the study was to determine whether the age of the child significantly influenced the scores on each of the three scales. The researchers believed that involvement in activities with and interest in pets were good measures of attachment to pets, who showed more interest in them, and who took more responsibility for them also were reported as having a greater affection for their pets than children who were scored lower on these scales. A higher score on each of these scales is interpreted as displaying more activities with the pets, showing more interest in the pets and more responsibility with the pets. The results of the pet-owning children on the three scales is given in Table C. The results of the nonpet-owning children on the three scales were discussed in Case Study 16.3.

Table C
Mean and Standard Deviations For Pet-Owning Children On Three Scales

	n	Mean	Std Dev
Activities Scale			
Ages			
0–5	120	27.12	6.32
Grade School	141	34.26	8.62
High School	137	35.12	7.98
Interest Scale			
Ages			
0–5	120	19.35	6.40
Grade School	141	24.28	7.71
High School	137	22.63	7.63
Responsibility Scale			
Ages			
0–5	120	2.32	2.17
Grade School	141	6.04	4.04
High School	137	4.48	3.79

Table D
One-Way Analysis of Variance of Activities With Pets, Interest In Pets, and Responsibility For Care of Pets: Scores of Children With Household Pets By Age Groups

Source of Variance	SS	df	MS	F
Activities With Pets				
SS Between Age Groups	140.61	2	70.31	3.34*
SS Within Groups	8314.75	395	21.05	
Interest In Pets				
SS Between Age Groups	88.02	1	86.02	4.53*
SS Within Groups	7702.46	396	19.45	
Responsibility For Care				
SS Between Age Groups	891.52	2	445.76	36.78†
SS Within Groups	4786.92	395	12.12	

*p0.05. †p 0.0005.

Using the information on the Activities Scale in Table C and Table D:
a) Identify the independent variable. State the different levels of this variable.
b) Which age group should the greatest activity with pets? the smallest activity with pets?
c) Which age group had the greatest variability in the activities scale with pets? the smallest variability?
d) Which age group had the greatest mean score on the Activities scale? the smallest mean score?
e) Determine the difference between the mean scores of part d? Do you think this difference is significant? If it is significant, state the alpha level and explain how you came to this conclusion.
f) What are the degrees of freedom for the within groups and the degrees of freedom for the between age groups?
g) Determine the Sums of Square due to Factor variance. Determine the Sums of Square due to Error variance.
h) Explain what the MS = 70.31 indicates about the between age groups for the activities scale.
i) Explain what the MS = 21.05 indicates about the within groups for the activities scale.
j) Identify the Mean Square population variance estimate for the variance Between the groups.
k) Identify the Mean Square population variance estimate for the variance within the groups.
l) Based upon the results of Table D, can you conclude that the age of a child is a significant factor in their activities with pets at an alpha level of 0.01? at an alpha level of 0.10? Explain.

Using the information on the Interest Scale in Table C and Table D:
m) Identify the independent variable. State the different levels of this variable.
n) Which age group should the greatest interest with pets? the smallest interest with pets?
o) Which age group had the greatest variability in the interest scale with pets? the smallest variability?
p) Which age group had the greatest mean score on the Interest scale? the smallest mean score?
q) Determine the difference between the mean scores of part p? Do you think this difference is significant? If it is significant, state the alpha level and explain how you came to this conclusion.
r) What are the degrees of freedom for the within groups and the degrees of freedom for the between age groups?
s) Determine the Sums of Square due to Factor variance. Determine the Sums of Square due to Error variance.
t) Explain what the MS = 86.02 indicates about the between age groups for the interest scale.
u) Explain what the MS = 19.45 indicates about the within groups for the interest scale.
v) Identify the Mean Square population variance estimate for the variance Between the groups.
w) Identify the Mean Square population variance estimate for the variance within the groups.
x) Based upon the results of Table D, can you conclude that the age of a child is a significant factor in their interest with pets at an alpha level of 0.01? at an alpha level of 0.10? Explain.

Using the information on the Responsibility Scale in Table C and Table D:
y) Identify the independent variable. State the different levels of this variable.
z) Which age group should the highest responsibility with pets? the lowest responsibility with pets?
aa) Which age group had the greatest variability in the Responsibility Scale with pets? the smallest variability?

bb) Which age group had the greatest mean score on the Responsibility Scale? the smallest mean score?
cc) Determine the difference between the mean scores of part bb? Do you think this difference is significant? If it is significant, state the alpha level and explain how you came to this conclusion.
dd) What are the degrees of freedom for the within groups and the degrees of freedom for the between age groups?
ee) Determine the Sums of Square due to Factor variance. Determine the Sums of Square due to Error variance.
ff) Explain what the MS = 445.76 indicates about the between age groups for the Responsibility Scale.
gg) Explain what the MS = 12.12 indicates about the within groups for the Responsibility Scale.
hh) Identify the Mean Square population variance estimate for the variance Between the groups.
ii) Identify the Mean Square population variance estimate for the variance within the groups.
jj) Based upon the results of Table D, can you conclude that the age of a child is a significant factor in their Responsibility with pets at an alpha level of 0.0001? at an alpha level of 0.01? Explain.

From the results of the previous tables, comment on the following statements.
kk) Why would you think that preschool children would have low activities scores? Comment.

Can we conclude that the Activities scores of preschool children are significantly lower than those of older children? What do you think?

ll) Why would you think that preschool children would have low interest scores? Comment.

Can we conclude that the Interest scores of preschool children are significantly lower than those of older children? What do you think?

mm) Why would you think that the grade school children would have a higher responsibility scores that either the high school or the preschool children? Comment.

Can we conclude that the Responsibility scores of grade school children were significantly higher than those of high school and preschool children? What do you think?

PROJECTS

138. Analysis of variance is often used as a statistical procedure in research scenarios in both the natural and social science arenas.
    a) Choose a professional journal from an area of interest and examine the statistical methods used in the analysis of the data to determine if an ANOVA model is employed.
    b) Read the research article and list the hypotheses that have been tested.
    c) Determine if significant results have been found. List these results with their corresponding p-value.
    d) Identify the statistical terms used in the article that are not familiar.

139. a) Develop a questionnaire to determine a student's class, (e.g. freshman, sophomore, junior, senior) and the number of hours the student dedicates to study.
    b) Randomly sample 100 students and administer the questionnaire.
    c) Use the one-way ANOVA procedure outlined in this chapter to test the hypothesis that the mean number of hours dedicated to study by each type student is equal at $\alpha = 0.05$.

140. a) Using a technological tool, input the following data.

C1	C2	C3	C4
82	71	70	65
65	70	65	62
73	64	72	72
71	65	63	64
84	79	59	65
62	68	77	60
67	77	73	68
78	83	65	65
79	78	63	59
79	63	69	72
81	71	71	69
76	72	68	64

b) Produce an ANOVA Table for this data.
Identify
c) the Sums of Square due to Factor variance.
d) the Sums of Square due to Between group variance.
e) the Sums of Square due to Error variance.
f) the Sums of Square due to Within group variance.
g) the degrees freedom for each sums of square.
h) the Mean Square population variance estimate for the variance Between the groups.

i) the Mean Square population variance estimate for the variance within the groups.
j) the calculated test statistic F.
k) the p-value for the test statistic F.
l) Are the group means significantly different if $\alpha = 0.05$?
m) Are the group means significantly different if $\alpha = 0.01$?

141. A researcher wishes to determine if three different diets result in different average weight losses. Subjects with similar weight problems are chosen for the study and randomly assigned one of the diets. The weight loss (in pounds) after one month are recorded and indicated as follows.

Diet A	Diet B	Diet C
3	24	21
15	12	24
16	10	9
7	11	15
13	5	19
	12	40
		12
		27

a) Input this data into a technological tool.
b) Produce an ANOVA Table for this data.
Identify
c) the Sums of Square due to Factor variance.
d) the Sums of Square due to Between group variance.
e) the Sums of Square due to Error variance.
f) the Sums of Square due to Within group variance.
g) the degrees freedom for each sums of square.
h) the Mean Square population variance estimate for the variance Between the groups.
i) the Mean Square population variance estimate for the variance within the groups.
j) the calculated test statistic F.
k) the p-value for the test statistic F.
l) Are the group means significantly different if $\alpha = 0.05$?
m) Are the group means significantly different if $\alpha = 0.01$?

142. Use a technological tool, enter the data for those problems in this chapter. Compare the output generated by the technological tool to the results obtained by pencil and paper calculation.

CHAPTER TEST

143. The F-Distribution is a symmetric and approximately bell shaped distribution. (T/F)

144. If $s_1^2 = 15.76$ and $s_2^2 = 13.85$, then F= _____.

145. When given population variances it is always true that Between Variance + Within Variance = Total Variance. (T/F)

146. Error Variance is often due to the individual differences for subjects within the treatment groups. (T/F)

147. Find the critical F-value if $dfs_1^2 = 22$, $dfs_2^2 = 24$ and $\alpha = 0.01$.

148. (a) Compute the test statistic F for the following sample variances.
$s_1^2 = 44 \quad s_2^2 = 58$
$n_1 = 33 \quad n_2 = 28$
(b) Conduct the test statistic F for the equality of the population variances using an $\alpha = 0.05$.

149. Harmony Music Associates is interested in the radio listening habits of men and women between the ages of 18 and 25. They found that although the mean number of minutes per day was similar for both men and women, the variances were not. Forty men and 48 women were sampled and the relevant statistics for the study is listed in the following table. Use the F-test to determine if there is statistical difference in the variation of the radio listening habits of men and women. (Use $\alpha = 0.10$).

Sex	Mean Time (MIN/day)	Variance	Sample Size
Men	98	23	8
Women	93	15	8

150. At the beginning of this chapter, Example 16.2 explored a problem that can be analyzed using the ANOVA procedure developed in this section. The example has been reprinted in this exercise. Apply ANOVA to answer the question posed by Jim.

Jim is interested in buying a highly rated computer. He checks a PC magazine and selects three manufacturers. From these manufacturers he draws a sample of computers rated by the magazine. Table 16.34 displays the samples with their respective means and variances.

TABLE 16.34
Ratings of Computers from a PC Magazine

Computers	Compost	Dovell	ACE Power
	29.2	33.6	24.4
	21.4	44.4	42.0
	47.1	25.2	25.2
	34.0	33.6	34.9
	37.1	46.4	52.9
$\bar{x}$	33.76	36.64	35.88
s^2	90.63	76.21	143.52

Perform a hypothesis test at $\alpha = 0.01$ to determine which manufacturer has a significantly higher rating.

CHAPTER 17

Nonparametric Statistics

Contents

17.1 Introduction
17.2 The Sign Test for Medians
17.3 The Mann-Whitney Rank-Sum Test
17.4 The Kruskal-Wallis H Test

17.1 INTRODUCTION

So far we have explored many elementary concepts of statistics. We studied descriptive statistics through the concepts of graphing, central tendency and variability. We introduced the concepts of probability and theoretical distributions such as the normal, binomial, t and F. The role of hypothesis testing and estimation as a branch of inferential statistics has been extensively examined. Linear correlation and regression was explored to measure the strength of the relationship between two variables and how such a relationship can be modelled.

Many of these statistical models were based on certain basic assumptions about the populations from which sample data was collected. Some of these assumptions were:

1. the random samples are selected from a *normal population.*
2. the population variances are *equal.*

Inferential statistics which are based on assumptions such as the shape of the population is normal or the population variances are equal are called **Parametric Statistics.** However, there are situations when it is **not** possible to assume that a population is normal or approximately normal or the population variances are approximately equal. Inferential statistics that make no assumptions about the shape of the population or the parameters of a population are referred to as **Nonparametric Statistics.**

DEFINITION 17.1
Nonparametric Statistics. Statistical tests that require few, if any, assumptions about the population from which sample data is collected.

There are many statistical tests that are classified as nonparametric. In fact we have already studied one such nonparametric test namely the *Chi-Square Test of Independence.* The Chi-Square Test of Independence is used to determine whether or not two distinct population classifications are independent or dependent. This test does not involve any population parameters. Such a test is a nonparametric test. The main objective of this chapter is to make you aware of the existence of nonparametric statistics and some nonparametric tests. Nonparametric tests will have the same basic characteristic,

Only available in the e-book.

that is, they will be statistical tests that do not require assumptions about the underlying population shapes or form. These statistical tests are also called *distribution-free* tests.

> **DEFINITION 17.2**
> **Distribution-free Tests.** Statistical tests that can be applied to many types of data regardless of the shape of the underlying population.

Nonparametric tests have many advantages over parametric tests. Let's examine some of these advantages.

> **The Advantages of Nonparametric Tests**
>
> (1) *Nonparametric tests are easy to apply.* Usually only ranking, counting, adding and subtraction of the data values are needed.
> (2) *Nonparametric tests are quick to apply for small samples.*
> (3) *Nonparametric tests can be applied to many applications.* Usually few, if any, assumptions are necessary for the applications.
> (4) *Nonparametric tests can be applied to data as long as the data values can be at least ranked.*

On the other hand, there are some drawbacks to their application. Some of the disadvantages of the nonparametric tests are:

> **The Disadvantages of Nonparametric Tests**
>
> (1) *Nonparametric tests are less powerful than a similar parametric test.* This means that the ability of a nonparametric test to detect a significant difference compared to its counterpart parametric test is not as efficient.
> (2) *Nonparametric tests do not use all the information contained in the sample data.*
> (3) *Nonparametric tests are difficult to use for large samples.*

Statisticians agree that if you are fairly sure that your data comes from a normal population, use the appropriate parametric test. This notwithstanding, in this chapter we will study three nonparametric statistical tests. They are:

The Sign Test for Medians
The Mann Whitney Rank Sum Test
The Kruskal-Wallis Test

17.2 THE SIGN TEST FOR MEDIANS

One of the easiest nonparametric tests, the **SIGN test for Medians** is the nonparametric counterpart to the **t test** about a population mean. Let's examine a situation where the Sign test can be applied.

Suppose the snack bar owner at a university claims that students drink, on the "average", four cups of coffee per day. A representative of the student government believe's that the owner's claim is too high. The representative randomly samples twenty students to determine the number of cups of coffee per day each student consumes. The results of the sample are as follows:

number of cups of coffee consumed each day
3 2 3 6 1 2 4 8 1 0 2 2
7 3 1 0 4 2 7 5 2 1 4 3

The student government representative decides to perform a hypothesis test using the sample data.

The student government representative is testing the hypothesis that students drink, on the "average", four cups of coffee per day. Thus, the null hypothesis is:

H_o: students drink four cups of coffee per day.

Since the student government representative believes the claim is *too high* the alternative hypothesis is:

H_a: students drink less than four cups of coffee per day.

To test the null hypothesis, the student representative randomly interviews 24 students.

Usually, when performing a hypothesis test, an appropriate distribution hypothesis testing model is chosen. The model is typically based on the *assumption that the population from which the sample data is drawn is **approximately normal**.* **But this is NOT the case here. In fact, the distribution of coffee consumption is known to be skewed. This violates an assumption that is required for parametric tests.**

If you were to construct a Stem-and-Leaf display as shown in Figure 17.1 for the sample data, you would notice a sample that is skewed right.

Stem-and-Leaf Display

```
                    0 | 00
                    1 | 0000
      number        2 | 000000
        of          3 | 0000
      cups of       4 | 0000    skewed right
     coffee per     5 |
        day         6 | 0
                    7 | 00
                    8 | 0
```

FIGURE 17.1

Computing the measures of Central Tendency for the sample data, the mean is 3.04, the median is 2.5 and the mode is 2 cups of coffee per day. Nonparametric tests are used *whenever the population from the which the data is sampled is not approximately normally distributed.* In this example we suspect from the examination of the stem-and-leaf display in Figure 17.1 that the distribution of coffee consumption is not normally distributed.

The Stem-and-Leaf display introduced in Chapter 2 is an exploratory data analysis technique that enables the experimenter to learn about a distribution. The Stem-and-Leaf display for the sample data in Figure 17.1 is skewed and the sample size, n = 24, is relatively small. In a skewed distribution, the mean may not be appropriate as a measure of central tendency. [Why?] For a skewed distribution the median is usually the measure of central tendency that best indicates the "typical value or average". Recall that the median is the value above which half or 50% of the data values lie and below which the other half or 50% lie.

To conduct an hypothesis test about the median, **the number of values above the median and the number of values below the median are counted. If a data value is equal to the median value, then the data value is omitted from the analysis.**

Thus, the claim made by the snack bar owner is a claim about a population *median*. The student government representative believes that the median value of 4 is too high. If we assume that the null hypothesis is correct, (i.e. H_o: median = 4) then the number of sample values above the median of 4, and the number of sample values below the median of 4 must be equal. [WHY?] The sample data indicates that there are 3 data values above the median and 13 data values less than the median. Since four data values are the same as the median, they are omitted from the analysis. *The first calculation in conducting the SIGN test about a Median is to determine the number of data values above the median value and the number of data values below the median value.* Table 17.1 illustrates the tabulation of the sample data values used in the SIGN Test. Notice that a "+" sign is listed next to a sample data value that is more than the median, while a "-" sign is listed next to a sample data value that is less than the median. The data values which are equal to the median have a "0" listed next to them. These data values are omitted from further analysis.

TABLE 17.1 Number of cups of coffee consumed each day

3	−	7	+
2	−	3	−
3	−	1	−
6	+	0	−
1	−	4	0
2	−	2	−
4	0	7	+
8	+	4	0
1	−	2	−
0	−	1	−
2	−	4	0
2	−	3	−

Thus, by counting the number of "+" signs and the number of "−" signs we see that there are 4 data values greater than the hypothesized median of 4 and 16 data values less than the hypothesized median of 4. The question we need to answer is: *"Do these sample results indicate enough evidence to reject the null hypothesis that the median number of cups of coffee is 4?"* Or could this *difference or imbalance* be due to sampling error?

If the null hypothesis about the median were true, then we would expect half the students drink less than 4 cups of coffee each day and half the students drink more than 4 cups of coffee each day. Thus, the probability that the number of cups of coffee consumed per day by a university student is less than the median would be 0.5 and, similarly, the probability that the number of cups of coffee consumed per day is greater than the median would also be 0.5.

To determine if the sample results indicate enough evidence to reject the null hypothesis, we must calculate the probability that for a sample of 20 coffee consumers, 3 students drink less than four cups per day. The sign test for medians satisfies the conditions of a binomial probability experiment. These conditions are:

1. there are *n identical trials,* n = 20.
2. there are *only two possible and independent* results: a "+" sign or a "−" sign.

3. the *probability for either a "+" sign or a "–" sign is ½* since the median is the middle value and we are assuming that half of the values are above the median and half the values are below the median.

Using this information, this probability question can be analyzed as a binomial probability experiment with a probability of success of p = 0.5, the number of identical trials is n = 16 and the outcome, O, is 3. Furthermore, for a binomial distribution where both np and n(1 – p) are at least than 10, the normal distribution can be used to approximate the binomial distribution. For this sample of n = 16 and p = 0.5, we have

$$np = (20)(0.5) = 10$$
and
$$n(1-p) = (20)(0.5) = 10$$

are both at least ten.

Thus, the normal distribution can be used to approximate the binomial distribution. To apply this approximation, we must compute the binomial mean, μ_s, and the binomial standard deviation, σ_s. For n = 16 and p = 0.5, the binomial mean and standard deviation are:

$$\mu_s = np = (20)(0.5) = 10$$

and

$$\sigma_s = \sqrt{np(1-p)}$$
$$= \sqrt{(20)(0.5)(0.5)}$$
$$\sigma_s \approx 2.24$$

When using the normal distribution to approximate binomial probability, the continuity correction must be applied. Therefore, to calculate the probability that **3 students or less** will drink more than 4 cups of coffee per day, you need to find the area under the normal curve to the left of 3.5 (i.e. the continuity correction factor). To determine this area, calculate the z-score for 3.5 and use TABLE II to find the area to the left of the z-score. This is illustrated in Figure 17.2.

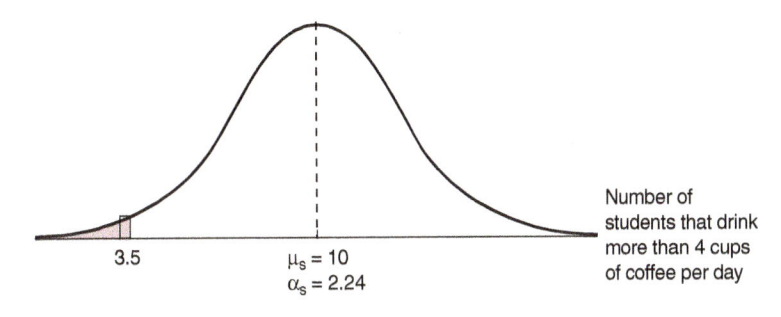

FIGURE 17.2

Using the z-score formula,

$$z = \frac{x - \mu_s}{\sigma_s} \text{ with } \mu_s = 10, \sigma_s = 2.24 \text{ and } x = 3.5$$

$$z = \frac{3.5 - 10}{2.24}$$

$$z \approx -2.90$$

870 *Chapter 17 Nonparametric Statistics*

Using TABLE II the portion of area to the left of z = –2.90 is 0.0018. Therefore, for a sample of sixteen students the probability of 3 students or less will drink 4 cups of coffee per day is 0.0018.

Remember, that this probability is based on the assumption that the median number of cups of coffee consumed per day is 4. Since the probability of this occurrence is so small (i.e. 0.0018) you might conclude that either the random sample was highly unusual or that the *null hypothesis of 4 cups per day* is **not likely.**

The student government representative decided that this unusual random sample was unlikely to occur just by chance and concluded the snack bar owner's claim was incorrect!

The above explanation is called the sign test for MEDIANS. It is a hypothesis testing procedure about a median. The sign test procedure will be developed. Within this procedure we will assume that the normal distribution can be used to approximate the binomial distribution.

The SIGN TEST for MEDIANS

Step 1 Formulate the two hypotheses, H_o and H_a.
Null Hypothesis:
In general, the null hypothesis for the sign test for a median has the following form.

Null Hypothesis Form:

H_o: The population median is claimed to be equal to a numerical value which will be symbolized as: M_o.

The null hypothesis can be symbolized as:

$$H_o: \text{population median} = M_o$$

Alternative Hypothesis:
Since the alternative hypothesis, H_a, can be stated as either *greater than, less than,* or *not equal* to the hypothesized median value, M_o, then the alternative hypothesis can have one of three forms.

Forms of the Alternative Hypothesis

GREATER THAN FORM FOR H_a

H_a: The population median is claimed to be **greater than** the hypothesized median value, M_o.

This is symbolized as:

$$H_a: \text{population median} > M_o$$

LESS THAN FORM FOR H_a

H_a: The population median is claimed to be **less than** the hypothesized median value, M_o.

This is symbolized as:

$$H_a: \text{population median} < M_o$$

Section 17.2   The Sign Test for Medians   871

> ### NOT EQUAL FORM FOR H$_a$
>
> *H$_a$: The population median is claimed to be **not equal** to the hypothesized median value, M$_o$.*
>
> This is symbolized as:
>
> **H$_a$: population median ≠ M$_o$**

Step 2 **Determine the model to test the null hypothesis, H$_o$.**

Under the assumption that H$_o$ is true, a binomial distribution is used as the hypothesis testing model. For this distribution, the usable sample size, n, is equal to the number of negative (−) signs, denoted as n(− signs), plus the number of plus (+) signs, symbolized as n(+ signs). Thus: n = n(− signs) + n(+ signs).

Remember the number of sample data values *not* equal to the median value represents the usable sample size, n. While the probability of a success is p = 0.5 since we are assuming there is a 1/2 chance that a data value within the sample is either below the median (assigned a "−" sign) or above the median (assigned a "+" sign). If both np and n(1 − p) are at least ten, the normal approximation to the binomial distribution can be used as the model distribution. When using the normal distribution to approximate the binomial distribution, the mean and standard deviation of the binomial distribution must be computed. The formulas to determine the binomial mean and standard deviation are:

$$\mu_s = np$$

and

$$\sigma_s = \sqrt{np(1-p)}$$

Step 3 **Formulate the Decision Rule.**
  a) Determine the type of alternative hypothesis: directional or nondirectional.
  b) Determine type of test: 1TT or 2TT.
  c) Identify the significance level: $\alpha = 0.01$ or $\alpha = 0.05$.
  d) Construct a normal distribution (used to approximate the binomial distribution) as the hypothesis testing model.
  e) Determine the critical z-score(s) using the following chart.

> ### Critical z-scores
>
> *For significance level of $\alpha = 0.05$:*
> For a 2TT: $z_{LC} = -1.96$ and $z_{RC} = +1.96$.
> For a 1TT test on left side: $z_c = -1.65$
> For a 1TT test on right side: $z_c = +1.65$
>
> *For significance level of $\alpha = 0.01$:*
> For a 2TT: $z_{LC} = -2.58$ and $z_{RC} = +2.58$.
> For a 1TT test on left side: $z_c = -2.33$
> For a 1TT test on right side: $z_c = +2.33$

f) Using the critical z-score(s), state the decision rule.

Step 4 **Analyze the sample data.**
Using the necessary sample data, tabulate the number of data values that are either less than the median or are greater than the median. This statistical result will be denoted by O. (**For a nondirectional alternative 2TT, use the number of the less frequent sign when computing the statistical result O.**) Calculate the z-score for the statistical result, O. This is referred to as the **test statistic.** Remember, when using the normal approximation to the binomial distribution, a continuity correction factor of 1/2 is either subtracted or added to the statistical result, O. If the statistical result O, is less than n/2, then the test statistic is:

$$z = \frac{(O + \frac{1}{2}) - \mu_s}{\sigma_s}$$

If the statistical result O, is greater than n/2, then the test statistic is:

$$z = \frac{(O - \frac{1}{2}) - \mu_s}{\sigma_s}$$

Step 5 **Determine the Conclusion.**
Compare the test statistic to the critical z-score and either:

a) **Reject H_o and accept H_a at α,**
OR
b) **Fail to reject H_o at α.**

Example 17.1 will illustrate the hypothesis testing procedure for the SIGN TEST for MEDIANS.

EXAMPLE 17.1

The corporate division of Real Low Eye wear believes that the median number of designer sunglasses purchases per customer per year is 3. The sales manager at one of the local stores does not believe this claim to be correct. She decides to use the Sign Test for Medians and conducts an experiment by randomly keeping track of the number of designer sunglasses purchases of 50 customers. For the 50 customers she finds that twenty five purchased more than 3 designer sunglasses, fifteen purchased less than 3 designer sunglasses, and ten purchased 3 designer sunglasses. Test the sales manager's claim using an $\alpha = 5\%$.

Solution

Step 1 **Formulate the two hypotheses, H_o and H_a.**
The null hypothesis has the general form:

H_o: The population median is claimed to be equal to a numerical value M_o.

For this example the corporate division of Real Low Eye wear believes that the median number of designer sunglasses purchases per customer per year is 3. The null hypothesis is:

H_o: The population median number of designer sunglasses purchases per customer per year is 3.

The null hypothesis can be symbolized as:

H₀: population median = 3

Since the sales manager believes the claim by the corporate division is **not correct**, then the alternative hypothesis, Hₐ, is a not equal statement and is nondirectional. Thus, the alternative hypothesis is:

Hₐ: The population median number of designer eye wear purchases per customer per year is not 3.

This alternative hypothesis is symbolized as:

Hₐ: population median ≠ 3

Step 2 Determine the model to test the null hypothesis, H₀.
Under the assumption that H₀ is true, a binomial distribution is used as the hypothesis testing model.

For the 50 customers, the sales manager determined the following information:

less than 3 sunglasses:	15 customers
3 sunglasses (the hypothesized median):	10 customers
more than 3 sunglasses:	25 customers

Converting the data to plus (+) signs, negative (−) signs and a zero (0), using the following rule: assign a negative (−) sign to each customer below the hypothesized value of the median, a plus (+) sign to the customers above the hypothesized value of the median, and a zero (0) to the customers who purchased exactly the same number of designer sunglasses equal to the hypothesized median value, we have:

15 negative (−) signs
25 plus (+) signs
10 zeros (0).

Since the sign test only uses negative and plus signs, the 10 zeros are discarded from the test. Consequently, the usable sample size, n, is: n = 50 − 10 = 40.

The number of negative (−) signs, symbolized as n(−), is n(−) = 15, while the number of plus (+) signs, symbolized as n(+), is n(+) = 25. Thus the usable sample size, n, is equal to the number of negative (−) signs plus the number of plus (+) signs. That is:

$$n = n(-) + n(+)$$
$$= 15 + 25$$
$$n = 40$$

Since the number of data values *not* equal to the median value is n = 40, and p = 0.5, then

$$np = (40)(0.5) = 20$$
and
$$n(1-p) = (40)(0.5) = 20$$

are both at least ten. Thus, the normal approximation to the binomial distribution can be used as the hypothesis testing model. Using the normal distribution to approximate the binomial distribution, we must compute the mean and standard deviation of the binomial distribution by the formulas:

$$\mu_s = np$$

and

$$\sigma_s = \sqrt{np(1-p)}$$

For p = 0.5 and n = 40:

$$\mu_s = np = (40)(0.5) = 20$$

and

$$\sigma_s = \sqrt{np(1-p)}$$
$$= \sqrt{(40)(0.5)(0.5)}$$
$$\sigma_s = \sqrt{10} \approx 3.16.$$

Step 3 **Formulate the Decision Rule.**
For this hypothesis test:
a) the type of alternative hypothesis is nondirectional.
b) the type of test is 2TT.
c) the significance level is $\alpha = 5\%$.
d) The hypothesis testing model is the normal distribution.
e) For a 2TT and a significance level of $\alpha = 5\%$, the critical z-score(s) are $z_{LC} = -1.96$ and $z_{RC} = +1.96$.
f) The decision rule is:
Reject the null hypothesis, H_o, if the test statistic is either less than –1.96 or greater than 1.96.
The decision rule is illustrated in Figure 17.3.

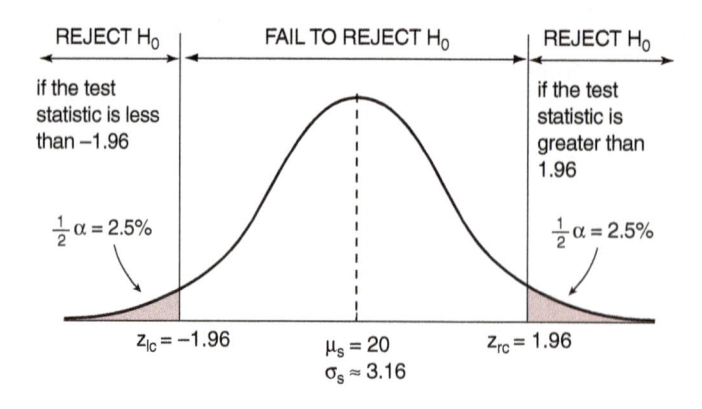

FIGURE 17.3

Step 4 **Analyze the sample data.**
Using the fifteen customers who purchased *less than* the hypothesized median number of 3 designer sunglasses as the statistical result, this statistical result is expressed as O = 15. To determine the test statistic z, we will use a continuity correction value of 15.5 since we are interested in the area under the normal curve that corresponds to 15 or less customers. Using the z score formula, $\mu_s = 20$ and $\sigma_s = 3.16$, the z-score of 15.5 is:

$$z = \frac{(O + \frac{1}{2}) - \mu_s}{\sigma_s}$$

$$z = \frac{15.5 - 20}{3.16}$$

$$z \approx -1.42$$

$z \approx -1.42$ is the test statistic.

Step 5 **Determine the Conclusion.**
Since the test statistic z = –1.42 is not less than –1.96 or greater than 1.96 as stated in the decision rule, we fail to reject H_o at α = 5%. Therefore, the sales manager **cannot** conclude that the corporate division of Real Low Eye wear is incorrect in their claim that the median number of designer sun glasses purchased by a customer per year is 3.

17.3 THE MANN-WHITNEY RANK-SUM TEST

As the Sign Test for Medians is comparable to the one sample t-test, the Mann-Whitney rank-sum test is the nonparametric alternative to the two-sample t test. *This test is designed to determine whether two independent samples came from the same population.* However, when samples are selected from two populations that are essentially the same, the test can be used to investigate if there is a difference between one specific characteristic of each population. For example, the test can be used to determine if the medians or means of two populations are the same. Besides being called the Mann-Whitney Rank-Sum test, the test is also referred to as the Wilcoxon Rank-Sum test as well as the Mann-Whitney-Wilcoxon U test.

To use the Mann-Whitney Rank-Sum test on two independent samples, the samples are combined into one larger sample and the data are ranked accordingly. The null hypothesis is that each sample comes from the same population, or, if there are two populations, they are the same. *If the null hypothesis is true, then the ranks would essentially be evenly distributed between the samples.* If the ranks are evenly distributed between the samples then the *sum* of the ranks for each sample would be the *same.*

On the other hand *if the alternative hypothesis is true, that is, that the samples come from different distributions, then the ranks would not be evenly distributed between the samples. That is, one of the samples would contain more of one type of rank. For example, one sample may contain more lower ranks.* In this case, the sum of the ranks for each sample *would not be the same.*

To determine if the difference in the sum of the ranks is statistically significant, we can use the standard normal distribution for the hypothesis testing model provided each sample contains at least ten observations. When the normal distribution can be applied, the test statistic, z, is computed using the formula:

$$z = \frac{W - \frac{n_1(n_1 + n_2 + 1)}{2}}{\sqrt{\frac{n_1 n_2 (n_1 + n_2 + 1)}{12}}}$$

where:
n_1 = the number of data values in sample one.
n_2 = the number of data values in sample two.
W = the **sum** of the **ranks** from **sample one.**

Let's examine a problem that can be analyzed using the Mann-Whitney Rank-Sum test.

Mr. Barry, the current registrar at a college has noticed that the number of withdrawals per class has increased for day students versus evening students. A random sample of 10 classes were chosen from all day classes while a random sample of 10 classes were selected from all the evening classes. The number of withdrawals per class was tabulated and are summarized in Table 17.2.

TABLE 17.2 Number of Withdrawals per class

Day Classes	Evening Classes
8	6
6	7
10	3
7	5
8	6
11	4
9	4
8	7
11	4
8	9

Can Mr. Barry conclude that there are significantly more withdrawals in the day versus the evening? (Use $\alpha = 5\%$).

In order to conclude this Mr. Barry would test the null hypothesis, H_0: the population number of day class withdrawals and the population number of evening class withdrawals are the same, and use the Mann-Whitney Rank-Sum test to determine if the null hypothesis can be rejected.

To reject H_0 would imply an acceptance of the alternative hypothesis, H_a: there is a significantly greater number of withdrawals from the day classes than from the evening classes.

To use the Mann-Whitney Rank-Sum test we must first combine the data of the two samples (sample one being the withdrawals in the day and sample two being the withdrawals in the evening) into **one** sample and **ASSIGN a rank** to each data value. Table 17.3 list the original data along with an assigned ranking that combines **all** the data values as if they belonged to **one** sample. The letter D is indicated for those data values from the day classes and E for those data values from the evening classes. Notice that if data values are equal, the *assigned* rank is the **mean** of the ranks they occupy. For example, data value 4 occupies the ranks 2, 3, and 4. The mean of the possible ranks 2, 3 and 4 is 3. The rank of data value 5 is assigned 5 since it **alone** occupies the fifth rank.

TABLE 17.3 Number of Withdrawals per class along with RANKing

Number of Withdrawals	Class	Possible Rank	Assigned Rank
3	E	1	1
4	E	2 — mean	3
4	E	3	3
4	E	4 — is 3	3
5	E	5	5
6	D	6 — mean	7
6	E	7	7
6	E	8 — is 7	7
7	D	9 — mean	10
7	E	10	10
7	E	11 — is 10	10
8	D	12 — mean	13.5
8	D	13	13.5
8	D	14	13.5
8	D	15 — is 13.5	13.5

Section 17.3 The Mann-Whitney Rank-Sum Test

Number of Withdrawals	Class	Possible Rank		Assigned Rank
9	D	16 ⎤ mean		16.5
9	E	17 ⎦ is 16.5		16.5
10	D	18		18
11	D	19 ⎤ mean		19.5
11	D	20 ⎦ is 19.5		19.5

To compute the test statistic, use the z formula for The Mann-Whitney Test:

$$z = \frac{W - \frac{n_1(n_1 + n_2 + 1)}{2}}{\sqrt{\frac{n_1 n_2 (n_1 + n_2 + 1)}{12}}}$$

Since n_1 is the number of data values in sample one (i.e. day classes), then, $n_1 = 10$. Similarly n_2 is the number of data values in sample two (i.e., evening classes), thus, $n_2 = 10$. To compute W which represents the **sum** of the **ranks** from **sample one**, it is helpful to list the **assigned ranks** for the day classes separately. Table 17.4 lists the data values for the number of day class withdrawals along with their assigned ranks.

TABLE 17.4

Day Class Data Values	Assigned Ranks
8	13.5
6	7
10	18
7	10
8	13.5
11	19.5
9	16.5
8	13.5
11	19.5
8	13.5

Thus, W, the sum of the assigned ranks for sample one is:

$$13.5 + 7 + 18 + 10 + 13.5 + 19.5 + 16.5 + 13.5 + 19.5 + 13.5 = 144.5$$

W = 144.5

Therefore, substituting the values of $n_1 = 10$, $n_2 = 10$ and $W = 144.5$ into the z statistic formula for the Mann-Whitney Test we have:

$$z = \frac{W - \frac{n_1(n_1+n_2+1)}{2}}{\sqrt{\frac{n_1 n_2(n_1+n_2+1)}{12}}}$$

$$z = \frac{144.5 - \frac{(10)(10+10+1)}{2}}{\sqrt{\frac{(10)(10)(10+10+1)}{12}}}$$

$$z = \frac{144.5 - \frac{(10)(21)}{2}}{\sqrt{\frac{(10)(10)(21)}{12}}}$$

$$z = \frac{144.5 - 105}{\sqrt{175}}$$

$$z \approx 2.99$$

$z \approx 2.99$ is the test statistic.

For a 1TT on the right, the critical z score at $\alpha = 5\%$ is 1.65. This is illustrated in Figure 17.4.

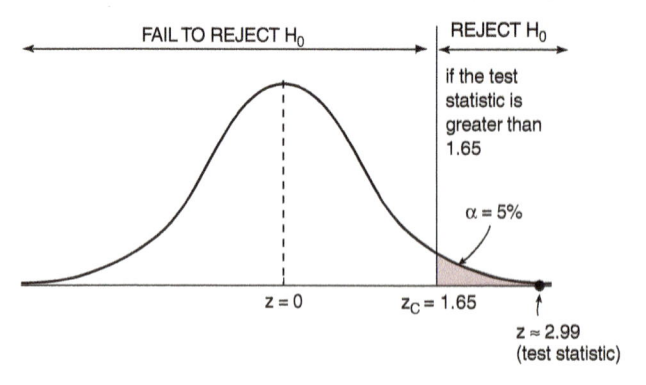

FIGURE 17.4

Since the test statistic $z \approx 2.99$ is greater than the critical z score 1.65, Mr. Barry can reject H_o and accept H_a at $\alpha = 0.05$. Therefore, the number of day class withdrawals is significantly greater than the number of evening class withdrawals.

Let's modify the hypothesis testing procedure to accommodate the Mann-Whitney Test. Remember the test can be used as an alternative to the two sample t test whenever the assumption about the normality of the populations or the assumption regarding the population variances being equal are not made. For our application of this test we will assume that the number of data values for each sample is at least ten. However, although we will not deal here with samples that are smaller than ten, there are statistical tables available that will enable you to perform the Mann-Whitney Test for these smaller samples.

The Mann-Whitney Test for Samples of Size of at Least 10

Step 1 *Formulate the two hypotheses, H_o and H_a.*
Null Hypothesis:
In general, the null hypothesis for the Mann-Whitney Test has the following form.

Section 17.3 The Mann-Whitney Rank-Sum Test

> **Null Hypothesis Form:**
>
> H_o: The two populations are the same

Alternative Hypothesis: The alternative hypothesis for the Mann-Whitney Test is stated as:

> **Alternative Hypothesis Form:**
>
> H_a: The two populations are NOT the same

The alternative hypothesis, H_a, could be that the population means, medians, or shapes are different.

Step 2 **Determine the model to test the null hypothesis, H_o.**

Under the assumption that H_o is true, when each sample contains at least ten data values, the *standard normal distribution* can be used as the hypothesis testing model. This is illustrated in Figure 17.5.

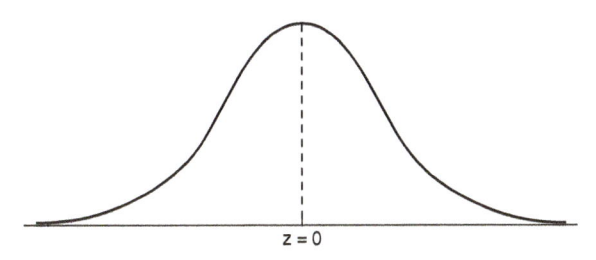

FIGURE 17.5

Step 3 **Formulate the Decision Rule.**
  a) Determine the type of alternative hypothesis: directional or nondirectional.
  b) Determine type of test: 1TT or 2TT.
  c) Identify the significance level: $\alpha = 0.01$ or $\alpha = 0.05$.
  d) Construct a standard normal distribution as the hypothesis testing model.
  e) Determine the critical z-score(s) using the following chart.

> **Critical z-scores**
>
> *For significance level of $\alpha = 0.05$:*
> For a 2TT: $z_{LC} = -1.96$ and $z_{RC} = +1.96$.
> For a 1TT test on left side: $z_c = -1.65$
> For a 1TT test on right side: $z_c = +1.65$
>
> *For significance level of $\alpha = 0.01$:*
> For a 2TT: $z_{LC} = -2.58$ and $z_{RC} = +2.58$.
> For a 1TT test on left side: $z_c = -2.33$
> For a 1TT test on right side: $z_c = +2.33$

f) Using the critical z-score(s), state the decision rule.

Step 4 **Analyze the sample data.**
  a) Collect the necessary sample data from each population and combine the data into one sample. Assign a rank to each data value. For data values that are equal, the rank assigned is the **mean** of the ranks they occupy.
  b) Calculate W, the sum of the assigned ranks from the first sample. The value of W represents the statistical result.
  c) Calculate the test statistic using the formula:

$$z = \frac{W - \frac{n_1(n_1+n_2+1)}{2}}{\sqrt{\frac{n_1 n_2 (n_1+n_2+1)}{12}}}$$

where:
n_1 = the number of data values in sample one.
n_2 = the number of data values in sample two.
W = the **sum** of the **ranks** from **sample one.**
z = test statistic

Step 5 **Determine the Conclusion.**
Compare the Mann-Whitney z statistic, test statistic, to the critical z-score and either:

  a) **Reject H_o and accept H_a at α,**

  OR

  b) **Fail to reject H_o at α.**

Example 17.2 will illustrate the hypothesis testing procedure for the Mann-Whitney Test.

EXAMPLE 17.2

The number of trains arriving at least 5 minutes late on Route A and Route B (two major commuter routes) were determined for randomly selected days in the month of December. The number of late arrivals for each of the selected days are listed in Table 17.5.

TABLE 17.5

Route A	Route B
5	18
34	23
11	19
12	32
19	8
24	25
28	22
17	16
24	14
31	10
29	

Use the Mann-Whitney Test at an $\alpha = 5\%$ to determine if there is a significant difference between the average number of late arrivals.

Solution

Step 1 **Formulate the two hypotheses, H_o and H_a.**

H_o: The population average number of late arrivals for trains using Route A and Route B are the same during the month of December.

H_a: The population average number of late arrivals for trains using Route A and Route B are NOT the same during the month of December.

Step 2 **Determine the model to test the null hypothesis, H_o.**

Under the assumption that H_o is true, when each sample contains at least ten data values, the *standard normal distribution* can be used as the hypothesis testing model. This is illustrated in Figure 17.6.

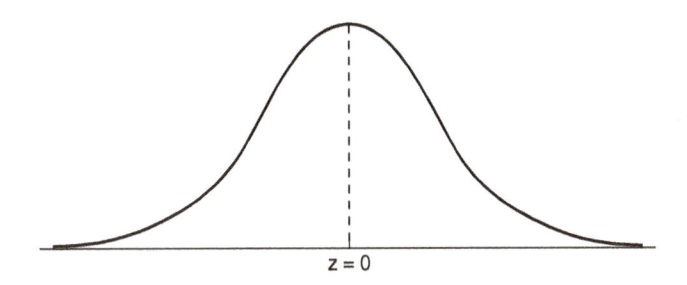

FIGURE 17.6

Step 3 **Formulate the Decision Rule.**

For this hypothesis test:
a) the type of alternative hypothesis, H_a, is nondirectional.
b) the type of test is 2TT.
c) the significance level is $\alpha = 0.05$.
d) The hypothesis testing model is the normal distribution.
d) For a 2TT and a significance level of $\alpha = 0.05$, the critical z-score(s) are $z_{LC} = -1.96$ and $z_{RC} = +1.96$.
f) The decision rule is: Reject H_o, if the Mann Whitney z statistic, test statistic, is either less than -1.96 or greater than 1.96. The decision rule is illustrated in Figure 17.7.

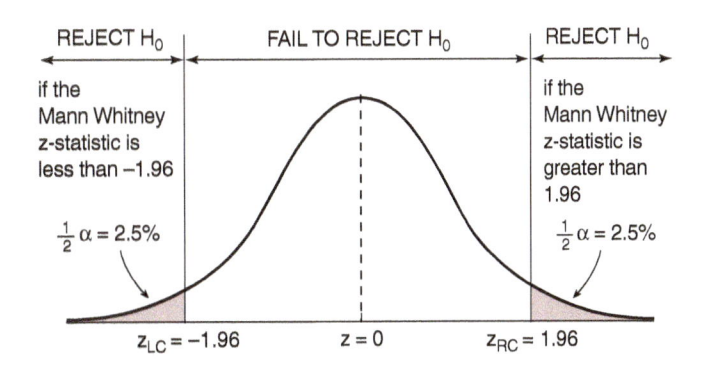

FIGURE 17.7

Chapter 17 Nonparametric Statistics

Step 4 Analyze the sample data.
a) Table 17.6 lists the sorted data values for the number of trains that were at least five minutes late along with their assigned ranks.

TABLE 17.6

Number of Late Trains	Route	Possible Rank	Assigned Rank
5	A	1	1
8	B	2	2
10	B	3	3
11	A	4	4
12	A	5	5
14	B	6	6
16	B	7	7
17	A	8	8
18	B	9	9
19	A	10	10.5
19	B	11	10.5
22	B	12	12
23	B	13	13
24	A	14	14.5
24	A	15	15.5
25	B	16	16
28	A	17	17
29	A	18	18
31	A	19	19
32	B	20	20
34	A	21	21

(mean is 10.5 for ranks 10, 11; mean is 14.5 for ranks 14, 15)

For Route A (sample 1) the Assigned Ranks are listed in Table 17.7.

TABLE 17.7

Route A Data Values	Assigned Ranks
5	1
34	21
11	4
12	5
19	10.5
24	14.5
28	17
17	8
24	14.5
31	19
29	18
34	21

b) W, the statistical result, is the sum of the assigned ranks for sample one is:

$$1 + 21 + 4 + 5 + 10.5 + 14.5 + 17 + 8 + 14.5 + 19 + 18 + 21 = 153.5$$

c) Substituting the values of $n_1 = 12$, $n_2 = 11$ and $W = 153.5$ into the test statistic formula z for the Mann-Whitney Test we have:

$$z = \frac{W - \frac{n_1(n_1+n_2+1)}{2}}{\sqrt{\frac{n_1 n_2 (n_1+n_2+1)}{12}}}$$

$$z = \frac{153.5 - \frac{(12)(12+11+1)}{2}}{\sqrt{\frac{(12)(11)(12+11+1)}{12}}}$$

$$z = \frac{153.5 - \frac{(12)(24)}{2}}{\sqrt{\frac{(12)(11)(24)}{12}}}$$

$$z = \frac{153.5 - 144}{\sqrt{16.25}}$$

$$z \approx 0.59$$

$z \approx 0.59$ is the test statistic.

Step 5 **Determine the Conclusion.**

Since the Mann-Whitney test z statistic, $z = 0.59$, falls between the critical z scores, we fail to reject the null hypothesis, H_o, at $\alpha = 5\%$. Therefore, although there is a difference for the average number of trains arriving late for Route A verses Route B, the difference is not statistically significant at $\alpha = 5\%$. The result is illustrated in Figure 17.8.

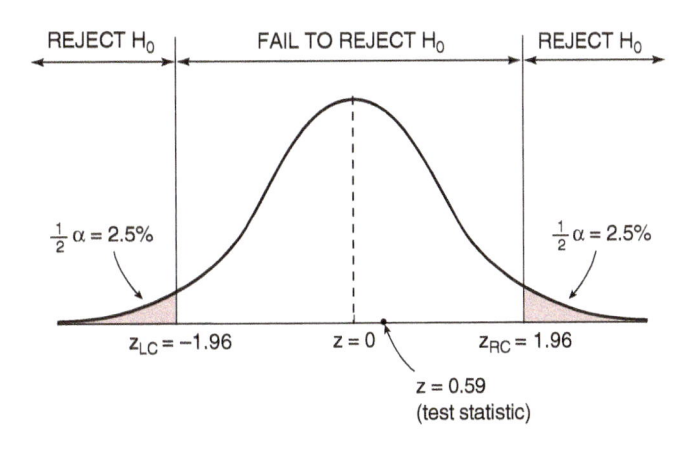

FIGURE 17.8

17.4 THE KRUSKAL-WALLIS H TEST

In the previous section the Mann-Whitney test was used to test whether or not two populations were the same. Recall that the Mann-Whitney test was the nonparametric counterpart to the two-sample t test studied in Chapter 14. A nonparametric test that is analogous to the Analysis of Variance test (ANOVA) covered in Chapter 16 is called the *Kruskal-Wallis H Test*. It is also referred to as simply the H test. Essentially, the Kruskal-Wallis tests whether three or more populations are the same.

The null hypothesis, H_o, for this test is that the populations are the same. While the alternative hypothesis, H_a, is that the populations are not the same. That is there are differences among the populations, in particular the populations may have different means or medians.

If each sample has at least five observations the sampling distribution of the data can be approximated by a Chi-Square, X^2, distribution. The test statistic for a Kruskal-Wallis test, denoted by H, is computed using the formula:

$$H = \frac{12}{N(N+1)}\left[\frac{(\Sigma R_1)^2}{n_1} + \frac{(\Sigma R_2)^2}{n_2} + \ldots + \frac{(\Sigma R_k)^2}{n_k}\right] - 3(N+1)$$

where $\Sigma R_1, \Sigma R_2, \ldots, \Sigma R_k$ are the sum of the ranks for samples $1, 2, \ldots, k$.
$n_1, n_2, \ldots, n_k$ are the sizes of samples $1, 2, \ldots, k$.
and N is the combined number of data values for all samples (i.e. $N = n_1 + n_2 + \ldots + n_k$)

After the Kruskal-Wallis test statistic, H, is computed its value is compared to a *critical value* approximated by using Table IV: the critical values for the Chi-Square Distributions in Appendix C. To determine the critical value for either $\alpha = 5\%$ or 1%, use k-1 degrees of freedom, where k represents the number of distinct random samples.

Let's apply the Kruskal-Wallis H test to a selection process problem for a college admissions officer.

For example, suppose the admissions officer at a prestigious university is interested in attempting to determine if there are differences between students transferring from three community colleges that are within one hundred miles of each other. Part of the selection process that the university uses for entrance is to administer an achievement test to prospective entrants. Table 17.8 contains the achievement test scores for recent prospective entrants from the three community colleges.

TABLE 17.8 Achievement Scores: Students from Three Community Colleges

1 NCC	2 WCC	3 TCC
30	65	55
75	25	65
65	35	85
90	20	95
95	45	75
85	40	80
85		75

If the administrator were able to assume that the population of achievement scores are normally distributed with their respective population variances equal, then a nonparametric test would not be necessary. [What test would the administrator use?] Since these assumptions cannot be made, the administrator decides to use the Kruskal-Wallis H test.

To apply the test, the three samples of achievement scores are combined into one group and are ranked as though they were scores from one population. (This is based on the assumption that the null hypothesis for the test is true.) The sorted scores and ranks are listed in Table 17.9.

TABLE 17.9 The combined Samples of Achievement Scores from three community colleges Sorted and Ranked based on the null hypothesis, H_0.

scores	sample	sorted scores	rank
30	1	20	1.0
75	1	25	2.0
65	1	30	3.0
90	1	35	4.0
95	1	40	5.0
85	1	45	6.0
85	1	55	7.0
65	2	65	9.0
25	2	65	9.0
35	2	65	9.0
20	2	75	12.0
45	2	75	12.0
40	2	75	12.0
55	3	80	14.0
65	3	85	16.0
85	3	85	16.0
95	3	85	16.0
75	3	90	18.0
80	3	95	19.5
75	3	95	19.5

In Table 17.9, notice the second column is *labelled sample* to identify the sample the ranked score belongs. For example, the score, 30, in the first row is from sample 1, NCC. Since it is the third lowest ranked score it has a rank of 3. Whereas, 65, the first score from sample 2, has a rank of 9, as well as the second score, 65, from sample 3.

Table 17.10 lists the individual scores, their ranks, the sum of their ranks, and the sample size, n, for each sample. The three samples have been separated as they were initially collected, with the overall rank shown alongside each data value.

TABLE 17.10 Students from Three Community Colleges Scores, Ranks and Sum of Ranks

NCC		WCC		TCC	
Score	Ranks from Sample 1	Score	Ranks from Sample 2	Score	Ranks from Sample 3
30	3	65	9	55	7
75	12	25	2	65	9
65	9	35	4	85	16
90	18	20	1	95	19.5
95	19.5	45	6	75	12
85	16	40	5	80	14
85	16			75	12
	$\Sigma R_1 = 93.5$		$\Sigma R_2 = 27$		$\Sigma R_3 = 89.5$
	$n_1 = 7$		$n_2 = 6$		$n_3 = 7$

To determine H, the Kruskal-Wallis H test statistic, we use the formula:

$$H = \frac{12}{N(N+1)} \left[\frac{(\Sigma R_1)^2}{n_1} + \frac{(\Sigma R_2)^2}{n_2} + \ldots + \frac{(\Sigma R_k)^2}{n_k} \right] - 3(N+1)$$

Since N is the combined number of observations for all the samples, N = 7 + 6 + 7 = 20. Substituting N and the values from Table 17.10 in the formula H, we get:

$$H = \frac{12}{20(20+1)} \left[\frac{(93.5)^2}{7} + \frac{(27)^2}{6} + + \frac{(89.5)^2}{7} \right] - 3(20+1)$$

$$H = (0.029)[2514.714] - 63$$

$H \approx 9.93$ is the test statistic.

To determine if the Kruskal-Wallis H test statistic, H = 9.93, is significant, compare H to the critical chi-square value χ_α^2 found in TABLE IV of Appendix C. For an $\alpha = 1\%$ and $df = k-1 = 3-1 = 2$, the critical value, χ_α^2, is 9.21. Since the test statistic H = 9.93 is greater than the critical value $\chi_\alpha^2 = 9.21$, the null hypothesis, H_o, is *rejected at an* $\alpha = 1\%$ and the alternative hypothesis, H_a, is accepted. Thus, we can conclude that there is a significant difference between the achievement scores of the prospective entrants from the three community colleges.

Figure 17.9 illustrates the results of the hypothesis test.

Chi-Square Distribution with $df = 2$

FAIL TO REJECT H_0 | REJECT H_0

$\alpha = 1\%$

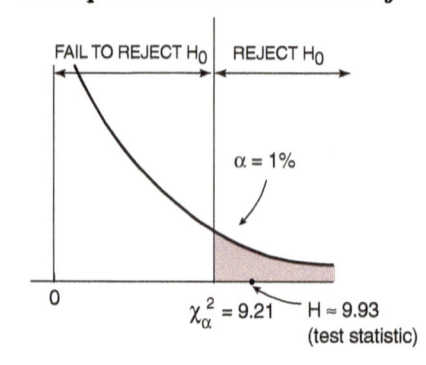

$\chi_\alpha^2 = 9.21$ $H \approx 9.93$ (test statistic)

FIGURE 17.9

Let's now summarize the hypothesis testing procedure for the Kruskal-Wallis H test.

The Kruskal-Wallis H Test

Step 1. Formulate the two hypotheses, H_0 and H_a.
Null Hypothesis:
In general, the null hypothesis for the Kruskal-Wallis H Test has the following form.

Null Hypothesis Form:

H_0: *The two populations are the same*

Alternative Hypothesis: The alternative hypothesis for the Kruskal-Wallis H Test is stated as:

Alternative Hypothesis Form:

H_a: The two populations are NOT the same

When it is assumed that certain characteristics of the distributions are the same the alternative hypothesis could be that the population means, medians, or shapes are different.

Step 2 **Determine the model to test the null hypothesis, H_o.**
Under the assumption that H_o is true, and when each sample contains at least five data values, a *Chi-Square Distribution* with df = k-1 degrees of freedom, where k represents the number of distinct samples, can be used as the hypothesis testing model. Figure 17.10 displays a typical chi-square distribution.

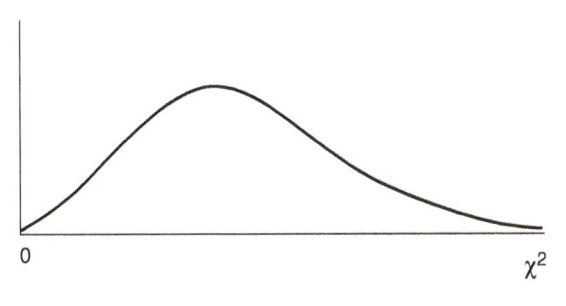

FIGURE 17.10

Step 3 **Formulate the Decision Rule.**
  a) Identify the significance level, α = 1% or α = 5%.
  b) Find the critical chi-square value, χ_α^2, in TABLE IV: Critical Values for the Chi-Square Distribution, in Appendix C: Statistical Tables.
  c) Construct a Chi Square Distribution with k – 1 degrees of freedom as the hypothesis testing model with the critical chi-square value χ_α^2 appropriately placed on the curve.
  d) The decision rule is: if the Kruskal-Wallis H test statistic is larger than the critical chi-square value, χ_α^2, then reject H_o and accept H_a.

Step 4 **Analyze the sample data.**
  a) Collect the necessary sample data from each population and combine the data into one sample. **Assign a rank** to each data value. For data values that are equal, the rank assigned is the **mean** of the ranks they occupy.
  b) Calculate the Kruskal-Wallis H test statistic using the formula:

$$H = \frac{12}{N(N+1)}\left[\frac{(\Sigma R_1)^2}{n_1} + \frac{(\Sigma R_2)^2}{n_2} + \ldots + \frac{(\Sigma R_k)^2}{n_k}\right] - 3(N+1)$$

where $\sum R_1, \sum R_2, \ldots, \sum R_k$ are the sum of the ranks for samples $1, 2, \ldots, k$.
$n_1, n_2, \ldots n_k$, are the sizes of samples $1, 2, \ldots, k$.
and N is the combined number of data values for all samples

Step 5 Determine the Conclusion.
Compare the Kruskal-Wallis H test statistic to the critical chi-square value χ_α^2 and either:
a) **Reject H_o and accept H_a at α,**

OR

b) **Fail to reject H_o at α.**

Example 17.3 will illustrate a Kruskal-Wallis H hypothesis test procedure.

EXAMPLE 17.2

Allison, a health and fitness coordinator, at a local racquet and fitness club wants to experiment with three different fitness routines for women over 35 years old. Since no experience exits regarding the three routines, she is unwilling to make any assumptions about the shape of the distribution of pulse rate recovery times. Use the Kruskal-Wallis H test to test if the participants in the fitness routines have different pulse rate recovery times after the fitness routines are given for a specified length of time. Use $\alpha = 5\%$. Table 17.11 represents the pulse rate recovery times for each participant along with their respective fitness routine type.

TABLE 17.11

Pulse rate recovery times in minutes

Routine A	Routine B	Routine C
3.2	7.5	3.5
2.1	3.8	4.2
4.4	5.6	5.5
8.3	6.4	8.4
5.7	4.8	6.1
	6.2	6.3

Solution

Step 1 Formulate the two hypotheses, H_o and H_a.
The null hypothesis is:

H_o: The populations of pulse rate recovery times are the same.

The alternative hypothesis is:

H_a: The populations of pulse rate recovery times are not the same.

Step 2 Determine the model to test the null hypothesis, H_o.
Since each of the three samples contain at least five data values, the Chi-Square Distribution will be used as the hypothesis testing model. For this Chi-Square Distribution, the degrees of freedom are determined by the formula: $df = k-1$. Since there are $k = 3$ distinct samples, then $df = 3-1 = 2$. Figure 17.11 shows the Chi-Square Distribution with 2 degrees of freedom.

Chi-Square Distribution with df = 2

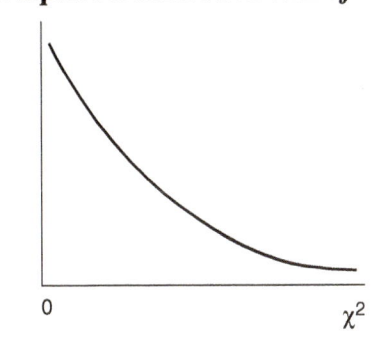

FIGURE 17.11

Step 3 **Formulate the Decision Rule.**
  a) The significance level is: $\alpha = 5\%$.
  b) Using TABLE IV: Critical Values for the Chi-Square Distribution, in Appendix C, the critical chi-square, χ_α^2, for $df = 2$ and $\alpha = 5\%$ is: $\chi_\alpha^2 = 5.99$.
  c) The Chi-Square Distribution with 2 degrees of freedom and $\alpha = 5\%$ serves as the hypothesis testing model and is shown in Figure 17.12 with the critical chi-square value $\chi_\alpha^2 = 5.99$.
  d) The decision rule is: if the Kruskal-Wallis H test statistic is larger than the critical chi-square value χ_α^2 of 5.99, then reject H_o and accept H_a. The decision rule is illustrated in Figure 17.12.

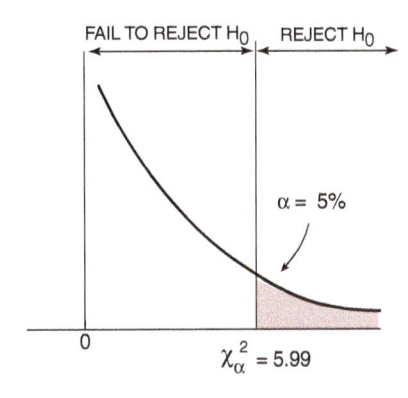

FIGURE 17.12

Step 4 **Analyze the sample data.**
  a) The sample data from Table 17.11, have been combined, labelled and sorted along with their assigned ranks. The results are indicated in Table 17.12.
  Table 17.13 lists the score, the ranks, the sum of the ranks, and the sample size, n, from each sample. The samples have been separated as they were initially collected, but this time the overall ranks are alongside each of the initial data values.
  b) To determine H, the Kruskal-Wallis H test statistic, use the formula:

$$H = \frac{12}{N(N+1)}\left[\frac{(\Sigma R_1)^2}{n_1} + \frac{(\Sigma R_2)^2}{n_2} + \ldots + \frac{(\Sigma R_k)^2}{n_k}\right] - 3(N+1)$$

TABLE 17.12 The combined Samples of Pulse Rate Recovery Times from three Fitness Routines Sorted and Ranked based on the null hypothesis, H_o.

rates	sample	sorted rates	rank
3.2	1	2.1	1
2.1	1	3.2	2
4.4	1	3.5	3
8.3	1	3.8	4
5.7	1	4.2	5
7.5	2	4.4	6
3.8	2	4.8	7
5.6	2	5.5	8
6.4	2	5.6	9
4.8	2	5.7	10
6.2	2	6.1	11
4.2	3	6.2	12
5.5	3	6.3	13
8.4	3	6.4	14
6.1	3	7.5	15
3.5	3	8.3	16
6.3	3	8.4	17

TABLE 17.13 Pulse Rate Recovery Times for Each Fitness Routine Rates, Ranks and Sum of Ranks

Routine A		Routine B		Routine C	
Rate	Ranks from Sample 1	Rate	Ranks from Sample 2	Rate	Ranks from Sample 3
3.2	2	7.5	15	3.5	3
2.1	1	3.8	4	4.2	5
4.4	6	5.6	9	5.5	8
8.3	16	6.4	14	8.4	17
5.7	10	4.8	7	6.1	11
		6.2	12	6.3	13
	$\Sigma R_1 = 35$		$\Sigma R_2 = 61$		$\Sigma R_3 = 57$
	$n_1 = 5$		$n_2 = 6$		$n_3 = 6$

Since N is the combined number of observations for all the samples, N = 19. Substituting N and the values from Table 17.13 in the test statistic formula H, we get:

$$H = \frac{12}{17(17+1)} \left[\frac{(35)^2}{5} + \frac{(61)^2}{6} + + \frac{(57)^2}{6} \right] - 3(17+1)$$

$$H = (0.039)[1406.667] - 54$$
$$H \approx 0.86 \text{ is the test statistic.}$$

Step 5 **Determine the Conclusion.**
Comparing the Kruskal-Wallis H test statistic to the critical chi-square value χ^2_α, you will find that test statistic H = 0.86 is less than the critical chi-square value, $\chi^2_\alpha = 5.99$. Consequently, we fail to reject the null hypothesis, H_o, at $\alpha = 5\%$. This is illustrated in Figure 17.13.

Section 17.4 The Kruskal-Wallis H Test

FAIL TO REJECT H₀ | REJECT H₀

$\alpha = 5\%$

$\chi_\alpha^2 = 5.99$

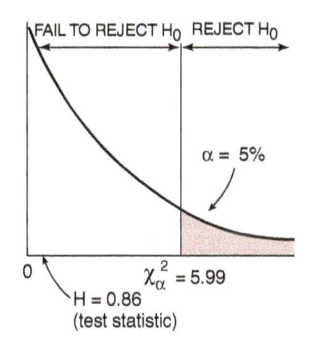

$H = 0.86$ (test statistic)

FIGURE 17.13

Thus, although there are differences between the samples, the differences are not statistically significant. Therefore, Allison cannot conclude that the different fitness routines for women over the age of 35 will yield different population recovery pulse rate times.

CASE STUDY 17.1

The following information is based on the 1990 U.S. Census Bureau survey data. The data represents housing costs (rent or mortgage plus taxes and utilities) for some randomly selected Nassau County communities on Long Island, New York.

Community	1990 monthly housing cost	Community	housing cost
Baldwin	$809	Levittown	$846
Bellmore	$950	Long Beach	$714
Carle Place	$680	Lynbrook	$727
East Meadow	$732	Malverne	$785
E. Rockaway	$799	Massapequa	$876
Elmont	$692	Merrick	$998
Garden City	$1,027	Oceanside	$847
Glen Cove	$661	Rockville Centre	$708
Great Neck	$801	Roosevelt	$894
Hempstead	$694	Uniondale	$835
Hewlett	$835	Wantagh	$909
Lawrence	$1,207	Westbury	$822

The Census Bureau states that the monthly median housing cost for Nassau County for 1990 is $801. Using this value, and the sign test for medians:

a) what sign would be assigned to the community of Elmont?
b) what sign would be assigned to the community of Merrick?
c) what sign would be assigned to the community of Great Neck?
d) what is the n(− signs)? n(+ signs)? usable sample size (n)?
e) if you are interested in determining whether the Census Bureau's statement about the median housing cost is reasonable, how would you state the alternative hypothesis?
f) how would you state the null hypothesis?
g) would it be acceptable to use the normal approximation to the binomial distribution?
h) what is the mean of the binomial distribution? the standard deviation?
i) what are the critical z score(s) for a significance level of $\alpha = 0.05$?
j) what is your statistical result? Explain why you chose this value for the statistical result?
k) can you reject the null hypothesis at $\alpha = 0.05$?
l) do you find the Census Bureau's statement about the median housing cost for all Nassau County Communities reasonable? Explain.

CASE STUDY 17.2

The article "Senior Nursing Students' and Professional Nurses' Perceptions of Effective Caring Behaviors: A Comparative Study" which appeared in the *Journal of Nursing Education* (volume 30, 1991) compared the ranks of a group of 30 senior baccalaureate students to 30 professional nurses on 50 behavioral items. Both groups of nurses were given 50 cards containing behavioral items which were ordered in six subscales of caring. These subscales of caring were:

1) Accessible - 6 items
2) Explains and facilitates - 6 items
3) Comforts—9 items
4) Anticipates—5 items
5) Trusting relationship—16 items
6) Monitors and follows through—8 items

All the nurses were given 50 cards containing these caring behavioral items and were instructed to sort the cards from most important to not important. The Mann-Whitney test was used to determine if there was a significant difference in the ranks of the caring behaviors of the nursing students and the professional nurses. Table 17.14 contains the sum of the ranks for the nursing students and the professional nurses on each of the six subscales of caring.

TABLE 17.14

Subscales	Nursing Group	Student Group	z*	p
Accessibility	906.0	924.0	−0.12	0.89
Explains	874.5	955.5	−0.59	0.54
Anticipates	903.5	926.5	−0.16	0.86
Comforts	820.5	1009.5	−1.39	0.16
Monitors	903.0	927.0	−0.17	0.86
Trusts	1041.5	788.5	1.86	0.06

* The Mann-Whitney test requires finding a z-value if one of the groups is greater than 20.

a) Describe the type of data used in this study.
b) Identify each population within this study.
c) The behavioral items in each subscale were rated from one for most important to N for the least important nursing care behavior. Using this information, determine which group rated the items on each subscale as more important.
d) For the accessibility subscale, state the null and alternative hypotheses for the Mann-Whitney test. At the 0.10 significance level, can the null hypothesis be rejected?
c) For the trusts subscale, state the null and alternative hypotheses for the Mann-Whitney test. At the 0.10 significance level, can the null hypothesis be rejected?
d) For what p-value can the null hypothesis be rejected on the trust subscale?
e) For what p-value can the null hypothesis be rejected on the comforts subscale?
f) For the Anticipates subscale, which group was identified as population one on the Mann-Whitney test? Explain.
g) Was there a significant difference at $\alpha = 0.20$ in the way that the professional nurses ranked those items in the comforts subscales than the student nurses? If it is significant, then which group ranked the items higher? Explain.
h) Is there a significant difference between the way the professional nurses and the student nurses perceive nursing care behavior?

KEY TERMS

Parametric Statistics 865
Nonparametric Statistics 865
Distribution Free Tests 866
n (– signs) 868
n (+ signs) 868

Sign Test for Medians 867
Mann-Whitney Rank Sum Test 875
Critical z scores 871
Usable sample size, n 871
Kruskal-Wallis H Test 884

Population Median, μ_o 870
Test statistic z 872

SECTION EXERCISES

Section 17.1

1. Inferential statistics which are based on assumptions such as the shape of the population is normal or the population variances are equal are called _____ statistics.

2. Inferential statistics that make no assumptions about the shape of the population or the parameters of a population are referred to as _____ statistics.

3. Nonparametric tests that do not require assumptions about the underlying population shape are also called _____ tests.

4. If you are not sure that your sample data comes from a normal population, you should
   a) use a parametric test
   b) use a nonparametric test
   c) increase your sample size
   d) use the normal distribution
   e) redesign your experiment

5. Nonparametric statistics are used when the population parameters are not known.
   (T/F)

6. You should use a nonparametric test whenever it is known that the distribution from which you are sampling is highly skewed.
   (T/F)

Section 17.2

7. a) Define Nonparametric Statistics.
   b) Why do statisticians frequently call nonparametric tests, distribution free tests?

8. What are some of the advantages of using nonparametric tests over parametric tests?

9. The sign test for medians requires the use of the binomial distribution where the probability of a success is p = ____.

10. In the sign test for medians, the data value is assigned a "+" if the data value is _____ the median, while a data value equal to the median is assigned a ____ and a data value below the median is assigned a ____.

11. The formulas for the binomial mean and standard deviation are:
    μ_s = ____ and σ_s = ____.

12. The null hypothesis for the sign test for a median is a statement about the _____ median.

13. In the sign test for a median, the usable sample size, denoted n, is equal to the number of _____ signs, plus the number of _____.

14. In the sign test for a median, if the $\alpha = 0.01$ and a 1TT on the right side, the critical z score is: ____. While a 1TT on the left side and $\alpha = 0.05$ has a critical z score of ____.

15. In the sign test for medians, if there are 18 values below the hypothesized median, 4 equal to the hypothesized median and 17 values above the hypothesized median then the usable sample size is:
    a) n = 39
    b) n = 18
    c) n = 17
    d) n = 35
    e) none of these

16. Consider the following sample of 23 measurements.
    18, 21, 8, 19, 22, 7, 5, 23, 11, 6, 13, 21
    23, 11, 19, 9, 25, 11, 22, 29, 24, 29, 25
    Use these data to conduct a Sign Test for Medians using the following hypotheses (use $\alpha = 0.05$):
    a) H_0: M = 12 versus H_a: M > 12
    b) H_0: M = 12 versus H_a: M ≠ 12
    c) H_0: M = 22 versus H_a: M < 22
    d) H_0: M = 22 versus H_a: M ≠ 22

17. Perform a sign test for medians using the following data where the median = 26 against the claim that the median does not equal 26. Use an 0.05 level of significance.
    30, 106, 44, 217, 19, 94, 21, 92, 22, 109, 11, 7, 90,
    55, 29, 44, 64, 4, 53, 59, 55, 89, 17, 19, 21, 294,
    26, 14, 15, 31, 18, 99, 6, 32, 12, 17, 10, 26, 65, 17,
    42, 50, 106, 102, 18, 74, 49, 14, 44, 40, 91, 94, 21,
    49, 17, 22, 42, 8, 52, 106, 100, 80, 42, 49, 35, 64,
    80, 30.
    a) Determine the sample size.
    b) How many data values are less than the hypothesized median?
    c) How many data values equal the hypothesized median?
    d) How many data values are greater than the hypothesized median?
    e) What is the median sample value?

f) Is the hypothesized median value rejected?

g) What is the p-value for the median sample value?

18. Perform a sign test for medians using the following data where the median = 10.6 against the claim that the median is greater than 10.6. Use an 0.05 level of significance.

11.6, 8.8, 10.4, 10.8, 11.2, 10.9, 10.4, 10.8, 10.1, 9.3, 10.7, 10.8, 10.2, 10.9, 10.5, 10.9, 11.5, 10.8, 10.9, 10.9, 11.3, 11.1.

a) Determine the sample size.
b) How many data values are less than the hypothesized median?
c) How many data values equal the hypothesized median?
d) How many data values are greater than the hypothesized median?
e) What is the median sample value?
f) Is the hypothesized median value rejected?
g) What is the p-value for the median sample value?

19. A manufacturer of DVD players has established that the median time to failure for its DVD players is 8340 hours of use. A sample 30 DVD players from the manufacturer's leading competitor is obtained. Each DVD player is tested until failure. For the sample of 30 DVD players, the 30 failure times ranged from 2 hours until 11,740 hours. Of the 30 failure times 20 was less than the 8340 hours. Do these results support the claim that the median failure time of the competitor differs from 8340 hours? Use $\alpha = 0.05$.

20. The following data represent the waiting time to play tennis at Salisbury Tennis Club on a typical Saturday morning.

Time (in minutes)

13, 14, 12, 13, 16, 17, 6, 4, 8, 6, 20, 8, 6, 1, 10, 7, 11, 9, 11, 13, 15, 10.

The club's manager claims that the median waiting time is 8 minutes. Aaron, the club's principal owner, believes that the claim is too low. Use the Sign Test for Medians to test Aaron's claim against the club manager's claim using a 0.05 level of significance.

21. Most nutritionists believe that a healthy diet contains no more than 30% of its total calories from fat. Dr. John believes that the claim is too high and determines the amount of calories from fat in diets of 23 healthy men that are on a 2000 calorie diet per day. The results of the sample are as follows:

Calories (from fat)

590, 450, 720, 600, 450, 320, 550, 490, 600, 340, 570, 400, 700, 800, 200, 445, 600, 590, 480, 730, 570, 520, 690

Do these results support Dr. John's claim at 0.05 level of significance?

22. The median hourly wage for the population of students between the ages of 18 and 21 is $8 per hour. Professor DePorto believes that the median hourly for students at his college is not equal to $8 per hour. To test his claim a sample of students in this age group was selected and the results are listed below:

Hourly Wages:

10.50, 7.40, 10.75, 9.05, 9.25, 7.00, 12.65, 6.05, 8.05, 10.90, 8.80, 5.85, 14.35, 10.10, 13.70, 12.15, 12.10, 5.55, 12.10, 8.10, 9.15, 10.05, 9.45, 5.85, 9.80

Using a 0.05 level of significance can Professor DePorto support his claim?

Section 17.3

23. State the assumptions for using the Mann-Whitney Rank Sum Test.

24. The null hypothesis of the Mann-Whitney Rank Sum test is

H_0: The two populations are the same.

State three different alternative hypotheses that may be tested.

25. The nonparametric alternative hypothesis test to the two-sample t test which determines whether two independent samples came from the same population is called the _____.

26. In the Mann-Whitney rank-sum test, the test statistic, z, is computed using the formula: z = _____ where n_1 = _____, n_2 = _____ and W = _____.

27. In the Mann-Whitney rank-sum test, all the data values combined into one group and are ranked. If four data values are equal then their rank is determined by finding the _____ of the ranks of the four data values.

28. In the Mann-Whitney Test, if the $\alpha = 0.01$ and a 1TT on the left side, the critical z score is: _____. While a 1TT on the right side and $\alpha = 0.05$ has a critical z score of _____.

29. In the Mann-Whitney Test, the statistical result is equal to the value of _____.

30. The Mann-Whitney test
   a) combines all the data values into one sample
   b) ranks the data values of the combined samples
   c) assumes the data values come from a normal distribution
   d) is also referred to as the Wilcoxon rank-sum test
   e) all of the above
   f) none of the above
   g) does only a and b
   h) does only a, b, and c
   i) does only a, b, and d

31. If the data values of two combined samples are:

20, 20, 21, 22, 22, 25, 27, 30, 31, 31, 33, 35

then the rank assigned to 22 in the Mann-Whitney test is:
   a) 4 for the both values of 22
   b) 4 for the first value of 22 and 5 for the second one
   c) 4.5 for both values of 22
   d) 4 but only one value of 22 is kept
   e) 0 since there is a tie in the rankings

32. The Wilcoxon Rank-Sum Test is also referred to as the Mann-Whitney Rank Sum Test.
   (T/F)

33. If it is known that the shapes of two populations are the same, then the Mann-Whitney Rank-Sum Test can be used to determine if the population means are different. (T/F)

34. Apply the Mann-Whitney Rank Sum Test to determine if the samples come from different distributions. Use a 0.05 level of significance.

Sample 1	Sample 2
85	57
67	80
56	73
61	61
76	75
64	86
54	64
65	72
	79

a) Identify the sample size of each sample.
b) What is the median for each sample?
c) Identify W.
d) Is W significant?
e) What is the p-value for W?

35. Apply the Mann-Whitney Rank Sum Test to determine if the median of distribution 1 is less than the median of distribution 2. Use a 0.05 level of significance.

Sample 1	Sample 2
28	26
16	42
42	65
29	38
31	29
22	32
50	59
42	42
23	27
25	41
	46
	18

a) Identify the sample size of each sample.
b) What is the median for each sample?
c) Identify W.
d) Is W significant?
e) What is the p-value for W?

36. The following collections of test scores have been randomly sampled from two distinct groups of students taking a reading placement exam.

Group 1	Group 2
78	86
29	65
74	78
66	80
82	65
65	95
66	88
84	94
73	75
55	89

a) Combine the two groups into one and assign a ranking of the data values.
b) Using the assigned ranking of the data values for the combined group, list the data values of group 1 and their respective assigned ranks.
c) Using the formula for the Mann-Whitney test, calculate the test statistic z.
d) If the Dean of Students believed that the reading placement exam scores are higher for group 2 as compared to group 1, perform a hypothesis test comparing the placement exam scores for the two groups. Use $\alpha = 0.01$.

37. Suppose you want to compare two treatments of physical therapy of the lower back. You wish to determine whether treatment A and treatment B are not the same by using the Mann-Whitney test.
a) State the null and alternative hypotheses you would test.
b) Suppose you were able to obtain for the following independent random samples measures as to the effectiveness of each treatment.

Treatment A: 45, 47, 25, 32, 54, 24, 18, 19, 42, 33
Treatment B: 23, 42, 31, 29, 24, 24, 19, 18, 17, 21
Calculate W

c) Using the formula for the Mann-Whitney test, calculate the test statistic z.
d) Do these data indicate that there is a difference in treatments at $\alpha = 0.05$?

38. Use the Mann-Whitney test to compare the number of sick days taken per year by male and female employees of a certain company. Use the 0.05 level of significance. The following data represents two randomly selected samples of the number of sick days taken for employees of the company.

Males: 11, 2, 4, 4, 4, 8, 3, 7, 8, 4
Females: 2, 4, 1, 6, 5, 3, 2, 3, 8, 4

39. The following data represents the result from a preference test of 100 individuals in a taste test involving Cokey versus Pipsi diet soft drink.

preference results:
Cokey 45   Pipsi 55

If H_0 is: There is no difference in preference of one product exists over the other (i.e. p = 0.50). and If H_a is: There is a difference in preference for one product over the other (i.e. p ≠ 0.50). Do the sample data indicate a significant difference in preference of one product over the other at $\alpha = 0.05$?

40. Ron's Japanese restaurant has noted an increase in the number of no-shows for dinner reservations when the reservations are made by a male caller than when made by a female caller. For the past 10 Saturdays the number of no-show reservations when the call was made by a male caller versus a female caller has been recorded.

896 Chapter 17 Nonparametric Statistics

Number of No-Shows for Dinner

male caller	female caller
11	13
15	14
10	10
18	9
11	10
12	8
10	11
14	12
20	15
11	7

At the 0.05 level of significance can Ron conclude that there are more no-shows for dinner when the caller is male versus when the caller is female?

41. Rich, the research director for a micro brewery, wants to know if there is a difference in the distribution of the number of liters consumed by men in the 22 to 35 age category for their Big Foot versus their Little Foot premium beer. A careful record of the total number of liters consumed by these men was collected over an eight day period. The results are as follows:

total number of liters consumed

Big Foot: 252, 263, 279, 273, 271, 265, 257, 280

Little Foot: 262, 242, 256, 260, 258, 243, 239, 265

Can Rich conclude that there is a difference between the two distributions? (use $\alpha = 0.05$)

42. Two groups are randomly selected. One group is to be given an experimental drug that is to enhance learning, the other a placebo. The time in minutes necessary to learn a specific task is recorded for each subject. The results are as follows:

Experimental Group: 41, 36, 42, 39, 36, 48, 49, 48, 52

Placebo Group: 61, 48, 55, 49, 70, 63, 74, 82

Does the drug enhance learning? Use a 0.05 level of significance.

Section 17.4

43. When using the Kruskal-Wallis H test state a typical null and alternative hypotheses.

44. If you able to apply either the Kruskal-Wallis H test or ANOVA in an experiment, which is more desirable? Why?

45. A nonparametric test that is analogous to the Analysis of Variance test (ANOVA) is called the _____.

46. In the Kruskal-Wallis H Test, the null hypothesis, H_o, is: the populations are _____. While the alternative hypothesis, H_a, is stated as the populations are _____.

47. The test statistic, for a Kruskal-Wallis H test is denoted by _____ and computed using the formula _____.

48. In the Kruskal-Wallis H test, the critical chi-square value, χ^2_α, is dependent upon the significance level, α, and the formula for the degrees of freedom $df = $ _____ where the number of distinct samples is denoted by _____.

49. The decision rule of the Kruskal-Wallis H test is: reject H_o and accept H_a if the _____ is larger than the _____.

50. The quantity ΣR_2 represents the sum of the _____ for sample _____, while n_3 equals the size of sample _____ and N equals the combined number of data values for _____ samples.

51. The Kruskal-Wallis H Test is a nonparametric statistic that is used to test whether populations are approximately normally distributed. (T/F)

52. The binomial approximation by the normal distribution is used to calculate the H statistic. (T/F)

53. The Kruskal-Wallis H test is use to test whether three or more populations are the same. (T/F)

54. Use the Kruskal-Wallis H test statistic to compute the value of H for each of the following samples:

a)
X	Y	C
93	84	88
86	81	66
90	78	71
73	83	75
86	60	68
96	71	

b)
A	B	C	D
8.3	9.8	13.2	11.4
10.1	13.2	7.3	10.8
10.3	11.1	8.1	9.0
13.8	7.6	9.5	12.1
14.1	13.4	11.4	8.9
7.8	12.6	8.6	11.1
79	12.8	7.9	

55. For the following data, use the the Kruskal-Wallis H Test to determine if the test statistic H is significant.

sample 1	sample 2	sample 3
30	68	56
75	25	77
65	35	64
90	22	84
98	43	96
95	36	71
88		79

a) Apply the Kruskal-Wallis H Test.
b) Identify the value of H.
c) Identify the df for the test.
d) Is H significant at 0.05? at 0.01?
e) If H is significant write the conclusion in terms of the null and alternative hypotheses.
f) What is the p-value for H?

56. The following are the final examination scores of samples of students who are taught statistics by three different methods:

Method A: traditional lecture with no emphasis on the use of technology, Method B: traditional lecture with emphasis on the use of technology with no laboratory requirement, and Method C: traditional lecture with an emphasis on the use of technology with a laboratory requirement.

Method A: 88, 66, 71, 75, 68
Method B: 84, 81, 78, 83, 60, 71, 80
Method C: 95, 88, 91, 75, 88, 97

Use the Kruskal-Wallis H test to test the null hypothesis that the three populations are the same against the alternative hypothesis that they are not the same. Use $\alpha = 0.05$.

57. Samples of annual income in thousands of dollars for police officers of patrolmen rank representing three major metropolitan cities are listed below. Use the Kruskal-Wallis H test to determine if the data indicate significant differences in annual income among the police officers representing the three cities at $\alpha = 0.05$.

NYC	LA	Boston
46	38	34
56	46	38
42	41	43
44	58	39
53	48	44

58. The following are the number of hours worked per week for a part-time job for students majoring in either Liberal Arts, Science or Business.

Liberal Arts: 23, 18, 35, 42, 22, 38
Science: 14, 9, 11, 25, 16
Business: 19, 22, 32, 39, 25, 48

Use the H test at a 0.05 level of significance to test the null hypothesis that there is no difference in the number of hours worked per week for a part-time job for the students majoring in Liberal Arts, Science or Business.

59. The number of patients treated in the emergency room of three area hospitals during the holiday week of July 4 are as follows:

Mercy Hospital: 157, 160, 178, 109, 99, 78, 77
Nassau Hospital: 180, 178, 190, 120, 102, 88, 90
Helpme Hospital: 140, 160, 128, 111, 98, 67, 40

Using an 0.01 level of significance, test the null hypothesis that the number of patients treated at the hospitals is the same.

CHAPTER REVIEW

60. Which one is *not* a nonparametric test?
    a) Sign Test
    b) Chi-Square
    c) Mann Whitney
    d) Kruskal-Wallis
    e) none of these

61. What are some of the disadvantages of using nonparametric tests over parametric tests?

62. The normal distribution can be used to approximate the binomial distribution if the quantities _____ and _____ are both at least 10.

63. The sign test for median
    a) stipulates how to assign each data value a "+", a "–" or a "0".
    b) uses the binomial distribution
    c) assumes equal variances
    d) assumes the data values come from a normal distribution
    e) does all of the above
    f) none of the above
    g) does a and b
    h) does a, b, and c
    i) does a, b, and d

64. For the Mann-Whitney Test, the null hypothesis is stated as:
    H_0: The two populations are _____ _____. While the alternative hypothesis is written as: H_a: The two populations are _____

65. When using the Mann-Whitney Rank Sum Test the normal distribution can be used as a model for the test if
    a) each sample contains at least 5 data values
    b) at least 80% of the samples contain 5 data values
    c) each sample contains at least 10 data values
    d both np and n(1 – p) are greater than 5
    e) at least 80% of the samples contain 10 data values

66. When using the Mann-Whitney Rank Sum Test, the null hypothesis is based on the assumption that for each sample the sum of the ranks are equal.
    (T/F)

67. Identify the conditions to use the Kruskal-Wallis H test.

68. In the Kruskal-Wallis H test, the value of the test statistic is compared to the critical _____ to determine if the populations are the same.

69. To reject the null hypothesis in a Kruskal-Wallis H Test means that
    a) H is larger than W
    b) W is significant at $\alpha = 0.05$
    c) H is larger than $\alpha = 0.05$
    d) the populations from which the samples were drawn are not the same
    e) W is greater than the critical χ_α^2 value at $\alpha = 0.05$

70. The Kruskal-Wallis H test
    a) uses the chi-square distribution
    b) uses the binomial distribution
    c) is also called the H test

d) is similar to the counterpart parametric ANOVA test
e) combines all the data values into one group
f) assumes the data values come from a normal distribution
g) all of the above
h) none of the above
i) does only a and b
j) does only a, b, and c
k) does only a, b, and d
l) does only a, c, and d
m) does only a, c, d, and e
n) does only b, c, d, e, and f

71. A medical researcher believes that the effects of a particular drug therapy will begin to appear approximately 12 days after regular use of the drug begins. The researcher randomly selects 24 patients that will be administered the drug therapy. The patients begin to report drug related effects for the following number of days after the drug therapy was administered: 10, 14, 13, 18, 11, 12, 9, 13, 10, 14, 17, 20, 17, 12, 16, 15, 9, 16, 17, 15, 18, 19, 11 and 14 days.

Use the Sign Test for Medians to test the null hypothesis that the population median number of days that the effects of the drug will begin to appear is 12 days against the claim that it is more than 12 days. Use $\alpha = 0.05$.

72. Big Joe's Pizza Palace believes families order more pies with toppings than without toppings. A sample of 20 orders are collected where the number of pies containing toppings versus without toppings is compared. The results are listed as follows:

number of pies ordered

with topping	without topping
3	2
4	1
3	4
2	1
1	3
5	2
3	1
2	3
1	2
2	1

At the 0.05 level of significance, can Joe conclude that more pies are order with toppings than without toppings?

73. A young new automobile driver suspects that the annual median cost for automobile insurance in his geographic region is approximately $2000. His parents believe that the annual cost is greater than $2000. He samples 23 insurance companies and finds the following insurance quotes for automobile insurance in his region:
$2500 $2350 $2350 $1920 $1936 $2008 $1995 $2300
$2100 $2000 $2200 $1900 $1924 $2550 $2320 $2036
$2108 $1945 $2000 $2040 $2010 $2440 $1850

Use the Sign Test for Medians to test the null hypothesis that the population median cost for automobile insurance is $2000 against the claim that it is more than $2000. Use $\alpha = 0.05$.

74. Bobby's music store has three mall locations. For the last ten weeks, Bobby has kept track of the number of customers who make a purchase of at least $100. The sample of those data are represented below. Use the Kruskal-Wallis H test to determine if there is a significant difference in the number of purchases of at least $100 for the three locations?
New York Mall: 98 63 100 84 78 87 98 94 90 99
Huntington Mall: 84 101 124 60 90 95 93 88 92 74
New Jersey Mall: 88 97 57 106 87 91 86 100 76 90

75. Two independent groups of nursing students have been instructed on the proper techniques of administering an upper arm injection. Group I was instructed using a preprogrammed video instruction tape while Group II was instructed by a traditional instructional lecture technique. The following data represents the number of minutes needed by each group's members to learn the procedure.

time to learn procedure in minutes
Group I 71 82 51 33 45 46 52 63 48 77
Group II 28 44 32 19 22 61 23 34 42 47

Use the Mann-Whitney Test to determine if the population mean time for learning the procedure is different between the groups. (Use $\alpha = 0.01$).

76. Use the Kruskal-Wallis H test to compare the probability distributions of three populations. The following three random samples were selected from each of the populations.
Pop I: 36, 58, 67, 61, 84, 72, 47
Pop II: 26, 20, 29, 43, 36, 44, 35
Pop III: 74, 103, 93, 78, 82, 76

Determine if there is a significant difference in the probability distributions for the three populations at $\alpha = 0.01$?

77. The human resource manager at a large municipality is interested in conducting a sexual awareness seminar to its employees. Before scheduling the seminar sessions, the manager is interested in finding out whether different groups of workers within the company are equally knowledgeable about sexual awareness. The manager randomly samples individuals from each group of company workers and administer's a test to each worker to determine if there are significant differences between the groups. The results of the tests are in the following table. Use these data and conduct a Kruskal-Wallis H test at $\alpha = 0.01$ to determine if the distributions of the sexual awareness scores for each group are equal.

Group 1	Group 2	Group 3	Group 4
42	71	43	29
51	74	58	45
50	69	46	49
61	80	50	56
68	81	38	52
38	75		46

WHAT DO YOU THINK?

78. The following information is based on the 1990 U.S. Census Bureau survey data. The data represents age of housing for some randomly selected Nassau County communities on Long Island, New York.

Community	age of housing (years)	Community	age of housing (yrs)
Baldwin	47.3	Levittown	41.2
Bellmore	35.7	Long Beach	35.0
Carle Place	39.5	Lynbrook	49.0
East Meadow	35.9	Malverne	48.0
E. Rockaway	42.6	Massapequa	35.3
Elmont	39.9	Merrick	35.7
Garden City	42.5	Oceanside	35.9
Glen Cove	35.8	Rockville Centre	48.8
Great Neck	42.5	Roosevelt	37.4
Hempstead	37.6	Uniondale	39.6
Hewlett	41.4	Wantagh	36.4
Lawrence	42.7	Westbury	38.5

The Census Bureau states that the median age of housing (years) for Nassau County in 1990 was 37.6. Using this value, and the sign test for medians:

a) what sign would be assigned to the community of Elmont?
b) what sign would be assigned to the community of Merrick?
c) what sign would be assigned to the community of Hempstead?
d) what is the n(– signs)? n(+ signs)? usable sample size (n)?
e) if you are interested in determining whether the Census Bureau's statement about the median age is reasonable, how would you state the alternative hypothesis?
f) how would you state the null hypothesis?
g) would it be acceptable to use the normal approximation to the binomial distribution?
h) what is the mean of the binomial distribution? the standard deviation?
i) what are the critical z score(s) for a significance level of $\alpha = 0.05$?
j) what is your statistical result? Explain why you chose this value for the statistical result?
k) can you reject the null hypothesis at $\alpha = 0.05$?
l) do you find the Census Bureau's statement about the median age for all Nassau County Communities reasonable? Explain.

79. In the Journal of Personality and Social Psychology (volume 54, 1988), a study entitled "*The Dark Side of Self- and Social Perception: Black Uniforms and Aggression in Professional Sports*" was conducted to determine if the color of a football team's uniform within the National Football League (NFL) has a significant effect on a person's perception of the team's behavior. Twenty-five subjects were selected to rate the uniforms of all the 28 teams in the NFL on several scales from one to seven. The researchers were interested in determining if there was a perception between black uniforms and aggressive behavior. Five of the 28 NFL teams have uniforms which are considered to be black. These teams are: Chicago Bears, Cincinnati Bengals, New Orleans Saints, Pittsburgh Steelers and the Los Angeles Raiders. Table 17.18 represents the average malevolence ratings on three semantic differential scales: nice/mean, timid/aggressive, and good/bad. Notice the top five teams within Table 17.18 all wear a uniform considered to be black.

TABLE 17.18

Football Team	Rating
Los Angeles Raiders	5.10
Pittsburgh Steelers	5.00
Cincinnati Bengals	4.97
New Orleans Saints	4.83
Chicago Bears	4.68
Kansas City Chiefs	4.58
Washington Redskins	4.40

900 Chapter 17 Nonparametric Statistics

Team	Rating
St. Louis Cardinals	4.27
New York Jets	4.12
Los Angeles Rams	4.10
Cleveland Browns	4.05
San Diego Chargers	4.05
Green Bay Packers	4.00
Philadelphia Eagles	3.97
Minnesota Vikings	3.90
Atlanta Falcons	3.87
San Francisco 49ers	3.83
Indianapolis Colts	3.83
Seattle Seahawks	3.82
Denver Broncos	3.80
Tampa Bay Buccaneers	3.77
New England Patriots	3.60
Buffalo Bills	3.53
Detroit Lions	3.38
New York Giants	3.27
Dallas Cowboys	3.15
Houston Oilers	2.88
Miami Dolphins	2.80

a) How would you interpret a higher malevolence rating? a lower malevolence rating?
b) Separate the 28 NFL teams into two groups: the teams who wear black uniforms and the teams not identified as wearing black uniforms. If you wanted to perform a large-sample Mann-Whitney test to determine whether teams wearing black uniforms are perceived as being more aggressive, mean or bad, then state a null and alternative hypothesis for such a test.
c) Identify the data type for this test.
d) If the null hypothesis is true, then how would you expect the ranks to be distributed within each of the two groups? Does this appear to be true for the information contained in Table 17.18?
e) What rank would you assign the Cleveland Browns? the San Francisco 49ers?
f) What is the value of W for this Mann-Whitney test?
g) Can you conclude at $\alpha = 0.05$ that malevolence rating of the NFL teams that wear black uniforms is significantly higher? Can you conclude the same conclusion at $\alpha = 0.01$? Explain. What do you think has contributed to this conclusion?

80. The following article entitled *Order of the Draft Drawing* which appeared in the New York Times shows the order in which birth dates were selected during the 1969 draft lottery. On December 1, 1969 the lottery was held in the Selective Service headquarters in Washington, D.C. to determine the draft status of all 19 year-old males for the Vietnam Conflict. The 366 possible birth dates were inserted within a small capsule. The capsules were placed in a stationary drum beginning with January, then February and so on until all the birth dates of December were placed in the drum last. The first capsule drawn would have the highest draft priority, and so on. Table 17.19 shows the actual order in which the 366 days were selected.

Theoretically, if the capsules were adequately mixed, then the resulting draft order should have been random with respect to each month. That is, if the lottery were fair then we would expect a birth date in December should have the same chance of being selected early in the draft as a birth date for any other month, such as January. Examine the list of birth dates within the Table 17.19 which represents the actual order of the 1969 draft drawing.
a) If the lottery was random, then how would you expect the assigned ranks to be distributed within each month?
b) If you wanted to conduct a Kruskal-Wallis test, then how would you state the null hypothesis? the alternative hypothesis?
c) How many groups would you have within this test? What does each group represent? In the Kruskal-Wallis test, what would the values of k and N represent for this test?
d) What would the value of the test statistic H for this Kruskal-Wallis test?
e) Is this value of the H statistic significant at $\alpha = 0.01$? at $\alpha = 0.0001$?
f) Do you think the 1969 Draft Lottery was fair? Explain your answer. What do you think might have contributed to this result?
g) If you were going to conduct a draft lottery how would you conduct the draft to insure that the lottery would be fair?

TABLE 17.19

Order of the Draft Drawing

Special to The New York Times

WASHINGTON, *Dec. 1—Following is the order in which birth dates were drawn tonight in the draft lottery.*

1	Sept. 14	48	Aug. 8	95	Dec. 24	142	Aug. 12	189	Feb. 17	236	Feb. 24	283	Oct. 8	325	Jan. 10
2	April 24	49	Sept. 3	96	Dec. 16	143	Nov. 17	190	July 18	237	Oct. 11	284	July 10	326	May 22
3	Dec. 30	50	July 7	97	Nov. 8	144	Feb. 2	191	April 29	238	Jan. 14	285	Feb. 29	327	July 6
4	Feb. 14	51	Nov. 7	98	July 17	145	Aug. 4	192	Oct. 20	239	Mar. 28	286	Aug. 25	328	Dec. 2
5	Oct. 18	52	Jan. 25	99	Nov. 29	146	Nov. 18	193	July 31	240	Dec. 19	287	July 30	329	Jan. 11
6	Sept. 6	53	Dec. 22	100	Dec. 31	147	Apr. 7	194	Jun. 9	241	Oct. 19	288	Oct. 17	330	May 1
7	Oct. 26	54	Aug. 5	101	Jan. 5	148	Apr. 16	195	Sept. 24	242	Sept. 12	289	July 27	331	July 14
8	Sept. 7	55	May 16	102	Aug. 15	149	Sept. 25	196	Oct. 24	243	Oct. 21	290	Feb. 22	332	Mar. 18
9	Nov. 22	56	Dec. 5	103	May 30	150	Feb. 11	197	May 9	244	Oct. 3	291	Aug. 21	333	Aug. 30
10	Dec. 6	57	Feb. 23	104	June 19	151	Sept. 29	198	Aug. 14	245	Aug. 26	292	Feb. 18	334	Mar. 21
11	Aug. 31	58	Jan. 19	105	Dec. 8	152	Feb. 13	199	Jan. 3	246	Sept. 18	293	Mar. 5	335	June 9
12	Dec. 7	59	Jan. 24	106	Aug. 9	153	July 22	200	Mar. 19	247	June 22	294	Oct. 14	336	April 19
13	July 8	60	June 21	107	Nov. 16	154	Aug. 17	201	Oct. 23	248	July 11	295	May 13	337	Jan. 22
14	April 11	61	Aug. 29	108	March 1	155	May 6	202	Oct. 4	249	June 1	296	May 27	338	Feb. 9
15	July 12	62	April 21	109	June 23	156	Nov. 21	203	Nov. 19	250	May 21	297	Feb. 3	339	Aug. 22
16	Dec. 29	63	Sept. 20	110	June 6	157	Dec. 3	204	Sept. 21	251	Jan. 3	298	May 2	340	April 26
17	Jan. 15	64	June 27	111	Aug. 1	158	Sept. 11	205	Feb. 27	252	April 23	299	Feb. 28	341	June 18
18	Sept. 26	65	May 10	112	May 17	159	Jan. 2	206	June 10	253	April 6	300	Mar. 12	342	Oct. 9
19	Nov. 1	66	Nov. 12	113	Sept. 15	160	Sept. 22	207	Sept. 16	254	Oct. 16	301	June 3	343	Mar. 25
20	June 4	67	July 25	114	Aug. 6	161	Sept. 2	208	April 30	255	Sept. 17	302	Feb. 20	344	Aug. 26
21	Aug. 10	68	Feb. 12	115	July 3	162	Dec. 23	209	June 30	256	Mar. 23	303	July 26	345	April 20
22	June 26	69	June 13	116	Aug. 23	163	Dec. 13	210	Feb. 4	257	Sept. 28	304	Dec. 17	346	April 12
23	July 24	70	Dec. 21	117	Oct. 22	164	Jan. 30	211	Jan. 31	258	Mar. 24	305	Jan. 1	347	Feb. 6
24	Oct. 5	71	Sept. 10	118	Jan. 23	165	Dec. 4	212	Feb. 16	259	Mar. 13	306	Jan. 7	348	Nov. 3
25	Feb. 19	72	Oct. 12	119	Sept. 23	166	Mar. 16	213	Mar. 8	260	April 17	307	Aug. 13	349	Jan. 29
26	Dec. 14	73	June 17	120	July 16	167	Aug. 28	214	Feb. 5	261	Aug. 3	308	May 28	350	July 2
27	July 21	74	April 27	121	Jan. 16	168	Aug. 7	215	Jan. 4	262	April 28	309	Nov. 26	351	April 25
28	June 5	75	May 19	122	Mar. 7	169	Mar. 15	216	Feb. 10	263	Sept. 9	310	Nov. 5	352	Aug. 27
29	Mar. 2	76	Nov. 6	123	Dec. 28	170	Mar. 26	217	Mar. 30	264	Oct. 27	311	Aug. 19	353	June 29
30	Mar. 21	77	Jan. 28	124	April 13	171	Oct. 15	218	April 10	265	Mar. 22	312	April 8	354	Mar. 14
31	May 24	78	Dec. 27	125	Oct. 2	172	July 23	219	April 9	266	Nov. 4	313	May 11	355	Jan. 27
32	April 1	79	Oct. 31	126	Nov. 13	173	Dec. 26	220	Oct. 10	267	Mar. 3	314	Dec. 12	356	June 14
33	Mar. 17	80	Nov. 9	127	Nov. 14	174	Nov. 30	221	Jan. 12	268	Mar. 27	315	Sept. 30	357	May 26
34	Nov. 2	81	April 4	128	Dec. 18	175	Sept. 13	222	Jan. 25	269	April 5	316	April 22	358	June 24
35	May 7	82	Sept. 5	129	Dec. 1	176	Oct. 25	223	Mar. 25	270	July 29	317	Mar. 9	359	Oct. 1
36	Aug. 24	83	April 3	130	May 15	177	Sept. 19	224	Jan. 6	271	April 2	318	Jan. 13	360	June 20
37	May 31	84	Dec. 23	131	Nov. 15	178	May 14	225	Sept. 1	272	June 12	319	May 23	361	May 25
38	Oct. 30	85	June 7	132	Nov. 25	179	Feb. 25	226	May 29	273	April 15	320	Dec. 18	362	Mar. 29
39	Dec. 11	86	Feb. 1	133	May 12	180	June 13	227	July 19	274	June 16	321	May 8	363	Feb. 21
40	May 3	87	Oct. 6	134	June 11	181	Feb. 8	228	June 2	275	Mar. 4	322	July 15	364	May 5
41	Dec. 10	88	July 28	135	Dec. 20	182	Nov. 23	229	Oct. 29	276	May 4	323	Mar. 10	365	Feb. 26
42	July 13	89	Feb. 18	136	Mar. 11	183	May 20	230	Nov. 24	277	July 9	324	Aug. 11	366	June 8
43	Dec. 9	90	April 18	137	June 25	184	Sept. 3	231	April 14	278	May 13				
44	Aug. 16	91	Feb. 7	138	Oct. 13	185	Nov. 20	232	Sept. 4	279	July 4				
45	Aug. 2	92	Jan. 26	139	Mar. 6	186	Jan. 21	233	Sept. 27	280	Jan. 20				
46	Nov. 11	93	July 1	140	Jan. 18	187	July 20	234	Oct. 7	281	Nov. 28				
47	Nov. 27	94	Oct. 28	141	Aug. 18	188	July 5	235	Jan. 17	282	Nov. 10				

Following is the order of the alphabet to be applied to the first letter of last names in determining the order of call for inductees with the same birth dates:
J, G, D, X, N, O, Z, T, W, P, Q, Y, U, C, F, I, K, H, S, L, M, A, R, E, B, V

PROJECTS

81. Find examples of the uses of nonparametric statistics in a professional journal.
  a) Identify the type of test that is demonstrated.
  b) Do the authors give a rationale for the use of the statistics? Explain.

82. a) Identify five quantitative variables that you are able to measure.
  b) For each variable randomly sample 250 data values.
  c) For each sample construct a Stem-and-Leaf display.
  d) Do the sample results indicate that the populations from which the data were sampled suggest the use of nonparametric statistics? Explain.

CHAPTER TEST

83. Which assumptions are necessary to use a parametric test?
  I. The distributions from which the samples are selected are not symmetric.
  II. The population variances must be equal.
  III. The populations are normally distributed.
  a) I only
  b) II only
  c) III only
  d) I and II
  e) II and III

84. The nonparametric counterpart test to the t-test about a population mean is called the _____.

85. The sign test for medians is analogous to the one sample t test. (T/F)

86. The Kruskal-Wallis H test is the nonparametric counterpart to the Analysis of Variance test. (T/F)

87. Everett Evergreen of Evergreen Entertainment hypothesized that the population median price a couple will spend on a wedding band is $4400. His partner Francis, believes that the claim is too high. They decide to randomly sample twenty-one recent wedding band contracts. The following data represent the result of the sample:

contract amount is dollars: 3850, 6000, 3600, 3900, 4200, 4700, 3900, 4100, 4400, 3750, 4200, 4500, 4600, 5500, 3800, 4200, 4800, 5200, 3900, 4000, 4100

Use the Sign Test for Medians to test Francis's claim against Everett's claim using a 0.01 level of significance.

88. The following collections of times have been randomly sampled from two distinct groups of runners performing in the 440m relay. Group 1 represents runners who have trained while under a specific high carbohydrate diet while Group 2 represents those runners who have trained using a balanced diet. The track coach believes that the runners that have been using the high carbohydrate diet will perform better in the next track meet.

Times in Seconds

Group 1	Group 2
54.2	64.1
53.8	55.6
54.9	58.1
60.1	53.4
63.4	55.6
51.9	53.2
56.4	58.4
54.2	64.1
53.7	52.2
55.6	55.6
60.5	

a) Combine the two groups into one and assign a ranking of the data values.
b) Using the assigned ranking of the data values for the combined group, list the data values of group 1 and their respective assigned ranks.
c) Using the formula for the Mann-Whitney test, calculate the test statistic z.
d) State a null and alternative hypothesis to test the coach's belief.
e) Do these data support the track coach's claim at $\alpha = 0.05$?

89. Frozen Yogurt may contain different amounts of calories for a 3 oz serving. The data in the table below show the calorie content from samples of three different brands of frozen yogurt. Test for significant differences in the calorie content of these three frozen yogurts. Are there significant differences at the 0.01 level of significance?

Yummy Frozen Yogurt: 105, 110, 102, 106, 104
Creamy Frozen Yogurt: 109, 115, 107, 109, 118
Dreamy Frozen Yogurt: 100, 104, 108, 112, 108

# CHAPTER 18

Bootstrapping Concepts & Methods: An Introduction

Contents

18.1 Introduction
18.2 Bootstrap Distribution
    Plug-In Principle
    Justification for the Bootstrap Procedure
    Bootstrap Procedure to Generate the
      Bootstrap Distribution
    Bootstrap Notation
    Using Bootstrapping to Estimate BIAS
    Comments about the Bootstrap Procedure
18.3 Estimating Standard Error Using Bootstrapping
    Revisiting the Bootstrap Idea
    Key Bootstrap Ideas
    Useful Characteristics of the Bootstrap Distribution
    Comparing the Bootstrap Distribution and
      the Sampling Distribution
18.4 Bootstrap Confidence Intervals
    Method 1: Bootstrap Percentile Confidence
      Interval Method
    Cautions about the Bootstrap Percentile
      Confidence Interval Method
    Method 2: Bootstrap Standard Error t
      Confidence Interval Method
    Method 3 Basic Bootstrap Confidence
      Interval Method (or the Reverse Bootstrap
      Percentile Confidence Interval)
    Important Bootstrap Concepts to Remember when
      Applying the Bootstrap Confidence Interval Methods
    Bootstrap Confidence Interval Accuracy

18.1 INTRODUCTION

As you recall from previous chapters, inferential statistics relies on the concept of the sampling distribution of a sample statistic. Previously, we examined the sampling distribution of the mean, the sampling distribution of the proportion, the sampling distribution of the difference between means, the sampling distribution of r, etc. to perform hypothesis testing.

In Chapter 8, the concept of a sampling distribution was first examined where samples of size n were randomly selected from the same population and a sample statistic was computed for each sample such as the sample mean or sample proportion. The theoretical distribution of all possible sample means, or sample proportions produced the Sampling Distribution of the Mean or Proportion respectively. The Theoretical Sampling Distribution model is illustrated in Figure 18.1.

Only available in the e-book.

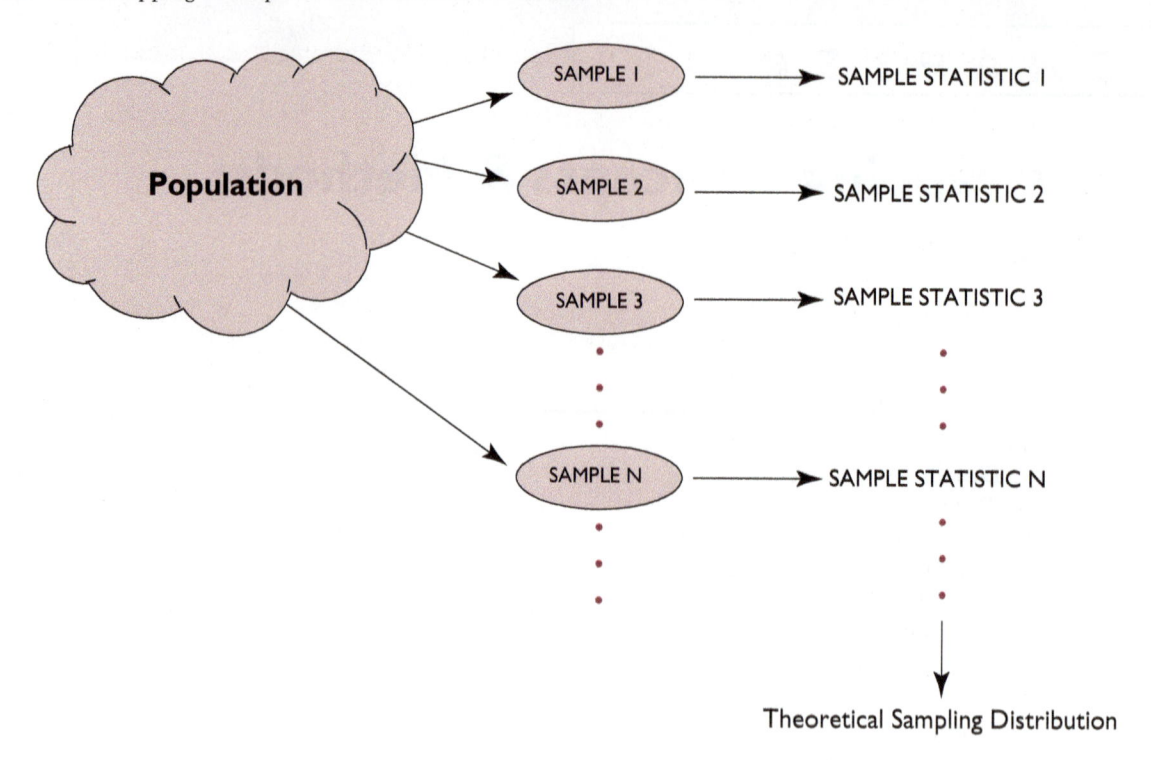

FIGURE 18.1

Specifically, before we could apply the central limit theorem to generate the theoretical sampling distribution of the mean, there were required assumptions about normality, or the size of the random samples selected from the population. If those assumptions are violated, the Chapter 8 methods pertaining to the sampling distribution of the mean may fail. Without knowledge of the sampling distribution of the mean, we are unable to conduct inferential statistics or generate confidence intervals pertaining to the mean.

There are real life situations when it is not possible to satisfy the required Chapter 8 assumptions about the population from which the random samples are selected to generate the characteristics of the theoretical sampling distribution. For example, the population might not be a normal distribution, or it is not possible to randomly select a sample of size greater than 30 when the population standard deviation is unknown. Furthermore, we may be limited to selecting only one small random sample of data due to the cost of sampling or required time to select the sample. Or what could one do if we wanted to construct a confidence interval for a median, or a correlation coefficient or a standard deviation for which there is no statistical theory to apply to generate such confidence intervals?

Consequently, in such instances, a statistical technique called *bootstrapping* has become very popular as an alternate approach that can be applied for a broad range of sample statistics to approximate the sampling distribution of a statistic, to compute the standard error of the sampling distribution of a statistic and to generate confidence intervals.

18.2 BOOTSTRAP DISTRIBUTION

The bootstrap technique was first introduced by Bradley Efron back in 1979 as a new technique to estimate the standard error of a sampling distribution when the usual statistical methods do not apply. The main benefit of the bootstrap technique is that it permits statisticians to not only compute estimates for the standard error of a sampling distribution but to also construct confidence intervals on parameters without having to make unreasonable assumptions about the population as well as to generate confidence intervals for a median, a correlation coefficient, standard deviation or other parameters for which there are no theoretical statistical methods.

The objective of this chapter will be to introduce you to this bootstrapping technique as a method of generating a sampling distribution from just one random sample and to examine two of the most common applications of bootstrapping: estimating the standard error of a sampling distribution and constructing a confidence interval for a statistic of interest.

Bootstrapping is an alternative statistical simulation technique to the traditional statistical technique of assuming a population has a particular probability distribution as discussed in Chapter 8 when the population was assumed to be approximately normally distributed. However, there may be instances where this is clearly not the case. In fact, there may not be any agreement on what probability distribution would be acceptable for the population. Thus, bootstrapping can be applied to generate an empirical sampling distribution without knowing the type of distribution from which a sample was selected. That is, when the standard parametric assumptions about the population are unknown. This type of bootstrap method that doesn't require any assumptions about the population is called *nonparametric bootstrapping*.

> **DEFINITION 18.1 NONPARAMETRIC BOOTSTRAPPING**
> Nonparametric Bootstrapping refers to a bootstrap method that doesn't make any assumptions about the population distribution.

Although the nonparametric bootstrap method makes no assumptions about the population distribution, it is a good method to use when you only have a limited random sample data selected from a population. However, this bootstrap method will require the use of a computer software package due to the numerous computations required to generate an empirical sampling distribution by mimicking the process of randomly selecting samples from an estimate of the population or a "proxy" population. The following definitions describe important concepts pertaining to the bootstrap method.

> **DEFINITION 18.2 BOOTSTRAPPING**
> *Bootstrapping* is a computational approach where samples are selected with replacement from an estimate of the population called the *original sample* to generate an empirical sampling distribution.

The random sample serving as the proxy population is called the original sample.

> **DEFINITION 18.3 ORIGINAL SAMPLE**
> Original sample is the only available sample information randomly selected from a population and serves as the best possible estimate of the population. It can be thought of as a "proxy population".

The sampling process is called resampling.

> **DEFINITION 18.4 RESAMPLING**
> The process of selecting *with replacement* a random sample from the original sample where the selected sample is the same size as the original sample is called *resampling*.

The resampling process generates bootstrap samples.

> **DEFINITION 18.5 BOOTSTRAP SAMPLE**
> A bootstrap sample is a random sample selected (or "bootstrapped") from the original sample having the same size as the original sample.

A bootstrap statistic is then computed for each resample.

> **DEFINITION 18.6 BOOTSTRAP STATISTIC**
> The sample statistic computed from a bootstrap sample is called a bootstrap statistic.

All the bootstrap sample statistics form an empirical sampling distribution.

> **DEFINITION 18.7 BOOTSTRAP DISTRIBUTION (OR BOOTSTRAP SAMPLING DISTRIBUTION)**
> The empirical sampling distribution of all the bootstrap sample statistics is called the *bootstrap distribution*.

Figure 18.2 illustrates the **bootstrap procedure** where n bootstrap random samples of the same size as the original sample are repeatedly selected, with replacement, from the data set of the original sample and then a bootstrap sample statistic for each selected resample is computed. The empirical sampling distribution of all the bootstrap statistics is called the bootstrap sampling distribution or just the bootstrap distribution.

Bootstrap Procedure

| Population | Original Random Sample Data Set (Selected from Population) (Serves as a "proxy population") | Bootstrap Samples (Selected from Original Sample) | Bootstrap Statistics |

POPULATION → ORIGINAL SAMPLE → Bootstrap sample 1 → Bootstrap Statistic 1
→ Bootstrap sample 2 → Bootstrap Statistic 2
→ Bootstrap sample 3 → Bootstrap Statistic 3
⋮
→ Bootstrap sample n → Bootstrap Statistic n
↓
Bootstrap Distribution
(an empirical sampling distribution)

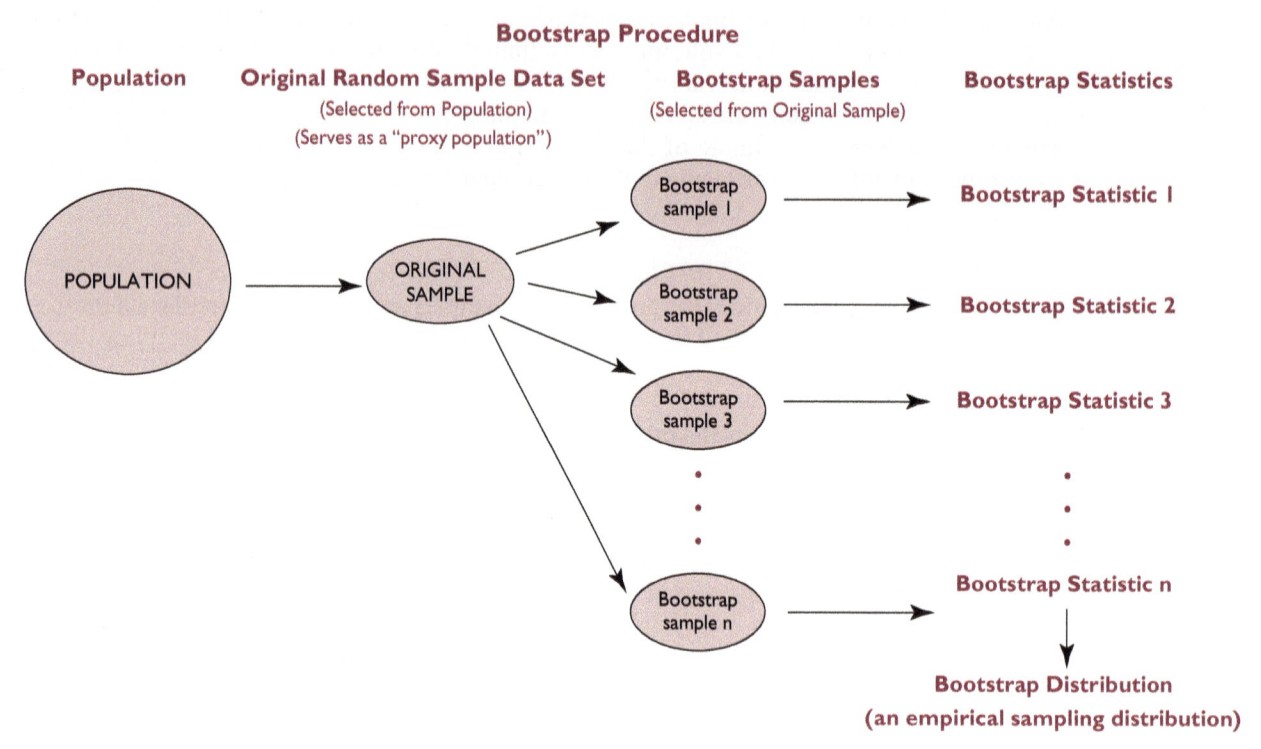

FIGURE 18.2

The bootstrap procedure illustrated in Figure 18.2 is a powerful resampling statistical technique that uses resampling with replacement to approximate the sampling distribution of a statistic such as the mean, median, correlation coefficient, regression coefficients, ... by selecting numerous resamples and calculating the sample statistic for each resample. Since this bootstrap procedure is very computationally intensive, it is necessary to rely on the use of computers to perform the numerous calculations to generate an empirical sampling distribution in place of the theoretical sampling distribution. The bootstrapping process substitutes intensive computational calculations to mimic the theoretical sampling distribution of a statistic.

In the bootstrap procedure, the original random sample serves as our best estimate of the population from which the resamples are actually randomly selected to generate an empirical bootstrap sampling distribution as opposed to hypothetically drawing samples from the true population to form the Theoretical Sampling Distribution of a statistic as presented in Chapter 8. It is important to know that a large (in size) original sample will reduce the variation in a bootstrap distribution. Bootstrapping cannot overcome the shortcoming of a small original sample.

Typically, when something is unknown, we substitute an estimate for it. This principle is very familiar to statisticians when they substitute the value of the standard deviation of a random sample for the unknown standard deviation value of the population. With the bootstrap procedure, we go one step farther—instead of plugging in an estimate for a single parameter, we plug in an estimate for the whole population.

Plug-In Principle

In previous chapters, when a population parameter was unknown, we substituted a sample statistic as an estimate for the population parameter within a statistical formula. However, in the bootstrap procedure, we take this plug-in principle one step beyond when an entire sample estimate called the original sample is substituted for an entire population.

As observed, the bootstrap procedure uses the original sample as a "proxy" population since the actual population distribution is unknown to generate the bootstrap distribution. That is, the bootstrap procedure is replacing the actual population with the original random sample. The original sample that serves as an estimate for the population is being substituted or *"plugged-in"* for the actual population. This *"plug-in principle"* concept has been applied throughout the text. The sample mean was used as an estimate of the population mean and the population proportion was estimated by the sample proportion. Furthermore, in Chapter 9 when the population standard deviation, σ, was unknown, we used the sample standard, s, as an estimate for the population standard deviation in the standard error of the sampling distribution of the mean formula. That is, the standard error of the mean is computed using the formula: $\frac{\sigma}{\sqrt{n}}$. However, when the population standard deviation is unknown, we substituted s for σ in the formula to obtain an estimate for the standard error of the mean to obtain: $\frac{s}{\sqrt{n}}$. This process of replacing the actual population parameter by the corresponding sample statistic is referred to as the *plug-in principle*.

> **DEFINITION 18.8 PLUG-IN PRINCIPLE**
> The plug-in principle refers to the process of using a sample statistic to estimate the corresponding population parameter.

The key idea behind the bootstrap procedure is that the original random sample is being used as the best possible estimate of the true population and can be thought of as a "proxy population". Thus, the "plug-in principle" is an essential component of the bootstrap procedure since the bootstrap procedure is based on substituting the original sample in for the true population. Since we are unable to select samples from the true population, the bootstrap procedure selects random samples from an estimate of the true population, the original sample, to generate the bootstrap distribution.

You might ask, "Does this bootstrap procedure generate accurate results for the empirical sampling distribution when you apply the plug-in principle to replace the true population with the original sample?"

Justification for the Bootstrap Procedure

The bootstrap procedure does work but the statistical proof to verify this procedure is beyond the scope of an introductory statistics text. In the bootstrap procedure, resamples are selected from the original sample to generate an empirical bootstrap sampling distribution. It is then necessary to take sufficient resamples to obtain a good estimate of the theoretical sampling distribution. As a general guideline, 1,000 resamples is generally considered adequate to obtain a good estimate. However, the more resamples selected will improve the accuracy of the bootstrap procedure. For more accurate results, 10,000 resamples or even 100,000 bootstrap samples should be used. The reason for selecting a large number of bootstrap samples is the bootstrap procedure is essentially based on the law of large numbers which states with enough data the empirical bootstrap distribution of a statistic will serve as an appropriate estimate of the theoretical sampling distribution of the statistic.

The bootstrap procedure has increased in usage over the years and various research have concluded that bootstrap distributions are good approximations of the true sampling distributions. Furthermore, the bootstrap procedure is easier to understand than the complex traditional statistical methods that rely on the assumption you know the true population from which the sample was selected and the theoretical formulas to generate sampling distributions. The bootstrap procedure works surprisingly well for a wide variety of unknown distributions and doesn't require complex formulas to generate the empirical sampling distribution.

Bootstrap Procedure to Generate the Bootstrap Distribution

The following four steps highlight the bootstrap procedure to generate a bootstrap distribution.

1. Start with the **original random sample** serving as the best possible estimate of the population. That is, it serves as a "proxy" population.
2. Select **n resamples** with replacement from the original sample and **having the same sample size** as the original sample to generate **n bootstrap samples**. The resamples have the same size as the original sample since the variability of the bootstrap statistic will depend on the size of the resamples. Using the same sample size as the original sample, the standard errors reflect the actual sample data rather than a hypothetical larger or smaller sample dataset.
3. Compute the sample statistic called a **bootstrap statistic** for each of the n bootstrap samples.
4. The empirical sampling distribution of all the bootstrap statistics represents the **bootstrap distribution**.

The term "Bootstrap" refers to the saying "to pull oneself by one's own bootstrap" which is thought to be based on one of the 18th Century famous *Adventures of Baron Munchausen* by Rudolph Erich Raspe. According to one of the adventures, the Baron had fallen into a swamp and saved himself by pulling himself up by his own bootstraps. In statistical applications, bootstrap means generating a bootstrap distribution by sampling from an existing original random sample itself without using any other external samples.

Bootstrap Notation

The following statistical notation is used to symbolize bootstrap samples and bootstrap statistics.

1. The number of bootstrap samples is typically symbolized by the capital letter B.
   For example, if 1000 resamples are selected from the original sample, then B = 1000.
2. The data values of a bootstrap sample are identified by an "asterisk notation" where an asterisk is placed in the exponent position of the symbol representing the data value.
   For example, if the original sample has n data values symbolized as: $x_1, x_2, x_3, \ldots, x_n$ then the n data values of the bootstrap sample would be identified as: $x_1^*, x_2^*, x_3^*, \ldots, x_n^*$.
3. An asterisk placed in the exponent position of a sample statistic will indicate the statistic was computed from a bootstrap sample. For example, the sample mean of a bootstrap sample would be symbolized as $\bar{x}^*$ and the sample standard deviation of a bootstrap sample is identified as s^*.

Before we apply the bootstrap procedure to generate a bootstrap distribution, it is important to clarify the resampling process. The resampling process with replacement requires that the randomly selected data value be placed back into the original sample before the next data value of the bootstrap sample is selected from the original sample. When the first data value has been selected from the original sample, it is *not* deleted from the original sample. It is *replaced* back into the original sample. The next data value of the bootstrap sample is then randomly selected from the original sample for the resample and then returned back to the original sample. This resampling process is repeated until the resample data set of size n is generated for the bootstrap sample. Consequently, the sample data value from the original sample may be included in the bootstrap sample data set once, twice, or more times or possibly not at all.

As an example, let's assume our original sample data set is:

$$x_1 = 72, x_2 = 57, x_3 = 46, x_4 = 64, x_5 = 55$$

and we decide to select bootstrap samples of size 5. It is possible to randomly select a bootstrap sample containing the following five numbers when *sampling with replacement*:

$$x_1^* = 55, x_2^* = 64, x_3^* = 72, x_4^* = 57, x_5^* = 57.$$

Notice, in this bootstrap sample, the data value 57 was randomly selected twice since the bootstrap sample was selected *with replacement* but the data value 46 was not selected for this resample. It would also be possible to select a bootstrap sample where all the data values randomly selected are the same value such as:

$$x_1^* = 55, x_2^* = 55, x_3^* = 55, x_4^* = 55, x_5^* = 55$$

since the selected data value is replaced back into the original sample before the next value is randomly selected.

Using Bootstrapping to Estimate BIAS

As previously mentioned, the bootstrap procedure will be used to generate the empirical bootstrap distribution of a statistic to estimate the true sampling distribution of the statistic to help us estimate standard errors and construct confidence intervals. In addition, we will examine the shape of the empirical bootstrap distributions to approximate the shape and spread of the true sampling distributions. However, the empirical bootstrap distribution cannot be used to determine the center of the true sampling distribution since the bootstrap distribution of a statistic is centered at the statistic of the original random sample and not the population parameter. But we can use the empirical bootstrap distribution to check for bias.

> **DEFINITION 18.9 BIAS OF A STATISTIC**
> The bias of a statistic is the mean of the sampling distribution minus the value of the population parameter.

When a statistic is used to estimate a population parameter and the sampling distribution is not centered at the value of the true population parameter then the statistic is biased. The bootstrap procedure can be used to check for bias by detecting whether the empirical bootstrap distribution of the statistic is centered at the statistic of the original random sample. Thus, we can use the bootstrap procedure to estimate the bias of an estimator by subtracting the statistic for the original data from the mean of the empirical bootstrap distribution.

Let's examine this idea further. Let Θ be the value of an unknown population parameter of interest. Let $\hat{\Theta}$ be the statistic of interest computed from the original sample where the original sample data set of size n is defined as $x_1, x_2, x_3, \ldots, x_n$. The resamples with replacement of size n selected from the original sample will be denoted as $x_1^*, x_2^*, x_3^*, \ldots, x_n^*$. Then we can estimate Θ by computing the bootstrap statistic for each resample. Let's denote each bootstrap statistic estimate of Θ by $\hat{\Theta}^*_i$, where the subscript i represents the ith resample. Then the empirical sampling distribution $\hat{\Theta}^*$ is the bootstrap distribution of B statistics denoted by: $\hat{\Theta}^*_1, \hat{\Theta}^*_2, \hat{\Theta}^*_3, \ldots, \hat{\Theta}^*_B$ and serves as an approximation of the true sampling distribution of the statistic $\hat{\Theta}$.

The bias of the estimator $\hat{\Theta}$ is defined as: Bias $(\hat{\Theta}) = E[\hat{\Theta} - \Theta] = E[\hat{\Theta}] - \Theta$. That is, the bias is a measure of a systematic error since $\hat{\Theta}$ tends to be either smaller or larger than Θ. To obtain the bootstrap estimates of bias, we plug the original sample statistic estimate $\hat{\Theta}$, which is a constant, in for the value of the unknown population Θ within the Bias formula. In addition, we replace $\hat{\Theta}$ by the empirical bootstrap distribution $\hat{\Theta}^*$. This leads us to the following approximation:

$$\text{Bias }(\hat{\Theta}) \approx \frac{1}{B}\Sigma_{i=1}^{B} \hat{\Theta}^*_i - \hat{\Theta} = \overline{\hat{\Theta}^*} - \hat{\Theta}$$

Thus, the bias is estimated by how much the original sample statistic estimate $\hat{\Theta}$ deviates from the mean of the bootstrap sample statistics.

> **DEFINITION 18.10 BOOTSTRAP ESTIMATE OF BIAS**
> The bootstrap estimate of bias is the mean of the bootstrap sample statistics minus the original sample statistic.

It is important to understand that the center of the empirical bootstrap distribution and the center of the true sampling distribution are different. The true sampling distribution of a statistic used to estimate a population parameter is centered at the actual value of the population parameter, plus any bias. However, the empirical bootstrap distribution that is generated by resampling from the original random sample without replacement is centered at the value of the original random sample statistic, plus any bias. Although the location of the two centers are not equal, their biases are similar.

Comments about the Bootstrap Procedure

1) Bootstrap technique uses the plug-in principle to generate an empirical sampling distribution (bootstrap distribution) to approximate the theoretical sampling distribution of the statistic of interest by using the original random sample as an estimate of the true population. The resampling procedure with replacement from the original sample mimics the process of randomly selecting samples from an infinite

population to reveal information about the sampling distribution of the sample statistic in question and about sampling error. It doesn't reveal any information about the true population.

2) Although the bootstrap distribution is a sampling distribution that will be used to estimate things about the theoretical sampling distribution, it is critical to understand that the bootstrap sampling distribution is centered at the original sample statistic and not the population parameter. The theoretical sampling distribution of the mean is centered at the mean of the population, µ. That is, a bootstrap distribution of the mean will be centered at the mean of the original sample, $\bar{x}$, and not the mean of the true population, µ. No matter how many resamples are selected to generate the bootstrap sampling distribution of the mean, the bootstrap procedure cannot be applied to improve on the original sample statistic estimate, $\bar{x}$, since the bootstrap sampling distribution is centered at $\bar{x}$.

3) Bootstrap sampling distributions are useful for estimating the shape, spread and bias of the true sampling distributions. The original sample should be reasonably large to generate reasonably accurate results. Since the bootstrap procedure generates results based on the number of resamples, you might get different results each time you run a simulation. Consequently, it is recommended to run as many resamples as is practical or until the simulations produce results that vary less than some acceptable level. As a general guideline, 1,000 resamples is generally considered adequate to obtain a good estimate. However, the more resamples selected will improve the accuracy of the bootstrap procedure. For more accurate results, 10,000 resamples or even 100,000 bootstrap samples should be used.

4) The bootstrap procedure can be applied to the median, correlation, standard deviation and other statistics to generate their sampling distributions that are difficult to describe theoretically.

5) The empirical bootstrap distributions approximate the shape, spread, and bias of the true sampling distribution of the statistic of interest.

18.3 ESTIMATING STANDARD ERROR USING BOOTSTRAPPING

In Chapter 8, we examined the theoretical sampling distribution of a statistic for both the mean and proportion. In general, the sampling distribution of a statistic is a probability distribution obtained by selecting all possible random samples of the same size from a specific population as shown in Figure 18.1. In reality, we are unable to select all possible random samples to construct a sampling distribution. We relied on statistical theory to help identify the sampling distribution model. For example, Theorem 8.1 states if the population being sampled is a normal distribution, then the sampling distribution of the mean is also a normal distribution. However, there are many situations when we don't have any information about the population so we must rely on the bootstrap procedure to generate the bootstrap distribution as a bootstrap estimate of the theoretical sampling distribution. Then we can use the bootstrap distribution to provide a bootstrap estimate of the standard error of the sampling distribution. The bootstrap procedure permits us to compute standard errors for sample statistics without needing to make any assumptions about the true population or using statistical formulas to compute standard errors. Rather we will use the bootstrap distribution to compute a bootstrap estimate of the standard error of the sampling distribution by computing the bootstrap standard deviation of the bootstrap distribution of the statistic of interest.

> **DEFINITION 18.11 BOOTSTRAP STANDARD ERROR OF A STATISTIC**
> The bootstrap standard error of a statistic denoted by $SE_b(statistic)$ is the standard deviation of the bootstrap distribution of that statistic.

If the statistic of interest is the sample mean, $\bar{x}$, then the bootstrap standard error for B number of resamples is computed by the formula:

$$SE_b(\bar{x}) = \sqrt{\frac{1}{B-1} * \sum (\bar{x}^* - \frac{1}{B}\sum \bar{x}^*)^2}$$

Where: $\bar{x}^*$ = the mean of a resample
B = number of resamples

Remark: The bootstrap standard error is just the standard deviation of the B resample means. Recall that the asterisk notation is used to distinguish the mean of a resample, $\bar{x}^*$, from the mean of the original sample, $\bar{x}$.

Section 18.3 Estimating Standard Error Using Bootstrapping

In Chapter 8, the sampling distribution of the mean helped to specify the degree to which means from different random samples can differ from each other and from the population mean. We were able to get a sense of how close a particular sample mean was likely to be to the population mean. The statistical measure used to determine how much sample means differ from each other was the **standard deviation of the sampling distribution of the mean**. This standard deviation was also called the **standard error of the mean**. Recall, if all the sample means of the sampling distribution were very close to the population mean, then the standard error of the mean would be a small value. On the other hand, if the sample means varied significantly, then the standard error of the mean would be a large value.

Let's apply the bootstrap procedure to generate a bootstrap distribution of the mean and determine a bootstrap estimate of the standard error of the sampling distribution of the mean by computing the standard error of the bootstrap distribution of the mean.

Suppose we are interested in generating a bootstrap distribution for the mean age of the Chief Executive Officer (CEO) of U.S. companies so we can determine a bootstrap estimate of the standard error of the mean. Consider that the only information we have about the age distribution of this CEO population is a random sample of 25 CEO ages displayed in Table 18.1. This random sample of 25 CEO ages represents the original sample and will serve as our best estimate of the true population of CEO ages.

TABLE 18.1 Original Random Sample of 25 CEO ages

$x_1 = 40$	$x_2 = 43$	$x_3 = 52$	$x_4 = 56$	$x_5 = 62$	$x_6 = 64$	$x_7 = 65$	$x_8 = 70$	$x_9 = 47$
$x_{10} = 50$	$x_{11} = 57$	$x_{12} = 61$	$x_{13} = 50$	$x_{14} = 72$	$x_{15} = 57$	$x_{16} = 46$	$x_{17} = 60$	$x_{18} = 55$
$x_{19} = 63$	$x_{20} = 42$	$x_{21} = 71$	$x_{22} = 47$	$x_{23} = 50$	$x_{24} = 57$	$x_{25} = 63$		

In addition, the mean CEO age of the original sample is $\bar{x} = 56.00$.

Since we have no other information about the true population distribution of CEO ages, we will apply the resampling nonparametric bootstrap procedure to construct the bootstrap distribution since this bootstrapping procedure doesn't require any assumptions or information about the population. Before we apply the resampling bootstrap procedure to generate the bootstrap distribution let's first just display only 7 randomly selected bootstrap samples from the original sample that can fit on the page to help observe and to provide a better understanding of the bootstrap procedure.

Table 18.2 lists an Excel output of 7 bootstrap samples with their means that was randomly selected from the original sample.

TABLE 18.2 7 Bootstrap Samples of size 25 and their Mean

x_1^*	60	50	43	70	46	65	72
x_2^*	42	42	42	47	64	57	60
x_3^*	55	63	71	43	61	72	72
x_4^*	46	50	50	55	63	72	57
x_5^*	60	56	71	47	46	55	65
x_6^*	70	52	46	50	65	47	43
x_7^*	64	42	57	40	70	62	50
x_8^*	46	63	72	46	46	50	50
x_9^*	63	63	57	43	40	57	57
x_{10}^*	57	57	43	56	61	61	52
x_{11}^*	57	61	63	64	64	60	65
x_{12}^*	50	50	63	72	50	57	64
x_{13}^*	50	40	55	52	57	61	63
x_{14}^*	57	46	46	61	57	63	63

Continued

TABLE 18.2 Continued

x_{15}^*	71	70	61	56	63	65	56
x_{16}^*	60	57	60	47	50	72	42
x_{17}^*	46	42	71	65	61	62	64
x_{18}^*	63	52	57	52	47	40	63
x_{19}^*	72	43	64	57	71	57	43
x_{20}^*	60	63	43	61	46	46	63
x_{21}^*	57	72	57	50	72	43	46
x_{22}^*	57	46	42	43	65	52	52
x_{23}^*	57	70	57	70	57	65	65
x_{24}^*	42	57	63	72	63	47	55
x_{25}^*	47	56	62	63	62	50	71
SAMPLE MEANS	$\bar{x}_1^* = 56.36$	$\bar{x}_2^* = 54.52$	$\bar{x}_3^* = 56.64$	$\bar{x}_4^* = 55.28$	$\bar{x}_5^* = 57.88$	$\bar{x}_6^* = 57.52$	$\bar{x}_7^* = 58.12$

Notice each bootstrap sample has a sample size n = 25 which is the same size as the original sample. For each bootstrap sample, the bootstrap sample mean was also computed. Observe the 7 bootstrap samples are different and have a different sample mean from the mean of the original sample. Note the bootstrap asterisk notation for the individual data values and the mean of each bootstrap sample.

Before an actual bootstrap distribution is generated, we need to decide on an appropriate number of resamples that will be selected from the original sample using the bootstrap procedure. As a general guideline, 1,000 resamples will be sufficient for our purposes to generate a bootstrap distribution. Recall, the more resamples selected should improve the accuracy of the bootstrap process. In real life, you probably want to take 10,000 or even 100,000 bootstrap samples.

In Example 18.1, we will use the given original sample of 25 CEO ages to generate a bootstrap distribution and to use the bootstrap distribution to determine a bootstrap estimate of the standard error of the mean.

EXAMPLE 18.1

Use the original sample of 25 ages of the Chief Executive Officers (CEO) of U.S. companies listed in in Table 18.3 to:

a) generate a bootstrap distribution of the mean using 1,000 resamples
b) construct a histogram of the bootstrap distribution of the mean
c) compute the mean of the bootstrap distribution and the bootstrap standard error of the mean
d) determine an estimate of the standard error of the sampling distribution of the mean
e) summarize the characteristics of the bootstrap distribution using the previous results

TABLE 18.3 Original Random Sample of 25 CEO Ages

$x_1 = 40$	$x_2 = 43$	$x_3 = 52$	$x_4 = 56$	$x_5 = 62$	$x_6 = 64$	$x_7 = 65$	$x_8 = 70$	$x_9 = 47$
$x_{10} = 50$	$x_{11} = 57$	$x_{12} = 61$	$x_{13} = 50$	$x_{14} = 72$	$x_{15} = 57$	$x_{16} = 46$	$x_{17} = 60$	$x_{18} = 55$
$x_{19} = 63$	$x_{20} = 42$	$x_{21} = 71$	$x_{22} = 47$	$x_{23} = 50$	$x_{24} = 57$	$x_{25} = 63$		

Solution

a) Following the general guideline to construct a bootstrap distribution, let's select 1,000 bootstrap samples of size 25 from the original sample of 25 CEO ages and calculate the sample mean of each bootstrap sample using Excel to generate the bootstrap distribution of the mean. The Excel results are listed in Table 18.4. Notice the 1,000 bootstrap sample means are listed in numerical order from lowest to highest. The sample means range from the smallest bootstrap sample mean value of 49.96 years to the largest sample bootstrap mean value of 61.96 years.

TABLE 18.4 Bootstrap Distribution of 1,000 Bootstrap Sample Means of CEO ages

49.96	53.68	54.44	54.96	55.52	56	56.48	57.04	57.56	58.24
51.4	53.68	54.44	54.96	55.52	56	56.48	57.04	57.56	58.24
51.64	53.72	54.44	55	55.52	56	56.52	57.04	57.6	58.24
51.68	53.72	54.44	55	55.52	56	56.52	57.04	57.6	58.28
51.72	53.72	54.44	55	55.56	56	56.52	57.04	57.6	58.28
51.76	53.76	54.44	55	55.56	56	56.52	57.04	57.64	58.28
51.88	53.76	54.44	55	55.56	56.04	56.52	57.04	57.64	58.32
51.92	53.8	54.44	55.04	55.56	56.04	56.52	57.04	57.64	58.32
51.92	53.8	54.48	55.04	55.56	56.04	56.56	57.04	57.64	58.32
51.92	53.84	54.48	55.04	55.56	56.04	56.56	57.04	57.64	58.32
51.96	53.84	54.48	55.04	55.56	56.04	56.56	57.08	57.64	58.32
52	53.84	54.48	55.04	55.56	56.04	56.56	57.08	57.64	58.4
52.16	53.84	54.48	55.04	55.6	56.04	56.56	57.08	57.64	58.4
52.2	53.84	54.48	55.08	55.6	56.04	56.6	57.08	57.64	58.44
52.2	53.84	54.48	55.08	55.6	56.04	56.6	57.08	57.68	58.44
52.24	53.88	54.48	55.08	55.6	56.04	56.6	57.08	57.68	58.44
52.24	53.88	54.52	55.08	55.6	56.04	56.6	57.08	57.68	58.48
52.32	53.88	54.52	55.08	55.6	56.08	56.6	57.08	57.68	58.48
52.36	53.88	54.52	55.08	55.6	56.08	56.6	57.08	57.68	58.48
52.44	53.88	54.52	55.08	55.6	56.08	56.64	57.12	57.68	58.52
52.48	53.88	54.52	55.08	55.6	56.08	56.64	57.12	57.72	58.52
52.52	53.92	54.52	55.08	55.6	56.08	56.64	57.12	57.72	58.52
52.6	53.92	54.56	55.12	55.6	56.08	56.64	57.12	57.72	58.52
52.6	53.92	54.56	55.12	55.6	56.08	56.64	57.12	57.72	58.52
52.6	53.92	54.56	55.12	55.6	56.08	56.64	57.12	57.72	58.56
52.64	53.92	54.56	55.12	55.6	56.08	56.68	57.12	57.72	58.56
52.64	53.92	54.56	55.12	55.64	56.08	56.68	57.12	57.76	58.56
52.72	53.96	54.6	55.12	55.64	56.08	56.68	57.12	57.76	58.56
52.72	53.96	54.6	55.12	55.64	56.08	56.68	57.16	57.76	58.56
52.76	53.96	54.6	55.16	55.64	56.12	56.68	57.16	57.76	58.6
52.76	53.96	54.6	55.16	55.64	56.12	56.68	57.16	57.76	58.6
52.76	53.96	54.6	55.16	55.64	56.12	56.68	57.16	57.76	58.64
52.8	53.96	54.6	55.16	55.64	56.12	56.68	57.16	57.76	58.64
52.84	53.96	54.6	55.16	55.64	56.12	56.68	57.16	57.76	58.64
52.84	54	54.64	55.16	55.64	56.12	56.68	57.16	57.76	58.68
52.88	54	54.68	55.16	55.68	56.12	56.68	57.16	57.76	58.68
52.88	54	54.68	55.16	55.68	56.12	56.68	57.16	57.76	58.72
52.88	54	54.68	55.16	55.68	56.16	56.68	57.2	57.76	58.72
52.92	54	54.68	55.2	55.68	56.16	56.72	57.2	57.76	58.76
52.92	54	54.68	55.2	55.68	56.16	56.72	57.2	57.8	58.76
52.92	54	54.68	55.2	55.68	56.16	56.72	57.2	57.8	58.84
52.96	54.04	54.72	55.2	55.68	56.16	56.72	57.2	57.8	58.84
52.96	54.04	54.72	55.2	55.68	56.16	56.72	57.24	57.8	58.84
52.96	54.04	54.72	55.2	55.68	56.16	56.72	57.24	57.8	58.84
52.96	54.04	54.72	55.2	55.68	56.16	56.76	57.24	57.8	58.84

Continued

TABLE 18.4 Continued

53.04	54.08	54.72	55.2	55.68	56.2	56.76	57.24	57.8	58.88
53.04	54.08	54.72	55.24	55.68	56.2	56.76	57.24	57.84	58.88
53.04	54.08	54.72	55.24	55.72	56.2	56.76	57.24	57.84	58.92
53.08	54.08	54.72	55.24	55.72	56.2	56.76	57.24	57.84	58.92
53.08	54.08	54.76	55.24	55.72	56.2	56.76	57.24	57.84	58.92
53.12	54.12	54.76	55.24	55.72	56.2	56.76	57.24	57.84	58.96
53.12	54.12	54.76	55.24	55.72	56.2	56.76	57.24	57.84	58.96
53.16	54.12	54.76	55.24	55.72	56.2	56.76	57.24	57.88	59
53.16	54.12	54.76	55.24	55.72	56.2	56.76	57.24	57.88	59
53.16	54.12	54.76	55.24	55.72	56.2	56.8	57.24	57.88	59
53.2	54.12	54.76	55.28	55.72	56.24	56.8	57.28	57.88	59.08
53.2	54.12	54.76	55.28	55.72	56.24	56.8	57.28	57.88	59.16
53.2	54.12	54.76	55.28	55.72	56.24	56.8	57.28	57.88	59.16
53.24	54.12	54.8	55.28	55.76	56.24	56.8	57.32	57.92	59.2
53.24	54.12	54.8	55.28	55.76	56.24	56.8	57.32	57.92	59.2
53.24	54.12	54.8	55.28	55.76	56.24	56.8	57.32	57.92	59.24
53.28	54.12	54.8	55.28	55.76	56.24	56.8	57.32	57.92	59.24
53.28	54.16	54.8	55.28	55.76	56.28	56.84	57.36	57.92	59.24
53.28	54.16	54.8	55.32	55.76	56.28	56.84	57.36	57.96	59.28
53.28	54.16	54.84	55.32	55.8	56.28	56.84	57.36	57.96	59.28
53.28	54.16	54.84	55.32	55.8	56.28	56.84	57.36	57.96	59.28
53.32	54.2	54.84	55.32	55.8	56.28	56.84	57.36	58	59.32
53.32	54.2	54.84	55.32	55.8	56.28	56.84	57.36	58	59.32
53.32	54.2	54.84	55.32	55.8	56.28	56.84	57.4	58	59.36
53.32	54.2	54.84	55.32	55.8	56.28	56.84	57.4	58	59.36
53.32	54.2	54.84	55.32	55.8	56.32	56.84	57.4	58	59.4
53.36	54.24	54.84	55.32	55.84	56.32	56.84	57.4	58	59.4
53.36	54.24	54.84	55.36	55.84	56.32	56.88	57.4	58	59.6
53.4	54.24	54.84	55.36	55.84	56.32	56.88	57.4	58	59.6
53.4	54.24	54.84	55.36	55.84	56.32	56.88	57.44	58	59.64
53.44	54.28	54.88	55.36	55.84	56.32	56.88	57.44	58	59.64
53.44	54.28	54.88	55.36	55.84	56.32	56.88	57.44	58.04	59.68
53.44	54.28	54.88	55.36	55.84	56.36	56.92	57.44	58.04	59.72
53.44	54.28	54.88	55.36	55.84	56.36	56.92	57.44	58.04	59.76
53.44	54.28	54.88	55.36	55.88	56.36	56.92	57.44	58.04	59.8
53.48	54.32	54.88	55.4	55.88	56.36	56.92	57.44	58.04	59.8
53.48	54.32	54.88	55.4	55.88	56.4	56.92	57.44	58.04	59.8
53.48	54.32	54.88	55.4	55.88	56.4	56.92	57.48	58.04	59.84
53.52	54.32	54.88	55.4	55.88	56.4	56.92	57.48	58.08	59.84
53.52	54.32	54.88	55.4	55.92	56.4	56.96	57.48	58.08	60.12
53.52	54.32	54.88	55.44	55.92	56.4	56.96	57.52	58.08	60.12
53.52	54.32	54.88	55.44	55.92	56.4	56.96	57.52	58.08	60.12
53.52	54.32	54.92	55.44	55.92	56.4	56.96	57.52	58.08	60.32
53.6	54.36	54.92	55.44	55.96	56.44	56.96	57.52	58.08	60.36
53.6	54.36	54.92	55.44	55.96	56.44	56.96	57.52	58.08	60.4

TABLE 18.4 Continued

53.6	54.36	54.92	55.44	55.96	56.44	56.96	57.52	58.08	60.44
53.64	54.36	54.92	55.44	55.96	56.44	56.96	57.52	58.12	60.6
53.64	54.36	54.92	55.44	55.96	56.44	56.96	57.56	58.12	60.76
53.64	54.4	54.92	55.48	55.96	56.48	56.96	57.56	58.16	60.84
53.64	54.4	54.96	55.48	55.96	56.48	57	57.56	58.16	60.84
53.64	54.4	54.96	55.48	55.96	56.48	57	57.56	58.2	60.88
53.64	54.4	54.96	55.48	56	56.48	57	57.56	58.2	60.88
53.68	54.4	54.96	55.48	56	56.48	57	57.56	58.2	61.04
53.68	54.44	54.96	55.48	56	56.48	57.04	57.56	58.2	61.72
53.68	54.44	54.96	55.48	56	56.48	57.04	57.56	58.2	61.96

b) The histogram of the bootstrap distribution of the mean in Figure 18.3 was generated using Excel.

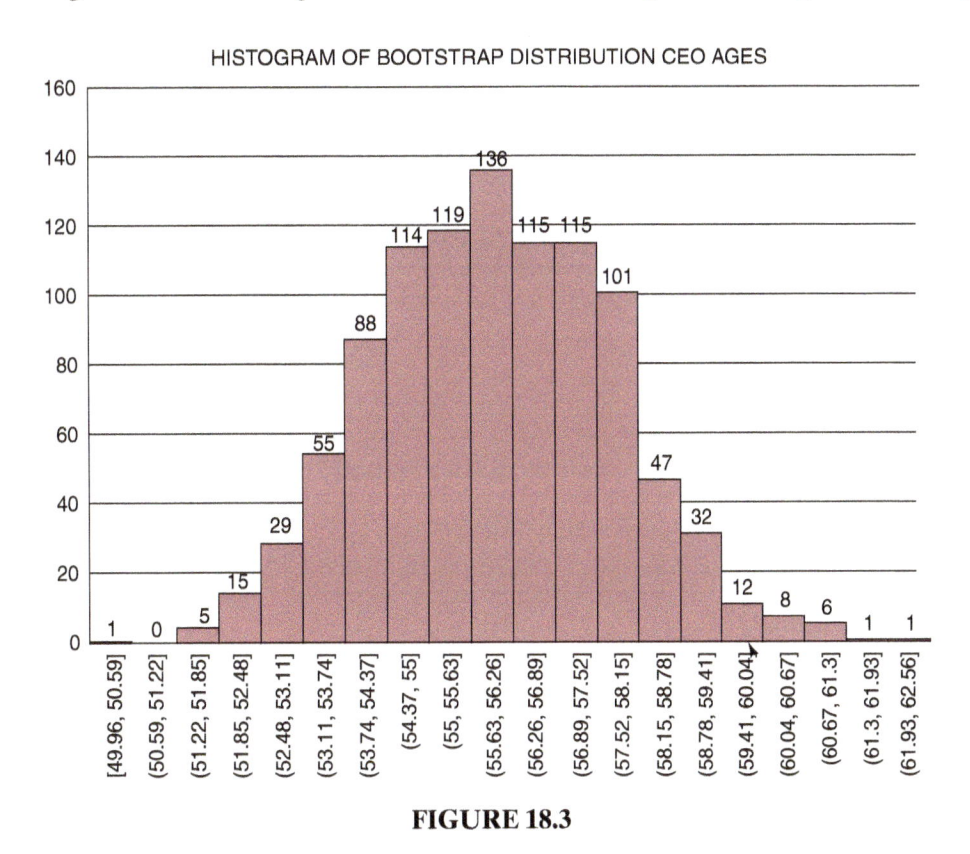

FIGURE 18.3

Notice the shape of the histogram of the bootstrap distribution is approximately a bell-shaped distribution.

c) The mean of the bootstrap distribution is equal to $\frac{1}{B}\sum \bar{x}^*$,

where B = 1,000 and $\bar{x}^*$ represents the individual values of the 1000 resample means. Using Excel, the mean of the bootstrap distribution is 56.00.

To compute the standard error of the bootstrap distribution, we can use the bootstrap standard error formula:

$$SE_b(\bar{x}) = \sqrt{\frac{1}{B-1} * \sum (\bar{x}^* - \frac{1}{B}\sum \bar{x}^*)^2}$$

An easier approach would be to use Excel again to compute the standard deviation of the 1,000 resample means using the sample standard deviation Excel function. Thus, the bootstrap standard error of the mean is: $SE_b(\bar{x}) = 1.81$.

d) Since the bootstrap standard error of the mean is an estimate of the standard error of the sampling distribution of the mean, then $SE_b(\bar{x}) = 1.81$ would serve as an estimate of the standard error of the sampling distribution of the mean.

e) In summary, the shape of the bootstrap distribution is approximately bell-shaped with its center at a mean of 56 with a standard error of 1.81. This information helps to give us an idea of some of the characteristics of the true sampling distribution of the mean. We would expect the sampling distribution of the mean to also be approximately bell-shaped with a standard error of the mean of approximately 1.81. However, we can't conclude what the mean of the sampling distribution of the mean would be based on the mean of the original sample. The mean of the true sampling distribution of the mean should be centered around the mean of the true population, μ, since the random samples of the sampling distribution are selected from the true population. However, we expect the mean of the bootstrap distribution to be close to the mean of the original sample since the bootstrap distribution was generated by taking resamples from the original sample. That is, we would expect the bootstrap distribution to be centered around the mean of the original sample, $\bar{x} = 56$, since the bootstrap distribution of the 1000 resamples were generated from the original sample.

Let's compute the mean and standard deviation of the original sample and then compare the statistical summary results of the bootstrap distribution of the mean in Example 18.1 to the statistical results of the original sample.

Table 18.5 lists the mean and standard deviation of the original sample while Table 18.6 lists the mean and standard error of the bootstrap distribution.

Table 18.5 Original Sample Statistical Results

Mean of the Original Sample,	56.00
Standard Deviation of the Original Sample	9.15
Sample Size of the Original Sample	25

Table 18.6 Bootstrap Distribution Statistical Results

Mean of the Bootstrap Distribution	56.00
Standard Error of the Bootstrap Distribution, $SE_b(\bar{x})$	1.81

In comparing Tables 18.5 and 18.6, notice the mean of the bootstrap distribution and the mean of the original sample are equal as one might expect since the sample size was reasonably large and the resamples were selected from the original sample. Since the histogram of the bootstrap distribution is nearly normal in shape, let's consider using the standard error formula for the mean presented in Chapter 8 to compute the standard error of the true sampling distribution of the mean.

That is, the standard error of the mean = $\dfrac{\text{population standard deviation}}{\sqrt{n}}$.

Since we are assuming the original sample is an estimate of the population, then we can use the standard deviation of the original sample, s, as an estimate of the population standard deviation, σ, within the standard error formula and the sample size of the original sample will be substituted for the value of n. Substituting these values into the standard error formula, we can compute an estimate for the standard error of the mean. Thus,

$$\text{estimate for the standard error of the mean} \approx \dfrac{\text{original sample standard deviation}}{\sqrt{n}} = \dfrac{9.15}{\sqrt{25}} = 1.83.$$

Notice that the bootstrap standard error of 1.81 is extremely close to the estimate for the standard error formula of 1.83 which is what was expected since the bootstrap distribution of the mean is an estimate of the true sampling distribution of the mean. Let's now compare the histogram and characteristics of the original sample in Figure 18.4a to the histogram and characteristics of the bootstrap distribution in Figure 18.4b.

FIGURE 18.4 a

FIGURE 18.4 b

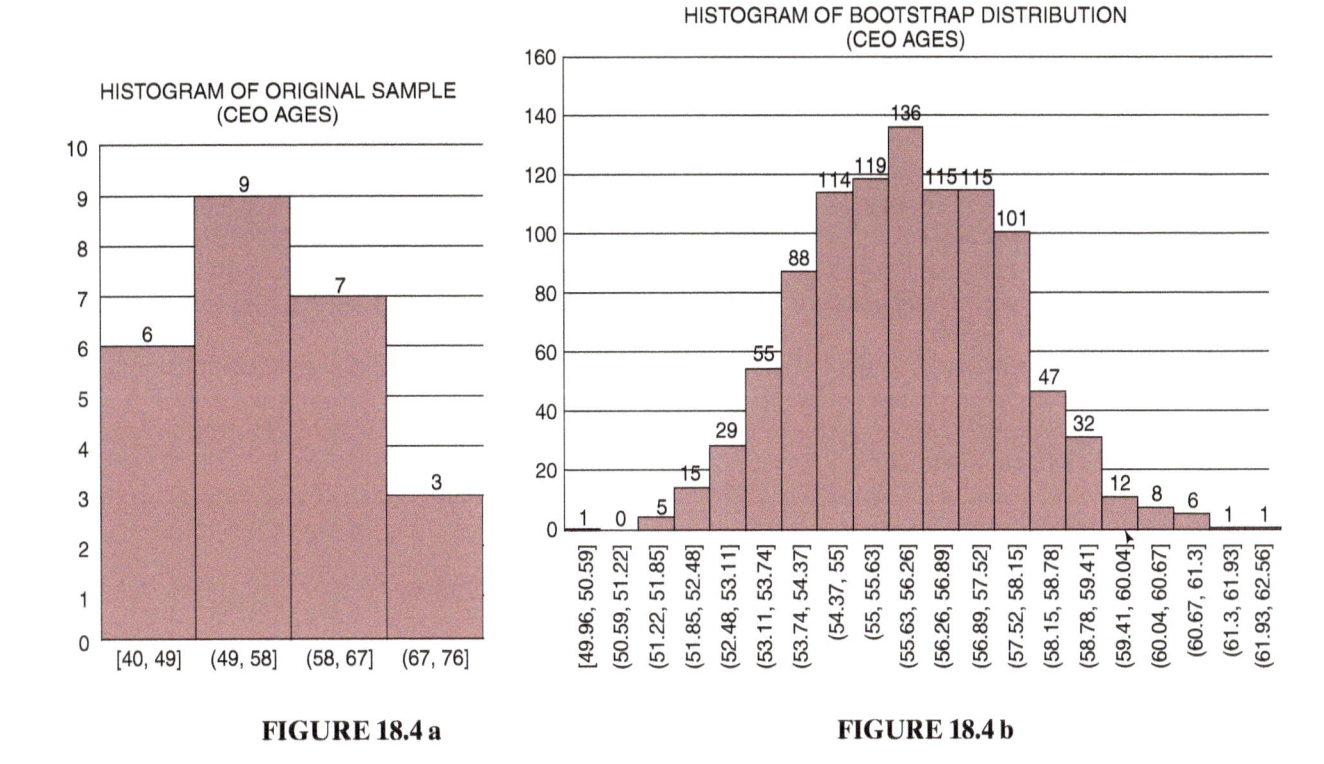

Original Random Sample Characteristics
Mean = 56.00 years
Standard Deviation = 9.15 years
Sample Size = 25

Bootstrap Distribution Characteristics
Mean = 56.00 years
Standard Error of the Bootstrap
Distribution, $SE_b(\bar{x})$ = 1.81 years

Notice that the histogram shape of the original sample is skewed right but the histogram shape of the bootstrap distribution is bell-shaped. With a sample size of the original sample close to 30, the shape of the bootstrap distribution is approximately a normal distribution and the mean of both the original sample and the bootstrap distribution are equal. In addition, the standard error of the bootstrap distribution is $SE_b(\bar{x})$ = 1.81 years which is approximately equal to the standard deviation of the original sample divided by the square root of n (that is, $\frac{original\ sample\ standard\ deviation}{\sqrt{n}} = \frac{9.15}{\sqrt{25}} = 1.83\ years$), the formula for the standard error of the mean.

Revisiting the Bootstrap Idea

This helpful comparison of the original sample to the bootstrap distribution helps to verify the bootstrap idea! That is, the bootstrap procedure is a computational resampling technique used to estimate standard errors of a sampling distribution of a statistic. The bootstrap idea is to mimic the procedure of randomly sampling from a population by treating the original sample as if it were the population. We then repeatedly take B samples (of the same size as the original sample), with replacement, from our original sample. For each of these B samples, we calculate the statistic of interest to generate the bootstrap distribution of these statistics. The standard error of the bootstrap distribution will represent an estimate of the standard error of the statistic. This is called a nonparametric bootstrap procedure since we are not making any assumptions about the population. The only information used is the original sample itself.

Key Bootstrap Ideas

1) The bootstrap distribution of a statistic of interest uses the original sample as an estimate of the population to estimate the sampling distribution of the statistic.

2) The bootstrap distribution estimates the variability between the sample estimates we would obtain if we selected samples from the true population. Thus, the bootstrap distribution is helpful for estimating standard errors.

3) The bootstrap distribution does not provide new information about the population since the resamples were generated from the original sample. For example, if we use the mean of the original sample, $\bar{x}$, as an estimate of the population mean, μ, the bootstrap sample mean, $\bar{x}^*$, computed for each resample would comprise the bootstrap distribution. If we compute the mean of all the sample means, $\bar{x}^*$, we would expect this overall mean of all the bootstrap sample means to be very close to the mean of the original sample. Thus, we would not gain any new information about the mean of the true population. Furthermore, the center of the histogram of the bootstrap distribution will be very close to the center of histogram of the original sample.

4) The bootstrap procedure relies on the very important assumption that the original sample provides a reasonable representation of the population from which the sample data was selected. The larger the size of our original random sample, the more reliable the original sample can be expected to be representative of the population.

5) The bootstrap distribution provides helpful information about the shape, center, and spread of the sampling distribution of the statistic.

Let's discuss the useful characteristics of the bootstrap distribution.

Useful Characteristics of the Bootstrap Distribution

As seen in Example 18.1, the original sample of CEO ages served as an estimate of the population of CEO ages. Then 1,000 resamples were randomly selected from the original sample to generate the bootstrap distribution. This resampling process is mimicking what we would expect to get if we took all random samples from the true population. Consequently, the resulting bootstrap distribution of a statistic, based on the 1,000 resamples, represents an estimate of the sampling distribution of the statistic. This is the importance of the bootstrap procedure to generate an estimate of the sampling distribution.

Once we generate the bootstrap distribution of a statistic, we will use the bootstrap distribution to provide valuable information about the shape, center and spread of the sampling distribution of that statistic.

Shape:

The histogram of the bootstrap distribution in Figure 18.3 was approximately bell-shaped. So, the bootstrap distribution shape is close to the shape one would expect the sampling distribution to have.

Center:

Comparing the mean of the original sample to the mean of the bootstrap distribution shown in Tables 18.4 and 18.5, the bootstrap distribution was centered at the mean of the original sample. But we expected such a result. Recall from Chapter 8, the true sampling distribution of the mean is centered at the population mean, μ. So, we would expect the bootstrap distribution to behave in a similar manner since the bootstrap distribution was generated from the original sample. Although, we expect the mean of the bootstrap distribution to be centered at the mean of the original sample, $\bar{x}$, we can check for bias using the bootstrap distribution. Recall, the bootstrap estimate of bias is the mean of the bootstrap distribution minus the statistic for the original sample data.

Spread:

The histogram of the bootstrap distribution in Figure 18.3 provides a graphical idea of the variability among the bootstrap sample means. We obtained a more accurate sense of the variability by computing the standard error of the bootstrap distribution to be $SE_b(\bar{x}) = 1.81$.

Since the bootstrap standard error is the standard deviation of the bootstrap distribution and serves as an estimate of the standard error of the true sampling distribution of the mean, then we use the bootstrap standard error of 1.81 as an estimate of the standard error of the true sampling distribution of the mean.

Comparing the Bootstrap Distribution and the Sampling Distribution

1) The bootstrap distribution has approximately a similar shape, spread and bias as the sampling distribution of the statistic. However, the bootstrap distribution is centered at the statistic of the original sample, but the sampling distribution is centered at the population parameter.
2) The bootstrap standard error is the standard deviation of the bootstrap distribution and it provides information about the spread of the true sampling distribution of the statistic. That is, the bootstrap distribution provides a way to estimate the variability in a statistic based on the original sample data.

For most statistics, the bootstrap distribution of the statistic will approximate the true sampling distribution of the statistic. Consequently, the bootstrap procedure is very useful in providing valuable information about the true sampling distribution of a statistic.

18.4 BOOTSTRAP CONFIDENCE INTERVALS

In Chapter 9, we discussed using a sample statistic as a point estimate of a population parameter. However, we explained that a point statistic does not allow us to estimate a population parameter with any certainty. Consequently, we examined confidence interval procedures with the objective to enclose the true population parameter value at a certain level of confidence.

In Sections 9.4 and 9.5, confidence intervals for a population mean or a population proportion were constructed using the sampling distribution of the mean or proportion respectively. However, when it is not possible to satisfy the population distribution conditions to use the theoretical sampling distribution, then bootstrapping should be considered to generate a confidence interval for statistics of interest. Bootstrap confidence intervals can be constructed using the bootstrap sampling distribution to approximate the classical confidence intervals discussed in Chapter 9.

Bootstrapping has become a very popular procedure to produce confidence intervals because the bootstrapping procedure will work in essentially the same way when applied to different population parameters. In addition, the bootstrap procedure can be used for a population parameter such as a median or standard deviation where there may be no available alternatives to constructing a confidence interval as general as bootstrapping.

There are multiple methods to generate bootstrap confidence intervals. Bootstrap confidence intervals can be classified into two categories: parametric and nonparametric, according to the assumptions that are applied to construct the confidence interval. The decision of which bootstrap confidence interval method to use is contingent on the particular statistical situation. None of these methods offers the best confidence interval in every situation, because the criteria for determining the quality of the results can vary widely. Although there are different bootstrap methods to construct a confidence interval, several basic nonparametric bootstrap methods will be examined using the bootstrap distribution to construct a confidence interval since these methods do not require any assumptions about the population or original sample.

> **DEFINITION 18.12: NONPARAMETRIC BOOTSTRAP METHOD**
> The nonparametric bootstrap method is a resampling process used to generate an empirical sampling distribution that doesn't require any assumptions about the population or the original sample.

Thus, the bootstrap sampling distribution can then be used to construct confidence intervals for a particular statistic. One nonparametric method to construct a confidence interval is referred to as the **bootstrap percentile confidence interval method**.

Method 1: Bootstrap Percentile Confidence Interval Method

Although the statistical techniques learned in Chapter 9 to construct confidence intervals required that we know the mean or standard deviation of the population, the bootstrap percentile confidence interval method does not require any information other than the original sample. The bootstrap percentile confidence interval method constructs a confidence interval for the population parameter using the bootstrap distribution without needing to compute the standard error (SE) of the sampling distribution as described in Chapter 9.

Using the bootstrap percentile confidence interval method, confidence intervals are constructed from the bootstrap distribution by finding two percentiles in the bootstrap distribution so that the proportion of bootstrap statistics between the percentiles matches the desired confidence level. With the bootstrap percentile method, a certain percentage, say 2.5% or 5%, is trimmed from the lower and upper ends of an ordered bootstrap distribution of the statistic. The percentage that is trimmed will depend on the confidence interval being constructed.

Let's consider the construction of a bootstrap percentile 95% confidence interval. A bootstrap percentile 95% confidence interval is constructed by taking the definition of a 95% Confidence Interval literally. After the distribution of bootstrap statistics is generated and ordered from the smallest to largest values, 2.5% of the bootstrap statistics is trimmed off the lower end of the bootstrap distribution as well as from the upper end of the bootstrap distribution of statistics. Thus, the 2.5th and 97.5th percentiles will serve as the endpoints of the 95% confidence interval. The reason these two percentiles are considered is that we split 100% − 95% = 5% in half so that the middle 95% of all of the bootstrap sample statistics will fall between the endpoints of the confidence interval. Thus, the middle 95% of the bootstrap statistics are sandwiched between the 2.5th percentile as the lower bound of the confidence interval and the 97.5th percentile as the upper bound of the 95% confidence interval as illustrated in Figure 18.5.

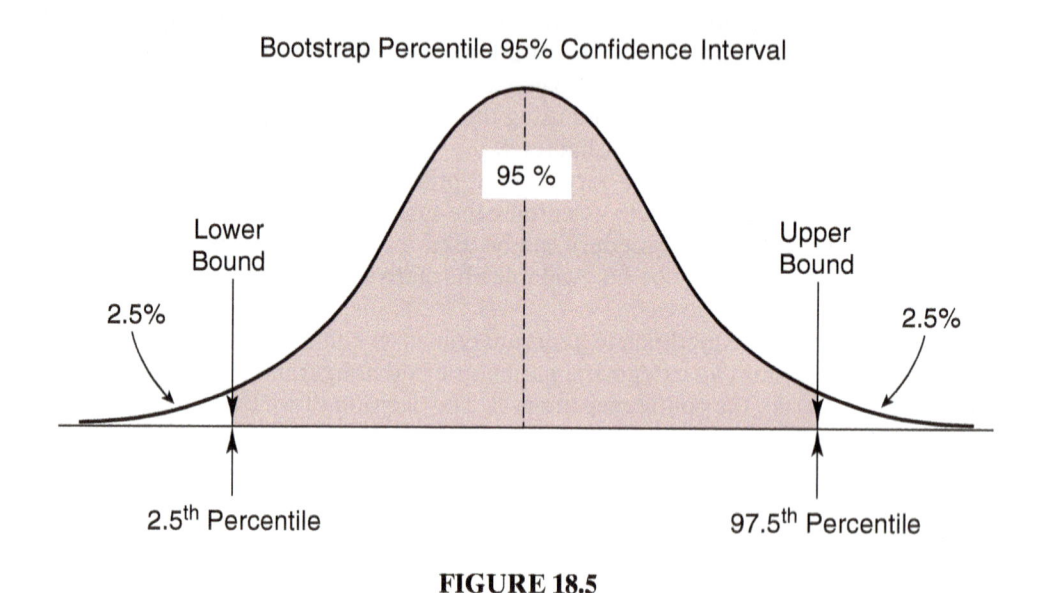

FIGURE 18.5

Notice, in terms of percentiles, the lower bound of the 95% confidence interval is the 2.5th percentile while the upper bound is the 97.5th percentile. Thus, the lower bound (2.5th percentile) of the 95% confidence interval has only 2.5% of the bootstrap statistics smaller than the lower bound of the confidence interval while the upper bound (97.5th percentile) of the 95% confidence interval has only 2.5% of the bootstrap statistics larger than the upper bound of the confidence interval. In a 95% confidence interval, the middle 95% of the bootstrap statistics are sandwiched between the lower and upper bound of the confidence interval, leaving 2.5% in each tail. Therefore, a 95% confidence interval would be denoted as: (2.5th percentile, 97.5th percentile) where the percentiles refer to the ordered bootstrap distribution.

In general, a bootstrap percentile P% confidence interval has the middle P% of the bootstrap statistics sandwiched between a lower and upper bound of the P% confidence interval as illustrated in Figure 18.6.

Bootstrap Percentile P% Confidence Interval

Bootstrap Percentile P% Confidence Interval

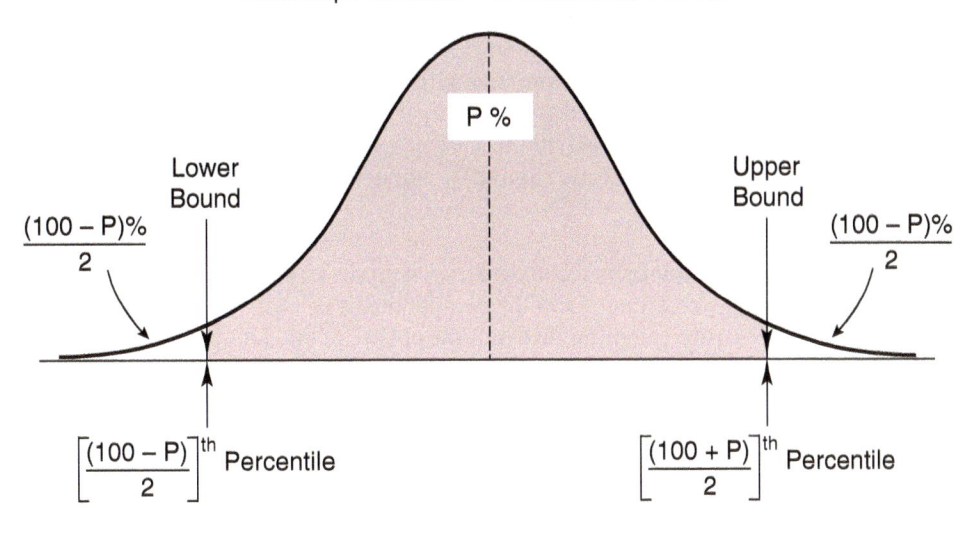

- Lower Bound: $\left[\dfrac{(100-P)}{2}\right]^{\text{th}}$ Percentile
- Upper Bound: $\left[\dfrac{(100+P)}{2}\right]^{\text{th}}$ Percentile
- Tails: $\dfrac{(100-P)\%}{2}$ on each side
- Center: $P\%$

FIGURE 18.6

DEFINITION 18.13: BOOTSTRAP PERCENTILE P% CONFIDENCE INTERVAL

The bootstrap percentile P% confidence interval is a nonparametric bootstrap confidence interval where P% of the ordered bootstrap sample statistics fall between the [(100 -P)/2]th percentile and the [(100 + P)/2th percentile.

Example 18.2 demonstrates how to apply the bootstrap percentile confidence interval method to construct a confidence interval of a population mean.

EXAMPLE 18.2

Suppose we are interested in constructing a 95% confidence interval to estimate the population mean body fat percentage for U.S. males aged 12 to 15 years. However, the only information that we have about this population is a random sample of 36 male body fat percentages for U.S. males aged 12 to 15 years listed in Table 18.7.

Use the bootstrap percentile confidence interval method to construct a 95% confidence interval to estimate the population mean body fat percentage for U.S. males aged 12 to 15 years.

Table 18.7 RANDOM SAMPLE of 36 Body Fat Percentages for U.S. Males aged 12 to 15 Years

15.1	14.4	13.6	15.7	16.1	17.2	15.9	14.8	16.9	16.5	17.1
15.8	18.6	17.4	18.8	18.9	23.0	21.7	20.9	23.6	31.1	28.5
32.8	31.9	24.5	25.8	29.8	36.1	30.9	37.7	36.9	31.9	37.1
40.6	39.2	38.4								

Solution

The random sample of 36 body fat percentages will serve as the original sample since it is our only best estimate of the population of body fat percentages for U.S. males aged 12 to 15 years.

To construct a 95% confidence interval using the bootstrap percentile confidence interval method, it is necessary to generate the bootstrap distribution of the mean. As a reminder, the steps are:

1. generate bootstrap samples by sampling with replacement from the original sample, using the same sample size n = 36 as the original sample.
2. compute the statistic of interest, in this case a bootstrap sample mean, for each of the bootstrap samples.
3. list the mean for each bootstrap sample in numerical order to form an ordered bootstrap distribution of the mean.

Following the general guideline, 1000 bootstrap samples of size 36 were randomly selected from the original sample of 36 body fat percentages and the sample mean of each bootstrap sample was computed using Excel to generate the bootstrap sampling distribution of the mean. The bootstrap sampling distribution is listed in Table 18.8. Notice the 1000 bootstrap sample means represent an ordered bootstrap distribution since the bootstrap sample means are listed in numerical order from lowest to highest. The sample means range from the smallest sample mean value of 19.65 to the largest sample mean value of 29.73.

TABLE 18.8   Bootstrap Distribution of Sample Mean Body Fat Percentages of U.S. Males 12–15 Years

19.65	22.73	23.37	23.86	24.30	24.63	25.00	25.36	25.86	26.49
20.82	22.73	23.38	23.86	24.30	24.63	25.00	25.36	25.86	26.50
20.88	22.73	23.38	23.87	24.30	24.64	25.01	25.37	25.86	26.55
20.89	22.73	23.39	23.87	24.30	24.64	25.01	25.37	25.86	26.55
20.90	22.73	23.40	23.89	24.30	24.64	25.01	25.37	25.87	26.57
20.92	22.73	23.40	23.90	24.31	24.65	25.02	25.39	25.88	26.57
20.92	22.73	23.40	23.90	24.31	24.65	25.02	25.40	25.88	26.59
20.96	22.73	23.40	23.90	24.32	24.65	25.03	25.41	25.88	26.59
21.00	22.73	23.43	23.91	24.32	24.65	25.03	25.41	25.89	26.59
21.09	22.73	23.43	23.92	24.32	24.65	25.04	25.41	25.89	26.60
21.19	22.73	23.43	23.92	24.32	24.66	25.04	25.42	25.89	26.61
21.22	22.73	23.44	23.93	24.32	24.66	25.04	25.42	25.90	26.61
21.23	22.73	23.44	23.93	24.32	24.66	25.05	25.42	25.91	26.61
21.33	22.73	23.45	23.95	24.33	24.67	25.05	25.42	25.91	26.62
21.40	22.73	23.45	23.96	24.33	24.68	25.05	25.43	25.92	26.64
21.40	22.73	23.46	23.96	24.33	24.68	25.06	25.43	25.93	26.64
21.42	22.73	23.46	23.96	24.33	24.68	25.06	25.43	25.93	26.64
21.55	22.73	23.47	23.96	24.33	24.69	25.06	25.43	25.93	26.65
21.60	22.73	23.48	23.97	24.34	24.69	25.06	25.44	25.94	26.68
21.64	22.73	23.49	23.98	24.34	24.71	25.06	25.44	25.94	26.70
21.65	22.73	23.49	23.98	24.35	24.71	25.06	25.44	25.94	26.71
21.75	22.73	23.49	23.99	24.35	24.71	25.06	25.45	25.95	26.74
21.80	22.73	23.50	23.99	24.35	24.71	25.07	25.46	25.95	26.75
21.81	22.73	23.50	24.00	24.35	24.72	25.08	25.46	25.96	26.76
21.83 *(25th Value)*	22.73	23.50	24.01	24.36	24.72	25.08	25.46	25.96	26.78
21.84	22.73	23.50	24.02	24.36	24.72	25.08	25.46	25.97	26.78
21.86	22.73	23.50	24.02	24.36	24.72	25.08	25.46	25.98	26.79
21.90	22.73	23.51	24.03	24.37	24.73	25.08	25.46	25.98	26.79
21.91	22.73	23.52	24.03	24.37	24.73	25.09	25.47	26.00	26.80

TABLE 18.8 Continued

21.92	22.73	23.52	24.04	24.37	24.73	25.09	25.48	26.01	26.81
21.93	22.73	23.52	24.04	24.38	24.73	25.09	25.49	26.01	26.81
21.96	22.73	23.54	24.04	24.39	24.74	25.09	25.49	26.02	26.81
21.97	22.73	23.54	24.04	24.39	24.74	25.09	25.49	26.02	26.84
21.97	22.73	23.54	24.05	24.39	24.74	25.09	25.50	26.03	26.84
21.99	22.73	23.55	24.06	24.41	24.74	25.09	25.51	26.04	26.85
22.05	22.73	23.56	24.06	24.41	24.74	25.10	25.51	26.05	26.86
22.06	22.73	23.57	24.06	24.42	24.75	25.11	25.51	26.05	26.86
22.07	22.73	23.58	24.07	24.42	24.75	25.11	25.52	26.05	26.87
22.11	22.73	23.58	24.07	24.42	24.76	25.11	25.52	26.06	26.90
22.11	22.73	23.59	24.07	24.43	24.77	25.11	25.52	26.07	26.91
22.14	22.73	23.59	24.07	24.43	24.77	25.12	25.53	26.08	26.91
22.15	22.73	23.59	24.07	24.43	24.77	25.12	25.53	26.09	26.93
22.18	22.73	23.60	24.07	24.43	24.77	25.12	25.53	26.10	26.93
22.18	22.73	23.61	24.08	24.44	24.77	25.13	25.53	26.11	26.94
22.23	22.73	23.61	24.08	24.44	24.78	25.13	25.53	26.12	26.94
22.24	22.73	23.61	24.09	24.44	24.78	25.13	25.53	26.13	26.94
22.24	22.73	23.61	24.09	24.44	24.78	25.13	25.53	26.14	26.96
22.26	22.73	23.62	24.09	24.44	24.78	25.14	25.53	26.17	26.96
22.29	22.73	23.63	24.09	24.44	24.79	25.15	25.54	26.17	26.99
22.30	22.73	23.63	24.10	24.45	24.79	25.15	25.54	26.18	27.01
22.31	22.73	23.64	24.10	24.45	24.79	25.16	25.56	26.18	27.01
22.32	22.73	23.65	24.10	24.45	24.80	25.18	25.56	26.18	27.03
22.33	22.73	23.65	24.11	24.46	24.80	25.18	25.57	26.18	27.03
22.34	22.73	23.65	24.11	24.47	24.81	25.19	25.58	26.19	27.04
22.36	22.73	23.65	24.11	24.48	24.81	25.19	25.58	26.19	27.04
22.37	22.73	23.66	24.11	24.48	24.81	25.20	25.59	26.20	27.11
22.41	22.73	23.66	24.12	24.48	24.81	25.20	25.59	26.21	27.12
22.41	22.73	23.66	24.13	24.48	24.81	25.21	25.60	26.21	27.17
22.41	22.73	23.66	24.14	24.48	24.82	25.21	25.60	26.21	27.19
22.41	22.73	23.67	24.14	24.49	24.82	25.22	25.60	26.22	27.19
22.43	22.73	23.67	24.14	24.49	24.82	25.22	25.61	26.23	27.26
22.47	22.73	23.67	24.15	24.50	24.82	25.22	25.61	26.23	27.28
22.48	22.73	23.68	24.16	24.50	24.83	25.23	25.62	26.24	27.33
22.50	22.73	23.68	24.16	24.50	24.83	25.23	25.63	26.24	27.33
22.50	22.73	23.68	24.16	24.50	24.83	25.23	25.63	26.26	27.34
22.51	22.73	23.68	24.16	24.50	24.83	25.24	25.63	26.27	27.36
22.51	22.73	23.68	24.17	24.51	24.84	25.24	25.64	26.28	27.37
22.53	22.73	23.69	24.17	24.51	24.84	25.24	25.64	26.28	27.44
22.54	22.73	23.69	24.17	24.51	24.85	25.24	25.65	26.28	27.46
22.54	22.73	23.70	24.18	24.51	24.86	25.25	25.66	26.30	27.46
22.54	22.73	23.70	24.18	24.51	24.87	25.25	25.68	26.30	27.47
22.55	22.73	23.72	24.18	24.52	24.88	25.26	25.68	26.31	27.60
22.56	22.73	23.72	24.19	24.52	24.88	25.26	25.68	26.32	27.64

Continued

TABLE 18.8 Continued

22.57	22.73	23.73	24.20	24.52	24.88	25.27	25.69	26.32	27.65	975th
22.58	22.73	23.73	24.20	24.53	24.89	25.28	25.69	26.32	27.71	Data Value
22.60	22.73	23.73	24.20	24.53	24.89	25.28	25.70	26.33	27.85	
22.61	22.73	23.73	24.20	24.53	24.90	25.28	25.70	26.34	27.92	
22.62	22.73	23.74	24.20	24.55	24.90	25.29	25.71	26.34	27.93	
22.62	22.73	23.74	24.20	24.55	24.90	25.29	25.71	26.35	27.99	
22.64	22.73	23.75	24.21	24.55	24.90	25.29	25.71	26.35	28.04	
22.64	22.73	23.75	24.21	24.55	24.91	25.29	25.71	26.35	28.05	
22.65	22.73	23.75	24.21	24.56	24.92	25.29	25.72	26.35	28.07	
22.65	22.73	23.75	24.22	24.57	24.93	25.29	25.73	26.35	28.07	
22.65	22.73	23.76	24.23	24.57	24.93	25.29	25.74	26.36	28.08	
22.66	22.73	23.76	24.23	24.58	24.94	25.32	25.74	26.36	28.12	
22.66	22.73	23.76	24.24	24.58	24.94	25.32	25.74	26.36	28.21	
22.66	22.73	23.76	24.24	24.59	24.94	25.32	25.75	26.36	28.22	
22.68	22.73	23.76	24.25	24.60	24.95	25.32	25.75	26.38	28.29	
22.68	22.73	23.78	24.25	24.60	24.96	25.33	25.76	26.39	28.29	
22.69	22.73	23.80	24.25	24.60	24.96	25.33	25.77	26.41	28.31	
22.69	22.73	23.81	24.26	24.61	24.96	25.33	25.78	26.42	28.32	
22.69	22.73	23.81	24.26	24.61	24.97	25.33	25.78	26.42	28.64	
22.70	22.73	23.81	24.26	24.62	24.97	25.34	25.79	26.44	28.68	
22.70	22.73	23.81	24.26	24.62	24.97	25.34	25.81	26.44	28.73	
22.70	22.73	23.82	24.27	24.62	24.97	25.34	25.81	26.45	28.73	
22.71	22.73	23.83	24.27	24.62	24.98	25.34	25.82	26.46	28.93	
22.72	22.73	23.83	24.27	24.62	24.98	25.34	25.84	26.47	29.21	
22.72	22.73	23.85	24.28	24.63	24.99	25.34	25.84	26.48	29.35	
22.73	22.73	23.85	24.28	24.63	24.99	25.36	25.84	26.48	29.46	
22.73	22.73	23.86	24.29	24.63	25.00	25.36	25.85	26.49	29.73	

Using the ordered bootstrap sampling distribution listed in Table 18.8, the 95% confidence interval can be constructed using this ordered list of 1,000 bootstrap sample means. To construct a 95% confidence interval, the 2.5th percentile will serve as the lower bound of the 95% confidence interval and the 97.5th percentile will serve as the upper bound of the 95% confidence interval. The reason these two percentiles serve as the endpoints of the confidence interval is that we split 100% −95% = 5% in half so that the confidence interval will include the middle 95% of all of the bootstrap sample means of the bootstrap sampling distribution.

To determine the bootstrap sample mean that represents the 2.5th percentile, it is necessary to find the sample mean of the bootstrap sampling distribution that lies at the 2.5th percentile after all the bootstrap sample means have been ordered from low to high values. Since there are 1,000 bootstrap sample means, 2.5% of 1,000 is 25. Thus, the bootstrap sample mean representing the 2.5th percentile is the 25th sample mean value will serve as the lower confidence interval value within the ordered 1000 values. Scanning the bootstrap sample means that are listed in numerical order within Table 18.8, the 25th bootstrap sample mean 21.83 represents the lower limit of the 95% confidence interval. That is, 21.83 = 2.5th percentile.

To determine the bootstrap sample mean that represents the 97.5th percentile, it is necessary to find the sample mean of the bootstrap sampling distribution that represents the 97.5th percentile after all the bootstrap sample means have been ordered from low to high values. Since there are 1,000 bootstrap sample means, 97.5% of 1,000 is 975. Thus, the bootstrap sample mean representing the 97.5th percentile is the 975th value as the upper confidence interval value within the ordered 1000 values. Scanning the bootstrap sample means that are listed in numerical order within Table 18.8, the 975th bootstrap sample mean 27.71 represents the upper limit of the 95% confidence interval. That is, 27.71 = 97.5th percentile. Therefore, 95% of the bootstrap sample means fall between 21.83 and 27.71. Then, the 95% bootstrap percentile confidence interval to estimate the population mean body fat percentage for U.S. males aged 12 to 15 years is (21.83, 27.71). Thus, we are 95% confident that the population mean body fat percentage of U.S. males is between 21.83 and 27.71. The 95% confidence interval (21.83, 27.71) is shown in Figure 18.7 on the histogram of the bootstrap distribution of body fat percentages.

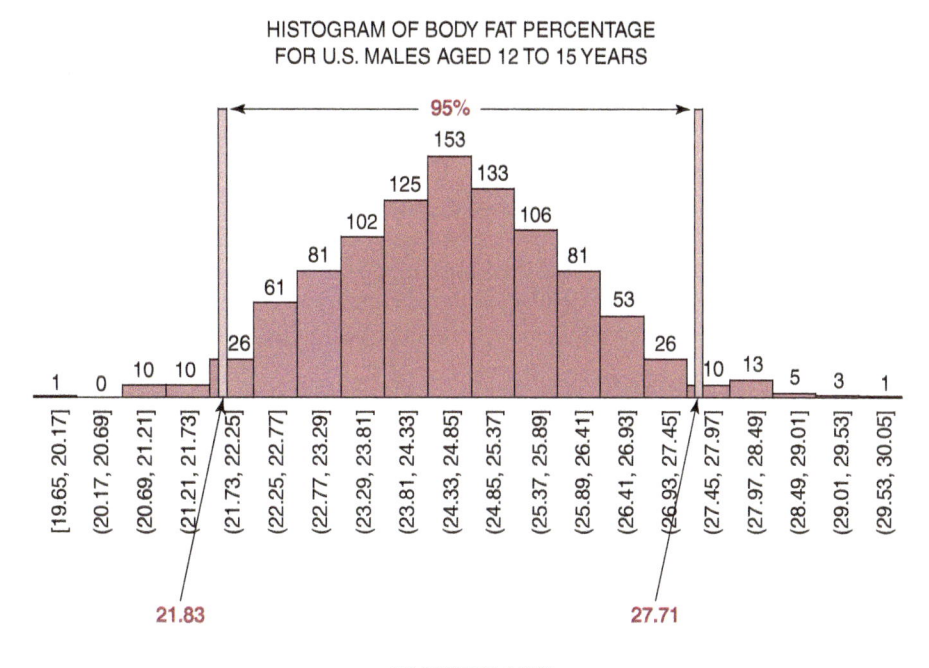

FIGURE 18.7

Cautions about the Bootstrap Percentile Confidence Interval Method

Although the bootstrap percentile method is very simple to use when constructing confidence intervals, the accuracy of the method depends on the original sample being a very good approximation to the true population. Furthermore, the bootstrap percentile confidence intervals are not very accurate for small sized original samples since the intervals tend to be too narrow.

Method 2: Bootstrap Standard Error t Confidence Interval Method

We previously discussed the bootstrap concept to construct a bootstrap percentile confidence interval. In Chapter 9, we constructed confidence intervals of the mean using an estimate of the standard error. In fact, standard errors are often used to construct confidence intervals based on the bootstrap distribution being approximately bell-shaped. Now we will examine constructing a bootstrap confidence interval following the general one sample t confidence interval mean formula presented in Section 9.4:

$$\bar{x} \pm (\text{t score}) * \frac{s}{\sqrt{n}}$$

This population mean confidence interval was based on the sampling distribution of the mean being normal and using the $\frac{s}{\sqrt{n}}$ formula to compute an estimate of the standard error of the mean.

When the bootstrap distribution of a statistic is approximately a normal distribution with small bias, we can follow this same process using the bootstrap standard error to generate a bootstrap confidence interval for any population parameter. This bootstrap confidence interval method is referred to as the Bootstrap Standard Error t Confidence Interval.

Chapter 18 Bootstrapping Concepts & Methods: An Introduction

> **DEFINITION 18.14 BOOTSTRAP STANDARD ERROR T CONFIDENCE INTERVAL**
>
> If the bootstrap distribution of a statistic generated from an original random sample of size n is approximately normal and has a small bias, then an approximate confidence interval of level C for the population parameter for this statistic is
>
> **statistic ± (t score) * (SE_b(statistic))**
>
> Where: t score is the appropriate critical t score that corresponds to the confidence level C and SE_b(statistic) is the bootstrap standard error of the statistic

This bootstrap standard error t confidence interval method is appropriate provided the bootstrap distribution is symmetric and bell-shaped.

In Section 18.3, we showed how to compute the standard deviation of the bootstrap distribution of the mean to estimate the standard error of the true sampling distribution for a mean. Recall, the standard error (SE) of a statistic is the standard deviation of the sampling distribution. After the bootstrap distribution of the mean was generated in Example 18.1 for CEO ages, we computed an estimate of the standard error of the mean by calculating the standard deviation of the bootstrap distribution using the bootstrap standard error formula:

$$SE_b(\bar{x}) = \sqrt{\frac{1}{B-1} * \sum (\bar{x}* - \frac{1}{B}\sum \bar{x}*)^2}$$

Let's use the bootstrap standard error t confidence interval method to construct a 95% confidence interval of the mean using the bootstrap distribution of the male body fat percentages presented in Example 18.2.

EXAMPLE 18.3

Construct a 95% confidence interval to estimate the population mean body fat percentage for U.S. males aged 12 to 15 years using the bootstrap standard error t confidence interval. The only information that we have about this population is a random sample of 36 male body fat percentages for U.S. males aged 12 to 15 years listed in Table 18.9.

TABLE 18.9 Random Sample of 36 Body Fat Percentages for U.S. Males aged 12 to 15 Years

15.1	14.4	13.6	15.7	16.1	17.2	15.9	14.8	16.9	16.5	17.1
15.8	18.6	17.4	18.8	18.9	23.0	21.7	20.9	23.6	31.1	28.5
32.8	31.9	24.5	25.8	29.8	36.1	30.9	37.7	36.9	31.9	37.1
40.6	39.2	38.4								

Solution

To construct a 95% confidence interval of the mean using the bootstrap standard error t confidence interval, we first need to generate the bootstrap distribution of the mean to determine if the bootstrap distribution is approximately normal before we can apply the bootstrap standard error t confidence interval formula. Using the original random sample of 36 male body fat percentages listed in Table 18.9, we can generate a bootstrap distribution of the mean. As a reminder, the steps are:

1. generate bootstrap samples by sampling with replacement from the original sample, using the same sample size n = 36 as the original sample.
2. compute the statistic of interest, in this case a bootstrap sample mean, for each of the bootstrap samples.
3. list the mean for each bootstrap sample in numerical order to form the bootstrap distribution of the mean.

Following the general guideline, 1000 bootstrap samples of size 36 were randomly selected from the original sample of 36 body fat percentages and the sample mean of each bootstrap sample was computed using Excel to generate the bootstrap sampling distribution of the mean. The bootstrap sampling distribution is listed in Table 18.10.

TABLE 18.10 Bootstrap Distribution of Sample Mean Body Fat Percentages of U.S. Males 12–15 Years

19.65	22.73	23.37	23.86	24.30	24.63	25.00	25.36	25.86	26.49
20.82	22.73	23.38	23.86	24.30	24.63	25.00	25.36	25.86	26.50
20.88	22.73	23.38	23.87	24.30	24.64	25.01	25.37	25.86	26.55
20.89	22.73	23.39	23.87	24.30	24.64	25.01	25.37	25.86	26.55
20.90	22.73	23.40	23.89	24.30	24.64	25.01	25.37	25.87	26.57
20.92	22.73	23.40	23.90	24.31	24.65	25.02	25.39	25.88	26.57
20.92	22.73	23.40	23.90	24.31	24.65	25.02	25.40	25.88	26.59
20.96	22.73	23.40	23.90	24.32	24.65	25.03	25.41	25.88	26.59
21.00	22.73	23.43	23.91	24.32	24.65	25.03	25.41	25.89	26.59
21.09	22.73	23.43	23.92	24.32	24.65	25.04	25.41	25.89	26.60
21.19	22.73	23.43	23.92	24.32	24.66	25.04	25.42	25.89	26.61
21.22	22.73	23.44	23.93	24.32	24.66	25.04	25.42	25.90	26.61
21.23	22.73	23.44	23.93	24.32	24.66	25.05	25.42	25.91	26.61
21.33	22.73	23.45	23.95	24.33	24.67	25.05	25.42	25.91	26.62
21.40	22.73	23.45	23.96	24.33	24.68	25.05	25.43	25.92	26.64
21.40	22.73	23.46	23.96	24.33	24.68	25.06	25.43	25.93	26.64
21.42	22.73	23.46	23.96	24.33	24.68	25.06	25.43	25.93	26.64
21.55	22.73	23.47	23.96	24.33	24.69	25.06	25.43	25.93	26.65
21.60	22.73	23.48	23.97	24.34	24.69	25.06	25.44	25.94	26.68
21.64	22.73	23.49	23.98	24.34	24.71	25.06	25.44	25.94	26.70
21.65	22.73	23.49	23.98	24.35	24.71	25.06	25.44	25.94	26.71
21.75	22.73	23.49	23.99	24.35	24.71	25.06	25.45	25.95	26.74
21.80	22.73	23.50	23.99	24.35	24.71	25.07	25.46	25.95	26.75
21.81	22.73	23.50	24.00	24.35	24.72	25.08	25.46	25.96	26.76
21.83	22.73	23.50	24.01	24.36	24.72	25.08	25.46	25.96	26.78
21.84	22.73	23.50	24.02	24.36	24.72	25.08	25.46	25.97	26.78
21.86	22.73	23.50	24.02	24.36	24.72	25.08	25.46	25.98	26.79
21.90	22.73	23.51	24.03	24.37	24.73	25.08	25.46	25.98	26.79
21.91	22.73	23.52	24.03	24.37	24.73	25.09	25.47	26.00	26.80
21.92	22.73	23.52	24.04	24.37	24.73	25.09	25.48	26.01	26.81
21.93	22.73	23.52	24.04	24.38	24.73	25.09	25.49	26.01	26.81
21.96	22.73	23.54	24.04	24.39	24.74	25.09	25.49	26.02	26.81
21.97	22.73	23.54	24.04	24.39	24.74	25.09	25.49	26.02	26.84
21.97	22.73	23.54	24.05	24.39	24.74	25.09	25.50	26.03	26.84
21.99	22.73	23.55	24.06	24.41	24.74	25.09	25.51	26.04	26.85
22.05	22.73	23.56	24.06	24.41	24.74	25.10	25.51	26.05	26.86
22.06	22.73	23.57	24.06	24.42	24.75	25.11	25.51	26.05	26.86
22.07	22.73	23.58	24.07	24.42	24.75	25.11	25.52	26.05	26.87
22.11	22.73	23.58	24.07	24.42	24.76	25.11	25.52	26.06	26.90
22.11	22.73	23.59	24.07	24.43	24.77	25.11	25.52	26.07	26.91
22.14	22.73	23.59	24.07	24.43	24.77	25.12	25.53	26.08	26.91
22.15	22.73	23.59	24.07	24.43	24.77	25.12	25.53	26.09	26.93
22.18	22.73	23.60	24.07	24.43	24.77	25.12	25.53	26.10	26.93
22.18	22.73	23.61	24.08	24.44	24.77	25.13	25.53	26.11	26.94

Continued

TABLE 18.10 Continued

22.23	22.73	23.61	24.08	24.44	24.78	25.13	25.53	26.12	26.94
22.24	22.73	23.61	24.09	24.44	24.78	25.13	25.53	26.13	26.94
22.24	22.73	23.61	24.09	24.44	24.78	25.13	25.53	26.14	26.96
22.26	22.73	23.62	24.09	24.44	24.78	25.14	25.53	26.17	26.96
22.29	22.73	23.63	24.09	24.44	24.79	25.15	25.54	26.17	26.99
22.30	22.73	23.63	24.10	24.45	24.79	25.15	25.54	26.18	27.01
22.31	22.73	23.64	24.10	24.45	24.79	25.16	25.56	26.18	27.01
22.32	22.73	23.65	24.10	24.45	24.80	25.18	25.56	26.18	27.03
22.33	22.73	23.65	24.11	24.46	24.80	25.18	25.57	26.18	27.03
22.34	22.73	23.65	24.11	24.47	24.81	25.19	25.58	26.19	27.04
22.36	22.73	23.65	24.11	24.48	24.81	25.19	25.58	26.19	27.04
22.37	22.73	23.66	24.11	24.48	24.81	25.20	25.59	26.20	27.11
22.41	22.73	23.66	24.12	24.48	24.81	25.20	25.59	26.21	27.12
22.41	22.73	23.66	24.13	24.48	24.81	25.21	25.60	26.21	27.17
22.41	22.73	23.66	24.14	24.48	24.82	25.21	25.60	26.21	27.19
22.41	22.73	23.67	24.14	24.49	24.82	25.22	25.60	26.22	27.19
22.43	22.73	23.67	24.14	24.49	24.82	25.22	25.61	26.23	27.26
22.47	22.73	23.67	24.15	24.50	24.82	25.22	25.61	26.23	27.28
22.48	22.73	23.68	24.16	24.50	24.83	25.23	25.62	26.24	27.33
22.50	22.73	23.68	24.16	24.50	24.83	25.23	25.63	26.24	27.33
22.50	22.73	23.68	24.16	24.50	24.83	25.23	25.63	26.26	27.34
22.51	22.73	23.68	24.16	24.50	24.83	25.24	25.63	26.27	27.36
22.51	22.73	23.68	24.17	24.51	24.84	25.24	25.64	26.28	27.37
22.53	22.73	23.69	24.17	24.51	24.84	25.24	25.64	26.28	27.44
22.54	22.73	23.69	24.17	24.51	24.85	25.24	25.65	26.28	27.46
22.54	22.73	23.70	24.18	24.51	24.86	25.25	25.66	26.30	27.46
22.54	22.73	23.70	24.18	24.51	24.87	25.25	25.68	26.30	27.47
22.55	22.73	23.72	24.18	24.52	24.88	25.26	25.68	26.31	27.60
22.56	22.73	23.72	24.19	24.52	24.88	25.26	25.68	26.32	27.64
22.57	22.73	23.73	24.20	24.52	24.88	25.27	25.69	26.32	27.65
22.58	22.73	23.73	24.20	24.53	24.89	25.28	25.69	26.32	27.71
22.60	22.73	23.73	24.20	24.53	24.89	25.28	25.70	26.33	27.85
22.61	22.73	23.73	24.20	24.53	24.90	25.28	25.70	26.34	27.92
22.62	22.73	23.74	24.20	24.55	24.90	25.29	25.71	26.34	27.93
22.62	22.73	23.74	24.20	24.55	24.90	25.29	25.71	26.35	27.99
22.64	22.73	23.75	24.21	24.55	24.90	25.29	25.71	26.35	28.04
22.64	22.73	23.75	24.21	24.55	24.91	25.29	25.71	26.35	28.05
22.65	22.73	23.75	24.21	24.56	24.92	25.29	25.72	26.35	28.07
22.65	22.73	23.75	24.22	24.57	24.93	25.29	25.73	26.35	28.07
22.65	22.73	23.76	24.23	24.57	24.93	25.29	25.74	26.36	28.08
22.66	22.73	23.76	24.23	24.58	24.94	25.32	25.74	26.36	28.12
22.66	22.73	23.76	24.24	24.58	24.94	25.32	25.74	26.36	28.21
22.66	22.73	23.76	24.24	24.59	24.94	25.32	25.75	26.36	28.22
22.68	22.73	23.76	24.25	24.60	24.95	25.32	25.75	26.38	28.29
22.68	22.73	23.78	24.25	24.60	24.96	25.33	25.76	26.39	28.29

TABLE 18.10 Continued

22.69	22.73	23.80	24.25	24.60	24.96	25.33	25.77	26.41	28.31
22.69	22.73	23.81	24.26	24.61	24.96	25.33	25.78	26.42	28.32
22.69	22.73	23.81	24.26	24.61	24.97	25.33	25.78	26.42	28.64
22.70	22.73	23.81	24.26	24.62	24.97	25.34	25.79	26.44	28.68
22.70	22.73	23.81	24.26	24.62	24.97	25.34	25.81	26.44	28.73
22.70	22.73	23.82	24.27	24.62	24.97	25.34	25.81	26.45	28.73
22.71	22.73	23.83	24.27	24.62	24.98	25.34	25.82	26.46	28.93
22.72	22.73	23.83	24.27	24.62	24.98	25.34	25.84	26.47	29.21
22.72	22.73	23.85	24.28	24.63	24.99	25.34	25.84	26.48	29.35
22.73	22.73	23.85	24.28	24.63	24.99	25.36	25.84	26.48	29.46
22.73	22.73	23.86	24.29	24.63	25.00	25.36	25.85	26.49	29.73

We need to construct a histogram of this bootstrap sampling distribution of the mean to verify that the histogram shape is approximately normal. Figure 18.8 represents the histogram of the bootstrap distribution.

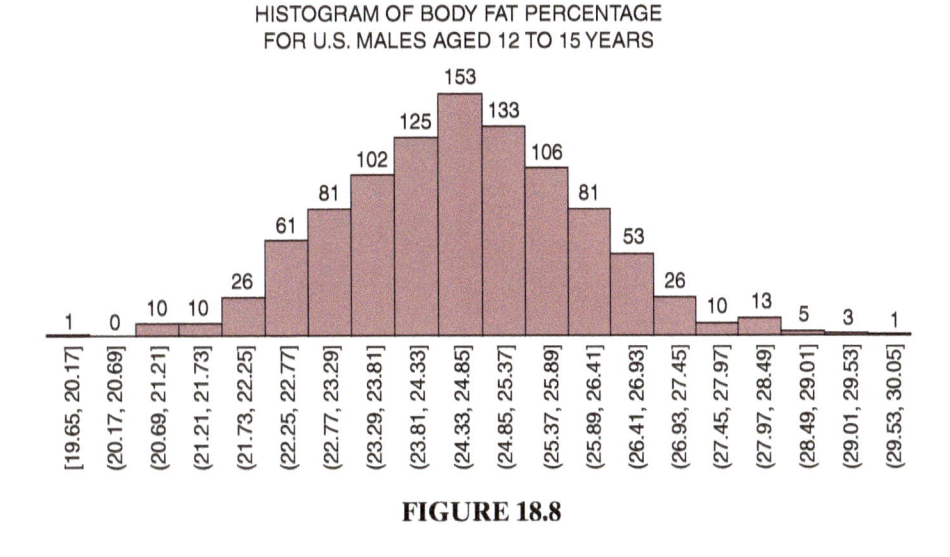

HISTOGRAM OF BODY FAT PERCENTAGE FOR U.S. MALES AGED 12 TO 15 YEARS

FIGURE 18.8

From Figure 18.8, the histogram of the bootstrap sampling distribution of the mean is symmetric and bell-shaped. So, let's proceed to apply the bootstrap standard error t confidence interval.

To apply the bootstrap standard error t confidence interval, we need to apply the bootstrap standard error t confidence interval formula:

$$\text{statistic} \pm (\text{t score}) * (SE_b(\text{statistic}))$$

Since the sample statistic of our bootstrap sampling distribution is the mean, then this general confidence interval formula can be rewritten where $\bar{x}$ replaces the term statistic as:

$$\bar{x} \pm (\text{t score}) * SE_b(\bar{x})$$

To determine the 95% confidence interval of the mean, we need to determine:
the value of $\bar{x}$,
the appropriate t score,
and the standard deviation of the bootstrap distribution using the bootstrap standard error formula:

$$SE_b(\bar{x}) = \sqrt{\frac{1}{B-1} * \Sigma(\bar{x}^* - \frac{1}{B}\Sigma\bar{x}^*)^2}$$

The value of $\bar{x}$ is the mean of the original sample. Computing the mean of the 36 original sample data values we get: $\bar{x} = 24.59$.

To determine the appropriate t score for a 95% confidence interval for a sample size of n = 36, we need to compute the degrees of freedom using the formula df = n – 1. Thus, df = 35.

Using Definition 9.12, the appropriate t score for a 95% confidence interval corresponds to the positive critical t score $t_{97.5\%}$ with 35 degrees of freedom. From Table III: Critical Values for the t Distribution, the t score = 2.04. (Note: Table III does not have an entry for 35 degrees of freedom, we selected t = 2.04, the entry for 30 degrees of freedom.)

Finally, we need to determine the standard deviation of the bootstrap distribution using the bootstrap standard error formula. Using Excel and the 1000 bootstrap sample means to compute the standard deviation of the bootstrap distribution, we get: $SE_b(\bar{x}) = 1.49$. Thus, the bootstrap standard error is $SE_b(\bar{x}) = 1.49$.

Substituting these values into the bootstrap standard error t confidence interval formula, we have:

$$\bar{x} \pm (\text{t score}) * SE_b(\bar{x}) = 24.59 \pm (2.04) * (1.49)$$
$$= 24.59 \pm 3.04$$

Therefore, the 95% bootstrap standard error t confidence interval to estimate the population mean body fat percentage for U.S. males aged 12 to 15 years is (21.55, 27.63). Figure 18.9 displays the 95% confidence interval (21.55, 27.63) on the histogram of the bootstrap distribution of body fat percentages. Thus, we are 95% confident that the population mean body fat percentage of U.S. males is between 21.55 and 27.63.

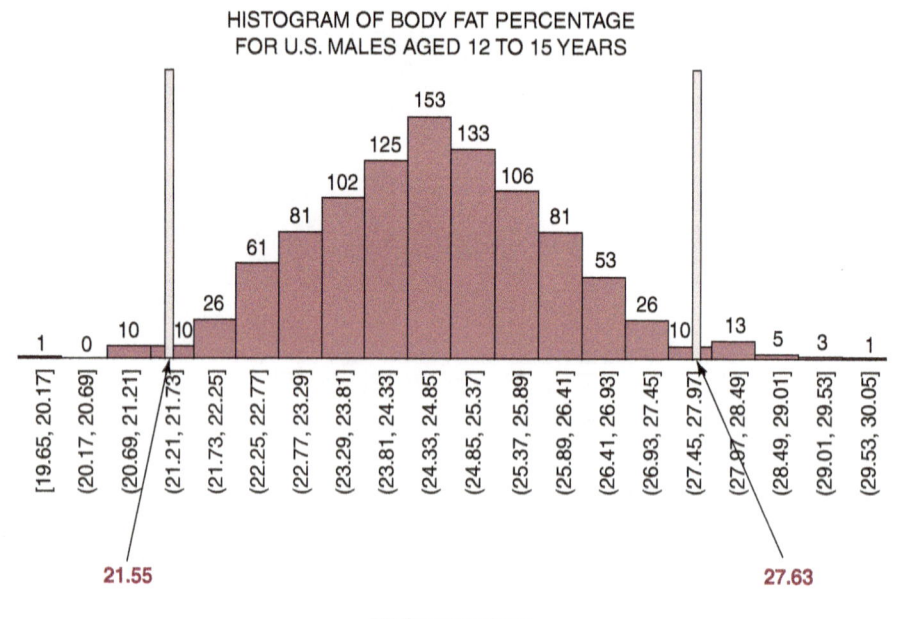

FIGURE 18.9

Method 3 Basic Bootstrap Confidence Interval Method (or the Reverse Bootstrap Percentile Confidence Interval)

Before considering the basic bootstrap method, we need to discuss notation. Let Θ be the value of an unknown population parameter of interest that we want to construct a confidence interval for. The only information we have about the population is the original sample that was randomly selected from the population.

Let $\hat{\Theta}$ be the statistic of interest computed from the original sample. Since the original random sample will serve as an estimate of this population parameter, then $\hat{\Theta}$ will serve as a point estimate of the true population parameter Θ.

The bootstrap procedure is then applied using the original random sample to generate B resamples selected from the original random sample and a bootstrap statistic for each resample is computed and will be denoted by $\hat{\Theta}^*_i$, where the subscript i represents the ith resample. The bootstrap distribution of B statistics is denoted by: $\hat{\Theta}^*_1, \hat{\Theta}^*_2, \hat{\Theta}^*_3, \ldots, \hat{\Theta}^*_B$ and serves as an estimate of the true sampling distribution of the statistic. If we took the average of all the B bootstrap statistics (i.e. $\frac{1}{B}\sum_{i=1}^{B}\hat{\Theta}^*_i$), we would expect the average to be very close to $\hat{\Theta}$. However, this result would not give us any new information about the population parameter Θ. Regardless of how many B resamples are selected, the true sampling distribution and the bootstrap sampling distribution will not be identical. Thus, there will be error in estimating the value of the population parameter Θ using the bootstrap sampling distribution. However, if the true sampling distribution and the bootstrap sampling distribution are reasonably close, then they should exhibit similar amounts of variation. Recall, the true sampling distribution of sample statistics $\hat{\Theta}$ is "centered" at Θ. However, the bootstrap sampling distribution of $\hat{\Theta}^*$ is "centered" at $\hat{\Theta}$. If there is a significant separation between $\hat{\Theta}$ and Θ, then these two sampling distributions will also differ significantly. So, then what we would like to know is how much the distribution of $\hat{\Theta}$ varies around the population parameter Θ. That is, the distribution of $\delta = \hat{\Theta} - \Theta$ describes the variation of $\hat{\Theta}$ about its center Θ. The bootstrap principle offers a practical approach to estimating the distribution of $\delta = \hat{\Theta} - \Theta$. It states that we can approximate $\delta = \hat{\Theta} - \Theta$ by the distribution of $\delta^* = \hat{\Theta}^* - \hat{\Theta}$ since the distribution of $\delta^* = \hat{\Theta}^* - \hat{\Theta}$ describes the variation about its center $\hat{\Theta}$. So, even if the centers are different, the two variations about the centers can be approximately equal.

The basic bootstrap confidence interval for Θ uses the bootstrap statistics: $\hat{\Theta}^*_1, \hat{\Theta}^*_2, \hat{\Theta}^*_3, \ldots, \hat{\Theta}^*_B$ to approximate the distribution $\hat{\Theta} - \Theta$ by $\hat{\Theta}^* - \hat{\Theta}$. We need to compute the $\alpha/2$ and $(1 - \alpha/2)$ quantiles of $\hat{\Theta}^*_1, \hat{\Theta}^*_2, \hat{\Theta}^*_3, \ldots, \hat{\Theta}^*_B$ and denote them by $q_{\alpha/2}$ and $q_{1-\alpha/2}$, and then contend that:

$$\begin{aligned} 1 - \alpha &= P(q_{\alpha/2} \leq \hat{\Theta}^* \leq q_{1-\alpha/2}) \\ &= P(q_{\alpha/2} - \hat{\Theta} \leq \hat{\Theta}^* - \hat{\Theta} \leq q_{1-\alpha/2} - \hat{\Theta}) \\ &\approx P(q_{\alpha/2} - \hat{\Theta} \leq \hat{\Theta} - \Theta \leq q_{1-\alpha/2} - \hat{\Theta}) \\ &= P(q_{\alpha/2} - 2\hat{\Theta} \leq -\Theta \leq q_{1-\alpha/2} - 2\hat{\Theta}) \\ &= P(2\hat{\Theta} - q_{1-\alpha/2} \leq \Theta \leq 2\hat{\Theta} - q_{\alpha/2}) \end{aligned}$$

Thus, the interval $[2\hat{\Theta} - q_{1-\alpha/2}, 2\hat{\Theta} - q_{\alpha/2}]$ serves as the Basic Bootstrap $(1 - \alpha)$ Confidence Interval for Θ. Note the lower limit of the $(1 - \alpha)$ confidence interval is $L = 2\hat{\Theta} - q_{1-\alpha/2}$, where $\hat{\Theta}$ is the value of the original random sample statistic and $q_{1-\alpha/2}$ is the $(1 - \alpha/2)$ quantile of the ordered bootstrap statistics distribution. The upper limit of the $(1 - \alpha)$ confidence interval is $U = 2\hat{\Theta} - q_{\alpha/2}$, where $\hat{\Theta}$ is the value of the original random sample statistic and $q_{1-\alpha/2}$ is the $(\alpha/2)$ quantile of the ordered bootstrap statistics distribution.

Note: It is important to state if *the distributions $\hat{\Theta} - \Theta$ and $\hat{\Theta}^* - \hat{\Theta}$ are not close, then the basic bootstrap confidence interval can be inaccurate.*

To construct the basic bootstrap confidence interval $[2\hat{\Theta} - q_{1-\alpha/2}, 2\hat{\Theta} - q_{\alpha/2}]$ it is necessary to first order all the bootstrap sample statistics from low to high values. Then you will need to identify the sample statistic of the ordered bootstrap sampling distribution that represents quantile $q_{1-\alpha/2}$ and the quantile $q_{\alpha/2}$ using the quantile q_α formula.

DEFINITION 18.15 QUANTILE q_α FORMULA

The quantile q_α is the bootstrap sample statistic representing the α quantile within the ordered bootstrap statistics distribution. The formula to determine the position of the bootstrap statistic within the ordered bootstrap statistics distribution associated to the quantile q_α is:

$$\text{quantile } q_\alpha = \text{Integer value of } [\alpha * (B + 1)],$$

where: B = the number of bootstrap sample statistics

Let's use the Basic Bootstrap Confidence Interval Method to construct a 95% confidence of the mean using the bootstrap distribution of the male body fat percentages presented in Example 18.2.

EXAMPLE 18.4

Use the basic bootstrap confidence interval method to construct a 95% confidence interval of the mean using the bootstrap distribution of the male body fat percentages presented in Example 18.2. The original sample data set is listed in Table 18.11

TABLE 18.11 Random Sample of 36 Body Fat Percentages for U.S. Males aged 12 to 15 Years

15.1	14.4	13.6	15.7	16.1	17.2	15.9	14.8	16.9	16.5	17.1
15.8	18.6	17.4	18.8	18.9	23.0	21.7	20.9	23.6	31.1	28.5
32.8	31.9	24.5	25.8	29.8	36.1	30.9	37.7	36.9	31.9	37.1
40.6	39.2	38.4								

Solution

The basic bootstrap $(1 - \alpha)$ confidence interval is defined by $[2\hat{\theta} - q_{1-\alpha/2}, 2\hat{\theta} - q_{\alpha/2}]$. For a 95% confidence interval, $\alpha = 0.05$. Substituting $\alpha = 0.05$ into the quantile $q_{1-\alpha/2}$, we get:

$q_{1-0.05/2} = q_{0.975}$ and the quantile $q_{\alpha/2}$ is $q_{0.05/2} = q_{0.025}$. Then we can express the basic bootstrap 95% confidence interval as: $[2\hat{\theta} - q_{0.975}, 2\hat{\theta} - q_{0.025}]$.

To construct a 95% basic bootstrap confidence interval, we need to determine the three quantities:

1) $\hat{\theta}$, the mean value of the original sample
2) the value of the quantile $q_{0.975}$.
3) the value of the quantile $q_{0.025}$.

The mean of the original sample data values listed in Table 18.11 is:

$\hat{\theta} = \frac{1}{36}$ (sum of the 36 body fat percentages) $= \frac{1}{36}$ (885.2) ≈ 24.59.

Now we need to determine the bootstrap sample mean within the ordered bootstrap distribution of bootstrap sample means listed in Table 18.8 that represents the quantile $q_{0.975}$. To determine the position of the bootstrap sample mean that represents the quantile $q_{0.975}$ after all the bootstrap sample means have been ordered from low to high values, we use the Quantile q_α Formula:

$$\text{quantile } q_\alpha = \text{Integer value of } [\alpha * (B + 1)]$$

Since there are B = 1,000 bootstrap sample means and $\alpha = 0.975$, then we have:

$$\text{quantile } q_{0.975} = \text{Integer value of } [0.975 * (1,001)]$$
$$= \text{Integer value of } [975.975]$$
$$= 975$$

Thus, the bootstrap sample mean representing quantile $q_{0.975}$ is the 975th value of the ordered bootstrap distribution of sample means within the ordered 1000 values. Using the bootstrap sample means that are listed in numerical order within Table 18.8, the 975th bootstrap sample mean 27.71 represents the quantile $q_{0.975}$. Thus, $q_{0.975} = 27.71$.

The position of the bootstrap sample mean representing the quantile $q_{0.025}$ after all the bootstrap sample means have been ordered from low to high values, we use the formula:

$$\text{quantile } q_\alpha = \text{Integer value of } [\alpha * (B + 1)]$$

Since there are B = 1,000 bootstrap sample means and q = 0.025, then we have:

$$\text{quantile } q_{0.025} = \text{Integer value of } [0.025 * (1,001)]$$
$$= \text{Integer value of } [25.025]$$
$$= 25$$

Thus, the bootstrap sample mean representing quantile $q_{0.025}$ is the 25th value of the ordered bootstrap distribution of sample means within the ordered 1000 values. Using the bootstrap sample means that are listed in numerical order within Table 18.8, the 25th bootstrap sample mean 21.83 represents the quantile $q_{0.025}$. Thus, $q_{0.025} = 21.83$.

Substituting the values: $\hat{\theta} \approx 24.59$, $q_{0.975} = 27.71$, and $q_{0.025} = 21.83$, into basic bootstrap 95% confidence interval: $[2\hat{\theta} - q_{0.975}, 2\hat{\theta} - q_{0.025}]$, we have:

$$[2\hat{\Theta} - q_{1-\alpha/2}, 2\hat{\Theta} - q_{\alpha/2}] = [2*(24.59) - 27.71, 2*(24.59) - 21.83]$$
$$= [49.18 - 27.71, 49.18 - 21.83]$$
$$= [21.47, 27.35]$$

Therefore, the 95% basic bootstrap confidence interval to estimate the population mean body fat percentage for U.S. males aged 12 to 15 years is (21.47, 27.35). Thus, we are 95% confident that the population mean body fat percentage of U.S. males is between 21.47 and 27.35.

Let's compare the 95% confidence bootstrap interval estimates of the population mean body fat percentage of U.S. males for the three examples Example 18.2, Example 18.3 and Example 18.4. The three 95% bootstrap confidence interval methods are listed in Table 18.12.

TABLE 18.12 Comparison of Bootstrap Confidence Intervals

Bootstrap Confidence Interval Method	95% Bootstrap Confidence Interval
Percentile Confidence Interval Method	(21.83, 27.71)
Standard Error t Confidence Interval Method	(21.55, 27.63)
Basic Confidence Interval Method	(21.47, 27.35)

The three bootstrap 95% confidence intervals are different even though they were constructed using the same bootstrap sampling distribution of the bootstrap sample means.

Note, the 95% Basic Confidence Interval and the 95% Percentile Confidence Interval are different, but they have the same width. The 95% Standard Error t confidence interval has the widest 95% confidence interval. Furthermore, the mean of the original sample, $\bar{x} = 24.59$, is centered within the 95% Standard Error t confidence interval. However, the mean of the original sample falls below the center of the 95% Percentile Confidence Interval, but the mean of the original sample falls above the center of the 95% Basic Confidence interval. Figure 18.10 graphically displays the three 95% Confidence Intervals with respect to the mean of the original sample.

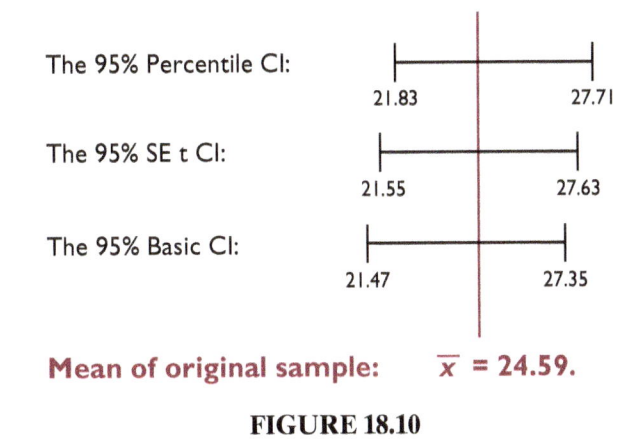

The 95% Percentile CI: 21.83 — 27.71

The 95% SE t CI: 21.55 — 27.63

The 95% Basic CI: 21.47 — 27.35

Mean of original sample: $\bar{x} = 24.59$.

FIGURE 18.10

Important Bootstrap Concepts to Remember when Applying the Bootstrap Confidence Interval Methods

- When a confidence interval of level $(1 - \alpha)$ is constructed using a bootstrap confidence interval method the coverage of the confidence interval or the true confidence interval level is not quite the desired confidence level $(1 - \alpha)$ because of the inability to work with the true sampling distribution. However, the better the bootstrap distribution approximates the true sampling distribution, the more accurate the confidence interval.
- The percentile method for a confidence interval works for any statistic, if the bootstrap distribution is

approximately symmetric and approximately continuous. The percentile bootstrap confidence intervals may not have the correct coverage when the bootstrap sampling distribution is skewed.
- The bootstrap standard error t confidence interval method also requires an approximately bell-shaped bootstrap distribution.
- The bootstrap standard error t method and the basic confidence method use the value of the original random sample statistic when computing a confidence interval. The percentile method relies only on the bootstrap sampling distribution of the bootstrap statistics.
- Once the bootstrap sampling distribution is generated, it is important to always examine a histogram of the bootstrap distribution to observe the shape and variability of the bootstrap sampling distribution.
- When using bootstrapping, you may get a slightly different confidence interval each time. This is fine. The more bootstrap samples you use, the more precise your answer will be. As a general guideline, 1,000 bootstrap samples are fine for rough approximation. In real life, you probably want to take 10,000 or even 100,000 bootstrap samples for more accurate results.
- The most important limitation of the bootstrap process is the assumption that the original sample data values are a reasonable estimate of the population. That is, the original sample must represent the data characteristics of the population in terms of diversity and range from which it was sampled. If the original sample does not reflect the population, then the random sampling performed in the bootstrap procedure may add another level of sampling error, resulting in invalid statistical estimations.
- The bootstrap distribution mirrors the original sample. So, if the original sample is narrower than the true population, the bootstrap distribution will be narrower than the sampling distribution. However, larger original samples of data usually represent the population very well. But, for small original samples of data, the original sample is not likely to represent the entire population well. Thus, the smaller the original sample, the more difficult it becomes to construct acceptable confidence intervals. The bootstrap method relies heavily on the tails of the bootstrap sampling distribution when computing confidence intervals and using small samples may jeopardize the validity of this computation.
- The bootstrap is an extremely useful procedure to estimate the standard error of a statistic.

Bootstrap Confidence Interval Accuracy

The accuracy of a P% confidence interval method is dependent upon whether the particular method produces intervals that capture the population parameter P% of the time as intended. For example, when we apply a confidence interval method to construct a 95% confidence interval the method may give intervals that only capture the true population parameter less than 95% such as 91% or 87% of the time. A confidence interval method for producing 95% confidence intervals is said to be accurate if, when applying the method, 95% of the time the method produces intervals that capture the population parameter and if the 5% misses are distributed equally on each side of the interval between the low misses and the high misses. No bootstrap confidence interval method will produce exactly the intended confidence level in practice and are never exactly accurate. However, the confidence interval methods will be more accurate for larger sample sizes. The three bootstrap confidence intervals that were presented are considered "first order" intervals. First order versus second order accuracy is not about the width of the confidence interval; it is about the coverage probability. That is, their accuracy improves as the original sample size increases. However, to improve accuracy for a first order confidence interval by a factor of 10, the sample size must increase 100 times. Thus, their accuracy is dependent on very large original sample sizes. The percentile method is a simple "first order" interval that uses percentiles of the bootstrap distribution to construct a confidence interval. This method has two limitations. First, the percentile method does not use the original random sample data estimate for the confidence interval. The percentile method is based solely on the bootstrap resamples. Second, the percentile method does not adjust for skewness in the bootstrap distribution. Consequently, it is critical to check the histogram of the bootstrap distribution when considering applying the percentile method.

There are "second order" bootstrap confidence intervals that are highly recommended over the first order intervals since their accuracy improves much faster, requiring only 10 times more data to improve accuracy by a factor of 10. Due to the complexity of these intervals, we will not address those intervals in an introductory statistics textbook. However, we would like to mention a popular second order bootstrap confidence interval called the bias-corrected and accelerated (BCa) bootstrap interval method that corrects for bias and skewness of the bootstrap distribution. The BCa bootstrap interval method is a second-order accurate interval procedure that addresses the bias and skewness issues by estimating two parameters. The bias-correction parameter, z_0, is related to the proportion of bootstrap sample statistic estimates that are less than the original sample statistic. And the acceleration parameter, a, is proportional to the skewness of the bootstrap distribution. The BCa interval method assumes the data are independent and identically distributed. The BCa

KEY TERMS

Basic Bootstrap Confidence Interval   930
Bias of a Statistic   909
Bootstrap Confidence Intervals   919
Bootstrap Distribution   906
Bootstrap Estimate of Bias   909
Bootstrap Idea   917
Bootstrap Percentile Confidence
  Interval   919
Bootstrapping   905
Bootstrap Procedure   906
Bootstrap Sample   905
Bootstrap Sampling Distribution   906
Bootstrap Standard Error   910
Bootstrap Standard Error of a Mean,
  $SE_b(\bar{x})$ 910

Bootstrap Standard Error of a Statistic,
  $SE_b(Statistic)$ 910
Bootstrap Standard Error t   925
Bootstrap Statistic   905
Central Limit Theorem   904
Confidence Interval   919
Empirical Sampling Distribution   905
Inferential Statistics   904
Nonparametric Bootstrap Methods   919
Nonparametric Bootstrapping   905
Original Random Sample   908
Parameter   904
Plug-In Principle   907
Population   903
Proxy Population   905

Random Sample   904
Resampling   905
Reverse Bootstrap Percentile Confidence
  Interval Method   930
Sample   905
Sample Statistic   903
Standard Error   904
Standard Error of a Sampling
  Distribution   904
Statistic   904
Theoretical Sampling Distribution   904

SECTION EXERCISES

Section 18.2

1. The mean of a sample is a:
   a) parameter
   b) statistic
   c) distribution
   d) error

2. An alternate approach to computing the standard error of a sampling distribution is called:
   a) Inferential Statistics
   b) Preferred Selection
   c) Bootstrapping
   d) none of these

3. A bootstrap method that doesn't require any assumptions about the population is called _____.

4. In bootstrapping, samples are randomly selected from an estimate of the population called the _____.

5. The original random sample has the following 5 data values: 20, 22, 23, 24, 31. Would the sample: 22, 20, 20, 31, 31 be a possible bootstrap sample?
   a) No   b) Yes   c) cannot be determined

6. The original random sample has the following 5 data values: 20, 22, 23, 24, 31. Would the sample: 22, 20, 30, 31, 31 be a possible bootstrap sample?
   a) No   b) Yes   c) cannot be determined

7. The original random sample has the following 5 data values: 20, 22, 23, 24, 31. Would the sample: 22, 20, 23, 31 be a possible bootstrap sample?
   a) No   b) Yes   c) cannot be determined

8. The original sample has a sample size of n = 40. The bootstrap procedure is used to generate 2000 resamples. What is the sample size of each bootstrap sample?
   a) 2000   b) 40   c) 40*2000   d) cannot be determined

9. The original sample has a sample size of n = 100. The bootstrap procedure is used to generate 1000 resamples. How many bootstrap statistics will you have?
   a) 100
   b) 1000
   c) 100*1000
   d) cannot be determined

10. Who invented a revolutionary new statistical procedure called the bootstrap?
    a) John Lennon
    b) Karl Pearson
    c) Brad Efron
    d) Albert Einstein

11. In the bootstrap procedure, the original sample is treated as an estimate of the:
    a) bootstrap sample
    b) mean
    c) bootstrap statistic
    d) population

12. In the bootstrap procedure, the samples of size n randomly selected from the original sample of size n are called:
    a) bootstrap statistics
    b) random samples
    c) secondary original samples
    d) resamples

13. The bootstrap sampling distribution of a statistic of interest is usually centered at:
    a) the original sample statistic
    b) the parameter of the underlying population
    c) the center of the theoretical sampling distribution
    d) the statistic of a bootstrap sample

14. The standard deviation of the bootstrap distribution represents an estimate of the:
    a) the standard deviation of the original sample
    b) the standard deviation of a bootstrap sample
    c) the standard error of the true sampling distribution
    d) the standard error of the original sample

15. Bootstrapping procedures that only use information about the original sample and make no assumption about the nature of the underlying population is called:
    a) parametric bootstrapping
    b) nonparametric bootstrapping
    c) original sample bootstrapping
    d) theoretical bootstrapping

16. Bootstrapping can be used to:
    a) estimating standard errors
    b) obtaining confidence intervals
    c) getting more information about population parameter
    d) only a and b
    e) all of these

17. If the original sample has n data values symbolized as: $x_1, x_2, x_3, \ldots, x_n$ then the n data values of the bootstrap sample would be identified as:
    a) $x_1^*, x_2^*, x_3^*, \ldots, x_n^*$
    b) $x_1', x_2', x_3', \ldots, x_n'$
    c) $x_1^b, x_2^b, x_3^b, \ldots, x_n^b$
    d) $x_1^\wedge, x_2^\wedge, x_3^\wedge, \ldots, x_n^\wedge$

18. If 2000 resamples are selected from the original sample, then:
    a) B = 1000   b) N = 2000   c) n = 2000   d) X = 2000

19. The notation for the sample mean of a bootstrap sample would be:
    a) x^* b) x   c) s   d) $\bar{x}^*$

20. The notation for the sample standard deviation of a bootstrap sample is:
    a) x^* b) s^* c) s   d) $\bar{x}^*$

21. The mean of the sampling distribution minus the value of the population parameter is:
    a) sample error
    b) standard error
    c) statistical error
    d) bias of a statistic

22. The mean of the bootstrap sample statistics minus the original sample statistic is:
    a) bootstrap estimate of bias
    b) statistical error
    c) sample error

Section 18.3

23. If Θ represents the parameter of a population, then $\hat{\Theta}$ represents the statistic of the:
    a) sample
    b) random sample
    c) Original sample
    d) bootstrap sample

24. The formula $SE_b(\bar{x}) = \sqrt{\frac{1}{B-1} * \sum(\bar{x}^* - \frac{1}{B}\sum \bar{x}^*)^2}$ is used to compute the:
    a) standard deviation of the original sample
    b) standard deviation of the bootstrap sample
    c) standard error of the bootstrap distribution
    d) standard deviation of the true sampling distribution

25. The standard deviation of the bootstrap distribution represents an estimate of the:
    a) standard deviation of the original sample
    b) population standard deviation
    c) standard deviation of the bootstrap sample
    d) standard error of the true sampling distribution

26. Consider the small original random sample of 6 data values: 22, 20, 31, 24, 40, 18
    a) Generate 20 resamples of size 6 using a fair die. Roll a fair die to sample of size 6 with replacement from this original sample. Use the following table to identify the members of the resample.

Value on Top face of Rolled Die	1	2	3	4	5	6
Member selected from the original sample	22	20	31	24	40	18

b) Compute the mean of each bootstrap sample.
c) Create a histogram of the 20 bootstrap sample means.
d) Compute the standard error of the bootstrap distribution using the 20 bootstrap sample means.
e) What would you use to estimate the standard error of the true population sampling distribution?
f) What is the bootstrap estimate of the bias from your bootstrap sample means of the bootstrap distribution? What does this tell us about the bias encountered in using the mean of the original sample to estimate the true population mean?

27. The following random sample represents the age at death of U.S. smokers.

81	65	89	70	87	61	94	98	81	70
62	86	89	75	76	69	69	61	72	77
90	88	74	88	90	87	95	66	72	84
79	72	92	77	45	75	70	91	73	67

Assuming this random sample is a good estimate of the population, then:
a) compute an estimate of the median age life expectancy of U.S. smokers assuming the given random sample is representative of all U.S. smokers.
b) define the original sample if we are interested in generating the bootstrap distribution of the median.
c) select 1,250 resamples and compute the median of each bootstrap sample to generate a bootstrap distribution of the median.
d) construct a histogram of the bootstrap distribution of the median.
e) compute the mean of the bootstrap distribution of the median. Compare this result to the median of the original sample.
f) compute the bootstrap standard error of the median. Did you have a formula to compute this standard error? Explain.
g) Why is bootstrapping an important statistical concept when generating the bootstrap distribution of the median?
h) What is the bootstrap estimate of the bias from your bootstrap sample medians of the bootstrap distribution? What does this tell us about the bias encountered in using the median of the original sample to estimate the median age life expectancy of U.S. smokers?
i) summarize the characteristics of the bootstrap distribution of the median using the previous results.

28. The following random sample represents the dollar amounts per order of 60 Amazon customers.

Amazon Customer Dollar Purchases Per Order

$39.88	$210.99	$16.85	$47.40	$89.99	$29.94	$23.77	$231.05	$95.97	$18.75
$23.26	$17.44	$35.98	$15.99	$51.94	$10.00	$107.95	$99.95	$89.99	$17.95
$22.72	$69.99	$39.15	$11.98	$29.99	$89.99	$38.24	$109.99	$139.99	$14.99
$26.65	$17.42	$43.97	$229.00	$11.99	$69.00	$78.95	$36.88	$89.90	$23.99
$40.91	$49.99	$19.95	$29.24	$245.99	$89.99	$55.37	$35.43	$89.99	$15.99
$64.35	$89.99	$37.30	$25.05	$155.30	$23.19	$33.95	$16.99	$89.99	$74.06

Assuming this sample is representative of the purchase orders of all Amazon customers, then:
a) construct a histogram of the sample data.
b) state the shape of the histogram.

c) define the original sample if we are interested in generating the bootstrap distribution of the mean.
d) select 1,000 resamples and compute the mean of each bootstrap sample to generate a bootstrap distribution of the mean.
e) construct a histogram of the bootstrap distribution of the mean
f) state the shape of the histogram of the bootstrap distribution of the mean.
g) the central limit theorem (chapter 8) states the true sampling distribution of the mean is approximately a normal distribution for a large sample size. Examine the shape of the histogram you constructed in part d and comment on what you expected the histogram shape to be?
h) compute the mean of the bootstrap distribution and compare it to the mean of the original sample. Did you expect these results? Explain.
i) compute the bootstrap standard error of the mean using the bootstrap distribution of the mean.
j) use the original sample data and the formula $\frac{s}{\sqrt{n}}$ to estimate the standard error of the true sampling distribution of the mean.
k) compare the result of part j to the bootstrap standard error of the mean. How close are the results?
l) Using the results of parts i and j, what is the relationship between the standard deviation of the bootstrap distribution of the mean to the standard error of the true sampling distribution of the mean?
m) What is the bootstrap estimate of the bias from your bootstrap sample means of the bootstrap distribution? What does this tell us about the bias encountered in using the mean of the original sample to estimate the mean dollar purchase per Amazon customer order?

29. The following random sample of the number of hours college students in Nassau County in Long Island, NY work in a week during the academic year.

Number of Hours College Students Work Weekly

10	16	36	27	0	28	40	35	27	29
45	19	24	34	37	42	40	20	0	0
25	26	18	32	40	0	20	42	40	0
22	26	40	38	45	38	0	40	35	50

Suppose one is interested in estimating the variability in the number of hours college students in Nassau County work per week during an academic year.

Assuming this sample is representative of the number of hours college students in Nassau County work per week during the academic year, then:

a) construct a histogram of the sample data.
b) state the shape of the histogram.
c) define the original sample if one is interested in generating the bootstrap distribution of the standard deviation of the weekly work hours of a college student.
d) select 1,000 resamples and compute the sample standard deviation of each bootstrap sample to generate a bootstrap distribution of the standard deviation.
e) construct a histogram of the bootstrap distribution of the standard deviation
f) state the shape of the histogram of the bootstrap distribution of the standard deviation

g) compute the mean of the bootstrap distribution and compare it to the standard deviation of the original sample. What did you expect when comparing these two results?
h) compute the bootstrap standard error of the standard deviation using the bootstrap distribution of the standard deviation. Did you have a formula to compute this standard error? Explain.
i) Why is the bootstrapping procedure essential when generating the bootstrap distribution of the standard deviation based on part h?
j) What is the bootstrap estimate of the bias from your bootstrap sample standard deviations of the bootstrap distribution? What does this tell us about the bias encountered in using the standard deviation of the original sample to estimate the standard deviation of the college student weekly work hours?
k) summarize the characteristics of the bootstrap distribution of the standard deviation using the previous results.

Section 18.4

30. A resampling process that is used to generate an empirical sampling distribution that doesn't require any assumptions about the population or the original sample is called _____ _____.

31. The lower bound of a bootstrap percentile 95% confidence interval is the:
   a) 5% percentile   b) 2.5% percentile
   c) 95% percentile   d) 97.5% percentile

32. If there are 2,000 bootstrap sample means within the bootstrap sampling distribution of means, then the bootstrap sample mean representing the 2.5^{th} percentile is the _____ sample value that serves as the lower 95% confidence interval value within the ordered 2,000 sample mean values.
   a) 25th b) 50th c) 2.5^{th} d) 95th

33. A statistician decides to use the bootstrap percentile confidence interval method to construct a confidence to estimate the population mean body fat percentage for U.S. males aged 12 to 15 years using the following random sample information.

RANDOM SAMPLE of 36 Body Fat Percentages for U.S. Males aged 12 to 15 Years

15.1	14.4	13.6	15.7	16.1	17.2	15.9	14.8	16.9	16.5	17.1
15.8	18.6	17.4	18.8	18.9	23.0	21.7	20.9	23.6	31.1	28.5
32.8	31.9	24.5	25.8	29.8	36.1	30.9	37.7	36.9	31.9	37.1
40.6	39.2	38.4								

a) Generate a bootstrap distribution of the mean using the random sample as the best estimate of the population and B = 2,000 bootstrap samples.
b) Construct a histogram of the bootstrap distribution to determine if the shape is approximately normal and check to see if the bias is small.
c) If the bootstrap distribution of the mean is normal with a small bias, then construct a 95% confidence interval using the bootstrap percentile confidence interval method and the B = 2,000 bootstrap samples. Write a statement explaining the meaning of the 95% confidence interval.
d) Generate a bootstrap distribution of the mean using the random sample as the best estimate of the population and B = 2,500 bootstrap samples.

e) Construct a histogram of the bootstrap distribution to determine if the shape is approximately normal and check to see if the bias is small.
f) If the bootstrap distribution of the mean is normal with a small bias, then construct a 95% confidence interval using the bootstrap percentile confidence interval method and the B = 2,500 bootstrap samples. Write a statement explaining the meaning of the 95% confidence interval.
g) Compare the confidence intervals from parts c and f. What do you notice about the width of the 95% confidence intervals when increasing the value of B?
h) Graphically, show where the mean of the original sample falls within the 95% confidence intervals of parts c and f?

34. The bootstrap standard error t confidence interval formula to construct an approximate confidence interval of level C for a population parameter for a statistic is

statistic ± (t score) * _____.

35. The notation for the bootstrap standard error of a statistic is:
a) $SE_b(\bar{x})$
b) $SE_b(\text{mean})$
c) $SE_b(\text{statistic})$
d) $SE_b(\text{parameter})$

36. A researcher is planning to construct a confidence interval to estimate the population mean body fat percentage for U.S. males aged 12 to 15 years using the bootstrap standard error t confidence interval based on the following random sample information.

RANDOM SAMPLE of 36 Body Fat Percentages for U.S. Males aged 12 to 15 Years

15.1	14.4	13.6	15.7	16.1	17.2	15.9	14.8	16.9	16.5	17.1
15.8	18.6	17.4	18.8	18.9	23.0	21.7	20.9	23.6	31.1	28.5
32.8	31.9	24.5	25.8	29.8	36.1	30.9	37.7	36.9	31.9	37.1
40.6	39.2	38.4								

a) Generate a bootstrap distribution of the mean using the random sample as the best estimate of the population and B = 1,500 bootstrap samples.
b) Construct a histogram of the bootstrap distribution to determine if the shape is approximately normal and check to see if the bias is small.
c) If the bootstrap distribution of the mean is normal with a small bias, then construct a 90% confidence interval using the bootstrap standard error t confidence interval method and the B = 1,500 bootstrap samples. Write a statement explaining the meaning of the 90% confidence interval.
d) Generate a bootstrap distribution of the mean using the random sample as the best estimate of the population and B = 2,500 bootstrap samples.
e) Construct a histogram of the bootstrap distribution to determine if the shape is approximately normal and check to see if the bias is small.
f) If the bootstrap distribution of the mean is normal with a small bias, then construct a 90% confidence interval using the bootstrap standard error t confidence interval method and the B = 2,500 bootstrap samples. Write a statement explaining the meaning of the 90% confidence interval.
g) Compare the confidence intervals from parts c to f. What do you notice about the width of the 90% confidence intervals and the value of the bootstrap standard error when increasing the value of B?
h) Graphically, show where the mean of the original sample falls within the 90% confidence intervals of parts c and f?

37. A statistician decides to the basic bootstrap confidence interval method to construct a confidence to estimate the population mean body fat percentage for U.S. males aged 12 to 15 years using the following random sample information.

RANDOM SAMPLE of 36 Body Fat Percentages for U.S. Males aged 12 to 15 Years

15.1	14.4	13.6	15.7	16.1	17.2	15.9	14.8	16.9	16.5	17.1
15.8	18.6	17.4	18.8	18.9	23.0	21.7	20.9	23.6	31.1	28.5
32.8	31.9	24.5	25.8	29.8	36.1	30.9	37.7	36.9	31.9	37.1
40.6	39.2	38.4								

a) Construct a histogram of the bootstrap sampling distribution of the mean to determine if it is reasonable to use the basic bootstrap confidence interval method for B = 1,500 bootstrap samples. If it is reasonable, then construct 90% confidence interval using the basic bootstrap confidence interval method and B = 1,500 bootstrap samples.
b) Construct a histogram of the bootstrap sampling distribution of the mean to determine if it is reasonable to use the basic bootstrap confidence interval method for B = 2,500 bootstrap samples. If it is reasonable, then construct 90% confidence interval using the basic bootstrap confidence interval method and B = 2,500 bootstrap samples.
c) Construct a histogram of the bootstrap sampling distribution of the mean to determine if it is reasonable to use the basic bootstrap confidence interval method for B = 1,500 bootstrap samples. If it is reasonable, then construct 95% confidence interval using the basic bootstrap confidence interval method and B = 1,500 bootstrap samples.
d) Construct a histogram of the bootstrap sampling distribution of the mean to determine if it is reasonable to use the basic bootstrap confidence interval method for B = 2,500 bootstrap samples. If it is reasonable, then construct 95% confidence interval using the basic bootstrap confidence interval method and B = 2,500 bootstrap samples.
e) Compare the confidence intervals from parts a to d. What do you notice about the width of the 90% and 95% confidence intervals when increasing the value of B?
f) Graphically, show where the mean of the original sample falls within the 90% and the 95% confidence intervals?

38. The following random sample represents the age at death of U.S. smokers.

81	65	89	70	87	61	94	98	81	70
62	86	89	75	76	69	69	61	72	77
90	88	74	88	90	87	95	66	72	84
79	72	92	77	45	75	70	91	73	67

Assuming this random sample is a good estimate of the population, then:

a) construct a histogram of the bootstrap sampling distribution of the median to determine if it is reasonable to use the bootstrap percentile confidence interval method for

B = 1,500 bootstrap samples. If it is reasonable, then construct a 95% confidence interval to estimate the median age life expectancy of U.S. smokers interval using the bootstrap percentile confidence interval method and B = 1,500 bootstrap samples.

b) construct a histogram of the bootstrap sampling distribution of the median to determine if it is reasonable to use the bootstrap percentile confidence interval method for B = 2,500 bootstrap samples. If it is reasonable, then construct a 95% confidence interval to estimate the median age life expectancy of U.S. smokers interval using the bootstrap percentile confidence interval method and B = 2,500 bootstrap samples.

c) construct a histogram of the bootstrap sampling distribution of the median to determine if it is reasonable to use the basic bootstrap confidence interval method for B = 1,500 bootstrap samples. If it is reasonable, then construct a 95% confidence interval to estimate the median age life expectancy of U.S. smokers interval using the basic bootstrap interval method and B = 1,500 bootstrap samples.

d) construct a histogram of the bootstrap sampling distribution of the median to determine if it is reasonable to use the bootstrap percentile confidence interval method for B = 1,500 bootstrap samples. If it is reasonable, then construct a 95% confidence interval to estimate the median age life expectancy of U.S. smokers interval using the basic bootstrap interval method and B = 2,500 bootstrap samples.

e) Compare the confidence intervals from parts a to d. What do you notice about the width of the 95% confidence intervals when increasing the value of B?

f) What do notice about the 95% confidence intervals using the different bootstrap confidence interval methods?

39. The following random sample represents the age at death of U.S. non-smokers.

Age at Death of U.S. Non-Smokers

81	65	89	70	87	61	94	98	81	70	62
86	89	75	76	69	69	61	72	77	90	88
74	88	90	87	95	66	72	84	79	72	92
77	45	75	70	91	73	67				

Assuming this random sample is a good estimate of the population, then:

a) construct a histogram of the bootstrap sampling distribution of the median to determine if it is reasonable to use the bootstrap percentile confidence interval for B = 1,500 bootstrap samples. If it is reasonable, then construct a 95% confidence interval to estimate the median age life expectancy of U.S. non-smokers interval using the bootstrap percentile confidence interval method and B = 1,500 bootstrap samples.

b) construct a histogram of the bootstrap sampling distribution of the median to determine if it is reasonable to use the bootstrap percentile confidence interval for B = 2,500 bootstrap samples. If it is reasonable, then construct a 95% confidence interval to estimate the median age life expectancy of U.S. non-smokers interval using the bootstrap percentile confidence interval method and B = 2,500 bootstrap samples.

c) construct a histogram of the bootstrap sampling distribution of the median to determine if it is reasonable to use the basic bootstrap confidence interval for B = 1,500 bootstrap samples. If it is reasonable, then construct a 95% confidence interval to estimate the median age life expectancy of U.S. non-smokers interval using the basic bootstrap interval method and B = 1,500 bootstrap samples.

d) construct a histogram of the bootstrap sampling distribution of the median to determine if it is reasonable to use the basic bootstrap confidence interval for B = 2,500 bootstrap samples. If it is reasonable, then construct a 95% confidence interval to estimate the median age life expectancy of U.S. non-smokers interval using the basic bootstrap interval method and B = 2,500 bootstrap samples.

e) Compare the confidence intervals from parts a to d. What do you notice about the width of the 95% confidence intervals when increasing the value of B?

f) What do notice about the 95% confidence intervals using the different bootstrap confidence interval methods?

40. The following random sample represents the dollar amounts per order by 60 Amazon customers.

Amazon Customer Dollar Purchases Per Order

$39.88	$210.99	$16.85	$47.40	$89.99	$29.94	$23.77	$231.05	$95.97	$18.75
$23.26	$17.44	$35.98	$15.99	$51.94	$10.00	$107.95	$99.95	$89.99	$17.95
$22.72	$69.99	$39.15	$11.98	$29.99	$89.99	$38.24	$109.99	$139.99	$14.99
$26.65	$17.42	$43.97	$229.00	$11.99	$69.00	$78.95	$36.88	$89.90	$23.99
$40.91	$49.99	$19.95	$29.24	$245.99	$89.99	$55.37	$35.43	$89.99	$15.99
$64.35	$89.99	$37.30	$25.05	$155.30	$23.19	$33.95	$16.99	$89.99	$74.06

Assuming this sample is representative of the purchase orders of all Amazon customers, then:

a) construct a histogram of the bootstrap sampling distribution of the mean to determine if it is reasonable to use the bootstrap percentile confidence interval method for B = 1,500 bootstrap samples. If it is reasonable, then construct a 90% confidence interval to estimate the population mean dollar purchase per order for Amazon customers using the bootstrap percentile confidence interval method and B = 1,500.

b) construct a histogram of the bootstrap sampling distribution of the mean to determine if it is reasonable to use the bootstrap standard error t confidence interval method for B = 1,500 bootstrap samples. If it is reasonable, then construct a 90% confidence interval to estimate the population mean dollar purchase per order for Amazon customers using the bootstrap standard error t confidence interval method where B = 1,500.

c) construct a histogram of the bootstrap sampling distribution of the mean to determine if it is reasonable to use the basic bootstrap confidence interval method for B = 1,500 bootstrap samples. If it is reasonable, then construct a 90% confidence interval to estimate the population mean dollar purchase per order for Amazon customers using the basic bootstrap confidence interval method and B = 1,500.

d) compare the confidence intervals from parts a and c. What do you notice about the width of the 90% confidence intervals for the different methods?

e) graphically, where does the mean of the original sample fall within the three confidence intervals? Explain.

41. The following random sample of the number of hours college students in Nassau County in Long Island, NY work in a week during the academic year.

Number of Hours College Students Work Weekly

10	16	36	27	0	28	40	35	27	29
45	19	24	34	37	42	40	20	0	0
25	26	18	32	40	0	20	42	40	0
22	26	40	38	45	38	0	40	35	50

Assuming this sample is representative of the number of hours college students in Nassau County work per week during the academic year and it is reasonable to use the bootstrap percentile and basic bootstrap confidence intervals to estimate the variability in the number of hours college students work, then:

a) construct a 95% confidence interval to estimate the population standard deviation of the college student weekly work hours using the bootstrap percentile confidence interval method and B = 1,500. Write a statement explaining the meaning of the 95% confidence interval.

b) construct a 95% confidence interval to estimate the population standard deviation of the college student weekly work hours using the basic bootstrap confidence interval method and B = 1,500. Write a statement explaining the meaning of the 95% confidence interval.

c) compare the confidence intervals from parts a and b. What do you notice about the width of the 95% confidence intervals for the different methods?

d) graphically, show the value of the standard deviation of the original sample falls within the previous confidence intervals.

CHAPTER REVIEW

42. A bootstrap method that doesn't require any assumptions about the population is called _____.

43. In bootstrapping, samples are randomly selected from a proxy population called the _____.

44. The original random sample has the following 5 data values: 40, 42, 43, 44, 51. Would the sample: 42, 40, 40, 51, 51 be a possible bootstrap sample?
    a) No b) Yes c) cannot be determined

45. The original random sample has the following 5 data values: 40, 42, 43, 44, 51. Would the sample: 42, 40, 50, 51, 51 be a possible bootstrap sample?
    a) No b) Yes c) cannot be determined

46. The original random sample has the following 5 data values: 40, 42, 43, 44, 51. Would the sample: 42, 40, 43, 51 be a possible bootstrap sample?
    a) No b) Yes c) cannot be determined

47. The original sample has a sample size of n = 50. The bootstrap procedure is used to generate 3000 resamples. What is the sample size of each bootstrap sample?
    a) 3000 b) 50 c) 50*3000 d) cannot be determined

48. The following random sample represents the age at death of U.S. non-smokers.

Age at Death of U.S. Non-Smokers

81	65	89	70	87	61	94	98	81	70	62
86	89	75	76	69	69	61	72	77	90	88
74	88	90	87	95	66	72	84	79	72	92
77	45	75	70	91	73	67				

Assuming this random sample is a good estimate of the population, then:

a) construct a histogram of the bootstrap sampling distribution of the median to determine if it is reasonable to use the bootstrap percentile confidence interval method for B = 1,500 bootstrap samples. If it is reasonable, then construct a 90% confidence interval to estimate the median age life expectancy of U.S. non-smokers interval using the bootstrap percentile confidence interval method and B = 1,000 bootstrap samples.

b) construct a 90% confidence interval to estimate the median age life expectancy of U.S. non-smokers interval using the bootstrap percentile confidence interval method and B = 3,000 bootstrap samples.

c) construct a 90% confidence interval to estimate the median age life expectancy of U.S. non-smokers interval using the basic bootstrap interval method and B = 1,000 bootstrap samples.

d) construct a 90% confidence interval to estimate the median age life expectancy of U.S. non-smokers interval using the basic bootstrap interval method and B = 3,000 bootstrap samples.

e) Compare the confidence intervals from parts a and d. What do you notice about the width of the 90% confidence intervals when increasing the value of B?

f) What do notice about the 90% confidence intervals using the different bootstrap confidence interval methods?

g) Graphically, show where the median value of the original sample falls within the previous confidence intervals.

49. A statistician wants to use the bootstrap percentile confidence interval method to construct a confidence to estimate the population mean body fat percentage for U.S. males aged 12 to 15 years using the following random sample information.

RANDOM SAMPLE of 36 Body Fat Percentages for U.S. Males aged 12 to 15 Years

15.1	14.4	13.6	15.7	16.1	17.2	15.9	14.8	16.9	16.5	17.1
15.8	18.6	17.4	18.8	18.9	23.0	21.7	20.9	23.6	31.1	28.5
32.8	31.9	24.5	25.8	29.8	36.1	30.9	37.7	36.9	31.9	37.1
40.6	39.2	38.4								

a) Construct a histogram of the bootstrap sampling distribution of the mean to determine if it is reasonable to use the bootstrap percentile confidence interval method for B = 1,500 bootstrap samples. If it is reasonable, then construct 90% confidence interval using the bootstrap percentile confidence interval method and B = 1,500 bootstrap samples.

b) Construct a histogram of the bootstrap sampling distribution of the mean to determine if it is reasonable to use the bootstrap percentile confidence interval method for B = 2,500 bootstrap samples. If it is reasonable, then construct 90% confidence interval using the bootstrap percentile confidence interval method and B = 2,500 bootstrap samples.

c) Construct a histogram of the bootstrap sampling distribution of the mean to determine if it is reasonable to use the bootstrap percentile confidence interval method for B = 1,500 bootstrap samples. If it is reasonable, then construct 95% confidence interval using the bootstrap percentile confidence interval method and B = 1,500 bootstrap samples.

d) Construct a histogram of the bootstrap sampling distribution of the mean to determine if it is reasonable to use the bootstrap percentile confidence interval method for B = 2,500 bootstrap samples. If it is reasonable, then construct 95% confidence interval using the bootstrap percentile confidence interval method and B = 2,500 bootstrap samples.

e) Compare the confidence intervals from parts a to d. What do you notice about the width of the confidence intervals when increasing the value of B?

50. The following random sample represents the dollar amounts per order by 60 Amazon customers.

Amazon Customer Dollar Purchases Per Order

$39.88	$210.99	$16.85	$47.40	$89.99	$29.94	$23.77	$231.05	$95.97	$18.75
$23.26	$17.44	$35.98	$15.99	$51.94	$10.00	$107.95	$99.95	$89.99	$17.95
$22.72	$69.99	$39.15	$11.98	$29.99	$89.99	$38.24	$109.99	$139.99	$14.99
$26.65	$17.42	$43.97	$229.00	$11.99	$69.00	$78.95	$36.88	$89.90	$23.99
$40.91	$49.99	$19.95	$29.24	$245.99	$89.99	$55.37	$35.43	$89.99	$15.99
$64.35	$89.99	$37.30	$25.05	$155.30	$23.19	$33.95	$16.99	$89.99	$74.06

Assuming this sample is representative of the purchase orders of all Amazon customers, then:

a) construct a histogram of the bootstrap sampling distribution of the mean to determine if it is reasonable to use the bootstrap percentile confidence interval method for B = 2,500 bootstrap samples. If it is reasonable, then construct a 95% confidence interval to estimate the population mean dollar purchase per order for Amazon customers using the bootstrap percentile confidence interval method and B = 2,500.

b) construct a histogram of the bootstrap sampling distribution of the mean to determine if it is reasonable to use the bootstrap standard error t confidence interval method for B = 2,500 bootstrap samples. If it is reasonable, then construct a 95% confidence interval to estimate the population mean dollar purchase per order for Amazon customers using the bootstrap standard error t confidence interval method where B = 2,500.

c) construct a histogram of the bootstrap sampling distribution of the mean to determine if it is reasonable to use the basic bootstrap confidence interval method for B = 2,500 bootstrap samples. If it is reasonable, then construct a 95% confidence interval to estimate the population mean dollar purchase per order for Amazon customers using the basic bootstrap confidence interval method and B = 2,500.

d) Compare the confidence intervals from parts a to c. What do you notice about the width of the 95% confidence intervals for the different methods?

e) Graphically, where does the mean of the original sample fall within the three confidence intervals? Explain.

51. A statistician prefers to use the bootstrap standard error t confidence interval to construct a confidence to estimate the population mean body fat percentage for U.S. males aged 12 to 15 years using the following random sample information.

RANDOM SAMPLE of 36 Body Fat Percentages for U.S. Males aged 12 to 15 Years

15.1	14.4	13.6	15.7	16.1	17.2	15.9	14.8	16.9	16.5	17.1
15.8	18.6	17.4	18.8	18.9	23.0	21.7	20.9	23.6	31.1	28.5
32.8	31.9	24.5	25.8	29.8	36.1	30.9	37.7	36.9	31.9	37.1
40.6	39.2	38.4								

Before the statistician can use this confidence method, she decides to generate the bootstrap distribution of the mean to determine if the bootstrap distribution is approximately normal.

a) Construct the bootstrap distribution of the mean using B = 2,000 bootstrap samples.

b) Construct a histogram of the bootstrap distribution of the mean to determine if the distribution is approximately normal.

c) If the distribution is approximately normal, then apply the bootstrap standard error t confidence interval formula to construct a 95% confidence interval.

d) Write a statement explaining the meaning of the 95% confidence interval.

WHAT DO YOU THINK?

52. **WHAT DO YOU THINK?**

Suppose a layperson who doesn't know much about statistics asks you to explain why bootstrapping works. Specifically, suppose the person said, "how is resampling from a sample rather than the population helping you learn something about the population rather than only the sample?" You comment that the bootstrapping procedure assumes the sample is the best estimate of the population. Furthermore, you state resampling is not done to provide an estimate of the true population distribution, it is done to provide an estimate of the true sampling distribution. Using this last statement as the key idea, explain how you would help the layperson understand how the bootstrapping procedure is used to learn about the standard error of the true sampling distribution of a statistic.

53. **WHAT DO YOU THINK?**

Bootstrap distributions mimic the shape, variability and bias of the true sampling distributions. What does a small bias mean tell us about the center of the bootstrap distribution and what would that suggest about the true sampling distribution?

PROJECTS

54. Consider the small original random sample of 6 data values: 12, 10, 21, 14, 30, 18
  a) Generate 20 resamples of size 6 using a fair die. Roll a fair die to sample of size 6 with replacement from this original sample. Use the following table to identify the members of the resample.

Value on Top face of Rolled Die	1	2	3	4	5	6
Member selected from the original sample	12	10	21	14	30	18

  b) Compute the mean of each bootstrap sample.
  c) Create a graph of the 20 bootstrap sample means.
  d) Compute the standard error of the bootstrap distribution using the 20 bootstrap sample means.
  e) What would you use to estimate the standard error of the true population sampling distribution?
  f) What is the bootstrap estimate of the bias from your bootstrap sample means of the bootstrap distribution? What does this tell us about the bias encountered in using the mean of the original sample to estimate the true population mean?

55. Consider the small original random sample of 6 data values: 32, 30, 41, 34, 50, 28
  a) Generate 20 resamples of size 6 using a fair die. Roll a fair die to sample of size 6 with replacement from this original sample. Use the following table to identify the members of the resample.

Value on Top face of Rolled Die	1	2	3	4	5	6
Member selected from the original sample	32	30	41	34	50	28

  b) Compute the mean of each bootstrap sample.
  c) Create a graph of the 20 bootstrap sample means.
  d) Compute the standard error of the bootstrap distribution using the 20 bootstrap sample means.
  e) What would you use to estimate the standard error of the true population sampling distribution?
  f) What is the bootstrap estimate of the bias from your bootstrap sample means of the bootstrap distribution? What does this tell us about the bias encountered in using the mean of the original sample to estimate the true population mean?

CHAPTER TEST

56. The original random sample has the following 5 data values: 70, 72, 73, 74, 91. Would the sample: 72, 70, 73, 91 be a possible bootstrap sample?
  a) No   b) Yes   c) cannot be determined

57. The original sample has a sample size of n = 90. The bootstrap procedure is used to generate 5000 resamples. What is the sample size of each bootstrap sample?
  a) 5000   b) 90   c) 90*3000   d) cannot be determined

58. The bootstrap standard error is the _____ _____ of the bootstrap distribution.

59. Bootstrap distribution of a statistic typically has approximately the same shape and spread as the _____ sampling distribution of the statistic.

60. The bootstrap distribution is centered at population parameter.
  a) True   b) False   c) Not enough information

61. The following random sample of the number of hours college students in Nassau County in Long Island, NY work in a week during the academic year.

  Number of Hours College Students Work Weekly

10	16	36	27	0	28	40	35	27	29
45	19	24	34	37	42	40	20	0	0
25	26	18	32	40	0	20	42	40	0
22	26	40	38	45	38	0	40	35	50

Suppose one is interested in estimating the variability in the number of hours college students in Nassau County work per week during an academic year.

Assuming this sample is representative of the number of hours college students in Nassau County work per week during the academic year, then:

  a) construct a 90% confidence interval to estimate the population standard deviation of the college student weekly work hours using the bootstrap percentile confidence interval method and B = 2,500 provided it is reasonable to use this method. Write a statement explaining the meaning of the 90% confidence interval.
  b) construct a 90% confidence interval to estimate the population standard deviation of the college student weekly work hours using the basic bootstrap confidence interval method and B = 2,500 provided it is reasonable to use this method. Write a statement explaining the meaning of the 90% confidence interval.
  c) compare the confidence intervals from parts a and b. What do you notice about the width of the 90% confidence intervals for the different methods?
  d) construct a histogram of the bootstrap distribution of the standard deviation
  e) state the shape of the histogram of the bootstrap distribution of the standard deviation
  f) compute the standard deviation of the bootstrap distribution and compare it to the standard deviation of the original sample. What did you expect when comparing these two results?
  g) compute the bootstrap standard error of the standard deviation using the bootstrap distribution of the standard deviation. Did you have a formula to compute this standard error? Explain.
  h) Why is bootstrapping an important statistical concept when generating the bootstrap distribution of the standard deviation based on part g?
  i) What is the bootstrap estimate of the bias from your bootstrap sample standard deviations of the bootstrap distribution? What does this tell us about the bias

encountered in using the standard deviation of the original sample to estimate the standard deviation of the college student weekly work hours?
  j) summarize the characteristics of the bootstrap distribution of the standard deviation using the previous results
62. A statistician wants to use the bootstrap percentile confidence interval method and the basic confidence interval method to construct a confidence to estimate the population mean body fat percentage for U.S. males aged 12 to 15 years using the following random sample information.

RANDOM SAMPLE of 36 Body Fat Percentages for U.S. Males aged 12 to 15 Years

15.1	14.4	13.6	15.7	16.1	17.2	15.9	14.8	16.9	16.5	17.1
15.8	18.6	17.4	18.8	18.9	23.0	21.7	20.9	23.6	31.1	28.5
32.8	31.9	24.5	25.8	29.8	36.1	30.9	37.7	36.9	31.9	37.1
40.6	39.2	38.4								

  a) Construct 90% confidence interval using the bootstrap percentile confidence interval method and B = 1,000 bootstrap samples provided it is reasonable to use this method.
  Write a statement explaining the meaning of the 90% confidence interval.
  b) Construct 90% confidence interval using the basic bootstrap confidence interval method and B = 1,000 bootstrap samples provided it is reasonable to use this method.
  Write a statement explaining the meaning of the 90% confidence interval.
  c) Construct 95% confidence interval using the bootstrap percentile confidence interval method and B = 3,000 bootstrap samples provided it is reasonable to use this method.
  Write a statement explaining the meaning of the 95% confidence interval.
  d) Construct 95% confidence interval using the basic bootstrap confidence interval method and B = 3,000 bootstrap samples provided it is reasonable to use this method. Write a statement explaining the meaning of the 95% confidence interval.
  e) Compare the confidence intervals from parts a to d. What do you notice about the width of the 90% and 95% confidence intervals when increasing the value of B?
63. A statistician prefers to use the bootstrap standard error t confidence interval to construct a confidence to estimate the population mean body fat percentage for U.S. males aged 12 to 15 years using the following random sample information.

RANDOM SAMPLE of 36 Body Fat Percentages for U.S. Males aged 12 to 15 Years

15.1	14.4	13.6	15.7	16.1	17.2	15.9	14.8	16.9	16.5	17.1
15.8	18.6	17.4	18.8	18.9	23.0	21.7	20.9	23.6	31.1	28.5
32.8	31.9	24.5	25.8	29.8	36.1	30.9	37.7	36.9	31.9	37.1
40.6	39.2	38.4								

Before the statistician can use this confidence method, she decides to generate the bootstrap distribution of the mean to determine if the bootstrap distribution is approximately normal.
  a) Construct the bootstrap distribution of the mean using B = 1,500 bootstrap samples.
  b) Construct a histogram of the bootstrap distribution of the mean to determine if the distribution is approximately normal.
  c) If the distribution is approximately normal, then apply the bootstrap standard error t confidence interval formula to construct a 90% confidence interval. Write a statement explaining the meaning of the 90% confidence interval.
64. The following random sample represents the age at death of U.S. non-smokers.

Age at Death of U.S. Non-Smokers

81	65	89	70	87	61	94	98	81	70	62
86	89	75	76	69	69	61	72	77	90	88
74	88	90	87	95	66	72	84	79	72	92
77	45	75	70	91	73	67				

Assuming this random sample is a good estimate of the population, then:
  a) construct a 95% confidence interval to estimate the median age life expectancy of U.S. non-smokers interval using the bootstrap percentile confidence interval method and B = 2,500 bootstrap samples provided it is reasonable to use this method. Write a statement explaining the meaning of the 95% confidence interval.
  b) construct a 95% confidence interval to estimate the median age life expectancy of U.S. non-smokers interval using the bootstrap percentile confidence interval method and B = 3,000 bootstrap samples provided it is reasonable to use this method. Write a statement explaining the meaning of the 95% confidence interval.
  c) construct a 95% confidence interval to estimate the median age life expectancy of U.S. non-smokers interval using the basic bootstrap interval method and B = 2,500 bootstrap samples provided it is reasonable to use this method. Write a statement explaining the meaning of the 95% confidence interval.
  d) construct a 95% confidence interval to estimate the median age life expectancy of U.S. non-smokers interval using the basic bootstrap interval method and B = 3,500 bootstrap samples provided it is reasonable to use this method. Write a statement explaining the meaning of the 95% confidence interval.
  e) Compare the confidence intervals from parts a and d. What do you notice about the width of the 95% confidence intervals when increasing the value of B?
  f) What do notice about the 95% confidence intervals using the different bootstrap confidence interval methods?
  g) Graphically, where does the mean of the original sample fall within the three confidence intervals? Explain.
65. The following random sample represents the age at death of U.S. non-smokers.

Age at Death of U.S. Non-Smokers

81	65	89	70	87	61	94	98	81	70	62
86	89	75	76	69	69	61	72	77	90	88
74	88	90	87	95	66	72	84	79	72	92
77	45	75	70	91	73	67				

Assuming this random sample is a good estimate of the population, then:
  a) compute an estimate of the median age life expectancy of U.S. smokers assuming the given random sample is representative of all U.S. smokers.

b) define the original sample if we are interested in generating the bootstrap distribution of the median.
c) select 1,500 resamples and compute the median of each bootstrap sample to generate a bootstrap distribution of the median.
d) construct a histogram of the bootstrap distribution of the median.
e) compute the mean of the bootstrap distribution of the median.
f) compute the bootstrap standard error of the median. Did you have a formula to compute this standard error? Explain.
g) Why is bootstrapping an important statistical concept when generating the bootstrap distribution of the median based on part h?
h) What is the bootstrap estimate of the bias from your bootstrap sample medians of the bootstrap distribution? What does this tell us about the bias encountered in using the median of the original sample to estimate the median age life expectancy of U.S. smokers?
i) summarize the characteristics of the bootstrap distribution of the median using the previous results

66. The following random sample of the number of hours college students in Nassau County in Long Island, NY work in a week during the academic year.

Number of Hours College Students Work Weekly

10	16	36	27	0	28	40	35	27	29
45	19	24	34	37	42	40	20	0	0
25	26	18	32	40	0	20	42	40	0
22	26	40	38	45	38	0	40	35	50

Suppose one is interested in estimating the number of hours college students in Nassau County work per week during the academic year.

Assuming this sample is representative of the number of hours college students in Nassau County work per week during the academic year, then:
a) construct a histogram of the sample data.
b) state the shape of the histogram.
c) define the original sample if one is interested in generating the bootstrap distribution of the standard deviation of the college student weekly work hours.
d) select 2,500 resamples and compute the sample standard deviation of each bootstrap sample to generate a bootstrap distribution of the standard deviation.
e) construct a histogram of the bootstrap distribution of the standard deviation.
f) state the shape of the histogram of the bootstrap distribution of the standard deviation.
g) compute the standard deviation of the bootstrap distribution and compare it to the standard deviation of the original sample. What did you expect when comparing these two results?
h) compute the bootstrap standard error of the standard deviation using the bootstrap distribution of the standard deviation. Did you have a formula to compute this standard error? Explain.
i) Why is bootstrapping an important statistical concept when generating the bootstrap distribution of the standard deviation?
j) What is the bootstrap estimate of the bias from your bootstrap sample standard deviations of the bootstrap distribution? What does this tell us about the bias encountered in using the standard deviation of the original sample to estimate the standard deviation of the college student weekly work hours?
k) summarize the characteristics of the bootstrap distribution of the standard deviation using the previous results.

67. The following random sample represents the dollar amounts per order by 60 Amazon customers.

Amazon Customer Dollar Purchases Per Order

$39.88	$210.99	$16.85	$47.40	$89.99	$29.94	$23.77	$231.05	$95.97	$18.75
$23.26	$17.44	$35.98	$15.99	$51.94	$10.00	$107.95	$99.95	$89.99	$17.95
$22.72	$69.99	$39.15	$11.98	$29.99	$89.99	$38.24	$109.99	$139.99	$14.99
$26.65	$17.42	$43.97	$229.00	$11.99	$69.00	$78.95	$36.88	$89.90	$23.99
$40.91	$49.99	$19.95	$29.24	$245.99	$89.99	$55.37	$35.43	$89.99	$15.99
$64.35	$89.99	$37.30	$25.05	$155.30	$23.19	$33.95	$16.99	$89.99	$74.06

Assuming this sample is representative of the purchase orders of all Amazon customers, then:
a) construct a histogram of the sample data.
b) state the shape of the histogram.
c) define the original sample if we are interested in generating the bootstrap distribution of the mean.
d) select 2,000 resamples and compute the mean of each bootstrap sample to generate a bootstrap distribution of the mean.
e) construct a histogram of the bootstrap distribution of the mean.
f) state the shape of the histogram of the bootstrap distribution of the mean.
g) the central limit theorem (chapter 8) states the true sampling distribution of the mean is approximately a normal distribution for a large sample size. Examine the shape of the histogram you constructed in part d and comment on what you expected the histogram shape to be?
h) compute the mean of the bootstrap distribution and compare it to the mean of the original sample. What did you expect when comparing these two results?
i) compute the bootstrap standard error of the mean using the bootstrap distribution of the mean.
j) Use the original sample data and the formula $\frac{s}{\sqrt{n}}$ to estimate the standard error of the mean.
k) Compare the results of part i to the bootstrap standard error of the mean. How close are the results?
l) Using the results of parts i and j, what is the relationship between the standard deviation of the bootstrap distribution of the mean to the standard error of the true sampling distribution of the mean?
m) What is the bootstrap estimate of the bias from your bootstrap sample means of the bootstrap distribution? What does this tell us about the bias encountered in using the mean of the original sample to estimate the mean dollar purchase per Amazon customer order?

68. The following random sample represents the age at death of U.S. smokers.

Age at Death of U.S. Non-Smokers

81	65	89	70	87	61	94	98	81	70
62	86	89	75	76	69	69	61	72	77
90	88	74	88	90	87	95	66	72	84
79	72	92	77	45	75	70	91	73	67

Assuming this random sample is a good estimate of the population, then:

a) construct a 95% confidence interval to estimate the median age life expectancy of U.S. smokers interval using the bootstrap percentile confidence interval method and B = 1,000 bootstrap samples provided it is reasonable to use this method. Write a statement explaining the meaning of the 95% confidence interval.

b) construct a 95% confidence interval to estimate the median age life expectancy of U.S. smokers interval using the bootstrap percentile confidence interval method and B = 3,000 bootstrap samples provided it is reasonable to use this method. Write a statement explaining the meaning of the 95% confidence interval.

c) construct a 95% confidence interval to estimate the median age life expectancy of U.S. smokers interval using the basic bootstrap interval method and B = 1,000 bootstrap samples provided it is reasonable to use this method. Write a statement explaining the meaning of the 95% confidence interval.

d) construct a 95% confidence interval to estimate the median age life expectancy of U.S. smokers interval using the basic bootstrap interval method and B = 3,000 bootstrap samples provided it is reasonable to use this method. Write a statement explaining the meaning of the 95% confidence interval.

e) Compare the confidence intervals from parts a and d. What do you notice about the width of the 95% confidence intervals when increasing the value of B?

f) What do notice about the 95% confidence intervals using the different bootstrap confidence interval methods?

ANSWER SECTION

CHAPTER 1

1. Raw
3. Descriptive; Inferential
5. Population
7. Statistic
9. b
11. b
13. false
15. **I.** NFL weigh-in illustrates descriptive statistics.
    II. a) U.S. mail vs. email illustrates inferential statistics
    b) population: USA households
    c) n = 1014
    d) answers vary; one possible solution is that the majority still trust the post office.
17. Inferential
19. census
21. population; sample
23. N
25. statistic
27. population
29. probability; sample
31. **a)** 511, 151, 135, 098, 083, 010, 200, 084, 283, 559, 572, 099, 197, 401, 266
    b) 010, 559, 462, 098, 200, 572, 401, 216, 637, 266, 447, 169, 511, 135, 084
33. 2487; 8467; 1898; 0402; 3469; 0784; 6202; 7791; 1277; 8596; 3891; 7965; 3610; 0010; 8314
35. Pictograph

37a.

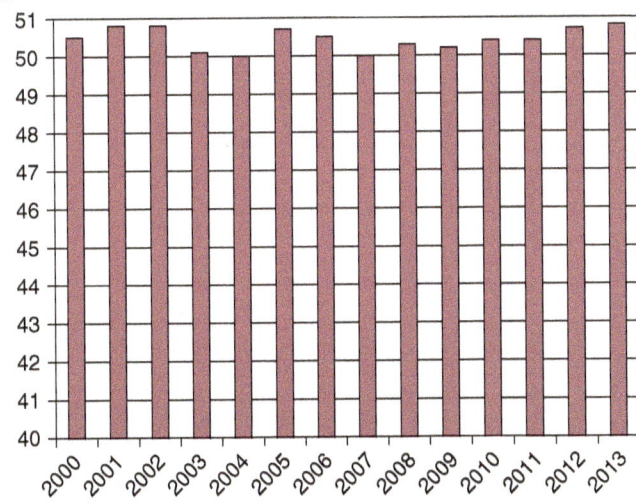

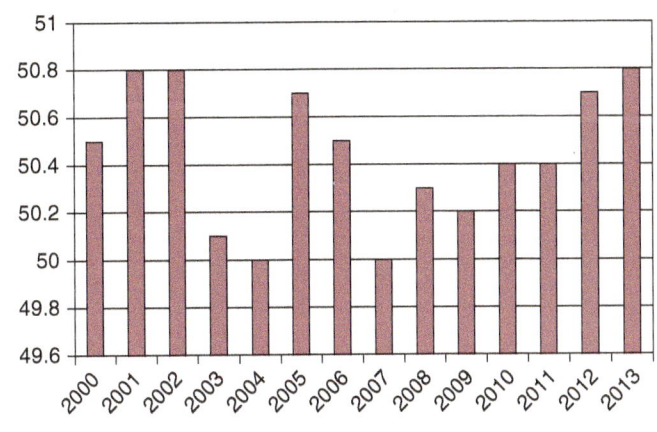

37b.

39. Discuss this with your instructor.
41. inferential; representative
43. inferential
45. **a)** all registered Democrats
    b) 629 registered Democrats
    c) sample
47. a
49. true
51. false
53. true
55. false
57. No, it would not be representative of all her constituents because everyone probably didn't return the survey. For example, lazy people might not have returned the survey.
59. no; only people interested in gun control will participate
61. 239, 422, 274, 523, 479, 157, 100, 505, 097, 542, 422, 019, 017, 532, 618, 041, 186, 187, 453, 494
63. 14,877; 10,924; 2,771; 5,697; 2,865; 4,237; 8,710; 4,284; 1,181; 4,230; 9,205

Ans1

Ans2 Answers

65.

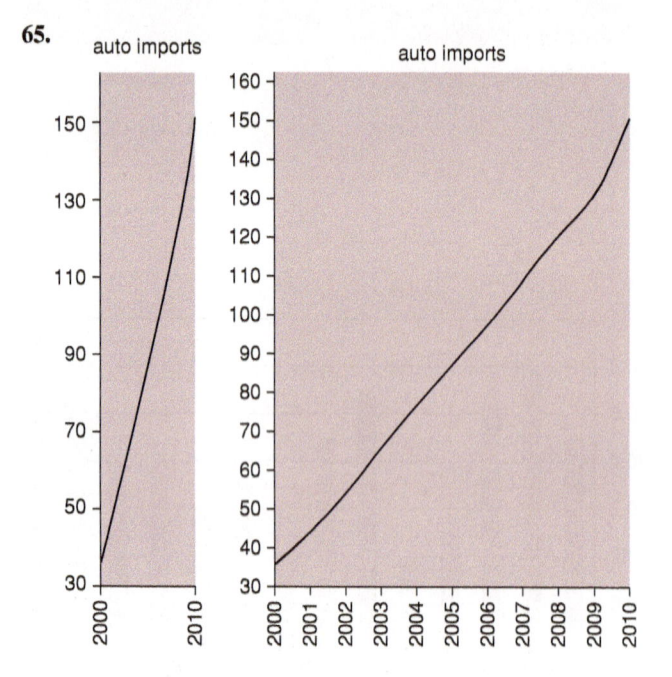

auto imports (2000–2010, two graphs)

67. Both are examples of inferential statistics
  i. Why don't we just do it?
   Population: assume US population
   Sample size: n = 2000
   Conclusion: answers vary
  ii. When pets get people food
   Population: pet owners
   Sample size: n = 1049
   Conclusion: answers vary

CHAPTER 2

1. variable
3. categorical; numerical
5. numerical; data
7. continuous
9. true
11. exploratory data analysis
13. stem-and-leaf
15. outlier
17. a
19. 10|2, 3|1, 2. |4 (or 102, 31, 2.4)
21. (a)

   18 | 89
   19 | 146
   20 | 00111223346666799
   21 | 001122222233344456778889
   22 | 0236

   (b) skewed to the left
23. classes; frequency
25. significant
27. midpoint
29. percent; 100; 100%
31. sum; greater than
33. b
35. b
37. true

39. a)

Class	Frequency
67–71	3
72–76	7
77–81	9
82–86	10
87–91	1

40. b) i) No
   ii) No
   iii) No
   iv) No
41. false
43. time
45. frequency or relative frequency or relative percentage
47. variable; gaps; breaks
49. ogive
51. true
53. a)

   30–34   364
   35–39   325
   40–44   169
   45–49   143
   50–54   104
   55–59   104
   60–64   91

   b) 5

   c)

Classes	Frequency	Class boundaries	Class Marks	Relative %
30–34	364	29.5–34.5	32	28%
35–39	325	34.5–39.5	37	25%
40–44	169	39.5–44.5	42	13%
45–49	143	44.5–49.5	47	11%
50–54	104	49.5–54.5	52	8%
55–59	104	54.5–59.5	57	8%
60–64	91	59.5–64.5	62	7%

   d) Histogram:

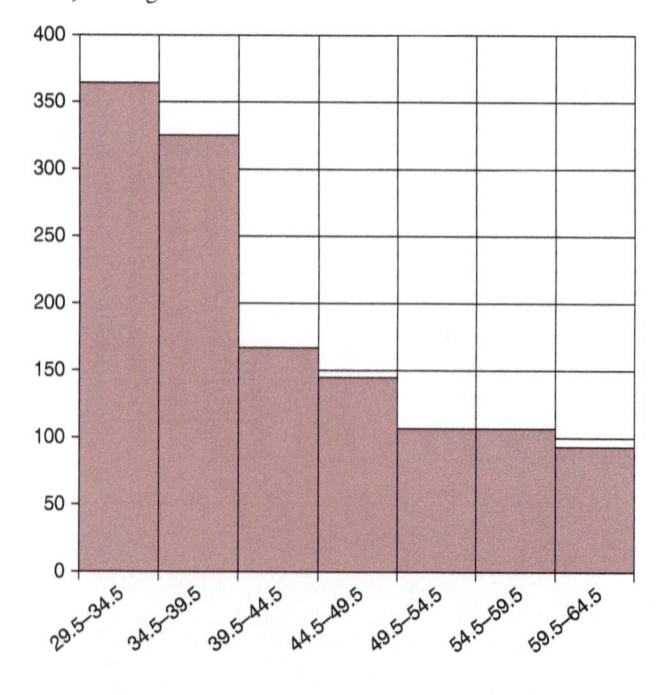

Frequency Polygon:

[Frequency polygon chart with x-axis from 27 to 67 and y-axis from 0 to 30, Series 2]

50 or more	299
55 or more	195
60 or more	91
65 or more	0

h) 66%
i) 53%
j) 77%

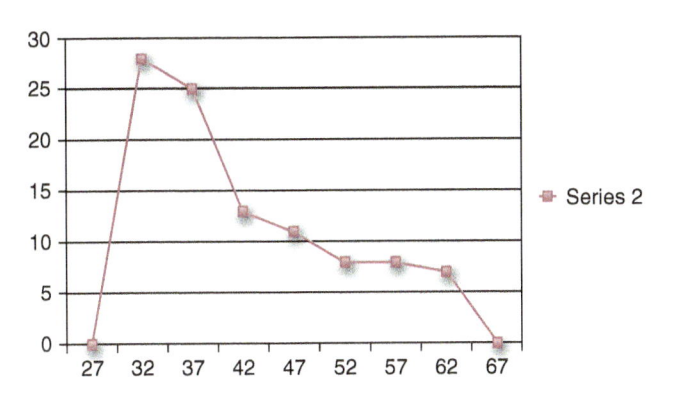

55. [Bar chart: Growth of iPad Ownership, Number of iPad Owners (in millions) vs YEAR: 2010: 11.5, 2011: 32.4, 2012: 58.1, 2013: 71, 2014: 68, 2015: 53.9, 2016: 45.6, 2017: 43.8]

e) Decreasing. Older people are less likely to end the relationship.

f)

Classes	Cumulative frequency
Less than 30	0
Less than 35	364
Less than 40	689
Less than 45	858
Less than 50	1001
Less than 55	1105
Less than 60	1209
Less than 65	1300

g)

Classes	Cumulative frequency
30 or more	1300
35 or more	936
40 or more	611
45 or more	442

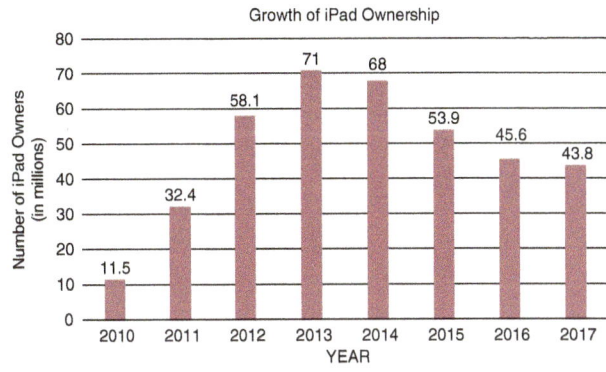

57. pictograph
59. center; decreasing
61. right; low; high
63. same
65. one; opposite
67. d
69. a
71. c
73. d
75. **a)** numerical; continuous
  b) numerical; continuous
  c) categorical
  d) numerical; discrete

77. a)

```
                        smoking        non-smoking
                    4 1 | 3 |
                        | 4 |
      12 12 9 9 7 5 4 3 3 2 2 | 5 | 9 11 11 13 15
      9 9 8 8 8 6 6 6 5 5 2 2 1 1 | 6 | 5 6 6 6 8 8 9 10 12 14 14 14
      8 8 7 6 6 5 5 5 4 4 3 2 2 0 | 7 | 0 1 1 2 3 3 5 6 6 8 9 9 10 11 12 14 15
                  3 3 2 2 0 0 | 8 | 1 2 4 4 5 6 8 9
                        3 2 1 | 9 | 0 3 3 4 8
                              | 10 | 1 5
                              | 11 | 2
```

b) skewed
c) yes

79. a)

Class	Class marks	Freq.	Class boundaries	Perce.	Rel. Freq
92–97	94.5	12	91.5–97.5	10.00%	0.10
98–103	100.5	11	97.5–103.5	9.17%	0.0917
104–109	106.5	17	103.5–109.5	14.17%	0.1417
110–115	112.5	19	109.5–115.5	15.83%	0.1583
116–121	118.5	22	115.5–121.5	18.33%	0.1833
122–127	124.5	19	121.5–127.5	15.83%	0.1583
128–133	130.5	8	127.5–133.5	6.67%	0.0667
134–139	136.5	4	133.5–139.5	3.33%	0.0333
140–145	142.5	5	139.5–145.5	4.17%	0.0417
146–151	148.5	3	145.5–151.5	2.50%	0.025

b) 1; 100%
c) 1) 49.17%, 2) 6.67%, 3) 32.50%, 4) 33.34%, 5) 10%.

81. a)

Finished shopping by	Number of Shoppers	Percentage
Nov 30	45	6.02%
Dec 15	389	52.01%
Dec 23	127	16.97%
Dec 25	187	25%

b) December 15
c) 75%
d) 25%

Answers

83. a) Income is numerical & treated as continuous.

b)

classes	freq	class boundaries	class marks	relative %
15–29	8	14.5–29.5	22	16%
30–44	14	29.5–44.5	37	28%
45–59	11	44.5–59.5	52	22%
60–74	17	59.5–74.5	67	34%

c)

classes	cumm. freq
less than 15	0
less than 30	8
less than 45	22
less than 60	33
less than 75	50

d) 66%
e) 84%
f) no
g) specific data values are lost; no

85. a)

```
                       Men          Stem   Women
                   6 6 6 6 5 4 4    10   4 4 5 5 6 7 8 8 8
     8 8 8 8 8 6 6 5 4 2 2 2 2 0 0 0 0 0 0  11  0 0 0 0 0 0 0 0 0 2 2 2 2 2 2 4 4 4 4 6 6 6 6 6 6 8 8 8 8 8 8 9
 8 8 6 6 6 6 6 4 2 0 0 0 0 0 0 0 0 0 0 0 0 0  12  0 0 0 0 0 2 2 2 4 4 5 6 6 6 6 6 8 8 8 8 8 9
                6 6 4 4 2 0 0 0 0 0 0 0 0 0  13  0 0 0 0 4 4 5 8 8 9
                         6 6 5 5 3 2 0  14   0
                                 6 6 0 0  15   6
```

b) *Men:*

Classes	Cumulative Frequency
Less than 104	0
Less than 115	18
Less than 126	43
Less than 137	64
Less than 148	71
Less than 159	75

Women:

Classes	Cumulative Frequency
Less than 104	0
Less than 115	28
Less than 126	52
Less than 137	70
Less than 148	74
Less than 159	75

c) 85.33%
d) 93.33%
e) Men:

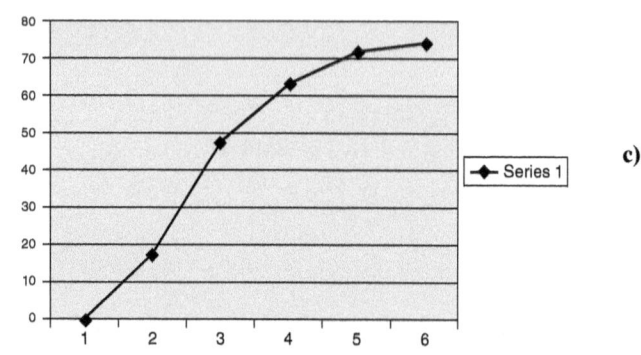

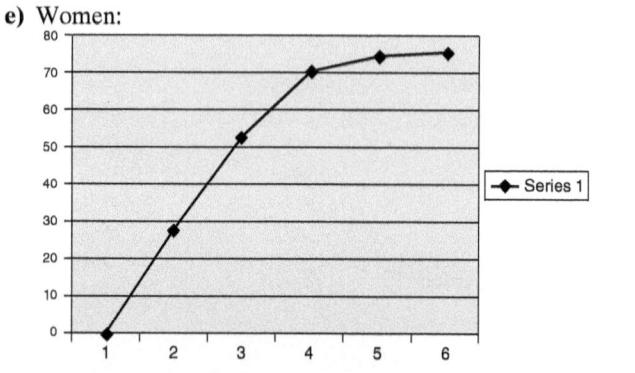

e) Women:

87. a)

classes (age)	freq.	class marks	class boundaries	relative frequency
18–22	27	20	17.5–22.5	0.13
23–27	54	25	22.5–27.5	0.26
28–32	55	30	27.5–32.5	0.26
33–37	31	35	32.5–37.5	0.15
38–42	21	40	37.5–42.5	0.10
43–47	10	45	42.5–47.5	0.05
48–52	7	50	47.5–52.5	0.03
53–57	3	55	52.5–57.5	0.01
58–62	2	60	57.5–62.5	0.01

b) shape: skewed to the right

Frequency Polygon
Age When Starting a Business

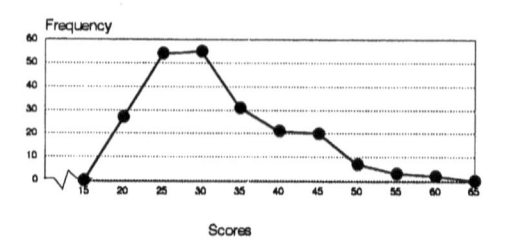

c)

classes	cumulative frequency
less than 18	0
less than 23	27
less than 28	81
less than 33	136
less than 38	167
less than 43	188
less than 48	198
less than 53	205
less than 58	208
less than 63	210

d)

classes	cumulative frequency
18 or more	210
23 or more	183
28 or more	129
33 or more	74
38 or more	43
43 or more	22

48 or more	12
53 or more	5
58 or more	2
63 or more	0

e) Ogive — Age When Starting a Business

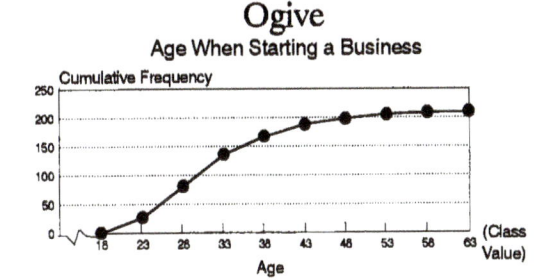

f) 65% g) 10% h) 52%

89. At Home:

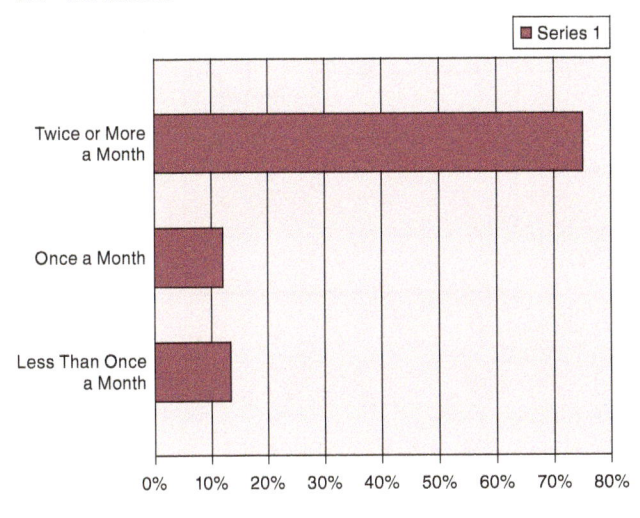

At Theater:

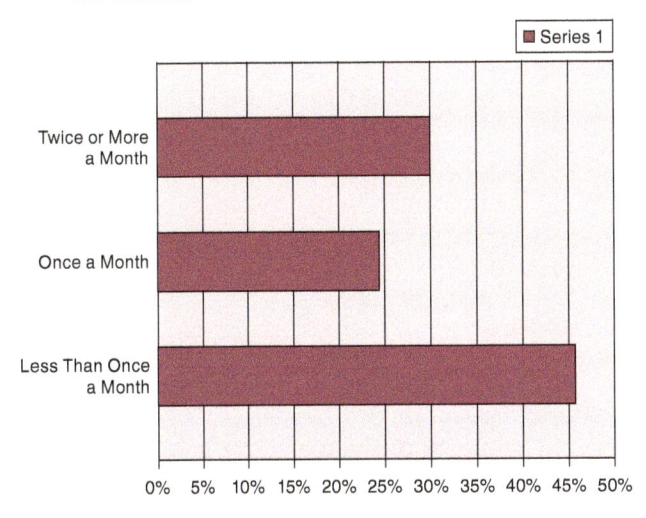

91. a) 18 b) 20% c) 20% d) 36% e) 36% f) yes g) 4%

CHAPTER 3

1. mean, median, mode
3. resistant
5. a) 45 b) 3 c) 5 d) 20
7. −5
9. mode
11. mode
13. (a)
15. resistant
17. false
19. a) mean, $\bar{x} = 516.83$; median = 473; mode = 560
    b) The data value 1334 appears to be an outlier. If we remove this data value and recalculate the 3 measures of center we get: mean, $\bar{x} = 488.66$; median = 471; mode = 560. Notice how the removal of the potential outlier has a big impact on the value of the mean, but little impact on the median. Because of this, the median would be a better indicator of the central tendency of the distribution.
21. a) $\mu = 89.5$
    b) The student is basing his remark on the mean. He is not taking the fact that approximately 46% of the grades are above 90 into consideration.
23. a) Monday — Fried Chicken
       Tuesday — Spag. & Meatballs
       Wednesday — Meatloaf
       Thursday — Beef Stew
       Friday — Pizza
    b) Spaghetti & Meatballs
    c) Spaghetti & Meatballs
25. b
27. 1) skewed left, median
    2) symmetric bell-shaped, mean = median
    3) skewed right, mean
    4) skewed right, mean
    5) symmetric bell-shaped, mean = median
29. dispersion or spread
31. 5
33. a) s   b) σ (sigma)
35. 1) below   2) above   3) $\bar{x}$
37. b
39. a
41. b
43. false
45. false
47. a) Men: Women:
       mean = 89   mean = 89
       sd = 5.31   sd = 4.94
    b) The golf scores of the women's team were more consistent since they had a smaller standard deviation.
    c) Men: Women:
       mean = 88   mean = 88
       sd = 3.52   sd = 3.97
    d) The 10 best golf scores of the men's team are more consistent. The extreme golf score 102 had a major influence on the standard deviation of the men's team.
49. a) $\bar{x} = 42$; $s = 12.96$ e) 30, 36
    b) 54.96   f) 16.08; 67.92
    c) 48, 54   g) 71.4%
    d) 29.04   h) 100%

51. a) day class because if the standard deviation is larger, then the grades are more spread apart
  b) evening class because if the standard deviation is smaller, then the grades are closer to the mean
53. standard deviations
55. a) 15.85%
  b) 99.85%
  c) 81.5%
  d) Approx. 63 students
  e) 500
  f) 575
  g) 425
57. b
59. a) $\mu = 136; \sigma = 11.2$
  b) 65%
  c) 92.5%
  d) 100%
  e) Yes, data adhere to Empirical Rule
  f) Bell-Shaped Distribution
61. a) 16%
  b) 2.5%
  c) 81.5%
  d) 48 cups
  e) 124 mg
  f) 0.15%
63. a) 11.1%
  b) yes, because it is 5 standard deviations above the mean
65. a) 61005.58
  b) 59020
67. z score
69. 1
71. mean
73. 3; greater
75. less
77. p
79. percentile
81. data value; 0; 100
83. a)

x	z
45	–1.67
165	4.50
590	1.27
830	0.80
1470	0

  b) 165
85. a) 2.75
  b) yes; since the z-score is close to 3
  c) approximately 81 ft/sec
  d) Yes, the person would travel 124.2 yards.
  e) Don't text and drive!
87. a) 3.13
  b) yes
  c) One would question who the father is.
89. a) 23
  b) 25
  c) 40%
  d) 4
  e) 13
  f) 21
  g) 31

91. The z score for you is 2.25; the z score for your friend is 2.67. Your friend's grade is more outstanding because his z score is larger.
93. a) 75%
  b) 75%
  c) 57%
  d) 30%
  e) 50%
  f) 50%
95. exploratory; center; spread; shape
97. whisker
99. symmetric
101. c
103. 23
105. true
107. true
109. a) 19, 64, 81, 89, 98
  b)

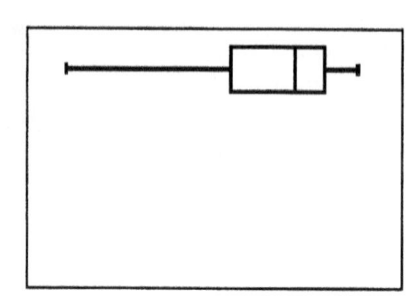

  c) skewed left
  d) IQR = 25
  e) outlier = 19
  f) z of 19 = –3.20
  g) lowers mean & no effect on median
111. male sample results:
  a)

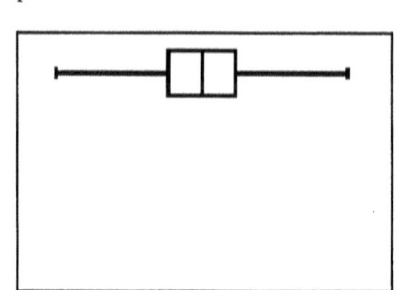

  b) symmetric
  c) Q1 = 23; med = 30; Q3 = 37
  d) yes: 0 & 60
  e) z of 0 = –2.05; z of 60 = 2.00
  f) very little effect on the mean and no effect on median
  female sample results:
  a)

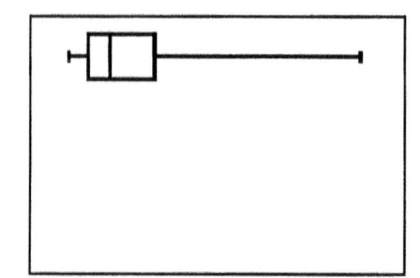

  b) skewed right

c) Q1=15.5; med = 20; Q3 = 28.5
d) yes: 69
e) z of 69 = 3.63
f) raises the mean and no effect on median
113. b
115. non-modal
117. sample mean: $\bar{x} = \dfrac{\Sigma x}{n}$

population mean: $\mu = \dfrac{\Sigma x}{N}$
119. sample
121. 75
123. false
125. a
127. mean = 14
mode = 14
median = 14
range = 0
variance = 0
standard deviation = 0
129. a) mean = 65.1; median = 80.5; mode = 82
b) μ
c) median b/c it is not influenced by the extreme data values
131. mode, because the data is categorical
133. No! This logic is not correct, since he doesn't take into consideration that each semester's avg is for a different number of credits.
135. a) 1) Brown
2) Brown
3) Brown Hair; Brown Eyes
4) Brown Hair; Blue Eyes
b) Mode, because the data is categorical
137. a) 414 mg.
b) −20 mg; Possibly since the sample mean is less than the mean stated by the manufacturer.
c) no
139. i) a. bell   **b.** mean and median are equal
ii) a. skewed left   **b.** median
iii) a. skewed right   **b.** mean
iv) a. skewed left   **b.** median
141. s
143. a) 18 **b)** 52 **c)** 7.21 **d)** 7.21 **e)** 6.45
145. true
147. true
149. a) Resort 1 s = 22.2; Resort 2 s = 6.1
b) Resort 2 has more consistent weather because the standard deviation is smaller.
151. a) Lucy $\mu = 6$ $\sigma = 4.5$
Allison $\mu = 10.4$ $\sigma = 6.0$
Lilia $\mu = 17$ $\sigma = 2.3$
Kathleen $\mu = 10$ $\sigma = 4.5$
Debbie $\mu = 20$ $\sigma = 4.4$
Ellen $\mu = 11$ $\sigma = 4.8$
Grace $\mu = 13$ $\sigma = 5.9$
b) Lilia because she has the lowest standard deviation therefore she is the most consistent
153. false
155. false
157. c
159. a) $\bar{x}$, the sample mean = 911 dollars per month and s, the sample standard deviation ≈ 215.68 dollars per month.
b) Data values that are within 2 standard deviations of the mean must be between 479.64 and 1342.36. Examining the data values in the sample we find that there are 14 out of 15 data values that are between 479.64 and 1342.36. This represents 93%.
c) According to Chebychev's Theorem the percent of data values that are within 2 standard deviations of the mean is at least 75%.
d) The results are consistent with Chebychev's Theorem.
161. program A (Group 1)
163. a) 68%
b) 95%
c) 99%–100%
d) Yes, 77 inches is an outlier because it is 5 standard deviations above the mean
165. a) 2.63
b) over-estimate is 2.8 and the under-estimate is 1.87. 2.63 is in that range.
c) 95%
d) 100%
e) using Chebyshev's Theorem, for 2 standard deviations from the mean, there should be at least 75% of the data. For 3 standard deviations from the mean, there should be at least 89% of the data.
167. c
169. c
171. true
173. c
175. 4
177. quartile
179. false
181. a) i) 1
ii) −3.5
iii) −1.33
iv) 4.25
v) 0
b) 875 because the z score is 4.25 which means that it is more than 4 standard deviations above the mean.
183. a) Condo A
b) Condo M
185. z score sprinter = −2
z score long jumper = 1.53846 the sprinter's record is more outstanding.
187. a) P_{10} **b)** P_{25} **c)** P_{50}
d) P_{70} **e)** P_{75} **f)** P_{81}
189. a) 20 seconds
b) 28 seconds
c) 8 seconds
d) 85%
e) 16 seconds
f) 12 seconds
g) 36 seconds
191. Barbara; Mike; Alice
193. b

195. a. 160 b. 185 c. 40% d. 160 e. 60 f. 260
g. 200 h. 50 i. 247.5 j. 272.5

197. outer; inner

199. d

201. false

203. a) Roger Maris   Babe Ruth
$Q_1 = 11$ $Q_1 = 34.5$
median = 19.5   median = 46
$Q_3 = 30.5$ $Q_3 = 51.5$

b) Roger Maris

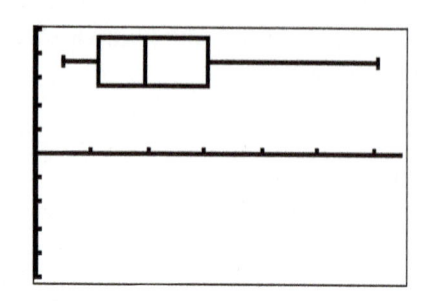

Babe Ruth

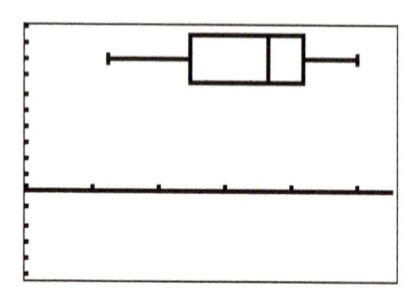

c) Babe Ruth's Q_1, median, Q_3 are much larger than Roger Maris'. Roger Maris' box and whisker is skewed to the right and Babe Ruth's box and whisker is skewed to the left.

d) Babe Ruth's box and whisker has a smaller variability and Roger Maris's box and whisker has a larger variability.

Roger Maris   Babe Ruth
IQR = 19.5   IQR = 17
s = 15.98   s = 11.49

Roger Maris's IQR and standard deviation is larger than Babe Ruth's. The larger the IQR, the larger the standard deviation, which means that numbers are more spread apart.

e) 61 is a potential outlier for Roger Maris because it is between the inner (59.75) and outer fences (89). Babe Ruth doesn't have any outliers. Roger Maris' 61 has a z score of 2.38, which means that it is more than 2 standard deviations above the mean.

f) Roger Maris' box and whisker is skewed to the right and Babe Ruth's box and whisker is skewed to the left.

205. a) & b)

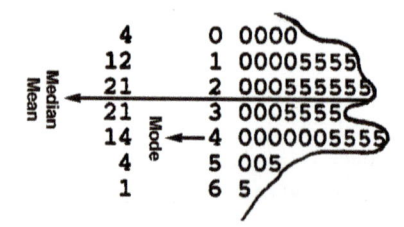

```
         4      0  0000
        12      1  00005555
Median  21      2  000555555
Mean    21      3  0005555
        14  ←4  0000005555   Mode
         4      5  005
         1      6  5
```

c) answers vary

207. a) cereal:

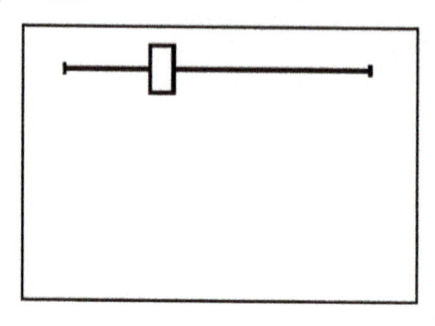

whole-grain:

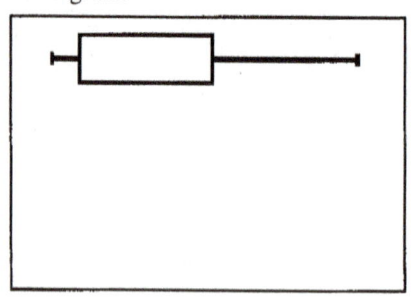

all-bran:

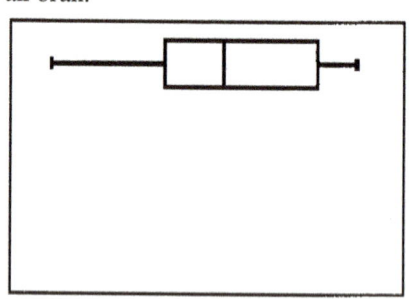

b) answers will vary

c) 5-number summary for cereal: 1.6, 2, 2, 2.1, 3
5-number summary for whole-grain cereal: 2.9, 3.1, 4, 4, 5
5-number summary for all-bran cereal: 5, 8, 9.5, 12, 13

d) potential outliers:
cereal: 1.6 & 3.0

e) answers vary

CHAPTER 4

1. measurements
3. scatter diagram
5. linear correlation
7. b
9. false
11. a) negative
    b) positive
    c) positive
    d) no linear correlation
13. Pearson's correlation coefficient, r.
15. strongest; perfect; upward
17. –1; +1; weak

19. true
21. −1.00, 0.95, −0.83, 0.50, 0.25, −0.10, 0
23. a) matches to III.
    b) matches to I.
    c) matches to IV.
    d) matches to II.
25. a)

x	y	x²	y²	xy
2	4	4	16	8
3	7	9	49	21
4	8	16	64	32
5	11	25	121	55
6	11	36	121	66

n =	5
$\sum x$	20
$\sum y$	41
$\sum x^2$	90
$\sum y^2$	371
$\sum xy$	182
r =	.9649

b) −0.7399
c) 1.0
d) 0.8370
e) −0.9538
f) 0.9317

27. c)
29. least squares, regression.
31. a) Scatterplot of Expenditures (thousands) vs Sales (millions)

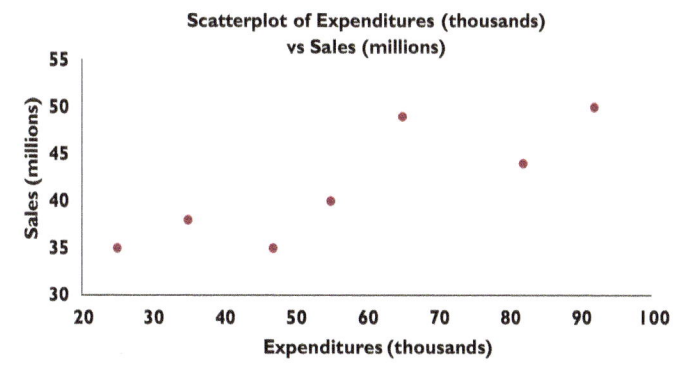

b) r = 0.8514
c) r² = 0.7249. This means about 72% of the variance in Sales (dependent variable) can be accounted for by the variance in expenditure (independent variable).
d) regression equation: y' = 29.01 + 0.22x
e) y'(50) = 0.22(50) + 29.01 ≈ 40.01 million dollars.
f) y' is an interpolated value since $50,000 is within the range of expenditure values given.

33. a) Scatterplot of Time on Social Media vs Extrovert Scale

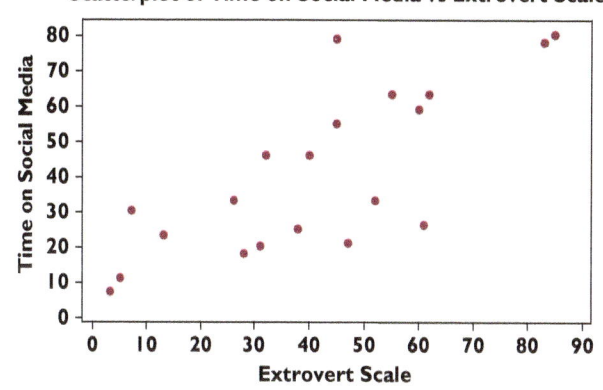

b) r = 0.762
c) r² = .58. This means about 58% of the variation in time spent online is accounted for by variation in score on extroversion scale.
d) y' = 9.95 + 0.75x
e) The predicted amount of time spent on a social network site online for a person who's extroversion scale score is 35 is 36 minutes.

35. a) Scatterplot of Finish Time (mins) vs Temp (F)

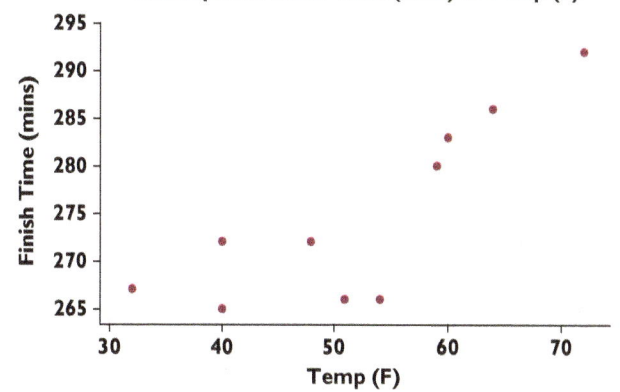

b) r = 0.8294
c) r² = 0.688. This means about 68.8% of the variation in finish time is accounted for by variation in temperature.
d) y' = 241.02 + 0.652x
e) For 68°, y' = 241.02 + 0.652(68) = 285.356 ≈ 285.4 minutes
f) For 80°, y' = 241.02 + 0.652(80) = 293.18 ≈ 293.2 minutes
g) It is extrapolated since 80° is not with in the given range of temperature values.

37. outliers, curvilinear data, heteroscedastic data.
39. False
41.

y	y'	Residual
45	47.5	−2.5
34	32.6	1.4
52	50.1	1.9
46	47.8	−1.8
35	36.4	−1.4

43. a)

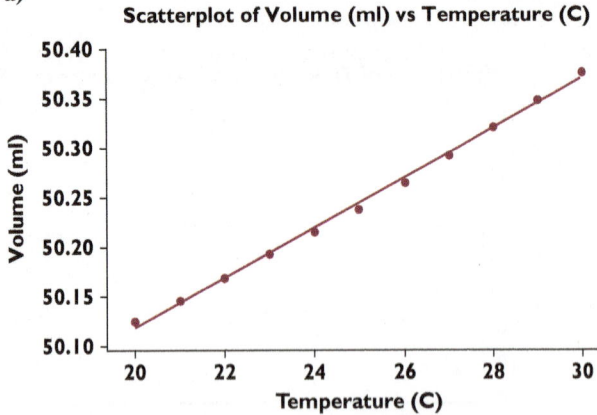

Scatterplot of Volume (ml) vs Temperature (C)

Yes, at first glance there does appear to be a linear relationship between volume and temperature.

b) regression equation: $y' = 49.61 + 0.0252x$, $r = 0.998$.

c)

Temperature (x)	Volume (y)	y'	residual (y – y')
20	50.124	50.114	0.01
21	50.146	50.1392	0.0068
22	50.169	50.1644	0.0046
23	50.193	50.1896	0.0034
24	50.215	50.2148	0.0002
25	50.237	50.24	–0.003
26	50.265	50.2652	–0.0002
27	50.292	50.2904	0.0016
28	50.321	50.3156	0.0054
29	50.349	50.3408	0.0082

d)

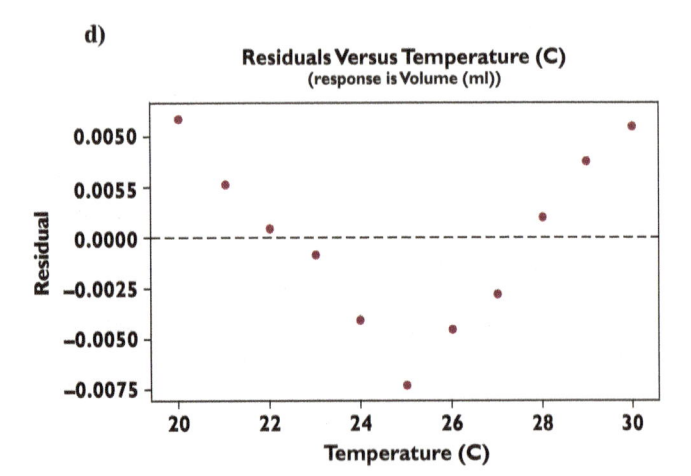

Residuals Versus Temperature (C) (response is Volume (ml))

Notice how the residual plot magnifies the curvilinear tendency of this data set. This makes us second-guess our original assumption of linearity.

45.
a) Not a linear relationship because the data is heteroscedastic.
b) Not a linear relationship because the data is curvilinear.
c) This is an example of a typical linear relationship between x & y.
d) Not a linear relationship because this data is heteroscedastic.

47. a)

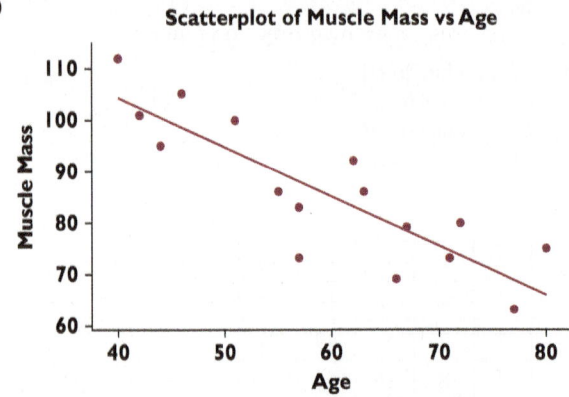

Scatterplot of Muscle Mass vs Age

The scatterplot shows a negative linear relationship between age and muscle mass. The data gives a sample correlation value of $r = -0.854$ showing a strong negative linear correlation.

b) *Regression Equation*: $y' = 142.6 - 0.9573x$. The y-intercept value of 142.6 tells us the muscle mass for a healthy female, and the negative slope indicates the amount by which muscle mass decreases each year. This regression equation is already in the scatterplot in part a.

c) For $x = 60$, $y' = 142.6 - 0.9573(60) = 85.162 \approx 85.2$

d) It is interpolated since 60 is within the given range of age values.

e) Solve the equation $58 = 142.6 - 0.9573x$ for x. The solution is $x \approx 88.4$ years

f) The prediction in part e should not be trusted since it is considered to be extrapolated.

g)

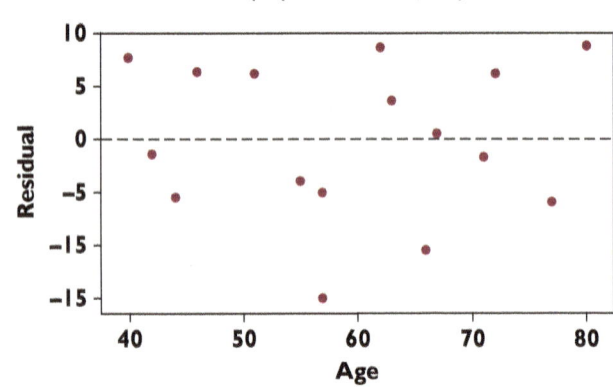

Residuals Versus Age (response is Muscle Mass)

Note how the scatter of the residual plot is random and shows no evidence of an outlier, curvilinear data or heteroscedastic data. Our original assumption of linearity has *not* been violated.

49. a)

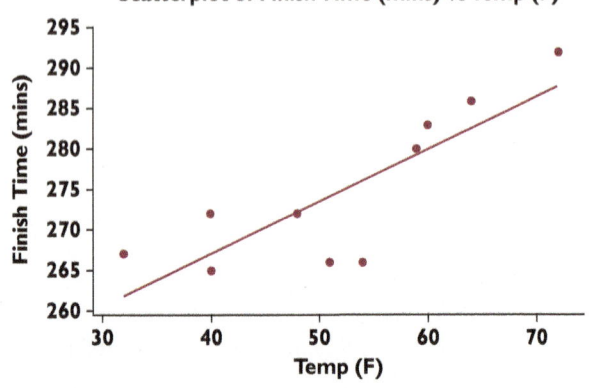

Scatterplot of Finish Time (mins) vs Temp (F)

b) *Regression Equation:* $y = 241.02 + 0.652x$; $r \approx 0.8294$ showing a strong positive correlation.

c)

Temp (x)	Finish Time (y)	y'	residual (y − y')
54	266	276.228	−10.228
51	266	274.272	−8.272
40	265	267.1	−2.1
72	292	287.964	4.036
64	286	282.748	3.252
32	267	261.884	5.116
60	283	280.14	2.86
59	280	279.488	0.512
40	272	267.1	4.9
48	272	272.316	−0.316

d)

Note how the scatter of the residual plot is random and shows no evidence of an outlier, curvilinear data or heteroscedastic data. Our original assumption of linearity has *not* been violated.

51. c

53. a) Negative
b) none
c) Positive
d) Negative
e) none
f) Positive
g) Positive
h) Positive
i) Positive

55. a) Moderate negative correlation
b) Strong positive correlation
c) Strong negative correlation
d) no linear correlation
e) no linear correlation

57. a) matches with IV.
b) matches with III.
c) matches with I.
d) matches with V.
e) matches with VI.
f) matches with II.

59. true

61. a)

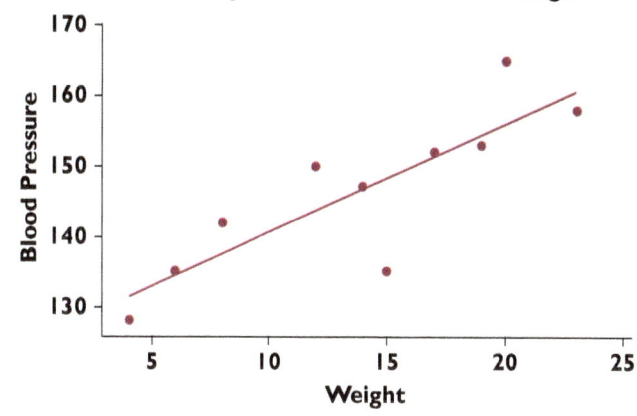

b) $r = 0.85$
c) 71.44% of the variance in blood pressure that can be accounted for by the variance in pounds above ideal weight.
d) $y' = 1.54x + 125.19$
e) $y' \approx 141$
f) The value is interpolated since 10 is within the given range of pounds above ideal weight.

63. a)

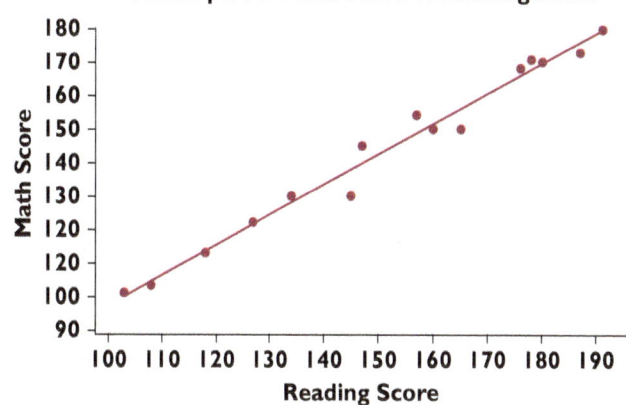

b) Based on the scatter diagram there appears to be a strong positive linear relationship between reading and math scores.
c) $r = 0.99$.
d) $r^2 \approx 0.981$. This means about 98.1% of the variation in Math Score is accounted for by variation in Reading Score.
e) $y' = 0.9053x + 6.632$
f) For $x = 182$, $y' = 0.9053(182) + 6.632 = 171.3966 \approx 171.4$
g) The value is interpolated since 182 is within the given range of reading scores.

CHAPTER 5

1. experiment
3. true
5. event
7. c
9. d

11. a) R, Y, B
    b) BB ** AA
       B* *B A*
       BA *A AB
    c) HHH
       HHT
       HTH
       HTT
       THH
       THT
       TTH
       TTT
    d) same as answer 1c
    e) BYE
       BEY
       EBY
       EYB
       YBE
       YEB
    f) GOOD OGOD
       GDOO OGDO
       OODG ODOG
       OOGD ODGO
       DOOG DOGO
       DGOO GODO
13. True
15. $_nP_n$; n!
17. a
19. 24
21. 120
23. a) 1. 5
       2. 20
       3. 60
       4. 120
       5. 120
    b) 1. 5
       2. 10
       3. 10
       4. 5
       5. 1
25. c
27. a) 453,600 b) 3780 c) 1680 d) 3360
29. $_nC_s$; $\dfrac{_nP_s}{s!}$ or $\dfrac{n!}{(n-s)!\,s!}$
31. 56
33. 75,287,520 = $_{100}C_5$
35. 150 = $_5C_3\,_6C_4$
37. a) 308,915,776 = 26^6
    b) 1,000,000 = 10^6
    c) 17,576,000 = $26^3 * 10^3$
    d) 165,765,600 = $_{26}P_6$
    e) 151,200 = $_{10}P_6$
39. a) 5,005 = $_{15}C_9$
    b) 1,890 = $_9C_5\,_6C_4$
41. Equally Likely
43. Subjective (or Personal)
45. a
47. False
49. Law of Averages
51. a) 1/12
    b) 1/4
    c) 0
53. b
55. a) BBBB BBBG BBGB BBGG
       BGBB BGBG BGGB BGGG
       GBBB GBBG GBGB GBGG
       GGBB GGBG GGGB GGGG
    b) 1/4
    c) 15/16
    d) 11/16
    e) 5/16
    f) 1/16
57. 0; 1
59. false
61. true
63. queen; red card; queen and a red card; queen; red card; queen and a red card; 4/52; 26/52; 2/52; 28/52
65. mutually exclusive
67. false
69. false
71. mutually exclusive; same; ace; king; 4/52; 4/52; 8/52
73. not mutually exclusive; 4; even number; 4
75. P(getting 4 heads)
77. independent; and; times; A and B; A; •; B

79.
outcome	2	3	4	5	6	7	8	9	10	11	12
probability	$\frac{1}{36}$	$\frac{2}{36}$	$\frac{3}{36}$	$\frac{4}{36}$	$\frac{5}{36}$	$\frac{6}{36}$	$\frac{5}{36}$	$\frac{4}{36}$	$\frac{3}{36}$	$\frac{2}{36}$	$\frac{1}{36}$

b) $\dfrac{8}{36}$ c) $\dfrac{28}{36}$ d) $\dfrac{26}{36}$ e) $\dfrac{18}{36}$

81. 39/45 = $(_6C_1)(_4C_1)/(_{10}C_2) + (_6C_2)(_4C_0)/(_{10}C_2)$
83. $\dfrac{4}{10}$
85. dependent; B; occurred; A and B; B | A
87. And; B; A | B; B
89. No two children have the same birthday; 81.4%
91. a) 58/100
    b) 29/53
    c) 18/42
93. a) $\dfrac{506}{7,893,600}$ = $(_1P_1)(_1P_1)(_1P_1)(_{23}P_2)/(_{26}P_5)$
    b) 1/7,893,600 = $1/(_{26}P_5)$
95. a) $\dfrac{1,069,263}{2,598,960}$ = $(_{13}C_1)(_{39}C_4)/(_{52}C_5)$
    b) P(at least one diamond) = 1 − P(no diamonds) = $1 - \dfrac{_{39}C_5}{_{52}C_5}$
       = 1 − 575757/2598960 = 7411/9520 ≈ .7785
97. a) 1/3
    b) 2/3
    c) 1/2

99. a) 0.75 b) manager should bring relief pitcher in to get batter out.
101. a) 1/13 b) 10000/99999 c) 1/52 d) 1/365 e) 1/5 f) 1/8
103. a) 144/2704 b) 132/2652 c) 400/2704 d) 380/2652
105. a) $\frac{21}{26}$ b) $\frac{5}{26}$ c) $\frac{15}{26}$ d) $\frac{2}{26}$ e) $\frac{18}{26}$
  f) $\frac{5}{26}$ g) $\frac{21}{26}$ h) $\frac{6}{26}$ i) $\frac{4}{26}$ j) $\frac{1}{26}$
107. a) classical
  b) subjective
  c) subjective
  d) relative frequency
  e) relative frequency
109. 8! = 40,320
111. $_7P_4$ = 840
113. 30
115. 4,200 = 10!/((4!)(3!)(3!))
117. a) 24
  b) 1/24
  c) 6
  d) 1/6
  e) 12
  f) 1/12
119. a) 720
  b) 120
  c) 1/720
121. a) 4
  b) 21
  c) 495
  d) 1,326
  e) 635,013,559,600
123. 45
125. a) 1/5
  b) 4/25
  c) 16/125
  d) 369/625

CHAPTER 6

1. chance
3. any value; uncountable number
5. b
7. a) continuous
  b) discrete
  c) discrete
  d) discrete
9. probability histogram; horizontal; vertical
11. d
13. true
15. a) probability distributions A & D
  b) probability distributions B: the sum of all the probabilities should equal 1
  probability distribution C: the probability of an event must be a non-negative value
17. probability distribution I:
  a) X: 0, 1, 2, 3
  b) shape: uniform

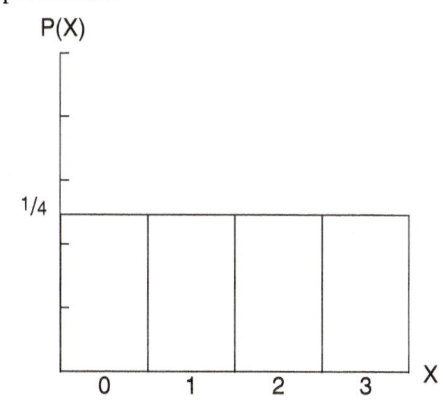

  c) 1; yes
  d) mean = 1.5
  e) sd = 1.12
  f) 0.5
  g) 1
  probability distribution II:
  a) X: 1, 2, 3, 4
  b) shape: skewed right

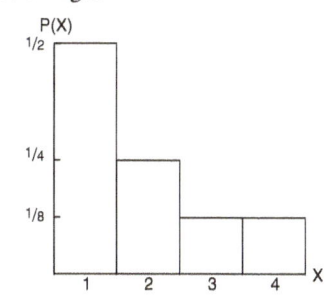

  c) 1; yes
  d) mean = 1.88
  e) sd = 1.05
  f) 3/4
  g) 7/8
  probability distribution III:
  a) X: –1, 0, 1
  b) shape: symmetric

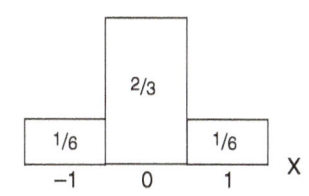

  c) 1; yes
  d) mean = 0
  e) sd = 0.58
  f) 2/3
  g) 1
  probability distribution IV:
  a) X: 2, 4, 6
  b) shape: skewed left

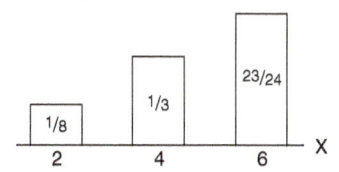

  c) 1; yes

      d) mean = 4.83
      e) sd = 1.4
      f) 7/8
      g) 7/8
19. a) S.S. = {0, 1, 2, 3}
    b) Using Y for earners with exactly 3 earners and N for otherwise;
      P(0) = P(NNN) = P(N)P(N)P(N) = 0.9 * 0.9 * 0.9 = 0.729
      P(1) = P(YNN or NYN or NNY) = P(YNN) + P(NYN) + P(NNY) = 0.1 * 0.9 * 0.9 + 0.9 * 0.1 * 0.9 + 0.9 * 0.9 * 0.1 = 0.081 + 0.081 + 0.081 = 0.243
      P(2) = P(YYN or YNY or NYY) = P(YYN) + P(YNY) + P(NYY) = 0.1 * 0.1 * 0.9 + 0.1 * 0.9 * 0.1 + 0.9 * 0.1 * 0.1 = 0.009 + 0.009 + 0.009 = 0.027
      P(3) = P(YYY) = P(Y)P(Y)P(Y) = 0.1 * 0.1 * 0.1 = 0.001
    c)

X	P(X)
0	$.9^3 = .729$
1	$3(.1 * .9^2) = .243$
2	$3(.1^2 * .9) = .027$
3	$.1^3 = .001$
	$\Sigma P(X) = 1$

    ** note that X is Binomial with n = 3 and p = .1 so we could have used the binomial distribution to work out the P(X)'s.
    d) 0 is most likely
    e) .271
    f) .028
    g) .729
    h) .972
21. $\sigma, \sqrt{\Sigma[(X-\mu)^2 * P(X)]}$
23. d)
25. F
27. The mean of this distribution is 8.02 > 8 so yes a traffic light is necessary.
29. a) identical
    b) independent
    c) success; failure
    d) success
31. true
33. 4; HHHT, HHTH, HTHH, THHH
35. 7; 5; 1/4; 3/4; 21; 5; 21
37. binomial probability formula
39. μ_s; $\mu_s = np$
41. acceptance sampling
43. e
45. false
47. true
49. a) $\dfrac{256}{10,000} = 0.0256$ b) $\dfrac{1296}{10,000} = 0.1296$ c) $\dfrac{1792}{10,000} = 0.1792$
51. a) 0.1074   b) 0.8791   c) 0.2013   d) 0.3222   e) 0.9999
53. a) 0.0106   b) 0.6331
55. a) $\dfrac{1}{625} = 0.0016$ b) $\dfrac{512}{625} = 0.8192$
    c) $\dfrac{608}{625} = 0.9728$ d) $\dfrac{81}{625} = 0.1296$
57. 0.1488
59. a)
X	P(X)
0	0.2373
1	0.3955
2	0.2637
3	0.0879
4	0.0146
5	0.0010

    b) skewed right

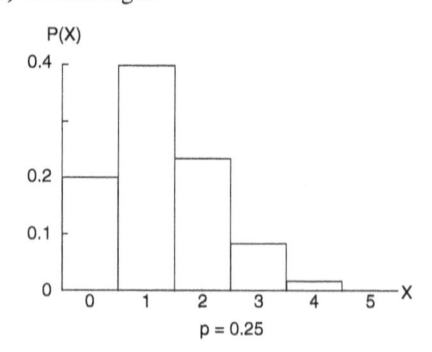

    c) mean = 1.25; sd = 0.9682

61. a)
X	P(X)
0	0.000
1	0.000
2	0.003
3	0.016
4	0.057
5	0.137
6	0.228
7	0.260
8	0.195
9	0.087
10	0.071

    b) skewed left

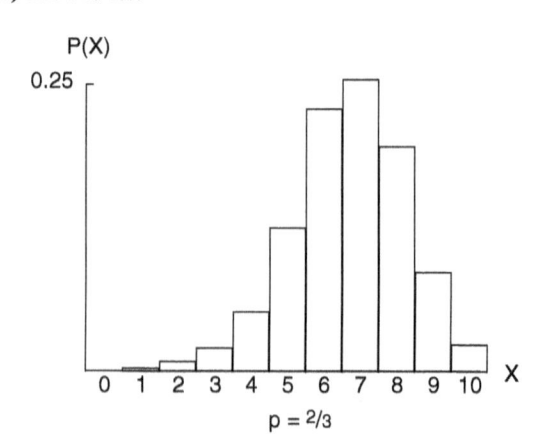

    c) mean = 6.667; sd = 1.491

63. a)

X	P(X)
0	0.0751
1	0.2253
2	0.3003
3	0.2336
4	0.1168
5	0.0389
6	0.0087
7	0.0012
8	0.0001
9	0.0000

b) skewed right

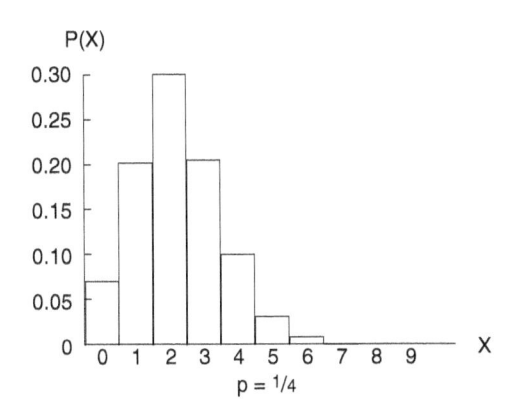

P(X) histogram, p = 1/4

c) mean = 2.25; sd = 1.3

65. a)

X	P(X)
0	0.7738
1	0.2036
2	0.0214
3	0.0011
4	0.0000
5	0.0000

b) skewed right

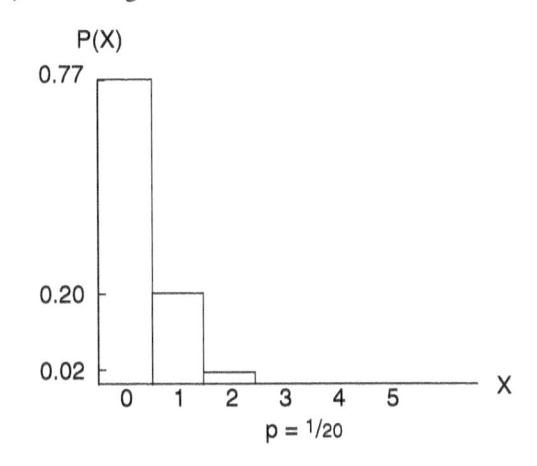

P(X) histogram, p = 1/20

c) mean = 0.25; sd = 0.4873

67. a)

X	P(X)
0	0.0003
1	0.0064
2	0.0512
3	0.2048
4	0.4096
5	0.3277

b) skewed left

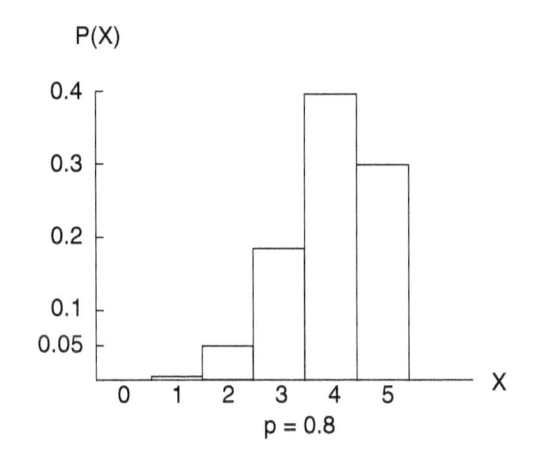

P(X) histogram, p = 0.8

c) mean = 4; sd = 0.8944

69. a)

X	P(X)
0	0.0000
1	0.0000
2	0.0000
3	0.0021
4	0.0305
5	0.2321
6	0.7351

b) skewed left

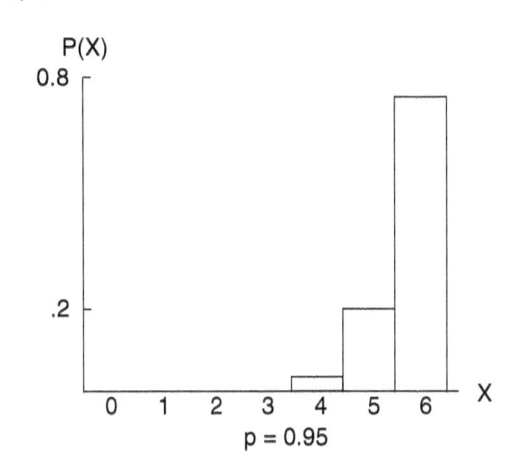

P(X) histogram, p = 0.95

c) mean = 5.7; sd = 0.5339

71. a. The shipment will be rejected if at least 25% of the balls are defective in the sample of 16. 25% of 16 is 4, so the entire shipment will NOT be rejected if the # of defective balls is less than 4.

p(defect) = .2; n = 16
P(X < 4) = P(X = 0, 1, 2 or 3) = binomcdf(16, .20, 3) ≈ .5981

b. The shipment will be rejected if the random variable X, the # of defective balls, is greater than or equal to 4.

p(defect) = .2; n = 16
P(X ≥ 4) = 1 − binomcdf(16, .20, 3) = .4019 (which is just 1 − part a.)

73. True

75. skewed right, symmetrical

77.

a.
X	P(X)
0	0.2865
1	0.3581
2	0.2238
3	0.0933

b.
X	P(X)
0	0.0001
1	0.0011
2	0.0050
3	0.0150

c.
X	P(X)
0	0.0183
1	0.0733
2	0.1465
3	0.1954

79. a. 0.1353 b. 0.0446 c. 0.5940
81. a. 0.0993 b. 0.0624
83. a. 300, 17.3 b. 20, 4.5 c. 0.0176
   d. 0.0214 e. 0.0888
85. failures; success
87. $\mu = p/q$ failures; rq/p failures
89. a) 20/38
   b) $P(X=5) = $ geometpdf(18/38,6) = .0191
   c) $P(X=4) = $ geometPDF(18/38, 4) = 0.9233
   d) $\mu = 1.1111$ losses
   e) $\sigma^2 = 2.3457$; $\sigma = 1.5316$ losses
91. a) 0.75
   b) $P(X=9) = $ geometpdf(0.25,10) = .0188
   c) $P(X=6) = $ geometPDF(0.25, 6) = 0.8220
   d) $\mu = 3$ incorrect guesses
   e) $\sigma^2 = 12$; $\sigma = 3.4641$ incorrect guesses
93. a) p = 18/38; q = 20/38
   b) $P(X=7) = 0.0428$
   c) $P(X=11) = 0.0157$
   d) $\mu = 22$ losses
95. a) p = 0.25; q = 0.75
   b) $P(X=30) = 0.0031$
   c) $P(X=7) = 0.0751$
   d) $\mu = 90$ incorrect guesses
97. discrete; continuous
99. a) continuous
   b) discrete
   c) discrete
   d) continuous
   e) discrete
   f) continuous
   g) discrete
   h) discree
   i) discrete
   j) continuous
   k) discrete
   l) discrete
   m) continuous
   n) discrete
   o) continuous
101. 1 (one)
103. a) 0.98
   b) 0.005
   c) 0.003
   d) unusual
   e) the production processes might be faulty
105. a) Average Risk is 2.95
   b) medium risk since the average risk is close to 3

107. a) 1,2,3
   b) Probability histogram shape both skewed right

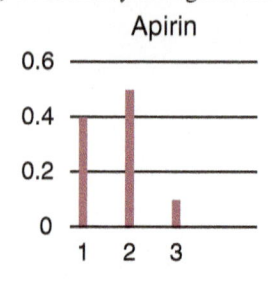

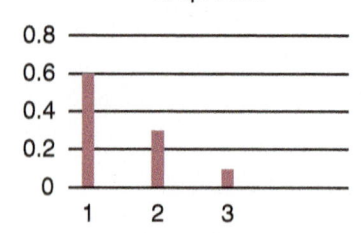

   c) 1; yes; Probability Distribution
   d) μ aspirin = 1.7 tablets; μ ibuprofen = 1.5 tablets
   e) σ aspirin = 0.64 tablets; σ ibuprofen = 0.67 tablets
   f) more than 2 aspirin. Larger mean number of tablets

109. a)

Craig
correct pick	P(correct pick)
0	0.32768
1	0.4096
2	0.2048
3	0.0512
4	0.0064
5	0.00032

Matthew
correct pick	P(correct pick)
0	0.031250
1	0.15625
2	0.3125
3	0.3125
4	0.15625
5	0.031252

Debbie
correct pick	P(correct pick)
0	0.00032
1	0.0064
2	0.0512
3	0.2048
4	0.4096
5	0.32768

b)

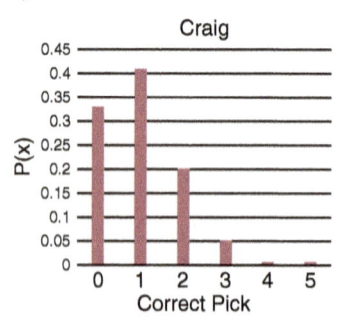

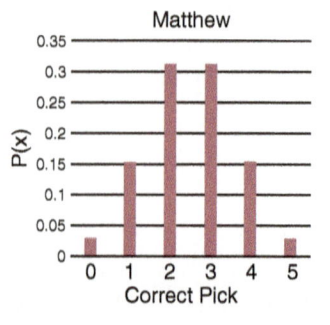

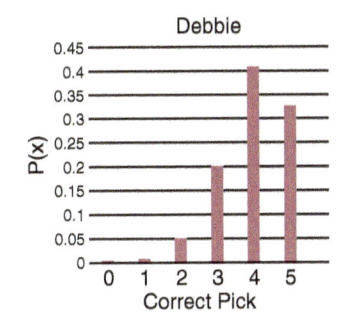

c) shape: Craig skewed right; Matthew symmetric; Debbie skewed left shape for probability histogram of 0.3 skewed right; shape for probability histogram of 0.7 skewed left

d) balance point for Craig histogram between 1 and 2 on horizontal balance point for Matthew histogram between 2 and 3 on horizontal balance point for Debbie histogram between 3 and 4 on horizontal

e) Craig $\mu_s = 1$ correct selection; Matthew $\mu_s = 2.5$ correct selections; Debbie $\mu_s = 4$ correct selections

f) Craig $\sigma_s = 0.894427$ correct selection; Matthew $\sigma_s = 1.118034$ correct selections; Debbie $\sigma_s = 0.894427$ correct selections

g) 1; They are all probability distributions

111. c
113. q = 1/5
115. e
117. e
119. a) $(1/3)^5 = 0.00415$
    b) 0.209877
    c) 0.045267
121. a) $(5/6)^4 = 0.482253$
    b) $(1/6)^4 = 7.716049\text{E}{-4}$
    c) 0.868055
123. a) 0.829847
    b) 0.170153
125. a) $\lambda = 8.35$

k	$P(x=k) = \dfrac{e^{-8.35}(8.35^k)}{k!}$
0	2.4E-4
1	0.00197
2	0.00824
3	0.02294
4	0.04788
5	0.07996

b) mean = 12 = λ

k	$P(x=k) = \dfrac{e^{-12}(12^k)}{k!}$
0	6.1E-6
1	7.4E-5
2	4.4E-4
3	0.00177
4	0.00531
5	0.01274

c) standard deviation = 9, thus $\lambda = 81$

k	$P(x=k) = \dfrac{e^{-81}(81^k)}{k!}$
0	7E-36
1	5E-34
2	2E-32
3	6E-31
4	1E-29
5	2E-28

127. a) $\lambda = 250$; standard deviation ≈ 15.8114
    b) $\lambda \approx 20.8333$; standard deviation ≈ 4.5644
    c) 0.004217
    d) if we assume that for a given week there are 7 twelve hour shifts then we might expect 250 for 7 days to be the number of students per week = 1750 students per week. Thus the probability that the number of students is less than a 1799 per week $\approx .8813$
    e) If for a given week we expect 1750 students, then $P(X=1800) \approx 0.0046$
129. a) 3/16
    b) 0.078125
    c) 0.031250
    d) 3

CHAPTER 7

1. true
3. true
5. true
7. b
9. answers vary
11. normal distribution properties are:
    - The normal curve is bell-shaped and has a single peak which is located at the center.
    - The mean, median and mode all have the same numerical value and are located at the center of the distribution.
    - The normal curve is symmetric about the mean.
    - The data values in the normal distribution are clustered about the mean.
    - In theory, the normal curve extends infinitely in both directions, always getting closer to the horizontal axis but never touching it.
    - The total area under the normal curve is 1 which can also be interpreted as probability.
13. a) 0.0013 b) 0.1587 c) 0.5 d) 0.9772
    e) 1 f) 0.5 g) 0.8413 h) 0.0228
    i) 0.9986 j) 0 k) 0.6827 l) 0.9545
    m) 0.9973 n) 0.9999 or approx. 1
15. center
17. 100
19. 15.87; 84.13
21. false

23. false
25. false
27. a. normalcdf(–e99, –1.65) ≈ .0495 = 4.95%
    b. normalcdf(–3, 3) ≈ .9973 = 99.73%
    c. normalcdf(2.33, e99) ≈ .0099 = .99%
29. a) –0.25   b) 2.33   c) –1.65 and 1.65
31. a) 0.8577   b) 0.6589   c) 0.2514
33. a) z of 8.5 = 0.91, percentage of users who spend less than 8.5 hours = 81.86%
    b) z of 4.5 = –1.52, percentage of users who spend at least 4.5 hours = 93.57%
    c) z of 4.5 = –1.52 and z of 10 = 1.82, percentage of users who spend either less than 8.5 hours or more than 10 hours = 6.42% + 3.43% = 9.85%
    d) z of 390 minutes = z of 6.5 hours = –0.30, percentage of users who spend at most 390 minutes on the site = 38.21%
35. The male height that cuts off the lower 90% from the upper 10% is found by invnorm(.90, 69, 4) ≈ 74.13 inches. Debbie wants the opening to have at least a 1 foot clearance, so adding 12" to 74.13 inches gives 86.13 inches.
37. The BMI that cuts off the lower 82% from the upper 18% is found by invnorm(.82, 28.7, 1.4) ≈ 29.98. The minimum BMI of a female to be considered obese is approximately 30.
39. mean, 0
41. a. normalcdf(–e99, 286, 250, 20) ≈ .9641 = 96
    b. normalcdf(–e99, 265, 250, 20) ≈ .7734 = 77
    c. normalcdf(–e99, 225, 250, 20) ≈ .1056 = 11
43. ans d.: invnorm(.77, 75, 15) = 86.1
45. a. P.R. of 163 = normalcdf(–e99, 163, 150, 10) ≈ .9032 = 90
    b. P.R. of 135 = normalcdf(–e99, 135, 150, 10) ≈ .0668 = 7
    c. invnorm(.98, 150, 10) ≈ 170.5 = 171
    d. 50/300 ≈ .1667; invnorm(.8333, 150, 10) ≈ 159.67 = 160
    e. invnorm(.60, 150, 10) ≈ 152.53 = 153
    f. invnorm(.25, 150, 10) ≈ 143.25 = 143
    g. invnorm(.75, 150, 10) ≈ 156.74 = 157
    h. invnorm(.15, 150, 10) ≈ 139.63 = 140
    i. You should guess the mean (150), since the largest area under a normal curve is centered around the mean.
       P(138 < X < 162) = normalcdf(138, 162, 150, 10) ≈ .7699 = 76.99%
47. a. P(X < 500) = normalcdf(–e99, 500, 900, 350) ≈ .1265
    b. P(800 < X < 1200) = normalcdf(800, 1200, 900, 350) ≈ .4168
    c. P(X > 1100) = normalcdf(1100, e99, 900, 350) ≈ .2839
49. a. P(X < 8) = normalcdf(–e99, 8, 7, 3.2) ≈ .6227 = 62.27%
    b. P(X > 10) = normalcdf(10, e99, 7, 3.2) ≈ .1743
    c. P(4 < X < 12) = normalcdf(4, 12, 7, 3.2) ≈ .7667 = 76.67%
    d. The number of job changes that cut off the middle 90% of the distribution is found by invnorm(.05, 7, 3.2) ≈ 1.74 and invnorm(.95, 7, 3.2) ≈ 12.26. The expected number of job changes is between 1.74 and 12.26.
51. a. P(X < 11.98) = normalcdf(–e99, 11.98, 12, .018) ≈ .1333
    b. P(11.96 < X < 12.01) = normalcdf(11.96, 12.01, 12, .018) ≈ .6976
    c. P(X > 12.03) = normalcdf(12.03, e99, 12, .018) ≈ .0478
    d. P(X > 12.05) = normalcdf(12.05, e99, 12, .018) ≈ .0027

53. a. P(X < 75) = normalcdf(–e99, 75, 55, 10) ≈ .9772
    b. P(X > 55) = normalcdf(55, e99, 55, 10) ≈ .5
    c. P(X ≤ 50) = normalcdf(–e99, 50, 55, 10) ≈ .3085
55. a. No, because n*p = 2 < 10 and n*(1 – p) = 8 < 10
    b. Yes, because n*p = 36 ≥ 10 and n*(1 – p) = 12 ≥ 10
57. a. $\mu_s = np$
    b. $\sigma_s = \sqrt{npq}$
59. a) 200*0.37 = 74
61. T
63. np = nq = 400 * 1/2 = 200 > 5, so we can use the normal approximation to the binomial . . . also $\sigma_s = \sqrt{npq} = 10$.
    a. P(s ≥ 217) = P(X ≥ 216.5) = normalcdf(216.5, e99, 200, 10) = .0495
    b. P(180 < s < 195) = P(179.5 < X < 195.5) = normalcdf(179.5, 195.5, 200, 10) = .3062
    c. P(s = 200) = P(199.5 < X < 200.5) = normalcdf(199.5, 200.5, 200, 10) = .0399
    d. P(s < 185 or s > 210) = P(X < 184.5 or X > 210.5) = normalcdf(–e99, 184.5, 200, 10) + normalcdf(210.5, e99, 200, 10) = .2074
65. np = 20 and nq = 80 which are both > 5 so the normal approximation to the binomial can be used. $\mu_s = np = 20$ and $\sigma_s = \sqrt{npq} = 4$.
    P(s ≥ 33) = P(X > 32.5) = normalcdf(32.5, e99, 20, 4) ≈ .0009
67. For Monday's: $\mu_s = np = 43$ and $\sigma_s = \sqrt{npq} = 4.951$.
    For Friday's: $\mu_s = np = 58$ and $\sigma_s = \sqrt{npq} = 4.936$.
    a. P(s ≥ 55) = P(X ≥ 54.5) = normalcdf(54.5, e99, 43, 4.951) ≈ .0101
    b. P(s ≤ 45) = P(X ≤ 45.5) = normalcdf(–e99, 45.5, 58, 4.936) ≈ .0057
    c. P(s > 40) = P(X > 40.5) = normalcdf(40.5, e99, 42, 4.936) ≈ .6194
    d. P(s < 60) = P(X < 59.5) = normalcdf(–e99, 59.5, 57, 4.951) ≈ .6932
69. $\mu_s = np = 80$ and $\sigma_s = \sqrt{npq} = 4$.
    a. P(s ≥ 90) = P(X > 89.5) = normalcdf(89.5, e99, 80, 4) ≈ .0088
    b. P(s ≤ 15) = P(X ≤ 15.5) = normalcdf(–e99, 15.5, 20, 4) ≈ .1303
71. $\mu_s = np = 80$ and $\sigma_s = \sqrt{npq} = 8$.
    a. P(s ≤ 60) = P(X ≤ 60.5) = normalcdf(–e99, 60.5, 80, 8) ≈ .0074
    b. P(s ≥ 90) = P(X > 89.5) = normalcdf(89.5, e99, 80, 8) ≈ .1175
73. $\mu_s = np = 4000$ and $\sigma_s = \sqrt{npq} = 60$
    P(s ≤ 4100) = P(X ≤ 4100.5) = normalcdf(–e99, 4100.5, 4000, 60) ≈ .9530
75. True
77. 68.26%
79. a. 68.26%
    b. 95.45%
    c. 99.73%
81. a) 0.0494   b) 0.1490   c) 0.3821
83. a) P(X < 165) = normalcdf(–e99, 165, 190, 20) ≈ .1056
    b) P(X > 235) = normalcdf(235, e99, 190, 20) ≈ .0122
    c) P(178 < X < 224) = normalcdf(178, 224, 190, 20) ≈ .6812
    d) invnorm(.75, 190, 20) ≈ 203.49 pounds

85. a) 9.68%
    b) approx. 27 games
    c) 0.3446
87. The 85th percentile for sit-ups is found by invnorm (.85, 27, 3) ≈ 30.11 ≈ 31 sit-ups. Therefore a student needs a minimum of 31 sit-ups to qualify for the award. Lylah, Hector, Hassan, Katie, and Michael failed to qualify for the award since they each did less than 31 sit-ups.
89. a
91. a) 360   b) 14.07
93. c
95. a) 0.0035
    b) 0.0007
97. a) 0.0301
    b) 0.1292
    c) 0.0838
99. a) 0.1003
    b) 0.9932

CHAPTER 8

1. nonsampling
3. μ
5. $\mu_{\bar{x}}$
7. c
9. a) $\mu = 121.2$
   b) i) 122   ii) 0.8
   c) i) 119.67   ii) –1.53
   d) i) 119   ii) –2.2
11. a) MA, MT, MH, ME, MW, AT, AH, AE, AW, TH, TE, TW, HE, HW, EW
    b) $\frac{1}{15}$ c) HM, WT
13. all
15. standard error
17. b)
19. mean = 850, 111.42
21. $\frac{\sigma}{\sqrt{n}} \times \sqrt{\frac{N-n}{N-1}}$
23. sample size is small relative to population size
25. a) 0.2523
    b) 0.9975
    c) The value of the ratio n/N is equal to .005. This ratio is less than 5% and tells us that the use of the finite correction factor is not necessary.
27. shape
29. normal
31. narrow; close
33. false
35. 2
37. a) 0.1056
    b) 0.7499
    c) 0.3385
39. 0.9332
41. 0.0171
43. 0.0148. This is not a likely event.
45. 0.9522
47. a) 0.0013
    b) maybe; the EPA rating doesn't indicate whether it is estimated based on highway or city driving conditions.
49. a) 1.52%
    b) 0.1587
    c) 18.14%
51. decreases
53. a)
55. 1.78
57. a) mean = 5′3″, standard error = 0.34″
    b) mean = 5.3″, standard error = 0.27″
    c) mean = 5.3″, standard error = 0.2″
    d) mean = 5.3″, standard error = 0.08″
    e) The standard error decreases.
59. $\dfrac{\hat{P} - \mu_{\hat{p}}}{\sigma_{\hat{p}}}$
61. $\dfrac{X}{N}$
63. $\mu_{\hat{p}}$
65. $\mu_{\hat{p}} = p$
67. $\hat{p}$, p
69. (c)
71. (b)
73. a) mean = 0.40, standard error = 0.0693, yes
    b) mean = 0.65, standard error = 0.0337, yes
    c) mean = 0.70, standard error = 0.0229, yes
75. a) 0.0839
    b) 0.0839
    c) 0.9615
77. a) 0.0742
    b) 0.7967
    c) 0.9258
79. mean = 1540, standard error = 20
81. a) mean = 2.7; standard error = 0.1968
    b) mean = 2.7; standard error = 0.0860
    c) mean = 2.7; standard error = 0.0353
    d) The larger the sample size, the smaller the value.
83. a) 896 dollars, 8 dollars
    b) 0.0228
    c) 0.6247
    d) 889.267 dollars
85. a) mean = 92.5, standard error = 2.5143, the sampling distribution of the mean will be a normal distribution with a mean of 92.5 and a standard error of 2.5143.
    b) mean = 92.5, standard error = 1.4667, the sampling distribution of the mean will be a normal distribution with a mean of 92.5 and a standard error of 1.4667.
87. a) mean = 0.10, standard error = 0.05477, cannot be approximated by normal distribution.
    b) mean = 0.80, standard error = 0.04, can be approximated by normal distribution.
    c) mean = 0.20, standard error = 0.0327, can be approximated by normal distribution.

Ans20 Answers

89. a) 0.1727
    b) 0.0973
    c) 0.9532
    d) –0.01 or –1%
    e) –0.09 or –9%
    f) yes, if $\hat{p} < p$

CHAPTER 9

1. parameter
3. estimation
5. sample proportion, $\hat{p}$
7. d
9. true
11. sampling
13. point
15. unbiasedness; efficiency; consistency; sufficiency
17. 164.12 to 175.88 miles
19. $\overline{X} - (2.58)\dfrac{\sigma}{\sqrt{n}}$; $\overline{X} + (2.58)\dfrac{\sigma}{\sqrt{n}}$
21. $\overline{X} - (t_{99.5\%})\dfrac{s}{\sqrt{n}}$; $\overline{X} + (t_{99.5\%})\dfrac{s}{\sqrt{n}}$
23. unknown; sample standard deviation, s; normal
25. d
27. true
29. d
31. 42.922 to 48.278 milligrams
33. 10.93 to 14.07 years
35. $35,047 to $44,953
37. $\sqrt{\dfrac{\hat{p}(1-\hat{p})}{n}}$
39. c
41. a) 90%:0.2244 to 0.3756; margin of error = 0.0756
       95%:0.2102 to 0.3898; margin of error = 0.0898
       99%:0.1818 to 0.4182; margin of error = 0.1182
    b) 90%:0.2810 to 0.3190; margin of error = 0.0190
       95%:0.2775 to 0.3225; margin of error = 0.0225
       99%:0.2703 to 0.3297; margin of error = 0.0297
    c) 90%:0.2977 to 0.3023; margin of error = 0.0023
       95%:0.2973 to 0.3027; margin of error = 0.0027
       99%:0.2964 to 0.3036; margin of error = 0.0036
    d) The width of each confidence interval and the margin of error at each confidence level decreases.
43. 0.0177
45. Either the true population proportion will be within 45% ± 3 percentage points or not, thus the chance is either 0% or 100%.
47. population parameter; margin
49. E; width
51. increased
53. $[\dfrac{z\sigma}{E}]^2$
55. false
57. n = 1849
59. 167
61. 46% to 52% Americans adults suffer from sleep related problems;
    49% to 55% American women had sleep difficulties;
    42% to 48% American men had sleep difficulties;
    37% to 43% American adults sometimes doze off when bored
63. a) study times of day students
    b) population mean
    c) sample mean
    d) 18.5 hrs/week
    e) 90%: margin of error = 0.33 hrs/wk
       95%: margin of error = 0.39 hrs/wk
       99%: margin of error = 0.52 hrs/wk
    f) 90%: 18.17 to 18.83 hrs/wk
    g) 95%: 18.11 to 18.89 hrs/wk
    h) 99%: 17.98 to 19.02 hrs/wk
65. a) 103.1 to 112.9
    b) 105.06 to 110.94
    c) 106.53 to 109.47
    d) 9.8; 5.88; 2.94
    e) the width decreases
67. 95%: 0.3822 to 0.4594; margin of error = 0.0386
69. a) 0.5124 to 0.6877
    b) 0.0877
    c) We are 95% confident that the proportion of students with a tattoo at Mike's college falls within the interval 0.5124 to 0.6877.
    d) Mike agreed to get a tattoo if more than 50% of the students at his college have one. Based on the 95% confidence interval, the population proportion of students with a tattoo is between 0.5124 and 0.6877 which is more than 50%. Therefore, Mike will get the tattoo.
71. a) 65.61 to 70.99
    b) 5.38
    c) 2.69
    d) n = 355
73. $(z)\left(\dfrac{\sigma}{\sqrt{n}}\right)$
75. $\dfrac{z^2(0.25)}{E^2}$

CHAPTER 10

1. true; reject H_0; H_0
3. II
5. true
7. a
9. I
11. one
13. –1.65; 1.65
15. fail to reject
17. c
19. true

21. H_o: The population mean weight loss is 12 lbs for the first ten days.

H_a: The population mean weight loss is less than 12 lbs for the first ten days.

This is a directional test.

23. H_o: The population proportion of asthmatic children whose mothers smoked during pregnancy is 10%.

H_a: The population proportion of asthmatic children whose mothers smoked during pregnancy is not 10%.

This is a non-directional test.

25. H_o: The population mean consumption of electricity per day of the energy saving refrigerator is 4.9 k.w.h.

H_a: The population mean consumption of electricity per day of the energy saving refrigerator is more than 4.9 k.w.h.

This is a directional test.

27. **a)** Reject H_o if the test statistic is either less than –1.96 or greater than 1.96

b) Yes

c) Yes

29. **a)** Reject H_o if the test statistic is less than –2.33.

b) No

c) No

d) Yes

e) Yes

31. **a)**

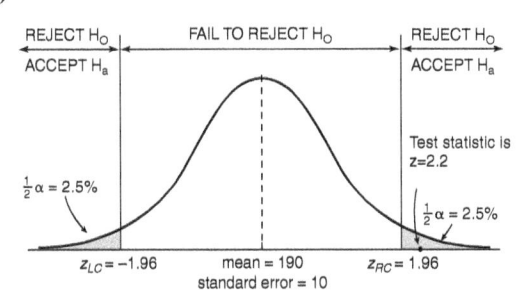

b) Reject H_o, accept H_a

c) Yes

33. **a)** For z = 2.33, chance of Type I error = 1%

b) 1%

35. **a)** For z = –1.65, chance of Type I error = 5%

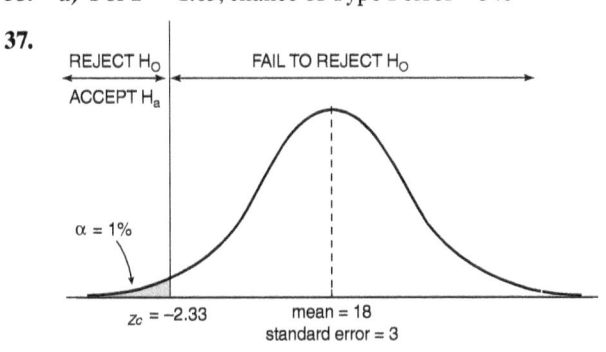

37.

Decision rule: Reject H_o if the test statistic is less than –2.33

a) 1) Fail to reject H_o
2) Fail to reject H_o
3) Reject H_o, accept H_a
4) Fail to reject H_o

b) –2.35

39.

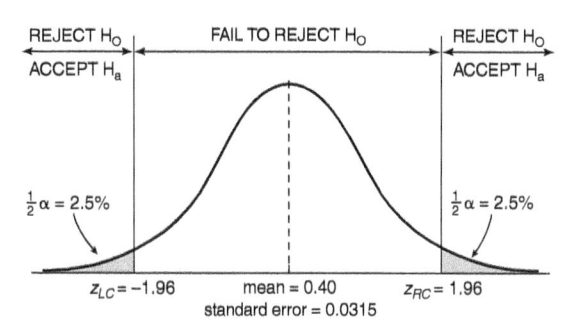

Decision rule: Reject H_o if the test statistic is less than –1.96 or greater than 1.96.

a) 1) Fail to reject H_o
2) Reject H_o, accept H_a
3) Reject H_o, accept H_a
4) Fail to reject H_o

b) 2.36; –2.49

41. very; significant; marginally or not; not

43. true

45. **a)** H_o: The population mean age of recipients of unemployment benefits is 37 years.

H_a: The population mean age of recipients of unemployment benefits is not 37 years.

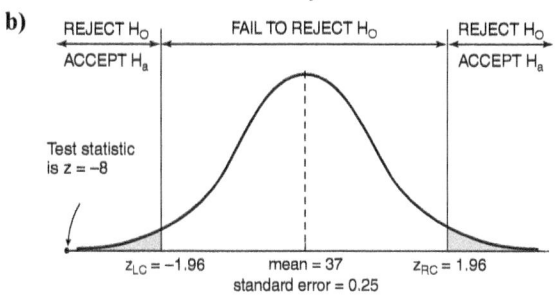

b)

c) p-value = 0.0000 or 0%

d) Reject H_o, accept H_a.

47. **a)** H_o: The population proportion of the babies delivered by Cesarean Section is 20%.

H_a: The population proportion of the babies delivered by Cesarean Section is more than 20%.

b) $Z_c = 1.65$

Decision rule: Reject H_o if the test statistic is more than $Z_c = 1.65$.

c)

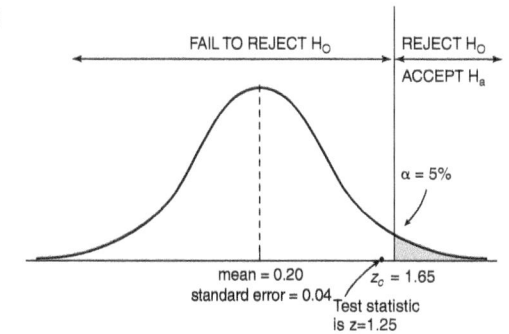

d) p-value = 0.1057 or 10.57%

e) *Conclusion*: Fail to reject H_o.

f) The test statistic is not statistically significant.

g) The research group cannot conclude that C-Sections have increased.

49. a) *Formulate the two hypotheses H_o and H_a.*

H_o: The population monthly mean number of hours spent on Facebook by an American Facebook user is 7 hours.

H_a: The population monthly mean number of hours spent on Facebook by an American Facebook user is greater than 7 hours.

b) $Z_c = 2.33$

Decision Rule: Reject H_o if the test statistic is more than the $Z_c = 2.33$

c)

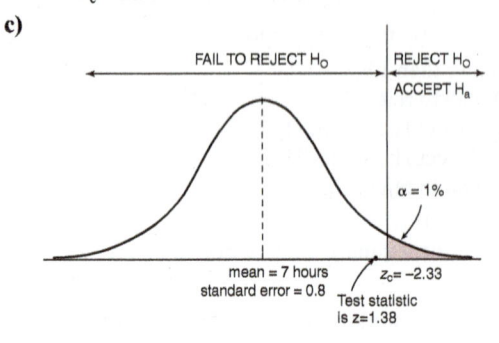

[Normal distribution curve: FAIL TO REJECT H_o | REJECT H_o ACCEPT H_a; $\alpha = 1\%$; mean = 7 hours, standard error = 0.8; $Z_c = -2.33$; Test statistic is $z = 1.38$]

d) p-value 0.0846 or 8.46%

e) *Conclusion*: Fail to reject H_o at $\alpha = 1\%$.

f) The test statistic is not statistically significant.

g) We cannot conclude that population monthly average spent on Facebook by an American Facebook user has increased this year at $\alpha = 1\%$.

51. a. 2.75E-7%
b. Reject H_o.
c. 2.8E-7%

53. b

55. two

57. −1.96, 1.96

59. a

61. a

63. a) yes; **b)** 3%

65. H_o: The population proportion of wives who hold full-time jobs earn more than their husbands is 20%.

H_a: The population proportion of wives who hold full-time jobs earn more than their husbands is less than 20% directional test.

67. a) Reject H_o if the test statistic is less than −1.65.
b) Yes
c) Yes
d) No
e) No

69. For $z = -1.65$, chance of Type I error = 5%

71.

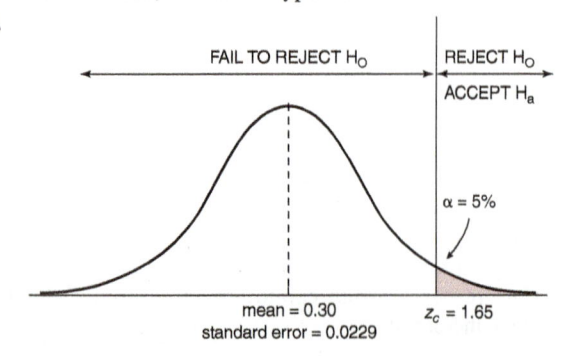

[Normal distribution curve: FAIL TO REJECT H_o | REJECT H_o ACCEPT H_a; $\alpha = 5\%$; mean = 0.30, standard error = 0.0229; $Z_c = 1.65$]

Decision rule: Reject H_o if the test statistic is greater $Z_c = 1.65$.

a) 1) Fail to reject H_o
2) Fail to reject H_o
3) Fail to reject H_o
4) Fail to reject H_o

b) none

73. a) H_o: The population mean reaction time is 0.20 seconds.

H_a: The population mean reaction time is less than 0.20 seconds.

b) $Z_c = -2.33$

Decision rule: Reject H_o if the test statistic is less $Z_c = -2.33$.

c)

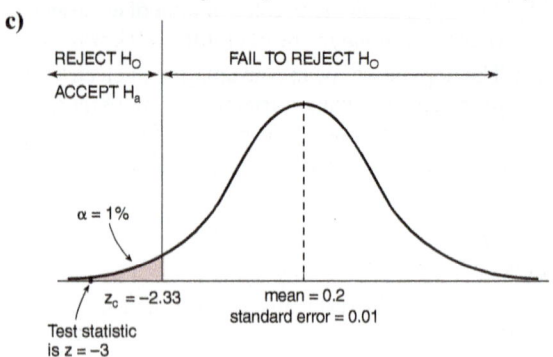

[Normal distribution curve: REJECT H_o ACCEPT H_a | FAIL TO REJECT H_o; $\alpha = 1\%$; $Z_c = -2.33$; mean = 0.2, standard error = 0.01; Test statistic is $z = -3$]

d) p-value = 0.0013 or 0.13%

e) *Conclusion*: Reject H_o, accept H_a.

f) The test statistic is statistically significant.

g) The sample result supports the psychiatrist's hypothesis.

CHAPTER 11

1. proportion

3. a) p; >
b) population; <
c) proportion; ≠

5. d

7. c

9. a) H_o: population proportion of cars with major defective parts is 10%: p = 0.1

H_a: population proportion of cars with major defective parts is greater than 10%: p > 0.1

b) $\mu_{\hat{p}} = 0.1$; $\sigma_{\hat{p}} = 0.03$; distribution of the test statistic is a normal distribution

c) Reject H_o if the test statistic is greater than 1.65

d) $\hat{p} = 0.16$ and $z = 2$

e) Reject H_o, accept H_a at $\alpha = 5\%$; Yes

11. a) H_o: population proportion of American people who say life in the U.S. is getting worse is 40%: p = 0.40

H_a: population proportion of American people who say life in the U.S. is getting worse is not 40%: p ≠ 0.40

b) $\mu_{\hat{p}} = 0.40$; $\sigma_{\hat{p}} = 0.0245$; distribution of the test statistic is a normal distribution

c) Reject H_o, if the test statistic is either less than −1.96 or greater than 1,96

d) $\hat{p} = 0.4375$ and $z = 1.53$

e) Fail to reject H_o at $\alpha = 5\%$; No

13. **a)** H₀: population proportion of LI shoppers being charged sales tax on non-taxable items in supermarkets is 25%: p = 0.25

   Hₐ: population proportion of LI shoppers being charged sales tax on non-taxable items in supermarkets is more than 25%: p > 0.25

   b) $\mu_{\hat{p}} = 0.25$; $\sigma_{\hat{p}} = 0.0306$; distribution of the test statistic is a normal distribution

   c) Reject H₀, if the test statistic is greater than 2.33

   d) $\hat{p} = 0.325$ and $z = 2.45$

   e) Reject H₀, accept Hₐ at α = 1%; Yes

15. **a)** H₀: population proportion of men who wed before age 22 were divorced within 20 years is 68%: p = 0.68

   Hₐ: population proportion of men who wed before age 22 were divorced within 20 years is not 68%: p ≠ 0.68

   b) $\mu_{\hat{p}} = 0.68$ $\sigma_{\hat{p}} = 0.0295$, distribution of the test statistic is a normal distribution

   c) Reject H₀, if the test statistic is either less than –1.96 or greater than 1.96

   d) $\hat{p} = 0.744$ and $z = 2.17$

   e) Reject H₀, accept Hₐ at α = 5%; Yes

17. **a)** H₀: population proportion of women who fear walking alone at night in their neighborhoods is 75%: p = 0.75

   Hₐ: population proportion of women who fear walking alone at night in their neighborhoods is less than 75%: p < 0.75

   b) $\mu_{\hat{p}} = 0.75$; $\sigma_{\hat{p}} = 0.0274$; distribution of the test statistic is a normal distribution

   c) Reject H₀, if the test statistic is less than –1.65

   d) $\hat{p} = 0.648$ and $z = -3.72$

   e) Reject H₀, accept Hₐ at α = 5%; Yes

19. directional
21. true
23. **a)** μ_o; >
    b) mean; μ_o; <
    c) population; ≠
25. b
27. a
29. false
31. **a)** H₀: The population mean breaking strength is 90 lbs using the new manufacturing process: μ = 90

   Hₐ: The population mean breaking strength is greater than 90 lbs using the new manufacturing process: μ > 90

   b) $\mu_{\bar{x}} = 90$; $\sigma_{\bar{x}} = 0.75$; the distribution of the test statistic is a normal distribution.

   c) reject H₀ if the test statistic is greater than 2.33.

   d) $\bar{x} = 93$ and $z = 4$

   e) reject H₀, accept Hₐ at α = 1%; yes

33. **a)** H₀: The population mean number of daily passengers is 125,000 during snow storms: μ = 125,000

   Hₐ: The population mean number of daily passengers is greater than 125,000 during snow storms: μ > 125,000

   b) $\mu_{\bar{x}} = 125{,}000$; $\sigma_{\bar{x}} = 666.67$; the distribution of the test statistic is a normal distribution.

   c) reject H₀ if the test statistic is greater than 1.65

   d) $\bar{x} = 126{,}000$ and $z = 1.5$

   e) fail to reject H₀ at α = 5%; no

35. **a)** H₀: The population mean time to harvest 1 acre of oranges is 5 hours under the new method: μ = 5

   Hₐ: The population mean time to harvest 1 acre of oranges is less than 5 hours under the new method: μ < 5

   b) $\mu_{\bar{x}} = 5$; $\sigma_{\bar{x}} = 0.075$ hours, the distribution of the test statistic is a normal distribution.

   c) reject H₀ if the test statistic is less than –2.33.

   d) $\bar{x} = 4.75$ and $z = -3.33$

   e) reject H₀, accept Hₐ at α = 1%; yes

37. **a)** H₀: The population mean saving passbook account balance is $1236.45: μ = 1236.45

   Hₐ: The population mean saving passbook account balance is less than $1236.45: μ < 1236.45

   b) $\mu_{\bar{x}} = 1236.45$; $\sigma_{\bar{x}} = 15.88$; the distribution of the test statistic is a normal distribution.

   c) reject H₀ if the test statistic is less than –1.65.

   d) $\bar{x} = 1217.69$ and $z = -1.18$

   e) fail to reject H₀ at α = 5%; no

39. **a.** a bell curve
    b. center at t = 0
    c. freedom

41. c
43. T
45. **a.** –1.80   **b.** 1.70   **c.** 2.65
    d. –2.57   **e.** –2.01, 2.01   **f.** –2.63, 2.63

47. **a)** H₀: The population mean growth for the new variety of oak tree is 14.3 inches: μ = 14.3 inches

   Hₐ: The population mean growth for the new variety of oak tree is greater than 14.3 inches: μ > 14.3 inches

   b) $\mu_{\bar{x}} = 14.3$; $S_{\bar{x}} = 0.3$; the distribution of the test statistic is a *t* distribution.

   c) Reject H₀ if the test statistic is greater than 1.68.

   d) $\bar{x} = 15.4$ inches, and t = 3.67

   e) Reject H₀ and Accept Hₐ at α = 5%.
   We can conclude that the biologist is correct α = 5%.

49. **a)** H₀: The population mean number of minutes per day spent on the site is 55. μ = 55 minutes per day.

   Hₐ: The population mean number of minutes per day spent on the site is more than μ > 55 minutes.

   b) $\mu_{\bar{x}} = 55$ minutes per day; $s_{\bar{x}} \approx 1.42$; the distribution of the test statistic is a *t* distribution.

   c) Reject H₀ if the test statistic is greater than 1.68.

   d) $\bar{x} = 58$ minutes per day and t = 2.1

   e) Reject H₀ and Accept Hₐ at α = 5%
   Yes, the findings support the professor's belief that Facebook.com claim is too low at α = 5%.

51. **a)** *Formulate the two hypotheses H₀ and Hₐ.*

   H₀: The population mean number of sleep hours per night for college students at the local college is 6 hours: μ = 6 hours

   Hₐ: population mean number of sleep hours per night for college students at the local college is less than 6 hours: μ < 6 hours

   b) $\mu_{\bar{x}} = 6$ hours; $s_{\bar{x}} = 0.2$; the distribution of the test statistic is a *t* distribution.

   c) Reject H₀ if the test statistic is less than –2.40

d) $\bar{x} = 5.5$ hours, $t = -2.5$ and p-value $= 0.0079$

e) Reject H_o and Accept H_a at $\alpha = 1\%$

We can conclude that the population mean number of sleep hours per night for college students at the local college has decreased at $\alpha = 1\%$.

53. a) H_o: The population mean rental for a 3 room apartment in a California County is $850: \mu = \$850$

H_a: The population mean rental for a 3 room apartment in a California County is not $850: \mu \neq \$850$

b) $\mu_{\bar{x}} = \$850$; $S_{\bar{x}} = \$7.50$; the distribution of the test statistic is a t distribution.

c) Reject H_o if the test statistic is less than -2.63 or greater than 2.63.

d) $\bar{x} = \$872$ and $t = 2.93$

e) Reject H_o and Accept H_a at $\alpha = 1\%$. The sample data indicates that the commerce department is not correct at 1%.

55. a) H_o: The population mean number of weeks that a new form of psychotherapy will help a patient to be ready for outpatient therapy is 12 weeks: $\mu = 12$ weeks

H_a: The population mean number of weeks that a new form of psychotherapy will help a patient to be ready for outpatient therapy is more than 12 weeks: $\mu > 12$ weeks

b) $\mu_{\bar{x}} = 12$ weeks; $S_{\bar{x}} = 0.50$ weeks; the distribution of the test statistic is a t distribution.

c) Reject H_o if the test statistic is greater than 2.42.

d) $\bar{x} = 12.70$ weeks and $t = 8.51$ (using calculations of part b)

e) Reject H_o and Accept H_a at $\alpha = 1\%$. The sample data supports the claim made by the resident psychologist at $\alpha = 1\%$.

57. a) H_o: The population proportion of lawn mowers that start up on the first attempt is 95%: $p = 0.95$

H_a: The population proportion of lawn mowers that start up on the first attempt is less than 95%: $p < 0.95$

b) $\mu_{\hat{p}} = 0.95$; $\sigma_{\hat{p}} = 0.0154$; the distribution of the test statistic is a normal distribution.

c) Reject H_o if the p-value of the test statistic is less than 1%.

d) $\hat{p} = 0.89$, $z = -3.89$ and p-value $= 4.95$ E -5

e) Reject H_o and Accept H_a at a p-value of 4.95 E –5. The sample data indicates that the manufacturer is exaggerating at $\alpha = 1\%$.

f) smallest alpha level $= 4.95$ E -5

59. a) H_o: The population proportion of high school students' dropout rate for the year 2013 is 7.9%: $p = 0.079$

H_a: The population proportion of high school students' dropout rate for the year 2013 is greater than 7.9%: $p > 0.079$

b) $\mu_{\hat{p}} = 0.079$; $\sigma_{\hat{p}} = 0.0064$; distribution of the test statistic is a normal distribution

c) Reject H_0, if the p-value of the test statistic is less than 5%

d) $\hat{p} = 0.0881$ and p-value $= 0.0771$

e) Fail to Reject H_0 at $\alpha = 5\%$; No

61. a) H_o: The population proportion of US college undergraduates receiving financial aid is 57%: $p = 0.57$

H_a: The population proportion of US college undergraduates receiving financial aid is greater than 57%: $p > 0.57$

b) $\mu_{\hat{p}} = 0.57$; $\sigma_{\hat{p}} = 0.0350$; distribution of the test statistic is a normal distribution

c) Reject H_0, if the p-value of the test statistic is less than 5%

d) $\hat{p} = 0.6550$ and p-value $= 0.0076$

e) Reject H_0 at $\alpha = 5\%$; Yes

63. a) H_o: The population proportion of users of 3D printers is 81%: $p = 0.81$

H_a: The population proportion of users of 3D printers is less than 81%: $p < 0.81$

b) $\mu_{\hat{p}} = 0.81$; $\sigma_{\hat{p}} = 0.0185$; distribution of the test statistic is a normal distribution

c) Reject H_0, if the p-value of the test statistic is less than 1%

d) $\hat{p} = 0.80$ and p-value $= 0.2943$

e) Reject H_0 at $\alpha = 1\%$; Yes

65. a) H_o: The population proportion of secretaries experiencing eye strain when using word processors with a CRT display screen is 60%: $p = 0.60$

H_a: The population proportion of secretaries experiencing eye strain when using word processors with a CRT display screen is not 60%: $p \neq 0.60$

b) $\mu_{\hat{p}} = 0.60$; $\sigma_{\hat{p}} = 0.0693$; the distribution of the test statistic is a normal distribution.

c) Reject H_o if the p-value of the test statistic is less than 1%.

d) $\hat{p} = 0.70$, $z = 1.44$ and p-value $= 0.1489$

e) Fail to Reject H_o at a p-value of 0.1489. Based upon the sample data, the executive secretary cannot reject the national claim at $\alpha = 1\%$.

f) smallest alpha level $= 0.1489$.

67. a) H_0: The population mean breaking strength of the tennis racket using the new manufacturing process is 90 lbs.: $\mu = 90$ lbs

H_a: The population mean breaking strength of the tennis racket using the new manufacturing process is greater than 90 lbs.: $\mu > 90$ lbs

b) $\mu_{\bar{x}} = 90$ lbs; $\sigma_{\bar{x}} = 0.75$ lbs; the distribution of the test statistic is a normal distribution.

c) Reject H_o if the p-value of the test statistic is less than 1%.

d) $\bar{x} = 93$ lbs, $z = 4.00$ and p-value $= 3.17$ E -5

e) Reject H_o and Accept H_a at a p-value of 3.17 E –5. Based upon the sample data, Larry K. can agree with the manufacturer's claim at $\alpha = 1\%$.

f) smallest alpha level $= 3.17$ E -5

69. a) H_0: The population mean number of hours first year full-time students at four year U.S. Colleges spend per week preparing for class is 14: $\mu = 14$ hours per week

H_a: The population mean number of hours first year full-time students at four year U.S. Colleges spend per week preparing for class is greater than 14: $\mu > 14$ hours per week

b) $\mu_{\bar{x}} = 14$ $s_{\bar{x}} = 0.39$; distribution of the test statistic is a t distribution

c) Reject H_o if the p-value of the test statistic is less than $\alpha = 1\%$.

d) $\bar{x} = 14.8$ and p-value $= 0.0257$

e) Fail to Reject H_0 at $\alpha = 1\%$; No

71. a) H_o: population of Facebook users post a mean of 49 photos of themselves on their profile: $\mu = 49$ photo posts

H_a: population of Facebook users post more than 49 photos of themselves on their profile: $\mu > 49$ photo posts

b) $\mu_{\bar{x}} = 49$; $s_{\bar{x}} = 1.43$; distribution of the test statistic is a t distribution

c) Reject H_0, if the p-value of the test statistic is less than to 5%

d) $\bar{x} = 50$, and p-value = 0.2439

e) Fail to Reject H_0 at $\alpha = 5\%$; No

73. a) H_o: The population mean life of the Louisiana shellfish is 27 months: $\mu = 27$ months

H_a: The population mean life of the Louisiana shellfish is greater than 27 months: $\mu > 27$ months

b) $\mu_{\bar{x}} = 27$; $s_{\bar{x}} = 0.7$; the distribution of the test statistic is a t distribution.

c) reject H_o if the p-value of the test statistic is less than 5%

d) $\bar{x} = 28.05$, t = 1.5 and p-value = 0.0772

e) fail to reject H_o at a p-value of 0.0772; The ecologist's claim is not too low.

f) smallest alpha level = 7.72%

75. a) H_o: The population mean effective period for the muscle relaxing tablets is 4.4 hours: $\mu = 4.4$ hours

H_a: The population mean effective period for the muscle relaxing tablets is less than 4.4 hours: $\mu < 4.4$ hours

b) $\mu_{\bar{x}} = 4.4$; $s_{\bar{x}} = 0.1$; the distribution of the test statistic is a t distribution.

c) reject H_o if the p-value of the test statistic is less than 1%.

d) $\bar{x} = 4.2$, t = −2 and p-value = 0.0239

e) fail to reject H_o at a p-value of 0.0239; The sample results do not support the researcher's suspicions.

f) smallest alpha level = 2.39%

77. a) H_o: The population mean decrease in blood pressure over a three week period is 20 points: $\mu = 20$ points

H_a: The population mean decrease in blood pressure over a three week period is less than 20 points: $\mu < 20$ points

b) $\mu_{\bar{x}} = 20$; $s_{\bar{x}} = 0.5$; the distribution of the test statistic is a t distribution.

c) reject H_o if the p-value of the test statistic is less than 5%.

d) $\bar{x} = 19.14$, t = −1.72 and p-value = 0.0468

e) reject H_o, accept H_a at a p-value of 0.0468; The sample data indicate that the pharmaceutical company's claim is supported.

f) smallest alpha level = 4.68%

79. directional

81. equal; =

83. true

85. false

87. d

89. false

91. a) H_o: The population mean norm for the mathematics anxiety test is 6 for the students at her school: $\mu = 6$

H_a: The population mean norm for the mathematics anxiety test is not 6 for the students at her school: $\mu \neq 6$

b) $\mu_{\bar{x}} = 6$; $\sigma_{\bar{x}} = 0.16$; the distribution of the test statistic is a normal distribution.

c) reject H_o if the test statistic is either less than −1.96 or greater than 1.96.

d) $\bar{x} = 5.73$ and $z = -1.69$.

e) fail to reject H_0 at $\alpha = 5\%$; no

93. a) H_o: population proportion of US adults 18 and older that smoke cigarettes is 18%: p = 0.18

H_a: population proportion of US adults 18 and older that smoke cigarettes is not 18%: p ≠ 0.18

b) $\mu_{\hat{p}} = 0.18$; $\sigma_{\hat{p}} = 0.0172$; distribution of the test statistic is a normal distribution

c) Reject H_0, if the p-value of the test statistic is less than 1%

d) $\hat{p} = 0.2220$ and p-value = 0.0145

e) Fail to Reject H_0 at $\alpha = 1\%$; No

95. a) H_o: population of the children age two have a mean vocabulary of one hundred words: $\mu = 100$ words

H_a: population of the children age two have a mean vocabulary of less than one hundred words: $\mu < 100$

b) $\mu_{\bar{x}} = 100$; $s_{\bar{x}} = 2.86$; distribution of the test statistic is a t distribution

c) Reject H_0, if the p-value of the test statistic is less than to 5%

d) $\bar{x} = 95$ and p-value = 0.0432

e) Reject H_0 at = 5%; Yes

97. a) H_o: The population mean time children aged 6 to 10 within the upper-middle class neighborhood watch TV is 21 hours per week: $\mu = 21$

H_a: The population mean time children aged 6 to 10 within the upper-middle class neighborhood watch TV is less than 21 hours per week: $\mu < 21$

b) $\mu_{\bar{x}} = 21$; $s_{\bar{x}} = 0.69$; the distribution of the test statistic is a t distribution.

c) reject H_o if the test statistic is less than −1.71.

d) $\bar{x} = 20$ and $t = -1.67$.

e) fail to reject H_o at $\alpha = 5\%$; no

CHAPTER 12

1. counts

3. p_2

5. proportion for population 1; size of sample 1; count possessing a particular characteristic for sample 1

7. $\hat{p}_1 - \hat{p}_2$

9. $p_1 - p_2$

11. normal distribution

13. $s_{\hat{p}_1 - \hat{p}_2}$; $\sqrt{\dfrac{\hat{p}_1(1-\hat{p}_1)}{n_1} + \dfrac{\hat{p}_2(1-\hat{p}_2)}{n_2}}$

15. sample proportion of men who never eat breakfast; $\hat{p}_1 = \dfrac{389}{1450} = 0.2683$;

sample proportion of women who never eat breakfast; $\hat{p}_2 = \dfrac{327}{1500} = 0.218$;

17. a) the mean is denoted $\mu_{\hat{p}_1-\hat{p}_2}$ and the formula is

$$\mu_{\hat{p}_1-\hat{p}_2} = p_1 - p_2$$

b) $\sigma_{\hat{p}_1-\hat{p}_2} = \sqrt{\dfrac{p_1(1-p_1)}{n_1} + \dfrac{p_2(1-p_2)}{n_2}}$

c) the standard error is denoted $\sigma_{\hat{p}_1-\hat{p}_2}$ and the formula is

$$\sigma_{\hat{p}_1-\hat{p}_2} = \sqrt{\dfrac{p_1(1-p_1)}{n_1} + \dfrac{p_2(1-p_2)}{n_2}}$$

d) the estimate of the standard error is denoted $s_{\hat{p}_1-\hat{p}_2}$ and the formula is

$$s_{\hat{p}_1-\hat{p}_2} = \sqrt{\dfrac{\hat{p}_1(1-\hat{p}_1)}{n_1} + \dfrac{\hat{p}_2(1-\hat{p}_2)}{n_2}}$$

19. a) the estimate of the standard error is denoted $s_{\hat{p}_1-\hat{p}_2}$; and the formula is

$$s_{\hat{p}_1-\hat{p}_2} = \sqrt{\dfrac{\hat{p}_1(1-\hat{p}_1)}{n_1} + \dfrac{\hat{p}_2(1-\hat{p}_2)}{n_2}}$$

b) when the independent sample selected from each population is a large sample

c) the estimate of the population standard deviation is denoted $s_{\hat{p}_1-\hat{p}_2}$, and the formula is

$$s_{\hat{p}_1-\hat{p}_2} = \sqrt{\dfrac{\hat{p}_1(1-\hat{p}_1)}{n_1} + \dfrac{\hat{p}_2(1-\hat{p}_2)}{n_2}}$$

21. a) $\hat{p}_1 = 0.24$; $\hat{p}_2 = 0.3104$
    b) $s_{\hat{p}_1-\hat{p}_2} \approx 0.0170$
    c) $s_{\hat{p}_1-\hat{p}_2} \approx 0.0170$

23. a) p_1 = proportion of all households watching the TV networks in 2001; p_2 = proportion of all households watching the TV networks in 2000
    b) $\hat{p}_1$ = proportion of all households surveyed watching the TV networks in 2001; $\hat{p}_2$ = proportion of all households surveyed watching the TV networks in 2000
    c) $\hat{p}_1 \approx 0.61$; $\hat{p}_2 \approx 0.70$
    d) $\sigma_{\hat{p}_1-\hat{p}_2} \approx 0.0163$
    e) $\sigma_{\hat{p}_1-\hat{p}_2} \approx 0.0163$

25. b
27. true
29. true
31. a) 0.4008
    b) 0.7189
    c) 0.5992
    d) 0.2811
33. a) 0.7325
    b) 0.6299
    c) yes

d) 0.1026
e) normal
f) 0.11
g) 0.0178
h) 0.0178

35. p; pooling

37. $\sqrt{\hat{p}(1-\hat{p})\left(\dfrac{1}{n_1} + \dfrac{1}{n_2}\right)}$

39. sample proportions; proportion; proportion
41. a) proportion of population 2; >
    b) proportion; population 2; <
    c) population; proportion; 2; ≠
43. z
45. a)
47. c)
49. a)
51. true
53. false
55. true
57. 0
59. a) i) p_1: married women
        p_2: unmarried women
    ii) H_o: $p_1 - p_2 = 0$, H_a: $p_1 - p_2 > 0$
    iii) 1TT
    b) i) p_1: males that believe wearing miniskirts is inappropriate attire for women, p_2: females that believe wearing miniskirts is inappropriate attire for women
    ii) H_o: $p_1 - p_2 = 0$, H_a: $p_1 - p_2 < 0$.
    iii) 1TT
    c) i) p_1: new homes with garage is large enough for 2 or more cars in 2000, p_2: new homes with garage is large enough for 2 or more cars in 1990.
    ii) H_o: $p_1 - p_2 = 0$, H_a: $p_1 - p_2 \neq 0$.
    iii) 2TT
    d) i) p_1: college graduate mothers who breast feed their babies in the hospital, p_2: non-college graduate mothers who breast feed their babies in the hospital
    ii) H_o: $p_1 - p_2 = 0$, H_a: $p_1 - p_2 > 0$.
    iii) 1TT
    e) i) p_1: US populations that attends the movies every week in 2001, p_2: US populations that attends the movies every week in 1990.
    ii) H_o: $p_1 - p_2 = 0$, H_a: $p_1 - p_2 < 0$.
    iii) 1TT
61. a) 0.4494
    b) 0.6283
    c) 0.5232
    d) 0.7163
    e) 0.6916
    f) 0.6425
    g) 0.8260
    h) 0.4126

63. a) p_1: population proportion of females who like to take chances, p_2: population proportion of males who like to take chances.
   H_o: There is no difference between the population proportion of females and males with regards to who likes to take chances: $p_1 - p_2 = 0$.
   H_a: There is a difference between the population proportion of females and males with regards to who likes to take chances: $p_1 - p_2 \neq 0$.
   b) the distribution of the test statistic is approximately a normal distribution; $\mu_{\hat{p}_1-\hat{p}_2} = 0$ and $s_{\hat{p}_1-\hat{p}_2} = 0.0242$
   c) Reject H_o if the test statistic of the difference between the sample proportions is either less than $Z_c = -2.576$ greater than $Z_c = 2.576$.
   d) the test statistic is $z = -2.4793$
   e) Fail to reject H_o at $\alpha = 1\%$. There is no difference between the population proportion of females and males with regards to who likes to take chances.

65. a) Pop$_1$: population proportion of divorced couples who lived together prior to marriage, Pop$_2$: population proportion of divorced couples who did not live together prior to marriage.
   H_o: There is no difference between the population proportion of divorced couples who lived together prior to marriage and population proportion of divorced couples who did not live together prior to marriage: $p_1 - p_2 = 0$.
   H_a: The difference between the population proportion of divorced couples who lived together prior to marriage and population proportion divorced couples who did not live together prior to marriage is greater than zero: $p_1 - p_2 > 0$.
   b) The distribution of the test statistic is approximately a normal distribution where: $\mu_{\hat{p}_1-\hat{p}_2} = 0$ and $s_{\hat{p}_1-\hat{p}_2} = 0.0207$ (using $\hat{p} = 0.323415$)
   c) Reject H_o if the test statistic of the difference between the sample proportions is greater than $z_c = 2.33$.
   d) Test statistic is: $z = 3.4397$;
   e) Reject H_o and Accept H_a at $\alpha = 1\%$. The marriage counselor can conclude that the proportion of couples that got divorced that lived together prior to marriage is significantly higher than the proportion of couples that got divorced that did not live together prior to marriage at $\alpha = 1\%$.

67. a) H_o: There is no difference between the population proportion of women in the Northwest and women in the Midwest who say they are attractive without makeup: $p_1 - p_2 = 0$
   H_a: The population proportion of women from the Northwest who say they are attractive without makeup is greater than the population proportion from the Midwest: $p_1 - p_2 > 0$
   b) the distribution of the test statistic is approximately a normal distribution; $\mu_{\hat{p}_1-\hat{p}_2} = 0$ and $s_{\hat{p}_1-\hat{p}_2} = 0.0203$
   c) Reject H_o if the test statistic of the difference between the sample proportions is greater than $z_c = 1.65$.
   d) The test statistic $z = 6.90$
   e) Reject H_o and accept H_a at $\alpha = 5\%$. The population proportion of women from the Northwest who say they are attractive without makeup is greater than the proportion from the Midwest.

69. a) H_o: There is no difference between the population proportion of smokers who strongly believe that nicotine is an addictive substance and the population proportion of non-smokers who believe that nicotine is an addictive substance: $p_1-p_2=0$.
   H_a: The population proportion of smokers who strongly believe that nicotine is an addictive substance is less than the population proportion of non-smokers who believe that nicotine is an addictive substance: $p_1-p_2 < 0$.
   b) the distribution of the test statistic is approximately a normal distribution, $\mu_{\hat{p}_1-\hat{p}_2} = 0$ and $s_{\hat{p}_1-\hat{p}_2} = 0.0174$
   c) Reject H_o if the test statistic of the difference between the population proportions is less than $Z_c = -1.645$.
   d) the test statistic is $z = -2.2988$
   e) Reject H_o at $\alpha = 5\%$. The population proportion of smokers who strongly believe that nicotine is an addictive substance is significantly less than the population proportion of non-smokers who believe that nicotine is an addictive substance.

71. a) H_o: There is no difference between the population proportion of men chosen from society A who had experienced some form of coronary heart disease, and main protein component of their diet is red meat and the population proportion of men chosen from society B who had experienced some form of coronary heart disease, and main protein component of their diet is fish: $p_1-p_2=0$.
   H_a: People chosen from society A had higher rate of some form of coronary heart disease than those chosen from society B: $p_1-p_2 > 0$.
   b) The distribution of the test statistic is approximately a normal distribution, $\mu_{\hat{p}_1-\hat{p}_2} = 0$ and $s_{\hat{p}_1-\hat{p}_2} = 0.0300$
   c) reject H_o if the test statistic is larger than 2.3263
   d) the test statistic $z = 2.0000$
   e) Fail to reject H_o at $\alpha = 1\%$. No.

73. a) Pop$_1$: population proportion of women who state they use social media to stay in touch with family and friends, Pop$_2$: population proportion of men who state they use social media to stay in touch with family and friends.
   H_o: There is no difference between the population proportion of women who state they use social media to stay in touch with family and friends and population proportion of men who state they use social media to stay in touch with family and friends: $p_1 - p_2 = 0$.
   H_a: The difference between the population proportion of women who state they use social media to stay in touch with family and friends and population proportion of men who state they use social media to stay in touch with family and friends is greater than zero: $p_1 - p_2 > 0$.
   b) The distribution of the test statistic is approximately a normal distribution where: $\mu_{\hat{p}_1-\hat{p}_2} = 0$ and $s_{\hat{p}_1-\hat{p}_2} = 0.01882$ (using $\hat{p} = 0.617778$)

c) Reject H_o if the test statistic of the difference between the sample proportions is greater than $z_c = 2.33$.
d) Test statistic is: $z = 2.66$;
e) Reject H_o and Accept H_a at $\alpha = 1\%$. The social researcher can conclude that more women use social media to stay in touch with family and friends at $\alpha = 1\%$

75. a) H_o: There is no difference between the population proportion of men who approve of mothers working when their children are less than 5 years old and the population proportion of women who approve of mothers working when their children are less than 5 years old: $p_1 - p_2 = 0$.
H_a: There is significant difference in the proportions of men and women who approve of mothers working when their children are less than 5 years old: $p_1 - p_2 \neq 0$.
b) The distribution of the test statistic is approximately a normal distribution, $\mu_{\hat{p}_1 - \hat{p}_2} = 0$ and $s_{\hat{p}_1 - \hat{p}_2} = 0.0451$
c) reject H_o if the test statistic is larger than 1.9600 or less than −1.9600
d) test statistic $z = -1.1086$
e) Fail to reject H_o at $\alpha = 5\%$. There is no difference between the population proportion of men who approve of mothers working when their children are less than 5 years old and the population proportion of women who approve of mothers working when their children are less than 5 years old. No.

77. a) H_o: The is no difference between the proportion of patients cured by Drug A and the proportion of patients cured by Drug B: $p_1 - p_2 = 0$.
H_a: The proportion of patients cured by Drug B is higher than the proportion of patients cured by Drug A: $p_1 - p_2 < 0$.
b) the distribution of the test statistic is approximately a normal distribution, $\mu_{\hat{p}_1 - \hat{p}_2} = 0$ and $s_{\hat{p}_1 - \hat{p}_2} = 0.0671$
c) reject H_o if the test statistic is less than −2.3263
d) test statistic $z = -0.5535$
e) Fail to reject H_o at $\alpha = 1\%$. The is no difference between the proportion of patients cured by Drug A and the proportion of patients cured by Drug B. No.

79. placebo
81. double-blind; experiment
83. An experimental study involves a treatment group and a control group. The only difference between the treatment and control groups is that the subjects of the treatment group are administered the treatment drug and the subjects of the control group are administered the placebo.
85. A treatment effect is the psychological effect of the treatment group feeling better not because of the actual treatment they received but because they feel they were selected for some special treatment.
87. a) treatment group = group receiving the surgery and the cancer vaccine treatment
b) control group = group receiving the surgery and the placebo vaccine
c) H_o: There is no difference between the population proportion of patients receiving the surgery and the cancer vaccine treatment and the population proportion of patients receiving the surgery and the placebo vaccine who remain disease-free after 3 years: $p_1 - p_2 = 0$.
H_a: The population proportion of patients receiving the surgery and the cancer vaccine treatment who remain disease-free after 3 years is greater than the population proportion of patients receiving the surgery and the placebo vaccine who remain disease-free after 3 years: $p_1 - p_2 > 0$.

89. a) H_o: There is no difference between the population proportion of patients receiving the experimental drug CentroRx and angioplasty and the population proportion of patients receiving the placebo drug and angioplasty after 1 year: $p_1 - p_2 = 0$.
H_a: The population proportion of patients receiving the experimental drug CentroRx and angioplasty is greater than the population proportion of patients receiving the placebo drug and angioplasty after 1 year: $p_1 - p_2 > 0$.
b) the distribution of the test statistic is approximately a normal distribution; $\mu_{\hat{p}_1 - \hat{p}_2} = 0$ and $s_{\hat{p}_1 - \hat{p}_2} = 0.0588$
c) Reject H_o if the test statistic of the difference between the sample proportions is greater than $z_c = 2.33$.
d) The test statistic $z = 2.83$.
e) Reject H_o and Accept H_a at $\alpha = 1\%$. The population proportion of patients receiving the experimental drug CentroRx and angioplasty is greater than the population proportion of patients receiving the placebo drug and angioplasty after 1 year.

91. a) H_o: There is no difference between the population proportion of thighs which were administered the treatment cream that lost an average of 0.5" of fat and the population proportion of thighs which were administered the placebo cream and lost an average of 0.5" of fat: $p_1 - p_2 = 0$.
H_a: The population proportion of thighs which were administered the treatment cream that lost an average of 0.5" of fat is greater than the population proportion of thighs which were administered the placebo cream and lost an average of 0.5" of fat: $p_1 - p_2 > 0$.
b) the distribution of the test statistic is approximately a normal distribution; $\mu_{\hat{p}_1 - \hat{p}_2} = 0$ and $s_{\hat{p}_1 - \hat{p}_2} = 0.0434$
c) Reject H_o if the test statistic of the difference between the sample proportions is greater than $z_c = 2.33$.
d) The test statistic $z = 6.35$.
e) Reject H_o and Accept H_a at $\alpha = 1\%$. The population proportion of thighs which were administered the treatment cream that lost an average of 0.5" of fat is greater than the population proportion of thighs which were administered the placebo cream and lost an average of 0.5" of fat.

93. Yes, since from the hypothesis test of problem 108 we can conclude that the population proportion of people administered Hismanal getting drowsy is not significantly different than the population proportion of people administered the placebo getting drowsy.

95. a) H_o: There is no difference between the population proportion of females and the population proportion of males who take chances: $p_1 - p_2 = 0$
H_a: There is a difference between the population proportions of females and the population proportion of males who take chances: $p_1 - p_2 \neq 0$
b) the distribution of the test statistic is approximately a normal distribution; $\mu_{\hat{p}_1 - \hat{p}_2} = 0$ and $s_{\hat{p}_1 - \hat{p}_2} = 0.0242$

c) Reject H_o if the p-value of the test statistic is less than 1%.

d) Test statistic: z = –2.48 and p-value = 0.0134

e) Fail to reject H_o at a p-value of 0.0134.

There is no difference between the population proportions of females and males who take chances.

f) Smallest alpha = 1.34%

97. Using p value approach:

a) Pop_1: population proportion of divorced couples who lived together prior to marriage, Pop_2: population proportion of divorced couples who did not live together prior to marriage.

H_o: There is no difference between the population proportion of divorced couples who lived together prior to marriage and population proportion of divorced couples who did not live together prior to marriage: $p_1 - p_2 = 0$.

H_a: The difference between the population proportion of divorced couples who lived together prior to marriage and population proportion divorced couples who did not live together prior to marriage is greater than zero: $p_1 - p_2 > 0$.

b) The distribution of the test statistic is approximately a normal distribution where:

$\mu_{\hat{p}_1-\hat{p}_2} = 0$ and $s_{\hat{p}_1-\hat{p}_2} = 0.0207$ (using $\hat{p} = 0.323415$)

c) Reject H_o if the p-value of the test statistic is less than 1%

d) Test statistic is: z = 3.4397 and p value = 2.9126 E-4

e) Reject H_o and Accept H_a at a p value of 2.9126 E –4. The marriage counselor can conclude that the proportion of couples that got divorced that lived together prior to marriage is significantly higher than the proportion of couples that got divorced that did not live together prior to marriage at a p value of 2.9126 E –4.

99. a) Population 1: proportion of women in the Northwest say they are attractive with no makeup.

Population 2: proportion of women in the Midwest say they are attractive with no makeup.

H_o: There is no difference between the population proportion of women in the Northwest and women in the Midwest who sat they are attractive without makeup: $p_1 - p_2 = 0$

H_a: The population proportion of women from the Northwest who say they are attractive without makeup is greater than the population proportion from the Midwest: $p_1 - p_2 > 0$

b) the distribution of the test statistic is approximately a normal distribution; $\mu_1 - \hat{p}_2 = 0$ and $s\hat{p}_1 - \hat{p}_2 = 0.0203$

c) Reject H_o if the p-value of the test statistic is less than 5%.

d) Test statistic: z = 6.90 and p-value = 2.74 · 10^{-24}

e) 1%: reject H_o. 5%: reject H_o.

The population proportion of women from the Northwest who say they are attractive without makeup is greater than the population proportion from the Midwest.

f) same as (d)

101. a) H_o: There is no difference between the population proportion of nonsmokers and the population proportion of smokers who strongly believe that nicotine is an addictive substance: $p_1 - p_2 = 0$

H_a: The difference between the population proportion of nonsmokers who strongly believe that nicotine is an addictive substance and the population proportion of smokers who strongly believe that nicotine is an addictive substance is less than zero: $p_1 - p_2 < 0$

b) the distribution of the test statistic is approximately a normal distribution; $\mu_{\hat{p}_1-\hat{p}_2} = 0$ and $s_{\hat{p}_1-\hat{p}_2} = 0.0174$

c) Reject H_o if the p-value of the test statistic is less than 5%.

d) Test statistic: z = –2.29 and p-value = 0.0109

e) Reject H_o and accept H_a at a p-value of 0.0109

The population proportion of nonsmokers who strongly believe that nicotine is an addictive substance is less than the population proportion of smokers who strongly believe that nicotine is an addictive substance.

f) Smallest alpha = 1.09%

103. a) H_o: There is no difference between the population proportion of coronary heart disease for Society A and Society B: $p_1 - p_2 = 0$

H_a: There is difference between the population proportion of coronary heart disease for Society A and Society B: $p_1 - p_2 \neq 0$

b) the distribution of the test statistic is approximately a normal distribution; $\mu_{\hat{p}_1-\hat{p}_2} = 0$ and $s_{\hat{p}_1-\hat{p}_2} = 0.03$

c) Reject H_o if the p-value of the test statistic is less than 1%.

d) Test statistic: z = 2 and p-value = 0.0455

e) Fail to reject H_o at a p-value of 0.0455.

There is no difference between the population proportion of coronary heart disease for Society A and Society B.

f) Smallest alpha = 4.55 %

105. Using p value approach:

a) Pop_1: population proportion of women who state they use social media to stay in touch with family and friends, Pop_2: population proportion of men who state they use social media to stay in touch with family and friends.

H_o: There is no difference between the population proportion of women who state they use social media to stay in touch with family and friends and population proportion of men who state they use social media to stay in touch with family and friends: $p_1 - p_2 = 0$.

H_a: The difference between the population proportion of women who state they use social media to stay in touch with family and friends and population proportion of men who state they use social media to stay in touch with family and friends is greater than zero: $p_1 - p_2 > 0$.

b) The distribution of the test statistic is approximately a normal distribution where:

$\mu_{\hat{p}_1-\hat{p}_2} = 0$ and $s_{\hat{p}_1-\hat{p}_2} = 0.01882$ (using $\hat{p} = 0.617778$)

c) Reject H_o if the p-value of the test statistic is less than 1%.

d) Test statistic: z = 2.66 and p-value = 0.0039

e) Reject H_o and Accept H_a at a p value = 0.0039. The social researcher can conclude that more women use social media to stay in touch with family and friends at a p value = 0.0039.

107. a) H_o: There is no difference between the population proportion of men and the population proportion of women who approve of mothers working when their children are less than five years old: $p_1 - p_2 = 0$

H_a: There is difference between the population proportion of men and the population proportion of women who approve of mothers working when their children are less than five years old: $p_1 - p_2 \neq 0$

b) the distribution of the test statistic is approximately a normal distribution; $\mu_{\hat{p}_1 - \hat{p}_2} = 0$ and $s_{\hat{p}_1 - \hat{p}_2} = 0.0451$

c) Reject H_o if the p-value of the test statistic is less than 5%.

d) Test statistic: $z = -1.11$ and p-value = 0.2671

e) Fail to reject H_o at a p-value of 0.2671.

There is no difference between the population proportion of men and the population proportion of women who approve of mothers working when their children are less than five years old.

f) Smallest alpha = 26.51 %

109. a) H_o: There is no difference between the population proportion of people cured using drug A and the population proportion of people cured using drug B: $p_1 - p_2 = 0$

H_a: The difference between the population proportion people cured using drug A and the population proportion of people cured using drug B is less than zero: $p_1 - p_2 < 0$

b) the distribution of the test statistic is approximately a normal distribution; $\mu_{\hat{p}_1 - \hat{p}_2} = 0$ and $s_{\hat{p}_1 - \hat{p}_2} = 0.0671$

c) Reject H_o if the p-value of the test statistic is less than 1%.

d) Test statistic: $z = -0.55$ and p-value = 0.2900

e) Fail to reject H_o at a p-value of 0.2900

There is no difference between the population proportion of people cured using drug A and the population proportion of people cured using drug B.

f) Smallest alpha = 29%

111. a) Population 1: proportion of those 250 subjects who were administered 2 or more aspirins a week.

Population 2: proportion of those 225 subjects who were administered fake pills

H_o: There is no difference between the population proportion of patients receiving 2 or more aspirins a week and the population proportion of patients receiving the fake pills: $p_1 - p_2 = 0$.

H_a: The population proportion of patients receiving 2 or more aspirins a week is less than the population proportion of the patients who received the fake pills: $p_1 - p_2 < 0$.

b) the distribution of the test statistic is approximately a normal distribution; $\mu_{p1-p2} = 0$ and $S_{p1-p2} = 0.0287$

c) Reject H_o if the p-value of the test statistic is less than 1%.

d) Test statistic: $z = -2.1680$ and p-value = 0.01506

e) 1%: fail to reject H_o. There is no difference between the population proportion of patients receiving 2 or more aspirins a week and the population proportion of patients receiving the fake pills.

5%: reject H_o. The population proportion of patients receiving 2 or more aspirins a week is less than the population proportion of the patients who received the fake pills

f) same as (d)

113. a) Population 1: proportion of those 228 elderly women who were administered one cup of cranberry juice.

Population 2: proportion of those 228 elderly women who were administered imitation liquid.

H_o: There is no difference between the population proportion of elderly women who were administered one cup of cranberry juice and elderly women who were administered imitation liquid: $p_1 - p_2 = 0$.

H_a: The population proportion of elderly women who were administered one cup of cranberry juice is less than the population proportion of elderly women who were administered imitation liquid: $p_1 - p_2 < 0$.

b) the distribution of the test statistic is approximately a normal distribution; $\mu_{p1-p2} = 0$ and $s_{p1-p2} = 0.0385$

c) Reject H_o if the p-value of the test statistic is less than 1%.

d) Test statistic: $z = -3.3766$ and p-value = $3.13 \cdot 10^{-4}$

e) 1%: reject H_o. 5%: reject H_o.

The population proportion of elderly women who were administered one cup of cranberry juice is less than the population proportion of elderly women who were administered imitation liquid

f) same as (d)

115. counts

117. normal distribution

119. false

121. a) The sampling distribution of $_{p1-p2}$ is a theoretical distribution of all the possible differences between the values of the sample proportions selected from each of the two populations.

b) np and n(1–p) are both larger than 5

c) $n_1 p_1, n_1(1 - p_1), n_2 p_2, n_2(1 - p_2)$ are all greater than 5

123. a) Population 1: consumer spending plans for the upcoming holiday season in the Green Acres Shopping Mall on Long Island.

Population 2: consumer spending plans for year 2000 holiday season in the Green Acres Shopping Mall on Long Island.

b) p_1: 650 out of 1952 people. p_2: 809 out of 1924 people.

c) $p_1 = 0.3330$, $p_2 = 0.4205$

d) 0.0155

e) 0.0155

125. a) H_o: There is no difference between the proportion of population 1, written p_1, and the proportion of population 2, written p_2.

H_o: $p_1 - p_2 = 0$

b) i) less than form:

H_a: the proportion of population 1, p_1, is less than the proportion of population 2, p_2.

H_a: $p_1 - p_2 < 0$

ii) greater than form:

H_a: the proportion of population 1, p_1, is greater than the proportion of population 2, p_2.

H_a: $p_1 - p_2 > 0$

iii) not equal to form:

H_a: the proportion of population 1, p_1, is not equal to the proportion of population 2, p_2

H_a: $p_1 - p_2 \neq 0$.

127. The sampling distribution for large sample size of $\hat{p}_1 - \hat{p}_2$ is approximately normal with mean zero and standard deviation

$$\sqrt{\frac{\hat{p}_1(1-\hat{p}_1)}{n_1} + \frac{\hat{p}_2(1-\hat{p}_2)}{n_2}}.$$

129. When the common hypothesized population proportion p is unknown, we need to use the following pooled formulas to estimate the standard error:

$$S_{\hat{p}1-\hat{p}2} = \sqrt{\frac{\hat{p}(1-\hat{p})}{n_1} + \frac{\hat{p}(1-\hat{p})}{n_2}} = \sqrt{\hat{p}(1-\hat{p})(\frac{1}{n_1} + \frac{1}{n_2})}$$

131. a) i) Population 1: proportion of no-show dinner reservations on the weekend, p_1. Population 2: proportion of no-show dinner reservations during the weekdays, p_2.

ii) H_o: There is no difference between the population proportion of no-show dinner reservations on the weekend and during the weekdays: $p_1 - p_2 = 0$.

H_a: The population proportion of no-show dinner reservations on the weekend is greater than during the weekdays: $p_1 - p_2 > 0$.

iii) 1TT

b) i) Population 1: proportion of males who pray, p_1.
Population 2: proportion of females who pray, p_2.

ii) H_o: There is no difference between the population proportion of males who pray and females who pray: $p_1 - p_2 = 0$

H_a: There is a significant difference between the population proportion of males who pray and females who pray: $p_1 - p_2 \neq 0$

iii) 2TT

133. a) 0.0220
b) 0.0155

135. a) Population 1: proportion of males said that people have very little influence on government, p_1.

Population 2: proportion of females said that people have very little influence on government, p_2.

H_o: There is no difference between the population proportion of males that said that people have very little influence on government and the population proportion of females that said that people have very little influence on government: $p_1 - p_2 = 0$

H_a: There is a significant difference between the population proportion of males that said that people have very little influence on government and the population proportion of females that said that people have very little influence on government: $p_1 - p_2 \neq 0$

b) Hypothesis testing model: sampling distribution of $\hat{p}_1 - \hat{p}_2$.
Population 1: $p_1 = 0.63$
Population 2: $p_2 = 0.52$
The distribution of the test statistic is approximately a normal distribution; $\mu_{\hat{p}1-\hat{p}2} = 0$
$\hat{p} = 0.5803$ and $S_{\hat{p}1-\hat{p}2} = 0.0206$

c) decision rule: Reject H_o if the test statistic of the difference between the sample proportion is either less than $Z_c = -2.575$ or greater than $Z_c = 2.575$.

d) The test statistic z = 5.3398

e) Reject H_o and accept H_a at α = 1%. There is a significant difference between the population proportion of males that said that people have very little influence on government and the population proportion of females that said that people have very little influence on government.

137. a) *Formulate the two hypotheses H_o and H_a.*

H_o: There is no difference between the population proportion of male wine drinkers and the population proportion of female wine drinkers who prefer red wine: $p_1 - p_2 = 0$.

H_a: The population proportions of male wine drinkers is significantly greater than the population proportion of female wine drinkers who prefer red wine: $p_1 - p_2 > 0$.

b) The distribution of the test statistic is approximately a normal distribution, $\mu_{\hat{p}1-\hat{p}2} = 0$ and $s_{\hat{p}1-\hat{p}2} = 0.0559$

c) Decision Rule: Reject H_o if the test statistic is greater than 2.33.

d) test statistic z = 2.39 and p-value = 0.0084

e) Reject H_o and Accept H_a at α = 1%.
The population proportion of male wine drinkers is significantly greater than the population proportion of female wine drinkers who prefer red wine.
Yes.

CHAPTER 13

1. The difference between two means
3. a) $\mu_{\bar{x}_1-\bar{x}_2}$; μ_1; μ_2; population; population

b) $\sqrt{\frac{(n_1-1)s_1^2 + (n_2-1)s_2^2}{n_1+n_2-2} \cdot (\frac{1}{n_1}+\frac{1}{n_2})}$

c) t; $n_1 + n_2 - 2$

5. No; mean; mean; $\mu_1 - \mu_2$
7. true
9. test statistic
11. c
13. false
15. a) Pop 1: male SAT math scores
Pop 2: female SAT math scores

b) *Hypotheses*
H_o: There is no difference between math SAT scores of male and female students: $\mu_1 - \mu_2 = 0$
H_a: There is a difference between math SAT scores of male and female students: $\mu_1 - \mu_2 \neq 0$

c) *Expected Results:*
$\mu_{\bar{x}_1-\bar{x}_2} = 0$
$s_{\bar{x}_1-\bar{x}_2} = 18.066$
Hypothesis testing model: the distribution of the test statistic is a *t* distribution.

d) *Decision rule:*
Reject H_0 if the test statistic is less than $t_c = -1.98$ or greater than $t_c = 1.98$

e) *test statistic:* t = 2.14

f) *Conclusion:*
Reject H_0 at α = 5%
The results support the high school teacher's claim.

17. a) Pop 1: 18 to 24 year old Internet users
Pop 2: 25 to 34 year old Internet users.

b) *Formulate the two hypotheses H_o and H_a.*

H_o: There is no difference in the population mean online hours of 18 to 24 year old Internet users and the population mean online hours of 25 to 34 year old Internet users: $\mu_1 - \mu_2 = 0$.

H_a: The difference between the population mean online hours of 18 to 24 year old Internet users and the population mean online hours of 25 to 34 year old Internet users is not zero: $\mu_1 - \mu_2 \neq 0$.

c) *Expected Results:*
$\mu_{\bar{x}_1} - \mu_{\bar{x}_2} = 0$
$s_{\bar{x}_1 - \bar{x}_2} = 0.8284$
Hypothesis test model: the distribution of the test statistic is a *t* distribution.

d) *Decision Rule:*
Reject H_o if the test statistic is either less than $t_{LC} = -1.99$ or greater than $t_{RC} = 1.99$.

e) *test statistic:* t = -2.46 and p-value = 0.01568

f) *Conclusion:*
Reject H_o and Accept H_a at $\alpha = 5\%$.
Answer:
The researcher can conclude that there is a significant difference in the population mean online usage between 18 to 24 year old and 25 to 34 year old Internet users at $\alpha = 5\%$.

19. a) Pop 1: married graduate students
Pop 2: unmarried graduate students

b) *Hypotheses:*
H_o: There is no difference between the population mean cumulative average of married and unmarried graduate students: $\mu_1 - \mu_2 = 0$
H_a: The difference between population mean cumulative average of married graduate students and the population mean cumulative average of unmarried graduate students is less than zero: $\mu_1 - \mu_2 < 0$

c) *Expected Results:*
$\mu_{\bar{x}_1 - \bar{x}_2} = 0$
$s_{\bar{x}_1 - \bar{x}_2} \approx 0.16$
Hypothesis test model: the distribution of the test statistic is a *t* distribution.

d) *Decision rule:*
Reject H_o if the test statistic is less than $t_c = -2.50$.

e) *test statistic:* t = -1.32

f) *Conclusion:*
Fail to reject H_o at $\alpha = 1\%$
Answer:
We cannot agree with the sociologist's claim.

21. a) Pop 1: male SAT math scores
Pop 2: female SAT math scores

b) *Hypotheses*
H_o: There is no difference between math SAT scores of male and female students: $\mu_1 - \mu_2 = 0$
H_a: There is a difference between math SAT scores of male and female students: $\mu_1 - \mu_2 \neq 0$

c) *Expected Results:*
$\mu_{\bar{x}_1 - \bar{x}_2} = 0$
$s_{\bar{x}_1 - \bar{x}_2} = 18.520$
Hypothesis testing model: the distribution of the test statistic is a *t* distribution.

d) *Decision rule:*
Reject H_0 if the test statistic is less than $t_c = -2.01$ or greater than $t_c = 2.01$

e) *test statistic:* t = 2.16

f) *Conclusion:*
Reject H_0, Accept H_a at $\alpha = 5\%$
The results support the high school teacher's claim.

23. a) Pop 1: 18 to 24-year old Internet users
Pop 2: 25 to 34-year-old Internet users.

b) *Hypotheses*
H_o: There is no difference in the population mean online hours of 18 to 24-year-old Internet users and the population mean online hours of 25 to 34-year-old Internet users: $\mu_1 - \mu_2 = 0$.
H_a: The difference between the population mean online hours of 18 to 24-year-old Internet users and the population mean online hours of 25 to 34-year-old Internet users is not zero: $\mu_1 - \mu_2 \neq 0$.

c) *Expected Results:*
$\mu_{\bar{x}_1 - \bar{x}_2} = 0$
$s_{\bar{x}_1 - \bar{x}_2} = 0.8283$
Hypothesis testing model: the distribution of the test statistic is a *t* distribution.

d) *Decision rule:*
Reject H_0 if the test statistic is less than $t_c = -2.02$ or greater than $t_c = 2.02$

e) *test statistic:* t = -2.41

f) *Conclusion:*
Reject H_0, Accept H_a at $\alpha = 5\%$
The results support the social researcher's claim.

25. a) Pop 1: married graduate students
Pop 2: unmarried graduate students

b) *Hypotheses:*
H_o: There is no difference between the population mean cumulative average of married and unmarried graduate students: $\mu_1 - \mu_2 = 0$
H_a: The difference between population mean cumulative average of married graduate students and the population mean cumulative average of unmarried graduate students is less than zero: $\mu_1 - \mu_2 < 0$

c) *Expected Results:*
$\mu_{\bar{x}_1 - \bar{x}_2} = 0$
$s_{\bar{x}_1 - \bar{x}_2} = 1828$
Hypothesis testing model: the distribution of the test statistic is a *t* distribution.

d) *Decision rule:*
Reject H_0 if the test statistic is less than $t_c = -2.90$.

e) *test statistic:* t = -1.14

f) *Conclusion:*
Fail to Reject H_0 at $\alpha = 1\%$
We cannot agree with the sociologist's claim.

27. B, A

29. a) Pop 1: control group
Pop 2: treatment group

b) *Hypotheses:*
H_o: There is no difference between the population mean amount of Vitamin A stored in mice between the control and treatment groups: $\mu_1 - \mu_2 = 0$
H_a: The difference between population mean amount of Vitamin A stored in the mice of the treatment group and

the population mean amount of Vitamin A stored in the mice of the control group is greater than zero: $\mu_1 - \mu_2 > 0$

c) *Expected Results:*
$\mu_{\bar{x}_1-\bar{x}_2} = 0$
$s_{\bar{x}_1-\bar{x}_2} = 139.63$
Hypothesis testing model: the distribution of the test statistic is a *t* distribution.

d) *Decision rule:*
Reject H_0 if the test statistic is greater than $t_c = 2.62$.

e) *test statistic:* t = 6.40

f) Reject H_0, Accept H_a at $\alpha = 1\%$
Answer:
The researcher can conclude that the population mean amount of Vitamin A stored in mice on the Vitamin E deficient diet is significantly less.

31. $\bar{x}_d, s_d$

33. estimate of the standard error

35. a) *Hypotheses:*
H_0: The population mean difference in LDL cholesterol in the before and after groups equals 0: $\mu_d = 0$
H_a: The population mean difference in LDL cholesterol in the before and after groups is significantly less than 0: $\mu_d < 0$

b) *Expected Results:*
The population mean difference of the paired observations is assumed to be 0: $\mu_d = 0$
The sample mean difference of the paired observations is: $\bar{x}_d = -2.5$
The standard deviation of the difference of the paired observations is: $s_d = 6.819$
Hypothesis testing model: the distribution of the test statistic is a *t* distribution with $df = n - 1 = 10 - 1 = 9$.

c) *Decision rule:*
Reject H_0 if the test statistic is less than $t_c = -1.83$.

d) *test statistic:* $t = \dfrac{\bar{x}_d - \mu_d}{\dfrac{s_d}{\sqrt{n}}} = -1.16$

e) Fail to Reject H_0, Reject H_a at $\alpha = 5\%$
Answer:
The medical researcher cannot conclude that the population mean difference in LDL cholesterol is significantly less for individuals using Co-X 15 daily.

37. a) *Hypotheses:*
H_0: The population mean interaction time in the before and after groups equals 0: $\mu_d = 0$
H_a: The population mean difference interaction time in the before and after groups is significantly greater than 0: $\mu_d > 0$

b) *Expected Results:*
The population mean difference of the paired observations is assumed to be 0: $\mu_d = 0$
The sample mean difference of the paired observations is: $\bar{x}_d = 0.48$
The standard deviation of the difference of the paired observations is: $s_d = 3.05$
Hypothesis testing model: the distribution of the test statistic is a *t* distribution with $df = n - 1 = 200 - 1 = 199$.

c) *Decision rule:*
Reject H_0 if the test statistic is greater than $t_c = 2.35$.

d) *test statistic:* $t = \dfrac{\bar{x}_d - \mu_d}{\dfrac{s_d}{\sqrt{n}}} = 2.23$

e) Fail to Reject H_0, Reject H_a at $\alpha = 1\%$
Answer:
The sociologist cannot conclude that the population mean interaction time is significantly greater when cell phones are turned off for at least five hours prior to 14-year-old females going to bed.

39. a) *Hypotheses:*
H_0: The population mean difference in mood in the pre and post medicine groups equals 0: $\mu_d = 0$
H_a: The population mean difference in mood in the pre and post medicine groups significantly greater than 0: $\mu_d > 0$

b) *Expected Results:*
The population mean difference of the paired observations is assumed to be 0: $\mu_d = 0$
The sample mean difference of the paired observations is: $\bar{x}_d = 1.1$
The standard deviation of the difference of the paired observations is: $s_d = 1.97$
Hypothesis testing model: the distribution of the test statistic is a *t* distribution with $df = n - 1 = 10 - 1 = 9$.

c) *Decision rule:*
Reject H_0 if the test statistic is less than $t_c = 2.82$.

d) *test statistic:* $t = \dfrac{\bar{x}_d - \mu_d}{\dfrac{s_d}{\sqrt{n}}} = 2.91$

e) Reject H_0, Accept H_a at $\alpha = 1\%$
Answer:
The psychologist can conclude that the population mean difference in mood for the pre and post medicine groups is significantly greater than 0.

41. a) Pop 1: 18 to 24-year old Internet users
Pop 2: 25 to 34-year-old Internet users.

b) *Hypotheses*
H_0: There is no difference in the population mean online hours of 18 to 24-year-old Internet users and the population mean online hours of 25 to 34-year-old Internet users: $\mu_1 - \mu_2 = 0$.
H_a: The difference between the population mean online hours of 18 to 24-year-old Internet users and the population mean online hours of 25 to 34-year-old Internet users is not zero: $\mu_1 - \mu_2 \neq 0$.

c) *Expected Results:*
$\mu_{\bar{x}_1-\bar{x}_2} = 0$
$s_{\bar{x}_1-\bar{x}_2} = 0.8284$
Hypothesis testing model: the distribution of the test statistic is a *t* distribution.

d) *Decision rule:*
Reject H_0 if the p-value is less than $\alpha = 5\%$

e) *p-value:* p = 0.0157

f) *Conclusion:*
Since the p-value is less than 0.05, Reject H_0 at $\alpha = 5\%$
The results support the social researcher's claim.

43. a) Pop 1: Group 1, the special fertilizer group.
Pop 2: Group 2, the usual fertilizer group.

b) *Hypotheses*
H_0: There is no difference in the population mean tomato crop between special and usual fertilizer groups: $\mu_1 - \mu_2 = 0$
H_a: The population mean tomato crop produced for the special fertilizer group is significantly greater than the usual fertilizer group: $\mu_1 - \mu_2 > 0$

c) *Expected Results:*
$\mu_{\bar{x}_1-\bar{x}_2} = 0$
$s_{\bar{x}_1-\bar{x}_2} = 5.381$
Hypothesis testing model: the distribution of the test statistic is a *t* distribution.

d) *Decision rule:*
Reject H_0 if the p-value is less than $\alpha = 1\%$

e) *p-value:* p = 3.1 E-6 (*essentially* 0)

f) *Conclusion:*
Since the p-value is less than 0.01, Reject H_0 at $\alpha = 1\%$
The results support the farmer's claim.

45. a) *Hypotheses:*
H_o: The population mean difference in the throwing distance between the before and after groups equals 0: $\mu_d = 0$
H_a: The population mean difference in the throwing distance between the before and after groups is significantly greater than 0: $\mu_d > 0$

b) *Expected Results:*
The population mean difference of the paired observations is assumed to be 0: $\mu_d = 0$
The sample mean difference of the paired observations is: $\bar{x}_d = 2.5$
The standard deviation of the difference of the paired observations is: $s_d = 6.82$
Hypothesis testing model: the distribution of the test statistic is a *t* distribution with $df = n - 1 = 10 - 1 = 9$.

c) *Decision rule:*
Reject H_0 if the p-value is less than $\alpha = 1\%$.

d) *p-value:* p = 0.138

e) Since the p-value is greater than 0.01,
Fail to Reject H_o, Reject H_a at $\alpha = 1\%$
Answer:
The softball coach cannot conclude that the population mean throwing distance is significantly greater when weight training is added to the player's exercise program.

47. b

49. $s_{\bar{x}_1-\bar{x}_2} = \sqrt{\frac{s_1^2}{n_1} + \frac{s_2^2}{n_2}} = \sqrt{\frac{3^2}{25} + \frac{5^2}{50}} \approx 0.927$

51. false
53. H_o
55. d
57. true
59. a) Pop 1: urban sixth grade students
Pop 2: rural sixth grade students

b) *Hypotheses*
H_o: There is no difference between the population mean score of the urban sixth grade students and the population mean score of the rural sixth grade students on the standardized reading comprehension test: $\mu_1 - \mu_2 = 0$
H_a: The difference between the population mean score of the urban sixth grade students and the population mean score of the rural sixth grade students on the standardized reading comprehension test is greater than zero: $\mu_1 - \mu_2 > 0$

c) *Expected Results:*
$\mu_{\bar{x}_1-\bar{x}_2} = 0$
$\sigma_{\bar{x}_1-\bar{x}_2} = 2.08$
Hypothesis testing model: the distribution of the test statistic is a *normal* distribution.

d) *Decision rule:*
Reject H_0 if the test statistic is greater than $z_c = 1.65$.

e) *test statistic:* z = 2.16; p-value = 0.0154

f) *Conclusion:*
Reject H_0, Accept H_a at $\alpha = 5\%$
The educational researcher can conclude that the urban sixth grade students performed significantly better on the standardized test

61. a) Pop 1: women
Pop 2: men

b) *Hypotheses*
H_o: There is no difference in population mean amount of debt between women and men: $\mu_1 - \mu_2 = 0$.
H_a: The population mean amount of debt for woman is less than men: $\mu_1 - \mu_2 < 0$.

c) *Expected Results:*
$\mu_{\bar{x}_1-\bar{x}_2} = 0$
$s_{\bar{x}_1-\bar{x}_2} = 1351.4$
Hypothesis testing model: the distribution of the test statistic is a *t* distribution.

d) *Decision rule:*
Reject H_0 if the test statistic is less than $t_c = -1.68$.

e) *test statistic:* t = −1.73, p-value = 0.044

f) *Conclusion:*
Reject H_0 and accept H_a at $\alpha = 5\%$
The results support the financial analyst's claim.

63. Part I.

a) Pop 1: Ivy League College
Pop 2: Military Academy

b) *Hypotheses:*
H_o: There is no difference between the population mean verbal S.A.T score of the IVY League College and the Military Academy: $\mu_1 - \mu_2 = 0$
H_a: The difference between the population mean verbal S.A.T score of the IVY League College and the population mean verbal S.A.T score of the Military Academy is greater than zero: $\mu_1 - \mu_2 > 0$

c) *Expected Results:*
$\mu_{\bar{x}_1-\bar{x}_2} = 0$
$s_{\bar{x}_1-\bar{x}_2} = 1.05$
Hypothesis test model: the distribution of the test statistic is a *t* distribution.

d) *Decision rule:*
Reject H_0 if the test statistic is greater than $t_c = 1.66$.

e) *test statistic:* t = 3.62; p-value: 0.0002

f) *Conclusion:*
Reject H_0, Accept H_a at $\alpha = 5\%$
Answer:
The educational researcher can conclude that the IVY League College has a higher mean verbal S.A.T.

63. Part II.

a) Pop 1: IVY League College
Pop 2: Military Academy

b) *Hypotheses:*
H_o: There is no difference between the population mean mathematics S.A.T score of the IVY League College and the Military Academy: $\mu_1 - \mu_2 = 0$
H_a: The difference between the population mean mathematics S.A.T score of the IVY League College and the

population mean mathematics S.A.T score of the Military Academy is less than zero: $\mu_1 - \mu_2 < 0$

c) *Expected Results:*
$\mu_{\bar{x}_1-\bar{x}_2} = 0$
$s_{\bar{x}_1-\bar{x}_2} \approx 12.17$
Hypothesis test model: the distribution of the test statistic is a *t* distribution.

d) *Decision rule:*
Reject H_o if the test statistic is less than $t_c = -1.66$.

e) *test statistic*: t = –2.88; p-value: 0.0025

f) *Conclusion:*
Reject H_o, Accept H_a at $\alpha = 5\%$
Answer:
The educational researcher can conclude the mean math S.A.T score is greater for the Military Academy.

63. Part III.
a) Pop 1: IVY League College
Pop 2: Military Academy

b) *Hypotheses:*
H_o: There is no difference between the population mean combined S.A.T score of the IVY League College and the Military Academy: $\mu_1 - \mu_2 = 0$
H_a: The difference between the population mean combined S.A.T score of the IVY League College and the Military Academy is not equal to zero: $\mu_1 - \mu_2 \neq 0$

c) *Expected Results:*
$\mu_{\bar{x}_1-\bar{x}_2} = 0$
$s_{\bar{x}_1-\bar{x}_2} \approx 16.42$
Hypothesis test model: the distribution of the test statistic is a *t* distribution.

d) *Decision rule:*
Reject H_o if the test statistic is less than tLC = –1.66 or greater than tRC = 1.66.

e) *test statistic*: t = 0.30; p-value: 0.7614

f) *Conclusion:*
Fail to reject H_o at $\alpha = 5\%$
Answer:
The educational researcher cannot conclude that the combined mean S.A.T score is significantly different for the IVY League College and the Military Academy.

65. a) *Hypotheses:*
H_o: The population mean difference in the lead levels between last year and this year equals 0: $\mu_d = 0$
H_a: The population mean difference in the lead levels between last year and this year is significantly greater than 0: $\mu_d > 0$

b) *Expected Results:*
The population mean difference of the paired observations is assumed to be 0: $\mu_d = 0$
The sample mean difference of the paired observations is: $\bar{x}_d \approx 0.667$
The standard deviation of the difference of the paired observations is: $s_d = 1.366$
Hypothesis testing model: the distribution of the test statistic is a *t* distribution with $df = n - 1 = 6 - 1 = 5$.

c) *Decision rule:*
Reject H_0 if the test statistic is greater than $t_c = 3.36$.

d) *test statistic*: $t = \dfrac{\bar{x}_d - \mu_d}{\dfrac{s_d}{\sqrt{n}}} = 1.20$

e) Fail to Reject H_o, Reject H_a at $\alpha = 1\%$
Answer:
The water authority cannot conclude that there is a Statistically significant increase in the amount of lead in this communities drinking water.

CHAPTER 14

1. classifications
3. 4; 3
5. c
7. non-negative
9. false
11. a) $X^2 = \sum \dfrac{(O-E)^2}{E}$

b) $X^2 = \sum \dfrac{O^2}{E} - n$

13. total; column; sample size
15. a
17. c
19. false
21. a) *Formulate the two hypotheses H_o and H_a.*
H_o: The population classifications gender and movie preference are independent.
H_a: The population classifications gender and movie preference are dependent.

b) For $\alpha = 5\%$ and $df = 4$, $\chi^2_\alpha = 9.49$.
Decision Rule: reject H_o if X^2 is greater than 9.49.

c) Test Statistic: $X^2 = 36.96$

d) *Conclusion:* reject H_o and accept H_a at $\alpha = 5\%$.
Yes, we can conclude that there is a relationship between gender and movie preference.

23. a) *Formulate the two hypotheses H_o and H_a.*
H_o: The population classifications drug treatment and result of the treatment are independent.
H_a: The population classifications drug treatment and result of the treatment are dependent.

b) For $\alpha = 5\%$ and $df = 4$, $\chi^2_\alpha = 9.49$.
Decision Rule: reject H_o if X^2 is greater than 9.49.

c) Test Statistic: $X^2 = 1.01$

d) *Conclusion:* fail to reject H_o at $\alpha = 5\%$.
Yes, Drug treatment and relief from pain are independent at $\alpha = 5\%$.

25. a) *Formulate the two hypotheses H_o and H_a.*
H_o: The population classifications educational level and job satisfaction are independent.
H_a: The population classifications educational level and job satisfaction are dependent.

b) For $\alpha = 5\%$ and $df = 4$, $X^2_\alpha = 9.49$. *Decision rule:* Reject H_o if X^2 is greater than 9.49.

c) Test Statistic: $X^2 = 5.49$

d) *Conclusion:* Fail to reject H_o at $\alpha = 5\%$.
No, educational level and job satisfaction are independent at $\alpha = 5\%$.

27. c

29. Sample
31. False
33. a) Formulate the two hypotheses H_o and H_a.
   H_o: The sample data fits a uniform distribution.
   H_a: The sample data does not fit a uniform distribution.
   b) For $\alpha = 5\%$ and $df = 4$, $X^2_\alpha = 9.49$.
   Decision rule: Reject H_o if X^2 is greater than 9.49.
   c) *Test Statistic:* $X^2 = 10.31$.
   d) *Conclusion:* Reject H_o and accept H_a at $\alpha = 5\%$.
   Yes, the marketing agency should recommend mass production of the new product to the manufacturer at $\alpha = 5\%$.

35. a) Formulate the two hypotheses H_o and H_a.
   H_o: The sample data fits the published distribution.
   H_a: The sample data does not fit the published distribution.
   b) For $\alpha = 1\%$ and $df = 2$, $X^2_\alpha = 9.21$.
   Decision rule: Reject H_o if X^2 is greater than 9.21.
   c) *Test Statistic:* $X^2 = 10.45$.
   d) *Conclusion:* Fail to reject H_o at $\alpha = 1\%$.
   No, we can conclude that the sample was randomly selected from the town at $\alpha = 1\%$.

37. a) Formulate the two hypotheses H_o and H_a.
   H_o: The dice are not "loaded."
   H_a: The dice are "loaded."
   b) For $\alpha = 5\%$ and $df = 10$, $X^2_\alpha = 18.31$.
   Decision rule: Reject H_o if X^2 is greater than 18.31.
   c) *Test Statistic:* $X^2 = 18.71$.
   d) *Conclusion:* Reject H_o and accept H_a at $\alpha = 5\%$.
   Yes, the Pit Boss can conclude that the dice are "loaded" at $\alpha = 5\%$.

39. a) Formulate the two hypotheses H_o and H_a.
   H_o: The population classifications membership in a fraternity group and GPA are independent.
   H_a: The population classifications membership in a fraternity group and GPA are dependent.
   b) *Decision rule:* Reject H_o if the p-value of the X^2 test statistic is less than 1%.
   c) *Test Statistic:* $X^2 = 9.56$; p-value = 0.008376
   d) *Conclusion:* Reject H_o and accept H_a at $\alpha = 1\%$.
   Yes, we can conclude that there is a relationship between membership in a fraternity group and GPA at $\alpha = 1\%$.
   e) Smallest alpha = 0.8376%

41. **Using p-value approach:**
   a) Formulate the two hypotheses H_o and H_a.
   H_o: The population classifications Caffeine Consumption Level and Weight Loss are independent.
   H_a: The population classifications Caffeine Consumption Level and Weight Loss are dependent.
   b) *Decision rule:* reject H_o if the p-value of the X^2 test statistic is less than 1%.
   c) **Test Statistic:** $X^2 = 13.996$ (or 14.00) with a **p-value** = 0.0029
   d) *Conclusion:* reject H_o and accept H_a at p-value of 0.0029.
   Yes, the researcher can conclude that caffeine consumption level and weight loss are dependent at $\alpha = 1\%$.
   e) Smallest alpha = 0.29%

43. **Using p-value approach:**
   a) Formulate the two hypotheses H_o and H_a
   H_o: The population classifications Classical Music Rating and IQ Level are independent.
   H_a: The population classifications Classical Music Rating and IQ Level are dependent.
   b) *Decision Rule:* reject H_o if the p-value of X^2 test statistic is less than 1%.
   c) **Test Statistic:** $X^2 = 13.42$ with a **p-value** = 0.0094
   d) *Conclusion:* reject H_o and accept H_a at a p-value of 0.0094.
   Yes, the PhD student researcher can conclude that an individual's IQ level and their Classical Music rating are dependent at $\alpha = 1\%$.
   e) smallest alpha = 0.94%

45. false
47. cell
49. count
51. true
53. true
55. false

57. **Not using p-value approach:**
   Part a)
   a) Formulate the two hypotheses H_o and H_a.
   H_o: The population classifications Relationship Status and Number of Friends are independent.
   H_a: The population classifications Relationship Status and Number of Friends are dependent.
   b) for $\alpha = 5\%$ and $df = 4$, $\chi^2_\alpha = 9.49$.
   Decision Rule: reject H_o if X^2 is greater than 9.49.
   c) Test statistic: $X^2 = 9.66$; p-value = 0.0465
   d) *Conclusion:* reject H_o and accept H_a at $\alpha = 5\%$.
   Yes, the PhD student researcher can conclude that an individual's IQ level and their Classical Music rating are dependent at $\alpha = 5\%$.
   Part b) No, since the assumption is that the sample size, n, be sufficiently large. That is, the smallest cell frequency is at least five.

59. a) Formulate the two hypotheses H_o and H_a.
   H_o: The population classifications educational background and the frequency of movie attendance per year are independent.
   H_a: The population classifications educational background and the frequency of movie attendance are dependent.
   b) For $\alpha = 5\%$ and $df = 6$, $X^2_\alpha = 12.59$.
   Decision Rule: reject H_o if X^2 is greater than 12.59.
   c) test statistic: $X^2 = 54.32$; p-value = 6.37 E –10
   d) *Conclusion:* reject H_o and accept H_a at $\alpha = 5\%$.
   Yes, educational background and the frequency of movie attendance per year are dependent at $\alpha = 5\%$.

61. **Not using p-value approach:**
   a) Formulate the two hypotheses H_o and H_a.
   H_o: The population classifications a mother's educational level and number of children are independent.
   H_a: The population classifications a mother's educational level and number of children are dependent.

b) For $\alpha = 5\%$ and $df = 6$, $\chi_\alpha^2 = 12.59$.
Decision Rule: reject H_o if X^2 is greater than 12.59.
c) Test statistic: $X^2 = 16.30$; p-value = 0.0122
d) *Conclusion:* reject H_o and accept H_a at $\alpha = 5\%$.
Yes, the researcher can conclude a mother's educational level and number of children are dependent at $\alpha = 5\%$.

63. a) Formulate the two hypotheses H_o and H_a.
H_o: The die is fair.
H_a: The die is not fair.
b) For $\alpha = 5\%$ and df = 5, $X_\alpha^2 = 11.07$.
Decision rule: Reject H_o if X^2 is greater than 11.07.
c) Test Statistic: $X^2 = 4.9$.
d) *Conclusion:* Fail to reject H_o at $\alpha = 5\%$.
Yes, we can conclude that the die is fair at $\alpha = 5\%$.

CHAPTER 15

1. rho: ρ
3. a) zero; no; 0
b) positive; >; negative; <; significant; ≠
5. r_α; n − 2
7. false
9. a) r = 0.86
b) H_o: $\rho = 0$; there is no relationship between seed density and grass density.
H_a: $\rho > 0$; the relationship between seed density and grass density is positive.
n = 7 so there are 5 degrees of freedom, and the critical value from Table V at $\alpha = 5\%$ is $r_\alpha = 0.67$.
Since the test statistic r = 0.86 lands in the critical region, we reject the null hypothesis and accept the alternative hypothesis and conclude that r is statistically significant at $\alpha = 5\%$. Therefore, there is a significant positive linear relationship between seed and grass density.

11. a) r = −0.89
b) H_o: $\rho = 0$; there is no relationship between number of sick days and % of salary increase.
H_a: $\rho < 0$; the relationship between number of sick days and % of salary increase is negative.
n = 20 so there are 18 degrees of freedom, and the critical value from Table V at $\alpha = 5\%$ is $r_\alpha = -0.38$.
Since the test statistic r = −0.89 lands in the critical region, we reject the null hypothesis and accept the alternative hypothesis and conclude that r is statistically significant at $\alpha = 5\%$. Therefore, there is a significant negative linear relationship between number of sick days and percent of salary increase.

13. a)

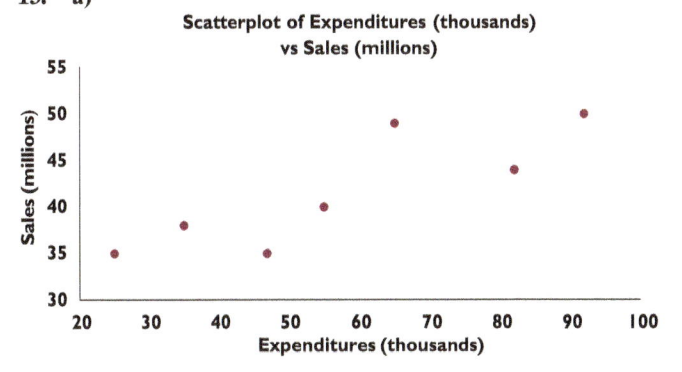

Scatterplot of Expenditures (thousands) vs Sales (millions)

b) r = 0.8514
c) H_o: $\rho = 0$; there is no relationship between expenditure and sales.
H_a: $\rho > 0$; the relationship between expenditure and sales is positive.
n = 7 so there are 5 degrees of freedom, and the critical value from Table V at $\alpha = 1\%$ is $r_\alpha = 0.83$.
Since the test statistic r = 0.85 lands in the critical region, we reject the null hypothesis and accept the alternative hypothesis and conclude that r is statistically significant at $\alpha = 1\%$. Therefore, there is a significant positive linear relationship between expenditure and sales.
d) $r^2 = 0.7249$; 72% of the variation in sales is accounted for by variation in advertising expenditure.
e) $y' = 29.01 + 0.22x$
f) Using the regression equation in part e, the predicted amount of sales for an expenditure of $50,000 is $40.01 million dollars.

15. a) r = −0.707
b) H_o: $\rho = 0$; there is no relationship between number of hours worked and final course grade.
H_a: $\rho < 0$; the relationship between number of hours worked and final course grade is negative.
n = 20 so there are 18 degrees of freedom, and the critical value from Table V at $\alpha = 5\%$ is $r_\alpha = -0.38$.
Since the test statistic r = −0.707 lands in the critical region, we reject the null hypothesis and accept the alternative hypothesis and conclude that r is statistically significant at $\alpha = 5\%$. Therefore, there is a significant negative relationship between number of hours worked and final course grade for these students.

17. a) r = 1.00
b) H_o: $\rho = 0$; there is no relationship between number of drinks consumed in 1 hour for a 100 lb. female and blood alcohol level.
H_a: $\rho > 0$; the relationship between number of drinks consumed in 1 hour for a 100 lb. female and blood alcohol level is positive.
n = 8 so there are 6 degrees of freedom, and the critical value from Table V at $\alpha = 5\%$ is $r_\alpha = 0.62$.
Since the test statistic r = 1.00 lands in the critical region, we reject the null hypothesis and accept the alternative hypothesis and conclude that r is statistically significant at $\alpha = 5\%$. Therefore, there is a significant positive linear relationship between number of drinks consumed in 1 hour and blood alcohol level for 100-lb. females.

19. a) r = 0.749
b) H_o: $\rho = 0$; there is no relationship between salary and years of education.
H_a: $\rho > 0$; the relationship between salary and years of education is positive.
n = 12 so there are 10 degrees of freedom, and the critical value from Table V at $\alpha = 1\%$ is $r_\alpha = 0.66$.
Since the test statistic r = 0.749 lands in the critical region, we reject the null hypothesis and accept the alternative hypothesis and conclude that r is statistically significant at $\alpha = 1\%$. Therefore, there is a significant positive linear relationship between salary and years of education.

21. a) *Formulate the two Hypotheses H_o and H_a*
H_o: The population linear correlation coefficient between a patient's score on an extrovert scale and time spent on a social networking site is 0, $\rho = 0$.

H_a: The population linear correlation coefficient between a patient's score on an extrovert scale and time spent on a social networking site is greater than 0, $\rho > 0$.

b) *Hypothesis testing model* is the sampling distribution of the correlation coefficients where $\rho = 0$.

c) *Decision Rule*: Reject H_o if the p-value for the correlation coefficient testing statistic is less than 1%; less than 5%.

d) *Test Statistic*: r = 0.76, p-value = 4.80 E-5.

e) *Conclusion*: Reject H_o and accept H_a at a p-value of 4.80 E-5.
There is a significant positive correlation between a patients score on an extrovert scale and time spent on a social networking site.

f) The smallest alpha level is 4.80 E-5.

23. a) *Formulate the two Hypotheses H_o and H_a*
H_o: The population linear correlation coefficient between the midterm grades and final exam grades for students taking Prof DePorto's introduction to statistics class is 0, $\rho = 0$.
H_a: The population linear correlation coefficient between the midterm grades and final exam grades for students taking Prof DePorto's introduction to statistics class is greater than 0, $\rho > 0$.

b) *Hypothesis testing model* is the sampling distribution of the correlation coefficients where $\rho = 0$.

c) *Decision Rule*: Reject H_o if the p-value for the correlation coefficient testing statistic is less than 1%; less than 5%.

d) *Test Statistic*: r = 0.27; p-value = 0.094

e) *Conclusion*: Fail to Reject H_o at a p-value of 0.094.
There is not a significant positive correlation between the midterm grades and final exam grades for students taking Prof DePorto's introduction to statistics class.

f) The smallest alpha level is 9.4%.

25. a) *Formulate the two Hypotheses H_o and H_a*
H_o: The population linear correlation coefficient between fat and cholesterol levels of fast food items is 0, $\rho = 0$.
H_a: The population linear correlation coefficient between fat and cholesterol levels of fast food items is greater than 0, $\rho > 0$.

b) *Hypothesis testing model* is the sampling distribution of the correlation coefficients where $\rho = 0$.

c) *Decision Rule*: Reject H_o if the p-value for the correlation coefficient testing statistic is less than 1%; less than 5%.

d) *Test Statistic*: r = 0.809; p-value = 0.0001

e) *Conclusion*: Reject H_o and accept H_a at a p-value of 0.0001.
There is a significant positive correlation between fat and cholesterol levels of fast food items.

f) The smallest alpha level is *essentially* 0.

27. a) *Formulate the two Hypotheses H_o and H_a*
H_o: The population linear correlation coefficient between the temperature outside and number of sunglasses sold by the Sunglass Shack Corp. is 0, $\rho = 0$.
H_a: The population linear correlation coefficient between the temperature outside and number of sunglasses sold by the Sunglass Shack Corp. is greater than 0, $\rho > 0$.

b) *Hypothesis testing model* is the sampling distribution of the correlation coefficients where $\rho = 0$.

c) *Decision Rule*: Reject H_o if the p-value for the correlation coefficient testing statistic is less than 1%; less than 5%.

d) *Test Statistic*: r = 0.169; p-value = 0.374

e) *Conclusion*: Fail to Reject H_o and reject H_a at a p-value of 0.374.
There is not a significant positive correlation between the temperature outside and number of sunglasses sold by the Sunglass Shack Corp.

f) The smallest alpha level is 37.4%.

29. a) *Formulate the two Hypotheses H_o and H_a*
H_o: The population linear correlation coefficient between temperature in degrees Fahrenheit and the finish time for the Chicago Marathon is 0, $\rho = 0$.
H_a: The population linear correlation coefficient between temperature in degrees Fahrenheit and the finish time for the Chicago Marathon is greater than 0, $\rho > 0$.

b) *Hypothesis testing model* is the sampling distribution of the correlation coefficients where $\rho = 0$.

c) *Decision Rule*: Reject H_o if the p-value for the correlation coefficient testing statistic is less than 1%; less than 5%.

d) *Test Statistic*: r = 0.829, p-value = 0.003.

e) *Conclusion*: Reject H_o and accept H_a at a p-value of 0.003.
There is a significant positive correlation between temperature in degrees Fahrenheit and the finish time for the Chicago Marathon.

f) The smallest alpha level is 0.003.

31. y'; a; $b_1, b_2, \ldots, b_k$; $x_1, x_2, \ldots x_k$

33. three

35. True

37. strong

39. R = 0.985

41. a) $R^2 = 0.985^2 = 0.970$

b) The value for R^2 is smaller. The amount of explained variance in the regression model is 0.985.

c) Residual Error = $1 - R^2 = 1 - 0.970 = 0.03$

43. a. $R^2 = 0.914^2 = 0.835$. The amount of explained variance in the regression model is 0.835.

b) The statistician may feel uncomfortable because the number of independent variables is 5. As the number of independent variables increase, so does the value of R^2. She should compute the value of the adjusted R^2.

c)
$$R^2_{adj} = 1 - \frac{(1-R^2)(n-1)}{n-k-1}$$
$$= 1 - \frac{(1-0.835)(75-1)}{75-5-1} = 0.823$$

45. a. R = 0.848, $R^2 = 0.719$, $R^2_{adj} = 0.687$. There appears to be a linear relationship among these variables.

b) The hypotheses are H_o: $\rho = 0$ and H_a: $\rho \neq 0$. The p-value for the testing statistic is p = 2.45 E-7 which is less than $\alpha = 0.01$. With such a small p-value, we would reject H_o and accept H_a at any alpha level and conclude there is a significant correlation between these variables.

c) y' = 0.0197 − 1.04(Price) + 0.003(income) + 0.003(temp)

d) y' ≈ 0.445

47. a) The hypotheses are H_o: $\rho = 0$ and H_a: $\rho \neq 0$. The p-value for the testing statistic is p = 1.28 E-21 which is less than $\alpha = 0.01$. With such a small p-value, we would reject H_o and accept H_a at any alpha level and conclude there is a significant correlation between these variables.

b) The reason for the p-value being so small and the test being so significant is due to the fact that the gender of an individual has a huge impact on that individual's height and weight.

49. False
51. c)
53. b) Negative
   b) none
   c) Positive
   d) Negative
   e) none
   f) Positive
   g) Positive
   h) Positive
   i) Positive
55. d)
57. 1.00, –0.95, 0.69 ~ –0.69, 0.50, 0.05, 0
59. False
61. False
63. True
65. a) r = –0.764
   b) H_o: The population linear correlation coefficient between number of practice hours and golf score is 0, $\rho = 0$. H_a: The population linear correlation coefficient between number of practice hours and golf score is less than 0, $\rho < 0$.
   n = 10 so there are 8 degrees of freedom, and the critical value from Table V at $\alpha = 1\%$ is $r_\alpha = -0.72$.
   Since the test statistic r = –0.764 lands in the critical region, we reject the null hypothesis and accept the alternative hypothesis and conclude that r is statistically significant at $\alpha = 1\%$. Therefore, there is a significant negative linear relationship between number of hours practicing golf and one's golf score.
67. a) r = –0.694
   b) H_o: The population linear correlation coefficient between average attractiveness and fine given by a judge is 0, $\rho = 0$.
   H_a: The population linear correlation coefficient between average attractiveness and fine given by a judge is less than 0, $\rho < 0$.
   n = 15 so there are 13 degrees of freedom, and the critical value from Table V at $\alpha = 1\%$ is $r_\alpha = 0.59$.
   Since the test statistic r = –0.694 lands in the critical region, we reject the null hypothesis and accept the alternative hypothesis and conclude that r is statistically significant at $\alpha = 1\%$. Therefore, there is a significant negative linear relationship for average attractiveness of an individual and the fine given to the individual by a judge for a first-time misdemeanor crime.

CHAPTER 16

1. Analysis of Variance
3. S_2^2
5. 1.53
7. F = 1
9. $F_{0.01}(10,20) = 3.37$
11. false
13. $F_{0.01}(8,9) = 5.47$
15. $F_{0.01}(13,20) = 3.23$
17. a) $F_{0.10}(30,40) = 1.54$
    b) $F_{0.05}(30,40) = 1.74$ (Place numbers on graph)
    c) $F_{0.025}(30,40) = 1.94$
    d) $F_{0.01}(30,40) = 2.20$
19. $\dfrac{\text{Larger Variance}}{\text{Smaller Variance}}$
21. a
23. c
25. a
27. false
29. a) F=1.58
    b) $H_0: \sigma_1^2 = \sigma_2^2$; $H_a: \sigma_1^2 \neq \sigma_2^2$; F-statistic = 1.58; $F_{0.05}(49,49)$ = 1.69 (use closest degrees of freedom)
    Conclusion: Since the calculated F, test statistic, 1.58, is not greater than the critical F-value, 1.69, then the population variances are assumed equal at $\alpha = 0.10$.
31. a) F = 2.72
    b) $H_0: \sigma_1^2 = \sigma_2^2$; $H_a: \sigma_1^2 \neq \sigma_2^2$; F-statistic = 2.72; $F_{0.025}(20,21)$ = 2.42 (use c closest degrees of freedom)
    Conclusion: Since the calculated F, test statistic, 2.72, is greater than the critical F-value, 2.42, then the population variances are assumed unequal at $\alpha = 0.05$.
33. a) F = 2.38
    b) $H_0: \sigma_1^2 = \sigma_2^2$; $H_a: \sigma_1^2 \neq \sigma_2^2$; F-statistic = 2.38; $F_{0.05}(99,99)$ = 1.35 (use closest degrees of freedom)
    Conclusion: Since the calculated F, test statistic, 2.38, is greater than the critical F-value, 1.35, then the population variances are assumed unequal at $\alpha = 0.10$.
35. a) $S_1^2 = 306.6$ $S_2^2 = 620.51$;   F = 2.02
    b) $H_0: \sigma_1^2 = \sigma_2^2$; $H_a: \sigma_1^2 \neq \sigma_2^2$; F-statistic = 2.02; $F_{0.01}(7,5)$ = 10.46
    Conclusion: Since the calculated F, test statistic, 2.02, is not greater than the critical F-value, 10.46, then the population variances are assumed equal at $\alpha = 0.02$.
37. $H_0: \sigma_1^2 = \sigma_2^2$; $H_a: \sigma_1^2 \neq \sigma_2^2$; F-statistic = 1.98; $F_{0.05}(30,40) = 1.74$
    Conclusion: Since the calculated F, test statistic, 1.98, is greater than the critical F-value, 1.74, then the sample variances are assumed unequal at $\alpha = 0.10$. No, Kathleen cannot pool the sample variances.
39. $S_1^2 = 71.4$ $S_2^2 = 37.7$; $H_0: \sigma_1^2 = \sigma_2^2$; $H_a: \sigma_1^2 \neq \sigma_2^2$; F-statistic = 1.89; $F_{0.01}(9,10) = 4.94$
    Conclusion: Since the calculated F, test statistic, 1.89, is less than the critical F-value, 4.94, then the sample variances are assumed equal at $\alpha = 0.02$. The results support the use of the two sample t-test with a pooled standard deviation.
41. between variance
43. within variance
45. sums of squares; $\sum (x - \bar{x})^2$
47. b
49. true
51. a) $\sigma_1^2 = 4$; $\sigma_2^2 = 4$; $\sigma_3^2 = 4$
    b) 4
    c) 40.82
    d) 44.82
    e) No, the total group variance is the sum of the within group variance, 4, and between group variance, 40.82.

53. a) 336
    b) 2744
    c) 3080
    d) yes
    e) 16; 16; 16
    f) 16
    g) 130.67
    h) 146.67
    i) No, the total variance is due to within variance, 16, and between variance, 130.67.
55. a) 270
    b) 208544
    c) 208824
    d) yes
    e) 18; 18; 18
    f) 18
    g) 14
    h) 32
    i) No, the total variance is due to within variance, 18, and between variance, 14.
57. a) Each sample selected from a normal population
    b) Each sample is selected from a population that has the same population variance, σ^2.
    c) When samples are drawn from different populations the samples must be random and independent.
59. true
61. false
63. false
65. increase; decrease
67. number of groups; total number of data values
69. population; equal; one population; group; different
71. d
73. a
75. reject the null hypothesis
77. $\left[\dfrac{(\sum T_1)^2}{n_1} + \dfrac{(\sum T_2)^2}{n_2} + \dfrac{(\sum T_3)^2}{n_3} + \dfrac{(\sum T_4)^2}{n_4}\right] - \dfrac{(\sum x)^2}{N}$
79. sample size of group 4; total number of data values; data values in group 3
81. 34; 36; 13.25; 0.94; 14.10
83. 2.98
85. 2.40
87. 2.12
89. 2.86
91. a) $F_{0.10}(20,20) = 1.79$
    b) $F_{0.05}(20,20) = 2.12$ (Place numbers on graph)
    c) $F_{0.025}(20,20) = 2.46$
    d) $F_{0.01}(20,20) = 2.94$
93.

Source of Variation	Sums of Squares	Degrees of Freedom	Mean Square	F-Stat
Between	39.2	1	39.2	1.80
Within	392.6	18	21.81	
Total	431.8	19		

$F_{0.01}(1,18) = 8.29$ No, since F, the test statistic, 1.80, is less than the critical F, 8.29, the test statistic is not statistically significant at $\alpha = 0.01$.

95. a) 596
    b) 9
    c) 228
    d) 15.56
    e) 14.65
    f) $F_{0.05}(2,9) = 4.26$ Yes, since F, the test statistic, 14.65, is greater than the critical F, 4.26, the test statistic is significantly significant at $\alpha = 0.05$.
97. a) $H_0: \mu_1 = \mu_2 = \mu_3$ There is no difference between treatment means
    $H_a: \mu_1 \neq \mu_2 \neq \mu_3$ There is a difference between treatment means
    b) F-distribution
    c) If the test statistic, F, is greater than the critical $F_{0.05}(2,9) = 4.26$, then Reject the Null Hypothesis
    d, e) Sums of Squares and df are indicated in the table.

Source of Variation	Sums of Squares	Degrees of Freedom	Mean Square	F-Stat
Between	72.67	2	36.34	36.34
Within	9	9	1	
Total	81.67	11		

   f) Reject H_0 and accept H_a, there is a difference between treatment means at $\alpha = 0.05$.
99. a) $H_0: \mu_1 = \mu_2 = \mu_3$ There is no difference between mean playing weights of female athletes in basketball, track and soccer.
    $H_a: \mu_1 \neq \mu_2 \neq \mu_3$ There is a difference between mean playing weights of female athletes in basketball, track and soccer.
    b) F-statistic = 5.29
    c) The $F_{0.01}(2,27) = 5.49$.
    d) Since the test statistic, F = 5.29, is less than the critical F, 5.49, the results are not significant at $\alpha = 0.01$. There is no difference between mean playing weights of female athletes in basketball, track and soccer.
101. a) $H_0: \mu_1 = \mu_2 = \mu_3$ There is no difference between mean water holding properties of the three brands of diapers.
    $H_a: \mu_1 \neq \mu_2 \neq \mu_3$ There is a difference between mean water holding properties of the three brands of diapers.
    b) F-distribution
    c) If the test statistic, F, is greater than the critical $F_{0.05}(2,12) = 3.89$, then Reject the Null Hypothesis
    d, e) Sums of Squares and df are indicated in the table.

Source of Variation	Sums of Squares	Degrees of Freedom	Mean Square	F-Stat
Between	2.70	2	1.35	13.5
Within	1.22	12	0.10	
Total	3.92	14		

f) Reject H_0 and accept H_a, there is a difference between mean water holding properties of the three brands of diapers at $\alpha = 0.05$.

103. a) $H_0: \mu_1 = \mu_2 = \mu_3$ There is no difference between mean mercury levels for the three age groups of loons.

$H_a: \mu_1 \neq \mu_2 \neq \mu_3$ There is a difference between mean mercury levels for the three age groups of loons.

b) F-distribution

c) If the test statistic, F, is greater than the critical $F_{0.01}(2,9) = 8.02$, then Reject the Null Hypothesis

d, e) Sums of Squares and df are indicated in the table.

*NOTE: Since the data values are small, the calculations in the source table use 5 decimal places.

Source of Variation	Sums of Squares	Degrees of Freedom	Mean Square	F-Stat
Between	0.2665	2	0.13325	17.17139
Within	0.06984	9	0.00776	
Total	0.33634	11		

f) Reject H_0 and accept H_a, there is a difference between mean mercury levels for the three age groups of loons at $\alpha = 0.01$.

105. 1.86
107. false
109. c
111. $F_{0.05}(7,10) = 3.01$
113. $F_{0.025}(15,17) = 2.72$
115. c
117. c
119. a) F = 2.69
  b) $F_{0.025}(18,21) = 2.42$
121. a) F = 3.21
  b) $H_0: \sigma_1^2 = \sigma_2^2; H_a: \sigma_1^2 \neq \sigma_2^2$; F-statistic = 3.21; $F_{0.025}(5,5) = 7.15$
  Conclusion: Since the calculated F, test statistic, 3.21, is not greater than the critical F-value, 7.15, then the population variances are assumed equal at $\alpha = 0.05$
123. between group variance
125. SS_{total}; $SS_{between\ treatments}$; $SS_{within\ treatments}$
127. a) Population variances Within are the same for each group, 25
  b) Average Within group variance = 25
  c) Between group variance = 0
  d) Total group variance = 25
  e) Total group variance is due entirely to Within group variance
129. a) 336   **b)** 2744   **c)** 3080   **d)** yes
  e) 16   **f)** 16   **g)** 130.65   **h)** 146.65
  i) No, total group variance is the sum of Within and Between group variance
131. sample size of group 6; total number of data values of all groups; data values for group 5

133. a)

Source of Variation	Sums of Squares	Degrees of Freedom	Mean Square	F-Stat
Between	150.51	4	37.63	11.19
Within	87.43	26	3.36	
Total	237.94	31		

b) $F_{0.05}(4,26) = 2.74$ Yes, since the test statistic, F = 11.19, is greater than the critical F, 2.74 at $\alpha = 0.05$.

135. a) $H_0: \mu_1 = \mu_2 = \mu_3$ There is no difference between mean number of hours worked P/T by students from the three regions.

$H_a: \mu_1 \neq \mu_2 \neq \mu_3$ There is a difference between mean number of hours worked P/T by students from the three regions.

b) F-distribution

c) If the test statistic, F, is greater than the critical $F_{0.05}(2,12) = 3.89$, then Reject the Null Hypothesis

d,e) Sums of Squares and df are indicated in the table.

Source of Variation	Sums of Squares	Degrees of Freedom	Mean Square	F-Stat
Between	7.09	2	3.55	12.24
Within	3.51	12	0.29	
Total	10.6	14		

f) Reject H_0 and accept H_a, there is a difference between mean number of hours worked P/T by students from the three regions at $\alpha = 0.05$.

CHAPTER 17

1. parametric
3. distribution free
5. false
7. Statistics test that require few assumptions about the population from which sample data is collected.
9. ½
11. np; $\sqrt{np(1-p)}$
13. negative; plus
15. d
17. a) 66   **b)** 23   **c)** 2   **d)** 43   **e)** 42
  f) 1. H_0: Population median = 26
       H_a: Population median ≠ 26
    2. Binomial distribution is the hypothesis testing model. Since np and n(1–p) are both at least 10, the normal distribution is used to approximate the binomial distribution. $\mu_s = 33$ and $\sigma_s = 4.06$.
    3. Reject H_0 if the test statistic, z, is either less than –1.96 or greater than 1.96.
    4. O = 23 and test statistic, z = –2.34
    5. Reject H_0 at $\alpha = 0.05$. The median is not equal to 26.
  g) p-value = 0.02

19. 1. H_0: Population median = 8340   Median time for DVD players to fail is 8,340 hours.
    H_a: Population median ≠ 8340   Median time for DVD players to fail is not 8,340 hours.
    2. Binomial distribution is the hypothesis testing model. Since np and n(1–p) are both at least 10, the normal distribution is used to approximate the binomial distribution. $\mu_s = 15$ and $\sigma_s = 2.74$.
    3. Reject H_0 if the test statistic, z, is either less than –1.96 or greater than 1.96.
    4. O = 10 and test statistic, z = –1.64   p-value = 0.10
    5. Fail to reject H_0 at $\alpha = 0.05$. There is no reason to believe the median time for DVD players to fail is different than 8340 hrs.

21. 1. H_0: Population median = 600   Median number of calories from fat is 600.
    H_a: Population median < 600   Median number of calories from fat is less than 600.
    2. Binomial distribution is the hypothesis testing model. Since np and n(1–p) are both at least 10, the normal distribution is used to approximate the binomial distribution. $\mu_s = 10$ and $\sigma_s = 2.24$.
    3. Reject H_0 if the test statistic, z, is less than –1.65.
    4. Test statistic, z = –2.01   p-value = 0.02
    5. Reject H_0 and Accept H_a at $\alpha = 0.05$. Yes, the median number of calories from fat is less than 600.

23. If either (or both) of the following is/are NOT met:
    a) The population(s) from which the samples are drawn is/are not normal
    b) The populations' variances are not equal

25. The Mann-Whitney Rank Sum Test
27. mean
29. W
31. c
33. true
35. H_0: The two populations are the same. The median of population 1 = median of population 2.
    H_a: The two populations are different. The median of population 1 < median of population 2.
    a) $n_1 = 10$; $n_2 = 12$
    b) median sample 1 = 28.5;   median sample 2 = 39.5
    c) W = 94.5
    d) No, since the test statistic, z = –1.35, is not less than –1.65
    e) p = 0.09

37. a) H_0: Treatment A and Treatment B are the same (or There is no difference between Treatments A and B)
    H_a: Treatment A and Treatment B are not the same (or There is a difference between Treatments A and B)
    b) W = 129.5
    c) z = 1.85
    d) No, since the test statistic, z = 1.65, is not less than z_{cl} = –1.96 or greater than z_{cr} = 1.96 where $\alpha = 0.05$.

39. No, since the test statistics, z = –0.93, is not less than $z_c = -1.65$. The difference is not statistically significant at $\alpha = 0.05$.

41. No, since the test statistics, z = 1.08, is not less than $z_{cl} = -1.96$ or greater than $z_{cr} = 1.96$.
    There is no significant difference between the two distributions at $\alpha = 0.05$.

43. H_0: The two populations are the same.
    H_a: The two populations are different.

45. Kruskal-Wallis H test

47. H;
$$\frac{12}{N(N+1)}\left[\frac{\left(\sum R_1\right)^2}{n_1} + \frac{\left(\sum R_2\right)^2}{n_2} + \ldots + \frac{\left(\sum R_k\right)^2}{n_k}\right] - 3(N+1)$$

49. H statistic; χ_α^2
51. false
53. true
55. a) H_0: The populations are the same.
    H_a: The populations are different
    Chi-Square distribution is used as the hypothesis testing model
    Reject H_0 if the test statistic, Kruskal-Wallis H statistic, is greater than χ_α^2
    b) H = 8.46
    c) df = 2
    d) At $\alpha = 0.05$, H, 8.46, is significant since it exceeds $\chi_{0.05}^2 = 5.99$.; At $\alpha = 0.01$, H, 8.46, is not significant since it is less than $\chi_{0.01}^2 = 9.21$.
    e) Since the test statistic H, 8.46, is greater than 5.99, reject H_0 and accept H_a at $\alpha = 0.05$.
    The distributions are not the same.
    f) p-value = 0.01

57. 1. H_0: The populations are the same. Annual income for police officers at the Patrolmen Rank in NYC, LA, and Boston is the same.
    H_a: populations are not the same. Annual income for police officers at the Patrolmen Rank in NYC, LA, and Boston is not the same.
    2. Chi-Square distribution is used as the hypothesis testing model
    3. Reject H_0 if the test statistic, Kruskal-Wallis H statistic, is greater than $\chi_{0.05}^2$, which equals 5.99.
    4. H = 4.13
    5. Fail to Reject H_0, since the test statistic H, 4.13, is less than 5.99 at $\alpha = 0.05$. The data do not indicate a significant difference between annual income for police officers at the Patrolmen Rank in NYC, LA and Boston.

59. 1. H_0: The populations are the same. There is no difference in the number of patients treated in the emergency room at Mercy, Nassau, and Helpme Hospitals.
    H_a: The populations are not the same. There is a difference in the number of patients treated in the emergency room at Mercy, Nassau, and Helpme Hospitals.
    2. Chi-Square distribution is used as the hypothesis testing model.
    3. Reject H_0 if the test statistic, Kruskal-Wallis H statistic, is greater than $\chi_{0.01}^2$, which equals 9.21.
    4. H = 1.37

5. Fail to Reject H_0, since the test statistic H, 1.37, is less than 9.21 at $\alpha = 0.01$. The data do not indicate a significant difference between the number of patients treated in the emergency room at Mercy, Nassau, and Helpme Hospitals.
61. less powerful; do not all sample data; difficult to use for large samples
63. g
65. c
67. tests whether 3 or more populations are the same; uses χ^2 distribution if each sample has at least 5 observations
69. d
71. 1. H_0: Population median = 12
    H_a: Population median > 12
    2. Binomial distribution is the hypothesis testing model. Since np and n(1–p) are both at least 10, the normal distribution is used to approximate the binomial distribution. $\mu_s = 11$ and $\sigma_s = 2.35$.
    3. Reject H_0 if the test statistic, z, is greater than 1.65.
    4. Test statistic, z = 1.91, p-value = 0.03
    5. Reject H_0 at $\alpha = 0.05$. The median number of days in which the effects of the drug will appear is more than 12.
73. 1. H_0: Population median = $2000   The median cost for automobile insurance is $2000.
    H_a: Population median > $2000   The median cost for automobile insurance is greater than $2000.
    2. Binomial distribution is the hypothesis testing model. Since np and n(1–p) are both at least 10, the normal distribution is used to approximate the binomial distribution. $\mu_s = 10.5$ and $\sigma_s = 2.29$.
    3. Reject H_0 if the test statistic, z, is greater than 1.65.
    4. O = 14; Test statistic, z = 1.31, p-value = 0.10
    5. Fail to Reject H_0, at $\alpha = 0.05$. There is no reason to believe that the median cost for automobile insurance is greater than $2000.
75. 1. H_0: The two populations are the same. The mean time for learning each procedure is the same.
    H_a: The two populations are different. The mean time for learning each procedure is different.
    2. Normal distribution is used as the hypothesis testing model.
    3. Reject H_0 if the test statistic, z, is either less than $z_{cl} = -2.58$ or greater than $z_{cr} = 2.58$.
    4. Test statistic z = 2.82, p-value = 3.49E-05
    5. Reject H_0, Accept H_a at $\alpha = 0.01$. The mean time for learning the procedures is significantly different between the groups of nursing students.
77. 1. H_0: The populations are the same. There is no difference in the sexual awareness scores of the groups.
    H_a: The populations are not the same. There is a difference in the sexual awareness scores of the groups.
    2. Chi-Square distribution is used as the hypothesis testing model.
    3. Reject H_0 if the test statistic, Kruskal-Wallis H statistic, is greater than $\chi^2_{0.01}$, which equals 11.34.
    4. H = 13.11
    5. Reject H_0, Accept H_a at $\alpha = 0.01$. There is a difference in the sexual awareness scores of the groups.

CHAPTER 18

1. b
3. Nonparametric Bootstrapping
5. b
7. a
9. b
11. d
13. a
15. b
17. a
19. d
21. a
23. c
25. d
27. ANSWERS WILL VARY SINCE ANSWER IS DEPENDENT ON THE RANDOM SELECTION OF THE RESAMPLES.
29. ANSWERS WILL VARY SINCE ANSWER IS DEPENDENT ON THE RANDOM SELECTION OF THE RESAMPLES.
31. b
33. ANSWERS WILL VARY SINCE ANSWER IS DEPENDENT ON THE RANDOM SELECTION OF THE RESAMPLES.
35. c
37. ANSWERS WILL VARY SINCE ANSWER IS DEPENDENT ON THE RANDOM SELECTION OF THE RESAMPLES.
39. ANSWERS WILL VARY SINCE ANSWER IS DEPENDENT ON THE RANDOM SELECTION OF THE RESAMPLES.
41. ANSWERS WILL VARY SINCE ANSWER IS DEPENDENT ON THE RANDOM SELECTION OF THE RESAMPLES.
43. original sample
45. a
47. b
49. ANSWERS WILL VARY SINCE ANSWER IS DEPENDENT ON THE RANDOM SELECTION OF THE RESAMPLES.
51. ANSWERS WILL VARY SINCE ANSWER IS DEPENDENT ON THE RANDOM SELECTION OF THE RESAMPLES.

APPENDICES

A Chapter Formulas   A47

B Summary of Tests   A53
Hypothesis Tests   A53
Confidence Intervals   A55

C Statistical Tables   A57
Table I: Random Number Table   A57
Table II: Standard Normal Curve Area Table   A58
Table III: Critical Values for the t Distributions   A60
Table IV: Critical Values for the Chi-Square Distributions, χ^2_α A61
Table V: Critical Values of the Correlation Coefficient, r_α A62

D TI-84 Plus Instructions   A63

E Databases   A111
Database I: Student Characteristics   A111
Database II: Student Interests   A113

APPENDIX A: CHAPTER FORMULAS

The Databases and End of Chapter Exercises pertaining to the Databases have been placed in MyStatLab under the Additional Resources Tab.

CHAPTER 2

Class Width = Range/ (# of classes) and roundup

CHAPTER 3

Population Mean: $\mu = \dfrac{\Sigma x}{N}$

Median is the middle value (after the data values have been arranged in numerical order)

position of median = $\dfrac{N+1}{2}$ th data value

Mode is most frequent data value

Sample Mean: $\bar{x} = \dfrac{\Sigma x}{n}$

Sample Variance = $\dfrac{\Sigma(x-\bar{x})^2}{(n-1)}$

or

$s^2 = \dfrac{\Sigma(x-\bar{x})^2}{(n-1)}$

Range = largest value − smallest value

Population Variance = $\dfrac{\Sigma(x-\mu)^2}{N}$

OR

$\sigma^2 = \dfrac{\Sigma(x-\mu)^2}{N}$

Definition Formula for the Population Standard Deviation:

$\sigma = \sqrt{variance}$

OR

$\sigma = \sqrt{\dfrac{\Sigma(x-\mu)^2}{N}}$

Computational Formula for the Population Standard Deviation:

$\sigma = \sqrt{\dfrac{\Sigma x^2 - \dfrac{(\Sigma x)^2}{N}}{N}}$

Definition Formula for the Sample Standard Deviation:

$s = \sqrt{\dfrac{\Sigma(x-\bar{x})^2}{n-1}}$

Computational Formula for the Sample Standard Deviation:

$s = \sqrt{\dfrac{\Sigma x^2 - \dfrac{(\Sigma x)^2}{n}}{n-1}}$

z Score Formula:

Population: $z = \dfrac{x-\mu}{\sigma}$

Sample: $z = \dfrac{x-\bar{x}}{s}$

Raw Score Formula:

Population: $x = \mu + z\sigma$
Sample: $x = \bar{x} + zs$

Percentile Rank Formula:

PR of $x = \dfrac{[B + (\frac{1}{2})E]}{N}(100)$

Coefficient of Variation

Coefficient of Variation = $\dfrac{standard\ deviation}{mean} = (100\%)$

Empirical Rule (or 68-95-99.7) Rule

The Empirical Rule states that for a mound or bell-shaped distribution:
- Approximately 68% of the data will lie within one standard deviation of the mean.
- Approximately 95% of the data will lie within two standard deviations of the mean.
- Approximately 99.7% of the data will lie within three standard deviations of the mean.

Chebyshev's Theorem

Chebyshev's Theorem states for any distribution, regardless of its shape:

at least $1 - \dfrac{1}{k^2}$ of the data values will lie within k standard deviations of the mean, where k is greater than one.

5-Number Summary

A **5-number summary** uses the following five numbers to describe a data set:
1. The smallest data value
2. The first quartile (Q_1)
3. The median
4. The third quartile (Q_3)
5. The largest data value

Inter-Quartile Range (IQR)

IQR = $Q_3 - Q_1$

CHAPTER 4

An Alternate Formula to compute Pearson's Correlation Coefficient, **the sample correlation coefficient, r**, is:

$r = \dfrac{1}{n-1}\Sigma z_x z_y$

CHAPTER 5

COUNTING RULE 1: Permutation Rule.
The number of permutations (arrangements) of n different objects taken altogether, denoted $_nP_n$, is:

$$_nP_n = n!$$

COUNTING RULE 2: Permutation Rule for n objects taken s at a time.
The number of permutations of *n different* objects taken *s* at a time, denoted $_nP_s$, is:

$$_nP_s = n(n-1)(n-2)\cdots(n-s+1)$$

COUNTING RULE 3: Permutation Rule of N objects with k alike objects
Given N objects where n_1 are alike, n_2 are alike, ... n_k are alike, then the number of permutations of these N objects is:

$$\frac{N!}{(n_1!)(n_2!)\ldots(n_k!)}$$

COUNTING RULE 4: Number of Combinations of n objects taken s at a time
The number of combinations of n objects taken s at a time, is:

$$_nC_s = \frac{_nP_s}{s!}$$

The formula for $_nC_s$ in factorial notation is written as:

$$_nC_s = \frac{n!}{s!(n-s)!}$$

Classical Probability Definition for Equally Likely Outcomes:

$$P(\text{Event E}) = \frac{\text{number of outcomes satisfying event E}}{\text{total \# of outcomes in sample space}}$$

Relative Frequency Approach to Calculating Probability:

$$P(\text{Event A}) = \frac{\text{number of times event A occurred}}{\text{total \# of times the experiment was repeated}}$$

The Addition Rule:
$$P(A \text{ or } B) = P(A) + P(B) - P(A \text{ and } B)$$

The Complement Rule:
$$P(E) + P(E') = 1$$

Multiplication Rule for Independent Events:
$$P(A \text{ and } B) = P(A) \cdot P(B)$$

Multiplication Rule for Dependent Events:
$$P(A \text{ and } B) = P(A) \cdot P(B \mid A)$$

Conditional Probability Formula

$$P(A \mid B) = \frac{P(A \text{ and } B)}{P(B)}$$

CHAPTER 6

The Mean of a Discrete Random Variable X

$$\mu = \sum_{\substack{\text{all possible} \\ X \text{ values}}} [X \cdot P(X)]$$

Standard Deviation of a Discrete Random Variable

$$\sigma = \sqrt{\sum_{\substack{\text{all possible} \\ X \text{ values}}} [(X-\mu)^2 \cdot P(X)]}$$

Binomial Probability Formula

$$P(s \text{ successes in n trials}) = {}_nC_s p^s q^{(n-s)}$$

Mean of a Binomial Distribution

$$\mu_s = np$$

Standard Deviation of a Binomial Distribution

$$\sigma_s = \sqrt{n \cdot p \cdot (1-p)}$$

Poisson Probability formula: $P(x-k) = \dfrac{e^{-\lambda}\lambda^k}{k!}$

Mean of a Poisson Distribution: $\mu = \lambda$

Standard deviation of a Poisson Distribution: $\sigma = \sqrt{\lambda}$

Geometric Probability Formula
$$P(X = k) = q^k p, \quad k = 0, 1, 2\ldots$$

Mean of a Geometric Distribution

$$\mu = E(X) = \frac{q}{p}$$

Variance and Standard Deviation of a Geometric Distribution

$$\sigma^2 = Var(X) = \frac{q}{p^2} \qquad \sigma = \sqrt{VarX} = \sqrt{\frac{q}{p^2}}$$

Negative Binomial Probability Formula

$$P(X = k) = \binom{r+k-1}{r-1} q^k p^r, \quad k = 0, 1, 2, 3\ldots$$

Mean of a Negative Binomial Distribution

$$\mu = E(X) = \frac{rq}{p}$$

Variance and Standard Deviation of a Negative Binomial Distribution

$$VarX = \frac{rq}{p^2} \qquad \sigma = \sqrt{VarX} = \sqrt{\frac{rq}{p^2}}$$

CHAPTER 7

z Score Formula:

$$z = \frac{X - \mu}{\sigma}$$

Raw Score Formula:

$$X = \mu + (z)\sigma$$

Mean of a Binomial Distribution

$$\mu_s = np$$

Standard Deviation of a Binomial Distribution

$$\sigma_s = \sqrt{np(1-p)}$$

CHAPTER 8

Mean of the Sampling Distribution of the Mean:

$$\mu_{\bar{x}} = \mu$$

Standard Error or Standard Deviation of the Sampling Distribution of the Mean:

$$\sigma_{\bar{x}} = \frac{\sigma}{\sqrt{n}}$$

The population proportion is: $p = \frac{X}{N}$

The sample proportion is: $\hat{p} = \frac{x}{n}$

Mean of the Sampling Distribution of The Proportion is:

$$\mu_{\hat{p}} = p$$

Standard Error or Standard Deviation of the Sampling Distribution of The Proportion is:

$$\sigma_{\hat{p}} = \sqrt{\frac{p(1-p)}{n}}$$

CHAPTER 9

A 90% Confidence Interval for the population mean, μ, (with known population standard deviation) is:

$$\bar{x} - (1.65)\frac{\sigma}{\sqrt{n}} \text{ to } \bar{x} + (1.65)\frac{\sigma}{\sqrt{n}}$$

A 95% Confidence Interval for the population mean, μ, (with known population standard deviation) is:

$$\bar{x} - (1.96)\frac{\sigma}{\sqrt{n}} \text{ to } \bar{x} + (1.96)\frac{\sigma}{\sqrt{n}}$$

A 99% Confidence Interval for the population mean, μ, (with known population standard deviation) is:

$$\bar{x} - (2.58)\frac{\sigma}{\sqrt{n}} \text{ to } \bar{x} + (2.58)\frac{\sigma}{\sqrt{n}}$$

A 90% Confidence Interval for the population mean, μ, when the population standard deviation is *unknown*, is:

$$\bar{x} - (t_{95\%})\frac{s}{\sqrt{n}} \text{ to } \bar{x} + (t_{95\%})\frac{s}{\sqrt{n}}$$

A 95% Confidence Interval for the population mean, μ, when the population standard deviation is *unknown*, is:

$$\bar{x} - (t_{97.5\%})\frac{s}{\sqrt{n}} \text{ to } \bar{x} + (t_{97.5\%})\frac{s}{\sqrt{n}}$$

A 99% Confidence Interval for the population mean, μ, when the population standard deviation is *unknown*, is:

$$\bar{x} - (t_{99.5\%})\frac{s}{\sqrt{n}} \text{ to } \bar{x} + (t_{99.5\%})\frac{s}{\sqrt{n}}$$

A 90% Confidence Interval for the Population Proportion, p, is:

$$\hat{p} - (1.65)s_{\hat{p}} \text{ to } \hat{p} + (1.65)s_{\hat{p}}$$

A 95% Confidence Interval for the Population Proportion, p, is:

$$\hat{p} - (1.96)s_{\hat{p}} \text{ to } \hat{p} + (1.96)s_{\hat{p}}$$

A 99% Confidence Interval for the Population Proportion, p, is:

$$\hat{p} - (2.58)s_{\hat{p}} \text{ to } \hat{p} + (2.58)s_{\hat{p}}$$

The margin of error for estimating a population mean, where the population standard deviation is known, is:

$$E = (z)\left(\frac{\sigma}{\sqrt{n}}\right)$$

The sample size formula when estimating the population mean is:

$$n = \left[\frac{z\sigma}{E}\right]^2$$

The margin of error for estimating a population proportion is:

$$E = (z)\left(\sqrt{\frac{\hat{p}(1-\hat{p})}{n}}\right)$$

The sample size formula when estimating the population proportion using $\hat{p}$, a point estimate of the population proportion, is:

$$n = \frac{z^2(\hat{p})(1-\hat{p})}{E^2}$$

The conservative sample size formula in estimating the population proportion is:

$$n = \frac{z^2(0.25)}{E^2}$$

Margin of Error for Estimating the Population Mean, where the Population Standard Deviation is Unknown. The formula to determine this margin of error is:

$$E = (t)\left(\frac{s}{\sqrt{n}}\right)$$

CHAPTER 10

z SCORE Formula:

$$z = \frac{X - \mu}{\sigma}$$

Test Statistic Formula:

$$z = \frac{\text{sample proportion} - \text{population proportion}}{\text{standard error}}$$

$$z = \frac{\text{sample mean} - \text{population mean}}{\text{standard error}}$$

CHAPTER 11

Population proportion is:

$$p = \frac{X}{N}$$

Sample proportion is:

$$\hat{p} = \frac{x}{n}$$

Mean of the Sampling Distribution of The Proportion is:

$$\mu_{\hat{p}} = p$$

Standard error or standard deviation of the Sampling Distribution of The Proportion is:

$$\sigma_{\hat{p}} = \sqrt{\frac{p(1-p)}{n}}$$

Test Statistic Formula is: $z = \frac{\hat{p} - \mu_{\hat{p}}}{\sigma_{\hat{p}}}$

Mean of the sampling distribution of the mean is:

$$\mu_{\bar{x}} = \mu$$

Standard error or standard deviation of the sampling distribution of the mean is:

$$\sigma_{\bar{x}} = \frac{\sigma}{\sqrt{n}}$$

Estimate of the standard error or standard deviation of the sampling distribution of the mean is:

$$s_{\bar{x}} = \frac{s}{\sqrt{n}}$$

Test Statistic Formula:
a) for normal distribution: $z = \frac{\bar{x} - \mu_{\bar{x}}}{\sigma_{\bar{x}}}$

b) for t distribution: $t = \frac{\bar{x} - \mu_{\bar{x}}}{s_{\bar{x}}}$

To find t_c, need to determine degrees of freedom using formula:

$$df = n - 1$$

CHAPTER 12

Sampling Distributions of the Difference Between Two Proportions

Mean: $\mu_{\hat{p}_1 - \hat{p}_2} = p_1 - p_2$

Standard Error:

$$s_{\hat{p}_1 - \hat{p}_2} = \sqrt{\frac{\hat{p}(1-\hat{p})}{n_1} + \frac{\hat{p}(1-\hat{p})}{n_2}} \quad \text{or}$$

$$s_{\hat{p}_1 - \hat{p}_2} = \sqrt{\hat{p}(1-\hat{p})\left(\frac{1}{n_1} + \frac{1}{n_2}\right)}$$

where $\hat{p} = \frac{x_1 + x_2}{n_1 + n_2}$ or $\hat{p} = \frac{n_1 \hat{p}_1 + n_2 \hat{p}_2}{n_1 + n_2}$

Test Statistic Formula is: $z = \frac{(\hat{p}_1 - \hat{p}_2) - \mu_{\hat{p}_1 - \hat{p}_2}}{s_{\hat{p}_1 - \hat{p}_2}}$

CHAPTER 13

Mean of the Sampling Distribution of the Difference Between Two Means is:

$$\mu_{\bar{x}_1 - \bar{x}_2} = \mu_1 - \mu_2$$

The Two Sample z Test

The expected results for the Two Sample z test are:

a) The mean of the sampling distribution of the difference between two means is:

$$\mu_{\bar{x}_1 - \bar{x}_2} = \mu_1 - \mu_2 = 0$$

b) The standard error of the difference between two means, written $\sigma_{\bar{x}_1 - \bar{x}_2}$, is:

$$\sigma_{\bar{x}_1 - \bar{x}_2} = \sqrt{\frac{\sigma_1^2}{n_1} + \frac{\sigma_2^2}{n_2}}$$

c) The test statistic z is:

$$z = \frac{(\bar{x}_1 - \bar{x}_2)}{\sigma_{\bar{x}_1 - \bar{x}_2}}$$

The Pooled Two Sample t Test

The expected results for the Pooled Two Sample t test are:

a) The mean of the sampling distribution of the difference between two means, denoted by $\mu_{\bar{x}_1 - \bar{x}_2}$, is

$$\mu_{\bar{x}_1 - \bar{x}_2} = \mu_1 - \mu_2 = 0$$

b) The population standard deviations are <u>unknown</u> and assumed to be <u>equal</u>, then the pooled t test estimate of the standard error is:

$$s_{\bar{x}_1-\bar{x}_2} = \sqrt{\frac{(n_1-1)s_1^2+(n_2-1)s_2^2}{n_1+n_2-2}\cdot\left(\frac{1}{n_1}+\frac{1}{n_2}\right)}$$

Note: $s_{\bar{x}_1-\bar{x}_2} = s_{x_p}\sqrt{\left(\frac{1}{n_1}+\frac{1}{n_2}\right)}$, where s_{x_p} is called the pooled standard deviation and is computed by the formula $s_{x_p} = \sqrt{\frac{(n_1-1)s_1^2+(n_2-1)s_2^2}{n_1+n_2-2}}$.

c) The test statistic t formula is:

$$t = \frac{\bar{x}_1-\bar{x}_2}{s_{x_p}\sqrt{\left(\frac{1}{n_1}+\frac{1}{n_2}\right)}}$$

where:

$$s_{x_p} = \sqrt{\frac{(n_1-1)s_1^2+(n_2-1)s_2^2}{n_1+n_2-2}}$$

d) The test statistic's distribution is a t distribution with $df = n_1 + n_2 - 2$.

Welch's Two Sample t Test

The expected results for Welch's t test are:

a) The test statistic hypothesis testing model is approximately a t distribution with degrees of freedom given by:

$$df = \text{smaller of } (n_1-1) \text{ and } (n_2-1).$$

b) The mean of the sampling distribution of the difference between two means is given by the formula:

$$\mu_{\bar{x}_1-\bar{x}_2} = \mu_1-\mu_2 = 0$$

c) Welch's approximation for the estimate of the standard error of the difference between two means, written $S_{\bar{x}_1-\bar{x}_2}$, is:

$$s_{\bar{x}_1-\bar{x}_2} = \sqrt{\frac{s_1^2}{n_1}+\frac{s_2^2}{n_2}}$$

d) The test statistic t formula is:

$$t = \frac{\bar{x}_1-\bar{x}_2}{\sqrt{\frac{s_1^2}{n_1}+\frac{s_2^2}{n_2}}}$$

The Paired t Test

The expected results for the Paired t test are:

a) The mean of the sampling distribution of the differences of the paired observations is equal to μ_d.

b) The estimate of the standard deviation of the differences of the paired observations referred to as the standard error of the differences of the paired observations is:

Standard Error of the difference of the paired observations equals $\frac{s_d}{\sqrt{n}}$,

where n is the number of data pairs.

c) The test statistic t formula is:

$$t = \frac{\bar{x}_d-\mu_d}{\frac{s_d}{\sqrt{n}}}$$

d) The test statistic's distribution is a t distribution with $df = n-1$, where n is the number of data pairs.

CHAPTER 14

Pearson's Chi-Square Test Statistic is:

$$X^2 = \sum_{\text{all cells}} \frac{(O-E)^2}{E}$$

or

$$X^2 = \sum_{\text{all cells}} \left(\frac{O^2}{E}\right) - n$$

Chi-Square Distribution:
degrees of freedom:

$$df = (r-1)(c-1)$$

$$\frac{\text{expected cell}}{\text{frequency}} = \frac{(RT)(CT)}{n}$$

CHAPTERS 4 & 15

Pearson's or Sample Correlation Coefficient, r (test statistic).

$$r = \frac{n\Sigma(xy)-(\Sigma x)(\Sigma y)}{\sqrt{n(\Sigma x^2)-(\Sigma x)^2}\sqrt{n(\Sigma y^2)-(\Sigma y)^2}}$$

degrees of freedom, df, for testing the correlation coefficient is:

$$df = n-2$$

Coefficient of Determination $= r^2$

Regression Line Formula: $y' = a + bx$

where:
y' is the predicted value of y, the dependent variable, given the value of x, the independent variable,

and

a and b are the regression coefficients obtained by the formulas:

$$b = \frac{n\Sigma(xy) - \Sigma x \Sigma y}{n(\Sigma x^2) - (\Sigma x)^2}$$

$$a = \frac{\Sigma y - b\Sigma x}{n}$$

CHAPTER 16: ANOVA

Sum of Squares $SS = \Sigma(x - \bar{x})^2$

$SS_{(between\ treatments)} + SS_{(within\ treatments)} = SS_{(total)}$

$$SS_{(total)} = \Sigma x^2 - \frac{(\Sigma x)^2}{N}$$

$SS_{(between\ treatments)} =$

$$\left(\frac{(\Sigma T_1)^2}{n_1} + \frac{(\Sigma T_2)^2}{n_2} + \ldots + \frac{(\Sigma T_i)^2}{n_i}\right) - \frac{(\Sigma x)^2}{N}$$

$SS_{(within\ treatments)} =$

$$\Sigma x^2 - \left(\frac{(\Sigma T_1)^2}{n_1} + \frac{(\Sigma T_2)^2}{n_2} + \ldots + \frac{(\Sigma T_i)^2}{n_i}\right)$$

Degrees of Freedom: $df_{(Between)} + df_{(within)} = df_{(Total)}$
Where: $df_{(Between)} = k - 1$
$df_{(within)} = N - k$
$df_{(Total)} = N - 1$

Mean Square, (denoted MS), formulas:

$$MS_{(Between)} = \frac{SS_{bet}}{df_{bet}}$$

$$MS_{(within)} = \frac{SS_{within}}{df_{within}}$$

Test statistic formula: $F = \dfrac{s_1^2}{s_2^2}$

Test statistic: F-statistic

$$F = \frac{treatment\ sample\ variance}{within\ sample\ variance}$$

CHAPTER 17
NONPARAMETRIC TESTS
SIGN TEST for MEDIANS

binomial mean: $\mu_s = np$

binomial std dev: $\sigma_s = \sqrt{np(1-p)}$

If the statistic result O, is less than n/2,

then test statistic is: $z = \dfrac{(O + \frac{1}{2}) - \mu_s}{\sigma_s}$

If the statistical result O, is greater than n/2,

then the test statistic is: $z = \dfrac{(O - \frac{1}{2}) - \mu_s}{\sigma_s}$

MANN-WHITNEY RANK-SUM TEST

When normal distribution can be applied, then test statistic, z, is:

$$z = \frac{w - \dfrac{n_1(n_1 + n_2 + 1)}{2}}{\sqrt{\dfrac{n_1 n_2 (n_1 + n_2 + 1)}{12}}}$$

KRUSKAL-WALLIS H TEST

Kruskal-Wallis H test statistic formula:

$$H = \frac{12}{N(N+1)}\left[\frac{(\Sigma R_1)^2}{n_1} + \frac{(\Sigma R_2)^2}{n_2} + \ldots + \frac{(\Sigma R_k)^2}{n_k}\right] - 3(N+1)$$

An Alternate Formula to compute Pearson's Correlation Coefficient, the **sample correlation coefficient**, r, is:

$$r = \frac{1}{n-1} \Sigma z_x z_y$$

CHAPTER 18: BOOTSTRAPPING

Bootstrap Estimate of Bias

$$Bias(\hat{\Theta}) \approx \frac{1}{B}\sum_{i=1}^{B} \hat{\Theta}_i^* - \hat{\Theta} = \hat{\Theta}^* - \hat{\Theta}$$

Mean of the bootstrap distribution

$$Mean = \frac{1}{B}\sum \bar{x}^*$$

Bootstrap Standard Error of a Statistic
Standard Error of the Bootstrap Distribution

$$SE_b(\bar{x}) = \sqrt{\frac{1}{B-1} * \sum \left(\bar{x}^* - \frac{1}{B}\sum \bar{x}^*\right)^2}$$

Bootstrap Estimate for the Standard Error of the Mean

$$\approx \frac{original\ sample\ standard\ deviation}{\sqrt{n}}$$

Bootstrap Standard Error t Confidence Interval Formula:

statistic $\pm$ (t score) * (SE_b(statistic))

When the statistic is a mean,
the Bootstrap Standard Error t Confidence Interval Formula:

$\bar{x} \pm$ (t score) * $SE_b(\bar{x})$

Basic Bootstrap Confidence Interval
$[2\hat{\Theta} - q_{1-\alpha/2}, 2\hat{\Theta} - q_{\alpha/2}]$

quantile q_α formula
quantile q_α = Integer value of $[\alpha * (B+1)]$

APPENDIX B: SUMMARY OF HYPOTHESIS TESTS

SUMMARY OF TEST STATISTIC APPROACH TO THE HYPOTHESIS TESTS FOR TESTING THE VALUE OF A POPULATION PARAMETER

FORM OF THE NULL HYPOTHESIS	CONDTIONS OF TEST	TEST STATISTIC FORMULA	TEST STATISTIC DISTRIBUTION IS A:
$\mu = \mu_o$	KNOWN σ and $n > 30$	$z = \dfrac{\bar{x} - \mu_{\bar{x}}}{\sigma_{\bar{x}}}$ where: $\sigma_{\bar{x}} = \dfrac{\sigma}{\sqrt{n}}$	Normal Distribution
$\mu = \mu_o$	UNKNOWN σ	$t = \dfrac{\bar{x} - \mu_{\bar{x}}}{\sigma_{\bar{x}}}$ where: $S_{\bar{x}} = \dfrac{S}{\sqrt{n}}$	t Distribution with: df = n − 1
$p = p_o$	$np \geq 10$ and $n(1-p) \geq 10$	$z = \dfrac{\hat{p} - \mu_{\hat{p}}}{\sigma_{\hat{p}}}$ where: $\sigma_{\hat{p}} = \sqrt{\dfrac{p(1-p)}{n}}$	Normal Distribution

SUMMARY OF TEST STATISTIC APPROACH TO THE HYPOTHESIS TESTS INVOLVING TWO POPULATION MEANS

TEST	FORM OF THE NULL HYPOTHESIS	CONDITIONS OF TEST	TEST STATISTIC FORMULA	TEST STATISTIC DISTRIBUTION
Two Sample z Test	$\mu_1 - \mu_2 = 0$	KNOWN: σ_1 and σ_2 and n_1 and n_2 are both greater than 30	$z = \dfrac{(\bar{x}_1 - \bar{x}_2)}{\sigma_{\bar{x}_1 - \bar{x}_2}}$ where: $\sigma_{\bar{x}_1 - \bar{x}_2} = \sqrt{\dfrac{\sigma_1^2}{n_1} + \dfrac{\sigma_2^2}{n_2}}$	Normal Distribution
Pooled Two Sample t Test	$\mu_1 - \mu_2 = 0$	UNKNOWN: σ_1 and σ_2 and $\sigma_1 = \sigma_2$	$t = \dfrac{\bar{x}_1 - \bar{x}_2}{s_{x_p}\sqrt{\left(\dfrac{1}{n_1} + \dfrac{1}{n_2}\right)}}$ where: $s_{x_p} = \sqrt{\dfrac{(n_1-1)S_1^2 + (n_2-1)S_2^2}{n_1 + n_2 - 2} \cdot \left(\dfrac{1}{n_1} + \dfrac{1}{n_2}\right)}$	t Distribution with: df = $n_1 + n_2 - 1$

TEST	FORM OF THE NULL HYPOTHESIS	CONDITIONS OF TEST	TEST STATISTIC FORMULA	TEST STATISTIC DISTRIBUTION
Welch's Two Sample t Test	$\mu_1 - \mu_2 = 0$	UNKNOWN: σ_1 and σ_2 and $\sigma_1 \neq \sigma_2$	$t = \dfrac{\bar{x}_1 - \bar{x}_2}{\sqrt{\dfrac{s_1^2}{n_1} + \dfrac{s_2^2}{n_2}}}$	t Distribution with: df = smaller of $(n_1 - 1)$ and $(n_2 - 1)$
Paired t Test	$\mu_d = 0$	UNKNOWN: σ_d	$t = \dfrac{\bar{x}_d - \mu_d}{\dfrac{s_d}{\sqrt{n}}}$ where: $\bar{x}_d$ = sample mean of the differences s_d = sample standard deviation of the differences	t Distribution with: $df = n - 1$, where: n is the number of data pairs.

SUMMARY OF CONFIDENCE INTERVALS

CONFIDENCE INTERVALS FORMULAS AND CONDITIONS

PARAMETER BEING ESTIMATED	CONDTIONS OF TEST	POINT ESTIMATE	CONFIDENCE INTERVAL	SAMPLING DISTRIBUTION IS A:
μ	KNOWN σ	$\bar{x}$	$\bar{x} - (z)\left(\dfrac{\sigma}{\sqrt{n}}\right)$ to $\bar{x} + (z)\left(\dfrac{\sigma}{\sqrt{n}}\right)$ where: $z = 1.65$ for 90% $z = 1.96$ for 95% $z = 2.58$ for 99%	Normal Distribution
μ	UNKNOWN σ	$\bar{x}$	$\bar{x} - (t)\left(\dfrac{S}{\sqrt{n}}\right)$ to $\bar{x} + (t)\left(\dfrac{S}{\sqrt{n}}\right)$ where: $t = t_{95\%}$ for 90% $t = t_{97.5\%}$ for 95% $t = t_{99.5\%}$ for 99%	t Distribution df = n - 1
p	$n\hat{p} \geq 10$ and $n(1 - \hat{p}) \geq 10$	$\hat{p}$	$\hat{p} - (z)(s_{\hat{p}})$ to $\hat{p} + (z)(s_{\hat{p}})$ where: $z = 1.65$ for 90% $z = 1.96$ for 95% $z = 2.58$ for 99%	Normal Distribution

APPENDIX C: STATISTICAL TABLES

TABLE I: RANDOM NUMBER TABLE

63271	59986	71744	51102	15141	80714	58683	93108	13554	79945
88547	09896	95436	79115	08303	01041	20030	63754	08459	28364
55957	57243	83865	09911	19761	66535	40102	26646	60147	15702
46276	87453	44790	67122	45573	84358	21625	16999	13385	22782
55363	07449	34835	15290	76616	67191	12777	21861	68689	03263
69393	92785	49902	58447	42048	30378	87618	26933	40640	16281
13186	29431	88190	04588	38733	81290	89541	70290	40113	08243
17726	28652	56836	78351	47327	18518	92222	55201	27340	10493
36520	64465	05550	30157	82242	29520	69753	72602	23756	54935
81628	36100	39254	56835	37636	02421	98063	89641	64953	99337
84649	48968	75215	75498	49539	74240	03466	49292	36401	45525
63291	11618	12613	75055	43915	26488	41116	64531	56827	30825
70502	53225	03655	05915	37140	57051	48393	91322	25653	06543
06426	24771	59935	49801	11082	66762	94477	02494	88215	27191
20711	55609	29430	70165	45406	78484	31639	52009	18873	96927
41990	70538	77191	25860	55204	73417	83920	69468	74972	38712
72452	36618	76298	26678	89334	33938	95567	29380	75906	91807
37042	40318	57099	10528	09925	89773	41335	96244	29002	46453
53766	52875	15987	46962	67342	77592	57651	95508	80033	69828
90585	58955	53122	16025	84299	53310	67380	84249	25348	04332
32001	96293	37203	64516	51530	37069	40261	61374	05815	06714
62606	64324	46354	72157	67248	20135	49804	09226	64419	29457
10078	28073	85389	50324	14500	15562	64165	06125	71353	77669
91561	46145	24177	15294	10061	98124	75732	00815	83452	97355
13091	98112	53959	79607	52244	63303	10413	63839	74762	50289
73864	83014	72457	22682	03033	61714	88173	90835	00634	85169
66668	25467	48894	51043	02365	91726	09365	63167	95264	45643
84745	41042	29493	01836	09044	51926	43630	63470	76508	14194
48068	26805	94595	47907	13357	38412	33318	26098	82782	42851
54310	96175	97594	88616	42035	38093	36745	56702	40644	83514
14877	33095	10924	58013	61439	21882	42059	24177	58739	60170
78295	23179	02771	43464	59061	71411	05697	67194	30495	21157
67524	02865	39593	54278	04237	92441	26602	63835	38032	94770
58268	57219	68124	73455	83236	08710	04284	55005	84171	42596
97158	28672	50685	01181	24262	19427	52106	34308	73685	74246
04230	16831	69085	30802	65559	09205	71829	06489	85650	38707
94879	56606	30401	02602	57658	70091	54986	41394	60437	03195
71446	15232	66715	26385	91518	70566	02888	79941	39684	54315
32886	05644	79316	09819	00813	88407	17461	73925	53037	91904
62048	33711	25290	21526	02223	75947	66466	06232	10913	75336
84534	42351	21628	53669	81352	95152	08107	98814	72743	12849
84707	15885	84710	35866	06446	86311	32648	88141	73902	69981
19409	40868	64220	80861	13860	68493	52908	26374	63297	45052
57978	48015	25973	66777	45924	56144	24742	96702	88200	66162
57295	98298	11199	96510	75228	41600	47192	43267	35973	23152
94044	83785	93388	07833	38216	31413	70555	03023	54147	06647
30,014	25879	71763	96679	90603	99396	74557	74224	18211	91637
07265	69563	64268	88802	72264	66540	01782	08396	19251	83613
84404	88642	30263	80310	11522	57810	27627	78376	36240	48952
21778	02085	27762	46097	43324	34354	09369	14966	10158	76089

Reprinted from page 44 of *A Million Random Digits with 100.000 Normal Deviates* by The Rand Corporation (New York: The Free Press, 1955). Copyright 1955 and 1983 by The Rand Corporation. Use by permission.

TABLE II: THE STANDARD NORMAL CURVE AREA TABLE

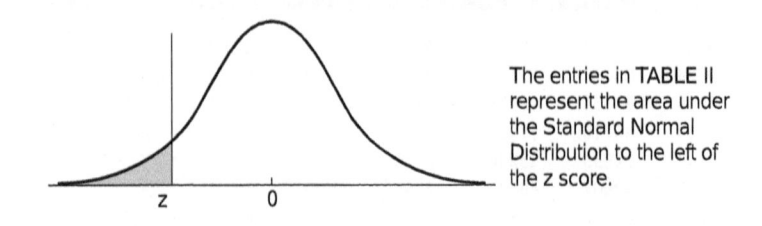

The entries in TABLE II represent the area under the Standard Normal Distribution to the left of the z score.

z	0	1	2	3	4	5	6	7	8	9
−3.	.0013	.0009	.0007	.0005	.0003	.0002	.0001	.0001	.0001	.0000[2]
−2.9	.0018	.0018	.0017	.0017	.0016	.0016	.0015	.0015	.0014	.0014
−2.8	.0025	.0024	.0024	.0023	.0023	.0022	.0021	.0021	.0020	.0019
−2.7	.0035	.0034	.0033	.0032	.0031	.0030	.0029	.0028	.0027	.0026
−2.6	.0047	.0045	.0044	.0043	.0041	.0040	.0039	.0038	.0037	.0036
−2.5	.0062	.0060	.0059	.0057	.0055	.0054	.0052	.0051	.0049	.0048
−2.4	.0082	.0080	.0078	.0075	.0073	.0071	.0069	.0068	.0066	.0064
−2.3	.0107	.0104	.0102	.0099	.0096	.0094	.0091	.0089	.0087	.0084
−2.2	.0139	.0136	.0132	.0129	.0125	.0122	.0119	.0116	.0113	.0110
−2.1	.0179	.0174	.0170	.0166	.0162	.0158	.0154	.0150	.0146	.0143
−2.0	.0228	.0222	.0217	.0212	.0207	.0202	.0197	.0192	.0188	.0183
−1.9	.0287	.0281	.0274	.0268	.0262	.0256	.0250	.0244	.0239	.0233
−1.8	.0359	.0352	.0344	.0336	.0329	.0322	.0314	.0307	.0301	.0294
−1.7	.0446	.0436	.0427	.0418	.0409	.0401	.0392	.0384	.0375	.0367
−1.6	.0548	.0537	.0526	.0516	.0505	.0495	.0485	.0475	.0465	.0455
−1.5	.0668	.0655	.0643	.0630	.0618	.0606	.0594	.0582	.0571	.0559
−1.4	.0808	.0793	.0778	.0764	.0749	.0735	.0722	.0708	.0694	.0681
−1.3	.0968	.0951	.0934	.0918	.0901	.0885	.0869	.0853	.0838	.0823
−1.2	.1151	.1131	.1112	.1093	.1075	.1056	.1038	.1020	.1003	.0985
−1.1	.1357	.1335	.1314	.1292	.1271	.1251	.1230	.1210	.1190	.1170
−1.0	.1587	.1562	.1539	.1515	.1492	.1469	.1446	.1423	.1401	.1379
−0.9	.1841	.1814	.1788	.1762	.1736	.1711	.1685	.1660	.1635	.1611
−0.8	.2119	.2090	.2061	.2033	.2005	.1977	.1949	.1922	.1894	.1867
−0.7	.2420	.2389	.2358	.2327	.2296	.2266	.2236	.2206	.2177	.2148
−0.6	.2743	.2709	.2676	.2643	.2611	.2578	.2546	.2514	.2483	.2451
−0.5	.3085	.3050	.3015	.2981	.2946	.2912	.2877	.2843	.2810	.2776
−0.4	.3446	.3409	.3372	.3337	.3300	.3264	.3228	.3192	.3156	.3121
−0.3	.3821	.3783	.3745	.3707	.3669	.3632	.3594	.3557	.3520	.3483
−0.2	.4207	.4168	.4129	.4090	.4052	.4013	.3974	.3936	.3897	.3859
−0.1	.4602	.4562	.4522	.4483	.4443	.4404	.4364	.4325	.4286	.4247
−0.0	.5000	.4960	.4920	.4880	.4840	.4801	.4761	.4721	.4681	.4641

[2] Please note: to four decimal places, the area to the left of z = −3.9 is *approximately* zero.

TABLE II: THE STANDARD NORMAL CURVE AREA TABLE

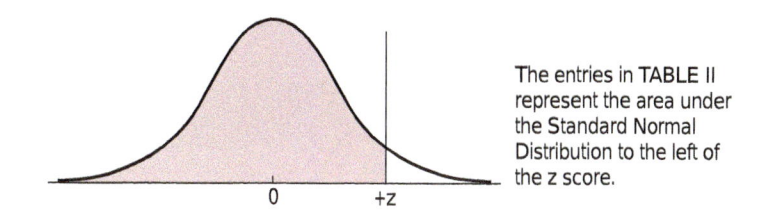

The entries in TABLE II represent the area under the Standard Normal Distribution to the left of the z score.

z	0	1	2	3	4	5	6	7	8	9
0.0	.5000	.5040	.5080	.5120	.5160	.5199	.5239	.5279	.5319	.5359
0.1	.5398	.5438	.5478	.5517	.5557	.5596	.5636	.5675	.5714	.5753
0.2	.5793	.5832	.5871	.5910	.5948	.5987	.6026	.6064	.6103	.6141
0.3	.6179	.6217	.6255	.6293	.6331	.6368	.6406	.6443	.6480	.6517
0.4	.6554	.6591	.6628	.6664	.6700	.6736	.6772	.6808	.6844	.6879
0.5	.6915	.6950	.6985	.7019	.7054	.7088	.7123	.7157	.7190	.7224
0.6	.7257	.7291	.7324	.7357	.7389	.7422	.7454	.7486	.7517	.7549
0.7	.7580	.7611	.7642	.7673	.7704	.7734	.7764	.7794	.7823	.7852
0.8	.7881	.7910	.7939	.7967	.7995	.8023	.8051	.8078	.8106	.8133
0.9	.8159	.8186	.8212	.8238	.8264	.8289	.8315	.8340	.8365	.8389
1.0	.8413	.8438	.8461	.8485	.8508	.8531	.8554	.8577	.8599	.8621
1.1	.8643	.8665	.8686	.8708	.8729	.8749	.8770	.8790	.8810	.8830
1.2	.8849	.8869	.8888	.8907	.8925	.8944	.8962	.8980	.8997	.9015
1.3	.9032	.9049	.9066	.9082	.9099	.9115	.9131	.9147	.9162	.9177
1.4	.9192	.9207	.9222	.9236	.9251	.9265	.9278	.9292	.9306	.9319
1.5	.9332	.9345	.9357	.9370	.9382	.9394	.9406	.9418	.9429	.9441
1.6	.9452	.9463	.9474	.9484	.9495	.9505	.9515	.9525	.9535	.9545
1.7	.9554	.9564	.9573	.9582	.9591	.9599	.9608	.9616	.9625	.9633
1.8	.9641	.9649	.9656	.9664	.9671	.9678	.9686	.9693	.9699	.9706
1.9	.9713	.9719	.9726	.9732	.9738	.9744	.9750	.9756	.9761	.9767
2.0	.9772	.9778	.9783	.9788	.9793	.9798	.9803	.9808	.9812	.9817
2.1	.9821	.9826	.9830	.9834	.9838	.9842	.9846	.9850	.9854	.9857
2.2	.9861	.9864	.9868	.9871	.9875	.9878	.9881	.9884	.9887	.9890
2.3	.9893	.9896	.9898	.9901	.9904	.9906	.9909	.9911	.9913	.9916
2.4	.9918	.9920	.9922	.9925	.9927	.9929	.9931	.9932	.9934	.9936
2.5	.9938	.9940	.9941	.9943	.9945	.9946	.9948	.9949	.9951	.9952
2.6	.9953	.9955	.9956	.9957	.9959	.9960	.9961	.9962	.9963	.9964
2.7	.9965	.9966	.9967	.9968	.9969	.9970	.9971	.9972	.9973	.9974
2.8	.9974	.9975	.9976	.9977	.9977	.9978	.9979	.9979	.9980	.9981
2.9	.9981	.9982	.9982	.9983	.9984	.9984	.9985	.9985	.9986	.9986
3.	.9987	.9990	.9993	.9995	.9997	.9998	.9998	.9999	.9999	1.0000[3]

[3]Please note: to four decimal places, the area to the left of z = 3.9 is *approximately* 1.

TABLE III: CRITICAL VALUES FOR THE t DISTRIBUTIONS
FOR HYPOTHESIS TESTING

The entries in the table give the critical t values for the specified number of degrees of freedom (df) and the level of significance (α) for a one tailed or a two tailed test.

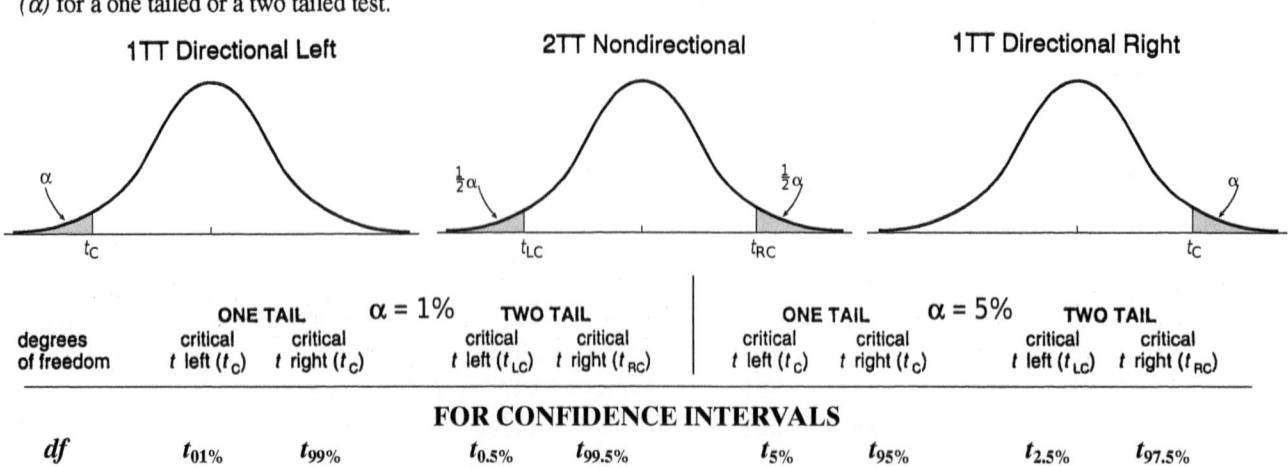

	$\alpha = 1\%$				$\alpha = 5\%$			
	ONE TAIL		TWO TAIL		ONE TAIL		TWO TAIL	
degrees of freedom	critical t left (t_C)	critical t right (t_C)	critical t left (t_{LC})	critical t right (t_{RC})	critical t left (t_C)	critical t right (t_C)	critical t left (t_{LC})	critical t right (t_{RC})

FOR CONFIDENCE INTERVALS

df	$t_{01\%}$	$t_{99\%}$	$t_{0.5\%}$	$t_{99.5\%}$	$t_{5\%}$	$t_{95\%}$	$t_{2.5\%}$	$t_{97.5\%}$
1	−31.82	31.82	−63.66	63.66	−6.31	6.31	−12.71	12.71
2	−6.96	6.96	−9.92	9.92	−2.92	2.92	−4.30	4.30
3	−4.54	4.54	−5.84	5.84	−2.35	2.35	−3.18	3.18
4	−3.75	3.75	−4.60	4.60	−2.13	2.13	−2.78	2.78
5	−3.36	3.36	−4.03	4.03	−2.02	2.02	−2.57	2.57
6	−3.14	3.14	−3.71	3.71	−1.94	1.94	−2.45	2.45
7	−3.00	3.00	−3.50	3.50	−1.90	1.90	−2.36	2.36
8	−2.90	2.90	−3.36	3.36	−1.86	1.86	−2.31	2.31
9	−2.82	2.82	−3.25	3.25	−1.83	1.83	−2.26	2.26
10	−2.76	2.76	−3.17	3.17	−1.81	1.81	−2.23	2.23
11	−2.72	2.72	−3.11	3.11	−1.80	1.80	−2.20	2.20
12	−2.68	2.68	−3.06	3.06	−1.78	1.78	−2.18	2.18
13	−2.65	2.65	−3.01	3.01	−1.77	1.77	−2.16	2.16
14	−2.62	2.62	−2.98	2.98	−1.76	1.76	−2.14	2.14
15	−2.60	2.60	−2.95	2.95	−1.75	1.75	−2.13	2.13
16	−2.58	2.58	−2.92	2.92	−1.75	1.75	−2.12	2.12
17	−2.57	2.57	−2.90	2.90	−1.74	1.74	−2.11	2.11
18	−2.55	2.55	−2.88	2.88	−1.73	1.73	−2.10	2.10
19	−2.54	2.54	−2.86	2.86	−1.73	1.73	−2.09	2.09
20	−2.53	2.53	−2.84	2.84	−1.72	1.72	−2.09	2.09
21	−2.52	2.52	−2.83	2.83	−1.72	1.72	−2.08	2.08
22	−2.51	2.51	−2.82	2.82	−1.72	1.72	−2.07	2.07
23	−2.50	2.50	−2.81	2.81	−1.71	1.71	−2.07	2.07
24	−2.49	2.49	−2.80	2.80	−1.71	1.71	−2.06	2.06
25	−2.48	2.48	−2.79	2.79	−1.71	1.71	−2.06	2.06
26	−2.48	2.48	−2.78	2.78	−1.71	1.71	−2.06	2.06
27	−2.47	2.47	−2.77	2.77	−1.70	1.70	−2.05	2.05
28	−2.47	2.47	−2.76	2.76	−1.70	1.70	−2.05	2.05
29	−2.46	2.46	−2.76	2.76	−1.70	1.70	−2.04	2.04
30	−2.46	2.46	−2.75	2.75	−1.70	1.70	−2.04	2.04
40	−2.42	2.42	−2.70	2.70	−1.68	1.68	−2.02	2.02
50	−2.40	2.40	−2.68	2.68	−1.68	1.68	−2.01	2.01
60	−2.39	2.39	−2.66	2.66	−1.67	1.67	−2.00	2.00
80	−2.37	2.37	−2.64	2.64	−1.66	1.66	−1.99	1.99
100	−2.36	2.36	−2.63	2.63	−1.66	1.66	−1.98	1.98
200	−2.34	2.34	−2.60	2.60	−1.65	1.65	−1.97	1.97
500	−2.33	2.33	−2.59	2.59	−1.65	1.65	−1.96	1.96
(normal distribution)	−2.33	2.33	−2.58	2.58	−1.65	1.65	−1.96	1.96

TABLE IV: CRITICAL VALUES FOR THE CHI-SQUARE DISTRIBUTIONS, χ^2_α

The entries in the table give the critical chi-square values (χ^2_α) for the specified number of degrees of freedom (*df*) and the level of significance (α).

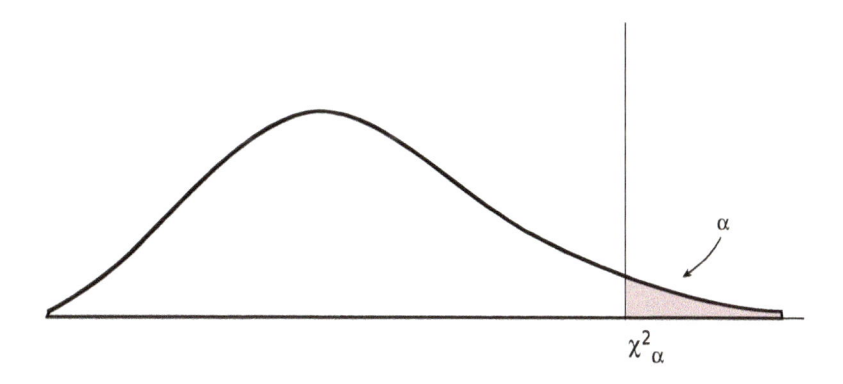

degrees of freedom	significance level	
df	$\alpha = 5\%$	$\alpha = 1\%$
1	3.84	6.64
2	5.99	9.21
3	7.82	11.34
4	9.49	13.28
5	11.07	15.09
6	12.59	16.81
7	14.07	18.48
8	15.51	20.09
9	16.92	21.67
10	18.31	23.21
11	19.68	24.72
12	21.03	26.22
13	22.36	27.69
14	23.68	29.14
15	25.00	30.58
16	26.30	32.00
17	27.59	33.41
18	28.87	34.80
19	30.14	36.19
20	31.41	37.57

TABLE V: CRITICAL VALUES OF THE CORRELATION COEFFICIENT, r_α

1TT Directional Left
use a negative critical r value, $-r_\alpha$

2TT Nondirectional
use a positive critical r value, r_α
and a negative critical r value, $-r_\alpha$

1TT Directional Right
use a positive critical r value, r_α

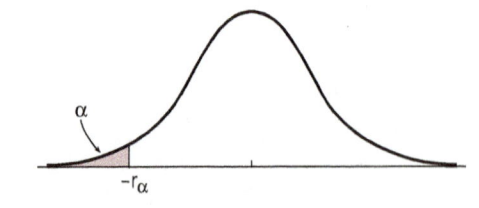

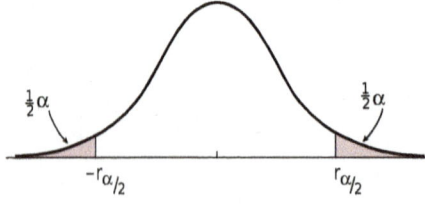

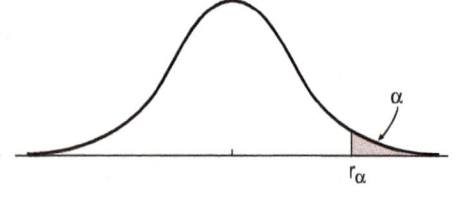

	$\alpha = 5\%$		$\alpha = 1\%$	
df	one tail	two tail	one tail	two tail
1	0.99	1.00	1.00	1.00
2	0.90	0.95	0.98	0.99
3	0.81	0.88	0.93	0.96
4	0.73	0.81	0.88	0.92
5	0.67	0.75	0.83	0.87
6	0.62	0.71	0.79	0.83
7	0.58	0.67	0.75	0.80
8	0.54	0.63	0.72	0.76
9	0.52	0.60	0.69	0.73
10	0.50	0.58	0.66	0.71
11	0.48	0.53	0.63	0.68
12	0.46	0.53	0.61	0.66
13	0.44	0.51	0.59	0.64
14	0.42	0.50	0.57	0.61
15	0.41	0.48	0.56	0.61
16	0.40	0.47	0.54	0.59
17	0.39	0.46	0.53	0.58
18	0.38	0.44	0.52	0.56
19	0.37	0.43	0.50	0.55
20	0.36	0.42	0.49	0.54
21	0.35	0.41	0.48	0.53
22	0.34	0.40	0.47	0.52
23	0.34	0.40	0.46	0.51
24	0.33	0.39	0.45	0.50
25	0.32	0.38	0.45	0.49
26	0.32	0.37	0.44	0.48
27	0.31	0.37	0.43	0.47
28	0.31	0.36	0.42	0.46

To determine the critical value of the correlation coefficient, r_α, when the *df* is greater than 28, use the formula:

$$r_\alpha = \frac{t_c}{\sqrt{t_c^2 + (n-2)}}$$

where:
   t_c is the corresponding critical value of *t* for (n – 2) degrees of freedom

APPENDIX D: TI-84 Plus INSTRUCTIONS

Instructions To Enter Data, Sort Data And Insert a New LIST.

TO ENTER DATA INTO A STAT LIST:
1) Press STAT to display EDIT menu shown in VIEW SCREEN 1.

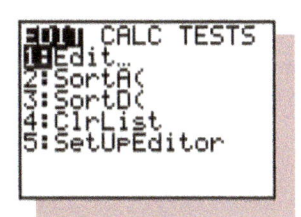

VIEW SCREEN 1

2) Press ENTER to select 1:EDIT
   This displays Lists L1, L2, and L3 shown in VIEW SCREEN 2.

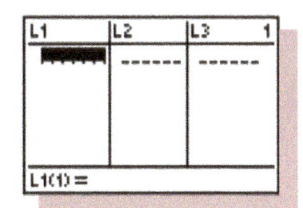

VIEW SCREEN 2

ENTERING DATA INTO L1:
1) Move cursor into List L1
2) Key data value & press ENTER
   Repeat until all data are entered

EXAMPLE:
Enter: 19, 16, 48, 42, 33 into L1. Notice the highlighted value of L1 is displayed on the bottom of the screen. Observe in VIEW SCREEN 3, the data value 33 is represented as L1(5) = 33.

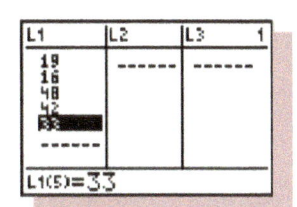

VIEW SCREEN 3

SORTING A LIST IN ASCENDING ORDER:
1) Press STAT to display EDIT menu shown in VIEW SCREEN 4.

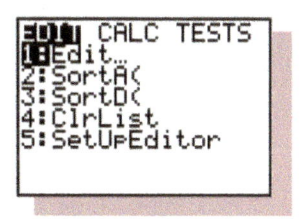

VIEW SCREEN 4

2) Select 2 for 2:SORTA(
   SORTA(appears on home screen
3) Press 2nd & 1 for List L1
4) Press ENTER & DONE appears on home screen indicating L1 has been sorted in ascending order.

To VIEW List L1:
1) Press STAT & ENTER to select 1:EDIT. List L1 appears in ascending order shown in VIEW SCREEN 5.

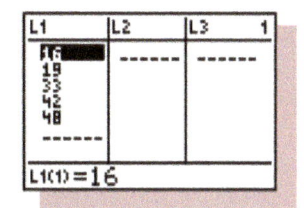

VIEW SCREEN 5

SORTING A LIST IN DESCENDING ORDER:
1) Press STAT
2) Select 3 for 3:SORTD(to sort data in descending order. SORTD(appears on home screen
3) Press 2nd & 1 for List L1
4) Press ENTER & DONE appears indicating L1 has been sorted in descending order.

To VIEW List L1:
Press STAT & ENTER to select 1:EDIT. List L1 appears in descending order shown in VIEW SCREEN 6.

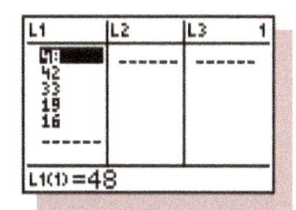

VIEW SCREEN 6

TO EDIT A VALUE WITHIN A LIST:
1) Press STAT & ENTER to select 1:EDIT
2) Highlight data value within List to be changed.
3) Key new value over old value and press ENTER

DELETING A VALUE FROM A LIST:
1) Highlight data value
2) Press DEL NOTE: CLEAR KEY WILL NOT DELETE A DATA VALUE!

TO CREATE A NEW LIST NAMED AVG:
EXAMPLE: To create a new List called *AVG between L1 and L2*:
1) Press STAT & ENTER to select 1:EDIT
2) Use up cursor ▲ to highlight List L2
3) Press 2nd & INS (DEL key) keys
A blank List will be displayed between L1 and L2 and at the bottom of the screen the NAME= prompt appears shown in VIEW SCREEN 7.

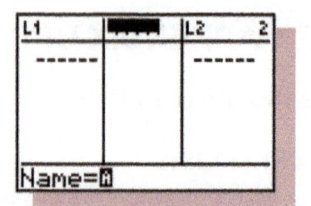

VIEW SCREEN 7

4) Input the name AVG
(Reminder: the first character of a List name must be a letter and cannot be more than 5 characters long)
5) Press ENTER
A new List named AVG will appear between Lists L1 and L2 as shown in VIEW SCREEN 8.

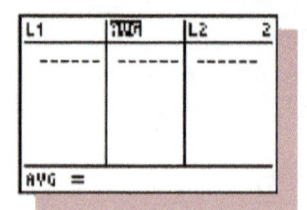

VIEW SCREEN 8

USING THE TI-84 Plus TO PLOT A HISTOGRAM

GRAPHS FOR ONE VARIABLE STATS:
I: HISTOGRAMS

TO CONSTRUCT A HISTOGRAM:

1) ENTER data values in L1
2) Press WINDOW to display the **WINDOW** variable values.
   VIEW SCREEN 9 shows the default values for: Xmin, Xmax, Xscl, Ymin, Ymax, Yscl, Xres.

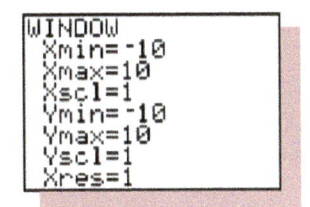

VIEW SCREEN 9

3) Set values for each window variable using the following guide for a **HISTOGRAM**.

   For a HISTOGRAM:
   set XMIN = smallest data value
   set XMAX > largest data value
   set XSCL = class width
   set YMIN = 0
   set YMAX > greatest class freq.
   set YSCL = 1 OR appropriate frequency scale
   set Xres = 1

4) Press 2nd & STAT PLOT (Y = key)
   The **STAT PLOTS Menu** is displayed with 1:PLOT1 . . . highlighted.

5) Press ENTER to display **stat plot editor** screen for Plot 1. This is shown in VIEW SCREEN 10.

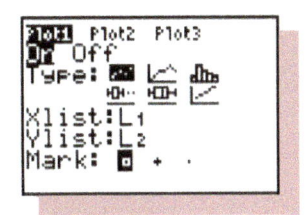

VIEW SCREEN 10

6) Highlight ON & press ENTER
7) Highlight **HISTOGRAM:** 3rd graph to right of TYPE and press ENTER
8) For Xlist: key in the name of data List. For this example, key in L1 and press ENTER
9) For Freq: press ENTER for 1

The previous inputs to generate a Histogram are shown in **VIEW SCREEN 11.**

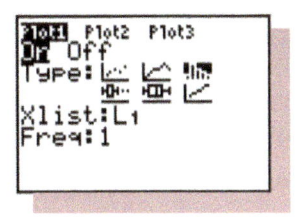

VIEW SCREEN 11

10) Press GRAPH to display the HISTOGRAM.

A HISTOGRAM for the data set in CHAPTER 2 PROBLEM #111 USING SIX CLASSES is shown in VIEW SCREEN 12 using the following WINDOW variable settings:
   Xmin = 40
   Xmax = 101
   Xscl = 10
   Ymin = 0
   Ymax = 12
   Yscl = 1
   Xres = 1

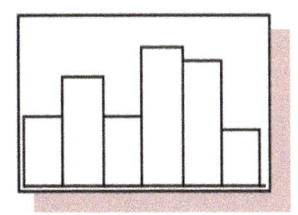

VIEW SCREEN 12

TO DISPLAY THE CLASS LIMITS AND CLASS FREQUENCIES OF A HISTOGRAM

1) Press TRACE : a blinking cursor appears at top center of first bar of the histogram. In lower left corner of the screen appear the quantities: Min, Max and n.
   These quantities represent:
   Min = lower class limit of bar highlighted by cursor
   Max < lower class limit of bar to the right of cursor
   n = frequency of bar highlighted by cursor

VIEW SCREEN 13 shows **TRACE information of 1st bar of HISTOGRAM** for data set in **CHAPTER 2: PROBLEM #111.**

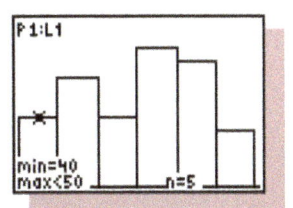

VIEW SCREEN 13

P1:L1 in the upper left corner identifies the HISTOGRAM as **STAT PLOT P1** where **L1** contains the data. Notice in the lower left screen are the values for the first bar: **Min = 40, Max < 50 & n = 5.** These values indicate that the *FIRST class* has a lower limit of 40 (**Mn = 40**) with a class frequency of 5 (**n = 5**) while the *SECOND class* has a lower limit of 50 (**Max < 50**).

2) Use cursor to display the class marks and frequencies of the other bars of the Histogram.

DISPLAYING CLASS BOUNDARIES

To display the class boundaries, you need to change only the Xmin value of the Window variables to the lower class boundary of the 1st class. For this problem, we ONLY change the Xmin value to 39.5. To accomplish this:

1) Press WINDOW
2) Key 39.5 (lower boundary of 1st class) for Xmin & press ENTER
3) Press GRAPH to display HISTOGRAM

4) Press TRACE to display the boundaries and frequency of the 1st class.

The values **Min = 39.5, Max < 49.5,** and **n = 5** for the 1st class are displayed in VIEW SCREEN 14.

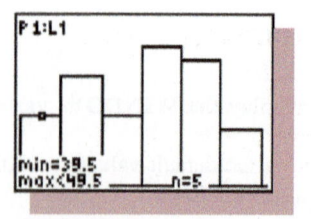

VIEW SCREEN 14

In VIEW SCREEN 14, the boundaries for the 1st class are 39.5 (**inclusive since Min = 39.5**) to 49.5 (**not inclusive since Max < 49.5**). Use right cursor to display the remaining class boundaries.

USING THE TI-84 Plus TO PLOT A FREQUENCY POLYGON

GRAPHS FOR ONE VARIABLE STATS:
II: FREQUENCY POLYGONS

TO CONSTRUCT A FREQUENCY POLYGON

1) class marks in L1
2) ENTER corresponding frequencies in L2

   REMEMBER TO INCLUDE AN ADDITIONAL CLASS MARK TO THE LEFT OF THE LOWEST CLASS AND TO THE RIGHT OF THE HIGHEST CLASS. THE FREQUENCY FOR EACH OF THESE ADDITIONAL CLASS MARKS IS ZERO!

3) Press WINDOW to set the **WINDOW** variable values using following guide for a Frequency Polygon.

   For a FREQUENCY POLYGON:
   set Xmin < smallest class mark
   set Xmax > largest class mark
   set Xscl = class width OR appropriate scale
   set Ymin = 0
   set Ymax > greatest class freq.
   set Yscl = 1 OR appropriate frequency scale
   set Xres = 1

4) Press 2nd & STAT PLOT (Y = key) **1:PLOT1. . . .** of the STAT PLOTS Menu is displayed

5) Press ENTER to display **stat plot editor** screen for Stat Plot 1 as shown in VIEW SCREEN 15.

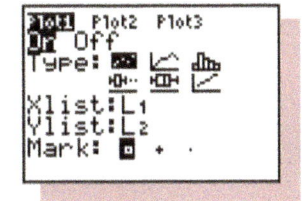

VIEW SCREEN 15

6) Select ON & press ENTER
7) Select **xyLine** graph: 2nd graph to right of TYPE and press ENTER. THIS GRAPH TYPE WILL PRODUCE A FREQUENCY POLYGON.
8) For Xlist: Input **L1** which contains the class marks and press ENTER. **THIS INDICATES THAT LIST L1 CONTAINS THE CLASS MARKS.**
9) For Ylist: Input **L2** which contains the frequencies and press ENTER. **THIS INDICATES THAT List L2 CONTAINS THE FREQUENCIES.**
10) Select ⁻ for MARK: and press ENTER

    The selections to generate a Frequency Polygon are shown in VIEW SCREEN 16.

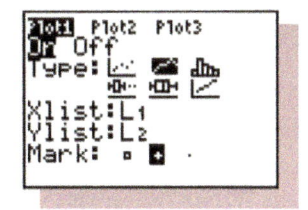

VIEW SCREEN 16

10) Press GRAPH to display FREQUENCY POLYGON. A FREQUENCY POLYGON for the data in Table 2.21 is shown in VIEW SCREEN 17 using the following WINDOW variable settings:

    Xmin = 2
    Xmax = 45
    Xscl = 5
    Ymin = 0
    Ymax = 40
    Yscl = 1
    Xres = 1

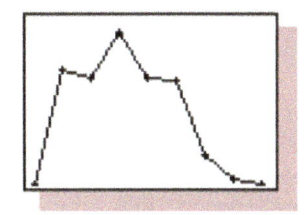

VIEW SCREEN 17

TO VERIFY THE CLASS MARKS AND THE CORRESPONDING FREQUENCIES

1) Press TRACE: a blinking cursor appears at the 1st class mark with its frequency as shown in VIEW SCREEN 18.

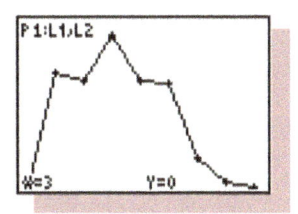

VIEW SCREEN 18

In the lower left corner of SCREEN 18 appears X = 3 & Y = 0.

FOR A FREQUENCY POLYGON, the X value is the class mark and the Y value is the class frequency. For this frequency polygon, the 1st class has a class mark of: X = 3 with a frequency of: Y = 0.

The **P1:L1,L2** in the upper left corner identifies the graph as Stat Plot 1 where L1 contains the class marks and L2 contains the frequencies.

2) Use cursor keys to identify the class marks and frequencies of the other classes.

USING THE TI-84 Plus TO PLOT AN OGIVE

GRAPHS FOR ONE VARIABLE STATS:
III: OGIVE

TO CONSTRUCT AN OGIVE:
1) ENTER class values in L1
2) ENTER cumulative freq into L2
3) Press WINDOW to set the WINDOW variable values using the following guide for an Ogive.

   For an OGIVE:
   set Xmin < smallest class value
   set Xmax > largest class value
   set Xscl = class width OR appropriate scale
   set Ymin = 0
   set Ymax > greatest freq
   set Yscl = 1 OR appropriate frequency scale
   set Xres = 1

4) Press 2nd & STAT PLOT (Y = key) 1:PLOT1 . . . of the STAT PLOTS Menu is displayed.
5) Press ENTER
6) Select ON & press ENTER
7) Select xyLine graph: 2nd graph to right of TYPE and press ENTER.
8) For Xlist: Input L1 which indicates that L1 contains the class values and press ENTER.
9) For Ylist: Input L2 which indicates that L2 contains the cumulative frequencies and press ENTER.
10) Select + for MARK: and press ENTER.

The selections to generate an ogive are shown in VIEW SCREEN 19.

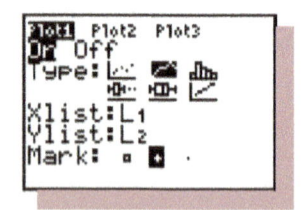

VIEW SCREEN 19

11) Press GRAPH to display OGIVE. An OGIVE for the data in Table 2.23 is shown in VIEW SCREEN 20 with the following WINDOW variable settings:

Xmin = 40
Xmax = 250
Xscl = 30
Ymin = 0
Ymax = 55
Yscl = 1
Xres = 1

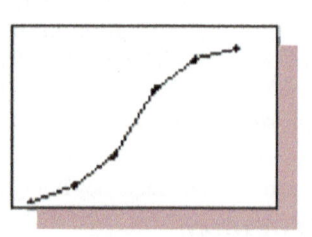

VIEW SCREEN 20

TO VERIFY THE CLASS VALUES AND THE CORRESPONDING CUMULATIVE FREQUENCIES

1) Press TRACE : a blinking cursor appears at the 1st class value with its cumulative frequency displayed in VIEW SCREEN 21.

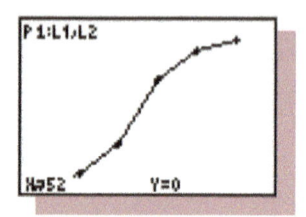

VIEW SCREEN 21
Notice in the lower left corner of VIEW SCREEN 21 appears X = 52 and Y = 0.

FOR AN OGIVE, the X value is the class value and the Y value is the cumulative frequency.
FOR THIS OGIVE, the 1st class has a class value of X = 52 with a cumulative frequency of Y = 0.
The P1:L1,L2 in the upper left corner identifies the graph as Stat Plot 1 where the class values are in List L1 and the cumulative frequencies are in List L2.

2) Use cursor key to identify the class values and cumulative frequencies of the other classes.

USING THE TI-84 Plus TO COMPUTE Σ EXPRESSIONS, MEAN, STANDARD DEVIATION AND 5–NUMBER SUMMARY

STAT ANALYSIS FOR 1-VARIABLE
1-VARIABLE STAT ANALYSIS THE MEAN, Σx^2, Σx, **POPULATION** and **SAMPLE STANDARD DEVIATIONS, n,** and the **5-Number Summary: MinX, Q_1, Median, Q_3 & MaxX**

1) Press STAT
2) Highlight CALC & 1:1-VAR STATS as shown in
3) VIEW SCREEN 22.

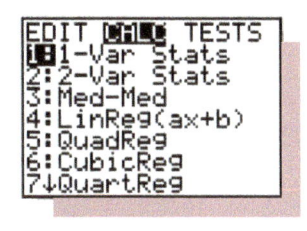

VIEW SCREEN 22

4) Press ENTER & **1-VAR STATS** will appear on the home screen
5) Input List name that contains the data values.
   For **L1:** Press 2nd and 1 as displayed in VIEW SCREEN 23.

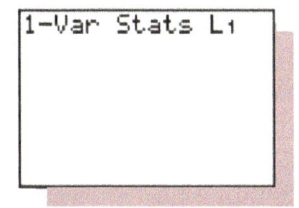

VIEW SCREEN 23

6) Press ENTER for 1-VAR STATS results

EXAMPLE:
The 1-VAR STATS results for the data set: *16, 19, 33, 42, 48* are displayed in VIEW SCREENs 24a & 24b.

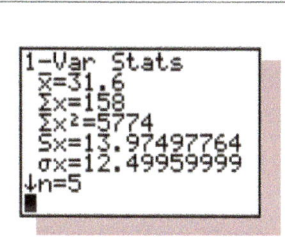

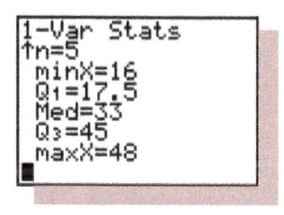

VIEW SCREEN 24a   VIEW SCREEN 24b

In VIEW SCREEN 24a, the **1-VAR STATS** results shown are:

mean: $\bar{x} = 31.6$
sum of data values: $\Sigma x = 158$
sum of squared values: $\Sigma x^2 = 5774$
sample std dev: $Sx = 13.97497764$
pop std dev: $\sigma x = 12.49959999$
number of data values: $n = 5$

In VIEW SCREEN 24a, the downward arrow appearing next to n = 5 indicates there are additional **1-VAR STATS** results. **To obtain the remaining results shown in VIEW SCREEN 24b, press the down cursor ▼ five times to display these results.** An upward arrow appears at the top of the VIEW SCREEN when you reached the final result of VIEW SCREEN 24b.

The results are:
number of data values: $n = 5$
minimum data value: $minX = 16$
1st quartile: $Q_1 = 17.5$
median: $med = 33$
3rd quartile: $Q_3 = 45$
maximum data value: $maxX = 48$

The last five results of VIEW SCREEN 24b:

MinX, Q_1, Median, Q_3 & MaxX are called the **5-Number Summary**.

USING THE TI-84 Plus TO COMPUTE POPULATION AND SAMPLE STANDARD DEVIATIONS

After computing the **1-VAR STATS** analysis, you can also retrieve the value of any one of these statistical variables: $\bar{x}$, Σx, Σx^2, Sx, σx, n, minX, Q_1, med, Q_3, and maxX using the VARS key.

One use of the VARS key is to compute the **variance of a sample and a population.**

PROCEDURE TO COMPUTE THE SAMPLE AND POPULATION VARIANCES

Before you can compute the VARIANCES OF A SAMPLE OR A POPULATION, you must first compute 1-VAR STATS on the List containing the sample or population data values. This step will store the appropriate values within the stat variables needed to compute the variance.

Computing the sample variance

1) Press VARS to display the VARS menu shown in VIEW SCREEN 25.

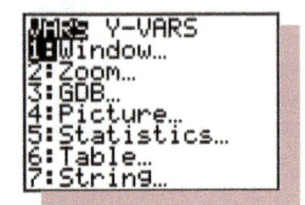

VIEW SCREEN 25

2) Press 5 for 5: STATISTICS to display the STATISTICS MENU shown in VIEW SCREEN 26.

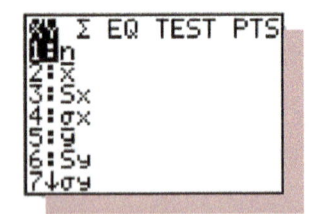

VIEW SCREEN 26

3) Press 3 to select 3: Sx The sample std deviation **Sx** appears on the home screen.

To compute the sample variance, square the sample deviation.

4) Press x^2 key and Enter

The sample variance Sx^2 is computed.

The sample variance for the data set: *16, 19, 33, 42, 48* is shown in VIEW SCREEN 27.

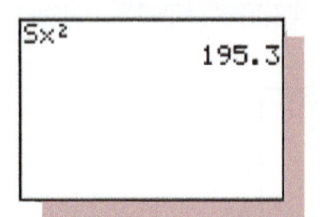

VIEW SCREEN 27

Computing the population variance

1) Press VARS
2) Press 5 for 5: STATISTICS to display the STATISTICS MENU shown in VIEW SCREEN 26.
3) Press 4 to select 4: σx

The population standard deviation **σx** appears on home screen.

To compute the population variance, square the population standard deviation.

4) Press x^2 key and Enter

The value of the population variance $σx^2$ is computed.

The population variance for the data set: *16, 19, 33, 42, 48* is shown in VIEW SCREEN 28.

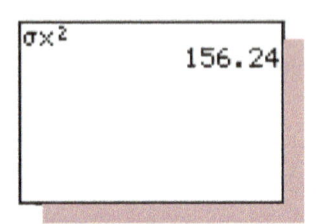

VIEW SCREEN 28

USING THE TI-84 Plus TO DETERMINE THE MODE

The TI-84 plus doesn't compute the mode directly for you. However you can use the sort feature to arrange the data values in numerical order and then scan the data set to identify the mode, if any.

Enter the following data into L1:

19, 18, 21, 20, 19, 20, 23, 21, 22, 18, 23,

21, 22, 23, 21, 19, 19, 21, 22, 23, 18

To sort L1:

1) Press STAT & Select 2 for 2: SortA(. SORTA(appears on home screen.
2) Press 2nd & 1 for List L1 containing data values to be sorted.

This is shown in VIEW SCREEN 29.

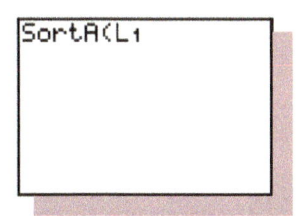

VIEW SCREEN 29

3) Press Enter & **Done** appears indicating List L1 has been sorted in ascending order.
4) Press STAT & Enter to display L1 in ascending order shown in VIEW SCREEN 30.

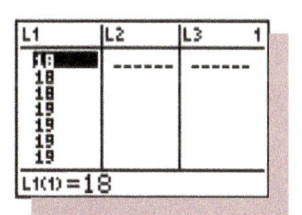

VIEW SCREEN 30

To determine the mode, scan L1 using the down cursor key to identify the most frequent data value. For this data set, the data value 21 is the mode since it has the greatest frequency.

USING THE TI-84 Plus TO CALCULATE 1-VAR STATS USING TWO LISTS

CALCULATING 1-VAR STATS USING TWO LISTS: L1 & L2 WHERE L1 CONTAINS THE DATA VALUES AND L2 HAS THE CORRESPONDING FREQUENCIES.

Enter the following data values into L1 and the corresponding frequencies into L2.

data values	frequency
90	6
87	4
83	2
79	5
76	6
73	6
69	9

1) Enter data values into L1.
2) Enter corresponding frequencies into L2. This is shown in VIEW SCREEN 31.

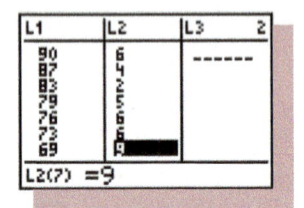

VIEW SCREEN 31

3) Press STAT
4) Highlight CALC & 1:1-VAR STATS
5) Press ENTER and **1-VAR STATS** appears on home screen
6) Key in **L1,L2**
where **L1** represents the List containing the data values followed by a comma (the key directly above the 7 key) and L2 which is the List containing the corresponding frequencies.
This is shown in VIEW SCREEN 32.

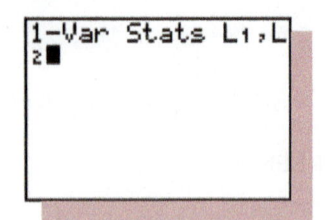

VIEW SCREEN 32

8) Press ENTER to display some 1-VAR STATS results shown in VIEW SCREEN 33.

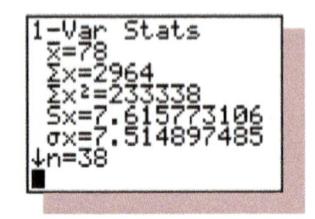

VIEW SCREEN 33

USING STAT LISTS TO COMPUTE THE SUM OF THE DEVIATIONS FROM THE MEAN: $\Sigma(x-\mu)$ or $\Sigma(x-\bar{x})$

TO COMPUTE THE SUM OF THE DEVIATIONS FROM THE MEAN:

$$\Sigma(x-\mu) \text{ or } \Sigma(x-\bar{x})$$

EXAMPLE:
Enter the data values into L1:
  6, 2, 4, 5, 3, 1, 7

1) Enter data values into L1
   To compute mean using 1-VAR STATS
2) Press STAT & Highlight CALC
3) Highlight 1:1-VAR STATS & ENTER
   1-VAR STATS appears on home screen
4) Input L1 for List name containing the data values and press ENTER

   Notice, the **mean = 4**.
   Now input the formula to compute the sum of the deviations from the mean. This is accomplished by the steps:
5) Press STAT & ENTER
6) Move cursor to L2
7) Press the up cursor ▲ key to highlight L2 as shown in VIEW SCREEN 34.

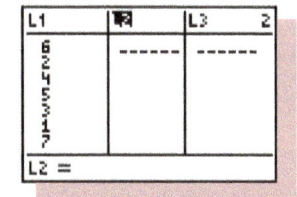

VIEW SCREEN 34
Notice at the bottom of the screen, **L2 =** is displayed. Once the name of a List is highlighted as shown in VIEW SCREEN 34, a formula can be inserted for the highlighted List.
We need to enter the formula for the deviations from the mean:

$$(x - \text{mean})$$

Instead of the letter x, we will input **L1** where the x or data values are listed followed by a minus then a 4 for the mean to obtain the expression:

$$L2 = (L1-4)$$

The exact key strokes to obtain this formula as shown at the bottom of VIEW SCREEN 35 are:

Press (key
Press 2nd & 1 to obtain **L1**
Press − (subtraction key)
Press 4 for the mean
Press) key
Press ENTER key

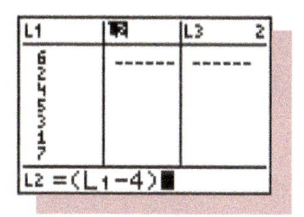

VIEW SCREEN 35
The deviations from the mean are now displayed in L2 shown in VIEW SCREEN 36.

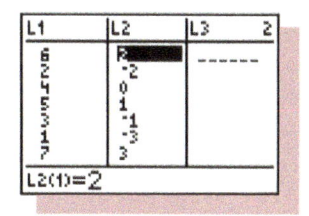

VIEW SCREEN 36
To compute the sum of the deviations in L2:

1) Press 2nd & QUIT (MODE key)
2) Press 2nd & List (STAT key)
3) Highlight MATH.
   This is shown in VIEW SCREEN 37.

VIEW SCREEN 37
4) Press 5 for 5: SUM
5) Input L2: the name of List to be summed.
6) Press ENTER to obtain the sum of L2.
   Notice, the **SUM** of **L2** is **zero** as shown in VIEW SCREEN 38 which represents the sum of the deviations from the mean.

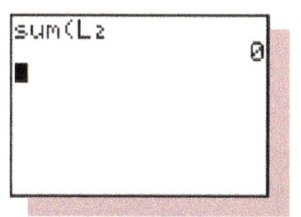

VIEW SCREEN 38

USING THE TI-84 Plus TO CONSTRUCT BOX PLOTS AND TRACING THE 5-NUMBER SUMMARY

GRAPHS FOR ONE VARIABLE STATS:

I. BOXPLOTS

To create a box plot:
1) ENTER data values into L1
2) Press WINDOW to set the WINDOW variable values using the following guide for a Box Plot.

For a BOX PLOT:

set Xmin < smallest data value
set Xmax > largest data value
The values for Xscl, Ymin, Ymax, Yscl and Xres are not used to construct a Boxplot so you can disregard these variables.

3) Press 2nd & STAT PLOT (Y = key) 1:PLOT1 ... of the STAT PLOTS Menu is displayed.
4) Press ENTER to display **stat plot editor** screen for Stat Plot 1 as shown in VIEW SCREEN 39.

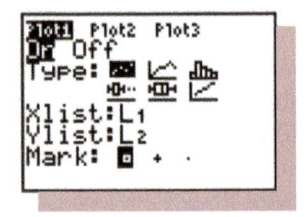

VIEW SCREEN 39

5) Select ON & press ENTER
6) For TYPE : select 5th graph: BOXPLOT to right of TYPE and press ENTER
7) Input **L1**: List name containing the data values and press ENTER
8) Press ENTER to select **1** for Freq:
   The selections to generate a BOXPLOT are shown in VIEW SCREEN 40.

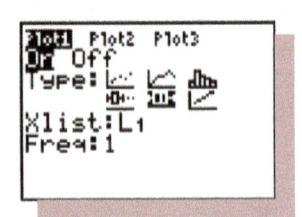

VIEW SCREEN 40

9) Press GRAPH
   BOXPLOT appears on display.
   A BOXPLOT for the data set:
   0, 10, 10, 20, 40, 54, 55,
   60, 65, 66, 66, 66, 67, 67,
   68, 68, 69, 70, 71, 71, 72,
   73, 75, 88, 96, 100
   is shown in VIEW SCREEN 41 using the following WINDOW variable settings:
   Xmin = 0
   Xmax = 105
   Xscl = 1
   Ymin = 0
   Ymax = 10
   Yscl = 1
   Xres = 1

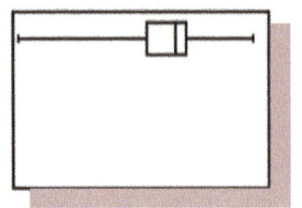

VIEW SCREEN 41

Tracing the 5-NUMBER SUMMARY

The 5-number summary values: minX, Q_1, Med, Q_3, & maxX can be traced from the BOXPLOT:

1) Press TRACE: a blinking cursor appears on the BOX-PLOT.
2) Use cursor keys to trace the values of the 5-Number Summary:
   minX, Q_1, Med, Q_3, & maxX on the BOXPLOT. Each value is displayed in the lower left corner of the screen.
   In the upper left corner of the screen appears **P1:L1** which indicates this BOXPLOT graph is STAT PLOT 1 and the data values are in LIST L1.
   VIEW SCREEN 42 displays the BOXPLOT with the TRACE positioned at the MEDIAN value: Med = 67.
   Verify the 5-Number Summary is:
   minX = 0, Q_1 = 55, Med = 67, Q_3 = 71, & maxX = 100.

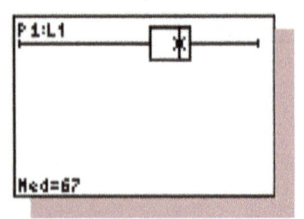

VIEW SCREEN 42

USING THE MODBOXPLOT TO DETECT OUTLIERS

GRAPHS FOR ONE VARIABLE STATS:

II. MODIFIED BOXPLOTS

A modified BOXPLOT (or ModBoxplot) graphs all the data values of a List that are within *1.5*IQR of the first (Q1) and third quartile (Q3)* like a typical BOXPLOT. Any data value **BEYOND** these **FENCES** is plotted individually using one of the Marks: □ **or + or *** you select and is defined as an **OUTLIER**. These **outliers** can be identified using the TRACE key and are prompted by: **x =**.

Using the DATA SET below, let's construct a ModBoxplot.

To Create a ModBoxplot:

1) ENTER the data into L1.
   Use the following data set:
   **0, 10, 10, 20, 40, 54, 55, 60,
   65, 66, 66, 66, 67, 67, 68, 68,
   69, 70, 71, 71, 72, 73, 75, 88,
   96, 100**

2) Press WINDOW to set the **WINDOW** variable values using the following guide for ModBoxplot.

For a BOXPLOT or MODBOXPLOT:
set Xmin < smallest data value
set Xmax > largest data value
The values for Xscl, Ymin, Ymax, Yscl and Xres are not used to construct a ModBoxplot so you can disregard these variables.

3) Press 2nd & STAT PLOT (Y = key) **1:PLOT1** . . . of the **STAT PLOTS Menu** is displayed.

4) Press ENTER to display the **stat plot editor** screen for **Stat Plot 1**.

5) Select ON & press ENTER

6) For TYPE : select 4th graph MODBOXPLOT to right of TYPE and press ENTER

7) For Xlist : Input **L1**, the List containing the data values, and press ENTER

8) For Freq : press ENTER to select **1** for Frequency

9) For Mark: select one symbol: □ **or + or *** to use to identify any outliers. For this example, select □ for the mark.

The selections to generate a MODBOXPLOT are shown in VIEW SCREEN 43.

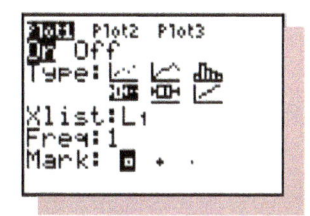

VIEW SCREEN 43

A MODBOXPLOT for the data set in L1 is shown in VIEW SCREEN 44 with the following WINDOW variable settings:

Xmin = 0
Xmax = 105
Xscl = 1
Ymin = 0
Ymax = 10
Yscl = 1
Xres = 1

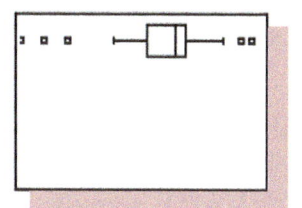

VIEW SCREEN 44

Notice in VIEW SCREEN 44 there are 5 outliers marked by 5 □ (boxes) beyond the whiskers of the MODBOXPLOT.

IDENTIFYING THE OUTLIERS

1) Press TRACE: a blinking cursor appears on MODBOXPLOT.

2) Use cursor keys to identify the **5 number summary** values: **minX, Q1, Med, Q3, maxX and the outliers**.

Notice in VIEW SCREEN 45, the value x = 20 is represented by a **box** □ beyond the left whisker indicating that 20 is an outlier.

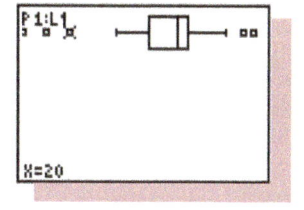

VIEW SCREEN 45

Notice if you move the cursor to the left you should verify that there are additional outliers. These outlier values are: x = 10(twice) and x = 0 (shown as minX = 0). Moving the cursor beyond the right whisker, you should verify that the value x = 96 is also an outlier. This is shown in VIEW SCREEN 46.

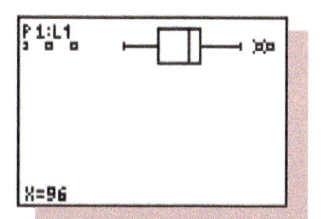

VIEW SCREEN 46

Move the cursor to the last box □ on the right to verify that x = 100 (shown as maxX = 100) is an outlier.

PLOTTING MORE THAN ONE BOXPLOT

The TI-84 Plus can plot 3 BOXPLOTS at once using the three STAT PLOTS. When 3 boxplots are displayed at once, PLOT 1 displays the boxplot on the top of the screen while the boxplot of PLOT 2 is shown in the middle and the boxplot of PLOT 3 is displayed at the bottom.

1) ENTER the data into L1.
   Use the following data set:
   0, 10, 10, 20, 40, 54, 55, 60,
   65, 66, 66, 66, 67, 67, 68, 68,
   69, 70, 71, 71, 72, 73, 75, 88,
   96, 100
2) Press WINDOW to set the WINDOW variable values using the following guide for BoxPlot and ModBoxplot.

For a BOXPLOT or MODBOXPLOT:
set Xmin < smallest data value
set Xmax > largest data value
Remember, the values for Xscl, Ymin, Ymax, Yscl and Xres are not used to construct a ModBoxplot so you can disregard these variables.

3) Press 2nd & STAT PLOT (Y = key) 1:PLOT1 . . . of the STAT PLOTS Menu is displayed.
4) Press ENTER to display stat plot editor screen.
   Input the following to set up the **stat plot editor** screen for **Stat Plot 1**.
5) Select ON & press ENTER
6) For TYPE : select 4th graph: MODBOXPLOT to right of TYPE and press ENTER
7) Input L1 and press ENTER
8) Press ENTER to select 1 for Freq:
9) For Mark : select □ for the mark.

The STAT PLOT 1 selections to generate a MODBOXPLOT are shown in VIEW SCREEN 47.

VIEW SCREEN 47

Now we will use STAT PLOT 2 and the same data in L1 to graph a BOXPLOT by performing the following steps:

10) Press 2nd & STAT PLOT (Y = key) 1:PLOT1 . . . of the STAT PLOTS Menu is displayed.
11) Press 2 & ENTER to display the **stat plot editor** screen for **STAT PLOT 2**.
12) Select ON & press ENTER
13) For TYPE: highlight 5th graph to right of TYPE for BOXPLOT and press ENTER
14) For Xlist: Input L1 for List name and press ENTER
15) For Freq: press ENTER to select 1 for Frequency
    VIEW SCREEN 48 shows the Stat Plot 2 selections to generate a BOXPLOT using List L1.

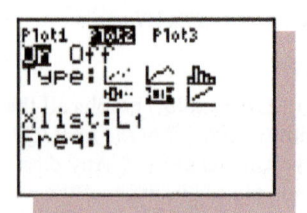

VIEW SCREEN 48

16) Press Graph to display two BOXPLOTS as shown in VIEW SCREEN 49.

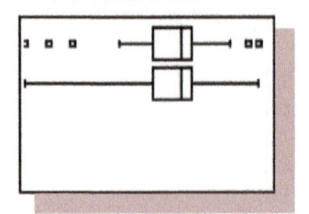

VIEW SCREEN 49
Notice MODBOXPLOT is on the top while the BOXPLOT is below.

17) Press TRACE to verify the MODBOXPLOT is STAT PLOT 1.
    VIEW SCREEN 50 shows STAT PLOT 1 as the MODBOXPLOT with Med = 67 highlighted by a blinking cursor.

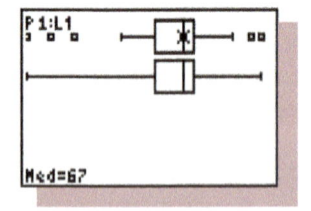

VIEW SCREEN 50
The **P1:L1** in the upper left corner indicates MODBOXPLOT as STAT PLOT 1 using LIST L1.

18) Press down cursor ▼ to highlight the BOXPLOT of STAT PLOT 2 with a blinking cursor at the Med = 67. This is shown in VIEW SCREEN 51.

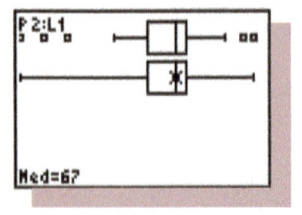

VIEW SCREEN 51

The **P2:L1** in the upper left corner identifies the BOXPLOT as STAT PLOT 2 using List L1.

Verify that the values for Q_1, Med, and Q_3 are the same for both BOXPLOTS. Explain why this is true and why the values at the end of the whiskers for both boxplots are not the same.

PROCEDURE TO COMPUTE BINOMIAL PROBABILITIES

There are two TI-84 Plus functions to compute binomial probabilities. They are: **binompdf(** and **binomcdf(**. We will first illustrate **binompdf(**. binompdf(n,p,s) computes the binomial probability for a specified success value, **s**, for **n** trials & prob of a success, **p**.

To illustrate binompdf(n,p,s) consider **EXAMPLE 6.18**.

For Example 6.18(a): we use: n = 5, p = 1/4 & s = 3 to compute P(3 correct answers) as follows:

1) Press **2nd** & **VARS** for **DISTR** to display the Distribution Menu shown in VIEW SCREEN 52.

VIEW SCREEN 52

2) Select **0** for **0:binompdf(**
    (0:binompdf(appears on the home screen shown in VIEW SCREEN 53.

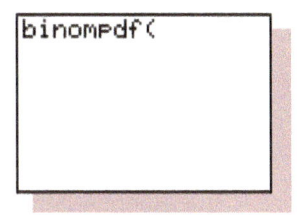

VIEW SCREEN 53

3) Input **5, 1/4, 3** for n, p, s to obtain binompdf (5, 1/4, 3) as shown in VIEW SCREEN 54.

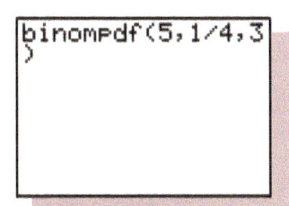

VIEW SCREEN 54

4) Press **ENTER** to obtain the result shown in VIEW SCREEN 55. Rounding off to 4 decimal places,
    P(3 correct answers) ≈ .0879

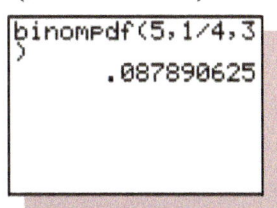

VIEW SCREEN 55

For Example 6.18(b): use: n = 5, p = 1/4 & s = 5 to compute P(5 correct answers) as follows:

1) Press **2nd** & **VARS** for **DISTR**
2) Select **0** for **0:binompdf(**
3) Input **5, 1/4, 5** for n,p,s to obtain binompdf(5, 1/4, 5) as shown in VIEW SCREEN 56.

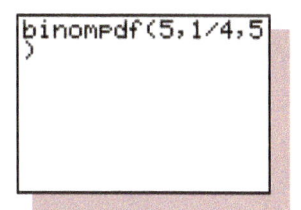

VIEW SCREEN 56

4) Press **ENTER** to obtain the result shown in VIEW SCREEN 57. Rounding off to 4 decimal places,
    P(5 correct answers) ≈ .0010

VIEW SCREEN 57

For Example 6.18(c): we need to compute: P(at most 2 correct). Remember P(at most 2 correct) equals the **sum** of the binomial probabilities for **s = 0 to s = 2**.
P(at most 2) = P(s = 0)+P(s = 1)+ P(s = 2)

The function: binomcdf (n, p, s) computes a **cumulative** probability at a specified "s" value. That is **binomcdf(n, p, s)** = P(at most s). So, **binomcdf (5, 1/4, 2)** = P(at most 2).

To compute this result:

1) Press **2nd** & **VARS** for **DISTR**
2) Press **ALPHA** & **A** (MATH key) for **A:binomcdf(**
3) Input **5, 1/4, 2** for n, p, s to obtain binomcdf(5, 1/4, 2) as shown in VIEW SCREEN 58.

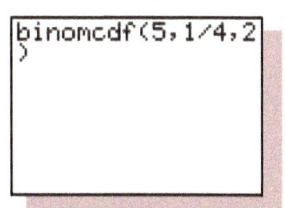

VIEW SCREEN 58

5) Press **ENTER** to obtain the result shown in VIEW SCREEN 59. Rounding off to 4 decimal places,
    P(at most 2 correct) ≈ .8965

Appendix D TI-84 Plus Instructions

```
binomcdf(5,1/4,2
)
           .896484375
```

VIEW SCREEN 59

```
binompdf(5,1/4,4
)+binompdf(5,1/4
,5)
              .015625
```

VIEW SCREEN 60

For Example 6.18(d): use binompdf(twice to compute the:

P(at least 4 correct) as shown in **VIEW SCREEN 60** to verify:

P(at least 4 correct) ≈ **.0156.**

PROCEDURE TO COMPUTE PROBABILITIES FOR THE NORMAL DISTRIBUTION

The TI-84Plus function normalcdf(can be used to compute normal distribution probabilities between two z scores.

normalcdf(z_1,z_2) computes the probability between the z scores, z_1 and z_2, where z_1 represents the lowerbound and z_2 represents the upperbound.

To illustrate normalcdf(consider **EXAMPLE 7.3**.

For Example 7.3(a): to find the area between z = 0 & z = 1.5, then z = 0 is the lowerbound of the area and z = 1.5 is the upperbound.

To compute this area:

1) Press 2nd & VARS for DISTR to display the Distribution Menu shown in VIEW SCREEN 61.

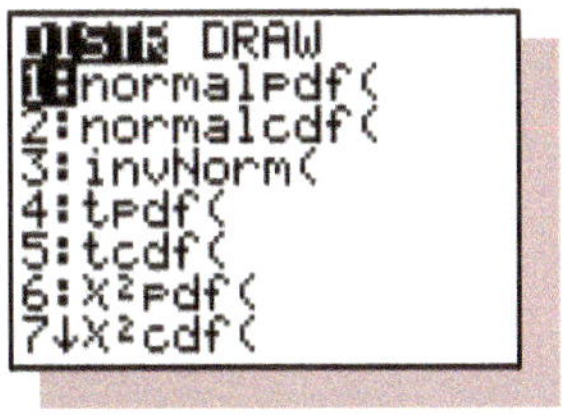

VIEW SCREEN 61v

2) Select **2** for 2:normalcdf

(2:normalcdf(appears on the home screen shown in VIEW SCREEN 62.

VIEW SCREEN 62

3) Input 0,1.5 for z_1,z_2 to obtain normalcdf(0,1.5) as shown in VIEW SCREEN 63.

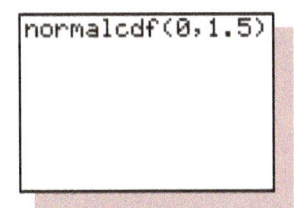

VIEW SCREEN 63

4) Press ENTER to obtain the result shown in VIEW SCREEN 64. Rounding off to 4 decimal places, **area between z = 0 & z = 1.5 ≈ .4332**

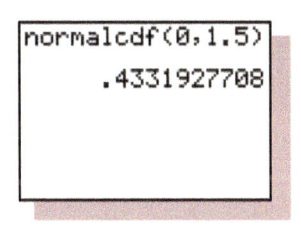

VIEW SCREEN 64

Consider EXAMPLE 7.1(a): we need to compute the area to the left of z = –0.53. Whenever we want to compute the area to the LEFT of a z score we will use –4 as the z score representing the LOWER bound.

For area to the left of -0.53:

1) Press 2nd & VARS for DISTR
2) Select **2** for 2:normalcdf(
3) Input -4,-.53 for z_1,z_2 to obtain normalcdf(-4,-.53) as shown in VIEW SCREEN 65.

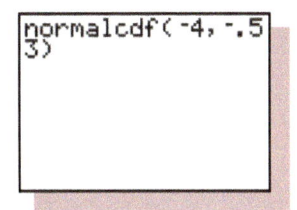

VIEW SCREEN 65

4) Press ENTER to obtain the result shown in VIEW SCREEN 66. Rounding off to 4 decimal places, **area to LEFT of z = –.53 ≈ .2980**

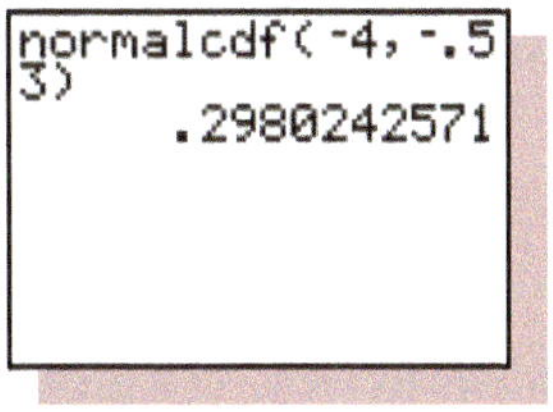

VIEW SCREEN 66

Notice this result is slightly different from the textbook. This difference is due to the roundoff errors associated with Table II which was used to compute the textbook answer.

Now consider EXAMPLE 7.2(a): we need to compute the area to the right of z = –1.37. To compute the area to the RIGHT of a z score we will use 4 as the z score representing the UPPER bound.

For area to the right of -1.37:
1) Press 2nd & VARS for DISTR
2) Select **2** for 2:normalcdf(
3) Input -1.37,4 for $z_1 z_2$ to obtain normalcdf(-1.37,4) as shown in VIEW SCREEN 67.

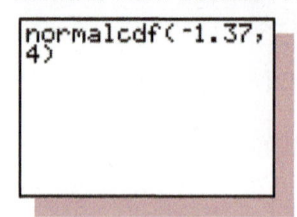

```
normalcdf(-1.37,
4)
```

VIEW SCREEN 67

4) Press ENTER to obtain the result shown in VIEW SCREEN 68. Rounding off to 4 decimal places, **area to RIGHT of z = –1.37 ≈ .9146**

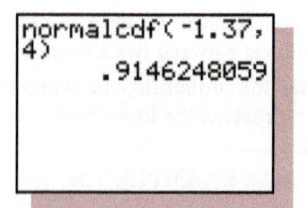

```
normalcdf(-1.37,
4)
       .9146248059
```

VIEW SCREEN 68

Again you will notice a slight discrepancy with the textbook result due to roundoff errors of Table II.

PROCEDURE TO DETERMINE THE Z SCORE GIVEN THE AREA TO THE LEFT OF A NORMAL DISTRIBUTION

The TI-84Plus function **invNorm(** determines the z score associated with a specified area to the left of the z score **invNorm(area to left as decimal)** computes the z score associated given the **area to left** as a **DECIMAL**.

To illustrate **invNorm(** consider **EXAMPLE 7.5**.

For Example 7.5(a), we need to determine the z score that cuts off the lowest 20% of area. **To determine this z score:**

1) Press **2nd** & **VARS** for **DISTR** to display the Distribution Menu shown in VIEW SCREEN 69.

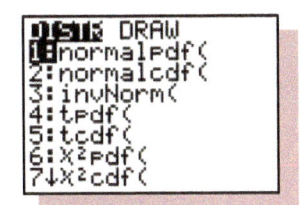

```
DISTR DRAW
1:normalpdf(
2:normalcdf(
3:invNorm(
4:tpdf(
5:tcdf(
6:X²pdf(
7↓X²cdf(
```

VIEW SCREEN 69

2) Select 3 for **3:invNorm(**

3:invNorm(appears on the home screen shown in VIEW SCREEN 70.

```
invNorm(
```

VIEW SCREEN 70

3) Input **.20** which represents 20% as a decimal to get **invNorm(.20)** as shown in VIEW SCREEN 71.

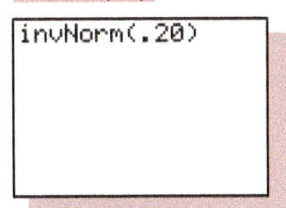

```
invNorm(.20)
```

VIEW SCREEN 71

4) Press **ENTER** to obtain the result shown in VIEW SCREEN 72. Rounding off to 2 decimal places, **the z score that cuts off the lowest 20% of area is z ≈ -.84**.

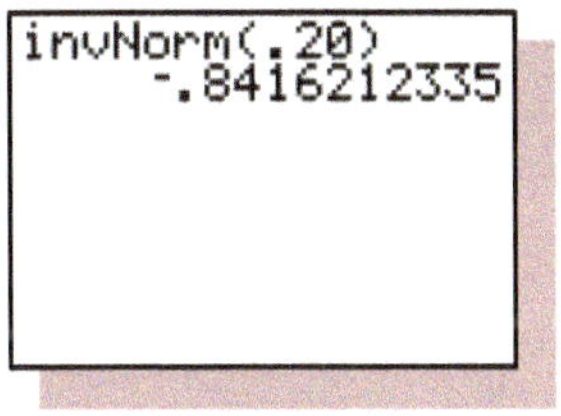

```
invNorm(.20)
       -.8416212335
```

VIEW SCREEN 72

For Example 7.5(b), we need to determine the z score that cuts off the highest 10% of area.

REMEMBER: to use the **invNorm(** function we need to input the area to the **LEFT** of the z score. The area to the left is 90% or 0.90. **So we need to determine the z score that cuts off the lowest 90%. To determine this z score:**

1) Press **2nd** & **VARS** for **DISTR**
2) Select 3 for **3:invNorm(**
3) Input **.90** which represents 90% as a decimal to get
4) **invNorm(.90)** as shown in VIEW SCREEN 73.

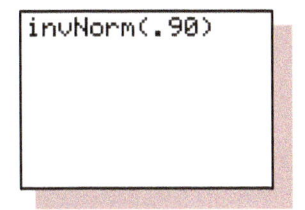

```
invNorm(.90)
```

VIEW SCREEN 73

5) Press **ENTER** to obtain the result shown in VIEW SCREEN 74. Rounding off to 2 decimal places, **the z score that cuts off the highest 10% of area is z ≈ 1.28**.

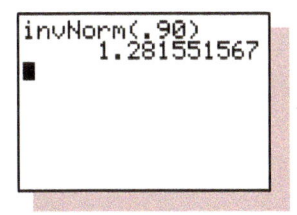

```
invNorm(.90)
       1.281551567
```

VIEW SCREEN 74

For Example 7.5(c), we need to determine the z scores that cutoff the middle 90% of the area. The middle 90% is between two z scores. The smaller z score has 5% of the area to its left. **To determine this z score:**

1) Press 2nd & VARS for DISTR
2) Select 3 for 3:invNorm(
3) Input .05 which represents the area to the left of the lower z score to obtain invNorm(.05) as shown in **VIEW SCREEN 75**.

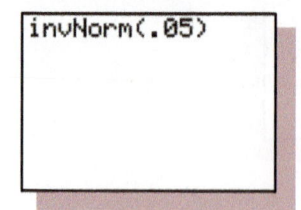

```
invNorm(.05)
```

VIEW SCREEN 75

4) Press ENTER to obtain the result shown in VIEW SCREEN 76. Rounding off to 2 decimal places, **the z score that cuts off the lowest 5% of area is z ≈ -1.64.**

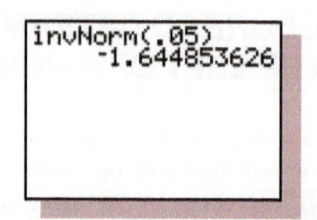

```
invNorm(.05)
        -1.644853626
```

VIEW SCREEN 76

Using the symmetry of the normal curve, the z scores that cutoff the middle 90% are:

z ≈ -1.64 and z ≈ 1.64.

Again you will notice a slight discrepancy with the textbook result due to roundoff errors of Table II.

Appendix D   TI-84 Plus Instructions   **A83**

PROCEDURE TO CONSTRUCT A CONFIDENCE INTERVAL FOR A POPULATION MEAN USING NORMAL DISTRIBUTION WITH STATS INPUT

Confidence Interval For Pop Mean Where Pop Std Dev is KNOWN and Sample Size is greater than 30

1) Press STAT & HIGHLIGHT TESTS
2) Select 7 for **7:ZInterval** . . . input screen for Z confidence interval for a population mean is displayed in VIEW SCREEN 77.

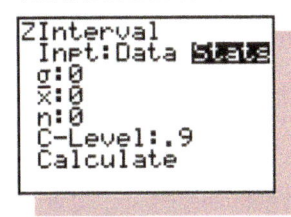

```
ZInterval
  Inpt:Data Stats
  σ:0
  x:0
  n:0
  C-Level:.9
  Calculate
```

VIEW SCREEN 77

3) INPUT FOR A ZInterval USING SAMPLE RESULTS:
   a) at **Inpt:**highlight Stats & ENTER
   b) at σ:input **pop std dev** & ENTER
   c) at $\bar{x}$:input **sample mean** & ENTER
   d) at n:input **sample size** & ENTER
   e) at C-Level:input **confidence level (as a decimal)** & ENTER
   f) highlight Calculate & ENTER to determine confidence interval

TO CONSTRUCT A 95% CONFIDENCE INTERVAL FOR THE POP MEAN WHERE:
σ = 18; $\bar{x}$ = 114 & n = 36.

The ZINTERVAL INPUT is:
   a) **Highlight** STATS & ENTER
   b) for σ: INPUT **18** & ENTER
   c) for $\bar{x}$: INPUT **114** & ENTER
   d) for n: INPUT **36** & ENTER
   e) for **C-Level:** INPUT **.95** & ENTER The **ZInterval** input screen should contain the info shown in VIEW SCREEN 78.

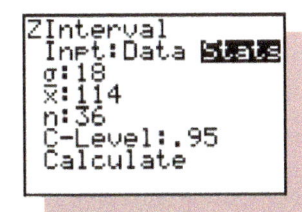

```
ZInterval
  Inpt:Data Stats
  σ:18
  x:114
  n:36
  C-Level:.95
  Calculate
```

VIEW SCREEN 78
   f) Highlight Calculate & ENTER
      The 95% ZInterval results are shown in VIEW SCREEN 79

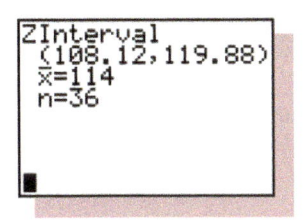

```
ZInterval
  (108.12,119.88)
  x=114
  n=36
```

VIEW SCREEN 79

The ZInterval results are:

(108.12,119.88) *is the 95% confidence interval for the population mean*

$\bar{x}$ = **114** is the sample mean

n = **36** is the sample size

TO CONSTRUCT A 99% CONFIDENCE INTERVAL FOR THE POP MEAN WHERE:
   σ = 18; $\bar{x}$ = 114 & n = 36.

1) Press STAT & HIGHLIGHT TESTS
2) Select 7 for **7:ZInterval** . . .

The ZINTERVAL INPUT is:
   a) **Highlight** STATS & ENTER
   b) for σ: INPUT **18** & ENTER
   c) for $\bar{x}$: INPUT **114** & ENTER
   d) for n: INPUT **36** & ENTER
   e) for **C-Level:** INPUT **.99** & ENTER
   f) Highlight Calculate & ENTER
      The 99% ZInterval results are shown in VIEW SCREEN 80

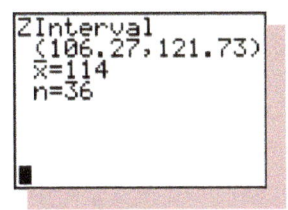

```
ZInterval
  (106.27,121.73)
  x=114
  n=36
```

VIEW SCREEN 80

The ZInterval results are:
ZInterval: (106.27,121.73) is the 99% confidence interval for the population mean

$\bar{x}$ = **114** *is the sample mean*

n = **36** *is the sample size*

Appendix D TI-84 Plus Instructions

PROCEDURE TO CONSTRUCT A CONFIDENCE INTERVAL FOR A POPULATION MEAN USING NORMAL DISTRIBUTION WITH DATA INPUT

Confidence Interval For Pop Mean Where Pop Std Dev is KNOWN and Sample Size is greater than 30

1) Press STAT & HIGHLIGHT TESTS
2) Select 7 for **7:ZInterval** . . . input screen for Z confidence interval for a population mean is displayed in VIEW SCREEN 81.

VIEW SCREEN 81

3) INPUT FOR A ZInterval USING SAMPLE DATA:
   a) at **Inpt:** highlight Data & ENTER
   b) at σ: input **pop std dev** & ENTER
   c) at **List:** input List name containing **sample data** & ENTER
   d) at **Freq:** input 1 or name of list containing frequencies & ENTER
   e) at **C-Level:** input **confidence level (as a decimal)** & ENTER
   f) highlight Calculate & ENTER to determine confidence interval

TO CONSTRUCT A 95% CONFIDENCE INTERVAL FOR THE POP MEAN WHERE σ = 9 AND THE SAMPLE DATA IS ENTERED INTO LIST L1.

SAMPLE DATA:
70 68 91 87 73 90 66 79 80 81
79 65 94 69 81 77 78 92 81 74
79 96 78 85 89 76 82 86 79 77 91
70 92 79 85 79 86 83 76 84

The **ZINTERVAL INPUT** is:
a) Highlight DATA & ENTER
b) for σ: INPUT 9 & ENTER
c) for **List:** INPUT L1 & ENTER
d) for **Freq:** INPUT 1 & ENTER
e) for **C-Level:** INPUT .95 & ENTER The **ZInterval** input screen should contain the info shown in VIEW SCREEN 82.

VIEW SCREEN 82

Highlight Calculate & ENTER
The 95% ZInterval results are shown in VIEW SCREEN 83

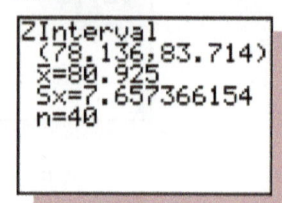

VIEW SCREEN 83
The ZInterval results are:

(78.136,83.714) is the 95% confidence interval for the **population mean**

$\bar{x}$ = **80.925** is the **sample mean**

Sx = **7.657366154** is the **sample standard deviation**

n = **40** is the **sample size**

TO CONSTRUCT A 99% CONFIDENCE INTERVAL FOR THE POPULATION MEAN USING THE SAME SAMPLE DATA IN LIST L1, PERFORM THE STEPS.

1) Press STAT & HIGHLIGHT TESTS
2) Select 7 for **7:ZInterval** . . .

The **ZINTERVAL INPUT** is:
a) Highlight DATA & ENTER
b) for σ: INPUT 9 & ENTER
c) for **List:** INPUT L1 & ENTER
d) for **Freq:** INPUT 1 & ENTER
e) for **C-Level:** INPUT .99 & ENTER
f) Highlight Calculate & ENTER
The 99% ZInterval results are shown in VIEW SCREEN 84

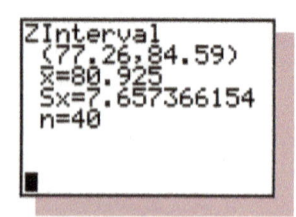

VIEW SCREEN 84
The ZInterval results are:

(77.26,84.59) is the 99% confidence interval for the **population mean**

$\bar{x}$ = **80.925** is the **sample mean**

Sx = **7.657366154** is the **sample std deviation**

n = **40** is the **sample size**

PROCEDURE TO CONSTRUCT A CONFIDENCE INTERVAL FOR A POPULATION MEAN USING *t* DISTRIBUTION WITH STATS INPUT

Confidence Interval For Pop Mean Where Pop Std Dev is UNKNOWN

1) Press STAT & HIGHLIGHT TESTS
2) Select 8 for **8:TInterval** . . . input screen for T confidence interval for a population mean is displayed in VIEW SCREEN 85.

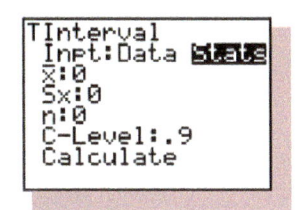

```
TInterval
 Inpt:Data Stats
 x:0
 Sx:0
 n:0
 C-Level:.9
 Calculate
```

VIEW SCREEN 85

3) **INPUT FOR A TInterval USING SAMPLE RESULTS:**
   a) at **Inpt:**highlight **Stats** & ENTER
   b) at $\bar{x}$:input **sample mean** & ENTER
   c) at **Sx:**input **sample std dev, s,** and ENTER
   d) at **n:** input **sample size** & ENTER
   e) at **C-Level:**input **confidence level (as a decimal)** & ENTER
   f) highlight Calculate & ENTER to determine confidence interval

TO CONSTRUCT A 95% CONFIDENCE INTERVAL FOR THE POP MEAN WHERE
$\bar{x} = 120$, s = 8 & n = 16.

FOR THIS 95% CONFIDENCE INTERVAL, INPUT THE RESULTS TO TInterval SCREEN:

1) Press STAT & HIGHLIGHT TESTS
2) Select 8 for **8:TInterval** . . .

The TINTERVAL INPUT is:
   a) **Highlight** STATS & ENTER
   b) at $\bar{x}$:input **120** & ENTER
   c) at **Sx:**input **8** & ENTER
   d) at **n:** input **16** & ENTER
   e) at **C-Level:**input **.95** & ENTER

The **TInterval** input screen should contain the info shown in VIEW SCREEN 86.

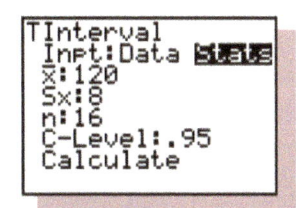

```
TInterval
 Inpt:Data Stats
 x:120
 Sx:8
 n:16
 C-Level:.95
 Calculate
```

VIEW SCREEN 86
f) Highlight Calculate & ENTER

The 95% TInterval results are shown in VIEW SCREEN 87

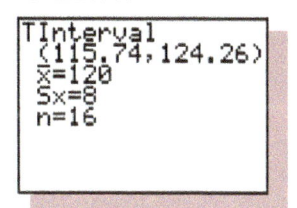

```
TInterval
 (115.74,124.26)
 x=120
 Sx=8
 n=16
```

VIEW SCREEN 87

The TInterval results are:

(115.74,124.26) *is the 95% confidence interval for the population mean*

$\bar{x} = 120$ *is the **sample mean***

Sx = 8 *is the **sample std dev***

n = 16 *is the **sample size***

TO CONSTRUCT A 99% CONFIDENCE INTERVAL FOR THE POP MEAN WHERE:

$\bar{x} = 120$, s = 8 and n = 16

1) Press STAT & HIGHLIGHT TESTS
2) Select 8 for **8:TInterval** . . .

The TINTERVAL INPUT is:
   a) **Highlight** STATS & ENTER
   b) at $\bar{x}$:input **120** & ENTER
   c) at **Sx:**input **8** & ENTER
   d) at **n:** input **16** & ENTER
   e) for **C-Level:** INPUT **.99** & ENTER
   f) Highlight Calculate & ENTER

The TInterval results are shown in VIEW SCREEN 88

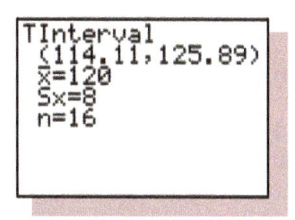

```
TInterval
 (114.11,125.89)
 x=120
 Sx=8
 n=16
```

VIEW SCREEN 88

The TInterval results are:

(114.11,125.89) – this is the 99% confidence interval for the population mean

$\bar{x} = 120$ *is the **sample mean***

Sx = 8 *is the **sample std dev***

n = 16 *is the **sample size***

PROCEDURE TO CONSTRUCT A CONFIDENCE INTERVAL FOR A POPULATION MEAN USING t DISTRIBUTION WITH DATA INPUT

Confidence Interval For Pop Mean Where Pop Std Dev is UNKNOWN

1) Press STAT & HIGHLIGHT TESTS
2) Select 8 for **8:TInterval** . . . input screen for T confidence interval for a population mean is displayed in VIEW SCREEN 89.

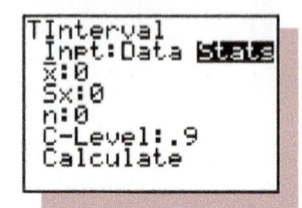

VIEW SCREEN 89

3) INPUT FOR A TInterval USING SAMPLE RESULTS:
   a) at **Inpt:**highlight Data & ENTER
   b) at **LIST:** input List name containing sample data & ENTER
   c) at **Freq:**input 1 or name of list containing frequencies & ENTER
   d) at **C-Level:**input **confidence level (as a decimal)** & ENTER
   e) highlight Calculate & ENTER to determine confidence interval

TO CONSTRUCT A 95% CONFIDENCE INTERVAL FOR THE POP MEAN WHERE THE SAMPLE DATA IS ENTERED INTO LIST L1.

SAMPLE DATA:
80 68 91 87 73 90 66 79 80 81
79 65 94 69 81 77 78 92 81 74

The TINTERVAL INPUT is:
   a) Highlight DATA & ENTER
   b) for **List:** INPUT L1 & ENTER
   c) for **Freq:** INPUT 1 & ENTER
   d) for **C-Level:** INPUT .95 & ENTER

The **TInterval** input screen should contain the info shown in VIEW SCREEN 90.

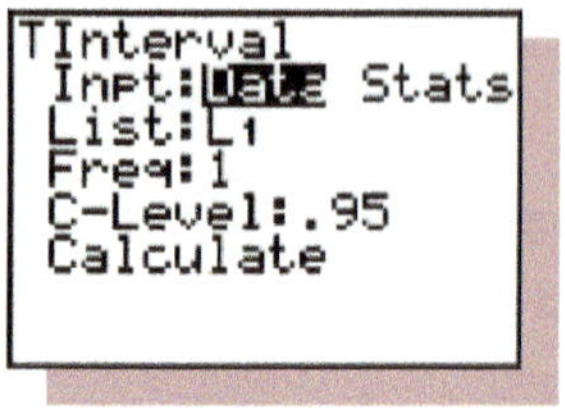

VIEW SCREEN 90
   e) Highlight Calculate & ENTER

The 95% TInterval results are shown in VIEW SCREEN 91

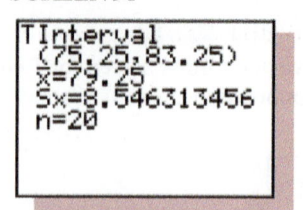

VIEW SCREEN 91

The TInterval results are:
(75.25,83.25) *is the 95% confidence interval for the population mean*
$\bar{x}$ = **79.25** *is the* **sample mean**
Sx = 8.546313456 *is the* **sample std dev**
n = 20 *is the* **sample size**

TO CONSTRUCT A 99% CONFIDENCE INTERVAL FOR THE POP MEAN USING THE SAME SAMPLE DATA IN LIST L1, PERFORM THE STEPS.

1) Press DATA & HIGHLIGHT TESTS
2) Select 8 for **8:ZInterval** . . .

The TINTERVAL INPUT is:
   a) Highlight DATA & ENTER
   b) for **List:** INPUT L1 & ENTER
   c) for **Freq:** INPUT 1 & ENTER
   d) for **C-Level:** INPUT .99 & ENTER
   f) Highlight Calculate & ENTER

The **TInterval** results are shown in VIEW SCREEN 92.

VIEW SCREEN 92

The TInterval results are:
(73.783,84.717) *is the 99% confidence interval for the population mean*
$\bar{x}$ = **79.25** *is the* **sample mean**
Sx = 8.546313456 *is the* **sample std dev**
n = 20 *is the* **sample size**

PROCEDURE TO CONSTRUCT A CONFIDENCE INTERVAL FOR A POPULATION PROPORTION USING NORMAL DISTRIBUTION

Confidence Interval For a Population Proportion

1) Press STAT & HIGHLIGHT TESTS
2) Select A for A: **1-PropZInt** . . . input screen for Z confidence interval for a pop proportion is displayed in VIEW SCREEN 93.

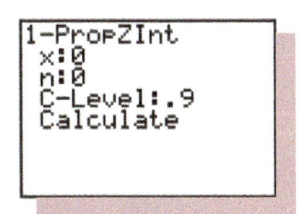

```
1-PropZInt
 x:0
 n:0
 C-Level:.9
 Calculate
```

VIEW SCREEN 93

3) **INPUT FOR 1-PropZInt USING SAMPLE RESULTS:**
   a) at **x:** input **number of successes** and ENTER
   b) at **n:** input **sample size and** ENTER
   c) at **C-Level:** input **confidence level (as a decimal)** & ENTER
   d) highlight Calculate & ENTER to determine confidence interval

TO CONSTRUCT A 95% CONFIDENCE INTERVAL FOR THE POPULATION PROPORTION WHERE:

  x = 918 and n = 1360

FOR THIS 95% CONFIDENCE INTERVAL, INPUT THE SAMPLE RESULTS TO 1-PropZInt SCREEN:

1) Press STAT & HIGHLIGHT TESTS
2) Select A for A: 1-PropZInt . . .

The 1-PropZINT INPUT is:
   a) for **x:** INPUT **918** & ENTER
   b) for **n:** INPUT **1360** & ENTER
   c) for **C-Level:** INPUT **.95** & ENTER

The **1-PropZINT** input screen should contain the info shown in VIEW SCREEN 94.

```
1-PropZInt
 x:918
 n:1360
 C-Level:.95
 Calculate
```

VIEW SCREEN 94
   e) Highlight Calculate & ENTER

The 95% 1-PropZInt results are shown in VIEW SCREEN 95

```
1-PropZInt
 (.65011,.69989)
 p=.675
 n=1360
```

VIEW SCREEN 95

The 1-PropZInt results are:
(.65011,.69989) *is the 95% confidence interval for the population proportion*
$\hat{p}$ = **.675** *is the **sample proportion***
n = 1360 *is the **sample size***

TO CONSTRUCT A 99% CONFIDENCE INTERVAL FOR THE POPULATION PROPORTION WHERE:

  x = 918 and n = 1360

1) Press STAT & HIGHLIGHT TESTS
2) Select A for A: 1-PropZInt . . .

The 1-PropZINT INPUT is:
   a) for **x:** INPUT **918** & ENTER
   b) for **n:** INPUT **1360** & ENTER
   c) for **C-Level:** INPUT **.99** & ENTER
   d) highlight Calculate & ENTER

The 99% 1-PropZInt results are shown in VIEW SCREEN 96

```
1-PropZInt
 (.64229,.70771)
 p=.675
 n=1360
```

VIEW SCREEN 96

The 1-PropZInt results are:
(.64229,.70771) *is the 99% confidence interval for the population proportion*
$\hat{p}$ = **.675** *is the **sample proportion***
n = 1360 *is the **sample size***

PROCEDURE TO PERFORM A HYPOTHESIS TEST INVOLVING A POPULATION PROPORTION USING A NORMAL DISTRIBUTION

ONE POP PROPORTION TEST USING THE NORMAL DISTRIBUTION

1) Press STAT & Highlight TESTS shown in VIEW SCREEN 97.

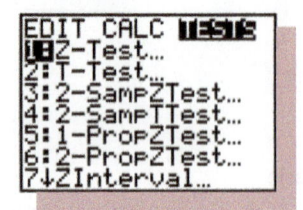

VIEW SCREEN 97

2) Press 5 for **5:1-PropZTest** . . . input screen as shown in VIEW SCREEN 98.

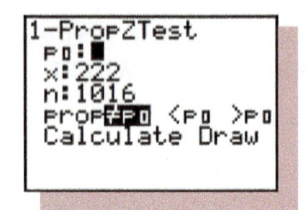

VIEW SCREEN 98

3) For a **1-PropZTEST, INPUT:**
   a) at p_0: input **pop proportion** as stated in H_0 & ENTER
   b) at **x**: input **number of successes for the sample** & ENTER
   c) at **n**: input **sample size** & ENTER
   d) at **prop**: highlight **form of H_a, alternative hypothesis,** & ENTER
   e) highlight Calculate & ENTER

LET'S PERFORM A TEST WHERE:

H_0: p = .20; H_a: p > .20; x = 222; n = 1016 and α = 5%.

For this 1-PropZTest, INPUT:
   a) at p_0: INPUT .2 & ENTER
   b) at **x**: INPUT 222 & ENTER
   c) at **n**: INPUT 1016 & ENTER
   d) at **prop**: HIGHLIGHT > p_0 & ENTER

The 1-PropZTest input screen should contain the info shown in VIEW SCREEN 99.

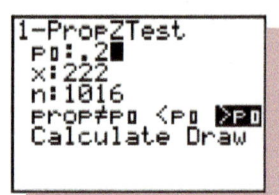

VIEW SCREEN 99

e) highlight Calculate & ENTER

The test results are shown in VIEW SCREEN 100.

VIEW SCREEN 100

The **1-PropZTest** test results are:

prop > .2 *is* H_a

z = 1.474521142 *is the z score of the sample proportion*

p = .0701706992 *is the p-value*

$\hat{p}$ = .218503937 *is the sample proportion*

n = 1016 *is the sample size*

We *FAIL TO REJECT* H_0 at α = 5% since the p-value > 5%.

What is the conclusion if α = 1%?

FOR A GRAPH OF THE TEST RESULTS:

1) press STAT & highlight TESTS
2) press 5 for **5:1-PropZTEST** . . .
3) highlight DRAW & ENTER

VIEW SCREEN 101 is displayed.

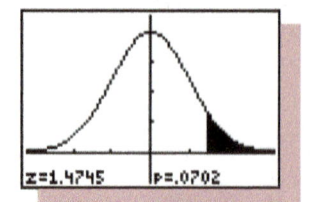

VIEW SCREEN 101

In VIEW SCREEN 101, the z score and the p-value are displayed. Since this is a 1TT where H_a is prop > .2, the right tail is shaded. This shaded area represents the p-value.

PROCEDURE TO PERFORM A HYPOTHESIS TEST INVOLVING A POPULATION MEAN USING NORMAL DISTRIBUTION WITH STATS INPUT

ONE POP MEAN HYPOTHESIS TEST USING THE NORMAL DISTRIBUTION WITH KNOWN POPULATION STD DEV

1) Press STAT & Highlight TESTS shown in VIEW SCREEN 102.

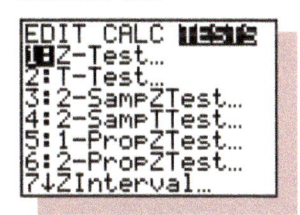

VIEW SCREEN 102

2) Press 1 for **1:Z-Test** . . . input screen as shown in VIEW SCREEN 103.

VIEW SCREEN 103

3) INPUT for a **Z-TEST USING SAMPLE RESULTS**:
   a) at **Inpt**:highlight Stats & ENTER
   b) at μ_0:input **population mean** as stated in null hypothesis & ENTER
   c) at σ:input **pop std dev** & ENTER
   d) at $\bar{x}$:input **sample mean** & ENTER
   e) at n:input **sample size** & ENTER
   f) at μ:highlight **form of H_a, alternative hypothesis**, & ENTER
   g) highlight Calculate & ENTER

LET'S PERFORM A TEST WHERE:

H_0: $\mu = 120$; H_a: $\mu < 120$
$\sigma = 18$, $\bar{x} = 114$, $n = 36$ & $\alpha = 1\%$.

For this Z-TEST, INPUT:
   a) Highlight STATS & ENTER
   b) for μ_0: INPUT 120 & ENTER
   c) for σ: INPUT 18 & ENTER
   d) for $\bar{x}$: INPUT 114 & ENTER
   e) for n: INPUT 36 & ENTER
   f) for μ: HIGHLIGHT $< \mu_0$ & ENTER

The Z-TEST input screen should contain the info shown in VIEW SCREEN 104.

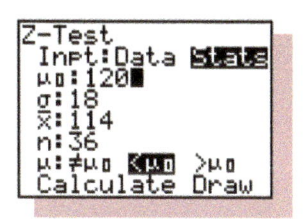

VIEW SCREEN 104

g) Highlight Calculate & ENTER

The test results are shown in VIEW SCREEN 105.

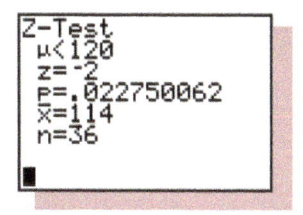

VIEW SCREEN 105

The Z-TEST results are:

$\mu < 120$: *is H_a*

$z = -2$: *is z score of sample mean*

$p = .022750062$: *is p-value of test*

$\bar{x} = 114$: *is sample mean*

$n = 36$: *is sample size*

We *FAIL TO REJECT H_0* at $\alpha = 1\%$ since the **p-value** $> 1\%$.

What is the conclusion if $\alpha = 5\%$?

FOR A GRAPH OF THE TEST RESULTS:

1) press STAT
2) highlight TESTS
3) press 1 for **1:Z-Test** . . .
   VIEW SCREEN 103 is displayed.
4) highlight Draw & ENTER
   VIEW SCREEN 106 is displayed.

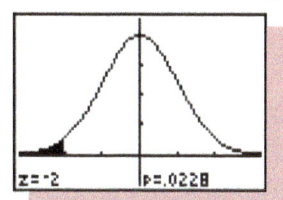

VIEW SCREEN 106

In VIEW SCREEN 106, the z score and the p-value are displayed. Since this is a 1TT where H_a is $\mu < 120$, the left tail is shaded. This shaded area represents the p-value of the test.

PROCEDURE TO PERFORM A HYPOTHESIS TEST INVOLVING A POPULATION MEAN USING NORMAL DISTRIBUTION WITH DATA INPUT

ONE POP MEAN HYPOTHESIS TEST USING THE NORMAL DISTRIBUTION WITH KNOWN POP STD DEVIATION

1) Press STAT & Highlight TESTS shown in VIEW SCREEN 107.

VIEW SCREEN 107

2) Press 1 for **1:Z-Test** . . . input screen as shown in VIEW SCREEN 108

VIEW SCREEN 108

3) INPUT for a Z-TEST USING SAMPLE DATA:
   a) at **Inpt:** highlight Data & ENTER
   b) at μ_0: input **population mean** as stated in null hypothesis & ENTER
   c) at σ: input **population std deviation** & ENTER
   d) at **List:** input List name containing **sample data** & ENTER
   e) at **Freq:** input 1 or name of list containing frequencies & ENTER
   f) at μ: highlight **form of Ha, alternative hypothesis,** & ENTER
   g) highlight Calculate & ENTER

LET'S PERFORM A Z-TEST WHERE THE FOLLOWING SAMPLE DATA HAS BEEN ENTERED INTO LIST L1:

SAMPLE DATA
750 765 780 710 720 715 730 735
740 745 750 760 680 675 690 695
650 600 615 620 630 640 590 585
560 555 490 485 475 465 460 695
650 610 575 450

AND WHERE:
H_0: μ = 610; Ha: μ > 610
σ = 120 and α = 5%.

For this Z-TEST, INPUT:
   a) Highlight DATA & ENTER
   b) for μ_0: INPUT 610 & ENTER
   c) for σ: INPUT 120 & ENTER
   d) for **List:** INPUT L1 & ENTER
   e) for **Freq:** INPUT 1 & ENTER
   f) for μ: HIGHLIGHT > μ_0 & ENTER

The Z-TEST input screen contains the info shown in VIEW SCREEN 109.

VIEW SCREEN 109

g) highlight Calculate & ENTER

The test results are shown in VIEW SCREEN 110.

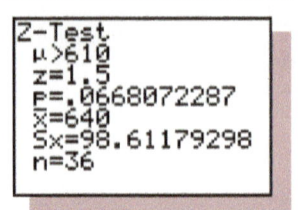

VIEW SCREEN 110

The Z-TEST results are:

μ > 610 *is* H_a

z = **1.5** is **z score** *of sample mean*

p = **.0668072287** *is the p-value*

$\bar{x}$ = **640** *is sample mean of L1*

Sx = **98.61179298** *is sample std dev. of List L1*

n = **36** *is the sample size*

We FAIL TO REJECT H_0 at α = 5% since the p-value > 5%. *What is the conclusion if α = 1%?*

FOR A GRAPH OF THE TEST RESULTS:

1) press STAT & highlight TESTS
2) press 1 for **1:Z-Test** . . .
3) highlight Draw & ENTER
   VIEW SCREEN 111 is displayed.

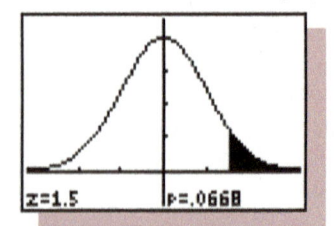

VIEW SCREEN 111

In VIEW SCREEN 111, the z score and the p-value are displayed. Since this is a 1TT where Ha is μ > 610, the right tail is shaded. This shaded area represents the p-value of the test.

PROCEDURE TO PERFORM A HYPOTHESIS TEST INVOLVING A POPULATION MEAN USING t DISTRIBUTION WITH STATS INPUT

ONE POPULATION MEAN HYPOTHESIS TEST USING THE t DISTRIBUTION WITH UNKNOWN POP STD DEVIATION

1) Press STAT & Highlight TESTS shown in VIEW SCREEN 112.

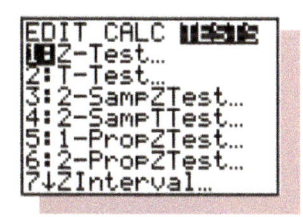

VIEW SCREEN 112

2) Press 2 for **2:T-Test** . . . input screen as shown in VIEW SCREEN 113

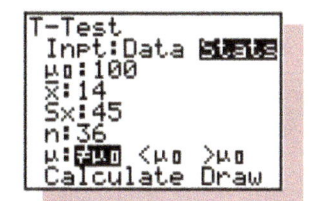

VIEW SCREEN 113

3) **INPUT for a T-TEST USING SAMPLE RESULTS:**
   a) at **Inpt:** highlight Stats & ENTER
   b) at μ_0: input **population mean** as stated in null hypothesis & ENTER
   c) at $\bar{x}$: input **sample mean** & ENTER
   d) at Sx: input **sample std dev, s,** and ENTER
   e) at n: input **sample size** & ENTER
   f) at μ: highlight **form of H_a, alternative hypothesis,** & ENTER
   g) highlight Calculate & ENTER

LET'S PERFORM A T-TEST WHERE:

H_0: μ = 45,000; Ha: μ < 45,000
$\bar{x}$ = 42,900; s = 3,590; n = 16 & α = 5%

For this T-TEST, INPUT:
   a) Highlight STATS & ENTER
   b) for μ_0: INPUT 45000 & ENTER
   c) for $\bar{x}$: INPUT 42900 & ENTER
   d) for Sx: INPUT 3590 & ENTER
   e) for n: INPUT 16 & ENTER
   f) for μ: HIGHLIGHT < μ_0 & ENTER

The T-TEST input screen should contain the info shown in VIEW SCREEN 114.

VIEW SCREEN 114

g) highlight Calculate & ENTER

The test results are shown in VIEW SCREEN 115.

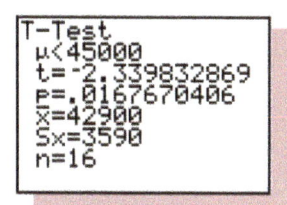

VIEW SCREEN 115

The **T-TEST** results are:

μ < 45000 is H_a

t = –2.339832869 is the *t score*

p = .0167670406 is the *p-value*

$\bar{x}$ = 42900 is the *sample mean*

Sx = 3590 is the *sample std dev*

n = 16 is the *sample size*

We REJECT H_0 at α = 5% since the p-value < 5%. What is the conclusion if α = 1%?

FOR A GRAPH OF THE TEST RESULTS:

1) press STAT & highlight TESTS
2) press 2 for **2:T-Test** . . .
3) highlight Draw & ENTER
   VIEW SCREEN 116 is displayed.

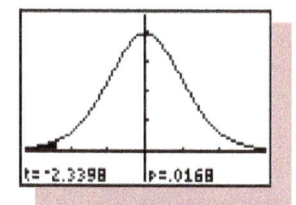

VIEW SCREEN 116

In VIEW SCREEN 116, the t score and the p-value are displayed. Since this is a 1TT where H_a is μ < 45000, the left tail is shaded. This shaded area represents the p-value.

Appendix D   TI-84 Plus Instructions

PROCEDURE TO PERFORM A HYPOTHESIS TEST INVOLVING TWO POPULATION PROPORTIONS USING A NORMAL DISTRIBUTION

1) Press STAT & Highlight TESTS shown in VIEW SCREEN 117.

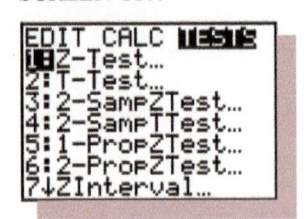

VIEW SCREEN 117

2) press 6 for **6:2-PropZTest** . . . input screen for a two sample proportion hypothesis test as displayed in VIEW SCREEN 118.

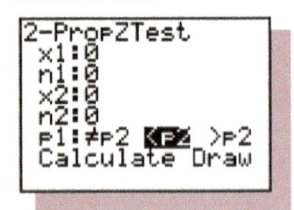

VIEW SCREEN 118

3) INPUT for a 2-PropZTEST USING SAMPLE RESULTS:
   a) for x1: input **the number of successes for sample one** & ENTER
   b) for n1: input the **sample size** for sample one & ENTER
   c) for x2: input **the number of successes for sample two** & ENTER
   d) for n2: input the **sample size** for sample two & ENTER
   e) for p1: highlight **form of Ha, alternative hypothesis**, & ENTER
   f) Highlight Calculate & ENTER

LET'S PERFORM A 2-PropZTEST WHERE
$\alpha = 5\%$, H_0: $p_1 - p_2 = 0$; H_a: $p_1 - p_2 > 0$
For sample 1: $x_1 = 870$ & $n_1 = 1500$
For sample 2: $x_2 = 702$ & $n_2 = 1300$

For this 2-PropZTEST, INPUT:
   a) For x1: INPUT 870 & ENTER
   b) For n1: INPUT 1500 & ENTER
   c) For x2: INPUT 702 & ENTER
   d) For n2: INPUT 1300 & ENTER
   e) For p1: HIGHLIGHT > p2 & ENTER
The 2-PropZTEST input screen should contain the info shown in VIEW SCREEN 119.

VIEW SCREEN 119
f) Highlight Calculate & ENTER
The test results are shown in VIEW SCREENS 120a & 120b.

VIEW SCREEN 120a   VIEW SCREEN 120b

The **2-PropZTEST** results are:

  p1 > p2 is H_a

  z = 2.127310376 *is the z score of the difference between two sample proportions*

  p = .0166970917 *is the p-value*

  $\hat{p}_1 = .58$ *is the proportion of sample 1*

  $\hat{p}_2 = .54$ *is the proportion of sample 2*

  $\hat{p} = .5614285714$ *is the pooled sample proportion*

  n1 = 1500 *is the size of sample 1*

  n2 = 1300 *is the size of sample 2*

We *REJECT H_0* and *ACCEPT H_a* at $\alpha = 5\%$ since the p-value < 5%.

What is the conclusion if $\alpha = 1\%$?

FOR A GRAPH OF THE TEST RESULTS:
1) Press STAT & highlight TESTS
2) press 2 for **2:PropZTest** . . .
3) highlight Draw & ENTER
   VIEW SCREEN 121 is displayed.

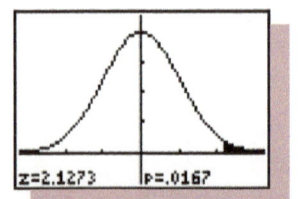

VIEW SCREEN 121
In VIEW SCREEN 121, the z score and the p-value are displayed. Since this is a 1TT where Ha is p1 > p2, the right tail is shaded. This shaded area represents the p-value.

PROCEDURE TO PERFORM A HYPOTHESIS TEST INVOLVING TWO POPULATION MEANS USING NORMAL DISTRIBUTION WITH STATS INPUT

TWO POP MEAN TEST WITH KNOWN POPULATION STD DEVIATIONS

1) Press STAT & Highlight TESTS shown in VIEW SCREEN 122.

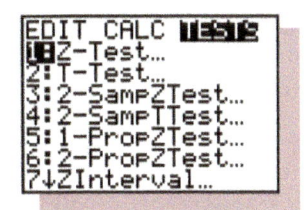

VIEW SCREEN 122

2) press 3 for **3:2-SampZTest**. . . . input screen for a 2 sample mean test with pop std deviations KNOWN as shown in VIEW SCREEN 123.

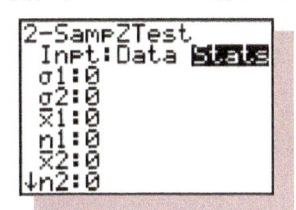

VIEW SCREEN 123

3) INPUT for a 2-SampZTEST USING SAMPLE RESULTS:
   a) at Inpt:Highlight Stats & ENTER
   b) at σ_1:key pop1 std dev & ENTER
   c) at σ_2:key pop2 std dev & ENTER
   d) at $\bar{x}$:key sample mean1 & ENTER
   e) ar n_1:key sample size1 & ENTER
   f) at $\bar{x}_2$:key sample mean2 & ENTER
   g) at n_2: key sample size2 & ENTER
   h) for μ_1 : highlight the form Ha, alternative hypothesis, & ENTER
   i) Highlight Calculate & ENTER

LET'S PERFORM A 2-SampZTEST WHERE

$\alpha = 1\%$, H_0: $\mu_1 - \mu_2 = 0$; H_a: $\mu_1 - \mu_2 > 0$
For pop 1: $\sigma_1 = 20$; $\bar{x}_1 = 120$ & $n_1 = 81$
For pop 2: $\sigma_2 = 30$; $\bar{x}_2 = 112$ & $n_2 = 64$

For this 2-SampZTEST, INPUT:
   a) at Inpt:Highlight Stats & ENTER
   b) for σ_1: INPUT **20** & ENTER
   c) for σ_2: INPUT **30** & ENTER
   d) for $\bar{x}1$: INPUT **120** & ENTER
   e) for n_1: INPUT **81** & ENTER
   f) for $\bar{x}2$: INPUT **112** & ENTER
   g) for n_2: INPUT **64** & ENTER
   h) for μ_1 : HIGHLIGHT > μ_2 & ENTER

The 2-SampZTest input screens should contain the info shown in VIEW SCREENS 124a and 124b.

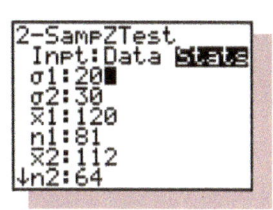

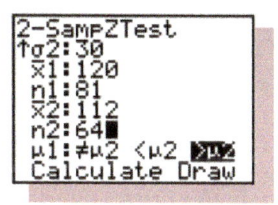

VIEW SCREEN 124a   VIEW SCREEN 124b

i) **Highlight** Calculate & ENTER

The 2-SampZTest results are shown in VIEW SCREENS 125a and 125b.

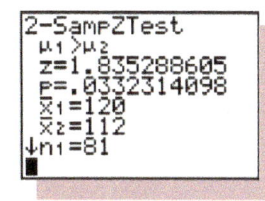

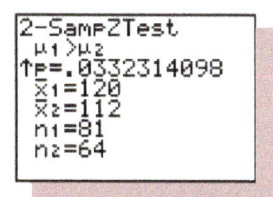

VIEW SCREEN 125a   VIEW SCREEN 125b

The 2-SampZTest test results are:

$\mu_1 > \mu_2$ is H_a: $\mu_1 - \mu_2 > 0$

$z = 1.835288605$ is the *z-score of the difference between the two sample means:* $\bar{x}_1 - \bar{x}_2$

$p = .0332314098$ is the *p-value*

$\bar{x}_1 = 120$ is the *mean of sample 1*

$\bar{x}_2 = 112$ is the *mean of sample 2*

$n_1 = 81$ is the *size of sample 1*

$n_2 = 64$ is the *size of sample 2*

We *FAIL TO REJECT H_0* at $\alpha = 1\%$ since the p-value > 1%. *What is the conclusion if $\alpha = 5\%$?*

FOR A GRAPH OF THE TEST RESULTS:

1) press STAT & highlight TESTS
2) press 3 for **3:2-SampZTest**. . . .
3) highlight Draw & ENTER

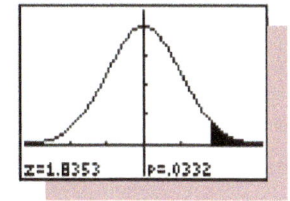

VIEW SCREEN 126

In VIEW SCREEN 126, the z score and the p-value are displayed. Since this is a 1TT where H_a is $\mu_1 > \mu_2$, the right tail is shaded. This shaded area represents the p-value.

PROCEDURE TO PERFORM A HYPOTHESIS TEST INVOLVING TWO POPULATION MEANS USING NORMAL DISTRIBUTION WITH DATA INPUT

TWO POP MEAN TEST WITH KNOWN POPULATION STD DEVIATIONS

1) Press STAT & Highlight TESTS shown in VIEW SCREEN 127.

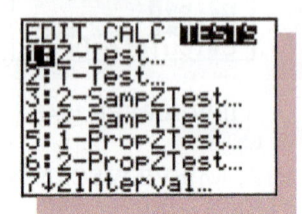

VIEW SCREEN 127

2) press 3 for **3:2-SampZTest**. . . . input screen for a 2 sample z test as shown in VIEW SCREEN 128.

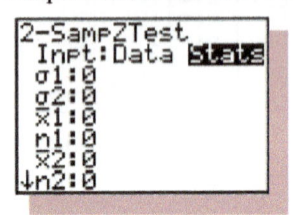

VIEW SCREEN 128

3) INPUT for a 2-SampZTEST USING DATA:
   a) at Inpt:Highlight Data & ENTER
   b) at σ_1:key pop1 std dev & ENTER
   c) at σ_2:key pop2 std dev & ENTER
   d) at List1: key List containing data for sample 1 & ENTER
   e) at List2: key List containing data for sample 2 & ENTER
   f) at Freq1: key 1 or List name containing frequencies for sample1 & ENTER
   g) at Freq2: key 1 or List name containing frequencies for sample2 & ENTER
   h) for μ_1: highlight the form Ha, alternative hypothesis, & ENTER
   i) Highlight Calculate & ENTER

LET'S PERFORM A 2-SampZTEST WHERE $\alpha = 5\%$ & THE SAMPLE DATA BELOW.
Enter Samples Into LISTS

INPUT SAMPLE 1 INTO L1
113 117 112 115 112 116 119
113 121 123 124 119 117 123
111 112 115 114 118 119 115
123 127 119 117 124 126 129
118 125 126 128 119 127 125
117 119 125 124 128 121 125

INPUT SAMPLE 2 INTO L2
124 126 128 113 118 117 114
117 113 120 125 120 127 125
121 115 117 117 128 124 126
127 129 129 127 126 122 126
128 129 123 129 126 128 127
127 129 122 123 126 123 126

LET'S PERFORM 2-SampZTEST USING THE SAMPLE DATA IN LISTS L1 & L2,
$H_0: \mu_1 - \mu_2 = 0$; $H_a: \mu_1 - \mu_2 < 0$ and
for pop 1: $\sigma_1 = 7$ & for pop 2: $\sigma_2 = 6$

For this 2-SampZTEST, INPUT:
   a) Highlight DATA & ENTER
   b) for σ_1: INPUT 7 & ENTER
   c) for σ_2: INPUT 6 & ENTER
   d) for List1: INPUT L1 & ENTER
   e) for List2: INPUT L2 & ENTER
   f) for Freq1: INPUT 1 & ENTER
   g) for Freq2: INPUT 1 & ENTER
   h) for μ_1: HIGHLIGHT $< \mu_2$ & ENTER

The 2-SampZTest input screens should contain the info shown in VIEW SCREENS 129a and 129b.

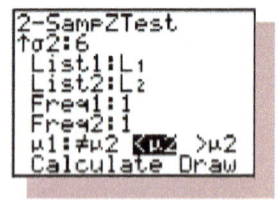

VIEW SCREEN 129a   VIEW SCREEN 129b

i) Highlight Calculate & ENTER 2-SampZTest results are shown in VIEW SCREENS 130a & 130b.

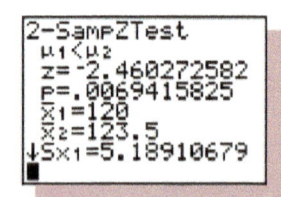

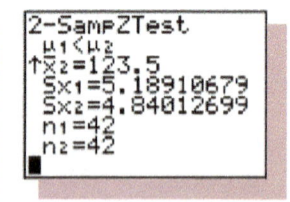

VIEW SCREEN 130a   VIEW SCREEN 130b

The 2-SampZTest test results are:
$\mu_1 > \mu_2$ is $H_a: \mu_1 - \mu_2 < 0$
$z = -2.460272582$ is the z-score of the difference between the two sample means: $\overline{x}_1 - \overline{x}_2$
$p = .0069415825$ is the *p-value*
$\overline{x}_1 = 120$ is the **mean of sample 1**
$\overline{x}_2 = 123.5$ is **mean of sample 2**

Sx1 = 5.1890679 *is the **std dev of sample 1***
Sx2 = 4.84012699 *is the **std dev of sample 2***
n1 = 42 *is the **size of sample 1***
n2 = 42 *is the **size of sample 2***

We *REJECT* H_0 and *ACCEPT* H_a at $\alpha = 5\%$ since the p-value < 5%.

What is the conclusion if $\alpha = 1\%$?

FOR A GRAPH OF THE TEST RESULTS:
1) press **STAT** & highlight **TESTS**
2) press 3 for **3:2-SampZTest. . . .**
3) highlight **Draw** & **ENTER**

VIEW SCREEN 131 is displayed.

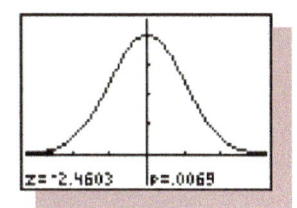

VIEW SCREEN 131

In VIEW SCREEN 131, the z score and the p-value are displayed. Since this is a 1TT where H_a is $\mu_1 < \mu_2$, the left tail is shaded. This shaded area represents the p-value. Note: since the p-value is very small, the shaded area is not noticeable.

PROCEDURE TO PERFORM A HYPOTHESIS TEST INVOLVING TWO POPULATION MEANS USING t DISTRIBUTION WITH STATS INPUT

TWO POP MEAN t TEST WITH UNKNOWN POPULATION STD DEVIATIONS

1) Press STAT & Highlight TESTS shown in VIEW SCREEN 132.

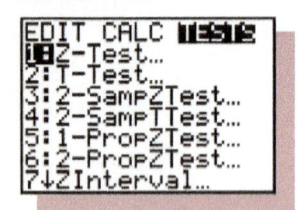

VIEW SCREEN 132

2) press 4 for **4:2-SampTTest**. . . . input screen for a 2 sample t test as shown in VIEW SCREEN 133.

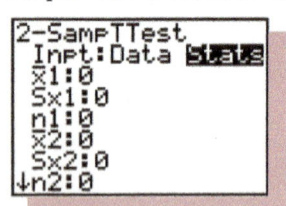

VIEW SCREEN 133

3) INPUT for a 2-SampTTEST USING SAMPLE RESULTS:
   a) at **Inpt:Highlight** Stats & ENTER
   b) at $\bar{x}_1$:mean of sample1 & ENTER
   c) at Sx1:std dev of sample1, s_1, & ENTER
   d) at n_1:size of sample1 & ENTER
   e) at $\bar{x}_2$:mean of sample2 & ENTER
   f) at Sx2:std dev of sample2, s_2, & ENTER
   g) at n_2:size of sample2 & ENTER
   h) at μ_1 : highlight the form Ha, alternative hypothesis, & ENTER
   i) at **Pooled:Highlight** YES and ENTER if the population standard deviations are unknown but assumed to be *equal*. In this case you'll be performing a pooled t-test. Highlight NO and ENTER if the population standard deviations are unknown but assumed to be *unequal*. In this case you'll be performing Welch's t-test.
   j) Highlight Calculate & ENTER

LET'S PERFORM A 2-SampTTEST WHERE THE POPULATION STANDARD DEVIATIONS ARE ASSUMED TO BE *EQUAL*,

α = 1% & H$_0$: $\mu_1 - \mu_2 = 0$ & H$_a$: $\mu_1 - \mu_2 < 0$
For sample1: $\bar{x}_1 = 116$; $s_1 = 6.7$ & $n_1 = 16$
For sample2: $\bar{x}_2 = 120$; $s_2 = 5.6$ & $n_2 = 20$

For this 2-SampTTEST, INPUT:
   a) at **Inpt:Highlight** Stats & ENTER
   b) at $\bar{x}1$: INPUT 116 & ENTER
   c) at Sx1: INPUT 6.7 & ENTER
   d) at n1: INPUT 16 & ENTER
   e) at $\bar{x}2$: INPUT 120 & ENTER
   f) at Sx2: INPUT 5.6 & ENTER
   g) at n2: INPUT 20 & ENTER
   h) at μ_1 : HIGHLIGHT $< \mu_2$ & ENTER
   i) at **Pooled:**HIGHLIGHT Yes & ENTER

The 2-SampTTest input screens should contain the info shown in VIEW SCREENS 134a and 134b.

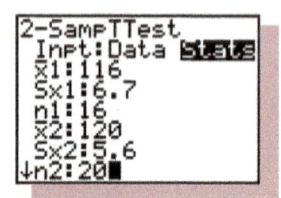

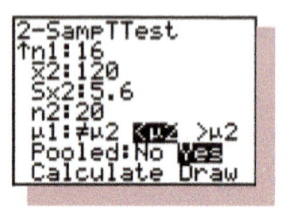

VIEW SCREEN 134a   VIEW SCREEN 134b

j) Highlight Calculate & ENTER 2-SampTTest results are shown in VIEW SCREENS 135a and 135b.

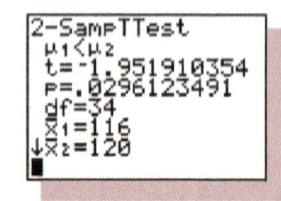

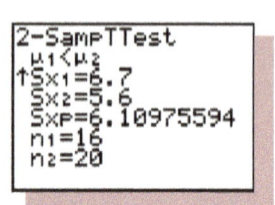

VIEW SCREEN 135a   VIEW SCREEN 135b

The pooled 2-SampTTest results are:

$\mu_1 < \mu_2$ is H$_a$: $\mu_1 - \mu_2 < 0$

$t = -1.951910354$ is the **t-score** of the difference between the two sample means: $\bar{x}_1 - \bar{x}_2$

p = .0296123491 is the **p-value**

df = 34 is the degrees of freedom

$\bar{x}1 = 116$ is the **mean of sample1**

$\bar{x}2 = 120$ is the **mean of sample2**

Sx1 = 6.7 is the std dev of sample1

Sx2 = 5.6 *is the std dev of sample2*

Sxp = 6.10975594 *is the pooled std deviation of both samples*

n1 = 16 *is the size of sample 1*

n2 = 20 *is the size of sample 2*

We *FAIL TO REJECT* H_0 $\alpha = 1\%$ since the p-value > 1%. *What is the conclusion if* $\alpha = 5\%$?

FOR A GRAPH OF THE TEST RESULTS:
1) press STAT & highlight TESTS
2) press 4 for **4:2-SampTTest.** . . .
3) highlight Draw & ENTER
   VIEW SCREEN 136 is displayed.

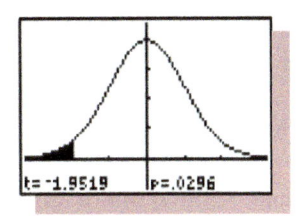

VIEW SCREEN 136

In VIEW SCREEN 136, the t score and the p-value are displayed. Since this is a 1TT where H_a is $\mu_1 < \mu_2$, the left tail is shaded. This shaded area represents the p-value.

PROCEDURE TO PERFORM A HYPOTHESIS TEST INVOLVING TWO POPULATION MEANS USING t DISTRIBUTION WITH DATA INPUT

TWO POP MEAN *t* TEST WITH UNKNOWN POPULATION STD DEVIATIONS

1) Press STAT & Highlight TESTS shown in VIEW SCREEN 137.

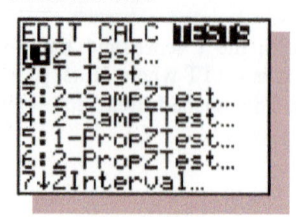

VIEW SCREEN 137

2) press 4 for **4:2-SampTTest**. . . . input screen for a 2 sample *t* test as shown in VIEW SCREEN 138.

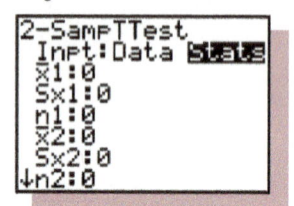

VIEW SCREEN 138

3) INPUT for a 2-SampTTEST USING SAMPLE DATA:
   a) at Inpt:Highlight Data & ENTER
   b) at List1: key List containing data for sample 1 & ENTER
   c) at List2: key List containing data for sample 2 & ENTER
   d) at Freq1: key 1 or List name containing frequencies for sample1 & ENTER
   e) at Freq2: key 1 or List name containing frequencies for sample2 & ENTER
   f) for μ_1: highlight the form H*a*, alternative hypothesis, & ENTER
   g) at Pooled:Highlight YES and ENTER if the population standard deviations are unknown but assumed to be *equal*. In this case you'll be performing a pooled t-test. Highlight NO and ENTER if the population standard deviations are unknown but assumed to be *unequal*. In this case you'll be performing Welch's t-test.
   h) Highlight Calculate & ENTER

LET'S PERFORM A 2-SampTTEST WHERE THE POPULATION STANDARD DEVIATIONS ARE ASSUMED TO BE *EQUAL*,

$\alpha = 5\%$ & H_0: $\mu_1 - \mu_2 = 0$ & H_a: $\mu_1 - \mu_2 > 0$

AND THE GIVEN SAMPLE DATA.
Enter Samples Into LISTS
INPUT SAMPLE 1 INTO L1
SAMPLE DATA FOR SAMPLE 1:
83 87 92 85 92 86 89 93
81 92 95 94 88 93 85

INPUT SAMPLE 2 INTO L2
SAMPLE DATA FOR SAMPLE 2:
74 96 78 83 86 87 94 77 83 80

For this 2-SampTTEST, INPUT:
   a) at Inpt:Highlight Data & ENTER
   b) at List1: INPUT L1 & ENTER
   c) at List2: INPUT L2 & ENTER
   d) at Freq1: INPUT 1 & ENTER
   e) at Freq2: INPUT 1 & ENTER
   f) at M1: HIGHLIGHT > M_2 & ENTER
   g) at Pooled:Highlight YES & ENTER

The 2-SampTTest input screen should contain the info shown in VIEW SCREEN 139.

VIEW SCREEN 139

h) Highlight Calculate & ENTER

The 2-SampTTest results are shown in VIEW SCREENS 140a and 140b.

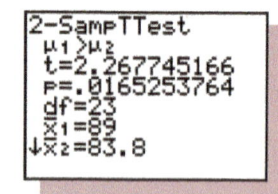

VIEW SCREEN 140a   VIEW SCREEN 140b

The pooled 2-SampTTest results are:

$\mu_1 > \mu_2$ is H_a: $\mu_1 - \mu_2 > 0$

t = 2.267745166 *is the t-score of the difference between the two sample means:* $\bar{x}_1 - \bar{x}_2$

p = .0165253764 *is the p-value*

df = 23 *is the degrees of freedom*

$\bar{x}1$ **= 89** *is the mean of sample 1*

$\bar{x}2$ **= 83.8** *is the mean of sample 2*

Sx1 = 4.35889894 *is the std dev of sample 1*

Sx2 = 7.1460945 *is the std dev of sample 2*

Sxp = 5.61674515 *is the pooled std dev of both samples*

n1 = 15 *is the size of sample 1*

n2 = 10 *is the size of sample 2*

We *REJECT* H_0 and ACCEPT H_a α = 5% since the p-value < 5%.

What is the conclusion if α = 1%?

FOR A GRAPH OF THE TEST RESULTS:

1) press STAT & highlight TESTS
2) press 4 for **4:2-SampTTest. . . .**
3) highlight Draw & ENTER
   VIEW SCREEN 141 is displayed.

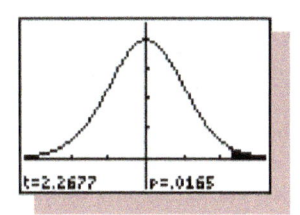

VIEW SCREEN 141

In VIEW SCREEN 141, the t score and the p-value are displayed. Since this is a 1TT with H_a is $\mu_1 > \mu_2$, the right tail is shaded. This shaded area represents the p-value.

THE PAIRED T-TEST COMPARING A SUBJECT PRE & POST SOME TREATMENT

Recall that the paired t-test compares observations that come in *matched pairs* (also called a *repeated measure* on the subject). This hypothesis test allows us to compare the population mean difference in these matched pairs of data. Prior to performing the paired t-test, we must input the data into the calculator.

To input the data into the TI-84:

1) Press **STAT** & **1: EDIT**. This allows you to enter the PRE data in L1 & the POST data in L2. This is shown in VIEW SCREEN 142a & 142b.

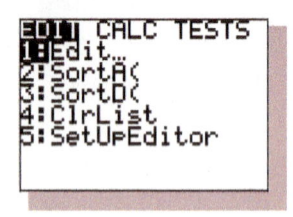

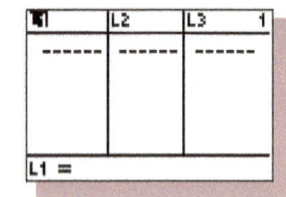

VIEW SCREEN 142a   VIEW SCREEN 142b

2) Once the data is input into L1 & L2, you will need to calculate the DIFFERENCES in these paired data values and place the differences in L3. For example, say the L1 data is 1, 2 & 3; and the L2 data is 7, 8 & 9 as shown in VIEW SCREEN 143.

VIEW SCREEN 143

To find the differences, use the arrow keys to place the cursor on top of L3 and then press:

2ND > L2 > – > 2ND > L1 > ENTER

The results are shown in VIEW SCREEN 144a & 144b.

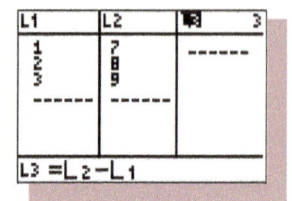

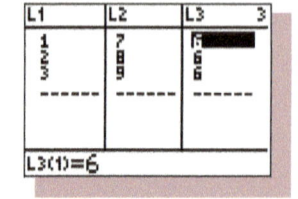

VIEW SCREEN 144a   VIEW SCREEN 144b

3) Recall, the paired t-test is merely a one sample t-test on the Differences in the paired data. To perform the t-test,

Press **STAT**, highlight **TESTS** and choose option **2: T-Test** as shown in VIEW SCREEN 145a & 145 b.

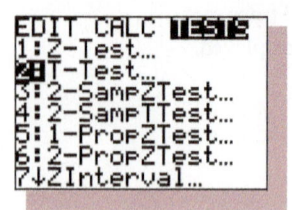

VIEW SCREEN 145a   VIEW SCREEN 145b

INPUTS for a paired t-TEST USING THE DIFFERENCE DATA:

a) at **Inpt**: highlight **Data** & press ENTER
b) at μ_0: input **population mean difference** as stated in null hypothesis which in this case *is always 0*, then press ENTER.
c) at List: press **2ND > 3** (assuming the difference data is stored in L3).
d) at **Freq**: input **1** or name of list containing frequencies & press ENTER
e) at μ: highlight **form of Ha, alternative hypothesis,** & ENTER
f) highlight Calculate & press ENTER

LET'S PERFORM A PAIRED T-TEST WHERE THE FOLLOWING SAMPLE WEIGHT DATA (for 6 individuals prior to a diet) HAS BEEN ENTERED INTO LISTS L1 & L2, AND THE DIFFERENCES HAVE ALSO BEEN CALCULATED AND PLACED IN L3:

L1(pre)	144	152	137	198	210	187
L2(post)	135	142	129	185	188	171

1) The question we would be interested in is, "Does this diet work?" In other words, do the participants have a significant amount of weight loss? Let's use $\alpha = 1\%$, and assume $H_a : \mu > 0$, indicating a significant positive weight loss for the participants. View Screen 146 shows the pre, post and difference data entered into the TI-84:

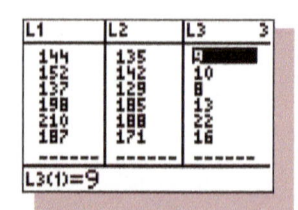

View Screen 146

2) INPUTS for a paired t-TEST USING THE DIFFERENCE DATA:

   a) at **Inpt**: highlight **Data** & press ENTER
   b) at μ_0: input *0*, then press ENTER.
   c) at **List**: press **2ND > 3**
   d) at **Freq**: input **1 & ENTER**
   f) at μ: highlight **> μ_0**, & press ENTER

   These inputs are show in VIEW SCREEN 147.

   View Screen 147

   g) highlight **Calculate** & press ENTER

   The test results are shown in VIEW SCREEN 148.

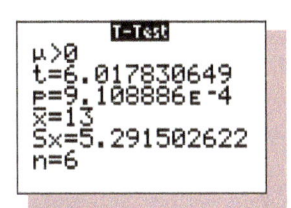

   View Screen 148

The test results are:

$\mu > 0$ is H_a
t = 6.02 is the t-score of the sample mean, also known as the test statistic.
p = 0.0009 is the p-value.
$\bar{x} = 13$ is the sample mean difference of the paired observations. We refer to this symbolically as $\bar{x}_d = 13$.
$S_x = 5.292$ is the sample standard deviation of the differences of the paired observations. We refer to this symbolically as $S_d = 5.292$.
n = 6 is the number of data pairs.

Since the p-value of 0.0009 is less than α = .01, we Reject H_o & Accept H_a at α = 1%

PROCEDURE TO INPUT OBSERVED VALUES OF A CHI-SQUARE TEST OF INDEPENDENCE INTO MATRIX A

1) INPUT OBSERVED VALUES INTO MATRIX A AS FOLLOWS.
   a) Press 2nd & MATRIX (x^{-1} key) to display matrix name menu as shown in VIEW SCREEN 149.

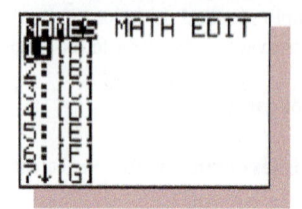

   VIEW SCREEN 149

   The first name: **1:[A]** refers to **MATRIX A** and is written **[A]**.

   b) HIGHLIGHT EDIT as shown in VIEW SCREEN 150.

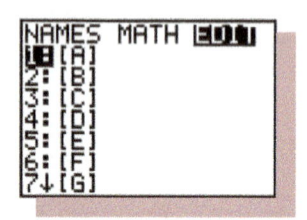

   VIEW SCREEN 150

   c) Select **1:[A]** for MATRIX[A] and its *size appears* as displayed in VIEW SCREEN 151.

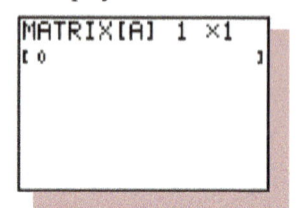

   VIEW SCREEN 151

   To define the size of MATRIX[A]:
   d) Input number of rows & ENTER
   e) Input number of columns & ENTER

   After **MATRIX[A] is defined,** you can enter the values of **[A]**.

WE WILL EXPLAIN HOW TO INPUT THE FOLLOWING OBSERVED VALUES LISTED IN THE 4x2 CONTINGENCY TABLE INTO MATRIX A.

Education Level		Grand Jury Selected		Not Selected
Elementary	a	24	b	36
Secondary	c	98	d	102
Some College	e	16	f	44
College Degree	g	42	h	38

TO INPUT THESE OBSERVED VALUES INTO A 4x2 MATRIX[A]:
a) Press 2nd & MATRIX (x^{-1} key)
b) HIGHLIGHT EDIT
c) Press 1 to select **1:[A]**
d) Press 4 ENTER 2 ENTER to define a 4x2 Matrix for **[A]**.

The cursor highlights the first element of the 4x2 MATRIX [A] as displayed in VIEW SCREEN 152.

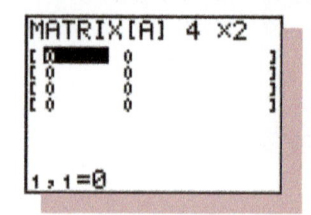

VIEW SCREEN 152

e) Input the first observed value **24** of **cell a** into [A] & ENTER.
f) Input the second observed value **36** of **cell b** into [A] & ENTER.

CONTINUE THIS PROCESS UNTIL ALL THE OBSERVED VALUES ARE ENTERED INTO MATRIX[A].

VIEW SCREEN 153 displays the final input screen for MATRIX[A].

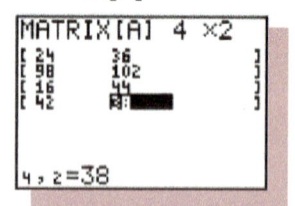

VIEW SCREEN 153

Notice the last cell value **38** in MATRIX[A] is highlighted. Its position within [A] is 4th row & 2nd column and is written at the bottom of the screen as 4,2 = 38. After all the observed results are entered into [A], you can conduct a chi-square test of independence.

Appendix D TI-84 Plus Instructions **A103**

PROCEDURE TO PERFORM A CHI-SQUARE HYPOTHESIS TEST OF INDEPENDENCE

TO PERFORM A CHI-SQUARE TEST AT $\alpha = 5\%$ TO DETERMINE IF A PERSON'S SELECTION TO SERVE ON THE GRAND JURY IS INDEPENDENT OF THEIR EDUCATIONAL LEVEL, WE WILL USE THE 4x2 CONTINGENCY TABLE.

Education Level	Grand Jury Selected	Not Selected
Elementary	a 24	b 36
Secondary	c 98	d 102
Some College	e 16	f 44
College Degree	g 42	h 38

To perform this chi-square test, input these observed values into a 4x2 matrix [A]:
  a) Press 2nd & MATRIX (x^{-1} key)
  b) HIGHLIGHT EDIT
  c) Press 1 to select 1:[A]
  d) Press 4 ENTER 2 ENTER to define a 4x2 Matrix for [A] shown in VIEW SCREEN 154.

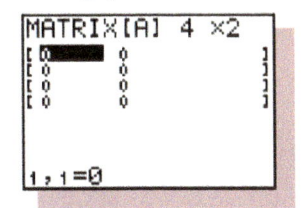

VIEW SCREEN 154

  e) Input all observed values into MATRIX [A]. VIEW SCREEN 155 shows the input screen for MATRIX[A].

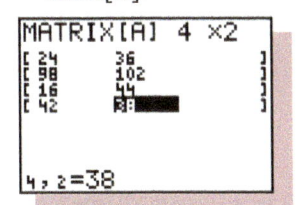

VIEW SCREEN 155

Notice the last cell value **38** in MATRIX[A] is highlighted. Its position is the 4th row & 2nd column of [A] and is written as 4,2 = 38 at the bottom of screen.

TO PERFORM THE CHI-SQUARE TEST:
1) Press STAT & Highlight TESTS
2) Highlight C: X^2-Test & ENTER
   The X^2- Test screen appears as shown in VIEW SCREEN 156.

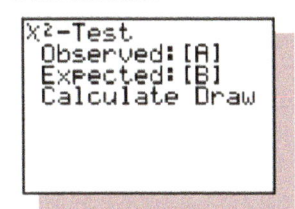

VIEW SCREEN 156

3) at **Observed:[A]** press ENTER to indicate the observed results are in MATRIX[A].
4) at **Expected:[B]** press ENTER to designate MATRIX[B] as the matrix where the expected results of the chi-square test will be stored.
5) HIGHLIGHT Calculate & ENTER
   The X^2-Test results are shown in VIEW SCREEN 157.

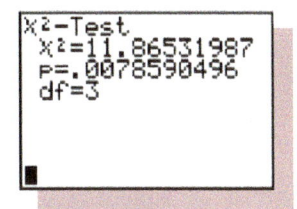

VIEW SCREEN 157

The X^2-Test results are:

$X^2 = 11.86531987$ *is Pearson's chi-square statistic,* X^2

$p = .0078590496$ *is the p-value*

$df = 3$ *is the degrees of freedom*

We *REJECT* H_0 & *ACCEPT* H_a at $\alpha = 5\%$ since the p-value $< 5\%$.

What is the conclusion if $\alpha = 1\%$?

TO RETRIEVE THE EXPECTED RESULTS:
1) Press 2nd & MATRIX (x^{-1} key)
2) press 2 for 2:[B] & ENTER to obtain the expected values of MATRIX[B] shown in VIEW SCREEN 158.

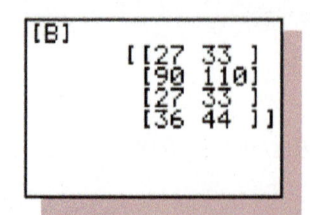

VIEW SCREEN 158

Notice the expected value in the 4th row & 1st column is 36.

FOR A GRAPH OF THE TEST RESULTS:

1) Press STAT & highlight TESTS
2) Highlight **C:** X^2 **-Test** & ENTER
3) Highlight Draw & ENTER

VIEW SCREEN 159 is displayed.

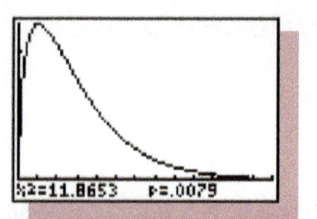

VIEW SCREEN 159

In VIEW SCREEN 152, *Pearson's chi-square statistic*, X^2 and the p-value are displayed. The right shaded tail area represents the p-value. Note: since the p-value is very small, the shaded area is not noticeable.

PROCEDURE TO CONSTRUCT A SCATTER DIAGRAM

CONSTRUCTING A SCATTER DIAGRAM

1) Enter the data values for each variable X and Y into a separate stat list.

   For this example, enter the X and Y values into **L1** and **L2** respectively.

   X: 23 24 25 26 27 28

   Y: 161 166 164 170 167 169

2) Press WINDOW to display the **window variables values.**

3) Set the window variables using the following guide for a **SCATTER DIAGRAM**.

For a SCATTER DIAGRAM:

set XMIN < smallest data value

set XMAX > largest data value

set XSCL = 1 OR appropriate frequency scale

set YMIN = < smallest data value

set YMAX > largest data value

set YSCL = 1 OR appropriate frequency scale

set Xres = 1

For this example, set the **window variables to:**

Xmin = 20   :less than the smallest X value

Xmax = 30   :greater than the largest X value

Xscl = 1   :set to one

Ymin = 155 :less than the smallest Y value

Ymax = 175 :greater than the largest Y value

Yscl = 1   :set to one

Xres = 1   :set to one

VIEW SCREEN 160. displays the WINDOW values for this scatter diagram.

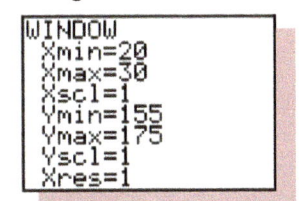

VIEW SCREEN 160

4) Press 2nd & STAT PLOT (Y = key)

   The **STAT PLOTS Menu** is displayed with **1:PLOT1 . . . .** highlighted.

5) Press ENTER to display **stat plot editor** screen for Plot 1.

6) Highlight ON & press ENTER

7) Highlight **SCATTER DIAGRAM**: 1st graph to right of TYPE & ENTER

8) For Xlist: : key in the name of data List. For this example, key in **L1** and press ENTER

9) For Ylist: : key in the name of data List. For this example, key in **L2** and press ENTER

10) For MARK: **select □ (1st item: box: □)** to designate how the points in the scatter diagram will be displayed & ENTER

The input screen for defining this scatter plot is shown in VIEW SCREEN 161.

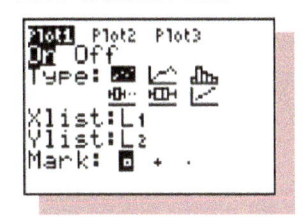

VIEW SCREEN 161

11) Press GRAPH to display the **scatter diagram** shown in VIEW SCREEN 162.

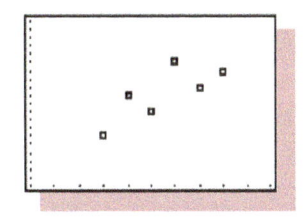

VIEW SCREEN 162

12) Press TRACE to identify the X and Y values of the points in the scatter diagram.

The X and Y values of the first point of the scatter diagram: **X = 23 and Y = 161** are displayed at the bottom of the screen as shown in VIEW SCREEN 163.

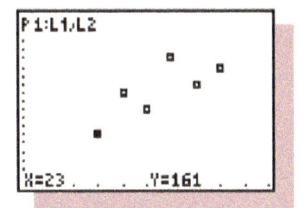

VIEW SCREEN 163

The **P1:L1,L2** in the upper left corner refers to stat plot P1 while the **L1,L2** indicates that the data values are stored in **Lists L1 and L2**.

13) Press cursor ▶ key to move to the next point on scatter diagram. This second point is identified by: **X = 24 and Y = 166.**

The 3rd point is **X = 25 and Y = 164.** And so on.

PROCEDURE TO PERFORM STATISTICAL ANALYSIS FOR 2-VARIABLES: TO DETERMINE THE REGRESSION COEFFICIENTS: a & b OF THE REGRESSION LINE y' = a + bx, PEARSON'S CORRELATION COEFFICIENT, r, AND THE COEFFICIENT OF DETERMINATION, r_2

To display the values of r and r^2, it is necessary to set the *Diagnostic display mode on.*

To set DiagnosticOn:

1) Press 2nd & CATALOG (0 KEY) and scroll to **DiagnosticOn** shown in VIEW SCREEN 164.

VIEW SCREEN 164

2) Press ENTER twice. DONE appears indicating mode is set to **DiagnosticOn**.

TO DISPLAY THE VALUES OF r, r^2 AND THE REGRESSION COEFFICIENTS

1) Input the X and Y values into **L1** and **L2** respectively.

   X: 23 24 25 26 27 28
   Y: 161 166 164 170 167 169

2) Press STAT & Highlight CALC
3) Press 8 for 8:LinReg(a+bx)
4) Input: XList name , YList name and press ENTER
   For this example, input: L1,L2

   The values of: **a, b, r and r^2** are displayed shown in VIEW SCREEN 165.

VIEW SCREEN 165

The **LinReg** results are:

y = a+bx

a = 130.4666667 *is the intercept of the regression line*

b = 1.4 *is the slope of the regression line*

r^2 = .6255319149 *is the coefficient of determination*

r = .790905756 *is Pearson's correlation coefficient*

PROCEDURE TO DISPLAY REGRESSION LINE AND TO COMPUTE THE PREDICTED Y VALUE FROM THE REGRESSION LINE

TO DISPLAY THE REGRESSION LINE

1) Input the X and Y values into **L1** and **L2** respectively.

   X: 23 24 25 26 27 28

   Y: 161 166 164 170 167 169

2) Press STAT & Highlight CALC
3) Press **8** for **8:LinReg(a+bx)**
4) Input L1,L2 & ENTER

   These 4 steps compute & store the regression variables required for the regression equation.

5) Press Y = . A blinking cursor appears to next to \Y1 = shown in VIEW SCREEN 166.

   VIEW SCREEN 166

6) Press VARS & press **5** for **5:Statistics. . .** . The **statistics vars menu** is displayed as shown in VIEW SCREEN 167.

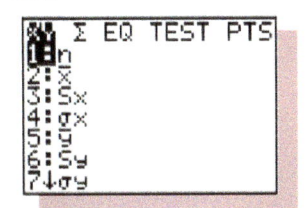

   VIEW SCREEN 167

7) HIGHLIGHT EQ (3rd COLUMN) shown in VIEW SCREEN 168.

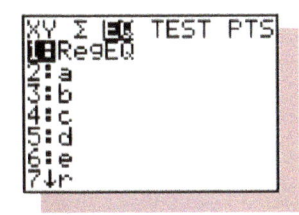

   VIEW SCREEN 168

8) Press ENTER to select 1:RegEQ . The **RegEQ: 130.4666666667 + 1.4X** appears next to \Y1 = shown in VIEW SCREEN 169.

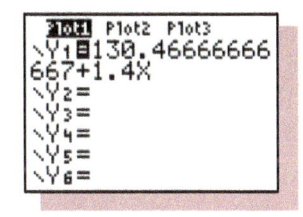

   VIEW SCREEN 169

9) Press GRAPH to draw regression line as shown in VIEW SCREEN 170.

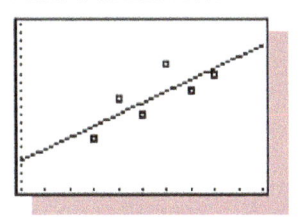

   VIEW SCREEN 170

10) Press TRACE to identify the X & Y values of the points on the REGRESSION LINE. Note the 1st point is: **X = 23 & Y = 161. P1:L1,L2** appears in left upper corner of the screen.

11) Press up cursor ▲ to place cursor ON THE REGRESSION LINE as shown in VIEW SCREEN 171.

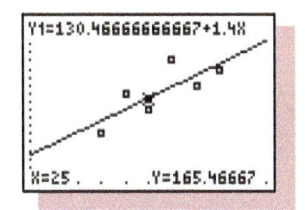

    VIEW SCREEN 171

The regression equation **Y1 = 130.46666666667 + 1.4X** appears in the upper left corner of the screen. Notice **THE POINT X = 25 and Y = 165.46667 IS DISPLAYED** at the bottom of the screen as shown in VIEW SCREEN 171. As the cursor moves along the regression line, the points of the regression line will be displayed at the bottom of the screen. *WHY IS THE Y VALUE: Y = 165.47 FOR X = 25 ON THE REGRESSION LINE DIFFERENT THAN THE Y VALUE: Y = 164 FOR X = 25 ON THE SCATTER DIAGRAM? EXPLAIN.*

TO PREDICT THE Y VALUE FOR A SPECIFIED X VALUE USING THE REGRESSION LINE

With the regression line graphed as shown in VIEW SCREEN 164, we can use the regression line to calculate a predicted y value(y') for a specified X value. To calculate a predicted y value for X = 26:

1) Press 2nd & CALC (TRACE key) to display the **CALCULATE menu** as shown in VIEW SCREEN 172.

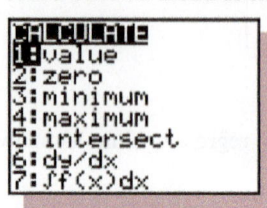

VIEW SCREEN 172

2) press ENTER to select 1:value. **X =** and a **blinking cursor** appears.

3) Input 26 for X value & ENTER The *predicted y value (y')*: $Y = 166.86667$ for $X = 26$ appears as shown in VIEW SCREEN 173.

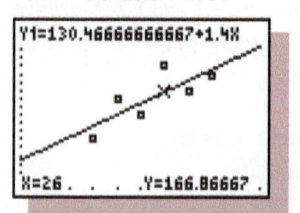

VIEW SCREEN 173

WHAT IS THE PREDICTED y VALUE (y') FOR x = 27?

Appendix D   TI-84 Plus Instructions   **A109**

PROCEDURE TO PERFORM A TWO-SAMPLE F-TEST

TO PERFORM A TWO-SAMPLE F-TEST
The TI–84 Plus 2-SampFTest performs an F-Test to compare two normal population standard deviations, S_1 and S_2.

Let's perform a TWO-SAMPLE F-TEST using the sample results of Example 16.4.

1) Press STAT & Highlight TESTS shown in VIEW SCREEN 174.

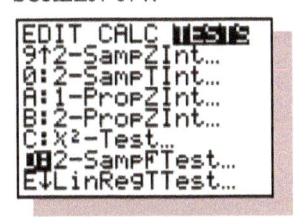

VIEW SCREEN 174

2) Press ALPHA & D (x^{-1} key) for **D:2-SampFTest.** . . . input screen for a 2 sample F test as shown in VIEW SCREEN 175.

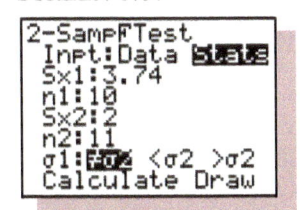

VIEW SCREEN 175

3) INPUT for a 2-SampFTEST USING SAMPLE RESULTS:
   a) at Inpt:Highlight Stats & ENTER
   b) at Sx1:std dev of sample1, s_1, & ENTER
   c) at n_1:size of sample1 & ENTER
   d) at Sx2:std dev of sample2, s_2, & ENTER
   e) at n_2:size of sample2 & ENTER
   h) at σ_1: highlight the form Ha, alternative hypothesis, & ENTER
   i) Highlight Calculate & ENTER

LET'S PERFORM A 2-SampFTEST USING THE INFO FROM EXAMPLE 16.4 WHERE
$\alpha = 0.10$ & H$_0$: $\sigma_1^2 = \sigma_2^2$ & H$_a$: $\sigma_1^2 \neq \sigma_2^2$

For sample1: $s_1 = 3.74$ & $n_1 = 10$
For sample2: $s_2 = 2$ & $n_2 = 11$

For this 2-SampFTEST, INPUT:
   a) at Inpt:Highlight Stats & ENTER
   b) Sx1: INPUT 3.74 & ENTER
   c) at n1: INPUT 10 & ENTER
   d) at Sx2: INPUT 2 & ENTER
   e) at n2: INPUT 11 & ENTER
   h) at σ_1: HIGHLIGHT $\neq S_2$ & ENTER

The 2-SampFTest input screen should contain the info shown in VIEW SCREEN 176.

VIEW SCREEN 176

j) Highlight Calculate & ENTER

2-SampTTest results are shown in VIEW SCREENS 177a and 177b.

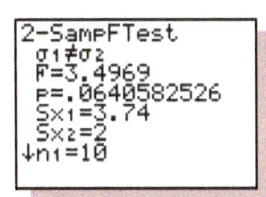

VIEW SCREEN 177a   VIEW SCREEN 177b

The 2-SampFTest results are:

2-SampFTest

$\sigma_1 \neq \sigma_2$

$F = 3.4969$ *is the test statistic F*

$p = .0640582526$ *is the p-value*

Sx1 = 3.74 *is std dev of sample 1*

Sx2 = 2 *is std dev of sample 2*

$n_1 = 10$ *is size of sample 1*

$n_2 = 11$ *is size of sample 2*

Since this 2-SampFTEST is being conducted at $\alpha = 0.10$ where H$_a$: $\sigma_1 \neq \sigma_2$, then using Table VII critical F-value is: $F_{0.05}(9,10) = 3.02$.

Since the test statistic, F = 3.5 (to 2 decimal places), is greater than the critical

F-value, $F_{0.05}(9,10) = 3.02$, reject H$_0$ and accept H$_a$. Thus, the population variances are significantly different at $\alpha = 0.10$.

PROCEDURE TO PERFORM ONE-WAY ANOVA

TO PERFORM A ONE-WAY ANALYSIS OF VARIANCE, ANOVA

The TI–84 Plus test function ANOVA(computes a one-way analysis of variance for comparing the means of 2 to 20 populations. The null hypothesis H_0:the population means are equal is tested against the alternative H_a: the population means are not all equal. To perform an ANOVA test, you must enter each sample data into a List.

Let's perform an ANOVA test using the information of Example 16.6.

1) INPUT the data of the 4 samples into Lists L1, L2, L3 & L4.

2) Press STAT & Highlight TESTS shown in VIEW SCREEN 178.

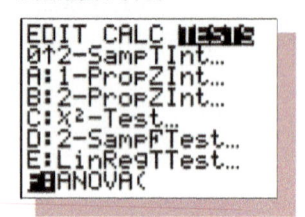

VIEW SCREEN 178

3) Select F : for F:ANOVA(
   ANOVA(appears on the home screen.

4) Input the 4 List names containing the sample data separating each List name by a comma to get: ANOVA(L1,L2,L3,L4) as shown in VIEW SCREEN 179.

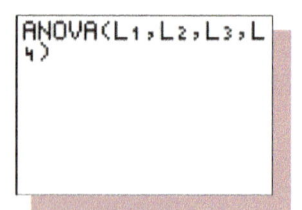

VIEW SCREEN 179

5) Press ENTER
   The ANOVA test results are shown in VIEW SCREENS 180a & 180b.

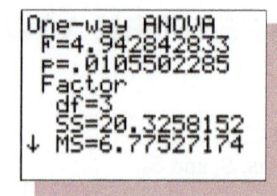

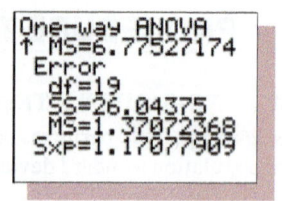

VIEW SCREEN 180a   VIEW SCREEN 180b

The One-way ANOVA results are:

F = 4.942842833 *is the test statistic*

p = .0105502285 *is the p-value*

Factor

df = 3 *is the degrees of freedom for the between treatments*

SS = 20.3258152 *is the sum of squares for between treatments*

MS = 6.77527174 *is the mean square for between treatments*

Error

df = 19 *is the degrees of freedom for within treatments*

SS = 26.04375 *is the sum of squares for within treatments*

MS = 1.37072368 *is the mean square for within treatments*

Sxp = 1.17077909 *is the pooled standard deviation*

Since this ANOVA test is being conducted at $\alpha = 0.05$, we need to look up the critical F-value, $F_{0.05}(3,19)$ in Table VII.

From Table VII, the critical F-value: $F_{0.05}(3,19) = 3.13$.
Since the calculated test statistic F, 4.94, is **greater than** the critical F-value (3.13), we **reject the null hypothesis, H_0, and accept the alternative hypothesis H_a** at $\alpha = 0.05$. Therefore, we conclude that the population means are **not** all equal.

APPENDIX E: DATABASES

DATABASE I: STUDENT CHARACTERISTICS

The following database represents a sample of 75 student records. The record layout for each student is as follows:

Column Label	Variable Name	Variable Description
C1	ID	identification number
C2	AGE	age in years
C3	SEX	gender
C4	MAJ	academic area of concentration
C5	AVE	high school average
C6	HR	hair color
C7	EYE	eye color
C8	WRK	hours worked per week
C9	STY	hours studied per week
C10	MAT	matriculation status
C11	GPA	grade point average
C12	NAP	hour of the day a nap is taken
C13	SHT	student height in inches
C14	PHT	same sex parent's height in inches
C15	AM	ideal age to marry
C16	MOB	month of birth
C17	LK	preference of soft drink: Coke, Pepsi, or neither
C18	HND	handedness
C19	GFT	amount in dollars spent on gift
C20	PH	number of hours spent on phone per week
C21	NB	response to question: Choose a number: 1, 2, 3, 4.
C22	AST	response to question: Do you believe in astrology?

Data Variable Values

Variable Name	Variable Value
ID	4 digit numeric
AGE	2 digit numeric
SEX	0=male, 1=female
MAJ	1=bus, 2=cmp, 3=fpa, 4=hth, 5=lib, 6=sci
AVE	2 digit numeric
HR	1=blonde 2=black, 3=brown, 4=red
EYE	1=blue, 2=brown, 3=green
WRK	2 digit numeric
STY	2 digit numeric
MAT	0=yes, 1=no
GPA	3 digit 2 decimal numeric
NAP	standard military hours
SHT	2 digit numeric
PHT	2 digit numeric
AM	2 digit numeric
MOB	2 digit numeric, 01 - 12, month
LK	1=n, 2=p 3=c
HND	0=left, 1=right
GFT	3 digit numeric
PH	2 digit numeric
NB	1 digit numeric
AST	0=yes, 1=no

Appendix E *Databases*

The database can be found at MyStatLab.com.

C1	C2	C3	C4	C5	C6	C7	C8	C9	C10	C11	C12	C13	C14	C15	C16	C17	C18	C19	C20	C21	C22
1618	19	0	5	81	3	1	16	11	0	1.87	1300	74	71	25	06	2	1	100	02	4	0
7290	20	1	5	87	1	1	12	19	0	3.05	2000	60	65	25	04	3	1	120	04	2	1
5506	21	0	1	85	3	3	15	10	1	3.62	1500	74	72	33	08	1	1	050	01	4	1
8496	19	1	2	88	3	3	14	09	0	3.67	1700	61	63	25	06	3	0	050	10	3	1
2832	18	1	1	72	3	3	08	23	0	2.25	0700	69	64	23	07	1	1	200	10	3	0
3407	22	0	5	83	3	1	14	20	0	2.70	1400	74	72	30	12	2	1	025	01	4	1
8573	19	1	5	78	3	2	15	17	1	2.30	1500	60	62	25	04	2	1	150	05	4	0
6265	20	0	6	76	3	2	22	16	1	2.51	0900	67	67	26	03	1	0	100	02	4	0
4821	19	0	1	74	1	1	18	22	1	2.10	2200	74	68	30	03	3	1	100	01	4	1
9029	19	0	1	79	1	1	10	12	0	1.92	1600	69	71	25	10	3	1	200	01	1	0
3001	19	1	5	83	3	1	10	20	1	3.15	1500	64	66	24	02	2	1	050	10	2	1
8010	23	1	1	78	3	2	20	11	0	2.00	1700	65	66	30	07	3	1	025	02	3	0
4699	19	1	5	74	1	3	26	09	0	1.95	1400	65	64	25	12	3	1	050	07	2	1
8298	24	0	1	88	3	3	12	19	1	2.88	1600	70	75	24	03	2	1	100	10	4	1
1896	18	1	5	81	3	1	18	09	1	2.65	1800	68	66	26	01	2	1	050	10	2	0
4789	19	1	5	84	3	1	17	08	0	1.98	0900	65	65	25	12	3	1	050	05	3	0
3592	27	1	1	83	3	1	15	19	1	3.21	2200	65	63	25	11	3	1	050	02	4	0
7679	21	0	1	87	1	1	16	07	0	1.95	1600	76	74	30	02	3	1	035	03	2	0
0209	18	1	6	80	3	1	22	10	1	2.96	1400	66	66	24	10	3	1	070	10	4	0
4489	21	0	5	84	1	3	14	08	0	2.95	1000	71	69	24	07	3	1	050	02	4	1
0880	18	0	1	90	3	2	08	18	0	2.00	1500	71	75	26	03	3	1	075	02	4	1
7168	37	1	5	87	3	2	11	16	1	2.82	2100	61	60	30	04	3	1	050	02	1	0
5709	28	1	1	85	2	2	07	10	1	1.95	1700	65	67	32	01	3	1	100	25	4	0
7520	20	1	5	77	2	2	25	12	1	3.01	1400	62	65	23	01	3	1	040	07	3	0
2757	20	1	4	72	1	1	27	15	0	2.70	1300	63	62	25	08	2	1	100	03	2	1
3016	19	1	5	83	3	3	20	11	0	2.67	1100	60	65	28	04	3	0	175	07	4	0
7298	21	0	2	77	3	3	12	09	0	2.81	1500	68	68	26	11	3	1	100	02	4	1
5677	18	0	6	85	3	1	16	27	1	3.76	1600	66	65	25	10	2	1	040	04	4	0
0406	21	1	2	70	3	2	14	10	1	2.67	1200	62	64	23	03	1	1	050	07	3	1
7673	19	1	4	77	3	2	16	07	0	2.32	1500	67	66	26	04	3	1	125	04	3	0
3480	20	1	1	83	4	3	24	08	0	2.09	1600	64	63	23	04	3	1	100	05	2	1
6994	18	1	5	86	3	3	19	09	0	2.42	1700	65	61	27	01	2	1	100	02	3	0
4009	18	1	5	84	3	2	10	10	1	1.90	1300	62	64	26	01	1	1	075	02	4	0
7879	20	0	6	77	2	2	08	07	1	1.98	1900	69	68	26	02	3	1	090	02	1	1
2021	19	1	5	82	1	3	20	12	0	3.15	1000	65	64	26	01	2	1	100	10	1	0
0044	19	1	4	90	1	2	00	14	0	3.40	1600	67	70	24	02	3	1	050	09	3	0
4508	19	1	1	84	1	3	15	10	1	2.87	2100	65	65	24	07	3	1	075	06	2	1
1085	23	1	5	91	3	2	10	28	1	3.51	1700	67	64	26	07	2	1	040	10	2	0
8289	20	1	1	78	2	2	22	08	1	1.95	1700	63	62	26	09	3	1	050	15	1	0
5664	18	0	5	75	3	1	26	06	0	2.43	1600	71	67	25	10	2	1	050	01	3	1
1284	19	1	5	81	2	2	12	08	0	2.47	1500	66	63	23	07	1	1	100	20	2	0
9406	19	0	6	79	2	2	14	10	0	2.66	1300	68	69	25	06	3	1	100	03	1	1
8022	19	0	1	88	1	3	00	12	1	3.28	1600	71	69	28	09	3	1	050	01	2	1
0954	25	1	5	87	3	2	09	15	1	3.41	1600	63	63	27	04	1	1	050	01	4	1
5012	20	0	5	75	2	2	10	21	1	2.20	2100	77	71	35	03	2	0	150	02	3	0
0013	19	1	5	80	3	1	30	11	1	2.32	1700	63	61	25	08	2	1	175	25	3	0
8809	22	1	4	88	1	2	08	20	1	2.97	1600	65	69	25	09	3	1	100	05	2	0
3697	20	1	1	93	3	2	16	29	1	3.75	1500	62	66	26	08	3	1	075	12	2	1
7807	18	1	5	77	3	2	18	20	0	2.78	1300	64	62	25	10	3	1	050	15	4	1
2058	30	1	4	82	3	2	20	15	1	2.84	1600	61	62	28	06	2	1	030	03	3	0
9801	18	1	4	79	3	2	08	10	1	3.10	1400	65	64	28	04	1	1	040	15	3	1
5464	18	1	5	75	1	1	24	06	0	2.49	1500	65	67	30	07	3	0	075	10	1	2
8002	24	1	4	88	3	2	10	10	1	3.24	1300	69	67	28	09	1	1	100	05	3	0
1919	18	0	1	81	3	1	09	21	1	2.95	1100	69	66	28	04	3	1	050	08	3	1
6728	19	0	1	86	3	2	17	23	1	2.81	1900	72	69	23	01	3	1	150	20	2	1
9603	26	1	4	79	2	2	16	10	0	2.36	1600	70	66	24	10	3	1	100	02	4	1
8871	22	0	2	88	3	2	08	17	1	3.10	2000	67	67	27	03	3	1	200	01	3	1
6587	19	1	5	86	3	2	15	28	1	3.78	2200	61	59	25	03	3	1	100	03	3	1
0250	19	0	5	90	3	2	12	15	0	3.08	1900	71	70	28	10	3	1	100	13	2	0
1010	19	0	1	81	3	2	20	11	1	3.07	1100	69	70	25	12	2	1	150	03	4	1

2721	22	1	5	92	1	2	17	12	0	3.15	1600	60	63	25	07	1	1	040	03	2	0
5256	18	1	5	95	1	2	12	15	1	3.65	1500	63	64	26	01	3	1	050	06	3	1
9847	18	0	1	89	3	2	08	14	1	3.70	1000	66	66	24	09	2	1	200	20	2	1
1617	20	1	5	81	2	1	16	21	1	2.73	1500	65	64	24	03	3	1	075	01	2	0
8077	20	0	6	88	3	1	10	17	1	3.75	1300	73	75	27	01	3	1	050	09	4	1
0001	19	1	5	90	3	2	10	20	1	3.17	1600	68	65	26	06	3	1	040	02	4	0
7755	31	0	1	87	3	2	37	25	0	3.56	1300	67	67	40	09	3	1	075	07	4	0
4607	18	1	3	84	3	2	16	30	0	3.75	1700	65	63	25	04	1	1	075	05	4	0
6930	20	1	1	76	3	2	28	23	1	2.05	1600	65	64	27	09	2	1	030	10	1	0
9788	19	0	5	89	3	2	17	28	1	3.85	1500	69	68	24	02	3	1	050	01	3	0
4697	22	0	5	74	3	2	16	09	0	2.74	1600	70	65	25	03	3	1	100	10	1	0
6796	18	1	1	86	3	2	17	09	0	2.64	1300	68	65	26	11	2	1	075	01	3	1
3529	23	1	3	73	3	2	15	12	0	2.93	1700	64	63	26	09	1	1	070	10	3	1
9244	19	0	5	79	1	1	12	24	0	2.47	1500	72	68	26	03	2	1	100	01	2	1
2581	24	0	3	82	3	2	25	08	1	2.18	1800	71	70	28	09	2	0	025	01	2	1

DATABASE II: STUDENT INTERESTS

The following survey was given to 75 community college students in the northeast.

The following are the names of the variables represented in each of the given questions.

1. Movie
2. Bike
3. Work
4. Gift
5. Study
6. Cell
7. Onlne
8. Drink
9. Gendr
10. Age

Survey: Please answer the following questions as honestly and accurately as possible.

1. Which movie category do you like **best**?   (*Choose One*)

   (a) Comedy   (b) Action Adventure   (c) Romance   (d) Drama   (e) Science Fiction/Horror   (f) Foreign Film
   (g) Other_____

2. To the best of your recollection, at what *age* did you learn how to ride a two-wheel bike?

3. Approximately how many *hours per week* do you work at a job for which you gain monetary reward?

4. If you were going to purchase a birthday gift for someone special, **approximately** *how much* would you spend? (nearest 10 dollars)

5. **Approximately how long (in hours per day)** do you put aside for studying your course work?

6. Who is your **carrier**?   (a) Verizon   (b) ATT / Cingular   (c) Nextel / Sprint   (d) T-Mobile   (g) Other

7. About how many **hours per day** do you spend online?

8. Do you drink alcoholic beverages on a regular basis?

   (a) Yes, *about* ONE alcoholic drink per day.
   (b) Yes, *about TWO or THREE* alcoholic drinks per day.
   (c) Yes, *more than THREE* alcoholic drinks per day.
   (d) NO, I DO NOT DRINK alcoholic beverages ON A REGULAR BASIS per day.

9. What is your age as of 4/1/2014?

10. What is your gender?

STUDENT INTEREST DATA

STDT	MOVIE	BIKE	WORK	GIFT	STUDY	CELL	ONLNE	DRINK	GENDR	AGE
2	6	5	21	50	3	2	3	4	1	18
4	7	7	0	100	20	3	5	4	1	40
5	2	8	0	20	2	4	4	4	1	19
6	2	6	24	100	5	1	1	4	1	21
8	3	7	0	120	1	1	3	4	1	18
9	7	5	0	70	1	1	1	1	1	19
10	2	6	24	100	5	5	3	4	1	20
11	2	7	21	40	1	4	4	4	1	19
13	3	7	0	20	5	2	3	4	1	20
14	2	9	20	100	1	2	3	4	1	19
16	1	5	10	100	2	3	3	4	1	17
17	2	13	30	80	2	1	12	4	1	18
18	1	5	0	100	0	2	2	4	1	20
20	7	7	35	250	1	4	2	1	1	22
27	4	7	0	100	2	3	1	4	1	20
29	1	5	30	150	1	2	1	2	1	19
31	3	5	20	150	2	4	4	2	1	21
32	2	6	20	50	0	1	2	1	1	19
35	1	5	30	100	4	2	4	4	1	25
36	4	6	20	100	3	2	8	4	1	18
37	1	6	0	40	1	1	1	2	1	18
40	1	10	30	60	2	2	5	4	1	18
41	1	5	12	150	2	4	1	4	1	23
42	3	6	27	100	1	5	2	4	1	20
43	1	8	20	20	3	2	4	4	1	18
46	2	10	4	30	3	3	9	4	1	19
47	5	5	40	90	2	1	2	4	1	24
49	1	8	40	40	2	4	1	1	1	20
50	4	6	35	100	1	4	2	4	1	19
53	7	10	7	100	2	2	2	4	1	19
54	1	4	20	100	5	3	5	2	1	21
55	3	7	40	50	1	4	1	4	1	21
58	1	5	24	100	5	4	1	4	1	19
60	4	5	50	100	3	5	2	4	1	22
61	1	7	0	100	0	1	3	4	1	20
62	2	6	0	100	1	2	2	4	1	23
66	2	10	40	150	4	1	5	4	1	19
68	4	7	30	30	5	4	5	4	1	20
75	4	5	40	150	1	2	6	4	1	19
1	1	5	0	40	2	5	1	4	2	18
3	1	5	40	20	1	1	3	4	2	21
7	1	7	10	90	2	2	3	4	2	20
12	1	7	20	20	4	2	5	4	2	20
15	2	4	35	50	2	2	1	4	2	19
19	7	7	40	300	0	3	1	4	2	19
21	7	7	8	40	1	4	2	4	2	18

(continues)

STUDENT INTEREST DATA (Continued)

STDT	MOVIE	BIKE	WORK	GIFT	STUDY	CELL	ONLNE	DRINK	GENDR	AGE
22	1	5	50	300	2	4	3	4	2	24
23	1	6	40	200	2	4	3	4	2	18
24	5	7	0	100	1	2	1	4	2	18
25	2	8	35	100	1	4	2	4	2	19
26	3	8	23	30	3	4	0	4	2	18
28	1	11	40	150	4	4	9	4	2	22
30	3	5	18	150	2	4	4	4	2	19
33	7	7	20	50	2	1	2	4	2	19
34	1	6	25	20	2	1	2	4	2	18
38	1	5	0	50	4	4	2	4	2	19
39	1	5	40	50	2	1	2	4	2	19
44	2	8	0	100	2	4	0	4	2	20
45	3	7	20	80	2	2	8	4	2	18
48	2	6	21	30	2	3	2	4	2	18
51	1	9	40	30	2	2	1	1	2	21
52	3	7	0	50	3	2	2	4	2	20
56	3	7	16	30	1	1	2	4	2	21
57	1	10	8	20	1	1	6	4	2	18
59	4	7	15	100	2	2	1	4	2	19
63	3	8	25	40	5	4	1	4	2	22
64	2	7	26	30	2	2	2	4	2	18
65	5	5	40	50	1	3	2	4	2	21
67	1	4	30	50	2	3	0	4	2	22
69	1	7	10	150	3	4	1	4	2	32
70	3	5	12	150	1	2	1	4	2	19
71	7	9	6	86	3	2	2	4	2	35
72	1	6	33	100	3	1	0	4	2	18
73	3	6	30	100	2	1	1	4	2	20
74	1	7	0	15	2	1	1	4	2	22

DIRECTIONS:

The following 78 exercises refer to DATABASES I & II listed in Appendix E. A survey was given to a sample of 75 community college students in the northeast from September 2008 to September 2010. DATABASE I contains 75 records on Student Characteristics and DATABASE II contains 75 records on Student Interests. The survey results as well as a copy of the surveys used are in Appendix E. The survey and the survey results can be downloaded from the **Additional Resources** link within MyStatLab. The downloadable files are in both Microsoft EXCEL and TI-84 format.

The following questions refer to **DATABASE I: Student Characteristics:**

#1 2 5 6 7 8 15 16 17 18 19 22 23 25 26 29
30 33 34 37 38 41 42 45 46 49 50 54 55 57 58
61 64 65 67 68 69 70 71 72 73 74 75 76 77 78

The following questions refer to **DATABASE II: Student Interests:**

#3 4 9 10 11 12 13 14 20 21 24 27 28 31 32 35
36 39 40 43 44 47 48 51 52 53 56 59 60 62 63 66

Using your technological tool, answer each exercise.

1. For the following variables, identify what each variable represents.
   a) AGE   b) AVE
   c) SEX   d) MAJ
   e) STY   f) GPA
   g) GFT

   For the variables SEX and MAJ identify what each data value represents.
   For the other variables except SEX and MAJ determine the lowest and highest data values.

2. Discuss the characteristics of the student whose ID number is:
   1) 8573   3) 8809
   2) 7679   4) 3529

3. By examining the survey, identify the questions that yield numerical versus non-numerical results.

4. Examine the survey and EXCEL file, and discuss the characteristics of the student responses for student numbers:
   a) 10   b) 75   c) 122

5. Construct a Stem-and-Leaf display for the variables:
   a) AGE   b) AVE
   c) WRK   d) STY
   Describe the general shape of each of the distributions using the terms developed in Section 2.7.

6. Construct a Histogram for the variables:
   a) GPA using increments of 0.5 by MAJ.
   b) Which majors have the highest GPAs?
   c) Which majors have the lowest GPAs?

7. Construct a Frequency Polygon for the variables:
   a) AGE
   b) AGE using increments of 5
   c) AGE by GENDER using increments of 5 on the same set of axes.
   How do the shapes of the graphs differ? Compare the shapes of the graphs for males versus females.

8. Construct a Pie Chart using the variables:
   a) LK   b) HND
   c) AST   d) MAJ
   For each of these variables, construct a Pie Chart by the variable GENDER.

9. Construct a stem-and-leaf display for the variables,
   a) WORK   b) ONLNE
   c) STUDY d) BIKE
   Describe the shape of each of the distributions using the terms developed in Section 2.7.

10. Construct a back to back stem-and-leaf plot for the variables
    a) GIFT separated by GENDR
    b) WORK separated by GENDR

11. Construct a frequency table and frequency histogram for the variable GIFT using a class width of
    a) 20
    b) 40
    Compare the shapes of the distributions for the histograms from parts a and b.

12. Construct a frequency table and bar graph for the categorical variable MOVIE.

13. Construct a frequency table and frequency polygon for the variable WORK. Use a class width of 5.

14. Construct a frequency table and bar graph for the variable MOVIE separated by GENDER.

15. Determine the measures of central tendency for the variables:
    a) AGE
    b) AVE
    c) STY
    d) GPA
    e) GFT

16. Display a table representing the effect of subtracting the mean of each variable from the data for the variables:
    a) AGE
    b) GPA
    c) GFT
    The table should include columns representing:
    1) the variable name along with the data values
    2) the data value minus the mean, as well as the *mean* of the variable and the *mean* of the column for the data value minus the mean.
    Examine the results within each column. Did you expect these results? Comment.

17. Determine the range and sample standard deviation for the variables:
    a) AGE   b) AVE
    c) STY   d) GPA
    e) GFT

18. Determine the mean and standard deviation of the variables:
    a) AVE   b) STY   c) GPA
    for
    1) the male students (Males are coded 0)
    2) the female students (Females are coded 1)

19. Numerically describe each of the given variables and answer the question using the statistical results:
    a) AGE
       What percent of the student ages are between 17 and 20?
    b) AVE
       What percent of the students have high school averages greater than 80?
    c) STY
       What percent of the students study more than 20 hours per week?
    d) GPA
       What percent of the students have grade point averages less than 3.0?
    e) GFT
       Determine the data value representing P_{90}.

20. For each of the following variables determine:
    i) Measures of Central Tendency,
    ii) Measures of Variability,
    iii) The 5-Number Summary and Box-and-Whisker Plot.
    a) WORK,
    b) GIFT,
    c) STUDY
    d) ONLNE

21. Using the variables in question 20, separate each variable by GENDR and determine for each gender:
    i) Measures of Central Tendency,
    ii) Measures of Variability,
    iii) The 5-Number Summary and Box-and-Whisker Plot.

22. a) For the variables: AVE and GPA, identify the independent and the dependent variable.
    b) Plot a scatter diagram for the variables AVE and GPA.
    c) Calculate the correlation between AVE and GPA. Determine the strength of the correlation between the two variables.
    d) Determine the regression equation.
    e) Calculate and interpret r^2.

23. a) For the variable names: SHT and PHT, identify the independent and the dependent variable.
    b) Plot a scatter diagram for the variables SHT and PHT.
    c) Calculate the correlation between SHT and PHT. Determine the strength of the correlation between the two variables.
    d) Determine the regression equation.
    e) Calculate and interpret r^2.

24. Using the variables WORK and STUDY,
    a) Construct a scatter diagram.
    b) By examining the scatter diagram estimate whether you believe there is a weak or strong linear relationship between the variables in part a. Why?
    c) Calculate the correlation coefficient for the variables WORK and STUDY. Is the value of the correlation coefficient consistent with your response to part b?

25. Count the number of *each type* of categorical value for the variables:
    a) SEX
    b) MAJ
    c) EYE
    d) HR
    e) MOB
    What is the probability of selecting each type of categorical value?

26. Construct a histogram for the variables:
    a) AGE
    b) GPA
    c) GFT
    Use the histogram to determine the probability that a student selected at random:
    1) will be at least 19 years of age.
    2) will have a grade point average of 2.5.
    3) will spend at most $100 for a gift.

27. Use the Relative Frequency Approach to calculate probabilities for the variables:
    a) MOVIE
    b) DRINK
    c) CELL

28. Construct a bar graph for the variables:
    a) MOVIE
    b) DRINK
    c) CELL
    For each variable, which outcome has the lowest and highest probability of occurring?

29. Using Technology, count the number of each discrete data value for the variable NB.
    a) Calculate the proportion of the sample that correspond to each data value.
    b) List the data values and their corresponding proportions and verify that this table is a discrete probability distribution.
    c) Find the mean, standard deviation and probability histogram of the probability distribution.
    d) Interpret the meaning of the values obtained for the mean and standard deviation in terms of the variable NB.

30. Using Technology, count the number of each discrete data value for the variable GFT.
    a) Calculate the proportion of the sample that correspond to each data value.
    b) List the data values and their corresponding proportions and verify that this table is a discrete probability distribution.
    c) Find the mean, standard deviation and probability histogram of the probability distribution.

d) Interpret the meaning of the values obtained for the mean and standard deviation in terms of the variable GFT.

e) What is the lower and upper dollar amounts that a student will spend on a gift if the value is to be within 1 standard deviation of the mean? What is the likelihood of this occurrence?

31. For the continuous variable BIKE, treat the data values as if they are measured discretely (measured to the nearest whole number).
    a. Construct a probability histogram.
    b. Determine the mean and standard deviation.

32. For the continuous variable ONLNE, treat the data values as if they are measured discretely (measured to the nearest whole number).
    c. Construct a probability histogram.
    d. Determine the mean and standard deviation.

33. Determine the mean and standard deviation for the variable AVE.
    a) Using a technological tool, obtain the probability of getting an average less than 78.
    b) Sort and display the data values for the variable AVE and find the proportion of data values below 78.
    c) Compare the results of parts (a) and (b).

34. Determine the mean and standard deviation for the variable GPA.
    a) Using a technological tool, obtain the probability of getting an average less than 3.00.
    b) Sort and display the data values for the variable GPA and find the proportion of data values less than 3.00.
    c) Compare the results of parts (a) and (b).

35. Use the variable BIKE to:
    a) Determine the mean and standard deviation for each variable.
    b) What percent of the data values are within one standard deviation from the mean?
    c) What percent of the data values are within two standard deviations from the mean?
    d) Using your answers from parts b and c, would you expect that the distribution of data values for the variable BIKE are normally distributed? Why?

36. Use the variable ONLNE to:
    a) Determine the mean and standard deviation.
    b) What percent of the data values are within one standard deviation from the mean?
    c) What percent of the data values are within two standard deviations from the mean?
    d) Using your answers from parts b and c, would you expect that the distribution of data values for the variable ONLNE are normally distributed? Why?

37. For each of the following variables:
    a) AGE   **b)** AVE
    c) STY   **d)** GPA
    e) GFT
    i) compute the mean and standard deviation.

    ii) select a random sample of 20 student records.
    iii) compute the mean and standard deviation of each of these random samples.
    iv) compute the sampling error for the mean.

38. For each of the following variables:
    a) AGE   **b)** AVE   **c)** WRK
    i) compute the mean and standard deviation.
    ii) construct a table of 10 random samples, where each sample contains 20 data values.
    iii) calculate the mean for each of the 10 random samples.
    iv) Form a distribution of the 10 sample means by recording the mean of each random sample within the table.
    Determine the mean and standard error for this distribution of sample means.
    v) Compare the results of part (iv) to the original mean and standard deviation computed in part (i). Explain any *differences* in the statistics in light of the theorems and properties discussed in this chapter.

39. For each of the following variables:
    a) AGE
    b) WORK
    c) BIKE
    i) Compute the mean and standard deviation.
    ii) Select a random sample of 10 student records.
    iii) Compute the mean and standard deviation of the sample.
    iv) Compute the sampling error for the mean.

40. For each of the following variables:
    a) AGE
    b) WORK
    c) BIKE
    i) Compute the mean and standard deviation.
    ii) Construct a table of 10 random samples where each sample contains 20 data values.
    iii) Calculate the mean for each of the 10 random samples.
    iv) Form a distribution of the 10 sample means by recording the mean of each random sample within the table. Determine the mean and standard error of this distribution of sample means.
    v) Compare the results of part (iv) to the original mean and standard deviation computed in part (i). Explain any differences in the statistics in light of the theorems and properties discussed in this chapter.

41. For the variable names: GPA, AVE, WRK, STY, GFT.
    a) Construct 90%, 95% and 99% confidence intervals for each of the variables.
    b) Interpret each of the confidence intervals in terms of the population means to which they are inferring.

42. For the variable names: SEX, EYE, HR, MAJ.
    a) Construct 90%, 95% and 99% confidence intervals for each categorical value of the population proportion contained within each variable.

b) Interpret each of the confidence intervals in terms of the population proportion to which they are inferring.

43. For the variables: WORK, ONLNE, GIFT, BIKE.
a) Construct 90%, 95% and 99% confidence intervals.
b) Interpret each of the confidence intervals in terms of the population means to which they are inferring.

44. For the variables: DRINK, MOVIE.
a) Construct 90%, 95% and 99% confidence intervals for each categorical value of the population proportion contained within each variable.
b) Interpret each of the confidence intervals in terms of the population proportions to which they are inferring.

45. For each variable, conduct a hypothesis test. Test the null hypothesis that the population mean is: $\mu = k$ against a nondirectional alternative hypothesis. Have your instructor help you interpret the output.
a) $k = 18$ for the variable AGE.
b) $k = 85$ for the variable AVE.
c) $k = 8$ for the variable STY.
d) $k = 2.75$ for the variable GPA.
e) $k = 100$ for the variable GFT.

46. For each variable, conduct a hypothesis test. Test the null hypothesis that the population mean is: $\mu = k$ against a directional alternative hypothesis. Have your instructor help you interpret the output.
a) $k = 18$ for the variable AGE and perform a greater than test.
b) $k = 85$ for the variable AVE and perform a less than test.
c) $k = 8$ for the variable STY and perform a greater than test.
d) $k = 2.75$ for the variable GPA and perform a greater than test.
e) $k = 100$ for the variable GFT and perform a less than test.
Have your instructor help you interpret the output.

47. For each variable, construct a hypothesis test. Test the null hypothesis that the population mean is: $\mu = k$ against a non-directional alternative hypothesis. Have your instructor help you interpret the output.
a) $k = 20$ for the variable WORK
b) $k = 2.5$ for the variable STUDY
c) $k = 100$ for the variable GIFT

48. For each variable, construct a hypothesis test. Test the null hypothesis that the population mean is: $\mu = k$ against a directional alternative hypothesis. Have your instructor help you interpret the output.
a) $k = 20$ for the variable WORK and perform a greater than test.
b) $k = 2.5$ for the variable STUDY and perform a less than test.
c) $k = 100$ for the variable GIFT and perform a less than test.

49. Use a 2TT to test the null hypothesis at $\alpha = 5\%$ and the population mean, $\mu = k$.
a) $k = 68$ for the variable SHT.
b) $k = 15$ for the variable WRK.
State the null and alternative hypothesis for each variable in words and symbols. Identify the sample mean, the z or t score, p-value and interpret. Is the null hypothesis rejected? State the conclusion in words with respect to the variable tested.

50. For the variable name: SEX
The chairman of the math department believes that the proportion of female students is 0.5.
Test the claim that the proportion of female students is *not* 0.5.
State the null and alternative hypotheses for this test. Is the null hypothesis rejected? Identify the p-value and interpret.
State the conclusion.

51. Use a 2TT to test the null hypothesis at $\alpha = 5\%$ and the population mean $\mu = k$.
a) $k = 20$ for the variable WORK.
b) $k = 2.7$ for the variable STUDY
c) $k = 100$ for the variable GIFT

State the null and alternative hypothesis for each variable in words and in symbols. Identify the sample mean, the z or t score, p-value and interpret. Is the null hypothesis rejected? State the conclusion in words with respect to the variable tested.

52. An advertisement claims that Verizon has captured 25% of the cell phone market in the northeast. You believe this claim is too high. Test the advertisements claim at $\alpha = 5\%$.

State the null and alternative hypothesis for each variable in words and in symbols. Identify the sample mean, the z or t score, p-value and interpret. Is the null hypothesis rejected? State the conclusion in words with respect to the variable tested.

53. Executives in Hollywood feel that 18% of college students feel that their preferred movie choice is an Action/Adventure film. You believe this claim is too low. Test their claim at $\alpha = 1\%$.

State the null and alternative hypothesis for each variable in words and in symbols. Identify the sample proportion, the z or t score, p-value and interpret. Is the null hypothesis rejected? State the conclusion in words with respect to the variable tested.

54. For the variables: SEX and AST.
a) Conduct a hypothesis test to determine if there is a difference between males and females regarding the question—"Do you believe in Astrology?"
b) State the null and alternative hypotheses.
c) Is the null hypothesis rejected? Why? What is the p-value?

55. For the variable names: SEX and HND.
   a) Conduct a hypothesis test to determine if there is a difference between males and females regarding the proportion of *left handedness* for each gender.
   b) State the null and alternative hypotheses.
   c) Is the null hypothesis rejected? Why? What is the p-value?
   d) Conduct a hypothesis test that compares *the proportion of right handedness* for each gender.

   A survey on Student Interests was given to a sample of community college students in the northeast from September 2008 to September 2010. The results of the survey as well as a copy of the survey used are located in Appendix A. The survey and the survey's results can be downloaded from the textbook's web site at www.MyStatLab.com. The downloadable files are in both Microsoft EXCEL and TI-84 format.

56. For the variables GENDR and MOVIE, conduct a hypothesis test to determine if there is a difference between males and females and their movie choice comedy. Use a 5% level of significance.

   State the null and alternative hypothesis for each variable in words and in symbols. Identify the sample proportions, the z or t score, p-value and interpret. Is the null hypothesis rejected? State the conclusion in words with respect to the variable tested.

57. For the variable names:
   a) AVE
   b) SEX

   Test the hypothesis that the mean high school average of females is not equal to the mean high school average of males.
   State the null and alternative hypotheses for this test. Identify the sample statistics, the t score, p-value and interpret. Is the null hypothesis rejected? State the conclusion in words with respect to the variable tested.

58. For the variable names:
   a) GFT
   b) SEX

   Test the hypothesis that the mean amount of dollars spent on a gift by males is more than the mean amount of dollars spent on a gift by females.
   State the null and alternative hypotheses for this test. Identify the sample statistics, the t score, p-value and interpret. Is the null hypothesis rejected? State the conclusion in words with respect to the variable tested.

59. For the variable names:
   a) WORK
   b) GENDER

   Test the hypothesis that the population mean number of work hours of females is not equal to the population mean work hours of males.
   State the null and alternative hypothesis for each variable in words and in symbols. Identify the sample mean, the z or t score, p-value and interpret. Is the null hypothesis rejected? State the conclusion in words with respect to the variable tested.

60. For the variable names:
   a) GIFT
   b) GENDER

   Test the hypothesis that the population mean amount of dollars spent on a gift for someone special is greater for males than it is for females.
   State the null and alternative hypothesis for each variable in words and in symbols. Identify the sample mean, the z or t score, p-value and interpret. Is the null hypothesis rejected? State the conclusion in words with respect to the variable tested.

61. For the variables Hair color and Eye color:
   a) construct a contingency table for the variables HR and EYE.
   b) calculate Pearson's chi-square test statistic along with the expected and the observed frequencies in each cell.
   c) test the significance at $\alpha = 5\%$.
   d) interpret the results in terms of the variables Hair color and Eye color.

62. For the variables GENDR and MAJ.
   a) construct a table for the variables GENDR and MAJ.
   b) calculate Pearson's chi-square test statistic along with the expected and the observed frequencies in each cell.
   c) test the significance at $\alpha = 5\%$.
   d) interpret the results in terms of the variables GENDR and MAJ.

63. For the variables GENDR and MOVIE:
   a) Construct a contingency table of GENDR by MOVIE.
   b) Calculate Pearson's chi-square statistic along with the expected frequencies.
   c) Test the significance at $\alpha = 5\%$. Interpret the results in terms of the variables GENDR and MOVIE.

64. a) For the variables: AVE and GPA, identify the independent and the dependent variable.
   b) Plot a scatter diagram for the variables AVE and GPA.
   c) Calculate the correlation between AVE and GPA. Determine the strength of the correlation between the two variables. Identify the p-value and determine if the correlation coefficient is significant at $\alpha = 5\%$.
   d) Determine the regression equation.
   e) Calculate and interpret r^2.

65. a) For the variable names: SHT and PHT, identify the independent and the dependent variable.
   b) Plot a scatter diagram for the variables SHT and PHT.
   c) Calculate the correlation between SHT and PHT. Determine the strength of the correlation between the two variables. Identify the p-value and determine if the correlation coefficient is significant at $\alpha = 5\%$.

d) Determine the regression equation.
e) Calculate and interpret r^2.

66. Using the variables WORK and STUDY,
a) Construct a scatter diagram.
b) By examining the scatter diagram estimate whether you believe there is a weak or strong linear relationship between the variables in part a. Why?
c) Calculate the correlation coefficient for the variables WORK and STUDY.
d) Identify the p-value of the correlation coefficient and determine if it is significant at $\alpha = 5\%$.

67. Using TECHNOLOGY,
a) on the data values for variable GPA and MAJ perform a one-way ANOVA to determine if there is a significant difference in the mean GPA for the indicated student majors at $\alpha = 0.05$.
b) Identify the independent and dependent variable.
c) What is the calculated test statistic F?
d) Identify the p-value for the test statistic.
e) Is the F-statistic significant?
f) Interpret the results in terms of the null and alternative hypotheses.

68. Using TECHNOLOGY,
a) on the data values for variable GFT and EYE to perform a one-way ANOVA to determine if there is a significant difference in the mean GFT for the indicated student eye color at $\alpha = 0.05$.
b) Identify the independent and dependent variable.
c) What is the calculated test statistic F?
d) Identify the p-value for the test statistic.
e) Is the F-statistic significant?
f) Interpret the results in terms of the null and alternative hypotheses.

69. Using TECHNOLOGY,
a) on the data values for variable AM and HR to perform a one-way ANOVA to determine if there is a significant difference in the mean AM for the indicated hair colors at $\alpha = 0.05$.
b) Identify the independent and dependent variable.
c) What is the calculated test statistic F?
d) Identify the p-value for the test statistic.
e) Is the F-statistic significant?
f) Interpret the results in terms of the null and alternative hypotheses.

70. Using TECHNOLOGY,
a) on the data values for variable STY and MAJ to perform a one-way ANOVA to determine if there is a significant difference in the mean STY for the indicated student majors at $\alpha = 0.05$.
b) Identify the independent and dependent variable.
c) What is the calculated test statistic F?
d) Identify the p-value for the test statistic.
e) Is the F-statistic significant?
f) Interpret the results in terms of the null and alternative hypotheses.

71. a) Construct a STEM-and-LEAF Display and a BOX PLOT for the variables: GFT and PH
b) Describe the shape of each variable.
c) Explain why it is reasonable to perform a nonparametric test on each variable.
d) For the variable GFT test that the population median amount spent on a gift is not equal to 100 dollars. Use $\alpha = 0.05$.
e) For the variable PH test that the population median number of hours spent on the phone per week is less than 5. Use $\alpha = 0.05$.

72. a) For the variable GFT, use the Mann-Whitney test to determine if there is a difference in the spending habits of males versus females. Use $\alpha = 0.05$.
b) By examining the output, determine
1. the sample size for each group.
2. W
3. the p-value.

73. a) For the variable PH, use the Mann-Whitney test to determine if there is a difference in the spending habits of males versus females. Use $\alpha = 0.05$.
b) By examining the output, determine
1. the sample size for each group.
2. W
3. the p-value.

74. a) For the variable WRK, use the Kruskal-Wallis H test to determine if there is a difference in the number of hours worked for the variable academic area of concentration (MAJ). Use $\alpha = 0.05$.
b) By examining the output, determine
1. the sample size for each major.
2. H
3. df
4. the p-value.

75. a) For the variable GPA, use the Kruskal-Wallis H test to determine if there is a difference in the number of hours worked for the variable academic area of concentration (MAJ). Use $\alpha = 0.05$.
b) By examining the output, determine
1. the sample size for each major.
2. H
3. df
4. the p-value.

76. For the variable STY, use the 75 data values listed in the database as the original random sample.
a) Compute the mean of the original sample.
b) Construct a bootstrap distribution of the mean using B = 2,000.
c) Construct a histogram of the bootstrap sampling distribution of the mean.
d) The central limit theorem (chapter 8) states the true sampling distribution of the mean is approximately a normal distribution for a large sample size. Examine the shape of the histogram you constructed in part d and comment on what you expected the histogram shape to be?
e) Compute the mean of the bootstrap distribution and compare it to the mean of the original sample. What did you expect to find when comparing these two results?

f) Compute the bootstrap standard error of the mean for the bootstrap sampling distribution of the mean.

g) Use the original sample data and the formula $\frac{s}{\sqrt{n}}$ to estimate the standard error of the mean for the true sampling distribution of the mean.

h) Compare the result of part g to the bootstrap standard error of the mean. How close are the results?

i) Using the results of parts f and g, what is the relationship between the standard deviation of the bootstrap distribution of the mean to the standard error of the true sampling distribution of the mean?

j) What is the bootstrap estimate of the bias from your bootstrap sample means of the bootstrap distribution? What does this tell us about the bias encountered in using the mean of the original sample to estimate the mean number of hours student's study per week?

k) If the histogram of the bootstrap sampling distribution of the mean is approximately normal, then apply the bootstrap standard error t confidence interval formula to construct a 95% confidence interval of the mean.

l) Construct a 95% confidence interval using the bootstrap percentile confidence interval method and the 2,000 bootstrap sample means provided it is reasonable to use this method.

m) Construct 95% confidence interval using the basic bootstrap confidence interval method and the 2,000 bootstrap samples provided it is reasonable to use this method.

n) Compare the 95% confidence intervals for the three different methods. What do you notice about the width of the 95% confidence intervals for the different methods?

77. For the variable GFT, use the 75 data values listed in the database as the original random sample.
  a) Compute the median of the original sample.
  b) Construct a bootstrap distribution of the median using B = 2,000.
  c) Construct a histogram of the bootstrap sampling distribution of the median.
  d) Compute the mean of the bootstrap sampling distribution of the median and compare it to the median of the original sample. What did you expect to find when comparing these two results?
  e) Compute the bootstrap standard error of the median using the bootstrap distribution of the median. Did you have a formula to compute this standard error? Explain.
  f) Why is the bootstrapping procedure essential when generating the bootstrap sampling distribution of the median?
  g) What is the bootstrap estimate of the bias from your bootstrap sample medians of the bootstrap samling distribution? What does this tell us about the bias encountered in using the median of the original sample to estimate the median dollar amount spent on a gift for a loved one?
  h) Construct a 95% confidence interval using the bootstrap percentile confidence interval method and the 2,000 bootstrap sample medians provided it is reasonable to use this method.
  i) Construct 95% confidence interval using the basic bootstrap confidence interval method and the 2,000 bootstrap sample medians provided it is reasonable to use this method.
  j) Compare the 95% confidence intervals for the two different methods. What do you notice about the width of the 95% confidence intervals for the different methods?

78. For the variable WRK, use the 75 data values listed in the database as the original random sample.
  a) Compute the standard deviation of the original sample.
  b) Construct a bootstrap distribution of the standard using B = 2,000.
  c) Construct a histogram of the bootstrap sampling distribution of the standard deviation.
  d) Compute the mean of the bootstrap sampling distribution of the standard deviation and compare it to the standard deviation of the original sample. What did you expect to find when comparing these two results?
  e) Compute the bootstrap standard error of the standard deviation using the bootstrap sampling distribution of the standard deviation. Did you have a formula to compute this standard error? Explain.
  f) Why is the bootstrapping procedure essential when generating the bootstrap sampling distribution of the standard deviation?
  g) What is the bootstrap estimate of the bias from your bootstrap sample standard deviations of the bootstrap sampling distribution? What does this tell us about the bias encountered in using the standard deviation of the original sample to estimate the standard deviation of the student work hours per week?
  h) Construct a 95% confidence interval using the bootstrap percentile confidence interval method and the 2,000 bootstrap sample standard deviations provided the histogram of the bootstrap distribution is approximately normal with small bias.
  i) Construct 95% confidence interval using the basic bootstrap confidence interval method and the 2,000 bootstrap sample standard deviations provided the histogram of the bootstrap distribution is approximately normal with small bias.
  j) Compare the 95% confidence intervals for the two different methods. What do you notice about the width of the 95% confidence intervals for the different methods?

INDEX

A

Absolute variation, 127
Acceptance sampling, 336–38, 545
Adjusted R^2, 784
Alternative hypothesis (H_a)
   analysis of variance, 824, 836
   difference between population means, 675–76, 681–82, 689, 702
   vs. hypothesis, 546–48
   Kruskal-Wallis H test, 887
   Mann-Whitney test, 879
   population correlation coefficient, 769–70
   population means, 594, 836
   population proportion, 585, 638
   sign test for median, 870–71
   variance of population, 810
Analysis of variance (ANOVA), 816, 822–49
   assumptions for, 837
   definition, 823
   formulas, 835–36
   hypothesis testing model and procedure, 826, 836
   table, 828
Approximately normal, 465
Assumptions
   for multiple regression, 782
   for paired t-test, 710
At least/at most, 239
Average, 88

B

Back-to-back stem-and-leaf display, 42–43
Bar graph, 19, 22, 54–57
Basic bootstrap confidence interval, 930–33
Bell-shaped distribution, 452
Best fitting line, 195
Between Group Variance, 816–17
Bias-corrected and accelerated (BCa) bootstrap interval method, 934–35
Biased sample, 11, 20–21
Bias of a statistic, 909
Bimodal distribution, 71
Binomial distribution
   definition, 422
   mean, 423
   normal approximation to, 428
   standard deviation, 423–25
Binomial experiment, 322
Binomial probability distribution, 320–40
   acceptance sampling, 336–38
   formula, 324
   mean, 335
   standard deviation, 335
Binomial probability formula, 324

Bootstrap confidence intervals, 919–35
   accuracy, 934
   basic method, 930–33
   bootstrap percentile confidence interval method, 919–25
   bootstrap standard error t confidence interval method, 925–30
   important concepts, 933–34
Bootstrap distribution, 904–10
   bootstrap procedure, 906
      comments about, 909–10
      to generating, 907–8
      justification for, 907
   characteristics of, 918
   definition, 906
   notation, 908
   plug-in principle, 907
   and sampling distribution, comparing, 919
   using bootstrapping to estimating bias, 909
Bootstrap estimate of bias, 909
Bootstrap idea
   key, 917
   revisiting, 917
Bootstrap percentile confidence interval method, 919
   cautions about, 925
   reverse, 930–33
Bootstrap percentile P% confidence interval, 921
Bootstrapping. *See also under* Bootstrap
   definition, 905
   estimating standard error using, 910–19
Bootstrap sample, definition of, 905
Bootstrap standard error, 910
   of a mean $SE_b\,(\bar{x})$, 910
   of a statistic, $SE_b\,(statistic)$, 910
   t confidence interval method, 925–30
Bootstrap statistic, 905, 908
Bootstrap technique, 904
Box-and-whisker plot (box plot), 138–48
   construction procedure, 144
   definition, 141
   interpretations, 145
Break-mark, 59

C

Categorical data, 43–44
Categorical variable, 36
Cell, 732
Census, 8, 10
Central limit theorem, 904
   definition, 463–64
   for sampling distribution of proportion, 486–87
Central tendency, measures, 88–105
Chance, 232
Chance of Type II error (β), 559

Chebyshev's Theorem, 118–20
Chi-square
   assumptions about, 745
   degrees of freedom, 736–41, 746–47
   distribution properties, 733
   goodness-of-fit test, 745–52
   hypothesis test of independence, 734–45
   introduction, 731–33
   Pearson's formulas, 735
   p-value approach, using graphing calculator, 752–57
Chi-square test of independence, 731, 865
Class boundary, 48–49, 428
Classes (class intervals), 43
Classical (a priori) probability, 252
Class mark, 47–48, 65
Class width, 46–47
Coefficient of determination (r^2), 193
Coefficient of variation, 126–27
Combinations, 239, 245–50, 273–76
Common population proportion value (p), 639
Commutative probability rule, 280
Complement of event E, 268
Compound or multi-stage experiment, 235
Computational formula for Pearson's chi-square test statistic (X^2), 735
Conditional probability, 277–86
   definition, 278
   formula, 282
   multiplication rule, 278
Confidence intervals, 919. *See also* Bootstrap confidence intervals
   90% for population mean, 512, 518
   90% for population proportion, 524
   95% for confidence intervals, 510
   95% for population mean, 510, 518
   95% for population proportion, 524
   99% for population mean, 512, 518
   99% for population proportion, 524
   confidence level, 511
   confidence limits, 511
   degrees of freedom, 516
   population standard deviation unknown, 515–21
   summary of, 536
   t distribution, 515–17
   width of, 515
Confidence level, 511
Confidence limits, 511
Conservative estimate of sample size, 534–36
Conservative method, 688
Contingency Table, 54, 731–32
Continuity correction, 427–28
Continuous data, 36–37
Continuous probability distributions, 371–76
   characteristics, 374
   probability statements associated with, 376
Continuous random variable, 302, 371
Continuous variable, 376

Control groups, 648–53, 693–96, 700
Cordan, Jerome, 232
Correlation
   calculating, 192
   cautions, 192
   coefficient of determination, 193–94
   coefficient of linear correlation, 181–92
   correlation coefficient, 192, 770
   definition, 176
   degrees of freedom, 770
   interpreting values of r, 181
   introduction, 173–74, 767–68
   linear, 173–203
   linear regression analysis, 194–203
   negative, 176, 178
   no linear, 176–77, 179
   ordered pair, 174
   Pearson's correlation coefficient, 181–92
   population correlation coefficient, 769
   positive, 175–76
   p-value testing using graphing calculator, 777
   regression analysis, 194–203
   scatter diagram and, 174–80, 193
   testing significance of correlation coefficient, 776–80
Count data, 37, 628
Counting Rules
   1: Permutation Rule, 240
   2: Permutation Rule for n object taken s at a time ($_nP_s$), 242
   3: Permutation Rule of N objects with k alike objects, 245
   4: Number of combinations of n objects taken s at a time ($_nC_s$), 246–48
Critical chi-square (χ_α^2), 734
Critical F-value, 808
Critical t score (t_c), 600–2
Critical value of correlation coefficient (r_α), 771
Critical z-score (z_c), 560–68, 603, 871, 879
Cross Tabulation Table. *See* Contingency Table
Cumulative frequency, 51–54

D

Data, 33–75
   categorical, 36, 43–44
   classification of, 35–37
   continuous, 36–37
   count, 628
   discrete, 36–37
   exploring, 38–43
   numerical, 36, 44–49
   outlier, 38–39
   raw, 2, 33–35
   set, 2, 35–36
Deciles, 135–36
Decision rule
   development of, 554–68
   formulation procedure, 564–65, 734

de Fermat, Pierre, 232
Definition formula for Pearson's chi-square
   test statistic (X^2), 732, 735
Degrees of freedom (*df*)
   calculations for, 844
   chi-square, 736–44, 746–47
   confidence intervals, 516–20
   definition, 770
   F-Distribution for, 807
   formulas for, 835, 839
   t distribution, 599
de Laplace, Pierre Simon, 232
Dependent, classifications, 731
Dependent events, 277
Dependent variable, 199
Descriptive statistics, 2–3
Destructive testing, 9
Deviations from the mean, 108
Difference between sample means, 673–74
Difference between sample proportions ($\hat{p}_1-\hat{p}_2$), 628–37
Difference of paired observations, 703
Directional alternative hypothesis, 591
Discrete data, 36–37
Discrete probability distribution, 305, 350
Discrete random variables
   definition, 301
   expected value, 312
   mean, 310–14
   probability distribution of, 304–10
   standard deviation, 314–18
   variance, 314
   weighted mean, 312
Distribution
   bell-shaped, 452
   bimodal, 71
   characteristics of, 38
   chi-square, 731–57
   definition, 38
   frequency, 43
   mean, median, and mode in, 102
   median's location within, 95
   outliers, 38–39, 40
   range of, 106–8
   reverse J-shaped, 71
   skewed, 38, 70, 103–4, 144
   spread of data within, 38
   symmetric, 38
   symmetric bell-shaped, 69–70, 103
   t, 599–602
   of test statistic, 602, 695, 703, 710
   uniform (rectangular), 71, 452, 746
   U-shaped, 71
Distribution-free tests, 866
Distribution of sample means (sampling distribution of
   the mean), 448
Double-blind experiment, 649
Double-blind study, 693

E

Efron, Bradley, 904
Empirical Rule, 120–24
Empirical sampling distribution, 905
Equally likely, 252
Error variance, 817
Estimated standard error, 516
Estimate/estimation
   conservative, of sample size, 534–36
   definition, 504
   interval, 506–26
   point, 504
   population standard deviation,
      117–18, 599
   sample size and margin of error,
      526–36
   standard error, 719
   standard error of difference between two means
      ($\mu_{\bar{x}_1-\bar{x}_2}$), 680–81, 688
   standard error of difference between two proportions
      ($\mu_{\hat{p}_1-\hat{p}_2}$), 634, 636
   standard error of mean, 596, 602
   standard error of sampling distribution,
      pooled formula ($\sigma_{\hat{p}_1-\hat{p}_2}$), 639–40
Estimator, 505
Event, definition, 233
Excel applications
   multiple regression analysis, 787–88
Expected cell frequency, 732, 734
Expected result (the mean), 552–54, 585, 640–48,
   645–46, 650, 654–56, 712
Expected value, 312
Experiment, 232–39
Experimental group, 700
Experimental study, 648
Experimental/treatment variance, 817
Exploratory data analysis, 17
   box-and-whisker plot, 138–48
   stem-and-leaf display, 38–43
Extrapolation, 199–200

F

Failure, 322
F-Distribution, 802–9
   assumptions of, 807
   characteristics of, 807–8
   critical values of, 808–9
Finite correction factor, 457–60
   formula for proportion, 484
   for mean, 457
5-number summary, 138–41
F-ratio, 803
Frequency, 38, 731
   cumulative, 51–54
   distribution, 43
   relative percentage, 50
   zero, 65

Frequency distribution tables, 43–54
   categorical data, 43–44
   class boundary, 48–49
   class mark, 47–48
   construction guidelines, 45
   construction procedure, 49
   cumulative frequency, 51–54
   definition, 43
   numerical data, 44–49
   relative frequency, 49–50
Frequency polygon, 64–66
F-Statistic, 810–12
F-Test, 809–15
   hypothesis testing procedure, 810–12
   for variances, 693
Fundamental counting principle, 236–39, 274

G

General multiplication rule, 281
Geometric and negative binomial distributions, 346–58
Geometric experiment, 347
Geometric probability distribution, 348
Geometric random variable, 354
Good estimator, 506
Goodness-of-Fit Test, 745–52
Gosset, William S., 599
Graphing calculator applications
   binomial probability distribution, 329–31
   box-and-whisker plot, 143–44
   combinations, 248
   confidence intervals for population mean, 513–14, 521
   confidence intervals for population proportion, 525
   correlation coefficient, 190–91
   factorials, 240–41
   5-number summary, 140
   frequency distribution tables/histograms, 61–63
   geometric distribution, 350–53
   histograms, 22
   Law of Large Numbers, 260
   mean and median, 97–98
   normal curve area, 385–89
   normal distribution, 402–3, 420
   permutations, 242–43
   Poisson distribution, 342–43
   probability distribution, 318–19
   p-value approach to chi-square hypothesis test of independence using, 752–57
   for p-values, 611–18
   p-values for hypothesis testing, 654–57, 659, 705–18
   p-values for testing correlation coefficient significance, 777–80
   regression analysis, 197–98
   regression line, 202–3
   sample correlation coefficient (r), 190–91
   sample standard deviation, 112
   sampling distribution of mean, 472
   sampling distribution of proportion, 490
   scatter diagram, 180
Graphs, 18–20
   bar, 19, 22, 54–57
   box-and-whisker plot (box plot), 138–48
   definition, 18
   frequency polygon, 64–66
   histograms, 58–64
   interpreting, 72–73
   ogive, 67–68
   pictograph, 19, 20, 69
   pie chart, 68
   scatter diagram, 174–80, 193
   shapes, 69–72
   stem-and-leaf, 38–43
   time-series, 57

H

Histograms, 58–64, 320
Horizontal bar graph, 55
H test. *See* Kruskal-Wallis H test
Hypothesis testing
   alternative hypothesis, 546–48
   critical z-score, 560–68
   decision rule, 554–68
   directional alternative hypothesis, 548
   introduction, 545
   involving one population, 583–618
   involving population mean, 594–611
   involving two population means using independent samples, 673–719
   vs. judicial system, 549
   level of significance, 554–60
   model, 586, 589, 595
   nondirectional alternative hypothesis, 548
   null hypothesis, 546–48
   one-tailed test, 561–67
   procedure, 549, 566, 574
   p-values, 569–74
   test statistic, 561
   two-tailed test, 563–67
   Type I error, 550–51
   Type II error, 550–51

I

Identical trials, 322
Independent, classifications, 731
Independent events, 271–73
Independent samples
   definition, 627
   hypothesis testing involving two population means, 673
   hypothesis testing involving two population proportions, 627–59
   random, 673
Independent trials, 322
Independent variable (x), 199, 767
Inferential statistics, 3–5, 8, 9, 446, 904
Inner fences, 146
Interpolation, 199–201
Interquartile Range (IQR), 146
Interval estimate/estimation, 506–26
   central limit theorem, 508
   confidence intervals, 507–26
   definition, 507

margin of error, 506, 526–36
population mean, 507–21
population proportion, 522–26

K
Kruskal-Wallis H test, 884–91

L
Left critical z score (z_{LC}), 562
"Less than" cumulative frequency distribution, 51–54, 67
Level of significance (α), 554–60
Linear correlation, 173–203. *See also* Correlation
Linear model, 195
Linear regression, 174
   assumptions for analysis of, 777
Lower confidence limit, 511

M
Mann-Whitney Rank-Sum test, 875–83
Marginally or not significant p-values, 571
Margin of error, 506, 526–36
Matched pairs, 700
Mean
   binomial distribution, 335, 423
   definition, 88
   deviations from, 93, 108
   of geometric random variable, 347
   of negative binomial random variable, 354
   Poisson distribution, 340
   population, 90–91
   property of, 94
   relationship with median and mode, 102–5
   sample, 89–90
   sampling distribution. *See* Sampling distribution of mean; Sampling distribution of proportion
   sampling error of, 447
   weighted, 312
Mean of sampling distribution of difference between two means ($\mu_{\bar{x}_1-\bar{x}_2}$), 680
Mean Square (MS), 833
   calculations for, 844
   formulas for, 836, 839
Mean value, of discrete random variable, 310–14
Measures of central tendency, 88–105
Measures of variability, 106–18
Median, 88
   definition, 94
   determining, 94
   location of, 95
   relationship with mean and mode, 102–5
Method of least squares, 195
MINITAB applications
   multiple regression analysis, 785–88
Mode
   bimodal, 99
   definition, 98
   nonmodal, 99
   relationship with mean and median, 102–5
"More than" cumulative frequency distribution, 51–54, 67
Multiple coefficient of determination (R^2), 784
Multiple correlation coefficient (R), 783–85
Multiple regression, 782–88
Multiplication rule
   dependent events, 278
   fundamental counting principle, 236
   general, 281
   independent events, 271–73
Mutually exclusive events, 266–68

N
n (– signs), 868
n (+ signs), 868
n factorial (n!), 240
n greater than 30 rule, 464
Negative binomial distribution, 353–58
   mean and variance of, 356
Negative binomial experiment, 354
Negative binomial random variable, 354
Negative correlation, 176, 178
Negatively skewed (skewed left) distribution, 70
No linear correlation, 176–77, 179
Nondirectional alternative hypothesis, 548, 589
Nonmodal, 99
Non-normal population, sampling from, 460–65
Non-numerical value, 2
Nonparametric bootstrap method, 919
Nonparametric bootstrapping, 905
Nonparametric statistics
   advantages/disadvantages, of tests, 866
   definition, 865
   Kruskal-Wallis H test, 884–91
   Mann-Whitney Rank-Sum test, 875–83
   SIGN test for Medians, 867–75
Nonparametric tests, 866
Nonrepresentative sample, 11, 20–21
Nonsampling errors, 447
Normal approximation, to binomial distribution, 422–33
Normal curve area table, using, 383–400
Normal distribution
   applications of, 401–6
   approximately binomial probability, 422–24
   continuity correlation, 427
   introduction, 376–78
   percentile/percentile rank, 406–17
   probability applications, 417–21
   properties of, 378–83
Normal population, sampling from, 457–60
Not significant p-values, 571
Null hypothesis (H_0)
   vs. alternative hypothesis, 546–48
   analysis of variance, 824–25, 836
   difference between population means, 675, 681, 689, 702
   Kruskal-Wallis H test, 886
   Mann-Whitney test, 878–79
   population correlation coefficient, 769
   population mean, 594, 836
   population proportion, 585, 638
   p-values and, 571
   sign test for median, 870–71
   variance of population, 810

Numerical data, 44–49
Numerical value, 2
Numerical variable, 36

O

Observed cell frequency, 732
Observed frequencies, 731
Ogive, 67–68
One-tailed test (1TT), 561–67, 588
One-way ANOVA, 822–49
Ordered pair, 174
Original random sample, 908
Original sample, 905
Outcome, of experiment, 232–34
Outer fences, 146
Outliers, 38–39, 40
   detecting, 146
   detecting using z scores, 132
   potential, 146

P

Paired t-test, 700–11
Parameter, 10, 904
   definition, 6
Parametric Statistics, 865
Pascal, Blaise, 232
Pearson's chi-square statistic (X^2), 732
Pearson's correlation coefficient (r), 181–92
Percentile
   deciles and quartiles, 135–36
   defined, 133
   meaning of, 135
   normal distribution, 406–17
Percentile rank, 133–37
   computation of, 134
   definition, 133
   meaning of, 135
   normal distribution, 406–17
Permutation Rule, 240
Permutations, 239–45, 250, 273–76
Pictograph, 19, 20, 69
Pie chart, 68
Placebo, 628, 694, 697
"Plug-in principle" concept, 907
Point estimate
   definition, 504
   of population mean, 504–6
   of population proportion (p), 505
Poisson distribution, 340–54
Pooled estimate of standard error, 633, 646–47, 650–51
Pooled formula for estimated standard error of sampling distribution ($\sigma_{\hat{p}_1-\hat{p}_2}$), 639
Pooled sample proportion ($\hat{p}$), 629
Pooled standard deviation, 680
Pooled t-test, 680–87
Pooling, 680
Pooling Variance, 818

Population, 4–6
   average, 10
   classifications, 731
   definition, 4
   size, 9
   standard deviation of, 113–17
Population correlation coefficient (ρ), 769–72
Population mean (μ)
   definition, 90–91
   estimate of, 507–21
   formula, 90
   hypothesis testing involving, 594–98
Population median, 870
Population multiple correlation coefficient (ρ), 785
Population proportion
   estimate of, 522–26
   hypothesis testing involving, 583–93
Population (true) proportion (p), 477, 628
Population standard deviation (σ), 114, 596
   estimate, 599
Population variance (σ^2), 113–17
Positive correlation, 175–76
Positively skewed (skewed right) distribution, 70
Potential outlier, 146
Precision, 526
Predicted value of y (y'), 195, 767
Probability
   addition rule, 264–66
   birthday problem, 284–85
   calculating, 258
   classical or a priori, 252
   combinations, 239, 245–50, 273–76
   commutative, 280
   complement of event E, 268–70
   conditional, 277–86
   conditional probability formula, 282
   definition, 251
   density curve, 373–75
   dependent events, 278–80
   equally likely, 252
   event, 233–39
   experiment, 232–39
   fundamental counting principle, 236–39
   fundamental rules and relationships, 261–76
   general multiplication rule, 281–82
   histogram, 305
   independent events, 271–73
   mutually exclusive events, 266–68
   permutations, 239–40, 250, 273–76
   relative frequency or posteriori, 258–59
   sample space, 234–35
   subjective or personal, 259–60
   using sampling distribution of mean, 468–74
   using sampling distribution of proportion, 488–92
Probability distribution
   definition, 304
   of discrete random variable, 304–10
Proportion, definition, 476

Proxy population, 905
p-values
  approach to hypothesis testing procedure, 574
  calculating procedure, 572
  chi-square independence test using graphing calculator, 752–57
  correlation coefficient testing using graphing calculator, 777–80
  defined, 570
  graphing calculator for, 611–18
  for hypothesis testing, 569–74
  hypothesis testing using graphing calculator, 654–57, 659, 705–18
  interpreting, 571
  paired t-test, 705
  rule to reject H_0, 573

Q

Qualitative data, 731
Quantile q_α, 931
Quartiles, 135–36

R

Random (probability) sampling, 11–15, 904
Random variable, 300, 448
Range, 22, 106–8
  definition, 106
  estimating standard deviation, 124–25
Range Rule, 124
Raspe, Rudolph Erich, 908
Raw, definition, 2
Raw data, 2, 33–35
Raw scores, 412–14
  converting to z scores, 132
  formula, 133
Real class limits, 48
Regression
  linear, 777
  multiple, 782–88
  technology, 786–88
Regression analysis, 194–203, 777
Regression coefficients (a and b), 195, 782
Regression line, 195–99
Relative frequency, 49–50
Relative frequency concept (posteriori) probability, 258–59
Relative percentage frequency, 50
Relative percentages, 50
Relative standing, measures, 127–38
Relative variation, 127
Repeated measure, 700
Representative sample, 5, 10–11
Resampling, 905
Residual error, 784
Resistant measure, 104
Reverse bootstrap percentile confidence interval method, 930–33
Reverse J-shaped distribution, 71
Right critical z score (z_{RC}), 562

S

Sample, 4–6, 905
  definition, 4
  representative, 5, 10–11
  size of, 9
Sample correlation coefficient (r), 181–92, 771
Sample information, 5
Sample mean ($\bar{x}$), 89–90, 446, 597
Sample proportion ($\hat{p}$), 478–81, 586, 628
Sample size, 9
  conservative estimate of, 534–36
  effect on standard error of mean, 474–76
  margin of error, 526–36
Sample space, 234–35
Sample standard deviation, 110
Sample statistic, 903
Sample variance (s^2), 109–10
Sampling, 8–22
  acceptance, 336–38
  biased, 11
  definition, 8
  from non-normal population, 463–68
  nonrepresentative, 11, 20–21
  from normal distribution, 460–63
  probability, 11–12
  random, 11–13
  reasons for, 8–10
  representative, 10–11
  simple random, 12
Sampling distribution
  of correlation coefficients, 770
  of difference between two means, 674
Sampling distribution of mean, 445–76, 596
  calculating probabilities using, 468–74
  central limit theorem, 463–64
  characteristics of, 461, 465, 467–68
  definition, 448
  difference between two means, 673–76
  effect of sample size on, 474–76
  mean of, 451, 453–57, 595
  shape of, 460–68
  standard deviation of, 451, 453–57
Sampling distribution of proportion
  calculating probabilities using, 488–90
  central limit theorem for, 486–87
  characteristics of, 487
  definition, 479
  difference between two proportions ($\mu_{\hat{p}_1-\hat{p}_2}$), 634
  estimate of standard error of proportion, 523, 586, 589, 591
  finite correction factor formula, 484
  interpretation of, 486
  mean of, 481–85, 489–90
  sample proportion, 478–81
  sampling error, 485
  shape of, 486
  standard deviation, 481–85
Sampling errors, 446–48, 485

Scatter diagram, 174–80, 193
Selection with/without replacement, 249
Significant p-values, 571
SIGN test for Medians, 867–75
Simple random sampling (random sampling), 12
Single-blind experiment, 649
Single-blind study, 693
68-95-99.7 Rule. *See* Empirical Rule
Skewed distribution, 70, 103–4, 144, 733
Standard deviation
   applications, 118–27
   binomial distribution, 335, 423–25
   of discrete random variable, 314–18
   estimate, 117–18
   estimate of sampling distribution of
      proportion (s_p), 523
   interpretation, 111
   Poisson distribution, 340
   population, 113–17
   sample, 110
   sampling distribution (p_1-p_2), 635
   sampling distribution of difference between two means
      ($\mu_{\bar{x}_1-\bar{x}_2}$), 674–75, 688
   sampling distribution of difference between two
      proportions ($\mu_{\hat{p}_1-\hat{p}_2}$), 634
   sampling distribution of mean ($\sigma_{\bar{x}}$), 449,
      452–56, 595, 911
   sampling distribution of proportion ($\sigma_{\hat{p}}$), 481–85
Standard error
   difference between two means ($\mu_{\bar{x}_1-\bar{x}_2}$), 680
   difference between two proportions ($\mu_{\hat{p}_1-\hat{p}_2}$), 634
   estimate of, 719
   estimating using bootstrapping, 910–19
   of mean, 449, 452–56, 595, 911, 916
   pooled estimate of, 643, 646–57, 650–51
   of proportion (σ_p), 482, 586
   of sampling distribution, 904
Standard normal curve, 380–83
Standard normal distribution, 382
Statistic, 6, 904
Statistical result, 551
Statistical significance, 552
Statistics
   definition, 2
   descriptive, 2–3
   inappropriate comparisons, 21–22
   inferential, 3–5, 8, 9, 446
   misuses of, 18–20
   uses of, 16–18
Stem-and-leaf display, 38–43
Student's t distribution, 599
Subjective (personal) probability, 259–60
Success, 322
Sums of squares, 822
   calculations for, 843
   formulas for, 835, 838
Symmetric bell-shaped distribution, 69–70, 103, 144
Symmetric distribution, 733

T

t distribution, 515–17, 599–602
   critical values for, 600–2
   degrees of freedom (*df*), 601
   properties of, 599
t score (t_c), 600
Test statistic (z)
   analysis of variance, 833, 837–38
   calculated, 836, 839, 844
   correlation coefficient significance, 777
   definition, 561
   distribution of, 602–9, 695, 703, 710
   F-Statistic, 811–12
   hypothesis testing, 586–87
   for Kruskal-Wallis test, 884
   paired t-test, 703
   Pearson's chi-square statistic, 732
   pooled t-test, 682–83
   population mean, 610
   population proportion, 593
   sample correlation coefficient, 771
   for SIGN test for Medians, 872
   two population means, 698–99
   two population proportions for independent
      samples, 658
   two sample z test, 676
Theoretical probability distribution, 632
Theoretical sampling distribution, 904
Three-stage experiment, 236
Time-series graph, 57
Total variance, 816
Treatment effect, 649
Treatment groups, 648–53, 693–96
Treatment variance, 817
Tree diagram, 234–35
Trials, 322
Two sample z test, 675–80
Two-stage experiment, 236
Two-tailed test (2TT), 518, 563–67
Two-way ANOVA, 849
Type I error, 550–51, 557–59
Type II error, 550–51, 557–59

U

Unbiased estimator, 506
Uniform (rectangular) distribution, 71, 452, 746
Upper confidence limit, 511
Usable sample size (n), 871
U-shaped distribution, 71

V

Value, of variable, 35
Variability, 10
   measures, 106–18
Variables
   categorical, 36
   definition, 35
   numerical, 36

Variance(s)
  of discrete random variable, 314–18
  of geometric random variable, 347
  of negative binomial random variable, 354
  one-way analysis of, 822–49
  population, 113–17
  testing, 809–15
  types of, 815–22
Vertical bar graph, 55, 56
Very significant p-values, 571

W

Weighted mean, 312
Welch's t-test, 687–93
Wilcoxon Rank-Sum test. *See* Mann-Whitney Rank-Sum test
Within Group Variance, 816–17
Without replacement, 448

Y

y'-intercept, 782

Z

z score
  converting to data value, 412
  converting to raw score, 132
  definition, 128
  detecting outliers using, 132
  formula, 129
  interpreting, 131
  normal distribution, 382
  proportion of area between two, 374, 391
  when proportion of area to left is known, 376
  proportion of area to left of, 384–85
  proportion of area to right of, 390
  relative standing, 128
Zero frequency, 65